Human
Anatomy

Text and Illustration Team

 **Frederic H. Martini** received his Ph.D. from Cornell University in Comparative and Functional Anatomy. He has broad interests in vertebrate biology, with special expertise in anatomy, physiology, histology, and embryology. Dr. Martini's publications include journal articles, technical reports, magazine articles, and a book for naturalists on the biology and geology of tropical islands. He is the author of *Fundamentals of Anatomy and Physiology* (third edition, 1995) and coauthor of *Essentials of Anatomy and Physiology* (first edition, 1997 by F. H. Martini and E. Bartholomew).

Dr. Martini has been involved in teaching undergraduate courses in anatomy and physiology (comparative and/or human) since 1970. During the 1980s he spent his winters teaching courses, including human anatomy and physiology, at Maui Community College, and his summers teaching an upper-level field course in vertebrate biology and evolution for Cornell University, at the Shoals Marine Laboratory (SML). Dr. Martini now teaches part-time in the winters, and devotes most of his attention to developing new approaches to A&P education. His primary interest is in the use of appropriate technologies in creating an integrated learning system. Dr. Martini is a member of the Human Anatomy and Physiology Society, the National Association of Biology Teachers, the American Society of Zoologists, the Society for College Science Teachers, the Western Society of Naturalists, and the National Association of Underwater Instructors.

 **Michael J. Timmons** received his degrees from Loyola University, Chicago. For more than two decades he has had a strong commitment to teaching human anatomy and physiology to nursing, and preprofessional students at Moraine Valley Community College. Early in his teaching career, Professor Timmons became very interested in publishing in the fields he taught in. He is coauthor of *Human Anatomy Laboratory Guide and Dissection Manual,* the lab manual that accompanies this text; three anatomy and physiology laboratory manuals; editor of the *Prentice Hall Laser Disk with Bar Code Manual* and author of *ATLAS,* the laboratory learning system, which are companions to accompany this text. His special areas of interest are biomedical photography, crafting illustration programs, and developing instructional technology learning systems. He has authored a series of titles on the dissection of the cat, *A Photographic Atlas of Cat Anatomy, Cat Anatomy Slides: A Visual Guide to Dissection,* numerous study guides, and is an active member of scientific and professional associations.

Dr. William C. Ober (art coordinator and illustrator) received his undergraduate degree from Washington and Lee University and his M.D. from the University of Virginia in Charlottesville. While in medical school he also studied in the Department of Art as Applied to Medicine at Johns Hopkins University. After graduation Dr. Ober completed a residency in family practice, and is currently on the faculty of the University of Virginia as a Clinical Assistant Professor in the Department of Family Medicine. He is also part of the Core Faculty at Shoals Marine Laboratory, where he teaches biological illustration in the summer program. Dr. Ober now devotes his full attention to medical and scientific illustration.

Claire W. Garrison, R.N., (illustrator) practiced pediatric and obstetric nursing for nearly 20 years before turning to medical illustration as a full-time career. Following a five-year apprenticeship, she has worked as Dr. Ober's associate since 1986. Ms. Garrison is also a Core Faculty member at Shoals.

Texts illustrated by Dr. Ober and Ms. Garrison have received national recognition and awards from the Association of Medical Illustrators (Award of Excellence), American Institute of Graphics Arts (Certificate of Excellence), Chicago Book Clinic (Award for Art and Design), Printing Industries of America (Award of Excellence), and Bookbuilders West. They are also recipients of the Art Directors Award.

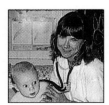 **Dr. Kathleen Welch** (clinical consultant) received her M.D. from the University of Washington in Seattle with her residency at the University of North Carolina at Chapel Hill. She served as Director of Maternal and Child Health at the LBJ Tropical Medical Center in American Samoa, and subsequently was a member of the Department of Family Practice at a Kaiser Permanente Clinic in Hawaii prior to entering private practice. She is a member of the American Academy of Family Practice, the Hawaii Medical Association, and the Human Anatomy and Physiology Society.

 **Ralph T. Hutchings** is a biomedical photographer who was associated with the Anatomy Department of the Royal College of Surgeons for 20 years. An engineer by training, Ralph has focused his attention for years now on photographing the structure of the human body. The result has been a series of color atlases, including the *Color Atlas of Human Anatomy, The Color Atlas of Surface Anatomy* and *The Human Skeleton* (all published by Mosby-Yearbook Publishing, St. Louis, Missouri, USA). Ralph makes his home in North London where he tries to balance the demands of his photographic assignments with his hobbies of early motor cars and airplanes.

Human Anatomy

Second Edition

Frederic H. Martini, Ph.D.

Michael J. Timmons, M.S.
Moraine Valley Community College
Palos Hills, Illinois

with

William C. Ober, M.D.
Art coordinator and illustrator

Claire W. Garrison, R.N.
Illustrator

Kathleen Welch, M.D.
Clinical consultant

Ralph T. Hutchings
Biomedical photographer

Prentice Hall
Upper Saddle River, New Jersey 07458

Library of Congress Cataloging-in-Publication Data

MARTINI, FREDERIC
  Human anatomy/Frederic H. Martini, Michael J. Timmons; with William C. Ober, art coordinator and illustrator,
Claire W. Garrison, illustrator, Kathleen Welch, clinical consultant; Ralph T. Hutchings, biomedical photographer.
    p.   cm.
  Includes index.
  ISBN 0-13-267691-5
  1. Human anatomy.   2. Human anatomy—Atlases.   I. Timmons, Michael J.
    II. Title.
  QM23.2.M356 1997
  611—dc20
        96-27166
        CIP

Executive Editor: *David Kendric Brake*
Editorial Director: *Tim Bozik*
Editor in Chief: *Paul Corey*
Vice President of Production and Manufacturing: *David W. Riccardi*
Executive Managing Editor: *Kathleen Schiaparelli*
Assistant Managing Editor: *Shari Toron*
Production Editor: *Karen M. Malley*
Creative Director: *Paula Maylahn*
Art Director: *Joseph Sengotta*
Interior Designer: *Lorraine Mullaney*
Cover Designer: *Lisa Jones*
Cover Illustration: *Vincent Perez*
Manufacturing Manager: *Trudy Pisciotti*
Page Layout: *Karen Noferi, Jeff Henn, Shari Toron,*
    Wanda Espana, Steven Greydanus
Art Manager: *Gus Vibal*
Technical Art: *Karen Noferi*
Photo Research: *Stuart Kenter*
Photo Editor: *Carolyn Gauntt*

 © 1997 by Frederic H. Martini, Inc. and Michael J. Timmons
Published by Prentice Hall
Simon & Schuster/A Viacom Company
Upper Saddle River, New Jersey 07458

First edition © 1995 by Prentice-Hall, Inc., Simon & Schuster/A Viacom Company

Printed in the United States of America
10 9 8 7 6 5 4 3

ISBN 0-13-267691-5

Prentice-Hall International (UK) Limited, *London*
Prentice-Hall of Australia Pty. Limited, *Sydney*
Prentice-Hall Canada Inc., *Toronto*
Prentice-Hall Hispanoamericana, S.A., *Mexico*
Prentice-Hall of India Private Limited, *New Delhi*
Prentice-Hall of Japan, Inc., *Tokyo*
Simon & Schuster Asia Pte. Ltd., *Singapore*
Editora Prentice-Hall do Brasil, Ltda., *Rio de Janeiro*

Contents in Brief

Dedication

To Kitty, Judy, P.K., Molly, Kelly, Patrick, and Katie:

We couldn't have done this without you.

Thank you for your encouragement and devotion.

Contents

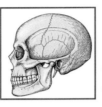

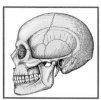
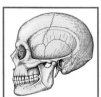
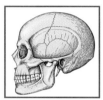

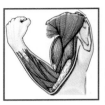
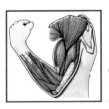

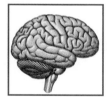
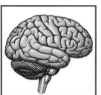
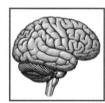

16

**The Nervous System: Pathways
and Higher-Order Functions** 409

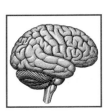

17

**The Nervous System: Autonomic
Division** 427

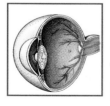

18

The Nervous System: General and Special Senses 445

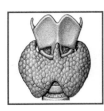
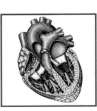
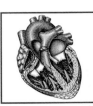

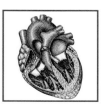

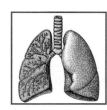

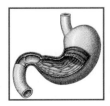

25

The Digestive System 625

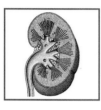
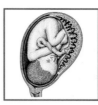

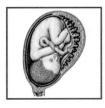

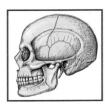

Preface

The second edition of Human Anatomy builds upon a reputation for excellence and utility established by our successful first edition. We are grateful for this success and pleased that so many instructors and students have taken the time to contact us and share their comments about this book. All of these comments have influenced us in our work on the second edition, and we hope that the addition of new line drawings and photographs, a Scanning Atlas in the appendix, a Human Anatomy presentational CD-ROM, and a dedicated World Wide Web site will help students better visualize the structures of the human body. From the very beginning, our goal has been to help users of this book "see" human anatomy and understand the basic principles of this discipline.

Many people find it enough of a challenge just to understand what is read or heard about the human body in the news. Others have a practical need to explore the field in greater detail. Students in the allied-health fields, as well as interested non-professionals, find that they must learn to "see" human anatomy as the first step toward understanding and applying anatomical knowledge. We intended this book to be a new kind of anatomy text, designed to improve the ability of students to *visualize* the human body, to *understand* the principles of human anatomy, and to *apply* the terms and concepts of the discipline to a variety of real-world and clinical situations.

Research-Based Approach to the Development of this Text

Before making the commitment to author the first edition of this text, we evaluated existing texts, and discussed our perceptions with numerous instructors, students, and allied-health professionals. We concluded that significant improvements could be made in the way human anatomy texts were illustrated and in the way anatomical structures and concepts were presented to the reader. Next, we asked more than forty human anatomy instructors to complete a comprehensive questionnaire in which they identified strengths and weaknesses of current anatomy texts and then commented on our approach to the presentation of anatomical information.

The results of this research strongly supported our conclusions. More importantly, we were pleased to receive additional suggestions that allowed us to enhance and improve our vision of what this book should accomplish. At that time, we formulated five basic goals for this book:

1. To take greater pedagogical advantage of the visual aspect of anatomy by
   - increasing the overall number and size of illustrations
   - increasing, wherever possible, the number of different perspectives shown in the illustrations
   - emphasizing the three-dimensional relationships between anatomical structures.

2. To improve the presentation of the most difficult topics covered in human anatomy as identified by written surveys and countless conversations with instructors, students, and allied-health professionals.

3. To give students a framework for the organization, interpretation, and application of anatomy-related information derived from the world around them.

4. To create a book that functions as a solid foundation and "information base" for students, providing them with the tools for tackling related or more advanced courses.

5. To develop an innovative format which encourages use of the text not only in the classroom but also in the laboratory.

The focus throughout has been to present essential information clearly, with emphasis on the concrete, visual aspects of each topic. We worked through many nights and time zones developing illustration concepts that would give students multiple views and perspectives of complex structures, enabling them to develop accurate mental images of these structures. The resulting figures should reduce the need for an anatomy instructor to juggle chalk, slides, photographic atlases, models, and a text to help students "see" human anatomy.

The larger format has enabled us to create what we feel is a pedagogically superior art and illustration program. In all cases, decisions regarding size, placement, perspective, and accuracy were reviewed by other instructors and clinicians, and their evaluations and recommendations taken into account.

During the production of this text, our primary concerns were (1) ensuring accuracy in every respect, from narrative text to labels and leaders on the illustrations, and (2) maintaining figure/text correlation, both in terms of figure callouts and placement. Beyond our own scrutiny, which we exercised on each element and at every stage, the finished artwork, galleys, and page proofs were subjected to review by specialists. These individuals provided a rigorous "safety net" prior to the actual printing.

Themes of this Text and Success in the Human Anatomy Course

Two primary themes are woven into the words and images of this text. We have already mentioned the first: students must be able to "see" human anatomy. The second theme follows from the first: students must be able to relate and apply anatomical concepts to the world around them. More specifically, they must be able to relate and apply anatomical concepts to real-life problems and situations.

For both teacher and student, a human anatomy course is an awesome undertaking. To succeed in this course, students must do more than develop a large technical vocabulary and retain a large volume of detailed information. They must relate the information learned in each chapter to data introduced in earlier chapters, and understand how these specific details affect the overall functioning of the body. In other words, anatomy students must develop a capacity for critical thinking, abstraction, and concept integration. These skills are important for everyone, but they are especially vital for those pursuing careers in the health sciences. Recent educational research has demonstrated that memorization of information does not improve diagnostic abilities *unless the individual learns the material in a logical framework that stresses concept organization.*

Unfortunately, introductory students are often unprepared for this type of course because their previous courses stressed rote memorization and recital of discrete blocks of information. An anatomy course thus gives the student a chance to develop or improve the ability to think clearly and logically. This text attempts to make this process as easy and enjoyable as possible.

General Features of this Text

As this project evolved, several innovative concepts were developed to provide new solutions to such problems as the diversity of students' learning types, backgrounds, degrees of preparation, and intellectual abilities, as well as limited instructional time and resources.

Larger Physical Format

The most conspicuous characteristic of our text is its larger format. In our opinion, it offers at least three distinct advantages: (1) Key illustrations throughout can be enlarged to provide greater detail. (2) A more visual presentation can be achieved through inclusion of additional illustrations, thereby improving the overall ratio of art and photos to narrative text. (3) The enlarged size of key illustrations eliminates the need for students to purchase a separate anatomical atlas for use in the laboratory.

Treatment of Difficult Topics

A conscious effort was made from the outset to identify the topics students typically find most difficult and adjust our coverage accordingly. In defining these areas, we relied on our classroom experience and the opinions of other instructors and students. In total, more than fifty instructors of human anatomy, carefully selected from a cross section of schools, were consulted. Everyone agreed that two systems—the muscular and nervous systems—are particularly difficult for students in this course.

We addressed this problem by reexamining the traditional scope and pace of presentation. Our conclusion was that three important modifications should be made. First, the space devoted to the most abstract topics should be expanded— not to allow for inclusion of greater detail, but to permit more careful explanation and description of key concepts. Second, the supporting illustrations should be increased in number and enlarged, to further clarify the narrative description. Third, the material should be broken down into smaller blocks of information, with shorter, less intimidating chapters. As a result, we have five chapters on the nervous system rather than four (the usual number in other texts) and three chapters on the muscular system (compared to one or two).

Compound Figures Integrating Macroscopic and Microscopic Structure

Introductory students are most familiar with (and most comfortable with) the highest levels of organization, those of the individual or organ system. They are much less familiar with, and considerably more apprehensive about, structures and events at the molecular or cellular level. Wherever possible, figures have been designed that bridge the gap between the familiar, macroscopic world and the unfamiliar world of cells and tissues.

An Emphasis on Applied Topics

This text makes frequent reference to concrete, real-life examples that drive home the importance of key topics. Supplementary material dealing with clinical details, sports-related information, or other applied topics may be found in small boxes scattered throughout the text. More detailed coverage of applied topics, especially clinical information, can be found in the Clinical Issues appendix.

A Focus on Pedagogical Aids

The pedagogical elements in this text are varied and unique. Realizing that not all students comprehend and internalize information in the same way, the pedagogical structure of each chapter now offers more help to the diversity of students taking this course. Examples include: (1) *integration of the chapter outline with the learning objectives* to provide perspective and aid in review; (2) *concept checkpoints,* which stimulate review of and reflection on small units of information; (3) *concept links,* which interconnect structures and concepts covered in different chapters; (4) *visual keys* to speed cross-referencing between text and art; and (5) *end of chapter questions* that stimulate analysis and integration of information.

Summaries of Embryological and Fetal Development

These summaries are provided for instructors who opt to cover basic morphogenesis on a system-by-system basis. For instructors who prefer to cover developmental topics exclusively in Chapter 28, that chapter contains a summary table that includes cross-references to appropriate portions of the text.

A Discussion of the Effects of Aging on Each System

The field of geriatrics is becoming increasingly important as the baby-boomer generation grows older. Today roughly 12% of the U.S. population is age 65 or older; a typical undergraduate anatomy student today will watch that percentage double. The resulting changes in population dynamics, politics, and—perhaps most significantly—the demand for medical care will profoundly affect our society.

As with our coverage of the more challenging topics in this course, many of these new features have been specifically designed to address common problems students encounter with this material. The macro-to-micro figures bring the microscopic world into perspective and make structural and functional relationships easier to understand. The emphasis on applied topics maintains student interest; the expanded pedagogy enables students to keep track of their progress. Cross-referencing of text and figures helps students develop an integrated perspective on the structure of the body.

Supplements

Prentice Hall has supported the development of a comprehensive supplement package to accompany our text.

- **Overhead Transparency Acetate Package** Instructors need a strong visual support package for the classroom. With this in mind, Prentice Hall has created a set of 310 transparencies and masters.

- **Laser Disk with Bar Code Manual** Prentice Hall has supported us in the development of a video disk containing more than 2000 still images as well as important animations and video segments. The laser disk is accompanied by a bar code manual with brief descriptions of each still, animation, and video segment. The laser disk can also be used with Prentice Hall's Anatomy and Physiology Multimedia Presenter software, an excellent tool for turning your lecture hall or laboratory into an electronic presentation gallery.

- **CD-ROM with Presentational Software** Designed for the instructor who wants to bring the latest technology into the classroom, this resource features images from the text as well as an assortment of histology and cadaver dissection stills. The presentational software embedded on the CD allows instructors to customize lectures or create mini student tutorials by selecting and sorting images in the desired order of presentation. The software also allows the user to import files from word processing and presentational programs, making this an even more effective and versatile teaching tool.

- **Instructor's Manual and Resource Guide** This supplement contains detailed lecture outlines and suggested classroom discussion questions, along with a collection of helpful, class-tested hints, useful analogies, and suggestions for conveying basic anatomical concepts to your students.

- **Test Item File** A collection of over 3000 test questions organized around a proven three-level method of question presentation. Level One focuses on a review of the fundamental terms and concepts of each chapter. Level Two questions focus on the basic synthesis and application of the concepts and terms from Level One. Level Three questions involve real-world and clinical problems that urge the student to move to the next cognitive level, or, as some instructors prefer to label it, the critical thinking mode. The test item file allows instructors the flexibility to create tests based upon the depth and breadth of coverage they deem appropriate for each section of the book.

- **Computerized Testing Programs** The same test package described above in IBM or Macintosh formats.

- **Student Study Guide** An excellent review and study tool for students. All questions and exercises are organized in accordance with the three-level system featured in the test item file. This study guide also features concept maps that review key terms, concepts, and their relationships. Many chapters also include important line drawings from the text to be used as labeling exercises.

■ **ATLAS** A supplement that organizes and directs students in the identification of anatomical structures. Designed to facilitate learning by linking illustrations in the *Human Anatomy* text to anatomical models and specimens seen in the laboratory, ATLAS will help students make key visual and tactile connections. Helpful icons direct students through each exercise module. Space is provided for students to take notes or make drawings, unlabeled drawings provide an opportunity to label and color significant structures, and printed bar codes allow students to access relevant structures found on the Prentice Hall Anatomy and Physiology Laser Disk.

■ **Human Anatomy Laboratory Guide and Dissection Manual** Written by Mike Timmons and Mike McKinley with 235 photographs by world-renown medical photographer, Ralph Hutchings. Combining the traditional features of a lab manual with those of a dissection guide and a pictorial anatomical atlas, this manual provides an innovative framework for organizing, interpreting, and applying anatomical information. Features many color illustrations and black and white photographs of prosected cadavers, anatomical laboratory models, and assorted anatomical specimens. The manual features separate instructions for the student dissector and key observation points for students who will study the prosected cadaver. Step-by-step procedures for accurately performing each dissection are provided for both human and cat. Provides page references to illustrations, specimen photos, radiographs, and MRI images in the Martini/Timmons text.

■ **Other Laboratory Manuals and Dissection Guides** Prentice Hall and its sister company Appleton & Lange publish several other laboratory manuals and dissection guides that would be appropriate for the undergraduate anatomy laboratory. We especially suggest that you consider the following publications:

• *Learning Human Anatomy: A Laboratory Text and Workbook* by Julia F. Guy (Appleton & Lange)

• *Manual for Human Dissection: Photographs with Clinical Applications* by Gerald Callas (Appleton & Lange)

• *Human Anatomy Laboratory Manual* by Gerard Tortora (Prentice Hall)

■ **New York Times Contemporary View** A program designed to enhance student access to current information of relevance in the classroom. Through this program, the core subject matter in the text is supplemented by a collection of time-sensitive articles from one of the world's most distinguished newspapers, *The New York Times*. These articles, selected and submitted by the authors, demonstrate the connection between what is learned in the classroom and what is happening in the world around us. Prentice Hall and *The New York Times* are proud to cosponsor A Contemporary View. We hope it will make the reading of both textbooks and newspapers a more dynamic, evolving process.

Acknowledgments

We would like to acknowledge the many users, reviewers, survey respondents, and focus group members whose advice, comments, and collective wisdom helped to shape this text into its final form. Their interest in the subject, their concern for accuracy and method of presentation, and their experience with students of widely varying abilities and backgrounds made the review process much more inspiring and useful than it might otherwise have been.

We would also like to thank the production staff whose efforts made all the difference, with special thanks to John J. Jordan, Karen Malley, Joanne Jimenez, Karen Noferi, and Richard Foster. Thanks are also due to our editor, David K. Brake, for his encouragement and support.

No two people could expect to produce a flawless textbook of this scope and complexity. Any errors or oversights are strictly our own rather than those of the reviewers, artists, or editors. In an effort to improve future editions, we would ask that readers with pertinent information, suggestions, or comments concerning the organization or content of this textbook send their remarks to: David Brake, Executive Editor, Prentice Hall Publishing, 1208 East Broadway, Suite 200, Tempe, Arizona, 85282. Any and all comments and suggestions will be deeply appreciated, and carefully considered in the preparation of the next edition.

Frederic H. Martini, Haiku, HI
Michael J. Timmons, Orland Park, IL

First Edition Survey Respondents

Robert Anthony, Triton College
Linda Banta, Sierra College
Cynthia Battie, University of Missouri
Christopher Bennett, University of Kansas
Berrit Blomquist, Salt Lake Community College
Suzanne Byrd, Eastern Kentucky University
Neil Cumberlidge, Northern Michigan
Randy Curtiss, Eastern Arizona College
Max Dalsing, Johnson County Community College
John A. Dorsch, University of Souther Colorado
Jim Gale, Sonoma State University
Larry Ganion, Ball State University
Lynda Gordon, Long Beach City College
Fay Hansen-Smith, Oakland University
Ruth L. Hays, Clemson University
Anne Johnson, Southern Illinois University
Michael Kesner, Indiana University of Pennsylvania
Raj Kilambi, University of Arkansas
Stan Kunigelis, University of Connecticut
Andrew Kuntzman, Wright State University
T. L. Maguder, University of Hartford
Anthony Mescher, Indiana University
Bryan Miller, Eastern Illinois University
John Morrow, Mississippi State University
Virginia Naples, Northern Illinois University
Stuart Neff, Temple University
Mark T. Neilson, University of Utah
Debra T. Palatinus, Roane State Community College
Nicholas Pantazis, University of Iowa
Gail R. Patt, Boston University
Ronald K. Plakke, University of Northern Colorado
David Rabuck, Oakton Community College
David J. Saxon, Morehead State University
Ronald A. Siders, University of Florida
Tom Sourisseau, Cabrillo College
Bernice Speer, Austin Community College-Rio Grande
Frank Sullivan, Front Rage Community College
Kathleen A. Tatum, Iowa State University
Virgiania Sparks Volker, University of Alabama-Birmingham
Frank Ray Voorhees, Central Missouri State University
Phil Walcott, Grand Valley State University

First Edition Focus Group Participants

Mary Jo Fourier, Johnson County Community College
Edward Lutsch, Northeastern Illinois University
Bryan Miller, Eastern Illinois University
Virginia Naples, Northern Illinois University
Carl Sievert, University of Wisconsin
Reuben Barrett, Chicago State University

First Edition Manuscript and Text Reviewers

Cynthia Battie, University of Missouri-Kansas City
Annalisa Berta, San Diego State University
Paul Biersuck, Nassau Community College
Leann Blem, Virginia Commonwealth University
Debbie Borosh, Mount San Antonio Community College
Neil Cumberlidge, Northern Michigan University
Mary Jo Fourier, Johnson County Community College
Glenn Gorelick, Citrus College
Edward Lutsch, Northeastern Illinois University
Bryan Miller, Eastern Illinois University
Sherwin Mizell, Indiana University
Virginia Naples, Northern Illinois University
Gail Platt, Boston University
Ron Plakke, University of Northern Colorado
Carl Sievert, University of Wisconsin

First Edition Technical Reviewers

Norman Lieska, University of Illinois School of Medicine
Lissa Little, Northwestern Medical School
Diane Merlos, Grossmont College
Larry Olver, Northwestern Medical School
Randy Perkins, Northwestern Medical School

Second Edition Manuscript and Text Reviewers

Joan Aloi, Saddleback College
Cynthia Battie, University of Missouri, Kansas City
Annatisa Berla, San Diego University
Paul Biersuck, Nassau Community College
Leann Blem, Virginia Commonwealth University
Debbie Borosh, Mount San Antonio Community College
Neil Cumberlidge, Northern Michigan University
Mary Jo Fourier, Johnson County Community College
Anthony J. Gaudin, Ph. D., California State University–Northridge
Glenn Gornick, Citrus College
Glenn E. Kietzmann, Wayne State College
Tom Linder, University of Washington
Edward Lutsch, Northeastern Illinois University
Bryan Miller, Eastern Illinois University
Sherwin Mizell, Indiana University
Virginia Naples, Northern Illinois University
Mary Oreluk, R.N., Little Company of Mary Hospital
Gail Platt, Boston University
Annie Peterman, University of Washington
Ron Plakke, University of Northern Colorado
Carl Sievert, University of Wisconsin
James Smith, California State University, Fullerton
Thomas M. Vollberg, Sr., Creighton University

Second Edition Technical Reviewers

Norman Lieska, University of Illinois School of Medicine
Lissa Little, Northwestern Medical School
Diane Merlos, Grossmont College
Larry Olver, Northwestern Medical School
Randy Perkins, Northwestern Medical School
Mike McKinley, Glendale Community College

To the Student

Dear Student,

This text was designed to help you master the terminology and basic concepts of human anatomy and to begin to apply what you learn to real-world and clinical situations. There are several learning aids built into the format of this book that should make your study of human anatomy more manageable as well as rewarding. We encourage you to examine this overview carefully and to consult your instructor if you have further questions about how to use this book.

Though we are no doubt biased, we believe that the study of human anatomy is one of the most interesting and rewarding undertakings available on campus. This book has been designed for you, and we hope that you can make full use of the many learning aids described in this overview.

Our Best Wishes,

Frederic H. Martini

Michael J. Timmons

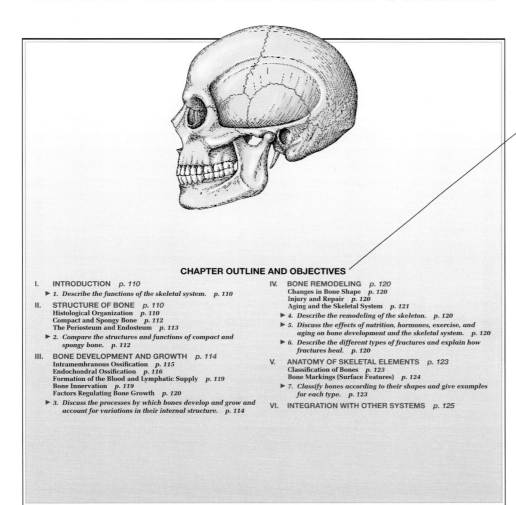

5

The Skeletal System: Osseous Tissue and Skeletal Structure

■ Chapter Outline and Objectives
Before you begin using a chapter, it is important to know where you are going and what is expected of you. With this in mind, every chapter opens with an outline of that chapter's content. Notice that each section of the outline corresponds to an actual heading in the text. Furthermore, the basic concepts (or learning objectives) are integrated into this outline. This valuable learning aid allows you to quickly preview the material for that chapter and to make note of the learning objectives. The addition of page references makes this feature even more valuable.

■ Macro-to-Micro Illustrations

One of the challenges of learning anatomy is learning to "see" anatomical structures. Throughout this book you will find illustrations that provide an orientation icon indicating where in the human body a particular organ or structure is located. The orientation icon is then followed by (1) a large, clear painting of that structure, (2) a sectional view, and (3) a photomicrograph. In this fashion, the illustration provides you with a "macro-to-micro" view that should aid greatly in your understanding of the structure.

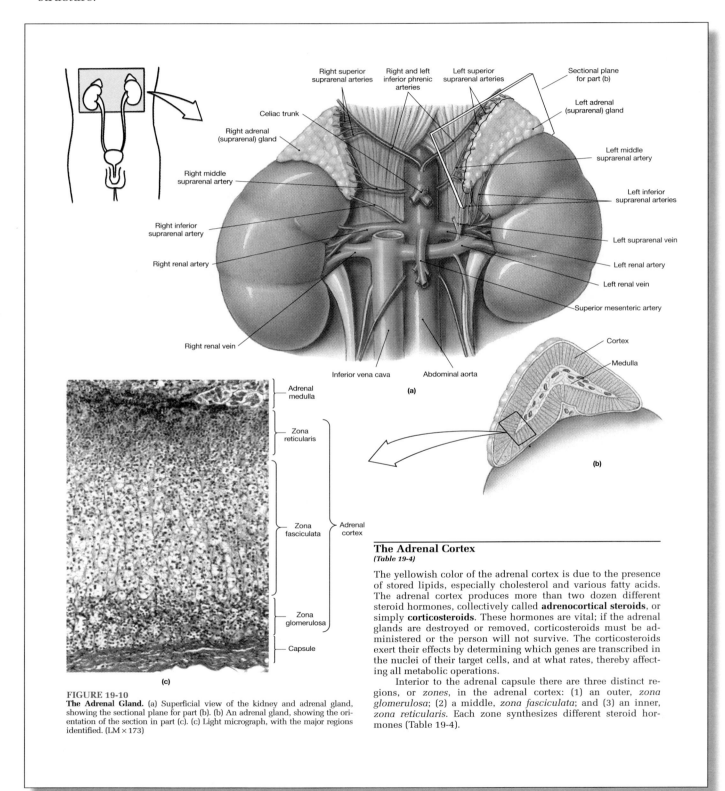

FIGURE 19-10
The Adrenal Gland. (a) Superficial view of the kidney and adrenal gland, showing the sectional plane for part (b). (b) An adrenal gland, showing the orientation of the section in part (c). (c) Light micrograph, with the major regions identified. (LM × 173)

The Adrenal Cortex
(Table 19-4)

The yellowish color of the adrenal cortex is due to the presence of stored lipids, especially cholesterol and various fatty acids. The adrenal cortex produces more than two dozen different steroid hormones, collectively called **adrenocortical steroids**, or simply **corticosteroids**. These hormones are vital; if the adrenal glands are destroyed or removed, corticosteroids must be administered or the person will not survive. The corticosteroids exert their effects by determining which genes are transcribed in the nuclei of their target cells, and at what rates, thereby affecting all metabolic operations.

Interior to the adrenal capsule there are three distinct regions, or *zones*, in the adrenal cortex: (1) an outer, *zona glomerulosa*; (2) a middle, *zona fasciculata*; and (3) an inner, *zona reticularis*. Each zone synthesizes different steroid hormones (Table 19-4).

■ Illustrations Combined with Cadaver Dissection Photos

An effectively rendered piece of art can communicate a lot, but we have tried to take your view of anatomy a step further by providing high quality cadaver photos, allowing you to compare a medical illustrator's interpretation with a photo of the actual structure. In addition to cadaver photos, some illustrations also include X-rays or MRI scans. Labels in plain type identify structures that are detailed in the accompanying text. Italicized labels identify anatomical landmarks that are not part of the system under discussion.

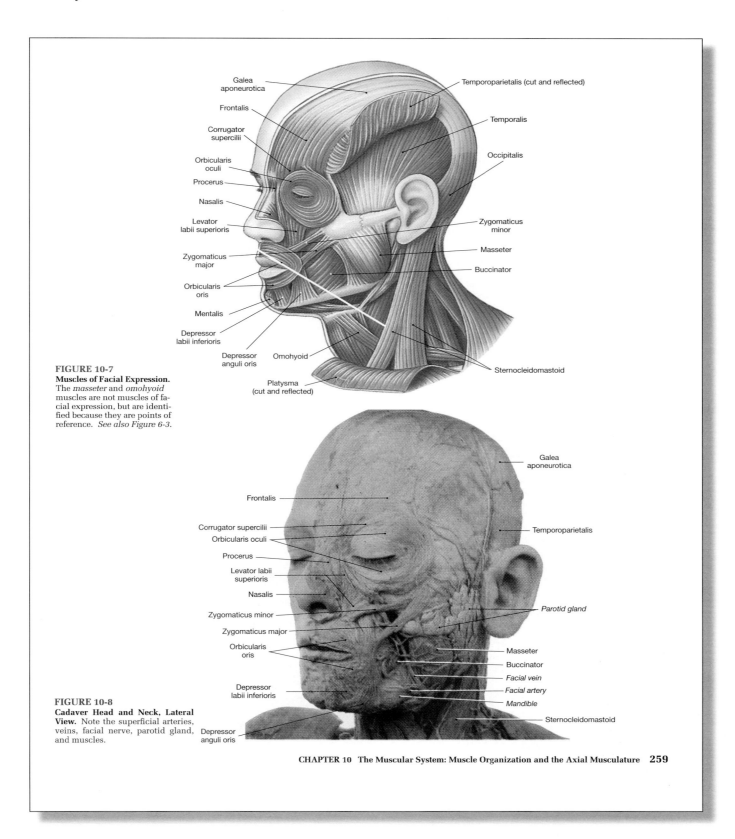

FIGURE 10-7
Muscles of Facial Expression. The *masseter* and *omohyoid* muscles are not muscles of facial expression, but are identified because they are points of reference. *See also Figure 6-3.*

FIGURE 10-8
Cadaver Head and Neck, Lateral View. Note the superficial arteries, veins, facial nerve, parotid gland, and muscles.

very living cell relies on the surrounding interstitial fluid for oxygen, nutrients, and waste disposal. Conditions in the interstitial fluid are kept stable through continuous exchange between the peripheral tissues and the circulating blood. The blood can help to maintain homeostasis only as long as it stays in motion. If blood remains stationary, its oxygen and nutrient supplies are quickly exhausted, its capacity to absorb wastes is soon saturated, and neither hormones nor white blood cells can reach their intended targets. Thus all of the functions of the cardiovascular system ultimately depend on the heart. This muscular organ beats approximately 100,000 times each day, forcing blood through the vessels of the circulatory system. Each year the heart pumps over 1.5 million gallons of blood, enough to fill 200 train tank cars.

For a practical demonstration of the heart's pumping abilities, turn on the faucet in the kitchen and open it all the way. To deliver an amount of water equal to the volume of blood pumped by the heart in an average lifetime, that faucet would have to be left on for at least 45 years. Equally remarkable, the volume of blood pumped by the heart can be varied over a wide range, from 5 to 30 liters per minute. The performance of the heart is closely monitored and finely regulated to ensure that conditions in the peripheral tissues remain within normal limits.

We begin this chapter by examining the structural features that enable the heart to perform so reliably, even in the face of widely varying physical demands. We will then consider the mechanisms that regulate cardiac activity to meet the body's ever-changing needs.

AN OVERVIEW OF THE CIRCULATORY SYSTEM
(Figure 21-1)

The circulatory system consists of a network of blood vessels that carry blood between the heart and peripheral tissues. This network can be subdivided into the **pulmonary circuit**, which carries blood to and from the exchange surfaces of the lungs, and the **systemic circuit**, which transports blood to and from the rest of the body. Each circuit begins and ends at the heart. **Arteries** transport blood away from the heart; **veins** return blood to the heart (Figure 21-1●). Blood travels through these circuits in sequence. For example, blood returning to the heart in the systemic veins must complete the pulmonary circuit before reentering the systemic arteries. **Capillaries** are small, thin-walled vessels that interconnect the smallest arteries and veins. Capillaries are called **exchange vessels** because their thin walls permit exchange of nutrients, dissolved gases, and waste products between the blood and surrounding tissues.

Despite its impressive workload, the heart is a small organ; your heart is roughly the size of your clenched fist. The heart contains four hollow muscular chambers, two associated with each circuit. The **right atrium** (Ā-trē-um; "chamber") receives blood from the systemic circuit, and the **right ventricle** (VEN-tri-KL; "little belly") discharges blood into the pulmonary circuit. The **left atrium** (plural *atria*) collects blood from the pulmonary circuit, and the **left ventricle** ejects blood into the systemic circuit. When the heart beats, the atria contract first, followed by the ventricles. The two ventricles contract at the same time and eject equal volumes of blood into the pulmonary and systemic circuits.

THE PERICARDIUM
(Figure 21-2)

The heart is located near the anterior chest wall, directly posterior to the sternum (Figure 21-2a●). The heart is surrounded by the **pericardial** (per-i-KAR-dē-al) **cavity**. The pericardial cavity is situated between the pleural cavities, in the mediastinum, which also contains the thymus, esophagus, and trachea. ∞ [p. 20] Figure 21-2b● is a sectional view that illustrates the position of the heart relative to other structures in the mediastinum.

The serous membrane lining the pericardial cavity is called the **pericardium** (Figure 21-2b–d●). To visualize the relationship between the heart and the pericardial cavity, imagine pushing your fist toward the center of a large balloon. The balloon represents the pericardium, and your fist is the heart. The pericardium can be subdivided into the *visceral pericardium* (the part of the balloon in contact with your fist) and the *parietal pericardium*. Your wrist, where the balloon folds back upon itself, corresponds to the base of the heart.

The **visceral pericardium**, or *epicardium*, covers the outer surface of the heart; the **parietal pericardium** lines the inner surface of the **pericardial sac** that surrounds the heart. Its outer layer, which contains abundant collagen fibers, is called the **fibrous pericardium**. The pericardial sac is further reinforced by a dense network of collagen fibers. At the base of the heart, they sta-

FIGURE 21-1
A Generalized View of the Pulmonary and Systemic Circuits. Blood flows through separate pulmonary and systemic circuits, driven by the pumping of the heart. Each circuit begins and ends at the heart and contains arteries, capillaries, and veins.

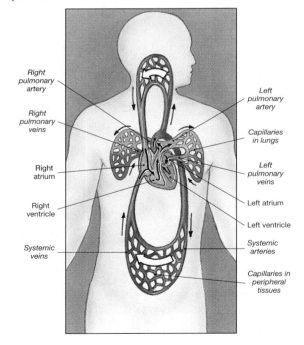

Right pulmonary artery

Left pulmonary artery

Right pulmonary veins

Capillaries in lungs

Right atrium

Left pulmonary veins

Left atrium

Right ventricle

Left ventricle

Systemic veins

Systemic arteries

Capillaries in peripheral tissues

■ **Tables**
Tables are used throughout the text to summarize concepts or categorize key information.

TABLE 11-3   Muscles That Move the Forearm and Hand

Muscle	Origin	Insertion	Action	Innervation
PRIMARY ACTION AT THE ELBOW				
Flexors				
Biceps brachii	*Short head* from the coracoid process; *long head* from the supraglenoid tubercle (both on the scapula)	Tuberosity of radius	Flexes and supinates forearm; flexes arm	Musculocutaneous nerve
Brachialis	Anterior, distal surface of humerus	Tuberosity of ulna	Flexes forearm	As above
Brachoradialis	Lateral epicondyle of humerus	Lateral aspect of styloid process of radius	As above	Radial nerve
Extensors				
Anconeus	Posterior surface of lateral epicondyle of humerus	Lateral margin of olecranon on ulna	Extends forearm, moves ulna laterally during pronation	As above
Triceps brachii				
lateral head	Superior, lateral margin of humerus	Olecranon process of ulna	Extends forearm	As above
long head	Infraglenoid tubercle of scapula	As above	As above	As above
medial head	Posterior surface of humerus inferior to radial groove	As above	As above	As above
PRONATORS/SUPINATORS				
Pronator				
quadratus	Medial surface of distal portion of ulna	Anterolateral surface of distal portion of radius	Pronates forearm	Median nerve
teres	Medial epicondyle of humerus and coronoid process of ulna	Distal lateral surface of radius	As above	As above
Supinator	Lateral epicondyle of humerus and ulna	Anterolateral surface of radius distal to the radial tuberosity	Supinates forearm	Radial nerve
PRIMARY ACTION AT THE HAND				
Flexors				
Flexor carpi radialis	Medial epicondyle of humerus	Bases of 2nd and 3rd metacarpal bones	Flexes and abducts hand	Median nerve
Flexor carpi ulnaris	Medial epicondyle of humerus; adjacent medial surface of olecranon and anteromedial portion of ulna	Pisiform, hamate, and base of 5th metacarpal bone	Flexes and adducts hand	Ulnar nerve
Palmaris longus	Medial epicondyle of humerus	Palmar aponeurosis and flexor retinaculum	Flexes hand	Median nerve
Extensors				
Extensor carpi radialis, longus	Lateral supracondylar ridge of humerus	Base of 2nd metacarpal	Extends and abducts hand	Radial nerve
brevis	Lateral epicondyle of humerus	Base of 3rd metacarpal	As above	As above
Extensor carpi ulnaris	Lateral epicondyle of humerus; adjacent dorsal surface of ulna	Base of 5th metacarpal	Extends and adducts hand	Deep radial nerve

The **pronator teres** and the **supinator** arise on both the humerus and forearm. They rotate the radius without flexing or extending the elbow. The **pronator quadratus** arises on the ulna and assists the pronator teres in opposing the actions of the supinator or biceps brachii. The muscles involved in pronation and supination can be seen in Figures 11-7c,d and 11-8d,e●, p. 289. Note the changes in orientation that occur as the pronator teres and pronator quadratus contract. During pronation the tendon of the biceps brachii rolls under the radius, and a bursa (Figure 8-10) prevents abrasion against the tendon. ↩ [p. 218]

As you study the muscles in Table 11-3, note that in general, extensor muscles lie along the dorsal and lateral surfaces of the forearm, and flexors are found on the ventral and medial surfaces. Many of the muscles that move the forearm and wrist can be seen from the body surface (Figures 11-7a and 11-8a●, p. 288).

Muscles That Move the Fingers
(Figures 11-7 to 11-11, Tables 11-4/11-5)

Several superficial and deep muscles of the forearm (Table 11-4, p. 290) perform flexion and extension of the fingers. Only the tendons of these muscles cross the articulation. These are relative-

■ **Concept Links**
Thoroughly understanding human anatomy requires you to remember and relate a number of terms and concepts to one another. Concept links provide a quick visual indicator and a page reference to signal you that it may be helpful to revisit a related concept from an earlier chapter or section of the book.

■ Embryology Summaries

Located throughout the text, these summaries highlight the developmental stages of significant organs, structures and systems. They are boxed and are designed to be relevant but isolated from the running text. Your instructor may or may not assign these summaries, but we wanted to make them available to you because they can function as an important resource to aid you in your study of human anatomy.

EMBRYOLOGY SUMMARY: INTRODUCTION TO THE DEVELOPMENT OF THE NERVOUS SYSTEM

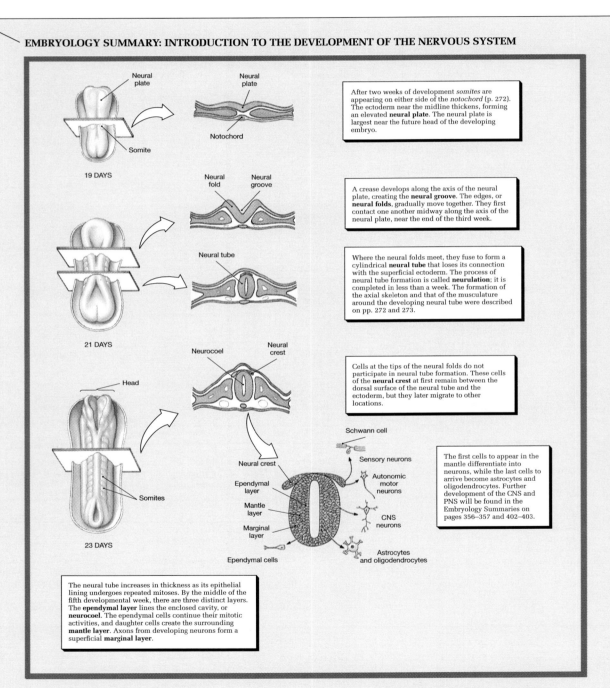

After two weeks of development *somites* are appearing on either side of the *notochord* (p. 272). The ectoderm near the midline thickens, forming an elevated **neural plate**. The neural plate is largest near the future head of the developing embryo.

A crease develops along the axis of the neural plate, creating the **neural groove**. The edges, or **neural folds**, gradually move together. They first contact one another midway along the axis of the neural plate, near the end of the third week.

Where the neural folds meet, they fuse to form a cylindrical **neural tube** that loses its connection with the superficial ectoderm. The process of neural tube formation is called **neurulation**; it is completed in less than a week. The formation of the axial skeleton and that of the musculature around the developing neural tube were described on pp. 272 and 273.

Cells at the tips of the neural folds do not participate in neural tube formation. These cells of the **neural crest** at first remain between the dorsal surface of the neural tube and the ectoderm, but they later migrate to other locations.

The first cells to appear in the mantle differentiate into neurons, while the last cells to arrive become astrocytes and oligodendrocytes. Further development of the CNS and PNS will be found in the Embryology Summaries on pages 356–357 and 402–403.

The neural tube increases in thickness as its epithelial lining undergoes repeated mitoses. By the middle of the fifth developmental week, there are three distinct layers. The **ependymal layer** lines the enclosed cavity, or **neurocoel**. The ependymal cells continue their mitotic activities, and daughter cells create the surrounding **mantle layer**. Axons from developing neurons form a superficial **marginal layer**.

The combination of pia mater and arachnoid is sometimes referred to as the *pia-arachnoid*. Along the length of the spinal cord, paired **denticulate ligaments** connect the pia-arachnoid to the dura mater. The denticulate ligaments originate along either side of the spinal cord, between the ventral and dorsal roots. These ligaments, which begin at the foramen magnum of the skull, help prevent side-to-side movement and inferior movement of the spinal cord. The connective tissue fibers of the pia mater continue from the inferior tip of the conus medullaris as the filum terminale. As noted above, the filum terminale blends into the coccygeal ligament; this arrangement prevents superior movement of the spinal cord.

■ CLINICAL BRIEF
Spinal Taps and Myelography

Tissue samples, or *biopsies*, are taken from many organs to assist in diagnosis. For example, when a liver or skin disorder is suspected, small plugs of tissue are removed and examined for signs of infection or cell damage or are used to identify the bacteria causing an infection. Unlike many other tissues, however, neural tissue consists largely of cells rather than extracellular fluids or fibers. Tissue samples are seldom removed for analysis because any extracted or damaged neurons will not be replaced. Instead, small volumes of cerebrospinal fluid (CSF) are collected and analyzed. CSF is intimately associated with the neural tissue of the CNS, and pathogens, cell debris, or metabolic wastes in the CNS will therefore be detectable in the CSF.

The withdrawal of cerebrospinal fluid, known as a **spinal tap**, must be done with care to avoid injuring the spinal cord. The adult spinal cord extends only as far as vertebra L_1 or L_2. Between vertebra L_2 and the sacrum, the meningeal layers remain intact, but they enclose only the relatively sturdy components of the cauda equina and a significant quantity of CSF. With the vertebral column flexed, a needle can be inserted between the lower lumbar vertebrae and into the subarachnoid spaces with minimal risk to the cauda equina. In this procedure, known as a **lumbar puncture**, 3–9 mℓ of fluid are taken from the subarachnoid space between vertebrae L_3 and L_4 (Figure 14-4a●). Spinal taps are performed when CNS infection is suspected or when diagnosing severe back pain, headaches, disc problems, and some types of strokes.

Myelography involves the introduction of radiopaque dyes into the CSF of the subarachnoid space. Because the dyes are opaque to X-rays, the CSF appears white on an X-ray photograph (Figure 14-4b●). Any tumors, inflammation, or adhesions that distort or divert CSF circulation will be shown in silhouette. In the event of severe infection, inflammation, or leukemia (cancer of the white blood cells), antibiotics, steroids, or anti-cancer drugs can be injected into the subarachnoid space.

FIGURE 14-4
Spinal Taps and Myelography. (a) The position of the lumbar puncture needle is in the subarachnoid space, between the nerves of the cauda equina. The needle has been inserted in the midline between the third and fourth lumbar vertebral spines, pointing at a superior angle toward the umbilicus. Once the needle correctly punctures the dura and enters the subarachnoid space, a sample of CSF may be obtained. (b) A myelogram—an X-ray photograph of the spinal cord after introduction of a radiopaque dye into the CSF—showing the cauda equina in the lower lumbar region.

The spinal meninges surround the dorsal and ventral roots within the intervertebral foramina. As seen in Figure 14-2●, the meningeal membranes are continuous with the connective tissues surrounding the spinal nerves and their peripheral branches.

√ Damage to which root of a spinal nerve would interfere with motor function?

√ Identify the location of the cerebrospinal fluid that surrounds the spinal cord.

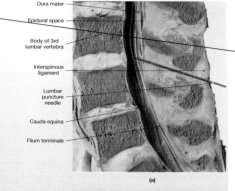

Dura mater
Epidural space
Body of 3rd lumbar vertebra
Interspinous ligament
Lumbar puncture needle
Cauda equina
Filum terminale

(a)

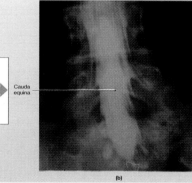

Cauda equina

(b)

■ Concept Checkpoints
This learning aid is designed to stop you for a moment and assess your understanding of the basic concepts just addressed in the previous few pages. Because concepts build upon one another, you need to master important material sequentially. If you are comfortable with the material, these checkpoints can be easily skipped. If you want to be sure that you know the material, we encourage you to pause and assess your understanding.

■ Clinical Briefs
Significant clinical considerations are titled and highlighted in a different typeface to distinguish them from the running narrative of the main text. Designed to enhance your understanding of normal anatomy by focusing upon disease, dysfunction, or injury, these Clinical Briefs are not only relevant but interesting, especially to those preparing for careers in the allied-health sciences.

■ Related Clinical Terms
A brief list of related clinical terms and their definitions is found toward the end of each chapter. Each of these terms is related to clinical information located within that chapter or in the Clinical Issues appendix. Page references or caduceus icons help you find the terms in context for further review or reference.

■ Clinical Appendix Reference
While not wanting to interrupt the flow of our text with too many clinical briefs but still wanting to provide you with ample clinical material, we chose to locate a substantial amount of the clinical material in the appendix. In an effort to help you find and relate this material, we have used the caduceus icon along with a title of the clinical material being referenced to signal you that relevant clinical information can be found in the appendix at the end of the book.

Related Clinical Terms

meningitis: An inflammation of the meningeal membranes. ⊤ *Spinal Meningitis [p. 759]*

epidural block: Regional anesthesia produced by the injection of an anesthetic into the epidural space near targeted spinal nerve roots. ⊤ *Spinal Anesthesia [p. 759]*

caudal anesthesia: Injection of anesthetics into the epidural space of the sacrum to paralyze lower abdominal and perineal structures. ⊤ *Spinal Anesthesia [p. 759]*

spinal tap: A procedure in which fluid is extracted from the subarachnoid space through a needle inserted between the vertebrae. *[p. 343]*

lumbar puncture: A spinal tap performed between adjacent lumbar vertebrae. *[p. 343]*

myelography: A diagnostic procedure in which a radiopaque dye is introduced into the cerebrospinal fluid in order to obtain an X-ray of the spinal cord. *[p. 343]*

spinal shock: A period of sensory and motor paralysis following any severe injury to the spinal cord. *[p. 345]*

quadriplegia: Paralysis involving loss of sensation and motor control of the upper and lower limbs. *[p. 345]*

paraplegia: Paralysis involving loss of motor control of the lower limbs. *[p. 345]*

nerve graft: Insertion of an intact section from a different peripheral nerve to bridge the gap between the cut ends of a damaged nerve and provide a route for axonal regeneration. ⊤ *Technology and Motor Paralysis [p. 759]*

functional electrical stimulation (FES): A technique for stimulating specific muscles and muscle groups using electrodes controlled by a computer. ⊤ *Technology and Motor Paralysis [p. 759]*

multiple sclerosis (skler-Ō-sis; *sklerosis*, hardness) **(MS):** A disease of the nervous system characterized by recurrent, often progressive incidents of demyelination affecting tracts in the brain and/or spinal cord. Common symptoms include partial loss of vision and problems with speech, balance, and general motor coordination. ⊤ *Multiple Sclerosis [p. 760]*

patellar reflex: The "knee jerk" reflex; often used to provide information about the related spinal segments. *[p. 360]*

Babinski sign (positive Babinski reflex): A spinal reflex in infants, consisting of a fanning of the toes, produced by stroking the foot on the side of the sole; in adults, a sign of CNS injury. ⊤ *Reflexes and Diagnostic Testing [p. 760]*

hyporeflexia: A condition where normal spinal reflexes are present but weak. ⊤ *Reflexes and Diagnostic Testing [p. 760]*

areflexia (ā-re-FLEK-sē-a; *a*-, without): The lack of normal reflex responses to stimuli. ⊤ *Reflexes and Diagnostic Testing [p. 760]*

hyperreflexia: Exaggerated reflex responses that may develop in some pathological states or following stimulation of spinal and cranial nuclei by higher centers. ⊤ *Reflexes and Diagnostic Testing [p. 760]*

CHAPTER SUMMARY AND REVIEW

STUDY OUTLINE   **Related Key Terms**

INTRODUCTION *[p. 339]*

1. The **central nervous system (CNS)** consists of the spinal cord and brain. In addition to relaying information to and from the brain, the spinal cord integrates and processes information on it...

Study Outline

The Chapter Summary and Review section of each chapter begins with a detailed Study Outline. This outline includes page and figure references to help you quickly find the material you want to review in more depth. In the column to the right of the outline you will find a list of Related Key Terms, terms that enjoy some association with the material found on the corresponding line(s) immediately to the left in the Study Outline.

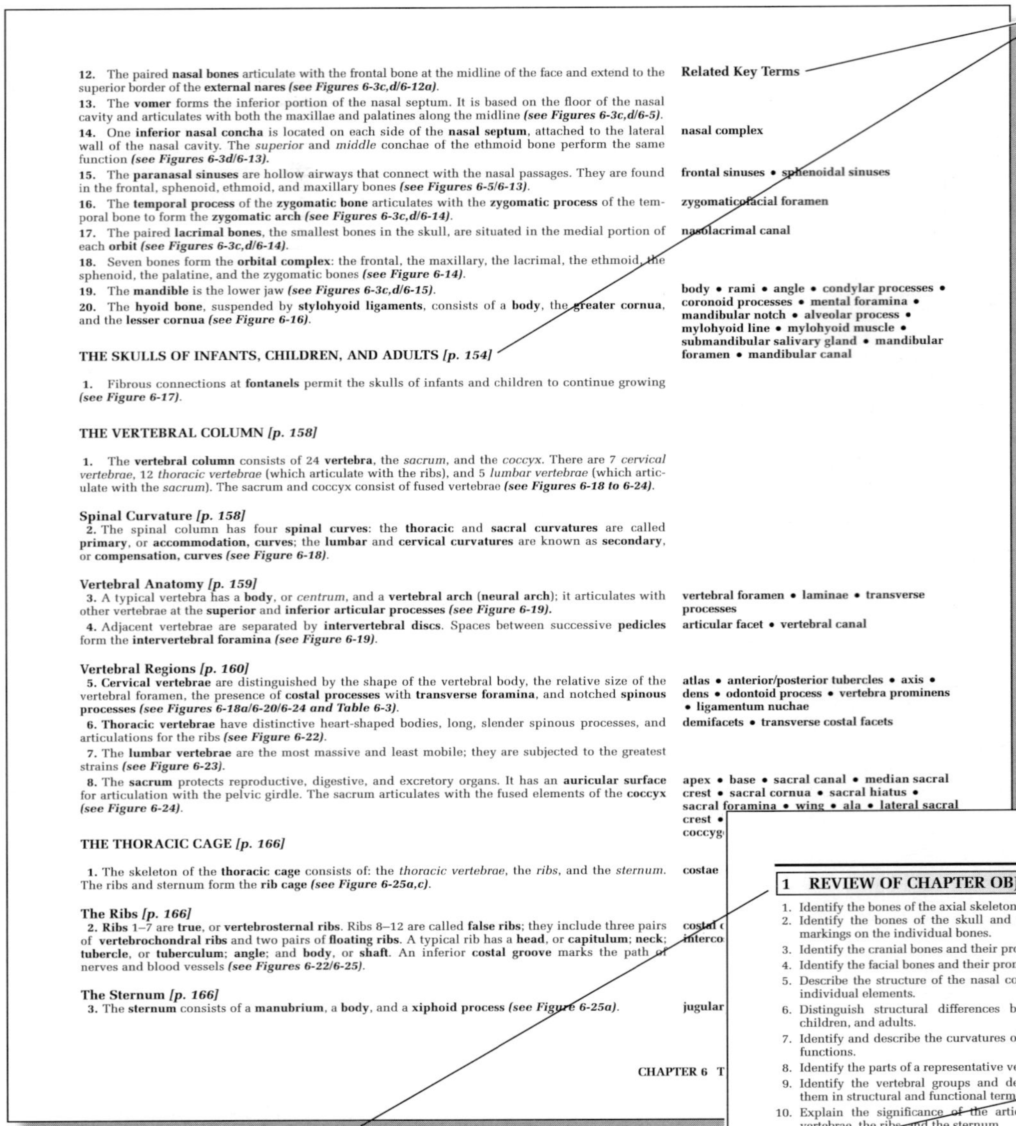

Related Key Terms

12. The paired **nasal bones** articulate with the frontal bone at the midline of the face and extend to the superior border of the **external nares** ***(see Figures 6-3c,d/6-12a)***.

13. The **vomer** forms the inferior portion of the nasal septum. It is based on the floor of the nasal cavity and articulates with both the maxillae and palatines along the midline ***(see Figures 6-3c,d/6-5)***.

14. One **inferior nasal concha** is located on each side of the **nasal septum**, attached to the lateral wall of the nasal cavity. The *superior* and *middle* conchae of the ethmoid bone perform the same function ***(see Figures 6-3d/6-13)***. — **nasal complex**

15. The **paranasal sinuses** are hollow airways that connect with the nasal passages. They are found in the frontal, sphenoid, ethmoid, and maxillary bones ***(see Figures 6-5/6-13)***. — **frontal sinuses • sphenoidal sinuses**

16. The **temporal process** of the **zygomatic bone** articulates with the **zygomatic process** of the temporal bone to form the **zygomatic arch** ***(see Figures 6-3c,d/6-14)***. — **zygomaticofacial foramen**

17. The paired **lacrimal bones**, the smallest bones in the skull, are situated in the medial portion of each **orbit** ***(see Figures 6-3c,d/6-14)***. — **nasolacrimal canal**

18. Seven bones form the **orbital complex**: the frontal, the maxillary, the lacrimal, the ethmoid, the sphenoid, the palatine, and the zygomatic bones ***(see Figure 6-14)***.

19. The **mandible** is the lower jaw ***(see Figures 6-3c,d/6-15)***. — **body • rami • angle • condylar processes • coronoid processes • mental foramina • mandibular notch • alveolar process • mylohyoid line • mylohyoid muscle • submandibular salivary gland • mandibular foramen • mandibular canal**

20. The **hyoid bone**, suspended by **stylohyoid ligaments**, consists of a **body**, the **greater cornua**, and the **lesser cornua** ***(see Figure 6-16)***.

THE SKULLS OF INFANTS, CHILDREN, AND ADULTS *[p. 154]*

1. Fibrous connections at **fontanels** permit the skulls of infants and children to continue growing ***(see Figure 6-17)***.

THE VERTEBRAL COLUMN *[p. 158]*

1. The **vertebral column** consists of 24 **vertebra**, the *sacrum*, and the *coccyx*. There are 7 *cervical vertebrae*, 12 *thoracic vertebrae* (which articulate with the ribs), and 5 *lumbar vertebrae* (which articulate with the *sacrum*). The sacrum and coccyx consist of fused vertebrae ***(see Figures 6-18 to 6-24)***.

Spinal Curvature *[p. 158]*

2. The spinal column has four **spinal curves**: the **thoracic** and **sacral curvatures** are called **primary**, or **accommodation, curves**; the **lumbar** and **cervical curvatures** are known as **secondary**, or **compensation, curves** ***(see Figure 6-18)***.

Vertebral Anatomy *[p. 159]*

3. A typical vertebra has a **body**, or *centrum*, and a **vertebral arch** (**neural arch**); it articulates with other vertebrae at the **superior** and **inferior articular processes** ***(see Figure 6-19)***. — **vertebral foramen • laminae • transverse processes**

4. Adjacent vertebrae are separated by **intervertebral discs**. Spaces between successive **pedicles** form the **intervertebral foramina** ***(see Figure 6-19)***. — **articular facet • vertebral canal**

Vertebral Regions *[p. 160]*

5. Cervical vertebrae are distinguished by the shape of the vertebral body, the relative size of the vertebral foramen, the presence of **costal processes** with **transverse foramina**, and notched **spinous processes** ***(see Figures 6-18a/6-20/6-24 and Table 6-3)***. — **atlas • anterior/posterior tubercles • axis • dens • odontoid process • vertebra prominens • ligamentum nuchae**

6. Thoracic vertebrae have distinctive heart-shaped bodies, long, slender spinous processes, and articulations for the ribs ***(see Figure 6-22)***. — **demifacets • transverse costal facets**

7. The **lumbar vertebrae** are the most massive and least mobile; they are subjected to the greatest strains ***(see Figure 6-23)***.

8. The **sacrum** protects reproductive, digestive, and excretory organs. It has an **auricular surface** for articulation with the pelvic girdle. The sacrum articulates with the fused elements of the **coccyx** ***(see Figure 6-24)***. — **apex • base • sacral canal • median sacral crest • sacral cornua • sacral hiatus • sacral foramina • wing • ala • lateral sacral crest • coccyg**

THE THORACIC CAGE *[p. 166]*

1. The skeleton of the **thoracic cage** consists of: the *thoracic vertebrae*, the *ribs*, and the *sternum*. The ribs and sternum form the **rib cage** ***(see Figure 6-25a,c)***. — **costae**

The Ribs *[p. 166]*

2. Ribs 1–7 are **true**, or **vertebrosternal ribs**. Ribs 8–12 are called **false ribs**; they include three pairs of **vertebrochondral ribs** and two pairs of **floating ribs**. A typical rib has a **head**, or **capitulum**; **neck**; **tubercle**, or **tuberculum**; **angle**; and **body**, or **shaft**. An inferior **costal groove** marks the path of nerves and blood vessels ***(see Figures 6-22/6-25)***. — **costal c interco**

The Sternum *[p. 166]*

3. The **sternum** consists of a **manubrium**, a **body**, and a **xiphoid process** ***(see Figure 6-25a)***. — **jugular**

CHAPTER 6 T

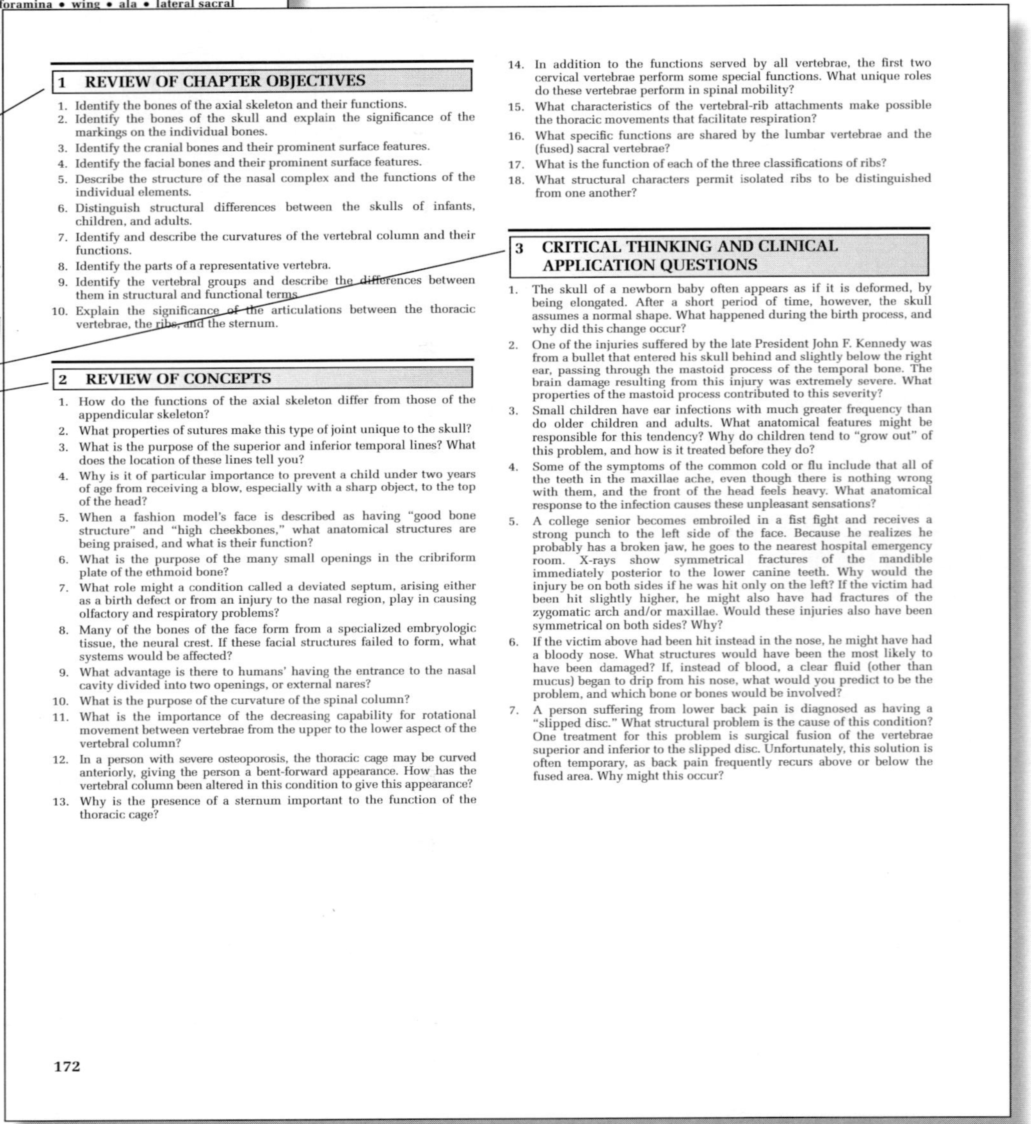

1 REVIEW OF CHAPTER OBJECTIVES

1. Identify the bones of the axial skeleton and their functions.
2. Identify the bones of the skull and explain the significance of the markings on the individual bones.
3. Identify the cranial bones and their prominent surface features.
4. Identify the facial bones and their prominent surface features.
5. Describe the structure of the nasal complex and the functions of the individual elements.
6. Distinguish structural differences between the skulls of infants, children, and adults.
7. Identify and describe the curvatures of the vertebral column and their functions.
8. Identify the parts of a representative vertebra.
9. Identify the vertebral groups and describe the differences between them in structural and functional terms.
10. Explain the significance of the articulations between the thoracic vertebrae, the ribs, and the sternum.

2 REVIEW OF CONCEPTS

1. How do the functions of the axial skeleton differ from those of the appendicular skeleton?
2. What properties of sutures make this type of joint unique to the skull?
3. What is the purpose of the superior and inferior temporal lines? What does the location of these lines tell you?
4. Why is it of particular importance to prevent a child under two years of age from receiving a blow, especially with a sharp object, to the top of the head?
5. When a fashion model's face is described as having "good bone structure" and "high cheekbones," what anatomical structures are being praised, and what is their function?
6. What is the purpose of the many small openings in the cribriform plate of the ethmoid bone?
7. What role might a condition called a deviated septum, arising either as a birth defect or from an injury to the nasal region, play in causing olfactory and respiratory problems?
8. Many of the bones of the face form from a specialized embryologic tissue, the neural crest. If these facial structures failed to form, what systems would be affected?
9. What advantage is there to humans' having the entrance to the nasal cavity divided into two openings, or external nares?
10. What is the purpose of the curvature of the spinal column?
11. What is the importance of the decreasing capability for rotational movement between vertebrae from the upper to the lower aspect of the vertebral column?
12. In a person with severe osteoporosis, the thoracic cage may be curved anteriorly, giving the person a bent-forward appearance. How has the vertebral column been altered in this condition to give this appearance?
13. Why is the presence of a sternum important to the function of the thoracic cage?
14. In addition to the functions served by all vertebrae, the first two cervical vertebrae perform some special functions. What unique roles do these vertebrae perform in spinal mobility?
15. What characteristics of the vertebral-rib attachments make possible the thoracic movements that facilitate respiration?
16. What specific functions are shared by the lumbar vertebrae and the (fused) sacral vertebrae?
17. What is the function of each of the three classifications of ribs?
18. What structural characters permit isolated ribs to be distinguished from one another?

3 CRITICAL THINKING AND CLINICAL APPLICATION QUESTIONS

1. The skull of a newborn baby often appears as if it is deformed, by being elongated. After a short period of time, however, the skull assumes a normal shape. What happened during the birth process, and why did this change occur?
2. One of the injuries suffered by the late President John F. Kennedy was from a bullet that entered his skull behind and slightly below the right ear, passing through the mastoid process of the temporal bone. The brain damage resulting from this injury was extremely severe. What properties of the mastoid process contributed to this severity?
3. Small children have ear infections with much greater frequency than do older children and adults. What anatomical features might be responsible for this tendency? Why do children tend to "grow out" of this problem, and how is it treated before they do?
4. Some of the symptoms of the common cold or flu include that all of the teeth in the maxillae ache, even though there is nothing wrong with them, and the front of the head feels heavy. What anatomical response to the infection causes these unpleasant sensations?
5. A college senior becomes embroiled in a fist fight and receives a strong punch to the left side of the face. Because he realizes he probably has a broken jaw, he goes to the nearest hospital emergency room. X-rays show symmetrical fractures of the mandible immediately posterior to the lower canine teeth. Why would the injury be on both sides if he was hit only on the left? If the victim had been hit slightly higher, he might also have had fractures of the zygomatic arch and/or maxillae. Would these injuries also have been symmetrical on both sides? Why?
6. If the victim above had been hit instead in the nose, he might have had a bloody nose. What structures would have been the most likely to have been damaged? If, instead of blood, a clear fluid (other than mucus) began to drip from his nose, what would you predict to be the problem, and which bone or bones would be involved?
7. A person suffering from lower back pain is diagnosed as having a "slipped disc." What structural problem is the cause of this condition? One treatment for this problem is surgical fusion of the vertebrae superior and inferior to the slipped disc. Unfortunately, this solution is often temporary, as back pain frequently recurs above or below the fused area. Why might this occur?

172

End of Chapter Three-Level Review System

Each chapter ends with a three-level questioning scheme. **Level One** is a basic **Review of Chapter Objectives** in which the objectives that were integrated into the opening chapter outline are restated. You should review and attempt to master these basic objectives. **Level Two, Review of Concepts,** requires slightly more advanced thinking skills since you are asked to synthesize or apply concepts by answering a list of questions. **Level Three** are **Critical Thinking and Clinical Application Questions** that ask you to consider and respond to real-world or clinical scenarios. The three-level review system has been designed to help you grow intellectually from a basic level in which you master terms and concepts to more advanced levels in which critical thinking skills are promoted. For those students who like this three-level review system but desire a greater number and variety of questions, we encourage you to obtain the Study Guide to accompany this text.

xxxii

1

An Introduction to Anatomy

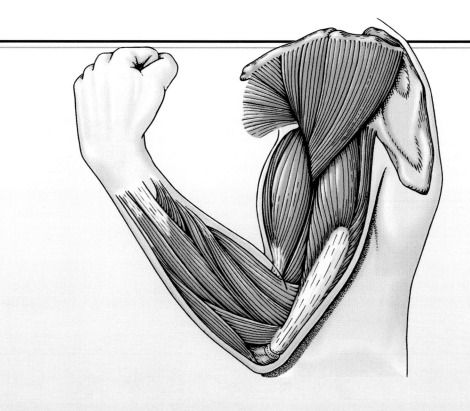

CHAPTER OUTLINE AND OBJECTIVES

The word *anatomy* has its origins in Greek, as do many other anatomical terms and phrases. A literal translation would be "to cut open." **Anatomy** is the study of internal and external structures and the physical relationships between body parts. Anatomical information provides clues about probable functions, and physiological mechanisms can be explained only in terms of the underlying anatomy. This observation leads to a very important concept: *All specific functions are performed by specific structures.*

The link between structure and function is always present, but not always understood. For example, the superficial anatomy of the heart was clearly described in the fifteenth century, but almost 200 years passed before the pumping action of the heart was demonstrated. On the other hand, many important cell functions were recognized decades before the electron microscope revealed the anatomical basis for those functions.

This text will discuss the anatomical structures and functions that make human life possible. This information will prepare you for more advanced courses in anatomy, physiology, and related subjects and will enable you to make informed decisions about your personal health.

MICROSCOPIC ANATOMY
(Figure 1-1)

Microscopic anatomy considers structures that cannot be seen without magnification. The boundaries of microscopic anatomy, or *fine anatomy*, are established by the limits of the equipment used (Figure 1-1●). A simple hand lens shows details that escape the naked eye, while an electron microscope demonstrates structural details that are only a few ten-thousandths of a millimeter across. As we proceed through the text, we will be considering details at all levels, from macroscopic to microscopic. (Readers unfamiliar with the terms used to describe measurements and weights over this size range should consult the reference tables in Appendix II.)

Microscopic anatomy can be subdivided into specialties that consider features within a characteristic range of sizes. **Cytology** (sī-TOL-ō-jē) analyzes the internal structure of individual cells, the smallest units of life. Living cells are composed of chemical substances in various combinations, and our lives depend on the chemical processes occurring in the trillions of cells that form our body.

Histology (his-TOL-ō-jē) takes a broader perspective and examines **tissues**, groups of specialized cells and cell products that work together to perform specific functions. The cells in the human body can be assigned to four major tissue types, and these tissues are the focus of Chapter 3. Tissues in combination form **organs** such as the heart, kidney, liver, or brain. Organs are anatomical units that have multiple functions. Many organs are easily examined without a microscope, and at the organ level we cross the boundary into gross anatomy.

FIGURE 1-1
The Study of Anatomy at Different Scales. The amount of detail recognized depends on the method of study and the degree of magnification.

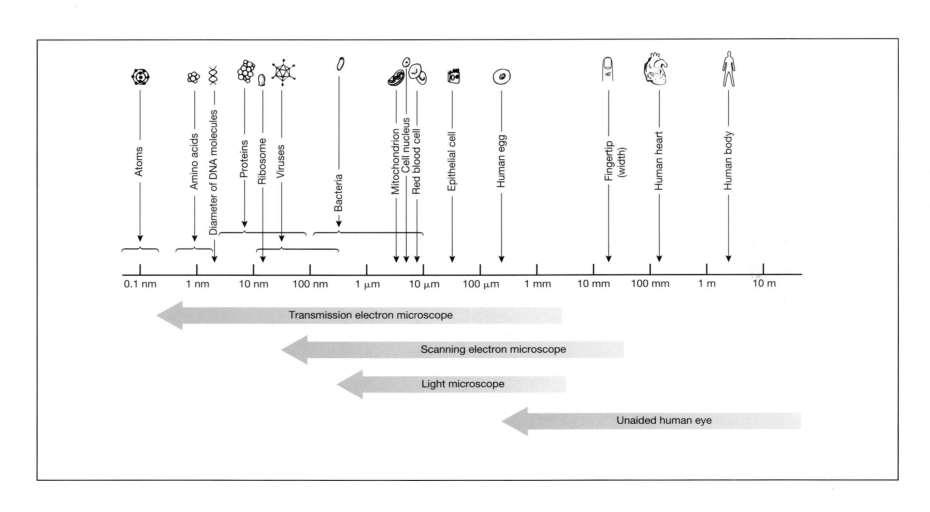

GROSS ANATOMY

Gross anatomy, or **macroscopic anatomy**, considers relatively large structures and features visible with the unaided eye. There are many ways to approach gross anatomy:

- **Surface anatomy** refers to the study of general form and superficial anatomical markings.
- **Regional anatomy** considers all of the superficial and internal features in a specific area of the body, such as the head, neck, or trunk. Advanced courses in anatomy often stress a regional approach because it emphasizes the spatial relationships between structures already familiar to the students.
- **Systemic anatomy** considers the structure of major *organ systems*, such as the skeletal or muscular system. Organ systems are groups of organs that function together to produce coordinated effects. For example, the heart, blood, and blood vessels form the *cardiovascular system*, which distributes oxygen and nutrients throughout the body. There are 11 organ systems in the human body, and they will be introduced later in the chapter. Intro-

ductory texts in anatomy, including this one, use a systemic approach because it provides a framework for organizing information about important structural and functional patterns.

OTHER PERSPECTIVES ON ANATOMY
(Figure 1-2)

There are other anatomical specialties that will be encountered in this text.

- **Developmental anatomy** examines the changes in form that occur during the period between conception and physical maturity. Because it considers anatomical structures over such a broad range of sizes (from a single cell to an adult human), developmental anatomy uses techniques that are similar to those used in both microscopic and gross anatomy. Developmental anatomy is important in medicine because many structural abnormalities can result from errors that occur during development. The most extensive structural changes occur during the first 2 months of development, and **embryology** (em-brē-OL-ō-jē) is the study of these early developmental processes.

FIGURE 1-2

Comparative Anatomy. Human beings are classified as *vertebrates*, a group that also includes animals as different in appearance as fish, snakes, and cats. Despite their differences, all vertebrates share a basic pattern of anatomical organization that differs from that of other animals.

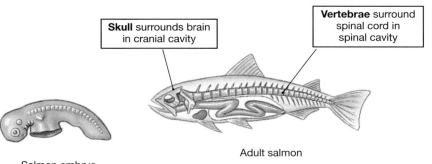

Skull surrounds brain in cranial cavity

Vertebrae surround spinal cord in spinal cavity

Adult salmon

Salmon embryo (bony fish)

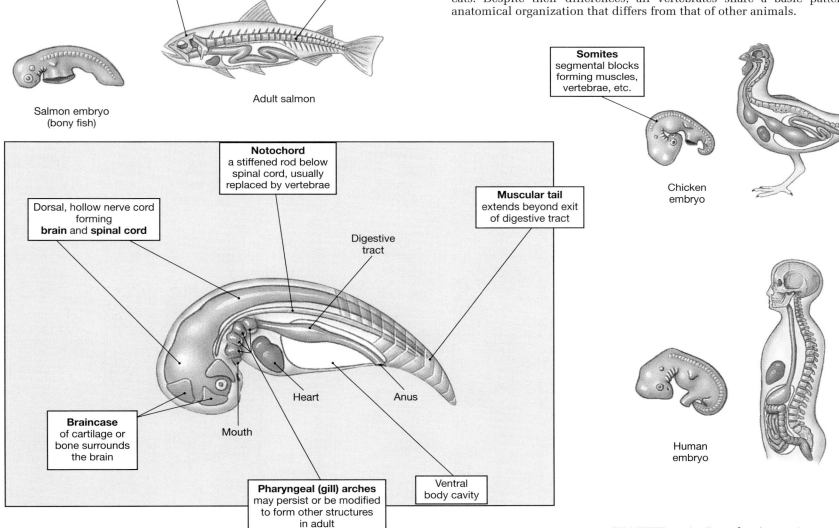

Somites — segmental blocks forming muscles, vertebrae, etc.

Chicken embryo

Notochord — a stiffened rod below spinal cord, usually replaced by vertebrae

Dorsal, hollow nerve cord forming **brain** and **spinal cord**

Muscular tail — extends beyond exit of digestive tract

Digestive tract

Braincase — of cartilage or bone surrounds the brain

Heart

Anus

Mouth

Pharyngeal (gill) arches — may persist or be modified to form other structures in adult

Ventral body cavity

Human embryo

- **Comparative anatomy** considers the anatomical organization of different types of animals. Observed similarities may reflect evolutionary relationships. Human beings, lizards, and sharks are all called *vertebrates* because they share a combination of anatomical features that is not found in any other group of animals. For example, each of the animals included in this group has a spinal column composed of individual elements, called *vertebrae*. Comparative anatomy uses techniques of gross and microscopic anatomy. Information on developmental anatomy is also very important, because closely related animals typically go through very similar developmental stages (Figure 1-2●).

Several other gross anatomical specialties are important in medical diagnosis.

- **Medical anatomy** focuses on anatomical features that may undergo characteristic changes during illness.
- **Radiographic anatomy** involves the study of anatomical structures as they are visualized by X-rays, ultrasound scans, or other specialized procedures performed on an intact body.
- **Surgical anatomy** studies anatomical landmarks important for surgical procedures.

LEVELS OF ORGANIZATION
(Figures 1-3/1-4)

Our study of the human body will begin with an overview of microscopic anatomy and then proceed to the gross and microscopic anatomy of each organ system. When considering events from the microscopic to macroscopic scales we are examining several interdependent *levels of organization*.

We begin at the *chemical* or *molecular level* of organization. The human body consists of over a dozen different elements, but four of them (hydrogen, oxygen, carbon, and nitrogen) account for more than 99% of the total number of atoms (Figure 1-3a●). At the chemical level, atoms interact to form compounds with distinctive properties. The major classes of compounds in the human body are indicated in Figure 1-3b●.

Figure 1-4● presents an example of the relationships between the chemical level and higher levels of organization. *Cells* are the smallest living units in the body. Each cell contains internal structures, called *organelles*, that are composed of complex chemicals. A discussion of cell structures and functions focuses attention on the *cellular level of organization*. Cell structure and the function of the major organelles will be presented in Chapter 2. In Figure 1-4, chemical interactions produce complex proteins within a *muscle cell* in the heart. Muscle cells are unusual because they can contract powerfully, shortening along their longitudinal axis.

Heart muscle cells are tied together to form a distinctive *muscle tissue*, an example of the *tissue level of organization*. Layers of muscle tissue form the bulk of the wall of the heart, a hollow, three-dimensional organ. We are now at the *organ level of organization*.

Normal functioning of the heart depends on events at the chemical, cellular, tissue, and organ levels of organization. Muscle contractions involve chemical interactions inside individual muscle cells. Because adjacent muscle cells are linked together in cardiac muscle tissue, their contractions are coordinated, producing a heartbeat. When that beat occurs, the internal anatomy of the organ enables it to function as a pump.

Each time it contracts, the heart pushes blood into the *circulatory system*, a network of blood vessels. Together the heart, blood, and circulatory system form an *organ system*, the *cardiovascular system*.

Each level of organization is totally dependent on the others. For example, damage at the cellular, tissue, or organ level may affect the entire system. A chemical change in heart muscle cells can cause abnormal contractions or even stop the heartbeat. Physical damage to the muscle tissue, as in a chest wound, can make the heart ineffective even when most of the heart muscle cells are intact and uninjured. An inherited abnormality in heart structure can make it an ineffective pump, although the muscle cells and muscle tissue are perfectly normal.

Finally, it should be noted that something that affects the *system* will ultimately affect all of its components. For example, the heart may not be able to pump blood effectively after a massive blood loss. If the heart cannot pump and blood cannot flow, oxygen and nutrients cannot be distributed. In a very short time, the tissue begins to break down as heart muscle cells die from oxygen and nutrient starvation.

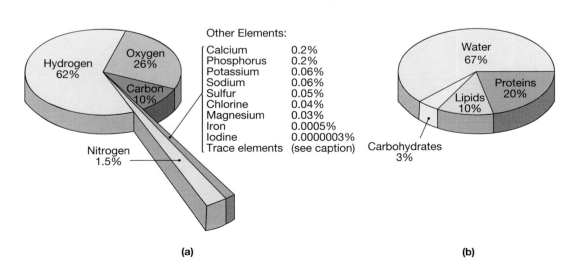

Other Elements:

Calcium	0.2%
Phosphorus	0.2%
Potassium	0.06%
Sodium	0.06%
Sulfur	0.05%
Chlorine	0.04%
Magnesium	0.03%
Iron	0.0005%
Iodine	0.0000003%
Trace elements	(see caption)

Hydrogen 62%
Oxygen 26%
Carbon 10%
Nitrogen 1.5%

Water 67%
Proteins 20%
Lipids 10%
Carbohydrates 3%

(a)

(b)

FIGURE 1-3

Composition of the Human Body at the Chemical Level of Organization. (a) Elemental composition of the human body. Trace elements include silicon, fluorine, copper, manganese, zinc, selenium, cobalt, molybdenum, cadmium, chromium, tin, aluminum, and boron. (b) Molecular composition of the human body.

FIGURE 1-4

Levels of Organization. Interacting atoms form molecules that combine in the protein fibers of heart muscle cells. These cells interlock, creating heart muscle tissue that constitutes most of the walls of a three-dimensional organ, the heart. The heart is one component of the cardiovascular system, which also includes the blood and blood vessels. All of the organ systems combine to create an organism, a living human being.

Of course these changes will not be restricted to the cardiovascular system; all of the cells, tissues, and organs in the body will be damaged. This observation brings us to another higher level of organization, that of the *organism*, in this case a human being. This level reflects the interactions among organ systems. All are vital; every system must be working properly and in harmony with every other system, or survival will be impossible. When those systems are functioning normally, the characteristics of the internal environment will be relatively stable at all levels. This vital state of affairs is called **homeostasis** (hō-mē-ō-STĀ-sis; *homeo*, unchanging + *stasis*, standing).

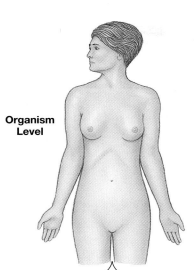

Organism Level

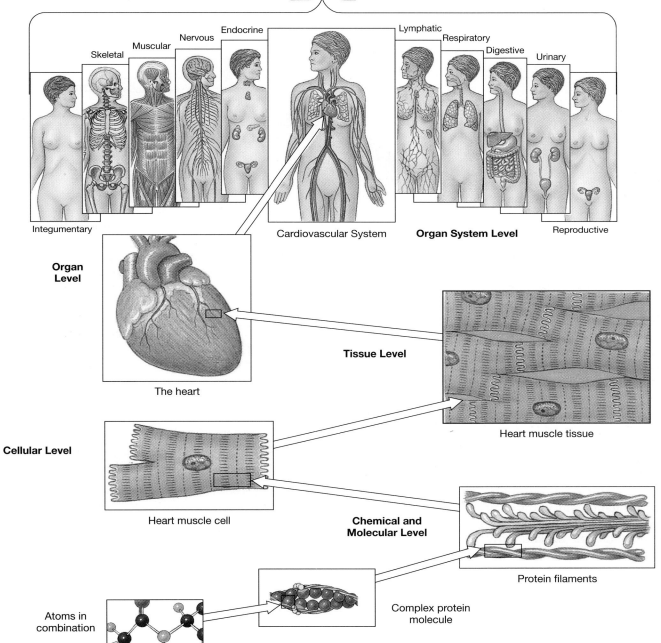

Skeletal   Muscular   Nervous   Endocrine   Lymphatic   Respiratory   Digestive   Urinary

Integumentary

Cardiovascular System

Organ System Level

Reproductive

Organ Level

The heart

Tissue Level

Heart muscle tissue

Cellular Level

Heart muscle cell

Chemical and Molecular Level

Protein filaments

Atoms in combination

Complex protein molecule

AN INTRODUCTION TO ORGAN SYSTEMS
(Figures 1-5/1-6)

All living organisms perform the same basic functions:

- **Responsiveness:** Organisms respond to changes in their immediate environment; this property is also called irritability. You move your hand away from a hot stove, your dog barks at approaching strangers, fish are scared by loud noises, and amoebas glide toward potential prey. Organisms also make longer-term changes as they adjust to their environments. For example, an animal may grow a heavier coat as winter approaches, or migrate to a warmer climate. The capacity to make such adjustments is termed *adaptability*.

- **Growth and Differentiation:** Over a lifetime, organisms grow larger, increasing in size through an increase in the size or number of their cells. In multicellular organisms, the individual cells become specialized to perform particular functions. This specialization is called **differentiation**. Growth and differentiation often produce changes in form and function. For example, the anatomical proportions and physiological capabilities of an adult human are quite different from those of an infant.

- **Reproduction:** Organisms reproduce, creating subsequent generations of similar organisms.

- **Movement:** Organisms are capable of producing movement, which may be internal (transporting food, blood, or other materials inside the body) or external (moving through the environment).

- **Metabolism and Excretion:** Organisms rely on complex chemical reactions to provide the energy for responsiveness, growth, reproduction, and movement. They must also synthesize complex chemicals, such as proteins. The term *metabolism* refers to all the chemical operations under way in the body. Normal metabolic operations require the *absorption* of materials from the environment. To generate energy efficiently, most cells require various nutrients, as well as oxygen, an atmospheric gas. The term **respiration** refers to the absorption, transport, and use of oxygen by cells. Metabolic operations often generate unneeded or potentially harmful waste products that must be eliminated through the process of **excretion**.

Several additional functions can be distinguished when you consider animals as complex as fish, cats, or human beings. For very small organisms, absorption, respiration, and excretion involve the movement of materials through exposed surfaces. But creatures larger than a few millimeters seldom absorb nutrients directly from their environment. For example, human beings cannot absorb steaks, apples, or ice cream without processing them first. That processing, called **digestion**, occurs in specialized areas where complex foods are broken down into simpler components that can be absorbed easily. Respiration and excretion are also more complicated for large organisms. Humans have specialized structures responsible for gas exchange (lungs) and waste elimination (kidneys). Finally, because absorption, respiration, and excretion are performed in different portions of the body, there must be an internal transportation system, or **cardiovascular system**.

The organ systems of the human body must perform these vital functions to maintain homeostasis. Figure 1-5● provides an overview of the 11 organ systems in the human body and indicates their size relative to the body as a whole. Figure 1-6● introduces the major organs in each system.

√ A histologist investigates structures at what level of organization?

√ Which level(s) of organization would a gross anatomist investigate?

FIGURE 1-5
An Introduction to Organ Systems

Organ System		Major Functions
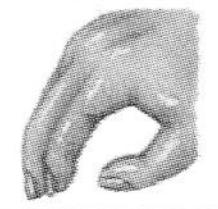	Integumentary system	Protection from environmental hazards, temperature control
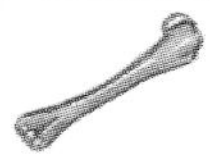	Skeletal system	Support, protection of soft tissues, mineral storage, blood formation
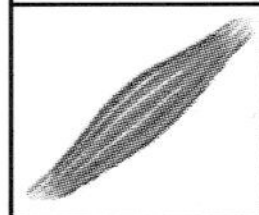	Muscular system	Locomotion, support, heat production
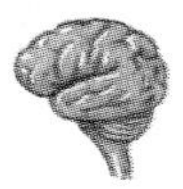	Nervous system	Directing immediate responses to stimuli, usually by coordinating the activities of other organ systems
	Endocrine system	Directing long-term changes in the activities of other organ systems
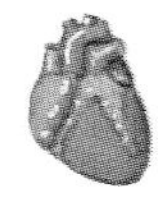	Cardiovascular system	Internal transport of cells and dissolved materials, including nutrients, wastes, and gases
	Lymphatic system	Defense against infection and disease
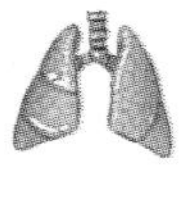	Respiratory system	Delivery of air to sites where gas exchange can occur between the air and circulating blood
	Digestive system	Processing of food and absorption of nutrients, minerals, vitamins, and water
	Urinary system	Elimination of excess water, salts, and waste products
	Reproductive system	Production of sex cells and hormones

FIGURE 1-6

The Organ Systems of the Human Body. (a) Integumentary system (skin). (b) Skeletal system. (c) Muscular system. (d) Nervous system. (e) Endocrine system. (f) Cardiovascular system. (g) Lymphatic system. (h) Respiratory system. (i) Digestive system. (j) Urinary system. (k) Male reproductive system. (l) Female reproductive system.

FIGURE 1-6a The Integumentary System

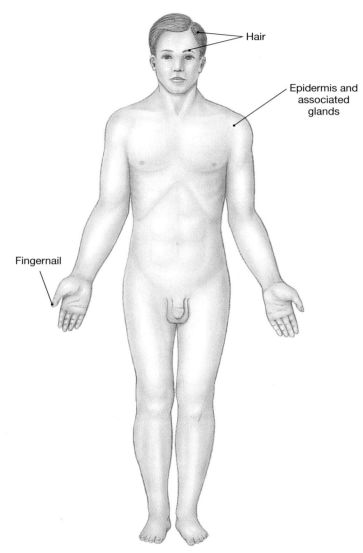

Organ	Primary Functions
CUTANEOUS MEMBRANE	
Epidermis	Protects underlying tissues
Dermis	Nourishes epidermis, provides strength
HAIR FOLLICLES	Produce hair
Hairs	Provide sensation, provide some protection for head
Sebaceous glands	Secrete lipid coating that lubricates hair shaft
SWEAT GLANDS	Produce perspiration for evaporative cooling
NAILS	Protect and stiffen distal tips of digits
SENSORY RECEPTORS	Provide sensations of touch, pressure, temperature, pain
SUBCUTANEOUS LAYER	Stores lipids, attaches skin to deeper structures

FIGURE 1-6b The Skeletal System

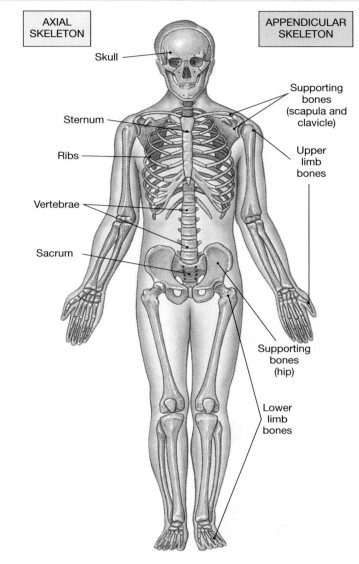

Organ	Primary Functions
BONES (206), CARTILAGES, AND LIGAMENTS	Support, protect soft tissues, store minerals
Axial skeleton	Protects brain, spinal cord, sense organs, and soft tissues of chest cavity; supports the body weight over the lower limbs
Appendicular skeleton	Internal support and positioning of the limbs, supporting and moving axial skeleton
Bone marrow	Primary site of blood cell production

FIGURE 1-6c   The Muscular System

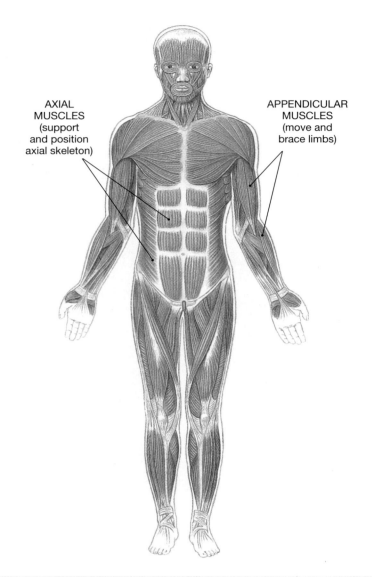

AXIAL MUSCLES (support and position axial skeleton)

APPENDICULAR MUSCLES (move and brace limbs)

Organ	Primary Functions
SKELETAL MUSCLES (700)	Provide skeletal movement, control entrances and exits of digestive tract, heat production, support skeletal position, protect soft tissues
TENDONS, APONEUROSES	Harness forces of contraction to perform specific tasks

FIGURE 1-6d   The Nervous System

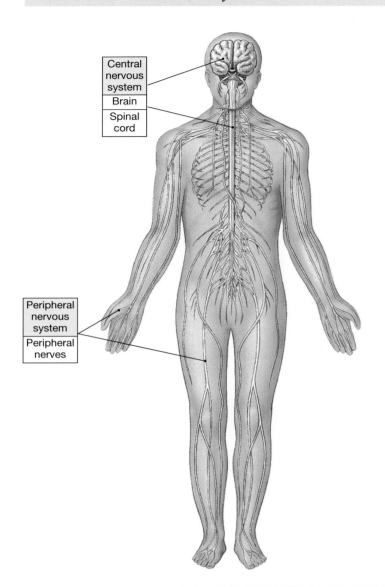

Central nervous system
Brain
Spinal cord

Peripheral nervous system
Peripheral nerves

Organ	Primary Functions
CENTRAL NERVOUS SYSTEM (CNS)	Control center for nervous system: processes information, provides short-term control over activities of other systems
Brain	Performs complex integrative functions, controls voluntary activities
Spinal cord	Relays information to the brain and performs less complex integrative functions; directs many simple involuntary activities
PERIPHERAL NERVOUS SYSTEM (PNS)	Links CNS with other systems and with sense organs
AUTONOMIC NERVOUS SYSTEM (ANS)	A functional system that includes components of the CNS and PNS; involved with the regulation of visceral functions, such as digestion, secretion, respiration, and cardiovascular activities

FIGURE 1-6e   The Endocrine System

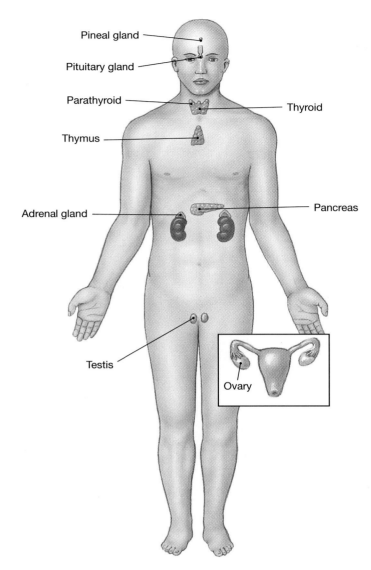

Organ	Primary Functions
PITUITARY GLAND	Controls other endocrine glands, regulates growth and fluid balance
THYROID GLAND	Controls tissue metabolic rate and regulates blood calcium levels
PARATHYROID GLAND	Regulates blood calcium levels (with thyroid)
THYMUS	Controls functional maturation of lymphocytes
ADRENAL GLANDS	Adjust water balance, tissue metabolism, cardiovascular and respiratory activity
KIDNEYS	Control red blood cell production and elevate blood pressure
PANCREAS	Regulates blood glucose levels
HEART	Regulates fluid balance
DIGESTIVE TRACT	Coordinates digestive activities
TESTES	Support male sexual characteristics and reproductive functions
OVARIES	Support female sexual characteristics and reproductive function
PINEAL GLAND	Controls the timing of reproduction

FIGURE 1-6f   The Cardiovascular System

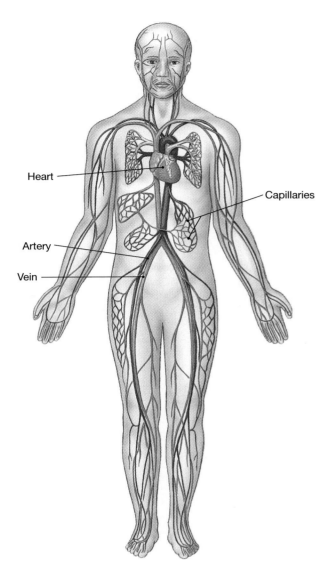

Organ	Primary Functions
HEART	Propels blood, maintains blood pressure
BLOOD VESSELS	Distribute blood around the body
Arteries	Carry blood from heart to capillaries
Capillaries	Site of diffusion between blood and interstitial fluids
Veins	Return blood from capillaries to the heart
BLOOD	Transports oxygen and carbon dioxide, delivers nutrients and hormones, removes waste products, assists in defense against disease

FIGURE 1-6g The Lymphatic System

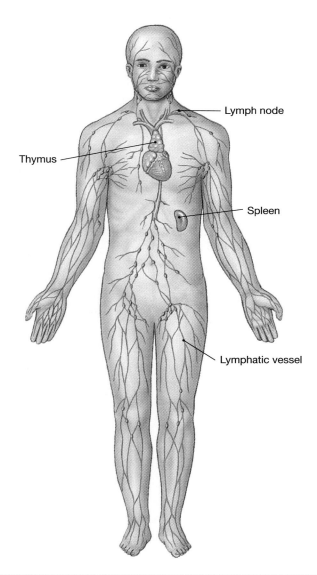

Organ	Primary Functions
LYMPHATIC VESSELS	Carry lymph (water and proteins) and lymphocytes from peripheral tissues to the veins of the cardiovascular system
LYMPH NODES	Monitor the composition of lymph, engulf pathogens, stimulate immune response
SPLEEN	Monitors circulating blood, engulfs pathogens, stimulates immune response
THYMUS	Controls development and maintenance of one important class of lymphocytes (T cells)

FIGURE 1-6h The Respiratory System

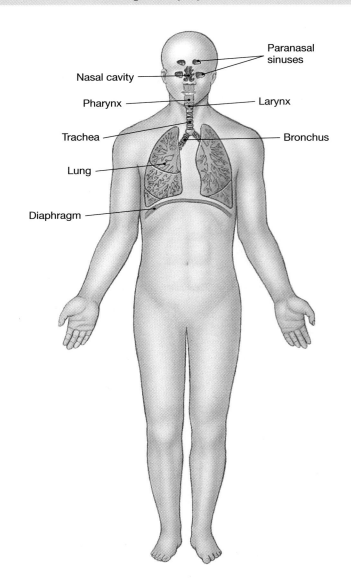

Organ	Primary Functions
NASAL CAVITIES	Filter, warm, humidify air, detect smells
PARANASAL SINUSES	Lighten skull, produce mucus that flushes walls of nasal cavities
PHARYNX	Chamber shared with digestive tract, conducts air to larynx
LARYNX	Protects opening to trachea and contains vocal cords
TRACHEA	Filters air, traps particles in mucus, cartilages keep airway open
LUNGS	Includes airways and alveoli; volume changes responsible for air movement
BRONCHI	Same as trachea
ALVEOLI	Sites of gas exchange between air and blood

FIGURE 1-6i   The Digestive System

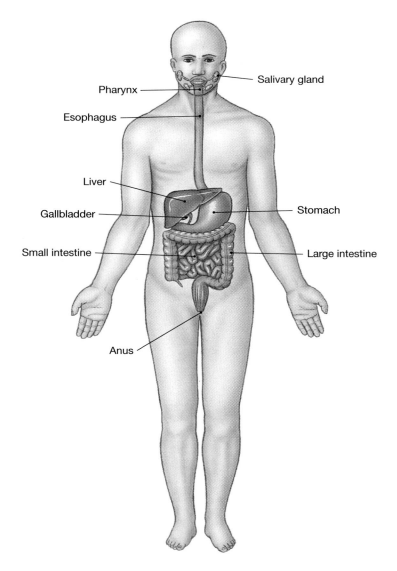

Organ	Primary Functions
MOUTH	Chewing mixes food with salivary secretions
SALIVARY GLANDS	Produce enzymes that begin digestion, provide buffers and lubrication
PHARYNX	Passageway shared with respiratory system, leads to esophagus
ESOPHAGUS	Delivers food to stomach
STOMACH	Secretes acids and digestive enzymes
SMALL INTESTINE	Absorbs nutrients, secretes buffers, enzymes, and hormones
LIVER	Secretes bile, regulates blood composition, stores lipids and carbohydrates
GALLBLADDER	Stores bile for release into small intestine
PANCREAS	Secretes digestive enzymes and buffers into small intestine; contains endocrine cells
LARGE INTESTINE	Removes water from fecal material, stores wastes
ANUS	Opening to exterior for discharge of feces

FIGURE 1-6j   The Urinary System

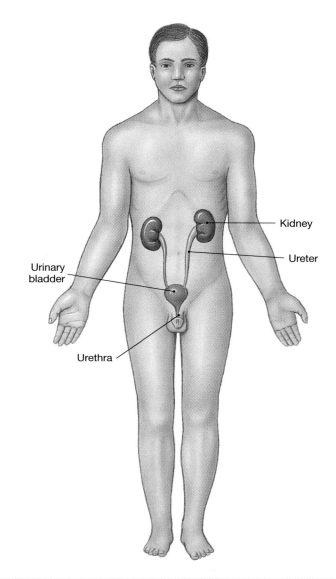

Organ	Primary Functions
KIDNEYS	Form and concentrate urine, regulate blood pH and ion concentrations; endocrine functions noted in Figure 1-6e
URETERS	Conduct urine from kidneys to urinary bladder
URINARY BLADDER	Stores urine for eventual elimination
URETHRA	Carries urine to exterior

FIGURE 1-6k   The Reproductive System of the Male

FIGURE 1-6l   The Reproductive System of the Female

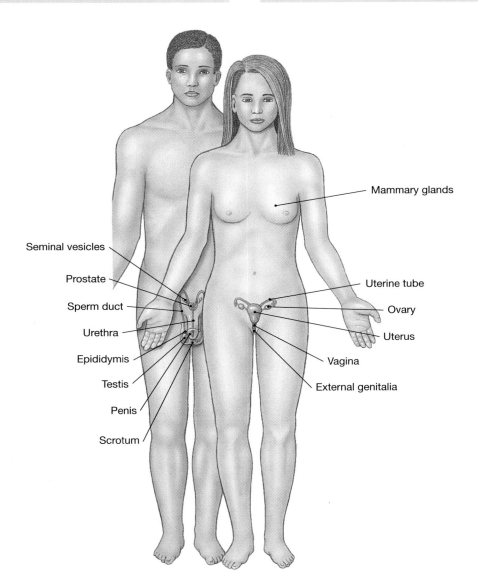

Organ	Primary Functions
TESTES	Produce sperm and hormones; see also Figure 1-6e
ACCESSORY ORGANS	
Epididymis	Site of sperm maturation
Ductus deferens (sperm duct)	Conducts sperm between epididymis and prostate
Seminal vesicles	Secrete fluid that makes up much of the volume of semen
Prostate	Secretes fluid and enzymes
EXTERNAL GENITALIA	
Penis	Erectile organ used to deposit sperm in the vagina of a female, produces pleasurable sensation during sexual act
Scrotum	Surrounds the testes, controls their temperature

Organ	Primary Functions
OVARIES	Produce ova (eggs) and hormones; see also Figure 1-6e
UTERINE TUBES	Deliver egg(s) or embryo(s) to uterus; site of normal fertilization
UTERUS	Site of embryonic development and diffusion between maternal and embryonic blood-streams
VAGINA	Site of sperm deposition; birth canal at delivery; provides passage of fluids at menses
EXTERNAL GENITALIA	
Clitoris	Erectile organ, produces pleasurable sensations during sexual act
Labia	Contain glands that lubricate entrance to vagina
MAMMARY GLANDS	Produce milk that nourishes newborn infant

THE LANGUAGE OF ANATOMY
(Figure 1-7)

If you discovered a new continent, how would you begin collecting information so that you could report your findings? You would have to construct a detailed map of the territory. The completed map would contain (1) prominent landmarks, such as mountains, valleys, or volcanoes; (2) the distance between them; and (3) the direction you traveled to get from one place to another. The distances might be recorded in miles, and the directions recorded as compass bearings (north, south, northeast, southwest, and so on). With such a map, anyone could go directly to a specific location on that continent.

Early anatomists faced similar communication problems. Stating that a bump is "on the back" does not give very precise information about its location. So anatomists created maps of the human body. The landmarks are prominent anatomical structures, and distances are measured in centimeters or inches. In effect, anatomy uses a special language that must be learned at the start. It takes time and effort to develop a working anatomical vocabulary, but it is absolutely essential if one is to avoid a situation like that depicted in Figure 1-7●.

New anatomical terms continue to appear as technology advances, but many of the older words and phrases remain in use. As a result, the vocabulary of this science represents a form of historical record. Latin and Greek words and phrases form the basis for an impressive number of anatomical terms. For example, many of the Latin names assigned to specific structures 2000 years ago are still in use today.

A familiarity with Latin roots and patterns makes anatomical terms more understandable, and the notes included on word derivation are intended to assist you in that regard. In English, when you want to indicate more than one of something, you usually add an *s* to the name—girl/girls or doll/dolls. Latin words change their endings. Those ending in *-us* convert to *-i*, and other conversions involve changing from *-um* to *-a*, and from *-a* to *-ae*. Additional information on foreign word roots, prefixes, suffixes, and combining forms can be found on the back endpapers.

Latin and Greek terms are not the only foreign terms imported into the anatomical vocabulary over the centuries, and the vocabulary continues to expand. Many anatomical structures and clinical conditions were initially named after either the discoverer or, in the case of diseases, the most famous victim. The major problem with this practice is that it is difficult for someone to remember a connection between the structure or disorder and the name. Over the last 100 years most of these commemorative names, or *eponyms*, have been replaced by more precise terms. For those interested in historical details, the section titled "Eponyms in Common Use" at the beginning of the Glossary provides information about the commemorative names in occasional use today.

Superficial Anatomy

A familiarity with major anatomical landmarks and directional references will make subsequent chapters more understandable because none of the organ systems except the integument can be seen from the body surface. You must create your own mental maps and extract information from the anatomical illustrations that accompany this discussion.

Anatomical Landmarks
(Figure 1-8)

Important anatomical landmarks are presented in Figure 1-8●. You should become familiar with the adjectival form as well as the anatomical term. Understanding the terms and their origins will help you to remember the location of a particular structure, as well as its name. For example, the term *brachium* refers to the arm, and later chapters discuss the *brachialis muscle* and branches of the *brachial artery*.

Standard anatomical illustrations show the human form in the **anatomical position**. In the anatomical position, the person stands with the legs together and the feet flat on the floor. The hands are at the sides, and the palms face forward. The individual shown in Figure 1-8 is in the anatomical position, as seen from the front (Figure 1-8a●) and back (Figure 1-8b●). Unless otherwise noted, all of the descriptions given in this text refer to the body in the anatomical position. A person lying down in the anatomical position is said to be **supine** (SŪ-pīn) when lying face up and **prone** when lying face down.

FIGURE 1-7
The Importance of a Precise Vocabulary.
Would you want to be this patient?

Drawing by Ed Fisher; ©1990 The New Yorker Magazine, Inc.

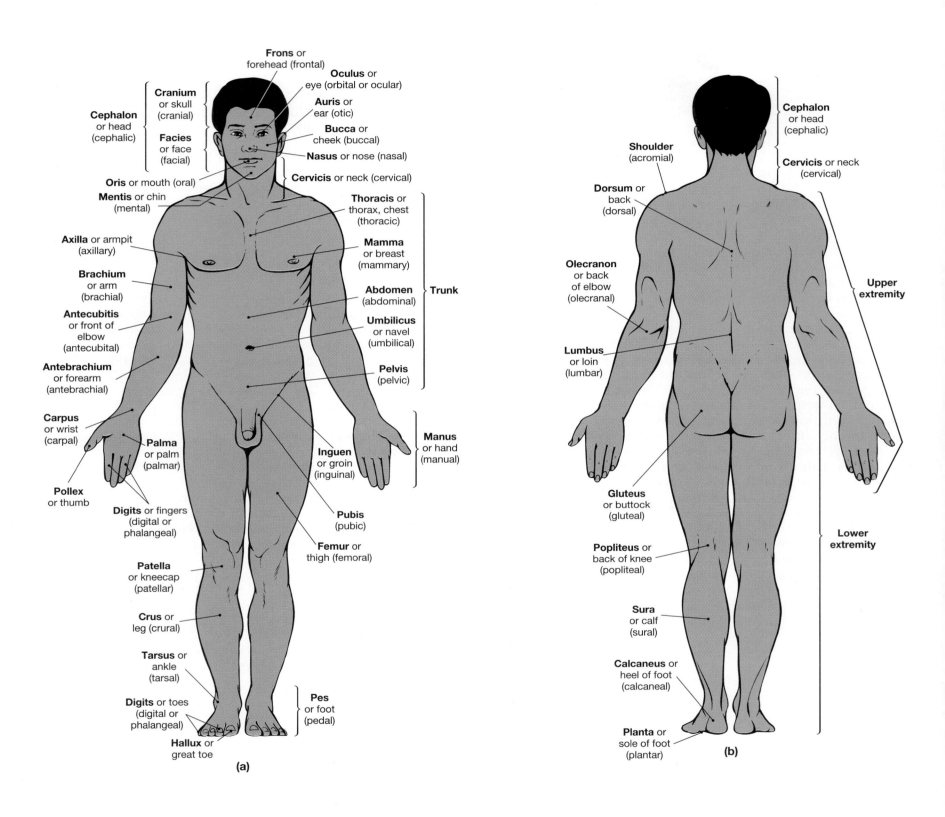

Frons or forehead (frontal)
Oculus or eye (orbital or ocular)
Auris or ear (otic)
Bucca or cheek (buccal)
Nasus or nose (nasal)
Cervicis or neck (cervical)
Cranium or skull (cranial)
Cephalon or head (cephalic)
Facies or face (facial)
Oris or mouth (oral)
Mentis or chin (mental)
Thoracis or thorax, chest (thoracic)
Axilla or armpit (axillary)
Mamma or breast (mammary)
Brachium or arm (brachial)
Abdomen (abdominal)
Antecubitis or front of elbow (antecubital)
Umbilicus or navel (umbilical)
Antebrachium or forearm (antebrachial)
Pelvis (pelvic)
Trunk
Carpus or wrist (carpal)
Manus or hand (manual)
Palma or palm (palmar)
Inguen or groin (inguinal)
Pollex or thumb
Digits or fingers (digital or phalangeal)
Pubis (pubic)
Femur or thigh (femoral)
Patella or kneecap (patellar)
Crus or leg (crural)
Tarsus or ankle (tarsal)
Digits or toes (digital or phalangeal)
Pes or foot (pedal)
Hallux or great toe
(a)

Cephalon or head (cephalic)
Shoulder (acromial)
Cervicis or neck (cervical)
Dorsum or back (dorsal)
Olecranon or back of elbow (olecranal)
Upper extremity
Lumbus or loin (lumbar)
Gluteus or buttock (gluteal)
Lower extremity
Popliteus or back of knee (popliteal)
Sura or calf (sural)
Calcaneus or heel of foot (calcaneal)
Planta or sole of foot (plantar)
(b)

FIGURE 1-8
Anatomical Landmarks. The anatomical terms are shown in boldface type, the common names are in plain type, and the anatomical adjectives are in parentheses.

Anatomical Regions
(Figures 1-8/1-9)

Major regions of the body are indicated in Table 1-1. These and additional regions and anatomical landmarks are noted in Figure 1-8●. Anatomists and clinicians often use specialized regional terms to indicate a specific area of the abdominal or pelvic regions. There are two different methods in use. Clinicians refer to the **abdominopelvic quadrants**. The region is divided into four segments using a pair of imaginary lines (one horizontal and one vertical) that intersect at the *umbilicus* (navel). This simple method, shown in Figure 1-9a●, provides useful references for the description of aches, pains, and injuries. The location can assist the doctor in deciding the possible cause; for example, tenderness in the right lower quadrant (RLQ) is a symptom of appendicitis, whereas tenderness in the right upper quadrant (RUQ) may indicate gallbladder or liver problems. ✝ *Anatomy and Observation [p. 749].*

Anatomists tend to use more precise regional distinctions to describe the location and orientation of internal organs. They recognize nine **abdominopelvic regions**, diagrammed in Figure 1-9b●. Figure 1-9c● shows the relationship between quadrants, regions, and internal organs.

TABLE 1-1	Regions of the Human Body*	
Anatomical Name	*Anatomical Region*	*Area Indicated*
Cephalon	Cephalic	Area of head
Cervicis	Cervical	Area of neck
Thoracis	Thoracic	The chest
Abdomen	Abdominal	The abdomen
Pelvis	Pelvic	The pelvis (in general)
Pubis	Pubic	The anterior pelvis
Inguen	Inguinal	The groin (crease between thigh and trunk)
Lumbus	Lumbar	The lower back
Gluteus	Gluteal	The buttock
Brachium	Brachial	The segment of the upper limb closest to the trunk; the arm
Antebrachium	Antebrachial	The forearm
Manus	Manual	The hand
Femur	Femoral	The thigh
Patella	Patellar	The kneecap
Crus	Crural	Anterior portion of leg
Sura	Sural	Posterior portion of leg
Pes	Pedal	The foot

*See Figures 1-8 and 1-9

FIGURE 1-9
Abdominopelvic Quadrants and Regions. (a) Abdominopelvic quadrants divide the area into four sections. These terms, or their abbreviations, are most often used in clinical discussions. (b) More precise anatomical descriptions are provided by reference to the appropriate abdominopelvic region. (c) Quadrants or regions are useful because there is a known relationship between superficial anatomical landmarks and underlying organs.

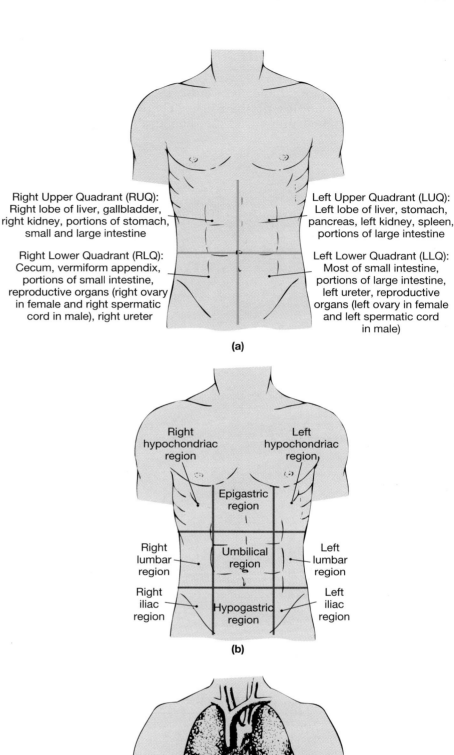

Right Upper Quadrant (RUQ): Right lobe of liver, gallbladder, right kidney, portions of stomach, small and large intestine

Left Upper Quadrant (LUQ): Left lobe of liver, stomach, pancreas, left kidney, spleen, portions of large intestine

Right Lower Quadrant (RLQ): Cecum, vermiform appendix, portions of small intestine, reproductive organs (right ovary in female and right spermatic cord in male), right ureter

Left Lower Quadrant (LLQ): Most of small intestine, portions of large intestine, left ureter, reproductive organs (left ovary in female and left spermatic cord in male)

(a)

Right hypochondriac region

Left hypochondriac region

Epigastric region

Right lumbar region

Umbilical region

Left lumbar region

Right iliac region

Hypogastric region

Left iliac region

(b)

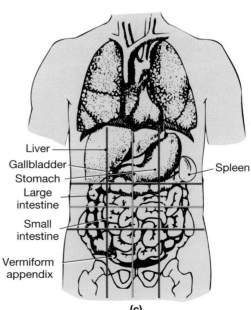

Liver
Gallbladder
Stomach
Large intestine
Small intestine
Vermiform appendix
Spleen

(c)

Anatomical Directions
(Figure 1-10)

Figure 1-10● and Table 1-2 show the principal directional terms and examples of their use. There are many different terms, and some can be used interchangeably. For example, *an-*terior refers to the front of the body, when viewed in the anatomical position; in human beings, this term is equivalent to *ventral*, which actually refers to the belly side. Although your instructor may have additional terminology, the terms that appear frequently in later chapters have been emphasized in Table 1-2.

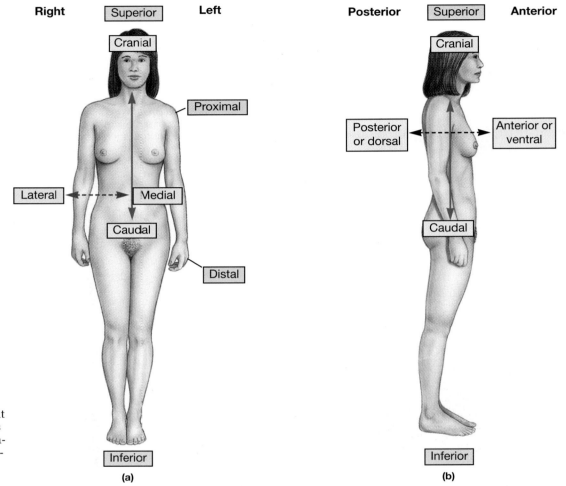

FIGURE 1-10
Directional References. Important directional references used in this text are indicated by arrows; definitions and descriptions are included in Table 1-2. (a) Anterior view. (b) Lateral view.

TABLE 1-2   Regional and Directional Terms*

Term	Region or Reference	Example
Anterior	The front; before	The navel is on the *ventral* (*anterior*) surface of the trunk.
Ventral	The belly side (equivalent to anterior when referring to human body)	
Posterior	The back; behind	The shoulder blade is located *posterior* to the rib cage.
Dorsal	The back (equivalent to posterior when referring to human body)	The *dorsal* body cavity encloses the brain and spinal cord.
Cranial	Toward the head	The *cranial*, or *cephalic*, border of the pelvis is *superior* to the thigh.
Cephalic	Same as cranial	
Superior	Above; at a higher level (in human body, toward the head)	
Caudal	The tail (coccyx in humans)	The hips are *caudal* to the waist.
Inferior	Below; at a lower level	The knees are *inferior* to the hips.
Medial	Toward the midline longitudinal axis of the body	The *medial* surfaces of the thighs may be in contact.
Lateral	Away from the midline longitudinal axis of the body	The thigh bone articulates with the *lateral* surface of the pelvis.
Proximal	Toward an attached base	The thigh is *proximal* to the foot.
Distal	Away from an attached base	The fingers are *distal* to the wrist.

*See Figure 1-10

16

FIGURE 1-11
Planes of Section. The three primary planes of section are indicated here. Table 1-3 defines and describes them.

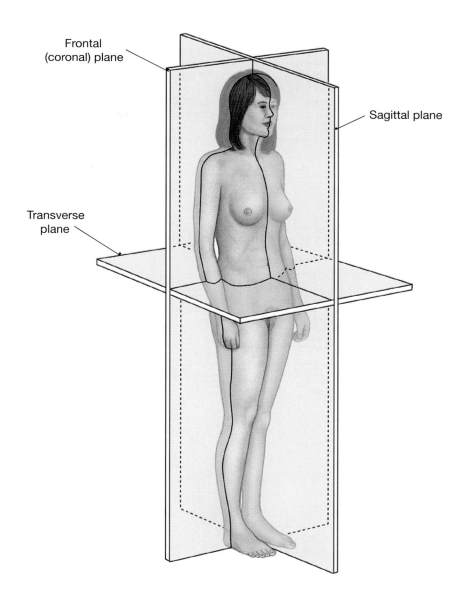

Frontal (coronal) plane

Sagittal plane

Transverse plane

When following anatomical descriptions, you may also find it useful to remember that the terms *left* and *right* always refer to the left and right sides of the subject, not the observer.

Sectional Anatomy

A presentation in sectional view is sometimes the only way to illustrate the relationships between the parts of a three-dimensional object. An understanding of sectional views has become increasingly important since the development of electronic imaging techniques that enable us to see inside the living body without resorting to surgery.

Planes and Sections
(Figures 1-11/1-12)

Any slice through a three-dimensional object can be described with reference to three **sectional planes**, indicated in Table 1-3 and Figure 1-11●. The **transverse plane** lies at right angles to the long axis of the body, dividing it into **superior** and **inferior** sections. A division along this plane is called a **transverse section**, or *cross section*. The **frontal plane**, or **coronal plane**, and the **sagittal plane** parallel the longitudinal axis of the body. The frontal plane extends from side to side, dividing the body into **anterior** and **posterior** sections. The sagittal plane extends from anterior to posterior, dividing the body into *left* and *right* sections. A section that passes along the midline and divides the body into left and right halves is a **midsagittal section**, or a **median sagittal section**; a section parallel to the midsagittal line is a **parasagittal section**.

Sometimes it is helpful to compare the information provided by sections made along different planes. You can experiment with this procedure by mentally sectioning this book, as in Figure 1-12a●. Each sectional plane provides a different perspective on the structure of the book; when combined with observations on the external anatomy, they create a reasonably complete picture.

A more accurate and detailed picture would entail choosing one sectional plane and making a series of sections at small intervals. This process, called **serial reconstruction**, permits the analysis of relatively complex structures. Figure 1-12b● shows the serial reconstruction of a single bent tube. The

---

TABLE 1-3   Terms That Indicate Planes of Section

Orientation of Plane	Adjective	Directional Term	Description
Parallel to long axis	Sagittal	Sagittally	A *sagittal* section separates right and left portions. You examine a sagittal section, but you section sagittally.
	Midsagittal		In a *midsagittal* section the plane passes through the midline, dividing the body in half and separating right and left sides.
	Parasagittal		A *parasagittal* section misses the midline, separating right and left portions of unequal size.
	Frontal or coronal	Frontally or coronally	A *frontal*, or *coronal*, section separates anterior and posterior portions of the body; coronal usually refers to sections passing through the skull.
Perpendicular to long axis	Transverse or horizontal or cross-section	Transversely or horizontally	A *transverse*, or *horizontal*, section separates superior and inferior portions of the body; sections typically pass through head and trunk regions.

CHAPTER 1 An Introduction to Anatomy   17

FIGURE 1-12

Sectional Planes and Visualization. (a) Visualizing three different sections through a book provides detailed information about its three-dimensional structure. (b) More complete pictures can be assembled by taking a series of sections at small intervals. This process is called serial reconstruction. Notice how the sectional views change; although a simple tube, the small intestine can look like a pair of tubes, a dumbbell, an oval, or a solid, depending on where the section was taken. The effects of sectional planes should be kept in mind when looking at slides under the microscope.

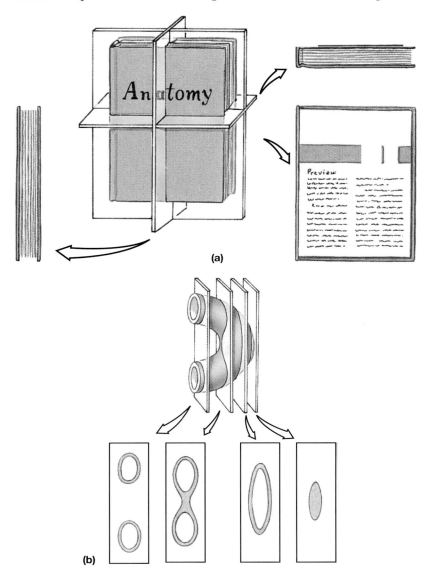

(a)

(b)

procedure could be used to visualize the path of a small blood vessel or to follow a loop of the intestine. Serial reconstruction is an important method for studying histological structure and for analyzing the images produced by sophisticated clinical procedures (see the Clinical Brief on p. 21).

Body Cavities
(Figure 1-13)

Viewed in sections, the human body is not a solid object, and many vital organs are suspended in internal chambers called **body cavities**. These cavities have two essential functions: (1) they protect delicate organs, such as the brain and spinal cord, from accidental shocks, and cushion them from the thumps and bumps that occur during walking, jumping, and

running; and (2) they permit significant changes in the size and shape of body (visceral) organs. For example, because they are situated within body cavities, the lungs, heart, stomach, intestines, urinary bladder, and many other organs can expand and contract without distorting surrounding tissues and disrupting the activities of nearby organs.

The **dorsal body cavity** surrounds the brain and spinal cord, and a much larger **ventral body cavity**, or **coelom** (SĒ-lom; *koila*, cavity), surrounds organs of the respiratory, cardiovascular, digestive, urinary, and reproductive systems. Relationships between the dorsal and ventral body cavities and their various subdivisions are diagrammed in Figure 1-13●.

DORSAL BODY CAVITY *(Figure 1-14a,b)*. The dorsal body cavity (Figure 1-14a●) is a fluid-filled space whose limits are established by the *cranium*, the bones of the skull that surround the brain, and the *spinal canal*, enclosed by the spinal vertebrae. The dorsal body cavity is subdivided into the **cranial cavity**, which encloses the brain, and the **spinal cavity**, which surrounds the spinal cord.

VENTRAL BODY CAVITY *(Figure 1-14a)*. As development proceeds, internal organs grow and change their relative positions. These changes lead to the subdivision of the ventral body cavity. The **diaphragm** (DĪ-a-fram), a flat muscular sheet, divides the ventral body cavity into a superior *thoracic cavity*, enclosed by the chest wall, and an inferior *abdominopelvic cavity*, enclosed by the abdomen and pelvis.

Many of the organs within these cavities change size and shape as they perform their functions. For example, the stomach swells at each meal, and the heart contracts and expands with each beat. These organs project into moist internal spaces that permit expansion and limited movement, but prevent friction. There are three such chambers in the thoracic cavity and one in the abdominopelvic cavity. The internal organs that project into these cavities are called **viscera** (VIS-e-ra).

The Thoracic Cavity. The walls of the **thoracic cavity** surround the lungs and heart, associated organs of the respiratory, cardiovascular, and lymphatic systems, the inferior portions of the esophagus, and the thymus gland. The thoracic cavity contains three internal chambers: a single *pericardial cavity* and a pair of *pleural cavities*. These cavities are lined by shiny, slippery, and delicate *serous membranes*.

- The heart projects into the **pericardial cavity**. The relationship between the heart and the cavity resembles that of a fist pushing into a balloon (Figure 1-14b●). The wrist corresponds to the base of the heart, and the balloon corresponds to the serous membrane lining the pericardial cavity. During each beat, the heart changes in size and shape. The pericardial cavity permits these changes, and the slippery lining of the cavity prevents friction between the heart and adjacent structures. The serous membrane is called the **pericardium** (*peri-*, around + *kardia*, heart). The layer covering the heart is the *visceral pericardium*, and the opposing surface is the *parietal pericardium*.

The pericardium lies within the **mediastinum** (mē-dē-as-TĪ-num or mē-dē-AS-ti-num). The mediastinum is the portion of the thoracic cavity that lies between the left and right *pleural cavities*. The connective tissue of the mediastinum surrounds the pericardial cavity and heart,

FIGURE 1-13
Relationships of the Various Body Cavities

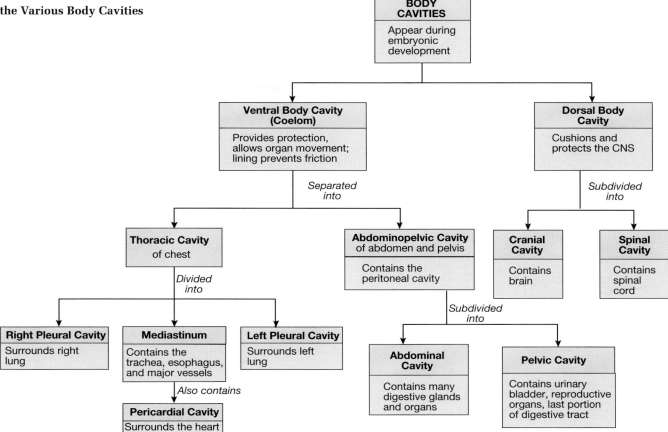

the large arteries and veins attached to the heart, and the thymus, trachea, and esophagus.

- There is one **pleural cavity** on each side of the mediastinum. Each pleural cavity encloses a lung, and the relationship between a lung and a pleural cavity is the same as that between the heart and the pericardial cavity. The serous membrane lining a pleural cavity is called a **pleura** (PLOO-ra). The outer surfaces of the lungs are covered by the *visceral pleura*, and the *parietal pleura* covers the opposing mediastinal surfaces and the inner body wall.

The Abdominopelvic Cavity. The **abdominopelvic cavity** can be divided into a superior *abdominal cavity* and an inferior *pelvic cavity*. The abdominopelvic cavity contains the **peritoneal** (per-i-tō-NĒ-al) cavity, an internal chamber lined by a serous membrane known as the **peritoneum** (per-i-tō-NĒ-um). The *parietal peritoneum* lines the body wall. A narrow, fluid-filled space separates the parietal peritoneum from the *visceral peritoneum* that covers the enclosed organs. Organs such as the stomach, small intestine, and portions of the large intestine are suspended within the peritoneal cavity by double sheets of peritoneum, called **mesenteries** (MES-en-ter-ēs). Mesenteries provide support and stability while permitting limited movement.

- The **abdominal cavity** extends from the inferior surface of the diaphragm to an imaginary plane extending from the inferior surface of the lowest spinal vertebra to the anterior and superior margin of the pelvic girdle. The abdominal cavity contains the liver, stomach, spleen, kidneys, pancreas, small intestine, and most of the large intestine. (The positions of these organs can be seen in Figure 1-9c●, p. 15). Many of these organs project par-

tially or completely into the peritoneal cavity, much as the heart or lungs project into the pericardial or pleural cavities.

- The portion of the ventral body cavity inferior to the abdominal cavity is the **pelvic cavity**. The pelvic cavity, enclosed by the bones of the pelvis, contains the last segments of the large intestine, the urinary bladder, and various reproductive organs. For example, the pelvic cavity of a female contains the ovaries, uterine tubes, and uterus; in a male, it contains the prostate gland and seminal vesicles. The inferior portion of the peritoneal cavity extends into the pelvic cavity. The superior portion of the urinary bladder in both sexes, and the uterine tubes, the ovaries, and the superior portion of the uterus in females are covered by peritoneal membrane.

√ **What type of section would separate the two eyes?**

√ **If a surgeon makes an incision just inferior to the diaphragm, what body cavity will be opened?**

√ **You fall and break your antebrachium. What part of the body is affected?**

This chapter provided an overview of the locations and functions of the major components of each organ system. It also introduced the anatomical vocabulary needed to follow more detailed anatomical descriptions in later chapters. Modern methods of visualizing anatomical structures in living individuals are summarized in the Clinical Brief on p. 21. Many of the figures in later chapters contain images produced by the procedures outlined in that section.

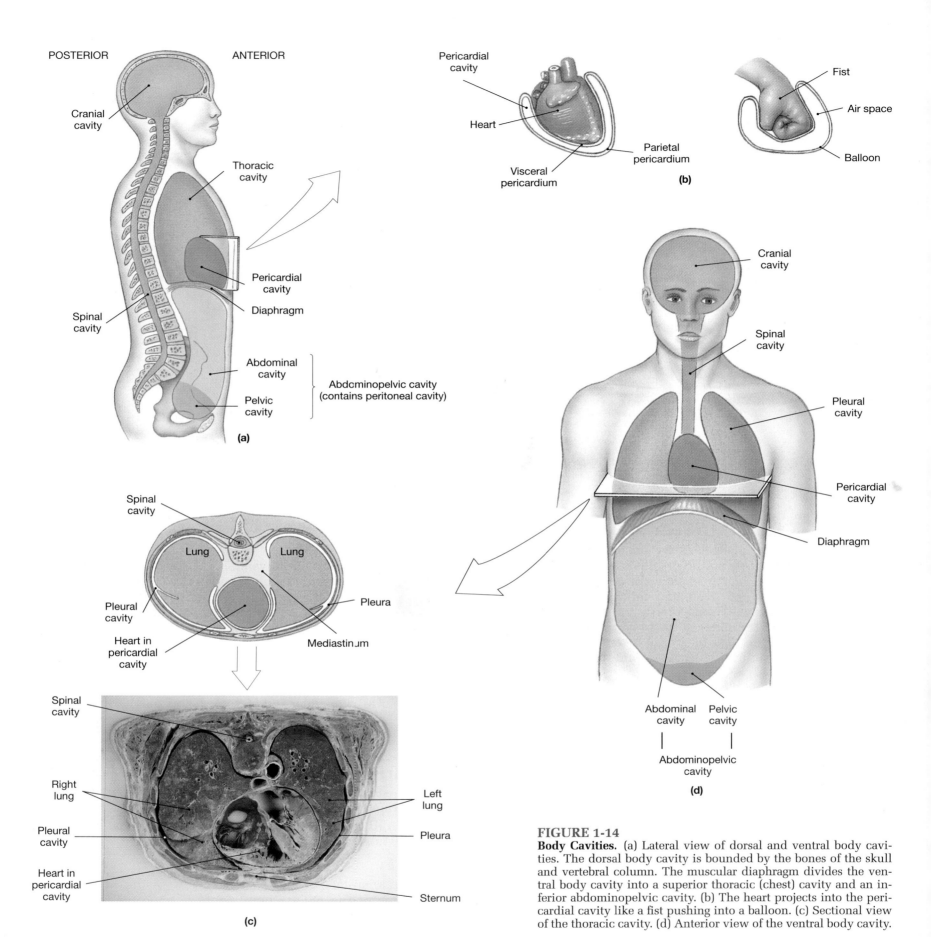

POSTERIOR   **ANTERIOR**

Cranial
cavity

Thoracic
cavity

Pericardial
cavity

Diaphragm

Spinal
cavity

Abdominal
cavity

Abdominopelvic cavity
(contains peritoneal cavity)

Pelvic
cavity

(a)

Pericardial
cavity

Heart

Visceral
pericardium

Parietal
pericardium

(b)

Fist

Air space

Balloon

Cranial
cavity

Spinal
cavity

Pleural
cavity

Pericardial
cavity

Diaphragm

Abdominal
cavity

Pelvic
cavity

Abdominopelvic
cavity

(d)

Spinal
cavity

Lung   Lung

Pleural
cavity

Heart in
pericardial
cavity

Pleura

Mediastinum

Spinal
cavity

Right
lung

Pleural
cavity

Heart in
pericardial
cavity

Left
lung

Pleura

Sternum

(c)

FIGURE 1-14
Body Cavities. (a) Lateral view of dorsal and ventral body cavities. The dorsal body cavity is bounded by the bones of the skull and vertebral column. The muscular diaphragm divides the ventral body cavity into a superior thoracic (chest) cavity and an inferior abdominopelvic cavity. (b) The heart projects into the pericardial cavity like a fist pushing into a balloon. (c) Sectional view of the thoracic cavity. (d) Anterior view of the ventral body cavity.

Sectional Anatomy and Clinical Technology

The term **radiological procedures** includes not only those scanning techniques that involve radioisotopes but also methods that employ radiation and other sources outside the body. Physicians who specialize in the performance and analysis of these procedures are called **radiologists.** Radiologi-cal procedures can provide detailed information about internal systems. Figures 1-15● and 1-16● compare the views provided by several different techniques; other examples will be found in later chapters and in Appendix I. Most of the procedures produce black and white images on film sheets. Colors can be added by computer to illustrate subtle variations in contrast and shading. Note that when anatomical diagrams or scans present cross-sectional views, the sections are presented as though the observer were standing at the feet and looking toward the head of the subject.

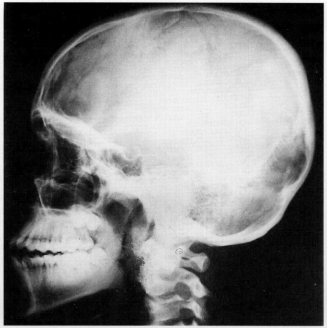
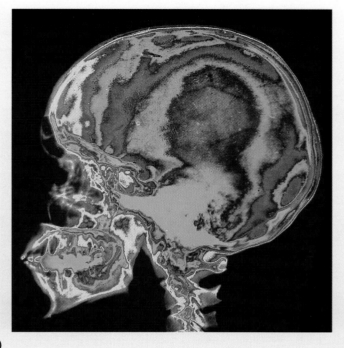

(a)

FIGURE 1-15

X-rays. (a) An X-ray of the skull and color-enhanced X-ray of the skull, taken from the left side. X-rays are a form of high-energy radiation that can penetrate living tissues. In the most familiar procedure, a beam of X-rays travels through the body and strikes a photographic plate. All of the projected X-rays do not arrive at the film; some are absorbed or deflected as they pass through the body. The resistance to X-ray penetration is called **radiodensity**. In the human body, the order of increasing radiodensity is as follows: air, fat, liver, blood, muscle, bone. The result is an image with radio-dense tissues, such as bone, appearing white, while less dense tissues are seen in shades of gray to black. The picture is a two-dimensional image of a three-dimensional object; in this image it is difficult to decide whether a particular feature is on the left side (toward the viewer) or on the right side (away from the viewer).

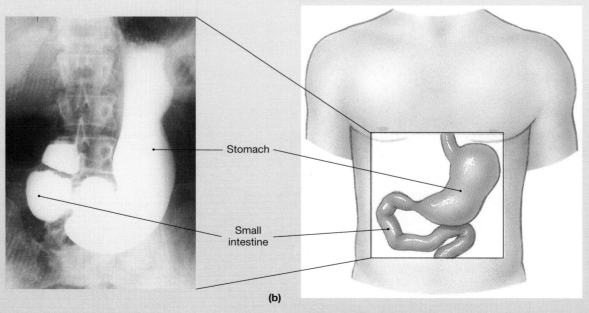

Stomach

Small
intestine

(b)

(b) A barium-contrast X-ray of the upper digestive tract. Barium is very radiodense, and the contours of the gastric and intestinal lining can be seen outlined against the white of the barium solution.

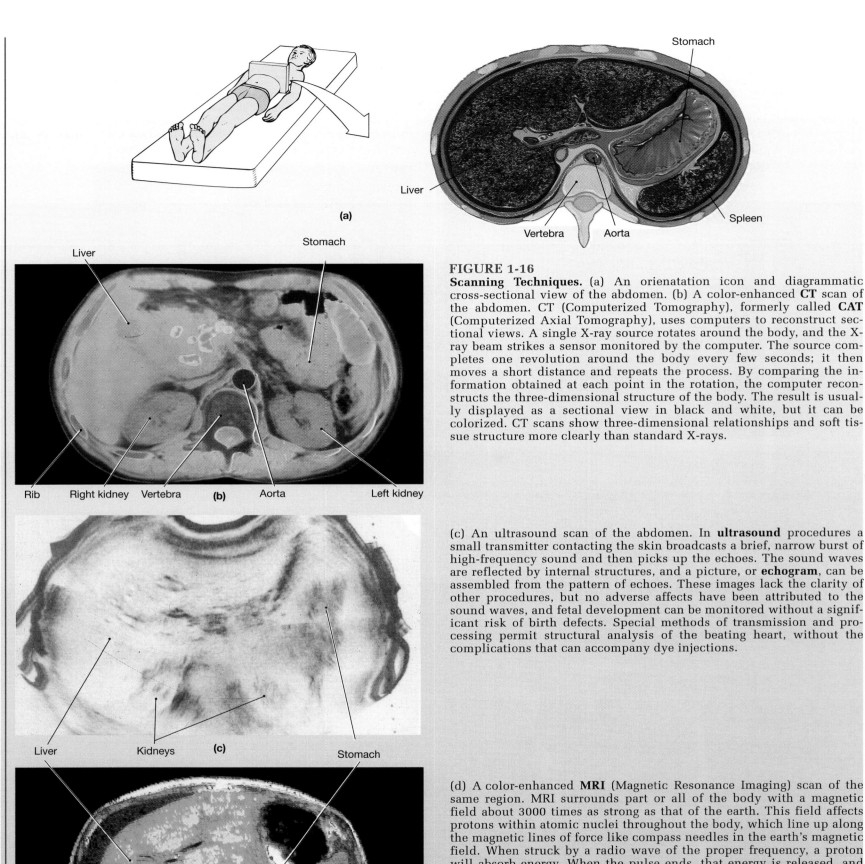

Stomach

Liver

Vertebra   Aorta

Spleen

(a)

Liver   Stomach

Rib   Right kidney   Vertebra   **(b)**   Aorta   Left kidney

Liver   Kidneys   **(c)**   Stomach

Vertebra   **(d)**   Kidney

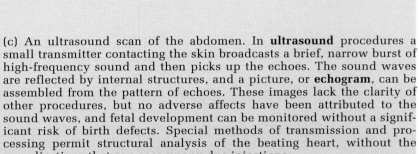

FIGURE 1-16

Scanning Techniques. (a) An orienatation icon and diagrammatic cross-sectional view of the abdomen. (b) A color-enhanced **CT** scan of the abdomen. CT (Computerized Tomography), formerly called **CAT** (Computerized Axial Tomography), uses computers to reconstruct sectional views. A single X-ray source rotates around the body, and the X-ray beam strikes a sensor monitored by the computer. The source completes one revolution around the body every few seconds; it then moves a short distance and repeats the process. By comparing the information obtained at each point in the rotation, the computer reconstructs the three-dimensional structure of the body. The result is usually displayed as a sectional view in black and white, but it can be colorized. CT scans show three-dimensional relationships and soft tissue structure more clearly than standard X-rays.

(c) An ultrasound scan of the abdomen. In **ultrasound** procedures a small transmitter contacting the skin broadcasts a brief, narrow burst of high-frequency sound and then picks up the echoes. The sound waves are reflected by internal structures, and a picture, or **echogram**, can be assembled from the pattern of echoes. These images lack the clarity of other procedures, but no adverse affects have been attributed to the sound waves, and fetal development can be monitored without a significant risk of birth defects. Special methods of transmission and processing permit structural analysis of the beating heart, without the complications that can accompany dye injections.

(d) A color-enhanced **MRI** (Magnetic Resonance Imaging) scan of the same region. MRI surrounds part or all of the body with a magnetic field about 3000 times as strong as that of the earth. This field affects protons within atomic nuclei throughout the body, which line up along the magnetic lines of force like compass needles in the earth's magnetic field. When struck by a radio wave of the proper frequency, a proton will absorb energy. When the pulse ends, that energy is released, and the energy source of the radiation is detected by the MRI computers. Each element differs in terms of the radio frequency required to affect its protons. Note the differences in detail between this image, the CT scan, and the X-rays in Figure 1-15.

Related Clinical Terms

abdominopelvic quadrant: One of four divisions of the anterior abdominal surface. *[p. 15]*

abdominopelvic region: One of nine divisions of the abdominal surface. *[p. 15]*

radiologist: A physician who specializes in performing and analyzing radiological procedures. *[p. 21]*

X-rays: High-energy radiation that can penetrate living tissues. *[p. 21]*

CT, CAT (computerized [axial] tomography): An imaging technique that reconstructs the three-dimensional structure of the body. *[p. 22]*

ultrasound: An imaging technique that uses brief bursts of high-frequency sound reflected by internal structures. *[p. 22]*

MRI (magnetic resonance imaging): An imaging technique that employs a magnetic field and radio waves to portray subtle structural differences. *[p. 22]*

CHAPTER SUMMARY AND REVIEW

STUDY OUTLINE

INTRODUCTION *[p. 2]*

1. **Anatomy** is the study of internal and external structures and the physical relationships between body parts. Specific functions are performed by specific structures.

MICROSCOPIC ANATOMY *[p. 2]*

1. The boundaries of **microscopic anatomy** are established by the equipment used. **Cytology** analyzes the internal structure of individual cells. **Histology** examines **tissues** (groups of cells that work together to perform specific functions). Tissues combine to form **organs**, anatomical units with multiple functions *(see Figure 1-1)*.

GROSS ANATOMY *[p. 3]*

1. **Gross (macroscopic) anatomy** considers features visible without a microscope. It includes **surface anatomy** (general form and superficial markings); **regional anatomy** (superficial and internal features in a specific area of the body); and **systemic anatomy** (structure of major organ systems).

OTHER PERSPECTIVES ON ANATOMY *[p. 3]*

1. **Developmental anatomy** examines the changes in form that occur between conception and physical maturity. **Embryology** studies processes during the first 2 months of development.
2. **Comparative anatomy** considers the anatomical organization of different types of animals *(see Figure 1-2)*.
3. Anatomical specialties important to clinical practice include **medical anatomy** (features that undergo characteristic changes during illness), **radiographic anatomy** (structures as they are visualized by specialized procedures performed on an intact body), and **surgical anatomy** (landmarks important for surgical procedures) *(see Figures 1-15/1-16)*.

LEVELS OF ORGANIZATION *[p. 4]*

1. Anatomical structures are arranged in a series of interacting levels of organization *(see Figures 1-3/1-4)*.

AN INTRODUCTION TO ORGAN SYSTEMS *[p. 5]*

1. All living organisms perform the same basic functions: They **respond** to changes in their environment; they show **adaptability** to their environment; they **grow** and **reproduce** to create future generations; they are capable of producing **movement**; and they **absorb** materials from the environment, and use them in **metabolism**. Organisms absorb and consume oxygen during **respiration**, and discharge waste products during **excretion**. Digestion occurs in specialized areas of the body to break down complex foods. The **circulation** forms an internal transportation system between areas of the body *(see Figures 1-5/1-6)*.
2. The organ systems of the human body perform these vital functions to maintain **homeostasis**.

THE LANGUAGE OF ANATOMY *[p. 13]*

1. Anatomy uses a special language that includes many terms that are imported from foreign languages, especially Latin and Greek *(see Figures 1-7 to 1-14)*.

Superficial Anatomy *[p. 13]*

2. Standard anatomical illustrations show the body in the **anatomical position**. In the anatomical position the person stands with the legs together and the feet flat on the floor. The hands are at the sides, and the palms face forward *(see Figures 1-8/1-10)*.
3. A person lying down in the anatomical position may be **supine** (face up) or **prone** (face down).

4. The terms **cephalic, cervical, thoracic, abdominal, pelvic, lumbar, gluteal, pubic, brachial, antebrachial, manual, femoral, patellar, crural, sural,** and **pedal** are applied to specific regions of the body *(see Figure 1-8 and Table 1-1).*

5. **Abdominopelvic quadrants** and **abdominopelvic regions** represent two different approaches to describing locations in the abdominal and pubic areas of the body *(see Figure 1-9).*

6. There are specific directional terms used to indicate relative location on the body: **anterior, posterior, dorsal, ventral, superior, inferior, medial, lateral, cranial, cephalic, caudal, proximal,** and **distal,** which will be encountered throughout the text *(see Figure 1-10 and Table 1-2).*

Sectional Anatomy *[p. 17]*

7. The three **sectional planes** (**frontal plane** or **coronal plane, sagittal plane,** and **transverse plane**) describe relationships between the parts of the three-dimensional human body *(see Figure 1-11).*

8. **Serial reconstruction** is an important technique for studying histological structure and analyzing images produced by radiological procedures *(see Figure 1-12).*

9. **Body cavities** protect delicate organs and permit changes in the size and shape of visceral organs. The **dorsal body cavity** includes the **cranial cavity** (enclosing the brain) and **spinal cavity** (surrounding the spinal cord). The **ventral body cavity,** or **coelom,** surrounds organs of the respiratory, cardiovascular, digestive, urinary, and reproductive systems *(see Figures 1-13/1-14).*

10. The **diaphragm** divides the ventral body cavity into the superior **thoracic** and inferior **abdominopelvic cavities** *(see Figures 1-13/1-14).*

11. The **abdominal cavity** extends from the inferior surface of the diaphragm to an imaginary line drawn from the inferior surface of the lowest spinal vertebra to the anterior and superior margin of the pelvic girdle. The portion of the ventral body cavity inferior to this imaginary line is the **pelvic cavity** *(see Figure 1-14).*

12. These cavities contain narrow fluid-filled spaces lined by a *serous membrane.* The thoracic cavity contains two **pleural cavities** (each surrounding a lung) separated by the **mediastinum** *(see Figure 1-14).*

13. The mediastinum contains the thymus, trachea, esophagus, blood vessels, and the **pericardial cavity,** which surrounds the heart. The membrane lining the pleural cavities is called the **pleura;** the membrane lining the pericardial cavity is called the **pericardium** *(see Figure 1-14).*

14. The abdominopelvic cavity or **peritoneal cavity** is lined by the **peritoneum.** Many digestive organs are supported and stabilized by **mesenteries.**

15. Important **radiological procedures** (which can provide detailed information about internal systems) include **X-rays, CT scans, MRI,** and **ultrasound.** Physicians who perform and analyze these procedures are called **radiologists** *(see Figures 1-15/1-16).*

1   REVIEW OF CHAPTER OBJECTIVES

1. Understand the reasons for studying anatomy and describe the relationship between structure and function.
2. Define the various specialties of anatomy.
3. Identify the major levels of organization in living organisms.
4. Describe the basic functions of living organisms.
5. Identify the organ systems of the human body, and understand their major functions.
6. Use anatomical terms to describe body sections, body regions, and relative positions.
7. Identify the major body cavities.

2   REVIEW OF CONCEPTS

1. Why is the study of anatomy important to an understanding of many aspects of biology?
2. What features are important in determining the subdiscipline of microscopic anatomy appropriate for a given topic of study?
3. How do different approaches to anatomy yield different types of information?
4. How can comparative anatomy contribute to a better understanding of anatomical structure and function in humans?
5. Why is it important to study the different levels of organization of organs and organ systems?

6. Does a problem in one system affect other systems, and if so, why?
7. Why do more complex organisms have specialized systems not present in simpler life forms?
8. What is the significance of using specialized anatomical terminology whenever discussing bodily structures or functions?
9. What is the functional significance of individual body cavities that contain internal organs?

3   CRITICAL THINKING AND CLINICAL APPLICATION QUESTIONS

1. If a person becomes ill and some of the symptoms indicate that he or she may be infected with a parasitic organism, treatment will depend upon the correct diagnosis of the problem. What level of anatomical study would be most appropriate to identify an infectious agent in the blood or muscle tissue? What other kinds of effects would either of these infections have?
2. A child born with a cleft palate requires surgery to construct a functional nasal septum, upper lip, and hard palate. Prior to correction of this defect, the function of several body systems is affected. What are these systems? Also, studies of animals that also have this type of defect have served as models to help understand how to solve these anatomical problems. What systems and anatomical approaches are necessary to solve this problem?

2

The Cell

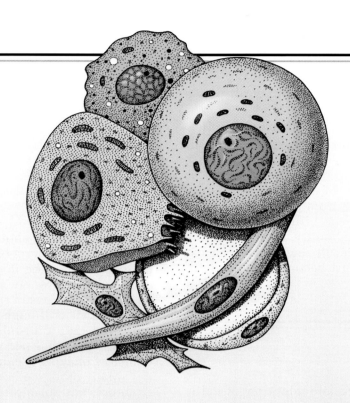

CHAPTER OUTLINE AND OBJECTIVES

Atoms are the building blocks of molecules; cells are the building blocks of the human body. Cells were first described by the English scientist Robert Hooke, around 1665. Hooke used an early light microscope to examine dried cork. He observed thousands of tiny empty chambers, which he named *cells*. Later that decade other scientists, observing the structure of living plants, realized that in life these spaces were filled with a gelatinous material. Research over the next 175 years led to the cell theory, the concept that cells are the fundamental units of all plant and animal tissues. Since that time the cell theory has been expanded to incorporate several basic concepts relevant to our discussion of the human body. The basic concepts of the cell theory are:

1. Cells are the structural "building blocks" of all plants and animals.

2. Cells are produced by the division of preexisting cells.

3. Cells are the smallest structural units that perform all vital functions.

The human body contains trillions of cells, and all our activities, from running to thinking, result from the combined and coordinated responses of millions or even billions of cells. Yet each cell also functions as an individual entity, responding to a variety of environmental cues. As a result, before we can explore human structure in detail we must become familiar with basic aspects of cell biology.

THE STUDY OF CELLS
(Figure 2-1)

Cytology is the study of the structure and function of cells. What we have learned over the past 40 years has given us new insights into cellular physiology and the mechanisms of homeostatic control. The two most common methods used to study cell and tissue structure are *light microscopy* and *electron microscopy*.

Before the 1950s most information was provided by light microscopy. A photograph taken through a light microscope is called a light micrograph (LM) (see Figure 2-1a●). Light microscopy can magnify cellular structures about 1000 times and show details as fine as 0.25 μm. (The symbol μm stands for micrometer, often shortened to micron; 1 μm = 0.001 mm, or 0.00004 in.) With a light microscope one can identify cell types, such as neurons or blood cells (Figure 2-1a, b●), and see large intracellular structures (Figure 2-1c●). The relative proportions of the cells in Figure 2-1c are correct, but all have been magnified roughly 500 times. Together, these and other types of cells create and maintain all anatomical structures and perform all vital functions.

Because individual cells are relatively transparent and difficult to distinguish from their neighbors, cells are treated with a variety of dyes that stain intracellular structures, thereby making them easier to see. Although special staining techniques can show the general distribution of protein, lipid, carbohydrate, or nucleic acids in the cell, many fine details of intracellular structure remained a mystery until investigators began using electron microscopy. This technique uses a focused beam of electrons, rather than a beam of light, to examine cell structure. In **transmission electron microscopy**, electrons penetrate an ultrathin

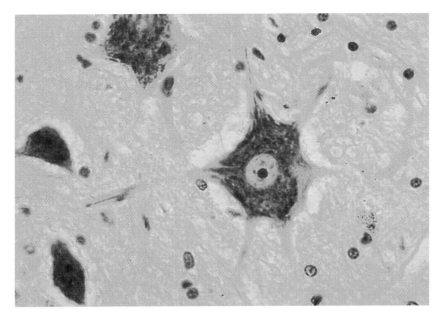

FIGURE 2-1a
A Light Micrograph of Neurons in the Spinal Cord

section to strike a photographic plate. The result is a transmission electron micrograph (TEM). Transmission electron microscopy shows the fine structure of cell membranes and intracellular structures. In **scanning electron microscopy**, electrons bouncing off exposed surfaces create a scanning electron micrograph (SEM). Although scanning microscopy provides less magnification than transmission electron microscopy, it provides a three-dimensional perspective on cell structure due to multiple planes of imaging (Figure 2-1b●).

Many other methods can be used to examine cell and tissue structure, and examples will be found in the pages that follow and throughout the book. This chapter describes the structure of a typical cell, some of the ways in which cells interact with their environment, and how cells reproduce.

FIGURE 2-1b
A Scanning Electron Micrograph of Human Red Blood Cells. (SEM × 1250)

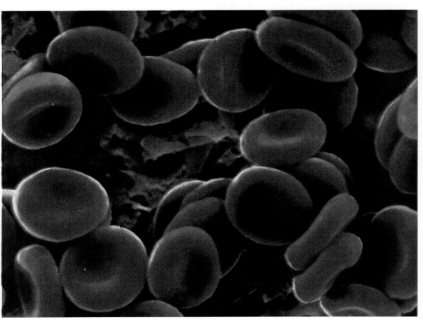

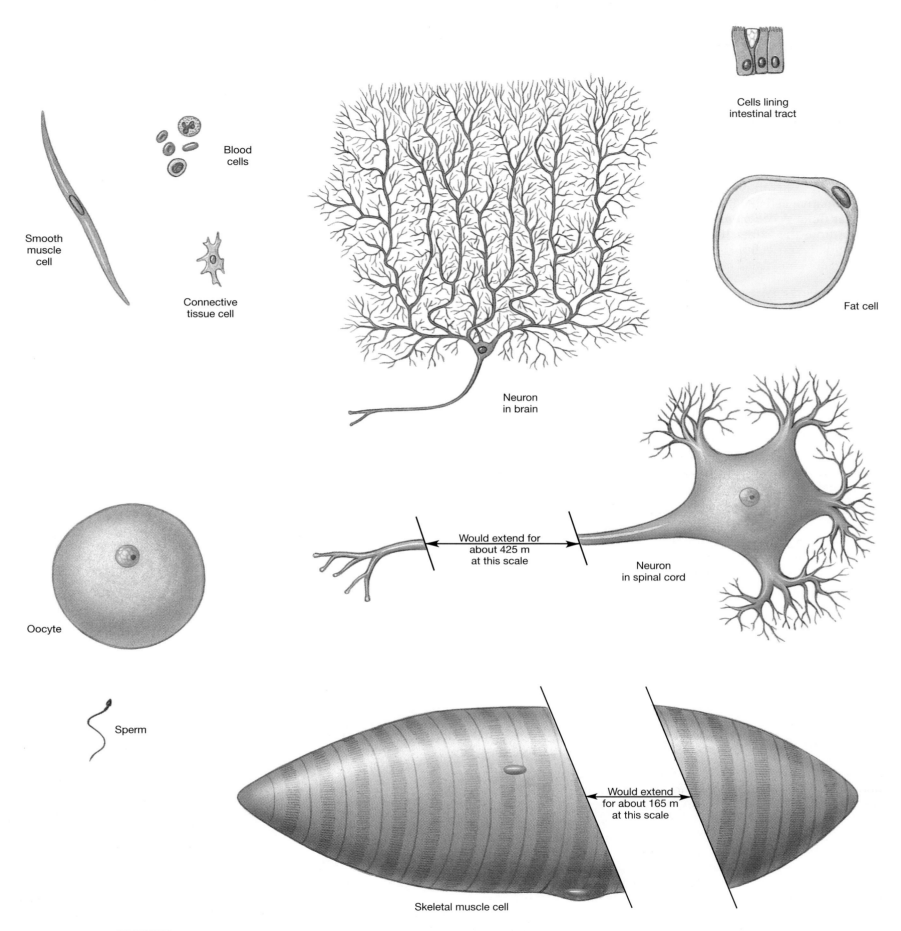

Blood cells

Smooth muscle cell

Connective tissue cell

Cells lining intestinal tract

Fat cell

Neuron in brain

Oocyte

Would extend for about 425 m at this scale

Neuron in spinal cord

Sperm

Would extend for about 165 m at this scale

Skeletal muscle cell

FIGURE 2-1c
The Diversity of Cells in the Human Body. The cells of the body have many different shapes and a variety of special functions. These examples give an indication of the range of forms and sizes; all of the cells are shown with the dimensions they would have if magnified approximately 500 times.

CELLULAR ANATOMY
(Figures 2-2/2-3)

The "typical" cell is like the "average" person. Any description can be thought of only in general terms because enormous individual variations occur. Our typical model cell will share features with most cells of the body without being identical to any. Figure 2-2● shows such a cell, and Table 2-1 summarizes the major structures and functions of its parts.

Figure 2-3● previews the organization of this chapter. Our representative cells float in a watery medium known as the **extracellular fluid**. A *cell membrane* separates the cell contents, or *cytoplasm*, from the extracellular fluid. The cytoplasm can be further subdivided into a fluid, the *cytosol*, and intracellular structures collectively known as *organelles* (or-gan-ELS, "little organs").

FIGURE 2-2
Anatomy of a Typical Cell. See Table 2-1 for a summary of the functions associated with the various cell structures.

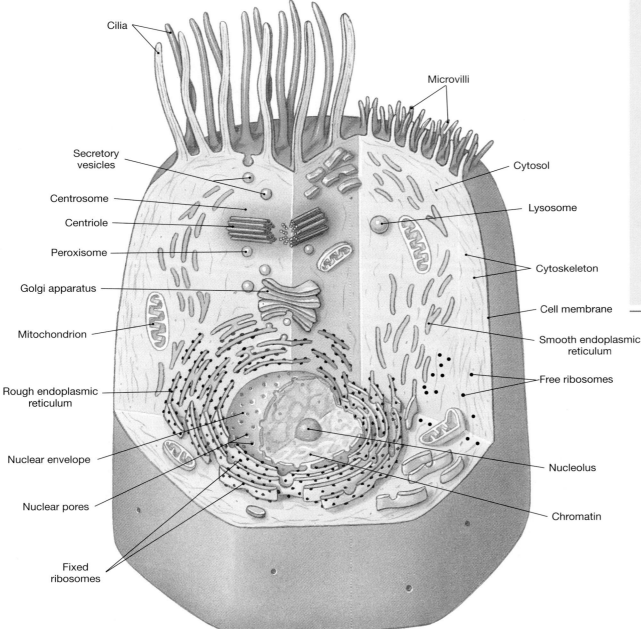

TABLE 2-1 Anatomy of a Representative Cell

Appearance

	Cell membrane
	Cytosol
	Cytoskeleton: microtubules, microfilaments
	Microvilli
	Cilia
	Centrioles
	Ribosomes
	Mitochondria
	Nucleus
	Nucleolus
	Endoplasmic reticulum (ER) Rough ER Smooth ER
	Golgi apparatus
	Lysosome
	Peroxisome

Structure	Composition	Function
CELL MEMBRANE	Lipid bilayer, containing phospholipids, steroids, and proteins	Isolation, protection, sensitivity, organization
CYTOSOL	Fluid component of cytoplasm	Distributes materials by diffusion
NONMEMBRANOUS ORGANELLES		
Cytoskeleton Microtubules, Microfilaments	Proteins organized in fine filaments or slender tubes	Strength and support, movement of cellular structures and materials
Microvilli	Membrane extensions containing microfilaments	Absorption of extracellular materials
Cilia	Membrane extensions containing microtubules in 9 + 2 arrangement	Movement of materials over cell surface
Centrioles	Two centrioles, at right angles within the centrosome; each centriole composed of microtubules in 9 × 3 array	Movement of chromosomes during cell division
Ribosomes	RNA + proteins; fixed ribosomes bound to endoplasmic reticulum, free ribosomes scattered in cytoplasm	Protein synthesis
MEMBRANOUS ORGANELLES **Mitochondria**	Double membrane, with inner membrane folds (cristae) enclosing important metabolic enzymes	Produce 95% of the ATP required by the cell
Nucleus	Nucleoplasm containing nucleotides, enzymes, and nucleoproteins; surrounded by double membrane or "nuclear envelope"	Control of metabolism; storage and processing of genetic information
Nucleolus	Dense region in nucleoplasm containing chromatin	Site of RNA synthesis and formation of ribosomal subunits
Endoplasmic reticulum	Network of membranous channels extending throughout the cytoplasm	Synthesis of secretory products; intracellular storage and transport
Rough ER	Ribosomes bound to membranes	Secretory protein synthesis
Smooth ER	Lacks attached ribosomes	Lipid and carbohydrate synthesis
Golgi apparatus	Series of stacked, flattened membranes (saccules) containing chambers (cisternae)	Storage, alteration, and packaging of secretory products and lysosomes
Lysosomes	Vesicles containing powerful digestive enzymes	Intracellular removal of damaged organelles or of pathogens
Peroxisomes	Vesicles containing degradative enzymes	Neutralization of toxic compounds

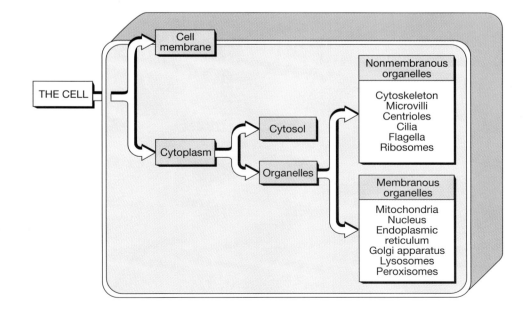

FIGURE 2-3
A Flow Chart for the Study of Cell Structure

The Cell Membrane
(Figure 2-4)

The **cell membrane**, also called the **plasma membrane** or *plasmalemma*, forms the outer boundary of the cell. It is extremely thin and delicate, ranging from 6 to 10 nm in thickness. Nevertheless, it has a complex structure composed of phospholipids, proteins, glycolipids, and cholesterol. The structure of the cell membrane is shown in Figure 2-4●.

The cell membrane is called a **phospholipid bilayer** because phospholipids form two distinct layers. In each layer the phospholipid molecules lie so that the heads are at the surface and the tails are on the inside. Dissolved ions and water-soluble compounds cannot cross the lipid portion of a cell membrane because the lipid tails will not associate with water molecules. This feature makes the membrane very effective in isolating the cytoplasm from the surrounding fluid environment. Such isolation is important because the composition of the cytoplasm is very different from that of the extracellular fluid, and that difference must be maintained.

Peripheral proteins are attached to the inner membrane surface; **integral proteins** are embedded in the membrane. Some of the integral proteins form **channels** that let water molecules, ions, and small water-soluble compounds into or out of the cell. Most of the communication between the interior and exterior of the cell occurs via these channels. Some of the channels are called **gated** because they can open or close to regulate the passage of materials.

The inner and outer surfaces of the cell membrane differ in protein and lipid composition. The glycolipids and glycoproteins that extend away from the outer surface of the cell membrane form a viscous, superficial coating known as the

glycocalyx (*calyx*, cup). Some of these molecules function as receptors: When bound to a specific molecule in the extracellular fluid, a membrane receptor can trigger a change in cellular activity. On the inner surface of the cell membrane, cytoplasmic enzymes may be bound to integral proteins, and these enzymes may be affected by events on the membrane surface.

The general functions of the cell membrane include:

1. *Physical isolation*: The cell membrane is a physical barrier that separates the inside of the cell from the surrounding extracellular fluid.

2. *Regulation of exchange with the environment*: The cell membrane controls the entry of ions and nutrients, the elimination of wastes, and the release of secretory products.

3. *Sensitivity*: The cell membrane is the first part of the cell affected by changes in the extracellular fluid. It also contains a variety of receptors that allow the cell to recognize and respond to specific molecules in its environment. Any alteration in the cell membrane may affect all cellular activities.

4. *Structural support*: Specialized connections between cell membranes or between membranes and extracellular materials give tissues a stable structure.

Membrane structure is not rigid, and integral proteins can drift from place to place across the surface of the membrane like ice cubes in a bowl of punch. In addition, the composition of the cell membrane can change over time, through the removal and replacement of membrane components.

FIGURE 2-4
The Cell Membrane

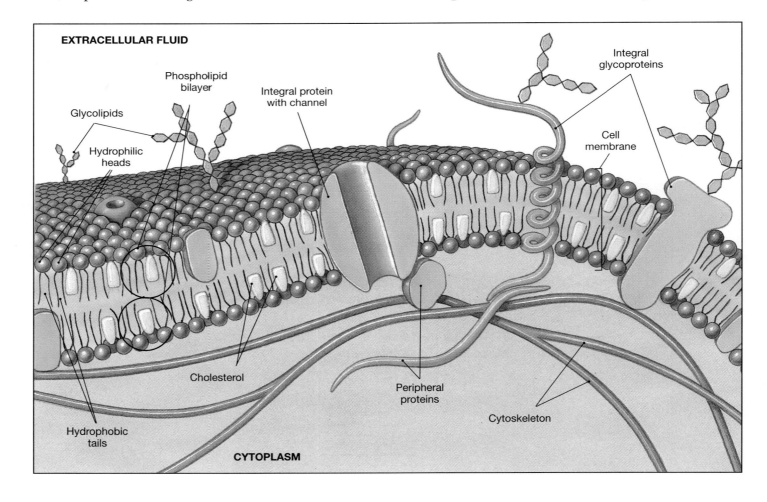

Membrane Permeability: Passive Processes

Precisely which substances can enter or leave the cytoplasm are determined by a property of the plasma membrane called its permeability. The **permeability** of a membrane is a property that determines its effectiveness as a barrier. If nothing can cross a membrane, it is described as **impermeable**. If any substance at all can cross without difficulty, the membrane is **freely permeable**. Cell membranes fall somewhere in between and are thus said to be **selectively permeable**. A selectively permeable membrane permits the free passage of some materials and restricts the passage of others. The distinction may be on the basis of size, electrical charge, molecular shape, solubility, or some combination of factors.

The permeability of a cell membrane varies depending on the organization and identity of membrane lipids and proteins. Passage across the membrane may be active or passive. Active processes, discussed later in this chapter, require that the cell draw on an energy source, usually ATP. Passive processes move ions or molecules across the cell membrane without any energy expenditure by the cell. Passive processes include *diffusion, osmosis, bulk flow, filtration,* and *facilitated diffusion.*

DIFFUSION *(Figure 2-5).* Ions and molecules in solution are in constant motion, bouncing off one another and colliding with water molecules. The result of the continual collisions and rebounds that occur is the process called diffusion. **Diffusion** can be defined as the net movement of material from an area where its concentration is relatively high to an area where its concentration is relatively low. The difference between the high and low concentrations represents a **concentration gradient**, and diffusion takes place until that gradient has been eliminated. Because diffusion occurs from a region of high concentration to one of relatively lower concentration, it is often described as proceeding "down a concentration gradient." When a concentration gradient has been eliminated, molecular motion contin-

ues, but it is random. An *equilibrium* now exists, and there is no longer a net movement in any direction.

Diffusion is important in body fluids because it tends to eliminate local concentration gradients. For example, an active cell generates carbon dioxide and absorbs oxygen. As a result, the extracellular fluid around the cell develops a relatively high concentration of CO_2 and a relatively low concentration of O_2. Diffusion then distributes the carbon dioxide through the tissue and into the bloodstream. At the same time, oxygen diffuses out of the blood and into the tissue.

In the extracellular fluids of the body, water and dissolved solutes diffuse freely. A cell membrane, however, acts as a barrier that selectively restricts diffusion. Some substances can pass through easily, whereas others cannot penetrate the membrane at all. There are only two ways for an ion or molecule to cross a cell membrane: diffuse through one of the membrane channels or diffuse across the lipid portion of the membrane. The size of the ion or molecule and any electrical charge it might carry determines its ability to pass through membrane channels. To cross the lipid portion of the membrane, the molecule must be lipid-soluble. These mechanisms are summarized in Figure 2-5●.

OSMOSIS. Cell membranes are very permeable to water molecules. This diffusion of water across a membrane in response to differences in concentration is so important that it is given a special name, **osmosis** (oz-MŌ-sis; *osmos,* thrust). For convenience, we will always use the term *osmosis* when considering water movement and restrict use of the term *diffusion* to the movement of solutes (substances dissolved in water).

Whenever an osmotic gradient exists, water molecules will diffuse rapidly across the cell membrane until the osmotic gradient is eliminated.

BULK FLOW. Osmosis eliminates solute concentration differences quickly. When water molecules cross a membrane they

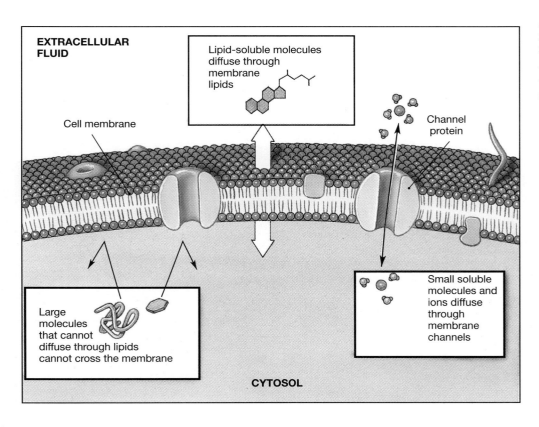

FIGURE 2-5
Diffusion across Cell Membranes. Small ions and water-soluble molecules diffuse through membrane channels. Lipid-soluble molecules can cross the membrane by diffusing through the phospholipid bilayer. Large molecules that are not lipid-soluble cannot diffuse through the membrane at all.

EXTRACELLULAR FLUID

Lipid-soluble molecules diffuse through membrane lipids

Cell membrane

Channel protein

Large molecules that cannot diffuse through lipids cannot cross the membrane

Small soluble molecules and ions diffuse through membrane channels

CYTOSOL

move in groups held together by hydrogen bonds. So, while solute molecules usually diffuse through membrane channels one at a time, water molecules move together in large clusters.

FILTRATION. In **filtration**, hydrostatic pressure forces water across a membrane, and solute molecules are selected on the basis of size. If the membrane pores are large enough, molecules of solute will be carried along with the water. We can see filtration in action in a coffee machine. Gravity forces hot water through the filter, and the water carries with it a variety of dissolved compounds. The large coffee grounds never reach the pot because they cannot fit through the fine pores in the filter. In the body, the heart pushes blood through the circulatory system and generates *hydrostatic pressure*. Filtration occurs across the walls of small blood vessels, pushing water and dissolved nutrients into the tissues of the body.

FACILITATED DIFFUSION. Many essential nutrients, such as glucose and amino acids, are insoluble in lipids but too large to fit through membrane channels. These compounds can be passively transported across the membrane by special **carrier proteins** in a process called **facilitated diffusion**. The molecule to be transported first binds to a **receptor site** on the protein. It is then moved to the inside of the cell membrane and released into the cytoplasm. No ATP is expended in facilitated diffusion or simple diffusion, and in each case molecules move from an area of higher concentration to one of lower concentration.

Membrane Permeability: Active Processes

All **active membrane processes** require energy. By spending energy, usually in the form of ATP, the cell can transport substances *against their concentration gradients*. We will consider two active processes: *active transport* and *endocytosis*.

ACTIVE TRANSPORT. In **active transport** the high-energy bond in ATP provides the energy needed to move ions or molecules across the membrane. The process is complex, and specific enzymes must be present in addition to carrier proteins. Although it has an energy cost, active transport offers one great advantage: It is not dependent on a concentration gradient. As a result the cell can import or export specific materials *regardless of their intracellular or extracellular concentrations*.

All living cells show active transport of sodium (Na^+), potassium (K^+), calcium (Ca^{2+}), and magnesium (Mg^{2+}). Specialized cells can transport additional ions such as iodide (I^-) or iron (Fe^{2+}). Many of these carrier mechanisms, known as **ion pumps**, move a specific cation or anion in one direction, either in or out of the cell. If one ion moves in one direction and another moves in the opposite direction, the carrier is called an **exchange pump**. The energy demands of these pumps are impressive; a resting cell may use up to 40% of the ATP it produces to power its exchange pumps.

ENDOCYTOSIS. Endocytosis (EN-dō-sī-TŌ-sis) is the packaging of extracellular materials into a vesicle at the cell surface for importation into the cell. This process,

which involves relatively large volumes of extracellular material, is sometimes called *bulk transport*. There are three major types of endocytosis: *pinocytosis, receptor-mediated endocytosis*, and *phagocytosis*. All three require energy in the form of ATP and so are classified as active processes. The mechanism is presumed to be the same in all three cases, but the mechanism itself remains unknown.

All forms of endocytosis produce cytoplasmic vesicles. Once an endocytotic vesicle has been formed, the contents do not necessarily enter the cytosol. They remain isolated within the vesicle unless they can pass through the vesicle wall. They may pass by means of active transport, simple or facilitated diffusion, or the destruction of the vesicle membrane.

Pinocytosis. Pinocytosis (PIN-ō-sī-TŌ-sis), or "cell drinking," is the formation of vesicles filled with extracellular fluid. In this process, a deep groove or pocket forms in the cell membrane and then pinches off. Nutrients, such as lipids, sugars, or amino acids, then enter the cytoplasm by diffusion or active transport. The membrane of the pinocytotic vesicle then pinches off and returns to the cell surface.

Virtually all cells perform pinocytosis in this manner. In a few specialized cells, the vesicles form on one side of the cell and travel through the cytoplasm to the opposite side. There they fuse with the cell membrane and discharge their contents. This method of bulk transport is found in cells lining capillaries, the most delicate blood vessels. These cells use pinocytosis to transfer fluid and solutes from the bloodstream to the surrounding tissues.

Receptor-Mediated Endocytosis (**Figure 2-6**). **Receptor-mediated endocytosis** resembles pinocytosis, but it is far more selec-

FIGURE 2-6
Receptor-Mediated Endocytosis

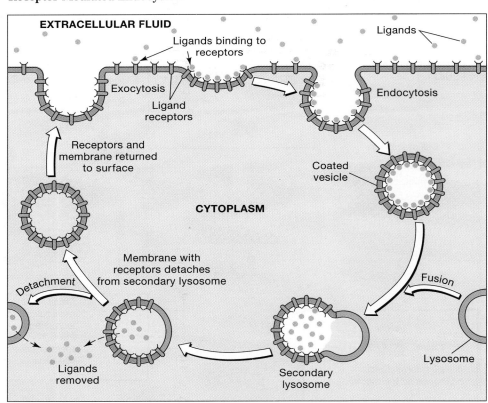

tive (Figure 2-6●). Pinocytosis produces vesicles filled with extracellular fluid; receptor-mediated endocytosis produces vesicles that contain a specific target molecule in high concentrations. The target substances are bound to receptors on the membrane surface. Many important substances, including cholesterol and iron ions (Fe^{2+}), are distributed through the body attached to special transport proteins. The proteins are too large to pass through membrane pores, but they can enter the cell through receptor-mediated endocytosis.

Phagocytosis (Figure 2-7). Phagocytosis (FA-gō-sī-TŌ-sis), or "cell eating," produces vesicles containing *solid objects* that may be as large as the cell itself. This process is shown in Figure 2-7●. Cytoplasmic extensions called **pseudopodia** (soo-dō-PŌ-dē-a; *pseudo-*, false + *podon*, foot) surround the object, and their membranes fuse to form a vesicle. The vesicle may then fuse with a lysosome, whereupon its contents are digested by lysosomal enzymes.

Most cells display pinocytosis, but phagocytosis, especially the entrapment of living or dead cells, is performed only by specialized cells of the immune system. The phagocytic activity of these cells will be considered in chapters dealing with blood cells (Chapter 20) and the lymphatic system (Chapter 23). A summary and comparison of the mechanisms involved in movement across cell membranes is presented in Table 2-2.

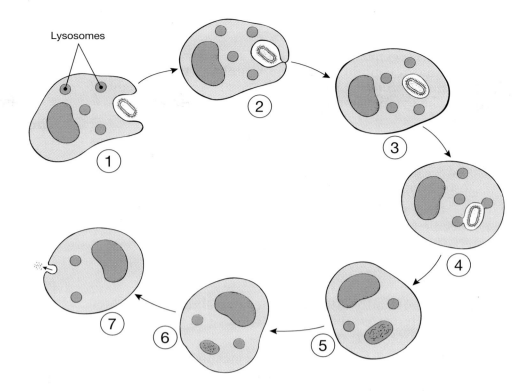

FIGURE 2-7
Phagocytosis. A phagocytic cell first comes in contact with the foreign object and sends cytoplasmic extensions around it (1). The extensions approach one another (2) and then fuse to trap the material within an endocytotic vesicle (3). Lysosomes fuse with this vesicle, activating digestive enzymes that gradually break down the structure of the phagocytized material (4–6). Undissolved residue can then be ejected by exocytosis (7).

TABLE 2-2 Summary of Mechanisms Involved in Movement across Cell Membranes

Mechanism	Process	Factors Affecting Rate	Substances Involved
PASSIVE			
Diffusion	Molecular movement of solutes; direction determined by relative concentrations	Size of gradient, molecular size, charge, lipid solubility	Small inorganic ions, lipid-soluble materials (all cells)
Osmosis	Movement of water (solvent) molecules toward high solute concentrations; requires membrane	Concentration gradient, opposing pressure	Water only (all cells)
Filtration	Movement of water, usually with solute, by hydrostatic pressure; requires membrane filter	Amount of pressure, size of pores	Water and small ions (blood vessels)
Facilitated diffusion	Carrier molecules transport down concentration gradient; requires membrane	As above, plus availability of carrier	Glucose and amino acids (all cells)
ACTIVE			
Active transport	Carrier molecules work despite opposing concentration gradients	Availability of carrier, substrate, and ATP	Na^+, K^+, Ca^{2+}, Mg^{2+} (all cells); probably other materials in special cases
Endocytosis	Creation of vesicles containing fluid or solid material	Stimulus and mechanics not understood; requires ATP	Fluids, nutrients (all cells); debris, pathogens (special cells)
Exocytosis	Fusion of vesicles with the cell membrane, releasing fluids and/or solids	Stimulus and mechanism incompletely understood; requires ATP and calcium ions	Fluid and wastes (all cells)

The Cytoplasm

Cytoplasm is a general term for all of the material inside the cell. Cytoplasm contains many more proteins than the extracellular fluid; proteins account for 15–30% of the weight of the cell. The cytoplasm includes two major subdivisions:

  1. **Cytosol**, or intracellular fluid. The cytosol contains dissolved nutrients, ions, soluble and insoluble proteins, and waste products. The cell membrane separates the cytosol from the surrounding extracellular fluid.

  2. **Organelles** (or-gan-ELS) are structures that perform specific functions within the cell.

The Cytosol

Cytosol is significantly different from extracellular fluid. Three important differences are:

  1. The cytosol contains a high concentration of potassium ions, whereas extracellular fluid contains a high concentration of sodium ions.

  2. The cytosol contains a relatively high concentration of dissolved and suspended proteins. Many of these proteins are enzymes that regulate metabolic operations, while others are associated with the various organelles. These proteins give the cytosol a consistency that varies between that of thin maple syrup and almost-set gelatin.

  3. The cytosol contains relatively small quantities of carbohydrates and large reserves of amino acids and lipids. The carbohydrates are broken down to provide energy, and the amino acids are used to manufacture proteins. The lipids stored in the cell are used primarily as an energy source when carbohydrates are unavailable.

The cytosol of cells contains masses of insoluble materials known as **inclusions**. Among the most common inclusions are stored nutrients: for example, glycogen granules in liver or skeletal muscle cells, and lipid droplets in fat cells.

Organelles
(Figure 2-2)

Organelles are found in all body cells (Figure 2-2●, p. 28). Each organelle performs specific functions that are essential to normal cell structure, maintenance, and metabolism. Cellular organelles can be divided into two broad categories: (1) **nonmembranous organelles**, which are always in contact with the cytosol; and (2) **membranous organelles** surrounded by membranes that isolate their contents from the cytosol, just as the cell membrane isolates the cytosol from the extracellular fluid. Table 2-1 on pp. 28 and 29 identifies those organelles which are categorized as nonmembranous and membranous.

Nonmembranous Organelles

Nonmembranous organelles include the *cytoskeleton, microvilli, centrioles, cilia, flagella,* and *ribosomes*.

The Cytoskeleton

The **cytoskeleton** is an internal protein framework that gives the cytoplasm strength and flexibility. It has four major components: *microfilaments, intermediate filaments, thick filaments,* and *microtubules*.

MICROFILAMENTS *(Figure 2-8a).*   **Microfilaments** are slender protein strands, composed primarily of the protein **actin**. In most cells microfilaments are scattered throughout the cytoplasm and form a dense network under the cell membrane. Figure 2-8a● shows the superficial layers of microfilaments in an intestinal cell.
  Microfilaments have two major functions:

  1. Microfilaments anchor the cytoskeleton to integral proteins of the cell membrane. This function stabilizes the position of the membrane proteins, provides additional mechanical strength to the cell, and firmly attaches the cell membrane to the underlying cytoplasm.

  2. Actin microfilaments can interact with microfilaments or thick filaments composed of the protein **myosin** to produce active movement of a portion of a cell, or a change in the shape of the entire cell. Such interactions are responsible for the contraction of muscle cells.

INTERMEDIATE FILAMENTS.   **Intermediate filaments** are defined chiefly by their size; their composition varies from one cell type to another. Intermediate filaments (1) provide strength, (2) stabilize the positions of organelles, and (3) transport materials within the cytoplasm. For example, specialized intermediate filaments, called **neurofilaments**, are found in neurons, where they provide structural support within long cellular processes, or axons, that may be up to a meter in length.

THICK FILAMENTS.   **Thick filaments**, not shown in Figure 2-8, are relatively massive filaments composed of myosin protein subunits. Thick filaments are abundant in muscle cells, where they interact with actin filaments to produce powerful contractions.

MICROTUBULES *(Figures 2-8b/2-9).*   **Microtubules**, found in all body cells, are hollow tubes built from the globular protein **tubulin**. Figure 2-8b● shows the microtubules in the cytoplasm of a representative cell. A microtubule forms through the aggregation of tubulin molecules; it persists for a time and then disassembles into individual tubulin molecules once again.

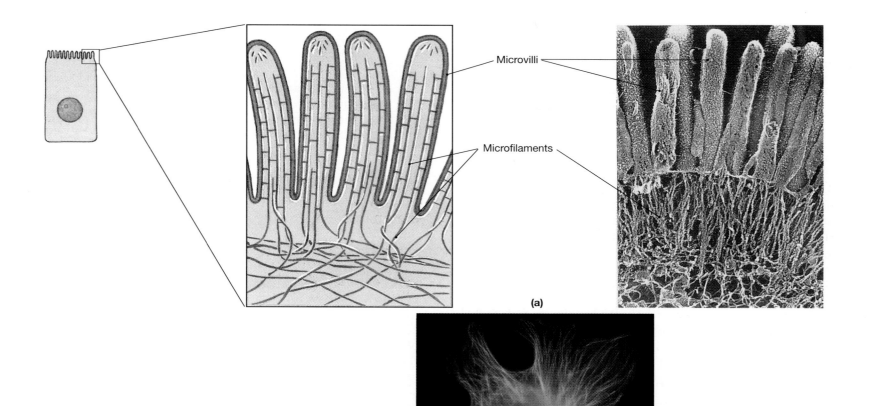

(a)

(b)

FIGURE 2-8
Microfilaments, Microvilli, and Microtubules. (a) Microfilaments form a network just under the cell membrane. They also extend into microvilli that project above the membrane surface. (SEM) (b) Microtubules in a living cell, as seen after special fluorescent labeling. (LM × 3200)

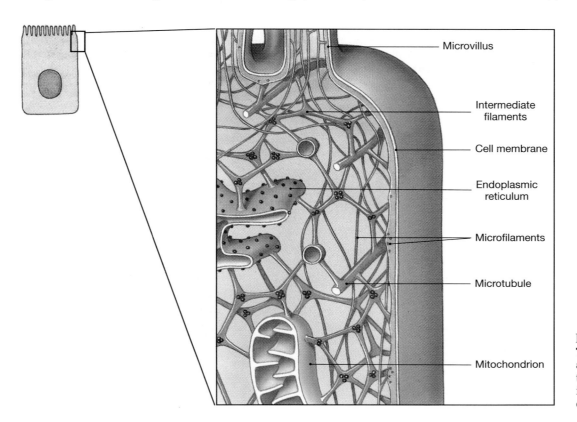

Microvillus

Intermediate filaments

Cell membrane

Endoplasmic reticulum

Microfilaments

Microtubule

Mitochondrion

FIGURE 2-9
The Cytoskeleton. The cytoskeleton provides strength and structural support for the cell and its organelles. Interactions between cytoskeletal components are also important in moving organelles and changing the shape of the cell.

Microtubules have a variety of functions:

1. Microtubules form the primary components of the cytoskeleton, giving the cell strength and rigidity, and anchoring the positions of major organelles.

2. Disassembly of microtubules provides a mechanism for changing the shape of the cell, perhaps assisting in cell movement.

3. Microtubules can attach to organelles and other intracellular materials and move them around within the cell.

4. During cell division, microtubules form the *spindle apparatus* that distributes the duplicated chromosomes to opposite ends of the dividing cell. This process will be considered in more detail in a later section.

5. Microtubules form structural components of organelles such as *centrioles*, *cilia*, and *flagella*.

The cytoskeleton as a whole incorporates microfilaments, intermediate filaments, and microtubules into a network that extends throughout the cytoplasm. The organizational details are as yet poorly understood, because the network is extremely delicate and thus hard to study in an intact state. Figure 2-9● is based on our current knowledge of cytoskeletal structure.

Microvilli
(Figure 2-8a)

Microvilli are small, finger-shaped projections of the cell membrane. They are found in cells that are actively engaged in absorbing materials from the extracellular fluid, such as the cells of the small intestine and kidneys. View Figure 2-8a● and note how a network of microfilaments stiffens each microvillus and anchors it to the underlying cytoskeleton. Interactions between these microfilaments and the cytoskeleton can produce a waving or bending action. Microvilli are important because they increase the surface area exposed to the extracellular environment for increased absorption. Their movements help to circulate fluid around the microvilli, bringing dissolved nutrients into contact with receptors on the membrane surface.

Centrioles, Cilia, and Flagella

The cytoskeleton contains numerous microtubules that function individually. Microtubules can also interact to form more complex structures known as *centrioles*, *cilia*, and *flagella*. These structures are summarized in Table 2-3.

CENTRIOLES *(Figures 2-2/2-10a).* A **centriole** (Figure 2-10a●) is a cylindrical structure composed of short microtubules. There are nine groups of microtubules, with three in each group. Body cells that are capable of reproducing themselves contain a pair of centrioles arranged as indicated in Figure 2-2●, p. 28. Cells that do not divide, such as mature red blood cells and skeletal muscle cells, lack centrioles. The **centrosome** is the cytoplasm surrounding this pair. Microtubules of the cytoskeleton usually begin within the centrosome and radiate through the cytoplasm. Centrioles direct the movement of chromosomes during cell division (discussed later in this chapter).

CILIA *(Figure 2-10b,c).* **Cilia** (singular *cilium*) contain nine pairs of microtubules surrounding a central pair (Figure 2-10b●). Cilia are anchored to a compact **basal body** situated just be-

FIGURE 2-10
Centrioles and Cilia. (a) The centrosome contains a pair of centrioles oriented at right angles to one another. (b) A cilium contains nine pairs of microtubules surrounding a central pair. (c) A single cilium swings forward and then returns to its original position. During the power stroke, the cilium is relatively stiff, but during the return stroke, it bends and moves parallel to the cell surface.

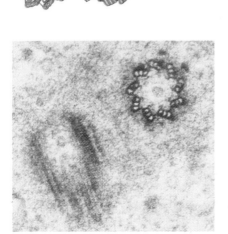

Microtubules

(a) Centrioles

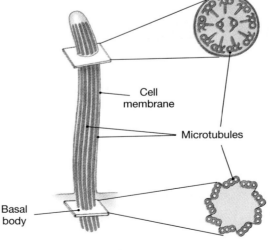

Cell membrane

Microtubules

Basal body

(b) Cilium

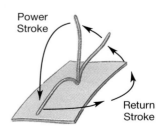

Power Stroke

Return Stroke

(c)

Structure	Microtubule Organization	Location	Function
Centriole	Nine groups of microtubules form a short cylinder	In centrosome near nucleus	Moves chromosomes during cell division
Cilium	Nine groups of long microtubules form a cylinder around a central pair	At cell surface	Moves fluids or solids across cell surface
Flagellum	Same as cilium	At cell surface	Moves sperm cells through fluid

TABLE 2-3   A Comparison of Centrioles, Cilia, and Flagella

neath the cell surface. The structure of the basal body resembles that of a centriole. The exposed portion of the cilium is completely covered by the cell membrane. Cilia "beat" rhythmically, as depicted in Figure 2-10c●, and their combined efforts move fluids or secretions across the cell surface. Cilia lining the respiratory tract beat in a synchronized manner to move sticky mucus and trapped dust particles toward the throat and away from delicate respiratory surfaces. If the cilia are damaged or immobilized by heavy smoking or some metabolic problem, the cleansing action is lost, and the irritants will no longer be removed. As a result, chronic respiratory infections develop.

FLAGELLA.   **Flagella** (fla-JEL-ah; singular *flagellum*, "whip") resemble cilia but they are much longer. Flagella move a cell through the surrounding fluid, rather than moving the fluid past a stationary cell. The sperm cell is the only human cell that has a flagellum, and it is used to move the cell along the female reproductive tract. If sperm flagella are paralyzed or otherwise abnormal, the individual will be sterile because immobile sperm cannot reach and fertilize an egg.

Ribosomes
(Figure 2-11)

Ribosomes are small, dense structures that cannot be seen with the light microscope. In an electron micrograph, ribosomes are dense granules roughly 25 nm in diameter (Figure 2-11a●). They are found in all cells, but their number varies depending on the type of cell and its activities. Each ribosome consists of roughly 60% RNA and 40% protein. At least 80 ribosomal proteins have been identified. These organelles are intracellular factories that manufacture proteins, using information provided by the DNA of the nucleus.

There are two major types of ribosomes: free ribosomes and fixed ribosomes as seen in Figure 2-11a●. **Free ribosomes** are scattered throughout the cytoplasm; the proteins they manufacture enter the cytosol. **Fixed ribosomes** are attached to the *endoplasmic reticulum (ER)*, a membranous organelle. Proteins manufactured by fixed ribosomes enter the *lumen*, or internal cavity, of the ER, where they are modified and packaged for export. These processes are described in the section on the ER later in this chapter.

√ **Cells lining the small intestine have numerous fingerlike projections on their free surfaces. What are these structures, and what is their function?**

√ **How would the absence of a flagellum affect a sperm cell?**

FIGURE 2-11
Ribosomes. (a) Both free and fixed ribosomes can be seen in the cytoplasm of this cell. (TEM × 73,600) (b) An individual ribosome, consisting of small and large subunits.

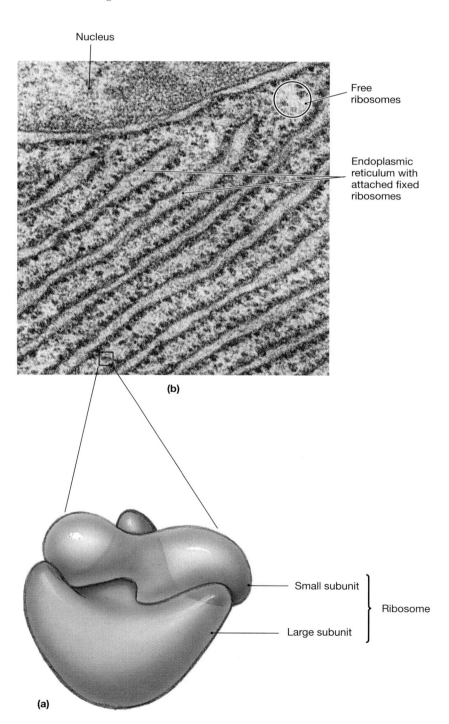

Membranous Organelles

A phospholipid bilayer membrane similar to the cell membrane surrounds each membranous organelle, isolating it from the cytosol. This isolation allows the organelle to manufacture or store secretions, enzymes, or toxins that could adversely affect the cytoplasm in general. Table 2-1 on pp. 28 and 29 includes six types of membranous organelles: *mitochondria, the nucleus, the endoplasmic reticulum, the Golgi apparatus, lysosomes,* and *peroxisomes.*

Mitochondria
(Figure 2-12)

Mitochondria (mī-tō-KON-drē-ah; singular *mitochondrion*; *mitos*, thread + *chondros*, cartilage) are organelles that have an unusual double membrane (Figure 2-12●). An outer membrane surrounds the entire organelle, and a second, inner membrane contains numerous folds, called **cristae**. Cristae increase the surface area exposed to the fluid contents, or **matrix**, of the mitochondrion. The matrix contains metabolic enzymes that perform the reactions that provide energy for cellular functions.

Enzymes attached to the cristae produce most of the ATP generated by mitochondria. Mitochondrial activity produces about 95% of the energy needed to keep a cell alive. Mitochondria produce ATP through the breakdown of organic molecules in a series of reactions that also consume oxygen (O_2) and generate carbon dioxide (CO_2).

Mitochondria have a variety of shapes, from long and slender to short and fat. Mitochondria control their own maintenance, growth, and reproduction. The number of mitochondria in a particular cell varies depending on the cell's energy demands. Red blood cells have none, but liver and skeletal muscle cells typically contain as many as 300 mitochondria. Muscle cells have high rates of energy consumption, and over time the mitochondria respond to the increased energy demands by reproducing. The increased numbers of mitochondria can provide energy faster and in greater amounts, improving muscular performance.

The Nucleus
(Figures 2-13/2-14)

The **nucleus** is the control center for cellular operations. A single nucleus stores all the information needed to control the synthesis of the approximately 100,000 different proteins in the human body. The nucleus determines the structural and functional characteristics of the cell by controlling what proteins are synthesized, and in what amounts. A cell without a nucleus could be compared to a car without a driver. However, a car can sit idle for years, but a cell without a nucleus will disintegrate within 3–4 months.

Most cells contain a single nucleus, but there are exceptions. For example, skeletal muscle cells have many nuclei, and mature red blood cells have none. Figure 2-13● details the structure of a typical nucleus. A **nuclear envelope** surrounds the nucleus and separates it from the cytosol. The nuclear envelope is a double membrane containing a narrow **perinuclear space** (*peri-*, around). At several locations, the nuclear envelope is connected to the rough endoplasmic reticulum, as shown in Figure 2-2●, p. 28.

The nucleus directs processes that take place in the cytosol and must in turn receive information about conditions and activities in the cytosol. Chemical communication between the nucleus and cytosol occurs through **nuclear pores**. These pores, which account for about 10% of the surface of the nucleus, are large enough to permit the movement of ions and small molecules, but too small for the passage of proteins or DNA.

The term **nucleoplasm** refers to the fluid contents of the nucleus. The nucleoplasm contains ions, enzymes, RNA and

FIGURE 2-12
Mitochondria. The sketch details the three-dimensional organization of a mitochondrion, and the false-color TEM shows a typical mitochondrion in section. (TEM × 61,776)

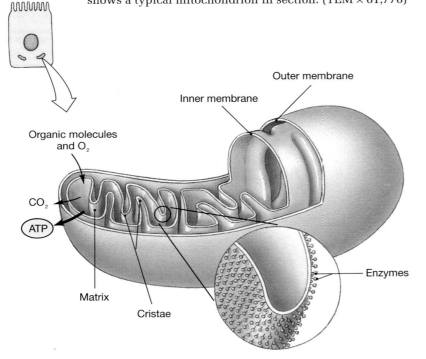

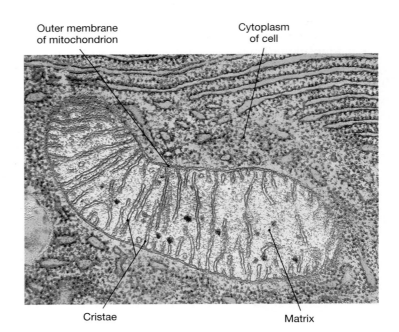

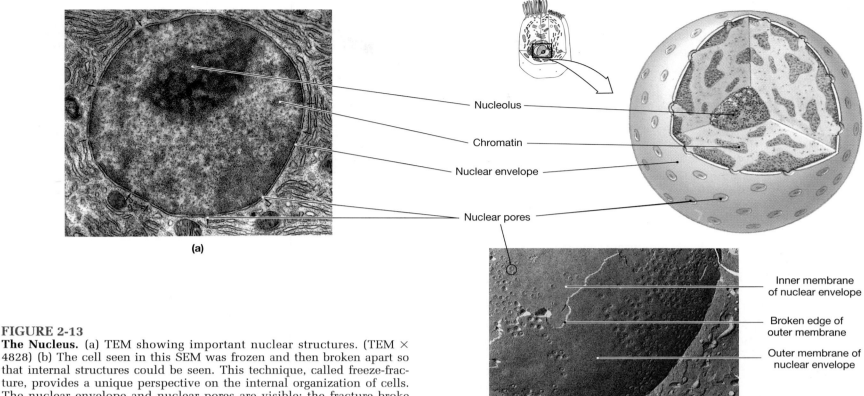

FIGURE 2-13
The Nucleus. (a) TEM showing important nuclear structures. (TEM ×
4828) (b) The cell seen in this SEM was frozen and then broken apart so
that internal structures could be seen. This technique, called freeze-frac-
ture, provides a unique perspective on the internal organization of cells.
The nuclear envelope and nuclear pores are visible; the fracture broke
away part of the outer membrane of the nuclear envelope, and the cut edge
can be seen crossing the center of the nucleus. (SEM × 9240)

DNA nucleotides, proteins, small amounts of RNA, and DNA.
The DNA strands form complex structures known as *chromosomes* (*chroma*, color). The nucleoplasm also contains a network of fine filaments, the **nuclear matrix**, that provides structural support and may be involved in the regulation of genetic
activity. Each **chromosome** contains DNA strands bound to special proteins called **histones**. Our cell nuclei contain 23 pairs of
chromosomes; one member of each pair is derived from our
mother and one from our father. The structure of a typical chromosome is diagrammed in Figure 2-14●.

At intervals the DNA strands wind around the histones,
forming a complex known as a **nucleosome**. The entire chain
of nucleosomes may coil around other histones. The degree
of coiling determines whether the chromosome is long and
thin or short and fat. Chromosomes in a dividing cell are very
tightly coiled, and so can be seen clearly as separate structures in light or electron micrographs. In cells that are not dividing, the chromosomes are loosely coiled, forming a tangle
of fine filaments known as **chromatin** (KRŌ-ma-tin). Each
chromosome may have some coiled regions, and only the
coiled areas stain clearly. As a result, the nucleus has a
clumped, grainy appearance.

The chromosomes also have direct control over the synthesis of RNA. Most nuclei contain one to four dark-staining areas
called **nucleoli** (noo-KLĒ-ō-lī; singular *nucleolus*). Nucleoli are
nuclear organelles that synthesize the components of ribosomes. A nucleolus contains histones and enzymes as well as
RNA, and it forms around a chromosomal region containing the
genetic instructions for producing ribosomal proteins and RNA.
Nucleoli are most prominent in cells that manufacture large
amounts of proteins, such as liver cells and muscle cells, because these cells need large numbers of ribosomes.

FIGURE 2-14
Chromosome Structure. DNA strands are coiled around histones to form nucleosomes. Nucleosomes form coils that may be very tight or rather loose. In
cells that are not dividing, the DNA is loosely coiled, forming a tangled network known as chromatin. When the coiling becomes tighter, as it does in
preparation for cell division, the DNA becomes visible as distinct structures
called chromosomes.

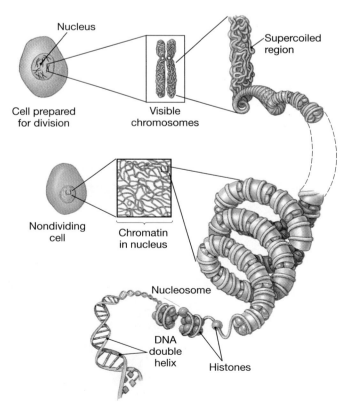

The Endoplasmic Reticulum
(Figure 2-15)

The **endoplasmic reticulum** (en-dō-PLAZ-mik re-TIK-ū-lum), or **ER,** is a network of intracellular membranes. It has three major functions:

1. *Synthesis*: The membrane of the endoplasmic reticulum manufactures proteins, carbohydrates, and lipids.
2. *Storage*: The ER can hold synthesized molecules or substances absorbed from the cytosol without affecting other cellular operations.
3. *Transport*: Substances can travel from place to place within the cell inside the endoplasmic reticulum.

The endoplasmic reticulum forms hollow tubes, flattened sheets, and round chambers (Figure 2-15●). The chambers are called **cisternae** (sis-TUR-nē; singular *cisterna*, a reservoir for water). There are two distinct types of endoplasmic reticulum, **rough endoplasmic reticulum (RER)** and **smooth endoplasmic reticulum (SER)**.

The RER functions as a combination workshop and shipping depot. It is where many newly synthesized proteins undergo chemical modification and where they are packaged for export to their next destination, the *Golgi apparatus*.

The outer surface of the rough endoplasmic reticulum contains fixed ribosomes. Ribosomes synthesize proteins using instructions provided by a strand of RNA. As the polypeptide chains grow, they enter the cisternae of the endoplasmic reticulum, where they may be further modified. Most of the proteins and glycoproteins produced by the RER are packaged into small membrane sacs that pinch off the tips of the cisternae. These **transport vesicles** deliver the proteins to the Golgi apparatus.

No ribosomes are associated with smooth endoplasmic reticulum. The SER has a variety of functions that center around the synthesis of lipids and carbohydrates, the storage of calcium ions, and the removal and inactivation of toxins.

The amount of endoplasmic reticulum and the proportion of RER to SER vary depending on the type of cell and its ongoing activities. For example, pancreatic cells that manufacture digestive enzymes contain an extensive RER, and the SER is relatively small. The situation is just the reverse in the cells that synthesize steroid hormones in the reproductive system.

The Golgi Apparatus
(Figure 2-16)

The **Golgi** (GOL-jē) **apparatus** consists of flattened membrane discs, called **saccules** (SAK-ūls). A typical Golgi apparatus, shown in Figure 2-16●, consists of five to six saccules; a single cell may contain several sets, each resembling a stack of dinner plates. Most often these stacks lie near the nucleus of the cell.

The major functions of the Golgi apparatus are:

1. Synthesis and packaging of secretions, such as mucus or enzymes.
2. Packaging of special enzymes for use in the cytosol.
3. Renewal or modification of the cell membrane.

The Golgi saccules communicate with the ER and with the cell surface. This communication involves the formation, movement, and fusion of vesicles.

VESICLES AND SECRETION (Figure 2-17). The role played by the Golgi apparatus in packaging secretions is illustrated in Figure 2-17a●. Protein and glycoprotein synthesis occurs in the RER, and transport vesicles (packages) then move these products to the Golgi apparatus. The vesicles usually arrive at a convex saccule known as the *cis* (sis) *saccule,* or *forming face.* The transport vesicles then fuse with the Golgi membrane, emptying their contents into the cisternae. Inside the Golgi, enzymes modify the arriving proteins and glycoproteins.

Material moves from saccule to saccule by means of small **transfer vesicles**. Ultimately the product arrives at the *trans saccule,* or *maturing face.* At the *trans* saccule, vesicles form that carry materials away from the Golgi. Vesicles containing secretions that will be discharged from the cell are called **secretory vesicles**. Secretion occurs as the membrane of a secretory vesicle fuses with the cell membrane. This ejection process (Figure 2-17b●) is called **exocytosis** (eks-ō-sī-TŌ-sis).

MEMBRANE TURNOVER. Because the Golgi apparatus continually adds new membrane to the cell surface in this way, it has the ability to change the properties of the cell membrane

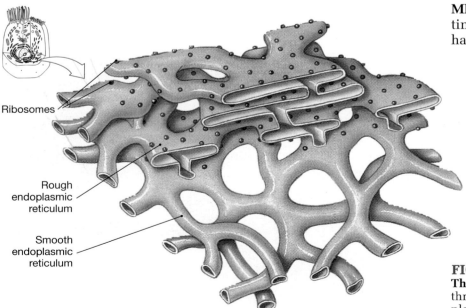

Ribosomes

Rough endoplasmic reticulum

Smooth endoplasmic reticulum

FIGURE 2-15
The Endoplasmic Reticulum. This diagrammatic sketch indicates the three-dimensional relationships between the rough and smooth endoplasmic reticulum.

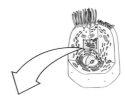

FIGURE 2-16

The Golgi Apparatus. (a) A sectional view of the Golgi apparatus of an active secretory cell. (TEM × 83,520) (b) A three-dimensional view of the Golgi apparatus with a cut edge corresponding to part (a).

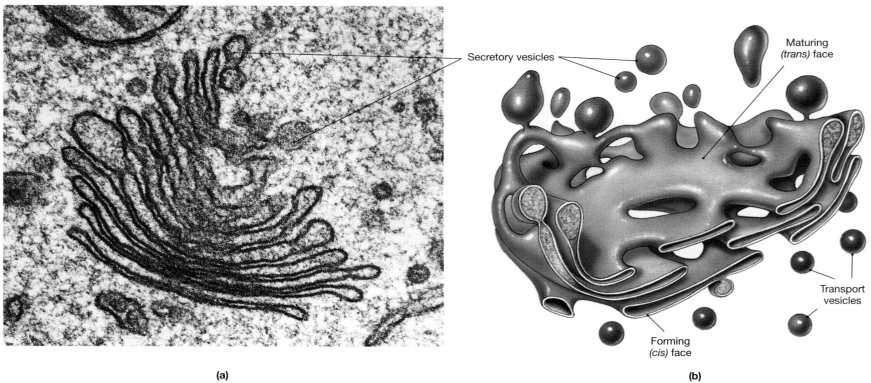

(a)

Secretory vesicles

Maturing (trans) face

Transport vesicles

Forming (cis) face

(b)

FIGURE 2-17

Golgi Function. (a) This diagram shows the functional link between the ER and the Golgi apparatus. Golgi structure has been simplified to clarify the relationships between the membranes. Transport vesicles carry the secretory product from the endoplasmic reticulum to the Golgi apparatus, and transfer vesicles move membrane and materials between the Golgi saccules. At the maturing face, three functional categories of vesicles develop. Secretory vesicles carry the secretion from the Golgi to the cell surface, where exocytosis releases the contents into the extracellular fluid. Other vesicles add surface area and integral proteins to the cell membrane. Lysosomes, which remain in the cytoplasm, are vesicles filled with enzymes. (b) Exocytosis at the surface of a cell.

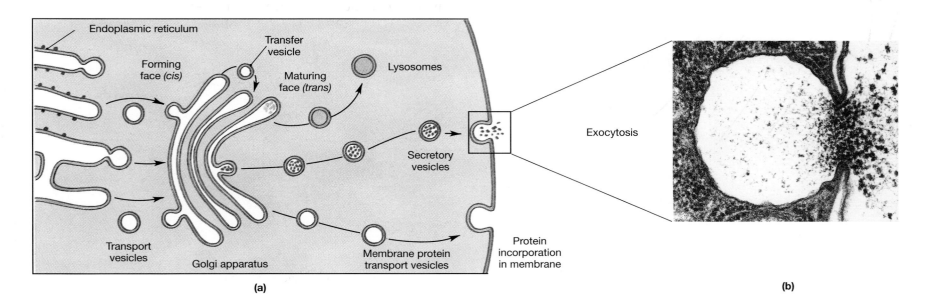

Endoplasmic reticulum

Transfer vesicle

Forming face (cis)

Maturing face (trans)

Lysosomes

Exocytosis

Transport vesicles

Golgi apparatus

Secretory vesicles

Membrane protein transport vesicles

Protein incorporation in membrane

(a)

(b)

over time. Such changes can profoundly alter the sensitivity and functions of the cell. In an actively secreting cell, the Golgi membranes may undergo a complete turnover every 40 minutes. The membrane lost from the Golgi is added to the cell surface, and that addition is balanced by the formation of vesicles at the membrane surface. As a result, an area equal to the entire membrane surface may be replaced each hour.

A third class of vesicles produced at the Golgi apparatus never leaves the cytoplasm. These vesicles contain digestive enzymes. The most important are *lysosomes*.

Lysosomes
(Figure 2-18)

Lysosomes (LĪ-sō-sōms; *lyso-*, dissolution + *soma*, body), are vesicles filled with digestive enzymes. Refer to Figure 2-18● as we describe the types of lysosomes and lysosomal functions. *Primary lysosomes* contain inactive enzymes. Activation occurs when the lysosome fuses with the membranes of damaged organelles, such as mitochondria or fragments of the endoplasmic reticulum. This fusion creates a *secondary lysosome*, which contains active enzymes. These enzymes then break down the lysosomal contents. Nutrients reenter the cytosol, and the remaining waste material is eliminated by exocytosis.

Lysosomes also function in the defense against disease. Cells may remove bacteria, as well as fluids and organic debris, from their surroundings in vesicles formed at the cell surface. Lysosomes may fuse with vesicles created in this way, and the digestive enzymes within the secondary lysosome then break down the contents and release usable substances such as sugars or amino acids. In this way the cell at once protects itself against pathogenic organisms and obtains valuable nutrients.

Lysosomes perform essential cleanup and recycling functions inside the cell. For example, when muscle cells are inactive, lysosomes gradually break down their contractile proteins; if the cells become active once again, this destruction ceases. This regulatory mechanism fails in a damaged or dead cell. Lysosomes then disintegrate, releasing active enzymes into the cytosol. These enzymes rapidly destroy the proteins and organelles of the cell, a process called **autolysis** (aw-TOL-i-sis; *auto-*, self). Because the breakdown of lysosomal membranes can destroy a cell, lysosomes have been called cellular "suicide packets." We do not know how to control lysosomal activities, or why the enclosed enzymes do not digest the lysosomal membranes unless the cell is damaged. Problems with lysosomal enzyme production cause more than 30 serious diseases affecting children. In these conditions, called *lysosomal storage diseases*, the lack of a specific lysosomal enzyme results in the buildup of waste products and debris normally removed and recycled by lysosomes. Affected individuals may die when vital cells, such as those of the heart, can no longer continue to function.

Peroxisomes

Peroxisomes are smaller than lysosomes and carry a different group of enzymes. Peroxisomes are thought to originate at the RER, whereas lysosomes are produced at the Golgi apparatus. Peroxisomes absorb and neutralize toxins, such as alcohol, which is absorbed from the extracellular fluid. Peroxisomes are most abundant in liver cells, which are responsible for removing and neutralizing toxins absorbed in the digestive tract.

Membrane Flow

With the exception of the mitochondria, all the membranous organelles in the cell are either interconnected or in communication through the movement of vesicles. The RER and SER are

FIGURE 2-18
Lysosomal Functions. Primary lysosomes, formed at the Golgi apparatus, contain inactive enzymes. Activation may occur under three basic conditions: (1) when the primary lysosome fuses with the membrane of another organelle, such as a mitochondrion; (2) when the primary lysosome fuses with an endocytotic vesicle containing fluid or solid materials from outside the cell; or (3) in autolysis, when the lysosomal membrane breaks down following death or injury to the cell.

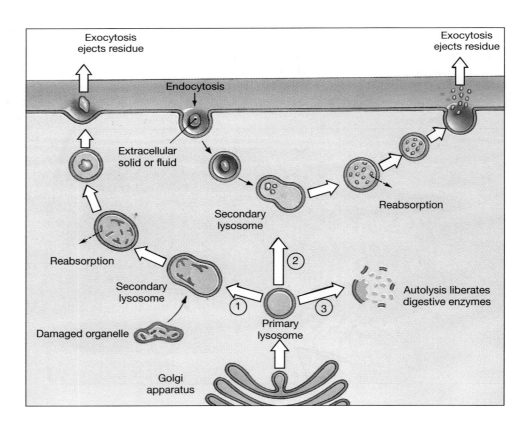

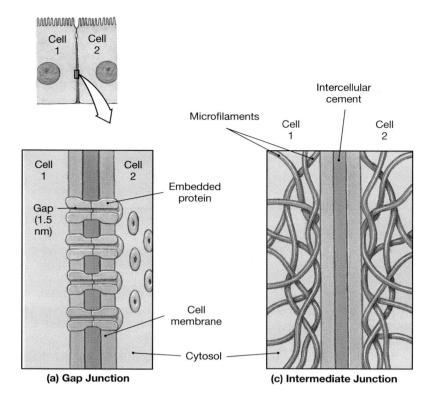

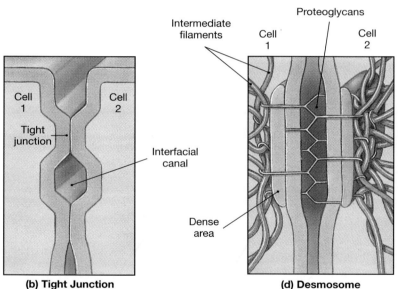

(a) Gap Junction

Microfilaments

Intercellular cement

Cell 1 — Cell 2

(c) Intermediate Junction

Embedded protein

Gap (1.5 nm)

Cell membrane

Cytosol

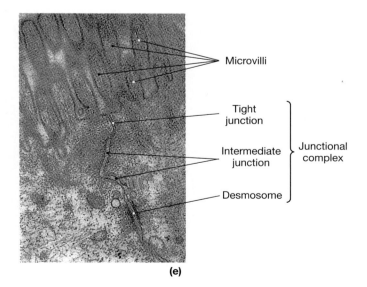

Intermediate filaments

Proteoglycans

Cell 1 — Cell 2

Tight junction

Interfacial canal

Dense area

(b) Tight Junction

(d) Desmosome

Microvilli

Tight junction

Intermediate junction

Desmosome

Junctional complex

(e)

continuous and connected to the nuclear envelope. Transport vesicles connect the ER with the Golgi apparatus, and secretory vesicles link the Golgi apparatus with the cell membrane. Finally, vesicles forming at the exposed surface of the cell remove and recycle segments of the cell membrane. This continual movement and exchange is called **membrane flow**. In an actively secreting cell, an area equal to the entire membrane surface may be replaced each hour.

Membrane flow is another example of the dynamic nature of cells. It provides a mechanism for cells to change the characteristics of their cell membranes—lipids, receptors, channels, anchors, and enzymes—as they grow, mature, or respond to a specific environmental stimulus.

√ **Microscopic examination of a cell reveals that it contains many mitochondria. What does this observation imply about the cell's energy requirements?**

√ **Cells in the ovaries and testes contain large amounts of smooth endoplasmic reticulum (SER). Why?**

INTERCELLULAR ATTACHMENT

Most cells in the body are firmly attached to other cells or to extracellular protein fibers. The attachments occur at four types of cell junctions: *gap junctions*, *tight junctions*, *intermediate junctions*, and *desmosomes*.

Gap Junctions
(Figure 2-19a)

In a **gap junction** (Figure 2-19a●), two cells are held together by an interlocking of membrane proteins. Because these are channel proteins, the result is a narrow passageway that lets small molecules and ions pass from cell to cell. Gap junctions are common among epithelial cells, where they help coordinate functions such as the beating of cilia. Gap junctions are most abundant in cardiac muscle and smooth muscle tissue.

Tight Junctions
(Figure 2-19b)

At a **tight junction**, shown in Figure 2-19b●, there is a partial fusion of the lipid portions of adjacent cell membranes. Because the membranes are fused together, tight junctions block the passage of water or solutes between the cells. For example, tight junctions found near the exposed surfaces of cells lining the digestive tract keep enzymes, acids, and wastes from damaging delicate underlying tissues.

FIGURE 2-19
Cell Attachments. (a) At a gap junction, binding of membrane proteins creates a cytoplasmic connection between two cells. (b) A tight junction is formed by fusion of the outer layers of two cell membranes. (c) At an intermediate junction, the membranes are held together by intercellular cement. A network of microfilaments strengthens the region of attachment. (d) A desmosome has a more organized network of microfilaments. Desmosomes attach one cell to another or attach a cell to extracellular structures, such as the protein fibers in connective tissues. (e) A junctional complex consists of a tight junction, an intermediate junction, and a desmosome. The tight junction is closest to the cell surface.

Intermediate Junctions
(Figure 2-19c)

At an **intermediate junction** (Figure 2-19c●), the opposing cell membranes, while not fused, are held together by a thick layer of proteoglycans. This proteoglycan layer is called **intercellular cement**. The polysaccharide **hyaluronic acid** is the most important component. The cytoplasm at an intermediate junction contains a dense network of microfilaments that anchor the junction to the cytoskeleton. This arrangement adds strength and helps to stabilize the shape of the cell.

Desmosomes
(Figure 2-19d)

At **desmosomes** (DEZ-mō-sōmz; *desmos*, ligament + *soma*, body), there is a very thin proteoglycan layer between the opposing cell membranes, reinforced by a network of intermediate filaments that lock the two cells together (Figure 2-19d●). Desmosomes are very strong and can resist stretching and twisting. These connections are most abundant between cells in the superficial layers of the skin. The desmosomes create links so strong that even dead skin cells are usually shed in thick sheets, rather than individually.

Junctional Complexes
(Figure 2-19e)

Cells lining the digestive tract, respiratory tract, or other passageways are held together by **junctional complexes**. A single junctional complex consists of a tight junction, an intermediate junction, and a desmosome, with the tight junction closest to the cell surface. Figure 2-19e● details the structure of a typical junctional complex.

THE CELL LIFE CYCLE
(Figure 2-20)

Between fertilization and physical maturity a human being goes from a single cell to roughly 75 trillion cells. This amazing increase in number occurs through a form of cellular reproduction called **cell division**. The division of a single cell produces a pair of *daughter cells*, each half the size of the original. Thus, two new cells have replaced the original one.

Even when development has been completed, cell division continues to be essential to survival. Although cells are highly adaptable, they can be damaged by physical wear and tear, toxic chemicals, temperature changes, or other environmental hazards. Cells are subject to aging. The life span of a cell varies from hours to decades, depending on the type of cell and the environmental stresses involved. A typical cell does not live nearly as long as a typical person, so over time cell populations must be maintained by cell division.

Central to cell reproduction is the accurate duplication of the cell's genetic material and its distribution to the two new daughter cells. This process is called **mitosis** (mī-TŌ-sis). Mitosis occurs during the division of **somatic** (*soma*, body) **cells**. Somatic cells include all of the cells in the body other than the **reproductive cells**, which give rise to sperm or eggs. Production of sperm and eggs involves a distinct process, *meiosis* (mī-Ō-sis), which will be described in Chapter 28. An overview of the life cycle of a typical somatic cell is presented in Figure 2-20●.

Interphase
(Figures 2-20/2-21)

Most cells spend only a small part of their time actively engaged in cell division. Somatic cells spend the majority of their functional lives in *interphase*. During **interphase** the cell performs all of its normal functions plus, if necessary, making preparations for division. In a cell preparing for division, interphase can be divided into the G_1, S, and G_2 phases. An interphase cell in the **G_0 phase** is not preparing for mitosis, but is performing all other normal cell functions. Some mature cells, such as skeletal muscle cells and many neurons, remain in G_0 indefinitely and may never undergo mitosis. In contrast, *stem cells*, which divide repeatedly with very brief interphase periods, never enter G_0.

FIGURE 2-20
The Cell Life Cycle

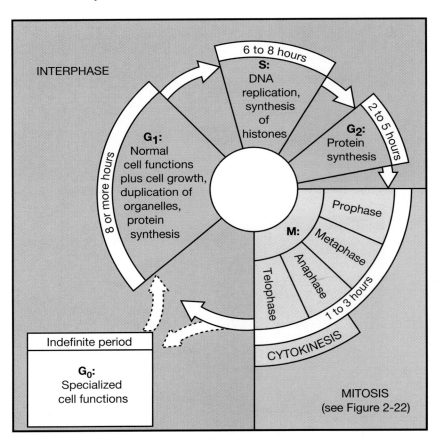

In the **G₁ phase** the cell manufactures enough mitochondria, centrioles, cytoskeletal elements, endoplasmic reticulum, ribosomes, Golgi membranes, and cytosol to make two functional cells. In cells dividing at top speed, G₁ may last as little as 8–12 hours. Such cells pour all of their energy into mitosis, and all other activities cease. If G₁ lasts for days, weeks, or months, preparation for mitosis occurs as the cells perform their normal functions. When preparations have been completed, the cell enters the **S phase**. Over the next 6–8 hours, the cell replicates its chromosomes.

Throughout the life of a cell, the DNA strands in the nucleus remain intact. DNA synthesis, or **DNA replication**, occurs in cells preparing to undergo mitosis or meiosis. The goal of replication is to copy the genetic information in the nucleus so that one set of chromosomes can be given to each of the two cells produced. Several different enzymes are needed for the process.

A DNA molecule consists of a pair of nucleotide strands held together by hydrogen bonding between complementary nitrogen bases. Figure 2-21● diagrams the process of DNA replication. It starts when the weak bonds between the nitrogenous bases are disrupted, and the strands unwind. As they do so, molecules of the enzyme **DNA polymerase** bind to the exposed nitrogenous bases. This enzyme promotes bonding between the nitrogenous bases of the DNA strand and complementary DNA nucleotides dissolved in the nucleoplasm.

Many molecules of DNA polymerase are working simultaneously, along different portions of each DNA strand. This process produces short complementary nucleotide chains that are then linked together by enzymes called **ligases** (LĪ-gās-ez; *liga*, to tie). The final result is a pair of identical DNA molecules.

Once DNA replication has been completed, there is a brief (2–5 hours) **G₂ phase** devoted to last-minute protein synthesis. The cell then enters the **G_M phase**, and mitosis begins.

FIGURE 2-21
DNA Replication. In replication the DNA strands unwind, and DNA polymerase begins attaching complementary DNA nucleotides along each strand. This process produces two identical copies of the original DNA molecule.

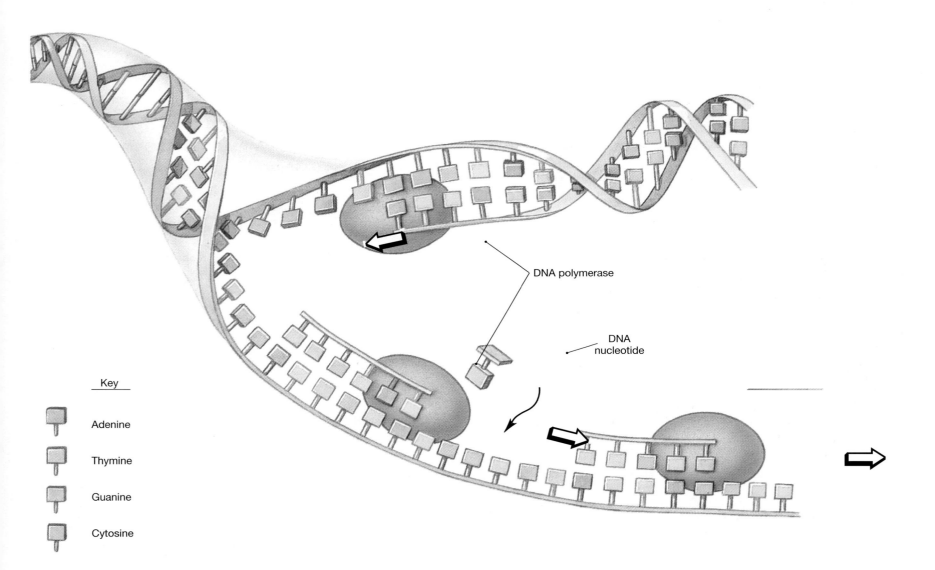

Key

Adenine

Thymine

Guanine

Cytosine

DNA polymerase

DNA nucleotide

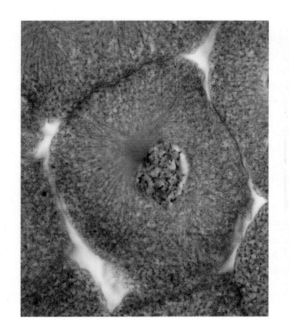
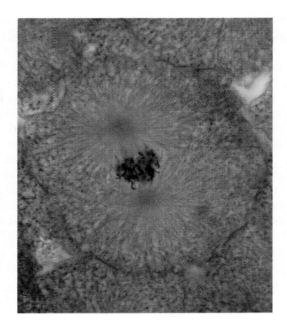
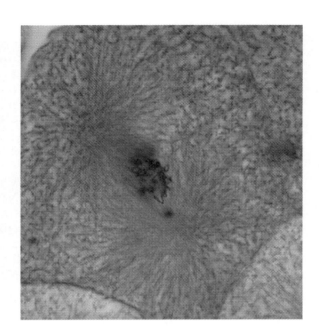

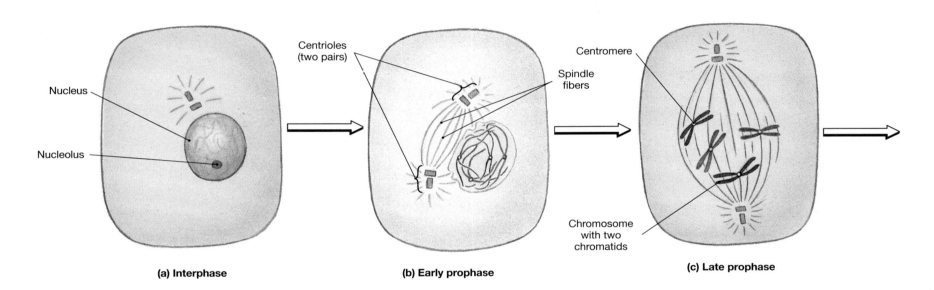

(a) Interphase

Nucleus

Nucleolus

Centrioles
(two pairs)

Spindle
fibers

(b) Early prophase

Centromere

Chromosome
with two
chromatids

(c) Late prophase

FIGURE 2-22
Mitosis. (a) Interphase. (b) Early prophase. (c) Late prophase. (d) Metaphase. (e) Anaphase. (f) Telophase. (LMs × 775)

Mitosis
(Figure 2-22)

Mitosis consists of four stages, detailed in Figure 2-22●.

STAGE 1: Prophase (PRŌ-fāz; *pro*, before). (See Figure 2-22b, c●.) Prophase begins when the chromosomes coil so tightly that they become visible as individual structures. As a result of DNA replication during the S phase, there are two copies of each chromosome, called **chromatids** (KRŌ-ma-tids), connected at a single point, the **centromere** (SEN-trō-mir). The two pairs of centrioles, duplicated in the G_1 phase, now move apart. **Spindle fibers** extend between the centriole pairs; smaller microtubules radiate into the surrounding cytosol. Prophase ends with the disappearance of the nuclear envelope.

STAGE 2: Metaphase (MET-a-fāz; *meta*, after). (See Figure 2-22d●.) The chromatids now move to a narrow central zone called the **metaphase plate**. A microtubule of the spindle apparatus attaches to each centromere.

STAGE 3: Anaphase (AN-uh-fāz; *ana*, back). (See Figure 2-22e●.) As if responding to a single command, the chromatid pairs separate, and the **daughter chromosomes** move toward opposite ends of the cell.

46

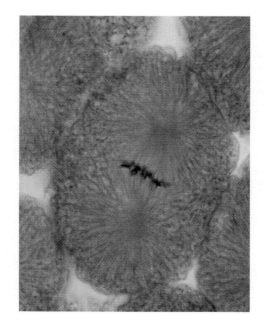 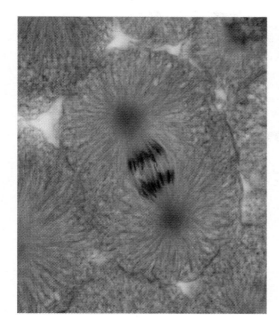 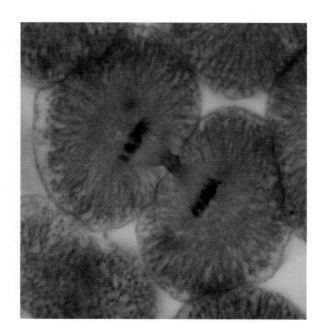

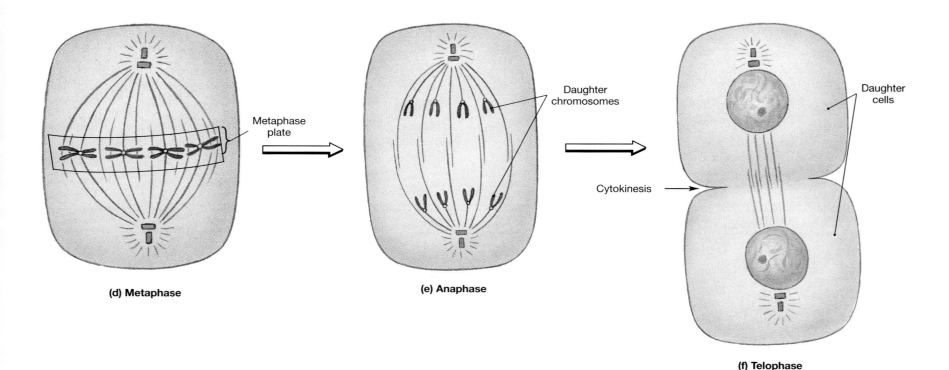

Metaphase plate

Daughter chromosomes

Daughter cells

Cytokinesis

(d) Metaphase

(e) Anaphase

(f) Telophase

STAGE 4: Telophase (TEL-ō-fāz; *telo*, end). (See Figure 2-22f●.) This stage is in many ways the reverse of prophase. The nuclear membranes form, the nuclei enlarge, and the chromosomes gradually uncoil. Once the chromosomes disappear, nucleoli reappear and the nuclei resemble those of interphase cells.

Telophase marks the end of mitosis proper, but the daughter cells have yet to complete their physical separation. This separation process, called **cytokinesis** (sī-tō-ki-NĒ-sis; *cyto-*, cell + *kinesis*, motion), usually begins in late anaphase. As the daughter chromosomes near the ends of the spindle apparatus, the cytoplasm constricts along the plane of the metaphase plate.

This process continues through telophase, and the completion of cytokinesis marks the end of cell division.

The frequency of cell division can be estimated by the number of cells in mitosis at any given time. As a result, the term **mitotic rate** is often used when discussing rates of cell division. In general, the longer the life expectancy of a cell type, the slower the mitotic rate. Relatively long-lived cells, such as muscle cells and neurons, either never divide or do so only under special circumstances. Other cells, like skin or the lining of the digestive tract, are constantly subjected to attack by chemicals, pathogens, and abrasion and survive only for days or even hours. Special cells called **stem cells** maintain these cell populations through repeated cycles of cell division.

Mitotic rates are usually well controlled, and in normal tissue the rate of cell division balances cell loss or destruction. When that balance breaks down, the tissue begins to enlarge. A **tumor**, or **neoplasm**, is a mass or swelling produced by abnormal cell growth and mitosis. In a **benign tumor** the cells remain within a connective tissue capsule. Such a tumor seldom threatens an individual's life. Surgery can usually remove the tumor if its size or position disturbs tissue function.

Cells in a **malignant tumor** are no longer responding to normal control mechanisms. These cells divide rapidly, spreading into the surrounding tissues, and they may also spread to other tissues and organs. This spread is called **metastasis** (me-TAS-ta-sis). Metastasis is dangerous and difficult to control. Once in a new location, the metastatic cells produce secondary tumors.

The term **cancer** refers to an illness characterized by malignant cells. Cancer cells gradually lose their resemblance to normal cells. They change size and shape, often becoming unusually large or abnormally small. Organ function begins to deteriorate as the number of cancer cells increases. The cancer cells may not perform their original functions at all, or they may perform normal functions in an unusual way. For example, endocrine cancer cells may produce normal hormones, but in abnormally large amounts. Cancer cells compete for space and nutrients with normal cells. They do not use energy very efficiently, and they grow and multiply at the expense of normal tissues. This activity accounts for the starved appearance of many patients in the late stages of cancer. ✝ *Causes of Cancer [p. 749]*

Related Clinical Terms

tumor (neoplasm): A mass or swelling produced by abnormal cell growth and division. *[p. 48]*

benign tumor: A mass or swelling in which the cells remain within a connective tissue capsule; rarely life-threatening. *[p. 48]*

malignant tumor: A mass or swelling in which the cells no longer respond to normal control mechanisms, but divide rapidly. *[p. 48]*

metastasis (me-TAS-ta-sis): The spread of malignant cells into surrounding tissues and organs. *[p. 48]*

cancer: An illness characterized by malignant cells. *[p. 48]*

hereditary predisposition: An individual born with genes that increase the likelihood of disease. ✝ *Causes of Cancer [p. 749]*

oncogene (ON-kō-jēn): A cancer-causing gene created by a somatic mutation in a normal gene (**proto-oncogene**) involved with growth, differentiation, or cell division. ✝ *Causes of Cancer [p. 749]*

tumor-suppressing genes (TSG) or **anti-oncogenes:** These genes suppress mitosis and growth in normal cells. ✝ *Causes of Cancer [p. 749]*

carcinogen (kar-SIN-ō-jen): An environmental factor that stimulates the conversion of a normal cell to a cancer cell. ✝ *Causes of Cancer [p. 749]*

mutagen (MŪ-ta-jen): A factor that can damage DNA strands and sometimes cause chromosomal breakage, stimulating the development of cancer cells. ✝ *Causes of Cancer [p. 749]*

CHAPTER SUMMARY AND REVIEW

STUDY OUTLINE

Related Key Terms

INTRODUCTION *[p. 26]*

1. Contemporary cell theory incorporates several basic concepts: (1) cells are the building blocks of all plants and animals; (2) cells are produced by the division of preexisting cells; (3) cells are the smallest units that perform all vital functions.

THE STUDY OF CELLS *[p. 26]*

1. *Cytology* analyzes the internal structure of individual cells.
2. *Light microscopy* permits the magnification and viewing of cellular structures about 1000 times their natural size *(see Figure 2-1)*.
3. The technique of *electron microscopy* uses a focused beam of electrons, rather than a beam of light, to examine cell structures.

**transmission electron microscopy •
scanning electron microscopy**

CELLULAR ANATOMY *[p. 28]*

1. A cell floats in the **extracellular fluid**. The cell's outer boundary is the **cell membrane**, or **plasma membrane**, which is a **phospholipid bilayer**. *Table 2-1* summarizes the anatomy of a typical cell *(see Figures 2-2/2-3)*.

The Cell Membrane *[p. 30]*

2. **Integral proteins** are part of the membrane itself, while **peripheral proteins** are attached but can separate from it. **Channels** allow water and ions to move across the membrane; some channels are called **gated** because they can open or close *(see Figures 2-4/2-5)*.

glycocalyx

3. Cell membranes are **selectively permeable**.

4. **Diffusion** is the net movement of material from an area where its concentration is relatively high to an area where its concentration is lower. Diffusion occurs until the **concentration gradient** is eliminated *(see Figure 2-5 and Table 2-2)*.

5. Diffusion of water across a membrane in response to differences in water concentration is called **osmosis** *(see Table 2-2)*.

6. In **filtration**, hydrostatic pressure forces water across a membrane; if membrane pores are large enough, molecules of solute will be carried along *(see Table 2-2)*.

7. **Facilitated diffusion** requires the presence of **carrier proteins** *(see Table 2-2)*.

8. All **active membrane processes** require energy. Two important active processes are active transport and endocytosis *(see Table 2-2)*.

9. **Active transport** mechanisms consume ATP and are independent of concentration gradients. Some **ion pumps** are **exchange pumps** *(see Table 2-2)*.

10. **Endocytosis** is an active process that can take three forms: **pinocytosis**, **receptor-mediated endocytosis**, and **phagocytosis**. A summary of mechanisms involved in movement of substances across cell membranes is presented in **Table 2-2** *(see Figures 2-6/2-7)*.

The Cytoplasm *[p. 34]*
11. The **cytoplasm** contains a fluid **cytosol** and surrounds **organelles** *(see Figure 2-2 and Table 2-1)*.

Nonmembranous Organelles *[p. 34]*
12. **Nonmembranous organelles** are always in contact with the cytosol. They include the cytoskeleton, microvilli, centrioles, cilia, flagella, and ribosomes *(see Figures 2-8 to 2-11 and Table 2-1)*.

13. The **cytoskeleton** gives the cytoplasm strength and flexibility. It has four components: **microfilaments**, **intermediate filaments**, **thick filaments**, and **microtubules** *(see Figures 2-8/2-9 and Table 2-1)*.

14. **Microvilli** are small projections of the cell membrane that increase the surface area exposed to the extracellular environment *(see Figure 2-8 and Table 2-1)*.

15. **Centrioles** direct the movement of DNA molecules during cell division *(see Figure 2-10 and Table 2-1)*.

16. **Cilia** beat rhythmically to move fluids or secretions across the cell surface *(see Figure 2-10 and Table 2-1)*.

17. **Flagella** move a cell through surrounding fluid, rather than moving fluid past a stationary cell. *Table 2-3* presents a comparison of centrioles, cilia, and flagella.

18. **Ribosomes** are intracellular factories that manufacture proteins. There are **free ribosomes** and **fixed ribosomes** *(see Figure 2-11 and Table 2-1)*.

Membranous Organelles *[p. 38]*
19. **Membranous organelles** are surrounded by lipid membranes that isolate them from the cytosol.

20. **Mitochondria** are responsible for 95% of the ATP production within a typical cell *(see Figure 2-12 and Table 2-1)*.

21. The **nucleus** is the control center for cellular operations. It is surrounded by a **nuclear envelope**, through which it communicates with the cytosol through **nuclear pores** *(see Figure 2-13 and Table 2-1)*.

22. The **endoplasmic reticulum (ER)** is a network of intracellular membranes. There are two types: rough and smooth. **Rough endoplasmic reticulum (RER)** contains ribosomes; **smooth endoplasmic reticulum (SER)** does not *(see Figure 2-15 and Table 2-1)*.

23. The **Golgi apparatus** packages **lysosomes**, **peroxisomes**, and **secretory vesicles**. Secretions are discharged from the cell in a process called **exocytosis** *(see Figures 2-16/2-17 and Table 2-1)*.

24. **Lysosomes** are vesicles filled with digestive enzymes. The process of *endocytosis* is important in ridding the cell of bacteria and debris *(see Figure 2-18 and Table 2-1)*.

25. **Peroxisomes** also carry enzymes; they absorb and neutralize toxins.

INTERCELLULAR ATTACHMENT *[p. 43]*

1. Cells can attach to other cells or to extracellular protein fibers in four ways: gap junctions, tight junctions, intermediate junctions, and desmosomes.

Gap Junctions *[p. 43]*
2. In a **gap junction** two cells are held together by interlocked membrane proteins, forming a narrow passageway *(see Figure 2-19a)*.

Tight Junctions *[p. 43]*
3. At a **tight junction** there is a partial fusion of the two cell membranes; these are the strongest intercellular connections *(see Figure 2-19b)*.

Intermediate Junctions *[p. 44]*
4. At an **intermediate junction** two cells are held together by a thick layer of proteoglycans called **intercellular cement** *(see Figure 2-19c)*.

Related Key Terms

permeability • impermeable • freely permeable

bulk flow

receptor site

pseudopodia

inclusions

actin • myosin • neurofilaments • tubulin

centrosome

basal body

cristae • matrix • respiratory enzymes

perinuclear space • nucleoplasm • chromosomes • histones • nucleosome • chromatin • nucleoli

cisternae • transport vesicles

saccules • transfer vesicles

autolysis

hyaluronic acid

Desmosomes *[p. 44]*

5. A **desmosome** has a very thin proteoglycan layer between the cell membranes, reinforced by a network of microfilaments *(see Figure 2-19d)*.

Junctional Complexes *[p. 44]*

6. Cells in some areas of the body are linked by **junctional complexes** *(see Figure 2-19e)*.

THE CELL LIFE CYCLE *[p. 44]*

Interphase *[p. 44]*

1. Mitosis refers to the nuclear division of **somatic cells. Reproductive cells** (sperm and eggs) are produced by *meiosis (see Figures 2-20 to 2-22)*.

2. Most somatic cells spend most of their time in **interphase** *(see Figure 2-20)*.

Mitosis *[p. 46]*

3. Mitosis proceeds in four stages: **prophase, metaphase, anaphase,** and **telophase** *(see Figure 2-22)*.

4. In general, the longer the life expectancy of a cell type, the slower the **mitotic rate. Stem cells** undergo frequent mitoses to replace other, more specialized cells.

Related Key Terms

cell division • G_0 phase • G_1 phase • S phase • DNA replication • DNA polymerase • ligases • G_2 phase • G_M phase

chromatids • centromere • spindle fibers • metaphase plate • daughter chromosomes • cytokinesis

---

1   REVIEW OF CHAPTER OBJECTIVES

1. Discuss the basic concepts of the cell theory.
2. Explain the structure and importance of the cell membrane.
3. Describe the ways in which materials move across the cell membrane.
4. Compare the fluid contents of a cell with the extracellular fluid.
5. Compare the structure and functions of the various cellular organelles.
6. Discuss the role of the nucleus as the cell's control center.
7. Discuss the way cells can be interconnected to maintain structural stability in body tissues.
8. Describe the cell life cycle and how cells divide through the process of mitosis.

2   REVIEW OF CONCEPTS

1. Why is the study of individual cells important to an understanding of the human body as a whole?
2. Describe the structure of a typical cell membrane.
3. What are protein channels and how do they work?
4. What are the general functions of cell membranes?
5. What is the function of a membrane receptor on the outer surface of a cell membrane?
6. How does membrane permeability affect the passage of substances into or out of the cell?
7. What passive processes allow materials to enter or leave a cell?
8. How does osmosis differ from diffusion?
9. How does filtration differ from osmosis?
10. How does facilitated diffusion differ from diffusion?
11. How do active processes differ from passive processes?
12. What are the requirements for active transport?
13. What is endocytosis?

14. Differentiate among pinocytosis, receptor-mediated endocytosis, and phagocytosis.
15. How does cytosol, the intracellular fluid, differ from extracellular fluid?
16. Other than presence or absence of membranes, how do membranous organelles differ from nonmembranous organelles?
17. What is the purpose of microvilli?
18. What type of structures are formed by combinations of microtubules?
19. How do mitochondria use oxygen?
20. What is the function of nuclear pores?
21. How do histones affect the structure of DNA?
22. Why are nucleoli prominent in cells that manufacture large amounts of proteins?
23. How do rough endoplasmic reticulum (RER) and smooth endoplasmic reticulum (SER) differ in function?
24. How do the Golgi apparatus and the endoplasmic reticulum interact?
25. What happens to cell contents when lysosomes break down?
26. How would a decrease in the production of peroxisomes affect liver function?
27. What type of cell-to-cell junction permits the least passage of materials between the cells?
28. What is the most important result of mitosis?
29. If normal activities during the S phase of interphase did not occur, how would mitosis be affected?

3   CRITICAL THINKING AND CLINICAL APPLICATION QUESTIONS

1. People who have high blood pressure often are put on a low-sodium (i.e., low-salt) diet, and their blood pressure decreases. Why does this occur?

3

The Tissue Level of Organization

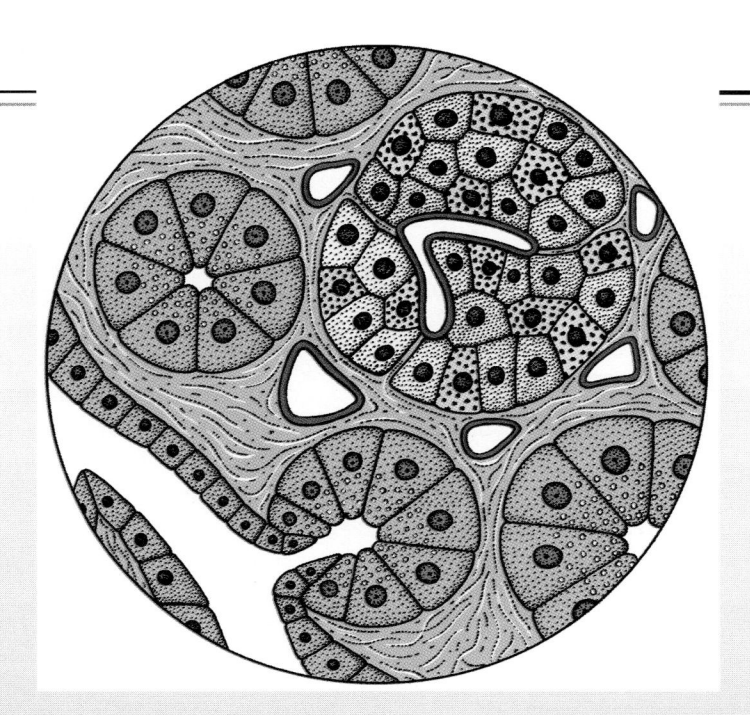

CHAPTER OUTLINE AND OBJECTIVES

No single cell contains the metabolic machinery and organelles needed to perform all the many functions of the human body. Instead, through the process of differentiation, each cell develops a characteristic set of structural features and a limited number of functions. These structures and functions can be quite distinct from those of nearby cells. Nevertheless, cells in a given location all work together. A detailed examination of the body reveals a number of patterns at the cellular level. Although the body contains trillions of cells, there are only about 200 types of cells. These cell types combine to form **tissues**, collections of specialized cells and cell products that perform a relatively limited number of functions. There are four **primary tissue types**: *epithelial tissue*, *connective tissue*, *muscle tissue*, and *neural tissue*. The basic functions of these tissue types are introduced in Figure 3-1.●

This chapter will discuss the characteristics of each major tissue type, focusing on the relationship between cellular organization and tissue function. Later chapters will consider the patterns of tissue interaction in various organs and systems in greater detail.

EPITHELIAL TISSUE

Epithelial tissue includes *epithelia* and *glands*, secretory structures derived from epithelia. An **epithelium** (e-pi-THĒ-lē-um) is a layer of cells that forms a barrier with specific properties. Epithelia cover every exposed body surface. The surface of the skin is a good example, but epithelia also line the digestive, respiratory, reproductive, and urinary tracts—passageways that communicate with the outside world. Epithelia also line internal cavities and passageways, such as the chest cavity, fluid-filled chambers in the brain, eye, inner ear, and the inner surfaces of blood vessels and the heart.

Important characteristics of epithelia include:

1. *Cellularity*: Epithelia are composed almost entirely of cells bound closely together by cell junctions of one or more types. In other tissue types the cells are often widely separated by extracellular materials.

2. *Polarity*: An epithelium always has an exposed surface, or *apical surface*, that faces the exterior of the body or some internal space. It also has an attached *basal surface* where the epithelium is attached to underlying tissues.

3. *Attachment*: The basal surface of an epithelium is bound to a thin *basement membrane*. The basement membrane is a complex structure produced by the basal surface of the epithelium and the underlying connective tissue.

4. *Avascularity*: Epithelia do not contain blood vessels. Because of this **avascular** (ā–VAS–kū–lar; *a*-, without + *vas*, vessel) condition, epithelial cells must obtain nutrients by diffusion or absorption across the apical or basal surfaces.

5. *Regeneration*: Epithelial cells damaged or lost at the apical surface are continually being replaced through the divisions of stem cells within the epithelium.

Functions of Epithelial Tissue

Epithelia perform essential functions that can be summarized as follows:

1. **Provide physical protection**: Epithelia protect exposed and internal surfaces from abrasion, dehydration, and destruction by chemical or biological agents.

2. **Control permeability**: Any substance that enters or leaves the body has to cross an epithelium. Some epithelia are relatively impermeable, whereas others are easily crossed by compounds as large as proteins. Many epithelia contain the molecular "machinery" needed for selective absorption or secretion. The epithelial barrier can be regulated and modified in response to various stimuli. For example, hormones can affect the transport of ions and nutrients through epithelial cells. Even physical stress can alter the structure and properties of epithelia—think of the calluses that form on your hands when you do rough work for a period of time.

3. **Provide sensations**: Most epithelia are extensively innervated by sensory nerves. Specialized epithelial cells can detect changes in the environment and convey information about such changes to the nervous system. For example, touch receptors in the deepest epithelial layers of the skin respond to pressure by stimulating adjacent sensory nerves. A **neuroepithelium** is an epithelium containing sensory cells providing sensations of smell, taste, sight, equilibrium, and hearing.

4. **Produce specialized secretions**: Epithelial cells that produce secretions are called **gland cells**. Individual gland cells are often scattered among other cell types in an epithelium. In a **glandular epithelium**, most or all of the epithelial cells produce secretions.

FIGURE 3-1
An Orientation to the Tissues of the Body

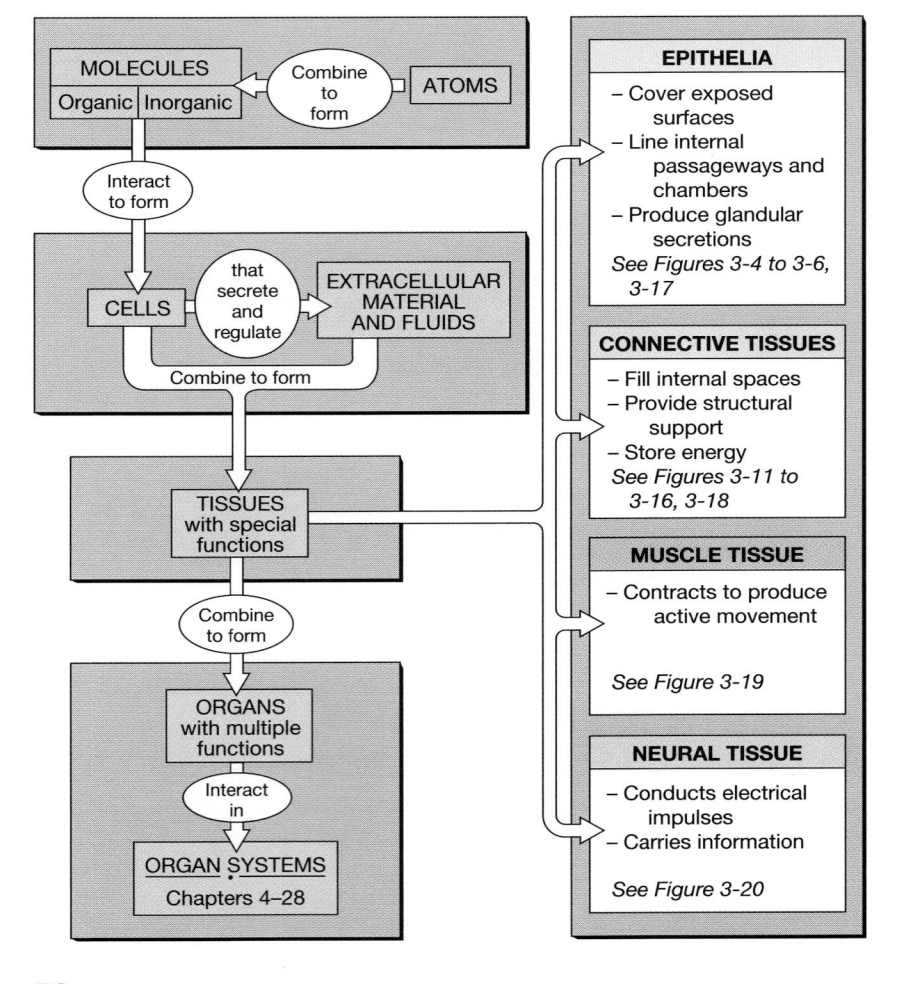

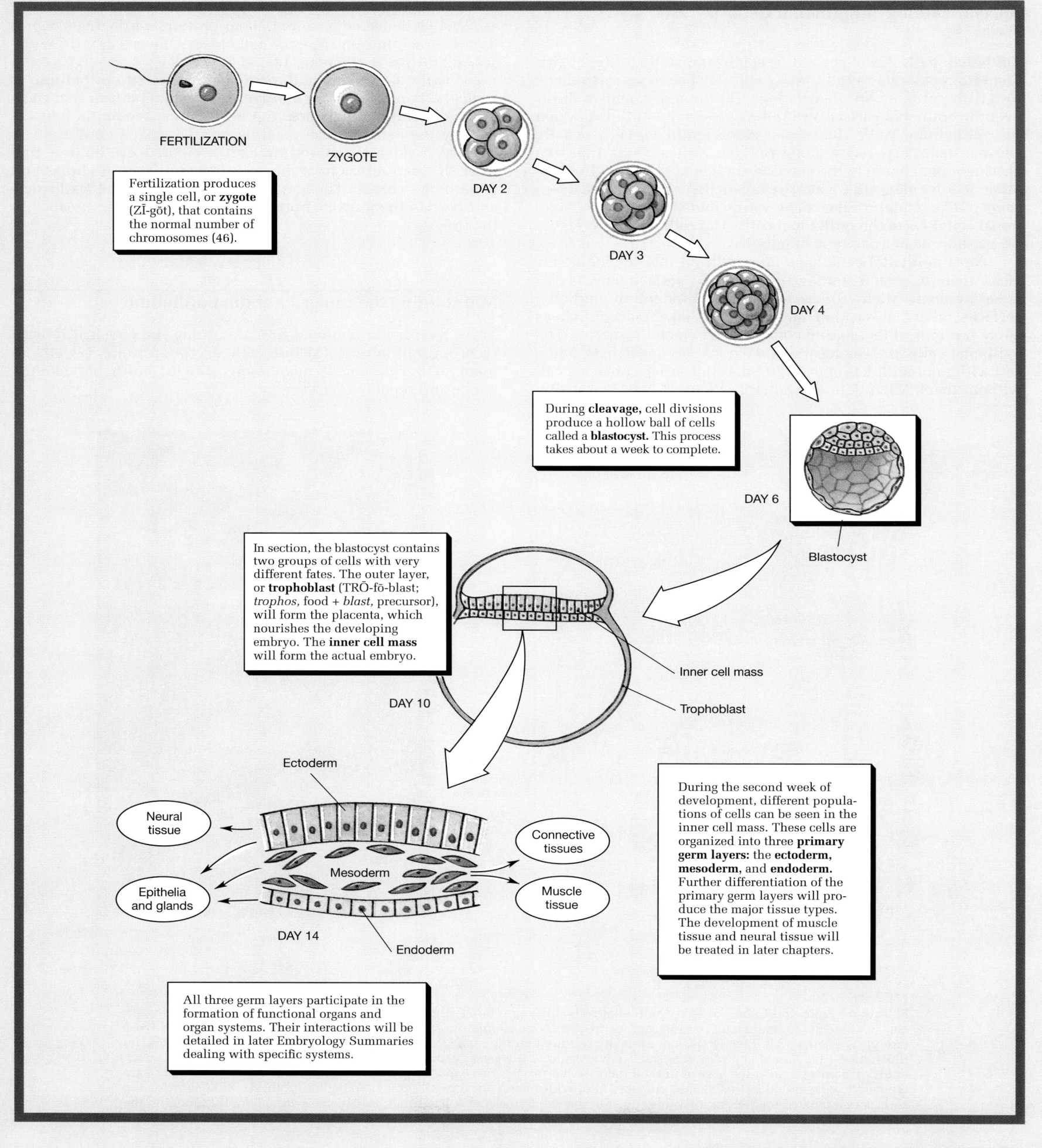

FERTILIZATION

ZYGOTE

DAY 2

DAY 3

DAY 4

Fertilization produces a single cell, or **zygote** (ZĪ-gōt), that contains the normal number of chromosomes (46).

During **cleavage,** cell divisions produce a hollow ball of cells called a **blastocyst.** This process takes about a week to complete.

DAY 6

Blastocyst

In section, the blastocyst contains two groups of cells with very different fates. The outer layer, or **trophoblast** (TRŌ-fō-blast; *trophos,* food + *blast,* precursor), will form the placenta, which nourishes the developing embryo. The **inner cell mass** will form the actual embryo.

Inner cell mass

Trophoblast

DAY 10

Ectoderm

Neural tissue

Connective tissues

Mesoderm

Muscle tissue

Epithelia and glands

DAY 14

Endoderm

During the second week of development, different populations of cells can be seen in the inner cell mass. These cells are organized into three **primary germ layers:** the **ectoderm, mesoderm,** and **endoderm.** Further differentiation of the primary germ layers will produce the major tissue types. The development of muscle tissue and neural tissue will be treated in later chapters.

All three germ layers participate in the formation of functional organs and organ systems. Their interactions will be detailed in later Embryology Summaries dealing with specific systems.

Specializations of Epithelial Cells
(Figure 3-2)

Epithelial cells have several specializations that distinguish them from other body cells. Many epithelial cells are specialized for (1) the production of secretions, (2) the movement of fluids over the epithelial surface, or (3) the movement of fluids through the epithelium itself. These specialized epithelial cells usually show a definite **polarity** along the axis that extends from the basement membrane to the exposed surface of the epithelium. In other words, along this axis the organelles are distributed unevenly. The actual arrangement varies depending on the functional activities of the individual cells. The cells shown in Figure 3-2a● show a common type of polarity.

Most epithelial cells have microvilli on their exposed surfaces; there may be just a few, or the entire surface may be carpeted by them. Microvilli are especially abundant on epithelial surfaces where absorption and secretion take place, such as along portions of the digestive and urinary tracts. ∞[p. 36] The epithelial cells in these locations are transport specialists, and a cell with microvilli has at least 20 times the surface area of a cell without them. Microvilli are shown in Figure 3-2b●. **Stereocilia**

are very long microvilli (up to 250 μm) that are incapable of movement. Stereocilia, found along portions of the male reproductive tract and on receptor cells of the inner ear, will be described further in Chapters 18 and 27.

Figure 3-2b● shows the surface of **ciliated epithelium**. A typical ciliated cell contains about 250 cilia that beat in a coordinated fashion. Substances are moved over the epithelial surface by the synchronized beating of cilia, like a continuously moving escalator. For example, the ciliated epithelium that lines the respiratory tract moves mucus up from the lungs and toward the throat. The mucus traps particles and pathogens and carries them away from more delicate surfaces deeper in the lungs.

Maintaining the Integrity of the Epithelium

Three factors are involved in maintaining the physical integrity of an epithelium: (1) intercellular connections, (2) attachment to the basement membrane, and (3) epithelial maintenance and repair.

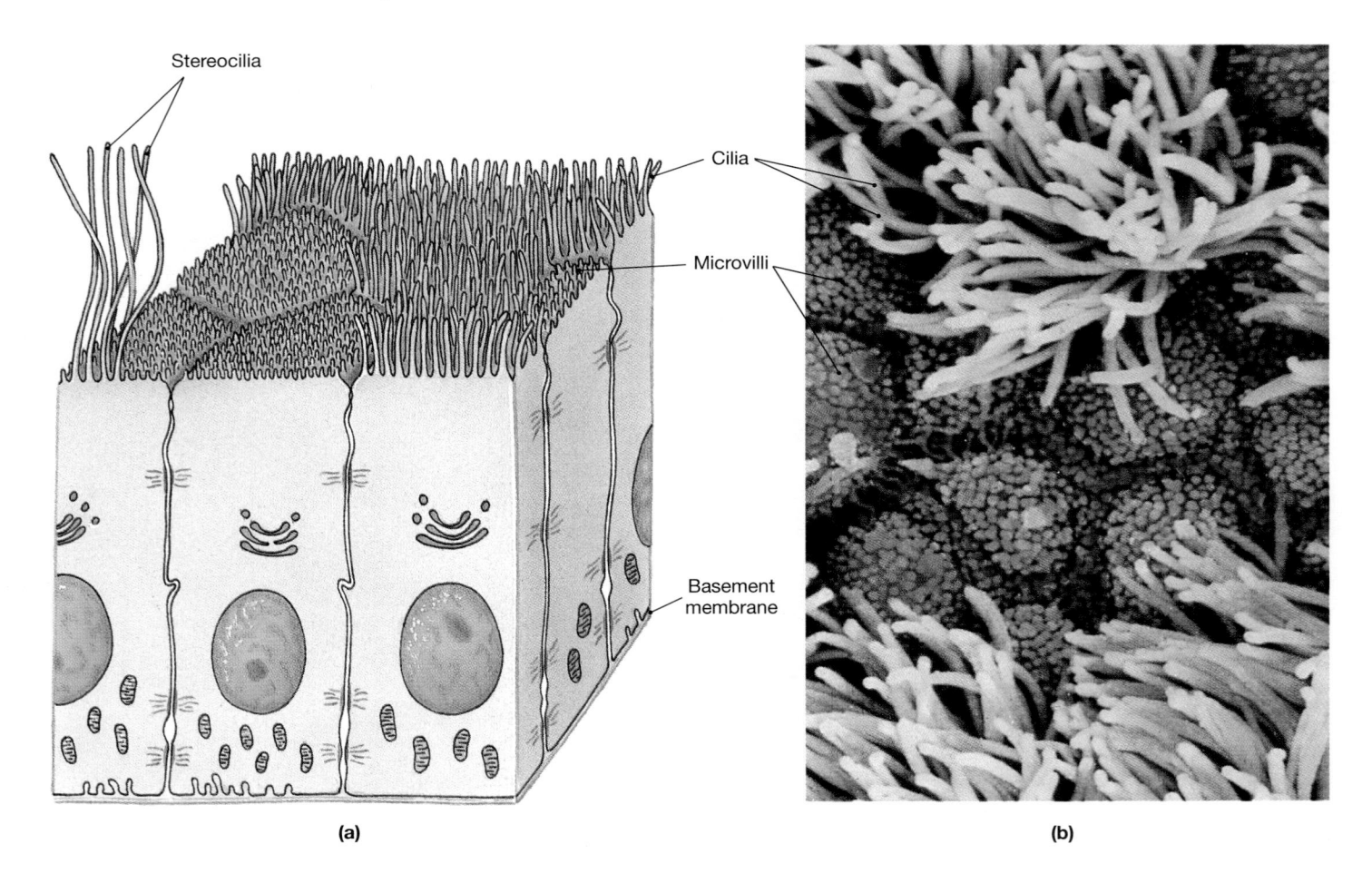

(a)

(b)

FIGURE 3-2
Polarity of Epithelial Cells. (a) Many epithelial cells differ in internal organization along an axis between the free surface and the basement membrane. The free surface frequently bears microvilli; less often, this surface may have cilia or (very rarely) stereocilia. (All three would not normally be found on the same group of cells but are depicted here for purposes of illustration.) Junctional complexes prevent movement of pathogens or diffusion of dissolved materials between the cells. Folds of membrane near the base of the cell increase the surface area exposed to the basement membrane. Mitochondria are typically concentrated in this region, probably to provide energy for the cell's transport activities. (b) An SEM showing the surface of a ciliated epithelium that lines most of the respiratory tract. The small, bristly areas are microvilli found on the exposed surfaces of mucus-producing cells that are scattered among the ciliated epithelial cells. (SEM × 15,846)

Intercellular Connections
(Figure 3-3)

Cells in epithelia are usually bound together by *junctional complexes* and by an extensive infolding of opposing cell membranes. ∞ [pp. 43, 44] Note the degree of interlocking that can exist between two cell membranes as seen in Figure 3-3●. The combination of junctional complexes, intercellular cement, and physical interlocking gives the epithelium strength and stability. The extensive connections between cells hold them together and may deny access to chemicals or pathogens that may cover their free surfaces.

Attachment to the Basement Membrane
(Figure 3-3b)

Epithelial cells not only hold onto one another, they also remain firmly connected to the rest of the body. The inner surface of each epithelium is attached to a special two-part **basement membrane** (Figure 3-3b●). The layer closest to the epithelium, called the **basal lamina** (LA-mi-na; *lamina*, thin layer), contains glycoproteins and a network of fine protein filaments. The basal lamina,

secreted by the adjacent layer of epithelial cells, provides a barrier that restricts the movement of proteins and other large molecules from the underlying connective tissue into the epithelium. The deeper portion of the basement membrane, the **reticular lamina**, contains bundles of coarse protein fibers produced by connective tissue cells. The reticular lamina gives the basement membrane its strength. Attachments between the fibers of the basal lamina and those of the reticular lamina hold the two together.

Epithelial Maintenance and Repair

An epithelium must continually repair and renew itself. Epithelial cells lead hard lives, for they may be exposed to disruptive enzymes, toxic chemicals, pathogenic bacteria, or mechanical abrasion. Under severe conditions, such as those encountered inside the small intestine, an epithelial cell may survive for just a day or two before it is lost or destroyed. The only way the epithelium can maintain its structure over time is through the continual division of stem cells. These stem cells, also known as **germinative cells**, are usually found in the deepest layers of the epithelium, close to the basement membrane. ∞ [p. 44]

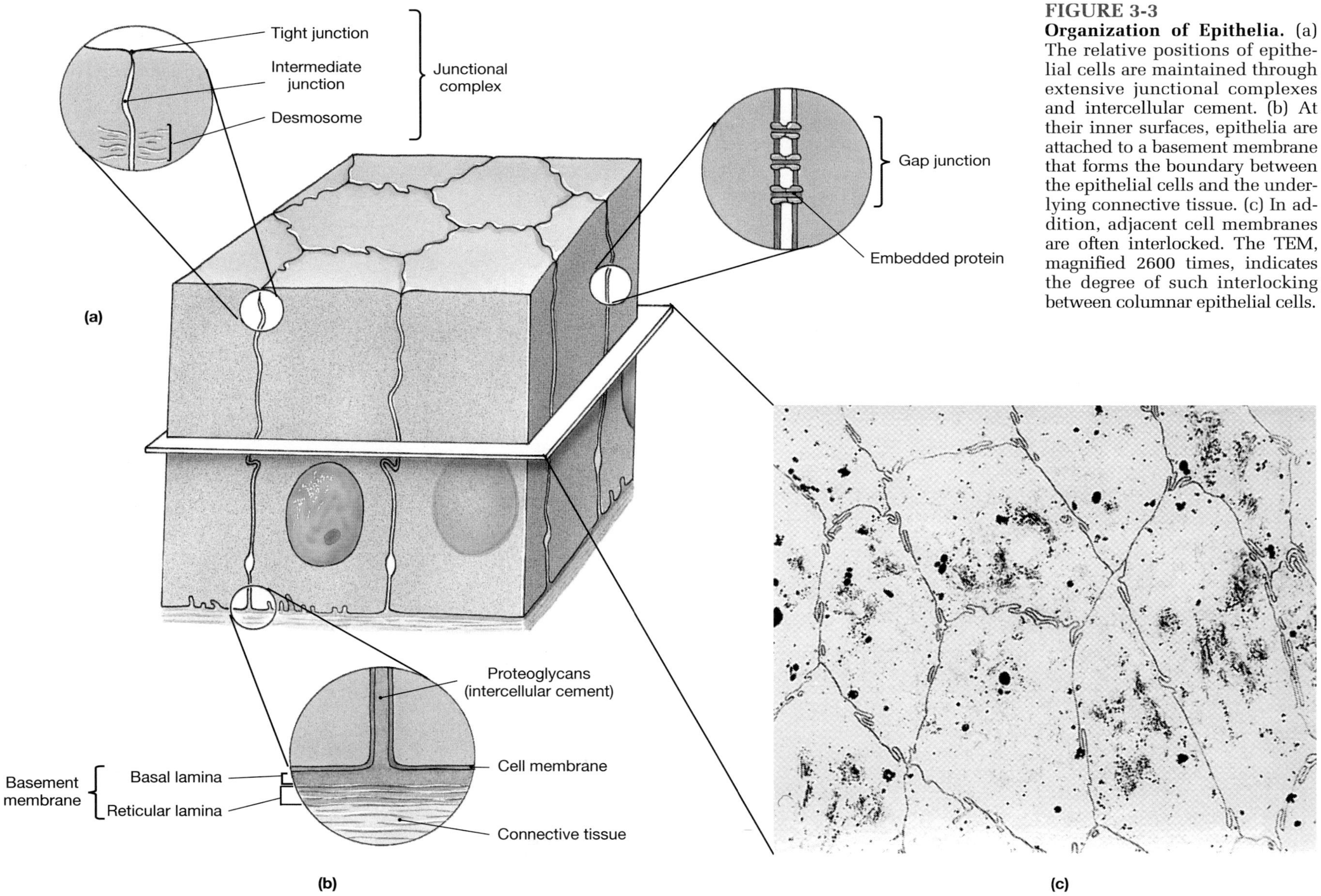

FIGURE 3-3
Organization of Epithelia. (a) The relative positions of epithelial cells are maintained through extensive junctional complexes and intercellular cement. (b) At their inner surfaces, epithelia are attached to a basement membrane that forms the boundary between the epithelial cells and the underlying connective tissue. (c) In addition, adjacent cell membranes are often interlocked. The TEM, magnified 2600 times, indicates the degree of such interlocking between columnar epithelial cells.

Classification of Epithelia

Epithelia are classified according to the number of cell layers and the shape of the exposed cells. The classification scheme recognizes two types of layering—*simple* and *stratified*—and three cell shapes—*squamous*, *cuboidal*, and *columnar*.

If there is only a single layer of cells covering the basement membrane, the epithelium is a **simple epithelium**. Simple epithelia are relatively thin, and because all the cells have the same polarity, the nuclei form a row above the basement membrane. Because they are so thin, simple epithelia are also relatively fragile. A single layer of cells cannot provide much mechanical protection, and simple epithelia are found only in protected areas inside the body. They line internal compartments and passageways, including the ventral body cavities, the chambers of the heart, and all blood vessels.

Simple epithelia are also characteristic of regions where secretion or absorption occurs, such as the lining of the intestines and the gas-exchange surfaces of the lungs. In these places the thin single layer of simple epithelia is an advantage, for it speeds the passage of materials through or across the epithelial barrier.

A **stratified epithelium** has several layers of cells above the basement membrane. Stratified epithelia are usually found in areas subject to mechanical or chemical stresses, such as the surface of the skin and the lining of the mouth. Combining the two basic epithelial layouts (simple and stratified) and the three possible cell shapes (squamous, cuboidal, and columnar) enables one to describe almost every epithelium in the body. Our discussion will now focus on the primary types of epithelial cells.

Squamous Epithelia
(Figure 3-4)

In a **squamous epithelium** (SKWĀ-mus; *squama*, plate or scale) the cells are thin, flat, and somewhat irregular in shape—like puzzle pieces, as seen in Figure 3-4a●. In a sectional view the nucleus occupies the thickest portion of each cell; from the surface, the cells look like fried eggs laid side by side. A **simple squamous epithelium** is the most delicate type of epithelium in the body. This type of epithelium is found in protected regions where absorption takes place or where a slick, slippery surface reduces friction. Examples include the respiratory exchange surfaces (*alveoli*) of the lungs, the lining of the ventral body cavities, and the inner surfaces of the circulatory system.

FIGURE 3-4
Squamous Epithelia. (a) A superficial view of the simple squamous epithelium (mesothelium) that lines the peritoneal cavity. The three-dimensional drawing shows the epithelium in superficial and sectional view. (b) A sectional view of the stratified squamous epithelium that covers the tongue.

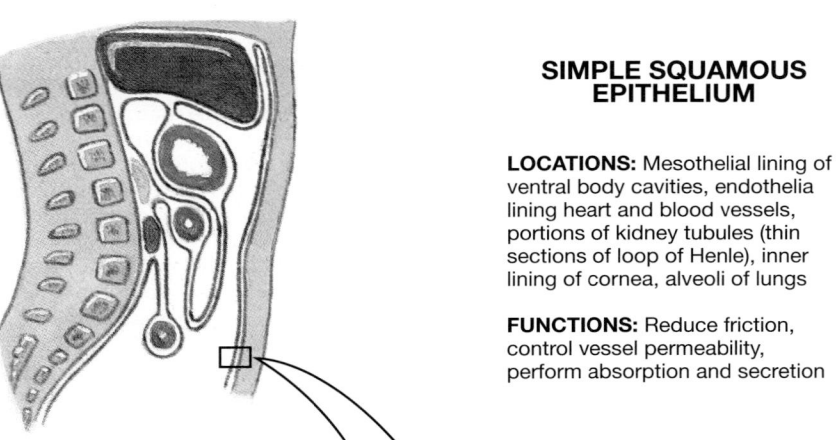

SIMPLE SQUAMOUS EPITHELIUM

LOCATIONS: Mesothelial lining of ventral body cavities, endothelia lining heart and blood vessels, portions of kidney tubules (thin sections of loop of Henle), inner lining of cornea, alveoli of lungs

FUNCTIONS: Reduce friction, control vessel permeability, perform absorption and secretion

(a)

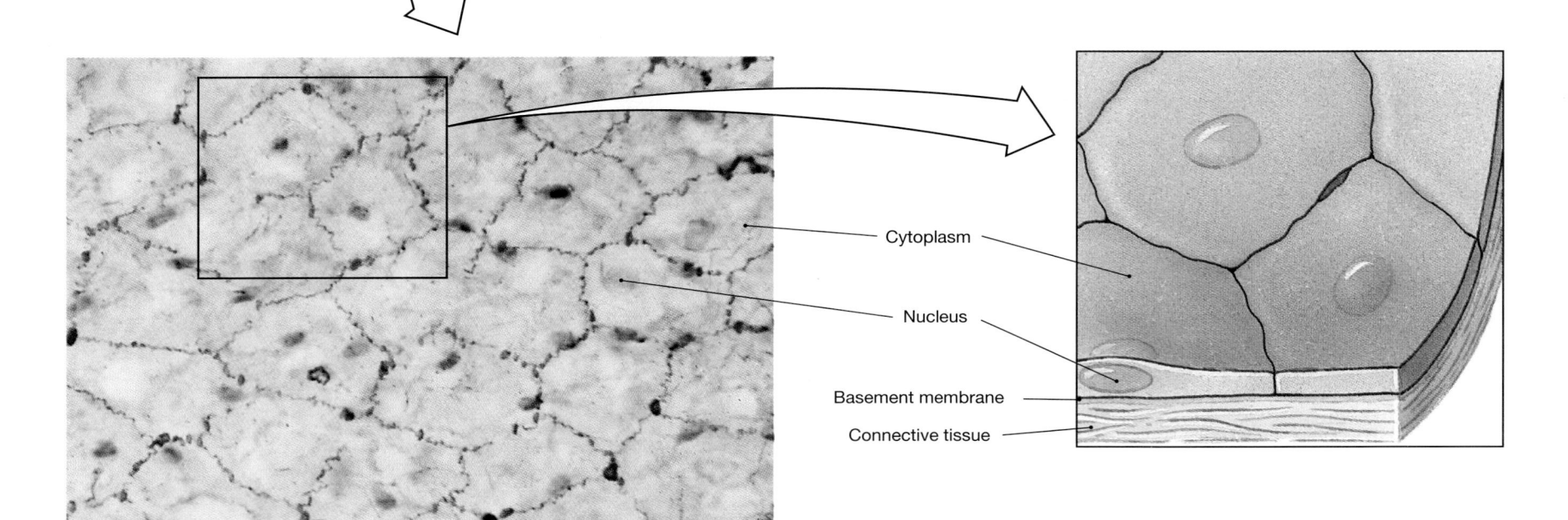

Mesothelium × 263

Cytoplasm

Nucleus

Basement membrane

Connective tissue

Special names have been given to simple squamous epithelia that line chambers and passageways that do not communicate with the outside world. The simple squamous epithelium that lines the ventral body cavities is known as a **mesothelium** (mez-ō-THĒ-lē-um; *mesos*, middle). The pleura, peritoneum, and pericardium each contain a superficial layer of mesothelium. The simple squamous epithelium lining the heart and all blood vessels is called an **endothelium** (en-dō-THĒ-lē-um).

A **stratified squamous epithelium** (Figure 3-4b●) is usually found where mechanical stresses are severe. Note how the cells form a series of layers, like a stack of plywood sheets. The surface of the skin and the lining of the mouth, throat, esophagus, rectum, vagina, and anus are areas where this epithelial type provides protection from physical and chemical attack. On exposed body surfaces, where mechanical stress and dehydration are potential problems, the apical layers of epithelial cells are packed with filaments of the protein *keratin*. As a result, the superficial layers are both tough and water-resistant, and the epithelium is said to be **keratinized**. A **nonkeratinized** stratified squamous epithelium provides resistance to abrasion, but will dry out and deteriorate unless kept moist. Nonkeratinized stratified squamous epithelia are found in the oral cavity, pharynx, esophagus, anus, and vagina.

Cuboidal Epithelia
(Figure 3-5)

The cells of a **cuboidal epithelium** resemble little hexagonal boxes; they appear square in typical sectional views. Each nucleus is near the center of the cell, with the distance between adjacent nuclei roughly equal to the height of the epithelium. **Simple cuboidal epithelium** provides limited protection and occurs in regions where secretion or absorption takes place. Such an epithelium lines portions of the kidney tubules, as seen in Figure 3-5a●. In the pancreas and salivary glands, simple cuboidal epithelia secrete enzymes and buffers and line the ducts that discharge those secretions. The thyroid gland contains chambers called *thyroid follicles* that are lined by a cuboidal secretory epithelium. Thyroid hormones, especially *thyroxine*, accumulate within the follicles before they are released into the bloodstream.

Stratified cuboidal epithelia are relatively rare; they are found along the ducts of sweat glands (Figure 3-5b●) and in the larger ducts of the mammary glands. A **transitional epithelium**, shown in Figure 3-5c,d●, lines the renal pelvis, the ureters, and the urinary bladder. This epithelium permits considerable stretching, and significant changes in volume occur at these locations. In an empty bladder (Figure 3-5c●), the epithelium seems to have many layers, and the outermost cells are typically plump cuboidal cells. The layered appearance results from overcrowding; the actual structure of the epithelium can be seen in the full bladder, when the pressure of the urine has stretched the lining (Figure 3-5d●).

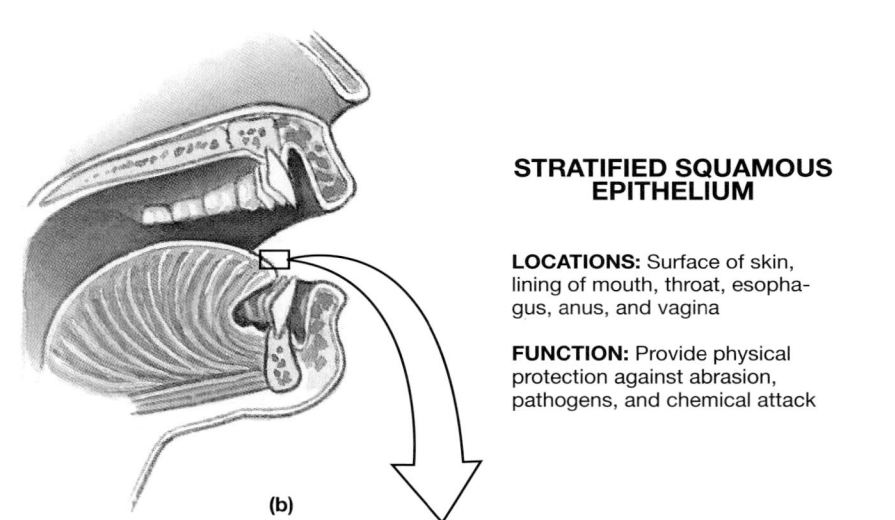

STRATIFIED SQUAMOUS EPITHELIUM

LOCATIONS: Surface of skin, lining of mouth, throat, esophagus, anus, and vagina

FUNCTION: Provide physical protection against abrasion, pathogens, and chemical attack

(b)

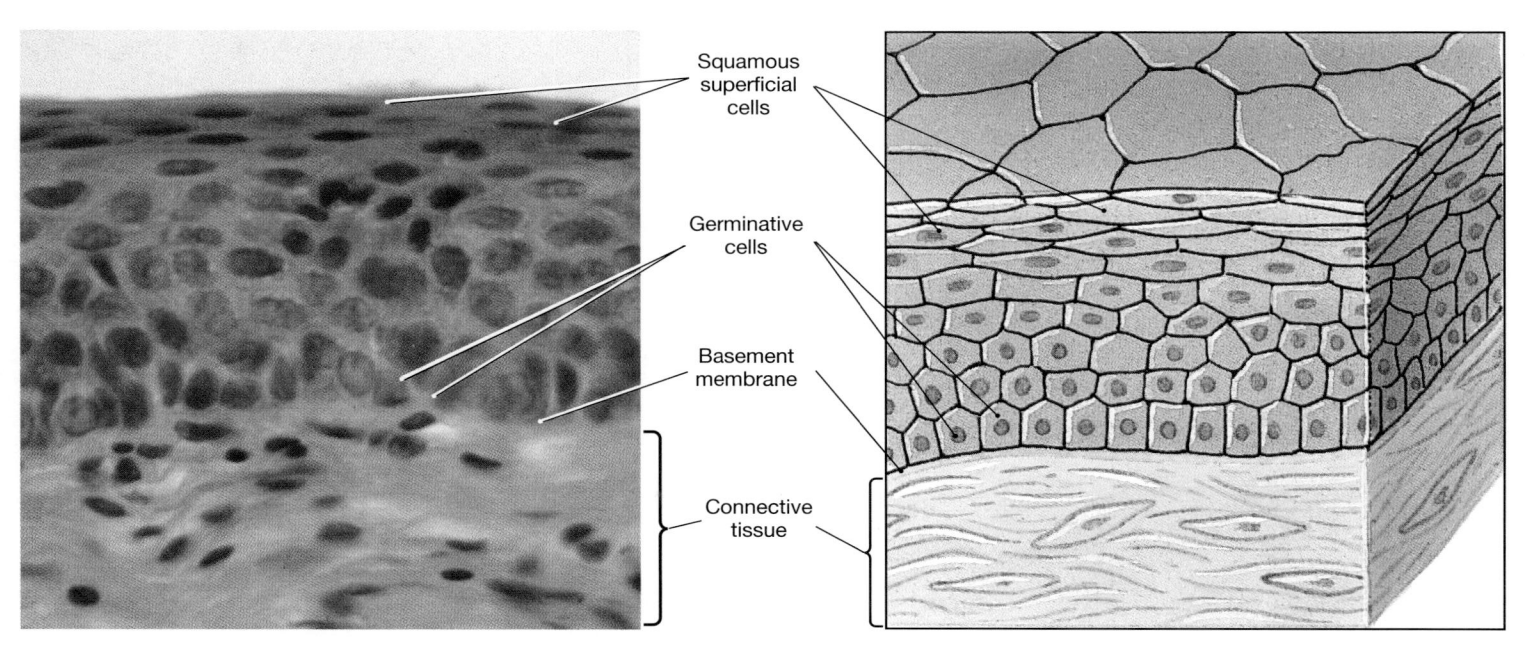

Squamous superficial cells

Germinative cells

Basement membrane

Connective tissue

Stratified squamous epithelium × 310

Columnar Epithelia
(Figure 3-6)

Columnar epithelial cells are also hexagonal in cross section, but they are taller and more slender than cuboidal epithelial cells. The nuclei are crowded into a narrow band close to the basement membrane, and the height of the epithelium is several times the distance between two nuclei (Figure 3-6a●). A **simple columnar epithelium** provides some protection and may also be encountered in areas where absorption or secretion occurs. This type of epithelium lines the stomach, intestinal tract, and many excretory ducts.

Portions of the respiratory tract contain a columnar epithelium that includes a mixture of cell types. Because their nuclei are situated at varying distances from the surface, the epithelium appears to be layered or stratified. But it is not truly stratified, because all of the cells contact the basement membrane. Since it looks stratified but isn't, it is known as a **pseudostratified columnar epithelium** (Figure 3-6b●). Pseudostratified columnar epithelial cells typically possess cilia. This epithelium lines most of the nasal cavity, the trachea (windpipe), bronchi, and portions of the male reproductive tract.

Stratified columnar epithelia are relatively rare, providing protection along portions of the pharynx, urethra, and anus, as well as along a few large excretory ducts. The epithelium may have two layers (Figure 3-6c●) or multiple layers; when multiple layers exist, only the superficial cells have the classic columnar shape.

Glandular Epithelia

Many epithelia contain gland cells that produce secretions. Exocrine glands may be classified by their *mode of secretion*, the *type of secretion*, and the *structure of the gland*.

Modes of Secretion
(Figure 3-7)

A glandular epithelial cell may use one of three methods to release its secretions: *merocrine secretion*, *apocrine secretion*, or *holocrine secretion*. In **merocrine secretion** (MER-ō-krin; *meros*,

FIGURE 3-5
(a) **Simple Cuboidal Epithelia.** A section through the cuboidal epithelial cells of a kidney tubule. The diagrammatic view emphasizes structural details that permit the classification of an epithelium as cuboidal. (b) **Stratified Cuboidal Epithelia.** Sectional view of the stratified cuboidal epithelium lining a sweat gland duct in the skin. **Transitional Epithelia.** (c) The lining of the empty urinary bladder, showing transitional epithelium in the contracted state. (d) The lining of the full bladder, showing the effects of stretching on the arrangement of cells in the epithelium.

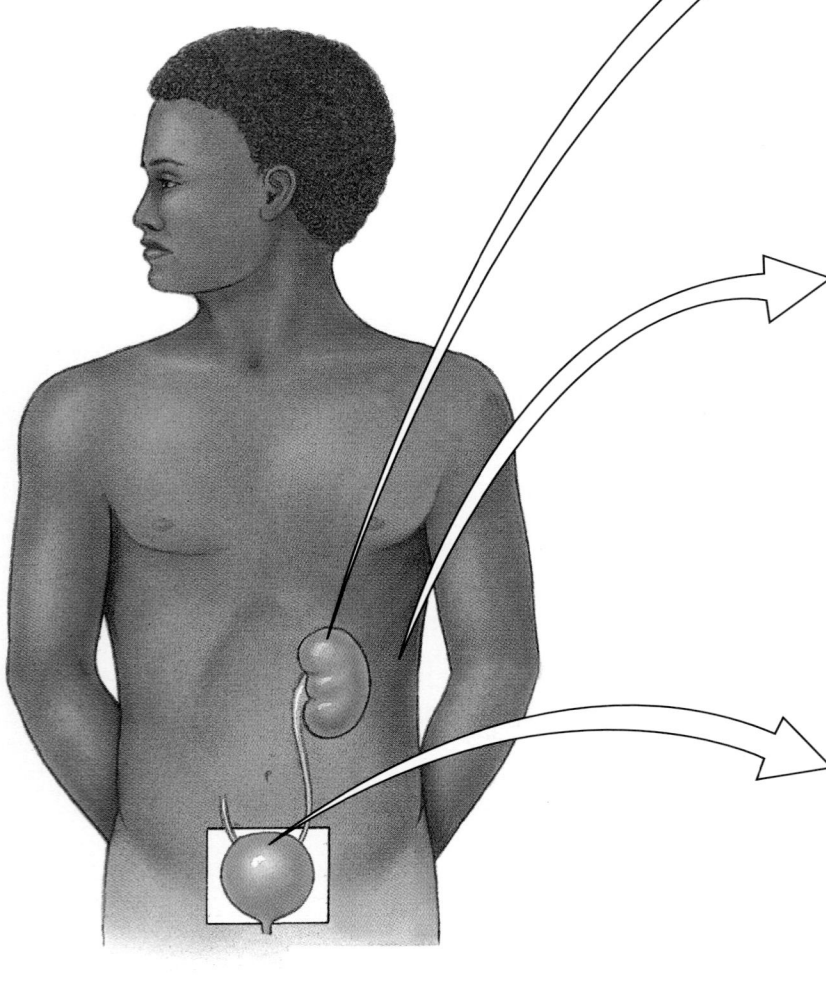

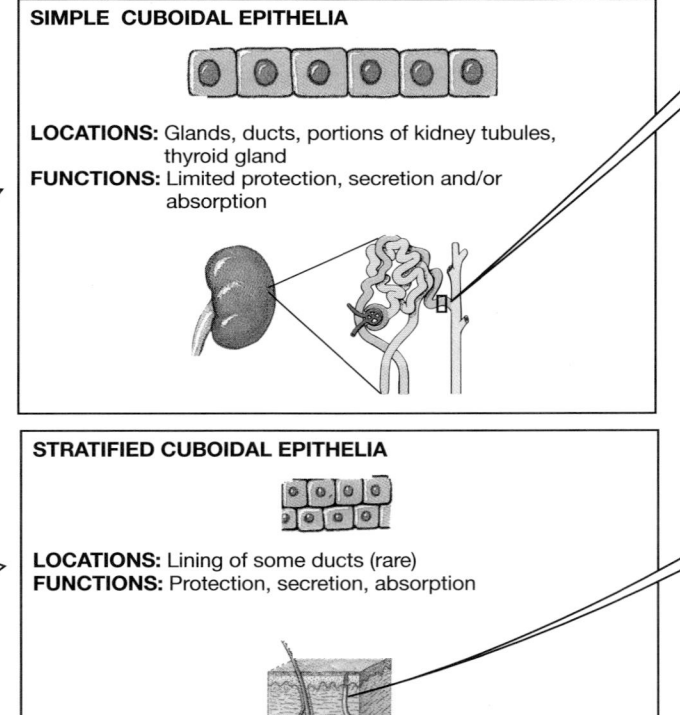

SIMPLE CUBOIDAL EPITHELIA

LOCATIONS: Glands, ducts, portions of kidney tubules, thyroid gland
FUNCTIONS: Limited protection, secretion and/or absorption

STRATIFIED CUBOIDAL EPITHELIA

LOCATIONS: Lining of some ducts (rare)
FUNCTIONS: Protection, secretion, absorption

TRANSITIONAL EPITHELIA

LOCATIONS: Urinary bladder, renal pelvis, ureters
FUNCTIONS: Permit expansion and recoil after stretching

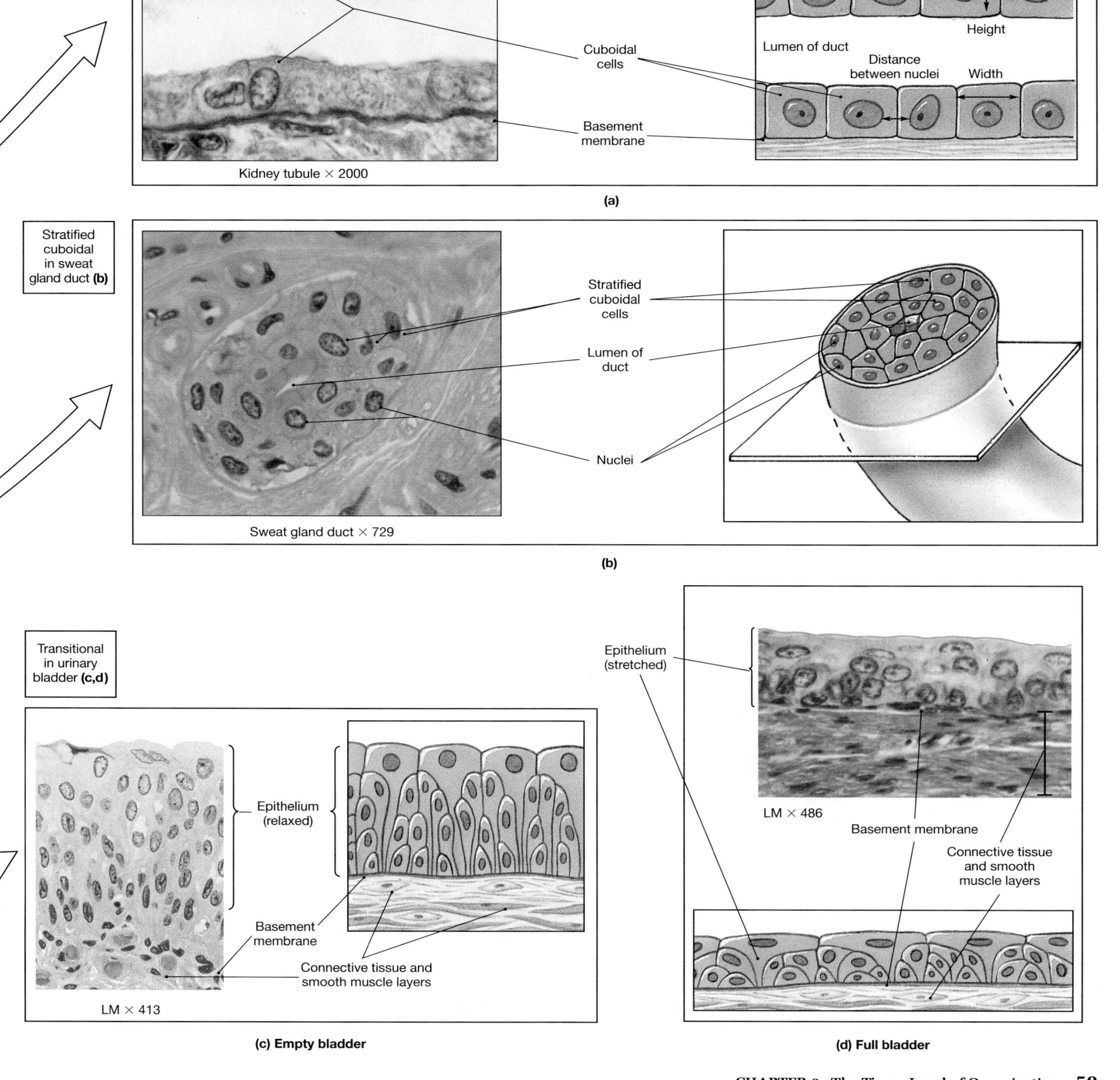

Simple cuboidal epithelium in kidney tubule **(a)**

Connective tissue

Cuboidal cells

Basement membrane

Kidney tubule × 2000

Height

Lumen of duct

Distance between nuclei

Width

(a)

Stratified cuboidal in sweat gland duct **(b)**

Stratified cuboidal cells

Lumen of duct

Nuclei

Sweat gland duct × 729

(b)

Transitional in urinary bladder **(c,d)**

Epithelium (relaxed)

Basement membrane

Connective tissue and smooth muscle layers

LM × 413

(c) Empty bladder

Epithelium (stretched)

LM × 486

Basement membrane

Connective tissue and smooth muscle layers

(d) Full bladder

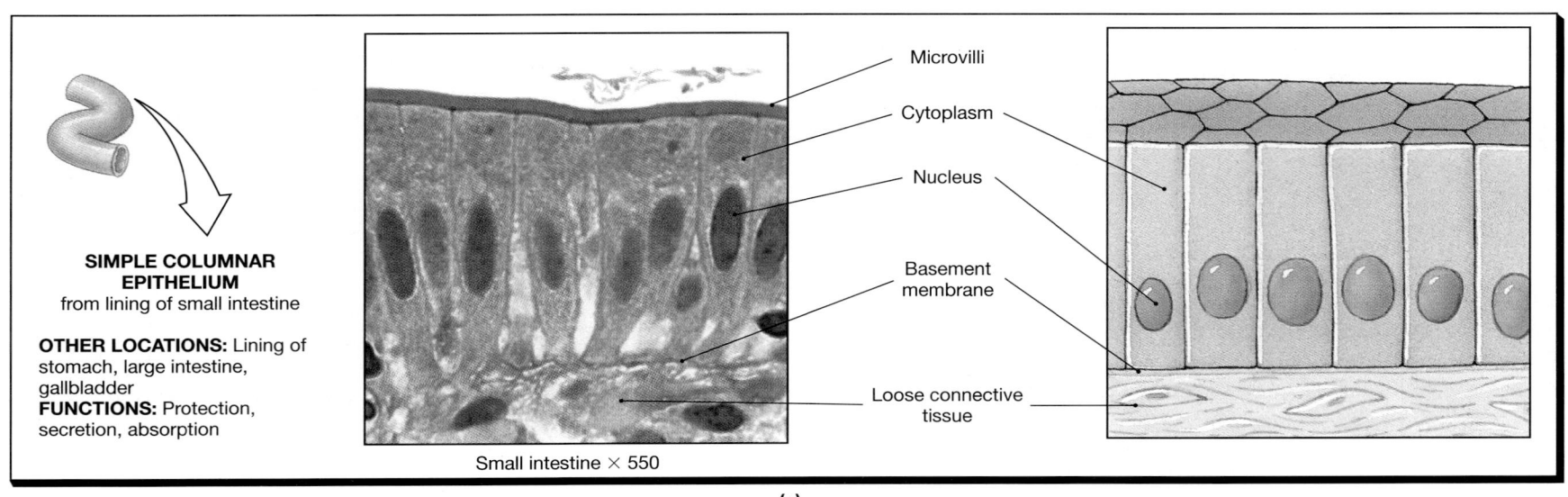

SIMPLE COLUMNAR EPITHELIUM
from lining of small intestine

OTHER LOCATIONS: Lining of stomach, large intestine, gallbladder
FUNCTIONS: Protection, secretion, absorption

Microvilli

Cytoplasm

Nucleus

Basement membrane

Loose connective tissue

Small intestine × 550

(a)

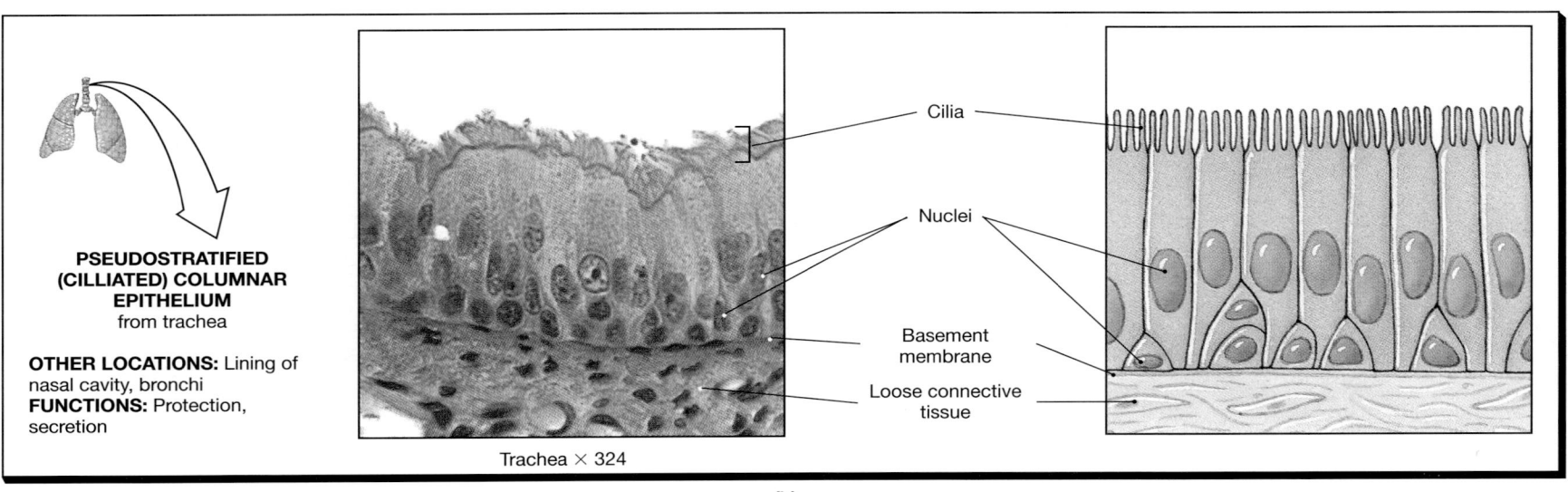

PSEUDOSTRATIFIED (CILLIATED) COLUMNAR EPITHELIUM
from trachea

OTHER LOCATIONS: Lining of nasal cavity, bronchi
FUNCTIONS: Protection, secretion

Cilia

Nuclei

Basement membrane

Loose connective tissue

Trachea × 324

(b)

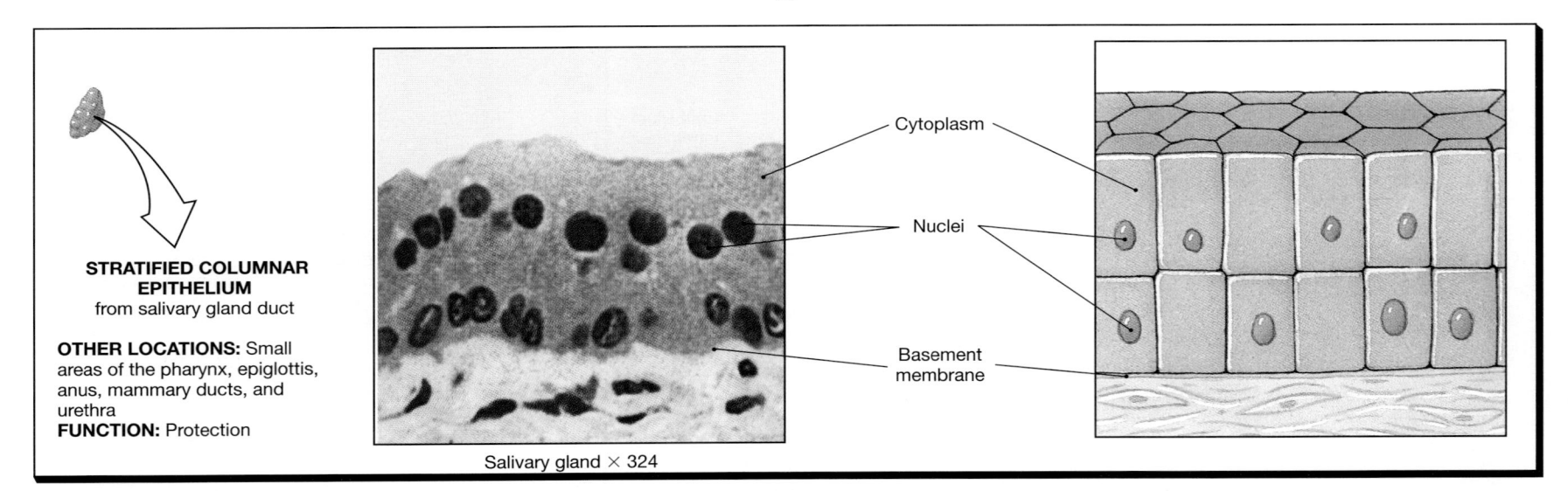

STRATIFIED COLUMNAR EPITHELIUM
from salivary gland duct

OTHER LOCATIONS: Small areas of the pharynx, epiglottis, anus, mammary ducts, and urethra
FUNCTION: Protection

Cytoplasm

Nuclei

Basement membrane

Salivary gland × 324

(c)

FIGURE 3-6
Columnar Epithelia. (a) Micrograph showing the characteristics of simple columnar epithelium. In the diagrammatic sketch, note the relationships between the height and width of each cell; the relative size, shape, and location of nuclei; and the distance between adjacent nuclei. Contrast these observations with the corresponding characteristics of simple cuboidal epithelia. (b) The pseudostratified, ciliated, columnar epithelium of the respiratory tract. Note the uneven layering of the nuclei. (c) A stratified columnar epithelium is sometimes found along large ducts, such as this salivary gland duct. Note the overall height of the epithelium and the location and orientation of the nuclei.

part + *krinein*, to separate), the product is released through exocytosis (Figure 3-7a●). This is the most common mode of secretion. One merocrine secretion, called **mucus**, is an effective lubricant, protective barrier, and a sticky trap for foreign particles and microorganisms. Mucus secretions coat the passageways of the digestive and respiratory tracts. **Apocrine secretion** (AP-ō-krin; *apo-*, off) involves the loss of cytoplasm as well as the secretory product (Figure 3-7b●). The outermost (apical) portion of the cytoplasm becomes packed with secretory vesicles before it is shed. Milk production by the lactiferous glands in the breasts involves a combination of merocrine and apocrine secretions; the viscous underarm perspiration targeted by the deodorant industry results from apocrine secretion.

Merocrine and apocrine secretions leave the nucleus and Golgi apparatus of the cell intact, so it can perform repairs and continue secreting. **Holocrine secretion** (HOL-ō-krin; *holos*, entire) destroys the gland cell. During holocrine secretion, the entire cell becomes packed with secretory products and then bursts apart (Figure 3-7c●). The secretion is released and the cell dies. Further secretion depends on gland cells being replaced by the division of stem cells. Sebaceous glands, associated with hair follicles, produce a waxy hair coating by means of holocrine secretion.

Types of Secretion

Exocrine (*exo-*, outside) secretions are discharged onto the surface of the skin or onto an epithelial surface lining one of the internal passageways that communicates with the exterior. There are many kinds of exocrine secretions, performing a variety of functions. Enzymes entering the digestive tract, perspiration on the skin, and the milk produced by mammary glands are examples of exocrine secretions. Exocrine cells often form pockets that are connected to the epithelial surface by tubes, called **ducts**.

Exocrine glands may be categorized according to the nature of the secretion produced:

- **serous glands** secrete a watery solution that usually contains enzymes;
- **mucous glands** secrete a viscous mucus; and
- **mixed exocrine glands** contain more than one type of gland cell and may produce two different exocrine secretions, one serous and the other mucous. The *submandibular gland*, one of the salivary glands, is an example of a mixed exocrine gland.

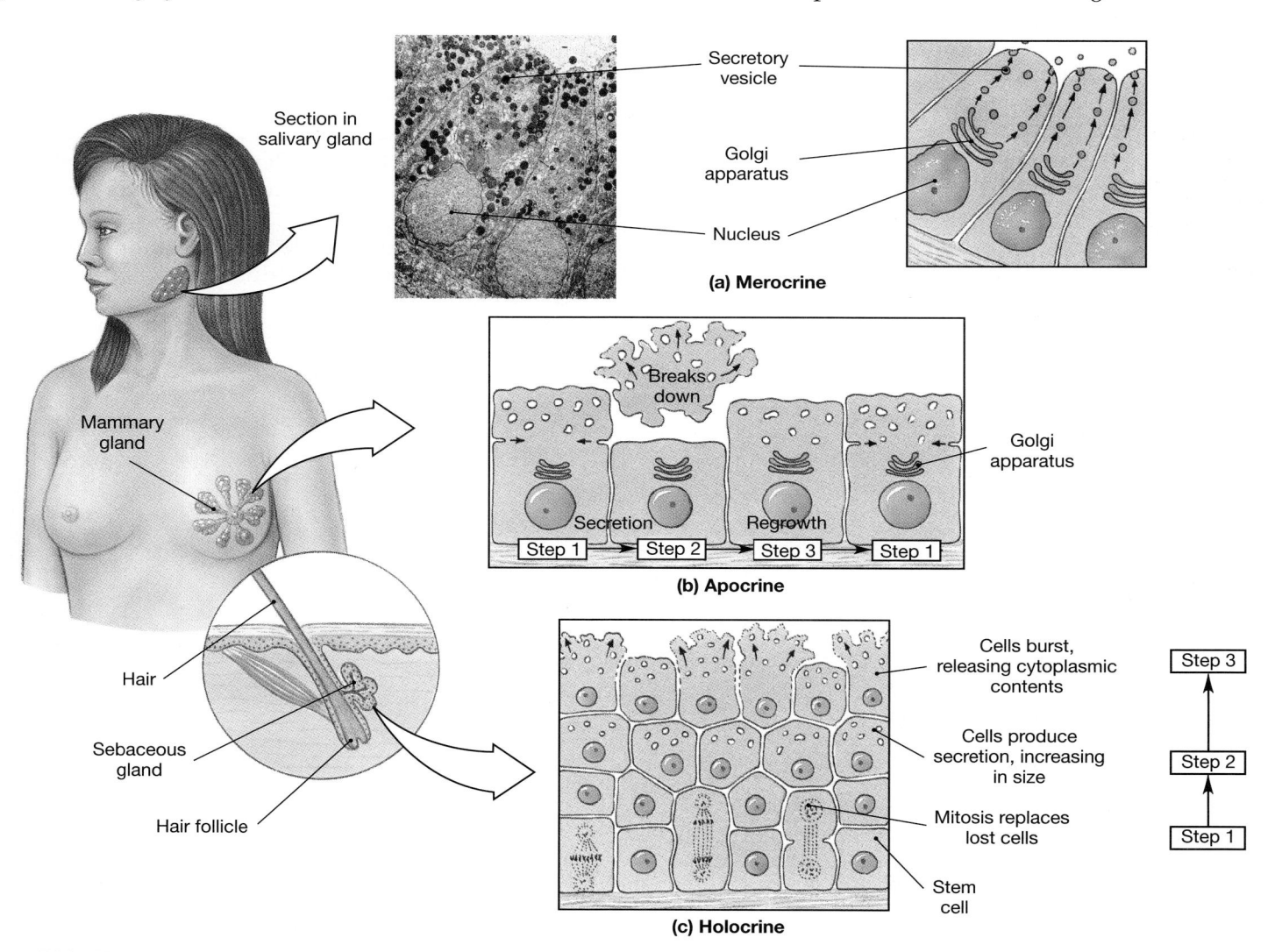

FIGURE 3-7
Mechanisms of Glandular Secretion. (a) In merocrine secretion, secretory vesicles are discharged at the surface of the gland cell through exocytosis. (b) Apocrine secretion involves the loss of cytoplasm. Inclusions, secretory vesicles, and other cytoplasmic components are shed at the apical surface of the cell. The gland cell then undergoes a period of growth and repair before releasing additional secretions. (c) Holocrine secretion occurs as superficial gland cells break apart. Continued secretion involves the replacement of these cells through the mitotic divisions of underlying stem cells.

Endocrine (*endo-*, inside) secretions are released by the gland cells into the interstitial fluid surrounding the cell. These secretions, called **hormones**, diffuse into the blood for distribution to other portions of the body, where they regulate or coordinate the activities of various tissues, organs, and organ systems. Endocrine cells may be part of an epithelial surface, such as the lining of the digestive tract, or they may be separate, as in the pancreas, thyroid gland, thymus, and pituitary gland. In either case, the gland cells release their hormones directly into body fluids, and there are no endocrine ducts. For this reason endocrine glands are often called *ductless glands*. Endocrine cells, organs, and hormones are considered further in Chapter 19. A few complex glands produce both exocrine and endocrine secretions. For example, the pancreas contains endocrine cells that secrete hormones, as well as exocrine cells and ducts responsible for the production of digestive enzymes and buffers.

Gland Structure
(Figures 3-8 to 3-10)

In epithelia that contain scattered gland cells, the individual secretory cells are called **unicellular glands. Multicellular glands** include glandular epithelia and aggregations of gland cells that produce exocrine or endocrine secretions.

Goblet cells are the only example of **unicellular exocrine glands** in the body. Goblet cells secrete *mucins*, glycoproteins that upon hydration form *mucus*, or slippery lubricant. Goblet cells are scattered among other epithelial cells (Figure 3-8●). For example, the pseudostratified columnar epithelium that lines the trachea and the columnar epithelium of the small and large intestines contain an abundance of goblet cells.

The simplest **multicellular exocrine gland** is called a **secretory sheet**. In a secretory sheet, glandular cells dominate the epithelium and release their secretions into an inner compartment (Figure 3-9a●). The mucus-secreting cells that line the stomach are an example. Their continual secretion protects the stomach from the acids and enzymes it contains. Most other multicellular glands are found in pockets set back from the epithelial surface; their secretory products travel through one or more ducts to reach the epithelial surface. Figure 3-9b● shows one example, a salivary gland that produces mucus and digestive enzymes.

Multicellular exocrine glands may be complex in structure. Two characteristics are used to describe the organization of a multicellular gland: (1) the shape of the secretory portion of the gland and (2) the branching pattern of the duct.

1. Glands made up of cells arranged in a tube are **tubular**; those made up of cells in a blind pocket are **alveolar** (al-VĒ-ō-lar; *alveolus*, sac), or **acinar** (A-si-nar; *acinus*, chamber). Glands that have a combination of the two arrangements are called **tubuloalveolar** or *tubuloacinar*.

2. A duct is called **simple** if it does not branch and **compound** if it branches repeatedly. Each glandular area may have its own duct; in the case of **branched** glands, several glands share a duct.

Figure 3-10●, p. 65, diagrams this method of classification based on gland structure. Specific examples of each gland type will be discussed in later chapters.

√ **You look at a tissue under a microscope and see a simple squamous epithelium. Can it be a sample of the skin surface?**

√ **The secretory cells of sebaceous glands fill with secretions and then rupture, releasing their contents. What kind of secretion is this?**

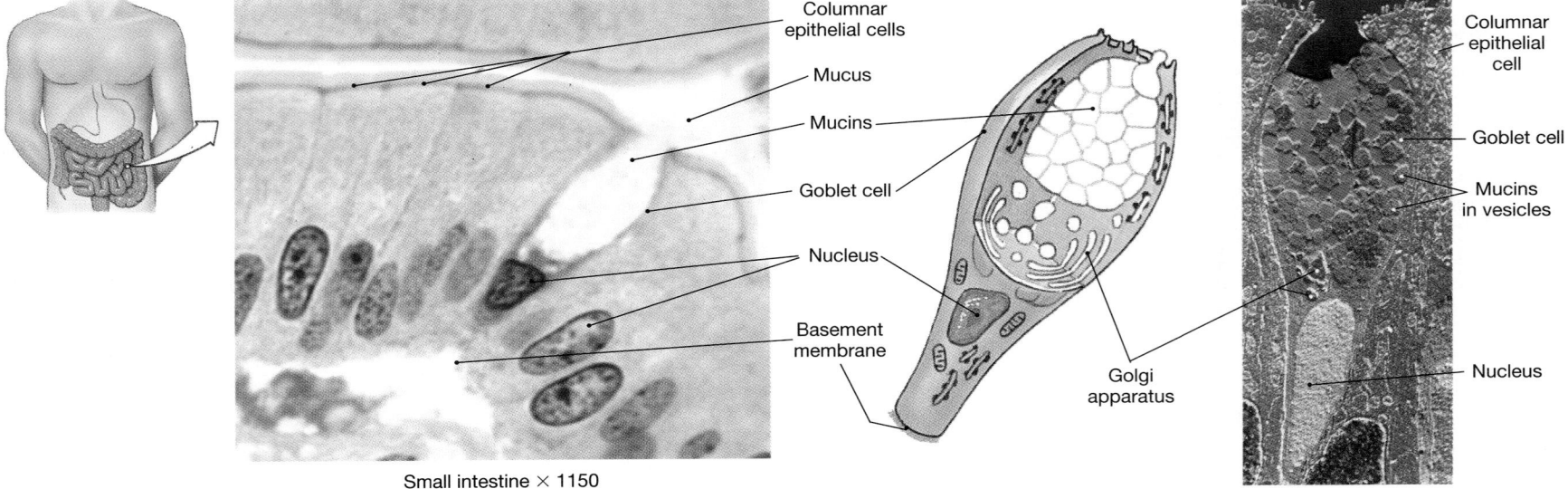

Small intestine × 1150

FIGURE 3-8
Goblet Cells. Goblet cells are unicellular exocrine glands. These secretory cells are often scattered among the simple or pseudostratified columnar epithelial cells of the digestive and respiratory tracts. The micrograph shows a goblet cell in the intestinal epithelium (simple columnar type). Figure 3-2b, p. 54, provides a superficial view of the microvilli that carpet goblet cells in the respiratory epithelium.

EMBRYOLOGY SUMMARY: DEVELOPMENT OF EPITHELIA

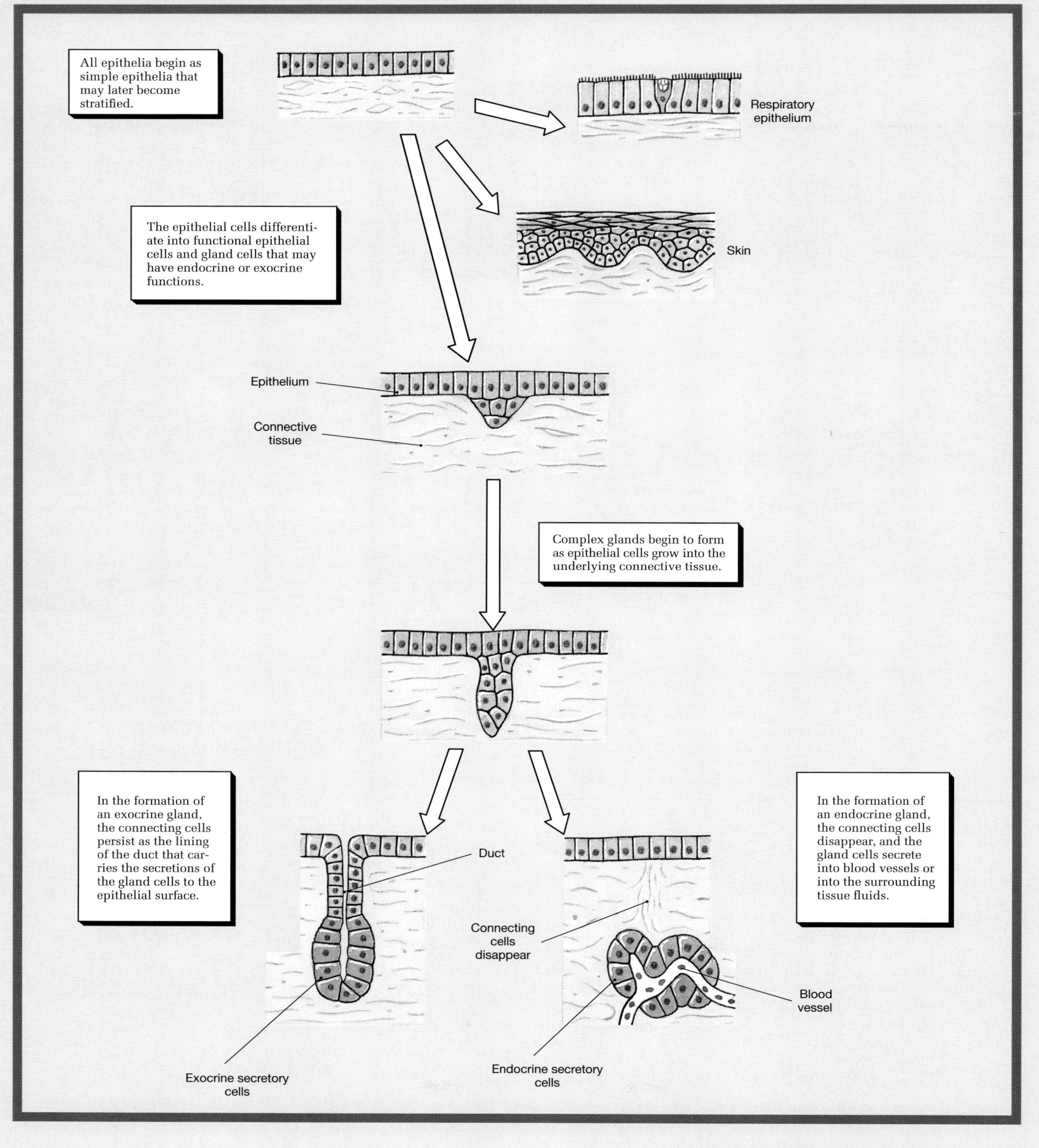

All epithelia begin as simple epithelia that may later become stratified.

Respiratory epithelium

The epithelial cells differentiate into functional epithelial cells and gland cells that may have endocrine or exocrine functions.

Skin

Epithelium

Connective tissue

Complex glands begin to form as epithelial cells grow into the underlying connective tissue.

In the formation of an exocrine gland, the connecting cells persist as the lining of the duct that carries the secretions of the gland cells to the epithelial surface.

In the formation of an endocrine gland, the connecting cells disappear, and the gland cells secrete into blood vessels or into the surrounding tissue fluids.

Duct

Connecting cells disappear

Blood vessel

Exocrine secretory cells

Endocrine secretory cells

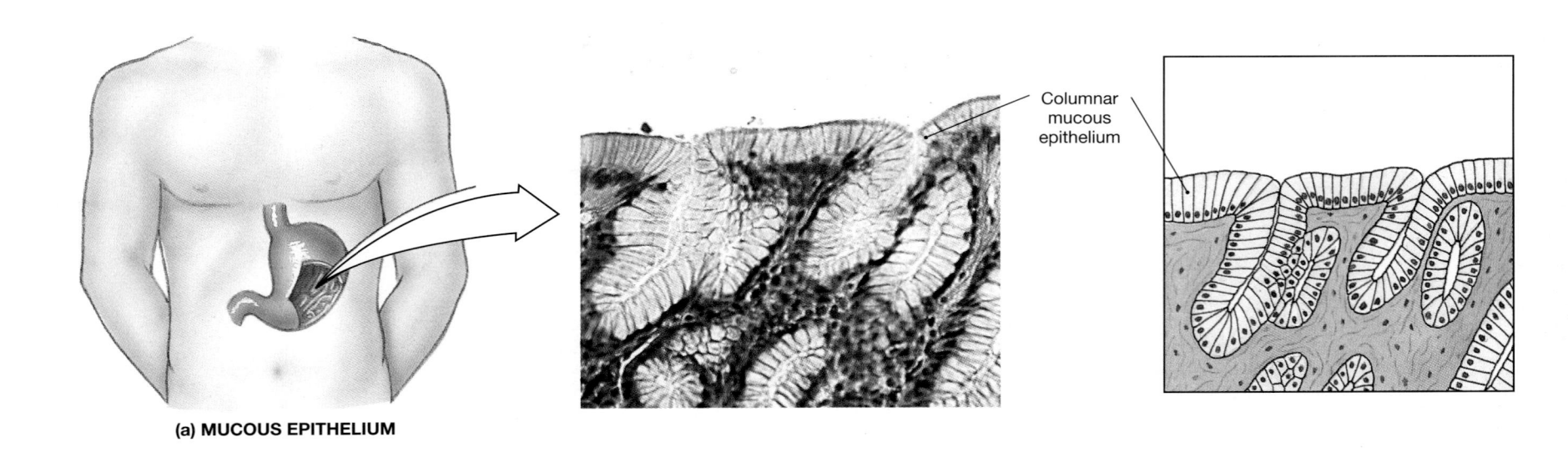

(a) MUCOUS EPITHELIUM

Columnar
mucous
epithelium

(b) MIXED GLANDULAR EPITHELIUM

Serous
cells

Mucous
cells

Duct

FIGURE 3-9
Mucous and Mixed Glandular Epithelia. (a) The interior of the stomach is lined by a mucous epithelium (a secretory sheet) whose secretions protect the walls from acids and enzymes. (The acids and enzymes are produced by glands that discharge their secretions onto the mucous epithelial surface.) (b) The submandibular salivary gland is a mixed gland containing cells that produce both serous and mucous secretions. The mucous cells contain large vesicles containing mucins, and they look pale and foamy. The serous cells secrete enzymes, and the proteins stain darkly. (LM × 252)

CONNECTIVE TISSUES

Connective tissues are found throughout the body but are never exposed to the environment outside the body. Connective tissues include bone, fat, and blood, tissues that are quite different in appearance and function. Nevertheless, all connective tissues have three basic components: (1) specialized cells, (2) extracellular protein fibers, and (3) a fluid known as the **ground substance**. The extracellular fibers and ground substance constitute the **matrix** that surrounds the cells. Whereas epithelial tissue consists almost entirely of cells, the extracellular matrix accounts for most of the volume of connective tissues.

The functions of connective tissues include:

1. establishing a structural framework for the body;

2. transporting fluids and dissolved materials from one region of the body to another;

3. providing protection for delicate organs;

4. supporting, surrounding, and interconnecting other tissue types;

5. storing energy reserves, especially in the form of lipids; and

6. defending the body from invasion by microorganisms.

Classification of Connective Tissues

Connective tissue can be classified into three categories: (1) *connective tissue proper*, (2) *fluid connective tissues*, and (3) *supporting connective tissues*.

1. **Connective tissue proper** refers to connective tissues with many types of cells and extracellular fibers in a syrupy

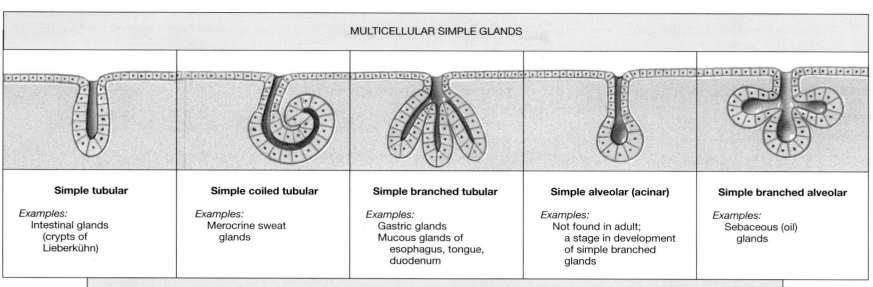

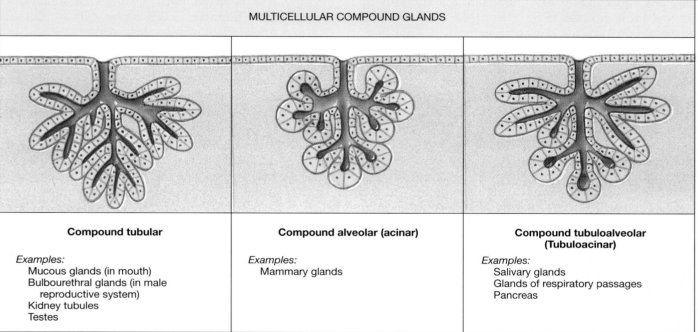

FIGURE 3-10
A Structural Classification of Exocrine Glands

ground substance. These connective tissues may differ in terms of the number of cell types they contain and the relative properties and proportions of fibers and ground substance. *Adipose* (fat) *tissue* and *tendons* are two examples.

2. **Fluid connective tissues** have a distinctive population of cells suspended in a watery matrix that contains dissolved proteins. There are two types of fluid connective tissues, *blood* and *lymph*.

3. **Supporting connective tissues** are of two types, *cartilage* and *bone*. These tissues have a less diverse cell population than connective tissue proper and a matrix that contains closely packed fibers. The matrix of cartilage is a gel whose characteristics vary depending on the predominant fiber type. The matrix of bone is said to be **calcified** because it contains mineral deposits, primarily calcium salts. These minerals give the bone strength and rigidity.

Connective Tissue Proper
(Figure 3-11)

Connective tissue proper contains extracellular fibers, a viscous ground substance, and two classes of cells. **Fixed cells** are stationary and are involved primarily with local maintenance, repair, and energy storage. **Wandering cells** are concerned primarily with the defense and repair of damaged tissues. The number of wandering cells at any given moment varies depending on local conditions. Refer to Figure 3-11● as we describe the cells and fibers of connective tissue proper.

Fixed Cells

Fixed cells include *fibroblasts*, *fixed macrophages*, *adipocytes*, *mesenchymal cells*, and, in a few locations, *melanocytes*.

■ **Fibroblasts** (FĪ-brō-blasts) are the most abundant fixed cells in connective tissue proper. These slender or *stel-*

late (star-shaped) cells are responsible for the production and maintenance of the connective tissue fibers. Each fibroblast manufactures and secretes protein subunits that interact to form large extracellular fibers. In addition, fibroblasts secrete hyaluronic acid that gives the ground substance its viscous consistency.

- **Fixed macrophages** (MAC-rō-fā-jez; *phagein*, to eat) are large, amoeboid cells that are scattered among the fibers. These cells engulf damaged cells or pathogens that enter the tissue. Although they are not abundant, they play an important role in mobilizing the body's defenses. When stimulated they release chemicals that activate the immune system and attract large numbers of wandering cells involved in the body's defense mechanisms.

- **Adipocytes** (AD-i-pō-sīts) are also known as fat cells, or *adipose cells*. A typical adipocyte contains a single, enormous lipid droplet. The nucleus and other organelles are squeezed to one side, making the cell in section resemble a class ring. The number of fat cells varies from one type of connective tissue to another, from one region of the body to another, and from individual to individual.

- **Mesenchymal** (MES-en-kī-mul) **cells** are stem cells that are present in many connective tissues. These cells respond to local injury or infection by dividing to produce daughter cells that differentiate into fibroblasts, macrophages, or other connective tissue cells.

- **Melanocytes** (MEL-an-ō-sīts) synthesize and store a brown pigment, **melanin** (MEL-a-nin), that gives the tissue a dark color. Melanocytes are common in the epithelium of the skin, where they play a major role in determining skin color. However, melanocytes are also abundant in connec-

tive tissues of the eye, and they are present in the dermis of the skin, although there are regional, individual, and racial differences in the number present.

Wandering Cells

Wandering cells include *free macrophages*, *mast cells*, *lymphocytes*, *plasma cells*, and *microphages*.

- **Free macrophages** are relatively large phagocytic cells that wander through the connective tissues of the body. When circulating within the blood, these cells are called **monocytes**. In effect, the few *fixed macrophages* in a tissue provide a "front-line" defense that will be reinforced by the arrival of free macrophages and other specialized cells.

- **Mast cells** are small, mobile connective tissue cells often found near blood vessels. The cytoplasm of a mast cell is filled with granules of **histamine** (HIS-ta-mēn) and **heparin** (HEP-a-rin). These chemicals, released after injury or infection, stimulate local inflammation.

- **Lymphocytes** (LIM-fō-sīts), like free macrophages, migrate throughout the body. Their numbers increase markedly wherever tissue damage occurs, and some of the lymphocytes may then develop into **plasma cells**. Plasma cells are responsible for the production of *antibodies*.

- **Microphages** are phagocytic blood cells that are smaller than monocytes. These cells, called *neutrophils* and *eosinophils*, migrate through connective tissues in small numbers. When an infection or injury occurs, chemicals released by macrophages and *mast cells* attract microphages in large numbers.

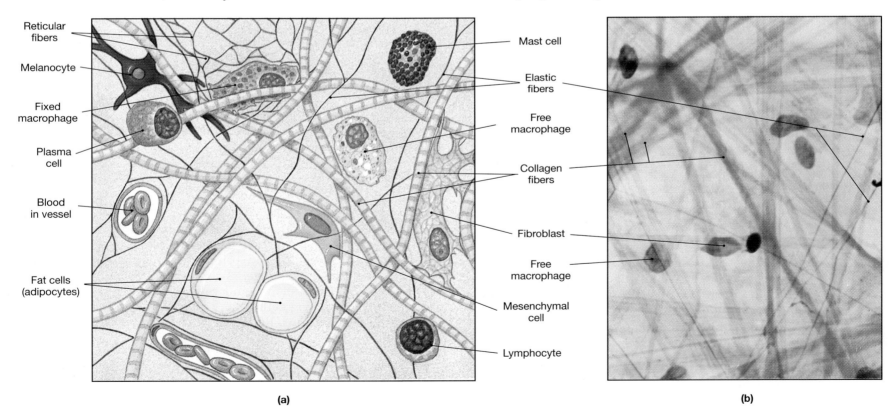

(a)

(b)

FIGURE 3-11
The Cells and Fibers of Connective Tissue Proper. (a) A summary of the cell types and fibers of connective tissue proper. (b) Loose (areolar) connective tissue underlies the mesothelium of the peritoneum. (LM × 502)

FIGURE 3-12
Embryonic Connective Tissues. (a) Mesenchyme, the first connective tissue to appear in the embryo. (LM × 1036) (b) Mucous connective tissue. This sample was taken from the umbilical cord of a fetus. Mucous connective tissue in this location is also known as *Wharton's jelly*. (LM × 650)

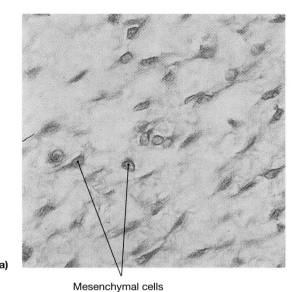

(a)

Mesenchymal cells

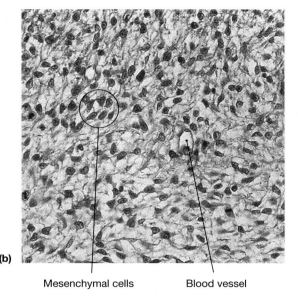

(b)

Mesenchymal cells   Blood vessel

Connective Tissue Fibers
(Figures 3-11 to 3-14)

Three types of fibers are found in connective tissue: *collagen*, *reticular*, and *elastic fibers*. Fibroblasts form all three types of fibers through the secretion of protein subunits that combine or aggregate within the matrix.

1. **Collagen fibers** are long, straight, and unbranched. These are the most common fibers in connective tissue proper. Each collagen fiber consists of three fibrous protein subunits wound together like the strands of a rope. Like a rope, a collagen fiber is flexible but very strong when pulled from either end. **Tendons** (Figure 3-14a●, p. 71) consist almost entirely of collagen fibers, and they connect skeletal muscles to bones. Typical **ligaments** (LIG-a-ments) are similar in structure to tendons, but they connect one bone to another. Tendons and ligaments can withstand tremendous forces; uncontrolled muscle contractions or skeletal movements are more likely to break a bone than snap a tendon or ligament.

2. **Reticular fibers** (*reticulum*, network) contain the same protein subunits as collagen fibers, but the protein subunits are combined in a different way. These fibers are thinner than collagen fibers and form a branching, interwoven framework that is tough but flexible. Reticular fibers are especially abundant in organs such as the spleen and liver, where they create a complex three-dimensional network, or *stroma*, that supports the *parenchyma* (pa-RENG-ki-ma), or distinctive functional cells, of these organs (Figures 3-12 and 3-13c●). Because they form a network, rather than sharing a common alignment, reticular fibers can resist forces applied from many different directions. This ability stabilizes the relative positions of the organ's cells, blood vessels, and other structures despite changing positions and the pull of gravity.

3. **Elastic fibers** contain the protein *elastin*. Elastic fibers are branched and wavy, and after stretching they return to their original length. **Elastic ligaments** are dominated by elastic fibers. They are relatively rare, but have important functions, such as interconnecting the vertebrae (Figures 3-12 and 3-14b●, p. 71).

Ground Substance
(Figures 3-11/3-12)

Ground substance fills the spaces between cells and surrounds the connective tissue fibers (Figure 3-12●). Ground substance in normal connective tissue proper is clear, colorless, and similar in consistency to maple syrup. In addition to hyaluronic acid, it contains a mixture of other proteoglycans and glycoproteins.

Connective tissue proper can be divided into *loose connective tissues* and *dense connective tissues* on the basis of the relative proportions of cells, fibers, and ground substance (Figure 3-11●).

Embryonic Connective Tissues
(Figure 3-12)

Mesenchyme is the first connective tissue to appear in the developing embryo. Mesenchyme contains star-shaped cells that are separated by a matrix that contains very fine protein filaments. This connective tissue (Figure 3-12a●) gives rise to all other connective tissues, including fluid connective tissues, cartilage, and bone. **Mucous connective tissue** (Figure 3-12b●) (*Wharton's jelly*) is a loose connective tissue found in many portions of the embryo, including the umbilical cord.

Neither of these embryonic connective tissues is found in the adult. However, many adult connective tissues contain scattered mesenchymal (stem) cells that can assist in tissue repairs after injury.

Loose Connective Tissues

Loose connective tissues are the "packing material" of the body. These tissues fill spaces between organs, provide cushioning, and support epithelia. Loose connective tissues also surround and support blood vessels and nerves, store lipids, and provide a route for the diffusion of materials. There are three types of loose connective tissues: *loose (areolar) connective tissue*, *adipose tissue*, and *reticular tissue*.

LOOSE (AREOLAR) CONNECTIVE TISSUE *(Figure 3-13a)*. **Loose connective tissue**, or **areolar tissue** (*areola*, little space), is the least specialized connective tissue in the adult body. This tissue, shown in Figure 3-13a●, contains all of the cells and fibers found in any connective tissue proper. Loose connective tissue has an open framework, and ground substance accounts for most of its volume. This viscous fluid cushions shocks, and because the fibers are loosely organized, loose connective tissue can be distorted without damage. The presence of elastic fibers makes it fairly resilient, so this tissue returns to its original shape after external pressure is relieved.

Loose connective tissue forms a layer that separates the skin from deeper structures. In addition to providing padding, the elastic properties of this layer allow a considerable amount of independent movement. Thus, pinching the skin of the arm does not affect the underlying muscle. Conversely, contractions of the underlying muscles do not pull against the skin—as the muscle bulges, the loose connective tissue stretches. Because this tissue has an extensive circulatory supply, drugs are typically injected into the loose connective tissue layer under the skin.

In addition to delivering oxygen and nutrients and removing carbon dioxide and waste products, the capillaries in loose connective tissue carry wandering cells to and from the tissue. Epithelia usually cover a layer of loose connective tissue, and fibroblasts are responsible for maintaining the reticular lamina of the basement membrane. The epithelial cells rely on diffusion across that membrane, and the capillaries in the underlying connective tissue provide the necessary oxygen and nutrients.

ADIPOSE TISSUE *(Figure 3-13b).*   The distinction between loose connective tissue and fat, or **adipose tissue**, is usually somewhat arbitrary. Adipocytes account for most of the volume of adipose tissue (Figure 3-13b●) but only a fraction of the volume of loose connective tissue. Adipose tissue provides padding, cushions shocks, acts as an insulator to slow heat loss through the skin, and serves as packing or filler around structures. Adipose tissue is common under the skin of the groin, sides, buttocks, and breasts. It fills the bony sockets behind the eyes, surrounds the kidneys, and dominates extensive areas of loose connective tissue in the pericardial and abdominal cavities. ⚕ *Infants and Brown Fat [p. 750]*

Adipocytes are metabolically active cells—their lipids are continually being broken down and replaced. Although adipocytes are incapable of dividing, an excess of nutrients can cause the division of mesenchymal cells, which then differentiate into additional fat cells. As a result, areas of loose connective tissue can become adipose tissue in times of nutritional plenty. When nutrients are scarce, as during a weight-loss program, adipocytes deflate like collapsing balloons. Because the cells are not killed, merely reduced in size, the lost weight can easily be regained in the same areas of the body. ⚕ *Liposuction [p. 750]*

RETICULAR TISSUE *(Figure 3-13c).*   **Reticular tissue** forms the stroma in many large and complex organs, including the liver (Figure 3-13c●), the spleen, lymph nodes, and bone marrow. Fixed macrophages and fibroblasts are associated with the reticular tissue, but these cells are seldom visible because the organs are dominated by parenchymal cells with other functions.

Dense Connective Tissues

Most of the volume of **dense connective tissues** is occupied by fibers. Dense connective tissues are often called **collagenous** (ko-LA-jin-us) **tissues** because collagen fibers are the dominant fiber type. Two types of dense connective tissue are found in the body: (1) *dense regular connective tissue* and (2) *dense irregular connective tissue.*

DENSE REGULAR CONNECTIVE TISSUE *(Figures 3-5c,d/3-14a,b).* In **dense regular connective tissue** the collagen fibers are

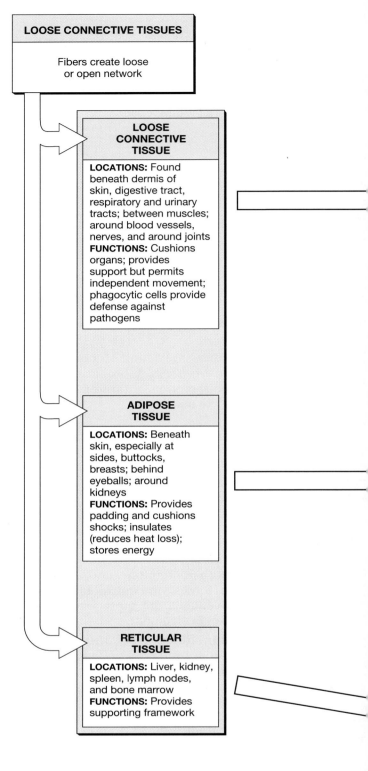

FIGURE 3-13
Loose Connective Tissues. (a) Note the open framework of loose connective tissue in this micrograph. All the cells of connective tissue proper are found in loose connective tissue. (b) Adipose tissue is a loose connective tissue that is dominated by adipocytes. In standard histological preparations, the tissue looks empty because the lipids in the fat cells dissolve during the sectioning and staining procedures. (c) Reticular tissue has an open framework of reticular fibers. These fibers are usually very difficult to see because of the large numbers of cells around them.

arranged parallel to each other, packed tightly, and aligned with the forces applied to the tissue. Four major examples of this tissue type are *tendons, aponeuroses, elastic tissue,* and *ligaments.*

Loose connective tissue from pleura (a)

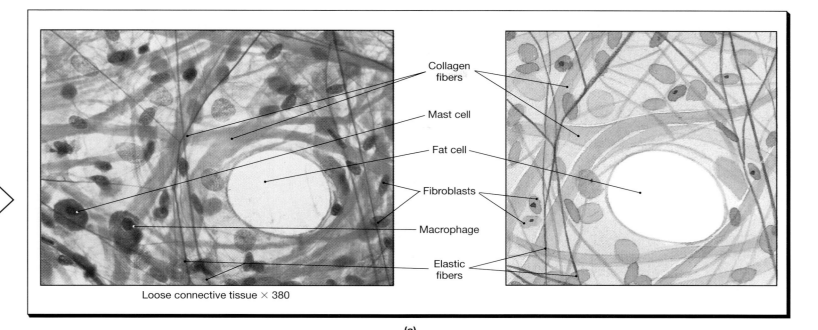

Collagen fibers

Mast cell

Fat cell

Fibroblasts

Macrophage

Elastic fibers

Loose connective tissue × 380

(a)

Adipose tissue from sub-cutaneous fat deposit (b)

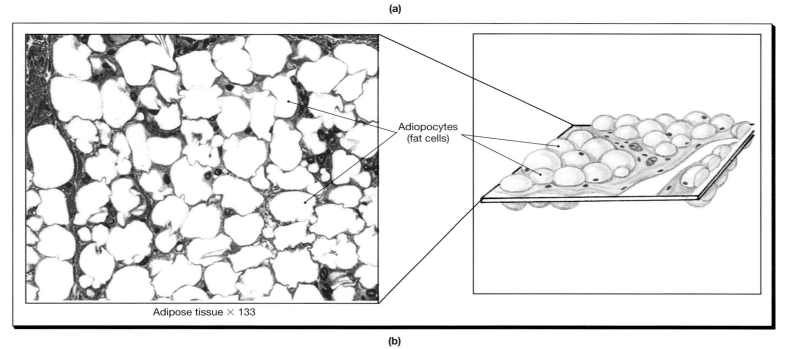

Adiopocytes (fat cells)

Adipose tissue × 133

(b)

Reticular tissue from liver (c)

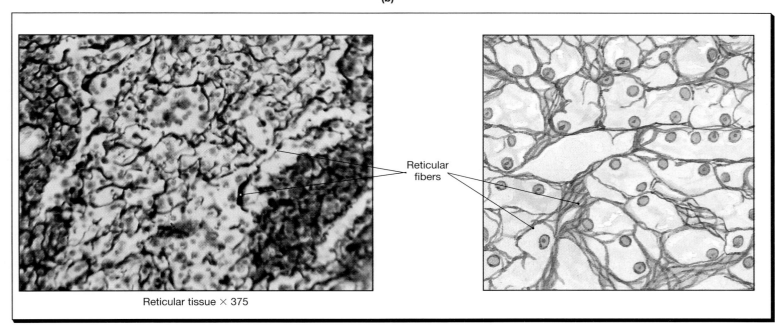

Reticular fibers

Reticular tissue × 375

(c)

1. **Tendons** (Figure 3-14a●) are cords of dense regular connective tissue that attach skeletal muscles to bones. The collagen fibers run along the longitudinal axis of the tendon and transfer the pull of the contracting muscle to the bone. Large numbers of fibroblasts are found between the collagen fibers.

2. **Aponeuroses** (ap-ō-nū-RŌ-sēz) are collagenous sheets or ribbons that resemble flat, broad tendons. Aponeuroses may cover the surface of a muscle and assist in attaching superficial muscles to another muscle or separate structure.

3. **Elastic tissue** contains large numbers of elastic fibers. Because elastic fibers outnumber collagen fibers, the tissue has a springy, resilient nature. This ability to stretch and rebound allows it to tolerate cycles of expansion and contraction. Elastic tissue often underlies transitional epithelia (Figure 3-5c,d●, p. 59); it is also found in the walls of blood vessels and surrounds the respiratory passageways.

4. **Ligaments** resemble tendons, but they connect one bone to another. Ligaments often contain significant numbers of elastic fibers as well as collagen fibers, and they can tolerate a modest amount of stretching. An even higher proportion of elastic fibers is found in **elastic ligaments**, which resemble tough rubber bands. Although uncommon elsewhere, elastic ligaments along the spinal column are very important in stabilizing the positions of the vertebrae (Figure 3-14b●).

DENSE IRREGULAR CONNECTIVE TISSUE *(Figure 3-14c)* The fibers in **dense irregular connective tissue** form an interwoven meshwork and do not show any consistent pattern (Figure 3-14c●). These tissues provide strength and support to areas subjected to stresses from many directions. A layer of dense irregular connective tissue, the *dermis*, gives skin its strength; a piece of cured leather (animal skin) provides an excellent illustration of the interwoven nature of this tissue. Except at joints, dense irregular connective tissue forms a sheath around cartilages (the perichondrium) and bones (the periosteum). Dense irregular connective tissue also forms the thick fibrous **capsule** that surrounds internal organs, such as the liver, kidneys, and spleen, and encloses the cavities of joints.

Fluid Connective Tissues

Blood and **lymph** are connective tissues that contain distinctive collections of cells in a fluid matrix. The watery matrix of blood and lymph contains cells and many different types of dissolved proteins that do not form insoluble fibers under normal conditions. We will discuss blood and lymph in Chapters 20 and 23.

Supporting Connective Tissues

Cartilage and bone are called **supporting connective tissues** because they provide a strong framework that supports the rest of the body. In these connective tissues, the matrix contains numerous fibers and, in some cases, deposits of insoluble calcium salts.

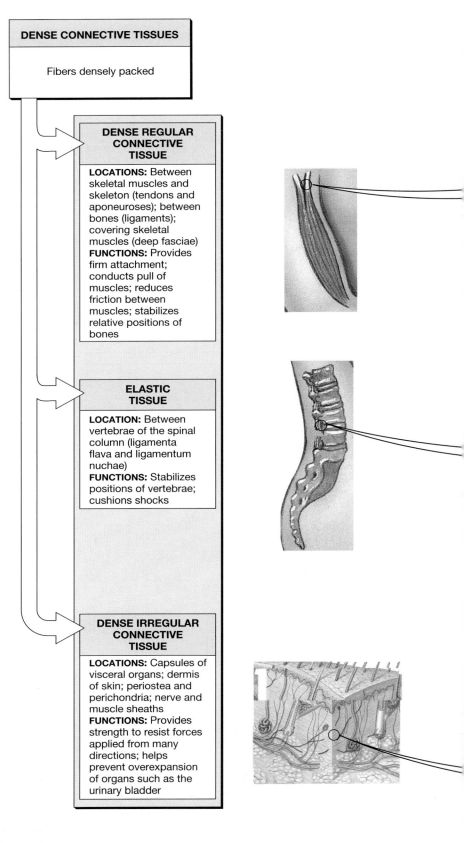

DENSE CONNECTIVE TISSUES

Fibers densely packed

DENSE REGULAR CONNECTIVE TISSUE

LOCATIONS: Between skeletal muscles and skeleton (tendons and aponeuroses); between bones (ligaments); covering skeletal muscles (deep fasciae)
FUNCTIONS: Provides firm attachment; conducts pull of muscles; reduces friction between muscles; stabilizes relative positions of bones

ELASTIC TISSUE

LOCATION: Between vertebrae of the spinal column (ligamenta flava and ligamentum nuchae)
FUNCTIONS: Stabilizes positions of vertebrae; cushions shocks

DENSE IRREGULAR CONNECTIVE TISSUE

LOCATIONS: Capsules of visceral organs; dermis of skin; periostea and perichondria; nerve and muscle sheaths
FUNCTIONS: Provides strength to resist forces applied from many directions; helps prevent overexpansion of organs such as the urinary bladder

FIGURE 3-14
Dense Connective Tissues. (a) The dense regular connective tissue in a tendon. Notice the densely packed, parallel bundles of collagen fibers. The fibroblast nuclei can be seen flattened between the bundles. (b) An elastic ligament from between the vertebrae of the spinal column. The bundles are fatter than those of a tendon or ligament composed of collagen. (c) The dermis of the skin contains a thick layer of dense irregular connective tissue.

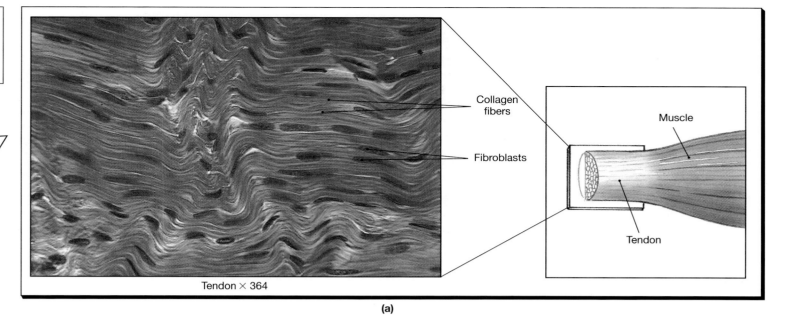

Collagen
fibers

Muscle

Fibroblasts

Tendon

Tendon × 364

(a)

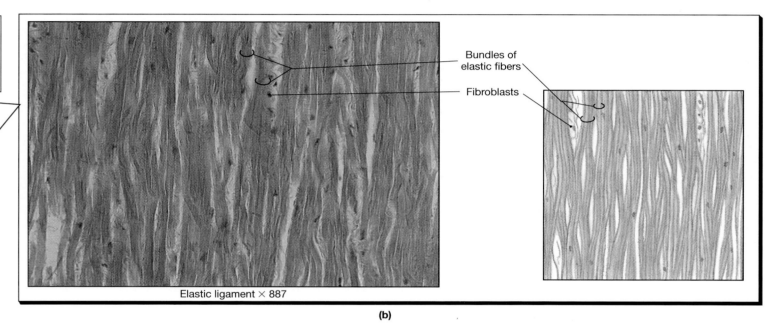

Bundles of
elastic fibers

Fibroblasts

Elastic ligament × 887

(b)

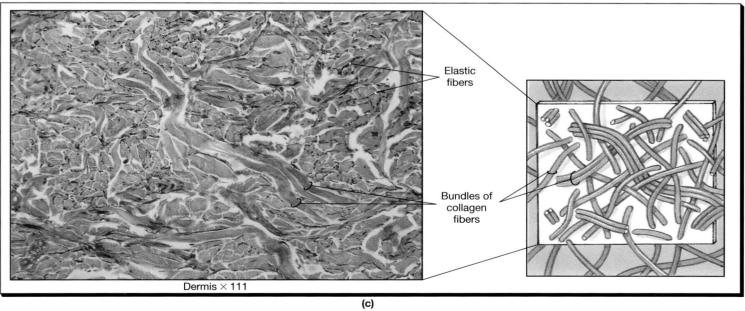

Elastic
fibers

Bundles of
collagen
fibers

Dermis × 111

(c)

CHAPTER 3 The Tissue Level of Organization   71

Cartilage

The matrix of **cartilage** is a firm gel that contains complex polysaccharides called **chondroitin sulfates** (kon-DROY-tin; *chondros*, cartilage). The chondroitin sulfates form complexes with proteins, forming proteoglycans. Cartilage cells, or **chondrocytes** (KON-drō-sīts), are the only cells found within the matrix. These cells live in small pockets known as **lacunae** (la-KOO-nē; *lacus*, pool). The physical properties of cartilage depend on the type and abundance of extracellular fibers, as well as the proteoglycan content.

Cartilage is avascular because chondrocytes produce a chemical that discourages the formation of blood vessels. All nutrient and waste-product exchange must occur by diffusion through the matrix. A cartilage is usually set apart from surrounding tissues by a fibrous **perichondrium** (per-ē-KON-drē-um; *peri-*, around). The perichondrium contains two distinct layers: an outer, fibrous region of dense irregular connective tissue and an inner, cellular layer. The fibrous layer provides mechanical support and protection, and attaches the cartilage to other structures. The cellular layer is important to the growth and maintenance of the cartilage.

Cartilages grow by two mechanisms. In **appositional growth**, cells of the inner layer of the perichondrium undergo repeated cycles of division. The innermost cells begin producing cartilage matrix and differentiate into chondrocytes. This differentiation gradually increases the size of the cartilage by adding to its surface. Chondrocytes within the cartilage matrix also undergo division, and the daughter cells produce additional matrix. This cycle enlarges the cartilage rather as a balloon is inflated; the process is called **interstitial growth**. Neither interstitial nor appositional growth occurs in adult cartilages, and most cartilages cannot repair themselves after a severe injury.

TYPES OF CARTILAGE *(Figure 3-15).* Three major types of cartilage are found in the body: (1) *hyaline cartilage*, (2) *elastic cartilage*, and (3) *fibrocartilage*.

1. **Hyaline cartilage** (HĪ-a-lin; *hyalos*, glass) is the most common type of cartilage. The matrix of hyaline cartilage contains closely packed collagen fibers, making it tough but somewhat flexible. Because the fibers do not stain well, they are not always apparent in light microscopy (Figure 3-15a●). Examples of this type of cartilage in the adult body include: (1) the connections between the ribs and the sternum, (2) the supporting cartilages along the conducting passageways of the respiratory tract, and (3) the cartilages covering articular surfaces within synovial joints, such as the elbow or knee.

2. **Elastic cartilage** contains numerous elastic fibers that make it extremely resilient and flexible. Elastic cartilage forms the external flap (*pinna*) of the outer ear (Figure 3-15b●), the epiglottis, and the tip of the nose.

3. **Fibrocartilage** has little ground substance, and the matrix is dominated by collagen fibers (Figure 3-15c●). The collagen fibers are densely interwoven, making this tissue extremely durable and tough. Fibrocartilaginous pads lie between the spinal vertebrae, between the pubic bones of the pelvis, and around or within a few joints and tendons. In these positions they resist compression, absorb shocks, and prevent damaging bone-to-bone contact. ✚ *Cartilages and Knee Injuries [p. 750]*

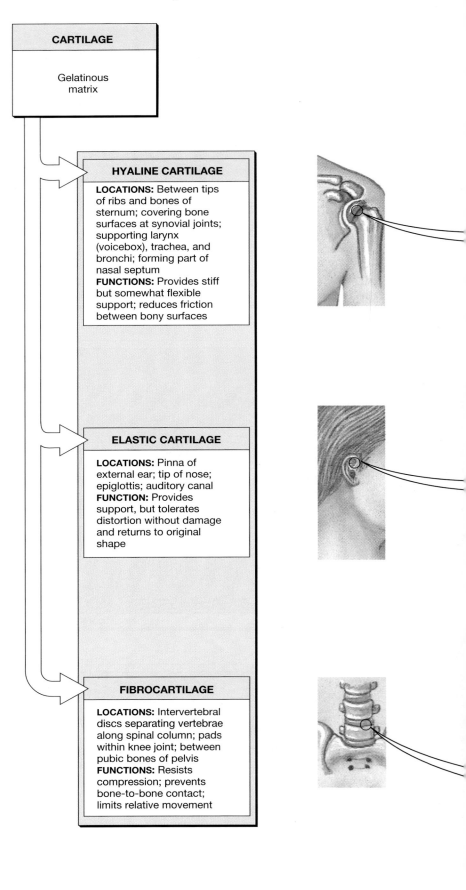

FIGURE 3-15
Types of Cartilage. (a) *Hyaline cartilage.* Note the translucent matrix and the absence of prominent fibers. (b) *Elastic cartilage.* The closely packed elastic fibers are visible between the chondrocytes. (c) *Fibrocartilage.* The collagen fibers are extremely dense, and the chondrocytes are relatively far apart.

Hyaline cartilage in shoulder joint (a)

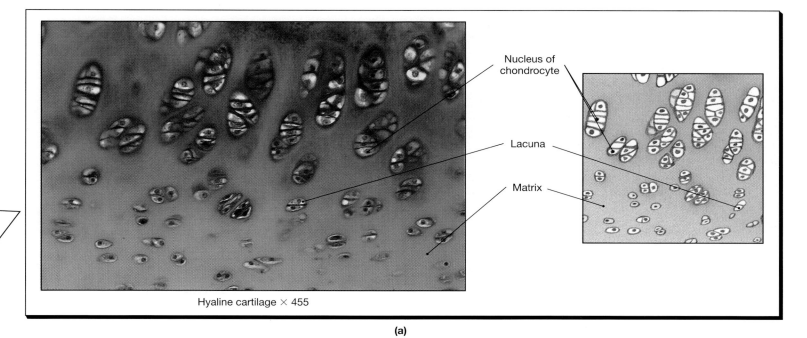

Nucleus of chondrocyte

Lacuna

Matrix

Hyaline cartilage × 455

(a)

Elastic cartilage supporting ear (b)

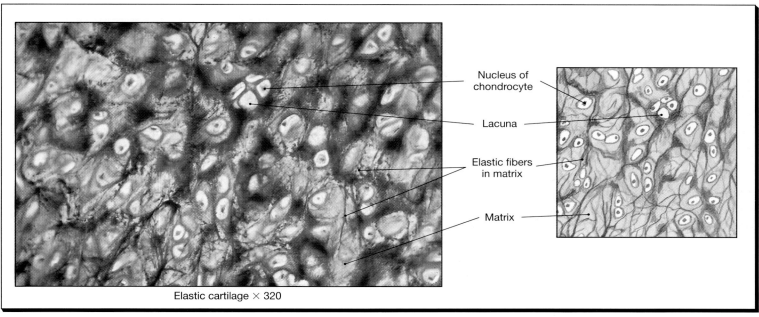

Nucleus of chondrocyte

Lacuna

Elastic fibers in matrix

Matrix

Elastic cartilage × 320

(b)

Fibrocartilage in intervertebral disc (c)

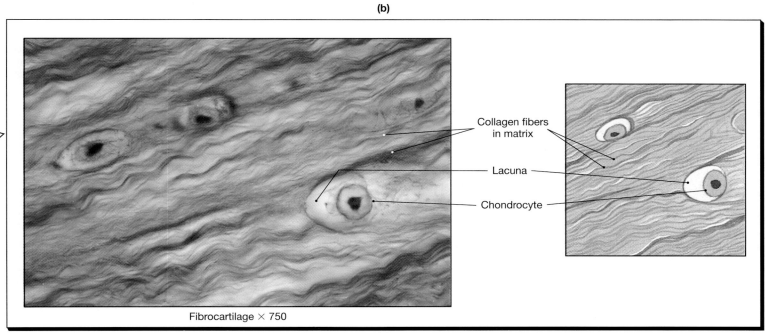

Collagen fibers in matrix

Lacuna

Chondrocyte

Fibrocartilage × 750

(c)

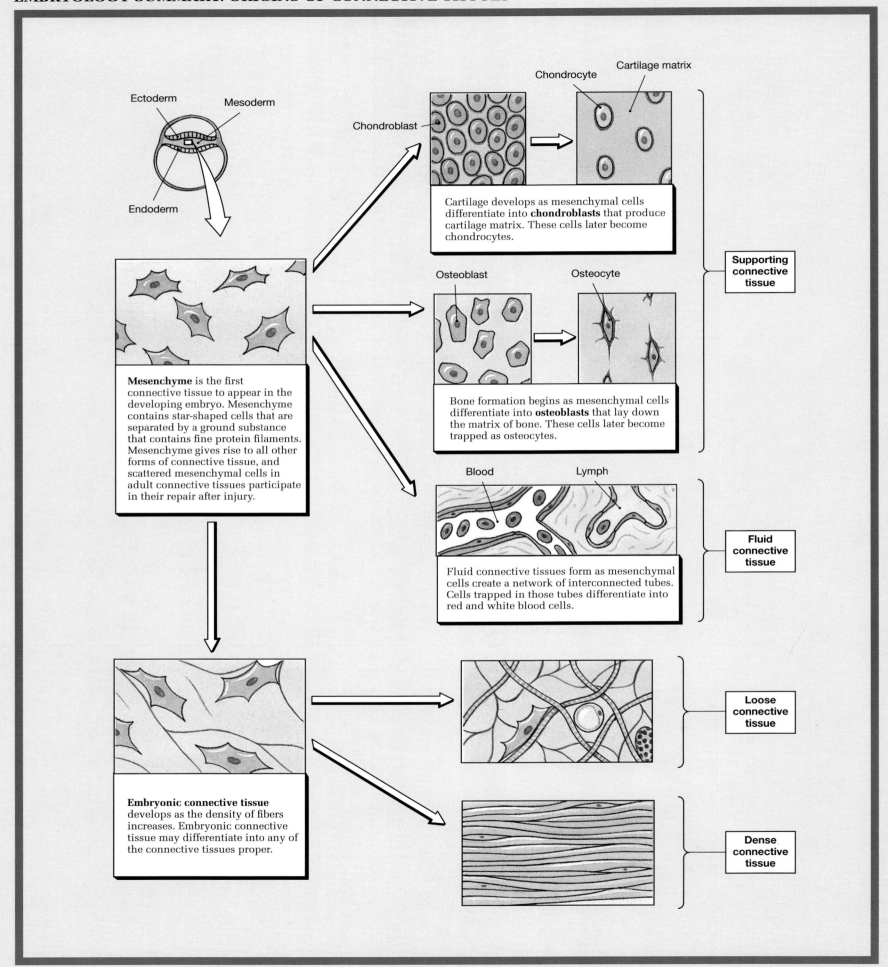

Ectoderm Mesoderm

Endoderm

Chondroblast

Chondrocyte Cartilage matrix

Cartilage develops as mesenchymal cells differentiate into **chondroblasts** that produce cartilage matrix. These cells later become chondrocytes.

Supporting connective tissue

Osteoblast Osteocyte

Bone formation begins as mesenchymal cells differentiate into **osteoblasts** that lay down the matrix of bone. These cells later become trapped as osteocytes.

Mesenchyme is the first connective tissue to appear in the developing embryo. Mesenchyme contains star-shaped cells that are separated by a ground substance that contains fine protein filaments. Mesenchyme gives rise to all other forms of connective tissue, and scattered mesenchymal cells in adult connective tissues participate in their repair after injury.

Blood Lymph

Fluid connective tissues form as mesenchymal cells create a network of interconnected tubes. Cells trapped in those tubes differentiate into red and white blood cells.

Fluid connective tissue

Loose connective tissue

Embryonic connective tissue develops as the density of fibers increases. Embryonic connective tissue may differentiate into any of the connective tissues proper.

Dense connective tissue

Bone
(Figure 3-16)

Because the detailed histology of bone, or **osseous tissue** (OS-ē-us; *os*, bone), will be considered in Chapter 5, this discussion will focus on significant differences between cartilage and bone. Roughly one-third of the matrix of bone consists of collagen fibers. The balance is a mixture of calcium salts, primarily calcium phosphate with lesser amounts of calcium carbonate. This combination gives bone truly remarkable properties. By themselves, calcium salts are strong but rather brittle. Collagen fibers are weaker, but relatively flexible. In bone, the minerals are organized around the collagen fibers. The result is a strong, somewhat flexible combination that is very resistant to shattering. In its overall properties, bone can compete with the best steel-reinforced concrete.

FIGURE 3-16
Bone. The osteocytes in bone are usually organized in groups around a central space that contains blood vessels. For the photomicrograph, a sample of bone was ground thin enough to become transparent. Bone dust produced during the grinding filled the lacunae and the central canal, making them appear dark.

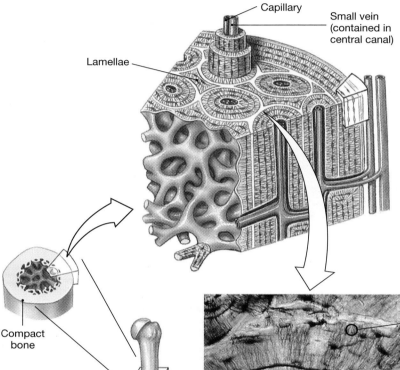

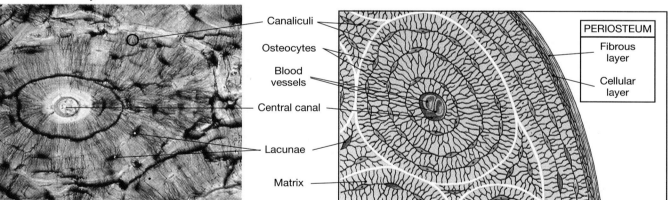

TABLE 3-1 A Comparison of Cartilage and Bone

Characteristic	Cartilage	Bone
STRUCTURAL FEATURES		
Cells	Chondrocytes in lacunae	Osteocytes in lacunae
Matrix	Chondroitin sulfates with proteins, forming hydrated proteoglycans	Insoluble crystals of calcium phosphate and calcium carbonate
Fibers	Collagen, elastic, reticular fibers (proportions vary)	Collagen fibers predominate
Vascularity	None	Extensive
Covering	Perichondrium with two layers	Periosteum with two layers
Strength	Limited: bends easily but hard to break	Strong: resists distortion until breaking point is reached
Growth	Interstitial and appositional	Appositional only
Repair capabilities	Limited ability	Extensive ability

The general organization of osseous tissue can be seen in Figure 3-16●. **Lacunae** within the matrix contain bone cells, or **osteocytes** (OS-tē-ō-sīts). The lacunae are often organized around blood vessels that branch through the bony matrix. Although diffusion cannot occur through the calcium salts, osteocytes communicate with the blood vessels and with one another through slender cytoplasmic extensions. These extensions run through long, slender passages in the matrix. These passageways, called **canaliculi** (kan-a-LIK-ū-lē; "little canals"), form a branching network for the exchange of materials between the blood vessels and the osteocytes.

Except on articular surfaces within synovial joints, where they are covered by a layer of hyaline cartilage, bone surfaces are sheathed by a **periosteum** (per-ē-OS-tē-um) composed of fibrous (outer) and cellular (inner) layers. The periosteum assists in the attachment of a bone to surrounding tissues and to associated tendons and ligaments. The cellular layer functions in bone growth and participates in repairs after an injury. Unlike cartilage, bone undergoes extensive remodeling on a regular basis, and complete repairs can be made even after severe damage has occurred. Bones also respond to the stresses placed upon them,

growing thicker and stronger with exercise, and thin and brittle with inactivity. Table 3-1 summarizes the similarities and differences between cartilage and bone.

MEMBRANES

Epithelia and connective tissues combine to form membranes that cover and protect other structures and tissues in the body. There are four such types of membranes: (1) *mucous membranes*, (2) *serous membranes*, (3) the *cutaneous membrane (skin)*, and (4) *synovial membranes.*

Mucous Membranes
(Figure 3-17a)

Mucous membranes line cavities that communicate with the exterior, including the digestive, respiratory, reproductive,

FIGURE 3-17
Membranes. (a) *Mucous membranes* are coated with the secretions of mucous glands. Mucous membranes line most of the digestive and respiratory tracts and portions of the urinary and reproductive tracts. (b) *Serous membranes* line the ventral body cavities (the peritoneal, pleural, and pericardial cavities). (c) The *cutaneous membrane* of the skin covers the outer surface of the body. (d) *Synovial membranes* line joint cavities and produce the fluid within the joint.

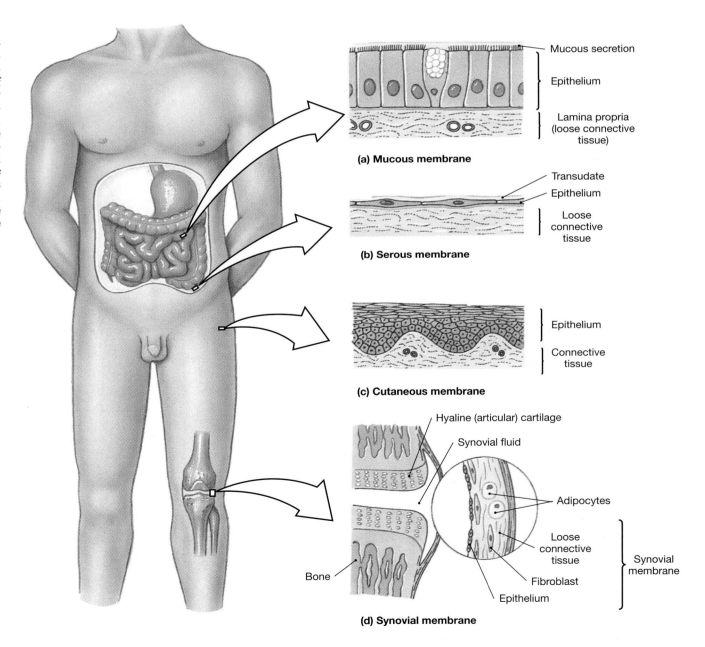

(a) **Mucous membrane**

- Mucous secretion
- Epithelium
- Lamina propria (loose connective tissue)

(b) **Serous membrane**

- Transudate
- Epithelium
- Loose connective tissue

(c) **Cutaneous membrane**

- Epithelium
- Connective tissue

(d) **Synovial membrane**

- Hyaline (articular) cartilage
- Synovial fluid
- Adipocytes
- Loose connective tissue
- Fibroblast
- Epithelium
- Bone
- Synovial membrane

and urinary tracts (Figure 3-17a●). The epithelial surfaces are kept moist at all times; they may be lubricated by mucus produced by goblet cells or multicellular glands or by exposure to fluids such as urine or semen. The loose connective tissue component of a mucous membrane is called the **lamina propria** (PRŌ-prē-a). We will consider the organization of specific mucous membranes in greater detail in later chapters.

Many mucous membranes are lined by simple epithelia that perform absorptive or secretory functions, such as the simple columnar epithelium of the digestive tract. However, other types of epithelia may be involved. For example, a stratified squamous epithelium is part of the mucous membrane of the mouth, and the mucous membrane along most of the urinary tract has a transitional epithelium.

Serous Membranes
(Figure 3-17b)

Serous membranes line the sealed, internal cavities of the body. There are three serous membranes, each consisting of a *mesothelium* (p. 57) supported by loose connective tissue (Figure 3-17b●). These membranes were introduced in Chapter 1 ∞ [p. 19]: (1) The *pleura* lines the pleural cavities and covers the lungs; (2) the *peritoneum* lines the peritoneal cavity and covers the surfaces of the enclosed organs; and (3) the *pericardium* lines the pericardial cavity and covers the heart. Serous membranes are very thin, and they are firmly attached to the body wall and to the organs they cover. When you are looking at an organ, such as the heart or stomach, you are really seeing the tissues of the organ through a transparent serous membrane.

The parietal and visceral portions of a serous membrane are in close contact at all times. Minimizing friction between these opposing surfaces is the primary function of serous membranes. Because the mesothelia are very thin, serous membranes are relatively permeable, and tissue fluids diffuse onto the exposed surface, keeping it moist and slippery.

The fluid formed on the surfaces of a serous membrane is called a **transudate** (TRANS-ū-dāt; *trans-*, across). Specific transudates are called **pleural fluid**, **peritoneal fluid**, or **pericardial fluid**, depending on their source. In normal healthy individuals, the total volume of transudate is extremely small, just enough to prevent friction between the walls of the cavities and the surfaces of internal organs. But after an injury or in certain disease states, the volume of transudate may increase dramatically, complicating existing medical problems or producing new ones. ☤ *Problems with Serous Membranes [p. 750]*

The Cutaneous Membrane
(Figure 3-17c)

The **cutaneous membrane** of the skin covers the surface of the body. It consists of a stratified squamous epithelium and an underlying layer of loose connective tissue reinforced by a layer of dense connective tissue (Figure 3-17c●). In contrast to serous or mucous membranes, the cutaneous membrane is thick, relatively waterproof, and usually dry.

Synovial Membranes
(Figure 3-17d)

A **synovial membrane** (sin-Ō-vē-al) consists of extensive areas of loose connective tissue bounded by a superficial layer of squamous or cuboidal cells (Figure 3-17d●). Although usually called an epithelium, it differs from other epithelia in two respects: (1) There is no basement membrane, and (2) the cellular layer is incomplete, and gaps exist between adjacent cells. Some of the lining cells are phagocytic and others are secretory. The secretory cells regulate the composition of the **synovial fluid** within the joint cavity.

THE CONNECTIVE TISSUE FRAMEWORK OF THE BODY
(Figure 3-18)

Connective tissues create the internal framework of the body. Layers of connective tissue connect the organs within the dorsal and ventral body cavities with the rest of the body. These layers (1) provide strength and stability, (2) maintain the relative positions of internal organs, and (3) provide a route for the distribution of blood vessels, lymphatics, and nerves. The connective tissue layers and wrappings can be divided into three major components: the *superficial fascia*, the *deep fascia*, and the *subserous fascia*. The functional anatomy of these layers is illustrated in Figure 3-18●.

- The **superficial fascia**, or **subcutaneous layer** (*sub*, below + *cutis*, skin) is also termed the **hypodermis** (*hypo*, below + *dermis*, skin). This layer of loose connective tissue separates the skin from underlying tissues and organs. It provides insulation and padding and lets the skin or underlying structures move independently.

- The **deep fascia** consists of dense connective tissue. The fiber organization resembles that of plywood: All the fibers in an individual layer run in the same direction, but the orientation of the fibers changes from one layer to another. This variation helps the tissue to resist forces applied from many different directions. The tough *capsules* that surround most organs, including the kidneys and the organs in the thoracic and peritoneal cavities, are components of the deep fascia. The perichondrium around cartilages, the periosteum around bones and the ligaments that interconnect them, and the connective tissues of muscle, including tendons and aponeuroses, are all part of the deep fascia. The dense connective tissue components are interwoven; for example, the deep fascia around a muscle blends into the tendon, whose fibers intermingle with those of the periosteum. This arrangement creates a strong, fibrous network for the body and ties structural elements together.

- The **subserous fascia** is a layer of loose connective tissue that lies between the deep fascia and the serous membranes that line body cavities. Because this layer separates the serous membranes from the deep fascia, movements of muscles or muscular organs do not severely distort the delicate lining.

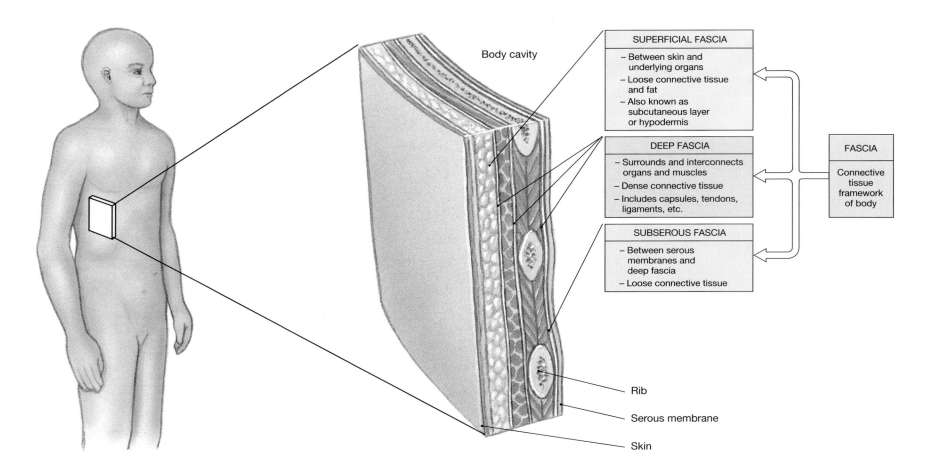

FIGURE 3-18
The Fasciae. The functional anatomy relationship of connective tissue elements in the body.

√ **A sheet of tissue has many layers of collagen fibers that run in different directions in successive layers. What type of tissue is this?**

√ **Lack of vitamin C in the diet interferes with the ability of fibroblasts to produce collagen. What effect might this limited ability to produce collagen have on connective tissue?**

MUSCLE TISSUE
(Figure 3-19)

Muscle tissue is specialized for contraction (Figure 3-19●). Muscle cells possess organelles and properties distinct from those of other cells. They are capable of powerful contractions that shorten the cell along its longitudinal axis. Because they are different from "typical" cells, the term **sarcoplasm** is used to refer to the cytoplasm of a muscle cell, and one refers to the **sarcolemma** rather than the cell membrane.

Three types of muscle tissue are found in the body: (1) *skeletal*, (2) *cardiac*, and (3) *smooth*. The contraction mechanism is similar in all three, but they differ in their internal organization. We will describe each muscle type in greater detail in later chapters (skeletal muscle in Chapter 9, cardiac muscle in Chapter 21, and smooth muscle in Chapter 25). This discussion will focus on general characteristics rather than specific details.

Skeletal Muscle Tissue
(Figure 3-19a)

Skeletal muscle tissue contains very large muscle fibers. Because individual skeletal muscle cells are relatively long and slender, they are usually called **muscle fibers**. Skeletal muscle fibers are very unusual because they may be a foot (0.3 meter) or more in length, and each cell is **multinucleated**, containing hundreds of nuclei. Nuclei lie just under the surface of the sarcolemma (Figure 3-19a●). Skeletal muscle fibers are incapable of dividing, but new muscle fibers can be produced through the division of **satellite cells**, embryonic stem cells that persist in adult skeletal muscle tissue. As a result, skeletal muscle tissue can at least partially repair itself after an injury.

Skeletal muscle fibers contain *actin* and *myosin* filaments arranged in organized groups. As a result, skeletal muscle fibers appear to have a banded or striated appearance (Figure 3-19a●). Skeletal muscle fibers will not usually contract unless stimulated by nerves, and the nervous system provides voluntary control over their activities. Thus, skeletal muscle is called **striated voluntary muscle**.

Skeletal muscle tissue is tied together by loose connective tissue. The collagen and elastic fibers surrounding each cell and group of cells blend into those of a tendon or aponeurosis that conducts the force of contraction, usually to a bone of the skeleton. When the muscle tissue contracts, it pulls on the bone, and the bone moves.

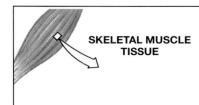

SKELETAL MUSCLE TISSUE

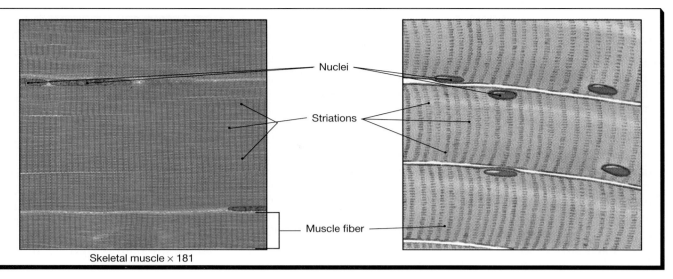

LOCATIONS: Combined with connective tissues and nervous tissue in skeletal muscles, organs such as the leg muscles or arm muscles

FUNCTIONS: Moves or stabilizes the position of the skeleton; guards entrances and exits to the digestive, respiratory, and urinary tracts; generates heat; protects internal organs

Nuclei

Striations

Muscle fiber

Skeletal muscle × 181

(a)

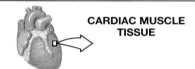

CARDIAC MUSCLE TISSUE

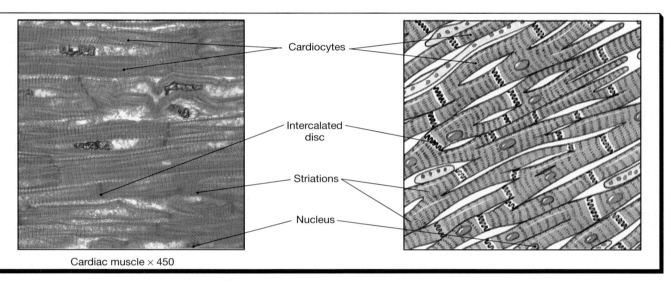

LOCATION: Heart

FUNCTIONS: Circulates blood; maintains blood (hydrostatic) pressure

Cardiocytes

Intercalated disc

Striations

Nucleus

Cardiac muscle × 450

(b)

SMOOTH MUSCLE TISSUE

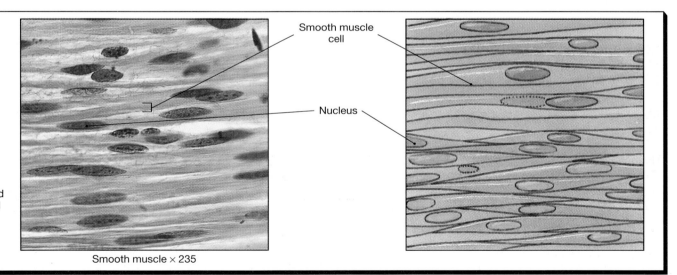

LOCATIONS: Encircles blood vessels; found in the walls of digestive, respiratory, urinary, and reproductive organs

FUNCTIONS: Moves food, urine, and reproductive tract secretions; controls diameters of respiratory passageways; regulates diameter of blood vessels; and contributes to regulation of tissue blood flow

Smooth muscle cell

Nucleus

Smooth muscle × 235

(c)

FIGURE 3-19

Muscle Tissue. (a) *Skeletal muscle fibers.* Note the large fiber size, prominent banding pattern, multiple nuclei, and un-branched arrangement. (b) *Cardiac muscle cells.* Cardiac muscle cells differ from skeletal muscle fibers in three major ways: size (cardiac muscle cells are smaller), organization (cardiac muscle cells branch), and number of nuclei (a typical cardiac muscle cell has one centrally placed nucleus). Both contain actin and myosin filaments in an organized array that produces the striations seen in both types of muscle cell. (c) *Smooth muscle cells.* Smooth muscle cells are small and spindle-shaped, with a central nucleus. They do not branch, and there are no striations.

Cardiac Muscle Tissue
(Figure 3-19b)

Cardiac muscle tissue is found only in the heart. A typical cardiac muscle cell, or **cardiocyte,** is smaller than a skeletal muscle fiber, and it usually has one centrally placed nucleus. The prominent striations, seen in Figure 3-19b●, resemble those of skeletal muscle. Cardiac muscle cells form extensive connections with one another. The connections occur at specialized regions known as **intercalated discs.** As a result, cardiac muscle tissue consists of a branching network of interconnected muscle cells. The interconnections help channel the forces of contraction, and gap junctions at the intercalated discs help coordinate the activities of individual cardiac muscle cells. Like skeletal muscle fibers, cardiac muscle cells are incapable of dividing, and because this tissue lacks satellite cells, cardiac muscle tissue damaged by injury or disease cannot regenerate.

Cardiac muscle cells do not rely on nerve activity to start a contraction. Instead, specialized cardiac muscle cells called **pacemaker cells** establish a regular rate of contraction. Although the nervous system can alter the rate of pacemaker activity, it does not provide voluntary control over individual cardiac muscle cells. Therefore, cardiac muscle is called **striated involuntary muscle.**

Smooth Muscle Tissue
(Figure 3-19c)

Smooth muscle tissue can be found in the walls of blood vessels; around hollow organs such as the urinary bladder; and in layers around the respiratory, circulatory, digestive, and reproductive tracts. A smooth muscle cell is a small cell with tapering ends, containing a single oval nucleus (Figure 3-19c●). Smooth muscle cells can divide, and smooth muscle tissue can regenerate after an injury. The actin and myosin filaments in smooth muscle cells are organized differently from those of skeletal and cardiac muscle, and there are no striations. Smooth muscle cells may contract on their own, or their contractions may be triggered by neural activity. The nervous system usually does not provide voluntary control over smooth muscle contractions, and smooth muscle is therefore known as **nonstriated involuntary muscle.**

NEURAL TISSUE
(Figure 3-20)

Neural tissue, also known as **nervous tissue** or *nerve tissue,* is specialized for the conduction of electrical impulses from one region of the body to another. Most of the neural tissue in the body (98%) is concentrated in the brain and spinal cord, the control centers for the nervous system. Neural tissue contains two basic types of cells: nerve cells, or **neurons** (NOO-rons; *neuro,* nerve), and several different kinds of supporting cells, collectively called **neuroglia** (noo-ROG-lē-a; *glia,* glue). Neurons transmit electrical impulses in the form of changes in the transmembrane potential. Neuroglia have different functions, such as to provide a supporting framework for neural tissue and to play a role in providing nutrients to neurons.

Neurons are the longest cells in the body, many reaching a meter in length. Most neurons are incapable of dividing under normal circumstances, and they have a very limited ability to repair themselves after injury. A typical neuron has a cell body, or **soma,** that contains a large prominent nucleus (Figure 3-20●). Extending from the soma are various branching processes termed **dendrites** (DEN-drīts; *dendron,* tree) and a single **axon.** Dendrites receive incoming messages, axons conduct outgoing messages. It is the length of the axon that can make a neuron so long; because axons are very slender, they are also called **nerve fibers.** Chapter 13 will discuss the properties of neural tissue and provide additional histological and cytological details.

√ **The lining of the nasal cavity is normally moist, contains numerous goblet cells, and rests on a layer of connective tissue called the lamina propria. What type of membrane is this?**

√ **What type of muscle tissue has small, tapering cells with single nuclei and no obvious striations?**

√ **Why do you find the same epithelial organization in the pharynx, esophagus, anus, and vagina?**

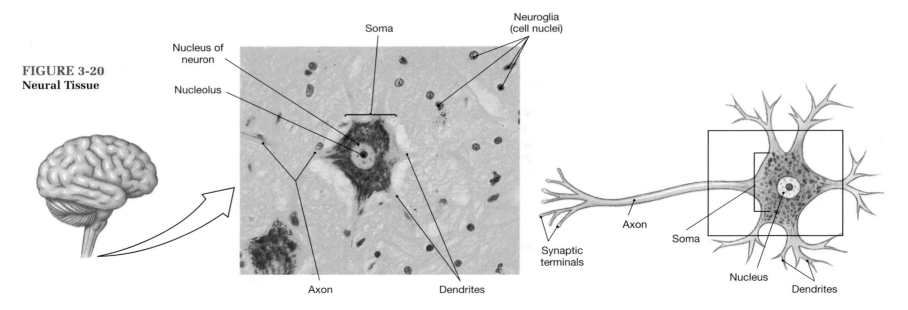

FIGURE 3-20
Neural Tissue

Tumor Formation and Growth

Physicians who specialize in the identification and treatment of cancers are called **oncologists** (on-KOL-ō-jists; *onkos*, mass). Pathologists and oncologists classify cancers according to their cellular appearance and their sites of origin. Over a hundred kinds have been described, but broad categories are usually used to indicate the location of the primary tumor. Table 3-2 summarizes information concerning benign and malignant tumors (cancers) associated with the tissues discussed in this chapter.

Cancer develops in a series of steps, diagrammed in Figure 3-21●. Initially the cancer cells are restricted to a single location, called the **primary tumor** or **primary neoplasm**. All of the cells in the tumor are usually the daughter cells of a single malignant cell. At first the growth of the primary tumor simply distorts the tissue, and the basic tissue organization remains intact. Metastasis begins as tumor cells "break out" of the primary tumor and invade the surrounding tissue. When this invasion is followed by penetration of nearby blood vessels, the cancer cells begin circulating throughout the body.

Responding to cues that are as yet unknown, these cells later escape from the circulatory system and establish **secondary tumors** at other sites. These tumors are extremely active metabolically, and their presence stimulates the growth of blood vessels into the area. The increased circulatory supply provides additional nutrients and further accelerates tumor growth and metastasis. Death may occur because vital organs have been compressed, because nonfunctional cancer cells have killed or replaced the normal cells in vital organs, or because the voracious cancer cells have starved normal tissues of essential nutrients. ♈ *Cancer Treatment and Statistics [p. 751]*

TABLE 3-2   Benign and Malignant Tumors in the Major Tissue Types

Tissue	Description
Epithelia	
Carcinomas	Any cancer of epithelial origin
Adenocarcinomas	Cancers of glandular epithelia
Angiosarcomas	Cancers of endothelial cells
Mesotheliomas	Cancers of mesothelial cells
Connective tissues	
Fibromas	Benign tumors of fibroblast origin
Lipomas	Benign tumors of adipose tissue
Liposarcomas	Cancers of adipose tissue
Leukemias, lymphomas	Cancers of blood-forming tissues
Chondromas	Benign tumors in cartilage
Chondrosarcomas	Cancers of cartilage
Osteomas	Benign tumors in bone
Osteosarcomas	Cancers of bone
Muscle tissues	
Myomas	Benign muscle tumors
Myosarcomas	Cancers of skeletal muscle tissue
Cardiac sarcomas	Cancers of cardiac muscle tissue
Leiomyomas	Benign tumors of smooth muscle tissue
Leiomyosarcomas	Cancers of smooth muscle tissue
Neural tissues	
Gliomas, neuromas	Cancers of neuroglial origin

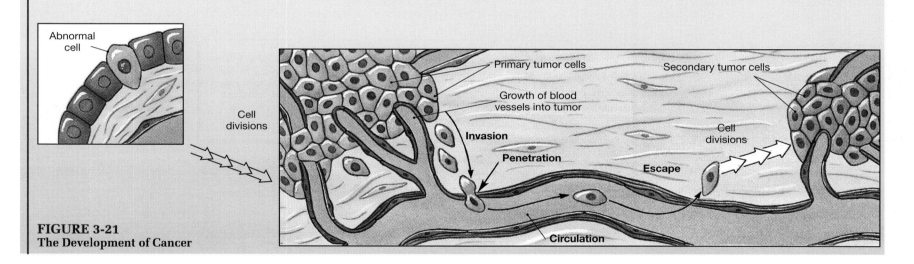

FIGURE 3-21
The Development of Cancer

TISSUES, NUTRITION, AND AGING

Tissues change with age. In general, repair and maintenance activities grow less efficient, and a combination of hormonal changes and alterations in lifestyle affect the structure and chemical composition of many tissues. Epithelia get thinner, and connective tissues more fragile. Individuals bruise easily and bones become brittle; joint pains and broken bones are common complaints. Because cardiac muscle fibers and neurons cannot be replaced, over time, cumulative losses from relatively minor damage can contribute to major health problems such as cardiovascular disease or deterioration in mental function.

In future chapters we will consider the effects of aging on specific organs and systems. Some of these changes are ge-netically programmed. For example, the chondrocytes of older individuals produce a slightly different form of proteoglycan than those of younger people. The difference probably accounts for the observed changes in the thickness and resilience of cartilage. In other cases the tissue degeneration may be temporarily slowed or even reversed. The age-related reduction in bone strength in women, a condition called *osteoporosis*, is often caused by a combination of inactivity, low dietary calcium levels, and a reduction in circulating *estrogens* (female sex hormones). A program of exercise, calcium supplements, and hormonal replacement therapies can usually maintain normal bone structure for many years.

In this chapter we have introduced the four basic types of tissue found in the human body. In combination these tissues form all of the organs and systems that will be discussed in subsequent chapters. ♈ *Tissue Structure and Disease [p. 750]*

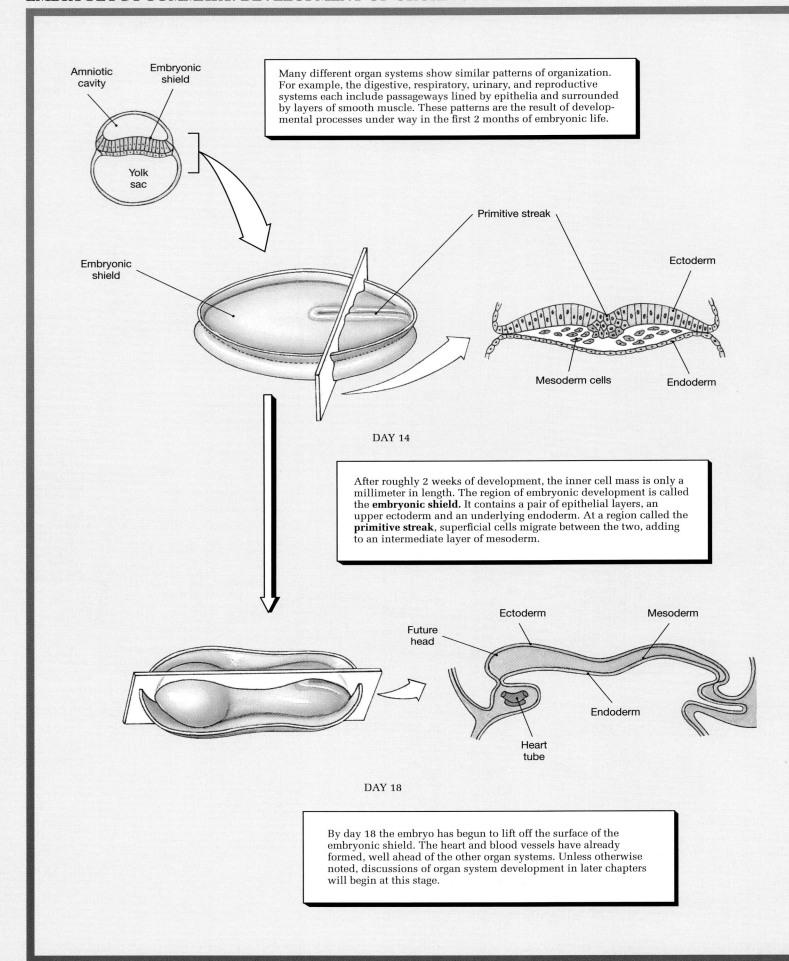

Many different organ systems show similar patterns of organization. For example, the digestive, respiratory, urinary, and reproductive systems each include passageways lined by epithelia and surrounded by layers of smooth muscle. These patterns are the result of developmental processes under way in the first 2 months of embryonic life.

Amniotic cavity

Embryonic shield

Yolk sac

Embryonic shield

Primitive streak

Ectoderm

Mesoderm cells

Endoderm

DAY 14

After roughly 2 weeks of development, the inner cell mass is only a millimeter in length. The region of embryonic development is called the **embryonic shield.** It contains a pair of epithelial layers, an upper ectoderm and an underlying endoderm. At a region called the **primitive streak**, superficial cells migrate between the two, adding to an intermediate layer of mesoderm.

Ectoderm

Mesoderm

Future head

Endoderm

Heart tube

DAY 18

By day 18 the embryo has begun to lift off the surface of the embryonic shield. The heart and blood vessels have already formed, well ahead of the other organ systems. Unless otherwise noted, discussions of organ system development in later chapters will begin at this stage.

DERIVATIVES OF PRIMARY GERM LAYERS	
Ectoderm Forms:	Epidermis and epidermal derivatives of the integumentary system, including hair follicles, nails, and glands communicating with the skin surface (sweat, milk, and sebum) Lining of the mouth, salivary glands, nasal passageways, and anus Nervous system, including brain and spinal cord Portions of endocrine system (pituitary and parts of adrenal glands) Portions of skull, pharyngeal arches, and associated musculature
Mesoderm Forms:	Lining of the body cavities (pleural, pericardial, peritoneal) Muscular, skeletal, cardiovascular, and lymphatic systems Kidneys and part of the urinary tract Gonads and most of the reproductive tracts Connective tissues supporting all organ systems Portions of endocrine system (parts of adrenal glands and endocrine tissues of the reproductive tracts)
Endoderm Forms:	Most of the digestive system: epithelium (except mouth and anus), exocrine glands (except salivary glands), the liver and pancreas Most of the respiratory system: epithelium (except nasal passageways) and mucous glands Portions of urinary and reproductive systems (ducts and the stem cells that produce gametes) Portions of endocrine system (thymus, thyroid, pancreas)

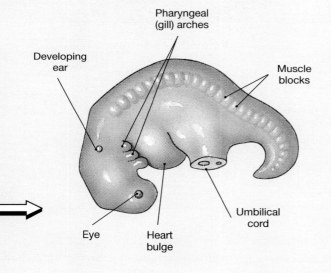

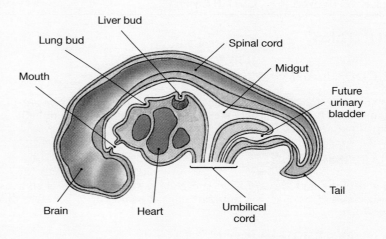

DAY 28

After 1 month you can find the beginnings of all major organ systems. The role of each of the primary germ layers in the formation of organs is summarized in the accompanying table; details will be found in later chapters.

Related Clinical Terms

liposuction: A surgical procedure to remove unwanted adipose tissue by sucking it through a tube. ☤ *Liposuction [p. 750]*

adhesions: Restrictive fibrous connections that can result from surgery, infection, or other injuries to serous membranes. ☤ *Problems with Serous Membranes [p. 750]*

pleuritis (pleurisy): An inflammation of the lining of the pleural cavities. ☤ *Problems with Serous Membranes [p. 750]*

effusion: The accumulation of fluid in body cavities. ☤ *Problems with Serous Membranes [p. 750]*

pericarditis: An inflammation of the pericardium. ☤ *Problems with Serous Membranes [p. 750]*

peritonitis: An inflammation of the peritoneum. ☤ *Problems with Serous Membranes [p. 750]*

ascites (a-SĪ-tēz): An accumulation of fluid that creates a characteristic abdominal swelling. ☤ *Problems with Serous Membranes [p. 750]*

pathologists (pa-THOL-ō-jists): Physicians who specialize in diagnosing disease processes. ☤ *Tissue Structure and Disease [p. 750]*

dysplasia (dis-PLĀ-zē-a): A change in the normal shape, size, and organization of tissue cells. ☤ *Tissue Structure and Disease [p. 750]*

metaplasia (me-ta-PLĀ-zē-a): A structural change that alters the character of a tissue. ☤ *Tissue Structure and Disease [p. 750]*

anaplasia (a-na-PLĀ-zē-a): An irreversible change in the size and shape of tissue cells. ☤ *Tissue Structure and Disease [p. 750]*

oncologists (on-KOL-ō-jists): Physicians who specialize in identifying and treating cancers. *[p. 81]*

primary tumor (primary neoplasm): The site at which a cancer initially develops. *[p. 81]*

secondary tumor: A colony of cancerous cells formed by metastasis, the spread of cells from a primary tumor. *[p. 81]*

remission: A stage in which a tumor stops growing or grows smaller; the goal of cancer treatment. ☤ *Cancer Treatment and Statistics [p. 751]*

chemotherapy: Administering drugs that either kill cancerous tissues or prevent mitotic divisions. ☤ *Cancer Treatment and Statistics [p. 751]*

immunotherapy: Administering drugs that help the immune system recognize and attack cancer cells. ☤ *Cancer Treatment and Statistics [p. 751]*

CHAPTER SUMMARY AND REVIEW

STUDY OUTLINE

Related Key Terms

INTRODUCTION *[p. 52]*

1. **Tissues** are collections of specialized cells and cell products that are organized to perform a relatively limited number of functions. There are four **primary tissue types**: epithelia, connective tissue, muscle tissue, and neural tissue *(see Figure 3-1)*.

EPITHELIAL TISSUE *[p. 52]*

1. **Epithelial tissue** includes *epithelia* and *glands*. An **epithelium** is an **avascular** layer of cells that forms a surface, lining, or covering. Epithelia consist mainly of cells, rather than extracellular materials *(see Figures 3-2 to 3-10)*.

Functions of Epithelial Tissue *[p. 52]*
2. Epithelia provide physical protection, control permeability, provide sensation, and produce specialized secretions. **Gland cells** are epithelial cells (or cells derived from epithelia) that produce secretions.

neuroepithelium • glandular epithelium

Specializations of Epithelial Cells *[p. 54]*
3. Epithelial cells are specialized to allow them to maintain the physical integrity of the epithelium and to perform secretory or transport functions.
4. Epithelial cells may show **polarity** and often have microvilli *(see Figure 3-2)*.

stereocilia

5. The coordinated beating of the cilia on a **ciliated epithelium** moves materials across the epithelial surface *(see Figure 3-2)*.

Maintaining the Integrity of the Epithelium *[p. 54]*
6. The inner surface of each epithelium is connected to a two-part **basement membrane** consisting of a **basal lamina** and a **reticular lamina**. In areas exposed to extreme chemical or mechanical stresses, divisions by **germinative cells** replace the short-lived epithelial cells *(see Figure 3-3)*.

Classification of Epithelia *[p. 56]*
7. Epithelia are classified on the basis of the number of cell layers and the shape of the exposed cells *(see Figures 3-4 to 3-6)*.

8. A **simple epithelium** has a single layer of cells covering the basement membrane. A **stratified epithelium** has several layers. In a **squamous epithelium** the cells are thin and flat; in a **cuboidal epithelium** the cells resemble short hexagonal boxes; in a **columnar epithelium** the cells are also hexagonal, but they are relatively tall and slender. **Transitional epithelium** is characterized by a mixture of cuboidal and squamous cells arranged to permit stretching, as in the urinary bladder. **Pseudostratified columnar epithelium** is columnar cells that possess cilia and goblet secreting cells that are arranged to appear stratified, but are not *(see Figures 3-4 to 3-6)*.

Glandular Epithelia [p. 58]

9. Glands may be classified by their mode of secretion, the type of secretion produced, or the structure of the gland *(see Figures 3-7/3-10)*.

10. A glandular epithelial cell may release its secretions through merocrine, apocrine, or holocrine mechanisms *(see Figure 3-7)*.

11. In **merocrine secretion**, the most common method of secretion, the product is released through exocytosis. **Apocrine secretion** involves the loss of both secretory product and some cytoplasm. Unlike the first two methods, **holocrine secretion** destroys the cell, which becomes packed with secretory product and finally bursts *(see Figure 3-7)*.

12. **Exocrine** secretions leave the body through **ducts**; **endocrine** secretions, known as **hormones**, are released by gland cells into the interstitial fluid surrounding the cell.

13. Exocrine glands may be classified as **serous** (producing a watery solution usually containing enzymes), **mucous** (producing a viscous, sticky *mucus*), or **mixed** (producing both types of secretions).

14. In epithelia that contain scattered gland cells, individual secretory cells are called **unicellular glands**. **Multicellular glands** are glandular epithelia or aggregations of gland cells that produce exocrine or endocrine secretions *(see Figures 3-8/3-9)*.

CONNECTIVE TISSUES [p. 64]

1. All connective tissues have three ingredients: specialized cells, extracellular protein fibers, and a **ground substance**. The extracellular fibers and ground substance constitute the **matrix** of the tissue *(see Figures 3-11 to 3-18)*.

2. Whereas epithelia consist almost entirely of cells, the matrix accounts for most of the volume of a connective tissue.

3. Connective tissues are internal tissues with many important functions: establishing a structural framework; transporting fluids and dissolved materials; protecting delicate organs; supporting, surrounding, and interconnecting tissues; storing energy reserves; and defending the body from microorganisms.

Classification of Connective Tissues [p. 64]

4. **Connective tissue proper** refers to connective tissues that contain varied cell populations and fiber types surrounded by a syrupy ground substance.

5. **Fluid connective tissues** have a distinctive population of cells suspended in a watery ground substance containing dissolved proteins. *Blood* and *lymph* are fluid connective tissues.

6. **Supporting connective tissues** have a less diverse cell population than connective tissue proper and a dense matrix that contains closely packed fibers. The two types of supporting connective tissues are *cartilage* and *bone*.

Connective Tissue Proper [p. 65]

7. Connective tissue proper contains extracellular fibers, a viscous ground substance, and two types of cells: **fixed cells** and **wandering cells** *(see Figure 3-11)*.

8. There are three types of fibers in connective tissue: **collagen fibers**, **reticular fibers**, and **elastic fibers** *(see Figures 3-11 to 3-14)*. All connective tissues are derived from embryonic **mesenchyme**.

9. Connective tissue proper is classified as **loose** or **dense connective tissues**. There are three types of loose connective tissues: **loose connective tissue**, or **areolar tissue**, **adipose tissue**, and **reticular tissue**. Most of the volume in dense connective tissue consists of extracellular protein fibers. There are two types of dense connective tissue: **dense regular connective tissue** and **dense irregular connective tissue** *(see Figures 3-13 to 3-14)*.

Fluid Connective Tissues [p. 70]

10. **Blood** and **lymph** are connective tissues that contain distinctive collections of cells in a fluid matrix. The watery matrix of blood is *plasma*. Both blood and lymph contain cells and many different types of proteins which are dissolved in the supporting cells, or matrix, but these proteins do not form insoluble fibers under normal conditions.

Supporting Connective Tissues [p. 70]

11. Cartilage and bone are called **supporting connective tissues** because they support the rest of the body *(see Figures 3-15/3-16)*.

12. The matrix of **cartilage** is a firm gel that contains **chondroitin sulfates** and cells called **chondrocytes**. A fibrous **perichondrium** separates cartilage from surrounding tissues. There are three types of cartilage: **hyaline cartilage**, **elastic cartilage**, and **fibrocartilage** *(see Figure 3-15 and Table 3-1)*.

13. Cartilage grows by two different mechanisms, **interstitial growth** and **appositional growth**.

14. **Bone (osseous tissue)** has a matrix consisting of collagen fibers and calcium salts, giving it unique properties *(see Figure 3-16)*.

mesothelium • endothelium

mucus

goblet cells • unicellular exocrine glands • multicellular exocrine glands • secretory sheet • tubular • alveolar • acinar • tubuloalveolar • simple • compound • branched

calcified • fibroblasts • fixed macrophages • adipocytes • mesenchymal cells • melanocytes • melanin • free macrophages • monocytes • mast cells • histamine • heparin • lymphocytes • plasma cells • microphages • tendons • ligaments • elastic ligaments • mucous connective tissue • collagenous tissue • aponeuroses • elastic tissue • capsule

lacunae

15. **Osteocytes** in **lacunae** depend on diffusion through **canaliculi** for nutrient intake *(see Figure 3-16 and Table 3-1).*

16. All bone surfaces except those inside joint cavities are covered by a **periosteum** that has fibrous and cellular layers. The periosteum assists in attaching the bone to surrounding tissues, tendons, and ligaments, and it participates in the repair of bone after an injury.

Related Key Terms

MEMBRANES *[p. 76]*

1. Membranes form a barrier or interface. Epithelia and connective tissues combine to form membranes that cover and protect other structures and tissues. There are four types of membranes: *mucous, serous, cutaneous,* and *synovial* **(see Figure 3-17).**

Mucous Membranes *[p. 76]*
2. **Mucous membranes** line cavities that communicate with the exterior, such as the digestive and respiratory tracts. These surfaces are usually moistened by mucous secretions *(see Figure 3-17a).*

lamina propria

Serous Membranes *[p. 77]*
3. **Serous membranes** line internal cavities and are delicate, moist, and very permeable. Examples include the peritoneal, pericardial, and pleural membranes *(see Figure 3-17b).*

transudate • pleural fluid • peritoneal fluid • pericardial fluid

The Cutaneous Membrane *[p. 77]*
4. The **cutaneous membrane** covers the body surface. Unlike serous and mucous membranes, it is relatively thick, waterproof, and usually dry *(see Figure 3-17c).*

Synovial Membranes *[p. 77]*
5. The **synovial membrane**, located at synovial joints, produces **synovial fluid** that fills joint cavities. Synovial fluid helps lubricate the joint and promotes smooth movement in joints such as the knee *(see Figure 3-17d).*

THE CONNECTIVE TISSUE FRAMEWORK OF THE BODY *[p. 77]*

1. Internal organs and systems are tied together by a network of connective tissue proper that includes the **superficial fascia** (separating the skin from underlying tissues and organs), the **deep fascia** (dense connective tissue), and the **subserous fascia** (the layer between the deep fascia and the serous membranes that line body cavities) *(see Figure 3-18).*

subcutaneous layer • hypodermis

MUSCLE TISSUE *[p. 78]*

1. **Muscle tissue** consists primarily of cells, termed **muscle fibers**, that are specialized for contraction along their longitudinal axes. There are three different types of muscle tissue: *skeletal muscle, cardiac muscle,* and *smooth muscle* **(see Figure 3-19).**

sarcoplasm • sarcolemma

Skeletal Muscle Tissue *[p. 78]*
2. **Skeletal muscle tissue** contains very large fibers tied together by collagen and elastic fibers. Skeletal muscle fibers have striations due to the organization of contractile proteins. Because we can control the contraction of skeletal muscle fibers through the nervous system, skeletal muscle is classified as **striated voluntary muscle** *(see Figure 3-19a).*

multinucleated • satellite cells

Cardiac Muscle Tissue *[p. 80]*
3. **Cardiac muscle tissue** is found only in the heart. The nervous system does not provide voluntary control over cardiac muscle cells, or **cardiocytes**. Thus, cardiac muscle is classified as **striated involuntary muscle** *(see Figure 3-19b).*

intercalated discs • pacemaker cells

Smooth Muscle Tissue *[p. 80]*
4. **Smooth muscle tissue** is found in the walls of blood vessels, around hollow organs, and in layers around various tracts. It is classified as **nonstriated involuntary muscle** *(see Figure 3-19c).*

NEURAL TISSUE *[p. 80]*

1. **Neural tissue** or **nervous tissue** (nerve tissue) is specialized to conduct electrical impulses from one area of the body to another.

2. Neural tissue consists of two cell types, neurons and neuroglia. **Neurons** transmit information as electrical impulses. There are different kinds of **neuroglia**, and among their other functions these cells provide a supporting framework for neural tissue and play a role in providing nutrients to neurons *(see Figure 3-20).*

3. Neurons have a cell body, or **soma**, that contains a large prominent nucleus. Extending from the soma are various branching processes termed **dendrites** and a single **axon**, or **nerve fiber**. Dendrites receive incoming messages, axons conduct outgoing messages *(see Figure 3-20).*

TISSUES, NUTRITION, AND AGING *[p. 81]*

1. Tissues change with age. Repair and maintenance grow less efficient, and the structure and chemical composition of many tissues are altered.

1 REVIEW OF CHAPTER OBJECTIVES

1. Classify the tissues of the body into four major categories.
2. Discuss the types and functions of epithelial tissues.
3. Describe the relationship between form and function for each epithelial type.
4. Compare the structure and functions of the various connective tissues.
5. Explain how epithelia and connective tissues combine to form four different types of membranes, and specify the functions of each.
6. Describe how connective tissue establishes the framework of the body.
7. Describe the three types of muscle tissue.
8. Discuss the basic structure and role of neural tissue.
9. Describe how nutrition and aging affect tissues.

2 REVIEW OF CONCEPTS

1. How does the role of a tissue in the body differ from that of a single cell?
2. How does the body distinguish the touch of a feather from that of the point of a pen?
3. What is the relationship between the thickness of epithelium and its function at a specific location in the body?
4. How do connective tissues differ from epithelial tissues?
5. What characteristics would you expect the epithelial lining of the intestinal tract to possess?
6. What are the differences between tendons and ligaments?
7. What mechanical requirements are shared by the pinnae of the ears and the tip of the nose, and what tissue type(s) allow these structures to function normally?
8. What are the main structural differences between bone and cartilage?
9. What types of tissues are lined by mucous membranes?
10. How does skeletal muscle tissue structure differ from that of a more typical tissue?
11. What properties of smooth muscle tissue are important to the tissues where they occur?
12. What properties of tissue(s) that make up the lungs facilitate the uptake of oxygen and the elimination of carbon dioxide?
13. How does a synovial membrane contribute to the mobility of a joint between two long bones?

14. What changes would occur when a blood vessel dilates to allow increased blood flow through it?
15. What is the significance of the perichondrium in the function and maintenance of cartilage?
16. How do appositional and interstitial growth mechanisms in cartilage differ from one another, and what effects do they have on cartilage size and structure?
17. How does the role of collagen fibers in bone differ from the role these fibers play in regular dense connective tissues?
18. What is the significance of the double-layered structure of serous membranes, and what advantages does this arrangement confer upon the organs they surround?
19. How does the superficial layer of a synovial membrane differ from a typical superficial epithelium?
20. What characteristics of a neuron facilitate its ability to conduct electrical signals?
21. On the basis of your knowledge of the materials that form bone, what changes would you consider to be responsible for causing bone to become brittle in an aged person?
22. What characteristics of a cancerous cell would distinguish it from a normal cell in the same tissue?
23. In what regions of the body would you expect to find dense, irregular connective tissue, and why would this tissue be located in these regions?
24. What characteristics would you expect to find in a cartilage that was located between the vertebral bodies or between the pubic bones of the pelvis? What is this type of cartilage called?
25. What is the significance of the rich blood supply that occurs in some body tissues, and how is this characteristic related to growth and tissue repair? Give an example of one of the most and least highly vascularized tissues.
26. How is the function of the hypodermis analogous to that of the subserous fascia?

3 CRITICAL THINKING AND CLINICAL APPLICATION QUESTIONS

1. Why does skin tend to become dry, scaly, and itchy in the winter?
2. How does a skin-softening lotion work?
3. A common injury for a football player is to have a "pulled muscle." What tissues are most likely to be damaged in this type of injury, and what kind of damage has occurred?

4
The Integumentary System

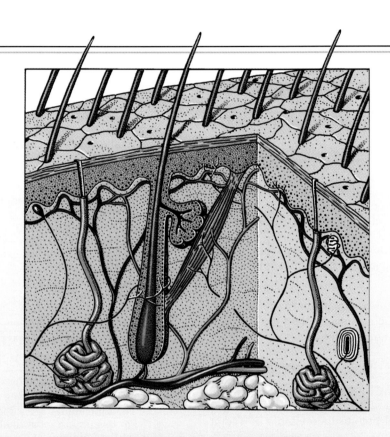

CHAPTER OUTLINE AND OBJECTIVES

The **integumentary system** or *integument*, is probably the most overexamined and underrated component of the human body. Of all the body systems, this is the only one we see every day, almost in its entirety. Because others see this system as well, we devote a lot of time to improving the appearance of the integument and associated structures. Washing the face, brushing and trimming hair, taking showers, and applying makeup are activities that modify the appearance or properties of the integument. And when something goes wrong with the integument, the effects are immediately apparent. Even a relatively minor condition or blemish will be noticed at once, whereas more serious problems in other systems are often ignored. In this chapter we will focus on the functional anatomy of the integumentary system.

INTEGUMENTARY STRUCTURE AND FUNCTION
(Figure 4-1)

The integument covers the entire body surface, including the anterior surfaces of the eyes and the tympanic membranes (eardrums) at the ends of the external auditory canals. At the nostrils, lips, anus, urethral opening, and vaginal opening the integument turns inward, meeting the mucous membranes lining the respiratory, digestive, urinary, and reproductive tracts. At these sites the transition is seamless, and the epithelial defenses remain intact and functional.

All four tissue types contribute to the structure of the integument. An epithelium covers its surface, and underlying connective tissues provide strength and resiliency. Blood vessels within the connective tissue nourish the epidermal cells. Smooth muscle tissue within the integument controls the diameters of the blood vessels and adjusts the positions of the hairs that project above the body surface. Neural tissue controls these smooth muscles and monitors sensory receptors providing sensations of touch, pressure, temperature, and pain.

Figure 4-1● shows the functional organization of the integumentary system. It has two major components, the *cutaneous membrane* and the *accessory structures*.

1. The **cutaneous membrane**, or skin, has two components, the superficial epithelium, termed the *epidermis* (*epi-*, above + *derma*, skin) and the underlying connective tissues of the *dermis*. Beneath the dermis, the loose connective tissue of the subcutaneous layer, also known as the *superficial fascia*, or *hypodermis*, separates the integument from the deep fascia around other organs, such as muscles or bones. ∞ [p. 77] Although it is not usually considered to be a part of the integument, we will consider the subcutaneous layer in this chapter because of its extensive interconnections with the dermis.

2. The **accessory structures** include hair, nails, and a variety of multicellular exocrine glands. These structures are located in the dermis and protrude through the epidermis to the surface.

FIGURE 4-1
Functional Organization of the Integumentary System

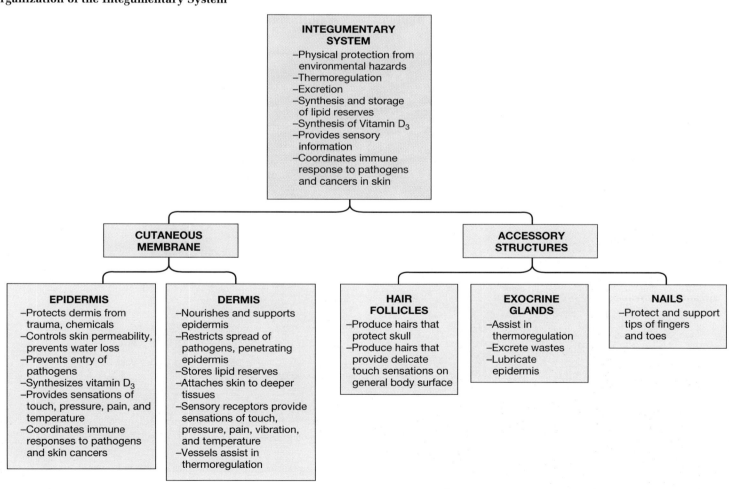

THE EPIDERMIS
(Figure 4-2)

The **epidermis** consists of a stratified squamous epithelium, as seen in Figure 4-2●. There are several different cell layers present, but the precise boundaries between them are often difficult to see in a light micrograph. In **thick skin**, found on the palms of the hands and soles of the feet, five layers can be distinguished. Only four layers can be distinguished in the **thin skin** that covers the rest of the body.

Layers of the Epidermis
(Figure 4-3)

Refer to Figure 4-3● as we describe the layers in a section of thick skin. Beginning at the basement membrane and traveling toward the outer surface, we find the *stratum germinativum*, the *stratum spinosum*, the *stratum granulosum*, the *stratum lucidum*, and the *stratum corneum*.

Stratum Germinativum

The innermost epidermal layer is the **stratum germinativum** (STRĀ-tum jer-mi-na-TĒ-vum), or *stratum basale*. This layer is firmly attached to the basement membrane that separates the epidermis from the loose connective tissue of the adjacent dermis. Large stem cells dominate the stratum germinativum. The divisions of stem cells replace the more superficial cells that are lost or shed at the epithelial surface. The brown tones of the skin result from the synthetic activities of *melanocytes*, pigment cells introduced in Chapter 3. ∞ [p. 66] These cells are scattered throughout the stratum germinativum, with cell processes extending into more superficial layers.

Skin surfaces that lack hair contain specialized epithelial cells known as **Merkel cells**. These cells are found among the deepest cells of the stratum germinativum. They are sensitive to touch, and when compressed, Merkel cells release chemicals that stimulate sensory nerve endings, providing information about objects touching the skin. (There are many other kinds of touch receptors, but they are located in the dermis and will be described in later sections.)

Stratum Spinosum

Each time a stem cell divides, one of the daughter cells is pushed above the stratum germinativum into the next layer, the **stratum spinosum** ("spiny layer"). The stratum spinosum is several cells thick, and the cells are bound together by desmosomes. Standard histological procedures shrink the cytoplasm, but the cytoskeletal elements and desmosomes remain intact, making the cells look like miniature pincushions. Some of the cells entering this layer from the stratum germinativum continue to divide, further increasing the thickness of the epithelium. Melanocytes and their processes are common in this layer. **Langerhans cells** are often present, but they cannot be seen in standard histological preparations. These cells play an important role in initiating an immune response against (1) pathogens that have penetrated the superficial layers of the epidermis and (2) epidermal cancer cells.

Stratum Granulosum

The layer of cells above the stratum spinosum is the **stratum granulosum** ("granular layer"). The stratum granulosum consists

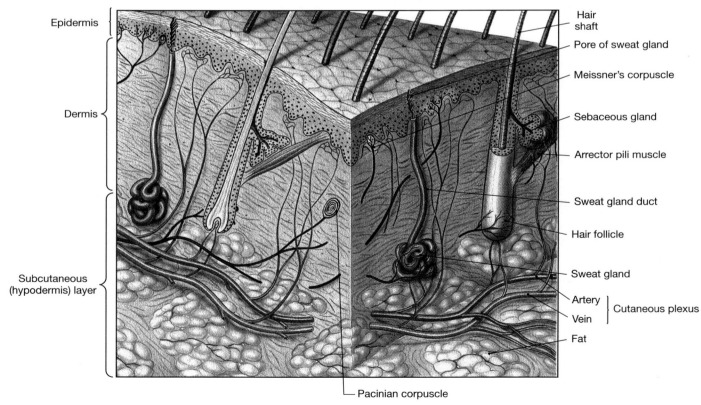

Epidermis

Dermis

Subcutaneous (hypodermis) layer

Hair shaft

Pore of sweat gland

Meissner's corpuscle

Sebaceous gland

Arrector pili muscle

Sweat gland duct

Hair follicle

Sweat gland

Artery

Vein

Cutaneous plexus

Fat

Pacinian corpuscle

FIGURE 4-2
Components of the Integumentary System. Relationships among the major components of the integumentary system (with the exception of nails, shown in Figure 4-15).

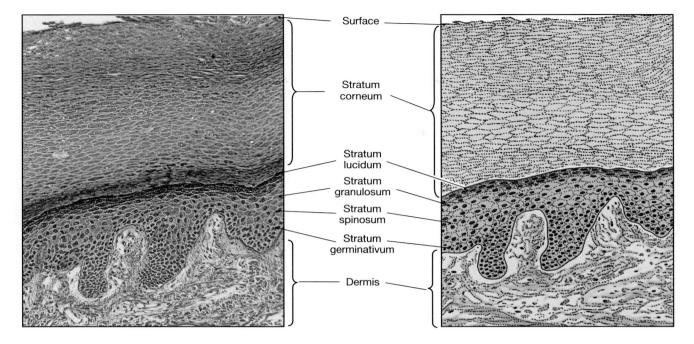

FIGURE 4-3
The Structure of the Epidermis. A light micrograph through a portion of the epidermis, showing the major stratified layers of epidermal cells. (LM × 200)

of cells displaced from the stratum spinosum. By the time cells reach this layer, most have stopped dividing, and they begin manufacturing large quantities of the proteins **keratohyalin** (ker-a-tō-HĪ-a-lin) and **keratin** (KER-a-tin; *keros*, horn). In the human body, keratin forms the basic structural component of hair and nails. As true keratin fibers are developing, the cells become thinner and flatter. The cell membranes thicken and become less permeable. The nuclei and other organelles disintegrate, the cells die, and their subsequent dehydration creates a tightly interlocked layer of keratin fibers surrounded by keratohyalin and sandwiched between phospholipid membranes.

Stratum Lucidum

In the thick skin of the palms and soles, a glassy **stratum lucidum** ("clear layer") covers the stratum granulosum. The cells in this layer are flattened, densely packed, and filled with keratin.

Stratum Corneum

The **stratum corneum** (KOR-nē-um; *cornu*, horn) is found at the surface of both thick and thin skin. It consists of multiple layers of flattened, dead, and interlocking cells.

An epithelium containing large amounts of keratin is said to be **keratinized** (KER-a-tin-īzed), or **cornified** (KOR-ni-fīd; *cornu*, horn + *facere*, to make). Normally the stratum corneum is relatively dry, which makes the surface unsuitable for the growth of many microorganisms. Maintenance of this barrier involves coating the surface with lipid secretions from integumentary glands (sebaceous and sweat). The process of *cornification* occurs everywhere on exposed skin surfaces except over the anterior surfaces of the eyes.

Although the stratum corneum is water-resistant, it is not waterproof, and water from the interstitial fluids slowly penetrates the surface, to be evaporated into the surrounding air. This process, called **insensible perspiration** accounts for a loss of roughly 500 m*ℓ*, (about 1 pt) of water per day. *Psoriasis and Xerosis [p. 752]*

It takes approximately 14 days for a cell to move from the stratum germinativum to the stratum corneum. The dead cells usually remain in the exposed stratum corneum layer for an additional 2 weeks before they are shed or washed away. This arrangement places the deeper portions of the epithelium and underlying tissues beneath a protective barrier composed of dead, durable, and expendable cells.

√ **Excessive shedding of cells from the outer layer of skin in the scalp causes dandruff. What is the name of this layer of skin?**

√ **As you pick up a piece of lumber, a splinter pierces the palm of your hand and lodges in the third layer of the epidermis. Identify this layer.**

▮ CLINICAL BRIEF
Transdermal Medication

Drugs in oils or other liquid-soluble carriers can penetrate the epidermis. The movement is slow, particularly through the layers of cell membranes in the stratum corneum, but once a drug reaches the underlying tissues, it will be absorbed into the circulation. A useful technique involves placing a sticky patch containing a drug over an area of thin skin. To overcome the relatively slow rate of diffusion, the patch must contain an extremely high concentration of the drug. This procedure, called *transdermal administration*, has the advantage that a single patch may work for several days, making daily pills unnecessary. *Scopolamine*, a drug that affects the nervous system, is administered transdermally to control the nausea associated with motion sickness. Transdermal *nitroglycerin* can be used to improve blood flow within heart muscle and prevent a heart attack.

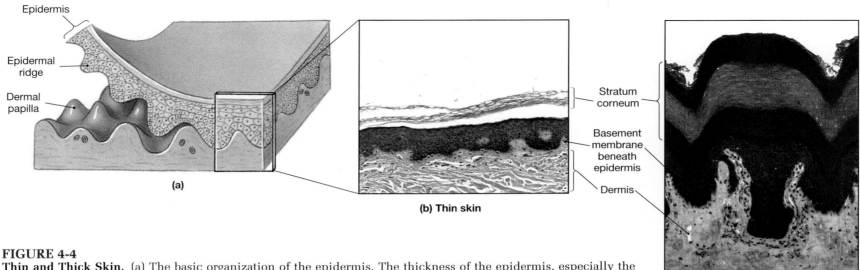

FIGURE 4-4
Thin and Thick Skin. (a) The basic organization of the epidermis. The thickness of the epidermis, especially the thickness of the stratum corneum, changes radically depending on the location sampled. (b) Thin skin covers most of the exposed body surface. (During sectioning the stratum corneum has pulled away from the rest of the epidermis.) (LM × 154) (c) Thick skin covers the surfaces of the palms and soles. (LM × 154)

Thick and Thin Skin
(Figure 4-4)

In descriptions of the skin, the terms *thick* and *thin* refer to the relative thickness of the epidermis, not to the integument as a whole. Most of the body is covered by thin skin. In a sample of thin skin, seen in Figure 4-4a,b●, the epidermis is a mere 0.08 mm thick, and the stratum corneum is only a few cell layers deep. Thick skin on the palms of the hands may be covered by 30 or more layers of cornified cells. As a result, the epidermis in these locations may be as much as six times thicker than the epidermis covering the general body surface (Figure 4-4c●).

Epidermal Ridges
(Figures 4-4/4-5)

The deeper layers of the epidermis form **epidermal ridges** that extend into the dermis, increasing the area of contact between the two regions. Dermal projections called **dermal papillae** (singular *papilla*, "nipple-shaped mound") extend between adjacent ridges, as indicated in Figure 4-4a●.

The contours of the skin surface follow the ridge patterns, which vary from small conical pegs (in thin skin) to the complex whorls seen on the thick skin of the palms and soles. Ridges on the palms and soles increase the surface area of the skin and increase friction, ensuring a secure grip. Ridge shapes are genetically determined: Those of each person are unique and do not change in the course of a lifetime. Fingerprint-ridge patterns on the tips of the fingers (Figure 4-5●) can therefore be used to identify individuals, and they have been so used in criminal investigation for over a century.

√ **What are the stages of keratin production?**

√ **Some criminals sand the tips of their fingers so as not to leave recognizable fingerprints. Would this practice permanently remove fingerprints? Why or why not?**

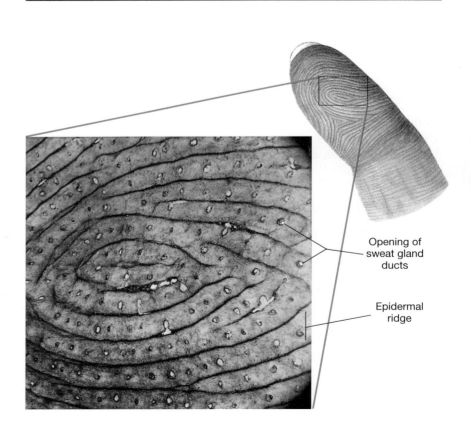

FIGURE 4-5
The Epidermal Ridges of Thick Skin. Fingerprints reveal the pattern of epidermal ridges in thick skin. This scanning electron micrograph shows the ridges on a fingertip. The pits are the openings of sweat gland ducts. (SEM × 25). [Reproduced from R. G. Kessel and R. H Kardon, *Tissues and Organs: A Text-Atlas of Scanning Electron Microscopy*, W. H. Freeman & Co., 1979.]

Skin Color
(Figure 4-6)

The color of the epidermis is due to an interaction between (1) the dermal blood supply and (2) pigment composition and concentration. Blood contains red blood cells filled with the pigment *hemoglobin*. When bound to oxygen, hemoglobin has a bright red color, giving blood vessels in the dermis a reddish tint that is most easily seen in lightly pigmented individuals. When those vessels are dilated, as during inflammation, the red tones become much more pronounced.

When the circulatory supply is temporarily reduced, the skin becomes relatively pale; a frightened Caucasian may "turn white" because of a sudden drop in blood supply to the skin. During a sustained reduction in circulatory supply, the blood in the superficial vessels loses oxygen and the hemoglobin changes color to a much darker red tone. Seen from the surface, the skin takes on a bluish coloration called **cyanosis** (sī-a-NŌ-sis; *kyanos*, blue). In individuals of any skin color, cyanosis is most apparent in areas of thin skin, such as the lips or beneath the nails. It can be a response to extreme cold or a result of circulatory or respiratory disorders, such as heart failure or severe asthma.

The epidermis contains variable quantities of two pigments, *carotene* and *melanin*. **Carotene** (KAR-ō-tēn) is an orange-yellow pigment that normally accumulates inside epidermal cells. Carotene pigment is found in a variety of orange-colored vegetables, such as carrots and squashes.

Melanocytes (me-LAN-ō-sīts) are pigment cells found in the stratum germinativum, squeezed between the epithelial cells. Melanocytes, shown in Figure 4-6●, manufacture and store **melanin** (MEL-a-nin), a yellow-brown, brown, or black pigment. Melanocytes inject melanin into the epithelial cells of the stratum germinativum and stratum spinosum, thereby coloring the entire epidermis. The rate at which melanin is synthesized by melanocytes is influenced by the blood level of **melanocyte-stimulating hormone (MSH)**, which is secreted by the pituitary gland. Observed individual and even racial differences in skin color do not reflect different *numbers* of melanocytes, merely different levels of melanocyte activity. Even the melanocytes of *albino* individuals are distributed normally. (**Albinism** is an inherited condition in which melanocytes are incapable of producing melanin; it affects approximately 1 person in 10,000.)

Melanin pigments protect epidermal cells from the **ultraviolet (UV) radiation** contained in sunlight. A little ultraviolet is useful because the skin needs UV to convert a steroid related to cholesterol into a member of the family of hormones collectively known as vitamin D.[1] Vitamin D is required for normal calcium and phosphorus absorption by the small intestine, and an inadequate supply of this vitamin leads to impaired bone maintenance and growth. But too much ultraviolet radiation produces immediate effects of mild or even serious burns. Long-term damage can result from repeated exposure, even though tanning occurs. Alterations in the underlying connective tissues lead to premature wrinkling, and skin cancers can result from chromosomal damage in germinative cells or melanocytes.

---

[1]Specifically, vitamin D₃, or *cholecalciferol*, which undergoes further modification in the liver and kidneys before circulating as the active hormone *calcitriol*.

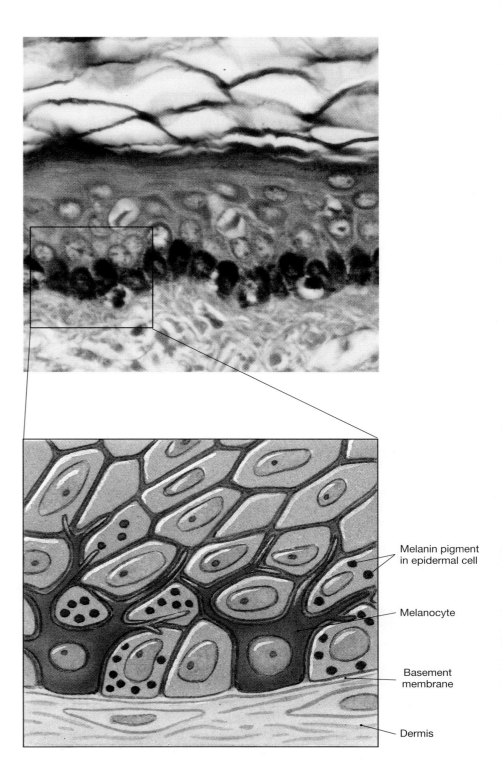

FIGURE 4-6
Melanocytes. The micrograph and accompanying drawing indicate the location and orientation of melanocytes in the stratum germinativum of a black person.

Melanin pigment in epidermal cell

Melanocyte

Basement membrane

Dermis

Melanin helps prevent skin damage by absorbing ultraviolet radiation before it reaches the deep layers of the epidermis and dermis. Within the epidermal cells, melanin concentrates around the outer wall of the nucleus, so it absorbs the UV before it can damage the nuclear DNA. Melanocytes respond to UV exposure by increasing their activity. ✝ *Skin Cancers [p. 752]*

The **dermis** lies beneath the epidermis (Figure 4-2●, p. 90). It has two major components, a superficial *papillary layer* and a deeper *reticular layer*.

Dermal Organization
(Figures 4-4/4-7)

The **papillary layer** consists of loose connective tissue (Figure 4-7a●). This region contains the capillaries and the sensory neu-

rons supplying the surface of the skin. The papillary layer derives its name from the dermal papillae that project between the epidermal ridges, as shown in Figure 4-4●.

The deeper **reticular layer** consists of an interwoven meshwork of dense irregular connective tissue (Figure 4-7b●). Bundles of collagen fibers leave the reticular layer to blend into those of the papillary layer above, so the boundary line between these layers is indistinct. Collagen fibers of the reticular layer also extend into the underlying subcutaneous layer (Figure 4-7c●). ⚕ *Dermatitis [p. 752]*

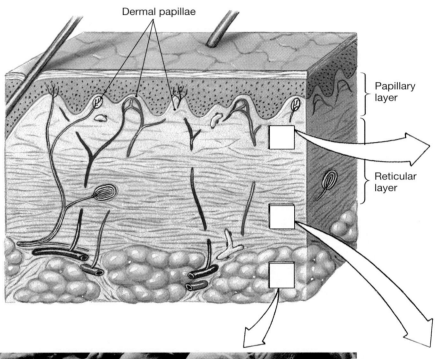

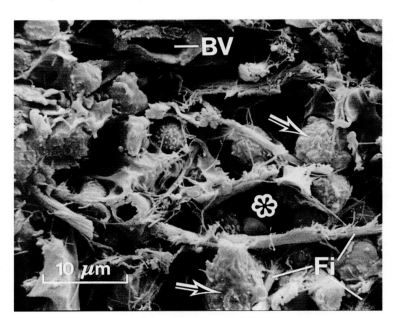

(a) Papillary layer of dermis

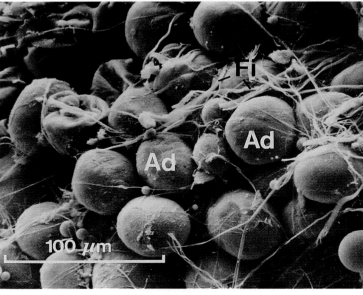

(c) Subcutaneous layer

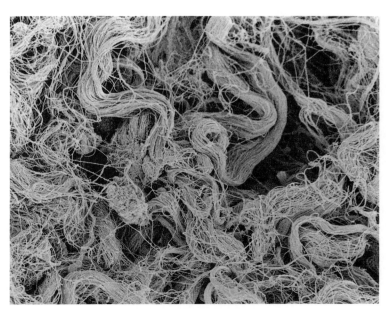

(b) Reticular layer of dermis

FIGURE 4-7
The Structure of the Dermis and Subcutaneous Layer. (a) The papillary layer of the dermis consists of loose connective tissue that contains numerous blood vessels (BV), fibers (Fi), and macrophages (arrows). Open spaces, such as the one marked by an asterisk, would be filled with fluid ground substance. (SEM × 649) (b) The reticular layer of the dermis contains dense, irregular connective tissue. (SEM × 1340) (c) The subcutaneous layer contains large numbers of adipocytes (Ad) in a framework of loose connective tissue fibers (Fi). (SEM × 268) [(a,c) Reproduced from R. G. Kessel and R. H. Kardon, *Tissues and Organs: A Text-Atlas of Scanning Electron Microscopy*, W. H. Freeman & Co., 1979.]

Wrinkles, Stretch Marks, and Lines of Cleavage
(Figure 4-8)

The interwoven collagen fibers of the reticular layer provide considerable tensile strength, and the extensive array of elastic fibers enables the dermis to stretch and contract repeatedly during normal movements. Age, hormones, and the destructive effects of ultraviolet radiation reduce the thickness and flexibility of the dermis, producing wrinkles and sagging skin. The extensive distortion of the dermis that occurs over the abdomen during pregnancy or after a substantial weight gain often exceeds the elastic capabilities of the skin. Although the skin stretches, it does not contract to its original size after delivery or a rigorous diet. The skin then wrinkles and creases, creating a network of **stretch marks**.

Tretinoin (*Retin-A*) is a derivative of vitamin A that can be applied to the skin as a cream or gel. This drug was originally developed to treat acne, but it also increases blood flow to the dermis and stimulates dermal repairs. As a result, the rate of wrinkle formation decreases, and existing wrinkles become smaller. The degree of improvement varies from individual to individual.

At any one location, the majority of the collagen and elastic fibers are arranged in parallel bundles. The orientation of these bundles depends on the stress placed on the skin during normal movement; the bundles are aligned to resist the applied forces. The resulting pattern of fiber bundles establishes the **lines of cleavage** of the skin. Lines of cleavage, shown in Figure 4-8●, are clinically significant because a cut parallel to a cleavage line will usually remain closed, whereas a cut at right angles to a cleavage line will be pulled open as cut elastic fibers recoil. Surgeons choose their incision patterns accordingly, for a parallel incision will heal faster and with less scarring than an incision at right angles.

Other Dermal Components
(Figures 4-2/4-7/4-9)

In addition to extracellular protein fibers, the dermis contains all of the cells of connective tissue proper. ∞ [p. 65] Accessory organs of epidermal origin, such as hair follicles and sweat glands, extend into the dermis (Figure 4-9●). In addition, the reticular and papillary layers of the dermis contain networks of blood vessels, lymph vessels, and nerve fibers (Figure 4-2●, p. 90).

The Blood Supply to the Skin
(Figure 4-7)

Arteries supplying the skin form a network in the subcutaneous layer along the border with the reticular layer. This network is called the *cutaneous plexus* (Figure 4-7●). Tributaries of these arteries supply the adipose tissues of the subcutaneous layer and the tissues of the integument. As small arteries travel toward the epidermis, branches supply the hair follicles, sweat glands, and other structures in the dermis. On reaching the papillary layer, these small arteries form another branching network, the *papillary plexus*, that provides arterial blood to capillary loops that follow the contours of the epidermal-dermal boundary. These capillaries empty into a network of venules that in turn form small veins that descend through the dermis to reach larger veins in the subcutaneous layer. ⚕ *Tumors in the Dermis [p. 752]*

The Nerve Supply to the Skin

Nerve fibers in the skin control blood flow, adjust gland secretion rates, and monitor sensory receptors in the dermis and the deeper layers of the epidermis. We have already noted the presence of Merkel cells in the deeper layers of the epidermis. These cells are monitored by sensory terminals known as *Merkel's discs*. The epidermis also contains the dendrites of sensory nerves that probably respond to pain and temperature. The dermis contains similar receptors as well as other, more specialized receptors. Examples discussed in Chapter 18 include receptors sensitive to light touch (*Meissner's corpuscles*, located in dermal papillae and the root hair plexus surrounding each hair follicle), stretch (*Ruffini corpuscles*, in the reticular layer), and deep pressure and vibration (*Pacinian corpuscles*, in the reticular layer).

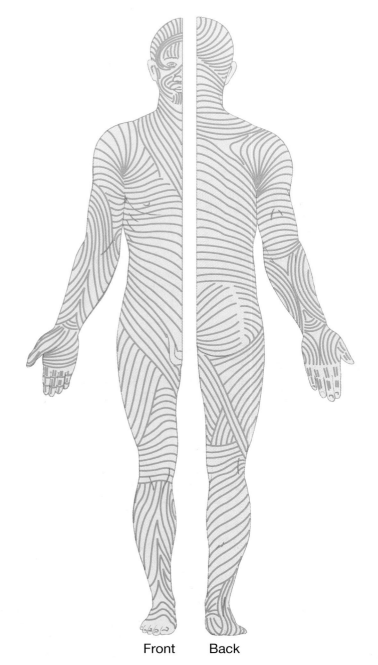

Front   Back

FIGURE 4-8
Lines of Cleavage of the Skin. Lines of cleavage follow lines of tension in the skin. They reflect the orientation of collagen fiber bundles in the dermis.

THE SUBCUTANEOUS LAYER
(Figures 4-2/4-7c)

The connective tissue fibers of the reticular layer are extensively interwoven with those of the **subcutaneous layer**, or **hypodermis**, and the boundary between the two is usually indistinct (Figure 4-2●, p. 90). Although the subcutaneous layer is not a part of the integument, it is important in stabilizing the position of the skin in relation to underlying tissues, such as skeletal muscles or other organs, while permitting independent movement.

The subcutaneous layer consists of loose connective tissue with abundant fat cells (Figure 4-7c●). Infants and small children usually have extensive "baby fat," which helps reduce heat loss. Subcutaneous fat also serves as a substantial energy reserve and a shock absorber for the rough-and-tumble activities of our early years.

As we grow, the distribution of subcutaneous fat changes. Men accumulate subcutaneous fat at the neck, upper arms, along the lower back, and over the buttocks. In women the breasts, buttocks, hips, and thighs are the primary sites of subcutaneous fat storage. In adults of either sex, the subcutaneous layer of the backs of the hands and the upper surfaces of the feet contain few fat cells, whereas distressing amounts of adipose tissue can accumulate in the abdominal region, producing a prominent "pot belly."

The subcutaneous layer is quite elastic. Only the superficial region contains large arteries and veins, and the rest contains a limited number of capillaries and no vital organs. This last characteristic makes **subcutaneous injection** a useful method for administering drugs. The familiar term **hypodermic needle** refers to the region targeted for injection.

ACCESSORY STRUCTURES
(Figure 4-2)

Hair follicles, sebaceous glands, sweat glands, and nails are considered accessory structures of the integument (Figure 4-2●, p. 90). During embryological development, these structures originate from the epidermis, but they are located in the dermis and may even project through the epidermis to the surface.

Hair Follicles and Hair

Hairs project above the surface of the skin almost everywhere except over the sides and soles of the feet, the palms of the hands, the sides of the fingers and toes, the lips, and portions of the external genitalia.[2] There are about 5 million hairs on the human body, and 98% of them are on the general body surface, not the head. Hairs originate in complex organs called **hair follicles**.

---

[2]The glans penis and prepuce of the male; the clitoris, labia minora, and inner surfaces of the labia majora in the female.

Hair Production
(Figures 4-9/4-10)

Hair follicles extend deep into the dermis, often projecting into the underlying subcutaneous layer (Figure 4-9●). The epithelium at the base of a hair follicle surrounds a small **hair papilla**, a peg of connective tissue containing capillaries and nerves. The **hair bulb** consists of epithelial cells that surround the papilla.

Hair production involves a specialization of the cornification process. The epithelial layer involved is called the **hair matrix**. Basal cells near the center of the matrix divide, producing daughter cells that are gradually pushed toward the surface. Those cells produced closest to the center of the matrix form the soft core, or **medulla**, of the hair, whereas cells closer to the edge of the developing hair form the relatively hard **cortex** (Figure 4-10●). The medulla contains flexible **soft keratin**. **Hard keratin** in the cortex gives the hair its stiffness. Cells at the surface of the hair differentiate to form the **cuticle**, a layer of hard keratin that coats the hair.

The **root** of the hair extends from the hair bulb to the point where the internal organization of the hair is complete. The **shaft** extends from this point, usually halfway to the skin surface, to the exposed tip of the hair. The size, shape, and color of the hair shaft are highly variable.

Follicle Structure
(Figure 4-10)

The cells of the follicle walls are organized into several concentric layers (Figure 4-10a●). Beginning at the hair cuticle, these layers include:

- The *internal root sheath:* This layer surrounds the hair root and the deeper portion of the shaft. It is produced by the cells at the periphery of the hair matrix. The cells of the internal root sheath disintegrate relatively quickly, and this layer does not extend the entire length of the follicle.
- The *external root sheath:* This layer extends from the skin surface to the hair matrix. Over most of that distance it has all of the cell layers found in the superficial epidermis. However, where the external root sheath joins the hair matrix, all of the cells resemble those of the stratum germinativum.
- The *glassy membrane*, a thickened basement membrane, is wrapped in a dense connective tissue sheath.

Functions of Hair
(Figure 4-9)

The 5 million hairs on the human body have important functions. The roughly 100,000 hairs on the head protect the scalp from ultraviolet light, cushion a blow to the head, and provide insulating benefits for the skull. The hairs guarding the entrances to the nostrils and external auditory canals help prevent the entry of foreign particles and insects, and eyelashes perform a similar function for the surface of the eye. A **root hair plexus** of sensory nerves surrounds the base of each hair follicle. As a result, the movement of the shaft of even a single hair can be felt at a conscious level. This sensitivity provides an early-warning system that may help to prevent injury. For example, you may be able to swat a mosquito before it reaches the skin surface.

SKIN OF SCALP   SKIN OF CHEEK   SKIN OF AXILLA

Old club hair being
extruded and
replaced by
new hair

Cortex ⎫
Medulla ⎬ of hair

Meissner's corpuscle

Sebaceous gland

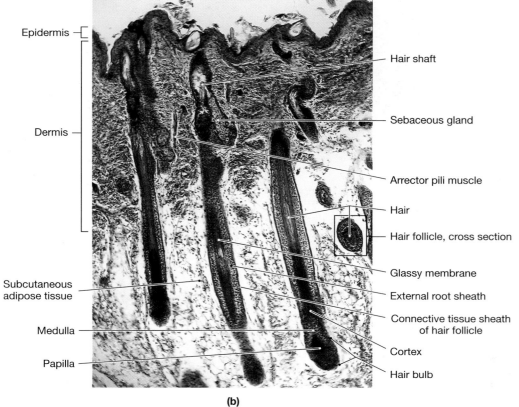

Bundles of
collagen
fibers in dermis

Subcutaneous
fat

Internal
root sheath

Sensory
nerve

External
root sheath

Hair
bulb

Hair
papilla

Hair
matrix

Merocrine
(eccrine)
sweat gland

Apocrine
sweat gland

(a)

Epidermis

Dermis

Subcutaneous
adipose tissue

Medulla

Papilla

Hair shaft

Sebaceous gland

Arrector pili muscle

Hair

Hair follicle, cross section

Glassy membrane

External root sheath

Connective tissue sheath
of hair follicle

Cortex

Hair bulb

(b)

FIGURE 4-9
Accessory Structures of the Skin. (a) A three-dimensional view of the skin in different parts of the body. Note the structure of the follicles and the orientation of the accessory structures. (b) A light micrograph showing the sectional appearance of the skin of the scalp. Note the abundance of hair follicles and the way they extend into the dermis. (LM × 66)

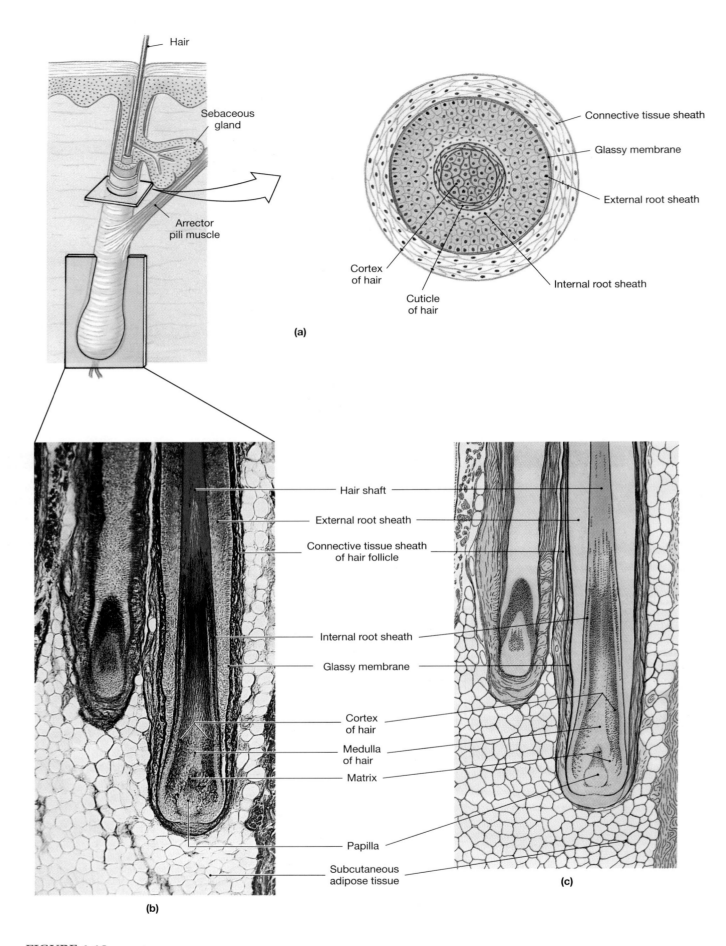

Hair

Sebaceous gland

Arrector pili muscle

Connective tissue sheath

Glassy membrane

External root sheath

Cortex of hair

Internal root sheath

Cuticle of hair

(a)

Hair shaft

External root sheath

Connective tissue sheath of hair follicle

Internal root sheath

Glassy membrane

Cortex of hair

Medulla of hair

Matrix

Papilla

Subcutaneous adipose tissue

(b)

(c)

FIGURE 4-10
Hair Follicles. (a) A longitudinal section and a cross section through a hair follicle. (b) and (c) A section along the longitudinal axis of a hair follicle. (LM × 60)

Ribbons of smooth muscle, called **arrector pili** (a-REK-tor PĪ-li) **muscles** (Figure 4-9a,b●) extend from the papillary dermis to the connective tissue sheath surrounding the hair follicle. When stimulated, the arrector pili pull on the follicles and elevate the hairs. Contraction may be caused by emotional states, such as fear or rage, or as a response to cold, producing characteristic "goose bumps." In a furry mammal, this action increases the thickness of the insulating coat, rather like putting on an extra sweater. Although we do not receive any comparable insulating benefits, the reflex persists.

Types of Hairs

There are three major types of hairs in the integument, *vellus hairs*, *terminal hairs*, and *intermediate hairs*.

- **Vellus hairs** are the fine "peach fuzz" hairs found over much of the body surface.

- **Terminal hairs** are heavy, more deeply pigmented, and sometimes curly. The hairs on your head, including your eyebrows and eyelashes, are examples of terminal hairs.

- **Intermediate hairs** are those hairs that change in their distribution, such as the hairs of the arms and legs.

Hair follicles may alter the structure of the hairs in response to circulating hormones, and this alteration accounts for many of the changes in hair distribution that begin at puberty.

Hair Color

Variations in hair color reflect differences in structure and variations in the pigment produced by melanocytes at the papilla. These characteristics are genetically determined, but the condition of your hair may be influenced by hormonal or environmental factors. As pigment production decreases with age, the hair color lightens toward gray. White hair results from the combination of a lack of pigment and the presence of air bubbles within the medulla of the hair shaft. Because the hair itself is dead and inert, changes in coloration are gradual.

Growth and Replacement of Hair
(Figures 4-9a/4-11)

A hair in the scalp grows for 2–5 years, at a rate of around 0.33 mm/day (about 1/64th inch). Variations in the hair growth rate and in the duration of the **hair growth cycle**, illustrated in Figure 4-11●, account for individual differences in uncut hair length.

While hair growth is under way, the root of the hair is firmly attached to the matrix of the follicle. At the end of the growth cycle, the follicle becomes inactive, and the hair is now termed a **club hair**. The follicle gets smaller, and over time the connections between the hair matrix and the root of the club hair break down. When another growth cycle begins, the follicle produces a new hair, and the old club hair gets pushed toward the surface (Figure 4-9a●).

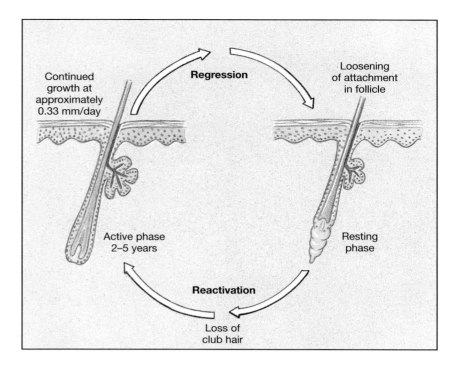

FIGURE 4-11
The Hair Growth Cycle

In healthy adults, about 50 hairs are lost each day, but several factors may affect this rate. Sustained losses of over 100 hairs per day usually indicate that something is wrong. Temporary increases in hair loss can result from drugs, dietary factors, radiation, high fever, stress, and hormonal factors related to pregnancy. In males, changes in the level of the sex hormones circulating in the blood can affect the scalp, causing a shift from terminal hair to vellus hair production. This alteration is called **male pattern baldness**. ⚕ *Baldness and Hirsutism [p. 752]*

√ **What happens to the dermis when it is excessively stretched, as in pregnancy or weight gain?**

√ **What condition is produced by the contraction of the arrector pili muscles?**

√ **A person suffers a burn on the forearm that destroys the epidermis and a portion of the dermis. When the injury heals, would you expect to find hair growing again in the area of the injury?**

Glands in the Skin
(Figure 4-12)

The skin contains two types of exocrine glands, *sebaceous glands* and *sweat glands*. Sebaceous glands produce an oily lipid that coats hair shafts and the epidermis. Sweat glands produce a watery solution and perform other special functions. Figure 4-12● summarizes the functional classification of the exocrine glands of the skin.

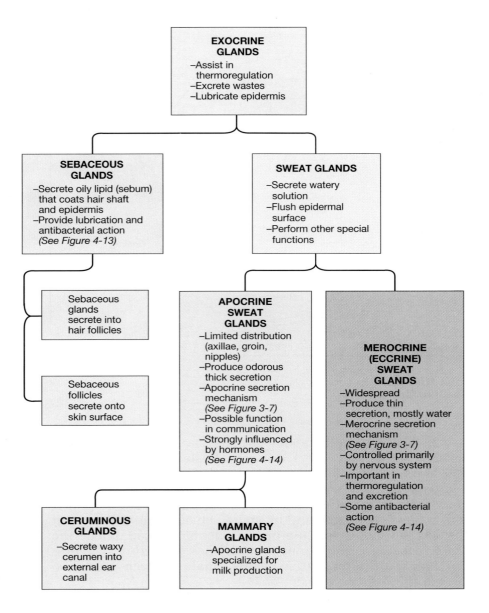

FIGURE 4-12
A Classification of Exocrine Glands in the Skin

Sebaceous Glands
(Figure 4-13)

Sebaceous (se-BĀ-shus) (oil) **glands** are holocrine glands that discharge a waxy, oily secretion into hair follicles (Figure 4-13●). Several sebaceous glands may communicate with a single follicle by means of one or more short ducts. Depending on whether the glands share a common duct, they may be classified as *simple alveolar glands* (each gland has its own duct) or *simple branched alveolar glands* (several glands empty into a single duct). The gland cells manufacture large quantities of lipids as they mature, and the lipid product is released through holocrine secretion. ∞ [p. 61]

The lipids released enter the open passageway, or **lumen**, of the gland. Contraction of the arrector pili muscle that elevates the hair squeezes the sebaceous gland, forcing the waxy secretions into the follicle and onto the surface of the skin. This secretion, called **sebum** (SĒ-bum), provides lubrication and inhibits the growth of bacteria. Keratin is a tough protein, but dead, cornified cells become dry and brittle once exposed to the environment. Sebum lubricates and protects the keratin

of the hair shaft and conditions the surrounding skin. Shampooing removes the natural oily coating, and excessive washing can make hairs stiff and brittle.

Sebaceous follicles are large sebaceous glands that communicate directly with the epidermis. These follicles, which never produce hairs, are found on the integument covering the face, back, chest, nipples, and male sex organs. Although sebum has bactericidal (bacteria-killing) properties, under some conditions bacteria can invade sebaceous glands or follicles. The presence of bacteria in glands or follicles can produce a local inflammation known as **folliculitis** (fo-lik-ū-LĪ-tis). If the duct of the gland becomes blocked, a distinctive abscess called a **furuncle** (FUR-ung-kl), or "boil," develops. The usual treatment for a furuncle is to cut it open, or "lance" it, so that normal drainage and healing can occur.

Sebaceous glands and sebaceous follicles are very sensitive to changes in the concentrations of sex hormones, and their secretory activities accelerate at puberty. For this reason an individual with large sebaceous glands may be especially prone to develop **acne** during adolescence. In acne, sebaceous ducts become blocked and secretions accumulate, causing inflammation and providing a fertile environment for bacterial infection. ✝ *Acne [p. 753]*

Sweat Glands
(Figures 4-12/4-14)

The skin contains two different groups of sweat glands, *apocrine sweat glands* and *merocrine sweat glands* (Figures 4-12 and 4-14●). These names refer to the mechanism of secretion, introduced in Chapter 3. ∞ [p. 58] Both gland types contain **myoepithelial cells** (*myo-*, muscle), specialized epithelial cells located between the gland cells and the underlying basement membrane. Myoepithelial cell contractions squeeze the gland and discharge the accumulated secretions. The secretory activities of the gland cells and the contractions of myoepithelial cells are controlled by the autonomic nervous system and by circulating hormones.

APOCRINE SWEAT GLANDS *(Figures 4-9a/4-14a).* In the armpits (axillae), around the nipples, and in the groin, **apocrine sweat glands** communicate with hair follicles (Figures 4-9a, p. 97, and 4-14a●). These are coiled tubular glands that produce a viscous, cloudy, and potentially odorous secretion. Apocrine sweat glands begin secreting at puberty, and the sweat produced is a nutrient source for bacteria that enhance its odor.

MEROCRINE (ECCRINE) SWEAT GLANDS *(Figures 4-9a/4-14b).* **Merocrine**, or **eccrine** (EK-rin), **sweat glands** are far more numerous and widely distributed than apocrine glands (Figures 4-9a, p. 97, and 4-14b●). The adult integument contains around 3 million eccrine glands. They are smaller than apocrine sweat

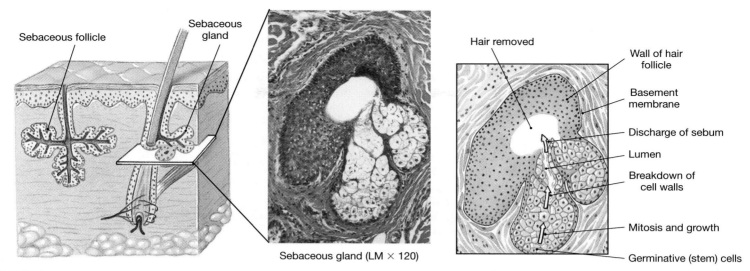

FIGURE 4-13
Sebaceous Glands and Follicles. The structure of sebaceous glands and sebaceous follicles in the skin.

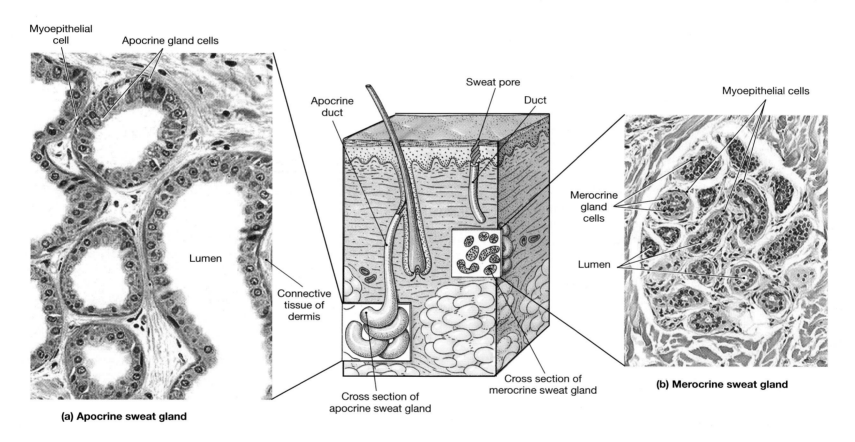

FIGURE 4-14
Sweat Glands. (a) Apocrine sweat glands are found in the axillae (armpits), groin, and nipples. They produce a thick, odorous fluid by apocrine secretion. (LM × 459) (b) Merocrine (eccrine) sweat glands produce a watery fluid by merocrine secretion. (LM × 243)

glands, and they do not extend as far into the dermis. Palms and soles have the highest numbers; estimates are that the palm of the hand has about 500 per cm² (3000 glands per square inch). Merocrine sweat glands are coiled tubular glands that discharge their secretions directly onto the surface of the skin.

The clear secretion produced by merocrine glands is termed **sweat**, or **sensible perspiration**. Sweat is chiefly water (99%), but it does contain some electrolytes (chiefly sodium chloride, NaCl), metabolites, and waste products. It is the presence of the sodium chloride that gives sweat a salty taste.

The functions of merocrine sweat gland activity include:

- Cooling the surface of the skin to reduce body temperature. Cooling is the primary function of sensible perspiration, and the degree of secretory activity is regulated by neural and hormonal mechanisms. When all of the merocrine sweat glands are working at maximum, the rate of perspiration may exceed a gallon per hour, and dangerous fluid and electrolyte losses can occur. For this reason athletes in endurance sports must pause frequently to drink fluids.

- Merocrine secretions can also provide a significant excretory route for water and electrolytes, as well as for a number of ingested drugs.
- Merocrine secretions provide protection from environmental hazards by diluting harmful chemicals and discouraging the growth of microorganisms.

Control of Glandular Secretions

Sebaceous and apocrine glands can be collectively turned on by the autonomic nervous system, but no regional control is possible. [p. 8] When one sebaceous or apocrine gland is activated, so are all the other glands of that type in the body. Merocrine sweat glands are much more precisely controlled, and the amount of secretion and the area of the body involved can be varied independently. For example, when you are nervously awaiting an anatomy exam, your palms may begin to sweat.

Other Integumentary Glands

Sebaceous glands and merocrine sweat glands are found over most of the body surface. Apocrine sweat glands are found in relatively restricted areas. The skin also contains a variety of specialized glands that are restricted to specific locations. Many will be encountered in later chapters; two important examples will be noted here.

1. The **mammary glands** of the breasts are anatomically related to apocrine sweat glands. A complex interaction between sexual and pituitary hormones controls their development and secretion. Mammary gland structure and function will be discussed in Chapter 27.
2. **Ceruminous** (se-ROO-mi-nus) **glands** are modified sweat glands located in the external auditory canal. Their secretions combine with those of nearby sebaceous glands, forming a mixture called **cerumen**, or simply "ear wax." Ear wax, together with tiny hairs along the ear canal, probably helps trap foreign particles or small insects and keeps them from reaching the eardrum.

√ **What are the functions of sebaceous secretions?**

√ **Deodorants are used to mask the effects of secretions from what type of skin gland?**

Nails
(Figure 4-15)

Nails form on the dorsal surfaces of the tips of the fingers and toes. The nails protect the exposed tips of the fingers and toes and help limit their distortion when they are subjected to mechanical stress—for example, in running or grasping objects. The structure of a nail can be seen in Figure 4-15●. The body of the nail covers the **nail bed,** but nail production occurs at the **nail root**, an epithelial fold not visible from the surface. The deepest portion of the nail root lies very close to the periosteum of the bone of the fingertip.

A portion of the stratum corneum of the fold extends over the exposed nail nearest the root, forming the **cuticle**, or **eponychium** (ep-ō-NIK-ē-um; *epi-*, over + *onyx*, nail). Underlying blood vessels give the nail its characteristic pink color, but near the root these vessels may be obscured, leaving a pale crescent known as the **lunula** (LOO-nu-la; *luna*, moon). The **nail body** is recessed beneath the level of the surrounding epithelium, and it is bounded by **nail grooves** and **nail folds**. The **free edge** of the nail extends over a thickened stratum corneum, the **hyponychium** (hī-pō-NIK-ē-um).

Changes in the shape, structure, or appearance of the nails are clinically significant. A change may indicate the existence of a disease process affecting metabolism throughout the body. For example, the nails may turn yellow in patients who have chronic respiratory disorders, thyroid gland disorders, or AIDS. They may become pitted and distorted in psoriasis and concave in some blood disorders.

LOCAL CONTROL OF INTEGUMENTARY FUNCTION

The integumentary system displays a significant degree of functional independence. It responds directly and automatically to local influences without the involvement of the nervous or endocrine systems. For example, when the skin is subjected to mechanical stresses, stem cells in the stratum germinativum divide more rapidly, and the depth of the epithelium increases. That is why calluses form on your palms when you perform manual labor. A more dramatic display of local regulation can be seen after an injury to the skin. The skin can regenerate effectively

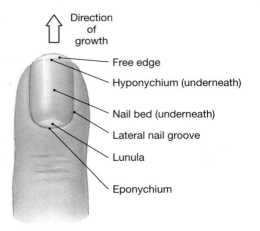

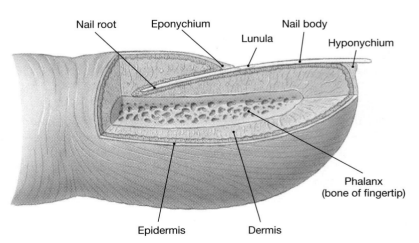

FIGURE 4-15
Structure of a Nail. These drawings illustrate the prominent features of a typical fingernail as viewed from the surface and in section.

even after considerable damage has occurred, because stem cells persist in both the epithelial and connective tissue components. Germinative cell divisions replace epidermal cells, and mesenchymal cell divisions replace lost dermal cells. This process can be slow, and when large surface areas are involved, problems of infection and fluid loss complicate the situation. ✝ *Inflammation of the Skin, Complications of Inflammation, Burns and Grafts, Scar Tissue Formation,* and *Synthetic Skin [pp. 753–754]*

After severe damage the repair process does not return the integument to its original condition. The injury site contains an abnormal density of collagen fibers and relatively few blood vessels. Damaged hair follicles, sebaceous or sweat glands, muscle cells, and nerves are seldom repaired, and they too are replaced by fibrous tissue. The formation of this rather inflexible, fibrous, noncellular **scar tissue** is a practical limit to the healing process.

Skin repairs proceed most rapidly in young, healthy individuals. For example, it takes 3–4 weeks to complete the repairs to a blister site in a young adult. The same repairs at age 65–75 take 6–8 weeks. However, this is just one example of the changes that occur in the integumentary system as a result of the aging process.

AGING AND THE INTEGUMENTARY SYSTEM
(Figure 4-16)

Aging affects all of the components of the integumentary system. These changes, summarized in Figure 4-16●, include:

1. The epidermis thins as germinative cell activity declines, making older people more prone to injury and skin infections.

2. The number of Langerhans cells decreases to around 50% of levels seen at maturity. This decrease may reduce the sensitivity of the immune system and further encourage skin damage and infection.

3. Vitamin D production declines by around 75%. The result can be muscle weakness and a reduction in bone strength.

4. Melanocyte activity declines, and in Caucasians the skin becomes very pale. With less melanin in the skin, older persons are more sensitive to sun exposure and more likely to experience sunburn.

5. Glandular activity declines. The skin becomes dry and often scaly because sebum production is reduced; sweat glands are also less active.

6. The blood supply to the dermis is reduced at the same time that sweat glands become less active. This combination makes the elderly less able to lose body heat, and overexertion or overexposure to warm temperatures can cause dangerously high body temperatures.

7. Hair follicles stop functioning or produce thinner, finer hairs. With decreased melanocyte activity, these hairs are gray or white.

8. The dermis becomes thinner, and the elastic fiber network decreases in size. The integument therefore becomes weaker and less resilient; sagging and wrinkling occurs. These effects are most pronounced in areas exposed to the sun.

9. With changes in levels of sex hormones, secondary sexual characteristics in hair and body fat distribution begin to fade. In consequence, people age 90–100 of both sexes and all races look very much alike.

10. Skin repairs proceed relatively slowly, and recurring infections may result.

FIGURE 4-16
Changes in the Skin during the Aging Process

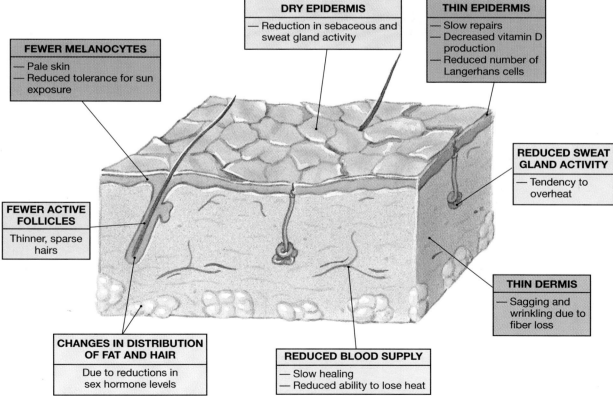

FEWER MELANOCYTES
— Pale skin
— Reduced tolerance for sun exposure

DRY EPIDERMIS
— Reduction in sebaceous and sweat gland activity

THIN EPIDERMIS
— Slow repairs
— Decreased vitamin D production
— Reduced number of Langerhans cells

REDUCED SWEAT GLAND ACTIVITY
— Tendency to overheat

FEWER ACTIVE FOLLICLES
Thinner, sparse hairs

THIN DERMIS
— Sagging and wrinkling due to fiber loss

CHANGES IN DISTRIBUTION OF FAT AND HAIR
Due to reductions in sex hormone levels

REDUCED BLOOD SUPPLY
— Slow healing
— Reduced ability to lose heat

EMBRYOLOGY SUMMARY: DEVELOPMENT OF THE INTEGUMENTARY SYSTEM

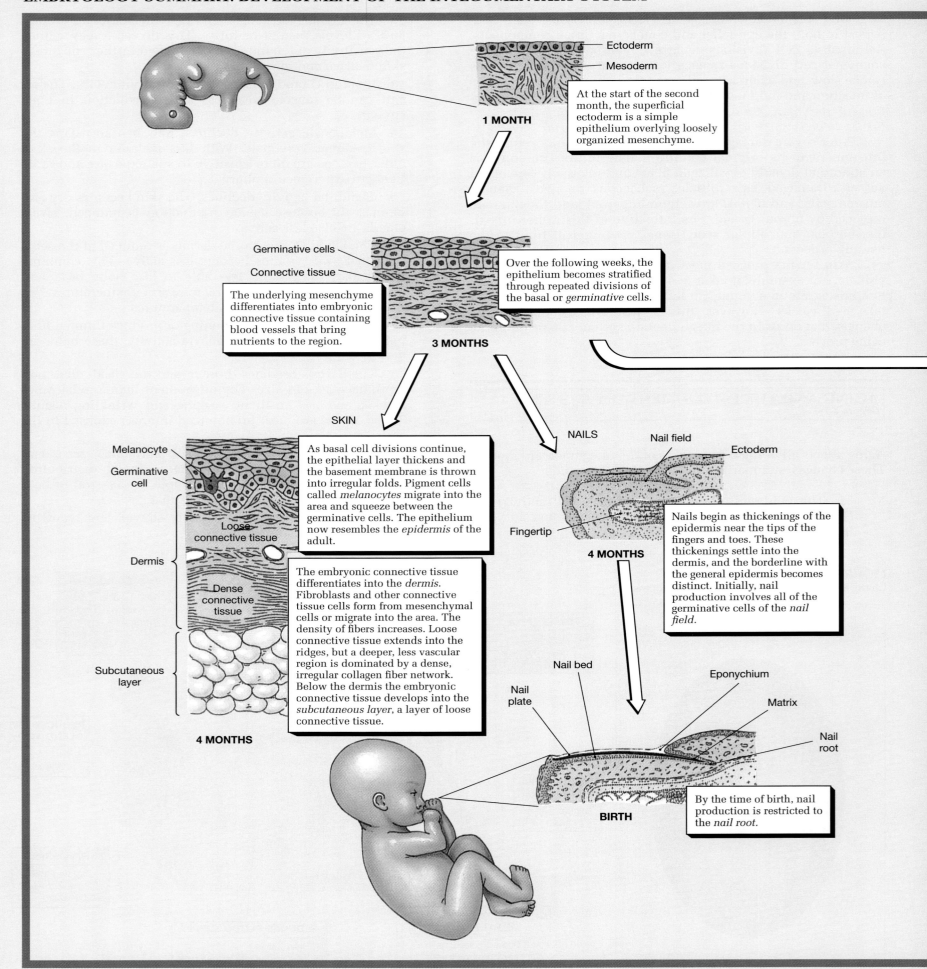

Ectoderm

Mesoderm

1 MONTH

At the start of the second month, the superficial ectoderm is a simple epithelium overlying loosely organized mesenchyme.

Germinative cells

Connective tissue

The underlying mesenchyme differentiates into embryonic connective tissue containing blood vessels that bring nutrients to the region.

Over the following weeks, the epithelium becomes stratified through repeated divisions of the basal or *germinative* cells.

3 MONTHS

SKIN

Melanocyte

Germinative cell

Loose connective tissue

Dermis

Dense connective tissue

Subcutaneous layer

4 MONTHS

As basal cell divisions continue, the epithelial layer thickens and the basement membrane is thrown into irregular folds. Pigment cells called *melanocytes* migrate into the area and squeeze between the germinative cells. The epithelium now resembles the *epidermis* of the adult.

The embryonic connective tissue differentiates into the *dermis*. Fibroblasts and other connective tissue cells form from mesenchymal cells or migrate into the area. The density of fibers increases. Loose connective tissue extends into the ridges, but a deeper, less vascular region is dominated by a dense, irregular collagen fiber network. Below the dermis the embryonic connective tissue develops into the *subcutaneous layer*, a layer of loose connective tissue.

NAILS

Nail field

Ectoderm

Fingertip

4 MONTHS

Nails begin as thickenings of the epidermis near the tips of the fingers and toes. These thickenings settle into the dermis, and the borderline with the general epidermis becomes distinct. Initially, nail production involves all of the germinative cells of the *nail field*.

Nail bed

Eponychium

Nail plate

Matrix

Nail root

BIRTH

By the time of birth, nail production is restricted to the *nail root*.

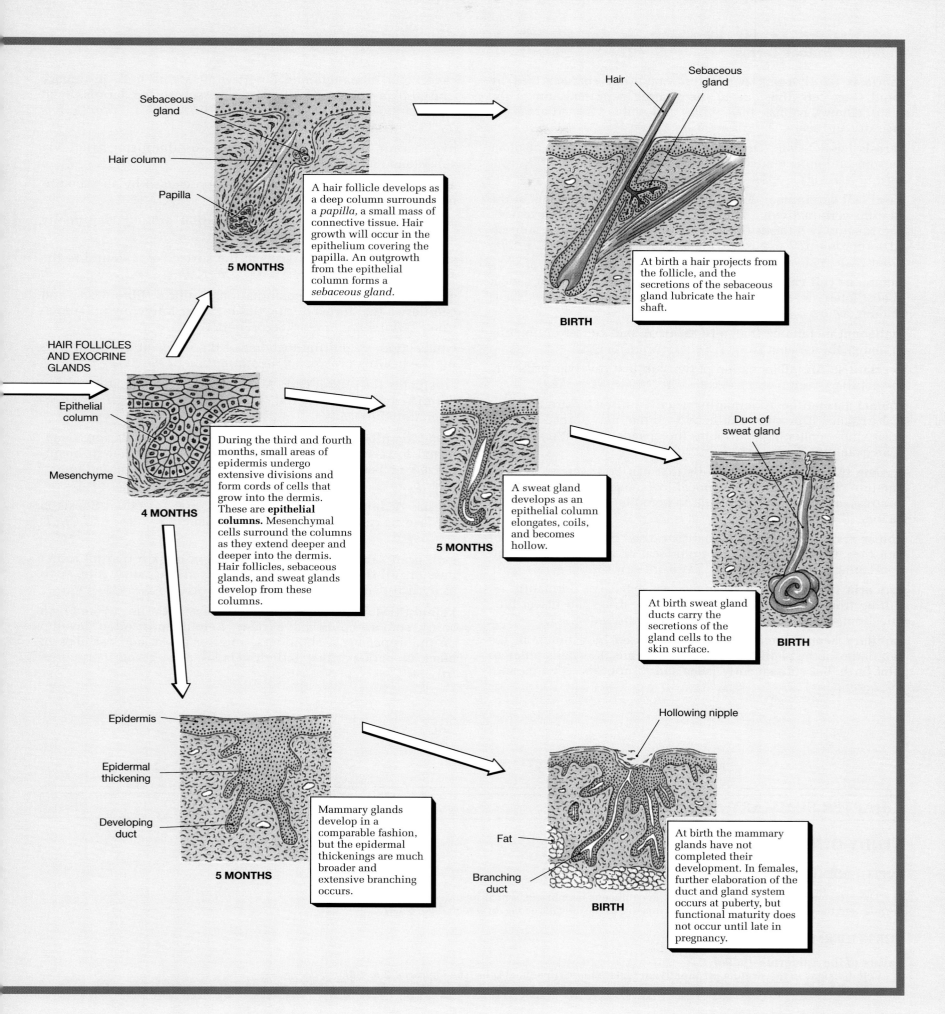

Sebaceous
gland

Hair column

Papilla

5 MONTHS

A hair follicle develops as
a deep column surrounds
a *papilla*, a small mass of
connective tissue. Hair
growth will occur in the
epithelium covering the
papilla. An outgrowth
from the epithelial
column forms a
sebaceous gland.

Hair   Sebaceous
        gland

At birth a hair projects from
the follicle, and the
secretions of the sebaceous
gland lubricate the hair
shaft.

BIRTH

HAIR FOLLICLES
AND EXOCRINE
GLANDS

Epithelial
column

Mesenchyme

4 MONTHS

During the third and fourth
months, small areas of
epidermis undergo
extensive divisions and
form cords of cells that
grow into the dermis.
These are **epithelial
columns.** Mesenchymal
cells surround the columns
as they extend deeper and
deeper into the dermis.
Hair follicles, sebaceous
glands, and sweat glands
develop from these
columns.

5 MONTHS

A sweat gland
develops as an
epithelial column
elongates, coils,
and becomes
hollow.

Duct of
sweat gland

At birth sweat gland
ducts carry the
secretions of the
gland cells to the
skin surface.

BIRTH

Epidermis

Epidermal
thickening

Developing
duct

5 MONTHS

Mammary glands
develop in a
comparable fashion,
but the epidermal
thickenings are much
broader and
extensive branching
occurs.

Hollowing nipple

Fat

Branching
duct

BIRTH

At birth the mammary
glands have not
completed their
development. In females,
further elaboration of the
duct and gland system
occurs at puberty, but
functional maturity does
not occur until late in
pregnancy.

Related Clinical Terms

psoriasis (sō-RĪ-a-sis): A painless condition characterized by rapid stem cell divisions in the stratum germinativum of the scalp, elbows, palms, soles, groin, and nails. Affected areas appear dry and scaly. ✝ *Psoriasis and Xerosis [p. 752]*

xerosis (zē-RŌ-sis): "Dry skin," a common complaint of older persons and almost anyone living in an arid climate. ✝ *Psoriasis and Xerosis [p. 752]*

basal cell carcinoma: A malignant cancer that originates in the stratum germinativum. This is the most common skin cancer, and roughly two-thirds of these cancers appear in areas subjected to chronic UV exposure. Metastasis seldom occurs. ✝ *Skin Cancers [p. 752]*

squamous cell carcinoma: A less common form of skin cancer almost totally restricted to areas of sun-exposed skin. Metastasis seldom occurs. ✝ *Skin Cancers [p. 752]*

malignant melanoma (mel-a-NŌ-ma): A skin cancer originating in malignant melanocytes. ✝ *Skin Cancers [p. 752]*

dermatitis: An inflammation of the skin that involves primarily the papillary region of the dermis. ✝ *Dermatitis [p. 752]*

contact dermatitis: A dermatitis usually caused by strong chemical irritants. It produces an itchy rash that may spread to other areas since scratching distributes the chemical agent; poison ivy is an example. ✝ *Dermatitis [p. 752]*

eczema (EK-se-ma): A dermatitis that can be triggered by temperature changes, fungus, chemical irritants, greases, detergents, or stress, and that can be related to hereditary or environmental factors. ✝ *Dermatitis [p. 752]*

diaper rash: A localized dermatitis caused by a combination of moisture, irritating chemicals from fecal or urinary wastes, and flourishing microorganisms. ✝ *Dermatitis [p. 752]*

urticaria (ur-ti-KAR-ē-a) or **hives:** An extensive dermatitis resulting from an allergic reaction to food, drugs, an insect bite, infection, stress, or other stimulus. ✝ *Dermatitis [p. 752]*

capillary hemangioma: A birthmark caused by a tumor in the capillaries of the papillary layer of the dermis. It usually enlarges after birth, but subsequently fades and disappears. ✝ *Tumors in the Dermis [p. 752]*

cavernous hemangiomas ("port-wine stains"): A birthmark caused by a tumor affecting larger vessels in the dermis. Such birthmarks usually last a lifetime. ✝ *Tumors in the Dermis [p. 752]*

hypodermic needle: A needle used to administer drugs via subcutaneous injection. *[p. 96]*

acne: A sebaceous gland inflammation caused by an accumulation of secretions. ✝ *Acne [p. 753]*

seborrheic dermatitis: An inflammation around abnormally active sebaceous glands. *[p. 100]*

scab: A fibrin clot that forms at the surface of a wound to the skin. ✝ *Inflammation of the Skin [p. 753]*

granulation tissue: A combination of fibrin, fibroblasts, and capillaries that forms during tissue repair following inflammation. ✝ *Inflammation of the Skin [p. 753]*

contraction: A pulling together of the edges of a wound during the healing process. ✝ *Inflammation of the Skin [p. 753]*

erysipelas (er-i-SIP-e-las): A widespread inflammation of the dermis caused by bacterial infection. ✝ *Complications of Inflammation [p. 753]*

decubitis ulcers ("bedsores"): Ulcers that occur in areas subjected to restricted circulation, especially common in hospitalized or bedridden persons. ✝ *Complications of Inflammation [p. 753]*

sepsis: A dangerous, widespread bacterial infection. Sepsis is the leading cause of death in burn patients. ✝ *Burns and Grafts [p. 754]*

skin graft: Transplantation of a section of skin (partial thickness or full thickness) to cover an extensive injury site, such as a third-degree burn. ✝ *Burns and Grafts [p. 754]*

keloid (KĒ-loyd): A thickened area of scar tissue covered by a shiny, smooth epidermal surface. Keloids most often develop on the upper back, shoulders, anterior chest, and earlobes in black or dark-skinned individuals. ✝ *Scar Tissue Formation [p. 754]*

CHAPTER SUMMARY AND REVIEW

STUDY OUTLINE

INTEGUMENTARY STRUCTURE AND FUNCTION *[p. 89]*

1. The **integumentary system** consists of the **cutaneous membrane**, which includes the **epidermis** and **dermis**, and the **accessory structures**. Underneath lies the *subcutaneous layer (see Figure 4-1)*.

THE EPIDERMIS *[p. 90]*

Layers of the Epidermis *[p. 90]*
1. Cell divisions in the **stratum germinativum** replace more superficial cells *(see Figures 4-2 to 4-6)*.

2. As new epidermal cells differentiate, they pass through the **stratum spinosum**, the **stratum granulosum**, the **stratum lucidum** (if thick skin), and the **stratum corneum**. In the process they accumulate large amounts of **keratin**. Ultimately the cells are shed or lost *(see Figure 4-3)*.

3. **Langerhans cells** (phagocytic cells) and **Merkel cells** are specialized cells found in the deeper layers of the epidermis.

Thick and Thin Skin *[p. 92]*
4. **Thin skin** covers most of the body; heavily abraded body surfaces of the palms and soles are covered by **thick skin** *(see Figure 4-4)*.

5. **Epidermal ridges**, such as those on the palms and soles, improve our gripping ability and increase the skin's sensitivity *(see Figures 4-4/4-5)*.

6. The color of the epidermis depends on two factors: blood supply and pigment composition and concentration. **Melanocytes** protect us from excessive amounts of **ultraviolet radiation** *(see Figure 4-6)*.

THE DERMIS *[p. 94]*

1. The **dermis** consists of the superficial **papillary layer** and the deeper **reticular layer** *(see Figures 4-2/4-4/4-7 to 4-9)*.

Dermal Organization *[p. 94]*
2. The papillary layer of the dermis contains blood vessels, lymphatics, and sensory nerves. This layer supports and nourishes the overlying epidermis *(see Figures 4-4/4-7)*.

3. The reticular layer consists of a meshwork of collagen and elastic fibers oriented to resist tension in the skin *(see Figure 4-8)*.

4. The extensive blood supply to the skin includes the *cutaneous* and *papillary* arterial plexuses. The papillary layer contains abundant capillaries *(see Figure 4-2)*.

Other Dermal Components *[p. 95]*
5. The skin is innervated by sensory nerves that monitor touch, temperature, pain, pressure, and vibration *(see Figures 4-2/4-9)*.

THE SUBCUTANEOUS LAYER *[p. 96]*

1. The **subcutaneous layer** or **hypodermis**, stabilizes the skin's position against underlying organs and tissues yet permits limited independent movement *(see Figures 4-2/4-7c)*.

ACCESSORY STRUCTURES *[p. 96]*

Hair Follicles and Hair *[p. 96]*
1. Hairs originate in complex organs called **hair follicles.** Each hair has a **bulb**, a **root**, and a **shaft**. Hair production involves cell specialization to form a soft core, or **medulla**, surrounded by a **cortex**. The **cuticle** is a hard layer that coats the hair *(see Figures 4-2/4-9 to 4-11)*.

2. The lumen of the follicle is lined by an internal root sheath, which is surrounded by the external root sheath, the glassy membrane, and a connective tissue layer. The internal root sheath extends only partway toward the surface of the skin *(see Figure 4-10)*.

3. A **root hair plexus** of sensory nerves surrounds the base of each hair follicle. Contraction of the **arrector pili muscles** elevates the hairs by pulling on the follicles *(see Figures 4-9/4-10a)*.

4. **Vellus hairs** ("peach fuzz"), heavy **terminal hairs**, and **intermediate hairs** (of transient character) make up the hair population on our bodies *(see Figure 4-9)*.

5. Our hairs grow and are shed according to the **hair growth cycle**. A single hair grows for 2–5 years and is subsequently shed *(see Figures 4-9/4-11)*.

Glands in the Skin *[p. 99]*
6. **Sebaceous glands** discharge the waxy **sebum** into hair follicles. **Sebaceous follicles** lack hairs but have large sebaceous glands *(see Figure 4-13)*.

7. **Apocrine sweat glands** produce an odorous secretion; the more numerous **merocrine**, or **eccrine**, **sweat glands** produce a watery secretion, known as **sensible perspiration**, or **sweat** *(see Figures 4-12/4-14)*.

8. **Ceruminous glands** in the ear canal produce a waxy **cerumen**.

Nails *[p. 102]*
9. The **nails** protect the exposed tips of the fingers and toes and help limit their distortion when they are subjected to mechanical stress.

10. The **nail body** covers the **nail bed,** with nail production occurring at the **nail root**. The **cuticle**, or **eponychium**, is formed by a fold of the stratum corneum extending from the nail root to the exposed nail *(see Figure 4-15)*.

LOCAL CONTROL OF INTEGUMENTARY FUNCTION *[p. 102]*

1. The skin can regenerate effectively even after considerable damage, as in the case of burns (*see Table A-3 in the Clinical Issues appendix*).

Related Key Terms
keratohyalin • eleidin • keratinized • cornified • insensible perspiration

dermal papillae

cyanosis • carotene • melanin • melanocyte-stimulating hormone (MSH) • albinism

stretch marks • lines of cleavage

subcutaneous injection

papilla • matrix • soft keratin • hard keratin

club hair • male pattern baldness

lumen • folliculitis • furuncle • acne

myoepithelial cells

mammary glands

lunula • nail grooves • nail folds • free edge • hyponychium

2. Severe damage to the dermis and accessory glands cannot be completely repaired, and fibrous **scar tissue** remains at the injury site.

AGING AND THE INTEGUMENTARY SYSTEM [p. 103]

1. Aging affects all layers and accessory structures of the integumentary system *(see Figure 4-16)*.

---

1   REVIEW OF CHAPTER OBJECTIVES

1. Describe and compare the structure and functions of the epidermis and dermis of the skin with the underlying hypodermis.
2. Explain what accounts for individual and racial differences in skin, such as skin color.
3. Discuss the effects of ultraviolet radiation on the skin and the role played by melanocyte cells in this regard.
4. Describe the structure of the subcutaneous layer (hypodermis) and its importance in stabilizing the skin.
5. Discuss the anatomy of the skin's accessory structures: hair, glands, and nails.
6. Describe the mechanisms that produce hair and determine hair texture and color.
7. Describe how the sweat glands of the integumentary system function in the regulation of body temperature.
8. Explain how the skin responds to injuries and repairs itself.
9. Summarize the effects of the aging process on the skin.

2   REVIEW OF CONCEPTS

1. What characteristics of the epidermis contribute to the formation of fingerprints?
2. What tissue types and structures permit movement of the skin over the underlying tissues in most areas of the body?
3. How does the protein keratin affect the appearance and function of the integument?
4. What types of sensations is the epidermis capable of discerning, and by what structures are these stimuli received?
5. What changes occur in the skin when a callus forms?
6. What skin changes occur in a person who is cold or frightened?
7. What is the importance of the secretion of sebum, an oily lipid that makes hair become limp and greasy?
8. Why do washing the skin and applying deodorant reduce the odor of apocrine sweat gland secretions?

9. What is happening to an individual who is cyanotic, and what body structures would show this condition most easily?
10. What is the arrangement of collagen and elastic fibers in the dermis, and how does the orientation of these fibers affect the lines of cleavage in the skin? Under what circumstances might lines of cleavage be important?
11. Why are the body contours of healthy infants typically rounder than those of older children and adults? Where is the tissue shaping these contours located, and why is having this tissue important to the health and well-being of the child?
12. What role might be played by the intermediate hairs on the arms and legs in sensing that a mosquito is walking along your skin?
13. What is the significance of increasing age in the changes seen in hair color and texture?
14. What causes the discoloration we call a bruise on the skin, and what layers or structures are involved?

3   CRITICAL THINKING AND CLINICAL APPLICATION QUESTIONS

1. People enjoy being at the beach and in the sun during warm weather. Exposure to the sun causes the skin to tan. What changes occur during this process?
2. Some people, men in particular, begin to lose the hair on their head, sometimes as early as in their twenties. What has influenced the hair loss?
3. Samples of hair can be tested to determine exposure to toxins, poisons, drugs, and levels of trace elements in the body over long periods of time. Why is this possible?
4. A person with diabetes must receive an injection of insulin each day. This injection is said to be subcutaneous. What layers of skin does the needle pass through, and into which tissue is the insulin introduced?
5. After successfully completing the Chicago marathon (a 26.2-mile race), on a pleasantly cool day in October, a runner complains of feelings of dizziness and exhaustion. Although she is not thirsty, she is given a drink and shortly begins to feel better. Why?

5

The Skeletal System: Osseous Tissue and Skeletal Structure

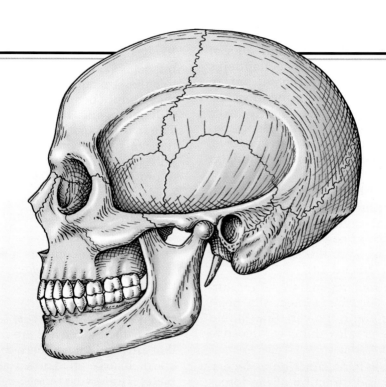

CHAPTER OUTLINE AND OBJECTIVES

The skeletal system includes the bones of the skeleton and the cartilages, ligaments, and other connective tissues that stabilize or connect them. Skeletal elements are more than just racks to hang muscles on; they have a great variety of vital functions. We begin our study of the skeletal system by identifying those functions:

1. **Support:** The skeletal system provides structural support for the entire body. Individual bones or groups of bones provide a framework for the attachment of soft tissues and organs.

2. **Storage of Minerals and Lipids:** The calcium salts of bone represent a valuable mineral reserve that maintains normal concentrations of calcium and phosphate ions in body fluids. Calcium is the most abundant mineral in the human body. A typical human body contains 1–2 kg (2.2–4.4 lb) of calcium, with more than 98% of it deposited in the skeleton. In addition to acting as a mineral reserve, the bones of the skeleton store energy reserves as lipids in areas of *yellow marrow*.

3. **Blood Cell Production:** Red blood cells, white blood cells, and other blood elements are produced within the *red marrow* that fills the internal cavities of many bones. The role of the bone marrow in blood cell formation will be described in later chapters dealing with the cardiovascular and lymphatic systems (Chapters 20 and 23).

4. **Protection:** Delicate tissues and organs are often surrounded by skeletal elements. The ribs protect the heart and lungs, the skull encloses the brain, the vertebrae shield the spinal cord, and the pelvis cradles delicate digestive and reproductive organs.

5. **Leverage:** The bones of the skeleton function as levers. They can change the magnitude and direction of the forces generated by skeletal muscles. The movements produced range from the delicate motion of a fingertip to powerful changes in the position of the entire body.

The bones of the skeleton are actually complex, dynamic organs that contain osseous tissue, other connective tissues, smooth muscle tissue, and neural tissue. We will now consider the internal organization of a typical bone.

STRUCTURE OF BONE

Bone tissue, or **osseous tissue**, is one of the supporting connective tissues. (You should consider reviewing the sections on dense connective tissues, cartilage, and bone at this time.) ∞ [pp. 68–76] Like other connective tissues, osseous tissue contains specialized cells and a **matrix** consisting of extracellular protein fibers and a ground substance. The matrix of bone tissue is solid and sturdy, because of the deposition of calcium salts around the protein fibers.

Histological Organization

The basic organization of bone tissue was introduced in Chapter 3. ∞ [p. 75] We will now take a closer look at the organization of the matrix and cells of bone.

The Matrix of Bone

Calcium phosphate, $Ca_3(PO_4)_2$, accounts for almost two-thirds of the weight of bone. The calcium phosphate interacts with calcium hydroxide $[Ca(OH)_2]$ to form crystals of **hydroxyapatite**, $Ca_{10}(PO_4)_6(OH)_2$. As they form, these crystals also incorporate other calcium salts, such as calcium carbonate, and ions such as sodium, magnesium, and fluoride. Roughly one-third of the weight of bone is contributed by collagen fibers. Osteocytes and other cell types account for only 2% of the mass of a typical bone.

Calcium phosphate crystals are very strong, but relatively inflexible. They can withstand compression, but the crystals are likely to shatter when exposed to bending, twisting, or sudden impacts. Collagen fibers are tough and flexible. They can easily tolerate stretching, twisting, and bending, but when compressed they simply bend out of the way. In bone, the collagen fibers provide an organic framework for the formation of mineral crystals. The hydroxyapatite crystals form small plates that lie alongside of the collagen fibers. The result is a protein-crystal combination with properties intermediate between those of collagen and those of pure mineral crystals.

Cells in Osseous Tissue
(Figure 5-1)

Osseous tissue contains a distinctive population of cells, including *osteocytes, osteoblasts, osteoclasts,* and *osteoprogenitor cells* (Figure 5-1a-c●).

OSTEOCYTES *(Figure 5-1b).* **Osteocytes** (*osteon*, bone) are mature bone cells. Osteocytes are found in small pockets, called **lacunae**, that are sandwiched between layers of calcified matrix. These layers of matrix are known as **lamellae** (lah-MEL-lē; *lamella*, thin plate). Hollow channels, called **canaliculi** (kan-a-LIK-ū-lē), penetrate the lamellae, radiating through the matrix and connecting lacunae with one another and with the nutrient sources (Figure 5-1b-d●). The canaliculi contain cytoplasmic extensions of the osteocytes, and the fluid that surrounds the osteocytes and their extensions provides a route for the diffusion of nutrients and waste products. The two types of osseous tissue, *compact bone* and *spongy bone*, have the same basic components but differ in their three-dimensional organization.

OSTEOBLASTS *(Figure 5-1a).* **Osteoblasts** (OS-tē-ō-blasts; *blast*, precursor) are cuboidal cells that synthesize the organic components of the bone matrix (Figure 5-1a●). This material, called **osteoid** (OS-tē-oyd), later becomes mineralized through an unknown mechanism. Osteoblasts are responsible for the production of new bone, a process called **osteogenesis** (os-tē-ō-JEN-e-sis; *gennan*, to produce).

OSTEOCLASTS *(Figure 5-1a).* **Osteoclasts** (OS-tē-ō-klasts; *clast*, break) are giant cells with 50 or more nuclei (Figure 5-1a●). Acids secreted by the lysosomes of osteoclasts dissolve the bony matrix and release the stored minerals of calcium and phosphate. This process, called **osteolysis** (os-tē-OL-ī-sis), is important in the regulation of calcium and phosphate concentrations in body fluids. Osteoclasts are always removing matrix, and osteoblasts are always adding to it. The balance between the activities of osteoblasts and osteoclasts is very important; when osteoclasts remove calcium salts faster than osteoblasts deposit them, bones become weaker; when osteoblast activity predominates, bones become stronger and more massive.

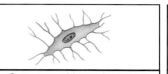

Osteocyte: Mature bone cell that turns over bone minerals and assists in repairs

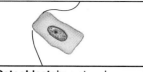

Osteoblast: Immature bone cell that secretes organic components of matrix

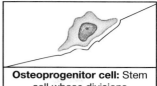

Osteoprogenitor cell: Stem cell whose divisions produce osteoblasts

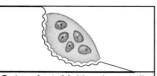

Osteoclast: Multinucleate cell that secretes acids and enzymes to dissolve bone matrix

(a) Cells of bones

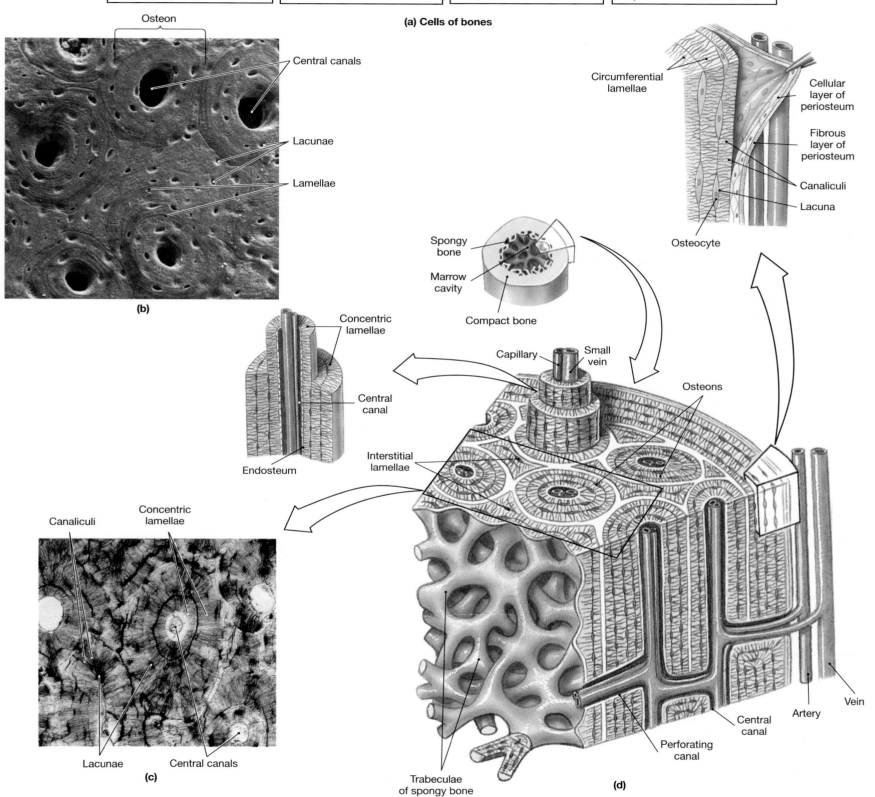

Osteon

Central canals

Lacunae

Lamellae

(b)

Circumferential lamellae

Cellular layer of periosteum

Fibrous layer of periosteum

Canaliculi

Lacuna

Osteocyte

Spongy bone

Marrow cavity

Compact bone

Concentric lamellae

Central canal

Endosteum

Interstitial lamellae

Capillary

Small vein

Osteons

Canaliculi

Concentric lamellae

Lacunae

Central canals

(c)

Trabeculae of spongy bone

Perforating canal

Central canal

Artery

Vein

(d)

FIGURE 5-1

Structure of a Typical Bone. (a) The cells bone. (b) A scanning electron micrograph of several osteons in compact bone. (SEM × 182) (c) A thin section through compact bone; in this procedure the intact matrix and central canals appear white, and the lacunae and canaliculi are shown in black. (LM × 272) (d) Diagrammatic view of the structure of a representative bone. [Reproduced from R. G. Kessel and R. H. Kardon, "Tissues and Organs: A Text–Atlas of Scanning Electron Microscopy," W. H. Freeman & Co., 1979.]

OSTEOPROGENITOR CELLS *(Figure 5-1a).* Bone tissue also contains small numbers of mesenchymal cells that can divide to produce daughter cells that differentiate into osteoblasts. These **osteoprogenitor cells** (os-tē-ō-prō-JEN-i-tor; *progenitor*, ancestor) (Figure 5-1a●) play an important role in fracture repair, a process discussed later in the chapter.

Compact and Spongy Bone
(Figure 5-1d)

There are two types of osseous tissue: *compact bone*, or *dense bone*, and *spongy bone*, or *cancellous* (KAN-sel-us) *bone*. **Compact bone** is relatively dense and solid, whereas **spongy bone** forms an open network of struts and plates. Both compact and spongy bone are present in a bone of the skeleton, such as a humerus or a vertebra. Compact bone tissue forms the walls, and an internal layer of spongy bone surrounds the **marrow cavity** (Figure 5-1d●). The marrow cavity contains **bone marrow**, a loose connective tissue that may be dominated by adipocytes (**yellow marrow**) or by a mixture of mature and immature red and white blood cells, and the stem cells that produce them (**red marrow**).

Structural Differences between Compact and Spongy Bone

The matrix composition in compact bone is the same as that of spongy bone, but they differ in the three-dimensional arrangement of osteocytes, canaliculi, and lamellae.

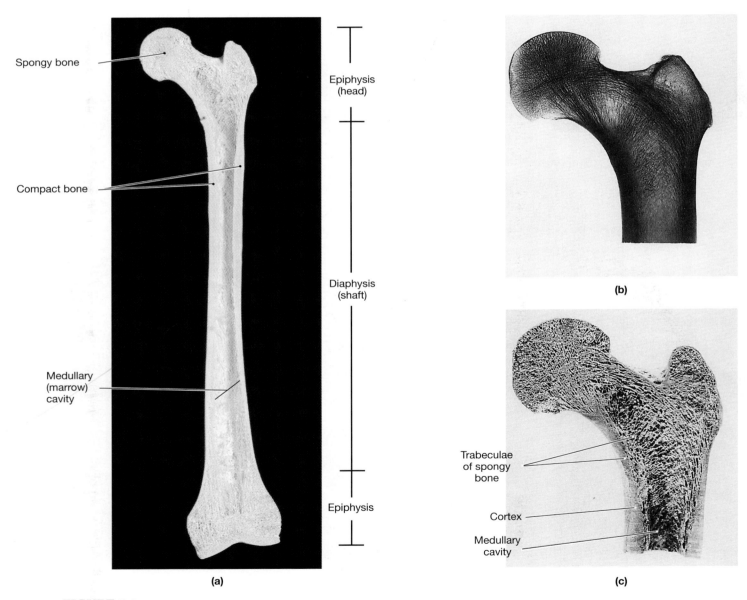

Spongy bone

Compact bone

Medullary (marrow) cavity

Epiphysis (head)

Diaphysis (shaft)

Epiphysis

(a)

(b)

Trabeculae of spongy bone

Cortex

Medullary cavity

(c)

FIGURE 5-2
Lamellar Organization in a Long Bone. (a) The femur, or thigh bone, has a diaphysis (shaft) with walls of compact bone and epiphyses (heads) filled with spongy bone. The body weight is transferred to the femur at the hip joint. Because the hip joint is off-center relative to the axis of the shaft, the body weight is distributed along the bone so that the medial portion of the shaft is compressed and the lateral portion is stretched. (b) An X-ray showing the orientation of the trabeculae in the epiphysis. (c) A photograph showing the epiphysis after sectioning.

COMPACT BONE *(Figure 5-1d).* The basic functional unit of mature compact bone is the **osteon** (OS-tē-on), or *Haversian system*. Within an osteon the osteocytes are arranged in concentric layers around a **central canal**, or *Haversian canal*, which contains one or more blood vessels that supply that osteon. Central canals usually run parallel to the surface of the bone. Other passageways, known as **perforating canals**, or the *canals of Volkmann*, extend roughly perpendicular to the surface. Blood vessels in the perforating canals deliver blood to osteons deeper in the bone and service the interior marrow cavity. The lamellae of each osteon are cylindrical and aligned parallel to the long axis of the central canal. These are known as *concentric lamellae.* Collectively the concentric lamellae form a series of concentric rings, resembling a "bull's-eye" target, around the central canal (Figure 5-1d●). Canaliculi radiating through the lamellae interconnect the lacunae of the osteon with one another and with the central canal. *Interstitial lamellae* fill in the spaces between the osteons in compact bone. Depending on their location, these lamellae may have been produced during the growth of the bone or they may represent remnants of osteons whose matrix components have been recycled by osteoclasts.

SPONGY BONE. There are no osteons in spongy bone, and the concentric lamellae form struts or plates called **trabeculae** (tra-BEK-ū-lē). The thin trabeculae often branch, creating an open network. Nutrients reach the osteocytes by diffusion along canaliculi that open onto the surfaces of the trabeculae.

Functional Differences between Compact and Spongy Bone
(Figure 5-2)

A layer of compact bone covers the surfaces of bones; the thickness of that layer varies from region to region and from one bone to another. This superficial layer of compact bone is in turn covered by the *periosteum*, a connective tissue component of the deep fascia, everywhere except inside joint capsules, where hyaline *articular cartilages* cover opposing surfaces.

Compact bone is thickest where stresses arrive from a limited range of directions. Osteons in compact bone are all lined up the same way, and such bones are very strong when stressed along that axis. You might envision a single osteon as a drinking straw with very thick walls. When you attempt to push the ends of a straw together, or to pull them apart, the straw is quite strong. However, if you hold the ends and push from the side, it will break easily.

Figure 5-2● shows the orientation of osteons and trabeculae in the femur, a typical long bone. The compact bone of the **cortex** surrounds the marrow cavity, also known as the **medullary cavity** (*medulla*, innermost part). Stresses are normally applied along the tubular **shaft**, or **diaphysis** (dī-A-fi-sis). The osteons are parallel to the long axis of the shaft. Although it does not bend when forces are applied to either end, an impact to the side of the shaft can lead to a femoral fracture.

The hip joint consists of the head of the femur and a corresponding socket on the lateral surface of the hip bone. The femoral head projects medially, and body weight compresses the medial side of the diaphysis. Because the force is applied off-center, the bone has a tendency to bend into a lateral bow. The other side of the shaft, which resists this bending, is placed under a stretching load, or *tension*. Bending does occur in disorders that reduce the amount of calcium salts in the skeleton, as in **rickets**. Rickets usually develops in children as a result of a vitamin D deficiency; this vitamin is essential to normal calcium absorption and deposition in the skeleton. In this disorder the bones are poorly mineralized, and they become very flexible. Affected individuals develop a bowlegged appearance as the thigh and leg bones bend under the weight of the body.

Spongy bone is found where bones are not heavily stressed, or where stresses arrive from many directions. It is present at the expanded ends of long bones, where they articulate with other skeletal elements. These expanded regions are the heads, or **epiphyses** (e-PIF-i-sēs), of the bones. Figure 5-2b,c● shows the trabecular alignment in the proximal epiphysis of the femur. The trabeculae are oriented along the stress lines, but with extensive cross-bracing. At the proximal epiphysis, the trabeculae transfer forces from the hip to the femoral shaft; at the distal epiphysis, the trabeculae direct the forces across the knee joint to the leg.

In addition to being able to withstand stresses applied from many directions, spongy bone is much lighter than compact bone. This reduces the weight of the skeleton and makes it easier for muscles to move the bones. Finally, the trabecular framework supports and protects the cells of the bone marrow. *Yellow marrow*, often found in the marrow cavity of the shaft, is an important energy reserve. Extensive areas of *red marrow*, such as that found in the spongy bone of the femoral epiphyses, are important sites of blood cell formation.

The Periosteum and Endosteum
(Figures 5-1d/5-3)

The outer surface of a bone is covered by a **periosteum** that consists of a fibrous outer layer and a cellular inner layer (Figure 5-1d●). The periosteum (1) isolates the bone from surrounding tissues, (2) provides a route for circulatory and nervous supply, and (3) actively participates in bone growth and repair.

Near joints, the periosteum becomes continuous with the connective tissues that bind the bones together. At a synovial joint, the periosteum is continuous with the joint capsule. The fibers of the periosteum are also interwoven with those of the tendons attached to the bone (Figure 5-3●). As the bone grows, these tendon fibers are cemented into the superficial lamellae by osteoblasts from the cellular layer of the periosteum. This cementing makes the tendon fibers part of the general structure of the bone, providing a much stronger bond than would otherwise be possible. An extremely powerful pull on a tendon or ligament will usually break a bone rather than snap the collagen fibers at the bone surface. The collagen fibers incorporated into bone tissue from tendons and from the superficial periosteum are called *Sharpey's fibers.*

Inside the bone, a cellular **endosteum** lines the marrow cavity as seen in Figure 5-3●. This layer covers the trabeculae of spongy bone and lines the inner surfaces of the central canals. The endosteum is active during the growth of bone

FIGURE 5-3
The Periosteum and Endosteum. (LM × 100)

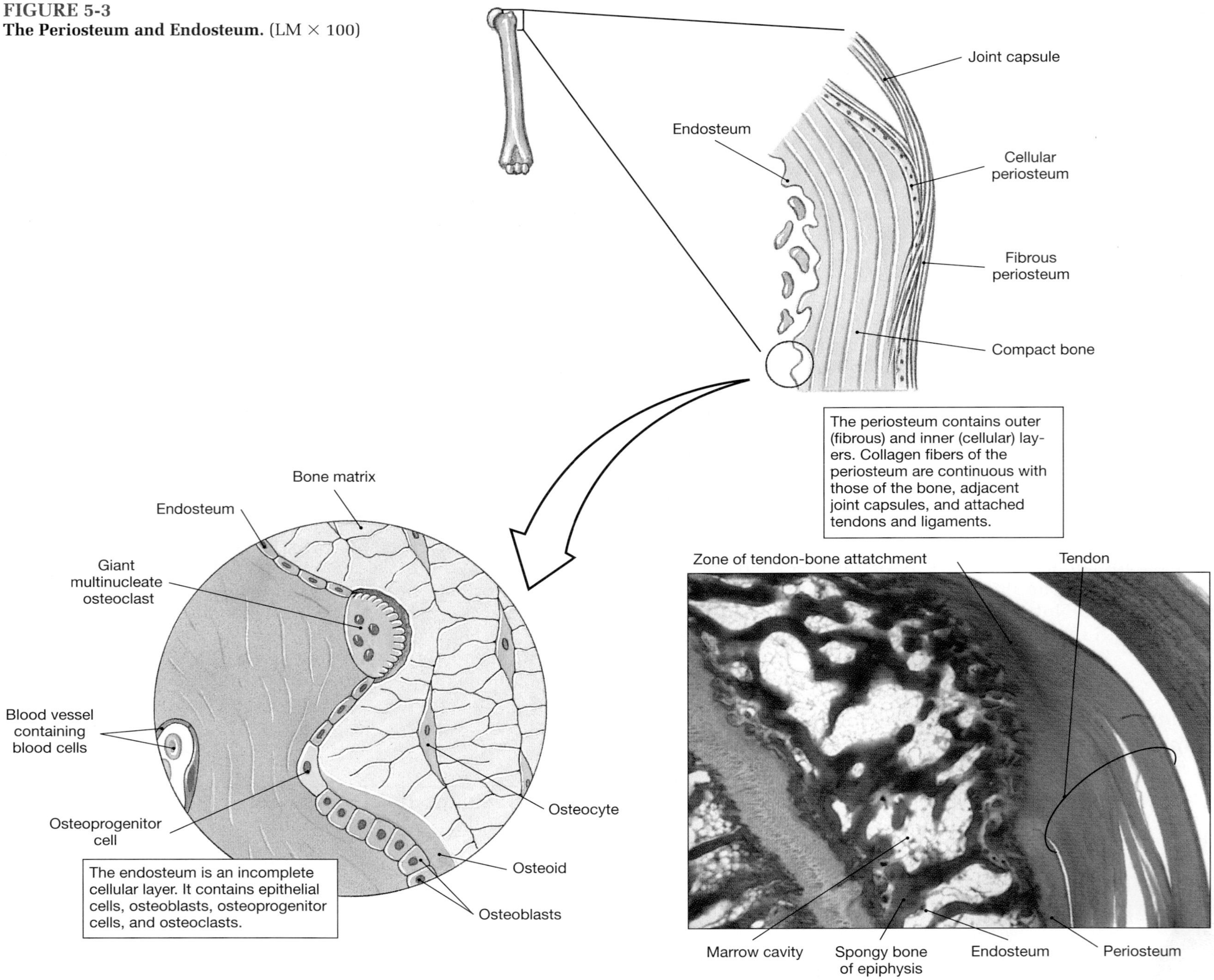

Joint capsule

Endosteum

Cellular periosteum

Fibrous periosteum

Compact bone

The periosteum contains outer (fibrous) and inner (cellular) layers. Collagen fibers of the periosteum are continuous with those of the bone, adjacent joint capsules, and attached tendons and ligaments.

Bone matrix

Endosteum

Giant multinucleate osteoclast

Blood vessel containing blood cells

Osteoprogenitor cell

Osteocyte

Osteoid

Osteoblasts

The endosteum is an incomplete cellular layer. It contains epithelial cells, osteoblasts, osteoprogenitor cells, and osteoclasts.

Zone of tendon-bone attatchment

Tendon

Marrow cavity

Spongy bone of epiphysis

Endosteum

Periosteum

and whenever repair or remodeling is under way. The endosteum is not a complete epithelial layer, and the matrix is occasionally exposed. At these exposed sites, osteoclasts and osteoblasts have access to the mineralized surfaces.

√ **How would the strength of a bone be affected if the ratio of collagen to calcium (hydroxyapatite) increased?**

√ **A sample of bone shows concentric lamellae surrounding a central canal. Is it from the cortex or the medullary cavity of a long bone?**

√ **If the activity of osteoclasts exceeds the activity of osteoblasts in a bone, how will the mass of the bone be affected?**

BONE DEVELOPMENT AND GROWTH

The growth of the skeleton determines the size and proportions of our body. The bony skeleton begins to form about 6 weeks after fertilization, when the embryo is approximately 12 mm (1/2 in.) long. (Before this time all of the skeletal elements are cartilaginous.) During subsequent development, the bones undergo a tremendous increase in size. Bone growth continues through adolescence, and portions of the skeleton usually do not stop growing until age 25. The entire process is carefully regulated, and a breakdown in regulation will ultimately affect all of the body systems. In this section we will consider the physical process of osteogenesis (bone formation) and bone growth. The next section will examine the maintenance and replacement of mineral reserves in the adult skeleton.

During development, mesenchyme or cartilage is replaced by bone. This process of replacing other tissues with bone is called **ossification**. Ossification refers specifically to the formation of bone. The process of **calcification** refers to the deposition of calcium salts within a tissue. There are two major forms of ossification. In *intramembranous ossification*, bone develops from mesenchyme or fibrous connective tissue. In *endochondral ossification*, bone replaces an existing cartilage model.

Intramembranous Ossification
(Figures 5-4/5-5)

Intramembranous (in-tra-MEM-bra-nus) **ossification**, also called *dermal ossification*, begins when osteoblasts differentiate within a mesenchymal or fibrous connective tissue. This type of ossification normally occurs in the deeper layers of the dermis, and the bones that result are often called **dermal bones.** Examples of dermal bones include the roofing bones of the skull, the mandible, and the clavicle. *Sesamoid bones* form within tendons; the *patella* (kneecap) is an example of a sesamoid bone. Intramembranous bone may also develop in other connective tissues subjected to chronic mechanical stresses. For example, cowboys in the nineteenth century sometimes developed small bony plates in the dermis on the insides of their thighs, from friction and impact against their saddles. In some disorders affecting calcium ion metablism or excretion, intramembranous

bone formation occurs in many areas of the dermis and deep fascia. Bones in abnormal locations are called *heterotopic bones* (*heteros*, different + *topos,* place).

The steps in the process of intramembranous ossification are illustrated in Figure 5-4● and may be summarized as:

STEP 1: Osteoblasts first cluster together and start to secrete the organic components of the matrix. The resulting mixture of dermal collagen fibers and osteoid then becomes mineralized through the crystallization of calcium salts. The location in a bone where ossification first occurs is called an **ossification center.** As ossification proceeds, it traps some osteoblasts inside bony pockets; these cells differentiate into osteocytes.

STEP 2: The developing bone grows outward from the ossification center in small struts, called **spicules.** Although osteoblasts are still being trapped in the expanding bone, mesenchymal cell divisions continue to produce additional osteoblasts. Bone growth is an active process, and osteoblasts require oxygen and a reliable supply of nutrients. Blood vessels that branch between the spicules meet these demands. As a result, the rate of bone growth actually accelerates.

STEP 3: Over time, the bone assumes the structure of spongy bone. Although initially the intramembranous bone resembles spongy bone, subsequent remodeling around the trapped blood vessels can produce compact bone.

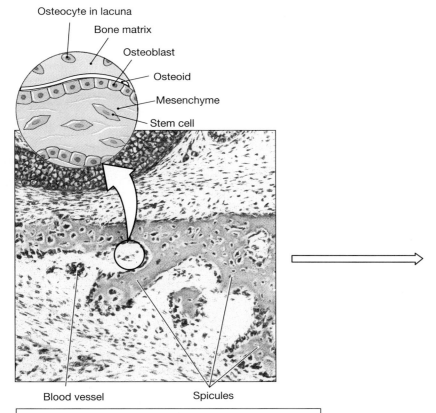

Step 1: Mesenchymal cells aggregate, differentiate, and begin the ossification process. The bone expands as a series of spicules that spread into surrounding tissues. (LM × 32)

FIGURE 5-4
A Three-Dimensional View of Intramembranous Ossification.
(Step 1, LM × 32; Step 2, LM × 32)

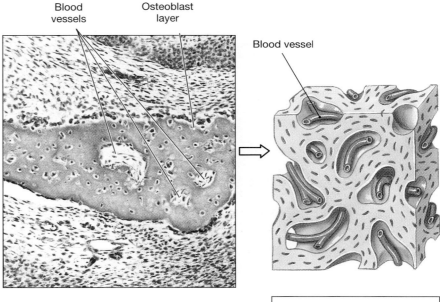

Step 2: As the spicules interconnect, they trap blood vessels within the bone. (LM × 32)

Step 3: Over time, the bone assumes the structure of spongy bone. Areas of spongy bone may later be removed, creating marrow cavities. Through remodeling, spongy bone formed in this way can be converted to compact bone.

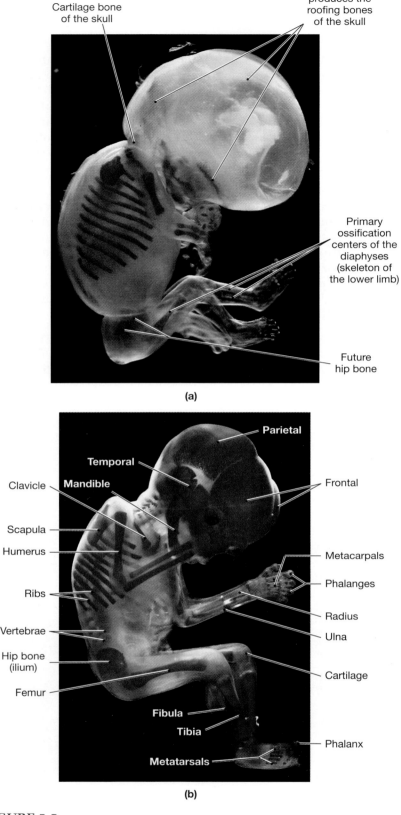

Cartilage bone of the skull

Intramembranous ossification produces the roofing bones of the skull

Primary ossification centers of the diaphyses (skeleton of the lower limb)

Future hip bone

(a)

Parietal

Temporal

Mandible

Clavicle

Scapula

Humerus

Ribs

Vertebrae

Hip bone (ilium)

Femur

Fibula

Tibia

Metatarsals

Frontal

Metacarpals

Phalanges

Radius

Ulna

Cartilage

Phalanx

(b)

FIGURE 5-5
Fetal Intramembranous and Endochondral Ossification. These 10- and 16-week human fetuses have been specially stained and cleared to show developing skeletal elements. (a) At 10 weeks the fetal skull clearly shows both membrane and cartilaginous bone, but the boundaries that indicate the limits of future skull bones have yet to be established. (b) At 16 weeks the fetal skull shows the irregular margins of the future skull bones. Most elements of the appendicular skeleton form through endochondral ossification. Note the appearance of the wrist and ankle bones at 16 weeks versus at 10 weeks.

Figure 5-5● shows skull bones forming through intramembranous ossification in the head of an embryo.

Endochondral Ossification
(Figures 5-5/5-6)

Endochondral ossification (en-dō-KON-dral; *endo*, inside + *chondros*, cartilage) begins with the formation of a hyaline cartilage model. Limb bone development is a good example of this process. By the time an embryo is 6 weeks old, the proximal bone of the limb, either the humerus or femur (thigh), is present, but it is composed entirely of cartilage. This model continues to grow by expansion of the cartilage matrix (interstitial growth) and the production of new cartilage at the outer surface (appositional growth). These growth mechanisms were introduced in Chapter 3. ∞ [p. 72] Steps in the growth and ossification of one of the limb bones are diagrammed in Figure 5-6●.

STEP 1: As the cartilage enlarges, chondrocytes near the center of the shaft increase greatly in size, and the surrounding matrix begins to calcify. Deprived of nutrients, these chondrocytes die and disintegrate.

STEP 2: Cells of the perichondrium surrounding this region of the cartilage develop into osteoblasts. The perichondrium has now been converted into a periosteum, and the inner **osteogenic layer** (os-tē-ō-JEN-ik) soon produces a thin layer of bone around the shaft of the cartilage.

STEP 3: While these changes are under way, the blood supply to the periosteum increases, and capillaries and osteoblasts migrate into the heart of the cartilage, invading the spaces left by the disintegrating chondrocytes. The calcified cartilaginous matrix breaks down, and osteoblasts replace it with spongy bone. Bone development proceeds from this **primary center of ossification**, located in the shaft, toward the ends of the cartilaginous model.

STEP 4: While the diameter is small, the entire diaphysis is filled with spongy bone, but as it enlarges, osteoclasts erode the central portion and create a marrow cavity. Further growth involves two distinct processes: an enlargement in *diameter* and an increase in *length*.

Figure 5-5● shows endochondral ossification occurring in the limb bones of an embryo.

Increasing the Diameter of a Developing Bone
(Figure 5-7)

The diameter of a bone enlarges through appositional growth at the outer surface. In this process, periosteal cells differentiate into osteoblasts and contribute to the growth of the bone matrix. Eventually they become surrounded by matrix and differentiate into osteocytes. As this process continues to add layers of bone tissue, blood vessels and collagen fibers of the periosteum become incorporated into the bony structure. Steps in appositional bone growth are described and illustrated in Figure 5-7●.

While bone matrix is being added to the outer surface of the growing bone, osteoclasts are removing bone matrix at the inner surface. As a result, the marrow cavity gradually enlarges as the bone increases in diameter.

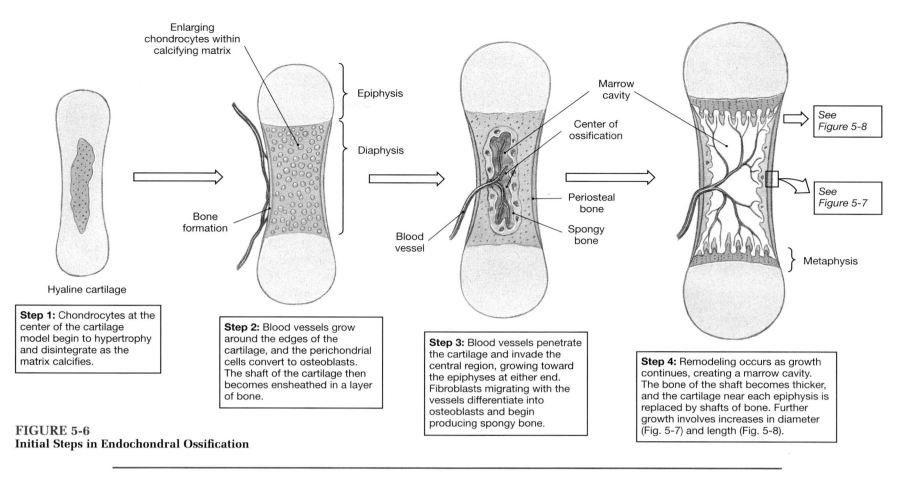

Enlarging chondrocytes within calcifying matrix

Epiphysis

Diaphysis

Bone formation

Hyaline cartilage

Step 1: Chondrocytes at the center of the cartilage model begin to hypertrophy and disintegrate as the matrix calcifies.

Step 2: Blood vessels grow around the edges of the cartilage, and the perichondrial cells convert to osteoblasts. The shaft of the cartilage then becomes ensheathed in a layer of bone.

Center of ossification

Blood vessel

Periosteal bone

Spongy bone

Step 3: Blood vessels penetrate the cartilage and invade the central region, growing toward the epiphyses at either end. Fibroblasts migrating with the vessels differentiate into osteoblasts and begin producing spongy bone.

Marrow cavity

See Figure 5-8

See Figure 5-7

Metaphysis

Step 4: Remodeling occurs as growth continues, creating a marrow cavity. The bone of the shaft becomes thicker, and the cartilage near each epiphysis is replaced by shafts of bone. Further growth involves increases in diameter (Fig. 5-7) and length (Fig. 5-8).

FIGURE 5-6
Initial Steps in Endochondral Ossification

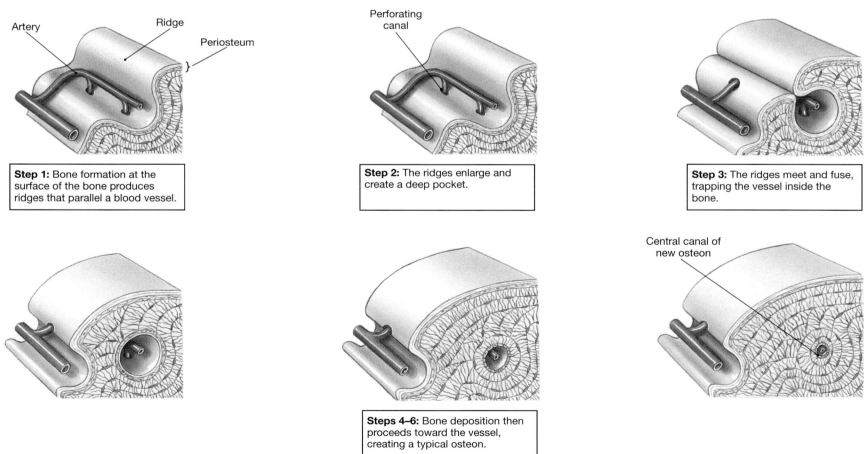

Artery

Ridge

Periosteum

Perforating canal

Step 1: Bone formation at the surface of the bone produces ridges that parallel a blood vessel.

Step 2: The ridges enlarge and create a deep pocket.

Step 3: The ridges meet and fuse, trapping the vessel inside the bone.

Central canal of new osteon

Steps 4–6: Bone deposition then proceeds toward the vessel, creating a typical osteon.

FIGURE 5-7
Appositional Bone Growth. Three-dimensional diagrams illustrating the mechanism responsible for increasing the diameter of a growing bone.

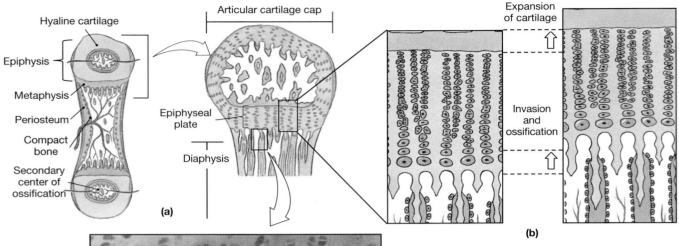

Capillaries and osteoblasts migrate into the epiphysis, creating a secondary ossification center.

Soon the epiphysis is filled with spongy bone. A thin cap of articular cartilage remains exposed to the joint cavity, and at the metaphysis an epiphyseal plate separates the epiphysis from the diaphysis.

Within the epiphyseal plate, chondrocytes near the epiphysis undergo division, thickening the plate and forcing the epiphysis farther from the shaft. As the daughter cells mature, they become enlarged, and the surrounding matrix becomes calcified. On the other side of the epiphyseal plate, osteoblasts and capillaries invade the lacunae and replace the dead cartilage with living bone. As long as the rate of cartilage growth keeps pace with the rate of osteoblast invasion, the shaft grows longer but the epiphyseal plate survives.

Hyaline cartilage

Epiphysis

Metaphysis

Periosteum

Compact bone

Secondary center of ossification

(a)

Articular cartilage cap

Epiphyseal plate

Diaphysis

Expansion of cartilage

Invasion and ossification

(b)

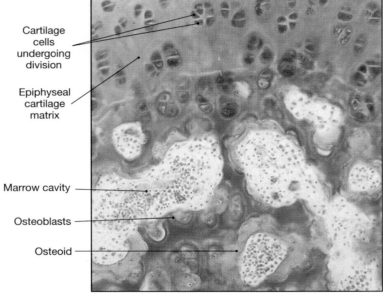

Cartilage cells undergoing division

Epiphyseal cartilage matrix

Marrow cavity

Osteoblasts

Osteoid

(c)

FIGURE 5-8
Bone Growth at the Epiphyseal Plate. (a), (b) Diagrammatic summary of growth mechanisms at the epiphyseal plate. (c) Light micrograph showing the degenerating cartilage and the advancing osteoblasts.

Increasing the Length of a Developing Bone
(Figures 5-8/5-9)

Growth in bone length involves a very different process, summarized and illustrated in Figure 5-8●. During the initial stages of osteogenesis, osteoblasts move away from the center of ossification toward the epiphyses. But they do not manage to complete the ossification of the model immediately, because the epiphyseal cartilages continue to enlarge.

STEP 1: The region where the cartilage is being replaced by bone lies at the junction between the diaphysis (shaft) and epiphyses (heads) of the bone (Figure 5-8a●). This region is known as the **metaphysis** (me-TAF-i-sis). On the shaft side of the metaphysis, osteoblasts are continually invading the cartilage and converting it to bone. But on the epiphyseal side, new cartilage is produced at the same rate. As a result, the osteoblasts never quite catch up with the epiphysis, although the skeletal element continues to enlarge.

STEP 2: Around the time of birth, the centers of some epiphyses begin to calcify. Capillaries and osteoblasts migrate into these areas, creating **secondary ossification centers**. Soon the epiphyses are filled with spongy bone. A thin cap of the original cartilage model remains exposed to the joint cavity as the **articular cartilage.** This cartilage prevents damaging bone-to-bone contact within the joint. At the metaphysis, a relatively narrow cartilaginous **epiphyseal plate** separates the epiphysis from the diaphysis.

STEP 3: Within the epiphyseal plate, the chondrocytes near the epiphysis continue to enlarge and divide, adding to the thickness of the plate. The light micrograph in Figure 5-8c● shows the interface between the degenerating cartilage and the advancing osteoblasts. The continual expansion of the epiphyseal cartilage forces the epiphysis farther from the shaft (Figure 5-8b●). As the daughter cells mature, they become enlarged, and the surrounding matrix becomes calcified. On the shaft side of the epiphyseal plate, osteoblasts and capillaries continue to invade these lacunae and replace the dead cartilage with living bone. As long as the rate of cartilage growth keeps pace with the rate of osteoblast invasion, the shaft grows longer but the epiphyseal plate survives. Figure 5-9a● shows an X-ray of epiphyseal plates in the hand of a young child. ♐*Inherited Abnormalities in Skeletal Development [p. 754]*

FIGURE 5-9
Epiphyseal Plates and Lines. (a) X-ray of the hand of a young child. The arrows indicate the locations of the epiphyseal plates. (b) X-ray of the hand of an adult. The arrows indicate the locations of epiphyseal lines.

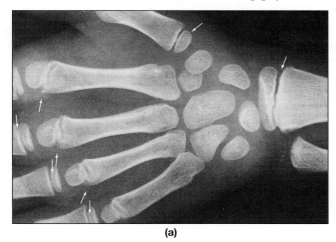

(a)

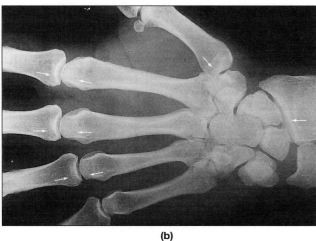

(b)

FIGURE 5-10
Circulatory Supply to a Mature Bone

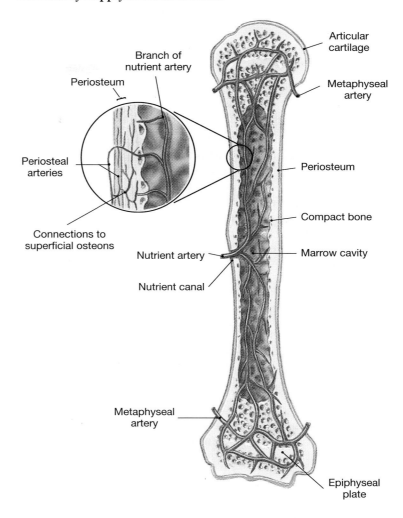

At maturity, the rate of epiphyseal cartilage production slows, and the rate of osteoblast activity accelerates. As a result, the epiphyseal plate gets narrower and narrower, until it ultimately disappears. The former location of the plate can still be detected in X-rays as a distinct **epiphyseal line** that remains after epiphyseal growth has ended. ✝ *Hyperostosis [p. 755]*

Formation of the Blood and Lymphatic Supply
(Figures 5-7/5-10)

Osseous tissue is very vascular, and the bones of the skeleton have an extensive blood supply. In a typical bone such as the humerus, three major sets of blood vessels develop (Figure 5-10●).

1. **The nutrient artery and vein:** These vessels form as blood vessels invade the cartilage model at the start of endochondral ossification. There is usually only one **nutrient artery** and one **nutrient vein** entering the diaphysis, although a few bones, including the femur, have more than one pair. Branches of these vessels extend along the length of the shaft and into the osteons of the surrounding cortex.
2. **Metaphyseal vessels:** These vessels supply blood to the inner (diaphyseal) surface of each epiphyseal plate, where bone is replacing the cartilage.

3. **Periosteal vessels:** Blood vessels from the periosteum are incorporated into the developing bone surface as described and illustrated in Figure 5-7●. These vessels provide blood to the superficial osteons of the shaft. During endochondral bone formation, branches of periosteal vessels also enter the epiphyses, providing blood to the secondary ossification centers. The periosteum also contains an extensive network of lymphatic vessels, and many of these have branches that enter the bone and reach individual osteons via the perforating canals.

Following the closure of the epiphyses, all three sets of vessels become extensively interconnected.

Bone Innervation

Bones are innervated by sensory nerves, and injuries to the skeleton can be very painful. Sensory nerve endings branch throughout the periosteum, and sensory nerves penetrate the cortex with the nutrient artery to innervate the endosteum, marrow cavity, and epiphyses.

√ **How could X-rays of the femur be used to determine whether a person had reached full height?**

Factors Regulating Bone Growth

Normal bone growth depends on a combination of nutritional and hormonal factors:

- Normal bone growth cannot occur without a constant dietary source of calcium and phosphate salts.

- **Vitamins A** and **C** are essential for normal bone growth and remodeling. These vitamins must be obtained from the diet.

- The group of related steroids collectively known as **vitamin D** plays an important role in normal calcium metabolism by stimulating the absorption and transport of calcium and phosphate ions into the blood. The active form of vitamin D, *Calcitriol*, is synthesized in the kidneys; this process ultimately depends on the availability of a related steroid that may be absorbed from the diet or synthesized in the skin in the presence of UV radiation. ∞ [p. 93] ✝ *Osteomalacia [p. 755]*

Hormones regulate the pattern of growth by changing the rates of osteoblast and osteoclast activity.

- The parathyroid glands release **parathyroid hormone**, which stimulates osteoclast activity, increases the rate of calcium absorption along the small intestine, and reduces the rate of calcium loss in the urine.

- The thyroid glands of children and pregnant women secrete the hormone **calcitonin** (kal-si-TO-nin), which inhibits osteoclasts and increases the rate of calcium loss in the urine. Calcitonin is of uncertain significance in the healthy nonpregnant adult.

- **Growth hormone**, produced by the pituitary gland, and **thyroxine**, from the thyroid gland, stimulate bone growth. In proper balance, these hormones maintain normal activity at the epiphyseal plates until roughly the time of puberty.

- At puberty, bone growth accelerates dramatically. The **sex hormones** (*estrogen* and *testosterone*) stimulate osteoblasts to produce bone faster than the rate of epiphyseal cartilage expansion. Over time, the epiphyseal plates narrow and eventually ossify or "close."

There are differences from bone to bone and individual to individual as to the timing of epiphyseal plate closure. The toes may complete their ossification by age 11, whereas portions of the pelvis or the wrist may continue to enlarge until age 25. Differences in the male and female sex hormones account for the variation between the sexes and for related variations in body size and proportions.

■ CLINICAL BRIEF
Abnormal Growth Hormone Production

Excessive or inadequate hormone production has a pronounced effect on activity at epiphyseal plates. **Gigantism** results from an overproduction of growth hormone before puberty (the world record is 8 feet 11 inches and a weight of 475 pounds). The opposite extreme is **pituitary growth failure** (*pituitary dwarfism*), in which inadequate growth hormone production leads to reduced epiphyseal plate activity and abnormally short bones. This form of growth failure is becoming rare in the United States because children can be treated with human growth hormone.

BONE REMODELING

Bone remodeling may involve a change in the shape or internal architecture of a bone or a change in the total amount of minerals deposited in the skeleton. In the adult, osteocytes are continually removing and replacing the surrounding calcium salts. But osteoclasts and osteoblasts also remain active, even after the epiphyseal plates have closed. Normally their activities are balanced. As one osteon forms through the activity of osteoblasts, another is destroyed by osteoclasts. The turnover rate for bone is quite high. Each year almost one-fifth of the adult skeleton is demolished and then rebuilt or replaced. Every part of every bone may not be affected, as there are regional and even local differences in the rate of turnover. For example, the spongy bone in the head of the femur may be replaced two or three times each year, whereas the compact bone along the shaft remains largely untouched.

Changes in Bone Shape

The turnover and recycling of minerals give each bone the ability to adapt to new stresses. Osteoblast sensitivity to electrical events has been theorized as the mechanism that controls the internal organization and structure of bone. Whenever a bone is stressed, the mineral crystals generate minute electrical fields. Osteoblasts are apparently attracted to these electrical fields, and once in the area they begin to produce bone. (Electrical fields may also be used to stimulate the repair of severe fractures.)

Because bones are adaptable, their shapes reflect the forces applied to them. For example, bumps and ridges on the surface of a bone mark the sites where tendons attach to the bone. If muscles become more powerful, the corresponding bumps and ridges enlarge to withstand the increased forces. Heavily stressed bones become thicker and stronger, whereas bones not subjected to ordinary stresses will become thin and brittle. Regular exercise is therefore important as a stimulus that maintains normal bone structure.

Degenerative changes in the skeleton occur after relatively brief periods of inactivity. For example, using a crutch while wearing a cast takes weight off the injured limb. After a few weeks, the unstressed bones will lose up to about a third of their mass. However, the bones rebuild just as quickly when normal loading resumes.

Injury and Repair
(Figure 5-11)

Despite its mineral strength, bone may crack or even break if subjected to extreme loads, sudden impacts, or stresses from unusual directions. The damage produced constitutes a **fracture**. Healing of a fracture usually occurs even after severe damage, provided the blood supply and the cellular components of the endosteum and periosteum survive. Steps in the repair of a fracture are illustrated in Figure 5-11●. The final repair will be slightly thicker and probably stronger than the original bone; under comparable stresses, a second fracture will usually occur at a different site.

Aging and the Skeletal System

The bones of the skeleton become thinner and relatively weaker as a normal part of the aging process. Inadequate ossification is called **osteopenia** (os-tē-ō-PĒ-nē-a; *penia*, lacking), and everyone becomes slightly osteopenic as they age. This reduction in bone mass occurs between the ages of 30 and 40. Over this period, osteoblast activity begins to decline while osteoclast activity continues at previous levels. Once the reduction begins, women lose roughly 8% of their skeletal mass every decade, the skeletons of men deteriorate at the slower rate of about 3% per decade. All parts of the skeleton are not equally affected. Epiphyses, vertebrae, and the jaws lose more than their fair share, resulting in fragile limbs, a reduction in height, and the loss of teeth. A significant percentage of older women and a smaller proportion of older men suffer from **osteoporosis** (os-tē-ō-por-Ō-sis; *porosus*, porous). This condition is characterized by a reduction in bone mass sufficient to compromise normal function. Osteoporosis is discussed further in the Clinical Issues appendix. ⚕ *Osteoporosis and Other Skeletal Abnormalities Associated with Aging [p. 755]*

√ **Would you expect to see any difference in the bones of an athlete before and after the addition of muscle mass? Why or why not?**

FIGURE 5-11
Steps in the Repair of a Fracture

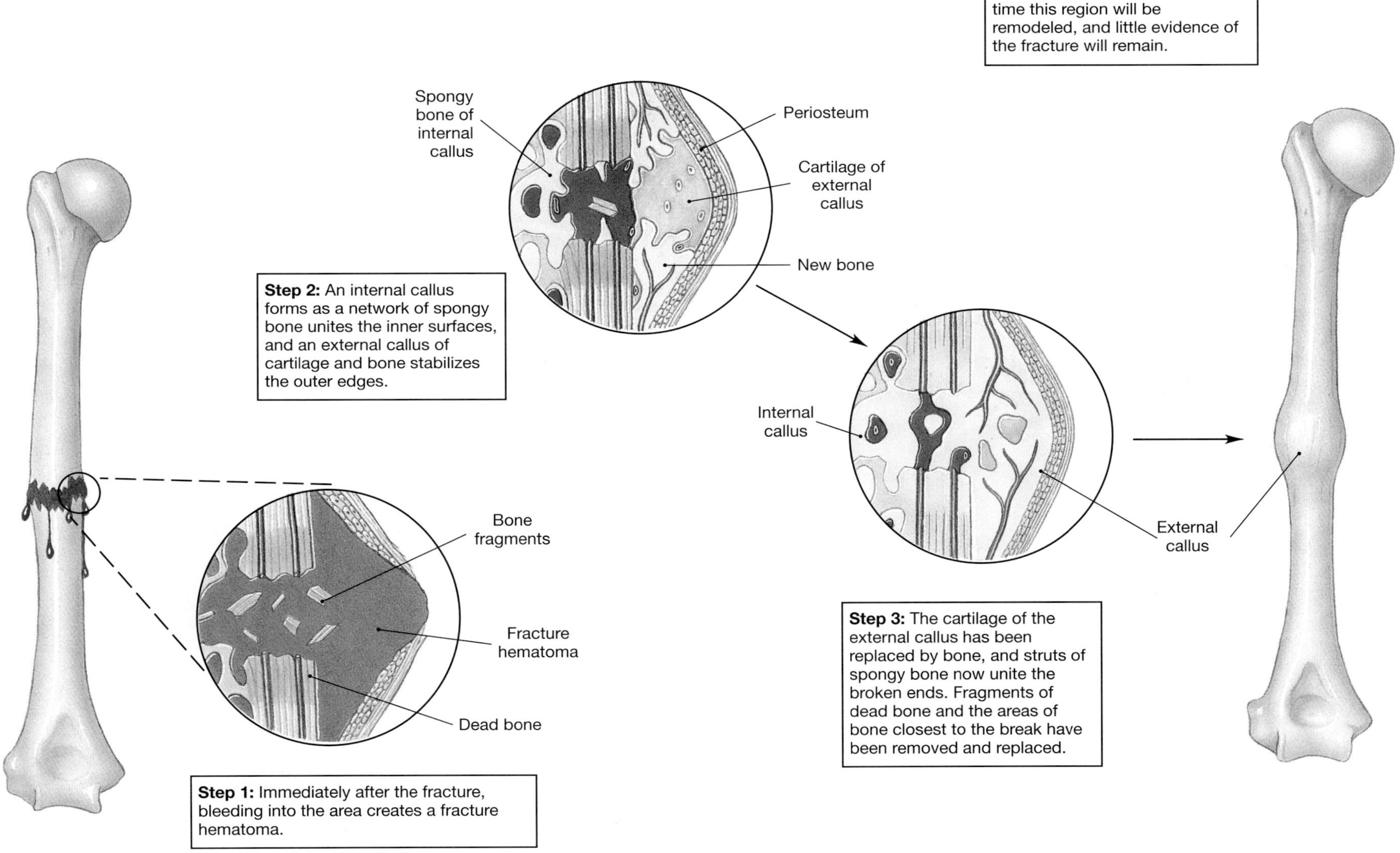

Step 4: A swelling initially marks the location of the fracture. Over time this region will be remodeled, and little evidence of the fracture will remain.

Spongy bone of internal callus

Periosteum

Cartilage of external callus

New bone

Step 2: An internal callus forms as a network of spongy bone unites the inner surfaces, and an external callus of cartilage and bone stabilizes the outer edges.

Internal callus

External callus

Bone fragments

Fracture hematoma

Dead bone

Step 3: The cartilage of the external callus has been replaced by bone, and struts of spongy bone now unite the broken ends. Fragments of dead bone and the areas of bone closest to the break have been removed and replaced.

Step 1: Immediately after the fracture, bleeding into the area creates a fracture hematoma.

Fractures are classified according to their external appearance, the site of the fracture, and the nature of the crack or break in the bone. Important fracture types are indicated below, with representative X-rays. Many fractures fall into more than one category. For example, a Colles' fracture is a transverse fracture, but depending on the injury, it may also be a comminuted fracture that can be either open or closed. **Closed,** or *simple,* fractures are completely internal; they do not involve a break in the skin. **Open,** or *compound,* fractures project through the skin; they are more dangerous because of the possibility of infection or uncontrolled bleeding.

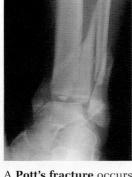

A **Pott's fracture** occurs at the ankle and affects both bones of the lower leg.

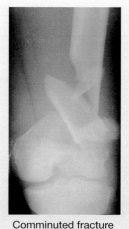

Comminuted fracture of distal femur

Comminuted fractures shatter the affected area into a multitude of bony fragments.

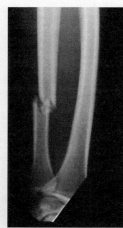

Transverse fractures break a shaft bone across its long axis.

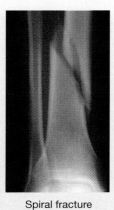

Spiral fracture of tibia

Spiral fractures, produced by twisting stresses, spread along the length of the bone.

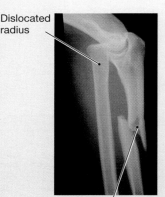

Dislocated radius

Displaced ulnar fracture

Displaced fractures produce new and abnormal arrangements of bony elements.

Nondisplaced fractures retain the normal alignment of the bone elements or fragments.

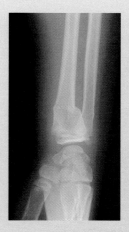

A **Colles' fracture** is a break in the distal portion of the radius, the slender bone of the forearm; it is often the result of reaching out to cushion a fall.

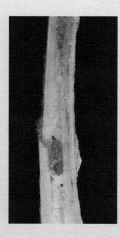

In a **greenstick fracture,** only one side of the shaft is broken, and the other is bent; this type usually occurs in children, whose long bones have yet to fully ossify.

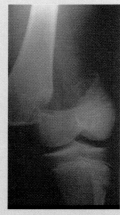

Epiphyseal fracture of femur

Epiphyseal fractures usually occur where the matrix is undergoing calcification and chondrocytes are dying. A clean transverse fracture along this line usually heals well. Fractures between the epiphysis and the epiphyseal plate can permanently halt further longitudinal growth unless carefully treated; often surgery is required.

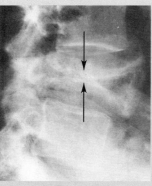

Compression fracture of vertebra

Compression fractures occur in vertebrae subjected to extreme stresses, as when landing on your seat after a fall.

ANATOMY OF SKELETAL ELEMENTS

The human skeleton contains more than 206 bones. We can divide these bones into six broad categories according to their individual shapes.

Classification of Bones
(Figure 5-12)

Refer to Figure 5-12● as we describe the anatomical classification of bones.

1. **Long bones** are characterized by diaphysis and epiphysis portions and contain a marrow cavity (Figure 5-12a●). Long bones are found in the upper arm and forearm, thigh and lower leg, palms, soles, fingers, and toes.
2. **Short bones** are boxlike in appearance (Figure 5-12b●). Their external surfaces are covered by compact bone, but the interior contains spongy bone. Examples of short bones include the carpals (wrists) and tarsals (ankles).
3. **Flat bones** have thin, roughly parallel surfaces of compact bone. In structure a flat bone resembles a spongy bone sandwich; such bones are strong but relatively light.

Flat bones form the roof of the skull (Figure 5-12c●), the sternum, the ribs, and the scapulae. They provide protection for underlying soft tissues and offer an extensive surface area for the attachment of skeletal muscles. Special terms are used when describing the flat bones of the skull, such as the parietals. Their relatively thick layers of compact bone are called the *internal* and *external* **tables**, and the layer of spongy bone between the tables is called the **diploë** (DIP-lō-ē).

4. **Irregular bones** have complex shapes with short, flat, notched, or ridged surfaces (Figure 5-12d●). Their internal structure is equally varied. The vertebrae that form the spinal column and several bones in the skull are examples of irregular bones.
5. **Sesamoid bones** are usually small, round, and flat (Figure 5-12e●). They develop inside tendons and are most often encountered near joints at the knee, the hands, and the feet. Few individuals have sesamoid bones at every possible location, but everyone has sesamoid **patellae** (pa-TEL-ē), or kneecaps.
6. **Sutural (Wormian) bones** are small, flat, oddly shaped bones found between the flat bones of the skull in the suture line (Figure 5-12f●). There are individual variations in the number, shape, and position of the sutural bones. Their borders are like puzzle pieces and may range in size from a grain of sand to the size of a quarter.

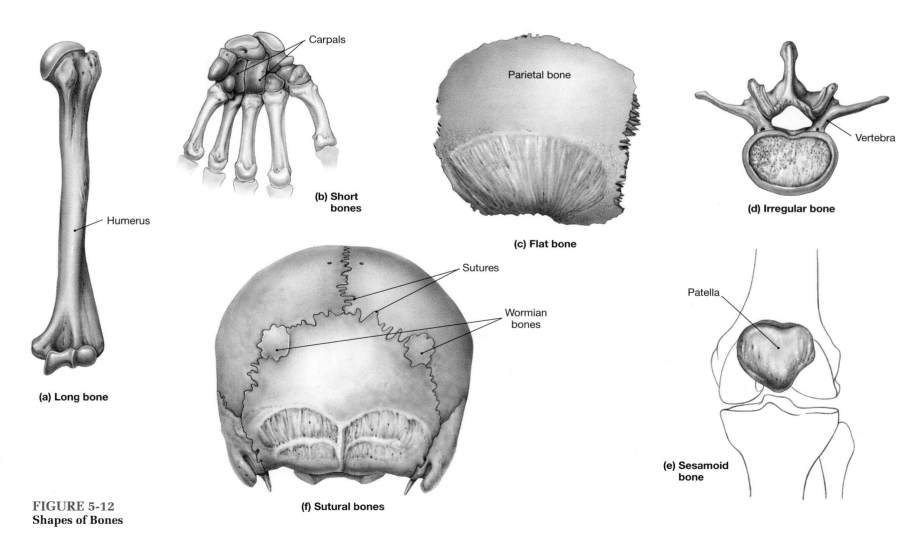

(a) Long bone — Humerus

(b) Short bones — Carpals

(c) Flat bone — Parietal bone

(d) Irregular bone — Vertebra

(e) Sesamoid bone — Patella

(f) Sutural bones — Sutures, Wormian bones

FIGURE 5-12
Shapes of Bones

Bone Markings (Surface Features)
(Figure 5-13)

Each bone in the body has a distinctive shape and characteristic external and internal features. Elevations or projections form where tendons and ligaments attach and where adjacent bones articulate. Depressions, grooves, and tunnels in bone indicate sites where blood vessels and nerves lie alongside or penetrate the bone. Detailed examination of these **bone markings** or *surface features* can yield an abundance of anatomical information. For example, anthropologists, criminologists, and pathologists can often determine the size, weight, sex, and general appearance of an individual on the basis of incomplete skeletal remains. (This topic will be discussed further in Chapter 6.)

We will ignore minor variations of individual bones and focus on prominent features that identify the bone. These markings are useful because they provide fixed landmarks that can help in determining the position of the soft tissue components of other systems. Specific anatomical terms are used to describe the various elevations and depressions. Bone marking terminology is presented in Table 5-1 and illustrated in Figure 5-13●.

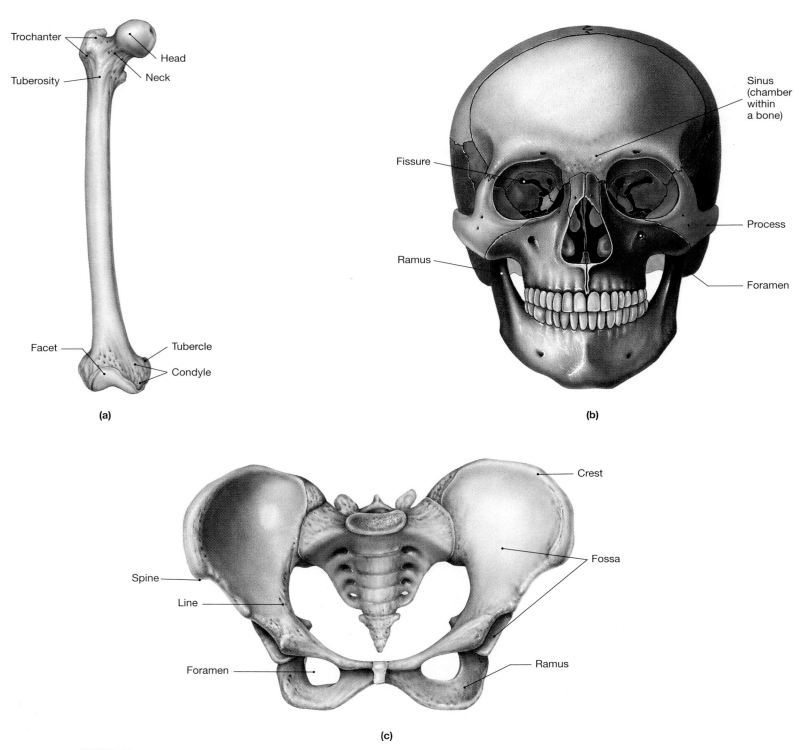

FIGURE 5-13
Examples of Bone Markings (Surface Features). (a) Femur, anterior view. (b) Skull, anterior view. (c) Hip, anterior view.

TABLE 5-1 Common Bone Marking Terminology

General Description	Anatomical Term	Definition	Example
Elevations and projections (general)	Process	Any projection or bump	See Figure 5-13b, c
	Ramus	An extension of a bone making an angle to the rest of the structure	
Processes formed where tendons or ligaments attach	Trochanter	A large, rough projection	See Figure 5-13a, c
	Tuberosity	A smaller, rough projection	
	Tubercle	A small, rounded projection	
	Crest	A prominent ridge	
	Line	A low ridge	
	Spine	A pointed process	
Processes formed for articulation with adjacent bones	Head	The expanded articular end of an epiphysis, separated from the shaft by a narrower **neck**	See Figure 5-13a, c
	Condyle	A smooth, rounded or articular process	
	Trochlea	A smooth, grooved articular process shaped like a pulley	
	Facet	A small, flat articular surface	
Depressions	Fossa	A shallow depression	See Figure 5-13a, c
Openings	Foramen	A rounded passageway for blood vessels and/or nerves	See Figure 5-13b
	Fissure	An elongate cleft	
	Meatus	The entrance to a canal leading through the substance of a bone	
	Sinus or antrum	A chamber within a bone, normally filled with air	

INTEGRATION WITH OTHER SYSTEMS

Although the bones may seem inert, you should now realize that they are quite dynamic structures. The entire skeletal system is intimately associated with other systems. Bones are attached to the muscular system, extensively connected to the cardiovascular and lymphatic systems, and largely under the physiological control of the endocrine system. Also the digestive and excretory systems play important roles in providing the calcium and phosphate minerals needed for bone growth. In return, the skeleton represents a reserve of calcium, phosphate, and other minerals that can compensate for changes in the dietary supply of these ions.

Related Clinical Terms

rickets: A disorder that reduces the amount of calcium salts in the skeleton; often characterized by a "bowlegged" appearance. [p. 113]

osteogenesis imperfecta (im-per-FEK-ta): An inherited condition affecting the organization of collagen fibers. Osteoblast function is impaired, growth is abnormal, and the bones are very fragile, leading to progressive skeletal deformation and repeated fractures. ⚕ *Inherited Abnormalities in Skeletal Development [p. 754]*

Marfan's syndrome: An inherited condition linked to defective production of a connective tissue glycoprotein. Extreme height and long, slender limbs are the most obvious physical indications of this disorder. The abnormal proportions result from excessive cartilage formation at epiphyseal plates. ⚕ *Inherited Abnormalities in Skeletal Development [p. 754]*

achondroplasia (a-kon-drō-PLĀ-sē-a): A condition resulting from abnormal epiphyseal plate activity; the epiphyseal plates grow unusually slowly, and the individual develops short, stocky limbs. The trunk is normal in size, and sexual and mental development remain unaffected. ⚕ *Inherited Abnormalities in Skeletal Development [p. 754]*

hyperostosis (hī-per-os-TŌ-sis): The excessive formation of bone tissue. ⚕ *Hyperostosis [p. 755]*

osteopetrosis (os-tē-ō-pe-TRŌ-sis): A condition caused by a decrease in osteoclast activity, causing increased bone mass and various skeletal deformities. ⚕ *Hyperostosis [p. 755]*

acromegaly: A condition caused by excessive secretion of growth hormone after puberty. Skeletal abnormalities develop, affecting the cartilages and various small bones. ⚕ *Hyperostosis [p. 755]*

osteomalacia (os-tē-ō-ma-LĀ-shē-ah): A softening of bone due to a decrease in the mineral content. ⚕ *Osteomalacia [p. 755]*

gigantism: A condition resulting from an overproduction of growth hormone before puberty. [p. 120]

pituitary growth failure: A type of dwarfism caused by inadequate growth hormone production. [p. 120]

fracture: A crack or break in a bone. [p. 120]

osteopenia (os-tē-ō-PĒ-nē-a): Inadequate ossification, leading to thinner, weaker bones. [p. 121]

osteoporosis (os-tē-ō-por-Ō-sis): A reduction in bone mass to a degree that compromises normal function. †*Osteoporosis and Other Skeletal Abnormalities Associated with Aging [p. 755]*

osteoclast-activating factor: A compound released by cancers of the bone marrow, breast, or other tissues. It produces a severe osteoporosis. †*Osteoporosis and Other Skeletal Abnormalities Associated with Aging [p. 755]*

osteomyelitis (os-tē-ō-mī-e-LĪ-tis): A painful infection in a bone, usually caused by bacteria. †*Osteoporosis and Other Skeletal Abnormalities Associated with Aging [p. 755]*

Paget's disease (osteitis deformans) (os-tē-Ī-tis de-FOR-mans): A condition characterized by gradual deformation of the skeleton.

†*Osteoporosis and Other Skeletal Abnormalities Associated with Aging [p. 755]*

external callus: A toughened layer of connective tissue that encircles and stabilizes a bone at a fracture site. *[p. 121]*

internal callus: A bridgework of trabecular bone that unites the broken ends of a bone on the marrow side of the fracture. *[p. 121]*

fracture hematoma: A large blood clot that closes off the injured vessels and leaves a fibrous meshwork in the damaged area. *[p. 121]*

CHAPTER SUMMARY AND REVIEW

STUDY OUTLINE

Related Key Terms

INTRODUCTION *[p. 110]*

1. The skeletal system includes the bones of the skeleton and the cartilages, ligaments, and other connective tissues that stabilize or interconnect bones. Its functions include: structural support, storage, blood cell production, protection, and leverage.

STRUCTURE OF BONE *[p. 110]*

1. **Osseous tissue** is a supporting connective tissue with a solid **matrix**.

Histological Organization *[p. 110]*

2. Crystals of calcium phosphate, $Ca_3(PO_4)_2$, account for almost two-thirds of the weight of bone. The remaining third is dominated by collagen fibers and small amounts of other calcium salts; cells contribute only around 2% to the volume of bone tissue.   **hydroxyapatite**

3. **Osteocytes** are mature bone cells surrounded by bone matrix. They reside in **lacunae** interconnected by **canaliculi**. **Lamellae** are layers of calcified matrix *(see Figure 5-1)*.

4. **Osteoclasts** dissolve the bony matrix through **osteolysis**. Other cells called **osteoblasts** synthesize the matrix in the process of **osteogenesis**. **Osteoprogenitor cells** are mesenchymal cells that play a role in fracture repair *(see Figure 5-1)*.   **osteoid**

Compact and Spongy Bone *[p. 112]*

5. There are two types of bone: **dense**, or **compact**, **bone** and **spongy bone**. The matrix composition in compact bone is the same as that of spongy bone, but they differ in the three-dimensional arrangement of osteocytes, canaliculi, and lamellae *(see Figures 5-1d/5-2)*.   **marrow cavity • bone marrow • yellow marrow • red marrow**

6. The basic functional unit of compact bone is the **osteon**, containing osteocytes arranged around a **central canal** *(see Figure 5-1d)*.   **perforating canals**

7. Spongy bone contains **trabeculae**, often in an open network *(see Figure 5-2)*.

8. Compact bone is found where stresses come from a limited range of directions; spongy bone is located where stresses are few or come from many different directions.   **cortex • medullary cavity • shaft diaphysis • epiphyses**

The Periosteum and Endosteum *[p. 113]*

9. A bone is covered by a **periosteum** and lined internally by an **endosteum** *(see Figures 5-1d/5-3)*.

BONE DEVELOPMENT AND GROWTH *[p. 114]*

1. **Ossification** is the process of converting other tissues to bone; **calcification** is the process of depositing calcium salts within a tissue.

Intramembranous Ossification *[p. 115]*

2. **Intramembranous ossification** begins when osteoblasts differentiate within fibrous connective tissue. This process can ultimately produce spongy or compact bone *(see Figures 5-4/5-5)*.   **dermal bones • ossification center • spicules**

Endochondral Ossification *[p. 116]*

3. **Endochondral ossification** begins by forming a cartilaginous model that is gradually replaced by bone *(see Figures 5-5/5-6)*.   **osteogenic layer • primary center of ossification**

4. The diameter of a bone enlarges through appositional growth at the outer surface. For the steps in appositional bone growth, see *Figure 5-7*.   **metaphysis • secondary ossification centers • articular cartilage • epiphyseal plate • epiphyseal line**

Formation of the Blood and Lymphatic Supply *[p. 119]*

5. A typical bone formed through endochondral ossification has three blood supplies: the **nutrient artery** and **vein**, **metaphyseal vessels**, and **periosteal vessels**. Lymphatic vessels are found in the periosteum and enter the osteons via the perforating canals *(see Figures 5-7/5-10)*.

Bone Innervation *[p. 119]*

6. Sensory nerve endings branch throughout the periosteum, and sensory nerves penetrate the cortex with the nutrient artery to innervate the endosteum, marrow cavity, and epiphyses.

Factors Regulating Bone Growth *[p. 120]*

7. Normal osteogenesis requires a reliable source of minerals, vitamins, and hormones.

8. Calcitonin, secreted by the thyroid gland, stimulates osteoblast activity, and **parathyroid hormone,** secreted by the parathyroid glands, stimulates osteoclasts. These hormones control the rate of mineral deposition in the skeleton and regulate the calcium ion concentrations in body fluids.

9. Growth hormone, **thyroxine**, and **sex hormones** stimulate bone growth.

10. There are differences between bones and between individuals regarding the timing of epiphyseal closure.

BONE REMODELING *[p. 120]*

1. The turnover rate for bone is quite high. Each year almost one-fifth of the adult skeleton is demolished and then rebuilt or replaced.

Changes in Bone Shape *[p. 120]*

2. Mineral turnover allows bone to adapt to new stresses.

3. Calcium is the most common mineral in the human body, with more than 98% of it located in the skeleton.

Injury and Repair *[p. 120]*

4. A **fracture** is a crack or break in a bone. Healing can usually occur if portions of the blood supply, endosteum, and periosteum remain intact *(see Figure 5-11)*. For a classification of fracture types, *see Clinical Brief on page 122*.

Aging and the Skeletal System *[p. 121]*

5. The bones of the skeleton become thinner and relatively weaker as a normal part of the aging process. This **osteopenia** affects everyone to some degree.

ANATOMY OF SKELETAL ELEMENTS *[p. 123]*

Classification of Bones *[p. 123]*

1. Categories of bones include: **long bones**, **short bones**, **flat bones**, **irregular bones**, **sesamoid bones**, and **sutural bones (Wormian bones)** *(see Figure 5-12)*.

Bone Markings (Surface Features) *[p. 124]*

2. Bone markings (or *surface features*) can be used to identify specific elevations, depressions, and openings of bones *(see Figure 5-13)*. Common bone marking terminology is presented in *Table 5-1*.

INTEGRATION WITH OTHER SYSTEMS *[p. 125]*

1. The skeletal system is anatomically and physiologically linked to other body systems. The skeleton represents a reservoir for calcium, phosphate, and other minerals that under the influence of the endocrine system can be supplied to meet the body's needs.

Related Key Terms

vitamin A • vitamin C • vitamin D

tables • diploë • patellae

---

1   REVIEW OF CHAPTER OBJECTIVES

1. Describe the functions of the skeletal system.
2. Compare the structures and functions of compact and spongy bone.
3. Discuss the processes by which bones develop and grow and account for variations in their internal structure.
4. Describe the remodeling of the skeleton.
5. Discuss the effects of nutrition, hormones, exercise, and aging on bone development and the skeletal system.
6. Describe the different types of fractures and explain how fractures heal.
7. Classify bones according to their shapes and give examples for each type.

2   REVIEW OF CONCEPTS

1. What bone cell type(s) is(are) involved in the healing of a fracture in a growing teenager, and how would this process differ in a 40-year-old adult?
2. What kind of bone would be most likely to be found in a sample taken from the area of a long bone immediately under the periosteum?
3. What effect(s) on bone structure occur when normal stresses are applied from a limited range of directions or along the bony shaft?
4. What are the advantages of spongy bone over compact bone in an area such as the expanded ends of long bones?
5. What is the most common injury that results from an extremely strong pull on a tendon or ligament, and why does this specific type of injury occur?
6. What factors determine the type of ossification that occurs in a specific bone?

7. What does the presence of cartilage signify regarding the development of bone in the same area?

8. What is the significance of epiphyseal cartilages at the shaft of a long bone?

9. How does the process of increasing in length of a long bone differ from the process of increasing in diameter in the same bone?

10. What is the meaning of a prominent epiphyseal line on the X-ray of a long bone?

11. What is the purpose of the small openings in the shafts of long bones?

12. Describe the role of the periosteum.

13. How does adequate exposure to the sun contribute to proper bone growth?

14. What events that occur at puberty affect the changes in the skeletal system that result in male and female adult skeletal characteristics?

15. What is the role of weight bearing in the maintenance of bone?

16. Why is a healed area of a bone less likely to fracture in the same place a second time from similar stresses?

17. Why are most cranial bones classified as "flat" bones?

18. What properties of a sesamoid bone allow it to be distinguished from a Wormian bone?

19. How does the area of bone where tendons or ligaments attach differ from an area midway along the shaft?

3 CRITICAL THINKING AND CLINICAL APPLICATION QUESTIONS

1. In leukemia, a disease of the blood, either nonfunctioning, impaired, or insufficient numbers of red blood cells are produced for normal oxygen uptake and delivery. One treatment for leukemia in children over the age of four is subjecting the bones to high-dosage radiation. What is the functional significance of this process? How can normal function be recovered?

2. Over several years, a person diagnosed with a severe form of osteo-arthritis gradually loses the ability to move joints. When X-rays are taken, it is discovered that excess bone deposited at the joints is restricting the range of motion. The control of bone remodeling is dependent on several bone cell types. If excess bone is being deposited, activity of which cells predominates? What hormonal treatment might improve this condition?

3. In a child with a vitamin D deficiency, rickets results in a bowlegged appearance. Once this condition is corrected, can the shape of the leg bones change? If so, what mechanism(s) might allow such a change?

4. An 11-year-old gymnast, falling off the balance beam, tries to save the dismount by doing a forward handspring. The result is a fracture to the forearm. What bone was most likely to be broken, and how would the fracture be classified? If, instead of landing on her arm, she had landed on her seat, what type of back injury would she most likely have incurred?

5. An otherwise healthy, but inactive 78-year-old woman falls when she gets up suddenly, and suffers a broken hip. What is the most likely diagnosis as to why the fracture occurred? Activity of what type(s) of bone cells is implicated in this result? How might these conditions be improved?

6. A person has suffered a comminuted fracture of the humerus. The bone must be "set" before the cast is applied. What is done to set this type of fracture, and what is likely to happen to the bone while it is immobilized during the healing process?

7. A small child falls off a bicycle and breaks an arm. The bone is set correctly and heals well. After the cast is removed, there remains an enlarged bony bump at the region of the fracture. After several months this enlargement disappears, and the arm is essentially normal in appearance. What happened during this healing process?

6

The Skeletal System: Axial Division

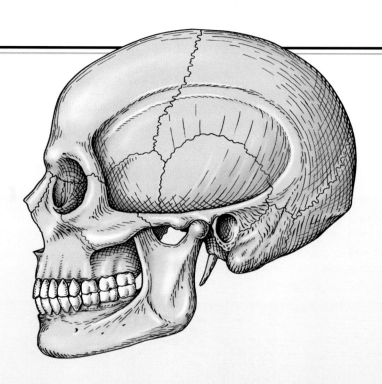

CHAPTER OUTLINE AND OBJECTIVES

The skeletal system is divided into *axial* and *appendicular divisions*; the axial division components are highlighted in Figure 6-1●. The skeletal system includes 206 separate bones and a number of associated cartilages. The **axial skeleton** consists of the bones of the skull, thorax, and vertebral column. These elements form the longitudinal axis of the body. There are 80 bones in the axial skeleton, roughly 40% of the bones in the human body. The axial components include:

- the **skull** (22 bones),
- bones associated with the skull (6 **auditory ossicles** + 1 **hyoid bone**),
- the **vertebral column** (26 bones), and
- the **thoracic cage** (24 **ribs** + 1 **sternum**).

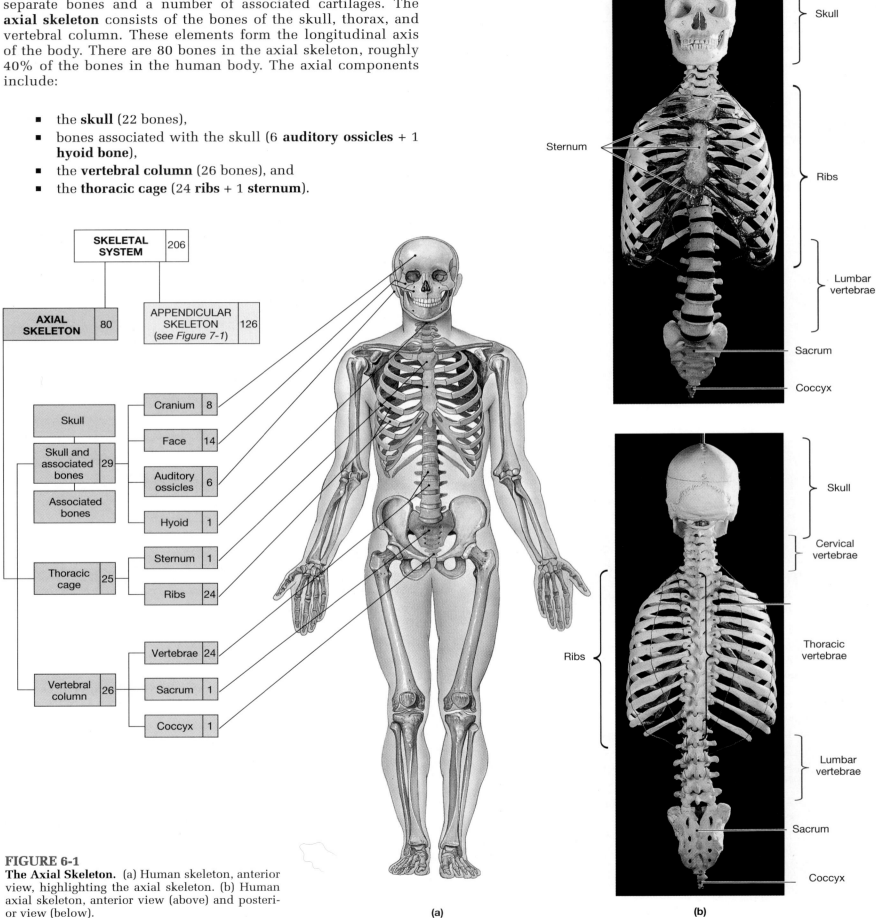

FIGURE 6-1
The Axial Skeleton. (a) Human skeleton, anterior view, highlighting the axial skeleton. (b) Human axial skeleton, anterior view (above) and posterior view (below).

(a)

(b)

The function of the axial skeleton is to create a framework that supports and protects organs in the dorsal and ventral body cavities. In addition, it provides an extensive surface area for the attachment of muscles that (1) adjust the positions of the head, neck, and trunk, (2) perform respiratory movements, and (3) stabilize or position structures of the appendicular skeleton. The joints of the axial skeleton permit limited movement, but they are very strong and heavily reinforced with ligaments.

The **appendicular skeleton** consists of 126 bones. This division includes the bones of the limbs and the **pectoral** and **pelvic girdles** that attach the limbs to the trunk. The appendicular skeleton will be examined in Chapter 7. This chapter describes the functional anatomy of the axial skeleton, and we will begin with the skull. Before proceeding, you may find it helpful to review the directional references included in Tables 1-1 and 1-2 and the terms introduced in Table 5-2. ∞[pp. 15, 16, 125]

THE SKULL
(Figures 6-2/6-3)

The bones of the skull protect the brain and guard the entrances to the digestive and respiratory systems. The skull contains 22 bones: 8 form the **cranium**, or *"braincase,"* and 14 are associated with the face (Figure 6-2●).

The cranium consists of the *occipital, parietal, frontal, temporal, sphenoid,* and *ethmoid* bones. Together they enclose the cranial cavity, a fluid-filled chamber that cushions and supports the brain. Blood vessels, nerves, and membranes that stabilize the position of the brain are attached to the inner surface of the cranium. Its outer surface provides an extensive area for the attachment of muscles that move the eyes, jaws, and head. A specialized joint between the occipital bone and the first spinal vertebra stabilizes the positions of the brain and spinal cord while permitting a considerable range of head movements.

If the cranium is the house where the brain resides, the *facial complex* is the front porch. **Facial bones** protect and support the entrances to the digestive and respiratory tracts. The superficial facial bones, the *maxillary, palatine, nasal, zygomatic, lacrimal,* and *mandible* (Figure 6-2●), provide areas for the attachment of muscles that control facial expressions and assist in the manipulation of food.

The boundaries between skull bones are immovable joints called **sutures**. At a suture, the bones are tied firmly together with dense fibrous connective tissue. Each of the sutures of the skull has a name, but you need to know only four major sutures at this time: the *lambdoidal, coronal, sagittal,* and *squamosal* sutures.

- **Lambdoidal** (lam-DOYD-al) **suture.** The lambdoidal suture arches across the posterior surface of the skull (Figure 6-3a●), separating the *occipital bone* from the *parietal bones.* One or more **sutural bones** (*Wormian bones*) may be found along this suture; they range from a bone the size of a grain of sand to one as large as a quarter.

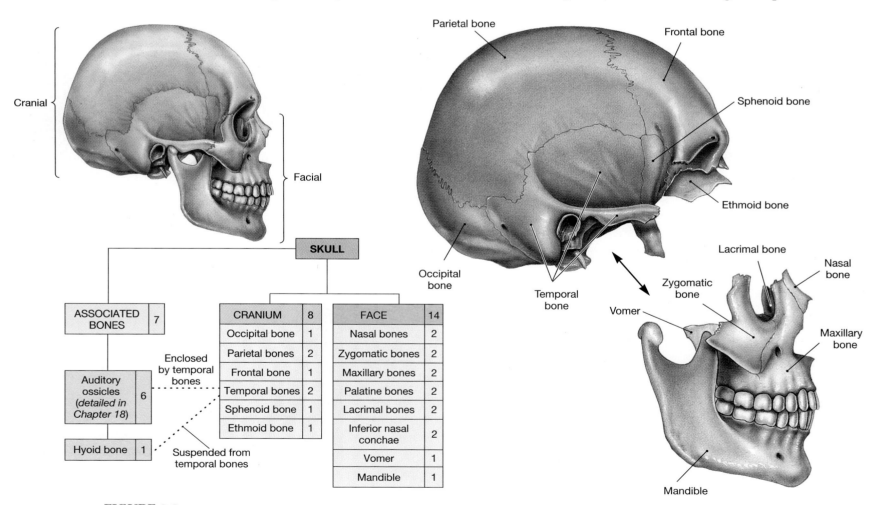

FIGURE 6-2
Cranial and Facial Subdivisions of the Skull. The skull can be divided into the cranial and the facial divisions. The inferior nasal conchae of the facial division are not visible from this aspect.

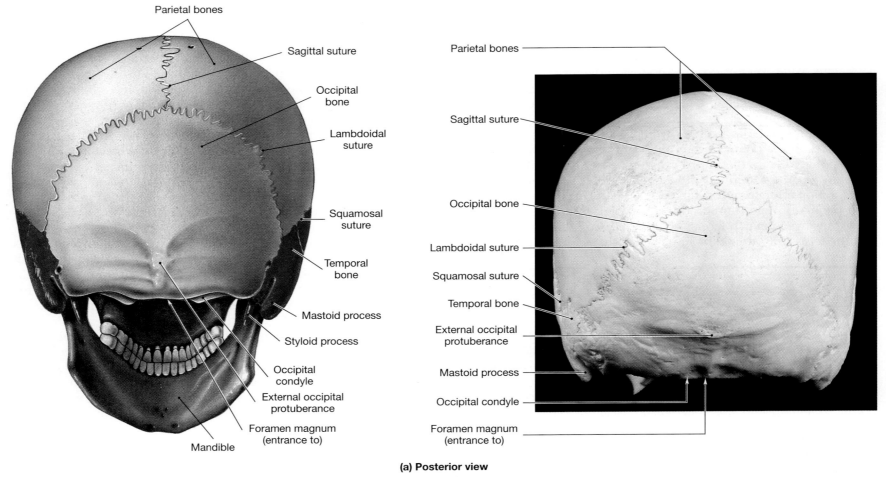

(a) Posterior view

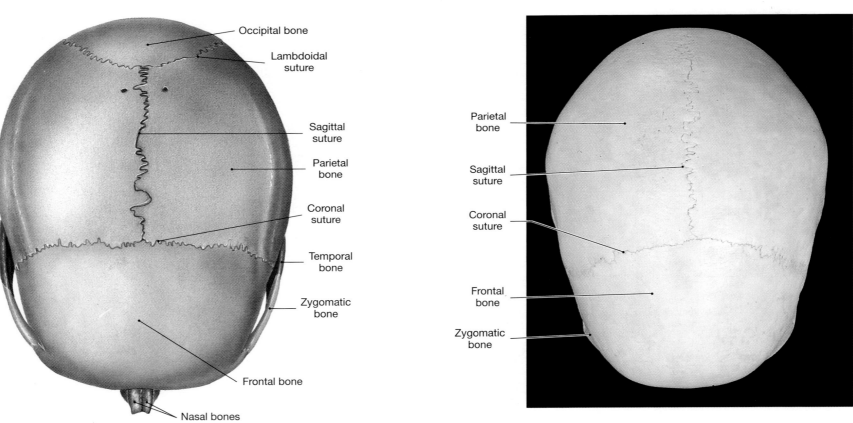

(b) Superior view

FIGURE 6-3
The Adult Skull. The adult skull is shown in (a) posterior view, (b) superior view, (c) lateral view, (d) anterior view, and (e) inferior view.

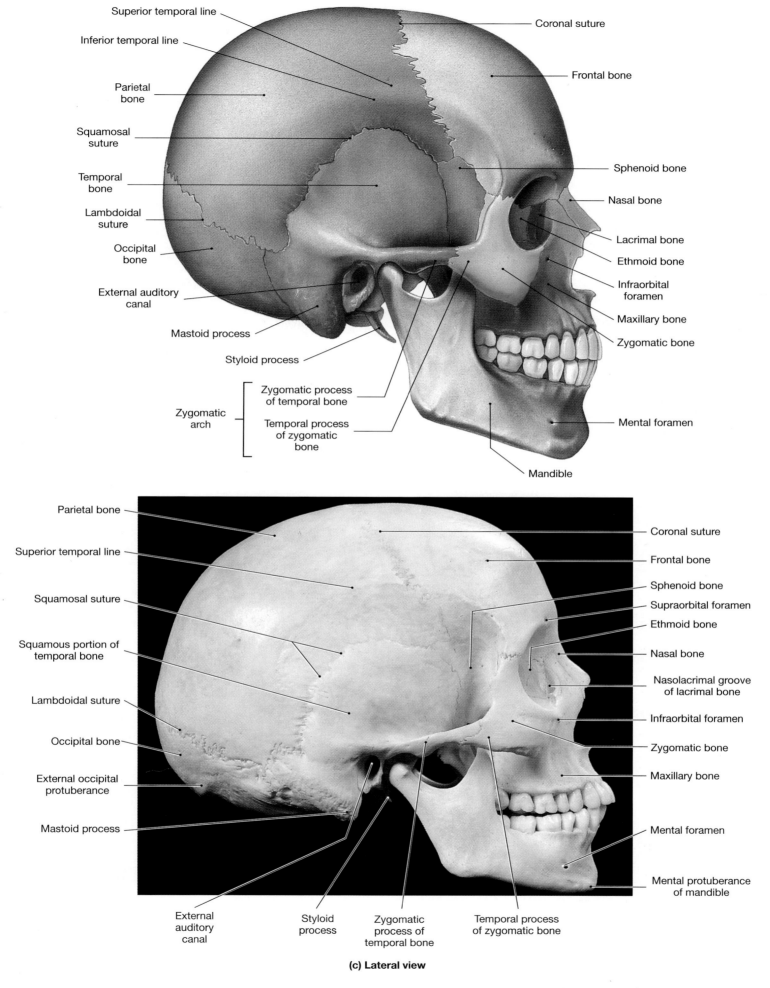

Superior temporal line

Inferior temporal line

Parietal bone

Squamosal suture

Temporal bone

Lambdoidal suture

Occipital bone

External auditory canal

Mastoid process

Styloid process

Zygomatic arch
- Zygomatic process of temporal bone
- Temporal process of zygomatic bone

Coronal suture

Frontal bone

Sphenoid bone

Nasal bone

Lacrimal bone

Ethmoid bone

Infraorbital foramen

Maxillary bone

Zygomatic bone

Mental foramen

Mandible

Parietal bone

Superior temporal line

Squamosal suture

Squamous portion of temporal bone

Lambdoidal suture

Occipital bone

External occipital protuberance

Mastoid process

External auditory canal

Styloid process

Zygomatic process of temporal bone

Temporal process of zygomatic bone

Coronal suture

Frontal bone

Sphenoid bone

Supraorbital foramen

Ethmoid bone

Nasal bone

Nasolacrimal groove of lacrimal bone

Infraorbital foramen

Zygomatic bone

Maxillary bone

Mental foramen

Mental protuberance of mandible

(c) Lateral view

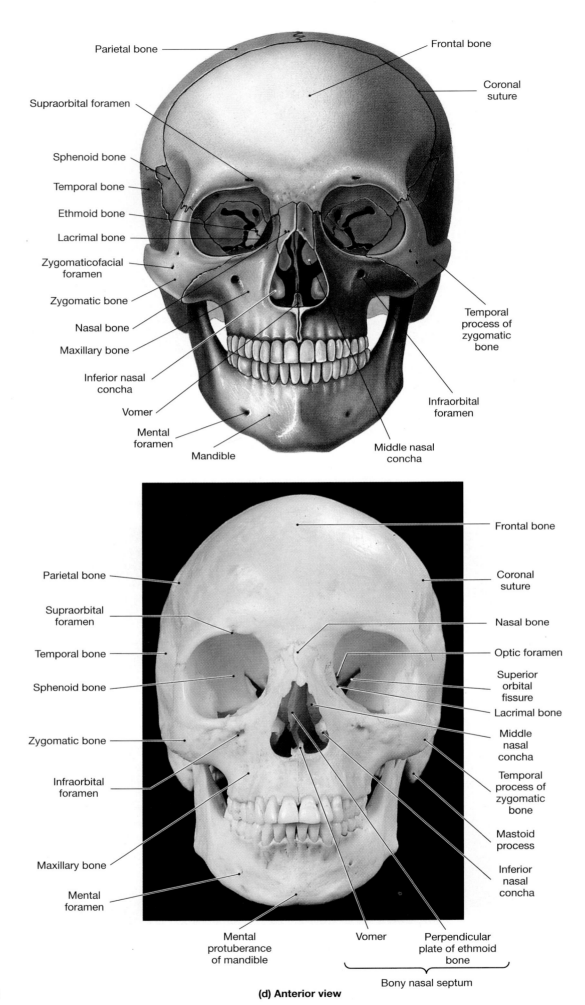

Parietal bone

Frontal bone

Supraorbital foramen

Coronal suture

Sphenoid bone

Temporal bone

Ethmoid bone

Lacrimal bone

Zygomaticofacial foramen

Zygomatic bone

Nasal bone

Maxillary bone

Inferior nasal concha

Temporal process of zygomatic bone

Vomer

Infraorbital foramen

Mental foramen

Mandible

Middle nasal concha

Frontal bone

Parietal bone

Coronal suture

Supraorbital foramen

Nasal bone

Temporal bone

Optic foramen

Sphenoid bone

Superior orbital fissure

Lacrimal bone

Middle nasal concha

Zygomatic bone

Temporal process of zygomatic bone

Infraorbital foramen

Mastoid process

Maxillary bone

Inferior nasal concha

Mental foramen

Mental protuberance of mandible

Vomer

Perpendicular plate of ethmoid bone

Bony nasal septum

FIGURE 6-3 continued

(d) Anterior view

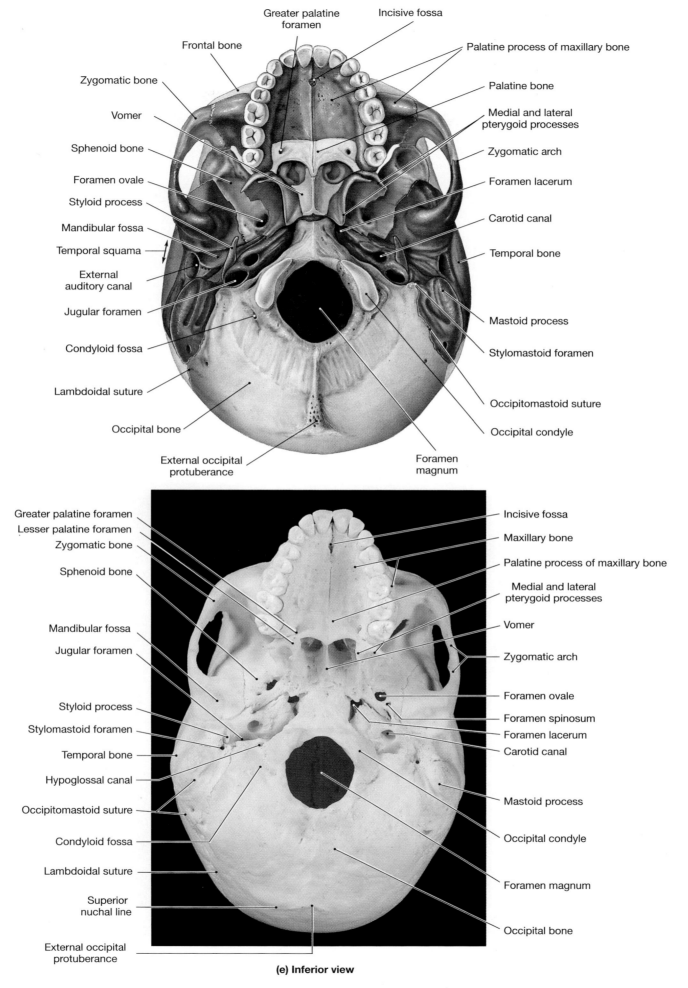

Greater palatine foramen

Incisive fossa

Frontal bone

Palatine process of maxillary bone

Zygomatic bone

Palatine bone

Vomer

Medial and lateral pterygoid processes

Sphenoid bone

Zygomatic arch

Foramen ovale

Foramen lacerum

Styloid process

Carotid canal

Mandibular fossa

Temporal squama

Temporal bone

External auditory canal

Jugular foramen

Condyloid fossa

Mastoid process

Lambdoidal suture

Stylomastoid foramen

Occipital bone

Occipitomastoid suture

Occipital condyle

External occipital protuberance

Foramen magnum

Greater palatine foramen

Incisive fossa

Lesser palatine foramen

Maxillary bone

Zygomatic bone

Palatine process of maxillary bone

Sphenoid bone

Medial and lateral pterygoid processes

Mandibular fossa

Vomer

Jugular foramen

Zygomatic arch

Styloid process

Foramen ovale

Stylomastoid foramen

Foramen spinosum

Temporal bone

Foramen lacerum

Hypoglossal canal

Carotid canal

Occipitomastoid suture

Condyloid fossa

Mastoid process

Lambdoidal suture

Superior nuchal line

Occipital condyle

External occipital protuberance

Foramen magnum

Occipital bone

(e) Inferior view

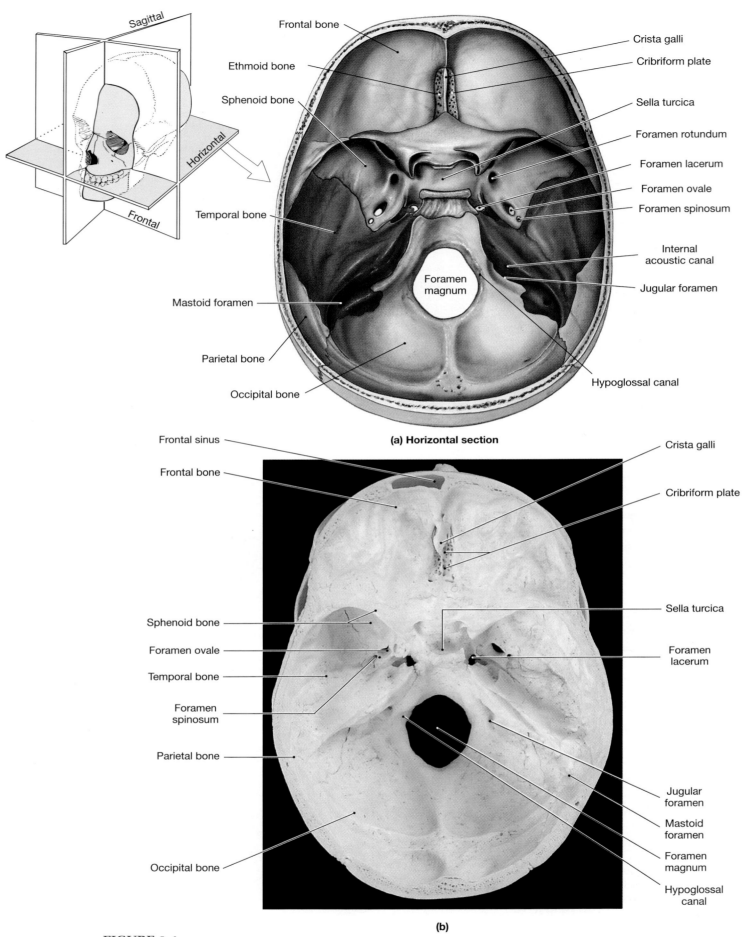

(a) Horizontal section

(b)

FIGURE 6-4
Sectional Anatomy of the Skull—Part I. (a,b) Horizontal section through the skull, showing the floor of the cranial cavity. Compare with part (b) and with Figure 6-3 and MRI SCANS 1a-c, p. 741.

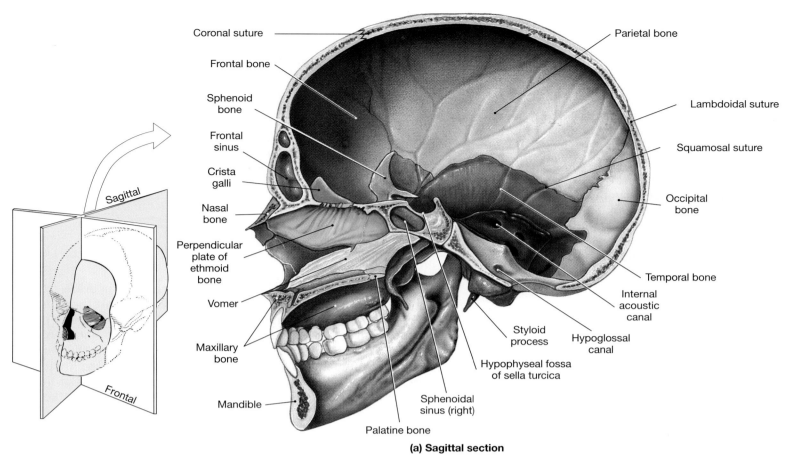

(a) Sagittal section

Coronal suture
Frontal bone
Sphenoid bone
Frontal sinus
Crista galli
Nasal bone
Perpendicular plate of ethmoid bone
Vomer
Maxillary bone
Mandible
Palatine bone
Sphenoidal sinus (right)
Hypophyseal fossa of sella turcica
Styloid process
Hypoglossal canal
Internal acoustic canal
Temporal bone
Occipital bone
Squamosal suture
Lambdoidal suture
Parietal bone

Sagittal
Frontal

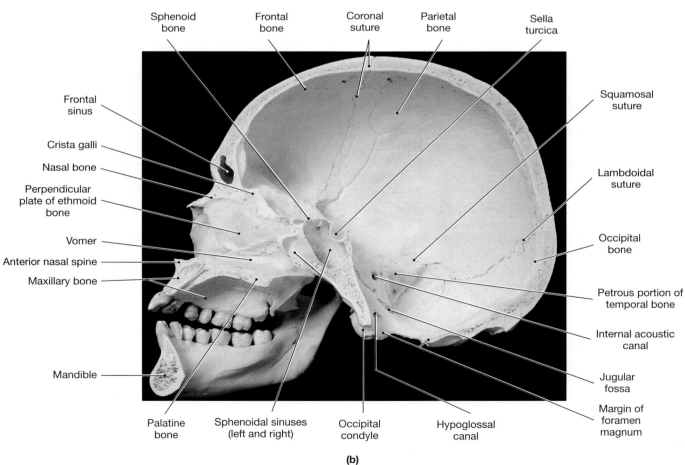

(b)

Sphenoid bone
Frontal bone
Coronal suture
Parietal bone
Sella turcica
Frontal sinus
Crista galli
Nasal bone
Perpendicular plate of ethmoid bone
Vomer
Anterior nasal spine
Maxillary bone
Mandible
Palatine bone
Sphenoidal sinuses (left and right)
Occipital condyle
Hypoglossal canal
Margin of foramen magnum
Jugular fossa
Internal acoustic canal
Petrous portion of temporal bone
Occipital bone
Lambdoidal suture
Squamosal suture

FIGURE 6-5
Sectional Anatomy of the Skull—Part II. (a,b) Sagittal section of the skull, showing detail of paranasal bones and cranial floor. See also MRI SCANS 1d-e, p. 741

- **Sagittal suture.** The sagittal suture begins at the superior midline of the lambdoidal suture and extends rostrally between the parietal bones to the coronal suture (Figure 6-3b●, p. 132).
- **Coronal suture.** Anteriorly, the sagittal suture ends when it intersects the coronal suture. The coronal suture crosses the superior surface of the skull, separating the anterior *frontal bone* from the more posterior parietals (Figure 6-3b●). The occipital, parietal, and frontal bones form the **calvaria** (kal-VAR-ē-a) or "skullcap."
- **Squamosal suture.** A squamosal suture on each side of the skull marks the boundary between the *temporal bone* and the parietal bone of that side. The squamosal sutures can be seen in Figure 6-3a●, p. 132, where they intersect the lambdoidal suture. The path of the squamosal suture on the right side of the skull can be seen in Figure 6-3c●, p. 133.

Bones of the Cranium
(Figures 6-3 to 6-5)

We will now consider each of the bones of the cranium. As we proceed, use the figures provided to develop a three-dimensional perspective on the individual bones. Ridges and foramina that are detailed here mark either the attachment of muscles or the passage of nerves and blood vessels that will be studied in later chapters. Figures 6-3, 6-4, and 6-5● present the adult skull in superficial and sectional views.

Occipital Bone
(Figures 6-3a,c,e/6-6a,b)

The **occipital bone** contributes to the posterior, lateral, and inferior surfaces of the cranium (Figure 6-3a,c,e●, pp. 132, 133, 135). The inferior surface of the occipital bone contains a large, circular opening, the **foramen magnum** (Figure 6-3e●), which connects the cranial cavity with the spinal cavity enclosed by the vertebral column. At the adjacent **occipital condyles**, the skull articulates with the first cervical vertebra. The posterior, external surface of the occipital bone (Figure 6-6a●) bears a number of prominent ridges. The **occipital crest** extends posteriorly from the foramen magnum, ending in a small midline bump called the **external occipital protuberance**. Two horizontal ridges intersect the crest, the **inferior** and **superior nuchal** (NOO-kal) **lines**. These mark the attachment of muscles and ligaments that stabilize the articulation at the occipital condyles and balance the weight of the head over the vertebrae of the neck. The occipital bone forms part of the wall of the large **jugular foramen** (Figure 6-3e●). The *internal jugular vein* passes through this foramen to drain venous blood from the brain. The **hypoglossal canals** begin at the lateral base of each occipital condyle, just superior to the condyles. The *hypoglossal nerves*, cranial nerves that control the tongue muscles, pass through these canals.

Inside the skull, the hypoglossal canals begin on the inner surface of the occipital bone near the foramen magnum (Figure 6-6b●). Note the concave internal surface of the occipital bone, which closely follows the contours of the brain. The grooves follow the path of major vessels, and the ridges mark the attachment site of membranes that stabilize the position of the brain.

Parietal Bones
(Figures 6-3b,c/6-5/6-6c)

The paired **parietal** (pa-RĪ-e-tal) **bones** contribute to the superior and lateral surfaces of the cranium (Figure 6-3b,c●, pp. 132, 133). The lateral surface of each parietal bone (Figure 6-6c●) bears a pair of low ridges, the **superior** and **inferior temporal lines**. These lines mark the attachment of the *temporalis muscle*, a large muscle that closes the mouth. The smooth parietal surface above these lines is called the **parietal eminence**. The inner surfaces of the parietal bones retain the impressions of cranial veins and arteries that branch inside the cranium (Figure 6-5●).

Frontal Bone
(Figures 6-3b,c,d/6-5a/6-7)

The **frontal bone** forms the forehead and roof of the orbits (Figure 6-3b,c,d●, pp. 132, 133, 134). During development, the bones of the cranium form through the fusion of separate centers of ossification, and at birth the fusions have not been completed. At this time there are two frontal bones that articulate along the **metopic suture**. Although the suture usually disappears by age 8 with the fusion of the bones, the adult skull often retains traces of the suture line. This suture, or what remains of it, runs down the center of the **frontal squama**, or *forehead* (Figure 6-7a●). To either side is the anterior continuation of the superior temporal line from the parietal surface.

The frontal squama ends at the **supraorbital margins**. Above each margin are raised ridges, the **superciliary arches**, which support the eyebrows. The center of each margin is perforated by a single **supraorbital foramen** or **notch**. The orbital surface is relatively smooth, but it contains small openings for blood vessels and nerves heading to or from structures in the orbit. The shallow **lacrimal fossa** marks the location of the *lacrimal* (tear) *gland* that lubricates the surface of the eye (Figure 6-7b●).

The **frontal sinuses** (Figure 6-5a●) are variable in size and time of appearance. They usually appear after age 6, but some people never develop them at all. The frontal sinuses and other sinuses will be described in a later section.

Temporal Bones
(Figures 6-3c,e/6-8)

The **squama**, or **squamous portion**, of the **temporal bone** is the convex surface inferior to the squamosal suture (Figure 6-3c●, p. 133). Inferior to the squama, the temporal bone has a prominent **zygomatic process** that curves laterally and anteriorly to meet the **temporal process** of the **zygomatic bone**. Together these processes form the **zygomatic arch**, or *cheekbone*. Medially, the **jugular foramen** is bounded by the temporal and occipital bones (Figure 6-3e●, p. 135). Anterior and slightly medial to the jugular foramen is the entrance to the **carotid canal**. The *internal carotid artery*, a major artery that supplies blood to the brain, penetrates the skull through this passageway. Anterior and medial to this opening, an elongated, jagged slit, the **foramen lacerum** (LA-se-rum; *lacerare*, to tear) extends between the occipital and temporal bones. In life this space contains hyaline cartilage and small arteries supplying the inner surface of the cranium.

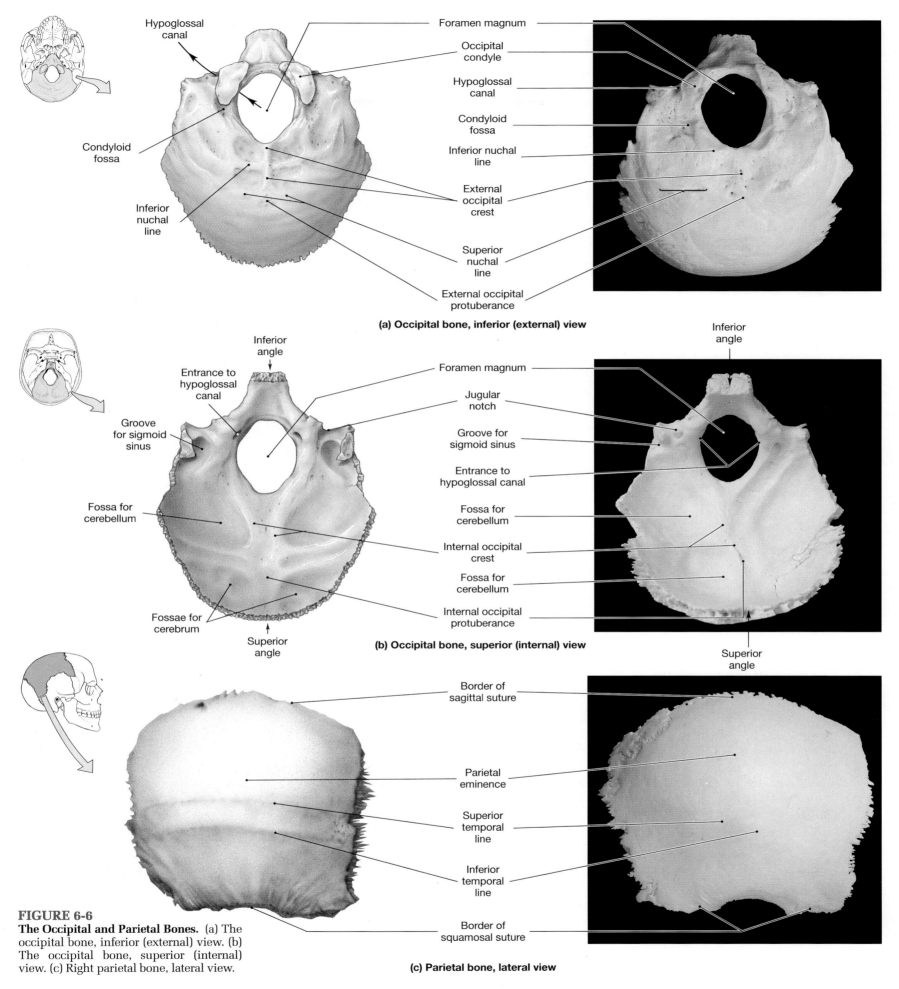

Hypoglossal canal

Foramen magnum

Occipital condyle

Hypoglossal canal

Condyloid fossa

Inferior nuchal line

External occipital crest

Condyloid fossa

Inferior nuchal line

Superior nuchal line

External occipital protuberance

(a) Occipital bone, inferior (external) view

Inferior angle

Inferior angle

Entrance to hypoglossal canal

Foramen magnum

Jugular notch

Groove for sigmoid sinus

Groove for sigmoid sinus

Entrance to hypoglossal canal

Fossa for cerebellum

Fossa for cerebellum

Internal occipital crest

Fossa for cerebellum

Fossae for cerebrum

Internal occipital protuberance

Superior angle

Superior angle

(b) Occipital bone, superior (internal) view

Border of sagittal suture

Parietal eminence

Superior temporal line

Inferior temporal line

Border of squamosal suture

FIGURE 6-6
The Occipital and Parietal Bones. (a) The occipital bone, inferior (external) view. (b) The occipital bone, superior (internal) view. (c) Right parietal bone, lateral view.

(c) Parietal bone, lateral view

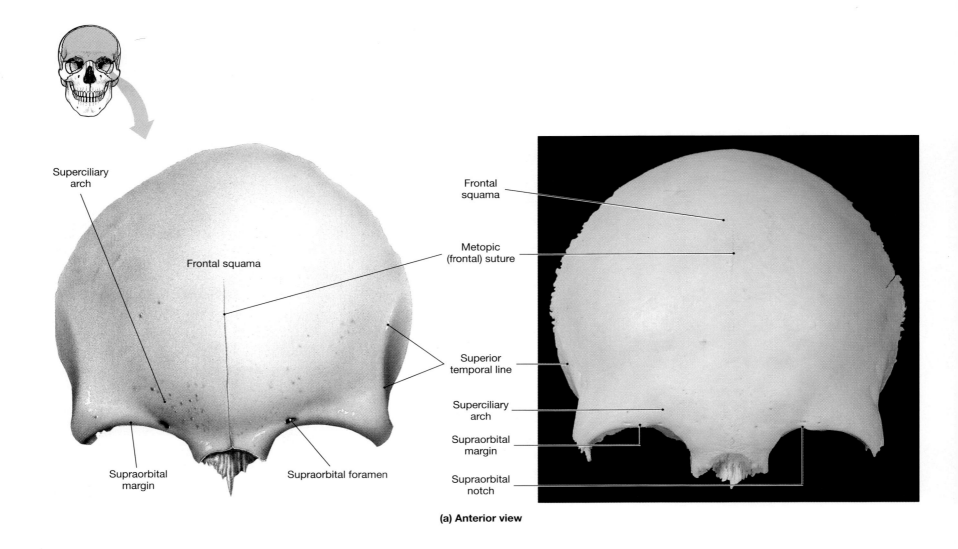

Superciliary arch

Frontal squama

Supraorbital margin

Supraorbital foramen

Frontal squama

Metopic (frontal) suture

Superior temporal line

Superciliary arch

Supraorbital margin

Supraorbital notch

(a) Anterior view

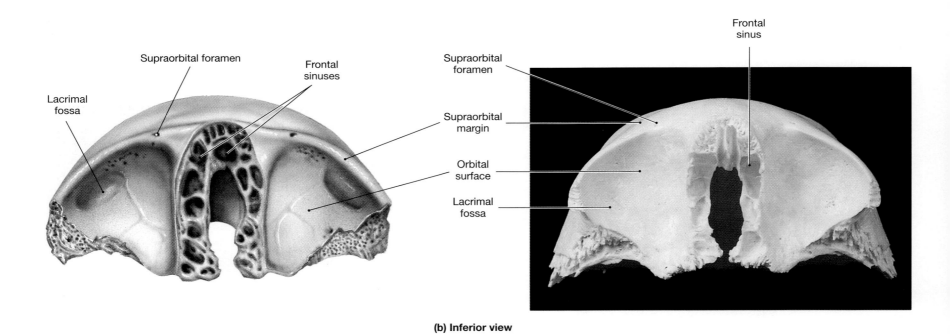

Supraorbital foramen

Frontal sinuses

Lacrimal fossa

Frontal sinus

Supraorbital foramen

Supraorbital margin

Orbital surface

Lacrimal fossa

(b) Inferior view

FIGURE 6-7
The Frontal Bone. (a) Anterior surface. (b) Inferior surface.

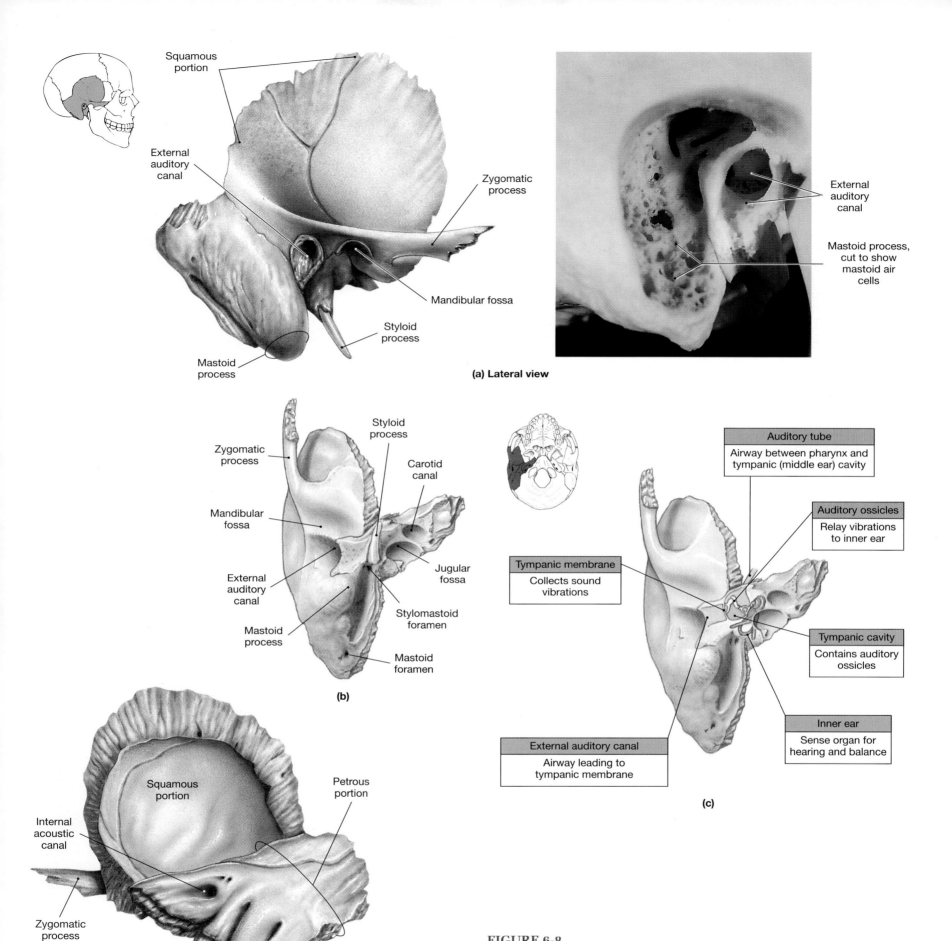

(a) Lateral view

(b)

(c)

(d) Medial view

Squamous portion

External auditory canal

Zygomatic process

Mandibular fossa

Styloid process

Mastoid process

External auditory canal

Mastoid process, cut to show mastoid air cells

Zygomatic process

Mandibular fossa

External auditory canal

Mastoid process

Styloid process

Carotid canal

Jugular fossa

Stylomastoid foramen

Mastoid foramen

| **Auditory tube** |
| Airway between pharynx and tympanic (middle ear) cavity |

| **Auditory ossicles** |
| Relay vibrations to inner ear |

| **Tympanic membrane** |
| Collects sound vibrations |

| **Tympanic cavity** |
| Contains auditory ossicles |

| **Inner ear** |
| Sense organ for hearing and balance |

| **External auditory canal** |
| Airway leading to tympanic membrane |

Internal acoustic canal

Squamous portion

Petrous portion

Zygomatic process

Styloid process

Mastoid process

FIGURE 6-8
The Temporal Bone. (a) Lateral view of the right temporal bone, showing major anatomical landmarks, and a cutaway view of the mastoid air cells. (b) Inferior view of the right temporal bone. (c) Inferior view of the right temporal bone, showing the orientation of the auditory canal and internal structures. (d) Medial view of the right temporal bone.

Inferior to the base of the zygomatic process, the temporal bone articulates with the mandible. A depression called the **mandibular fossa** and an elevated **articular tubercle** mark this site (Figure 6-8a,b●). Immediately posterior and lateral to that mandibular fossa is the round entrance to the **external auditory canal.** In life, this canal ends at the delicate **tympanic membrane**, or *eardrum*, but this membrane disintegrates during the preparation of a dried skull. The prominent bulge just posterior and inferior to the meatus is the **mastoid process**. This process provides an attachment site for muscles that rotate or extend the head. Numerous interconnected mastoid sinuses, termed *mastoid air cells* are contained within the mastoid process (Figure 6-8a●).

Many of the important features of the temporal bone can be seen on its inferior surface (Figure 6-8b●). Near the base of the mastoid process, the **mastoid foramen** penetrates the temporal bone. Blood vessels travel through this passageway to reach the membranes surrounding the brain. Ligaments that support the hyoid bone attach to the sharp **styloid process** (STĪ-loyd; *stylos*, pillar), as do some of the tongue muscles. The **stylomastoid foramen** lies posterior to the base of the styloid process. The *facial nerve* passes through this foramen to control the facial muscles.

Lateral and anterior to the carotid foramen, the temporal bone articulates with the sphenoid. A small canal begins at that articulation and ends inside the mass of the temporal bone. In life it contains an air-filled passageway that begins at the pharynx and ends at the **tympanic cavity**, a chamber inside the temporal bone (Figure 6-8c●). This passageway has long been known as the *Eustachian* (ū-STĀ-kē-an) *tube*, although the terms **auditory tube**, or *pharyngotympanic tube*, are now used to designate this structure.

The tympanic cavity, or *middle ear*, contains the **auditory ossicles**, or *ear bones*. These tiny bones, three on each side of the skull, transfer sound vibrations from the eardrum to the hearing receptors of the *inner ear*. (Details concerning the individual bones and their role in hearing will be discussed in Chapter 18.) The auditory ossicles and inner ear are completely enclosed by the temporal bone.

A medial view of the temporal bone (Figure 6-8d●) shows several additional features. The thick **petrous portion** of the temporal bone houses the inner ear structures that provide information about hearing and balance. The **internal acoustic canal** carries blood vessels and nerves to the inner ear and the facial nerve to the stylomastoid foramen.

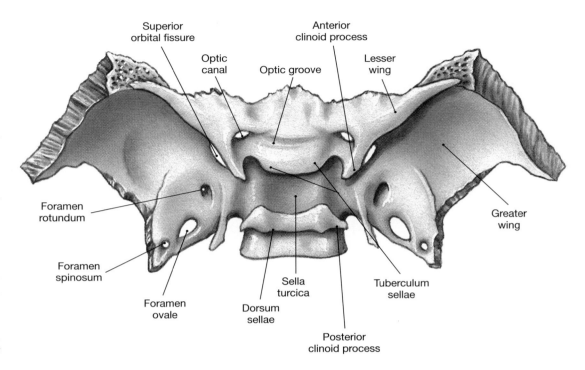

(a) Superior surface

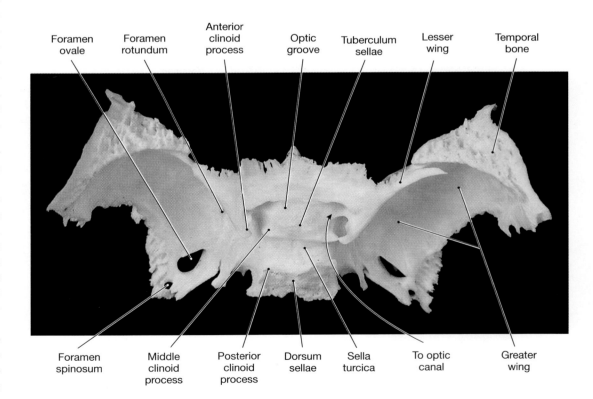

(c) Superior surface

FIGURE 6-9
The Sphenoid Bone. (a,c) Superior surface. (b,d) Anterior surface.

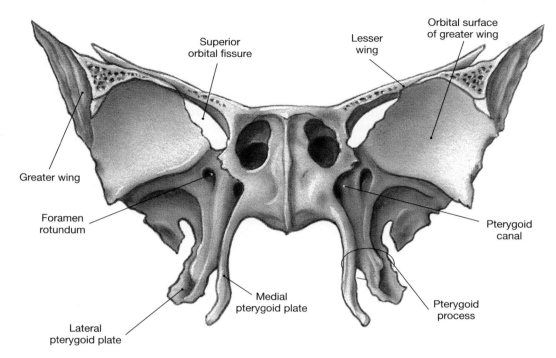

Superior orbital fissure

Lesser wing

Orbital surface of greater wing

Greater wing

Foramen rotundum

Medial pterygoid plate

Lateral pterygoid plate

Pterygoid canal

Pterygoid process

(b) Anterior surface

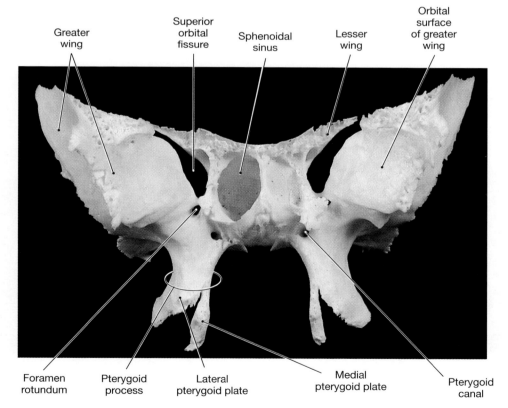

Greater wing

Superior orbital fissure

Sphenoidal sinus

Lesser wing

Orbital surface of greater wing

Foramen rotundum

Pterygoid process

Lateral pterygoid plate

Medial pterygoid plate

Pterygoid canal

(d) Anterior surface

Sphenoid Bone
(Figures 6-3c,d,e/6-4/6-9)

The **sphenoid bone** contributes to the floor of the cranium. Although it is relatively large, much of it is hidden by more superficial bones. The sphenoid bone is a bridge uniting the cranial and facial bones; it articulates with the frontal, occipital, parietal, ethmoid, and temporal bones of the cranium, and the palatine bones, zygomatic bones, maxillary bones, and vomer of the facial complex (Figure 6-3c,d,e●, pp. 133, 134, 135). It also acts as a brace, strengthening the sides of the skull.

The general shape of the sphenoid bone has been compared to a giant bat, with its wings extended. The wings can be seen most clearly on the superior surface (Figures 6-4, p. 136, and 6-9a-c●). A prominent central depression between the wings cradles the pituitary gland below the brain. This recess is called the **hypophyseal** (hī-pō-FIZ-ē-al) **fossa**, and the bony enclosure is called the **sella turcica** (TUR-si-ka) because it supposedly resembles a "Turkish saddle." If the Turk rode facing forward, his back would rest against the **posterior clinoid** (KLĪ-noid) **process** and he could reach forward to grasp the **anterior clinoid processes** on either side. The anterior clinoid processes are posterior projections of the **lesser wings** of the sphenoid. The **tuberculum sellae** forms the anterior border of the sella turcica.

The transverse groove that crosses the front of the saddle, above the level of the seat, is the **optic groove**. At either end of this groove is an **optic canal**. The optic nerves that carry visual information from the eyes to the brain travel through these foramina. On either side of the sella turcica, the **superior orbital fissure**, the **foramen rotundum**, the **foramen ovale** (ō-VAH-lā), and the **foramen spinosum** penetrate the sphenoid bone. These passages carry blood vessels and cranial nerves to structures of the orbit, face, and jaws. These foramina penetrate the **greater wings** of the sphenoid bone. The superior orbital fissures can also be seen in an anterior view.

The **pterygoid processes** (TER-i-goyd; *pterygion*, wing) of the sphenoid bone are vertical projections that begin at the boundary between the greater and lesser wings. Each process forms a pair of *pterygoid plates* that are important sites for the attachment of muscles that move the lower jaw and soft palate. At the base of each process, the **pterygoid canal** provides a route for a small nerve and an artery that supply the soft palate and adjacent structures.

Ethmoid Bone
(Figures 6-3d/6-4/6-5/6-10)

The **ethmoid** is an irregularly shaped bone that forms part of the orbital wall (Figure 6-3d●, p. 134), the anteromedial floor of the cranium (Figure 6-4●, p. 136), and the roof of the nasal cavity and part of the nasal septum (Figures 6-5●, p. 137). The ethmoid has three parts: the *cribriform plate*, the paired *lateral masses*, and the *perpendicular plate* (Figure 6-10●).

The superior surface of the ethmoid (Figure 6-10a●) contains a prominent ridge, the **crista galli** (*crista*, crest + *gallus*, chicken; "cock's comb"). The *falx cerebri*, a membrane that stabilizes the position of the brain, attaches to this bony ridge. The foramina in the adjacent **cribriform plate** permit passage of the *olfactory nerves* providing the sense of smell.

The **lateral masses**, the **superior nasal conchae** (KONG-ke; singular *concha*, "a snail shell"), and the **middle nasal conchae** are best viewed from the anterior and posterior surfaces of the ethmoid bone (Figure 6-10b,c●).

- The lateral masses contain the **ethmoidal sinuses**, which open into the nasal cavity on each side. Mucous secretions from these sinuses flush the surfaces of the nasal cavities.

- The conchae are thin scrolls of bone that project into the nasal cavity. The projecting conchae break up the airflow, creating swirls and eddies. This mechanism slows air movement, but provides additional time for warming, humidification, and dust removal before the air reaches more delicate portions of the respiratory tract.

- The **perpendicular plate** forms part of the *nasal septum*, along with the vomer and a piece of hyaline cartilage. Olfactory receptors are located in the epithelium covering the inferior surfaces of the cribriform plate, the me-

dial surfaces of the superior nasal conchae, and the superior portion of the perpendicular plate.

The Cranial Fossae
(Figure 6-11)

The contours of the cranium closely follow the shape of the brain. Proceeding from anterior to posterior, the floor of the cranium is not horizontal; it descends in two steps (Figure 6-11a●). Viewed from the superior surface (Figure 6-11b,c●), the cranial floor at each level forms a curving depression known as a **cranial fossa**. The **anterior cranial fossa** is formed by the frontal bone, the ethmoid bone, and the lesser wings of the sphenoid bone. The anterior cranial fossa cradles the frontal lobes of the cerebral hemispheres. The **middle cranial fossa** extends from the "step" at the lesser wings to the petrous portion of the temporal bone. The sphenoid, temporal, and parietal bones form this fossa, which cradles the temporal lobes of the cerebral hemispheres, the *diencephalon*, and the anterior portion of the brain stem (*mesencephalon*). The more inferior **posterior cranial fossa** extends from the petrous portion of the temporal bones to the posterior surface of the skull. The posterior fossa is formed primarily by the occipital bone, with contributions from the temporal and parietal bones. The posterior cranial fossa supports the occipital lobes of the cerebral hemispheres, the cerebellum, and the posterior brain stem (*pons* and *medulla*).

√ **The internal jugular veins are important blood vessels of the head. What bones do these blood vessels pass through?**

√ **What bone contains the depression called the sella turcica? What is located in the depression?**

√ **Which of the five senses would be affected if the cribriform plate of the ethmoid bone failed to form?**

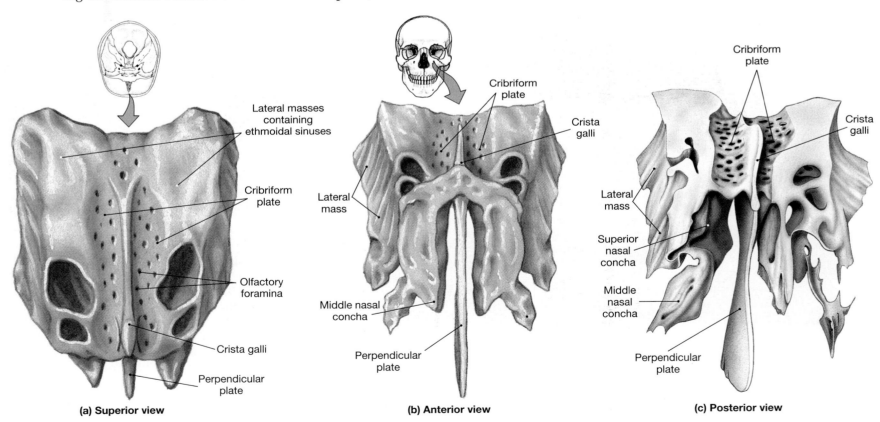

(a) Superior view   **(b) Anterior view**   **(c) Posterior view**

FIGURE 6-10
The Ethmoid Bone. (a) Superior surface. (b) Anterior surface. (c) Posterior surface.

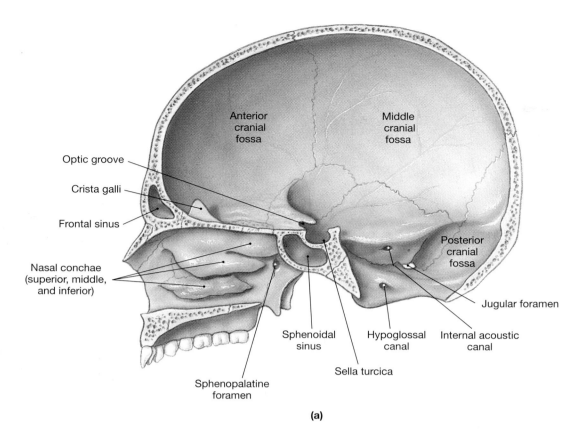

(a)

Bones of the Face

The facial bones are the *maxillary bones*, the *palatine bones*, the *nasal bones*, the *mandible*, the *zygomatic bones*, the *lacrimal bones*, the *inferior conchae*, and the *vomer*. Facial bones and cranial bones together form the *nasal complex* that surrounds the nasal cavities and the *orbital complex* surrounding each eye.

The Maxillary Bones
(Figures 6-3d/6-12a,b)

The left and right **maxillary bones** (*maxillae*), are the largest facial bones, and form the upper jaw. They articulate with all other facial bones except the mandible (Figure 6-3d●, p. 134). The **orbital rim** (Figure 6-12a●) provides protection for the eye and other structures in the orbit. A large **infraorbital foramen** marks the path of a major sensory nerve from the face. In the orbit, it runs along the *infraorbital groove* before passing through the inferior orbital fissure and the *foramen rotundum* to reach the brain stem. An elongated **in-**

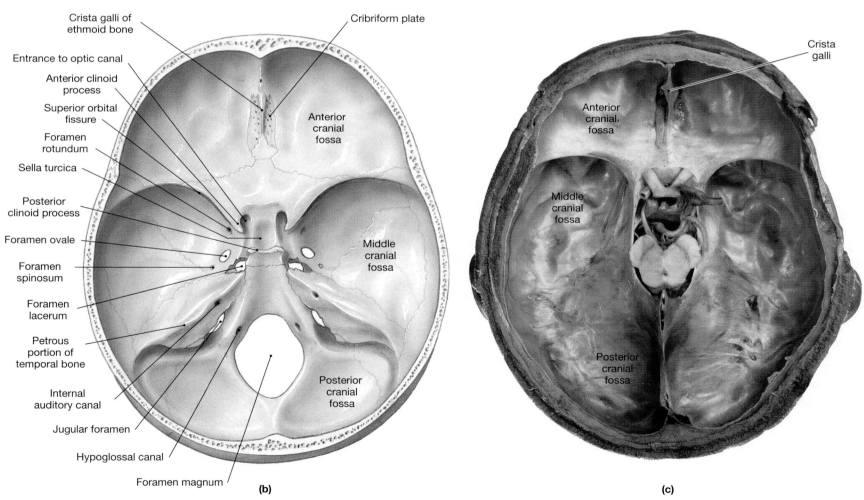

(b)

(c)

FIGURE 6-11
The Cranial Fossae. (a) A sagittal section through the skull, showing the relative positions of the cranial fossae. (b,c) Superior views of the cranial floor. See also MRI SCANS 1a-e, p. 741.

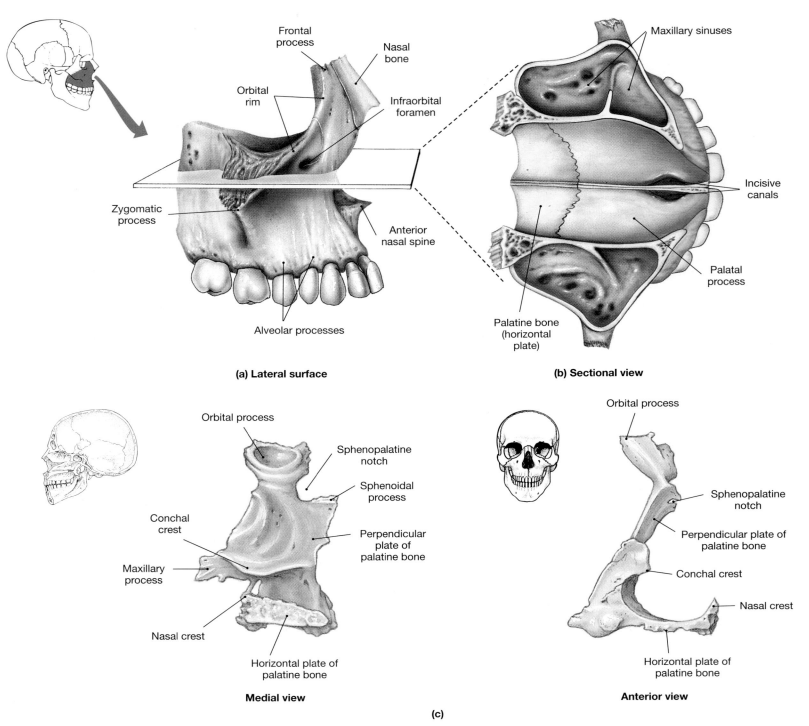

Frontal process

Nasal bone

Orbital rim

Infraorbital foramen

Zygomatic process

Anterior nasal spine

Alveolar processes

(a) Lateral surface

Maxillary sinuses

Incisive canals

Palatal process

Palatine bone (horizontal plate)

(b) Sectional view

Orbital process

Sphenopalatine notch

Sphenoidal process

Conchal crest

Perpendicular plate of palatine bone

Maxillary process

Nasal crest

Horizontal plate of palatine bone

Medial view

Orbital process

Sphenopalatine notch

Perpendicular plate of palatine bone

Conchal crest

Nasal crest

Horizontal plate of palatine bone

Anterior view

(c)

FIGURE 6-12
The Maxillary and Palatine Bones. (a) Lateral surface of right maxilla, showing superficial landmarks and the sectional plane for part (b). (b) Horizontal section through both maxillary bones, showing the size and orientation of the maxillary sinuses. (c) The right palatine bone.

ferior orbital fissure** within each orbit lies between the maxilla and the sphenoid. The oral margins of the maxillary bones form the **alveolar processes** that contain the upper teeth.

The large **maxillary sinuses** are evident in horizontal section (Figure 6-12b●). These are the largest sinuses in the skull; they lighten the portion of the maxillary bones above the teeth and produce mucous secretions that flush the inferior surfaces of the nasal cavities. The sectional view also shows the extent of the **palatal processes** that form most of the bony roof, or **hard palate**, of the mouth.

The Palatine Bones
(Figure 6-3e/6-12b,c)

The **palatine bones** are small, L-shaped bones. Their *horizontal plates* articulate with the maxillary bones to form the posterior portions of the hard palate (Figure 6-3e●, p. 135). The inferior surface of each horizontal plate has a prominent foramen, the *greater palatine foramen*, and usually one or more *lesser palatine foramina*. The *perpendicular plate* of the palatine bone extends superiorly (Figure 6-12b,c●), where it forms a small portion of the floor of the orbit.

146

The Nasal Bones
(Figures 6-3c,d/6-12a)

The paired **nasal bones** articulate with the frontal bone at the midline of the face (Figure 6-3d●, p. 134). The nasal bones extend to the superior border of the **external nares** (NA-rēz), or nasal openings. Cartilage attached to the nasal bones forms the flexible portion of the nose. The lateral surfaces of the nasal bones articulate with the maxillary bones on either side (Figures 6-3c, p. 133, and 6-12a●).

The Vomer
(Figures 6-3d,e/6-5)

The **vomer** forms the inferior portion of the nasal septum (Figure 6-5●, p. 137). It is based on the floor of the nasal cavity and articulates with both the maxillary bones and palatine bones along the midline. The vertical portion of the vomer is thin. Its curving superior surface articulates with the sphenoid bone and the perpendicular plate of the ethmoid bone, forming a bony **nasal septum** (*septum*, wall) that separates the right and left nasal cavities (Figure 6-3d,e●, pp. 134, 135). Anteriorly, the vomer supports a cartilaginous extension of the nasal septum that continues into the fleshy portion of the nose and separates the external nares.

The Inferior Nasal Conchae
(Figures 6-3d/6-13)

The **inferior nasal conchae** are paired scroll-like bones that resemble the *superior* and *middle conchae of the ethmoid bone*. One inferior concha is located on each side of the nasal septum, attached to the lateral wall of the nasal cavity (Figures 6-3d, p. 134, and 6-13●). They perform the same functions as the superior and middle conchae of the ethmoid bone.

The Nasal Complex
(Figures 6-5/6-13a,b)

The **nasal complex** includes the bones that enclose the nasal cavities and the *paranasal sinuses*, airspaces connected to the nasal cavities (Figure 6-5●, p. 137). The frontal, sphenoid, and ethmoid bones form the superior wall of the nasal cavities; the lateral walls are formed by the maxillary bones, the lacrimal bones, and the ethmoidal and inferior conchae (Figure 6-13a,b●). Much of the anterior margin of the nasal cavity is formed by the soft tissues of the nose, but the bridge of the nose is supported by the maxillary and nasal bones.

PARANASAL SINUSES *(Figure 6-13b,c,d).* The frontal, sphenoid, ethmoid, and maxillary bones contain the **paranasal sinuses**, air-filled chambers that open into the nasal cavities. Figure 6-13b● shows the location of the **frontal** and **sphenoidal sinuses**. The **ethmoidal** and **maxillary sinuses** are shown in Figure 6-13c,d●. Sinuses make skull bones lighter, produce mucus, and resonate during sound production. The mucous secretions are released into the nasal cavities, and the ciliated epithelium passes the mucus back toward the throat, where it is eventually swallowed. Incoming air is humidified and warmed as it flows across this carpet of mucus. Foreign particulate matter, such as dust or microorganisms, becomes trapped in this sticky mucus and then is swallowed. This mechanism helps protect the delicate exchange surfaces of the fragile lung tissue portions of the respiratory tract.

The Zygomatic Bones
(Figures 6-3c,d/6-14)

As noted above, the temporal process of the **zygomatic bone** articulates with the zygomatic process of the temporal bone to form the zygomatic arch (Figure 6-3c,d●, pp. 133, 134). A **zygomaticofacial foramen** on the anterior surface of each zygomatic bone carries a sensory nerve innervating the cheek. The zygomatic bone also forms the lateral rim of the orbit (Figure 6-14●) and contributes to the interior orbital wall.

The Lacrimal Bones
(Figures 6-3c,d/6-14)

The paired **lacrimal bones** (*lacrima*, tear) are the smallest bones in the skull. One is situated in the medial portion of each orbit, where it articulates with the frontal, maxillary, and ethmoid bones (Figures 6-3c,d, pp. 133, 134, and 6-14●). A small passageway, the **nasolacrimal canal** surrounds the tear duct as it passes toward the nasal cavity.

The Orbital Complex
(Figure 6-14)

The **orbits** are the bony recesses that contain the eyes. Seven bones form the **orbital complex** that forms each orbit (Figure 6-14●). The frontal bone forms the roof, and the maxillary bone provides most of the orbital floor. Proceeding from medial to lateral, the orbital rim and the first portion of the wall are contributed by the maxillary bone, the lacrimal bone, and the lateral mass of the ethmoid bone, which articulates with the sphenoid bone and a small process of the palatine bone. Several prominent foramina and fissures penetrate the sphenoid bone or lie between the sphenoid and maxillary bone. Laterally, the sphenoid bone and maxillary bone articulate with the zygomatic bone, which forms the lateral wall and rim of the orbit.

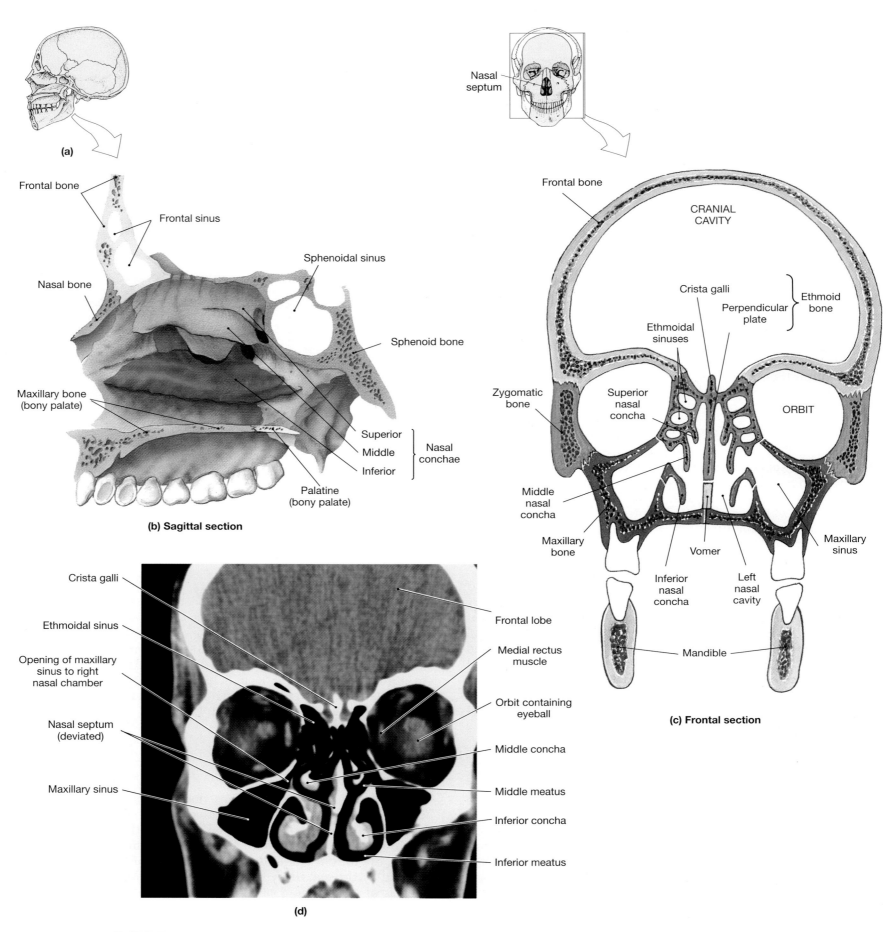

(a)

Frontal bone

Frontal sinus

Nasal bone

Sphenoidal sinus

Sphenoid bone

Maxillary bone (bony palate)

Superior
Middle } **Nasal conchae**
Inferior

Palatine (bony palate)

(b) Sagittal section

Nasal septum

Frontal bone

CRANIAL CAVITY

Crista galli

Perpendicular plate } **Ethmoid bone**

Ethmoidal sinuses

Zygomatic bone

Superior nasal concha

ORBIT

Middle nasal concha

Maxillary sinus

Maxillary bone

Vomer

Inferior nasal concha

Left nasal cavity

Mandible

(c) Frontal section

Crista galli

Ethmoidal sinus

Opening of maxillary sinus to right nasal chamber

Nasal septum (deviated)

Maxillary sinus

Frontal lobe

Medial rectus muscle

Orbit containing eyeball

Middle concha

Middle meatus

Inferior concha

Inferior meatus

(d)

FIGURE 6-13
The Nasal Complex. (a) Sagittal section through the skull, with the nasal septum in place. (b) Sagittal section with the nasal septum removed to show major features of the wall of the right nasal cavity. (c) A frontal section showing the positions of the paranasal sinuses. (d) An MRI scan showing the location of the sinuses; compare with (c). See also MRI SCAN 2a, p. 742.

148

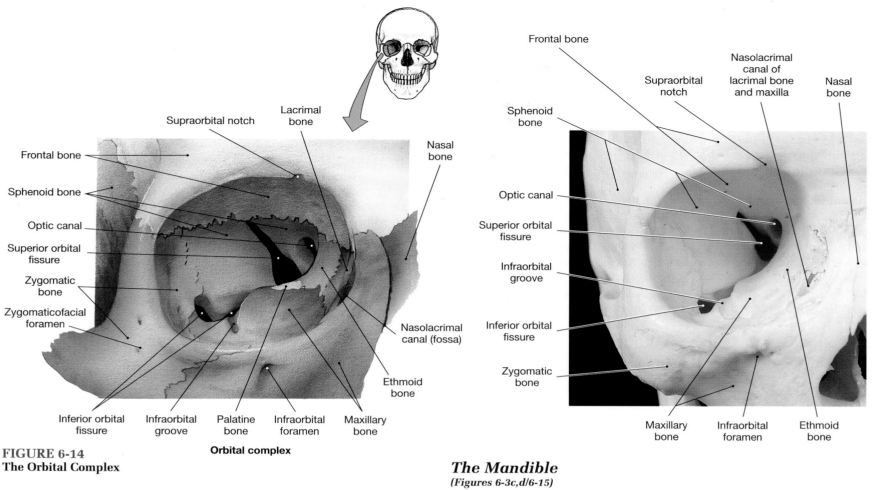

FIGURE 6-14
The Orbital Complex

Orbital complex

Frontal bone

Supraorbital notch

Lacrimal bone

Nasal bone

Sphenoid bone

Optic canal

Superior orbital fissure

Zygomatic bone

Zygomaticofacial foramen

Inferior orbital fissure

Infraorbital groove

Palatine bone

Infraorbital foramen

Maxillary bone

Ethmoid bone

Nasolacrimal canal (fossa)

Frontal bone

Supraorbital notch

Nasolacrimal canal of lacrimal bone and maxilla

Nasal bone

Sphenoid bone

Optic canal

Superior orbital fissure

Infraorbital groove

Inferior orbital fissure

Zygomatic bone

Maxillary bone

Infraorbital foramen

Ethmoid bone

The Mandible
(Figures 6-3c,d/6-15)

The **mandible** forms the entire lower jaw (Figures 6-3c,d, pp. 133, 134, and 6-15●). This bone can be subdivided into the horizontal **body** and the ascending **rami** (singular *ramus, "branch"*). The teeth are supported by the mandibular body. Each ramus meets the body at the mandibular **angle**. The **condylar processes** articulate with the mandibular fossae of the temporal bone at the *temporomandibular joint (TMJ)*. This joint is quite mobile, as evidenced by the jaw movements during chewing or talking. The disadvantage of such mobility is that the jaw can easily be dislocated by forceful protraction or lateral displacement.

Temporomandibular joint

Teeth (molar)

Coronoid process

Condylar process

Mandibular notch

Alveolar margin

Mandibular notch

Ramus   Angle   Body   Mental foramen

Mental protuberance

(a) Lateral view

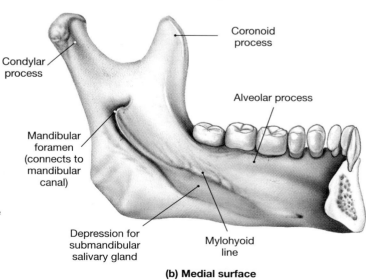

Coronoid process

Condylar process

Alveolar process

Mandibular foramen (connects to mandibular canal)

Depression for submandibular salivary gland

Mylohyoid line

(b) Medial surface

FIGURE 6-15
The Mandible. (a) Lateral view. (b) Medial view of the left side of the mandible.

At the **coronoid** (kor-Ō-noid) **processes**, the *temporalis* muscle inserts onto the mandible. This is one of the most forceful muscles involved in closing the mouth. Anteriorly, the **mental foramina** (*mentalis*, chin) penetrate the body on each side of the chin. Nerves pass through these foramina carrying sensory information from the lips and chin back to the brain. The **mandibular notch** is the depression that lies between the condylar and coronoid processes.

Medially (Figure 6-15b●), an **alveolar process** covers the alveoli and the roots of the teeth in the lower jaw. A **mylohyoid line** lies on the medial aspect of each ramus. It marks the insertion of the *mylohyoid muscle* that supports the floor of the mouth and tongue. The *submandibular salivary gland* nestles in a depression inferior to the mylohyoid line. Near the posterior, superior end of the mylohyoid line, a prominent **mandibular foramen** leads into the **mandibular canal**. This is a passageway for blood vessels to service the lower teeth and for nerves. The nerve that uses this passage carries sensory information from the teeth and gums; dentists typically anesthetize this nerve before working on the lower teeth.

The Hyoid Bone
(Figure 6-16)

The **hyoid bone** lies inferior to the skull, suspended by the **stylohyoid ligaments** (Figure 6-16●). The **body** of the hyoid serves as a base for several muscles concerned with movements of the tongue and larynx. Because muscles and ligaments form the only connections between the hyoid and other skeletal structures, the entire complex is quite mobile. The larger spinous processes on the hyoid are the **greater cornua** (singular *cornu*, "horns"), which help support the larynx and serve as the base for muscles that move the tongue. The **lesser cornua** are connected to the stylohyoid ligaments, and from these ligaments the hyoid and larynx hang beneath the skull like a swing from the limb of a tree.

Many superficial bumps and ridges are associated with the skeletal muscles described in Chapter 10; learning the names now will help you organize the material in that chapter. Tables 6-1 and 6-2 summarize information concerning the foramina and fissures introduced thus far. Table 6-1 is intended as a reference that will be especially important in later chapters dealing with the nervous and cardiovascular systems.

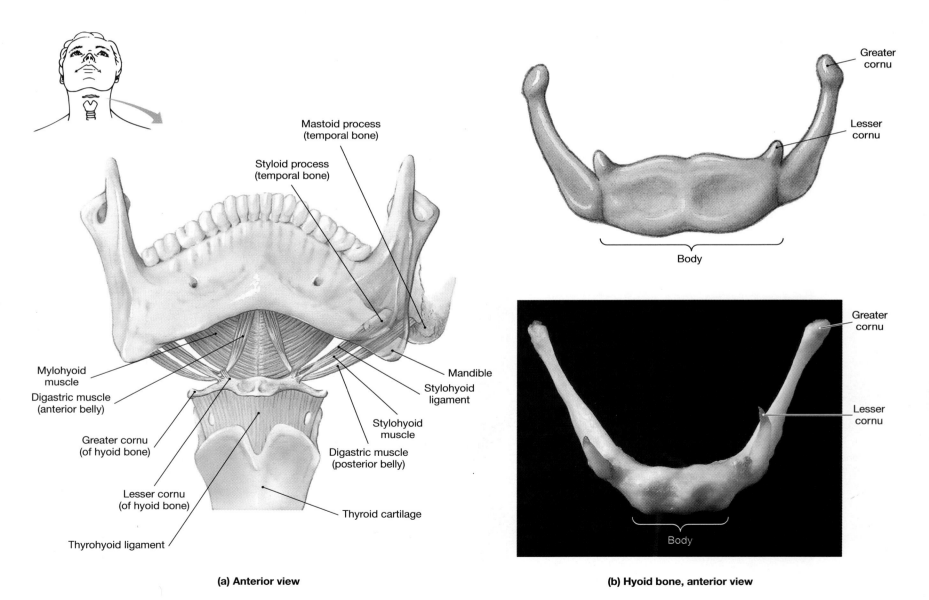

(a) Anterior view

(b) Hyoid bone, anterior view

FIGURE 6-16
The Hyoid Bone. (a) Hyoid bone shown in relation to muscles of the mandible, tongue, and the larynx. (b) Hyoid bone, isolated, anterior view.

150

TABLE 6-1   A Key to the Foramina and Fissures of the Skull

| Bone | Foramen/Fissure | Major Structures Using Passageway | |
		Neural Tissue	Vessels and Other Structures
OCCIPITAL BONE	Foramen magnum	Medulla oblongata (last portion of brain) and accessory nerve (XI).	Vertebral arteries and supporting membranes around CNS.
	Hypoglossal canal	Hypoglossal nerve (XII) provides motor control to muscles of the tongue.	
With temporal bone	Jugular foramen	Glossopharyngeal nerve (IX), vagus nerve (X), accessory nerve (XI). Nerve IX provides taste sensation; X is important for visceral functions; XI innervates important muscles of the back and neck.	Internal jugular vein, important vein returning blood from brain to heart.
FRONTAL BONE	Supraorbital foramen	Supraorbital nerve, sensory branch of the ophthalmic nerve, innervating the eyebrow, eyelid, and frontal sinus.	Supraorbital artery delivers blood to same region.
LACRIMAL BONE	Lacrimal foramen		Tear duct drains into the nasal chamber.
TEMPORAL BONE	Mastoid foramen	Facial nerve (VII) provides motor control of facial muscles.	Vessels to membranes around CNS.
	Stylomastoid foramen		
	Carotid foramen		Internal carotid artery, major arterial supply to the brain.
	External auditory canal		Air conducts sound to eardrum.
	Internal acoustic canal	Vestibulocochlear nerve (VIII) goes to sense organs for hearing and balance. Facial nerve (VII) enters here, exits at stylomastoid foramen.	Internal acoustic artery to inner ear.
SPHENOID BONE	Optic canal	Optic nerve (II) brings information from the eye to the brain.	Ophthalmic artery brings blood into orbit.
	Superior orbital fissure	Oculomotor nerve (III), trochlear nerve (IV), ophthalmic branch of trigeminal nerve (V), abducens nerve (VI). Ophthalmic nerve provides sensory information about eye and orbit; other nerves control muscles that move the eye.	Ophthalmic vein returns blood from orbit.
	Foramen rotundum	Maxillary branch of trigeminal nerve (V) provides sensation to the face.	
	Foramen ovale	Mandibular branch of trigeminal nerve (V) controls the muscles that move the lower jaw and provides sensory information from that area.	
With temporal and occipital bones	Foramen spinosum		Vessels to membranes around CNS.
	Foramen lacerum		Internal carotid artery leaves carotid canal, enters cranium via foramen lacerum.
With maxillary bone	Inferior orbital fissure	Maxillary branch of trigeminal nerve (V). See *Foramen rotundum.*	
ETHMOID BONE	Cribriform plate	Olfactory nerve (I) provides sense of smell.	
MAXILLARY BONE	Infraorbital foramen	Infraorbital nerve, maxillary branch of trigeminal nerve (V) from the inferior orbital fissure to face.	Infraorbital artery with the same distribution.
ZYGOMATIC BONE	Zygomaticofacial foramen	Zygomaticofacial nerve, sensory branch of mandibular nerve to cheek.	Zygomaticofacial nerve.
MANDIBLE	Mental foramen	Mental nerve, sensory nerve branch of the mandibular nerve, provides sensation from the chin and lips.	Mental vessels to chin and lips.
	Mandibular foramen	Inferior alveolar nerve, sensory branch of the mandibular nerve, provides sensation from the gums, teeth.	Inferior alveolar vessels supply same region.

TABLE 6-2   Surface Features and Foramina of the Skull

Region	Bone	Articulates with:	Surface Features	Functions	Foramina	Functions
CRANIUM (8)	**Occipital bone (1) (Figure 6-6)**	Parietal, temporal, sphenoid bones	**External:** Occipital condyles	Articulate with first cervical vertebra	Jugular foramen (with temporal)	Carries blood from smaller veins in the cranial cavity
			Occipital crest, external occipital protuberance, and inferior and superior nuchal lines	Attachment of muscles and ligaments that move the head and stabilize the occipital joint	Hypoglossal canal	Passageway for hypoglossal nerve that controls tongue muscles
			Internal: Internal occipital crest	Attachment of membranes that stabilize position of the brain		
	Parietal bones (2) (Figure 6-6)	Occipital, frontal, temporal, sphenoid bones	**External:** Superior and inferior temporal lines	Attachment of major jaw-closing muscle		
			Parietal eminence	Attachment of scalp to skull		
	Frontal bone (1) (Figure 6-7)	Parietal, sphenoid, ethmoid, nasal, maxillary, zygomatic bones	Metopic suture	Marks fusion of frontal bones in development	Supraorbital foramina	Passageways for sensory branch of ophthalmic nerve and supraorbital artery to the eyebrow and eyelid
			Frontal squama	Attachment of muscles of scalp		
			Supraorbital margin Lacrimal fossa	Protects eye Recesses enclose lacrimal glands		
			Frontal sinuses	Lighten bone and produce mucous secretions		
	Temporal bones (2) (Figure 6-8)	Occipital, parietal, frontal, sphenoid, zygomatic bones, and mandible; encloses auditory ossicles and suspends hyoid by stylohyoid ligaments	**External:** Mastoid process	Attachment of muscles that extend or rotate head	**External:** External auditory canal	Entrance and passage to tympanum
			Styloid process	Attachment of stylohyoid ligament and muscles attached to hyoid	Mastoid foramen	Passage for blood vessels to membranes of brain
			Mandibular fossa and articular tubercle	Form articulation with mandible	Stylomastoid foramen	Exit for nerve that controls facial muscles
			Zygomatic process	Articulates with zygomatic bone	Jugular foramen (with occipital bone)	Carries blood from smaller veins in the cranial cavity
			Squama	Attachment of jaw muscles	Carotid foramen	Entryway for carotid artery bringing blood to the brain
Internal:			Mastoid air cells	Lighten mastoid process	Foramen lacerum between temporal and occipital bones	Cartilage
			Petrous portion	Protects middle and inner ear	Auditory (pharyngotympanic) tube	Connects airspace of middle ear with pharynx
					Internal: Internal acoustic canal	Passage for blood vessels and nerves to the inner ear and stylomastoid foramen

TABLE 6-2   Surface Features and Foramina of the Skull (Continued)

Region	Bone	Articulates with:	Surface Features	Functions	Foramina	Functions
FACE (14)	**Sphenoid bone (1)** (Figure 6-9)	Occipital, frontal, temporal, ethmoid, zygomatic, maxillary, palatine bones, and vomer	**Internal:** Sella turcica	Protects pituitary gland	Optic foramen	Passage of optic nerve
			Anterior and posterior clinoid processes, optic groove	Protect pituitary gland and optic nerve	Superior orbital fissure	Entrance for nerves that control eye movements
			External: Pterygoid processes and spines	Attachment of jaw muscles	Foramen rotundum	Passage for sensory nerves from face
					Foramen ovale	Passage for nerves that control jaw movement
					Foramen spinosum	Passage of vessels to membranes around brain
	Ethmoid bone (1) (Figure 6-10)	Frontal, sphenoid, nasal, maxillary, lacrimal bones, and vomer	Crista galli	Attachment of membranes that stabilize position of brain	Cribriform plate	Passage of olfactory nerves
			Ethmoidal sinuses	Lighten bone and site of mucus production		
			Superior and middle conchae	Create turbulent airflow		
			Perpendicular plate	Separates nasal cavities (with vomer and nasal cartilage)		
	Zygomatic bones (2) (Figure 6-3c,d)	Frontal, temporal, sphenoid, maxillary bones	Temporal process	With zygomatic process of temporal, completes zygomatic arch for attachment of jaw muscles		
	Maxillary bones (2) (Figure 6-12)	Frontal, sphenoid, ethmoid, zygomatic, palatine, lacrimal bones, and inferior nasal concha	Orbital margin	Protects eye	Inferior orbital fissure	Exit for nerves entering skull at foramen rotundum
			Palatal process	Forms most of the bony palate	Infraorbital foramen	Passage of sensory nerves from face
			Maxillary sinus	Lightens bone, secretes mucus		
			Alveolar process	Surrounds articulations with teeth		
	Mandible (1) (Figure 6-15)	Temporal bones	Condylar process	Articulates with temporal	Mental foramen	Passage for sensory nerve from chin and lips
			Coronoid process	Attachment of temporalis muscle from parietal surface	Mandibular foramen	Passage of sensory nerve from teeth and gums
			Alveolar process	Protects articulations with teeth		
			Mylohyoid line	Attachment of muscle supporting floor of mouth		
			Mandibular groove	Protects salivary gland		

TABLE 6-2 Surface Features and Foramina of the Skull (Continued)

Region	Bone	Articulates with:	Surface Features	Functions	Foramina	Functions
	Nasal bones (2) (Figure 6-3c,d)	Frontal, ethmoid, maxillary bones		Support bridge of nose		
	Palatine bones (2) (Figure 6-12)	Sphenoid, maxillary bones, and vomer		Complete bony palate		
	Lacrimal bone (2) (Figure 6-3c,d)				Lacrimal foramen	Drains tears from orbit to nasal cavity
	Inferior nasal conchae (2) (Figure 6-3d)	Maxillary bone		Create turbulent airflow		
	Vomer (1) (Figures 6-3d,e, 6-5)	Ethmoid, maxillary, palatine bones		Forms inferior and posterior part of nasal septum		
ASSOCIATED BONES	Hyoid bone (1) (Figure 6-16)	Suspended by ligaments from styloid process of temporal bone; connected by ligaments to larynx	Greater cornua Lesser cornua	Attachment of tongue muscles and ligaments to larynx Attachment of stylohyoid ligaments		
	Auditory ossicles (6) (Figure 6-8c)	3 are enclosed by the petrous portion of each temporal bone		Conduct sound vibrations from tympanum to inner ear		

THE SKULLS OF INFANTS, CHILDREN, AND ADULTS
(Figure 6-17)

Many different centers of ossification are involved in the formation of the skull, but as development proceeds, fusion of the centers produces a smaller number of composite bones. For example, the sphenoid bone begins as 14 separate ossification centers. At birth fusion has not been completed, and there are two frontal bones, four occipital bones, and a number of sphenoid and temporal elements.

The skull organizes around the developing brain, and as the time of birth approaches, the brain enlarges rapidly. Although the bones of the skull are also growing, they fail to keep pace, and at birth the cranial bones are connected by areas of fibrous connective tissue. These connections are quite flexible, and the skull can be distorted without damage. Such distortion normally occurs during delivery and eases the passage of the infant along the birth canal. The fibrous areas between the cranial bones are known as **fontanels** (fon-tah-NELS; sometimes spelled *fontanelles*) (Figure 6-17a,b,c●).

The skulls of infants and adults differ in terms of the shape and structure of cranial elements, and this difference accounts for variations in proportions as well as in size. The most significant growth in the skull occurs before age 5, for at that time the brain stops growing and the cranial sutures develop. As a result, when compared with the skull as a whole, the cranium of a young child is relatively larger than that of an adult (Figure 6-17●).

■ CLINICAL BRIEF
Problems with Growth of the Skull

The growth of the cranium is usually coordinated with the expansion of the brain. Unusual distortions of the skull result from the premature closure of one or more fontanels, a condition called **craniostenosis** (krā-nē-ō-sten-Ō-sis; *stenosis*, narrowing). As the brain continues to enlarge, the rest of the skull accommodates it. A long and narrow head will be produced by early closure of the sagittal suture, whereas a very broad skull results if the coronal suture forms prematurely. Closure of *all* of the cranial sutures restricts the development of the brain, and surgery must be performed to prevent brain damage. However, if the brain enlargement stops because of genetic or developmental abnormalities, skull growth ceases as well. This condition, which results in a very undersized head, is called **microcephaly** (mī-krō-SEF-a-lē).

√ Identify the functions of the paranasal sinuses.

√ Why would a fracture of the coronoid process of the mandible make it difficult to close the mouth?

√ What symptoms would you expect to see in a person suffering from a fractured hyoid bone?

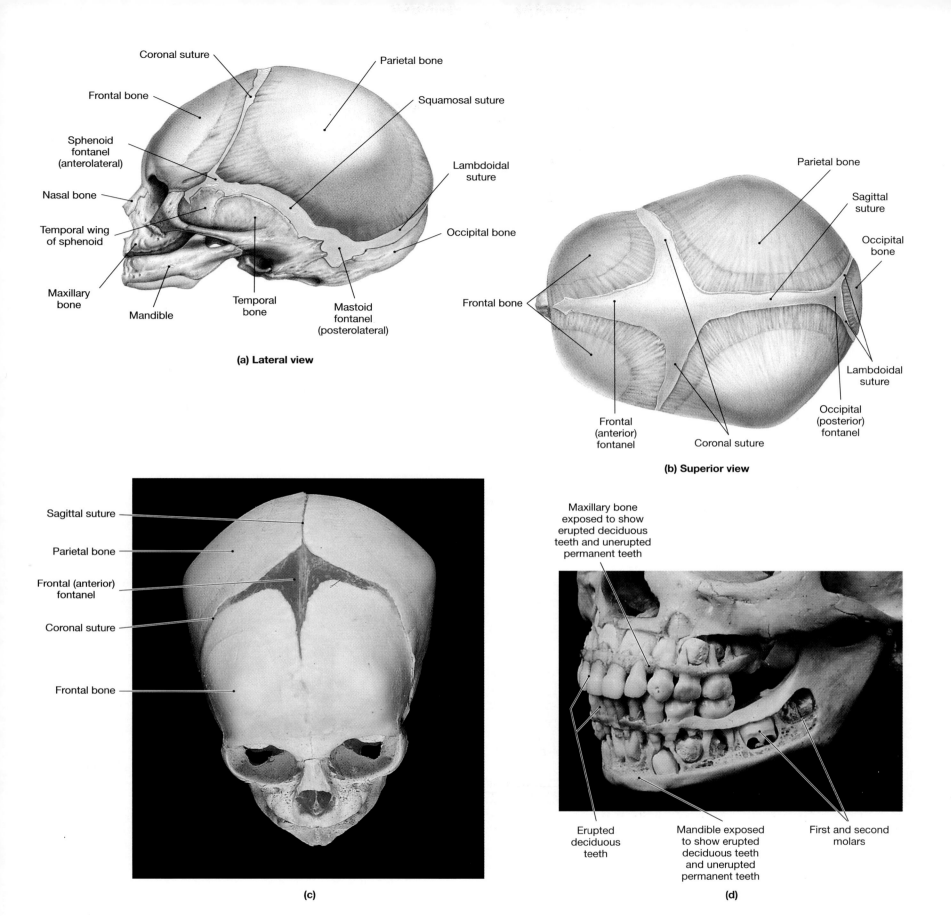

(a) Lateral view

Coronal suture
Parietal bone
Frontal bone
Squamosal suture
Sphenoid fontanel (anterolateral)
Lambdoidal suture
Nasal bone
Temporal wing of sphenoid
Occipital bone
Maxillary bone
Mandible
Temporal bone
Mastoid fontanel (posterolateral)

(b) Superior view

Parietal bone
Sagittal suture
Occipital bone
Frontal bone
Lambdoidal suture
Frontal (anterior) fontanel
Coronal suture
Occipital (posterior) fontanel

Sagittal suture
Parietal bone
Frontal (anterior) fontanel
Coronal suture
Frontal bone

(c)

Maxillary bone exposed to show erupted deciduous teeth and unerupted permanent teeth

Erupted deciduous teeth
Mandible exposed to show erupted deciduous teeth and unerupted permanent teeth
First and second molars

(d)

FIGURE 6-17

The Skull of an Infant. (a) Lateral view. The skull of an infant contains a greater number of individual bones than that of an adult. Many of the bones will eventually fuse; thus there will be fewer bones in the adult skull. The flat bones of the skull are separated by areas of fibrous connective tissue, allowing for cranial expansion and the distortion of the skull during birth. The large fibrous areas are called fontanels. By about age 4 these areas will disappear, and skull growth will be completed. (b) Superior view. (c) Infant skull, anterior view. (d) Skull of a young child, with part of the maxillary bone and mandible cut away to show the developing teeth.

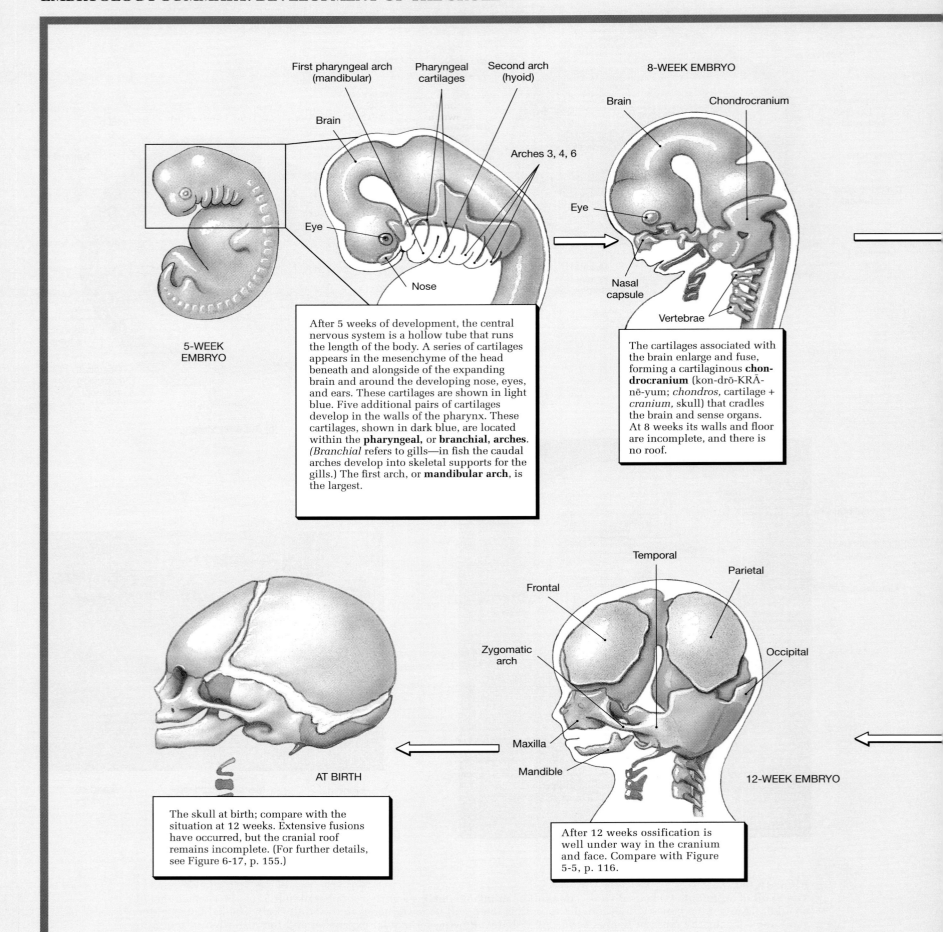

First pharyngeal arch (mandibular)

Pharyngeal cartilages

Second arch (hyoid)

Brain

Eye

Nose

Arches 3, 4, 6

5-WEEK EMBRYO

After 5 weeks of development, the central nervous system is a hollow tube that runs the length of the body. A series of cartilages appears in the mesenchyme of the head beneath and alongside of the expanding brain and around the developing nose, eyes, and ears. These cartilages are shown in light blue. Five additional pairs of cartilages develop in the walls of the pharynx. These cartilages, shown in dark blue, are located within the **pharyngeal**, or **branchial, arches.** *(Branchial* refers to gills—in fish the caudal arches develop into skeletal supports for the gills.) The first arch, or **mandibular arch,** is the largest.

8-WEEK EMBRYO

Brain

Chondrocranium

Eye

Nasal capsule

Vertebrae

The cartilages associated with the brain enlarge and fuse, forming a cartilaginous **chondrocranium** (kon-drō-KRĀ-nē-yum; *chondros,* cartilage + *cranium,* skull) that cradles the brain and sense organs. At 8 weeks its walls and floor are incomplete, and there is no roof.

Temporal

Parietal

Frontal

Zygomatic arch

Occipital

Maxilla

Mandible

12-WEEK EMBRYO

After 12 weeks ossification is well under way in the cranium and face. Compare with Figure 5-5, p. 116.

AT BIRTH

The skull at birth; compare with the situation at 12 weeks. Extensive fusions have occurred, but the cranial roof remains incomplete. (For further details, see Figure 6-17, p. 155.)

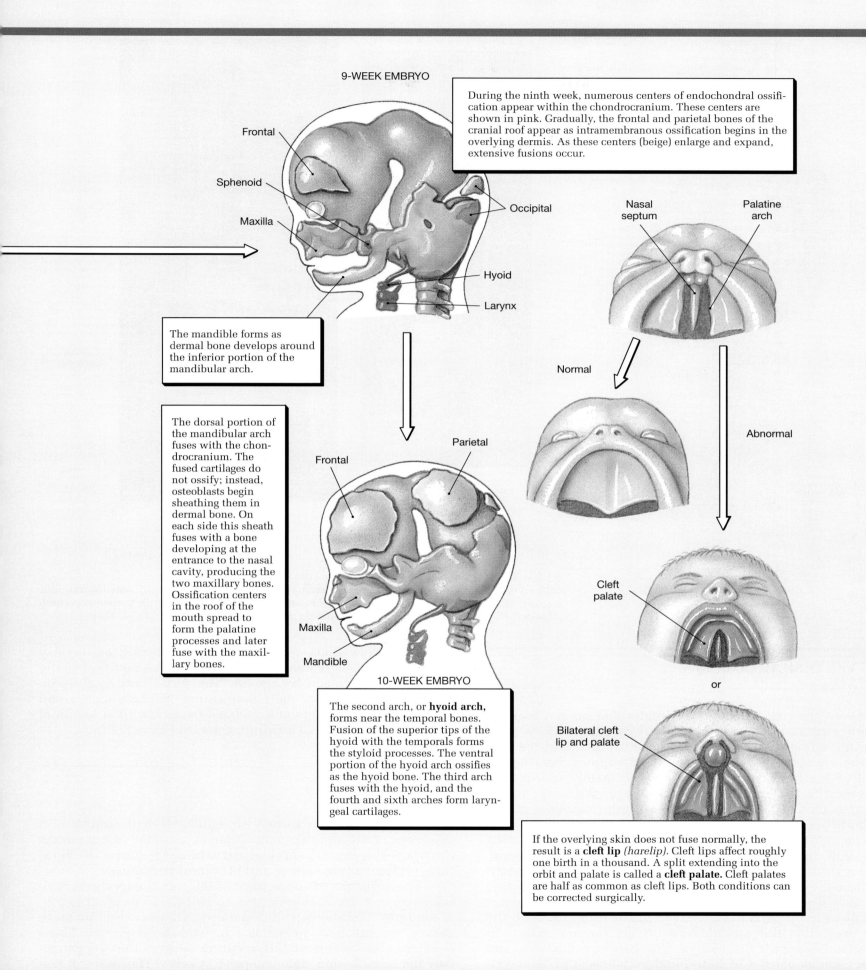

9-WEEK EMBRYO

Frontal

Sphenoid

Maxilla

Occipital

Hyoid

Larynx

During the ninth week, numerous centers of endochondral ossification appear within the chondrocranium. These centers are shown in pink. Gradually, the frontal and parietal bones of the cranial roof appear as intramembranous ossification begins in the overlying dermis. As these centers (beige) enlarge and expand, extensive fusions occur.

Nasal septum

Palatine arch

The mandible forms as dermal bone develops around the inferior portion of the mandibular arch.

Normal

The dorsal portion of the mandibular arch fuses with the chondrocranium. The fused cartilages do not ossify; instead, osteoblasts begin sheathing them in dermal bone. On each side this sheath fuses with a bone developing at the entrance to the nasal cavity, producing the two maxillary bones. Ossification centers in the roof of the mouth spread to form the palatine processes and later fuse with the maxillary bones.

Abnormal

Parietal

Frontal

Cleft palate

Maxilla

Mandible

10-WEEK EMBRYO

The second arch, or **hyoid arch**, forms near the temporal bones. Fusion of the superior tips of the hyoid with the temporals forms the styloid processes. The ventral portion of the hyoid arch ossifies as the hyoid bone. The third arch fuses with the hyoid, and the fourth and sixth arches form laryngeal cartilages.

or

Bilateral cleft lip and palate

If the overlying skin does not fuse normally, the result is a **cleft lip** *(harelip).* Cleft lips affect roughly one birth in a thousand. A split extending into the orbit and palate is called a **cleft palate.** Cleft palates are half as common as cleft lips. Both conditions can be corrected surgically.

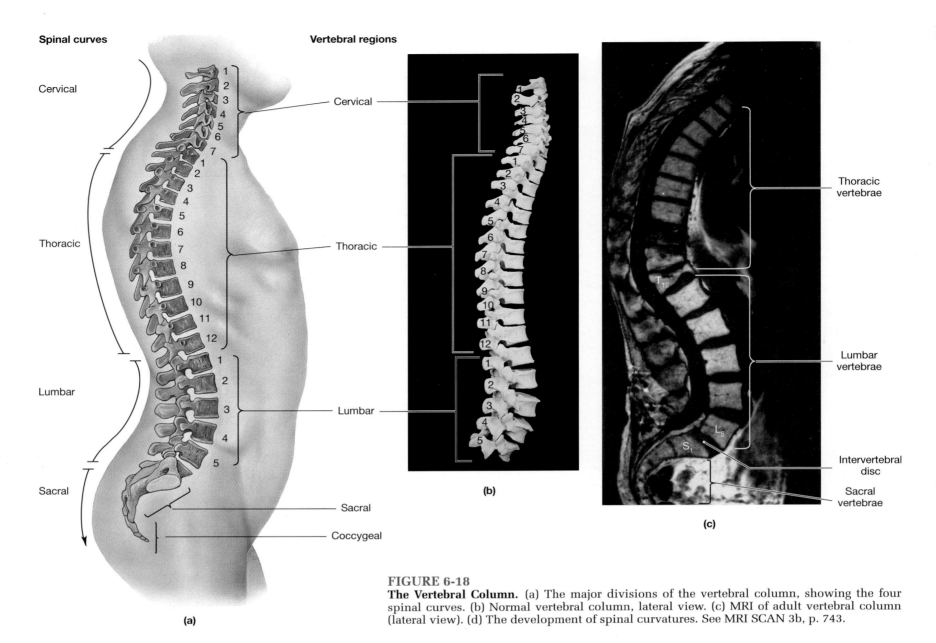

Spinal curves

Cervical

Thoracic

Lumbar

Sacral

Vertebral regions

Cervical

Thoracic

Lumbar

Sacral

Coccygeal

(a)

(b)

Thoracic vertebrae

Lumbar vertebrae

Intervertebral disc

Sacral vertebrae

(c)

FIGURE 6-18
The Vertebral Column. (a) The major divisions of the vertebral column, showing the four spinal curves. (b) Normal vertebral column, lateral view. (c) MRI of adult vertebral column (lateral view). (d) The development of spinal curvatures. See MRI SCAN 3b, p. 743.

THE VERTEBRAL COLUMN
(Figure 6-18)

The rest of the axial skeleton is subdivided on the basis of vertebral structure. The adult **vertebral column** consists of 26 bones, including the **vertebrae** (24), the *sacrum*, and the *coccyx*. The vertebrae provide a column of support, bearing the weight of the head, neck, trunk, and ultimately transferring that weight to the appendicular skeleton of the lower limbs. They also protect the spinal cord and help maintain an upright body position, as in sitting or standing.

The vertrebral column is divided into regions. Beginning at the skull, the regions are: *cervical, thoracic, lumbar, sacral,* and *coccygeal* (Figure 6-18●). Seven *cervical vertebrae* constitute the neck and extend inferiorly to the trunk. Twelve *thoracic vertebrae* form the midback regions, and each articulates with one or more pairs of ribs. Five *lumbar vertebrae* form the lower back; the fifth articulates with the sacrum, which in turn articulates with the coccyx. The cervical, thoracic, and lumbar regions consist of individual vertebrae. During development, the *sacrum* originates as a group of five vertebrae,

and the *coccyx* (KOK-siks), or "tailbone," begins as three to five very small vertebrae. The vertebrae of the sacrum usually complete their fusion by age 25. The distal coccygeal vertebrae do not complete their ossification before puberty, and thereafter fusion occurs at a variable pace. The total length of the vertebral column of an adult averages 71 cm (28 in.).

Spinal Curvature
(Figure 6-18)

The vertebrae do not form a straight and rigid structure. A side view of the adult spinal column shows four **spinal curves** (Figure 6-18a-c●): (1) **cervical curvature**, (2) **thoracic curvature**, (3) **lumbar curvature**, and (4) **sacral curvature**.

The sequence of appearance of the spinal curvatures is illustrated from fetus, to newborn, to child, and to adult in Figure 6-18d●. The thoracic and sacral curves are called **primary curves** because they appear late in fetal development. These are also called **accommodation curves** because they accommodate the thoracic and abdominopelvic viscera. The vertebral column in the newborn is C-shaped in contrast to the reversed S-shape

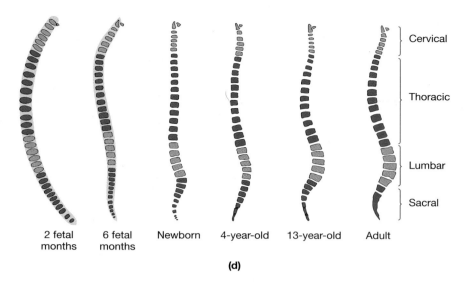

Cervical

Thoracic

Lumbar

Sacral

| 2 fetal months | 6 fetal months | Newborn | 4-year-old | 13-year-old | Adult |

(d)

of the adult, because only the primary curves are present. The lumbar and cervical curves, known as **secondary curves**, do not appear until several months after birth. These are also called **compensation curves** because they help shift the trunk weight over the legs. They become accentuated as the toddler learns to walk and run. All four curves are fully developed by the time a child is 10 years old.

When standing, the weight of the body must be transmitted through the vertebral column to the hips and ultimately to the lower extremities. Yet most of the body weight lies in front of the vertebral column. The various curves bring that weight in line with the body axis. Consider what people do automatically when they stand holding a heavy object. To avoid toppling forward, they exaggerate the lumbar curvature, bringing the weight closer to the body axis. This posture can lead to discomfort at the base of the spinal column. Similarly, women in the last 3 months of pregnancy often develop chronic back pain from the changes in lumbar curvature that adjust for the increasing weight of the fetus. No doubt you have seen pictures of African or South American people carrying heavy objects balanced on their heads. Such a practice increases the load on the vertebral column, but because the weight is aligned with the axis of the spine, the spinal curves are not affected and strain is minimized.

■ CLINICAL BRIEF
Kyphosis, Lordosis, Scoliosis

There are several abnormal distortions of the normal spinal curvature. In **kyphosis** (kī-FŌ-sis), the normal thoracic curvature becomes exaggerated posteriorly, producing a "roundback" appearance. This can be caused by (1) osteoporosis or a compression fracture affecting the anterior portions of vertebral centra, (2) chronic contractions in muscles that insert on the vertebrae, or (3) abnormal vertebral growth. In **lordosis** (lor-DŌ-sis), or "swayback," the abdomen and buttocks protrude because of an anterior exaggeration of the lumbar curvature.

Scoliosis (skō-lē-Ō-sis) involves an abnormal lateral curvature. This lateral deviation may occur in one or more of the movable vertebrae. Scoliosis is the most common distortion of the spinal curvature. Scoliosis may result from developmental problems, such as incomplete vertebral formation, or from muscular paralysis affecting one side of the back. In four out of five cases, it is impossible to determine the structural or functional cause of the abnormal spinal curvature. Scoliosis usually appears in girls during adolescence, when periods of growth are most rapid. Treatment consists of a combination of exercises, braces, and sometimes surgical modifications of the affected vertebrae. Early detection greatly improves the chances for successful treatment.

Vertebral Anatomy
(Figure 6-19)

Each vertebra consists of three basic parts: (1) a *body* (*centrum*), (2) a *vertebral arch*, and (3) *articular processes* (Figure 6-19●).

The Vertebral Body
(Figure 6-19)

The **body**, or *centrum* (plural *centra*) is the part of a vertebra that transfers weight along the axis of the vertebral column (Figure 6-19●). Each vertebra articulates with neighboring vertebrae; the bodies are interconnected by ligaments and separated by pads of fibrocartilage, the **intervertebral discs**.

The Vertebral Arch
(Figure 6-19)

The **vertebral arch** (Figure 6-19●), also called the **neural arch**, encloses the **vertebral foramen** that in life surrounds a portion of the spinal cord. The vertebral arch has a floor (the posterior surface of the body), walls (the *pedicles*) (PED-i-kls), and a roof (the *laminae*) (LA-mi-nē; singular *lamina*, "a thin plate"). The **pedicles** arise along the *posterolateral* (posterior and lateral) margins of the body. The **laminae** on either side extend *dorsomedially* (dorsally and medially) to complete the roof. From the fusion of the laminae, a **spinous process**, also known as a *spinal process*, projects dorsally and posteriorly from the midline. These processes can be seen and felt through the skin of the back. **Transverse processes** project laterally or dorsolaterally on both sides from the point where the laminae join the pedicles. These processes are sites of muscle attachment, and they may also articulate with the ribs.

The Articular Processes
(Figure 6-19)

The **articular processes** also arise at the junction between the pedicles and laminae. There is a superior and inferior articular process on each side of the vertebra. The **superior articular processes** project cranially; the **inferior articular processes** project caudally (Figure 6-19●).

Vertebral Articulation
(Figure 6-19)

The inferior articular processes of one vertebra articulate with the superior articular processes of the more caudal vertebra. Each articular process has a polished surface, called an **articular facet**. The superior processes have articular facets on their dorsal surfaces, whereas the inferior processes articulate along their ventral surfaces.

The vertebral arches of the vertebral column together form the **vertebral canal** that encloses the spinal cord. However, the spinal cord is not completely encased in bone. The vertebral bodies are separated by the intervertebral discs, and there are gaps between the pedicles of successive vertebrae. These **intervertebral foramina** (Figure 6-19●) permit the passage of nerves running to or from the enclosed spinal cord.

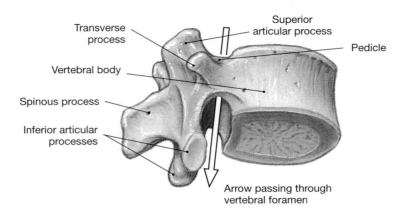

Transverse process

Superior articular process

Pedicle

Vertebral body

Spinous process

Inferior articular processes

Arrow passing through vertebral foramen

Lateral and inferior view

Vertebral Regions
(Figures 6-18a)

In references to the vertebrae, a capital letter indicates the vertebral region, and a number indicates the vertebra in question, starting with the cervical vertebra closest to the skull. For example, C_3 refers to the third cervical vertebra, with C_1 in contact with the skull; L_4 is the fourth lumbar vertebra, with L_1 in contact with the last thoracic vertebra (Figure 6-18a●). This shorthand will be used throughout the text.

Although each vertebra bears characteristic markings and articulations, focus on the general characteristics of each region and how the regional variations determine the vertebral group's basic function. Table 6-3 compares typical vertebrae from each region of the vertebral column.

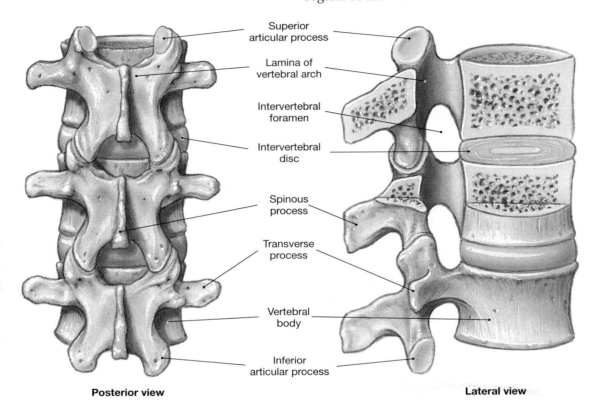

Superior articular process

Lamina of vertebral arch

Intervertebral foramen

Intervertebral disc

Spinous process

Transverse process

Vertebral body

Inferior articular process

Posterior view

Lateral view

FIGURE 6-19
Vertebral Anatomy. The anatomy of a typical vertebra from posterior and lateral views, and the arrangement of articulations between vertebrae.

TABLE 6-3   Regional Differences in Vertebral Structure and Function

Type (Number)	Location	Centrum	Vertebral Foramen	Spinous Process	Transverse Process	Functions
Cervical vertebrae (7) (*see Figure 6-20*)	Neck	Small, oval, curved faces	Large	Long, split, tip points caudally	Has transverse foramen	Support skull, stabilize relative positions of brain and spinal cord, allow controlled head movement
Thoracic vertebrae (12) (*see Figure 6-22*)	Chest	Medium, heart-shaped, flat faces, facets for rib articulations	Smallest	Long, slender, not split, tip points caudally	All but two (T_{11}, T_{12}) have facets for rib articulations	Support weight of head, neck, upper limbs, organs of thoracic cavity; articulations with ribs allow changes in volume of thoracic cage
Lumbar vertebrae (5) (*see Figure 6-23*)	Lower back	Massive, oval, flat faces	Smaller	Blunt, broad, tip points posteriorly	Short, no articular facets or transverse foramina	Support weight of head, neck, upper limbs, organs of thoracic and abdominal cavities

Cervical Vertebrae
(Figure 6-20)

The seven **cervical vertebrae** are the smallest of the vertebrae (Figure 6-20●). They extend from the occipital bone of the skull to the thorax. Notice that the body of a cervical vertebra is relatively small as compared with the size of the vertebral foramen. At this level the spinal cord still contains most of the nerves that connect the brain to the rest of the body. As you continue along the vertebral canal, the diameter of the spinal cord decreases, and so does the diameter of the vertebral arch. On the other hand, cervical vertebrae support only the weight of the head, so the vertebral bodies can be relatively small and light. As you continue caudally along the column, the loading increases and the centra gradually enlarge.

In a cervical vertebra, the superior surface of the body is concave from side to side, and it slopes, with the anterior edge inferior to the posterior edge. The spinous process is relatively stumpy, usually shorter than the diameter of the vertebral foramen, and the tip of the process bears a prominent notch. Laterally, the transverse processes are fused to the **costal processes** that originate near the ventrolateral portion of the body. *Costal* refers to rib, and these processes represent the fused remnants of cervical ribs. The costal and transverse processes encircle prominent, round, **transverse foramina**. In life these passageways protect the *vertebral arteries and veins*, important blood vessels to the brain.

This description would be adequate to identify all but the first two cervical vertebrae. When cervical vertebrae C_3–C_7 articulate, their interlocking centra permit a relatively greater degree of flexibility than do those of other regions. The first two cervical vertebrae are unique and the seventh is modified. Table 6-3 summarizes the features of cervical vertebrae.

THE ATLAS (C_1) *(Figure 6-21a,b).* The **atlas** (C_1) holds up the head, articulating with the occipital condyles of the skull (Figure 6-21a,b●). It is named after Atlas, a figure in Greek mythology who held up the world. The articulation between the occipital condyles and the atlas is a joint that permits nodding (as when indicating "yes"), but prevents twisting. The atlas can be distinguished from the other vertebrae by the following features: (1) the lack of a body, (2) possession of semicircular **anterior** and **posterior vertebral arches**, each containing **anterior** and **posterior tubercles**; and (3) presence of oval **superior facets** and round **inferior articular facets**.

The atlas articulates with the second cervical vertebra, the *axis*. This articulation permits rotation (as when shaking the head to indicate "no").

THE AXIS (C_2) *(Figure 6-21c,d).* During development, the body of the atlas fuses to the body of the second cervical vertebra, called the **axis** (C_2) (Figure 6-21c,d●). This fusion creates the prominent **dens** (*denz*, tooth), or **odontoid process** (ō-DON-toid; *odontos*, tooth) of the axis. A transverse ligament binds the dens to the inner surface of the atlas, forming a pivot for rotation of the atlas and skull. Important muscles controlling the position of the head and neck attach to the especially robust spinous process of the axis.

In a child the fusion between the dens and axis is incomplete, and impacts or even severe shaking can cause dislocation of the dens and severe damage to the spinal cord. In the adult, a blow to the base of the skull can be equally dangerous because a dislocation of the axis-atlas joint can force the dens into the base of the brain, with fatal results.

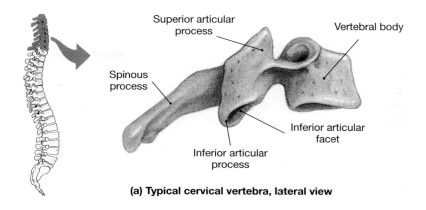

(a) Typical cervical vertebra, lateral view

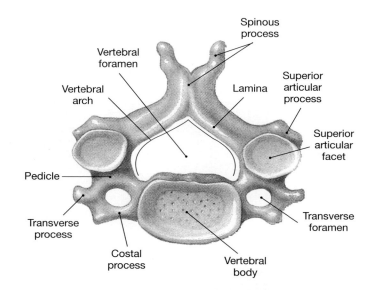

(b) Typical cervical vertebra, superior view

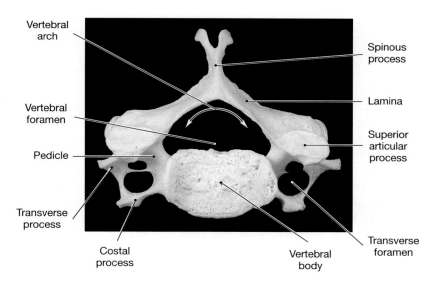

(c) Cervical vertebra, superior view

FIGURE 6-20
Cervical Vertebrae. (a) Lateral view of a typical cervical vertebra (C_3–C_6). (b) Superior view of the same vertebra. Note the characteristic features listed in Table 6-3. See also MRI SCAN 3a, p. 743.

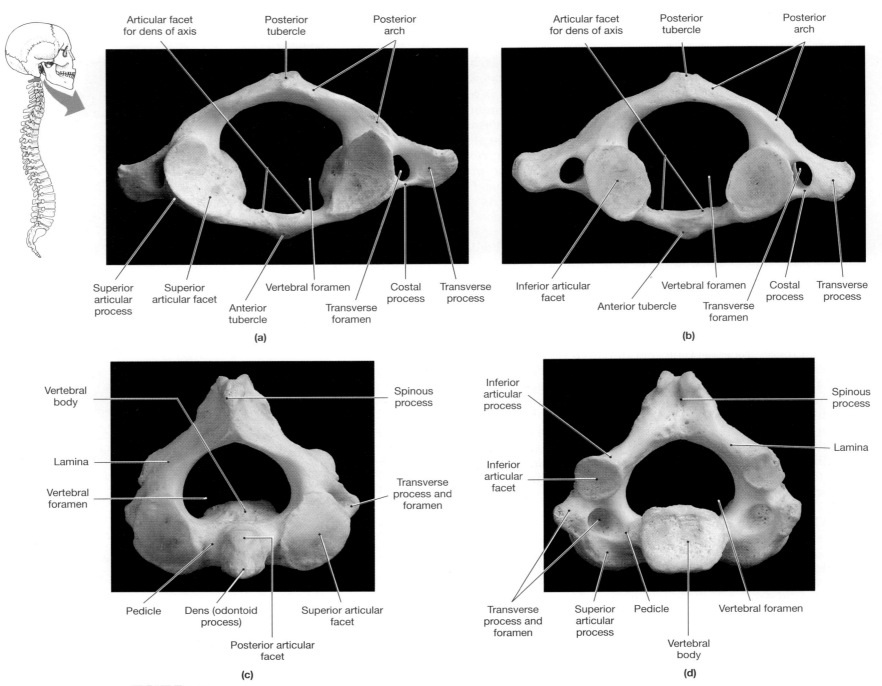

Articular facet for dens of axis   Posterior tubercle   Posterior arch

Superior articular process   Superior articular facet   Anterior tubercle   Vertebral foramen   Transverse foramen   Costal process   Transverse process

(a)

Articular facet for dens of axis   Posterior tubercle   Posterior arch

Inferior articular facet   Anterior tubercle   Vertebral foramen   Transverse foramen   Costal process   Transverse process

(b)

Vertebral body   Spinous process

Lamina

Vertebral foramen   Transverse process and foramen

Pedicle   Dens (odontoid process)   Posterior articular facet   Superior articular facet

(c)

Inferior articular process   Spinous process

Inferior articular facet   Lamina

Transverse process and foramen   Superior articular process   Pedicle   Vertebral foramen   Vertebral body

(d)

FIGURE 6-21
Atlas and Axis. (a) Atlas, superior view. (b) Atlas, inferior view. (c) Axis, superior view. (d) Axis, inferior view.

VERTEBRA PROMINENS (C₇) *(Figure 6-22e).* The transition from one vertebral region to another is not abrupt, and the last vertebra of one region usually resembles the first vertebra of the next. The **vertebra prominens** (C₇) has a long, slender spinous process that ends in a broad tubercle that can be felt beneath the skin at the base of the neck. This vertebra, shown in Figure 6-22e●, is the interface between the cervical curve, which arches forward, and the thoracic curve, which arches backward. The transverse processes are large, providing additional surface area for muscle attachment, and the transverse foramina are either reduced or absent. A large elastic ligament, the **ligamentum nuchae** (li-ga-MEN-tum NOO-kē; *nucha*, nape) begins at the vertebra prominens and extends cranially to an insertion along the external occipital crest. Along the way, it attaches to the spinous processes of the other cervical vertebrae. When the head is upright, this liga-

ment acts like the string on a bow, maintaining the cervical curvature without muscular effort. If the neck has been bent forward, the elasticity in this ligament helps return the head to an upright position.

The head is relatively massive, and it sits atop the cervical vertebrae like a soup bowl on the tip of a finger. With this arrangement, small muscles can produce significant effects by tipping the balance one way or another. But if the body suddenly changes position, as in a fall or during rapid acceleration (a jet taking off) or deceleration (a car crash), the balancing muscles are not strong enough to stabilize the head. A dangerous partial or complete dislocation of the cervical vertebrae can result, with injury to muscles and ligaments and potential injury to the spinal cord. The term **whiplash** is used to describe such an injury, because the movement of the head resembles the cracking of a whip.

162

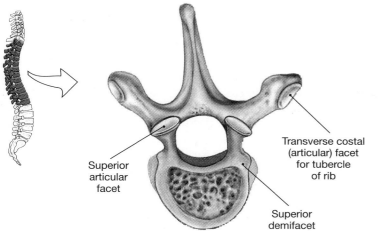

Thoracic Vertebrae
(Figure 6-22)

There are 12 **thoracic vertebrae**. A typical thoracic vertebra (Figure 6-22●) has a distinctive heart-shaped body that is more massive than that of a cervical vertebra. The vertebral foramen is relatively smaller, and the long, slender spinous process projects posterocaudally. The spinous processes of T_{10}, T_{11}, and T_{12} increasingly resemble those of the lumbar series, as the transition between the thoracic and lumbar curvatures approaches. Because of the weight carried by the lower thoracic and lumbar vertebrae, it is difficult to stabilize the transition between the thoracic and lumbar curves. As a result, compression fractures or compression-dislocation fractures after a hard fall most often involve the last thoracic and first two lumbar vertebrae.

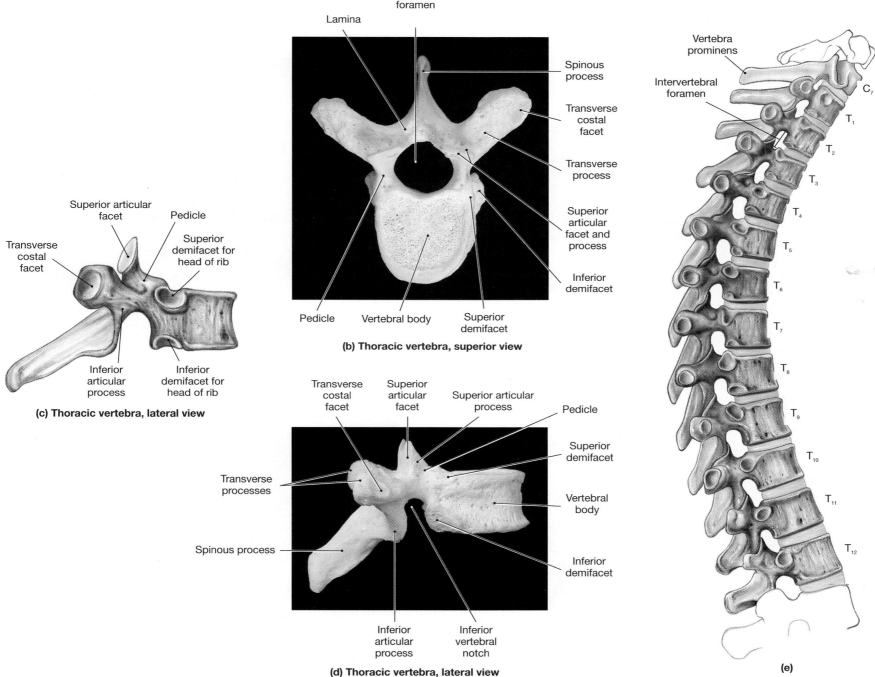

(a) Thoracic vertebra, superior view

(b) Thoracic vertebra, superior view

(c) Thoracic vertebra, lateral view

(d) Thoracic vertebra, lateral view

(e)

FIGURE 6-22
Thoracic Vertebrae. (a,b) Thoracic vertebra, superior views. (c,d) Thoracic vertebra, lateral views. Note the characteristic features listed in Table 6-3. (e) Lateral view of the thoracic region of the spinal column. The vertebra prominens (C_7) resembles T_1, but it lacks facets for rib articulation. Vertebra T_{12} resembles the first lumbar vertebra (L_1), but it has a facet for rib articulation. See MRI SCAN 3b, p. 743.

Each thoracic vertebra articulates with ribs along the dorso-lateral surfaces of the body. The location and structure of the articulations vary somewhat from vertebra to vertebra (Figure 6-22e●). Rib pairs 2 through 8 originate between adjacent vertebrae, so vertebrae T_2 to T_8 have **superior** and **inferior demifacets** on each side. The first rib originates at the body of T_1, so that the vertebra has an articular facet and an inferior demifacet on each side, but no superior demifacets. Vertebra T_9 has only a pair of superior demifacets, whereas T_{10}, T_{11}, and T_{12} have a single *whole facet* on either side.

The transverse processes of vertebrae T_1 to T_{10} are relatively thick, and they contain **transverse costal facets** for articulation with the tubercles of ribs. Thus, ribs 1 through 10 contact their vertebrae at two points, at an articular facet (ribs 1 and 10) or demifacet (ribs 2–9) and at a transverse costal facet. Table 6-3, p. 160, summarizes the features of the thoracic vertebrae.

Lumbar Vertebrae
(Figure 6-23)

The **lumbar vertebrae** are the largest of the vertebrae. The body of a typical lumbar vertebra (Figure 6-23●) is thicker than that of a thoracic vertebra, and the superior and inferior surfaces are oval rather than heart-shaped. There are no articular facets on either the body or the transverse processes, and the vertebral foramen is triangular. The transverse processes are slender and project dorsolaterally, and the stumpy spinous processes project dorsally.

The lumbar vertebrae bear the most weight, and their massive spinous processes provide surface area for the attachment of lower back muscles that reinforce or adjust the lumbar curvature. Table 6-3, p. 160, summarizes the characteristics of lumbar vertebrae.

FIGURE 6-23
Lumbar Vertebrae. (a,b) Lateral view of a typical lumbar vertebra. (c,d) Superior view. See MRI SCAN 3b, p. 743.

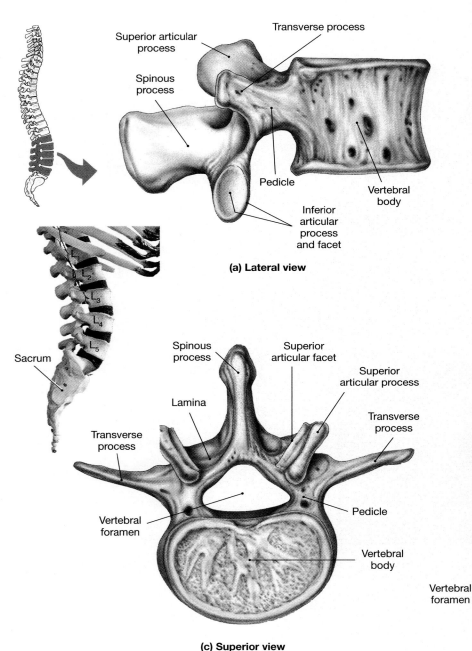

(a) Lateral view

(c) Superior view

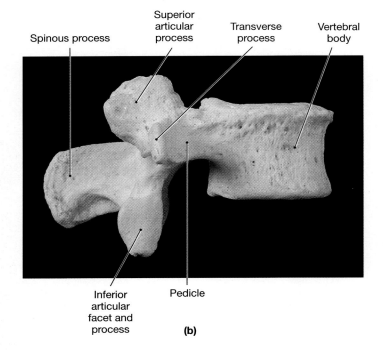

(b)

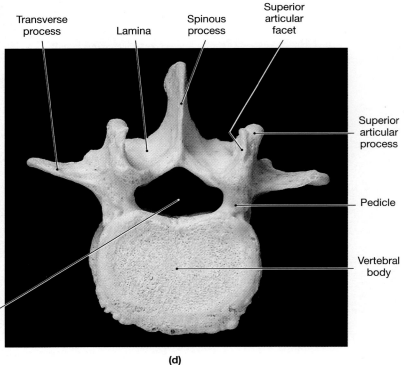

(d)

The Sacrum
(Figure 6-24)

The **sacrum** consists of the fused components of five sacral vertebrae. These vertebrae begin fusing shortly after puberty and are usually completely fused between ages 25 and 30. This composite structure affords protection for reproductive, digestive, and excretory organs, and, via paired articulations, attaches the axial skeleton to the pelvic girdle of the appendicular skeleton. The broad surface area of the sacrum provides an extensive area for the attachment of muscles, especially those responsible for movement of the thigh. Figure 6-24● shows the posterior, lateral, and anterior surfaces of the sacrum.

The sacrum is curved, with a convex dorsal surface (Figure 6-24a●). The narrow, caudal portion is the sacral **apex**, whereas the broad superior surface forms the **base**. The **articular processes** form synovial articulations with the last lumbar vertebra. The **sacral canal** begins between those processes and extends the length of the sacrum. Nerves and membranes that line the vertebral canal in the spinal cord continue into the sacral canal.

The spinous processes of the five fused sacral vertebrae form a series of elevations along the **median sacral crest**. The laminae of the fifth sacral vertebra fail to contact one another at the midline, and they form the **sacral cornua**. These ridges establish the margins of the **sacral hiatus** (hī-Ā-tus), the end of the sacral canal. In life this opening is covered by connective tissues. On either side of the median sacral crest, the **sacral foramina** represent the intervertebral foramina, now enclosed by the fused sacral bones. A broad sacral **wing**, or **ala**, extends laterally from each **lateral sacral crest**. The median and lateral sacral crests provide surface area for the attachment of muscles of the lower back and hip.

Viewed laterally (Figure 6-24b●), the *sacral curvature* is more apparent. The degree of curvature is greater in males than in females (see Table 7-1, p. 202). Laterally, the **auricular surface** marks the site of articulation with the pelvic girdle, the **sacroiliac joint**. Dorsal to the auricular surface is a roughened area, the **sacral tuberosity**, marking the attachment of a ligament that stabilizes this articulation. The anterior surface of the sacrum is concave (Figure 6-24c●), and after fusion is completed, prominent *transverse lines* mark the former boundaries of individual vertebrae. At the apex, a flattened area marks the site of articulation with the *coccyx*.

The Coccyx
(Figure 6-24)

The small **coccyx** consists of three to five (most often four) coccygeal vertebrae that have usually begun fusing by age 26 (Figure 6-24●). The coccyx provides an attachment site for a number of ligaments and for a muscle that constricts the anal opening. The first two coccygeal vertebrae have transverse processes and unfused vertebral arches. The prominent laminae of the first are known as the **coccygeal cornua**, and they curve to meet the cornua of the sacrum. The coccygeal vertebrae do not complete their fusion until late in adulthood. In very old people, the coccyx may fuse with the sacrum.

■ CLINICAL BRIEF
Spina Bifida

Spina bifida (SPĪ-na BI-fi-da) results when the vertebral laminae fail to unite during development. The vertebral arch is incomplete, and the membranes that line the dorsal body cavity bulge outward. In mild cases, most often involving the sacral and lumbar regions, the condition may pass unnoticed. In severe cases, the entire spinal column and skull are affected. This condition is often associated with developmental abnormalities of the brain and spinal cord. See Embryology Summary: Development of Spinal Cord and Nerves, Chapter 14.

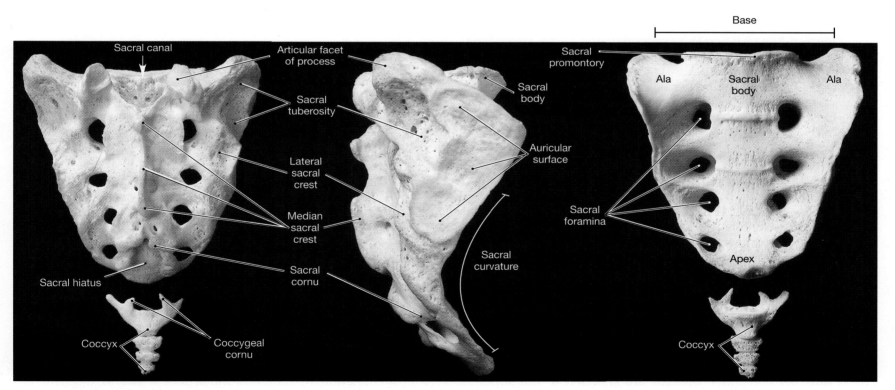

(a) Posterior surface   (b) Lateral surface   (c) Anterior surface

FIGURE 6-24
The Sacrum and Coccyx. (a) Posterior view. (b) Lateral view from the right side. (c) Anterior view.

THE THORACIC CAGE
(Figure 6-25a,c)

The skeleton of the chest, or **thoracic cage**, consists of the thoracic vertebrae, the *ribs*, and the *sternum* (Figure 6-25a,c●). The *ribs*, or **costae**, and the sternum form the **rib cage** and support the walls of the thoracic cavity. The thoracic cage serves two functions:

- It protects the heart, lungs, thymus, and other structures in the thoracic cavity; and
- It serves as an attachment point for muscles involved with (1) respiration, (2) the position of the vertebral column, and (3) movements of the pectoral girdle and upper extremity.

The Ribs
(Figures 6-22/6-25)

Ribs are elongated, curved, flattened bones that (1) originate on or between thoracic vertebrae and (2) end in the wall of the thoracic cavity. There are 12 pairs of ribs (Figure 6-25●). The first seven pairs are called **true**, or **vertebrosternal**, **ribs**. They reach the anterior body wall and are connected to the sternum by separate cartilaginous extensions, the **costal cartilages**. Beginning with the first rib, the vertebrosternal ribs gradually increase in length and in the radius of curvature.

Ribs 8–12 are called **false ribs** because they do not attach directly to the sternum. The costal cartilages of ribs 8–10, the **vertebrochondral ribs**, fuse together before reaching the sternum (Figure 6-25a●). The last two pairs of ribs are called **floating ribs** because they have no connection with the sternum.

Figure 6-25b● shows the superior surface of a typical rib. The *vertebral end* of the rib articulates with the vertebral column at the **head**, or **capitulum** (ka-PIT-ū-lum). When the rib articulates between adjacent vertebrae, the articular surface is divided into **superior** and **inferior articular facets** by the **interarticular crest**. After a short **neck**, the **tubercle**, or **tuberculum** (tū-BER-kū-lum), projects dorsally. The inferior portion of the tubercle contains an articular facet that contacts the transverse process of the thoracic vertebra.

Ribs 1 and 10 originate at whole facets on vertebrae T_1 and T_{10}, and their tubercular facets articulate with their respective vertebrae. Ribs 2–9 originate at demifacets, and their tubercular facets articulate with the inferior member of the vertebral pair. Ribs 11 and 12 originate at whole facets on T_{11} and T_{12}. These ribs do not have tubercular facets. The difference in rib orientation can be seen by comparing Figures 6-22, p. 163, and 6-25c●.

The bend, or **angle**, of the rib indicates the site where the tubular **body**, or **shaft**, begins curving toward the sternum. The internal rib surface is concave, and a prominent **costal groove** along its inferior border marks the path of nerves and blood vessels. The superficial surface is convex and provides an attachment site for muscles of the pectoral girdle and trunk. The **intercostal muscles** that move the ribs are attached to the superior and inferior surfaces.

With their complex musculature, dual articulations at the vertebrae, and flexible connection to the sternum, the ribs are quite mobile. Note how the ribs curve away from the vertebral column to angle downward (Figure 6-25d●). Functionally, a typical rib acts as if it were the handle on a bucket, lying just below the horizontal plane. Pushing it down forces it inward; pulling it up swings it outward. In addition, because of the curvature of the ribs, the same movements change the position of the sternum. Depressing the ribs moves the sternum posteriorly (inward), whereas elevation moves it anteriorly (outward). As a result, movements of the ribs affect both the width and the depth of the thoracic cage, increasing or decreasing its volume accordingly.
✝ *The Thoracic Cage and Surgical Procedures [p. 755]*

The Sternum
(Figure 6-25a)

The adult **sternum** is a flat bone that forms in the anterior midline of the thoracic wall (Figure 6-25a●). The sternum has three components:

- The broad, triangular **manubrium** (ma-NŪ-brē-um) articulates with the clavicles and the cartilages of the first pair of ribs. This is the widest and most superior portion of the sternum. Only the first pair of ribs is attached by cartilage to this portion of the sternum. The **jugular notch** is the shallow indentation on the superior surface of the manubrium. It is located between the clavicular articulations.
- The tongue-shaped **body** attaches to the inferior surface of the manubrium and extends caudally along the midline. Individual costal cartilages from rib pairs 2–7 are attached to this portion of the sternum. The rib pairs 8–10 are also attached to the body, but by a single pair of cartilages shared with rib pair 7.
- The **xiphoid** (ZĪ-foid) **process**, the smallest part of the sternum, is attached to the inferior surface of the body. The muscular *diaphragm* and *rectus abdominis muscles* attach to the xiphoid process.

Ossification of the sternum begins at six to ten different centers, and fusion is not completed until at least age 25. Before age 25, the sternal body consists of four separate bones. Their boundaries can be detected as a series of transverse lines crossing the adult sternum. The xiphoid process is usually the last of the sternal components to undergo ossification and fusion. Its connection to the body of the sternum can be broken by an impact or strong pressure, creating a spear of bone that can severely damage the liver. To reduce the chances of that happening, strong emphasis is placed on the proper positioning of the hand during cardiopulmonary resuscitation (CPR) training.

√ **Joe suffered a hairline fracture at the base of the odontoid process. What bone is fractured, and where would you find it?**

√ **Improper administration of CPR (cardiopulmonary resuscitation) could result in a fracture of what bone?**

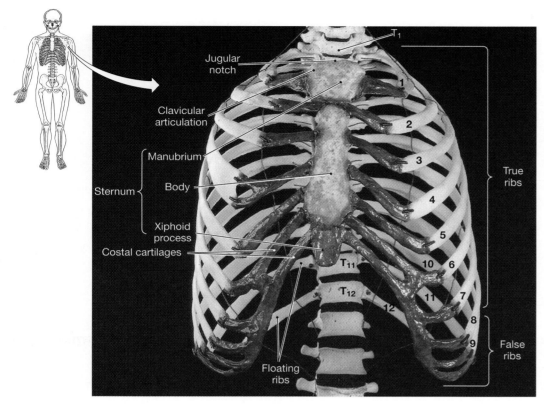

(a) Anterior view

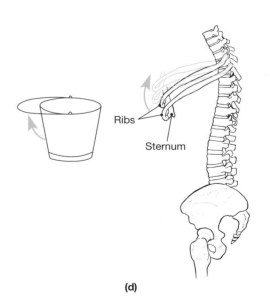

(b) Posterior view

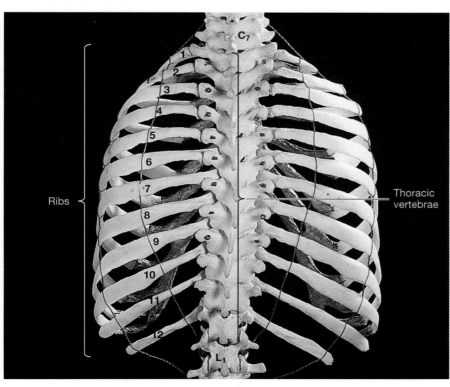

(c) Posterior view

(d)

FIGURE 6-25
The Thoracic Cage. (a) Anterior view of the rib cage and sternum. (b) Details of rib structure and the articulations between the ribs and thoracic vertebrae. (c) Posterior view of the rib cage. (d) Effect of rib elevation on the thoracic cavity. These movements are important in respiration.

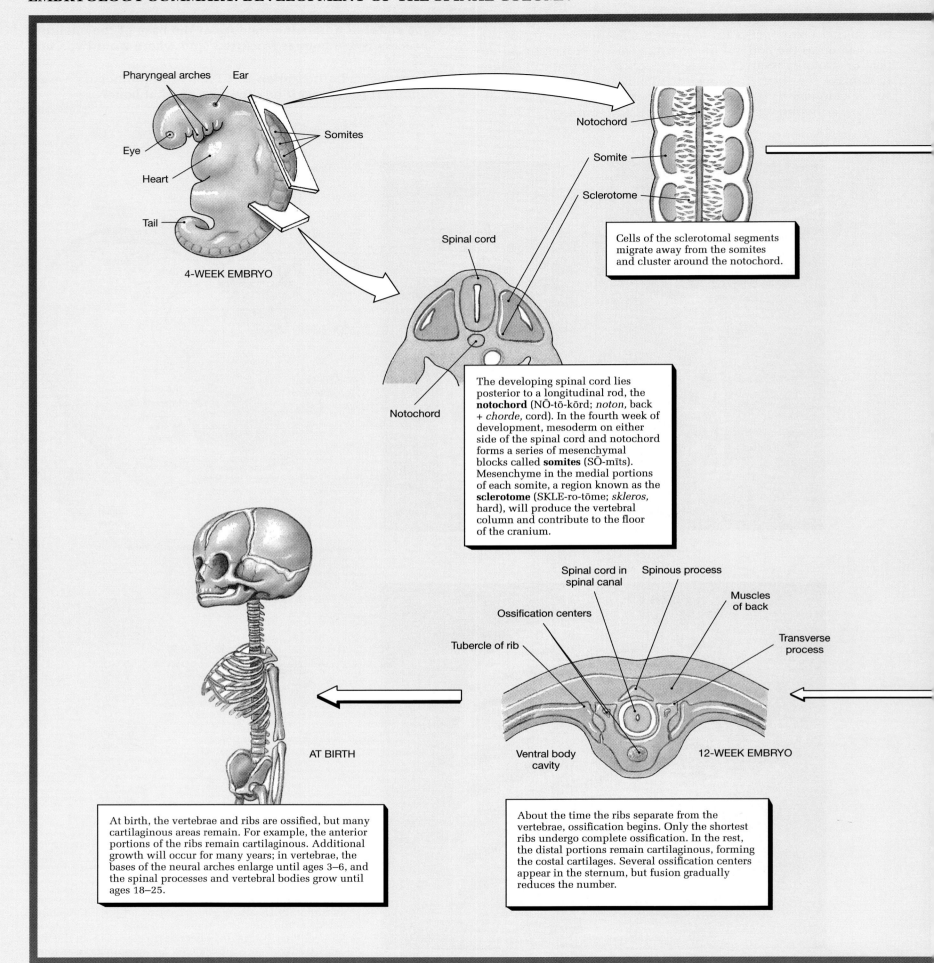

Pharyngeal arches **Ear**

Eye

Heart

Tail

Somites

4-WEEK EMBRYO

Notochord

Somite

Sclerotome

Cells of the sclerotomal segments migrate away from the somites and cluster around the notochord.

Spinal cord

Notochord

The developing spinal cord lies posterior to a longitudinal rod, the **notochord** (NŌ-tō-kōrd; *noton,* back + *chorde,* cord). In the fourth week of development, mesoderm on either side of the spinal cord and notochord forms a series of mesenchymal blocks called **somites** (SŌ-mīts). Mesenchyme in the medial portions of each somite, a region known as the **sclerotome** (SKLE-ro-tōme; *skleros,* hard), will produce the vertebral column and contribute to the floor of the cranium.

Spinal cord in spinal canal **Spinous process**

Ossification centers **Muscles of back**

Tubercle of rib **Transverse process**

Ventral body cavity **12-WEEK EMBRYO**

AT BIRTH

At birth, the vertebrae and ribs are ossified, but many cartilaginous areas remain. For example, the anterior portions of the ribs remain cartilaginous. Additional growth will occur for many years; in vertebrae, the bases of the neural arches enlarge until ages 3–6, and the spinal processes and vertebral bodies grow until ages 18–25.

About the time the ribs separate from the vertebrae, ossification begins. Only the shortest ribs undergo complete ossification. In the rest, the distal portions remain cartilaginous, forming the costal cartilages. Several ossification centers appear in the sternum, but fusion gradually reduces the number.

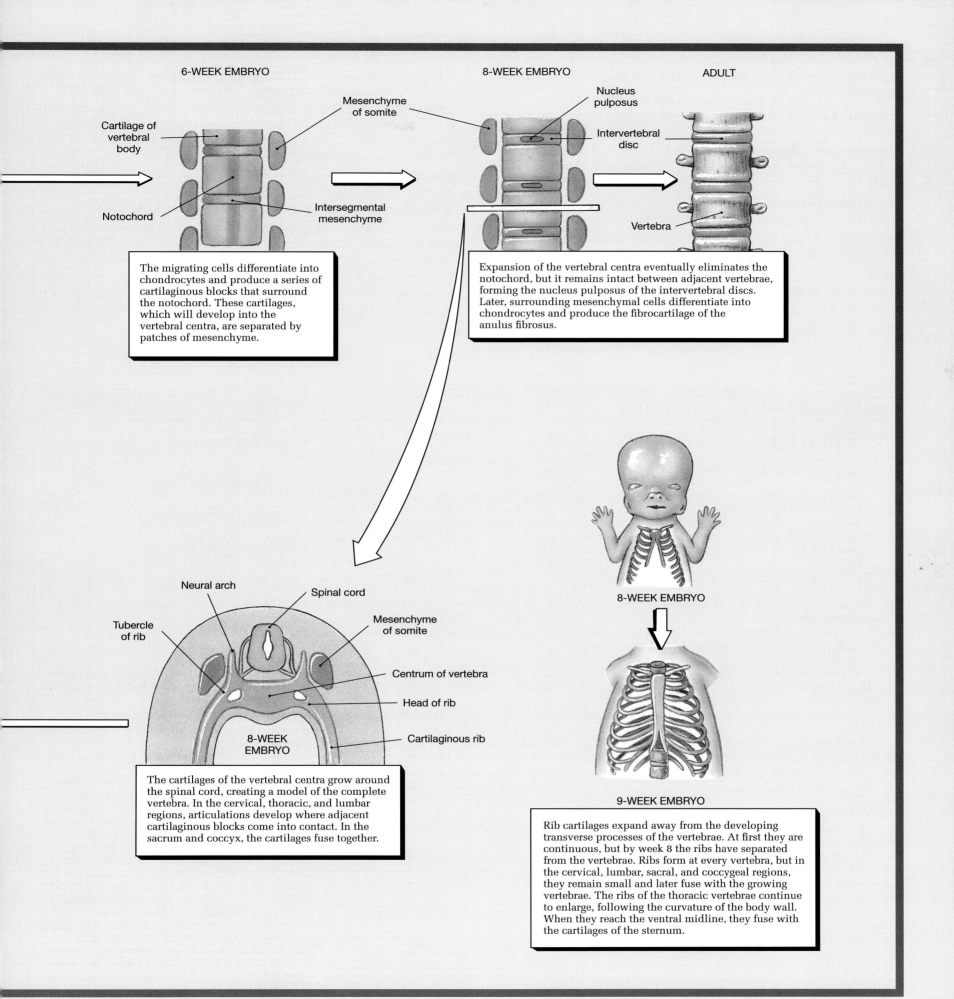

6-WEEK EMBRYO

Cartilage of vertebral body

Mesenchyme of somite

Notochord

Intersegmental mesenchyme

The migrating cells differentiate into chondrocytes and produce a series of cartilaginous blocks that surround the notochord. These cartilages, which will develop into the vertebral centra, are separated by patches of mesenchyme.

8-WEEK EMBRYO

Nucleus pulposus

Intervertebral disc

ADULT

Vertebra

Expansion of the vertebral centra eventually eliminates the notochord, but it remains intact between adjacent vertebrae, forming the nucleus pulposus of the intervertebral discs. Later, surrounding mesenchymal cells differentiate into chondrocytes and produce the fibrocartilage of the anulus fibrosus.

Neural arch

Tubercle of rib

Spinal cord

Mesenchyme of somite

Centrum of vertebra

Head of rib

Cartilaginous rib

8-WEEK EMBRYO

The cartilages of the vertebral centra grow around the spinal cord, creating a model of the complete vertebra. In the cervical, thoracic, and lumbar regions, articulations develop where adjacent cartilaginous blocks come into contact. In the sacrum and coccyx, the cartilages fuse together.

8-WEEK EMBRYO

9-WEEK EMBRYO

Rib cartilages expand away from the developing transverse processes of the vertebrae. At first they are continuous, but by week 8 the ribs have separated from the vertebrae. Ribs form at every vertebra, but in the cervical, lumbar, sacral, and coccygeal regions, they remain small and later fuse with the growing vertebrae. The ribs of the thoracic vertebrae continue to enlarge, following the curvature of the body wall. When they reach the ventral midline, they fuse with the cartilages of the sternum.

Related Clinical Terms

sinusitis: Inflammation and congestion of the paranasal sinuses. *[p. 147]*

deviated nasal septum: A bent nasal septum that slows or prevents sinus drainage. *[p. 147]*

craniostenosis (krā-nē-ō-sten-Ō-sis): Premature closure of one or more fontanels, which can lead to unusual distortions of the skull. *[p. 154]*

microcephaly (mī-krō-SEF-a-lē): An undersized head resulting from genetic or developmental abnormalities. *[p. 154]*

kyphosis (kī-FŌ-sis): Abnormal exaggeration of the thoracic curvature that produces a "roundback" appearance. *[p. 159]*

lordosis (lōr-DŌ-sis): Abnormal lumbar curvature giving a "swayback" appearance. *[p. 159]*

scoliosis (skō-lē-Ō-sis): Abnormal lateral curvature of the spine. *[p. 159]*

whiplash: An injury resulting from a sudden change in body position that can injure the cervical vertebrae. *[p. 162]*

spina bifida (SPĪ-na BI-fi-da): A condition resulting from failure of the vertebral laminae to unite during development; it is often associated with developmental abnormalities of the brain and spinal cord. *[p. 165]*

pneumothorax (nū-mō-THŌR-aks): The entry of air into a pleural cavity. *[p. 166]*

hemothorax: Bleeding into the thoracic cavity. *[p. 166]*

thoracentesis (thō-ra-sen-TĒ-sis) or **thoracocentesis:** The penetration of the thoracic wall along the superior border of one of the ribs. ☩ *The Thoracic Cage and Surgical Procedures [p. 755]*

chest tube: A drain installed after thoracic surgery to permit removal of blood and pleural fluid. ☩ *The Thoracic Cage and Surgical Procedures [p. 755]*

CHAPTER SUMMARY AND REVIEW

STUDY OUTLINE

Related Key Terms

INTRODUCTION *[p. 130]*

1. The skeletal system consists of the axial skeleton and the appendicular skeleton. The **axial skeleton** can be subdivided into the **skull** and associated bones (the **auditory ossicles** and **hyoid bone**), the **vertebral column**, and the **thoracic cage** composed of the **ribs** and **sternum** *(see Figure 6-1)*.

2. The **appendicular skeleton** includes the **pectoral** and **pelvic girdles** that support the upper and lower limbs *(see Figure 6-1)*.

THE SKULL *[p. 131]*

1. The **cranium** encloses the **cranial cavity**, a division of the dorsal body cavity *(see Figures 6-2 to 6-15 and Tables 6-1/6-2)*.

2. Prominent superficial landmarks on the skull include the **lambdoidal**, **sagittal**, **squamosal**, and **coronal sutures** *(see Figure 6-3a,b,c,d and Tables 6-1/6-2)*.

facial bones • sutural bones • calvaria

Bones of the Cranium *[p. 138]*

3. The **occipital bone** surrounds the **foramen magnum** and articulates with the sphenoid, temporal, and parietal bones *(see Figures 6-3a,c,e/6-6a,b)*.

4. The **parietal bones** articulate with the occipital, the temporal, and the frontal bones *(see Figures 6-3b,c/6-5/6-6c)*.

5. The **frontal bone** articulates with the parietal, the temporal, the sphenoid, the ethmoid, the nasal, the lacrimal, the **zygomatic**, and the maxillary bones *(see Figures 6-3b,c,d/6-5a/6-7)*.

6. The **temporal bone** articulates with the parietal, sphenoid, and occipital bones *(see Figures 6-3c,e/6-8)*.

7. The **sphenoid bone** articulates with all the cranial bones as well as the maxillary bones, the zygomatics, and the palatines. The maxillary bones articulate with every bone in the face except the mandible *(see Figures 6-3c,d,e/6-4/6-9)*.

8. The **cribriform plate** of the **ethmoid bone** contains perforations for olfactory nerves. The **perpendicular plate** forms part of the nasal septum *(see Figures 6-3d/6-4/6-5/6-10)*.

9. The **anterior cranial fossa** is formed by the frontal bone, the ethmoid, and the **lesser wings** of the sphenoid. The **middle cranial fossa** is created by the sphenoid, temporal, and parietal bones. The **posterior cranial fossa** is primarily formed by the occipital bone, with contributions from the temporal and parietal bones *(see Figure 6-11)*.

occipital condyles • occipital crest • external occipital protuberance • nuchal lines • jugular foramen • hypoglossal canals

temporal lines • parietal eminence

metopic suture • frontal squama • supraorbital margins • superciliary arches • supraorbital foramen • supraorbital notch • lacrimal fossa • frontal sinuses

squama (squamous portion) • carotid canal • foramen lacerum • mandibular fossa • articular tubercle • external auditory canal • tympanic membrane • mastoid process • mastoid foramen • styloid process • stylomastoid foramen • tympanic cavity • auditory tube • auditory ossicles • petrous portion • internal acoustic canal • hypophyseal fossa • sella turcica • clinoid processes • tuberculum sellae • optic groove • optic canal • superior orbital fissure • foramen rotundum • foramen ovale • foramen spinosum • greater wings • pterygoid processes • pterygoid canal • crista galli • lateral masses • conchae • ethmoidal sinuses • orbital rim • infraorbital foramen • inferior orbital fissure • alveolar processes • maxillary sinuses • palatal processes

Bones of the Face *[p. 145]*

10. The left and right **maxillary bones**, or *maxillae*, are the largest facial bones and form the upper jaw. They articulate with all other facial bones except the mandible *(see Figures 6-3d/6-12a,b)*.

11. The **palatine bones** are small, L-shaped bones that articulate with the maxillary bones, forming the posterior portions of the **hard palate**, and contributing to the floor of the orbital cavities *(see Figures 6-3e/6-12b,c)*.

12. The paired **nasal bones** articulate with the frontal bone at the midline of the face and extend to the superior border of the **external nares** *(see Figures 6-3c,d/6-12a)*.

13. The **vomer** forms the inferior portion of the nasal septum. It is based on the floor of the nasal cavity and articulates with both the maxillae and palatines along the midline *(see Figures 6-3c,d/6-5)*.

14. One **inferior nasal concha** is located on each side of the **nasal septum**, attached to the lateral wall of the nasal cavity. The *superior* and *middle* conchae of the ethmoid bone perform the same function *(see Figures 6-3d/6-13)*.

15. The **paranasal sinuses** are hollow airways that connect with the nasal passages. They are found in the frontal, sphenoid, ethmoid, and maxillary bones *(see Figures 6-5/6-13)*.

16. The **temporal process** of the **zygomatic bone** articulates with the **zygomatic process** of the temporal bone to form the **zygomatic arch** *(see Figures 6-3c,d/6-14)*.

17. The paired **lacrimal bones**, the smallest bones in the skull, are situated in the medial portion of each **orbit** *(see Figures 6-3c,d/6-14)*.

18. Seven bones form the **orbital complex**: the frontal, the maxillary, the lacrimal, the ethmoid, the sphenoid, the palatine, and the zygomatic bones *(see Figure 6-14)*.

19. The **mandible** is the lower jaw *(see Figures 6-3c,d/6-15)*.

20. The **hyoid bone**, suspended by **stylohyoid ligaments**, consists of a **body**, the **greater cornua**, and the **lesser cornua** *(see Figure 6-16)*.

THE SKULLS OF INFANTS, CHILDREN, AND ADULTS [p. 154]

1. Fibrous connections at **fontanels** permit the skulls of infants and children to continue growing *(see Figure 6-17)*.

THE VERTEBRAL COLUMN [p. 158]

1. The **vertebral column** consists of 24 **vertebra**, the *sacrum*, and the *coccyx*. There are 7 *cervical vertebrae*, 12 *thoracic vertebrae* (which articulate with the ribs), and 5 *lumbar vertebrae* (which articulate with the *sacrum*). The sacrum and coccyx consist of fused vertebrae *(see Figures 6-18 to 6-24)*.

Spinal Curvature [p. 158]
2. The spinal column has four **spinal curves**: the **thoracic** and **sacral curvatures** are called **primary**, or **accommodation, curves**; the **lumbar** and **cervical curvatures** are known as **secondary**, or **compensation, curves** *(see Figure 6-18)*.

Vertebral Anatomy [p. 159]
3. A typical vertebra has a **body**, or *centrum*, and a **vertebral arch** (**neural arch**); it articulates with other vertebrae at the **superior** and **inferior articular processes** *(see Figure 6-19)*.

4. Adjacent vertebrae are separated by **intervertebral discs**. Spaces between successive **pedicles** form the **intervertebral foramina** *(see Figure 6-19)*.

Vertebral Regions [p. 160]
5. Cervical vertebrae are distinguished by the shape of the vertebral body, the relative size of the vertebral foramen, the presence of **costal processes** with **transverse foramina**, and notched **spinous processes** *(see Figures 6-18a/6-20/6-24 and Table 6-3)*.

6. Thoracic vertebrae have distinctive heart-shaped bodies, long, slender spinous processes, and articulations for the ribs *(see Figure 6-22)*.

7. The **lumbar vertebrae** are the most massive and least mobile; they are subjected to the greatest strains *(see Figure 6-23)*.

8. The **sacrum** protects reproductive, digestive, and excretory organs. It has an **auricular surface** for articulation with the pelvic girdle. The sacrum articulates with the fused elements of the **coccyx** *(see Figure 6-24)*.

THE THORACIC CAGE [p. 166]

1. The skeleton of the **thoracic cage** consists of: the *thoracic vertebrae*, the *ribs*, and the *sternum*. The ribs and sternum form the **rib cage** *(see Figure 6-25a,c)*.

The Ribs [p. 166]
2. Ribs 1–7 are **true**, or **vertebrosternal ribs**. Ribs 8–12 are called **false ribs**; they include three pairs of **vertebrochondral ribs** and two pairs of **floating ribs**. A typical rib has a **head**, or **capitulum**; **neck**; **tubercle**, or **tuberculum**; **angle**; and **body**, or **shaft**. An inferior **costal groove** marks the path of nerves and blood vessels *(see Figures 6-22/6-25)*.

The Sternum [p. 166]
3. The **sternum** consists of a **manubrium**, a **body**, and a **xiphoid process** *(see Figure 6-25a)*.

Related Key Terms

nasal complex

frontal sinuses • sphenoidal sinuses

zygomaticofacial foramen

nasolacrimal canal

body • rami • angle • condylar processes • coronoid processes • mental foramina • mandibular notch • alveolar process • mylohyoid line • mylohyoid muscle • submandibular salivary gland • mandibular foramen • mandibular canal

vertebral foramen • laminae • transverse processes

articular facet • vertebral canal

atlas • anterior/posterior tubercles • axis • dens • odontoid process • vertebra prominens • ligamentum nuchae

demifacets • transverse costal facets

apex • base • sacral canal • median sacral crest • sacral cornua • sacral hiatus • sacral foramina • wing • ala • lateral sacral crest • sacroiliac joint • sacral tuberosity • coccygeal cornua

costae

costal cartilages • interarticular crest • intercostal muscles

jugular notch

1   REVIEW OF CHAPTER OBJECTIVES

1. Identify the bones of the axial skeleton and their functions.
2. Identify the bones of the skull and explain the significance of the markings on the individual bones.
3. Identify the cranial bones and their prominent surface features.
4. Identify the facial bones and their prominent surface features.
5. Describe the structure of the nasal complex and the functions of the individual elements.
6. Distinguish structural differences between the skulls of infants, children, and adults.
7. Identify and describe the curvatures of the vertebral column and their functions.
8. Identify the parts of a representative vertebra.
9. Identify the vertebral groups and describe the differences between them in structural and functional terms.
10. Explain the significance of the articulations between the thoracic vertebrae, the ribs, and the sternum.

2   REVIEW OF CONCEPTS

1. How do the functions of the axial skeleton differ from those of the appendicular skeleton?
2. What properties of sutures make this type of joint unique to the skull?
3. What is the purpose of the superior and inferior temporal lines? What does the location of these lines tell you?
4. Why is it of particular importance to prevent a child under two years of age from receiving a blow, especially with a sharp object, to the top of the head?
5. When a fashion model's face is described as having "good bone structure" and "high cheekbones," what anatomical structures are being praised, and what is their function?
6. What is the purpose of the many small openings in the cribriform plate of the ethmoid bone?
7. What role might a condition called a deviated septum, arising either as a birth defect or from an injury to the nasal region, play in causing olfactory and respiratory problems?
8. Many of the bones of the face form from a specialized embryologic tissue, the neural crest. If these facial structures failed to form, what systems would be affected?
9. What advantage is there to humans' having the entrance to the nasal cavity divided into two openings, or external nares?
10. What is the purpose of the curvature of the spinal column?
11. What is the importance of the decreasing capability for rotational movement between vertebrae from the upper to the lower aspect of the vertebral column?
12. In a person with severe osteoporosis, the thoracic cage may be curved anteriorly, giving the person a bent-forward appearance. How has the vertebral column been altered in this condition to give this appearance?
13. Why is the presence of a sternum important to the function of the thoracic cage?

14. In addition to the functions served by all vertebrae, the first two cervical vertebrae perform some special functions. What unique roles do these vertebrae perform in spinal mobility?
15. What characteristics of the vertebral-rib attachments make possible the thoracic movements that facilitate respiration?
16. What specific functions are shared by the lumbar vertebrae and the (fused) sacral vertebrae?
17. What is the function of each of the three classifications of ribs?
18. What structural characters permit isolated ribs to be distinguished from one another?

3   CRITICAL THINKING AND CLINICAL APPLICATION QUESTIONS

1. The skull of a newborn baby often appears as if it is deformed, by being elongated. After a short period of time, however, the skull assumes a normal shape. What happened during the birth process, and why did this change occur?
2. One of the injuries suffered by the late President John F. Kennedy was from a bullet that entered his skull behind and slightly below the right ear, passing through the mastoid process of the temporal bone. The brain damage resulting from this injury was extremely severe. What properties of the mastoid process contributed to this severity?
3. Small children have ear infections with much greater frequency than do older children and adults. What anatomical features might be responsible for this tendency? Why do children tend to "grow out" of this problem, and how is it treated before they do?
4. Some of the symptoms of the common cold or flu include that all of the teeth in the maxillae ache, even though there is nothing wrong with them, and the front of the head feels heavy. What anatomical response to the infection causes these unpleasant sensations?
5. A college senior becomes embroiled in a fist fight and receives a strong punch to the left side of the face. Because he realizes he probably has a broken jaw, he goes to the nearest hospital emergency room. X-rays show symmetrical fractures of the mandible immediately posterior to the lower canine teeth. Why would the injury be on both sides if he was hit only on the left? If the victim had been hit slightly higher, he might also have had fractures of the zygomatic arch and/or maxillae. Would these injuries also have been symmetrical on both sides? Why?
6. If the victim above had been hit instead in the nose, he might have had a bloody nose. What structures would have been the most likely to have been damaged? If, instead of blood, a clear fluid (other than mucus) began to drip from his nose, what would you predict to be the problem, and which bone or bones would be involved?
7. A person suffering from lower back pain is diagnosed as having a "slipped disc." What structural problem is the cause of this condition? One treatment for this problem is surgical fusion of the vertebrae superior and inferior to the slipped disc. Unfortunately, this solution is often temporary, as back pain frequently recurs above or below the fused area. Why might this occur?

7

The Skeletal System:
Appendicular Division

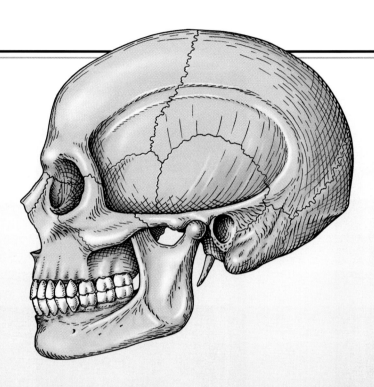

CHAPTER OUTLINE AND OBJECTIVES

The **appendicular skeleton** includes the bones of the upper and lower extremities and the supporting elements, or *girdles*, that connect them to the trunk (Figure 7-1●). This chapter describes the bones of the appendicular skeleton; as in Chapter 6, the descriptions emphasize surface features that have functional importance and highlight the interactions among the skeletal system and other systems. For example, many of the anatomical features noted in this chapter are attachment sites for skeletal muscles or openings for nerves and blood vessels that supply the bones or other organs of the body.

There are direct anatomical connections between the skeletal and muscular systems. As noted in Chapter 5, the connective tissue of the deep fascia that surrounds a skeletal muscle is continuous with that of its tendon, which continues into the periosteum and becomes part of the bone matrix at its attachment site. Muscles and bones are also physiologically linked, because muscle contractions can occur only when the extracellular concentration of calcium remains within relatively narrow limits. The skeleton contains most of the body's calcium, and these reserves are vital to calcium homeostasis.

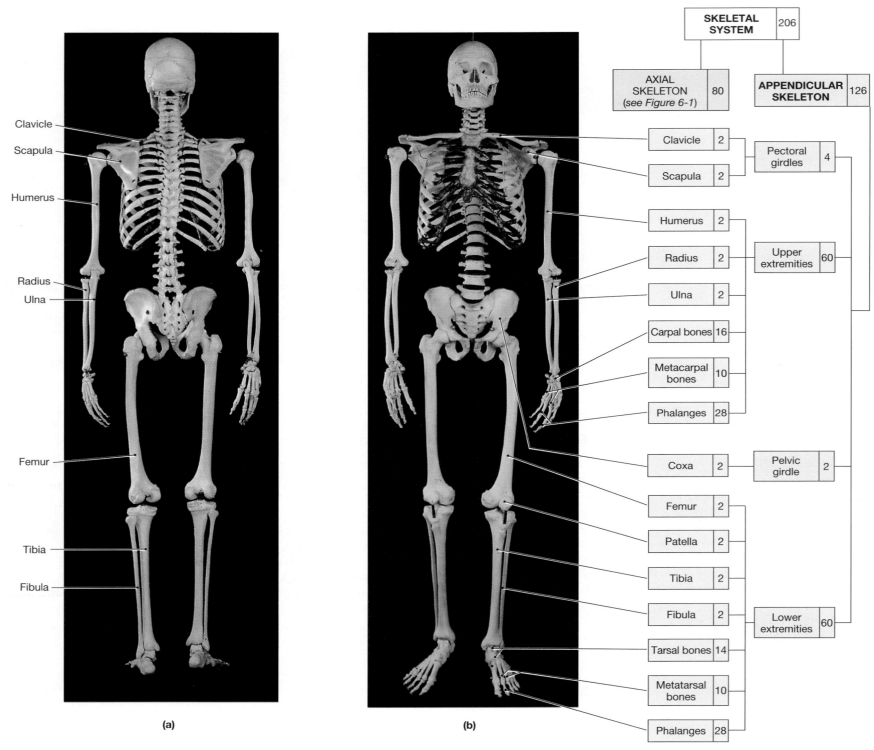

FIGURE 7-1
The Appendicular Skeleton. (a) Posterior view of the skeleton. (b) Anterior view of the skeleton, highlighting the appendicular components.

Each arm articulates with the trunk at the **shoulder**, or **pectoral girdle**. The shoulder girdle consists of the S-shaped *clavicle* (*collarbone*) and a broad, flat *scapula* (*shoulder blade*) as seen in Figure 7-2●. The clavicle articulates with the manubrium of the sternum, and this is the *only* direct connection between the pectoral girdle and the axial skeleton. Skeletal muscles support and position the scapula, which has no direct bony or ligamentous connections to the thoracic cage.

The Pectoral Girdle

Movements of the clavicle and scapula position the shoulder joint and provide a base for arm movement. Once the shoulder joint is in position, muscles that originate on the pectoral girdle help to move the upper extremity. The surfaces of the scapula and clavicle are therefore extremely important as sites for muscle

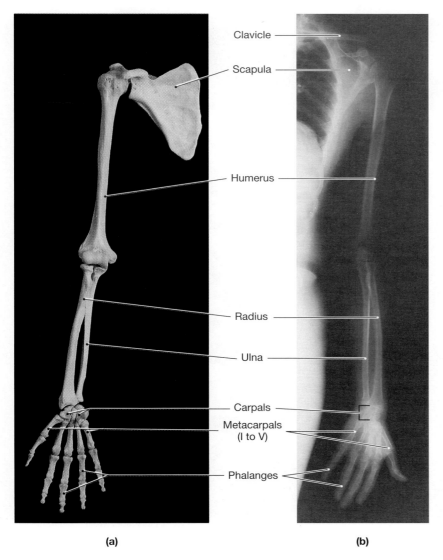

Clavicle
Scapula
Humerus
Radius
Ulna
Carpals
Metacarpals (I to V)
Phalanges

(a)   (b)

FIGURE 7-2
The Pectoral Girdle and Upper Extremity. (a) Right upper limb, anterior view. (b) X-ray of right pectoral girdle and upper limb, posterior view.

attachment. Where major muscles attach, they leave their marks, creating bony ridges and flanges. Other bone markings, such as foramina, indicate the position of nerves or blood vessels that control the muscles and nourish the muscles and bones.

The Clavicle
(Figures 7-3/7-4)

The **clavicle** (KLAV-i-kul) is an S-shaped bone that originates at the craniolateral border of the manubrium of the sternum, lateral to the jugular notch (see Figures 6-25a and 7-3●). ∞ [p. 167] From the roughly pyramidal **sternal end,** the clavicle curves laterally and dorsally until it articulates with the acromion of the scapula. The **acromial end** is broader than the sternal end.

The smooth superior surface of the clavicle lies just beneath the skin; the rough inferior surface of the acromial end is marked by prominent lines and tubercles that indicate the attachment sites for muscles and ligaments. The **conoid tubercle** is on the inferior surface at the acromial end, and the **costal tuberosity** is at the sternal end. These are attachment sites for ligaments of the shoulder.

You can explore the interaction between scapulae and clavicles. With your fingers in the jugular notch, locate the clavicle to either side. When you move your shoulders you can feel the clavicles change their positions. Because the clavicles are so close to the skin, you can trace one laterally until it articulates with the scapula. Shoulder movements are limited by the position of the clavicle, as shown in Figure 7-4●. Fractures of the medial portion of the clavicle are common because a fall on the palm of the hand of an outstretched arm produces compressive forces that are conducted to the clavicle and its articulation with the manubrium. Fortunately, these fractures usually heal rapidly without a cast.

The Scapula
(Figure 7-5)

The anterior aspect of the **body** of the **scapula** (SCAP-ū-le) forms a broad triangle (Figure 7-5a,d●). The three sides of that triangle are the **superior border;** the **medial border,** or **vertebral border;** and the **lateral border,** or **axillary border** (*axilla,* armpit). Muscles that position the scapula attach along these edges. The corners of the triangle are called the **superior angle,** the **inferior angle,** and the **lateral angle.** The lateral angle, or *head,* of the scapula, forms a broad process that supports the cup-shaped **glenoid cavity,** or *glenoid fossa.* The lateral angle is separated from the body of the scapula by the rounded **neck.** At the glenoid cavity, the scapula articulates with the proximal end of the *humerus,* the bone of the arm. This articulation is the **shoulder joint,** also known as the **scapulohumeral joint.** The relatively smooth, concave **subscapular fossa** forms most of the anterior surface of the scapula.

The articular head of the humerus is smooth, round, and several times the diameter of the glenoid cavity (Figure 7-5b,e●). Two large scapular processes extend over the superior margin of the glenoid cavity, superior to the head of the humerus. The smaller, anterior projection is the **coracoid** (KOR-a-koid) **process.** The **acromion** (a-KRŌ-mē-on) is the larger posterior process. If you run your fingers along the superior surface of the shoulder joint, you will feel this process. The acromion articulates with the clavicle at the **acromioclavicular joint.** Both the acromion and the coracoid process are attached to ligaments and tendons associated with the shoulder joint, which will be considered further in Chapter 8.

Most of the surface markings of the scapula represent the attachment sites for muscles that position the shoulder and arm. For example, the **supraglenoid tubercle** marks the origin of one portion of the *biceps brachii*, a prominent muscle on the anterior surface of the arm. The **infraglenoid tubercle** marks the origin of a portion of the *triceps brachii*, an equally prominent muscle on the posterior surface of the arm. The acromion is continuous with the **scapular spine** (Figure 7-5c,f●). This ridge crosses the scapular body before ending at the medial border. The scapular spine divides the convex dorsal surface of the body into two regions. The area superior to the spine constitutes the **supraspinous fossa** (*supra*, above); the region inferior to the spine is the **infraspinous fossa** (*infra*, beneath). The spine is an attachment site for two large muscles (the *supraspinatus* and the *infraspinatus*), and the entire dorsal surface is marked by small ridges and lines where smaller muscles attach to the scapula.

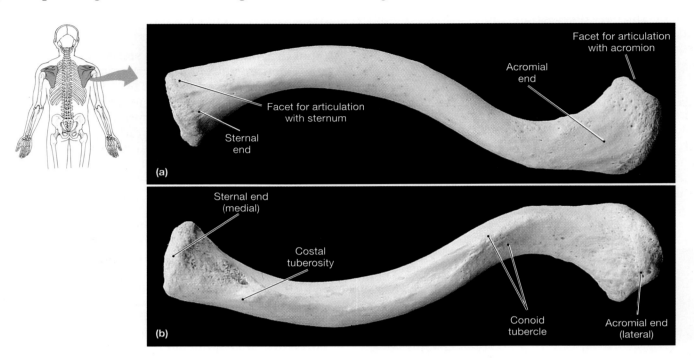

FIGURE 7-3
The Clavicle. Superior (a) and inferior (b) views of the right clavicle.

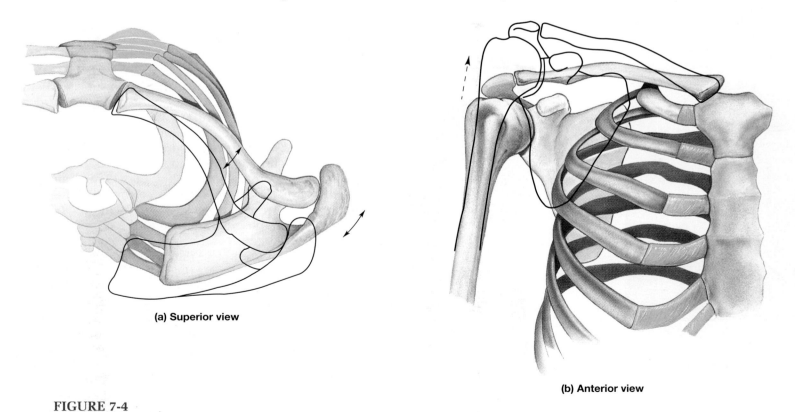

(a) Superior view

(b) Anterior view

FIGURE 7-4
Mobility of the Pectoral Girdle. (a) Alterations in position that occur during protraction and retraction of the right shoulder. (b) Alterations in position that occur during elevation and depression of the right shoulder. In each instance note that the clavicle is responsible for limiting the range of motion.

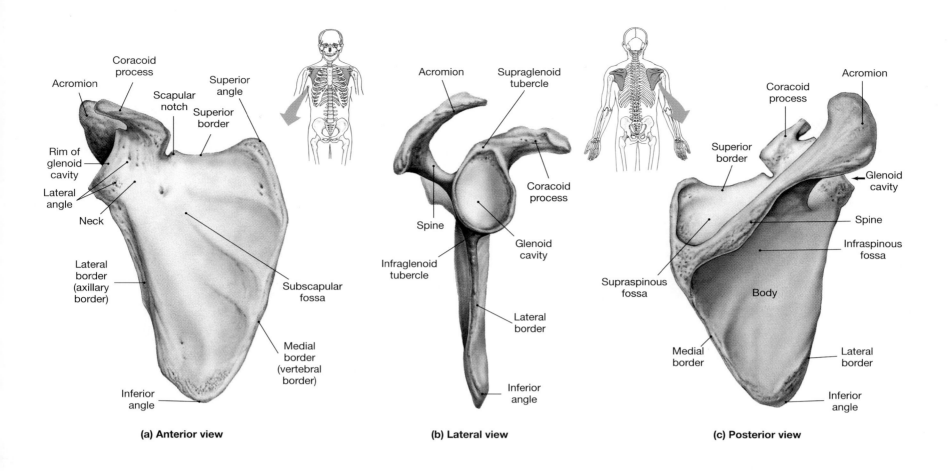

(a) Anterior view   (b) Lateral view   (c) Posterior view

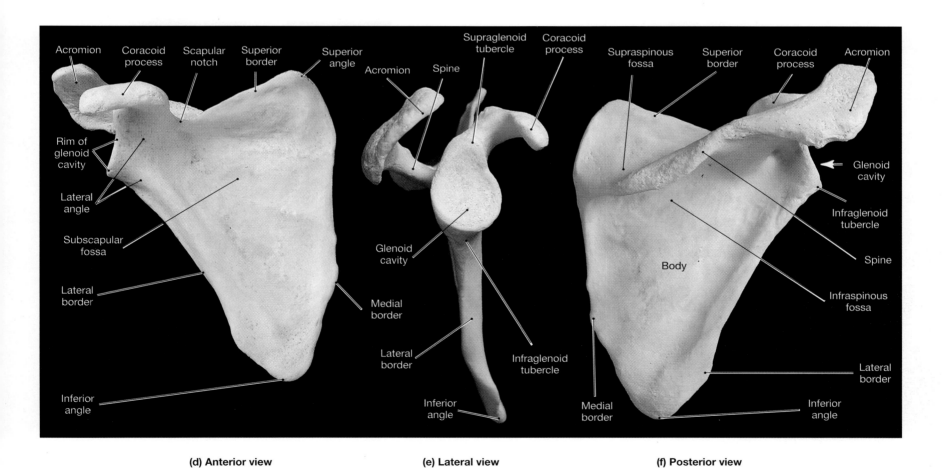

(d) Anterior view   (e) Lateral view   (f) Posterior view

FIGURE 7-5
The Scapula. Anterior (a,d), lateral (b,e), and posterior (c,f) views of the right scapula.

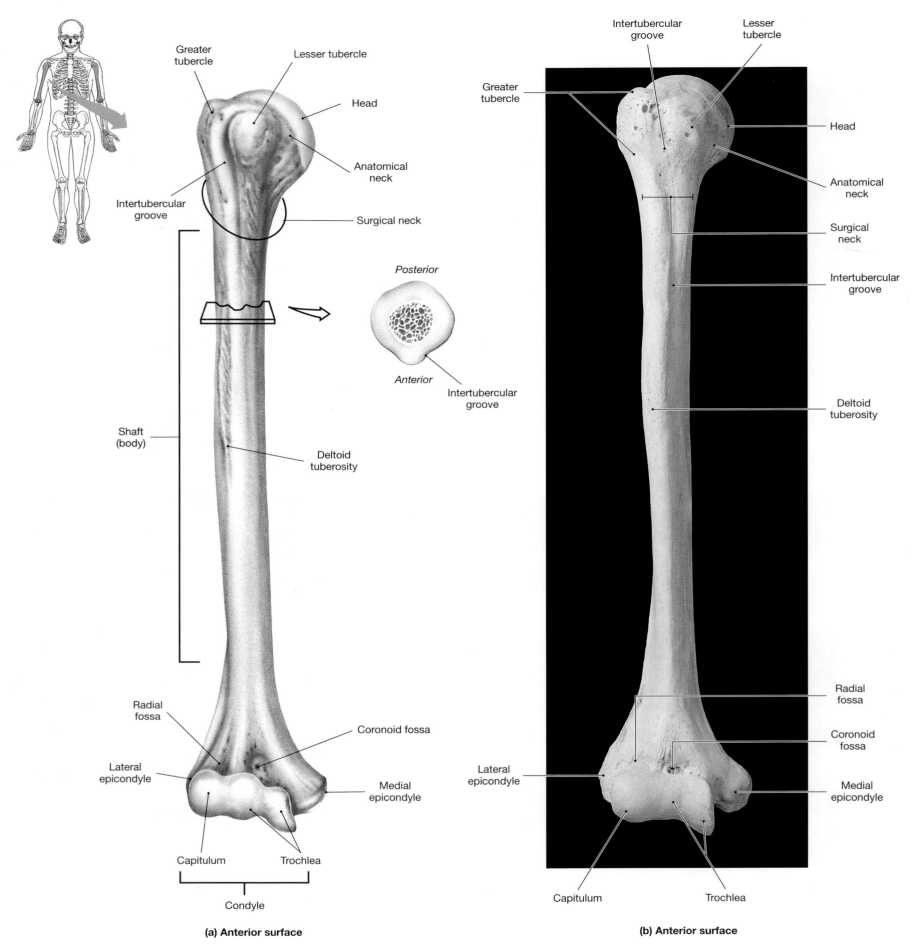

(a) Anterior surface

Greater tubercle

Lesser tubercle

Head

Anatomical neck

Intertubercular groove

Surgical neck

Shaft (body)

Posterior

Anterior

Intertubercular groove

Deltoid tuberosity

Radial fossa

Coronoid fossa

Lateral epicondyle

Medial epicondyle

Capitulum

Trochlea

Condyle

(b) Anterior surface

Intertubercular groove

Lesser tubercle

Greater tubercle

Head

Anatomical neck

Surgical neck

Intertubercular groove

Deltoid tuberosity

Radial fossa

Coronoid fossa

Lateral epicondyle

Medial epicondyle

Capitulum

Trochlea

FIGURE 7-6

Humerus. Major landmarks on the anterior (a,b,e) and the posterior (c,d,f) surface of the right humerus.

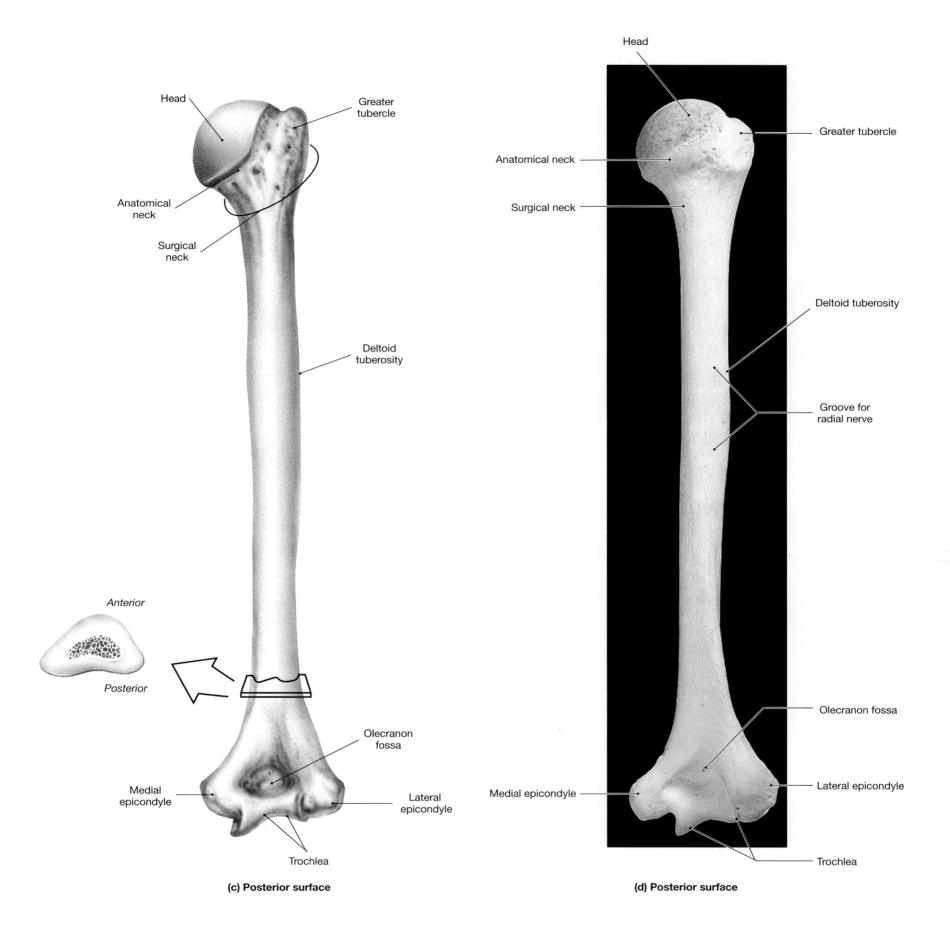

Head

Greater tubercle

Anatomical neck

Surgical neck

Deltoid tuberosity

Anterior

Posterior

Olecranon fossa

Medial epicondyle

Lateral epicondyle

Trochlea

(c) Posterior surface

Head

Greater tubercle

Anatomical neck

Surgical neck

Deltoid tuberosity

Groove for radial nerve

Olecranon fossa

Medial epicondyle

Lateral epicondyle

Trochlea

(d) Posterior surface

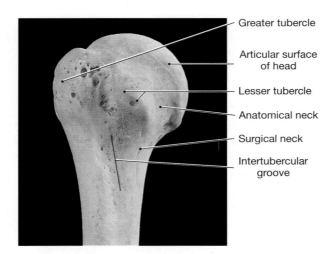

Greater tubercle

Articular surface of head

Lesser tubercle

Anatomical neck

Surgical neck

Intertubercular groove

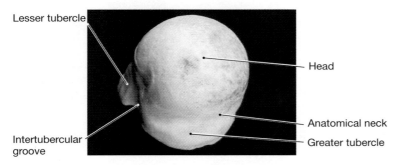

Lesser tubercle

Head

Intertubercular groove

Anatomical neck

Greater tubercle

(e) Proximal end, anterior and superior surfaces

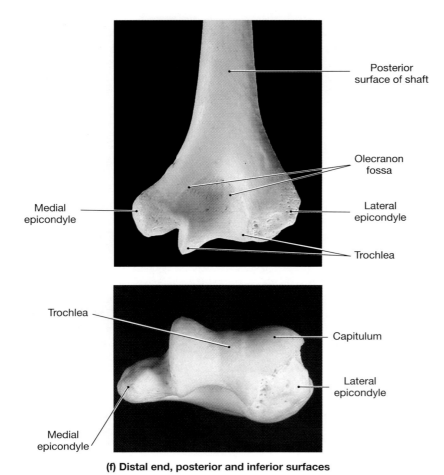

Posterior surface of shaft

Olecranon fossa

Medial epicondyle

Lateral epicondyle

Trochlea

Trochlea

Capitulum

Lateral epicondyle

Medial epicondyle

(f) Distal end, posterior and inferior surfaces

FIGURE 7-6 continued

The Upper Extremity
(Figure 7-2)

The arm, or *brachium*, contains a single bone, the **humerus**. The humerus extends from the scapula to the elbow. At its proximal end, the round **head** of the humerus articulates with the scapula. At its distal end, it articulates with the bones of the forearm, or *antebrachium*, the *radius* and *ulna* (Figure 7-2●, p. 175).

The Humerus
(Figure 7-6)

The prominent **greater tubercle** of the humerus is located near the head, on the lateral surface of the epiphysis (Figure 7-6a,b,e●). The greater tubercle establishes the lateral contour of the shoulder. You can verify its position by feeling for a bump situated a few centimeters anterior and inferior to the tip of the acromion. Three muscles that originate on the scapula (the *supraspinatus, infraspinatus,* and *teres minor*) are attached to the humerus at the greater tubercle. The **lesser tubercle** lies on the anterior and medial surface of the epiphysis. The lesser tubercle marks the insertion point of another scapular muscle, the *subscapularis.* The lesser tubercle and greater tubercle are separated by the **intertubercular groove**, or *intertubercular sulcus*. A tendon of the *biceps brachii* muscle runs along this groove from its origin at the supraglenoid tubercle of the scapula. The **anatomical neck** marks the distal limit of the articular capsule. It lies between the tubercles and the articular surface of the head. Distal to the tubercles, the narrow **surgical neck** corresponds to the metaphysis of the growing bone. This name reflects the fact that fractures often occur at this site.

The proximal **shaft** of the humerus is round in section. The elevated **deltoid tuberosity** runs along the lateral border of the shaft, extending more than halfway down its length. It is named after the *deltoid muscle* that attaches to it. On the anterior surface of the shaft, the intertubercular groove continues alongside the deltoid tuberosity.

On the posterior surface (Figure 7-6c,d,f●), the deltoid tuberosity ends at the **radial groove**. This depression marks the path of the *radial nerve*, a large nerve that provides sensory information from the back of the hand and motor control over large muscles that extend to the forearm. Distal to the radial groove, the posterior surface of the humerus is relatively flat. Near the distal end, the shaft expands to either side, forming a broad triangle. The **medial** and **lateral epicondyles** that project to either side provide additional surface area for muscle attachment. The articular **condyle** dominates the distal, inferior surface of the humerus.

A low ridge crosses the condyle, dividing it into two distinct articular regions (Figure 7-6a,b,c,d,f●). The **trochlea** (*trochlea*, pulley) is the spool-shaped medial portion that extends from the base of the **coronoid fossa** (KOR-ō-noyd; *corona*, crown) on the anterior surface (Figure 7-6a,b●) to the **olecranon fossa** on the posterior surface (Figure 7-6c,d,f●). These depressions accept projections from the surface of the ulna as the elbow approaches full flexion or extension. The rounded **capitulum** forms the lateral surface of the condyle. A shallow **radial fossa** proximal to the capitulum accommodates a small projection on the radius (Figure 7-6a,b●).

The Ulna
(Figures 7-2/7-7)

The **ulna** and *radius* are parallel bones that support the forearm (Figure 7-2●, p. 175). In the anatomical position, the ulna lies medial to the radius.

The **olecranon** (ō-LEK-ra-non), or *olecranon process*, is the point of the elbow. This process forms the superior and posterior portions of the proximal epiphysis of the ulna (Figure 7-7a,b●). On the anterior surface of the proximal epiphysis (Figure 7-7c,d●), the **trochlear notch** (or *semilunar notch*) articulates with the trochlea of the humerus at the **olecranal (elbow) joint**, also called the *humeroulnar joint*. The olecranon forms the superior lip of the trochlear notch, and the **coronoid process** forms its inferior lip. During extreme extension, the olecranon swings into the olecranon fossa on the posterior surface of the humerus. In extreme flexion, the coronoid process projects into the coronoid fossa on the anterior humeral surface. Lateral to the coronoid process, a smooth **radial notch** accommodates the head of the radius at the *proximal radioulnar joint*.

When viewed in cross section, the shaft of the ulna is roughly triangular, with the smooth medial surface at the base of the triangle and the lateral margin at the apex. A fibrous sheet, the **antebrachial interosseous membrane**, connects the lateral margin of the ulna to the medial margin of the radius (Figure 7-7a,c●). Distally, the ulnar shaft narrows before ending at a disc-shaped **ulnar head** whose posterior margin supports a short **styloid process** (*styloid*, long and pointed). A triangular **articular cartilage** attaches to the styloid process, isolating the ulnar head from the bones of the wrist. The *distal radioulnar joint* lies near the lateral border of the ulnar head.

The Radius
(Figures 7-2/7-7)

The **radius** is the lateral bone of the forearm (Figure 7-2●, p. 175). View Figure 7-7● as we describe the prominent features of the radius. The disc-shaped **radial head** articulates with the capitulum of the humerus. A narrow *neck* extends from the radial head to a prominent **radial tuberosity** that marks the attachment site of a muscle that flexes the forearm. The shaft of the radius curves along its length, and the **distal extremity** is considerably larger than the distal portion of the ulna. Because the articular cartilage and an articulating disc separate the ulna from the wrist, only the distal extremity of the radius participates in the wrist joint. The **styloid process** on the lateral surface of the distal extremity assists in the stabilization of the joint.

The medial surface of the distal extremity articulates with the ulnar head at the **ulnar notch**. The proximal radioulnar articulation permits rotation; when this movement occurs, the ulnar notch rolls across the rounded surface of the ulnar head. This movement is called **pronation**; the reverse movement, which returns the radius and ulna to the anatomical position, is called **supination**.

The Carpal Bones
(Figure 7-8)

The eight **carpal bones** of the wrist, or *carpus*, form two rows; there are four **proximal carpal bones** and four **distal carpal**

bones. The proximal carpals are the *scaphoid, lunate, triquetrum,* and *pisiform* (PĪ-si-form) *bones*. The distal carpals are the *trapezium, trapezoid, capitate,* and *hamate bones* (Figure 7-8●, p. 184).

THE PROXIMAL CARPAL BONES

- The **scaphoid bone** is the proximal carpal bone located on the lateral border of the wrist adjacent to the styloid process of the radius.

- The comma-shaped **lunate** (*luna*, moon) **bone** lies medial to the scaphoid bone. Like the scaphoid bone, the lunate bone articulates with the radius.

- The **triquetrum** (*triangular*) is medial to the lunate bone. It has the shape of a small pyramid. The triangular bone articulates with the cartilage that separates the ulnar head from the wrist.

- The small, pear-shaped **pisiform bone** lies anterior to the triangular bone and extends farther medially than any other carpal bone in the proximal or distal rows.

THE DISTAL CARPAL BONES

- The **trapezium** is the lateral bone of the distal row. It forms a proximal articulation with the scaphoid bone.

- The wedge-shaped **trapezoid bone** lies medial to the trapezium; it is the smallest distal carpal bone. Like the trapezium, it has a proximal articulation with the scaphoid bone.

- The **capitate bone** is the largest carpal bone. It sits between the trapezoid and the hamate bone.

- The **hamate** (*hamatum*, hooked) **bone** is a hook-shaped bone that is the medial distal carpal bone.

The Hand
(Figure 7-8)

Five **metacarpal** (met-a-KAR-pal) **bones** articulate with the distal carpal bones and support the palm (Figure 7-8●, p. 184). Roman numerals I–V are used to identify the metacarpal bones, beginning with the lateral metacarpal bone that articulates with the trapezium. Each metacarpal bone has a wide, concave, proximal *base*, a small *body*, and a distal *head*. Distally, the metacarpal bones articulate with the finger bones, or **phalanges** (fa-LAN-jēz; singular *phalanx*). There are 14 phalangeal bones in each hand. The thumb, or **pollex** (POL-eks), has two phalanges (proximal and distal), and each of the other fingers has three phalanges (proximal, middle, and distal).

√ Why would a broken clavicle affect the mobility of the scapula?

√ The rounded projections on either side of the elbow are parts of what bone?

√ Which antebrachial bone is lateral in the anatomical position?

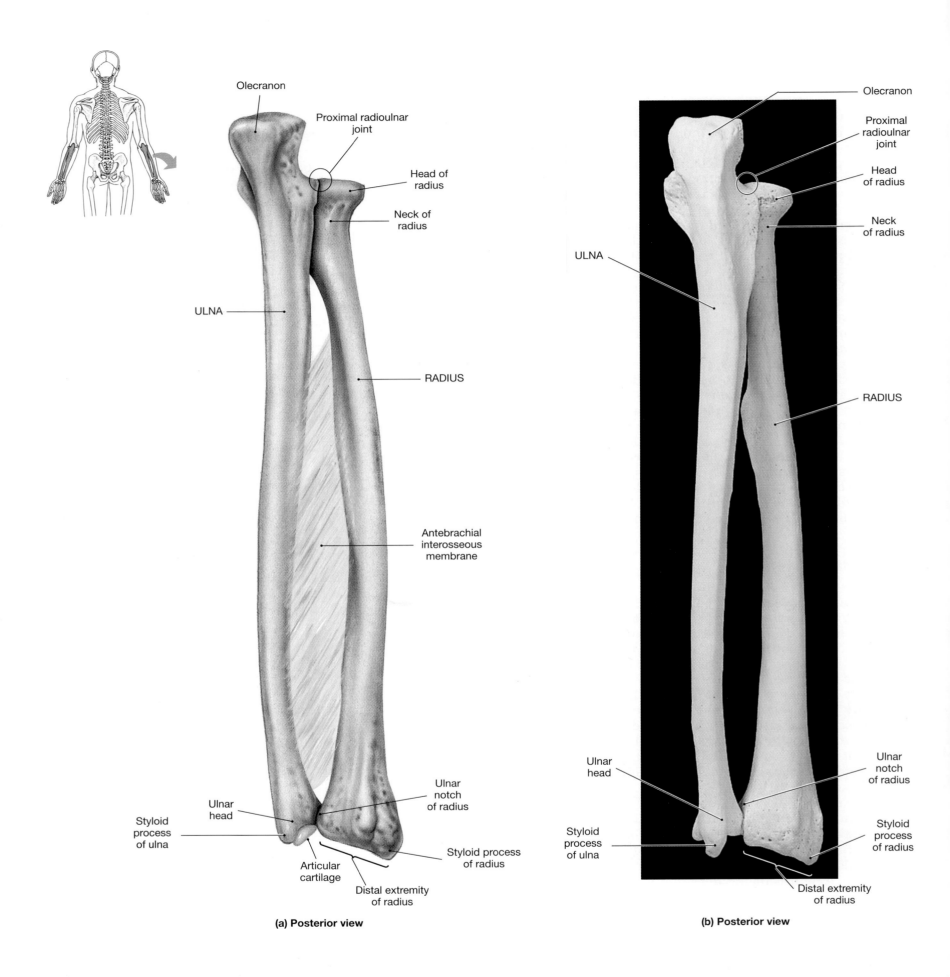

Olecranon

Proximal radioulnar joint

Head of radius

Neck of radius

ULNA

RADIUS

Antebrachial interosseous membrane

Ulnar head

Ulnar notch of radius

Styloid process of ulna

Articular cartilage

Styloid process of radius

Distal extremity of radius

(a) Posterior view

Olecranon

Proximal radioulnar joint

Head of radius

Neck of radius

ULNA

RADIUS

Ulnar head

Ulnar notch of radius

Styloid process of ulna

Styloid process of radius

Distal extremity of radius

(b) Posterior view

FIGURE 7-7
Radius and Ulna. The radius and ulna of the right forearm are shown in posterior view (a,b) and anterior view (c,d).

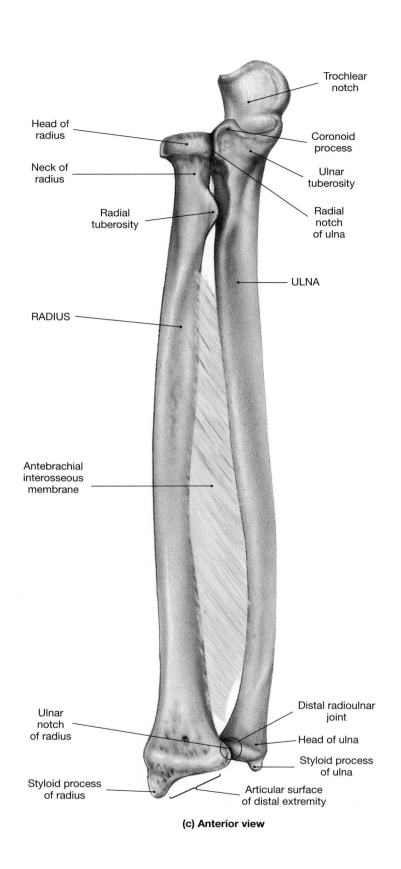

Trochlear
notch

Head of
radius

Coronoid
process

Neck of
radius

Ulnar
tuberosity

Radial
tuberosity

Radial
notch
of ulna

RADIUS

ULNA

Antebrachial
interosseous
membrane

Ulnar
notch
of radius

Distal radioulnar
joint

Head of ulna

Styloid process
of ulna

Styloid process
of radius

Articular surface
of distal extremity

(c) Anterior view

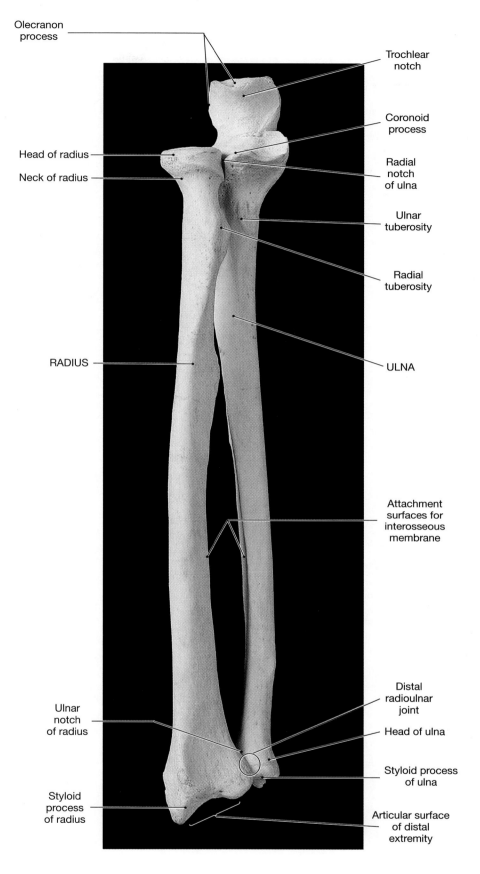

Olecranon
process

Trochlear
notch

Coronoid
process

Head of radius

Radial
notch
of ulna

Neck of radius

Ulnar
tuberosity

Radial
tuberosity

RADIUS

ULNA

Attachment
surfaces for
interosseous
membrane

Ulnar
notch
of radius

Distal
radioulnar
joint

Head of ulna

Styloid process
of ulna

Styloid
process
of radius

Articular surface
of distal
extremity

(d) Anterior view

CHAPTER 7 The Skeletal System: Appendicular Division **183**

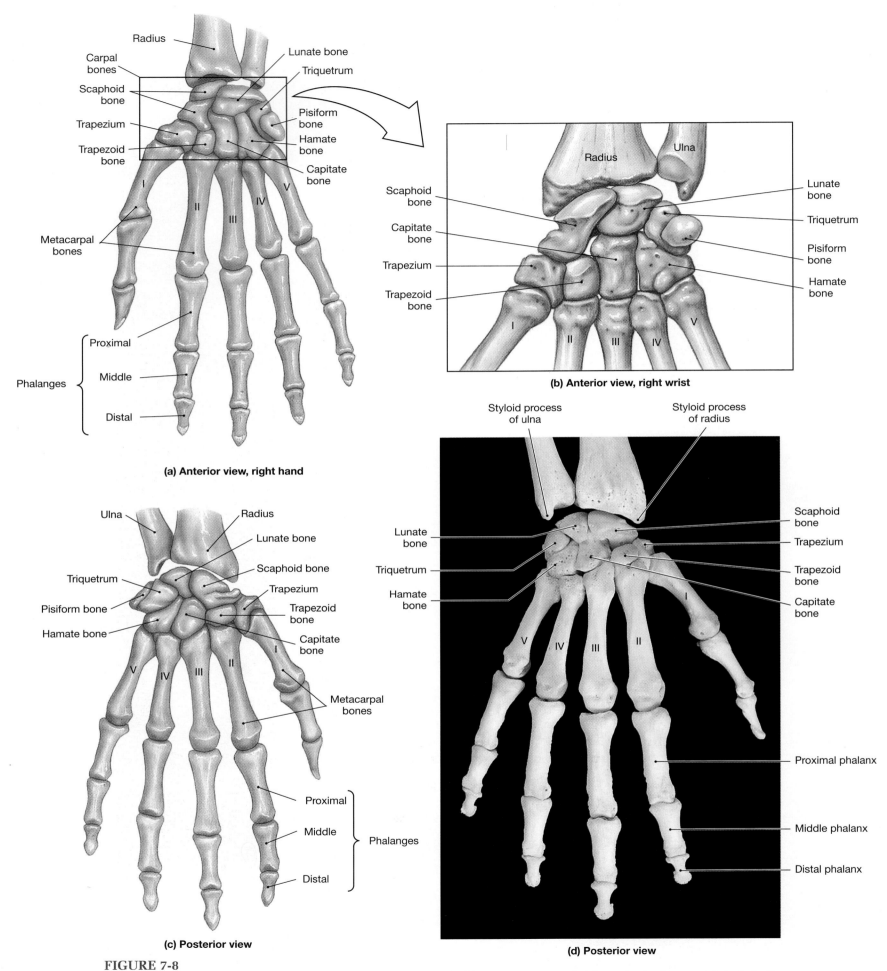

(a) Anterior view, right hand

(b) Anterior view, right wrist

(c) Posterior view

(d) Posterior view

FIGURE 7-8

Bones of the Wrist and Hand. (a) Anterior view of right hand. (b) Anterior view of right wrist (radiocarpal joint). (c,d) Posterior views of right hand; pisiform bone not visible in (d).

THE PELVIC GIRDLE AND LOWER EXTREMITY
(Figure 7-9)

The bones of the **pelvic girdle** are more massive than those of the pectoral girdle because of the stresses involved in weight bearing and locomotion. The bones of the lower extremities are more massive than those of the upper extremities, for similar reasons. The pelvic girdle consists of the two fused **coxae**, or *innominate bones*. The *pelvis* is a composite structure that includes the coxae of the appendicular skeleton and the sacrum and coccyx of the axial skeleton. Each lower extremity consists of the *femur* (thigh), the *tibia* and *fibula* (leg), and the bones of the ankle (*tarsals*) and foot (Figure 7-9●).

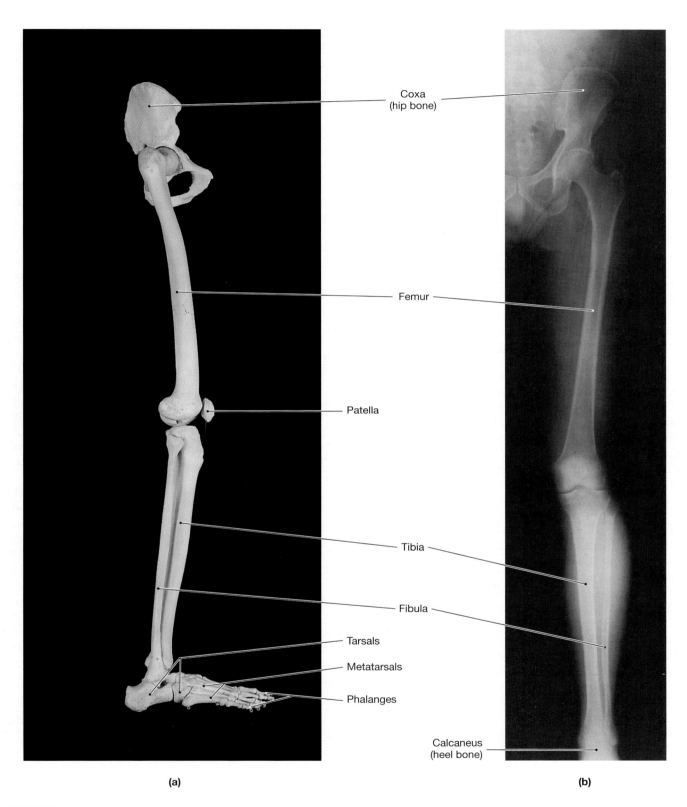

Coxa
(hip bone)

Femur

Patella

Tibia

Fibula

Tarsals

Metatarsals

Phalanges

Calcaneus
(heel bone)

(a)   (b)

FIGURE 7-9
The Pelvic Girdle and Lower Extremity. (a) Right leg, anterior/lateral view. (b) X-ray, pelvic girdle and lower extremity, anterior/posterior projection.

The Pelvic Girdle
(Figure 7-10)

Each coxa forms through the fusion of three bones, an *ilium* (IL-ē-um), an *ischium* (IS-kē-um), and a *pubis* (PYŪ-bis) (Figure 7-10●). The articulation between the coxa and the auricular surfaces of the sacrum occurs at the posterior and medial aspect of the ilium. Ventrally the coxae are connected by a pad of fibrocartilage at the *pubic symphysis*. On the lateral surface of each coxa, the head of the femur articulates with the curved surface of the **acetabulum** (as-e-TAB-ū-lum; *acetabulum*, vinegar cup).

The acetabulum lies inferior and anterior to the center of the coxa (Figure 7-10a●). The space enclosed by the walls of the acetabulum is the **acetabular fossa**, which has a diameter of approximately 5 cm (2 in.). The acetabulum contains a smooth curved surface that forms the shape of the letter *C*. This is the **lunate surface**, which articulates with the head of the femur.

The Coxa
(Figures 7-10/7-11a)

The three coxal bones meet inside the acetabular fossa, as if it were a pie sliced into three pieces. The **ilium** (IL-ē-um) the largest coxal bone, provides the superior slice that includes around two-fifths of the acetabular surface. Superior to the acetabulum, the ilium forms a broad, curved surface that provides an extensive area for the attachment of muscles, tendons, and ligaments (Figure 7-10a●). The **anterior, posterior, and inferior gluteal lines** mark the attachment sites for the *gluteal muscles* that insert on the femur. The iliac expansion begins superior to the **arcuate** (AR-kū-āt) **line**. The anterior border includes the **anterior inferior iliac spine**, above the **inferior iliac notch**, and continues anteriorly to the **anterior superior iliac spine**. Curving posteriorly, the superior border supports the **iliac crest**, a ridge marking the attachments of both ligaments and muscles. The iliac crest ends at the **posterior superior iliac spine**. Inferior to the spine, the ilial margin continues inferiorly to the rounded **posterior inferior iliac spine** that is superior to the **greater sciatic** (sī-AT-ik) **notch**.

Near the superior and posterior margin of the acetabulum, the ilium fuses with the **ischium**, which accounts for the posterior two-fifths of the acetabular surface. Posterior to the acetabulum, the prominent **ischial spine** projects above the **lesser sciatic notch**. The rest of the ischium forms a sturdy process that turns medially and inferiorly. A roughened projection, the **ischial tuberosity**, forms its posterolateral border. When seated, the body weight is borne by the ischial tuberosities. The narrow **ramus** (branch) of the ischium continues toward its anterior fusion with the **pubis** (PYŪ-bis).

At the point of fusion, the ramus of the ischium meets the **inferior ramus** of the pubis. Anteriorly, the inferior ramus ends at the **pubic tubercle**, where it meets the **superior ramus** of the pubis. The anterior, superior surface of the superior ramus bears a roughened ridge, the **pubic crest**, that ends at the pubic tubercle. The pubic and ischial rami encircle the **obturator** (OB-tū-rā-tor) **foramen**. In life, this space is closed by a sheet of collagen fibers whose inner and outer surfaces provide a firm base for the attachment of muscles and visceral structures. The superior ramus originates at the anterior margin of the acetabulum. Inside the acetabulum, the pubis contacts the ilium and ischium.

The concave medial surface of the **iliac fossa** helps support the abdominal organs and provides additional surface area for muscle attachment (Figure 7-10b●). The arcuate line marks the inferior border of the iliac fossa. The anterior and medial surface of the pubis contains a roughened area that marks the site of articulation with the pubis of the opposite side. At this articulation, the **pubic symphysis**, the two pubic bones are attached to a median fibrocartilage pad (Figure 7-11a●, p. 189). The **iliopectineal** (il-ē-ō-pek-TIN-ē-al) **line** begins near the symphysis and extends diagonally across the pubis to merge with the arcuate line, which continues toward the **auricular surface** of the ilium. Ligaments arising at the **iliac tuberosity** stabilize the joint. On the medial surface of the superior ramus of the pubis lies the **obturator groove**, for the obturator blood vessels and nerves.

The Pelvis
(Figures 7-11 to 7-13)

Figure 7-11●, p. 189, shows anterior and posterior views of the pelvis, which consists of the coxae, the sacrum, and the coccyx. An extensive network of ligaments connects the lateral borders of the sacrum with the iliac crest, the ischial tuberosity, the ischial spine, and the iliopectineal line. Other ligaments tie the ilia to the posterior lumbar vertebrae. These interconnections increase the stability of the pelvis.

The pelvis may be subdivided into the **false (greater) pelvis** and the **true (lesser) pelvis**. The boundaries of each are indicated in Figure 7-12●, p. 191. The greater pelvis consists of the expanded, bladelike portions of each ilium superior to the iliopectineal line. Structures inferior to that line, including the inferior portions of each ilium, both pubic bones, the ischia, the sacrum, and the coccyx, form the true pelvis.

The true pelvis encloses the *pelvic cavity*, a subdivision of the abdominopelvic cavity. ∞ [p. 18] In lateral view (Figure 7-12b●), the superior limit of the true pelvis is a line that extends from either side of the base of the sacrum, along the iliopectineal lines to the superior margin of the pubic symphysis. The bony edge of the true pelvis is called the **pelvic brim**, and the enclosed space is the **pelvic inlet**.

The **pelvic outlet** is the opening bounded by the inferior edges of the pelvis (Figure 7-12b,c●). In life, this region, called the **perineum** (per-i-NĒ-um), is bounded by the coccyx, the ischial tuberosities, and the inferior border of the pubic symphysis. Perineal muscles form the floor of the pelvic cavity and support the enclosed organs. Figure 7-12d● shows the appearance of the pelvis in anterior view.

The shape of the female pelvis is somewhat different from that of the male pelvis (Figure 7-13●, p. 192). Some of these differences are the result of variations in body size and muscle mass. For example, the female pelvis is usually smoother, lighter, and has less prominent markings. Other differences are adaptations for childbearing, including:

- an enlarged pelvic outlet,
- less curvature on the sacrum and coccyx, which in the male arc into the pelvic outlet,
- a wider, more circular pelvic inlet,
- a relatively broad, low pelvis,
- ilia that project farther laterally, but do not extend as far superior to the sacrum, and
- a broader pubic arch, with the inferior angle between the pubic bones greater than 100°.

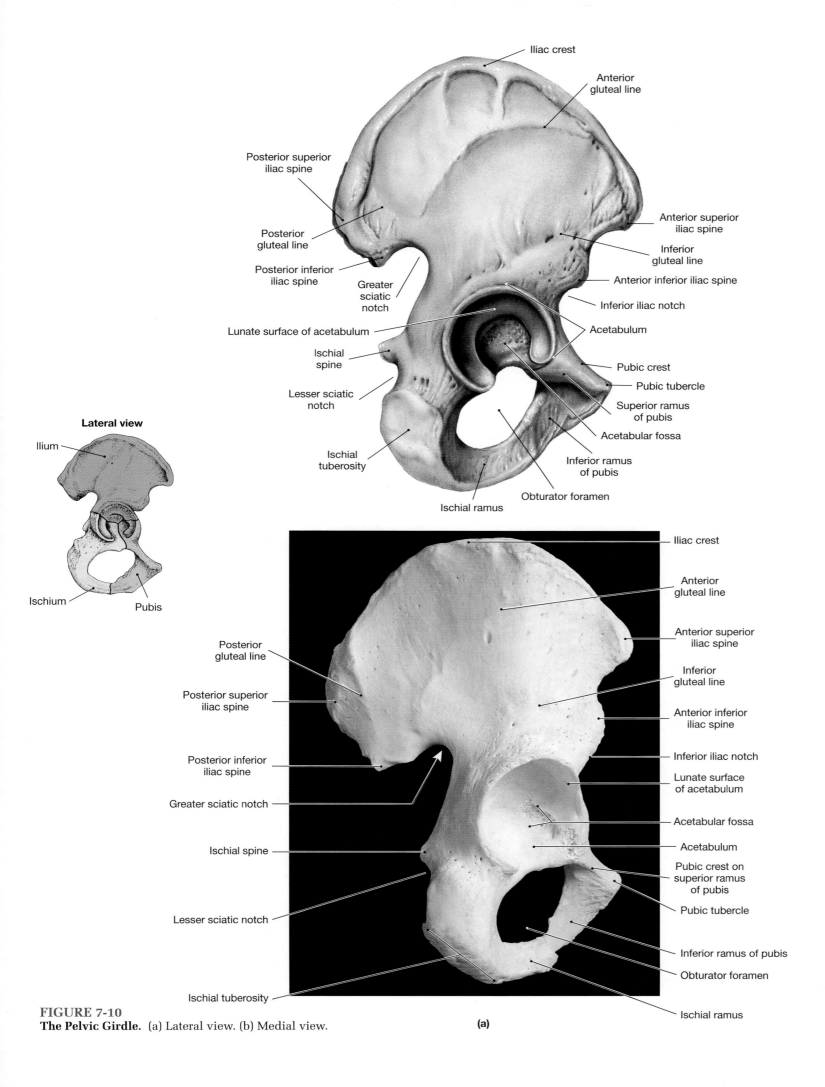

Iliac crest

Anterior
gluteal line

Posterior superior
iliac spine

Posterior
gluteal line

Anterior superior
iliac spine

Inferior
gluteal line

Posterior inferior
iliac spine

Anterior inferior iliac spine

Greater
sciatic
notch

Inferior iliac notch

Lunate surface of acetabulum

Acetabulum

Ischial
spine

Pubic crest

Lesser sciatic
notch

Pubic tubercle

Superior ramus
of pubis

Acetabular fossa

Ischial
tuberosity

Inferior ramus
of pubis

Obturator foramen

Ischial ramus

Lateral view

Ilium

Ischium

Pubis

Iliac crest

Anterior
gluteal line

Posterior
gluteal line

Anterior superior
iliac spine

Posterior superior
iliac spine

Inferior
gluteal line

Posterior inferior
iliac spine

Anterior inferior
iliac spine

Greater sciatic notch

Inferior iliac notch

Lunate surface
of acetabulum

Ischial spine

Acetabular fossa

Acetabulum

Lesser sciatic notch

Pubic crest on
superior ramus
of pubis

Pubic tubercle

Inferior ramus of pubis

Obturator foramen

Ischial tuberosity

Ischial ramus

FIGURE 7-10
The Pelvic Girdle. (a) Lateral view. (b) Medial view.

(a)

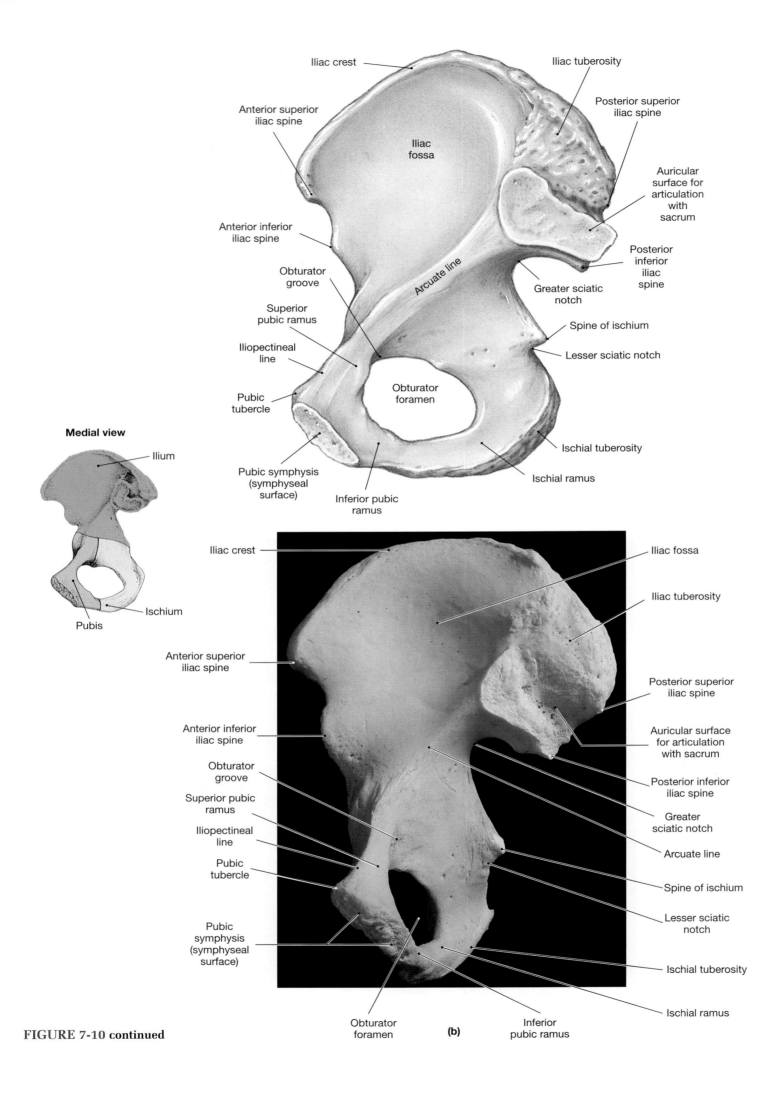

Iliac crest

Iliac tuberosity

Anterior superior
iliac spine

Posterior superior
iliac spine

Iliac
fossa

Auricular
surface for
articulation
with
sacrum

Anterior inferior
iliac spine

Posterior
inferior
iliac
spine

Obturator
groove

Arcuate line

Greater sciatic
notch

Superior
pubic ramus

Spine of ischium

Iliopectineal
line

Lesser sciatic notch

Pubic
tubercle

Obturator
foramen

Medial view

Ilium

Pubic symphysis
(symphyseal
surface)

Ischial tuberosity

Ischium

Ischial ramus

Pubis

Inferior pubic
ramus

Iliac crest

Iliac fossa

Iliac tuberosity

Anterior superior
iliac spine

Posterior superior
iliac spine

Anterior inferior
iliac spine

Auricular surface
for articulation
with sacrum

Obturator
groove

Posterior inferior
iliac spine

Superior pubic
ramus

Greater
sciatic notch

Iliopectineal
line

Arcuate line

Pubic
tubercle

Spine of ischium

Pubic
symphysis
(symphyseal
surface)

Lesser sciatic
notch

Ischial tuberosity

Ischial ramus

Obturator
foramen

(b)

Inferior
pubic ramus

FIGURE 7-10 continued

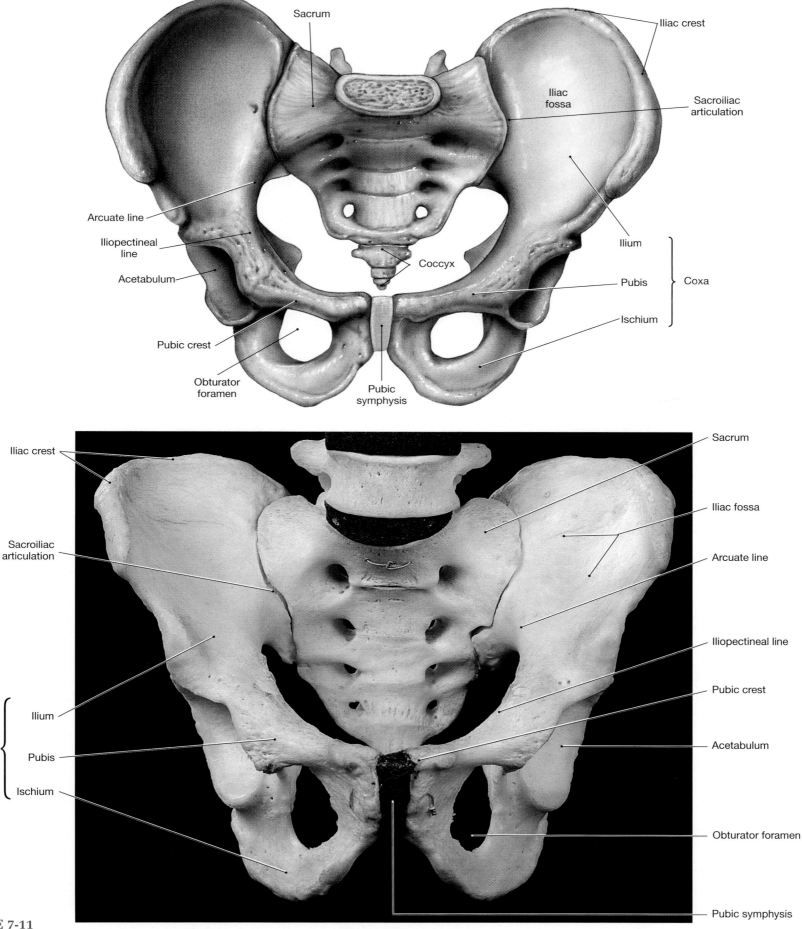

FIGURE 7-11
The Male Pelvis. (a) Anterior view. (b) Posterior view.

(a) Anterior view

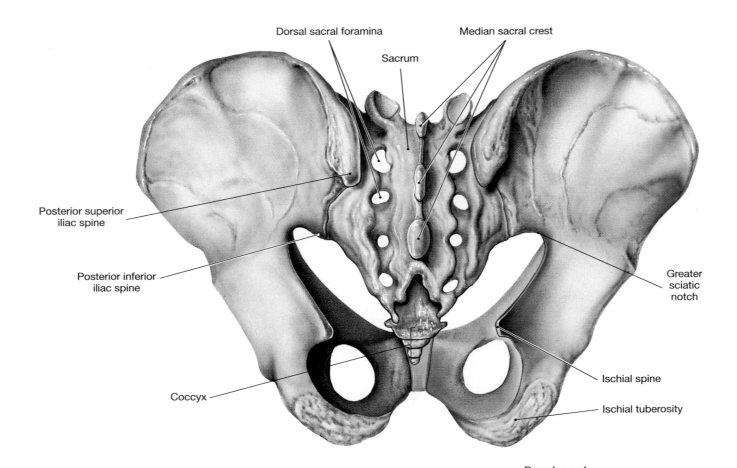

Dorsal sacral foramina

Sacrum

Median sacral crest

Posterior superior iliac spine

Posterior inferior iliac spine

Coccyx

Greater sciatic notch

Ischial spine

Ischial tuberosity

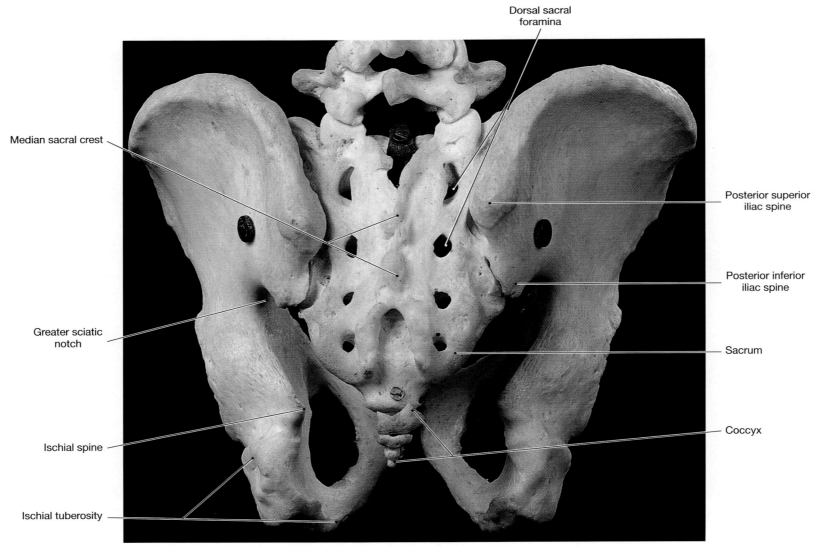

Dorsal sacral foramina

Median sacral crest

Posterior superior iliac spine

Posterior inferior iliac spine

Greater sciatic notch

Sacrum

Ischial spine

Coccyx

Ischial tuberosity

FIGURE 7-11 continued

(b) Posterior view

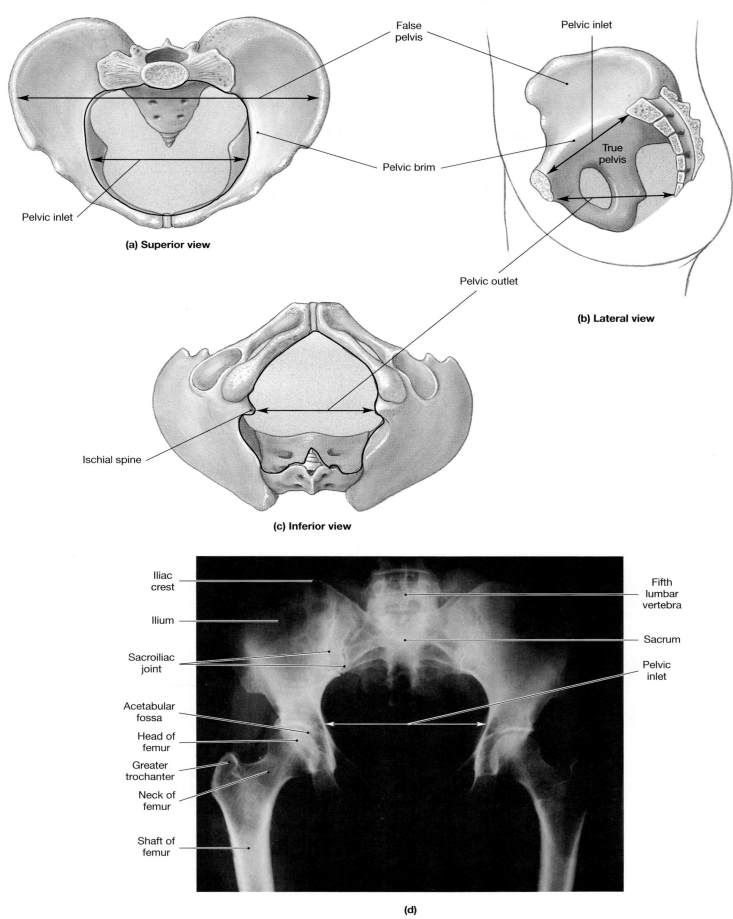

(a) Superior view

False pelvis

Pelvic brim

Pelvic inlet

(b) Lateral view

Pelvic inlet

True pelvis

Pelvic outlet

Ischial spine

(c) Inferior view

Iliac crest

Ilium

Sacroiliac joint

Acetabular fossa

Head of femur

Greater trochanter

Neck of femur

Shaft of femur

Fifth lumbar vertebra

Sacrum

Pelvic inlet

(d)

FIGURE 7-12
Divisions of the Pelvis. (a) Superior view, showing the pelvic brim and pelvic inlet of a male. (b) Lateral view, showing the boundaries of the true (lesser) and false (greater) pelvis. (c) Inferior view, showing the limits of the pelvic outlet. (d) X-ray of pelvis and femora.

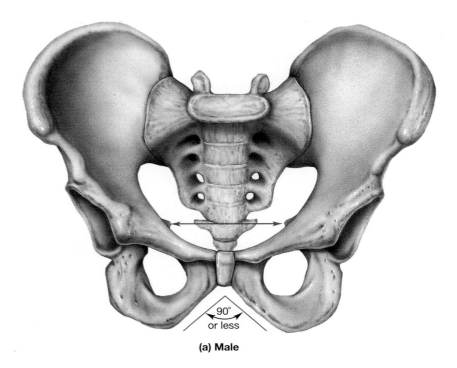

(a) Male

90° or less

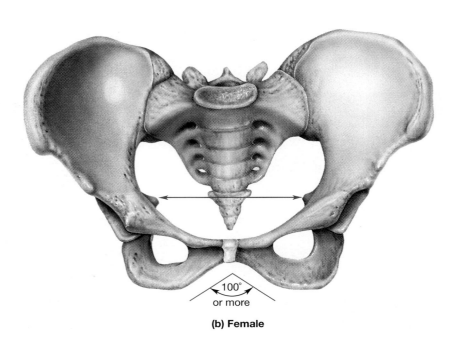

(b) Female

100° or more

FIGURE 7-13
Anatomical Differences in the Male and Female Pelvis. Note the much sharper pubic angle in the pelvis of a male (a) than in that of a female (b).

These adaptations are related to (1) the support of the weight of the developing fetus and uterus and (2) the passage of the newborn through the pelvic outlet at the time of delivery. In addition, a hormone produced during pregnancy loosens the pubic symphysis, allowing relative movement between the coxae that can further increase the size of the pelvic inlet and outlet.

The Lower Extremity
(Figure 7-9)

The lower extremity consists of the *femur*, the *patella* (kneecap), the *tibia* and *fibula*, and the bones of the ankle and foot (Figure 7-9●, p. 185). The functional anatomy of the lower extremity is very different from that of the upper extremity, primarily because the lower extremity must transfer the body weight to the ground.

The Femur
(Figures 7-9/7-12d/7-14)

The **femur** is the longest and heaviest bone in the body. Distally, the femur articulates with the *tibia* of the lower leg at the knee joint. The rounded epiphysis, or **head** of the femur, articulates with the pelvis at the acetabulum (Figures 7-9, p. 185, and 7-12d●). A stabilizing ligament attaches to the head at a depression, the **fovea capitis**. Distal to the head, the **neck** joins the **shaft** at an angle of about 125° (Figure 7-14a●). The **greater trochanter** (trō-KAN-ter) projects laterally from the junction of the neck and shaft. The **lesser trochanter** originates on the posteromedial surface of the femur. Both trochanters develop where large tendons attach to the femur. On the anterior surface of the femur, the raised **intertrochanteric** (in-ter-trō-kan-TER-ik) **line** marks the distal edge of the articular capsule. This line continues around to the posterior surface, passing inferior to the trochanters as the **intertrochanteric crest** (Figure 7-14b●). The **pectineal line,** inferior to the intertrochanteric crest, marks the attachment of the *pectineus muscle.*

A prominent elevation, the **linea aspera** (*aspera*, rough), runs along the center of the posterior surface, marking the attachment site of powerful muscles that adduct the femur (Figure 7-14b●). Distally, the linea aspera divides into a **medial** and **lateral supracondylar ridge** to form a flattened triangular area, the **popliteal surface**. The medial supracondylar ridge terminates in a raised, rough projection, the **adductor tubercle** on the **medial epicondyle**. The lateral ridge ends at the **lateral epicondyle**. The smoothly rounded **medial** and **lateral condyles** are primarily distal and posterior to the epicondyles. On the posterior surface, the two condyles are separated by a deep **intercondylar fossa**.

The condyles continue across the inferior surface of the femur to the anterior surface, but the intercondylar fossa does not. As a result, the smooth articular faces merge, producing an articular surface with elevated lateral borders (Figure 7-14a●). This is the **patellar surface** over which the patella glides.

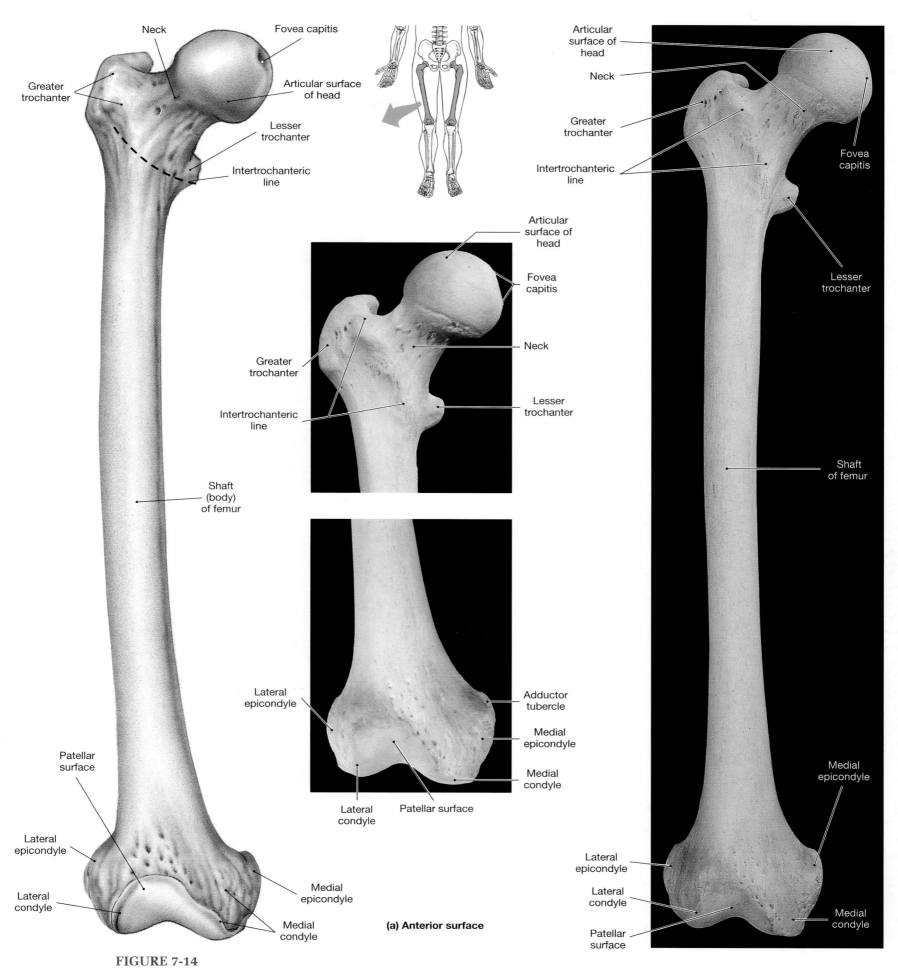

Neck

Fovea capitis

Greater trochanter

Articular surface of head

Lesser trochanter

Intertrochanteric line

Shaft (body) of femur

Patellar surface

Lateral epicondyle

Lateral condyle

Medial epicondyle

Medial condyle

Articular surface of head

Fovea capitis

Greater trochanter

Intertrochanteric line

Neck

Lesser trochanter

Lateral epicondyle

Medial condyle

Lateral condyle

Patellar surface

Adductor tubercle

Medial epicondyle

Medial condyle

(a) Anterior surface

Articular surface of head

Neck

Greater trochanter

Intertrochanteric line

Fovea capitis

Lesser trochanter

Shaft of femur

Medial epicondyle

Lateral epicondyle

Lateral condyle

Patellar surface

Medial condyle

FIGURE 7-14
Femur. Bone markings on the right femur are presented as seen from (a) the anterior and (b) the posterior surface.

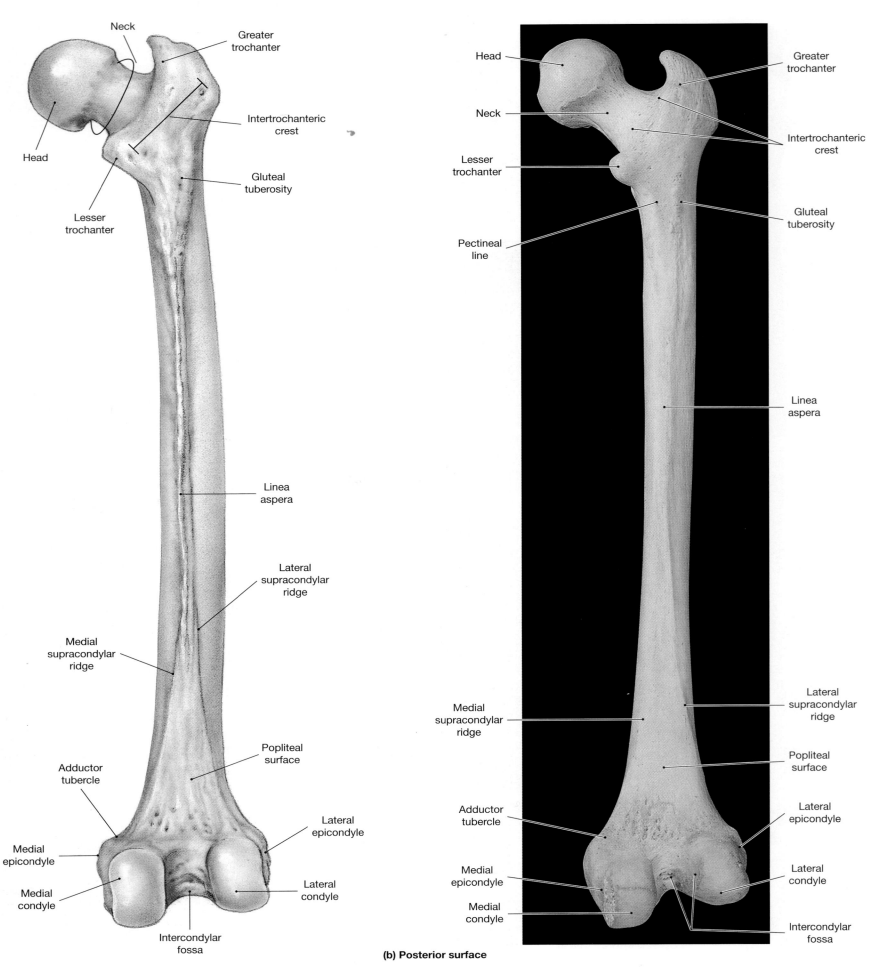

Neck

Greater
trochanter

Head

Intertrochanteric
crest

Lesser
trochanter

Gluteal
tuberosity

Linea
aspera

Lateral
supracondylar
ridge

Medial
supracondylar
ridge

Popliteal
surface

Adductor
tubercle

Lateral
epicondyle

Medial
epicondyle

Medial
condyle

Lateral
condyle

Intercondylar
fossa

Head

Greater
trochanter

Neck

Intertrochanteric
crest

Lesser
trochanter

Gluteal
tuberosity

Pectineal
line

Linea
aspera

Medial
supracondylar
ridge

Lateral
supracondylar
ridge

Popliteal
surface

Adductor
tubercle

Lateral
epicondyle

Medial
epicondyle

Medial
condyle

Lateral
condyle

Intercondylar
fossa

(b) Posterior surface

FIGURE 7-14 continued

194

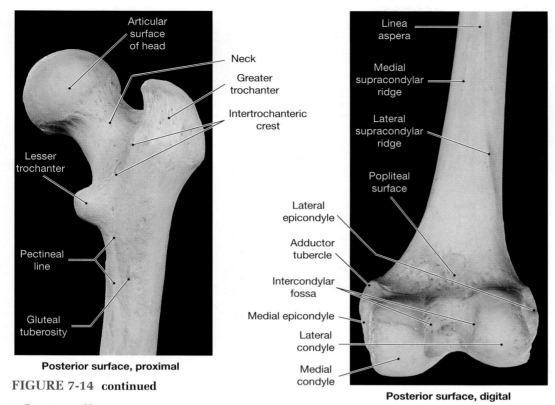

Posterior surface, proximal

FIGURE 7-14 continued

The Patella
(Figure 7-15)

The **patella** (pa-TEL-a) is a large sesamoid bone that forms within the tendon of the *quadriceps femoris*, a group of muscles that extend the leg. The roughly triangular patella has a rough, convex anterior surface (Figure 7-15a●). The roughened surface and broad **base** reflect the attachment of the quadriceps tendon (anterior and superior surfaces) and the *patellar ligament* (anterior and inferior surfaces). The patellar ligament continues from the **apex** of the patella to the tibia. The posterior surface (Figure 7-15b●) presents two concave **facets** (**medial** and **lateral**) for articulation with the medial and lateral condyles of the femur.

The Tibia
(Figure 7-16a,b)

The **tibia** (TIB-ē-a) is the large medial bone (Figure 7-16a,b●) of the leg. The medial and lateral condyles of the femur articulate with the **medial** and **lateral tibial condyles** at the proximal end of the tibia. The anterior surface of the tibia near the condyles bears a prominent, rough **tibial tuberosity** that can easily be felt beneath the skin of the leg. This tuberosity marks the attachment of the stout patellar ligament.

Posterior surface, digital

The **anterior crest** is a ridge that begins at the tibial tuberosity and extends distally along the anterior tibial surface. The anterior crest of the tibia can be felt through the skin. Distally, the tibia narrows, and the medial border ends in a large process, the **medial malleolus** (ma-LĒ-ō-lus; *malleolus*, hammer). The inferior surface of the tibia forms a hinge joint with the proximal bone of the ankle; the medial malleolus provides medial support for the joint. The posterior surface of the tibia bears a prominent **popliteal line,** or *soleal line,* which marks the attachment of several leg muscles, including the *popliteus* and *soleus.*

A ridge, the **intercondylar eminence,** separates the medial and lateral condyles of the tibia (Figure 7-16b●). There are two *tubercles* (*medial* and *lateral*) on the intercondylar eminence.

The Fibula
(Figure 7-16)

The slender **fibula** (FIB-yoo-la) parallels the lateral border of the tibia (Figure 7-16a,b●). The **fibular head** articulates along the lateral margin of the tibia, inferior and slightly posterior to the lateral tibial condyle. The medial border of the thin shaft is bound to the tibia by the **crural interosseous membrane**, which extends from the **interosseous border** of the fibula to that of the tibia. A sectional view through the shafts of the tibia and fibula (Figure 7-16c●) shows the locations of the tibial and fibular interosseous borders and the fibrous *crural interosseous membrane* that extends between them. This membrane helps stabilize the positions of the two bones and provides additional surface area for muscle attachment.

The fibula is excluded from the knee joint and does not transfer weight to the ankle and foot. However, it is an important site for muscle attachment. In addition, the distal tip of the fibula extends laterally to the ankle joint. This fibular process, the **lateral malleolus**, provides lateral stability to the ankle.

FIGURE 7-15
Right Patella. (a) Anterior surface. (b) Posterior surface.

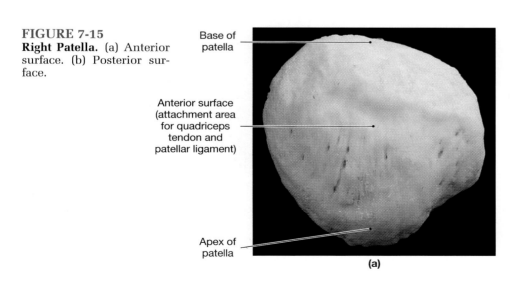

(a)

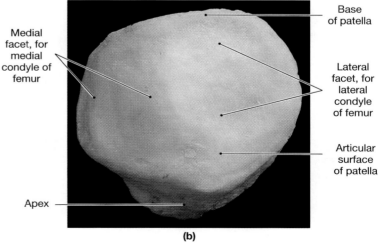

(b)

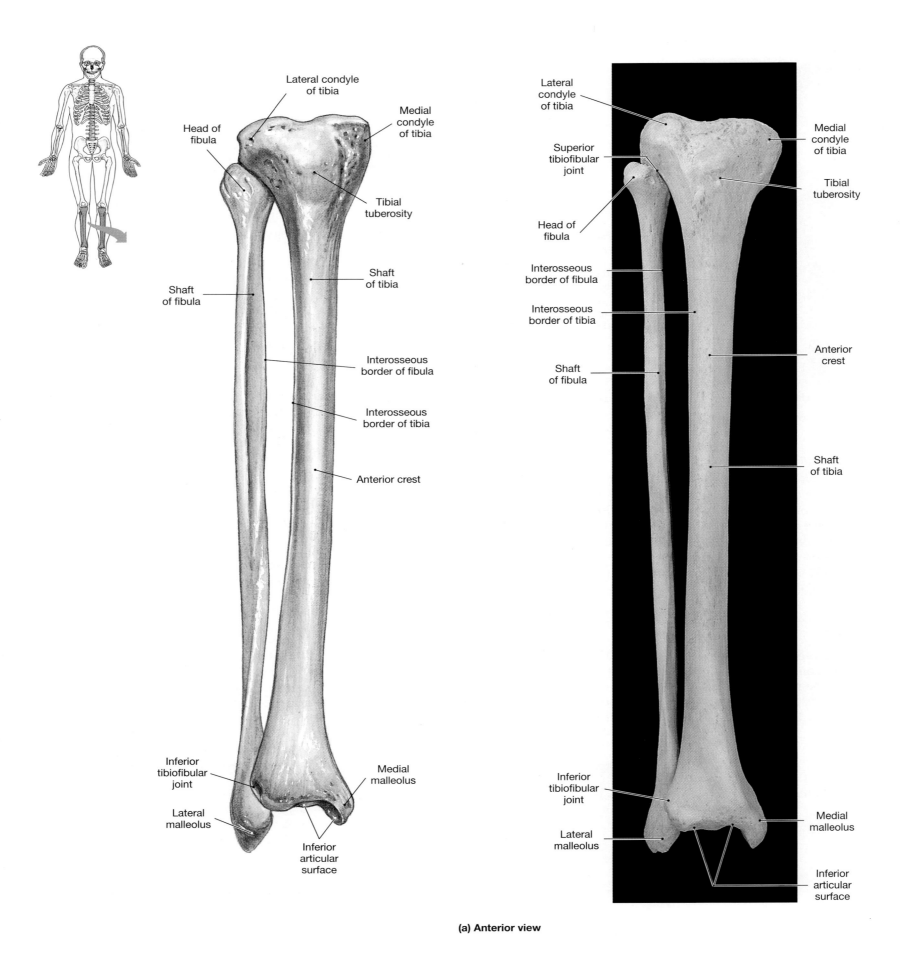

(a) Anterior view

FIGURE 7-16
Tibia and Fibula. (a) Anterior view of the right tibia and fibula. (b) Posterior view. (c) Sectional view at the level indicated in (b).

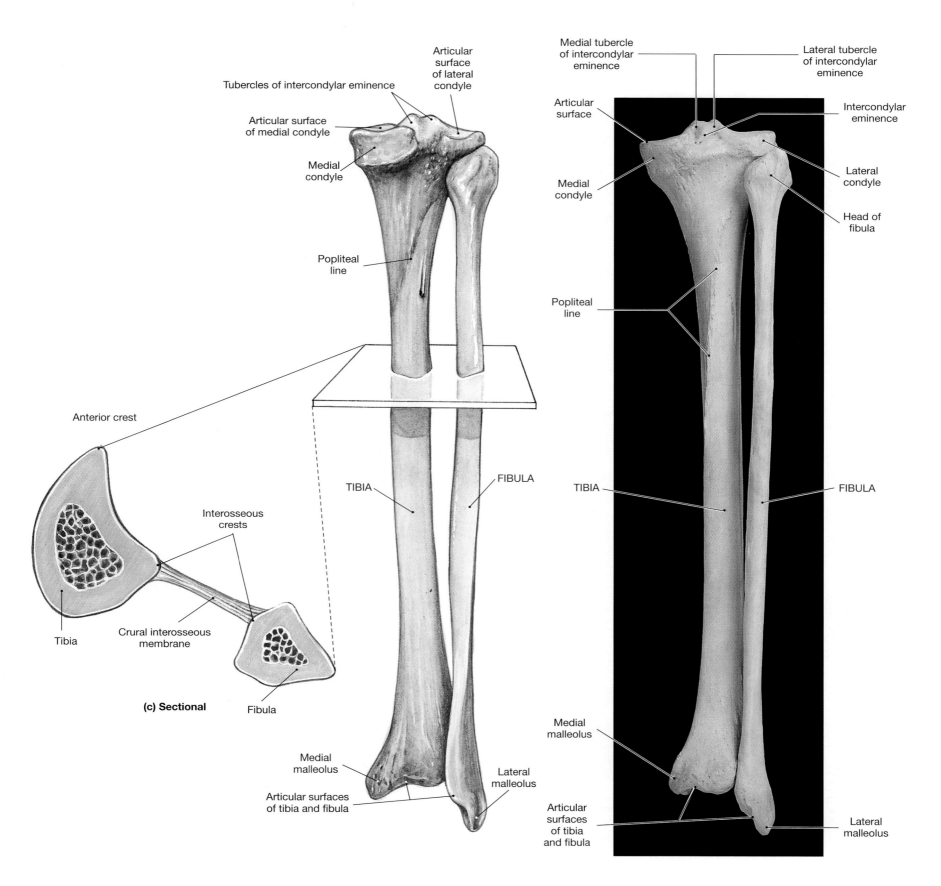

Tubercles of intercondylar eminence

Articular surface of medial condyle

Medial condyle

Popliteal line

Anterior crest

Tibia

Interosseous crests

Crural interosseous membrane

(c) Sectional

Fibula

Articular surface of lateral condyle

TIBIA

FIBULA

Medial malleolus

Articular surfaces of tibia and fibula

Lateral malleolus

(b) Posterior view

Medial tubercle of intercondylar eminence

Lateral tubercle of intercondylar eminence

Articular surface

Intercondylar eminence

Medial condyle

Lateral condyle

Head of fibula

Popliteal line

TIBIA

FIBULA

Medial malleolus

Articular surfaces of tibia and fibula

Lateral malleolus

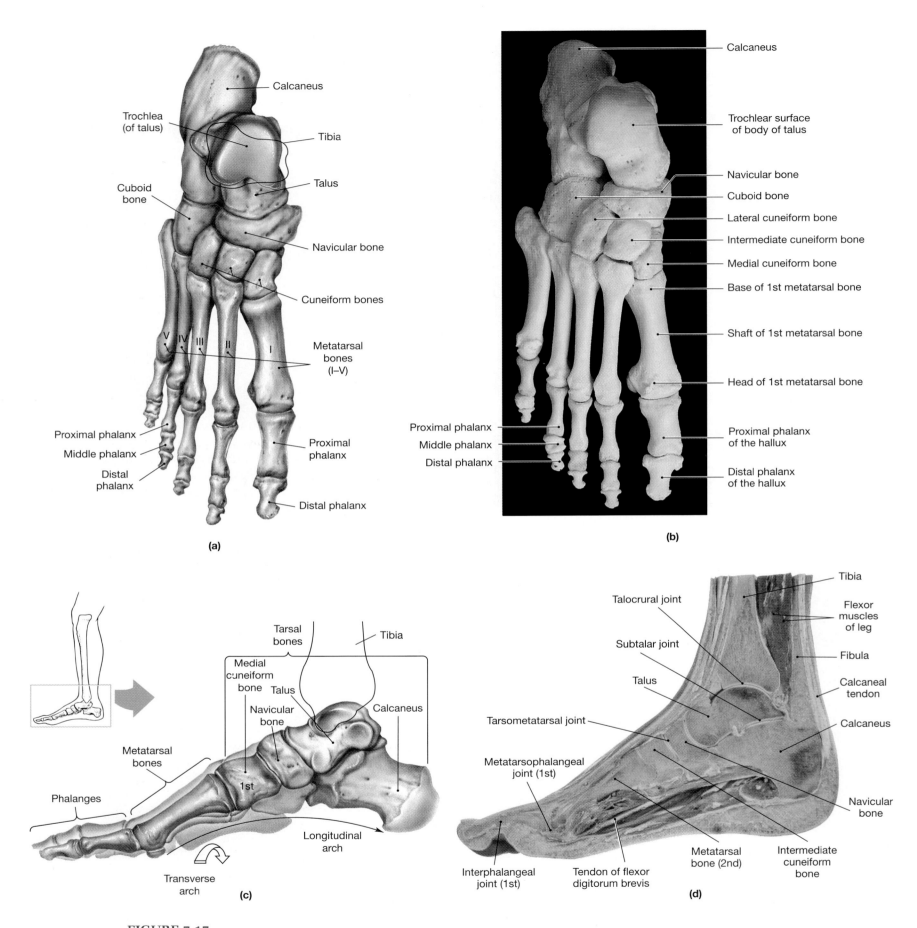

(a)

Calcaneus

Trochlea (of talus)

Tibia

Cuboid bone

Talus

Navicular bone

Cuneiform bones

V IV III II I

Metatarsal bones (I–V)

Proximal phalanx

Middle phalanx

Distal phalanx

Proximal phalanx

Distal phalanx

(b)

Calcaneus

Trochlear surface of body of talus

Navicular bone

Cuboid bone

Lateral cuneiform bone

Intermediate cuneiform bone

Medial cuneiform bone

Base of 1st metatarsal bone

Shaft of 1st metatarsal bone

Head of 1st metatarsal bone

Proximal phalanx

Middle phalanx

Distal phalanx

Proximal phalanx of the hallux

Distal phalanx of the hallux

(c)

Tarsal bones

Tibia

Medial cuneiform bone

Talus

Navicular bone

Calcaneus

Metatarsal bones

1st

Phalanges

Longitudinal arch

Transverse arch

(d)

Talocrural joint

Tibia

Flexor muscles of leg

Subtalar joint

Fibula

Talus

Calcaneal tendon

Tarsometatarsal joint

Calcaneus

Metatarsophalangeal joint (1st)

Navicular bone

Interphalangeal joint (1st)

Tendon of flexor digitorum brevis

Metatarsal bone (2nd)

Intermediate cuneiform bone

FIGURE 7-17

Bones of the Ankle and Foot. (a) Superior view of the right foot. Note the orientation of the tarsal bones that convey the weight of the body to the heel and the plantar surfaces of the foot. (b) Dorsal view of bones of the foot. (c) Medial view, showing the relative positions of the tarsal bones and the orientation of the transverse and longitudinal arches. (d) Sectional view of the right foot. Note the muscles and tendons within the longitudinal arch. See MRI SCAN 8a, p. 745.

The Tarsus
(Figure 7-17)

The ankle, or **tarsus** (Figure 7-17●), consists of seven **tarsal bones**: *talus, calcaneus, cuboid, navicular,* and three *cuneiforms.*

- The **talus** transmits from the tibia the weight of the body anterior toward the toes. The talus is the second largest foot bone. The primary tibial articulation occurs across the smooth superior surface of the **trochlea** of the talus. The trochlea has lateral and medial extensions that articulate with the lateral malleolus (fibula) and medial malleolus (tibia). The lateral surfaces of the talus are roughened where ligaments connect it to the tibia and fibula, further stabilizing the ankle joint.

- The **calcaneus** (kal-KĀ-nē-us), or heel bone, is the largest of the tarsal bones and may be easily palpated. When standing normally (Figure 7-17c●), most of your weight is transmitted from the tibia to the talus to the calcaneus, and then to the ground. The posterior portion of the calcaneus is a rough, knob-shaped projection. This is the attachment site for the **calcaneal tendon (tendon of Achilles)** that arises at strong calf muscles. These muscles raise the heel and depress the sole of the foot, as when standing on tiptoe. The superior and anterior surfaces of the calcaneus bear smooth facets for articulation with other tarsal bones.

- The **cuboid bone** articulates with the anterior, lateral surface of the calcaneus.

- The **navicular bone** is located anterior to the talus, on the medial side of the ankle. Its proximal surface articulates with the talus, its distal surface with the cuneiform bones.

- The three **cuneiform bones** are wedge-shaped bones arranged in a row, with articulations between them, located anterior to the navicular bone. They are named according to their position: **medial cuneiform, intermediate cuneiform**, and **lateral cuneiform bones**. Proximally the cuneiform bones articulate with the anterior surface of the navicular bone. The lateral cuneiform bone also articulates with the medial surface of the cuboid bone.

The distal surfaces of the cuboid bone and the cuneiform bones articulate with the bones of the foot.

The Foot
(Figure 7-17)

The **metatarsal bones** are five long bones that form the sole of the foot (Figure 7-17●). The metatarsal bones are identified with Roman numerals I–V, proceeding from medial to lateral across the sole. Proximally, the first three metatarsal bones articulate with the three cuneiform bones, and the last two articulate with the cuboid bone. Distally, each metatarsal bone articulates with a proximal phalanx.

The **phalanges**, or toes (Figure 7-17●), have the same anatomical organization as the fingers. The toes contain 14 phalanges. The great toe, or **hallux**, has two phalanges (**proximal** and **distal**), and the other four toes have three phalanges apiece (proximal, **middle**, and distal).

ARCHES OF THE FOOT *(Figure 7-17c,d).* Weight transfer occurs along the **longitudinal arch** of the foot (Figure 7-17c,d●). Ligaments and tendons maintain this arch by tying the calcaneus to the distal portions of the metatarsal bones. The lateral, calcaneal side of the foot carries most of the weight of the body while standing normally. This portion of the arch has much less curvature than the medial, talar portion. The talar arch also has considerably more elasticity than the calcaneal arch. As a result, the medial, plantar surface remains elevated, and the muscles, nerves, and blood vessels that supply the inferior surface of the foot are not squeezed between the metatarsals and the ground. The elasticity of the talar arch absorbs the shocks that accompany sudden changes in weight loading. For example, the stresses involved with running or ballet dancing on the toes are cushioned by the elasticity of this portion of the arch. Because the degree of curvature changes from the medial to the lateral borders of the foot, a **transverse arch** also exists.

The amount of weight transferred forward depends on the position of the foot and the placement of body weight. During *dorsiflexion* of the foot, as when "digging in the heels," all of the body weight rests on the calcaneus. During plantar flexion and "standing on tiptoe," the talus and calcaneus transfer the weight to the metatarsal bones and phalanges through more anterior tarsal bones.

√ **What three bones make up the coxa?**

√ **The fibula does not participate in the knee joint nor does it bend, but when it is fractured, it is difficult to walk. Why?**

√ **While jumping off the back steps at his house, 10-year-old Mark lands on his right heel and breaks his foot. What foot bone is most likely broken?**

INDIVIDUAL VARIATION IN THE SKELETAL SYSTEM

A comprehensive study of a human skeleton can reveal important information about the individual. For example, there are characteristic racial differences in portions of the skeleton, especially the skull and pelvis, and the development of various ridges and general bone mass can permit an estimation of muscular development and body weight. Details such as the condition of the teeth or the presence of healed fractures can provide information about the individual's medical history. Two important details, sex and age, can be determined or closely estimated on the basis of measurements indicated in Tables 7-1 and 7-2 on p. 202. Table 7-1 identifies characteristic differences between the skeletons of males and females, but not every skeleton shows every feature in classic detail. Many differences, including markings on the skull, cranial capacity, and general skeletal features, reflect differences in average body size, muscle mass, and muscular strength. The general changes in the skeletal system that take place with age are summarized in Table 7-2. Note how these changes begin at age 3 and continue throughout life. For example, fusion of the epiphyseal plates begins about age 3, while degenerative changes in the normal skeletal system, such as a reduction in mineral content in the bony matrix, do not begin until age 45.

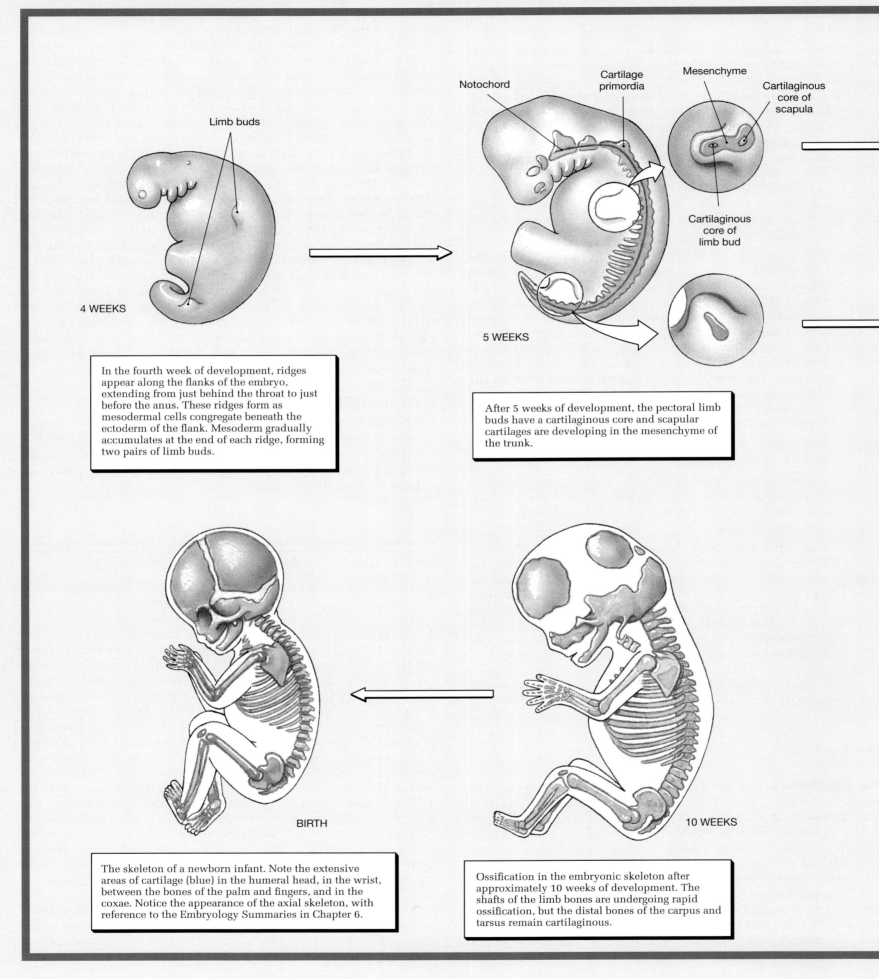

Limb buds

4 WEEKS

In the fourth week of development, ridges appear along the flanks of the embryo, extending from just behind the throat to just before the anus. These ridges form as mesodermal cells congregate beneath the ectoderm of the flank. Mesoderm gradually accumulates at the end of each ridge, forming two pairs of limb buds.

Notochord

Cartilage primordia

Mesenchyme

Cartilaginous core of scapula

Cartilaginous core of limb bud

5 WEEKS

After 5 weeks of development, the pectoral limb buds have a cartilaginous core and scapular cartilages are developing in the mesenchyme of the trunk.

BIRTH

The skeleton of a newborn infant. Note the extensive areas of cartilage (blue) in the humeral head, in the wrist, between the bones of the palm and fingers, and in the coxae. Notice the appearance of the axial skeleton, with reference to the Embryology Summaries in Chapter 6.

10 WEEKS

Ossification in the embryonic skeleton after approximately 10 weeks of development. The shafts of the limb bones are undergoing rapid ossification, but the distal bones of the carpus and tarsus remain cartilaginous.

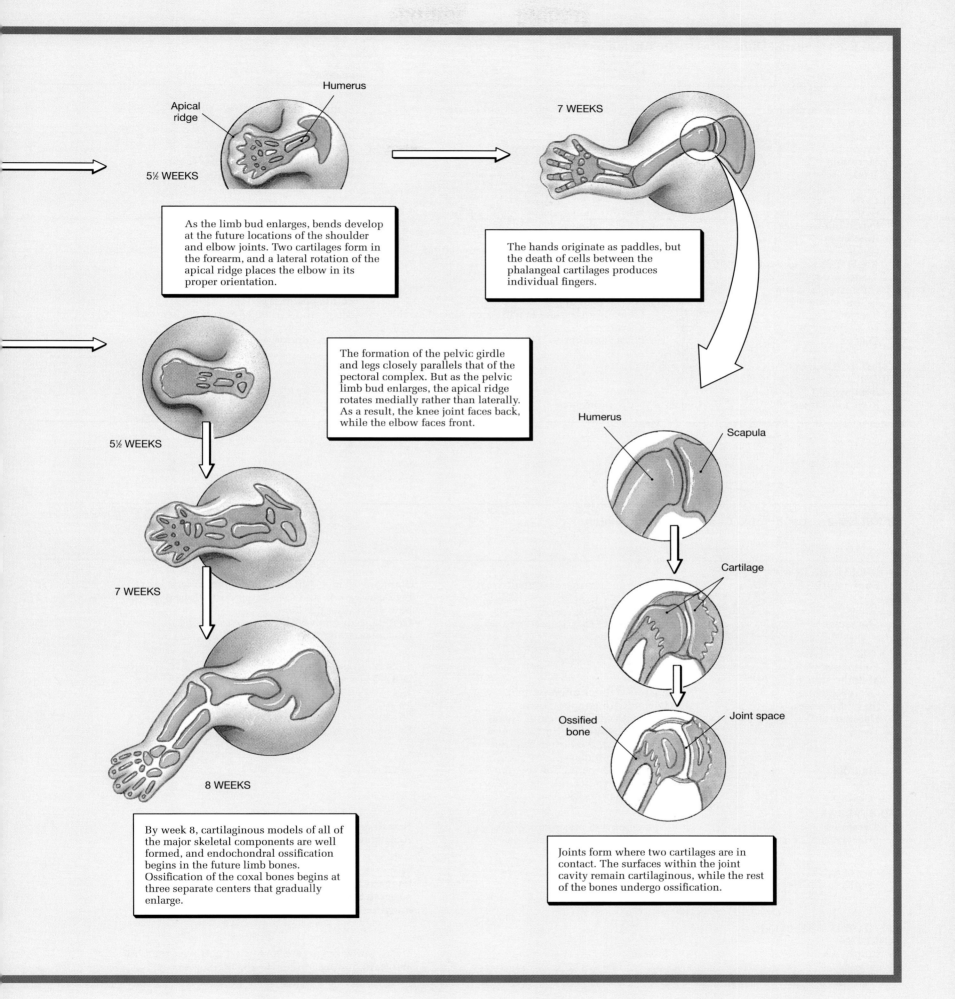

Apical ridge

Humerus

5½ WEEKS

As the limb bud enlarges, bends develop at the future locations of the shoulder and elbow joints. Two cartilages form in the forearm, and a lateral rotation of the apical ridge places the elbow in its proper orientation.

7 WEEKS

The hands originate as paddles, but the death of cells between the phalangeal cartilages produces individual fingers.

5½ WEEKS

The formation of the pelvic girdle and legs closely parallels that of the pectoral complex. But as the pelvic limb bud enlarges, the apical ridge rotates medially rather than laterally. As a result, the knee joint faces back, while the elbow faces front.

7 WEEKS

8 WEEKS

By week 8, cartilaginous models of all of the major skeletal components are well formed, and endochondral ossification begins in the future limb bones. Ossification of the coxal bones begins at three separate centers that gradually enlarge.

Humerus

Scapula

Cartilage

Ossified bone

Joint space

Joints form where two cartilages are in contact. The surfaces within the joint cavity remain cartilaginous, while the rest of the bones undergo ossification.

TABLE 7-1 Sexual Differences in the Human Skeleton

Region/Feature	Male	Female
SKULL		
General appearance	Heavier, rougher surface than female	Lighter, smoother surface than male
Forehead	Sloping	More vertical
Sinuses	Larger	Smaller
Cranium	About 10% larger (average)	About 10% smaller
Mandible	Larger, robust	Lighter, smaller
Teeth	Larger	Smaller
PELVIS		
General appearance	Narrow, robust, heavy, rough surface	Broad, light, smoother surface
Pelvic inlet	Heart-shaped	Oval to round
Iliac fossa	Relatively deep	Relatively shallow
Ilium	Extends farther above sacral articulation	More vertical; less extension above sacroiliac joint
Angle inferior to pubic symphysis	Less than 90°	100° or more
Acetabulum	Directed laterally	Faces slightly anteriorly as well as laterally
Obturator foramen	Oval	Triangular
Ischial spine	Points medially	Points posteriorly
Sacrum	Long, narrow triangle with pronounced sacral curvature	Broad, short triangle with less curvature
Coccyx	Points anteriorly	Points inferiorly
OTHER SKELETAL ELEMENTS		
Bone weight	Heavier	Lighter
Bone markings	More prominent	Less prominent

TABLE 7-2 Age-Related Changes in the Skeleton

Region/Structure	Event(s)	Age (Years)
GENERAL SKELETON		
Bony matrix	Reduction in mineral content	Standard values differ for males versus females between ages 45 and 65; similar reductions occur in both sexes after age 65.
Markings	Reduction in size, roughness	Gradual reduction with increasing age and decreasing muscular strength and mass.
SKULL		
Fontanels	Closure	1–24 months
Metopic suture	Fusion	2–8
Occipital bone	Fusion of ossification centers	1–4
Styloid process	Fusion with temporal bone	12–16
Hyoid bone	Complete ossification and fusion	25–30
Teeth	Loss of "baby teeth"; appearance of secondary dentition; eruption of posterior molars	Detailed in Chapter 25 (Digestive System)
Mandible	Loss of teeth; reduction in bone mass; change in angle at mandibular notch	Accelerates in later years (60+)
VERTEBRAE		
Curvature	Appearance of major curves	Described in Figure 6-18
Intervertebral discs	Reduction in size, percentage contribution to height	Accelerates in later years (60+)
LONG BONES		
Epiphyseal plates	Fusion	Ranges vary according to specific bone under discussion, but general analysis permits determination of approximate age (3–7, 15–22, etc.).
PECTORAL AND PELVIC GIRDLES		
Epiphyses	Fusion	Overlapping ranges are somewhat narrower than the above, including 14–16, 16–18, 22–25 years.

STUDY OUTLINE

INTRODUCTION [p. 174]

1. The **appendicular skeleton** includes the bones of the upper and lower extremities and the pectoral and pelvic girdles that connect the limbs to the trunk *(see Figure 7-1)*.

THE PECTORAL GIRDLE AND UPPER EXTREMITY [p. 175]

1. Each upper limb articulates with the trunk via the **shoulder**, or **pectoral girdle**, which consists of the **scapula** and **clavicle** *(see Figures 7-2 to 7-5)*.

The Pectoral Girdle [p. 175]

2. The clavicle and scapula position the shoulder joint, help move the upper extremity, and provide a base for muscle attachment *(see Figures 7-3/7-4)*.

3. The scapula articulates with the humerus at the **shoulder joint (scapulohumeral joint)**. Both the **coracoid process** and the **acromion process** are attached to ligaments and tendons. The acromion articulates with the clavicle at the **acromioclavicular joint**. The **scapular spine** crosses the scapular body *(see Figure 7-5)*.

sternal end • acromial end • conoid tubercle • costal tuberosity

body of scapula • borders • angles • supraglenoid tubercle • infraglenoid tubercle • glenoid cavity • neck • subscapular fossa • supraspinous fossa • infraspinous fossa

The Upper Extremity [p. 180]

4. The capsule of the shoulder attaches to the **humerus** at the **anatomical neck**; the **greater tubercle** and **lesser tubercle** are important sites for muscle attachment. Other prominent surface features include the **deltoid tuberosity**, the **radial groove**, the **medial** and **lateral epicondyles**, and the articular **condyle** *(see Figures 7-2/7-6 to 7-8)*.

5. Distally the humerus articulates with the radius and ulna. The medial **trochlea** extends from the **coronoid fossa** to the **olecranon fossa** *(see Figure 7-6)*.

6. The **ulna** and **radius** are the bones of the forearm. The olecranon fossa accommodates the **olecranon** during extension of the **olecranal joint** *(humeroulnar joint)*. The coronoid and radial fossae accommodate the **coronoid process** *(see Figures 7-2/7-7)*.

7. The bones of the wrist, or *carpus,* form two rows of **carpal bones**, **proximal** and **distal**. Four of the fingers contain three **phalanges**; the **pollex** (thumb) has only two *(see Figure 7-8)*.

8. Five **metacarpal bones** articulate with the distal carpal bones. Distally, the metacarpal bones articulate with the phalanges.

head • intertubercular groove • surgical neck • shaft

capitulum • radial fossa

trochlear notch • radial notch • antebrachial interosseous membrane • ulnar head • styloid process • articular cartilage • radial head • radial tuberosity • distal extremity • ulnar notch • pronation • supination scaphoid bone • lunate bone • triquetrum • pisiform bone • trapezium • trapezoid bone • capitate bone • hamate bone

THE PELVIC GIRDLE AND LOWER EXTREMITY [p. 185]

The Pelvic Girdle [p. 186]

1. The **pelvic girdle** consists of two **coxae**; each coxa forms through the fusion of an ilium, an ischium, and a pubis *(see Figures 7-9/7-10)*.

2. The largest coxal bone is the **ilium**. Inside the **acetabulum** the ilium is fused to the **ischium** (posteriorly) and the **pubis** (anteriorly). The **pubic symphysis** limits movement between the pubic bones of the left and right coxae *(see Figures 7-11/7-13)*.

3. The **pelvis** consists of the coxae, sacrum, and coccyx. It may be subdivided into the **false (greater) pelvis** and the **true (lesser) pelvis** *(see Figures 7-11 to 7-13)*.

acetabular fossa • lunate surface • gluteal line • arcuate line • iliac spines • inferior iliac notch • iliac crest • greater sciatic notch • ischial spine • lesser sciatic notch • ischial tuberosity • ischial ramus • pubic rami • pubic tubercle • pubic crest • obturator foramen • iliac fossa • iliopectineal line • auricular surface of ilium • iliac tuberosity • obturator groove pelvic brim • pelvic inlet • pelvic outlet • perineum

The Lower Extremity [p. 192]

4. The **femur** is the longest bone in the body. It articulates with the pelvis at the acetabulum and with the tibia at the knee joint *(see Figures 7-9/7-12d/7-14)*.

5. The **patella** is a large sesamoid bone that forms within the tendon of the quadriceps muscle. The patellar ligament extends from the patella to the **tibial tuberosity** *(see Figure 7-15)*.

6. The **tibia** is the largest medial bone of the lower leg. Other tibial landmarks include the **anterior crest**, the **interosseous border**, the **crural interosseous membrane**, which connects to the **fibula**, and the **medial malleolus**. The **fibular head** articulates with the tibia below the knee, and the **lateral malleolus** stabilizes the ankle *(see Figure 7-16)*.

7. The **tarsus**, or ankle, includes seven **tarsal bones**; only the **trochlea** portion of the **talus** articulates with the tibia and fibula. When standing normally, most of our weight is transferred to the **calcaneus**, and the rest is passed on to the **metatarsal bones**. Weight transfer occurs along the **longitudinal arch** and **transverse arches** *(see Figure 7-17)*.

8. The basic organizational pattern of the metatarsal bones and phalanges of the foot resembles that of the hand *(see Figure 7-17)*.

head • fovea capitis • neck • shaft • trochanters • intertrochanteric line • intertrochanteric crest • pectineal line • pectineus muscle • linea aspera • supracondylar ridges • popliteal surface • adductor tubercle • epicondyles • condyles • intercondylar fossa • patellar surface medial/lateral facets • medial/lateral tibial condyles • popliteal line • intercondylar eminence

calcaneal tendon (tendon of Achilles) • cuboid bone • navicular bone • cuneiform bones • phalanges • hallux

INDIVIDUAL VARIATION IN THE SKELETAL SYSTEM [p. 199]

1. Studying a human skeleton can reveal important information such as race, medical history, weight, sex, body size, muscle mass, and age *(see Tables 7-1/7-2)*.

2. A number of age-related changes and events take place in the skeletal system. These changes begin about age 3 and continue throughout life *(see Tables 7-1/7-2)*.

REVIEW OF CHAPTER OBJECTIVES

1. Identify each bone of the appendicular skeleton and note prominent surface features.
2. Identify the bones of the pectoral girdle and upper extremity and their prominent surface features.
3. Identify the bones that form the pelvic girdle and lower extremity and their prominent surface features.
4. Discuss the structural and functional differences between the pelvis of a female and that of a male.
5. Explain how study of the skeleton can reveal important information about an individual.
6. Summarize the skeletal differences between males and females.
7. Briefly describe how the aging process affects the skeletal system.

REVIEW OF CONCEPTS

1. The pectoral girdle attaches to the axial skeleton only at the sternoclavicular joint, while the pelvic girdle is firmly attached at the sacroiliac joint. Of what functional significance is the differing attachment of these girdles to the axial skeleton?
2. What properties of the glenohumeral articulation permit a human to circumduct the humerus in a 360° circle?
3. What deviation from the normal movement pattern would you predict to occur if the clavicle of the human pectoral girdle failed to develop?
4. What change in the position of the scapula would you expect to see when the arm is raised over the head?
5. Upon what structures of the pectoral girdle does the scapular spine have an effect, and how does this structure affect the function of the pectoral girdle?
6. What bony structures contribute to the "funny bone," and why do you get a tingling sensation when this region is struck?
7. Why is it possible to turn the hand through at least 180° of rotation even though the humerus is held stationary?
8. What is the function of the olecranon process of the ulna?
9. What is the function of an interosseous membrane?
10. How would the functional capability of the pelvic girdle be affected if it had as little bone-bone contact among the elements as does the pectoral girdle?
11. How do the shape, orientation, and movement capability of the hip joint compare with those of the scapulohumeral joint?
12. A fibrocartilage pad separates the rami of the pubis at the pubic symphysis, allowing this joint to have some flexibility. How would this attribute be advantageous during locomotion? How is this joint important in women during the birth process?
13. What contribution does the patella make in the function of the knee? How does the bone type classification of the patella affect this role?
14. How do movement capabilities of the knee joint compare with those of the elbow?

15. What is the importance of maintaining the correct amount of curvature of the longitudinal arch of the foot?
16. "Turn out" is important in performing many ballet movements correctly. This orientation involves the entire lower limb, not merely the distal segments of it. Which joint(s) is (are) involved in creating the position?
17. It is possible to discern "handedness" (i.e., being left- or right-handed) from skeletal features. What would these features be, and which bones would be involved to the greatest extent?
18. If two skeletal elements are approximately the same size and age, why does a heavier skeletal element with larger, rougher, and more prominent features such as ridges, crests, and processes suggest that the person to whom it belonged was a male?
19. Women tend to be "knock-kneed" more often than men. What features of the lower limbs and pelvic girdle contribute to this condition?

CRITICAL THINKING AND CLINICAL APPLICATION QUESTIONS

1. Muscular activity has a marked and clearly definable effect on skeletal structures. What specific effects would you predict would occur on the humerus of a person who works in a restaurant carrying large trays of dishes and food using a single hand, perhaps balancing the tray against one shoulder?
2. One of the most common sports injuries is called a "shoulder separation." What structures are actually damaged in this injury? What is the effect on the pectoral girdle?
3. People who perform repetitive tasks, such as working a cash register or a computer keyboard for extended periods of time, often develop a condition called carpal tunnel syndrome. What anatomical structures are being irritated in this condition? One method of alleviating pain from this problem involves cutting what structures?
4. When a clinician taps the anterior of the knee distal to the patella of a person sitting with one leg crossed over the other and the feet dangling in the air, what is being tested? If the person tenses the muscles of the leg before the tap, what will happen to the response? Why does the response change?
5. If a person breaks the "heel" of the foot, which structures are damaged and what problems will this cause for walking or running? Is this break likely to cause a significant or minor problem for the person? Why?
6. Although they may be in excellent health, recruits are still not permitted to serve in the army if they have a condition called "flat feet." Anatomically speaking, what does it mean to say a person has this condition, and why would this be a problem for an infantry soldier?
7. A person has the misfortune to break a great toe, which necessitates placing a stiff bandage around the toe and the metatarsophalangeal joint to immobilize the toe. If the person had a normal walking gait before the injury, will the bandaged toe and anterior metatarsal region change the gait? If so, how will the movement be affected?

8

The Skeletal System: Articulations

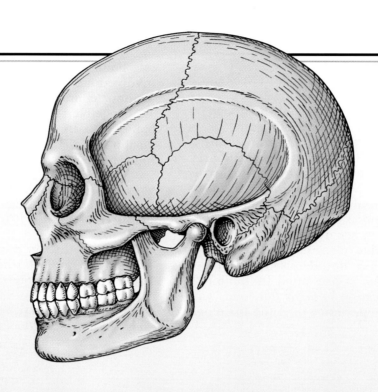

CHAPTER OUTLINE AND OBJECTIVES

We depend upon our bones for support, but support without mobility would leave us little better than statues. In this chapter we will focus on how bones are linked together to give us freedom of movement. Joints, or **articulations** (ar-tik-ū-LĀ-shuns), exist wherever two bones meet. The function of each joint depends on its anatomical design. Some joints permit extensive movement, others permit slight movement, and there are interlocking joints that completely prohibit movement.

CLASSIFICATION OF JOINTS

Three major categories of joints are based on the range of motion permitted (Table 8-1). An immovable joint is a **synarthrosis** (sin-ar-THRŌ-sis; *syn,* together + *arthros,* joint); a slightly movable joint is an **amphiarthrosis** (am-fē-ar-THRŌ-sis; *amphi,* on both sides); and a freely movable joint is a **diarthrosis** (dī-ar-THRŌ-sis; *dia,* through). Subdivisions within each major category indicate significant structural differences. Synarthrotic or amphiarthrotic joints are classified as fibrous or cartilaginous, and diarthrotic joints are subdivided according to the degree of movement permitted. An alternative classification scheme is based on joint structure (bony fusion, fibrous, cartilaginous, or synovial). This classification scheme is presented in Table 8-2. We will use the functional classification here, as our focus will be on the degree of motion permitted, rather than the histological structure of the articulation.

Immovable Joints (Synarthroses)

At a synarthrosis the bony edges are quite close together and may even interlock. A **suture** (*sutura,* a sewing together) is a synarthrotic joint found only between the bones of the skull. The edges of the bones are interlocked and bound together at the suture by dense connective tissue. A different type of synarthrosis binds each tooth to the surrounding bony socket. This fibrous connection is the **periodontal ligament** (per-ē-ō-DON-tal; *peri,* around + *odontos,* tooth), and the articulation is a **gomphosis** (gom-FŌ-sis; *gomphosis,* a bolting together).

An epiphyseal plate is an articulation between two separate bones, the shaft and the epiphysis, even though the bones involved are part of the same skeletal element. This rigid, cartilaginous connection is called a **synchondrosis** (sin-kon-DRŌ-sis; *syn,* together + *chondros,* cartilage). Sometimes two separate bones actually fuse together, and the boundary between them disappears. This creates a **synostosis** (sin-os-TŌ-sis), a totally rigid immovable joint.

Slightly Movable Joints (Amphiarthroses)

An amphiarthrosis permits very limited movement, and the bones are usually farther apart than they are at a synarthrosis. The bones may be connected by collagen fibers or cartilage. At a **syndesmosis** (sin-dez-MŌ-sis; *desmo,* band or ligament) bones are connected by a ligament. Examples include the distal articulation between the tibia and fibula. At a **symphysis**

TABLE 8-1   A Functional Classification of Articulations

Functional Category	Structural Category	Description	Example
SYNARTHROSIS (no movement)	**Fibrous**		
	Suture	Fibrous connections plus interdigitation	Between the bones of the skull
	Gomphosis	Fibrous connections plus insertion in alveolus	Periodontal ligaments between the teeth and jaws
	Cartilaginous		
	Synchondrosis	Interposition of cartilage plate	Epiphyseal plates
	Bony fusion		
	Synostosis	Conversion of other articular form to solid mass of bone	Portions of the skull
AMPHIARTHROSIS (little movement)	**Fibrous**		
	Syndesmosis	Ligamentous connection	Between the bones of the forearm or leg
	Cartilaginous		
	Symphysis	Connection by a fibrocartilage pad	Between right and left halves of pelvis; between adjacent vertebrae of spinal column
DIARTHROSIS (free movement)	**Synovial**	Complex joint bounded by joint capsule and containing synovial fluid	Numerous; subdivided by range of movement (see Figure 8-6)
	Monaxial	Permits movement in one plane	Elbow, ankle
	Biaxial	Permits movement in two planes	Ribs, wrist
	Triaxial	Permits movement in all three planes	Shoulder, hip

TABLE 8-2   A Structural Classification of Articulations

Example*	Structure	Type	Functional Category
Metopic suture (fusion) — **Frontal bone**	Bony fusion	Synostosis	Synarthrosis
Lambdoidal suture — **Skull**	Fibrous joint	Suture Gomphosis Syndesmosis	Synarthrosis Synarthrosis Amphiarthrosis
Symphysis — **Symphysis pubis**	Cartilaginous joint	Synchondrosis Symphysis	Synarthrosis Amphiarthrosis
Synovial joint	Synovial joint	Monaxial Biaxial Triaxial	All diarthroses

*For other examples, see Table 8-1.

the bones are separated by a wedge or pad of fibrocartilage. The articulations between adjacent vertebral bodies (via the *intervertebral disc*) and the anterior connection between the two pubic bones (the *pubic symphysis*) are examples of this type of joint.

Freely Movable Joints (Diarthroses)
(Figure 8-1)

Diarthroses, or **synovial** (si-NŌ-vē-al) **joints**, permit a wide range of motion. The basic structure of a synovial joint was introduced in Chapter 3 during the discussion of synovial membranes. ∞ [p. 77] Synovial joints are typically found at the ends of long bones, such as those of the upper and lower extremities. Under normal conditions, the bony surfaces cannot contact one another, because the articulating surfaces are covered by special **articular cartilages**. Articular cartilages resemble hyaline cartilages elsewhere in the body, but they have no perichondrium, and the matrix contains more water than other cartilages have. View Figure 8-1● as we describe the structure of synovial joints.

Synovial Fluid

A synovial joint is surrounded by a fibrous **joint capsule**, and a synovial membrane lines the joint cavity. ∞ [p. 76] Synovial membranes produce the **synovial fluid** that fills the joint cavity. Synovial fluid serves three functions.

1. *It provides lubrication:* When a portion of the cartilage is compressed, some of the water is squeezed out of the cartilage and into the space between the opposing surfaces. This thin layer of fluid reduces friction between moving surfaces in a joint to around one-fifth of that between two pieces of ice. The articular cartilages act like sponges, for when the compression stops, fluid is sucked back into them.

2. *It nourishes the chondrocytes:* The total quantity of synovial fluid in a joint is normally less than 3 mℓ, even in a large joint such as the knee. This relatively small volume of fluid must be circulated continually to provide nutrients and a route for waste disposal for the chondrocytes of the articular cartilages. The synovial fluid circulates whenever the joint moves, and the compression and reexpansion of the articular cartilages pump synovial fluid into and out of the cartilage matrix.

3. *It acts as a shock absorber:* Synovial fluid cushions shocks in joints that are subjected to compression. For example, the hip, knee, and ankle joints are compressed during walking, and they are severely compressed during jogging or running. When the pressure suddenly increases, the synovial fluid absorbs the shock and distributes it evenly across the articular surfaces. ♆ *Rheumatism, Arthritis, and Synovial Function [p. 755]*

Accessory Structures

Synovial joints may have a variety of accessory structures, including pads of cartilage or fat, ligaments, tendons, and bursae.

CARTILAGES AND FAT PADS *(Figure 8-1).* In complex joints such as the knee (Figure 8-1●), accessory structures may lie between the opposing articular surfaces. These include:

- **Menisci** (men-IS-kē; *meniscus*, crescent), or **articular discs**, may subdivide a synovial cavity, channel the flow of synovial fluid, or allow for variations in the shapes of the articular surfaces.
- **Fat pads** are often found around the periphery of the joint, lightly covered by a layer of synovial membrane. Fat pads provide protection for the articular cartilages. They also act as packing material for the joint; when the bones move, the fat pads fill in the spaces created as the joint cavity changes shape.

LIGAMENTS *(Figure 8-1b).* The joint capsule that surrounds the entire joint is continuous with the periostea of the articulating bones. **Accessory ligaments** are localized thickenings of the capsule. **Extracapsular ligaments** can be seen on the outside of the capsule; **intracapsular ligaments** are found inside the capsule (Figure 8-1b●).

TENDONS *(Figure 8-1b).* While not part of the articulation itself, tendons (Figure 8-1b●) passing across or around a joint may limit the range of motion and provide mechanical support.

BURSAE *(Figure 8-1b).* **Bursae** (Figure 8-1b●) are small, fluid-filled pockets in connective tissue. They are filled with synovial fluid and lined by synovial membrane. Bursae may be connected to the joint cavity, or they may be completely separate from it. Bursae form where a tendon or ligament rubs against other tissues. Their function is to reduce friction and act as a shock absorber. Bursae are found around most synovial joints, such as the shoulder joint. **Synovial tendon sheaths** are tubular bursae that surround tendons where they pass across bony surfaces. Bursae may also appear beneath the skin covering a bone or within other connective tissues exposed to friction or pressure. Bursae that develop in abnormal locations, or due to abnormal stresses, are called *adventitious bursae.*

Strength versus Mobility

A joint cannot be both highly mobile and very strong. The greater the range of motion at a joint, the weaker it becomes. A synarthrosis, the strongest type of joint, does not permit any movement. Any mobile diarthrosis may be damaged by movement beyond its normal range of motion. Several factors are responsible for limiting that range of motion and reducing the frequency of injury:

- the presence of accessory ligaments and the collagen fibers of the joint capsule;
- the shapes of the articulating surfaces that prevent movement in specific directions;
- the presence of other bones, skeletal muscles, or fat pads around the joint;
- tension in tendons attached to the articulating bones. When a skeletal muscle contracts and pulls on a tendon, it may either encourage or oppose movement in a specific direction.

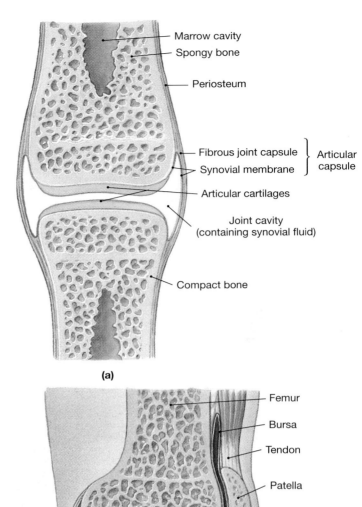

(a)

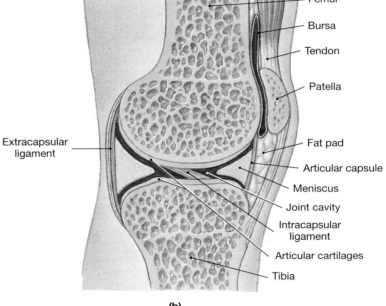

(b)

FIGURE 8-1
Structure of a Synovial Joint. (a) Diagrammatic view of a simple articulation. (b) A simplified sectional view of the knee joint.

■ CLINICAL BRIEF
Dislocation of a Synovial Joint

When a **dislocation**, or **luxation** (luks-Ā-shun), occurs, the articulating surfaces are forced out of position. This displacement can damage the articular cartilages, tear ligaments, or distort the joint capsule. Although the *inside* of a joint has no pain receptors, nerves that monitor the capsule, ligaments, and tendons are quite sensitive, and dislocations are very painful. The damage accompanying a partial dislocation, or **subluxation** (sub-luks-Ā-shun), is less severe. People who are "double-jointed" have joints that are weakly stabilized. Although their joints permit a greater range of motion than those of other individuals, they are more likely to suffer partial or complete dislocations.

ARTICULAR FORM AND FUNCTION

To *understand* human movement you must become aware of the relationship between structure and function at each articulation. To *describe* human movement you need a frame of reference that permits accurate and precise communication. The synovial joints can be classified according to their anatomical and functional properties. To demonstrate the basis for that classification, we will describe the movements that can occur at a typical synovial joint, using a simplified model.

Describing Dynamic Motion
(Figure 8-2)

Take a pencil (or pen) as your model, and stand it upright on the surface of a desk or table, as shown in Figure 8-2a●. The pencil represents a bone, and the desk is an articular surface. A little imagination and a lot of twisting, pushing, and pulling will demonstrate that there are only three ways to move the model. Considering them one at a time will provide a frame of reference for analyzing any complex movement.

Possible Movement 1: The point can move.

If you hold the pencil upright but do not secure the point, you can push the pencil across the surface. This kind of motion is called *gliding* (Figure 8-2b●), and it is an example of **linear motion**. You could slide the point forward or backward, from one side to the other, or diagonally. However you choose to move the pencil, the motion can be described using two lines of reference. One line represents forward/backward motion, and the other left/right movement. For example, a simple movement along one axis could be described as "forward 1 cm" or "left 2 cm." A diagonal movement could be described using both axes, as in "backward 1 cm and to the right 2.5 cm."

Possible Movement 2: The shaft can change its angle with the surface.

While holding the tip in position, you can still move the free (eraser) end forward and backward or from side to side. These movements, which change the angle between the shaft and the articular surface, are examples of **angular motion** (Figure 8-2c●).

Any angular movement can be described with reference to the same two axes (forward/backward, left/right) and the angular change (in degrees). However, in one instance a special term is used to describe a complex angular movement. Grasp the free end of the pencil, and move it until the shaft is no longer vertical. Now move that end through a complete circle (Figure 8-2d●). This movement is very difficult to describe. Anatomists avoid the problem entirely by using a special term, **circumduction** (sir-kum-DUK-shun; *circum*, around), for this type of angular motion.

Possible Movement 3: The shaft can rotate.

If you prevent movement of the base and keep the shaft vertical, you can still spin the shaft around its longitudinal axis. This movement is called **rotation** (Figure 8-2e●). Several articulations will permit partial rotation, but none can rotate freely; such a movement would hopelessly tangle the blood vessels, nerves, and muscles that cross the joint.

If an articulation permits movement along only one axis, it is called **monaxial** (mon-AKS-ē-al). In the above model, if an ar-

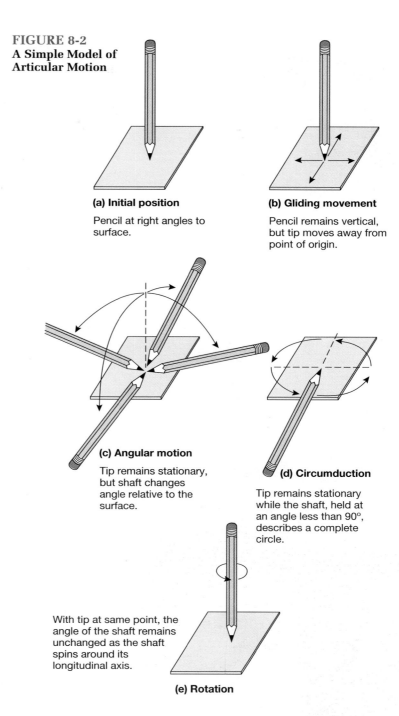

FIGURE 8-2
A Simple Model of Articular Motion

(a) Initial position
Pencil at right angles to surface.

(b) Gliding movement
Pencil remains vertical, but tip moves away from point of origin.

(c) Angular motion
Tip remains stationary, but shaft changes angle relative to the surface.

(d) Circumduction
Tip remains stationary while the shaft, held at an angle less than 90°, describes a complete circle.

(e) Rotation
With tip at same point, the angle of the shaft remains unchanged as the shaft spins around its longitudinal axis.

ticulation permits angular movement only in the forward/backward plane, or prevents any movement other than rotation around its longitudinal axis, it is monaxial. If movement can occur along two axes, the articulation is **biaxial** (bī-AKS-ē-al). If the pencil could undergo angular motion in the forward/backward or left/right plane, but not in some combination of the two, it would be biaxial. **Triaxial** (trī-AKS-ē-al) joints permit rotation and angular motion in multiple planes.

Types of Movements

In descriptions of motion at synovial joints, phrases such as "bend the leg" or "raise the arm" are not sufficiently precise. Anatomists use descriptive terms that have specific meanings. We will consider these movements with regard to the basic categories of movement considered in the previous section.

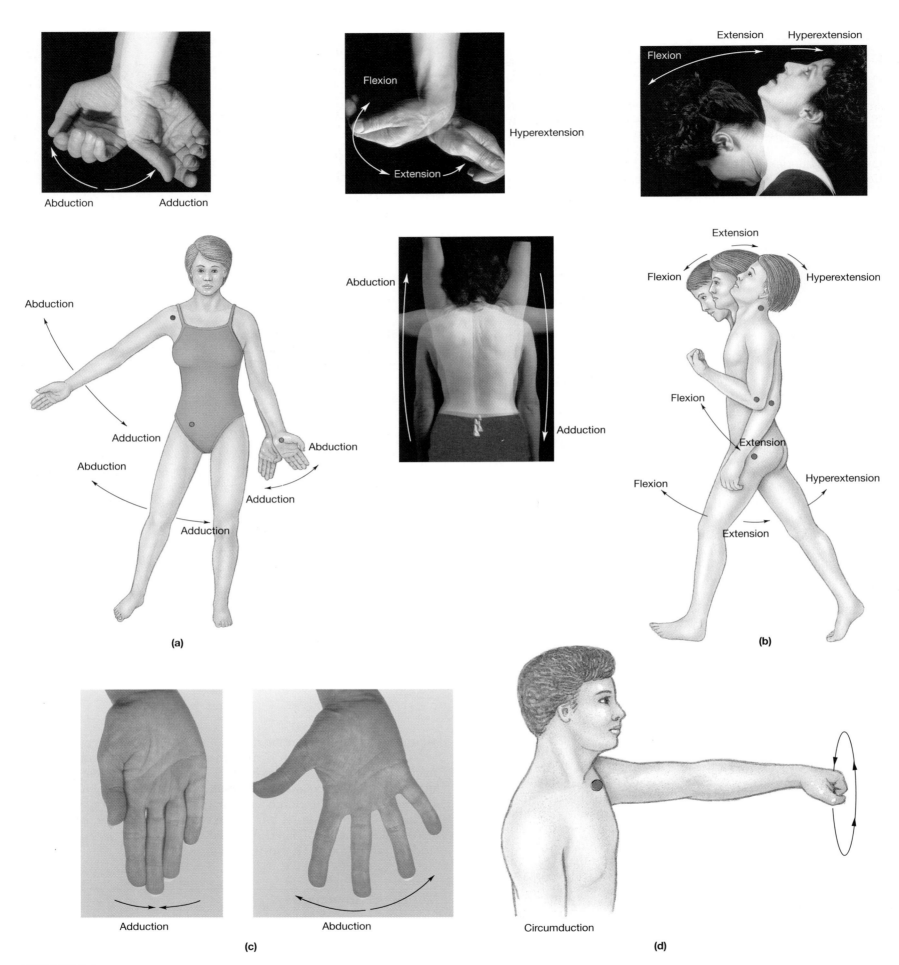

Abduction   Adduction

Flexion

Extension   Hyperextension

Hyperextension

Flexion

Extension

Abduction

Abduction

Adduction

Abduction

Adduction

Abduction

Adduction

Adduction

Extension

Flexion   Hyperextension

Flexion

Flexion   Extension

Flexion   Hyperextension

Extension

(a)

(b)

Adduction   Abduction

Circumduction

(c)   (d)

FIGURE 8-3
Angular Movements

210

Gliding

In **gliding**, two opposing surfaces slide past one another, as in possible movement 1. Gliding occurs between the surfaces of articulating carpals and tarsals and between the clavicles and the sternum. The movement can occur in almost any direction, but the amount of movement is slight, and rotation is usually prevented by the capsule and associated ligaments.

Angular Motion
(Figure 8-3)

Examples of angular motion include *flexion, extension, adduction*, and *abduction*. The description of each movement is based on reference to an individual in the anatomical position. View Figure 8-3● as we describe examples of angular motion.

- **Flexion** (FLEK-shun) can be defined as movement in the anterior-posterior plane that *reduces the angle between the articulating elements*. **Extension** occurs in the same plane, but it *increases the angle between articulating elements* (Figure 8-3b●). When you bring your head toward your chest, you flex the head. When you bend down to touch your toes, you flex the spine. Extension reverses these movements.

  Flexion at the shoulder or hip moves the limbs forward, whereas extension moves them back. Flexion of the wrist moves the palm forward, and extension moves it back. In each of these examples, extension can be continued past the anatomical position, in which case **hyperextension** occurs. You can also hyperextend the head, a movement that allows you to gaze at the ceiling (Figure 8-3b●). Hyperextension of other joints is usually prevented by ligaments, bony processes, or soft tissues.

- **Abduction** (*ab*, from) is movement *away from the longitudinal axis of the body* in the frontal plane. For example, swinging the upper limb to the side is abduction of the limb; moving it back constitutes **adduction** (*ad*, to). Adduction of the wrist moves the heel of the hand *toward* the body, whereas abduction moves it farther away. Spreading the fingers or toes apart abducts them, because they move *away* from a central digit (finger or toe). Bringing them together constitutes adduction. Abduction and adduction always refer to movements of the appendicular skeleton (Figure 8-3a,c●).

- A special type of angular motion, **circumduction**, (Figure 8-3d●) was introduced in our model. A familiar example of circumduction is moving your arm in a loop, as when drawing a large circle on a chalkboard.

Rotation
(Figure 8-4)

Rotational movements are described with reference to a figure in the anatomical position. Rotation of the head may involve **left rotation** or **right rotation**. In analyses of movements of the limbs, if the anterior aspect of the limb rotates *inward*, toward the ventral surface of the body, you have **internal rotation**, or **medial rotation**. If it turns outward, you have **external rotation**, or **lateral rotation**. These rotational movements are illustrated in Figure 8-4●.

The articulations between the radius and ulna permit the rotation of the distal end of the radius across the anterior surface of the ulna. This moves the wrist and hand from palm-facing-front to palm-facing-back. This motion is called **pronation** (prō-NĀ-shun). The opposing movement, in which the palm is turned forward, is **supination** (sū-pi-NĀ-shun).

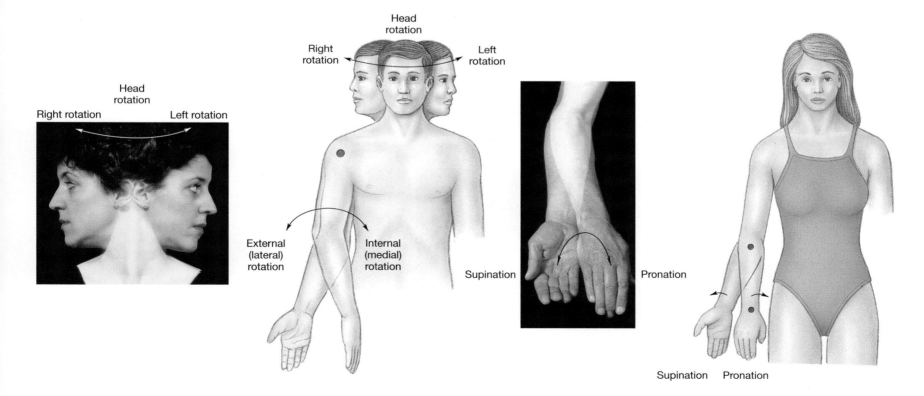

FIGURE 8-4
Rotational Movements

Special Movements
(Figure 8-5)

There are a number of special terms that apply to specific articulations or unusual types of movement (Figure 8-5●).

- **Inversion** (*in*, into + *vertere*, to turn) is a twisting motion of the foot that turns the sole inward. The opposite movement is called **eversion** (ē-VER-shun; *e*, out).
- **Dorsiflexion** and **plantar flexion** (*planta*, sole) also refer to movements of the foot. Dorsiflexion is flexion of the ankle and elevation of the sole, as when "digging in the heels." Plantar flexion, the opposite movement, extends the ankle and elevates the heel, as when standing on tiptoes.
- **Opposition** is the special movement of the thumb that enables it to grasp and hold an object.
- **Protraction** entails moving a part of the body anteriorly in the horizontal plane. **Retraction** is the reverse movement. You protract your jaw when you grasp your upper lip with your lower teeth, and you protract your clavicles when you cross your arms.
- **Elevation** and **depression** occur when a structure moves in a superior or inferior direction. You depress your mandible when you open your mouth and elevate it as you close it. Another familiar elevation occurs when you shrug your shoulders.
- **Lateral flexion** occurs when the vertebral column bends to the side. This movement is most pronounced in the cervical and thoracic regions.

A Structural Classification of Synovial Joints
(Figure 8-6)

- **Gliding joints:** Gliding joints (Figure 8-6c●), also called *planar joints*, have flattened or slightly curved faces. The relatively flat articular surfaces slide across one another, but the amount of movement is very slight. Ligaments usually prevent or restrict rotation. Gliding joints are found at the ends of the clavicles, between the carpals, between the tarsals, and between the articular facets of adjacent spinal vertebrae. Gliding joints may be classified as *monaxial*, because they permit only small sliding movements along one axis, or *multiaxial*, because that sliding may occur in any direction.

- **Hinge joints:** The elbow (Figure 8-6a●) and knee are called hinge joints because they permit angular movement in a single plane, like the opening and closing of a door. A hinge joint is an example of a monaxial joint.

- **Pivot joints:** Pivot joints (Figure 8-6b●) are also monaxial, but they permit only rotation. A pivot joint between the atlas and axis allows you to turn your head to either side.

- **Ellipsoidal joints:** In an ellipsoidal joint (Figure 8-6d●), an oval articular face nestles within a depression on the opposing surface. With such an arrangement, angular motion occurs in two planes, along or across the length of the oval. It is thus an example of a biaxial joint. Ellipsoidal joints connect the fingers and toes with the metacarpals and metatarsals.

- **Saddle joints:** Saddle joints (Figure 8-6e●) have complex articular faces. Each one resembles a saddle because it is concave on one axis and convex on the other. Saddle joints are extremely mobile, allowing extensive angular motion without rotation. They are usually classified as biaxial joints. Moving the saddle joint at the base of your thumb is an excellent demonstration that also provides an excuse for twiddling your thumbs during a lecture.

- **Ball-and-socket joints:** In a ball-and-socket joint (Figure 8-6f●), the round head of one bone rests within a cup-shaped depression in another. All combinations of movements, including rotation, can be performed at ball-and-socket joints. These are triaxial joints, and examples include the shoulder and hip joints.

√ In a newborn infant, the large bones of the skull are joined by fibrous connective tissue. What type of joint is this? These bones later grow, interlock, and form immovable joints. What type of joints are these?

√ Give the proper term for each of the following types of motion: (a) moving the humerus away from the midline of the body, (b) turning the palms so that they face forward, (c) bending the elbow.

FIGURE 8-5
Special Movements

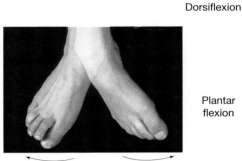

Eversion Inversion

Dorsiflexion

Plantar flexion

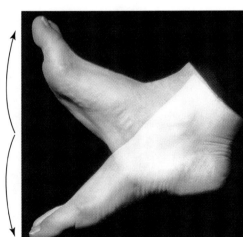
Lateral flexion

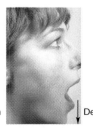
Opposition

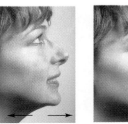
Retraction Protraction

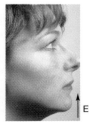
Elevation Depression

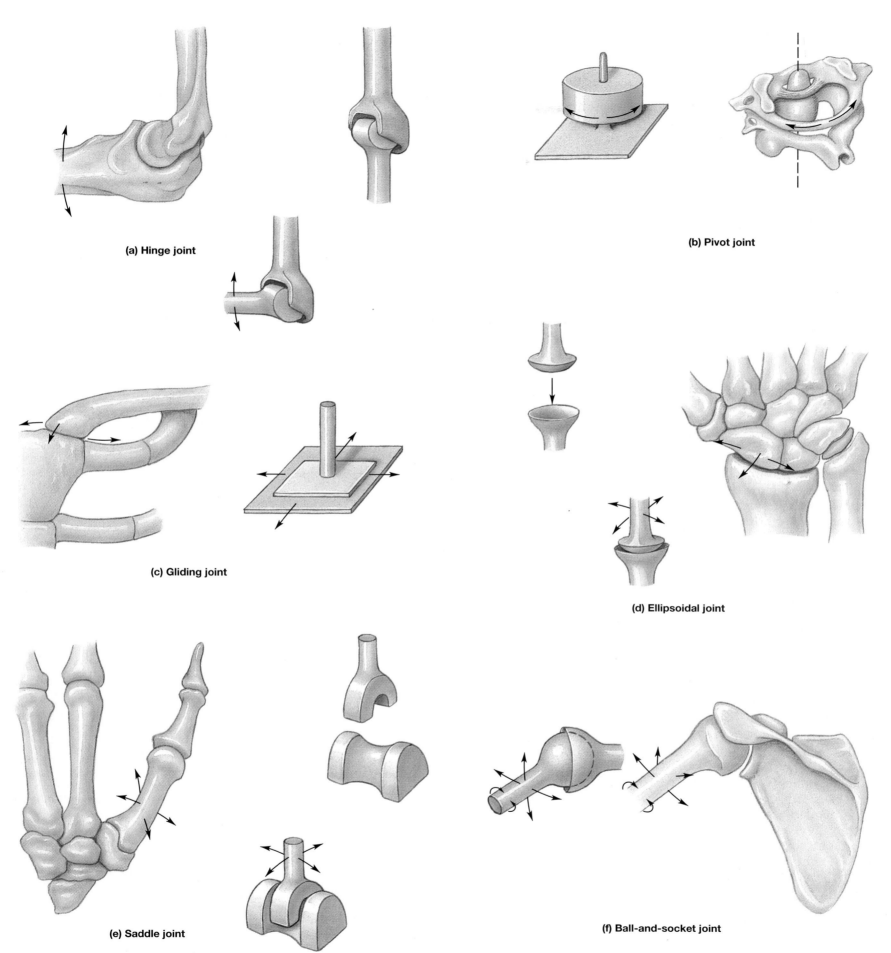

(a) Hinge joint

(b) Pivot joint

(c) Gliding joint

(d) Ellipsoidal joint

(e) Saddle joint

(f) Ball-and-socket joint

FIGURE 8-6
A Structural Classification of Synovial Joints

REPRESENTATIVE ARTICULATIONS

This section considers examples of articulations that demonstrate important functional principles. We will first consider the *intervertebral articulations* of the axial skeleton. The articulations between adjacent vertebrae include gliding articulations and cartilaginous symphyses. We will then proceed to a discussion of the synovial articulations of the appendicular skeleton. The shoulder has great mobility, the elbow has great strength, and the wrist makes fine adjustments in the orientation of the palm and fingers. The functional requirements of the joints in the lower limb are very different from those of the upper limb. Articulations at the hip, knee, and ankle must transfer the body weight to the ground, and during movements such as running, jumping, or twisting, the applied forces are considerably greater than the weight of the body. Although this section considers representative articulations, Tables 8-3, 8-4, and 8-5 summarize information concerning the majority of articulations in the body.

Intervertebral Articulations
(Figure 8-7)

The articulations between the superior and inferior articular processes of adjacent vertebrae are gliding joints that permit small movements associated with flexion and rotation of the vertebral column. Little gliding occurs between adjacent vertebral bodies, and Figure 8-7● illustrates the structure of this joint. From axis to sacrum, the vertebrae are separated and cushioned by pads of fibrocartilage called **intervertebral discs**. Intervertebral discs are not found in the sacrum and coccyx, where vertebrae have fused, nor are they found between the first and second cervical vertebrae.

The Intervertebral Discs
(Figures 8-7/8-8)

Each intervertebral disc (Figures 8-7 and 8-8●) has a tough outer layer of fibrocartilage, the **anulus fibrosus** (AN-ū-lus fī-BRŌ-sus). The collagen fibers of the anulus fibrosus attach the disc to the bodies of adjacent vertebrae. The anulus also surrounds a soft, elastic, and gelatinous core, the **nucleus pulposus** (pul-PŌ-sus). The nucleus pulposus is composed primarily of water (about 75%) with scattered reticular and elastic fibers. The nucleus pulposus gives the disc resiliency and enables it to act as a shock absorber.

Movements of the vertebral column compress the nucleus pulposus substance and displace it in the opposite direction. This displacement permits smooth gliding movements by each vertebra while still maintaining the alignment of all the vertebrae. The discs make a significant contribution to an individual's height; they account for roughly one-quarter of the length of the vertebral column above the sacrum. As we grow older, the water content of the nucleus pulposus within each disc decreases. The discs gradually become less effective as a cushion, and the chances for vertebral injury increase. Loss of water by the discs also causes shortening of the vertebral column; this shortening accounts for the characteristic decrease in height with advanced age.

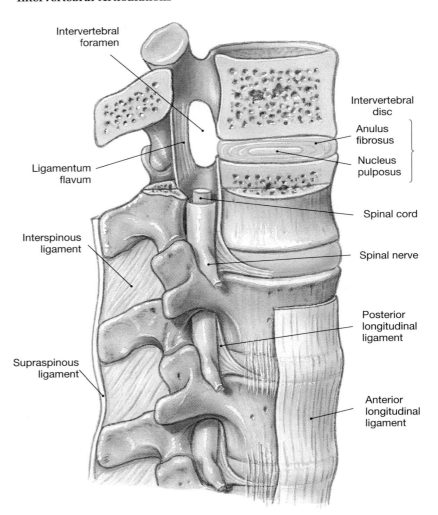

FIGURE 8-7
Intervertebral Articulations

Intervertebral foramen

Intervertebral disc

Anulus fibrosus

Ligamentum flavum

Nucleus pulposus

Interspinous ligament

Spinal cord

Spinal nerve

Posterior longitudinal ligament

Supraspinous ligament

Anterior longitudinal ligament

■ CLINICAL BRIEF
Problems with the Intervertebral Discs

An intervertebral disc compressed beyond its normal limits may become temporarily or permanently damaged. If the posterior longitudinal ligaments are weakened, as often occurs with advancing age, the compressed nucleus pulposus may distort the anulus fibrosus, forcing it partway into the vertebral canal. This condition is often called a **slipped disc** (Figure 8-8a●), although disc slippage does not actually occur. The most common sites for disc problems are at C_5–C_6, L_4–L_5, and at L_5–S_1.

Under severe compression the nucleus pulposus may break through the anulus and enter the vertebral canal. This condition is called a **herniated disc** (Figure 8-8b●). When a disc herniates, sensory nerves are distorted, producing pain, and the protruding mass can also compress the nerves passing through the intervertebral foramen. **Sciatica** (sī-AT-i-ka) is the painful result of compression of the roots of the sciatic nerve. The acute initial pain in the lower back is sometimes called **lumbago** (lum-BĀ-gō).

Most lumbar disc problems can be treated successfully with some combination of rest, back braces, analgesic (pain-killing) drugs, and physical therapy. Surgery to relieve the symptoms is required in only about 10% of cases involving lumbar disc herniation. In this procedure, the disc is removed, and the vertebral bodies fuse together to prevent relative movement. To access the offending disc, the surgeon must remove the nearest vertebral arch by shaving away the laminae. For this reason the procedure is known as a **laminectomy** (la-mi-NEK-tō-mē).

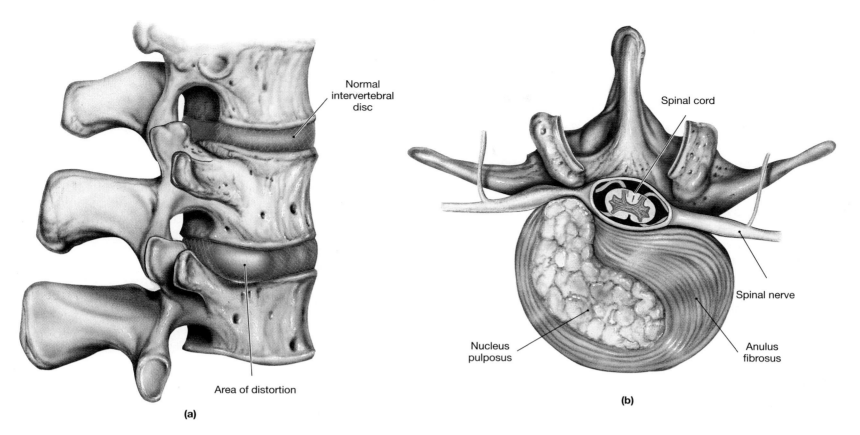

FIGURE 8-8
Damage to the Intervertebral Discs. (a) Lateral view of the lumbar region of the spinal column, showing a distorted ("slipped") intervertebral disc. (b) Sectional view through a herniated disc, showing displacement of the nucleus pulposus and its effect on the spinal cord and adjacent nerves.

TABLE 8-3   Articulations of the Axial Skeleton

Element	Joint	Type of Articulation	Movements
Cranial and facial bones of skull	Various	Synarthroses (suture or synostosis)	None
Maxilla/teeth Mandible/teeth		Synarthrosis (gomphosis)	None
Temporal bone/mandible	Temporomandibular	Hinge diarthrosis	Elevation/depression, lateral gliding
Occipital bone/atlas	Atlanto-occipital	Ellipsoidal diarthrosis	Flexion/extension
Atlas/axis	Atlanto-axial	Pivot diarthrosis	Rotation
Other vertebral elements	Intervertebral (between centra) Intervertebral (between articular processes)	Amphiarthrosis (symphysis) Gliding diarthrosis	Slight movement Slight rotation and flexion/ extension
Thoracic vertebrae/ribs	Vertebrocostal	Gliding diarthrosis	Elevation/depression
Ribs/sternum	Sternocostal	Synarthrosis (synchondrosis)	No movement
Sternum/clavicle	Sternoclavicular	Gliding diarthrosis	Protraction/retraction, depression/ elevation
L₅/sacrum	Between body of L₅ and sacral body	Amphiarthrosis (symphysis)	Slight movement
	Between inferior articular processes of L₅ and articular processes of sacrum	Gliding diarthrosis	Slight flexion/extension
Sacrum/coxae	Sacroiliac	Gliding diarthrosis	Slight movement
Sacrum/coccyx	Sacrococcygeal	Gliding diarthrosis (may become fused)	Slight movement
Coccygeal bones		Synarthrosis (synostosis)	No movement

Intervertebral Ligaments
(Figure 8-7)

Numerous ligaments are attached to the bodies and processes of all vertebrae to bind them together and stabilize the vertebral column (Figure 8-7●). Ligaments interconnecting adjacent vertebrae include:

- The *anterior longitudinal ligament* connects the anterior surfaces of each vertebral body;
- The *posterior longitudinal ligament* parallels the anterior longitudinal ligament but passes across the posterior surfaces of each body;
- The *ligamentum flavum* (plural *ligamenta flava*) connects the laminae of adjacent vertebrae;
- The *interspinous ligament* connects the spinous processes of adjacent vertebrae;
- The *supraspinous ligament* interconnects the tips of the spinous processes from C₇ to the sacrum. The *ligamentum nuchae*, discussed in Chapter 6, is a supraspinous ligament that extends from C₇ to the base of the skull. ∞ [p. 162]

Vertebral Movements

The following movements of the vertebral column are possible: (1) **anterior flexion**, bending forward; (2) **extension**, bending backward; (3) **lateral flexion**, bending to the side; and (4) **rotation**.

Table 8-3 summarizes information concerning other articulations of the axial skeleton.

The Glenohumeral (Shoulder) Joint
(Figure 8-9)

The **glenohumeral joint** (Figure 8-9●) permits the greatest range of motion of any joint in the body. Because it is also the most frequently dislocated joint, it provides an excellent demonstration of the principle that strength and stability must be sacrificed to obtain mobility.

This joint is a ball-and-socket type, formed by the articulation of the head of the humerus with the glenoid fossa of the scapula. In life, the surface of the glenoid fossa is covered by a fibrocartilaginous **glenoid labrum** (*labrum*, lip or edge), that encloses the glenoid cavity. The relatively loose articular capsule extends from the scapular neck to the humerus. It is a relatively oversized capsule that permits an extensive range of motion. The bones of the pectoral girdle provide some stability to the superior surface, because the acromion and coracoid processes project laterally superior to the humeral head. However, most of the stability at this joint is provided by (1) ligaments and (2) surrounding skeletal muscles and their associated tendons.

Ligaments
(Figure 8-9b,c,d)

Major ligaments involved with stabilizing the scapulohumeral joint are shown in Figure 8-9b,c,d● and described below.

- The capsule surrounding the shoulder joint is relatively thin, but it thickens anteriorly in regions known as the **glenohumeral ligaments**. Because the capsular fibers are usually loose, these ligaments participate in joint stabilization only as the humerus approaches, or exceeds, the limits of normal motion.
- The large **coracohumeral ligament** originates at the base of the coracoid process and inserts on the head of the humerus.
- The **coracoacromial ligament** spans the gap between the coracoid process and the acromion, just superior to the capsule. This ligament provides additional support to the superior surface of the capsule. The strong **acromioclavicular ligament** binds the acromion to the clavicle, thereby restricting clavicular movement at the acromion end. A **shoulder separation** is a relatively common injury involving partial or complete dislocation of the acromioclavicular joint. This injury can result from a blow to the upper surface of the shoulder. The acromion is forcibly depressed, but the clavicle is held back by strong muscles.
- The **coracoclavicular ligaments** tie the clavicle to the coracoid process and help to limit the relative motion between the clavicle and scapula.

Skeletal Muscles and Tendons

Muscles that move the humerus do more to stabilize the shoulder joint than all the ligaments and capsular fibers combined. Muscles originating on the trunk, shoulder girdle, and humerus cover the anterior, superior, and posterior surfaces of the capsule. Tendons passing across the joint reinforce the anterior and superior portions of the capsule.

Bursae
(Figure 8-9b,c)

As at other joints, *bursae* at the shoulder reduce friction where large muscles and tendons pass across the joint capsule. ∞ [p. 208] The shoulder has a relatively large number of important bursae. The **subacromial bursa** and the **subcoracoid bursa** (Figure 8-9a,b●) prevent contact between the acromial and coracoid processes and the capsule. The **subdeltoid bursa** and the **subscapular bursa** (Figure 8-9b,c●) lie between large muscles and the capsular wall. Inflammation of one or more of these bursae can restrict motion and produce the painful symptoms of **bursitis**. ✝ *Bursitis [p. 755]*

■ CLINICAL BRIEF
Shoulder Injuries

When a head-on charge leads to a collision, such as a block (in football) or check (in hockey), the shoulder usually lies in the impact zone. The clavicle provides the only fixed support for the pectoral girdle, and it cannot resist large forces. Because the inferior surface of the shoulder capsule is poorly reinforced, a dislocation caused by an impact or violent muscle contraction most often occurs at this site. Such a dislocation can tear the inferior capsular wall and the glenoid labrum. The healing process often leaves a weakness that increases the chances for future dislocations.

In a **shoulder separation**, the acromioclavicular joint undergoes partial or complete dislocation. This injury can result from a blow to the upper surface of the shoulder. The acromion is forcibly depressed, but the clavicle is held back by strong muscles.

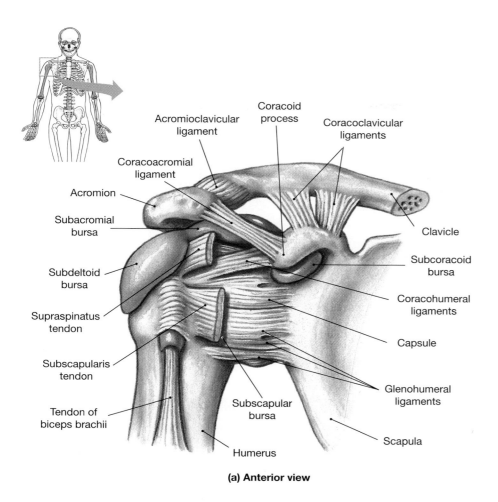

(a) Anterior view

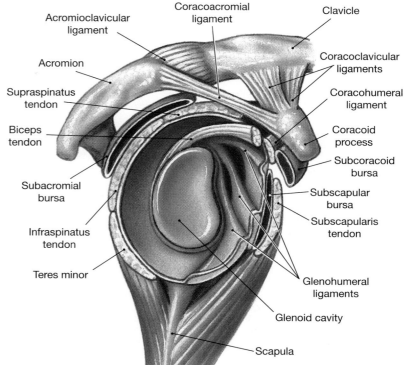

(b) Lateral view of pectoral girdle

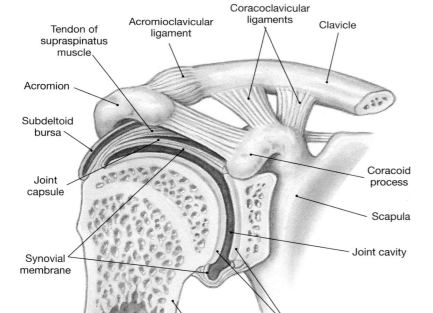

(c) Anterior view, frontal section

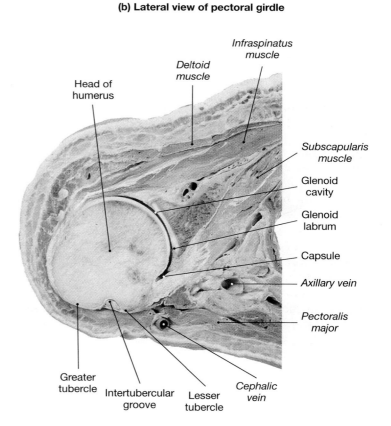

(d) Superior view, horizontal section

FIGURE 8-9
The Shoulder Joint. (a) Anterior view of the right shoulder joint. (b) Lateral view, right shoulder joint (humerus removed). (c) A diagrammatic frontal section through the right shoulder joint. (d) Horizontal section of right shoulder joint, superior view.

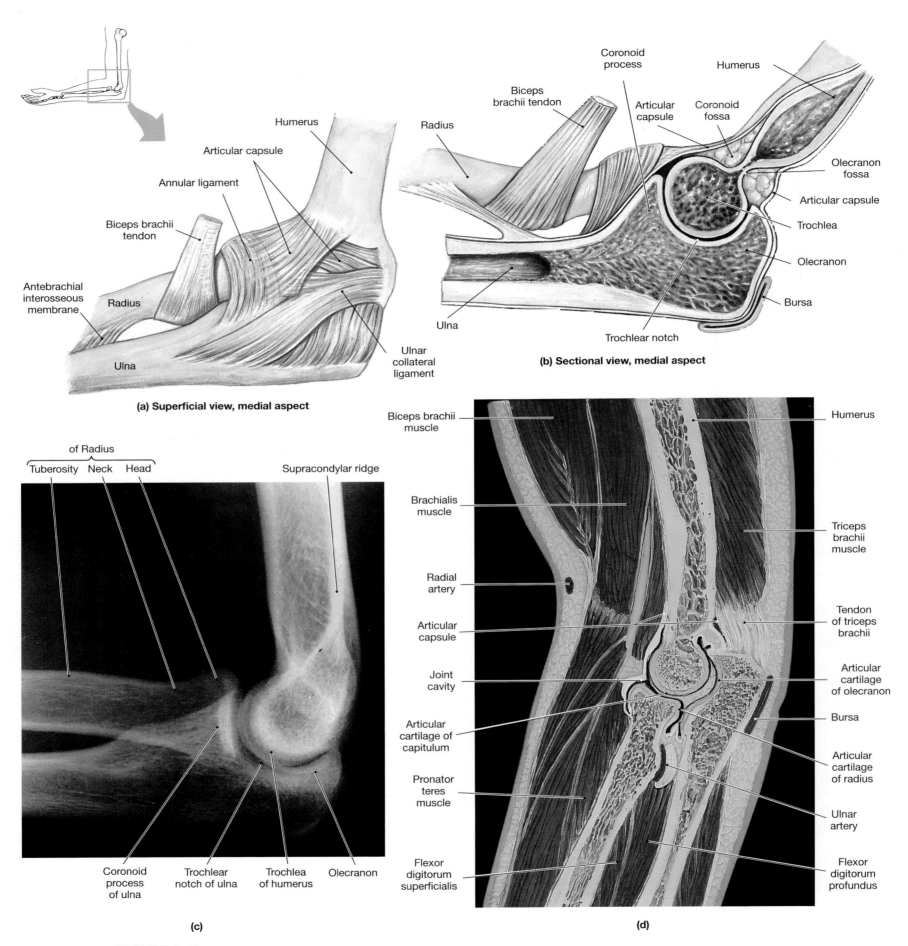

(a) Superficial view, medial aspect

Humerus

Articular capsule

Annular ligament

Biceps brachii
tendon

Antebrachial
interosseous
membrane

Radius

Ulna

Ulnar
collateral
ligament

(b) Sectional view, medial aspect

Coronoid
process

Humerus

Biceps
brachii tendon

Articular
capsule

Coronoid
fossa

Radius

Olecranon
fossa

Articular capsule

Trochlea

Olecranon

Bursa

Ulna

Trochlear notch

(c)

of Radius

Tuberosity Neck Head

Supracondylar ridge

Coronoid
process
of ulna

Trochlear
notch of ulna

Trochlea
of humerus

Olecranon

(d)

Biceps brachii
muscle

Humerus

Brachialis
muscle

Triceps
brachii
muscle

Radial
artery

Articular
capsule

Tendon
of triceps
brachii

Joint
cavity

Articular
cartilage
of olecranon

Articular
cartilage of
capitulum

Bursa

Pronator
teres
muscle

Articular
cartilage
of radius

Flexor
digitorum
superficialis

Ulnar
artery

Flexor
digitorum
profundus

FIGURE 8-10
The Elbow Joint. (a) Medial view, right elbow joint. (b) Sectional view of the same joint. (c) X-ray of the right elbow joint,
lateral view. (d) Coronal section of elbow joint cast from cadaver specimen.

218

The Humeroulnar (Elbow) Joint
(Figure 8-10)

The **humeroulnar joint**, or **olecranal joint**, is a hinge joint. The trochlea of the humerus articulates with the trochlear notch of the ulna, and the capitulum of the humerus articulates with the head of the radius (Figure 8-10●).

Muscles that extend the elbow attach to the rough surface of the olecranon process. These muscles are primarily under the control of the radial nerve, which passes along the *radial groove* of the humerus. ∞ [p. 180] The large *biceps brachii* muscle covers the anterior surface of the arm. Its tendon is attached to the radius at the **radial tuberosity**. Contraction of this muscle produces supination of the forearm and flexion of the elbow.

The elbow joint is extremely stable because: (1) the bony surfaces of the humerus and ulna interlock; (2) the articular capsule is very thick; and (3) the capsule is reinforced by strong ligaments. The **radial collateral ligament** stabilizes the lateral surface of the joint. It extends between the lateral epicondyle of the humerus and the **annular ligament** that binds the proximal radial head to the ulna. The medial surface of the joint is stabilized by the **ulnar collateral ligament**. This ligament extends from the medial epicondyle of the humerus anteriorly to the coronoid processes of the ulna, and posteriorly to the olecranon (Figure 8-10a●).

Despite the strength of the capsule and ligaments, the elbow joint can be damaged by severe impacts or unusual stresses. For example, when you fall on a hand with a partially flexed elbow, contractions of muscles that extend the elbow may break the ulna at the center of the trochlear notch. Less violent stresses can produce dislocations or other injuries to the elbow, especially if epiphyseal growth has not been completed. For example, parents in a hurry may drag a toddler along behind them, exerting an upward, twisting pull on the elbow joint that can result in a partial dislocation known as a "nursemaid's elbow."

The Carpus
(Figure 8-11)

The **carpus**, or wrist, contains the **wrist joint** (Figure 8-11●). The wrist joint consists of the **distal radioulnar articulation**, the **radiocarpal articulation**, and the **intercarpal articulations**. The distal radioulnar articulation permits pronation and supination. The radiocarpal articulation involves the distal articular surface of the radius and three proximal carpals, the scaphoid, lunate, and triquetrum bones. The radiocarpal joint is an ellipsoidal articulation that permits flexion/extension, adduction/abduction, and circumduction. The intercarpal articulations are gliding joints that permit sliding and slight twisting movements.

Stability of the Carpus
(Figure 8-11b,c)

Carpal surfaces that do not participate in articulations are roughened by the attachment of ligaments and for the passage of tendons. A tough connective tissue capsule, reinforced by broad ligaments, surrounds the wrist complex and stabilizes the positions of the individual carpal bones (Figure 8-11b,c●). The major ligaments include:

- the *palmar radiocarpal ligament*, which connects the distal radius to the anterior surfaces of the scaphoid, lunate, and triquetrum bones;

- the *dorsal radiocarpal ligament*, which connects the distal radius to the posterior surfaces of the same carpal bones (not seen from palmar surface);

- the *ulnar collateral ligament*, which extends from the styloid process of the ulna to the medial surface of the triquetrum bone; and

- the *radial collateral ligament*, which extends from the styloid process of the radius to the lateral surface of the scaphoid bone.

In addition to these prominent ligaments, a variety of intercarpal ligaments interconnect the distal carpal bones to the metacarpal bones.

Tendons producing flexion of the palm and fingers pass over the anterior surface of the wrist joint superficial to these ligaments. Tendons producing extension and hyperextension pass across the posterior surface in a similar fashion. A pair of broad transverse ligaments arch across the anterior and posterior surfaces of the wrist superficial to these tendons. These bands are attached to the distal portions of the radius and ulna, as well as the carpal bones. The posterior band is called the **extensor retinaculum**; the anterior band is the **flexor retinaculum**. Tendon sheaths of the flexor muscles are trapped in the *carpal tunnel*, the region between the flexor retinaculum and the carpal bones. Inflammation of the tissues beneath the flexor retinaculum can compress the flexor tendons and adjacent sensory nerves, producing pain and a loss of wrist mobility. This condition is called **carpal tunnel syndrome**.

The Hand
(Figure 8-11b,d)

The carpal bones articulate with the metacarpal bones of the palm. The first metacarpal bone has a saddle-type articulation at the wrist, the **carpometacarpal joint** of the thumb. All other carpal/metacarpal articulations are gliding joints. An **intercarpal joint** is formed by carpal/carpal articulation. The articulations between the metacarpal bones and the proximal phalanges (**metacarpophalangeal joints**) are ellipsoidal, permitting flexion/extension and adduction/abduction. The **interphalangeal joints** are "hinge" joints that allow flexion and extension (Figure 8-11b,d●).

Table 8-4 summarizes the characteristics of the articulations of the upper extremity.

√ **Would a tennis player or a jogger be more likely to develop inflammation of the subscapular bursa? Why?**

√ **Mary falls on the palm of her hands with her elbows slightly flexed. After the fall, she can't move her left arm at the elbow. If a fracture exists, what bone is most likely broken?**

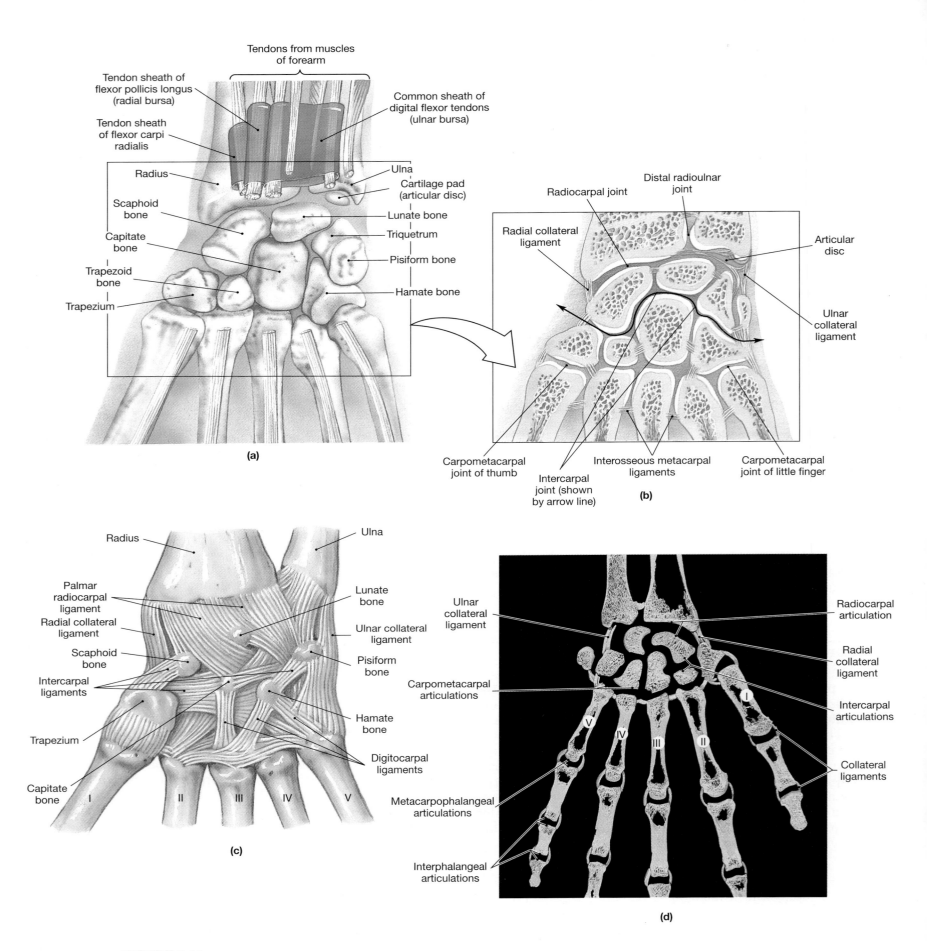

FIGURE 8-11

Joints of the Wrist and Hand. (a) Anterior view of the right wrist, identifying the structural components of the wrist joint. (b) Sectional view through the wrist, showing the radiocarpal, intercarpal, and carpometacarpal joints. (c) Stabilizing ligaments on the anterior (palmar) surface of the wrist. (d) Sectional view of the bones that form the wrist and hand.

220

TABLE 8-4   Articulations of the Upper Extremity

Element	Joint	Type of Articulation	Movements
Clavicle/sternum	Sternoclavicular	Gliding diarthrosis	Gliding
Scapula/clavicle	Acromioclavicular	Gliding diarthrosis	Gliding
Scapula/humerus	Glenohumeral, or shoulder	Ball-and-socket diarthrosis	Flexion/extension, adduction/abduction, circumduction, rotation
Humerus/ulna and radius	Olecranal, or elbow	Hinge diarthrosis	Flexion/extension
Radius/ulna	Proximal radioulnar	Pivot diarthrosis	Pronation/supination
	Distal radioulnar	Pivot diarthrosis	Pronation/supination
Radius/carpals	Radiocarpal, or wrist	Ellipsoidal diarthrosis	Flexion/extension, adduction/abduction, circumduction
Carpal/carpal	Intercarpal	Gliding diarthrosis	Gliding
Carpal/metacarpal (first)	Carpometacarpal of thumb	Saddle diarthrosis	Flexion/extension, adduction/abduction, circumduction, opposition
Carpal/metacarpal (2–5)	Carpometacarpal	Gliding diarthrosis	Slight flexion/extension, adduction/abduction
Metacarpals/phalanges	Metacarpophalangeal	Ellipsoidal diarthrosis	Flexion/extension, adduction/abduction, circumduction
Phalanges/phalanges	Interphalangeal	Hinge diarthrosis	Flexion/extension

The Hip Joint
(Figure 8-12)

Figure 8-12 introduces the structure of the **hip joint**. A fibro-cartilage pad covers the articular surface of the acetabulum and extends like a horseshoe along the sides of the **acetabular notch** (Figure 8-12a●). A fat pad covered by synovial membrane covers the central portion of the acetabulum. This pad acts as a shock absorber, and the adipose tissue stretches and distorts without damage.

The Articular Capsule
(Figure 8-13b,c)

The articular capsule of the hip joint is extremely dense and strong (Figure 8-13b,c●). It extends from the lateral and inferior surfaces of the pelvic girdle to the intertrochanteric line and trochanteric crest of the femur, enclosing both the femoral head and neck. This arrangement helps keep the head from moving away from the acetabulum.

Stabilization of the Hip
(Figures 8-12b,c/8-13)

Four broad ligaments reinforce the articular capsule (Figure 8-12b,c●). Three of them are regional thickenings of the capsule: the **iliofemoral**, **pubofemoral**, and **ischiofemoral ligaments**. The **transverse acetabular ligament** crosses the acetabular notch and completes the inferior border of the acetabular fossa. A fifth ligament, the **ligament of the femoral head**, or *ligamentum teres* (*teres*, long and round) originates along the transverse acetabular ligament and attaches to the center of the femoral head. This ligament tenses only when the thigh is flexed and undergoing external rotation. Much more important stabilization is provided by the bulk of the surrounding muscles (Figure 8-13b●). Although flexion, exten-

sion, adduction, abduction, and rotation are permitted, hip flexion is the most important normal movement, and the primary limits are imposed by the surrounding muscles. Other directions of movement are restricted by ligaments and capsular fibers.

The combination of an almost complete bony socket, a strong articular capsule, supporting ligaments, and muscular padding makes this an extremely stable joint. Fractures of the femoral neck or between the trochanters are actually more common than hip dislocations. ✝ *Hip Fractures and the Aging Process [p. 756]*, and *A Case Study: Avascular Necrosis [p. 756]*

The Knee Joint

Although the knee functions as a hinge joint, the articulation is far more complex than that of the elbow or even the ankle. The rounded femoral condyles roll across the top of the tibia, so the points of contact are constantly changing.

Structurally the knee resembles three separate joints, two between the femur and tibia (medial condyle to medial condyle and lateral condyle to lateral condyle) and one between the patella and the patellar surface of the femur.

The Articular Capsule
(Figures 8-14/8-15a,b)

There is no single unified capsule in the knee, nor is there a common synovial cavity (Figure 8-14●). A pair of fibrocartilage pads, the **medial** and **lateral menisci**, lie between the femoral and tibial surfaces (Figure 8-15a,b●, p. 224). The menisci (1) act as cushions, (2) conform to the shape of the articulating surfaces as the femur changes position, and (3) provide some lateral stability to the joint. Prominent **fat pads** provide padding around the margins of the joint and assist the bursae in reducing friction between the patella and other tissues (Figure 8-14●).

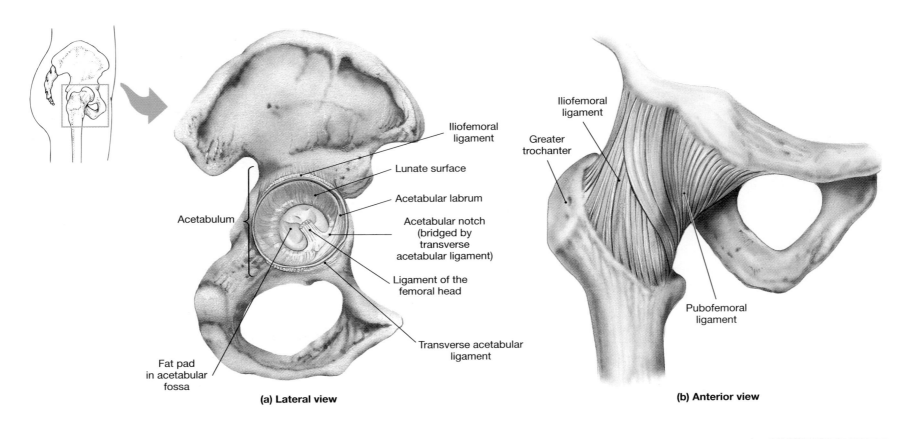

Iliofemoral
ligament

Lunate surface

Acetabular labrum

Acetabular notch
(bridged by
transverse
acetabular ligament)

Ligament of the
femoral head

Transverse acetabular
ligament

Acetabulum

Fat pad
in acetabular
fossa

(a) Lateral view

Iliofemoral
ligament

Greater
trochanter

Pubofemoral
ligament

(b) Anterior view

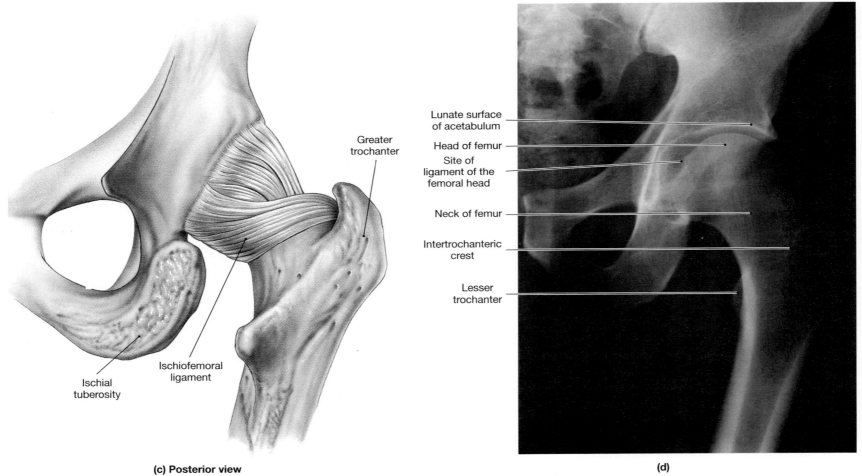

Greater
trochanter

Ischiofemoral
ligament

Ischial
tuberosity

(c) Posterior view

Lunate surface
of acetabulum

Head of femur

Site of
ligament of the
femoral head

Neck of femur

Intertrochanteric
crest

Lesser
trochanter

(d)

FIGURE 8-12

The Hip Joint. (a) Lateral view of the right hip joint with the femur removed. (b) Anterior view of the right hip joint. This joint is extremely strong and stable, in part because of the massive capsule. (c) Posterior view of the right hip joint, showing additional ligaments that add strength to the capsule. (d) X-ray of right hip joint, anterior/posterior view. See MRI SCAN 4, p. 743.

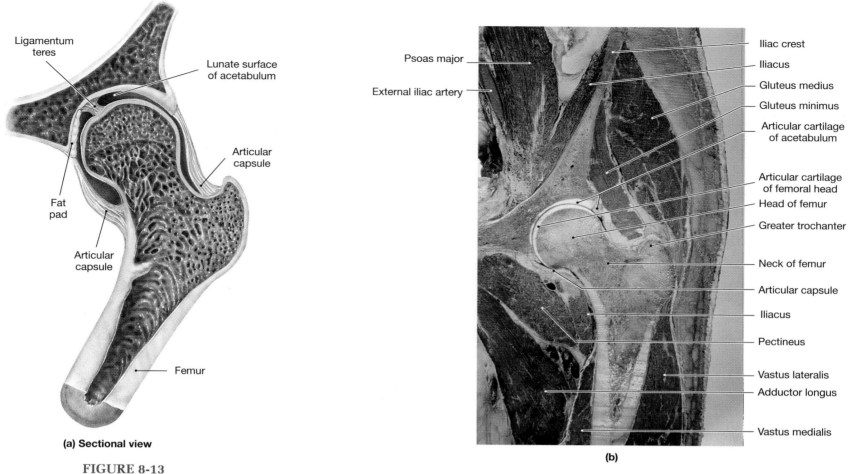

(a) Sectional view

Ligamentum teres

Lunate surface of acetabulum

Fat pad

Articular capsule

Articular capsule

Femur

Psoas major

External iliac artery

Iliac crest

Iliacus

Gluteus medius

Gluteus minimus

Articular cartilage of acetabulum

Articular cartilage of femoral head

Head of femur

Greater trochanter

Neck of femur

Articular capsule

Iliacus

Pectineus

Vastus lateralis

Adductor longus

Vastus medialis

(b)

FIGURE 8-13
Articular Capsule of the Hip Joint. (a) Coronal sectional view of the hip, showing the position and orientation of the ligamentum teres. (b) Coronal section of the hip joint from cadaver, showing the arrangement of major joint elements. See MRI SCAN 4, p. 743.

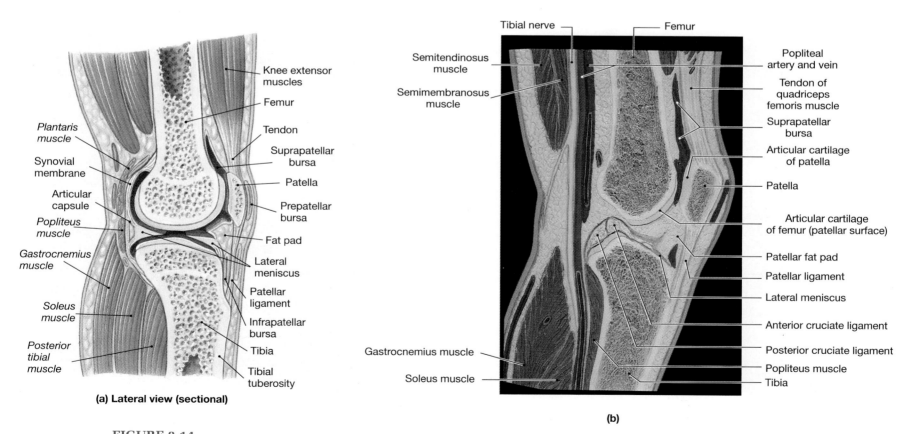

Knee extensor muscles

Femur

Tendon

Suprapatellar bursa

Patella

Prepatellar bursa

Fat pad

Lateral meniscus

Patellar ligament

Infrapatellar bursa

Tibia

Tibial tuberosity

Plantaris muscle

Synovial membrane

Articular capsule

Popliteus muscle

Gastrocnemius muscle

Soleus muscle

Posterior tibial muscle

(a) Lateral view (sectional)

Tibial nerve

Femur

Semitendinosus muscle

Semimembranosus muscle

Gastrocnemius muscle

Soleus muscle

Popliteal artery and vein

Tendon of quadriceps femoris muscle

Suprapatellar bursa

Articular cartilage of patella

Patella

Articular cartilage of femur (patellar surface)

Patellar fat pad

Patellar ligament

Lateral meniscus

Anterior cruciate ligament

Posterior cruciate ligament

Popliteus muscle

Tibia

(b)

FIGURE 8-14
The Knee Joint (Sectional Views). (a) A lateral view of the extended right knee as seen in sagittal section, showing major anatomical features. (b) Sagittal section of knee from cadaver specimen. See MRI SCANS 5a,b and 6a,b, p. 744.

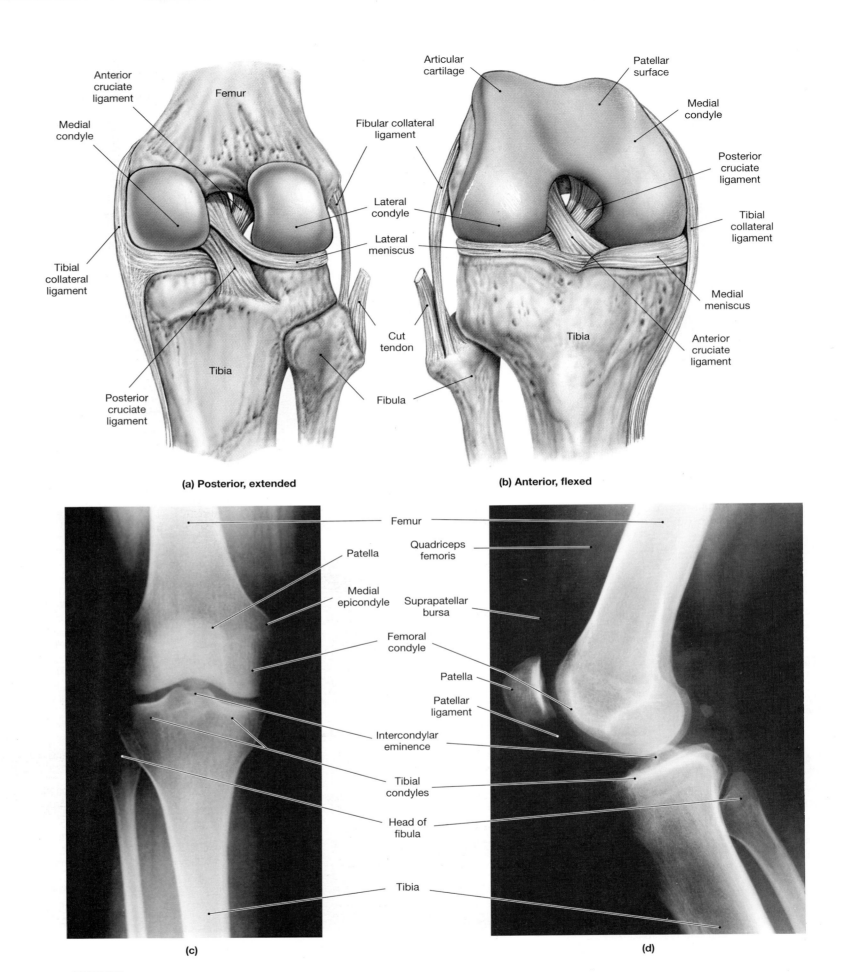

(a) Posterior, extended

Anterior cruciate ligament

Femur

Medial condyle

Fibular collateral ligament

Lateral condyle

Lateral meniscus

Tibial collateral ligament

Tibia

Posterior cruciate ligament

Cut tendon

Fibula

(b) Anterior, flexed

Articular cartilage

Patellar surface

Medial condyle

Posterior cruciate ligament

Tibial collateral ligament

Medial meniscus

Anterior cruciate ligament

Tibia

Femur

Quadriceps femoris

Patella

Suprapatellar bursa

Medial epicondyle

Femoral condyle

Patella

Patellar ligament

Intercondylar eminence

Tibial condyles

Head of fibula

Tibia

(c)

(d)

FIGURE 8-15
The Knee Joint. (a) Posterior view of the right knee at full extension. (b) Anterior view of the right knee at full flexion. (c) X-ray of the right knee in extension, anteroposterior projection. (d) X-ray of the right knee joint in flexion, lateral projection. See MRI SCANS 5a,b and 7a,b, pp. 744, 745.

Supporting Ligaments
(Figures 8-14/8-15a,b)

Seven major ligaments stabilize the knee joint, and a complete dislocation of the knee is an extremely rare event.

- The tendon from the muscles responsible for extending the knee passes over the anterior surface of the joint. The patella is embedded within this tendon, and the **patellar ligament** continues to its attachment on the anterior surface of the tibia. The patellar ligament provides support to the anterior surface of the knee joint (Figure 8-14●).
- Two superficial **popliteal ligaments** (not shown) extend between the femur and the heads of the tibia and fibula. These ligaments reinforce the back of the knee joint.
- Inside the joint capsule, the **anterior cruciate** and **posterior cruciate ligaments** attach the intercondylar area of the tibia to the condyles of the femur. *Anterior* and *posterior* refer to their sites of origin on the tibia, and they cross one another as they proceed to their destinations on the femur (Figure 8-15a,b●). (The term *cruciate* is derived from the Latin word *crucialis*, meaning "a cross.") These ligaments limit the anterior and posterior movement of the femur and maintain the alignment of the femoral and tibial condyles.
- The **tibial collateral ligament** reinforces the medial surface of the knee joint, and the **fibular collateral ligament** reinforces the lateral surface (Figure 8-15a,b●). These ligaments tighten only at full extension, and in this position they act to stabilize the joint.

Locking of the Knee
(Figure 8-15c,d)

The knee joint can "lock" in the extended position. At full extension a slight external rotation of the tibia tightens the anterior cruciate ligament and jams the meniscus between the tibia and femur (Figure 8-15c,d●). This mechanism allows you to stand for prolonged periods without using (and tiring) the extensor muscles of the leg. Unlocking the joint requires muscular contractions that produce internal rotation of the tibia or external rotation of the femur. ⚕ *Knee Injuries [p. 756]*

✓ **Where would you find the following ligaments: iliofemoral ligament, pubofemoral ligament, and ischiofemoral ligament?**

✓ **What symptoms would you expect to see in an individual who has damaged the menisci of the knee joint?**

The Ankle and Foot

The **tibiotalar joint**, or ankle joint, involves the distal articular surface of the tibia, including the medial malleolus, the lateral malleolus of the fibula, and the trochlea and lateral articular facets of the talus. The tibiotalar joint is a hinge joint that permits limited dorsiflexion (sole elevated) and plantar flexion (sole depressed).

Stabilization of the Ankle Joint
(Figures 8-16b,d/8-17)

The articular capsule of the ankle joint extends from the distal surfaces of the tibia and medial malleolus, the lateral malleolus, and the talus. The anterior and posterior portions of the capsule are thin, but the lateral and medial surfaces are strong and reinforced by stout ligaments. The major ligaments are the medial **deltoid ligament** and the three **lateral ligaments** (Figure 8-16b,d●). The malleoli, supported by these ligaments and associated fat pads, prevent the ankle bones from sliding from side to side (Figure 8-17●).

Joints of the Foot
(Figures 8-16d/8-17)

Four types of synovial joints are found in the foot (Figures 8-16d and 8-17●):

1. Tarsal to tarsal (**intertarsal joints**). These are gliding joints that permit limited sliding and twisting movements. The articulations between the tarsals are comparable to those between the carpals of the wrist.
2. Tarsal to metatarsal (**tarsometatarsal joints**). These are gliding articulations that allow limited sliding and twisting movements. The first three metatarsals articulate with the medial, intermediate, and lateral cuneiforms. The fourth and fifth metatarsals articulate with the cuboid.
3. Metatarsal to phalanx (**metatarsophalangeal joints**). These are ellipsoidal joints that permit flexion/extension and adduction/abduction. Joints between the metatarsal bones and phalanges resemble those between the metacarpal bones and phalanges of the hand. Because the first metatarsophalangeal joint is ellipsoidal, rather than saddle-shaped, the hallux lacks the mobility of the thumb. A pair of sesamoid bones often forms in the tendons that cross the inferior surface of this joint, and their presence further restricts movement.
4. Phalanx to phalanx (**interphalangeal joints**). These are hinge joints that permit flexion and extension. ⚕ *Problems with the Ankle and Foot [p. 757]*

Table 8-5 summarizes information about the articulations of the lower extremity.

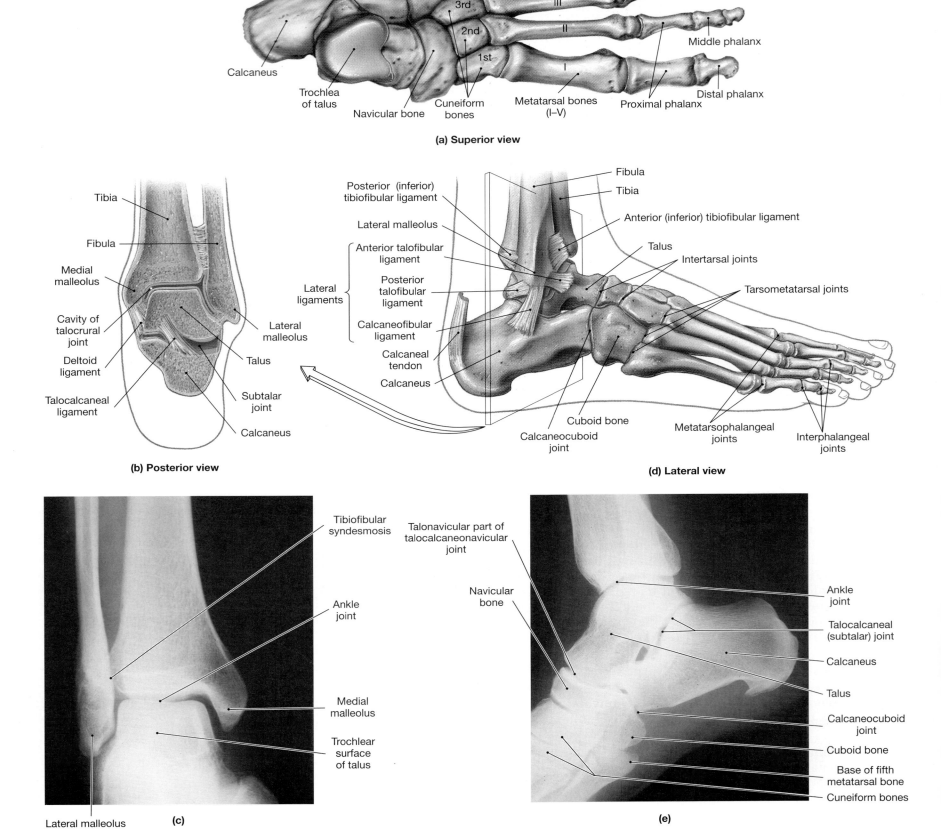

(a) Superior view

(b) Posterior view

(d) Lateral view

(c)

(e)

FIGURE 8-16
Joints of the Ankle and Foot. (a) Superior view of bones of the foot. (b) Posterior view showing the orientation of the tibio-talar joint and the placement of the medial and lateral malleoli. (c) X-ray of right foot, anteroposterior view. (d) Lateral view of the ankle joint, showing the ligamentous and bony props that help stabilize the joint. (e) X-ray of right foot, lateral view. See MRI SCAN 8b, p. 745.

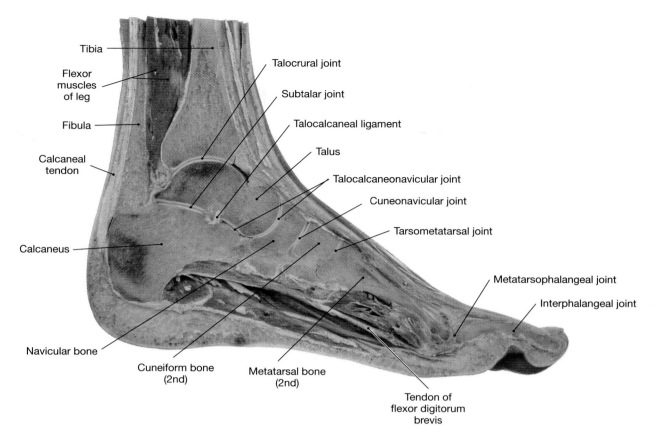

FIGURE 8-17
The Foot. Longitudinal section of the foot from a cadaver. See MRI SCAN 8a, p. 745.

TABLE 8-5   Articulations of the Lower Extremity

Element	Joint	Type of Articulation	Movements
Sacrum/ilium of coxa	Sacroiliac	Gliding diarthrosis	Gliding
Coxa/coxa	Symphysis pubis	Amphiarthrotic symphysis	Slight
Coxa/femur	Hip	Ball-and-socket diarthrosis	Flexion/extension, adduction/abduction, circumduction, rotation
Femur/tibia	Knee	Complex, functions as hinge	Flexion/extension, limited rotation
Tibia/fibula	Tibiofibular (proximal) Tibiofibular (distal)	Gliding diarthrosis Gliding diarthrosis and amphiarthrotic syndesmosis	Gliding Slight gliding
Tibia and fibula with talus	Ankle, or tibiotalar	Hinge diarthrosis	Dorsiflexion/plantar flexion
Tarsal/tarsal	Intertarsal	Gliding diarthrosis	Gliding
Tarsal/metatarsal	Tarsometatarsal	Gliding diarthrosis	Gliding
Metatarsal/phalanges	Metartarsophalangeal	Ellipsoidal diarthrosis	Flexion/extension, adduction/abduction
Phalanges/phalanges	Interphalangeal	Hinge diarthrosis	Flexion/extension

BONES AND MUSCLES

The skeletal and muscular systems are structurally and functionally interdependent; their interactions are so extensive that they are often considered to be parts of a single *musculoskeletal system*.

There are direct physical connections, for the connective tissues that surround the individual muscle fibers are continuous with those that establish the tissue framework of an attached bone. Muscles and bones are also physiologically linked, because muscle contractions can occur only when the extracellular concentration of calcium remains within relatively narrow limits, and most of the body's calcium reserves are held within the skeleton. The next three chapters will examine the structure and function of the muscular system and discuss how muscular contractions perform specific movements.

Related Clinical Terms

ankylosis (an-ke-LŌ-sis): An abnormal fusion between articulating bones in response to trauma and friction within a joint. ⚕️*Rheumatism, Arthritis, and Synovial Function [p. 755]*

continuous passive motion (CPM): A therapeutic procedure involving passive movement of an injured joint to stimulate circulation of synovial fluid. The goal is to prevent degeneration of the articular cartilages. ⚕️*Rheumatism, Arthritis, and Synovial Function [p. 755]*

rheumatism (ROO-ma-tizm): A general term that indicates pain and stiffness affecting the skeletal system, the muscular system, or both. ⚕️*Rheumatism, Arthritis, and Synovial Function [p. 755]*

arthritis (ar-THRĪ-tis): Rheumatic diseases that affect synovial joints. Arthritis always involves damage to the articular cartilages, but the specific cause may vary. The diseases of arthritis are usually classified as either **degenerative** or **inflammatory** in nature. ⚕️*Rheumatism, Arthritis, and Synovial Function [p. 755]*

osteoarthritis (os-tē-ō-ar-THRĪ-tis), (**degenerative arthritis**, or **degenerative joint disease (DJD)):** An arthritic condition resulting from (1) cumulative wear and tear on joint surfaces or (2) genetic predisposition. In the U.S. population, 25% of women and 15% of men over 60 years of age show signs of this disease. ⚕️*Rheumatism, Arthritis, and Synovial Function [p. 756]*

rheumatoid arthritis: An inflammatory arthritis that affects roughly 2.5% of the adult population. The cause is uncertain, although allergies, bacteria, viruses, and genetic factors have all been proposed. The primary symptom is **synovitis** (sin-ō-VĪ-tis), swelling and inflammation of the synovial membrane. ⚕️*Rheumatism, Arthritis, and Synovial Function [p. 756]*

luxation (luks-Ā-shun): A dislocation; a condition in which the articulating surfaces are forced out of position. *[p. 208]*

subluxation (sub-luks-Ā-shun): A partial dislocation; displacement of articulating surfaces sufficient to cause discomfort, but resulting in less physical damage to the joint than during a complete dislocation. *[p. 208]*

slipped disc: A common name for a condition caused by distortion of an intervertebral disc. The distortion applies pressure to spinal nerves, causing pain and limiting range of motion. *[p. 214]*

herniated disc: A condition caused by intervertebral compression severe enough to rupture the anulus fibrosus and displace the nucleus pulposus. *[p. 214]*

sciatica (sī-AT-i-ka): The painful result of compression of the roots of the sciatic nerve. The acute initial pain in the lower back is sometimes called **lumbago** (lum-BĀ-gō). *[p. 214]*

laminectomy (la-mi-NEK-tō-mē): Removal of vertebral laminae; may be performed to access the vertebral canal and relieve symptoms of a herniated disc. *[p. 214]*

bursitis: Inflammation of a bursa that causes pain whenever the associated tendon or ligament moves. ⚕️*Bursitis [p. 756]*

bunion: The most common pressure-related bursitis, involving a tender nodule formed around bursae over the base of the great toe. ⚕️*Bursitis [p. 756]*

bursectomy: The surgical removal of inflamed bursae. ⚕️*Bursitis [p. 756]*

shoulder separation: The partial or complete dislocation of the acromioclavicular joint. *[p. 216]*

carpal tunnel syndrome: Inflammation of the tissues beneath the flexor retinaculum, causing compression of the flexor tendons and adjacent sensory nerves. Symptoms are pain and a loss of wrist mobility. *[p. 219]*

arthroscope: An instrument that uses fiber optics to explore a joint without major surgery. ⚕️*Knee Injuries [p. 756]*

arthroscopic surgery: The surgical modification of a joint using an arthroscope. ⚕️*Knee Injuries [p. 756]*

meniscectomy: The surgical removal of an injured meniscus. ⚕️*Knee Injuries [p. 756]*

sprain: Condition caused when a ligament is stretched to the point where some of the collagen fibers are torn. The ligament remains functional, and the structure of the joint is not affected. ⚕️*Problems with the Ankle and Foot [p. 757]*

dancer's fracture: A fracture of the fifth metatarsal, usually near its proximal articulation. ⚕️*Problems with the Ankle and Foot [p. 757]*

flat feet: The loss or absence of a longitudinal arch. ⚕️*Problems with the Ankle and Foot [p. 757]*

congenital talipes equinovarus (clubfoot): A congenital deformity affecting one or both feet. It develops secondary to abnormalities in muscular development. ⚕️*Problems with the Ankle and Foot [p. 757]*

STUDY OUTLINE   Related Key Terms

INTRODUCTION *[p. 206]*

CLASSIFICATION OF JOINTS *[p. 206]*

1.   **Articulations** (joints) exist wherever two bones interact. Immovable joints are **synarthroses**, slightly movable joints are **amphiarthroses**, and those that are freely movable are called **diarthroses**. Joints may be classified by function *(see Table 8-1)* and by structure *(see Table 8-2)*.

Immovable Joints (Synarthroses) *[p. 206]*
2.   Examples of synarthroses include a **suture**, a **gomphosis**, a **synchondrosis**, and a **synostosis**.   periodontal ligament

Slightly Movable Joints (Amphiarthroses) *[p. 206]*
3.   Examples of amphiarthroses are a **syndesmosis** and a **symphysis**.

Freely Movable Joints (Diarthroses) *[p. 207]*
4.   The bony surfaces at diarthroses are covered by **articular cartilages**, lubricated by **synovial fluid**, and enclosed within a **joint capsule**. Other synovial structures can include **menisci**, or **articular discs**; **fat pads**; and **accessory ligaments** *(see Figure 8-1)*.   synovial joints • extracapsular ligaments • intracapsular ligaments • bursae • synovial tendon sheaths

ARTICULAR FORM AND FUNCTION *[p. 209]*

Describing Dynamic Motion *[p. 209]*
1.   Possible movements can be classified as **linear motion**, **angular motion**, and **rotation** *(see Figure 8-2)*.

Types of Movements *[p. 209]*
2.   Important terms that describe dynamic motion are **gliding**, **flexion**, **extension**, **hyperextension**, **rotation**, **circumduction**, **abduction**, and **adduction** *(see Figures 8-3 to 8-5)*.   left rotation • right rotation • internal (medial) rotation • external (lateral) rotation
3.   The ankle undergoes **dorsiflexion** and **plantar flexion**. Movements of the foot include **inversion** and **eversion**. **Opposition** is the thumb movement that enables us to grasp objects *(see Figure 8-5)*.
4.   The bones in the forearm permit **pronation** and **supination** *(see Figure 8-4)*.
5.   **Protraction** involves moving a structure anteriorly; **retraction** involves moving it posteriorly. **Depression** and **elevation** occur when we move a structure inferiorly or superiorly *(see Figure 8-5)*.
6.   Joints are called **monaxial**, **biaxial**, or **triaxial** depending on the degree of movement they allow. *(see Figure 8-6)*.

A Structural Classification of Synovial Joints *[p. 212]*
7.   **Gliding joints** permit limited movement, usually in a single plane *(see Figure 8-6 and Table 8-2)*.
8.   **Hinge joints** and **pivot joints** are monaxial joints *(see Figure 8-6 and Table 8-2)*.
9.   Biaxial joints include **ellipsoidal joints** and **saddle joints** *(see Figure 8-6 and Table 8-2)*.
10. Triaxial, or **ball-and-socket joints**, permit rotation as well as other movements *(see Figure 8-6 and Table 8-2)*.

REPRESENTATIVE ARTICULATIONS *[p. 214]*

Intervertebral Articulations *[p. 214]*
1.   The articular processes form gliding joints with those of adjacent vertebrae. The bodies form symphyseal joints. They are separated by **intervertebral discs** containing an inner **nucleus pulposus** and an outer **anulus fibrosus** *(see Figures 8-7/8-8)*.
2.   Numerous ligaments bind together the bodies and processes of all vertebrae *(see Figure 8-7)*.
3.   The articulations of the vertebral column permit **anterior flexion**, **lateral flexion**, **extension**, and **rotation** movements.
4.   Articulations of the axial skeleton are summarized in *Table 8-3*.

The Glenohumeral (Shoulder) Joint *[p. 216]*
5.   The **glenohumeral** (shoulder) **joint**, formed by the glenoid fossa and the head of the humerus, permits the greatest range of motion of any joint in the body. It is a ball-and-socket diarthrosis. Strength and stability are sacrificed to obtain mobility. Ligaments and the surrounding muscle provide strength and stability. The shoulder has a large number of *bursae* that reduce friction as large muscles and tendons pass across the joint capsule *(see Figure 8-9)*.   glenoid labrum • ligaments and bursae of the glenohumeral joint

The Humeroulnar (Elbow) Joint *[p. 219]*

6. The **humeroulnar** or **olecranal** (elbow) **joint** permits only flexion/extension movements. It is a "hinge" diarthrosis. **Radial** and **ulnar collateral** and **annular ligaments** aid in stabilizing this joint *(see Figure 8-10)*.

The Carpus *[p. 219]*

7. The **carpus** or **wrist joint** is formed by the distal articular surface of the radius and the proximal row of carpals (scaphoid, lunate, and triquetral). The wrist joint consists of the **distal radioulnar articulation**, the **radiocarpal articulation**, and the **intercarpal articulations**. The joint permits flexion/extension, adduction/abduction, and circumduction movements. A connective tissue capsule and broad ligaments stabilize the positions of the individual carpals *(see Figure 8-11)*.

The Hand *[p. 219]*

8. Five types of diarthrotic joints are found in the hand: (1) carpal/carpal (**intercarpal**); gliding diarthrosis, permitting gliding movements; (2) carpal/metacarpal (first) (**carpometacarpal of thumb**); saddle diarthrosis, permitting flexion/extension, adduction/abduction, circumduction, opposition movements; (3) carpal/metacarpal (2–5) (**carpometacarpal**); gliding diarthrosis, permitting slight flexion/extension, adduction/abduction movements; (4) metacarpal/phalanx (**metacarpophalangeal**); ellipsoidal diarthrosis, permitting flexion/extension, adduction/abduction, and circumduction; and (5) phalanx/phalanx (**interphalangeal**); hinge diarthrosis, permitting flexion/extension movements *(see Figure 8-11b,d)*.

The Hip Joint *[p. 221]*

9. The **hip joint** is formed by the union of the acetabulum with the head of the femur. The joint permits flexion/extension, adduction/abduction, circumduction, and rotational movements *(see Figures 8-12/8-13)*.

10. The articular capsule of the hip joint is reinforced and stabilized by ligaments: **iliofemoral** and **transverse acetabular ligaments** and the **ligament of the femoral head** *(see Figures 8-12b,c/8-13)*.

The Knee Joint *[p. 221]*

11. The knee joint is a hinge joint formed by the union of condyles of the femur with the superior condylar surfaces of the tibia. The joint permits flexion/extension and limited rotational movements *(see Figures 8-14/8-15)*.

12. Seven ligaments bind and stabilize this joint: **patellar**, **popliteal** (two), **anterior** and **posterior cruciate**, **tibial collateral**, and **fibular collateral ligaments** *(see Figures 8-14/8-15)*.

The Ankle and Foot *[p. 225]*

13. The ankle (**tibiotalar**) joint, is a hinge joint formed by the medial malleolus of the tibia, the lateral malleolus of the fibula, and the trochlea of the talus. The joint permits dorsiflexion/plantar flexion movements. The medial **deltoid ligament** and three **lateral ligaments** stabilize the ankle joint *(see Figures 8-16/8-17)*.

14. Four types of diarthrotic joints are found in the foot: (1) tarsal/tarsal (**intertarsal joint**); gliding diarthrosis, permitting gliding movements; (2) tarsal/metatarsal (**tarsometatarsal joint**); gliding diarthrosis, permitting gliding movements; (3) metatarsal/phalanx (**metatarsophalangeal joint**); ellipsoidal diarthrosis, permitting flexion/extension, adduction/abduction movements; and (4) phalanx/phalanx (**interphalangeal joint**); hinge diarthrosis, permitting flexion/extension *(see Figures 8-16/8-17 and Table 8-5)*.

BONES AND MUSCLES *[p. 227]*

1. The skeletal and muscular systems are structurally and functionally interdependent and comprise the *musculoskeletal system*.

Related Key Terms
radial tuberosity

extensor retinaculum • flexor retinaculum

acetabular notch

medial meniscus • lateral meniscus • fat pads

---

1   REVIEW OF CHAPTER OBJECTIVES

1. Distinguish between different types of joints and link anatomical design to joint functions.
2. Describe the dynamic movements of the skeleton.
3. Describe the articulations between the vertebrae of the vertebral column.
4. Describe the structure and function of the joints of the upper extremity: shoulder, elbow, wrist, and hand.
5. Describe the structure and function of the joints of the lower extremity: hip, knee, ankle, and foot.

2   REVIEW OF CONCEPTS

1. How does the structure of a synarthrotic or amphiarthrotic joint affect the range of motion shown by those joints?
2. What is the common mechanism holding together immovable joints such as skull sutures and the gomphoses holding teeth in their alveoli?
3. What type of articular structures do the intervertebral joints (those between the vertebral bodies) and the pubic symphysis have in common? How do these structures affect the amount of movement of these joints?
4. How does the production of either too much or too little synovial fluid impair the proper function of a synovial joint?

5. What roles do accessory structures such as fat pads, bursae, and menisci play in complex synovial joints?

6. In what ways can the morphology of the articular surface of a joint, in conjunction with the accessory structures, affect the range through which that joint can move?

7. Does the term "double-jointed" imply that a person's joints are structurally different from "normal" joints, or is there some other explanation for this appellation?

8. What components of movement make up the special movement of the thumb toward the base of the little finger?

9. If an intervertebral disc is compressed to the point of rupture, which structures will be damaged, and what will most likely be the effect on movements of the vertebral column at this location?

10. Movements of which joints of the upper limb distal to the elbow are responsible for permitting you to hoist a large and heavy glass of liquid to your lips and drain it to the last drop?

11. How would the action of the arm and forearm muscles with tendons inserting on the carpal, metacarpal, and phalangeal bones change if the flexor retinaculum were absent or cut in a proximal-distal orientation?

12. A sixth grader is standing still with his knees in the locked position when a bully sneaks up behind him and, using his own knee, pushes the victim's knee forward, which nearly causes him to fall before being able to "catch" himself. Why was the victim barely able to prevent himself from falling?

13. How do the malleoli of the tibia and fibula function to retain the correct positioning of the tibiotalar joint?

14. In a hip injury, why is the neck of the femur more susceptible to damage than the head?

15. What anatomical reason is there for knee injuries to be among the most common injuries that occur in contact sports?

16. How can rotation be distinguished from circumduction of a skeletal element?

17. Of what importance are capsular ligaments to a complex synovial joint? Think of the scapulohumeral joint as an illustration for your answer.

18. Luxation of the head of the humerus from the glenoid cavity usually occurs in a specific direction. Which direction is involved, and how does the structure of the joint contribute to this tendency?

19. Particularly in a flexed joint, the capsular ligament may appear to be loose and may even sag, wrinkle, and show extra bulges. How do loose joint capsules affect the function of a joint?

20. Which of the joints are chiefly responsible for causing the actions of eversion and inversion of the foot?

3 CRITICAL THINKING AND CLINICAL APPLICATION QUESTIONS

1. When a person involved in an automobile accident suffers from "whiplash," what structures have been affected and what movements could be responsible for this injury?

2. Little League pitchers who play too often and attempt to throw the ball with too much force, snapping the forearm to full extension immediately before releasing the ball, may suffer from a condition called "pitcher's elbow." What structures associated with the elbow joint are involved? When the young pitcher grows to full adulthood, some aspects of this problem no longer occur. Why?

3. A person complains of pain in the sacroiliac joint. Symptoms include lower back pain and aches that might become sharp pains upon movement. What structures are being irritated here? Is the sacroiliac joint really the cause?

4. An athlete suffers an injury to the knee from twisting the leg unexpectedly. This injury results in severe pain, swelling, and a greater-than-usual ability for anterior-posterior movement of the femur on the tibia. When the knee is "tapped" to relieve some of the swelling, a watery fluid tinged with blood is withdrawn. What is this fluid, and from what location was it removed? Arthroscopy is then used to visualize the joint. What structures are examined, and which ones are most likely to have been injured?

5. "Clipping" is an illegal move in football. It is defined as a blow to the lateral side of the knee joint. Clipping can cause serious knee damage. Why is damage from this type of blow so serious, and what structures are most likely to be damaged?

6. A jogger tripped while trying to negotiate a pothole. She fell to her side, twisting her ankle severely. After being examined, she was told the ankle was severely sprained; not broken. The ankle will probably take longer to heal than would a broken bone. Which structures were damaged, and why would they take so long to heal?

7. In an injury described as a "shoulder separation," which structures are most likely to be damaged, and what effect does this have on the movement capacity or stability of the scapulohumeral joint?

9

The Muscular System: Skeletal Muscle Tissue

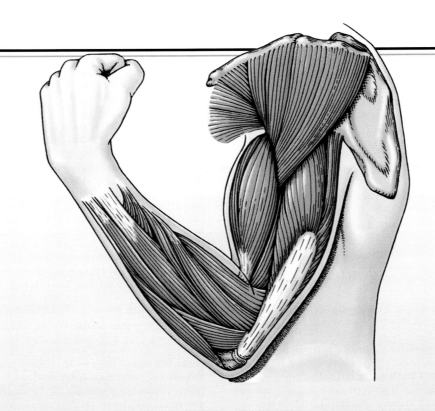

CHAPTER OUTLINE AND OBJECTIVES

Muscle tissue, one of the four primary tissue types, consists chiefly of *muscle fibers*—elongate cells, each capable of contracting along its longitudinal axis. Muscle tissue also includes the connective tissue fibers that harness those contractions to perform useful work. There are three types of muscle tissue: *skeletal muscle*, *cardiac muscle*, and *smooth muscle*. **Skeletal muscle tissue** moves the body by pulling on bones of the skeleton, making it possible for us to walk, dance, or play a musical instrument. **Cardiac muscle tissue** pushes blood through the arteries and veins of the circulatory system, and **smooth muscle tissues** push fluids and solids along the digestive tract and perform varied functions in other systems.

Skeletal muscles are organs that include all four basic tissue types, but that consist primarily of skeletal muscle tissue. The **muscular system** of the human body consists of over 700 skeletal muscles. The muscular system will be the focus of the next three chapters. This chapter considers the function, gross structure, and microstructure of skeletal muscle tissue. Chapters 10 and 11 describe the anatomical organization and functions of the major skeletal muscles. The structure and function of cardiac muscle tissue will be discussed in Chapter 21, and smooth muscle tissue will be considered in Chapter 23.

FUNCTIONS OF SKELETAL MUSCLE

Skeletal muscles are contractile organs directly or indirectly attached to bones of the skeleton. Skeletal muscles perform the following functions:

1. **Produce skeletal movement**: Muscle contractions pull on tendons and move the bones of the skeleton. The effects range from simple motions such as extending the arm to the highly coordinated movements of swimming, skiing, or typing.

2. **Maintain posture and body position**: Contraction of specific muscles also maintains body posture—for example, holding the head in position when reading a book or balancing the weight of the body above the feet when walking. Without constant muscular contraction, we could not sit upright without collapsing or stand without toppling over.

3. **Support soft tissues**: The abdominal wall and the floor of the pelvic cavity consist of layers of skeletal muscle. These muscles support the weight of visceral organs and protect internal tissues from injury.

4. **Guard entrances and exits**: Openings or **orifices** of the digestive and urinary tracts are encircled by skeletal muscles. These muscles provide voluntary control over swallowing, defecation, and urination.

5. **Maintain body temperature**: Muscle contractions require energy, and whenever energy is used in the body, some of it is converted to heat. The heat lost by muscles contracting keeps our body temperature in the range required for normal functioning.

ANATOMY OF SKELETAL MUSCLES

When naming structural features of muscles and their components, anatomists often used the Greek words *sarkos* (flesh) and *mys* (muscle). These root words should be kept in mind as our discussion proceeds. First, we will discuss the gross anatomy of skeletal muscle and then describe the microstructure that makes contraction possible.

Gross Anatomy
(Figure 9-1)

Figure 9-1● illustrates the appearance and organization of a typical skeletal muscle. Notice how connective tissue plays a major role in the structural design of the muscle. We begin our study of the gross anatomy of muscle with a description of the connective tissues that bind and attach skeletal muscles to other structures.

Connective Tissue of Muscle
(Figure 9-1)

Each skeletal muscle has three concentric layers, or wrappings, of connective tissue: an outer *epimysium*, a central *perimysium*, and an inner *endomysium* (Figure 9-1●).

- The **epimysium** (ep-i-MĪZ-ē-um; *epi*, on + *mys*, muscle) is a dense irregular connective tissue layer that surrounds the entire skeletal muscle. The epimysium, which separates the muscle from surrounding tissues and organs, is a component of the *deep fascia*. ∞ [p. 77]

- The connective tissue fibers of the **perimysium** (per-i-MĪZ-ē-um; *peri-*, around) divide the muscle into a series of internal compartments, each containing a bundle of muscle fibers called a **fascicle** (FAS-i-kul; *fasciculus*, bundle). In addition to collagen and elastic fibers, the perimysium contains numerous blood vessels and nerves that branch to supply each individual fascicle.

- The **endomysium** (en-dō-MĪZ-ē-um; *endo*, inside + *mys*, muscle) surrounds each skeletal muscle fiber and binds each muscle fiber to its neighbor. The endomysium consists of a delicate network of reticular fibers. Scattered **satellite cells** lie between the endomysium and the muscle fibers. These cells function in the repair of damaged muscle tissue.

TENDONS AND APONEUROSES. The connective tissue fibers of the endomysium and perimysium are interwoven, and those of the perimysium blend into those of the epimysium. At each end of the muscle, the collagen fibers of the epimysium, perimysium, and endomysium often converge to form a fibrous **tendon**. Tendons often resemble thick cords or cables. Tendons that form thick, flattened sheets are called *aponeuroses*. The structural characteristics of tendons and aponeuroses were considered in Chapter 3. ∞ [p. 70]

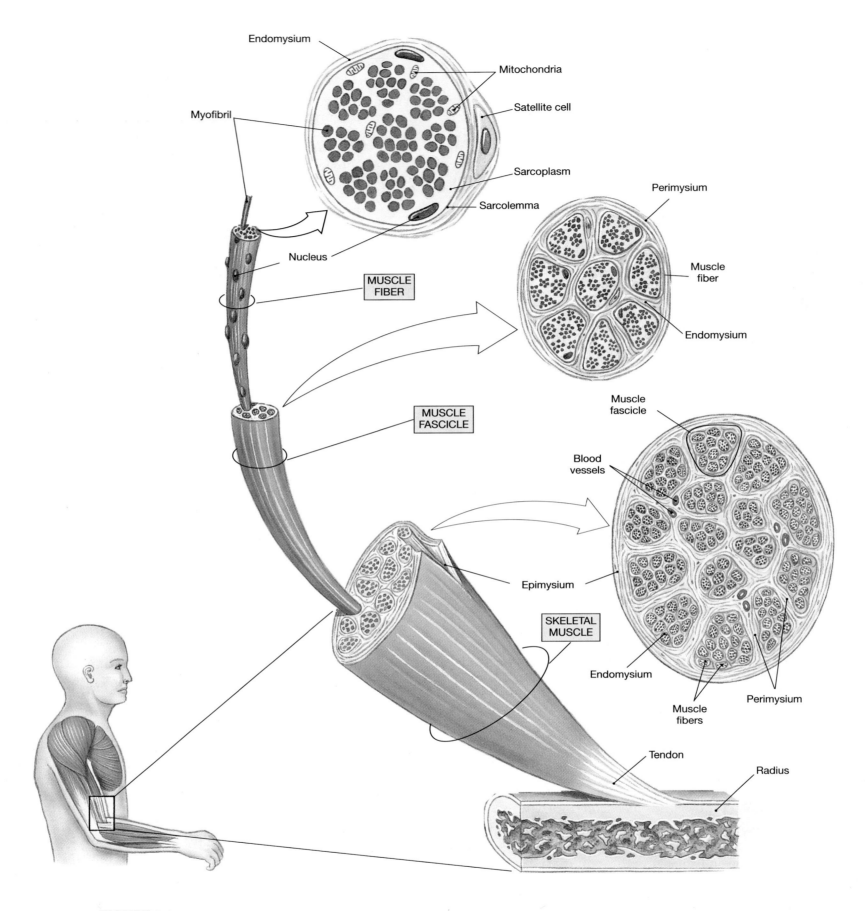

FIGURE 9-1

Organization of Skeletal Muscles. A skeletal muscle consists of fascicles (bundles of muscle fibers) enclosed by the epimysium. The bundles are separated by connective tissue fibers of the perimysium, and within each bundle the muscle fibers are surrounded by the endomysium. Each muscle fiber has many nuclei. The cytoplasm contains mitochondria and other organelles seen in this figure and in Figure 9-4.

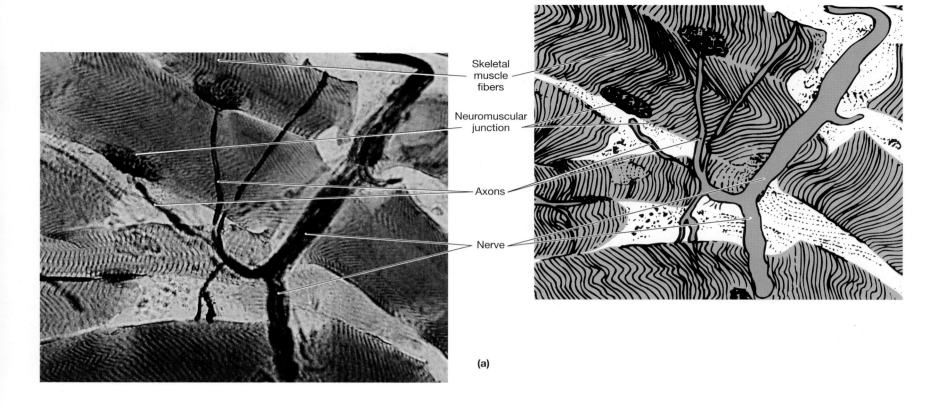

Skeletal
muscle
fibers

Neuromuscular
junction

Axons

Nerve

(a)

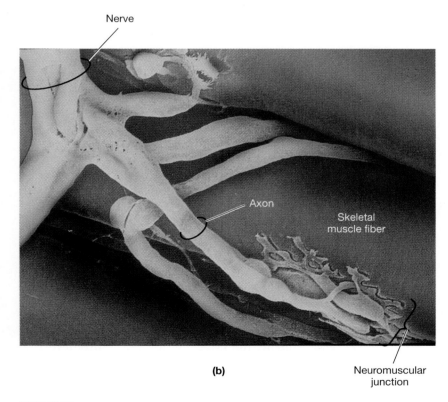

Nerve

Axon

Skeletal
muscle fiber

(b)

Neuromuscular
junction

FIGURE 9-2
Skeletal Muscle Innervation. (a) Several neuromuscular junctions are seen on the muscle fibers of this fascicle. (LM × 230) (b) SEM of neuromuscular junction.

The tendon fibers are interwoven into the periosteum and matrix of the associated bone. This meshwork provides an extremely strong bond. As a result, any contraction of the muscle will exert a pull on the attached bone.

Nerves and Blood Vessels
(Figures 9-2/9-3)

The connective tissues of the epimysium and perimysium contain the nerves and blood vessels that supply the muscle fibers. Skeletal muscles are often called *voluntary muscles* because they contract when stimulated by motor neurons of the central nervous system. Nerves, which are bundles of axons, penetrate the epimysium and branch through the perimysium. Individual axons, or *nerve fibers*, enter the endomysium to innervate individual muscle fibers. Chemical communication between the synaptic terminal of the neuron and the *motor end plate* of the skeletal muscle fiber occurs at a site called the **neuromuscular junction**, or *myoneural junction*. Several of these junctions are shown in Figure 9-2●.

An extensive vascular supply delivers the oxygen and nutrients needed for the production of ATP in active skeletal muscles. These blood vessels often enter the muscle alongside the associated nerves, and the vessels and nerves follow the same branching pattern through the perimysium. Once within the endomysium, the arteries supply an extensive capillary network around each muscle fiber. Capillaries surrounding a portion of a single muscle fiber can be seen in Figure 9-3●. Because these capillaries are coiled, rather than straight, they are able to tolerate changes in the length of the muscle fiber.

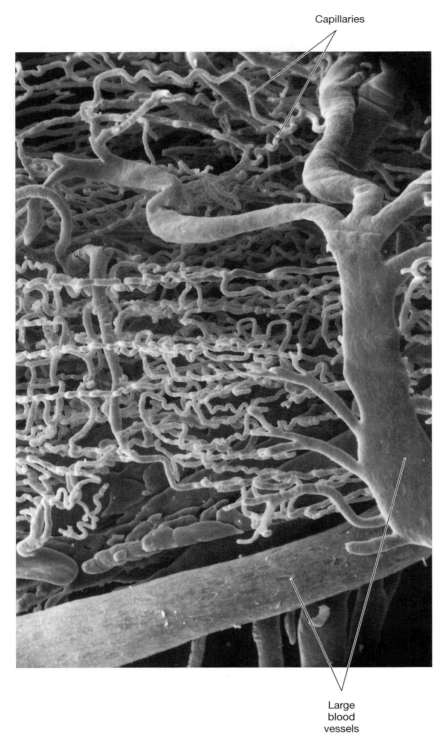

Capillaries

Large
blood
vessels

FIGURE 9-3
Capillary Supply to Skeletal Muscle. Capillaries along the axis of a skeletal muscle fiber. Note the looping and curling of the capillaries that allow cycles of stretching and contraction. (SEM × 168) [Reproduced from R. G. Kessel and R. H. Kardon, *Tissues and Organs: A Text-Atlas of Scanning Electron Microscopy,* W. H. Freeman & Co., 1979.]

Microanatomy of Skeletal Muscle Fibers
(Figures 9-1/9-4)

The cell membrane, or **sarcolemma** (sar-cō-LEM-a; *sarkos*, flesh + *lemma*, husk) of a skeletal muscle fiber surrounds the cytoplasm, or **sarcoplasm** (SAR-kō-plazm). In several other respects, skeletal muscle fibers are quite different from the "typical" cell described in Chapter 2:

- Skeletal muscle fibers are very large. A fiber from a leg muscle could have a diameter of 100 μm and a length equal to that of the entire muscle (30–40 cm, or 10–16 in.).

- Skeletal muscle fibers are *multinucleated.* Each skeletal muscle fiber contains hundreds of nuclei just beneath the sarcolemma (Figure 9-4b●). This characteristic distinguishes skeletal muscle fibers from cardiac and smooth muscle fibers. During development, groups of embryonic cells called **myoblasts** fuse together to create individual skeletal muscle fibers (Figure 9-4a●). Each nucleus in a skeletal muscle fiber reflects the contribution of a single myoblast. Some myoblasts do not fuse with developing muscle fibers, but remain in adult skeletal muscle tissue as *satellite cells* (Figure 9-1●). When a skeletal muscle is injured, satellite cells may differentiate and assist in the repair and regeneration of the muscle.

- Deep depressions in the sarcolemmal surface form a network of narrow tubules that extend into the sarcoplasm. Electrical impulses conducted by the sarcolemma and these **transverse tubules**, or **T-tubules**, help stimulate and coordinate muscle contractions.

Myofibrils and Myofilaments
(Figure 9-4b)

The sarcoplasm of a skeletal muscle fiber contains hundreds to thousands of **myofibrils**. Each myofibril is a cylindrical structure 1–2 μm in diameter and as long as the entire cell (Figure 9-4b●). Myofibrils can shorten, and these are the structures responsible for skeletal muscle fiber contraction. Because the myofibrils are attached to the sarcolemma at each end of the cell, their contraction shortens the entire cell.

Surrounding each myofibril is a sleeve made up of membranes of the **sarcoplasmic reticulum (SR)**, a membrane complex (Figure 9-4b●) similar to the smooth endoplasmic reticulum of other cells. This membrane network, which is closely associated with the transverse tubules, plays a key role in controlling the contraction of individual myofibrils. Scattered between the myofibrils are mitochondria and glycogen granules. The breakdown of glycogen and the activity of mitochondria provide the ATP needed to power muscular contractions. A typical skeletal muscle fiber has about 300 mitochondria, more than most other cells in the body.

Myofibrils consist of bundles of **myofilaments**, protein filaments consisting primarily of the proteins *actin* and *myosin*. The actin filaments are found in *thin filaments* and the myosin filaments are found in *thick filaments*. ∞ [p. 34]

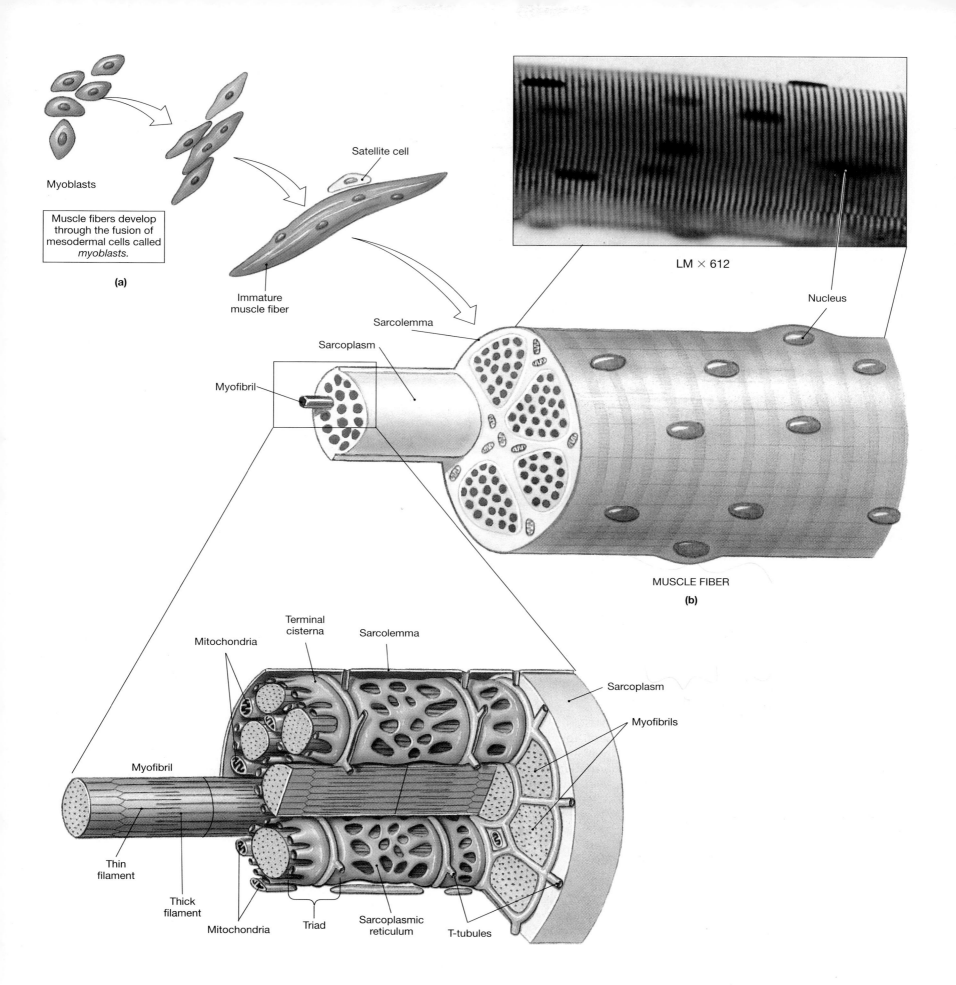

(a)

Myoblasts

Muscle fibers develop through the fusion of mesodermal cells called *myoblasts*.

Satellite cell

Immature muscle fiber

Sarcolemma

Sarcoplasm

Myofibril

LM × 612

Nucleus

MUSCLE FIBER

(b)

Mitochondria

Terminal cisterna

Sarcolemma

Sarcoplasm

Myofibrils

Myofibril

Thin filament

Thick filament

Mitochondria

Triad

Sarcoplasmic reticulum

T-tubules

FIGURE 9-4
Organization of Muscle Tissue. (a) Formation of skeletal muscle fiber during development. (b) Muscle fiber microanatomy.

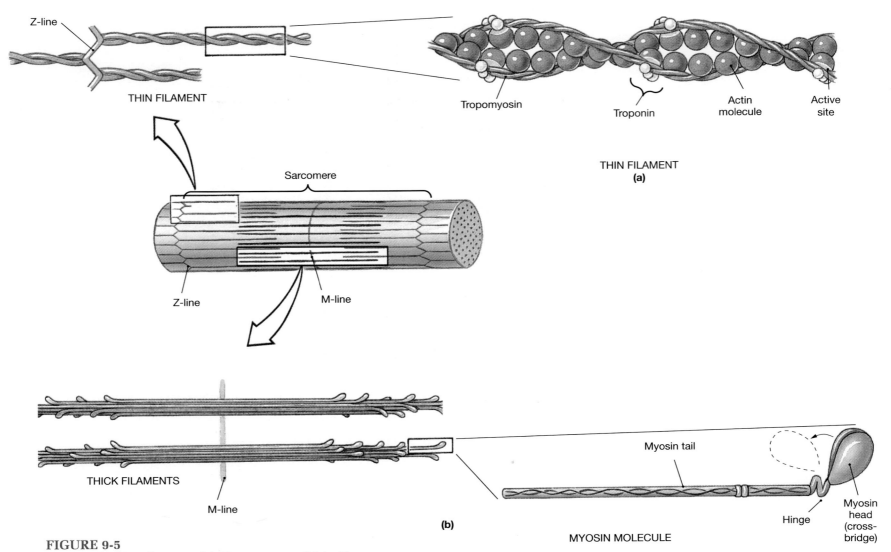

THIN FILAMENT

Tropomyosin

Troponin

Actin molecule

Active site

THIN FILAMENT
(a)

Z-line

Sarcomere

Z-line

M-line

THICK FILAMENTS

M-line

(b)

Myosin tail

Myosin head (cross-bridge)

Hinge

MYOSIN MOLECULE

FIGURE 9-5
Thick and Thin Filaments. (a) The structure of thin filaments.
(b) The structure of thick filaments.

Thin and Thick Filaments
(Figure 9-5)

Each **thin filament** (Figure 9-5a●) consists of a twisted strand composed of 300–400 globular actin molecules and the associated proteins *tropomyosin* and *troponin.* **Thick filaments** are composed of a bundle of myosin molecules. Each myosin molecule consists of a double myosin strand with an attached, elongate *tail* and a free globular *head.* The myosin molecules are oriented away from the center of the thick filament, with the heads projecting outward toward the surrounding thin filaments (Figure 9-5b●). Because during a contraction the myosin heads connect thick filaments and thin filaments, they are also known as **cross-bridges.**

THICK AND THIN FILAMENT INTERACTIONS *(Figure 9-5a).* Contraction involves an interaction between the myosin heads and **active sites** on actin molecules along the thin filaments. At rest, these interactions are prevented by two other proteins,**tropomyosin** (trō-pō-MĪ-ō-sin) and **troponin** (TRŌ-pō-nin; *trope,* turning). Tropomyosin molecules form a long chain that covers the active sites, blocking the actin-myosin interaction (Figure 9-5a●). Troponin holds the tropomyosin strand in place. Before a contraction

can begin, the troponin molecules must change position, moving the tropomyosin molecules and exposing the active sites. This happens when calcium ion concentrations rise in the sarcoplasm. The calcium ions bind to troponin, causing a conformational change that moves tropomyosin away from the active sites. Cross-bridge binding can then occur, and a contraction begins. The mechanism will be detailed in a later section.

Sarcomere Organization
(Figures 9-2/9-4/9-6)

Thick and thin filaments within a myofibril are organized in repeating functional units called **sarcomeres** (SAR-kō-mērs; *sarkos,* flesh + *meros,* part). The arrangement of thick and thin filaments within the sarcomere gives it a banded appearance. All of the myofibrils are arranged parallel to the long axis of the cell, with their sarcomeres lying side by side. As a result, the entire muscle fiber has a banded appearance corresponding to the bands of the individual sarcomeres (see Figures 9-2, p. 235, and 9-4●).

Each myofibril consists of a linear series of approximately 10,000 sarcomeres. *Sarcomeres are the smallest functional*

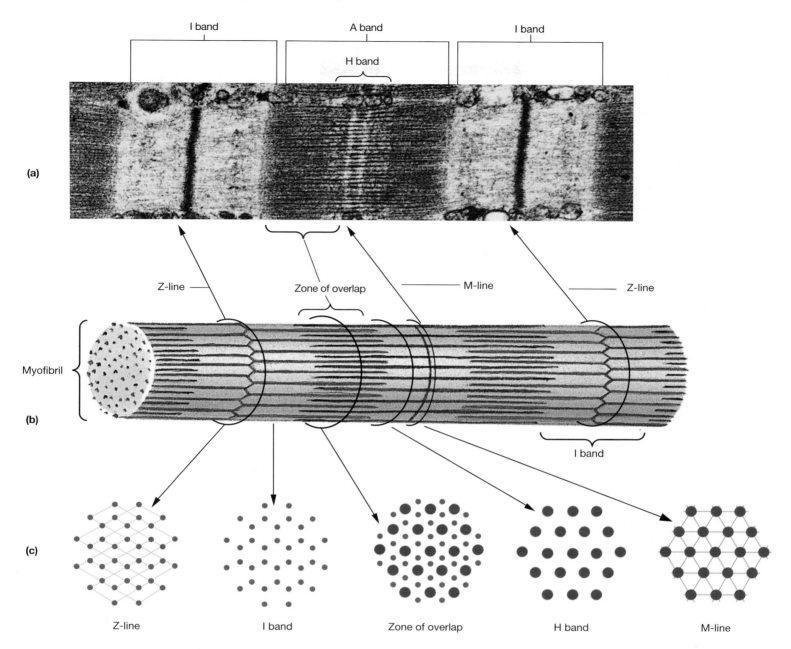

FIGURE 9-6
Sarcomere Structure. (a) A sarcomere in a myofibril from a muscle fiber in the *gastrocnemius* muscle of the calf. (TEM × 64,000) (b) Organization of thick and thin filaments in a sarcomere. (c) Cross-sectional views of different portions of the sarcomere.

units of the muscle fiber—interactions between the thick and thin filaments of sarcomeres are responsible for skeletal muscle fiber contractions.

Figure 9-6a● diagrams the structure of an individual sarcomere. The thick filaments lie in the center of the sarcomere, linked by filaments of the **M-line**. Thin filaments at either end of the sarcomere, attached to interconnecting filaments that make up the **Z-lines**, extend toward the M-line. In the **zone of overlap**, the thin filaments pass between the thick filaments. Figure 9-6b● is a cross section through the zone of overlap, showing the relative sizes and arrangement of thick and thin filaments. Each thin filament sits in a triangle formed by three thick filaments, and each thick filament is surrounded by six thin filaments.

The differences in the size and density of thick filaments and thin filaments account for the banded appearance of the sar-

comere. The **A band** is the area containing thick filaments. The A band includes the M-line, the **H band** (thick filaments only), and the zone of overlap (thick and thin filaments). Between the A band and the Z-line is the **I band**, which contains only thin filaments. The terms *A band* and *I band* are derived from *anisotropic* and *isotropic*, which refer to the appearance of these bands when viewed under polarized light.

√ **How would severing a skeletal muscle's tendon affect the ability of the muscle to move a body part?**
√ **Why does skeletal muscle appear striated when viewed with a microscope?**

MUSCLE CONTRACTION

A contracting muscle fiber exerts a pull, or **tension**, and shortens in length. Muscle fiber contraction results from interactions between the thick and thin filaments in each sarcomere. The mechanism for muscle contraction is explained by the *sliding filament theory*.

The Sliding Filament Theory
(Figures 9-7/9-8)

When a sarcomere contracts, the H band and I band get smaller, the zone of overlap gets larger, and the Z-lines move closer together (Figure 9-7●). The width of the A band remains constant throughout the contraction. The explanation for these observations is known as the **sliding filament theory**, which is illustrated in Figure 9-8●. The sliding filament theory explains the physical changes that occur during contraction, but not why it begins, how it uses energy, or what stops the contraction.

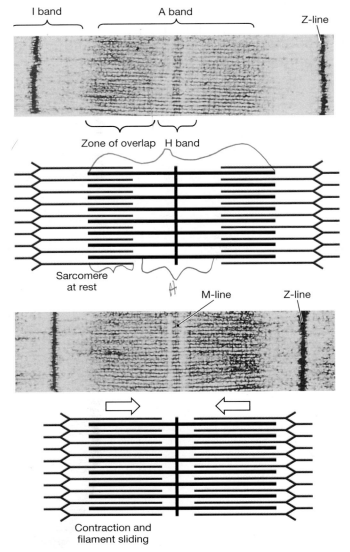

FIGURE 9-7
Changes in the Appearance of a Sarcomere during Contraction of a Skeletal Muscle Fiber. During a contraction, the A band stays the same width, but the Z-lines move closer together and the I band gets smaller.

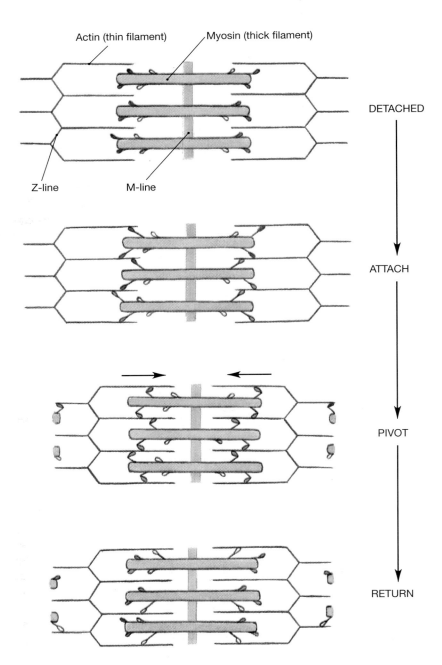

FIGURE 9-8
The Sliding Filament Theory. During a contraction, binding occurs between the myosin heads of the thick filaments and active sites on the actin molecules along thin filaments. The heads pivot, pulling the thin filaments toward the center of the sarcomere. Repeated cycles of attachment, pivoting, and release move the Z-lines toward the ends of the thick filaments.

The Sliding Mechanism
(Figures 9-8/9-9)

Sliding occurs when the myosin heads of thick filaments bind to active sites on thin filaments. When cross-bridge binding occurs, the myosin head pivots toward the M-line, pulling the thin filament toward the center of the sarcomere. The cross-bridge then detaches and returns to its original position, ready to repeat the cycle of "attach, pivot, detach, and return." The net effect is diagrammed in Figure 9-8●. Each time a cross-bridge detaches, an ATP molecule is broken down. A muscle fiber contraction consumes large amounts of energy. Although contraction is an active process, the return to resting length is entirely passive and may be due to elastic forces or the pull of antagonistic muscles.

When many people are pulling on a rope, the amount of tension produced is proportional to the number of people involved. In a muscle fiber, the amount of tension generated during a contraction depends on the number of cross-bridge interactions that occur in the sarcomeres of the myofibrils. The number of cross-bridges is in turn determined by the degree of overlap between thick and thin filaments. Only myosin heads within the zone of overlap can bind to active sites and produce tension. The tension produced by the muscle fiber can therefore be related directly to the structure of an individual sarcomere (Figure 9-9●). The normal range of sarcomere lengths is from 75 to 130% of the optimal length. During normal movements our muscle fibers perform over a broad range of intermediate lengths, and the tension produced therefore varies from moment to moment. During an activity such as walking, in which muscles contract and relax in a cyclical fashion, muscle fibers are stretched to a length very close to optimal before they are stimulated to contract.

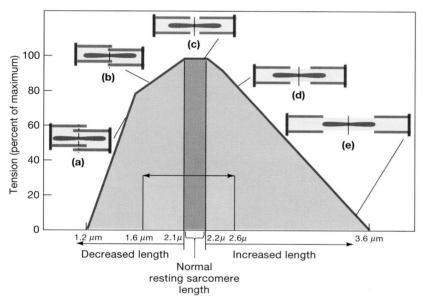

FIGURE 9-9

The Effect of Sarcomere Length on Tension. When the sarcomeres are too short (a,b), contraction cannot occur, because the thick filaments come in contact with the Z-lines. If the sarcomeres are stretched too far, the zone of overlap is reduced (d) or disappears (e), and cross-bridge interactions are reduced or cannot occur. The tension produced reaches a maximum when the zone of overlap is large, but the thin filaments do not extend across the center of the sarcomere (c).

The Role of the Transverse Tubules and the Sarcoplasmic Reticulum
(Figures 9-4/9-10)

Electrical events at the sarcolemmal surface cause a contraction by altering the chemical environment around every sarcomere in the muscle fiber. The electrical "message" is distributed by the transverse tubules that extend deep into the sarcoplasm of the muscle fiber. A transverse tubule begins at the sarcolemma and travels inward at right angles to the membrane surface. Along the way, branches from the transverse tubule encircle each of the individual sarcomeres at the boundary between the A band and the I band (see Figures 9-4, p. 237, and 9-10●).

At the zones of overlap between the thin and thick filaments, the individual tubules of the sarcoplasmic reticulum enlarge, fuse, and form expanded chambers called **terminal cisternae** (Figure 9-10●). A transverse tubule lies sandwiched between each pair of terminal cisternae, forming a complex known as a **triad**. The membranes of the triad do not connect with each other, so their fluid contents remain separate and distinct.

Calcium ions are continually transported out of the sarcoplasm and into the sarcoplasmic reticulum. These ions are released from the sarcoplasmic reticulum when electrical changes occur in the membranes of the T-tubules. The duration of the contraction depends on the duration of the electrical events. When the electrical stimulation ceases, the sarcoplasmic reticulum recaptures the calcium ions, the troponin-tropomyosin complex covers the active sites, and the contraction ends.

■ CLINICAL BRIEF
Rigor Mortis

When death occurs, circulation ceases and the skeletal muscles are deprived of nutrients and oxygen. Within a few hours, the skeletal muscle fibers have run out of ATP, and the sarcoplasmic reticulum becomes unable to remove calcium ions from the sarcoplasm. Calcium ions diffusing into the sarcoplasm from the extracellular fluid or leaking out of the sarcoplasmic reticulum then trigger a sustained contraction. Without ATP, the cross-bridges cannot detach from the active sites, and the muscle locks in the contracted position. All of the body's skeletal muscles are involved, and the individual becomes "stiff as a board." This physical state, called **rigor mortis**, lasts until the lysosomal enzymes released by autolysis break down the myofilaments 15–25 hours later.

FIGURE 9-10

The Orientation of the Sarcoplasmic Reticulum, T-Tubules, and Individual Sarcomeres. A triad occurs where a T-tubule encircles a sarcomere between two terminal cisternae. Compare with Figure 9-4b, p. 237; note that triads occur in the zones of overlap.

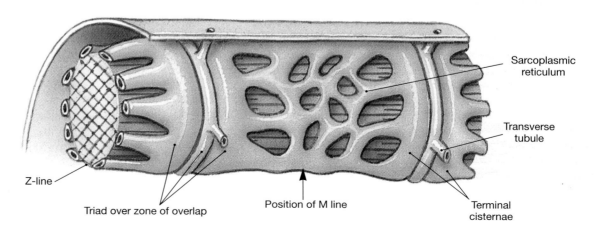

FIGURE 9-11
The Neuromuscular Junction.
See also Figure 9-2.

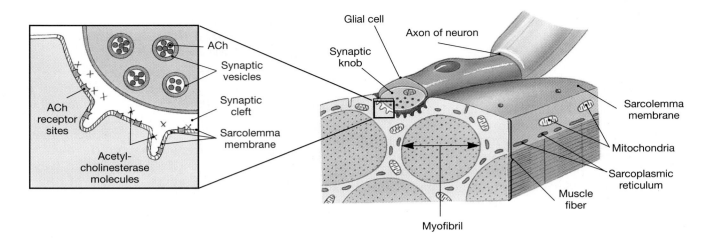

FIGURE 9-12
Summary of the Events in Muscle Contraction

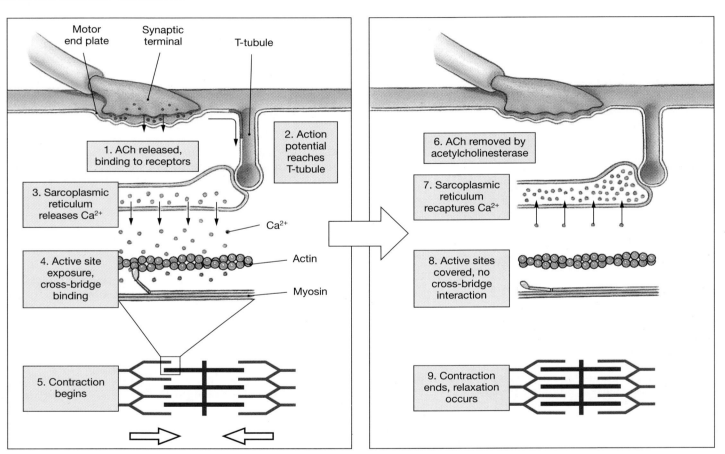

1. At the neuromuscular junction, ACh released by the synaptic knob binds to receptors on the sarcolemma.

2. The resulting change in the transmembrane potential of the muscle fiber leads to the production of an action potential that spreads across its entire surface and reaches the triads via the transverse tubules.

3. The sarcoplasmic reticulum releases stored calcium ions, increasing the calcium concentration of the sarcoplasm in and around the sarcomeres.

4. Calcium ions bind to troponin, producing a change in the orientation of the troponin-tropomyosin complex that exposes active sites on the thin (actin) filaments.

5. Repeated cycles of cross-bridge binding, pivoting, and detachment occur, powered by the breakdown of ATP.

These events produce filament sliding, and the muscle fiber shortens.

6. Action potential generation ceases as ACh is removed by acetylcholinesterase.

7. The sarcoplasmic reticulum reabsorbs calcium ions, and the concentration of calcium ions in the sarcoplasm declines.

8. When calcium ion concentrations approach normal resting levels, the troponin-tropomyosin complex returns to its normal position. This change covers the active sites and prevents further cross-bridge interaction.

9. Without cross-bridge interactions, further sliding will not take place; the contraction ends, and the muscle relaxes.

Neural Control of Muscle Fiber Contraction

The basic sequence of events involved in the neural control of skeletal muscle function can be summarized as follows:

1. Chemicals released by the motor neuron at the neuromuscular junction alter the transmembrane potential of the sarcolemma. This change sweeps across the surface of the sarcolemma and into the transverse tubules.

2. The change in the transmembrane potential of the T-tubules triggers the release of calcium ions by the sarcoplasmic reticulum. This release initiates the contraction, as detailed above.

The Neuromuscular Junction
(Figures 9-2/9-11)

Each skeletal muscle fiber is controlled by a motor neuron at a single neuromuscular junction midway along its length (Figures 9-2, p. 235, and 9-11●). The **synaptic knob**, the expanded tip of an axonal branch, faces a specialized region of the sarcolemma. A narrow space, the **synaptic cleft**, separates the two. The cytoplasm of the synaptic knob contains numerous mitochondria and **synaptic vesicles** filled with molecules of the neurotransmitter **acetylcholine (ACh)** (as-e-til-KŌ-lēn). The synaptic cleft contains the enzyme **acetylcholinesterase (AChE)**, or *cholinesterase*, which breaks down molecules of ACh.

When a nerve impulse arrives at the synaptic knob, ACh is released into the synaptic cleft. The ACh released then binds to receptors on the sarcolemma, initiating a change in the local transmembrane potential. This change results in the generation of electrical signals that sweep over the surface of the sarcolemma and into each T-tubule. These signals persist until acetylcholinesterase removes the bound ACh. ✝*Duchenne's Muscular Dystrophy* and *Myasthenia Gravis* [p. 757]

Muscle Contraction: A Summary
(Figure 9-12)

The entire sequence of events from neural activation to the completion of a contraction is visually summarized in Figure 9-12●.

✓ **What happens to the A bands and I bands of a myofibril during a contraction?**

✓ **How does stimulation of a motor neuron produce a contraction?**

MOTOR UNITS AND MUSCLE CONTROL
(Figure 9-13)

All of the muscle fibers controlled by a single motor neuron constitute a **motor unit.** A typical skeletal muscle contains thousands of muscle fibers. Although some motor neurons control a single muscle fiber, most control hundreds. The size of a motor unit is an indication of how fine the control of

movement can be. In the muscles of the eye, where precise control is extremely important, a motor neuron may control two or three muscle fibers. We have much less precise control over power-generating muscles, such as our leg muscles, where up to 2000 muscle fibers may be controlled by a single motor neuron.

A skeletal muscle contracts when its motor units are stimulated. The amount of tension produced depends on two factors: (1) the frequency of stimulation and (2) the number of motor units involved. As the rate of stimulation increases, tension production will rise to a peak and plateau at maximal levels. Most muscle contractions involve this type of stimulation.

Each muscle fiber either contracts completely or does not contract at all. This characteristic is called the *all or none principle.* All of the fibers in a motor unit contract at the same time, and the amount of force exerted by the muscle as a whole thus depends on how many motor units are activated. By varying the number of motor units activated at any one time, the nervous system provides precise control over the pull exerted by a muscle.

When a decision is made to perform a movement, specific groups of motor neurons are stimulated. The stimulated neurons do not respond simultaneously, and over time the number of activated motor units gradually increases. Figure 9-13● shows how the muscle fibers of each motor unit are intermingled with those of other units. Because of this intermingling, the direction of pull exerted on the tendon does not change as more motor units are activated, but the total amount of force steadily increases. The smooth but steady increase in muscular tension produced by increasing the number of active motor units is called **recruitment**, or **multiple motor unit summation**.

Peak tension production occurs when all of the motor units in the muscle are contracting at the maximal rate of stimulation. However, such powerful contractions cannot last

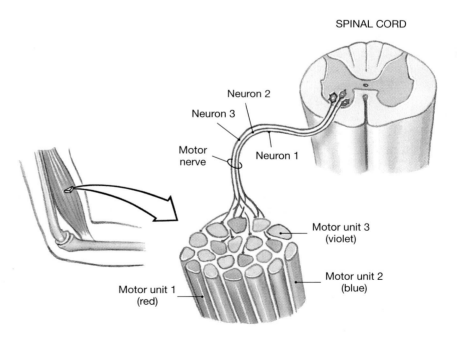

FIGURE 9-13
Arrangement of Motor Units in a Skeletal Muscle. Muscle fibers of different motor units are intermingled, so that the net distribution of force applied to the tendon remains constant even when individual muscle groups cycle between contraction and relaxation.

long, because the individual muscle fibers soon use up their available energy reserves. To lessen the onset of fatigue during periods of sustained contraction, motor units are activated on a rotating basis, so that some of them are resting and recovering while others are actively contracting.

Muscle Tone

Even when a muscle is at rest, some motor units are always active. Their contractions do not produce enough tension to cause movement, but they do tense the muscle. This resting tension in a skeletal muscle is called **muscle tone**. Motor units are randomly being stimulated, so that a constant tension in the attached tendon is maintained but individual muscle fibers can relax. Resting muscle tone stabilizes the position of bones and joints. For example, in muscles involved with balance and posture, enough motor units are stimulated to produce the isometric tension needed to maintain body position. Specialized muscle cells called **muscle spindles** are monitored by sensory nerves that control the muscle tone in the surrounding muscle tissue. Reflexes triggered by activity in these sensory nerves play an important role in the reflex control of position and posture, a topic that we will discuss in Chapter 18.

Muscle Hypertrophy

Exercise increases the activity of muscle spindles and may enhance muscle tone. As a result of repeated, exhaustive stimulation, muscle fibers develop a larger number of mitochondria, a higher concentration of glycolytic enzymes, and larger glycogen reserves. These muscle fibers have more myofibrils, and each myofibril contains a larger number of thick and thin filaments. The net effect is an enlargement, or **hypertrophy** (hī-PER-trō-fē), of the stimulated muscle. Hypertrophy occurs in muscles that have been repeatedly stimulated to produce near-maximal tension; the intracellular changes that occur increase the amount of tension produced when these muscles contract. A champion weight lifter or body builder is an excellent example of hypertrophied muscular development.

Muscle Atrophy

When a skeletal muscle is not stimulated by a motor neuron on a regular basis, it loses muscle tone and mass. The muscle becomes flaccid, and the muscle fibers become smaller and weaker. This reduction in muscle size, tone, and power is called **atrophy**. Individuals paralyzed by spinal injuries or other damage to the nervous system will gradually lose muscle tone and size in the areas affected. Even a temporary reduction in muscle use can lead to muscular atrophy; this loss of tone and size may easily be seen by comparing limb muscles before and after a cast has been worn. Muscle atrophy is initially reversible, but dying muscle fibers are not replaced, and in extreme atrophy the functional losses are permanent. That is why physical therapy is crucial in cases where people are temporarily unable to move normally. ✝ *Polio [p. 757]*

√ **A motor unit from a skeletal muscle contains 1500 muscle fibers. Would this muscle be involved in fine, delicate movements or powerful movements? Explain.**

TYPES OF SKELETAL MUSCLE FIBERS
(Figure 9-14)

Skeletal muscles are designed for various actions. The types of fibers that comprise a muscle will, in part, determine its action. There are three major types of skeletal muscle fibers in the body: *fast*, *slow*, and *intermediate*.

Fast fibers are large in diameter; they contain densely packed myofibrils, large glycogen reserves, and relatively few mitochondria. Most of the skeletal muscle fibers in the body are called fast fibers because they can contract in 0.01 seconds or less following stimulation. The tension produced by a muscle fiber is directly proportional to the number of sarcomeres, so fast-fiber muscles produce powerful contractions. However, because these contractions use enormous amounts of ATP, prolonged activity is supported primarily by anaerobic glycolysis, and fast fibers fatigue rapidly.

Slow fibers are only about half the diameter of fast fibers, and they take three times as long to contract after stimulation. Slow fibers are specialized to enable them to continue contracting for extended periods, long after a fast muscle would have become fatigued. Slow muscle tissue contains a more extensive network of capillaries than fast muscle tissue, so oxygen supply is dramatically increased. In addition, slow muscle fibers contain the red pigment **myoglobin** (MĪ-ō-glō-bin). This globular protein is structurally related to hemoglobin, the oxygen-binding pigment found in red blood cells. Myoglobin also binds oxygen molecules; thus resting slow muscle fibers contain substantial oxygen reserves that can be mobilized during a contraction.

The mitochondria of slow muscle fibers are able to contribute a greater amount of ATP than fast-fiber muscles while contractions are under way, making the cell less dependent on anaerobic metabolism. Some of the mitochondrial energy production involves the breakdown of stored lipids rather than glycogen reserves, which are smaller than those of fast muscle cells. Slow muscles also contain a relatively larger number of mitochondria than do fast muscle fibers. Fast and slow muscle fibers are shown in Figure 9-14●.

Intermediate fibers have properties intermediate between those of fast fibers and slow fibers. Histologically, intermediate fibers are very similar to fast fibers, although they have a greater resistance to fatigue.

Distribution of Fast, Slow, and Intermediate Fibers

The percentage of fast, slow, and intermediate muscle fibers varies from one skeletal muscle to another. Most muscles contain a mixture of fiber types, although all of the fibers within one motor unit are of the same type. However, there are no slow fibers in muscles of the eye and hand, where swift but brief contractions are required. Many back and calf muscles are dominated by slow fibers; these muscles contract almost continually to maintain an upright posture.

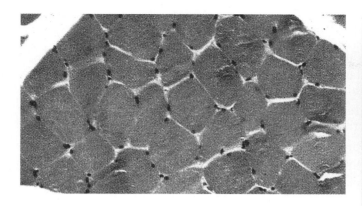

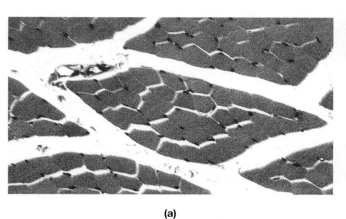

(a)

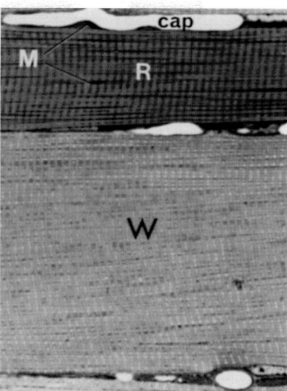

(b)

FIGURE 9-14
Fast and Slow Muscle Fibers. (a) Note the difference in the size of fast muscle fibers, above, and slow muscle fibers, below. (LMs × 229) (b) The relatively slender slow muscle fiber (R) has more mitochondria (M) and a more extensive capillary supply (cap) than the fast muscle fiber (W). (LM × 1044)

The percentage of fast versus slow fibers in each muscle is genetically determined, and there are significant individual differences. These variations have an effect on endurance. A person with more slow muscle fibers in a particular muscle will be better able to perform repeated contractions under aerobic conditions. For example, marathon runners with high proportions of slow muscle fibers in their leg muscles outperform those with more fast muscle fibers. For brief periods of intense activity, such as a sprint or a weight-lifting event, the individual with a higher percentage of fast muscle fibers will have the advantage.

The proportion of intermediate fibers changes with physical conditioning; if used repeatedly for endurance events, fast fibers can develop the appearance and functional capabilities of intermediate fibers. Thus, physical conditioning enables athletes to improve both strength and endurance.

AGING AND THE MUSCULAR SYSTEM

As the body ages, there is a general reduction in the size and power of all muscle tissues. The effects of aging on the muscular system can be summarized as follows:

1. *Skeletal muscle fibers become smaller in diameter.* This reduction in size reflects primarily a decrease in the number of myofibrils. In addition, the muscle fibers contain less ATP, glycogen reserves, and myoglobin. The overall effect is a reduction in muscle strength and endurance and a tendency to fatigue rapidly. Because cardiovascular performance also decreases with age, blood flow to active muscles does not increase with exercise as rapidly as it does in younger people.

2. *Skeletal muscles become smaller and less elastic.* Aging skeletal muscles develop increasing amounts of fibrous connective tissue, a process called **fibrosis**. Fibrosis makes the muscle less flexible, and the collagen fibers can restrict movement and circulation.

3. *Tolerance for exercise decreases.* A lower tolerance for exercise results in part from the tendency for rapid fatigue and in part from the reduction in the ability to eliminate the heat generated during muscular contraction.

4. *Ability to recover from muscular injuries decreases.* The number of satellite cells steadily decreases with age, and the amount of fibrous tissue increases. As a result, when an injury occurs, repair capabilities are limited, and scar tissue formation is the usual result.

The rate of decline in muscular performance is the same in all individuals, regardless of their exercise patterns or lifestyle. Therefore to be in good shape late in life, one must be in very good shape early in life. Regular exercise helps control body weight, strengthens bones, and generally improves the quality of life at all ages. Extremely demanding exercise is not as important as regular exercise. In fact, extreme exercise in the elderly may lead to problems with tendons, bones, and joints. Although it has obvious effects on the quality of life, there is no clear evidence that exercise prolongs life expectancy.

√ **Why does a sprinter experience muscle fatigue after a few minutes, while a marathon runner can run for hours?**

√ **What type of muscle fibers would you expect to predominate in the large leg muscles of someone who excels at endurance activities such as cycling or long-distance running?**

Related Clinical Terms

rigor mortis: A state following death during which muscles are locked in the contracted position, making the body extremely stiff. *[p. 241]*

muscular dystrophy: A condition characterized by a generalized muscular weakness most evident in the arms, head, and chest; caused by a reduction in the number of ACh receptors on motor end plates. ☘ *Duchenne's Muscular Dystrophy [p. 757]*

myasthenia gravis: A congenital disease that produces progressive muscular weakness, and that often develops at an early age. ☘ *Myasthenia Gravis [p. 757]*

polio: Progressive paralysis due to destruction of CNS motor neurons by the polio virus. ☘ *Polio [p. 757]*

fibrosis: A process in which increasing amounts of fibrous connective tissue develop, making muscles less flexible. *[p. 245]*

CHAPTER SUMMARY AND REVIEW

STUDY OUTLINE

Related Key Terms

INTRODUCTION *[p. 233]*

1. There are three types of muscle tissue: **skeletal muscle**, **cardiac muscle**, and **smooth muscle**.

muscular system

FUNCTIONS OF SKELETAL MUSCLE *[p. 233]*

1. **Skeletal muscles** attach to bones directly or indirectly and perform these functions: (1) produce skeletal movement; (2) maintain posture and body position; (3) support soft tissues; (4) guard entrances and exits; (5) maintain body temperature.

ANATOMY OF SKELETAL MUSCLES *[p. 233]*

Gross Anatomy *[p. 233]*
1. Each muscle fiber is surrounded by an **epimysium**, a **perimysium**, and an **endomysium**. At the ends of the muscle are **tendons** or *aponeuroses* that attach the muscle to other structures *(see Figure 9-1)*.
2. Communication between a neuron and a muscle fiber occurs across the **neuromuscular** (*myoneural*) **junction** *(see Figure 9-2)*.

fascicle • satellite cells

Microanatomy of Skeletal Muscle Fibers *[p. 236]*
3. A muscle cell has a cell membrane, or **sarcolemma**, **sarcoplasm** (cytoplasm), and **sarcoplasmic reticulum (SR)**, similar to the endoplasmic reticulum of other cells. Filaments in a **myofibril** are organized into repeating functional units called **sarcomeres** *(see Figure 9-4)*.
4. Myofilaments consist of **thin filaments** and **thick filaments** *(see Figures 9-4/9-5)*.

myoblasts • transverse tubules • T-tubules • myofilaments • M-line • Z-line • zone of overlap • A band • H band • I band

MUSCLE CONTRACTION *[p. 240]*

The Sliding Filament Theory *[p. 240]*
1. The **sliding filament theory** explains contraction, during which a muscle fiber exerts **tension** (a pull) and shortens *(see Figure 9-7)*.
2. Sliding involves a cycle of "attach, pivot, detach, and return." The four-step contraction process involves **active sites** on thin filaments and **cross-bridges** of the thick filaments. At rest, the necessary interactions are prevented by two proteins, **tropomyosin** and **troponin**, on the thin filaments *(see Figures 9-5/9-8)*.
3. Contraction is active, but elongation of a muscle fiber is a passive process that can occur either through elastic forces or through the movement of other, opposing muscles.
4. The amount of tension produced during a contraction is proportional to the degree of overlap between thick and thin filaments *(see Figure 9-9)*.

terminal cisternae • triad

Neural Control of Muscle Fiber Contraction *[p. 243]*
5. Neural control of muscle function involves a link between electrical activity in the sarcolemma and the initiation of a contraction.
6. Each fiber is controlled by a neuron at a *neuromuscular (myoneural) junction*; the junction includes the **synaptic knob**, **synaptic vesicles**, and the **synaptic cleft**. **Acetylcholine (ACh)** release leads to the stimulation of the motor end plate and the generation of electrical impulses that spread across the sarcolemma. **Acetylcholinesterase (AChE)** breaks down ACh and limits the duration of stimulation *(see Figures 9-2/9-11)*.

Muscle Contraction: A Summary *[p. 243]*
7. The steps involved in contraction are: ACh release → binding of ACh to the motor end plate → generation of an electrical impulse in the sarcolemma → conduction of the impulse along T-tubules → release of calcium ions by the SR → exposure of active sites on thin filaments → cross-bridge formation and contraction *(see Figure 9-12)*.

MOTOR UNITS AND MUSCLE CONTROL [p. 243]

Related Key Terms

1. The number and size of a muscle's **motor units** indicate how precisely controlled its movements are *(see Figure 9-13).*

recruitment • multiple motor unit summation • muscle spindles

Muscle Tone [p. 244]

2. Even when a muscle is at rest, motor units are always randomly being stimulated, so that a constant tension in the attached tendon is maintained. This resting tension in a skeletal muscle is called **muscle tone**. Resting muscle tone stabilizes bones and joints.

Muscle Hypertrophy [p. 244]

3. Excessive repeated stimulation to produce near-maximal tension in skeletal muscle can lead to **hypertrophy** (enlargement) of the stimulated muscles.

Muscle Atrophy [p. 244]

4. Inadequate stimulation to maintain resting muscle tone causes muscles to become flaccid and undergo **atrophy**.

TYPES OF SKELETAL MUSCLE FIBERS [p. 244]

1. The three types of skeletal muscle fibers are **fast fibers**, **slow fibers**, and **intermediate fibers** *(see Figure 9-14).*
2. Fast fibers are large in diameter; contain densely packed myofibrils, large glycogen reserves, and relatively few mitochondria. They produce rapid and powerful contractions of relatively brief duration.

myoglobin

3. Slow fibers are only about half the diameter of fast fibers, and they take three times as long to contract after stimulation. Slow fibers are specialized to enable them to continue contracting for extended periods.
4. Intermediate fibers are very similar to fast fibers, although they have a greater resistance to fatigue.

Distribution of Fast, Slow, and Intermediate Fibers [p. 244]

5. The percentage of fast, slow, and intermediate fibers varies from one skeletal muscle to another. Muscles contain a mixture of fiber types, but the fibers that comprise one motor unit are of the same type. The percentage of fast versus slow fibers in each muscle is genetically determined.

AGING AND THE MUSCULAR SYSTEM [p. 245]

1. The aging process reduces the size, elasticity, and power of all muscle tissues. Exercise tolerance and the ability to recover from muscular injuries both decrease.

1   REVIEW OF CHAPTER OBJECTIVES

1. Describe the characteristics of muscle tissue.
2. Describe the functions of skeletal muscle tissue.
3. Discuss the organization of skeletal muscle from the gross level to the molecular level.
4. Identify the unique characteristics of skeletal muscle fibers.
5. Summarize the process of muscular contraction.
6. Describe a motor unit and the control of muscle fibers.
7. Relate types of skeletal muscle fibers to muscular performance.
8. Describe the effects of exercise and aging on skeletal muscle.

2   REVIEW OF CONCEPTS

1. How does the structure of smooth muscle differ from that of striated muscle?
2. What are the main functions of skeletal muscle?
3. What is the role of connective tissue in the organization of skeletal muscle?
4. How are tendons and aponeuroses related to the connective tissues that form part of the muscles?
5. Why are some muscles connected to origins and insertions by tendons and others by aponeuroses?
6. What is the function of a neuromuscular junction?
7. How is an electrical signal transmitted along muscle fibers?
8. How do the actin and myosin filaments interact to cause a muscle fiber to contract?
9. What is the role of the "zone of overlap" in the structure of skeletal muscle?
10. What is the role of ATP in allowing muscle fibers to contract?
11. How does the electrical stimulation of a muscle affect the distribution of calcium?

12. What is the role of the T-tubules in propagating the electrical impulse in the muscle?
13. What is the relationship between a motor neuron and the muscle fibers of a motor unit?
14. What is the role of muscle spindles in controlling tone in skeletal muscles?
15. Why can slow muscle fibers sustain activity for a longer period of time than fast fibers?
16. What is the relationship between muscle fiber type and muscle action?

3   CRITICAL THINKING AND CLINICAL APPLICATION QUESTIONS

1. Recently, the parents of a three-year-old boy have noticed that he is slow at developing motor skills, and he appears to have weak and atrophying skeletal muscles. The child did not sit, stand, or walk until later than normal; he is clumsy and falls frequently. What is a possible diagnosis for this condition, and what anatomical mechanism is responsible?
2. Several anatomy students decide to take up weight lifting and body building. After several months, they notice many physical changes, including an increase in muscle mass, lean body weight, and greater muscular strength. What anatomical mechanism is responsible for these changes?
3. One of the body-building anatomy students tripped over some barbells and fell; he broke his left radius and ulna. His arm was placed in a cast, and for six weeks he could neither move it nor lift weights. Once the cast was removed, the student realized his left forearm had become dramatically smaller than his right. Why did this change occur?
4. Many people of all ages train for and participate in athletic competitions. However, in most sports the competitors are divided into age classes. The best performances in sports such as running, swimming, and gymnastics are turned in by the more youthful competitors. How does aging affect the activity of muscles?

10

The Muscular System: Muscle Organization and the Axial Musculature

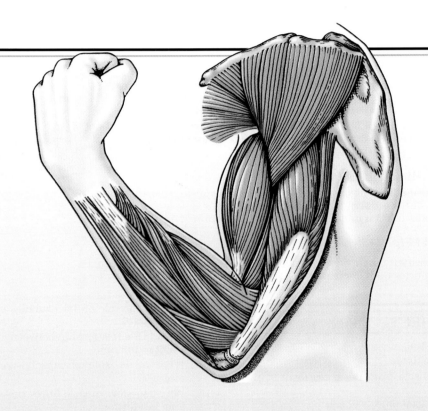

CHAPTER OUTLINE AND OBJECTIVES

The **muscular system** includes all of the skeletal muscles that can be controlled voluntarily. There are approximately 700 skeletal muscles, and most of the muscle tissue in the body is part of this system. Some skeletal muscles are attached to bony processes, others to broad sheets of connective tissue, but all are directly or indirectly associated with the skeletal system. We will not attempt to examine all 700 skeletal muscles, and the focus of this chapter will be the major skeletal muscles. For clarity, these muscles, roughly 20% of the total number, have been organized into anatomical and functional groups.

The shape or appearance of each muscle provides clues to its primary function. Muscles involved with locomotion and posture work across joints, producing skeletal movement. Those that support soft tissue form slings or sheets between relatively stable bony elements, whereas those that guard an entrance or exit completely encircle the opening.

At the level of the individual skeletal muscle, two factors interact to determine the effects of its contraction: (1) the arrangement of muscle fibers and (2) the way the muscle attaches to the skeletal system. Although most skeletal muscle fibers contract at comparable speeds, and shorten to the same degree, variations in microscopic and macroscopic organization can dramatically affect the tension, range, and speed of movement produced when a muscle contracts. The performance of muscles in the body can be understood in terms of basic mechanical laws. The analysis of biological systems in mechanical terms is the study of **biomechanics**. This chapter considers the organization of skeletal muscles, muscle terminology, and the gross anatomy of the *axial musculature*, skeletal muscles associated with the *axial skeleton*. ∞[p. 78] In the process we will introduce basic principles of biomechanics.

ORGANIZATION OF SKELETAL MUSCLE FIBERS
(Figure 10-1)

Muscle fibers within a skeletal muscle form bundles called *fascicles*. ∞[p. 233] The muscle fibers of each fascicle lie parallel to one another, but the organization of the fascicles in the skeletal muscle can vary, as can the relationship between the fascicles and the associated tendon. Four different patterns of fascicle organization produce *parallel muscles*, *convergent muscles*, *pennate muscles*, and *circular muscles*. Figure 10-1● illustrates the fascicle organization of skeletal muscle fibers.

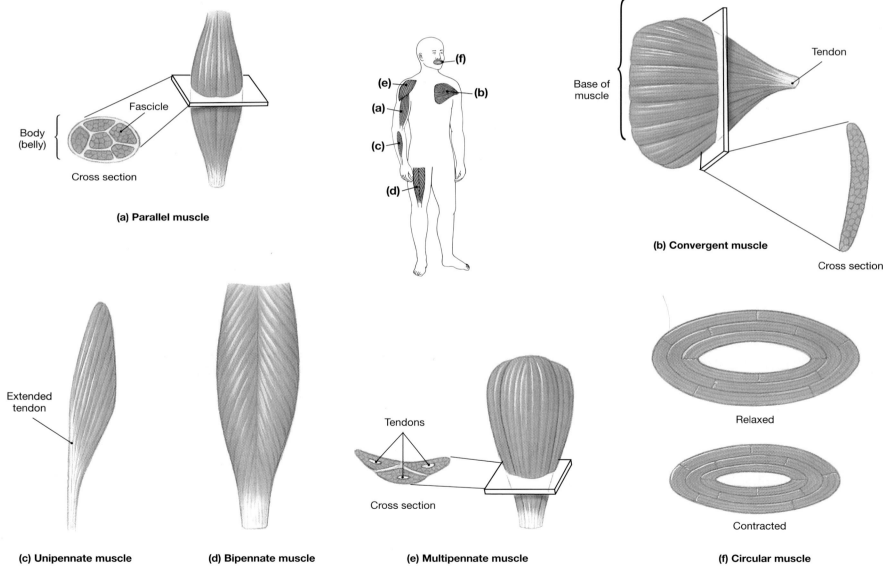

(a) Parallel muscle

(b) Convergent muscle

(c) Unipennate muscle

(d) Bipennate muscle

(e) Multipennate muscle

(f) Circular muscle

FIGURE 10-1
Different Arrangements of Skeletal Muscle Fibers

Parallel Muscles
(Figure 10-1a)

In a **parallel muscle**, the fascicles are parallel to the long axis of the muscle. The functional characteristics of a parallel muscle resemble those of an individual muscle fiber. Consider the skeletal muscle shown in Figure 10-1a●. It has a firm attachment and a tendon that extends from the free tip to a movable bone of the skeleton. Most of the skeletal muscles in the body are parallel muscles. Some form flat bands, with broad aponeuroses at each end; others are spindle-shaped, with cordlike tendons at one or both ends (Figure 10-1a●). Such a muscle has a central **body**, also known as the *belly*, or *gaster* (GAS-ter; *gaster*, stomach). When this muscle contracts, it gets shorter and the body increases in diameter. The *biceps brachii* muscle of the arm is an example of a parallel muscle with a central body. The bulge of the contracting biceps can be seen on the front of the arm when the forearm is flexed.

A skeletal muscle cell can contract effectively until it has been shortened by roughly 30%. Because the muscle fibers are parallel to the long axis of the muscle, when they contract together, the entire muscle shortens by the same amount. For example, if the skeletal muscle is 10 cm long, the end of the tendon will move 3 cm when the muscle contracts. The tension developed by the muscle during this contraction depends on the total number of myofibrils it contains. Because the myofibrils are distributed evenly through the sarcoplasm of each cell, the tension can be estimated on the basis of the cross-sectional area of the resting muscle. A parallel skeletal muscle 6.45 cm^2 (1 in.^{2}) in cross-sectional area can develop approximately 23 kg (50 lb) of tension.

Convergent Muscles
(Figure 10-1b)

In a **convergent muscle**, the muscle fibers are based over a broad area, but all the fibers come together at a common attachment site. They may pull on a tendon, a tendinous sheet, or a slender band of collagen fibers known as a **raphe** (RAY-fē; seam). The muscle fibers often spread out, like a fan or a broad triangle, with a tendon at the tip as shown in Figure 10-1b●. The prominent chest muscles of the *pectoralis group* have this shape. Such a muscle has versatility, for the direction of pull can be changed by stimulating only one group of muscle cells at any one time. But when they all contract at once, they do not pull as hard on the tendon as a parallel muscle of the same size. The reason is that the muscle fibers on opposite sides of the tendon are pulling in different directions, rather than working together.

Pennate Muscles
(Figure 10-1c,d,e)

In a **pennate muscle** (*penna*, feather), one or more tendons run through the body of the muscle, and the fascicles form an oblique angle to the tendon. Because the muscle cells pull at an angle, contracting pennate muscles do not move their tendons as far as parallel muscles do. But a pennate muscle will contain more muscle fibers than a parallel muscle of the same size, and as a result the contraction of the pennate muscle generates more tension than that of a parallel muscle of the same size.

If all of the muscle cells are found on the same side of the tendon, the muscle is **unipennate** (Figure 10-1c●). A long muscle that extends the fingers, the *extensor digitorum* is an example of a unipennate muscle. More commonly, there are muscle fibers on both sides of the tendon. A prominent muscle of the thigh, the *rectus femoris* is a **bipennate** muscle (Figure 10-1d●) that helps to extend the knee. If the tendon branches within the muscle, the muscle is **multipennate** (Figure 10-1e●). The triangular *deltoid* muscle that covers the superior surface of the shoulder joint is an example of a multipennate muscle.

Circular Muscles
(Figure 10-1f)

In a **circular muscle**, or **sphincter** (SFINK-ter), the fibers are concentrically arranged around an opening or recess (Figure 10-1f●). When the muscle contracts, the diameter of the opening decreases. Circular muscles guard entrances and exits of internal passageways such as the digestive and urinary tracts. An example is the *orbicularis oris* muscle of the mouth.

LEVERS: A SYSTEMS DESIGN FOR MOVEMENT

Skeletal muscles do not work in isolation. When a muscle is attached to the skeleton, the nature and site of the connection will determine the force, speed, and range of the movement produced. These characteristics are interdependent, and the relationships can explain a great deal about the general organization of the muscular and skeletal systems.

The force, speed, or direction of movement produced by contraction of a muscle can be modified by attaching the muscle to a lever. A **lever** is a rigid structure—such as a board, a crowbar, or a bone—that moves on a fixed point, called the **fulcrum**. In the body, each bone is a lever and each joint a fulcrum. The teeter-totter, or seesaw, at the park provides a more familiar example of lever action. Levers can change (1) the direction of an applied force, (2) the distance and speed of movement produced by a force, and (3) the strength of a force.

Classes of Levers
(Figure 10-2)

Three classes of levers are found in the human body:

1. **First-Class Levers**: The seesaw is an example of a **first-class lever**: one in which the fulcrum lies between the

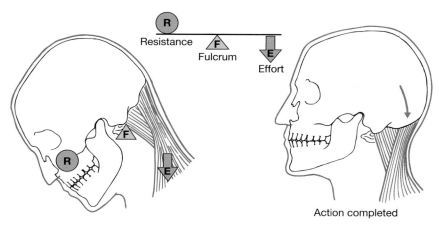

(a) First-class lever

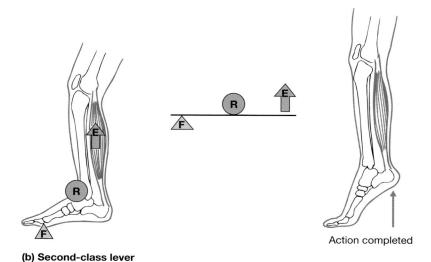

(b) Second-class lever

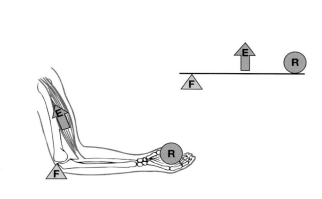

(c) Third-class lever

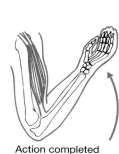

FIGURE 10-2
The Three Classes of Levers. (a) In a first-class lever, the applied force and the resistance are on opposite sides of the fulcrum. First-class levers can change the amount of force transmitted to the resistance and alter the direction and speed of movement. (b) In a second-class lever, the resistance lies between the applied force and the fulcrum. This arrangement magnifies force at the expense of distance and speed; the direction of movement remains unchanged. (c) In a third-class lever, the force is applied between the resistance and the fulcrum. This arrangement increases speed and distance moved but requires a larger applied force. The example shows the force and distance relationships involved in the contraction of the biceps brachii muscle that flexes the forearm.

applied force and the resistance as seen in Figure 10-2a●. There are not many examples of first-class levers in the body. One, involving the muscles that extend the neck, is shown in this figure.

2. Second-Class Levers: In a **second-class lever**, the resistance is located between the applied force and the fulcrum. A familiar example of such a lever is a loaded wheelbarrow. The weight of the load is the resistance, and the upward lift on the handle is the applied force. Because in this arrangement the force is always farther from the fulcrum than the resistance is, a small force can balance a larger weight. In other words, the force is magnified. Notice, however, that when a force moves the handle, the resistance moves more slowly and covers a shorter distance. There are few examples of second-class levers in the body. In performing plantar flexion, the calf muscles act across a second-class lever (Figure 10-2b●).

3. Third-Class Levers: In a **third-class lever** system, a force is applied between the resistance and the fulcrum (Figure 10-2c●). Third-class levers are the most common levers in

the body. The effect of this arrangement is the reverse of that produced by a second-class lever: Speed and distance traveled are increased at the expense of force. In the example illustrated (the biceps brachii, which flexes the forearm), the resistance is six times farther away from the fulcrum than the applied force. The biceps can develop an effective force of 180 kg, which now will be reduced from 180 kg to 30 kg. However, the distance traveled and the speed of movement are *increased* by the same ratio (6×): the resistance travels 45 cm while the insertion point moves only 7.5 cm.

Although every muscle does not operate as part of a lever system, the presence of levers provides speed and versatility far in excess of what we would predict on the basis of muscle physiology alone. Skeletal muscle cells resemble one another closely, and their abilities to contract and generate tension are quite similar. Consider a skeletal muscle that can contract in 500 ms and shorten 1 cm while exerting a 10-kg pull. Without using a lever, this muscle would be performing efficiently only when moving a 10-kg weight a distance of 1 cm. But by using a lever, the same muscle operating at the same efficiency could move 20 kg a distance of 0.5 cm, 5 kg a distance of

2 cm, or 1 kg a distance of 10 cm. Thus, the lever system design produces the maximum movements with the greatest efficiency.

MUSCLE TERMINOLOGY
(Table 10-1)

Each muscle begins at an **origin**, ends at an **insertion**, and contracts to produce a specific **action**. Terms indicating the actions of muscles, specific regions of the body, and structural characteristics of muscle are presented in Table 10-1.

Origins and Insertions

Typically the origin remains stationary and the insertion moves, or the origin is proximal to the insertion. For example, the triceps inserts on the olecranon process and originates closer to the shoulder. Such determinations are made during normal movement with the individual in the anatomical position. Part of the fun of studying the muscular system is that you can actually do the movements and think about the muscles involved. (Laboratory discussions of the muscular system often resemble a poorly organized aerobics class.)

When the origins and insertions cannot be determined easily on the basis of movement or position, other rules are used. If a muscle extends between a broad aponeurosis and a narrow tendon, the aponeurosis is considered to be the origin, and the tendon is attached to the insertion. If there are several tendons at one end and just one at the other, there are multiple origins and a single insertion. These simple rules cannot cover every situation, and knowing which end is the origin and which is the insertion is ultimately less important than knowing where the two ends attach and what the muscle does when it contracts.

Actions

Almost all skeletal muscles either originate or insert upon the skeleton. When a muscle moves a portion of the skeleton, that movement may involve *flexion, extension, adduction, abduction, protraction, retraction, elevation, depression, rotation, circumduction, pronation, supination, inversion,* or *eversion.* Before proceeding, consider reviewing the discussion of planes of motion and Figures 8-3 to 8-5. ∞[pp. 210, 211, 212]

Muscles can be grouped according to their **primary actions** into three types:

1. **Prime Movers (Agonists):** A **prime mover,** or **agonist**, is a muscle whose contraction is chiefly responsible for producing a particular movement, such as flexion of the forearm. The biceps brachii is an example of a prime mover or agonist of the forearm into flexion.

2. **Synergists:** When a **synergist** (*syn-*, together + *ergon*, work) contracts, it assists the prime mover in performing that action. Synergists may provide additional pull near the insertion or stabilize the point of origin. Their importance in assisting a particular movement may change as the movement progresses; in many cases they are most useful at the start, when the prime mover is stretched and its power is relatively low. For example, the *latissimus dorsi* and the *teres major* pull the arm inferiorly. With the arm pointed at the ceiling, the muscle fibers of the massive latissimus are at maximum stretch, and they are aligned parallel to the humerus. The latissimus dorsi cannot develop much tension in this position. However, the orientation of the teres major, which originates on the scapula, can contract more efficiently, and it assists the latissimus in starting an inferior movement. The importance of this smaller "assistant" decreases as the inferior movement proceeds. In this example, the latissimus dorsi is the agonist and the teres major the synergist. Synergists may also assist an agonist by *preventing* movement at a joint and thereby stabilizing the origin of the agonist. These muscles are called *fixators.*

3. **Antagonists:** **Antagonists** are muscles whose actions oppose that of the agonist; if the agonist produces flexion, the antagonist will produce extension. When an agonist contracts to produce a particular movement, the corresponding antagonist will be stretched, but it will usually not relax completely. Instead, its tension will be adjusted to control the speed of the movement and ensure its smoothness. For example, the biceps brachii acts as an agonist when it contracts, thereby producing flexion of the forearm. The triceps brachii muscle, located on the opposite side of the humerus, acts as an antagonist to stabilize the flexion movement and to produce the opposing action, extension of the forearm.

Names of Skeletal Muscles

You will not need to learn every one of the nearly 700 muscles in the human body, but you will have to become familiar with the most important ones. Fortunately, the names of most skeletal muscles provide clues to their identification. The name may indicate a specific region (Table 10-1), such as the *brachialis* of the arm, the shape of the muscle (*trapezius, piriformis*), or some combination of the two (*biceps femoris*).

Some names include reference to the orientation of the muscle fibers within a particular skeletal muscle. For example, **rectus** means "straight," and rectus muscles are parallel muscles whose fibers generally run along the long axis of the body. Because there are several rectus muscles, the name usually includes a second term that refers to a precise region of the body. The *rectus abdominis* is found on the abdomen, and the *rectus femoris* on the thigh. Other directional indicators include **transversus** and **obliquus** for muscles whose fibers run across or at an oblique angle to the longitudinal axis of the body.

Other muscles were named after specific and unusual structural features. The *biceps* muscle has two tendons of origin (*bi-*, two + *caput*, head), the *triceps* has three, and the *quadriceps* four. Shape is sometimes an important clue to the name of a muscle. For example, the *trapezius* (tra-PĒ-zē-us), *deltoid, rhomboideus* (rom-BOY-dē-us), and *orbicularis* (or-bik-ū-LAR-is) refer to prominent muscles that look like a trapezoid, a triangle, a rhomboid, and a circle, respectively. Long muscles are called **longus** (long) or **longissimus** (longest), and **teres** muscles are both long and round. Short

TABLE 10-1 Muscle Terminology

Terms Indicating Direction Relative to Axes of the Body	Terms Indicating Specific Regions of the Body*	Terms Indicating Structural Characteristics of the Muscle	Terms Indicating Actions
Anterior (front)	Abdominis (abdomen)	**Origin**	**General**
Externus (superficial)	Anconeus (elbow)	Biceps (two heads)	Abductor
Extrinsic (outside)	Auricularis (auricle of ear)	Triceps (three heads)	Adductor
Inferioris (inferior)	Brachialis (brachium)	Quadriceps (four heads)	Depressor
Internus (deep, internal)	Capitis (head)		Extensor
Intrinsic (inside)	Carpi (wrist)	**Shape**	Flexor
Lateralis (lateral)	Cervicis (neck)	Deltoid (triangle)	Levator
Medialis/medius (medial, middle)	Cleido/clavius (clavicle)	Orbicularis (circle)	Pronator
Obliquus (oblique)	Coccygeus (coccyx)	Pectinate (comblike)	Rotator
Posterior (back)	Costalis (ribs)	Piriformis (pear-shaped)	Supinator
Profundus (deep)	Cutaneous (skin)	Platys- (flat)	Tensor
Rectus (straight, parallel)	Femoris (femur)	Pyramidal (pyramid)	
Superficialis (superficial)	Genio- (chin)	Rhomboideus (rhomboid)	**Specific**
Superioris (superior)	Glosso/glossal (tongue)	Serratus (serrated)	Buccinator (trumpeter)
Transversus (transverse)	Hallucis (great toe)	Splenius (bandage)	Risorius (laugher)
	Ilio- (ilium)	Teres (long and round)	Sartorius (like a tailor)
	Inguinal (groin)	Trapezius (trapezoid)	
	Lumborum (lumbar region)		
	Nasalis (nose)	**Other Striking Features**	
	Nuchal (back of neck)	Alba (white)	
	Oculo- (eye)	Brevis (short)	
	Oris (mouth)	Gracilis (slender)	
	Palpebrae (eyelid)	Lata (wide)	
	Pollicis (thumb)	Latissimus (widest)	
	Popliteus (behind knee)	Longissimus (longest)	
	Psoas (loin)	Longus (long)	
	Radialis (radius)	Magnus (large)	
	Scapularis (scapula)	Major (larger)	
	Temporalis (temples)	Maximus (largest)	
	Thoracis (thoracic region)	Minimus (smallest)	
	Tibialis (tibia)	Minor (smaller)	
	Ulnaris (ulna)	-tendinosus (tendinous)	
	Uro- (urinary)	Vastus (great)	

*For other regional terms, refer to Figure 1-8, p. 14, which deals with anatomical landmarks.

muscles are called **brevis**; large ones are called **magnus** (big), **major** (bigger), or **maximus** (biggest); and small ones are called **minor** (smaller) or **minimus** (smallest).

Muscles visible at the body surface are external and often called **externus** or **superficialis** (superficial), whereas those lying beneath are internal, termed **internus** or **profundus**. Superficial muscles that position or stabilize an organ are called **extrinsic** muscles; those that operate within the organ are called **intrinsic** muscles.

The names of many muscles identify their origins and insertions. In such cases, the first part of the name indicates the origin and the second part the insertion. For example, the *genioglossus* originates at the chin (*geneion*) and inserts in the tongue (*glossus*).

Names that include flexor, extensor, retractor, and so on, indicate the primary function of the muscle. These are such common actions that the names almost always include other clues concerning the appearance or location of the muscle. For example, the *extensor carpi radialis longus* is a long muscle found along the radial (lateral) border of the forearm. When it contracts, its primary function is extension of the carpus (wrist).

A few muscles are named after the specific movements associated with special occupations or habits. For example, the *sartorius* (sar-TOR-ē-us) muscle is active when crossing the legs. Before sewing machines were invented, a tailor would sit on the floor cross-legged, and the name of the muscle was derived from *sartor*, the Latin word for "tailor." On the face, the *buccinator* (BUK-si-nā-tor) muscle compresses the cheeks, as when pursing the lips and blowing forcefully. *Buccinator* translates as "trumpet player." Finally, another facial muscle, the *risorius* (ri-SOR-ē-us), was supposedly named after the mood expressed. However, the Latin term *risor* means "laughter," while a more appropriate description for the effect would be "grimace."

√ **What type of muscle would you expect to find guarding the opening between the stomach and the small intestine?**

√ **What muscle would be the antagonist of the biceps brachii?**

√ **What does the name *flexor digitorum longus* tell you about this muscle?**

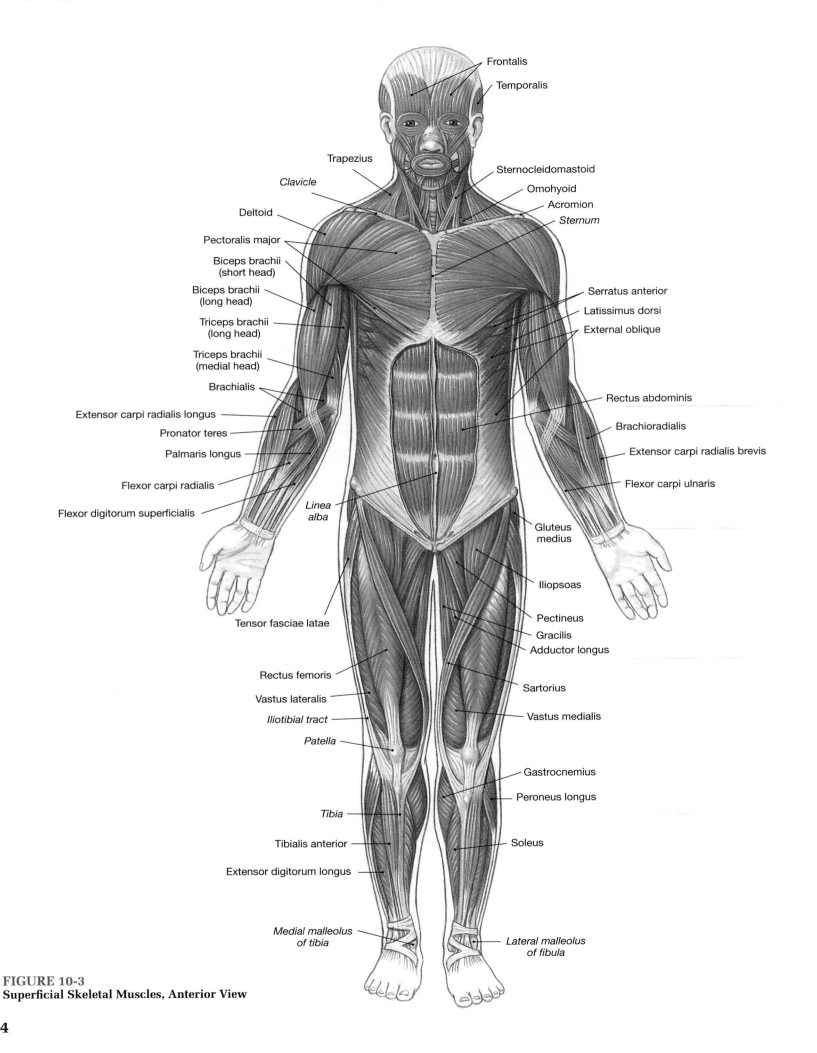

Frontalis
Temporalis
Trapezius
Sternocleidomastoid
Clavicle
Omohyoid
Acromion
Deltoid
Sternum
Pectoralis major
Biceps brachii (short head)
Serratus anterior
Biceps brachii (long head)
Latissimus dorsi
Triceps brachii (long head)
External oblique
Triceps brachii (medial head)
Brachialis
Rectus abdominis
Extensor carpi radialis longus
Brachioradialis
Pronator teres
Extensor carpi radialis brevis
Palmaris longus
Flexor carpi ulnaris
Flexor carpi radialis
Flexor digitorum superficialis
Linea alba
Gluteus medius
Tensor fasciae latae
Iliopsoas
Pectineus
Gracilis
Adductor longus
Rectus femoris
Vastus lateralis
Sartorius
Iliotibial tract
Vastus medialis
Patella
Gastrocnemius
Peroneus longus
Tibia
Tibialis anterior
Soleus
Extensor digitorum longus
Medial malleolus of tibia
Lateral malleolus of fibula

FIGURE 10-3
Superficial Skeletal Muscles, Anterior View

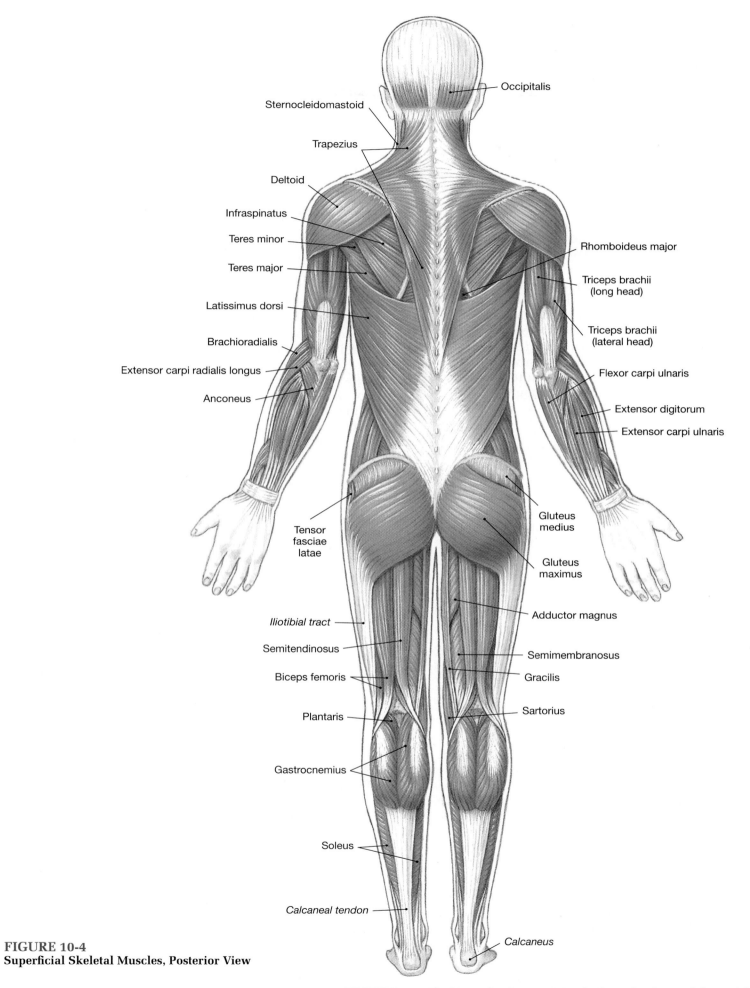

Occipitalis

Sternocleidomastoid

Trapezius

Deltoid

Infraspinatus

Teres minor

Teres major

Latissimus dorsi

Brachioradialis

Extensor carpi radialis longus

Anconeus

Rhomboideus major

Triceps brachii
(long head)

Triceps brachii
(lateral head)

Flexor carpi ulnaris

Extensor digitorum

Extensor carpi ulnaris

Tensor
fasciae
latae

Gluteus
medius

Gluteus
maximus

Adductor magnus

Iliotibial tract

Semitendinosus

Biceps femoris

Plantaris

Gastrocnemius

Semimembranosus

Gracilis

Sartorius

Soleus

Calcaneal tendon

Calcaneus

FIGURE 10-4
Superficial Skeletal Muscles, Posterior View

The separation of the skeletal system into axial and appendicular divisions provides a useful guideline for subdividing the muscular system as well. The **axial musculature** arises and inserts on the axial skeleton. It positions the head and spinal column and also moves the rib cage, assisting in the movements that make breathing possible. The axial muscles do not play a role in the movement or stabilization of the pectoral or pelvic girdles or the limbs. Roughly 60% of the skeletal muscles in the body are axial muscles. The **appendicular musculature** stabilizes or moves components of the appendicular skeleton, and these muscles will be considered in Chapter 11. The major axial and appendicular muscles are illustrated in Figures 10-3 and 10-4●.

Since our discussion relies heavily on an understanding of skeletal anatomy, you may find it helpful to review appropriate figures in Chapters 6 and 7 as we proceed. The relevant figures in those chapters are noted in the captions below.

The axial muscles fall into four logical groups based on location, function, or both. The groups do not always have distinct anatomical boundaries. For example, a function such as the extension of the spine involves muscles along the entire length of the spinal column.

1. The first group includes the *muscles of the head and neck* that are not associated with the spinal column. These muscles include those that move the face, tongue, and larynx. They are therefore responsible for verbal and nonverbal communication—laughing, talking, frowning, smiling, and whistling are examples. This group of muscles also performs movements associated with feeding, such as sucking, chewing, or swallowing, as well as contractions of the eye muscles that help us look around for something to eat.

2. The second group, the *muscles of the spine*, includes numerous flexors and extensors of the head, neck, and spinal column.

3. The third group, the *oblique* and *rectus* muscles, form the muscular walls of the thoracic and abdominopelvic cavities between the first thoracic vertebra and the pelvis. In the thoracic area, these muscles are partitioned by the ribs, but over the abdominal surface, they form broad muscular sheets. There are also oblique and rectus muscles in the neck. Although they do not form a complete muscular wall, they are included in this group because they share a common developmental origin.

4. The fourth group, the *muscles of the pelvic floor*, extend between the sacrum and pelvic girdle, forming the muscular *perineum* that closes the pelvic outlet.

To facilitate the review process, information concerning the origin, insertion, and action of each muscle has been summarized in tables. These tables also contain information about the *innervation* of individual muscles. **Innervation** refers to the identity of the nerve that controls the muscle. The names of the nerves provide clues to the distribution of the nerve or the site at which the nerve leaves the cranial or spinal cavities. ∞[p. 151] For example, the *facial nerve* innervates the facial musculature, and the various *spinal nerves* leave the vertebral canal via intervertebral foramina. ∞[p. 160] Figures 10-3 and 10-4● provide an overview of the superficial skeletal muscles of the human body.

Muscles of the Head and Neck

The muscles of the head and neck can be divided into several groups. The *muscles of facial expression*, the *muscles of mastication*, the *muscles of the tongue*, and the *muscles of the pharynx* originate on the skull or hyoid bone. Muscles involved with sight and hearing are also based on the skull. The *extrinsic eye muscles*—those associated with movements of the eye—will be considered here. The intrinsic eye muscles, nonstriated muscles located inside the eyeball, control the diameter of the pupil and the shape of the lens. They are discussed in Chapter 18 (general and special senses), along with muscles associated with the ear and hearing. The *anterior muscles of the neck* are concerned primarily with adjusting the position of the hyoid bone and larynx. These muscles include the *extrinsic laryngeal muscles*; the intrinsic laryngeal muscles, including those of the vocal cords, will be discussed in Chapter 24 (respiratory system).

Muscles of Facial Expression
(Figures 10-5 to 10-8, Table 10-2)

The muscles of facial expression originate on the surface of the skull. View Figures 10-5 to 10-8● as we describe their structure. Table 10-2 provides a detailed summary of their characteristics. At their insertions the epimysial fibers are woven into those of the superficial fascia and the dermis of the skin; when they contract, the skin moves. These muscles are innervated by the seventh cranial nerve, the *facial nerve*.

The largest group of facial muscles is associated with the mouth (Figure 10-5●). The **orbicularis oris** (OR-is) constricts the opening, while other muscles move the lips or the corners of the mouth. The **buccinator**, one of the muscles associated with the mouth, has two functions related to feeding (in addition to its importance to musicians). During chewing it cooperates with the masticatory muscles by moving food back across the teeth from the space inside the cheeks. In infants, the buccinator is responsible for producing the suction required for suckling at the breast.

Smaller groups of muscles control movements of the eyebrows and eyelids, the scalp, the nose, and the external ear. The *epicranius* (ep-i-KRĀ-nē-us; *epi-*, on + *kranion*, skull), or scalp, contains two muscles, the **frontalis** and the **occipitalis**, separated by a collagenous sheet, the **galea aponeurotica** (GĀ-lē-a ap-ō-nū-RŌ-ti-ka; *galea*, helmet, + *aponeurosis*) (Figure 10-7●). The superficial **platysma** (pla-TIZ-ma; *platys*, flat) covers the ventral surface of the neck, extending from the base of the neck to the periosteum of the mandible and the fascia at the corners of the mouth (Figures 10-5 and 10-6●).

TABLE 10-2 Muscles of Facial Expression

Region/Muscle	Origin	Insertion	Action	Innervation
Mouth				
Buccinator	Alveolar processes of maxilla and mandible	Blends into fibers of orbicularis oris	Compresses cheeks	Facial nerve (N VII)
Depressor labii inferioris	Mandible between the anterior midline and the mental foramen	Skin of lower lip	Depresses lip	As above
Levator labii superioris	Lower margin of orbit, superior to the infraorbital foramen	Orbicularis oris	Raises upper lip	As above
Mentalis	Incisive fossa of mandible	Skin of chin	Elevates and protrudes lower lip	As above
Orbicularis oris	Maxilla and mandible	Lips	Compresses, purses lips	As above
Risorius	Fascia surrounding parotid salivary gland	Angle of mouth	Draws corner of mouth to the side	As above
Depressor anguli oris	Anterolateral surface of mandibular body	Skin at angle of mouth	Depresses corner of mouth	As above
Zygomaticus major	Zygomatic bone near the zygomatic maxillary suture	Angle of mouth	Retracts and elevates corner of mouth	As above
Zygomaticus minor	Zygomatic bone posterior to zygomaticotemporal suture	Upper lip	Retracts and elevates corner of mouth	As above
Eye				
Corrugator supercilii	Orbital rim of frontal bone near nasal suture	Eyebrow	Pulls skin inferiorly and anteriorly; wrinkles brow	As above
Levator palpebrae superioris	Tendinous band around optic foramen	Upper eyelid	Raises upper lid	Oculomotor nerve (N III)[a]
Orbicularis oculi	Medial margin of orbit	Skin around eyelids	Closes eye	Facial nerve (N VII)
Nose				
Procerus	Nasal bone and lateral nasal cartilages	Aponeurosis at bridge of nose and skin of forehead	Moves nose, changes position, shape of nostrils	As above
Nasalis	Maxilla and alar cartilage of nose	Bridge of nose	Compresses bridge, depresses tip of nose; elevates corners of nostrils	As above
Ear (extrinsic)				
Temporoparietalis	Fascia around external ear	Galea aponeurotica	Tenses scalp, moves pinna of ear	As above
Scalp (Epicranius)				
Frontalis	Galea aponeurotica	Skin of eyebrow and bridge of nose	Raises eyebrows, wrinkles forehead	As above
Occipitalis	Superior nuchal line	Galea aponeurotica	Tenses and retracts scalp	As above
Neck				
Platysma	Upper thorax between cartilage of second rib and acromion of scapula	Mandible and skin of cheek	Tenses skin of neck, depresses mandible	As above

[a]This muscle originates in association with the extrinsic oculomotor muscles, so its innervation is unusual.

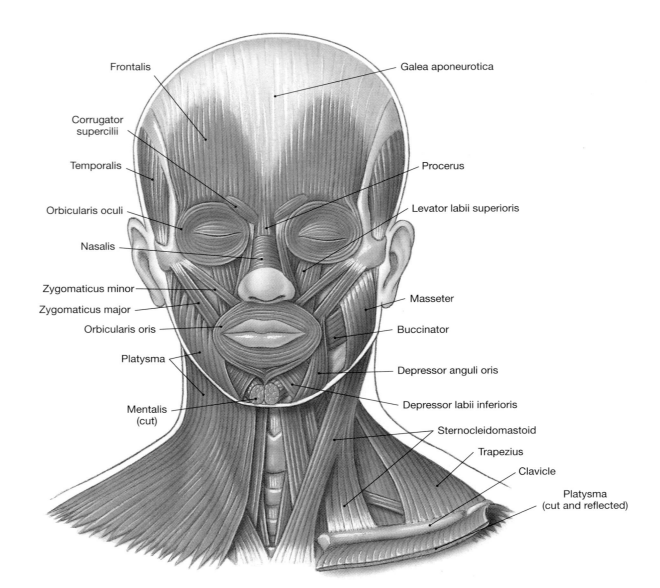

Frontalis

Corrugator
supercilii

Temporalis

Orbicularis oculi

Nasalis

Zygomaticus minor

Zygomaticus major

Orbicularis oris

Platysma

Mentalis
(cut)

Galea aponeurotica

Procerus

Levator labii superioris

Masseter

Buccinator

Depressor anguli oris

Depressor labii inferioris

Sternocleidomastoid

Trapezius

Clavicle

Platysma
(cut and reflected)

FIGURE 10-5
**Head and Neck Muscles,
Anterior View**

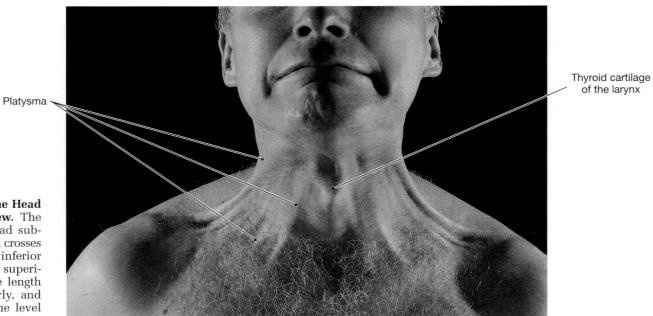

Platysma

Thyroid cartilage
of the larynx

FIGURE 10-6
**Surface Anatomy of the Head
and Neck, Anterior View.** The
platysma muscle, spread subcutaneously like a sheet, crosses
the whole length of the inferior
border of the mandible superiorly, crosses the whole length
of the clavicle inferiorly, and
extends inferiorly to the level
of the 2nd rib to the acromion.

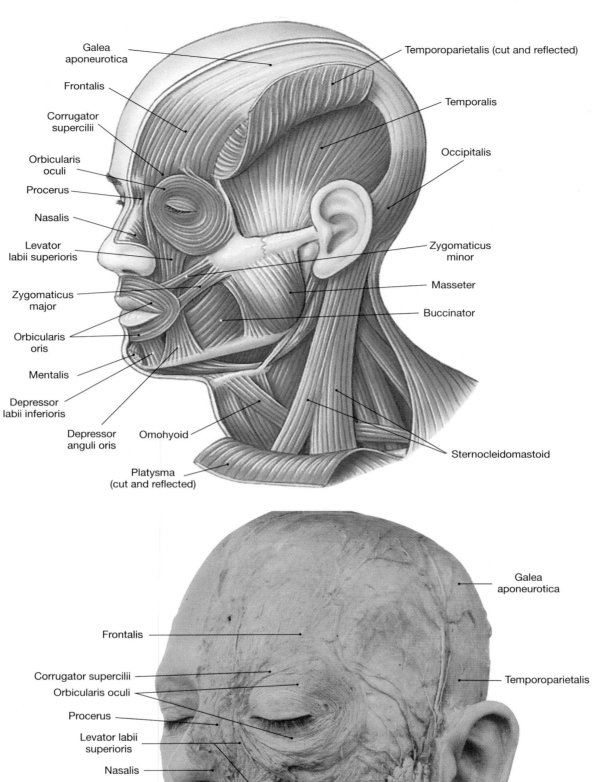

Galea
aponeurotica

Temporoparietalis (cut and reflected)

Frontalis

Temporalis

Corrugator
supercilii

Occipitalis

Orbicularis
oculi

Procerus

Nasalis

Levator
labii superioris

Zygomaticus
minor

Zygomaticus
major

Masseter

Buccinator

Orbicularis
oris

Mentalis

Depressor
labii inferioris

Depressor
anguli oris

Omohyoid

Platysma
(cut and reflected)

Sternocleidomastoid

FIGURE 10-7
Muscles of Facial Expression.
The *masseter* and *omohyoid*
muscles are not muscles of facial expression, but are identified because they are points of
reference. *See also Figure 6-3.*

Galea
aponeurotica

Frontalis

Corrugator supercilii

Temporoparietalis

Orbicularis oculi

Procerus

Levator labii
superioris

Nasalis

Zygomaticus minor

Parotid gland

Zygomaticus major

Orbicularis
oris

Masseter

Buccinator

Facial vein

Facial artery

Depressor
labii inferioris

Mandible

Sternocleidomastoid

FIGURE 10-8
**Cadaver Head and Neck, Lateral
View.** Note the superficial arteries,
veins, facial nerve, parotid gland,
and muscles.

Depressor
anguli oris

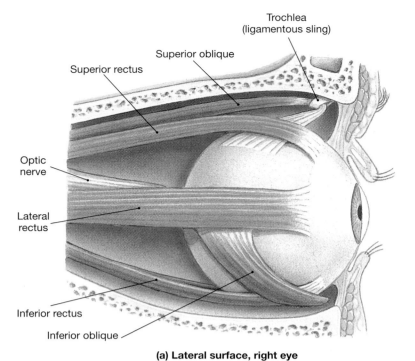

(a) Lateral surface, right eye

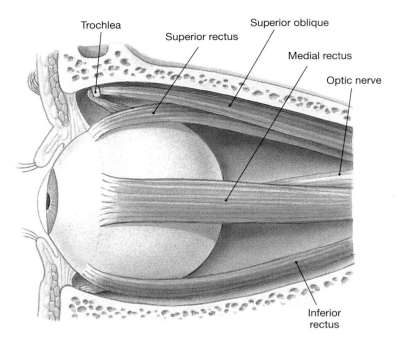

(b) Medial surface, right eye

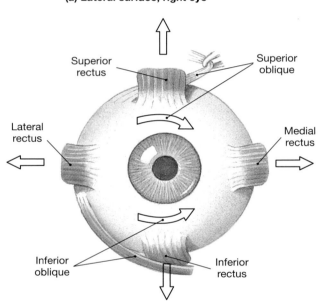

(c) Anterior view, right eye

Extrinsic Eye Muscles
(Figure 10-9, Table 10-3)

Six extrinsic eye muscles, sometimes called **oculomotor** (ok-ū-lō-MŌ-ter) **muscles**, originate on the surface of the orbit and control the position of each eye. These muscles are the **inferior rectus**, **lateral rectus**, **medial rectus**, **superior rectus**, **inferior oblique**, and **superior oblique** (Figure 10-9● and Table 10-3). The extrinsic eye muscles are innervated by the third (oculomotor), fourth (trochlear), and sixth (abducens) cranial nerves.

FIGURE 10-9
Extrinsic Eye Muscles. *See also Figure 6-3.* (a) Muscles on the lateral surface of the right eye. (b) Muscles on the medial surface of the right eye. (c) Anterior view of the right eye, showing the orientation of the oculomotor muscles and the directions of eye movement produced by contractions of the individual muscles.

TABLE 10-3 Extrinsic Eye Muscles

Muscle	Origin	Insertion	Action	Innervation
Inferior rectus	Sphenoid bone around optic foramen	Inferior, medial surface of eyeball	Eye looks down	Oculomotor nerve (N III)
Lateral rectus	As above	Lateral surface of eyeball	Eye rotates laterally	Abducens nerve (N VI)
Medial rectus	As above	Medial surface of eyeball	Eye rotates medially	Oculomotor nerve (N III)
Superior rectus	As above	Superior surface of eyeball	Eye looks up	As above
Inferior oblique	Maxilla at anterior portion of orbit	Inferior, lateral surface of eyeball	Eye rolls, looks up and to the side	As above
Superior oblique	Sphenoid bone around optic foramen	Superior, lateral surface of eyeball	Eye rolls, looks down and to the side	Trochlear nerve (N IV)

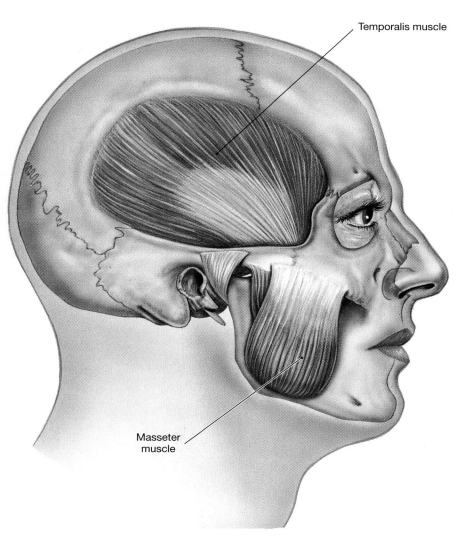

Temporalis muscle

Masseter muscle

(a) Lateral view

Muscles of Mastication
(Figure 10-10, Table 10-4)

The muscles of mastication (Figure 10-10● and Table 10-4) move the lower jaw at the temporomandibular joint. ∞[p. 149] The large **masseter** (ma-SĒ-ter) is the most powerful and important of the masticatory muscles. The **temporalis** (tem-po-RĀ-lis) assists in elevation of the mandible, whereas the **pterygoid** (TER-i-goyd) **muscles** used in various combinations can elevate or protract the mandible or slide it from side to side. These movements are important in maximizing the efficient use of the teeth while chewing foods of various consistencies. The muscles of mastication are innervated by the fifth cranial nerve, the *trigeminal nerve.*

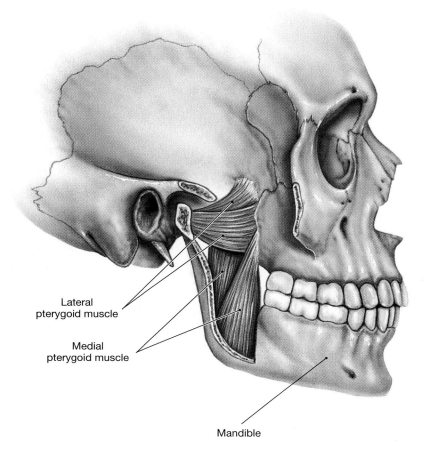

Lateral pterygoid muscle

Medial pterygoid muscle

Mandible

(b) Lateral view, pterygoid muscle exposed

FIGURE 10-10
Muscles of Mastication. *See also Figures 6-3 and 6-12.* (a) The temporalis and masseter are prominent muscles on the lateral surface of the skull. The temporalis passes medial to the zygomatic arch to insert on the coronoid process of the mandible. The masseter inserts on the angle and lateral surface of the mandible. (b) The location and orientation of the pterygoid muscles can be seen after removing the overlying muscles, along with a portion of the mandible.

TABLE 10-4   Muscles of Mastication

Muscle	Origin	Insertion	Action	Innervation
Masseter	Zygomatic arch	Lateral surface of mandibular ramus	Elevates mandible	Trigeminal nerve (N V), mandibular branch
Temporalis	Along temporal lines of skull	Coronoid process of mandible	Elevates mandible	As above
Pterygoids (medial and lateral)	Lateral pterygoid plate	Medial surface of mandibular ramus	*Medial*: Elevates the mandible and closes the jaws, or moves mandible from side to side	As above
			Lateral: Opens jaws, protudes mandible, or moves mandible from side to side	As above

Muscles of the Tongue
(Figure 10-11, Table 10-5)

These muscles have names ending in *glossus*, meaning "tongue." Once you can recall the structures referred to by *palato-*, *stylo-*, *genio-*, and *hyo-*, you shouldn't have much trouble with this group. The **palatoglossus** originates at the palate, the **styloglossus** at the styloid process, the **genioglossus** at the chin, and the **hyoglossus** at the hyoid bone (Figure 10-11●). These muscles, used in various combinations, move the tongue in the delicate and complex patterns necessary for speech and manipulate food within the mouth in preparation for swallowing. Most of these muscles are innervated by the *hypoglossal nerve*, a cranial nerve whose name indicates its function as well as its location (Table 10-5).

FIGURE 10-11
Muscles of the Tongue

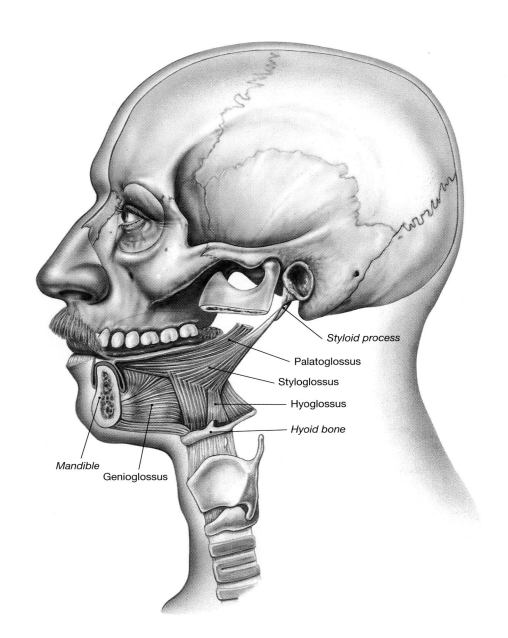

Styloid process

Palatoglossus

Styloglossus

Hyoglossus

Hyoid bone

Mandible

Genioglossus

TABLE 10-5	Muscles of the Tongue			
Muscle	*Origin*	*Insertion*	*Action*	*Innervation*
Genioglossus	Medial surface of mandible around chin	Body of tongue, hyoid bone	Depresses and protracts tongue	Hypoglossal nerve (N XII)
Hyoglossus	Body and greater cornu of hyoid bone	Side of tongue	Depresses and retracts tongue	As above
Palatoglossus	Anterior surface of soft palate	As above	Elevates tongue, depresses fleshy palate	Cranial branch of accessory nerve (N XI) via vagus nerve (N X)
Styloglossus	Styloid process of temporal bone	Via side to the tip and base of tongue	Retracts tongue, elevates sides	Hypoglossal nerve (N XII)

Muscles of the Pharynx
(Figure 10-12, Table 10-6)

The pharyngeal muscles are important in the initiation of swallowing. The **pharyngeal constrictors** begin the process of moving a mass of food into the esophagus. The **palatopharyngeus** (pal-āt-ō-fār-IN-jē-us) and **stylopharyngeus** (stī-lō-far-IN-jē-us) elevate the larynx, and the **palatal muscles** raise the soft palate and adjacent portions of the pharyngeal wall. The latter muscles also pull open the entrance to the auditory tube. As a result, swallowing repeatedly can help one adjust to pressure changes when flying or diving. These muscles are illustrated in Figure 10-12●, and additional information can be found in Table 10-6.

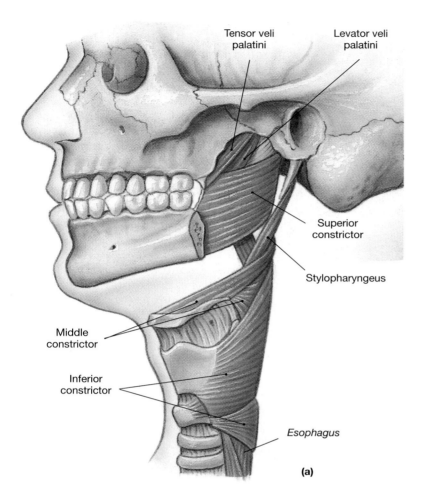

(a)

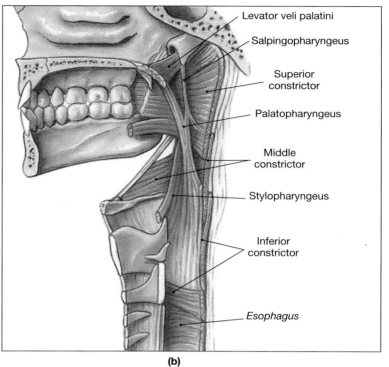

(b)

FIGURE 10-12
Muscles of the Pharynx. (a) Lateral view. (b) Midsagittal view.

TABLE 10-6   Muscles of the Pharynx

Muscle	Origin	Insertion	Action	Innervation
PHARYNGEAL CONSTRICTORS			Constrict pharynx to propel bolus into esophagus	Branches of pharyngeal plexus (N X)
Superior constrictor	Pterygoid processes of sphenoid, medial surfaces of mandible	Median raphe attached to occipital bone		
Middle constrictor	Cornua of hyoid	Median raphe		
Inferior constrictor	Cricoid and thyroid cartilages of larynx	Median raphe		
LARYNGEAL ELEVATORS[a]			Elevate larynx	Branches of pharyngeal plexus (N IX & X)
Palatopharyngeus	Soft palate	Thyroid cartilage		N X
Salpingopharyngeus	Cartilage around the inferior portion of the auditory tube	Thyroid cartilage		N X
Stylopharyngeus	Styloid process of temporal bone	Thyroid cartilage		N IX
PALATAL MUSCLES			Elevate soft palate	
Levator veli palatini	Petrous portion of temporal bone and tissues around the auditory tube	Soft palate		Cranial branch of accessory nerve (N XI) via vagus (N X)
Tensor veli palatini	Spine of sphenoid, tissues around the auditory tube	Soft palate		Mandibular branch of trigeminal nerve (N V)

[a]Assisted by the thyrohyoid, geniohyoid, stylohyoid, and hyoglossal muscles.

Anterior Muscles of the Neck
(Figures 10-5/10-7/10-13, Table 10-7)

The anterior muscles of the neck (Figure 10-13● and Table 10-7) control the position of the larynx, depress the mandible, tense the floor of the mouth, and provide a stable foundation for muscles of the tongue and pharynx. The **digastric** (dī-GAS-trik) has two bellies, as the name implies (*di-*, two + *gaster*, stomach). One belly extends from the chin to the hyoid, and the other continues from the hyoid to the mastoid portion of the temporal bone. This muscle opens the mouth by depressing the mandible. It overlies the broad, flat **mylohyoid** (mī-lō-HĪ-oyd), which provides a muscular floor to the mouth. The **stylohyoid** (stī-lō-HĪ-oyd) forms a muscular connection between the hyoid apparatus and the styloid process of the skull. The **sternocleidomastoid** (ster-nō-klī-dō-MAS-toid) extends from the clavicle and the sternum to the mastoid region of the skull (Figures 10-5, p. 258, and 10-7●, p. 259). As indicated in Table 10-7, these extensive muscles are innervated by more than one nerve, and specific regions can be made to contract independently. As a result, their actions are quite varied. The other members of this group are straplike muscles that run between the sternum and the chin.

(a) Anterior view

FIGURE 10-13
Anterior Muscles of the Neck.
(a) Anterior view of neck muscles. (b) Lateral view of neck, showing neck muscles and adjacent structures. *See also Figures 6-3 and 6-16.*

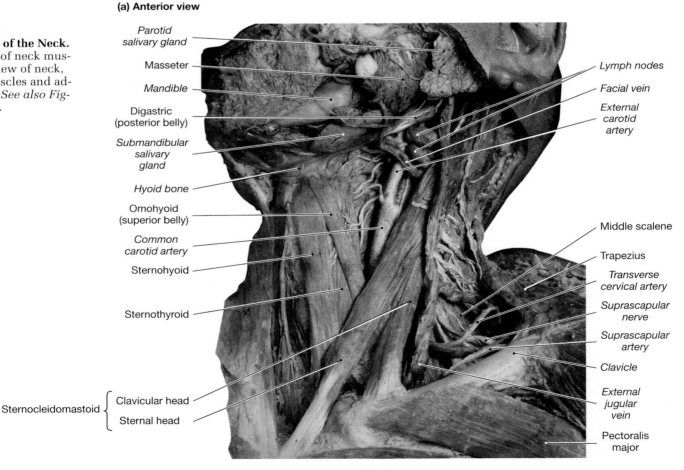

(b) Lateral view

TABLE 10-7 Anterior Muscles of the Neck

Muscle	Origin	Insertion	Action	Innervation
Digastric	Body and greater cornu of hyoid bone	Two bellies: *posterior* to mastoid region of temporal; *anterior* to inferior surface of mandible at chin	Depresses mandible and/or elevates larynx	Facial nerve (N VII) to posterior belly; trigeminal nerve (N V), mandibular branch, to anterior belly
Geniohyoid	Medial surface of mandible at chin	Body of hyoid bone	As above and pulls hyoid anteriorly	Cervical spinal nerves, via hypoglossal nerve (N XII)
Mylohyoid	Mylohyoid line of mandible	Median connective tissue band (raphe) that runs to hyoid bone	Elevates floor of mouth, elevates hyoid, and/or depresses mandible	Trigeminal nerve (N V), mandibular branch
Omohyoid	Central tendon attaches to clavicle and 1st rib	Two bellies: *superior* attaches to body of hyoid bone; *inferior* to superior margin of scapula	Depresses hyoid and larynx	Cervical spinal nerves (ansa cervicalis, C_1–C_3)
Sternohyoid	Clavicle and manubrium	Inferior surface of the body of the hyoid bone	As above	As above
Sternothyroid	Dorsal surface of manubrium and 1st costal cartilage	Thyroid cartilage of larynx	As above	As above
Stylohyoid	Styloid process of temporal bone	Lesser cornu of hyoid bone	Elevates larynx	Facial nerve (N VII)
Thyrohyoid	Thyroid cartilage of larynx	Inferior border of the greater cornu of the hyoid bone	Elevates thyroid, depresses hyoid	Cervical spinal nerves via hypoglossal nerve (N XII)
Sternocleidomastoid	Superior margins of manubrium and clavicle Two bellies: *clavicular head* attaches to sternal end of clavicle; *sternal head* attaches to manubrium of sternum	Mastoid process of skull	Together they flex the neck; alone one side bends head toward shoulder and turns face to opposite side	Accessory nerve (N XI) and cervical spinal nerves

Muscles of the Spine
(Figures 10-4/10-14, Table 10-8)

The muscles of the spine (Table 10-8) are covered by more superficial back muscles, such as the trapezius and latissimus dorsi (see Figure 10-4●, p. 255). The underlying muscles of the back can be divided into a *superficial layer* and a *deep layer*. The superficial layer includes the *splenius muscles* and the spinal extensors, or **erector spinae**. The names of the individual muscles within these groups provide useful information about their insertions. For example, a muscle with the name *capitis* inserts on the skull, whereas *cervicis* indicates an insertion on the upper cervical vertebrae and *thoracis* an insertion on the lower cervical and upper thoracic vertebrae. The erector spinae are divided into **spinalis, longissimus,** and **iliocostalis** divisions (Figure 10-14a●). These divisions are based on proximity to the vertebral column, with the spinalis group being the closest and the iliocostalis the farthest away. In the lower lumbar and sacral regions, the distinction between the longissimus and iliocostalis muscles becomes indistinct, and they are sometimes known as the **sacrospinalis** muscles. When contracting together, the erector spinae extend the spinal column. When the muscles on only one side contract, the spine is bent laterally.

Beneath the spinalis muscles, deep muscles of the spine interconnect and stabilize the vertebrae. These muscles, some-

times called the *transversospinalis group,* include the **semispinalis,** and the **multifidus, interspinales, intertransversarii, and rotatores** (Figure 10-14b●). By working in various combinations, they produce slight extension or rotation of the spinal column. They are also important in making delicate adjustments in the positions of individual vertebrae, and stabilizing adjacent vertebrae. If injured, these muscles can start a cycle of pain → muscle stimulation → contraction → pain. This can lead to pressure on adjacent spinal nerves, leading to sensory losses as well as limiting mobility. Many of the warmup and stretching exercises recommended before athletic events are intended to prepare these small but very important muscles for their supporting roles.

The muscles of the spine include many dorsal extensors, but few ventral flexors. The spinal column does not need a massive series of flexor muscles because (1) many of the large trunk muscles flex the spine when they contract, and (2) most of the body weight lies anterior to the spinal column, and gravity tends to flex the spine. However, there are a few spinal flexors associated with the anterior surface of the spinal column. In the neck (Figure 10-14c●) the **longus capitis** and the **longus colli** rotate or flex the neck, depending on whether the muscles of one or both sides are contracting. In the lumbar region, the large **quadratus lumborum** muscles flex the spine and depress the ribs.

TABLE 10-8 Muscles of the Spine

Group/Muscle	Origin	Insertion	Action	Innervation
SUPERFICIAL LAYER				
Splenius (Splenius capitis, splenius cervicis)	Spinous processes and ligaments connecting lower cervical and upper thoracic vertebrae	Mastoid process, occipital bone of skull, and upper cervical vertebra	The two sides act together to extend head; either alone rotates and tilts head to that side	Cervical spinal nerves
Erector Spinae				
Spinalis group				
Spinalis cervicis	Inferior portion of ligamentum nuchae and spinous process of C_7	Spinous process of axis	Extends neck	Cervical spinal nerves
Spinalis thoracis	Spinous processes of lower thoracic and upper lumbar vertebrae	Spinous processes of upper thoracic vertebrae	Extends spinal column	Thoracic and lumbar spinal nerves
Longissimus group				
Longissimus capitis	Processes of lower cervical and upper thoracic vertebrae	Mastoid process of temporal bone	The two sides act together to extend head; either alone rotates and tilts head to that side	Cervical and thoracic spinal nerves
Longissimus cervicis	Transverse processes of upper thoracic vertebrae	Transverse processes of middle and upper cervical vertebrae	As above	As above
Longissimus thoracis	Broad aponeurosis and at transverse processes of lower thoracic and upper lumbar vertebrae; joins iliocostalis to form "sacrospinalis"	Transverse processes of higher vertebrae and inferior surfaces of ribs	Extends and/or bends spine to the side	As above
Iliocostalis group				
Iliocostalis cervicis	Superior borders of vertebrosternal ribs near the angles	Transverse processes of middle and lower cervical vertebrae	Extends or bends neck, elevates ribs	As above
Iliocostalis thoracis	Superior borders of lower 7 ribs medial to the angles	Upper ribs and transverse process of last cervical vertebra	Stabilizes thoracic vertebrae in extension	Thoracic spinal nerves
Iliocostalis lumborum	Sacrospinal aponeurosis and iliac crest	Inferior surfaces of lower 7 ribs near their angles	Extends spine, depresses ribs	Lumbar spinal nerves
DEEP LAYER (TRANSVERSOSPINALIS)				
Semispinalis group				
Semispinalis capitis	Processes of lower cervical and upper thoracic vertebrae	Occipital bone, between nuchal lines	The two sides act together to extend head; either alone extends and tilts head to that side	Cervical spinal nerves
Semispinalis cervicis	Transverse processes of T_1–T_5 / T_6	Spinous processes of C_2–C_5	Extends vertebral column and rotates toward opposite side	As above
Semispinalis thoracis	Transverse processes of T_6–T_{10}	Spinous processes of C_5–T_4	As above	Thoracic spinal nerves
Multifidus	Sacrum and transverse processes of each vertebra	Spinous processes of the third or fourth more superior vertebrae	Extend vertebral column and rotate toward opposite side	Cervical, thoracic, and lumbar spinal nerves
Rotatores	Transverse processes of each vertebra	Spinous process of adjacent, more superior vertebra	Extend vertebral column and rotate toward opposite side	As above
Interspinales	Spinous processes of each vertebra	Spinous processes of preceding vertebra	Extend vertebral column	As above
Intertransversarii	Transverse processes of each vertebra	Transverse process of preceding vertebra	Bend the vertebral column laterally	As above
SPINAL FLEXORS				
Longus capitis	Transverse processes of cervical vertebrae	Base of the occipital bone	The two sides act together to bend head forward; either alone rotates head to that side	Cervical spinal nerves
Longus olli	Anterior surfaces of cervical and upper thoracic vertebrae	Transverse processes of upper cervical vertebrae	Flexes and/or rotates neck; limits hyperextension	As above
Quadratus lumborum	Iliac crest and iliolumbar ligament	Last rib and transverse processes of lumbar vertebrae	Together they depress ribs; one side alone flexes spine laterally	Thoracic and lumbar spinal nerves

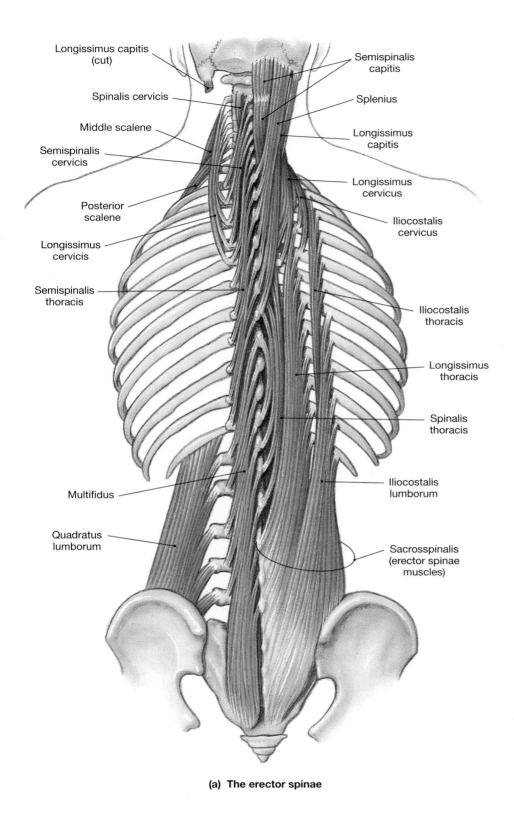

Longissimus capitis (cut)

Spinalis cervicis

Middle scalene

Semispinalis cervicis

Posterior scalene

Longissimus cervicis

Semispinalis thoracis

Multifidus

Quadratus lumborum

Semispinalis capitis

Splenius

Longissimus capitis

Longissimus cervicus

Iliocostalis cervicus

Iliocostalis thoracis

Longissimus thoracis

Spinalis thoracis

Iliocostalis lumborum

Sacrosspinalis (erector spinae muscles)

(a) The erector spinae

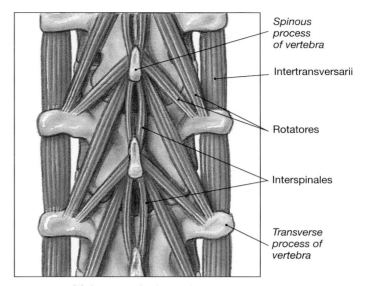

Spinous process of vertebra

Intertransversarii

Rotatores

Interspinales

Transverse process of vertebra

(b) Intervertebral muscles

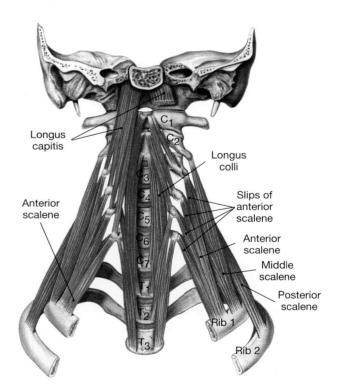

Longus capitis

Anterior scalene

C_1

C_2

C_3

C_4

C_5

C_6

C_7

T_1

T_2

T_3

Longus colli

Slips of anterior scalene

Anterior scalene

Middle scalene

Posterior scalene

Rib 1

Rib 2

(c) Muscles arising from the anterior surfaces of the superior vertebrae

FIGURE 10-14

Muscles of the Spine. *See also Figures 6-18 and 6-19.* (a) Superficial (right) and deep (left) muscles of the erector spinae group. (b) Intervertebral muscles. (c) Muscles arising from the anterior surfaces of the superior vertebrae.

TABLE 10-9 Oblique and Rectus Muscles

Group/Muscle	Origin	Insertion	Action	Innervation
OBLIQUE GROUP **Cervical region** Scalenes (anterior, middle, and posterior)	Transverse and costal processes of cervical vertebrae C_2–C_7	Superior surfaces of first two ribs	Elevate ribs, and/or flex neck; one side bends neck and rotates head and neck toward opposite side	Cervical spinal nerves
Thoracic region External intercostals	Inferior border of each rib	Superior border of the next rib, caudally	Elevate ribs	Intercostal nerves (branches of thoracic spinal nerves)
Internal intercostals	Superior border of each rib	Inferior border of the previous rib, cranially	Depress ribs	As above
Transversus thoracis	Posterior surface of sternum	Cartilages of ribs	As above	As above
Abdominal region External oblique	External and inferior borders of ribs 5–12	Linea alba and iliac crest	Compresses abdomen; depresses ribs; flexes, bends to side, or rotates spine	Intercostal iliohypogastric and ilioinguinal nerves
Internal oblique	Lumbodorsal fascia and iliac crest	Inferior surfaces of ribs 9–12, costal cartilages 8–10, linea alba, and pubis	As above	As above
Transversus abdominis	Cartilages of lower ribs, iliac crest, and lumbodorsal fascia	Linea alba and pubis	Compresses abdomen	As above
Serratus posterior (inferior)	Aponeurosis from spinous processes T_{10}–L_3	Inferior borders of ribs 8–12	Pulls ribs inferiorly; also pulls outward, opposing diaphragm	Thoracic nerves T_9–T_{12}
RECTUS GROUP **Cervical region**	*See muscles in Table 10-7 (except sternocleidomastoid)*			
Thoracic region Diaphragm	Xiphoid process, ribs 7–12 and associated costal cartilages, and anterior surfaces of lumbar vertebrae	Central tendinous sheet	Contraction expands thoracic cavity, compresses abdominopelvic cavity	Phrenic nerves
Abdominal region Rectus abdominis	Superior surface of pubis around symphysis	Inferior surfaces of costal cartilages (ribs 5–7) and xiphoid process of sternum	Depresses ribs, flexes vertebral column	Thoracic spinal nerves (T_7–T_{12})

Oblique and Rectus Muscles
(Figures 10-3/10-14/10-15, Table 10-9)

The muscles of the oblique and rectus groups (Table 10-9) lie between the vertebral column and the ventral midline (see Figure 10-3●, p. 254). The oblique muscles can compress underlying structures or rotate the spinal column, depending on whether one or both sides are contracting. The rectus muscles are important flexors of the spinal column, acting in opposition to the erector spinae. The oblique and rectus muscles are united by their common embryological origins. They can be divided into cervical, thoracic, and abdominal groups.

The oblique series includes the **scalenes** (SKĀ-lēnz) of the neck and the **intercostal** (in-ter-KOS-tul) and **transversus** muscles of the thoracic region. The scalenes include three muscles, *anterior*, *posterior*, and *middle*, that elevate the first two ribs and assist in flexion of the head (Figures 10-14a,c●). In the thorax, the oblique muscles lie between the ribs, and the **external intercostals** cover the **internal intercostals** (Figure 10-15a●). Both intercostal muscles are important in respiratory movements of the ribs. A small **transversus thoracis** crosses the inner surface of the rib cage and is covered by the serous membrane (*pleura*) that lines the pleural cavities. The sternum occupies the place where one might expect to find thoracic rectus muscles.

The same basic pattern of musculature extends unbroken across the abdominopelvic surface. Here the muscles are known as the **external obliques**, the **internal obliques**, the **transversus abdominis** (ab-DOM-i-nus), and the **rectus abdominis** (Figure 10-15a,c●). Collectively these muscles form the abdominal wall. An excellent way to observe the relationship of these muscles is to view them in horizontal section (Figure 10-15b●). The rectus abdominis begins at the xiphoid process and ends near the pubic symphysis. This muscle is longitudinally divided by a median collagenous partition, the **linea alba** (white line). The transverse **tendinous inscriptions** divide this muscle into repeated segments (Figure 10-15a,c●). The surface anatomy of the oblique and rectus muscles of the thorax and abdomen is shown in Figure 10-15d●.

The Diaphragm
(Figure 10-16)

The term *diaphragm* refers to any muscular sheet that forms a wall. When used without a modifier, however, the **diaphragm,**

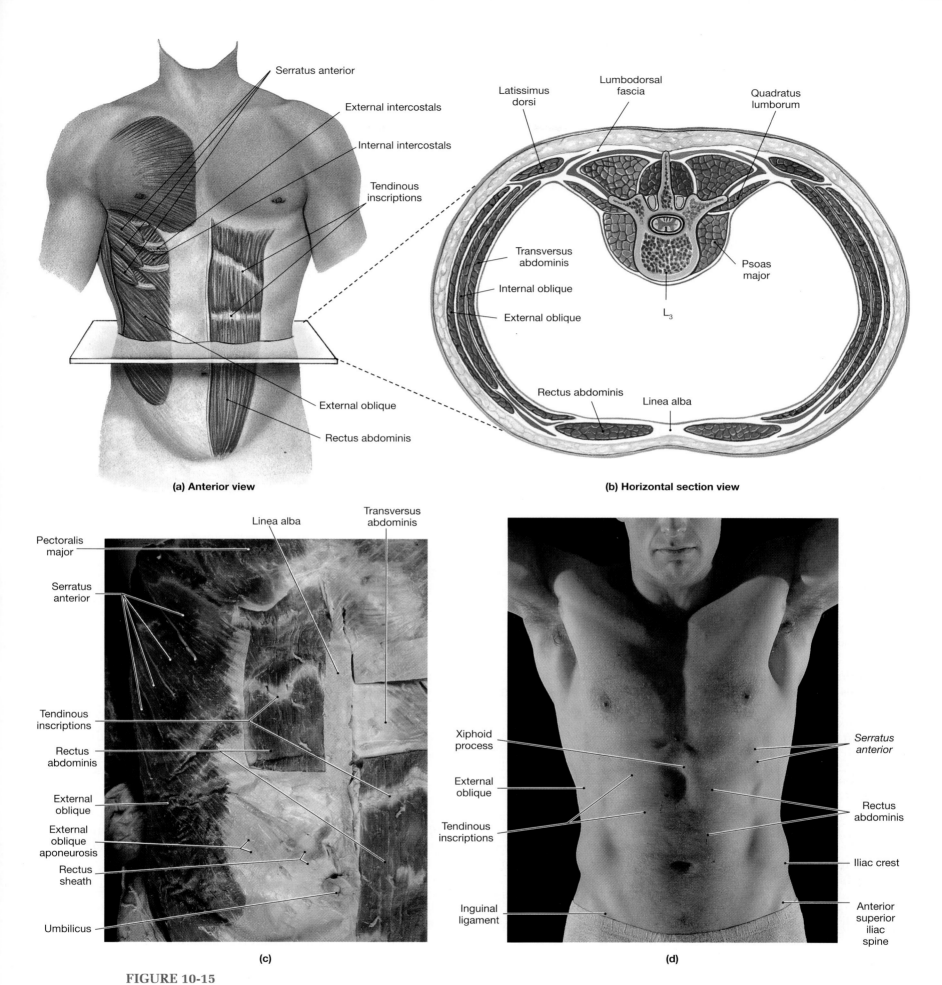

Serratus anterior
External intercostals
Internal intercostals
Tendinous inscriptions
External oblique
Rectus abdominis

(a) Anterior view

Latissimus dorsi
Lumbodorsal fascia
Quadratus lumborum
Transversus abdominis
Internal oblique
External oblique
Psoas major
L_3
Rectus abdominis
Linea alba

(b) Horizontal section view

Linea alba
Transversus abdominis
Pectoralis major
Serratus anterior
Tendinous inscriptions
Rectus abdominis
External oblique
External oblique aponeurosis
Rectus sheath
Umbilicus

(c)

Xiphoid process
External oblique
Tendinous inscriptions
Inguinal ligament
Serratus anterior
Rectus abdominis
Iliac crest
Anterior superior iliac spine

(d)

FIGURE 10-15
Oblique and Rectus Muscles. *See also Figures 6-14, 6-23, and 6-24.* (a) Anterior view of the trunk, showing superficial and deep members of the oblique and rectus groups, and the sectional plane shown in part (b). (b) Diagrammatic horizontal section through the abdominal region. (c) Cadaver, anterior superficial view of the abdominal wall. (d) Surface anatomy of the abdominal wall, anterior view. The *serratus anterior*, seen in parts (a) and (c), is an appendicular muscle detailed in Chapter 11.

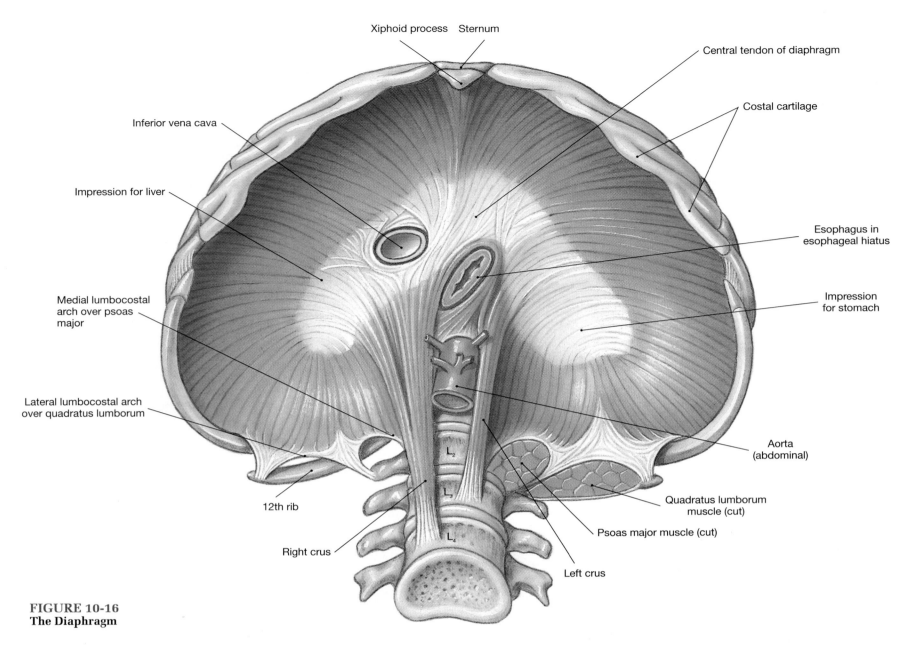

Xiphoid process   Sternum

Central tendon of diaphragm

Inferior vena cava

Costal cartilage

Impression for liver

Esophagus in esophageal hiatus

Medial lumbocostal arch over psoas major

Impression for stomach

Lateral lumbocostal arch over quadratus lumborum

Aorta (abdominal)

12th rib

Quadratus lumborum muscle (cut)

Right crus

Psoas major muscle (cut)

Left crus

L₂

L₃

L₄

FIGURE 10-16
The Diaphragm

or *diaphragmatic muscle*, specifies the muscular partition that separates the abdominopelvic and thoracic cavities (Figure 10-16●). This muscle is included here because it is developmentally linked to the other muscles of the chest wall. The diaphragm is a major muscle of respiration. ☤ *Hernias [p. 757]*

Muscles of the Pelvic Floor
(Figure 10-17, Table 10-10)

The muscles of the pelvic floor extend from the sacrum and coccyx to the ischium and pubis. These muscles (Figure 10-17● and Table 10-10, p. 274) (1) support the organs of the pelvic cavity, (2) flex the sacrum and coccyx, and (3) control the movement of materials through the urethra and anus.

The boundaries of the **perineum** (the pelvic floor and associated structures) are established by the inferior margins of the pelvis. If you draw a line between the ischial tuberosities, you will divide the perineum into two triangles, an **anterior** or **urogenital triangle** and a **posterior** or **anal triangle** (Figure 10-17b●). The superficial muscles of the anterior triangle are the muscles of the external genitalia. They overlie deeper muscles that strengthen the pelvic floor and encircle the urethra. These muscles

constitute the **urogenital diaphragm**, a muscular layer that extends between the pubic bones.

An even more extensive muscular sheet, the **pelvic diaphragm**, forms the muscular foundation of the anal triangle. This layer extends anteriorly superior to the urogenital diaphragm as far as the pubic symphysis.

The urogenital and pelvic diaphragms do not completely close the pelvic outlet, because the urethra, vagina, and anus pass through them to open on the external surface. Muscular sphincters surround their openings and permit voluntary control of urination and defecation. Muscles, nerves, and blood vessels also pass through the pelvic outlet as they travel to or from the legs.

√ **If you were contracting and relaxing your masseter muscle, what would you probably be doing?**

√ **Damage to the external intercostal muscles would interfere with what important process?**

√ **If someone hit you in your rectus abdominis muscle, how would your body position change?**

270

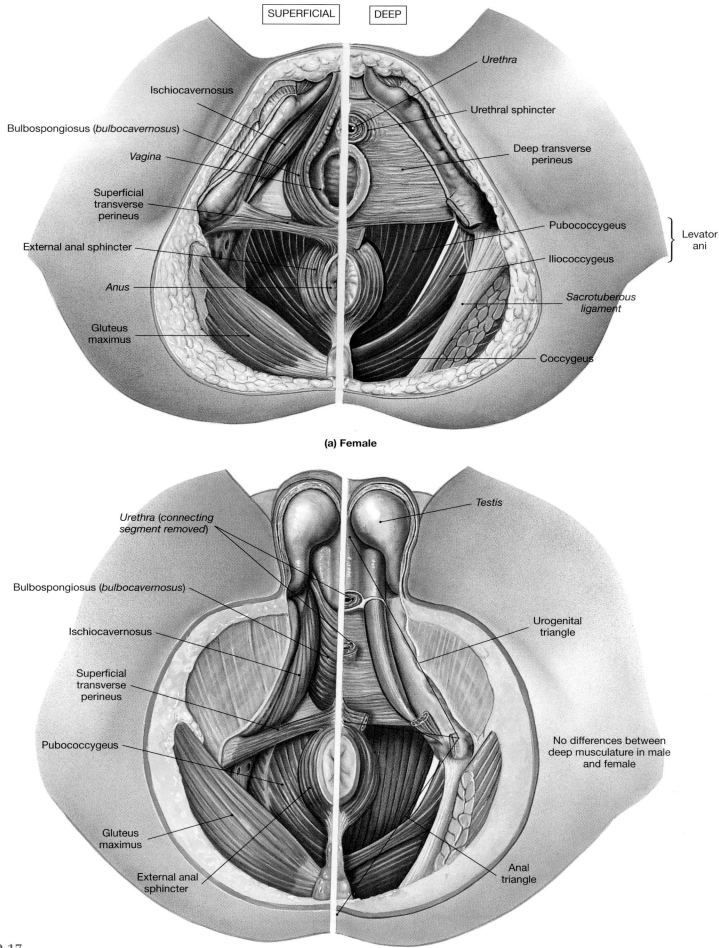

SUPERFICIAL   DEEP

Ischiocavernosus

Bulbospongiosus (*bulbocavernosus*)

Vagina

Superficial
transverse
perineus

External anal sphincter

Anus

Gluteus
maximus

Urethra

Urethral sphincter

Deep transverse
perineus

Pubococcygeus

Iliococcygeus

} Levator
ani

*Sacrotuberous
ligament*

Coccygeus

(a) Female

*Urethra (connecting
segment removed)*

Bulbospongiosus (*bulbocavernosus*)

Ischiocavernosus

Superficial
transverse
perineus

Pubococcygeus

Gluteus
maximus

External anal
sphincter

Testis

Urogenital
triangle

No differences between
deep musculature in male
and female

Anal
triangle

(b) Male

FIGURE 10-17
Muscles of the Pelvic Floor. *See also Figures 7-10 and 7-11.*

EMBRYOLOGY SUMMARY: DEVELOPMENT OF THE MUSCULAR SYSTEM

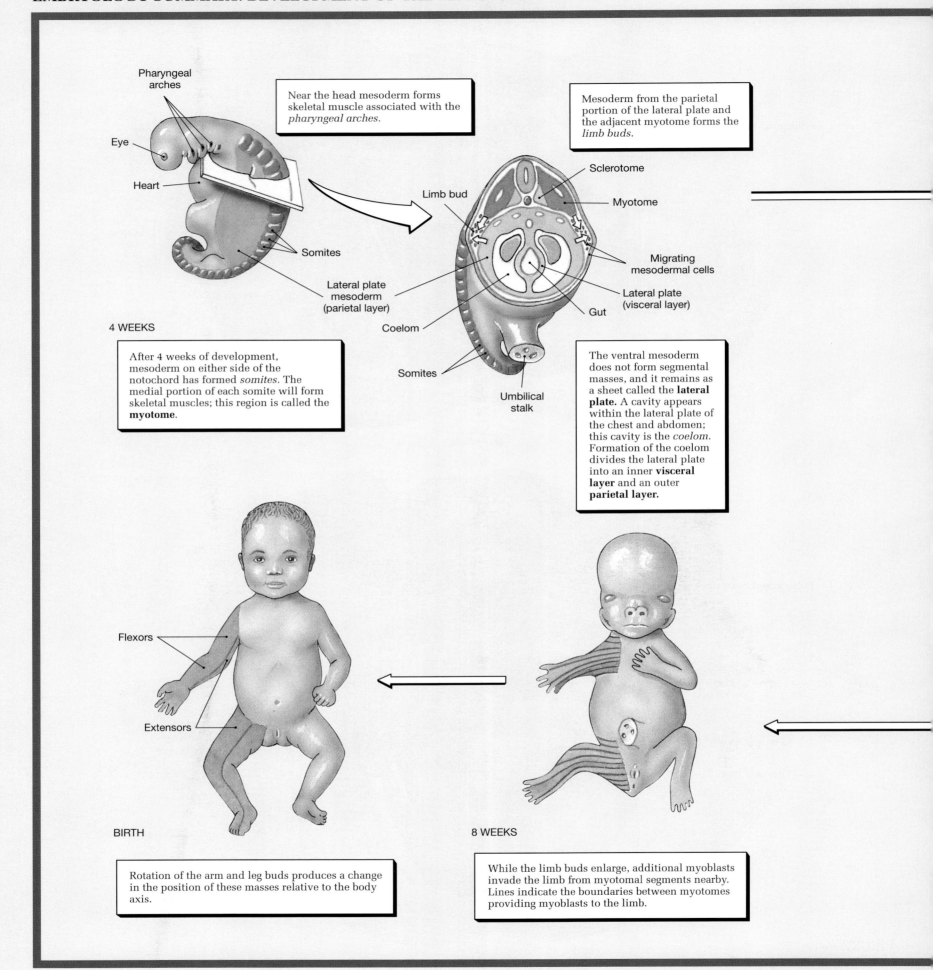

Pharyngeal arches

Near the head mesoderm forms skeletal muscle associated with the *pharyngeal arches*.

Mesoderm from the parietal portion of the lateral plate and the adjacent myotome forms the *limb buds*.

Eye

Heart

Limb bud

Sclerotome

Myotome

Somites

Migrating mesodermal cells

Lateral plate mesoderm (parietal layer)

Lateral plate (visceral layer)

Coelom

Gut

4 WEEKS

Somites

After 4 weeks of development, mesoderm on either side of the notochord has formed *somites*. The medial portion of each somite will form skeletal muscles; this region is called the **myotome.**

Umbilical stalk

The ventral mesoderm does not form segmental masses, and it remains as a sheet called the **lateral plate.** A cavity appears within the lateral plate of the chest and abdomen; this cavity is the *coelom.* Formation of the coelom divides the lateral plate into an inner **visceral layer** and an outer **parietal layer.**

Flexors

Extensors

BIRTH

8 WEEKS

Rotation of the arm and leg buds produces a change in the position of these masses relative to the body axis.

While the limb buds enlarge, additional myoblasts invade the limb from myotomal segments nearby. Lines indicate the boundaries between myotomes providing myoblasts to the limb.

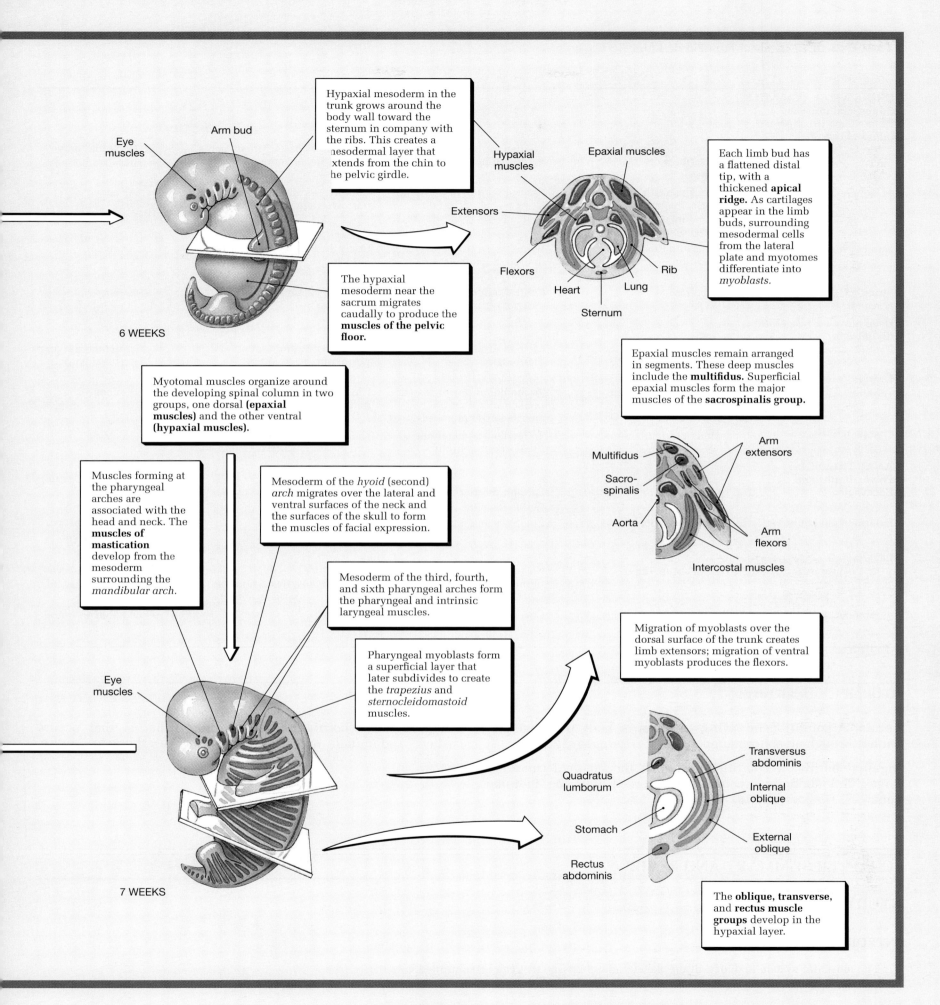

Eye muscles

Arm bud

Hypaxial mesoderm in the trunk grows around the body wall toward the sternum in company with the ribs. This creates a mesodermal layer that extends from the chin to the pelvic girdle.

Hypaxial muscles

Epaxial muscles

Each limb bud has a flattened distal tip, with a thickened **apical ridge.** As cartilages appear in the limb buds, surrounding mesodermal cells from the lateral plate and myotomes differentiate into *myoblasts.*

Extensors

Flexors

Rib

Heart

Lung

Sternum

The hypaxial mesoderm near the sacrum migrates caudally to produce the **muscles of the pelvic floor.**

6 WEEKS

Epaxial muscles remain arranged in segments. These deep muscles include the **multifidus.** Superficial epaxial muscles form the major muscles of the **sacrospinalis group.**

Myotomal muscles organize around the developing spinal column in two groups, one dorsal **(epaxial muscles)** and the other ventral **(hypaxial muscles).**

Multifidus

Arm extensors

Sacro-spinalis

Aorta

Arm flexors

Muscles forming at the pharyngeal arches are associated with the head and neck. The **muscles of mastication** develop from the mesoderm surrounding the *mandibular arch.*

Mesoderm of the *hyoid* (second) *arch* migrates over the lateral and ventral surfaces of the neck and the surfaces of the skull to form the muscles of facial expression.

Intercostal muscles

Mesoderm of the third, fourth, and sixth pharyngeal arches form the pharyngeal and intrinsic laryngeal muscles.

Migration of myoblasts over the dorsal surface of the trunk creates limb extensors; migration of ventral myoblasts produces the flexors.

Pharyngeal myoblasts form a superficial layer that later subdivides to create the *trapezius* and *sternocleidomastoid* muscles.

Eye muscles

Transversus abdominis

Quadratus lumborum

Internal oblique

Stomach

External oblique

Rectus abdominis

7 WEEKS

The **oblique, transverse, and rectus muscle groups** develop in the hypaxial layer.

CHAPTER 10 **The Muscular System: Muscle Organization and the Axial Musculature** 273

TABLE 10-10   Muscles of the Pelvic Floor

Group/Muscle	Origin	Insertion	Action	Innervation
UROGENITAL TRIANGLE **Superficial muscles** Bulbospongiosus (or *bulbocavernosus*):				
Male	Collagen sheath at base of penis; fibers cross over urethra	Median raphe and central tendon of perineum	Compresses base, stiffens penis, ejects urine or semen	Pudendal nerve, perineal branch
Female	Collagen sheath at base of clitoris; fibers run on either side of urethral and vaginal openings	Central tendon of perineum	Compresses and stiffens clitoris, narrows vaginal opening	As above
Ischiocavernosus	Ramus and tuberosity of ischium	Symphysis pubis anterior to base of penis or clitoris	Compresses and stiffens penis or clitoris	As above
Superficial transverse perineus	Ischial ramus	Central tendon of perineum	Stabilizes central tendon of perineum	As above
Deep muscles: urogenital diaphragm Deep transverse perineus	Ischial ramus	Median raphe of urogenital diaphragm	Stabilizes central tendon of perineum	Pudendal nerve, perineal branch
Urethral sphincter: Male	Ischial and pubic rami	To median raphe at base of penis; inner fibers encircle urethra	Closes urethra, compresses prostate and bulbourethral glands	As above
Female	Ischial and pubic rami	To median raphe; inner fibers encircle urethra	Closes urethra, compresses vagina and greater vestibular glands	As above
ANAL TRIANGLE **Pelvic diaphragm** Coccygeus	Ischial spine	Lateral, inferior borders of the sacrum	Flexes coccyx and coccygeal vertebrae	Lower sacral nerves (S_4–S_5)
External anal sphincter	Via tendon from coccyx	Encircles anal opening	Closes anal opening	Pudendal nerve, inferior rectal branch
Levator ani: Iliococcygeus	Ischial spine, pubis	Coccyx and median raphe	Tenses floor of pelvis, supports pelvic organs, flexes coccyx, elevates and retracts anus	Pudendal nerve and lower sacral nerves
Pubococcygeus	Inner margins of pubis	Coccyx and median raphe	As above	As above

Related Clinical Terms

hernia: A condition involving an organ or body part that protrudes through an abnormal opening. ☤ *Hernias [p. 758]*

inguinal hernia: A condition in which the inguinal canal enlarges and abdominal contents are forced into the inguinal canal. ☤ *Hernias [p. 758]*

diaphragmatic hernia (hiatal hernia): A hernia that occurs when abdominal organs slide into the thoracic cavity. ☤ *Hernias [p. 758]*

CHAPTER SUMMARY AND REVIEW

STUDY OUTLINE

INTRODUCTION *[p. 249]*

1.   The **muscular system** includes all the skeletal muscle tissue that can be controlled voluntarily. The analysis of biological systems in mechanical terms is the study of **biomechanics**.

ORGANIZATION OF SKELETAL MUSCLE FIBERS *[p. 249]*

Related Key Terms

1. A muscle can be classified according to the arrangement of fibers and fascicles as a parallel muscle, convergent muscle, pennate muscle, or circular muscle, or sphincter.

Parallel Muscles *[p. 250]*
2. In a **parallel muscle**, the fascicles are parallel to the long axis of the muscle. Most of the skeletal muscles in the body are parallel muscles, such as the *biceps brachii (see Figure 10-1)*.

body

Convergent Muscles *[p. 250]*
3. In a **convergent muscle**, the muscle fibers are based over a broad area, but all the fibers come together at a common attachment site. The *pectoralis group* of the chest is a good example of this type of muscle *(see Figure 10-1)*.

raphe

Pennate Muscles *[p. 250]*
4. In a **pennate muscle**, one or more tendons run through the body of the muscle, and the fascicles form an oblique angle to the tendon. Contraction of pennate muscles generates more tension than that of parallel muscles of the same size. A pennate muscle may be **unipennate, bipennate,** or **multipennate** *(see Figure 10-1)*.

Circular Muscles *[p. 250]*
5. In a **circular muscle (sphincter)**, the fibers are concentrically arranged around an opening or recess *(see Figure 10-1)*.

LEVERS: A SYSTEMS DESIGN FOR MOVEMENT *[p. 250]*

1. A **lever** can change the direction, speed, or distance of muscle movements, and it can modify the force applied to them.

fulcrum

Classes of Levers *[p. 250]*
2. Levers may be classified as **first-class, second-class,** or **third-class levers**; the last are the most common type of lever in the body *(see Figure 10-2)*.

MUSCLE TERMINOLOGY *[p. 252]*

Origins and Insertions *[p. 252]*
1. Typically, the origin remains stationary and the insertion moves, or the origin is proximal to the insertion.

Actions *[p. 252]*
2. Each muscle may be identified by its **origin, insertion,** and **primary action**. A muscle may be classified as a **prime mover**, or **agonist**, a **synergist**, or an **antagonist**.

Names of Skeletal Muscles *[p. 252]*
3. The names of muscles often provide clues to their location, orientation, or function *(see Table 10-1)*.

THE AXIAL MUSCULATURE *[p. 256]*

1. The **axial musculature** arises on the axial skeleton; it positions the head and spinal column and moves the rib cage. The **appendicular musculature** stabilizes or moves components of the appendicular skeleton *(see Figures 10-3/10-4)*.
2. The axial muscles fall into logical groups based on location and/or function.
3. **Innervation** refers to the identity of the nerve that controls a muscle.

Muscles of the Head and Neck *[p. 256]*
4. The muscles of facial expression are: **orbicularis oris, buccinator,** *epicranius* (**frontalis** and **occipitalis** muscles), and **platysma** *(see Figures 10-5 to 10-8 and Table 10-2)*.

galea aponeurotica

5. The six **oculomotor muscles**, or extrinsic eye muscles, control external eye movements: **inferior** and **superior recti; lateral** and **medial recti; superior** and **inferior obliques** *(see Figure 10-9 and Table 10-3)*.
6. The muscles of mastication (chewing) are: **masseter, temporalis, pterygoid muscles** *(see Figure 10-10 and Table 10-4)*.
7. The muscles of the tongue are necessary for speech and swallowing, and they assist in mastication; they are the **genioglossus, hyoglossus, palatoglossus,** and **styloglossus** *(see Figure 10-11 and Table 10-5)*.
8. Muscles of the pharynx initiate the swallowing process. These muscles include the **pharyngeal constrictors, palatopharyngeus, stylopharyngeus,** and **palatal muscles** *(see Figure 10-12 and Table 10-6)*.
9. The anterior muscles of the neck control the position of the larynx, depress the mandible, and provide a foundation for the muscles of the tongue and pharynx. The extrinsic muscles of the larynx are: **digastric, geniohyoid, mylohyoid, omohyoid, sternohyoid, sternothyroid, stylohyoid, thyrohyoid** and **sternocleidomastoid** *(see Figures 10-5/10-7/10-13 and Table 10-7)*.

Muscles of the Spine [p. 265]

10. The superficial muscles of the spine can be classified into the **spinalis**, **longissimus**, and **iliocostal-is** groups. In the lower lumbar and sacral regions, the longissimus and iliocostalis are sometimes called the **sacrospinalis** muscles *(see Figure 10-14 and Table 10-8)*.

11. Other muscles of the spine include the **transverse muscles**, the **longus capitis**, the **longus colli**, and the **quadratus lumborum** muscles *(see Figure 10-14 and Table 10-8)*.

Oblique and Rectus Muscles [p. 268]

12. The oblique muscles include the **scalenes** and the **intercostal** and **transversus** muscles. The **external intercostals** and the **internal intercostals** are important in respiratory movements of the ribs *(see Figure 10-15 and Table 10-9)*.

13. Also important to respiration is the **diaphragm** (*diaphragmatic muscle*), which separates the abdominopelvic and thoracic cavities *(see Figure 10-16)*.

Muscles of the Pelvic Floor [p. 270]

14. The **perineum** can be divided into an **anterior** or **urogenital triangle** and a **posterior** or **anal triangle**. The pelvic floor consists of the **urogenital diaphragm** and the **pelvic diaphragm** *(see Figure 10-17 and Table 10-10)*.

Related Key Terms

erector spinae • semispinalis • multifidus • interspinales • intertransversarii • rotatores

transversus thoracis • external obliques • internal obliques • transversus abdominis • rectus abdominis • linea alba • tendinous inscriptions

---

1 REVIEW OF CHAPTER OBJECTIVES

1. Describe the arrangement of fascicles in the various types of muscles, and explain the resulting functional differences.
2. Describe the different classes of levers and how they make muscles more efficient.
3. Predict the actions of a muscle on the basis of its origin and insertion.
4. Explain how muscles interact to produce or oppose movements.
5. Use the name of a muscle to help identify and remember its location, appearance, and function.
6. Identify and locate the principal axial muscles of the body, together with their origins and insertions, and describe their actions.

2 REVIEW OF CONCEPTS

1. How does the structure of smooth muscle differ from that of striated muscle?
2. What are the main functions of skeletal muscle?
3. What is the role of connective tissue in the organization of skeletal muscle?
4. How are tendons and aponeuroses related to the connective tissues that form part of the muscles?
5. Why are some muscles connected to origins and insertions by tendons and others by aponeuroses?
6. What is the function of a neuromuscular junction?
7. How is an electrical signal transmitted along muscle fibers?
8. How do the actin and myosin filaments interact to cause a muscle fiber to contract?
9. What is the role of the "zone of overlap" in the structure of skeletal muscle?
10. What is the role of ATP in allowing muscle fibers to contract?

11. How does the electrical stimulation of a muscle affect the distribution of calcium?
12. What is the role of the T-tubules in propagating the electrical impulse in the muscle?
13. What is the relationship between a motor neuron and the muscle fibers of a motor unit?
14. What is the role of muscle spindles in controlling tone in skeletal muscles?
15. Why can slow muscle fibers sustain activity for a longer period of time than fast fibers?
16. What is the relationship between muscle fiber type and muscle action?

3 CRITICAL THINKING AND CLINICAL APPLICATION QUESTIONS

1. A man had surgery to remove a stone from a salivary gland duct on the right side of his face. Unfortunately, during the surgery, branches of the facial nerve (N VII) were damaged. What effects would this have on the facial musculature?
2. At the beginning of the semester, a generous anatomy student not only moves all of his belongings into his new apartment, but assists several of the women moving in next door. Although he is merely tired at the end of the day, the next morning he has an extremely stiff and sore back. Which muscles are the sore ones, and what is the anatomical basis for the soreness?
3. If the student in the question above had experienced severe pain in the groin area upon lifting a heavy weight, what injury might he have incurred?
4. A woman in her forties has been experiencing recurring heartburn, sometimes has difficulty in swallowing, and often has a feeling of fullness or pressure in the substernal region. What condition might she have, and how can it be diagnosed?

11

The Muscular System: The Appendicular Musculature

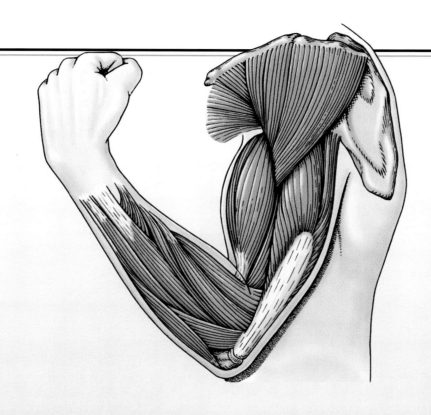

CHAPTER OUTLINE AND OBJECTIVES

In this chapter we will describe the **appendicular musculature**. These muscles are responsible for stabilizing the pectoral and pelvic girdles and for moving the upper and lower limbs. Appendicular muscles account for roughly 40% of the skeletal muscles in the body.

This discussion assumes an understanding of skeletal anatomy, and you may find it helpful to review figures in Chapters 6 and 7 as you proceed. The appropriate figures are referenced in the figure captions throughout this chapter.

The appendicular musculature stabilizes and moves the appendicular skeleton. There are two major groups of appendicular muscles: (1) the muscles of the shoulders and upper extremities and (2) the muscles of the pelvic girdle and lower extremities. The functions and required ranges of motion are very different from one group to another. In addition to increasing the mobility of the upper limb, the muscular connections between the pectoral girdle and the axial skeleton must act as shock absorbers. For example, people who are jogging can still perform delicate hand movements because the muscular con-

nections between the axial and appendicular skeleton smooth out the bounces in their stride. In contrast, the pelvic girdle has evolved to transfer weight from the axial to the appendicular skeleton. A muscular connection would reduce the efficiency of the transfer. The emphasis is on strength rather than versatility.

MUSCLES OF THE SHOULDER AND UPPER EXTREMITY
(Figures 11-1 to 11-11)

Muscles associated with the shoulders and upper extremities can be divided into four groups: (1) muscles that position the pectoral girdle (Figures 11-2 and 11-3●), (2) muscles that move the arm (Figure 11-5●), (3) muscles that move the forearm and hand (Figures 11-7, 11-8, and 11-9●), and (4) muscles that move the hand and fingers (Figures 11-10 and 11-11●). As we describe

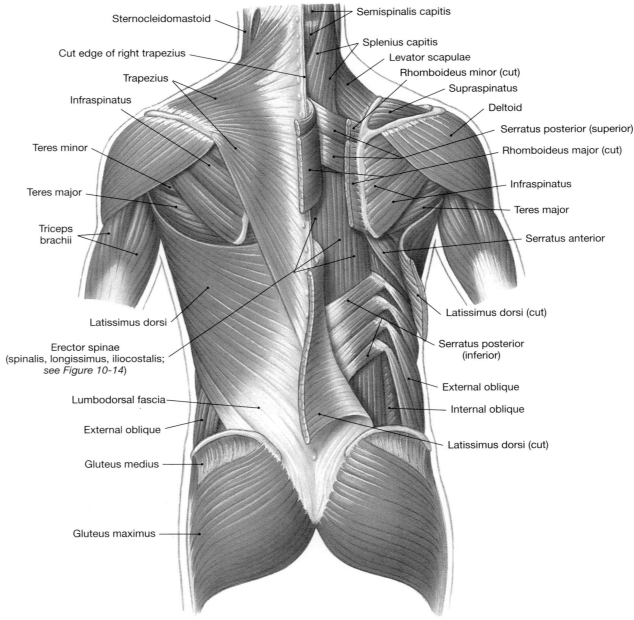

FIGURE 11-1
Superficial and Deep Muscles of the Neck, Shoulder, Back, and Gluteal Regions, Posterior View

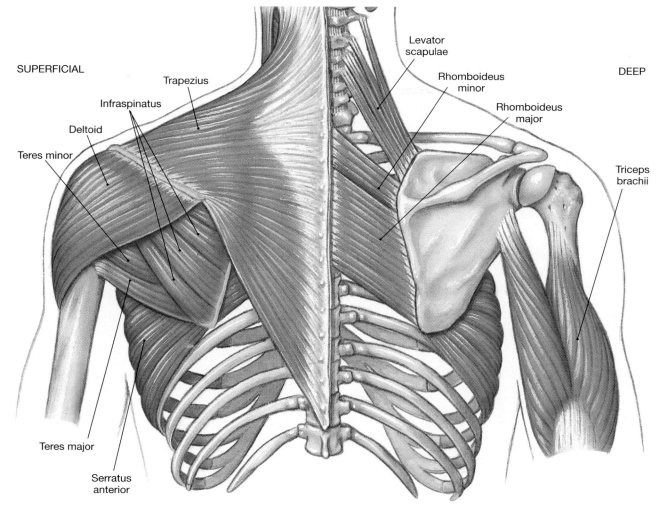

SUPERFICIAL

Trapezius

Infraspinatus

Deltoid

Teres minor

Levator scapulae

Rhomboideus minor

Rhomboideus major

DEEP

Triceps brachii

Teres major

Serratus anterior

FIGURE 11-2

Muscles of the Shoulder Girdle. *See also Figures 6-25, 7-4, and 8-9.* Posterior view, showing superficial muscles and deep muscles of the pectoral girdle.

the various muscles of the shoulders and upper extremities, refer first to Figure 11-1●, then to Figure 11-4● for the general location of the muscle under study with respect to the surrounding musculature.

Muscles That Position the Pectoral Girdle
(Figures 11-1 to 11-4, Table 11-1)

The large **trapezius** (tra-PĒ-zē-us) muscles cover the back and portions of the neck, extending to the base of the skull. These muscles originate along the middle of the neck and back and insert upon the clavicles and the scapular spines. Together, these triangular muscles form a broad diamond (Figures 11-1 and 11-2●). The trapezius muscles are innervated by more than one nerve (Table 11-1), and specific regions can be made to contract independently. As a result, their actions are quite varied.

Removing the trapezius reveals the **rhomboideus** (rom-BOY-dē-us) muscles and the **levator scapulae** (SKAP-yoo-lē) (Figures 11-1 and 11-2●). These muscles are attached to the dorsal surfaces of the cervical and thoracic vertebrae. They insert along the vertebral border of each scapula, between the superior and inferior angles. Contraction of the rhomboideus muscles adducts (retracts) the scapula, pulling it toward the center of the back. The levator scapulae, as the name implies, elevates the scapula, as in shrugging the shoulders.

On the chest, the **serratus** (se-RĀ-tus) **anterior** originates along the anterior surfaces of several ribs (Figures 11-3 and 11-4●). This fan-shaped muscle inserts along the anterior margin of the vertebral border of the scapula. When the serratus anterior contracts, it abducts (protracts) the scapula and swings the shoulder forward.

Two other deep chest muscles arise along the ventral surfaces of the ribs. The **subclavius** (sub-KLĀ-vē-us; *sub*, below + *clavius*, clavicle) inserts upon the inferior border of the clavicle (Figures 11-3 and 11-4●). When it contracts, it depresses and protracts the scapular end of the clavicle. Because ligaments connect this end to the shoulder joint and scapula, those structures move as well. The **pectoralis minor** (pek-tō-RĀ-lis) attaches to the coracoid process of the scapula (Figures 11-3 and 11-4●). Its contraction usually complements that of the subclavius. Table 11-1 identifies the muscles that move the shoulder girdle and the nerves that innervate those muscles.

Muscles That Move the Arm
(Figures 11-1/11-4 to 11-6, Table 11-2)

The muscles that move the arm are easiest to remember when grouped by primary actions. Some of these muscles are best

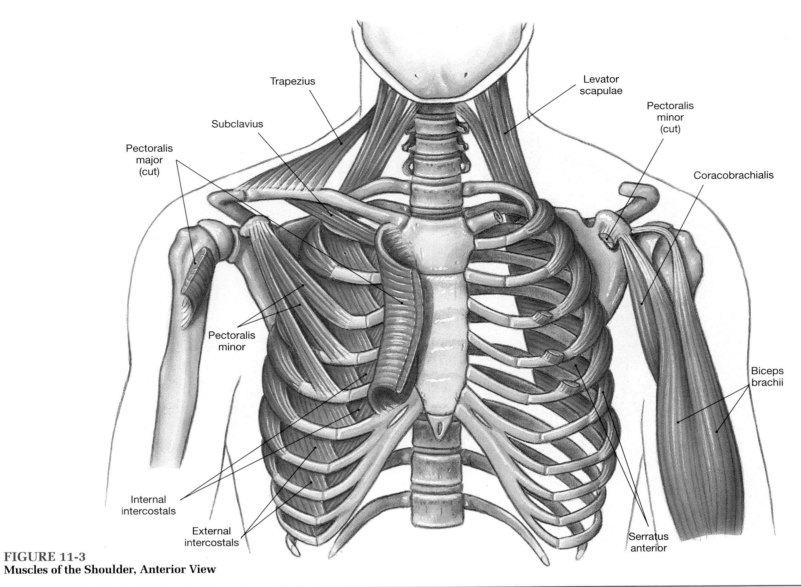

FIGURE 11-3
Muscles of the Shoulder, Anterior View

TABLE 11-1   Muscles That Move the Pectoral Girdle

Muscle	Origin	Insertion	Action	Innervation
Levator scapulae	Transverse processes of first 4 cervical vertebrae	Vertebral border of scapula near superior angle	Elevates scapula	Dorsal scapular nerve
Pectoralis minor	Ventral surfaces of ribs 3–5	Coracoid process of scapula	Depresses and protracts shoulder; rotates scapula so glenoid cavity moves inferiorly (downward rotation); elevates ribs if scapula is stationary	Medial pectoral nerve
Rhomboideus major	Spinous processes of upper thoracic vertebrae	Vertebral border of scapula from spine to inferior angle	Adducts and performs downward rotation	Dorsal scapular nerve
Rhomboideus minor	Spinous processes of vertebrae C_7–T_1	Vertebral border near spine	As above	As above
Serratus anterior	Ventral and superior margins of ribs 1–9	Ventral surface of vertebral border of scapula	Protracts shoulder, rotates scapula so glenoid cavity moves superiorly (upward rotation)	Long thoracic nerve
Subclavius	First rib	Clavicle	Depresses and protracts shoulder	Subclavian nerve
Trapezius	Occipital bone, ligamentum nuchae, and spinous processes of thoracic vertebrae	Clavicle and scapula (acromion and scapular spine)	Depends on active region and state of other muscles; may elevate, retract, depress, or rotate scapula upward and/or elevate clavicle; can also extend head and neck	Accessory nerve (N XI) and cervical spinal nerves

280

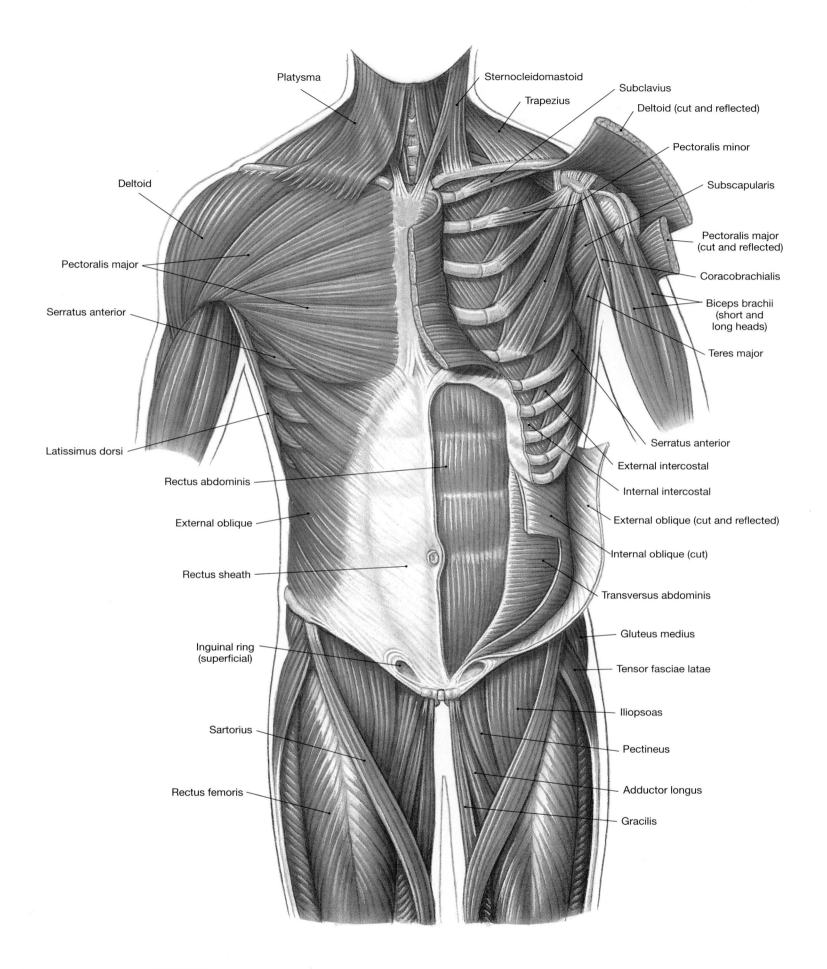

Platysma

Sternocleidomastoid

Subclavius

Trapezius

Deltoid (cut and reflected)

Pectoralis minor

Deltoid

Subscapularis

Pectoralis major

Pectoralis major
(cut and reflected)

Serratus anterior

Coracobrachialis

Biceps brachii
(short and
long heads)

Teres major

Latissimus dorsi

Serratus anterior

External intercostal

Rectus abdominis

Internal intercostal

External oblique

External oblique (cut and reflected)

Internal oblique (cut)

Rectus sheath

Transversus abdominis

Gluteus medius

Inguinal ring
(superficial)

Tensor fasciae latae

Iliopsoas

Sartorius

Pectineus

Rectus femoris

Adductor longus

Gracilis

FIGURE 11-4
Superficial and Deep Muscles of the Neck, Shoulder, Abdomen, and Upper Thigh, Anterior View

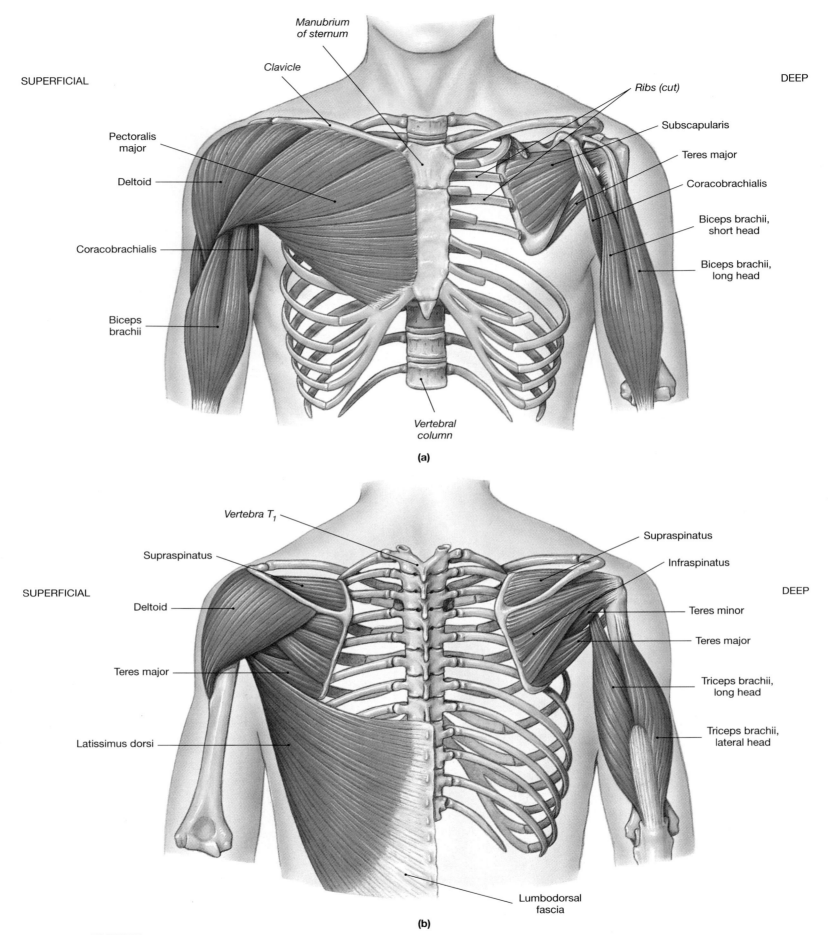

SUPERFICIAL

Manubrium
of sternum

Clavicle

DEEP

Ribs (cut)

Subscapularis

Pectoralis
major

Teres major

Deltoid

Coracobrachialis

Biceps brachii,
short head

Coracobrachialis

Biceps brachii,
long head

Biceps
brachii

Vertebral
column

(a)

Vertebra T₁

Supraspinatus

SUPERFICIAL

Supraspinatus

Infraspinatus

DEEP

Deltoid

Teres minor

Teres major

Teres major

Triceps brachii,
long head

Latissimus dorsi

Triceps brachii,
lateral head

Lumbodorsal
fascia

(b)

FIGURE 11-5
Muscles That Move the Arm. *See also Figures 7-4, 7-5, 7-6, and 8-9.* (a) Anterior view, showing superficial muscles and deep muscles. (b) Posterior view, showing superficial muscles and deep muscles.

TABLE 11-2 Muscles That Move the Arm

Muscle	Origin	Insertion	Action	Innervation
Coracobrachialis	Coracoid process	Medial margin of shaft of humerus	Adducts and flexes humerus	Musculocutaneous nerve
Deltoid	Clavicle and scapula (acromion and adjacent scapular spine)	Deltoid tuberosity of humerus	Abducts arm	Axillary nerve
Supraspinatus	Supraspinous fossa of scapula	Greater tubercle of humerus	Abducts arm	Suprascapular nerve
Infraspinatus	Infraspinous fossa of scapula	Greater tubercle of humerus	Lateral rotation of humerus	Suprascapular nerve
Subscapularis	Subscapular fossa of scapula	Lesser tubercle of humerus	Medial rotation of humerus	Subscapular nerves
Teres major	Inferior angle of scapula	Medial lip of the intertubercular groove of the humerus	Extends, adducts, and medially rotates humerus	Lower subscapular nerve
Teres minor	Lateral (axillary) border of scapula	Greater tubercle of humerus	Lateral rotation of humerus	Axillary nerve
Triceps brachii (long head)	See Table 11-3			
Latissimus dorsi	Spinous processes of lower thoracic vertebrae, ribs 8–12, the spines of lumbar vertebrae, and the lumbodorsal fascia	Floor of the intertubercular groove of the humerus	Extends, adducts, and medially rotates humerus	Thoracodorsal nerve
Pectoralis major	Cartilages of ribs 2–6, body of sternum, and inferior, medial portion of clavicle	Crest of greater tubercle of humerus (lateral lip of intertubercular groove)	Flexes, adducts, and medially rotates humerus	Pectoral nerves

seen in posterior view (Figure 11-1●, p. 278) and others in anterior view (Figure 11-4●). Information on the muscles that move the arm is summarized in Table 11-2. The **deltoid** is the major abductor of the arm, but the **supraspinatus** (soo-pra-spī-NĀ-tus) assists at the start of this movement. The **subscapularis** and **teres** (TER-ēz) **major** rotate the arm medially, whereas the **infraspinatus** (in-fra-spī-NĀ-tus) and the **teres minor** perform lateral rotation. All of these muscles originate on the scapula. The small **coracobrachialis** (kor-a-kō-brā-kē-AL-is) (Figure 11-5a●) is the only muscle attached to the scapula that flexes and adducts the humerus.

The **pectoralis major** extends between the anterior chest and the crest of the greater tubercle of the humerus. The **latissimus dorsi** (la-TIS-i-mus DOR-sē) extends between the thoracic vertebrae at the posterior midline and the floor of the intertubercular groove of the humerus (Figures 11-1, p. 278, 11-4, and 11-5a,b●). The pectoralis major flexes the arm, and the latissimus dorsi extends it. These two muscles can also work together to produce adduction and medial rotation of the humerus. The surface anatomy of many of the muscles that move the arm is shown in Figure 11-6●.

These muscles provide substantial support for the loosely-built glenohumeral joint. ∞ [p. 216] The tendons of the supraspinatus, infraspinatus, subscapularis, and teres minor support the joint capsule and limit the range of movement. They are known as the muscles of the **rotator cuff**, a frequent site of sports injuries, sustained especially by baseball players. Powerful, repetitive arm movements (such as pitching a fastball at 96 mph for nine innings) can place intolerable strains on the muscles of the rotator cuff, leading to muscle strains, bursitis, and other painful injuries. ☤ *Sports Injuries [p. 758]*

Muscles That Move the Forearm and Hand
(Figures 11-4/11-7/11-8, Table 11-3)

Although most of the muscles that insert upon the forearm and wrist originate on the humerus, there are two noteworthy exceptions. The **biceps** (BĪ-seps) **brachii** (BRĀ-kē-ī) and the long head of the **triceps brachii** originate on the scapula and insert upon the bones of the forearm (Figures 11-4, 11-7, pp. 286, 287, and 11-8●, pp. 288, 289). Although their contractions can have a secondary effect on the shoulder, their primary actions are at the elbow joint. The triceps brachii extends the forearm when, for example, we do push-ups. The biceps brachii both flexes and supinates the forearm. With the forearm pronated (in the anatomical position), the biceps brachii cannot function effectively due to the position of its muscular insertion. As a result, we are strongest when flexing the supinated forearm; the biceps brachii then makes a prominent bulge. Muscles that move the forearm and wrist along with their nerve innervations are detailed in Table 11-3.

The **brachialis** (brā-kē-Ā-lis) and **brachioradialis** (brā-kē-ō-rā-dē-Ā-lis) also flex the forearm, opposed by the **anconeus** (an-KŌ-nē-us) and the triceps. The **flexor carpi ulnaris**, the **flexor carpi radialis**, and the **palmaris longus** are superficial muscles that work together to produce flexion of the wrist (Figures 11-7b,c and 11-8b●, pp. 286, 287, 288). Because of differences in their sites of origin, the flexor carpi radialis flexes and abducts while the flexor carpi ulnaris flexes and adducts. The **extensor carpi radialis** muscles and the **extensor carpi ulnaris** have a similar relationship; the former produces extension and abduction, the latter extension and adduction (Figure 11-8●, pp. 288, 289).

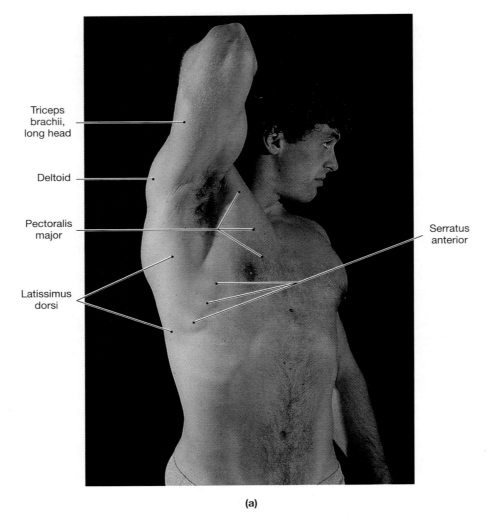

Triceps
brachii,
long head

Deltoid

Pectoralis
major

Serratus
anterior

Latissimus
dorsi

(a)

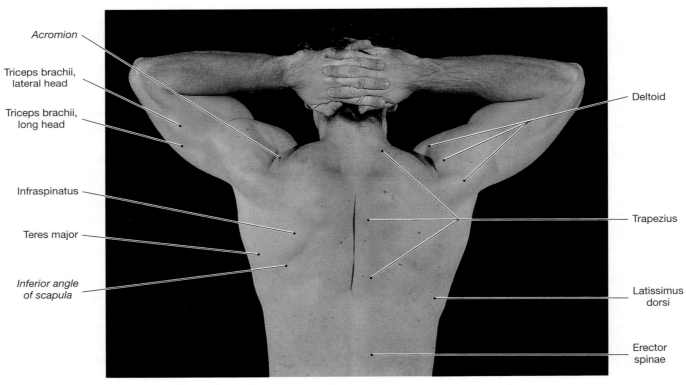

Acromion

Triceps brachii,
lateral head

Triceps brachii,
long head

Infraspinatus

Teres major

Inferior angle
of scapula

Deltoid

Trapezius

Latissimus
dorsi

Erector
spinae

(b)

FIGURE 11-6
Surface Anatomy of the Trunk, Showing Muscles That Move the Arm. (a) Anterolateral view. (b) Posterior view. *See also Figures 12-2 and 12-4, pp. 313 and 315.*

TABLE 11-3 Muscles That Move the Forearm and Hand

Muscle	Origin	Insertion	Action	Innervation
PRIMARY ACTION AT THE ELBOW				
Flexors				
Biceps brachii	*Short head* from the coracoid process; *long head* from the supraglenoid tubercle (both on the scapula)	Tuberosity of radius	Flexes and supinates forearm; flexes arm	Musculocutaneous nerve
Brachialis	Anterior, distal surface of humerus	Tuberosity of ulna	Flexes forearm	As above
Brachioradialis	Lateral epicondyle of humerus	Lateral aspect of styloid process of radius	As above	Radial nerve
Extensors				
Anconeus	Posterior surface of lateral epicondyle of humerus	Lateral margin of olecranon on ulna	Extends forearm, moves ulna laterally during pronation	As above
Triceps brachii				
lateral head	Superior, lateral margin of humerus	Olecranon process of ulna	Extends forearm	As above
long head	Infraglenoid tubercle of scapula	As above	As above	As above
medial head	Posterior surface of humerus inferior to radial groove	As above	As above	As above
PRONATORS/SUPINATORS				
Pronator				
quadratus	Medial surface of distal portion of ulna	Anterolateral surface of distal portion of radius	Pronates forearm	Median nerve
teres	Medial epicondyle of humerus and coronoid process of ulna	Distal lateral surface of radius	As above	As above
Supinator	Lateral epicondyle of humerus and ulna	Anterolateral surface of radius distal to the radial tuberosity	Supinates forearm	Radial nerve
PRIMARY ACTION AT THE HAND				
Flexors				
Flexor carpi radialis	Medial epicondyle of humerus	Bases of 2nd and 3rd metacarpal bones	Flexes and abducts hand	Median nerve
Flexor carpi ulnaris	Medial epicondyle of humerus; adjacent medial surface of olecranon and anteromedial portion of ulna	Pisiform, hamate, and base of 5th metacarpal bone	Flexes and adducts hand	Ulnar nerve
Palmaris longus	Medial epicondyle of humerus	Palmar aponeurosis and flexor retinaculum	Flexes hand	Median nerve
Extensors				
Extensor carpi radialis, longus	Lateral supracondylar ridge of humerus	Base of 2nd metacarpal	Extends and abducts hand	Radial nerve
brevis	Lateral epicondyle of humerus	Base of 3rd metacarpal	As above	As above
Extensor carpi ulnaris	Lateral epicondyle of humerus; adjacent dorsal surface of ulna	Base of 5th metacarpal	Extends and adducts hand	Deep radial nerve

The **pronator teres** and the **supinator** arise on both the humerus and forearm. They rotate the radius without flexing or extending the elbow. The **pronator quadratus** arises on the ulna and assists the pronator teres in opposing the actions of the supinator or biceps brachii. The muscles involved in pronation and supination can be seen in Figures 11-7c,d and 11-8d,e●, p. 289. Note the changes in orientation that occur as the pronator teres and pronator quadratus contract. During pronation the tendon of the biceps brachii rolls under the radius, and a bursa (Figure 8-10) prevents abrasion against the tendon. ∞ [p. 218]

As you study the muscles in Table 11-3, note that in general, extensor muscles lie along the dorsal and lateral surfaces of

the forearm, and flexors are found on the ventral and medial surfaces. Many of the muscles that move the forearm and wrist can be seen from the body surface (Figures 11-7a and 11-8a●, p. 288).

Muscles That Move the Fingers
(Figures 11-7 to 11-11, Tables 11-4/11-5)

Several superficial and deep muscles of the forearm (Table 11-4, p. 290) perform flexion and extension of the fingers. Only the tendons of these muscles cross the articulation. These are relative-

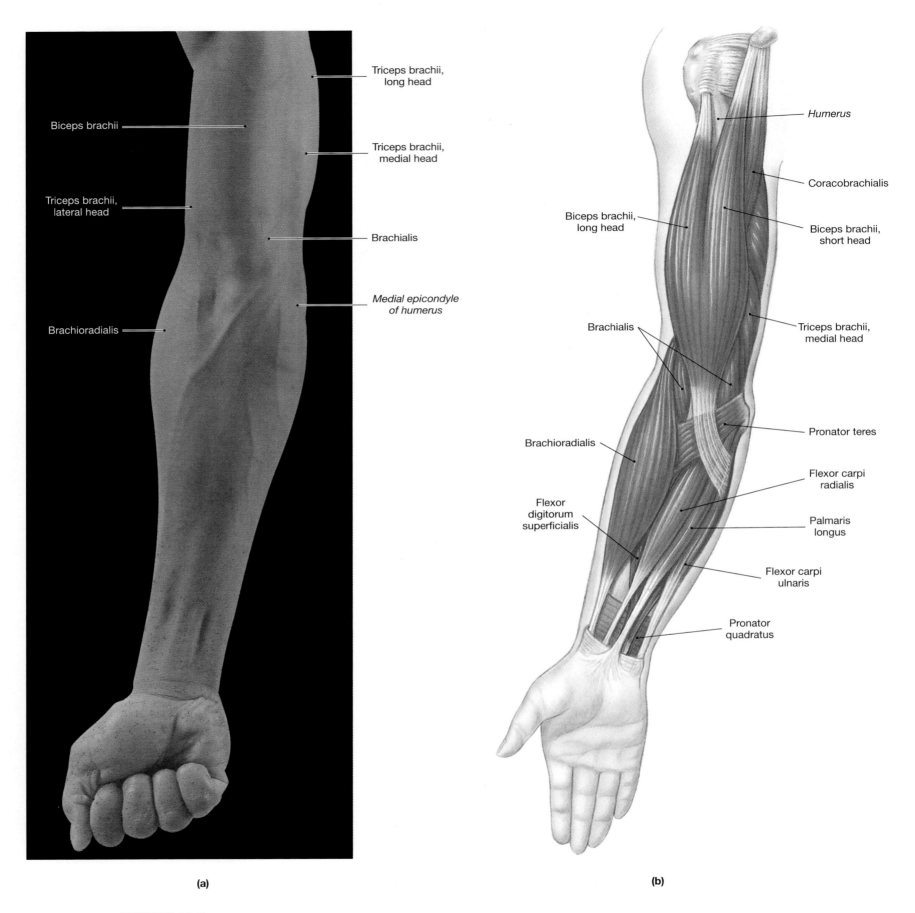

Triceps brachii,
long head

Biceps brachii

Triceps brachii,
medial head

Triceps brachii,
lateral head

Brachialis

Medial epicondyle
of humerus

Brachioradialis

(a)

Humerus

Coracobrachialis

Biceps brachii,
long head

Biceps brachii,
short head

Brachialis

Triceps brachii,
medial head

Brachioradialis

Pronator teres

Flexor carpi
radialis

Flexor
digitorum
superficialis

Palmaris
longus

Flexor carpi
ulnaris

Pronator
quadratus

(b)

FIGURE 11-7
Muscles That Move the Forearm and Hand (part I). *See also Figures 7-6, 7-7, and 7-8.* (a) Surface anatomy of the right arm, anterior view. (b) Muscles on the anterior aspect of the arm. (c) Forearm muscles of a cadaver, anterior view of the middle layer. The palmaris longus, flexor carpi radialis and ulnaris have been partly removed, and the flexor retinaculum has been divided. The relationships among the deeper muscles are best seen in the sectional view. (d) Anterior view of the supine forearm (deep muscles).

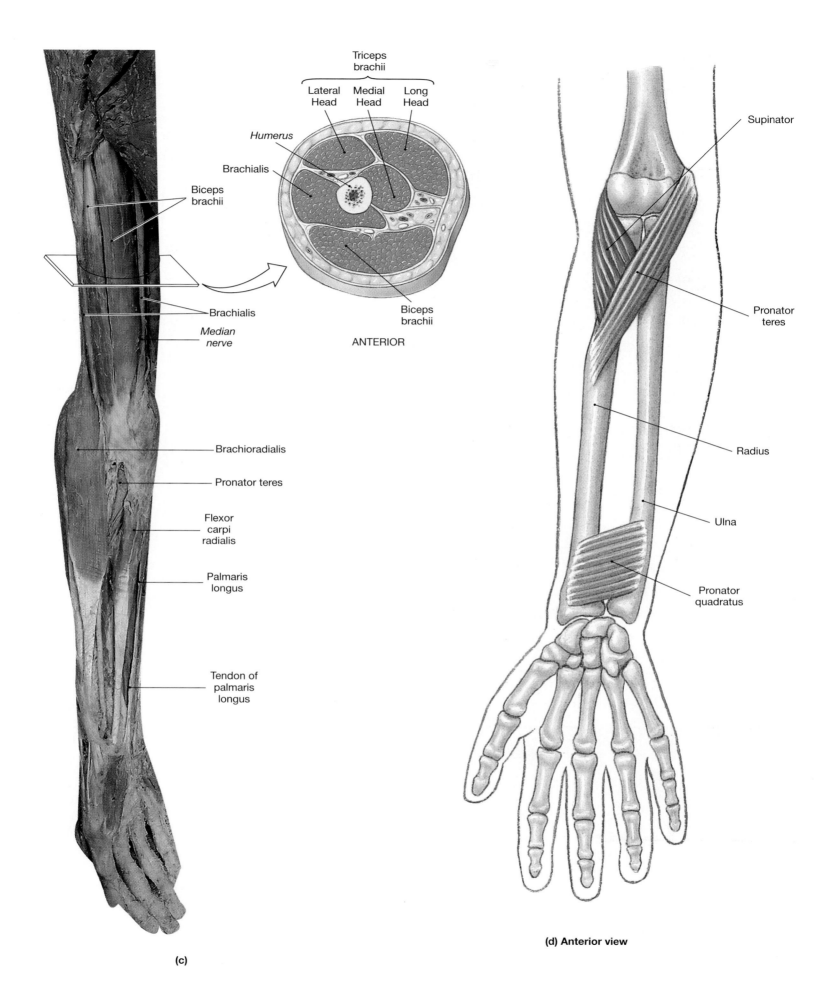

Triceps
brachii

Lateral   Medial   Long
Head   Head   Head

Humerus

Brachialis

Biceps
brachii

ANTERIOR

Biceps
brachii

Brachialis

*Median
nerve*

Brachioradialis

Pronator teres

Flexor
carpi
radialis

Palmaris
longus

Tendon of
palmaris
longus

Supinator

Pronator
teres

Radius

Ulna

Pronator
quadratus

(c)

(d) Anterior view

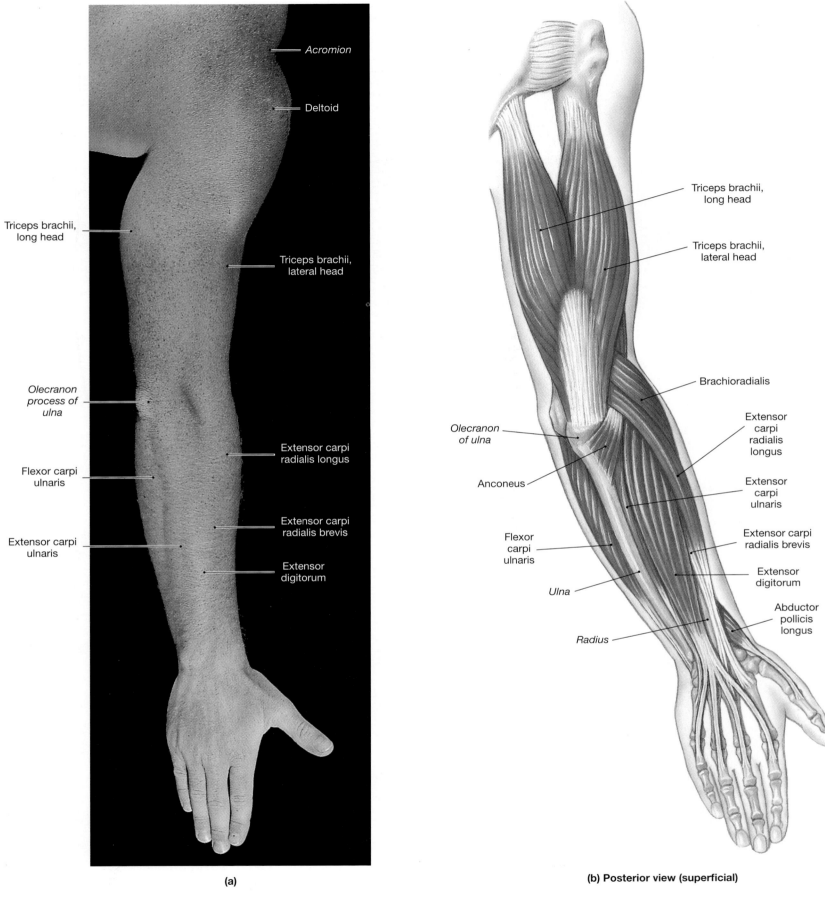

(a)

(b) Posterior view (superficial)

FIGURE 11-8
Muscles That Move the Forearm and Hand (part II). *See also Figure 7-7.* (a) Surface anatomy of the right arm, posterior view. (b) Muscles of the posterior aspect of the arm. (c) Forearm and hand muscles of the cadaver, superficial layer, posterior view. Relationships among deeper muscles are best seen in the sectional view. (d) Pronation of the hand. (e) Sectional view in pronation at the level of the biceps tendon insertion.

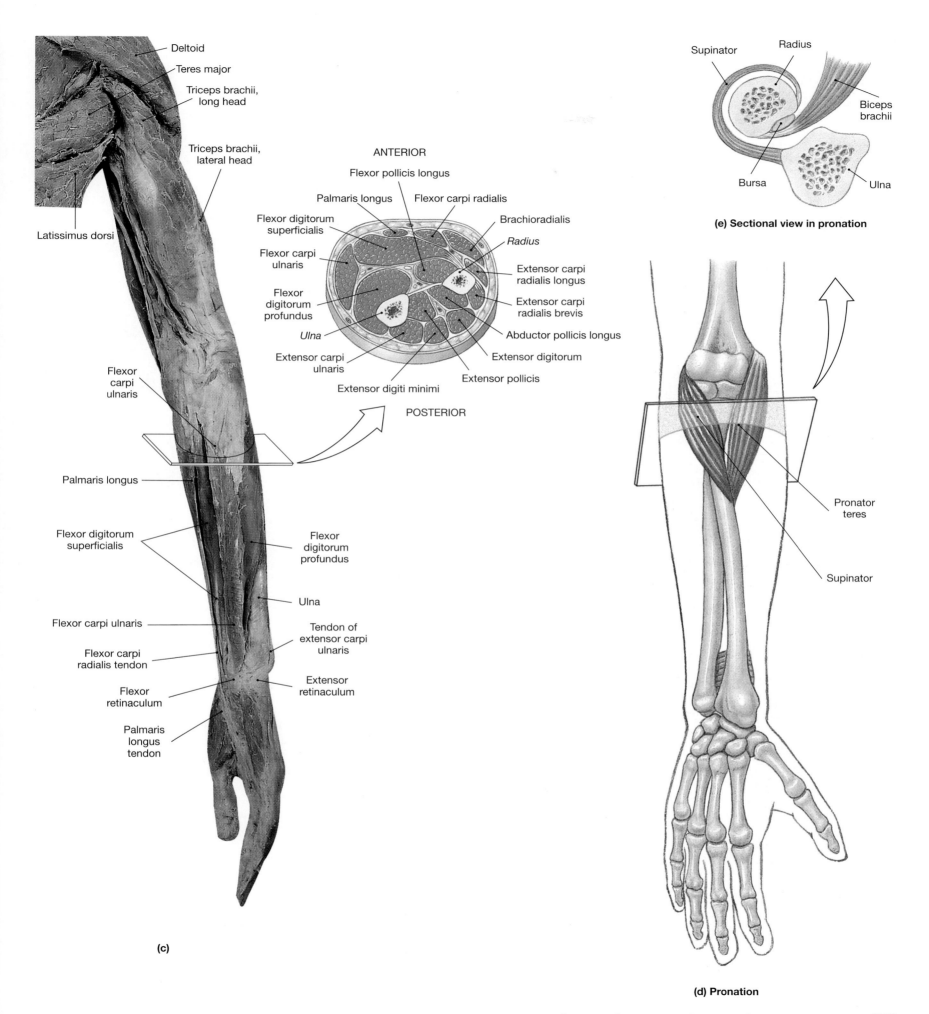

Deltoid
Teres major
Triceps brachii, long head
Triceps brachii, lateral head
Latissimus dorsi
Flexor carpi ulnaris
Palmaris longus
Flexor digitorum superficialis
Flexor carpi ulnaris
Flexor carpi radialis tendon
Flexor retinaculum
Palmaris longus tendon

Flexor digitorum profundus
Ulna
Tendon of extensor carpi ulnaris
Extensor retinaculum

(c)

ANTERIOR

Flexor pollicis longus
Palmaris longus
Flexor digitorum superficialis
Flexor carpi radialis
Flexor carpi ulnaris
Flexor digitorum profundus
Brachioradialis
Radius
Extensor carpi radialis longus
Extensor carpi radialis brevis
Ulna
Abductor pollicis longus
Extensor carpi ulnaris
Extensor digitorum
Extensor digiti minimi
Extensor pollicis

POSTERIOR

Supinator
Radius
Biceps brachii
Bursa
Ulna

(e) Sectional view in pronation

Pronator teres
Supinator

(d) Pronation

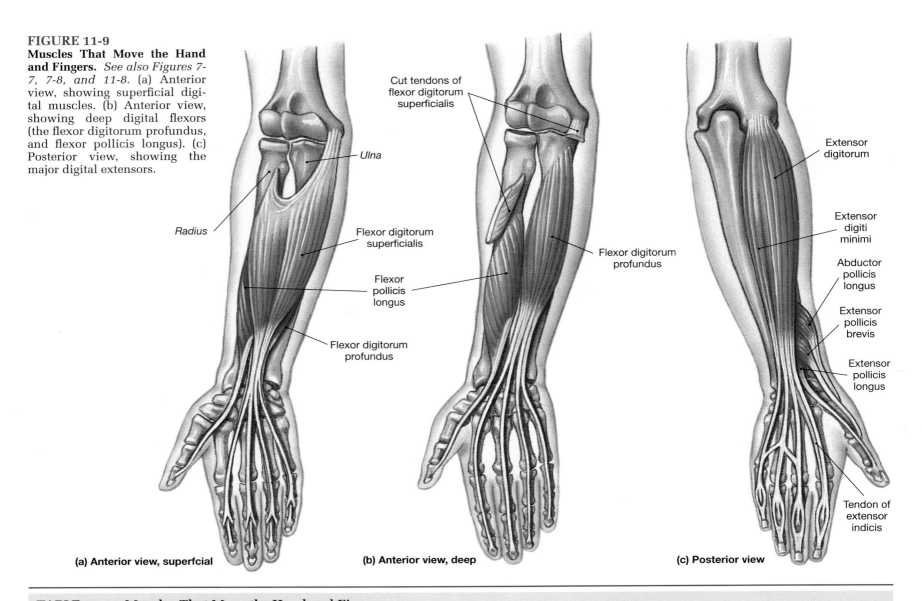

FIGURE 11-9
Muscles That Move the Hand and Fingers. *See also Figures 7-7, 7-8, and 11-8.* (a) Anterior view, showing superficial digital muscles. (b) Anterior view, showing deep digital flexors (the flexor digitorum profundus, and flexor pollicis longus). (c) Posterior view, showing the major digital extensors.

Ulna

Radius

Flexor digitorum superficialis

Flexor pollicis longus

Flexor digitorum profundus

(a) Anterior view, superfcial

Cut tendons of flexor digitorum superficialis

Flexor digitorum profundus

(b) Anterior view, deep

Extensor digitorum

Extensor digiti minimi

Abductor pollicis longus

Extensor pollicis brevis

Extensor pollicis longus

Tendon of extensor indicis

(c) Posterior view

TABLE 11-4   Muscles That Move the Hand and Fingers

Muscle	Origin	Insertion	Action	Innervation
Abductor pollicis longus	Proximal dorsal surfaces of ulna and radius	Lateral margin of 1st metacarpal bone	Abducts thumb	Deep radial nerve
Extensor digitorum	Lateral epicondyle of humerus	Posterior surfaces of the phalanges, fingers 2-5	Extends fingers and hand	As above
Extensor pollicis brevis	Shaft of radius distal to origin of adductor pollicis longus	Base of proximal phalanx of thumb	Extends thumb, abducts hand	As above
Extensor pollicis longus	Posterior and lateral surfaces of ulna and interosseous membrane	Base of distal phalanx of thumb	Extends thumb, abducts hand	As above
Extensor indicis	Posterior surface of ulna and interosseous membrane	Posterior surface of phalanges of little (5th) finger, with tendon of extensor digitorum	Extends and adducts little finger	As above
Extensor digiti minimi	Via extensor tendon to lateral epicondyle of humerus, and from intermuscular septa	Posterior surface of proximal phalanx of little finger	Extends little finger	As above
Flexor digitorum superficialis	Medial epicondyle of humerus; adjacent anterior surfaces of ulna and radius	Midlateral surfaces of middle phalanges of fingers 2-5	Flexes fingers, specifically middle phalanx on proximal; flexes hand	As above
Flexor digitorum profundus	Medial and posterior surfaces of ulna, medial surface of coronoid process, and interosseous membrane	Bases of distal phalanges of fingers 2-5	Flexes distal phalanges and to a lesser degree the other phalanges and hand	Palmar interosseous nerve, from median nerve and ulnar nerve
Flexor pollicis longus	Anterior shaft of radius and interosseous membrane	Base of distal phalanx of thumb	Flexes thumb	As above

290

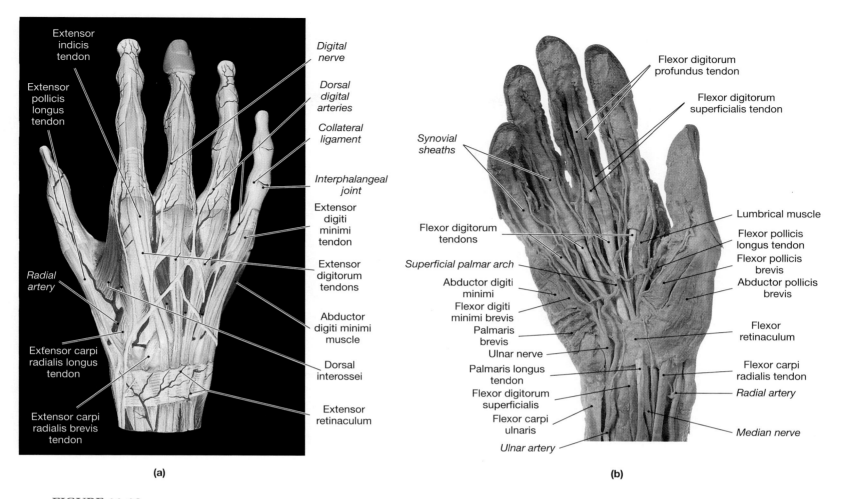

FIGURE 11-10
Muscles, Tendons, and Ligaments of the Right Wrist and Hand. (a) Cadaver cast model of right wrist and hand (dorsal surface). (b) Right wrist and hand (palmar surface).

ly large muscles (Figures 11-8 and 11-9●), and keeping them clear of the joints ensures maximum mobility at both the wrist and hand. The tendons that cross the dorsal and ventral surfaces of the wrist pass through **tendon sheaths**, elongated bursae that reduce friction. These sheaths were seen in Figure 7-10. ∞ [p. 187] These muscles and their tendons are shown in anterior view in Figures 11-7b,c, pp. 286, 287, and 11-9a●, and in posterior view in Figure 11-8b,c●, pp. 288, 289. The fascia of the forearm thickens on the posterior surface of the wrist to form a wide band of connective tissue, the **extensor retinaculum** (ret-i-NAK-ū-lum). The extensor retinaculum holds all the tendons of the extensor muscles in place. On the anterior surface, the fascia also thickens to form another wide band of connective tissue, the **flexor retinaculum,** which retains the tendons of the flexor muscles (Figure 11-10●). Inflammation of the retinacula and tendon sheaths can restrict movement and irritate the median nerve. This condition, known as *carpal tunnel syndrome*, causes chronic pain.

The muscles of the forearm provide strength and crude control of the hand and fingers. They are known as the *extrinsic muscles of the hand*. Fine control of the hand involves small *intrinsic muscles* that originate on the carpals and metacarpals (Figures 11-10 and 11-11●). No muscles originate on the phalanges, and only tendons extend across the distal joints of the fingers. The intrinsic muscles of the hand are detailed in Table 11-5.

√ **What muscle are you using when you shrug your shoulders?**
√ **Sometimes baseball pitchers will suffer from rotator cuff injuries. What muscles are involved in this type of injury?**
√ **Injury to the flexor carpi ulnaris would impair what two movements?**

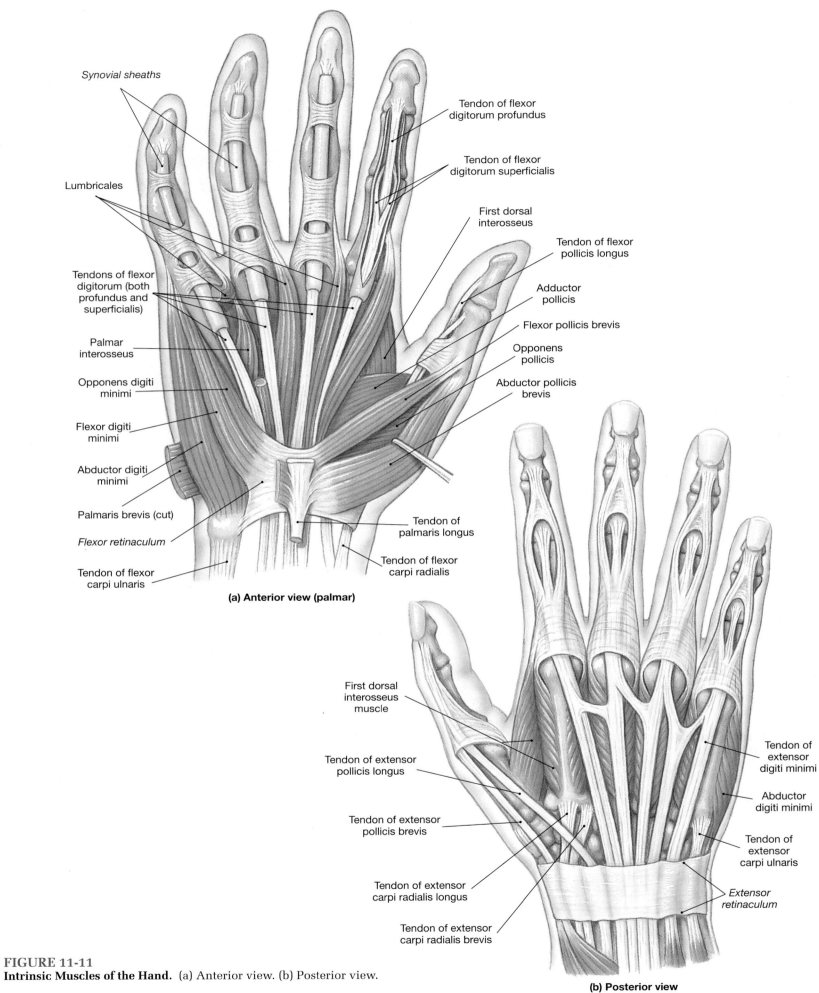

Synovial sheaths

Lumbricales

Tendons of flexor
digitorum (both
profundus and
superficialis)

Palmar
interosseus

Opponens digiti
minimi

Flexor digiti
minimi

Abductor digiti
minimi

Palmaris brevis (cut)

Flexor retinaculum

Tendon of flexor
carpi ulnaris

Tendon of flexor
digitorum profundus

Tendon of flexor
digitorum superficialis

First dorsal
interosseus

Tendon of flexor
pollicis longus

Adductor
pollicis

Flexor pollicis brevis

Opponens
pollicis

Abductor pollicis
brevis

Tendon of
palmaris longus

Tendon of flexor
carpi radialis

(a) Anterior view (palmar)

First dorsal
interosseus
muscle

Tendon of extensor
pollicis longus

Tendon of extensor
pollicis brevis

Tendon of extensor
carpi radialis longus

Tendon of extensor
carpi radialis brevis

Tendon of
extensor digiti minimi

Abductor
digiti minimi

Tendon of
extensor
carpi ulnaris

*Extensor
retinaculum*

(b) Posterior view

FIGURE 11-11
Intrinsic Muscles of the Hand. (a) Anterior view. (b) Posterior view.

TABLE 11-5 Intrinsic Muscles of the Hand

Muscle	Origin	Insertion	Action	Innervation
Adductor pollicis	Metacarpal and carpal bones	Proximal phalanx of thumb	Adducts thumb	Ulnar nerve, deep branch
Opponens pollicis	Trapezium	First metacarpal bone	Opposition of thumb	Median nerve
Palmaris brevis	Palmar aponeurosis	Skin of medial border of hand	Moves skin on medial border toward midline of palm	Ulnar nerve, superficial branch
Abductor digiti minimi	Pisiform bone	Proximal phalanx of little finger	Abducts little finger and flexes its proximal phalanx	Ulnar nerve, deep branch
Abductor pollicis brevis	Transverse carpal ligament, scaphoid bone, and trapezium	Radial side of the base of the proximal phalanx of the thumb	Abducts the thumb	Median nerve
Flexor pollicis brevis*	Flexor retinaculum, trapezium, capitate bone, and ulnar side of first metacarpal bone	Ulnar side of the proximal phalanx of the thumb	Flexes and adducts the thumb	Branches of median and ulnar nerves
Flexor digiti minimi brevis	Hamate bone	Proximal phalanx of little finger	Flexes little finger	Ulnar nerve, deep branch
Opponens digiti minimi	Hamate bone	Fifth metacarpal	Opposition of fifth metacarpal bone	Ulnar nerve, deep branch
Lumbricals (4)	Tendons of flexor digitorum profundus	Tendons of extensor digitorum	Flexes metacarpophalangeal joints, extends middle and distal phalanges	#1 and #2 by median nerve, #3 and #4 by ulnar nerve, deep branch
Dorsal interossei (4)	Each originates from opposing faces of two metacarpal bones (I and II, II and III, III and IV, IV and V)	Bases of proximal phalanges of fingers 2–4	Abduct fingers 2-4 away from the midline axis of the middle finger (3), flex metacarpophalangeal joints, extend fingertips	Ulnar nerve, deep branch
Palmar interossei (3-4)*	Sides of metacarpal bones II, IV, and V	Bases of proximal phalanges of fingers 2, 4, and 5	Adduct fingers 2, 4, and 5 toward the midline axis of the middle finger (3), flex metacarpophalangeal joints, extend fingertips	Ulnar nerve, deep branch

* The portion of the flexor pollicis brevis originating on the first metacarpal is sometimes called the *first palmar interosseus* muscle.

MUSCLES OF THE LOWER EXTREMITY

The pelvic girdle is tightly bound to the axial skeleton, and relatively little movement is permitted. The few muscles that can influence the position of the pelvis were considered in Chapter 10, during the discussion of the axial musculature. ∞ [p. 270] The muscles of the lower extremities can be divided into three groups: (1) muscles that move the thigh, (2) muscles that move the leg, and (3) muscles that move the foot and toes.

Muscles That Move the Thigh
(Figures 11-1/11-4/11-12 to 11-14, Table 11-6)

One method for organizing the diverse muscles that originate on the pelvis considers the orientation of each muscle around the hip joint. Figure 11-12● presents the organization of the attachment of muscles that originate on the pelvis. Muscles originating on the surface of the pelvis and inserting on the femur will produce characteristic movements determined by their position relative to the acetabulum. Information concerning the origins of the thigh muscles is summarized in Table 11-6.

Gluteal muscles cover the lateral surface of the ilium (Figures 11-1, p. 278, 11-4, p. 281, and 11-13●). The **gluteus maximus** is the largest and most posterior of the gluteal muscles. It originates along the edge of the posterior superior iliac spine and the ligaments that bind the sacrum to the ilium. Acting alone, this massive muscle extends and laterally rotates the thigh. The gluteus maximus shares an insertion with the **tensor** (TEN-sor) **fasciae latae** (FASH-i-ē LĀ-tē), a muscle that originates on the

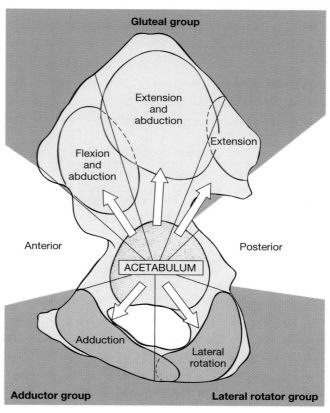

FIGURE 11-12
The Attachment of Muscles to the Pelvic Girdle. *See also Figures 11-13 and 11-14.* The attachment of muscles that originate on the pelvis can be roughly predicted on the basis of their positions relative to the acetabulum. For each major muscle group the direction of the pull exerted on the femur is indicated by an arrow. Specific muscles are detailed in *Figures 11-13 and 11-14.*

FIGURE 11-13

The Gluteal Muscles and Lateral Rotators of the Thigh. *See also Figures 7-10, 7-11, and 7-14.* (a) Lateral view of the gluteal muscle group. (b) Posterior view of pelvis, showing deep disection of the gluteal muscles and lateral rotators. (c) Posterior view of the gluteal and lateral rotator muscles. (d) Anterior view of the obturator and piriformis, the most important members of the lateral rotator group.

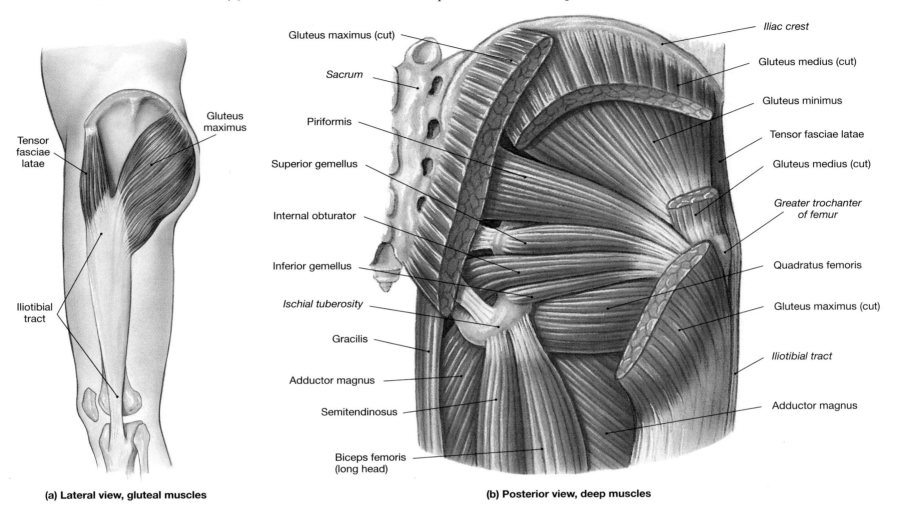

(a) Lateral view, gluteal muscles

(b) Posterior view, deep muscles

TABLE 11-6 Muscles That Move the Thigh

Group/Muscle	Origin	Insertion	Action	Innervation
Gluteal group				
Gluteus maximus	Iliac crest of ilium, sacrum, coccyx, and lumbodorsal fascia	Iliotibial tract and gluteal tuberosity of femur	Extends and laterally rotates thigh	Inferior gluteal nerve
Gluteus medius	Anterior iliac crest of ilium, lateral surface between posterior and anterior gluteal lines	Greater trochanter of femur	Abducts and medially rotates thigh	Superior gluteal nerve
Gluteus minimus	Lateral surface of ilium between inferior and anterior gluteal lines	Greater trochanter of femur	Abducts and medially rotates thigh	As above
Tensor fasciae latae	Iliac crest and lateral surface of anterior superior iliac spine	Iliotibial tract	Flexes, abducts, and medially rotates thigh; tenses fascia lata, which laterally supports the knee	As above
Lateral Rotator Group				
Obturators (externus and internus)	Lateral and medial margins of obturator foramen	Trochanteric fossa of femur (externus); medial surface of greater trochanter (internus)	Laterally rotate thigh	Obturator nerve (externus) and special nerve from sacral plexus (internus)
Piriformis	Anterolateral surface of sacrum	Greater trochanter of femur	Laterally rotates and abducts thigh	Branches of sacral nerves
Gemelli (superior and inferior)	Ischial spine and tuberosity	Medial surface of greater trochanter	Laterally rotates thigh	Nerves to obturator internus and quadratus femoris
Quadratus femoris	Lateral border of ischial tuberosity	Intertrochanteric crest of femur	Laterally rotates thigh	Special nerve from sacral plexus

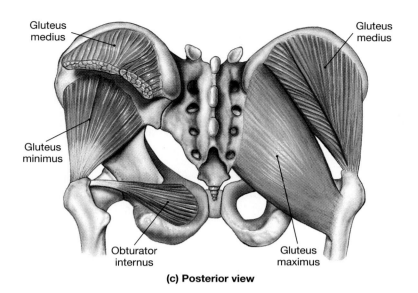

Gluteus medius

Gluteus medius

Gluteus minimus

Obturator internus

Gluteus maximus

(c) Posterior view

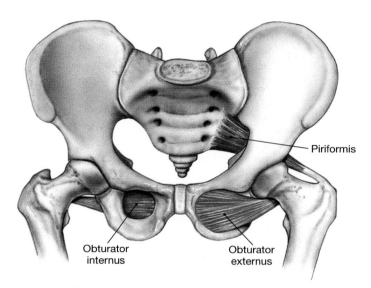

Piriformis

Obturator internus

Obturator externus

(d) Anterior view, major lateral rotator group

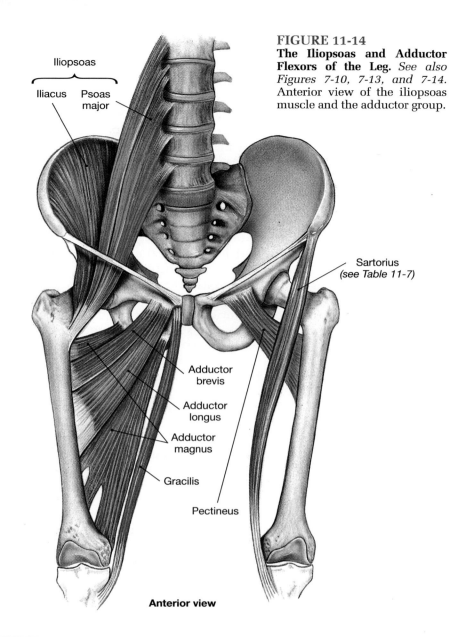

FIGURE 11-14
The Iliopsoas and Adductor Flexors of the Leg. *See also Figures 7-10, 7-13, and 7-14.* Anterior view of the iliopsoas muscle and the adductor group.

Iliopsoas

Iliacus   Psoas major

Sartorius *(see Table 11-7)*

Adductor brevis

Adductor longus

Adductor magnus

Gracilis

Pectineus

Anterior view

TABLE 11-6   Muscles That Move the Thigh (Cont.)

Group/Muscle	Origin	Insertion	Action	Innervation
Adductor Group				
Adductor brevis	Inferior ramus of pubis	Linea aspera of femur	Adducts, medially rotates, and flexes thigh	Obturator nerve
Adductor longus	Inferior ramus of pubis anterior to brevis	As above	Adducts, flexes, and medially rotates thigh	As above
Adductor magnus	Inferior ramus of pubis posterior to adductor brevis and ischial tuberosity	Linear aspera and adductor tubercle of femur	Adducts thigh; superior portion flexes and medially rotates thigh, inferior portion extends and laterally rotates thigh	Obturator and sciatic nerves
Pectineus	Superior ramus of pubis	Pectineal line inferior to lesser trochanter of femur	Flexes, medially rotates, and adducts thigh	Femoral nerve
Gracilis	Inferior ramus of pubis	Medial surface of tibia inferior to medial condyle	Flexes leg, adducts, and medially rotates thigh	Obturator nerve
Iliopsoas Group				
Iliacus	Iliac fossa of ilium	Femur distal to lesser trochanter; tendon fused with that of psoas	Flexes hip and/or lumbar spine	Femoral nerve
Psoas major	Anterior surfaces and transverse processes of vertebrae $T_{12}-L_5$	Lesser trochanter in company with iliacus	As above	Branches of the lumbar plexus

anterior superior iliac spine. Together these muscles pull on the **iliotibial** (il-ē-ō-TIB-ē-al) **tract**, a band of collagen fibers that extends along the lateral surface of the thigh and inserts upon the tibia. This tract provides a lateral brace for the knee that becomes particularly important when a person balances on one foot.

The **gluteus medius** and **gluteus minimus** (Figure 11-13b, c●) originate anterior to that of the gluteus maximus and insert upon the greater trochanter of the femur. The anterior gluteal line on the lateral surface of the ilium marks the boundary between these muscles. ∞ [p. 187]

The **lateral rotators** (Figure 11-13b-d●) arise at or inferior to the horizontal axis of the acetabulum. There are six muscles in all, of which the **piriformis** (pir-i-FOR-mis) and the **obturator** muscles are dominant.

The **adductors** are located inferior to the acetabular surface. This muscle group includes the **adductor magnus**, the **adductor brevis**, the **adductor longus**, the **pectineus** (pek-TIN-ē-us), and the **gracilis** (GRAS-i-lis) (Figure 11-14●). All but the magnus are found both anterior and inferior to the joint, so they can contribute to hip flexion as well as adduction. The magnus can produce either adduction and flexion or adduction and extension, depending on the region stimulated. These muscles insert upon low ridges along the posterior surface of the femur. The adductor magnus may also produce medial or lateral rotation. The other muscles in this group produce medial rotation.

The medial surface of the pelvis is dominated by a single pair of muscles. The large **psoas** (SŌ-us) **major** arises alongside the lower thoracic and lumbar vertebrae, and its insertion lies on the lesser trochanter of the femur. Before reaching this insertion, its tendon merges with that of the **iliacus** (i-LĒ-ak-us), which lies nestled within the iliac fossa. These two muscles are powerful flexors of the thigh and are often referred to as the **iliopsoas** (i-lē-ō-SŌ-us) muscle (Figure 11-14●).

Muscles That Move the Leg
(Figures 11-15 to 11-18, Table 11-7)

Muscles that move the leg are detailed in Table 11-7. Extensors are found along the anterior and lateral surfaces of the leg (Figures 11-15 and 11-16●, pp. 298, 299), and flexors lie along the posterior and medial surfaces (Figures 11-15b and 11-17●, p. 300). Although the flexors and adductors originate on the pelvic girdle, most of the extensors originate on the femoral surface.

The flexors of the leg include: the **biceps femoris**, the **semimembranosus** (sem-ē-mem-bra-NŌ-sus), the **semitendinosus** (sem-ē-ten-di-NŌ-sus), and the **sartorius** (sar-TOR-ē-us) (Figures 11-15 and 11-17●, p. 300). These muscles arise along the edges of the pelvis and insert upon the tibia and fibula, and their contractions produce flexion of the knee. Because the biceps femoris, semimembranosus, and semitendinosus originate on the pelvic surface inferior and posterior to the acetabulum, their contractions also produce extension of the hip. These muscles are often called the **hamstrings**. The sartorius is the only knee flexor that originates superior to the acetabulum, and its insertion lies along the medial aspect of the tibia. When it contracts, it flexes and laterally rotates the thigh, as when crossing the legs. In Chapter 8 we noted that the knee joint can be locked at full extension by a slight lateral rotation of the tibia. ∞ [p. 225] The small **popliteus** (pop-LI-tē-us) muscle arises on the femur near the lateral condyle and inserts on the posterior tibial shaft (Figure 11-18a●, p. 302). When flexion is initiated, this muscle contracts to produce a slight medial rotation of the tibia that unlocks the joint. Figure 11-17c●, p. 300, shows the surface anatomy of the leg, revealing the flexors of the posterior surface of the thigh.

Collectively the *knee extensors* (Figures 11-15 and 11-16●) are known as the **quadriceps femoris**. The three **vastus** mus-

TABLE 11-7	Muscles That Move the Leg			
Muscle	*Origin*	*Insertion*	*Action*	*Innervation*
Flexors of the Leg				
Biceps femoris	Ischial tuberosity and linea aspera of femur	Head of fibula, lateral condyle of tibia	Flexes leg, extends and laterally rotates thigh	Sciatic nerve; tibial portion (to long head) and common peroneal branch (to short head)
Semimembranosus	Ischial tuberosity	Posterior surface of medial condyle of tibia	Flexes and medially rotates leg, extends thigh	Sciatic nerve (tibial portion)
Semitendinosus	Ischial tuberosity	Proximal, medial surface of tibia near insertion of gracilis	As above	As above
Sartorius	Anterior superior iliac spine	Medial surface of tibia near tibial tuberosity	Flexes leg, flexes and laterally rotates thigh	Femoral nerve
Popliteus	Lateral condyle of femur	Posterior surface of proximal tibial shaft	Medially rotates tibia (or laterally rotates femur)	Tibial nerve
Extensors of the Leg				
Rectus femoris	Anterior inferior iliac spine and superior acetabular rim of ilium	Tibial tuberosity via patellar ligament	Extends leg, flexes thigh	Femoral nerve
Vastus intermedius	Anterolateral surface of femur and linea aspera (distal half)	As above	Extends leg	As above
Vastus lateralis	Anterior and inferior to greater trochanter of femur and along linea aspera (proximal half)	As above	As above	As above
Vastus medialis	Entire length of linea aspera of femur	As above	As above	As above

cles originate along the body of the femur, and they cradle the **rectus femoris** the way a bun surrounds a hot dog. All four muscles insert upon the patella and reach the tibial tuberosity via the patellar ligament. The rectus femoris originates on the anterior inferior iliac spine, so in addition to producing extension of the knee, it can assist in flexion of the thigh.

Muscles That Move the Foot and Toes
(Figures 11-18 to 11-22, Tables 11-8/11-9)

Extrinsic and intrinsic muscles that move the foot and toes (Figures 11-18 to 11-22●) are detailed in Tables 11-8 and 11-9. Most of the muscles that move the ankle produce the plantar flexion involved with walking and running movements.

The large **gastrocnemius** (gas-trok-NĒ-mē-us; *gaster,* stomach + *kneme,* knee) of the calf is an important plantar flexor, but the slow muscle fibers of the underlying **soleus** (SŌ-lē-us) are more powerful. These muscles are best seen from the posterior and lateral views (Figures 11-18 and 11-20●, pp. 302, 304). The gastrocnemius arises from two heads located on the medial and lateral epicondyles of the femur just proximal to the knee. A

sesamoid bone, the **fabella,** is usually found within the gastrocnemius. The gastrocnemius and soleus muscles share a common tendon, the **calcaneal,** or *Achilles,* **tendon.**

Deep to the gastrocnemius and soleus lie the two **peroneus muscles** (Figure 11-19●, p. 303). These muscles produce eversion as well as plantar flexion of the foot. Inversion of the foot is caused by contraction of the **tibialis** (tib-ē-Ā-lis) muscles; the large **tibialis anterior** opposes the gastrocnemius and dorsiflexes the foot (Figures 11-20 and 11-21●, pp. 304, 305).

Important muscles that move the toes originate on the surface of the tibia, the fibula, or both (Figures 11-19, 11-20, and 11-21●, pp. 303, 304, 305). Large tendon sheaths surround the tendons of the tibialis anterior, extensor digitorum longus, and extensor hallucis longus, where they cross the ankle joint. The positions of these sheaths are stabilized by the **superior** and **inferior extensor retinacula** (Figures 11-20a,c, 11-21a, and 11-22a,c●, pp. 304, 305, 306).

The small intrinsic muscles that move the toes originate on the bones of the tarsus and foot (Figure 11-22●, p. 306, and Table 11-9, p. 301). Some of the flexor muscles originate at the anterior border of the calcaneus; their muscle tone contributes to maintenance of the longitudinal arch of the foot. As in the hand, small **interosseus muscles** (plural *interossei*) originate on the lateral and medial surfaces of the metatarsals.

TABLE 11-8   Extrinsic Muscles That Move the Foot and Toes

Muscle	Origin	Insertion	Action	Innervation
PRIMARY ACTION AT THE ANKLE **Dorsiflexors**				
Tibialis anterior	Lateral condyle and proximal shaft of tibia	Base of 1st metatarsal bone and medial cuneiform bone	Dorsiflexes and inverts foot	Deep peroneal nerve
Plantar flexors				
Gastrocnemius	Femoral condyles	Calcaneus via calcaneal tendon	Plantar flexes, inverts, and adducts foot; flexes leg	Tibial nerve
Peroneus brevis	Midlateral margin of fibula	Base of 5th metatarsal bone	Everts and plantar flexes foot	Superficial peroneal nerve
longus	Lateral condyle of tibia, head and and proximal shaft of fibula	Base of 1st metatarsal bone and medial cuneiform bone	Everts and plantar flexes foot; supports longitudinal arch	As above
Plantaris	Lateral supracondylar ridge	Posterior portion of calcaneus	Plantar flexes foot, flexes leg	Tibial nerve
Soleus	Head and proximal shaft of fibula, and adjacent posteromedial shaft of tibia	Calcaneus via calcaneal tendon (with gastrocnemius)	Plantar flexes, inverts, and adducts foot	Sciatic nerve, tibial branch
Tibialis posterior	Interosseous membrane and adjacent shafts of tibia and fibula	Tarsals and metatarsal bones	Adducts, inverts, and plantar flexes foot	As above
PRIMARY ACTION AT THE TOES **Flexors**				
Flexor digitorum longus	Posteromedial surface of tibia	Inferior surfaces of distal phalanges, toes 2–5	Plantar flexes toes 2–5	As above
Flexor hallucis longus	Posterior surface of fibula	Inferior surface, distal phalanx of great toe	Plantar flexes great toe	As above
Extensors				
Extensor digitorum longus	Lateral condyle of tibia, anterior surface of fibula	Superior surfaces of phalanges, toes 2–5	Extends toes 2–5	Deep peroneal nerve
Extensor hallucis longus	Anterior surface of fibula	Superior surface, distal phalanx of great toe	Extends great toe	As above

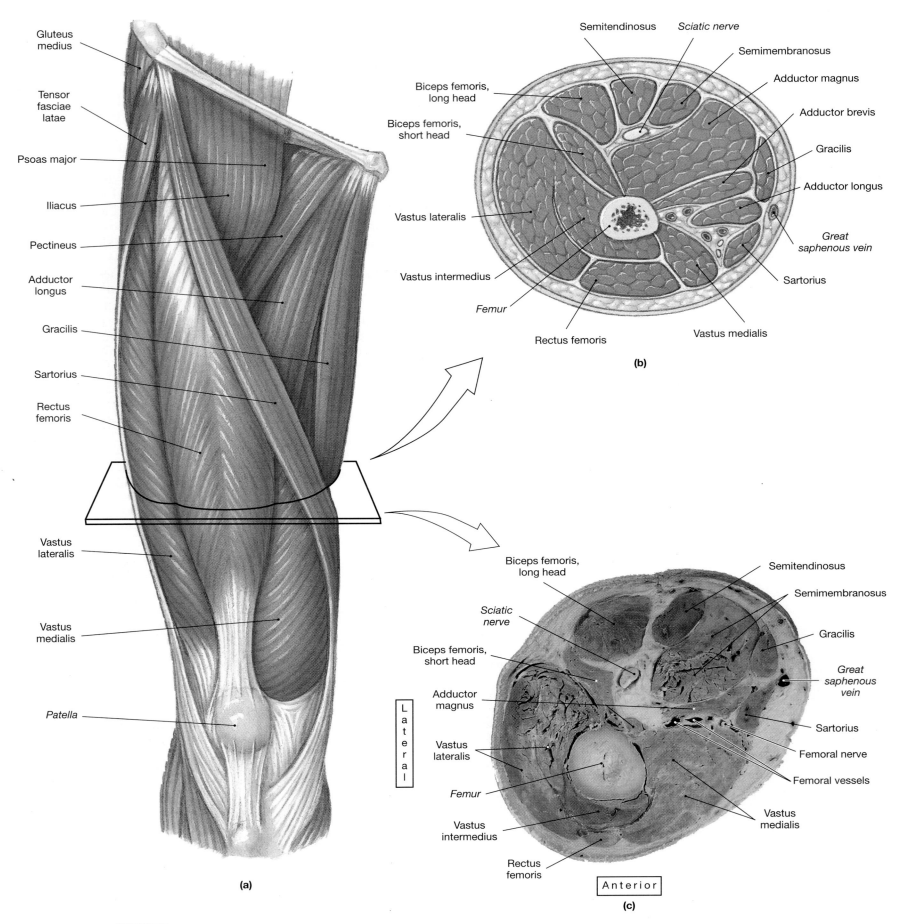

Gluteus medius

Tensor fasciae latae

Psoas major

Iliacus

Pectineus

Adductor longus

Gracilis

Sartorius

Rectus femoris

Vastus lateralis

Vastus medialis

Patella

(a)

Semitendinosus

Sciatic nerve

Semimembranosus

Adductor magnus

Adductor brevis

Gracilis

Adductor longus

Great saphenous vein

Sartorius

Vastus medialis

Rectus femoris

Vastus intermedius

Femur

Vastus lateralis

Biceps femoris, short head

Biceps femoris, long head

(b)

Biceps femoris, long head

Semitendinosus

Semimembranosus

Sciatic nerve

Gracilis

Biceps femoris, short head

Great saphenous vein

Adductor magnus

Sartorius

Vastus lateralis

Femoral nerve

Femur

Femoral vessels

Vastus intermedius

Vastus medialis

Rectus femoris

L a t e r a l

A n t e r i o r

(c)

FIGURE 11-15
The Quadriceps Group. *See also Figures 7-10, 7-11, 7-14, and 8-12.* (a) Quadriceps and thigh muscles, anterior view. The orientation of the quadriceps muscles and the anatomical relationships between the thigh muscles. (b) Transverse section through the distal end of the thigh. (c) Transverse section of the thigh of a cadaver. See MRI SCANS 5-7, pp. 744–745.

298

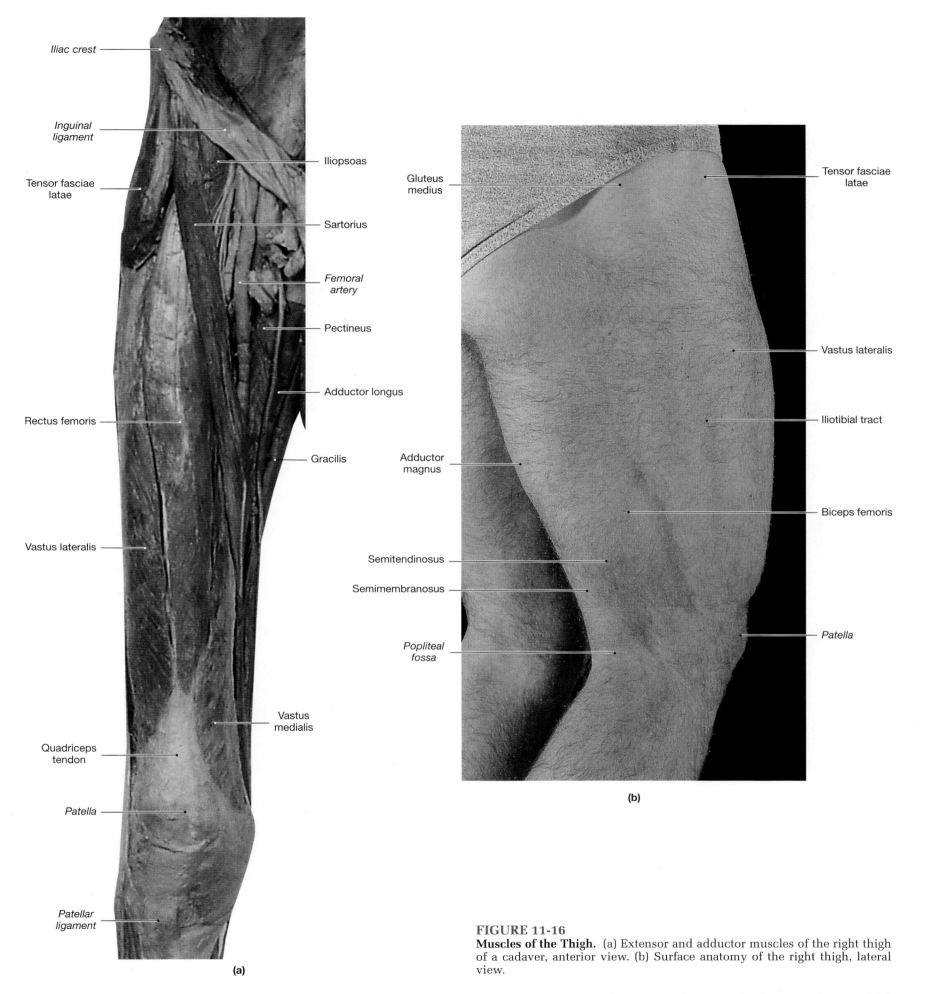

(a)

(b)

FIGURE 11-16
Muscles of the Thigh. (a) Extensor and adductor muscles of the right thigh of a cadaver, anterior view. (b) Surface anatomy of the right thigh, lateral view.

CHAPTER 11 The Muscular System: The Appendicular Musculature 299

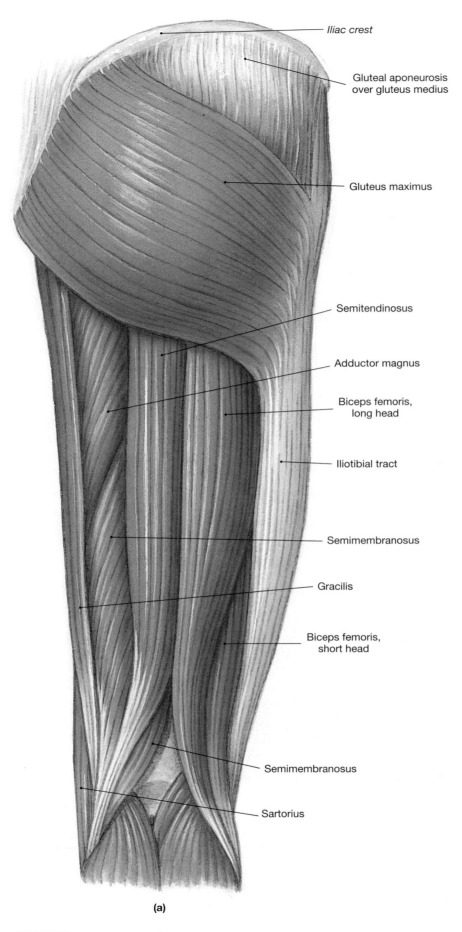

(a)

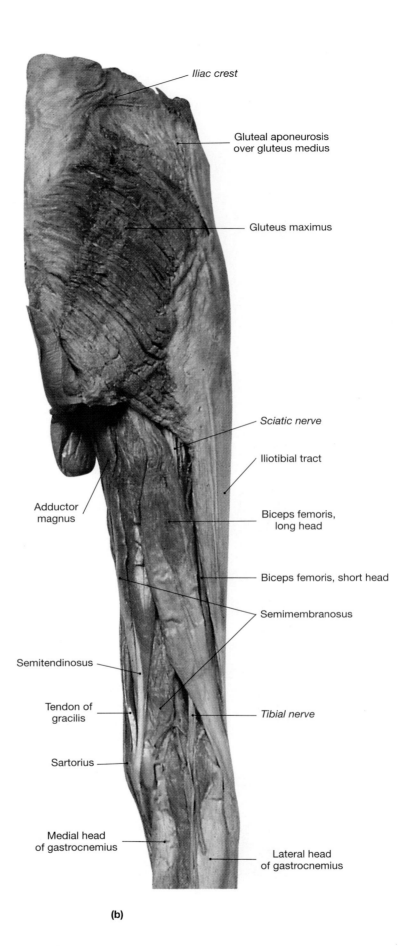

(b)

FIGURE 11-17
Thigh Muscles. (a) Superficial thigh muscles, posterior view. (b) Superficial muscles of the right thigh, posterior view of a cadaver. (c) Surface anatomy of the right thigh, posterior view. See MRI SCANS 4-7, pp. 743–745.

TABLE 11-9 Intrinsic Muscles of the Foot

Muscle	Origin	Insertion	Action	Innervation
Extensor digitorum brevis	Calcaneus (superior and lateral surfaces)	Dorsal surfaces of toes I–IV	Extends proximal phalanges of toes 1–4	Deep peroneal nerve
Abductor hallucis	Calcaneus (tuberosity on inferior surface)	Medial side of proximal phalanx of great toe	Abducts great toe	Medial plantar nerve
Flexor digitorum brevis	As above	Sides of middle phalanges, toes 2–5	Flexes the middle phalanx of toes 2–5	As above
Abductor digiti minimi	As above	Lateral side of proximal phalanx, little toe	Abducts the little toe	Lateral plantar nerve
Quadratus plantae	Calcaneus (medial surface and lateral border of inferior surface)	Tendon of flexor digitorum longus	Flexes toes 2–5	As above
Lumbricals (4)	Tendons of flexor digitorum longus	Insertions of extensor digitorum longus	Flexes proximal phalanges, extends middle phalanges, toes 2–5	Medial plantar nerve (1), lateral plantar nerve (2–4)
Flexor hallucis brevis	Cuboid and lateral cuneiform bones	Proximal phalanx of great toe	Flexes proximal phalanx of great toe	Medial plantar nerve
Adductor hallucis	Bases of metatarsals II–IV and plantar ligaments	Proximal phalanx of great toe	Adducts the great toe	Lateral plantar nerve
Flexor digiti minimi brevis	Base of 5th metatarsal bone	Lateral side of proximal phalanx of little (5th) toe	Flexes proximal phalanx of little toe	As above

Interossei
  dorsal (4)
  plantar (3)
	Sides of metatarsal bones
	Bases and medial sides of metatarsal bones
		Sides of toes 2–4
		Sides of toes 3–5
			Abducts the toes
			Adducts the toes
				As above
				As above

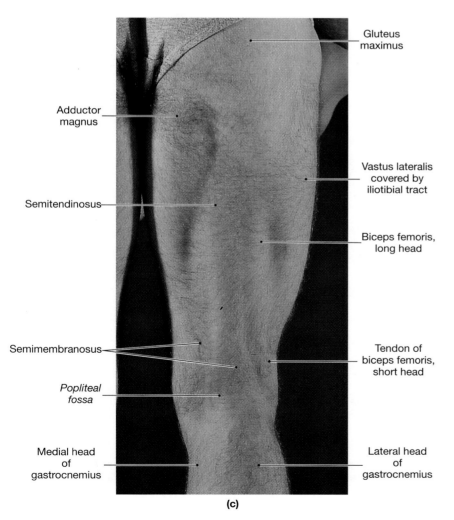

(c)

√ **What leg movement would be impaired by injury to the obturator muscle?**

√ **One often hears of athletes suffering a "pulled hamstring." To what does this phrase refer?**

√ **How would you expect a torn calcaneal tendon to affect movement of the foot?**

■ CLINICAL BRIEF
Intramuscular Injections

Drugs are frequently injected into tissues, rather than directly into the circulation. This method enables the physician to introduce a large amount of a drug at one time, yet have it enter the circulation gradually. In an **intramuscular (IM) injection**, the drug is introduced into the mass of a large skeletal muscle. Uptake is usually faster and accompanied by less tissue irritation than when drugs are administered *intradermally* or *subcutaneously* (injected into the dermis or subcutaneous layers). ∞ [p. 96] Up to 5 mℓ of fluid may be injected at one time, and multiple injections are possible.

The most common complications involve accidental injection into a blood vessel or piercing of a nerve. The sudden entry of massive quantities of a drug into the bloodstream can have unpleasant or even fatal consequences, while damage to a nerve can cause motor paralysis or sensory loss. As a result, the site of injection must be selected with care. Bulky muscles that contain few large vessels or nerves make ideal targets, and the gluteus medius or the posterior, lateral, superior portion of the gluteus maximus is often selected. The deltoid muscle of the arm, about 2.5 cm (1 in.) distal to the acromion, is another popular site. Probably the most satisfactory from an anatomical point of view is the vastus lateralis of the thigh, for an injection into this thick muscle will not encounter vessels or nerves. This is the preferred site in infants and young children whose gluteal and deltoid muscles are relatively small.

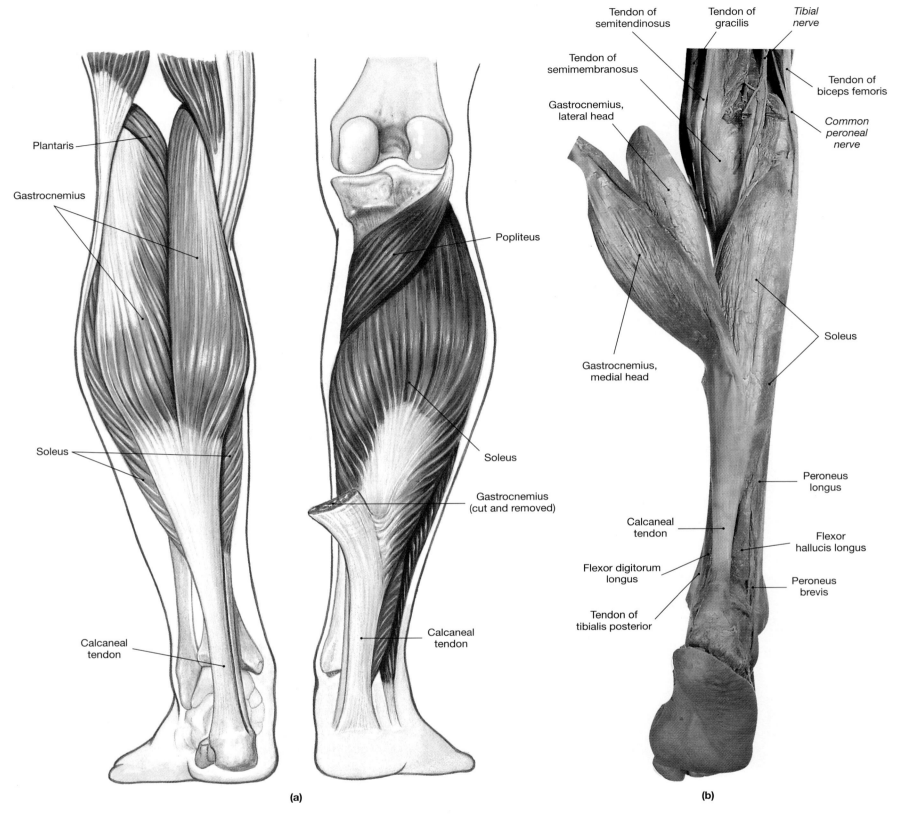

Plantaris

Gastrocnemius

Soleus

Calcaneal
tendon

Popliteus

Soleus

Gastrocnemius
(cut and removed)

Calcaneal
tendon

(a)

Tendon of
semitendinosus

Tendon of
gracilis

*Tibial
nerve*

Tendon of
semimembranosus

Tendon of
biceps femoris

Gastrocnemius,
lateral head

*Common
peroneal
nerve*

Soleus

Gastrocnemius,
medial head

Peroneus
longus

Calcaneal
tendon

Flexor
hallucis longus

Flexor digitorum
longus

Peroneus
brevis

Tendon of
tibialis posterior

(b)

Posterior view, superficial muscles

FIGURE 11-18
Superficial Muscles That Move the Foot and Toes, Posterior View. *See also Figures 7-16 and 7-17.* (a) Superficial muscles of the posterior surface of the leg; these large muscles are primarily responsible for plantar flexion. (b) Superficial muscles of the right leg of a cadaver. See MRI SCANS 5-7, pp. 744–745.

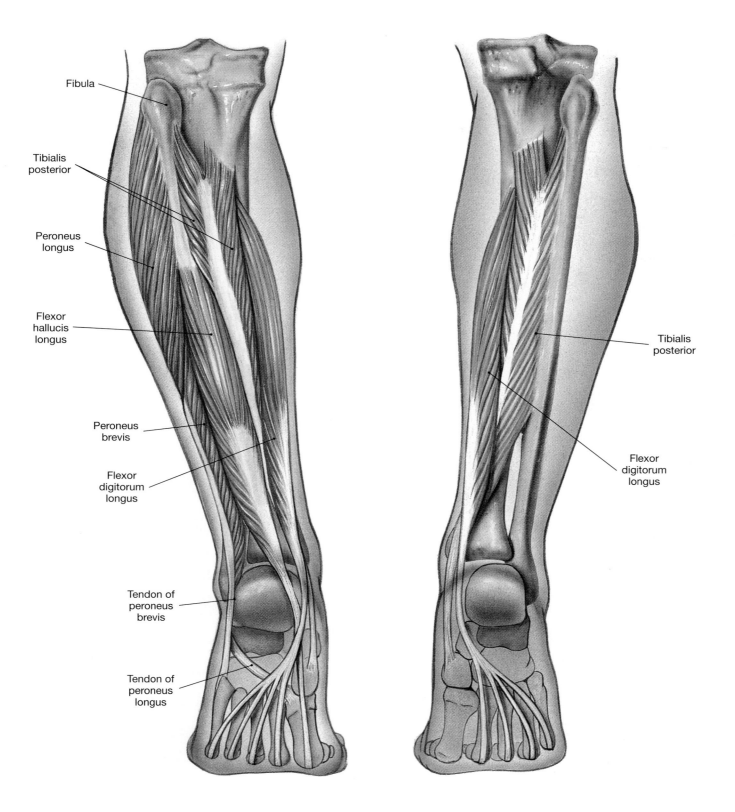

Fibula

Tibialis
posterior

Peroneus
longus

Flexor
hallucis
longus

Peroneus
brevis

Flexor
digitorum
longus

Tendon of
peroneus
brevis

Tendon of
peroneus
longus

Tibialis
posterior

Flexor
digitorum
longus

Posterior view, Deeper view

FIGURE 11-19
Deep Muscles That Move the Foot and Toes, Posterior View. These smaller muscles are primarily concerned with plantar flexion of the foot and flexion of the toes. See MRI SCANS 5–7, pp. 744–745.

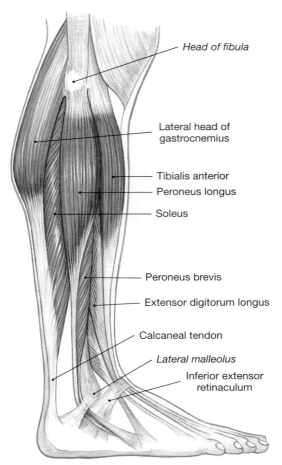

Head of fibula

Lateral head of
gastrocnemius

Tibialis anterior
Peroneus longus

Soleus

Peroneus brevis

Extensor digitorum longus

Calcaneal tendon

Lateral malleolus

Inferior extensor
retinaculum

(a) Lateral view

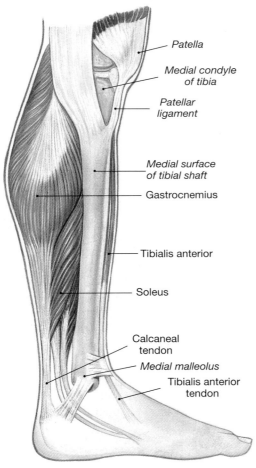

Patella

*Medial condyle
of tibia*

*Patellar
ligament*

*Medial surface
of tibial shaft*

Gastrocnemius

Tibialis anterior

Soleus

Calcaneal
tendon

Medial malleolus

Tibialis anterior
tendon

(b) Medial view

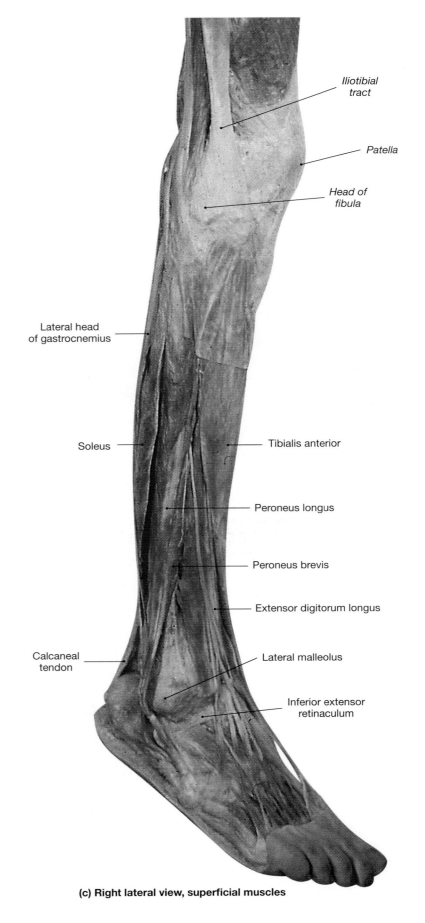

*Iliotibial
tract*

Patella

*Head of
fibula*

Lateral head
of gastrocnemius

Soleus

Tibialis anterior

Peroneus longus

Peroneus brevis

Extensor digitorum longus

Lateral malleolus

Calcaneal
tendon

Inferior extensor
retinaculum

(c) Right lateral view, superficial muscles

FIGURE 11-20
Muscles That Move the Foot and Toes, Medial and Lateral Views. (a) Lateral view of leg. (b) Medial view of leg. (c) Lateral view of the right leg of a cadaver.

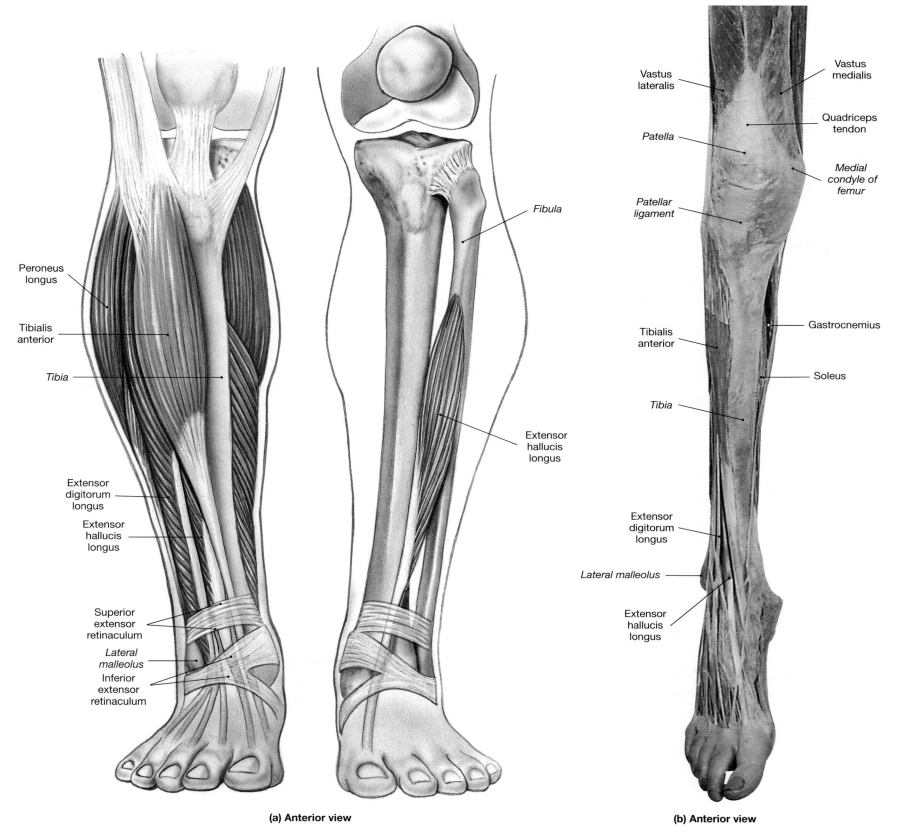

(a) Anterior view

(b) Anterior view

Figure labels (left diagram, a):
- Peroneus longus
- Tibialis anterior
- *Tibia*
- Extensor digitorum longus
- Extensor hallucis longus
- Superior extensor retinaculum
- *Lateral malleolus*
- Inferior extensor retinaculum

Figure labels (middle diagram, a):
- *Fibula*
- Extensor hallucis longus

Figure labels (right photo, b):
- Vastus lateralis
- Vastus medialis
- Quadriceps tendon
- *Patella*
- *Medial condyle of femur*
- *Patellar ligament*
- Tibialis anterior
- Gastrocnemius
- Soleus
- *Tibia*
- Extensor digitorum longus
- *Lateral malleolus*
- Extensor hallucis longus

FIGURE 11-21
Muscles That Move the Foot and Toes, Anterior View. (a) Superficial muscles of the legs. These muscles on the anterior surface of the leg are primarily concerned with dorsiflexion of the foot and extension of the toes. (b) Superficial muscles of the right leg of a cadaver, anterior view.

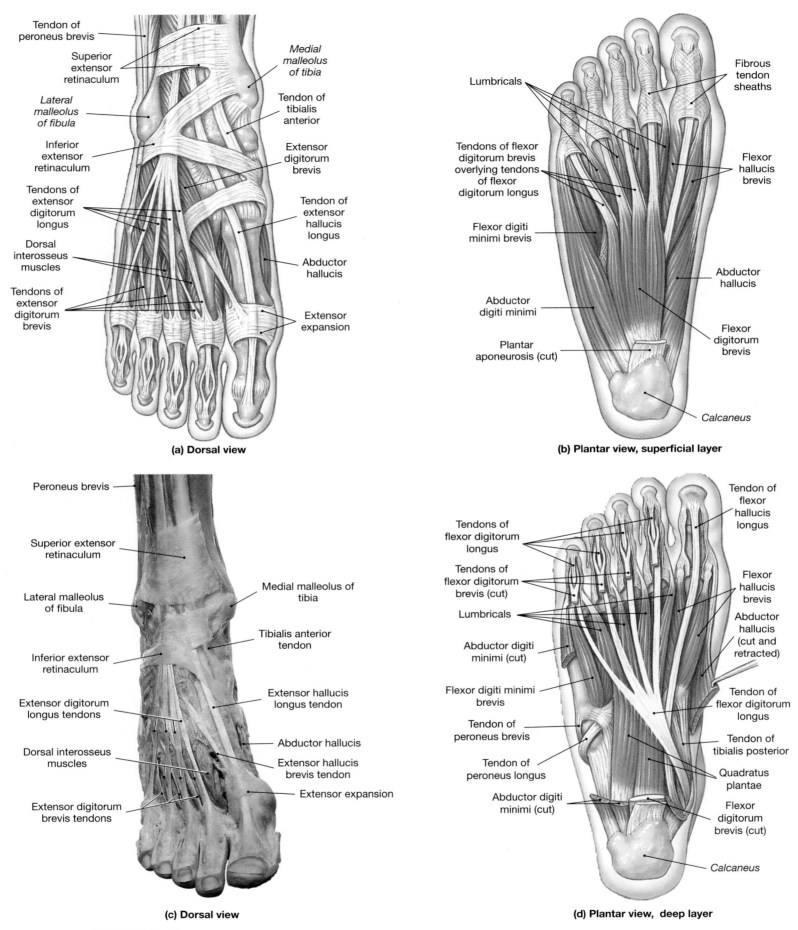

(a) Dorsal view

Tendon of peroneus brevis
Superior extensor retinaculum
Lateral malleolus of fibula
Inferior extensor retinaculum
Tendons of extensor digitorum longus
Dorsal interosseus muscles
Tendons of extensor digitorum brevis

Medial malleolus of tibia
Tendon of tibialis anterior
Extensor digitorum brevis
Tendon of extensor hallucis longus
Abductor hallucis
Extensor expansion

(b) Plantar view, superficial layer

Lumbricals
Tendons of flexor digitorum brevis overlying tendons of flexor digitorum longus
Flexor digiti minimi brevis
Abductor digiti minimi
Plantar aponeurosis (cut)

Fibrous tendon sheaths
Flexor hallucis brevis
Abductor hallucis
Flexor digitorum brevis
Calcaneus

(c) Dorsal view

Peroneus brevis
Superior extensor retinaculum
Lateral malleolus of fibula
Inferior extensor retinaculum
Extensor digitorum longus tendons
Dorsal interosseus muscles
Extensor digitorum brevis tendons

Medial malleolus of tibia
Tibialis anterior tendon
Extensor hallucis longus tendon
Abductor hallucis
Extensor hallucis brevis tendon
Extensor expansion

(d) Plantar view, deep layer

Tendons of flexor digitorum longus
Tendons of flexor digitorum brevis (cut)
Lumbricals
Abductor digiti minimi (cut)
Flexor digiti minimi brevis
Tendon of peroneus brevis
Tendon of peroneus longus
Abductor digiti minimi (cut)

Tendon of flexor hallucis longus
Flexor hallucis brevis
Abductor hallucis (cut and retracted)
Tendon of flexor digitorum longus
Tendon of tibialis posterior
Quadratus plantae
Flexor digitorum brevis (cut)
Calcaneus

FIGURE 11-22
Muscles That Move the Foot and Toes, Dorsal and Plantar Views. (a) Dorsal view of right foot. (b) Plantar view of right foot, superficial layer. (c) Muscles, tendons, and nerves of the right foot (dorsal view). (d) Plantar view of right foot, deep layer. See MRI SCAN 8, p. 745.

306

Compartment Syndrome

In the arms and legs, the interconnections between the superficial fascia, the deep fascia of the muscles, and the periostea of the appendicular skeleton are quite substantial. The muscles within a limb are effectively isolated in **compartments** formed by dense collagenous sheets, as shown in Figure 11-23●. Blood vessels and nerves traveling to specific muscles within the limb enter and branch within the appropriate compartments. When a crushing injury, severe contusion, or strain occurs, the blood vessels within one or more compartments may be damaged. When damaged, these compartments become swollen with tissue, fluid, and blood that has leaked from damaged blood vessels. Because the connective tissue partitions are very strong, the accumulated fluid cannot escape, and pressures rise within the affected compartments. Eventually compartment pressures may become so high that they compress the regional blood vessels and eliminate the circulatory supply to the muscles and nerves of the compartment. This compression produces a condition of **ischemia** (is-KĒ-mē-a), or "blood starvation," known as the **compartment syndrome.** Slicing into the compartment along its longitudinal axis or implanting a drain are emergency measures used to relieve the pressure. If such steps are not taken, the contents of the compartment will suffer severe damage. Nerves in the affected compartment will be destroyed after 2–4 hours of ischemia, although they can regenerate to some degree if the circulation is restored. After 6 hours or more, the muscle tissue will also be destroyed, and no regeneration will occur. The muscles will be replaced by scar tissue, and shortening of the connective tissue fibers may result in *contracture.*

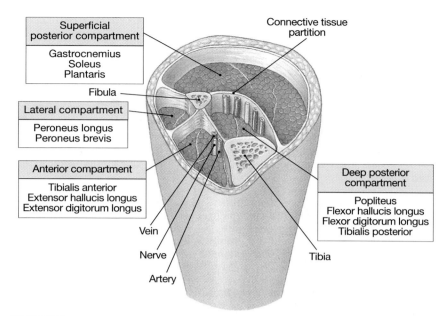

FIGURE 11-23
Musculoskeletal Compartments. A diagrammatic section through the right leg, with the muscles removed to show the arrangement of the compartments. A section through the thigh or arm would show a comparable arrangement of dense connective tissue partitions. The anterior and lateral compartments of the leg contain muscles of the extensor/flexor series, while the posterior compartments ensheathe the flexor/plantar flexor muscles.

Related Clinical Terms

rotator cuff: The muscles that surround the shoulder joint; a frequent site of sports injuries. *[p. 283]*

bone bruise: Bleeding within the periosteum of a bone. † *Sports Injuries [p. 758]*

bursitis: Inflammation of the bursae around one or more joints. † *Sports Injuries [p. 758]*

stress fractures: Cracks or breaks in bones subjected to repeated stress or trauma. † *Sports Injuries [p. 758]*

muscle cramps: Prolonged, involuntary, painful muscular contractions. † *Sports Injuries [p. 758]*

sprains: Tears or breaks in ligaments or tendons. † *Sports Injuries [p. 758]*

strains: Tears or breaks in muscles. † *Sports Injuries [p. 758]*

tendinitis: Inflammation of the connective tissue surrounding a tendon. † *Sports Injuries [p. 758]*

carpal tunnel syndrome: An inflammation within the sheath surrounding the flexor tendons of the palm. *[p. 291]*

intramuscular (IM) injection: Administering a drug by injecting it into the mass of a large skeletal muscle. *[p. 301]*

ischemia (is-KĒ-mē-a)**:** A condition of "blood starvation" resulting from compression of regional blood vessels. *[p. 307]*

compartment syndrome: Ischemia resulting from accumulated blood and fluid trapped within a musculoskeletal compartment. *[p. 307]*

CHAPTER SUMMARY AND REVIEW

STUDY OUTLINE

INTRODUCTION *[p. 278]*

1. The **appendicular musculature** is responsible for stabilizing the pectoral and pelvic girdles and for moving the upper and lower limbs.

MUSCLES OF THE SHOULDER AND UPPER EXTREMITY *[p. 278]*

Muscles That Position the Pectoral Girdle *[p. 279]*
1. The **trapezius** affects the positions of the shoulder girdle, head, and neck. Other muscles inserting on the scapula include the **rhomboideus**, the **levator scapulae**, the **serratus anterior**, the **subclavius**, and the **pectoralis minor** *(see Figures 11-1 to 11-4 and Table 11-1).*

Muscles That Move the Arm *[p. 279]*
2. The **deltoid** and the **supraspinatus** are important abductors. The **subscapularis** and the **teres major** rotate the arm medially; the **infraspinatus** and **teres minor** perform lateral rotation; and the **coraco-brachialis** flexes and adducts the humerus *(see Figures 11-1/11-4/11-5 and Table 11-2).*

3. The **pectoralis major** flexes the arm, while the **latissimus dorsi** extends it *(see Figures 11-1/11-5 and Table 11-2).*

rotator cuff

Muscles That Move the Forearm and Hand *[p. 283]*
4. The primary actions of the **biceps brachii** and the **triceps brachii** (long head) affect the elbow joint. The **brachialis** and **brachioradialis** flex the forearm, opposed by the **anconeus**. The **flexor carpi ulnaris**, the **flexor carpi radialis**, and the **palmaris longus** cooperate to flex the wrist. They are opposed by the **extensor carpi radialis** and the **extensor carpi ulnaris**. The **pronator teres** and **pronator quadratus** pronate the forearm, opposed by the **supinator** *(see Figures 11-4/11-7/11-8 and Table 11-3).*

Muscles That Move the Fingers *[p. 285]*
5. Muscles that perform flexion and extension of the fingers are illustrated in *Figures 11-7 to 11-11* and detailed in *Tables 11-4/11-5.*

tendon sheaths • extensor retinaculum • flexor retinaculum

MUSCLES OF THE LOWER EXTREMITY *[p. 293]*

Muscles That Move the Thigh *[p. 293]*
1. **Gluteal muscles** cover the lateral surface of the ilium. The largest is the **gluteus maximus**, which shares an insertion with the **tensor fasciae latae**; together these muscles pull on the **iliotibial tract** *(see Figures 11-1/11-4/11-13 and Table 11-6).*

gluteus medius • gluteus minimus

2. The **piriformis** and the **obturator** muscles are the most important **lateral rotators**. The **adductors** can produce a variety of movements *(see Figure 11-13 and Table 11-6).*

adductor magnus • adductor brevis • adductor longus • pectineus • gracilis

3. The **psoas major** and the **iliacus** merge to form the **iliopsoas** muscle, a powerful flexor of the thigh *(see Figure 11-14 and Table 11-6).*

Muscles That Move the Leg *[p. 296]*
4. The flexors of the leg include the **biceps femoris**, **semimembranosus**, and **semitendinosus** (the three **hamstrings**), and the **sartorius**. The **popliteus** unlocks the knee joint *(see Figures 11-15 to 11-18 and Table 11-7).*

5. Collectively the *knee extensors* are known as the **quadriceps femoris**. This group includes the three **vastus** muscles and the **rectus femoris** *(see Figures 11-15/11-16 and Table 11-7).*

Muscles That Move the Foot and Toes *[p. 297]*
6. The **gastrocnemius** and **soleus** muscles produce plantar flexion. A pair of **peroneus muscles** produce eversion as well as plantar flexion *(see Figures 11-18/11-20 and Table 11-8).*

fabella • calcaneal tendon • tibialis anterior

7. Smaller muscles of the calf and shin position the foot and move the toes. Precise control of the phalanges is provided by muscles originating at the tarsals and metatarsals *(see Figures 11-19 to 11-22 and Table 11-9).*

interosseus muscles • superior/inferior extensor retinacula

---

1   REVIEW OF CHAPTER OBJECTIVES

1. Describe the functions of the appendicular musculature.
2. Identify and locate the principal appendicular muscles of the body, together with their origins and insertions, and describe their actions.
3. Compare the major muscle groups of the upper and lower extremities, and relate their differences to their functional roles.

2   REVIEW OF CONCEPTS

1. How do the functions of the pectoral muscles differ from those of the pelvic muscles?
2. Why is the trapezius muscle able to participate in a wide variety of movements of the shoulder girdle?
3. How do the combined (synergistic) actions of the subclavius and pectoralis minor muscles affect the position of the clavicle and the shoulder joint?

4. Why are the muscles of the rotator cuff important in controlling the position of the head of the humerus?
5. If the pectoralis major flexes the arm while the latissimus dorsi extends it, how is it possible that these two muscles can together produce adduction and medial rotation of the humerus?
6. In addition to being the prime flexor of the forearm, why is the biceps brachii also capable of strong supination?
7. Why is it that muscles, such as the flexor carpi ulnaris and flexor carpi radialis, which produce the opposite actions of adduction and abduction, can flex the wrist when they act synergistically?
8. Why are the pronator teres and supinator muscles considered to be antagonists, and what actions do they affect?
9. Why are the extrinsic muscles of the hand located on the forearm?
10. What is the function of a synovial sheath?
11. How does the tensor fasciae latae muscle act synergistically with the gluteus maximus muscle?
12. Why are the thigh adductor muscles all located inferior to the acetabular surface?

11. How does the tensor fasciae latae muscle act synergistically with the gluteus maximus muscle?

12. Why are the thigh adductor muscles all located inferior to the acetabular surface?

13. How does the location of the origin of the "hamstring" muscle group affect the movements they make?

14. What is the significance of the popliteus muscle with regard to the knee joint "lock" mechanism?

15. What is the function of the patella?

16. How does the extensor retinaculum affect the orientation of the muscles that move the toes?

3 CRITICAL THINKING AND CLINICAL APPLICATION QUESTIONS

1. Among the most common shoulder injuries is injury to the rotator cuff. It results from excessive force generated by repeated movement of the humerus to the limits of its range. What structures are injured in this situation, and what could be done to treat this problem?

2. A cashier in a very busy department store begins to experience numbness in the palmar region and pain upon flexion of the hands as well as weakness of the abductor pollicis muscle. What could cause this problem?

3. An anatomy student wisely decides to be immunized against the hepatitis B virus and is given the first of a series of injections. Where is the injection most likely to be given, and why is this site chosen?

4. A player on the tennis team is training diligently for a series of important matches, but after several hours she begins to feel pain in the anterior tibial region of both legs. The areas are swollen and tender, so she decides to consult the team physician. If the diagnosis is compartment syndrome, what is the problem, and what can be done to ease the pain and swelling?

12
Surface Anatomy

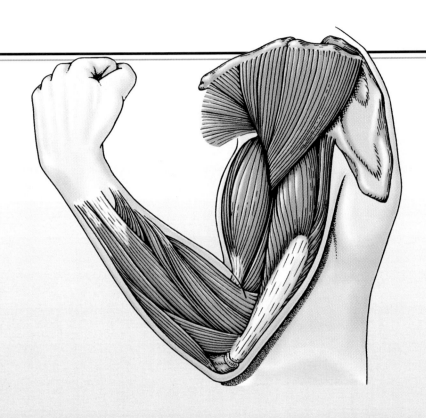

CHAPTER OUTLINE AND OBJECTIVES

This chapter focuses attention on anatomical structures that can be identified from the body surface. **Surface anatomy** is the study of anatomical landmarks on the exterior of the human body. The photographs in this chapter survey the entire body, providing a visual tour that highlights skeletal landmarks and muscle contours. Chapter 1 provided an overview of surface anatomy (Figures 1-8 and 1-9). ∞ [pp. 14, 15] Now that you are familiar with the basic anatomy of the skeletal and muscular systems, a detailed examination of surface anatomy will help demonstrate the structural and functional relationships between those systems. Many of the figures in earlier chapters included views of surface anatomy; those figures will be referenced throughout this chapter.

Surface anatomy has many practical applications. For example, an understanding of surface anatomy is crucial to medical examination in a clinical setting. In the laboratory, a familiarity with surface anatomy is essential for both invasive and noninvasive laboratory procedures. ✝ *Anatomy and Observation [p. 749]*

REGIONAL APPROACH TO SURFACE ANATOMY

Surface anatomy is best studied using a regional approach. The regions are: *head and neck, thorax, abdomen, upper extremity*, and *lower extremity*. This chapter presents the information in pictorial fashion, using photographs of the living human body. These models have well-developed muscles and very little body fat. Because many anatomical landmarks can be hidden by a layer of subcutaneous fat, you may not find it as easy to locate these structures on your own body. In practice, anatomical observation often involves estimating the location and then palpating for specific structures. In the sections that follow, identify through visual observation and palpation the surface anatomy of the regions of the body, using the labeled photographs for reference.

Head and Neck
(Figure 12-1)

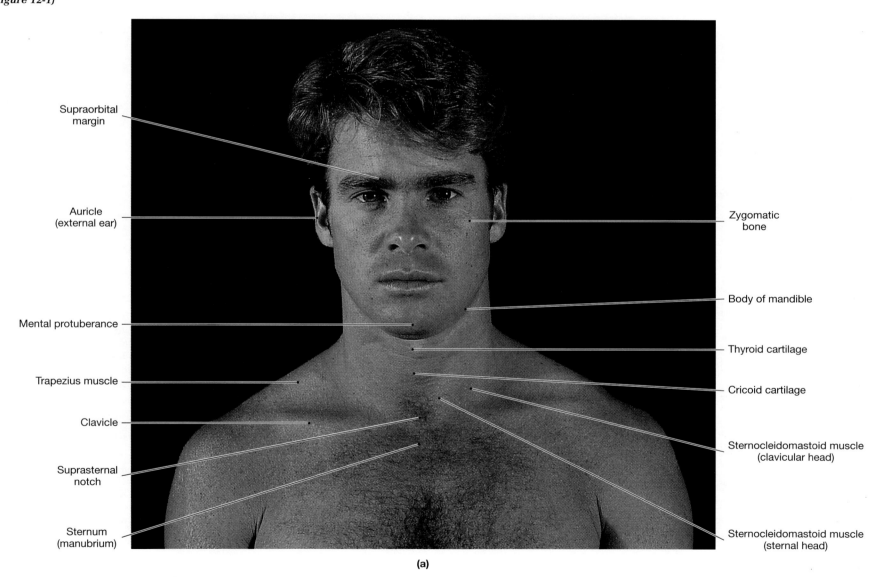

(a)

FIGURE 12-1
Head and Neck. (a) Anterior view. *For details concerning the musculature of this region, see Figures 10-5 and 10-6.* (b) Posterior cervical triangle. (c) Detail of anterior and posterior cervical triangles. The **anterior triangle** of the neck (b,c) extends from the anterior midline to the anterior border of the sternocleidomastoid muscle. The **posterior triangle** (b) extends between the posterior border of the sternocleidomastoid and the anterior border of the trapezius.

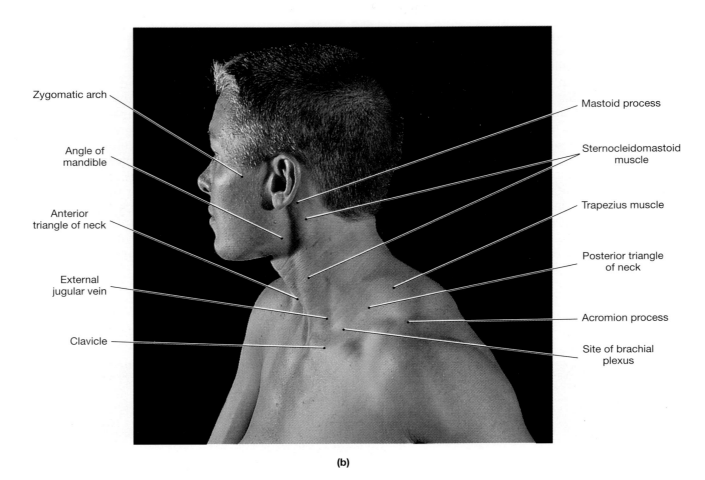

Zygomatic arch

Angle of mandible

Anterior triangle of neck

External jugular vein

Clavicle

Mastoid process

Sternocleidomastoid muscle

Trapezius muscle

Posterior triangle of neck

Acromion process

Site of brachial plexus

(b)

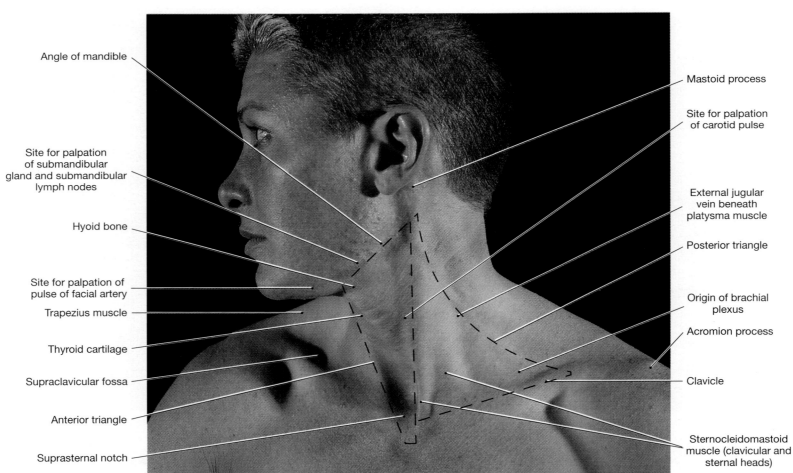

Angle of mandible

Site for palpation of submandibular gland and submandibular lymph nodes

Hyoid bone

Site for palpation of pulse of facial artery

Trapezius muscle

Thyroid cartilage

Supraclavicular fossa

Anterior triangle

Suprasternal notch

Mastoid process

Site for palpation of carotid pulse

External jugular vein beneath platysma muscle

Posterior triangle

Origin of brachial plexus

Acromion process

Clavicle

Sternocleidomastoid muscle (clavicular and sternal heads)

FIGURE 12-1 continued

(c)

Thorax

The Anterior Thorax
(Figure 12-2a)

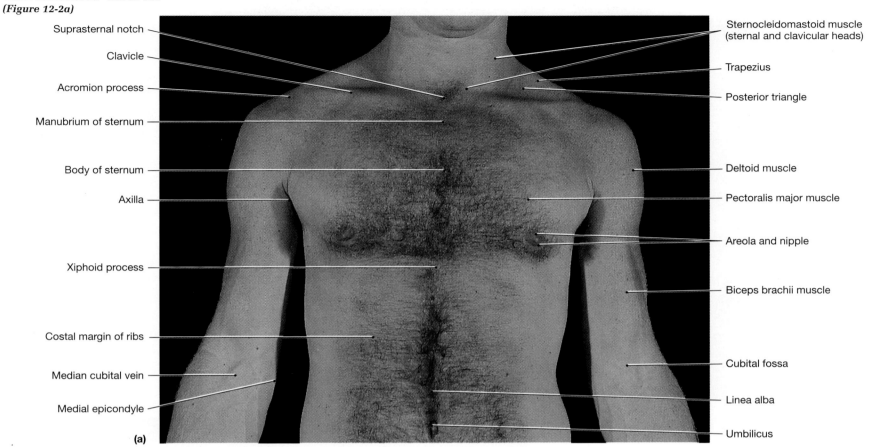

Suprasternal notch

Clavicle

Acromion process

Manubrium of sternum

Body of sternum

Axilla

Xiphoid process

Costal margin of ribs

Median cubital vein

Medial epicondyle

Sternocleidomastoid muscle (sternal and clavicular heads)

Trapezius

Posterior triangle

Deltoid muscle

Pectoralis major muscle

Areola and nipple

Biceps brachii muscle

Cubital fossa

Linea alba

Umbilicus

(a)

Back and Shoulder Regions
(Figure 12-2b)

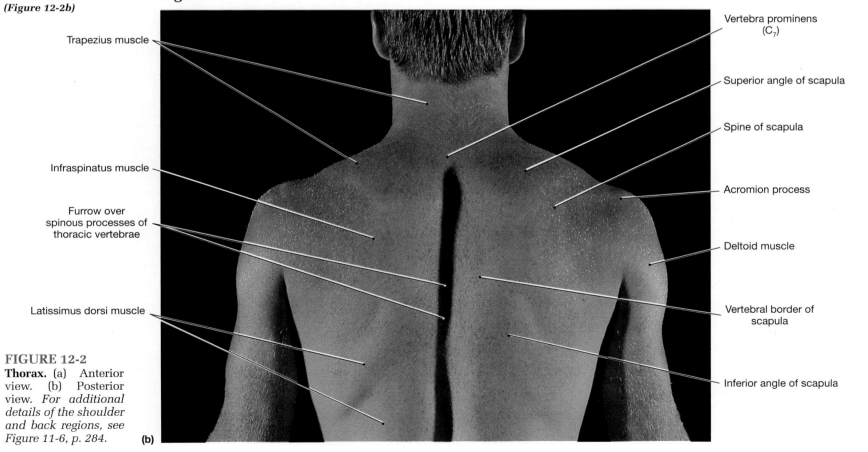

Trapezius muscle

Infraspinatus muscle

Furrow over spinous processes of thoracic vertebrae

Latissimus dorsi muscle

Vertebra prominens (C₇)

Superior angle of scapula

Spine of scapula

Acromion process

Deltoid muscle

Vertebral border of scapula

Inferior angle of scapula

FIGURE 12-2
Thorax. (a) Anterior view. (b) Posterior view. *For additional details of the shoulder and back regions, see Figure 11-6, p. 284.*

(b)

Abdomen
(Figure 12-3)

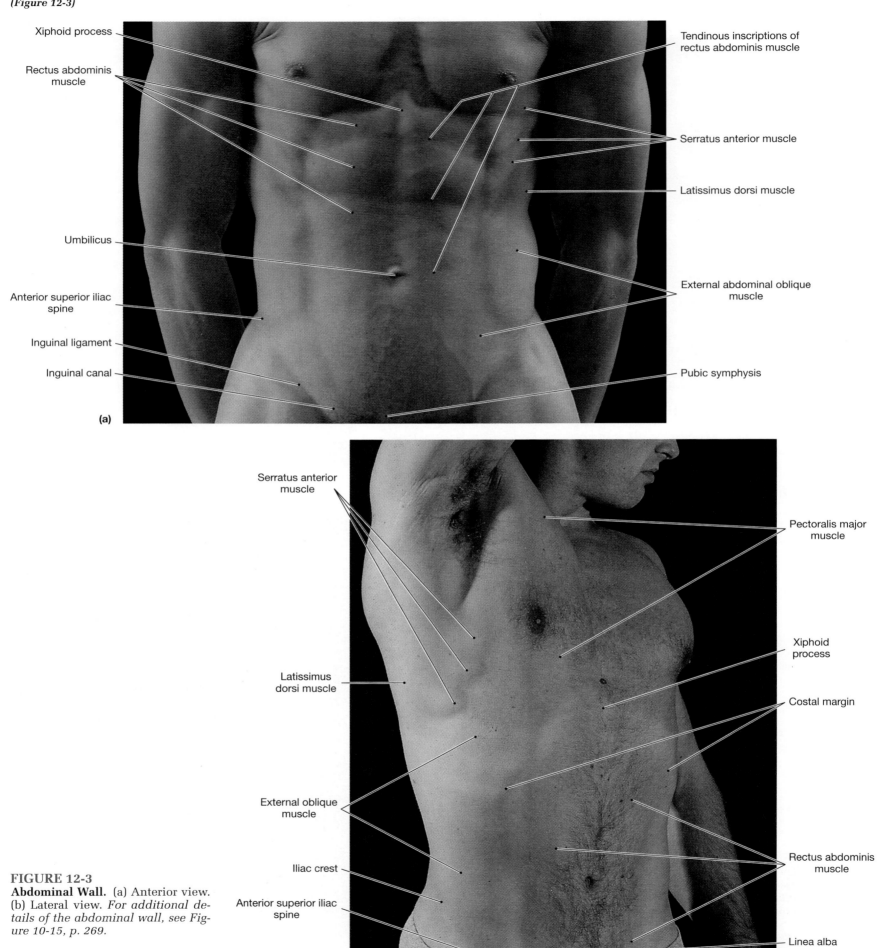

Xiphoid process

Rectus abdominis muscle

Umbilicus

Anterior superior iliac spine

Inguinal ligament

Inguinal canal

(a)

Tendinous inscriptions of rectus abdominis muscle

Serratus anterior muscle

Latissimus dorsi muscle

External abdominal oblique muscle

Pubic symphysis

Serratus anterior muscle

Latissimus dorsi muscle

External oblique muscle

Iliac crest

Anterior superior iliac spine

Pectoralis major muscle

Xiphoid process

Costal margin

Rectus abdominis muscle

Linea alba

(b)

FIGURE 12-3
Abdominal Wall. (a) Anterior view.
(b) Lateral view. *For additional details of the abdominal wall, see Figure 10-15, p. 269.*

314

Upper Extremity
(Figures 12-4/12-5)

Acromial end of clavicle

Deltoid muscle

Teres major muscle

Triceps brachii muscle, long head

Biceps brachii muscle

Triceps brachii muscle, lateral head

Brachialis muscle

Tendon of biceps brachii

Lateral epicondyle of humerus

Brachioradialis muscle

Extensor carpi radialis longus muscle

Olecranon process

Anconeus muscle

Extensor carpi radialis brevis muscle

Extensor digitorum muscle

Styloid process of radius

Head of ulna

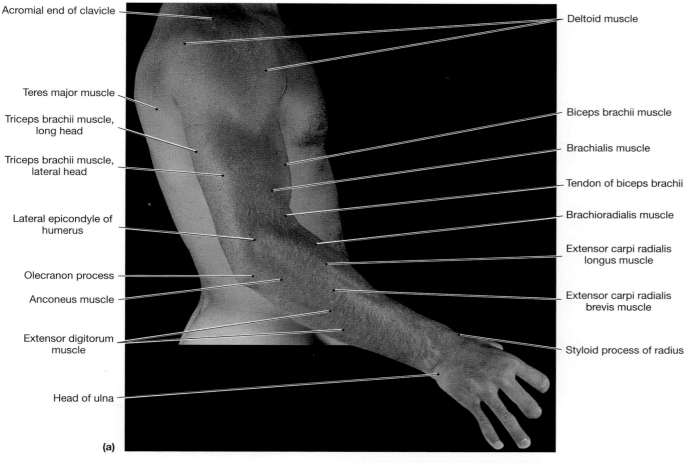

(a)

Spine of scapula

Vertebral border of scapula

Infraspinatus muscle

Teres major muscle

Inferior angle of scapula

Site of axillary nerve

Triceps brachii muscle, long head

Triceps brachii muscle, lateral head

Triceps brachii muscle, medial head

Tendon of insertion of triceps brachii

Medial epicondyle of humerus

Brachioradialis muscle

Extensor carpi radialis longus muscle

Site of palpation for ulnar nerve

Extensor carpi radialis brevis muscle

Olecranon process

Extensor digitorum muscle

Anconeus muscle

Extensor carpi ulnaris muscle

Flexor carpi ulnaris muscle

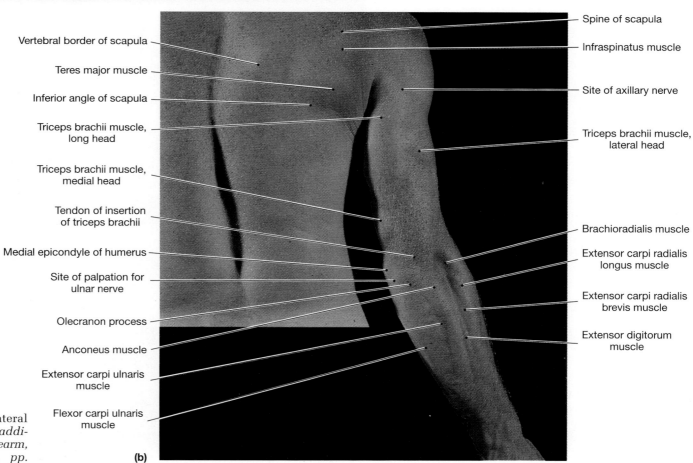

(b)

FIGURE 12-4
Right Upper Extremity. (a) Lateral view. (b) Posterior view. *For additional details of the arm and forearm, see Figures 11-7 and 11-8, pp. 286–289.*

Forearm and Wrist
(Figure 12-5)

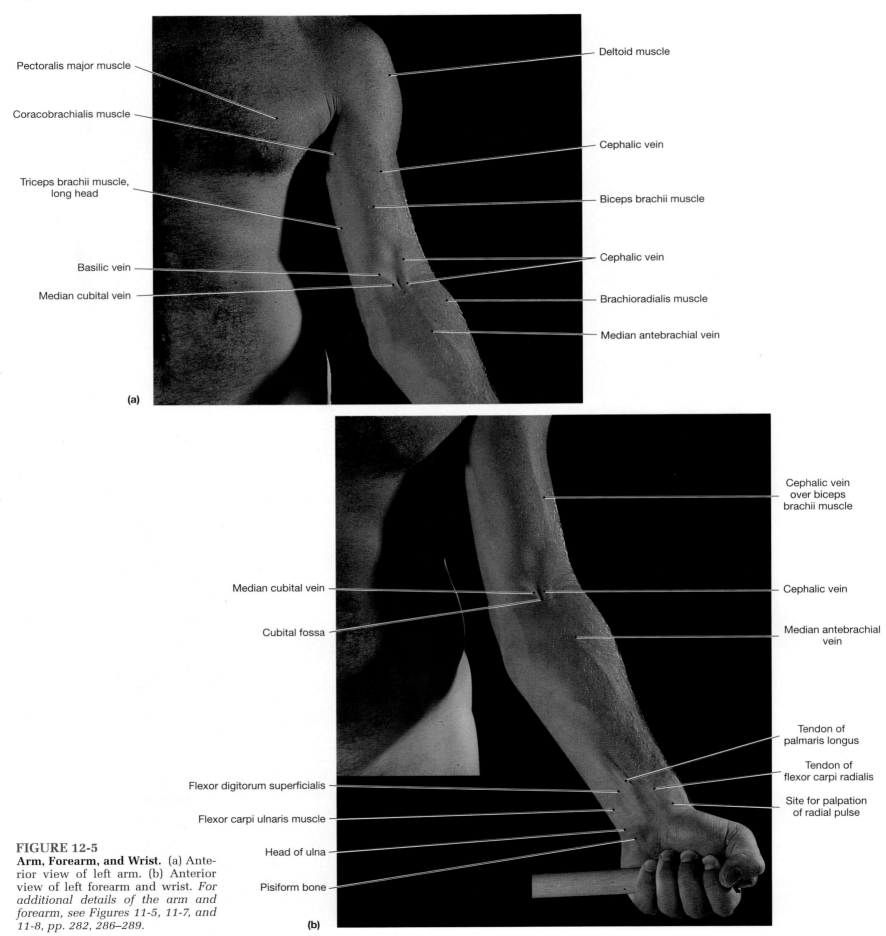

Pectoralis major muscle

Coracobrachialis muscle

Triceps brachii muscle, long head

Basilic vein

Median cubital vein

Deltoid muscle

Cephalic vein

Biceps brachii muscle

Cephalic vein

Brachioradialis muscle

Median antebrachial vein

(a)

Median cubital vein

Cubital fossa

Flexor digitorum superficialis

Flexor carpi ulnaris muscle

Head of ulna

Pisiform bone

Cephalic vein over biceps brachii muscle

Cephalic vein

Median antebrachial vein

Tendon of palmaris longus

Tendon of flexor carpi radialis

Site for palpation of radial pulse

(b)

FIGURE 12-5
Arm, Forearm, and Wrist. (a) Anterior view of left arm. (b) Anterior view of left forearm and wrist. *For additional details of the arm and forearm, see Figures 11-5, 11-7, and 11-8, pp. 282, 286–289.*

Pelvis and Lower Extremity
(Figures 12-6/12-7)

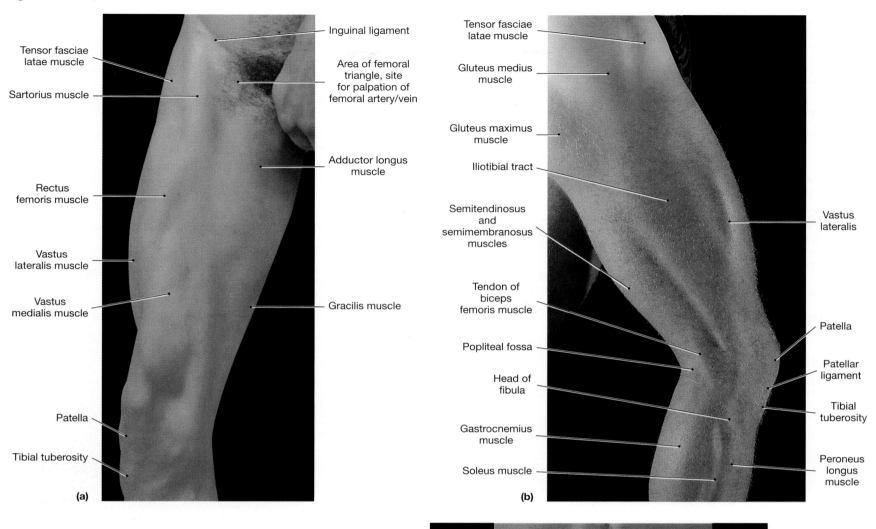

Tensor fasciae latae muscle

Sartorius muscle

Rectus femoris muscle

Vastus lateralis muscle

Vastus medialis muscle

Patella

Tibial tuberosity

Inguinal ligament

Area of femoral triangle, site for palpation of femoral artery/vein

Adductor longus muscle

Gracilis muscle

(a)

Tensor fasciae latae muscle

Gluteus medius muscle

Gluteus maximus muscle

Iliotibial tract

Semitendinosus and semimembranosus muscles

Tendon of biceps femoris muscle

Popliteal fossa

Head of fibula

Gastrocnemius muscle

Soleus muscle

Vastus lateralis

Patella

Patellar ligament

Tibial tuberosity

Peroneus longus muscle

(b)

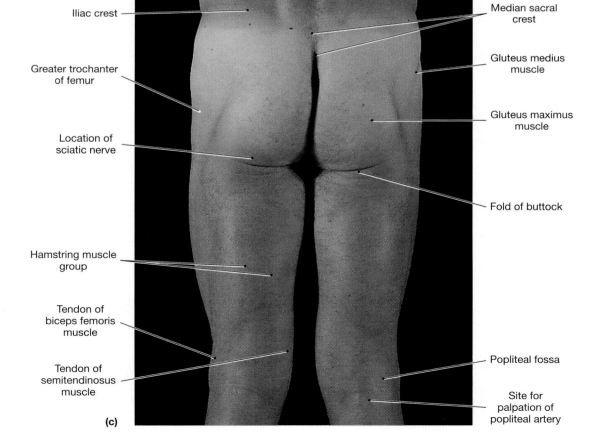

Iliac crest

Greater trochanter of femur

Location of sciatic nerve

Hamstring muscle group

Tendon of biceps femoris muscle

Tendon of semitendinosus muscle

Median sacral crest

Gluteus medius muscle

Gluteus maximus muscle

Fold of buttock

Popliteal fossa

Site for palpation of popliteal artery

(c)

FIGURE 12-6
The Pelvis and Lower Extremity. (a) Anterior surface of right thigh showing femoral region and adductor muscles. (b) Lateral view showing right gluteal region and thigh. (c) Posterior view of gluteal region and posterior surface of thigh. *For additional details of the thigh, see Figures 11-15, 11-16, and 11-17, pp. 298–301.*

Knee, Leg, Ankle, and Foot
(Figures 12-6/12-7)

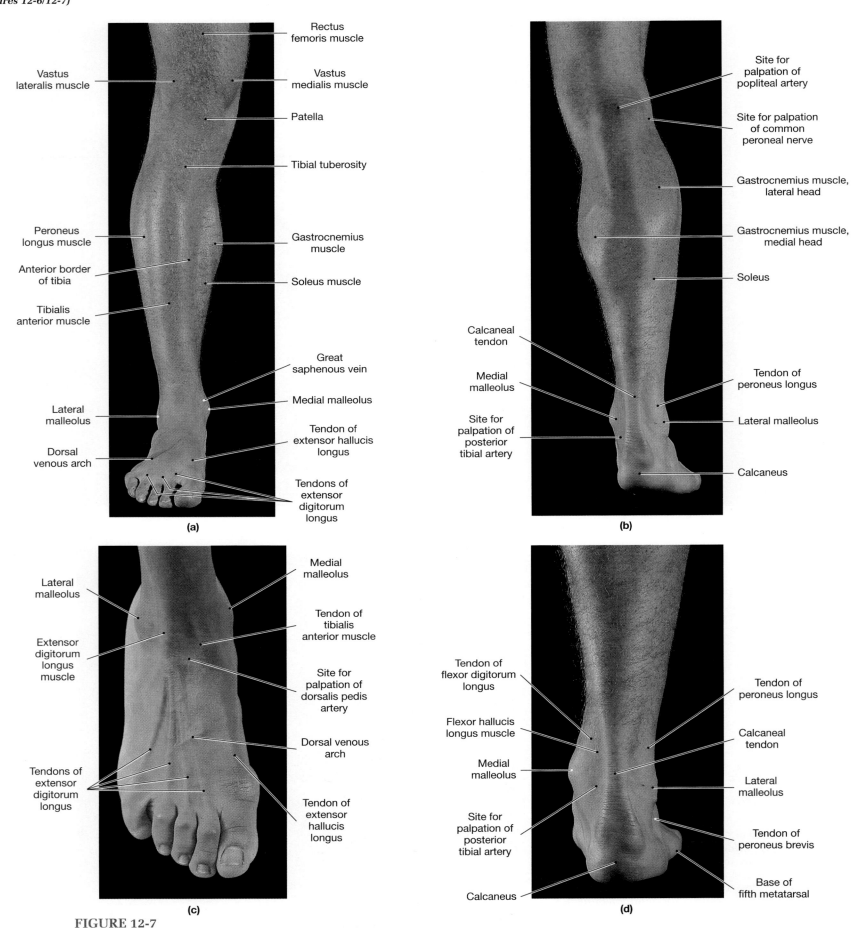

(a)

Rectus femoris muscle

Vastus lateralis muscle

Vastus medialis muscle

Patella

Tibial tuberosity

Peroneus longus muscle

Gastrocnemius muscle

Anterior border of tibia

Soleus muscle

Tibialis anterior muscle

Great saphenous vein

Lateral malleolus

Medial malleolus

Dorsal venous arch

Tendon of extensor hallucis longus

Tendons of extensor digitorum longus

(b)

Site for palpation of popliteal artery

Site for palpation of common peroneal nerve

Gastrocnemius muscle, lateral head

Gastrocnemius muscle, medial head

Soleus

Calcaneal tendon

Medial malleolus

Tendon of peroneus longus

Site for palpation of posterior tibial artery

Lateral malleolus

Calcaneus

(c)

Medial malleolus

Lateral malleolus

Tendon of tibialis anterior muscle

Extensor digitorum longus muscle

Site for palpation of dorsalis pedis artery

Dorsal venous arch

Tendons of extensor digitorum longus

Tendon of extensor hallucis longus

(d)

Tendon of flexor digitorum longus

Tendon of peroneus longus

Flexor hallucis longus muscle

Calcaneal tendon

Medial malleolus

Lateral malleolus

Site for palpation of posterior tibial artery

Tendon of peroneus brevis

Base of fifth metatarsal

Calcaneus

FIGURE 12-7
The Lower Extremity: Right Leg and Foot. (a) Knee and leg, anterior view. (b) Knee and leg, posterior view. (c) Ankle and foot, anterior view. (d) Ankle and foot, posterior view. *For a lateral view of the foot, see Figure 7-17, p. 198. For additional details of the foot, see Figure 11-22, p. 306.*

13

The Nervous System:
Neural Tissue

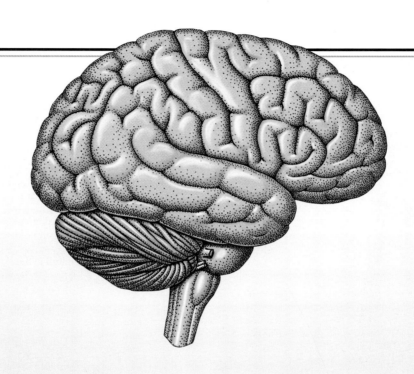

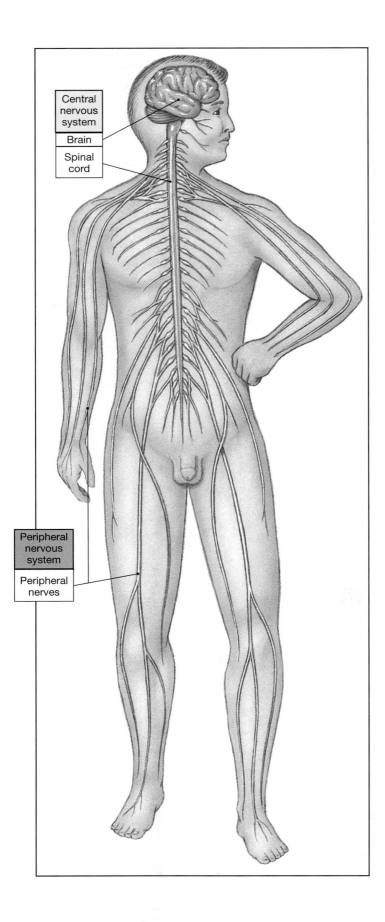

Central nervous system

Brain

Spinal cord

Peripheral nervous system

Peripheral nerves

FIGURE 13-1

The Nervous System. The nervous system includes all of the neural tissue in the body. The components of the nervous system include the brain, the spinal cord, sense organs such as the eye and ear, and the nerves that interconnect those organs and link the nervous system with other systems.

TABLE 13-1   An Introductory Glossary for the Nervous System

Major Anatomical and Functional Divisions:	
CENTRAL NERVOUS SYSTEM (CNS):	The brain and spinal cord, which contain control centers responsible for processing and integrating sensory information, planning and coordinating responses to stimuli, and providing short-term control over the activities of other systems.
PERIPHERAL NERVOUS SYSTEM (PNS):	Neural tissue outside of the CNS, whose function is to link the CNS with sense organs and other systems.
AUTONOMIC NERVOUS SYSTEM (ANS):	Components of the CNS and PNS that are concerned with the control of visceral functions.
Histology:	
NEURON:	The basic functional unit of the nervous system; a highly specialized cell (pp. 322, 326).
SENSORY NEURON:	A neuron whose axon carries sensory information from the PNS toward the CNS (p. 329).
MOTOR NEURON:	A neuron whose axon carries motor commands from the CNS toward effectors in the PNS (p. 330).
SOMA:	The cell body of a neuron (p. 322).
DENDRITES:	Neuronal processes that are specialized to respond to specific stimuli in the extracellular environment (p. 322).
AXON:	A long, slender cytoplasmic process of a neuron; axons are capable of conducting nerve impulses (action potentials) (p. 322).
ACTION POTENTIALS:	Sudden, transient changes in the membrane potential that are propagated along the surface of an axon or sarcolemma (p. 330).
MYELIN:	A membranous wrapping, produced by glial cells, that coats axons and increases the speed of action potential propagation; axons coated with myelin are said to be *myelinated* (p. 324).
NEUROGLIA OR GLIAL CELLS:	Supporting cells that interact with neurons and regulate the extracellular environment, provide defense against pathogens, and perform repairs within neural tissue (p. 322).
GRAY MATTER:	Neural tissue dominated by neuron cell bodies (p. 324).
WHITE MATTER:	Neural tissue dominated by myelinated axons (p. 324).
NEURAL CORTEX:	A layer of gray matter on the surface of the brain (p. 335).
Functional Categories:	
RECEPTOR:	A specialized cell, dendrite, or organ that responds to specific stimuli in the extracellular environment and whose stimulation alters the level of activity in a sensory neuron (pp. 321, 329).
EFFECTOR:	A muscle, gland, or other specialized cell or organ that responds to neural stimulation by altering its activity and producing a specific effect (p. 321).
SOMATIC:	Pertaining to the control of skeletal muscle activity (*somatic motor*) or sensory information from skeletal muscles, tendons, and joints (*somatic sensory*) (pp. 329–330).
VISCERAL:	Pertaining to the control of visceral functions, such as digestion, circulation, etc. (*visceral motor*) or sensory information from visceral organs (*visceral sensory*) (pp. 329–330).
VOLUNTARY:	Under direct conscious control (p. 321).
INVOLUNTARY:	Not under direct conscious control (p. 321).
Gross Anatomy:	
NUCLEUS:	A CNS center with discrete anatomical boundaries (p. 335).
CENTER:	A group of neuron cell bodies with a common function (p. 335).
TRACT:	A bundle of axons within the CNS that share a common origin, destination, and function (p. 335).
COLUMN:	A group of tracts found within a specific region of the spinal cord (p. 335).
PATHWAYS:	Centers and tracts that connect the brain with other organs and systems in the body (p. 335).
GANGLIA:	An anatomically distinct collection of sensory or motor neuron cell bodies within the PNS (p. 326).
NERVE:	A bundle of axons in the PNS (p. 326).

We now shift our attention to the structure of two systems that control and adjust the activities of other systems. These systems, the *nervous system* and the *endocrine system*, share important structural and functional characteristics. For example, they both rely on some form of chemical communication with targeted tissues and organs. These two systems often act in a complementary fashion. The nervous system usually provides relatively swift but brief responses to stimuli by temporarily modifying the activities of other organ systems. The response may appear almost immediately—in a few milliseconds—but the effects disappear soon after neural activity ceases. Endocrine responses are typically slower to develop than neural responses, but they often last much longer. The endocrine system, detailed in Chapter 19, adjusts the metabolic activity of other systems in response to changes in nutrient availability and energy demands. It also coordinates processes that continue for extended periods (months to years), such as growth. In this chapter we will introduce the structure of neurons and identify the basic principles of neural function.

AN OVERVIEW OF THE NERVOUS SYSTEM
(Figures 13-1/13-2)

The **nervous system** includes all of the **neural tissue** in the body. ∞[p. 80] Table 13-1 provides an overview of the most important concepts and terms introduced in this chapter.

There are two anatomical subdivisions of the nervous system: the *central nervous system* and the *peripheral nervous system*. The **central nervous system (CNS)** consists of the *brain* and *spinal cord*. The CNS is responsible for integrating, processing, and coordinating sensory data and motor commands. It is also the seat of higher functions, such as intelligence, memory, learning, and emotion. The CNS begins as a mass of neural tissue organized into a hollow tube. As development proceeds, the central cavity decreases in relative size, but the thickness of the walls and the diameter of the enclosed space vary from one region to another. The narrow central cavity that persists within the spinal cord is called the *central canal*; the *ventricles* are expanded chambers found in portions of the brain. *Cerebrospinal fluid* (CSF) fills the central canal and ventricles and surrounds the CNS, which is suspended within the dorsal body cavity. ∞[p. 17]

The **peripheral nervous system (PNS)** includes all of the neural tissue outside the CNS. The PNS provides sensory information to the CNS and carries motor commands to peripheral tissues and systems. The anatomical and functional divisions of the nervous system are included in Figure 13-1● and Table 13-1.

The functions of the anatomical subdivisions are diagrammed in Figure 13-2●. The **afferent division** of the PNS brings sensory information to the CNS, and the **efferent division** carries motor commands to muscles and glands. The afferent division begins at **receptors** that monitor specific characteristics of the environment. A receptor may be a *dendrite* (a sensory process of a neuron), a specialized cell or cluster of cells, or a complex sense organ such as the eye. Whatever its structure, the stimulation of a receptor provides information that may be carried to the CNS. The efferent division begins inside the CNS and ends at an **effector,** a muscle cell, gland cell, or other cell specialized to perform specific functions. The efferent division has *somatic* and *visceral* components. The **somatic nervous system (SNS)** controls skeletal muscle contractions. Those contractions may be *voluntary* or *involuntary*. Voluntary contractions are under conscious control; you exert voluntary con-

trol over your arm muscles as you raise a full glass of water to your lips. Involuntary contractions are directed outside of your conscious awareness; if you accidentally place your hand on a hot stove, it will be withdrawn immediately, usually before you even notice the pain. The *visceral motor system*, or **autonomic nervous system (ANS)**, regulates smooth muscle, cardiac muscle, and glandular activity, usually outside of our conscious awareness or control.

The organs of the CNS and PNS are complex, with numerous blood vessels and layers of connective tissue that provide physical protection and mechanical support. Nevertheless, all of the varied and essential functions of the nervous system are performed by individual neurons that must be kept safe, secure, and fully functional. Our discussion of the nervous system will begin at the cellular level, with the histology of neural tissue.

CELLULAR ORGANIZATION IN NEURAL TISSUE
(Figure 13-3)

Neural tissue contains two distinct cell types: nerve cells, or *neurons*, and supporting cells, or *neuroglia*. **Neurons** (*neuro*, nerve) are responsible for the transfer and processing of information in the nervous system.

FIGURE 13-2
Functional Overview of the Nervous System

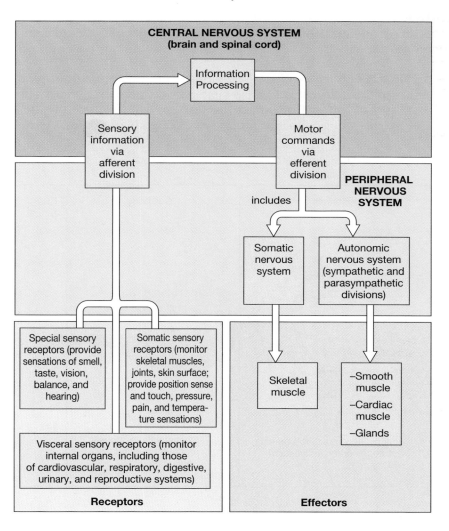

Neuron structure was introduced in Chapter 3. ∞[p. 80] A representative neuron (Figure 13-3●) has a **cell body**, or **soma**, several branching **dendrites**, and an elongate **axon** that ends at one or more **synaptic terminals**. At each synaptic terminal, the neuron communicates with another cell. The soma contains the organelles responsible for energy production and the synthesis of organic molecules, such as enzymes.

Supporting cells, or **neuroglia** (noo-RŌ-glē-a; *glia*, glue), isolate the neurons, provide a supporting framework for the neural tissue, and act as phagocytes. The neural tissue of the body contains approximately 100 billion neuroglia, or **glial cells**, which is roughly 20 times the number of neurons. Together, neuroglia account for roughly half of the volume of the nervous system. There are significant organizational differences between the neural tissue of the CNS and that of the PNS, primarily due to differences in the glial cell populations.

Neuroglia

The greatest variety of glial cells is found within the central nervous system. Table 13-2 summarizes information concerning the major glial cell populations in the CNS and PNS.

Neuroglia of the CNS

Four types of glial cells are found in the central nervous system: *astrocytes, oligodendrocytes, microglia,* and *ependymal cells.* These cell types can be distinguished on the basis of size, intracellular organization, and the presence of specific cytoplasmic processes.

ASTROCYTES *(Figure 13-4).* The largest and most numerous glial cells are the **astrocytes** (AS-trō-sīts; *astro-,* star + *cyte,* cell). The processes of astrocytes contact the surfaces of neuron cell bodies and axons, along with the processes of oligodendrocytes. Astro-

TABLE 13-2   Glial Cells in the CNS and PNS

Cell Type	Functions
CENTRAL NERVOUS SYSTEM	
Astrocytes	Maintain blood-brain barrier; provide structural support; regulate ion, nutrient, and dissolved gas concentrations; absorb and recycle neurotransmitters; form scar tissue after injury
Oligodendrocytes	Myelinate CNS axons; provide structural framework
Microglia	Remove cell debris, wastes, and pathogens by phagocytosis
Ependymal cells	Line ventricles (brain) and central canal (spinal cavity); assist in producing, circulating, and monitoring of cerebrospinal fluid
PERIPHERAL NERVOUS SYSTEM	
Satellite cells	Surround neuron cell bodies in ganglia
Schwann cells	Surround all axons in PNS; responsible for myelination of peripheral axons; participate in repair process after injury

cytes shield the neurons from direct contact with other neurons and limit their exposure to the surrounding interstitial fluid. Astrocytes (Figure 13-4●) have a variety of functions, but many are poorly understood. These functions can be summarized as:

- **Maintaining the blood-brain barrier:** Neural tissue must be isolated from the general circulation because hormones or other chemicals normally present in the blood could have disruptive effects on neuron function. The endothelial cells lining CNS capillaries control the chemical exchange between blood and in-

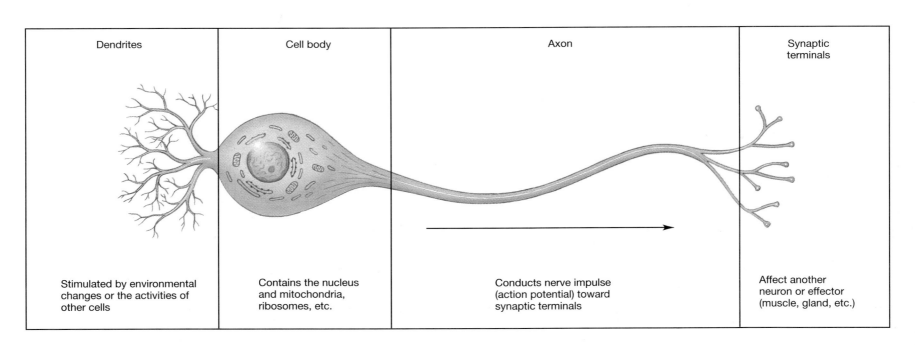

Dendrites	Cell body	Axon	Synaptic terminals
Stimulated by environmental changes or the activities of other cells	Contains the nucleus and mitochondria, ribosomes, etc.	Conducts nerve impulse (action potential) toward synaptic terminals	Affect another neuron or effector (muscle, gland, etc.)

FIGURE 13-3
A Review of Neuron Structure

terstitial fluid. These cells have very restricted permeability characteristics, and so create a **blood-brain barrier** that isolates the CNS from the general circulation. Chemicals secreted by astrocytes are essential for the maintenance of the blood-brain barrier. The slender cytoplasmic extensions of astrocytes end in expanded "feet" that wrap around capillaries. There are so many astrocytes, and so many astrocyte feet, that they form a complete cytoplasmic blanket around the capillaries, interrupted only where other glial cells contact the capillary walls. (The blood-brain barrier will be discussed further in Chapter 15.)

- **Creating a three-dimensional framework for the CNS:** Astrocytes are packed with microfilaments that extend across the breadth of the cell. This reinforcement assists them in providing a structural framework for the neurons of the brain and spinal cord.

- **Performing repairs in damaged neural tissue:** After damage to the CNS, astrocytes make structural repairs that stabilize the tissue and prevent further injury by producing scar tissue at the injury site.

- **Guiding neuron development:** Astrocytes in the embryonic brain appear to be involved in directing the growth and interconnection of developing neurons.

- **Controlling the interstitial environment:** Evidence suggests that astrocytes adjust the interstitial fluid composition by providing a rapid-transit system for the transport of nutrients, ions, and dissolved gases between capillaries and neurons.

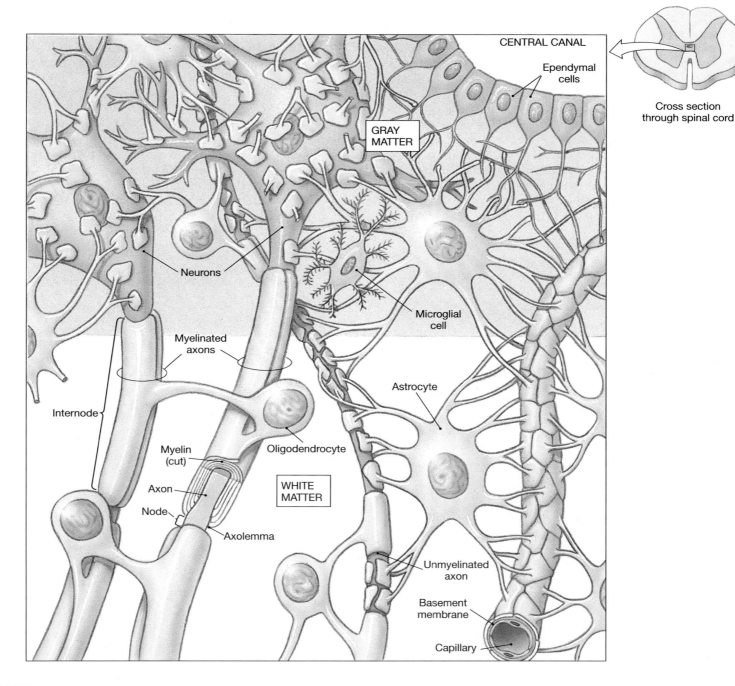

FIGURE 13-4

Histology of Neural Tissue in the CNS. A diagrammatic view of neural tissue in the CNS, showing relationships between major glial elements and neurons.

OLIGODENDROCYTES *(Figure 13-4)*. **Oligodendrocytes** (ō-li-gō-DEN-drō-sīts; *oligo*, few) resemble astrocytes only in that they both possess slender cytoplasmic extensions. However, the cell bodies of oligodendrocytes are smaller, and they have fewer processes (Figure 13-4●). Oligodendrocyte processes usually contact the axons or cell bodies of neurons, but the functions of processes ending at the cell bodies have yet to be determined. Oligodendrocytes (1) tie clusters of axons together and (2) improve the functional performance of neurons by wrapping axons in *myelin*, a material with insulating properties.

Many axons in the CNS are completely sheathed by the processes of oligodendrocytes. Near the tip of each process, the cytoplasm becomes very thin, but the cell membrane expands to form an enormous membranous pad. This flattened pancake is wrapped around the axon (Figure 13-4●), creating a multilayered sheath composed primarily of phospholipids. This membranous coating is called **myelin** (MĪ-e-lin), and the axon is said to be **myelinated**. Myelin improves the speed at which an action potential, or *nerve impulse*, is conducted along an axon. Not all axons in the CNS are myelinated, and **unmyelinated** axons may not be completely covered by glial cell processes.

Many oligodendrocytes cooperate in the formation of the myelin sheath along the entire length of a myelinated axon. Small gaps, called **nodes**, or the *nodes of Ranvier* (rahn-vē-Ā), exist between the myelin sheaths produced by individual oligodendrocytes. The relatively large areas wrapped in myelin are called **internodes** (*inter*, between). When dissected, myelinated axons appear a glossy white, primarily because of the lipids present. Regions dominated by myelinated axons constitute the **white matter** of the CNS. In contrast, regions dominated by neuron cell bodies and unmyelinated axons are called **gray matter** because of their dusky gray color.

MICROGLIA *(Figure 13-4)*. Roughly 5% of the CNS glial cells are **microglia** (mī-KRŌ-glē-a). Microglia are smaller than the other glial cells, and their slender cytoplasmic processes have many fine branches (Figure 13-4●). Microglia appear early in embryonic development, through the division of mesodermal stem cells. The stem cells that produce microglia are related to those that produce tissue macrophages and monocytes of the blood. The microglia migrate into the CNS as it forms, and thereafter they remain isolated within the neural tissue, acting as a roving security force. Microglia engulf cellular debris, waste products, and pathogens. In times of infection or injury, the number of microglia increases dramatically.

EPENDYMAL CELLS *(Figure 13-5)*. The ventricles of the brain and central canal of the spinal cord are lined by a cellular layer called the **ependyma** (ep-EN-di-mah). These chambers and passageways are filled with **cerebrospinal fluid (CSF)**. This fluid, which also surrounds the brain and spinal cord, provides a protective cushion and transports dissolved gases, nutrients, wastes, and other materials.

Ependymal cells are cuboidal to columnar in form. Unlike typical epithelial cells, ependymal cells have slender processes that branch extensively and make direct contact with glial cells in the surrounding neural tissue (Figure 13-5●). Experimental evidence suggests that ependymal cells may act as receptors that monitor the composition of the CSF. During development and early childhood the free surfaces of ependymal cells are covered with cilia. In the adult, cilia may persist on ependymal cells lining the ventricles of the brain, but the ependyma elsewhere usually has only scattered microvilli. The ciliated regions of the ependyma may assist in the circulation of CSF. In a few parts of the brain, specialized ependymal cells participate in the secretion of cerebrospinal fluid.

FIGURE 13-5
The Ependyma. (a) Light micrograph showing the ependymal lining of the central canal of the spinal cord. (LM × 257) The diagrammatic view of ependymal organization is based on information provided by electron microscopy. (b) An SEM of the ciliated surface of the ependyma from one of the ventricles. (Ci = cilia; SEM × 1825) [Reproduced from R. G. Kessel and R. H. Kardon, "Tissues and Organs: A Text–Atlas of Scanning Electron Microscopy," W. H. Freeman & Co., 1979.]

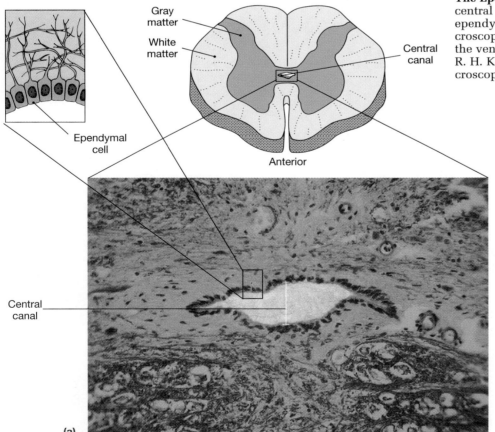

(a)

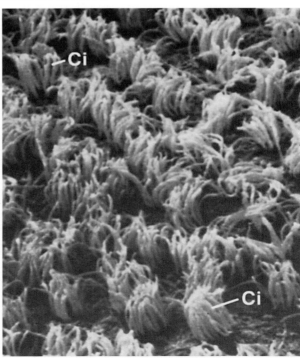

(b)

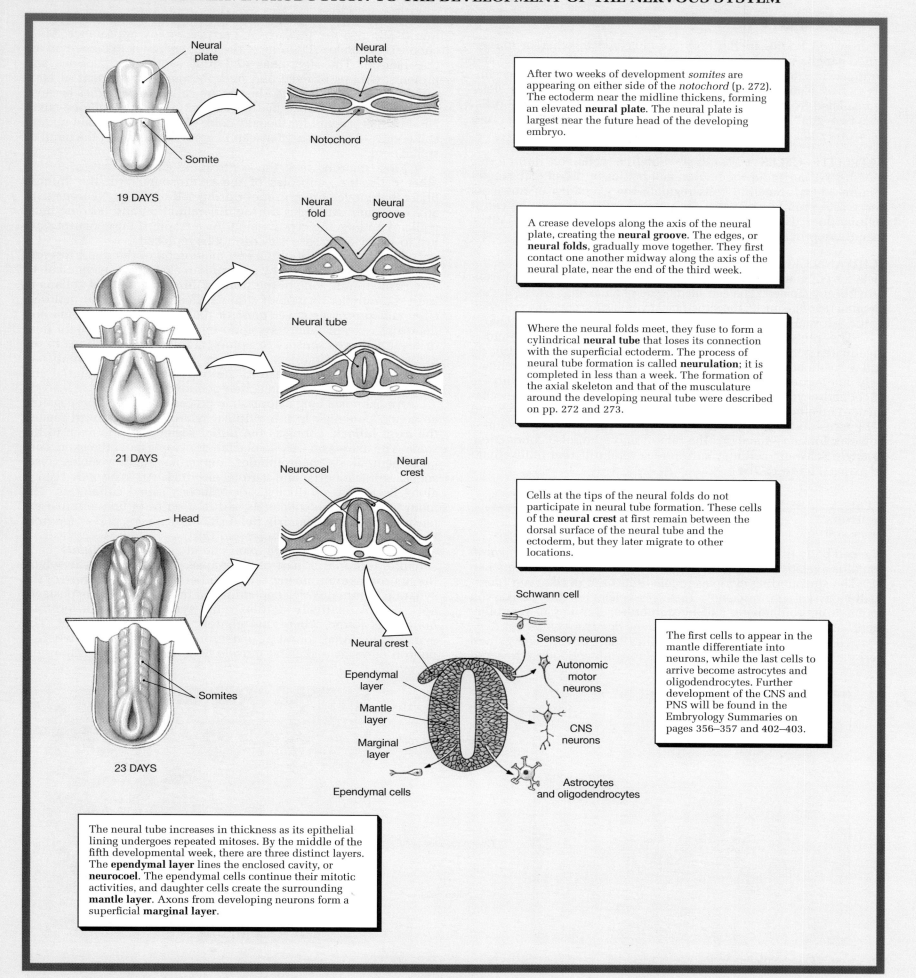

Neural plate · **Neural plate** · **Notochord** · **Somite**

19 DAYS

After two weeks of development *somites* are appearing on either side of the *notochord* (p. 272). The ectoderm near the midline thickens, forming an elevated **neural plate**. The neural plate is largest near the future head of the developing embryo.

Neural fold · **Neural groove**

21 DAYS · **Neural tube**

A crease develops along the axis of the neural plate, creating the **neural groove**. The edges, or **neural folds**, gradually move together. They first contact one another midway along the axis of the neural plate, near the end of the third week.

Where the neural folds meet, they fuse to form a cylindrical **neural tube** that loses its connection with the superficial ectoderm. The process of neural tube formation is called **neurulation**; it is completed in less than a week. The formation of the axial skeleton and that of the musculature around the developing neural tube were described on pp. 272 and 273.

Neurocoel · **Neural crest**

Head · **Somites**

23 DAYS

Cells at the tips of the neural folds do not participate in neural tube formation. These cells of the **neural crest** at first remain between the dorsal surface of the neural tube and the ectoderm, but they later migrate to other locations.

Neural crest · **Ependymal layer** · **Mantle layer** · **Marginal layer** · **Ependymal cells**

Schwann cell · **Sensory neurons** · **Autonomic motor neurons** · **CNS neurons** · **Astrocytes and oligodendrocytes**

The first cells to appear in the mantle differentiate into neurons, while the last cells to arrive become astrocytes and oligodendrocytes. Further development of the CNS and PNS will be found in the Embryology Summaries on pages 356–357 and 402–403.

The neural tube increases in thickness as its epithelial lining undergoes repeated mitoses. By the middle of the fifth developmental week, there are three distinct layers. The **ependymal layer** lines the enclosed cavity, or **neurocoel**. The ependymal cells continue their mitotic activities, and daughter cells create the surrounding **mantle layer**. Axons from developing neurons form a superficial **marginal layer**.

Neuroglia of the PNS

Neuron cell bodies in the PNS are clustered together in masses called **ganglia** (singular *ganglion*). Axons are bundled together and wrapped in connective tissue, forming **peripheral nerves**, or simply *nerves*. Neuron cell bodies and axons in the PNS are completely insulated from their surroundings by the processes of glial cells. The two glial cell types involved are called *satellite cells* and *Schwann cells*.

SATELLITE CELLS *(Figure 13-6)*. **Satellite cells**, or *amphicytes* (AM-fi-sīts), surround the neuron cell bodies in peripheral ganglia (Figure 13-6●). Satellite cells regulate the exchange of nutrients and waste products between the neuron cell body and the extracellular fluid. They also help isolate the neuron from stimuli other than those provided at synapses.

SCHWANN CELLS *(Figures 13-4/13-7)*. **Schwann cells** produce a complete covering around every peripheral axon, whether it is unmyelinated or myelinated. The cell membrane of an axon is called the **axolemma** (*lemma*, husk); the superficial cytoplasmic covering provided by the Schwann cells is known as the **neurilemma** (noo-ri-LEM-ma).

The physical relationship between a Schwann cell and a myelinated peripheral axon differs from that of an oligodendrocyte and a myelinated axon in the CNS. A Schwann cell can myelinate only one portion of a single axon (Figure 13-7a●; compare with the oligodendrocytes in Figure 13-4●, p. 323). Although the mechanism of myelination differs, myelinated axons in both the CNS and PNS have nodes and internodes, and the presence of myelin—however formed—increases the rate of nerve impulse conduction. A single Schwann cell may surround several different unmyelinated axons (Figure 13-7b●).

Neurons
(Figure 13-8)

The cell body (Figure 13-8●) of a representative *multipolar neuron* contains a relatively large, round nucleus with a prominent nucleolus. The surrounding cytoplasm constitutes the **perikaryon** (per-i-KAR-ē-on; *karyon*, nucleus). The cytoskeleton of the perikaryon contains **neurofilaments** and **neurotubules**. Bundles of neurofilaments, called **neurofibrils**, extend into the dendrites and axon.

The perikaryon contains organelles that provide energy and perform synthetic activities. The numerous mitochondria, free and fixed ribosomes, and membranes of the rough endoplasmic reticulum (RER) give the perikaryon a coarse, grainy appearance. Mitochondria generate ATP to meet the high energy demands of an active neuron. The ribosomes and RER synthesize peptides and proteins. Groups of fixed and free ribosomes are present in large numbers. These ribosomal clusters are called *Nissl bodies* because they were first described by the German microscopist Franz Nissl. Nissl bodies account for the gray color of areas that contain neuron cell bodies—the *gray matter* seen in gross dissection of the brain or spinal cord.

Most neurons lack the *centrosome* complex. ∞[p. 36] In other cells, the centrioles of the centrosome form the spindle fibers that move chromosomes during cell division. Neurons usually lose their centrioles during differentiation and become incapable of undergoing cell division. As a result, if these neurons are lost to injury or disease they cannot be replaced.

The permeability of the cell membrane of the dendrites and cell body can be changed by exposure to chemical, mechanical, or electrical stimuli. One of the primary functions of glial cells is to limit the number or type of stimuli affecting individual neurons. Glial cell processes cover most of the surfaces of the soma and dendrites, except where synaptic terminals exist or where dendrites function as sensory receptors, monitoring conditions in the extracellular environment. Exposure to appropriate stimuli can produce a localized change in the transmembrane potential and lead to the generation of an action potential at the axon.

An axon is a long cytoplasmic process capable of propagating an action potential. In a multipolar neuron, a specialized region, the **axon hillock**, connects the **initial segment** of the axon to the soma. The **axoplasm** (AK-so-plazm), or cytoplasm of the axon, contains neurofibrils, neurotubules, numerous small vesicles, lysosomes, mitochondria, and various enzymes. An axon may branch along its length, producing side-branches called **collaterals**. The main trunk and the collaterals end in a series of fine terminal extensions, called **telodendria** (tel-ō-DEN-drē-a; *telo-*, end + *dendron*, tree). ⚕ *Axoplasmic Transport and Disease [p. 758]*

The telodendria of an axon end at *synaptic terminals*. Each synaptic terminal is part of a **synapse**, a specialized site where the neuron communicates with another cell. The structure of the synaptic terminal varies depending on the identity of the postsynaptic cell. A relatively simple, round **synaptic knob,** or *terminal bouton*, is found where one neuron synapses on another. The synaptic terminal found at a *neuromuscular junction*, where a neuron contacts a skeletal muscle fiber, is much more complex.

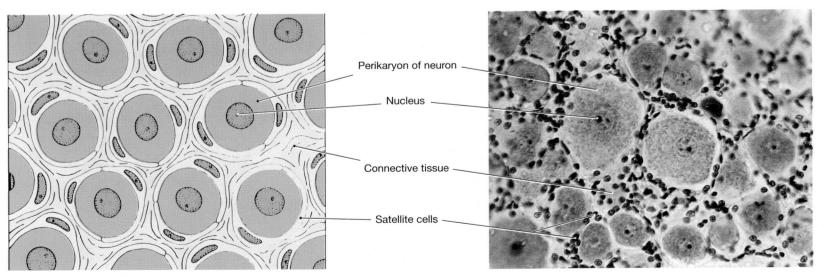

Perikaryon of neuron

Nucleus

Connective tissue

Satellite cells

FIGURE 13-6
Satellite Cells and Peripheral Neurons. Satellite cells surround neuron cell bodies in peripheral ganglia. (LM × 120)

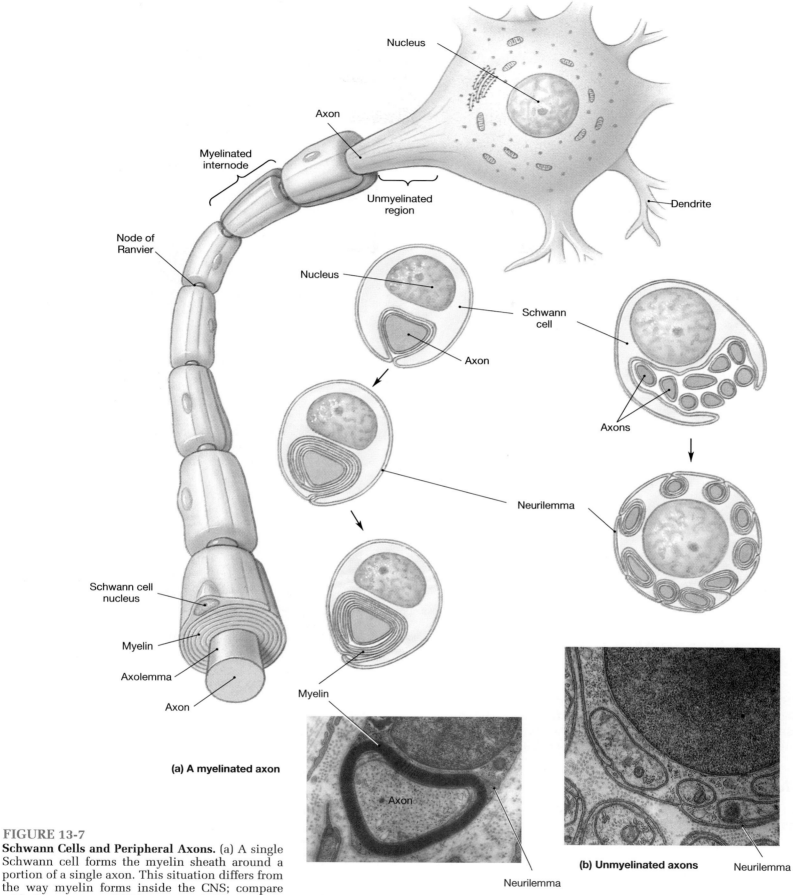

(a) A myelinated axon

(b) Unmyelinated axons

Neurilemma

Neurilemma

FIGURE 13-7

Schwann Cells and Peripheral Axons. (a) A single Schwann cell forms the myelin sheath around a portion of a single axon. This situation differs from the way myelin forms inside the CNS; compare with Figure 13-4. (TEM × 20,603) (b) A single Schwann cell can encircle several unmyelinated axons. Unlike the situation inside the CNS (Figure 13-4), every axon in the PNS has a complete neurilemmal sheath. (TEM × 27,627)

FIGURE 13-8
Anatomy of a Representative Neuron. (a) A multipolar neuron which may innervate other neurons (1), skeletal muscle fibers (2), or gland cells (3). (A single neuron would not innervate all three.) (b) Detailed organization of the cell body. (LM × 1600)

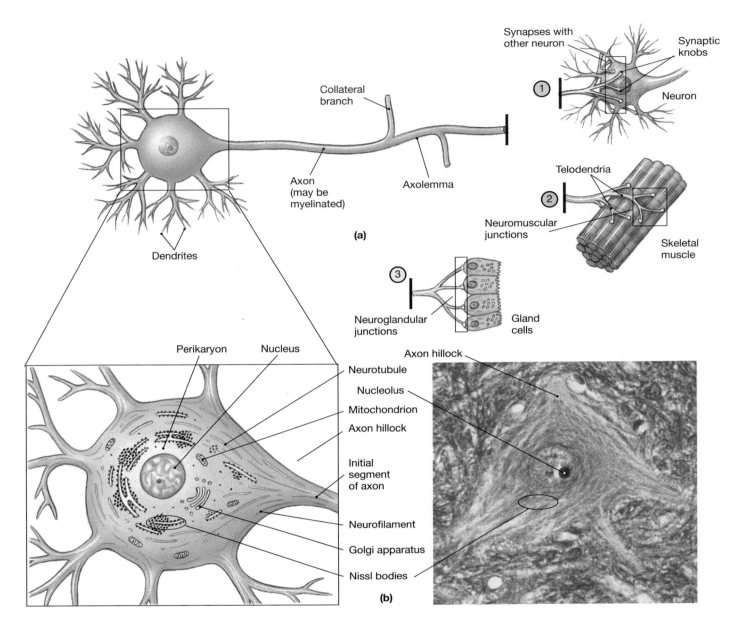

[p. 243] Synaptic communication most often involves the release of specific chemicals, called **neurotransmitters**. The release of these chemicals is triggered by the arrival of a nerve impulse; additional details are provided in a later section.

Neuron Classification

The billions of neurons in the nervous system are quite variable in form. Neurons are classified in two ways: (1) on the basis of structure and (2) on the basis of function.

STRUCTURAL CLASSIFICATION *(Figure 13-9).* The structural classification is based on the number of processes that project from the cell body (Figure 13-9●).

- **Anaxonic** (an-ak-SON-ik) neurons are small, and there are no anatomical clues to distinguish dendrites from axons. Anaxonic neurons (Figure 13-9a●) are found in

the CNS and in special sense organs, but their functions are poorly understood.

- In a **unipolar neuron,** or *pseudounipolar neuron,* the dendritic and axonal processes are continuous, and the cell body lies off to one side (Figure 13-9b●). In a unipolar neuron the initial segment lies at the base of the dendritic branches, and the rest of the process is considered an axon on both structural and functional grounds. Sensory neurons of the peripheral nervous system are usually unipolar, and their axons may be myelinated.

- **Bipolar neurons** (Figure 13-9c●) have one dendrite and one axon, with the cell body between them. Bipolar neurons are relatively rare but play an important role in relaying information concerning sight, smell, and hearing. Their axons are not myelinated.

- **Multipolar neurons** have several dendrites and a single axon that may have one or more branches. Multipolar neurons (Figure 13-9d●) are the most common type of neuron in the CNS. For example, all of the motor neurons that control skeletal muscles are multipolar neurons with myelinated axons.

FUNCTIONAL CLASSIFICATION *(Figure 13-10).* Neurons can be categorized into three functional groups: (1) *sensory neurons*, (2) *motor neurons*, and (3) *interneurons*, or *association neurons*. Their relationships are diagrammed in Figure 13-10●.

Sensory Neurons. **Sensory neurons** or *afferent neurons,* form the *afferent division* of the PNS. Their function is to deliver information to the CNS. The cell bodies of sensory neurons are found in peripheral *sensory ganglia*. Sensory neurons are unipolar neurons, and their processes, known as **afferent fibers,** extend between a sensory receptor and the spinal cord or brain. Sensory neurons collect information concerning the external or internal environment. There are about 10 million sensory neurons. **Somatic sensory neurons** monitor the outside world and our position within it. **Visceral sensory neurons** monitor internal conditions and the status of other organ systems.

Receptors may be the processes of specialized sensory neurons or cells monitored by sensory neurons. Receptors are broadly categorized as:

- **Exteroceptors** (*extero-,* outside) provide information about the external environment in the form of touch, temperature, and pressure sensations and the more complex senses of sight, smell, hearing, and taste.

- **Proprioceptors** (prō-prē-ō-SEP-torz) monitor the position and movement of skeletal muscles and joints.

- **Interoceptors** (*intero-,* inside) monitor the digestive, respiratory, cardiovascular, urinary, and reproductive systems and provide sensations of taste, deep pressure, and pain.

Data from exteroceptors and proprioceptors are carried by somatic sensory neurons. Interoceptive information is carried by visceral sensory neurons.

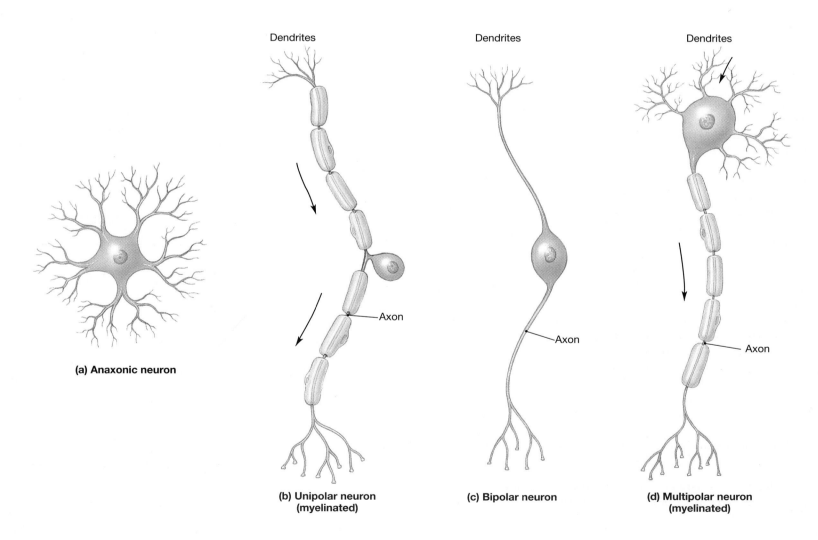

(a) Anaxonic neuron

(b) Unipolar neuron (myelinated)

(c) Bipolar neuron

(d) Multipolar neuron (myelinated)

FIGURE 13-9
A Structural Classification of Neurons. Arrows indicate the direction that messages travel along the neuron.

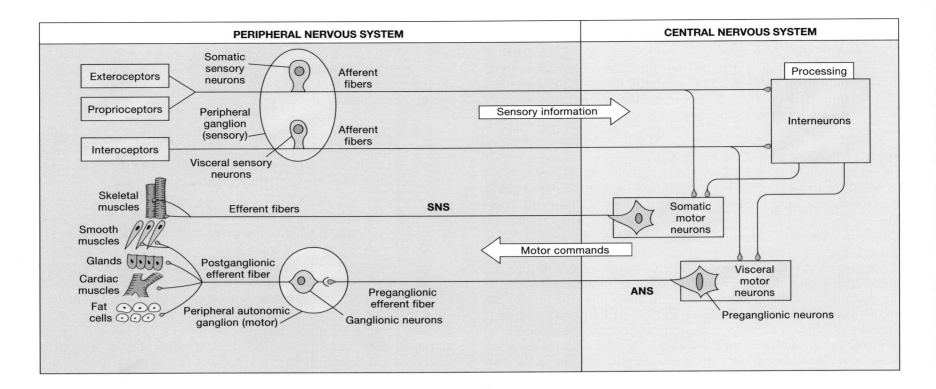

FIGURE 13-10
A Functional Classification of Neurons

Motor Neurons. **Motor neurons,** or *efferent neurons,* form the *efferent division* (*ex,* from) carrying instructions from the CNS to peripheral effectors. A motor neuron stimulates or modifies the activity of a peripheral tissue, organ, or organ system. About a half a million motor neurons are found in the body. Axons traveling away from the CNS are called **efferent fibers**. There are two major efferent divisions in the PNS. The *somatic nervous system* includes all of the **somatic motor neurons** that innervate skeletal muscles. We have conscious control over many of the activities of the somatic nervous system. The cell body of a somatic motor neuron lies inside the CNS, and its axon extends into the periphery to end at neuromuscular junctions.

The activities of the *autonomic nervous system* are controlled primarily outside of our conscious awareness. **Visceral motor neurons** innervate all peripheral effectors other than skeletal muscles. The axons of visceral motor neurons inside the CNS synapse on neurons in peripheral ganglia, and the ganglion cells control peripheral effectors. Axons extending from the CNS to a ganglion are called **preganglionic fibers**. Axons connecting the ganglion cells with the peripheral effectors are known as **postganglionic fibers**. This arrangement clearly distinguishes the autonomic (visceral motor) system from the somatic motor system.

Interneurons. **Interneurons,** or **association neurons,** may be situated between sensory and motor neurons. Interneurons are located entirely within the brain and spinal cord. There are 20 billion interneurons, outnumbering all other neurons combined. Interneurons are responsible for the analysis of sensory inputs and the coordination of motor outputs. The more complex the response to a given stimulus, the greater the number of interneurons involved. Interneurons can be classified as **excitatory** or **inhibitory** on the basis of their effects on the postsynaptic membranes of other neurons.

√ **Examination of a tissue sample shows unipolar neurons. Are these more likely to be sensory neurons or motor neurons?**

√ **What type of glial cell would you expect to find in large numbers in brain tissue from a person suffering from a CNS infection?**

THE NERVE IMPULSE

Excitability is the ability of a cell membrane to conduct electrical impulses. The cell membranes of skeletal muscle fibers and most neurons (including all multipolar and unipolar neurons) are excitable. The conducted changes in the transmembrane potential are called **action potentials**. The action potentials that trigger the contraction of a skeletal muscle fiber begin at the neuromuscular junction and sweep across the entire surface of the sarcolemma. ∞ [p. 235] In the nervous system, action potentials travel from one part of the body to another along axons. An action potential traveling along an axon is called a **nerve impulse**.

Before a nerve impulse can occur, a stimulus of sufficient strength must be applied to the membrane of the neuron. Once initiated, the rate of impulse conduction depends on the properties of the axon, specifically:

1. *The presence or absence of a myelin sheath*: A myelinated axon conducts impulses five to seven times faster than an unmyelinated axon.
2. *The diameter of the axon*: The larger the diameter, the more rapidly the impulse will be conducted.

The largest myelinated axons, with diameters ranging from 4 to 20 μm, conduct nerve impulses at speeds of up to 140 m/s (300 mph). In contrast, small unmyelinated fibers (less than 2 μm in diameter) conduct impulses at speeds below 1 m/s (2 mph). ✝*Demyelination Disorders [p. 759]*

√ **What effect would cutting the axon have in transmitting the action potential?**

√ **Two axons are tested for conduction velocities. One conducts action potentials at 50 m/s, the other at 1 m/s. Which axon is myelinated?**

SYNAPTIC COMMUNICATION

A synapse may be located on a dendrite, on the soma, or along the length of an axon. A synapse may also permit communication between a neuron and another cell type; such synapses are called **neuroeffector junctions**. The *neuromuscular junction* described in Chapter 9 was an example of a neuroeffector junction. ∞[p. 235] Neuroeffector junctions involving other cell types are less common and will be described in later chapters.

At the end of an axon, a nerve impulse triggers events at a synapse that transfers the information to another neuron or effector cell. A synapse may be **chemical**, involving a neurotransmitter substance, or **electrical**, with gap junctions providing direct physical contact between the cells.

Chemical Synapses
(Figure 13-11)

Chemical synapses are by far the most abundant, and there are several different types. Most interactions between neurons and all communications between neurons and peripheral effectors involve chemical synapses. At a chemical synapse involving two neurons (Figure 13-11b,c●), a neurotransmitter released at the *presynaptic membrane* binds to receptor proteins on the *postsynaptic membrane* and triggers a transient change in the transmembrane potential of the receptive cell. Only the presynaptic membrane releases a neurotransmitter, and communication can occur in only one direction: from the presynaptic neuron to the postsynaptic neuron.

The neuromuscular junction described in Chapter 9 is an example of a chemical synapse using the neurotransmitter acetylcholine (ACh). ∞[p. 243] Over 50 different neurotransmitters have been identified, but ACh is probably the best known. All neuromuscular junctions utilize ACh as a neurotransmitter; ACh is also released at many chemical synapses in the CNS and PNS. The general sequence of events is similar, regardless of the location of the synapse or the nature of the neurotransmitter.

- Arrival of a nerve impulse at the synaptic knob triggers release of neurotransmitter from secretory vesicles, through exocytosis at the presynaptic membrane.
- The neurotransmitter diffuses across the synaptic cleft and binds to receptors on the postsynaptic membrane.

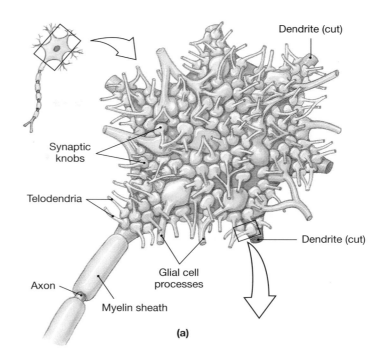

(a)

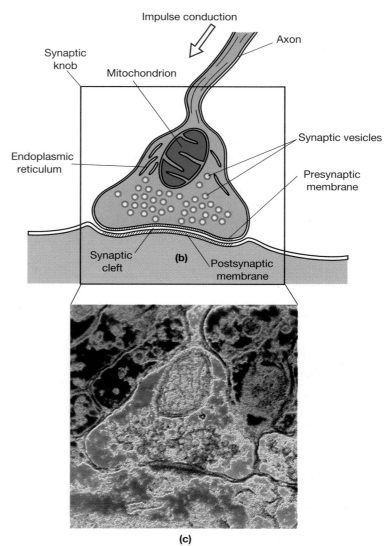

(b)

(c)

FIGURE 13-11
Structure of a Synapse. (a) There may be thousands of chemical synapses on the surface of a single neuron. Many of these synapses may be active at any one moment. (b) Diagrammatic view of a chemical synapse between two neurons. (c) Color-enhanced TEM of a chemical synapse. (TEM × 186,480)

- Receptor binding results in a change in the permeability of the postsynaptic cell membrane. Depending on the identity and abundance of the receptor proteins on the postsynaptic membrane, the result may be *excitatory* or *inhibitory*. In general, excitatory effects promote the generation of nerve impulses, whereas inhibitory effects oppose this process.

- If the degree of excitation is sufficient, receptor binding may lead to the generation of an action potential in the axon (if the postsynaptic cell is a neuron) or sarcolemma (if the postsynaptic cell is a skeletal muscle fiber).

- The effects on the postsynaptic membrane are short-lived, because the neurotransmitter molecules are either enzymatically removed or reabsorbed. To prolong the effects, additional nerve impulses must arrive at the synaptic terminal, and more neurotransmitter molecules must be released.

Examples of neurotransmitters other than ACh will be presented in later chapters. There may be thousands of synapses on the cell body of a single neuron (Figure 13-11a●). Many of these will be active at any given moment, releasing a variety of different neurotransmitters. Some will have excitatory effects, others inhibitory effects. The activity of the receptive neuron depends on the sum of all of the excitatory and inhibitory stimuli influencing the axon hillock at any given moment.

Electrical Synapses

Chemical synapses dominate the nervous system. **Electrical synapses** are found between neurons in the CNS and PNS, but they are relatively rare. At an electrical synapse, the presynaptic and postsynaptic cell membranes are tightly bound together, and *gap junctions* permit the passage of ions between the cells. ∞[p. 43] Because the two cells are linked in this way, they function as if they shared a common cell membrane, and the nerve impulse crosses from one cell membrane to the next without delay.

NEURAL REGENERATION
(Figure 13-12)

A neuron responds to injury in a very limited fashion. Within the cell body, the Nissl bodies disappear and the nucleus moves away from its centralized location. If the neuron recovers its functional abilities, it will return to a normal appearance. If the problem is impaired oxygen or nutrient supply, as in a stroke, or local pressure, as in spinal cord or nerve injuries, the affected neurons will recover if the problem is resolved within a period of minutes to hours. If the stresses continue for a longer period, the neuron may be permanently damaged or destroyed.

In the peripheral nervous system, Schwann cells participate in the repair of damaged nerves. In the process known as **Wallerian degeneration** (Figure 13-12●), the axon distal to the injury site deteriorates, and macrophages migrate in to phagocytize the debris. The Schwann cells in the area divide and form a solid cellular cord that follows the path of the original axon. As the neuron recovers, its axon grows into the injury site, and the Schwann cells wrap around it.

If the axon continues to grow into the periphery alongside the appropriate cord of Schwann cells, it may eventually reestablish its

normal synaptic contacts. If it stops growing, or wanders off in some new direction, normal function will not return. The growing axon is most likely to arrive at its appropriate destination if the cut proximal and distal stumps remain in contact after the injury. When an entire peripheral nerve is severed, only a relatively small number of axons will successfully reestablish normal synaptic contacts. As a result, nerve function will be permanently impaired.

Limited regeneration can occur inside the central nervous system, but the situation is more complicated because (1) many more axons are likely to be involved, (2) astrocytes produce scar tissue that can prevent axon growth across the damaged area, and (3) astrocytes release chemicals that block the regrowth of axons. Additional information concerning neural regeneration and surgical repairs will be presented in the next chapter.

NEURON ORGANIZATION AND PROCESSING
(Figure 13-13)

Interneurons within the CNS are organized into **neuronal pools**. A neuronal pool is a group of interconnected neurons with specific functions. Neuronal pools are defined on the basis of function rather than on anatomical grounds. A pool may be diffused, involving neurons in several different regions of the brain, or localized, with neurons restricted to one specific location, such as a single *nucleus*. (The anatomical organization of the nervous system is detailed in the next section.) Estimates concerning the actual number of neuronal pools vary, but range between a few hundred and a few thousand. Each pool has a limited number of input sources and output destinations, and the pool may contain both excitatory and inhibitory neurons.

The pattern of interaction between neurons provides clues to the functional characteristics of a neuronal pool. Five patterns may be distinguished.

1. **Divergence** is the spread of information from one neuron to several neurons, as in Figure 13-13a●, or from one pool to multiple pools. Divergence permits the broad distribution of a specific input. Considerable divergence occurs when sensory neurons bring information into the CNS, for the information is distributed to neuronal pools throughout the spinal cord and brain.

2. In **convergence**, several neurons synapse on the same postsynaptic neuron (Figure 13-13b●). Several different patterns of activity in the presynaptic neurons can have the same effect on the postsynaptic neuron. Convergence permits the variable control of motor neurons by providing a mechanism for their voluntary and involuntary control. For example, the movements of your diaphragm and ribs are now being controlled by respiratory centers in the brain that operate outside of your conscious awareness. But the same motor neurons can also be controlled voluntarily, as when you take a deep breath and hold it. Two different neuronal pools are involved, both synapsing on the same motor neurons.

3. Information may be relayed in a stepwise sequence, from one neuron to another or from one neuronal pool to the next. This pattern, called **serial processing,** is shown in Figure 13-13c●. Serial processing occurs as sensory information is relayed from one processing center to another in the brain.

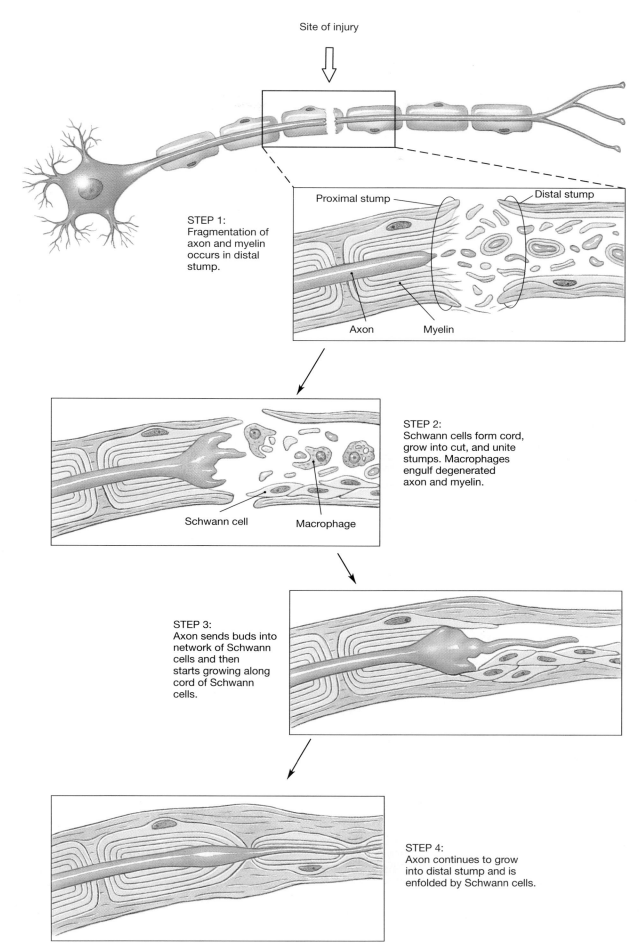

Site of injury

Proximal stump

Distal stump

STEP 1:
Fragmentation of axon and myelin occurs in distal stump.

Axon

Myelin

STEP 2:
Schwann cells form cord, grow into cut, and unite stumps. Macrophages engulf degenerated axon and myelin.

Schwann cell

Macrophage

STEP 3:
Axon sends buds into network of Schwann cells and then starts growing along cord of Schwann cells.

STEP 4:
Axon continues to grow into distal stump and is enfolded by Schwann cells.

FIGURE 13-12
Nerve Regeneration after Injury

FIGURE 13-13

Organization of Neuronal Pools. (a) Divergence, a mechanism for spreading stimulation to multiple neurons or neuronal pools in the CNS. (b) Convergence, a mechanism providing input to a single neuron from multiple sources. (c) Serial processing, in which neurons or pools work in a sequential manner. (d) Parallel processing, in which individual neurons or neuronal pools process information simultaneously. (e) Reverberation, a feedback mechanism that may be excitatory or inhibitory.

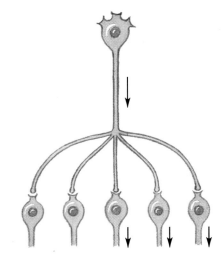
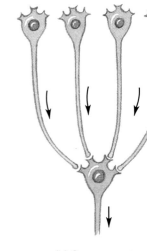

(a) Divergence

(b) Convergence

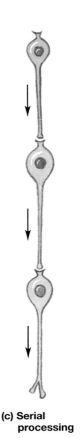
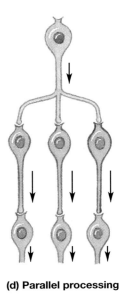
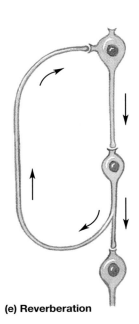

(c) Serial processing

(d) Parallel processing

(e) Reverberation

4. Parallel processing occurs when several neurons or neuronal pools are processing the same information at one time (Figure 13-13d●). Thanks to parallel processing, many different responses occur simultaneously. For example, stepping on a sharp object stimulates sensory neurons that distribute the information to a number of neuronal pools. As a result of parallel processing, you might withdraw your foot, shift your weight, move your arms, feel the pain, and shout "Ouch!" all at the same time.

5. Some neural circuits utilize positive feedback to produce **reverberation**. In this arrangement, collateral axons extend back toward the source of an impulse and further stimulate the presynaptic neurons. Once a reverberating circuit has been activated, it will continue to function until synaptic fatigue or inhibitory stimuli break the cycle. As with convergence or divergence, reverberation can occur within a single neuronal pool, or it may involve a series of interconnected pools. An example of reverberation is shown in Figure 13-13e●; much more complicated examples of reverberation between neuronal pools in the brain may be involved in the maintenance of conscious-

ness, muscular coordination, and normal breathing patterns. We will discuss these and other "wiring patterns" as we consider the organization of the spinal cord and brain in subsequent chapters.

ANATOMICAL ORGANIZATION OF THE NERVOUS SYSTEM
(Figure 13-14)

The functions of the nervous system depend on the interactions between neurons in neuronal pools, with the most complex neural processing steps occurring in the spinal cord and brain (CNS). The arriving sensory information and the outgoing motor commands are carried by the peripheral nervous system (PNS). Axons and cell bodies in the CNS and PNS are not randomly scattered. Instead, they form masses or bundles with distinct anatomical boundaries. The anatomical organization of the nervous system is depicted in Figure 13-14●.

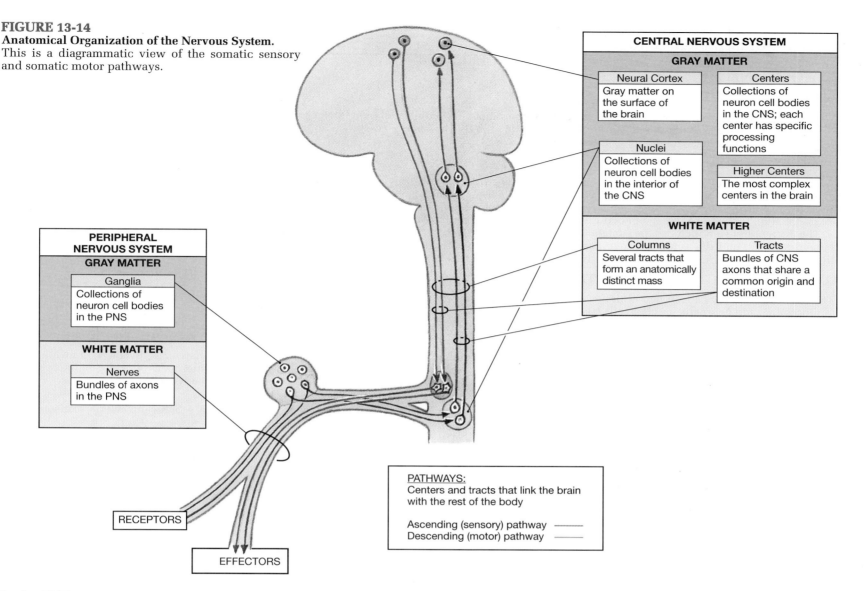

FIGURE 13-14
Anatomical Organization of the Nervous System.
This is a diagrammatic view of the somatic sensory and somatic motor pathways.

CENTRAL NERVOUS SYSTEM

GRAY MATTER

Neural Cortex	Centers
Gray matter on the surface of the brain	Collections of neuron cell bodies in the CNS; each center has specific processing functions

Nuclei	Higher Centers
Collections of neuron cell bodies in the interior of the CNS	The most complex centers in the brain

WHITE MATTER

Columns	Tracts
Several tracts that form an anatomically distinct mass	Bundles of CNS axons that share a common origin and destination

PERIPHERAL NERVOUS SYSTEM

GRAY MATTER

Ganglia
Collections of neuron cell bodies in the PNS

WHITE MATTER

Nerves
Bundles of axons in the PNS

RECEPTORS

EFFECTORS

PATHWAYS:
Centers and tracts that link the brain with the rest of the body

Ascending (sensory) pathway ———
Descending (motor) pathway ———

In the PNS:

- The cell bodies of sensory neurons and visceral motor neurons are found in *ganglia.*

- Axons are bundled together in nerves, with *spinal nerves* connected to the spinal cord, and *cranial nerves* connected to the brain.

In the CNS:

- A collection of neuron cell bodies with a common function is called a **center.** A center with a discrete anatomical boundary is called a **nucleus.** Portions of the brain surface are covered by a thick layer of gray matter, called the **neural cortex.** The term *higher centers* refers to the most complex integration centers, nuclei, and cortical areas of the brain.

- The white matter of the CNS contains bundles of axons that share common origins, destinations, and functions. These bundles are called **tracts.** Tracts in the spinal cord form larger groups, called **columns.**

- The centers and tracts that link the brain with the rest of the body are called **pathways.** For example, **sensory pathways,** or *ascending pathways,* distribute information from peripheral receptors to processing centers in the brain, and **motor pathways**, or *descending pathways,* begin at CNS centers concerned with motor control and end at the effectors they control. ⚕ *Growth and Myelination of the Nervous System [p. 759]*

Related Clinical Terms

rabies: An acute viral disease of the central nervous system usually transmitted by the bite of an infected mammal. The virus reaches the CNS by transport along the axons of neurons innervating the region of the bite. ⚕ *Axoplasmic Transport and Disease [p. 758]*

demyelination: The progressive destruction of myelin sheaths in the CNS and PNS, leading to a loss of sensation and motor control. Demyelination is associated with *heavy metal poisoning, diphtheria, multiple sclerosis,* and *Guillain-Barré syndrome.* ⚕ *Demyelination Disorders [p. 759]*

CHAPTER SUMMARY AND REVIEW

STUDY OUTLINE

Related Key Terms

INTRODUCTION *[p. 321]*

1. Two organ systems, the nervous and endocrine systems, coordinate the activities of other organ systems. The nervous system provides swift but brief responses to stimuli; the endocrine system adjusts metabolic operations and directs long-term changes.

AN OVERVIEW OF THE NERVOUS SYSTEM *[p. 321]*

1. The **nervous system** includes all the **neural tissue** in the body. Its anatomical divisions include the **central nervous system (CNS)** (the brain and spinal cord) and the **peripheral nervous system (PNS)** (all of the neural tissue outside the CNS). Functionally it can be divided into an **afferent division**, which brings sensory information to the CNS and an **efferent division**, which carries motor commands to muscles and glands. The efferent division includes the **somatic nervous system (SNS)** (voluntary control over skeletal muscle contractions), and the **autonomic nervous system (ANS)** (automatic, involuntary regulation of smooth muscle, cardiac muscle, and glandular activity) *(see Figures 13-1/13-2 and Table 13-1)*.

receptors

CELLULAR ORGANIZATION IN NEURAL TISSUE *[p. 321]*

1. There are two types of cells in neural tissue: **neurons**, which are responsible for information transfer and processing, and **neuroglia**, or **glial cells** *(see Table 13-2)*.

synaptic terminals

2. A typical neuron has a **cell body**, or **soma**, an **axon**, and several **dendrites** *(see Figure 13-3)*.

Neuroglia *[p. 322]*

3. There are four types of neuroglia: (1) **astrocytes** (largest and most numerous); (2) **oligodendrocytes**, which are responsible for the **myelination** of CNS axons; (3) **microglia** (phagocytic white blood cells); and (4) **ependymal cells** (with functions related to the **cerebrospinal fluid (CSF)** *(see Figures 13-4 to 13-7 and Table 13-2)*.

blood-brain barrier • myelin • unmyelinated • nodes • internodes • white matter • gray matter • ependyma

4. Nerve cell bodies in the PNS are clustered into **ganglia** (singular *ganglion*), and their axons form **peripheral nerves** *(see Figures 13-6/13-7)*.

satellite cells • Schwann cells • axolemma • neurilemma •

Neurons *[p. 326]*

5. The **perikaryon** of a neuron contains organelles, including **neurofilaments**, **neurotubules**, and **neurofibrils**. The **axon hillock** connects the **initial segment** of the axon to the cell body, and the **axoplasm** contains numerous organelles *(see Figure 13-8)*.

synaptic knobs • neurotransmitters

6. **Collaterals** may branch from an axon, with **telodendria** branching from the axon's tip *(see Figure 13-8)*.

7. Neurons may be described as **anaxonic**, **unipolar**, **bipolar**, or **multipolar** *(see Figure 13-9)*.

8. A **synapse** is a site of intercellular communication. A synapse where a neuron communicates with another cell type is a **neuroeffector junction** *(see Figure 13-11)*.

9. There are three functional categories of neurons: *sensory neurons*, *motor neurons*, and *interneurons* (association neurons) *(see Figure 13-10)*.

afferent fibers • somatic sensory neurons • visceral sensory neurons (interoceptors) • exteroceptors • proprioceptors • interoceptors • efferent fibers • somatic motor neurons • visceral motor neurons • preganglionic fibers • postganglionic fibers • excitatory • inhibitory

10. **Sensory neurons** form the afferent division of the PNS, and deliver information to the CNS.

11. **Motor neurons** stimulate or modify the activity of a peripheral tissue, organ, or organ system.

12. **Interneurons (association neurons)** may be located between sensory and motor neurons; they analyze sensory inputs and coordinate motor outputs.

THE NERVE IMPULSE *[p. 330]*

1. **Excitability** is the ability of a cell membrane to conduct electrical impulses; the cell membranes of skeletal muscle fibers and most neurons are excitable.

2. The conducted changes in the transmembrane potential are called **action potentials**. An action potential traveling along an axon is called a **nerve impulse**.

3. The rate of impulse conduction depends on the properties of the axon, specifically the presence or absence of a myelin sheath (a myelinated axon conducts impulses five to seven times faster than an unmyelinated axon) and the diameter of the axon (the larger the diameter, the faster the rate of conduction).

SYNAPTIC COMMUNICATION *[p. 331]*

1. A synapse may be **chemical** (involving a neurotransmitter) or **electrical** (with direct physical contact between cells). **Chemical synapses** are more common *(see Figure 13-11)*.

neuroeffector junctions

Chemical Synapses *[p. 331]*

2. At a chemical synapse involving two neurons, a neurotransmitter released at the *presynaptic membrane* binds to receptor proteins on the *postsynaptic membrane* and triggers a change in the transmembrane potential of the receptive cell. Only the presynaptic membrane releases a neurotransmitter, and communication can occur in only one direction: from the presynaptic neuron to the postsynaptic neuron *(see Figure 13-11)*.

3. There are more than 50 neurotransmitters known. All neuromuscular junctions utilize ACh as a neurotransmitter; ACh is also released at many synapses in the CNS and PNS.

4. The general sequence of events is as follows: (1) arrival of a nerve impulse at the synaptic knob triggers release of the neurotransmitter at the presynaptic membrane, (2) the neurotransmitter diffuses across the synaptic cleft and binds to receptors on the postsynaptic membrane, (3) receptor binding results in a change in the permeability of the postsynaptic cell membrane, which may lead to generation of an action potential, and (4) the effects on the postsynaptic membrane fade rapidly as the neurotransmitter molecules are enzymatically removed.

5. Neurotransmitters may have *excitatory* or *inhibitory* effects on the postsynaptic membrane depending on the nature of the receptor involved.

6. There may be thousands of synapses on the cell body of a single neuron. The activity of the neuron depends on the sum of all of the excitatory and inhibitory stimuli arriving at any given moment.

Electrical Synapses [p. 332]

7. Electrical synapses are found between neurons in the CNS and PNS, but they are relatively rare. At an electrical synapse, the presynaptic and postsynaptic cell membranes are tightly bound together, and they function as if they shared a common cell membrane.

NEURAL REGENERATION [p. 332]

1. Neurons have a very limited ability to regenerate after an injury. When an entire peripheral nerve is severed, only a relatively small number of axons will successfully reestablish normal synaptic contacts. As a result, nerve function is permanently impaired *(see Figure 13-12)*.

2. Limited regeneration can occur inside the central nervous system, but the situation is more complicated because (1) many more axons are likely to be involved, (2) astrocytes produce scar tissue that can prevent axon growth across the damaged area, and (3) astrocytes release chemicals that block the regrowth of axons *(see Figure 13-12)*.

NEURON ORGANIZATION AND PROCESSING [p. 332]

1. The roughly 20 billion interneurons can be classified into **neuronal pools** (groups of interconnected neurons with specific functions) *(see Figure 13-13)*.

2. Divergence is the spread of information from one neuron to several, or from one pool to several pools. In **convergence**, several neurons synapse on the same postsynaptic neuron. Neuronal pools may also function in sequence (**serial processing**) or may process the same information at one time (**parallel processing**). In **reverberation**, collateral axons establish a circuit that further stimulates presynaptic neurons *(see Figure 13-13)*.

ANATOMICAL ORGANIZATION OF THE NERVOUS SYSTEM [p. 334]

1. The functions of the nervous system as a whole depend on interactions between neurons in neuronal pools. *Spinal nerves* communicate with the spinal cord, and *cranial nerves* are connected to the brain *(see Figure 13-14)*.

2. Sensory *(ascending)* **pathways** carry information from peripheral receptors to the brain; **motor** *(descending)* **pathways** extend from CNS centers concerned with motor control to the associated skeletal muscles *(see Figure 13-14)*.

Related Key Terms

Wallerian degeneration

center • nucleus • neural cortex

tracts • columns • pathways

---

1   REVIEW OF CHAPTER OBJECTIVES

1. Discuss the anatomical organization and general functions of the nervous system.

2. Distinguish between neurons and neuroglia and compare their structures and functions.

3. Describe the structure of a typical neuron.

4. Understand the basis for the structural and functional classification of neurons.

5. Describe the significance of excitability in muscle and nerve cell membranes.

6. Identify the factors that determine the speed of nerve impulse conduction.

7. Summarize the events that occur during synaptic transmission and describe the effects of a typical neurotransmitter, ACh.

8. Describe the process of peripheral nerve regeneration after injury to an axon.

9. Discuss the possible methods of interaction between individual neurons or groups of neurons in neuronal pools.

10. Describe the anatomical organization of the nervous system.

---

2   REVIEW OF CONCEPTS

1. How does nervous control of an effector differ from endocrine control?

2. How does convergence function in the modulation of a nervous response?

3. How do CNS and PNS functions differ?

4. What roles do glial cells play in the nervous system?

5. What is the purpose of the blood-brain barrier?

6. Compare the functions of astrocytes with those of oligodendrocytes.

7. How does myelin improve the speed of electrical conduction along an axon?

8. Identify the location, function, and route of cerebrospinal fluid (CSF) within the CNS.

9. How do the ependymal cells and the choroid plexus function to produce and regulate cerebrospinal fluid (CSF)?

10. Distinguish between the autonomic (visceral motor) system and the somatic motor system?

11. How are action potentials involved in skeletal muscle contraction?

12. What is the role of neurotransmitter substances in chemical synapses? How does an electrical synapse differ from a chemical synapse?

---

3   CRITICAL THINKING AND CLINICAL APPLICATION QUESTIONS

1. A baby appears normal at birth, but shortly thereafter begins to show swelling of the cranial vault. What is this condition, what is the most common anatomical cause, and how can it be treated?

2. An 8-year-old girl was cut on the elbow by flying glass from a window shattered by a baseball. This injury caused little muscle damage, but partially severed the ulnar nerve. What is likely to happen to the severed axons, and will the little girl regain normal function of the nerve?

3. An engineering student discovers that a heated soldering iron looks exactly like a cold soldering iron. Her hand immediately pulls away from the hot iron. What neural mechanism controlled this motor response?

14

The Nervous System:
The Spinal Cord and Spinal Nerves

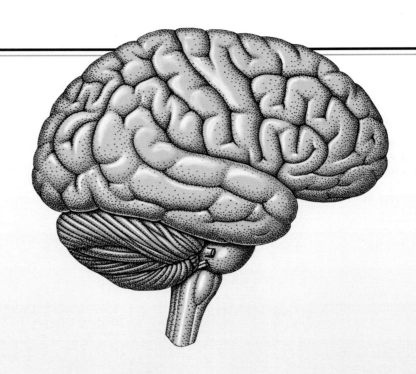

CHAPTER OUTLINE AND OBJECTIVES

The **central nervous system (CNS)** consists of the *spinal cord* and *brain*. Despite the fact that the two are anatomically connected, the brain and spinal cord show significant degrees of functional independence. The spinal cord is far more than just a highway for information traveling to or from the brain. Although most sensory data will be relayed to the brain, the spinal cord also integrates and processes information on its own. This chapter describes the anatomy of the spinal cord and examines the integrative activities that occur in this portion of the CNS.

GROSS ANATOMY OF THE SPINAL CORD
(Figures 14-1 to 14-3)

The adult spinal cord (Figure 14-1a●) measures approximately 45 cm (18 in.) in length. The dorsal surface of the spinal cord bears a shallow longitudinal groove, the **posterior median sulcus**. The deep crease along the ventral surface is the **anterior median fissure**. The cervical spinal cord contains all of the ascending and descending tracts linking the spinal cord with the brain. The thoracic spinal cord includes all of the tracts involved with thoracic, lumbar, and sacral segments. The lumbar spinal cord carries tracts for the lumbar and sacral sections, while the sacral spinal cord, the narrowest of all, consists only of tracts that begin or end in that region. Figure 14-1c● provides a series of sectional views that demonstrate the variations in the relative mass of gray matter in the cervical, thoracic, lumbar, and sacral regions of the spinal cord.

The amount of gray matter is substantially increased in segments of the spinal cord concerned with the sensory and motor innervation of the limbs. These areas are expanded, forming the **enlargements** of the spinal cord seen in Figure 14-1a●. The **cervical enlargement** supplies nerves to the shoulder girdle and upper extremities; the **lumbar enlargement** provides innervation to structures of the pelvis and lower extremities. Below the lumbar enlargement, the spinal cord becomes tapered and conical; this region is known as the **conus medullaris**. A slender strand of fibrous tissue, the **filum terminale** ("terminal thread"), extends from the inferior tip of the conus medullaris. This filamentous extension continues along the length of the vertebral canal as far as the second sacral vertebra (Figure 14-1a●). There it provides longitudinal support to the spinal cord as a component of the *coccygeal ligament*.

The entire spinal cord can be divided into 31 segments. Each segment is identified by a letter and number designation. For example, C_3, the segment in the uppermost section in Figure 14-1a●, is the third cervical segment.

Every spinal segment is associated with a pair of **dorsal root ganglia** that contain the cell bodies of sensory neurons. These sensory ganglia lie between the pedicles of adjacent vertebrae. On either side of the spinal cord, a **dorsal root**, which contains the axons of the neurons in the dorsal root ganglion, brings sensory information to the spinal cord. Anterior to the dorsal root, a **ventral root** leaves the spinal cord. The ventral root contains the axons of somatic and visceral motor neurons that control peripheral effectors. The dorsal roots of each segment enter the vertebral canal between adjacent vertebrae at the *intervertebral foramina*. ∞ [p. 160] The ventral roots that origi-

nate at the same segment leave the vertebral canal at the same location. Distal to each dorsal root ganglion, the sensory and motor fibers are bound together into a single **spinal nerve** (Figure 14-2●). Spinal nerves are classified as **mixed nerves**, because they contain both afferent (sensory) and efferent (motor) fibers. Figure 14-3●, p. 342, shows the spinal nerves as they emerge from intervertebral foramina.

The spinal cord continues to enlarge and elongate until an individual is approximately 4 years old. Up to that time, enlargement of the spinal cord keeps pace with the growth of the vertebral column (Figure 14-1b●). Throughout this period the ventral and dorsal roots are short, and they leave the vertebral canal through the adjacent intervertebral foramina. After age 4 the vertebral column continues to grow, but the spinal cord does not. This vertebral growth carries the dorsal root ganglia and spinal nerves farther and farther away from their original position relative to the spinal cord. As a result, the dorsal and ventral roots gradually elongate. The adult spinal cord extends only to the level of the first or second lumbar vertebra.

When seen in gross dissection, the filum terminale and the long ventral and dorsal roots that extend caudal to the conus medullaris reminded early anatomists of a horse's tail. With this in mind the complex was called the **cauda equina** (KAW-da ek-WĪ-na; *cauda*, tail + *equus*, horse) (Figure 14-1a●).

SPINAL MENINGES
(Figure 14-2)

The vertebral column and its surrounding ligaments, tendons, and muscles isolate the spinal cord from the external environment. ∞ [p. 160] The delicate neural tissues must also be protected against damaging contacts with the surrounding bony walls of the vertebral canal. A series of specialized membranes, the **spinal meninges** (men-IN-jēz), provide the necessary protection, physical stability, and shock absorption (Figure 14-2●). Blood vessels branching within these layers also deliver oxygen and nutrients to the spinal cord. There are three meningeal layers: the *dura mater*, the *arachnoid*, and the *pia mater*. At the foramen magnum of the skull, the spinal meninges are continuous with the **cranial meninges** that surround the brain. (The cranial meninges, which have the same three layers, will be described in Chapter 15.)

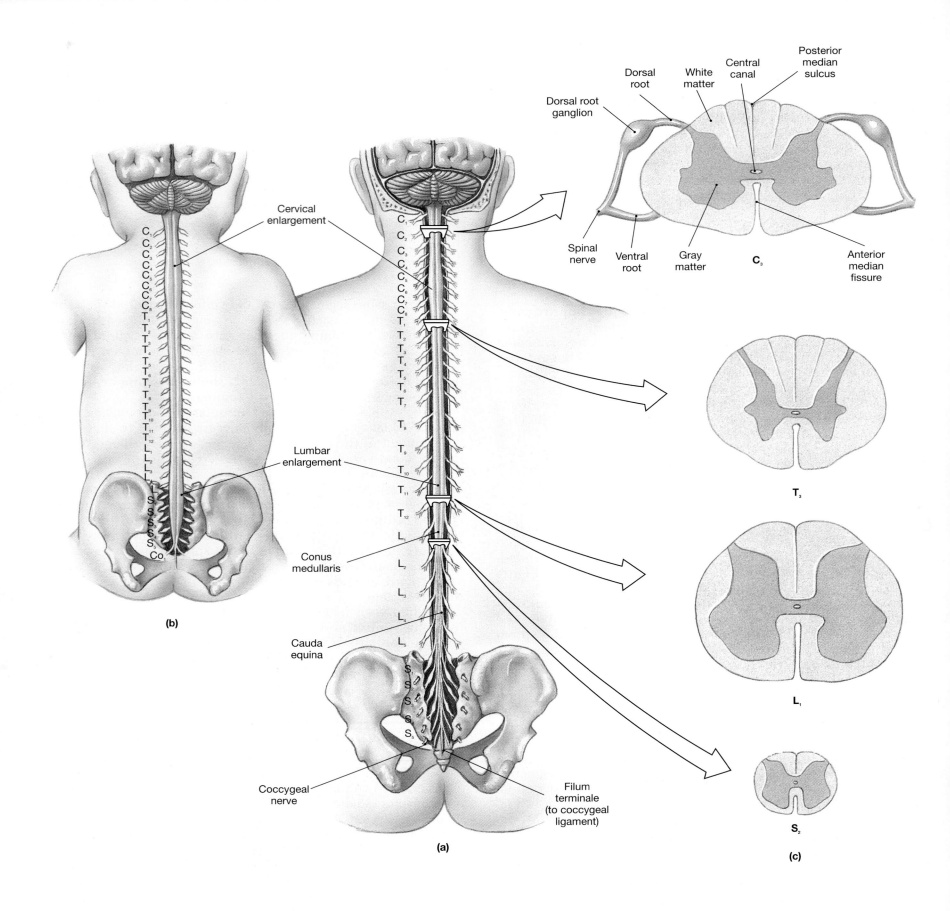

FIGURE 14-1
Gross Anatomy of the Spinal Cord. (a) Superficial anatomy and orientation of the adult spinal cord. (b) Position and extent of the spinal cord in a 4-year-old child; compare with part (a) and note the change in the relative positions of the spinal cord and vertebrae. (c) Inferior views of cross sections through representative regions of the spinal cord, showing the arrangement of gray and white matter.

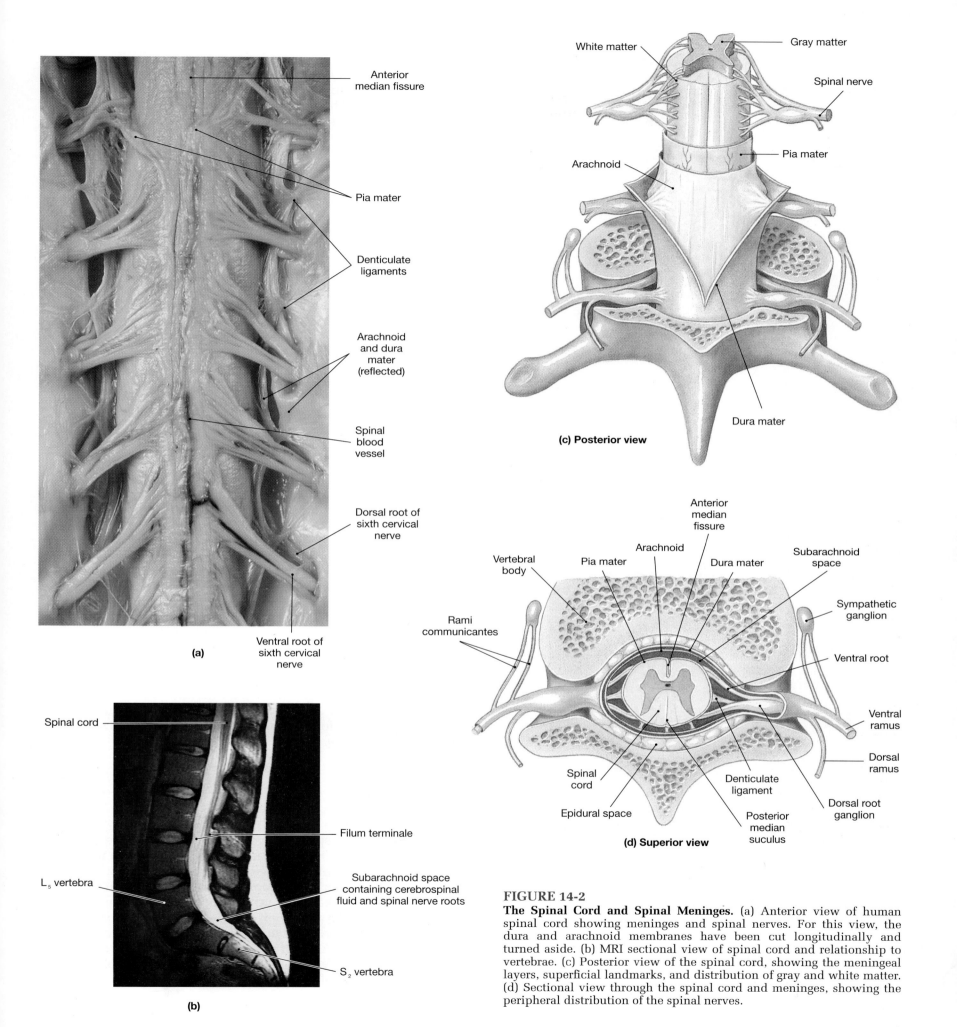

(a)

Anterior median fissure

Pia mater

Denticulate ligaments

Arachnoid and dura mater (reflected)

Spinal blood vessel

Dorsal root of sixth cervical nerve

Ventral root of sixth cervical nerve

(b)

Spinal cord

L₅ vertebra

Filum terminale

Subarachnoid space containing cerebrospinal fluid and spinal nerve roots

S₂ vertebra

(c) Posterior view

White matter

Gray matter

Spinal nerve

Arachnoid

Pia mater

Dura mater

(d) Superior view

Vertebral body

Pia mater

Arachnoid

Anterior median fissure

Dura mater

Subarachnoid space

Rami communicantes

Sympathetic ganglion

Ventral root

Ventral ramus

Dorsal ramus

Spinal cord

Denticulate ligament

Dorsal root ganglion

Epidural space

Posterior median suculus

FIGURE 14-2

The Spinal Cord and Spinal Meninges. (a) Anterior view of human spinal cord showing meninges and spinal nerves. For this view, the dura and arachnoid membranes have been cut longitudinally and turned aside. (b) MRI sectional view of spinal cord and relationship to vertebrae. (c) Posterior view of the spinal cord, showing the meningeal layers, superficial landmarks, and distribution of gray and white matter. (d) Sectional view through the spinal cord and meninges, showing the peripheral distribution of the spinal nerves.

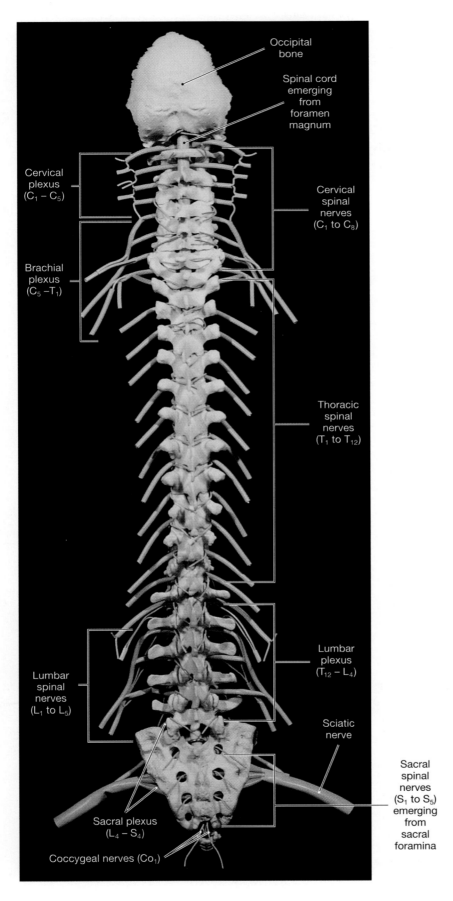

The Dura Mater
(Figure 14-2)

The tough, fibrous **dura mater** (DŪ-ra MĀ-ter; *dura*, hard + *mater*, mother) forms the outermost covering of the spinal cord and brain. The dense irregular collagen fibers of the dura are oriented along the longitudinal axis of the cord. Between the dura mater and the walls of the vertebral canal lies the **epidural space**, which contains loose connective tissue, blood vessels, and adipose tissue (Figure 14-2d●).

Longitudinal stability for the dura mater is provided by localized attachment sites at either end of the vertebral canal. Cranially, the dura mater fuses with the periosteum lining the cranial cavity at the margins of the foramen magnum. Caudally, the spinal dura mater tapers from a sheath to a dense cord of collagen fibers that ultimately blend with components of the filum terminale to form the **coccygeal ligament**. The coccygeal ligament extends along the sacral canal and is interwoven into the periosteum of the sacrum and coccyx. Lateral support for the dura mater is provided by the connective tissues within the epidural space. In addition, the dura mater extends between adjacent vertebrae at each intervertebral foramen, fusing with the connective tissues that surround the spinal nerves (Figure 14-2c,d●).

The Arachnoid
(Figure 14-2a,c)

In most anatomical and histological preparations, a narrow **subdural space** separates the dura from deeper meningeal layers. It is likely, however, that in life no such space exists, and the inner surface of the dura is in contact with the outer surface of the **arachnoid** (a-RAK-noyd; *arachne*, spider). The inner surface of the dura and the outer surface of the arachnoid are lined by simple squamous epithelia. The arachnoid (Figure 14-2a,c●) includes the squamous epithelium and the underlying **subarachnoid space**, which contains modified fibroblasts distributed among the *arachnoid trabeculae*, a delicate network of collagen and elastic fibers. The subarachnoid space also contains **cerebrospinal fluid** that acts as a shock absorber as well as a diffusion medium for dissolved gases, nutrients, chemical messengers, and waste products. The subarachnoid space and the role of cerebrospinal fluid will be discussed in Chapter 15. The subarachnoid space can be easily tapped between L_3 and L_4 for the clinical examination of cerebrospinal fluid or for the administration of anesthetics. *Spinal Meningitis* and *Spinal Anesthesia [p. 759]*

The Pia Mater
(Figure 14-2)

The subarachnoid space bridges the gap between the arachnoid epithelium and the innermost meningeal layer, the **pia mater** (*pia*, delicate + *mater*, mother) as seen in Figure 14-2a,c,d●. The elastic and collagen fibers of the pia are interwoven with those of the subarachnoid space. The blood vessels supplying the spinal cord are found here. Unlike the outer meninges, the pia mater is firmly bound to the underlying neural tissue, conforming to its bulges and fissures. The surface of the spinal cord consists of a thin layer of astrocytes, and cytoplasmic extensions of these glial cells lock the collagen fibers of the pia mater in place.

FIGURE 14-3
Posterior View of Vertebral Column and Spinal Nerves. This view, which extends from the occipital bone to the coccyx, shows the spinal nerves as they emerge from intervertebral foramina.

The combination of pia mater and arachnoid is sometimes referred to as the *pia-arachnoid*. Along the length of the spinal cord, paired **denticulate ligaments** connect the pia-arachnoid to the dura mater. The denticulate ligaments originate along either side of the spinal cord, between the ventral and dorsal roots. These ligaments, which begin at the foramen magnum of the skull, help prevent side-to-side movement and inferior movement of the spinal cord. The connective tissue fibers of the pia mater continue from the inferior tip of the conus medullaris as the filum terminale. As noted above, the filum terminale blends into the coccygeal ligament; this arrangement prevents superior movement of the spinal cord.

The spinal meninges surround the dorsal and ventral roots within the intervertebral foramina. As seen in Figure 14-2●, the meningeal membranes are continuous with the connective tissues surrounding the spinal nerves and their peripheral branches.

√ **Damage to which root of a spinal nerve would interfere with motor function?**

√ **Identify the location of the cerebrospinal fluid that surrounds the spinal cord.**

CLINICAL BRIEF
Spinal Taps and Myelography

Tissue samples, or *biopsies*, are taken from many organs to assist in diagnosis. For example, when a liver or skin disorder is suspected, small plugs of tissue are removed and examined for signs of infection or cell damage or are used to identify the bacteria causing an infection. Unlike many other tissues, however, neural tissue consists largely of cells rather than extracellular fluids or fibers. Tissue samples are seldom removed for analysis because any extracted or damaged neurons will not be replaced. Instead, small volumes of cerebrospinal fluid (CSF) are collected and analyzed. CSF is intimately associated with the neural tissue of the CNS, and pathogens, cell debris, or metabolic wastes in the CNS will therefore be detectable in the CSF.

The withdrawal of cerebrospinal fluid, known as a **spinal tap**, must be done with care to avoid injuring the spinal cord. The adult spinal cord extends only as far as vertebra L_1 or L_2. Between vertebra L_2 and the sacrum, the meningeal layers remain intact, but they enclose only the relatively sturdy components of the cauda equina and a significant quantity of CSF. With the vertebral column flexed, a needle can be inserted between the lower lumbar vertebrae and into the subarachnoid spaces with minimal risk to the cauda equina. In this procedure, known as a **lumbar puncture**, 3–9 mℓ of fluid are taken from the subarachnoid space between vertebrae L_3 and L_4 (Figure 14-4a●). Spinal taps are performed when CNS infection is suspected or when diagnosing severe back pain, headaches, disc problems, and some types of strokes.

Myelography involves the introduction of radiopaque dyes into the CSF of the subarachnoid space. Because the dyes are opaque to X-rays, the CSF appears white on an X-ray photograph (Figure 14-4b●). Any tumors, inflammation, or adhesions that distort or divert CSF circulation will be shown in silhouette. In the event of severe infection, inflammation, or leukemia (cancer of the white blood cells), antibiotics, steroids, or anticancer drugs can be injected into the subarachnoid space.

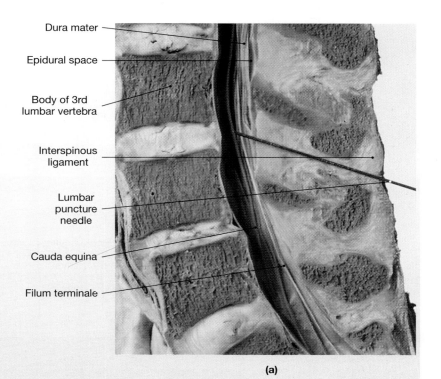

Dura mater
Epidural space
Body of 3rd lumbar vertebra
Interspinous ligament
Lumbar puncture needle
Cauda equina
Filum terminale

(a)

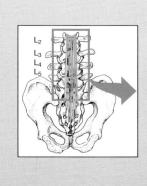

L₂
L₃
L₄
L₅

FIGURE 14-4
Spinal Taps and Myelography. (a) The position of the lumbar puncture needle is in the subarachnoid space, between the nerves of the cauda equina. The needle has been inserted in the midline between the third and fourth lumbar vertebral spines, pointing at a superior angle toward the umbilicus. Once the needle correctly punctures the dura and enters the subarachnoid space, a sample of CSF may be obtained. (b) A myelogram—an X-ray photograph of the spinal cord after introduction of a radiopaque dye into the CSF—showing the cauda equina in the lower lumbar region.

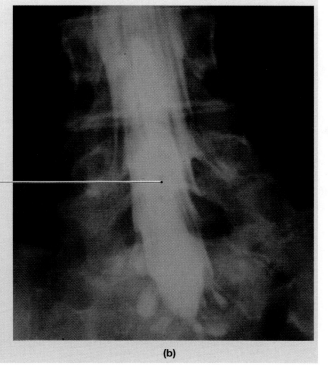

Cauda equina

(b)

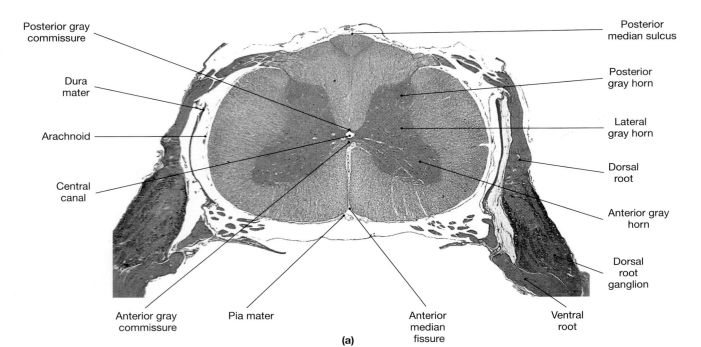

Posterior gray commissure

Dura mater

Arachnoid

Central canal

Anterior gray commissure

Pia mater

Anterior median fissure

Posterior median sulcus

Posterior gray horn

Lateral gray horn

Dorsal root

Anterior gray horn

Dorsal root ganglion

Ventral root

(a)

FIGURE 14-5

Sectional Organization of the Spinal Cord. (a) Transverse section of human spinal cord. (b) The left half of this sectional view shows important anatomical landmarks; the right half indicates the functional organization of the gray matter in the anterior and posterior gray horns. (c) The left half shows the major regions of white matter. The right half indicates the anatomical organization of sensory tracts in the posterior white column and motor nuclei in the anterior gray horn. Note that both sensory and motor components of the spinal cord have a definite regional organization.

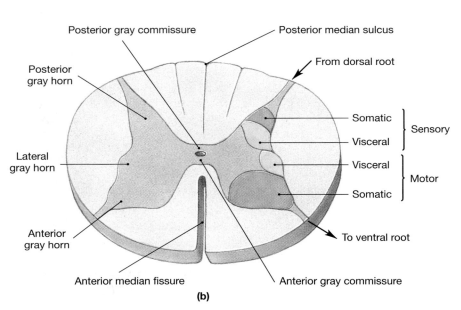

Posterior gray commissure

Posterior median sulcus

From dorsal root

Posterior gray horn

Somatic

Visceral

} Sensory

Lateral gray horn

Visceral

Somatic

} Motor

Anterior gray horn

To ventral root

Anterior median fissure

Anterior gray commissure

(b)

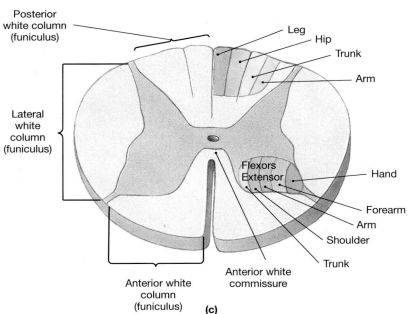

Posterior white column (funiculus)

Leg

Hip

Trunk

Arm

Lateral white column (funiculus)

Flexors

Extensor

Hand

Forearm

Arm

Shoulder

Trunk

Anterior white column (funiculus)

Anterior white commissure

(c)

SECTIONAL ANATOMY OF THE SPINAL CORD
(Figure 14-5)

The *anterior median fissure* and the *posterior median sulcus* are longitudinal landmarks that follow the division between the left and right sides of the spinal cord (Figure 14-5●). The peripherally situated **white matter** contains large numbers of myelinated and unmyelinated axons organized in *tracts* and *columns.* ∞ [p. 324] The **gray matter**, dominated by the cell bodies of neurons and glial cells, surrounds the narrow **central canal.** The projections of gray matter toward the outer surface of the spinal cord are called **horns** (Figure 14-5a,b●).

Organization of Gray Matter
(Figure 14-5b,c)

The cell bodies of neurons in the gray matter of the spinal cord are organized into groups, called *nuclei*, with specific functions. **Sensory nuclei** receive and relay sensory information from peripheral receptors, such as touch receptors located in the skin. **Motor nuclei** issue motor commands to peripheral effectors, such as skeletal muscles (Figure 14-5b●). Sensory and motor nuclei may extend for a considerable distance along the length of the spinal cord. A frontal section along the axis of the central canal separates the sensory (dorsal) nuclei from the motor (ventral) nuclei. The **posterior gray horns** contain somatic and visceral sensory nuclei, whereas the **anterior gray horns** are concerned with somatic motor control. The **lateral gray horns**, found in most segments, contain visceral motor neurons. The **gray commissures** (*commissura*, a joining together) posterior to and anterior to the central canal contain axons crossing from one side of the cord to the other before reaching a destination within the gray matter (Figure 14-5b●).

Figure 14-5b● shows the relationship between the function of a particular nucleus (sensory or motor) and its relative position within the gray matter of the spinal cord. The nuclei within each gray horn are also highly organized. Figure 14-5b,c● illustrates the distribution of somatic motor nuclei in the anterior gray horns of the cervical enlargement.

Organization of White Matter
(Figure 14-5)

The white matter can be divided into regions, or **columns** (*funiculi*, singular *funiculus*) (Figure 14-5c●). The **posterior white columns** lie between the posterior gray horns and the posterior median sulcus. The **anterior white columns** lie between the anterior gray horns and the anterior median fissure; they are interconnected by the **anterior white commissure**. The white matter on either side between the anterior and posterior columns represents the **lateral white columns**.

Each column contains **tracts**, or *fasciculi*, whose axons share functional and structural characteristics. A specific tract conveys either sensory data or motor commands, and the axons are relatively uniform with respect to diameter, myelination, and conduction speed. All of the axons within a tract relay information in the same direction. Small tracts carry sensory or motor signals between segments of the spinal cord, and larger tracts connect the spinal cord with the brain. **Ascending tracts** carry sensory information up toward the brain, and **descending tracts** convey motor commands down into the spinal cord. Within each column the tracts show a regional organization comparable to that found in the nuclei of the gray matter (Figure 14-5b,c●). The identities of the major CNS

■ **CLINICAL BRIEF**
Spinal Cord Injuries

Injuries affecting the spinal cord produce symptoms of sensory loss or motor paralysis that reflect the specific nuclei and tracts involved. At the outset, any severe injury to the spinal cord produces a period of sensory and motor paralysis termed **spinal shock**. The skeletal muscles become flaccid; neither somatic nor visceral reflexes function; and the brain no longer receives sensations of touch, pain, heat, or cold. The location and severity of the injury determine how long these symptoms persist and how completely the individual recovers.

Violent jolts, such as those associated with blows or gunshot wounds, may cause **spinal concussion** without visibly damaging the spinal cord. Spinal concussion produces a period of spinal shock, but the symptoms are only temporary and recovery may be complete in a matter of hours. More serious injuries, such as whiplash or falls, usually involve physical damage to the spinal cord. In a **spinal contusion**, hemorrhages occur in the meninges, pressure rises in the cerebrospinal fluid, and the white matter of the spinal cord may degenerate at the site of injury. Gradual recovery over a period of weeks may leave some functional losses. Recovery from a **spinal laceration** by vertebral fragments or other foreign bodies will usually be far slower and less complete. **Spinal compression** occurs when the spinal cord becomes physically squeezed or distorted within the vertebral canal. In a **spinal transection** the spinal cord is completely severed. Current surgical procedures cannot repair a severed spinal cord, but experimental techniques may restore partial function. ♈ *Technology and Motor Paralysis [p. 759]*

Spinal injuries often involve some combination of compression, laceration, contusion, and partial transection. Relieving pressure and stabilizing the affected area through surgery may prevent further damage and allow the injured spinal cord to recover as much as possible.

Extensive damage at the fourth or fifth cervical vertebra will eliminate sensation and motor control of the arms and legs. The extensive paralysis produced is called **quadriplegia**. If the damage extends from C_3 to C_5, the motor paralysis will include all of the major respiratory muscles, and the patient will usually need mechanical assistance in breathing. **Paraplegia**, the loss of motor control of the legs, may follow damage to the thoracic vertebrae. Injuries to the lower lumbar vertebrae may compress or distort the elements of the cauda equina, causing problems with peripheral nerve function.

tracts will be discussed when we consider sensory and motor pathways in Chapter 16.

SPINAL NERVES
(Figures 14-1/14-3/14-6)

There are 31 pairs of spinal nerves, each of which can be identified by its association with adjacent vertebrae (Figure 14-1a●, p. 340). Each spinal nerve has a regional number, as indicated in Figure 14-1●, p. 340. Spinal nerves caudal to the first thoracic vertebra take their names from the vertebra immediately preceding them. Thus the spinal nerve T_1 emerges immediately caudal to vertebra T_1, spinal nerve T_2 follows vertebra T_2, and so forth, as seen in Figure 14-3●, p. 342.

The arrangement differs in the cervical region because the first pair of spinal nerves, C_1, exits between the skull and the first cervical vertebra. For this reason, cervical nerves take their names from the vertebra immediately *following* them. In other words, cervical nerve C_2 *precedes* vertebra C_2, and the same system is used for the rest of the cervical series. The transition from this identification method occurs between the last cervical and first thoracic vertebrae. The spinal nerve lying between these two vertebrae has been designated C_8. Thus there are seven cervical vertebrae but *eight* cervical nerves.

A series of connective tissue layers surrounds each peripheral nerve and continues along all of its branches. These layers, shown in Figure 14-6●, are comparable to the connective tissue layers associated with skeletal muscles. ∞ [p. 233] The outermost layer, or **epineurium**, consists of a dense network of collagen fibers. The fibers of the **perineurium** divide the nerve into a series of compartments that contain bundles of axons. A single bundle of axons is known as a **fascicle**, or **fasciculus**. Arteries and veins penetrate the epineurium and branch within the perineurium. The **endoneurium** consists of delicate connective tissue fibers that surround individual axons. Capillaries leaving the perineurium branch in the endoneurium and provide oxygen and nutrients to the axons and Schwann cells of the nerve. ♈ *Multiple Sclerosis [p. 760]*

Peripheral Distribution of Spinal Nerves
(Figures 14-2d/14-3/14-7/14-8)

In the thoracic and upper lumbar regions (Figure 14-3●, p. 342), the first branch of each spinal nerve carries visceral motor fibers to a nearby **autonomic ganglion** (Figure 14-7a●). Because preganglionic axons are myelinated, this branch has a light color and is known as the **white ramus** (*ramus*, branch). Two groups of unmyelinated postganglionic fibers leave the ganglion. Those innervating glands and smooth muscles in the body wall or limbs form the **gray ramus** that rejoins the spinal nerve. The gray and white rami are collectively termed the **rami communicantes** (Figure 14-2d●, p. 341), or "communicating branches." Preganglionic or postganglionic fibers headed for internal organs do not rejoin the spinal nerves, but form a series of separate autonomic nerves, such as the *splanchnic nerves,* preganglionic autonomic nerves involved with the regulation of organs in the abdominopelvic cavity.

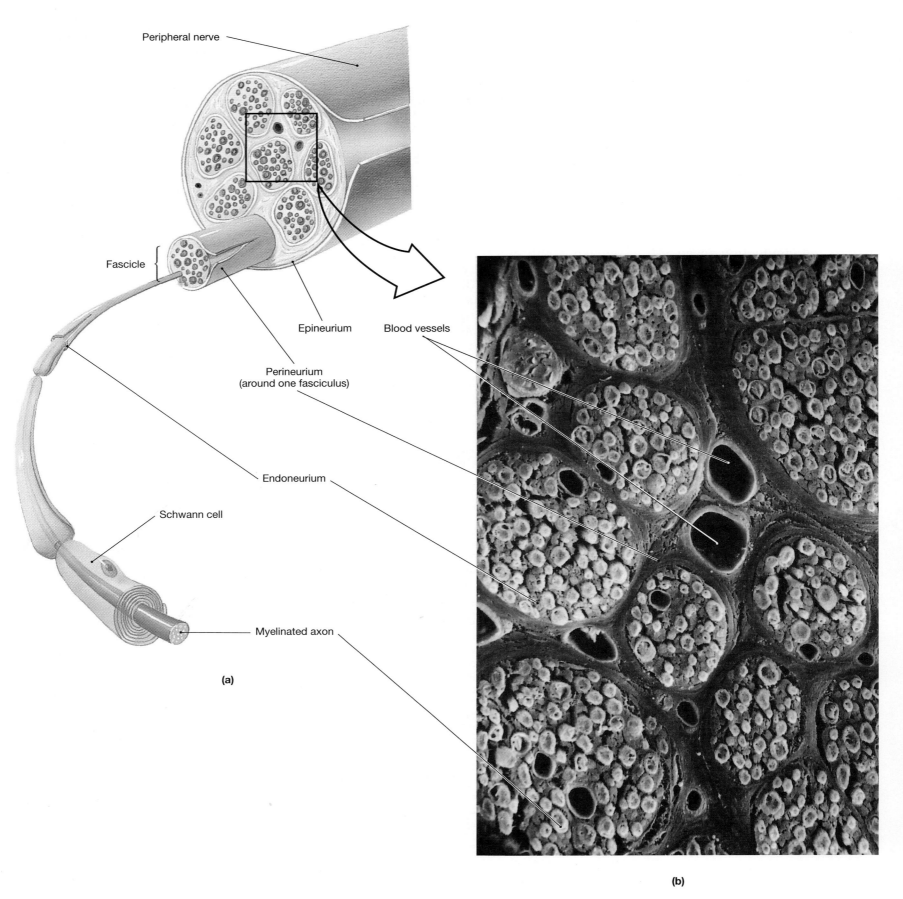

(a)

Peripheral nerve

Fascicle

Epineurium

Perineurium
(around one fasciculus)

Endoneurium

Schwann cell

Myelinated axon

Blood vessels

(b)

FIGURE 14-6

Peripheral Nerves. (a) A typical peripheral nerve and its connective tissue wrappings. (b) A scanning electron micrograph showing the various layers in great detail. (SEM × 425) [Reproduced from R. G. Kessel and R. H. Kardon, *Tissues and Organs: A Text-Atlas of Scanning Electron Microscopy*, W. H. Freeman & Co., 1979.]

The **dorsal ramus** of each spinal nerve provides sensory and motor innervation to the skin and muscles of the back. The relatively large **ventral ramus** supplies the ventrolateral body surface, structures in the body wall, and the limbs. The distribution of the dorsal rami illustrates the segmental division of labor along the length of the spinal cord. Note the distribution of sensory fibers in a branch of a typical spinal nerve (Figure 14-7b●). Each pair of spinal nerves monitors a specific region of the body surface, an area known as a **dermatome** (Figure 14-8●). Dermatomes are clinically important because damage to a spinal nerve or dorsal root ganglion will produce a characteristic loss of sensation in specific patches of the skin.

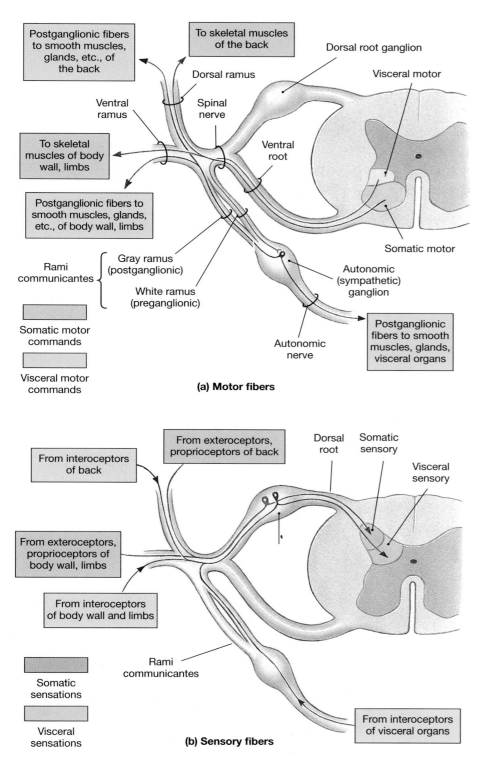

(a) Motor fibers

(b) Sensory fibers

FIGURE 14-7
Peripheral Distribution of Spinal Nerves. (a) Diagrammatic view indicating the distribution of motor fibers in the major branches of a typical spinal nerve. (b) A comparable view, detailing the distribution of sensory fibers.

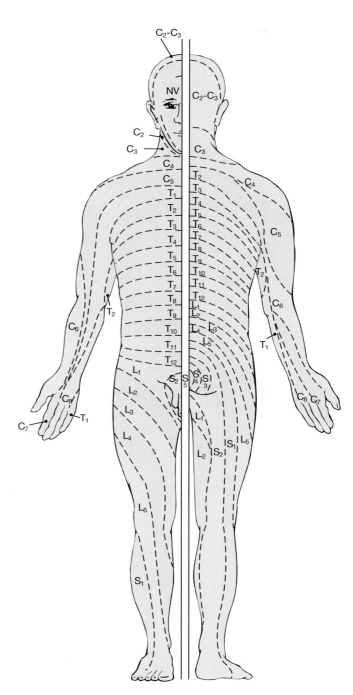

FIGURE 14-8
Dermatomes. Anterior and posterior distribution of dermatomes.

Nerve Plexuses
(Figures 14-3/14-7/14-9)

The relatively simple distribution pattern illustrated in Figure 14-7● applies to spinal nerves T_2–T_{12}. But in segments controlling the skeletal musculature of the neck and the upper and lower extremities, the situation becomes more complicated. During development, skeletal muscles fuse with their neighbors to form larger muscles with compound origins. Although the anatomical distinctions may disappear, ventral rami from the associated spinal segments continue to provide innervation and motor control. As they converge, the ventral rami of adjacent spinal nerves blend their fibers to produce a series of compound nerve trunks. Such a complex interwoven network of nerves is called a **nerve plexus** (PLEK-sus, "braid"). The three major nerve plexuses are the *cervical plexus*, *brachial plexus*, and *lumbosacral plexus* (Figures 14-3, p. 342, and 14-9●).

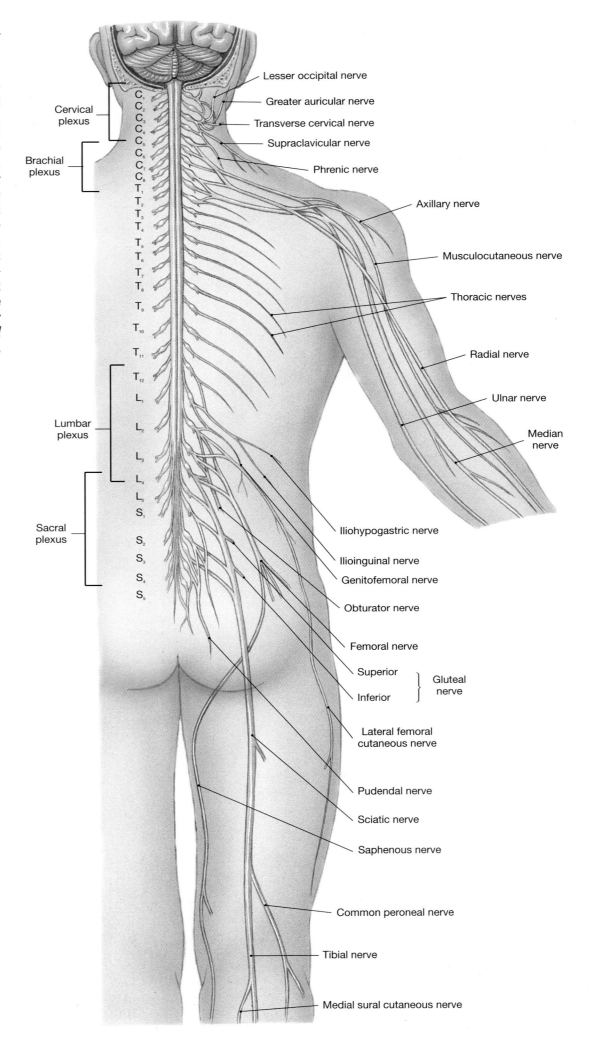

Cervical plexus

Brachial plexus

Lumbar plexus

Sacral plexus

Lesser occipital nerve
Greater auricular nerve
Transverse cervical nerve
Supraclavicular nerve
Phrenic nerve
Axillary nerve
Musculocutaneous nerve
Thoracic nerves
Radial nerve
Ulnar nerve
Median nerve
Iliohypogastric nerve
Ilioinguinal nerve
Genitofemoral nerve
Obturator nerve
Femoral nerve
Superior
Inferior
Gluteal nerve
Lateral femoral cutaneous nerve
Pudendal nerve
Sciatic nerve
Saphenous nerve
Common peroneal nerve
Tibial nerve
Medial sural cutaneous nerve

FIGURE 14-9
Peripheral Nerves and Nerve Plexuses

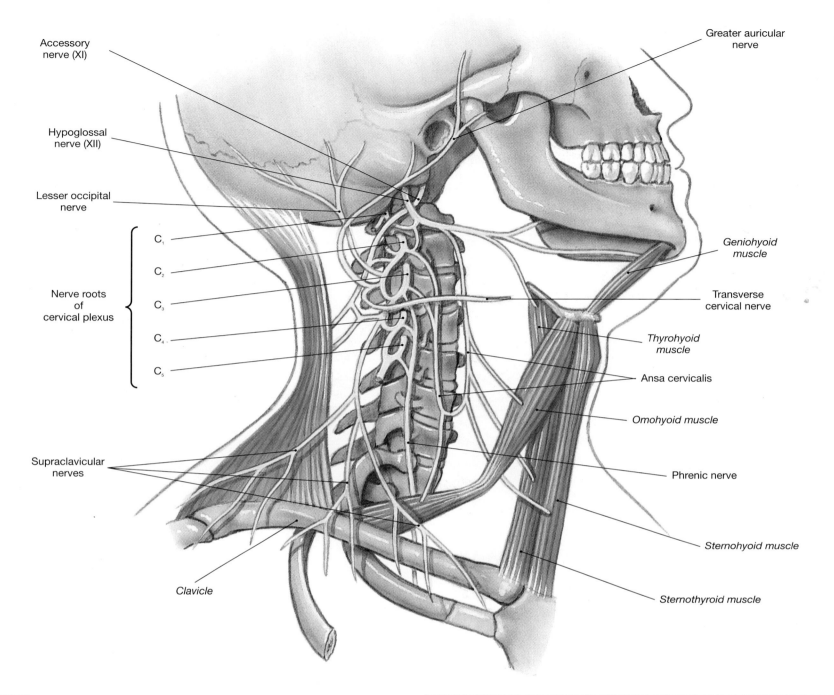

Accessory
nerve (XI)

Hypoglossal
nerve (XII)

Lesser occipital
nerve

C_1

C_2

Nerve roots
of
cervical plexus

C_3

C_4

C_5

Supraclavicular
nerves

Clavicle

Greater auricular
nerve

*Geniohyoid
muscle*

Transverse
cervical nerve

*Thyrohyoid
muscle*

Ansa cervicalis

Omohyoid muscle

Phrenic nerve

Sternohyoid muscle

Sternothyroid muscle

FIGURE 14-10
The Cervical Plexus

The Cervical Plexus
(Figures 14-9/14-10, Table 14-1)

The **cervical plexus** (Figures 14-9 and 14-10●) consists of the
ventral rami of spinal nerves C_1–C_4 and some nerve fibers
from C_5. Branches from the cervical plexus innervate the muscles of the neck and extend into the thoracic cavity to control
the diaphragmatic muscles (Table 14-1). The **phrenic nerve**,
the major nerve of this plexus, provides the entire nerve supply to the diaphragm. Other branches are distributed to the
skin of the neck, shoulder, and upper portion of the breast.

TABLE 14-1   The Cervical Plexus

Spinal Segment	Nerves	Distribution
C_1–C_4	Ansa cervicalis (superior and inferior branches)	Five of the extrinsic laryngeal muscles (sternothyroid, sternohyoid, omohyoid; geniohyoid and thyrohyoid via N XII)
C_2–C_3	Lesser occipital, transverse cervical, supraclavicular, and greater auricular nerves	Skin of upper chest, shoulder, neck, and ear
C_3–C_5	Phrenic nerve	Diaphragm
C_1–C_5	Cervical nerves	Levator scapulae, scalenes, sternocleidomastoid, and trapezius (with N XI)

The Brachial Plexus
(Figures 14-9/14-11/14-12, Table 14-2)

The **brachial plexus** (Table 14-2) innervates the shoulder girdle and upper extremity, with contributions from the ventral rami of spinal nerves C_5–T_1 (Figures 14-9, 14-11, and 14-12●). The nerves that form this plexus originate from one or more cords or trunks named according to their location. These are the *lateral cord*, the *medial cord*, and the *posterior cord*. The major nerves of the lateral cord are the **musculocutaneous nerve** and the **median nerve**. The **ulnar nerve** is the major nerve of the medial cord. The **axillary (circumflex) nerve** and **radial nerve** are the major nerves of the posterior cord.

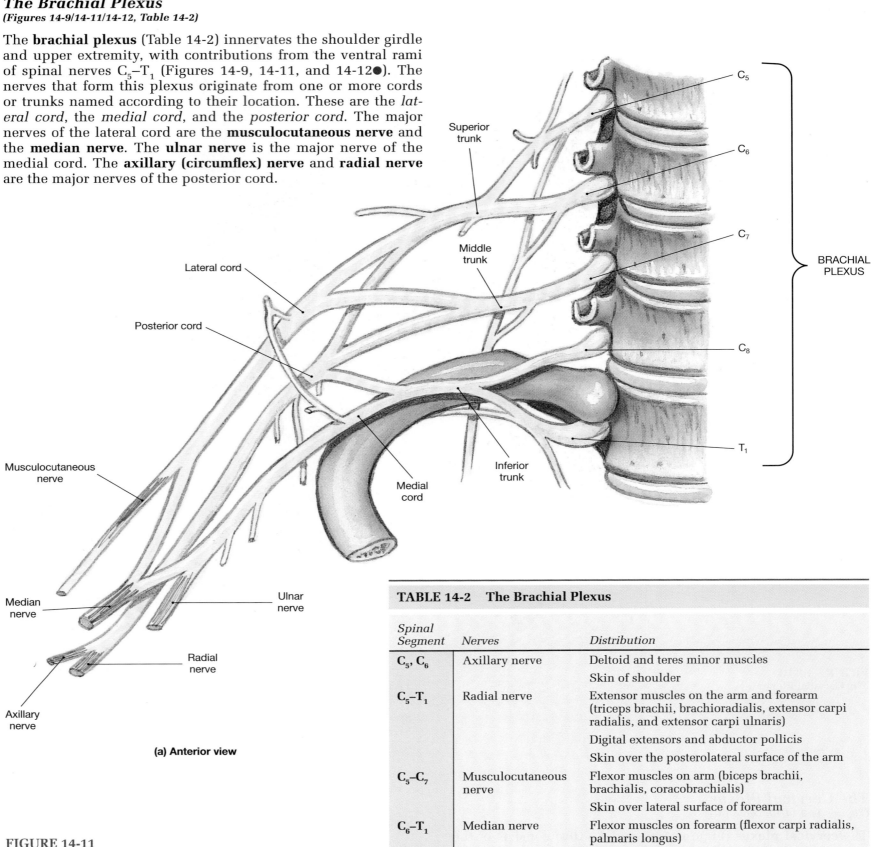

(a) Anterior view

FIGURE 14-11
The Brachial Plexus

TABLE 14-2   The Brachial Plexus

Spinal Segment	Nerves	Distribution
C_5, C_6	Axillary nerve	Deltoid and teres minor muscles
		Skin of shoulder
C_5–T_1	Radial nerve	Extensor muscles on the arm and forearm (triceps brachii, brachioradialis, extensor carpi radialis, and extensor carpi ulnaris)
		Digital extensors and abductor pollicis
		Skin over the posterolateral surface of the arm
C_5–C_7	Musculocutaneous nerve	Flexor muscles on arm (biceps brachii, brachialis, coracobrachialis)
		Skin over lateral surface of forearm
C_6–T_1	Median nerve	Flexor muscles on forearm (flexor carpi radialis, palmaris longus)
		Pronators (p. quadratus and p. teres)
		Digital flexors
		Skin over anterolateral surface of hand
C_8, T_1	Ulnar nerve	Flexor muscle on forearm (flexor carpi ulnaris)
		Adductor pollicis and small digital muscles
		Skin over medial surface of hand

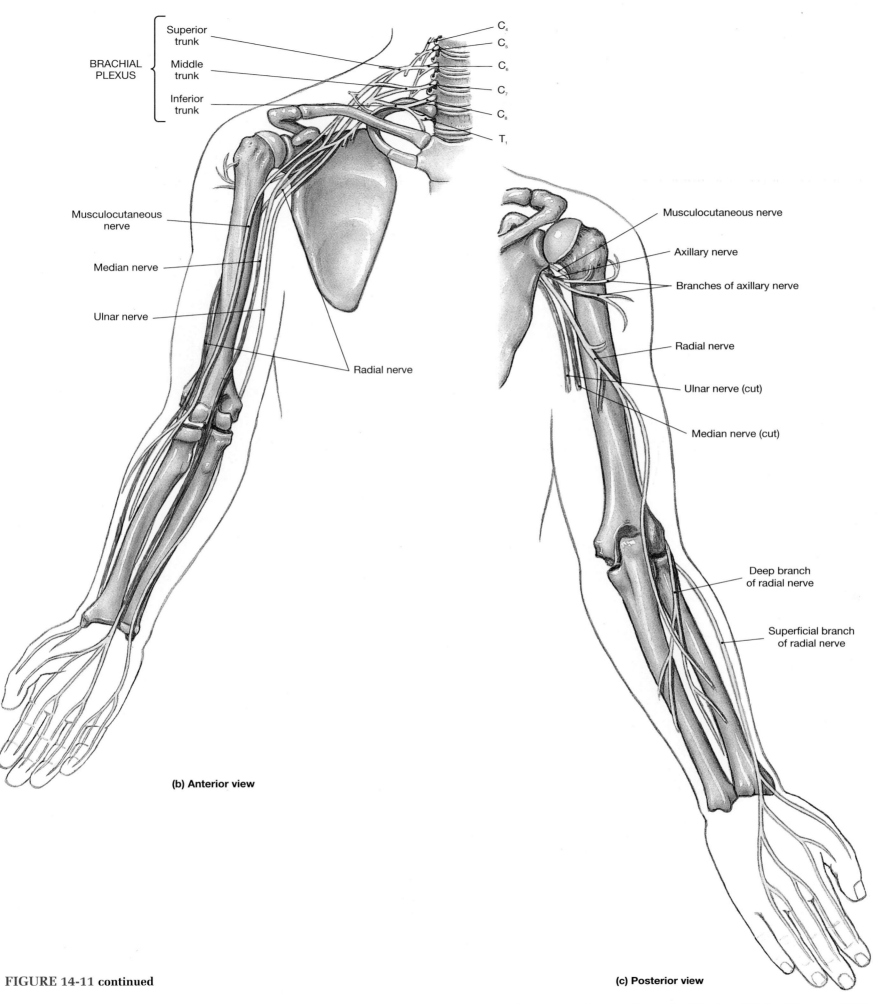

BRACHIAL PLEXUS

Superior trunk

Middle trunk

Inferior trunk

C₄
C₅
C₆
C₇
C₈
T₁

Musculocutaneous nerve

Median nerve

Ulnar nerve

Radial nerve

(b) Anterior view

Musculocutaneous nerve

Axillary nerve

Branches of axillary nerve

Radial nerve

Ulnar nerve (cut)

Median nerve (cut)

Deep branch of radial nerve

Superficial branch of radial nerve

(c) Posterior view

FIGURE 14-11 continued

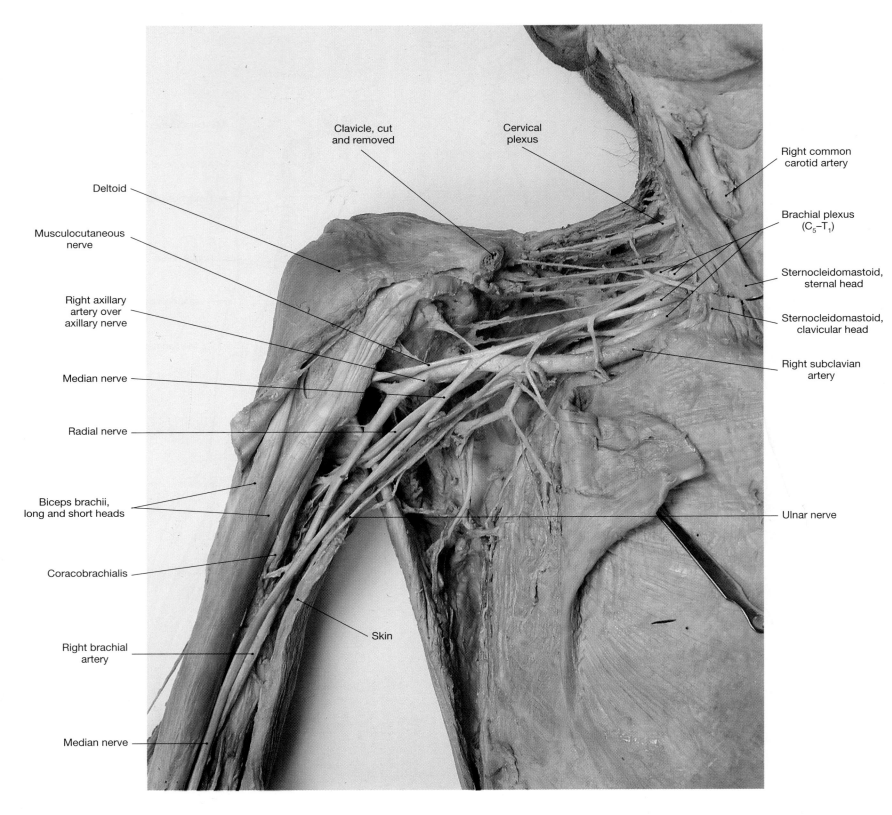

Deltoid

Musculocutaneous
nerve

Right axillary
artery over
axillary nerve

Median nerve

Radial nerve

Biceps brachii,
long and short heads

Coracobrachialis

Right brachial
artery

Median nerve

Clavicle, cut
and removed

Cervical
plexus

Right common
carotid artery

Brachial plexus
(C_5-T_1)

Sternocleidomastoid,
sternal head

Sternocleidomastoid,
clavicular head

Right subclavian
artery

Ulnar nerve

Skin

FIGURE 14-12
The Cervical and Brachial Plexuses. This cadaver specimen shows the main branches of the cervical and brachial plexuses.

TABLE 14-3   The Lumbosacral Plexus

Spinal Segment	Nerves	Distribution
THE LUMBAR PLEXUS		
T₁₂, L₁	Iliohypogastric nerve	Abdominal muscles (external and internal obliques, transversus abdominis) Skin over lower abdomen and buttocks
L₁	Ilioinguinal nerve	Abdominal muscles (with iliohypogastric) Skin over medial upper thigh and portions of external genitalia
L₁, L₂	Genitofemoral nerve	Skin over anteromedial surface of thigh and portions of external genitalia
L₂, L₃	Lateral femoral cutaneous nerve	Skin over anterior, lateral, and posterior surfaces of thigh
L₂–L₄	Femoral nerve	Anterior muscles of thigh (sartorius and quadriceps) Adductors of thigh (pectineus and iliopsoas) Skin over anteromedial surface of thigh, medial surface of leg and foot
L₂–L₄	Obturator nerve	Adductors of thigh (adductor magnus, brevis, longus) Gracilis muscle Skin over medial surface of thigh
L₂–L₄	Saphenous nerve	Skin over medial surface of leg
THE SACRAL PLEXUS		

L₄–S₂	Gluteal nerves:
	    Superior
	    Inferior
		
		Abductors of thigh (gluteus minimus, gluteus medius, and tensor fasciae latae)
		Extensor of thigh (gluteus maximus)

L₄–S₃	Sciatic nerve:	Two of the hamstrings (semimembranosus, semitendinosus) Adductor magnus (with obturator nerve)
	Tibial branch	Flexors of leg and plantar flexors of foot (popliteus, gastrocnemius, soleus, tibialis posterior, biceps femoris (long head)) Flexors of toes Skin over posterior surface of leg, plantar surface of foot
	Peroneal branch	Biceps femoris of hamstrings (short head) Peroneus (brevis and longus) and tibialis anterior Extensors of toes Skin over anterior surface of leg and dorsal surface of foot
S₂–S₄	Pudendal nerve	Skin and muscles of perineum, including urogenital diaphragm and external anal and urethral sphincters Skin of external genitalia and related skeletal muscles (bulbospongiosus and ischiocavernosus)

The Lumbosacral Plexus
(Figures 14-9/14-13/14-14, Table 14-3)

The **lumbosacral plexus** (Figures 14-9, p. 348, and 14-13●) arises from the posterior abdominal wall, and the ventral rami of these nerves supply the pelvic girdle and leg. This plexus can be subdivided into a **lumbar plexus** (T₁₂–L₄) and a **sacral plexus** (L₄–S₄). The individual nerves that form the lumbosacral plexus and their distribution are detailed in Table 14-3.

The major nerves of the lumbar plexus are the **genitofemoral nerve**, **lateral femoral cutaneous nerve**, and **femoral nerve**. The major nerves of the sacral plexus are the **sciatic nerve** and the **pudendal nerve**. The sciatic nerve passes posterior to the femur and deep to the long head of the biceps femoris muscle. As it approaches the popliteal fossa, the sciatic nerve divides into two branches: the **common peroneal nerve** and the **tibial nerve** (Figure 14-14●).

Chapters 10 and 11 introduced the peripheral nerves that control the major axial and appendicular muscles. At this point you should return to the tables in those chapters to review the innervation of the skeletal muscle groups. ∞ [pp. 255–274, 279–301]

Although dermatomes can provide clues to the location of injuries along the spinal cord, the loss of sensation at the skin does not provide precise information concerning the site of injury, because the boundaries of dermatomes are not precise, clearly defined lines. More exact conclusions can be drawn from the loss of motor control on the basis of the origin and distribution of the peripheral nerves originating at nerve plexuses. In the assessment of motor performance, a distinction is made between the conscious ability to control motor activities and the performance of automatic, involuntary motor responses. These latter, programmed motor patterns, called *reflexes*, will now be described.

√ **A patient suffering from polio has lost the use of his leg muscles. In what area of the spinal cord would you expect to locate the virally infected motor neurons in this individual?**

√ **An anesthetic blocks the function of the dorsal rami of the cervical spinal nerves. What area of the body would be affected?**

√ **Injury to which of the nerve plexuses would interfere with the ability to breathe?**

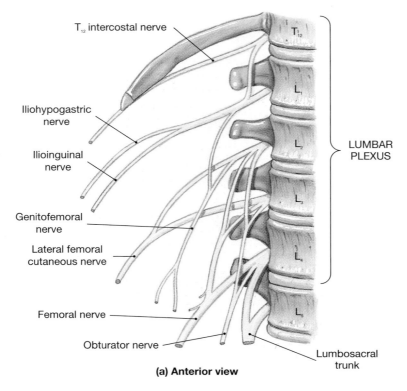

T_12 intercostal nerve

Iliohypogastric
nerve

Ilioinguinal
nerve

Genitofemoral
nerve

Lateral femoral
cutaneous nerve

Femoral nerve

Obturator nerve

T_12

L_1

L_2

L_3

L_4

L_5

LUMBAR
PLEXUS

Lumbosacral
trunk

(a) Anterior view

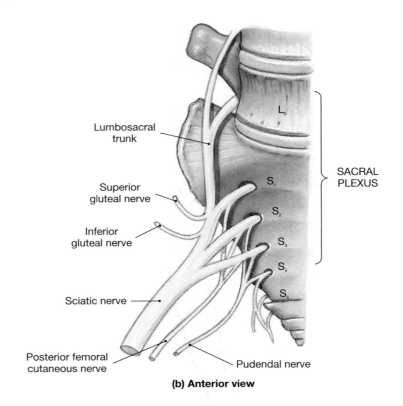

Lumbosacral
trunk

Superior
gluteal nerve

Inferior
gluteal nerve

Sciatic nerve

Posterior femoral
cutaneous nerve

L_5

S_1

S_2

S_3

S_4

S_5

SACRAL
PLEXUS

Pudendal nerve

(b) Anterior view

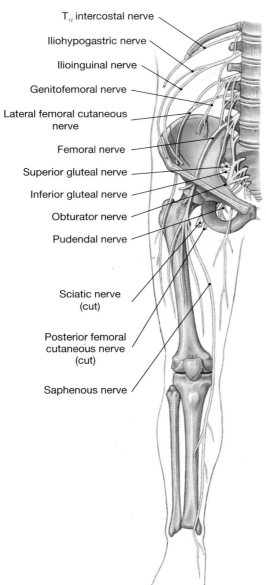

T_12 intercostal nerve

Iliohypogastric nerve

Ilioinguinal nerve

Genitofemoral nerve

Lateral femoral cutaneous
nerve

Femoral nerve

Superior gluteal nerve

Inferior gluteal nerve

Obturator nerve

Pudendal nerve

Sciatic nerve
(cut)

Posterior femoral
cutaneous nerve
(cut)

Saphenous nerve

(c) Anterior view

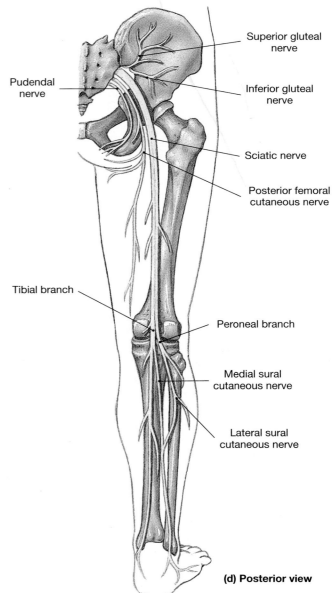

Pudendal
nerve

Superior gluteal
nerve

Inferior gluteal
nerve

Sciatic nerve

Posterior femoral
cutaneous nerve

Tibial branch

Peroneal branch

Medial sural
cutaneous nerve

Lateral sural
cutaneous nerve

(d) Posterior view

FIGURE 14-13
The Lumbosacral Plexus

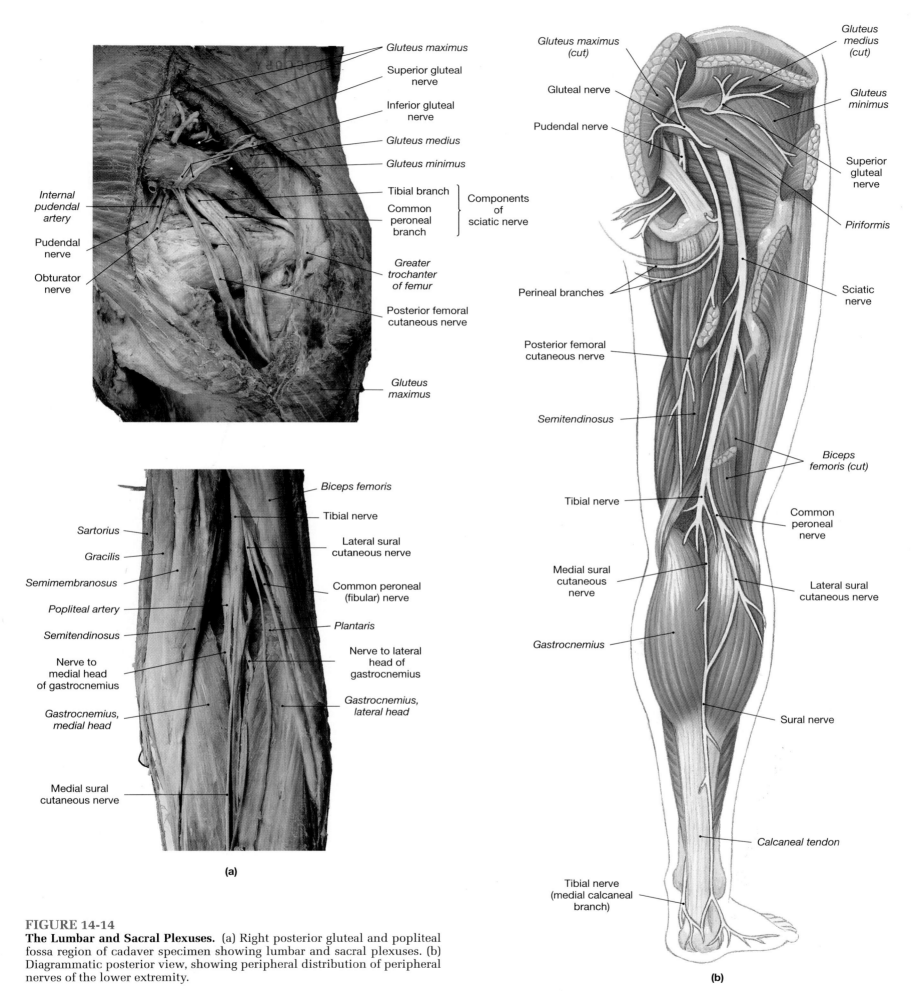

(a)

Gluteus maximus
Superior gluteal nerve
Inferior gluteal nerve
Gluteus medius
Gluteus minimus
Tibial branch
Common peroneal branch
} Components of sciatic nerve
Internal pudendal artery
Pudendal nerve
Obturator nerve
Greater trochanter of femur
Posterior femoral cutaneous nerve
Gluteus maximus

Biceps femoris
Tibial nerve
Lateral sural cutaneous nerve
Common peroneal (fibular) nerve
Plantaris
Nerve to lateral head of gastrocnemius
Gastrocnemius, lateral head
Sartorius
Gracilis
Semimembranosus
Popliteal artery
Semitendinosus
Nerve to medial head of gastrocnemius
Gastrocnemius, medial head
Medial sural cutaneous nerve

(b)

Gluteus maximus (cut)
Gluteus medius (cut)
Gluteal nerve
Gluteus minimus
Pudendal nerve
Superior gluteal nerve
Piriformis
Perineal branches
Sciatic nerve
Posterior femoral cutaneous nerve
Semitendinosus
Biceps femoris (cut)
Tibial nerve
Common peroneal nerve
Medial sural cutaneous nerve
Lateral sural cutaneous nerve
Gastrocnemius
Sural nerve
Calcaneal tendon
Tibial nerve (medial calcaneal branch)

FIGURE 14-14
The Lumbar and Sacral Plexuses. (a) Right posterior gluteal and popliteal fossa region of cadaver specimen showing lumbar and sacral plexuses. (b) Diagrammatic posterior view, showing peripheral distribution of peripheral nerves of the lower extremity.

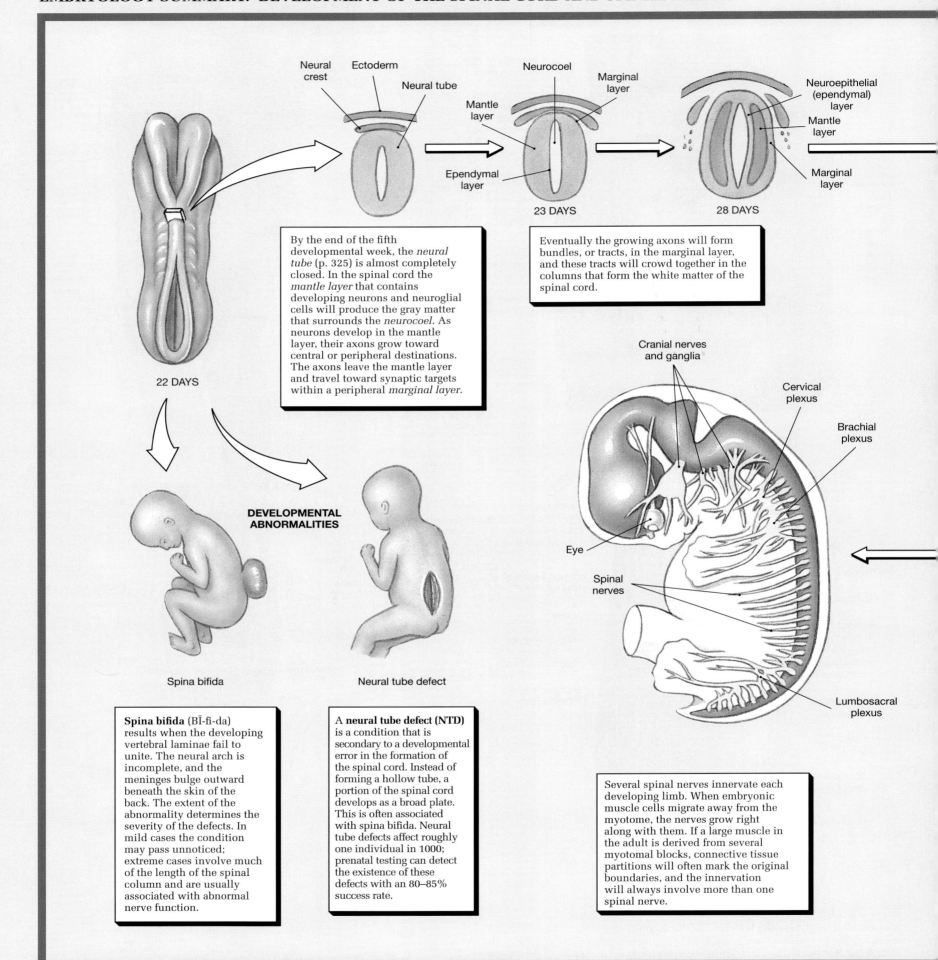

Neural crest · **Ectoderm** · **Neural tube**

Neurocoel · **Marginal layer** · **Mantle layer** · **Ependymal layer**

23 DAYS

Neuroepithelial (ependymal) layer · **Mantle layer** · **Marginal layer**

28 DAYS

22 DAYS

By the end of the fifth developmental week, the *neural tube* (p. 325) is almost completely closed. In the spinal cord the *mantle layer* that contains developing neurons and neuroglial cells will produce the gray matter that surrounds the *neurocoel*. As neurons develop in the mantle layer, their axons grow toward central or peripheral destinations. The axons leave the mantle layer and travel toward synaptic targets within a peripheral *marginal layer*.

Eventually the growing axons will form bundles, or tracts, in the marginal layer, and these tracts will crowd together in the columns that form the white matter of the spinal cord.

DEVELOPMENTAL ABNORMALITIES

Spina bifida

Neural tube defect

Cranial nerves and ganglia

Cervical plexus

Brachial plexus

Eye

Spinal nerves

Lumbosacral plexus

Spina bifida (BĪ-fi-da) results when the developing vertebral laminae fail to unite. The neural arch is incomplete, and the meninges bulge outward beneath the skin of the back. The extent of the abnormality determines the severity of the defects. In mild cases the condition may pass unnoticed; extreme cases involve much of the length of the spinal column and are usually associated with abnormal nerve function.

A **neural tube defect (NTD)** is a condition that is secondary to a developmental error in the formation of the spinal cord. Instead of forming a hollow tube, a portion of the spinal cord develops as a broad plate. This is often associated with spina bifida. Neural tube defects affect roughly one individual in 1000; prenatal testing can detect the existence of these defects with an 80–85% success rate.

Several spinal nerves innervate each developing limb. When embryonic muscle cells migrate away from the myotome, the nerves grow right along with them. If a large muscle in the adult is derived from several myotomal blocks, connective tissue partitions will often mark the original boundaries, and the innervation will always involve more than one spinal nerve.

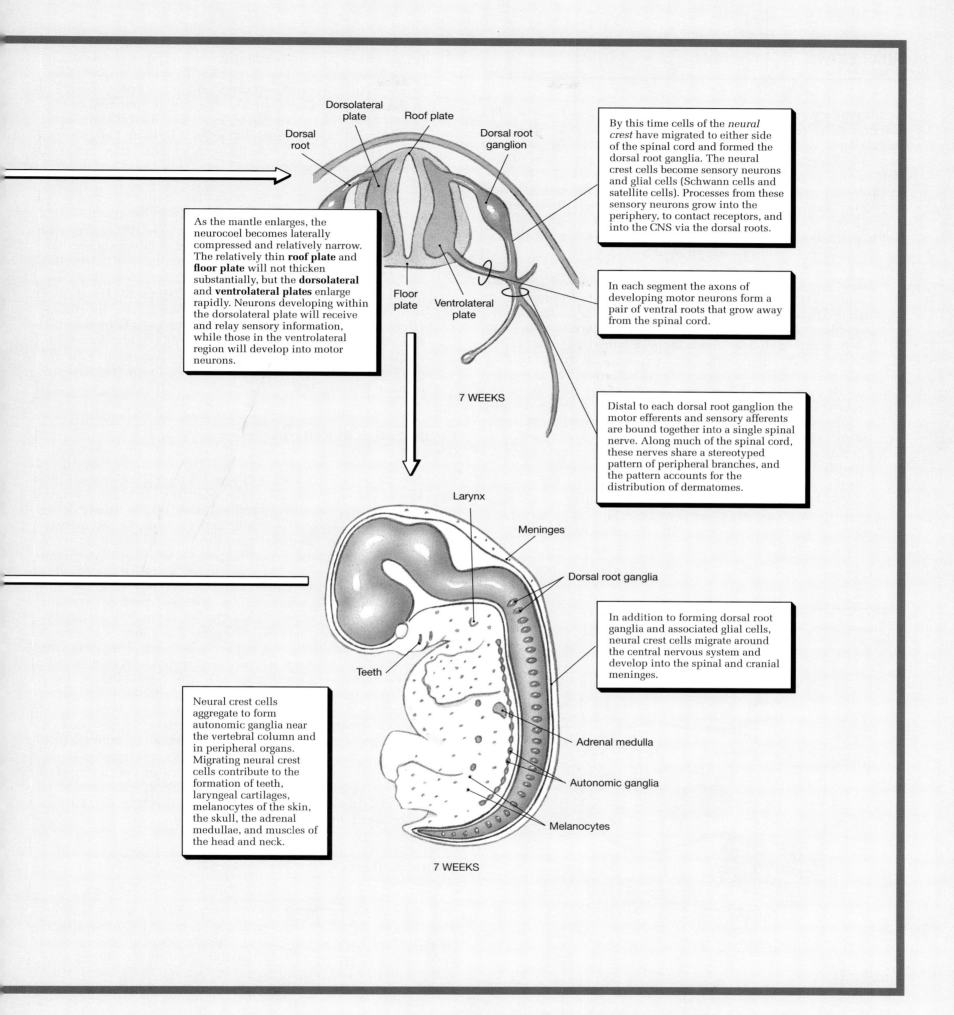

Dorsolateral
plate

Roof plate

Dorsal
root

Dorsal root
ganglion

By this time cells of the *neural crest* have migrated to either side of the spinal cord and formed the dorsal root ganglia. The neural crest cells become sensory neurons and glial cells (Schwann cells and satellite cells). Processes from these sensory neurons grow into the periphery, to contact receptors, and into the CNS via the dorsal roots.

As the mantle enlarges, the neurocoel becomes laterally compressed and relatively narrow. The relatively thin **roof plate** and **floor plate** will not thicken substantially, but the **dorsolateral** and **ventrolateral plates** enlarge rapidly. Neurons developing within the dorsolateral plate will receive and relay sensory information, while those in the ventrolateral region will develop into motor neurons.

Floor
plate

Ventrolateral
plate

In each segment the axons of developing motor neurons form a pair of ventral roots that grow away from the spinal cord.

7 WEEKS

Distal to each dorsal root ganglion the motor efferents and sensory afferents are bound together into a single spinal nerve. Along much of the spinal cord, these nerves share a stereotyped pattern of peripheral branches, and the pattern accounts for the distribution of dermatomes.

Larynx

Meninges

Dorsal root ganglia

In addition to forming dorsal root ganglia and associated glial cells, neural crest cells migrate around the central nervous system and develop into the spinal and cranial meninges.

Teeth

Neural crest cells aggregate to form autonomic ganglia near the vertebral column and in peripheral organs. Migrating neural crest cells contribute to the formation of teeth, laryngeal cartilages, melanocytes of the skin, the skull, the adrenal medullae, and muscles of the head and neck.

Adrenal medulla

Autonomic ganglia

Melanocytes

7 WEEKS

Conditions inside or outside the body can change rapidly and unexpectedly. A **reflex** is an immediate involuntary motor response to a specific stimulus. Neural reflexes help preserve homeostasis by making rapid adjustments in the function of organs or organ systems. The response shows little variability—activation of a particular reflex always produces the same motor response. The neural "wiring" of a single reflex is called a **reflex arc**. A reflex arc begins at a receptor and ends at a peripheral effector, such as a muscle or gland cell. Figure 14-15● illustrates the five steps involved in a neural reflex:

STEP 1 *Arrival of a Stimulus and Activation of a Receptor.* There are many types of sensory receptors, and general categories introduced in Chapter 13 include *exteroceptors*, *interoceptors*, and *proprioceptors*. ∞ [p. 330] Each receptor has a characteristic range of sensitivity, some responding to almost any stimulus. In this example, a painful stimulus activates a pain receptor. These receptors, the dendrites of sensory neurons, are stimulated by pressure, temperature extremes, physical damage, or exposure to abnormal chemicals. Other receptors, such as those providing visual, auditory, or taste sensations, are specialized cells that respond to only a limited range of stimuli.

STEP 2 *Relay of Information to the CNS.* Information is carried in the form of action potentials along an afferent fiber. In this case the axon conducts the action potentials into the spinal cord via one of the dorsal roots.

STEP 3 *Information Processing.* Information processing begins when a neurotransmitter released by synaptic terminals of the sensory neuron reaches the postsynaptic membrane of either a motor neuron or an interneuron. ∞ [p. 331] In the simplest reflexes, such as the one diagrammed in Figure 14-15●, this processing is performed by the motor neuron that controls peripheral effectors. In more complex reflexes, several pools of interneurons are interposed between the sensory and motor neurons, and both serial and parallel processing occur. ∞ [pp. 332, 334] The goal of this information processing is the selection of an appropriate motor response through the activation of specific motor neurons.

STEP 4 *Activation of a Motor Neuron.* Once stimulated to threshold, the axon of a motor neuron conducts action potentials into the periphery, in this example over the ventral root of a spinal nerve.

STEP 5 *Response of a Peripheral Effector.* Activation of the motor neuron then leads to a response by a peripheral effector, such as a skeletal muscle or gland. In general, this response is aimed at removing or counteracting the original stimulus. Reflexes play an important role in opposing potentially harmful changes in the internal or external environment.

Classification of Reflexes
(Figures 14-16/14-17)

Reflexes can be classified according to (1) their development (**innate** and **acquired reflexes**), (2) the site where information processing occurs (**spinal** and **cranial reflexes**), (3) the nature of the resulting motor response (**somatic** and **visceral**, or **autonomic, reflexes**), or (4) the complexity of the neural circuit involved (*monosynaptic* and *polysynaptic reflexes*). These categories, diagrammed in Figure 14-16●, are not mutually exclusive; they represent different ways of describing a single reflex.

In the simplest reflex arc, a sensory neuron synapses directly on a motor neuron. Such a reflex is termed a **monosynaptic reflex** (Figure 14-17a●). Transmission across a chemical synapse always involves a synaptic delay, but with only one synapse, the delay between stimulus and response is minimized.

Polysynaptic reflexes (Figure 14-17b●) have a longer delay between stimulus and response, the length of the delay being proportional to the number of synapses involved. Polysynaptic reflexes can produce far more complicated responses because the interneurons can control several different muscle groups. Many of the motor responses are extremely complicated; for example, stepping on a sharp object not only causes withdrawal of the foot, but triggers all of the muscular adjustments needed to prevent a fall. Such complicated responses result from the interactions between multiple interneuron pools.

Spinal Reflexes
(Figure 14-18)

The neurons of the spinal cord participate in a variety of reflex arcs. These *spinal reflexes* range in complexity from simple monosynaptic reflexes involving a single segment of the spinal cord to polysynaptic reflexes that integrate motor output from many different spinal cord segments to produce a coordinated motor response.

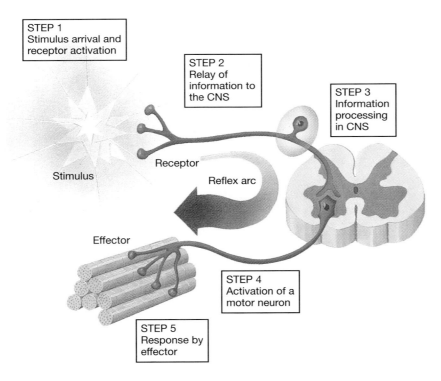

STEP 1
Stimulus arrival and receptor activation

STEP 2
Relay of information to the CNS

STEP 3
Information processing in CNS

Receptor

Stimulus

Reflex arc

Effector

STEP 4
Activation of a motor neuron

STEP 5
Response by effector

FIGURE 14-15
Components of a Reflex Arc

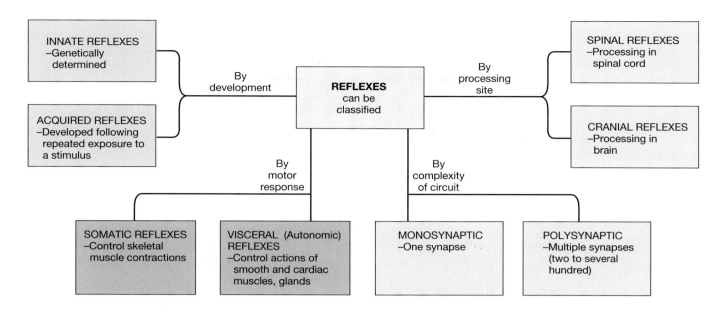

FIGURE 14-16
Four Different Methods of Classifying Reflexes

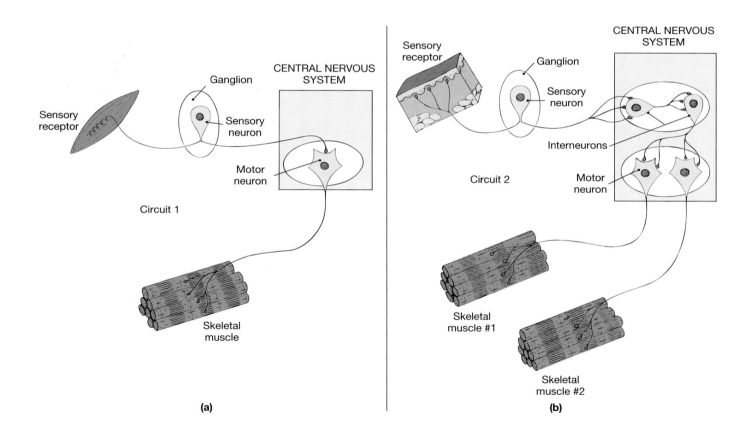

(a)

(b)

FIGURE 14-17
Neural Organization and Simple Reflexes. (a) A monosynaptic reflex involves a peripheral sensory neuron and a central motor neuron. In this example, stimulation of the receptor will lead to a reflexive contraction in a skeletal muscle. (b) A polysynaptic reflex involves a sensory neuron, interneurons, and motor neurons. In this example the stimulation of the receptor leads to the coordinated contractions of two different skeletal muscles.

The best-known example is a simple monosynaptic reflex, the *stretch reflex*. The **stretch reflex** (Figure 14-18a●) provides automatic regulation of skeletal muscle length. The stimulus, stretching of a relaxed muscle, activates a sensory neuron that triggers the contraction of that muscle. The stretch reflex also provides for the automatic adjustment of muscle tone, increasing or decreasing it in response to information provided by the stretch receptors of *muscle spindles*. Muscle spindles, which will be considered in Chapter 18, consist of specialized muscle fibers whose lengths are monitored by sensory neurons.

The most familiar stretch reflex is probably the *knee jerk*, or **patellar reflex**. In this reflex, a sharp rap on the patellar ligament stretches muscle spindles in the quadriceps muscles (Figure 14-18b●). With so brief a stimulus, the reflexive contraction occurs unopposed and produces a noticeable kick. Physicians often test this reflex to check the status of the lower segments of the spinal cord. A normal patellar reflex indicates that spinal nerves and spinal segments L_2–L_4 are undamaged.

The stretch reflex is an example of a **postural reflex**, a re-

flex that maintains normal upright posture. Postural muscles usually have a firm muscle tone and extremely sensitive stretch receptors. As a result, very fine adjustments are continually being made, and you are not aware of the cycles of contraction and relaxation that occur.

Higher Centers and Integration of Reflexes

Reflexive motor activities occur automatically, without instructions from higher centers in the brain. However, higher centers can have a profound effect on reflex performance. For example, processing centers in the brain can enhance or suppress spinal reflexes via descending tracts that synapse on interneurons and motor neurons throughout the spinal cord. Motor control therefore involves a series of interacting levels. At the lowest level are monosynaptic reflexes that are rapid but stereotyped and relatively inflexible. At the highest level are centers in the brain that can modulate or build upon reflexive motor patterns. ⸙*Reflexes and Diagnostic Testing [p. 760]*

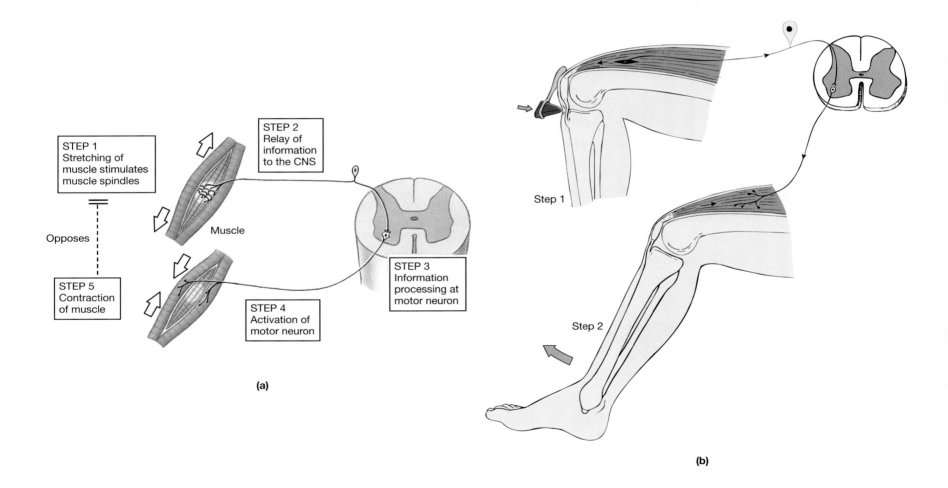

(a)

(b)

FIGURE 14-18
Components of the Stretch Reflex. (a) Diagram of the activities in a stretch reflex. (b) The patellar reflex is controlled by muscle spindles in the quadriceps group. In Step 1 a reflex hammer strikes the muscle tendon, stretching the spindle fibers. This results in a sudden increase in the activity of the sensory neurons, which synapse on spinal motor neurons. In Step 2 the activation of motor units in the quadriceps produces an immediate increase in muscle tone and a reflexive kick.

Related Clinical Terms

meningitis: An inflammation of the meningeal membranes. ⚡ *Spinal Meningitis [p. 759]*

epidural block: Regional anesthesia produced by the injection of an anesthetic into the epidural space near targeted spinal nerve roots. ⚡ *Spinal Anesthesia [p. 759]*

caudal anesthesia: Injection of anesthetics into the epidural space of the sacrum to paralyze lower abdominal and perineal structures. ⚡ *Spinal Anesthesia [p. 759]*

spinal tap: A procedure in which fluid is extracted from the subarachnoid space through a needle inserted between the vertebrae. *[p. 343]*

lumbar puncture: A spinal tap performed between adjacent lumbar vertebrae. *[p. 343]*

myelography: A diagnostic procedure in which a radiopaque dye is introduced into the cerebrospinal fluid in order to obtain an X-ray of the spinal cord. *[p. 343]*

spinal shock: A period of sensory and motor paralysis following any severe injury to the spinal cord. *[p. 345]*

quadriplegia: Paralysis involving loss of sensation and motor control of the upper and lower limbs. *[p. 345]*

paraplegia: Paralysis involving loss of motor control of the lower limbs. *[p. 345]*

nerve graft: Insertion of an intact section from a different peripheral nerve to bridge the gap between the cut ends of a damaged nerve and provide a route for axonal regeneration. ⚡ *Technology and Motor Paralysis [p. 759]*

functional electrical stimulation (FES): A technique for stimulating specific muscles and muscle groups using electrodes controlled by a computer. ⚡ *Technology and Motor Paralysis [p. 759]*

multiple sclerosis (skler-Ō-sis; *sklerosis,* hardness) **(MS):** A disease of the nervous system characterized by recurrent, often progressive incidents of demyelination affecting tracts in the brain and/or spinal cord. Common symptoms include partial loss of vision and problems with speech, balance, and general motor coordination. ⚡ *Multiple Sclerosis [p. 760]*

patellar reflex: The "knee jerk" reflex; often used to provide information about the related spinal segments. *[p. 360]*

Babinski sign (positive Babinski reflex): A spinal reflex in infants, consisting of a fanning of the toes, produced by stroking the foot on the side of the sole; in adults, a sign of CNS injury. ⚡ *Reflexes and Diagnostic Testing [p. 760]*

hyporeflexia: A condition where normal spinal reflexes are present but weak. ⚡ *Reflexes and Diagnostic Testing [p. 760]*

areflexia (ā-re-FLEK-sē-a; *a-,* without): The lack of normal reflex responses to stimuli. ⚡ *Reflexes and Diagnostic Testing [p. 760]*

hyperreflexia: Exaggerated reflex responses that may develop in some pathological states or following stimulation of spinal and cranial nuclei by higher centers. ⚡ *Reflexes and Diagnostic Testing [p. 760]*

CHAPTER SUMMARY AND REVIEW

STUDY OUTLINE

Related Key Terms

INTRODUCTION *[p. 339]*

1. The **central nervous system (CNS)** consists of the spinal cord and brain. In addition to relaying information to and from the brain, the spinal cord integrates and processes information on its own.

GROSS ANATOMY OF THE SPINAL CORD *[p. 339]*

1. The adult spinal cord includes localized **enlargements** that provide innervation to the limbs. The spinal cord has 31 segments, each associated with a pair of **dorsal roots** and a pair of **ventral roots** *(see Figures 14-1 to 14-3).*

2. The **filum terminale** (a strand of fibrous tissue) that originates at the **conus medullaris** ultimately becomes part of the *coccygeal ligament (see Figures 14-1a/14-2a/14-4a).*

posterior median sulcus ● **anterior median fissure** ● **cervical enlargement** ● **lumbar enlargement**

dorsal root ganglia ● **spinal nerve** ● **mixed nerves** ● **cauda equina**

SPINAL MENINGES *[p. 339]*

1. The **spinal meninges** provide physical stability and shock absorption for neural tissues of the spinal cord; the **cranial meninges** surround the brain *(see Figure 14-2).*

The Dura Mater *[p. 342]*
2. The **dura mater** covers the spinal cord; caudally it tapers into the **coccygeal ligament**. The **epidural space** separates the dura mater from the walls of the vertebral canal *(see Figures 14-2/14-4a).*

The Arachnoid *[p. 342]*
3. Beneath the inner surface of the dura mater are the **subdural space**, the **arachnoid** (the second meningeal layer), and the **subarachnoid space**. The latter contains **cerebrospinal fluid**, which acts as a shock absorber and a diffusion medium for dissolved gases, nutrients, chemical messengers, and waste products in a meshwork of elastic and collagen fibers *(see Figure 14-2).*

The Pia Mater [p. 342]

4. The **pia mater** is the innermost meningeal layer. Unlike more superficial meninges, it is bound to the underlying neural tissue (*see Figure 14-2*).

SECTIONAL ANATOMY OF THE SPINAL CORD [p. 344]

1. The **white matter** contains myelinated and unmyelinated axons, while the **gray matter** contains cell bodies of neurons and glial cells. The projections of gray matter toward the outer surface of the spinal cord are called **horns** (*see Figure 14-5*).

Organization of Gray Matter [p. 344]

2. The **posterior gray horns** contain somatic and visceral sensory nuclei, while nuclei in the **anterior gray horns** are concerned with somatic motor control. The **lateral gray horns** contain visceral motor neurons. The **gray commissures** contain the axons of interneurons that cross from one side of the cord to the other (*see Figure 14-5*).

Organization of White Matter [p. 345]

3. The white matter can be divided into six **columns**, each of which contains **tracts** (*fasciculi*). **Ascending tracts** relay information from the spinal cord to the brain, and **descending tracts** carry information from the brain to the spinal cord (*see Figure 14-5*).

SPINAL NERVES [p. 345]

1. There are 31 pairs of spinal nerves. Each has an **epineurium** (outermost layer), **perineurium**, and **endoneurium** (innermost layer) (*see Figures 14-1/14-3/14-6*).

Peripheral Distribution of Spinal Nerves [p. 345]

2. A typical spinal nerve has a **white ramus** (which contains myelinated axons), a **gray ramus** (containing unmyelinated fibers that innervate glands and smooth muscles in the body wall or limbs), a **dorsal ramus** (providing sensory and motor innervation to the skin and muscles of the back), and a **ventral ramus** (supplying the ventrolateral body surface, structures in the body wall, and the limbs). Each pair of spinal nerves monitors a region of the body surface called a **dermatome** (*see Figures 14-2/14-3/14-7/14-8*).

Nerve Plexuses [p. 348]

3. A complex, interwoven network of nerves is called a **nerve plexus**. The three large plexuses are the **cervical plexus**, the **brachial plexus**, and the **lumbosacral plexus**. The latter can be divided into the **lumbar plexus** and the **sacral plexus** (*see Figures 14-7/14-9 to 14-14 and Tables 14-1 to 14-3*).

REFLEXES [p. 358]

1. A neural **reflex** is an automatic, involuntary motor response that helps preserve homeostasis by rapidly adjusting the functions of organs or organ systems (*see Figure 14-15*).

2. A **reflex arc** is the neural "wiring" of a single reflex.

3. A receptor is a specialized cell that monitors conditions in the body or external environment. Each receptor has a characteristic range of sensitivity.

4. There are five steps involved in a neural reflex: (1) arrival of a stimulus and activation of a receptor; (2) activation of a sensory neuron; (3) information processing; (4) activation of a motor neuron; (5) response by an effector (*see Figure 14-15*).

Classification of Reflexes [p. 358]

5. **Innate reflexes** result from the connections that form between neurons during development. **Acquired reflexes** are learned, and often are more complex (*see Figure 14-16*).

6. Reflexes processed in the brain are **cranial reflexes**. In a **spinal reflex** the important interconnections and processing occur inside the spinal cord (*see Figure 14-16*).

7. **Somatic reflexes** control skeletal muscles, and **visceral (autonomic) reflexes** control the activities of other systems (*see Figure 14-16*).

8. A **monosynaptic reflex** is the simplest reflex arc, in which a sensory neuron synapses directly on a motor neuron that acts as the processing center. **Polysynaptic reflexes**, which have at least one interneuron placed between the sensory afferent and the motor efferent, have a longer delay between stimulus and response (*see Figures 14-16/14-17*).

Spinal Reflexes [p. 358]

9. Spinal reflexes range from simple monosynaptic reflexes to more complex polysynaptic reflexes in which many segments interact to produce a coordinated motor response (*see Figure 14-17*).

10. The **stretch reflex** is a monosynaptic reflex that automatically regulates skeletal muscle length and muscle tone. The sensory receptors involved are muscle spindles (*see Figure 14-18*).

11. A **postural reflex** maintains normal upright posture.

Higher Centers and Integration of Reflexes [p. 360]

12. Higher centers in the brain can enhance or inhibit reflex motor patterns based in the spinal cord.

1   REVIEW OF CHAPTER OBJECTIVES

1. Discuss the structure and functions of the spinal cord.
2. Locate and describe the spinal meninges.
3. Discuss the structure, location, and role of white matter and gray matter in processing and relaying sensory and motor information.
4. Describe the major components of a spinal nerve.
5. Identify spinal nerves to the anatomical regions or structures that they innervate.
6. Describe the structures involved in a neural reflex.
7. Classify reflexes and identify their structural components.
8. Give examples of the types of motor responses produced by spinal reflexes.

2   REVIEW OF CONCEPTS

1. How is the size of the spinal cord in different regions correlated with regional functions?
2. How does the relative length of the spinal cord differ from child to adult?
3. What is the role of the meninges in protecting the spinal cord?
4. What role does the dura mater play in stabilizing the spinal cord?
5. How does the relationship of the pia mater to the spinal cord differ from that of the other meninges?
6. How do gray matter and white matter differ, and where are these regions located in the spinal cord?
7. Why are there eight cervical spinal nerves but only seven cervical vertebrae?
8. How is sensation from a given dermatome related to the positions of spinal nerves?

9. Why are many skeletal muscles of the neck and limbs innervated by more than one nerve?
10. How does a reflex differ from a voluntary muscle movement?
11. How does information processing differ between simple and complex reflexes?
12. What is the role of an interneuron in a reflex?

3   CRITICAL THINKING AND CLINICAL APPLICATION QUESTIONS

1. Over a period of several days, a scientist in his seventies develops painful skin eruptions on the side of his face and on his chest. He also has a fever and headache. The skin eruptions appear to follow the path of the trigeminal and thoracic nerves, which leads him to suspect he has a specific disorder. His suspicions are confirmed when he visits his doctor, who asks if he has ever had the chicken pox. What condition does this scientist have, and what is the anatomical cause?
2. A young child living in one of the public housing projects in the city of Chicago is shot while playing in a nearby park. The bullet lodges near the spinal canal, but does not penetrate it. Nevertheless, the child shows evidence of both sensory and motor paralysis inferior to the location of the wound. What specific symptoms will the child show, and will it be likely that the damage is permanent?
3. Several tests are administered to determine the condition of the wounded child in question 2. Describe two tests that would be the most useful in determining what damage the child's spinal cord has sustained.
4. During a routine physical examination, a doctor, using a rubber hammer, taps the patellar tendon of the person he is examining. What response is he looking for, and what anatomical mechanism is involved in producing the response?

15

The Nervous System:
The Brain and Cranial Nerves

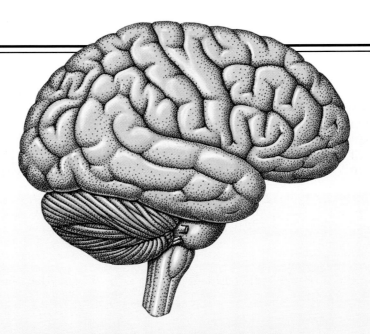

CHAPTER OUTLINE AND OBJECTIVES

The brain is probably the most fascinating organ in the body, because all of our dreams, passions, plans, and memories result from brain activity. The brain contains tens of billions of neurons organized into hundreds of neuronal pools; it has a complex three-dimensional structure and performs a bewildering array of functions.

The brain is far more complex than the spinal cord, and it can respond to stimuli with greater versatility. That versatility results from (1) the tremendous number of neurons and neuronal pools in the brain and (2) the complexity of the interconnections between those neurons and neuronal pools. The brain contains roughly 35 billion neurons, each of which may receive information across as many as 80,000 synapses at one time. The neurons are organized into pools with extensive interconnections. Excitatory and inhibitory interactions between these pools ensure that the response can vary to meet changing circumstances. But adaptability has a price. A response cannot be immediate, precise, and adaptable all at the same time. Adaptability requires multiple processing steps, and every synapse adds to the delay between stimulus and response. One of the major functions of spinal reflexes is to provide an *immediate* response that can be fine-tuned or elaborated on by more versatile but slower processing centers in the brain.

We now begin a detailed examination of the brain. This chapter focuses attention on the major structures of the brain and their relationships with the cranial nerves.

AN INTRODUCTION TO THE ORGANIZATION OF THE BRAIN
(Figure 15-1)

The adult human brain (Figure 15-1●) contains almost 98% of the neural tissue in the body. A "typical" adult brain weighs 1.4 kg (3 lb) and has a volume of 1200 cc (71 in.^{3}). There is considerable individual variation, and the brains of males average about 10% larger than those of females, owing to differences in average body size. No correlation exists between brain size and intelligence, and individuals with the smallest (750 cc) and largest (2100 cc) brains are functionally normal.

Major Regions and Landmarks
(Figure 15-1)

There are five major divisions in the adult brain: (1) the *telencephalon*, or *cerebrum*, (2) the *diencephalon*, (3) the *mesencephalon*, or *midbrain*, (4) the *metencephalon*, which includes the *cerebellum* and the *pons*, and (5) the *myelencephalon*, or *medulla oblongata*. During embryonic development these regions differentiate from three primary divisions; the process is detailed in the Embryology Summary on p. 402.

Telencephalon (Cerebrum)

Viewed from the superior surface (Figure 15-1●), the **cerebrum** (ser-Ē-brum) is divided into large, paired *cerebral hemispheres* separated by the **longitudinal fissure**. Conscious thought processes, intellectual functions, memory storage and retrieval, and complex motor patterns originate in the cerebrum. The cerebellum automatically adjusts motor activities on the basis of sensory information and memories of learned patterns of movement.

Diencephalon

The portion of the brain attached to the cerebrum is the **diencephalon** (dī-en-SEF-a-lon; *dia*, through). The walls of the diencephalon are composed of the **left** and **right thalamus**. Each thalamus contains relay and processing centers for sensory information. A narrow stalk connects the floor of the diencephalon, or **hypothalamus** (*hypo-*, below), to the **pituitary gland**, or *hypophysis* (*phyein*, to generate). The hypothalamus contains centers involved with emotions, autonomic function, and hormone production. It is the primary link between the nervous and endocrine systems.

The remaining regions of the brain are collectively referred to as the *brain stem*. The **brain stem** consists of three major structures: (1) mesencephalon, (2) pons, and (3) medulla oblongata.[1] The brain stem contains important processing centers and also relays information to or from the cerebrum or cerebellum. View Figure 15-1● as we describe the structure of the brain stem.

Mesencephalon (Midbrain)

Nuclei in the **mesencephalon** (mez-en-SEF-a-lon; *meso*, middle), or **midbrain**, process visual and auditory information and generate involuntary somatic motor responses to these stimuli. This region also contains centers involved with the maintenance of consciousness.

Metencephalon (Cerebellum and Pons)

Immediately inferior to the cerebrum are the somewhat smaller hemispheres of the **cerebellum** (ser-e-BEL-um) (Figure 15-1●). The cerebellum automatically adjusts motor activities on the basis of sensory information and memories of learned patterns of movement. The term **pons** refers to a bridge, and the pons of the brain connects the cerebellum to the brain stem. In addition to tracts and relay centers, this region of the brain also contains nuclei involved with somatic and visceral motor control.

Myelencephalon (Medulla Oblongata)

The spinal cord connects to the brain at the **myelencephalon**, also known as the **medulla oblongata**, or simply the *medulla*. The superior portion of the medulla oblongata has a thin, membranous roof, but the caudal portion resembles the spinal cord. The medulla relays sensory information to the thalamus and other brain stem centers; it also contains major centers concerned with the regulation of autonomic function, such as heart rate, blood pressure, and digestive activities.

Gray Matter and White Matter Organization

The distribution of gray matter in the brain stem resembles that in the spinal cord. *Nuclei* are clustered around the fluid-filled *ventricles* that correspond to the central canal of the spinal cord. *Tracts* of white matter surround this central region of gray matter. However, the arrangement of tracts and nuclei is not as regular as that in the spinal cord. The nuclei are of many different

---

[1] Some references include the diencephalon in the term *brain stem*; we will use the narrower definition here.

sizes, and tracts may begin, end, merge, or branch as they pass around or through nuclei in their path. In the cerebrum and cerebellum the white matter is covered by a superficial layer of **neural cortex** (*cortex*, rind).

The term *higher centers* refers to nuclei, centers, and cortical areas of the cerebrum, cerebellum, diencephalon, and mesencephalon. Output from these processing centers modifies the activities of nuclei and centers in the lower brain stem and spinal cord. The nuclei and cortical areas of the brain can receive sensory information and issue motor commands to peripheral effectors indirectly, via the spinal cord and spinal nerves, or directly, via the cranial nerves.

Embryology of the Brain

The development of the brain is detailed in the Embryology Summary on pages 402–403. The central nervous system begins as a hollow *neural tube*, with a fluid-filled internal cavity called the *neurocoel*. In the cephalic portion of the neural tube, three areas enlarge rapidly through expansion of the neurocoel. This enlargement creates three prominent **primary brain vesicles** named for their relative positions: the **prosencephalon** (prō-zen-SEF-a-lon; *proso,* forward + *enkephalos,* brain), or "forebrain"; the **mesencephalon** (*mesos,* middle), or "midbrain"; and the **rhombencephalon** (romben-SEF-a-lon), or "hindbrain."

The fate of the three primary divisions of the brain is summarized in Table 15-1. The prosencephalon and rhombencephalon are subdivided further, forming **secondary brain vesicles.** The prosencephalon forms the **telencephalon** (tel-en-SEF-a-lon; *telos,* end) and the diencephalon (dī-en-SEF-a-lon; *dia,* through). The telencephalon will form the *cerebrum,* including the paired *cerebral hemispheres* that dominate the superior and lateral surfaces of the adult brain. The diencephalon forms the *thalamus,* the *hypothalamus,* and the *pineal gland.* The mesencephalon thickens, and the neurocoel becomes a relatively narrow passageway with a diameter comparable to the central canal of the spinal cord. The portion of the rhombencephalon adjacent to the mesencephalon forms the **metencephalon** (met-en-SEF-a-lon; *meta,* after). The dorsal portion of the metencephalon will become the *cerebellum,* and the ventral portion will develop into the *pons.* The portion of the rhombencephalon closer to the spinal cord forms the **myelencephalon** (mī-el-en-SEF-a-lon; *myelon,* spinal cord), which will become the *medulla oblongata.*

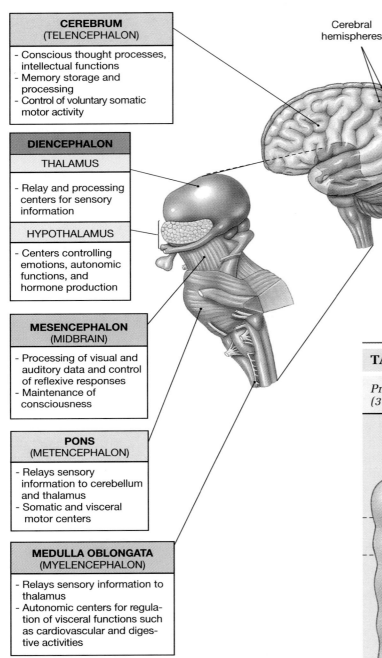

CEREBRUM
(TELENCEPHALON)
- Conscious thought processes, intellectual functions
- Memory storage and processing
- Control of voluntary somatic motor activity

DIENCEPHALON

THALAMUS
- Relay and processing centers for sensory information

HYPOTHALAMUS
- Centers controlling emotions, autonomic functions, and hormone production

MESENCEPHALON
(MIDBRAIN)
- Processing of visual and auditory data and control of reflexive responses
- Maintenance of consciousness

PONS
(METENCEPHALON)
- Relays sensory information to cerebellum and thalamus
- Somatic and visceral motor centers

MEDULLA OBLONGATA
(MYELENCEPHALON)
- Relays sensory information to thalamus
- Autonomic centers for regulation of visceral functions such as cardiovascular and digestive activities

Cerebral hemispheres

CEREBELLUM
(METENCEPHALON)
- Helps coordinate complex somatic motor patterns

FIGURE 15-1
An Introduction to Brain Functions

TABLE 15-1	Development of the Human Brain	
Primary Brain Vesicles (3 weeks)	*Secondary Brain Vesicles (6 weeks)*	*Brain Regions at Birth*
Prosencephalon	Telencephalon	Cerebrum
	Diencephalon	Diencephalon
Mesencephalon	Mesencephalon	Mesencephalon
Rhombencephalon	Metencephalon	Cerebellum
		Pons
	Myelencephalon	Medulla oblongata

The Ventricles of the Brain
(Figures 15-2/15-17b)

The brain is hollow, like the spinal cord, and has a central passageway filled with cerebrospinal fluid. Inside the telencephalon, diencephalon, metencephalon, and the superior portion of the myelencephalon, this passageway expands to form fluid-filled chambers called **ventricles** (VEN-tri-kls). Figure 15-2● shows the position and orientation of the ventricles.

Each cerebral hemisphere contains an enlarged ventricular chamber (Figure 15-2a,b,c●). A thin medial partition, the **septum pellucidum**, separates this pair of **lateral ventricles**. There is no direct connection between the two lateral ventricles, but each communicates with the ventricle of the diencephalon through an **interventricular foramen** (*foramen of Munro*) (Figure 15-2a●). Because there are two lateral ventricles (first and second), the diencephalic chamber is called the **third ventricle**.

The mesencephalon has a slender canal known as the **mesencephalic aqueduct** (*aqueduct of Sylvius* or *cerebral aqueduct*) (Figure 15-2a,c●), which connects the third ventricle with the **fourth ventricle** of the metencephalon and medulla oblongata (Figures 15-2d and 15-17b●, p. 388). In the inferior half of the medulla, the fourth ventricle narrows and becomes continuous with the central canal of the spinal cord.

The ventricles are filled with cerebrospinal fluid and lined by cells of the *ependyma*. ∞ [p. 324] There is a continuous circulation of cerebrospinal fluid from the ventricles and central canal into the *subarachnoid space* of the meninges that surround the CNS via foramina near the base of the cerebellum.

FIGURE 15-2

Ventricles of the Brain. (a) Orientation and extent of the ventricles as if seen through a transparent brain. (b) The two lateral ventricles of the cerebral hemispheres and the third ventricle in the diencephalon. (c) Anterior view of the ventricles. (d) Diagrammatic coronal section, showing the interconnections between the ventricles. See MRI SCANS 1 and 2, pp. 741–742.

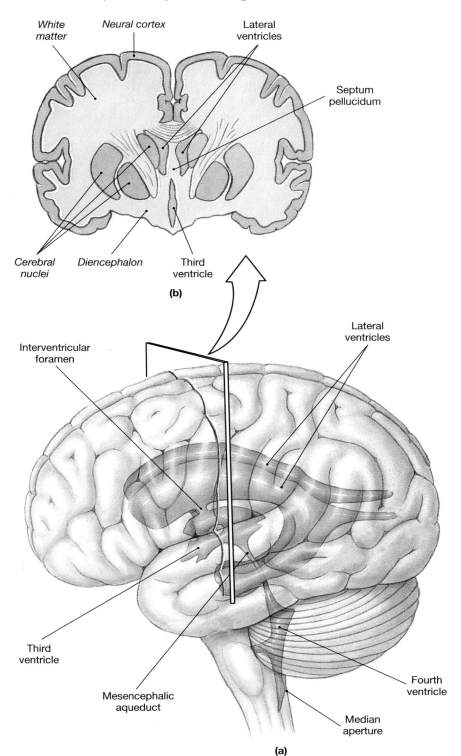

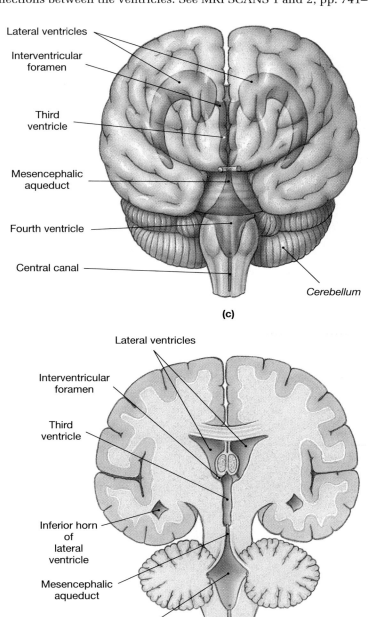

PROTECTION AND SUPPORT OF THE BRAIN

The human brain is extremely delicate, and it has a high demand for nutrients and oxygen. At the same time, it must be isolated from a variety of compounds in the blood that could interfere with its complex operations. There are several anatomical features that provide protection and support for the tissues of the brain.

The Cranial Meninges
(Figure 15-3)

The brain lies cradled within the cranium of the skull, and there is an obvious correspondence between the shape of the brain and that of the cranial cavity (Figure 15-3●). The massive cranial bones provide mechanical protection, but they also pose a threat. The brain is like a person driving a car. If the car hits a tree, the car protects the driver from contact with the tree, but serious injury will occur unless a seat belt or airbag protects the driver from contact with the car. ♈ *Cranial Trauma [p. 761]*

Within the cranial cavity, the **cranial meninges** that surround the brain provide this protection, acting as shock absorbers that prevent contact with surrounding bones (Figure 15-3a●). The membrane layers that make up the cranial meninges, *dura mater*, *arachnoid*, and *pia mater* are continuous with those of the spinal cord. ∞ [p. 339] However, the cranial meninges have distinctive anatomical features and functional roles.

Dura Mater
(Figures 15-3/15-4a)

Within the skull, the **dura mater** consists of two fibrous layers. The outermost, or *endosteal* layer, is fused to the periosteum of the cranial bones (Figure 15-3a●). The inner, or *meningeal*, and outer layers in many areas are separated by a slender gap that contains tissue fluids and blood vessels, including the large veins known as **dural sinuses**. The veins of the brain open into these sinuses, which return blood from the brain to the internal jugular vein of the neck.

At four locations the innermost layer of the dura mater extends deep into the cranial cavity, providing additional stabilization and support to the brain (Figure 15-3b,c●):

- The **falx** (falks) **cerebri** (ser-Ē-brē; *falx*, curving, or sickle-shaped) is a fold of dura mater that projects between the cerebral hemispheres in the longitudinal fissure. Its inferior portions attach to the crista galli (anteriorly) and the internal occipital crest and *tentorium cerebelli* (posteriorly). Two large venous sinuses, the **superior sagittal sinus** and the **inferior sagittal sinus**, travel within this dural fold (Figures 15-3b,c and 15-4a●).
- The **tentorium cerebelli** (ten-TOR-ē-um ser-e-BEL-ē; *tentorium*, covering), separates and protects the cerebellar hemispheres from those of the cerebrum. It extends across the cranium at right angles to the falx cerebri. The **transverse sinus** lies within the tentorium cerebelli.
- The **falx cerebelli** extends in the midsagittal line inferior to the tentorium cerebelli, dividing the two cerebellar hemispheres.
- The **diaphragma sellae** is a continuation of the dural sheet that lines the *sella turcica* of the sphenoid bone. ∞ [p. 143] The diaphragma sellae anchors the dura mater to the sphenoid bone and ensheathes the base of the pituitary gland.

Arachnoid
(Figure 15-4)

In most anatomical preparations a narrow **subdural space** separates the opposing epithelia of the dura mater and the arachnoid (Figure 15-4a●). The **arachnoid membrane** covers the brain, providing a smooth surface that does not follow the underlying neural convolutions or sulci. Deep to the arachnoid membrane is the **subarachnoid space**, which contains a delicate meshwork of collagen and elastic fibers that link the arachnoid membrane to the underlying pia mater. Cerebrospinal fluid percolates through the bundles of fibers, which are known as **arachnoid trabeculae**. The arachnoid membrane acts as a roof over the cranial blood vessels, and the underlying pia mater forms a floor. Cerebral arteries and veins are supported by the arachnoid trabeculae and surrounded by cerebrospinal fluid (Figure 15-4b●). Blood vessels, surrounded and suspended by arachnoid trabeculae, penetrate the substance of the brain within channels lined by pia mater.

Along the axis of the superior sagittal sinus, fingerlike extensions of the arachnoid penetrate the dura mater and form **arachnoid granulations**. Cerebrospinal fluid is absorbed into the venous circulation at the arachnoid granulations.

Pia Mater
(Figure 15-4b)

The **pia mater** closely adheres to the surface contours of the brain, anchored by the processes of astrocytes. The pia mater is a highly vascular membrane composed primarily of areolar connective tissue. It acts as a floor to support the large cerebral blood vessels as they branch over the surface of the brain, invading the neural contours to supply superficial areas of neural cortex (Figure 15-4b●). An extensive circulatory supply is vital, because the brain requires a constant supply of nutrients and oxygen.

■ CLINICAL BRIEF
Epidural and Subdural Hemorrhages

A severe head injury may damage meningeal vessels and cause bleeding into the epidural or subdural spaces. The most common cases of epidural bleeding, or **epidural hemorrhage**, involve an arterial break. The arterial blood pressure usually forces considerable quantities of blood into the epidural space, distorting the underlying soft tissues of the brain. The individual loses consciousness from minutes to hours after the injury, and death follows in untreated cases.

An epidural hemorrhage involving a damaged vein does not produce massive symptoms immediately, and the individual may become unconscious from several hours to several days or even weeks after the original incident. Consequently, the problem may not be noticed until the nervous tissue has been severely damaged by distortion, compression, and secondary hemorrhaging. Epidural hemorrhages are rare, occurring in fewer than 1% of head injuries. This rarity is rather fortunate, for the mortality rate is 100% in untreated cases and over 50% even after removal of the blood pool and closure of the damaged vessels.

The term **subdural hemorrhage** is somewhat misleading, because blood actually enters the inner layer of the dura, flowing beneath the epithelium that contacts the arachnoid membrane. Subdural hemorrhages are roughly twice as common as epidural hemorrhages. The most common source of blood is a small vein or one of the dural sinuses. Because the blood pressure is somewhat lower, the extent and effects of the condition may be quite variable.

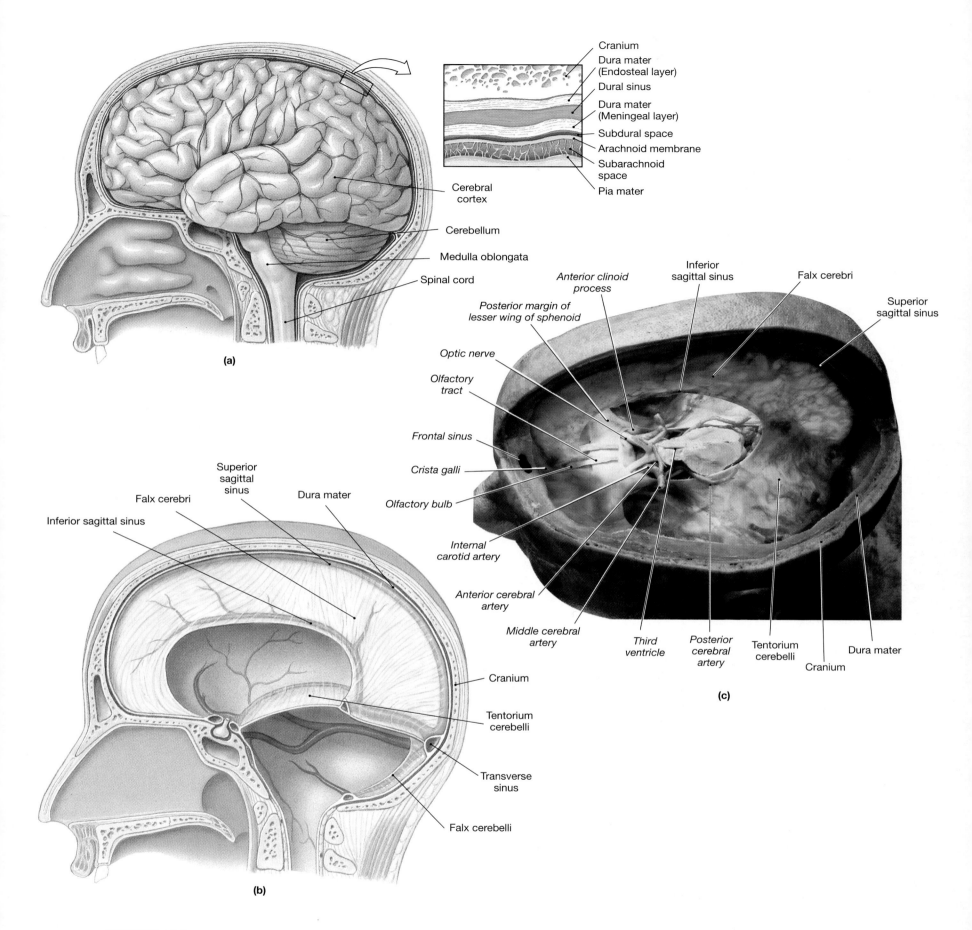

(a)

(b)

(c)

Cranium
Dura mater (Endosteal layer)
Dural sinus
Dura mater (Meningeal layer)
Subdural space
Arachnoid membrane
Subarachnoid space
Pia mater

Cerebral cortex

Cerebellum

Medulla oblongata

Spinal cord

Inferior sagittal sinus

Falx cerebri

Superior sagittal sinus

Anterior clinoid process

Posterior margin of lesser wing of sphenoid

Optic nerve

Olfactory tract

Frontal sinus

Crista galli

Olfactory bulb

Internal carotid artery

Anterior cerebral artery

Middle cerebral artery

Third ventricle

Posterior cerebral artery

Tentorium cerebelli

Cranium

Dura mater

Superior sagittal sinus

Falx cerebri

Dura mater

Inferior sagittal sinus

Cranium

Tentorium cerebelli

Transverse sinus

Falx cerebelli

FIGURE 15-3
Relationship between the Brain, Cranium, and Meninges. (a) Lateral view of the brain, showing its position in the cranium and the organization of the meningeal coverings. (b) A view of the cranial cavity with the brain removed, showing the orientation and extent of the falx cerebri and tentorium cerebelli. (c) Superior view of the open cranial cavity with the cerebrum and diencephalon removed. Note the relationships among the falx cerebri, tentorium cerebelli, and the cranial nerves and blood vessels. See MRI SCAN 2, p. 742.

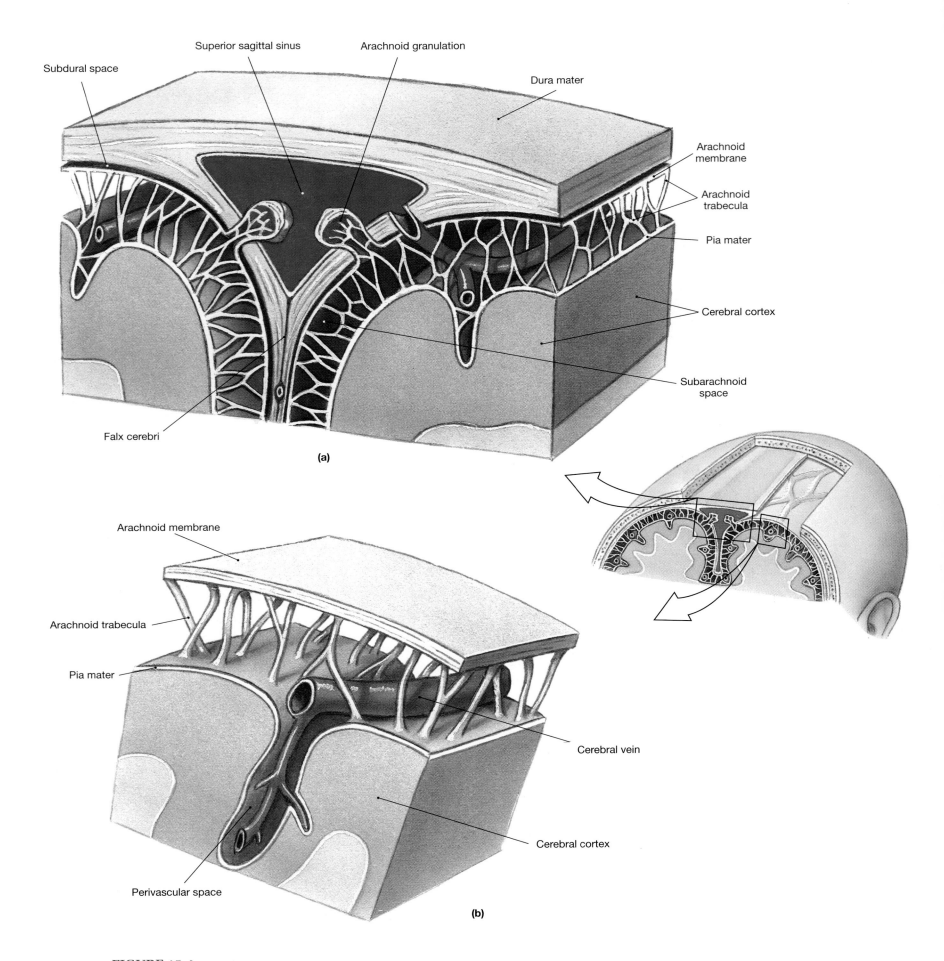

Subdural space

Superior sagittal sinus

Arachnoid granulation

Dura mater

Arachnoid membrane

Arachnoid trabecula

Pia mater

Cerebral cortex

Subarachnoid space

Falx cerebri

(a)

Arachnoid membrane

Arachnoid trabecula

Pia mater

Cerebral vein

Cerebral cortex

Perivascular space

(b)

FIGURE 15-4
The Cranial Meninges. (a) Organization and relationship of the cranial meninges to the brain. (b) A detailed view of the pia-arachnoid membrane and the relationship between a cerebral vein and the subarachnoid space.

The Blood-Brain Barrier
(Figure 15-5)

Neural tissue in the CNS is isolated from the general circulation by the **blood-brain barrier**. This barrier exists because the endothelial cells lining the capillaries of the CNS (Figure 15-5●) are extensively interconnected by tight junctions. ∞ [p. 43] These junctional complexes prevent the diffusion of materials between adjacent endothelial cells. In general, only lipid-soluble compounds can diffuse into the interstitial fluid of the brain and spinal cord. They do so by diffusing across the cell membranes of the endothelial cells. Water-soluble compounds can cross the capillary walls only through passive or active transport mechanisms. ∞ [p. 31] Many different transport proteins are involved, and their activities are quite specific. For example, the transport system that handles glucose is different from those transporting large amino acids. The restricted permeability characteristics of the endothelial lining of brain capillaries is in some way dependent on chemicals secreted by astrocytes. These cells, which are in close contact with CNS capillaries, were described in Chapter 13. ∞ [p. 323]

Endothelial transport across the blood-brain barrier is selective and directional. Neurons have a constant need for glucose that must be met regardless of the relative concentrations in the blood and interstitial fluid. Even when circulating glucose levels are low, endothelial cells continue to transport glucose from the blood to the interstitial fluid of the brain. In contrast, the amino acid *glycine* is a neurotransmitter, and its concentration in neural tissue must be kept much lower than that in the circulating blood. Endothelial cells actively absorb this compound from the interstitial fluid of the brain and secrete it into the blood.

The blood-brain barrier remains intact throughout the CNS, with three noteworthy exceptions:

1. In portions of the hypothalamus, the capillary endothelium is extremely permeable. This permeability exposes hypothalamic nuclei in the anterior and tuberal regions to circulating hormones, and permits the diffusion of hypothalamic hormones into the circulation.

2. Capillaries in the *pineal gland* are also very permeable. The pineal gland, an endocrine structure, is located in the roof of the diencephalon. The capillary permeability allows pineal secretions into the general circulation.

3. In the membranous roof of the third and fourth ventricles, the pia mater supports extensive capillary networks that project into the ventricles of the brain. These capillaries are unusually permeable. Substances do not have free access to the CNS, however, because the capillaries are covered by modified ependymal cells that are interconnected by tight junctions. This complex, the *choroid plexus*, is the site of cerebrospinal fluid production.

Cerebrospinal Fluid

Cerebrospinal fluid completely surrounds and bathes the exposed surfaces of the central nervous system. It has several important functions, including:

1. *Cushioning* delicate neural structures.

2. *Supporting the brain*: In essence, the brain is suspended inside the cranium floating in the cerebrospinal fluid. A human brain weighs about 1400 g in air, but only about 50 g when supported by the cerebrospinal fluid.

3. *Transporting nutrients, chemical messengers, and waste products*: Except at the choroid plexus, the ependymal lining is freely permeable, and the CSF is in constant chemical communication with the interstitial fluid of the CNS.

Because free exchange occurs between the interstitial fluid and CSF, changes in CNS function may produce changes in the composition of the CSF. As noted in Chapter 14, a spinal tap can provide useful clinical information concerning CNS injury, infection, or disease. ∞ [p. 343]

Formation of CSF
(Figure 15-5)

The **choroid plexus** (*choroid*, vascular coat; *plexus*, network) consists of a combination of specialized ependymal cells and permeable capillaries for the production of cerebrospinal fluid (CSF). Two extensive folds of the choroid plexus originate in the roof of the third ventricle and extend through the interventricular foramina. These folds cover the floors of the lateral ventricles (Figure 15-5●). In the lower brain stem, a region of the choroid plexus in the roof of the fourth ventricle projects between the cerebellum and the pons.

Large ependymal cells cover the capillaries of the choroid plexus and contact the CSF of the ventricles. Through a combination of active and passive transport, these cells secrete cerebrospinal fluid into the ventricles. The regulation of CSF composition involves transport in both directions, and the choroid plexus removes waste products from the CSF and fine-tunes its composition over time. There are many differences in composition between cerebrospinal fluid and blood plasma (blood with the cellular elements removed). For example, the blood contains high concentrations of suspended proteins, but the CSF does not. There are also differences in the concentrations of individual ions and in the levels of amino acids, lipids, and waste products.

Circulation of CSF
(Figures 15-4a/15-6)

The choroid plexus produces CSF at a rate of about 500 mℓ/day. The total volume of CSF at any given moment is approximately 150 mℓ; this means that the entire volume of CSF is replaced roughly every 8 hours. Despite this rapid turnover, the composition of CSF is closely regulated, and the rate of removal normally keeps pace with the rate of production.

Most of the CSF reaching the fourth ventricle enters the subarachnoid space by passing through the paired **lateral apertures** and a single **median aperture** in its membranous roof. (A relatively small quantity of cerebrospinal fluid circulates between the fourth ventricle and the central canal of the spinal cord.) CSF flows through the subarachnoid space surrounding the brain, spinal cord, and cauda equina. It eventually reenters the circulation via the *arachnoid granulations* (Figures 15-4a and 15-6●). If the normal circulation of CSF is interrupted, a variety of clinical problems may appear.

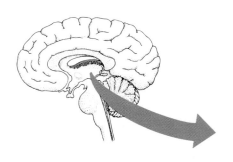

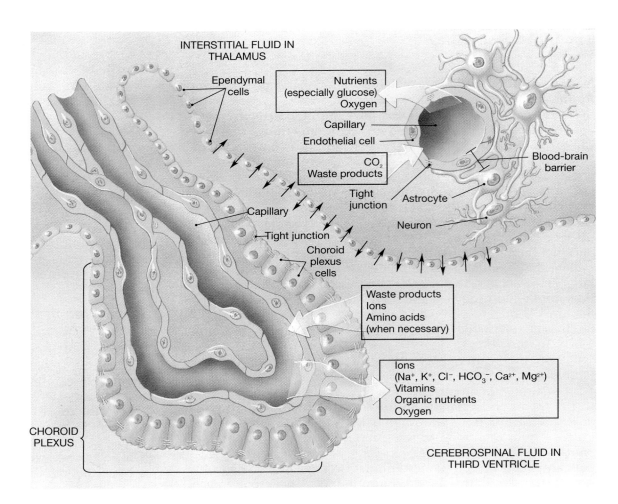

INTERSTITIAL FLUID IN
THALAMUS

Ependymal cells

Nutrients
(especially glucose)
Oxygen

Capillary

Endothelial cell

Blood-brain barrier

CO₂
Waste products

Tight junction

Astrocyte

Capillary

Tight junction

Neuron

Choroid plexus cells

Waste products
Ions
Amino acids
(when necessary)

Ions
(Na⁺, K⁺, Cl⁻, HCO₃⁻, Ca²⁺, Mg²⁺)
Vitamins
Organic nutrients
Oxygen

CHOROID PLEXUS

CEREBROSPINAL FLUID IN
THIRD VENTRICLE

FIGURE 15-5

The Choroid Plexus and Blood-Brain Barrier.
The choroid plexus consists of a combination of specialized ependymal cells and permeable capillaries. The ependymal cells act as the selective barrier, actively transporting nutrients, vitamins, and ions into the CSF. When necessary, these cells also actively remove ions or compounds from the CSF to stabilize its composition.

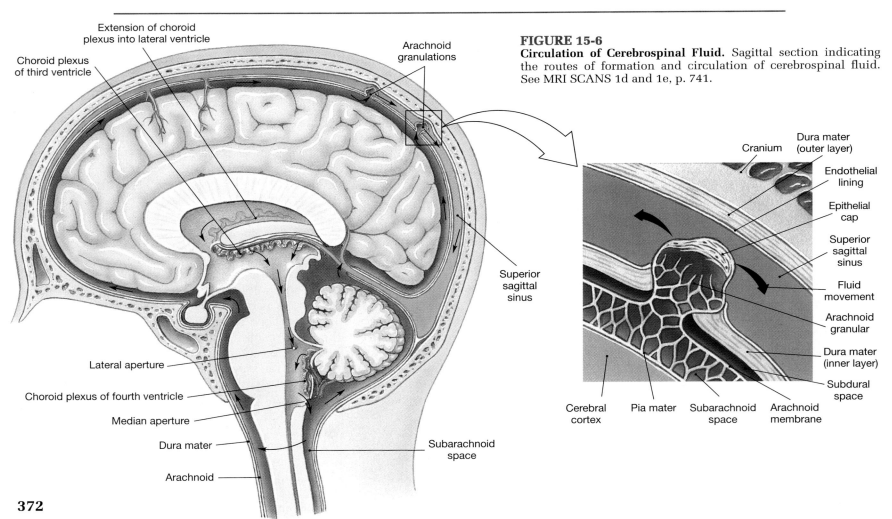

Extension of choroid plexus into lateral ventricle

Choroid plexus of third ventricle

Arachnoid granulations

FIGURE 15-6
Circulation of Cerebrospinal Fluid. Sagittal section indicating the routes of formation and circulation of cerebrospinal fluid. See MRI SCANS 1d and 1e, p. 741.

Cranium

Dura mater (outer layer)

Endothelial lining

Epithelial cap

Superior sagittal sinus

Fluid movement

Arachnoid granular

Dura mater (inner layer)

Subdural space

Cerebral cortex

Pia mater

Subarachnoid space

Arachnoid membrane

Superior sagittal sinus

Lateral aperture

Choroid plexus of fourth ventricle

Median aperture

Dura mater

Arachnoid

Subarachnoid space

The adult brain is surrounded by the inflexible bones of the cranium. The cranial cavity contains two fluids, blood and cerebrospinal fluid, and the relatively firm tissues of the brain. Because the total volume cannot change, when the volume of blood or CSF increases, the volume of the brain must decrease. In a subdural or epidural hemorrhage, the fluid volume increases as blood collects within the cranial cavity. ∞ [p. 368] The rising intracranial pressure compresses the brain, leading to neural dysfunction that often ends in unconsciousness and death.

Any alteration in the rate of cerebrospinal fluid production is normally matched by an increase in the rate of removal at the arachnoid granulations. If this equilibrium is disturbed, clinical problems appear as the intracranial pressure changes. The volume of cerebrospinal fluid will increase if the rate of formation accelerates or the rate of removal decreases. In either event the increased fluid volume leads to compression and distortion of the brain. Increased rates of formation may accompany head injuries, but the most common problems arise from masses, such as tumors or abscesses, or from developmental abnormalities. These conditions have the same effect: They restrict the normal circulation and reabsorption of CSF. Because CSF production continues, the ventricles gradually expand, distorting the surrounding neural tissues and causing the deterioration of brain function.

Infants are especially sensitive to alterations in intracranial pressure, because the arachnoid granulations do not appear until roughly 3 years of age. (Over the interim, CSF is reabsorbed into small vessels within the subarachnoid space and underlying the ependyma.) As in an adult, if intracranial pressure becomes abnormally high, the ventricles will expand. But in an infant, the cranial sutures have yet to fuse, and the skull can enlarge to accommodate the extra fluid volume. This enlargement produces an enormously expanded skull, a condition called **hydrocephalus**, or "water on the brain." Infant hydrocephalus (Figure 15-7●) often results from some interference with normal CSF circulation, such as blockage of the mesencephalic aqueduct or constriction of the connection between the subarachnoid spaces of the cranial and spinal meninges. Untreated infants often suffer some degree of mental retardation. Successful treatment usually involves the installation of a **shunt**, a bypass that either bypasses the blockage site or drains the excess cerebrospinal fluid. In either case, the goal is reduction of the intracranial pressure. The shunt may be removed if (1) further growth of the brain eliminates the blockage or (2) the intracranial pressure decreases following the development of the arachnoid granulations at 3 years of age.

FIGURE 15-7
Hydrocephalus. This infant suffers from hydrocephalus, a condition usually caused by impaired circulation and removal of cerebrospinal fluid. CSF buildup leads to distortion of the brain and enlargement of the cranium.

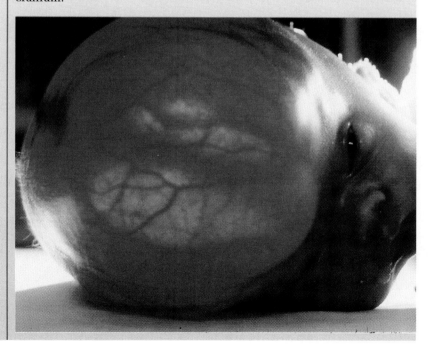

THE CEREBRUM
(Figures 15-1/15-8/15-9)

The cerebrum is the largest region of the brain. Conscious thought processes and all intellectual functions originate in the cerebral hemispheres. Much of the cerebrum is involved in the processing of somatic sensory and motor information. Somatic sensory information relayed to the cerebrum reaches our conscious awareness, and cerebral neurons exert direct (voluntary) or indirect (involuntary) control over somatic motor neurons. Most visceral sensory processing and visceral motor (autonomic) control occur at centers elsewhere in the brain, usually outside our conscious awareness. Figures 15-1, p. 366, 15-8, and 15-9● provide additional perspective on the cerebrum and its relationships with other regions of the brain.

Gray matter in the cerebrum is found in a superficial layer of neural cortex and in deeper *cerebral nuclei*. The *central white matter* lies beneath the neural cortex and surrounds the cerebral nuclei.

The Cerebral Cortex
(Figure 15-9)

A blanket of neural cortex (superficial gray matter) covers the paired **cerebral hemispheres** that form the superior and lateral surfaces of the cerebrum (Figure 15-9●). The cortical surface forms a series of elevated ridges, or **gyri** (JĪ-rī), separated by shallow depressions, called **sulci** (SUL-sī), or deeper grooves, called **fissures**. Gyri increase the surface area of the cerebral hemispheres and provide space for additional cortical neurons. The total surface area of the cerebral hemispheres is roughly equivalent to 2200 cm² (2.5 ft²) of flat surface, and that large an area can be packed into the skull only when folded, like a crumpled piece of paper. Complex analytical and integrative functions require large numbers of neurons. The entire brain has enlarged in the course of human evolution, but the cerebral hemispheres have enlarged at a much faster rate than the rest of the brain, even with extensive folding of the cerebral hemispheres.

Boundaries between Cerebral Lobes
(Figures 15-1/15-8/15-9)

The two cerebral hemispheres are separated by a deep **longitudinal fissure** (Figures 15-1, p. 366, and 15-8●), and each hemisphere can be divided into **lobes** named after the overlying bones of the skull. There are individual differences in the appearance of the sulci and gyri of each brain, but the boundaries between lobes are reliable landmarks. A deep groove, the **central sulcus**, extends laterally from the longitudinal fissure. The area anterior to the central sulcus is the **frontal lobe**, and the **lateral sulcus** marks its inferior border (Figures 15-8 and 15-9a●). The region inferior to the lateral sulcus is the **temporal lobe**. Reflecting this lobe to the side (Figure 15-9b●) exposes the **insula** (IN-sū-la), an "island" of cortex that is otherwise invisible. The **parietal lobe** extends posteriorly from the central sulcus to the **parieto-occipital sulcus**. The region posterior to the parieto-occipital sulcus is the **occipital lobe**.

Each lobe contains functional regions whose boundaries are less clearly defined. Some of these regions are concerned with sensory information and others with motor commands.

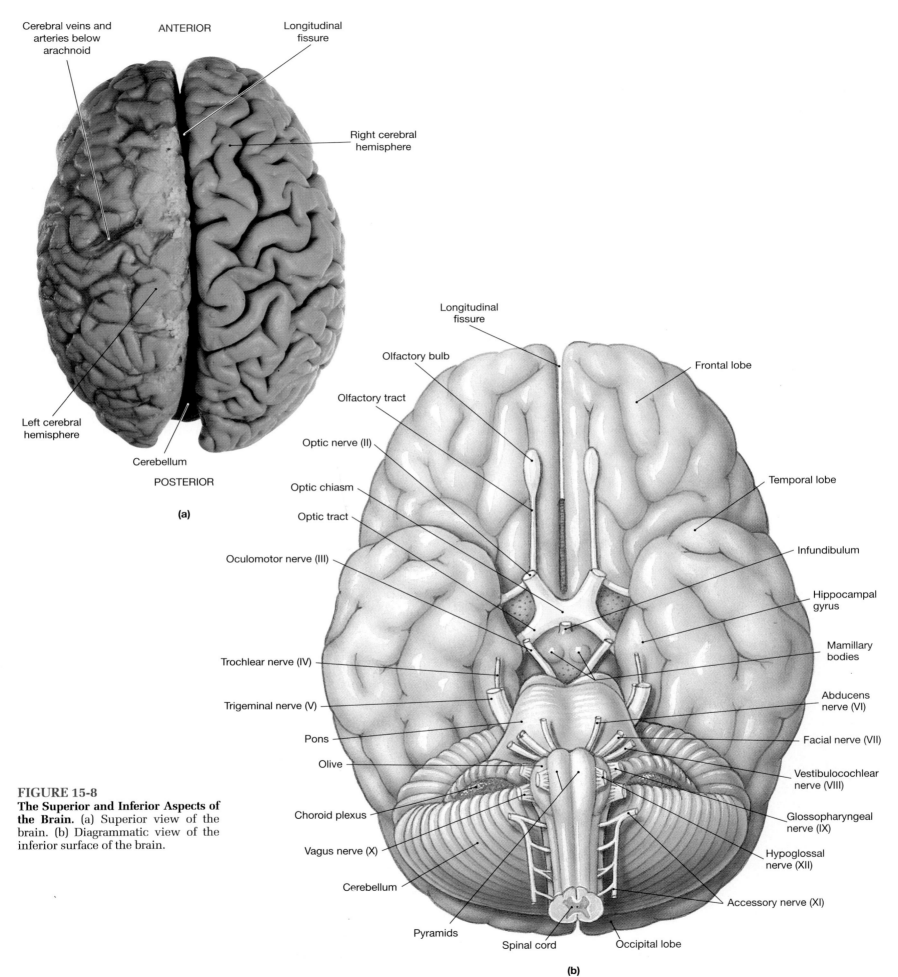

Cerebral veins and arteries below arachnoid

ANTERIOR

Longitudinal fissure

Right cerebral hemisphere

Left cerebral hemisphere

Cerebellum

POSTERIOR

(a)

FIGURE 15-8
The Superior and Inferior Aspects of the Brain. (a) Superior view of the brain. (b) Diagrammatic view of the inferior surface of the brain.

Longitudinal fissure

Olfactory bulb

Olfactory tract

Optic nerve (II)

Optic chiasm

Optic tract

Oculomotor nerve (III)

Trochlear nerve (IV)

Trigeminal nerve (V)

Pons

Olive

Choroid plexus

Vagus nerve (X)

Cerebellum

Pyramids

Spinal cord

Frontal lobe

Temporal lobe

Infundibulum

Hippocampal gyrus

Mamillary bodies

Abducens nerve (VI)

Facial nerve (VII)

Vestibulocochlear nerve (VIII)

Glossopharyngeal nerve (IX)

Hypoglossal nerve (XII)

Accessory nerve (XI)

Occipital lobe

(b)

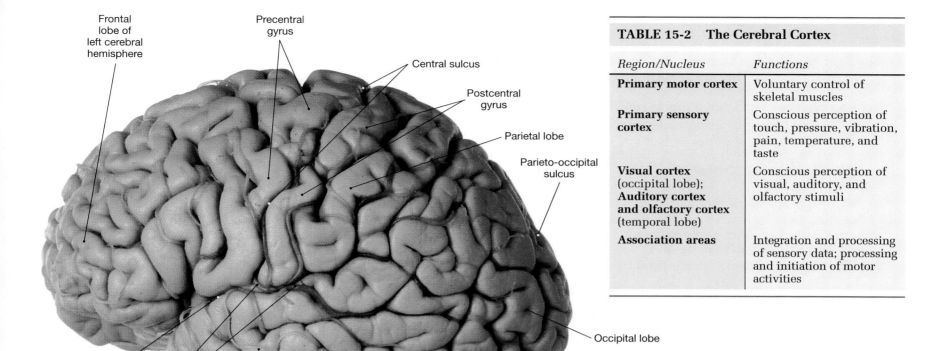

Frontal lobe of left cerebral hemisphere

Precentral gyrus

Central sulcus

Postcentral gyrus

Parietal lobe

Parieto-occipital sulcus

Occipital lobe

Lateral sulcus

Branches of middle cerebral artery emerging from lateral sulcus

Pons

Temporal lobe

Medulla oblongata

Cerebellum

(a)

TABLE 15-2	The Cerebral Cortex
Region/Nucleus	Functions
Primary motor cortex	Voluntary control of skeletal muscles
Primary sensory cortex	Conscious perception of touch, pressure, vibration, pain, temperature, and taste
Visual cortex (occipital lobe); **Auditory cortex and olfactory cortex** (temporal lobe)	Conscious perception of visual, auditory, and olfactory stimuli
Association areas	Integration and processing of sensory data; processing and initiation of motor activities

FIGURE 15-9
The Cerebral Hemispheres. (a) Lateral view of intact brain showing superficial surface anatomy of the left hemisphere. (b) Major anatomical landmarks on the surface of the left cerebral hemisphere. Association areas are colored. To expose the insula, the lateral sulcus has been opened.

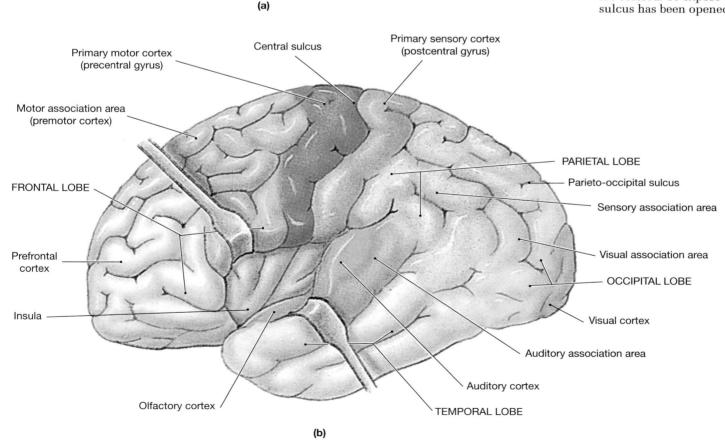

Primary motor cortex (precentral gyrus)

Central sulcus

Primary sensory cortex (postcentral gyrus)

Motor association area (premotor cortex)

PARIETAL LOBE

Parieto-occipital sulcus

FRONTAL LOBE

Sensory association area

Prefrontal cortex

Visual association area

OCCIPITAL LOBE

Insula

Visual cortex

Auditory association area

Olfactory cortex

Auditory cortex

TEMPORAL LOBE

(b)

Three points about the cerebral lobes should be kept in mind:

1. *Each cerebral hemisphere receives sensory information and generates motor commands that concern the opposite side of the body.* The left hemisphere controls the right side, and the right hemisphere controls the left side. This crossing over has no known functional significance.

2. *The two hemispheres have some functional differences, although anatomically they appear the same.* These differences affect primarily higher-order functions, a topic which will be discussed in Chapter 16.

3. *The assignment of a specific function to a specific region of the cerebral cortex is imprecise.* Because the boundaries are indistinct, with considerable overlap, any one region may have several different functions. Some aspects of cortical function, such as consciousness, cannot easily be assigned to any single region.

Our understanding of brain function is still incomplete, and every anatomical feature does not have a known function. However, it is clear that all portions of the brain are used in a normal individual.

Motor and Sensory Areas of the Cortex
(Figure 15-9b, Table 15-2)

The major motor and sensory regions of the cerebral cortex are detailed in Figure 15-9b● and Table 15-2. The central sulcus separates the motor and sensory portions of the cortex. The **precentral gyrus** of the frontal lobe forms the anterior margin of the central sulcus. The surface of this gyrus is the **primary motor cortex**. Neurons of the primary motor cortex direct voluntary movements by controlling somatic motor neurons in the brain stem and spinal cord. The neurons of the primary motor cortex are called **pyramidal cells**, and the pathway that provides voluntary motor control is known as the **pyramidal system**.

The **postcentral gyrus** of the parietal lobe forms the posterior margin of the central sulcus, and its surface contains the **primary sensory cortex**. Neurons in this region receive somatic sensory information from touch, pressure, pain, taste, and temperature receptors. We are consciously aware of these sensations because the sensory information has been relayed to the primary sensory cortex. At the same time, collaterals deliver information to the cerebral nuclei and other centers. As a result, sensory information is monitored at both conscious and unconscious levels.

Sensory information concerning sensations of sight, sound, and smell arrive at other portions of the cerebral cortex. The **visual cortex** of the occipital lobe receives visual information, and the **auditory cortex** and **olfactory cortex** of the temporal lobe receive information concerned with hearing and smelling, respectively. These regions are shown on Figure 15-9b●.

Association Areas
(Figure 15-9b)

The sensory and motor regions of the cortex are connected to nearby **association areas** that interpret incoming data or coordinate a motor response (Figure 15-9b●). The **somatic motor association area**, or **premotor cortex**, is responsible for the coordination of learned motor responses. The functional distinctions between the sensory and motor association areas are most evident after localized brain damage has occurred. For example, an individual with a damaged **visual association area** may see letters quite clearly, but be unable to recognize or interpret them. This person would scan the lines of a printed page and see rows of clear symbols that convey no meaning. Someone with damage to the area of the premotor cortex concerned with coordination of eye movements can understand written letters and words but cannot read, because his or her eyes cannot follow the lines on a printed page.

Integrative Centers
(Figure 15-9b)

Integrative centers receive and process information from many different association areas. These regions direct extremely complex motor activities and perform complicated analytical functions. For example, the **prefrontal cortex** of the frontal lobe (Figure 15-9b●) integrates information from sensory association areas and performs abstract intellectual functions, such as predicting the consequences of possible responses.

These lobes and cortical areas are found on both cerebral hemispheres. Higher-order integrative centers concerned with complex processes, such as speech, writing, mathematical computation, or understanding spatial relationships, are restricted to the left or right hemisphere. These centers and their functions are described in Chapter 16.

The Central White Matter
(Figure 15-10, Table 15-3)

The **central white matter** (Figure 15-10●) contains three major groups of axons: (1) *association fibers*, tracts that interconnect areas of neural cortex within a single cerebral hemisphere; (2) *commissural fibers*, tracts that connect the two cerebral hemispheres; and (3) *projection fibers*, tracts that link the cerebrum with other regions of the brain and the spinal cord. The functions of these groups are summarized in Table 15-3.

Association Fibers

Association fibers interconnect portions of the cerebral cortex. The shortest association fibers are called **arcuate** (AR-kū-āt) **fibers** because they curve in an arc to pass from one gyrus to another. The longer association fibers are organized into discrete bundles. The **longitudinal fasciculi** connect the frontal lobe to the other lobes of the same hemisphere.

Commissural Fibers

A dense band of **commissural fibers** (kom-MIS-ū-ral; *commissura*, a crossing over) permit communication between the two hemispheres. Prominent commissural bundles linking the cerebral hemispheres include the **corpus callosum** and the **anterior commissure**.

Projection Fibers

Projection fibers link the cerebral cortex to the diencephalon, brain stem, cerebellum, and spinal cord. All ascending and descending axons must pass through the diencephalon on their way to or from sensory, motor, or association areas of the cerebral cortex. In gross dissection the afferent fibers and efferent fibers look alike, and the entire collection of fibers is known as the **internal capsule**.

FIGURE 15-10

The Central White Matter. (a) Lateral aspect of the brain, showing arcuate fibers and longitudinal fasciculi. (b) Anterior view of the brain, showing orientation of the commissural and projection fibers.

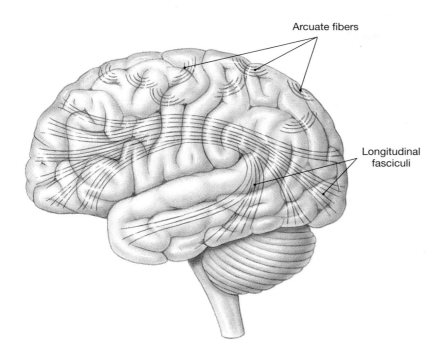

Arcuate fibers

Longitudinal fasciculi

(a) Lateral view

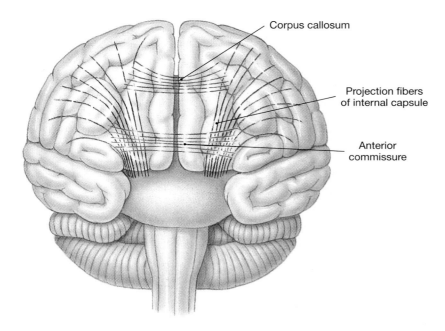

Corpus callosum

Projection fibers of internal capsule

Anterior commissure

(b) Anterior view

TABLE 15-3 White Matter of the Cerebrum

Region/Nucleus	Functions
Arcuate fibers	Interconnect gyri
Association fibers	Interconnect lobes of same hemisphere
Commissural fibers (anterior commissure and corpus callosum)	Interconnect corresponding lobes of different hemispheres
Projection fibers	Connect cerebral cortex with brain stem

The Cerebral Nuclei
(Figure 15-11, Table 15-4)

The **cerebral nuclei** are masses of gray matter within the cerebral hemispheres. The cerebral nuclei are also known as the **basal nuclei**, or *basal ganglia*. The latter term, although inappropriate, has persisted despite the fact that ganglia are otherwise restricted to the peripheral nervous system. The cerebral nuclei (Figure 15-11●) lie within each hemisphere inferior to the floor of the lateral ventricle. They are embedded within the central white matter, and the radiating projection and commissural fibers travel around or between these nuclei.

The **caudate nucleus** has a massive head and a slender, curving tail that follows the curve of the lateral ventricle. At the tip of the tail there is a separate nucleus, the **amygdaloid body** (ah-MIG-da-loyd; *amygdale*, almond). Three masses of gray matter lie between the bulging surface of the insula and the lateral wall of the diencephalon. These are the **claustrum** (KLAWS-trum), the **putamen** (pū-TĀ-men), and the **globus pallidus** (GLŌ-bus PAL-i-dus; pale globe).

Several additional terms are used to designate specific anatomical or functional subdivisions of the cerebral nuclei. The putamen and globus pallidus are often considered as subdivisions of a larger **lentiform** (lens-shaped) **nucleus**, for when exposed on gross dissection, they form a rather compact, rounded mass (Figure 15-11b,d,e●). The term **corpus striatum** (striated body) encompasses the caudate *and* lentiform nuclei. The name refers to the striated (striped) appearance of the internal capsule as it passes between the two nuclei. Table 15-4 summarizes these relationships and the functions of the cerebral nuclei.

Functions of the Cerebral Nuclei

The cerebral nuclei are important components of the **extrapyramidal system**, a pathway that controls muscle tone and coordinates learned movement patterns and other somatic motor activities. (This system will be discussed further in Chapter 16.) Some functions assigned to specific cerebral nuclei are detailed below.

CAUDATE NUCLEUS. Under normal conditions the cerebral nuclei do not initiate a particular movement. But once a movement is under way, they provide the general pattern and rhythm. For example, when walking, the caudate nucleus and putamen control the cycles of arm and leg movements that occur between the time the decision is made to "start walking" and the time the "stop" order is given.

GLOBUS PALLIDUS. The globus pallidus controls and adjusts muscle tone, particularly in the appendicular muscles, to set body position in preparation for a voluntary movement. For example, when you decide to pick up a pencil, the globus pallidus positions the shoulder and stabilizes the arm as you consciously reach and grasp with the forearm, wrist, and hand.

CLAUSTRUM AND AMYGDALOID BODY. The claustrum appears to be involved in the unconscious processing of visual information. The amygdaloid body is an important component of the *limbic system*, and will be considered in the next section. The functions of other cerebral nuclei are poorly understood.

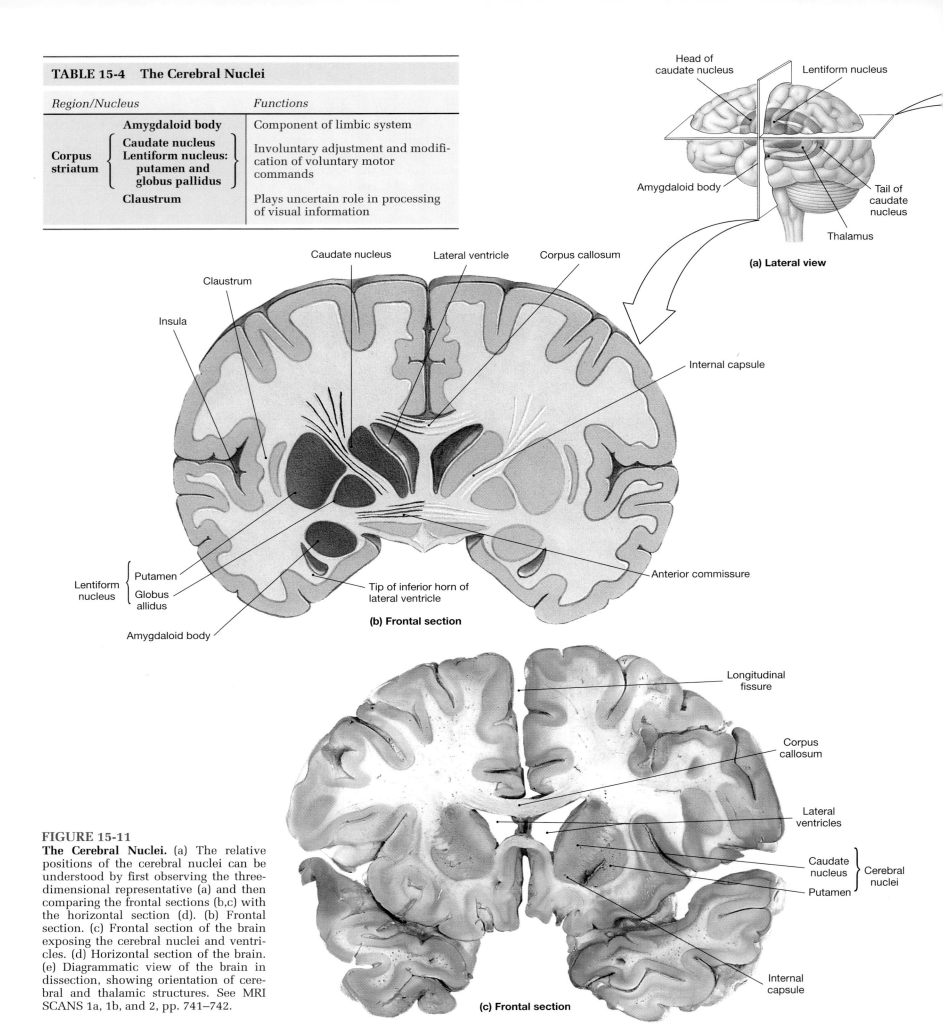

TABLE 15-4   The Cerebral Nuclei

Region/Nucleus		Functions
Corpus striatum	**Amygdaloid body**	Component of limbic system
	Caudate nucleus **Lentiform nucleus: putamen and globus pallidus**	Involuntary adjustment and modification of voluntary motor commands
	Claustrum	Plays uncertain role in processing of visual information

(a) Lateral view

- Head of caudate nucleus
- Lentiform nucleus
- Amygdaloid body
- Tail of caudate nucleus
- Thalamus

- Claustrum
- Caudate nucleus
- Lateral ventricle
- Corpus callosum
- Insula
- Internal capsule
- Lentiform nucleus { Putamen, Globus allidus }
- Tip of inferior horn of lateral ventricle
- Anterior commissure
- Amygdaloid body

(b) Frontal section

- Longitudinal fissure
- Corpus callosum
- Lateral ventricles
- Caudate nucleus } Cerebral nuclei
- Putamen
- Internal capsule

(c) Frontal section

FIGURE 15-11
The Cerebral Nuclei. (a) The relative positions of the cerebral nuclei can be understood by first observing the three-dimensional representative (a) and then comparing the frontal sections (b,c) with the horizontal section (d). (b) Frontal section. (c) Frontal section of the brain exposing the cerebral nuclei and ventricles. (d) Horizontal section of the brain. (e) Diagrammatic view of the brain in dissection, showing orientation of cerebral and thalamic structures. See MRI SCANS 1a, 1b, and 2, pp. 741–742.

378

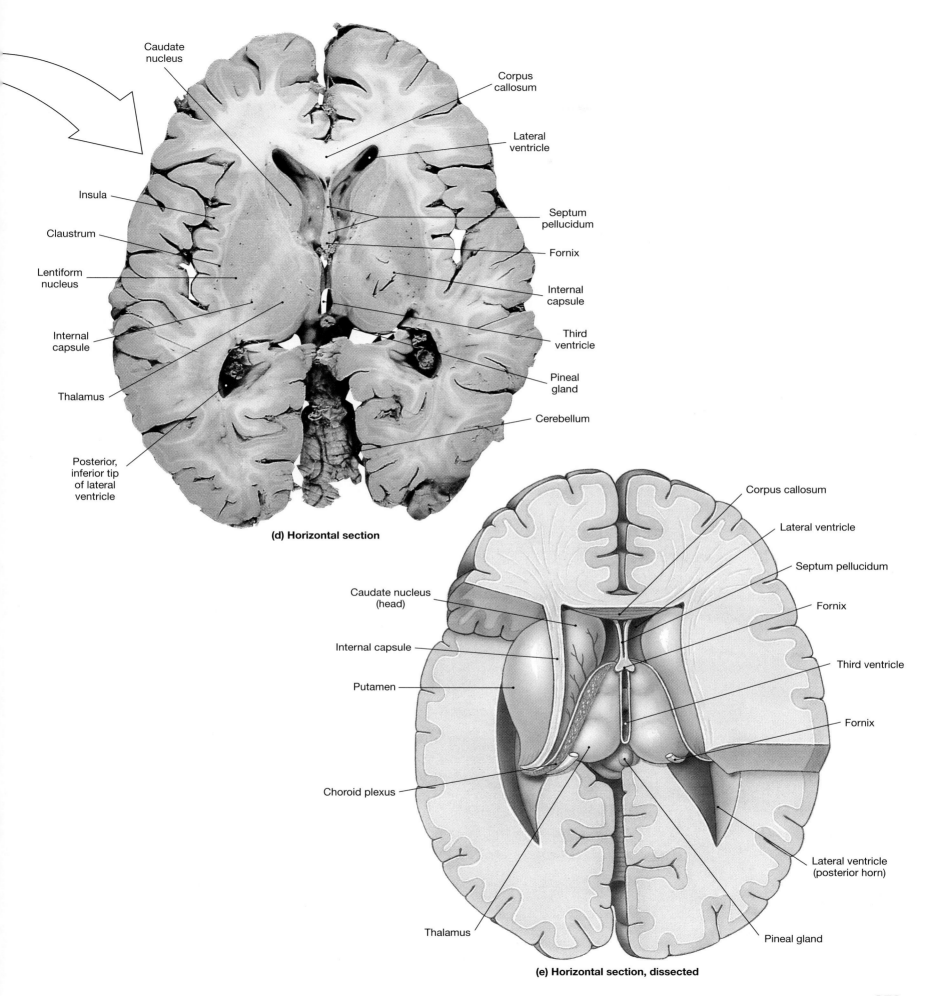

(d) Horizontal section

Caudate nucleus

Corpus callosum

Lateral ventricle

Insula

Septum pellucidum

Claustrum

Fornix

Lentiform nucleus

Internal capsule

Internal capsule

Third ventricle

Thalamus

Pineal gland

Posterior, inferior tip of lateral ventricle

Cerebellum

(e) Horizontal section, dissected

Corpus callosum

Lateral ventricle

Septum pellucidum

Caudate nucleus (head)

Fornix

Internal capsule

Third ventricle

Putamen

Fornix

Choroid plexus

Lateral ventricle (posterior horn)

Thalamus

Pineal gland

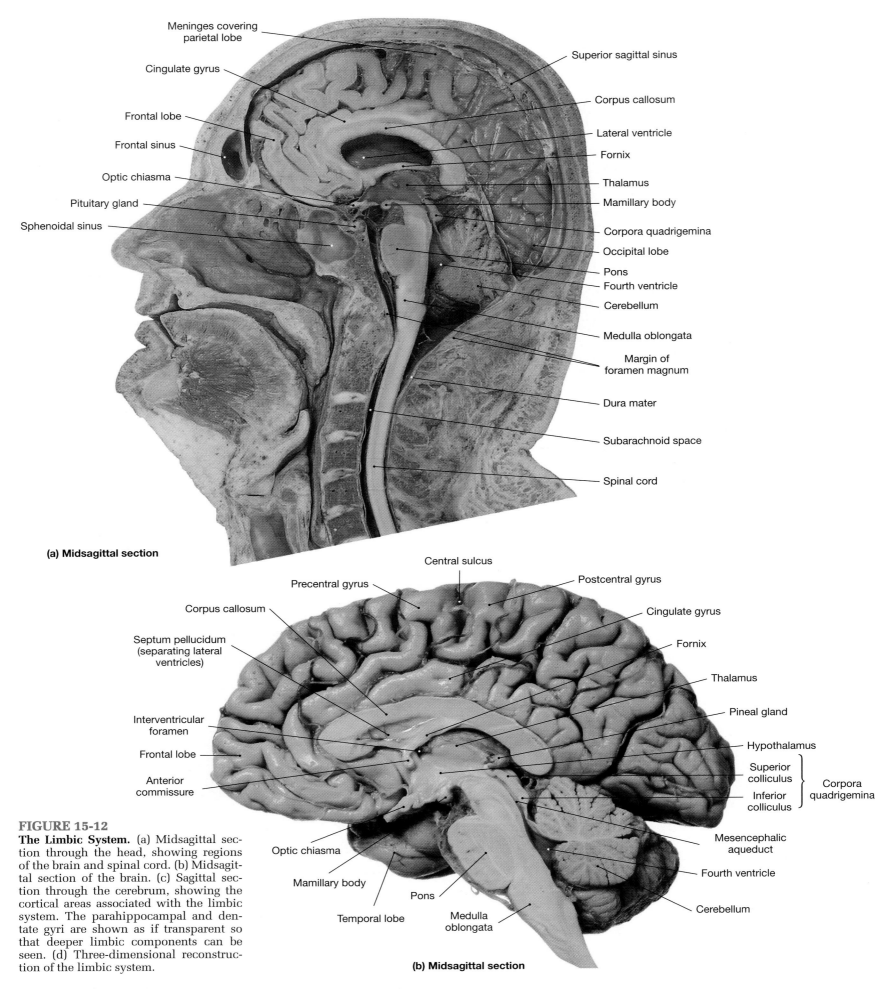

Meninges covering parietal lobe

Cingulate gyrus

Frontal lobe

Frontal sinus

Optic chiasma

Pituitary gland

Sphenoidal sinus

Superior sagittal sinus

Corpus callosum

Lateral ventricle

Fornix

Thalamus

Mamillary body

Corpora quadrigemina

Occipital lobe

Pons

Fourth ventricle

Cerebellum

Medulla oblongata

Margin of foramen magnum

Dura mater

Subarachnoid space

Spinal cord

(a) Midsagittal section

Central sulcus

Precentral gyrus

Postcentral gyrus

Corpus callosum

Cingulate gyrus

Septum pellucidum (separating lateral ventricles)

Fornix

Thalamus

Pineal gland

Interventricular foramen

Hypothalamus

Superior colliculus

Corpora quadrigemina

Frontal lobe

Anterior commissure

Inferior colliculus

Mesencephalic aqueduct

Optic chiasma

Fourth ventricle

Mamillary body

Pons

Cerebellum

Temporal lobe

Medulla oblongata

(b) Midsagittal section

FIGURE 15-12

The Limbic System. (a) Midsagittal section through the head, showing regions of the brain and spinal cord. (b) Midsagittal section of the brain. (c) Sagittal section through the cerebrum, showing the cortical areas associated with the limbic system. The parahippocampal and dentate gyri are shown as if transparent so that deeper limbic components can be seen. (d) Three-dimensional reconstruction of the limbic system.

THE LIMBIC SYSTEM
(Figures 15-11/15-12, Table 15-5)

The **limbic system** (LIM-bik; *limbus*, border) includes nuclei and tracts along the border between the cerebrum and diencephalon. The functions of the limbic system include: (1) establishment of emotional states and related behavioral drives, (2) linking the conscious, intellectual functions of the cerebral cortex with the unconscious and autonomic functions of other portions of the brain, and (3) facilitating memory storage and retrieval. This system is a functional grouping, rather than an anatomical one, and the limbic system includes components of the cerebrum, diencephalon, and mesencephalon (Table 15-5).

The amygdaloid body (Figure 15-11a,b,● p. 378) appears to act as an interface between the limbic system, the cerebrum, and various sensory systems. The **limbic lobe** of the cerebral hemisphere consists of two gyri that curve along the corpus callosum and onto the medial surface of the temporal lobe (Figure 15-12c●). The **cingulate gyrus** (SIN-gū-lāt; *cingulum*, girdle or belt) sits superior to the corpus callosum (Figure 15-12c●). The **dentate gyrus** and the adjacent **parahippocampal** (pa-ra-hip-ō-KAM-pal) **gyrus** conceal an underlying nucleus, the **hippocampus**, which lies beneath the floor of the lateral ventricle (see Figures 15-11e and 15-12c●). Early anatomists thought this nucleus resembled a sea horse (*hippocampus*); it appears to be important in learning and the storage of long-term memories.

The **fornix** (FŌR-niks) is a tract of white matter that connects the hippocampus with the hypothalamus. From the hippocampus the fornix curves medially and proceeds anteriorly and inferiorly to the corpus callosum, before arching inferiorly to the hypothalamus. Many of the fibers end in the **mamillary bodies** (MAM-i-lar-ē; *mamilla*, breast), prominent nuclei in the floor of the hypothalamus (Figure 15-12a,b●). The mamillary bodies contain motor nuclei that control reflex movements associated with eating, such as chewing, licking, and swallowing.

Several other nuclei in the wall (thalamus) and floor (hypothalamus) of the diencephalon are components of the limbic system. The **anterior nucleus** of the thalamus relays visceral sensations from the hypothalamus to the cerebral components of the limbic system. Experimental stimulation of the hypothalamus has outlined a number of important centers responsible for the emotions of rage, fear, pain, sexual arousal, and pleasure.

Stimulation of the hypothalamus can also produce heightened alertness and a generalized excitement. This response is caused by widespread stimulation of the **reticular formation**, an interconnected network of brain stem nuclei whose headquarters lies in the mesencephalon. Stimulation of adjacent portions of the hypothalamus or thalamus will depress reticular activity, resulting in generalized lethargy or actual sleep.

√ **What would happen if an interventricular foramen became blocked?**

√ **A patient suffers a head injury that damages the primary motor cortex. Where is this area located?**

TABLE 15-5   The Limbic System

FUNCTION	Processing of memories, creation of emotional states, drives, and associated behaviors
CEREBRAL COMPONENTS Cortical areas	Limbic lobe (cingulate gyrus and parahippocampal gyrus)
Nuclei	Hippocampus, amygdaloid body
Tracts	Fornix
DIENCEPHALIC COMPONENTS Thalamus	Anterior nuclear group
Hypothalamus	Centers concerned with emotions, appetites (thirst, hunger), and related behaviors (see Table 15-7)
OTHER COMPONENTS Reticular formation	Network of interconnected nuclei through brain stem

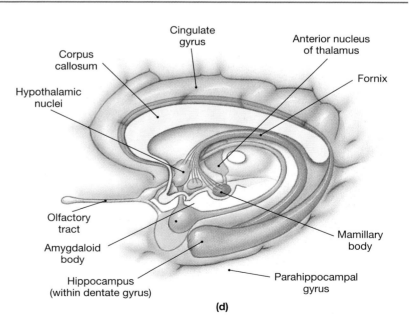

Central sulcus

Corpus callosum

Fornix

Cingulate gyrus (limbic lobe)

Pineal gland

Temporal lobe

Intermediate mass

Mamillary body

Parahippocampal gyrus (limbic lobe)

Hippocampus (within dentate gyrus)

(c) Sagittal section

Cingulate gyrus

Anterior nucleus of thalamus

Corpus callosum

Fornix

Hypothalamic nuclei

Olfactory tract

Amygdaloid body

Mamillary body

Hippocampus (within dentate gyrus)

Parahippocampal gyrus

(d)

FIGURE 15-13

The Brain Stem. (a) Diagrammatic view of the diencephalon and brain stem, seen from the left side. (b) Lateral view of the brain stem, cerebellum sectioned and removed. (c) Posterior diagrammatic view of the diencephalon and brain stem. (d) Brain stem, posterior view.

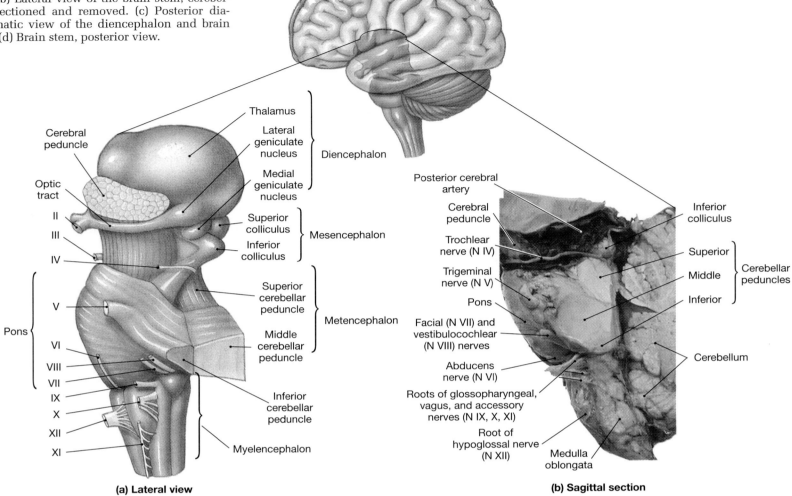

(a) Lateral view

(b) Sagittal section

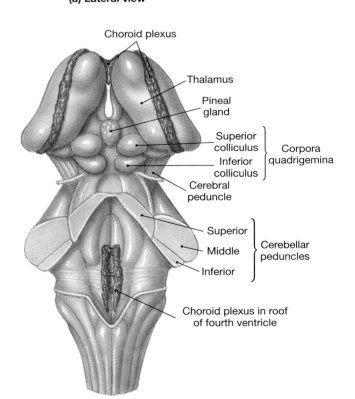

(c) Posterior view

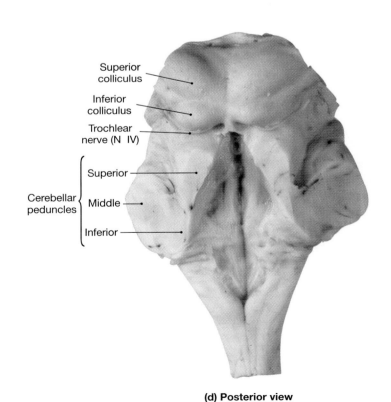

(d) Posterior view

THE DIENCEPHALON
(Figures 15-8/15-12a,b/15-13)

The diencephalon provides the switching and relay centers for both sensory and motor pathways. Figures 15-1, p. 366, 15-12a,b, and 15-13● show the position of the diencephalon and its relationship to other landmarks in the brain.

The diencephalic roof, or **epithalamus**, is membranous anteriorly. The region, which forms the roof of the third ventricle, contains an extensive area of choroid plexus that extends through the interventricular foramina into the lateral ventricles. The posterior epithalamus contains the **pineal gland**, an endocrine structure that secretes the hormone **melatonin**. Melatonin is important in the regulation of day-night cycles, with secondary effects on reproductive function. (The role of melatonin will be described in Chapter 19.)

Most of the neural tissue in the diencephalon is concentrated in two structures, the *thalami* (walls of the diencephalon surround the third ventricle) and the *hypothalamus* (floor of the third ventricle). Ascending sensory information from the spinal cord and cranial nerves (other than the olfactory tract) is processed in a nucleus in the left or right thalamus. The hypothalamus contains centers involved with emotions and visceral processes that affect the cerebrum as well as other components of the brain stem. It also controls a variety of autonomic functions and forms the link between the nervous and endocrine systems.

The Thalamus
(Figures 15-11e/15-12a,b/15-13)

The thalamus is the final relay point for ascending sensory information that will be projected to the primary sensory cortex. It acts as a filter, passing on only a small portion of the arriving sensory information. The thalamus also coordinates the activities of the pyramidal and extrapyramidal systems.

The left and right thalami are separated by the third ventricle. Viewed in midsagittal section (Figure 15-12a,b●, p. 380), the thalamus extends from the anterior commissure to the inferior base of the pineal gland. A round projection of gray matter, an **intermediate mass**, extends into the ventricle from the thal-

amus on either side. In roughly 70% of the population, the two fuse in the midline, interconnecting the two thalami.

The thalamus on each side bulges laterally, away from the third ventricle (Figures 15-11e, p. 379, and 15-13●), and anteriorly toward the cerebrum. The lateral border of the thalamus is established by the fibers of the *internal capsule*. ∞ [p. 376] Embedded within each thalamus is a rounded mass composed of several interconnected *thalamic nuclei*.

Functions of Thalamic Nuclei
(Figure 15-14, Table 15-6)

The thalamic nuclei are concerned primarily with the relay of sensory information to the cerebral nuclei and cerebral cortex. The four major groups of thalamic nuclei, detailed in Figure 15-14● and Table 15-6, are (1) the *anterior group*, (2) the *medial group*, (3) the *ventral group*, and (4) the *posterior group*.

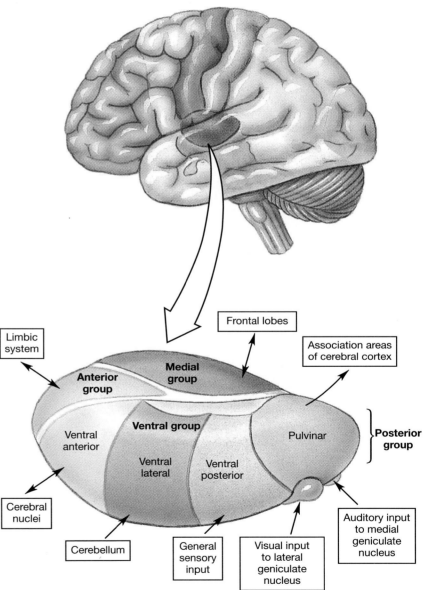

FIGURE 15-14
The Thalamus. (Above) Lateral view of the brain, showing the positions of the major thalamic structures. Functional areas of cerebral cortex are also indicated. (Below) Enlarged view of the thalamic nuclei of the left side. The color of each nucleus or group of nuclei matches the color of the associated cortical region. Arrows show sensory "input" or projection ("output") to the cerebral nuclei and cerebral cortex.

TABLE 15-6   The Thalamus

Structure/Nuclei	Functions
Anterior group	Part of the limbic system
Medial group	Integrates sensory information and other data arriving at the thalamus and hypothalamus for projection to the frontal lobes of the cerebral hemispheres
Ventral group	Projects sensory information to the primary sensory cortex of the parietal lobe; relays information from cerebellum and cerebral nuclei to motor areas of cerebral cortex
Posterior group Pulvinar	Integrates sensory information for projection to association areas of cerebral cortex
Lateral geniculate nuclei	Project visual information to the visual cortex of occipital lobe
Medial geniculate nuclei	Project auditory information to the visual cortex of temporal lobe

1. The **anterior nuclei** are part of the limbic system, discussed on p. 381.

2. The **medial nuclei** provide a conscious awareness of emotional states by connecting emotional centers in the hypothalamus with the frontal lobes of the cerebrum. These nuclei also integrate sensory information arriving at other portions of the thalamus for relay to the frontal lobes.

3. The **ventral nuclei** relay information from the cerebral nuclei and cerebellum, specifically to the primary motor cortex and the motor association area of the frontal lobe. They also relay sensory data concerning touch, pressure, pain, temperature, proprioception, and taste to the primary sensory cortex of the parietal lobe.

4. The **posterior nuclei** include the *pulvinar* and the *geniculates*. The **pulvinar** integrates sensory information for projection to the association areas of the cerebral cortex. The **lateral geniculate** (je-NIK-ū-lāt; *genicula*, little knee) of each thalamus receives visual information from the eyes, brought by the **optic tract**. Efferent fibers project to the visual cortex and descend to the mesencephalon. The **medial geniculate** relays auditory information to the auditory cortex from the specialized receptors of the inner ear.

The Hypothalamus
(Figures 15-8/15-12a/15-15)

The hypothalamus, which forms the floor of the third ventricle, extends from the area superior to the **optic chiasm**, where the *optic tracts* from the eyes arrive at the brain, to the posterior margins of the *mamillary bodies* (Figure 15-15●). (The mamillary bodies were introduced in the discussion of the limbic system on p. 381.) Posterior to the optic chiasm, the **infundibulum** (in-fun-DIB-ū-lum; *infundibulum*, funnel) extends inferiorly, connecting the hypothalamus to the pituitary gland (Figure 15-8●, p. 374). In life, the *diaphragma sellae* (p. 368) surrounds the infundibulum as it enters the hypophyseal fossa of the sphenoid bone.

Viewed in midsagittal section (Figures 15-12b, p. 380, and 15-15●), the floor of the hypothalamus between the infundibulum and the mamillary bodies is the **tuber cinereum** (sin-Ē-rē-um; *tuber*, swelling + *cinereus*, ashen color). The tuber cinereum is a mass of gray matter whose neurons are involved with the control of pituitary gland function. The rostral portion of the tuber cinereum adjacent to the infundibulum is thickened and slightly elevated; this region is called the **median eminence**.

Functions of the Hypothalamus
(Figure 15-15c, Table 15-7)

The hypothalamus contains a variety of important control and integrative centers, in addition to those associated with the limbic system. These centers and their functions are summarized in Figure 15-15c● and Table 15-7. Hypothalamic centers are continually receiving sensory information from the cerebrum, brain stem, and spinal cord. Hypothalamic neurons also detect and respond to changes in the CSF and interstitial fluid composition; they also respond to stimuli in the circulating blood, because of the high permeability of capillaries in this region (p. 371). Hypothalamic functions include:

1. *Control of involuntary somatic motor activities*: By stimulation of appropriate centers in other portions of the brain, hypothalamic nuclei direct somatic motor patterns associated with the emotions of rage, pleasure, pain, and sexual arousal.

2. *Control of autonomic function*: Hypothalamic centers adjust and coordinate the activities of autonomic centers in other parts of the brain stem concerned with regulating heart rate, blood pressure, respiration, and digestive functions.

3. *Coordination of activities of the nervous and endocrine systems*: Much of the regulatory control is exerted through inhibition or stimulation of portions of the pituitary gland.

4. *Secretion of hormones*: The hypothalamus secretes two hormones: (1) *antidiuretic hormone*, produced by the **supraoptic nucleus**, restricts water loss at the kidneys, and (2) *oxytocin*, produced by the **paraventricular nucleus**, stimulates smooth muscle contractions in the uterus and prostate gland, and myoepithelial cell contractions in the mammary glands. Both hormones are transported along axons down the infundibulum for release into the circulation at the posterior portion of the pituitary gland.

5. *Production of emotions and behavioral drives*: Specific hypothalamic centers produce sensations that lead to changes in voluntary or involuntary behavior patterns. For example, stimulation of the **feeding center** produces the desire to eat, and stimulation of the **thirst center** produces the desire to drink.

6. *Coordination between voluntary and autonomic functions*: When facing a stressful situation, your heart rate and respiratory rate go up and your body prepares for an emergency. These autonomic adjustments are made because cerebral activities are monitored by the hypothalamus.

7. *Regulation of body temperature*: The **preoptic area** of the hypothalamus controls the physiological responses to changes in body temperature. In doing so, it coordinates the activities of other CNS centers and regulates other physiological systems.

Table 15-7 lists the major nuclei and centers of the hypothalamus and provides greater detail regarding their known functions.

THE MESENCEPHALON (MIDBRAIN)
(Figures 15-13/15-16, Table 15-8)

The external anatomy of the mesencephalon can be seen in Figure 15-13●, and the major nuclei are detailed in Figure 15-16● and Table 15-8. The roof, or **tectum**, of the mesencephalon contains two pairs of sensory nuclei known collectively as the **corpora quadrigemina** (KOR-pō-ra quad-ri-JEM-i-na). These nuclei, the *colliculi*, are relay stations concerned with the processing of visual and auditory sensations. Each **superior colliculus** (kol-IK-ū-lus; *colliculus*, small hill) receives visual inputs from the lateral geniculate of the thalamus on that side. The **inferior colliculus** receives auditory data from nuclei in the medulla oblongata; some of this information may be forwarded to the medial geniculate on the same side.

The mesencephalon also contains the headquarters of the reticular formation. Specific patterns of stimulation in this region can produce a variety of involuntary motor responses. Each side of the reticular formation contains a pair of nuclei, the **red nucleus** and the *substantia nigra*. The red nucleus is provided

FIGURE 15-15
The Hypothalamus. (a) Sagittal section through the brain, showing the location of the hypothalamus and its relationship to other brain stem components. (b) Midsagittal section through the diencephalon of human brain. (c) Enlarged view of the hypothalamus, showing the locations of major nuclei and centers. Functions for these centers are summarized in Table 15-7. See MRI SCAN 1e, p. 741.

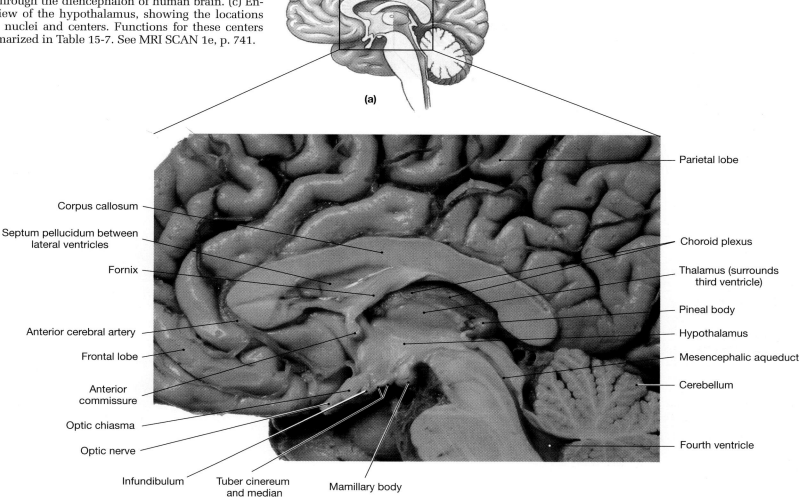

(a)

Parietal lobe

Corpus callosum

Septum pellucidum between lateral ventricles

Fornix

Choroid plexus

Thalamus (surrounds third ventricle)

Anterior cerebral artery

Pineal body

Frontal lobe

Hypothalamus

Anterior commissure

Mesencephalic aqueduct

Cerebellum

Optic chiasma

Optic nerve

Fourth ventricle

Infundibulum

Tuber cinereum and median eminence

Mamillary body

(b)

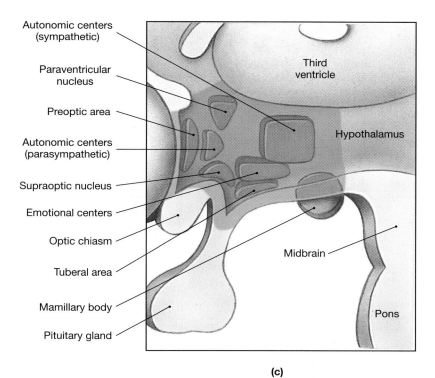

Autonomic centers (sympathetic)

Paraventricular nucleus

Preoptic area

Autonomic centers (parasympathetic)

Supraoptic nucleus

Emotional centers

Optic chiasm

Tuberal area

Mamillary body

Pituitary gland

Third ventricle

Hypothalamus

Midbrain

Pons

(c)

TABLE 15-7	Components and Functions of the Hypothalamus
Region/Nucleus	*Functions*
Hypothalamus in general	Controls autonomic functions; sets appetitive drives (thirst, hunger, sexual desire) and behaviors; sets emotional states (with limbic system); integrates with endocrine system (see Chapter 19)
Supraoptic nucleus	Secretes antidiuretic hormone, restricting water loss at the kidneys
Paraventricular nucleus	Secretes oxytocin, stimulating smooth muscle contractions in uterus and mammary glands
Preoptic area	Regulates body temperature via control of autonomic centers in the medulla
Tuberal area	Produces inhibitory and releasing hormones that target endocrine cells of the anterior pituitary
Autonomic centers	Control heart rate and blood pressure via regulation of autonomic centers in the medulla
Mamillary bodies	Control feeding reflexes (licking, swallowing, etc.)

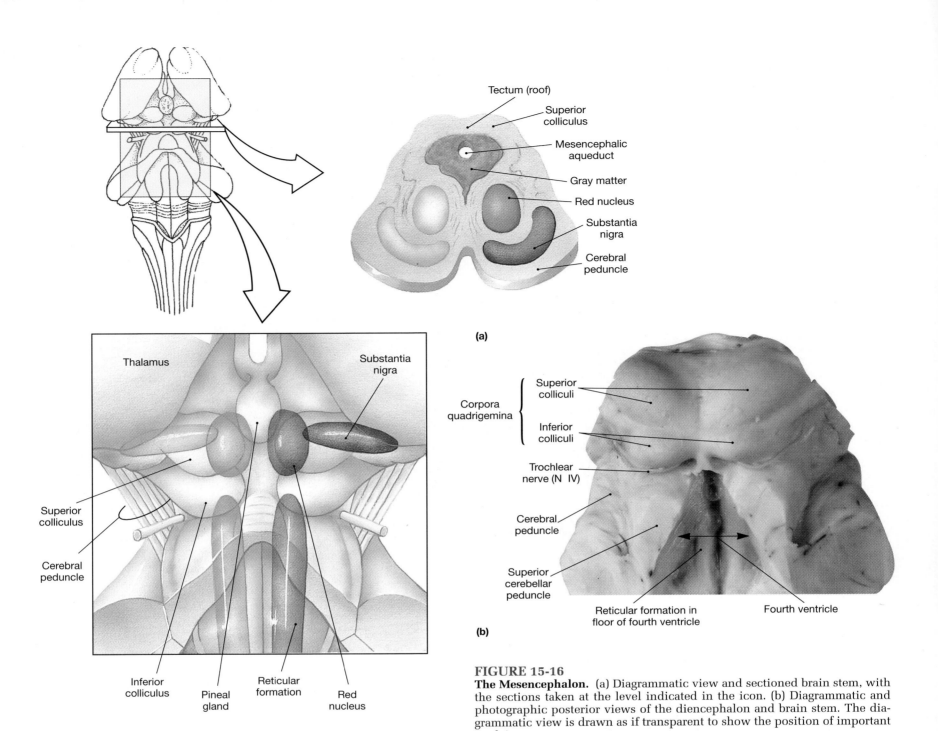

FIGURE 15-16
The Mesencephalon. (a) Diagrammatic view and sectioned brain stem, with the sections taken at the level indicated in the icon. (b) Diagrammatic and photographic posterior views of the diencephalon and brain stem. The diagrammatic view is drawn as if transparent to show the position of important nuclei.

TABLE 15-8 Components and Functions of the Mesencephalon (Midbrain)

Subdivision	Region/Nucleus	Functions
GRAY MATTER **Tectum (roof)**	Superior colliculus	Integrates visual information with other sensory inputs; initiates involuntary motor responses
	Inferior colliculus	Relays auditory information to medial geniculate
Walls and floor	Red nucleus	Involuntary control of muscle tone and posture
	Substantia nigra	Regulates activity in the cerebral nuclei
	Reticular formation (headquarters)	Automatic processing of incoming sensations and outgoing motor commands; can initiate involuntary motor responses to stimuli, maintenance of consciousness
	Other nuclei/centers	Nuclei associated with two cranial nerves (N III, IV)
WHITE MATTER	Cerebral peduncles	Connect primary motor cortex with motor neurons in brain and spinal cord; carry ascending sensory information to thalamus

with numerous blood vessels, giving it a rich red coloration. This nucleus integrates information from the cerebrum and cerebellum and issues involuntary motor commands concerned with the maintenance of muscle tone and posture. The **substantia nigra** (NĪ-grah; "black") lies lateral to the red nucleus. The gray matter in this region contains darkly pigmented cells, giving it a black color. The substantia nigra plays an important role in regulating the motor output of the cerebral nuclei.

The nerve fiber bundles on the ventrolateral surfaces of the mesencephalon (Figures 15-13a,b, p. 382, and 15-16b●) are the **cerebral peduncles** (*peduncles*, little feet). They contain ascending fibers headed for thalamic nuclei and the descending fibers of the pyramidal system that carry voluntary motor commands from the primary motor cortex of each cerebral hemisphere.

■ CLINICAL BRIEF
The Cerebral Nuclei and Parkinson's Disease

The cerebral nuclei contain two discrete populations of neurons. One group stimulates motor neurons by releasing acetylcholine (ACh), and the other inhibits motor neurons by the release of the neurotransmitter *gamma-aminobutyric acid*, or *GABA*. Under normal conditions, the excitatory neurons remain inactive, and the descending tracts are responsible primarily for inhibiting motor neuron activity. The excitatory neurons are quiet because they are continually exposed to the inhibitory effects of the neurotransmitter *dopamine*. This compound is manufactured by neurons in the substantia nigra and transported along axons to synapses in the cerebral nuclei. If the ascending tract or the dopamine-producing neurons are damaged, this inhibition is lost, and the excitatory neurons become increasingly active. This increased activity produces the motor symptoms of **Parkinson's disease**, or *paralysis agitans*.

Parkinson's disease is characterized by a pronounced increase in muscle tone. Voluntary movements become hesitant and jerky, for a movement cannot occur until one muscle group manages to overpower its antagonists. Individuals with Parkinson's disease show spasticity during voluntary movement and a continual **tremor** when at rest. A tremor represents a tug of war between antagonistic muscle groups that produces a background shaking of the limbs. Individuals with Parkinson's disease also have difficulty starting voluntary movements. Even changing one's facial expression requires intense concentration, and the individual acquires a blank, static expression. Finally, the positioning and preparatory adjustments normally performed automatically no longer occur. Every aspect of each movement must be voluntarily controlled, and the extra effort requires intense concentration that may prove tiring and extremely frustrating. In the late stages of this condition, other CNS effects, such as depression and hallucinations, often appear.

Providing the cerebral nuclei with dopamine can significantly reduce the symptoms for two-thirds of Parkinson's patients. Dopamine cannot cross the blood-brain barrier, and the most common treatment involves the oral administration of the drug L-DOPA (*levodopa*), a related compound that crosses the cerebral capillaries and is converted to dopamine. Surgery to control Parkinson's symptoms focuses on the destruction of large areas within the cerebral nuclei or thalamus to control the motor symptoms of tremor and rigidity. Transplantation of tissues that produce dopamine or related compounds directly into the cerebral nuclei is one method attempted as a cure. The transplantation of fetal brain cells into the cerebral nuclei of adult brains has slowed or even reversed the course of the disease in a significant number of patients.

THE METENCEPHALON

The metencephalon has two components: the *cerebellum* and the *pons*. The cerebellum, which is not part of the brain stem, will be described first.

The Cerebellum
(Figures 15-3a/15-8/15-17, Table 15-9)

The cerebellum (Figures 15-3a, p. 369, 15-8, p. 374, and 15-17●) has a complex, highly convoluted surface composed of neural cortex. The folds, or **folia** (FŌ-lē-a), of the surface are less prominent than the gyri of the cerebral hemispheres. The **anterior** and **posterior lobes** are separated by the **primary fissure**. Along the midline a narrow band of cortex known as the **vermis** (VER-mis; "worm") separates the **cerebellar hemispheres** of the posterior lobe. Slender **flocculonodular** (flok-ū-lō-NOD-ū-lar) **lobes** lie anterior and inferior to the cerebellar hemisphere.

The cerebellar cortex contains huge, highly branched **Purkinje** (pur-KIN-jē) **cells** (Figure 15-17b●). Purkinje cells have large pear-shaped somas, which have large numerous dendrites fanning out into the gray matter (neural cortex) of the cerebellar cortex. Axons project from the basal portion of the cell into the white matter to reach the cerebellar nuclei. Internally, the white matter of the cerebellum forms a branching array that in sectional view resembles a tree. Anatomists call it the **arbor vitae**, or "tree of life." The cerebellum receives proprioceptive information (position sense) from the spinal cord, and monitors all proprioceptive, visual, tactile, balance, and auditory sensations received by the brain. Information concerning motor commands issued by the cerebral cortex reaches the cerebellum indirectly, via nuclei in the pons. A relatively small portion of the afferent fibers synapse within **cerebellar nuclei** before projecting to the cerebellar cortex. Most axons carrying sensory information do not synapse in the cerebellar nuclei, but pass through the deeper layers of the cerebellar cortex to end near the cortical surface. There they synapse with the dendritic processes of the Purkinje cells. Tracts containing the axons of Purkinje cells then relay motor commands to nuclei within the cerebrum and brain stem. ▼*Cerebellar Dysfunction [p. 761]*

Tracts that link the cerebellum with the brain stem, cerebrum, and spinal cord leave the cerebellar hemispheres as the *superior*, *middle*, and *inferior cerebellar peduncles*. The **superior cerebellar peduncles** link the cerebellum with nuclei in the midbrain, diencephalon, and cerebrum. The **middle cerebellar peduncles** are connected to a broad band of fibers that cross the ventral surface of the pons at right angles to the axis of the brain stem. The middle cerebellar peduncles also connect the cerebellar hemispheres with sensory and motor nuclei in the pons. The **inferior cerebellar peduncles** permit communication between the cerebellum and nuclei in the medulla oblongata and carry ascending and descending cerebellar tracts from the spinal cord.

The cerebellum is an automatic processing center that has two primary functions:

- *Adjusting the postural muscles of the body*: The cerebellum coordinates rapid, automatic adjustments that maintain balance and equilibrium. These alterations in muscle tone and position are made by modifying the activity of the red nucleus.

- *Programming and fine-tuning voluntary and involuntary movements*: The cerebellum stores memories of learned movement patterns. These functions are performed indirectly, by regulating activity along both pyramidal and extrapyramidal motor pathways at the cerebral cortex, cerebral nuclei, and motor centers in the brain stem.

FIGURE 15-17

The Cerebellum. (a) Posterior and superior surface of the cerebellum, showing major anatomical landmarks and regions. (b) Sagittal view of the cerebellum, showing the arrangement of gray matter and white matter. Purkinje cells are seen in the photograph; these large neurons are found in the cerebellar cortex. (LM × 120)

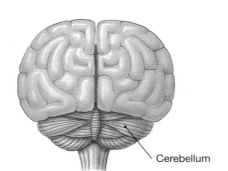

Cerebellum

Posterior view of brain

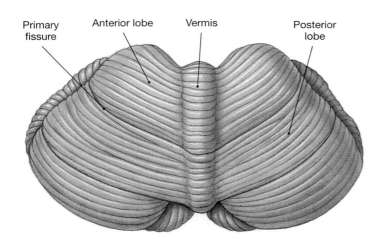

Primary fissure — Anterior lobe — Vermis — Posterior lobe

Superior view of cerebellum

Soma of Purkinje cell

Dendrites projecting into the gray matter of the cerebellum

Axons of Purkinje cells projecting into the white matter of the cerebellum

Purkinje cells

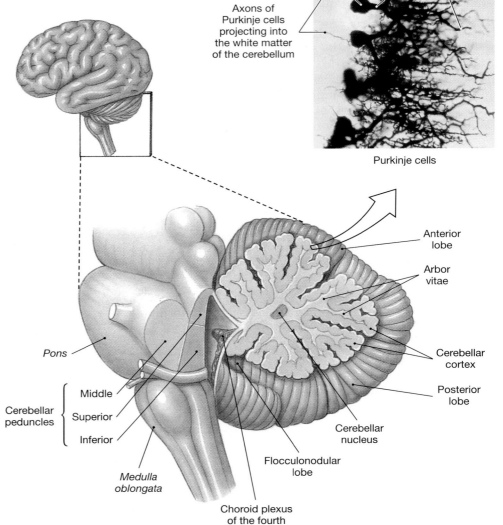

Anterior lobe

Arbor vitae

Cerebellar cortex

Posterior lobe

Pons

Cerebellar peduncles { Middle / Superior / Inferior }

Cerebellar nucleus

Flocculonodular lobe

Medulla oblongata

Choroid plexus of the fourth ventricle

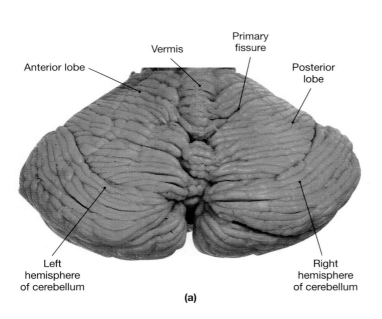

Anterior lobe — Vermis — Primary fissure — Posterior lobe

Left hemisphere of cerebellum

Right hemisphere of cerebellum

(a)

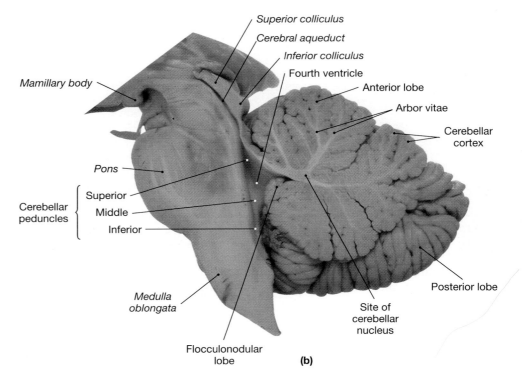

Superior colliculus

Cerebral aqueduct

Inferior colliculus

Fourth ventricle

Anterior lobe

Arbor vitae

Cerebellar cortex

Mamillary body

Pons

Cerebellar peduncles { Superior / Middle / Inferior }

Posterior lobe

Site of cerebellar nucleus

Medulla oblongata

Flocculonodular lobe

(b)

TABLE 15-9 Components of the Cerebellum

Subdivision	Region/Nucleus	Functions
Gray matter	Cerebellar cortex	Involuntary coordination and control of ongoing movements of body parts
	Cerebellar nuclei	As above
White matter	Arbor vitae	Connects cerebellar cortex and nuclei with cerebellar peduncles
	Cerebellar peduncles Superior	Link the cerebellum with mesencephalon, diencephalon, and cerebrum
	Middle	Contain transverse fibers and carry communications between the cerebellum and pons
	Inferior	Link the cerebellum with the medulla and spinal cord

The structures of the cerebellum and their functions are summarized in Table 15-9.

The Pons
(Figures 15-12/15-13/15-18, Table 15-10)

The pons links the cerebellar hemispheres with the mesencephalon, diencephalon, cerebrum, and spinal cord. Important features and regions are indicated in Figures 15-12, p. 380, 15-13, p. 382, and 15-18●; structures are detailed in Table 15-10. The pons contains:

- *Sensory and motor nuclei for four cranial nerves* (N V, N VI, N VII, and N VIII). These cranial nerves innervate the jaw muscles, the anterior surface of the face, one of the extrinsic eye muscles (the lateral rectus), and the sense organs of the inner ear.

- *Nuclei concerned with the involuntary control of respiration.* On each side of the brain, the reticular formation in this region contains two respiratory centers, the *apneustic center* and the *pneumotaxic center*. These centers modify the activity of the *respiratory rhythmicity center* in the medulla oblongata.

- *Nuclei that process and relay cerebellar commands arriving over the middle cerebellar peduncles.*

- *Ascending, descending, and transverse tracts.* The longitudinal tracts interconnect other portions of the CNS. The middle cerebellar peduncles are connected to the **transverse fibers** of the pons that cross its anterior surface. These fibers permit communication between the cerebellar hemispheres of opposite sides.

THE MEDULLA OBLONGATA (MYELENCEPHALON)
(Figures 15-1/15-8/15-12b/15-13/15-19, Table 15-11)

The medulla oblongata, or *medulla*, is continuous with the spinal cord. The external appearance of the medulla oblongata should be examined in Figures 15-1, p. 366, 15-8, p. 374, and

15-13●, p. 382. The important nuclei and centers are diagrammed in Figure 15-19● and detailed in Table 15-11.

Figure 15-12b●, p. 380, shows the medulla oblongata in midsagittal section. The caudal portion of the medulla resembles the spinal cord in having a rounded shape and a narrow central canal. Closer to the pons, the central canal becomes enlarged and continuous with the fourth ventricle.

The medulla oblongata physically connects the brain with the spinal cord, and many of its functions are directly related to this connection. For example, all communication between the brain and spinal cord involves tracts that ascend or descend through the medulla oblongata.

Nuclei in the medulla oblongata may be (1) relay stations along sensory or motor pathways, (2) sensory or motor nuclei associated with cranial nerves connected to the medulla oblongata, or (3) nuclei associated with the autonomic control of visceral activities.

1. *Relay stations*: Ascending tracts may synapse in sensory or motor nuclei that act as relay stations and processing centers. For example, the **nucleus gracilis** and the **nucleus cuneatus** pass somatic sensory information to the thalamus, and the **olivary nuclei** relay information from the spinal cord, the cerebral cortex, diencephalon, and brain stem to the cerebellar cortex. The bulk of these nuclei create the **olives**, prominent bulges along the ventrolateral surface of the medulla oblongata (Figure 15-19●).

2. *Nuclei of cranial nerves*: The medulla oblongata contains sensory and motor nuclei associated with five of the cranial nerves (N VIII, N IX, N X, N XI, and N XII). These cranial nerves innervate muscles of the pharynx, neck, and back, as well as visceral organs of the thoracic and peritoneal cavities.

3. *Autonomic nuclei*: The reticular formation in the medulla oblongata contains nuclei and centers responsible for the regulation of vital autonomic functions. These **reflex centers** receive inputs from cranial nerves, the cerebral cortex, the diencephalon, and the brain stem, and their output controls or adjusts the activities of one or more peripheral systems. Major centers include:

- The **cardiovascular centers** adjust heart rate, the strength of cardiac contractions, and the flow of blood through peripheral tissues. On functional grounds, the cardiovascular centers may be subdivided into **cardiac** (*kardia*, heart) and **vasomotor** (*vas*, canal) centers, but their anatomical boundaries are difficult to determine.

- The **respiratory rhythmicity centers** set the basic pace for respiratory movements, and their activity is regulated by inputs from the apneustic and pneumotaxic centers of the pons.

√ **What area of the diencephalon would be stimulated by changes in body temperature?**

√ **In what part of the brain would you find a worm (vermis) and a tree (arbor vitae)?**

√ **The medulla oblongata is one of the smallest sections of the brain, yet damage there can cause death, whereas similar damage in the cerebrum might go unnoticed. Why?**

FIGURE 15-18
The Pons

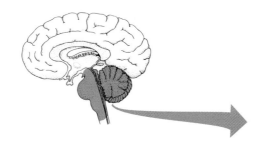

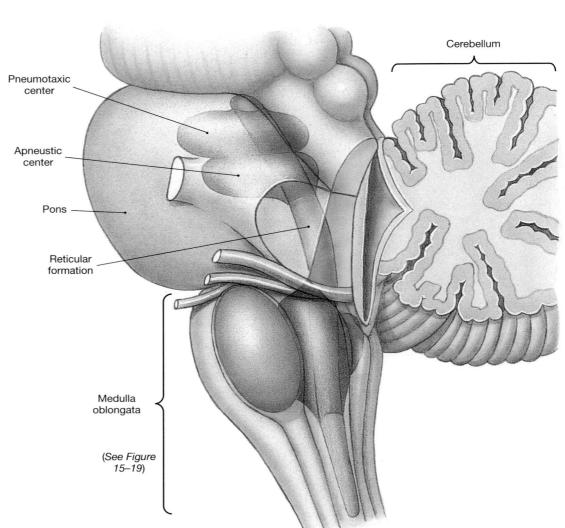

Cerebellum

Pneumotaxic center

Apneustic center

Pons

Reticular formation

Medulla oblongata

(See Figure 15–19)

TABLE 15-10 Components of the Pons

Subdivisions	Region/Nucleus	Functions
Gray matter	Respiratory centers	Modify output of respiratory centers in the medulla
	Other nuclei/centers	Nuclei associated with four cranial nerves and cerebellum
White matter	Transverse fibers	Interconnect cerebellar hemispheres
	Ascending and descending tracts	Interconnect other portions of CNS

TABLE 15-11 Components and Functions of the Medulla Oblongata

Subdivision	Region/Nucleus	Functions
Gray matter	Nucleus gracilis Nucleus cuneatus	Relay somatic sensory information to the ventral posterior nuclei of the thalamus
	Olivary nuclei	Relay information from the red nucleus, other midbrain centers, and the cerebral cortex to the vermis of cerebellum
	Reflex centers Cardiac centers	Regulate heart rate and force of contraction
	Vasomotor center	Regulates distribution of blood flow
	Respiratory rhythmicity center	Sets the pace of respiratory movements
	Other nuclei/centers	Sensory and motor nuclei of five cranial nerves Nuclei relaying ascending sensory information from the spinal cord to higher centers
White matter	Ascending and descending tracts	Link the brain with the spinal cord

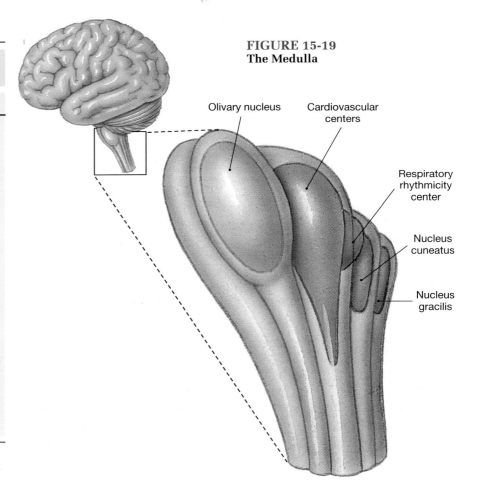

FIGURE 15-19
The Medulla

Olivary nucleus

Cardiovascular centers

Respiratory rhythmicity center

Nucleus cuneatus

Nucleus gracilis

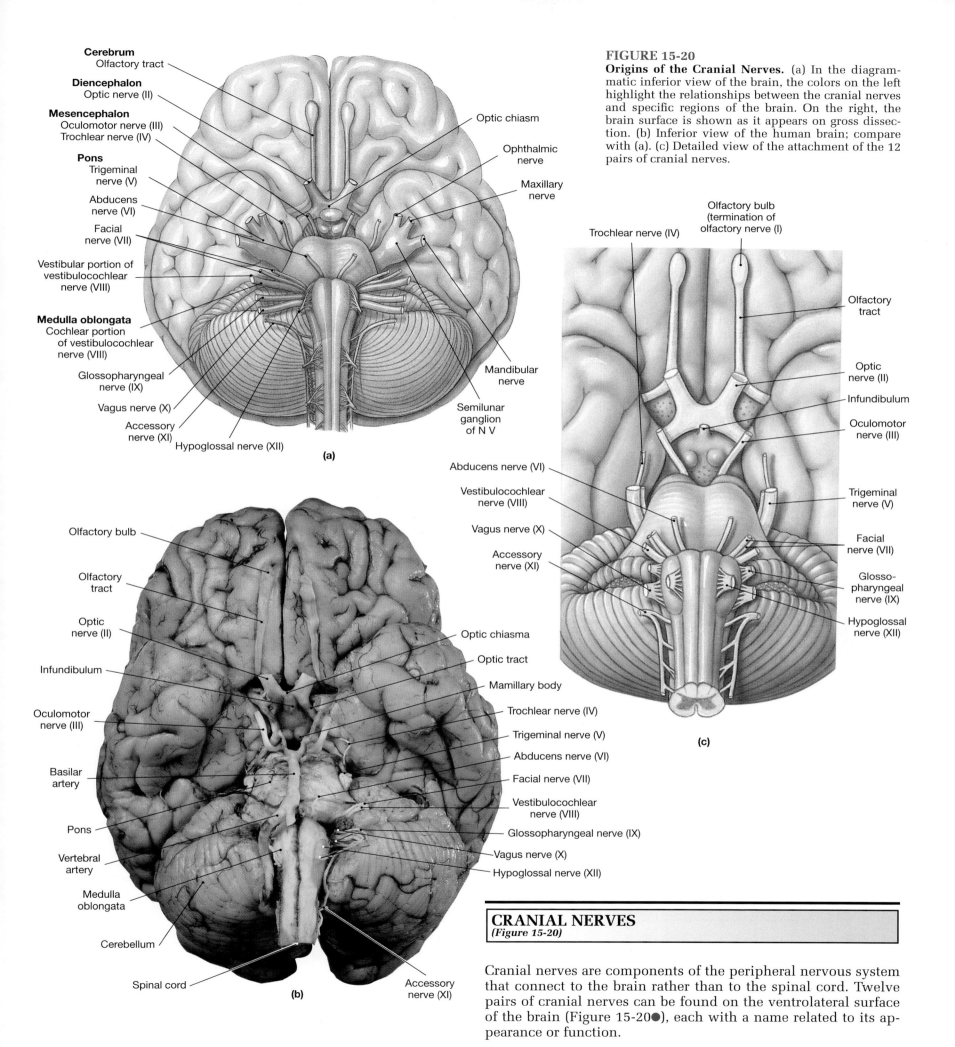

Cerebrum
Olfactory tract

Diencephalon
Optic nerve (II)

Mesencephalon
Oculomotor nerve (III)
Trochlear nerve (IV)

Pons
Trigeminal nerve (V)

Abducens nerve (VI)

Facial nerve (VII)

Vestibular portion of vestibulocochlear nerve (VIII)

Medulla oblongata
Cochlear portion of vestibulocochlear nerve (VIII)

Glossopharyngeal nerve (IX)

Vagus nerve (X)

Accessory nerve (XI)

Hypoglossal nerve (XII)

Optic chiasm

Ophthalmic nerve

Maxillary nerve

Mandibular nerve

Semilunar ganglion of N V

(a)

Olfactory bulb

Olfactory tract

Optic nerve (II)

Infundibulum

Oculomotor nerve (III)

Basilar artery

Pons

Vertebral artery

Medulla oblongata

Cerebellum

Spinal cord

(b)

Optic chiasma

Optic tract

Mamillary body

Trochlear nerve (IV)

Trigeminal nerve (V)

Abducens nerve (VI)

Facial nerve (VII)

Vestibulocochlear nerve (VIII)

Glossopharyngeal nerve (IX)

Vagus nerve (X)

Hypoglossal nerve (XII)

Accessory nerve (XI)

Trochlear nerve (IV)

Olfactory bulb (termination of olfactory nerve (I)

Olfactory tract

Optic nerve (II)

Infundibulum

Oculomotor nerve (III)

Abducens nerve (VI)

Vestibulocochlear nerve (VIII)

Vagus nerve (X)

Accessory nerve (XI)

Trigeminal nerve (V)

Facial nerve (VII)

Glosso-pharyngeal nerve (IX)

Hypoglossal nerve (XII)

(c)

FIGURE 15-20
Origins of the Cranial Nerves. (a) In the diagrammatic inferior view of the brain, the colors on the left highlight the relationships between the cranial nerves and specific regions of the brain. On the right, the brain surface is shown as it appears on gross dissection. (b) Inferior view of the human brain; compare with (a). (c) Detailed view of the attachment of the 12 pairs of cranial nerves.

CRANIAL NERVES
(Figure 15-20)

Cranial nerves are components of the peripheral nervous system that connect to the brain rather than to the spinal cord. Twelve pairs of cranial nerves can be found on the ventrolateral surface of the brain (Figure 15-20●), each with a name related to its appearance or function.

Cranial nerves are numbered according to their position along the longitudinal axis of the brain, beginning at the cerebrum. Roman numerals are usually used, either alone or with the prefix N or CN. We will use the abbreviation N, which is generally preferred by neuroanatomists and clinical neurologists. CN, an equally valid abbreviation, is preferred by comparative anatomists.

Each cranial nerve attaches to the brain near the associated sensory or motor nuclei. The sensory nuclei act as switching centers, with the postsynaptic neurons relaying the information to other nuclei or to processing centers within the cerebral or cerebellar cortex. In a similar fashion, the motor nuclei receive convergent inputs from higher centers or from other nuclei along the brain stem.

The next section classifies cranial nerves as primarily sensory, special sensory, motor, or mixed (sensory and motor). This is a useful method of classification, but it is based on the primary function, and a cranial nerve can have important secondary functions. Two examples are worth noting:

1. As elsewhere in the PNS, a nerve containing tens of thousands of motor fibers to a skeletal muscle will also contain sensory fibers from proprioceptors in that muscle. These sensory fibers are assumed to be present, but are ignored in the primary classification of the nerve.

2. Regardless of their other functions, several cranial nerves (N III, N VII, N IX, and N X) distribute autonomic fibers to peripheral ganglia, just as spinal nerves deliver them to ganglia along the spinal cord. The presence of small numbers of autonomic fibers will be noted (and discussed further in Chapter 17) but ignored in the classification of the nerve.

The Olfactory Nerve (N I)
(Figures 15-20/15-21)

> ***Primary function:*** Special sensory (smell)
> ***Origin:*** Receptors of olfactory epithelium
> ***Passes through:*** Cribriform plate of ethmoid bone ∞ [p. 144]
> ***Destination:*** Olfactory bulbs

The first pair of cranial nerves (Figure 15-21●) carries special sensory information responsible for the sense of smell. The olfactory receptors are specialized neurons in the epithelium covering the roof of the nasal cavity, the superior nasal conchae, and the superior portions of the nasal septum. Axons from these sensory neurons come together to form 20 or more bundles that penetrate the cribriform plate of the ethmoid. Almost at once they enter the **olfactory bulbs**, neural masses that lie on either side of the crista galli. The olfactory afferents synapse within the olfactory bulbs, and the axons of the postsynaptic neurons proceed to the cerebrum along the slender **olfactory tracts** (Figures 15-20 and 15-21●).

Because the olfactory nerves looked like a typical peripheral nerve, anatomists a hundred years ago called the olfactory tracts the first cranial nerve. Later studies demonstrated that the olfactory tracts and bulbs are actually part of the cerebrum, but by then the numbering system was already firmly established, and anatomists were left with a forest of tiny olfactory nerve bundles lumped together as N I.

The olfactory nerves are the only cranial nerves attached directly to the cerebrum. The rest originate or terminate within nuclei of the diencephalon or brain stem. As a result, all sensory information except olfactory sensations must be relayed across synapses in the thalamus before reaching the cerebrum.

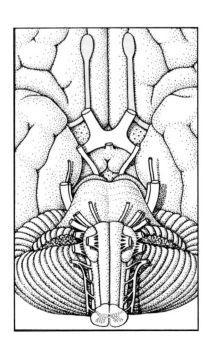

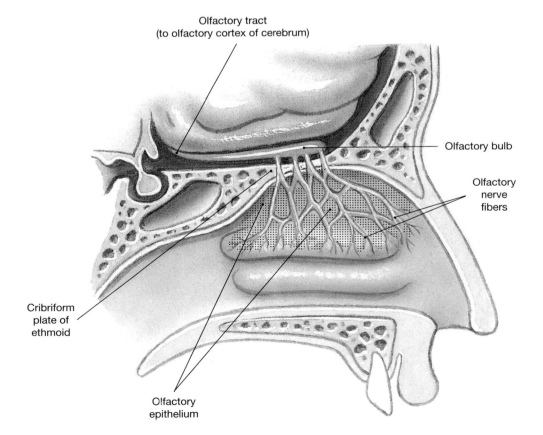

FIGURE 15-21
The Olfactory Nerve

The Optic Nerve (N II)
(Figures 15-20/15-22)

Primary function: Special sensory (vision)
Origin: Retina of eye
Passes through: Optic foramen of sphenoid bone ∞ [p. 143]
Destination: Diencephalon via optic chiasm

The **optic nerves** (N II) carry visual information from special sensory ganglia in the eyes. These nerves, diagrammed in Figure 15-22●, contain about 1 million sensory nerve fibers. These nerves pass through the optic foramina of the sphenoid before converging at the ventral and anterior margin of the dien-

cephalon, at the **optic chiasm** (*chiasma*, a crossing). At the optic chiasm, some of the fibers from each optic nerve cross over to the opposite side of the brain. The reorganized axons continue toward the lateral geniculate nuclei of the thalamus as the **optic tracts** (Figures 15-20 and 15-22●). After synapsing in the lateral geniculates, projection fibers deliver the information to the occipital lobe of the brain. This arrangement results in each cerebral hemisphere receiving visual information from the lateral half of the retina of the eye on that side and from the medial half of the retina of the eye on the opposite side. A relatively small number of axons in the optic tracts bypass the lateral geniculates and synapse in the superior colliculus of the midbrain. This pathway will be considered in Chapter 18.

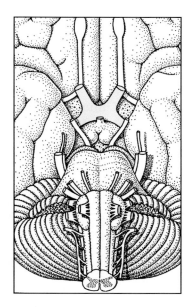

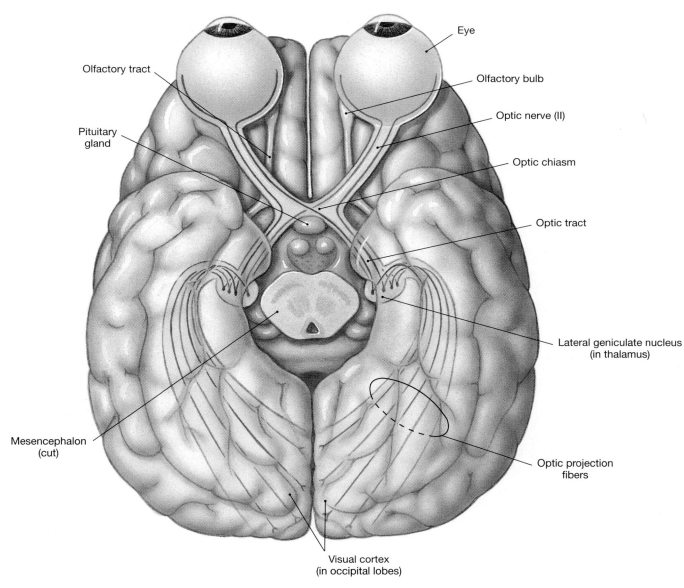

FIGURE 15-22
The Optic Nerve

The Oculomotor Nerve (N III)
(Figures 15-20/15-23)

> **Primary function:**   Motor, eye movements
>
> **Origin:**   Mesencephalon
>
> **Passes through:**   Superior orbital fissure of sphenoid bone ∞ [p. 143]
>
> **Destination:**   *Somatic motor*: superior, inferior, and medial rectus muscles; the inferior oblique muscle; the levator palpebrae superioris muscle ∞ [p. 255]; *Visceral motor*: intrinsic eye muscles

The mesencephalon contains the motor nuclei controlling the third and fourth cranial nerves. The **oculomotor nerves** (N III) innervate all but two of the extrinsic eye muscles. ∞ [p. 260] These cranial nerves emerge from the ventral surface of the mesencephalon (Figure 15-20●, p. 391) and penetrate the posterior orbital wall at the superior orbital fissure. ∞ [p. 143] The oculomotor nerve (Figure 15-23●) controls four of the six extrinsic eye muscles and the levator palpebrae superioris, the muscle that raises the upper eyelid.

The oculomotor nerve also delivers preganglionic autonomic fibers to neurons of the **ciliary ganglion**. The ganglionic neurons control intrinsic eye muscles. These muscles change the diameter of the pupil, adjusting the amount of light entering the eye, and change the shape of the lens to focus images on the retina.

The Trochlear Nerve (N IV)
(Figures 15-20/15-23)

> **Primary function:**   Motor, eye movements
>
> **Origin:**   Mesencephalon
>
> **Passes through:**   Superior orbital fissure of sphenoid bone ∞ [p. 143]
>
> **Destination:**   Superior oblique muscle ∞ [p. 260]

The **trochlear nerve** (TRŌK-lē-ar; *trochlea*, pulley), smallest of the cranial nerves, innervates the superior oblique muscle of the eye (Figure 15-23●). The motor nucleus lies in the ventrolateral portion of the mesencephalon, but the fibers emerge from the surface of the tectum to enter the orbit through the superior orbital fissure (Figure 15-20●, p. 391). The name "trochlear nerve" should remind you that the innervated muscle passes through a ligamentous sling, or *trochlea*, on its way to its insertion on the surface of the eye. ∞ [p. 260]

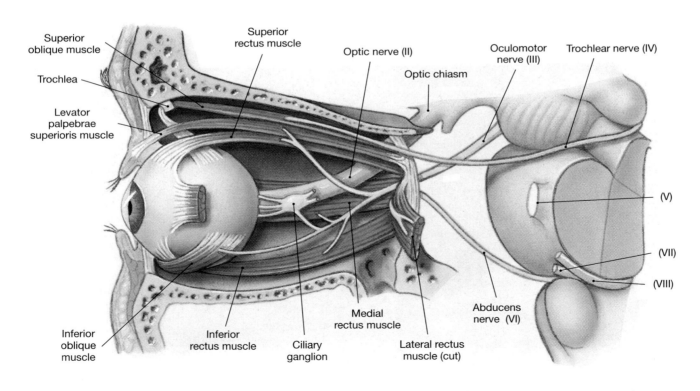
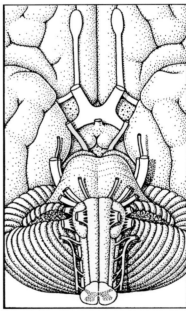

FIGURE 15-23
Cranial Nerves Controlling the Extrinsic Eye Muscles

The Trigeminal Nerve (N V)
(Figures 15-20/15-24)

Primary function: Mixed (sensory and motor); ophthalmic and maxillary branches sensory, mandibular branch mixed

Origin: *Ophthalmic branch* (sensory): orbital structures, nasal cavity, skin of forehead, upper eyelid, eyebrow, nose (part)

Maxillary branch (sensory): lower eyelid, upper lip, gums, and teeth; cheek; nose, palate, and pharynx (part)

Mandibular branch (mixed): sensory from lower gums, teeth, and lips; palate and tongue (part); motor from motor nuclei of pons (Figure 15-20●, p. 391)

Passes through: Ophthalmic branch via superior orbital fissure, maxillary branch via foramen rotundum, mandibular branch via foramen ovale ∞ [p. 143]

Destination: Ophthalmic and maxillary branches to sensory nuclei in pons; mandibular branch innervates muscles of mastication ∞ [p. 255]

The pons contains the nuclei associated with three cranial nerves (N V, N VI, and N VII) and contributes to the control of a fourth (N VIII). The **trigeminal** (trī-JEM-i-nal) **nerve** (Figure 15-24●) is the largest cranial nerve. This mixed nerve provides sensory information from the head and face and motor control over the muscles of mastication. Sensory (dorsal) and motor (ventral) roots originate on the lateral surface of the pons. The sensory branch is larger, and the enormous **semilunar ganglion** (*trigeminal ganglion*) contains the cell bodies of the sensory neurons. As the name implies, the trigeminal has three major branches; the relatively small motor root contributes to only one of the three.

Branch 1. The **ophthalmic branch** of the trigeminal nerve is purely sensory. This nerve innervates orbital structures, the nasal cavity and sinuses, and the skin of the forehead, eyebrows, eyelids, and nose. It leaves the cranium via the superior orbital fissure, then branches within the orbit.

Branch 2. The **maxillary branch** of the trigeminal nerve is also purely sensory. It supplies the lower eyelid, upper lip, cheek, and nose. Deeper sensory structures of the upper gums and teeth, the palate, and portions of the pharynx are also innervated by the maxillary nerve branch. The maxillary branch leaves the cranium at the foramen rotundum, entering the floor of the orbit through the inferior orbital fissure. A major branch of the maxillary, the *infraorbital nerve*, passes through the infraorbital foramen to supply adjacent portions of the face.

Branch 3. The **mandibular branch** is the largest branch of the trigeminal nerve, and it carries all of the fibers of the motor root. This branch exits the cranium through the foramen ovale. The motor components of the mandibular nerve innervate the mus-

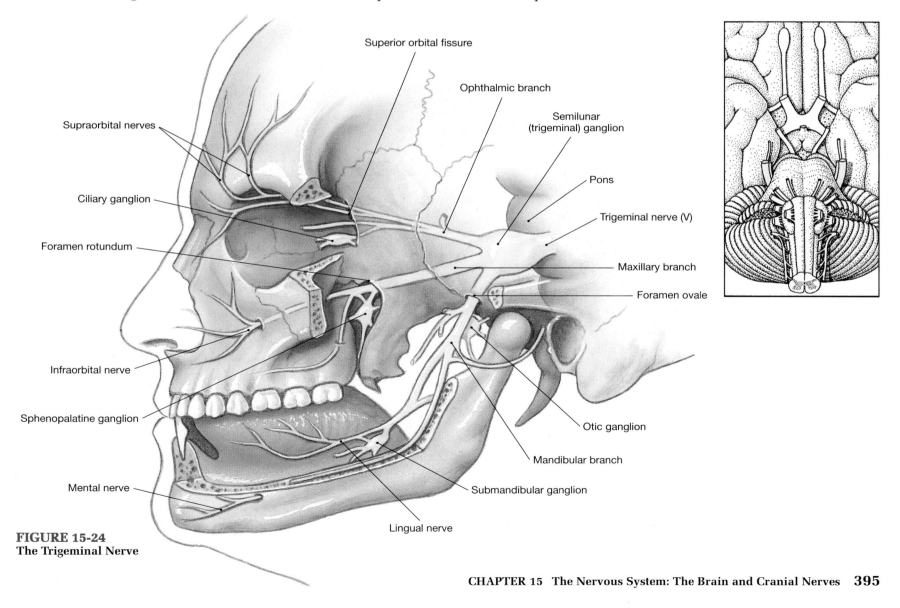

FIGURE 15-24
The Trigeminal Nerve

cles of mastication. The sensory fibers carry proprioceptive information from those muscles and monitor (1) the skin of the temples; (2) the lateral surfaces, gums, and teeth of the mandible; (3) the salivary glands; and (4) the anterior portions of the tongue.

The trigeminal nerve branches are associated with the *ciliary*, *sphenopalatine*, *submandibular*, and *otic* ganglia. These are autonomic ganglia whose neurons innervate structures of the face. The trigeminal nerve does not contain visceral motor fibers; the connections between the trigeminal nerve and these ganglia occur (1) where trigeminal fibers are passing through the ganglia without synapsing to reach other destinations, (2) where postganglionic autonomic fibers leave the ganglion and travel with the trigeminal nerve before continuing to peripheral structures, or (3) where branches of another cranial nerve, such as the *facial nerve*, travel with the trigeminal nerve on their way to a ganglion. The ciliary ganglion was discussed earlier (p. 394), and the other ganglia will be detailed below, with the branches of the *facial nerve* (N VII).

The Abducens Nerve (N VI)
(Figures 15-20/15-23)

Primary function: Motor, eye movements
Origin: Pons
Passes through: Superior orbital fissure of sphenoid bone ∞ [p. 143]
Destination: Lateral rectus muscle ∞ [p. 260]

The **abducens** (ab-DŪ-senz) **nerve** innervates the lateral rectus, the sixth of the extrinsic eye muscles. Innervation of this muscle makes lateral movements of the eyeball possible. The nerve emerges from the inferior surface of the brain at the border between the pons and the medulla (Figure 15-20●, p. 391). It reaches the orbit through the superior orbital fissure in company with the oculomotor and trochlear nerves (Figure 15-23●).

The Facial Nerve (N VII)
(Figures 15-20/15-25)

Primary function: Mixed (sensory and motor)
Origin: Sensory from taste receptors on anterior two-thirds of tongue; motor from motor nuclei of pons

Passes through: Internal acoustic meatus of temporal bone, along internal acoustic canal and facial canal to reach stylomastoid foramen ∞ [p. 142]
Destination: Sensory to sensory nuclei of pons; *Somatic motor*: muscles of facial expression ∞ [p. 255]; *Visceral motor*: lacrimal (tear) gland and nasal mucous glands via sphenopalatine ganglion; submandibular and sublingual salivary glands via submandibular ganglion

The **facial nerve** is a mixed nerve. The cell bodies of the sensory neurons are located in the **geniculate ganglion**, and the motor nuclei are in the pons (Figure 15-20●, p. 391). The sensory and motor roots combine to form a large nerve that passes through the internal acoustic canal of the temporal bone (Figure 15-25●). The nerve then passes through the facial canal to reach the face via the stylomastoid foramen. ∞ [p. 142] The sensory neurons monitor proprioceptors in the facial muscles, provide deep pressure sensations over the face, and receive taste information from receptors along the anterior two-thirds of the tongue. Somatic motor fibers control the superficial muscles of the scalp and face and deep muscles near the ear.

The facial nerve carries preganglionic autonomic fibers to the sphenopalatine and submandibular ganglia.

- **Sphenopalatine ganglion:** The *greater petrosal nerve* innervates the sphenopalatine ganglion. Postganglionic fibers from this ganglion innervate the lacrimal gland and small glands of the nasal cavity and pharynx.
- **Submandibular ganglion:** To reach the submandibular ganglion, autonomic fibers leave the facial nerve and travel along the mandibular branch of the trigeminal nerve. Postganglionic fibers from this ganglion innervate the *submandibular* and *sublingual* (*sub*, under + *lingua*, tongue) salivary glands.

The Vestibulocochlear Nerve (N VIII)
(Figures 15-20/15-26)

Primary function: Special sensory: balance and equilibrium (vestibular branch) and hearing (cochlear branch)
Origin: Monitors receptors of the inner ear (vestibule and cochlea)
Passes through: Internal acoustic canal and meatus of temporal bone ∞ [p. 142]
Destination: Vestibular and cochlear nuclei of pons and medulla (Figure 15-20●, p. 391)

The **vestibulocochlear nerve** is also known as the *acoustic nerve*, the *auditory nerve*, and the *statoacoustic nerve*. We will use the term vestibulocochlear because it indicates the names of

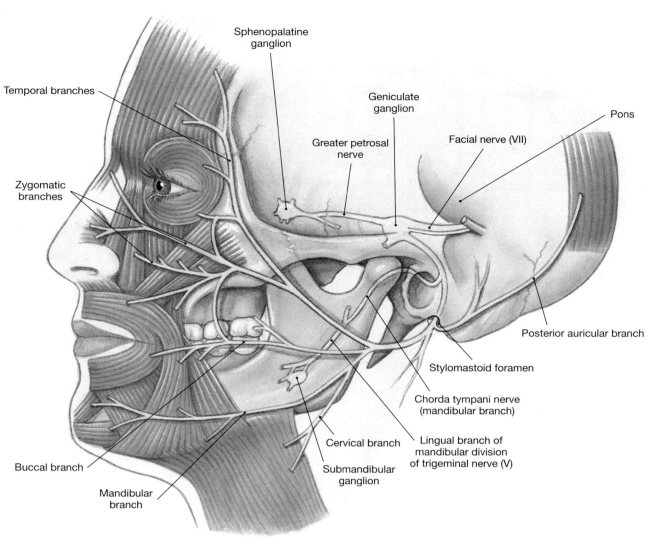

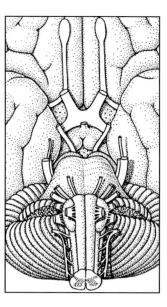

FIGURE 15-25
The Facial Nerve

The following labels appear in Figure 15-25:

Sphenopalatine ganglion

Geniculate ganglion

Temporal branches

Greater petrosal nerve

Facial nerve (VII)

Pons

Zygomatic branches

Posterior auricular branch

Stylomastoid foramen

Chorda tympani nerve (mandibular branch)

Lingual branch of mandibular division of trigeminal nerve (V)

Buccal branch

Cervical branch

Submandibular ganglion

Mandibular branch

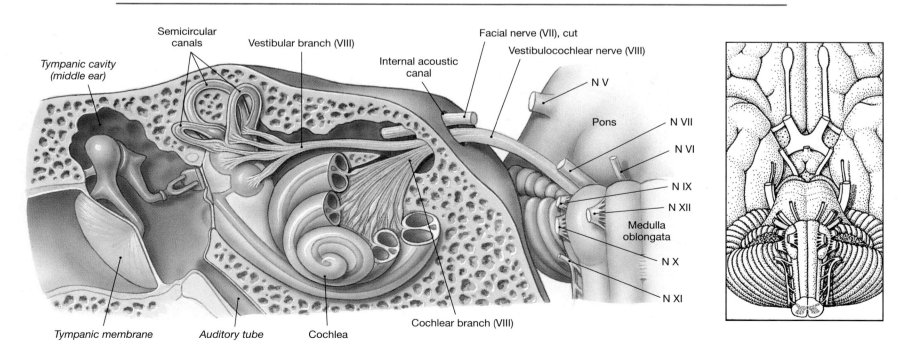

FIGURE 15-26
The Vestibulocochlear Nerve

The following labels appear in Figure 15-26:

Semicircular canals

Vestibular branch (VIII)

Facial nerve (VII), cut

Vestibulocochlear nerve (VIII)

Tympanic cavity (middle ear)

Internal acoustic canal

N V

Pons

N VII

N VI

N IX

N XII

Medulla oblongata

N X

N XI

Tympanic membrane

Auditory tube

Cochlea

Cochlear branch (VIII)

its two major branches: the *vestibular branch* and the *cochlear branch*. The vestibulocochlear nerve lies posterior to the origin of the facial nerve, straddling the boundary between the pons and the medulla (Figure 15-26●). This nerve reaches the sensory receptors of the inner ear by entering the internal acoustic canal in company with the facial nerve. There are two distinct bundles of sensory fibers within the vestibulocochlear nerve. The **vestibular nerve** (*vestibulum*, cavity) originates at the receptors of the *vestibule*, the portion of the inner ear concerned with balance sensations. The sensory neurons are located within an adjacent sensory ganglion, and their axons target the **vestibular nuclei** of the medulla. These afferents convey information concerning position, movement, and balance. The **cochlear nerve** (KOK-lē-ar; *cochlea*, snail shell) monitors the receptors in the cochlea that provide the sense of hearing. The nerve cells are located within a peripheral ganglion, and their axons synapse within the **cochlear nuclei** of the medulla. Axons leaving the vestibular and cochlear nuclei relay the sensory information to other centers or initiate reflexive motor responses. Balance and the sense of hearing will be discussed in Chapter 18.

The Glossopharyngeal Nerve (N IX)
(Figures 15-20/15-27)

Primary function: Mixed (sensory and motor)

Origin: Sensory from posterior one-third of the tongue, part of the pharynx and palate, the carotid arteries of the neck; motor from motor nuclei of medulla oblongata

Passes through: Jugular foramen between occipital and temporal bones ∞ [p. 138]

Destination: Sensory to sensory nuclei of medulla (Figure 15-20●, p. 391); *Somatic motor*: pharyngeal muscles involved in swallowing; *Visceral motor*: parotid salivary gland, via otic ganglion

In addition to the vestibular nucleus of N VIII, the medulla contains the sensory and motor nuclei for the ninth, tenth, eleventh, and twelfth cranial nerves. The **glossopharyngeal nerve** (glos-ō-fah-RIN-jē-al; *glossum*, tongue) innervates the tongue and pharynx. The glossopharyngeal nerve passes through the cranium via the jugular foramen in company with N X and N XI (Figure 15-27●).

The glossopharyngeal is a mixed nerve, but sensory fibers are most abundant. The sensory neurons are in the **superior ganglion** and the **inferior (petrosal) ganglion**. The afferent fibers carry general sensory information from the lining of the pharynx and the soft palate to a nucleus in the medulla. The glossopharyngeal nerve also provides taste sensations from the posterior third of the tongue and has special receptors monitoring the blood pressure and dissolved gas concentrations within major blood vessels.

The somatic motor fibers control the pharyngeal muscles involved in swallowing. Visceral motor fibers synapse in the otic ganglion, and postganglionic fibers innervate the parotid salivary gland of the cheek.

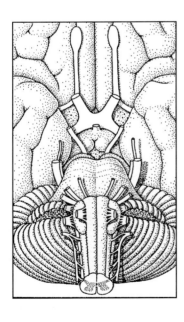

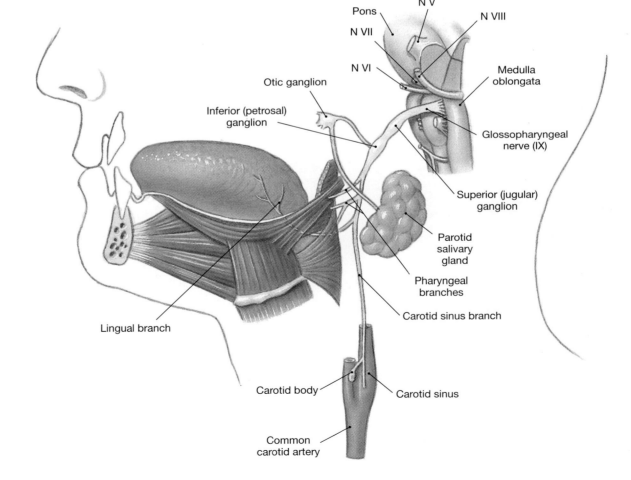

FIGURE 15-27
The Glossopharyngeal Nerve

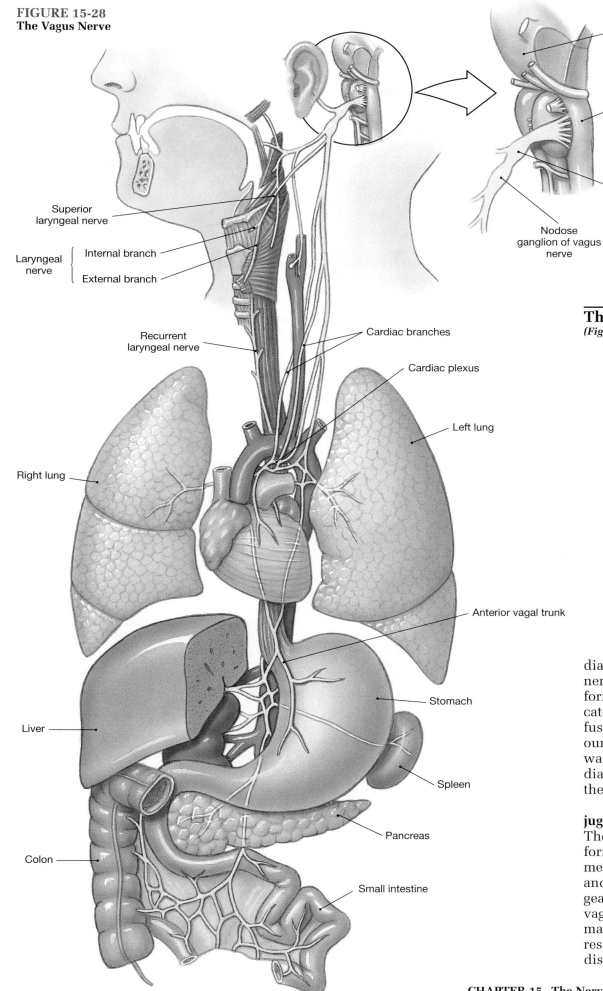

FIGURE 15-28
The Vagus Nerve

Superior laryngeal nerve

Laryngeal nerve {
  Internal branch
  External branch
}

Recurrent laryngeal nerve

Cardiac branches

Cardiac plexus

Left lung

Right lung

Anterior vagal trunk

Liver

Stomach

Spleen

Pancreas

Colon

Small intestine

Pons

Medulla oblongata

Jugular ganglion of vagus nerve

Nodose ganglion of vagus nerve

The Vagus Nerve (N X)
(Figures 15-20/15-28)

Primary function: Mixed (sensory and motor)

Origin: Visceral sensory from pharynx (part), pinna, external auditory canal, diaphragm, and visceral organs in thoracic and abdominopelvic cavities. Visceral motor from motor nuclei in medulla oblongata

Passes through: Jugular foramen between occipital and temporal bones ∞ [p. 138]

Destination: Sensory fibers to sensory nuclei and autonomic centers of medulla (Figure 15-20●, p. 391); *Visceral motor* fibers to muscles of the palate, pharynx, digestive, respiratory, and cardiovascular systems in the thoracic and abdominal cavities

The **vagus** (VĀ-gus) **nerve** arises immediately posterior to the glossopharyngeal nerve. Many small rootlets contribute to its formation, and developmental studies indicate that this nerve probably represents the fusion of several smaller cranial nerves during our evolution. As its name suggests (*vagus*, wanderer), the vagus nerve branches and radiates extensively. Figure 15-28● shows only the general pattern of distribution.

Sensory neurons are located within the **jugular** and **nodose ganglia** (NŌ-dos; knot). The vagus nerve provides somatic sensory information concerning the external auditory meatus, a portion of the ear, and the diaphragm, and special sensory information from pharyngeal taste receptors. But the majority of the vagal afferents provide visceral sensory information from receptors along the esophagus, respiratory tract, and abdominal viscera as distant as the terminal segments of the large intestine. Vagal afferents are vital to the autonomic control of visceral function, but because the information often fails to reach the cerebral cortex, we are not consciously aware of the sensations they provide.

The motor components of the vagus nerve are equally diverse. The vagus nerve carries preganglionic autonomic fibers that affect the heart and control smooth muscles and glands within the areas monitored by its sensory fibers, including the stomach, intestines, and gallbladder. The vagus nerve also distributes somatic motor fibers to muscles of the palate and pharynx, but these are actually branches of the *accessory nerve*, described below.

The Accessory Nerve (N XI)
(Figures 15-20/15-29)

> **Primary function:**   Motor
> **Origin:**   Motor nuclei of spinal cord and medulla oblongata
> **Passes through:**   Jugular foramen between occipital and temporal bones ∞ [p. 138]
> **Destination:**   Medullary branch innervates voluntary muscles of palate, pharynx, and larynx; spinal branch controls sternocleidomastoid and trapezius muscles

The **accessory nerve** differs from other cranial nerves in that some of its motor fibers originate in the lateral gray horns of the first five cervical segments of the spinal cord (Figures 15-20, p. 391, and 15-29●). These fibers enter the cranium through the foramen magnum, unite with the motor fibers from a nucleus in the medulla oblongata, and leave the cranium through the jugular foramen. The accessory nerve consists of two branches:

1.   The **medullary (cranial) branch** joins the vagus nerve and innervates the voluntary swallowing muscles of the soft palate and pharynx and the intrinsic muscles that control the vocal cords.
2.   The **spinal branch** controls the sternocleidomastoid and trapezius muscles of the neck and back. ∞ [p. 265] The motor fibers of this branch originate in the lateral gray horns of C_1 to C_5.

The Hypoglossal Nerve (N XII)
(Figures 15-20/15-29)

> **Primary function:**   Motor, tongue movements
> **Origin:**   Motor nuclei of medulla (Figure 15-20●, p. 391)
> **Passes through:**   Hypoglossal canal of occipital bone ∞ [p. 138]
> **Destination:**   Muscles of the tongue ∞ [p. 262]

The **hypoglossal** (hī-pō-GLOS-al) **nerve** leaves the cranium through the hypoglossal canal of the occipital bone. It then curves

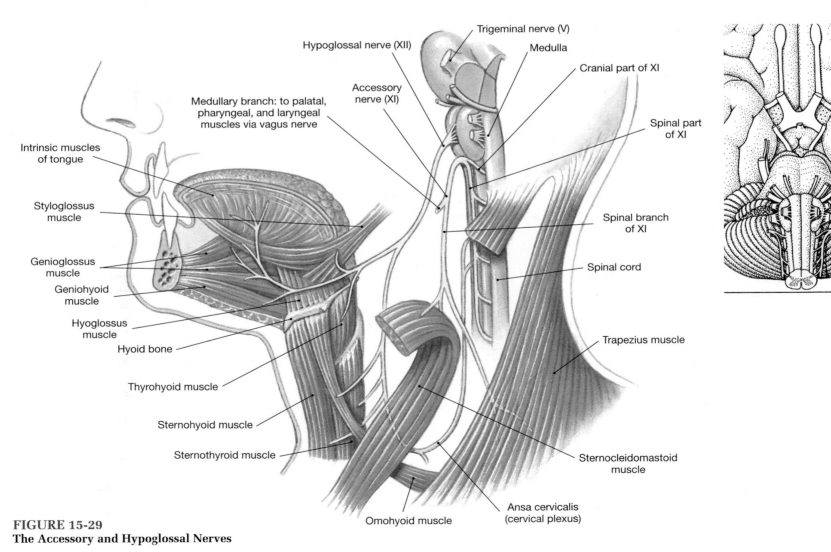

FIGURE 15-29
The Accessory and Hypoglossal Nerves

inferiorly, anteriorly, and then superiorly to reach the skeletal muscles of the tongue (Figure 15-29●). This nerve provides voluntary motor control over movements of the tongue.

√ **Damage to which of the cranial nerves can result in death?**

√ **John is experiencing problems in moving his tongue. His doctor tells him the problems are due to pressure on a cranial nerve. Which cranial nerve is involved?**

√ **What symptoms would you associate with damage to the abducens nerve (N VI)?**

A Summary of Cranial Nerve Branches and Functions

Few people are able to remember the names, numbers, and functions of the cranial nerves without a struggle. Mnemonic devices may prove useful. The most famous and oft-repeated is *On Old Olympus's Towering Top A Finn And German Viewed Some Hops.* (The *And* refers to the acoustic nerve, an alternative name for N VIII, the vestibulocochlear nerve.) A more modern one, *Oh, Once One Takes The Anatomy Final, Very Good Vacations Seem Heavenly,* may be a bit easier to remember. A summary of the basic distribution and function of each cranial nerve is detailed in Table 15-12.

TABLE 15-12   The Cranial Nerves

Cranial Nerve (No.)	Sensory Ganglion	Branch	Primary Function	Foramen	Innervation
Olfactory (I)			Special sensory	Cribriform plate of ethmoid bone	Olfactory epithelium
Optic (II)			Special sensory	Optic foramen	Retina of eye
Oculomotor (III)			Motor	Superior orbital fissure	Inferior, medial, superior rectus, inferior oblique, and levator palpebrae muscles; intrinsic muscles of eye
Trochlear (IV)			Motor	Superior orbital fissure	Superior oblique muscle
Trigeminal (V)	Semilunar		Mixed		Areas associated with the jaws
		Ophthalmic	Sensory	Superior orbital fissure	Orbital structures, nasal cavity, skin of forehead, upper eyelid, eyebrows, nose (part)
		Maxillary	Sensory	Foramen rotundum	Lower eyelid; upper lip, gums, and teeth; cheek, nose (part), palate and pharynx (part)
		Mandibular	Mixed	Foramen ovale	*Sensory* to lower gums, teeth, lips; palate (part) and tongue (part); *motor* to muscles of mastication
Abducens (VI)			Motor	Superior orbital fissure	Lateral rectus muscle
Facial (VII)	Geniculate		Mixed	Internal acoustic canal to facial canal; exits at stylomastoid foramen	*Sensory* to taste receptors on anterior 2/3 of tongue; *motor* to muscles of facial expression, lacrimal gland, submandibular salivary gland, sublingual salivary glands
Vestibulocochlear (Acoustic) (VIII)		Cochlear	Special sensory	Internal acoustic canal	Cochlea (receptors for hearing)
		Vestibular	Special sensory	As above	Vestibule (receptors for motion and balance)
Glossopharyngeal (IX)	Superior and inferior (petrosal)		Mixed	Jugular foramen	*Sensory* from posterior 1/3 of tongue; pharynx and palate (part); carotid body (monitors blood pressure, pH, and levels of respiratory gases); *motor* to pharyngeal muscles, parotid salivary gland
Vagus (X)	Jugular and nodose		Mixed	Jugular foramen	*Sensory* from pharynx; pinna and external meatus; diaphragm; visceral organs in thoracic and abdominopelvic cavities; *motor* to palatal and pharyngeal muscles, and visceral organs in thoracic and abdominopelvic cavities
Accessory (XI)		Medullary (cranial) portion	Motor	Jugular foramen	Voluntary muscles of palate, pharynx and larynx (with branches of the vagus nerve)
		Spinal portion			Sternocleidomastoid and trapezius muscles
Hypoglossal (XII)			Motor	Hypoglossal canal	Tongue musculature

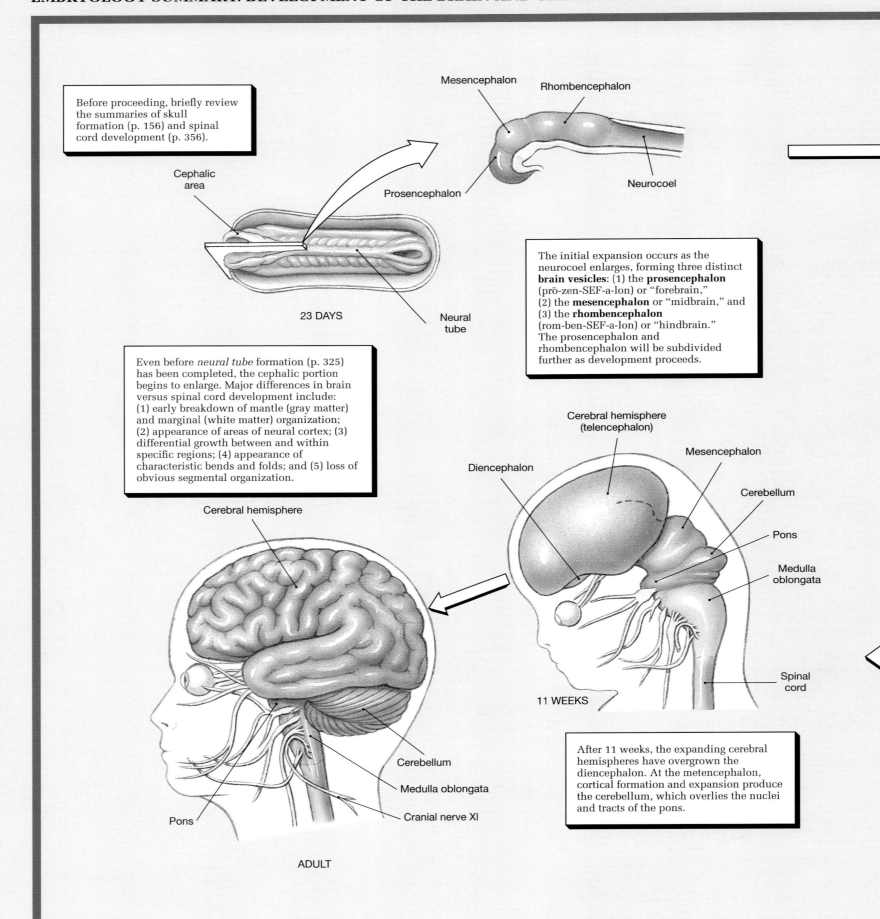

Before proceeding, briefly review the summaries of skull formation (p. 156) and spinal cord development (p. 356).

Cephalic area

Prosencephalon

Mesencephalon

Rhombencephalon

Neurocoel

23 DAYS

Neural tube

The initial expansion occurs as the neurocoel enlarges, forming three distinct **brain vesicles**: (1) the **prosencephalon** (prō-zen-SEF-a-lon) or "forebrain," (2) the **mesencephalon** or "midbrain," and (3) the **rhombencephalon** (rom-ben-SEF-a-lon) or "hindbrain." The prosencephalon and rhombencephalon will be subdivided further as development proceeds.

Even before *neural tube* formation (p. 325) has been completed, the cephalic portion begins to enlarge. Major differences in brain versus spinal cord development include: (1) early breakdown of mantle (gray matter) and marginal (white matter) organization; (2) appearance of areas of neural cortex; (3) differential growth between and within specific regions; (4) appearance of characteristic bends and folds; and (5) loss of obvious segmental organization.

Cerebral hemisphere (telencephalon)

Diencephalon

Mesencephalon

Cerebellum

Pons

Medulla oblongata

Spinal cord

11 WEEKS

Cerebral hemisphere

After 11 weeks, the expanding cerebral hemispheres have overgrown the diencephalon. At the metencephalon, cortical formation and expansion produce the cerebellum, which overlies the nuclei and tracts of the pons.

Cerebellum

Medulla oblongata

Cranial nerve XI

Pons

ADULT

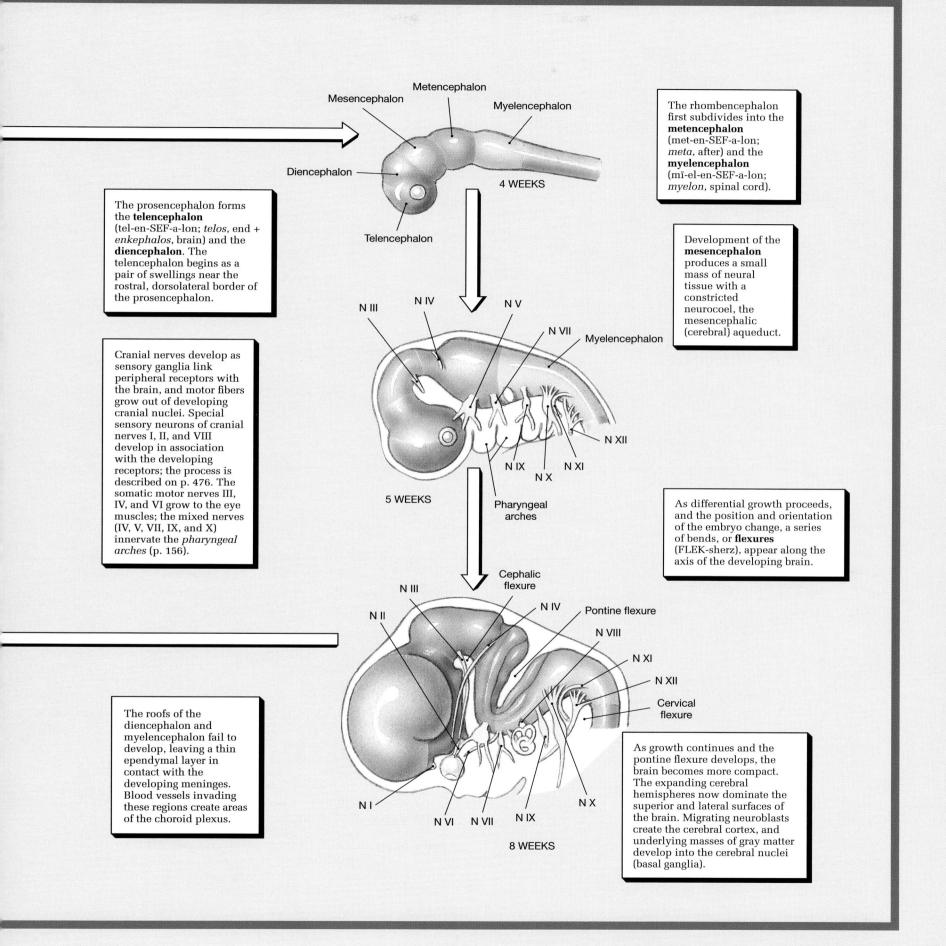

Mesencephalon

Metencephalon

Myelencephalon

Diencephalon

4 WEEKS

Telencephalon

The rhombencephalon first subdivides into the **metencephalon** (met-en-SEF-a-lon; *meta*, after) and the **myelencephalon** (mī-el-en-SEF-a-lon; *myelon*, spinal cord).

The prosencephalon forms the **telencephalon** (tel-en-SEF-a-lon; *telos*, end + *enkephalos*, brain) and the **diencephalon**. The telencephalon begins as a pair of swellings near the rostral, dorsolateral border of the prosencephalon.

Development of the **mesencephalon** produces a small mass of neural tissue with a constricted neurocoel, the mesencephalic (cerebral) aqueduct.

Cranial nerves develop as sensory ganglia link peripheral receptors with the brain, and motor fibers grow out of developing cranial nuclei. Special sensory neurons of cranial nerves I, II, and VIII develop in association with the developing receptors; the process is described on p. 476. The somatic motor nerves III, IV, and VI grow to the eye muscles; the mixed nerves (IV, V, VII, IX, and X) innervate the *pharyngeal arches* (p. 156).

N III   N IV   N V

N VII

Myelencephalon

N XII

N IX   N XI
N X

5 WEEKS

Pharyngeal arches

As differential growth proceeds, and the position and orientation of the embryo change, a series of bends, or **flexures** (FLEK-sherz), appear along the axis of the developing brain.

Cephalic flexure

N III

N II

N IV   Pontine flexure

N VIII

N XI

N XII

Cervical flexure

The roofs of the diencephalon and myelencephalon fail to develop, leaving a thin ependymal layer in contact with the developing meninges. Blood vessels invading these regions create areas of the choroid plexus.

N I

N VI   N VII   N IX

N X

8 WEEKS

As growth continues and the pontine flexure develops, the brain becomes more compact. The expanding cerebral hemispheres now dominate the superior and lateral surfaces of the brain. Migrating neuroblasts create the cerebral cortex, and underlying masses of gray matter develop into the cerebral nuclei (basal ganglia).

Cranial Reflexes

Cranial reflexes are reflex arcs that involve the sensory and motor fibers of cranial nerves. Examples of cranial reflexes are discussed in later chapters, and this section will simply provide an overview and general introduction.

Table 15-13 lists representative examples of cranial reflexes and their functions. These reflexes are clinically important because they provide a quick and easy method for observing the condition of cranial nerves and specific nuclei and tracts in the brain.

Cranial somatic reflexes are seldom more complex than the somatic reflexes of the spinal cord. This table includes four somatic reflexes: the *corneal reflex*, the *tympanic reflex*, the *auditory reflex*, and the *vestibulo-ocular reflex*. These reflexes are often used to check for damage to the cranial nerves or processing centers involved. The brain stem contains many reflex centers that control visceral motor activity. Many of these reflex centers are in the medulla, and they can direct very complex visceral motor responses to stimuli. These reflex responses are essential to the control of respiratory, digestive, and cardiovascular functions.

TABLE 15-13 Cranial Reflexes

Reflex	Stimulus	Afferents	Central Synapse	Efferents	Response
Corneal reflex	Contact with corneal surface	N V (trigeminal)	Motor nucleus for N VII (facial)	N VII	Blinking of eye
Tympanic reflex	Loud noise	N VIII (vestibulocochlear)	Inferior colliculus (midbrain)	N VII	Reduced movement of auditory ossicles
Auditory reflexes	Loud noise	N VIII	Motor nuclei of brain stem and spinal cord	N III, IV, VI, VII, X, cervical nerves	Movements triggered by sudden sounds
Vestibulo-ocular reflexes	Rotation of head	N VIII	Motor nuclei controlling eye muscles	N III, IV, VI	Opposite movement of eyes to stabilize field of vision
Direct light reflex	Light striking photoreceptors	N II (optic)	Superior colliculus (midbrain)	N III (oculomotor)	Constriction of ipsilateral pupil
Consensual light reflex	Light striking photoreceptors	N II	Superior colliculus	N III	Constriction of contralateral pupil

Related Clinical Terms

cranial trauma: A head injury resulting from violent contact with another object. Cranial trauma may cause a **concussion**, a condition characterized by a temporary loss of consciousness and a variable period of amnesia. ⚕ *Cranial Trauma [p. 761]*

epidural hemorrhage: A condition involving bleeding into the epidural spaces. *[p. 368]*

subdural hemorrhage: A condition in which blood accumulates between the dura and the arachnoid. *[p. 368]*

hydrocephalus: Also known as "water on the brain"; a condition in which the skull expands to accommodate extra fluid. *[p. 373]*

decerebrate rigidity: A generalized state of muscular contraction resulting from loss of CNS inhibitory control. *[p. 387]*

Parkinson's disease (paralysis agitans): A condition characterized by a pronounced increase in muscle tone, resulting from loss of inhibitory control over neurons in the cerebral nuclei. *[p. 387]*

spasticity: A condition characterized by hesitant, jerky voluntary movements and increased muscle tone. *[p. 387]*

tremor: A background shaking of the limbs resulting from a "tug of war" between antagonistic muscle groups. *[p. 387]*

ataxia: A disturbance of balance that in severe cases leaves the individual unable to stand without assistance; caused by problems affecting the cerebellum. ⚕ *Cerebellar Dysfunction [p. 761]*

dysmetria (dis-MET-rē-a): An inability to stop a movement at a precise, predetermined position; it often leads to an **intention tremor** in the affected individual; usually reflects cerebellar dysfunction. ⚕ *Cerebellar Dysfunction [p. 761]*

tic douloureux (doo-loo-ROO), or **trigeminal neuralgia:** A disorder of the maxillary and mandibular branches of N V characterized by severe, almost totally debilitating pain triggered by contact with the lip, tongue, or gums. *[p. 396]*

Bell's palsy: A condition resulting from an inflammation of the facial nerve; symptoms include paralysis of facial muscles on the affected side and loss of taste sensations from the anterior two-thirds of the tongue. *[p. 396]*

STUDY OUTLINE

Related Key Terms

INTRODUCTION *[p. 365]*

1. The brain is far more complex than the spinal cord; its complexity makes it adaptable but slower in response than spinal reflexes.

AN INTRODUCTION TO THE ORGANIZATION OF THE BRAIN *[p. 365]*

Major Regions and Landmarks *[p. 365]*

1. There are five regions in the adult brain: telencephalon (cerebrum), diencephalon, mesencephalon (midbrain), metencephalon (cerebellum and pons), and myelencephalon (medulla oblongata) *(see Figure 15-1)*.

2. Conscious thought, intellectual functions, memory, and complex involuntary motor patterns originate in the **cerebrum** (**telencephalon**) *(see Figure 15-1)*.

3. The walls of the **diencephalon** form the **thalamus**, which contains relay and processing centers for sensory data. The **hypothalamus** contains centers involved with emotions, autonomic function, and hormone production *(see Figure 15-1)*.

4. The **midbrain** (**mesencephalon**) processes visual and auditory information and generates involuntary somatic motor responses *(see Figure 15-1)*.

5. The **metencephalon** consists of the *cerebellum* and *pons*. The **cerebellum** adjusts voluntary and involuntary motor activities on the basis of sensory data and stored memories *(see Figure 15-1)*. The **pons** connects the cerebellum to the brain stem, and is involved with somatic and visceral motor control. The spinal cord connects to the brain at the **medulla oblongata** (**myelencephalon**), which relays sensory information and regulates autonomic functions *(see Figure 15-1)*.

longitudinal fissure • pituitary gland • brain stem

Gray Matter and White Matter Organization *[p. 365]*

6. The brain contains extensive areas of **neural cortex**, a layer of gray matter on the surfaces of the cerebrum and cerebellum.

Embryology of the Brain *[p. 366]*

7. The brain forms from three swellings at the superior tip of the developing neural tube: the **prosencephalon**, **mesencephalon**, and **rhombencephalon** *(see Table 15-1 and Embryology Summary, pp. 402–403)*.

Primary brain vesicles • seconday brain vesicles

The Ventricles of the Brain *[p. 367]*

8. The central passageway of the brain expands to form chambers called **ventricles**. Cerebrospinal fluid continually circulates from the ventricles and central canal of the spinal cord into the subarachnoid space of the meninges that surround the CNS *(see Figures 15-2/15-17b)*.

septum pellucidum • lateral ventricles • interventricular foramen • third ventricle • mesencephalic aqueduct • fourth ventricle

PROTECTION AND SUPPORT OF THE BRAIN *[p. 368]*

The Cranial Meninges *[p. 368]*

1. The **cranial meninges**—the **dura mater**, **arachnoid membrane**, and **pia mater** are continuous with those of the spinal cord, but have anatomical and functional differences *(see Figures 15-3/15-4)*.

dural sinuses

2. Folds of dura mater stabilize the position of the brain. The **falx cerebri**, **tentorium cerebelli**, **falx cerebelli**, and **diaphragma sellae** are examples *(see Figures 15-3/15-4)*.

superior/inferior sagittal sinus • transverse sinus • subdural space • subarachnoid space • arachnoid trabeculae

The Blood-Brain Barrier *[p. 371]*

3. The **blood-brain barrier** isolates neural tissue from the general circulation *(see Figure 15-5)*.

4. The blood-brain barrier remains intact throughout the CNS except in portions of the hypothalamus, the pineal gland, and at the choroid plexus in the membranous roof of the diencephalon and medulla.

Cerebrospinal Fluid *[p. 371]*

5. The **choroid plexus** is the site of cerebrospinal fluid production *(see Figure 15-5)*.

6. Cerebrospinal fluid (CSF): (1) cushions delicate neural structures, (2) supports the brain, and (3) transports nutrients, chemical messengers, and waste products. Cerebrospinal fluid reaches the subarachnoid space via the **lateral apertures** and a **median aperture**. Diffusion across the **arachnoid granulations** into the superior sagittal sinus returns CSF to the venous circulation *(see Figures 15-4/15-6)*.

THE CEREBRUM *[p. 373]*

The Cerebral Cortex *[p. 373]*

1. The cortical surface contains **gyri** (elevated ridges) separated by **sulci** (shallow depressions) or deeper grooves (**fissures**). The **longitudinal fissure** separates the two **cerebral hemispheres**. The **central sulcus** marks the boundary between the **frontal lobe** and the **parietal lobe**. Other sulci form the boundaries of the **temporal lobe** and the **occipital lobe** *(see Figures 15-1/15-8/15-9)*.

lateral sulcus • insula • parieto-occipital sulcus

2. Each cerebral hemisphere receives sensory information and generates motor commands that concern the opposite side of the body. There are significant functional differences between the two hemispheres.

3. The **primary motor cortex** of the **precentral gyrus** directs voluntary movements. The **primary sensory cortex** of the **postcentral gyrus** receives somatic sensory information from touch, pressure, pain, taste, and temperature receptors *(see Figure 15-9 and Table 15-2)*.

4. **Association areas**, such as the **visual association area** and **somatic motor association area** (**premotor cortex**), control our ability to understand sensory information and coordinate a motor response. "Higher-order" integrative centers receive information from many different association areas and direct complex motor activities and analytical functions *(see Figure 15-9 and Table 15-2)*.

The Central White Matter *[p. 376]*
5. The **central white matter** contains three major groups of axons: (1) **association fibers** (tracts that interconnect areas of neural cortex within a single cerebral hemisphere); (2) **commissural fibers** (tracts connecting the two cerebral hemispheres); and (3) **projection fibers** (tracts that link the cerebrum with other regions of the brain and spinal cord) *(see Figure 15-10 and Table 15-3)*.

The Cerebral Nucle i *[p. 377]*
6. The **cerebral nuclei** (**basal nuclei**) within the central white matter include the **caudate nucleus**, **amygdaloid body**, **claustrum**, **globus pallidus**, and **putamen**. The cerebral nuclei are part of the **extrapyramidal system**, which controls muscle tone and coordinates learned movement patterns and other somatic motor activities *(see Figure 15-11 and Table 15-4)*.

THE LIMBIC SYSTEM *[p. 381]*

1. The **limbic system** includes the amygdaloid body, **cingulate gyrus**, **dentate gyrus**, **parahippocampal gyrus**, **hippocampus**, and **fornix**. The **mamillary bodies** control reflex movements associated with eating. The functions of the limbic system involve emotional states and related behavioral drives *(see Figures 15-11/15-12 and Table 15-5)*.

2. The **anterior nucleus** relays visceral sensations, and stimulating the **reticular formation** produces heightened awareness and a generalized excitement.

THE DIENCEPHALON *[p. 383]*

1. The diencephalon provides the switching and relay centers necessary to integrate the conscious and unconscious sensory and motor pathways. The diencephalic roof (**epithalamus**) contains the **pineal gland** and a vascular network that produces cerebrospinal fluid *(see Figures 15-12/15-13)*.

The Thalamus *[p. 383]*
2. The thalamus is the final relay point for ascending sensory information and coordinates the pyramidal and extrapyramidal systems *(see Figures 15-11e/15-12 to 15-14 and Table 15-6)*.

The Hypothalamus *[p. 384]*
3. The hypothalamus contains important control and integrative centers. It can: (1) control involuntary somatic motor activities; (2) control autonomic function; (3) coordinate activities of the nervous and endocrine systems; (4) secrete hormones; (5) produce emotions and behavioral drives; (6) coordinate voluntary and autonomic functions; and (7) regulate body temperature *(see Figures 15-12/15-15 and Table 15-7)*.

THE MESENCEPHALON (MIDBRAIN) *[p. 384]*

1. The **tectum** (roof) of the mesencephalon contains two pairs of nuclei, the **corpora quadrigemina**. On each side, the **superior colliculus** receives visual inputs from the thalamus, and the **inferior colliculus** receives auditory data from the medulla. The **red nucleus** integrates information from the cerebrum and issues involuntary motor commands related to muscle tone and posture. The **substantia nigra** regulates the motor output of the cerebral nuclei. The **cerebral peduncles** contain ascending fibers headed for thalamic nuclei and descending fibers of the pyramidal system that carry voluntary motor commands from the primary motor cortex of each cerebral hemisphere *(see Figures 15-13/15-16 and Table 15-8)*.

THE METENCEPHALON *[p. 387]*

The Cerebellum *[p. 387]*
1. The cerebellum oversees the body's postural muscles, and programs and tunes voluntary and involuntary movements. The **cerebellar hemispheres** consist of neural cortex formed into folds, or **folia**. The surface can be divided into the **anterior** and **posterior lobes**, the **vermis**, and the **flocculonodular lobes** *(see Figures 15-1/15-3/15-8/15-17 and Table 15-9)*.

The Pons *[p. 389]*
2. The pons contains: (1) sensory and motor nuclei for four cranial nerves; (2) nuclei concerned with involuntary control of respiration; (3) tracts linking the cerebellum with the brain stem, cerebrum, and spinal cord; and (4) ascending and descending tracts *(see Figures 15-1/15-8/15-13/15-18 and Table 15-10)*.

Related Key Terms

pyramidal cells • pyramidal system • visual cortex • auditory cortex • olfactory cortex

prefrontal cortex

arcuate fibers • longitudinal fasciculi • corpus callosum • anterior commissure • internal capsule

lentiform nucleus • corpus striatum

limbic lobe

melatonin

intermediate mass • anterior nuclei • medial nuclei • ventral nuclei • posterior nuclei • pulvinar • lateral geniculate • optic tract • medial geniculate
optic chiasm • infundibulum • tuber cinereum • median eminence • supraoptic nucleus • paraventricular nucleus • feeding center • thirst center • preoptic area

primary fissure • Purkinje cells • arbor vitae • cerebellar nuclei • superior cerebellar peduncles • middle cerebellar peduncles • inferior cerebellar peduncles

transverse fibers

THE MEDULLA OBLONGATA (MYELENCEPHALON) *[p. 389]*

Related Key Terms

1. The medulla oblongata connects the brain to the spinal cord. It contains **olivary nuclei**, which relay information from the spinal cord, the cerebral cortex, and brain stem to the cerebellar cortex. Its **reflex centers**, including the **cardiovascular centers** and the **respiratory rhythmicity center**, control or adjust the activities of peripheral systems *(see Figures 15-1/15-8/15-12/15-13/15-19 and Table 15-11)*.

nucleus gracilis • nucleus cuneatus • olives • cardiac center • vasomotor center

CRANIAL NERVES *[p. 391]*

1. There are 12 pairs of cranial nerves. Each nerve attaches to the brain near the associated sensory or motor nuclei on the ventrolateral surface of the brain *(see Figure 15-20)*.

The Olfactory Nerve (N I) *[p. 392]*
2. The **olfactory tract** (nerve) (N I) carries sensory information responsible for the sense of smell. The olfactory afferents synapse within the **olfactory bulbs** *(see Figure 15-21)*.

The Optic Nerve (N II) *[p. 393]*
3. The **optic nerve** (N II) carries visual information from special sensory receptors in the eyes *(see Figure 15-22)*.

optic chiasm • optic tract

The Oculomotor Nerve (N III) *[p. 394]*
4. The **oculomotor nerve** (N III) is the primary source of innervation for the extrinsic oculomotor muscles that move the eyeball *(see Figure 15-23)*.

ciliary ganglion

The Trochlear Nerve (N IV) *[p. 394]*
5. The **trochlear nerve** (N IV), the smallest cranial nerve, innervates the superior oblique muscle of the eye *(see Figure 15-23)*.

The Trigeminal Nerve (N V) *[p. 395]*
6. The **trigeminal nerve** (N V), the largest cranial nerve, is a mixed nerve with **ophthalmic**, **maxillary**, and **mandibular branches** *(see Figure 15-24)*.

semilunar ganglion

The Abducens Nerve (N VI) *[p. 396]*
7. The **abducens nerve** (N VI) innervates the sixth extrinsic oculomotor muscle, the lateral rectus *(see Figure 15-23)*.

The Facial Nerve (N VII) *[p. 396]*
8. The **facial nerve** (N VII) is a mixed nerve that controls muscles of the scalp and face. It provides pressure sensations over the face and receives taste information from the tongue *(see Figure 15-25)*.

geniculate ganglion • sphenopalatine ganglion • submandibular ganglion

The Vestibulocochlear Nerve (N VIII) *[p. 396]*
9. The **vestibulocochlear nerve** (N VIII) contains the **vestibular nerve**, which monitors sensations of balance, position, and movement, and the **cochlear nerve**, which monitors hearing receptors *(see Figure 15-26)*.

vestibular nuclei • cochlear nuclei

The Glossopharyngeal Nerve (N IX) *[p. 398]*
10. The **glossopharyngeal nerve** (N IX) is a mixed nerve that innervates the tongue and pharynx and controls the action of swallowing *(see Figure 15-27)*.

superior ganglion • inferior (petrosal) ganglion

The Vagus Nerve (N X) *[p. 399]*
11. The **vagus nerve** (N X) is a mixed nerve that is vital to the autonomic control of visceral function and has a variety of motor components *(see Figure 15-28)*.

jugular ganglia • nodose ganglia

The Accessory Nerve (N XI) *[p. 400]*
12. The **accessory nerve** (N XI) has a **medullary cranial branch**, which innervates voluntary swallowing muscles of the soft palate and pharynx, and a **spinal branch**, which controls muscles associated with the pectoral girdle *(see Figure 15-29)*.

The Hypoglossal Nerve (N XII) *[p. 400]*
13. The **hypoglossal nerve** (N XII) provides voluntary motor control over tongue movements *(see Figure 15-29)*.

A Summary of Cranial Nerve Branches and Functions *[p. 401]*
14. See *Table 15-12*.

---

1 REVIEW OF CHAPTER OBJECTIVES

1. Name the major regions of the brain, and describe their functions.

2. Identify and describe the structures that protect and support the brain.

3. Describe the structures that constitute the blood-brain barrier and their functions.

4. Describe the structure and function of the choroid plexus.

5. Discuss the origin, function, and circulation of cerebrospinal fluid.

6. Identify the anatomical structures of the cerebrum, and list their functions.

7. Distinguish between motor, sensory, and association areas of the cerebral cortex.

8. Identify the anatomical structures that make up the limbic system, and identify its function.

9. Identify the anatomical structures that form the thalamus, and list their functions.
10. Identify and describe the structure and function of the hypothalamus.
11. Identify and describe the structure and function of the mesencephalon.
12. Identify and describe the structure and function of the metencephalon.
13. Identify the anatomical structures of the medulla oblongata and their functions.
14. Identify the 12 cranial nerves, and relate each pair of cranial nerves to its principal functions.

2 REVIEW OF CONCEPTS

1. What is the nature of the interaction between spinal reflexes and the brain?
2. What are the main functions of the cerebrum?
3. How do the cranial meninges support and protect the brain from injury?
4. Of what significance are the arachnoid granulations in cranial circulation?
5. How do the endothelial cells lining the CNS capillaries maintain the blood-brain barrier?

6. What are the primary functions of the limbic system?
7. What functions do N I, II, VII, VIII, IX, and X have in common?
8. Which cranial nerves are involved in controlling the functions of the eye?

3 CRITICAL THINKING AND CLINICAL APPLICATION QUESTIONS

1. While sitting and reading quietly in his living room one evening, a man in his middle seventies loses consciousness. When he awakens later in a hospital, he realizes that his left arm and leg are paralyzed. What may have happened to him, and what is the likely anatomical cause of the symptoms?
2. A man in his sixties has been experiencing tremors in the skeletal muscles, increased muscle tone, and is finding it increasingly difficult to perform normal motor movements. His movement patterns have become jerky, and it is getting harder for him to initiate movement. When he is at rest, he has a constant tremor. What is one possible cause of these symptoms?
3. A young man awakens one morning and discovers that the muscles on the right side of his face have become paralyzed. He does not feel any pain associated with this condition, but decides to consult his doctor. What cranial nerve(s) may be involved?

16
The Nervous System:
Pathways and Higher-Order Functions

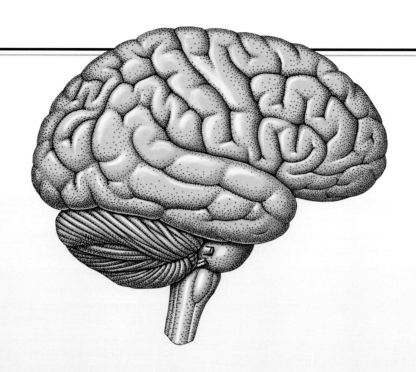

CHAPTER OUTLINE AND OBJECTIVES

There is a continuous flow of information between the brain, spinal cord, and peripheral nerves. At any given moment, millions of sensory neurons are delivering information to processing centers in the CNS, and millions of motor neurons are controlling or adjusting the activities of peripheral effectors. This process continues whether we are awake or asleep. Your conscious mind may sleep soundly, but many brain stem centers are active throughout our lives, performing vital autonomic functions.

Many subtle forms of interaction, feedback, and regulation link higher centers with the various components of the brain stem. Only a few are understood in any detail. In the following sections, we will focus on the anatomical design that allows neural structures to perform sensory, motor, and higher-order brain functions, such as learning and memory.

SENSORY AND MOTOR PATHWAYS

Communication between the CNS, the PNS (peripheral nervous system), and peripheral organs and systems involves pathways that relay sensory and motor information between the periphery and higher centers of the brain. Each ascending (sensory) or descending (motor) pathway consists of a chain of tracts and associated nuclei. There are usually several successive levels of processing along the sensory and motor pathways. Our attention will focus on pathways that utilize the major ascending and descending tracts of the spinal cord. When discussing tract names, be aware that:

1. The tract names indicate the origins and destinations of the axons. If the name of a tract begins with *spino-*, it must *start* in the spinal cord and *end* in the brain, and it therefore carries sensory information. If the name of a tract ends in *-spinal*, its axons must *start* in the brain and *end* in the spinal cord, bearing motor commands.

2. The rest of the tract name indicates the associated nucleus or cortical area of the brain.

Sensory Pathways
(Figure 16-1)

Sensory receptors monitor conditions in the body or the external environment. When stimulated, a receptor passes information to the central nervous system. This information, called a **sensation**, arrives in the form of action potentials in an afferent (sensory) fiber. Chapter 17 describes the distribution of visceral sensory information and considers reflexive responses to visceral sensations. Chapter 18 examines the origins of sensations and the pathways involved in relaying special sensory information, such as olfaction (smell) or vision, to conscious and unconscious processing centers in the CNS. Most of the processing occurs in centers along the sensory pathways in the spinal cord or brain stem; only about 1% of the information provided by afferent fibers reaches the cerebral cortex and our conscious awareness.

This section describes the pathways that carry somatic sensory information to the primary sensory cortex of the cerebral hemispheres. The sensory neuron that delivers the sensations to the CNS is often called a **first-order neuron**. Inside the CNS the axon of the first-order neuron synapses on an in-

terneuron known as a **second-order neuron**. The second-order neuron, which may be located in the spinal cord or brain stem, in turn synapses on a **third-order neuron** in the thalamus. Somewhere along its length, the axon of either the first-order or second-order neuron crosses over to the opposite side of the CNS. As a result of this crossing over, a process called **decussation**, the right side of the thalamus receives sensory information from the left side of the body. The axons of the third-order neurons synapse on neurons of the primary sensory cortex of the cerebral hemisphere. Because the axons of third-order neurons do not cross over, the right cerebral hemisphere receives sensory information from the left side of the body, and vice versa.

Table 16-1 identifies and summarizes the three major somatic sensory pathways, or **somatosensory pathways**: (1) *posterior column pathway*, (2) *spinothalamic pathway*, and (3) *spinocerebellar pathway*. Figure 16-1● indicates their relative positions in the spinal cord. For clarity, the figures dealing with spinal pathways show how sensations originating on one side of the body are relayed to the cerebral cortex. Keep in mind, however, that these pathways are present on *both* sides of the body.

The Posterior Column Pathway
(Figure 16-2a)

The **posterior column pathway** (Figure 16-2a●) carries highly localized "fine" touch, pressure, vibration, and proprioceptive (position) sensations. The axons of the first-order neurons reach the CNS via the dorsal roots of spinal nerves and the sensory roots of cranial nerves. Axons from spinal nerves ascend within the **fasciculus gracilis** and the **fasciculus cuneatus**, synapsing at the nucleus gracilis and the nucleus cuneatus of the medulla oblongata. The postsynaptic neurons then relay the information to the thalamus of the opposite side of the brain along a tract called the **medial lemniscus** (*lemniskos*, ribbon). The decussation occurs as the axons of the second-order neurons leave the nuclei to enter the medial lemniscus. As it travels toward the thalamus, the medial lemniscus incorporates the same classes of sensory information collected by cranial nerves V, VII, IX, and X.

Ascending sensory information maintains a strict regional organization along the pathway from center to center. ∞[p. 345] Data arriving over the posterior column pathway are integrated by the ventroposterolateral nucleus of the thalamus, which sorts them according to the region of the body involved and projects them to specific regions of the primary sensory cortex. The sensations arrive with sensory information from the toes at one end of the primary sensory cortex and information from the head at the other.

The individual "knows" the nature of the stimulus and its location because the information has been projected to a specific portion of the primary sensory cortex. If the sensation is relayed to the wrong part of the sensory cortex, the individual will believe that the sensation originated in a different part of the body. For example, the pain of a heart attack is often felt in the left arm. (This phenomenon, called *referred pain* will be considered in Chapter 18.) Our perception of a given sensation as touch, rather than as temperature or pain, depends on processing in the thalamus. If the cerebral cortex were damaged or the projection fibers cut, one could still be aware of a light touch because the thalamic nuclei remain intact. The individual, however, would be unable to determine its source, because localization is provided by the primary sensory cortex.

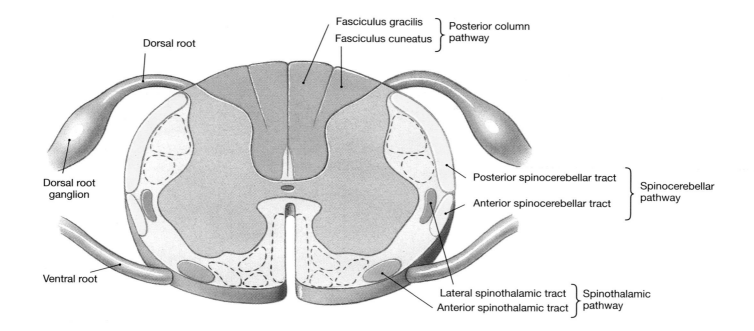

FIGURE 16-1
Ascending Tracts in the Spinal Cord. A cross-sectional view indicating the locations of
the major ascending sensory tracts in the spinal cord. For information about these tracts,
see Table 16-1. Descending motor tracts are shown in outline; these tracts are identified
in Figure 16-4.

TABLE 16-1   Principal Ascending (Sensory) Tracts in the Spinal Cord and a Summary of the Functional Roles of the Associated Nuclei in the Brain

Tract	Origin	Destination[a]	Sensations	Comments
THE POSTERIOR COLUMN PATHWAY				
Fasciculus gracilis	Proprioceptors and fine touch, pressure, and vibration receptors of lower body	Nucleus gracilis (medulla oblongata)	Position, fine touch, pressure, vibration	Ascends on same side as stimulus
Fasciculus cuneatus	Proprioceptors and fine touch, pressure, and vibration receptors of upper body	Nucleus cuneatus (medulla oblongata)	Position, fine touch, pressure, vibration	Ascends on same side as stimulus
THE SPINOTHALAMIC PATHWAY				
Lateral spinothalamic tract	Interneurons relaying information from pain and temperature receptors	Thalamus (ventral nuclei)	Pain and temperature	Crosses to ascend on side opposite stimulus
Anterior spinothalamic tract	Interneurons relaying information from crude touch and pressure receptors	Thalamus (ventral nuclei)	Crude touch and pressure	Crosses to ascend on side opposite stimulus
THE SPINOCEREBELLAR PATHWAY				
Posterior spinocerebellar tract	Interneurons relaying information from proprioceptors	Cerebellum	Proprioception	Ascends on same side as stimulus
Anterior spinocerebellar tract	Interneurons relaying information from proprioceptors	Cerebellum	Proprioception	Crosses to ascend on on side opposite stimulus

[a] Location of first synapse.

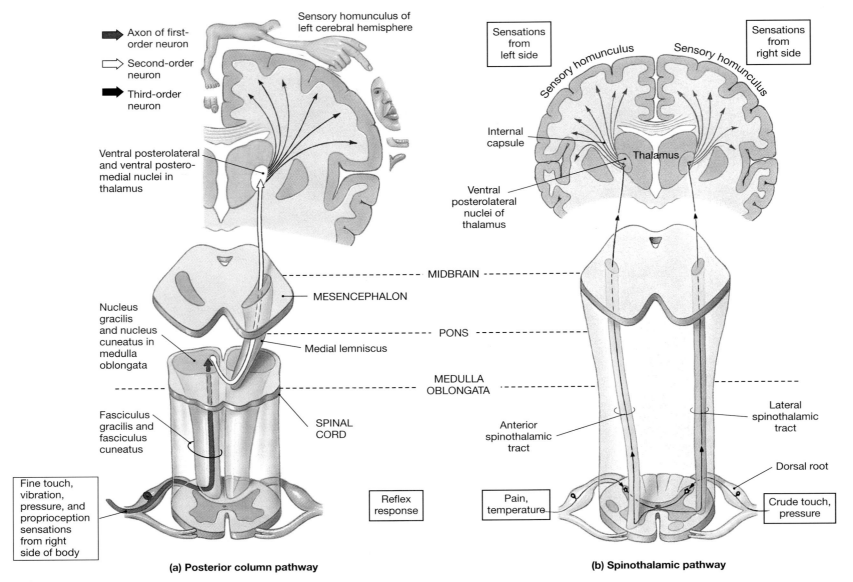

Legend (top left):
- ➡ Axon of first-order neuron
- ⇨ Second-order neuron
- ➡ Third-order neuron

Sensory homunculus of left cerebral hemisphere

Ventral posterolateral and ventral postero-medial nuclei in thalamus

Nucleus gracilis and nucleus cuneatus in medulla oblongata

Medial lemniscus

Fasciculus gracilis and fasciculus cuneatus

SPINAL CORD

Fine touch, vibration, pressure, and proprioception sensations from right side of body

(a) Posterior column pathway

Sensations from left side

Sensations from right side

Sensory homunculus

Sensory homunculus

Internal capsule

Thalamus

Ventral posterolateral nuclei of thalamus

MIDBRAIN

MESENCEPHALON

PONS

MEDULLA OBLONGATA

Lateral spinothalamic tract

Anterior spinothalamic tract

Dorsal root

Reflex response

Pain, temperature

Crude touch, pressure

(b) Spinothalamic pathway

FIGURE 16-2
The Posterior Column, Spinothalamic, and Spinocerebellar Sensory Pathways. (a) The posterior column pathway delivers fine touch, vibration, and proprioception information to the primary sensory cortex of the cerebral hemisphere on the opposite side of the body. The crossover occurs in the medulla, after a synapse in the nucleus gracilis or nucleus cuneatus. (For clarity, this figure shows only the pathway for sensations originating on the left side of the body.) (b) The spinothalamic pathways carry sensations of pain and temperature (lateral spinothalamic tract) and crude touch and pressure (anterior spinothalamic tract) to the sensory cortex on the opposite side. The crossover occurs in the spinal cord, at the level of entry. (Only one tract is detailed on each side, although each side has both tracts.) (c) The spinocerebellar pathway.

If a site on the primary sensory cortex is electrically stimulated, the individual reports feeling sensations in a specific part of the body. By electrically stimulating the cortical surface, experimenters have been able to create a functional map of the primary sensory cortex. The results are shown in Figure 16-2a●. This sensory map is called a **sensory homunculus** ("little man"). The proportions of the homunculus are obviously very different from those of the individual. For example, the face is huge and distorted, with enormous lips and tongue, whereas the back is relatively tiny. These distortions occur because the area of sensory cortex devoted to a particular region is proportional not to its absolute size, but rather to the *number of sensory receptors* it contains. In other words, it takes many more cortical neurons to process sensory information arriving from the tongue, which has tens of thousands of taste and touch receptors, than it does to analyze sensations originating on the back, where touch receptors are few and far between.

The Spinothalamic Pathway
(Figure 16-2b)

The **spinothalamic pathway** (Figure 16-2b●) begins as axons carrying "crude" sensations of touch, pressure, pain, and temperature enter the spinal cord and synapse within the posterior gray horns. Interneurons send the information to the opposite side, to ascend within the **anterior** and **lateral spinothalamic tracts**. These tracts and the medial lemniscus converge on the ventral posterolateral nuclei of the thalamus. Projection fibers then carry the information to the primary sensory cortex. Table 16-1 summarizes the origin and destination of these tracts and the associated sensations. For clarity, Figure 16-2b● shows the distribution route for pain and temperature sensations from the right side of the body and crude touch and pressure from the left side. However, both sides of the spinal cord have anterior and lateral spinothalamic tracts.

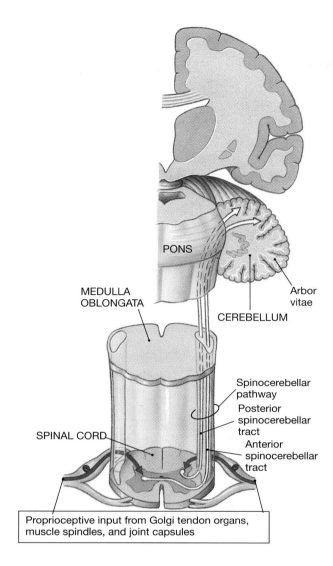

PONS

MEDULLA
OBLONGATA

Arbor
vitae

CEREBELLUM

Spinocerebellar
pathway

Posterior
spinocerebellar
tract

SPINAL CORD

Anterior
spinocerebellar
tract

Proprioceptive input from Golgi tendon organs,
muscle spindles, and joint capsules

(c) Spinocerebellar pathway

FIGURE 16-2 continued

The Spinocerebellar Pathway
(Figure 16-2c)

The **spinocerebellar pathway** includes the **posterior** and **anterior spinocerebellar tracts** (Figure 16-2c●). These tracts carry proprioceptive information concerning the position of muscles, tendons, and joints to the cerebellum for processing. Table 16-1 summarizes the origin and destination of these tracts and the associated sensations.

Motor Pathways
(Figure 16-3)

The central nervous system issues motor commands in response to information provided by sensory systems. These commands are distributed by the somatic nervous system and the autonomic nervous system. The *somatic nervous system (SNS)* issues somatic motor commands that direct the contractions of skeletal muscles. The *autonomic nervous system (ANS)*, also termed *vis-*

ceral motor system, innervates visceral effectors, such as smooth muscles, cardiac muscles, and glands.

The motor neurons of the SNS and ANS are organized in different ways. Somatic motor pathways (Figure 16-3a●) always involve at least two motor neurons: an **upper motor neuron**, whose cell body lies in a CNS processing center, and a **lower motor neuron** located in a motor nucleus of the brain stem or spinal cord. Activity in the upper motor neuron can facilitate or inhibit the lower motor neuron, but only the axon of the lower motor neuron extends outside of the CNS and contacts skeletal muscle fibers. Destruction or damage to a lower motor neuron produces a flaccid paralysis of the innervated motor unit. Damage to an upper motor neuron may produce muscle rigidity, flaccidity, or uncoordinated contractions.

At least two neurons are involved in the ANS, and one of them is always located in the periphery (Figure 16-3b●). Autonomic motor control involves a **preganglionic neuron** whose cell body lies within the CNS, and a **ganglionic neuron** in a peripheral ganglion. Higher centers in the hypothalamus and elsewhere in the brain stem may stimulate or inhibit the preganglionic neuron. Motor pathways of the ANS will be described in Chapter 17.

The Pyramidal System
(Figure 16-4)

Voluntary and involuntary somatic motor commands issued by the brain reach peripheral targets by traveling over two integrated motor pathways, the *pyramidal system* and the *extrapyramidal system*. The **pyramidal system** provides voluntary control of skeletal muscles. The pyramidal system owes its name to the **pyramidal cells** of the primary motor cortex. These neurons, which have cell bodies shaped like miniature pyramids, have axons that extend into the brain stem and spinal cord to synapse on lower motor neurons.

Components of the pyramidal system involved in controlling the right side of the body are illustrated in Figure 16-4●. If direct electrical stimulation is applied to the primary motor cortex, peripheral muscular contractions will occur. The muscle group involved will depend on the area stimulated. Mapping these areas produces a **motor homunculus** (Figure 16-4a●). The proportions of the motor homunculus are different from those of the body. The differences exist because the area of cortex devoted to a specific region depends on the number of motor units controlled. As a result, the motor homunculus gives an indication of the degree of fine motor control available. The hands, face, and tongue are very large, and the trunk relatively small; these proportions are similar to those of the sensory homunculus. The sensory and motor images differ in other respects because some highly sensitive regions, such as the sole of the foot, contain few motor units, and some areas with very small motor units, such as the eye muscles, are not equally sensitive.

The pyramidal system consists of three pairs of descending motor tracts: (1) the *corticobulbar tracts*, (2) the *lateral corticospinal tracts*, and (3) the *anterior corticospinal tracts*. These tracts enter the internal capsule, descend, and emerge briefly on either side of the mesencephalon as the *cerebral peduncles*. ∞ [p. 376]

THE CORTICOBULBAR TRACTS *(Figure 16-4a)*. The **corticobulbar tracts** (kor-ti-kō-BUL-bar; *bulbar*, brain stem) (Figure 16-4a●) terminate at the motor nuclei of cranial nerves controlling eye movements, facial muscles, tongue muscles, and superficial muscles of the neck and back (N III, N IV, N VI, N VII, N XI, and N XII). ∞ [pp. 396–402]

FIGURE 16-3

Organization of the Somatic and Autonomic Nervous Systems. (a) In the SNS, an upper motor neuron in the CNS controls a lower motor neuron in the brain stem or spinal cord. The axon of the lower motor neuron has direct control over skeletal muscle fibers. Stimulation of the lower motor neuron always has an excitatory effect on the skeletal muscle fibers. (b) In the ANS, the axon of a preganglionic neuron in the CNS controls ganglionic neurons in the periphery. Stimulation of the ganglionic neurons may lead to excitation or inhibition of the visceral effector innervated.

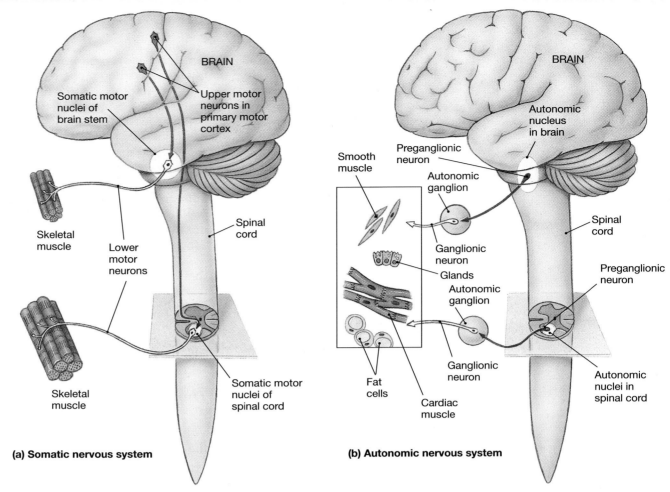

(a) Somatic nervous system

(b) Autonomic nervous system

THE CORTICOSPINAL TRACTS *(Figure 16-4b).*

The **corticospinal tracts** (Figure 16-4b●) synapse on motor neurons in the anterior gray horns of the spinal cord. As they descend, these tracts are visible along the ventral surface of the medulla as a pair of thick bands, the **pyramids**. Along the length of the pyramids, roughly 85% of the axons cross the midline to enter the descending **later-al corticospinal tracts** on the opposite side of the spinal cord. The remaining axons continue uncrossed along the spinal cord as the **anterior corticospinal tracts**, but these fibers too will cross over within the anterior white commissure before synapsing on motor neurons in the anterior gray horns. Information concerning these tracts and their associated functions is summarized in Table 16-2.

TABLE 16-2   Principal Descending (Motor) Tracts in the Spinal Cord and a Summary of the Functional Roles of the Associated Nuclei in the Brain

Tract	Origin	Destination[a]	Actions	Comments
PYRAMIDAL TRACTS				
Corticobulbar tracts	Primary motor cortex (cerebral hemispheres)	Motor neurons of cranial nerve nuclei in brain stem	Voluntary motor control of skeletal muscles	Crosses to opposite side within brain stem
Lateral corticospinal tract	As above	Motor neurons of anterior gray horns of spinal cord	As above	Crosses to opposite side before entering spinal cord
Anterior corticospinal tract	As above	As above	As above	Descends uncrossed but crosses to opposite side before synapsing
EXTRAPYRAMIDAL TRACTS				
Vestibulospinal tract	Vestibular nucleus (near superior border of medulla oblongata)	Motor neurons of anterior gray horns	Involuntary regulation of balance in response to sensations from inner ear	Descends without crossing to opposite side
Tectospinal tract	Tectum (midbrain)	Motor neurons of anterior gray horns of cervical spinal cord	Involuntary regulation of eye, head, neck, and arm position in response to visual and auditory stimuli	Crosses to opposite side before entering spinal cord
Rubrospinal tract	Red nucleus (midbrain)	Motor neurons of anterior gray horns	Involuntary regulation of posture and muscle tone	Crosses to opposite side before entering spinal cord
Reticulospinal tract	Reticular formation (network of nuclei in brain stem)	Somatic and visceral motor neurons of anterior and lateral gray horns	Involuntary regulation of reflex activity and autonomic functions	Descends without crossing to opposite side

[a]Location of first synapse.

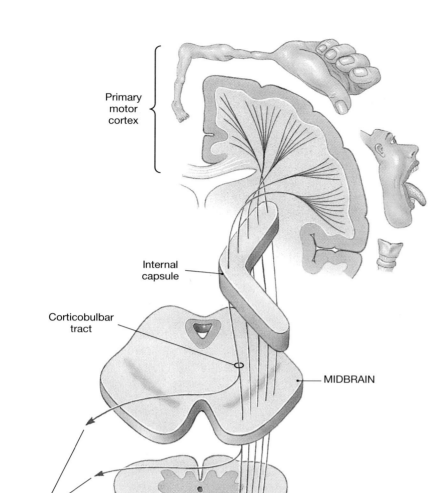

■ CLINICAL BRIEF
Cerebral Palsy

The term **cerebral palsy** refers to a number of disorders affecting voluntary motor performance that appear during infancy or childhood and persist throughout the life of the affected individual. The cause may be trauma associated with premature or unusually stressful birth, maternal exposure to drugs, including alcohol, or a genetic defect that causes improper development of the motor pathways. Prematurity has become less of a factor as maintenance and support procedures have improved, but difficult labor and deliveries still pose a problem. If the oxygen concentration in the fetal blood declines significantly for as little as 5–10 minutes, CNS function may be permanently impaired. Because of their high rates of oxygen consumption, the cerebral cortex, cerebellum, cerebral nuclei, hippocampus, and thalamus are likely targets. Damage to these sites may produce abnormalities in motor performance, involuntary control of posture and balance, memory, speech, and learning abilities. ⬧ *Tay-Sachs Disease [p. 761]*

The Extrapyramidal System
(Figures 16-4b/16-5)

The axons of the pyramidal cells of the motor cortex descend to synapse on motor neurons in the spinal cord and brain. Because there are no intervening synapses, the pyramidal system provides a rapid and direct mechanism for the control of skeletal muscles. Several other centers may issue motor commands as a result of processing performed at an unconscious level. These centers and their associated tracts constitute the **extrapyramidal system (EPS)**. By altering the sensitivity of motor neurons controlling muscles or groups of muscles, and by feedback loops to the primary motor cortex, the extrapyramidal system can modify or direct somatic motor patterns.

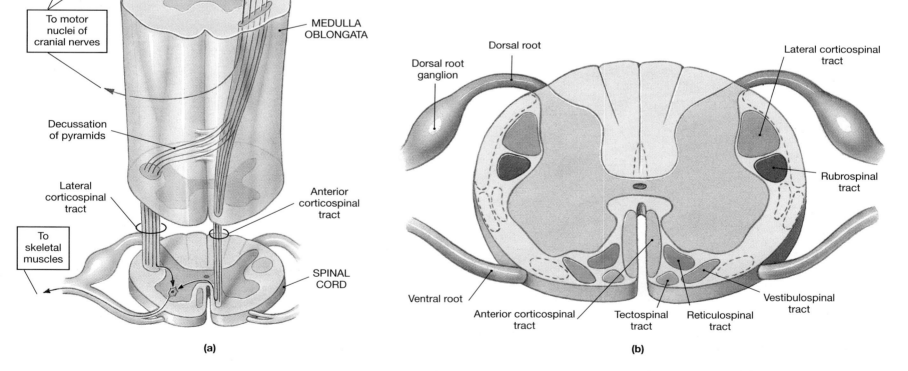

(a)

(b)

FIGURE 16-4
The Pyramidal System. (a) The pyramidal system originates at the primary motor cortex. Axons of the pyramidal cells descend in the internal capsule. The corticobulbar tracts end at the motor nuclei of cranial nerves. Most of the remaining pyramidal fibers cross over in the medulla before descending into the spinal cord. (b) A cross-sectional view of the spinal cord indicating the locations of the descending pyramidal and extrapyramidal tracts. Sensory tracts, detailed in Figure 16-1, are outlined.

Processing centers of the EPS include the vestibular nucleus, the superior colliculus, the red nucleus, the reticular formation, and the cerebral nuclei. Figure 16-5● provides an overview of the locations of these components. For additional details, review appropriate sections of Chapter 15. ∞ [pp. 375–382] Their outputs may target the motor nuclei of cranial nerves or descend into the spinal cord through: (1) the *vestibulospinal tracts*, (2) the *tectospinal tracts*, (3) the *rubrospinal tracts*, and (4) the *reticulospinal tracts* (Figure 16-4b●).

1. *The Vestibular Nuclei and the Vestibulospinal Tracts*: The vestibular nuclei receive information, via N VIII, from receptors in the inner ear that monitor position and movement. These nuclei respond to changes in the orientation of the head, sending motor commands that alter the position of the neck, eyes, head, and limbs via the contraction or relaxation of appropriate muscles. The primary goal is the maintenance of posture and balance. The descending fibers within the spinal cord constitute the **vestibulospinal tracts**.

2. *The Tectum and the Tectospinal Tracts*: Commands carried by the **tectospinal tracts** change the position of the eyes, head, neck, and arms in response to bright lights, sudden movements, or loud noises. These tracts originate in the roof (tectum) of the midbrain, in the superior and inferior colliculi, and cross to the opposite side. The motor commands carried by the tectospinal tracts are triggered by visual and auditory stimuli; the colliculi also process inputs from the cerebrum, the cerebellum, the reticular formation, and other nuclei in the brain stem.

3. *The Red Nucleus and the Rubrospinal Tracts*: The red nucleus receives extensive inputs from cerebral nuclei, the cerebellum, and the reticular formation. The **rubrospinal** tracts (*ruber*, red) carry the motor responses to spinal motor neurons. Direct stimulation of the red nucleus increases muscle tone and may produce either flexion or extension of the axial skeleton, depending on the area affected.

4. *The Reticular Formation and Reticulospinal Tracts*: The reticular formation receives collateral inputs from almost every ascending and descending pathway, as well as via extensive interconnections with the cerebrum, the cerebellum, and nuclei within the brain stem. The motor commands carried by the **reticulospinal tracts** vary depending on the region stimulated. For example, stimulation of one region produces involuntary eye movements, whereas stimulation of a different area affects the respiratory muscles.

THE CEREBRAL NUCLEI. The **cerebral nuclei** are the most important and complex components of the extrapyramidal system. ∞ [p. 377] These nuclei are processing centers whose function is to provide the background patterns of movement involved in the performance of voluntary motor activities. These nuclei adjust the motor commands issued in other processing centers, rather than exerting direct control over lower motor neurons. Although they do not initiate specific movements, once a movement is under way, the cerebral nuclei provide a background pattern and rhythm.

For example, consider what happens when you begin to walk; the basic rhythm and the motor patterns involved in moving your legs, shifting your weight, swinging your arms, and so forth are directed by the cerebral nuclei. When you make the conscious decision to start the movement, branches from the pyramidal tracts synapse in the cerebral nuclei, which respond by issuing the characteristic stereotyped motor commands that continue until you decide to stop walking.

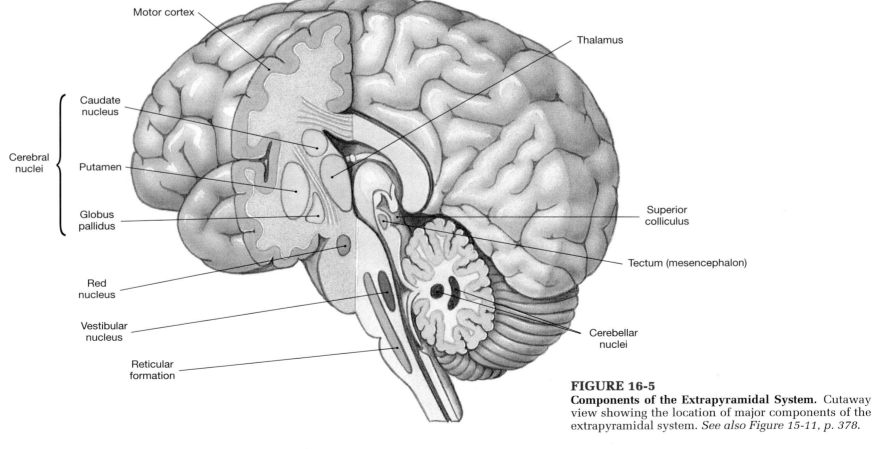

FIGURE 16-5
Components of the Extrapyramidal System. Cutaway view showing the location of major components of the extrapyramidal system. *See also Figure 15-11, p. 378.*

The Cerebellum

The pyramidal and extrapyramidal systems could not function effectively if their outputs were not continually readjusted as movements occurred. The complex processing and integration of neural information from peripheral structures, visual information from the eyes, and equilibrium-related sensations from the inner ear are not performed by either the cerebral cortex or the cerebral nuclei. Instead, these correlations occur within the **cerebellum**, and the cerebellar output regulates the activity along both pyramidal and extrapyramidal motor pathways.

The cerebellum adjusts muscle tone in the postural muscles of the body and coordinates complex motor patterns. ∞ [p. 387] The cerebellum receives proprioceptive information from the spinal cord along the spinocerebellar tracts. Additional sensory information arrives from the reticular formation and the vestibular nuclei. Some of the afferents pass through the arbor vitae to reach the superficial layers of the cerebellar cortex, while others synapse in cerebellar nuclei.

Both pyramidal and extrapyramidal centers send information to the cerebellum when motor commands are issued. As the movement proceeds, the cerebellum monitors proprioceptive (position) and vestibular (balance) information and adjusts the activities of the motor centers involved. In general, any voluntary movement begins with the activation of far more motor units than are actually required or even desirable. The cerebellum provides the necessary inhibition, reducing the number of motor units to the pattern and degree of inhibition required to produce the desired result. Extremely complex behaviors require considerable effort and practice until the cerebellum learns the proper motor pattern. For example, the cerebellum controls the smooth, rhythmic, and automatic movements of dancers and tennis players.

Levels of Somatic Motor Control
(Figure 16-6)

Ascending sensory information is relayed from one nucleus or center to another in a series of steps. For example, somatic sensory information from the spinal cord goes from a nucleus in the medulla to a nucleus in the thalamus before it reaches the primary sensory cortex. Information processing occurs at each step along the way, with results that may block, reduce, or heighten conscious awareness of the stimulus.

These processing steps are important, but they take time. Every synapse means another synaptic delay. Between the conduction time and the synaptic delays, several milliseconds can pass before the primary sensory cortex receives information from a peripheral receptor. Additional time will pass before the primary motor cortex orders a voluntary motor response.

This delay is not dangerous, because interim motor commands are issued by relay stations in the spinal cord and brain stem. While the conscious mind is still processing the information, neural reflexes provide an immediate response that can later be "fine-tuned." For example, when someone touches a hot stove top, the response (withdrawing the hand) occurs before the individual is aware of the injury. Voluntary motor responses, such as shaking the hand, stepping back, and crying out, occur somewhat later. In this case the initial reflexive response, directed by neurons in the spinal cord, was supplemented by a voluntary response controlled by the cerebral cortex. The spinal reflex provided a rapid, automatic, preprogrammed response that preserved homeostasis. The cortical response was more complex, but it required more time to prepare and execute.

Nuclei in the brain stem are also involved in a variety of complex reflexes. Some of these nuclei receive sensory information and generate appropriate motor responses. These motor responses may involve direct control over motor neurons or the regulation of reflex centers in other parts of the brain. This pattern can be seen in Figure 16-6●, which diagrams the levels of somatic motor control.

Motor neurons controlled by simple reflexes are at the bottom level. These reflexes are directed by nuclei in the spinal cord and brain stem. Higher levels perform more elaborate processing; as one moves from the medulla to the cerebral cortex, the motor patterns become increasingly complex and variable. For example, the respiratory rhythmicity center of the medulla sets a basic breathing rate. Centers in the pons adjust that rate in response to commands received from the hypothalamus (involuntary) or cerebral cortex (voluntary).

The cerebral nuclei, cerebellum, midbrain, and hypothalamus control the most complicated involuntary motor patterns. Examples include motor patterns associated with eating and reproduction (hypothalamus), walking and body positioning (cerebral nuclei), learned movement patterns (cerebellum), and movements in response to sudden visual or auditory stimuli (midbrain).

At the highest level are the complex, variable, and voluntary motor patterns dictated by the cerebral cortex. Motor commands may be given to specific motor neurons directly, or they may be given indirectly by altering the activity of a reflex control center. Figure 16-6b,c● provides a simple diagram of the steps involved in the planning and execution of a voluntary movement.

During development the control levels appear in sequence, beginning with the spinal reflexes. More complex reflexes develop as the neurons grow and interconnect. The process proceeds relatively slowly, as billions of neurons establish trillions of synaptic connections. At birth neither the cerebral nor the cerebellar cortex is fully functional, and their abilities take years to mature. A number of anatomical factors, noted in earlier chapters, contribute to this maturation:

1. Cortical neurons continue to increase in number until at least age 1;

2. The brain grows in size and complexity until at least age 4;

3. Myelination of CNS axons continues at least until age 1–2, and peripheral myelination may continue through puberty.

As these events occur, cortical neurons continue to establish new interconnections that will have a long-term effect on the functional capabilities of the individual.

√ **As a result of pressure on her spinal cord, Jill cannot feel touch or pressure on her legs. What spinal tract is being compressed?**

√ **What is the anatomical reason for the left side of the brain controlling motor function on the right side of the body?**

√ **An injury to the superior portion of the motor cortex would affect what part of the body?**

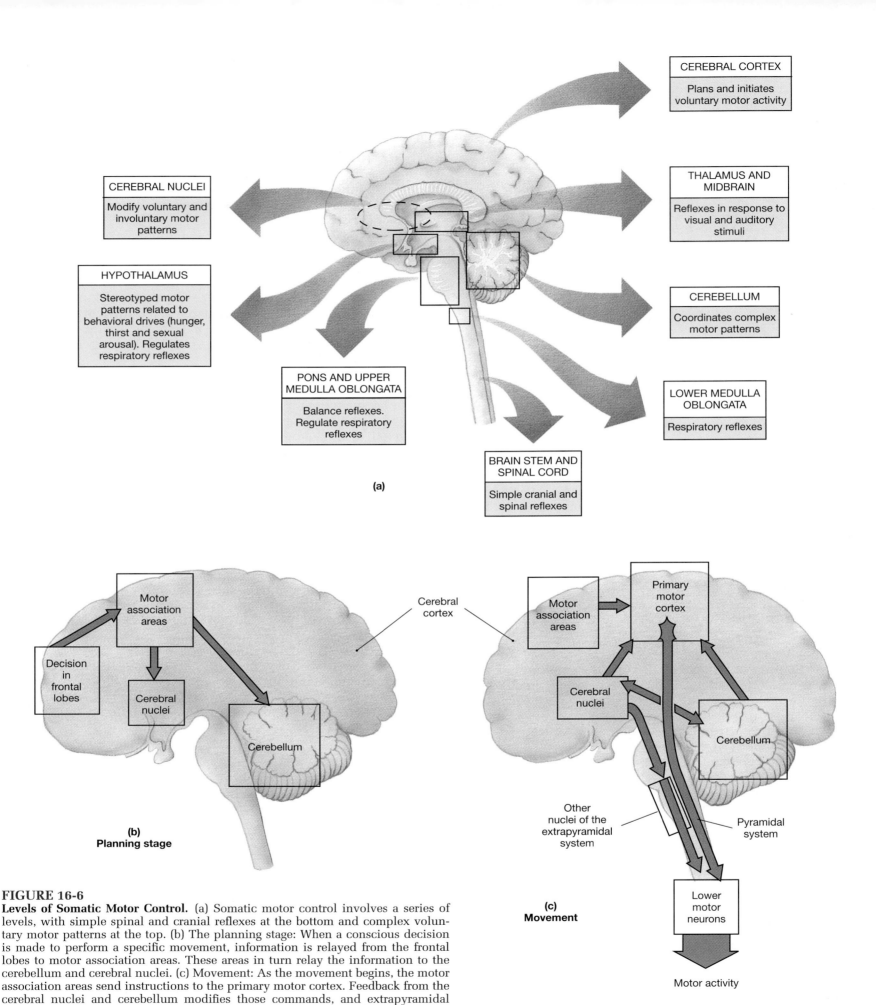

CEREBRAL CORTEX

Plans and initiates voluntary motor activity

THALAMUS AND MIDBRAIN

Reflexes in response to visual and auditory stimuli

CEREBELLUM

Coordinates complex motor patterns

LOWER MEDULLA OBLONGATA

Respiratory reflexes

CEREBRAL NUCLEI

Modify voluntary and involuntary motor patterns

HYPOTHALAMUS

Stereotyped motor patterns related to behavioral drives (hunger, thirst and sexual arousal). Regulates respiratory reflexes

PONS AND UPPER MEDULLA OBLONGATA

Balance reflexes. Regulate respiratory reflexes

BRAIN STEM AND SPINAL CORD

Simple cranial and spinal reflexes

(a)

Motor association areas

Decision in frontal lobes

Cerebral nuclei

Cerebellum

Cerebral cortex

(b)
Planning stage

Motor association areas

Primary motor cortex

Cerebral nuclei

Cerebellum

Other nuclei of the extrapyramidal system

Pyramidal system

Lower motor neurons

Motor activity

(c)
Movement

FIGURE 16-6
Levels of Somatic Motor Control. (a) Somatic motor control involves a series of levels, with simple spinal and cranial reflexes at the bottom and complex voluntary motor patterns at the top. (b) The planning stage: When a conscious decision is made to perform a specific movement, information is relayed from the frontal lobes to motor association areas. These areas in turn relay the information to the cerebellum and cerebral nuclei. (c) Movement: As the movement begins, the motor association areas send instructions to the primary motor cortex. Feedback from the cerebral nuclei and cerebellum modifies those commands, and extrapyramidal motor output directs involuntary adjustments in position and muscle tone.

HIGHER-ORDER FUNCTIONS

Higher-order functions have the following characteristics:

1. They are performed by the cerebral cortex.
2. They involve complex interactions between areas within the cortex and between the cerebral cortex and other areas of the brain.
3. They involve both conscious and unconscious information processing.
4. They are not part of the programmed "wiring" of the brain; therefore, the functions are subject to modification and adjustment over time.

Our discussion of higher-order function begins by identifying the cortical areas involved and considering functional differences between the left and right cerebral hemispheres. We will then briefly consider the mechanisms of memory, learning, and consciousness.

Integrative Regions of the Cerebral Cortex
(Figure 16-7)

The sensory, motor, and association areas of the cerebral hemispheres were introduced in Chapter 15. ∞ [p. 376] Figure 16-7a● reviews the major cortical regions of the left cerebral hemisphere. Scanning data, electrical monitoring, and clinical observation have shown that several cortical areas act as higher-order integrative centers for complex sensory stimuli and motor responses. These centers (Figure 16-7b●) include the *general interpretive area*, the *speech center*, and the *prefrontal cortex*.

The General Interpretive Area

The **general interpretive area**, or *gnostic area*, receives information from all the sensory association areas. This analytical center is present in only one hemisphere, usually the left. Damage to the general interpretive area affects the ability to interpret what is read or heard, even though the words are understood as individual entities. For example, an individual might understand the meaning of the words "sit" and "here" but be totally bewildered by the request "sit here."

The Speech Center

Efferents from the general interpretive area target the **speech center**, or *Broca's area*. This center lies along the edge of the premotor cortex in the same hemisphere as the general interpretive area. The speech center is a motor center that regulates the patterns of breathing and vocalization needed for normal speech. The corresponding regions on the opposite hemisphere are not "inactive," but their functions are less well defined. Damage to the speech center can manifest itself in various ways. Some individuals have difficulty speaking, although they know exactly what words to use; others talk constantly but use all the wrong words.

The Prefrontal Cortex

The **prefrontal cortex** of the frontal lobe coordinates information from the secondary and special association areas of the entire cortex. In doing so it performs such abstract intellectual functions as predicting the future consequences of events or actions. Damage to the prefrontal cortex leads to difficulties in estimating the temporal relationships between events; questions such as "How long ago did this happen?" or "What happened first?" become difficult to answer. ⊤ *Huntington's Disease [p. 761]*

The prefrontal cortex has extensive connections with other cortical areas and with other portions of the brain, such as the limbic system. Feelings of frustration, tension, and anxiety are generated at the prefrontal cortex as it interprets ongoing events and makes predictions about future situations or consequences. If the connections between the prefrontal cortex and other brain regions are severed, the tensions, frustrations, and anxieties are removed. Earlier in this century, this rather drastic procedure, called a **prefrontal lobotomy**, was used to "cure" a variety of mental illnesses, especially those associated with violent or antisocial behavior.

After a lobotomy, the patient would no longer be concerned about what had previously been a major problem, whether psychological (hallucinations) or physical (severe pain). However, the individual was often equally unconcerned about tact, decorum, and toilet training. Now that drugs have been developed to target specific pathways and regions of the CNS, lobotomies are no longer used to control behavior.

Hemispheric Specialization
(Figures 16-7a/16-8)

The regions seen in Figure 16-7a● are present on both hemispheres, but the higher-order functions are not equally distributed. Figure 16-8● indicates the major functional differences between the hemispheres. Higher-order centers in the left and right hemispheres have different but complementary func-

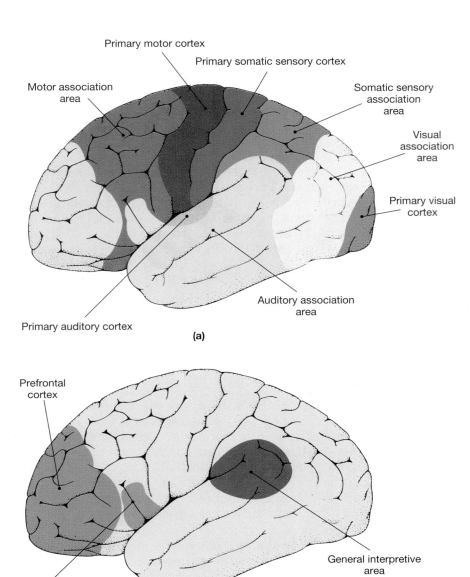

Primary motor cortex

Primary somatic sensory cortex

Motor association area

Somatic sensory association area

Visual association area

Primary visual cortex

Auditory association area

Primary auditory cortex

(a)

Prefrontal cortex

Speech center (Broca's area)

General interpretive area

(b)

FIGURE 16-7
Functional Areas of the Cerebral Cortex. (a) Centers responsible for higher-order functions are not found in both hemispheres. (b) The left hemisphere usually contains the general interpretive area and the speech center. Specializations of the two cerebral hemispheres are shown in Figure 16-8.

tions. In the majority of people in the United States, the left hemisphere contains the general interpretive and speech centers. This is the hemisphere responsible for language-based skills. For example, reading, writing, and speaking are dependent on processing done in the left cerebral hemisphere. This hemisphere is also important in performing analytical tasks, such as mathematical calculations and logical decision making. Although the left hemisphere was formerly called the *dominant hemisphere*, a more appropriate term is the **categorical hemisphere**, because the right hemisphere has many important functions.

The right cerebral hemisphere analyzes sensory information and relates the body to the sensory environment. Interpretive centers in this hemisphere permit the identification of familiar objects by touch, smell, taste, or feel. Because it is concerned with spatial relationships and analyses, the term **representational hemisphere** is used to refer to the right hemisphere.

The designation of a hemisphere as categorical, rather than representational, depends on the location of the major functional centers, especially the general interpretive and speech centers. Because the left hemisphere is categorical in the majority of both left-handed and right-handed individuals, there is probably a genetic basis for the distribution of functions. An estimated 90% of the population have an enlarged left hemisphere at birth. (The remaining 10% may have hemispheres of equal size or an enlarged right hemisphere.)

The functional significance (if any) of having the categorical hemisphere on the left side is unknown. Interestingly, there may be a link between being right- or left-handed and sensory and spatial abilities. An unusually high percentage of musicians and artists are left-handed; the complex motor activities performed by these individuals are directed by the primary motor cortex and association areas on the right (representational) hemisphere.

Hemispheric specialization does not mean that the two hemispheres are independent, merely that certain centers have evolved to process information gathered by the system as a whole. The intercommunication occurs over commissural fibers, especially those of the corpus callosum. The corpus callosum alone contains over 200 million axons, carrying an estimated 4 billion impulses per second!

■ **CLINICAL BRIEF**
Damage to the Integrative Centers

Aphasia (*a-*, without + *phasia*, speech) is a disorder affecting the ability to speak or read. Extreme, or **global**, **aphasia**, results from extensive damage to the general interpretive area or to the associated sensory tracts. Affected individuals are totally unable to speak, to read, or to understand or interpret the speech of others. Global aphasia often accompanies a severe stroke or tumor that affects a large area of cortex including the speech and language areas. Recovery is possible when the condition results from edema or hemorrhage, but the process often takes months or even years.

Dyslexia (*lexis*, diction) is a disorder affecting the comprehension and use of words. **Developmental dyslexia** affects children; there are estimates that up to 15% of children in the United States suffer from some degree of dyslexia. These children have difficulty reading and writing, although their other intellectual functions may be normal or above normal. Their writing looks uneven and disorganized; letters are often reversed or written in the wrong order. Recent evidence suggests that at least some forms of dyslexia result from problems in processing and sorting visual information.

Memory

Higher-order functions such as memory involve considerable interaction between the cerebral cortex and other areas of the brain. *Memory* is accessing stored bits of information gathered through experience. Some memories can be voluntarily retrieved and verbally expressed, such as recalling a phone number. Other memories are involuntarily accessible; for example, when hungry you may salivate at the smell of food. There are *short-term memories* that last seconds to hours, and *long-term memories* that can last for years. Each type of memory involves different anatomical structures. The conversion from a short-term to a long-term memory is called *memory consolidation*.

Two components of the limbic system (Figure 15-12●), the amygdaloid body and the hippocampus, are essential to memory

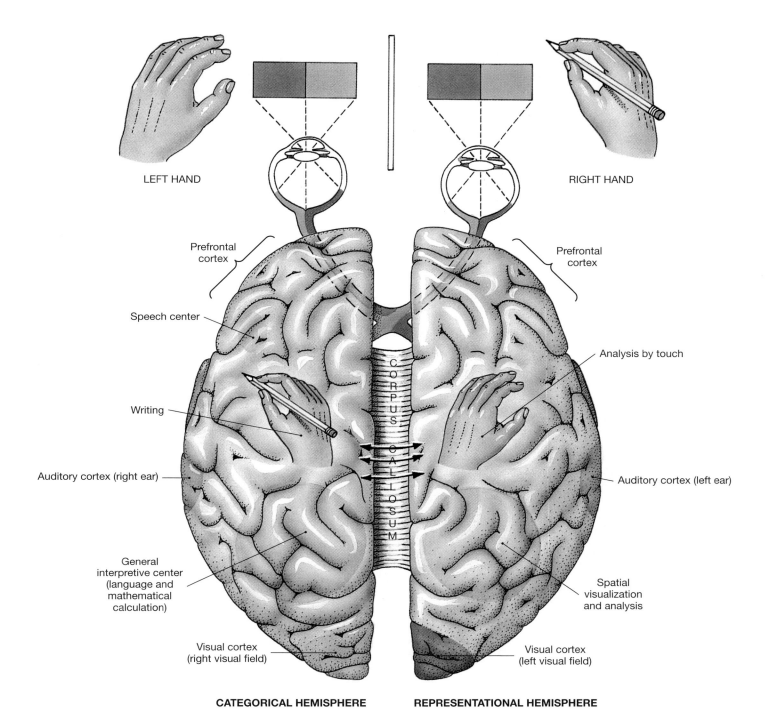

LEFT HAND

RIGHT HAND

Prefrontal cortex

Prefrontal cortex

Speech center

Analysis by touch

Writing

Auditory cortex (right ear)

Auditory cortex (left ear)

CORPUS CALLOSUM

General interpretive center (language and mathematical calculation)

Spatial visualization and analysis

Visual cortex (right visual field)

Visual cortex (left visual field)

CATEGORICAL HEMISPHERE

REPRESENTATIONAL HEMISPHERE

FIGURE 16-8
Hemispheric Specialization. Functional differences between the left and right cerebral hemispheres. Notice that special sensory information is relayed to the cerebral hemisphere on the opposite side of the body. Chapter 18 will provide additional details on these pathways.

■ CLINICAL BRIEF
Disconnection Syndrome

Communication across the corpus callosum permits the integration of sensory information and motor commands. Yet the two hemispheres are significantly different in terms of their ongoing processing activities. Otherwise untreatable seizures may sometimes be "cured" by severing the corpus callosum. This surgery produces symptoms of **disconnection syndrome**. In this condition the two hemispheres function independently, each remaining "unaware" of stimuli or motor commands involving their counterpart. The result is a number of rather interesting changes in the individual's abilities. For example, objects touched by the left hand can be recognized but not verbally identified, because the sensory information arrives at the right hemisphere and the speech center is on the left. The object can be verbally identified if felt with the right hand, but the person will not be able to say whether or not it is the same object previously touched with the left hand. This problem with cross-referencing sensory information applies to all incoming sensations.

Two years after a surgical sectioning of the corpus callosum, the most striking behavioral abnormalities have disappeared, and the individual may test normally. In addition, individuals born without a functional corpus callosum do not show obvious sensory or motor deficits. In some way the CNS adapts to the situation, probably by increasing the amount of information transferred across the anterior commissure.

consolidation. Damage to either of those areas will interfere with normal memory consolidation. ∞ [p. 381] Damage to the hippocampus leads to an immediate loss of short-term memory, although long-term memories remain intact and accessible. Tracts leading from the amygdaloid body to the hypothalamus may link memories to specific emotions. A cerebral nucleus near the diencephalon, the *nucleus basalis*, plays an uncertain role in memory storage and retrieval. Tracts connect this nucleus with the hippocampus, amygdaloid body, and all areas of the cerebral cortex. Damage to this nucleus is associated with changes in emotional states, memory, and intellectual function (see the discussion of Alzheimer's disease later in this chapter).

Long-term memories are stored in the cerebral cortex. Conscious motor and sensory memories are referred to the appropriate association areas. For example, visual memories are stored in the visual association area, and memories of voluntary motor activity in the premotor cortex. Special portions of the occipital and temporal lobes retain the memories of faces, voices, and words.

■ CLINICAL BRIEF
Amnesia

Amnesia refers to the loss of memory from disease or trauma. The type of memory loss depends upon the specific regions of the brain affected. Damage to sensory association areas produces memory loss of sensations arriving at the adjacent sensory cortex. Damage to thalamic and limbic structures, especially the hippocampus, will affect memory storage and consolidation. Amnesia may occur suddenly or progressively, and recovery may be complete, partial, or nonexistent, depending on the nature of the problem.

In **retrograde amnesia** (*retro-*, behind), the individual loses memories of past events. Some degree of retrograde amnesia often follows a head injury, and accident victims are frequently unable to remember the moments preceding a car wreck. In **anterograde amnesia** (*antero-*, ahead), an individual may be unable to store additional memories, but earlier memories are intact and accessible. The problem appears to involve an inability to generate long-term memories. At least two drugs—diazepam (*Valium*) and *Halcion*—have been known to cause brief periods of anterograde amnesia. A person with permanent anterograde amnesia lives in surroundings that are always new. Magazines can be read, chuckled over, and then reread a few minutes later with equal pleasure, as if they had never been seen before. Physicians and nurses must introduce themselves at every meeting, even if they have been treating the patient for years.

Post-traumatic amnesia (PTA) often develops after a head injury. The duration of the amnesia varies depending on the severity of the injury. PTA combines the characteristics of retrograde and anterograde amnesia; the individual can neither remember the past nor consolidate memories of the present.

√ **After suffering a head injury in an automobile accident, David has difficulty comprehending what he hears or reads. This problem might indicate damage to what portion of the brain?**

Consciousness: The Reticular Activating System
(Figure 16-9)

A conscious individual is alert and attentive; an unconscious individual is not. The difference is obvious, but there are many gradations of both the conscious and unconscious states. The state of consciousness experienced by an individ-

ual is determined by complex interactions between the brain stem and the cerebral cortex. One of the most important brain stem components is the **reticular activating system (RAS)**, a poorly defined network in the reticular formation. The RAS extends from the mesencephalon to the medulla oblongata (Figure 16-9●). The output of the RAS projects throughout the cerebral cortex. When the RAS is inactive, so is the cerebral cortex; stimulation of the RAS produces a widespread activation of the cerebral cortex.

The mesencephalic portion of the RAS appears to be the center of the system, and stimulation of this area produces the most pronounced and long-lasting effects on the cerebral cortex. Stimulating other portions of the RAS seems to have an effect only insofar as it changes the activity of this region. The greater the stimulation to the mesencephalic region of the RAS, the more alert and attentive the individual will be to incoming sensory information. Associated nuclei in the thalamus play a supporting role by focusing attention on specific mental processes.

AGING AND THE NERVOUS SYSTEM

The aging process affects all bodily systems, and the nervous system is no exception. Anatomical changes begin shortly after maturity (probably by age 30) and accumulate over time. Although an estimated 85% of the elderly (above age 65) lead relatively normal lives, there are noticeable changes in mental performance and CNS functioning.

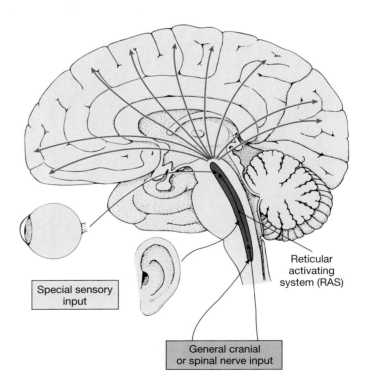

Special sensory input

Reticular activating system (RAS)

General cranial or spinal nerve input

FIGURE 16-9
The Reticular Activating System. The mesencephalic headquarters of the reticular formation receives collateral inputs from a variety of sensory pathways. Stimulation of this region produces arousal and heightened states of attentiveness.

Normal individuals cycle between the alert, conscious state and the asleep state each day. For reference purposes, Table 16-3 indicates the entire range of conscious and unconscious states, ranging from **delirium** through *coma*. It is important to realize that these states are external indications of the level of ongoing CNS activity. When CNS function becomes abnormal, the state of consciousness can be affected. As a result, clinicians are quick to note any abnormalities in the state of consciousness in their patients.

TABLE 16-3   States of Awareness

Level or State	Description
CONSCIOUS STATES	
Delirium	Disorientation, restlessness, confusion, hallucinations, agitation, alternating with other conscious states
Dementia	Difficulties with spatial orientation, memory, language, changes in personality
Confusion	Reduced awareness, easily distracted, easily startled by sensory stimuli, alternates between drowsiness and excitability; resembles minor form of delirium state
Normal consciousness	Aware of self and external environment, well-oriented, responsive
Somnolence	Extreme drowsiness, but will respond normally to stimuli
Chronic vegetative state	Conscious but unresponsive, no evidence of cortical function
UNCONSCIOUS STATES	
Asleep	Can be aroused by normal stimuli (light touch, sound, etc.)
Stupor	Can be aroused by extreme and/or repeated stimuli
Coma	Cannot be aroused and does not respond to stimuli (coma states can be further subdivided according to the effect on reflex responses to stimuli)

Age-related anatomical changes in the nervous system that are commonly seen include:

1.   A reduction in brain size and weight. This reduction results primarily from a decrease in the volume of the cerebral cortex. The brains of elderly individuals have narrower gyri and wider sulci than those of young persons, and the subarachnoid space and ventricles are enlarged.

2.   A reduction in the number of neurons. Brain shrinkage has been linked to a loss of cortical neurons, although evidence exists that neuronal loss does not occur to the same degree in all individuals nor in all brain stem nuclei.

3.   A decrease in blood flow to the brain. With age, fatty deposits gradually accumulate in the walls of blood vessels. These deposits reduce the rate of blood flow through arteries. (This process, called *atherosclerosis*, may affect arteries throughout the body; it is described further in Chapter 22.) The reduction in blood flow does not cause a cerebral crisis, but it does increase the chances that the individual will suffer a stroke.

4.   Changes in synaptic organization of the brain. The number of dendritic branches and interconnections appears to decrease. As synaptic connections are lost, the rate of neurotransmitter production declines.

5.   Intracellular and extracellular changes in CNS neurons. Many neurons in the brain begin accumulating abnormal intracellular deposits, such as *plaques* or *neurofibrillary tangles*. *Plaques* are accumulations of an unusual fibrillar protein, *amyloid*, surrounded by abnormal dendrites and axons. *Neurofibrillary tangles* are masses of neurofibrils that form dense mats inside the soma. The significance of these cellular and extracellular abnormalities remains to be determined. There is evidence that they appear in all aging brains (see the discussion of Alzheimer's disease on the next page), but when present in excess they seem to be associated with clinical abnormalities.

These anatomical changes are linked to a series of functional alterations. In general, neural processing becomes less efficient. For example, memory consolidation often becomes more difficult, and the sensory systems of the elderly, notably hearing, balance, vision, smell, and taste, become less acute. Light must be brighter, sounds louder, and smells stronger before they are perceived. Reaction times are slowed, and reflexes—even some monosynaptic reflexes—become weaker or even disappear. There is a decrease in the precision of motor control, and it takes longer to perform a given motor pattern than it did 20 years earlier. For the majority of the elderly population, these changes do not interfere with their abilities to function in society. But for as yet unknown reasons, many elderly individuals become incapacitated by progressive CNS changes. By far the most common such incapacitating condition is *Alzheimer's disease*.

√ **What would happen to a sleeping individual if his or her reticular activating system (RAS) were suddenly stimulated?**

√ **Which major anatomical structures in the CNS are affected by aging?**

Cerebrovascular diseases are circulatory disorders that interfere with the normal circulatory supply to the brain. The particular distribution of the vessel involved will determine the symptoms, and the degree of oxygen or nutrient starvation will determine their severity. A stroke, or **cerebrovascular accident (CVA)**, occurs when the blood supply to a portion of the brain is shut off by a vascular blockage or hemorrhage. Affected neurons begin to die in a matter of minutes.

The symptoms of a stroke provide an indication of the vessel and region of the brain involved. For example, the carotid artery enters the skull via the carotid foramen. One major branch of the carotid, the *middle cerebral artery*, is the most common site of a stroke. Superficial branches deliver blood to the temporal lobe and to large portions of the frontal and parietal lobes; deep branches supply the cerebral nuclei and portions of the thalamus. If a stroke blocks the middle cerebral artery on the left side of the brain, aphasia and a sensory and motor paralysis of the right side result. In a stroke affecting the middle cerebral artery on the right side, the individual experiences a loss of sensation and motor control over the left side and has difficulty drawing or interpreting spatial relationships. Strokes affecting vessels supplying the brain stem also produce distinctive symptoms; those affecting the lower brain stem are often fatal. (Further information on the causes, diagnosis, and treatment of strokes will be found in Chapter 22.)

Alzheimer's disease is a progressive disorder characterized by the loss of higher cerebral functions. It is the most common cause of **senile dementia**, commonly termed "senility." The first symptoms usually appear at 50–60 years of age. Alzheimer's disease affects an estimated 2 million people over age 65 in the United States, who suffer from some form of the condition. This disease causes approximately 100,000 deaths each year.

Alzheimer's disease produces a gradual deterioration of mental organization. The afflicted individual loses memories, verbal and reading skills, and emotional control. Initial symptoms are subtle—moodiness, irritability, depression, and a general lack of energy. As the condition progresses, however, it becomes more difficult to ignore or accommodate. The victim has difficulty making decisions, even minor ones. Mistakes—sometimes dangerous ones—are made, through either bad judgment or simple forgetfulness.

The memory losses continue, and the problems become more severe. The affected person may forget relatives or how to use the telephone. The memory loss often starts with an inability to store long-term memories, followed by the loss of recently stored memories, and eventually the loss of basic long-term memories, such as the sound of the victim's own name. The loss of memory affects both intellectual and motor abilities, and a patient with severe Alzheimer's disease has difficulty in performing even the simplest motor tasks.

Individuals with Alzheimer's disease show a pronounced decrease in the number of cortical neurons, especially in the frontal and temporal lobes. There are also unusually large concentrations of plaques and neurofibrillary tangles in the nucleus basalis, hippocampus, and parahippocampal gyrus. In addition, an abnormal protein, called **Alzheimer's disease-associated protein** (ADAP) appears in brain regions, such as the hippocampus, specifically associated with memory processing. Because this protein also appears in small quantities in the cerebrospinal fluid of many Alzheimer's patients, a blood screening test is now being developed to detect the condition before mental deterioration becomes pronounced. There is no treatment for the condition, and the exact cause—genetic, environmental, or pathogenic—remains a mystery.

Related Clinical Terms

cerebral palsy: A number of disorders affecting voluntary motor performance that appear during infancy or childhood and persist throughout the life of the affected individual. *[p. 415]*

Tay-Sachs' disease: A disease resulting from a genetic abnormality involving the metabolism of *gangliosides*. Affected infants seem normal at birth, but the progress of symptoms typically includes muscular weakness, blindness, seizures, and death, usually before age 4. ✝ *Tay-Sachs' Disease [p. 761]*

anencephaly (an-en-SEF-a-lē): A rare condition in which the brain fails to develop at levels above the mesencephalon or lower diencephalon. *[p. 419]*

aphasia: A disorder affecting the ability to speak or read. *[p. 420]*

dyslexia: A disorder affecting the comprehension and use of words. *[p. 420]*

disconnection syndrome: Syndrome caused by severing the corpus callosum and separating the two cerebral hemispheres. Each hemisphere continues to function independently, and the right hand literally doesn't know what the left hand is doing. *[p. 421]*

amnesia: The loss of memory from disease or trauma. *[p. 422]*

delirium: A conscious state involving confusion and wild oscillations in the level of consciousness. *[p. 423]*

dementia: A chronic state of consciousness characterized by deficits in memory, spatial orientation, language, or personality. *[p. 423]*

cerebrovascular diseases: Circulatory disorders that interfere with the normal circulatory supply to the brain. *[p. 423]*

cerebrovascular accident (CVA): Also known as "stroke"; occurs when the blood supply to a portion of the brain is shut off. *[p. 423]*

Alzheimer's disease: A progressive disorder marked by the loss of higher cerebral functions. *[p. 424]*

CHAPTER SUMMARY AND REVIEW

STUDY OUTLINE

Related Key Terms

INTRODUCTION *[p. 410]*

1. There is continual communication between the brain, spinal cord, and peripheral nerves.

SENSORY AND MOTOR PATHWAYS *[p. 410]*

Sensory Pathways *[p. 410]*
1. A **sensation** arrives in the form of an action potential in an afferent fiber. The **posterior column pathway** carries fine touch, pressure, and proprioceptive sensations. The axons ascend within the **fasciculus gracilis** and **fasciculus cuneatus**, synapsing in the nucleus gracilis and nucleus cuneatus. Information is then relayed to the thalamus via the **medial lemniscus**. As the axons enter the medial lemniscus, they **decussate** (cross over) to the opposite side of the brain stem *(see Figures 16-1/16-2 and Table 16-1)*.

first-order neuron • second-order neuron • third-order neuron • somatosensory pathways • sensory homunculus

424

2. The **spinothalamic pathway** carries poorly localized sensations of touch, pressure, pain, and temperature. The axons decussate in the spinal cord and ascend via the **anterior** and **lateral spinothalamic tracts** to the ventral posterolateral nuclei of the thalamus *(see Figure 16-2 and Table 16-1)*.

3. The **spinocerebellar pathway**, including the **posterior** and **anterior spinocerebellar tracts**, carries sensations to the cerebellum concerning the position of muscles, tendons, and joints *(see Figures 16-1/16-2 and Table 16-1)*.

Motor Pathways [p. 413]
4. Somatic motor pathways always involve an **upper motor neuron** (whose soma lies in a CNS processing center) and a **lower motor neuron** (located in a motor nucleus of the brain stem or spinal cord). Autonomic motor control requires a **preganglionic neuron** (in the CNS) and a **ganglionic neuron** (in a peripheral ganglion) *(see Figures 16-3 to 16-5)*.

5. The neurons of the primary motor cortex are **pyramidal cells**; the **pyramidal system** provides voluntary skeletal muscle control. The **corticobulbar tracts** terminate at the cranial nerves, while the **corticospinal tracts** synapse on motor neurons in the anterior gray horns of the spinal cord. The corticospinal tracts are visible along the medulla as a pair of thick bands, the **pyramids**, where most of the axons decussate to enter the descending **lateral corticospinal tracts** or the **anterior corticospinal tracts**. The pyramidal system provides a rapid, direct mechanism for controlling skeletal muscles *(see Figure 16-4 and Table 16-2)*.

6. The **extrapyramidal system (EPS)** consists of several other centers that may issue motor commands as a result of processing performed at an unconscious, involuntary level. Its outputs may descend via the vestibulospinal, tectospinal, rubrospinal, or reticulospinal tracts *(see Figures 16-4/16-5 and Table 16-2)*.

7. The **vestibulospinal tracts** carry information related to maintaining balance and posture. Commands carried by the **tectospinal tracts** change the position of the eyes, head, neck, and arms in response to bright lights, sudden movements, or loud noises. The **rubrospinal tracts** carry motor responses to spinal motor neurons. Motor commands carried by the **reticulospinal tracts** vary according to the region stimulated *(see Figure 16-5 and Table 16-2)*.

8. The **cerebral nuclei** use three major pathways to adjust the motor commands issued in other processing centers: (1) One group of axons synapses with thalamic neurons and creates a feedback loop that changes the sensitivity of pyramidal cells and alters the instructions carried by the corticospinal tracts; (2) another group innervates the red nucleus and alters activity in the rubrospinal tracts; (3) the third group travels through the thalamus to reach centers in the reticular formation where they adjust the output of the reticulospinal tracts *(see Figure 16-5 and Table 16-2)*.

9. The **cerebellum** regulates the activity along both pyramidal and extrapyramidal motor pathways. The integrative activities performed by neurons in the cerebellar cortex and cerebellar nuclei are essential for precise control of voluntary and involuntary movements *(see Figure 16-6)*.

Levels of Somatic Motor Control [p. 417]
10. Ascending sensory information is relayed from one nucleus or center to another in a series of steps. Information processing occurs at each step *(see Figure 16-6)*.

HIGHER-ORDER FUNCTIONS [p. 419]

1. Higher-order functions (1) are performed by the cerebral cortex; (2) involve complex interactions between areas of the cerebral cortex and between the cortex and other areas of the brain; (3) involve conscious and unconscious information processing; and (4) are subject to modification and adjustment over time *(see Figures 16-7 to 16-9)*.

Integrative Regions of the Cerebral Cortex [p. 419]
2. The **general interpretive area** receives information from all the sensory association areas. It is present in only one hemisphere, usually the left *(see Figures 16-7b/16-8)*.

3. The **speech center** regulates the patterns of breathing and vocalization needed for normal speech *(see Figure 16-7b)*.

4. The **prefrontal cortex** coordinates information from the secondary and special association areas of the entire cortex and performs abstract intellectual functions *(see Figure 16-7b)*.

Hemispheric Specialization [p. 419]
5. The left hemisphere is usually the **categorical hemisphere**; it contains the general interpretive and speech centers and is responsible for language-based skills. The right hemisphere, or **representational hemisphere**, is concerned with spatial relationships and analyses *(see Figures 16-7/16-8)*.

Memory [p. 420]
6. The amygdaloid body and the hippocampus are essential to memory consolidation *(see Figure 15-12)*.

Consciousness: The Reticular Activating System [p. 422]
7. Consciousness is determined by interactions between the brain stem and cerebral cortex; one of the most important brain stem components is a network in the reticular formation called the **reticular activating system (RAS)** *(see Figure 16-9)*.

Related Key Terms

motor homunculus

prefrontal lobotomy

AGING AND THE NERVOUS SYSTEM [p. 422]

1. Age-related changes in the nervous system include: (1) reduction in brain size and weight; (2) reduction in number of neurons; (3) decrease in blood flow to the brain; (4) changes in synaptic organization of the brain; and (5) intracellular and extracellular changes in CNS neurons.

1 REVIEW OF CHAPTER OBJECTIVES

1 Identify the principal sensory and motor pathways.

2. Identify the anatomical structures and functions of the pyramidal and extrapyramidal systems.

3. Describe the anatomical structures that allow us to distinguish between sensations that originate in different areas of the body.

4. Identify the major brain components that interact to determine somatic motor output.

5. Identify the integrative areas of the cerebral cortex and give their general functions.

6. Indicate significant functional differences between the left and right cerebral hemispheres.

7. Describe the regions and structures of the brain involved in memory storage and recall.

8. Explain the structure of the reticular activating system and how it maintains consciousness.

9. Summarize the effects of aging on the nervous system.

2 REVIEW OF CONCEPTS

1. What is the function of first-order neurons in the CNS?

2. What is the functional significance of decussation in transmitting information from the body to the brain or vice versa?

3. How are the abilities to discriminate the type and location of a sensation related?

4. Why is the sensory map (homunculus) distorted in comparison with the appearance of the structures in the body?

5. How do somatic motor pathways and visceral motor pathways differ?

6. How is the map of the motor homunculus related to the fineness of the motor control of any given area?

7. What is the role of the extrapyramidal system in processing motor commands in the CNS?

8. What is the role of the cerebral nuclei in the function of the extrapyramidal system?

9. How does the cerebellum regulate activity on the pyramidal and extrapyramidal motor pathways?

10. Why do students who "cram" for an exam retain the information for only a few days?

3 CRITICAL THINKING AND CLINICAL APPLICATION QUESTIONS

1. The skull of a baby that appears normal at birth seems to be translucent when a light is shined on it. But activities that a newborn usually engages in are normal. What is likely to be the reason for this observation, and what is the prognosis for this infant?

2. In an episode of a popular television science fiction series, an unknown virus infected space station crew members. The people suddenly began to speak unintelligibly, and they could not understand one another. What condition is similar to this, what is the usual cause, and how can it be treated?

17

The Nervous System: Autonomic Division

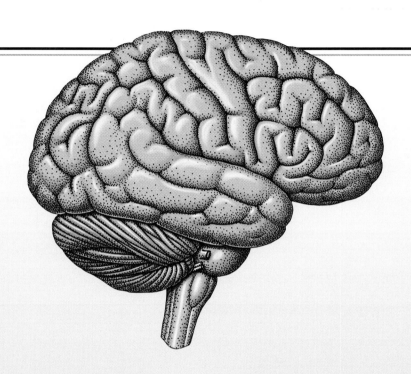

CHAPTER OUTLINE AND OBJECTIVES

Our conscious thoughts, plans, and actions represent only a tiny fraction of the activities of the nervous system. If all consciousness were eliminated, vital physiological processes would continue virtually unchanged, because routine adjustments in physiological systems are made by the autonomic nervous system (ANS). It is the ANS that regulates body temperature and coordinates cardiovascular, respiratory, digestive, excretory, and reproductive functions. In doing so, the ANS adjusts internal water, electrolyte, nutrient, and dissolved gas concentrations in body fluids without instructions or interference from the conscious mind.

This chapter examines the anatomical divisions and structure of the autonomic nervous system. Each of these divisions has a characteristic anatomical and functional organization. Our examination of the ANS begins with a description of the sympathetic and parasympathetic divisions. Then, we will briefly describe the way these divisions maintain and adjust various organ systems to meet the body's ever-changing physiological needs.

AN OVERVIEW OF THE AUTONOMIC NERVOUS SYSTEM (ANS)

It will be useful to compare the organization of the ANS with that of the somatic nervous system (SNS), which was discussed in Chapter 16. We will focus on (1) the pathways involved in visceral motor output and (2) the subdivisions of the ANS, based on structural and functional patterns of peripheral innervation.

Pathways for Visceral Motor Output
(Figure 17-1)

Lower motor neurons of the SNS exert direct control over skeletal muscles. ∞ [p. 413] In the ANS (Figure 17-1●), there is always a second neuron interposed between the visceral motor neuron in the CNS and the peripheral effector. Visceral motor neurons in the CNS, known as **first-order (preganglionic) neurons**, send their axons, called *preganglionic fibers*, to synapse on **second-order (ganglionic) neurons**, whose cell bodies are located in ganglia outside the CNS. Axons leaving the autonomic ganglia are relatively small and unmyelinated. The axon of a ganglionic neuron is called a *postganglionic fiber* because it carries impulses away from the ganglion. Postganglionic fibers innervate peripheral organs, such as cardiac muscle, smooth muscles, glands, and adipose tissues.

Subdivisions of the ANS
(Figure 17-2)

The ANS contains two subdivisions, the *sympathetic division* and the *parasympathetic division* (Figure 17-2●).

Sympathetic (Thoracolumbar) Division
(Figure 17-2)

Preganglionic fibers from the thoracic and upper lumbar spinal segments synapse in ganglia near the spinal cord.

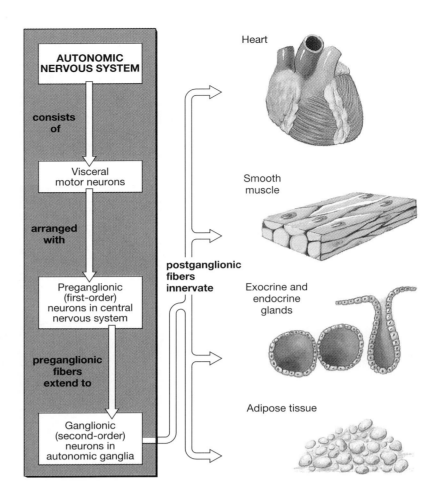

FIGURE 17-1
An Overview of the Autonomic Nervous System

These axons and ganglia are part of the **thoracolumbar** (thor-a-kō-LUM-bar) **division,** or **sympathetic division,** of the ANS (Figure 17-2●). The sympathetic division is often called the "fight or flight" system because an increase in sympathetic activity generally stimulates tissue metabolism, increases alertness, and prepares the body to deal with emergencies.

Parasympathetic (Craniosacral) Division
(Figure 17-2)

Preganglionic fibers originating in the brain and the sacral spinal cord segments are part of the **craniosacral** (krā-nē-ō-SĀ-kral) **division,** or **parasympathetic division,** of the ANS (Figure 17-2●). The preganglionic fibers synapse on neurons of **terminal ganglia** or **intramural ganglia** (*murus*, wall) located near or within the tissues of visceral organs. The parasympathetic division is often called the "rest and repose" system because it conserves energy and promotes sedentary type activities, such as digestion.

Innervation Patterns

The sympathetic and parasympathetic divisions affect target organs through the controlled release of neurotransmitters by postganglionic fibers. Whether the result is a stimulation or inhibition of activity depends on the response of the membrane

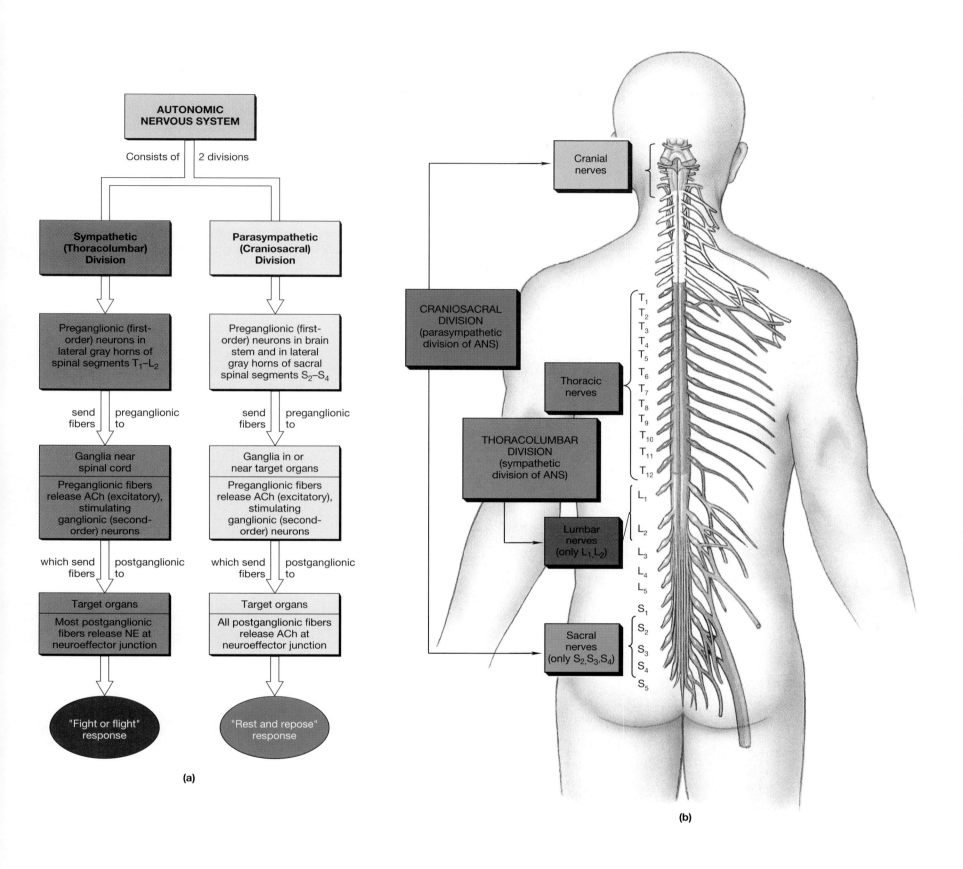

FIGURE 17-2
The Autonomic Nervous System. (a) Components of the autonomic nervous system (ANS). (b) Anatomical divisions of the autonomic nervous system. At the cranial and sacral levels, the visceral efferent fibers from the CNS constitute the parasympathetic or craniosacral division. At the thoracic and lumbar levels, the visceral efferent fibers that emerge form the sympathetic, or thoracolumbar, division.

receptor to the presence of the neurotransmitter. Three general statements can be made regarding the neurotransmitters and their effects:

1. All preganglionic autonomic fibers release acetylcholine (ACh) at their synaptic terminals. The effects are always excitatory.

2. Postganglionic parasympathetic fibers also release ACh, but the effects may be excitatory or inhibitory, depending on the nature of the receptor.

3. Most postganglionic sympathetic terminals release the neurotransmitter **norepinephrine (NE)**. The effects are usually excitatory.

THE SYMPATHETIC DIVISION
(Figure 17-3)

The sympathetic division (Figure 17-3●) consists of:

1. *Preganglionic (first-order) neurons located between segments T_1 and L_2 of the spinal cord.* These neurons are situated in the lateral gray horns, and their axons enter the ventral roots of those segments.

2. *Ganglionic (second-order) neurons in ganglia near the vertebral column.* There are two types of sympathetic ganglia:

- **Sympathetic chain ganglia**, also called *paravertebral ganglia*, or *lateral ganglia*, lie on either side of the vertebral column. Neurons in these ganglia control effectors in the body wall and inside the thoracic cavity.
- **Collateral ganglia**, also known as *prevertebral ganglia*, are found anterior to the vertebral centra. Neurons within the collateral ganglia innervate tissues and organs in the abdominopelvic cavity.

3. *Specialized second-order neurons in the interior of the adrenal gland.* The central core of each adrenal gland, an area known as the *adrenal medulla*, is a modified sympathetic ganglion. The ganglionic neurons of the medulla have very short axons, and when stimulated they release neurotransmitters into the bloodstream.

The Sympathetic Chain
(Figures 17-2a/17-4)

The ventral roots of spinal segments T_1 to L_2 contain sympathetic preganglionic fibers. The basic pattern of sympathetic innervation in these regions was described in Figure 14-2a●. ∞ [p. 341] After passing through the intervertebral foramen, each ventral root gives rise to a *white ramus*, or *white ramus communicans*, that carries preganglionic fibers into a nearby sympathetic chain ganglion. These fibers may synapse within the sympathetic

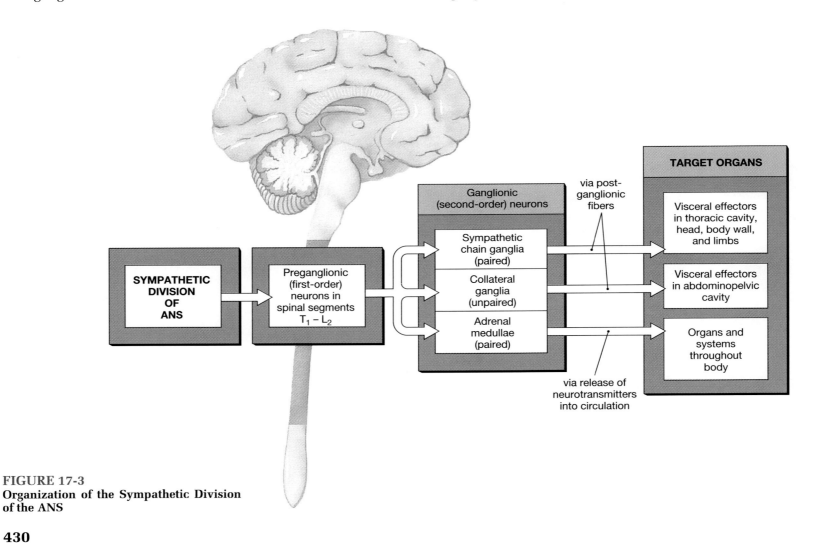

FIGURE 17-3
Organization of the Sympathetic Division of the ANS

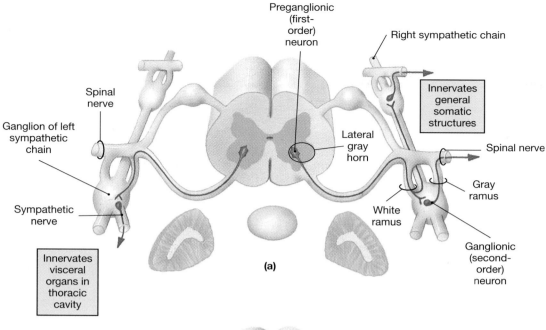

Preganglionic (first-order) neuron

Right sympathetic chain

Spinal nerve

Ganglion of left sympathetic chain

Innervates general somatic structures

Lateral gray horn

Spinal nerve

Gray ramus

Sympathetic nerve

White ramus

Ganglionic (second-order) neuron

Innervates visceral organs in thoracic cavity

(a)

Major effects produced by sympathetic postganglionic fibers in spinal nerves (Figure 17-4a):
- Constriction of cutaneous blood vessels and reduction in circulation to the skin as well as to most other organs in the body wall
- Acceleration of blood flow to skeletal muscles
- Stimulation of energy production and use by skeletal muscle tissue
- Release of stored lipids from subcutaneous adipose tissue
- Stimulation of secretion by sweat glands
- Dilation of the pupils to allow more light into the eyes and focusing of the eyes for viewing distant objects
- Stimulation of arrector pili muscles, producing "goose-bumps"

Major effects produced by postganglionic fibers entering the thoracic cavity in sympathetic nerves:
- Acceleration of the heart rate and increase in the strength of cardiac contractions. (The heart works harder, circulating blood at a higher rate and under increased pressures.)
- Dilation of the respiratory passageways. (Airflow becomes more efficient, improving the delivery of oxygen and the elimination of carbon dioxide.)

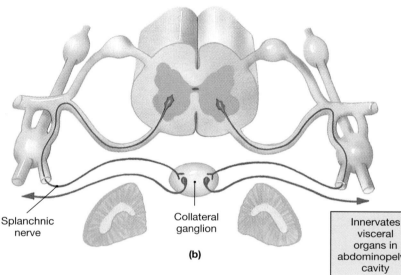

Splanchnic nerve

Collateral ganglion

Innervates visceral organs in abdominopelvic cavity

(b)

Major effects produced by preganglionic fibers innervating the collateral ganglia (Figure 17-4b):
- Constriction of small arteries and reduction in blood flow to visceral organs
- Decrease in the activity of digestive glands and organs, including the stomach, intestines, pancreas, and gall-bladder
- Stimulation of the release of glucose from glycogen reserves in the liver
- Stimulation of the release of lipids from adipose tissue
- Relaxation of the smooth muscle in the wall of the urinary bladder
- Reduction in the rate of urine formation at the kidneys
- Control of some aspects of sexual function, such as ejaculation in the male

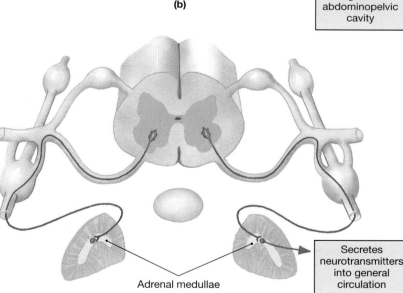

Adrenal medullae

Secretes neurotransmitters into general circulation

(c)

Major effects produced by preganglionic fibers innervating the adrenal medullae (Figure 17-4c):
- Release of epinephrine and norepinephrine into the general circulation

FIGURE 17-4
Sympathetic Pathways and Their General Functions. Preganglionic fibers leave the spinal cord in the ventral roots of spinal nerves. They synapse on ganglionic neurons in sympathetic chain ganglia (a), in collateral ganglia (b), or in the adrenal medullae (c). These sections are seen in inferior view, the standard format for radiographic scans and sectional views of the nervous system. ∞ [pp. 21, 340]

chain ganglia (Figure 17-4a●), at one of the collateral ganglia (Figure 17-4b●), or in the adrenal medullae (Figure 17-4c●). Extensive divergence occurs, with one preganglionic fiber synapsing on two dozen or more second-order (ganglionic) neurons. Preganglionic fibers running between the sympathetic chain ganglia interconnect them, making the chain resemble a string of beads. Each ganglion in the sympathetic chain innervates a particular body segment or group of segments.

If a preganglionic fiber carries motor commands that target structures in the body wall or the thoracic cavity, it will synapse in one or more of the sympathetic chain ganglia. Postganglionic fibers that control visceral effectors in the body wall, such as the sweat glands of the skin or the smooth muscles in superficial blood vessels, enter the *gray ramus* (*gray ramus communicans*) and return to the spinal nerve for subsequent distribution. However, spinal nerves do not innervate structures in the ventral body cavities. Postganglionic fibers targeting structures in the thoracic cavity, such as the heart and lungs, proceed directly to their peripheral targets as sympathetic nerves. These nerves are usually named after their primary targets; examples include the *cardiac* and *esophageal nerves*.

Functions of the Sympathetic Chain
(Figure 17-4a)

Postganglionic fibers leaving the sympathetic chain reach their peripheral targets via spinal nerves and sympathetic nerves. The primary results of increased activity in the postganglionic fibers leaving the sympathetic chain ganglia via spinal nerves and sympathetic nerves are summarized in Figure 17-4a●. In general, these responses help prepare the individual for a crisis that will require sudden, intensive physical activity. How these effects are brought about will be the focus of a later section.

Anatomy of the Sympathetic Chain
(Figure 17-5)

There are 3 cervical, 11–12 thoracic, 2–5 lumbar, 4–5 sacral, and 1 coccygeal sympathetic ganglia in each sympathetic chain. The numbers are variable because fusion of adjacent ganglia often occurs. The coccygeal ganglia from either side usually fuse to form a single median ganglion known as the *ganglion impar*. First-order sympathetic neurons are limited to segments T_1–L_2, and these spinal nerves have both white rami (preganglionic fibers) and gray rami (postganglionic fibers). The neurons in the cervical, lower lumbar, and sacral sympathetic chain ganglia are innervated by preganglionic fibers running along the axis of the chain. In turn, these chain ganglia provide postganglionic fibers, via gray rami, to the cervical, lumbar, and sacral spinal nerves. *Every spinal nerve has a gray ramus that carries sympathetic postganglionic fibers.* About 8% of the axons in each spinal nerve are sympathetic postganglionic fibers. As a result, the dorsal and ventral rami of the spinal nerves provide extensive sympathetic innervation to structures in the body wall and limbs. In the head, postganglionic fibers leaving the cervical chain ganglia supply the regions and structures innervated by cranial nerves N III, N VII, N IX, and N X (Figure 17-5●).

In summary: (1) *only the thoracic and upper lumbar ganglia receive preganglionic fibers via white rami*; (2) *the cervical, lower lumbar, and sacral chain ganglia receive preganglionic innervation via collateral fibers from first-order sympathetic neurons*; and (3) *every spinal nerve receives a gray ramus from a ganglion of the sympathetic chain.*

This anatomical arrangement has interesting functional consequences. If the ventral roots of thoracic spinal nerves are damaged, there will be no sympathetic motor function on the affected side of the head, neck, and trunk. Yet damage to the ventral roots of cervical spinal nerves will produce voluntary muscle paralysis on the affected side, *but leave sympathetic function intact* because the preganglionic fibers innervating the cervical ganglia originate in the white rami of thoracic segments, which are undamaged.

Collateral Ganglia
(Figures 17-4b/17-5)

The abdominopelvic viscera receive sympathetic innervation via preganglionic fibers that pass through the sympathetic chain without synapsing. These fibers originate at first-order neurons in the lower thoracic and upper lumbar segments. They synapse within separate *collateral ganglia* (Figures 17-4b and 17-5●). Preganglionic fibers that innervate the collateral ganglia form the **greater, lesser,** and **lumbar splanchnic** (SPLANK-nik) **nerves** in the dorsal wall of the abdominal cavity. Splanchnic nerves from both sides of the body converge on these ganglia. Although there are two sympathetic chains, one on each side of the vertebral column, most collateral ganglia are single, rather than paired.

Functions of the Collateral Ganglia
(Figure 17-4b)

Postganglionic fibers leaving the collateral ganglia extend throughout the abdominopelvic cavity, innervating a variety of visceral tissues and organs. A summary of the effects of increased sympathetic activity along these postganglionic fibers is included in Figure 17-4b●. The general pattern is (1) a reduction of blood flow and energy use by visceral organs that are not important to short-term survival, such as the digestive tract, and (2) the release of stored energy reserves.

Anatomy of the Collateral Ganglia
(Figures 17-5/17-11)

The splanchnic nerves innervate three collateral ganglia. Preganglionic fibers from the seven lower thoracic segments end at the **celiac** (SĒ-lē-ak) **ganglion** and the **superior mesenteric ganglion.** These ganglia are embedded in an extensive autonomic plexus that will be discussed in a later section. Preganglionic fibers from the lumbar segments form splanchnic nerves that end at the **inferior mesenteric ganglion**. These ganglia are diagrammed in Figure 17-5● and detailed in Figure 17-11●, p. 440.

THE CELIAC GANGLION. The celiac ganglion is variable in appearance. It most often consists of a pair of interconnected masses of gray matter situated at the base of the *celiac artery.* The celiac ganglion may also form a single mass, or many small, interwoven masses. Postganglionic fibers from the celiac ganglion innervate the stomach, duodenum, liver, gallbladder, pancreas, and spleen.

THE SUPERIOR MESENTERIC GANGLION. The superior mesenteric ganglion sits near the base of the *superior mesenteric artery.* Postganglionic fibers leaving the superior mesenteric ganglion innervate the small intestine and the initial segments of the large intestine.

THE INFERIOR MESENTERIC GANGLION. The inferior mesenteric ganglion is located near the base of the *inferior mesenteric artery.* Postganglionic fibers from this ganglion provide sympathetic innervation to the terminal portions of the large intestine, the kidney and bladder, and the sex organs.

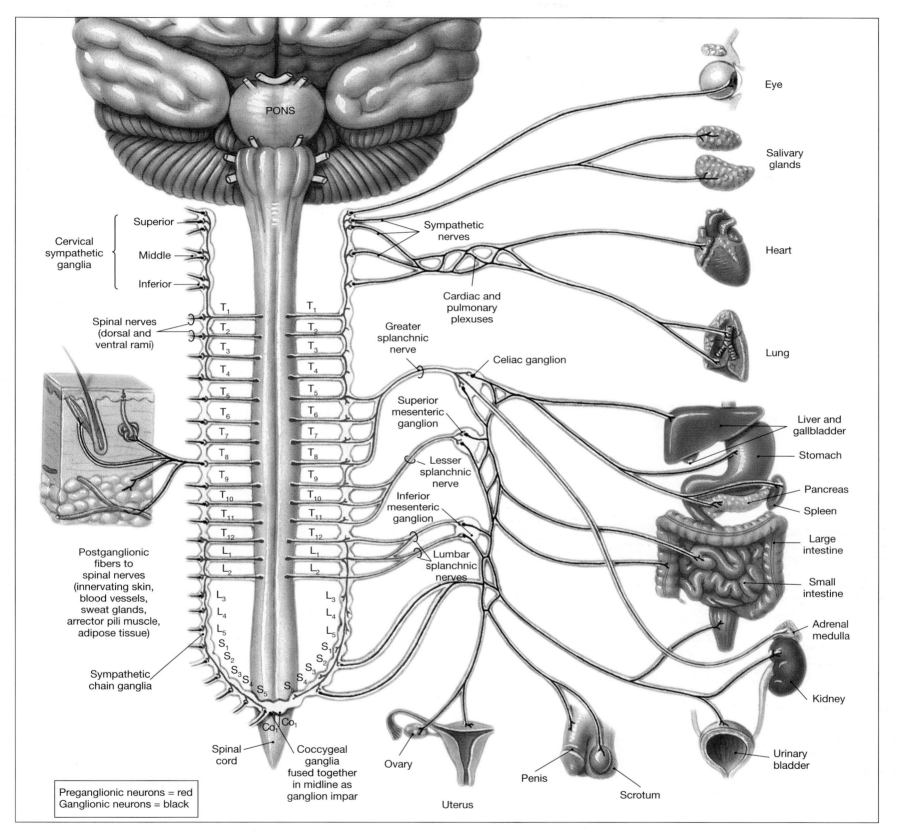

FIGURE 17-5
Anatomical Distribution of Sympathetic Postganglionic Fibers. The left side of this figure shows the distribution of sympathetic postganglionic fibers via the gray rami and spinal nerves. The right side shows the distribution of preganglionic and postganglionic fibers innervating visceral organs. However, *both* innervation patterns are found on *each* side of the body.

The Adrenal Medullae
(Figures 17-4c/17-5/17-6)

Preganglionic fibers entering each adrenal gland proceed to its center, to the region called the **adrenal medulla** (Figures 17-4c, p. 431, 17-5, and 17-6●). They synapse on modified neurons that perform an endocrine function. ∞ [p. 326] These neurons have short axons that end on an extensive network of capillaries (Figure 17-6●). When stimulated, these cells release the neurotransmitters *norepinephrine* (NE) and *epinephrine* (E) into the circulation. Epinephrine, also called *adrenaline*, accounts for 75–80% of the secretory output; the rest is norepinephrine.

Circulating blood then carries the neurotransmitters throughout the body, causing changes in the metabolic activities of many different cells. In general, these effects resemble those produced by the stimulation of sympathetic postganglionic fibers. But they differ in two respects: (1) cells not innervated by sympathetic post-ganglionic fibers are affected as well, and (2) the effects last much longer than those produced by direct sympathetic innervation.

Effects of Sympathetic Stimulation

The sympathetic division can change tissue and organ activities by releasing norepinephrine at peripheral synapses and distributing norepinephrine and epinephrine throughout the body in the bloodstream. The motor fibers that target specific effectors, such as smooth muscle fibers in blood vessels of the skin, can be activated in reflexes that do not involve other peripheral effectors. In a crisis, however, the entire division responds. This event is called **sympathetic activation**. Sympathetic activation is controlled by sympathetic centers in the hypothalamus. The effects are not limited to

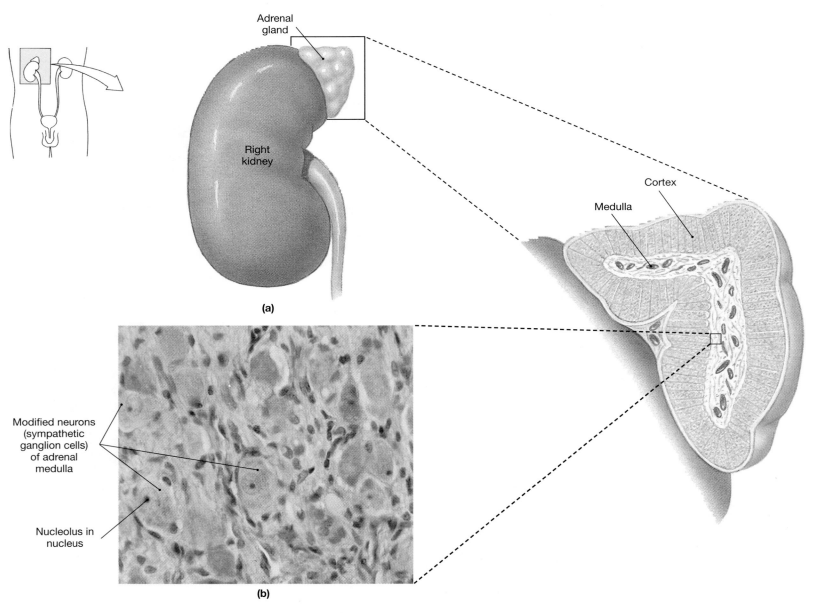

FIGURE 17-6
Adrenal Medulla. (a) Relationship of an adrenal gland to a kidney. (b) The adrenal medulla, a modified sympathetic ganglion. (LM × 426)

peripheral tissues, and sympathetic activation also alters CNS activity.

When sympathetic activation occurs, an individual experiences:

1. Increased alertness, via stimulation of the reticular activating system, causing the individual to feel "on edge."

2. A feeling of energy and euphoria, often associated with a disregard for danger and a temporary insensitivity to painful stimuli.

3. Increased activity in the cardiovascular and respiratory centers of the pons and medulla oblongata, leading to elevations in blood pressure, heart rate, breathing rate, and depth of respiration.

4. A general elevation in muscle tone through stimulation of the extrapyramidal system, so that the person *looks* tense, and may even begin to shiver.

5. The mobilization of energy reserves, through the accelerated breakdown of glycogen in muscle and liver cells and the release of lipids by adipose tissues.

These changes, coupled with the peripheral changes already noted, complete the preparations necessary for the individual to cope with stressful and potentially dangerous situations. We have examined the distribution of sympathetic impulses and the general effects of sympathetic activation. We will now consider the cellular basis for these effects on peripheral organs.

■ CLINICAL BRIEF
Horner's Syndrome

In **Horner's syndrome**, the sympathetic postganglionic innervation to one side of the face becomes interrupted, possibly as the result of an injury, a tumor, or some progressive condition such as multiple sclerosis. ∞ [p. 398] The affected side of the face becomes flushed because in the absence of sympathetic tone, the blood vessels dilate. Sweating stops in the affected region, and the pupil on that side becomes markedly constricted. Other symptoms include a drooping eyelid and an apparent retreat of the eye into the orbit.

The elimination of sympathetic innervation can have additional consequences that appear more gradually. Under normal conditions, sympathetic innervation provides the effectors with a background level of stimulation. After the disappearance of sympathetic stimulation, the effectors may become extremely sensitive to norepinephrine and epinephrine. This hypersensitivity can produce changes in facial blood flow and other functions when the adrenal medulla is stimulated.

Sympathetic Activation and Neurotransmitter Release
(Figure 17-7)

On stimulation, sympathetic preganglionic fibers release ACh at synapses with ganglionic neurons. Synapses using ACh as a transmitter are called *cholinergic*. The effect on the ganglionic neurons is always excitatory.

Stimulation of ganglionic neurons usually leads to the release of norepinephrine at neuroeffector junctions. These synapses are called *adrenergic*. The sympathetic division also contains a small but significant number of ganglionic neurons that release ACh, rather than NE, at neuroeffector junctions in the body wall, the skin, and in skeletal muscles.

Figure 17-7● details a representative sympathetic neuroeffector junction. Rather than ending at a single synaptic terminal, the telodendria form an extensive branching network. Each branch resembles a string of pop beads, and each pop bead, or **varicosity**, is packed with mitochondria and neurotransmitter vesicles. These varicosities pass along or near the surfaces of many effector cells. A single axon may supply 20,000 varicosities that can affect dozens of target cells. Receptor proteins are scattered across the opposing surfaces, but there are no specialized postsynaptic membranes.

The effects of neurotransmitter released by varicosities persist for at most a few seconds before the neurotransmitter is reabsorbed, broken down by enzymes, or removed by diffusion into the bloodstream. The effects of the E and NE secreted by the adrenal medullae are considerably longer in duration because (1) the bloodstream does not contain the enzymes that break down epinephrine or norepinephrine, and (2) most tissues contain relatively low concentrations of such enzymes. As a result, the effects of adrenal stimulation are widespread, and they continue for a relatively long time. For example, tissue concentrations of epinephrine may remain elevated for as long as 30 seconds, and the effects may persist for several minutes.

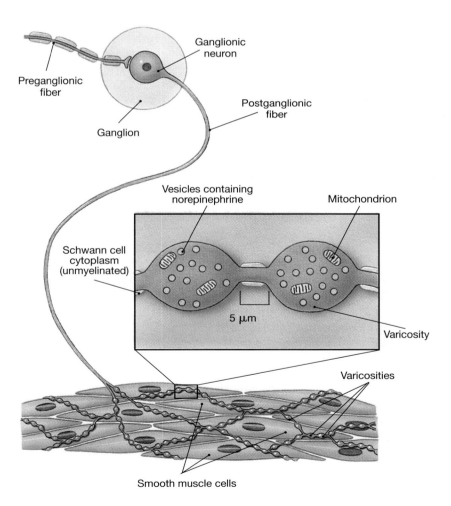

FIGURE 17-7
Sympathetic Postganglionic Nerve Endings

Membrane Receptors and Sympathetic Function

The effects of sympathetic stimulation result primarily from interactions with membrane receptors sensitive to norepinephrine and epinephrine. There are two classes of sympathetic receptors, *alpha receptors* and *beta receptors*. In general, norepinephrine stimulates alpha receptors more than it does beta receptors; however, epinephrine stimulates both classes of receptors.

Alpha and Beta Receptors

Stimulation of **alpha receptors** on the surfaces of smooth muscle cells causes constriction of peripheral blood vessels and the closure of sphincters along the digestive tract. **Beta receptors** are found in many organs, including skeletal muscles, smooth muscle surrounding airways in the lungs, the heart, and the liver. Stimulation of beta receptors at those sites triggers changes in the metabolic activity of the target cells. The result depends on which enzymes are involved. The most common response is an increase in metabolic activity; skeletal muscles use energy at a faster rate, and the heart rate accelerates. However, inhibition may also occur, and beta receptor–induced relaxation of smooth muscles causes a dilation of blood vessels supplying skeletal muscles and the enlargement of the respiratory passageways.

Sympathetic Stimulation and ACh

Although the vast majority of sympathetic postganglionic fibers are adrenergic, releasing norepinephrine, a few postganglionic fibers are cholinergic. These postganglionic fibers innervate sweat glands of the skin and the blood vessels to skeletal muscles. Activation of these sympathetic fibers stimulates sweat gland secretion and dilates the blood vessels. Neither the body wall nor skeletal muscles are innervated by the parasympathetic division of the ANS; the distribution of cholinergic fibers via the sympathetic division provides a method of regulating sweat gland secretion and selectively controlling blood flow to skeletal muscles while norepinephrine released by adrenergic terminals reduces the blood flow to other tissues in the body wall.

A Summary of the Sympathetic Division

1. The sympathetic division of the ANS includes two segmentally arranged sympathetic chains, one on each side of the spinal column; three collateral ganglia anterior to the spinal column; and two adrenal medullae.

2. The preganglionic fibers are short, because the ganglia are close to the spinal cord. The postganglionic fibers are relatively long and extend a considerable distance before reaching their target organs. (In the case of the adrenal medullae, very short axons end at capillaries that carry their secretions to the bloodstream.)

3. The sympathetic division shows extensive divergence, and a single preganglionic fiber may innervate as many as 32 second-order neurons in different ganglia. As a result, a single sympathetic motor neuron inside the CNS can control a variety of peripheral effectors and produce a complex and coordinated response.

4. All preganglionic neurons release ACh at their synapses with ganglionic neurons. Most of the postganglionic fibers release norepinephrine, but a few release ACh.

5. The effector response depends on the nature of the channels or enzymes activated when norepinephrine or epinephrine binds to alpha or beta receptors.

Table 17-1 (p. 440) summarizes the characteristics of the sympathetic division of the ANS.

√ **Where do the nerves that synapse in the collateral ganglia originate?**

√ **Individuals with high blood pressure may be given a medication that blocks beta receptors. How would this medication help their condition?**

THE PARASYMPATHETIC DIVISION
(Figure 17-8)

The parasympathetic division of the ANS (Figure 17-8●) includes:

1. *Preganglionic (first-order) neurons in the brain stem and in sacral segments of the spinal cord.* In the brain, the mesencephalon, pons, and medulla contain autonomic nuclei associated with cranial nerves III, VII, IX, and X. In the sacral segments of the spinal cord, the autonomic nuclei lie in the lateral gray horns of spinal segments S_2–S_4.

2. *Ganglionic (second-order) neurons in peripheral ganglia located within or adjacent to the target organs.* **Intramural ganglia** are parasympathetic ganglia located within the walls of target organs. The preganglionic fibers of the parasympathetic division do not diverge as extensively as do those of the sympathetic division. A typical preganglionic fiber synapses on 6–8 ganglionic neurons. These second-order neurons are all located in the same ganglion, and their postganglionic fibers influence the same target organ. As a result, *the effects of parasympathetic stimulation are more specific and localized than those of the sympathetic division.*

Organization and Anatomy of the Parasympathetic Division
(Figure 17-9)

Parasympathetic preganglionic fibers leave the brain in cranial nerves III (oculomotor), VII (facial), IX (glossopharyngeal), and X (vagus) (Figure 17-9●). Parasympathetic fibers in the oculomotor, facial, and glossopharyngeal nerves are concerned with the control of visceral structures in the head. These fibers synapse in the **ciliary, sphenopalatine, submandibular**, and **otic ganglia**. ∞ [p. 397] Short postganglionic fibers then continue to their peripheral targets. The vagus nerve provides preganglionic parasympathetic innervation to intramural ganglia within structures in the thoracic and abdominopelvic cavity as distant as the last segments of the large intestine. The vagus nerve alone provides roughly 75% of all parasympathetic outflow.

The sacral parasympathetic outflow does not join the ventral roots of the spinal nerves. Instead, the preganglionic fibers form distinct **pelvic nerves** that innervate intramural ganglia in the kidney and bladder, the terminal portions of the large intestine, and the sex organs.

FIGURE 17-8
Organization of the Parasympathetic Division of the ANS

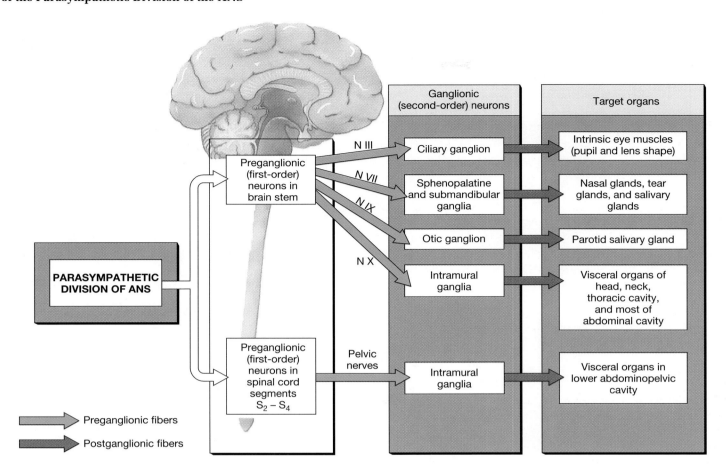

General Functions of the Parasympathetic Division

A partial listing of the major effects produced by the parasympathetic division includes:

1. Constriction of the pupils to restrict the amount of light entering the eyes, and focusing on nearby objects.

2. Secretion by digestive glands, including salivary glands, gastric glands, duodenal glands, intestinal glands, pancreas, and liver.

3. Secretion of hormones that promote nutrient absorption by peripheral cells.

4. Increased smooth muscle activity along the digestive tract.

5. Stimulation and coordination of defecation.

6. Contraction of the urinary bladder during urination.

7. Constriction of the respiratory passageways.

8. Reduction in heart rate and force of contraction.

9. Sexual arousal and stimulation of sexual glands in both sexes.

These functions center on relaxation, food processing, and energy absorption. The parasympathetic division has been called the *anabolic system* because stimulation leads to a general increase in the nutrient content of the blood. Cells throughout the body respond to this increase by absorbing nutrients and using them to support growth and other anabolic activities.

Parasympathetic Activation and Neurotransmitter Release

All of the preganglionic and postganglionic fibers in the parasympathetic division release ACh at synapses and neuroeffector junctions. The neuroeffector junctions are small, with narrow synaptic clefts. The effects of stimulation are short-lived, because most of the ACh released is inactivated by acetylcholinesterase within the synapse. Any ACh diffusing into the surrounding tissues will be deactivated by the enzyme *tissue cholinesterase.* As a result, the effects of parasympathetic stimulation are quite localized, and they last a few seconds at most.

Membrane Receptors and Responses

Although all the synapses (neuron-to-neuron) and neuroeffector junctions (neuron-to-effector) of the parasympathetic division use the same transmitter, acetylcholine, two different types of ACh receptors are found on the postsynaptic membranes.

1. **Nicotinic** (nik-ō-TIN-ik) **receptors** *are found on the surfaces of ganglion cells of both the parasympathetic and sympathetic divisions, as well as at neuromuscular junctions of the SNS.* Exposure to ACh always causes excitation of the ganglionic (second-order) neuron or muscle fiber via the opening of membrane ion channels.

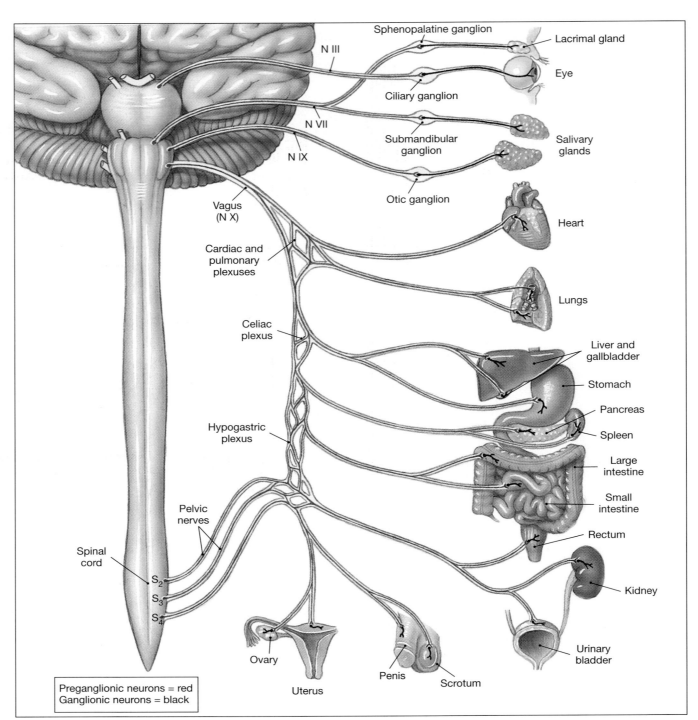

FIGURE 17-9
Anatomical Distribution of the Parasympathetic Output

Labels in figure:

Sphenopalatine ganglion
Lacrimal gland
N III
Eye
Ciliary ganglion
N VII
Submandibular ganglion
Salivary glands
N IX
Otic ganglion
Vagus (N X)
Heart
Cardiac and pulmonary plexuses
Lungs
Celiac plexus
Liver and gallbladder
Stomach
Pancreas
Spleen
Hypogastric plexus
Large intestine
Small intestine
Pelvic nerves
Rectum
Spinal cord
S_2
S_3
S_4
Kidney
Ovary
Uterus
Penis
Scrotum
Urinary bladder

Preganglionic neurons = red
Ganglionic neurons = black

2. Muscarinic (mus-kar-IN-ik) **receptors** *are found at cholinergic neuroeffector junctions in the parasympathetic division, as well as at the few cholinergic neuroeffector junctions in the sympathetic division.* Stimulation of muscarinic receptors produces longer-lasting effects than does stimulation of nicotinic receptors. The response, which reflects the activation or inactivation of specific enzymes, may be excitation or inhibition.

The names *nicotinic* and *muscarinic* indicate the chemical compounds that stimulate these receptor sites. Nicotinic receptors bind *nicotine*, a powerful component of tobacco smoke. Muscarinic receptors are stimulated by *muscarine*, a toxin produced by some poisonous mushrooms.

A Summary of the Parasympathetic Division

1. The parasympathetic division includes visceral motor nuclei associated with four cranial nerves (III, VII, IX, and X) and in sacral segments S_2–S_4.

2. The second-order neurons are situated in intramural ganglia or in ganglia closely associated with their target organs.

3. The parasympathetic division innervates areas serviced by the cranial nerves and organs in the thoracic and abdominopelvic cavities.

4. All parasympathetic neurons are cholinergic. Ganglionic neurons have nicotinic receptors that are excited by ACh. Muscarinic receptors present at neuroeffector junctions may produce either excitation or inhibition, depending on the nature of the enzymes activated when ACh binds to the receptor.

5. The effects of parasympathetic stimulation are usually brief and restricted to specific organs and sites.

Table 17-1 summarizes the characteristics of the parasympathetic division of the ANS.

RELATIONSHIPS BETWEEN THE SYMPATHETIC AND PARASYMPATHETIC DIVISIONS

The sympathetic division has widespread impact, reaching visceral organs and tissues throughout the body. The parasympathetic division innervates only visceral structures serviced by the cranial nerves or lying within the abdominopelvic cavity. Although some organs are innervated by one division or the other, most vital organs receive **dual innervation**—that is, they receive instructions from both the sympathetic and parasympathetic divisions. Where dual innervation exists, the two divisions often have opposing effects. Dual innervation is most prominent in the digestive tract, the heart, and the lungs. Secretory control of the salivary glands or the sexual functions of the male reproductive tract are also examples.

Anatomy of Dual Innervation
(Figure 17-11)

In the head, parasympathetic postganglionic fibers from the ciliary, sphenopalatine, submandibular, and otic ganglia accompany the cranial nerves to their peripheral destinations. The sympathetic innervation reaches the same structures by traveling directly from the superior cervical ganglia of the sympathetic chain.

In the thoracic and abdominopelvic cavities, the sympathetic postganglionic fibers mingle with parasympathetic preganglionic fibers at a series of plexuses (Figure 17-11●). These are the *cardiac plexus*, the *pulmonary plexus*, the *esophageal plexus*, the *celiac plexus*, the *inferior mesenteric plexus*, and the *hypogastric plexus*. Nerves leaving these plexuses travel with the blood vessels and lymphatics that supply visceral organs.

Autonomic fibers entering the thoracic cavity intersect at the **cardiac plexus** and the **pulmonary plexus**. These plexuses contain sympathetic and parasympathetic fibers bound for the heart and lungs, respectively, as well as the parasympathetic ganglia whose output affects those organs. The **esophageal plexus** contains descending branches of the vagus nerve and splanchnic nerves leaving the sympathetic chain on either side.

Parasympathetic preganglionic fibers of the vagus nerve enter the abdominopelvic cavity with the esophagus. There they join the network of the **celiac plexus**, also known as the *solar plexus*. The celiac plexus and associated smaller plexuses, such as the **inferior mesenteric plexus**, innervate viscera down to the initial segments of the large intestine. The **hypogastric plexus** contains the parasympathetic outflow of the

pelvic nerves and sympathetic postganglionic fibers from the inferior mesenteric ganglion and of the splanchnic nerves from the sacral sympathetic chain. This plexus innervates the digestive, urinary, and reproductive organs of the pelvic cavity.

A Comparison of the Sympathetic and Parasympathetic Divisions

Figure 17-10● and Table 17-1 compare key features of the sympathetic and parasympathetic divisions of the ANS.

INTEGRATION AND CONTROL OF AUTONOMIC FUNCTIONS

The ANS, like the somatic nervous system (SNS), is organized into a series of interacting levels. At the bottom are the visceral motor neurons that participate in cranial and spinal *visceral reflexes*. The cell bodies of these motor neurons are located in the spinal cord and lower brain stem.

Visceral Reflexes
(Figure 17-12)

Visceral reflexes are the simplest functional units in the autonomic nervous system. Visceral reflexes provide automatic motor responses that can be modified, facilitated, or inhibited by higher centers, especially those of the hypothalamus. All visceral reflexes are polysynaptic. Each visceral reflex arc (Figure 17-12●) consists of a receptor, a sensory nerve, a

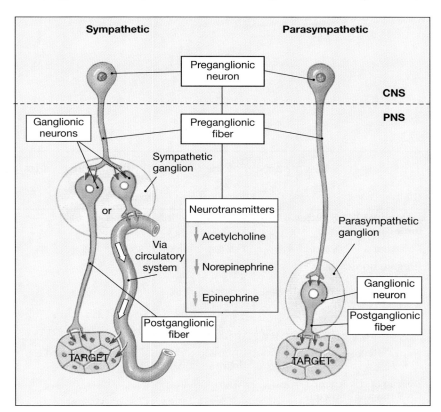

FIGURE 17-10
A Comparison of the Sympathetic and Parasympathetic Divisions

FIGURE 17-11
The Peripheral Autonomic Plexuses

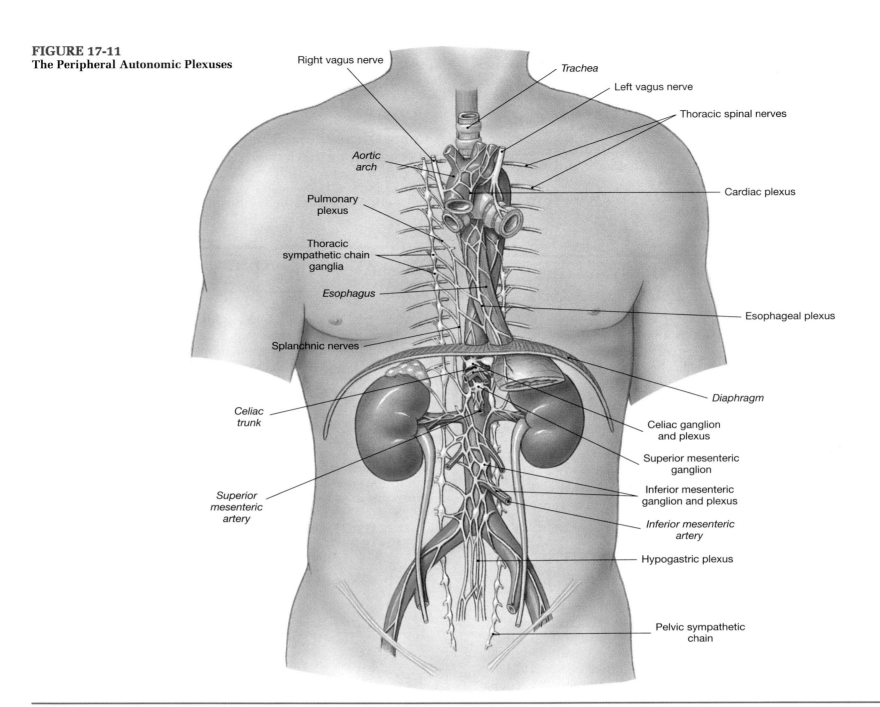

Right vagus nerve

Trachea

Left vagus nerve

Thoracic spinal nerves

Aortic arch

Cardiac plexus

Pulmonary plexus

Thoracic sympathetic chain ganglia

Esophagus

Esophageal plexus

Splanchnic nerves

Diaphragm

Celiac trunk

Celiac ganglion and plexus

Superior mesenteric ganglion

Inferior mesenteric ganglion and plexus

Superior mesenteric artery

Inferior mesenteric artery

Hypogastric plexus

Pelvic sympathetic chain

TABLE 17-1	A Comparison of the Sympathetic and Parasympathetic Divisions of the ANS	
Characteristic	*Sympathetic Division*	*Parasympathetic Division*
Location of CNS visceral motor neuron	Lateral gray horns, spinal segments T_1-L_2	Brain stem and spinal segments S_2-S_4
Location of PNS ganglia	Near spinal column	Typically intramural
Preganglionic fibers Length Neurotransmitter released	 Relatively short Acetylcholine	 Relatively long Acetylcholine
Postganglionic fibers Length Neurotransmitter released	 Relatively long Usually norepinephrine	 Relatively short Acetylcholine
Neuroeffector junction	Enlarged terminal knobs that release transmitter near target cells	Neuroeffector junctions that release transmitter to special receptor surface
Degree of divergence from CNS to ganglion cells	Approximately 1:32	Approximately 1:6
General function	Stimulate metabolism, increase alertness, prepare for emergency "fight or flight"	Promote relaxation, nutrient uptake, energy storage, "rest and repose"

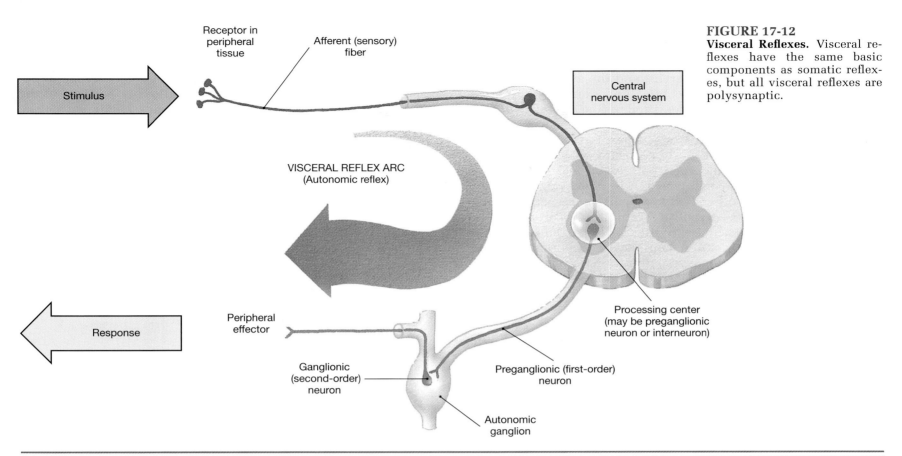

FIGURE 17-12
Visceral Reflexes. Visceral reflexes have the same basic components as somatic reflexes, but all visceral reflexes are polysynaptic.

Receptor in peripheral tissue

Afferent (sensory) fiber

Stimulus

Central nervous system

VISCERAL REFLEX ARC
(Autonomic reflex)

Response

Peripheral effector

Processing center (may be preganglionic neuron or interneuron)

Ganglionic (second-order) neuron

Preganglionic (first-order) neuron

Autonomic ganglion

TABLE 17-2 Representative Visceral Reflexes

Reflex	Stimulus	Response	Comments
PARASYMPATHETIC REFLEXES			
Gastric and intestinal reflexes	Pressure and physical contact with materials	Smooth muscle contractions that propel materials and mix with secretions	Via vagus nerve
Defecation	Distention of rectum	Relaxation of internal anal sphincter	Requires voluntary relaxation of external sphincter
Urination	Distention of urinary bladder	Contraction of bladder walls, relaxation of internal urethral sphincter	Requires voluntary relaxation of external sphincter
Light and consensual light reflexes	Bright light shining in eye(s)	Constriction of pupils of both eyes	
Swallowing reflex	Movement of material into upper pharynx	Smooth muscle contractions moving material to stomach	Coordinated by medullary swallowing center
Vomiting reflex	Irritation of digestive tract lining	Reversal of normal smooth muscle action to eject contents	Coordinated by medullary vomiting center
Coughing reflex	Irritation of respiratory tract lining	Sudden explosive ejection of air	Coordinated by medullary coughing center
Cardioinhibitory reflex	Sudden rise in blood pressure in carotid artery	Reduction in heart rate and force of contraction	Coordinated in cardiac center in medulla oblongata
Sexual arousal	Erotic stimuli (visual or tactile)	Increased glandular secretions, sensitivity, erection of the penis or clitoris	
SYMPATHETIC REFLEXES			
Cardioacceleratory reflex	Sudden decline in blood pressure in carotid artery	Increase in heart rate and force of contraction	Coordinated in cardiac center in medulla oblongata
Vasomotor reflexes	Changes in blood pressure in major arteries	Changes in diameter of peripheral vessels to maintain normal range of blood pressures	Coordinated in vasomotor center in medulla oblongata
Pupillary reflex	Low light level reaching visual receptors	Dilation of pupil	
Emission and ejaculation (in male)	Erotic stimuli (tactile)	Smooth and skeletal muscle contractions involved in the ejection of glandular secretions and the expulsion of semen	

processing center (interneuron or motor neuron), and two visceral motor neurons, preganglionic (first-order) and ganglionic (second-order). Sensory nerves deliver information to the CNS along spinal nerves, cranial nerves, and the autonomic nerves that innervate peripheral effectors. For example, shining a light in the eye triggers a visceral reflex (the *consensual light reflex*) that constricts the pupils of both eyes. ∞ [p. 358] In total darkness, the pupils dilate. But the motor nuclei directing pupillary constriction or dilation are also controlled by hypothalamic centers concerned with emotional states. When you are queasy or nauseated, your pupils constrict.

As we examine other body systems, many examples of autonomic reflexes involved in respiration, cardiovascular function, and other visceral activities will be examined. Table 17-2 summarizes information concerning important visceral reflexes. Note that the parasympathetic division participates in reflexes that affect individual organs and systems, reflecting the relatively specific and restricted pattern of innervation. In contrast, there are fewer sympathetic reflexes. This division is typically activated as a whole, in part because of the degree of divergence and in part because the release of hormones by the adrenal medullae produces widespread peripheral effects.

Higher Levels of Autonomic Control
(Figure 17-13)

The levels of activity in the sympathetic and parasympathetic divisions are controlled by centers in the brain stem concerned with specific visceral functions. Figure 17-13● diagrams the levels of autonomic control. As in the SNS, simple reflexes based in the spinal cord provide relatively rapid and automatic responses to stimuli. More complex sympathetic and parasympathetic reflexes are coordinated by processing centers in the medulla oblongata. In addition to the cardiovascular centers, the medulla oblongata contains centers and nuclei involved with respiration, digestive secretions, peristalsis, and urinary function. These medullary centers are in turn subject to regulation by the hypothalamus. In general, centers in the posterior and lateral hypothalamus are concerned with the coordination and regulation of sympathetic function, and portions of the anterior and medial hypothalamus control the parasympathetic division. ∞ [p. 384]

The term *autonomic* was originally applied to the visceral motor system because it was thought that the regulatory centers functioned without regard for other CNS activities. This view has been drastically revised in light of subsequent research. Because the hypothalamus interacts with all other portions of the brain, activity in the limbic system (memories, emotional states), thalamus (sensory information), or cerebral cortex (conscious thought processes) can have dramatic effects on autonomic function. For example, when you become angry, your heart rate accelerates and your blood pressure rises.

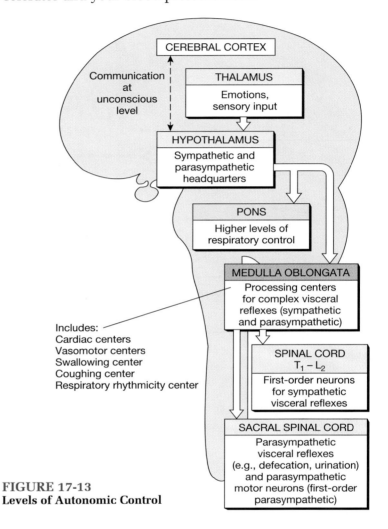

FIGURE 17-13
Levels of Autonomic Control

Related Clinical Terms

Horner's syndrome: A condition characterized by loss of sympathetic innervation to the face. *[p. 435]*

CHAPTER SUMMARY AND REVIEW

STUDY OUTLINE

INTRODUCTION *[p. 428]*

1. The autonomic nervous system (ANS) coordinates cardiovascular, respiratory, digestive, excretory, and reproductive functions.

AN OVERVIEW OF THE AUTONOMIC NERVOUS SYSTEM (ANS) *[p. 428]*

Pathways for Visceral Motor Output *[p. 428]*
1. **First-order (preganglionic) neurons** in the CNS send axons to synapse on **second-order (ganglionic) neurons** in autonomic ganglia outside the CNS *(see Figure 17-1)*.

Subdivisions of the ANS *[p. 428]*
2. Visceral efferents from the thoracic and lumbar segments form the **thoracolumbar (sympathetic) division** ("fight or flight" system) of the ANS. Visceral efferents leaving the brain and sacral segments form the **craniosacral (parasympathetic) division** ("rest and repose" system) *(see Figure 17-2)*.

terminal ganglia • intramural ganglia • norepinephrine (NE)

THE SYMPATHETIC DIVISION *[p. 430]*

1. The sympathetic division consists of preganglionic (first-order) neurons between segments T_1 and L_2, ganglionic (second-order) neurons in ganglia near the vertebral column, and specialized neurons inside the adrenal gland *(see Figures 17-2a to 17-5/17-10)*.
2. There are two types of sympathetic ganglia: **sympathetic chain ganglia** (*paravertebral ganglia*) and **collateral ganglia** (*prevertebral ganglia*).

The Sympathetic Chain *[p. 430]*
3. Postganglionic fibers targeting thoracic cavity structures form autonomic nerves that go directly to their visceral destination. Preganglionic fibers run between the sympathetic chain ganglia and interconnect them *(see Figures 17-4a,b/17-5)*.

Collateral Ganglia *[p. 432]*
4. The abdominopelvic viscera receive sympathetic innervation via preganglionic fibers that synapse within collateral ganglia. The preganglionic fibers that innervate the collateral ganglia form the **splanchnic nerves** *(see Figures 17-4b/17-5)*.
5. The **celiac ganglion** innervates the stomach, liver, pancreas, and spleen; the **superior mesenteric ganglion** innervates the small intestine and initial segments of the large intestine; and the **inferior mesenteric ganglion** innervates the kidney, bladder, sex organs, and terminal portions of the large intestine *(see Figures 17-4b/17-5/17-10)*.

The Adrenal Medullae *[p. 434]*
6. Preganglionic fibers entering the adrenal gland synapse within the **adrenal medulla** *(see Figures 17-4c/17-5/17-6)*.

Effects of Sympathetic Stimulation *[p. 434]*
7. In a crisis, the entire division responds, an event called **sympathetic activation**. Its effects include: increased alertness, a feeling of energy and euphoria, increased cardiovascular and respiratory activity, and general elevation in muscle tone.

Sympathetic Activation and Neurotransmitter Release *[p. 435]*
8. Stimulation of the sympathetic division has two distinctive results: the release of norepinephrine (or in some cases acetylcholine) at neuroeffector junctions, and the secretion of epinephrine and norepinephrine into the general circulation *(see Figure 17-7)*.

varicosities

Membrane Receptors and Sympathetic Function *[p. 435]*
9. There are two types of sympathetic receptors: **alpha receptors** (which respond to norepinephrine or epinephrine by depolarizing the membrane) and **beta receptors** (which are particularly sensitive to epinephrine).
10. Most postganglionic fibers are adrenergic, but a few are cholinergic. Postganglionic fibers innervating sweat glands of the skin and blood vessels to skeletal muscles release ACh.

A Summary of the Sympathetic Division *[p. 436]*
11. Preganglionic sympathetic fibers are relatively short. Except for those of the adrenal medullae, postganglionic fibers are quite long. Extensive divergence typically occurs, with a single preganglionic fiber synapsing with many ganglionic neurons in different ganglia *(see Table 17-1)*.

THE PARASYMPATHETIC DIVISION *[p. 436]*

1. The parasympathetic division includes preganglionic (first-order) neurons in the brain stem and sacral segments of the spinal cord, and ganglionic (second-order) neurons in peripheral ganglia located within or next to target organs *(see Figures 17-8/17-9 and Table 17-1)*.

intramural ganglia

Organization and Anatomy of the Parasympathetic Division *[p. 436]*
2. Preganglionic fibers leaving the sacral segments form **pelvic nerves** *(see Figure 17-9)*.

ciliary ganglion • sphenopalatine ganglion • submandibular ganglion • otic ganglion

General Functions of the Parasympathetic Division *[p. 437]*
3. The effects produced by the parasympathetic division center on relaxation, food processing, and energy absorption.

Parasympathetic Activation and Neurotransmitter Release *[p. 437]*
4. All the parasympathetic preganglionic and postganglionic fibers release ACh at synapses and neuroeffector junctions. The effects are short-lived because of the actions of enzymes at the postsynaptic membrane and in the surrounding tissues.

5. Two different ACh receptors are found in postsynaptic membranes. Stimulation of **muscarinic receptors** produces a longer-lasting effect than does stimulation of **nicotinic receptors**.

A Summary of the Parasympathetic Division *[p. 438]*
6. The parasympathetic division innervates areas serviced by cranial nerves and organs in the thoracic and abdominopelvic cavities. All preganglionic and postganglionic parasympathetic neurons are cholinergic, and the effects of stimulation are usually brief and restricted to specific sites *(see Figure 17-10 and Table 17-1)*.

RELATIONSHIPS BETWEEN THE SYMPATHETIC AND PARASYMPATHETIC DIVISIONS *[p. 439]*
1. The sympathetic division has widespread influence, reaching visceral and somatic structures throughout the body *(see Figure 17-5 and Table 17-1)*.
2. The parasympathetic division innervates only visceral structures serviced by cranial nerves or lying within the abdominopelvic cavity. Organs with **dual innervation** receive instructions from both divisions *(see Figure 17-11 and Table 17-1)*.

Anatomy of Dual Innervation *[p. 439]*
3. In body cavities the parasympathetic and sympathetic nerves intermingle to form a series of characteristic nerve plexuses (nerve networks), which include the **cardiac, pulmonary, esophageal, celiac, inferior mesenteric**, and **hypogastric plexuses** *(see Figure 17-11)*.

A Comparison of the Sympathetic and Parasympathetic Divisions *[p. 439]*
4. See *Figure 17-10 and Table 17-1*.

INTEGRATION AND CONTROL OF AUTONOMIC FUNCTIONS *[p. 439]*

Visceral Reflexes *[p. 439]*
1. **Visceral reflexes** are the simplest functional units in the ANS *(see Figure 17-12 and Table 17-2)*.

Higher Levels of Autonomic Control *[p. 442]*
2. In general, higher brain centers in the posterior and lateral hypothalamus are concerned with the coordination and regulation of sympathetic function, and portions of the anterior and medial hypothalamus control the parasympathetic division *(see Figure 17-13)*.

---

1   REVIEW OF CHAPTER OBJECTIVES

1. Compare the autonomic nervous system with the other divisions of the nervous system.
2. Identify the divisions of the autonomic nervous system.
3. Describe the structures that constitute the sympathetic division of the autonomic nervous system and the functions associated with them.
4. Discuss the mechanisms of neurotransmitter release by the sympathetic nervous system.
5. Compare the effects of norepinephrine and epinephrine on target organs and tissues.
6. Describe the structures that constitute the parasympathetic division of the autonomic nervous system and the functions associated with them.
7. Discuss the relationship between the sympathetic and parasympathetic divisions, and explain the implications of dual innervation.
8. Describe the levels of integration and control of the autonomic system.

2   REVIEW OF CONCEPTS

1. How do conscious thoughts, plans, and actions affect the function of the autonomic nervous system?
2. How does the sympathetic division of the autonomic nervous system differ in function and distribution from the parasympathetic division?
3. How is the response of an effector organ controlled by the sympathetic nervous system?
4. How do sympathetic chain ganglia differ from collateral ganglia?
5. How are the sympathetic preganglionic motor fibers that innervate body segments or groups of segments routed from the spinal cord to their effector organs?
6. How do the white and gray rami affect voluntary motor function in the cervical and thoracic regions?

7. What is the relationship between the adrenal medulla and the sympathetic division of the autonomic nervous system?
8. How does the sympathetic nervous system affect tissue and organ activities in the body?
9. What functional classes of membrane receptors are stimulated by sympathetic neurotransmitters?
10. How is sweat gland function regulated by sympathetic activity?
11. What is the role of the vagus nerve (N X) in the innervation of visceral structures?
12. What are the main functions of the parasympathetic division of the autonomic nervous system?
13. How does the organization of the parasympathetic division differ from that of the sympathetic division?
14. What is the function of a visceral reflex?

3   CRITICAL THINKING AND CLINICAL APPLICATION QUESTIONS

1. A student in anatomy class is going to present a discussion of the autonomic nervous system to the class. He is experiencing many of the symptoms of autonomic response to public speaking. What physical sensations is he most likely feeling, and how are they mediated by the autonomic system? How will these autonomic activities change after he successfully completes the presentation?
2. If the student becomes visibly embarrassed during his presentation, what physical signs will be present, and what causes these responses?

444

18

The Nervous System: General and Special Senses

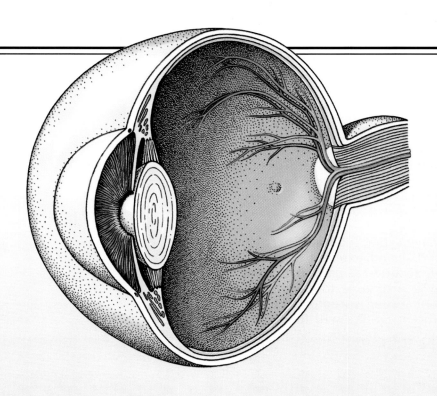

CHAPTER OUTLINE AND OBJECTIVES

Every cell membrane functions as a receptor for the cell, because it responds to changes in the extracellular environment. Cell membranes differ in their sensitivities to specific electrical, chemical, and mechanical stimuli. For example, a hormone that stimulates a neuron may have no effect on an osteocyte because the cell membranes of neurons and osteocytes contain different receptor proteins. A **sensory receptor** is a specialized cell or cell process that monitors conditions in the body or the external environment. Stimulation of the receptor directly or indirectly alters the production of action potentials in a sensory neuron. The sensory information arriving at the CNS is called a **sensation**. The term **general senses** refers to sensations of temperature, pain, touch, pressure, vibration, and proprioception (body position). General sensory receptors are distributed throughout the body. These sensations arrive at the primary sensory cortex, or *somatosensory cortex*, via pathways previously described. ∞ [p. 410]

The **special senses** are smell (*olfaction*), taste (*gustation*), balance (*equilibrium*), hearing, and vision. The sensations are provided by specialized receptor cells that are structurally more complex than those of the general senses. These receptors are found within complex **sense organs**, such as the eye or ear. The information provided by these receptors is distributed to specific areas of the cerebral cortex (such as the visual cortex) and to centers throughout the brain stem.

Sensory receptors represent the interface between the nervous system and the internal and external environments. The nervous system relies on accurate sensory data to control and coordinate relatively swift responses to specific stimuli. This chapter begins by summarizing receptor function and basic concepts in sensory processing. We will then apply this information to each of the general and special senses as we discuss their structure.

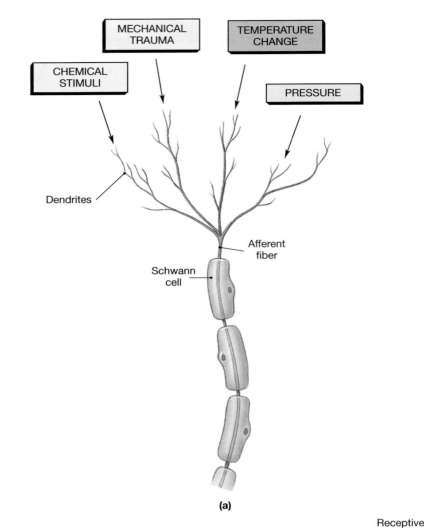

(a)

RECEPTORS
(Figure 18-1)

Each receptor has a characteristic sensitivity. For example, a touch receptor is very sensitive to pressure but relatively insensitive to chemical stimuli. This concept is called **receptor specificity**. Specificity results from the structure of the receptor cell itself or from the presence of accessory cells or structures that shield it from other stimuli. The simplest receptors are the dendrites of sensory neurons, called **free nerve endings** (Figure 18-1a●). They can be stimulated by many different stimuli. For example, free nerve endings that provide the sensation of pain may be responding to chemical stimulation, pressure, temperature changes, or physical damage. In contrast, the receptor cells of the eye are surrounded by accessory cells that normally prevent their stimulation by anything other than light. The highly specialized receptor cells in this complex organ communicate with sensory neurons over chemical synapses.

The area monitored by a single receptor cell is its **receptive field** (Figure 18-1b●). Whenever a sufficiently strong stimulus arrives in the receptive field, the CNS receives the information. The larger the receptive field, the poorer our ability to localize a stimulus. For example, a touch receptor on the general body surface may have a receptive field 7 cm (2.5 in.) in diameter. As a result, a light touch can be described only as affecting a general area of this size. On the tongue, where the receptive fields are

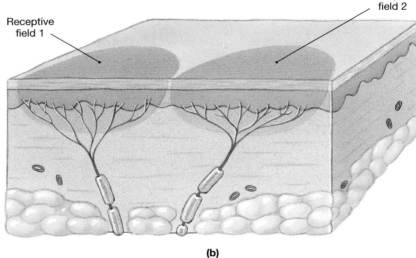

(b)

FIGURE 18-1
Receptors and Receptive Fields. (a) A free nerve ending consists of sensory dendrites that may be stimulated by a variety of stimuli. (b) Each receptor monitors a specific area known as the receptive field.

less than a millimeter in diameter, we can be very precise about the location of a stimulus.

An arriving stimulus can take many different forms; it may be a physical force, such as pressure, a dissolved chemical, a sound, or a beam of light. Regardless of the nature of the stimulus, however, sensory information must be sent to the CNS in the form of action potentials, which are electrical events. The arriving information is processed and interpreted by the CNS at the conscious and unconscious levels.

Interpretation of Sensory Information

When sensory information arrives at the CNS, it is routed according to the location and nature of the stimulus. Along sensory pathways, axons relay information from point A (the receptor) to point B (a neuron at a specific site in the cerebral cortex). The connection between receptor and cortical neuron is called a **labeled line**. Each labeled line carries information concerning a specific sensation (touch, pressure, vision, and so forth) from receptors in a specific part of the body. The identity of the active labeled line indicates the location and nature of the stimulus. *All other characteristics of the stimulus are conveyed by the pattern of action potentials in the afferent fibers.* This **sensory coding** provides information about the strength, duration, variation, and movement of the stimulus.

Some sensory neurons, called **tonic receptors**, are always active; the receptors involved in vision and hearing are examples. Other receptors are normally inactive, but become active for a short time whenever there is a change in the conditions they are monitoring. These are **phasic receptors** and provide information on the intensity and rate of change of a stimulus. Many touch receptors in the skin are examples of phasic receptors. Receptors that combine phasic and tonic coding convey extremely complicated sensory information.

Central Processing and Adaptation

Adaptation is a reduction in sensitivity in the presence of a constant stimulus. **Peripheral (sensory) adaptation** occurs when the receptors or sensory neurons alter their levels of activity. The receptor responds strongly at first, but thereafter the activity along the afferent fiber gradually declines, in part because of synaptic fatigue. This response is characteristic of phasic receptors, which are also called **fast-adapting receptors**. Tonic receptors show little peripheral adaptation, and so are called **slow-adapting receptors**.

Adaptation also occurs inside the CNS along the sensory pathways. For example, a few seconds after exposure to a new smell, conscious awareness of the stimulus virtually disappears, although the sensory neurons are still quite active. This process is known as **central adaptation**. Central adaptation usually involves the inhibition of nuclei along a sensory pathway. At the unconscious level, central adaptation further restricts the amount of detail arriving at the cerebral cortex. Most of the incoming sensory information is processed in centers along the spinal cord or brain stem, potentially triggering involuntary reflexes. Only about 1% of the information provided by afferent fibers reaches the cerebral cortex and our conscious awareness.

Sensory Limitations

Our sensory receptors provide a constant detailed picture of our bodies and our surroundings. This picture is, however, incomplete for several reasons:

1. Humans do not have receptors for every possible stimulus.
2. Our receptors have characteristic ranges of sensitivity.
3. A stimulus must be interpreted by the CNS. Our perception of a particular stimulus is an interpretation, and not always a reality.

This discussion has introduced basic concepts of receptor function and sensory processing. We will now discuss the structure and function of the general senses.

THE GENERAL SENSES

Receptors for the general senses are scattered throughout the body and are relatively simple in structure. A simple classification scheme divides them into *exteroceptors* and *interoceptors*. **Exteroceptors** provide information about the external environment; **interoceptors** monitor conditions inside the body.

A more detailed classification system divides the general sensory receptors into four types according to the nature of the stimulus that excites them:

1. **Nociceptors** (nō-sē-SEP-torz; *noceo*, hurt) respond to a variety of stimuli usually associated with tissue damage. Receptor activation causes the sensation of pain.
2. **Thermoreceptors** respond to changes in temperature.
3. **Mechanoreceptors** are stimulated or inhibited by physical distortion, contact, or pressure on their cell membranes.
4. **Chemoreceptors** monitor the chemical composition of body fluids and respond to the presence of specific molecules.

Each class of receptors has distinct structural and functional characteristics. You will find that some tactile receptors and mechanoreceptors are indentified by eponyms (commemorative names). Several authors have proposed differing alternatives for these names, and as yet no standardization or consensus exists. More significantly, *none* of the alternative names appear in the primary literature (professional, technical, or clinical journals or reports). To avoid later confusion, this chapter will use eponyms whenever there is no generally accepted or widely used alternative.

Nociceptors
(Figures 18-2/18-3a)

Nociceptors, or pain receptors, are especially common in the superficial portions of the skin (look ahead to Figure 18-3a●), in joint capsules, within the periostea of bones, and around the walls of blood vessels. There are few nociceptors in other deep tissues or in most visceral organs. Pain receptors are free nerve endings with large receptive fields. As a consequence it is often difficult to determine the exact origin of a painful sensation.

There are three types of nociceptors: (1) those sensitive to extremes of temperature, (2) those sensitive to mechanical damage, and (3) those sensitive to dissolved chemicals, such as those released by injured cells. However, very strong stimuli will excite all three receptor types.

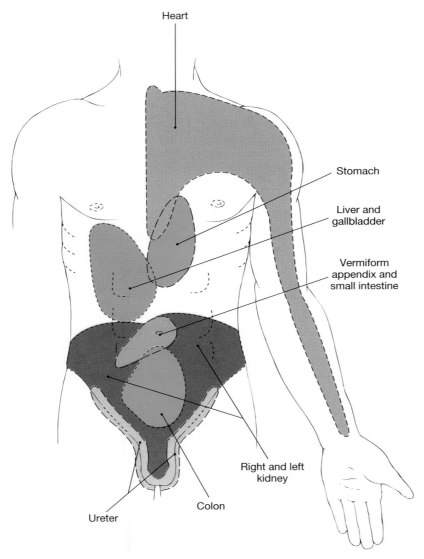

FIGURE 18-2
Referred Pain. Pain sensations originating in visceral organs are often perceived as involving specific regions of the body surface innervated by the same spinal nerves.

Sensations of **fast pain**, or **prickling pain**, are produced by deep cuts or similar injuries. These sensations reach the CNS very quickly, where they often trigger somatic reflexes. They are also relayed to the primary sensory cortex, and so receive conscious attention. Painful sensations cease only after tissue damage has ended. However, central adaptation may reduce *perception* of the pain while the pain receptors are still stimulated. ♱ *The Control of Pain [p. 761]*

Sensations of **slow pain**, or **burning and aching pain**, result from the same types of injuries as fast pain sensations. However, sensations of slow pain begin later and persist longer. These sensations cause a generalized activation of the reticular formation and thalamus. The individual is aware of the pain, but has only a general idea of the area affected. A person experiencing slow pain sensations will often palpate the area in an attempt to locate the source of the pain.

Pain sensations from visceral organs are often perceived as originating in more superficial regions, generally regions innervated by the same respective spinal nerves. The precise mechanism responsible for this **referred pain** remains to be determined, but several clinical examples are shown in Figure 18-2●. Cardiac pain, for example, is often perceived as originating in the upper chest and left arm.

Thermoreceptors

Temperature receptors are found in the dermis of the skin, in skeletal muscles, in the liver, and in the hypothalamus. Cold receptors are three or four times more numerous than warm receptors. The receptors are free nerve endings, and there are no known structural differences between warm and cold thermoreceptors.

Temperature sensations are conducted along the same pathways that carry pain sensations. They are sent to the reticular formation, the thalamus, and (to a lesser extent) the primary sensory cortex. Thermoreceptors are phasic receptors. They are very active when the temperature is changing, but they quickly adapt to a stable temperature. When you enter an air-conditioned classroom on a hot summer day, or a warm lecture hall on a brisk fall evening, the temperature seems unpleasant at first, but the discomfort fades as adaptation occurs.

Mechanoreceptors

Mechanoreceptors are sensitive to stimuli that stretch, compress, twist, or distort their cell membranes. There are three classes of mechanoreceptors: (1) **tactile receptors** provide sensations of touch, pressure, and vibration; (2) **baroreceptors** (bar-ō-rē-SEP-torz; *baro-*, pressure) detect pressure changes in the walls of blood vessels and in portions of the digestive, reproductive, and urinary tracts; (3) **proprioceptors** monitor the positions of joints and muscles and are the most complex of the general sensory receptors.

Tactile Receptors
(Figure 18-3)

Tactile receptors range in complexity from the simple *free nerve endings* to specialized sensory complexes with accessory cells and supporting structures. **Fine touch and pressure receptors** provide detailed information about a source of stimulation, including its exact location, shape, size, texture, and movement. These receptors are extremely sensitive and have relatively narrow receptive fields. **Crude touch and pressure receptors** provide poor localization and little additional information about the stimulus.

Figure 18-3● shows six different types of tactile receptors in the skin. **Free nerve endings** are common in the papillary layer of the dermis (Figure 18-3a●). In sensitive areas, the dendritic branches penetrate the epidermis and contact *Merkel cells* in the stratum germinativum. ∞ [p. 90] Each Merkel cell communicates with a sensory neuron across a chemical synapse that involves an expanded nerve terminal known as a **Merkel's** (MER-kelz) **disc** (Figure 18-3b●). Merkel cells are sensitive to fine touch and pressure. They are tonically active, extremely sensitive, and have narrow receptive fields.

Free nerve endings are also associated with hair follicles. The free nerve endings of the **root hair plexus** monitor distortions and movements across the body surface (Figure 18-3f●). When the hair is displaced, the movement of the follicle distorts the sensory dendrites and produces action potentials in the afferent fiber. These receptors adapt rapidly, so they are best at detecting initial contact and subsequent movements. For example, most people feel their clothing only when they move or when they consciously focus attention on tactile sensations from their skin.

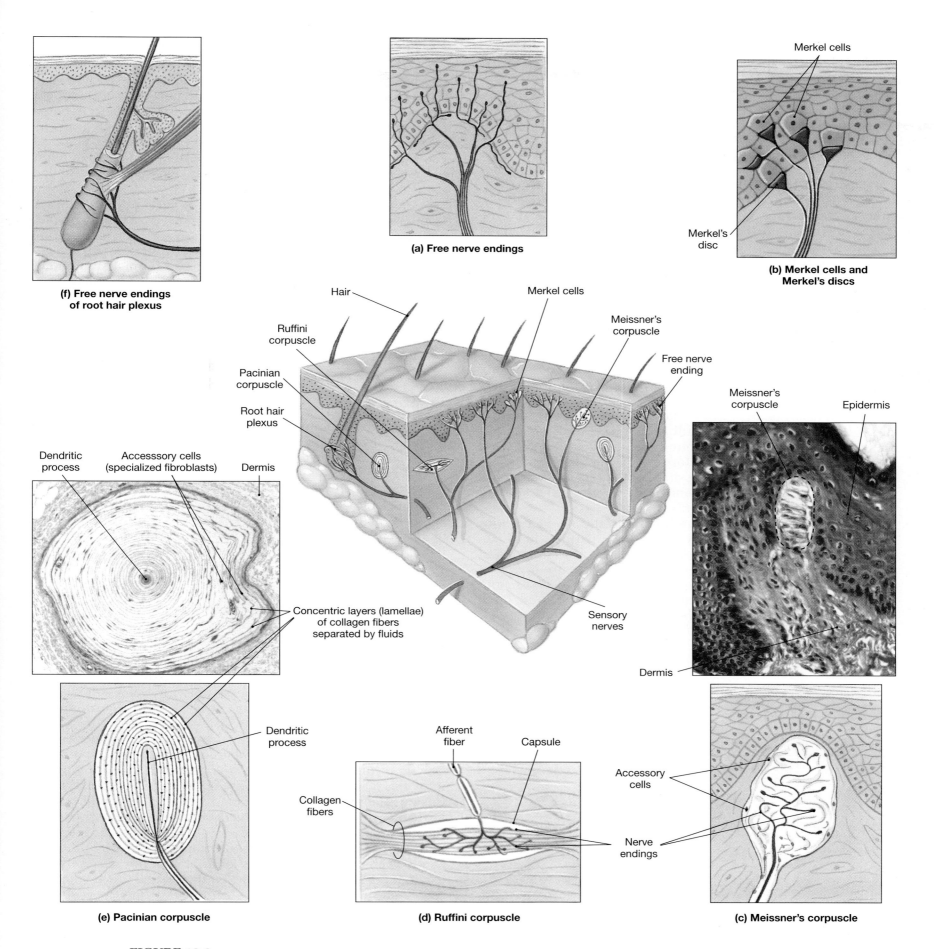

(f) Free nerve endings of root hair plexus

(a) Free nerve endings

(b) Merkel cells and Merkel's discs

Merkel cells

Merkel's disc

Hair

Ruffini corpuscle

Pacinian corpuscle

Root hair plexus

Merkel cells

Meissner's corpuscle

Free nerve ending

Meissner's corpuscle

Epidermis

Dendritic process

Accesssory cells (specialized fibroblasts)

Dermis

Concentric layers (lamellae) of collagen fibers separated by fluids

Dendritic process

Sensory nerves

Dermis

(e) Pacinian corpuscle

Afferent fiber

Capsule

Collagen fibers

Nerve endings

Accessory cells

(d) Ruffini corpuscle

(c) Meissner's corpuscle

FIGURE 18-3
Tactile Receptors in the Skin. The location and general appearance of six important tactile receptors. (a) Free nerve endings. (b) Merkel cells and Merkel's discs. (c) Meissner's corpuscle. (LM × 550) (d) Ruffini corpuscle. (e) Pacinian corpuscle. (LM × 125) (f) Free nerve endings of root hair plexus.

Large oval **Meissner's** (MĪS-nerz) **corpuscles** (Figure 18-3c●) are found where tactile sensitivities are extremely well developed. They are especially common at the eyelids, lips, fingertips, nipples, and external genitalia. The dendrites are highly coiled and interwoven, and they are surrounded by modified Schwann cells. A fibrous capsule surrounds the entire complex and anchors it within the dermis. Meissner's corpuscles detect light touch, movement, and vibration; they adapt to stimulation within a second after contact.

Pacinian (pa-SIN-ē-an) **corpuscles** are considerably larger encapsulated receptors (Figure 18-3e●). The dendritic process lies within a series of concentric cellular layers. These layers shield the dendrite from virtually every source of stimulation other than direct pressure. Pacinian corpuscles respond to heavy pressure but are most sensitive to pulsing or vibrating stimuli. These receptors are scattered throughout the dermis, notably in the fingers, breasts, and external genitalia. They are also encountered in the superficial and deep fasciae, in periostea and joint capsules, in mesenteries, in the pancreas and mammary glands, and in the walls of the urethra and urinary bladder.

Ruffini (ru-FĒ-nē) **corpuscles**, located in the dermis, are also sensitive to pressure and distortion of the skin, but they are tonically active and show little if any adaptation. The capsule surrounds a core of collagen fibers that are continuous with those of the surrounding dermis. Dendrites within the capsule are interwoven around the collagen fibers (Figure 18-3d●). Any tension or distortion of the dermis tugs or twists the fibers within the capsule, and this change stretches or compresses the attached dendrites and alters the activity in the myelinated afferent fiber.

Table 18-1 summarizes the functions and characteristics of the six tactile receptors discussed. The distribution of tactile sensations inside the CNS is via the posterior column and spinothalamic pathways. ∞ [p. 410] Tactile sensitivities may be altered by peripheral infection, disease processes, and damage to sensory afferents or central pathways, and there are important clinical tests that evaluate tactile sensitivity. ✝*Assessment of Tactile Sensitivities [p. 762]*

TABLE 18-1   Touch and Pressure Receptors

Sensation	Receptor	Responds to
Fine touch	Free nerve ending	Light contact with skin
	Merkel's disc	As above
	Root hair plexus	Initial contact with hair shaft
Pressure and vibration	Meissner's corpuscle	Initial contact and low-frequency vibrations
	Pacinian corpuscle	Initial contact (deep) and high-frequency vibrations
Deep pressure	Ruffini corpuscle	Stretching and distortion of the dermis

Baroreceptors
(Figure 18-4)

Baroreceptors monitor changes in pressure. The receptor consists of free nerve endings that branch within the elastic tissues in the wall of a hollow organ, a blood vessel or a portion of the respiratory, digestive, or urinary tracts. When the pressure changes, the elastic walls of these tubes or organs contract or expand. This movement distorts the dendritic branches and alters the rate of action potential generation. Baroreceptors respond immediately to a change in pressure.

Baroreceptors monitor blood pressure in the walls of major vessels, including the carotid artery (at the *carotid sinus*) and the aorta (at the *aortic sinus*). The information plays a major role in regulating cardiac function and adjusting blood flow to vital tissues. Baroreceptors in the lungs monitor the degree of lung expansion. This information is relayed to the respiratory rhythmicity center, which sets the pace of respiration. Baroreceptors in the urinary and digestive tracts trigger a variety of visceral reflexes, such as urination. Figure 18-4● provides examples of baroreceptor locations and functions.

Proprioceptors

Proprioceptors monitor the position of joints, the tension in tendons and ligaments, and the state of muscular contraction. Generally, proprioceptors do not adapt to constant stimulation. *Muscle spindles*, which monitor the length of a skeletal muscle, were considered in an earlier chapter. ∞ [p. 244] **Golgi tendon organs** monitor the tension in tendons. If tension becomes dangerously high, these receptors trigger the reflexive relaxation of the contracting muscle to prevent permanent damage to the muscle, tendon, or bone.

Chemoreceptors
(Figure 18-5)

Specialized chemoreceptive neurons can detect small changes in the concentration of specific chemicals or compounds. In general, chemoreceptors respond only to water-soluble and lipid-soluble substances that are dissolved in the surrounding fluid.

The locations of important chemosensory receptors are indicated in Figure 18-5●. Neurons within the respiratory centers of the brain respond to the concentration of hydrogen ions (pH) and carbon dioxide in the cerebrospinal fluid. Chemoreceptive neurons are found within the **carotid bodies**, near the origin of the internal carotid arteries on each side of the neck, and in the **aortic bodies** between the major branches of the aortic arch. These receptors monitor the carbon dioxide and oxygen concentration of arterial blood. The afferent fibers leaving the carotid and aortic bodies reach the respiratory centers by traveling along the ninth (glossopharyngeal) and tenth (vagus) cranial nerves. These chemoreceptors play an important role in the reflexive control of respiration and cardiovascular function.

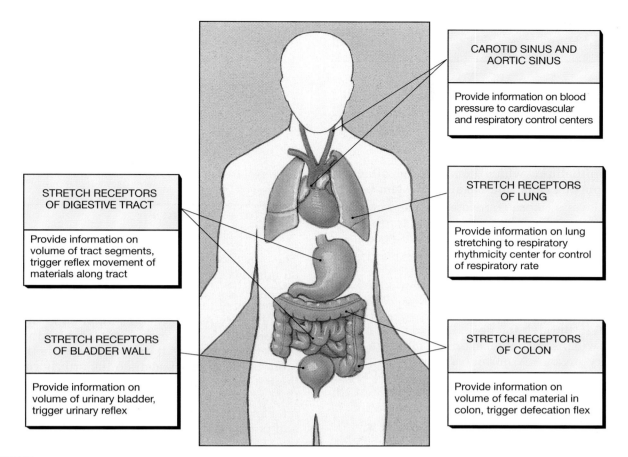

FIGURE 18-4

Baroreceptors and the Regulation of Autonomic Functions. Baroreceptors provide information essential to the regulation of autonomic activities, including respiration, digestion, urination, and defecation.

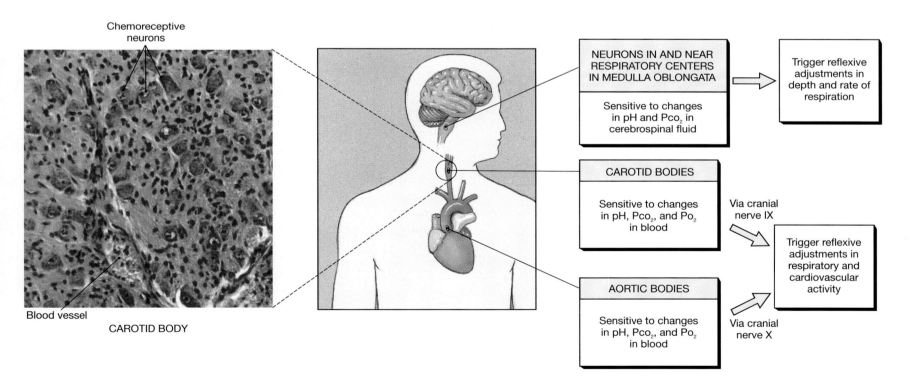

FIGURE 18-5

Chemoreceptors. Chemoreceptors are found inside the CNS, on the ventrolateral surfaces of the medulla oblongata, and in the aortic and carotid bodies. These receptors are involved in the autonomic regulation of respiratory and cardiovascular function. The micrograph shows the histological appearance of the chemoreceptive neurons in the carotid body. (LM × 1500)

OLFACTION (SMELL)
(Figure 18-6)

The sense of smell, more precisely called **olfaction**, is provided by paired **olfactory organs**. These organs are located in the nasal cavity on either side of the nasal septum. The olfactory organs (Figure 18-6●) consist of:

- an **olfactory epithelium**, which contains the **olfactory receptors**, **supporting cells**, and **basal cells (stem cells)**;
- an underlying layer of loose connective tissue known as the *lamina propria*. This layer contains (1) **olfactory glands**, also called *Bowman's glands*, which produce a thick, pigmented mucus, (2) blood vessels, and (3) nerves.

The olfactory epithelium covers the inferior surface of the cribriform plate, the superior portion of the nasal septum, and the superior nasal conchae. ∞ [p. 144]

When air is drawn in through the nose, it swirls and eddies within the nasal cavity. This turbulence brings airborne compounds to the olfactory organs. A normal, relaxed inspiration provides a small sample of the inspired air (around 2%) to the olfactory organs. Sniffing repeatedly increases the flow of air across the olfactory epithelium, intensifying the stimulation of the receptors. Once compounds have reached the olfactory organs, water-soluble and lipid-soluble materials must diffuse into the mucus before they can stimulate the olfactory receptors.

Olfactory Receptors
(Figure 18-6b)

The olfactory receptors are highly modified neurons. The apical portion of each receptor forms a prominent knob that pro-
jects above the epithelial surface and into the nasal cavity (Figure 18-6b●). That projection provides a base for up to 20 cilia that extend into the surrounding mucus, exposing their considerable surface area to the dissolved chemical compounds. Somewhere between 10 and 20 million olfactory receptors are packed into an area of roughly 5 cm². Olfactory reception occurs on the surface of an olfactory cilium, but how the dissolved chemicals interact with the receptor surface remains to be determined.

Olfactory Pathways
(Figure 18-6b)

The olfactory system is very sensitive. As few as four molecules of an odorous substance can activate an olfactory receptor. However, the activation of an afferent fiber does not guarantee a conscious awareness of the stimulus. Considerable convergence occurs along the olfactory pathway, and inhibition at the intervening synapses can prevent the sensations from reaching the cerebral cortex.

Axons leaving the olfactory epithelium collect into 20 or more bundles that penetrate the cribriform plate of the ethmoid bone to reach the olfactory bulbs where the first synapse occurs (Figure 18-6b●). Axons leaving the olfactory bulb travel along the olfactory tract to reach the olfactory cortex, the hypothalamus, and portions of the limbic system.

Olfactory sensations are the only sensations that reach the cerebral cortex without first synapsing in the thalamus. The extensive limbic and hypothalamic connections help to explain the profound emotional and behavioral responses that can be produced by certain smells, such as perfumes.

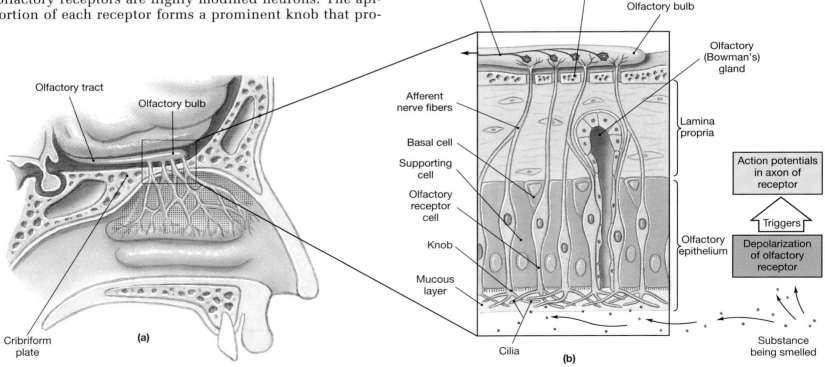

FIGURE 18-6
The Olfactory Organs. (a) The structure of the olfactory organ on the left side of the nasal septum. (b) Detail of olfactory epithelium and origin of olfactory tract.

Olfactory Discrimination

The olfactory system can make subtle distinctions between thousands of chemical stimuli. We know that there are at least 50 different "primary smells." No apparent structural differences exist among the olfactory cells, but the epithelium as a whole contains receptor populations with distinctly different sensitivities. The CNS interprets the smell on the basis of the particular pattern of receptor activity.

The olfactory receptor cells are the only known example of neuronal replacement in the adult human. Despite this process, the total number of receptors declines with age, and the remaining receptors become less sensitive. As a result, elderly individuals have difficulty detecting odors in low concentrations. This decline in the number of receptors accounts for Grandmother's tendency to apply perfume in excessive quantities and explains why Grandfather's aftershave seems so overdone; they must apply more to be able to smell it themselves.

Gustation, or taste, provides information about the foods and liquids that we consume. The **gustatory** (GUS-ta-tō-rē) **(taste) receptors** are distributed over the dorsal surface of the tongue (Figure 18-7a●) and adjacent portions of the pharynx and larynx. By adulthood the taste receptors on the epithelium of the pharynx and larynx have decreased in importance, and the *taste buds* of the tongue are the primary gustatory receptors.

Taste buds lie along the sides of epithelial projections called **papillae** (pa-PIL-lē; *papilla*, nipple-shaped mound). There are three types of papillae on the human tongue: **filiform** (*filum*, thread), **fungiform** (*fungus*, mushroom), and **circumvallate** (sir-kum-VAL-āt; *circum-*, around + *vallum*, wall). There are regional differences in the distribution of the papillae (Figure 18-7a●).

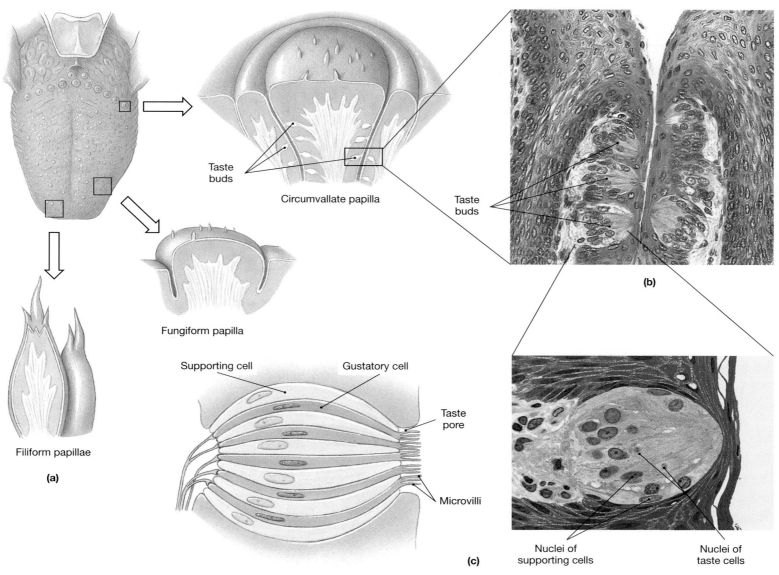

Taste buds

Circumvallate papilla

Fungiform papilla

Filiform papillae

(a)

Supporting cell Gustatory cell

Taste pore

Microvilli

(b)

Taste buds

(c)

Nuclei of supporting cells Nuclei of taste cells

FIGURE 18-7
Gustatory Reception. (a) Gustatory receptors are found in taste buds that form pockets in the epithelium of the fungiform and circumvallate papillae. (b) Photomicrograph of taste buds in a circumvallate papilla. (LM × 280) (c) A taste bud, showing receptor cells and supporting cells. The diagrammatic view shows details of the taste pore not visible in the light micrograph. (LM × 650)

Gustatory Receptors
(Figure 18-7b,c)

The taste receptors are clustered in individual **taste buds** (Figure 18-7b,c●). Each taste bud contains around 40 slender receptors, called **gustatory cells**, and a number of supporting cells. Taste buds are recessed into the surrounding epithelium and isolated from the relatively unprocessed oral contents. Each gustatory cell extends slender microvilli, sometimes called *taste hairs*, into the surrounding fluids through a narrow opening, the **taste pore**.

The small fungiform papillae each contain about five taste buds, while the large circumvallate papillae, which form a V shape near the posterior margin of the tongue, contain as many as 100 taste buds per papilla. The typical adult has over 10,000 taste buds.

The mechanism behind gustatory reception appears to parallel that of olfaction. Dissolved chemicals contacting the taste hairs provide the stimulus that produces a change in the transmembrane potential of the taste cell. Stimulation of the gustatory cell creates action potentials in the afferent fiber.

Gustatory Pathways
(Figure 18-8)

Taste buds are monitored by cranial nerves VII (facial), IX (glossopharyngeal), and X (vagus) (Figure 18-8●). The sensory afferents synapse within the **nucleus solitarius** of the medulla, and the axons of the postsynaptic neurons enter the medial lemniscus. After another synapse in the thalamus, the information is projected to the appropriate portions of the primary sensory cortex.

A conscious perception of taste involves correlating the information received from the taste buds with other sensory data. Information concerning the general texture of the food, together with taste-related sensations of "peppery" or "burning hot," is provided by sensory afferents in the trigeminal nerve (N V). In addition, the level of stimulation from the olfactory receptors plays an overwhelming role in taste perception. We are several thousand times more sensitive to "tastes" when our olfactory organs are fully functional. When you have a cold, airborne molecules cannot reach the olfactory receptors, and meals taste dull and unappealing. This reduction in taste perception will occur even though the taste buds may be responding normally.

Gustatory Discrimination
(Figure 18-8)

There are four **primary taste sensations**: sweet, salt, sour, and bitter. Each taste bud shows a particular sensitivity to one of these tastes, and a sensory map of the tongue indicates that there are regional differences in primary sensitivity (Figure

FIGURE 18-8
Gustatory Pathways. Three cranial nerves (VII, IX, and X) carry gustatory information to the primary sensory cortex of the cerebrum.

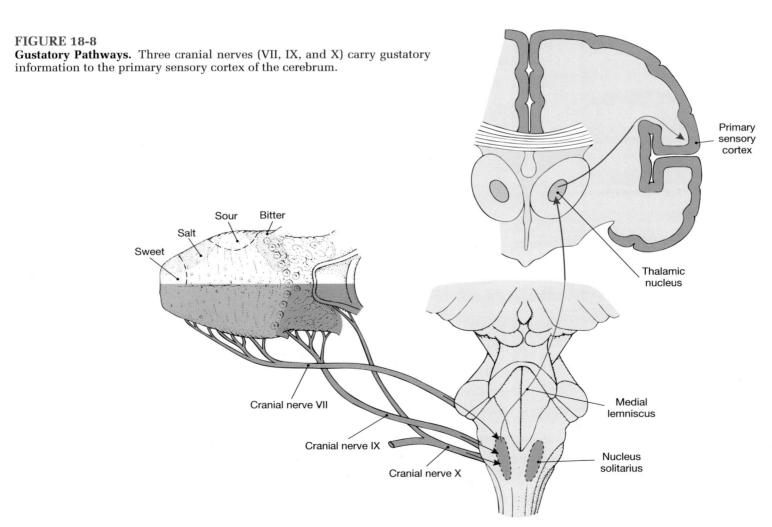

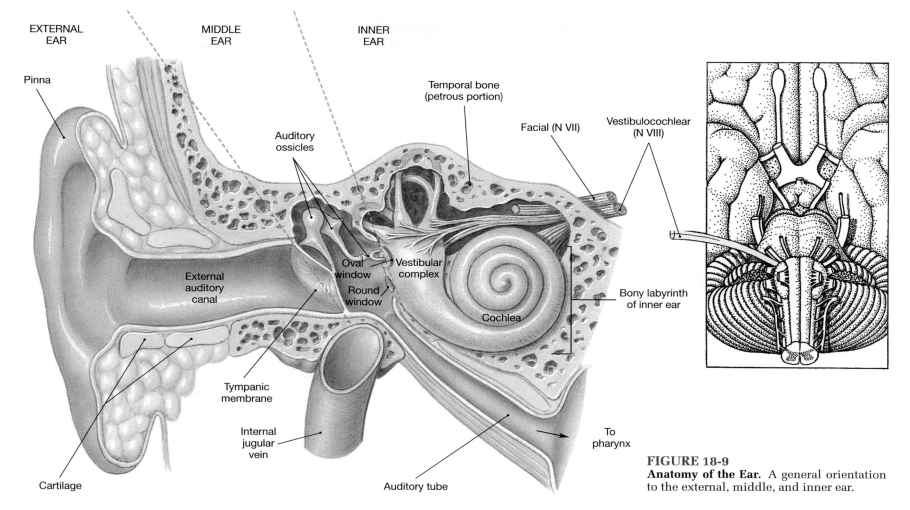

EXTERNAL EAR   MIDDLE EAR   INNER EAR

Pinna

Auditory ossicles

Temporal bone (petrous portion)

Facial (N VII)

Vestibulocochlear (N VIII)

External auditory canal

Oval window

Vestibular complex

Round window

Cochlea

Bony labyrinth of inner ear

Tympanic membrane

Internal jugular vein

To pharynx

Cartilage

Auditory tube

FIGURE 18-9
Anatomy of the Ear. A general orientation to the external, middle, and inner ear.

18-8●). The threshold for receptor stimulation varies for each of the primary taste sensations, and the taste receptors respond most readily to unpleasant rather than pleasant stimuli. For example, we are almost a thousand times more sensitive to acids, which give a sour taste, than to either sweet or salty chemicals, and we are a hundred times more sensitive to bitter compounds than to acids. This sensitivity has survival value, for acids can damage the mucous membranes of the mouth and pharynx, and many potent biological toxins produce an extremely bitter taste.

There are significant individual differences in taste sensitivity. Many of these conditions are inherited. The best-known example involves sensitivity to the compound phenylthiourea, also known as phenylthiocarbamide, or PTC. In the Caucasian population, roughly 70% can taste this substance; the rest are unable to detect it.

Our tasting abilities change with age. We begin life with over 10,000 taste buds, but the number begins declining dramatically by age 50. The sensory loss becomes especially significant because aging individuals also experience a decline in the population of olfactory receptors. As a result, many elderly people find that their food tastes bland and unappetizing, whereas children often find the same foods too spicy.

√ **When the nociceptors in your hand are stimulated, what sensation do you perceive?**

√ **What would happen to an individual if the information from proprioceptors in the legs were blocked from reaching the CNS? Identify the structures of the sensory pathway involved.**

EQUILIBRIUM AND HEARING
(Figure 18-9)

The ear is divided into three anatomical regions; the *external ear*, the *middle ear*, and the *inner ear* (Figure 18-9●). The external ear is the visible portion of the ear, and it collects and directs sound waves to the *eardrum*. The *middle ear* is a chamber located within the petrous portion of the temporal bone. Structures within the middle ear amplify sound waves and transmit them to an appropriate portion of the inner ear. The *inner ear* contains the sensory organs for equilibrium and hearing.

The External Ear
(Figures 18-9/18-10a,b)

The **external ear** includes the fleshy flap and cartilaginous **pinna**, or *auricle*, which surrounds the **external auditory canal**. The pinna protects and provides directional sensitivity to the ear by blocking or facilitating the passage of sound into the external auditory canal. The external auditory canal is a passageway that ends at the eardrum, also called the **tympanic membrane**, or **tympanum** (Figures 18-9 and 18-10a●). The tympanum is a thin, semitransparent sheet (Figure 18-10b●) that separates the external ear from the middle ear.

The tympanum is very delicate. The pinna and the narrow external auditory canal provide some protection from accidental injury to the typanum. In addition, **ceruminous glands** along the external auditory canal secrete a waxy material, and many small, outwardly projecting hairs help deny access to foreign objects or insects. The waxy secretion of the ceruminous glands, called **cerumen**, also slows the growth of microorganisms in the external auditory canal and reduces the chances of infection.

The Middle Ear
(Figures 18-9/18-10)

The **middle ear** includes the **tympanic cavity** and the *auditory ossicles*. The tympanic cavity is filled with air. It is separated from the external auditory canal by the tympanum, but it communicates with the nasopharynx and with the mastoid sinuses via a number of small and variable connections. ∞ [p. 141] The connection with the nasopharynx is the **auditory tube** (Figures 18-9 and 18-10●), also called the *pharyngotympanic tube* or the *Eustachian tube.* This tube is about 4.0 cm in length, and it con-

sists of two portions. The connection to the tympanic cavity is relatively narrow and supported by cartilage. The opening into the nasopharynx is relatively broad and funnel-shaped. The auditory tube serves to equalize the pressure in the middle ear cavity with external, atmospheric pressure. Unfortunately, it can also allow microorganisms to travel from the nasopharynx into the tympanic cavity.

The Auditory Ossicles
(Figures 18-9/18-10)

The tympanic cavity contains three tiny ear bones, collectively called **auditory ossicles**. The ear bones connect the tympanic membrane with the receptor complex of the inner ear (Figures 18-9 and 18-10●). The three auditory ossicles are the *malleus*, the *incus*, and the *stapes*. These bones act as levers that transfer sound vibrations from the tympanum to a fluid-filled chamber within the inner ear.

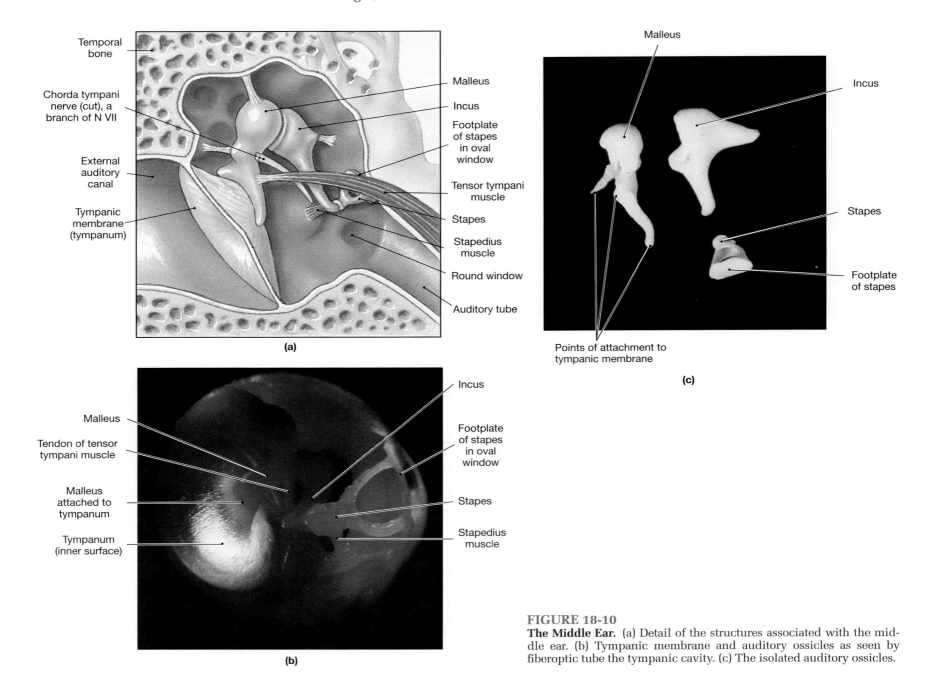

FIGURE 18-10
The Middle Ear. (a) Detail of the structures associated with the middle ear. (b) Tympanic membrane and auditory ossicles as seen by fiberoptic tube the tympanic cavity. (c) The isolated auditory ossicles.

The **malleus** (*malleus*, a hammer) attaches at three points to the interior surface of the tympanum. The middle bone, the **incus** (*incus*, anvil), attaches the malleus to the inner **stapes** (STĀ-pēz; *stapes*, stirrup). The *base* of the stapes attaches to the membranous *oval window* of the inner ear. Movement of the stapes sets up vibrations in the fluid contents of the inner ear.

Vibration of the tympanum converts arriving sound waves into mechanical movements. The auditory ossicles then conduct those vibrations to the inner ear. Because of the way these ossicles are connected, an in-out movement of the tympanum produces a rocking motion at the stapes. The amount of movement increases markedly from tympanic membrane to oval window. The tympanic membrane is 22 times larger and heavier than the oval window, and a 1-μm movement of the tympanic membrane produces a 22-μm deflection of the oval window.

Because this magnification occurs, we can hear very faint sounds. But that degree of magnification can be a problem when we are exposed to very loud noises. Within the tympanic cavity, two small muscles serve to protect the eardrum and ossicles from violent movements under very noisy conditions.

- The **tensor tympani** (TEN-sor tim-PAN-ē) **muscle** is a short ribbon of muscle whose origin is the petrous portion of the temporal bone and the auditory tube and whose insertion is on the "handle" of the malleus (Figure 18-10a●). When the tensor tympani contracts, the malleus is pulled medially, stiffening the tympanum. This increased stiffness reduces the amount of possible movement. The tympani muscle is innervated by motor fibers of the mandibular branch of N V.

- The **stapedius** (stā-PĒ-dē-us) **muscle**, innervated by the facial nerve (N VII), originates from the posterior wall of the tympanic cavity and inserts on the stapes (Figure 18-10a,b●). Contraction of the stapedius pulls the stapes, reducing movement of the stapes at the oval window.

The Inner Ear
(Figures 18-9 to 18-12)

The senses of equilibrium and hearing are provided by the receptors of the **inner ear** (Figures 18-9 and 18-11●). The receptors lie within a collection of fluid-filled tubes and chambers known as the **membranous labyrinth** (*labyrinthos*, network of canals). The membranous labyrinth contains a fluid called **endolymph** (EN-dō-limf). The receptor cells of the inner ear can function only when exposed to the unique ionic composition of the endolymph. Endolymph is quite distinctive; it has a relatively high potassium ion concentration, and a relatively low sodium ion concentration, whereas typical extracellular fluids have high sodium and low potassium ion concentrations.

The **bony labyrinth** is a shell of dense bone that surrounds and protects the membranous labyrinth. Its inner contours closely follow the contours of the membranous labyrinth (Figure 18-12b●), while its outer walls are fused with the surrounding temporal bone. ∞ [p. 142] Between the bony and membranous labyrinths flows the **perilymph** (PER-i-limf), a liquid whose properties closely resemble those of cerebrospinal fluid.

The bony labyrinth can be subdivided into the **vestibule** (VES-ti-būl), the **semicircular canals**, and the **cochlea** (KOK-lē-a; *cochlea*, snail shell) as seen in Figures 18-9 and 18-12b,c●. Receptors in the membranous labyrinth of the vestibule and semicircular canals provide the sense of equilibrium; those in the cochlea provide the sense of hearing. The structures and airspaces of the external ear and middle ear function in the capture and transmission of sound to the cochlea.

FIGURE 18-11
Structural Relationships of the Inner Ear

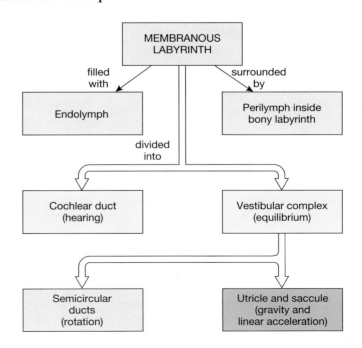

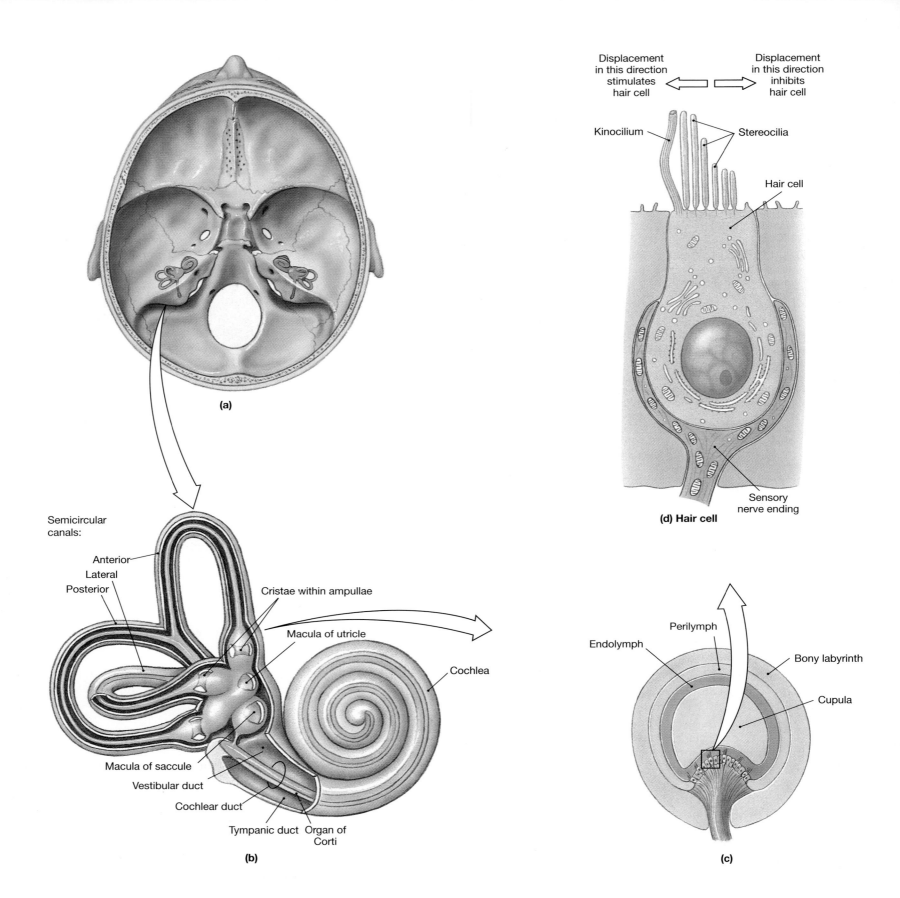

Displacement in this direction stimulates hair cell ⟸ ⟹ **Displacement in this direction inhibits hair cell**

Kinocilium — Stereocilia

Hair cell

Sensory nerve ending

(d) Hair cell

(a)

Semicircular canals:

Anterior
Lateral
Posterior

Cristae within ampullae

Macula of utricle

Cochlea

Macula of saccule

Vestibular duct

Cochlear duct

Tympanic duct Organ of Corti

(b)

Perilymph

Endolymph

Bony labyrinth

Cupula

(c)

FIGURE 18-12

Receptor Hair Cells of the Inner Ear. (a) Bony labyrinth projected on petrous portion of temporal bone to show location. (b) Anterior view of the bony labyrinth, showing the enclosed membranous labyrinth. (c) A section through the ampulla of a semicircular canal. Displacement of the cupula stimulates the hair cells. (d) Structure of a typical hair cell, showing details revealed by electron microscopy. An individual hair cell has a kinocilium and numerous stereocilia. Each cell contacts a sensory neuron. Bending the stereocilia toward the kinocilium depolarizes the cell and stimulates the sensory neuron. Displacement in the opposite direction inhibits the sensory neuron.

The combination of vestibule and semicircular canals is called the *vestibular complex*, because the fluid-filled chambers of the vestibule are broadly continuous with those of the semicircular canals. The vestibule includes a pair of membranous sacs, the **saccule** (SAK-ūl) and the **utricle** (Ū-tre-kl), or the *sacculus* and *utriculus*. Receptors in the saccule and utricle provide sensations of gravity and linear acceleration. Those in the semicircular canals are stimulated by rotation of the head.

The cochlea contains a slender, elongated portion of the membranous labyrinth known as the **cochlear duct** (Figure 18-12c,d●). The cochlear duct sits sandwiched between a pair of perilymph-filled chambers, and the entire complex makes 2½ turns around a central bony hub. In sectional view the spiral arrangement resembles that of a snail shell, *cochlea* in Latin.

The outer walls of the perilymphatic chambers consist of dense bone everywhere except at two small areas near the base of the cochlear spiral. The **round window** is a thin, membranous partition that separates the perilymph of the cochlear chambers from the air spaces of the middle ear (Figure 18-9●). The **oval window** has a similar structure, but it is firmly bound to the stapes (Figure 18-10a,b●). When a sound vibrates the tympanum, the movements are conducted over the auditory ossicles to the surface of the oval window. This process ultimately leads to the stimulation of receptors within the cochlear duct, and we "hear" the sound.

Receptor Function in the Inner Ear
(Figure 18-12d)

The sensory receptors of the inner ear are called **hair cells** (Figure 18-12d●). These receptor cells are surrounded by **supporting cells** and are monitored by sensory afferent fibers. The free surface of the hair cell supports 80–100 long **stereocilia**. Each hair cell in the vestibule also contains a **kinocilium**, a single large cilium. Hair cells do not actively move their kinocilia and stereocilia. However, when an external force pushes against these processes, the distortion of the cell membrane alters the rate of chemical transmitter released by the hair cell.

Hair cells provide information concerning the direction and strength of mechanical stimuli. The stimuli involved, however, are quite varied: gravity or acceleration in the vestibule, rotation in the semicircular canals, and sound in the cochlea activate the respective hair cells. The sensitivities of the hair cells differ because in each of these regions, there are different accessory structures that determine the source of stimulation. The importance of these accessory structures will become apparent as we consider the semicircular canals, utricle, and saccule.

The Vestibular Complex

The vestibular complex consists of the semicircular canals, the saccule, and the utricle.

THE SEMICIRCULAR CANALS *(Figures 18-12b,c/18-13).* Receptors in the semicircular canals respond to rotational movements of the head. The **anterior, posterior**, and **lateral** semicircular canals are continuous with the utricle (Figure 18-12b●). Each semicircular canal contains a swollen region, the **ampulla**, which contains the sensory receptors. Hair cells attached to the wall of the ampulla form a raised structure known as a **crista** (Figure 18-12b,c●). The kinocilia and stereocilia of the hair cells are embedded in a gelatinous structure, the **cupula** (KŪ-pū-la). Because the cupula has a density very close to that of the sur-

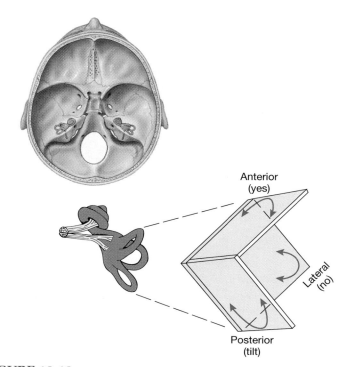

FIGURE 18-13
Function of the Semicircular Ducts. A superior view showing the planes of sensitivity for the semicircular ducts.

■ CLINICAL BRIEF
Nystagmus

Automatic eye movements occur in response to sensations of motion (whether real or illusory) under the direction of the *superior colliculus*. ∞ [p. 384] These movements attempt to keep the gaze focused on a specific point in space. When you spin around, your eyes fix on one point for a moment, then jump ahead to another, in a series of short, jerky movements. These eye movements may appear in stationary individuals after damage to the brain stem or inner ear. This condition is called **nystagmus**. Physicians often check for nystagmus by asking the subject to watch a small pen-light as it is moved across the field of vision.

rounding endolymph, it essentially "floats" above the receptor surface, nearly filling the ampulla. When the head rotates in the plane of the canal, movement of the endolymph along the canal axis pushes the cupula and distorts the receptor processes. Fluid movement in one direction stimulates the hair cells, and movement in the opposite direction inhibits them. When the endolymph stops moving, the elastic nature of the cupula makes it "bounce back" to its normal position.

Even the most complex movement can be analyzed in terms of motion in three rotational planes. Each semicircular canal responds to one of these rotational movements (Figure 18-13●). A horizontal rotation, as in shaking the head "no," stimulates the hair cells of the lateral semicircular canal. Nodding "yes" excites the anterior canal, while tilting the head from side to side activates the receptors in the posterior canal.

THE UTRICLE AND SACCULE *(Figures 18-12b/18-14).* The utricle and saccule (Figures 18-12b and 18-14●) are connected by a slender passageway continuous with the narrow **endolymphatic duct**. The endolymphatic duct ends in a blind pouch, the **endolymphatic sac**, that projects through the dura mater lining the temporal bone and into the subdural space. Portions of the

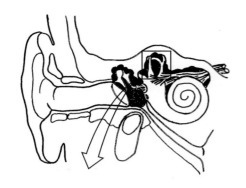

cochlear duct continually secrete endolymph, and at the endolymphatic sac, excess fluids return to the general circulation.

The hair cells of the utricle and saccule are clustered in the oval **maculae** (MAK-ū-lē; *macula*, spot) (Figures 18-12b and 18-14a●). As in the ampullae, the hair cell processes are embedded in a gelatinous mass, but the macular receptors lie under a thin layer containing densely packed mineral crystals. These **otoliths** (Ō-tō-liths; *oto-*, ear + *lithos*, stone), sometimes called **otoconia** (Ō-tō-KŌ-nē-a; *conia*, dust), can be seen in Figure 18-14b●. When the head is in the normal, upright position, the otoliths sit atop the maculae. Their weight presses down on the macular surfaces, pushing the sensory hairs down rather than to one side or another. When the head is tilted, the pull of gravity on the otoliths shifts the mass to the side. This shift distorts the sensory hairs, and the change in receptor activity tells the CNS that the head is no longer level (Figure 18-14c●).

For example, when an elevator starts its downward plunge, we are immediately aware of it because the otoliths no longer push so forcefully against the surface of the receptor cells. Once

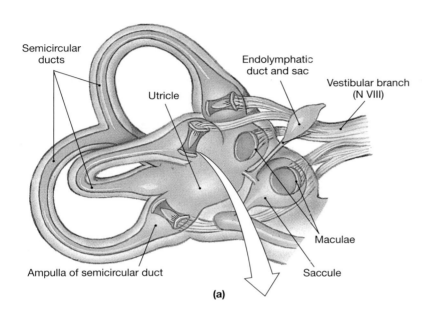

(a)

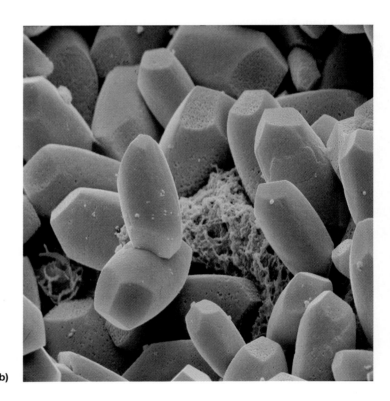

(b)

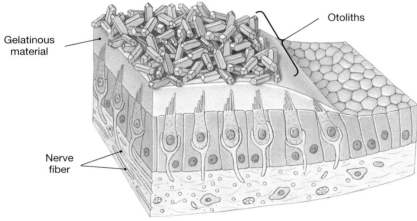

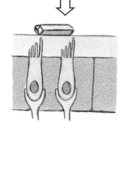

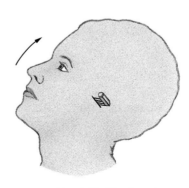

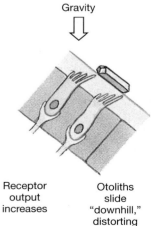

Gravity

Gravity

Head upright

(c)

Head tilted back

Receptor output increases

Otoliths slide "downhill," distorting hair cell processes

FIGURE 18-14
The Maculae of the Vestibule. (a) Location of the maculae within the utricle and saccule. (b) Detailed structure of a sensory macula and a scanning electron micrograph showing the crystalline structure of otoliths. (c) Diagrammatic view of changes in otoliths position during tilting of the head.

they catch up, we are no longer aware of any movement until the elevator brakes to a halt. As the body slows down, the otoliths press harder against the hair cells, and we "feel" the force of gravity increase. A similar mechanism accounts for our perception of linear acceleration in a car that speeds up suddenly. The otoliths lag behind, distorting the sensory hairs and changing the activity in the sensory neurons.

PATHWAYS FOR VESTIBULAR SENSATIONS *(Figure 18-15).* Hair cells of the vestibule and semicircular canals are monitored by sensory neurons located in adjacent **vestibular ganglia**. Sensory fibers from each ganglion form the **vestibular branch** of the vestibulocochlear nerve (N VIII). These fibers synapse on neurons within the **vestibular nuclei** at the boundary between the pons and medulla oblongata. The two vestibular nuclei:

1.   integrate the sensory information concerning balance and equilibrium arriving from each side of the head,

2.   relay information from the vestibular apparatus to the cerebellum,

3.   relay information from the vestibular apparatus to the cerebral cortex, providing a conscious sense of position and movement, and

4.   send commands to motor nuclei in the brain stem and spinal cord.

The reflexive motor commands issued by the vestibular nucleus are distributed to the motor nuclei for cranial nerves involved with eye, head, and neck movements (N III, N IV, N VI,

and N XI). Descending instructions along the **vestibulospinal tracts** of the spinal cord adjust peripheral muscle tone to complement the reflexive movements of the head or neck. These pathways are illustrated in Figure 18-15●. ⫟ *Vertigo, Dizziness, and Motion Sickness [p. 762]*

The Cochlea
(Figure 18-16)

The bony cochlea (Figure 18-16●) coils around a central hub, or **modiolus** (mō-DI-ō-lus). There are usually 2.5 turns in the cochlear spiral. The modiolus encloses the **spiral ganglion** containing the cell bodies of the sensory neurons that monitor the receptors in the cochlear duct. In sectional view, the **cochlear duct**, or *scala media*, lies between a pair of perilymphatic chambers; the **vestibular duct** (*scala vestibuli*) and the **tympanic duct** (*scala tympani*) are interconnected at the tip of the cochlear spiral. The oval window forms a portion of the wall of the vestibular duct, and the round window is a part of the tympanic duct.

THE ORGAN OF CORTI *(Figure 18-16b–e).*   The hair cells of the cochlear duct are found in the **organ of Corti** (Figure 18-16b–e●). This sensory structure sits above the **basilar membrane** that separates the cochlear duct from the tympanic duct. The hair cells are arranged in a series of longitudinal rows. They lack kinocilia, and their stereocilia are in contact with the overlying **tectorial membrane** (tek-TOR-ē-al; *tectum*, roof). This membrane is firmly attached to the inner wall of the cochlear duct. When a portion of the basilar membrane bounces up and down, the stereocilia of the hair cells are distorted.

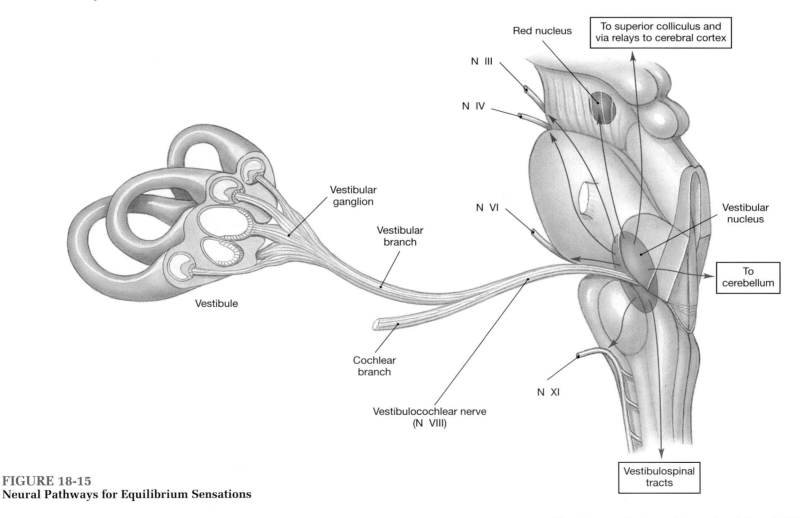

FIGURE 18-15
Neural Pathways for Equilibrium Sensations

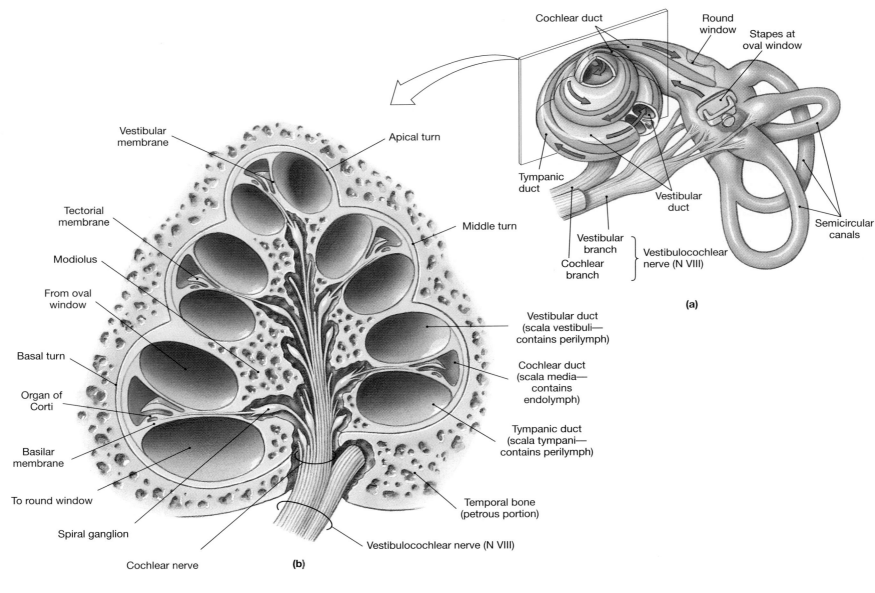

Cochlear duct

Round window

Stapes at oval window

Tympanic duct

Vestibular duct

Semicircular canals

Vestibular branch

Cochlear branch

Vestibulocochlear nerve (N VIII)

(a)

Vestibular membrane

Apical turn

Tectorial membrane

Middle turn

Modiolus

From oval window

Basal turn

Vestibular duct (scala vestibuli— contains perilymph)

Organ of Corti

Cochlear duct (scala media— contains endolymph)

Basilar membrane

Tympanic duct (scala tympani— contains perilymph)

To round window

Spiral ganglion

Temporal bone (petrous portion)

Cochlear nerve

Vestibulocochlear nerve (N VIII)

(b)

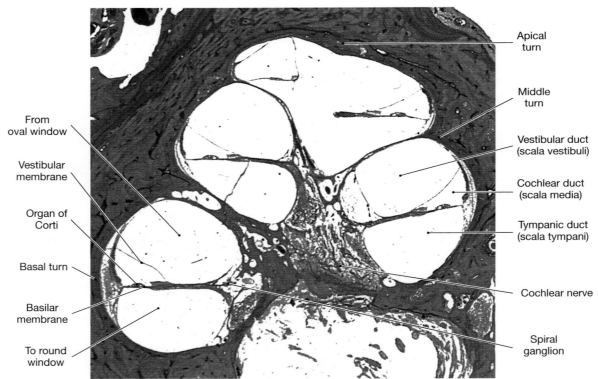

From oval window

Apical turn

Middle turn

Vestibular membrane

Vestibular duct (scala vestibuli)

Organ of Corti

Cochlear duct (scala media)

Basal turn

Tympanic duct (scala tympani)

Basilar membrane

Cochlear nerve

To round window

Spiral ganglion

(c)

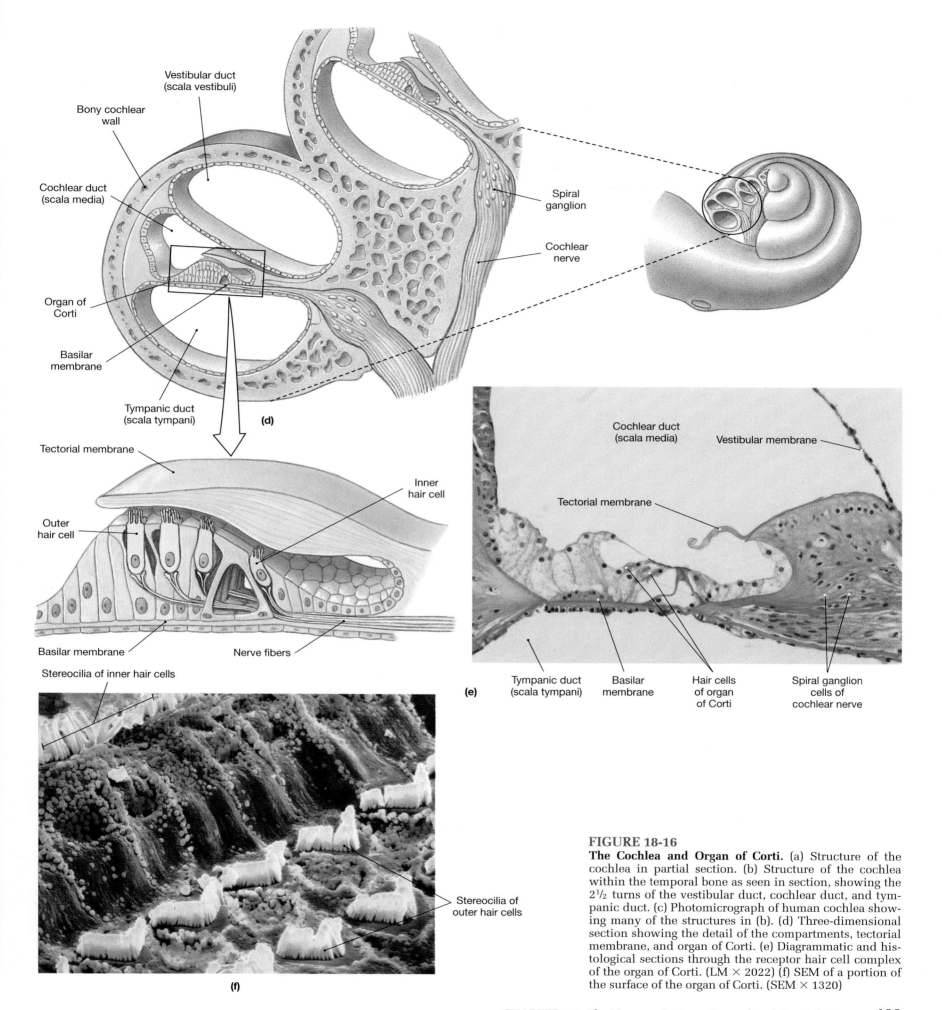

Vestibular duct
(scala vestibuli)

Bony cochlear
wall

Cochlear duct
(scala media)

Spiral
ganglion

Cochlear
nerve

Organ of
Corti

Basilar
membrane

Tympanic duct
(scala tympani)

(d)

Tectorial membrane

Inner
hair cell

Outer
hair cell

Basilar membrane

Nerve fibers

Stereocilia of inner hair cells

Cochlear duct
(scala media)

Vestibular membrane

Tectorial membrane

(e)

Tympanic duct
(scala tympani)

Basilar
membrane

Hair cells
of organ
of Corti

Spiral ganglion
cells of
cochlear nerve

Stereocilia of
outer hair cells

(f)

FIGURE 18-16
The Cochlea and Organ of Corti. (a) Structure of the
cochlea in partial section. (b) Structure of the cochlea
within the temporal bone as seen in section, showing the
$2^{1}/_{2}$ turns of the vestibular duct, cochlear duct, and tympanic duct. (c) Photomicrograph of human cochlea showing many of the structures in (b). (d) Three-dimensional
section showing the detail of the compartments, tectorial
membrane, and organ of Corti. (e) Diagrammatic and histological sections through the receptor hair cell complex
of the organ of Corti. (LM × 2022) (f) SEM of a portion of
the surface of the organ of Corti. (SEM × 1320)

Hearing

Hearing is the detection of sound, which consists of pressure waves conducted through air or water. Sound waves enter the external auditory meatus and travel along the auditory canal toward the tympanum. The tympanum provides the surface for sound collection, and it vibrates in response to sound waves with frequencies between approximately 20 and 20,000 Hz (in a young child). The auditory ossicles transfer these vibrations in modified form to the oval window.

Movement of the stapes at the oval window applies pressure to the perilymph of the vestibular duct. A property of liquids is their inability to be compressed. For example, when you sit on a waterbed, you know that when you push down *here* the waterbed bulges over *there*. Because the rest of the cochlea is sheathed in bone, pressure applied at the oval window can be relieved only at the round window. When the oval window moves inward, the round window bulges outward.

Movement of the stapes sets up pressure waves in the perilymph. These waves distort the cochlear duct and the organ of Corti, stimulating the hair cells. The location of maximum stimulation varies depending on the frequency (pitch) of the sound. High-frequency sounds affect the basilar membrane near the oval window; the lower the frequency of the sound, the farther away from the oval window the distortion will be.

The actual amount of movement at a given location will depend on the amount of force applied to the oval window. This relationship provides a mechanism for detecting the intensity (volume) of the sound. Very high-intensity sounds can produce hearing losses by breaking the stereocilia off the surfaces of the hair cells. The reflex contraction of the tensor tympani and stapedius in response to a dangerously loud noise occurs in less than 0.1 s, but this may not be fast enough to prevent damage and related hearing loss. Table 18-2 sum-

FIGURE 18-17
Pathways for Auditory Sensations. Auditory sensations are carried by the cochlear branch of N VIII to the cochlear nucleus of the medulla. From there the information is relayed to the inferior colliculus, a center that directs a variety of unconscious motor responses to sounds. Ascending acoustic information goes to the medial geniculate before being forwarded to the auditory cortex of the temporal lobe.

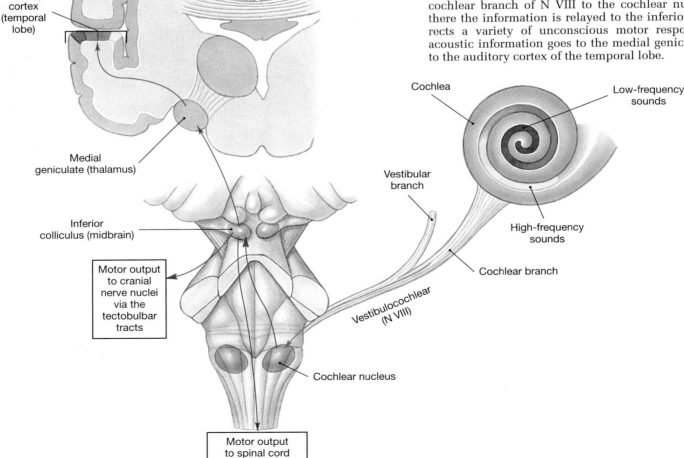

464

TABLE 18-2   Steps in the Production of an Auditory Sensation

1. Sound waves arrive at the tympanic membrane.
2. Movement of the tympanic membrane causes displacement of the auditory ossicles.
3. Movement of the stapes at the oval window establishes pressure waves in the perilymph of the vestibular duct.
4. The pressure waves distort the basilar membrane on their way to the round window of the tympanic duct.
5. Vibration of the basilar membrane causes vibration of hair cells against the tectorial membrane.
6. Information concerning the region and intensity of stimulation is relayed to the CNS over the cochlear branch of N VIII.

marizes the steps involved in translating a sound wave into an auditory sensation.

Auditory Pathways
(Figure 18-17)

Hair cell stimulation activates sensory neurons whose cell bodies are in the adjacent spiral ganglion. Their afferent fibers form the **cochlear branch** of the vestibulocochlear nerve (N VIII). These axons enter the medulla and synapse at the **cochlear nucleus** of the medulla. From here the information crosses to the opposite side of the brain and ascends to the inferior colliculus of the midbrain (Figure 18-17●). This processing center coordinates a number of responses to acoustic stimuli, including auditory reflexes involving skeletal muscles of the head, face, and trunk. These reflexes automatically change the position of the head in response to a sudden loud noise.

Before reaching the cerebral cortex and our conscious awareness, ascending auditory sensations synapse in the thalamus. Projection fibers then deliver the information to the auditory cortex of the temporal lobe. In effect, the auditory cortex contains a map of the organ of Corti. High-frequency sounds activate one portion of the cortex, and low-frequency sounds affect another. If the auditory cortex is damaged, the individual will respond to sounds and have normal acoustic reflexes, but sound interpretation and pattern recognition will be difficult or impossible. Damage to the adjacent association area does not affect the ability to detect the tones and patterns, but produces an inability to comprehend their meaning.

✓ **If the round window were not able to bulge out with increased pressure in the perilymph, how would sound perception be affected?**

✓ **How would loss of stereocilia from the hair cells of the organ of Corti affect hearing?**

VISION
(Figure 18-18)

Humans rely more on vision than on any other special sense. Our visual receptors are contained in elaborate structures, the eyes, which enable us not only to detect light but to create de-

tailed visual images. We will begin our discussion with the *accessory structures* of the eye that provide protection, lubrication, and support. The superficial anatomy of the eye and the major accessory structures are illustrated in Figure 18-18●.

Accessory Structures of the Eye

The **accessory structures** of the eye include the eyelids, the superficial epithelium of the eye, and the structures associated with the production, secretion, and removal of tears.

Eyelids
(Figures 18-18/18-19b)

The eyelids, or **palpebrae** (pal-PE-brē) are a continuation of the skin. The eyelids act like windshield wipers; their continual blinking movements keep the surface lubricated and free from dust and debris. They can also close firmly to protect the delicate surface of the eye. The free margins of the upper and lower eyelids are separated by the **palpebral fissure**, but the two are connected at the **medial canthus** (KAN-thus) and the **lateral canthus** (Figure 18-18●). The **eyelashes** along the palpebral margins are very robust hairs. These help to prevent foreign matter and insects from reaching the surface of the eye.

The eyelashes are associated with large sebaceous glands, the *glands of Zeis* (ZĪS). Along the inner margin of the lid, **Meibomian** (mī-BŌ-mē-an) **glands** secrete a lipid-rich product that helps to keep the eyelids from sticking together. At the medial canthus the **lacrimal caruncle** (KAR-unk-ul) (Figure 18-18a●) contains glands producing the thick secretions that contribute to the gritty deposits occasionally found after a good night's sleep. These various glands are subject to occasional invasion and infection by bacteria. A cyst, or **chalazion** (kah-LĀ-zē-on; "small lump"), usually results from the infection of a Meibomian gland. An infection in a sebaceous gland of one of the eyelashes, a Meibomian gland, or one of the many sweat glands that open to the surface between the follicles, produces a painful localized swelling known as a **sty**.

The visible surface of the eyelid is covered by a thin layer of stratified squamous epithelium. Deep to the subcutaneous layer, the eyelids are supported and strengthened by broad sheets of connective tissue, collectively called the **tarsal plate**. The muscle fibers of the *orbicularis oculi* and the *levator palpebrae superioris* lie between the tarsal plate and the skin. These skeletal muscles are responsible for closing the eyelids (orbicularis oculi) and raising the upper eyelid (levator palpebrae superioris). ∞ [pp. 255, 259]

The epithelium covering the inner surface of the eyelids and the outer surface of the eye is called the **conjunctiva** (kon-junk-TĪ-va) (Figure 18-19b●, p. 468). It is a mucous membrane covered by a specialized stratified squamous epithelium. The **palpebral conjunctiva** covers the inner surface of the eyelids, and the **ocular conjunctiva**, or **bulbar conjunctiva**, covers the anterior surface of the eye. A continuous supply of fluid washes over the surface of the eyeball, keeping the conjunctiva moist and clean. Goblet cells within the epithelium assist the various accessory glands in providing a superficial lubricant that prevents friction and drying of the opposing conjunctival surfaces.

Over the transparent **cornea** (KOR-nē-a) of the eye, the conjunctival epithelium changes to a very thin and delicate squa-

FIGURE 18-18

Accessory Structures of the Eye. (a) Superficial anatomy of the accessory structures. (b) Surface anatomy of the eye, with the margins of the orbit indicated for reference. (c) Anterior view of the eye dissected to show the accessory structures, emphasizing the lacrimal apparatus.

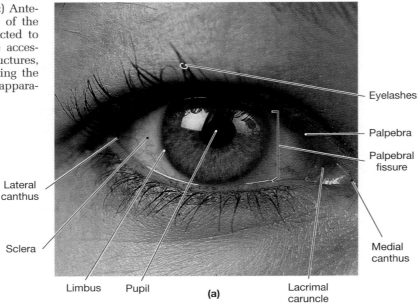

Eyelashes

Palpebra

Palpebral fissure

Medial canthus

Lateral canthus

Sclera

Limbus   Pupil

Lacrimal caruncle

(a)

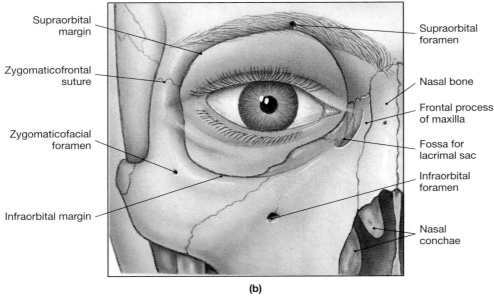

Supraorbital margin

Zygomaticofrontal suture

Zygomaticofacial foramen

Infraorbital margin

Supraorbital foramen

Nasal bone

Frontal process of maxilla

Fossa for lacrimal sac

Infraorbital foramen

Nasal conchae

(b)

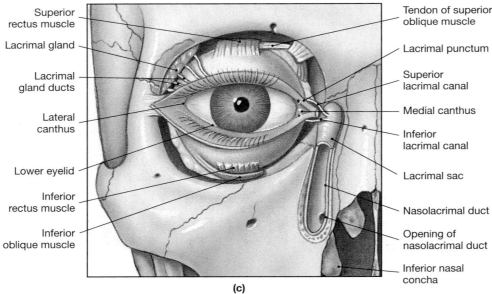

Superior rectus muscle

Lacrimal gland

Lacrimal gland ducts

Lateral canthus

Lower eyelid

Inferior rectus muscle

Inferior oblique muscle

Tendon of superior oblique muscle

Lacrimal punctum

Superior lacrimal canal

Medial canthus

Inferior lacrimal canal

Lacrimal sac

Nasolacrimal duct

Opening of nasolacrimal duct

Inferior nasal concha

(c)

mous epithelium 5–7 cells thick. Around the edges of the lids, the conjunctiva converts to the more robust stratified squamous epithelium characteristic of exposed bodily surfaces. Although there are no specialized sensory receptors beneath the conjunctival surface of the eye, the conjunctiva itself contains abundant free nerve endings with very broad sensitivities.

The Lacrimal Apparatus
(Figure 18-18b,c)

A constant flow of tears keeps conjunctival surfaces moist and clean. Tears reduce friction, remove debris, prevent bacterial infection, and provide nutrients and oxygen to portions of the conjunctival epithelium. The **lacrimal apparatus** produces, distributes, and removes tears. The lacrimal apparatus of each eye consists of: (1) a *lacrimal gland*, (2) *superior and inferior lacrimal canals*, (3) a *lacrimal sac*, and (4) a *nasolacrimal duct* (Figure 18-18b,c●).

The pocket created where the conjunctiva of the eyelid connects with that of the eye is known as the **fornix** (FOR-niks). The lateral portion of the superior fornix receives 10–12 ducts from the **lacrimal gland**, or tear gland. The lacrimal gland is about the size and shape of an almond, measuring roughly 12–20 mm (0.5–0.75 in.). It nestles within a depression in the frontal bone, just inside the orbit and superior and lateral to the eyeball. ∞ [p. 147] The lacrimal gland normally provides the key ingredients and most of the volume of the tears that bathe the conjunctival surfaces. Its secretions are watery, slightly alkaline, and contain the enzyme **lysozyme**, which attacks microorganisms.

The lacrimal gland produces tears at a rate of around 1 mℓ/day. Once the lacrimal secretions have reached the ocular surface, they mix with the products of accessory glands and the oily secretions of the Meibomian glands and glands of Zeis. The latter contributions produce a superficial "oil slick" that assists in lubrication and slows evaporation.

The blinking of the eye sweeps the tears across the ocular surface, and they accumulate at the medial canthus in an area known as the **lacus lacrimalis**, or "lake of tears." Two small pores, the **lacrimal puncta**, drain the lacrymal lake, emptying into the **lacrimal canals** that run along grooves in the surface of the lacrimal bone. These passageways lead to the **lacrimal sac**, and from there the **nasolacrimal duct** passes through the nasolacrimal groove of the lacrimal bone and maxilla, and delivers the tears to the inferior meatus of the nasal cavity. ∞ [p. 147]

Blockage of the lacrimal puncta or oversecretion of the lacrimal glands can produce "watery eyes" that are constantly overflowing. The condition of "dry eyes" is more common and is due to inadequate tear production. Lubricating "artificial tears" in the form of eye drops are the usual answer, but more serious cases may be treated by surgically closing the lacrimal puncta.

■ CLINICAL BRIEF
Conjunctivitis

Conjunctivitis, or "pink-eye," results from damage to and irritation of the conjunctival surface. The most obvious symptom results from dilation of the blood vessels beneath the conjunctival epithelium. The term *conjunctivitis* is more useful as the description of a symptom than as a name for a specific disease. A great variety of pathogens, including bacteria, viruses, and fungi can cause conjunctivitis, and a temporary form of the condition may be produced by chemical or physical irritation (including even such mundane experiences as prolonged crying or peeling an onion).

Chronic conjunctivitis, or **trachoma**, results from bacterial or viral invasion of the conjunctiva. Many of these infections are highly contagious, and severe cases may disrupt the corneal surface and affect vision. The pathogen most often involved is *Chlamydia trachomatis*. Trachoma is a relatively common problem in southwestern North America, North Africa, and the Middle East. The condition must be treated with topical and systemic antibiotics to prevent corneal damage and vision loss.

The Eye
(Figure 18-19a,e–g)

The eyes are slightly irregular spheroids with an average diameter of 24 mm (almost 1 in.), slightly smaller than a Ping-Pong ball. Each eye weighs around 8 g (0.28 oz). The eyeball shares space within the orbit with the extrinsic eye muscles, the lacrimal gland, and the cranial nerves and blood vessels that supply the eye and adjacent portions of the orbit and face (Figure 18-19e–g●). A mass of **orbital fat** provides padding and insulation.

The wall of the eye contains three distinct layers, or tunics (Figure 18-19a●): an outer *fibrous tunic*, an intermediate *vascular tunic*, and an inner *neural tunic*. The eyeball is hollow and the interior can be divided into two *cavities*. The large **posterior cavity** is also called the *vitreous chamber* because it contains the gelatinous *vitreous body*. The smaller **anterior cavity** is subdivided into two chambers, anterior and posterior. The shape of the eye is stabilized in part by the vitreous body and the clear *aqueous humor* that fills the anterior cavity.

The Fibrous Tunic
(Figure 18-19d)

The **fibrous tunic**, the outermost layer of the eye, consists of the *sclera* and the *cornea* (Figure 18-19d●). The fibrous tunic: (1) provides mechanical support and some degree of physical protection, (2) serves as an attachment site for the extrinsic eye muscles, and (3) contains structures that assist in the focusing process.

Most of the ocular surface is covered by the **sclera** (SKLER-a). The sclera, or "white of the eye," consists of a dense fibrous connective tissue containing both collagen and elastic fibers. This layer is thickest at the posterior portion of the eye, near the exit of the optic nerve, and thinnest over the anterior surface. The six extrinsic eye muscles insert upon the sclera, blending their collagenous fibers with those of the outer tunic. ∞ [p. 260]

The surface of the sclera contains small blood vessels and nerves that penetrate the sclera to reach internal structures. The network of small vessels that lie beneath the ocular conjunctiva usually does not carry enough blood to lend an obvious color to the sclera, but they are visible, on close inspection, as red lines against the white background of collagen fibers.

The transparent cornea of the eye is a clear area of the sclera. The corneal surface is covered by a delicate stratified squamous epithelium continuous with the ocular conjunctiva. Beneath that epithelium, the cornea consists primarily of a dense matrix containing multiple layers of collagen fibers. The transparency of the cornea results from the precise alignment of the collagen fibers within these layers.

The cornea is structurally continuous with the sclera; the **limbus** is the border between the two. The cornea is avascular, and there are no blood vessels between the cornea and the overlying conjunctiva. As a result, the superficial epithelial cells must obtain oxygen and nutrients from the tears that flow across their free surfaces. There are numerous free nerve endings in the cornea, and this is the most sensitive portion of the eye. This sensitivity is important because corneal damage will cause blindness even though the rest of the eye—photoreceptors included—is perfectly normal.

The Vascular Tunic (Uvea)
(Figures 18-19b,d,e/18-20)

The **vascular tunic**, or **uvea**, contains numerous blood vessels, lymphatics, and the intrinsic eye muscles. The functions of this layer include: (1) providing a route for blood vessels and lymphatics that supply tissues of the eye, (2) regulating the amount of light entering the eye, (3) secreting and reabsorbing the *aqueous humor* that circulates within the eye, and (4) controlling the shape of the lens, an essential part of the focusing process. The vascular tunic includes the *iris*, the *ciliary body*, and the *choroid* (Figures 18-19b,d,e● and 18-20●).

THE IRIS *(Figures 18-19/18-20).* The **iris** can be seen through the transparent corneal surface. The iris contains blood vessels, pigment cells, and two layers of smooth muscle fibers. When these muscles contract they change the diameter of the central opening, or **pupil**, of the iris. One group of smooth muscle fibers forms a series of concentric circles around the pupil (Figure 18-19b,c,e●). When these **pupillary constrictor muscles** contract, the diameter of the pupil decreases. A second group of smooth muscles extends radially away from the edge of the pupil. Contraction of these **pupillary dilator muscles** enlarges the pupil. Both muscle groups are controlled by the autonomic nervous system; parasympathetic activation causes pupillary constriction, and sympathetic activation causes pupillary dilation. ∞ [p. 436]

The body of the iris consists of a connective tissue whose posterior surface is covered by an epithelium containing pigment cells. Pigment cells may also be present in the connective tissue of the iris and in the epithelium covering its anterior surface. Eye color is determined by the density and distribution of pigment cells. When there are no pigment cells in the body of the iris, light passes through it and bounces off the inner surface of the pigmented epithelium. The eye then appears blue. Individuals with gray, brown, and black eyes have more pigment cells, respectively, in the body and surface of the iris.

THE CILIARY BODY *(Figures 18-19b,d,e/18-20).* At its periphery the iris attaches to the anterior portion of the ciliary body. The **cil-**

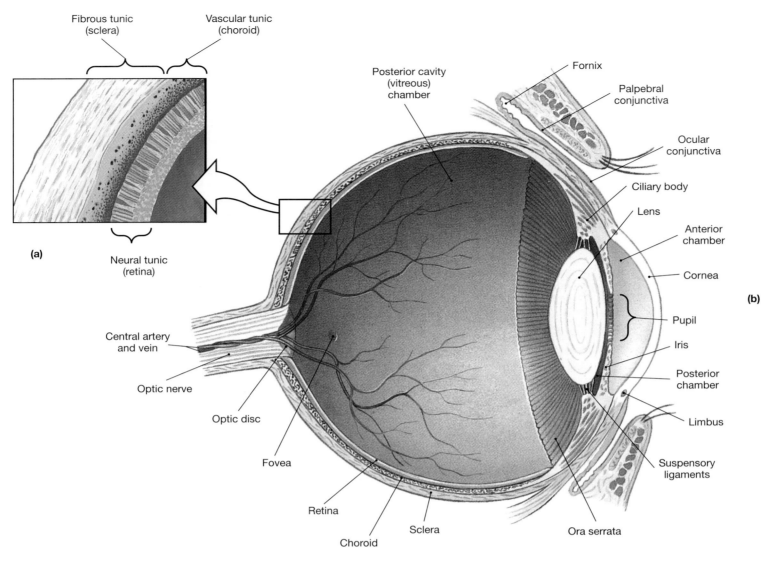

Fibrous tunic
(sclera)

Vascular tunic
(choroid)

(a)

Neural tunic
(retina)

Posterior cavity
(vitreous)
chamber

Fornix

Palpebral
conjunctiva

Ocular
conjunctiva

Ciliary body

Lens

Anterior
chamber

Cornea

(b)

Pupil

Iris

Posterior
chamber

Central artery
and vein

Optic nerve

Optic disc

Fovea

Retina

Choroid

Sclera

Ora serrata

Limbus

Suspensory
ligaments

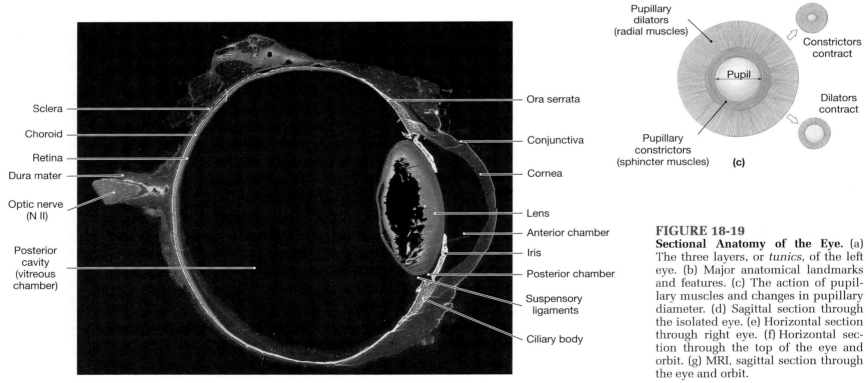

Sclera

Choroid

Retina

Dura mater

Optic nerve
(N II)

Posterior
cavity
(vitreous
chamber)

Ora serrata

Conjunctiva

Cornea

Lens

Anterior chamber

Iris

Posterior chamber

Suspensory
ligaments

Ciliary body

(d)

Pupillary
dilators
(radial muscles)

Constrictors
contract

Pupil

Dilators
contract

Pupillary
constrictors
(sphincter muscles)

(c)

FIGURE 18-19
Sectional Anatomy of the Eye. (a)
The three layers, or *tunics,* of the left
eye. (b) Major anatomical landmarks
and features. (c) The action of pupillary muscles and changes in pupillary
diameter. (d) Sagittal section through
the isolated eye. (e) Horizontal section
through right eye. (f) Horizontal section through the top of the eye and
orbit. (g) MRI, sagittal section through
the eye and orbit.

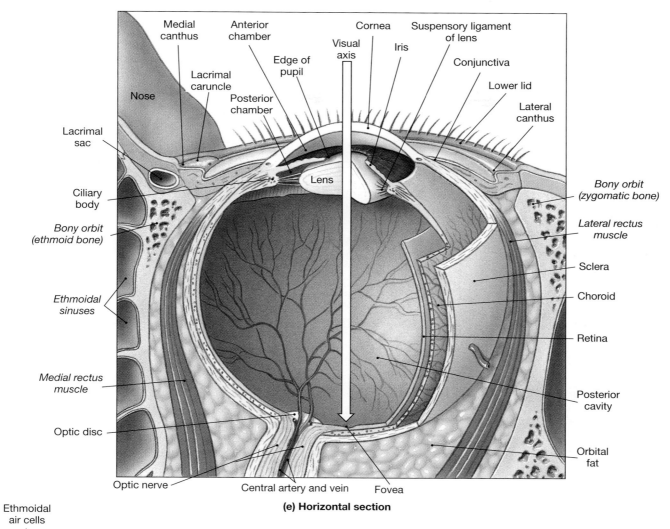

(e) Horizontal section

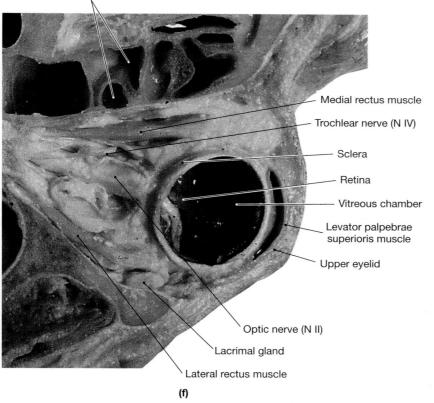

(f)

FIGURE 18-19 continued

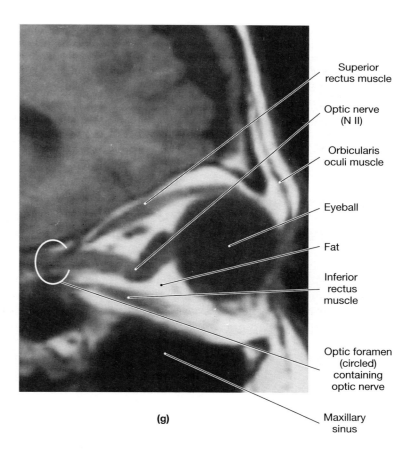

(g)

CHAPTER 18 The Nervous System: General and Special Senses 469

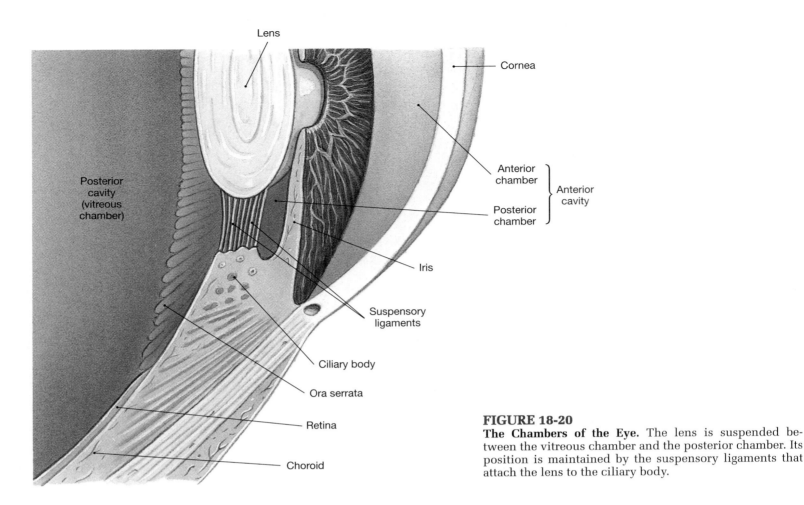

Lens

Cornea

Posterior
cavity
(vitreous
chamber)

Anterior
chamber

Anterior
cavity

Posterior
chamber

Iris

Suspensory
ligaments

Ciliary body

Ora serrata

Retina

Choroid

FIGURE 18-20
The Chambers of the Eye. The lens is suspended between the vitreous chamber and the posterior chamber. Its position is maintained by the suspensory ligaments that attach the lens to the ciliary body.

iary body begins at the junction between the cornea and sclera and extends posteriorly to the **ora serrata** (Ō-ra ser-RĀ-ta; "serrated mouth") (Figures 18-19b,d,e and 18-20●). The bulk of the ciliary body consists of the **ciliary muscle**, a muscular ring that projects into the interior of the eye. The epithelium is thrown into numerous folds, called **ciliary processes**. The **suspensory ligaments** of the lens attach to these processes. These connective tissue fibers hold the lens posterior to the iris and centered on the pupil. As a result, any light passing through the pupil and headed for the photoreceptors will pass through the lens.

THE CHOROID *(Figure 18-19).* The **choroid** contains an extensive capillary network that delivers oxygen and nutrients to the outer portion of the retina. It also contains scattered melanocytes, which are especially dense in the outermost portion of the choroid adjacent to the sclera (Figure 18-19a,b,d,e●). The innermost portion of the choroid attaches to the outer retinal layer.

The Neural Tunic
(Figures 18-19 to 18-21)

The **neural tunic**, or **retina**, consists of two distinct layers, an outer **pigmented layer** and an inner **neural retina** that contains the visual receptors and associated neurons (Figures 18-19 to 18-21●). The pigment layer absorbs light after it passes through the retina and has important biochemical interactions with retinal photoreceptors. The neural retina contains (1) the photoreceptors that respond to light, (2) supporting cells and neurons that perform preliminary processing and integration of visual information, and (3) blood vessels supplying tissues lining the posterior cavity.

The neural retina and pigmented layers are normally very close together, but not tightly interconnected. The pigmented layer continues over the ciliary body and iris, although the neural retina extends anteriorly only as far as the ora serrata. The neural retina thus forms a cup that establishes the posterior and lateral boundaries of the posterior cavity (Figures 18-19b,d,e and 18-20●).

RETINAL ORGANIZATION *(Figures 18-19b,e/18-21).* There are approximately 130 million photoreceptors in the retina, each monitoring a specific location. A visual image results from the processing of information provided by the entire receptor population. In sectional view, the retina contains several layers of cells (Figure 18-21a,b●). The outermost layer, closest to the pigmented layer, contains the visual receptors. There are two types of **photoreceptors**, **rods** and **cones**. Rods do not discriminate between different colors of light. They are very light-sensitive and enable us to see in dimly lit rooms, at twilight, or in pale moonlight. Cones provide us with color vision. There are three types of cones, and their stimulation in various combina-

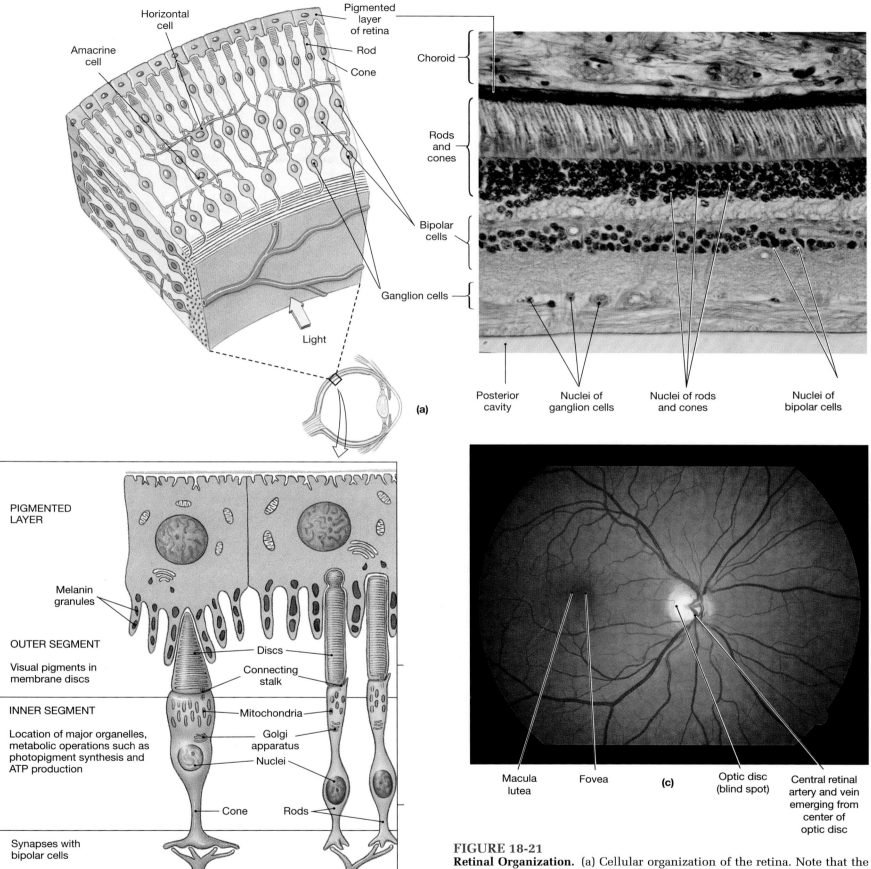

(a)

Horizontal cell

Amacrine cell

Pigmented layer of retina

Rod

Cone

Bipolar cells

Ganglion cells

Light

Choroid

Rods and cones

Bipolar cells

Ganglion cells

Posterior cavity

Nuclei of ganglion cells

Nuclei of rods and cones

Nuclei of bipolar cells

PIGMENTED LAYER

Melanin granules

OUTER SEGMENT

Visual pigments in membrane discs

INNER SEGMENT

Location of major organelles, metabolic operations such as photopigment synthesis and ATP production

Discs

Connecting stalk

Mitochondria

Golgi apparatus

Nuclei

Cone

Rods

Synapses with bipolar cells

(b)

LIGHT

Macula lutea

Fovea

(c)

Optic disc (blind spot)

Central retinal artery and vein emerging from center of optic disc

FIGURE 18-21

Retinal Organization. (a) Cellular organization of the retina. Note that the photoreceptors are located closest to the choroid rather than near the vitre-ous chamber. (LM × 73) (b) Diagrammatic view of the fine structure of rods and cones, based on data from electron microscopy. (c) A photograph taken through the pupil of the eye, showing the retinal blood vessels, the origin of the optic nerve, and the optic disc, or "blind spot."

tions provides the perception of different colors. Cones give us sharper, clearer images, but they require more intense light than rods. If you sit outside at sunset (or sunrise), you will probably be able to tell when your visual system shifts from cone-based vision (clear images in full color) to rod-based vision (relatively grainy images in black and white).

Rods and cones are not evenly distributed across the outer surface of the retina. Approximately 125 million rods form a broad band around the periphery of the retina. The posterior retinal surface is dominated by the presence of roughly 6 million cones. Most of these are concentrated in the area where a visual image arrives after passing through the cornea and lens. There are no rods in this region, which is known as the **macula lutea** (LOO-tē-a; "yellow spot"). The highest concentration of cones is found in the central portion of the macula lutea, called the **fovea** (FŌ-vē-a; "shallow depression"), or *fovea centralis*. The fovea is the site of sharpest vision; when you look directly at an object, its image falls upon this portion of the retina (Figures 18-19b,e and 18-21c●).

The rods and cones synapse with roughly 6 million **bipolar cells. Horizontal cells** at this level form a network that inhibits or facilitates communication between the visual receptors and bipolar cells. Bipolar cells in turn synapse within the layer of **ganglion cells** that faces the vitreous chamber. **Amacrine** (AM-a-krīn) **cells** at this level modulate communication between bipolar and ganglion cells.

Axons from an estimated 1 million ganglion cells converge on the **optic disc**, penetrate the wall of the eye, and proceed toward the diencephalon as the optic nerve (N II) (Figure 18-19b,e●). The *central retinal artery* and *central retinal vein* that supply the retina pass through the center of the optic nerve and emerge on the surface of the optic disc (Figure 18-21c●). There are no photoreceptors or other retinal structures at the optic disc. Because light striking this area goes unnoticed, it is commonly called the **blind spot.** You do not "notice" a blank spot in the visual field because involuntary eye movements keep the visual image moving and allow the brain to fill in the missing information. ♈*Scotomas and Floaters [p. 762]*

The Chambers of the Eye

The chambers of the eye are the *anterior, posterior,* and *vitreous chambers.* The anterior and posterior chambers are filled with *aqueous humor.*

AQUEOUS HUMOR *(Figure 18-22).* **Aqueous humor** forms as interstitial fluids pass between the epithelial cells of the ciliary processes and enter the posterior chamber (Figure 18-22●). The epithelial cells appear to regulate its composition, which resembles that of cerebrospinal fluid. The aqueous humor circulates so that in addition to forming a fluid cushion, it provides an important route for nutrient and waste transport.

Aqueous humor returns to the circulation in the anterior chamber near the edge of the iris. After diffusing through the local epithelium, it passes into the **canal of Schlemm,** a passageway that communicates with the veins of the eye.

The **lens** lies posterior to the cornea, held in place by the suspensory ligaments that originate on the ciliary body of the choroid (Figure 18-22●). The lens and its attached suspensory ligaments form the anterior boundary of the vitreous chamber. This chamber contains the **vitreous body,** a gelatinous mass sometimes called the *vitreous humor.* The vitreous body helps to maintain the shape of the eye and gives physical support to the retina. Aqueous humor produced in the posterior chamber freely diffuses through the vitreous body and across the retinal surface.

The Lens
(Figures 18-19/18-22)

The primary function of the lens is to focus the visual image on the retinal photoreceptors. It accomplishes this by changing its shape. The lens consists of concentric layers of cells that are precisely organized (Figures 18-19b,d,e and 18-22●). A dense fibrous capsule covers the entire lens. Many of the capsular fibers are elastic, and unless an outside force is applied,

FIGURE 18-22
The Circulation of Aqueous Humor

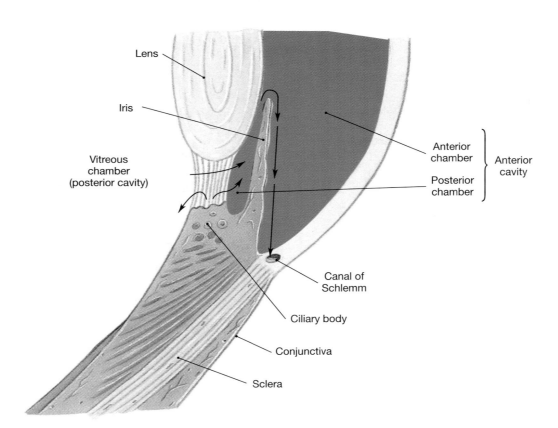

Lens

Iris

Vitreous chamber (posterior cavity)

Anterior chamber

Posterior chamber

Anterior cavity

Canal of Schlemm

Ciliary body

Conjunctiva

Sclera

they will contract and make the lens spherical. Around the edges of the lens, the capsular fibers intermingle with those of the suspensory ligaments.

At rest, tension in the suspensory ligaments overpowers the elastic capsule and flattens the lens. In this position the eye is focused for distant vision. When the ciliary muscles contract, the ciliary body moves toward the lens. This movement reduces the tension in the suspensory ligaments, and the elastic lens assumes a more spherical shape. This shape focuses the eye on nearby objects.

■ CLINICAL BRIEF
Glaucoma

Glaucoma affects roughly 2% of the population over 40. In this condition the aqueous humor no longer has free access to the canal of Schlemm. The primary factors responsible cannot be determined in 90% of all cases. Although drainage is impaired, production of aqueous humor continues unabated, and the intraocular pressure begins to rise. The fibrous scleral coat cannot expand significantly, so the increasing pressures begin to push against the surrounding soft tissues.

The optic nerve is not wrapped in connective tissue, for it penetrates all three tunics. When intraocular pressures have risen to roughly twice normal levels, distortion of the nerve fibers begins to affect visual perception. If this condition is not corrected, blindness eventually results.

Most eye exams include a glaucoma test. Intraocular pressure is tested by bouncing a tiny blast of air off the surface of the eye and measuring the deflection produced. Glaucoma may be treated by the application of drugs that constrict the pupil and tense the edge of the iris, making the surface more permeable to aqueous humor. Surgical correction involves perforating the wall of the anterior chamber to encourage drainage and is now performed by laser surgery on an outpatient basis.

Visual Pathways
(Figures 18-23/18-24)

Each rod and cone cell monitors a specific receptive field. A visual image results from the processing of information provided by the entire receptor population. A significant amount of processing occurs in the retina before the information is sent to the brain, because of interactions between the various cell types.

The two optic nerves, one from each eye, reach the diencephalon at the optic chiasm (Figure 18-23●). From this point approximately one-half of the fibers proceed toward the lateral geniculate of the same side of the brain, while the other half cross over to reach the lateral geniculate of the opposite side (Figure 18-24●). Visual information from the left half of each retina arrives at the lateral geniculate of the left side; information from the right half of each retina goes to the right side. The lateral geniculate nuclei act as switching and processing centers that relay visual information to reflex centers in the brain stem as well as to the cerebral cortex. For example, the pupillary reflexes and reflexes that control eye movement are triggered by information relayed by the lateral geniculate nuclei.

Cortical Integration
(Figure 18-24)

The sensation of vision arises from the integration of information arriving at the visual cortex of the cerebrum. The vi-sual cortex contains a sensory map of the entire field of vision. As in the case of the primary sensory cortex, the map does not faithfully duplicate the relative areas within the sensory field.

Each eye also receives a slightly different image, because (1) their foveae are 2–3 inches apart, and (2) the nose and eye socket block the view of the opposite side. The association and integrative areas of the cortex compare the two perspectives (Figure 18-24●) and use them to provide us with depth perception. The partial crossover that occurs at the optic chiasm ensures that the visual cortex receives a *composite* picture of the entire visual field.

The Brain Stem and Visual Processing

Many centers in the brain stem receive visual information, either from the lateral geniculate nuclei or via collaterals from the optic tracts. Collaterals that bypass the lateral geniculate nuclei synapse in the superior colliculus or hypothalamus. The superior colliculus of the midbrain issues motor commands controlling unconscious eye, head, or neck movements in response to visual stimuli. Visual inputs to the **suprachiasmatic** (soo-pra-kī-az-MA-tic) **nucleus** of the hypothalamus and the epithelium of the *pineal gland* affect the function of other brain stem nuclei. These nuclei establish a daily pattern of visceral activity that is tied to the day-night cycle. This **circadian rhythm** (*circa*, about + *dies*, day) affects metabolic rate, endocrine function, blood pressure, digestive activities, the awake-asleep cycle, and other physiological processes.

√ **What layer of the eye would be the first to be affected by inadequate tear production?**

√ **If the intraocular pressure becomes abnormally high, which structures of the eye are affected, and how are they affected?**

√ **Would a person born without cones in her eyes be able to see? Explain.**

■ CLINICAL BRIEF
Cataracts

The transparency of the lens depends on a precise combination of structural and biochemical characteristics. When that balance becomes disturbed, the lens loses its transparency, and the abnormal lens is known as a **cataract**. Cataracts may result from drug reactions, injuries, or radiation, but **senile cataracts** are the most common form. As aging proceeds, the lens becomes less elastic, and the individual has difficulty focusing on nearby objects. (The person becomes "farsighted.")

Over time, the lens takes on a yellowish hue, and eventually it begins to lose its transparency. As the lens becomes "cloudy," the individual needs brighter and brighter reading lights, and visual clarity begins to fade. If the lens becomes completely opaque, the person will be functionally blind, even though the retinal receptors are normal. Modern surgical procedures involve removing the lens, either intact or in pieces, after shattering it with high-frequency sound. The missing lens can be replaced by an artificial substitute, and vision can then be fine-tuned with glasses or contact lenses.

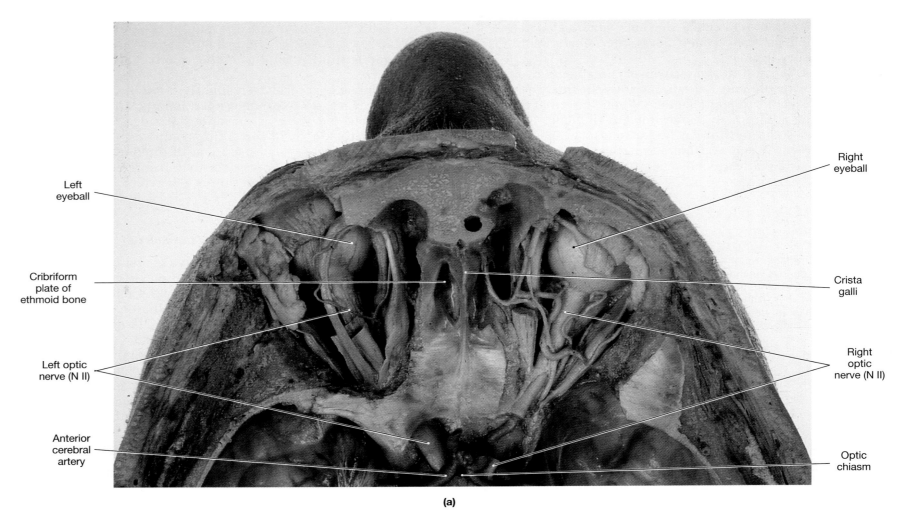

(a)

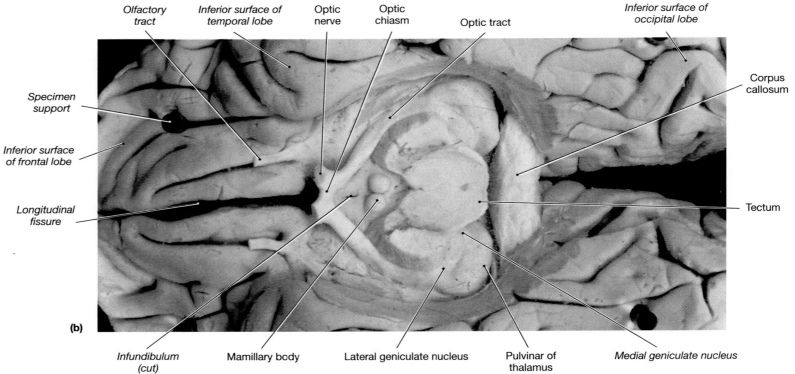

(b)

FIGURE 18-23

Anatomy of the Visual Pathways. (a) A horizontal section through the human head at the level of the optic chiasm as seen from the superior aspect. Note the relationship of the eye in the orbits and emergence of the optic nerves to form the optic chiasm and optic tract. (b) Inferior view of the brain. The brainstem has been mostly removed, leaving only the upper part of the midbrain. Portions of the cerebral hemispheres have been dissected away to show the optic tract.

FIGURE 18-24

An Overview of the Visual Pathways.
At the optic chiasm, a partial cross-over of nerve fibers occurs. As a result, each hemisphere receives visual information from the lateral half of the retina of the eye on that side and from the medial half of the retina of the eye on the opposite side. Visual association areas integrate this information to develop a composite picture of the entire visual field.

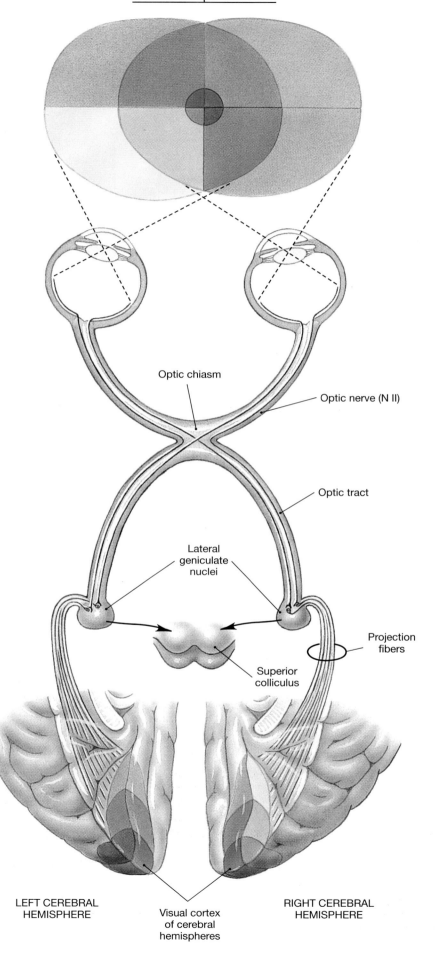

Optic chiasm

Optic nerve (N II)

Optic tract

Lateral geniculate nuclei

Superior colliculus

Projection fibers

LEFT CEREBRAL HEMISPHERE

RIGHT CEREBRAL HEMISPHERE

Visual cortex of cerebral hemispheres

CHAPTER 18 The Nervous System: General and Special Senses   475

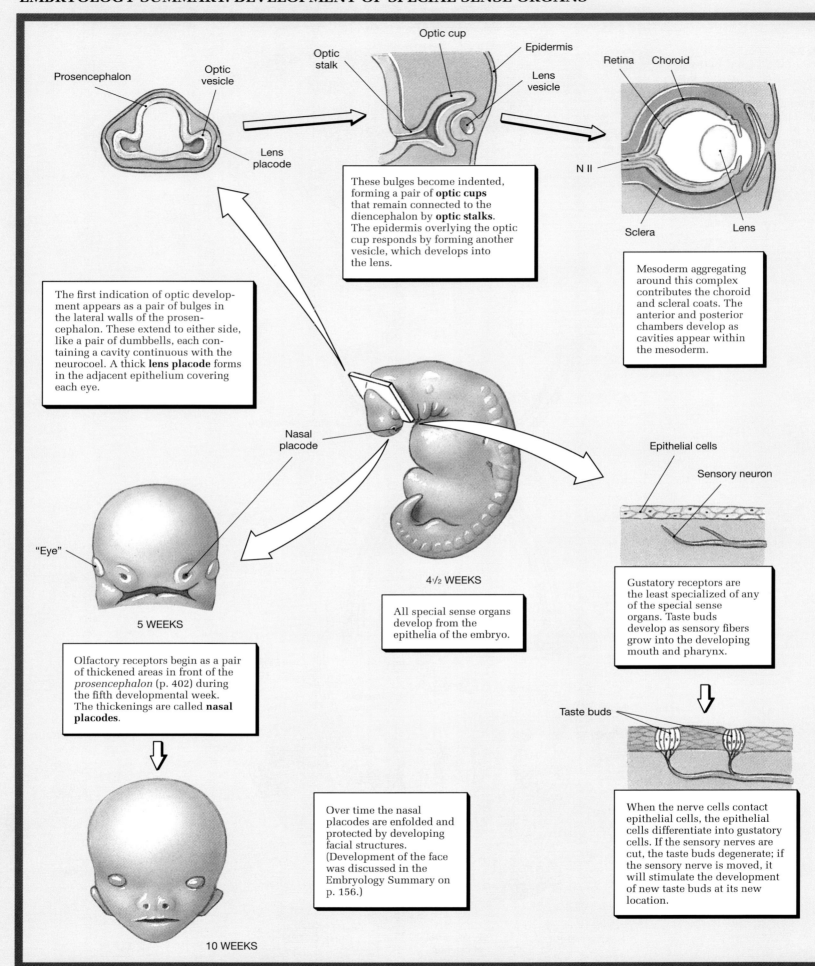

Prosencephalon

Optic vesicle

Lens placode

Optic stalk

Optic cup

Epidermis

Lens vesicle

Retina   Choroid

N II

Sclera   Lens

These bulges become indented, forming a pair of **optic cups** that remain connected to the diencephalon by **optic stalks**. The epidermis overlying the optic cup responds by forming another vesicle, which develops into the lens.

The first indication of optic development appears as a pair of bulges in the lateral walls of the prosencephalon. These extend to either side, like a pair of dumbbells, each containing a cavity continuous with the neurocoel. A thick **lens placode** forms in the adjacent epithelium covering each eye.

Mesoderm aggregating around this complex contributes the choroid and scleral coats. The anterior and posterior chambers develop as cavities appear within the mesoderm.

Nasal placode

Epithelial cells

Sensory neuron

"Eye"

5 WEEKS

4½ WEEKS

All special sense organs develop from the epithelia of the embryo.

Gustatory receptors are the least specialized of any of the special sense organs. Taste buds develop as sensory fibers grow into the developing mouth and pharynx.

Olfactory receptors begin as a pair of thickened areas in front of the *prosencephalon* (p. 402) during the fifth developmental week. The thickenings are called **nasal placodes**.

Taste buds

Over time the nasal placodes are enfolded and protected by developing facial structures. (Development of the face was discussed in the Embryology Summary on p. 156.)

When the nerve cells contact epithelial cells, the epithelial cells differentiate into gustatory cells. If the sensory nerves are cut, the taste buds degenerate; if the sensory nerve is moved, it will stimulate the development of new taste buds at its new location.

10 WEEKS

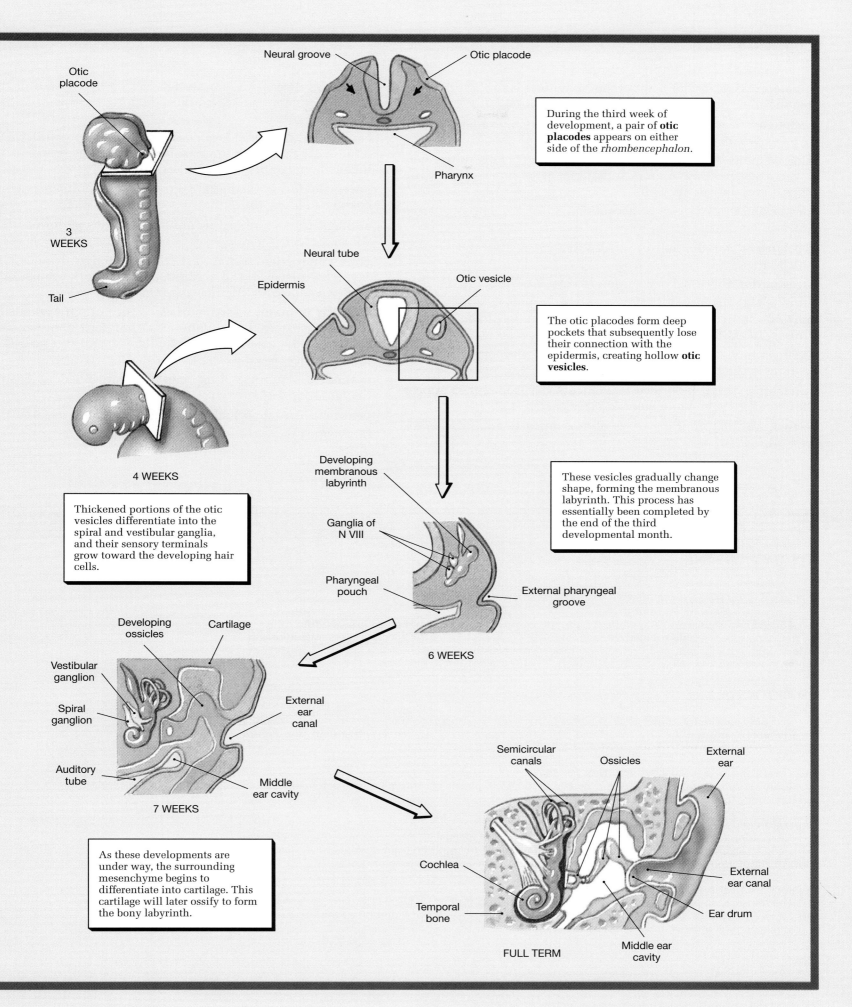

Otic
placode

Neural groove

Otic placode

During the third week of
development, a pair of **otic
placodes** appears on either
side of the *rhombencephalon*.

Pharynx

3
WEEKS

Tail

Neural tube

Epidermis

Otic vesicle

The otic placodes form deep
pockets that subsequently lose
their connection with the
epidermis, creating hollow **otic
vesicles**.

4 WEEKS

Thickened portions of the otic
vesicles differentiate into the
spiral and vestibular ganglia,
and their sensory terminals
grow toward the developing hair
cells.

Developing
membranous
labyrinth

These vesicles gradually change
shape, forming the membranous
labyrinth. This process has
essentially been completed by
the end of the third
developmental month.

Ganglia of
N VIII

Pharyngeal
pouch

External pharyngeal
groove

6 WEEKS

Developing
ossicles

Cartilage

Vestibular
ganglion

External
ear
canal

Spiral
ganglion

Auditory
tube

Middle
ear cavity

7 WEEKS

As these developments are
under way, the surrounding
mesenchyme begins to
differentiate into cartilage. This
cartilage will later ossify to form
the bony labyrinth.

Semicircular
canals

Ossicles

External
ear

Cochlea

External
ear canal

Temporal
bone

Ear drum

FULL TERM

Middle ear
cavity

Related Clinical Terms

rhizotomy: Cutting the dorsal roots providing sensation from an area to relieve pain. ✝ *The Control of Pain [p. 761]*

tractotomy: Severing ascending spinal tracts to relieve pain. ✝ *The Control of Pain [p. 761]*

referred pain: Pain sensations from visceral organs, often perceived as originating in more superficial areas innervated by the same spinal nerves. *[p. 448]*

anesthesia: A total loss of sensation. ✝ *Assessment of Tactile Sensitivities [p. 762]*

hypesthesia: A reduction in sensitivity. ✝ *Assessment of Tactile Sensitivities [p. 762]*

paresthesia: Abnormal sensations. ✝ *Assessment of Tactile Sensitivities [p. 762]*

mastoidectomy: Surgical opening and draining of the mastoid sinuses. *[p. 457]*

myringotomy: Drainage of the middle ear through a surgical opening in the tympanum. *[p. 457]*

nystagmus: Short, jerky eye movements that sometimes appear after damage to the brain stem or inner ear. *[p. 459]*

vertigo: An inappropriate sense of motion. ✝ *Vertigo, Dizziness, and Motion Sickness [p. 762]*

Ménière's disease: Acute vertigo caused by rupture of the wall of the membranous labyrinth. ✝ *Vertigo, Dizziness, and Motion Sickness [p. 762]*

conductive deafness: Deafness resulting from conditions in the middle ear that block the transfer of vibrations from the tympanic membrane to the oval window. *[p. 464]*

nerve deafness: Deafness resulting from problems within the cochlea or along the auditory pathway. *[p. 464]*

audiogram: A graphical record of a subject's performance during a hearing test. ✝ *Testing and Treating Hearing Deficits [p. 762]*

bone conduction test: A test for conductive deafness, usually involving placement of a vibrating tuning fork against the skull. ✝ *Testing and Treating Hearing Deficits [p. 762]*

cochlear implant: Insertion of electrodes into the cochlear nerve to provide external stimulation that provides some sensitivity to sounds in the absence of a functional organ of Corti. ✝ *Testing and Treating Hearing Deficits [p. 762]*

scotomas: Abnormal blind spots that are fixed in position. ✝ *Scotomas and Floaters [p. 762]*

cataract: An abnormal lens that has lost its transparency. *[p. 473]*

CHAPTER SUMMARY AND REVIEW

STUDY OUTLINE

Related Key Terms

INTRODUCTION *[p. 446]*

1. The **general senses** are temperature, pain, touch, pressure, vibration, and proprioception; receptors for these sensations are distributed throughout the body. Receptors for the **special senses (olfaction, gustation**, sight, equilibrium, and hearing) are located in specialized areas, or **sense organs**. A **sensory receptor** is a specialized cell that, when stimulated, sends a **sensation** to the CNS.

RECEPTORS *[p. 446]*

1. **Receptor specificity** allows each receptor to respond to particular stimuli. The simplest receptors are **free nerve endings**; the area monitored by a single receptor cell is the **receptive field** *(see Figure 18-1)*.

Interpretation of Sensory Information *[p. 447]*
2. **Tonic receptors** are always sending signals to the CNS; **phasic receptors** become active only when the conditions that they monitor change.

labeled line • sensory coding

Central Processing and Adaptation *[p. 447]*
3. **Adaptation** (a reduction in sensitivity in the presence of a constant stimulus) may involve changes in receptor sensitivity (**peripheral**, or **sensory**, **adaptation**) or inhibition along the sensory pathways (**central adaptation**). **Fast-adapting receptors** are phasic; **slow-adapting receptors** are tonic.

Sensory Limitations *[p. 447]*
4. The information provided by our sensory receptors is incomplete because: (1) we do not have receptors for every stimulus; (2) our receptors have limited ranges of sensitivity; and (3) a stimulus produces a neural event that must be interpreted by the CNS.

THE GENERAL SENSES *[p. 447]*

Related Key Terms

Nociceptors *[p. 447]*
1. **Nociceptors** respond to a variety of stimuli usually associated with tissue damage. There are two types of these painful sensations: **fast (prickling) pain** and **slow (burning and aching) pain** *(see Figures 18-2/18-3a).*

exteroceptors • interoceptors

Thermoreceptors *[p. 448]*
2. **Thermoreceptors** respond to changes in temperature.

Mechanoreceptors *[p. 448]*
3. **Mechanoreceptors** respond to physical distortion, contact, or pressure on their cell membranes: **tactile receptors** to touch, pressure, and vibration; **baroreceptors** to pressure changes in the walls of blood vessels and the digestive, reproductive, and urinary tracts; and **proprioceptors** (*muscle spindles*) to positions of joints and muscles *(see Figures 18-3/18-4).*
4. **Fine touch and pressure receptors** provide detailed information about a source of stimulation; **crude touch and pressure receptors** are poorly localized. Important tactile receptors include **free nerve endings**, the **root hair plexus**, **Merkel's discs**, **Meissner's corpuscles**, **Pacinian corpuscles**, and **Ruffini corpuscles** *(see Figure 18-3).*
5. Baroreceptors monitor changes in pressure; they respond immediately but adapt rapidly. Baroreceptors in the walls of major arteries and veins respond to changes in blood pressure. Receptors along the digestive tract help coordinate reflex activities of digestion *(see Figure 18-4).*
6. Proprioceptors monitor the position of joints, tension in tendons and ligaments, and the state of muscular contraction.

Golgi tendon organs

Chemoreceptors *[p. 450]*
7. In general, **chemoreceptors** respond to water-soluble and lipid-soluble substances that are dissolved in the surrounding fluid. They monitor the chemical composition of body fluids *(see Figure 18-5).*

carotid bodies • aortic bodies

OLFACTION (SMELL) *[p. 450]*

1. The **olfactory organs** contain the **olfactory epithelium** with **olfactory receptors** (neurons sensitive to chemicals dissolved in the overlying mucus), **supporting cells,** and **basal (stem) cells.** Their surfaces are coated with the secretions of the **olfactory glands** *(see Figure 18-6).*

Olfactory Receptors *[p. 452]*
2. The olfactory receptors are modified neurons *(see Figure 18-6b).*

Olfactory Pathways *[p. 452]*
3. The olfactory system has extensive limbic and hypothalamic connections that help explain the emotional and behavioral responses that can be produced by certain smells *(see Figure 18-6b).*

Olfactory Discrimination *[p. 453]*
4. The olfactory system can make subtle distinctions between thousands of chemical stimuli; the CNS interprets the smell on the basis of the particular pattern of receptor activity.
5. The olfactory receptor population shows considerable turnover and is the only known example of neuronal replacement in the adult human. The total number of receptors, however, declines with age.

GUSTATION (TASTE) *[p. 453]*

Gustatory Receptors *[p. 454]*
1. **Gustatory (taste) receptors** are clustered in **taste buds**, each of which contains **gustatory cells,** which extend taste hairs through a narrow **taste pore** *(see Figure 18-7).*
2. Taste buds are associated with epithelial projections (**papillae**) *(see Figure 18-7a).*

filiform • fungiform • circumvallate

Gustatory Pathways *[p. 454]*
3. The taste buds are monitored by cranial nerves VII, IX, and X. The afferent fibers synapse within the **nucleus solitarius** before proceeding to the thalamus and cerebral cortex *(see Figure 18-8).*

Gustatory Discrimination *[p. 454]*
4. The **primary taste sensations** are sweet, salt, sour, and bitter.
5. There are individual differences in the sensitivity to specific tastes. The number of taste buds and their sensitivity decline with age *(see Figure 18-8).*

EQUILIBRIUM AND HEARING *[p. 455]*

The External Ear *[p. 455]*
1. The **external ear** includes the **pinna**, which surrounds the entrance to the **external auditory canal** that ends at the **tympanic membrane (tympanum)**, or eardrum *(see Figures 18-9/18-10).*

ceruminous glands • cerumen

The Middle Ear [p. 456]

2.   In the **middle ear** the **tympanic cavity** encloses and protects the **auditory ossicles**, which connect the tympanic membrane with the receptor complex of the inner ear. The tympanic cavity communicates with the nasopharynx via the **auditory tube** *(see Figures 18-9/18-10)*.

3.   The **tensor tympani** and **stapedius muscles** contract to reduce the amount of motion of the tympanum when very loud sounds arrive *(see Figures 18-9/18-10)*.

Related Key Terms
malleus • incus • stapes

The Inner Ear [p. 457]

4.   The senses of equilibrium and hearing are provided by the receptors of the **inner ear** (also known as the **membranous labyrinth**). Its chambers and canals contain **endolymph**. The **bony labyrinth** surrounds and protects the membranous labyrinth. The bony labyrinth can be subdivided into the **vestibule** and **semicircular canals** (providing the sense of equilibrium) and the **cochlea** (providing the sense of hearing) *(see Figures 18-9/18-11 to 18-17)*.

perilymph • round window • oval window • modiolus

5.   The vestibule includes a pair of membranous sacs, the **saccule** and **utricle**, whose receptors provide sensations of gravity and linear acceleration. The cochlea contains the **cochlear duct**, an elongated portion of the membranous labyrinth *(see Figure 18-12)*.

6.   The basic receptors of the inner ear are **hair cells** whose surfaces support **stereocilia**. Hair cells provide information about the direction and strength of varied mechanical stimuli *(see Figure 18-12d)*.

supporting cells • kinocilium

7.   The **anterior**, **posterior**, and **lateral** semicircular canals are continuous with the utricle. Each contains an **ampulla** with sensory receptors. Here the cilia contact a gelatinous cupula *(see Figures 18-12/18-13)*.

crista • cupula

8.   The saccule and utricle are connected by a passageway continuous with the **endolymphatic duct**, which terminates in the **endolymphatic sac**. In the saccule and utricle, hair cells cluster within **maculae**, where their cilia contact **otoliths** (densely packed mineral crystals). When the head tilts, the mass of otoconia shifts, and the resulting distortion in the sensory hairs signals the CNS *(see Figures 18-12b/18-14)*.

otoconia • vestibulospinal tracts

9.   The vestibular receptors activate sensory neurons of the **vestibular ganglia**. The axons form the **vestibular branch** of the vestibulocochlear nerve (N VIII), synapsing within the **vestibular nuclei** *(see Figure 18-15)*.

Hearing [p. 464]

10.   Sound waves travel toward the tympanum, which vibrates; the auditory ossicles conduct the vibrations to the oval window. Movement at the oval window applies pressure to the perilymph of the **vestibular duct** and the **tympanic duct** *(see Figure 18-16)*.

11.   Pressure waves distort the **basilar membrane** and push the hair cells of the **organ of Corti** against the **tectorial membrane** *(see Figure 18-16 and Table 18-2)*.

Auditory Pathways [p. 465]

12.   The sensory neurons are located in the **spiral ganglion** of the cochlea. Their afferent fibers form the **cochlear branch** of the vestibulocochlear nerve (N VIII), synapsing at the **cochlear nucleus** *(see Figure 18-17)*.

VISION [p. 465]

Accessory Structures of the Eye [p. 465]

1.   The **accessory structures** of the eye include the **palpebrae** (eyelids), which are separated by the **palpebral fissure**. The **eyelashes** line the palpebral margins. Along the inner margin of the lid are **Meibomian glands**, which secrete a lipid-rich product. Glands at the **lacrimal caruncle** produce other secretions *(see Figures 18-18/18-19)*.

medial canthus • lateral canthus • chalazion • sty • tarsal plate

2.   An epithelium called the **conjunctiva** covers most of the exposed surface of the eye; the **bulbar**, or **ocular**, **conjunctiva** covers the anterior surface of the eye, and the **palpebral conjunctiva** lines the inner surface of the eyelids. The **cornea** is transparent *(see Figure 18-19)*.

3.   The secretions of the **lacrimal gland** bathe the conjunctiva; these secretions are slightly alkaline and contain **lysozymes** (enzymes that attack bacteria). Tears collect in the **lacus lacrimalis**. The tears reach the inferior meatus of the nose after passing through the **lacrimal puncta**, the **lacrimal canals**, the **lacrimal sac**, and the **nasolacrimal duct**. Collectively these structures constitute the **lacrimal apparatus** *(see Figures 18-18/18-19)*.

fornix • orbital fat

The Eye [p. 467]

4.   The eye has three layers: an outer fibrous tunic, a vascular tunic, and an inner neural tunic.

posterior cavity • anterior cavity

5.   The **fibrous tunic** includes most of the ocular surface, which is covered by the **sclera** (a dense fibrous connective tissue of the fibrous tunic); the **limbus** is the border between the sclera and the cornea *(see Figure 18-19)*.

6.   The **vascular tunic**, or **uvea**, includes the **iris**, the **ciliary body**, and the **choroid**. The iris forms the boundary between the anterior and posterior chambers. The ciliary body contains the **ciliary muscle** and the **ciliary processes**, which attach to the **suspensory ligaments** of the lens *(see Figures 18-19/18-20)*.

pupil • pupillary constrictor muscles • pupillary dilator muscles • ora serratta

7.   The **neural tunic (retina)** consists of an outer **pigmented layer** and an inner **neural retina**; the latter contains visual receptors and associated neurons *(see Figures 18-19 to 18-21)*.

8.   There are two types of **photoreceptors** (visual receptors of the retina): **rods** and **cones**. Rods provide black and white vision in dim light; cones provide color vision in bright light. Cones are concentrated in the **macula lutea**; the **fovea** is the area of sharpest vision *(see Figures 18-19/18-21)*.

9. The direct line to the CNS proceeds from the photoreceptors to **bipolar cells**, then to **ganglion cells**, and to the brain via the optic nerve. **Horizontal cells** and **amacrine cells** modify the signals passed between other retinal components *(see Figure 18-21).*

10. The fluid **aqueous humor** circulates within the eye and reenters the circulation after diffusing through the walls of the anterior chamber and into the **canal of Schlemm** *(see Figure 18-22).*

11. The **lens**, held in place by the suspensory ligaments, lies behind the cornea and forms the anterior boundary of the vitreous chamber. This chamber contains the **vitreous body**, a gelatinous mass that helps stabilize the shape of the eye and support the retina *(see Figures 18-19/18-22).*

12. The lens focuses a visual image on the retinal receptors.

Visual Pathways [p. 473]

13. Each photoreceptor monitors a specific receptive field. The axons of ganglion cells converge on the **optic disc** and proceed along the optic tract to the optic chiasm *(see Figures 18-21/18-23).*

14. From the optic chiasm, after a partial decussation, visual information is relayed to the lateral geniculate nuclei. From there the information is sent to the visual cortex of the occipital lobes *(see Figure 18-24).*

15. Visual inputs to the **suprachiasmatic nucleus** and the pineal gland affect the function of other brain stem nuclei. These nuclei establish a visceral **circadian rhythm** that is tied to the day-night cycle and affects other metabolic processes.

Related Key Terms

blind spot

---

1 REVIEW OF CHAPTER OBJECTIVES

1. Distinguish between the general and special senses.
2. Explain why receptors respond to specific stimuli and how the structure of a receptor affects its sensitivity.
3. Identify the receptors for the general senses, and briefly describe how they function.
4. Identify and discuss the receptors and neural pathways involved in the sense of smell.
5. Identify and discuss the receptors and neural pathways involved in the sense of taste.
6. Identify and describe the structures of the ear and their roles in the processing of equilibrium sensations.
7. Identify and describe those structures of the ear that collect, amplify, and conduct sound and the structures along the auditory pathway.
8. Identify and describe the layers of the eye and the function of each structure.
9. Identify the structures of the visual pathway.

---

2 REVIEW OF CONCEPTS

1. How does the cell membrane influence cellular responses to changes in the extracellular environment?
2. What is receptor specificity? What causes it?
3. What is the relationship between the size of a receptive field and an individual's ability to locate a stimulus?
4. How do tonic receptors differ from phasic receptors?
5. What is the functional role of sensory adaptation?
6. How does a baroreceptor monitor changes in pressure?
7. What are the functional relationships between gustatory and olfactory senses?
8. What is the function of the auditory tube?
9. How is a sound wave entering the external auditory canal converted to a nerve impulse at the inner ear?
10. What is at the structural relationship between the bony labyrinth and the membranous labyrinth?

11. What are the functions of hair cells in the inner ear?
12. Trace the path of an auditory sensation from the inner ear to the brain.
13. Why is the lubrication system an important component of the visual system?
14. How does the eye focus?
15. How does the iris respond to changes in light intensity? What neural pathways are involved?

---

3 CRITICAL THINKING AND CLINICAL APPLICATION QUESTIONS

1. On Thanksgiving one of the best experiences when one arrives home for dinner is the smell of the roasting turkey. The people preparing the dinner often do not notice the smell. But if they leave the house for more than a few minutes, they will be able to smell the turkey when they return. Why does this happen?

2. If a person goes from a relaxed state to a sudden sprint, the rate of breathing increases without conscious effort. What reflex mechanism is responsible for altering the respiratory rate?

3. Why can new perfumes be introduced onto the market, even though there are so many already available?

4. Babies and small children often dislike the vegetables and other foods that their parents want them to eat. However, later, when they become adults, these individuals may enjoy vegetables. In addition, as adults they often prefer foods with stronger spices than they did when they were younger. What has happened, and why have these changes occurred?

5. Persons suffering from a severe cold or throat infection often complain that they cannot hear sounds as well as when they are in normal health. A similar condition often occurs after swimming or when a person has gotten water in the ear during a shower. But some people suffer from hearing loss that is not associated with health or activity considerations. How do the causes of deafness differ, and what is responsible for each?

6. The classic response to peeling an onion is to cry, even though this is usually not a sad event. If the person cutting the onion rubs his eyes, they become reddened and irritated. What is happening, and why is this irritation occurring?

19

The Endocrine System

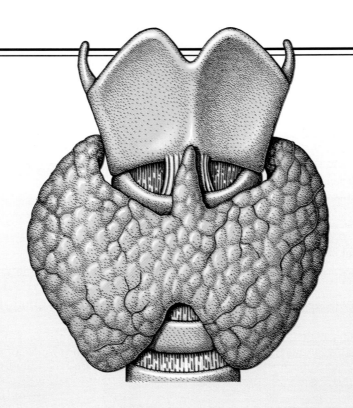

CHAPTER OUTLINE AND OBJECTIVES

Homeostatic regulation involves coordinating the activities of organs and systems throughout the body. The cells of the nervous and endocrine systems work together to monitor and adjust the physiological activities under way at any given moment. Their activities are closely coordinated, and their effects are typically complementary. In general, the nervous system produces short-term (usually a few seconds) but very specific responses to environmental stimuli. Endocrine cells, in contrast, release chemicals into the bloodstream for distribution throughout the body. These chemicals, called **hormones**, alter the metabolic activities of many different tissues and organs simultaneously. The hormonal effects may not be immediately apparent, but when they do appear, they often persist for days. This response pattern makes the endocrine system particularly effective in regulating ongoing processes such as growth and development.

The nervous and endocrine systems are compared in Table 19-1. From this general perspective, the two systems are easily distinguished. Yet when viewed in detail, the two systems are difficult to separate on either anatomical or functional grounds. For example, the adrenal medulla is a modified sympathetic ganglion whose neurons secrete epinephrine and norepinephrine into the blood. Here we have an endocrine structure that is functionally and developmentally part of the nervous system.

This chapter describes the components and functions of the endocrine system and considers some important interactions between the nervous and endocrine systems.

AN OVERVIEW OF THE ENDOCRINE SYSTEM
(Figures 19-1/19-2)

The endocrine system (Figure 19-1●) includes all of the endocrine cells and tissues of the body. Endocrine cells are glandular secretory cells that release their hormones into the interstitial fluids, rather than onto an epithelial surface. ∞ [p. 62]

Hormones can be divided into three groups on the basis of chemical structure:

- **Amino acid derivatives:** These are relatively small molecules that are structurally similar to amino acids. Examples include (1) derivatives of tyrosine, such as the **thyroid hormones** released by the thyroid gland and the **catecholamines** (epinephrine, norepinephrine) released by the adrenal medullae, and (2) derivatives of tryptophan, such as **melatonin** synthesized by the pineal gland.

- **Peptide hormones:** Peptide hormones are chains of amino acids. This is the largest group of hormones; all of the hormones of the pituitary gland are peptide hormones.

- **Steroid hormones:** Steroid hormones, derived from cholesterol, are released by the reproductive organs and the adrenal glands.

All cellular activities and metabolic reactions are controlled by enzymes. Hormones alter cellular operations by changing the *types, activities, or quantities* of key cytoplasmic enzymes. In this way a hormone can regulate the metabolic operations of **target cells**, peripheral cells sensitive to its presence. The first step in the process involves the release of the hormone and interaction between the hormone and a specific receptor protein in the cell membrane or cytoplasm. The presence or absence of the necessary receptor determines each cell's sensitivities to a hormone. Binding of a hormone to its receptor starts a biochemical chain of events that changes the pattern of enzymatic activity within the cell. The different classes of hormones work by different mechanisms to achieve this result (Figure 19-2●).

The catecholamines (epinephrine, norepinephrine, dopamine) and peptide hormones target receptors on the cell membrane because they cannot diffuse through a cell membrane. As a result, they do not exert their effects directly by entering a cell and begin building a protein or catalyzing a specific reaction. The hormone acts as a *first messenger* that causes the appearance of a *second messenger* in the cytoplasm of the target cell. The second messenger may function as an enzyme activator, inhibitor, or cofactor to change the direction and rate of the cell's metabolic reactions. The specific response of the target cell depends on the identity of the enzymes affected. Many hormones discussed later in this chapter, including calcitonin, parathyroid hormone, ADH, ACTH, FSH, LH, TSH, and glucagon, produce their effects in this way.

The thyroid hormones and steroid hormones cross the cell membrane and bind to intracellular receptors. The lipid-based steroid hormones diffuse rapidly through the lipid portion of the cell membrane and bind to receptors in the cytoplasm or nucleus. The hormone-receptor complex then binds to DNA segments and triggers the activation or inactivation of specific genes. Thyroid hormones may diffuse or be passively transported across the cell membrane. They then bind to chromatin in the nucleus, triggering the activation of specific genes.

This mechanism enables these hormones to alter the rate of mRNA transcription in the nucleus, thereby changing the pattern of enzyme synthesis and metabolic activity in their target cells. This may alter cell structure or, if enzymes are involved, change the metabolic activity of the target cell. For example, the hormone *testosterone* stimulates the production of enzymes and proteins in skeletal muscle fibers, causing an increase in muscle size and strength.

Endocrine activity is usually regulated by some form of negative feedback. In direct negative feedback control: (1) the endocrine cell responds to a change in the composition of the extracellular

TABLE 19-1 A Comparison of the Nervous and Endocrine Systems

	Nervous System	Endocrine System
METHOD OF INFORMATION TRANSFER	Release of neurotransmitters at specialized synapses at specific sites in the body	Release of hormones into general circulation for distribution throughout the body
EFFECTS		
Scope	Localized	Widespread/systemic
Primary targets	Nerve cells, gland cells, muscle cells, fat cells	All tissues
Onset	Immediate (milliseconds)	Gradual (seconds to hours)
Duration	Short-term (milliseconds to minutes)	Long-term (minutes to days)
Recovery	Immediately after stimulation ends	Slow, persisting after hormonal secretion stops

FIGURE 19-1
The Endocrine System

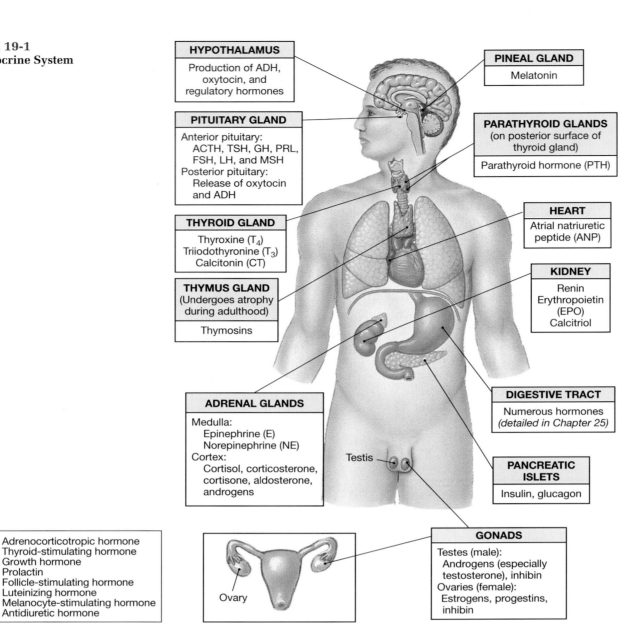

HYPOTHALAMUS
Production of ADH, oxytocin, and regulatory hormones

PITUITARY GLAND
Anterior pituitary: ACTH, TSH, GH, PRL, FSH, LH, and MSH
Posterior pituitary: Release of oxytocin and ADH

THYROID GLAND
Thyroxine (T$_4$)
Triiodothyronine (T$_3$)
Calcitonin (CT)

THYMUS GLAND
(Undergoes atrophy during adulthood)
Thymosins

ADRENAL GLANDS
Medulla:
Epinephrine (E)
Norepinephrine (NE)
Cortex:
Cortisol, corticosterone, cortisone, aldosterone, androgens

PINEAL GLAND
Melatonin

PARATHYROID GLANDS
(on posterior surface of thyroid gland)
Parathyroid hormone (PTH)

HEART
Atrial natriuretic peptide (ANP)

KIDNEY
Renin
Erythropoietin (EPO)
Calcitriol

DIGESTIVE TRACT
Numerous hormones
(detailed in Chapter 25)

Testis

PANCREATIC ISLETS
Insulin, glucagon

GONADS
Testes (male):
Androgens (especially testosterone), inhibin
Ovaries (female):
Estrogens, progestins, inhibin

Ovary

ACTH	Adrenocorticotropic hormone
TSH	Thyroid-stimulating hormone
GH	Growth hormone
PRL	Prolactin
FSH	Follicle-stimulating hormone
LH	Luteinizing hormone
MSH	Melanocyte-stimulating hormone
ADH	Antidiuretic hormone

fluid by releasing its hormone into the circulatory system, (2) the released hormone stimulates a target cell, and (3) the target cell response restores homeostasis. This response eliminates the source of stimulation of the endocrine cell. One example noted in Chapter 5 was the control of calcium levels by parathyroid hormone. When circulating calcium levels decline, parathyroid hormone is released, and the responses of target cells (osteoclasts) elevate blood calcium levels. As calcium levels climb, the level of parathyroid stimulation declines, and so does its rate of hormone secretion.

Endocrine activity can also be controlled (1) by neural activity, (2) by positive feedback, or (3) by complex negative feedback loops. An example of neural control is the regulation of secretion by the adrenal medullae, as considered in Chapter 17. Hormone regulation through positive feedback is very rare. In these instances, secretion of the hormone produces an effect that further stimulates hormone release. An example, the release of oxytocin during labor and delivery, was introduced in Chapter 1. Complex negative feedback loops are far more common. In these instances the secretion of one hormone, such as thyroid-stimulating hormone from the anterior pituitary gland, triggers the secretion of a second hormone, such as the thyroid hormones produced by the thyroid gland. The second hormone may have many effects, but one of them is suppression of the release of the first hormone.

The Hypothalamus and Endocrine Regulation
(Figure 19-3)

Coordinating centers in the hypothalamus regulate the activities of the nervous and endocrine systems via three different mechanisms (Figure 19-3●):

1. The hypothalamus contains autonomic centers that exert direct neural control over the endocrine cells of the adrenal medullae. ∞ [pp. 430, 431] When the sympathetic division is activated, the adrenal medullae release hormones into the bloodstream.

2. The hypothalamus acts as an endocrine organ, releasing the hormones *ADH* and *oxytocin* into the circulation at the posterior pituitary.

3. The hypothalamus secretes **regulatory hormones**, or *regulatory factors*. These are special hormones that regulate the activities of endocrine cells in the pituitary gland. There are two classes of regulatory hormones: (1) A **releasing hormone (RH)** stimulates production of one or more hormones at the anterior pituitary, and (2) an **inhibiting hormone (IH)** prevents the synthesis and secretion of pituitary hormones.

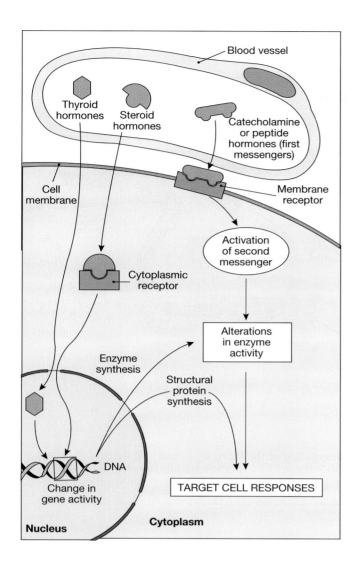

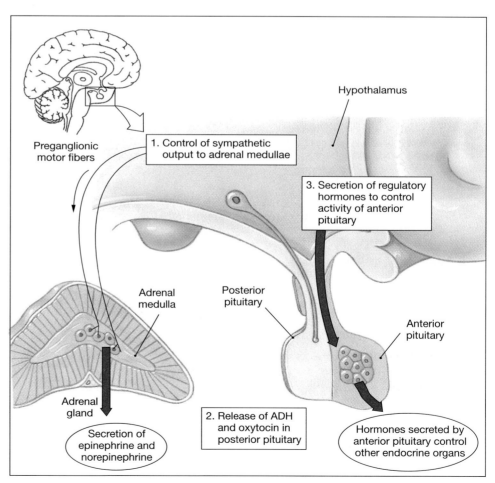

FIGURE 19-2
A Summary of the Mechanisms of Hormone Action. Catecholamines and peptide hormones act via second messengers released when the hormones bind to receptors at the membrane surface. Steroid hormones bind to receptors in the cytoplasm that then enter the nucleus. Thyroid hormones proceed directly to the nucleus to reach hormonal receptors.

THE PITUITARY GLAND
(Figures 19-4/19-5)

The **pituitary gland**, or **hypophysis** (hī-POF-i-sis), weighing one-fifth of an ounce, is the most compact chemical factory in the body. This small, oval gland, about the size and weight of a small grape, lies nestled within a depression in the sphenoid bone, the *sella turcica.* ∞ [p. 143] It lies inferior to the hypothalamus, connected to it by the infundibulum. The base of the infundibulum lies between the optic chiasma and the mamillary bodies (Figure 19-4a●). The gland is cradled by the sella turcica and held in position by the *diaphragma sellae* that encircles the stalk of the infundibulum. ∞ [p. 368]

The pituitary gland can be divided into *posterior* and *anterior* divisions on anatomical and developmental grounds (Figure 19-4●). Nine important peptide hormones are released by the pituitary gland, two by the posterior pituitary and seven by the anterior pituitary. Table 19-2 summarizes information about the hormones of the pituitary gland. The pituitary hormones and their targets are diagrammed in Figure 19-5●.

FIGURE 19-3
Three Types of Hypothalamic Control over Endocrine Organs.
(1) The hypothalamus exerts direct neural control over the secretory activity of the adrenal medullae. (2) Hypothalamic neurons secrete ADH and oxytocin, hormones that produce specific responses in peripheral target organs. (3) Hypothalamic neurons release regulatory hormones that control the secretory activity of the pituitary gland.

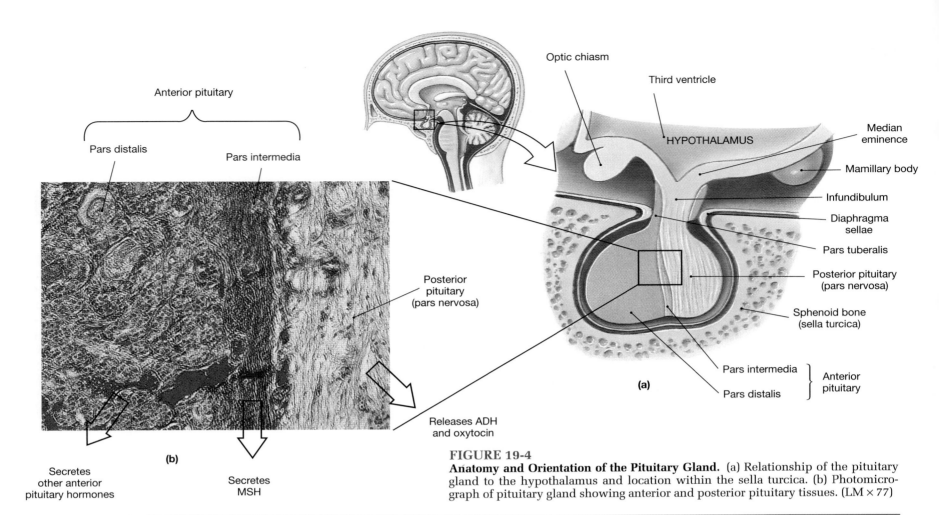

Anterior pituitary

Pars distalis

Pars intermedia

Optic chiasm

Third ventricle

HYPOTHALAMUS

Median eminence

Mamillary body

Infundibulum

Diaphragma sellae

Pars tuberalis

Posterior pituitary (pars nervosa)

Sphenoid bone (sella turcica)

Pars intermedia } Anterior pituitary
Pars distalis

(a)

Posterior pituitary (pars nervosa)

Releases ADH and oxytocin

(b)

Secretes other anterior pituitary hormones

Secretes MSH

FIGURE 19-4
Anatomy and Orientation of the Pituitary Gland. (a) Relationship of the pituitary gland to the hypothalamus and location within the sella turcica. (b) Photomicrograph of pituitary gland showing anterior and posterior pituitary tissues. (LM × 77)

TABLE 19-2   The Pituitary Hormones

Region/Area	Hormone	Targets	Hormonal Effects
ANTERIOR PITUITARY (Adenohypophysis, pars distalis)	Thyroid-stimulating hormone (TSH)	Thyroid gland	Secretion of thyroid hormones
	Adrenocorticotropic hormone (ACTH)	Adrenal cortex (zona fasciculata)	Glucocorticoid secretion
	Gonadotropic hormones: Follicle-stimulating hormone (FSH)	Follicle cells of ovaries in female	Estrogen secretion, follicle development
		Sustentacular cells of testes in male	Sperm maturation
	Luteinizing hormone (LH) or interstitial cell-stimulating hormone (ICSH)	Follicle cells before and after ovulation in female	Ovulation, formation of corpus luteum, and progesterone secretion
		Interstitial cells of testes in male	Androgen secretion
	Prolactin (PRL)	Mammary glands	Production of milk
	Growth hormone (GH)	All cells	Growth, protein synthesis, lipid mobilization and catabolism
Pars intermedia	Melanocyte-stimulating hormone (MSH)	Melanocytes of skin	Increased melanin synthesis, dispersion in epidermis
POSTERIOR PITUITARY (Neurohypophysis, pars nervosa)	Antidiuretic hormone (ADH)	Kidneys	Reabsorption of water; elevation of blood volume and pressure
	Oxytocin (OT)	Uterus; mammary glands in female	Labor contractions; milk ejection
		Prostate gland in male	Smooth muscle contractions, ejection of secretions

FIGURE 19-5
Pituitary Hormones and Their Targets

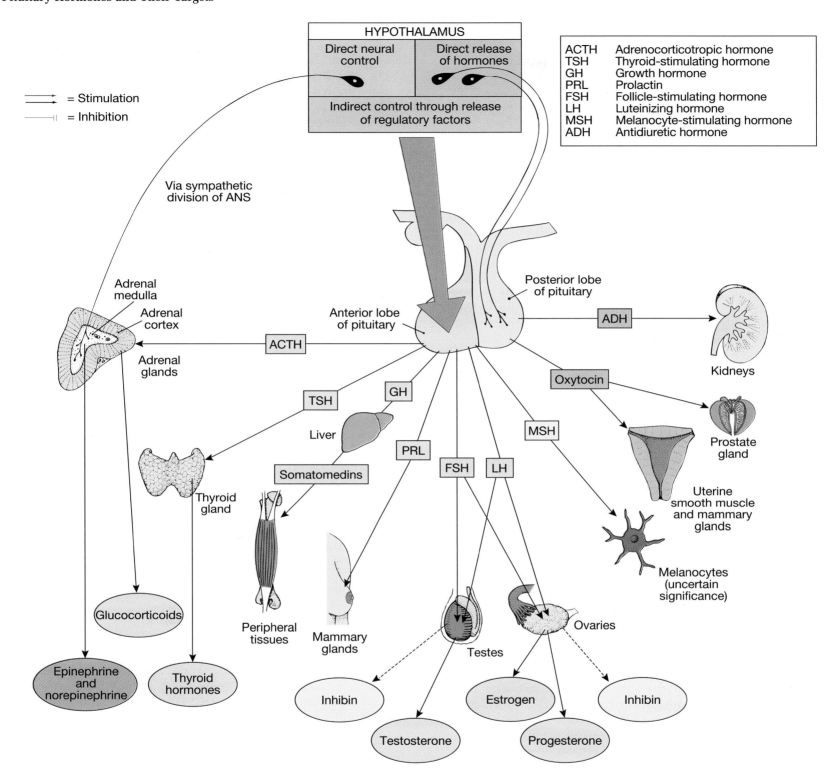

The Posterior Pituitary
(Figures 19-4 to 19-6)

The **posterior pituitary** (Figure 19-4●) is also called the **neu-rohypophysis** (noo-rō-hī-POF-i-sis) or *pars nervosa* ("nervous part"). The posterior pituitary contains the axons of roughly 50,000 hypothalamic neurons in the **supraoptic** and **paraventricular nuclei** (Figure 19-6●). The axons extend from these

nuclei through the infundibulum and end in synaptic terminals in the posterior pituitary. The hypothalamic neurons manufacture ADH (supraoptic nuclei) and oxytocin (paraventricular nuclei). ADH and oxytocin are called *neurosecretions* because they are produced and secreted by neurons. Once released, these hormones enter local capillaries supplied by the **inferior hypophyseal artery**. From there they will be transported into the general circulation.

Hormones released by the posterior pituitary (Figure 19-5●) include the following.

1. **Antidiuretic hormone (ADH)** is released in response to a variety of stimuli, most notably to a rise in the concentration of electrolytes in the blood or a fall in blood volume or pressure. The primary function of ADH is to decrease the amount of water lost at the kidneys. ADH also causes the constriction of peripheral blood vessels, which helps to elevate blood pressure. ✝ *Diabetes Insipidus [p. 762]*

2. The functions of **oxytocin** (ok-sē-TŌ-sin; *oxy-*, quick + *tokos*, childbirth) are best known in women, where it stimulates the contractions of smooth muscle cells in the uterus and contractile (myoepithelial) cells surrounding the secretory cells of the mammary glands. The stimulation of uterine muscles by oxytocin in the last stage of pregnancy is required for normal labor and childbirth. After birth, the sucking of an infant at the breast stimulates the release of oxytocin into the blood. Oxytocin then stimulates the smooth muscle cells in the mammary glands, causing discharge of milk from the nipple. In the human male, oxytocin causes smooth muscle contractions in the prostate gland. Animal studies suggest that its release may also trigger sexual behaviors by stimulating centers in the hypothalamus.

The Anterior Pituitary
(Figure 19-4)

The **anterior pituitary**, or **adenohypophysis** (ad-e-nō-hī-POF-i-sis), contains a variety of endocrine cell types. The adenohypophysis can be subdivided into two regions: (1) a large **pars distalis** (dis-TAL-is; "distal part"), which represents the major portion of the pituitary gland, and (2) a slender **pars intermedia** (in-ter-MĒ-dē-a; "intermediate part"), which forms a narrow band adjacent to the neurohypophysis (Figure 19-4●). The entire adenohypophysis is richly vascularized with an extensive capillary network.

The Hypophyseal Portal System
(Figure 19-6)

The production of hormones in the anterior pituitary is controlled by the hypothalamus through the secretion of specific regulatory factors. At the **median eminence**, a swelling near the attachment of the **infundibulum** (in-fun-DIB-ū-lum), neurons release regulatory factors into the surrounding interstitial fluids. ∞ [p. 384] Their secretions enter the circulation quite easily because the capillaries in this region have a "Swiss cheese" appearance, with open spaces between adjacent endothelial cells. Such capillaries are called **fenestrated** (FEN-es-trāt-ed; *fenestra*, window), and they are found only where relatively large molecules enter or leave the circulatory system. The capillary networks in the median eminence are supplied by the **superior hypophyseal artery** (Figure 19-6●).

Before leaving the hypothalamus, the capillary network unites to form a series of larger vessels that spiral around the infundibulum to reach the anterior pituitary gland. Once within the anterior pituitary, these vessels form a second capillary network that branches among the endocrine cells (Figure 19-6●).

This is an unusual vascular arrangement. Typically, an artery conducts blood from the heart to a capillary network, and a vein carries blood from a capillary network back to the heart. The vessels between the median eminence and the anterior pituitary, however, carry blood from one capillary network to another. Blood vessels that link two capillary networks are called **portal vessels**, and the entire complex is termed a **portal system**. Portal systems provide an efficient means of chemical communication by ensuring that all of the blood entering the portal vessels will reach the intended target cells before returning to the general circulation. The communication is strictly one-way, however, because any chemicals released by the cells "downstream" must do a complete tour of the circulatory system before reaching the capillaries at the start of the portal system. Portal vessels are named after their destinations, so this particular network of vessels is the **hypophyseal portal system**.

Hormones of the Anterior Pituitary
(Figure 19-5)

We will restrict our discussion to seven hormones whose functions and control mechanisms are reasonably well understood. All but one of these hormones are produced by the pars distalis, and four of them regulate the production of hormones by other endocrine glands. The names of these hormones indicate their activities; details are summarized in Table 19-2 and Figure 19-5●.

1. **Thyroid-stimulating hormone (TSH)** targets the thyroid gland and triggers the release of thyroid hormones.

2. **Adrenocorticotropic hormone (ACTH)** stimulates the release of steroid hormones by the adrenal gland. ACTH specifically targets cells producing hormones called **glucocorticoids (GC)** (gloo-kō-KOR-ti-koyds) that affect glucose metabolism.

3. **Follicle-stimulating hormone (FSH)** promotes follicle development in women and stimulates the secretion of **estrogens** (ES-trō-jens) by ovarian cells. Estrogens, which are steroids, are female sex hormones; *estradiol* is the most important estrogen. In men, FSH secretion supports sperm production in the testes.

4. **Luteinizing** (LOO-tē-in-ī-zing) **hormone (LH)** induces ovulation in women and promotes the ovarian secretion of **progestins** (prō-JES-tins), steroid hormones that prepare the body for possible pregnancy. *Progesterone* is the most important progestin. In men, LH is often called **interstitial cell-stimulating hormone (ICSH)**, because it stimulates the production of male sex hormones called **androgens** (AN-drō-jenz; *andros*, man) by the interstitial cells of the testes. *Testosterone* is the most important androgen. FSH and LH are called **gonadotropins** (gō-nad-ō-TRŌ-pinz; *tropos*, turning) because they regulate the activities of the male and female sex organs (gonads).

5. **Prolactin** (prō-LAK-tin; *pro-*, before + *lac*, milk), or **PRL**, stimulates the development of the mammary glands and the production of milk. PRL exerts the dominant effect on the glandular cells, but the mammary glands are regulated by the interaction of a number of other hormones, including *growth hormone* and hormones produced by the placenta. Prolactin has no known effects in the human male.

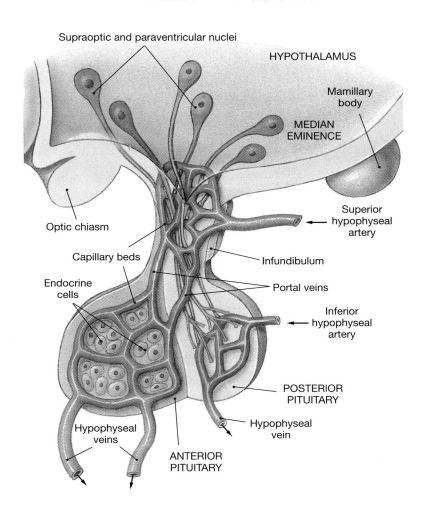

FIGURE 19-6
The Pituitary Gland and the Hypophyseal Portal System. Circulatory arrangement that forms the hypophyseal portal system.

Labels in figure:
Supraoptic and paraventricular nuclei
HYPOTHALAMUS
Mamillary body
MEDIAN EMINENCE
Optic chiasm
Superior hypophyseal artery
Capillary beds
Infundibulum
Endocrine cells
Portal veins
Inferior hypophyseal artery
POSTERIOR PITUITARY
Hypophyseal veins
Hypophyseal vein
ANTERIOR PITUITARY

6.   **Growth hormone (GH)**, also called *human growth hormone* (HGH) or **somatotropin** (*soma*, body), stimulates cell growth and replication by accelerating the rate of protein synthesis. Although virtually every tissue responds to some degree, growth hormone has a particularly strong effect on skeletal and muscular development.

Children unable to produce adequate concentrations of growth hormone suffer from *pituitary growth failure*, sometimes called *pituitary dwarfism*, a condition described in Chapter 5. ∞ [p. 120] The steady growth and maturation that normally precede and accompany puberty do not occur in these individuals.

7.   **Melanocyte-stimulating hormone (MSH)** is the only hormone released by the pars intermedia. As the name indicates, MSH stimulates the melanocytes of the skin, increasing their rates of melanin production and distribution. MSH is secreted only during fetal development, in young children, in pregnant women, and in some disease states.

THE THYROID GLAND
(Figure 19-7)

The **thyroid gland** curves across the anterior surface of the trachea, just inferior to the **thyroid** ("shield-shaped") **cartilage** that dominates the anterior surface of the larynx (Figure 19-7a●). Because of its location, the thyroid gland can be easily felt with the fingers; when something goes wrong with the gland, it may even become prominent. The size of the thyroid gland is quite variable, depending on heredity, environment, and nutritional

factors, but the average weight is about 34 g (1.2 oz). The gland has a deep red coloration because of the large number of blood vessels supplying the glandular cells. Blood supply to the gland is from two sources: (1) a *superior thyroid artery*, which is a branch from the external carotid artery, and (2) an *inferior thyroid artery*, a branch of the thyrocervical trunk. Venous drainage of the gland is via the *superior* and *middle thyroid veins*, which end in the internal jugular vein, and the *inferior thyroid veins*, which terminate at the brachiocephalic veins.

The thyroid gland has a "butterfly-like" appearance and consists of two main **lobes**. The superior portion of each lobe extends over the lateral surface of the trachea toward the inferior border of the thyroid cartilage. Inferiorly, the lobes of the thyroid gland extend to the level of the second or third tracheal ring. The two lobes are united by a slender connection, the **isthmus** (IS-mus). The thyroid gland is anchored to the tracheal rings by a thin capsule that is continuous with connective tissue partitions that segment the glandular tissue and surround the *thyroid follicles*.

Thyroid Follicles and Thyroid Hormones
(Figures 19-7b,c,d/19-8)

Thyroid follicles manufacture, store, and secrete thyroid hormones. Individual follicles are spherical, resembling miniature tennis balls, and they are lined by a simple cuboidal epithelium (Figure 19-7b,c●). The follicle cells surround a **follicle cavity**, which contains a **colloid**, a viscous fluid containing large quantities of suspended proteins. A network of capillaries surrounds each follicle, delivering nutrients and regulatory hormones to the glandular cells and accepting their secretory products and metabolic wastes.

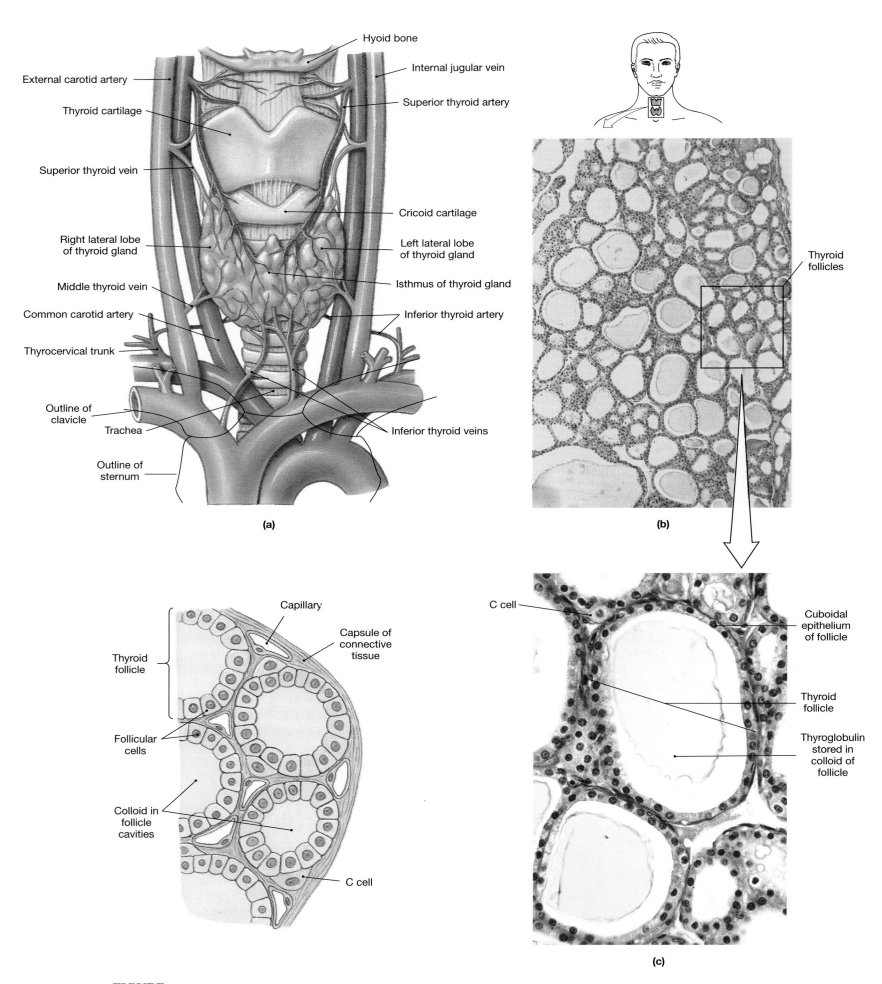

FIGURE 19-7
The Thyroid Gland. (a) Location and anatomy of the thyroid. (b) Histological organization of the thyroid. (LM × 122)
(c) Histological details of the thyroid gland showing thyroid follicles. (LM × 260)

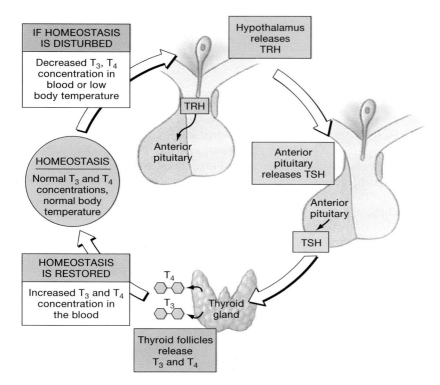

FIGURE 19-8
The Regulation of Thyroid Activity

Follicular cells synthesize a globular protein called **thyroglobulin** (thī-rō-GLOB-ū-lin) and secrete it into the colloid of the thyroid follicles. Thyroglobulin contains molecules of tyrosine, and these amino acids will be converted to thyroid hormones inside the follicle, through the attachment of iodine. The follicle cells absorb iodine as I^- through active transport from the interstitial fluid. The iodine is converted to a special ionized form (I^+) and attached to the tyrosine molecules of thyroglobulin by enzymes at the lumenal surfaces of the follicle cells. The hormone **thyroxine** (thī-ROKS-ēn), or **TX**, consists of two tyrosine molecules and four atoms of iodine; it is also known as tetraiodothyronine, or simply T_4. **Triiodothyronine**, or T_3, is a related molecule containing three iodine atoms. Eventually each molecule of thyroglobulin stored within a thyroid follicle will contain 4–8 molecules of T_3 and/or T_4 hormones.

The major factor controlling the rate of thyroid hormone release is the concentration of TSH in the circulating blood (Figure 19-8●). Under the influence of TSH, follicle cells remove thyroglobulin from the follicles, break the protein down, and release molecules of T_3 and T_4, which then enter the circulation.

Thyroxine (T_4) accounts for roughly 90% of all thyroid secretions. The thyroid hormones, which have complementary effects, increase the rate of cellular metabolism and oxygen consumption in almost every cell in the body. These hormones are included in Table 19-3. ☂ *Thyroid Gland Disorders [p. 763]*

The C Cells of the Thyroid Gland
(Figure 19-7c)

A second group of endocrine cells lies sandwiched between the cuboidal follicle cells and their basement membrane. These cells are larger than those of the follicular epithelium, and they do not stain as clearly (Figure 19-7c●). These are the **C cells**, or *parafollicular cells*, that produce the hormone **calcitonin** (kal-si-TŌ-nin) **(CT)**. Calcitonin assists in the regulation of calcium ion concentrations in body fluids, especially under stresses such as starvation or pregnancy. Calcitonin lowers calcium ion concentrations by (1) inhibiting osteoclasts and (2) stimulating calcium ion excretion at the kidneys. The actions of calcitonin are opposed by those of *parathyroid hormone* produced by the parathyroid glands.

THE PARATHYROID GLANDS
(Figures 19-7a/19-9)

Four pea-sized, reddish brown **parathyroid glands** are located on the posterior surfaces of the thyroid gland (Figure 19-9●). The glands are usually attached at the surface of the thyroid gland by the thyroid capsule. Like the thyroid gland, the parathyroids are surrounded by a connective tissue capsule that invades the interior of the gland, forming separations and small irregular *lobules*. Blood is supplied to the superior pair via the *superior thyroid artery* and to the inferior pair via the *inferior thyroid artery* (Figure 19-7a●). The venous drainage is the same as that of the thyroid. All together the four parathyroid glands weigh a mere 1.6 g. There are two types of glandular cells in the parathyroid gland. The **principal cells**, or **chief cells**, are small cells (Figure 19-9b,c●) that produce the hormone **parathyroid hormone (PTH)**; the functions of the other major cell type (oxyphil cells) are unknown.

Like the C cells of the thyroid, the chief cells of the parathyroids monitor the circulating concentration of calcium ions. When the calcium concentration falls below normal, the chief cells secrete parathyroid hormone. PTH stimulates osteoclasts, inhibits osteoblasts, promotes intestinal absorption of calcium, and reduces urinary excretion of calcium ions until blood concentrations return to normal. ☂ *Disorders of Parathyroid Function [p. 763]*

TABLE 19-3	Hormones of the Thyroid, Parathyroids, and Thymus		
Gland/Cells	*Hormones*	*Targets*	*Effects*
THYROID **Follicular epithelium**	Thyroxine (TX, T_4), triiodothyronine (T_3)	Most cells	Increase energy utilization, oxygen consumption, growth, and development
C cells	Calcitonin (CT)	Bone, kidneys	Decreases calcium ion concentrations in body fluids; uncertain significance in healthy nonpregnant adults
PARATHYROIDS **Chief cells**	Parathyroid hormone (PTH)	Bone, kidneys, intestines	Increases calcium ion concentrations in body fluids
THYMUS	"Thymosin" (see Chapter 23)	Lymphocytes	Maturation and functional competence of immune system

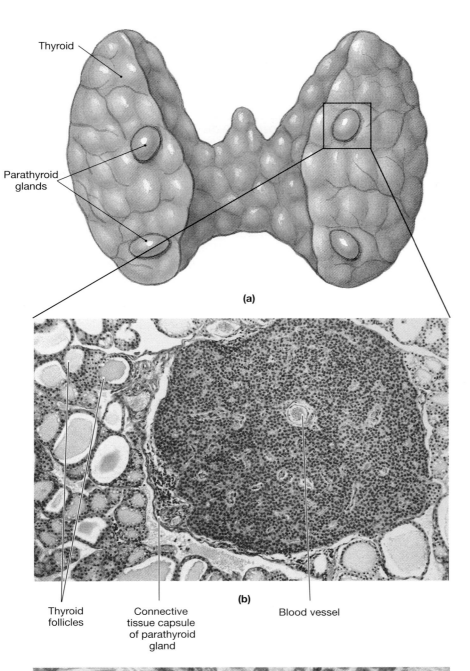

(a)

(b)

Thyroid

Parathyroid
glands

Thyroid
follicles

Connective
tissue capsule
of parathyroid
gland

Blood vessel

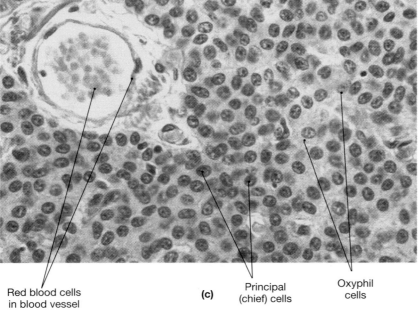

(c)

Red blood cells
in blood vessel

Principal
(chief) cells

Oxyphil
cells

THE THYMUS
(Figure 19-1)

The **thymus** is embedded in a mass of connective tissue inside the thoracic cavity, usually just posterior to the sternum (Figure 19-1●, p. 484). In newborn infants and young children, the thymus is relatively large, often extending from the base of the neck to the superior border of the heart. Although its relative size decreases as a child grows, the thymus continues to enlarge slowly, reaching its maximum size just before puberty, at a weight of around 40 g. After puberty it gradually diminishes in size; by age 50 the thymus may weigh less than 12 g.

The thymus produces several hormones important to the development and maintenance of normal immunological defenses (Table 19-3). **Thymosin** (thī-MŌ-sin) is a thymic extract that promotes the development and maturation of lymphocytes. It is becoming apparent that this process is complex, and "thymosin" is a blend of several different, complementary hormones (thymosin-1, thymopoietin, thymopentin, thymulin, thymic humoral factor, and IGF-1).

It has been suggested that the gradual decrease in the size and secretory abilities of the thymus may make the elderly more susceptible to disease. The histological organization of the thymus and the functions of thymosin will be discussed further in Chapter 23.

THE ADRENAL GLAND
(Figure 19-10)

A yellow, pyramid-shaped **adrenal gland**, or **suprarenal gland** (soo-pra-RĒ-nal; *supra-*, above + *renes*, kidneys), sits on the superior border of each kidney. Each gland lies at the level of the twelfth thoracic rib and is firmly attached to the superior portion of each kidney by a dense fibrous **capsule** (Figure 19-10a●). The adrenal gland on each side nestles between the kidney, the diaphragm, and the major arteries and veins running along the dorsal wall of the abdominopelvic cavity. The adrenal glands project into the abdominopelvic cavity, but they remain separated from it by the peritoneal lining. Like the other endocrine glands, the adrenal glands are highly vascularized. Branches of the *renal artery*, the *inferior phrenic artery*, and a direct branch from the aorta (the *middle suprarenal artery*) supply blood to each adrenal gland. The *suprarenal veins* carry blood away from the adrenal glands.

A typical adrenal gland weighs about 7.5 g. It is generally heavier in men than in women, but adrenal size can vary greatly as secretory demands change. Structurally and functionally, the adrenal gland can be divided into two regions: a superficial **adrenal cortex** and an inner **adrenal medulla** (Figure 19-10b,c●).

FIGURE 19-9
The Parathyroid Glands. (a) Location of parathyroids on posterior surface of thyroid lobes. (b) Photomicrograph showing both parathyroid and thyroid tissues. (LM × 116) (c) Photomicrograph of parathyroid cells. (LM × 850)

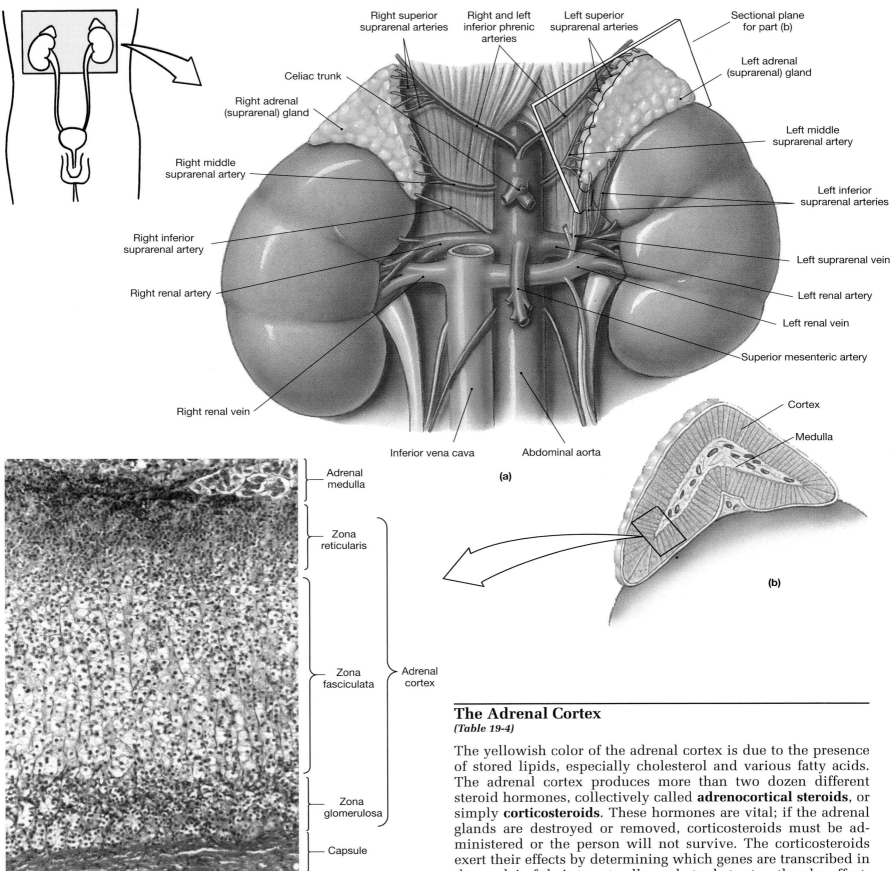

Right superior suprarenal arteries

Right and left inferior phrenic arteries

Left superior suprarenal arteries

Sectional plane for part (b)

Celiac trunk

Right adrenal (suprarenal) gland

Left adrenal (suprarenal) gland

Right middle suprarenal artery

Left middle suprarenal artery

Left inferior suprarenal arteries

Right inferior suprarenal artery

Right renal artery

Left suprarenal vein

Left renal artery

Left renal vein

Superior mesenteric artery

Right renal vein

Inferior vena cava

Abdominal aorta

(a)

Cortex

Medulla

(b)

Adrenal medulla

Zona reticularis

Zona fasciculata

Adrenal cortex

Zona glomerulosa

Capsule

(c)

FIGURE 19-10
The Adrenal Gland. (a) Superficial view of the kidney and adrenal gland, showing the sectional plane for part (b). (b) An adrenal gland, showing the orientation of the section in part (c). (c) Light micrograph, with the major regions identified. (LM × 173)

The Adrenal Cortex
(Table 19-4)

The yellowish color of the adrenal cortex is due to the presence of stored lipids, especially cholesterol and various fatty acids. The adrenal cortex produces more than two dozen different steroid hormones, collectively called **adrenocortical steroids**, or simply **corticosteroids**. These hormones are vital; if the adrenal glands are destroyed or removed, corticosteroids must be administered or the person will not survive. The corticosteroids exert their effects by determining which genes are transcribed in the nuclei of their target cells, and at what rates, thereby affecting all metabolic operations.

Interior to the adrenal capsule there are three distinct regions, or *zones*, in the adrenal cortex: (1) an outer, *zona glomerulosa*; (2) a middle, *zona fasciculata*; and (3) an inner, *zona reticularis*. Each zone synthesizes different steroid hormones (Table 19-4).

The Zona Glomerulosa
(Figure 19-10c)

The **zona glomerulosa** (glō-mer-ū-LŌ-sa), the outermost cortical region, accounts for about 15% of the cortical volume (Figure 19-10c●). This zone extends from the capsule to the radiating cords of the underlying zona fasciculata. A *glomerulus* is a little ball or knot, and here the endocrine cells form densely packed clusters.

The zona glomerulosa produces **mineralocorticoids (MC)**, steroid hormones that affect the electrolyte composition of body fluids. **Aldosterone** (al-DOS-ter-ōn) is the principal mineralocorticoid and targets kidney cells that regulate the ionic composition of the urine. It causes the retention of sodium ions and water, thereby reducing fluid losses in the urine. Aldosterone also reduces sodium and water losses at the sweat glands, salivary glands, and along the digestive tract. Aldosterone secretion occurs when the zona glomerulosa is stimulated by the hormone *angiotensin II* (*angeion*, vessel + *teinein*, to stretch). ✝*Disorders of the Adrenal Cortex [p. 763]*

The Zona Fasciculata
(Figure 19-10c)

The **zona fasciculata** (fa-sik-ū-LA-ta; *fasciculus*, little bundle) begins at the inner border of the zona glomerulosa and extends toward the medulla (Figure 19-10c●). It contributes about 78% of the cortical volume. The cells are larger and contain more lipids than those of the zona glomerulosa, and the lipid droplets give the cytoplasm a pale, foamy appearance. The cells of the zona fasciculata form cords that radiate like a sunburst from the zona reticularis. Adjacent cords are separated by flattened vessels with fenestrated walls.

Steroid production in the zona fasciculata is stimulated by ACTH from the anterior pituitary. This zone produces steroid hormones collectively known as **glucocorticoids (GC)** because of their effects on glucose metabolism. **Cortisol** (KOR-ti-sol; also called *hydrocortisone*), **corticosterone** (kor-ti-KOS-te-rōn), and **cortisone** are the three most important glucocorticoids. These hormones speed up the rates of glucose synthesis and glycogen formation, especially within the liver.

The Zona Reticularis
(Figure 19-10c)

The **zona reticularis** (re-tik-ū-LAR-is; *reticulum*, network) forms a narrow band bordering the adrenal medulla (Figure 19-10c●).

In total, the zona reticularis accounts for only around 7% of the total cell volume of the adrenal cortex. The endocrine cells of the zona reticularis are much smaller than those of the medulla, and they form a folded, branching network. Fenestrated blood vessels wind between the cells. The zona reticularis normally secretes small amounts of sex hormones called *androgens*. Androgens are produced in large quantities by the testes, and the significance of the small adrenal androgen contribution—in both males *and* females—remains uncertain.

The Adrenal Medulla
(Figure 19-10b,c)

The boundary between the adrenal cortex and medulla does not form a straight line (Figure 19-10b,c●), and the supporting connective tissues and blood vessels are extensively interconnected. The adrenal medulla has a reddish brown coloration due in part to the many blood vessels in this area. The medulla contains clumps of large, rounded cells, called **pheochromocytes**, or **chromaffin cells**, that resemble the neurons in sympathetic ganglia. These cells are innervated by preganglionic sympathetic fibers, and sympathetic activation triggers their secretory activity. ∞ [p. 434]

The adrenal medulla contains two populations of secretory cells, one producing epinephrine (adrenaline) and the other norepinephrine (noradrenaline). Epinephrine makes up 75–80% of the secretions from the adrenal medulla, and the rest is norepinephrine. The effects of these hormones were described in Chapter 17. ∞ [p. 434] Catecholamine secretion triggers cellular energy utilization and the mobilization of energy reserves. This combination increases muscular strength and endurance. The metabolic changes that follow catecholamine release are at their peak 30 seconds after adrenal stimulation, and they linger for several minutes thereafter. As a result, the effects produced by stimulation of the adrenal medulla outlast the other signs of sympathetic activation. ✝*Disorders of the Adrenal Medulla [p.763]*

√ **When a person's thyroid gland is removed, signs of decreased thyroid hormone concentration do not appear until about one week later. Why?**

√ **Removal of the parathyroid glands would result in a decrease in the blood of what important mineral?**

TABLE 19-4	The Adrenal Hormones			
Region/Zone	Hormones		Targets	Effects
CORTEX **Zona glomerulosa**	Mineralocorticoids (MC), primarily aldosterone		Kidneys	Increases renal reabsorption of sodium ions and water and accelerates renal potassium ion loss
Zona fasciculata	Glucocorticoids (GC): cortisol (hydrocortisone), corticosterone, cortisone		Most cells	Releases amino acids from skeletal muscles, lipids from adipose tissues; promotes liver glycogen and glucose formation; promotes peripheral utilization of lipids (glucose-sparing); anti-inflammatory effects
Zona reticularis	Androgens			Uncertain significance under normal conditions
MEDULLA	Epinephrine (adrenaline, E), norepinephrine (noradrenaline, NE)		Most cells	Increased cardiac activity, blood pressure, glycogen breakdown, blood glucose; release of lipids by adipose tissue (see Chapter 17)

ENDOCRINE FUNCTIONS OF THE KIDNEYS AND HEART

The kidneys and heart produce several hormones, most involved with the regulation of blood pressure and blood volume. The kidneys produce two peptide hormones, *renin* and *erythropoietin*, and one steroid hormone, *calcitriol*. **Renin** functions as an enzyme. It converts circulating **angiotensinogen**, an inactive protein produced by the liver, to **angiotensin I**. In capillaries of the lungs, this compound is converted to **angiotensin II**, the hormone that stimulates the adrenal production of aldosterone. **Erythropoietin (EPO)** (e-rith-rō-POY-e-tin) stimulates red blood cell production by the bone marrow. This hormone is released when blood pressure or blood oxygen levels in the kidneys decline. EPO stimulates red blood cell production and maturation, thus increasing the blood volume and its oxygen-carrying capacity.

Calcitriol is a steroid hormone secreted by the kidney in response to the presence of parathyroid hormone (PTH). Calcitriol synthesis is dependent on the availability of a related steroid, *cholecalciferol* (vitamin D_3), which may be synthesized in the skin or absorbed from the diet. Cholecalciferol from either source is absorbed from the bloodstream by the liver and converted to an intermediary product that is released into the circulation and absorbed by the kidneys for conversion to calcitriol. The term *vitamin D* is used to indicate the entire group of related steroids, including calcitriol, cholecacliferol, and various intermediaries.

The best-known function of calcitriol is stimulation of calcium and phosphate ion absorption along the digestive tract. PTH stimulates the release of calcitriol, and in this way PTH has an indirect effect on intestinal calcium absorption. The effects of calcitriol on the skeletal system and kidney are not well understood.

Cardiac muscle cells in the *right atrium* of the heart produce **atrial natriuretic peptide (ANP)** when blood pressure or blood volume becomes excessive. ANP suppresses the release of ADH and aldosterone and stimulates water and sodium ion loss at the kidneys. This effect gradually reduces blood volume.

THE PANCREAS AND OTHER ENDOCRINE TISSUES OF THE DIGESTIVE SYSTEM

The pancreas, the lining of the digestive tract, and the liver produce a variety of exocrine secretions that are essential to the normal digestion of food. Although the pace of digestive activities can be affected by the autonomic nervous system, most digestive processes are controlled locally by the individual organs. The various digestive organs communicate with one another via hormones, detailed in Chapter 25. This section focuses attention on one digestive organ, the pancreas, which produces hormones that affect metabolic operations throughout the body.

The Pancreas
(Figure 19-11)

The **pancreas** lies within the abdominopelvic cavity in the J-shaped loop between the stomach and small intestine (Figure 19-11a●, p. 498). It is a slender, pink organ with a nodu-

lar or lumpy consistency. The adult pancreas ranges between 20 and 25 cm (8–10 in.) in length and weighs about 80 g (2.8 oz). Chapter 25 considers the detailed anatomy of the pancreas, because it is primarily a digestive organ that manufactures digestive enzymes. The **exocrine pancreas**, roughly 99% of the pancreatic volume, produces large quantities of an enzyme-rich fluid that enters the digestive tract through a prominent secretory duct.

The **endocrine pancreas** consists of small groups of cells peppered throughout the gland, each group surrounded by the exocrine cells, which make up the digestive portion. The groups, known as **pancreatic islets**, or the *islets of Langerhans* (LAN-ger-hanz), account for only about 1% of the pancreatic cell population (Figure 19-11b●, p. 498). Nevertheless, there are roughly 2 million islets in the normal pancreas.

Like other endocrine tissue, the islets are surrounded by an extensive, fenestrated capillary network that carries its hormones into the circulation. Two major arteries supply blood to the pancreas, the *pancreaticoduodenal arteries* and *pancreatic arteries*. Venous blood returns to the *hepatic portal vein*. The autonomic nervous system innervates the islets by branches from the *celiac plexus*. ∞ [p. 439]

Each islet contains four cell types.

1. **Alpha cells** produce the hormone **glucagon** (GLOO-ka-gon), which raises blood glucose levels by increasing the rates of glycogen breakdown and glucose release by the liver (Figure 19-11c●, p. 498).

2. **Beta cells** (Figure 19-11c●, p. 498) produce **insulin** (IN-su-lin), which lowers blood glucose by increasing the rate of glucose uptake and utilization by most body cells.

3. **Delta cells** produce the hormone **somatostatin (growth-hormone-inhibiting factor)**, which inhibits the production and secretion of glucagon and insulin and slows the rates of food absorption and enzyme secretion along the digestive tract.

4. **F-cells** produce the hormone **pancreatic polypeptide**, which regulates the production of certain digestive enzymes.

Pancreatic alpha and beta cells are sensitive to blood glucose concentrations, and their regulatory activities are not under the direct control of other endocrine or nervous components. Yet because the islet cells are extremely sensitive to variations in blood glucose levels, any hormone that affects blood glucose concentrations will indirectly affect the production of insulin and glucagon. The major hormones of the pancreas are summarized in Table 19-5, p. 499. ✝*Diabetes Mellitus [p. 764]*

ENDOCRINE TISSUES OF THE REPRODUCTIVE SYSTEM

The endocrine tissues of the reproductive system are restricted primarily to the male and female gonads, the testes and ovaries. Details of the anatomy of the reproductive organs will be described in detail and visualized in the reproductive system (Chapter 27).

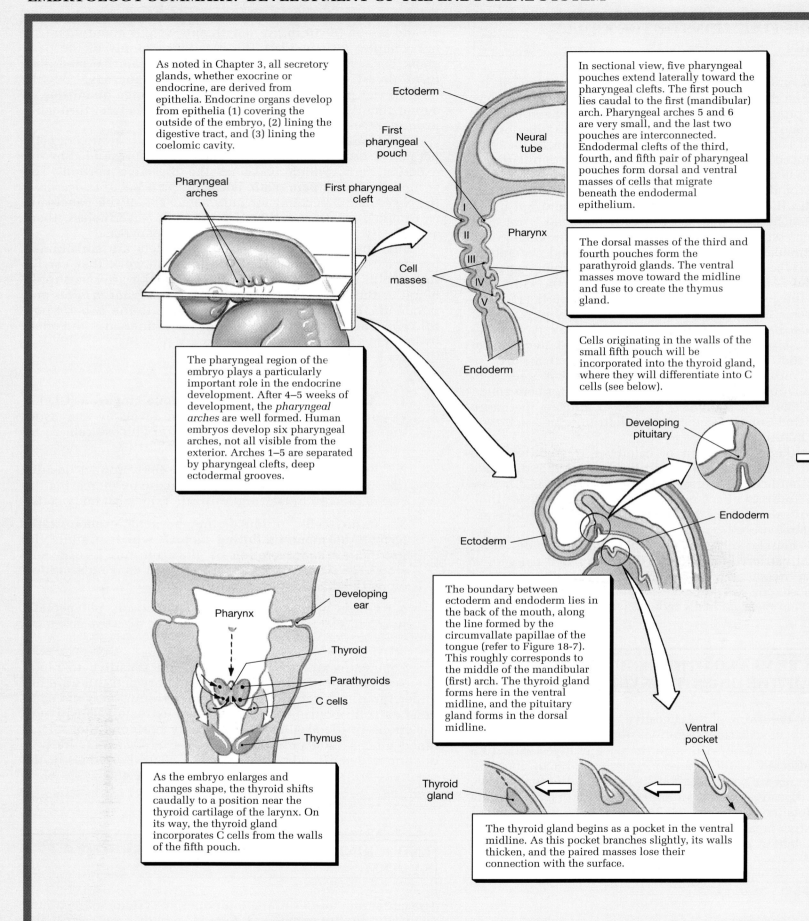

As noted in Chapter 3, all secretory glands, whether exocrine or endocrine, are derived from epithelia. Endocrine organs develop from epithelia (1) covering the outside of the embryo, (2) lining the digestive tract, and (3) lining the coelomic cavity.

In sectional view, five pharyngeal pouches extend laterally toward the pharyngeal clefts. The first pouch lies caudal to the first (mandibular) arch. Pharyngeal arches 5 and 6 are very small, and the last two pouches are interconnected. Endodermal clefts of the third, fourth, and fifth pair of pharyngeal pouches form dorsal and ventral masses of cells that migrate beneath the endodermal epithelium.

Ectoderm

First pharyngeal pouch

Neural tube

Pharyngeal arches

First pharyngeal cleft

Pharynx

Cell masses

The dorsal masses of the third and fourth pouches form the parathyroid glands. The ventral masses move toward the midline and fuse to create the thymus gland.

Endoderm

Cells originating in the walls of the small fifth pouch will be incorporated into the thyroid gland, where they will differentiate into C cells (see below).

The pharyngeal region of the embryo plays a particularly important role in the endocrine development. After 4–5 weeks of development, the *pharyngeal arches* are well formed. Human embryos develop six pharyngeal arches, not all visible from the exterior. Arches 1–5 are separated by pharyngeal clefts, deep ectodermal grooves.

Developing pituitary

Endoderm

Ectoderm

The boundary between ectoderm and endoderm lies in the back of the mouth, along the line formed by the circumvallate papillae of the tongue (refer to Figure 18-7). This roughly corresponds to the middle of the mandibular (first) arch. The thyroid gland forms here in the ventral midline, and the pituitary gland forms in the dorsal midline.

Developing ear

Pharynx

Thyroid

Parathyroids

C cells

Thymus

As the embryo enlarges and changes shape, the thyroid shifts caudally to a position near the thyroid cartilage of the larynx. On its way, the thyroid gland incorporates C cells from the walls of the fifth pouch.

Ventral pocket

Thyroid gland

The thyroid gland begins as a pocket in the ventral midline. As this pocket branches slightly, its walls thicken, and the paired masses lose their connection with the surface.

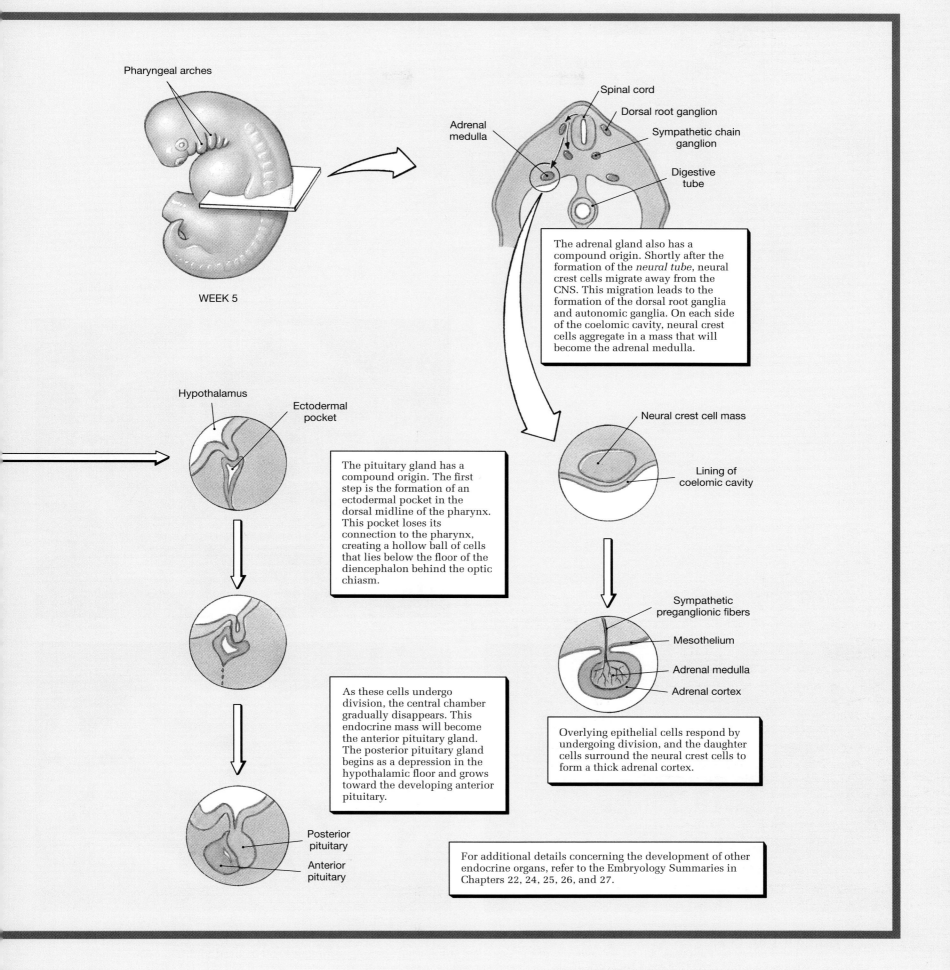

Pharyngeal arches

WEEK 5

Spinal cord

Dorsal root ganglion

Adrenal medulla

Sympathetic chain ganglion

Digestive tube

The adrenal gland also has a compound origin. Shortly after the formation of the *neural tube*, neural crest cells migrate away from the CNS. This migration leads to the formation of the dorsal root ganglia and autonomic ganglia. On each side of the coelomic cavity, neural crest cells aggregate in a mass that will become the adrenal medulla.

Hypothalamus

Ectodermal pocket

The pituitary gland has a compound origin. The first step is the formation of an ectodermal pocket in the dorsal midline of the pharynx. This pocket loses its connection to the pharynx, creating a hollow ball of cells that lies below the floor of the diencephalon behind the optic chiasm.

Neural crest cell mass

Lining of coelomic cavity

As these cells undergo division, the central chamber gradually disappears. This endocrine mass will become the anterior pituitary gland. The posterior pituitary gland begins as a depression in the hypothalamic floor and grows toward the developing anterior pituitary.

Sympathetic preganglionic fibers

Mesothelium

Adrenal medulla

Adrenal cortex

Overlying epithelial cells respond by undergoing division, and the daughter cells surround the neural crest cells to form a thick adrenal cortex.

Posterior pituitary

Anterior pituitary

For additional details concerning the development of other endocrine organs, refer to the Embryology Summaries in Chapters 22, 24, 25, 26, and 27.

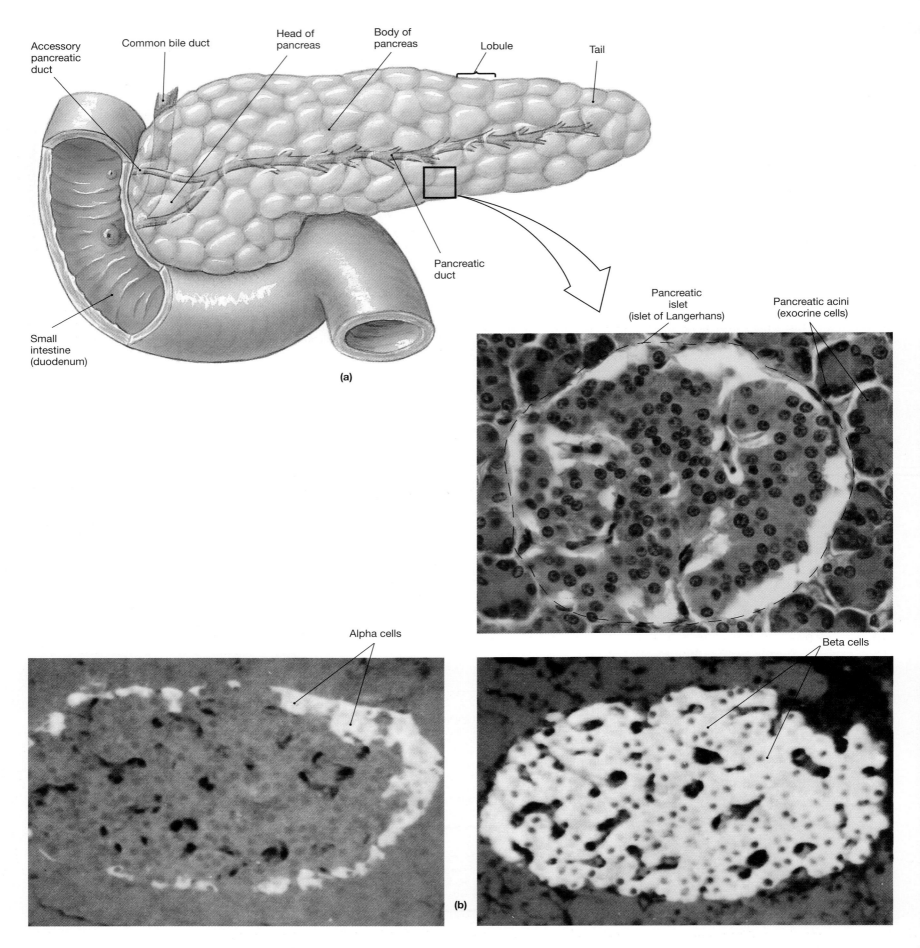

FIGURE 19-11
The Endocrine Pancreas. (a) Gross anatomy of the pancreas. (b) Special staining techniques can be used to differentiate between alpha cells (left) and beta cells (right) in pancreatic islets. (LMs × 184)

TABLE 19-5   Hormones of the Pancreas

Structure/Cells	Hormones	Primary Targets	Effects
PANCREATIC ISLETS			
Alpha cells	Glucagon	Liver, adipose tissues	Mobilization of lipid reserves, glucose synthesis and glycogen breakdown in liver, elevation of blood glucose concentrations
Beta cells	Insulin	All cells except those of brain, kidneys, digestive tract epithelium, and RBCs	Facilitation of uptake of glucose by cells, stimulation of lipid and glycogen formation and storage, decrease in blood glucose concentrations
Delta cells	Somatostatin	Alpha and beta cells	Inhibition of secretion
F-cells	Pancreatic polypeptides (PP)	Exocrine pancreas	Secretion of digestive enzymes

Testes

In the male, the **interstitial cells** of the testis produce male hormones known as *androgens*. **Testosterone** (tes-TOS-ter-ōn) is the most important androgen. This hormone promotes the production of functional sperm, maintains the secretory glands of the male reproductive tract, determines secondary sexual characteristics, and stimulates muscle growth. During embryonic development, the production of testosterone affects the anatomical development of the hypothalamic nuclei of the CNS.

Cells directly associated with the formation of functional sperm secrete an additional hormone, called **inhibin** (in-HIB-in). Inhibin production, which occurs under FSH stimulation, depresses the secretion of FSH by the anterior pituitary. Throughout adult life, these two hormones interact to maintain sperm production at normal levels.

Ovaries

In the ovaries, oocytes develop in specialized structures called **follicles**, under stimulation by FSH. Follicle cells surrounding the oocytes produce estrogens, especially the hormone **estradiol**. These steroid hormones support the maturation of the oocytes and stimulate the growth of the uterine lining. Under FSH stimulation, active follicles secrete inhibin, which suppresses FSH release through a feedback mechanism comparable to that described above for males.

After ovulation has occurred, the remaining follicular cells reorganize into a **corpus luteum** (LOO-tē-um) that releases a mixture of estrogens and progestins, especially **progesterone** (prō-JES-ter-ōn). Progesterone accelerates the movement of the fertilized egg along the uterine tube and prepares the uterus for the arrival of the developing embryo. A summary of information concerning the reproductive hormones can be found in Table 19-6.

THE PINEAL GLAND

The small, red, pinecone-shaped **pineal gland**, or **epiphysis** (e-PIF-e-sys) **cerebri**, is part of the epithalamus. ∞ [p. 382] The pineal gland contains neurons, glial cells, and special secretory cells called **pinealocytes** (PIN-ē-a-lō-sīts). Pinealocytes synthesize the hormone **melatonin** (mel-a-TŌ-nin) from molecules of the neurotransmitter *serotonin*. Melatonin slows the maturation of sperm, eggs, and reproductive organs by inhibiting the production of a hypothalamic releasing factor that stimulates FSH and LH secretion. Collaterals from the visual pathways enter the pineal gland and affect the rate of melatonin production. Melatonin production rises at night and declines during the day. This cycle is apparently important in regulating *circadian rhythms*, our natural awake-asleep cycles. This hormone is also a powerful antioxidant that may help protect CNS tissues from toxins generated by active neurons and glial cells.

TABLE 19-6   Hormones of the Reproductive System

Structure/Cells	Hormones	Primary Targets	Effects
TESTES **Interstitial cells**	Testosterone	Most cells	Supports functional maturation of sperm, protein synthesis in skeletal muscles, male secondary sex characteristics, and associated behaviors
	Inhibin	Anterior pituitary	Inhibits secretion of FSH
OVARIES **Follicle cells**	Estrogens (especially estradiol)	Most cells	Support follicle maturation, female secondary sex characteristics, and associated behaviors
	Inhibin	Anterior pituitary	Inhibits secretion of FSH
Corpus luteum	Progestins (especially progesterone)	Uterus, mammary glands	Prepare uterus for implantation; prepare mammary glands for secretory functions

The endocrine system shows relatively few functional changes with age. The most dramatic exception is the decline in the concen-tration of reproductive hormones. It is interesting to note that age-related changes in other tissues affect their abilities to respond to hormonal stimulation. As a result, most tissues may become less responsive to circulating hormones, even though hormone concentrations remain normal. ☩ *Endocrine Disorders [p. 764]*

■ **CLINICAL BRIEF**
Endocrine Disorders

Endocrine disorders fall into two basic categories: They reflect either abnormal hormone production or abnormal cellular sensitivity. The symptoms are interesting because they highlight the significance of normally "silent" hormonal contributions. The characteristic features of many of these conditions are illustrated in Figure 19-12 ●.

FIGURE 19-12
Endocrine Abnormalities

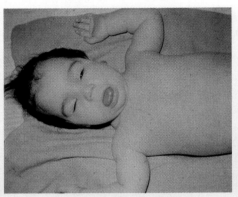

(a) *Acromegaly* results from the over-production of growth hormone after the epiphyseal plates have fused. Bone shapes change and cartilaginous areas of the skeleton enlarge. Note the broad facial features and the enlarged lower jaw.

(b) *Cretinism* results from thyroid hormone insufficiency in infancy.

(c) An enlarged thyroid gland, or *goiter*, is usually associated with thyroid hypo-secretion due to iodine insufficiency.

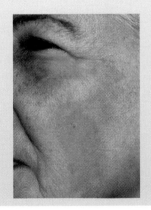

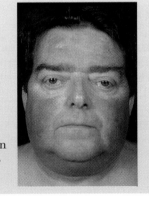

(d) *Addison's disease* is caused by hyposecre-tion of corticosteroids, especially glucocorti-coids. Pigment changes result from stimulation of melanocytes by ACTH, which is structurally similar to MSH.

(e) *Cushing's disease* is caused by hypersecretion of glucocorticoids. Lipid reserves are mobilized, and adipose tissue accumulates in the cheeks and at the base of the neck.

Related Clinical Terms

diabetes insipidus: A disorder that develops when the posterior pituitary no longer releases adequate amounts of ADH. ☩ *Dia-betes Insipidus [p. 762]*

myxedema (miks-e-DĒ-ma): Symptoms of hypothyroidism, which include subcutaneous swelling, dry skin, hair loss, low body tem-perature, muscular weakness, and slowed reflexes. ☩ *Thyroid Gland Disorders [p. 763]*

goiter: An enlargement of the thyroid gland. ☩ *Thyroid Gland Disorders [p. 763]*

exophthalmos (eks-ahf-THAL-mos): Protrusion of the eyes, a symptom of hyperthyroidism. ☩ *Thyroid Gland Disorders [p. 763]*

thyrotoxic crisis: A period when a subject with acute hyperthy-roidism experiences an extremely high fever, rapid heart rate, and the malfunctioning of a variety of physiological systems. ☩ *Thy-roid Gland Disorders [p. 763]*

diabetes mellitus (mel-LĪ-tus): A disorder characterized by glu-cose concentrations high enough to overwhelm the kidneys' re-absorption capabilities. ☩ *Diabetes Mellitus [p. 763]*

insulin-dependent diabetes mellitus (IDDM); (also known as **Type I diabetes** or **juvenile-onset diabetes**): A type of diabetes mellitus; the primary cause is inadequate insulin production by the beta cells of the pancreatic islets. †*Diabetes Mellitus [p. 764]*

ketoacidosis: A condition in which large numbers of ketone bodies in the blood lead to a dangerously low blood pH. †*Diabetes Mellitus [p. 764]*

non-insulin-dependent diabetes mellitus (NIDDM); (also known as **Type II diabetes** or **maturity-onset diabetes**): A type of diabetes mellitus in which insulin levels are normal or elevated, but peripheral tissues no longer respond normally. †*Diabetes Mellitus [p. 764]*

CHAPTER SUMMARY AND REVIEW

STUDY OUTLINE

Related Key Terms

INTRODUCTION *[p. 483]*

1. In general, the nervous system performs short-term "crisis management," while the endocrine system regulates longer-term, ongoing metabolic processes. Endocrine cells release chemicals called **hormones** that alter the metabolic activities of many different tissues and organs simultaneously. *(see Figure 19-1 and Table 19-1).*

AN OVERVIEW OF THE ENDOCRINE SYSTEM *[p. 483]*

1. Hormones can be divided into three groups based on chemical structure: *amino acid derivatives*, *peptide hormones*, and *steroids*.
2. **Amino acid derivatives** are structurally similar to amino acids; they include **catecholamines**, **thyroid hormones**, and **melatonin**.
3. **Peptide hormones** are chains of amino acids; **steroid hormones** are lipids structurally similar to cholesterol.
4. Hormones exert their effects by modifying the activities of **target cells** (peripheral cells that are sensitive to that particular hormone). Receptors for catecholamine and peptide hormones are located on the cell membranes of target cells; in this case the hormone acts as a first messenger that causes a second messenger to appear in the cytoplasm. Thyroid and steroid hormones cross the cell membrane and bind to receptors in the cytoplasm or nucleus *(see Figure 19-2).*

The Hypothalamus and Endocrine Regulation *[p. 484]*

releasing hormone (RH) • inhibiting hormone (IH)

5. The hypothalamus coordinates endocrine and neural activities. It (1) produces two hormones of its own (ADH and oxytocin), (2) controls the activity of the anterior pituitary through the production of **regulatory hormones**, and (3) controls the output of the adrenal medulla, an endocrine component of the sympathetic division of the ANS *(see Figure 19-3).*

THE PITUITARY GLAND *[p. 485]*

1. The **pituitary gland (hypophysis)** releases nine important peptide hormones. Seven are synthesized in the anterior pituitary, and two are synthesized in the hypothalamus and released at the posterior pituitary *(see Figures 19-4/19-5 and Table 19-2).*

The Posterior Pituitary *[p. 487]*

inferior hypophyseal artery

2. The **posterior pituitary (neurohypophysis)** contains the axons of hypothalamic neurons. Neurons of the **supraoptic** and **paraventricular nuclei** manufacture **antidiuretic hormone (ADH)** and oxytocin, respectively. ADH decreases the amount of water lost at the kidneys. In women, **oxytocin** stimulates smooth muscle cells in the uterus and contractile cells in the mammary glands. In men, it stimulates prostatic smooth muscle contractions *(see Figures 19-4 to 19-6 and Table 19-2).*

The Anterior Pituitary *[p. 488]*

gonadotropins

3. The **anterior pituitary (adenohypophysis)** can be subdivided into the large **pars distalis** and the slender **pars intermedia.** Important hormones released by the pars distalis include: (1) **thyroid-stimulating hormone (TSH)** (triggers the release of thyroid hormones); (2) **adrenocorticotropic hormone (ACTH)** (stimulates the release of **glucocorticoids** by the adrenal gland); (3) **follicle-stimulating hormone (FSH)** (stimulates **estrogen** secretion and egg development in women and sperm production in men); (4) **luteinizing hormone (LH),** or **interstitial cell-stimulating hormone (ICSH)** (causes ovulation and production of **progestins** in women and **androgens** in men); (5) **prolactin (PRL)** (stimulates the development of the mammary glands and the production of milk); and (6) **growth hormone (GH,** or **somatotropin)** (stimulates cells' growth and replication) *(see Figures 19-4/19-5).*
4. **Melanocyte-stimulating hormone (MSH),** released by the pars intermedia, stimulates melanocytes to produce melanin *(see Figure 19-5).*
5. At the **median eminence,** neurons release regulatory factors into the surrounding interstitial fluids. Their secretions enter the circulation easily, since the capillaries in this area are **fenestrated** *(see Figures 19-4/19-6).*

infundibulum • superior hypophyseal artery • portal system

6. The **hypophyseal portal system** ensures that all of the blood entering the **portal vessels** will reach the intended target cells before returning to the general circulation *(see Figure 19-6).*

7. Endocrine cells in the anterior pituitary are controlled by releasing hormones, inhibiting hormones, or some combination of the two.

THE THYROID GLAND *[p. 489]*

1. The **thyroid gland** lies near the **thyroid cartilage** of the larynx and consists of two **lobes** connected by a narrow **isthmus** *(see Figure 19-7).*

Thyroid Follicles and Thyroid Hormones *[p. 489]*
2. The thyroid gland contains numerous **thyroid follicles**, which release several hormones, including **thyroxine (TX or T$_4$)** and **triiodothyronine (T$_3$)** *(see Figure 19-7b,c and Table 19-3).*
3. The follicle cells manufacture **thyroglobulin** and store it as a **colloid** (a viscous fluid containing suspended proteins). The cells also transport iodine from the extracellular fluids into the **follicle cavity**, where they complex with the thyroglobulin molecules *(see Figure 19-7b,c).*
4. When stimulated by TSH the follicle cells reabsorb the thyroglobulin, break it apart, and release the thyroid hormones into the circulation *(see Figure 19-8).*

The C Cells of the Thyroid Gland *[p. 491]*
5. The **C cells** of the follicles produce **calcitonin (CT),** which helps to lower calcium ion concentrations in body fluids *(see Figure 19-7c).*

THE PARATHYROID GLANDS *[p. 491]*

1. Four **parathyroid glands** are embedded in the posterior surface of the thyroid gland. The **principal (chief) cells** of the parathyroid produce **parathyroid hormone (PTH)** in response to lower-than-normal concentrations of calcium ions. These and the C cells of the thyroid gland maintain calcium ion levels within relatively narrow limits *(see Figure 19-9 and Table 19-3).*

THE THYMUS *[p. 492]*

1. The **thymus** produces several hormones that stimulate the development and maintenance of normal immunological defenses *(see Figure 19-1).*
2. **Thymosin** promotes the development and maturation of lymphocytes.

THE ADRENAL GLAND *[p. 492]*

1. A single **adrenal (suprarenal) gland** lies along the superior border of each kidney. Each gland is surrounded by a fibrous **capsule**. The adrenal gland can be subdivided into the superficial **adrenal cortex** and the inner **adrenal medulla** *(see Figure 19-10).*

The Adrenal Cortex *[p. 493]*
2. The adrenal cortex manufactures steroid hormones called **adrenocortical steroids (corticosteroids)**. The cortex can be subdivided into three areas. (1) The **zona glomerulosa** releases **mineralocorticoids (MC)**, principally **aldosterone**, which restrict sodium and water losses at the kidneys, sweat glands, digestive tract, and salivary glands. The zona glomerulosa responds to the presence of angiotensin II, which appears after the enzyme renin has been secreted by kidney cells exposed to a decline in blood volume and/or blood pressure. (2) The **zona fasciculata** produces **glucocorticoids (GC)**, notably **cortisol**, **corticosterone**, and **cortisone** (all of which exert a glucose-sparing effect on peripheral tissues). (3) The **zona reticularis** produces androgens of uncertain significance *(see Figure 19-10c and Table 19-4).*

The Adrenal Medulla *[p. 494]*
3. The adrenal medulla produces epinephrine (75–80%) and norepinephrine (20–25%) *(see Figure 19-10b,c and Table 19-4).*

ENDOCRINE FUNCTIONS OF THE KIDNEYS AND HEART *[p. 495]*

1. Endocrine cells in the kidneys and heart produce hormones important for the regulation of blood volume and blood pressure. **Renin** converts **angiotensinogen** to **angiotensin I**. In the lungs this compound is converted to **angiotensin II**, the hormone that stimulates the adrenal production of aldosterone.

2. The kidney hormone **erythropoietin (EPO)** stimulates red blood cell production by the bone marrow.
3. Specialized muscle cells in the heart produce **atrial natriuretic peptide (ANP)** when blood pressure or blood volume becomes excessive.

THE PANCREAS AND OTHER ENDOCRINE TISSUES OF THE DIGESTIVE SYSTEM *[p. 495]*

1. The lining of the digestive tract, the liver, and the pancreas produce exocrine secretions that are essential to the normal breakdown and absorption of food.

The Pancreas *[p. 495]*

2. The **pancreas** contains both exocrine and endocrine cells. The **exocrine pancreas** secretes an enzyme-rich fluid that travels to the lumen of the digestive tract. Cells of the **endocrine pancreas** form clusters called **pancreatic islets** (*islets of Langerhans*), containing four cell types: **alpha cells,** which produce the hormone **glucagon**; **beta cells**, which secrete **insulin**; **delta cells**, which secrete **somatostatin (growth-hormone-inhibiting factor)**; and **F-cells**, which secrete **pancreatic polypeptide** *(see Figure 19-11 and Table 19-5).*

3. Insulin lowers blood glucose by increasing the rate of glucose uptake and utilization; glucagon raises blood glucose by increasing the rates of glycogen breakdown and glucose manufacture in the liver. Somatostatin slows the rates of food absorption and secretion of enzymes along the digestive tract. F-cells enhance the production of certain digestive enzymes *(see Table 19-5).*

ENDOCRINE TISSUES OF THE REPRODUCTIVE SYSTEM *[p. 495]*

Testes *[p. 499]*
1. The **interstitial cells** of the male testis produce androgens. **Testosterone** is the most important male sex hormone *(see Table 19-6).*

Ovaries *[p. 499]*
2. In women, eggs develop in **follicles**; follicle cells surrounding the eggs produce estrogens. After ovulation, the cells reorganize into a **corpus luteum** that releases a mixture of estrogens and **progesterone** *(see Table 19-6).*

THE PINEAL GLAND *[p. 499]*

1. The **pineal gland (epiphysis cerebri)** contains **pinealocytes**, which synthesize **melatonin**. Melatonin slows the maturation of sperm, eggs, and reproductive organs and may establish circadian rhythms.

HORMONES AND AGING *[p. 500]*

1. The endocrine system shows relatively few functional changes with advanced age. The most dramatic endocrine change is the decline in the concentration of reproductive hormones.

Related Key Terms

inhibin

estradiol

---

1 REVIEW OF CHAPTER OBJECTIVES

1. Compare the endocrine and nervous systems.
2. Define a hormone.
3. Explain how hormones act to exert their effects.
4. List and describe the location, structure, and hormones secreted by the endocrine organs and tissues.
5. Briefly describe the effects of aging in relation to hormones.
6. Discuss the results of abnormal hormone production.

2 REVIEW OF CONCEPTS

1. What are the functional differences between the endocrine and the nervous system?
2. How do hormones regulate cellular activity?
3. How do the coordinating centers in the hypothalamus regulate nervous and endocrine activities?
4. How do the anterior and posterior parts of the pituitary gland differ in function?
5. What is the significance of the capillary network within the hypophysis?
6. What is the role of iodine in thyroid gland function?
7. What effect do thyroid secretions have on body tissues?
8. How would the impairment of thyroid C cell function interfere with healing and remodeling of a broken bone?
9. Why is normal parathyroid function essential in maintaining normal calcium ion levels?
10. Even though too high a level of cholesterol in the blood can cause serious or fatal health problems, too low a level can also cause problems. Why?
11. How does aldosterone affect organs throughout the body?
12. What is the role of the pancreas in the regulation of the metabolism of glucose?
13. What are the functions of testosterone?
14. What is the function of the corpus luteum?
15. What is the role of melatonin in regulating reproductive function?

3 CRITICAL THINKING AND CLINICAL APPLICATION QUESTIONS

1. A child, physically normal until the age of puberty, fails to undergo the typical growth spurt, although the body proportions are normal, and sexual maturation occurs. What could be the cause of this problem, and how can it be corrected?
2. Many people, particularly women, who had their portraits painted during the Renaissance, show a smooth, rounded swelling of the anterior portion of the neck. Although this depiction has sometimes been referred to as an artistic convention only, it is more likely indicative of a physiological condition that was evidently common during that period. What was this condition, and what was the cause?
3. A woman in her forties appears at her doctor's office complaining of swelling of the anterior portion of her throat, dry skin, loss of hair, constantly feeling cold, a decrease in normal muscular strength, and being unable to react as quickly as she normally would. What could be the problem, and how would it be treated to eliminate these symptoms?

20

The Cardiovascular System: Blood

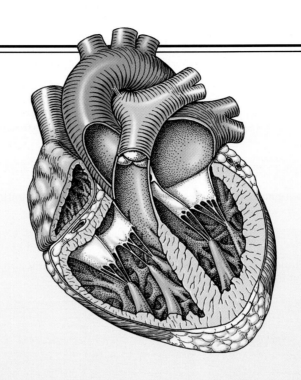

CHAPTER OUTLINE AND OBJECTIVES

The living body is in constant chemical communication with its external environment. Nutrients are absorbed through the lining of the digestive tract, gases move across the delicate epithelium of the lungs, and wastes are excreted in the feces and urine, as well as saliva, bile, sweat, and other exocrine secretions. These chemical exchanges occur at specialized sites or organs because all parts of the body are linked by a transport network, the *cardiovascular system* (*CVS*). The cardiovascular system can be compared to the cooling system of a car. The basic components include a circulating fluid (blood), a pump (the heart), and an assortment of conducting pipes (the circulatory system). The first three chapters of this unit examine those components individually: Chapter 20 discusses the nature of the circulating blood; Chapter 21 considers the structure and function of the heart; and Chapter 22 examines the network of blood vessels and the integrated functioning of the cardiovascular system. Chapter 23 details the components of the *lymphatic system*, which is intimately associated with the CVS.

FUNCTIONS OF THE BLOOD
(Table 20-1)

Blood is a specialized connective tissue that (1) distributes nutrients, oxygen, and hormones to each of the roughly 75 trillion cells in the human body, (2) carries metabolic wastes to the kidneys for excretion, and (3) transports specialized cells that defend peripheral tissues from infection and disease. Table 20-1 contains a detailed listing of the functions of the blood. These services are absolutely essential, and a region completely deprived of circulation may die in a matter of minutes.

COMPOSITION OF THE BLOOD
(Figure 20-1)

Blood is a fluid connective tissue normally confined to the circulatory system. It has a characteristic and unique composition (Figure 20-1●). Blood consists of two components:

1. **Plasma** (PLAZ-mah), the liquid matrix of blood, has a density only slightly greater than that of water. It contains dissolved proteins rather than the network of insoluble fibers found in loose connective tissues or cartilage.

2. **Formed elements** are blood cells and cell fragments that are suspended in the plasma. These elements are present in great abundance and are highly specialized. **Red blood cells (RBCs)** transport oxygen and carbon dioxide. The less numerous **white blood cells (WBCs)** are components of the immune system. **Platelets** (PLĀT-lets)

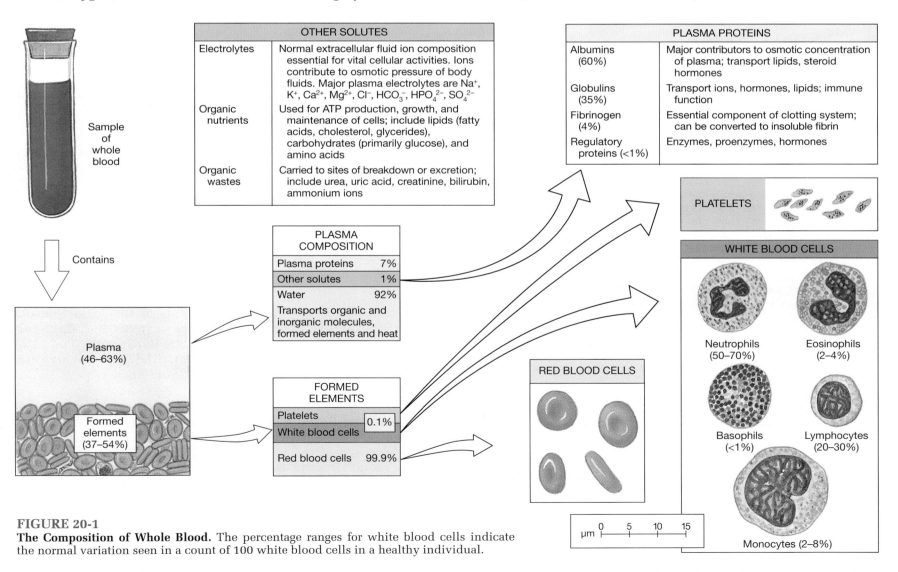

FIGURE 20-1
The Composition of Whole Blood. The percentage ranges for white blood cells indicate the normal variation seen in a count of 100 white blood cells in a healthy individual.

are small, membrane-enclosed packets of cytoplasm that contain enzymes and other *clotting factors* involved in blood coagulation.

Whole blood is a mixture of plasma and formed elements, but the components may be separated, or **fractionated**, for clinical purposes. Whole blood is sticky, cohesive, and resistant to flow, characteristics that determine the **viscosity** of a solution. Solutions are usually compared with pure water, which has a viscosity of 1.0. Plasma has a viscosity of 1.5, but the viscosity of whole blood is much greater (about 5.0) because of interactions between water molecules and the formed elements.

On average there are 5–6 liters of whole blood in the cardiovascular system of an adult man, and 4–5 liters in an adult woman. Clinicians use the terms **hypovolemic** (hī-pō-vō-LĒ-mik), **normovolemic** (nor-mō-vō-LĒ-mik), and **hypervolemic** (hī-per-vō-LĒ-mik) to refer to low, normal, or excessive blood volumes, respectively. These conditions are potentially dangerous because variations in blood volume affect other components of the cardiovascular system. For example, an abnormally large blood volume can place a severe stress on the heart (e.g., hypertension or "high blood pressure"), which must push the extra fluid around the circulatory system.

TABLE 20-1   Functions of the Blood

1. *Transportation of dissolved gases,* bringing oxygen from the lungs to the tissues and carrying carbon dioxide from the tissues to the lungs.

2. *Distribution of nutrients* absorbed at the digestive tract or released from storage in adipose tissue or the liver.

3. *Transportation of metabolic wastes* from peripheral tissues to sites of excretion, especially the kidneys.

4. *Delivery of enzymes and hormones* to specific target tissues.

5. *Stabilization of the pH and electrolyte composition of interstitial fluids* throughout the body. By absorbing, transporting, and releasing ions as it circulates, blood helps prevent regional variations in the ion concentrations of body tissues. An extensive array of buffers enables the bloodstream to deal with the acids generated by tissues, such as the lactic acid produced by skeletal muscles.

6. *Prevention of fluid losses* through damaged vessels or at other injury sites. The **clotting reaction** seals the breaks in the vessel walls, preventing changes in blood volume that could seriously affect blood pressure and cardiovascular function.

7. *Defense against toxins and pathogens.* Blood it transports *white blood cells,* specialized cells that migrate into peripheral tissues to fight infections or remove debris, and delivers *antibodies,* special proteins that attack invading organisms or foreign compounds. The blood also collects toxins, such as those produced by infection, and delivers them to the liver and kidneys, where they can be inactivated or excreted.

8. *Stabilization of body temperature* by absorbing and redistributing heat. Active skeletal muscles and other tissues generate heat, and the bloodstream carries it away. When body temperature is already high, that heat is lost across the surface of the skin. When body temperature is too low, warm blood is directed to the most temperature-sensitive organs. These changes in circulatory flow are controlled and coordinated by the *cardiovascular centers* in the medulla. ∞ [p. 389]

Plasma
(Figure 20-1, Table 20-2)

Plasma contributes approximately 55% of the volume of whole blood, and water accounts for 92% of the plasma volume. These are average values, and the actual concentrations vary depending on the area sampled and the ongoing activity within that particular region. Information concerning the composition of plasma is summarized in Figure 20-1● and Table 20-2.

Differences between Plasma and Interstitial Fluid

In many respects, the composition of the plasma resembles that of tissue fluid. The concentrations of the major plasma ions, for example, are similar to those of interstitial fluid and differ markedly from those found inside living cells. The chief differences between plasma and interstitial fluid involve the concentrations of dissolved gases and proteins.

1. **Concentrations of dissolved oxygen and carbon dioxide**: Because the dissolved oxygen concentration of plasma is higher than that of interstitial fluid, oxygen diffuses out of the bloodstream and into peripheral tissues. Carbon dioxide concentrations in the tissues are much higher than those in plasma, so carbon dioxide diffuses out of the tissues and into the bloodstream.

2. **Concentration of dissolved proteins**: The plasma contains significant quantities of dissolved proteins. The large size and globular shapes of most blood proteins pre-

TABLE 20-2   Composition of Whole Blood

Component	Significance
PLASMA	
Water	Dissolves and transports organic and inorganic molecules, distributes blood cells, and transfers heat
Electrolytes	Normal extracellular fluid ion composition essential for vital cellular activities
Nutrients	Used for energy production, growth, and maintenance of cells
Organic wastes	Travel to sites of breakdown or excretion
Proteins	
Albumins	Major contributor to osmotic concentration of plasma; transport lipids
Globulins	Transport ions, hormones, lipids
Fibrinogen	Essential component of clotting system; can be converted to insoluble fibrin
FORMED ELEMENTS	
Red blood cells	Transport oxygen and carbon dioxide
White blood cells	
Neutrophils	Phagocytic cells; engulf debris and pathogens
Eosinophils	Attack antigens, such as foreign proteins or pathogens, that are coated in antibodies, primarily through release of toxic chemicals; suppress inflammation
Basophils	Stimulate inflammation in tissues by releasing histamine
Monocytes	In tissues these become phagocytic cells that engulf debris and pathogens
Lymphocytes	Immune defense against specific pathogens, toxins, or foreign proteins
Platelets	Participate in clotting response

vent them from crossing capillary walls, and they remain trapped within the circulatory system.

The Plasma Proteins

One hundred milliliters of human plasma normally contains 6–7.8 g of soluble proteins. There are three primary classes of plasma proteins: *albumins* (al-BŪ-minz), *globulins* (GLOB-ū-linz), and *fibrinogen* (fī-BRIN-ō-jen).

1.   **Albumins** constitute roughly 60% of the plasma proteins. As the most abundant proteins, they are major contributors to the osmotic pressure of the plasma. They are also important in the transport of fatty acids, steroid hormones, and other substances. Albumins are the smallest of the major plasma proteins; they are cigar-shaped, and roughly 10 nm long.

2.   **Globulins** account for approximately 35% of the plasma protein population. Globulins include *imunoglobulins* and *transport proteins*. **Immunoglobulins** (i-mū-nō-GLOB-ū-linz), also called **antibodies**, attack foreign proteins and pathogens. **Transport proteins** bind small ions, hormones, or compounds that are either insoluble or that might be filtered out of the blood at the kidneys.

3.   **Fibrinogen,** which accounts for roughly 4% of the plasma proteins, functions in the clotting reaction. Under certain conditions fibrinogen molecules interact, forming large, insoluble strands of *fibrin* (FĪ-brin). These fibers pro-

vide the basic framework for a blood clot. If steps are not taken to prevent clotting, the conversion of fibrinogen to fibrin will occur in a sample of plasma. This conversion removes the clotting proteins, leaving a fluid known as **serum.** Fibrinogen is the largest of the plasma proteins. Each molecule is long and slender, roughly 70 nm × 4 nm. It would take over 100 fibrinogen molecules placed end to end to equal the diameter of a single red blood cell.

Both albumins and globulins can become attached to lipids, such as triglycerides, fatty acids, or cholesterol, that are not water-soluble. These protein-lipid combinations, called **lipoproteins** (lī-pō-PRŌ-tēnz), readily dissolve in plasma, and in this way the circulatory system transports insoluble lipids to peripheral tissues. A representative lipoprotein would be a sphere roughly 20 nm in diameter.

The liver synthesizes and releases more than 90% of the plasma proteins. Because the liver is the primary source of plasma proteins, liver disorders can alter the composition and functional properties of the blood. For example, some forms of liver disease can lead to uncontrolled bleeding, caused by inadequate synthesis of fibrinogen and other plasma proteins involved in the clotting response.

■ CLINICAL BRIEF
Plasma Expanders

Plasma expanders are solutions that can be used to increase blood volume temporarily, over a period of hours, while preparing for a transfusion of whole blood. Plasma expanders contain large carbohydrate molecules, rather than dissolved proteins, to maintain proper osmolarity. Although these carbohydrates are not metabolized, they are gradually removed from circulation by phagocytes, and the blood volume steadily declines. Plasma expanders are easily stored, and their sterile preparation ensures that there are no problems with viral or bacterial contamination. Although they provide a temporary solution to hypovolemia (low blood volume), plasma expanders fail to increase the amount of oxygen delivered to peripheral tissues.

FORMED ELEMENTS
(Table 20-3)

The major cellular components of blood are *red blood cells* and *white blood cells.* ∞ [p. 70] There are two major classes of white blood cells: *granular* (with granules) and *agranular* (without granules). In addition, blood contains the noncellular formed elements called *platelets*, small membrane-enclosed packets of cytoplasm that function in the clotting response. Table 20-3 summarizes information concerning the formed elements of the blood. The sections that follow will consider each of these components individually.

Red Blood Cells (RBCs)

Red blood cells (RBCs), or **erythrocytes** (e-RITH-rō-sīts; *erythros*, red) account for slightly less than half of the total blood volume. The **hematocrit** (hē-MA-tō-krit) value indicates the percentage of whole blood occupied by cellular elements. The normal hematocrit in adult men averages 46 (range: 40–54); the average for adult women is 42 (range: 37–47). Because whole blood contains roughly 1000 red blood cells for each white blood cell, the hematocrit closely approximates the vol-

TABLE 20-3   A Review of the Formed Elements of the Blood

Cell	Abundance (per μℓ)ᵃ	Characteristics	Functions	Remarks
RED BLOOD CELLS	5.2 million (range: 4.4–6.0 million)	Biconcave disc without a nucleus, mitochondria, or ribosomes; red color due to presence of hemoglobin molecules	Transport oxygen from lungs to tissues, and carbon dioxide from tissues to lungs	120-day life expectancy; amino acids and iron recycled; produced in bone marrow
WHITE BLOOD CELLS	7000 (range: 6000–9000)			
Granulocytes				
Neutrophils	4150 (range: 1800–7300) differential count: 57%	Round cell, nucleus resembles a series of beads; cytoplasm contains large, pale inclusions	Phagocytic; engulf pathogens or debris in tissues	Survive minutes to days, depending on activity; produced in bone marrow
Eosinophils	165 (range: 0–700) differential count: 2.4%	Round cell, nucleus usually in two lobes; cytoplasm contains large granules that stain bright orange-red with acid dyes	Attack anything that is labeled with antibodies; important in fighting parasitic infections; suppress inflammation	Produced in bone marrow
Basophils	44 (range: 0–150) differential count: 0.6%	Round cell, nucleus usually cannot be seen, because of dense, purple-blue granules in cytoplasm	Enter damaged tissues and release histamine and other chemicals	Assist mast cells of tissues in producing inflammation; produced in bone marrow
Agranulocytes				
Monocytes	456 (range: 200–950) differential count: 6.5%	Very large, kidney bean-shaped nucleus, abundant pale cytoplasm	Enter tissues to become free macrophages; engulf pathogens or debris	Primarily produced in bone marrow
Lymphocytes	2185 (range: 1500–4000) differential count: 31%	Slightly larger than RBC, round nucleus, very little cytoplasm	Cells of lymphatic system, providing defense against specific pathogens or toxins	T cells attack directly; B cells form plasma cells that produce antibodies; produced in bone marrow and lymphatic tissues
Platelets	350,000 (range: 150,000–500,000)	Cytoplasmic fragments, contain enzymes and proenzymes; no nucleus	Hemostasis: clump together and stick to vessel wall (platelet phase); activate intrinsic pathway of coagulation phase	Produced by megakaryocytes in bone marrow

ᵃDifferential count: percentage of circulating white blood cells.

■CLINICAL BRIEF
Anemia and Polycythemia

Anemia (a-NĒ-mē-a) exists when the oxygen-carrying capacity of the blood is reduced, diminishing the delivery of oxygen to peripheral tissues. Such a reduction causes a variety of symptoms, including premature muscle fatigue, weakness, lethargy, and a general lack of energy. Anemia may exist because the hematocrit is abnormally low or because the amount of hemoglobin in the RBCs is reduced. Standard laboratory tests can be used to differentiate between the various forms of anemia on the basis of the number, size, shape, and hemoglobin content of red blood cells. For a more detailed discussion of blood tests, see the Clinical Issues appendix. ⫙ *Blood Tests and RBCs* [p. 764]

An elevated hematocrit with a normal blood volume constitutes **polycythemia** (po-lē-sī-THĒ-mē-ah). There are several different types of polycythemia. *Erythrocytosis* (e-rith-rō-sī-TŌ-sis), a polycythemia affecting only red blood cells, will be considered later in the chapter. **Polycythemia vera** ("true polycythemia") results from an increase in the numbers of all blood cells. The hematocrit may reach 80–90, at which point the tissues become oxygen-starved because red blood cells are blocking the smaller vessels. This condition seldom strikes young people; most cases involve persons aged 60–80. There are several treatment options, but none cures the condition. The cause of polycythemia vera is unknown, although there is some evidence that the condition is linked to radiation exposure.

ume of erythrocytes. As a result, hematocrit values are often reported as the **volume of packed red cells (VPRC)** or simply the **packed cell volume (PCV)**.

The number of erythrocytes in the blood of a normal individual staggers the imagination. One microliter, or cubic millimeter (mm³), of whole blood from a man contains roughly 5.4 million erythrocytes; a microliter of blood from a woman contains about 4.8 million erythrocytes. There are approximately 260 million red blood cells in a single drop of whole blood, and 25 trillion (2.5×10^{13}) RBCs in the blood of an average adult.

Structure of RBCs
(Figure 20-2)

Erythrocytes transport oxygen and carbon dioxide within the bloodstream. They are among the most specialized cells of the body, and their anatomical specializations are apparent when RBCs are compared with "typical" body cells. (Refer to Figure 2-1.) ∞ [p. 26] Figure 20-2a,c● indicates the significant differences seen using light and electron microscopy. Each red blood cell has a thin central region and a thick outer margin. A typi-

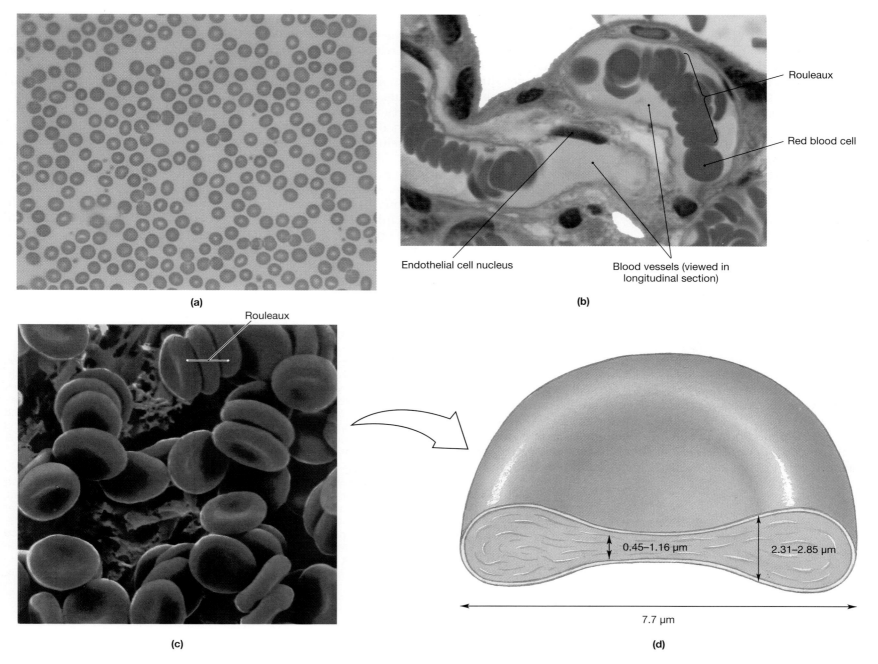

FIGURE 20-2
Anatomy of Red Blood Cells. (a) When viewed in a standard blood smear, red blood cells appear as two-dimensional objects because they are flattened against the surface of the slide. (LM × 477) (b) When traveling through relatively narrow capillaries, erythrocytes may stack like dinner plates, forming rouleaux. (LM × 1430) (c) A scanning electron micrograph of red blood cells reveals their three-dimensional structure quite clearly. (SEM × 1838) (d) A sectional view of a red blood cell.

cal erythrocyte has a diameter of 7.7 μm and a maximum thickness of 2.6 μm, although the center narrows to about 0.8 μm.

This unusual shape gives each RBC a relatively large surface area, allowing rapid diffusion between the cytoplasm and surrounding plasma. The total surface area of the red blood cells in the blood of a typical adult is roughly 3800 m², 2000 times the total surface area of the body. The flattened shape also enables them to form stacks, like dinner plates. These stacks, called **rouleaux** (roo-LŌ; "little rolls," singular *rouleau*), form and dissociate repeatedly without affecting the cells involved. An entire rouleau can pass along a blood vessel (Figure 20-2b●) little larger than the diameter of a single erythrocyte, whereas individual cells would bump the walls, band together, and form log jams that could block the circulatory passageway. Finally, the slender profile of an erythrocyte gives the cell considerable

flexibility, and erythrocytes can bend and flex with apparent ease. By changing their shape, individual red blood cells can squeeze through capillaries as seen in Figure 20-2b●.

RBC Life Span and Circulation

During their differentiation and maturation, red blood cells lose most of their organelles, retaining only the cytoskeletal elements. As a result, circulating RBCs lack mitochondria, ribosomes, and nuclei. (The process of RBC formation will be described in a later section.) Without mitochondria, these cells can obtain energy only through anaerobic metabolism, and they rely on glucose obtained from the surrounding plasma. This mechanism ensures that absorbed oxygen will be carried to peripheral tissues, not "stolen" by mitochondria in the RBCs. Without a nucleus or ribosomes, protein synthesis cannot occur, so as cellular enzymes and structural proteins age and deteriorate, they cannot be replaced.

An erythrocyte is exposed to severe stresses. A single circuit of the circulatory system usually takes less than 30 seconds. In that time it stacks in rouleaux, contorts and squeezes through tiny capillaries, and then joins its comrades in a headlong rush back to the heart for another round. With all this wear and tear and no repair mechanisms, a typical red blood cell has a relatively short life span of about 120 days. After traveling about 700 miles in 120 days, either the cell membrane ruptures or the aged cell is detected and destroyed by phagocytic cells. About 1% of the circulating erythrocytes are replaced each day, and in the process approximately 3 million new erythrocytes enter the circulation *each second*!

RBCs and Hemoglobin
(Figure 20-3)

A developing erythrocyte loses all intracellular components not directly associated with its primary functions: the transport of oxygen and carbon dioxide. A mature red blood cell consists of a cell membrane surrounding cytoplasm containing water (66%) and proteins (about 33%). Molecules of **hemoglobin** (HĒ-mō-glō-bin) (**Hb**) account for over 95% of the erythrocyte's proteins. Hemoglobin is responsible for the cell's ability to transport oxygen and carbon dioxide. Hemoglobin is a red pigment, and its presence gives blood its characteristic red color.

Four globular protein subunits combine to form a single molecule of hemoglobin. Each of the subunits contains a single molecule of **heme** (Figure 20-3●). Each heme unit holds an iron ion in

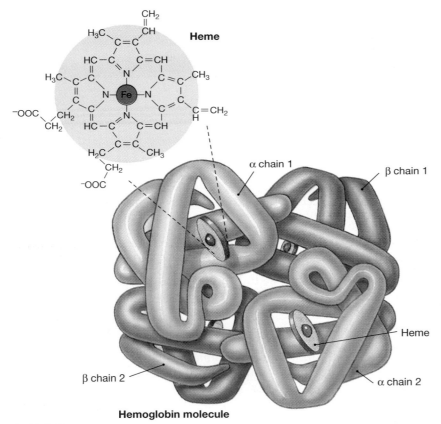

Hemoglobin molecule

FIGURE 20-3
The Structure of Hemoglobin. Hemoglobin consists of four globular protein subunits. Each subunit contains a single molecule of heme, a porphyrin ring surrounding a single ion of iron. It is the iron ion that reversibly binds to an oxygen molecule.

such a way that it can interact with an oxygen molecule. The iron-oxygen interaction is very weak, and the two can be easily separated without damage to either the hemoglobin or the oxygen molecule. There are approximately 280 million molecules of hemoglobin in each red blood cell, and because a hemoglobin molecule contains four heme units, each erythrocyte can potentially carry more than a billion molecules of oxygen. The binding of carbon dioxide to the globin portion of the molecule is just as reversible as the binding of oxygen to heme.

As red blood cells circulate through capillaries in the lungs, diffusion brings oxygen into the plasma and removes carbon dioxide. The hemoglobin molecules in red blood cells respond by absorbing oxygen and releasing carbon dioxide. In the peripheral tissues, the situation is reversed because active cells are consuming oxygen and producing carbon dioxide. As blood flows through these tissues, oxygen diffuses out of the blood and carbon dioxide diffuses in. ✝ *Sickle Cell Anemia [p. 766]*

■ CLINICAL BRIEF
Erythrocytosis

In **erythrocytosis** the blood contains abnormally large numbers of red blood cells. Erythrocytosis usually results from the massive release of erythropoietin by tissues deprived of oxygen. People moving to high altitudes usually experience erythrocytosis after their arrival, because the air contains less oxygen than it does at sea level. The increased number of red blood cells compensates for the fact that individually each RBC is carrying less oxygen than it would at sea level. Mountaineers and those living at altitudes of 10,000–12,000 feet may have hematocrits as high as 65.

Individuals whose hearts or lungs are functioning inadequately may also develop erythrocytosis. For example, this condition is often seen in heart failure and emphysema, two conditions discussed in later chapters. Whether the blood fails to circulate efficiently or the lungs do not deliver enough oxygen to the blood, peripheral tissues remain oxygen-poor despite the rising hematocrit. Having a higher concentration of red blood cells increases the oxygen-carrying capacity of the blood, but it also makes the blood thicker and harder to push around the circulatory system. This change increases the workload on the heart, making a bad situation even worse.

Blood Types
(Figure 20-4)

An individual's **blood type** is determined by the presence or absence of specific components in erythrocyte cell membranes. The cell membrane of a typical red blood cell contains a number of **agglutinogens** (a-gloo-TIN-ō-jenz), or *surface antigens*, exposed to the plasma. Agglutinogens are glycoproteins or glycolipids whose characteristics are genetically determined. There are at least 50 different kinds of agglutinogens on the surfaces of RBCs. Three of particular importance have been designated agglutinogens **A**, **B**, and **D (Rh)**.

The red blood cells of a particular individual may have any combination of these agglutinogens on their surfaces (Figure 20-4●). For example, **Type A** blood has agglutinogen A, **Type B** has agglutinogen B, **Type AB** has both, and **Type O** has neither. The average values for the U.S. population are: Type O 46%, Type A 40%, Type B 10%, and Type AB 4%. These values may differ among different racial and ethnic groups (Table 20-4).

The presence of the Rh agglutinogen, sometimes called the *Rh factor*, is indicated by the terms **Rh-positive** (present) or **Rh-negative** (absent). In recording the complete blood type, the term *Rh* is usually omitted, and the data are reported as O-negative, A-positive, and so forth.

TYPE A	TYPE B	TYPE AB	TYPE O
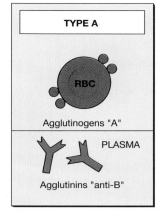	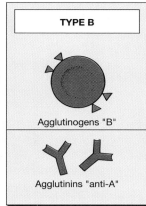	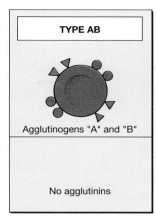	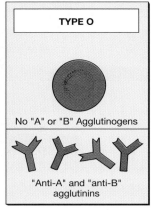
Agglutinogens "A"	Agglutinogens "B"	Agglutinogens "A" and "B"	No "A" or "B" Agglutinogens
PLASMA			
Agglutinins "anti-B"	Agglutinins "anti-A"	No agglutinins	"Anti-A" and "anti-B" agglutinins

FIGURE 20-4

Blood Typing. The blood type depends on the presence of agglutinogens on RBC surfaces. The plasma contains agglutinins, antibodies that will react with foreign agglutinogens. The relative frequencies of each blood type in the U.S. population are indicated in Table 20-4.

■ CLINICAL BRIEF
Elevating Hemoglobin Levels

Genetic engineering procedures are now being used to synthesize one of the subunits of normal human hemoglobin, which can be introduced into the circulation to increase oxygen transport and total blood volume. An alternative strategy involves removing hemoglobin molecules from red cells and attaching them to inert carrier molecules that will prevent their loss at the kidneys. Still a third approach draws on natural but nonhuman sources. The FDA recently approved testing of a blood substitute called *Hemopure* that contains purified hemoglobin from cattle. Because the hemoglobin is removed from erythrocytes and infused without plasma, *cross-reactions* are not expected to occur.

TABLE 20-4 Differences in Blood Group Distribution

Population	Percentage with Each Blood Type				
	O	A	B	AB	Rh⁺
U.S. (average)	46	40	10	4	85
Caucasian	45	40	11	4	85
African-American	49	27	20	4	95
Chinese	42	27	25	6	100
Japanese	31	39	21	10	100
Korean	32	28	30	10	100
Filipino	44	22	29	6	100
Hawaiian	46	46	5	3	100
Native North American	79	16	4	<1	100
Native South American	100	0	0	0	100
Australian Aborigines	44	56	0	0	100

AGGLUTININS AND CROSS-REACTIONS *(Figure 20-4).* You probably know that your blood type must be checked before you give or receive blood. The agglutinogens on your own red blood cells are ignored by your immune system. (This ability to recognize normal body cells will be examined in Chapter 23.) However, your plasma contains immunoglobulins called **agglutinins** (a-GLOO-ti-ninz) that will attack "foreign" agglutinogens. For example, if you have Type A blood, your plasma holds circulating anti-B agglutinins that will attack Type B erythrocytes (Figure 20-4●). If you are Type B, your plasma contains anti-A agglutinins. The red blood cells of an individual with Type O blood lack A and B agglutinogens, so the plasma of such an individual contains both anti-A and anti-B agglutinins. At the other extreme, Type AB individuals lack anti-A and anti-B agglutinins.

The blood of a person with Type A, Type B, or Type O agglutinogens always contains complementary agglutinins. Even if a Type A person has never been exposed to Type B blood, the individual will still have anti-B agglutinins in the plasma. In contrast, the plasma of an Rh-negative individual does not always contain anti-Rh agglutinins. These agglutinins are present only if the individual has been **sensitized** by previous exposure to Rh-positive erythrocytes. Such exposure may occur accidentally, during a transfusion, but it may also accompany a seemingly normal pregnancy involving an Rh-negative mother and an Rh-positive fetus. ✝ *Hemolytic Disease of the Newborn [p. 766]*

When an agglutinin meets its specific agglutinogen, a **cross-reaction** occurs. Initially the red blood cells clump together, a process called **agglutination** (a-gloo-ti-NĀ-shun), and they may also **hemolyze**, or rupture. Clumps and fragments of red blood cells under attack form drifting masses that can plug small vessels in the kidneys, lungs, heart, or brain, damaging or destroying the tissues deprived of circulation. Such reactions can be avoided by ensuring that the blood types of the donor and the recipient are **compatible**. In practice, this procedure involves choosing a donor whose blood cells will not undergo cross-reaction with the plasma of the recipient.

■ CLINICAL BRIEF
Testing for Compatibility

Testing for compatibility normally involves two steps: a determination of blood type and a cross-match test. At least 50 agglutinogens have been identified on red cell surfaces, but the standard test for blood type categorizes a blood sample on the basis of the three most likely to produce dangerous cross-reactions. The test involves taking drops of blood and mixing them with solutions containing anti-A, anti-B, *and* anti-Rh (anti-D) agglutinins. Any cross-reactions are then recorded. For example, if the red blood cells clump together when exposed to anti-A *and* anti-B, the individual has Type AB blood. If no reactions occur, the person must be Type O. The presence or absence of the Rh agglutinogen is also noted, and the individual is classified as Rh-positive or Rh-negative on that basis. Type O-positive is the most common blood type. The red blood cells of these individuals do not have agglutinogens A and B, but they do have agglutinogen D.

Standard blood typing can be completed in a matter of minutes, and Type O-negative blood can be safely administered in an emergency. However, with at least 48 other possible agglutinogens on the cell surface, cross-reactions can occur, even to Type O blood. Whenever time and facilities permit, further testing is performed to ensure complete compatibility.

Cross-match testing involves exposing the donor's red blood cells to a sample of the recipient's plasma under controlled conditions. This procedure reveals the presence of significant cross-reactions involving other agglutinogens and agglutinins.

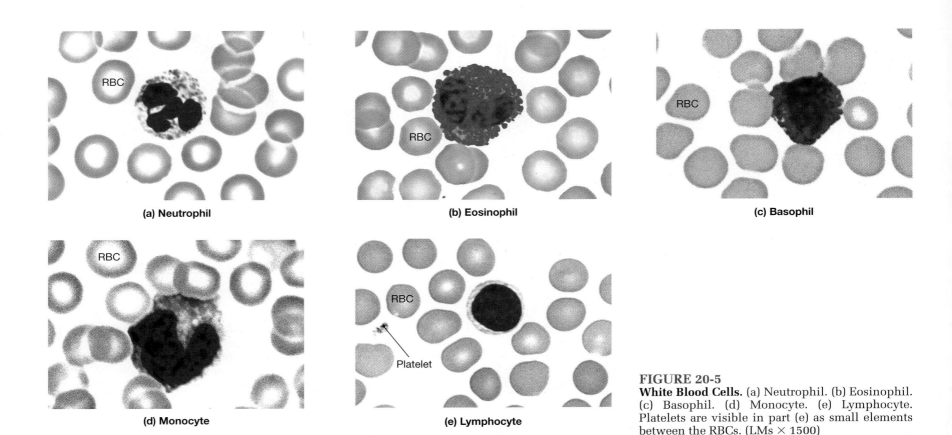

(a) Neutrophil

(b) Eosinophil

(c) Basophil

(d) Monocyte

(e) Lymphocyte

Platelet

RBC

FIGURE 20-5
White Blood Cells. (a) Neutrophil. (b) Eosinophil. (c) Basophil. (d) Monocyte. (e) Lymphocyte. Platelets are visible in part (e) as small elements between the RBCs. (LMs × 1500)

White Blood Cells (WBCs)
(Figure 20-5)

White blood cells (WBCs), or **leukocytes** (LOO-kō-sīts; *leukos,* white), are scattered throughout peripheral tissues. Most of the white blood cells in the body are found in peripheral tissues, and circulating leukocytes represent only a small fraction of the total population. White blood cells help defend the body against invasion by pathogens and remove toxins, wastes, and abnormal or damaged cells. WBCs contain nuclei of characteristic sizes and shapes (Figure 20-5●). Most white blood cells are as large as or larger than RBCs. There are two major classes of white blood cells: (1) **granular leukocytes**, or **granulocytes** (GRAN-ū-lō-sīts), which have large granular inclusions in their cytoplasm, and (2) **agranular leukocytes**, or **agranulocytes**, which do not possess the comparable cytoplasmic granules. Typical granular and agranular leukocytes are shown in Figure 20-5●.

A typical microliter of blood contains 6000–9000 leukocytes. The term **leukopenia** (loo-kō-PĒ-nē-a; *penia,* poverty) indicates inadequate numbers of white blood cells; a count of below 2500 per μl usually indicates a serious disorder. **Leukocytosis** (loo-kō-sī-TŌ-sis) refers to excessive numbers of white blood cells; a count of over 30,000 per μl usually indicates a serious disorder. A stained blood smear provides a **differential count** of the white blood cell population. The values obtained indicate the number of each type of cell encountered in a sample of 100 white blood cells. The normal range for each cell type is indicated in Table 20-3. The endings *-penia* and *-osis* can also be used to indicate low or high numbers, respectively, of specific types of white blood cells. For example, *lymphopenia* means too few lymphocytes, and *lymphocytosis* means an unusually high number.

Leukocytes have a very short life span, typically only a few days. The bloodstream provides rapid transportation of WBCs to areas of invasion or injury, where they are attracted to the chemical signs of inflammation or infection in the adjacent interstitial fluids. WBCs are attracted to specific chemical stimuli, and this **chemotaxis** (kē-mō-TAK-sis) draws them to invading pathogens, damaged tissues, and white blood cells already in damaged tissues. A leukocyte can migrate across the endothelial lining of a capillary by squeezing between adjacent endothelial cells, a process known as **diapedesis** (dī-a-ped-Ē-sis).

Granular Leukocytes

Granular leukocytes are subdivided into *neutrophils, eosinophils,* and *basophils* on the basis of their staining characteristics. Neutrophils and eosinophils are important phagocytic cells that participate in the immune response. Neutrophils and eosinophils are sometimes called *microphages,* so that they will not be confused with the monocytes of the blood and the fixed and free macrophages, phagocytes found in peripheral tissues.

NEUTROPHILS *(Figure 20-5a).* Up to 70% of the circulating white blood cells are **neutrophils** (NOO-trō-fils). A mature neutrophil (Figure 20-5a●) has a diameter of 12–15 μm, nearly twice that of a red blood cell. A neutrophil has a very dense, contorted nucleus that may be condensed into a series of lobes like beads on a string. This attribute has given these cells another name, **polymorphonuclear leukocytes** (pol-ē-mor-fō-NŪ-klē-ar; *poly,* many + *morphe,* form), or **PMNs.** Their cytoplasm is packed with pale granules containing lysosomal enzymes and bactericidal (bacteria-killing) compounds.

512

Neutrophils are highly mobile and are usually the first of the WBCs to arrive at an injury site. They are very active phagocytes, specializing in attacking and digesting bacteria. Neutrophils usually have a short life span, surviving for about 12 hours. After actively engulfing debris or pathogens, a neutrophil dies, but its breakdown releases chemicals that attract other neutrophils to the site.

EOSINOPHILS *(Figure 20-5b).* **Eosinophils** (ē-ō-SIN-ō-fils), also called **acidophils** (a-SID-ō-fils), are so named because their granules stain with *eosin*, an acidic red dye. Eosinophils represent 2–4% of the circulating WBCs. These cells are similar in size to neutrophils, but the combination of deep red granules and a bilobed (two-lobed) nucleus makes an eosinophil relatively easy to identify (Figure 20-5b●). Eosinophils are phagocytic cells attracted to foreign compounds that have reacted with circulating antibodies. Eosinophil numbers increase dramatically during an allergic reaction or a parasitic infection. Eosinophils are also attracted to injury sites, where they release enzymes that reduce the degree of inflammation and control its spread to adjacent tissues.

BASOPHILS *(Figure 20-5c).* **Basophils** (BĀ-sō-fils) have numerous granules that stain with basic dyes, and in a standard blood smear these inclusions are a deep purple or blue (Figure 20-5c●). They are relatively rare, accounting for less than 1% of the leukocyte population. Basophils migrate to sites of injury and cross the capillary endothelium to accumulate within the damaged tissues, where they discharge their granules into the interstitial fluids. The granules contain histamine, and its release exaggerates the inflammation response at the injury site. Similar compounds are produced by the mast cells of damaged connective tissues. Other chemicals released by stimulated basophils attract eosinophils and other basophils to the area.

Agranular Leukocytes

Circulating blood contains two types of agranular leukocytes: *monocytes* and *lymphocytes*.

MONOCYTES *(Figure 20-5d).* **Monocytes** (MON-ō-sīts) are 16–20 μm in diameter, 2–3 times the diameter of a typical RBC. These cells account for 2–8% of the WBC population. They are normally almost spherical, and when flattened in a blood smear they appear even larger. Their size makes monocytes relatively easy to identify. Each cell has a large oval or kidney bean-shaped nucleus (Figure 20-5d●). Monocytes remain in circulation for just a few days before entering peripheral tissues. Once outside the bloodstream, monocytes are called *free macrophages*, to distinguish them from the immobile *fixed macrophages* found in many connective tissues. ∞ [p. 66] Free macrophages are highly mobile, phagocytic cells. They usually arrive at the injury site almost immediately after the incident. While phagocytizing, they release chemicals that attract and stimulate other monocytes and other phagocytic cells. Free macrophages also secrete substances that lure fibroblasts into the region. The fibroblasts then begin producing a dense network of collagen fibers around the site. This *scar tissue* may eventually wall off the injured area. Monocytes are one component of the **monocyte-macrophage system** that includes related cell types, such as the microglia of the CNS, Langerhans cells of the skin, and phagocytic cells in the liver, spleen, and lymph nodes. This system will be discussed further in Chapter 23.

LYMPHOCYTES *(Figure 20-5e).* Typical **lymphocytes** (LIM-fō-sīts) have very little cytoplasm, just a thin halo around a relatively large, round, purple-staining nucleus, and they lack granules in their cytoplasm (Figure 20-5e●). Lymphocytes are usually slightly larger than RBCs and account for 20–30% of the WBC population. Blood lymphocytes represent a minute segment of the entire lymphocyte population, for lymphocytes are the primary cells of the **lymphatic system**, a network of special vessels and organs distinct from, but connected to those of the circulatory system.

Lymphocytes are the cells responsible for *specific immunity*: the ability of the body to mount a counterattack against invading pathogens or foreign proteins *on an individual basis*. Lymphocytes respond to such threats in three ways. Lymphocytes called **T cells** enter peripheral tissues and attack foreign cells directly. Another group of lymphocytes, the **B cells**, differentiate into plasma cells that secrete antibodies that attack foreign cells or proteins in distant portions of the body. T cells and B cells cannot be distinguished with the light microscope. **NK cells,** sometimes known as *large granular lymphocytes,* are responsible for *immune surveillance,* the destruction of abnormal tissue cells. These cells are important in preventing cancer. (The lymphatic system and immunity are discussed in Chapter 23.) ⟨ *The Leukemias [p. 766]*

Platelets
(Figures 20-5e/20-6/20-7)

Platelets are flattened discs, round when viewed from above (Figure 20-5e●) and spindle-shaped in section. Platelets were once thought to be cells that had lost their nuclei, because similar functions in vertebrates other than mammals are performed by small nucleated blood cells. Histologists called all of these cells **thrombocytes** (THROM-bō-sīts; *thrombos,* clot). The term is still in use, although in mammals the term *platelet* is more suitable because these are membrane-enclosed enzyme packets, not individual cells.

Normal bone marrow contains a number of very unusual cells, called **megakaryocytes** (meg-a-KAR-ē-ō-sīts; *mega-,* big + *karyon,* nucleus + *-cyte,* cell). As the name suggests, these are enormous cells (up to 160 μm in diameter) with large nuclei. The dense nucleus may be lobed or ring-shaped, and the surrounding cytoplasm contains Golgi apparatus, ribosomes, and mitochondria in abundance. The cell membrane communicates with an extensive membrane network that radiates throughout the peripheral cytoplasm (Figure 20-6●).

During their development and growth, megakaryocytes manufacture structural proteins, enzymes, and membranes. They then begin shedding cytoplasm in small membrane-enclosed packets. These packets are the platelets that enter the circulation. A mature megakaryocyte gradually loses all of its cytoplasm, producing around 4000 platelets before the nucleus is engulfed by phagocytes and broken down for recycling.

Platelets are continually replaced, and an individual platelet circulates for 10–12 days before being removed by

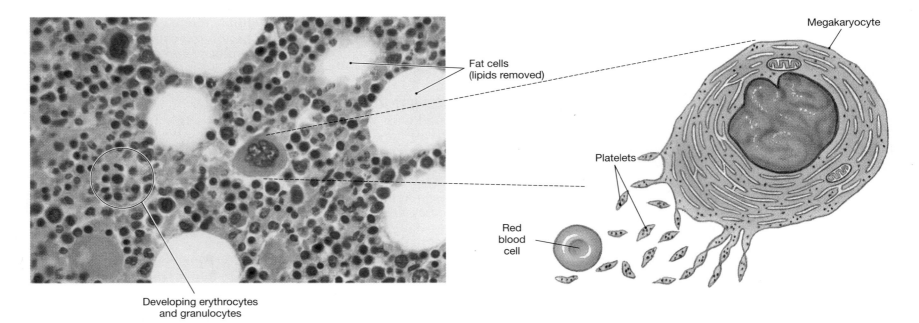

Fat cells
(lipids removed)

Megakaryocyte

Platelets

Red
blood
cell

Developing erythrocytes
and granulocytes

FIGURE 20-6
Megakaryocytes and Platelet Formation. Megakaryocytes stand out in bone marrow sections because of their enormous size and the unusual shape of their nuclei. These cells are continually shedding chunks of cytoplasm that enter the circulation as platelets. (LM × 673)

phagocytes. A microliter of circulating blood contains an average of 350,000 platelets. An abnormally low platelet count (80,000 per µℓ or less) is known as **thrombocytopenia** (throm-bō-sī-tō-PĒ-nē-a) and indicates excessive platelet destruction or inadequate platelet production. Symptoms include bleeding along the digestive tract, within the skin, and occasionally inside the CNS. Platelet counts in **thrombocytosis** (throm-bō-sī-TŌ-sis) may exceed 1,000,000 per µℓ, which usually results from accelerated platelet formation in response to infection, inflammation, or cancer.

Platelets are one participant in a vascular *clotting system* that also includes plasma proteins and the cells and tissues of the circulatory network. The process of **hemostasis** (*haima*, blood + *stasis*, halt) prevents the loss of blood through the walls of damaged vessels. In doing so it not only restricts blood loss but establishes a framework for tissue repairs. A portion of a blood clot is shown in Figure 20-7●.

Hemostasis involves a complex chain of events, and a disorder that affects a single step can disrupt the entire process. In addition, there are general requirements—a deficiency of calcium ions or vitamin K will interfere with virtually all aspects of hemostasis.

The functions of platelets include:

1. *Transport of chemicals important to the clotting process.* By releasing enzymes and other factors at the appropriate times, platelets help initiate and control the clotting process.

2. *Formation of a temporary patch in the walls of damaged blood vessels.* Platelets clump together at an injury site, forming a *platelet plug* that can slow the rate of blood loss while clotting occurs.

3. *Active contraction after clot formation has occurred.* Platelets contain filaments of actin and myosin. After a blood clot has formed, the contraction of platelets re-

duces the size of the clot and pulls together the cut edges of the vessel wall. T *Problems with the Clotting Process [p. 767]*

√ **Why can't a person with Type A blood receive blood from a person with Type B blood?**

√ **What type of white blood cell would you expect to find in the greatest numbers in an infected cut?**

FIGURE 20-7
Structure of a Blood Clot. A colorized scanning electron micrograph showing the network of fibers that forms the framework of the clot. Red blood cells trapped in the clot add to its mass and give it a red color. (SEM × 4625)

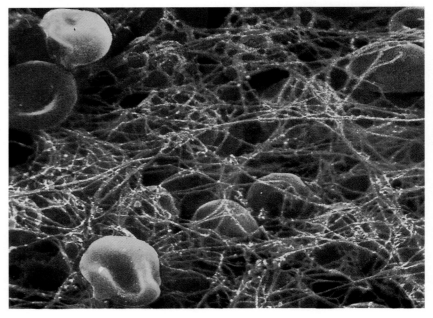

HEMOPOIESIS
(Figure 20-8)

The process of blood cell formation is called **hemopoiesis** (hēm-ō-poi-Ē-sis). Blood cells appear in the circulation during the third week of embryonic development. These cells divide repeatedly, increasing their numbers. As other organ systems appear, some of the embryonic blood cells move out of the circulation and into the liver, spleen, thymus, and bone marrow (Figure 20-8●). These embryonic cells differentiate into **stem cells**, which produce blood cells by their divisions. As the skeleton enlarges, the bone marrow becomes increasingly important; in the adult it is the primary site of blood cell formation.

Stem cells called **hemocytoblasts** produce all of the blood cells, but the process occurs in a series of steps. Hemocytoblast divisions produce at least four different types of stem cells with relatively restricted functions. These cells, called **progenitor cells**, remain capable of division, but their daughter cells will differentiate only into specific types of blood cells. For example, one type of progenitor cell produces daughter cells that mature into red blood cells; another gives rise to granulocytes or monocytes. Figure 20-8● includes important details concerning the formation of the various cellular components of the blood.

Erythropoiesis
(Figure 20-8)

Erythropoiesis (e-rith-rō-poi-Ē-sis) refers specifically to the formation of erythrocytes. The bone marrow, or **myeloid tissue** (MĪ-e-loyd; *myelos*, marrow), is the primary site of blood cell formation in the adult. The types of marrow were introduced in Chapter 5. ∞ [p. 112] Blood cells are produced in areas of **red marrow**. Red marrow is found in portions of the vertebrae, sternum, ribs, skull, scapulae, pelvis, and prox-

imal limb bones. Under extreme conditions the fatty **yellow marrow** found in other bones can be converted to red marrow. For example, this conversion may occur after a severe and sustained blood loss, thereby increasing the rate of red blood cell formation. For erythropoiesis to proceed normally, the myeloid tissues must receive adequate supplies of amino acids, iron, and **vitamin B$_{12}$**, a vitamin obtained from dairy products and meat.

Erythropoiesis is regulated by the hormone **erythropoiesis-stimulating hormone**, or **erythropoietin (EPO)**, introduced in Chapter 19. ∞ [p. 495] Erythropoietin is produced and secreted under hypoxic (low-oxygen) conditions, primarily in the kidneys. Erythropoietin has two major effects (Figure 20-8●):

- It stimulates increased rates of cell division in erythroblasts and in the stem cells that produce erythroblasts; and

- It speeds up the maturation of RBCs, primarily by accelerating the rate of hemoglobin synthesis. Under maximum EPO stimulation, the bone marrow can increase the rate of red blood cell formation tenfold, to around 30 million per second.

Stages in RBC Maturation
(Figure 20-8)

During its maturation, a red blood cell passes through a series of stages. **Hematologists** (hē-ma-TOL-ō-jists), specialists in blood formation and function, have given specific names to key stages (Figure 20-8●). **Erythroblasts** are very immature red blood cells that are actively synthesizing hemoglobin. **Reticulocytes** (re-TIK-ū-lō-sīts) represent the last step in the maturation process. After shedding their nuclei, these cells will develop the appearance of mature erythrocytes and enter the circulation. These immature red blood cells normally account for about 0.8% of the RBC population.

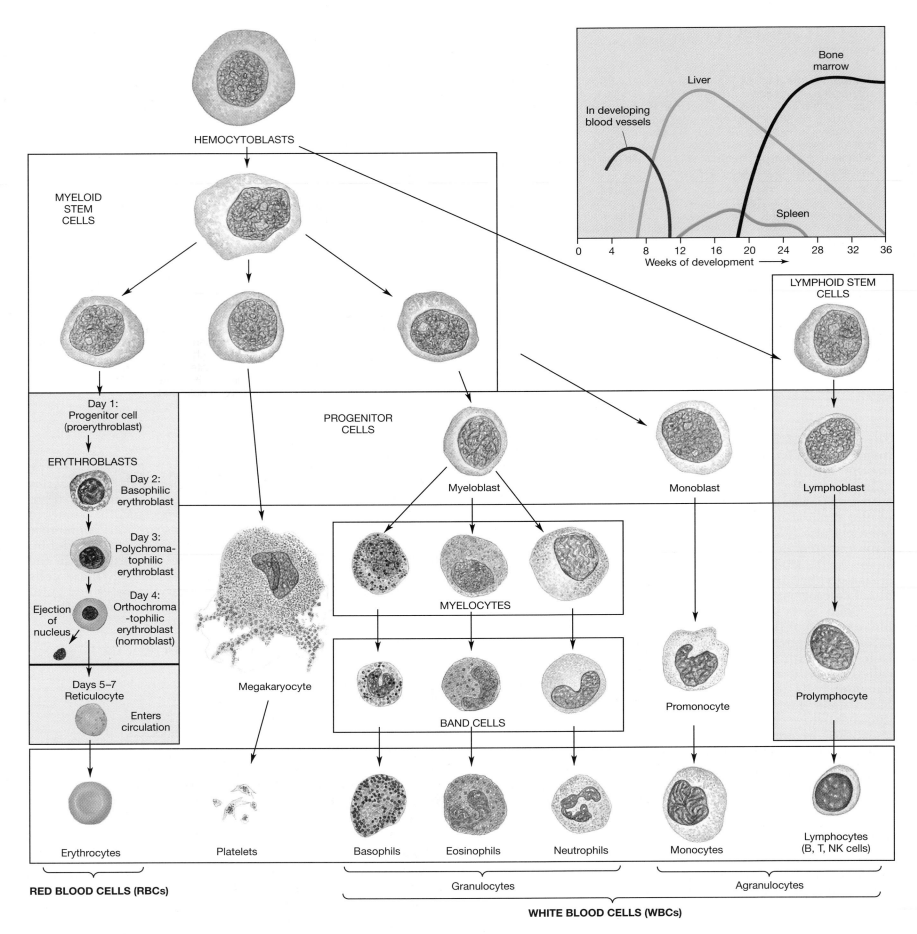

FIGURE 20-8
The Origins and Differentiation of Blood Cells. Hemocytoblast divisions give rise to myeloid and lymphoid stem cells. Myeloid stem cells with more restricted fates produce progenitor cells, which divide to produce the various classes of blood cells. The graph indicates the primary locations of blood cell formation during development.

Leukopoiesis
(Figure 20-8)

Stem cells responsible for the production of white blood cells, a process termed **leukopoiesis,** originate in the bone marrow (Figure 20-8●). Granulocytes complete their development in myeloid tissue; monocytes begin their differentiation in the bone marrow, enter the circulation, and complete their development when they become free macrophages in peripheral tissues. Stem cells responsible for the production of lymphocytes, a process called **lymphopoiesis**, also originate in the bone marrow, but many of them subsequently migrate to the thymus. The bone marrow and thymus are called **primary lymphoid organs** because the divisions of undifferentiated stem cells at these sites produce daughter cells destined to become specialized lymphocytes. Immature B cells and NK cells are produced in the bone marrow, and immature T cells are produced in the thymus. These cells may subsequently migrate to **secondary lymphoid organs**, such as the spleen, tonsils, or lymph nodes. Although they retain the ability to divide, their divisions always produce cells of the same type—a dividing T cell produces daughter T cells, and not NK or B cells. We will consider formation of lymphocytes in more detail in Chapter 23.

Factors that regulate lymphocyte maturation are as yet incompletely understood; however, prior to maturity, hormones of the thymus gland promote the differentiation and maintenance of T cell populations. Several hormones called **colony-stimulating factors (CSFs)** are involved in the regulation of other white blood cell populations. Commercially available CSFs are now used to stimulate the production of WBCs in individuals undergoing cancer chemotherapy.

■ CLINICAL BRIEF
Synthetic Blood

Despite improvements in transfusion technology, shortages of blood and anxieties over the safety of existing stockpiles persist. In addition, some people may be unable or unwilling to accept transfusions for medical or religious reasons. Thus there has been widespread interest in a number of recent attempts to develop synthetic blood components. Plasma expanders and hemoglobin additives were considered earlier (pp. 507, 511).

Whole blood substitutes are highly experimental solutions still undergoing clinical evaluation. In addition to the osmotic agents found in plasma expanders, these solutions contain small clusters of synthetic molecules built of carbon and fluorine atoms. The mixtures, known as **perfluorochemical (PFC) emulsions**, can carry roughly 70% of the oxygen of whole blood. Animals have been kept alive after an exchange transfusion that completely replaced their circulating blood with a PFC emulsion.

PFC solutions have the same advantages of other plasma expanders, plus the added benefits of transporting oxygen. Because there are no RBCs involved, the PFC emulsions can carry oxygen to regions whose capillaries have been partially blocked by fatty deposits or blood clots. Unfortunately, PFCs do not absorb oxygen as effectively as normal blood. To ensure that they deliver adequate oxygen to peripheral tissues, the individual must breathe air rich in oxygen, usually through an oxygen mask. In addition, phagocytes appear to engulf the PFC clusters. These problems have limited the use of PFC emulsions on humans. However, one PFC solution, *Fluosol*, has been used to enhance oxygen delivery to cardiac muscle during heart surgery.

Another approach involves the manufacture of miniature erythrocytes by enclosing small bundles of hemoglobin in a lipid membrane. These **neohematocytes** (nē-ō-hēm-AT-ō-sīts) are spherical, with a diameter of under 1 μm, and they can easily pass through narrow or partially blocked vessels. The major problem with this technique is that phagocytes treat neohematocytes like fragments of normal erythrocytes, so they remain in circulation for only about 5 hours.

Related Clinical Terms

normovolemic (nor-mo-vo-LĒ-mik): The condition of having normal blood volume. *[p. 506]*

transfusion: A procedure in which blood components are given to someone whose blood volume has been reduced or whose blood is defective. *[p. 506]*

packed red blood cells (PRBCs): Red blood cells from which most of the plasma has been removed. *[p. 506]*

anemia (a-NĒ-mē-ah): A condition in which the oxygen-carrying capacity of the blood is reduced, because of low hematocrit or low blood hemoglobin concentrations. *[p. 508]*

polycythemia (po-lē-sī-THĒ-mē-ah): A blood condition showing an elevated hematocrit with a normal blood volume. *[p. 508]*

normochromic: The condition in which red blood cells contain normal amounts of hemoglobin. ⚕*Blood Tests and RBCs* *[p. 764]*

normocytic: A term referring to red blood cells of normal size. ⚕*Blood Tests and RBCs* *[p. 764]*

hemorrhagic anemia: Anemia caused by severe bleeding, typified by a low hematocrit and a low hemoglobin but normal RBCs. ⚕*Blood Tests and RBCs* *[p. 766]*

aplastic anemia: Anemia caused by failure of the bone marrow, leading to a low hematocrit and a low reticulocyte count. ⚕*Blood Tests and RBCs* *[p. 766]*

hemolytic disease of the newborn (HDN): An anemia in the newborn caused by an incompatibility between the maternal (Rh⁻) and fetal (Rh⁺) blood types. ⚕*Hemolytic Disease of the Newborn* *[p. 766]*

bone marrow transplant: A transfusion of bone marrow cells, including stem cells, that can be used to repopulate the bone marrow after radiation exposure or chemotherapy. ⚕*The Leukemias* *[p. 766]*

hemophilia: One of many inherited disorders characterized by inadequate production of clotting factors. *[p. 515]*

embolism: A condition in which a drifting blood clot becomes stuck in a blood vessel, blocking circulation to the area downstream. ⚕*Problems with the Clotting Process* *[p. 767]*

thrombus: A blood clot. ⚕*Problems with the Clotting Process* *[p. 767]*

plaque: An abnormal area within a blood vessel where large quantities of lipids accumulate. ⚕*Problems with the Clotting Process* *[p. 767]*

STUDY OUTLINE

Related Key Terms

INTRODUCTION *[p. 505]*

1.   The cardiovascular system provides a mechanism for the rapid transport of nutrients, waste products, and cells within the body.

FUNCTIONS OF THE BLOOD *[p. 505]*

1.   Blood is a specialized connective tissue. Its functions include: (1) transporting dissolved gases; (2) distributing nutrients; (3) transporting metabolic wastes; (4) delivering enzymes and hormones; (5) stabilizing the pH and electrolyte composition of interstitial fluids; (6) restricting fluid losses through damaged vessels or injuries via the **clotting reaction**; (7) defending the body against toxins and pathogens; and (8) stabilizing body temperature by absorbing and redistributing heat *(see Table 20-1)*.

COMPOSITION OF THE BLOOD *[p. 505]*

1.   Blood contains **plasma, red blood cells (RBCs), white blood cells (WBCs)**, and **platelets**. The plasma and **formed elements** constitute **whole blood**, which can be **fractionated** for analytical or clinical purposes *(see Figure 20-1 and Table 20-2)*.

viscosity • hypovolemic • hypervolemic

Plasma *[p. 506]*
2.   Plasma accounts for about 55% of the volume of blood; roughly 92% of plasma is water *(see Figure 20-1 and Table 20-2)*.

3.   Plasma differs from interstitial fluid because it has a higher dissolved oxygen concentration and large numbers of dissolved proteins. There are three classes of plasma proteins: *albumins, globulins*, and *fibrinogen (see Table 20-2)*.

4.   **Albumins** constitute about 60% of plasma proteins. **Globulins** constitute roughly 33% of plasma proteins: they include **immunoglobulins (antibodies)**, which attach foreign proteins and pathogens, and **transport proteins**, which bind ions, hormones, and other compounds. **Fibrinogen** molecules function in the clotting reaction by interacting to form *fibrin*: removing fibrinogen from plasma leaves a fluid called **serum** *(see Table 20-2)*.

lipoproteins

FORMED ELEMENTS *[p. 507]*

Red Blood Cells (RBCs) *[p. 507]*
1.   Red blood cells (RBCs), or **erythrocytes**, account for slightly less than half the blood volume. The **hematocrit** value indicates the percentage of whole blood occupied by cellular elements *(see Figures 20-2 to 20-4)*.

volume of packed red cells (VPRC) • packed cell volume (PCV)

2.   RBCs transport oxygen and carbon dioxide within the bloodstream. They are highly specialized cells with large surface-to-volume ratios. Because RBCs lack mitochondria, ribosomes, and nuclei, they are unable to perform normal maintenance operations, so they usually degenerate after about 120 days in the circulation *(see Table 20-3)*.

rouleaux

3.   Molecules of **hemoglobin (Hb)** account for over 95% of the RBCs' proteins. Hemoglobin is a globular protein formed from four subunits. Each subunit contains a single molecule of **heme** and can reversibly bind an oxygen molecule. Damaged or dead RBCs are recycled by phagocytes *(see Figure 20-3 and Table 20-3)*.

4.   One's **blood type** is determined by the presence or absence of specific **agglutinogens** in the RBC cell membranes: agglutinogens A, B, and **D (Rh)**. **Agglutinins** within the plasma will react with RBCs bearing different agglutinogens *(see Figure 20-4 and Table 20-4)*.

Type A • Type B • Type AB • Type O • Rh-positive • Rh-negative • sensitization • cross-reaction • agglutination • hemolyze • compatible

White Blood Cells (WBCs) *[p. 512]*
5.   White blood cells (WBCs), or **leukocytes**, defend the body against pathogens and remove toxins, wastes, and abnormal or damaged cells *(see Figure 20-5)*.

leukopenia • leukocytosis • differential count

6.   Leukocytes show **chemotaxis** (are attracted to specific chemicals) and **diapedesis** (the ability to move through vessel walls).

7.   **Granular leukocytes (granulocytes)** are subdivided into **neutrophils, eosinophils (acidophils)**, and **basophils**. Fifty to 70% of circulating WBCs are neutrophils, which are highly mobile phagocytes. The much less common eosinophils are attracted to foreign compounds that have reacted with circulating antibodies. The relatively rare basophils migrate to damaged tissues and release histamines, aiding the inflammation response *(see Figure 20-5 and Table 20-3)*.

polymorphonuclear leukocytes (PMNs)

8.   **Agranular leukocytes** are subdivided into **monocytes** and **lymphocytes**. Monocytes migrating into peripheral tissues become free macrophages. Lymphocytes, the primary cells of the **lymphatic system**, include **T cells** (which enter peripheral tissues and attack foreign cells directly) and **B cells** (which produce antibodies) *(see Figure 20-5 and Table 20-3)*.

monocyte-macrophage system • NK cells

Platelets *[p. 513]*

9. Megakaryocytes in the bone marrow release packets of cytoplasm (platelets) into the circulating blood. The functions of platelets include: (1) transporting chemicals important to the clotting process; (2) forming a temporary patch in the walls of damaged blood vessels; and (3) contracting after a clot has formed in order to reduce the size of the break in the vessel wall *(see Figures 20-6/20-7 and Table 20-3).*

HEMOPOIESIS *[p. 515]*

1. Hemopoiesis is the process of blood cell formation. Circulating **stem cells** called **hemocytoblasts** divide to form all of the blood cells *(see Figure 20-8).*

Erythropoiesis *[p. 515]*

2. Erythropoiesis, the formation of erythrocytes, occurs mainly within the **myeloid tissue** (bone marrow) in adults. RBC formation increases under **erythropoiesis-stimulating hormone (erythropoietin, EPO)** stimulation. Stages in RBC development include **erythroblasts** and **reticulocytes** *(see Figure 20-8).*

Leukopoiesis *[p. 517]*

3. Leukopoiesis, the formation of white blood cells, occurs in bone marrow. Granulocytes and monocytes are produced by stem cells in the bone marrow. Stem cells responsible for **lymphopoiesis** (production of lymphocytes) also originate in the bone marrow, but many migrate to peripheral lymphoid tissues *(see Figure 20-8).*

4. Factors that regulate lymphocyte maturation are not completely understood. Several **colony-stimulating factors (CSFs)** are involved in regulating other WBC populations and coordinating RBC and WBC production.

Related Key Terms
thrombocytes • thrombocytopenia • thrombocytosis • hemostasis

progenitor cells

red marrow • yellow marrow • vitamin B$_{12}$ • hematologist

primary lymphoid organs • secondary lymphoid organs

---

1 REVIEW OF CHAPTER OBJECTIVES

1. List and describe the functions of the blood.
2. Discuss the composition of blood.
3. List the characteristics and functions of red blood cells.
4. Explain what determines a person's blood type and why blood types are important.
5. Categorize the various white blood cells on the basis of their structures and functions.
6. Describe how white blood cells fight infection.
7. Discuss the function of platelets.
8. Describe the life cycles of blood cells.
9. Identify the locations where the components of blood are produced, and discuss the factors that regulate their production.

2 REVIEW OF CONCEPTS

1. What is the role of the blood in the regulation of body temperature?
2. How does the composition of the blood in an anemic person differ from that of a healthy individual?
3. How does the composition of blood plasma differ from that of the interstitial fluid?
4. How do red blood cells differ in structure from more typical body cells?
5. How do gender and altitude affect the hematocrit of an individual?
6. What is the significance of the types of agglutinogens present on the cell membranes of red blood cells?
7. How do the immunoglobulins called agglutinins respond to different types of blood?
8. How do the leukocytes assist in the maintenance of tissues?
9. What is the role of the monocytes in the repair of injured tissues?

10. What is the role of the megakaryocyte in blood clotting?
11. How do platelets function in the process of clotting?
12. How do hemocytoblasts differ from progenitor cells and their daughter cells?
13. How does the production of red blood cells differ from that of white blood cells?
14. What are the main functions of the blood?

3 CRITICAL THINKING AND CLINICAL APPLICATION QUESTIONS

1. A person who loses too much blood from an injury or during surgery will need to receive additional blood or blood products to survive. At present, it is possible to store blood and to provide the specific fractions needed by any individual. How does this process work?

2. One of the factors that has become important in recent years as an indicator of circulatory system health and a predictor for such serious or fatal crises as heart attacks and stroke has been the levels of cholesterol, triglycerides, and fatty acids in the blood. Why are these compounds implicated in health problems?

3. In emergency situations it is often impossible to provide ideal medical care. People suffering injuries involving a large blood loss are often given solutions that temporarily raise the volume of blood, until a transfusion becomes available. What are these solutions, and how do they function?

4. From the time he was a baby, a small child has been experiencing bouts of unexplained bleeding, and the numerous falls he had while learning to walk resulted in unusually large amounts of bruising. He often has swelling of the joints of the limbs and acts as if he is in pain. What might explain these symptoms?

21

The Cardiovascular System: The Heart

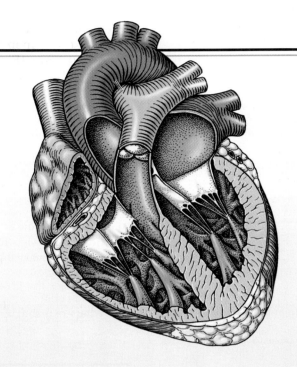

CHAPTER OUTLINE AND OBJECTIVES

Every living cell relies on the surrounding interstitial fluid for oxygen, nutrients, and waste disposal. Conditions in the interstitial fluid are kept stable through continuous exchange between the peripheral tissues and the circulating blood. The blood can help to maintain homeostasis only as long as it stays in motion. If blood remains stationary, its oxygen and nutrient supplies are quickly exhausted, its capacity to absorb wastes is soon saturated, and neither hormones nor white blood cells can reach their intended targets. Thus all of the functions of the cardiovascular system ultimately depend on the heart. This muscular organ beats approximately 100,000 times each day, forcing blood through the vessels of the circulatory system. Each year the heart pumps over 1.5 million gallons of blood, enough to fill 200 train tank cars.

For a practical demonstration of the heart's pumping abilities, turn on the faucet in the kitchen and open it all the way. To deliver an amount of water equal to the volume of blood pumped by the heart in an average lifetime, that faucet would have to be left on for at least 45 years. Equally remarkable, the volume of blood pumped by the heart can be varied over a wide range, from 5 to 30 liters per minute. The performance of the heart is closely monitored and finely regulated to ensure that conditions in the peripheral tissues remain within normal limits.

We begin this chapter by examining the structural features that enable the heart to perform so reliably, even in the face of widely varying physical demands. We will then consider the mechanisms that regulate cardiac activity to meet the body's ever-changing needs.

AN OVERVIEW OF THE CIRCULATORY SYSTEM
(Figure 21-1)

The circulatory system consists of a network of blood vessels that carry blood between the heart and peripheral tissues. This network can be subdivided into the **pulmonary circuit**, which carries blood to and from the exchange surfaces of the lungs, and the **systemic circuit**, which transports blood to and from the rest of the body. Each circuit begins and ends at the heart. **Arteries** transport blood away from the heart; **veins** return blood to the heart (Figure 21-1●). Blood travels through these circuits in sequence. For example, blood returning to the heart in the systemic veins must complete the pulmonary circuit before reentering the systemic arteries. **Capillaries** are small, thin-walled vessels that interconnect the smallest arteries and veins. Capillaries are called **exchange vessels** because their thin walls permit exchange of nutrients, dissolved gases, and waste products between the blood and surrounding tissues.

Despite its impressive workload, the heart is a small organ; your heart is roughly the size of your clenched fist. The heart contains four hollow muscular chambers, two associated with each circuit. The **right atrium** (Ā-trē-um; "chamber") receives blood from the systemic circuit, and the **right ventricle** (VEN-tri-KL; "little belly") discharges blood into the pulmonary circuit. The **left atrium** (plural *atria*) collects blood from the pulmonary circuit, and the **left ventricle** ejects blood into the systemic circuit. When the heart beats, the atria contract first, followed by the ventricles. The two ventricles contract at the same time and eject equal volumes of blood into the pulmonary and systemic circuits.

THE PERICARDIUM
(Figure 21-2)

The heart is located near the anterior chest wall, directly posterior to the sternum (Figure 21-2a●). The heart is surrounded by the **pericardial** (per-i-KAR-dē-al) **cavity**. The pericardial cavity is situated between the pleural cavities, in the mediastinum, which also contains the thymus, esophagus, and trachea. ∞ [p. 20] Figure 21-2b● is a sectional view that illustrates the position of the heart relative to other structures in the mediastinum.

The serous membrane lining the pericardial cavity is called the **pericardium** (Figure 21-2b–d●). To visualize the relationship between the heart and the pericardial cavity, imagine pushing your fist toward the center of a large balloon. The balloon represents the pericardium, and your fist is the heart. The pericardium can be subdivided into the *visceral pericardium* (the part of the balloon in contact with your fist) and the *parietal pericardium*. Your wrist, where the balloon folds back upon itself, corresponds to the base of the heart.

The **visceral pericardium**, or *epicardium*, covers the outer surface of the heart; the **parietal pericardium** lines the inner surface of the **pericardial sac** that surrounds the heart. Its outer layer, which contains abundant collagen fibers, is called the **fibrous pericardium**. The pericardial sac is further reinforced by a dense network of collagen fibers. At the base of the heart, they sta-

FIGURE 21-1
A Generalized View of the Pulmonary and Systemic Circuits. Blood flows through separate pulmonary and systemic circuits, driven by the pumping of the heart. Each circuit begins and ends at the heart and contains arteries, capillaries, and veins.

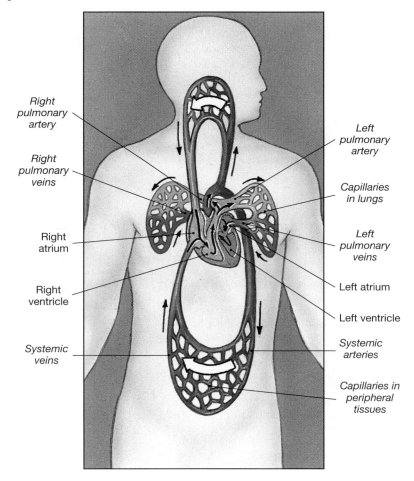

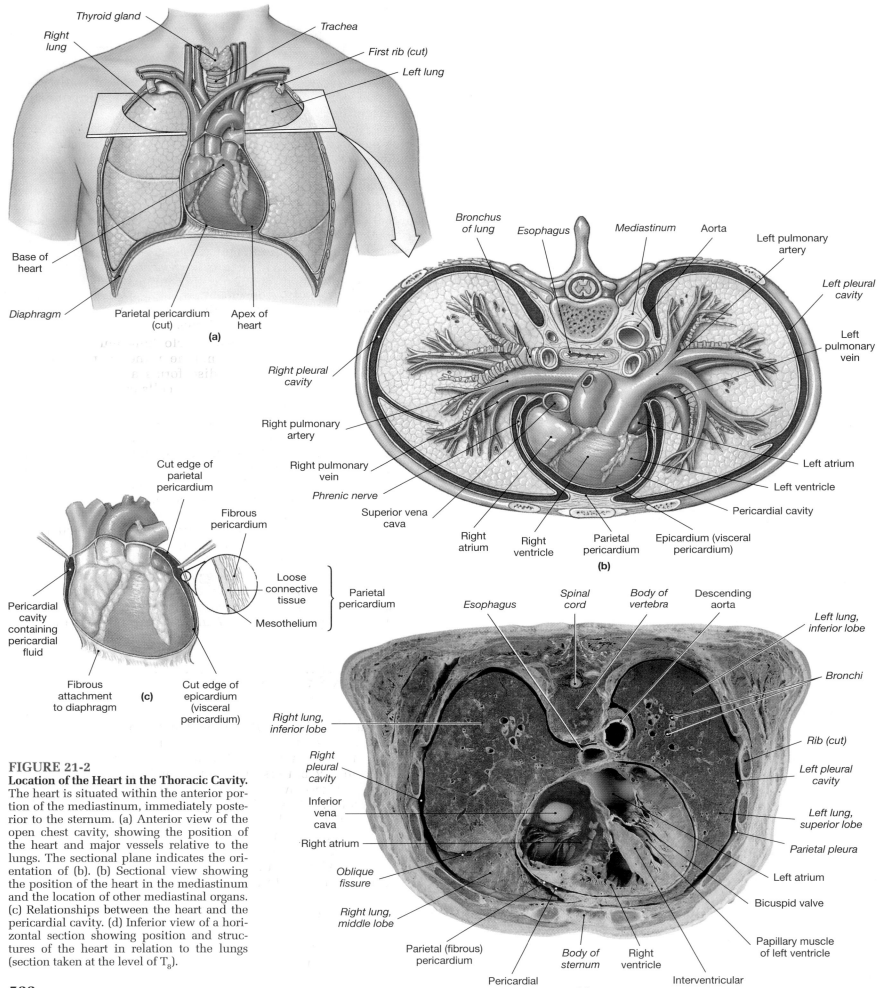

Thyroid gland

Trachea

Right lung

First rib (cut)

Left lung

Base of heart

Diaphragm

Parietal pericardium (cut)

Apex of heart

(a)

Bronchus of lung

Esophagus

Mediastinum

Aorta

Left pulmonary artery

Left pleural cavity

Left pulmonary vein

Right pleural cavity

Right pulmonary artery

Right pulmonary vein

Phrenic nerve

Superior vena cava

Right atrium

Right ventricle

Parietal pericardium

Epicardium (visceral pericardium)

Left atrium

Left ventricle

Pericardial cavity

(b)

Cut edge of parietal pericardium

Fibrous pericardium

Loose connective tissue

Parietal pericardium

Mesothelium

Pericardial cavity containing pericardial fluid

Fibrous attachment to diaphragm

(c)

Cut edge of epicardium (visceral pericardium)

Esophagus

Spinal cord

Body of vertebra

Descending aorta

Left lung, inferior lobe

Bronchi

Right lung, inferior lobe

Rib (cut)

Right pleural cavity

Left pleural cavity

Inferior vena cava

Left lung, superior lobe

Right atrium

Parietal pleura

Oblique fissure

Left atrium

Right lung, middle lobe

Bicuspid valve

Papillary muscle of left ventricle

Parietal (fibrous) pericardium

Body of sternum

Right ventricle

Interventricular septum

Pericardial cavity

(d)

FIGURE 21-2

Location of the Heart in the Thoracic Cavity. The heart is situated within the anterior portion of the mediastinum, immediately posterior to the sternum. (a) Anterior view of the open chest cavity, showing the position of the heart and major vessels relative to the lungs. The sectional plane indicates the orientation of (b). (b) Sectional view showing the position of the heart in the mediastinum and the location of other mediastinal organs. (c) Relationships between the heart and the pericardial cavity. (d) Inferior view of a horizontal section showing position and structures of the heart in relation to the lungs (section taken at the level of T_8).

522

bilize the positions of the pericardium, heart, and associated vessels in the mediastinum. The slender gap between the opposing parietal and visceral surfaces is the pericardial cavity. This cavity normally contains 10–20 mℓ of **pericardial fluid** secreted by the pericardial membranes. Pericardial fluid acts as a lubricant, reducing friction between the opposing surfaces. The moist pericardial lining prevents friction as the heart beats, and the collagen fibers binding the base of the heart to the mediastinum limit movement of the major vessels during a contraction.

STRUCTURE OF THE HEART WALL
(Figure 21-3)

A section through the wall of the heart (Figure 21-3a●) reveals three distinct layers: (1) an outer *epicardium* (visceral pericardium), (2) a middle *myocardium*, and (3) an inner *endocardium*.

1. The **epicardium** is the visceral pericardium, which covers the outer surface of the heart. This serous membrane consists of an exposed mesothelium and an underlying layer of loose connective tissue that is attached to the myocardium.

2. The **myocardium**, or muscular wall of the heart, forms both atria and ventricles. The myocardium contains cardiac muscle tissue and associated connective tissues, blood vessels, and nerves. The myocardium consists of concentric layers of cardiac muscle tissue. The atrial myocardium contains layers that wrap around the atria and form figure eights through the interatrial septum (Figure 21-3b●). Superficial ventricular muscles wrap around both ventricles; deeper muscle layers spiral around and between the ventricles toward the apex.

3. The inner surfaces of the heart, including the valves, are covered by a simple squamous epithelium, the **endocardium** (en-dō-KAR-dē-um; *endo-*, inside), which is continuous with the endothelium of the attached blood vessels (Figure 21-3a●).

Cardiac Muscle Tissue
(Figure 21-3c)

The unusual histological characteristics of cardiac muscle tissue give the myocardium its unique functional properties. Cardiac muscle tissue was introduced in Chapter 3, and its properties were briefly compared with those of other muscle types. ∞ [p. 80] Cardiac muscle cells, or **cardiocytes**, are relatively small, averaging 10–20 µm in diameter and 50–100 µm in length. A typical cardiocyte has a single, centrally placed nucleus (Figure 21-3c●).

Cardiac muscle cells are much smaller than skeletal muscle cells. Like skeletal muscle fibers, each cardiac muscle cell contains organized myofibrils, and the alignment of their sarcomeres gives the cardiocyte a striated appearance. However, there are significant differences between skeletal and cardiac muscle cells with regard to metabolism, membrane characteristics, T-tubule structure, and the organization of the sarcoplasmic reticulum.

Cardiac muscle cells are almost totally dependent on aerobic respiration to obtain the energy needed to continue contracting. The sarcoplasm of a cardiac muscle cell thus contains hundreds of mitochondria and abundant reserves of myoglobin (to store oxygen). Energy reserves are maintained in the form of glycogen and lipid inclusions.

The Intercalated Discs
(Figure 21-3d)

Cardiac muscle cells are connected to neighboring cells at specialized junctional sites known as **intercalated** (in-TER-ka-lā-ted) **discs**. Intercalated discs (Figure 21-3d●) are unique to cardiac muscle tissue. The jagged appearance is due to the interdigitation of opposing sarcolemmal membranes. At an intercalated disc:

1. The cell membranes of two cardiac muscle cells are bound together by desmosomes. This connection helps to stabilize their relative positions and maintain the three-dimensional structure of the tissue.

2. Myofibrils in the two interlocking muscle cells are firmly anchored to the membrane at the intercalated disc. Because the intercalated disc forms a bridge that links their myofibrils, the two muscle cells can "pull together" with maximum efficiency.

3. Cardiac muscle cells at an intercalated disc are also connected by gap junctions. Ions and small molecules can move from one cell to another at gap junctions, thereby creating a direct electrical connection between the two muscle cells; an action potential can move from one cardiac muscle cell to another as if the membranes were continuous.

Because cardiac muscle cells are mechanically, chemically, and electrically connected to one another, cardiac muscle tissue functions like a single, enormous muscle cell. The contraction of any one cell will trigger the contraction of several others, and the contraction will spread throughout the myocardium. For this reason cardiac muscle has been called a *functional syncytium* (sin-SIT-ē-um; "fused mass of cells").

The Fibrous Skeleton
(Figure 21-3a)

The connective tissues of the heart include large numbers of collagen and elastic fibers. Each cardiac muscle cell is wrapped in a strong but elastic sheath, and adjacent cells are tied together by fibrous cross-links, or "struts." These fibers are in turn interwoven in sheets to separate the superficial and deep muscle layers. Dense bands of fibroelastic tissue encircle the bases of the pulmonary trunk and aorta and the valves of the heart, almost completely separating the atria and ventricles. This internal connective tissue network is called the **fibrous skeleton** of the heart (Figure 21-3a●).

The fibrous skeleton performs the following functions:

1. Stabilizing the positions of the muscle cells and valves in the heart.

2. Providing physical support for the cardiac muscle cells and for the blood vessels and nerves in the myocardium.

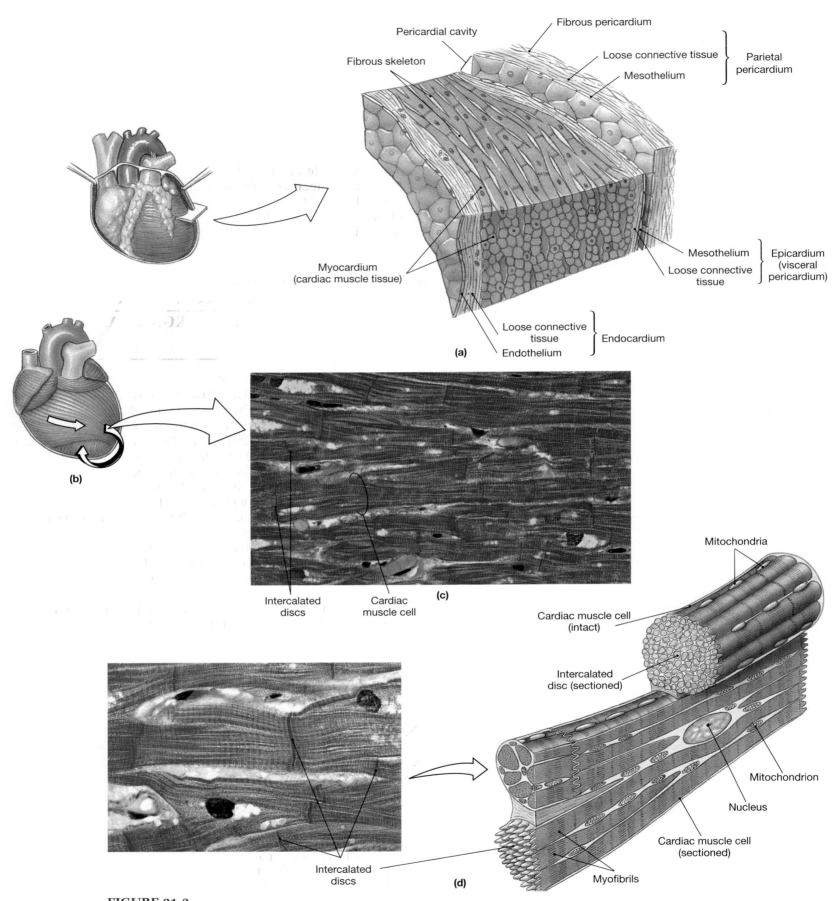

FIGURE 21-3

Organization of Muscle Tissue in the Heart. (a) A diagrammatic section through the heart wall showing the relative positions of the epicardium, myocardium, and endocardium. (b) Cardiac muscle tissue in the heart forms concentric layers that wrap around the atria and spiral within the walls of the ventricles. (c,d) Sectional and diagrammatic views of cardiac muscle tissue. Distinguishing characteristics of cardiac muscle cells include: (1) small size, (2) single, centrally placed nucleus, (3) branching interconnections between cells, and (4) presence of intercalated discs. (LMs × 350, × 763)

3. Distributing the forces of contraction.

4. Adding strength and helping to prevent overexpansion of the heart.

5. Providing elasticity that helps return the heart to its original shape after each contraction.

6. Physically isolating the muscle fibers of the atria from those in the ventricles.

✓ **How could you distinguish a sample of cardiac muscle tissue from a sample of skeletal muscle tissue?**

SUPERFICIAL ANATOMY OF THE HEART
(Figure 21-4)

Several external features distinguish the atria from the ventricles (Figure 21-4●). The atria have relatively thin muscular walls, and they are highly distensible. When not filled with blood, the outer portion of the atrium deflates and becomes a rather lumpy and wrinkled flap. This expandable extension of the atrium is called an **auricle** (AW-ri-kel; *auris*, ear) because it reminded early anatomists of the external ear. The auricle is also known as an **atrial appendage**. A deep groove, the **coronary sulcus**, marks the border between the atria and the ventricles. Shallower depressions, the **anterior interventricular sulcus** and the **posterior interventricular sulcus**, mark the boundary line between the left and right ventricles. The connective tissue of the epicardium at the coronary and interventricular sulci usually contains substantial amounts of fat, and in fresh or preserved hearts, this fat must be stripped away to expose the underlying grooves (Figure 21-4b,c●). These sulci also contain the arteries and veins that supply blood to the cardiac muscle of the heart.

A midsagittal section of the body would not separate the left and right sides of the heart because the heart (1) lies slightly to the left of the midline, (2) sits at an angle to the longitudinal axis of the body, and (3) is rotated toward the left side.

The Base and Apex
(Figures 21-2a,d/21-4a)

The heart lies slightly to the left of the midline. The great veins and arteries of the circulatory system are connected to the superior end of the heart, at the **base**. The base sits posterior to the sternum at the level of the third costal cartilage, centered about 1.2 cm (0.5 in.) to the left side (Figure 21-2a,d●, p. 522). The inferior, pointed tip of the heart is the **apex** (Ā-peks). A typical adult heart measures approximately 12.5 cm (5 in.) from the attached base to the apex. The apex reaches the fifth intercostal space approximately 7.5 cm (3 in.) to the left of the midline (Figure 21-4a●).

Borders of the Heart
(Figure 21-4b)

The heart sits at an angle to the longitudinal axis of the body. The **superior border** of the heart includes the bases of the major vessels and the two atria. The right atrium forms the **right border** of the heart; the **left border** is formed by the left ventricle and a small portion of the left atrium. The left border extends to the apex, where it meets the **inferior border**. The wall of the right ventricle forms most of the inferior border (Figure 21-4b●).

Surfaces of the Heart
(Figure 21-4c)

The heart is rotated slightly toward the left, so that the **anterior surface**, or **sternocostal** (ster-nō-KOS-tal) **surface**, consists primarily of the right atrium and right ventricle (Figure 21-4b●). The heart sits in a notch within the left lung, which comes in contact with the **pulmonary** (left) **surface** of the heart. The posterior wall of the left ventricle forms much of the sloping **posterior surface**, or **diaphragmatic surface**, between the base and the apex of the heart (Figure 21-4c●).

INTERNAL ANATOMY AND ORGANIZATION OF THE HEART
(Figure 21-5)

View Figure 21-5● as we describe the internal anatomy and organization of the heart. The two atria are separated by the *interatrial septum* (*septum*, wall), and the *interventricular septum* divides the two ventricles (Figure 21-5a,c●). Each atrium communicates with the ventricle of the same side. **Valves** are folds of fibrous tissue that extend into the openings between the atria and ventricles. These valves open and close to prevent backflow, thereby maintaining a one-way flow of blood from the atria into the ventricles. (Valve structure and function will be described under a separate heading.)

The function of an atrium is to collect blood returning to the heart and deliver it to the attached ventricle. The functional demands placed on the right and left atria are very similar, and the two chambers look almost identical. The demands placed on the right and left ventricles are very different, and there are significant structural differences between the two.

The Right Atrium
(Figures 21-4c/21-5a,c/21-8b)

The right atrium (Figure 21-5a,c●) receives blood from the systemic circuit via the two great veins, the **superior vena cava** (VĒ-na CĀ-va) and the **inferior vena cava**. The superior vena cava delivers blood to the right atrium from the head, neck, upper extremities, and chest. The superior vena cava opens into the posterior and superior portion of the right atrium. The inferior vena cava carries blood to the right atrium from the rest of the trunk, the viscera, and the lower extremities. The inferior vena cava opens into the posterior and inferior portion of the right atrium.

Prominent muscular ridges, the **pectinate muscles** (*pectin*, comb), or *musculi pectinati*, run along the inner surface of the auricle and across the adjacent anterior atrial wall. The *coronary veins* of the heart return blood to the **coronary sinus** (Figures 21-4c and 21-8b●, p. 532), which opens into the right atrium

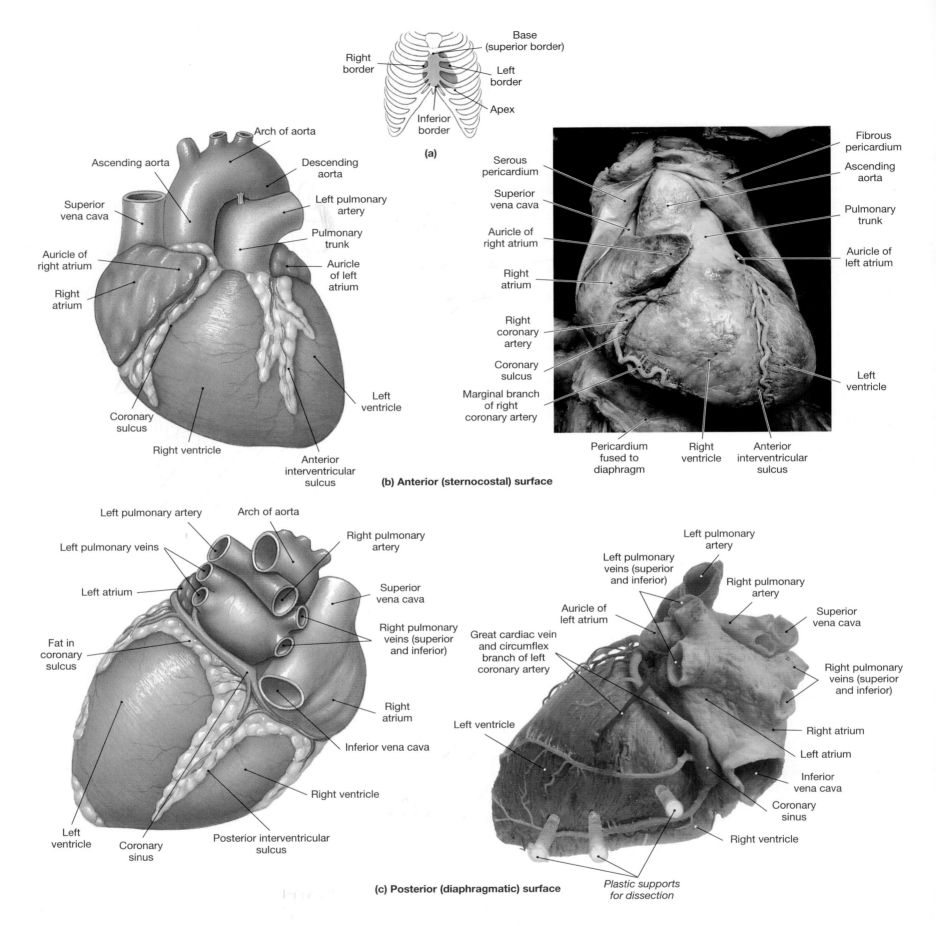

(a)

Base (superior border)
Right border
Left border
Apex
Inferior border

Arch of aorta
Ascending aorta
Descending aorta
Superior vena cava
Left pulmonary artery
Pulmonary trunk
Auricle of right atrium
Auricle of left atrium
Right atrium
Coronary sulcus
Left ventricle
Right ventricle
Anterior interventricular sulcus

Fibrous pericardium
Serous pericardium
Ascending aorta
Superior vena cava
Pulmonary trunk
Auricle of right atrium
Auricle of left atrium
Right atrium
Right coronary artery
Coronary sulcus
Left ventricle
Marginal branch of right coronary artery
Pericardium fused to diaphragm
Right ventricle
Anterior interventricular sulcus

(b) Anterior (sternocostal) surface

Left pulmonary artery
Arch of aorta
Right pulmonary artery
Left pulmonary veins
Left atrium
Superior vena cava
Right pulmonary veins (superior and inferior)
Fat in coronary sulcus
Right atrium
Left ventricle
Inferior vena cava
Coronary sinus
Right ventricle
Posterior interventricular sulcus

Left pulmonary artery
Left pulmonary veins (superior and inferior)
Right pulmonary artery
Auricle of left atrium
Superior vena cava
Great cardiac vein and circumflex branch of left coronary artery
Right pulmonary veins (superior and inferior)
Left ventricle
Right atrium
Left atrium
Inferior vena cava
Coronary sinus
Right ventricle
Plastic supports for dissection

(c) Posterior (diaphragmatic) surface

FIGURE 21-4
Superficial Anatomy of the Heart. (a) Anterior view of the chest, showing the position of the heart relative to the chest wall and the anatomical borders of the heart. (b) Anterior (sternocostal) view of the heart, showing major anatomical features. (c) The posterior (diaphragmatic) surface of the heart.

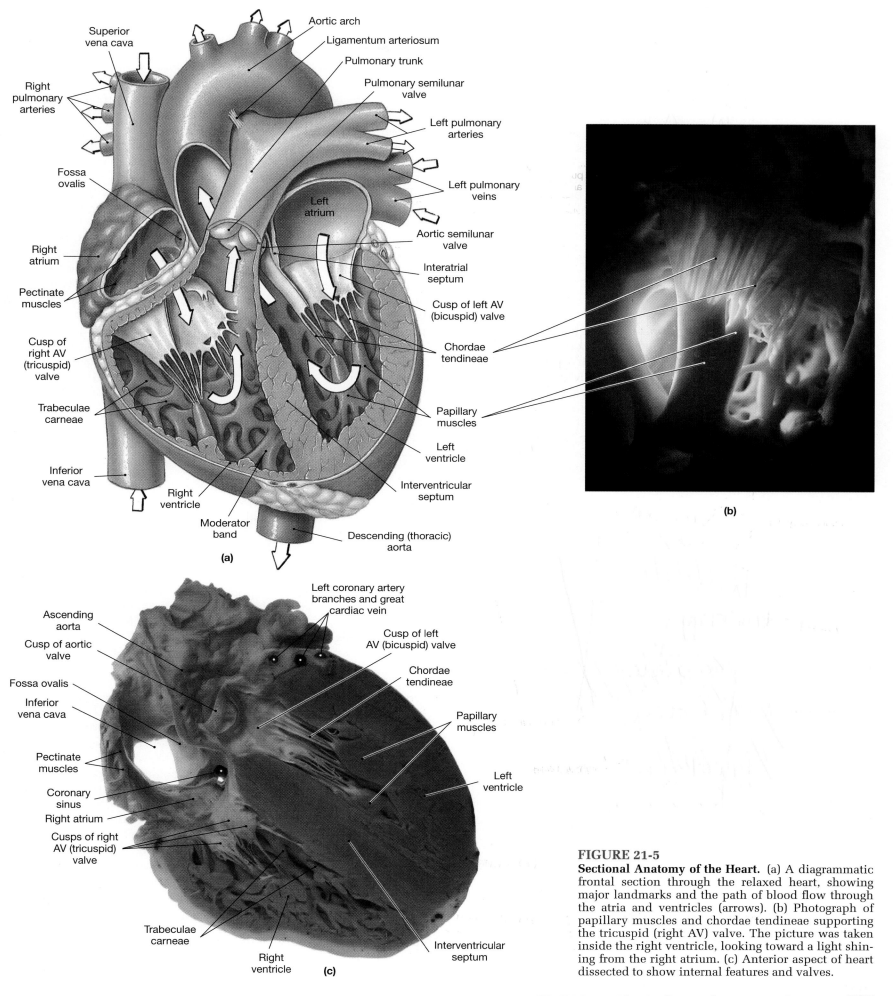

Superior
vena cava

Aortic arch

Ligamentum arteriosum

Pulmonary trunk

Right
pulmonary
arteries

Pulmonary semilunar
valve

Left pulmonary
arteries

Left pulmonary
veins

Fossa
ovalis

Left
atrium

Aortic semilunar
valve

Right
atrium

Interatrial
septum

Pectinate
muscles

Cusp of left AV
(bicuspid) valve

Chordae
tendineae

Cusp of
right AV
(tricuspid)
valve

Papillary
muscles

Trabeculae
carneae

Left
ventricle

Inferior
vena cava

Interventricular
septum

Right
ventricle

Moderator
band

Descending (thoracic)
aorta

(a)

(b)

Left coronary artery
branches and great
cardiac vein

Ascending
aorta

Cusp of left
AV (bicuspid) valve

Cusp of aortic
valve

Chordae
tendineae

Fossa ovalis

Inferior
vena cava

Papillary
muscles

Pectinate
muscles

Left
ventricle

Coronary
sinus

Right atrium

Cusps of right
AV (tricuspid)
valve

Trabeculae
carneae

Right
ventricle

Interventricular
septum

(c)

FIGURE 21-5

Sectional Anatomy of the Heart. (a) A diagrammatic
frontal section through the relaxed heart, showing
major landmarks and the path of blood flow through
the atria and ventricles (arrows). (b) Photograph of
papillary muscles and chordae tendineae supporting
the tricuspid (right AV) valve. The picture was taken
inside the right ventricle, looking toward a light shining from the right atrium. (c) Anterior aspect of heart
dissected to show internal features and valves.

slightly inferior to the opening of the inferior vena cava. The **interatrial septum** divides the right atrium from the left. From the fifth week of embryonic development until birth, an oval opening, the **foramen ovale**, penetrates the septum. The foramen ovale permits blood flow from the right atrium to the left atrium while the lungs are developing. At birth the foramen ovale closes, and after 48 hours the opening is permanently sealed. A small depression, the **fossa ovalis**, persists at this site in the adult heart. Occasionally this sealing process does not occur, and the foramen ovale remains *patent* (open). As a result, the circulation to the lungs is reduced, and blood oxygenation is low. The tissues of the newborn infant soon become starved for oxygen, and the *cyanosis* that develops makes the infant appear to be a "blue baby." ∞ [p. 93]

The Right Ventricle
(Figure 21-5a,c)

Blood travels from the right atrium into the right ventricle (Figure 21-5a,c●) through a broad opening bounded by three fibrous flaps. These flaps, or **cusps**, are part of the **right atrioventricular (AV) valve**, also known as the **tricuspid valve** (trī-KUS-pid; *tri*, three). Each cusp is braced by tendinous connective tissue fibers, the **chordae tendineae** (KOR-dē TEN-di-neē; "tendinous cords"), that are connected to cone-shaped muscular projections, the **papillary** (PAP-i-ler-ē) **muscles**, that arise from the inner surface of the right ventricle. The internal surface of the ventricle also contains a series of deep grooves and folds, the **trabeculae carneae** (tra-BEK-ū-lē CAR-neē; *carneus*, fleshy). The **moderator band** is a muscular ridge that extends horizontally from the inferior portion of the **interventricular septum,** a thick, muscular partition that separates the right ventricle from the left.

The superior end of the right ventricle tapers to a cone-shaped pouch, the **conus arteriosus**, which ends at the **pulmonary semilunar valve**. The pulmonary semilunar valve consists of three semilunar (half-moon-shaped) cusps of thick connective tissue. Blood flowing from the right ventricle passes through this valve to enter the **pulmonary trunk**, the start of the pulmonary circuit, and the arrangement of cusps prevents backflow when the right ventricle relaxes. Once within the pulmonary trunk, blood flows into the **left** and **right pulmonary arteries**. These vessels branch repeatedly within the lungs before supplying capillaries, where gas exchange occurs.

The Left Atrium
(Figure 21-5a,c)

From the respiratory capillaries, blood collects into small veins that ultimately unite to form the four *pulmonary veins*. The posterior wall of the left atrium (Figure 21-5a,c●) receives blood from two **left** and two **right pulmonary veins**. Like the right atrium, the left atrium has an auricle and a valve, the **left atrioventricular (AV) valve**, or **bicuspid** (bī-KUS-pid) **valve**. As the name *bicuspid* implies, the left AV valve contains a pair of cusps rather than a trio. Clinicians often use the term **mitral** (MĪ-tral; *mitre*, bishop's hat) when referring to this valve. The left atrioventricular valve permits the flow of blood from the left atrium into the left ventricle.

The Left Ventricle
(Figure 21-5a,c)

The left ventricle is the largest and thickest of the heart chambers. Its large size and extra thick myocardium enable it to pump blood to the entire body, whereas the right ventricle pumps blood only about 15 cm (6 in.) to the lungs. The internal organization of the left ventricle resembles that of the right ventricle (Figure 21-5a,c●). The trabeculae carneae are prominent, and a pair of large papillary muscles tense the chordae tendineae that brace the cusps of the bicuspid valve.

Blood leaves the left ventricle by passing through the **aortic semilunar valve** into the **ascending aorta**. The arrangement of cusps in the aortic semilunar valve is the same as in the pulmonary semilunar valve. Saclike dilations of the base of the ascending aorta occur adjacent to each cusp. These sacs, called **aortic sinuses**, prevent the individual cusps from sticking to the wall of the aorta when the valve opens. The aortic semilunar valve prevents the backflow of blood into the left ventricle once it has been pumped out of the heart and into the systemic circuit. From the ascending aorta, blood flows on through the **aortic arch** and into the **descending aorta** (Figure 21-5a,c●). The pulmonary trunk is attached to the aortic arch by the *ligamentum arteriosum*, which marks the path of an important fetal blood vessel. Circulatory changes that occur at birth are described in Chapter 22 (p. 571).

Structural Differences between the Left and Right Ventricles
(Figure 21-6)

Anatomical differences between left and right ventricles are best seen in a three-dimensional view (Figure 21-6●). The lungs are close to the heart, and the pulmonary arteries and veins are relatively short and wide. Thus the right ventricle normally does not need to push very hard to propel blood through the pulmonary circuit. The wall of the right ventricle is relatively thin, and in sectional view it resembles a pouch attached to the massive wall of the left ventricle. When the right ventricle contracts, it acts like a bellows pump, squeezing the blood against the mass of the left ventricle. This mechanism moves blood very efficiently with minimal effort, but it develops relatively low pressures. Functionally this is important because the pulmonary capillaries are very delicate. Pressures as high as those found in systemic capillaries would both damage the pulmonary vessels and force fluid into the alveoli of the lungs.

A comparable pumping arrangement would not be suitable for the left ventricle, because six to seven times as much force must be exerted to push blood around the systemic circuit. The left ventricle has an extremely thick muscular wall, and it is round in cross section. When this ventricle contracts, two things happen: The distance between the base and apex decreases, and the diameter of the ventricular chamber decreases. If you imagine the effects of simultaneously squeezing and rolling up the end of a toothpaste tube you will get the idea. The forces generated are quite powerful, more than enough to open the semilunar valve and eject blood into the ascending aorta. As the powerful left ventricle contracts, it also bulges into the right ventricular cavity. This intrusion improves the efficiency of the right ventricle's efforts. Individuals whose right ventricular musculature has been severely damaged may continue to survive because of the extra push provided by the contraction of the left ventricle. ✝ *The Cardiomyopathies [p. 767]*

FIGURE 22-1

A Comparison of Typical Arteries and Veins (a) Light micrograph of artery and vein. (LM × 95) (b) Light micrograph of artery. (LM × 35) (c) Light micrograph of vein. (LM × 35)

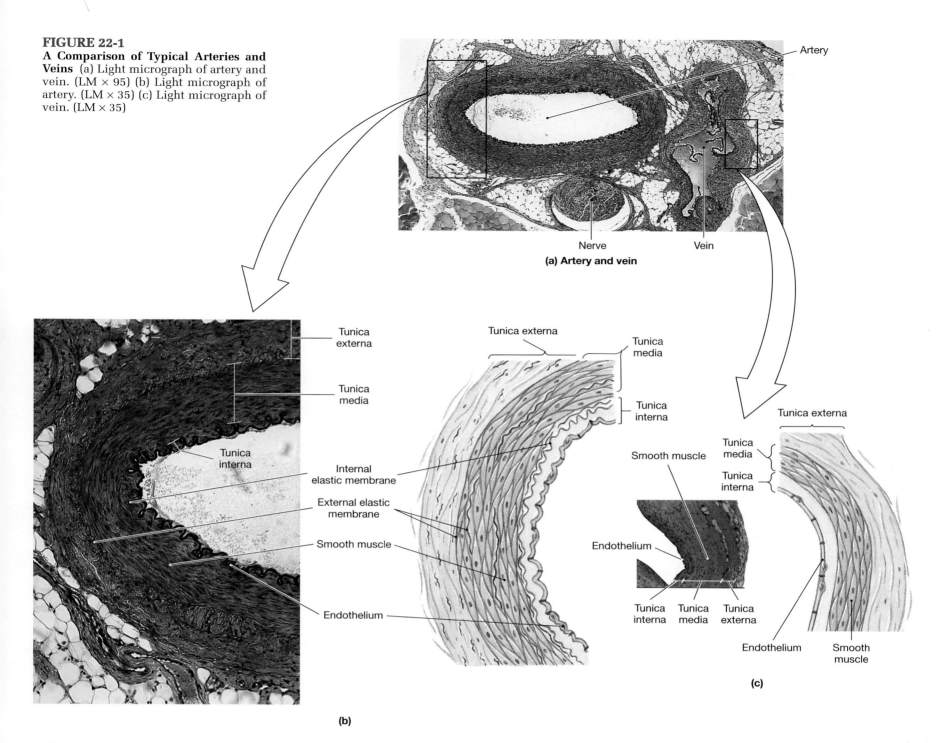

(a) Artery and vein

(b)

(c)

their original dimensions. Their expansion cushions the sudden rise in pressure during ventricular systole, and their contraction slows the decline in pressure during ventricular diastole.

Muscular Arteries
(Figures 22-1b/22-2)

Medium-sized arteries, or *distribution arteries* (also known as **muscular arteries**), distribute blood to the body's skeletal muscle and internal organs. A typical muscular artery has a diameter of approximately 0.4 cm (0.15 in.). Muscular arteries are characterized by a thick tunica media containing a greater amount of smooth muscle fibers than elastic arteries have (Figures 22-1b and 22-2●). The *external carotid artery* of the neck, the *brachial arteries* of the arms, and the *femoral arteries* of the thighs are examples of muscular arteries.

Arterioles
(Figure 22-2)

Arterioles (ar-TĒ-rē-ōlz) are considerably smaller than medium-sized arteries, with an average diameter of about 30 μm. Arterioles have a poorly defined tunica externa, and the tunica media consists of scattered smooth muscle fibers that do not form a complete layer (Figure 22-2●). The smaller muscular arteries and arterioles change their diameter in response to local conditions or to sympathetic or endocrine stimulation. For example, arterioles in most tissues vasoconstrict under sympathetic stimulation. ∞ [p. 434]

Elastic and muscular arteries are interconnected, and vessel characteristics change gradually as the vessels get farther away from the heart. For example, the largest of the muscular arteries contain a considerable amount of elastic tissue, while the smallest resemble heavily muscled arterioles.

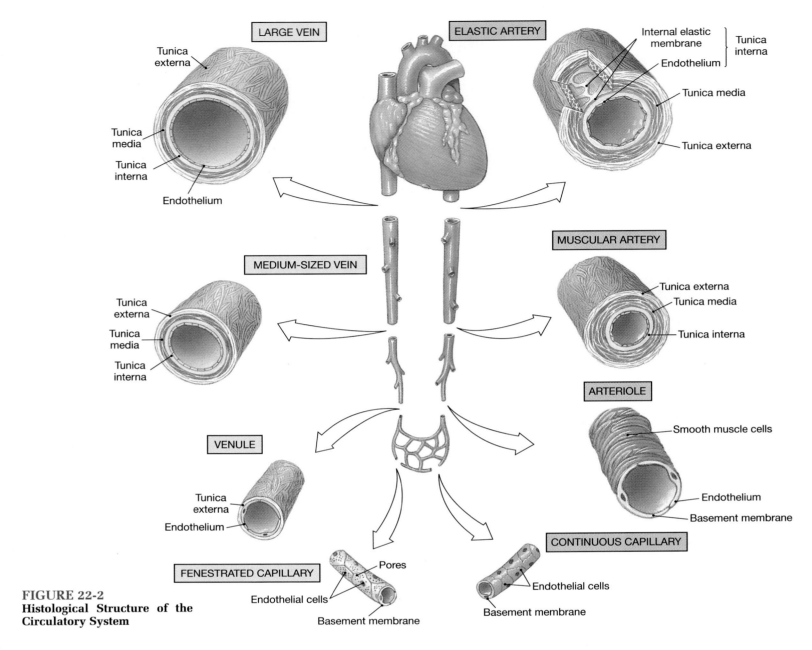

FIGURE 22-2
Histological Structure of the Circulatory System

Capillaries
(Figure 22-3)

Capillaries are the only blood vessels whose walls permit exchange between the blood and the surrounding interstitial fluids. Because the walls are relatively thin (Figure 22-3a,b●), the diffusion distances are small, and exchange can occur quickly. In addition, blood flows through capillaries slowly, allowing sufficient time for diffusion or active transport of materials across the capillary walls. Thus, the histological structure of capillaries permits a two-way exchange of substances between blood and body cells.

A typical capillary consists of an endothelial tube inside a delicate basement membrane. The average diameter of a capillary is a mere 8 μm, very close to that of a single red blood cell. In most regions of the body the capillary endothelium is a complete lining, with the endothelial cells connected by tight junctions and desmosomes. These vessels are called **continuous capillaries** (Figure 22-3a,b●). **Fenestrated capillaries** (FEN-es-trā-ted; *fenestra*,

window) are capillaries that contain "windows," or pores, due to an incomplete or perforated endothelial lining (Figure 22-3c●).

The walls of a fenestrated capillary have a "Swiss cheese" appearance, and the pores allow molecules as large as peptides and proteins to pass in or out of the circulation. This type of capillary permits very rapid exchange of fluids and solutes. Examples of fenestrated capillaries noted in earlier chapters include the choroid plexus of the brain (Chapter 15), and the capillaries in a variety of endocrine organs, including the hypothalamus, the pituitary, the pineal, the adrenals, and the thyroid gland (Chapter 19). ∞ [p. 488] Fenestrated capillaries are also found at filtration sites in the kidneys. ∞ [p. 665]

Sinusoids (SĪ-nus-oidz) resemble fenestrated capillaries but have larger pores and a thinner basement membrane. (In some organs, such as the liver, there is no basement membrane.) Sinusoids are flattened and irregular, and follow the internal contours of complex organs. Blood moves through sinusoids relatively slowly, maximizing the time available for absorption and secretion across the sinusoidal walls. Sinusoids are found in the liver, bone marrow, and adrenal glands.

Capillary Beds
(Figure 22-4)

Capillaries do not function as individual units, but as part of an interconnected network called a **capillary plexus**, or **capillary bed** (Figure 22-4●). A single arteriole usually gives rise to dozens of capillaries that empty into several venules. The entrance to each capillary is guarded by a band of smooth muscle, termed a **precapillary sphincter**. Contraction of the smooth muscle fibers constricts and narrows the diameter of the capillary entrance, thereby reducing the flow of blood. Relaxation of the sphincter dilates the opening, allowing blood to enter the capillary at a faster rate.

Within the capillary bed, **central** or **preferred channels** provide a relatively direct means of arteriole-venule communication. The arteriolar segment of the channel contains smooth muscles capable of altering its diameter, and this region is often called a **metarteriole** (met-ar-TĒ-rē-ōl) (Figure 22-4a●). The rest of the central channel resembles a typical capillary.

Blood usually flows from the arterioles to the venules at a constant rate, but the blood flow within a single capillary can be quite variable. Each precapillary sphincter goes through cycles of alternately contracting and relaxing, perhaps a dozen times each minute. As a result, the blood flow within any one capillary occurs in a series of pulses rather than as a steady and constant stream. The net effect is that blood may reach the venules by one route now, and by a quite different route later. This process, which is controlled at the tissue level, is called capillary **autoregulation**.

There are also mechanisms to modify the circulatory supply to the entire capillary complex. The capillary networks within an area are often supplied by more than one artery. The arteries, called **collaterals**, enter the region and fuse together, rather than ending in a series of arterioles. The interconnection is an **arterial anastomosis**. Such an arrangement guarantees a reliable blood supply to the tissues, for if one arterial supply should become blocked, the other will supply blood to the capillary bed. **Arteriovenous** (ar-tē-rē-ō-VĒ-nus)

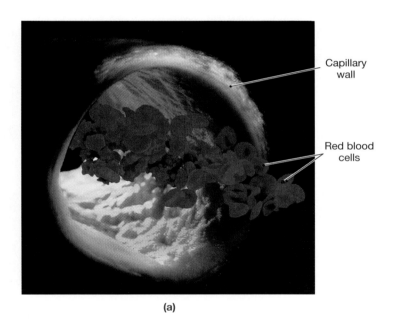

(a)

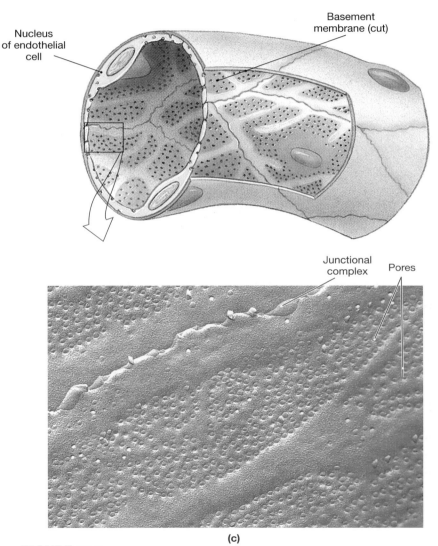

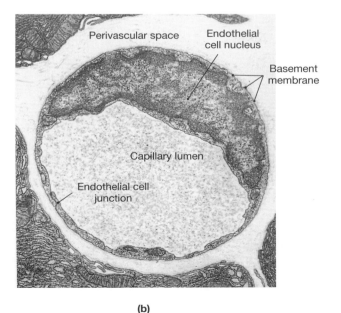

(b)

FIGURE 22-3

Structure of Capillaries. (a) Continuous capillary showing red blood cells passing through the vessel. (b) TEM of cross section through a continuous capillary. A single endothelial cell forms a complete lining around this portion of the capillary. (c) Longitudinal section (above) and SEM (below) showing the wall of a fenestrated capillary. The pores are gaps in the endothelial lining that permit the passage of large volumes of fluid and solutes. (SEM × 12,425)

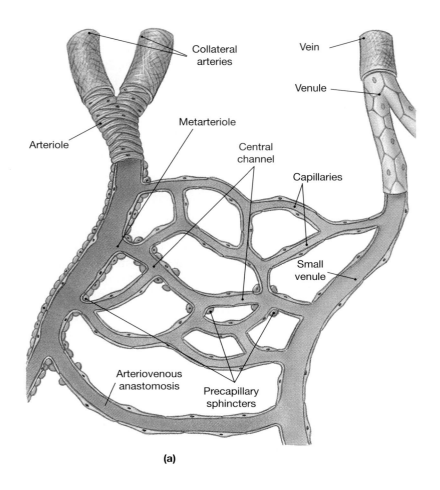

Collateral arteries

Vein

Venule

Metarteriole

Arteriole

Central channel

Capillaries

Small venule

Arteriovenous anastomosis

Precapillary sphincters

(a)

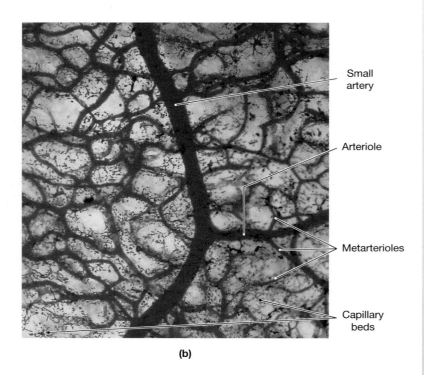

Small artery

Arteriole

Metarterioles

Capillary beds

(b)

FIGURE 22-4

Organization of a Capillary Bed. (a) Basic features of a typical capillary bed. The pattern of blood flow changes continually in response to regional alterations in tissue oxygen demand. (b) Capillary bed as seen in a living specimen.

■ CLINICAL BRIEF
Arteriosclerosis

Arteriosclerosis (ar-tē-rē-ō-skle-RŌ-sis) is a thickening and toughening of arterial walls. Complications related to arteriosclerosis account for roughly one-half of all deaths in the United States. There are many different forms of arteriosclerosis; one example is coronary artery disease (CAD), which was described in Chapter 21. ∞ [p. 533]

Atherosclerosis (ath-er-ō-skle-RŌ-sis) is a type of arteriosclerosis characterized by changes in the endothelial lining. Evidence now indicates that this condition begins when smooth muscle fibers near the tunica interna begin dividing repeatedly. Monocytes then invade the area, migrating between the endothelial cells, and both monocytes and smooth muscle fibers begin phagocytizing circulating lipids, primarily cholesterol. Even the endothelial cells begin to accumulate lipids. As the cells become engorged, gaps soon appear in the endothelial lining, and platelets begin sticking to the exposed collagen fibers. The result is a **plaque**, a fatty mass of tissue that projects into the lumen of the vessel (Figure 22-5●).

Potential treatments for atherosclerotic plaques include *catheterization* and *balloon angioplasty*. ∞ [p. 533] However, the best approach is to try to avoid atherosclerosis by eliminating or reducing risk factors. Suggestions include: (1) reducing the amount of dietary cholesterol and saturated fats, by restricting consumption of fatty meats (such as beef, lamb, and pork), egg yolks, and cream; (2) giving up smoking; (3) checking your blood pressure, and taking steps to lower it if necessary; (4) having your blood cholesterol levels checked at annual physical examinations; (5) controlling your weight; and (6) exercising regularly.

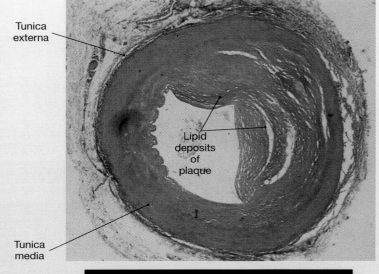

Tunica externa

Lipid deposits of plaque

Tunica media

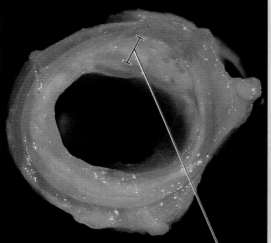

Plaque deposit

FIGURE 22-5
A Plaque Blocking a Peripheral Artery (LM × 28)

anastomoses are direct connections between arterioles and venules (Figure 22-4a●). Smooth muscles in the walls of these vessels can contract or relax to regulate the amount of blood reaching the capillary bed. For example, when the arteriovenous anastomoses are dilated, blood will bypass the capillary bed and flow directly into the venous circulation. Collateral blood flow is regulated primarily by sympathetic innervation, under the control of cardiovascular centers in the medulla oblongata. ∞ [p. 389]

Veins
(Figures 22-1/22-2)

Veins collect blood from all tissues and organs and return it to the heart. The walls of veins are thinner than those of corresponding arteries because the blood pressure in veins is lower than that in arteries. Veins are classified on the basis of their size, and in general veins are larger in diameter than their corresponding arteries. Review Figures 22-1, p. 545, and 22-2●, p. 546, to compare typical arteries and veins.

Venules

Venules vary widely in size and character, but an average venule has an internal diameter of roughly 20 μm. The smallest venules resemble expanded capillaries, and venules smaller than 50 μm lack a tunica media altogether.

Medium-Sized Veins

Medium-sized veins range from 2 to 9 mm in internal diameter and correspond in general size to medium-sized arteries. In these veins, the tunica media is thin, and it contains relatively few smooth muscle fibers. The thickest layer of a medium-sized vein is the tunica externa, which contains longitudinal bundles of elastic and collagen fibers.

Large Veins

Large veins include the superior and inferior venae cavae and their tributaries within the abdominopelvic and thoracic cavities. All the tunica layers are thickest in large veins. The slender tunica media is surrounded by a thick tunica externa composed of a mixture of elastic and collagenous fibers.

Venous Valves
(Figure 22-6)

The blood pressure in venules and medium-sized veins is too low to oppose the force of gravity. In the extremities, veins of this size contain one-way **valves** (Figure 22-6●). These valves act like the valves in the heart, preventing the backflow of blood. As long as the valves function normally, any movement that distorts or compresses a vein will push blood toward the heart. For example, when you are standing, blood returning from the foot must overcome the pull of gravity to ascend to the heart. Valves compartmentalize the blood within the veins, thereby dividing the weight of the blood between the compartments. Any movement in the

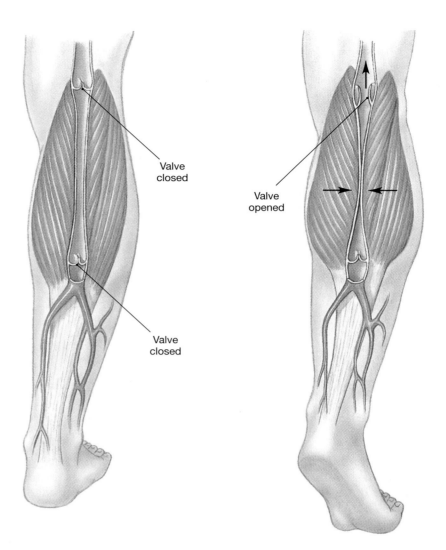

FIGURE 22-6
Function of Valves in the Venous System. Valves in the walls of medium-sized veins prevent the backflow of blood. Venous compression caused by the contraction of adjacent skeletal muscles assists in maintaining venous blood flow. Changes in body position and the thoracoabdominal pump may provide additional assistance.

surrounding skeletal muscles squeezes the blood toward the heart. Large veins such as the vena cava do not have valves, but changes of pressure in the thoracic cavity assist in moving blood toward the heart. ✝*Problems with Venous Valve Function [p. 768]*

The Distribution of Blood
(Figure 22-7)

The total blood volume is unevenly distributed among arteries, veins, and capillaries (Figure 22-7●). The heart, arteries, and capillaries normally contain 30–35% of the blood volume (roughly 1.5 ℓ of whole blood), and the venous system contains the rest (65–70% or around 3.5 ℓ).

Because their walls are thinner and contain a lower proportion of smooth muscle, veins are much more distensible than arteries. For a given rise in pressure, a typical vein will stretch about eight times as much as a corresponding artery. If the blood volume rises or falls, the elastic walls stretch or recoil, changing the volume of blood in the venous system.

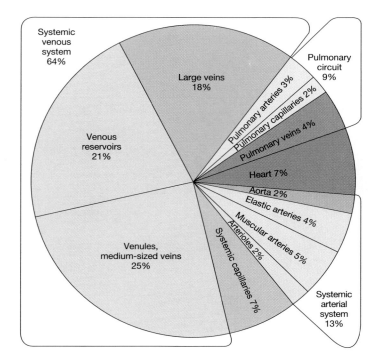

FIGURE 22-7
The Distribution of Blood in the Cardiovascular System

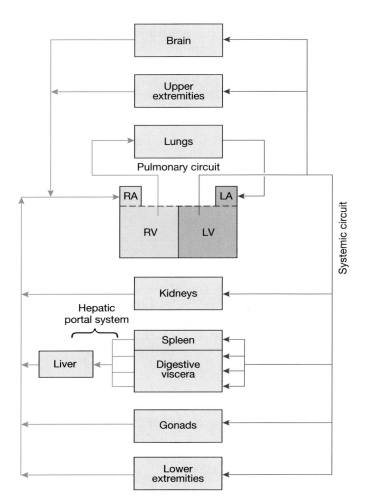

FIGURE 22-8
An Overview of the Pattern of Circulation

If serious hemorrhaging occurs, the vasomotor center of the medulla oblongata stimulates sympathetic nerves innervating smooth muscle cells in the walls of medium-sized veins. When the smooth muscles in the walls contract, this **venoconstriction** (vē-nō-kon-STRIK-shun) further reduces the volume of the venous system. In addition, blood enters the general circulation from venous networks in the liver, bone marrow, and skin. Reducing the amount of blood in the venous system enables the volume within the arterial system to be maintained at near-normal levels, despite a significant blood loss. The venous system therefore acts as a **blood reservoir**, and the change in volume constitutes the **venous reserve**. The venous reserve normally amounts to just over 1 ℓ, 21% of the total blood volume.

√ **Examination of a cross section of tissue shows several small, thin-walled vessels with very little smooth muscle tissue in the tunica media. What type of vessels are these?**

√ **Why are valves found in veins but not in arteries?**

THE CIRCULATORY SYSTEM
(Figure 22-8)

The circulatory system is divided into the pulmonary circuit and the systemic circuit. The pulmonary circuit is composed of arteries and veins that transport blood between the heart and the lungs. From the heart, the arteries of the systemic circuit transport oxygenated blood and nutrients to all organs and tissues. Figure 22-8● summarizes the primary circulatory routes within the pulmonary and systemic circuits.

Three important functional patterns will emerge from the tables and figures that follow:

1. The peripheral distribution of arteries and veins on the left and right sides is usually identical except near the heart, where the largest vessels connect to the atria or ventricles.

2. A single vessel may have several different names as it crosses specific anatomical boundaries, making accurate anatomical descriptions possible when the vessel extends far into the periphery.

3. Arteries and veins often make anastomotic connections that reduce the impact of a temporary or even permanent occlusion (blockage) of a single vessel.

The Pulmonary Circulation
(Figure 22-9)

Blood entering the right atrium has just returned from the peripheral capillary beds, where oxygen was released and carbon dioxide absorbed. After traveling through the right atrium and ventricle, blood enters the pulmonary trunk, the start of the pulmonary circuit. In this circuit (Figure 22-9a●), oxygen will be replenished, carbon dioxide excreted, and the oxygenated

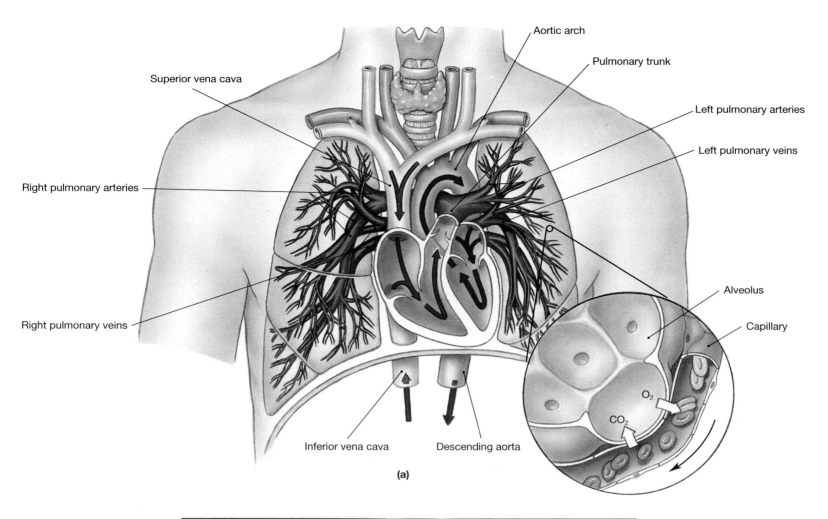

(a)

Aortic arch

Pulmonary trunk

Left pulmonary arteries

Left pulmonary veins

Superior vena cava

Right pulmonary arteries

Right pulmonary veins

Alveolus

Capillary

O_2

CO_2

Inferior vena cava

Descending aorta

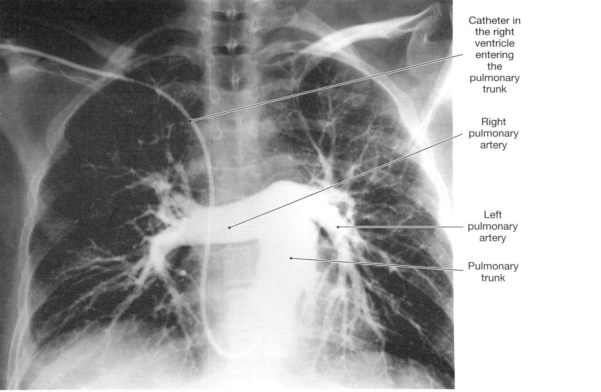

(b)

Catheter in the right ventricle entering the pulmonary trunk

Right pulmonary artery

Left pulmonary artery

Pulmonary trunk

FIGURE 22-9
The Pulmonary Circuit. (a) Anatomy of the pulmonary circuit. (b) Coronary angiogram, showing heart, vessels of the pulmonary circuit, diaphragm, and lungs.

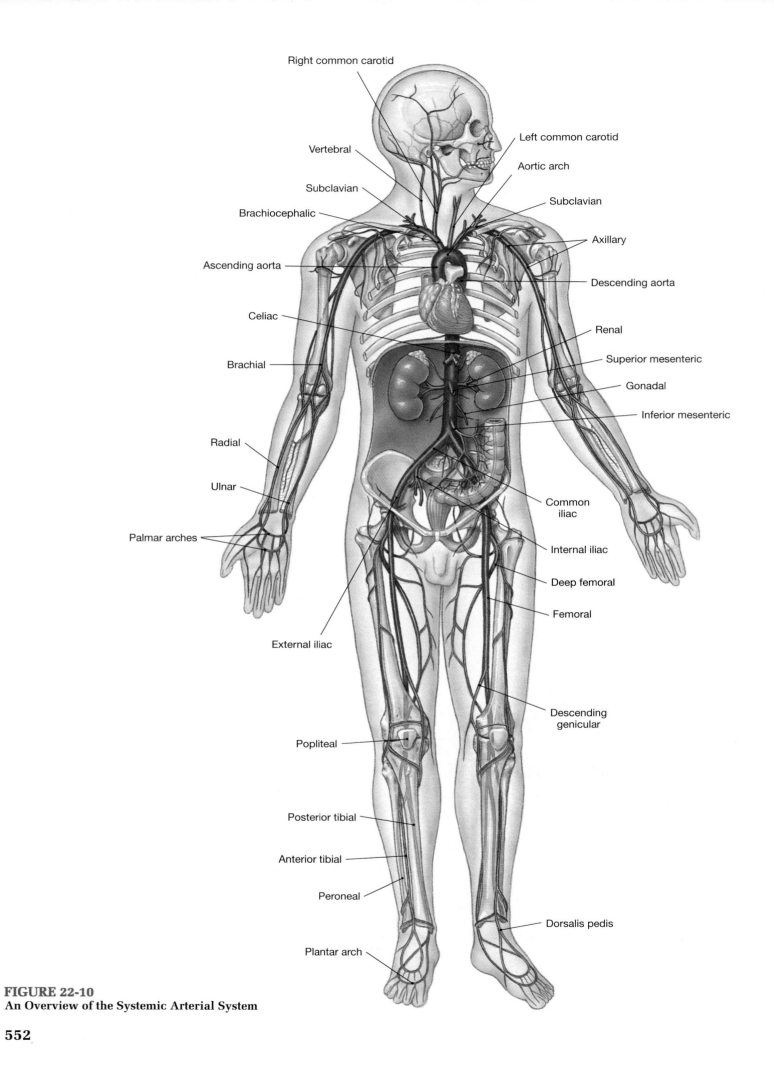

Right common carotid

Vertebral

Subclavian

Brachiocephalic

Ascending aorta

Celiac

Brachial

Radial

Ulnar

Palmar arches

External iliac

Popliteal

Posterior tibial

Anterior tibial

Peroneal

Plantar arch

Left common carotid

Aortic arch

Subclavian

Axillary

Descending aorta

Renal

Superior mesenteric

Gonadal

Inferior mesenteric

Common iliac

Internal iliac

Deep femoral

Femoral

Descending genicular

Dorsalis pedis

FIGURE 22-10
An Overview of the Systemic Arterial System

552

blood returned to the heart for distribution via the systemic circuit. Compared with the systemic circuit, the pulmonary circuit is relatively short; the base of the pulmonary trunk and lungs are only about 15 cm (6 in.) apart.

The arteries of the pulmonary circuit differ from those of the systemic circuit in that they carry deoxygenated blood. (For this reason, color-coded diagrams usually show the pulmonary arteries in blue, the same color as systemic veins.) As the pulmonary trunk curves over the superior border of the heart, it gives rise to the **left** and **right pulmonary arteries**. These large arteries enter the lungs before branching repeatedly, giving rise to smaller and smaller arteries. The smallest branches, the *pulmonary arterioles*, provide blood to capillary networks that surround small air pockets, or **alveoli** (al-VĒ-ō-lī; *alveolus*, sac). The walls of alveoli are thin enough for gas exchange to occur between the capillary blood and inspired air. (Alveolar structure is described in Chapter 24.) As it leaves the alveolar capillaries, oxygenated blood enters venules that in turn unite to form larger vessels carrying blood toward the **pulmonary veins**. These four veins, two from each lung, empty into the left atrium, completing the pulmonary circuit. Figure 22-9b● is a coronary angiogram showing the vessels of the pulmonary circuit and their relationship to the heart and lungs in a living subject.

The Systemic Circulation

The systemic circulation supplies the capillary beds in all other parts of the body. This circuit, which at any given moment contains about 84% of the total blood volume, begins at the left ventricle and ends at the right atrium.

Systemic Arteries
(Figures 22-10 to 22-20)

Figure 22-10● is an overview of the arterial system. This figure indicates the relative locations of major systemic arteries. The detailed distribution of these vessels and their branches will be found in Figures 22-11 to 22-20●.

THE ASCENDING AORTA *(Figures 21-5a/22-11).* The **ascending aorta** begins at the aortic semilunar valve of the left ventricle (Figures 21-5a and 22-11●). ∞ [p. 527] The left and right coronary arteries originate at the base of the ascending aorta, just superior to the aortic semilunar valve. The distribution of coronary vessels was described in Chapter 21 and illustrated in Figure 21-8. ∞ [p. 532]

THE AORTIC ARCH *(Figures 22-10 to 22-13).* The **aortic arch** curves like a cane handle across the superior surface of the heart, connecting the ascending aorta with the *descending aorta*. Three elastic arteries originate along the aortic arch (Figures 22-10, 22-11, and 22-12●). These arteries, the **brachiocephalic** (brā-kē-ō-se-FAL-ik) **trunk**, the **left common carotid**, and the **left subclavian** (sub-CLĀ-vē-an), deliver blood to the head, neck, shoulders, and upper extremities. The brachiocephalic trunk, also called the *innominate artery* (i-NOM-i-nāt; "unnamed"), ascends for a short distance before branching to form the **right subclavian artery** and the **right common carotid artery**. There is only one brachiocephalic artery, and the left common carotid and left subclavian arteries arise separately from the aortic arch. However, in terms of their peripheral distribution, the vessels on the left side are mirror images of those on the right side. Because the descriptions that follow focus attention on major branches found on both sides of

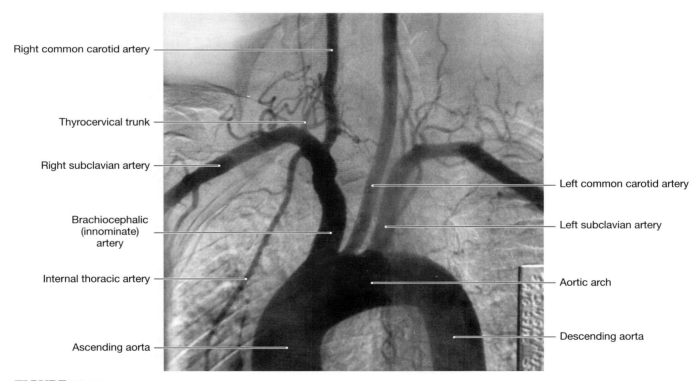

FIGURE 22-11
Aortic Angiogram. Observe the ascending aorta, the arch of the aorta, the descending aorta, the brachiocephalic trunk branching into the right subclavian and right common carotid arteries, and the left subclavian and left common carotid arteries arising directly from the aorta.

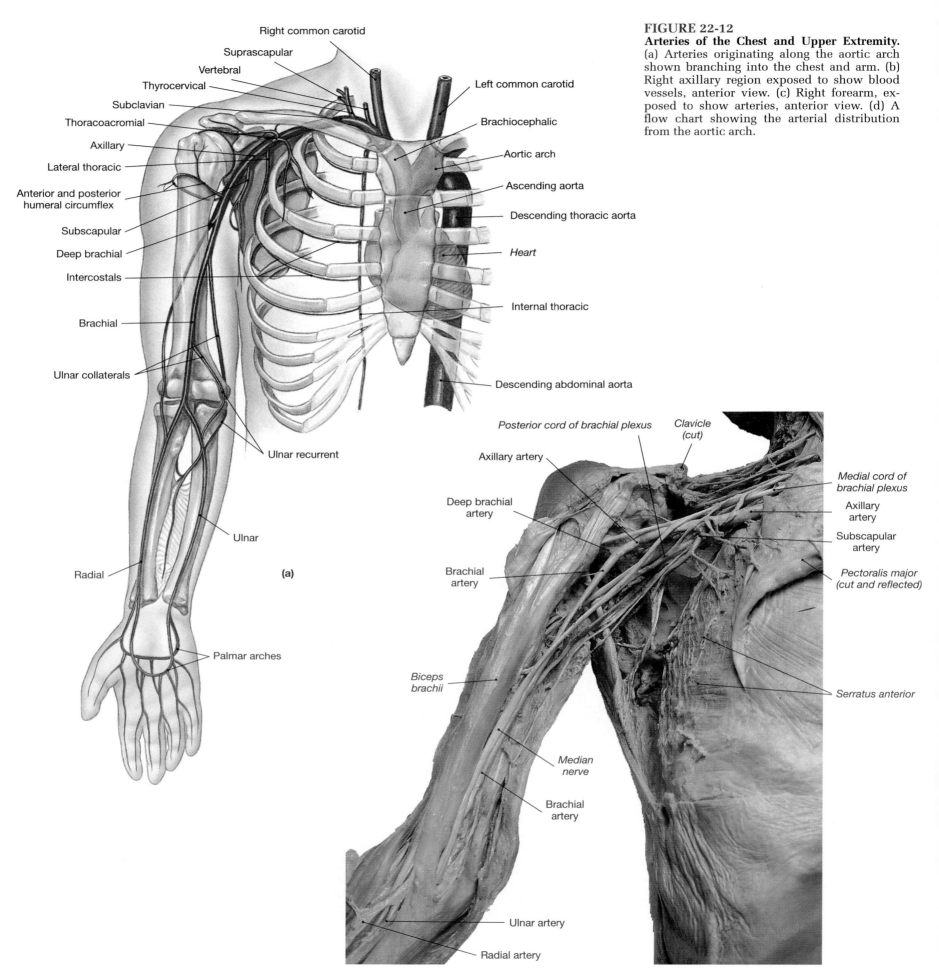

FIGURE 22-12
Arteries of the Chest and Upper Extremity.
(a) Arteries originating along the aortic arch shown branching into the chest and arm. (b) Right axillary region exposed to show blood vessels, anterior view. (c) Right forearm, exposed to show arteries, anterior view. (d) A flow chart showing the arterial distribution from the aortic arch.

Right common carotid
Suprascapular
Vertebral
Thyrocervical
Subclavian
Thoracoacromial
Axillary
Lateral thoracic
Anterior and posterior humeral circumflex
Subscapular
Deep brachial
Intercostals
Brachial
Ulnar collaterals
Ulnar recurrent
Ulnar
Radial
Palmar arches

Left common carotid
Brachiocephalic
Aortic arch
Ascending aorta
Descending thoracic aorta
Heart
Internal thoracic
Descending abdominal aorta

(a)

Posterior cord of brachial plexus
Clavicle (cut)
Axillary artery
Deep brachial artery
Brachial artery
Biceps brachii
Median nerve
Brachial artery
Ulnar artery
Radial artery

Medial cord of brachial plexus
Axillary artery
Subscapular artery
Pectoralis major (cut and reflected)
Serratus anterior

(b)

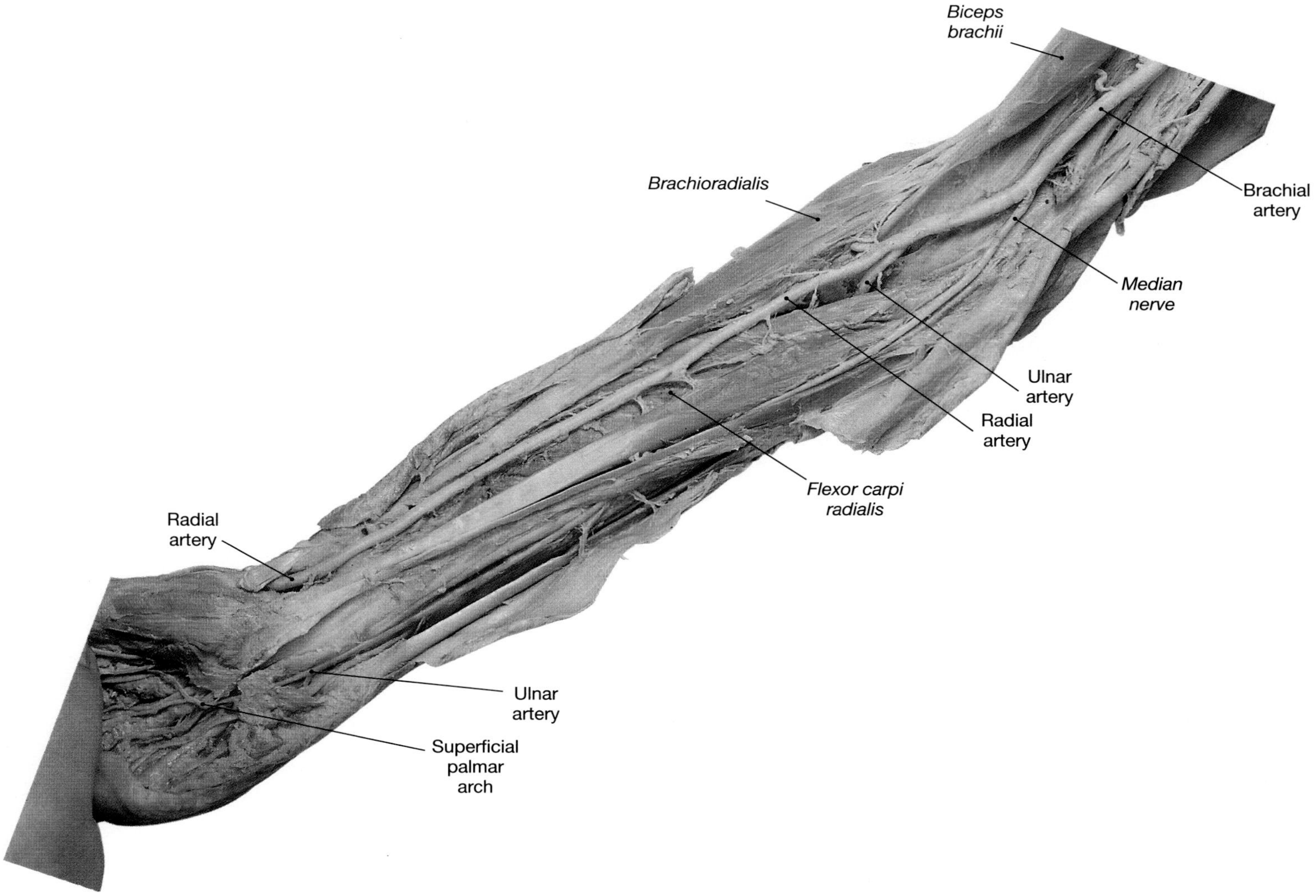
Biceps brachii
Brachioradialis
Brachial artery
Median nerve
Ulnar artery
Radial artery
Flexor carpi radialis
Radial artery
Ulnar artery
Superficial palmar arch

(c)

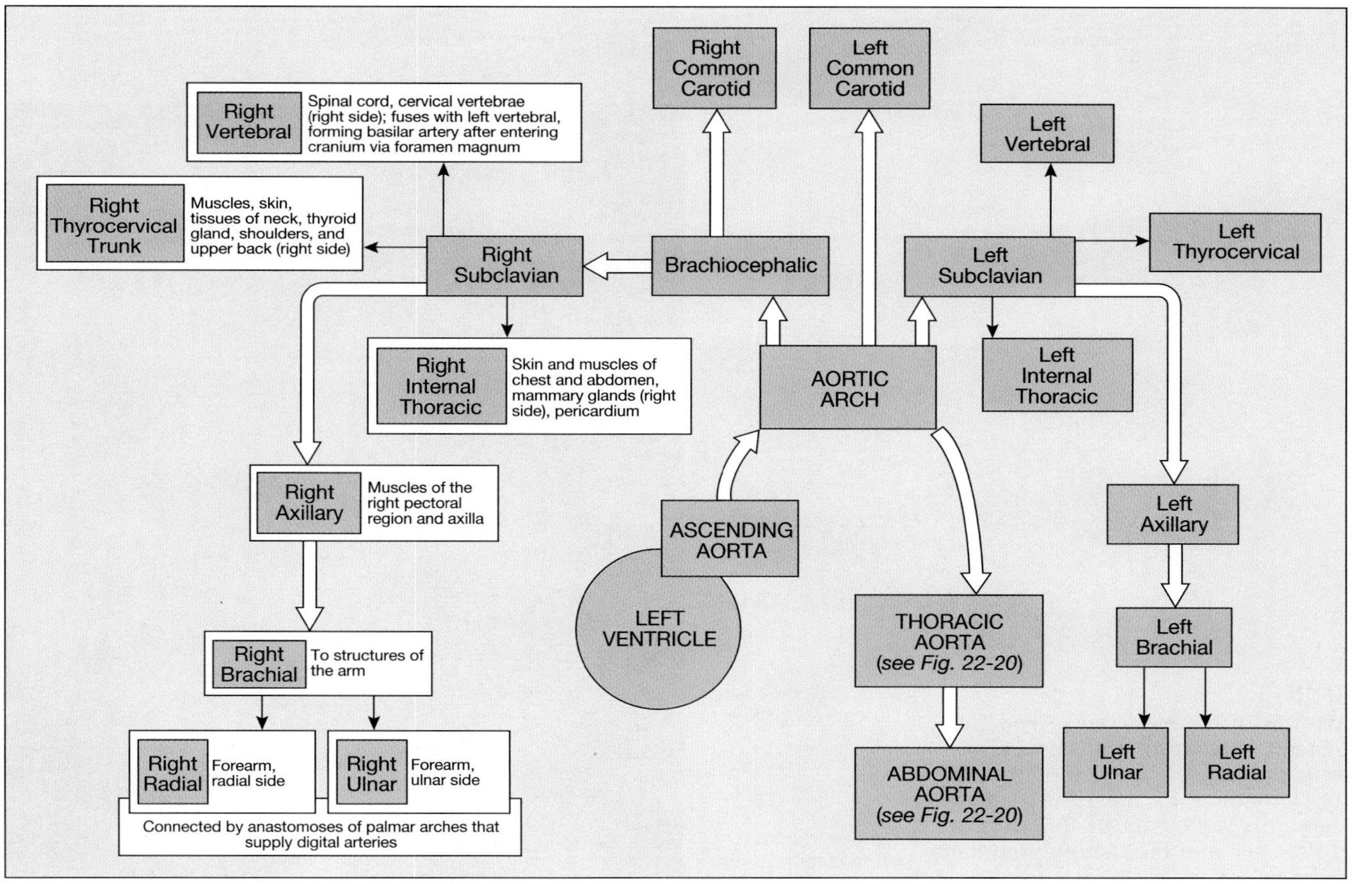
Right Common Carotid
Left Common Carotid
Right Vertebral
Spinal cord, cervical vertebrae (right side); fuses with left vertebral, forming basilar artery after entering cranium via foramen magnum
Left Vertebral
Right Thyrocervical Trunk
Muscles, skin, tissues of neck, thyroid gland, shoulders, and upper back (right side)
Right Subclavian
Brachiocephalic
Left Subclavian
Left Thyrocervical
Right Internal Thoracic
Skin and muscles of chest and abdomen, mammary glands (right side), pericardium
AORTIC ARCH
Left Internal Thoracic
Right Axillary
Muscles of the right pectoral region and axilla
ASCENDING AORTA
LEFT VENTRICLE
THORACIC AORTA (see Fig. 22-20)
Left Axillary
Right Brachial
To structures of the arm
Left Brachial
Right Radial
Forearm, radial side
Right Ulnar
Forearm, ulnar side
Connected by anastomoses of palmar arches that supply digital arteries
ABDOMINAL AORTA (see Fig. 22-20)
Left Ulnar
Left Radial

(d)

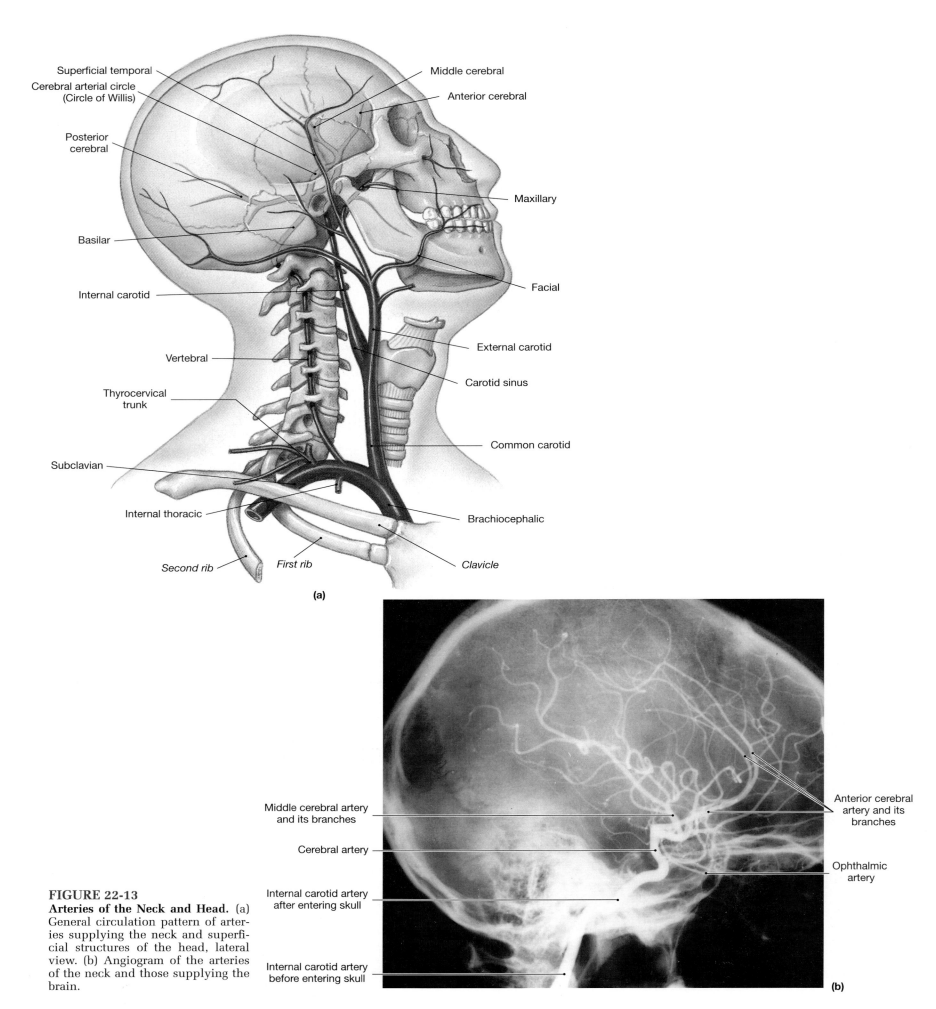

FIGURE 22-13
Arteries of the Neck and Head. (a) General circulation pattern of arteries supplying the neck and superficial structures of the head, lateral view. (b) Angiogram of the arteries of the neck and those supplying the brain.

(a)

Superficial temporal
Cerebral arterial circle (Circle of Willis)
Posterior cerebral
Basilar
Internal carotid
Vertebral
Thyrocervical trunk
Subclavian
Internal thoracic
Second rib
First rib

Middle cerebral
Anterior cerebral
Maxillary
Facial
External carotid
Carotid sinus
Common carotid
Brachiocephalic
Clavicle

(b)

Middle cerebral artery and its branches
Cerebral artery
Internal carotid artery after entering skull
Internal carotid artery before entering skull

Anterior cerebral artery and its branches
Ophthalmic artery

the body, for clarity the terms "right" or "left" will not be used. Figures 22-12 and 22-13● illustrate the major branches of these arteries.

The Subclavian Arteries *(Figures 22-10/22-12).*

The subclavian arteries supply blood to the arms, chest wall, shoulders, back, and central nervous system (Figures 22-10, p. 552, and 22-12●, pp. 554, 555). Three major branches arise before a subclavian artery leaves the thoracic cavity: (1) the **thyrocervical trunk**, which provides blood to muscles and other tissues of the neck, shoulder, and upper back; (2) an **internal thoracic artery**, supplying the pericardium and anterior wall of the chest; and (3) a **vertebral artery**, which provides blood to the brain and spinal cord.

After leaving the thoracic cavity and passing over the outer border of the first rib, the subclavian artery becomes the **axillary artery**. The axillary artery crosses the axilla to enter the arm, where it becomes the **brachial artery**. The brachial artery supplies blood to the upper extremity. At the cubital fossa, the brachial artery divides into the **radial artery**, which follows the radius, and the **ulnar artery**, which follows the ulna to the wrist. These arteries supply blood to the forearm. At the wrist, these arteries anastomose to form a **superficial palmar arch** and a **deep palmar arch** that respectively supply blood to the palm and to the **digital arteries** of the thumb and fingers.

The Carotid Arteries and the Blood Supply to the Brain *(Figures 22-13/22-14).*

The common carotid arteries ascend deep in the tissues of the neck. A carotid artery can usually be located by pressing gently along either side of the windpipe (trachea) until a strong pulse is felt. Each common carotid artery divides into an **external carotid** and an **internal carotid artery** at an expanded chamber, the **carotid sinus** (Figure 22-13●). (This sinus, which contains baroreceptors involved in cardiovascular regulation, was introduced in Chapter 18.) ∞ [p. 448] The external carotids supply blood to the structures of the neck, pharynx, esophagus, larynx, lower jaw, and face. The internal carotids enter the skull through the *carotid foramina* of the temporal bones, delivering blood to the brain. ∞ [p. 138] The internal carotids ascend to the level of the optic nerves, where each divides into three branches: (1) an **ophthalmic artery**, which supplies the eyes; (2) an **anterior cerebral artery**, which supplies the frontal and parietal lobes of the brain; and (3) a **middle cerebral artery**, which supplies the midbrain and lateral surfaces of the cerebral hemispheres (Figures 22-13 and 22-14●).

The brain is extremely sensitive to changes in its circulatory supply. An interruption of circulation for several seconds will produce unconsciousness, and after 4 minutes there may be some permanent neural damage. Such circulatory crises are rare, because blood reaches the brain through the vertebral arteries as well as by way of the internal carotids. The left and right vertebral arteries arise from the subclavian arteries and ascend within the *transverse foramina* of the cervical vertebrae. ∞ [p. 161] The vertebral arteries enter the cranium at the foramen magnum, where they fuse along the ventral surface of the medulla oblongata to form the **basilar artery**. The basilar artery continues on the ventral surface along the pons, branching many times before dividing into the **posterior cerebral arteries**. The **posterior communicating arteries** branch off the posterior cerebral arteries (Figures 22-13a and 22-14a●).

The internal carotids normally supply the arteries of the anterior half of the cerebrum, and the rest of the brain receives blood from the vertebral arteries. But this circulatory pattern can easily change, because the internal carotids and the basilar artery are interconnected in a ring-shaped anastomosis, the **cerebral arterial circle**, or *circle of Willis*, that encircles the infundibulum of the pituitary gland. With this arrangement, the brain can receive blood from either the carotids or the vertebrals, and the chances for a serious interruption of circulation are reduced.

THE DESCENDING AORTA *(Figures 22-10/22-20).*

The **descending aorta** is continuous with the aortic arch. The diaphragm divides the descending aorta into a superior **thoracic aorta** and an inferior **abdominal aorta** (Figure 22-10●, p. 552). A summary of the distribution of blood from the descending aorta is presented in Figure 22-20●, p. 564.

The Thoracic Aorta *(Figure 22-15).*

The thoracic aorta begins at the level of vertebra T_5 and penetrates the diaphragm at the level of vertebra T_{12}. The thoracic aorta travels within the mediastinum, on the dorsal thoracic wall, just slightly to the left of the vertebral column. It supplies blood to branches servicing the viscera of the thorax, the muscles of the chest and the diaphragm, and the thoracic spinal cord. The branches of the thoracic aorta are anatomically grouped as either *visceral branches* or *parietal branches*. Visceral branches supply the organs of the chest: the **bronchial arteries** supply the nonrespiratory tissues of the lungs, **pericardial arteries** supply the pericardium, **mediastinal arteries** supply general mediastinal structures, and **esophageal arteries** supply the esophagus. The parietal branches supply the chest wall: the **intercostal arteries** supply the chest muscles and the vertebral column area, and the **superior phrenic** (FREN-ik) **arteries** deliver blood to the superior surface of the muscular diaphragm that separates the thoracic and abdominopelvic cavities. The branches of the thoracic aorta are detailed in Figure 22-15●.

The Abdominal Aorta *(Figures 22-15 to 22-17).*

The abdominal aorta is a continuation of the thoracic aorta, immediately inferior to the diaphragm (Figures 22-15 and 22-16●, p. 560). The abdominal aorta descends slightly to the left of the vertebral column, but posterior to the peritoneal cavity, and is often surrounded by a cushion of adipose tissue. At the level of vertebra L_4, the abdominal aorta splits into two major arteries that supply deep pelvic structures and the lower extremities. The region where the aorta splits is called the **terminal segment of the aorta**.

The abdominal aorta delivers blood to all of the abdominopelvic organs and structures. The major branches to visceral organs are unpaired, and they arise on the anterior surface of the abdominal aorta and extend into the mesenteries. Branches to the body wall, the kidneys, and other structures outside of the abdominopelvic cavity are paired, and they originate along the lateral surfaces of the abdominal aorta. Figures 22-15 and 22-16●, p. 560, show the major arteries of the trunk, with the thoracic and abdominal organs removed.

There are three unpaired arteries: (1) the *celiac artery*, (2) the *superior mesenteric artery*, and (3) the *inferior mesenteric artery* (Figures 19-10a, 22-15, and 22-17●, p. 561). ∞ [p. 493]

  1. The **celiac** (SĒ-lē-ak) **artery** delivers blood to the liver, stomach, and spleen. The celiac artery divides into three branches:

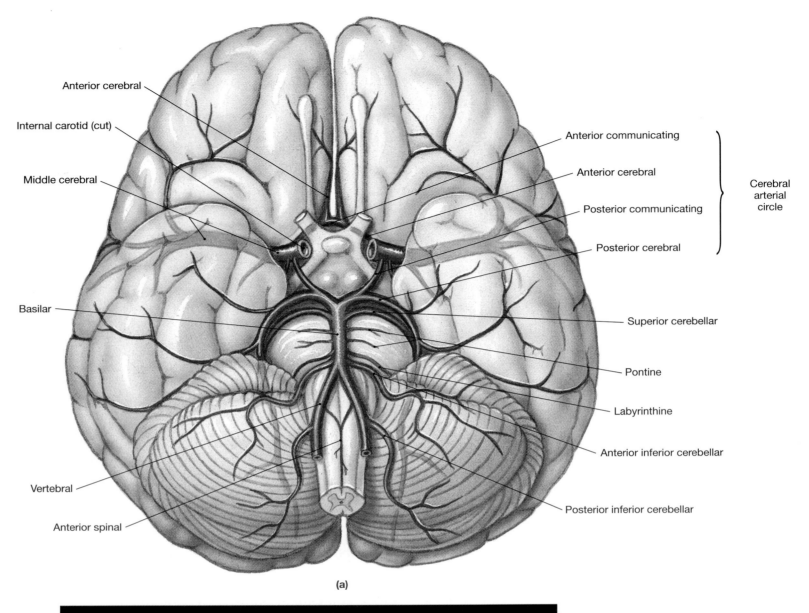

Anterior cerebral

Internal carotid (cut)

Middle cerebral

Basilar

Vertebral

Anterior spinal

Anterior communicating

Anterior cerebral

Posterior communicating

Posterior cerebral

Cerebral arterial circle

Superior cerebellar

Pontine

Labyrinthine

Anterior inferior cerebellar

Posterior inferior cerebellar

(a)

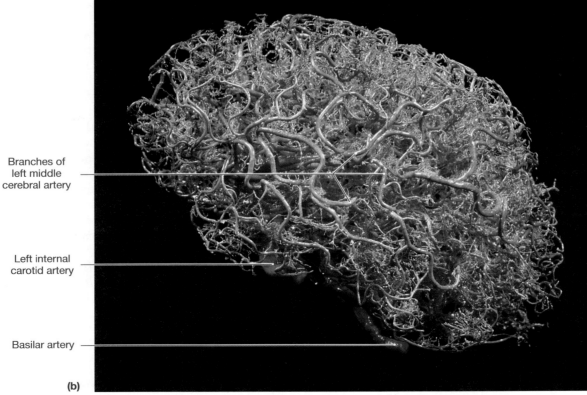

Branches of left middle cerebral artery

Left internal carotid artery

Basilar artery

(b)

FIGURE 22-14

The Arterial Supply to the Brain. (a) Distribution of arteries to the brain, inferior view. (b) Corrosion cast of brain circulatory network, showing distribution of cerebral arteries of the left hemisphere. *See Figure 22-22b* to compare the veins of the brain.

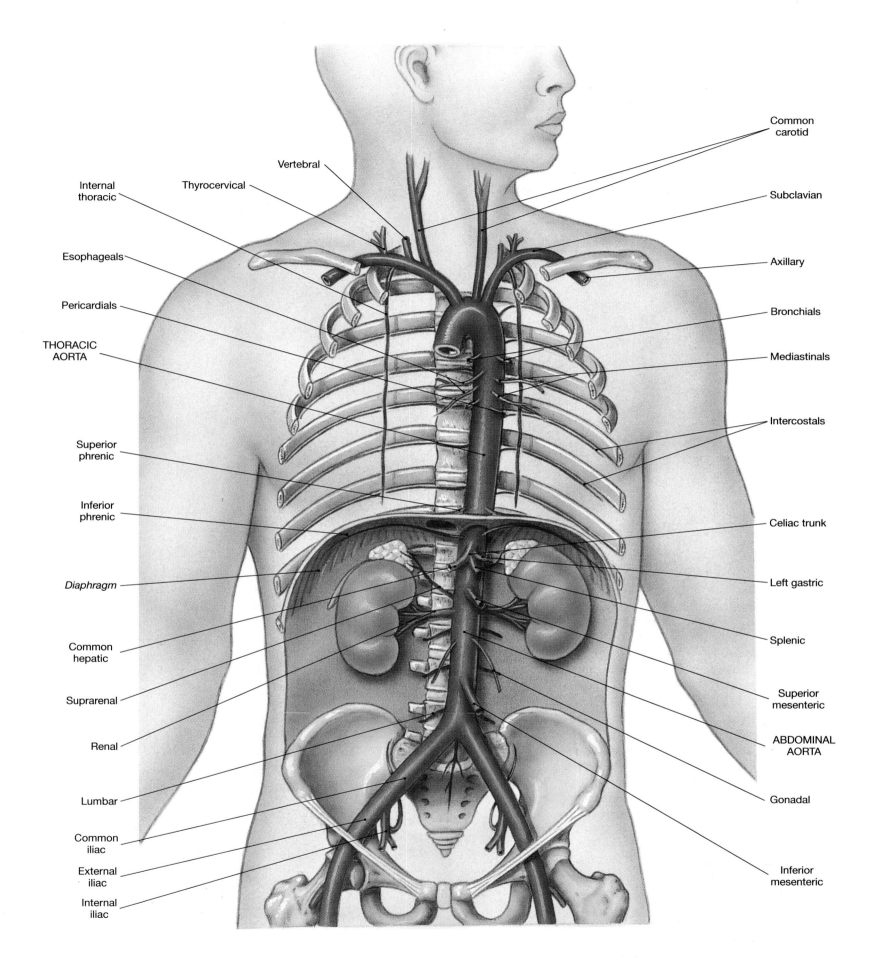

Common carotid

Vertebral

Internal thoracic

Thyrocervical

Subclavian

Esophageals

Axillary

Pericardials

Bronchials

THORACIC AORTA

Mediastinals

Superior phrenic

Intercostals

Inferior phrenic

Celiac trunk

Diaphragm

Left gastric

Common hepatic

Splenic

Suprarenal

Superior mesenteric

Renal

ABDOMINAL AORTA

Lumbar

Gonadal

Common iliac

External iliac

Inferior mesenteric

Internal iliac

FIGURE 22-15
Major Arteries of the Trunk. Diagrammatic view.

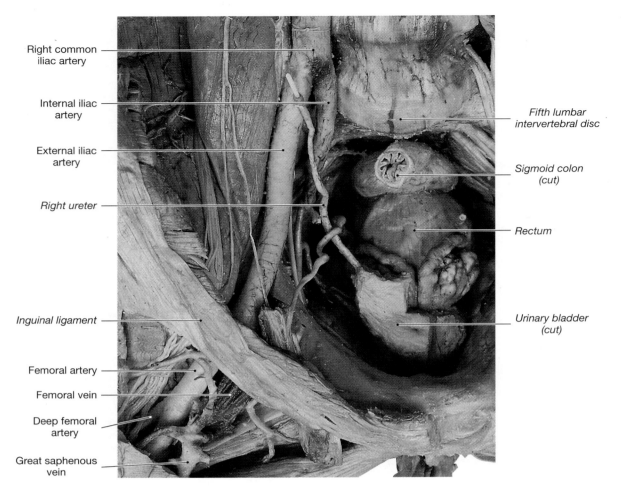

Right common
iliac artery

Internal iliac
artery

External iliac
artery

Right ureter

Inguinal ligament

Femoral artery

Femoral vein

Deep femoral
artery

Great saphenous
vein

*Fifth lumbar
intervertebral disc*

*Sigmoid colon
(cut)*

Rectum

*Urinary bladder
(cut)*

FIGURE 22-16
Major Arteries of the Pelvis. Most of the abdominal viscera have been removed to expose these vessels.

- the **left gastric artery**, which supplies the stomach and esophagus;
- the **common hepatic artery**, which supplies arteries to the liver (*hepatic*), stomach (*right gastric*), gallbladder (*cystic*), and duodenal area (*gastroduodenal* and *right gastroepiploic*);
- the **splenic artery**, which supplies the spleen and arteries to the stomach (*left gastroepiploic*) and pancreas (*pancreatics*).

2. The **superior mesenteric** (mez-en-TER-ik) **artery** arises about 2.5 cm inferior to the celiac trunk to supply arteries to the pancreas and duodenum (*pancreaticoduodenal*), small intestine (*intestinal*), and most of the large intestine (*right* and *middle colic* and *ileocolic*).

3. The **inferior mesenteric artery** arises about 5 cm superior to the terminal segment of the aorta and delivers blood to the terminal portions of the colon (*left colic* and *sigmoid*) and the rectum (*rectal*).

Paired arteries include (1) the *suprarenals*, (2) the *renals*, (3) the *lumbars*, and (4) the *gonadals*.

1. The **suprarenal arteries** originate in the same area as the superior mesenteric artery. The suprarenal arteries supply each of the adrenal glands, which cap the superior portion of the kidneys.

2. The short (about 7.5 cm) **renal arteries** arise along the posterolateral surface of the abdominal aorta, about 2.5 cm in-

ferior to the superior mesenteric artery, and travel posterior to the peritoneal lining to reach the adrenal glands and kidneys.

3. Small **lumbar arteries** arise on the posterior surface of the aorta and supply the spinal cord and the abdominal wall.

4. Paired **gonadal** (gō-NAD-al) **arteries** originate between the superior and inferior mesenteric arteries. In males, they are called *testicular arteries* and are long, thin arteries that supply blood to the testes and scrotum. In females, they are termed *ovarian arteries* and supply blood to the ovaries, uterine tubes, and uterus. The distribution of gonadal vessels (both arteries and veins) differs in males and females; the differences will be described in Chapter 27.

ARTERIES OF THE PELVIS AND LOWER EXTREMITIES
(Figures 22-15 to 22-17). Near the level of vertebra L$_4$, the terminal segment of the abdominal aorta divides to form a pair of muscular arteries, the **right** and **left common iliac** (IL-ē-ak) **arteries**. These arteries carry blood to the pelvis and lower extremities (Figures 22-15, 22-16, and 22-17●). As these arteries travel along the inner surface of the ilium, they descend posterior to the cecum and sigmoid colon, and, at the level of the lumbosacral joint, each common iliac divides to form an **internal iliac artery** and an **external iliac artery**. The internal iliac arteries enter the pelvic cavity to supply the urinary bladder, the internal and external walls of the pelvis, the external genitalia, and the medial side of the thigh. In females, these vessels also supply the uterus and vagina. The external iliac arteries supply blood to the lower extremities, and they are much larger in diameter than the internal iliac arteries.

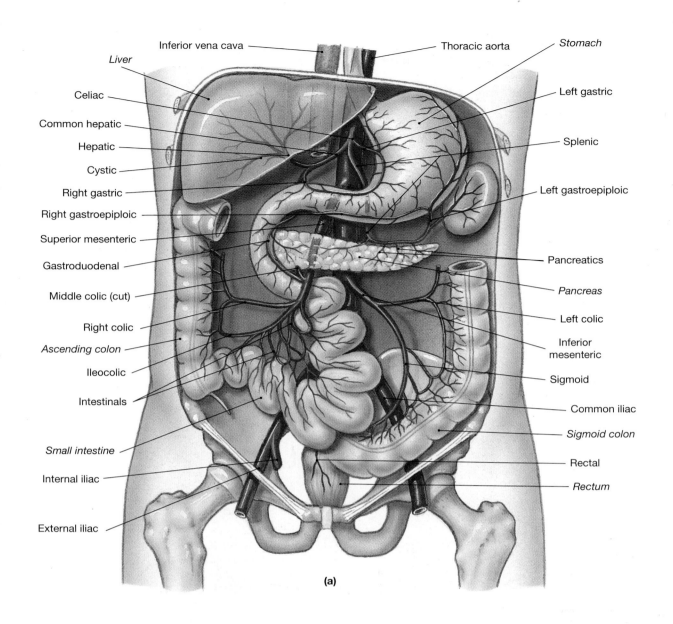

Inferior vena cava

Liver

Celiac

Common hepatic

Hepatic

Cystic

Right gastric

Right gastroepiploic

Superior mesenteric

Gastroduodenal

Middle colic (cut)

Right colic

Ascending colon

Ileocolic

Intestinals

Small intestine

Internal iliac

External iliac

Thoracic aorta

Stomach

Left gastric

Splenic

Left gastroepiploic

Pancreatics

Pancreas

Left colic

Inferior mesenteric

Sigmoid

Common iliac

Sigmoid colon

Rectal

Rectum

(a)

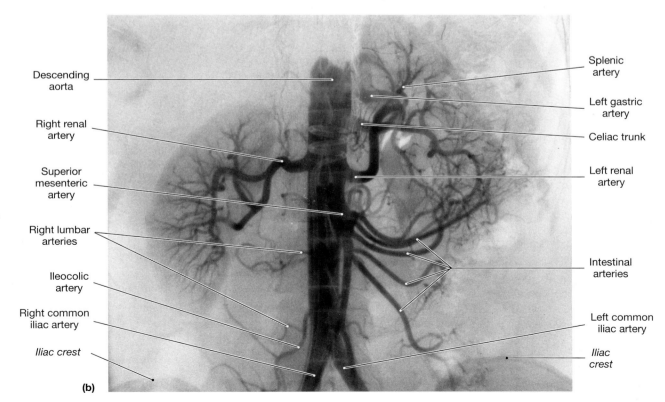

Descending aorta

Right renal artery

Superior mesenteric artery

Right lumbar arteries

Ileocolic artery

Right common iliac artery

Iliac crest

Splenic artery

Left gastric artery

Celiac trunk

Left renal artery

Intestinal arteries

Left common iliac artery

Iliac crest

(b)

FIGURE 22-17
Arteries Supplying the Abdominal Viscera. (a) Major arteries supplying the abdominal organs. (b) Angiogram of the abdominal aorta.

Arteries of the Thigh and Leg *(Figures 22-18/22-19).* The external iliac artery crosses the surface of the iliopsoas muscle and penetrates the abdominal wall midway between the anterior superior iliac spine and the symphysis pubis. It emerges on the anteromedial surface of the thigh as the **femoral artery**, and roughly 5 cm distal to its emergence, the **deep femoral artery** branches off its lateral surface (Figure 22-18●). The deep femoral artery supplies blood to the ventral and lateral regions of the skin and deep muscles of the thigh.

The femoral artery continues inferiorly and posterior to the femur. As it reaches the *popliteal fossa,* it passes through a hiatus in the adductor magnus muscle and becomes the **popliteal artery** (Figure 22-19●). The popliteal artery crosses the popliteal fossa before branching to form the **posterior tibial artery** and the **anterior tibial artery**. The posterior tibial artery gives rise to the **peroneal artery** and continues inferiorly along the posterior surface of the tibia. The anterior tibial artery passes between the tibia and fibula, emerging on the anterior surface of the tibia. As it descends toward the foot, the anterior tibial artery provides blood to the skin and muscles of the anterior portion of the leg.

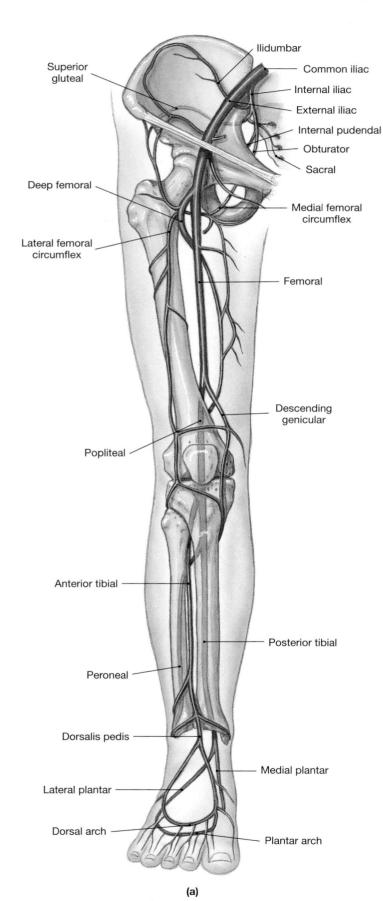

(a)

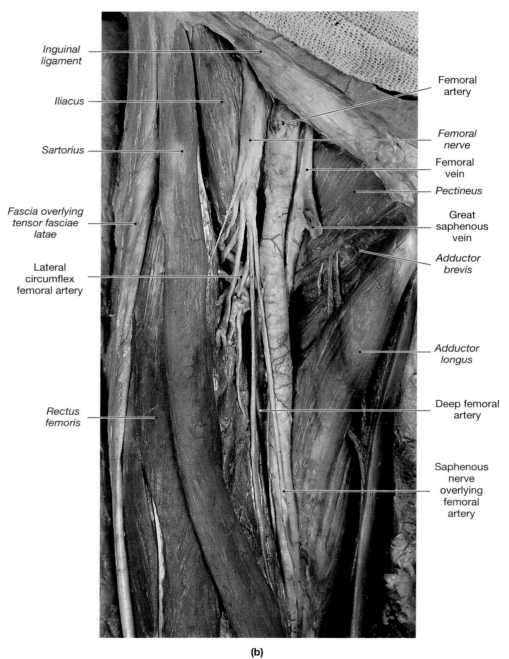

(b)

FIGURE 22-18
Major Arteries of the Lower Extremity, Anterior View. (a) Arteries of the lower extremity. (b) Major arteries of the thigh. See MRI SCANS 5, 6b, and 7, pp. 744 and 745.

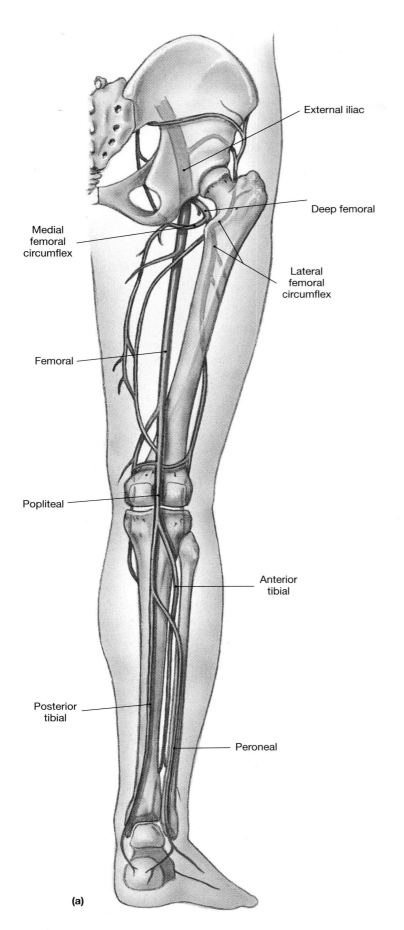

External iliac

Deep femoral

Medial femoral circumflex

Lateral femoral circumflex

Femoral

Popliteal

Anterior tibial

Posterior tibial

Peroneal

(a)

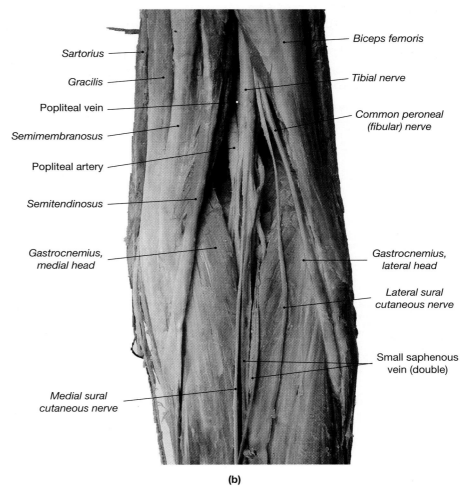

Sartorius

Gracilis

Popliteal vein

Semimembranosus

Popliteal artery

Semitendinosus

Gastrocnemius, medial head

Medial sural cutaneous nerve

Biceps femoris

Tibial nerve

Common peroneal (fibular) nerve

Gastrocnemius, lateral head

Lateral sural cutaneous nerve

Small saphenous vein (double)

(b)

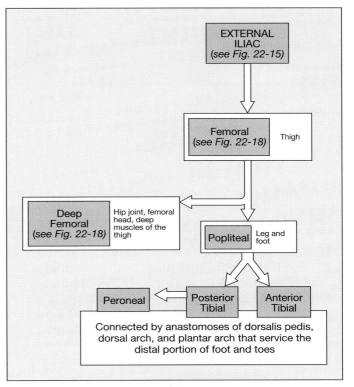

EXTERNAL ILIAC
(see Fig. 22-15)

Femoral
(see Fig. 22-18)   Thigh

Deep Femoral
(see Fig. 22-18)   Hip joint, femoral head, deep muscles of the thigh

Popliteal   Leg and foot

Peroneal   Posterior Tibial   Anterior Tibial

Connected by anastomoses of dorsalis pedis, dorsal arch, and plantar arch that service the distal portion of foot and toes

(c)

FIGURE 22-19

Major Arteries of the Lower Extremity, Posterior View. (a) Arteries of the lower extremity. (b) Vessels of the right popliteal fossa. (c) A summary of the arteries of the lower extremity. See MRI SCANS 5, 6b, and 7, pp. 744 and 745.

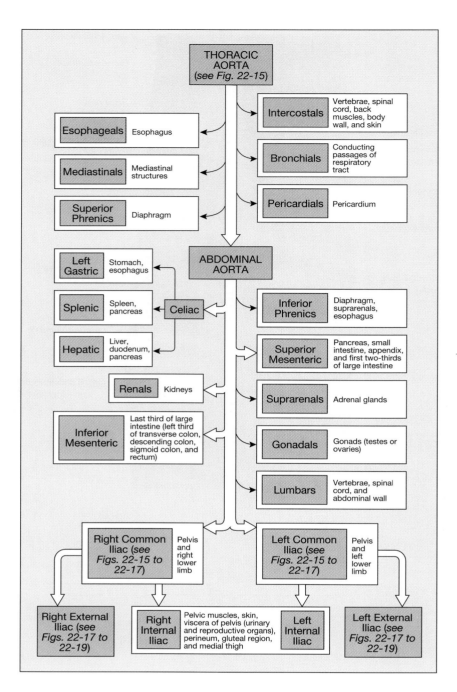

FIGURE 22-20
A Summary of the Distribution of Blood from the Aorta

Systemic Veins
(Figures 22-10/22-21)

Veins collect blood from the body's tissues and organs via an elaborate venous network that drains into the right atrium of the heart via the *superior* and *inferior vena cavae* (Figure 22-21●). A comparison between Figure 22-21 and Figure 22-10●, p. 552, reveals that complementary arteries and veins often run side by side, and in many cases they have comparable names. For example, the axillary arteries run alongside the axillary veins. In addition, arteries and veins often travel in the company of peripheral nerves that have the same names and innervate the same structures. ∞ [p. 351]

One significant difference between the arterial and venous systems concerns the distribution of major veins in the neck and extremities. Arteries in these areas are not found at the body surface; instead, they are deep beneath the skin, protected by bones and surrounding soft tissues. In contrast, the neck and extremities usually have two sets of peripheral veins, one superficial and the other deep. The superficial veins are so close to the surface that they can be seen quite easily. Because they are so close to the surface, they are easy targets for obtaining blood samples, and most blood tests are performed on venous blood collected from the superficial veins of the upper extremity (usually the antecubital surface).

This dual venous drainage plays an important role in the control of body temperature. When body temperature becomes abnormally low, the arterial blood supply to the skin is reduced, and the superficial veins are bypassed. Blood entering the limbs then returns to the trunk in the deep veins. When overheating occurs, the blood supply to the skin increases, and the superficial veins dilate. This mechanism is one reason why superficial veins in the arms and legs become prominent during periods of heavy exercise or when sitting in a sauna, hot tub, or steam bath.

The branching pattern of peripheral veins is much more variable than that of arteries. Arterial pathways are usually direct, because developing arteries grow toward active tissues. By the time blood reaches the venous system, pressures are low, and routing variations make little functional difference. The discussion that follows is based on the most common arrangement of veins.

THE SUPERIOR VENA CAVA *(Figure 22-21)*. All of the systemic veins (except the cardiac veins) drain into either the *superior vena cava* or the *inferior vena cava*. The **superior vena cava (SVC)** receives blood from the tissues and organs of the head, neck, chest, shoulders, and upper extremities (Figure 22-21●).

Venous Return from the Cranium (Figure 22-22). Numerous *superficial cerebral veins* and *internal cerebral veins* drain the cerebral hemispheres. The **superficial cerebral veins** empty into a network of dural sinuses, including the *superior* and *inferior sagittal sinuses*, the *left* and *right transverse sinuses*, and the *straight sinus* (Figure 22-22●). The largest sinus, the **superior sagittal sinus**, is in the falx cerebri. ∞ [p. 368] The majority of the **internal cerebral veins** collect inside the brain to form the **great cerebral vein**, which collects blood from the interior of the cerebral hemispheres and the choroid plexus, and delivers it to the **straight sinus**. Other cerebral veins drain into the **cavernous sinus** in company with numerous small veins from the orbit.

Arteries of the Foot (Figures 22-18/22-20). When the anterior tibial artery reaches the ankle, it becomes the **dorsalis pedis artery**. The dorsalis pedis branches repeatedly, supplying the ankle and dorsal portion of the foot (Figure 22-18●).

As it reaches the ankle, the posterior tibial artery divides to form the **medial** and **lateral plantar arteries**, which supply blood to the plantar surface of the foot. The medial and lateral plantar arteries are connected to the dorsalis pedis artery via a pair of anastomoses. This connection links the **dorsal arch** (*arcuate arch*) to the **plantar arch**. Small arteries branching off these arches supply the distal portions of the foot and the toes.

Before proceeding, review Figure 22-20●, which summarizes the distribution of blood from the abdominal aorta.

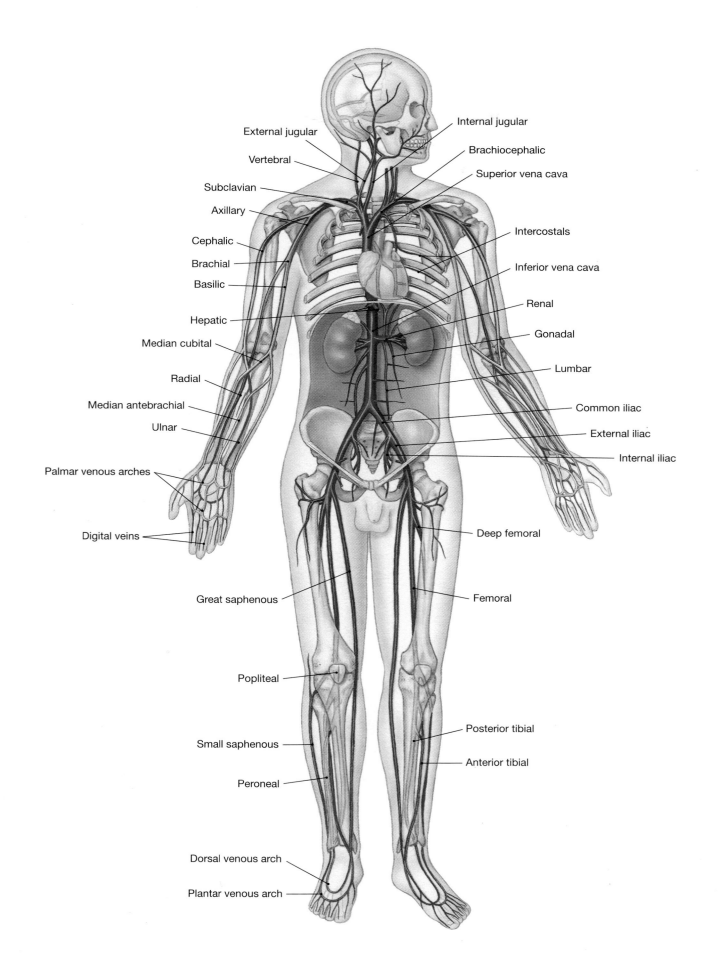

External jugular

Vertebral

Subclavian

Axillary

Cephalic

Brachial

Basilic

Hepatic

Median cubital

Radial

Median antebrachial

Ulnar

Palmar venous arches

Digital veins

Great saphenous

Popliteal

Small saphenous

Peroneal

Dorsal venous arch

Plantar venous arch

Internal jugular

Brachiocephalic

Superior vena cava

Intercostals

Inferior vena cava

Renal

Gonadal

Lumbar

Common iliac

External iliac

Internal iliac

Deep femoral

Femoral

Posterior tibial

Anterior tibial

FIGURE 22-21
An Overview of the Systemic Venous System

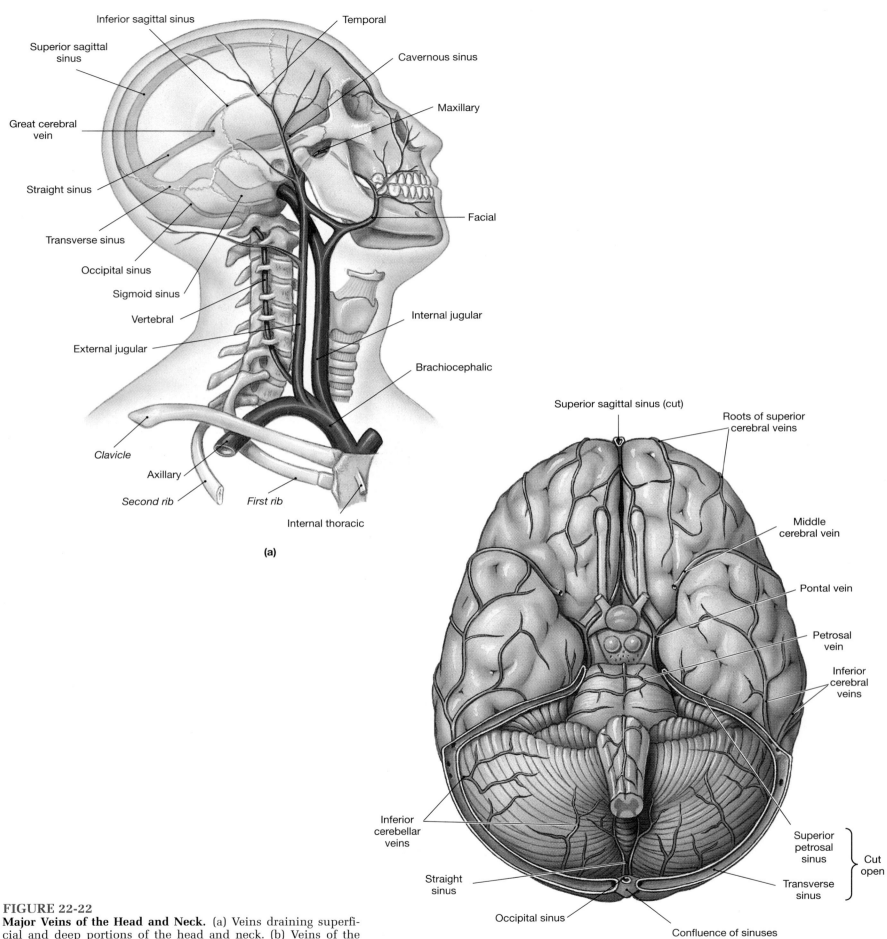

FIGURE 22-22

Major Veins of the Head and Neck. (a) Veins draining superficial and deep portions of the head and neck. (b) Veins of the brain, inferior view. *Compare with figure 22-14a (the arterial supply to the brain).*

566

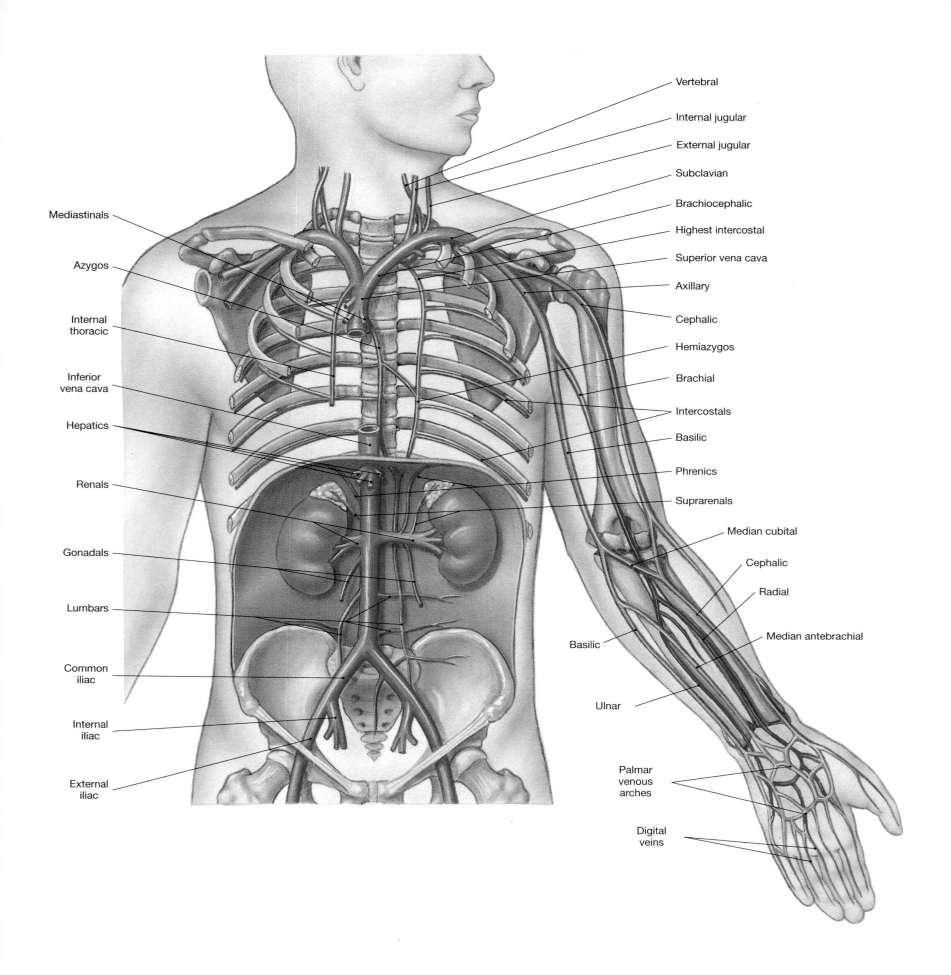

Vertebral

Internal jugular

External jugular

Subclavian

Brachiocephalic

Highest intercostal

Superior vena cava

Axillary

Cephalic

Hemiazygos

Brachial

Intercostals

Basilic

Phrenics

Suprarenals

Median cubital

Cephalic

Radial

Median antebrachial

Basilic

Ulnar

Palmar
venous
arches

Digital
veins

Mediastinals

Azygos

Internal
thoracic

Inferior
vena cava

Hepatics

Renals

Gonadals

Lumbars

Common
iliac

Internal
iliac

External
iliac

FIGURE 22-23
The Venous Drainage of the Upper Extremities, Chest, and Abdomen

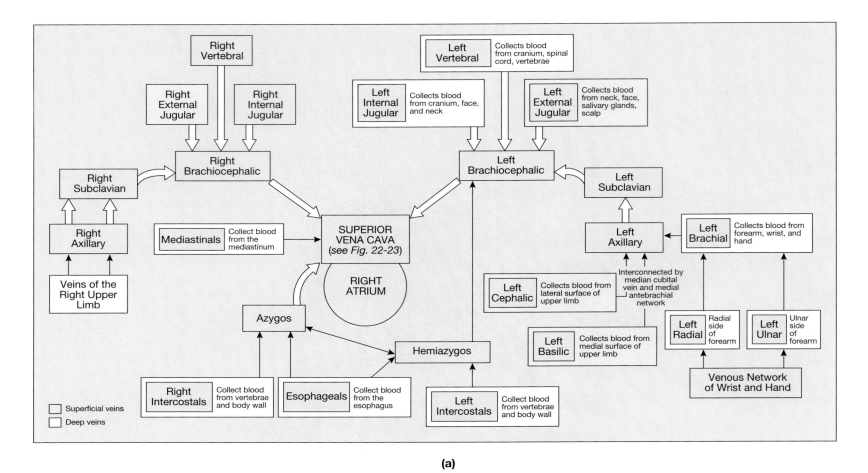

FIGURE 22-24
Summary of the Venous Tributaries of the (a) Superior and (b) Inferior Venae Cavae

The venous sinuses converge within the dura mater in the region of the lambdoidal suture. The left and right transverse sinuses converge at the base of the petrous portion of the temporal bone, forming the **sigmoid sinus**, which penetrates the jugular foramen and leaves the skull as the **internal jugular vein**. The internal jugular vein descends parallel to the common carotid artery in the neck. ∞ [p. 556]

Vertebral veins drain the cervical spinal cord and the posterior surface of the skull. These vessels descend within the transverse foramina of the cervical vertebrae, in company with the vertebral arteries. The vertebral veins empty into the *brachiocephalic veins* of the chest.

Superficial Veins of the Head and Neck *(Figure 22-22)*. Superficial veins of the head collect to form the **temporal**, **facial**, and **maxillary veins** (Figure 22-22●). The temporal and maxillary veins drain into the **external jugular vein**. The facial vein drains into the internal jugular vein; a broad anastomosis between the external and internal jugular veins at the angle of the mandible provides dual venous drainage of the face, scalp, and cranium. The external jugular vein descends toward the chest just beneath the skin, on the outer surface of the sternocleidomastoid muscle. Posterior to the clavicle, the external jugular empties into the *subclavian vein*. In healthy individuals, the external jugular vein is easily palpable, and a *jugular venous pulse* (*JVP*) can sometimes be seen at the base of the neck.

Venous Return from the Upper Extremity *(Figure 22-23)*. The **digital veins** empty into the **superficial** and **deep palmar veins** of the hand, which are interconnected to form the **palmar venous arches** (Figure 22-23●). The superficial arch empties into the **cephalic vein**, which ascends along the radial side of the forearm, the **median antebrachial vein**, and the **basilic vein**, which ascends on the ulnar side. Anterior to the elbow is the superficial **median cubital vein**. This vein passes from the cephalic vein, medially and at an oblique angle, to connect to the basilic vein. Venous blood samples are typically collected from the median cubital vein.

From the elbow, the basilic vein passes superiorly along the median surface of the biceps brachii. As it approaches the axilla, the basilic vein joins the brachial vein to form the **axillary vein** (Figure 22-23●).

The deep palmar veins drain into the **radial vein** and the **ulnar vein**. After crossing the elbow, these veins fuse to form the **brachial vein**. The brachial vein lies parallel to the brachial artery. As the brachial vein continues toward the trunk, it receives blood from the basilic vein before entering the axilla as the axillary vein.

The Formation of the Superior Vena Cava *(Figures 22-23/22-24)*. The cephalic vein joins the axillary vein on the outer surface of the first rib, forming the **subclavian vein**, which continues into the chest. The subclavian vein passes over the surface of the first rib, along the clavicle, and into the thoracic cavity. After traveling a short distance inside the thoracic cavity, the subclavian meets and merges with the external and internal

568

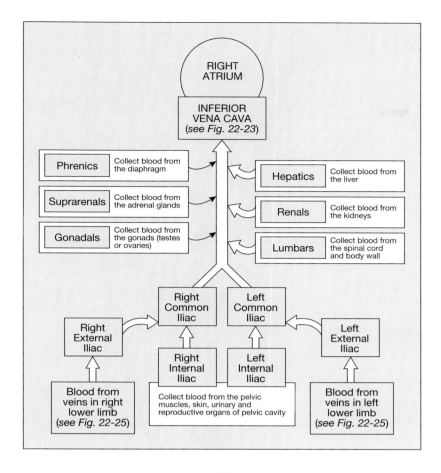

RIGHT
ATRIUM

INFERIOR
VENA CAVA
(see Fig. 22-23)

Phrenics	Collect blood from the diaphragm
Suprarenals	Collect blood from the adrenal glands
Gonadals	Collect blood from the gonads (testes or ovaries)

Hepatics	Collect blood from the liver
Renals	Collect blood from the kidneys
Lumbars	Collect blood from the spinal cord and body wall

Right Common Iliac

Left Common Iliac

| Right External Iliac | | Left External Iliac |

| Right Internal Iliac | Left Internal Iliac |

| Blood from veins in right lower limb (see Fig. 22-25) | Collect blood from the pelvic muscles, skin, urinary and reproductive organs of pelvic cavity | Blood from veins in left lower limb (see Fig. 22-25) |

(b)

jugular veins of that side. This fusion creates the **brachiocephalic vein**, or *innominate vein* (Figure 22-23●).

The brachiocephalic veins receive blood from the left and right **vertebral veins** draining the back of the skull and spinal cord. Near the heart, at the level of the first and second ribs, the left and right brachiocephalic veins combine, creating the superior vena cava (SVC). Close to the point of fusion, the **internal thoracic vein** empties into the brachiocephalic.

The **azygos** (AZ-i-gos) **vein** is the major tributary of the superior vena cava. This vessel ascends from the lumbar region over the right side of the vertebral column to invade the thoracic cavity through the diaphragm. The azygos joins the superior vena cava at the level of vertebra T_2. The azygos receives blood from the smaller **hemiazygos vein**. The hemiazygos vein may also drain into the **highest intercostal vein**, a tributary of the left brachiocephalic vein.

The azygos and hemiazygos veins are the chief collecting vessels of the thorax. They receive blood from: (1) numerous **intercostal veins**, which receive blood from the chest muscles; (2) **esophageal veins**, which drain blood from the esophagus; and (3) smaller veins draining other mediastinal structures.

Figure 22-24● summarizes the venous tributaries of the superior vena cava.

THE INFERIOR VENA CAVA. The **inferior vena cava (IVC)** collects most of the venous blood from organs below the diaphragm (a small amount reaches the superior vena cava via the azygos and hemiazygos veins).

Veins Draining the Lower Extremity (Figure 22-25). Blood leaving capillaries in the sole of each foot collects into a network

of **plantar veins**. The plantar network provides blood to the deep veins of the leg: the **anterior tibial vein**, the **posterior tibial vein**, and the **peroneal vein** (Figure 22-25a●). The **dorsal venous arch** collects blood from capillaries on the superior surface of the foot. There are extensive interconnections between the plantar arch and the dorsal arch, and the path of blood flow can easily shift from superficial to deep veins.

The dorsal venous arch is drained by two superficial veins, the **great saphenous vein** (sa-FĒ-nus; *saphenes*, prominent) and the **small saphenous vein**. The great saphenous vein ascends along the medial aspect of the leg and thigh, draining into the *femoral vein* near the hip joint. The small saphenous vein arises from the dorsal venous arch and ascends along the posterior and lateral aspect of the calf. It then enters the popliteal fossa, where it meets the **popliteal vein** formed by the union of the tibial and peroneal veins. The popliteal vein may be easily palpated in the popliteal fossa adjacent to the adductor magnus muscle (Figure 22-25b●). ∞ [p. 296]. Once it reaches the femur, the popliteal vein becomes the **femoral vein**, which ascends along the thigh next to the femoral artery. Immediately before penetrating the abdominal wall, the femoral vein receives blood from the great saphenous vein and the **deep femoral vein**. The large vein that results penetrates the body wall and emerges in the pelvic cavity as the **external iliac vein**.

Veins Draining the Pelvis (Figure 22-23). The external iliac veins receive blood from the lower extremities, pelvis, and lower abdomen. As the left and right external iliacs travel across the inner surface of the ilium, they are joined by the **internal iliac veins**, which drain the pelvic organs. The union of external and internal iliac veins results in the **common iliac vein**. The right and left common iliacs ascend at an oblique angle, and anterior to vertebra L_5 they unite to form the **inferior vena cava (IVC)** (Figure 22-23●).

Veins Draining the Abdomen (Figures 22-23/22-24). The inferior vena cava ascends posterior to the peritoneal cavity, parallel to the aorta. Blood from both inferior and superior vena cavae flow into the right atrium to be pumped out the right ventricle to the lungs for oxygenation. The abdominal portion of the inferior vena cava collects blood from six major veins (Figures 22-23 and 22-24●).

1. **Lumbar veins**, which drain the lumbar portion of the abdomen. Superior branches of these veins are connected to the azygos vein (right side) and hemiazygos vein (left side), which empty into the superior vena cava.

2. **Gonadal** (ovarian or testicular) **veins**, which drain the ovaries or testes. The right gonadal vein empties into the inferior vena cava; the left gonadal usually drains into the left renal vein.

3. **Renal veins**, which collect blood from the kidneys. These are the largest tributaries of the inferior vena cava.

4. **Suprarenal veins**, which drain the adrenal glands. Usually only the right suprarenal vein drains into the inferior vena cava, and the left drains into the left renal vein.

5. **Phrenic veins**, which drain the diaphragm. Only the *right phrenic vein* drains into the inferior vena cava; the *left* drains into the left renal vein.

6. **Hepatic veins**, which leave the liver and empty into the inferior vena cava at the level of vertebra T_{10}.

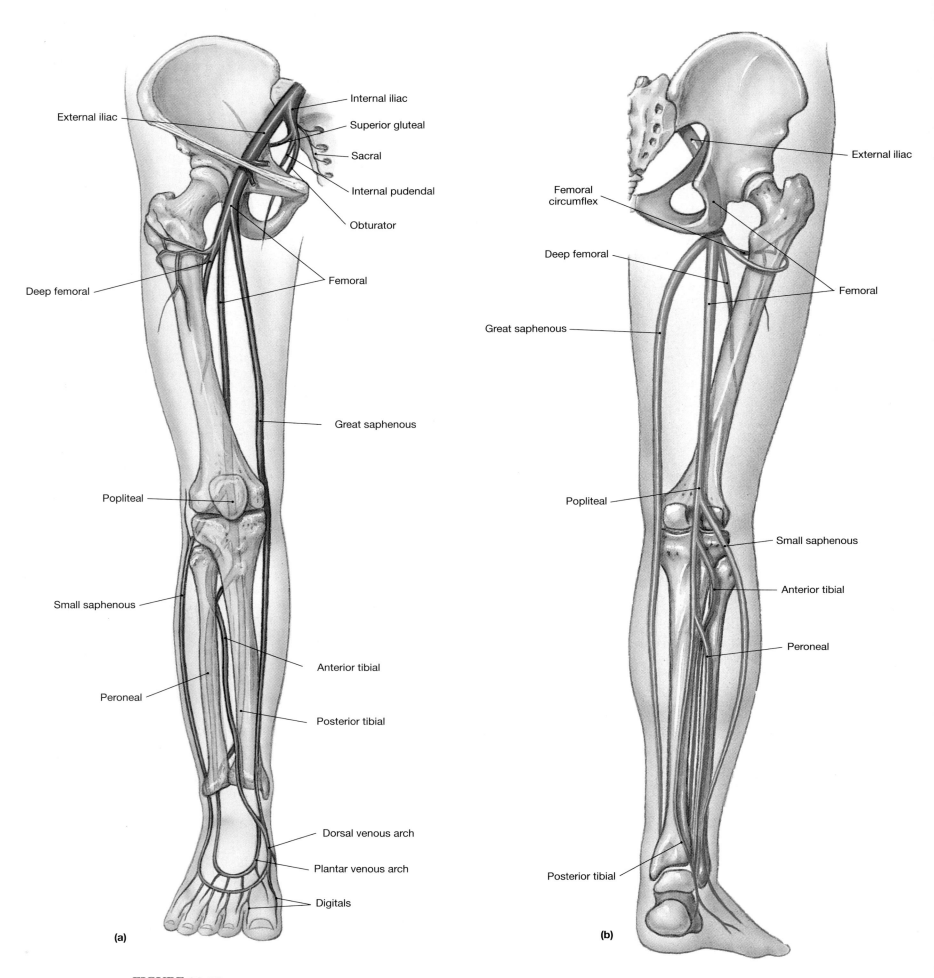

Internal iliac
Superior gluteal
Sacral
Internal pudendal
Obturator
External iliac
Deep femoral
Femoral
Great saphenous
Popliteal
Small saphenous
Anterior tibial
Peroneal
Posterior tibial
Dorsal venous arch
Plantar venous arch
Digitals

(a)

External iliac
Femoral circumflex
Deep femoral
Femoral
Great saphenous
Popliteal
Small saphenous
Anterior tibial
Peroneal
Posterior tibial

(b)

FIGURE 22-25
The Venous Drainage of the Lower Extremity and Foot. (a) Veins of the right lower limb, anterior view. (b) Veins of the right lower limb, posterior view. (c) A summary of the veins of the lower extremity. See MRI SCANS 5, 6, and 7, pp. 744 and 745.

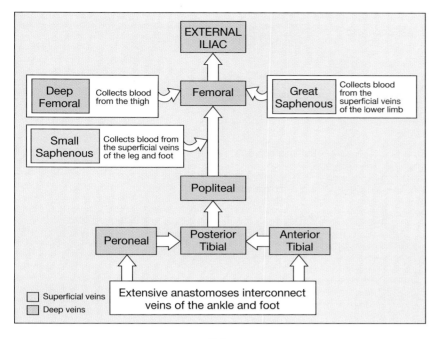

(c)

THE HEPATIC PORTAL SYSTEM *(Figure 22-26).* The liver is the only digestive organ drained by the inferior vena cava. Instead of traveling directly to the inferior vena cava, blood leaving the capillaries supplied by the celiac, superior, and inferior mesenteric arteries flows into the **hepatic portal system**. As noted in Chapter 19, a blood vessel connecting two capillary beds is called a *portal vessel*, and the network is a *portal system.* ∞ [p. 488]

Blood flowing in the hepatic portal system is quite different from that in other systemic veins, because the hepatic portal vessels contain substances absorbed by the stomach and intestines. For example, levels of blood glucose and amino acids in the hepatic portal vein often exceed those found anywhere else in the circulatory system.

The liver regulates the concentrations of nutrients, such as glucose or amino acids, in the circulating blood. During digestion, the stomach and intestines absorb high concentrations of nutrients, along with various wastes and even toxins. The hepatic portal system delivers these compounds directly to the liver for storage, metabolic conversion, or excretion by liver cells. After passing through the liver sinusoids, blood collects into the hepatic veins, which empty into the inferior vena cava (Figure 22-26●). Because blood goes to the liver first, the composition of the blood in the general systemic circuit remains relatively stable, regardless of the digestive activities under way.

The hepatic portal system begins in the capillaries of the digestive organs and ends as the hepatic portal vein discharges blood into sinusoids in the liver. The tributaries of the hepatic portal vein (Figure 22-26●) include:

1. The **inferior mesenteric vein,** which collects blood from capillaries along the lower portion of the large intestine.

2. The **left colic vein** and the **superior rectal vein,** which drain the descending colon, sigmoid colon, and rectum.

3. The **splenic vein,** formed by the union of the inferior mesenteric vein and veins from the spleen, the lateral border of the stomach (*left gastroepiploic*), and the pancreas (*pancreatic*).

4. The **superior mesenteric vein,** which collects blood from veins draining the stomach (*right gastroepiploic*), the small intestine (*intestinal*), and two-thirds of the large intestine (*ileocolic, right colic, and middle colic*).

The hepatic portal vein forms through the fusion of the superior mesenteric and splenic veins. Of the two, the superior mesenteric normally contributes the greater volume of blood and most of the nutrients. As it proceeds toward the liver, the hepatic portal receives blood from the **gastric vein**, which drains the medial border of the stomach and the **cystic vein** from the gallbladder.

√ **Blockage of which branch of the aortic arch would interfere with the blood flow to the left upper extremity?**

√ **Diane is in an automobile accident and ruptures her celiac artery. What organs would be affected most directly by this injury?**

CIRCULATORY CHANGES AT BIRTH
(Figure 22-27)

There are significant differences between the fetal and adult circulatory systems that reflect differing sources of respiratory and nutritional support. The embryonic lungs are collapsed and nonfunctional, and the digestive tract has nothing to digest. All of the embryonic nutritional and respiratory needs are provided by diffusion across the placenta. Two **umbilical arteries** leave the internal iliac arteries, enter the umbilical cord, and deliver blood to the placenta. Blood returns in the single **umbilical vein**, bringing oxygen and nutrients to the developing fetus. The umbilical vein drains into the **ductus venosus**, which is connected to an intricate network of veins within the developing liver. The ductus venosus collects the blood from the veins of the liver and from the umbilical vein, and dumps it into the inferior vena cava (Figure 22-27●, p. 576). When the placental connection is broken at birth, blood flow ceases along the umbilical vessels, and they soon degenerate.

One of the most interesting aspects of circulatory development reflects the differences between the life of an embryo or fetus and that of an infant. Throughout embryonic and fetal life, the lungs are collapsed; yet after delivery, the newborn infant must be able to extract oxygen from inspired air rather than across the placenta.

Although the interatrial and interventricular septa develop early in fetal life, the interatrial partition remains functionally incomplete up to the time of birth. The interatrial opening, or **foramen ovale**, is associated with an elongated flap that acts as a valve. Blood can flow freely from the right atrium to the left atrium, but any backflow will close the valve and isolate the two chambers. Thus blood can enter the heart at the right atrium and bypass the pulmonary circuit altogether. A second short circuit exists between the pulmonary and aortic trunks. This connection, the **ductus arteriosus**, consists of a short, muscular vessel.

With the lungs collapsed, the capillaries are compressed and little blood flows through the lungs. During diastole, blood enters the right atrium and flows into the right ventricle, but also passes into the left atrium via the foramen ovale. About 25% of the blood arriving at the right atrium bypasses

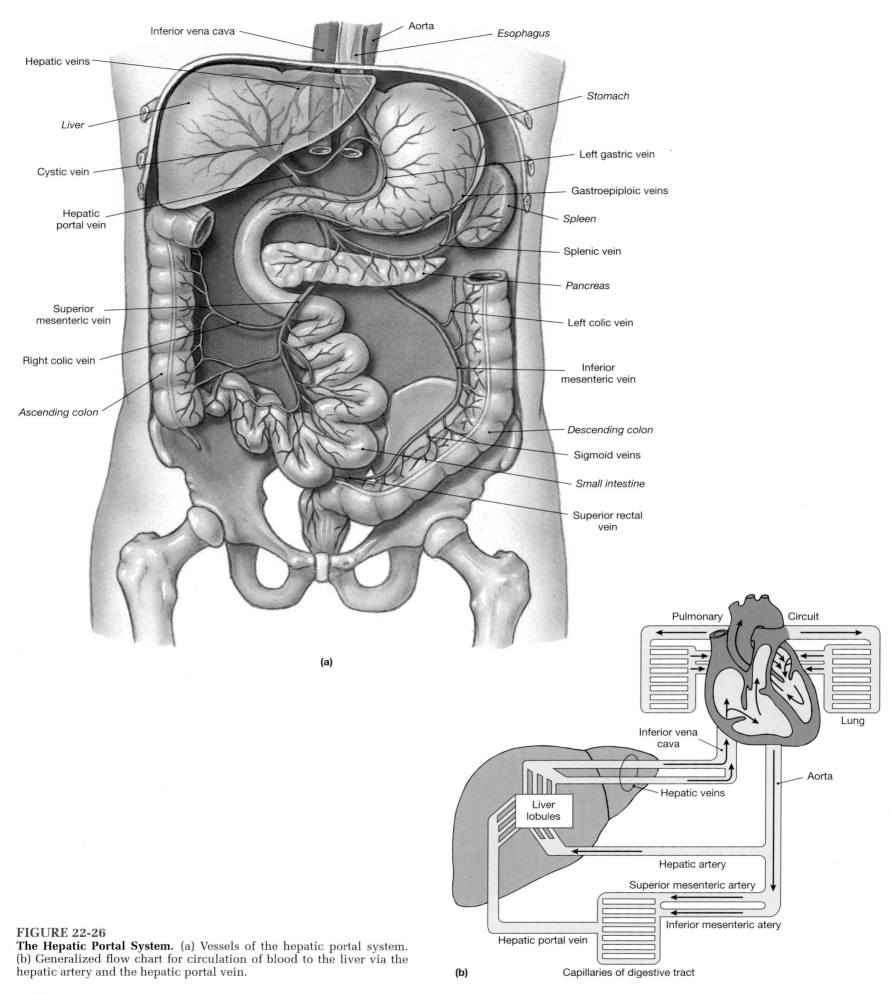

(a)

FIGURE 22-26
The Hepatic Portal System. (a) Vessels of the hepatic portal system.
(b) Generalized flow chart for circulation of blood to the liver via the
hepatic artery and the hepatic portal vein.

(b)

the pulmonary circuit in this way. In addition, over 90% of the blood leaving the right ventricle passes through the ductus arteriosus and enters the systemic circuit, rather than continuing to the lungs.

At birth, dramatic changes occur. When the infant takes its first breath, the lungs expand, and so do the pulmonary vessels. The smooth muscles in the ductus arteriosus contract, isolating the pulmonary and aortic trunks, and blood begins flowing through the pulmonary circuit. As pressures rise in the left atrium, the valvular flap closes the foramen ovale and completes the circulatory remodeling. These alterations are diagrammed and summarized in Figure 22-27a,b●, p. 576. In the adult, the interatrial septum bears a shallow depression, the **fossa ovale**, that marks the site of the original foramen ovale. ∞ [p. 528] The remnants of the ductus arteriosus persist as a fibrous cord, the **ligamentum arteriosum**.

If the proper circulatory changes do not occur at birth or shortly thereafter, problems will eventually develop. The severity of the problem varies depending on which connection remains open and the size of the opening. Treatment may involve surgical closure of the foramen ovale or the ductus arteriosus or both. Other forms of congenital heart defects result from abnormal cardiac development or inappropriate connections between the heart and major arteries and veins. (For further discussion of circulatory changes during development, see the Embryology Summary on the next page.)

EXERCISE AND CARDIOVASCULAR DISEASE

Regular exercise has several beneficial effects. Even a moderate exercise routine (power walking or bicycling, for example) can lower total blood cholesterol levels. High cholesterol is one of the major risk factors for atherosclerosis, leading to cardiovascular disease and strokes. In addition, a healthy lifestyle with regular exercise, a balanced diet, weight control, and no smoking reduces stress, lowers blood pressure, and slows plaque formation.

Large-scale statistical studies indicate that regular moderate exercise may cut the incidence of heart attacks almost in half. Exercise is also beneficial in accelerating recovery after a heart attack. Regular light to moderate exercise, cou-

pled with a low-fat diet and low-stress lifestyle, not only reduces symptoms, such as angina, but also improves both mood and the overall quality of life. However, exercise does not remove the underlying medical problem.

There is no evidence that intense athletic training lowers the incidence of cardiovascular disease. On the contrary, the strains placed on the cardiovascular system during intense athletic training can be severe. Healthy individuals can develop acute disorders, such as kidney failure, after extreme exercise.

AGING AND THE CARDIOVASCULAR SYSTEM

The capabilities of the cardiovascular system gradually decline with age. The major changes are listed and summarized below, in the same sequence as the cardiovascular chapters: blood, heart, and vessels.

1. *Age-related changes in the blood* may include: (1) decreased hematocrit; (2) constriction or blockage of peripheral veins by a **thrombus** (stationary blood clot); the thrombus can become detached, pass through the heart, and become wedged in a small artery, most often in the lungs, causing a **pulmonary embolism**; and (3) pooling of blood in the veins of the legs because valves are not working effectively.

2. *Age-related changes in the heart* include: (1) a reduction in the maximum cardiac output; (2) changes in the activities of the nodal and conducting fibers; (3) a reduction in the elasticity of the fibrous skeleton; (4) a progressive atherosclerosis that can restrict coronary circulation; and (5) replacement of damaged cardiac muscle fibers by scar tissue.

3. *Age-related changes in blood vessels* are often related to arteriosclerosis and include: (1) inelastic walls of arteries become less tolerant of sudden increases in pressure, which may lead to an **aneurysm** (AN-ū-rizm) causing a stroke, infarct, or massive blood loss, depending on the vessel involved; (2) calcium salts can deposit on weakened vascular walls, increasing the risk of a stroke or infarct; and (3) thrombi can form at atherosclerotic plaques.

Related Clinical Terms

aneurysm (AN-ū-rizm): A bulge in the weakened wall of a blood vessel, usually an artery. ☨ *Aneurysms [p. 768]*

arteriosclerosis (ar-tē-rē-ō-skle-RŌ-sis): A thickening and toughening of arterial walls. *[p. 548]*

atherosclerosis (ath-er-ō-skle-RŌ-sis): A type of arteriosclerosis characterized by changes in the endothelial lining. *[p. 548]*

varicose (VAR-i-kōs) **veins:** Sagging, swollen veins distorted by gravity and the failure of the venous valves. ☨ *Problems with Venous Valve Function [p. 768]*

hemorrhoids (HEM-ō-roidz): Varicose veins in the walls of the rectum and/or anus, often associated with frequent straining to force bowel movements. ☨ *Problems with Venous Valve Function [p. 768]*

thrombus: A stationary blood clot within a blood vessel. *[p. 573]*

pulmonary embolism: Circulatory blockage caused by the trapping of a freed thrombus in a pulmonary artery. *[p. 573]*

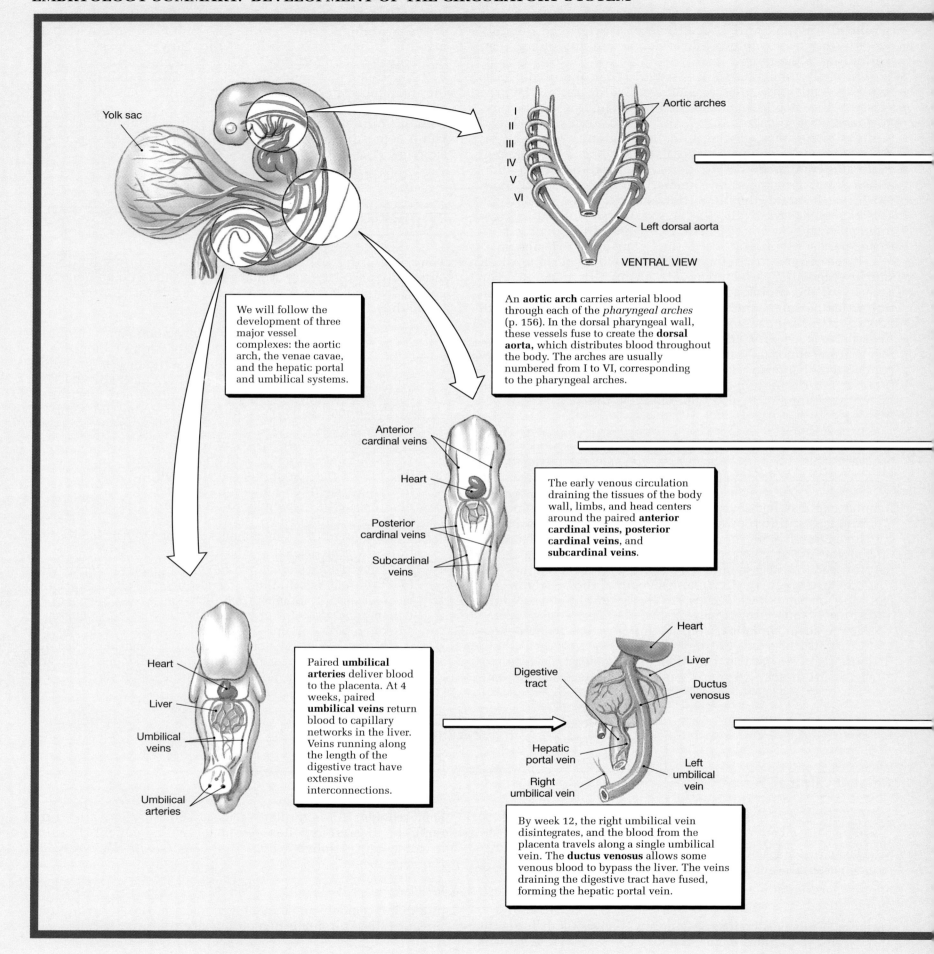

Yolk sac

Aortic arches

I
II
III
IV
V
VI

Left dorsal aorta

VENTRAL VIEW

We will follow the development of three major vessel complexes: the aortic arch, the venae cavae, and the hepatic portal and umbilical systems.

An **aortic arch** carries arterial blood through each of the *pharyngeal arches* (p. 156). In the dorsal pharyngeal wall, these vessels fuse to create the **dorsal aorta,** which distributes blood throughout the body. The arches are usually numbered from I to VI, corresponding to the pharyngeal arches.

Anterior cardinal veins

Heart

Posterior cardinal veins

Subcardinal veins

The early venous circulation draining the tissues of the body wall, limbs, and head centers around the paired **anterior cardinal veins, posterior cardinal veins**, and **subcardinal veins**.

Heart

Liver

Umbilical veins

Umbilical arteries

Paired **umbilical arteries** deliver blood to the placenta. At 4 weeks, paired **umbilical veins** return blood to capillary networks in the liver. Veins running along the length of the digestive tract have extensive interconnections.

Heart

Liver

Ductus venosus

Digestive tract

Hepatic portal vein

Right umbilical vein

Left umbilical vein

By week 12, the right umbilical vein disintegrates, and the blood from the placenta travels along a single umbilical vein. The **ductus venosus** allows some venous blood to bypass the liver. The veins draining the digestive tract have fused, forming the hepatic portal vein.

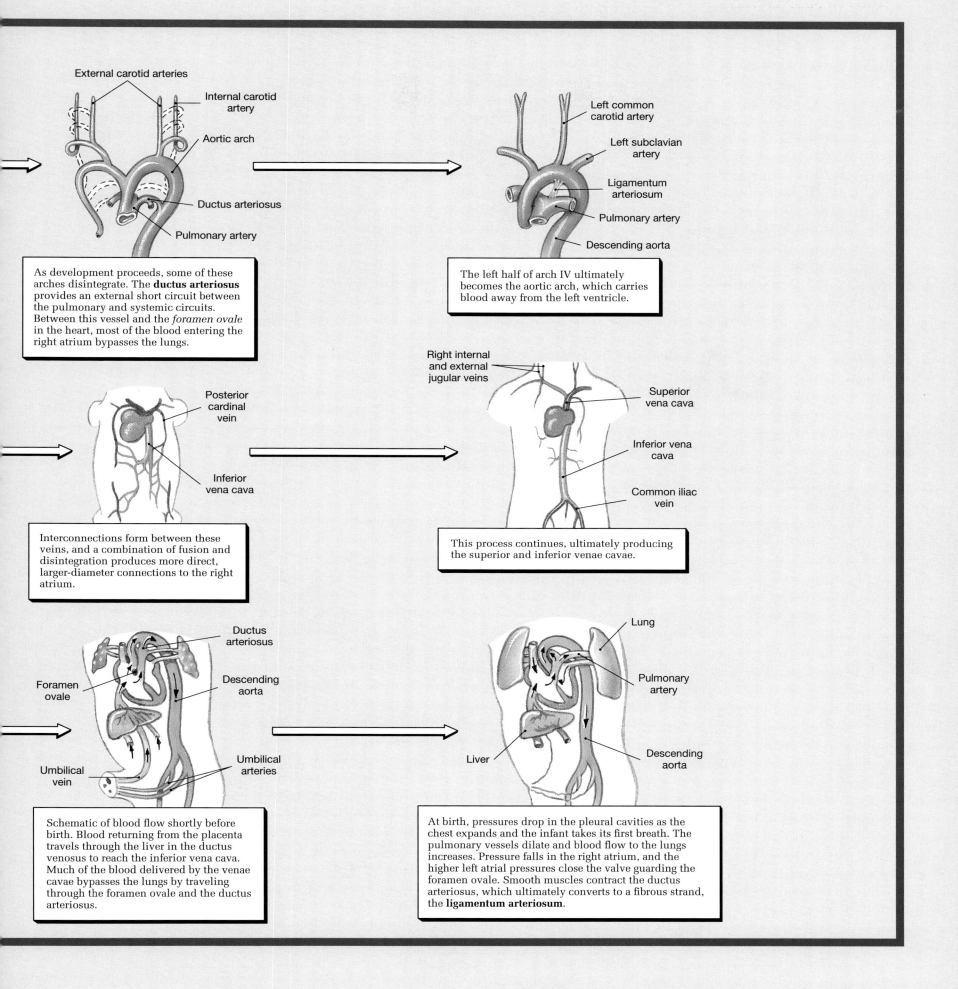

External carotid arteries

Internal carotid artery

Aortic arch

Ductus arteriosus

Pulmonary artery

As development proceeds, some of these arches disintegrate. The **ductus arteriosus** provides an external short circuit between the pulmonary and systemic circuits. Between this vessel and the *foramen ovale* in the heart, most of the blood entering the right atrium bypasses the lungs.

Left common carotid artery

Left subclavian artery

Ligamentum arteriosum

Pulmonary artery

Descending aorta

The left half of arch IV ultimately becomes the aortic arch, which carries blood away from the left ventricle.

Posterior cardinal vein

Inferior vena cava

Interconnections form between these veins, and a combination of fusion and disintegration produces more direct, larger-diameter connections to the right atrium.

Right internal and external jugular veins

Superior vena cava

Inferior vena cava

Common iliac vein

This process continues, ultimately producing the superior and inferior venae cavae.

Ductus arteriosus

Descending aorta

Foramen ovale

Umbilical vein

Umbilical arteries

Schematic of blood flow shortly before birth. Blood returning from the placenta travels through the liver in the ductus venosus to reach the inferior vena cava. Much of the blood delivered by the venae cavae bypasses the lungs by traveling through the foramen ovale and the ductus arteriosus.

Lung

Pulmonary artery

Descending aorta

Liver

At birth, pressures drop in the pleural cavities as the chest expands and the infant takes its first breath. The pulmonary vessels dilate and blood flow to the lungs increases. Pressure falls in the right atrium, and the higher left atrial pressures close the valve guarding the foramen ovale. Smooth muscles contract the ductus arteriosus, which ultimately converts to a fibrous strand, the **ligamentum arteriosum**.

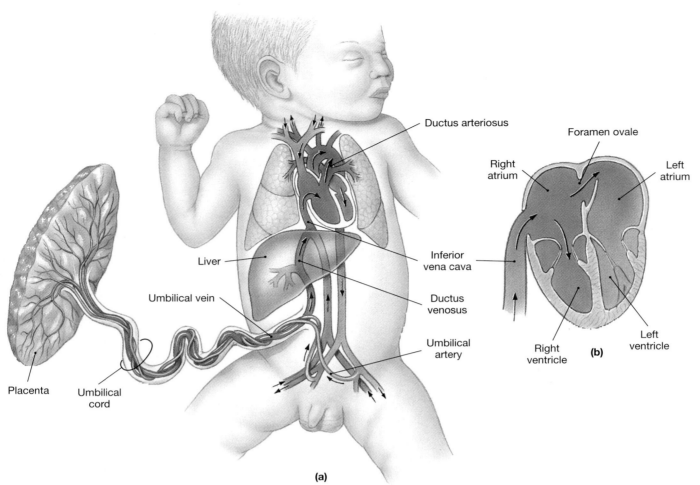

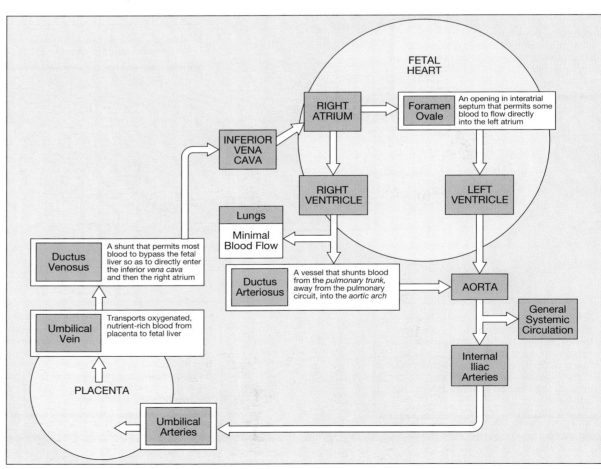

FIGURE 22-27
Changes in Fetal Circulation at Birth. (a) Circulation pathways in a full-term fetus. (b) Blood flow through the heart of the fetus. (c) Flow chart for circulatory patterns in the fetus and newborn infant.

Congenital Circulatory Problems

Congenital circulatory problems serious enough to represent a threat to homeostasis are relatively rare. They usually reflect abnormal formation of the heart or problems with the interconnections between the heart and the great vessels. Several examples of congenital circulatory defects are illustrated in Figure 22-28●. All of these conditions can be surgically corrected, although multiple surgeries may be required.

The incomplete closure of the foramen ovale or ductus arteriosus (Figure 22-28a●) results in the bypassing of the lungs and the recirculation of blood into the pulmonary circuit. Because normal blood oxygenation does not occur, the circulating blood has a deep red color. The skin then develops the blue tones typical of cyanosis, a condition noted in Chapter 4, and the infant is known as a "blue baby."

Ventricular septal defects (Figure 22-28b●) are the most common congenital heart problems, affecting 0.12% of newborn infants. The open-ing between the left and right ventricles has the reverse effect of a connection between the atria: When it beats, the more powerful left ventricle ejects blood into the right ventricle and pulmonary circuit. Pulmonary hypertension, pulmonary edema, and cardiac enlargement are the results.

The *tetralogy of Fallot* (fa-LŌ) (Figure 22-28c●) is a complex group of heart and circulatory defects that affect 0.10% of newborn infants. In this condition (1) the pulmonary trunk is abnormally narrow, (2) the interventricular septum is incomplete, (3) the aorta originates where the interventricular septum normally ends, and (4) the right ventricle is enlarged.

In the *transposition of great vessels* (Figure 22-28d●) the aorta is connected to the right ventricle and the pulmonary artery is connected to the left ventricle. This malformation affects 0.05% of newborn infants.

In an *atrioventricular septal defect* (Figure 22-28e●) the atria and ventricles are incompletely separated. The results are quite variable, depending on the extent of the defect and the effects on the atrioventricular valves. This type of defect most often affects infants with *Down syndrome*, a disorder caused by the presence of an extra copy of chromosome 21.

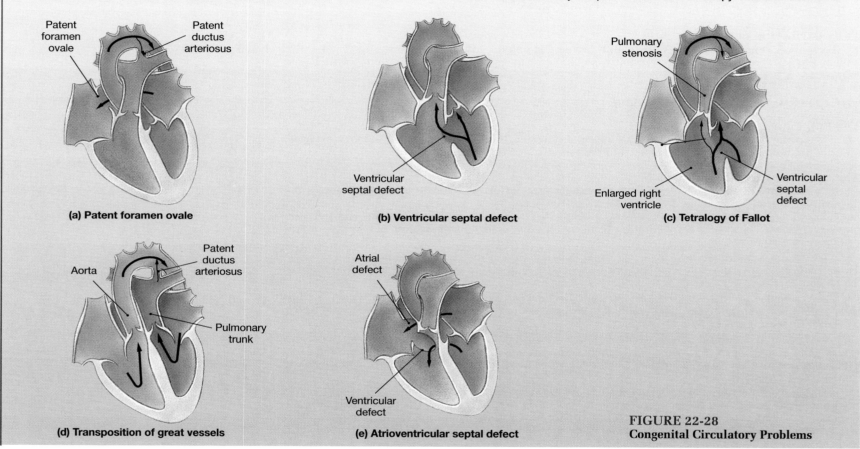

(a) Patent foramen ovale

(b) Ventricular septal defect

(c) Tetralogy of Fallot

(d) Transposition of great vessels

(e) Atrioventricular septal defect

FIGURE 22-28
Congenital Circulatory Problems

CHAPTER SUMMARY AND REVIEW

STUDY OUTLINE

Related Key Terms

INTRODUCTION *[p. 544]*

1. Blood flows through a network of arteries, veins, and capillaries. All chemical and gaseous exchange between the blood and interstitial fluid takes place across capillary walls.

HISTOLOGICAL ORGANIZATION OF BLOOD VESSELS *[p. 544]*

1. The walls of arteries and veins contain three layers: the **tunica interna**, **tunica media**, and outermost **tunica externa** *(see Figure 22-1).*

internal elastic membrane ● **external elastic membrane** ● **vasoconstriction** ● **vasodilation** ● **vasa vasorum**

Distinguishing Arteries from Veins *[p. 544]*
2. In general, the walls of arteries are thicker than those of veins. The endothelial lining of an artery cannot contract, so it is thrown into folds. Arteries constrict when blood pressure does not distend them; veins constrict very little *(see Figure 22-1).*

Arteries [p. 544]

3. The arterial system includes the large **elastic arteries**, **medium-sized** or **muscular arteries**, and smaller **arterioles**. As we proceed toward the capillaries, the number of vessels increase, but the diameters of the individual vessels decrease and the walls become thinner *(see Figures 22-1 and 22-2)*.

Capillaries [p. 546]

4. Capillaries are the only blood vessels whose walls permit exchange between blood and interstitial fluid. Capillaries may be **continuous** or **fenestrated**. **Sinusoids** are specialized fenestrated capillaries found in certain tissues that allow very slow blood flow *(see Figure 22-3)*.

5. Capillaries form interconnected networks called **capillary plexuses (capillary beds)**. A **precapillary sphincter** (a band of smooth muscle) adjusts the blood flow into each capillary. The entire capillary plexus may be bypassed by blood flow through **arteriovenous anastomoses** or via **central (preferred) channels** within the capillary plexus *(see Figure 22-4)*.

Veins [p. 549]

6. **Venules** collect blood from the capillaries and merge into **medium-sized veins** and then **large veins**. The arterial system is a high-pressure system; blood pressure in veins is much lower. **Valves** in these vessels prevent the backflow of blood *(see Figures 22-1/22-2/22-6)*.

The Distribution of Blood [p. 549]

7. Peripheral **venoconstriction** helps maintain adequate blood volume in the arterial system after a hemorrhage. The **venous reserve** normally accounts for up to 20% of the total blood volume *(see Figure 22-7)*.

THE CIRCULATORY SYSTEM [p. 550]

The Pulmonary Circulation [p. 550]

1. The pulmonary circuit includes the pulmonary trunk, the **left** and **right pulmonary arteries**, and the **pulmonary veins**, which empty into the left atrium *(see Figure 22-9)*.

The Systemic Circulation [p. 553]

2. The **ascending aorta** gives rise to the coronary circulation. The **aortic arch** continues as the **descending aorta**. Three large arteries arise from the aortic arch to collectively supply the head, neck, shoulder, and upper extremities: **brachiocephalic trunk**, **left common carotid**, and **left subclavian** arteries. The detailed distribution of these blood vessels and their branches will be found in *Figures 22-10 to 22-20*.

3. The brachiocephalic gives rise to the **right subclavian** and the **right common carotid** arteries. These arteries supply the right arm and portions of the right shoulder, neck, and head *(see Figures 22-12 and 22-13)*.

4. Each subclavian exits the thoracic cavity to become the **axillary artery**, which enters the arm to become the **brachial artery**. The brachial arteries and their branches supply blood to the upper extremities *(see Figure 22-12)*.

5. Each common carotid artery divides into an **external carotid artery** and **internal carotid artery**. The external carotids and their branches supply blood to structures in the neck and face. The internal carotids and their branches enter the skull to supply blood to the brain and eyes *(see Figures 22-13 and 22-14)*.

6. The descending aorta superior to the diaphragm is termed the **thoracic aorta** and inferior to it the **abdominal aorta**. The thoracic aorta and its branches supply blood to the thorax and thoracic viscera. The abdominal aorta and its branches supply blood to the abdominal wall, abdominal viscera, pelvic structures, and lower extremities. The detailed distribution of these blood vessels and their branches will be found in *Figures 22-10 to 22-17/22-20*.

7. Arteries in the neck, arms, and legs are deep beneath the skin; in contrast, there are usually two sets of peripheral veins, one superficial and one deep. This dual venous drainage is important for controlling body temperature *(see Figures 22-10/22-21)*.

8. The **superior vena cava (SVC)** receives blood from the head, neck, chest, shoulders, and arms. The detailed distribution of these collecting vessels and their branches may be seen in *Figures 22-21 to 22-24*. The **inferior vena cava** collects most of the venous blood from organs below the diaphragm. The detailed distribution of these collecting vessels and their branches may be seen in *Figures 22-23 to 22-26*.

9. Any blood vessel connecting two capillary beds is called a *portal vessel* and the network of blood vessels comprises a *portal system*.

10. Blood leaving the capillaries supplied by the celiac, superior, and inferior mesenteric arteries flows into the **hepatic portal system**. Blood in the hepatic portal system is unique, compared to that of the other systemic veins, because portal blood contains high concentrations of nutrients and waste products. These substances are collected from the digestive organs, via the vessels of the portal system and transported directly to the liver for processing.

11. The vessels that form the hepatic portal system are shown in *Figure 22-26*.

CIRCULATORY CHANGES AT BIRTH [p. 571]

1. During fetal development, the **umbilical arteries** carry blood to the placenta. It returns via the **umbilical vein** and enters a network of vascular sinuses in the liver. The **ductus venosus** collects this blood and returns it to the inferior vena cava *(see Figure 22-27)*.

Related Key Terms

metarteriole • autoregulation • collaterals • arterial anastomosis

blood reservoir

alveoli

2. At this time, the interatrial septum is incomplete and the **foramen ovale** allows the passage of blood from the right atrium to the left atrium. The **ductus arteriosus** also permits the flow of blood between the pulmonary trunk and the aortic arch. These connections are normally closed at birth or shortly thereafter as the pulmonary circuit becomes functional (*see Figure 22-27*).

EXERCISE AND CARDIOVASCULAR DISEASE [p. 573]

1. Regular moderate exercise can cut the risk of heart attack almost in half and lower total blood cholesterol levels (one of the major risk factors for atherosclerosis, leading to cardiovascular disease and strokes). In addition, regular exercise reduces stress, lowers blood pressure, and slows plaque formation.

AGING AND THE CARDIOVASCULAR SYSTEM [p. 573]

1. Age-related changes in the blood can include: (1) decreased hematocrit; (2) constriction or blockage of peripheral veins by a thrombus; and (3) pooling of blood in the veins of the lower legs because valves are not working effectively.

2. Age-related changes in the heart include: (1) a reduction in the maximum cardiac output; (2) changes in the activities of the nodal and conducting fibers; (3) a reduction in the elasticity of the fibrous skeleton; (4) a progressive atherosclerosis that can restrict coronary circulation; and (5) replacement of damaged cardiac muscle fibers by scar tissue.

3. Age-related changes in blood vessels are often related to arteriosclerosis and include: (1) inelastic walls of arteries being less tolerant of sudden increases in pressure, which may lead to an aneurysm; (2) calcium salts, which can deposit on weakened vascular walls, increasing the risk of a stroke or infarct; and (3) thrombi that form at atherosclerotic plaques.

1   REVIEW OF CHAPTER OBJECTIVES

1. Describe the general anatomical organization of the circulatory system and its relationship to the heart.
2. Identify the various types of blood vessels on the basis of their histological characteristics.
3. Describe how histological structure influences the function of blood vessels.
4. Describe the structure of two types of capillaries and discuss their permeability characteristics.
5. Explain the distribution of blood in arteries, veins, and capillaries, and discuss the function of blood reservoirs.
6. Identify and describe the vessels of the pulmonary circuit.
7. Identify the major vessels of the systemic circuit and the areas and organs supplied by each vessel.
8. Describe the major circulatory changes that occur at birth, and explain their functional significance.
9. Identify the benefits of regular exercise and its influence on cardiovascular disease.
10. Describe the age-related changes that occur in the cardiovascular system.

2   REVIEW OF CONCEPTS

1. How do the walls of arteries differ from those of veins?
2. Where do the vasa vasorum occur, and why are they present?
3. What are the structural differences of the three types of arteries?
4. What characteristics make capillaries different from other types of blood vessels?
5. What is the significance of a fenestrated capillary?
6. How is blood flow through capillaries regulated?
7. What is the significance of collateral arteries?
8. What is the function of an arteriovenous anastomosis?
9. Arteries carry oxygenated blood and veins carry deoxygenated blood through all body systems, but with what exception?
10. What are the structural differences of the three types of veins?

11. What is the function of venous valves?
12. How does the function of the pulmonary circuit of the heart differ from that of the systemic circuit?
13. What is the significance of the close relationship between the alveoli of the lungs and the capillary networks that branch from the pulmonary arterioles?
14. Why is it possible to take a pulse only from arteries and not from veins, and why can it be taken only in some locations on the body?
15. What are the significance and function of the circle of Willis?
16. What is the role of arterial anastomoses in supplying blood to the stomach and spleen?
17. What unusual circulation role is played by the superior mesenteric vein?
18. How does the path of the femoral artery differ from that of the femoral vein after they exit the inguinal region?
19. What is the role of the dual venous drainage (superficial and deep) in the control of body temperature?

3   CRITICAL THINKING AND CLINICAL APPLICATION QUESTIONS

1. An injured person may experience a loss of orientation or even consciousness, low blood pressure, and circulatory problems, and may be said to be "in shock." What physiological mechanism governs this response?
2. A blood test revealed that a man in his sixties has an elevated cholesterol level. Is this a health concern, and by what mechanism is a high cholesterol level correlated with circulatory problems?
3. In most injuries involving cuts or tears in body tissues, even profuse bleeding eventually slows down or stops. What mechanism is involved in minimizing blood loss?
4. A 32-year-old man with disturbance in the visual system, headaches, dizziness, and the sound of blood pounding in his ears has been diagnosed as having an aneurysm in the circle of Willis. What is this condition, and what will happen if it worsens?
5. A woman in her late fifties is concerned about the unsightly varicose veins in the area of her calves. What is the cause of varicose veins, and what anatomical complications might arise from them?

23

The Lymphatic System

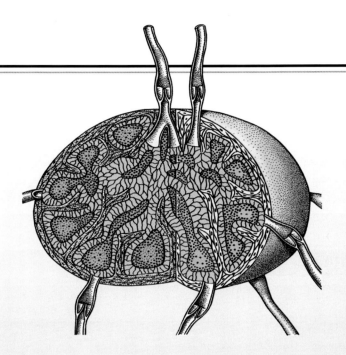

CHAPTER OUTLINE AND OBJECTIVES

The world is not always kind to the human body. Accidental collisions and other interactions with objects in our environment produce bumps, cuts, and burns. Making matters worse, an assortment of viruses, bacteria, and other microorganisms thrive in the environment. Some of these microorganisms normally live on the surface of and inside our bodies, but all have the potential to cause us great harm. Staying alive and healthy involves a massive, combined effort involving many different organs and systems. In this ongoing struggle, the lymphatic system plays the primary role.

In this chapter we will describe the anatomical organization of the lymphatic system and the anatomy of the structures of which it is composed. We will then consider how the lymphatic system interacts with other systems and tissues to defend the body against infection and disease.

ORGANIZATION OF THE LYMPHATIC SYSTEM
(Figure 23-1)

The **lymphatic system** (Figure 23-1●) includes the following structures:

1. *Lymphatic vessels*, which begin in peripheral tissues and empty into the venous system.

2. **Lymph**, the fluid transported by the lymphatic vessels. It is similar in composition to blood plasma, but contains a lower concentration of proteins. ∞ [p. 506]

3. *Lymphoid tissues* and *lymphoid organs* that are connected to the lymphatic vessels. These structures contain large numbers of lymphocytes and numerous macrophages.

Functions of the Lymphatic System

The primary function of the lymphatic system is *the production, maintenance, and distribution of lymphocytes.* Lymphocytes are essential to the normal defense of the body. These cells are produced and stored within lymphoid organs, such as the spleen, thymus, and bone marrow. Lymphoid structures can be classified as *primary* or *secondary*. In *primary lymphoid structures*, stem cells divide to produce daughter cells that differentiate into B, T, or NK cells. The bone marrow and thymus of the adult are primary lymphoid structures. In *secondary lymphoid structures*, immature or activated lymphocytes divide to produce additional lymphocytes of the same type. For example, the divisions of activated B cells within lymph nodes can produce the additional B cells needed to fight off an infection.

Two additional functions include *the maintenance of normal blood volume* and *the elimination of local variations in the chemical composition of the interstitial fluid.* The blood pressure at the start of a systemic capillary is approximately 35 mm Hg. This outward pressure is opposed by the *colloid osmotic pressure* of the blood, the osmotic pressure resulting from the presence of suspended plasma proteins, which is usually about 25 mm Hg. There is thus a net outward pressure of roughly 10 mm Hg that forces water and solutes out of the plasma and into the interstitial fluid. Some of that water will be reabsorbed as the blood pressure declines along the length of the capillary, but a significant amount—approximately 3.6l a day, or 72% of the total blood volume—will

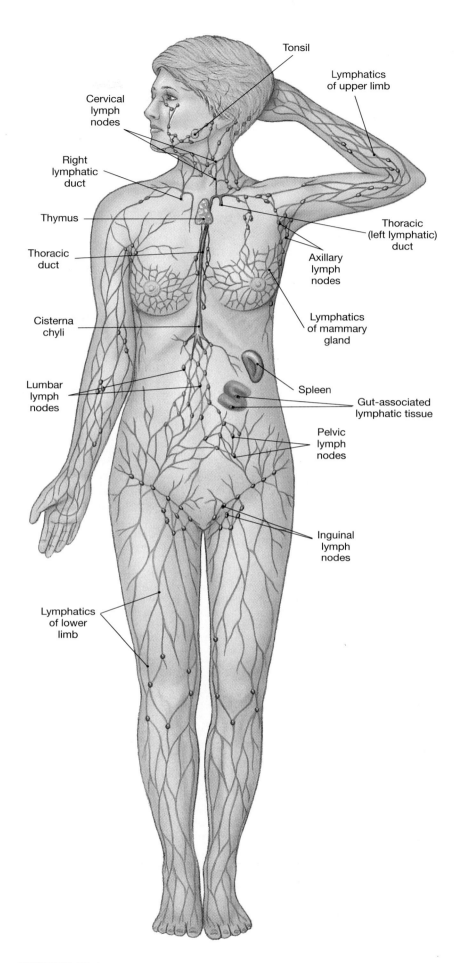

FIGURE 23-1
Components of the Lymphatic System

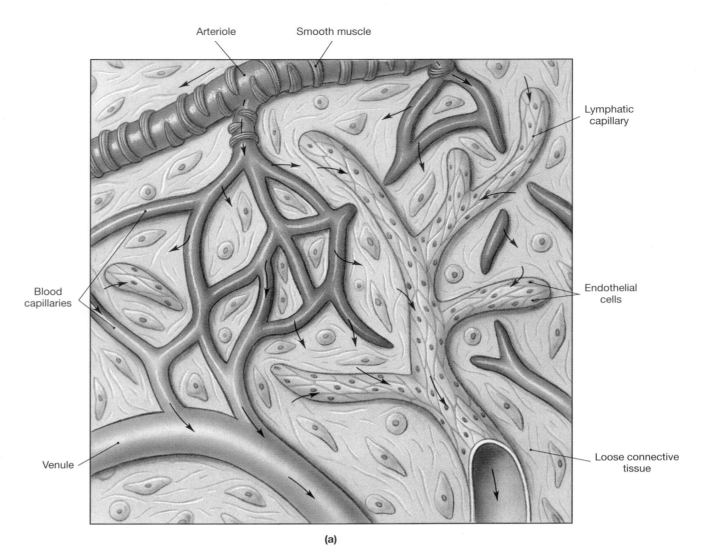

(a)

FIGURE 23-2
Lymphatic Capillaries. A three-dimensional view of the association of blood capillaries, tissue, interstitial fluid, and lymphatic capillaries. Lymphatic capillaries are blind vessels in areas of loose connective tissue. Interstitial fluid enters these capillaries by passing between adjacent endothelial cells. From the lymphatic capillaries, this fluid, now called lymph, moves into larger lymphatic vessels and eventually into the venous system. Arrows show the direction of interstitial fluid and lymph fluid movement. (b) Sectional view of this association.

be returned to the circulation only by vessels of the lymphatic system. There is thus a continual movement of fluid from the bloodstream into the tissues and then back to the bloodstream via lymphatic vessels. Because so much fluid moves through the lymphatic system each day, a break in a major lymphatic vessel can cause a rapid and potentially fatal decline in blood volume.

This circulation of fluid helps to eliminate regional differences in the composition of interstitial fluid. It also provides an alternative route for the distribution of hormones, nutrients, and waste products. For example, certain lipids absorbed by the digestive tract reach the bloodstream via passage along lymphatic vessels.

Structure of Lymphatic Vessels

Lymphatic vessels, often called *lymphatics*, carry lymph from peripheral tissues to the venous system. The smallest lymphatic vessels are called *lymphatic capillaries*.

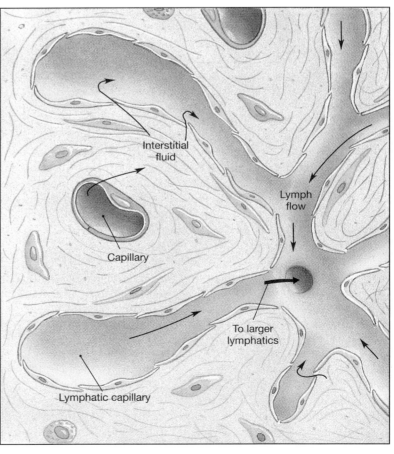

(b)

FIGURE 23-3

Lymphatic Vessels and Valves. (a) A diagrammatic view of loose connective tissue, showing small blood vessels and a lymphatic. (b) Lymphatic valves resemble those of the venous system. Each valve consists of a pair of flaps that permit fluid movement in only one direction. (LM × 63) (c) The cross-sectional view emphasizes the structural differences between them.

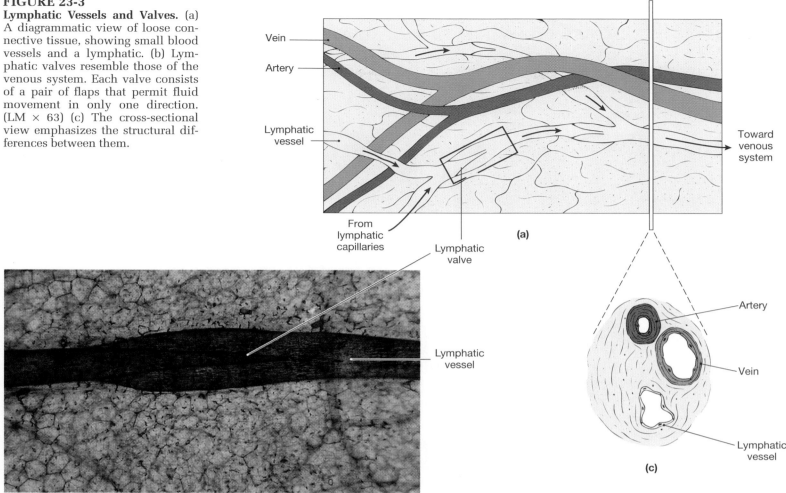

Vein

Artery

Lymphatic vessel

Toward venous system

From lymphatic capillaries

Lymphatic valve

Lymphatic vessel

(a)

Lymphatic vessel

(b)

Artery

Vein

Lymphatic vessel

(c)

Lymphatic Capillaries
(Figure 23-2)

The lymphatic network begins with the **lymphatic capillaries**, or *terminal lymphatics*, which radiate and branch through peripheral tissues. Lymphatic capillaries are larger in diameter than vascular capillaries, and in sectional view they often have a flattened or irregular outline (Figure 23-2a●). Lymphatic capillaries also have thinner walls than capillaries containing blood. Although lymphatic capillaries are lined by endothelial cells, there is no underlying basement membrane. The endothelial cells of a lymphatic capillary are not tightly bound together, but they do overlap. The region of overlap acts as a one-way valve, permitting fluid to enter the lymphatic, but preventing its return to the intercellular spaces (Figure 23-2b●).

Lymphatic capillaries are present in almost every tissue and organ in the body. Prominent lymphatic capillaries in the small intestine, called *lacteals*, are important in the transport of lipids absorbed by the digestive tract. Lymphatic capillaries are absent in areas that lack a blood supply, such as the cornea of the eye, and there are no lymphatics in the central nervous system. (The CNS is bathed in cerebrospinal fluid, which circulates from its production at the choroid plexus to its reabsorption into the venous system at the superior sagittal sinus. ∞ [p. 371]).

Valves of Lymphatic Vessels
(Figure 23-3)

From the lymphatic capillaries, lymph flows into larger lymphatic vessels that lead toward the trunk. The walls of these lymphatics contain layers comparable to those of veins, and like veins, the larger lymphatics contain valves. The valves are quite close together, and at each valve the lymphatic vessel bulges noticeably. This configuration gives large lymphatics a beaded appearance (Figure 23-3●). The valves prevent the backflow of lymph within lymph vessels, especially those of the extremities. Pressures within the lymphatic system are minimal, and the valves are essential to maintaining normal lymph flow toward the thoracic cavity.

If a lymphatic vessel has damaged valves, or if the vessel is compressed or blocked, lymph drainage slows or ceases in the affected area. Fluid continues to leave the capillaries in that region, but because the lymphatic system is no longer able to remove it at the same rate, the interstitial fluid volume and pressure gradually increase. The affected tissues become distended and swollen, and this condition is called *lymphedema*.

Lymphatic vessels are often found in association with blood vessels. Note the differences in relative size, general ap-

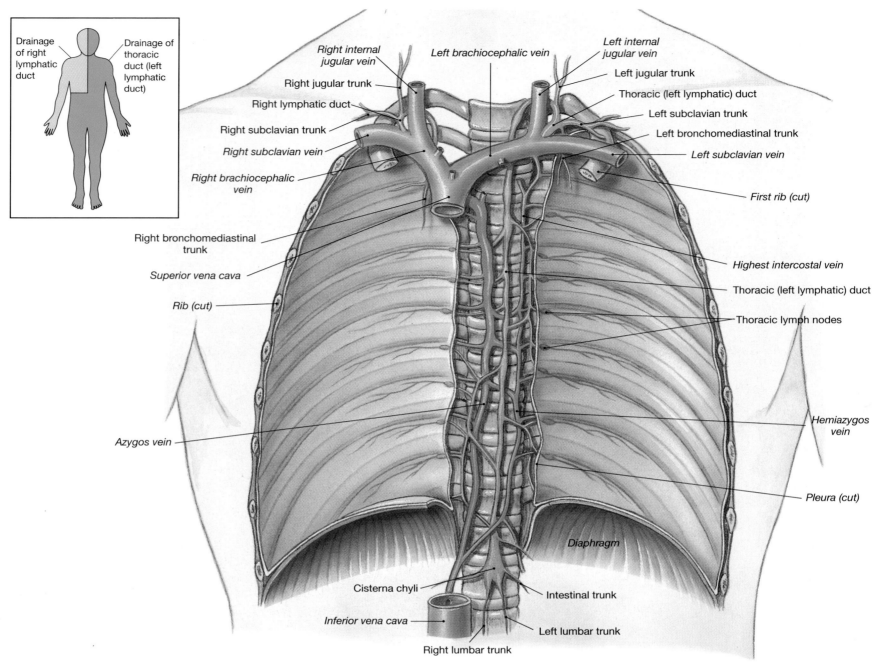

FIGURE 23-4

Relationship of Lymphatic Ducts and Circulatory System. The collecting system of lymph vessels, nodes, major lymphatic collecting ducts, and their relationship to the great veins of the circulatory system. The thoracic duct carries lymph originating in tissues inferior to the diaphragm and from the left side of the upper body. The right lymphatic duct drains the right half of the body superior to the diaphragm.

pearance, and branching pattern that distinguish the lymphatics from arteries and veins (Figure 23-3a●). There are also characteristic color differences that are apparent on examining living tissues. Arteries are usually a bright red, veins a dark red, and lymphatics a pale golden color.

Major Lymph-Collecting Vessels
(Figure 23-4)

Two sets of lymphatic vessels collect blood from the lymphatic capillaries. **Superficial lymphatics** are found:

- In the subcutaneous layer beneath the skin.
- In the loose connective tissues of the mucous mem-

branes lining the digestive, respiratory, urinary, and reproductive tracts.

- In the loose connective tissues of the serous membranes lining the pleural, pericardial, and peritoneal cavities.

Deep lymphatics are larger lymph vessels that accompany deep arteries and veins supplying skeletal muscles and other organs of the neck, limbs, and trunk, and the walls of visceral organs.

Superficial and deep lymphatics converge to form trunks that empty into two large collecting vessels, the *thoracic duct* and the *right lymphatic duct* (Figure 23-4●). The **thoracic duct** collects lymph from the body inferior to the diaphragm and

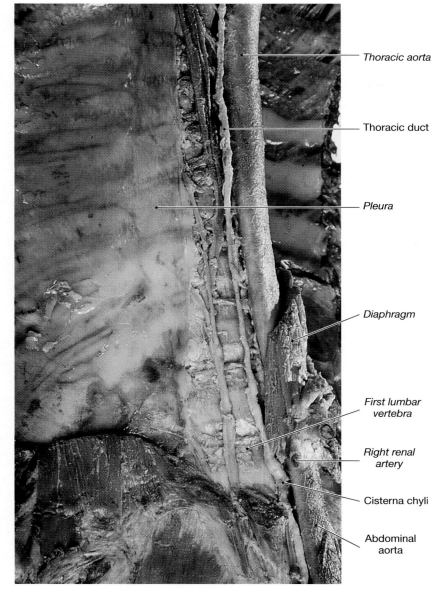

FIGURE 23-5
Major Lymphatic Vessels of the Trunk. Thoracic and abdominal organs have been removed to show vessels and their association with the circulatory system.

Image labels (top to bottom):
Thoracic aorta
Thoracic duct
Pleura
Diaphragm
First lumbar vertebra
Right renal artery
Cisterna chyli
Abdominal aorta

from the left side of the body superior to the diaphragm. The relatively small **right lymphatic duct** collects lymph from the right side of the body superior to the diaphragm.

THE THORACIC DUCT *(Figures 23-4/23-5).*   The thoracic duct is formed inferior to the diaphragm at the level of vertebra L_2. The base of the thoracic duct is an expanded, saclike chamber, the **cisterna chyli** (Figures 23-4 and 23-5●). The cisterna chyli receives lymph from the lower abdomen, pelvis, and lower limbs via the *right* and *left lumbar trunks* and the *intestinal trunk.*

The inferior segment of the thoracic duct lies anterior to the vertebral column. From the second lumbar vertebra, it penetrates the diaphragm in company with the aorta, at the *aortic hiatus,* and ascends along the left side of the vertebral column to the level of the left clavicle. After collecting lymph from the *left bronchomediastinal trunk,* the *left subclavian trunk,* and the *left jugular trunk*, it empties into the left subclavian vein

near the left internal jugular vein (Figure 23-4●). Lymph collected from the left side of the head, neck, and thorax as well as lymph from the entire body inferior to the diaphragm reenters the venous system in this way.

THE RIGHT LYMPHATIC DUCT *(Figure 23-4).*   The right lymphatic duct is formed by the merging of many smaller lymphatic vessels in the region of the right clavicle. This duct empties into the right subclavian vein, delivering lymph from the right side of the body superior to the diaphragm (Figure 23-4●). The distribution of the larger lymphatic vessels, lymph nodes, and collecting ducts can also be seen in Figure 23-4●. ⊤ *Lymphedema [p. 769]*

Lymphocytes

Lymphocytes, introduced in Chapters 3 and 20, are the primary cells of the lymphatic system. ∞ [pp. 66, 513] They respond to the presence of: (1) invading organisms, such as bacteria or viruses; (2) abnormal body cells, such as virus-infected cells or cancer cells; and (3) foreign proteins, such as the toxins released by some bacteria. Lymphocytes attempt to eliminate these threats or render them harmless by a combination of physical and chemical attack.

Types of Lymphocytes

There are three different classes of lymphocytes in the blood: **T cells** (thymus-dependent), **B cells** (bone marrow–derived), and **NK cells** (natural killer). Each type has distinctive biochemical and functional characteristics.

T CELLS.   Approximately 80% of circulating lymphocytes are classified as T cells. **Cytotoxic T cells** attack foreign cells or body cells infected by viruses. Their attack often involves direct contact. These lymphocytes are responsible for providing **cell-mediated immunity. Helper T cells** and **suppressor T cells** assist in the regulation and coordination of the immune response; for this reason they are also called *regulatory T cells*. Regulatory T cells control the activation and activity of B cells. **Memory T cells** are produced by the division of activated T cells when an individual is exposed to a particular antigen. They are called memory cells because they remain "on reserve," becoming activated only if the same antigen appears in the body at a later date. This is not a complete list, as there are several other types of specialized T cells in the body.

B CELLS.   B cells account for 10–15% of circulating lymphocytes. When stimulated, B cells can differentiate into **plasma cells**. Plasma cells, introduced in Chapter 3, are responsible for the production and secretion of *antibodies*. ∞ [p. 66] These proteins react with specific chemical targets, called **antigens**. Antigens are usually associated with pathogens, parts or products of pathogens, or other foreign compounds. Typical antigens are short peptide chains or short amino acid sequences along a complex protein, but some lipids, polysaccharides, and nucleic acids can also stimulate antibody production. When an antigen-antibody complex forms, it starts a chain of events leading to the destruction, neutralization, or elimination of the antigen. Antibodies are known as **immunoglobulins** (i-mū-nō-GLOB-ū-lins). Because the blood is the primary distribution route for immunoglobulins, B cells are said to be responsible for **antibody-mediated immunity,**

or *humoral* ("liquid") *immunity*. **Memory B cells** are produced by the division of activated B cells when an individual is exposed to a particular antigen. These cells will become activated only if the antigen appears in the body at a later date.

Helper T cells promote the differentiation of plasma cells and accelerate their production of antibodies. Suppressor T cells inhibit the formation of plasma cells and reduce the production of antibodies by existing plasma cells.

NK CELLS.   The remaining 5–10% of circulating lymphocytes are NK cells, also known as *large granular lymphocytes*. These lymphocytes will attack foreign cells, normal cells infected with viruses, and cancer cells that appear in normal tissues. The continual policing of peripheral tissues by NK cells and activated macrophages has been called **immunological surveillance**.

Lymphocytes and the Immune Response
(Figure 23-6)

The goal of the **immune response** is the destruction or inactivation of pathogens, abnormal cells, and foreign molecules such as toxins. The body has two different ways to do this:

1. Direct attack by activated T cells (*cell-mediated immunity*);
2. Attack by circulating antibodies released by *plasma cells* that are derived from activated B cells (*antibody-mediated immunity*).

Figure 23-6 provides an overview of the immune response to bacterial infection (Figure 23-6a●) and viral infection (Figure 23-6b●). The process begins with the appearance of an antigen. In many cases, the first step is the phagocytosis of the antigen by a macrophage. The macrophage then "presents" the antigen to T cells that are *preprogrammed to respond to that particular antigen*. These lymphocytes respond because their cell membranes contain receptors capable of binding that antigen. Some of the activated T cells differentiate into cytotoxic T cells, others into helper T cells that will in turn activate B cells.

Your immune system has no way of anticipating which antigens it will actually encounter. Its protective strategy is to prepare for *any* antigen that might appear. During development, differentiation of cells in the lymphatic system produces an enormous number of lymphocytes with varied antigen sensitivities. Among the trillion or so lymphocytes in the human body are millions of different lymphocyte populations. Each population consists of several thousand cells that are prepared to recognize a specific antigen. When activated, these cells divide and proliferate, *producing more lymphocytes sensitive to that particular antigen.* Some of the lymphocytes function immediately, to eliminate the antigen, and others (the memory cells) will be ready if the antigen reappears at a later date. This mechanism provides an immediate defense and ensures that there will be an even more massive and rapid response if the antigen appears in the body at some later date.

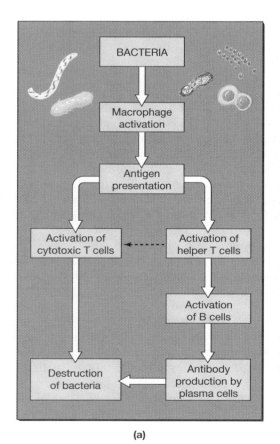

(a)

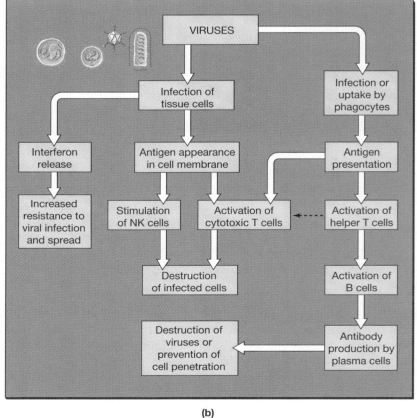

(b)

FIGURE 23-6
Lymphocytes and the Immune Response. (a) Defenses against bacterial pathogens are usually initiated by active macrophages. (b) Defenses against viruses are usually activated after infection of normal cells. In each instance, B cells and T cells cooperate to produce a coordinated chemical and physical attack.

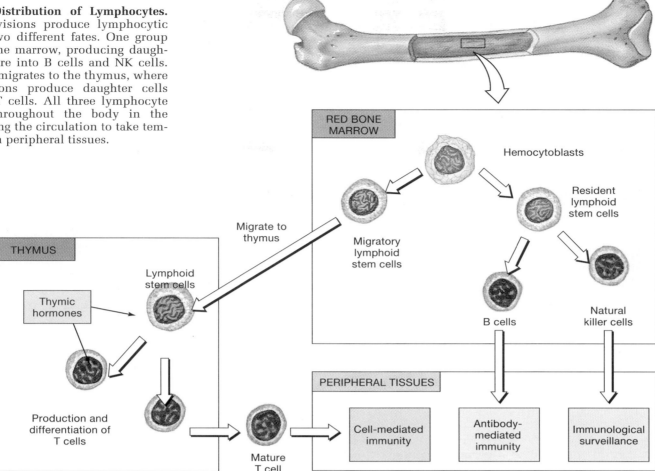

FIGURE 23-7

Derivation and Distribution of Lymphocytes. Hemocytoblast divisions produce lymphocytic stem cells with two different fates. One group remains in the bone marrow, producing daughter cells that mature into B cells and NK cells. The second group migrates to the thymus, where subsequent divisions produce daughter cells that mature into T cells. All three lymphocyte types circulate throughout the body in the bloodstream, leaving the circulation to take temporary residence in peripheral tissues.

Distribution and Life Span of Lymphocytes

The ratio of B cells to T cells varies depending on the tissue or organ considered. For example, B cells are seldom found in the thymus, and in the blood, T cells outnumber B cells by a ratio of 8 to 1. This ratio changes to 1:1 in the spleen and 1:3 in the bone marrow.

The lymphocytes within these organs are visitors, not residents. Lymphocytes move throughout the body; they wander through a tissue and then enter a blood vessel or lymphatic vessel for transport to another site. T cells move relatively quickly. For example, a wandering T cell may spend about 30 minutes in the blood and 15–20 hours in a lymph node. B cells move more slowly; a typical B cell spends around 30 hours in a lymph node before moving to another location.

In general, lymphocytes have relatively long life spans. Roughly 80% survive for 4 years, and some last 20 years or more. Throughout life, normal lymphocyte populations are maintained through lymphopoiesis in the bone marrow and lymphatic tissues.

Lymphopoiesis: Lymphocyte Production
(Figure 23-7)

Erythropoiesis in the adult is normally confined to the bone marrow, but **lymphopoiesis** involves the bone marrow and thymus. The relationships are diagrammed in Figure 23-7●.

Hemocytoblasts in the bone marrow produce lymphocytic stem cells with two distinct fates. One group remains in the bone marrow and generates NK cells and B cells that migrate into peripheral tissues. The B cells take up residence in lymph nodes, the spleen, and other lymphatic tissues. The second group of stem cells migrates to the thymus. Under the influence of the thymic hormones (thymosin-1, thymopentin, thymulin, and others) these stem cells divide repeatedly, producing daughter cells that undergo functional maturation into T cells. These T cells subsequently migrate to the spleen, other lymphoid organs, and the bone marrow.

As lymphocytes migrate through peripheral tissues, they retain the ability to divide. These divisions produce daughter cells of the same type. For example, a dividing B cell produces other B cells, not T cells or NK cells. The ability to increase the number of lymphocytes of a specific type is important to the success of the immune response. If that ability is compromised, the individual will become unable to mount an effective defense against infection and disease. For example, the disease *AIDS* (*acquired immune deficiency syndrome*) results from infection with a virus that selectively destroys T cells. Individuals with AIDS are likely to be killed by bacterial or viral infections that would be easily overcome by a normal immune system. For a detailed discussion of AIDS, see the Clinical Issues appendix. ✝ *SCID and AIDS [p. 769]*

Lymphoid Tissues
(Figure 23-8)

Lymphoid tissues are connective tissues dominated by lymphocytes. In a **lymphoid nodule,** the lymphocytes are densely packed in an area of loose connective tissue (Figure 23-8●). Typical nodules average around a millimeter in diameter, but the boundaries are indistinct because no fibrous capsule surrounds them. They

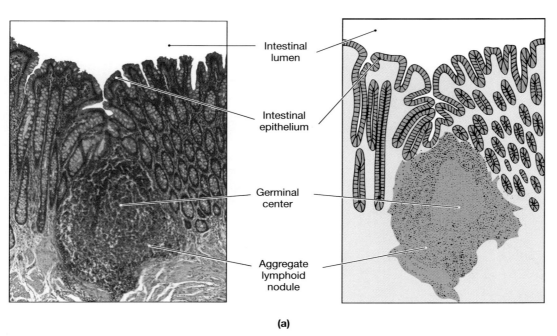

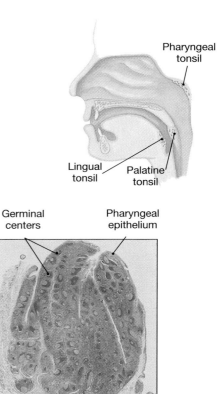

(a)

(b)

FIGURE 23-8
Lymphoid Nodules. (a) Appearance of a typical nodule (Peyer's patch from small intestine) as seen with the light microscope. Note the relatively pale (pinkish) germinal center, where lymphocyte cell divisions occur. (LM × 51) (b) The positions of the tonsils in sagittal section and a histological section of a single tonsil, showing the germinal centers.

often have a pale, central zone, called a **germinal center**, which contains dividing lymphocytes (Figure 23-8a●).

Lymphoid nodules are found beneath the epithelia lining the respiratory, digestive, and urinary tracts. Clusters of lymphoid nodules beneath the epithelial lining of the intestine are known as **aggregate lymphoid nodules,** or **Peyer's patches** (Figure 23-8a●). In addition, the walls of the appendix, a blind pouch that originates near the junction between the small and large intestines, contain a mass of fused lymphoid nodules.

Large nodules in the walls of the pharynx are called **tonsils** (Figure 23-8b●). There are usually five tonsils:

- A single **pharyngeal tonsil**, often called the *adenoids,* located in the posterior superior wall of the nasopharynx;
- A pair of **palatine tonsils**, located at the posterior margin of the oral cavity along the boundary with the pharynx to the soft palate; and
- A pair of **lingual tonsils**, which are not visible because they are located at the base of the tongue.

The extensive array of lymphoid tissue associated with the digestive system is called **gut-associated lymphatic tissue (GALT).** Lymphoid nodules are not always able to destroy bacterial or viral invaders that have crossed the adjacent epithelium. An infection may then develop, and familiar examples include *tonsillitis* and *appendicitis.* ✝ *Infected Lymphoid Nodules [p. 769]*

Lymphoid Organs

Lymphoid organs are separated from surrounding tissues by a fibrous connective tissue capsule. Important lymphoid organs include the *lymph nodes*, the *thymus*, and the *spleen*.

Lymph Nodes
(Figures 23-1/23-4/23-9 to 23-14)

Lymph nodes are small, oval lymphoid organs ranging in diameter from 1 to 25 mm (up to around 1 in.). The general pattern of lymph node distribution in the body can be seen in Figure 23-1●, p. 581. Each lymph node is covered by a dense fibrous connective tissue capsule. Fibrous extensions from the capsule extend partway into the interior of the node (Figure 23-9●). These fibrous partitions are called **trabeculae** (tra-BEK-ū-lē).

The shape of a typical lymph node resembles that of a lima bean. Blood vessels and nerves attach to the lymph node at the indentation, or **hilus** (Figure 23-9●). Two sets of lymphatic vessels are connected to each lymph node: *afferent lymphatics* and *efferent lymphatics*. The afferent lymphatics, which bring lymph from peripheral tissues, penetrate the capsule on the side opposite the hilus. Lymph then flows through the lymph node within a network of sinuses, open passageways with incomplete walls. Lymph first enters a *subcapsular sinus* that contains a meshwork of branching reticular fibers, macrophages, and *dendritic cells*. Dendritic cells collect and present antigens in their cell membranes. T cells encountering these bound antigens become activated, thus initiating an immune response. After passing through the subcapsular sinus, lymph flows through the **outer cortex** of the node. The outer cortex contains B cells within germinal centers resembling those of lymphoid nodules.

Lymph then continues through lymph sinuses in the **deep cortex** (*paracortical area*). Lymphocytes leave the circulation and enter the lymph node by crossing the walls of blood vessels within the deep cortex. The deep cortical area is dominated by T cells.

After flowing through the sinuses of the deep cortex, lymph continues into the core, or **medulla**, of the lymph node. The medulla contains B cells and plasma cells organized into elon-

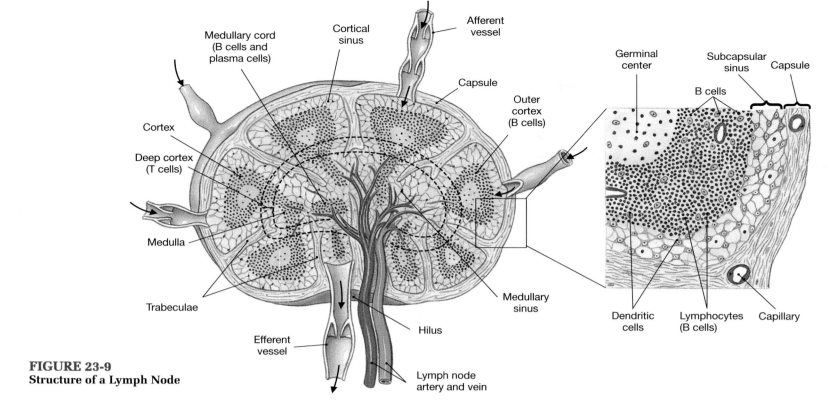

FIGURE 23-9
Structure of a Lymph Node

gate masses known as **medullary cords**. Lymph enters the efferent lymphatics at the hilus after passing through a network of sinuses in the medulla.

A lymph node functions like a kitchen water filter: It filters and purifies lymph before it reaches the venous system. As lymph flows through a lymph node, at least 99 percent of the antigens present in the arriving lymph will be removed. Fixed macrophages in the walls of the lymphatic sinuses engulf debris or pathogens in the lymph as it flows past. Antigens removed in this way are then processed by the macrophages and "presented" to nearby T cells. Other antigens stick to the surfaces of dendritic cells, where they can stimulate T cell activity. This process, called *antigen presentation*, is usually the first step in the activation of the immune response.

The largest lymph nodes are found where peripheral lymphatics connect with the trunk (Figure 23-4●, p. 584), in regions such as

■ CLINICAL BRIEF
Lymphadenopathy and Metastatic Cancer

A minor injury normally produces a slight enlargement of the nodes along the lymphatic draining the region. The enlargement usually results from an increase in the number of lymphocytes and phagocytes in the node, in response to a minor, localized infection. Chronic or excessive enlargement of lymph nodes constitutes **lymphadenopathy** (lim-fad-e-NOP-a-thē). This condition may occur in response to bacterial or viral infections, endocrine disorders, or cancer.

Lymphatics are found in most portions of the body, and lymphatic capillaries offer little resistance to the passage of cancer cells. As a result, metastasizing cancer cells often spread along the lymphatics. Under these circumstances the lymph nodes serve as way-stations for migrating cancer cells. Thus an analysis of lymph nodes can provide information on the spread of the cancer cells, and such information has a direct influence on the selection of appropriate therapies. One example is the classification of breast cancer or lymphomas by the degree of lymph node involvement.
✝ *Lymphomas* and *Breast Cancer* [p. 770]

the base of the neck, the axillae, and the groin (Figures 23-10 to 23-14●, pp. 590–591). These nodes are often called *lymph glands*. "Swollen glands" usually indicate inflammation or infection of peripheral structures. Dense collections of lymph nodes also exist within the mesenteries of the gut, near the trachea and passageways leading to the lungs, and in association with the thoracic duct.

Distribution of Lymphoid Tissues and Lymph Nodes
(Figures 23-1/23-4/23-5/23-10 to 23-15)

Lymphatic tissues and lymph nodes are distributed in areas particularly susceptible to injury or invasion. If we wanted to protect a house against intrusion, we might guard all doors and windows and perhaps keep a pitbull indoors. The distribution of lymphatic tissues and lymph nodes is based on a similar strategy.

1. The **cervical lymph nodes** monitor lymph originating in the head and neck (Figure 23-10●).

2. The **axillary lymph nodes** filter lymph arriving from the upper extremities (Figure 23-11a●). In women, the axillary nodes also drain lymph from the mammary glands (Figure 23-11b●).

3. The **popliteal lymph nodes** and the **inguinal lymph nodes** monitor lymph arriving from the lower extremities (Figures 23-12 to 23-14●).

4. The **thoracic lymph nodes** receive lymph from the lungs, respiratory passageways, and mediastinal structures (Figure 23-4●, p. 584).

5. The **abdominal lymph nodes** filter lymph arriving from the urinary and reproductive systems.

6. The lymphatic tissue of *Peyer's patches*, the **intestinal lymph nodes**, and the **mesenterial lymph nodes** receive lymph originating from the digestive tract (Figure 23-15●).

FIGURE 23-10
Lymphatic Drainage of the Head and Neck

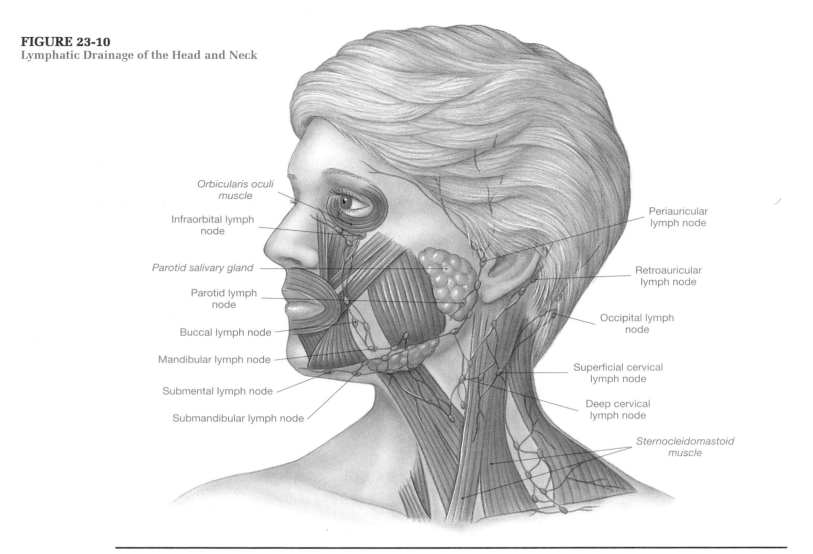

Orbicularis oculi
muscle

Infraorbital lymph
node

Parotid salivary gland

Parotid lymph
node

Buccal lymph node

Mandibular lymph node

Submental lymph node

Submandibular lymph node

Periauricular
lymph node

Retroauricular
lymph node

Occipital lymph
node

Superficial cervical
lymph node

Deep cervical
lymph node

Sternocleidomastoid
muscle

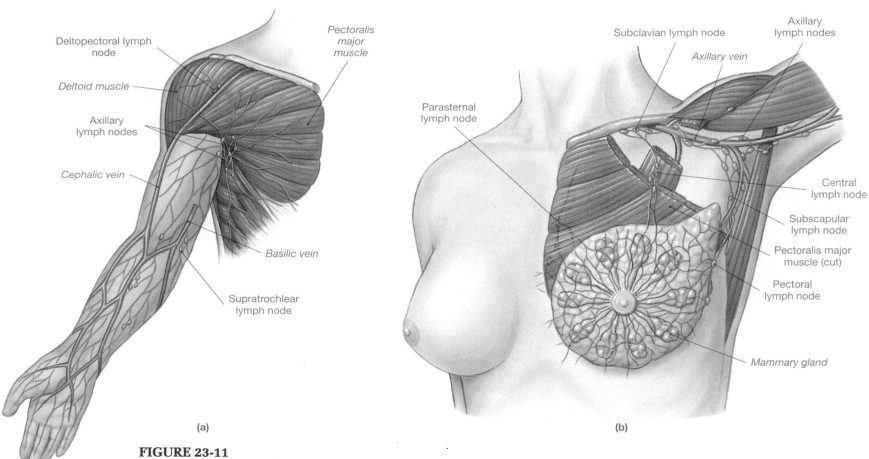

Deltopectoral lymph
node

Deltoid muscle

Axillary
lymph nodes

Cephalic vein

Basilic vein

Supratrochlear
lymph node

Pectoralis
major
muscle

(a)

Parasternal
lymph node

Subclavian lymph node

Axillary vein

Axillary
lymph nodes

Central
lymph node

Subscapular
lymph node

Pectoralis major
muscle (cut)

Pectoral
lymph node

Mammary gland

(b)

FIGURE 23-11
Lymphatic Drainage of the Upper Extremity. (a) In the male, anterior view. (b) In the female, anterior view.

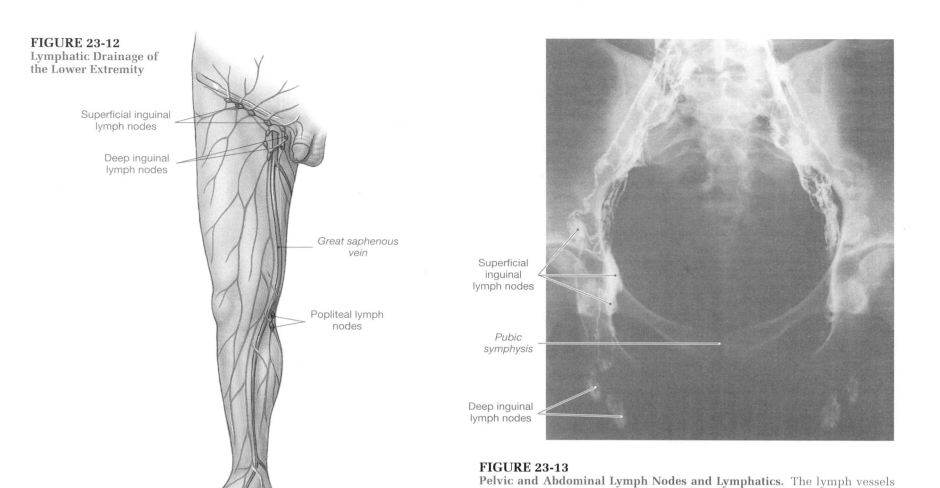

FIGURE 23-12
Lymphatic Drainage of the Lower Extremity

Superficial inguinal lymph nodes

Deep inguinal lymph nodes

Great saphenous vein

Popliteal lymph nodes

FIGURE 23-13
Pelvic and Abdominal Lymph Nodes and Lymphatics. The lymph vessels and nodes can be visualized in a *lymphangiogram*, an X-ray taken after the introduction of a radiopaque dye into the lymphatic system.

Superficial inguinal lymph nodes

Pubic symphysis

Deep inguinal lymph nodes

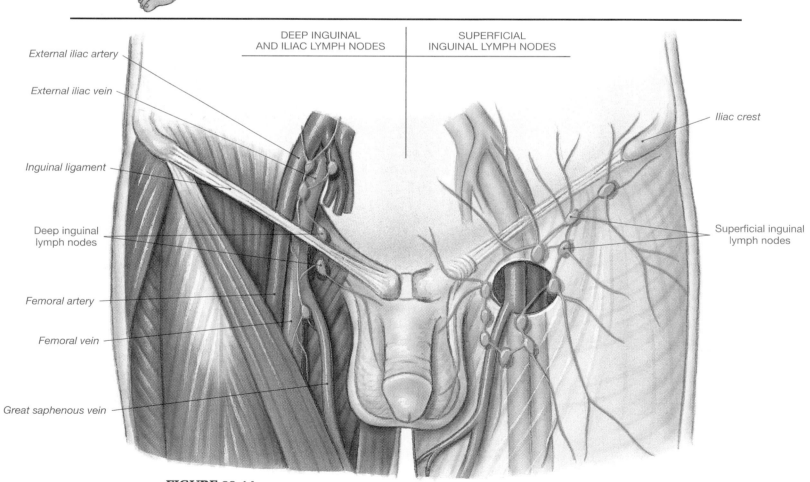

DEEP INGUINAL AND ILIAC LYMPH NODES

SUPERFICIAL INGUINAL LYMPH NODES

External iliac artery

External iliac vein

Inguinal ligament

Deep inguinal lymph nodes

Femoral artery

Femoral vein

Great saphenous vein

Iliac crest

Superficial inguinal lymph nodes

FIGURE 23-14
Lymphatic Drainage of the Inguinal Region. Inguinal region of the male, anterior view.

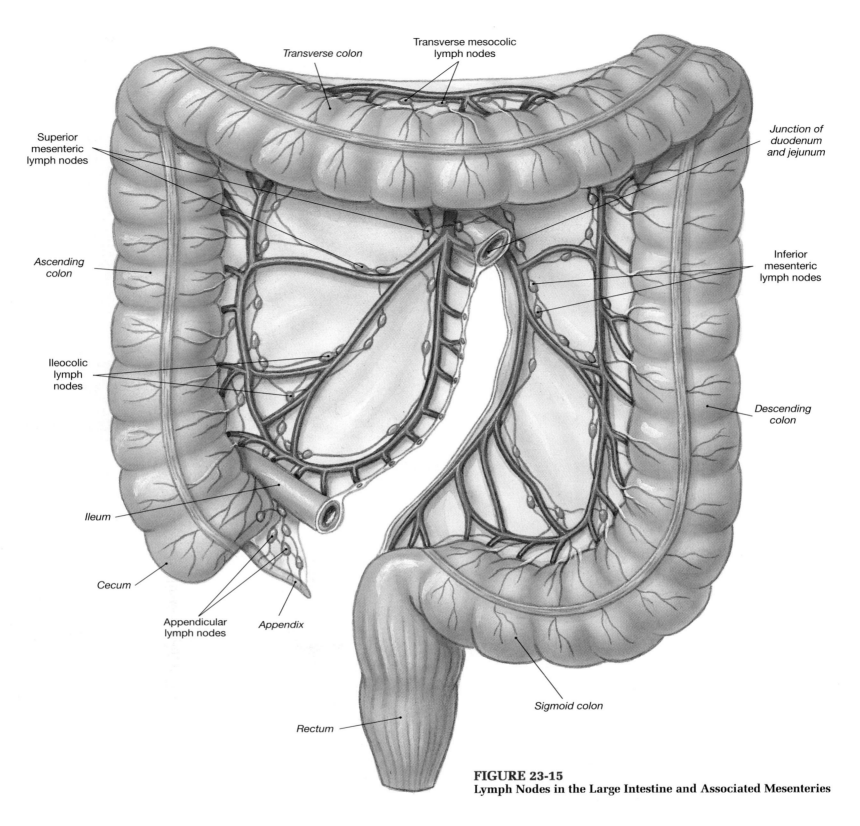

Transverse colon

Transverse mesocolic
lymph nodes

Junction of
duodenum
and jejunum

Superior
mesenteric
lymph nodes

Ascending
colon

Inferior
mesenteric
lymph nodes

Ileocolic
lymph
nodes

Descending
colon

Ileum

Cecum

Appendicular
lymph nodes

Appendix

Sigmoid colon

Rectum

FIGURE 23-15
Lymph Nodes in the Large Intestine and Associated Mesenteries

The Thymus
(Figure 23-16)

The **thymus** lies posterior to the sternum, in the anterior portion of the mediastinum. It has a pinkish coloration and a nodular consistency. The thymus reaches its greatest size (relative to body size) in the first year or two after birth, and its maximum absolute size during puberty, when it weighs between 30 and 40 g. Thereafter the thymus gradually decreases in size and becomes increasingly fibrous, a process called **involution**.

The capsule that covers the thymus divides it into two **thymic lobes** (Figure 23-16a,b●). Fibrous partitions, or **septae**, from the capsule divide the lobes into **lobules** averaging 2 mm in width (Figure 23-16b,c●). Each lobule consists of a densely packed outer **cortex** and a paler central **medulla**. Lymphoid stem cells in the cortex undergo mitosis, and as the T cells mature they migrate into the medulla. During the maturation process, any T cells that are sensitive to normal tissue antigens are destroyed. The surviving T cells eventually enter one of the specialized blood vessels in that region. Scattered among the lymphocytes are **epithelial cells** responsible for the production of thymic hormones. In the medullary region, the epithelial cells cluster together in concentric layers, forming distinctive structures known as **Hassall's corpuscles** (Figure 23-16d●) whose function remains unknown.

Cells produced in the thymus through the differentiation of lymphocytic stem cells are exposed to a mixture of thymic hormones. These hormones, produced by the epithelial cells of the thymus, promote the differentiation of functionally competent T cells. However, the T cells residing within the thymus do not participate in the immune response; they remain inactive until they leave the thymus and enter the general circulation. The capillaries of the thymus resemble those of the CNS in that they do not permit free exchange between the interstitial fluid and the circulation. This **blood-thymus barrier** prevents premature stimulation of the developing T cells by circulating antigens.

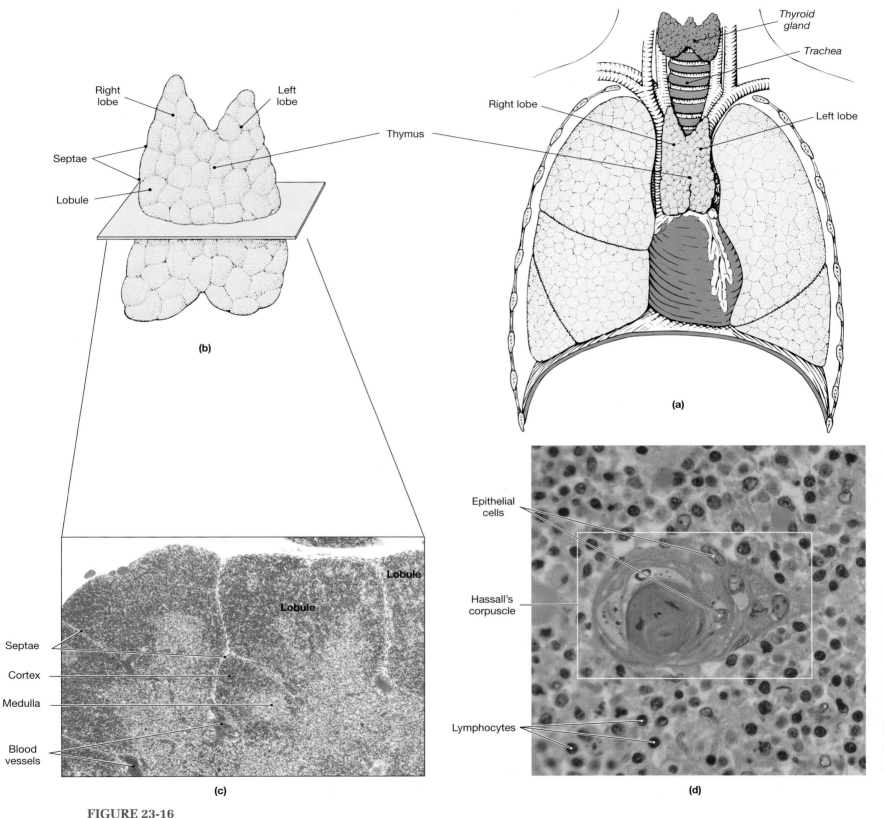

FIGURE 23-16
The Thymus. (a) Appearance and position of the thymus on gross dissection; note its relationship to other organs in the chest. (b) Anatomical landmarks on the thymus. (c) A low-power light micrograph of the thymus. Note the fibrous septae that divide the thymic tissue into lobules resembling interconnected lymphoid nodules. (LM × 43) (d) At higher magnification the unusual structure of Hassall's corpuscles can be examined. The small cells in view are lymphocytes in various stages of development. (LM × 700)

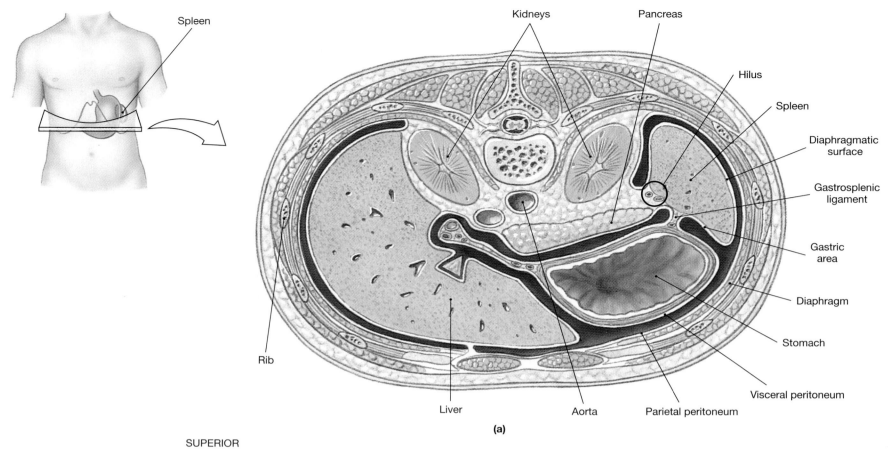

Kidneys Pancreas

Hilus

Spleen

Diaphragmatic surface

Gastrosplenic ligament

Gastric area

Diaphragm

Stomach

Visceral peritoneum

Rib

Liver   Aorta   Parietal peritoneum

(a)

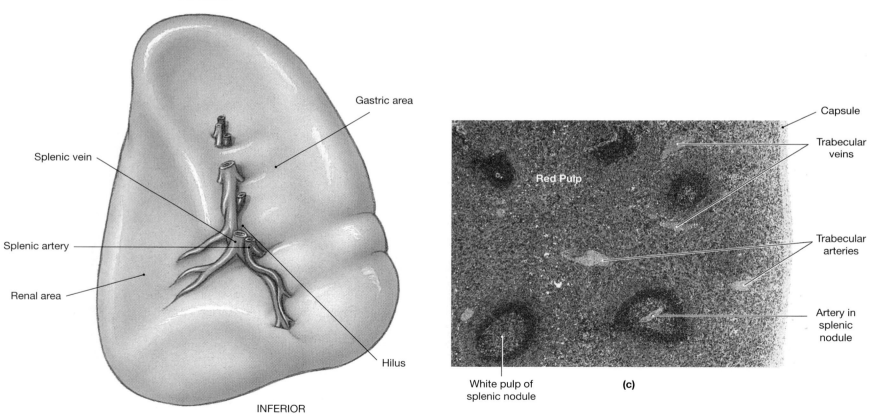

SUPERIOR

Gastric area

Splenic vein

Splenic artery

Renal area

Hilus

INFERIOR

(b) Visceral surface of spleen

Capsule

Trabecular veins

Red Pulp

Trabecular arteries

Artery in splenic nodule

White pulp of splenic nodule

(c)

FIGURE 23-17

The Spleen. (a) The shape of the spleen roughly conforms to the shapes of adjacent organs. This transverse section through the trunk shows the typical position of the spleen within the abdominopelvic cavity. (b) External appearance of the visceral surface of the intact spleen, showing major anatomical landmarks. This view should be compared with that of (a). (c) Histological appearance of the spleen. Areas of white pulp are dominated by lymphocytes; they appear blue because the nuclei of lymphocytes stain very darkly. Areas of red pulp contain a preponderance of red blood cells. (LM × 45)

The Spleen
(Figure 23-17a)

The **spleen** is the largest lymphoid organ in the body. It is around 12 cm (5 in.) long and weighs an average of up to 160 g. The spleen lies along the curving lateral border of the stomach, extending between the ninth and eleventh ribs on the left side. It is attached to the lateral border of the stomach by a broad mesenterial band, the **gastrosplenic ligament** (Figure 23-17a●).

On gross dissection the spleen has a deep red color because of the blood it contains. In essence, the spleen performs the same functions for the blood that lymph nodes perform for lymph. Splenic functions can be summarized as (1) the removal of abnormal blood cells and other blood components through phagocytosis, (2) the storage of iron from recycled red blood cells, and (3) the initiation of immune responses by B cells and T cells in response to antigens in the circulating blood.

SURFACES OF THE SPLEEN *(Figure 23-17).* The spleen has a soft consistency, and its shape primarily reflects its association with the structures around it. It lies wedged between the stomach, the left kidney, and the muscular diaphragm. The **diaphragmatic surface** (Figure 23-17a●) is smooth and convex, conforming to the shape of the diaphragm and body wall. The **visceral surface** (Figure 23-17b●) contains indentations that record the shapes of the stomach (the **gastric area**) and kidney (the **renal area**). Splenic blood vessels and lymphatics communicate with the spleen on the visceral surface at the **hilus**, a groove marking the border between the gastric and renal depressions. The **splenic artery**, **splenic vein**, and the lymphatics draining the spleen are attached at the hilus.

HISTOLOGY OF THE SPLEEN *(Figure 23-17c).* The spleen is surrounded by a capsule containing collagen and elastic fibers. The cellular components within constitute the **pulp** of the spleen (Figure 23-17c●). Areas of **red pulp** contain large quantities of red blood cells, whereas areas of **white pulp** resemble lymphatic nodules. The splenic artery enters at the hilus and branches to produce a number of arteries that radiate outward toward the capsule. These **trabecular arteries** branch extensively, and their finer branches are surrounded by areas of white pulp. Capillaries then discharge the blood into the red pulp.

The cell population of the red pulp includes all of the normal components of the circulating blood, plus fixed and free macrophages. The structural framework of the red pulp consists of a network of reticular fibers. The blood passes through this meshwork and enters large sinusoids, also lined by fixed macrophages. The sinusoids empty into small veins, and these ultimately collect into **trabecular veins** that continue toward the hilus.

This circulatory arrangement gives the phagocytes of the spleen an opportunity to identify and engulf any damaged or infected cells in the circulating blood. Lymphocytes are scattered throughout the red pulp, and the region surrounding each area of white pulp has a high concentration of macrophages. Thus any microorganisms or abnormal plasma components will quickly come to the attention of the splenic lymphocytes. ⊤ *Disorders of the Spleen, Immune Disorders,* and *Systemic Lupus Erythematosus [p. 771]*

√ **How would blockage of the thoracic duct affect the circulation of lymph?**

√ **Why do certain lymph nodes enlarge during some infections?**

AGE AND THE LYMPHATIC SYSTEM

With advancing age the lymphatic system becomes less effective at combating disease. T cells become less responsive to antigens; as a result, fewer cytotoxic T cells respond to an infection. Because the number of helper T cells is also reduced, B cells are less responsive, and antibody levels do not rise as quickly after antigen exposure. The net result is an increased susceptibility to viral and bacterial infection. For this reason, vaccinations for acute viral diseases, such as the flu (influenza), are strongly recommended for elderly individuals. The increased incidence of cancer in the elderly reflects the fact that surveillance by the lymphatic system declines, and tumor cells are not eliminated as effectively.

Related Clinical Terms

severe combined immunodeficiency disease (SCID): A congenital condition in which an individual fails to develop either cellular or humoral immunity because of a lack of normal B and T cells. ⊤ *SCID [p. 769]*

acquired immune deficiency syndrome (AIDS): A disorder that develops after HIV infection, characterized by reduced circulating antibody levels and depressed cellular immunity. ⊤ *AIDS [p. 769]*

tonsillectomy: The removal of an infected tonsil to remove symptoms of **tonsillitis**. ⊤ *Infected Lymphoid Nodules [p. 769]*

appendicitis: Inflammation of the appendix, often requiring treatment in the form of an **appendectomy**. ⊤ *Infected Lymphoid Nodules [p. 770]*

lymphadenopathy (lim-fad-e-NOP-a-thē): Chronic or excessive enlargement of lymph nodes. *[p. 589]*

lymphomas: Malignant cancers consisting of abnormal lymphocytes or lymphocytic stem cells; includes *Hodgkin's disease* and *non-Hodgkin's lymphoma.* ⊤ *Lymphomas [p. 770]*

splenectomy (splē-NEK-to-mē): Surgical removal of the spleen, usually after splenic rupture. ⊤ *Disorders of the Spleen [p. 771]*

splenomegaly (splē-nō-MEG-a-lē): Enlargement of the spleen; often caused by infection, inflammation, or cancer. ⊤ *Disorders of the Spleen [p. 771]*

immunodeficiency disease: A disease in which either the immune system fails to develop normally or the immune response is somehow blocked. ☦ *Immune Disorders [p. 771]*

autoimmune disorder: A disorder that develops when the immune response mistakenly targets normal body cells and tissues.☦ *Immune Disorders [p. 771]*

allergy: An inappropriate or excessive immune response to antigens. ☦ *Immune Disorders [p. 771]*

systemic lupus erythematosus (LOO-pus e-rith-ē-ma-TŌ -sis) **(SLE):** A condition resulting from a generalized breakdown in the antigen recognition mechanism. ☦ *Systemic Lupus Erythematosus [p. 771]*

CHAPTER SUMMARY AND REVIEW

STUDY OUTLINE

Related Key Terms

INTRODUCTION *[p. 581]*

1. The cells, tissues, and organs of the lymphatic system play a central role in the body's defenses.

ORGANIZATION OF THE LYMPHATIC SYSTEM *[p. 581]*

1. The **lymphatic system** includes a network of lymphatic vessels that carry **lymph** (a fluid similar to plasma but with a lower concentration of proteins). A series of lymphoid organs and lymphoid tissues are connected to the lymphatic vessels *(see Figure 23-1).*

Functions of the Lymphatic System *[p. 581]*
2. The lymphatic system produces, maintains, and distributes lymphocytes (cells that attack invading organisms, abnormal cells, and foreign proteins). The system also helps maintain blood volume and eliminate local variations in the composition of the interstitial fluid.

Structure of Lymphatic Vessels *[p. 582]*
3. Lymph flows along a network of lymphatics that originate in the **lymphatic capillaries** (*terminal lymphatics*). The **lymphatic vessels** empty into the **thoracic duct** and the **right lymphatic duct** *(see Figures 23-2 to 23-5).*

superficial lymphatics • deep lymphatics • cisterna chyli

Lymphocytes *[p. 585]*
4. There are three different classes of lymphocytes: **T cells** (thymus-dependent), **B cells** (bone marrow–derived), and **NK cells** (natural killer) *(see Figures 23-6/23-7).*
5. **Cytotoxic T cells** attack foreign cells or body cells infected by viruses; they provide **cell-mediated immunity**. *Regulatory* (**helper** and **suppressor**) *T cells* regulate and coordinate the immune response *(see Figure 23-6).*

memory T cells

6. B cells can differentiate into **plasma cells**, which produce and secrete *antibodies* that react with specific chemical targets, or **antigens**. Antibodies in body fluids are called **immunoglobulins**. B cells are responsible for **antibody-mediated immunity** *(see Figure 23-6).*

memory B cells

7. NK cells (also called *large granular lymphocytes*) attack foreign cells, normal cells infected with viruses, and cancer cells. They provide **immunological surveillance**.
8. The goal of the **immune response** is the destruction or inactivation of pathogens, abnormal cells, and foreign molecules such as toxins *(see Figure 23-6).*
9. Lymphocytes continually migrate in and out of the blood through the lymphoid tissues and organs. **Lymphopoiesis** (lymphocyte production) involves the bone marrow, thymus, and peripheral lymphatic tissues *(see Figure 23-7).*

Lymphoid Tissues *[p. 587]*
10. **Lymphoid tissues** are connective tissues dominated by lymphocytes. In a **lymphoid nodule**, the lymphocytes are densely packed in an area of loose connective tissue *(see Figure 23-8).*

germinal center • aggregate lymphoid nodules (Peyer's patches) • pharyngeal tonsil • palatine tonsils • lingual tonsils • gut-associated lymphatic tissue (GALT)

Lymphoid Organs *[p. 588]*
11. Important **lymphoid organs** include the *lymph nodes*, the *thymus*, and the *spleen*. Lymphoid tissues and organs are distributed in areas especially vulnerable to injury or invasion *(see Figures 23-1/23-9 to 23-15).*

12. **Lymph nodes** are encapsulated masses of lymphoid tissue. The **deep cortex** is dominated by T cells; the **outer cortex** and **medulla** contain B cells arranged into **medullary cords** *(see Figures 23-1/23-9 to 23-15).*

trabeculae • hilus

13. Lymphoid tissues and nodes are located in areas particularly susceptible to injury or invasion by microorganisms.
14. The **cervical lymph nodes, axillary lymph nodes, inguinal lymph nodes, popliteal lymph nodes, thoracic lymph nodes, abdominal lymph nodes, intestinal lymph nodes**, and **mesenterial lymph nodes** serve to protect the vulnerable areas of the body *(see Figures 23-4/23-5/23-10 to 23-15).*

The Thymus *[p. 592]*

15. The **thymus** lies behind the sternum, in the anterior mediastinum. **Epithelial cells** scattered among the lymphocytes produce thymic hormones *(see Figure 23-16).*

The Spleen *[p. 595]*

16. The adult **spleen** contains the largest mass of lymphoid tissue in the body. The cellular components form the **pulp** of the spleen. **Red pulp** contains large numbers of red blood cells, and areas of **white pulp** resemble lymphatic nodules *(see Figure 23-17).*

AGE AND THE LYMPHATIC SYSTEM *[p. 595]*

1. With aging, the immune system becomes less effective at combating disease.

Related Key Terms

involution • thymic lobes • septae • lobules • coretx • medulla • Hassall's corpuscles • blood-thymus barrier

gastrosplenic ligament • diaphragmatic surface • visceral surface • gastric area • renal area • hilus • splenic artery • splenic vein • trabecular arteries • trabecular veins

---

1 REVIEW OF CHAPTER OBJECTIVES

1. Describe the role played by the lymphatic system in the body's defenses.
2. Identify the major components of the lymphatic system.
3. Locate the lymphatic vessels and tissues and describe their structure and function.
4. Discuss the importance of lymphocytes and describe where they are found in the body.
5. Briefly describe the role of the various lymphatic system components in the immune response.
6. Describe the anatomical and functional relationship between the lymphatic and circulatory systems.
7. Locate the principal lymph nodes and and describe their structure and function.
8. Locate the thymus and describe its structure and function.
9. Locate the spleen and describe its structure and function.
10. Describe the changes in the immune system that occur with aging.

2 REVIEW OF CONCEPTS

1. What is the role played by lymphatic vessels in the maintenance of normal blood volume?
2. How does the structure of a lymphatic capillary differ from that of a vascular capillary?
3. How does the structure of a lymphatic vessel compare with that of a vein?
4. When located in a bundle of vessels, how can a lymphatic vessel be distinguished from an artery or a vein?
5. What is the role of the regulatory T cells in control and coordination of the immune response?
6. What is the difference between cellular immunity and humoral immunity?
7. Why does an immune response to a bacterial or viral infection increase with repeated exposure to the pathogen?
8. How does erythropoiesis differ from lymphopoiesis?
9. Why is there a risk of contracting an HIV infection during sexual activity even if a condom is worn?
10. How does AIDS-related complex (ARC) differ from a full-blown AIDS infection?

11. Why are lymph nodes considered to be an "early warning system" of the onset of infection or presence of abnormalities in peripheral tissues?
12. Why is the growth of the thymus so rapid in the first year to two after birth?
13. How is the function of the spleen analogous to that of the lymph nodes?
14. How do autoimmune disorders differ from those such as AIDS and severe combined immunodeficiency (SCID)?
15. How does aging affect the immune response of the lymphatic system?

3 CRITICAL THINKING AND CLINICAL APPLICATION QUESTIONS

1. An 8-year-old girl complains of a severe sore throat, which is red and irritated. She has a fever of 39°C, a headache, an earache, difficulty in swallowing, and enlarged and tender lymph nodes in the neck. She has had these symptoms several times a year for the last few years, and although she responded well to antibiotics each time, the problem has recurred. Assuming a diagnosis of tonsillitis, what structures have traditionally been blamed, and what remedy has historically been taken to treat it?

2. In addition to symptoms of nasal stuffiness, sore throat, malaise, headache, fever, and other symptoms of a cold or influenza, people often complain of having "swollen glands." What structures are involved, and why do they swell?

3. In women with diagnosed breast cancer, not only the breast tissue may be removed surgically, but the lymph nodes in the axillary region as well. Why would this procedure be followed, and how would involvement of the lymph nodes in the spread of breast cancer be a serious condition?

4. People often have adverse reactions to such components of the environment as dust, pollens, molds, animal dander, and hair. These effects may cause symptoms similar to those of a cold, but do not disappear in the same length of time as would the cold symptoms. These effects are often seasonal, becoming worse rapidly when there are high levels of the irritant present, and the victim is said to be allergic to the substance. What is an allergy, and why does the body respond in a particular fashion to these environmental factors?

24

The Respiratory System

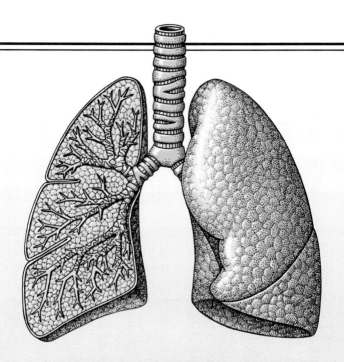

CHAPTER OUTLINE AND OBJECTIVES

Cells obtain energy primarily through aerobic metabolism, a process that requires oxygen and produces carbon dioxide. For cells to survive, they must have a way to obtain that oxygen and eliminate the carbon dioxide. The cardiovascular system provides the link between the interstitial fluids and the exchange surfaces of the lungs. The respiratory system facilitates the exchange of gases between the air and the blood. As it circulates, blood carries oxygen from the lungs to peripheral tissues; it also accepts the carbon dioxide produced by these tissues and transports it to the lungs.

We begin our discussion of the respiratory system by describing the anatomical structures that air will pass through as it travels toward the alveoli. We will then discuss the mechanics of breathing and the neural control of respiration.

FUNCTIONS OF THE RESPIRATORY SYSTEM

The functions of the respiratory system are:

1. Providing an extensive area for gas exchange between air and circulating blood;

2. Moving air to and from the exchange surfaces of the lungs;

3. Protecting respiratory surfaces from dehydration, temperature changes, or other environmental variations;

4. Defending the respiratory system and other tissues from invasion by pathogenic microorganisms;

5. Producing sounds involved in speaking, singing, or nonverbal communication;

6. Assisting in the regulation of blood volume, blood pressure, and the control of body fluid pH.

The respiratory system performs these functions with the cooperation of the circulatory system, selected skeletal muscles, and the nervous system.

ORGANIZATION OF THE RESPIRATORY SYSTEM
(Figure 24-1)

The **respiratory system** includes the nose, nasal cavity and sinuses, the pharynx, the larynx (voice box), the trachea (windpipe), and smaller conducting passageways leading to the exchange surfaces of the lungs. These structures are illustrated in Figure 24-1●. The **upper respiratory system** consists of the nose, nasal cavity, paranasal sinuses, and pharynx. These passageways

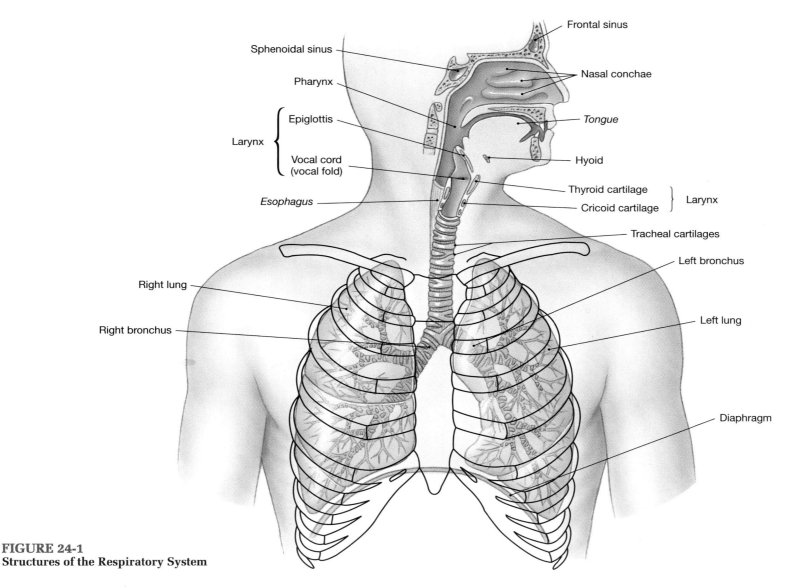

FIGURE 24-1
Structures of the Respiratory System

filter, warm, and humidify the air, protecting the more delicate conduction and exchange surfaces of the **lower respiratory system** from debris, pathogens, and environmental extremes. The lower respiratory system includes the larynx, trachea, bronchi, and lungs.

The **respiratory tract** consists of the airways that carry air to and from the exchange surfaces of the lungs. The respiratory tract can be divided into a *conducting portion* and a *respiratory portion*. The conducting portion extends from the entrance to the nasal cavity to the smallest *bronchioles* of the lungs. The respiratory portion of the tract includes the delicate *respiratory bronchioles* and the delicate air sacs, or **alveoli** (al-VĒ-ō-lī), where gas exchange occurs.

Filtering, warming, and humidification of the inspired air begin at the entrance to the upper respiratory system and continue throughout the rest of the conducting system. By the time the air reaches lung alveoli, most foreign particles and pathogens have been removed, and the humidity and temperature are within acceptable limits. The success of this "conditioning process" is due primarily to the properties of the *respiratory epithelium*.

THE RESPIRATORY EPITHELIUM
(Figure 24-2)

The **respiratory epithelium** consists of pseudostratified, ciliated, columnar epithelium with numerous goblet cells (Figure 24-2●). The respiratory epithelium lines the entire respiratory tract except for the finest conducting portions and the alveoli. Goblet cells and mucous glands beneath the epithelium produce a sticky mucus that bathes the exposed surfaces. In the nasal cavity, cilia sweep any debris trapped in mucus or microorganisms toward the pharynx, where it will be swallowed and exposed to the acids and enzymes of the stomach. In lower portions of the respiratory tract, the cilia also beat toward the pharynx, creating a *mucus escalator* that cleans the respiratory passageways. The delicate surfaces of the respiratory system can be severely damaged if the inspired air becomes contaminated with debris or pathogens. The respiratory mechanisms for filtration are responsible for the **respiratory defense system**. ⇑ *Overloading the Respiratory Defenses [p. 772]*

In the nasal cavity, this filtering mechanism removes virtually all particles larger than around 10 μm from the inspired air. Smaller particles may be trapped by the mucus of the nasopharynx or secretions of the pharynx before proceeding farther along the conducting system. Exposure to unpleasant stimuli, such as noxious vapors, large quantities of dust and debris, allergens, or pathogens, usually causes a rapid increase in the rate of mucus production. (The familiar symptoms of the "common cold" result from the invasion of the respiratory epithelium by one of over 200 viruses.)

Much of the filtration, warming, and humidification of inhaled air occurs in the nasal cavity. Breathing through the mouth eliminates much of the preliminary filtration, heating, and humidifying of the inspired air. Patients breathing on a respirator, which utilizes a tube to provide air directly into the trachea, must receive air that has been externally filtered and humidified, or risk alveolar damage.

■ CLINICAL BRIEF
Cystic Fibrosis

Cystic fibrosis (CF) is the most common lethal inherited disease in the Caucasian population, occurring at a frequency of 1 birth in 1600. Each year approximately 2000 babies are born with this condition in the U.S. alone. Individuals with CF seldom survive past age 30; death is usually the result of a massive bacterial infection of the lungs and associated heart failure.

The underlying problem involves a membrane protein responsible for the active transport of chloride ions. This membrane protein is abundant in exocrine cells that produce watery secretions. In persons with CF, these cells cannot transport salts and water effectively, and the secretions produced are thick and gooey. Mucous glands of the respiratory tract and secretory cells of the pancreas, salivary glands, and digestive tract are affected.

The most serious symptoms appear because the respiratory defense system cannot transport such dense mucus. The mucus escalator stops working, and mucus plugs block the smaller respiratory passageways. This blockage reduces the diameter of the airways, and the inactivation of the normal respiratory defenses leads to frequent bacterial infections.

The gene responsible for CF has been identified, and the structure of the membrane protein determined. Now that the structure of the gene is understood, research continues with the goal of correcting the defect by the insertion of normal genes.

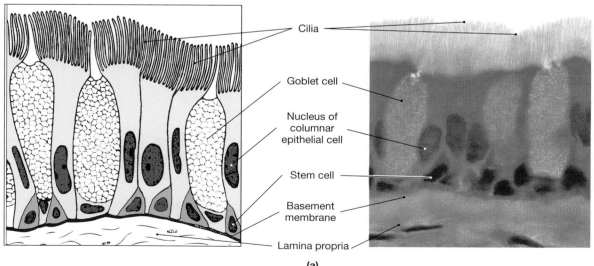

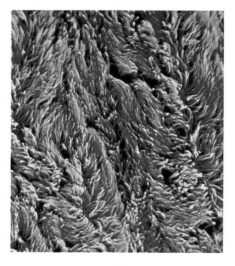

Cilia

Goblet cell

Nucleus of columnar epithelial cell

Stem cell

Basement membrane

Lamina propria

(a)

(b)

FIGURE 24-2
The Respiratory Epithelium. (a) Sketch and light micrograph showing the sectional appearance of the respiratory epithelium. (LM × 932) (b) A surface view of the epithelium, as seen with the scanning electron microscope. The cilia of the epithelial cells form a dense layer that resembles a shag carpet. The movement of these cilia propels mucus across the epithelial surface. (SEM × 1647)

THE UPPER RESPIRATORY SYSTEM

The Nose and Nasal Cavity
(Figure 24-3)

The nose is the primary passageway for air entering the respiratory system. The bones, cartilages, and sinuses associated with the nose were introduced in Chapter 6. ∞ [pp. 134, 144, 147] Air normally enters the respiratory system via the paired **external nares** that open into the **nasal cavity**. The **vestibule** (VES-ti-būl) is the portion of the nasal cavity contained within the flexible tissues of the external nose (Figure 24-3a,d●). The epithelium of the vestibule contains coarse hairs that extend across the external nares. Large airborne particles such as sand, sawdust, or even insects are trapped in these hairs and are prevented from entering the nasal cavity.

The *nasal septum* divides the nasal cavity into left and right portions. The bony portion of the nasal septum is formed by the fusion of the perpendicular plate of the ethmoid and the plate of the vomer. The anterior portion of the nasal septum is formed of hyaline cartilage. This cartilaginous plate supports the bridge, or **dorsum nasi** (DOR-sum NĀ-zī), and **apex** (tip) of the nose.

The maxillary, nasal, frontal, ethmoid, and sphenoid bones form the lateral and superior walls of the nasal cavity. ∞ [p. 148] The superior portion, or *olfactory region*, of the nasal cavity includes the area lined by olfactory epithelium: the inferior surface of the cribriform plate, the superior portion of the nasal septum, and the superior nasal conchae. ∞ [p. 144]

The *superior*, *middle*, and *inferior nasal conchae* project toward the nasal septum from the lateral walls of the nasal cavity. To pass from the vestibule to the internal nares, air tends to flow between adjacent conchae, through the **superior**, **middle**, or **inferior meatuses** (mē-Ā-tus-es; *meatus*, passage) (Figure 24-3b,d●). These are narrow grooves rather than open passageways, and the incoming air bounces off the conchal surfaces and churns around like water flowing over rapids. This turbulence serves a purpose: As the air eddies and swirls, small airborne particles are likely to come in contact with the mucus that coats the lining of the nasal cavity. For this reason the conchae are also called the **turbinate bones**. In addition to promoting filtration, the turbulence allows extra time for warming and humidifying the incoming air. ⊤ *Nosebleeds [p. 772]*

A bony **hard palate**, formed by the maxillary and palatine bones, forms the floor of the nasal cavity and separates the oral and nasal cavities. ∞ [p. 146] A fleshy **soft palate** extends posterior to the hard palate, marking the boundary line between the superior *nasopharynx* and the rest of the pharynx (Figure 24-3c●). The nasal cavity opens into the nasopharynx at the **internal nares** (NAR-ēz).

The Pharynx
(Figure 24-3c,d)

The nose, mouth, and throat connect to each other by a common passageway or chamber called the **pharynx** (FAR-inks). The pharynx is shared by the digestive and respiratory systems. It extends between the internal nares and the entrances to the larynx and esophagus. The curving superior and posterior walls are closely bound to the axial skeleton, but the lateral walls are quite flexible and muscular. The pharynx is divided into three regions (Figure 24-3c,d●): the *nasopharynx*, the *oropharynx*, and the *laryngopharynx*.

The Nasopharynx
(Figure 24-3c,d)

The **nasopharynx** (nā-zō-FAR-inks) is the superior portion of the pharynx. It is connected to the posterior portion of the nasal cavity via the internal nares and is separated from the oral cavity by the soft palate (Figure 24-3c,d●).

The nasopharynx is lined by typical respiratory epithelium. The *pharyngeal* (adenoid) *tonsil* is located on the posterior wall of the nasopharynx; the lateral walls contain the openings of the *auditory tubes* (see Figure 25-5a●). ∞ [p. 633]

The Oropharynx
(Figure 24-3c,d)

The **oropharynx** (ōr-ō-FAR-inks, *oris*, mouth) extends between the soft palate and the base of the tongue at the level of the hyoid bone. The posterior portion of the oral cavity communicates directly with the oropharynx, as do the posterior and inferior portions of the nasopharynx (Figure 24-3c,d●). At the boundary between the nasopharynx and oropharynx, the epithelium changes from a typical respiratory epithelium to a stratified squamous epithelium similar to that of the oral cavity.

The posterior margin of the soft palate supports the dangling **uvula** (Ū-vū-la) and two pairs of muscular **pharyngeal arches**. On either side a palatine tonsil lies between an anterior **palatoglossal** (pal-a-tō-GLOS-al) **arch** and a posterior **palatopharyngeal** (pal-a-tō-fa-RIN-jē-al) **arch** (see Figure 25-5a●). ∞ [p. 633] A curving line that connects the palatoglossal arches and uvula forms the boundaries of the **fauces** (FAW-sēz), the passageway between the oral cavity and the oropharynx.

The Laryngopharynx
(Figure 24-3c,d)

The narrow **laryngopharynx** (la-RING-gō-far-inks) includes that portion of the pharynx lying between the hyoid bone and the entrance to the esophagus (Figure 24-3c,d●). The laryngopharynx is the most inferior portion of the pharynx, and like the oropharynx it is lined by a stratified squamous epithelium that can resist mechanical abrasion, chemical attack, and pathogenic invasion.

THE LARYNX
(Figures 24-3d/24-4)

Inspired (inhaled) air leaves the pharynx by passing through a narrow opening, the **glottis** (GLOT-is) (Figure 24-3d●). The **larynx** (LAR-inks) surrounds and protects the glottis. The larynx begins at the level of vertebra C_4 or C_5 and ends at the level of vertebra C_7. The larynx is essentially a cylinder whose cartilaginous walls are stabilized by ligaments or skeletal muscles or both.

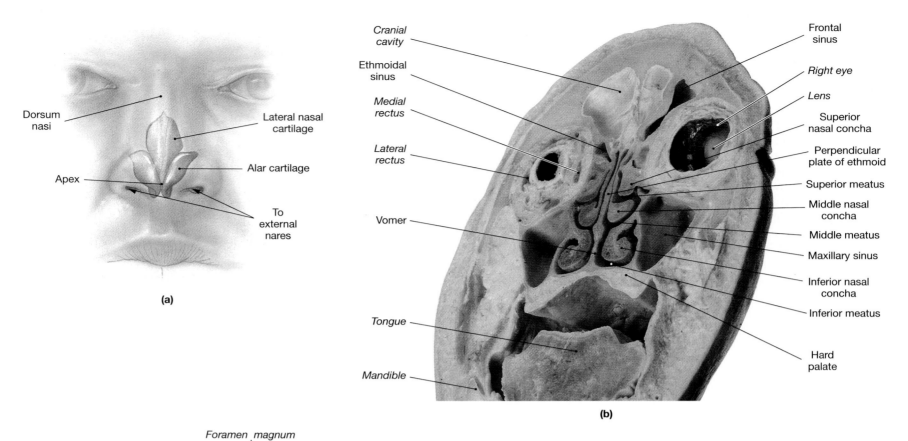

(a)

Dorsum nasi
Apex
Lateral nasal cartilage
Alar cartilage
To external nares

Cranial cavity
Ethmoidal sinus
Medial rectus
Lateral rectus
Vomer
Tongue
Mandible

Frontal sinus
Right eye
Lens
Superior nasal concha
Perpendicular plate of ethmoid
Superior meatus
Middle nasal concha
Middle meatus
Maxillary sinus
Inferior nasal concha
Inferior meatus
Hard palate

(b)

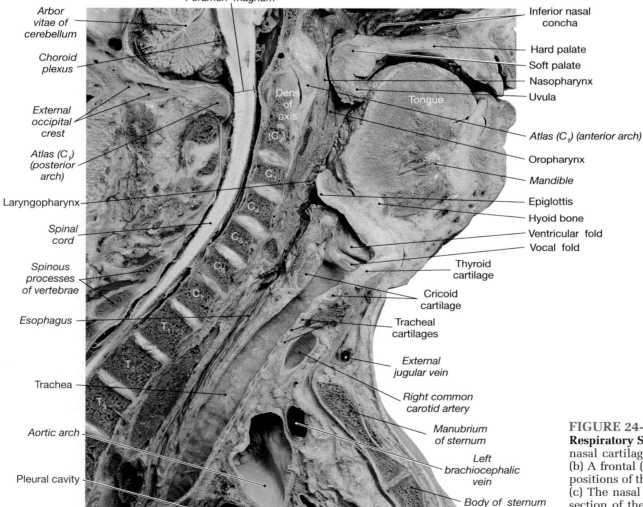

Arbor vitae of cerebellum
Choroid plexus
External occipital crest
Atlas (C₁) (posterior arch)
Laryngopharynx
Spinal cord
Spinous processes of vertebrae
Esophagus
Trachea
Aortic arch
Pleural cavity

Foramen magnum

Inferior nasal concha
Hard palate
Soft palate
Nasopharynx
Uvula
Atlas (C₁) (anterior arch)
Oropharynx
Mandible
Epiglottis
Hyoid bone
Ventricular fold
Vocal fold
Thyroid cartilage
Cricoid cartilage
Tracheal cartilages
External jugular vein
Right common carotid artery
Manubrium of sternum
Left brachiocephalic vein
Body of sternum

Dens of axis
Tongue

(c)

FIGURE 24-3

Respiratory Structures in the Head and Neck. (a) The nasal cartilages and external landmarks on the nose. (b) A frontal (coronal) section of the head showing the positions of the paranasal sinuses and nasal structures. (c) The nasal cavity and pharynx as seen in a sagittal section of the head and neck. (d) Diagrammatic view of the head and neck in sagittal section, for comparison with (c).

Cartilages of the Larynx
(Figure 24-4)

Three large cartilages form the body of the larynx: the *thyroid cartilage*, the *cricoid cartilage*, and the *epiglottis*. View Figure 24-4● as we describe the cartilages of the larynx.

The Thyroid Cartilage
(Figure 24-4a,b)

The **thyroid** ("shield-shaped") **cartilage** is the largest laryngeal cartilage, and it forms most of the anterior and lateral walls of the larynx. The thyroid cartilage, when viewed in sagittal section, has the form of an apostrophe (ʼ); it is incomplete posteriorly. The anterior surface of this cartilage is easily seen and felt, and the thyroid cartilage is commonly called the *Adam's apple*. During embryological development, the thyroid cartilage is formed by two pieces of cartilage that meet in the anterior midline to form the **laryngeal prominence** of the thyroid cartilage (Figure 24-4a,b●).

The inferior surface of the thyroid cartilage articulates with the cricoid cartilage; the superior surface has ligamentous attachments to the epiglottis and smaller laryngeal cartilages.

The Cricoid Cartilage
(Figure 24-4a,c)

The thyroid cartilage sits superior to the **cricoid cartilage** (KRĪ-koyd; "ring-shaped"). The posterior portion of the cricoid is greatly expanded, providing support in the absence of the thyroid cartilage. The cricoid and thyroid cartilages protect the glottis and the entrance to the trachea, and their broad surfaces provide sites for the attachment of important laryngeal muscles and ligaments. Ligaments attach the inferior surface of the

cricoid cartilage to the first cartilage of the trachea (Figure 24-4a,c●). The superior surface of the cricoid cartilage articulates with the small *arytenoid cartilages*, described below.

The Epiglottis
(Figures 24-3c,d/24-4b,c)

The shoehorn-shaped **epiglottis** (ep-i-GLOT-is) projects above the glottis (Figures 24-3c,d and 24-4b,c●). This structure, composed of elastic cartilage, has ligamentous attachments to the anterior and superior borders of the thyroid cartilage and the hyoid bone. During swallowing, the larynx is elevated, and the epiglottis folds back over the glottis, preventing the entry of liquids or solid food into the respiratory passageways.

Paired Laryngeal Cartilages
(Figures 24-4c,d/24-5)

The larynx also contains three pairs of smaller cartilages: the *arytenoid*, *corniculate*, and *cuneiform cartilages*.

- The paired **arytenoid cartilages** (ar-i-TĒ-noyd; "ladle-shaped") articulate with the superior border of the enlarged portion of the cricoid cartilage (Figure 24-4c●).
- The **corniculate cartilages** (kor-NIK-ū-lāt; "horn-shaped") articulate with the arytenoid cartilages (Figure 24-4c●). The corniculate and arytenoid cartilages are involved with the opening and closing of the glottis and the production of sound.
- Elongated, curving **cuneiform cartilages** (kū-NĒ-i-form; "wedge-shaped") lie within the *aryepiglottic fold* that extend between the lateral aspect of each arytenoid cartilage and the epiglottis (Figures 24-4d and 24-5●).

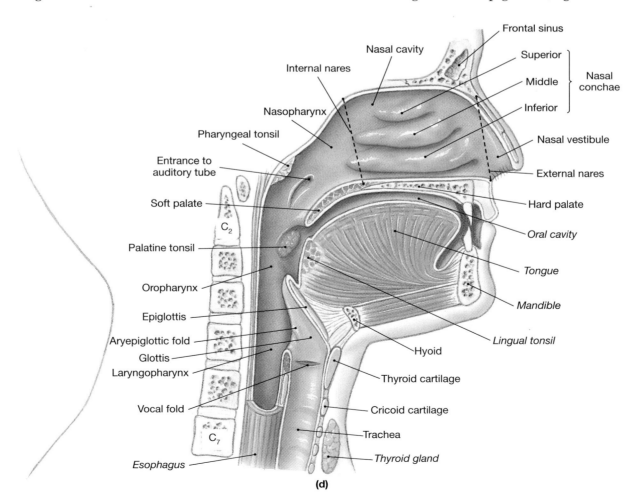

(d)

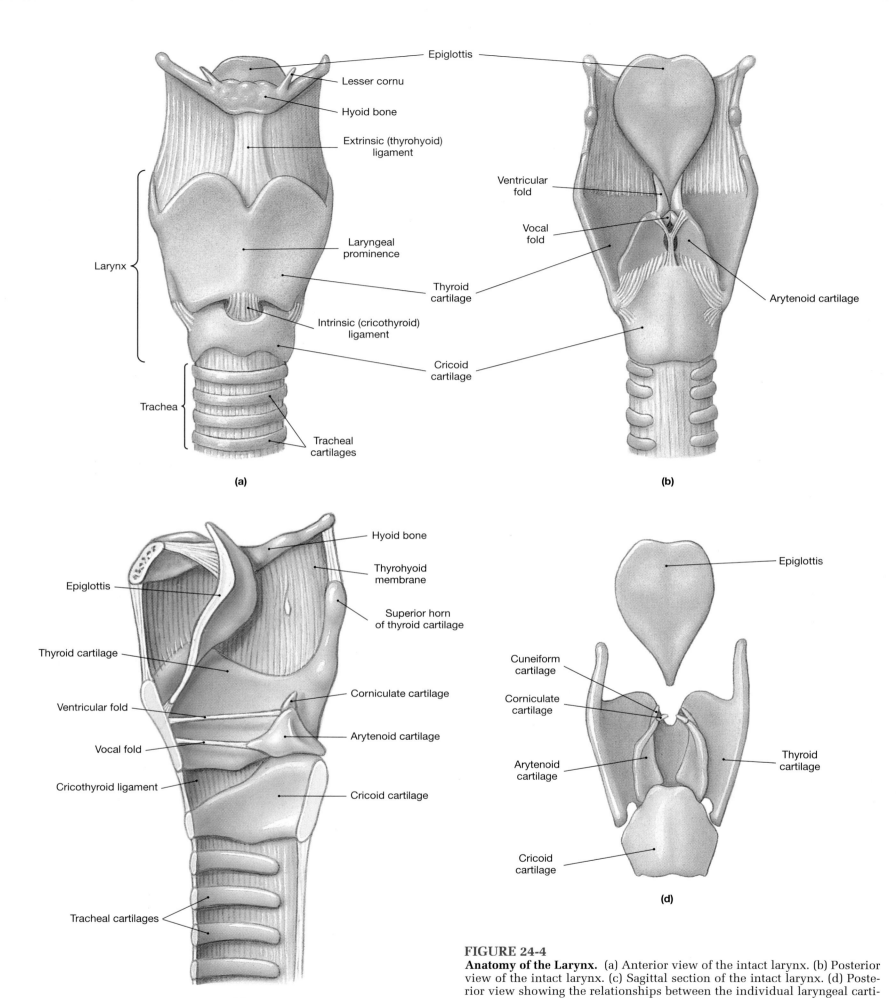

FIGURE 24-4
Anatomy of the Larynx. (a) Anterior view of the intact larynx. (b) Posterior view of the intact larynx. (c) Sagittal section of the intact larynx. (d) Posterior view showing the relationships between the individual laryngeal cartilages.

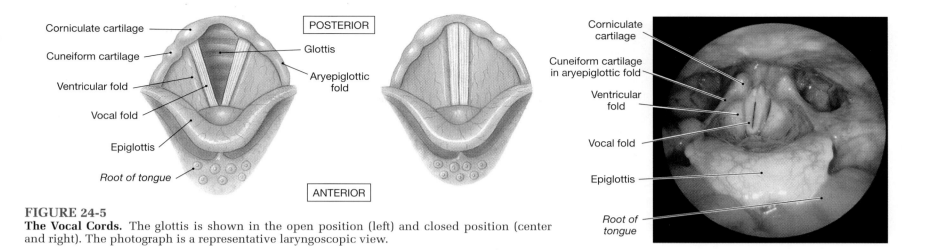

FIGURE 24-5
The Vocal Cords. The glottis is shown in the open position (left) and closed position (center and right). The photograph is a representative laryngoscopic view.

Laryngeal Ligaments
(Figures 24-4a,b/24-5)

A series of **intrinsic ligaments** binds all nine cartilages together to form the larynx (Figure 24-4a,b●). **Extrinsic ligaments** attach the thyroid cartilage to the hyoid bone and the cricoid cartilage to the trachea. The **ventricular ligaments** and the **vocal ligaments** extend between the thyroid cartilage and the arytenoids.

The ventricular and vocal ligaments are covered by folds of laryngeal epithelium that project into the glottis. The ventricular ligaments lie within the superior pair of folds, known as the **ventricular folds** (Figures 24-4b and 24-5●). The ventricular folds, which are relatively inelastic, help to prevent foreign objects from entering the glottis and provide protection for the more delicate **vocal folds**.

The vocal folds are highly elastic, because the vocal ligament is a band of elastic tissue. The vocal folds are involved with the production of sounds, and for this reason they are known as the **true vocal cords**. Because the ventricular folds play no part in sound production, they are often called the **false vocal cords**.

Sound Production

Air passing through the glottis vibrates the vocal folds and produces sound waves. The pitch of the sound produced depends on the diameter, length, and tension in the vocal folds. The diameter and length are directly related to the size of the larynx. The tension is controlled by the contraction of voluntary muscles that change the relative positions of the thyroid and arytenoid cartilages. When the distance increases, the vocal folds tense and the pitch rises; when the distance decreases, the vocal folds relax and the pitch falls.

Anatomically, children of both sexes have slender, short vocal folds, and their voices tend to be high-pitched. At puberty the larynx of a male enlarges considerably more than that of a female. The true vocal cords of an adult male are thicker and longer, and they produce lower tones than those of an adult female.

The entire larynx is involved in sound production, because its walls vibrate, creating a composite sound. Amplification and echoing of the sound occur within the pharynx, the oral cavity, the nasal cavity, and the paranasal sinuses. The final production of distinct sounds depends on voluntary movements of the tongue, lips, and cheeks.

The Laryngeal Musculature
(Figure 24-6)

The larynx is associated with two different groups of muscles, the *intrinsic laryngeal muscles* and the *extrinsic laryngeal muscles*. The **intrinsic laryngeal muscles** have two major functions. One group regulates tension in the vocal folds, while a second set opens and closes the glottis. Those involved with the vocal folds insert upon the thyroid, arytenoid, and corniculate cartilages. Opening or closing the glottis involves rotational movements of the arytenoids that move the vocal folds apart or together.

The **extrinsic laryngeal musculature** positions and stabilizes the larynx. These muscles were considered in Chapter 10. ∞ [p. 263]

During swallowing, both extrinsic and intrinsic muscles cooperate to prevent food or drink from entering the glottis. Before you swallow, the material is crushed and chewed into a pasty mass known as a **bolus**. Extrinsic muscles then elevate the larynx, bending the epiglottis over the entrance to the glottis, so that the bolus can glide across the epiglottis, rather than falling into the larynx (Figure 24-6●). While this movement is under way, intrinsic muscles close the glottis. Should any food particles or liquids manage to touch the surfaces of the ventricular or vocal folds, the coughing reflex will be triggered. Coughing usually prevents the material from entering the glottis. ✝ *Disorders of the Larynx [p. 772]*

THE TRACHEA
(Figures 24-2a/24-3c/24-7)

The epithelium of the larynx is continuous with that of the **trachea** (TRĀ-kē-a), or "windpipe." The trachea is a tough, flexible tube with a diameter of around 2.5 cm (1 in.) and a length of approximately 11 cm (4.25 in.) (Figures 24-3c, p. 602 and 24-7●). The trachea begins anterior to vertebra C_6 in a ligamentous attachment to the cricoid cartilage; it ends in the mediastinum, at the level of vertebra T_5, where it branches to form the *right* and *left primary bronchi*.

The lining of the trachea consists of respiratory epithelium overlying a layer of loose connective tissue called the **lamina propria** (LA-mi-na PRŌ-prē-a) (Figure 24-2a●, p. 600). The lamina propria separates the respiratory epithelium from underlying

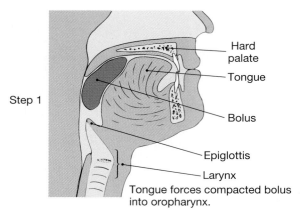

Step 1

Hard palate

Tongue

Bolus

Epiglottis

Larynx

Tongue forces compacted bolus into oropharynx.

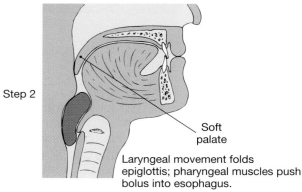

Step 2

Soft palate

Laryngeal movement folds epiglottis; pharyngeal muscles push bolus into esophagus.

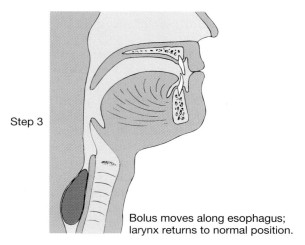

Step 3

Bolus moves along esophagus; larynx returns to normal position.

FIGURE 24-6
Movements of the Larynx during Swallowing. During swallowing the elevation of the larynx folds the epiglottis over the glottis, steering materials into the esophagus.

cartilages. The epithelium and lamina propria are interdependent, and the combination is an example of a *mucous membrane*, or **mucosa** (mū-KŌ-sa). ∞ [p. 77]

A thick layer of connective tissue, the **submucosa** (sub-mū-KŌ-sa), surrounds the mucosa. The submucosa contains mucous glands that communicate with the epithelial surface through a number of secretory ducts. The trachea contains 15–20 **tracheal cartilages** (Figure 24-7●). Each tracheal cartilage is bound to neighboring cartilages by elastic **annular ligaments**. The tracheal cartilages stiffen the tracheal walls and protect the airway. They also prevent its collapse or overexpansion as pressures change in the respiratory system.

Each tracheal cartilage is C-shaped. The closed portion of the C protects the anterior and lateral surfaces of the tra-

chea. The open portions of the C-shaped tracheal cartilages face posteriorly, toward the esophagus (Figure 24-7b●). Because the cartilages do not continue around the trachea, the posterior tracheal wall can easily distort during swallowing, permitting the passage of large masses of food.

An elastic ligament and a band of smooth muscle, the **trachealis**, connect the ends of each tracheal cartilage (Figure 24-7b●). Contraction of the trachealis muscle alters the diameter of the tracheal lumen, changing the resistance to airflow. The normal diameter of the trachea changes from moment to moment, primarily under the control of the sympathetic division of the autonomic nervous system. Sympathetic stimulation increases the diameter of the trachea and makes it easier to move large volumes of air along the respiratory passageways.

■ CLINICAL BRIEF
Tracheal Blockage

Foreign objects that become lodged in the larynx or trachea are usually expelled by coughing. If the individual can speak or make a sound, the airway is still open, and no emergency measures should be taken. If the victim can neither breathe nor speak, an immediate threat to life exists. Unfortunately many victims become acutely embarrassed by this situation, and rather than seek assistance, they run to the nearest restroom and quietly expire.

In the **Heimlich** (HĪM-lik) **maneuver**, or *abdominal thrust*, a rescuer applies compression to the abdomen just beneath the diaphragm. Compression elevates the diaphragm forcefully and may generate enough pressure to remove the blockage. This maneuver must be performed properly to avoid damage to internal organs. Organizations such as the Red Cross, the local fire department, or other charitable groups usually hold brief training sessions throughout the year.

If the blockage remains, professionally qualified rescuers may perform a **tracheostomy** (trā-kē-OS-tō-mē). In this procedure an incision is made through the anterior tracheal wall, and a tube is inserted. The tube bypasses the larynx and permits air to flow directly into the trachea. A tracheostomy may also be required when the larynx becomes blocked by a foreign object, inflammation, or sustained laryngeal spasms, or when a portion of the trachea has been crushed.

THE PRIMARY BRONCHI
(Figures 24-7a)

The trachea branches within the mediastinum, giving rise to the **right** and **left primary bronchi** (BRONG-kī). A ridge, the **carina** (ka-RĪ-na), marks the line of separation between the two bronchi (Figure 24-7a●). The histological organization of the primary bronchi is the same as the trachea, with cartilaginous C-shaped supporting rings. The right primary bronchus supplies the right lung, and the left supplies the left lung. The right primary bronchus has a larger diameter than the left, and it descends toward the lung at a steeper angle. For these reasons foreign objects that enter the trachea usually become lodged in the right bronchus rather than the left.

Each primary bronchus travels to a groove along the medial surface of its lung before branching further. This groove, the **hilus** (HĪ-lus), also provides access for entry to pulmonary vessels and nerves. The entire array is firmly anchored in a meshwork of dense connective tissue. This complex, known as the **root** of the lung, attaches it to the mediastinum and fixes the positions of the major nerves, vessels, and lymphatics. The roots of the lungs are located anterior to vertebrae T_5 (right) and T_6 (left).

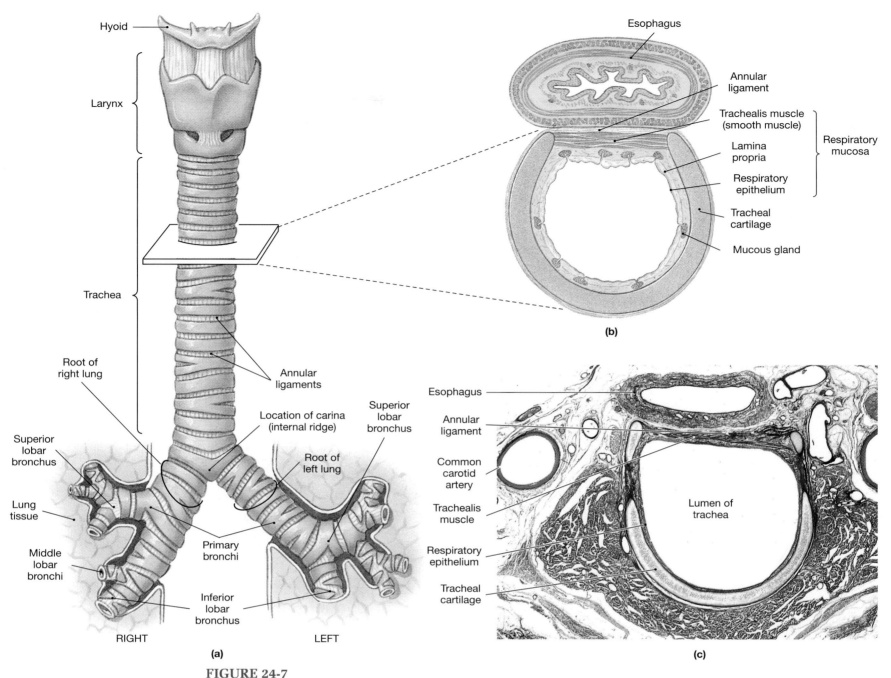

FIGURE 24-7
Anatomy of the Trachea and Primary Bronchi. (a) Anterior view on dissection, showing the plane of section for (b). (b,c) Cross-sectional views of the trachea. (LM × 60)

THE LUNGS
(Figure 24-8)

The left and right lungs (Figure 24-8●) are situated in the left and right pleural cavities. Each lung is a blunt cone with the tip, or **apex**, pointing superiorly. The apex on each side extends into the base of the neck above the first rib. The broad concave inferior portion, or **base**, of each lung rests on the superior surface of the diaphragm.

Lobes of the Lungs
(Figure 24-8)

The lungs have distinct **lobes** separated by deep fissures. The **right lung** has three lobes, **superior**, **middle**, and **inferior**. The *horizontal fissure* separates the superior and middle lobes. The *oblique fissure* separates the superior and inferior lobes. The **left lung** has only two lobes, **superior** and **inferior** separated by the *oblique fissure* (Figure 24-8●). The right lung is broader than the left because most of the heart and great vessels project into the left pleural cavity. However, the left lung is longer than the right lung, because the diaphragm rises on the right side to accommodate the mass of the liver.

Lung Surfaces
(Figure 24-8)

The curving anterior portion of the lung that follows the inner contours of the rib cage is the **costal surface** (Figure 24-8a●). The **mediastinal surface** containing the hilus has a more irregular shape (Figure 24-8b●). The mediastinal surfaces of both lungs bear grooves that mark the passage of vessels traveling to and from the heart. The

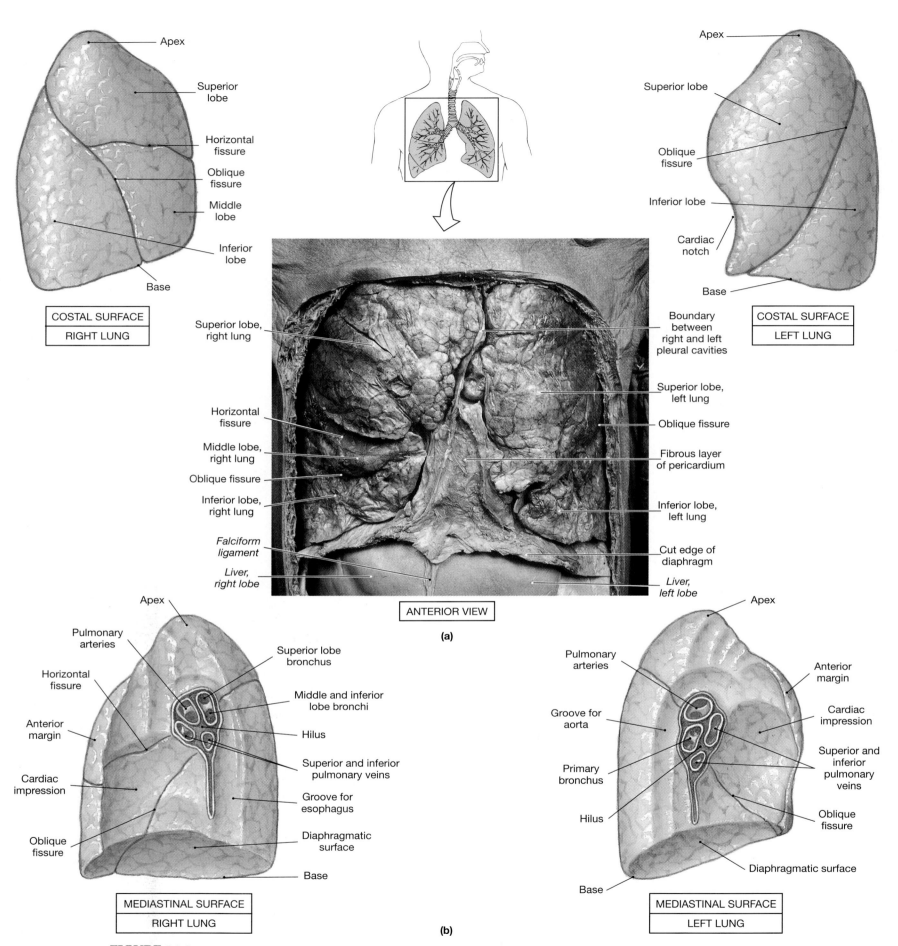

COSTAL SURFACE
RIGHT LUNG

Apex
Superior lobe
Horizontal fissure
Oblique fissure
Middle lobe
Inferior lobe
Base

COSTAL SURFACE
LEFT LUNG

Apex
Superior lobe
Oblique fissure
Inferior lobe
Cardiac notch
Base

Superior lobe, right lung
Horizontal fissure
Middle lobe, right lung
Oblique fissure
Inferior lobe, right lung
Falciform ligament
Liver, right lobe

Boundary between right and left pleural cavities
Superior lobe, left lung
Oblique fissure
Fibrous layer of pericardium
Inferior lobe, left lung
Cut edge of diaphragm
Liver, left lobe

ANTERIOR VIEW

(a)

MEDIASTINAL SURFACE
RIGHT LUNG

Apex
Pulmonary arteries
Horizontal fissure
Anterior margin
Cardiac impression
Oblique fissure

Superior lobe bronchus
Middle and inferior lobe bronchi
Hilus
Superior and inferior pulmonary veins
Groove for esophagus
Diaphragmatic surface
Base

(b)

MEDIASTINAL SURFACE
LEFT LUNG

Apex
Pulmonary arteries
Groove for aorta
Primary bronchus
Hilus
Base

Anterior margin
Cardiac impression
Superior and inferior pulmonary veins
Oblique fissure
Diaphragmatic surface

FIGURE 24-8

Superficial Anatomy of the Lungs. (a) Anterior view of open chest, showing relative positions of lungs and heart, and diagrammatic views of the lateral surfaces of the isolated right and left lungs. (b) Diagrammatic views of the mediastinal (medial) surfaces of the isolated right and left lungs.

608

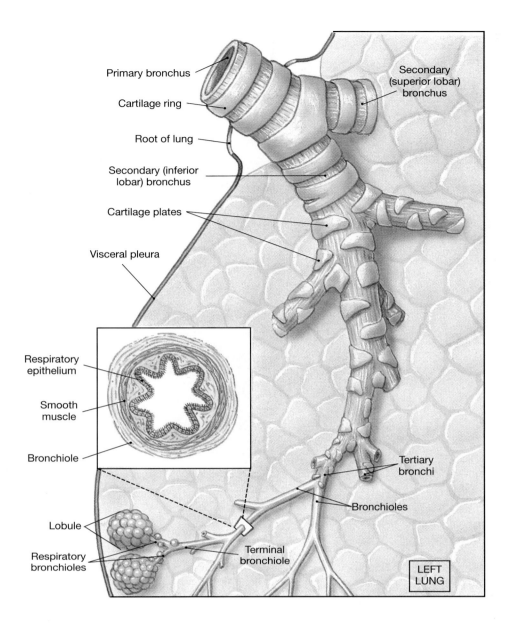

Labels for the figure:

Primary bronchus
Cartilage ring
Root of lung
Secondary (inferior lobar) bronchus
Cartilage plates
Visceral pleura
Respiratory epithelium
Smooth muscle
Bronchiole
Lobule
Respiratory bronchioles
Secondary (superior lobar) bronchus
Tertiary bronchi
Bronchioles
Terminal bronchiole
LEFT LUNG

FIGURE 24-9
Bronchi and Bronchioles. For clarity, the degree of branching has been reduced; an airway branches approximately 23 times before reaching the level of a lobule.

branches are collectively called the **intrapulmonary bronchi**.

Each primary bronchus divides to form **secondary bronchi**, also known as **lobar bronchi**. Secondary bronchi in turn branch to form **tertiary bronchi**, or **segmental bronchi**. The branching pattern differs depending on the lung considered; details are provided below. Each tertiary bronchus supplies air to a single *bronchopulmonary segment*, a specific region of one lung. There are 10 tertiary bronchi (and 10 bronchopulmonary segments) in the right lung. The left lung also has 10 segments during development, but subsequent fusion usually reduces that number to eight or nine. The walls of the primary, secondary, and tertiary bronchi contain progressively lesser amounts of cartilage. The walls of secondary and tertiary bronchi contain cartilage plates arranged around the lumen. These cartilages serve the same purpose as the rings of cartilage in the trachea and primary bronchi. *Bronchitis [p. 772]*

Branches of the Right Primary Bronchus
(Figures 24-7/24-10)

The right lung has three lobes, and the right primary bronchus divides into three secondary bronchi: a **superior lobar bronchus**, a **middle lobar bronchus**, and an **inferior lobar bronchus**. The middle and inferior lobar bronchi branch from the right primary bronchus almost as soon as it enters the lung at the hilus (Figure 24-7●). Each lobar branch delivers air to one of the lobes of the right lung (Figure 24-10●).

Branches of the Left Primary Bronchus
(Figures 24-7/24-9/24-10)

The left lung has two lobes, and the left primary bronchus divides into two secondary bronchi: a **superior lobar bronchus** and an **inferior lobar bronchus** (Figures 24-7, 24-9, and 24-10●).

Branches of the Secondary Bronchi
(Figure 24-10a)

The secondary bronchi in each lung divide to form tertiary bronchi. In the right lung three tertiary bronchi supply the superior lobe, two enter the middle lobe, and five supply the inferior lobe. The superior lobe of the left lung usually contains four tertiary bronchi, whereas the inferior lobe has five (Figure 24-10a●). The tertiary bronchi deliver air to the *bronchopulmonary segments* of the lungs.

The Bronchopulmonary Segments
(Figure 24-10a,b,d)

Each lobe of the lung can be divided into smaller units called **bronchopulmonary segments**. Each segment consists of the lung tissue associated with a single tertiary bronchus. The bronchopulmonary segments have names that corre-

mediastinal surface of the left lung bears a concavity, the **cardiac notch**, that conforms to the shape of the pericardium.

The connective tissues of the root of each lung extend into its substance, or **parenchyma** (par-ENG-ki-ma). These fibrous partitions, or **trabeculae**, contain elastic fibers, smooth muscles, and lymphatics. They branch repeatedly, dividing the lobes into smaller and smaller compartments, and the branches of the conducting passageways, pulmonary vessels, and nerves of the lungs follow these trabeculae to reach their peripheral destinations. The terminal partitions, or **septa**, divide the lung into **lobules** (LOB-ūlz), each supplied by tributaries of the pulmonary arteries, pulmonary veins, and respiratory passageways. The connective tissues of the septa are in turn continuous with those of the visceral pleura. We will now follow the branching pattern of the bronchi from the hilus to the alveoli of each lung.

The Bronchi
(Figures 24-7/24-9/24-10)

The left and right primary bronchi are outside the lungs and are called the **extrapulmonary bronchi**. As the primary bronchi enter the lungs, they divide to form smaller passageways (Figures 24-7, 24-9, and 24-10●). Those

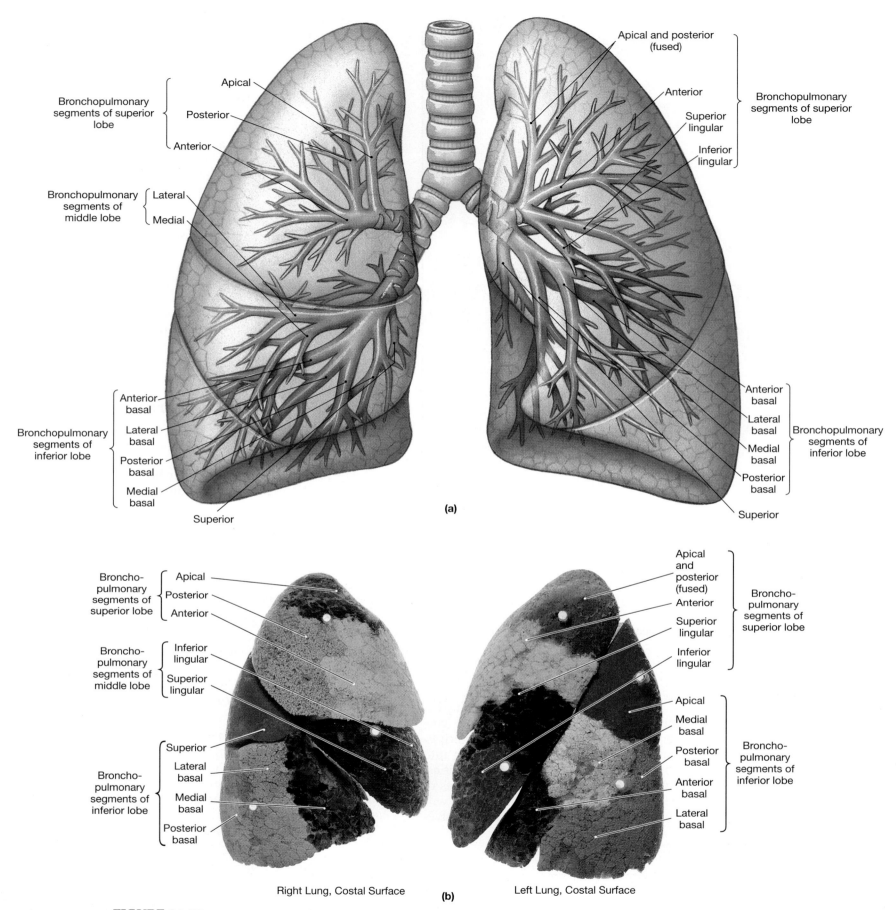

Bronchopulmonary segments of superior lobe
Apical
Posterior
Anterior

Apical and posterior (fused)
Anterior
Superior lingular
Inferior lingular
Bronchopulmonary segments of superior lobe

Bronchopulmonary segments of middle lobe
Lateral
Medial

Bronchopulmonary segments of inferior lobe
Anterior basal
Lateral basal
Posterior basal
Medial basal
Superior

Anterior basal
Lateral basal
Medial basal
Posterior basal
Bronchopulmonary segments of inferior lobe
Superior

(a)

Broncho-pulmonary segments of superior lobe
Apical
Posterior
Anterior

Apical and posterior (fused)
Anterior
Superior lingular
Inferior lingular
Broncho-pulmonary segments of superior lobe

Broncho-pulmonary segments of middle lobe
Inferior lingular
Superior lingular

Broncho-pulmonary segments of inferior lobe
Superior
Lateral basal
Medial basal
Posterior basal

Apical
Medial basal
Posterior basal
Anterior basal
Lateral basal
Broncho-pulmonary segments of inferior lobe

Right Lung, Costal Surface   (b)   Left Lung, Costal Surface

FIGURE 24-10

The Bronchial Tree and Divisions of the Lungs, Anterior View. (a) Gross anatomy of the lungs, showing the bronchial tree and its divisions. (b) The distribution of the bronchopulmonary segments. (c) Bronchogram of the right and left bronchial tree, slightly oblique, posteroanterior view. (d) Plastic cast of the adult bronchial tree. All of the branches in a given segment have been painted the same color.

610

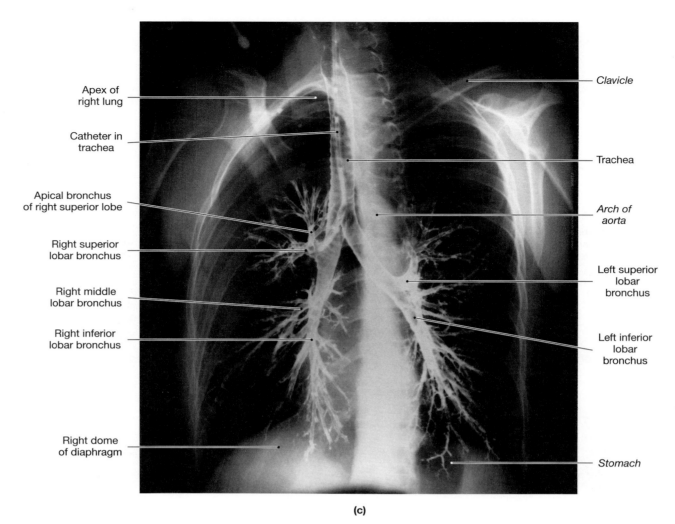

Apex of
right lung

Catheter in
trachea

Apical bronchus
of right superior lobe

Right superior
lobar bronchus

Right middle
lobar bronchus

Right inferior
lobar bronchus

Right dome
of diaphragm

Clavicle

Trachea

*Arch of
aorta*

Left superior
lobar
bronchus

Left inferior
lobar
bronchus

Stomach

(c)

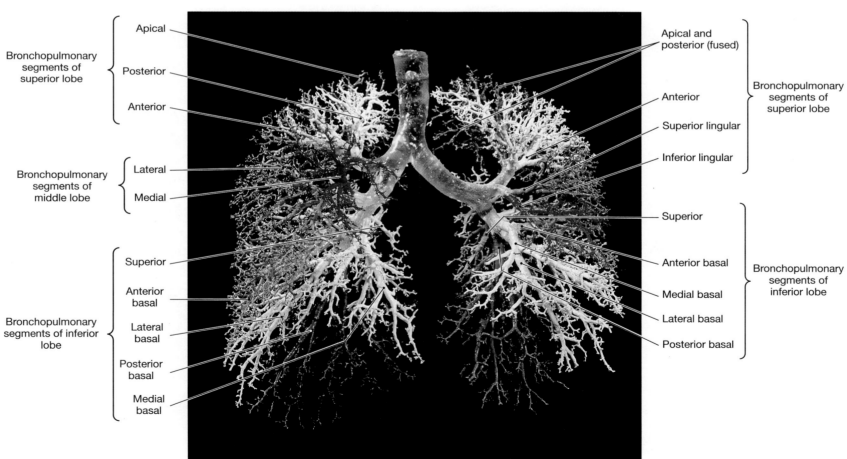

Bronchopulmonary
segments of
superior lobe
{ Apical
Posterior
Anterior

Bronchopulmonary
segments of
middle lobe
{ Lateral
Medial

Bronchopulmonary
segments of inferior
lobe
{ Superior
Anterior
basal
Lateral
basal
Posterior
basal
Medial
basal

Apical and
posterior (fused)

Anterior

Superior lingular

Inferior lingular

Bronchopulmonary
segments of
superior lobe

Superior

Anterior basal

Medial basal

Lateral basal

Posterior basal

Bronchopulmonary
segments of inferior
lobe

(d)

FIGURE 24-11

Bronchi and Bronchioles. (a) The structure of a single lobule, one portion of a bronchopulmomary segment. (b) Light micrograph of a section of the lung. (LM × 62)

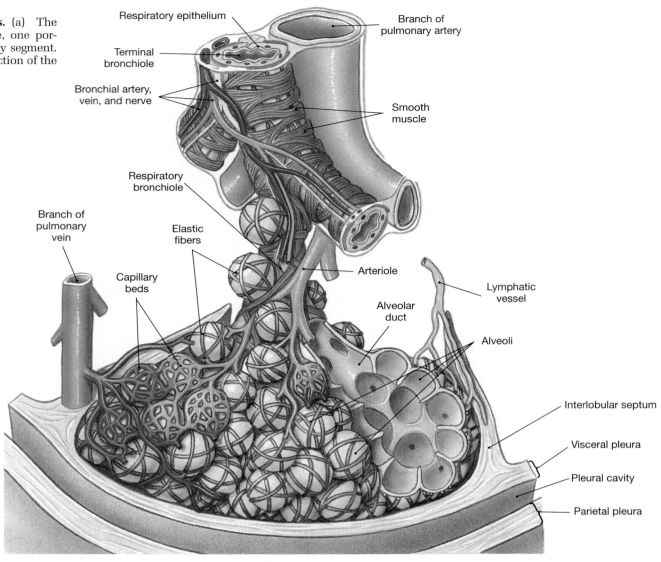

Respiratory epithelium

Terminal bronchiole

Bronchial artery, vein, and nerve

Branch of pulmonary artery

Smooth muscle

Respiratory bronchiole

Branch of pulmonary vein

Elastic fibers

Capillary beds

Arteriole

Lymphatic vessel

Alveolar duct

Alveoli

Interlobular septum

Visceral pleura

Pleural cavity

Parietal pleura

(a)

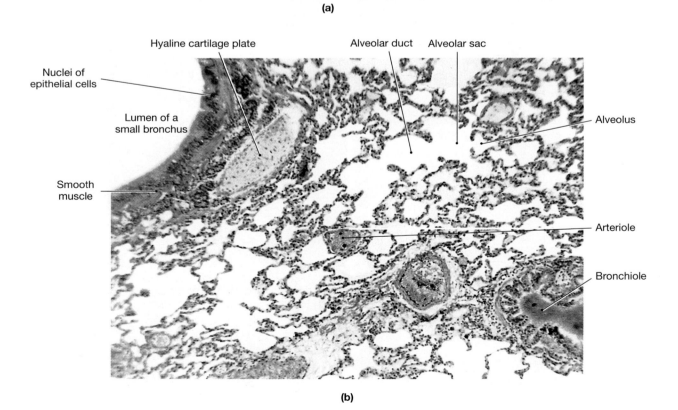

Hyaline cartilage plate

Alveolar duct

Alveolar sac

Nuclei of epithelial cells

Lumen of a small bronchus

Smooth muscle

Alveolus

Arteriole

Bronchiole

(b)

spond to those of the associated tertiary bronchi (Figure 24-10a,b,d●). ⍓*Examining the Living Lung [p. 772]*

The Bronchioles
(Figures 24-9/24-11b)

Each tertiary bronchus branches several times within the bronchopulmonary segment, ultimately giving rise to 6500 smaller **terminal bronchioles**. Terminal bronchioles have a lumenal diameter of 0.3–0.5 mm. The walls of terminal bronchioles, which lack cartilaginous supports, are dominated by smooth muscle tissue (Figures 24-9 and 24-11b●). The autonomic nervous system regulates the activity in the smooth muscle layer of the terminal bronchioles and thereby controls the diameter. Sympathetic activation leads to enlargement of the airway, or **bronchodilation**. Parasympathetic stimulation leads to **bronchoconstriction**. These changes alter the resistance to airflow toward or away from the respiratory exchange surfaces. Tension in the smooth muscles often throws the bronchiolar mucosa into a series of folds, and excessive stimulation, as in *asthma*, can almost completely prevent airflow along the terminal bronchioles.

Each terminal bronchiole delivers air to a single pulmonary lobule. Within the lobule, the terminal bronchiole branches to form several **respiratory bronchioles**. These are the thinnest and most delicate branches of the bronchial tree, and they deliver air to the exchange surfaces of the lungs. The preliminary filtration and humidification of the incoming air are completed before air leaves the terminal bronchioles. The epithelial cells of the respiratory bronchioles and the smaller terminal bronchioles are cuboidal. Cilia are rare, and there are no goblet cells or underlying mucous glands.

Alveolar Ducts and Alveoli
(Figure 24-11a/24-12)

Respiratory bronchioles are connected to individual alveoli and to multiple alveoli along regions called **alveolar ducts**. These passageways end at **alveolar sacs**, common chambers connected to multiple individual alveoli (Figure 24-11 and 24-12a,c●). Each lung contains approximately 150 million alveoli, and their abundance gives the lung an open, spongy appearance. An extensive network of capillaries is associated with each alveolus (Figure 24-12b●); the capillaries are surrounded by a network of elastic fibers. This elastic tissue helps maintain the relative positions of the alveoli and respiratory bronchioles. Recoil of these fibers during expiration reduces the size of the alveoli and assists in the process of expiration.

The Alveolus and the Respiratory Membrane
(Figure 24-12d,e)

The alveolar epithelium consists primarily of simple squamous epithelium (Figure 24-12d,e●). The squamous epithelial cells, called *Type I cells*, are unusually thin and delicate. **Septal cells**, also called **surfactant** (sur-FAK-tant) **cells** or *Type II cells*, are scattered among the squamous cells. These large cells produce an oily secretion containing a mixture of phospholipids. This secretion, termed **surfactant**, coats the inner surface of each alveolus. Surfactant reduces surface tension in the fluid coating the alveolar surface; without surfactant, the alveoli would collapse. Roaming **alveolar macrophages** (*dust cells*) patrol the epithelium, phagocytizing any particulate matter that has eluded the respiratory defenses and reached the alveolar surfaces. ⍓*Respiratory Distress Syndrome [p. 772]*

Gas exchange occurs in areas where the basement membranes of the alveolar epithelium and adjacent capillaries have fused (Figure 24-12e●). In these areas the total distance separating the respiratory and circulatory systems can be as little as 0.1 μm. Diffusion across this **respiratory membrane** proceeds very rapidly, because (1) the distance is small and (2) the gases are lipid-soluble. The membranes of the epithelial and endothelial cells thus do not pose a barrier to the movement of oxygen and carbon dioxide between the blood and alveolar airspaces. ⍓*Tuberculosis* and *Pneumonia [p. 772]*

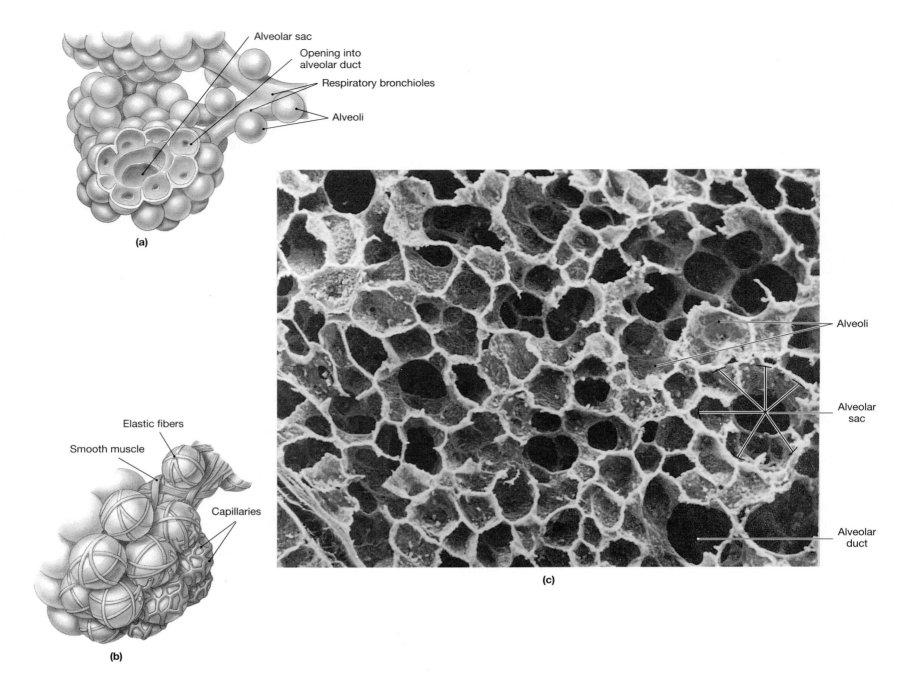

(a)

Alveolar sac
Opening into
alveolar duct
Respiratory bronchioles
Alveoli

Elastic fibers
Smooth muscle
Capillaries

(b)

Alveoli
Alveolar
sac
Alveolar
duct

(c)

FIGURE 24-12
Alveolar Organization. (a) Basic structure of a lobule, cut to reveal the arrangement between the alveolar ducts and alveoli. (b) Connective tissue layers and vascular supply to the alveoli. A network of capillaries surrounds each alveolus.
These capillaries are surrounded by elastic fibers. (c) SEM of lung tissue, showing the appearance and organization of the
alveoli. (d) Diagrammatic view of alveolar structure and the respiratory membrane. (e) The respiratory membrane.

The Blood Supply to the Lungs

The respiratory exchange surfaces receive blood from arteries of
the pulmonary circuit. ∞ [p. 550] The pulmonary arteries enter
the lungs at the hilus and branch with the bronchi as they approach the lobules. Each lobule receives an arteriole and a venule,
and a network of capillaries surrounds each alveolus directly beneath the respiratory membrane. In addition to providing a mechanism for gas exchange, the alveolar capillaries are the primary
source of *converting enzyme*, and the endothelial cells convert circulating angiotensin I to angiotensin II, a hormone involved with
the regulation of blood volume and blood pressure. ∞ [p. 495]

Blood from the alveolar capillaries passes through the
pulmonary venules, and then enters the pulmonary veins that
deliver it to the left atrium. The conducting portions of the
respiratory tract receive blood from the *external carotid arteries* (nasal passages and larynx), the *thyrocervical trunk*
(branches of the subclavian arteries that supply the inferior
larynx and trachea), and the *bronchial arteries*. ∞ [p. 557] The
capillaries supplied by the bronchial arteries provide oxygen
and nutrients to the conducting passageways of the lungs. The
venous blood flows into the pulmonary veins, bypassing the
rest of the systemic circuit and diluting the oxygenated blood
leaving the alveoli. ⸸*Pulmonary Embolism [p. 773]*

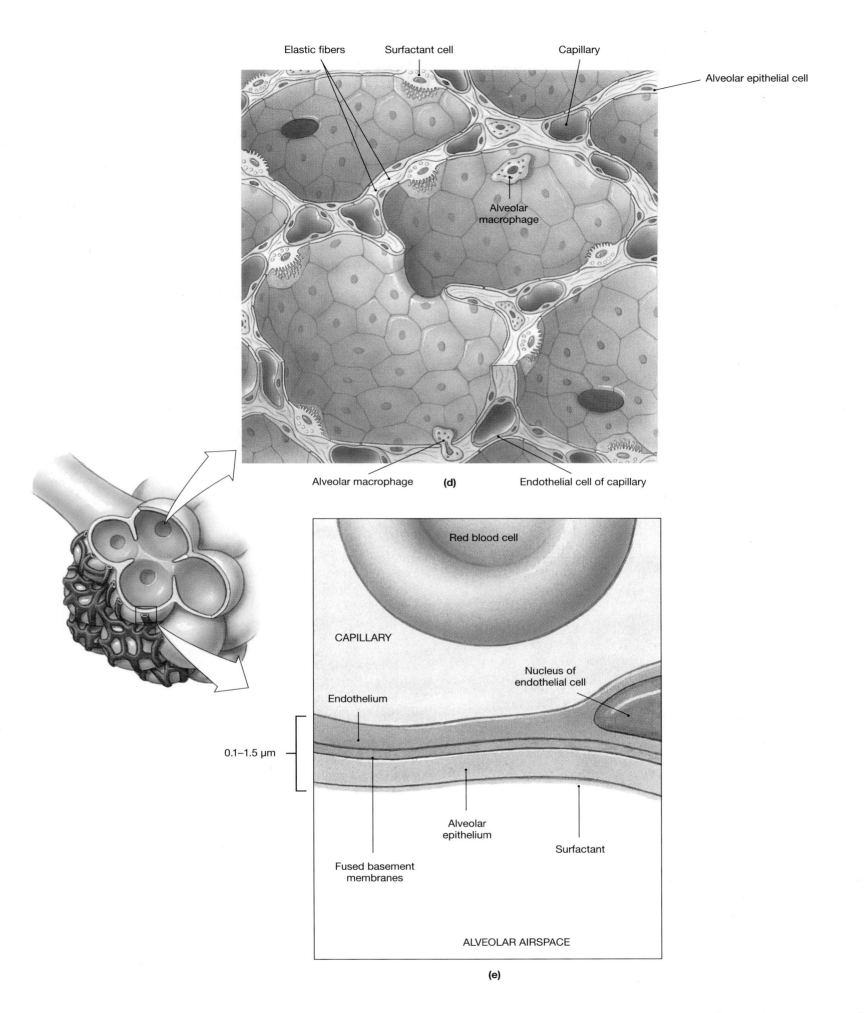

Elastic fibers   Surfactant cell   Capillary

Alveolar epithelial cell

Alveolar
macrophage

Alveolar macrophage   **(d)**   Endothelial cell of capillary

Red blood cell

CAPILLARY

Nucleus of
endothelial cell

Endothelium

0.1–1.5 μm

Alveolar
epithelium

Surfactant

Fused basement
membranes

ALVEOLAR AIRSPACE

(e)

THE PLEURAL CAVITIES AND MEMBRANES
(Figure 24-13)

The thoracic cavity has the shape of a broad cone. Its walls are the rib cage, and the muscular diaphragm forms the floor. The two pleural cavities are separated by the mediastinum (Figure 24-13●). Each lung occupies a single pleural cavity, lined by a serous membrane, or **pleura** (PLOO-ra). ∞ [p. 77] The **parietal pleura** covers the inner surface of the thoracic wall and extends over the diaphragm and mediastinum. The **visceral pleura** covers the outer surfaces of the lungs, extending into the fissures between the lobes. A small amount of **pleural fluid** is secreted by both pleural membranes. Pleural fluid gives a moist, slippery coating that provides lubrication, thereby reducing friction between the parietal and visceral surfaces during breathing. ⸸*Pneumothorax [p. 773]*

The **pleural cavity** actually represents a potential space rather than an open chamber, for the parietal and visceral layers are usually in close contact. Inflammation of the pleurae, a condition called **pleurisy,** may cause the membranes to produce and secrete excess amounts of pleural fluid, or the inflamed pleura may adhere to one another, limiting relative movement. In either form of this disorder, breathing becomes difficult, and prompt medical attention is required. ⸸*Thoracentesis [p. 773]*

√ **Why are the cartilages that reinforce the trachea C-shaped instead of complete rings?**

√ **Why do chronic smokers develop a hacking "smoker's" cough?**

RESPIRATORY MUSCLES AND PULMONARY VENTILATION

Pulmonary ventilation, or breathing, refers to the physical movement of air into and out of the bronchial tree. The function of pulmonary ventilation is to maintain adequate *alveolar*

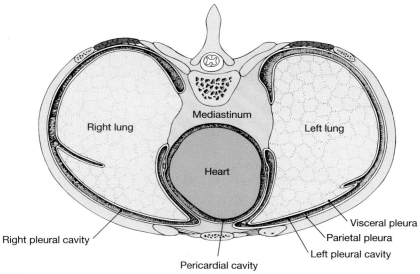

FIGURE 24-13
Anatomical Relationships in the Thoracic Cavity. Those interested in the appearance of other structures at this level should refer to *Figures 1-14a,c*, p. 20, and *Figure 21-2b*, p. 522.

616

ventilation, the movement of air into and out of the alveoli. Alveolar ventilation prevents the buildup of carbon dioxide in the alveoli and ensures a continual supply of oxygen that keeps pace with absorption by the bloodstream.

Respiratory Muscles
(Figure 24-14)

The skeletal muscles involved in respiratory movements were introduced in Chapters 10 and 11. ∞ [p. 268] Of these, the most important are the **diaphragm** and the **external** and **internal intercostals**. Contraction of the diaphragm increases the volume of the thoracic cavity by tensing and flattening its floor, and this increase draws air into the lungs. The external intercostals may assist in inspiration by elevating the ribs. When the ribs are elevated they swing anteriorly, and the width of the thoracic cage increases along its anterior-posterior axis. The internal intercostals depress the ribs and reduce the width of the thoracic cavity, thereby contributing to expiration. These muscles and their actions are diagrammed in Figure 24-14●.

The **accessory muscles** become active when the depth and frequency of respiration must be increased markedly. The sternocleidomastoid, the serratus anterior, the pectoralis minor, and the scalenes assist the external intercostals in elevating the ribs and performing inspiration. The transversus thoracis, abdominal obliques, and rectus abdominis assist the internal intercostals in expiration by compressing the abdominal contents, forcing the diaphragm upward, and further reducing the volume of the thoracic cavity.

Respiratory Movements

The respiratory muscles may be used in various combinations depending on the volume of air that must be moved in or out of the lungs. Respiratory movements may be classified as *eupnea* or *hyperpnea* on the basis of whether expiration is passive or active.

EUPNEA. In **eupnea** (ūp-NĒ-a), or **quiet breathing**, inspiration involves muscular contractions, but expiration is a passive process. During quiet breathing, expansion of the lungs stretches their elastic fibers. In addition, elevation of the rib cage stretches opposing skeletal muscles and elastic fibers in the connective tissues of the body wall. When the inspiratory muscles relax, these elastic structures contract, returning the diaphragm or rib cage or both to their original positions.

Eupnea may involve *diaphragmatic breathing* or *costal breathing*.

- During **diaphragmatic breathing**, or **deep breathing**, contraction of the diaphragm provides the necessary change in thoracic volume. Air is drawn into the lungs as the diaphragm contracts, and exhalation occurs when the diaphragm relaxes.

- In **costal breathing**, or **shallow breathing**, the thoracic volume changes because the rib cage changes shape. Inhalation occurs when contraction of the external intercostals elevates the ribs and enlarges the thoracic cavity. Exhalation occurs when these muscles relax. During pregnancy women increasingly rely on costal breathing as the uterus enlarges and pushes the abdominal viscera against the diaphragm.

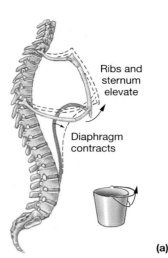

FIGURE 24-14
Respiratory Muscles. The primary muscles are shown in plain type, and the accessory muscles in italics. (a) As the ribs are elevated or the diaphragm depressed, the volume of the thoracic cavity increases and air moves into the lungs. The outward movement of the ribs as they are elevated resembles the outward swing of a raised bucket handle. (b) Lateral view at rest, with no air movement. (c) **Inhalation**, showing the primary and accessory respiratory muscles that elevate the ribs. (d) **Exhalation**, showing the primary and accessory respiratory muscles that depress the ribs.

Ribs and sternum elevate

Diaphragm contracts

(a)

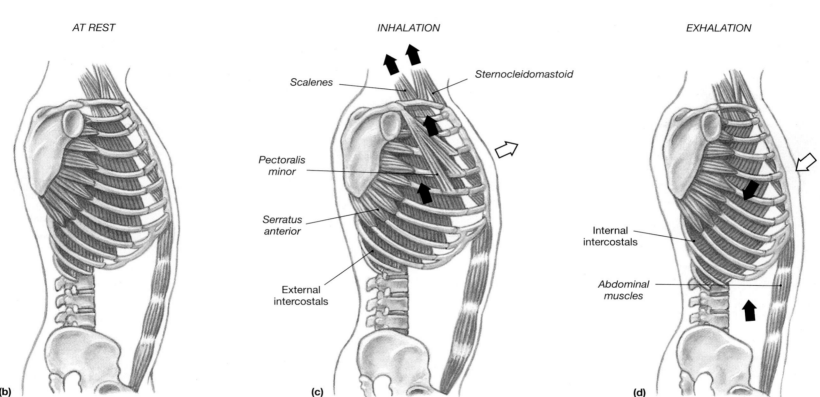

AT REST

INHALATION

Scalenes

Sternocleidomastoid

Pectoralis minor

Serratus anterior

External intercostals

EXHALATION

Internal intercostals

Abdominal muscles

(b)   (c)   (d)

HYPERPNEA. **Hyperpnea** (hī-perp-NĒ-a), or **forced breathing**, involves active inspiratory and expiratory movements. Forced breathing calls upon the accessory muscles to assist with inspiration, and expiration involves contraction of the transversus thoracis and the internal intercostals. When a person is breathing at absolute maximum levels, such as during vigorous exercise, the abdominal muscles are used in exhalation. Their contraction compresses the abdominal contents, pushing them up against the diaphragm and further reducing the volume of the thoracic cavity.

Respiratory Changes at Birth

There are several important differences between the respiratory system of a fetus and that of a newborn infant. Before delivery, pulmonary arterial resistance is high, because the pulmonary vessels are collapsed. The rib cage is compressed, and the lungs and conducting passageways contain only small amounts of fluid and no air. At birth the newborn infant takes a truly heroic first breath through powerful contractions of the diaphragmatic and external intercostal muscles. The inspired air enters the passageways with enough force to push the contained fluids out of the way and inflate the entire bronchial tree and most of the alveoli. The same drop in pressure that pulls air into the lungs pulls blood into the pulmonary circulation; the changes in blood flow that occur lead to the closure of the *foramen ovale*, an interatrial connection, and the *ductus arteriosus*, the fetal connection between the pulmonary trunk and the aorta. ∞ [p. 528]

The exhalation that follows fails to completely empty the lungs, for the rib cage does not return to its former, fully compressed state. Cartilages and connective tissues keep the conducting passageways open, and the surfactant covering the alveolar surfaces prevents their collapse. Subsequent breaths complete the inflation of the alveoli.

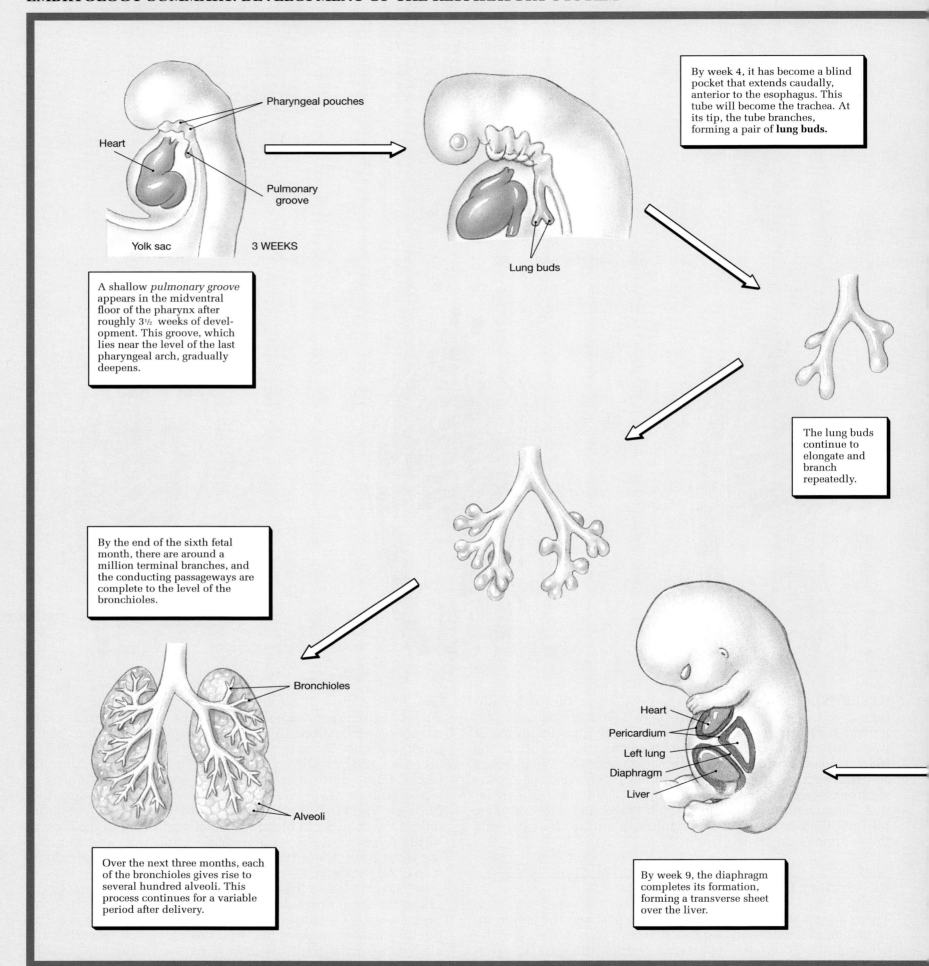

Pharyngeal pouches

Heart

Pulmonary groove

Yolk sac

3 WEEKS

Lung buds

By week 4, it has become a blind pocket that extends caudally, anterior to the esophagus. This tube will become the trachea. At its tip, the tube branches, forming a pair of **lung buds.**

A shallow *pulmonary groove* appears in the midventral floor of the pharynx after roughly 3½ weeks of development. This groove, which lies near the level of the last pharyngeal arch, gradually deepens.

The lung buds continue to elongate and branch repeatedly.

By the end of the sixth fetal month, there are around a million terminal branches, and the conducting passageways are complete to the level of the bronchioles.

Bronchioles

Alveoli

Heart

Pericardium

Left lung

Diaphragm

Liver

Over the next three months, each of the bronchioles gives rise to several hundred alveoli. This process continues for a variable period after delivery.

By week 9, the diaphragm completes its formation, forming a transverse sheet over the liver.

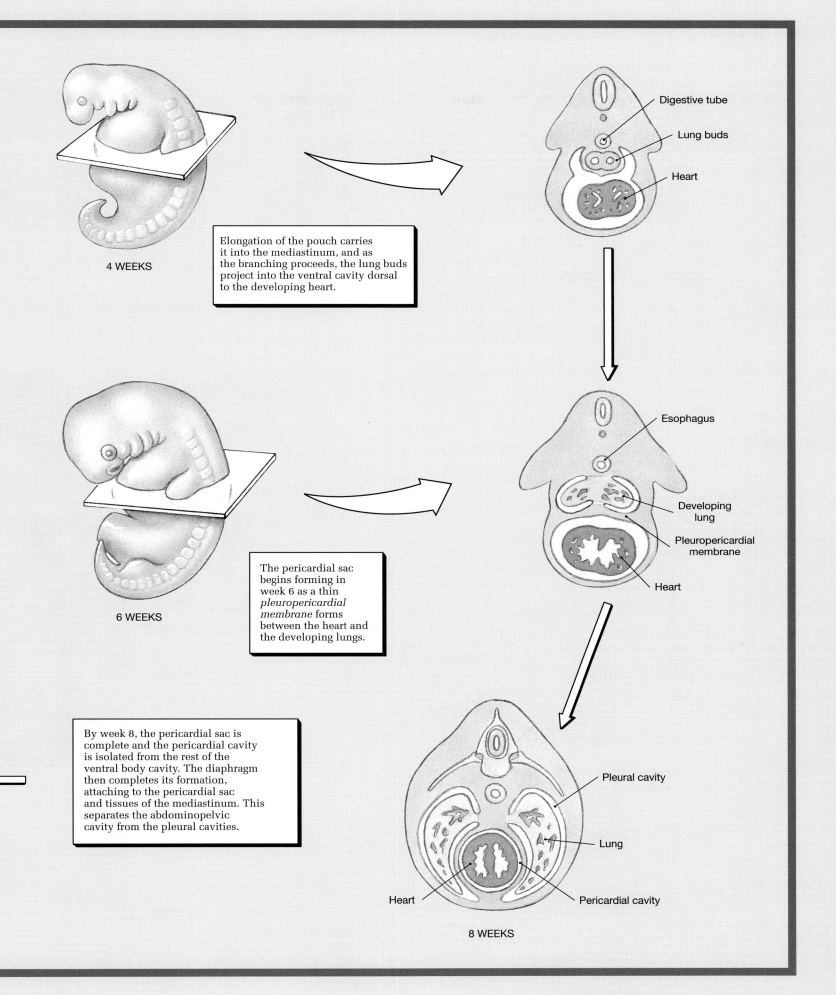

4 WEEKS

Digestive tube

Lung buds

Heart

Elongation of the pouch carries it into the mediastinum, and as the branching proceeds, the lung buds project into the ventral cavity dorsal to the developing heart.

6 WEEKS

The pericardial sac begins forming in week 6 as a thin *pleuropericardial membrane* forms between the heart and the developing lungs.

Esophagus

Developing lung

Pleuropericardial membrane

Heart

By week 8, the pericardial sac is complete and the pericardial cavity is isolated from the rest of the ventral body cavity. The diaphragm then completes its formation, attaching to the pericardial sac and tissues of the mediastinum. This separates the abdominopelvic cavity from the pleural cavities.

Pleural cavity

Lung

Heart

Pericardial cavity

8 WEEKS

√ John breaks a rib that punctures the thoracic cavity on his left side. Which structures are potentially damaged, and what do you predict will happen to the lung as a result?

√ In emphysema, alveoli are replaced by large air spaces and elastic fibrous connective tissue. How do these changes affect the lungs?

√ In pneumonia, fluid accumulates in the alveoli of the lungs. How does this fluid affect the lungs?

Respiratory Centers of the Brain
(Figure 24-15)

Under normal conditions cellular rates of absorption and generation are matched by the capillary rates of delivery and removal. When adjustments by the cardiovascular and respiratory systems are needed to meet the body's ever-changing demands for oxygen, these systems must be coordinated. The regulatory centers that integrate the responses by these systems are located in the pons and medulla oblongata.

The **respiratory centers** include three pairs of loosely organized nuclei in the reticular formation of the pons and medulla oblongata. These nuclei regulate the activities of the respiratory muscles by adjusting the frequency and depth of pulmonary ventilation.

The **respiratory rhythmicity center** sets the basic pace and depth of respiration. It can be subdivided into a **dorsal respiratory group (DRG)** and a **ventral respiratory group (VRG)**. The dorsal respiratory group, or *inspiratory center,* controls motor neurons innervating the external intercostal muscles and the diaphragm. This group functions in every respiratory cycle, whether quiet or forced. The ventral respiratory group functions only during forced respiration. It innervates motor neurons controlling accessory muscles involved in active exhalation and maximal inhalation. The neurons involved with active exhalation are sometimes said to form an *expiratory center.*

The **apneustic** (ap-NŪ-stik) and **pneumotaxic** (nū-mō-TAKS-ik) **centers** of the pons adjust the output of the rhythmicity center, thereby modifying the pace of respiration. Their activities adjust the respiratory rate and the depth of respiration in response to sensory stimuli or instructions from higher centers. The locations of the apneustic, pneumotaxic, and respiratory rhythmicity centers are reviewed in Figure 24-15●.

Normal breathing occurs automatically, without conscious control. Three different reflexes are involved in the regulation of respiration: (1) *mechanoreceptor reflexes* that respond to changes in the volume of the lungs or to changes in arterial blood pressure, (2) *chemoreceptor reflexes* that respond to changes in the P_{CO_2}, pH, and P_{O_2} of the blood and cerebrospinal fluid, and (3) *protective reflexes* that respond to physical injury or irritation of the respiratory tract. ∞ [p. 450]

Higher centers influence respiration via inputs to the pneumotaxic center and by their direct influence on respiratory muscles. The higher centers involved are found in the cerebrum, especially the cerebral cortex, and in the hypothalamus. Although pyramidal output provides conscious control over the respiratory muscles, these muscles most often receive instructions via extrapyramidal pathways. In addition, the respiratory centers are embedded in the reticular formation, and almost every sensory and motor nucleus has some connection with this complex. As a result, emotional and autonomic activities often affect the pace and depth of respiration.

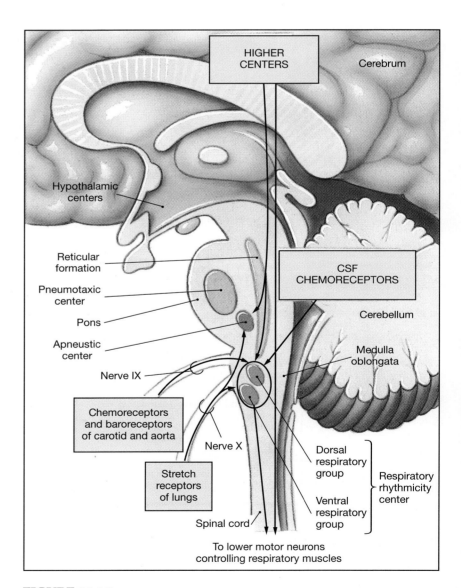

FIGURE 24-15
Respiratory Centers and Reflex Controls. The positions and relationships between the major respiratory centers and factors important to respiratory control.

AGING AND THE RESPIRATORY SYSTEM

Many factors interact to reduce the efficiency of the respiratory system in elderly individuals. Three examples are particularly noteworthy:

    1. With increasing age, elastic tissue deteriorates throughout the body. This deterioration reduces the lungs' ability to inflate and deflate.

    2. Movements of the rib cage are restricted by arthritic changes in the rib articulations and by decreased flexibility at the costal cartilages. In combination with the changes noted in 1, the stiffening and reduction in chest movement effectively limit the respiratory volume. This restriction contributes to the reduction in exercise performance and capabilities with increasing age.

    3. Some degree of emphysema is normally found in individuals aged 50–70. On average, roughly 1 square foot of respiratory membrane is lost each year after age 30. However, the extent varies widely depending on the lifetime exposure to cigarette smoke and other respiratory irritants. ☨*Lung Cancer, Smoking, and Diet [p. 773]*

Related Clinical Terms

silicosis (sil-i-KŌ-sis), **asbestosis** (as-bes-TŌ-sis), and **anthracosis** (an-thra-KŌ-sis): Serious clinical conditions caused by the inhalation of dust or other particulate matter in sufficient quantities to overload the respiratory defense system. ☨*Overloading the Respiratory Defenses [p. 772]*

cystic fibrosis (CF): A relatively common lethal inherited disease in which mucous secretions become too thick to be transported easily. *[p. 600]*

epistaxis (ep-i-STAK-sis): A nosebleed caused by trauma, infection, allergies, hypertension, or other factors. ☨*Nosebleeds [p. 772]*

laryngitis (lar-in-JĪ-tis): Infection or inflammation of the larynx. ☨*Disorders of the Larynx [p. 772]*

Heimlich (HĪM-lik) **maneuver:** A method of applying abdominal pressure to force the expulsion of foreign objects lodged in the trachea or larynx. *[p. 606]*

tracheostomy (trā-kē-OS-tō-mē): Insertion of a tube through an incision in the anterior tracheal wall, to bypass a foreign body or damaged larynx. *[p. 606]*

bronchitis (brong-KĪ-tis): An inflammation of the bronchial lining. ☨*Bronchitis [p. 772]*

asthma (AZ-ma): A condition characterized by unusually sensitive, irritable conducting passageways. *[p. 613]*

respiratory distress syndrome (RDS): A condition resulting from inadequate surfactant production; characterized by collapse of the alveoli and an inability to maintain adequate levels of gas exchange at the lungs. ☨*Respiratory Distress Syndrome (RDS) [p. 772]*

tuberculosis (tū-ber-kū-LŌ-sis) **(TB):** Infection of the lungs by the bacteria *Mycobacterium tuberculosis.* Symptoms are variable, but usually include coughing and chest pain, with fever, night sweats, fatigue, and weight loss. ☨*Tuberculosis [p. 772]*

pneumonia (nū-MŌ-nē-a): A condition caused by an infection of the lobules of the lung and characterized by a decline in respiratory function due to fluid leakage into the alveoli and/or swelling and constriction of the respiratory bronchioles. ☨*Pneumonia [p. 773]*

emphysema (em-fi-ZĒ-ma): A chronic, progressive condition characterized by shortness of breath and resulting from the destruction of respiratory exchange surfaces. *[p. 613]*

pulmonary embolism: Blockage of a pulmonary artery by a blood clot, fat mass, or air bubble. ☨*Pulmonary Embolism [p. 773]*

pneumothorax (nū-mō-THŌ-raks): The entry of air into the pleural cavity. ☨*Pneumothorax [p. 773]*

atelectasis (at-e-LEK-ta-sis): A collapsed lung. ☨*Pneumothorax [p. 773]*

thoracentesis: Removal of a sample of pleural fluid for diagnostic evaluation. ☨*Thoracentesis [p. 773]*

pleural effusion: An abnormal accumulation of fluid within the pleural cavities. ☨*Thoracentesis [p. 773]*

cardiopulmonary resuscitation (CPR): Applying cycles of compression to the rib cage and mouth-to-mouth breathing to maintain circulatory and respiratory function. *[p. 620]*

lung cancer (pleuropulmonary neoplasm): A class of aggressive malignancies originating in the bronchial passageways or alveoli. ☨*Lung Cancer, Smoking, and Diet [p. 773]*

CHAPTER SUMMARY AND REVIEW

STUDY OUTLINE

INTRODUCTION *[p. 599]*

FUNCTIONS OF THE RESPIRATORY SYSTEM *[p. 599]*

1. The respiratory system: (1) provides an area for gas exchange between air and circulating blood; (2) moves air to and from exchange surfaces; (3) protects respiratory surfaces; (4) defends the respiratory system and other tissues from pathogens; (5) permits vocal communication; and (6) helps regulate blood volume and pressure and body fluid pH.

ORGANIZATION OF THE RESPIRATORY SYSTEM *[p. 599]*

Related Key Terms

1. The **respiratory system** includes the nose, nasal cavity and sinuses, pharynx, larynx, trachea, and conducting passageways leading to the exchange surfaces of the lungs *(see Figure 24-1)*.

2. The **upper respiratory system** includes the nose, nasal cavity, and pharynx. These structures begin the process of filtration and humidification of the incoming air. The **lower respiratory system** includes the larynx, trachea, bronchi, bronchioles, and **alveoli**.

3. The **respiratory tract** consists of a *conducting portion*, which extends from the entrance to the nasal cavity to the bronchioles; and a *respiratory portion*, which includes the *respiratory bronchioles* and the **alveoli** *(see Figure 24-1)*.

THE RESPIRATORY EPITHELIUM *[p. 600]*

1. The **respiratory epithelium** lines the conducting portions of the respiratory system down to the level of the smallest bronchioles.

2. The respiratory epithelium consists of a pseudostratified, ciliated, columnar epithelium with goblet cells *(see Figure 24-2)*.

3. The respiratory epithelium produces mucus that traps incoming particles. Underneath is the **lamina propria** (a layer of connective tissue); the combined respiratory epithelium and lamina propria form a **mucosa** (mucous membrane) *(see Figure 24-2)*.

4. The **respiratory defense system** includes the *mucus escalator* (which washes particles toward the stomach), alveolar macrophages, hairs, and cilia.

THE UPPER RESPIRATORY SYSTEM *[p. 601]*

The Nose and Nasal Cavity *[p. 601]*
1. Air normally enters the respiratory system via the **external nares**, which open into the **nasal cavity**. The **vestibule** of the nose is guarded by hairs that screen out large particles *(see Figure 24-3)*.

dorsum nasi • apex

2. Incoming air flows through the **superior**, **middle**, or **inferior meatuses** (narrow grooves) and bounces off the conchal surfaces *(see Figure 24-3b,d)*.

turbinate bones

3. The **hard palate** separates the oral and nasal cavities. The **soft palate** separates the superior **nasopharynx** from the oral cavity. The connections between the nasal cavity and nasopharynx represent the **internal nares** *(see Figure 24-3)*.

The Pharynx *[p. 601]*
4. The **pharynx** is a chamber shared by the digestive and respiratory systems. The **oropharynx** is continuous with the oral cavity; the **laryngopharynx** includes the narrow zone between the hyoid and the entrance to the esophagus *(see Figure 24-3c)*.

uvula • pharyngeal arches • palatoglossal arch • palatopharyngeal arch • fauces

THE LARYNX *[p. 601]*

Cartilages of the Larynx *[p. 603]*
1. Inspired air passes through the **glottis** en route to the lungs; the **larynx** surrounds and protects the glottis. The **epiglottis** projects into the pharynx *(see Figures 24-3c,d/24-4/24-5)*.

2. Two pairs of folds span the glottal opening; the relatively inelastic **ventricular folds** and the more delicate **vocal folds**. Air passing through the glottis vibrates the vocal folds and produces sound *(see Figure 24-5)*.

thyroid cartilage • laryngeal prominence • cricoid cartilage • arytenoid cartilages • corniculate cartilages • cuneiform cartilages intrinsic ligaments • extrinsic ligaments • ventricular ligaments • vocal ligaments • true vocal cords • false vocal cords

The Laryngeal Musculature *[p. 605]*
3. The **intrinsic laryngeal muscles** regulate tension in the vocal folds and open and close the glottis. The **extrinsic laryngeal musculature** positions and stabilizes the larynx. During swallowing, both sets of muscles help to prevent particles from entering the glottis *(see Figure 24-6)*.

bolus

THE TRACHEA *[p. 605]*

1. The **trachea** ("windpipe") extends from the sixth cervical vertebra to the fifth thoracic vertebra. The **submucosa** contains C-shaped **tracheal cartilages** that stiffen the tracheal walls and protect the airway. The posterior tracheal wall can distort to permit large masses of food to move along the esophagus *(see Figures 24-2a/24-3c/24-7)*.

annular ligaments • trachealis

THE PRIMARY BRONCHI *[p. 606]*

1. The trachea branches within the mediastinum to form the **right** and **left primary bronchi**. Each bronchus enters a lung at the **hilus**. The **root** of the lung is a connective tissue mass including the bronchus, pulmonary vessels, and nerves *(see Figures 24-7/24-8)*.

carina

THE LUNGS [p. 607]

Lobes of the Lungs [p. 607]

1. The **lobes** of the lungs are separated by fissures; the **right lung** has three lobes and three secondary bronchi: **superior lobar, middle lobar,** and **inferior lobar bronchi.** The **left lung** has two lobes and two secondary bronchi: **superior lobar** and **inferior lobar bronchi** *(see Figure 24-8).*

Lung Surfaces [p. 607]

2. The anterior portion of the lung, termed the **costal surface,** follows the inner contours of the rib cage. The **mediastinal surface** contains a hilus, and the left lung bears the **cardiac notch** *(see Figure 24-8).*

3. The connective tissues of the root extend into the **parenchyma** of the lung as a series of **trabeculae** (partitions). These branches form **septa** that divide the lung into **lobules** *(see Figure 24-8).*

The Bronchi [p. 609]

4. **Extrapulmonary bronchi** (left and right primary bronchi) are outside of the lung tissue. **Intrapulmonary bronchi** (branches within the lung) are surrounded by bands of smooth muscle *(see Figures 24-7/24-9/24-10).*

5. Each lung may be further divided into smaller units called **bronchopulmonary segments.** These segments are named according to the associated tertiary bronchi. The right lung contains 10 and the left lung usually contains 8–9 bronchopulmonary segments *(see Figures 24-7/24-9/24-10).*

The Bronchioles [p. 613]

6. Each tertiary bronchus ultimately gives rise to 50–80 **terminal bronchioles** that supply individual lobules *(see Figures 24-9/24-11).*

Alveolar Ducts and Alveoli [p. 613]

7. The **respiratory bronchioles** open into **alveolar ducts**; many alveoli are interconnected at each duct *(see Figures 24-11/24-12).*

8. The **respiratory membrane** (alveolar lining) consists of a simple squamous epithelium of *Type I cells*; **septal cells** (**surfactant cells,** *Type II cells*) scattered in it produce an oily secretion (**surfactant**) that keeps the alveoli from collapsing. **Alveolar macrophages** (*dust cells*) patrol the epithelium and engulf foreign particles *(see Figure 24-12).*

The Blood Supply to the Lungs [p. 614]

9. The respiratory exchange surfaces are extensively connected to the circulatory system via the vessels of the pulmonary circuit.

THE PLEURAL CAVITIES AND MEMBRANES [p. 616]

1. Each lung occupies a single **pleural cavity** lined by a **pleura** (serous membrane) *(see Figure 24-13).*

RESPIRATORY MUSCLES AND PULMONARY VENTILATION [p. 616]

Respiratory Muscles [p. 616]

1. **Pulmonary ventilation** is the movement of air into and out of the lungs. The most important respiratory muscles are the **diaphragm** and the **external** and **internal intercostals.** Contraction of the diaphragm increases the volume of the thoracic cavity; the external intercostals may assist in inspiration by elevating the ribs; the internal intercostals depress the ribs and reduce the width of the thoracic cavity, thereby contributing to expiration. The **accessory muscles** become active when the depth and frequency of respiration must be increased markedly. Accessory muscles include the sternocleidomastoid, serratus anterior, transversus thoracis, scalenes, pectoralis minor, abdominal obliques, and rectus abdominis *(see Figure 24-14).*

Respiratory Changes at Birth [p. 617]

2. Before delivery the fetal lungs are fluid-filled and collapsed. After the first breath, they remain partially inflated, even after maximum exhalation.

Respiratory Centers of the Brain [p. 620]

3. The **respiratory centers** include three pairs of nuclei in the reticular formation of the pons and medulla. The **respiratory rhythmicity center** sets the pace for respiration. The **apneustic center** causes strong, sustained inspiratory movements, and the **pneumotaxic center** inhibits the apneustic center and the inspiratory center in the medulla *(see Figure 24-15).*

4. Three different reflexes are involved in the regulation of respiration: (1) *mechanoreceptor reflexes* respond to changes in the volume of the lungs or to changes in arterial blood pressure; (2) *chemoreceptor reflexes* respond to changes in the P_{CO_2}, pH, and P_{O_2} of the blood and cerebrospinal fluid; and (3) *protective reflexes* respond to physical injury or irritation of the respiratory tract *(see Figure 24-15).*

5. Conscious and unconscious thought processes can also control respiratory activity by affecting the respiratory centers or controlling the respiratory muscles.

AGING AND THE RESPIRATORY SYSTEM [p. 621]

1. The respiratory system is generally less efficient in the elderly because: (1) elastic tissue deteriorates, lowering the vital capacity of the lungs; (2) movements of the chest cage are restricted by arthritic changes and decreased flexibility of costal cartilages; and (3) some degree of emphysema is normal in the elderly.

1   REVIEW OF CHAPTER OBJECTIVES

1. Describe the primary functions of the respiratory system.
2. Describe the structural organization of the respiratory system and its major organs.
3. Describe the histology and function of the respiratory epithelium.
4. Describe the functional anatomy of the organs of the upper respiratory system.
5. Describe the histological specializations of each component of the respiratory tract.
6. Describe the functional anatomy of the bronchial tree and bronchopulmonary segments.
7. Describe the structure and function of the respiratory membrane.
8. Describe the pleural cavities and pleural membranes.
9. Identify the muscles of respiration, and discuss the movements responsible for pulmonary ventilation.
10. Describe the changes that occur in the respiratory system at birth.
11. Identify the respiratory control centers and how they interact.
12. Identify the reflexes that regulate respiration.
13. Describe the changes that occur in the respiratory system with age.

2   REVIEW OF CONCEPTS

1. What are the functional differences between the upper and lower respiratory tracts?
2. What effect does the respiratory epithelium have on the air traveling toward the alveoli?
3. What role do cilia play in the respiratory tract?
4. Why are the air passages within the nose convoluted and complex?
5. What role does the palate (hard and soft) have in respiratory function?
6. What features of the pharyngeal region contribute to a tendency to choke on food or liquid or to have items go "down the wrong" passageway?
7. What are the functions of the uvula and pharyngeal arches?
8. Why is the attachment of the hyoid bone to laryngeal structures important to respiration and vocalization?
9. Why does a person's voice sound different when he or she has a cold than it normally does?
10. What is the function of the ventricular folds?
11. How does the respiratory epithelium differ from the stratified squamous epithelium of the oral cavity and laryngopharynx?

12. How do the functions of the extrinsic and intrinsic laryngeal muscles differ?
13. What is the mediastinum, and of what importance is it to the respiratory system?
14. How can the autonomic nervous system alter the amount of alveolar ventilation?
15. What is the function of the surfactant secreted in the lungs?
16. What aspects of alveolar microstructure facilitate gas exchange?
17. What is unique about the blood transported in the pulmonary arteries and veins?
18. How do respiratory muscles and other structures involved in hyperpnea complement the action of muscles in active eupnea?
19. Why is it necessary for the respiratory system to undergo dramatic changes at birth?

3   CRITICAL THINKING AND CLINICAL APPLICATION QUESTIONS

1. A person with cystic fibrosis (CF) may show varied symptoms, but usually the most serious are those involving the respiratory system. Most victims die of respiratory ailments before the age of 30. What anatomical mechanism causes these problems?
2. A person with a blocked or crushed trachea is unable to breathe and will die unless a way is found to allow air to pass to the lungs. What can be done to save this person's life?
3. People often claim that they do not recognize or like the sound of their own voice when they hear a recording of it, even though other people assure them the recording is accurate. Why do people sound so different to themselves in contrast to the way other people hear them?
4. A man has suffered a chest injury from falling onto the tines of a rake. He has several punctures between the third and sixth rib on the right side, with air and bloody fluid intermittently bubbling out of the injury sites, and he is gasping for breath. What damage has occurred?
5. An 11-year-old girl has difficulty breathing in the spring and summer, and after taking antibiotics for a bacterial infection, she developed severe breathing problems. What is one possible cause for her condition?
6. A 63-year-old man who has smoked since he was 13 has a persistent cough and shortness of breath. What would be a likely diagnosis for these symptoms, and what factors are responsible?

25

The Digestive System

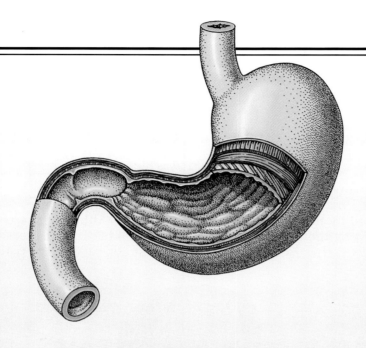

CHAPTER OUTLINE AND OBJECTIVES

Few of us give any serious thought to the digestive system unless it malfunctions. Yet each day we spend hours of conscious effort filling and emptying it. The digestive system consists of a muscular tube, called the **digestive tract**, and various **accessory organs**. The digestive tract and accessory organs work together to perform the following functions:

1. *Ingestion:* Ingestion occurs when foods and liquids enter the digestive tract via the mouth.

2. *Mechanical processing:* Most ingested solids must undergo mechanical processing before they are swallowed. Squashing with the tongue and tearing and crushing with the teeth are examples of mechanical processing that occur before ingestion. Swirling, mixing, and churning motions of the digestive tract provide mechanical processing after ingestion.

3. *Digestion:* Digestion is the chemical and enzymatic breakdown of complex sugars, lipids, and proteins into small organic molecules that can be absorbed by the digestive epithelium.

4. *Secretion:* Digestion usually involves the action of acids, enzymes, and buffers produced by active secretion. Some of these secretions are produced by the lining of the digestive tract, but most are provided by the accessory organs, such as the pancreas.

5. *Absorption:* Absorption is the movement of organic substances, electrolytes, vitamins, and water across the digestive epithelium and into the interstitial fluid of the digestive tract.

6. *Compaction:* Compaction is the progressive dehydration of indigestible materials and organic wastes prior to elimination from the body.

7. *Excretion:* Waste products are secreted into the digestive tract, primarily by the accessory glands (especially the liver). **Defecation** (def-e-KĀ-shun) is the elimination of fecal material from the body.

The lining of the digestive tract also plays a defensive role by protecting surrounding tissues against (1) the corrosive effects of digestive acids and enzymes, (2) mechanical stresses, such as abrasion, and (3) pathogens that are either swallowed with food or residing within the digestive tract.

AN OVERVIEW OF THE DIGESTIVE SYSTEM
(Figure 25-1)

The major components of the digestive system are shown in Figure 25-1●. Although these structures have overlapping functions, each has certain areas of specialization and shows distinctive histological characteristics.

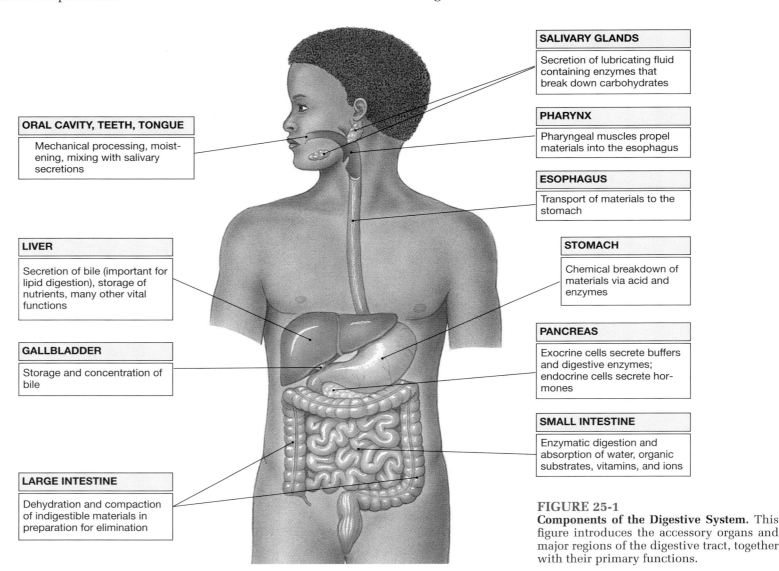

SALIVARY GLANDS

Secretion of lubricating fluid containing enzymes that break down carbohydrates

PHARYNX

Pharyngeal muscles propel materials into the esophagus

ESOPHAGUS

Transport of materials to the stomach

STOMACH

Chemical breakdown of materials via acid and enzymes

PANCREAS

Exocrine cells secrete buffers and digestive enzymes; endocrine cells secrete hormones

SMALL INTESTINE

Enzymatic digestion and absorption of water, organic substrates, vitamins, and ions

ORAL CAVITY, TEETH, TONGUE

Mechanical processing, moistening, mixing with salivary secretions

LIVER

Secretion of bile (important for lipid digestion), storage of nutrients, many other vital functions

GALLBLADDER

Storage and concentration of bile

LARGE INTESTINE

Dehydration and compaction of indigestible materials in preparation for elimination

FIGURE 25-1
Components of the Digestive System. This figure introduces the accessory organs and major regions of the digestive tract, together with their primary functions.

FIGURE 25-2

Histological Structure of the Digestive Tract. (a) Three-dimensional view of the histological organization of the digestive tube. (b) Photomicrograph of ileum showing general histological organization. (LM × 160)

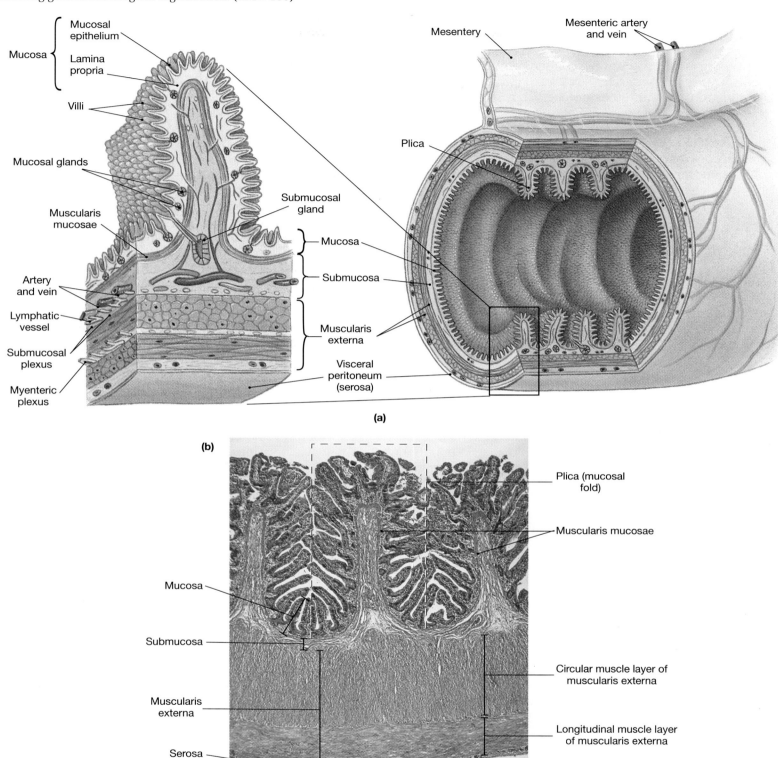

(a)

(b)

Histological Organization of the Digestive Tract
(Figure 25-2)

Sectional and diagrammatic views of the digestive tract are presented in Figure 25-2●. The major layers of the digestive tract are described in detail below, and include: (1) the *mucosa*, (2) the *submucosa*, (3) the *muscularis externa*, and (4) the *serosa*.

The Mucosa
(Figure 25-2)

The inner lining, or **mucosa**, of the digestive tract is an example of a **mucous membrane**. Mucous membranes, introduced in Chapter 3, consist of a layer of loose connective tissue covered by an epithelium moistened by glandular secretions. ∞ [p. 77]

The **mucosal epithelium** may be simple or stratified, depending on the location and the stresses involved. For example, the oral cavity and esophagus are lined by a stratified squamous epithelium, whereas the stomach, small intestine, and almost the entire large intestine have a simple columnar epithelium. The lining of the digestive tract is often organized in transverse or longitudinal folds (Figure 25-2●). The folds, which resemble pleats, serve to (1) permit expansion of the lumen after a large meal and (2) dramatically increase the surface area available for absorption. Ducts opening onto the epithelial surfaces carry the secretions of gland cells located in the mucosa and submucosa or within accessory organs.

The underlying layer of loose connective tissue is called the **lamina propria**. The lamina propria contains blood vessels, sensory nerve endings, lymphatic vessels, smooth muscle fibers, and scattered areas of lymphatic tissue. In most regions of the digestive tract the outer portion of the lamina propria is a narrow band of smooth muscle and elastic fibers. This band is called the **muscularis** (mus-kū-LAR-is) **mucosae**. The smooth muscle fibers in the muscularis mucosae are arranged in two thin concentric layers (Figure 25-2a●). The inner layer encircles the lumen (the *circular muscle*), and the outer layer contains muscle fibers oriented parallel to the long axis of the tract (the *longitudinal layer*). Contraction of these layers alters the shape of the lumen and moves the epithelial pleats and folds.

The Submucosa
(Figure 25-2a)

The **submucosa** (sub-myū-KŌ-sa) is a layer of loose connective tissue that surrounds the muscularis mucosae. Large blood vessels and lymphatics are found in this layer, and in some regions the submucosa also contains exocrine glands that secrete buffers and enzymes into the lumen of the digestive tract. Along its outer margin, the submucosa contains a network of nerve fibers and scattered nerve cells. This **submucosal plexus** (*plexus of Meissner*) contains sensory nerve cells, parasympathetic ganglia, and sympathetic postganglionic fibers (Figure 25-2a●).

The Muscularis Externa
(Figure 25-2)

The submucosal plexus lies along the inner border of the **muscularis externa**, a region dominated by smooth muscle fibers. The smooth muscle fibers of the muscularis externa are arranged in an inner, circular layer and an outer, longitudinal layer (Figure 25-2●). These layers play an essential role in mechanical processing and in the movement of materials along the digestive tract. These movements are coordinated primarily by neurons of the **myenteric plexus** (mī-en-TER-ik; *mys*, muscle + *enteron*, intestine), or *plexus of Auerbach*. This network of parasympathetic ganglia and sympathetic postganglionic fibers lies sandwiched between the circular and longitudinal muscle layers. Parasympathetic stimulation increases muscular tone and activity, and sympathetic stimulation promotes muscular inhibition and relaxation.

The Serosa
(Figure 25-2)

Along most portions of the digestive tract inside the peritoneal cavity, the muscularis externa is covered by a *serous membrane*

known as the **serosa** (Figure 25-2●). The muscularis externa of the oral cavity, pharynx, esophagus, and rectum is surrounded by a dense network of collagen fibers that firmly attaches the digestive tract to adjacent structures. This connective tissue layer is called the **adventitia** (ad-ven-TISH-a).

Muscularis Layers and the Movement of Digestive Materials

Smooth muscle cells are surrounded by connective tissue, but the collagen fibers do not form tendons or aponeuroses. ∞ [p. 70] A single smooth muscle cell ranges from 5 to 10 μm in diameter and from 30 to 200 μm in length. The contractile proteins are not organized in sarcomeres, so the muscle cells are not striated, but their contractions are as strong as those of skeletal or cardiac muscle cells.

Because the contractile filaments of smooth muscle cells are not rigidly organized, a stretched smooth muscle cell soon adapts to its new length and retains the ability to contract on demand. This ability to tolerate extreme stretching is called **plasticity**. Plasticity is especially important for digestive organs that undergo great changes in volume, such as the stomach.

The digestive system contains **visceral smooth muscle tissue**. In visceral smooth muscle tissue, many of the muscle cells have no motor innervation. The muscle cells are arranged in sheets or layers, and adjacent muscle cells are electrically connected by gap junctions. When one visceral smooth muscle cell contracts, the contraction spreads in a wave that travels throughout the tissue. The initial stimulus may be the activation of a motor neuron that contacts one of the muscle cells in the region. It may also be a local response to chemicals, hormones, the concentrations of oxygen or carbon dioxide, or physical factors such as extreme stretching or irritation.

Smooth muscle in the digestive tract shows rhythmic cycles of activity because of the presence of **pacesetter cells**. These smooth muscle cells undergo spontaneous depolarization, and their contraction triggers a wave of contraction that spreads through the entire muscular sheet. Pacesetter cells are found in the muscularis mucosae and muscularis externa.

Peristalsis
(Figure 25-3a)

The muscularis externa propels materials from one portion of the digestive tract to another through the contractions of **peristalsis** (per-i-STAL-sis). Peristalsis consists of waves of muscular contractions that move along the length of the digestive tract. During a **peristaltic wave**, the circular muscles contract behind the digestive contents. Longitudinal muscles contract next, shortening adjacent segments. A wave of contraction in the circular muscles then forces the materials in the desired direction (Figure 25-3a●).

Segmentation
(Figure 25-3b)

Most areas of the small intestine and some portions of the large intestine undergo **segmentation** (Figure 25-3b●). These movements churn and fragment the digestive materials, mixing the contents with intestinal secretions.

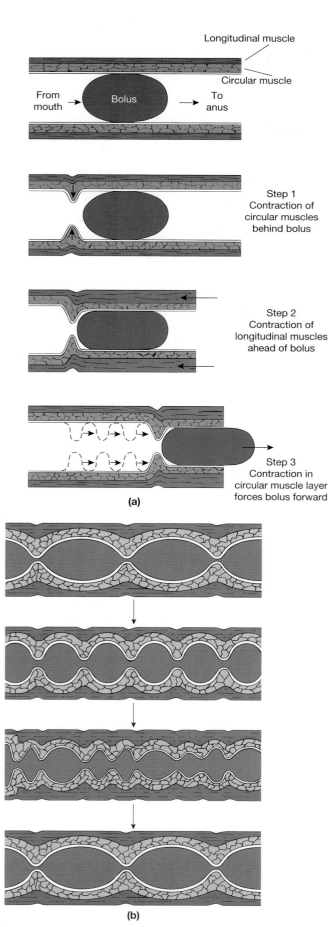

FIGURE 25-3
Peristalsis and Segmentation. (a) Peristalsis propels materials along the length of the digestive tract by coordinated contractions of the circular and longitudinal layers. (b) Segmentation movements involve primarily the circular muscle layers. These activities churn and mix the contents of the digestive tract but do not produce net movement in a particular direction.

Within image (a):

Longitudinal muscle

Circular muscle

From mouth — Bolus — To anus

Step 1
Contraction of circular muscles behind bolus

Step 2
Contraction of longitudinal muscles ahead of bolus

Step 3
Contraction in circular muscle layer forces bolus forward

(a)

(b)

Segmentation and peristalsis may be triggered by pacesetter cells, hormones, chemicals, and physical stimulation. Peristaltic waves can also be initiated by afferent and efferent fibers within the glossopharyngeal, vagus, or pelvic nerves. Local peristaltic movements limited to a few centimeters of the digestive tract are triggered by sensory receptors in the walls of the digestive tract. These afferent fibers synapse within the myenteric plexus to produce localized **myenteric reflexes**.

The Peritoneum

The serosa, or **visceral peritoneum**, is continuous with the **parietal peritoneum** that lines the inner surfaces of the body wall. ∞ [p. 77] The peritoneal lining continually produces peritoneal fluid that lubricates the peritoneal surfaces. About 7ℓ of fluid are secreted and reabsorbed each day, although the volume within the peritoneal cavity at any one time is very small. Under unusual conditions, the volume can increase markedly, reducing blood volume and distorting visceral organs.

■ **CLINICAL BRIEF**
Peritonitis

Inflammation of the peritoneal membrane produces symptoms of **peritonitis** (per-i-tō-NĪ-tis), a painful condition that interferes with the normal functioning of the affected organs. Physical damage, chemical irritation, or bacterial invasion of the peritoneum can lead to severe and even fatal cases of peritonitis. Peritonitis due to bacterial infection is a potential complication of any surgery that involves opening the peritoneal cavity. Liver disease, kidney disease, or heart failure can cause an increase in the rate of fluid movement through the peritoneal lining. The accumulation of fluid, called **ascites** (a-SĪ-tēz) creates a characteristic abdominal swelling. Distortion of internal organs by the contained fluid can result in a variety of symptoms; heartburn, indigestion, and low back pain are common complaints.

Mesenteries
(Figures 25-2a/25-4/25-10a/25-11b)

Portions of the digestive tract are suspended within the peritoneal cavity by sheets of serous membrane that connect the parietal peritoneum with the visceral peritoneum. These **mesenteries** (MEZ-en-ter-ēz) are double sheets of peritoneal membrane (Figure 25-2a●). The loose connective tissue between the mesothelial surfaces provides an access route for the passage of the blood vessels, nerves, and lymphatics to and from the digestive tract. Mesenteries also stabilize the positions of the attached organs and prevent the intestines from becoming entangled during digestive movements or sudden changes in body position.

During development, the digestive tract and accessory organs are suspended within the peritoneal cavity by dorsal and ventral mesenteries (Figure 25-4a,b●, p. 632). The ventral mesentery later disappears along most of the digestive tract, persisting only on the ventral surface of the stomach, between the stomach and liver (the **lesser omentum**), and between the liver and the anterior abdominal wall and diaphragm (the *falciform ligament*). (For additional information concerning the development of the digestive tract, accessory organs, and associated mesenteries, see the Embryology Summary on the next page.)

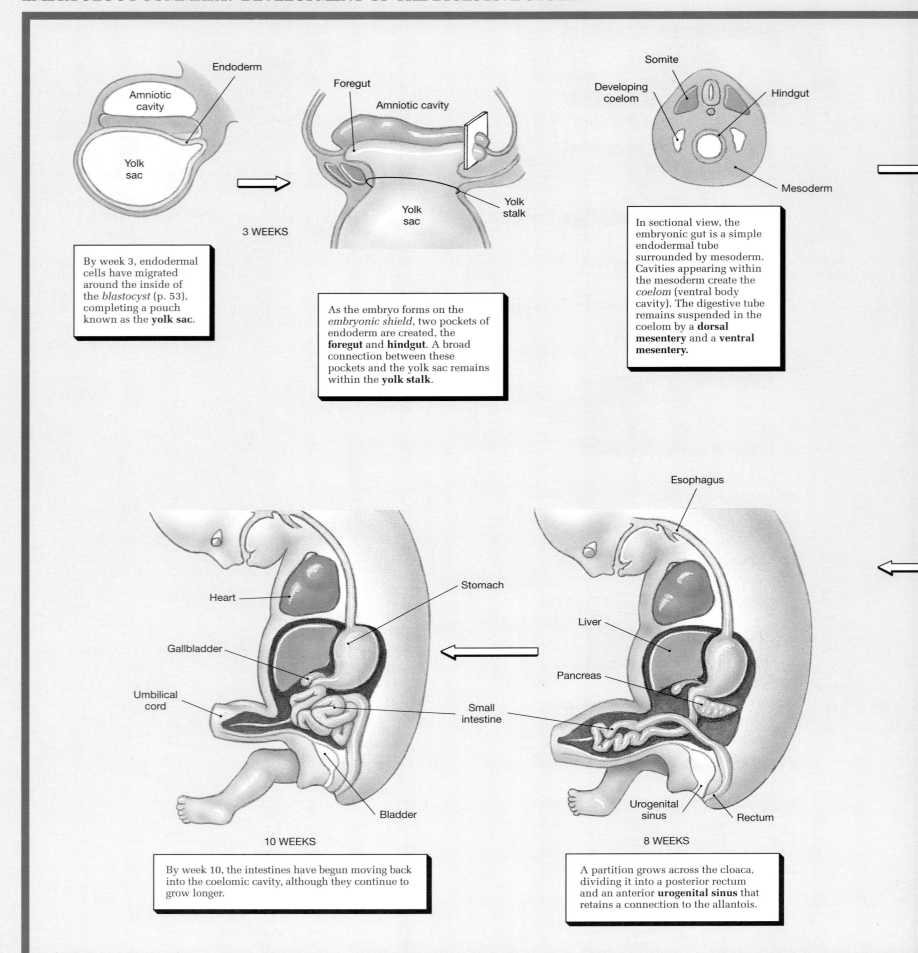

Endoderm

Amniotic cavity

Yolk sac

3 WEEKS

By week 3, endodermal cells have migrated around the inside of the *blastocyst* (p. 53), completing a pouch known as the **yolk sac**.

Foregut

Amniotic cavity

Yolk sac

Yolk stalk

As the embryo forms on the *embryonic shield*, two pockets of endoderm are created, the **foregut** and **hindgut**. A broad connection between these pockets and the yolk sac remains within the **yolk stalk**.

Somite

Developing coelom

Hindgut

Mesoderm

In sectional view, the embryonic gut is a simple endodermal tube surrounded by mesoderm. Cavities appearing within the mesoderm create the *coelom* (ventral body cavity). The digestive tube remains suspended in the coelom by a **dorsal mesentery** and a **ventral mesentery.**

Esophagus

Heart

Stomach

Gallbladder

Liver

Umbilical cord

Pancreas

Small intestine

Bladder

Urogenital sinus

Rectum

10 WEEKS

8 WEEKS

By week 10, the intestines have begun moving back into the coelomic cavity, although they continue to grow longer.

A partition grows across the cloaca, dividing it into a posterior rectum and an anterior **urogenital sinus** that retains a connection to the allantois.

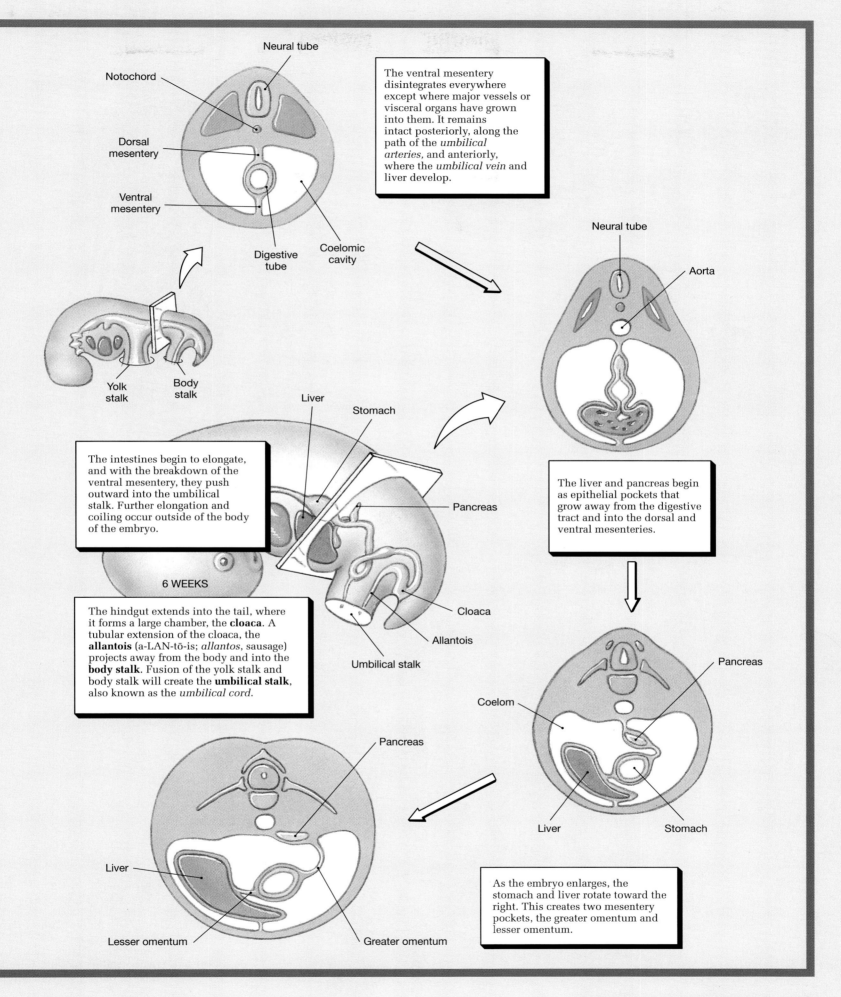

Neural tube

Notochord

Dorsal mesentery

Ventral mesentery

Digestive tube

Coelomic cavity

The ventral mesentery disintegrates everywhere except where major vessels or visceral organs have grown into them. It remains intact posteriorly, along the path of the *umbilical arteries,* and anteriorly, where the *umbilical vein* and liver develop.

Yolk stalk

Body stalk

Neural tube

Aorta

The intestines begin to elongate, and with the breakdown of the ventral mesentery, they push outward into the umbilical stalk. Further elongation and coiling occur outside of the body of the embryo.

Liver

Stomach

Pancreas

6 WEEKS

The liver and pancreas begin as epithelial pockets that grow away from the digestive tract and into the dorsal and ventral mesenteries.

The hindgut extends into the tail, where it forms a large chamber, the **cloaca**. A tubular extension of the cloaca, the **allantois** (a-LAN-tō-is; *allantos,* sausage) projects away from the body and into the **body stalk**. Fusion of the yolk stalk and body stalk will create the **umbilical stalk**, also known as the *umbilical cord.*

Cloaca

Allantois

Umbilical stalk

Pancreas

Coelom

Liver

Stomach

Pancreas

As the embryo enlarges, the stomach and liver rotate toward the right. This creates two mesentery pockets, the greater omentum and lesser omentum.

Liver

Lesser omentum

Greater omentum

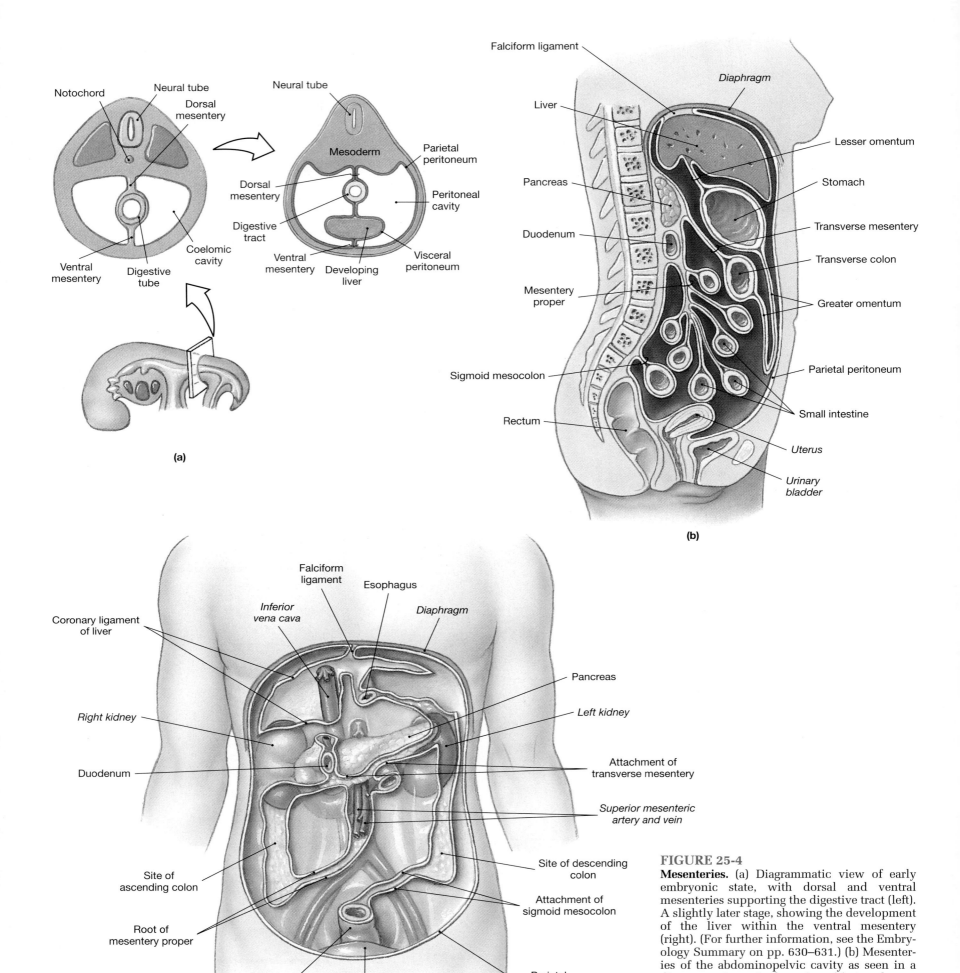

(a)

(b)

(c)

FIGURE 25-4

Mesenteries. (a) Diagrammatic view of early embryonic state, with dorsal and ventral mesenteries supporting the digestive tract (left). A slightly later stage, showing the development of the liver within the ventral mesentery (right). (For further information, see the Embryology Summary on pp. 630–631.) (b) Mesenteries of the abdominopelvic cavity as seen in a diagrammatic sagittal section. (c) Attachment of visceral organs to the peritoneal wall.

As the digestive tract elongates, it twists and turns within the relatively crowded peritoneal cavity. The dorsal mesentery of the stomach becomes greatly enlarged, and it forms a pouch that extends inferiorly, between the body wall and the anterior surface of the small intestine. This pouch is the **greater omentum** (ō-MEN-tum; *omentum*, fat skin) (see Figure 25-4b● and look ahead to Figures 25-10a, p. 640, and 25-11b●, p. 641). All but the first 25 cm of the small intestine is suspended by a thick **mesenterial sheet,** termed the **mesentery proper,** that provides stability but permits a degree of independent movement. The mesentery associated with the initial portion of the small intestine (the *duodenum*) and the pancreas fuse with the posterior abdominal wall, locking these structures in position. Their anterior surfaces are covered by peritoneum, but the rest of the organs lie outside of the peritoneal cavity. This position is called **retroperitoneal** (re-trō-per-i-tō-NĒ-al; *retro*, behind).

A **mesocolon** is a mesentery associated with the large intestine. The middle portion of the large intestine (the *transverse colon*) is suspended by a portion of the mesocolon known as the **transverse mesentery**. The *sigmoid colon*, which leads to the rectum and anus, is suspended by the **sigmoid mesocolon**. The mesocolon of the *ascending colon*, the *descending colon*, and the *rectum* of the large intestine usually fuse to the body wall, fixing them in position. These organs are now retroperitoneal, and visceral peritoneum covers only their anterior surfaces and portions of their lateral surfaces (Figure 25-4b,c●).

√ **What are the components and functions of the mucosa of the digestive tract?**

√ **What are the functions of mesenteries?**

√ **What is the functional difference between peristalsis and segmentation?**

THE ORAL CAVITY

Our exploration of the digestive tract will follow the path of food from the mouth to the anus. The mouth opens into the *oral cavity.* The functions of the oral cavity may be summarized as: (1) **analysis** of material before swallowing; (2) **mechanical processing** through the actions of the teeth, tongue, and palatal surfaces; (3) **lubrication** by mixing with mucous and salivary secretions; and (4) limited **digestion** of carbohydrates by a salivary enzyme.

Anatomy of the Oral Cavity
(Figure 25-5)

The oral cavity, or **buccal** (BUK-al) **cavity** (Figure 25-5●), is lined by the **oral mucosa**, which has a stratified squamous

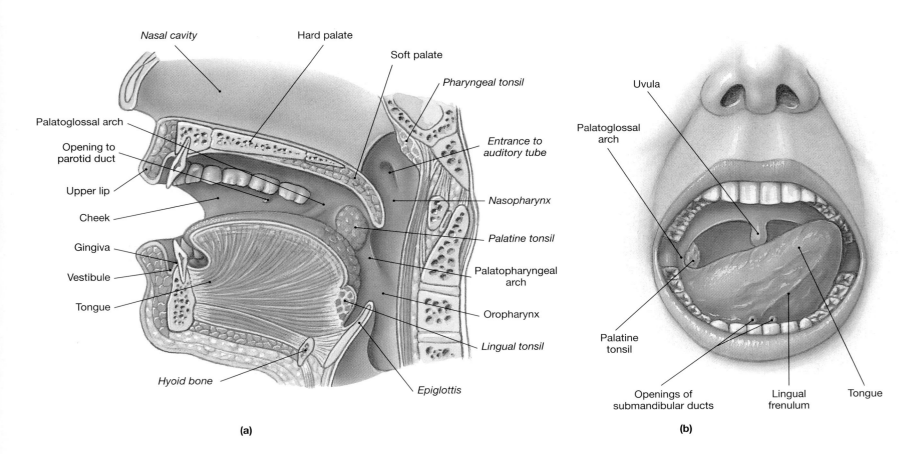

FIGURE 25-5
The Oral Cavity. (a) The oral cavity as seen in sagittal section. (b) An anterior view of the oral cavity, as seen through the open mouth.

epithelium. Unlike the stratified squamous epithelium of the skin, the oral epithelium does not undergo cornification. The mucosa of the **cheeks**, or lateral walls of the oral cavity, are supported and formed by **buccal fat pads** and the *buccinator muscles.* ` Anteriorly, the mucosa of the cheeks are continuous with the lips, or **labia** (LĀ-bē-a). The **vestibule** is the space between the cheeks, lips, and the teeth. A ridge of oral mucosa, the gums, or **gingivae** (JIN-ji-vē), surrounds the base of each tooth on the alveolar surfaces of the maxilla and mandible. ∞ [p. 150]

The roof of the oral cavity is formed by the **hard** and **soft palates**, while the tongue dominates its floor. Inferior to the tongue, the floor receives additional support from the *mylohyoid muscle.* ∞ [p. 264] The hard palate is formed by the palatine process of the maxillary bone and the palatine bone. The soft palate lies posterior to the hard palate. The posterior margin of the soft palate supports the dangling uvula and two pairs of muscular *pharyngeal arches*. The *palatopharyngeal arches*, the uvula, and the base of the tongue frame the *fauces*, the entrance to the pharynx. The **uvula** (Ū-vū-la), located at the center of the fauces, helps to prevent food from entering the pharynx prematurely. ∞ [p. 601]

The Tongue
(Figures 18-7/25-5)

The tongue (Figure 25-5●) manipulates materials inside the mouth and may occasionally be used to bring foods (such as ice cream) into the oral cavity. The primary functions of the tongue are: (1) mechanical processing by compression, abrasion, and distortion; (2) manipulation to assist in chewing and prepare the material for swallowing; and (3) sensory analysis by touch, temperature, and taste receptors.

The tongue can be divided into an anterior **body**, or *oral portion*, and a posterior **root**, or *pharyngeal portion*. The superior surface, or **dorsum**, of the body contains a forest of fine *papillae*. The thickened epithelium covering each papilla assists in the movement of materials by the tongue. Surface features and histological details of the tongue can be found in Figure 18-7●. ∞ [p. 453] A **V**-shaped line of circumvallate papillae roughly indicates the boundary between the body and the root of the tongue, which is situated in the pharynx.

The epithelium covering the inferior surface of the tongue is thinner and more delicate than that of the dorsum. Along the inferior midline is a thin fold of mucous membrane, the **lingual frenulum** (FREN-ū-lum; *frenulum*, small bridle), which connects the body of the tongue to the mucosa of the oral floor. Ducts from one of the salivary glands are visible as they open on either side of the lingual frenulum.

The lingual frenulum prevents extreme movements of the tongue. However, if the lingual frenulum is *too* restrictive, the individual cannot eat or speak normally. When properly diagnosed, this condition, called **ankyloglossia** (ang-ki-lō-GLOS-ē-a), can be corrected surgically.

The tongue contains two different types of muscles, **intrinsic tongue muscles** and **extrinsic tongue muscles**. Both intrinsic and extrinsic tongue muscles are under the control of the hypoglossal nerve (N XII). The extrinsic muscles, discussed in Chapter 10, include the *hyoglossus, styloglossus, genioglossus*, and *palatoglossus* muscles. ∞ [p. 262] All gross movements of the tongue are performed by the extrinsic muscles. The less massive intrinsic muscles alter the shape of the tongue and assist the extrinsic muscles during precise movements, as in speech.

Salivary Glands
(Figures 25-5/25-6)

Each salivary gland is covered by a fibrous capsule. As saliva is produced by the secretory cells of the gland, a network of ducts carries the saliva to a main drainage duct that penetrates the capsule and proceeds to the surface of the oral mucosa. Three pairs of salivary glands (Figure 25-6●) secrete into the oral cavity:

1. The large **parotid** (pa-ROT-id) **salivary glands** lie inferior to the zygomatic arch beneath the skin covering the lateral and posterior surface of the mandible. Each parotid gland has an irregular shape, extending from the mastoid process of the temporal bone across the outer surface of the masseter muscle. The secretions of each gland are drained by a **parotid duct**, which empties into the vestibule at the level of the second upper molar (Figure 25-5a●).

2. The **sublingual** (sub-LING-gwal) **salivary glands** are located beneath the mucous membrane of the floor of the mouth. Numerous **sublingual ducts** open along either side of the lingual frenulum.

3. The **submandibular salivary glands** are found in the floor of the mouth along the inner surfaces of the mandible within the *mandibular groove*. ∞ [p. 149] The **submandibular ducts** open into the mouth on either side of the lingual frenulum immediately posterior to the teeth (Figure 25-5b●). The histological appearance of the submandibular gland can be seen in Figure 25-6b●.

Each of the salivary glands has a distinctive cellular organization and produces saliva with slightly different properties. For example, the parotid salivary glands produce a thick, serous secretion containing large amounts of the digestive enzyme **salivary amylase**, which begins the chemical breakdown of complex carbohydrates. The saliva in the mouth is a mixture of glandular secretions; about 70% of the saliva originates in the submandibular salivary glands, 25% from the parotid salivary glands, and 5% from the sublingual salivary glands. Collectively the salivary glands produce 1.0–1.5 ℓ of saliva each day, with a composition of 99.4% water, plus an assortment of ions, buffers, metabolites, and enzymes.

At mealtimes the production of large quantities of saliva lubricates the mouth and dissolves chemicals that stimulate the taste buds. A continual background level of secretion flushes the oral surfaces and helps to control populations of oral bacteria. A reduction or elimination of salivary secretions triggers a bacterial population explosion in the oral cavity. This proliferation rapidly leads to recurring infections and the progressive erosion of the teeth and gums.

Salivary secretions are usually controlled by the autonomic nervous system. Each salivary gland receives parasympathetic and sympathetic innervation. Any object placed within the mouth can trigger a salivary reflex by stimulating receptors monitored by the trigeminal nerve or by stimulating taste buds innervated by N VII, N IX, or N X. Parasympathetic stimulation accelerates secretion by all of the salivary glands, resulting in the production of large amounts of saliva.

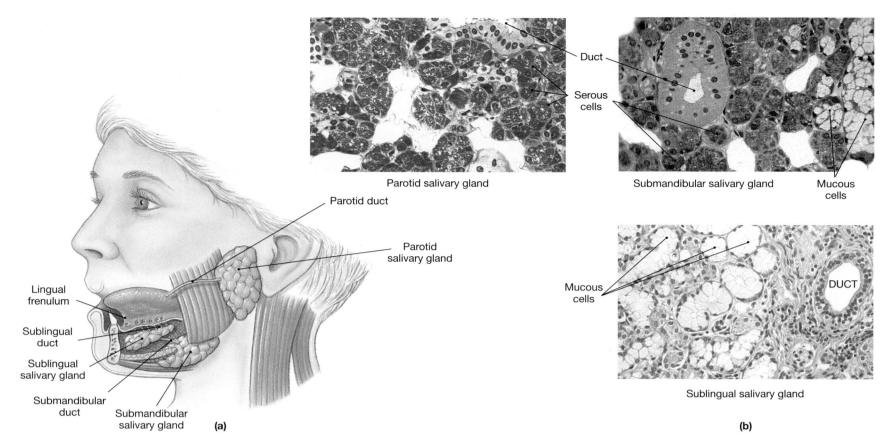

FIGURE 25-6
The Salivary Glands. (a) Lateral view, showing the relative positions of the salivary glands and ducts on the left side of the head. For the positions of the ducts inside the oral cavity, *see Figure 25-5.* (b) Photomicrographs showing histological detail of the parotid, submandibular, and sublingual salivary glands. The parotid salivary gland produces saliva rich in enzymes. The gland is dominated by serous secretory cells. The submandibular salivary gland produces saliva containing enzymes and mucins, and it contains both serous and mucous secretory cells. The sublingual salivary gland produces saliva rich in mucins. This gland is dominated by mucous secretory cells.

The Teeth
(Figure 25-7a)

The movements of the tongue are important in passing food across the surfaces of the teeth. Teeth perform chewing, or **mastication** (mas-ti-KĀ-shun), of food. Mastication breaks down tough connective tissues and plant fibers and helps saturate the materials with salivary secretions and enzymes.

Figure 25-7a● is a sectional view through an adult tooth. The bulk of each tooth consists of a mineralized matrix similar to that of bone. This material, called **dentin** (DEN-tin), differs from bone in that it does not contain living cells. Instead, cytoplasmic processes extend into the dentin from cells in the central **pulp cavity**. The pulp cavity is spongy and highly vascular. It receives blood vessels and nerves via a narrow tunnel, the **root canal**, which is located at the base or **root** portion of the tooth. The **dental artery**, **dental vein**, and **dental nerve** enter the root canal through the **apical foramen** to service the pulp cavity.

The root of the tooth sits within a bony socket, or *alveolus*. Collagen fibers of the **periodontal** (per-ē-ō-DON-tal) **ligament** extend from the dentin of the root to the alveolar bone, creating a strong articulation known as a *gomphosis*. ∞ [p. 206] A layer of **cementum** (se-MEN-tum) covers the dentin of the root, providing protection and firmly anchoring the periodontal ligament. Cementum is very similar in histological structure to bone, and less resistant to erosion than dentin.

The **neck** of the tooth marks the boundary between the root and the **crown**. The crown is the visible portion of the tooth. Epithelial cells of the **gingival** (JIN-ji-val) **sulcus** form tight attachments to the tooth above the neck, preventing bacterial access to the lamina propria of the gingiva or the relatively soft cementum of the root.

The dentin of the crown is covered by a layer of **enamel**. Enamel contains calcium phosphate in a crystalline form and is the hardest biologically manufactured substance. Adequate amounts of calcium, phosphates, and vitamin D during childhood are essential if the enamel coating is to be complete and resistant to decay.

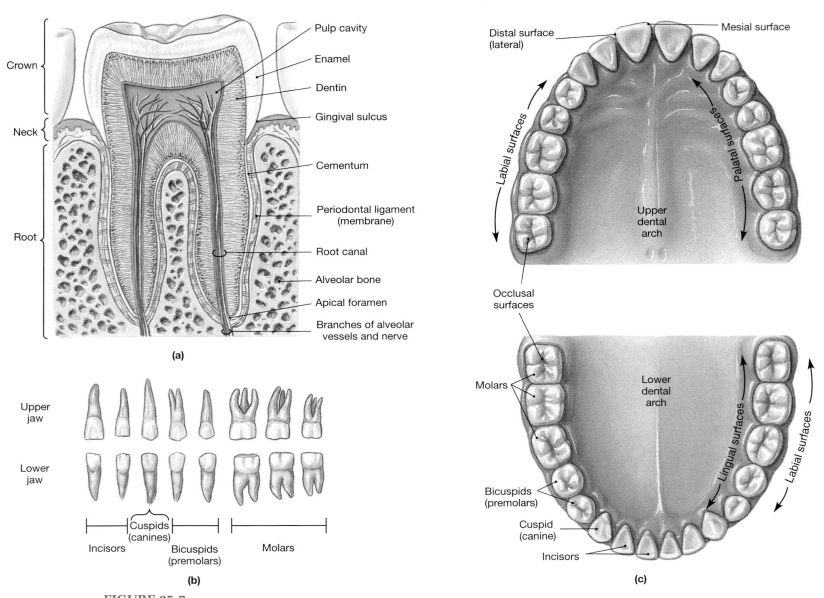

FIGURE 25-7
Teeth. (a) Diagrammatic section through a typical adult tooth. (b) The adult teeth. (c) Dental reference terms and the normal orientation of adult teeth.

TYPES OF TEETH *(Figure 25-7b,c).* There are four types of teeth, each with specific functions (Figure 25-7b,c●):

1. Incisors (in-SĪ-zerz) are blade-shaped teeth found at the front of the mouth. Incisors are useful for clipping or cutting, as when nipping off the tip of a carrot stick.

2. The **cuspids** (KUS-pidz), or *canines*, are conical with a sharp ridgeline and a pointed tip. They are used for tearing or slashing. A tough piece of celery might be weakened by the clipping action of the incisors, but then moved to one side to take advantage of the shearing action provided by the cuspids. Incisors and cuspids each have a single root.

3. Bicuspids (bī-KUS-pidz), or *premolars*, have one or two roots. Premolars have flattened crowns with prominent ridges. They are used for crushing, mashing, and grinding.

4. Molars have very large flattened crowns with prominent ridges and typically have three or more roots. Molars also have flattened crowns for crushing and grinding.

DENTAL SUCCESSION. During development, two sets of teeth begin to form. The first to appear are the **deciduous teeth** (dē-SID-ū-us; *deciduus*, falling off), also known as *primary teeth*, *milk teeth*, or *baby teeth* (Figure 6-17d●, p. 155). There are usually 20 deciduous teeth, five on each side of the upper and lower jaws. These teeth will later be replaced by the adult **secondary dentition**, or *permanent dentition*. The larger adult jaws can accommodate more than 20 permanent teeth, and three additional molars appear on each side of the upper and lower jaws as the individual ages. These teeth extend the length of the tooth rows posteriorly and bring the permanent tooth count to 32.

On each side of the upper or lower jaw, the primary dentition consists of two incisors, one cuspid, and a pair of deciduous molars. These are gradually replaced by the permanent dentition.

Table 25-1 presents the sequence of eruption of the primary dentition and the approximate ages at their replacement. In this process the periodontal ligaments and roots of the primary teeth are eroded away, until they fall out or are pushed aside by the emergence, or **eruption**, of the secondary teeth. The adult premolars take the place of the deciduous molars, and the definitive

TABLE 25-1   Tooth Eruption and Replacement

Teeth	Age at Eruption of Primary Teeth (in months)		Age at Eruption of Permanent Teeth (in years)	
	Lower	Upper	Lower	Upper
Central incisors	6	7.5	6–7	7–8
Lateral incisors	7	9	7–8	8–9
Cuspids (canines)	16	18	9–10	11–12
Primary first molar	12	14		
First bicuspids (premolars)			10–12	10–11
Primary second molar	20	24		
Second bicuspids (premolars)			11–12	10–12
First molars			6–7	6–7
Second molars			11–13	12–13
Third molars (wisdom teeth)			17–21	17–21

adult molars extend the tooth row as the jaw enlarges. The last molars, or *wisdom teeth*, may not erupt before age 21, if they appear at all. Wisdom teeth often develop in inappropriate positions, and they may be unable to erupt properly.

A DENTAL FRAME OF REFERENCE (*Figure 25-7b,c*). The upper and lower rows of teeth form a curving **dental arch**. Relative positions along the arch are indicated by the use of special terms (Figure 25-7b,c●). The terms **labial** or **buccal** refer to the outer surface of the dental arch, adjacent to the lips or cheeks. **Palatal** (upper) or **lingual** (lower) refers to the inner surface of the dental arch. **Mesial** (MĒ-zē-al) or **lateral** refers to the opposing surfaces between the teeth in a single dental arch. Mesial surfaces face forward or medially, and lateral surfaces face backward or laterally. For example, the mesial surface of each canine faces the lateral surface of the second incisor. The **occlusal surfaces** (o-KLOO-zal; *occlusio*,

■ **CLINICAL BRIEF**
Dental Problems

Most common dental problems result from the action of oral bacteria. Bacteria adhering to the surfaces of the teeth produce a sticky matrix that traps food particles and creates deposits of **plaque**. The mass of the plaque deposit protects the bacteria from salivary secretions, and as they digest nutrients, the bacteria generate acids that erode the structure of the tooth. The results are *dental caries*, otherwise known as cavities. Brushing the exposed surfaces of the teeth after meals helps to prevent the settling of bacteria and the entrapment of food particles, but bacteria between the teeth and within the gingival sulcus may elude the brush. Dentists therefore recommend the daily use of dental floss to clean these spaces.

If the bacteria remain within the gingival sulcus, the acids generated begin eroding the connections between the neck of the tooth and the gingiva. The gums appear to recede from the teeth, and **periodontal disease** develops. As it progresses, the bacteria attack the cementum, progressively destroying the periodontal ligament and eroding the alveolar bone. This deterioration loosens the tooth, and periodontal disease is the most common cause for the loss of teeth.

If teeth are broken or they must be removed because of disease, the usual treatment involves replacing them with "false teeth" attached to a plate or frame inserted into the mouth. Over the past 10 years, an alternative has been developed, using dental implants. A ridged titanium cylinder is inserted into the alveolus, and osteoblasts lock the ridges into the surrounding bone. After 4–6 months, an artificial tooth is screwed into the cylinder. Roughly 42% of individuals over age 65 have lost all of their teeth; the rest have lost an average of 10 teeth.

closed) of the teeth face their counterparts on the opposing dental arch. The occlusal surfaces perform the actual clipping, tearing, crushing, and grinding actions of the teeth.

MASTICATION. The *muscles of mastication* close the jaws and slide or rock the lower jaw from side to side. ∞ [p. 255] During mastication food is forced back and forth between the vestibule and the rest of the oral cavity, crossing and recrossing the occlusal surfaces. This movement results in part from the action of the masticatory muscles, but control would be impossible without the aid of the buccal, labial, and lingual muscles. Once the material has been shredded or torn to a satisfactory consistency and moistened with salivary secretions, the tongue begins compacting the debris into a small oval mass, or **bolus**, that can be swallowed relatively easily.

√ **What type of epithelium lines the oral cavity?**
√ **What are the functions of saliva?**

THE PHARYNX

The pharynx serves as a common passageway for food, liquids, and air. The epithelial lining and divisions of the pharynx, the nasopharynx, the oropharynx, and the laryngopharynx were described and illustrated in Chapter 24. ∞ [p. 601] Beneath the lamina propria lies a dense layer of elastic fibers, bound to the underlying skeletal muscles. The specific pharyngeal muscles involved in swallowing, summarized below, were detailed in Chapter 10. ∞ [p. 263]

- The *pharyngeal constrictors* provide the impetus for bolus movement.
- The *palatopharyngeus* and *stylopharyngeus* elevate the larynx.
- The *palatal muscles* raise the soft palate and adjacent portions of the pharyngeal wall.

The pharyngeal muscles cooperate with muscles of the oral cavity and esophagus to initiate the swallowing process, or **deglutition** (dē-gloo-TISH-un).

The Swallowing Process
(*Figure 25-8*)

Swallowing is a complex process whose initiation is voluntarily controlled, but it proceeds involuntarily once it begins. Swallowing can be divided into *buccal*, *pharyngeal*, and *esophageal phases*. Key aspects of each stage are illustrated in Figure 25-8●.

1. The **buccal phase** begins with the compression of the bolus against the hard palate. Subsequent retraction of the tongue then forces the bolus into the pharynx and assists in the elevation of the soft palate, thereby isolating the nasopharynx (Figure 25-8a,b●). The buccal phase is strictly voluntary; once the bolus enters the oropharynx, involuntary reflexes are initiated, and the bolus is moved toward the stomach.

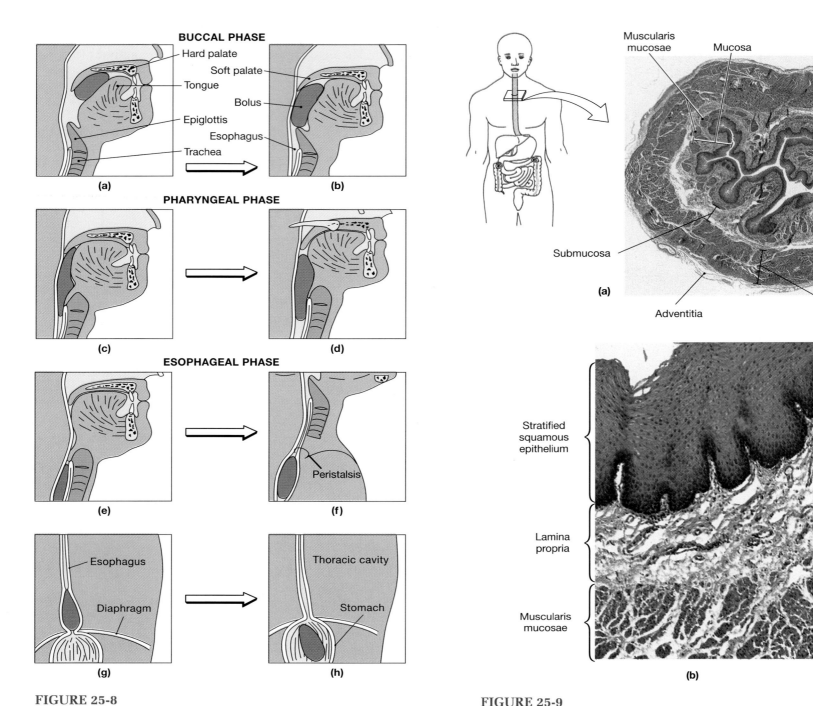

BUCCAL PHASE

- Hard palate
- Soft palate
- Tongue
- Bolus
- Epiglottis
- Esophagus
- Trachea

(a)　(b)

PHARYNGEAL PHASE

(c)　(d)

ESOPHAGEAL PHASE

Peristalsis

(e)　(f)

- Esophagus
- Diaphragm
- Thoracic cavity
- Stomach

(g)　(h)

FIGURE 25-8
The Swallowing Process. This sequence, based on a series of X-rays, shows the stages of swallowing and the movement of materials from the mouth to the stomach.

- Muscularis mucosae
- Mucosa
- Submucosa
- Adventitia
- Muscularis externa

(a)

- Stratified squamous epithelium
- Lamina propria
- Muscularis mucosae

(b)

FIGURE 25-9
The Esophagus. (a) Low-power view of a section through the esophagus. (b) The esophageal mucosa. (LM × 77)

2. The **pharyngeal phase** begins as the bolus comes in contact with the palatal arches or the posterior pharyngeal wall or both (Figure 25-8c,d●). Elevation of the larynx and folding of the epiglottis direct the bolus past the closed glottis, and in less than a second, the pharyngeal muscles have propelled the bolus into the esophagus. During the time it takes for the bolus to travel through the pharynx and into the esophagus, the respiratory centers are inhibited and breathing ceases.

3. The **esophageal phase** of swallowing (Figure 25-8e–g●) starts with the opening of the **upper esophageal sphincter**. After passing through the open sphincter, the bolus is pushed along the length of the esophagus by peristaltic waves. The approach of the bolus triggers the opening of the **lower esophageal sphincter**, and the bolus then continues into the stomach (Figure 25-8g,h●). ✝*Achalasia and Esophagitis [p. 774]*

THE ESOPHAGUS
(Figures 25-1/25-8)

The **esophagus** (Figure 25-1●, p. 626) is a hollow muscular tube that carries foods and liquids to the stomach. It is located posterior to the trachea (Figure 25-8●). It passes along the dorsal wall of the mediastinum in the thoracic cavity and enters the peritoneal cavity through an opening in the diaphragm, the **esophageal hiatus** (hī-Ā-tus), before emptying into the stomach. The esophagus is approximately 25 cm (1 ft) long and about 2 cm (0.75 in.) in diameter. It begins at the level of the cricoid cartilage anterior to vertebra C_6 and ends anterior to vertebra T_7.

The esophagus receives blood from the esophageal arteries and branches of (1) the *thyrocervical trunk* and *external carotid arteries* of the neck, (2) the *bronchial arteries* of the mediastinum, and (3) the *celiac artery* and *inferior phrenic artery* of the abdomen. Blood from the esophageal capillaries collects into the *inferior thyroid*, *azygos*, and *gastric* veins. These blood vessels were detailed in Chapter 22. ∞ [p. 557] The esophagus is innervated by the vagus and sympathetic trunks via the *esophageal plexus.* ∞ [p. 439]

Histology of the Esophageal Wall
(Figures 25-2/25-9)

The wall of the esophagus contains mucosal, submucosal, and muscularis layers comparable to those described above (Figure 25-2●, p. 627). Distinctive features of the esophageal wall, shown in Figure 25-9●, include:

- The mucosa of the esophagus contains a stratified squamous epithelium.

- The mucosa and submucosa are thrown into large folds that run the length of the esophagus. These folds allow for expansion during the passage of a large bolus; except during swallowing, muscle tone in the walls keeps the lumen closed.

- The smooth muscle layer of the muscularis mucosae may be very thin or absent near the pharynx, but it gradually thickens to 200–400 μm as it approaches the stomach.

- The submucosa contains scattered *esophageal glands* that produce a mucous secretion.

- The muscularis externa has inner circular and outer longitudinal layers. In the upper third of the esophagus these layers contain skeletal muscle fibers; in the middle third there is a mixture of skeletal and smooth muscle tissue; along the lower third only smooth muscles are found.

- There is no serosa, and a layer of connective tissue outside of the muscularis externa anchors the esophagus in position against the dorsal body wall. This outer fibrous layer is called the *adventitia.*

√ **What effect would a drug that blocks parasympathetic stimulation of the digestive tract have on peristalsis?**

√ **Where is the fauces?**

√ **What is occurring when the soft palate and larynx are elevated and the glottis closes?**

THE STOMACH

The stomach performs three major functions: (1) the bulk storage of ingested food, (2) the mechanical breakdown of ingested food, and (3) the disruption of chemical bonds through the action of acids and enzymes. The mixing of ingested substances with the gastric juices secreted by the glands of the stomach produces a viscous, soupy mixture called **chyme** (KĪM).

Anatomy of the Stomach
(Figures 25-10/25-11/25-12a/25-13)

The stomach has the shape of an expanded J (Figures 25-10 and 25-11●). The stomach occupies the *left hypochondriac*, *epigastric*, and portions of the *umbilical* and *left lumbar regions* (Figure 25-13●, p. 644). Those reference terms were introduced in Chapter 1. ∞ [p. 15] The shape and size of the stomach are extremely variable from individual to individual and from one meal to the next. The stomach typically extends between the levels of vertebrae T_7 and L_3.

The J-shaped stomach has a short **lesser curvature**, forming the **medial surface** of the organ, and a long **greater curvature**, which forms the **lateral surface**. The **anterior** and **posterior surfaces** are smoothly rounded.

The stomach is divided into four regions (Figures 25-10 and 25-11●):

1. The esophagus contacts the medial surface of the stomach at the **cardia** (KAR-dē-a). The esophageal lumen opens into the cardia at the **cardiac orifice**.

2. The portion of the stomach superior to the gastroesophageal junction is the **fundus** (FUN-dus). The fundus contacts the inferior and posterior surface of the diaphragm.

3. The area between the fundus and the curve of the J is the **body** of the stomach. The body is the largest region of the stomach, and it functions as a mixing tank for ingested food and gastric secretions.

4. The **pylorus** (pī-LŌR-us) is the curve of the J. This region is connected to the *duodenum*, the proximal segment of the small intestine. As mixing movements occur during digestion, the pylorus frequently changes shape. A muscular **pyloric sphincter** regulates the release of chyme from the **pyloric outlet** into the duodenum.

The volume of the stomach increases at mealtimes and then decreases as chyme enters the small intestine. In the relaxed (empty) stomach, the mucosa is thrown into a number of prominent longitudinal folds, called **rugae** (ROO-gē; "wrinkles") (Figures 25-11a and 25-12a●, p. 643). Rugae permit expansion of the gastric lumen. As expansion occurs, the epithelial lining, which cannot stretch, flattens out, and the rugae become less prominent. In a full stomach, the rugae almost disappear.

Mesenteries of the Stomach
(Figures 25-4/25-10a)

The visceral peritoneum covering the outer surface of the stomach is continuous with a pair of prominent mesenteries. The **greater omentum** forms an enormous pouch that hangs like an apron from the greater curvature of the stomach. The greater omentum lies posterior to the abdominal wall and anterior to the abdominal viscera (Figures 25-4, p. 632, and 25-10a●). Adipose tissue in the greater omentum conforms to the shapes of the surrounding organs, providing padding and protection across the anterior and lateral surfaces of the abdomen. The lipids in the adipose tissue represent an important energy reserve, and the greater omentum provides insulation that reduces heat loss across the anterior abdominal

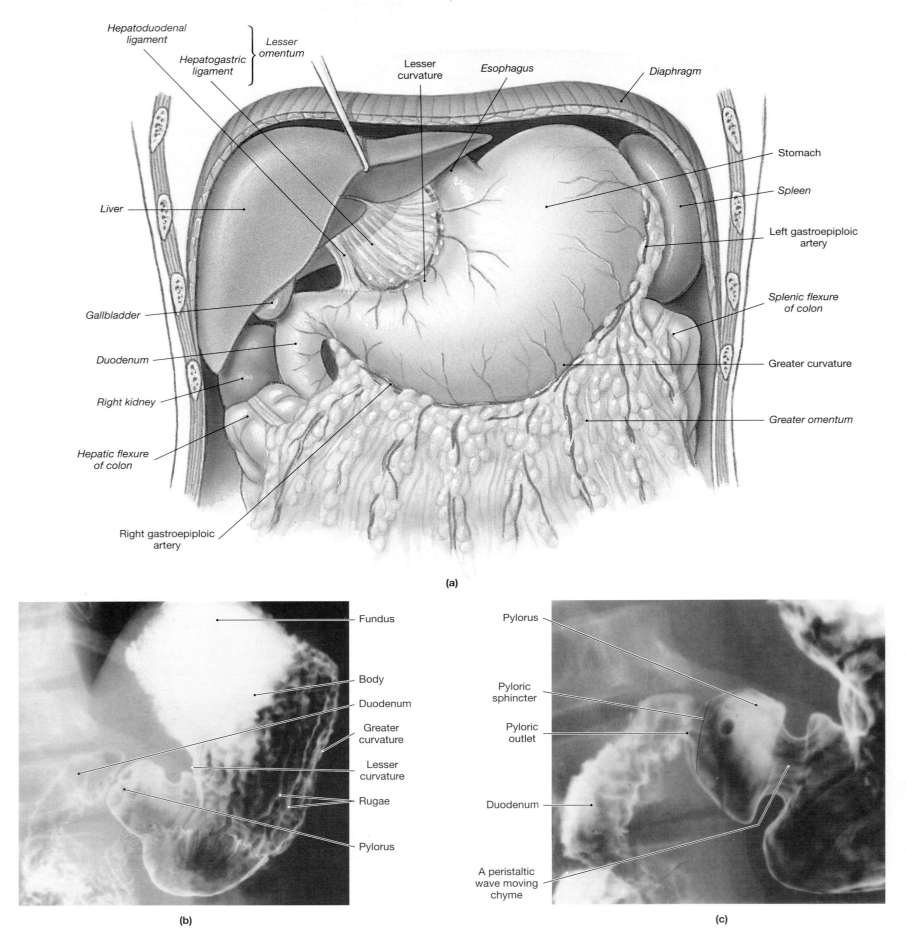

FIGURE 25-10

The Stomach and Omenta. (a) Surface anatomy of the stomach, showing blood vessels and relation to liver and intestines. (b) Radiograph of the stomach and duodenum, showing a barium meal. (c) Radiograph of the pyloric region, pyloric valve, and relationship to duodenum.

640

FIGURE 25-11
Gross Anatomy of the Stomach. (a) External and internal anatomy of the stomach. (b) The natural position of the stomach may be viewed after the left lobe of the liver and lesser omentum have been removed; ventral view.

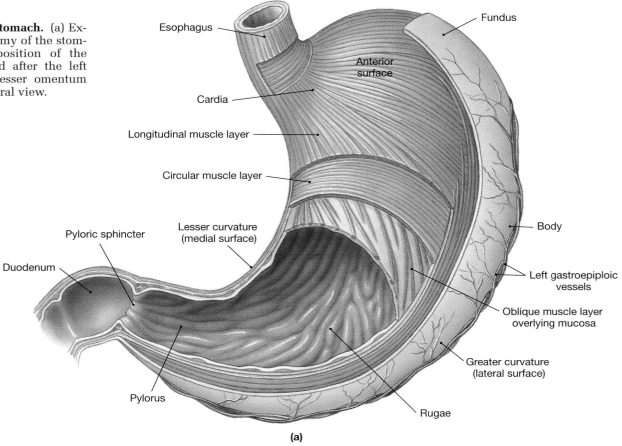

Esophagus

Cardia

Longitudinal muscle layer

Circular muscle layer

Pyloric sphincter

Duodenum

Lesser curvature (medial surface)

Pylorus

Rugae

Fundus

Anterior surface

Body

Left gastroepiploic vessels

Oblique muscle layer overlying mucosa

Greater curvature (lateral surface)

(a)

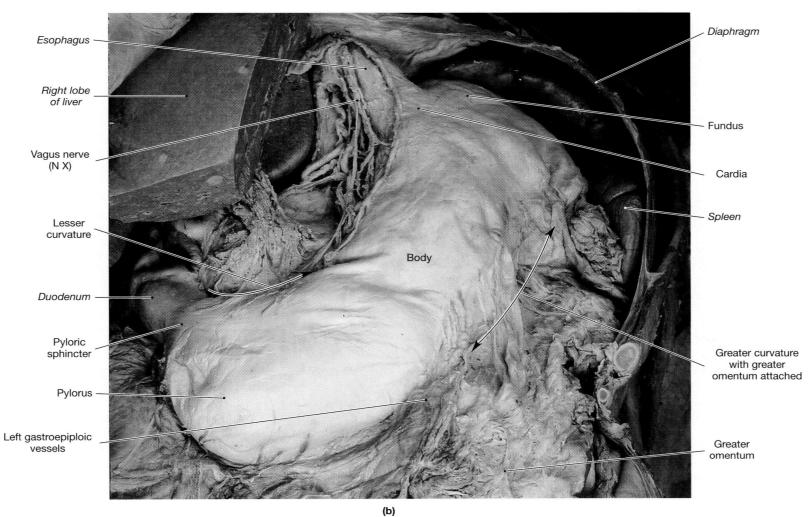

Esophagus

Right lobe of liver

Vagus nerve (N X)

Lesser curvature

Duodenum

Pyloric sphincter

Pylorus

Left gastroepiploic vessels

Diaphragm

Fundus

Cardia

Spleen

Body

Greater curvature with greater omentum attached

Greater omentum

(b)

wall. The *lesser omentum* is a much smaller pocket in the ventral mesentery between the lesser curvature of the stomach and the liver. The lesser omentum stabilizes the position of the stomach and provides an access route for blood vessels and other structures entering or leaving the liver.

Blood Supply to the Stomach
(Figures 22-17/22-26/25-10a)

The three branches of the celiac artery supply blood to the stomach:

- The *left gastric artery* supplies blood to the lesser curvature and cardia.
- The *splenic artery* supplies the fundus and greater curvature via the *left gastroepiploic artery* (Figure 25-10a●).
- The *common hepatic artery* supplies blood to the lesser and greater curvatures of the pylorus via the *right gastroepiploic artery* and the *gastroduodenal artery* (Figures 22-17 and 25-10a●). ∞ [pp. 561] Corresponding veins drain blood from the stomach into the hepatic portal vein (Figure 22-26●). ∞ [p. 572]

Musculature of the Stomach
(Figure 25-11a)

The muscularis mucosae and muscularis externa of the stomach contain extra layers of smooth muscle in addition to the usual *circular* and *longitudinal layers*. The muscularis mucosae usually contains an additional outer, circular layer of muscle fibers. The muscularis externa has an extra inner, *oblique layer* of smooth muscle (Figure 25-11a●). The extra layers of smooth muscle strengthen the stomach wall, and they perform the mixing and churning activities essential to the formation of chyme.

Histology of the Stomach
(Figure 25-12)

A simple columnar epithelium lines all portions of the stomach. The epithelium is a *secretory sheet* that produces a carpet of mucus that covers the interior surfaces of the stomach. ∞ [p. 77] The mucus layer provides protection against the acids and enzymes in the gastric lumen. Shallow depressions, called **gastric pits**, open onto the gastric surface (Figure 25-12●). The mucous cells at the base, or *neck*, of each gastric pit are actively dividing, replacing superficial cells that are shed into the chyme. The continual replacement of epithelial cells provides an additional defense against the gastric contents. If stomach acid and digestive enzymes penetrate the mucous layers, any damaged epithelial cells are quickly replaced.

In the fundus and body of the stomach, each gastric pit communicates with several **gastric glands** that extend deep into the underlying lamina propria. Gastric glands (Figure 25-12b●) are simple branched tubular glands dominated by two types of secretory cells: *parietal cells* and *chief cells*. Together they secrete about 1500 m*ℓ* of **gastric juice** each day.

PARIETAL CELLS.   **Parietal cells** are especially common along the proximal portions of each gastric gland. These cells secrete *intrinsic factor* and *hydrochloric acid* (HCl). **Intrinsic factor** facilitates the absorption of vitamin B_{12} across the intestinal lining. Hydrochloric acid lowers the pH of the gastric juice, kills

microorganisms, breaks down cell walls and connective tissues in food, and activates the secretions of the chief cells.

CHIEF CELLS.   **Chief cells** are most abundant near the base of a gastric gland. These cells secrete **pepsinogen** (pep-SIN-ō-jen), which is converted by the acids in the gastric lumen to an active proteolytic enzyme, **pepsin**. The stomachs of newborn infants (but not adults) also produce **rennin** and **gastric lipase**, enzymes important for the digestion of milk. Rennin coagulates milk proteins, and gastric lipase initiates the digestion of milk fats.

ENTEROENDOCRINE CELLS.   **Enteroendocrine** (en-ter-ō-EN-dō-krin) **cells** are scattered among the parietal and chief cells. These cells produce at least seven different secretions. The best-known is the hormone **gastrin** (GAS-trin), which stimulates the secretion of both parietal and chief cells. Gastrin release occurs when food enters the stomach.

■ CLINICAL BRIEF
Gastritis and Peptic Ulcers

Inflammation of the gastric mucosa causes **gastritis** (gas-TRĪ-tis). This condition may develop after swallowing drugs, including alcohol and aspirin. Gastritis may also appear after severe emotional or physical stress, bacterial infection of the gastric wall, or the ingestion of strong acid or alkaline chemicals. Infections involving the bacterium *Helicobacter pylori* are another important factor in ulcer formation. These bacteria are able to survive long enough to penetrate the mucus that coats the epithelium. Once within the protective layer of mucus, they are safe from the action of gastric acids and enzymes. Over time, the infection damages the epithelial lining, with two major results: (1) erosion of the lamina propria by gastric juices and (2) entry and spread of the bacteria through the gastric wall and into the bloodstream.

A **peptic ulcer** develops when the digestive acids and enzymes manage to erode their way through the defenses of the stomach lining or proximal portions of the small intestine. The locations may be indicated by using the terms **gastric ulcer** (stomach) or **duodenal ulcer** (duodenum). Peptic ulcers result from the excessive production of acid or the inadequate production of the alkaline mucus that poses an epithelial defense.

Once gastric juices have destroyed the epithelial layers, the virtually defenseless lamina propria will be exposed to digestive attack. Sharp abdominal pain results, and bleeding can develop. The administration of antacids can often control peptic ulcers by neutralizing the acids and allowing time for the mucosa to regenerate. The drug *cimetidine* (sī-MET-i-dēn), or *Tagamet*, inhibits the secretion of acid by the parietal cells. Dietary restrictions limit the intake of acidic beverages and eliminate foods that promote acid production (caffeine) or damage unprotected mucosal cells (alcohol). In severe cases the damage may provoke significant bleeding, and the acids may even erode their way through the wall of the digestive tract and enter the peritoneal cavity. This condition, called a **perforated ulcer**, requires immediate surgical correction.

Regulation of the Stomach

The production of acid and enzymes by the gastric mucosa can be directly controlled by the central nervous system and indirectly regulated by local hormones. CNS regulation involves the vagus nerve (parasympathetic innervation) and branches of the celiac plexus (sympathetic innervation). The sight or thought of food triggers motor output in the vagus nerve. Postganglionic parasympathetic fibers innervate parietal cells, chief cells, and mucous cells of the stomach. Stimulation causes an increase in the production of acids, enzymes, and mucus. The arrival of food in the stomach leads to stimu-

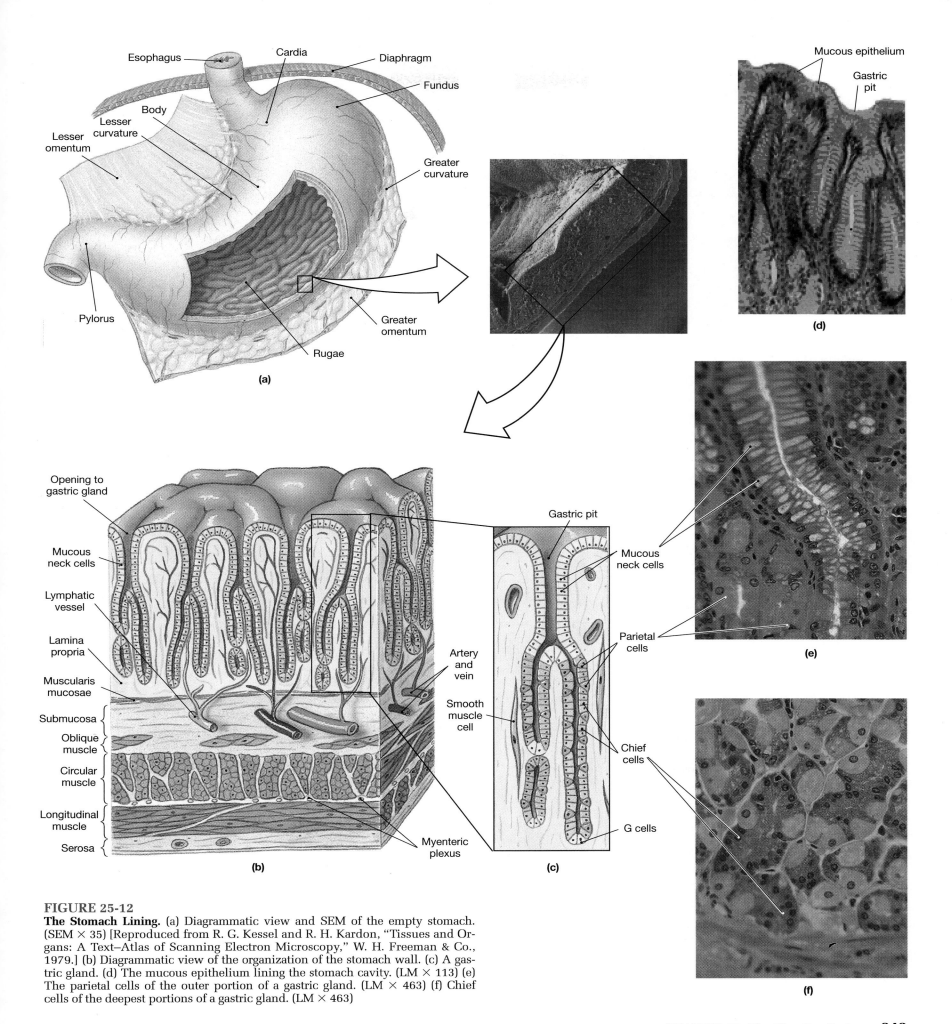

FIGURE 25-12
The Stomach Lining. (a) Diagrammatic view and SEM of the empty stomach. (SEM × 35) [Reproduced from R. G. Kessel and R. H. Kardon, "Tissues and Organs: A Text–Atlas of Scanning Electron Microscopy," W. H. Freeman & Co., 1979.] (b) Diagrammatic view of the organization of the stomach wall. (c) A gastric gland. (d) The mucous epithelium lining the stomach cavity. (LM × 113) (e) The parietal cells of the outer portion of a gastric gland. (LM × 463) (f) Chief cells of the deepest portions of a gastric gland. (LM × 463)

FIGURE 25-13
Abdominal Regions and Planes

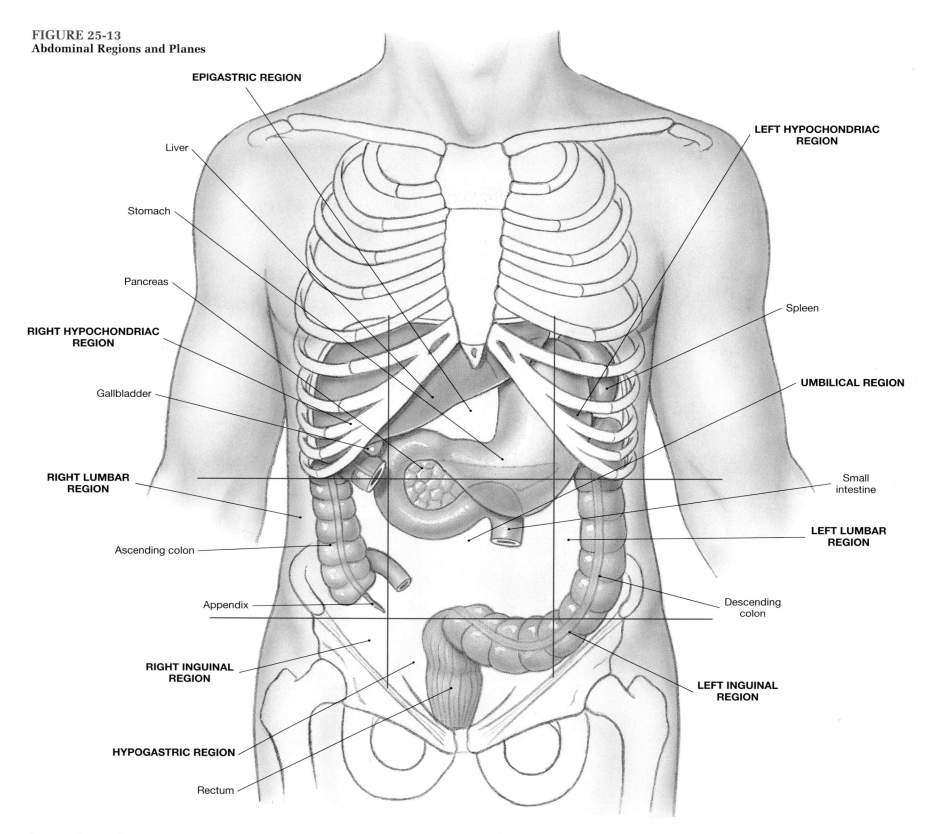

EPIGASTRIC REGION

LEFT HYPOCHONDRIAC REGION

Liver

Stomach

Pancreas

Spleen

RIGHT HYPOCHONDRIAC REGION

UMBILICAL REGION

Gallbladder

RIGHT LUMBAR REGION

Small intestine

LEFT LUMBAR REGION

Ascending colon

Appendix

Descending colon

RIGHT INGUINAL REGION

LEFT INGUINAL REGION

HYPOGASTRIC REGION

Rectum

lation of stretch receptors in the stomach wall and chemoreceptors in the mucosa. Reflexive contractions occur in the muscularis layers of the stomach wall, and gastrin is released by enteroendocrine cells. Both parietal and chief cells respond to the presence of gastrin by accelerating their secretory activities, but the parietal cells are especially sensitive, and the rate of acid production accelerates markedly.

Sympathetic activation leads to the inhibition of gastric activity. In addition, two hormones released by the small intestine inhibit gastric secretion. The release of these hormones, **secretin** (se-

KRĒ-tin) and **cholecystokinin** (kō-lē-sis-tō-KĪ-nin), stimulates secretion by the pancreas and liver; the depression of gastric activity is a secondary but complementary effect. ⊤ *Stomach Cancer [p. 774]*

√ **When a person suffers from chronic ulcers in the stomach, the branches of the vagus nerve that serve the stomach are sometimes severed surgically. Why?**

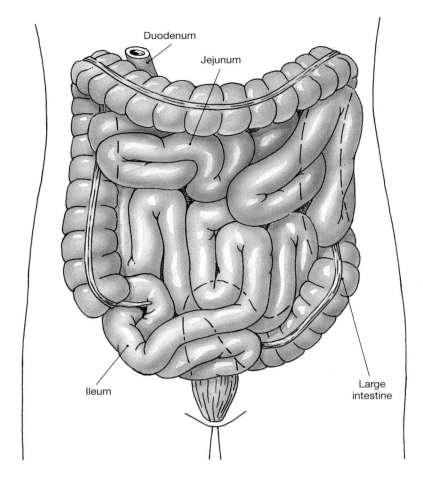

FIGURE 25-14
Regions of the Small Intestine. The color coding indicates the relative sizes and positions of the duodenum, jejunum, and ileum.

THE SMALL INTESTINE
(Figures 25-2b/25-4b/25-10/25-13/25-14/25-15a)

The small intestine plays the primary role in the digestion and absorption of nutrients. The small intestine averages 6 m (20 ft) in length (range, 15–25 ft) and has a diameter ranging from 4 cm at the stomach to about 2.5 cm at the junction with the large intestine. It occupies all abdominal regions except the hypochondriac left and epigastric regions (Figure 25-13●). Ninety percent of nutrient absorption occurs in the small intestine, and most of the rest occurs in the large intestine.

The small intestine fills much of the peritoneal cavity, and its position is stabilized by mesenteries attached to the dorsal body wall (Figure 25-4b●, p. 632). Movement of the small intestine during digestion is restricted by the stomach, the large intestine, the abdominal wall, and the pelvic girdle. Figures 25-10, p. 640, 25-13, and 25-14● show the position of the small intestine relative to the other segments of the digestive tract.

The intestinal lining bears a series of transverse folds called **plicae** (PLĪ-kē) **circulares** (Figures 25-2b, p. 627, and 25-15a●). Unlike the rugae in the stomach, each *plica* (PLĪ-ka) is a permanent feature of the intestinal lining; plicae do not disappear when the small intestine fills with chyme. Roughly 800 plicae are found along the length of the small intestine, and their presence greatly increases the surface area available for absorption.

Regions of the Small Intestine
(Figure 25-14)

The small intestine has three anatomical subdivisions: the *duodenum*, the *jejunum*, and the *ileum* (Figure 25-14●).

The Duodenum
(Figures 25-4b/25-10b,c/25-14)

The **duodenum** (dū-ō-DĒ-num) (Figures 25-10b,c, p. 640, and 25-14●) is the shortest and widest segment of the small intestine; it is approximately 25 cm (10 in.) long. The duodenum is connected to the pylorus of the stomach, and the passageway is guarded by the pyloric sphincter. The duodenum is a "mixing bowl" that receives chyme from the stomach and digestive secretions from the pancreas and liver. From its connection with the stomach, the duodenum curves in a C that encloses the pancreas. Except for the proximal 2.5 cm (1 in.), the duodenum is in a retroperitoneal position between vertebrae L_1 and L_4 (Figure 25-4b●, p. 632).

The Jejunum

A rather abrupt bend marks the boundary between the duodenum and the **jejunum** (je-JOO-num). At this junction the small intestine reenters the peritoneal cavity, supported by a sheet of mesentery. The jejunum is about 2.5 m (8 ft) long. The bulk of chemical digestion and nutrient absorption occurs in the jejunum.

The Ileum

The **ileum** (IL-ē-um) is the third and last segment of the small intestine. It is also the longest, averaging 3.5 m (12 ft) long. The ileum ends in a sphincter, the *ileocecal valve*, which controls the flow of materials from the ileum into the *cecum* of the large intestine. The ileocecal valve protrudes into the cecum.

Support of the Small Intestine
(Figure 25-4)

The duodenum has no supporting mesentery. The proximal 2.5 cm is movable, but the rest is retroperitoneal and fixed in position. The jejunum and ileum are supported by an extensive, fan-shaped mesentery known as the **mesentery proper** (Figure 25-4●, p. 632). Blood vessels, lymphatics, and nerves reach these segments of the small intestine via the connective tissue of the mesentery. The blood vessels involved are branches of the *superior mesenteric artery* and *vein* (Figures 22-17 and 22-26●). ∞ [pp. 561, 572] Parasympathetic innervation is provided by the vagus nerve; sympathetic innervation involves fibers from the superior mesenteric ganglion. ∞ [pp. 401, 432]

Histology of the Small Intestine

The Intestinal Epithelium
(Figures 25-15/25-16)

The mucosa of the small intestine is thrown into a series of finger-like projections, the **intestinal villi** (Figures 25-15 and 25-16●). The intestinal villi are covered by a simple columnar epithelium that is carpeted with microvilli. If the small intestine were a simple tube with smooth walls, it would have a total absorptive area of roughly 0.33 m². Instead, the epithelium contains plicae circulares, each plica supports a forest of villi, and each villus is covered by epithelial cells whose exposed surfaces contain microvilli. This arrangement increases the total area for absorption to over 200 m².

Intestinal Crypts
(Figure 25-15b,d)

Between the columnar epithelial cells, goblet cells eject mucus onto the intestinal surfaces. At the bases of the villi are found the entrances to the **intestinal crypts** (KRIPTS). These pockets extend deep into the underlying lamina propria (Figure 25-15b,d●). Near the base of each crypt, stem cell divisions continually produce new generations of epithelial cells. These new cells are continually displaced toward the intestinal surface, and within a few days they will have reached the tip of a villus, where they are shed into the intestinal lumen. This ongoing process renews the epithelial surface and adds intracellular enzymes to the chyme.

Intestinal crypts also contain enteroendocrine cells responsible for the production of several intestinal hormones, including cholecystokinin and secretin.

The Lamina Propria
(Figure 25-15b,c,e)

The lamina propria of each villus contains an extensive network of capillaries that carries absorbed nutrients to the hepatic portal circulation. In addition to capillaries and nerve endings, each villus contains a terminal lymphatic called a **lacteal** (LAK-tē-al; *lacteus*, milky) (Figure 25-15b,c,e●). Lacteals transport materials that cannot enter local capillaries. These materials, such as large lipid-protein complexes, ultimately reach the venous circulation via the thoracic duct. The name *lacteal* refers to the pale, cloudy appearance of lymph containing large quantities of lipids.

Regional Specializations
(Figures 25-14/25-15c–e/25-16)

The regions of the small intestine have histological specializations related to their primary functions. Representative sections from each region of the small intestine are presented in Figure 25-16● and should be viewed for each region along with Figures 25-14 and 25-15c–e●.

THE DUODENUM *(Figures 25-16/25-22b).* The duodenum contains numerous mucous glands, both within the epithelium and beneath it. In addition to the duodenal crypts, the submucosa contains **submucosal glands**, also known as *Brunner's glands*, that produce copious quantities of mucus (Figure 25-16a●). The mucus produced by the crypts and the submucosal glands protects the epithelium from the acid chyme arriving from the stomach. The mucus also contains buffers that help elevate the pH of the chyme. Submucosal glands are most abundant in the proximal portion of the duodenum, and their numbers decrease approaching the jejunum. Over this distance the pH of the chyme goes from 1–2 to 7–8; by the start of the jejunum, the extra mucus production is no longer needed.

Roughly halfway along its length, the duodenum receives buffers and enzymes from the pancreas and bile from the liver. Within the duodenal wall, the *common bile duct* from the liver and *pancreatic duct* from the pancreas come together at a muscular chamber called the **duodenal ampulla** (am-PYŪL-la). This chamber opens into the duodenal lumen at a small mound, the **duodenal papilla** (Figures 25-16 and 25-22b, p. 655●).

THE JEJUNUM AND ILEUM *(Figures 25-15d,e/25-16).* Plicae and villi remain prominent over the proximal half of the jejunum (Figure 25-15d,e●). As one approaches the ileum, the plicae and villi become smaller and continue to diminish in size to the end of the ileum. This reduction parallels the reduction in absorptive activity; most nutrient absorption has occurred before materials reach the terminal portion of the ileum. The region adjacent to the large intestine lacks plicae altogether, and the scattered villi are stumpy and conical.

Bacteria are normal inhabitants within the lumen of the large intestine. These bacteria are nourished by the surrounding mucosa. The epithelial barriers (cells, mucus, and digestive juices) and underlying cells of the immune system protect the small intestine from bacteria migrating from the large intestine. Small, isolated lymphoid nodules are present in the lamina propria of the jejunum. In the ileum the lymphoid nodules become more numerous, and they fuse together to form large masses of lymphoid tissue. These lymphoid centers, called **aggregate lymphoid nodules**, or *Peyer's patches*, may reach the size of cherries. ∞ [p. 588] They are most abundant in the terminal portion of the ileum, near the entrance to the large intestine (Figure 25-16●).

Regulation of the Small Intestine

As absorption occurs, weak peristaltic contractions slowly move materials along the length of the small intestine. Movements of the small intestine are controlled primarily by neural reflexes involving the submucosal and myenteric plexuses. Stimulation of the parasympathetic system increases the sensitivity of these reflexes and accelerates peristaltic contractions and segmentation movements. These contractions are limited to within a few centimeters of the site of the original stimulus. Coordinated intestinal move-

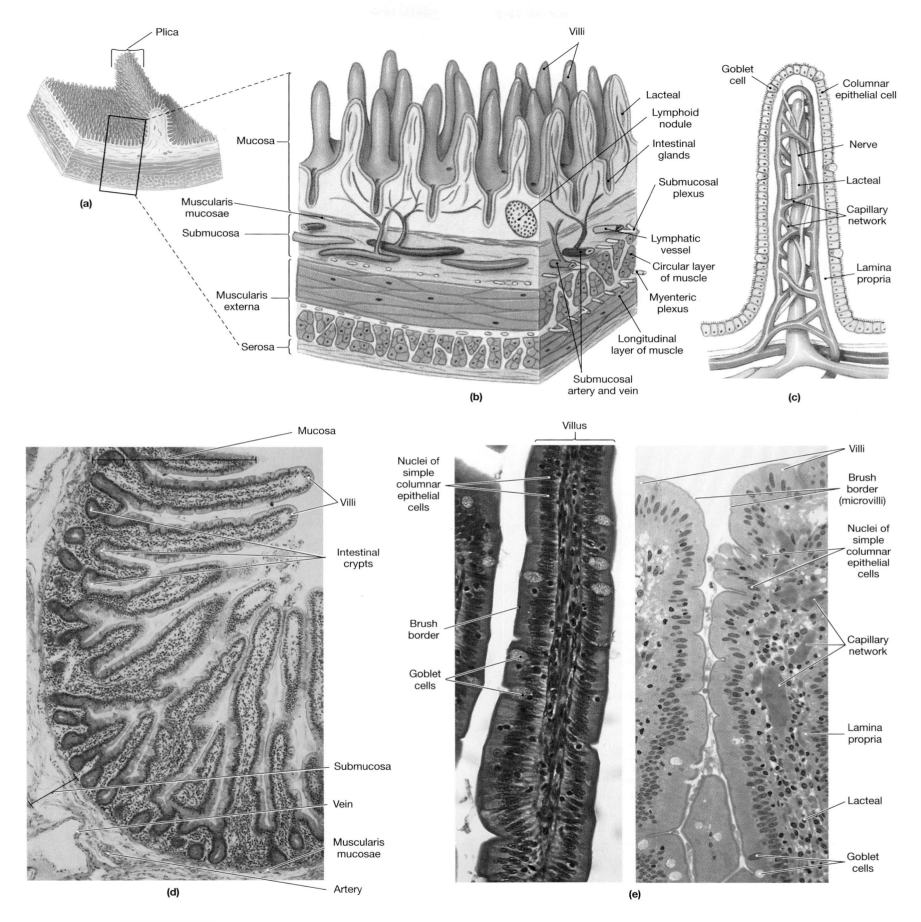

(a)

Plica

Mucosa

Muscularis mucosae

Submucosa

Muscularis externa

Serosa

(b)

Villi

Lacteal

Lymphoid nodule

Intestinal glands

Submucosal plexus

Lymphatic vessel

Circular layer of muscle

Myenteric plexus

Longitudinal layer of muscle

Submucosal artery and vein

(c)

Goblet cell

Columnar epithelial cell

Nerve

Lacteal

Capillary network

Lamina propria

(d)

Mucosa

Villi

Intestinal crypts

Submucosa

Vein

Muscularis mucosae

Artery

(e)

Villus

Nuclei of simple columnar epithelial cells

Brush border

Goblet cells

Villi

Brush border (microvilli)

Nuclei of simple columnar epithelial cells

Capillary network

Lamina propria

Lacteal

Goblet cells

FIGURE 25-15
The Intestinal Wall. (a) Characteristic features of the intestinal lining. (b) The organization of villi and the intestinal crypts. (c) Diagrammatic view of a single villus, showing the capillary and lymphatic supply. (d) Panoramic view of the wall of the small intestine showing mucosa with characteristic villi, plica, submucosa, and muscularis layers. (LM × 49) (e) Photomicrographs of villi from the jejunum. (LM × 360, LM × 620)

FIGURE 25-16

Regions of the Small Intestine. (a) Diagrammatic view highlighting the distinguishing features of each region of the small intestine. The detail shows a cross section through the ampulla. (b) The photographs show the gross anatomy of the intestinal lining in each region.

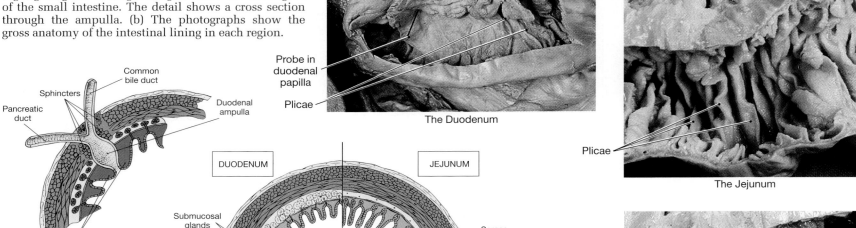

The Duodenum

The Jejunum

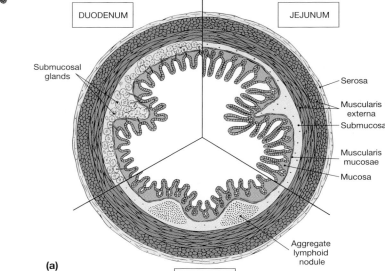

(a)

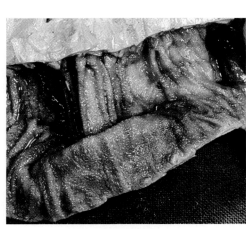

(b)

The Ileum

ments occur when food enters the stomach; these movements tend to move materials away from the duodenum and toward the large intestine. During these periods the ileocecal valve permits the passage of material into the large intestine. ⊤*Gastroenteritis [p. 774]*

Hormonal and CNS controls regulate the secretory output of the small intestine and accessory glands. The secretions of the small intestine are collectively called **intestinal juice**. Secretory activities are triggered by local reflexes and parasympathetic (vagal) stimulation. Sympathetic stimulation inhibits their activation. Duodenal enteroendocrine cells produce secretin and cholecystokinin, hormones that coordinate the secretory activities of the stomach, duodenum, liver, and pancreas.

√ **Which histological features of the small intestine facilitate the digestion and absorption of nutrients?**

√ **A narrowing (stenosis) of the ileocecal valve would interfere with the movement of chyme between what two organs?**

THE LARGE INTESTINE
(Figures 25-13/25-14/25-17)

The horseshoe-shaped large intestine begins at the end of the ileum and ends at the anus. The large intestine lies inferior to the stomach and liver, and almost completely frames the small intestine (Figures 25-13, p. 644, and 25-14●, p. 645). The major functions of the large intestine are (1) the resorption of water

and compaction of chyme into feces, (2) the absorption of important vitamins liberated by bacterial action, and (3) the storing of fecal material before defecation.

The large intestine, often called the **large bowel**, has an average length of about 1.5 m (5 ft) and a width of 7.5 cm (3 in.). It can be divided into three parts: (1) the *cecum*, the first portion of the large intestine, which appears as a pouch; (2) the *colon*, the largest portion of the large intestine; and (3) the *rectum*, the last 15 cm (6 in.) of the large intestine and the end of the digestive tract (Figure 25-17●).

The Cecum
(Figure 25-17)

Materials arriving from the ileum first enter an expanded pouch called the **cecum** (SĒ-kum). The ileum attaches to the medial surface of the cecum and opens into the cecum at the **ileocecal** (il-ē-ō-SĒ-kal) **valve** (Figure 25-17●), which regulates the passage of materials into the large intestine. The cecum collects and stores the arriving materials and begins the process of compaction. The slender, hollow **vermiform appendix** is attached to the posteromedial surface of the cecum. The appendix usually is approximately 9 cm (3.5 in.) long, but its size and shape are quite variable. A band of mesentery, the **mesoappendix**, connects the appendix to the ileum and cecum. The mucosa and submucosa of the appendix are dominated by lymphoid nodules, and its primary function is as an organ of the lymphatic system. Inflammation of the appendix produces the symptoms of *appendicitis.* ⚭ [p. 595]

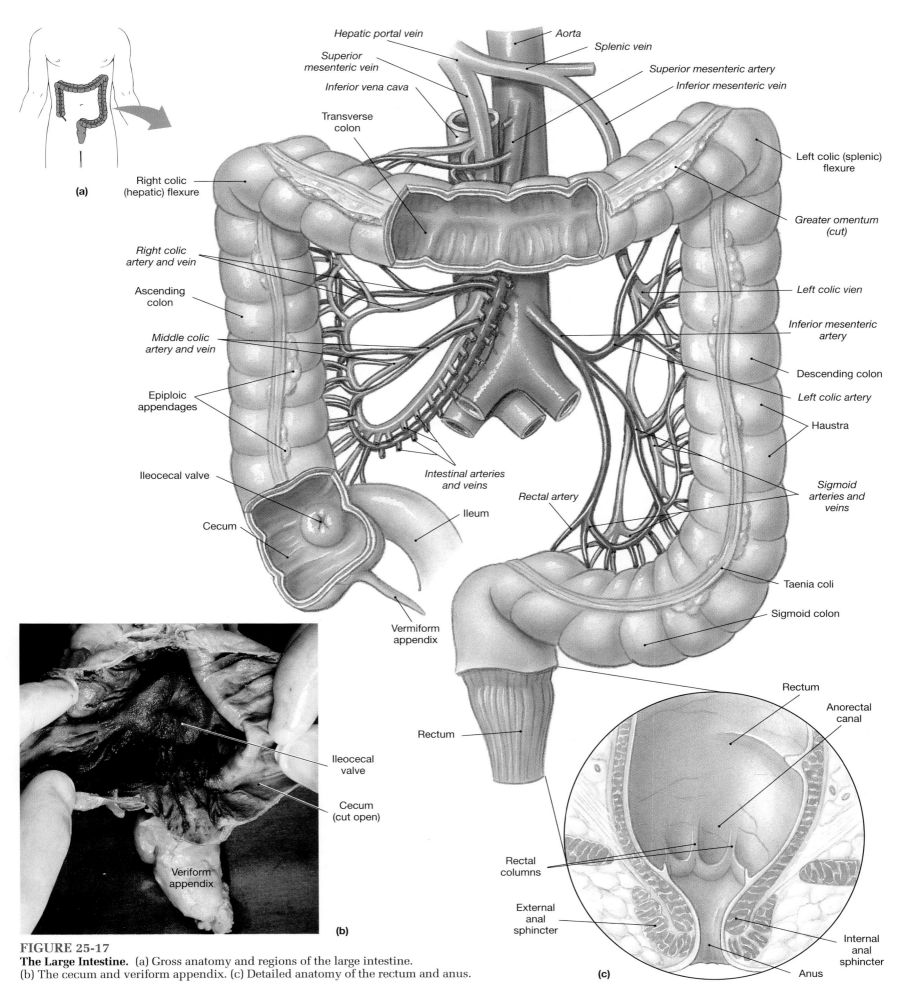

(a)

Hepatic portal vein

Aorta

Splenic vein

Superior mesenteric vein

Superior mesenteric artery

Inferior vena cava

Inferior mesenteric vein

Transverse colon

Right colic (hepatic) flexure

Left colic (splenic) flexure

Greater omentum (cut)

Right colic artery and vein

Ascending colon

Left colic vien

Middle colic artery and vein

Inferior mesenteric artery

Descending colon

Left colic artery

Epiploic appendages

Haustra

Ileocecal valve

Intestinal arteries and veins

Rectal artery

Cecum

Ileum

Vermiform appendix

Sigmoid arteries and veins

Taenia coli

Sigmoid colon

Rectum

Ileocecal valve

Cecum (cut open)

Veriform appendix

(b)

Rectum

Anorectal canal

Rectal columns

External anal sphincter

Internal anal sphincter

Anus

(c)

FIGURE 25-17
The Large Intestine. (a) Gross anatomy and regions of the large intestine.
(b) The cecum and veriform appendix. (c) Detailed anatomy of the rectum and anus.

The Colon
(Figure 25-17)

The **colon** has a larger diameter and a thinner wall than the small intestine. Refer to Figure 25-17● as we describe the colon. The distinctive features of the colon include:

1.   The wall of the colon forms a series of pouches, or **haustra** (HAWS-tra; singular *haustrum*), that permit considerable distension and elongation. Cutting into the intestinal lumen reveals that the creases between the haustra extend into the mucosal lining, producing a series of internal folds.

2.   Three separate longitudinal ribbons of smooth muscle, the **taeniae coli** (TĒ-nē-ā KŌ-lī), are visible on the outer surfaces of the colon just beneath the serosa. Muscle tone within these bands creates the haustra.

3.   The serosa of the colon contains numerous teardrop-shaped sacs of fat, called **epiploic appendages** (Figure 25-17●).

Regions of the Colon
(Figures 25-17/25-18)

The colon can be subdivided into four regions: the *ascending colon*, the *transverse colon*, the *descending colon*, and the *sigmoid colon* (Figure 25-17●). The regions of the colon may be clearly seen in the radiograph in Figure 25-18●.

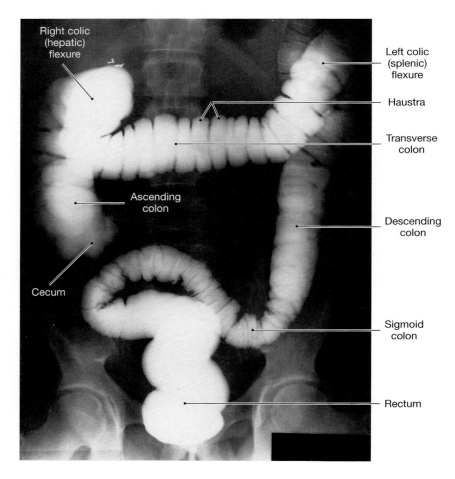

FIGURE 25-18
Anterior/Posterior Radiograph of the Colon

THE ASCENDING COLON *(Figure 25-4)*.   The **ascending colon** begins at the superior border of the cecum and ascends along the right lateral and posterior wall of the peritoneal cavity to the inferior surface of the liver. At this point the colon makes a sharp right bend to the left at the **right colic flexure**, or *hepatic flexure*. The right colic flexure marks the end of the ascending colon and the beginning of the *transverse colon*. The lateral and anterior surfaces of the ascending colon are covered by visceral peritoneum. There is normally no mesentery, and the ascending colon is retroperitoneal (Figure 25-4●, p. 632).

THE TRANSVERSE COLON.   The **transverse colon** begins at the right colic flexure. It curves anteriorly and crosses the abdomen from right to left. It is supported by the *transverse mesentery* and is separated from the anterior abdominal wall by the layers of the greater omentum. As the transverse colon reaches the left side, it passes inferior to the greater curvature of the stomach. The **gastrocolic ligament** attaches the transverse colon to the greater curvature of the stomach. Near the spleen, the colon makes a right-angle bend, termed the **left colic flexure**, or *splenic flexure*, and then proceeds caudally.

THE DESCENDING COLON.   The **descending colon** proceeds inferiorly along the left side until reaching the iliac fossa. The descending colon, which is retroperitoneal, is firmly attached to the abdominal wall. At the iliac fossa, the descending colon enters an S-shaped segment, the *sigmoid colon,* at the **sigmoid flexure** (SIG-moid; *sigmoides,* the Greek letter *S*).

THE SIGMOID COLON *(Figure 25-4)*.   The sigmoid flexure begins the **sigmoid colon**, an S-shaped segment that is only about 15 cm (6 in.) long. It lies posterior to the urinary bladder, suspended from the *sigmoid mesocolon* (Figure 25-4●, p. 632). The sigmoid colon empties into the *rectum*.

The Rectum
(Figures 25-13/25-17/25-18)

The **rectum** (REK-tum) forms the last 15 cm (6 in.) of the digestive tract (Figures 25-13, p. 644, 25-17, and 25-18●). The rectum is an expandable organ for the temporary storage of fecal material. Movement of fecal materials into the rectum triggers the urge to defecate.

The last portion of the rectum, the **anorectal** (ā-nō-REK-tal) **canal**, contains small longitudinal folds, the **rectal columns**. The distal margins of the rectal columns are joined by transverse folds that mark the boundary between the columnar epithelium of the proximal rectum and a stratified squamous epithelium similar to that found in the oral cavity. Very close to the **anus,** the epidermis becomes keratinized and identical to the surface of the skin.

Veins in the lamina propria and submucosa of the anorectal canal occasionally become distended, producing *hemorrhoids*. The circular muscle layer of the muscularis externa in this region forms the **internal anal sphincter**. The smooth muscle fibers of the internal anal sphincter are not under voluntary control. The **external anal sphincter** guards the exit of the anorectal canal, termed **anal orifice**. This sphincter, which consists of a ring of skeletal muscle fibers, is under voluntary control.

Histology of the Large Intestine
(Figure 25-19)

Although the diameter of the colon is roughly three times that of the small intestine, the wall is much thinner. The major characteristics of the colon are the lack of villi, the abundance of goblet cells, and the presence of distinctive intestinal glands (Figure 25-19●). The crypts of the large intestine are deeper than those of the small intestine, and they are dominated by goblet cells. These mucus-secreting pockets are known as the **intestinal glands**, or *crypts of Lieberkühn*. Secretion of the intestinal glands occurs as local stimuli trigger reflexes involving the local nerve plexuses, resulting in the production of copious amounts of mucus. Large lymphoid nodules (not seen in Figure 25-19) are scattered throughout the lamina propria and extend into the submucosa.

The muscularis externa differs from that of other intestinal regions because the longitudinal layer has been reduced to the muscular bands of the taeniae coli. However, the mixing and propulsive contractions of the colon resemble those of the small intestine. ✝*Diverticulitis and Colitis [p. 774]*

Regulation of the Large Intestine

Movement from the cecum to the transverse colon occurs very slowly, allowing hours for the fecal material to be converted into a sludgy paste. Movement from the transverse colon through the rest of the large intestine results from powerful peristaltic contractions, called **mass movements**, that occur a few times each day. The stimulus is distension of the stomach and duodenum, and the commands are relayed over the intestinal nerve plexuses. The contractions force fecal materials into the rectum and produce the conscious urge to defecate.

The rectal chamber is usually empty except when one of those powerful mass movements force fecal materials out of the sigmoid colon into the rectum. Distension of the rectal wall then stimulates the conscious urge to defecate. It also leads to the relaxation of the internal sphincter, and fecal material moves into the anorectal canal. When the external anal sphincter is voluntarily relaxed, defecation occurs.

ACCESSORY DIGESTIVE ORGANS

The major accessory organs are the *salivary glands*, the *liver*, the *gallbladder*, and the *pancreas*. The salivary glands were described on p. 634. The accessory digestive organs produce and store enzymes and buffers that are essential to normal digestive function. The liver and pancreas have other vital functions in addition to their roles in digestion.

The Liver
(Figure 25-13)

The **liver** is the largest visceral organ, and it is one of the most versatile organs in the body. Most of its mass lies within the right hypochondriac and epigastric regions (Figure 25-13●, p. 644). The liver weighs about 1.5 kg (3.3 lb). This large, firm, reddish-brown organ provides essential metabolic and synthetic services that fall into three basic categories: *metabolic regulation*, *hematological regulation*, and *bile production*.

1. *Metabolic Regulation:* The liver represents the central clearinghouse for metabolic regulation in the body. All blood leaving the absorptive surfaces of the digestive tract enters the hepatic portal system and flows into the liver. This arrangement gives liver cells the opportunity to ex-

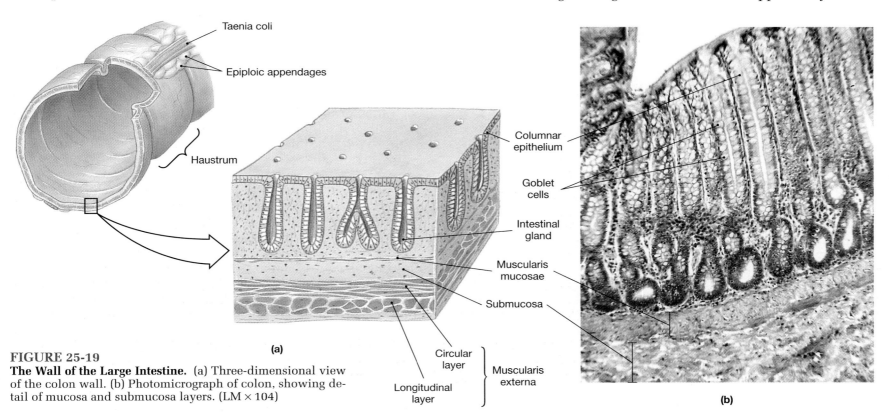

FIGURE 25-19
The Wall of the Large Intestine. (a) Three-dimensional view of the colon wall. (b) Photomicrograph of colon, showing detail of mucosa and submucosa layers. (LM × 104)

tract absorbed nutrients or toxins from the blood before it reaches the systemic circulation via the hepatic veins. Liver cells, or *hepatocytes* (he-PAT-ō-sīts), monitor the circulating levels of metabolites and adjust them as necessary. Excess nutrients are removed and stored, and deficiencies are corrected by mobilizing stored reserves or performing appropriate synthetic activities. Circulating toxins and metabolic waste products are also removed, for subsequent inactivation, storage, or excretion. Finally, fat-soluble vitamins (A, D, K, and E) are absorbed and stored in the liver.

2. *Hematological Regulation:* The liver is the largest blood reservoir in the body, and it receives about 25% of the cardiac output. As blood passes by, (1) phagocytic cells in the liver remove old or damaged RBCs, cellular debris, and pathogens from the circulation; and (2) liver cells synthesize plasma proteins that contribute to the osmotic concentration of the blood, transport nutrients, and establish the clotting and complement systems.

3. *Synthesis and Secretion of Bile:* **Bile** is synthesized by liver cells, stored in the gallbladder, and excreted into the lumen of the duodenum. Bile consists mostly of water, with minor amounts of ions, *bilirubin* (a pigment derived from hemoglobin), and an assortment of lipids collectively known as the **bile salts**. The water and ions assist in the dilution and buffering of acids in chyme as it enters the small intestine. Bile salts associate with lipids in the chyme and make it possible for enzymes to break down those lipids into fatty acids suitable for absorption.

To date, over 200 different functions have been assigned to the liver. A partial listing of these functions is presented in Table 25-2. Any condition that severely damages the liver represents a serious threat to life. The liver has a limited ability to regenerate after injury, but liver function will not fully recover unless the normal vascular pattern returns.

TABLE 25-2   Major Functions of the Liver

Digestive and Metabolic Functions
Synthesis and secretion of bile

Storage of glycogen and lipid reserves

Maintenance of normal blood glucose, amino acid, and fatty acid concentrations

Synthesis and interconversion of nutrient types (e.g., transamination of amino acids, or conversion of carbohydrates to lipids)

Synthesis and release of cholesterol bound to transport proteins

Inactivation of toxins

Storage of iron reserves

Storage of fat-soluble vitamins

Other Major Functions
Synthesis of plasma proteins

Synthesis of clotting factors

Synthesis of the inactive hormone, angiotensinogen

Phagocytosis of damaged red blood cells (by Kupffer cells)

Blood storage (major contributor to venous reserve)

Absorption and breakdown of circulating hormones (insulin, epinephrine) and immunoglobulins

Absorption and inactivation of lipid-soluble drugs

Anatomy of the Liver
(Figure 25-20)

The liver is wrapped in a tough fibrous capsule and covered by a layer of visceral peritoneum. On the anterior surface, a ventral mesentery, the **falciform** (FAL-si-form) **ligament**, marks the division between the **left lobe** and **right lobe** of the liver (Figure 25-20a-c●). A thickening in the inferior margin of the falciform ligament is the **round ligament,** or *ligamentum teres*, a fibrous band that marks the path of the degenerated fetal umbilical vein. The liver is suspended from the inferior surface of the diaphragm by the **coronary ligament**.

The shape of the liver conforms to its surroundings. The **anterior surface**, or *parietal surface*, follows the smooth curve of the body wall (Figure 25-20c●). The **posterior surface**, or *visceral surface*, bears the impressions of the stomach, small intestine, right kidney, and large intestine (Figure 25-20d●). The impression left by the inferior vena cava marks the division between the right lobe and the small **caudate** (KAW-dāt) **lobe**. Below the caudate lies the **quadrate lobe** sandwiched between the left lobe and the gallbladder. Afferent blood vessels and other structures reach the liver by traveling within the connective tissue of the lesser omentum. They converge at the **hilus** of the liver, a region known as the **porta hepatis** ("doorway to the liver").

The *gallbladder* is a muscular sac that stores and concentrates bile before its excretion into the small intestine. The gallbladder is located in a recess, or fossa, in the visceral surface of the right lobe. The gallbladder and associated structures are described in a later section.

THE BLOOD SUPPLY TO THE LIVER *(Figures 25-17a/25-20d)*.   The circulation to the liver was detailed in Chapter 22 and summarized in Figures 22-17 and 22-26●. ∞ [pp. 561, 572] Two blood vessels deliver blood to the liver, the **hepatic artery** and the **hepatic portal vein** (Figures 25-17a, p. 649, and 25-20d●). Roughly one-third of the normal hepatic blood flow arrives via the hepatic artery, and the rest is provided by the hepatic portal vein. Blood returns to the systemic circuit via the **hepatic veins** that open into the inferior vena cava.

Histological Organization of the Liver
(Figure 25-21)

Each lobe of the liver is divided by connective tissue into approximately 100,000 **liver lobules**, the basic functional units of the liver. The histological organization and structure of a typical liver lobule are shown in Figure 25-21●.

THE LIVER LOBULE *(Figure 25-21)*.   The liver cells, or **hepatocytes**, in a liver lobule form a series of irregular plates arranged like the spokes of a wheel (Figure 25-21a,c●). The plates are only one cell thick, and exposed hepatocyte surfaces are covered with short microvilli. *Sinusoids* between adjacent plates empty into the **central vein** (Figure 25-21b●). The walls of the sinusoids contain large openings that allow substances to pass out of the circulation and into the spaces surrounding the hepatocytes. In addition to typical endothelial cells, the sinusoidal lining includes a large number of **Kupffer** (KOOP-fer) **cells**, also known as *stellate reticuloendothelial cells*. These phagocytic cells are part of the monocyte-macrophage system, and they engulf pathogens, cell debris, and damaged blood cells. Kupffer cells are also responsible for engulfing and retaining any heavy metals, such as tin or mercury, that are absorbed by the digestive tract.

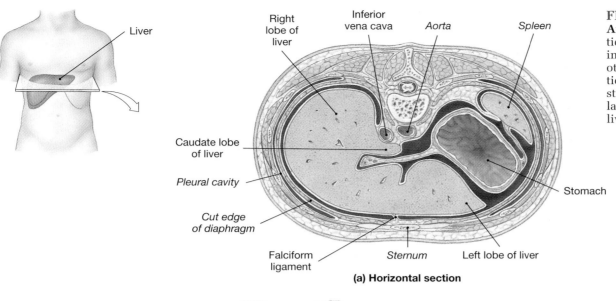

Liver

Right lobe of liver | Inferior vena cava | *Aorta* | *Spleen*

Caudate lobe of liver

Pleural cavity

Cut edge of diaphragm

Falciform ligament | *Sternum* | Left lobe of liver

Stomach

(a) Horizontal section

FIGURE 25-20

Anatomy of the Liver. (a) Horizontal section through the upper abdomen, showing the position of the liver relative to other visceral organs. (b) Horizontal section through the upper abdomen showing structures illustrated in (a). (c) Anatomical landmarks on the anterior surface of the liver. (d) The posterior surface of the liver.

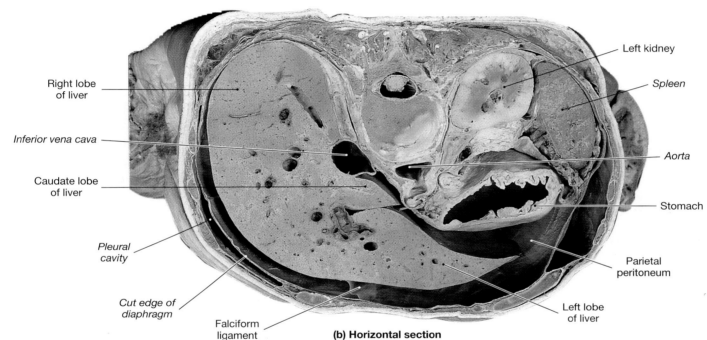

Right lobe of liver

Inferior vena cava

Caudate lobe of liver

Pleural cavity

Cut edge of diaphragm

Falciform ligament

Left kidney

Spleen

Aorta

Stomach

Parietal peritoneum

Left lobe of liver

(b) Horizontal section

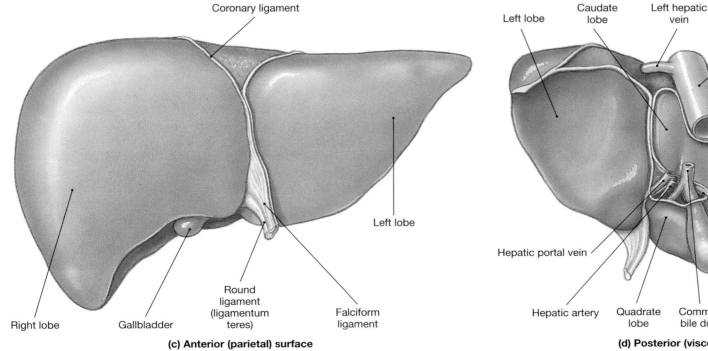

Coronary ligament

Right lobe | Gallbladder | Round ligament (ligamentum teres) | Falciform ligament

Left lobe

(c) Anterior (parietal) surface

Left lobe | Caudate lobe | Left hepatic vein | Inferior vena cava | Right lobe

Hepatic portal vein

Hepatic artery | Quadrate lobe | Common bile duct | Gallbladder | Porta hepatis

(d) Posterior (visceral) surface

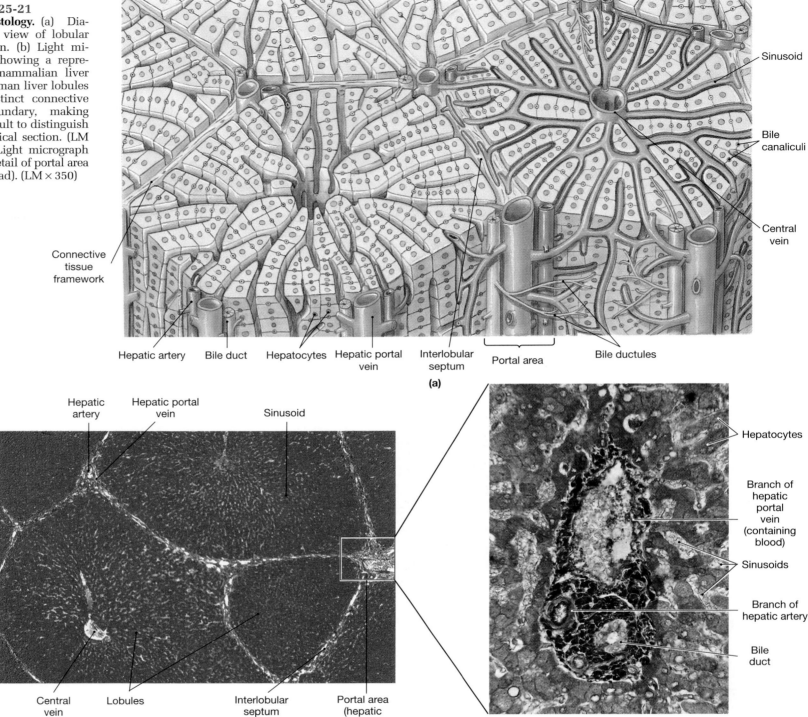

FIGURE 25-21
Liver Histology. (a) Diagrammatic view of lobular organization. (b) Light micrograph showing a representative mammalian liver lobule. Human liver lobules lack a distinct connective tissue boundary, making them difficult to distinguish in histological section. (LM × 47) (c) Light micrograph showing detail of portal area (hepatic triad). (LM × 350)

Connective tissue framework

Sinusoid

Bile canaliculi

Central vein

Hepatic artery   Bile duct   Hepatocytes   Hepatic portal vein   Interlobular septum   Portal area   Bile ductules

(a)

Hepatic artery   Hepatic portal vein   Sinusoid

Hepatocytes

Branch of hepatic portal vein (containing blood)

Sinusoids

Branch of hepatic artery

Bile duct

Central vein   Lobules   Interlobular septum   Portal area (hepatic triad)

(b)

(c)

Blood enters the liver sinusoids from small branches of the portal vein and hepatic artery. A typical lobule is hexagonal in cross section (Figure 25-21a,b●). There are six **portal areas**, or *hepatic triads*, one at each of the six corners of the lobule. A portal area (Figure 25-21c●) contains three structures: (1) a branch of the hepatic portal vein, (2) a branch of the hepatic artery, and (3) a small branch of the bile duct.

Branches from the arteries and veins deliver blood to the sinusoids of adjacent lobules (Figure 25-21a●). As blood flows through the sinusoids, hepatocytes absorb and secrete materials into the bloodstream across their exposed surfaces. Blood then leaves the sinusoids and enters the central vein of the lobule. The central veins ultimately merge to form the hepatic veins that empty into the inferior vena cava. ♈*Cirrhosis [p. 774]*

BILE SECRETION AND TRANSPORT *(Figure 25-22).* Bile is secreted into a network of narrow channels between the opposing membranes of adjacent liver cells. These passageways are called **bile canaliculi**. The bile canaliculi extend outward, away from the central vein. These eventually connect with fine **bile ductules** (DUK-tūlz) that carry bile to a bile duct in the nearest portal area. The **right** and **left hepatic ducts** collect bile from all of the bile ducts of the liver lobes. These ducts unite to form the **common hepatic duct** that leaves the liver. The bile within the common hepatic duct may either (1) flow into the *common bile duct* that empties into the duodenum or (2) enter the *cystic duct* that leads to the gallbladder. These structures are illustrated and shown in a radio-graph in Figure 25-22●.

The Gallbladder
(Figures 25-16a/25-20d/25-22)

The **gallbladder** is a hollow, pear-shaped muscular organ. It is divided into three regions: the **fundus**, the **body**, and the **neck** (Figure 25-22●). The **cystic duct** leads from the gallbladder toward the hilus of the liver. At the porta hepatis, the common hepatic duct and the cystic duct unite to create the **common bile duct** (Figure 25-20d●). At the duodenum, a muscular **pancreaticohepatic sphincter**, or *sphincter of Oddi*, surrounds the lumen of the common bile duct and the duodenal ampulla (Figures 25-16, p. 648, and 25-22b●). Contraction of this sphincter seals off the passageway and prevents bile from entering the small intestine.

The gallbladder has two major functions, *bile storage* and *bile modification*. When the pancreaticohepatic sphincter is closed, bile enters the cystic duct. Over the interim, when bile cannot flow along the common bile duct, it enters the cystic duct for storage within the expandable gallbladder. When filled to capacity, the gallbladder contains 40–70 mℓ of bile. As bile remains in the gallbladder, its composition gradually changes. Water is absorbed, and the bile salts and other components of bile become increasingly concentrated.

Bile ejection occurs under stimulation of the hormone *cholecystokinin*, or *CCK*. Cholecystokinin is released into the bloodstream at the duodenum, when chyme arrives containing large amounts of lipids and partially digested proteins. CCK causes relaxation of the pancreaticohepatic sphincter and contraction of the gallbladder.

■ CLINICAL BRIEF
Problems with Bile Storage and Secretion

If bile becomes too concentrated, crystals of insoluble minerals and salts begin to appear. These deposits are called **gallstones**. Merely having them, a condition termed **cholelithiasis** (kō-lē-li-THĪ-a-sis; *chole*, bile), does not represent a problem as long as the stones remain small. Small stones are normally flushed down the bile duct and excreted.

If gallstones enter and jam in the cystic or bile duct, the painful symptoms of **cholecystitis** (kō-lē-sis-TĪ-tis) appear. The gallbladder becomes swollen and inflamed, infections may develop, and if the blockage does not work its way down the common bile duct to the duodenum, it must be removed or destroyed. Small gallstones can be chemically dissolved. One chemical now under clinical review is *methyl tert-butyl ether* (MTBE). When introduced into the gallbladder, it dissolves gallstones in a matter of hours. Surgery is usually required to remove large gallstones, and the gallbladder is often removed at the same time, to prevent recurrence. The procedure can be performed with a laparoscope inserted through a very small incision.

A recent therapy for cholecystitis involves immersing the individual in water and shattering the stones with focused sound waves. The apparatus used is called a *lithotripter*. The particles produced are then small enough to pass through the bile duct without difficulty.

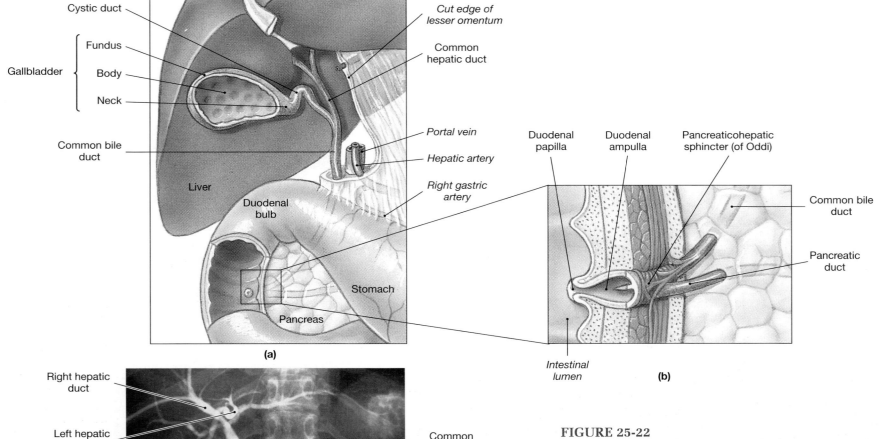

(a)

(c)

FIGURE 25-22
The Gallbladder and Associated Bile Ducts. (a) A view of the inferior surface of the liver, showing the position of the gallbladder and ducts that transport bile from the liver to the gallbladder and duodenum. (b) A portion of the lesser omentum has been cut away to make it easier to see the relationship between the common bile duct, the hepatic duct, and the cystic duct. (c) Radiograph (cholangiopancreatogram) of the gallbladder, biliary ducts, and pancreatic ducts, anterior-posterior view.

The Pancreas
(Figures 25-22a/25-23)

The **pancreas** lies behind the stomach, extending laterally from the duodenum toward the spleen (Figures 25-22a and 25-23●). The pancreas is an elongate, pinkish-gray organ, approximately 15 cm (6 in.) long and around 80 g (3 oz). The broad **head** of the pancreas lies within the loop formed by the duodenum as it leaves the pylorus. The slender **body** extends transversely toward the spleen, and the **tail** is short and bluntly rounded. The pancreas is retroperitoneal, and it is firmly bound to the posterior wall of the abdominal cavity.

The surface of the pancreas has a lumpy, nodular texture. A thin, transparent connective tissue capsule wraps the pancreas.

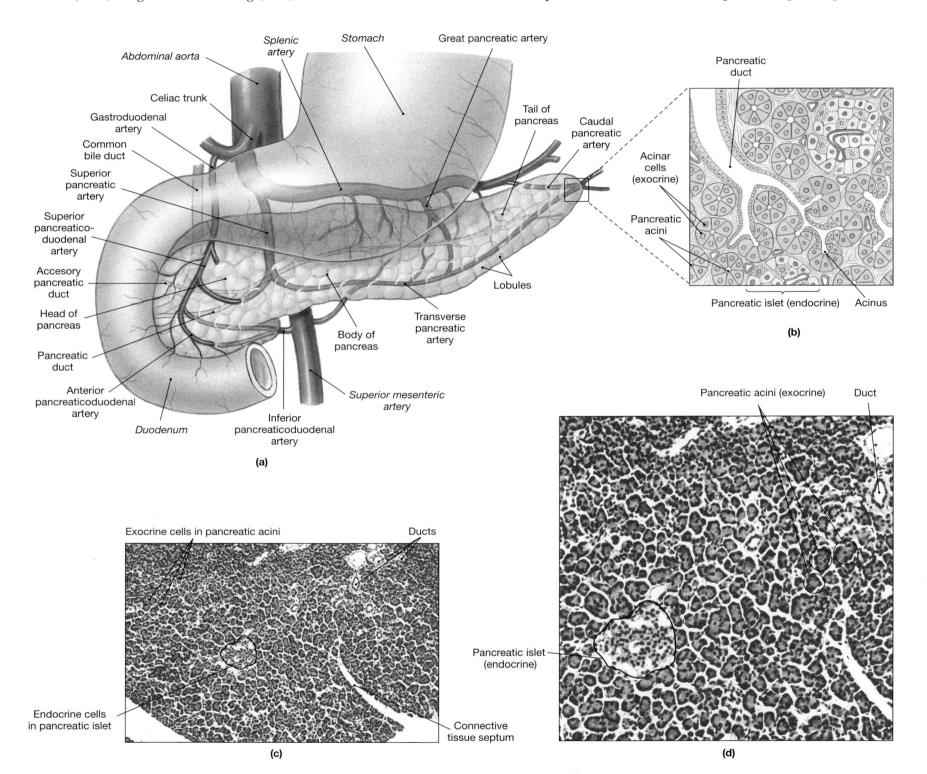

FIGURE 25-23
The Pancreas. (a) Gross anatomy of the pancreas. The head of the pancreas is tucked into a curve of the duodenum that begins at the pylorus of the stomach. (b) Diagrammatic view of the cellular organization of the pancreas, showing exocrine and endocrine regions. (c) Photomicrograph of the pancreas. This image shows the connective tissue septae that separate pancreatic lobules, a branch of the pancreatic duct, and exocrine and endocrine tissues. (LM × 70) (d) A view of pancreatic acini at higher magnification. (LM × 120)

The pancreatic lobules, associated blood vessels, and excretory ducts can be seen through the anterior capsule and the overlying layer of peritoneum.

The pancreas is primarily an exocrine organ producing digestive enzymes and buffers. The large **pancreatic duct** (*duct of Wirsung*) delivers these secretions to the duodenal ampulla. A small **accessory duct**, or *duct of Santorini*, may drain into the pancreatic duct before it leaves the pancreas (Figure 25-23a●).

Arterial blood reaches the pancreas via branches of the *splenic, superior mesenteric,* and *common hepatic arteries* (Figures 22-15 and 22-17●). ∞ [pp. 559, 561] The **pancreatic arteries** and **superior** and **inferior pancreaticoduodenal arteries** are the major branches from these vessels (Figure 19-11a●). ∞ [p. 498] The *splenic vein* and its branches drain the pancreas (Figures 22-15 and 22-26●). ∞ [pp. 559, 572]

Histological Organization of the Pancreas
(Figure 25-23b,c)

Partitions of connective tissue divide the pancreatic tissue into distinct lobules (Figure 25-23b,c●). The blood vessels and tributaries of the pancreatic ducts are found within these connective tissue septa. The pancreas is an example of a *compound tubuloacinar gland.* ∞ [p. 62] Within each lobule, the ducts branch repeatedly before ending in blind pockets, the **pancreatic acini** (AS-i-nī). Each pancreatic acinus is lined by a simple cuboidal epithelium. *Pancreatic islets* are scattered between the acini, but they account for only around 1% of the cellular population of the pancreas. ∞ [p. 495]

The pancreatic acini secrete a mixture of water, ions, and pancreatic digestive enzymes into the duodenum. Pancreatic enzymes do most of the digestive work in the small intestine, breaking down ingested materials into small molecules suitable for absorption. The pancreatic ducts secrete buffers (primarily sodium bicarbonate) in a watery solution. These secretions are important in neutralizing the acid in chyme and stabilizing the pH of the intestinal contents. The **pancreatic juice** produced by the pancreas contains the secretions of acinar cells and the epithelial cells of the pancreatic ducts.

Pancreatic Enzymes

Pancreatic enzymes are classified according to their intended targets. **Lipases** (LĪ-pā-zez) attack lipids, **carbohydrases** (kar-bō-HĪ-drā-zez) digest sugars and starches, **nucleases** attack nucleic acids, and **proteolytic** (prō-tē-ō-LIT-ik) **enzymes** break proteins apart. Proteolytic enzymes include **proteinases** and **peptidases**; proteinases break apart large protein complexes, whereas peptidases break small peptide chains into individual amino acids.

The Regulation of Pancreatic Secretion

Secretion of pancreatic juice occurs primarily in response to hormonal instructions from the duodenum. When acid chyme arrives in the small intestine, secretin is released. This hormone triggers the production of watery pancreatic juice containing buffers, especially sodium bicarbonate. Another duodenal hormone, cholecystokinin, stimulates the production and secretion of pancreatic enzymes.

√ **What structures are involved when the gallbladder is removed?**

√ **In cystic fibrosis, the pancreatic duct can become blocked with thickened secretions. This condition would interfere with the proper digestion of which group(s) of nutrients?**

AGING AND THE DIGESTIVE SYSTEM

Essentially normal digestion and absorption occur in elderly individuals. However, there are many changes in the digestive system that parallel age-related changes already described for other systems.

1. *The rate of epithelial stem cell division declines.* The digestive epithelium becomes more susceptible to damage by abrasion, acids, or enzymes. Peptic ulcers therefore become more likely. In the mouth, esophagus, and anus the stratified epithelium becomes thinner and more fragile.

2. *Smooth muscle tone decreases.* General motility decreases, and peristaltic contractions are weaker. This change slows the rate of chyme movement and promotes constipation. Sagging of the walls of haustra in the colon can produce symptoms of *diverticulitis.* Straining to eliminate compacted fecal materials can stress the less resilient walls of blood vessels, producing *hemorrhoids.* Weakening of the cardiac sphincter can lead to *esophageal reflux* and frequent bouts of "heartburn."

3. *The effects of cumulative damage become apparent.* A familiar example would be the gradual loss of teeth because of tooth decay or gingivitis. Cumulative damage can involve internal organs as well. Toxins such as alcohol, heavy metals, and other injurious chemicals that are absorbed by the digestive tract are transported to the liver for processing or storage. The liver cells are not immune to these compounds, and chronic exposure can lead to *cirrhosis* or other types of liver disease.

4. *Cancer rates increase.* Cancers are most common in organs where stem cells divide to maintain epithelial cell populations. ∞ [p. 81] Rates of colon cancer and stomach cancer rise in the elderly; oral and pharyngeal cancers are particularly common in elderly smokers.

5. *Changes in other systems have direct or indirect effects on the digestive system.* For example, the reduction in bone mass and calcium content in the skeleton is associated with erosion of the tooth sockets and eventual tooth loss. The decline in olfactory and gustatory sensitivity with age can lead to dietary changes that affect the entire body.

Related Clinical Terms

peritonitis (per-i-tō-NĪ-tis): A painful condition resulting from inflammation of the peritoneal membrane. *[p. 629]*

mumps: A viral infection that most often affects the parotid salivary glands at ages 5–9. *[p. 635]*

plaque: A dense deposit of food particles and bacterial secretions on the surfaces of teeth. *[p. 637]*

periodontal disease: A progressive condition resulting from erosion of the connections between the necks of teeth and the gingiva. *[p. 637]*

achalasia (ak-a-LĀ-zē-a): Blockage of the lower part of the esophagus due to weak peristalsis and malfunction of the lower esophageal sphincter. ✝*Achalasia and Esophagitis [p. 774]*

esophagitis (ē-sof-a-JĪ-tis): Inflammation of the esophagus due to erosion by gastric juices. ✝*Achalasia and Esophagitis [p. 774]*

gastritis (gas-TRĪ-tis): Inflammation of the gastric mucosa. *[p. 642]*

peptic ulcer: Erosion of the gastric or duodenal mucosa by acids and enzymes in chyme. *[p. 642]*

gastroscope: A fiberoptic instrument used to visualize the interior of the stomach. ✝*Stomach Cancer [p. 774]*

gastrectomy: Surgical removal of the stomach; a potential treatment for stomach cancer. ✝*Stomach Cancer [p. 774]*

gastric stapling: A surgical procedure to promote weight loss by blocking off a large portion of the gastric lumen. *[p. 645]*

enteritis (en-ter-Ī-tis): Irritation of the small intestine by toxins or other irritants; causes diarrhea due to frequent peristalsis along the small intestine. ✝*Gastroenteritis [p. 774]*

gastroenteritis: Vomiting and diarrhea caused by an extremely powerful irritating stimulus. ✝*Gastroenteritis [p. 774]*

diverticulosis (dī-ver-tik-ū-LŌ-sis): A condition in which pockets (*diverticula*) form in the mucosa of the colon, usually in the sigmoid colon. ✝*Diverticulitis and Colitis [p. 774]*

colitis: Irritation of the colon, leading to abnormal bowel function. ✝*Diverticulitis and Colitis [p. 774]*

colostomy: Attachment of the colon to the abdominal wall, bypassing the distal portion of the large intestine. ✝*Diverticulitis and Colitis [p. 774]*

cirrhosis: A condition caused by destruction of hepatocytes by drug exposure, viral infection, ischemia, or other factors. ✝*Cirrhosis [p. 774]*

cholelithiasis (kō-lē-li-THĪ-a-sis): Presence of gallstones in the gallbladder. *[p. 655]*

cholecystitis (kō-lē-sis-TĪ-tis): A painful condition caused by blockage of the cystic or common bile duct by gallstones. *[p. 655]*

pancreatitis (pan-krē-a-TĪ-tis): An inflammation of the pancreas due to blockage of the pancreatic ducts, bacterial or viral infections, circulatory ischemia, or drug reactions. *[p. 657]*

CHAPTER SUMMARY AND REVIEW

STUDY OUTLINE

Related Key Terms

INTRODUCTION *[p. 626]*

1. The digestive system consists of the muscular **digestive tract** and various **accessory organs** *(see Figure 25-1).*

2. Digestive functions include: ingestion, mechanical processing, digestion, secretion, absorption, compaction, and excretion.

defecation

AN OVERVIEW OF THE DIGESTIVE SYSTEM *[p. 626]*

Histological Organization of the Digestive Tract *[p. 627]*
1. The digestive tract is lined by a **mucosal epithelium** moistened by glandular secretions of the epithelial and accessory organs *(see Figure 25-2).*

muscularis mucosae

2. The **lamina propria** and epithelium form the **mucosa** (a **mucous membrane**) of the digestive tract. Proceeding outward, the layers are the **submucosa**, the **muscularis externa**, and (in the peritoneal cavity) a serous membrane called the **serosa** *(see Figure 25-2).*

submucosal plexus • myenteric plexus

Muscularis Layers and the Movement of Digestive Materials *[p. 628]*
3. The muscularis externa propels materials through the digestive tract through the contractions of **peristalsis. Segmentation** movements in areas of the small intestine churn digestive materials *(see Figure 25-3).*

plasticity • visceral smooth muscle tissue • pacesetter cells • peristaltic wave • myenteric reflexes

The Peritoneum *[p. 629]*
4. The serosa, also known as the **visceral peritoneum**, is continuous with the **parietal peritoneum** that lines the inner surfaces of the body wall.

5. Double sheets of peritoneal membrane called **mesenteries** suspend portions of the digestive tract *(see Figure 25-2a).*

6. Important **mesenteries** include the **greater omentum**, **lesser omentum**, **mesentery proper**, **transverse mesentery**, and **sigmoid mesocolon**. The ascending colon, descending colon, duodenum, and pancreas are attached to the posterior wall of the abdominopelvic cavity; they are **retroperitoneal** *(see Figures 25-4/25-10/25-11).*

Related Key Terms

mesenterial sheet • mesocolon

THE ORAL CAVITY *[p. 633]*

1. The functions of the oral cavity include: (1) **analysis** of potential foods; (2) **mechanical processing** using the teeth, tongue, and palatal surfaces; (3) **lubrication** by mixing with mucus and salivary secretions; and (4) **digestion** by salivary enzymes. Structures of the oral cavity include the tongue, salivary glands, and teeth *(see Figure 25-5).*

buccal cavity • oral mucosa • cheeks • buccal fat pads • labia • vestibule • gingivae hard and soft palates • uvula

Anatomy of the Oral Cavity *[p. 633]*
2. **Intrinsic** and **extrinsic tongue muscles** are controlled by the hypoglossal nerve *(see Figure 25-5).*
3. Saliva lubricates the mouth, dissolves chemicals, flushes the oral surfaces, and helps control bacteria. Salivation is usually controlled by the autonomic nervous system.

body • root • dorsum • lingual frenulum • ankyloglossia

4. The **parotid**, **sublingual**, and **submandibular salivary glands** discharge their secretions into the oral cavity *(see Figures 25-5/25-6).*
5. **Dentin** forms the basic structure of a tooth. The **crown** is coated with **enamel**, and the **root** with **cementum**. The **neck** marks the boundary between the root and the crown. The **periodontal ligament** anchors the tooth in an alveolar socket. **Mastication** (chewing) occurs through the contact of the opposing **occlusal surfaces** of the teeth *(see Figure 25-7).*

parotid duct • sublingual ducts • submandibular ducts • salivary amylase pulp cavity • root canal • dental artery • dental vein • dental nerve • apical foramen • gingival sulcus

6. There are four types of teeth, each with specific functions: **incisors**, for cutting; **cuspids** (canines) for tearing; **bicuspids** (premolars), for crushing; and **molars**, for grinding *(see Figure 25-7b,c).*
7. The sequence of tooth eruption is presented in ***Table 25-1.***

deciduous teeth • secondary dentition • eruption • dental arch

THE PHARYNX *[p. 637]*

1. Skeletal muscles involved with swallowing include the *pharyngeal constrictors* and the *palatopharyngeus, stylopharyngeus,* and *palatal muscles.*

The Swallowing Process *[p. 637]*
2. **Deglutition** (swallowing) begins with the compaction of a **bolus** and its movement into the pharynx, followed by the elevation of the larynx, reflection of the epiglottis, and closure of the glottis. After the opening of the **upper esophageal sphincter**, peristalsis moves the bolus down the esophagus to the **lower esophageal sphincter**. Swallowing, once initiated, proceeds automatically through the **buccal**, **pharyngeal**, and **esophageal phases** *(see Figure 25-8).*

THE ESOPHAGUS *[p. 638]*

1. The **esophagus** is a hollow muscular tube that transports food and liquid to the stomach.

esophageal hiatus

Histology of the Esophageal Wall *[p. 639]*
2. The wall of the esophagus is formed by *mucosa, submucosa, muscularis,* and *adventitia* layers *(see Figures 25-1/25-2/25-9).*

THE STOMACH *[p. 639]*

1. The stomach has three major functions: (1) bulk storage of ingested matter; (2) mechanical breakdown of resistant materials, and (3) disruption of chemical bonds using acids and enzymes.

chyme

Anatomy of the Stomach *[p. 639]*
2. The stomach wall is formed by mucosa, submucosa, muscularis, and serosa layers. The mucosa and submucosa are thrown into longitudinal folds, called **rugae**. The **pyloric sphincter** guards the exit from the stomach. The muscularis layer is formed of three bands of smooth muscle: a *longitudinal layer*, a *circular layer*, and an inner *oblique layer (see Figures 25-4/25-10 to 25-13).*

lesser curvature • greater curvature • medial/lateral/anterior/posterior surfaces • cardia • cardiac orifice • fundus • body • pylorus • pyloric outlet • greater omentum

3. Three branches of the celiac artery supply blood to the stomach: the *left gastric artery, splenic artery,* and the *common hepatic artery (see Figures 22-17/22-26/25-10a).*

Histology of the Stomach *[p. 642]*
4. Simpler columnar epithelium lines all portions off the stomach. Within the **gastric glands** of the fundus and body, **parietal cells** secrete **intrinsic factor** and hydrochloric acid. **Chief cells** secrete **pepsinogen**, which acids in the gastric lumen convert to the enzyme **pepsin**. **Enteroendocrine cells** of the stomach secrete several compounds, notably the hormone **gastrin** *(see Figure 25-12).*

gastric pits • rennin • gastric lipase

Regulation of the Stomach *[p. 642]*
5. The production and secretion of **gastric juices** are directly controlled by the CNS (the vagus nerve, parasympathetic innervation, and the celiac plexus, sympathetic innervation). The release of the local hormones **secretin** and **cholecystokinin** inhibits gastric secretion, but stimulates secretion by the pancreas and liver.

THE SMALL INTESTINE *[p. 645]*

Related Key Terms

Regions of the Small Intestine *[p. 645]*
1. The small intestine includes the **duodenum,** the **jejunum,** and the **ileum.** The intestinal mucosa bears transverse folds, called **plicae circulares,** and small projections, called **intestinal villi,** that increase the surface area for absorption. Each villus contains a terminal lymphatic called a **lacteal.** Pockets called **intestinal crypts** house enteroendocrine, goblet, and stem cells *(see Figures 25-4/25-10/25-13 to 25-16).*

Support of the Small Intestine *[p. 646]*
2. The *superior mesenteric artery* and *vein* supply numerous branches to the segments of the small intestine *(see Figures 22-17/22-26).*

3. The **mesentery proper** supports the branches of the superior mesenteric artery and vein, lymphatics, and nerves that supply the jejunum and ileum portions of the small intestine *(see Figure 25-4).*

Histology of the Small Intestine *[p. 646]*
4. The regions of the small intestine have histological specializations that determine their primary functions. The duodenum (1) contains **submucosal glands** that aid the crypts in producing mucus and (2) receives the secretions of the common bile duct and pancreatic duct. The jejunum and ileum contain large groups of **aggregate lymphoid nodules** (*Peyer's patches*) within the lamina propria *(see Figures 25-15/ 25-16).*

duodenal ampulla • duodenal papilla

Regulation of the Small Intestine *[p. 646]*
5. **Intestinal juice** moistens the chyme, helps to buffer acids, and dissolves digestive enzymes and the products of digestion.

6. Secretin and cholecystokinin are two hormones important in the coordination of digestive activities. Parasympathetic (vagal) innervation stimulates digestive function; sympathetic stimulation inhibits activity along the digestive tract.

THE LARGE INTESTINE *[p. 648]*

1. The large intestine, or **large bowel,** begins as a pouch inferior to the terminal portion of the ileum and ends at the anus. The main functions of the large intestine are to: (1) reabsorb water and compact the feces; (2) absorb vitamins liberated by bacteria; and (3) store fecal material before defecation *(see Figures 25-4/25-13/25-14/25-17 to 25-19).*

2. The large intestine is divided into three parts: the **cecum,** the **colon,** and the **rectum.**

The Cecum *[p. 648]*
3. The cecum collects and stores chyme. The **ileocecal valve** opens into the cecum *(see Figure 25-17).*

vermiform appendix • mesoappendix

The Colon *[p. 650]*
4. The colon has a larger diameter and a thinner wall than the small intestine. It bears **haustra** (pouches), the **taeniae coli** (longitudinal bands of muscle), and **epiploic appendages** (sacs of fat within the serosa) *(see Figure 25-17/25-18).*

5. The colon is subdivided into four regions: **ascending, transverse, descending,** and **sigmoid** *(see Figures 25-4/25-17/25-18).*

right/left colic flexure • gastrocolic ligament • sigmoid flexure

The Rectum *[p. 650]*
6. The rectum terminates in the **anorectal canal** leading to the **anus.** Muscular sphincters control the passage of fecal material to the anus. Distension of the rectal wall triggers the defecation reflex *(see Figures 25-13/25-17/25-18).*

rectal columns • internal anal sphincter • external anal sphincter • anal orifice

Histology of the Large Intestine *[p. 651]*
7. The major histological features of the colon are: lack of villi, abundance of goblet cells, distinctive intestinal glands, and deep mucus-secreting pockets, called the **intestinal glands,** or *crypts of Lieberkühn (see Figure 25-19).*

Regulation of the Large Intestine *[p. 651]*
8. Movement from the cecum to the transverse colon occurs slowly via peristalsis. Movement from the transverse to the sigmoid colon occurs several times each day via **mass movements.**

9. Distention of the rectal wall from a mass movement stimulates the conscious desire to relax internal and external anal sphincters to defecate.

ACCESSORY DIGESTIVE ORGANS *[p. 651]*

The Liver *[p. 651]*
1. The **liver** performs metabolic and hematological regulation, and produces **bile.** Liver cells are specialized epithelial cells, termed **hepatocytes** *(see Figures 25-13/25-20/25-21 and Table 25-2).*

bile salts

2. The liver is divided into four lobes: **left, right, quadrate,** and **caudate.** The gallbladder is located in a fossa within the **posterior surface** of the right lobe *(see Figure 25-20).*

falciform ligament • round ligament • coronary ligament • anterior surface • hilus • porta hepatis

3. The **liver lobule** is the basic functional unit of the liver. Each lobule is hexagonal in cross section and contains six **portal areas,** or *hepatic triads.* A portal area consists of a branch of the hepatic portal vein, a branch of the hepatic artery, and a branch of the hepatic (bile) duct. **Bile canaliculi** carry bile to **bile ductules** that lead to portal areas. The bile ducts from each lobule unite to form the **left** and **right hepatic ducts,** which merge to form the **common hepatic duct** *(see Figures 25-20 to 25-22).*

central vein • Kupffer cells

4. The **hepatic artery** and **hepatic portal vein** supply blood to the liver. **Hepatic veins** drain blood from the liver and return it to the systemic circuit via the inferior vena cava *(see Figures 22-17/*

The Gallbladder [p. 655]

5. The **gallbladder** is a hollow muscular organ that stores and concentrates bile. Bile salts break apart large drops of lipids and make them accessible to digestive enzymes.

6. The gallbladder is divided into: **fundus**, **body**, and **neck** regions. The **cystic duct** leads from the gallbladder to merge with the common hepatic duct to form the **common bile duct** *(see Figures 25-16a/ 25-20d/25-22).*

pancreaticohepatic sphincter

The Pancreas [p. 656]

7. The **pancreas** is divided into **head**, **body**, and **tail** regions. The **pancreatic duct** penetrates the wall of the duodenum. Within each lobule, ducts branch repeatedly before ending in the **pancreatic acini** (blind pockets). The **accessory duct** and pancreatic duct perforate the wall of the duodenum to discharge **pancreatic juice** *(see Figures 25-22/25-23).*

proteinases • peptidases

8. Pancreatic tissue includes endocrine and exocrine components. The bulk of the organ (99%) is exocrine in function, secreting water, ions, and digestive enzymes into the small intestine. Pancreatic enzymes include **lipases**, **carbohydrases**, **nucleases**, and **proteolytic enzymes**.

pancreatic arteries • posterior superior pancreaticoduodenal artery

9. Regulation of the production of pancreatic juice is via the hormones cholecystokinin and secretin.

AGING AND THE DIGESTIVE SYSTEM [p. 657]

1. Normal digestion and absorption occur in elderly individuals; however, changes in the digestive system reflect age-related changes in other body systems.

1 REVIEW OF CHAPTER OBJECTIVES

1. Summarize the functions of the digestive system.
2. Locate the components of the digestive tract and identify the functions of each.
3. Describe the histological organization and characteristics of each segment of the digestive tract in relation to its function.
4. Explain how ingested food is propelled via the actions of smooth muscle through the digestive tract.
5. Describe the anatomical structure of the tongue, teeth, salivary glands, and pharynx.
6. Describe the events of the swallowing process.
7. Describe the gross and microscopic anatomy of the esophagus, stomach, small intestine, and large intestine.
8. Identify the structures and hormones involved in the regulation of the stomach and the small and large intestines.
9. Describe the gross and microscopic anatomy of the accessory organs and the roles they play in digestion.
10. Identify the structures and hormones involved in the regulation of the liver, gallbladder, and pancreas.
11. Describe the changes in the digestive system that occur with aging.

2 REVIEW OF CONCEPTS

1. There are two basic types of mucosal epithelium (simple and stratified) in the digestive tract in different parts of the system. Why do the locations of the two types differ?
2. What is the function of the muscularis mucosae?
3. What is the significance of the muscularis externa?
4. What unique properties of smooth muscle fibers are vital to the role of smooth muscle in the digestive system?
5. What is the importance of the process of torsion, which the digestive tract undergoes during development?
6. How does a retroperitoneal structure differ from a peritoneal structure?
7. The process of swallowing is governed by both voluntary and autonomic actions. What is the significance of this dual control, and where does the change occur?

8. How does the structure of the esophagus differ from that of the trachea? How are these differences related to the primary functions of each passageway?
9. What is the greater omentum, and what are its functions?
10. Why do gastric and duodenal ulcers form, and are there different reasons for the two types?
11. What are the roles of the duodenum and jejunum in the digestive process?
12. How does the function of the large intestine differ from that of the small intestine?
13. How does the process of moving materials through the colon differ from that in the small intestine?
14. What is the role of the liver in digestion?
15. What is the role of the gallbladder in digestion?
16. What is the significance of the pancreas in the digestive process?

3 CRITICAL THINKING AND CLINICAL APPLICATION QUESTIONS

1. Many older people have all of their teeth removed as a result of dental decay or periodontal disease and are fitted with dentures. Initially, the false teeth work well and fit the mouth, but within years the dentures no longer fit, and the wearer complains of pain, particularly from the anterior mandibular rami. What changes have occurred that affect the fit of the dentures?

2. Some of the newest products available for improved dental hygiene are oral rinses that claim to prevent periodontal disease and dental caries. What do these rinses do, and how do they work?

3. A 45-year-old woman experiences frequent heartburn, occasional difficulty in swallowing, and sharp pains in the substernal region. Sometimes at night she experiences gastric reflux, or a regurgitation of stomach acid into the esophagus, a condition which is extremely painful. What could produce these symptoms, and how can they be treated?

26

The Urinary System

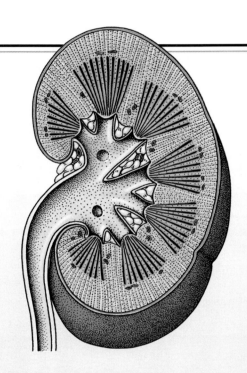

CHAPTER OUTLINE AND OBJECTIVES

The urinary system performs vital excretory functions and eliminates the organic waste products generated by cells throughout the body. But it also has a number of other essential functions that are often overlooked. A more complete list of urinary system functions includes:

1. Regulating plasma concentrations of sodium, potassium, chloride, calcium, and other ions by controlling the quantities lost in the urine;

2. Regulating blood volume and blood pressure by (a) adjusting the volume of water lost in the urine, (b) releasing erythropoietin, and (c) releasing renin; ∞ [p. 495]

3. Contributing to the stabilization of blood pH;

4. Conserving valuable nutrients by preventing their excretion in the urine;

5. Eliminating organic waste products, especially nitrogenous wastes such as *urea* and *uric acid*, toxic substances, and drugs; and

6. Synthesizing *calcitriol*, a hormone derivative of vitamin D_3 that stimulates calcium ion absorption by the intestinal epithelium.

These activities are carefully regulated to keep the composition of the blood within acceptable limits. A disruption of any one of these functions will have immediate and potentially fatal consequences.

This chapter considers the functional organization of the urinary system and describes the major regulatory mechanisms that control urine production and concentration.

ORGANIZATION OF THE URINARY SYSTEM
(Figure 26-1)

The urinary system (Figure 26-1●) includes the *kidneys*, *ureters*, *urinary bladder*, and *urethra*. The excretory functions of the urinary system are performed by a pair of **kidneys**. These organs produce **urine**, a fluid containing water, ions, and small soluble compounds. Urine leaving the kidneys travels along the paired **ureters** (yū-RĒ-terz) to the **urinary bladder** for temporary storage. When **urination**, or **micturition** (mik-tū-RISH-un) occurs, contraction of the muscular urinary bladder forces urine through the **urethra** (yū-RĒTH-ra) and out of the body.

THE KIDNEYS
(Figures 26-1 to 26-3)

The kidneys are located lateral to the vertebral column between the last thoracic and third lumbar vertebrae on each side (Figure 26-2a●). The superior surface of the right kidney is often situated inferior to the superior surface of the left kidney (Figures 26-1 and 26-2a●).

On gross dissection, the anterior surface of the right kidney is covered by the liver, the hepatic flexure of the colon, and the duodenum. The anterior surface of the left kidney is covered by the spleen, stomach, pancreas, jejunum, and splenic flexure of the colon. The superior surface of each kidney is capped by an adrenal gland (Figures 26-2a and 26-3●, pp. 666, 667). The kidneys, adrenal glands, and ureters lie between the muscles of the dorsal body wall and the parietal peritoneum in a retroperitoneal position (Figures 26-2b and 26-3●).

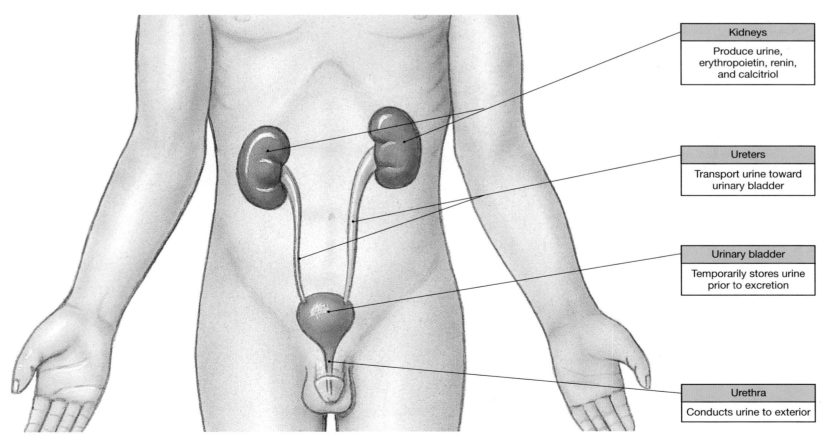

Kidneys
Produce urine, erythropoietin, renin, and calcitriol

Ureters
Transport urine toward urinary bladder

Urinary bladder
Temporarily stores urine prior to excretion

Urethra
Conducts urine to exterior

FIGURE 26-1
Components of the Urinary System

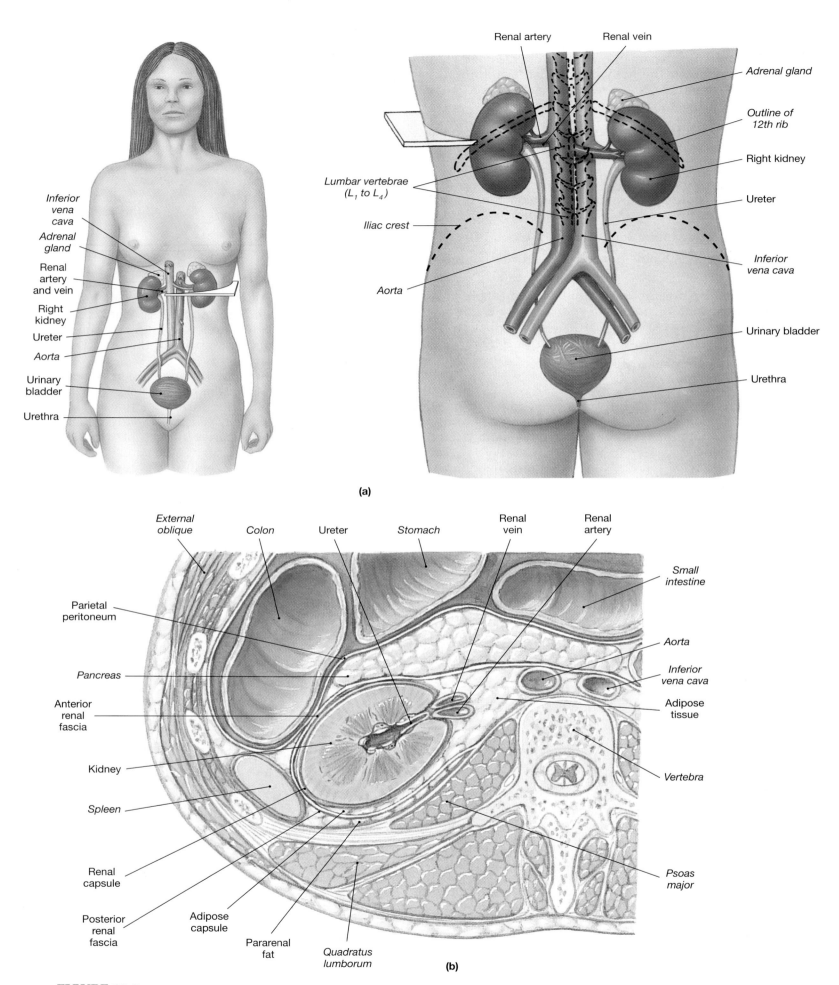

FIGURE 26-2

An Introduction to Renal Anatomy. (a) Anterior and posterior views of the trunk, showing the positions of the kidneys and other components of the urinary system. (b) Sectional view at the level indicated in (a).

The position of the kidneys in the abdominal cavity is maintained by: (1) the overlying peritoneum, (2) contact with adjacent visceral organs, and (3) supporting connective tissues. Each kidney is protected and stabilized by three concentric layers of connective tissue (Figure 26-2b●):

1. A layer of collagen fibers covers the outer surface of the entire organ. This layer, the **renal capsule**, is also known as the *fibrous tunic* of the kidney.

2. A layer of adipose tissue, the **adipose capsule**, or *perirenal fat*, surrounds the renal capsule. This layer can be quite thick, and on dissection the adipose capsule usually obscures the outline of the kidney.

3. Collagen fibers extend outward from the renal capsule through the adipose capsule to a dense outer layer known as the **renal fascia**. The renal fascia anchors the kidney to surrounding structures. Posteriorly the renal fascia is bound to the deep fascia surrounding the muscles of the body wall. A layer of *pararenal fat* separates the posterior and lateral portions of the renal fascia from the body wall. Anteriorly the renal fascia ia attached to the peritoneum and to the anterior renal fascia of the opposite side.

In effect, the kidney hangs suspended by collagen fibers from the renal fascia and is packed in a soft cushion of adipose tissue. This arrangement prevents the jolts and shocks of day-to-day existence from disturbing normal kidney function. If the suspensory fibers break or become detached, a slight bump or jar may displace the kidney and stress the attached vessels and ureter. This condition, called a **floating kidney**, can be especially dangerous because the ureters or renal blood vessels may become twisted or kinked during movement.

Superficial Anatomy of the Kidney
(Figures 26-3/26-4)

Each brownish-red kidney has the shape of a kidney bean. A typical adult kidney (Figures 26-3 and 26-4●, p. 668) measures approximately 10 cm (4 in.) in length, 5.5 cm (2.2 in.) in width, and 3 cm (1.2 in.) in thickness. A single kidney averages around 150 g (5.25 oz). A prominent medial indentation, the **hilus**, is the point of entry for the *renal artery* and the exit for the *renal vein* and *ureter*.

The fibrous renal capsule has inner and outer layers. In sectional view (Figure 26-4a,c●, p. 668), the inner layer folds inward at the hilus and lines the **renal sinus**. Renal blood vessels and the ureter draining the kidney pass through the hilus and branch within the renal sinus. The thick, outer layer of the capsule extends across the hilus and stabilizes the position of these structures.

Sectional Anatomy of the Kidney
(Figure 26-4)

The renal **cortex** is the outer layer of the kidney in contact with the capsule (Figure 26-4a,c●, p. 668). The cortex is reddish-brown and granular. The **medulla** consists of 6 to 18 distinct conical or triangular structures, called **renal pyramids**. The base of each pyramid faces the cortex, and the tip, or **renal papilla**, projects into the renal sinus. Each pyramid has a series of fine grooves that converge at the papilla. Adjacent renal pyramids are separated by bands of cortical tissue, called **renal columns**. The columns have a distinctly granular texture, similar to that of the cortex. A **renal lobe** contains a renal pyramid, the overlying area of renal cortex, and adjacent tissues of the renal columns.

Urine production occurs in the renal lobes, and ducts within each renal papilla discharge urine into a cup-shaped drain, called a **minor calyx** (KĀ-liks). Four or five minor calyces (KĀL-i-sēz) merge to form a **major calyx**, and the major calyces combine to form a large, funnel-shaped chamber, the **renal pelvis**. The renal pelvis, which fills most of the renal sinus, is connected to the ureter at the hilus of the kidney.

Histology of the Kidney
(Figure 26-5)

The **nephron** (NEF-ron), the basic structural and functional unit of the kidney, can be viewed only with a microscope. There are roughly 1.25 million nephrons in each kidney, with a combined length of around 145 km (85 miles). Each nephron consists of a long **renal tubule** that has straight and convoluted segments; for clarity, the nephron diagrammed in Figure 26-5●, p. 669, has been shortened and straightened out.

An Introduction to the Structure and Function of the Nephron
(Figures 26-5 to 26-7)

The renal tubule begins at an expanded chamber, called **Bowman's capsule**, containing a capillary knot, or **glomerulus** (glō-MER-ū-lus), as seen in Figures 26-5, p. 669, and 26-6●, p. 670. Together Bowman's capsule and the glomerulus form the **renal corpuscle** (KOR-pusl). A typical glomerulus consists of around 50 intertwining capillaries. Filtration across the walls of the glomerulus produces a protein-free solution known as a **filtrate**. From the renal corpuscle the filtrate enters a long tubular passageway that is subdivided into regions with varied structural and functional characteristics. Major subdivisions include the **proximal convoluted tubule (PCT)**, the **loop of Henle** (HEN-lē), and the **distal convoluted tubule (DCT)**.

Each nephron empties into the **collecting system**. A **collecting tubule** connected to the distal convoluted tubule carries the filtrate toward a nearby **collecting duct**. The collecting duct leaves the cortex and descends into the medulla, carrying fluid toward a **papillary duct** that drains into the renal pelvis.

There are slight differences in nephron structure depending on their location. Roughly 85% of the nephrons are called **cortical nephrons** because they are found in the superficial cortex of the kidney. The loops of Henle of cortical nephrons do not extend deeply into the medulla. The remaining 15% of nephrons, termed **juxtamedullary nephrons** (juks-ta-MED-ū-lar-ē; *juxta*, near) are located closer to the medulla, and they extend deep into the renal pyramids (Figure 26-7●, p. 671). Although relatively small in number, juxtamedullary nephrons play a key role in the formation and concentration of urine.

The urine arriving at the renal pelvis is very different from the filtrate produced at the renal corpuscle. Filtration is a passive process that permits or prevents movement across a barrier solely on the basis of solute size. A filter with pores large enough to permit the passage of organic waste products is unable to prevent the passage of water, ions, and other organic molecules, such as glucose, fatty acids, or amino acids.

FIGURE 26-3

The Urinary System in Gross Dissection. (a) Diagrammatic view and dissection (b) of the abdominopelvic cavity showing the kidneys, adrenal glands, ureters, urinary bladder, and the blood and nerve supply to the urinary structures. (c) Anterior view of the male urogenital system with abdominal organs removed. The urinary bladder has been cut to expose the ureteral openings and the internal urethral orifice.

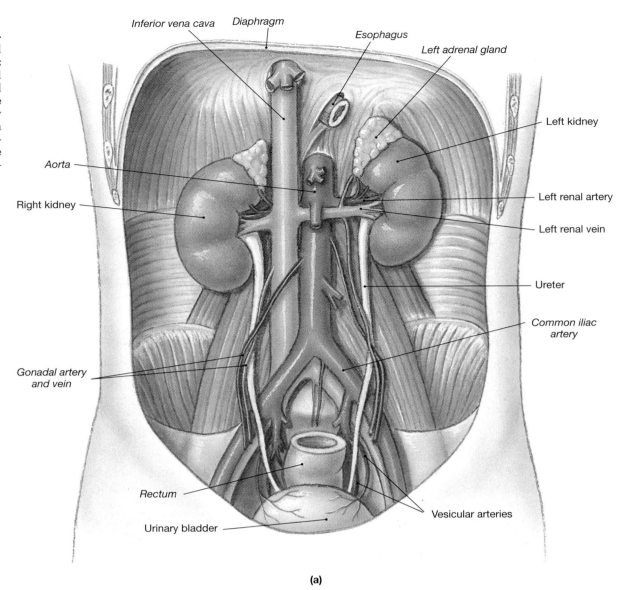

(a)

The other distal segments of the nephron are responsible for:

- Reabsorbing all of the useful organic substrates from the filtrate,
- Reabsorbing over 80% of the water in the filtrate, and
- Secreting into the filtrate waste products that were missed by the filtration process.

We will now examine each of the segments of a juxtamedullary nephron in greater detail.

The Renal Corpuscle
(Figure 26-6c-e)

The renal corpuscle has a diameter averaging 150–250 μm. It includes: (1) the capillary knot of the glomerulus and (2) the expanded initial segment of the renal tubule, a region known as *Bowman's capsule.* The glomerulus projects into Bowman's capsule like the heart projects into the pericardial cavity (Figure 26-6c●, p. 670). The outer wall of the capsule is lined by a simple squamous **capsular epithelium**, and this layer is continuous with the **glomerular epithelium** that covers the glomerular capillaries. The glomerular epithelium consists of large cells with complex processes, or "feet," that wrap around the glomerular capillaries. These specialized cells are called **podocytes** (POD-ō-sīts; *podos*, foot + *-cyte*, cell) as seen in Figure 26-6c-e●, p. 670. The **capsular space** separates the capsular

and glomerular epithelia. The connection between the capsular and glomerular epithelia lies at the **vascular pole** of the renal corpuscle. At the vascular pole, the glomerular capillaries are connected to the bloodstream. Blood arrives via the **afferent arteriole** and departs in the **efferent arteriole** (Figure 26-6c●). (This unusual circulatory arrangement will be discussed further in a later section.)

Filtration occurs as blood pressure forces fluid and dissolved solutes out of the glomerulus and into the capsular space. The filtrate produced in this way is very similar to plasma with the blood proteins removed. The filtration process involves passage across three physical barriers (Figure 26-6d●, p. 670):

1. *The Capillary Endothelium*: The glomerular capillaries are *fenestrated capillaries* with pores ranging from 60 to 100 nm (0.06–0.1 μm) in diameter. ∞ [p. 546] These openings are small enough to prevent the passage of blood cells, but they are too large to restrict the diffusion of solutes, even those the size of plasma proteins.

2. *The Basement Membrane*: The basement membrane that surrounds the capillary endothelium has several times the density and thickness of a typical basement membrane. This layer, called the **lamina densa**, restricts the passage of the larger plasma proteins, but permits the movement of smaller plasma proteins, nutrients, and ions.

666

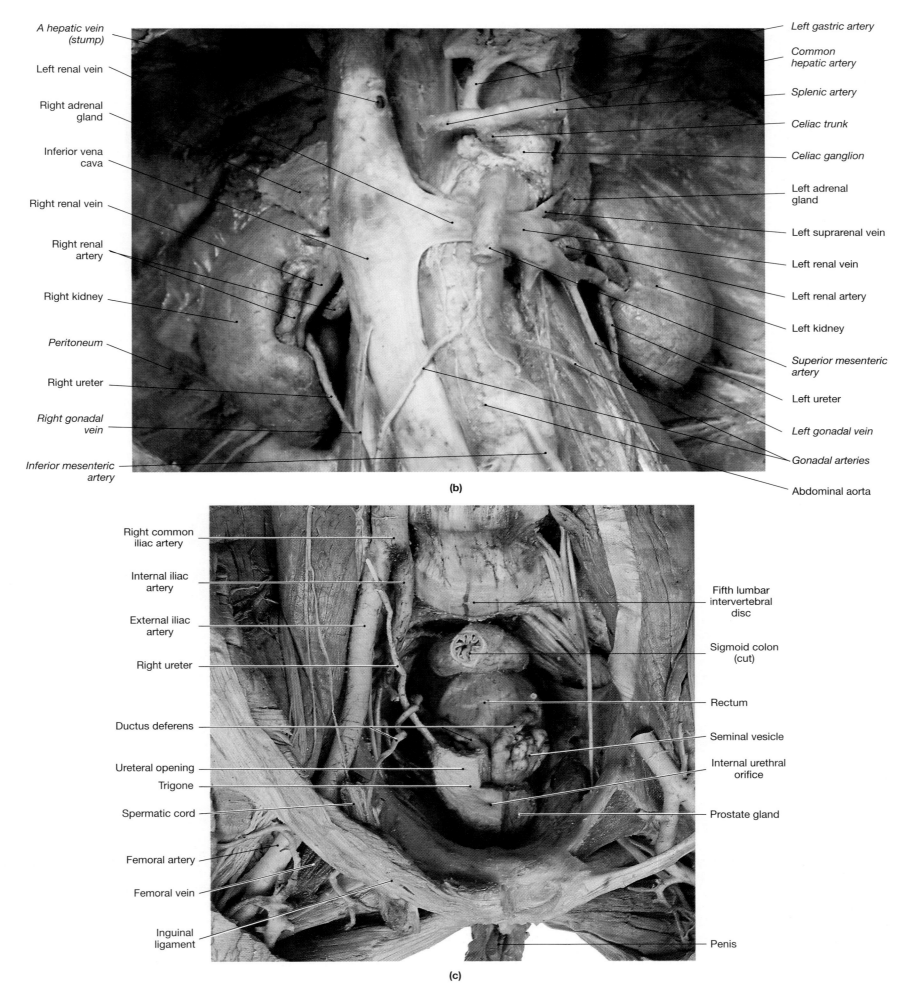

A hepatic vein (stump)

Left renal vein

Right adrenal gland

Inferior vena cava

Right renal vein

Right renal artery

Right kidney

Peritoneum

Right ureter

Right gonadal vein

Inferior mesenteric artery

Left gastric artery

Common hepatic artery

Splenic artery

Celiac trunk

Celiac ganglion

Left adrenal gland

Left suprarenal vein

Left renal vein

Left renal artery

Left kidney

Superior mesenteric artery

Left ureter

Left gonadal vein

Gonadal arteries

Abdominal aorta

(b)

Right common iliac artery

Internal iliac artery

External iliac artery

Right ureter

Ductus deferens

Ureteral opening

Trigone

Spermatic cord

Femoral artery

Femoral vein

Inguinal ligament

Fifth lumbar intervertebral disc

Sigmoid colon (cut)

Rectum

Seminal vesicle

Internal urethral orifice

Prostate gland

Penis

(c)

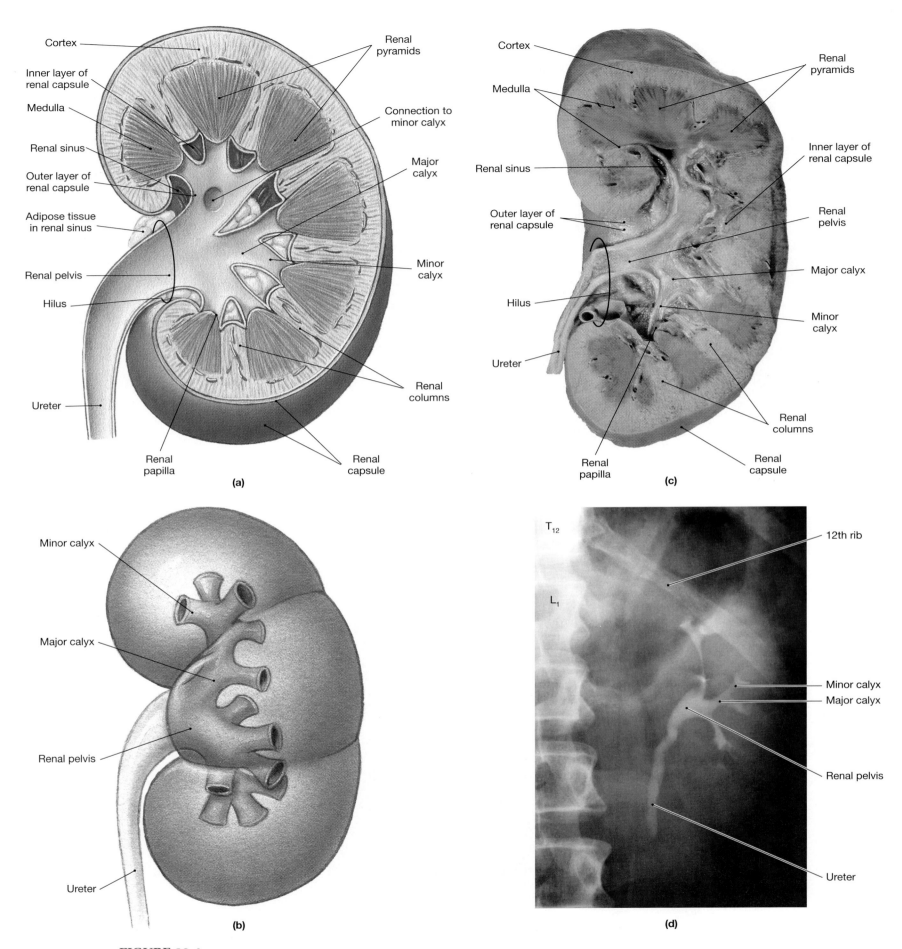

FIGURE 26-4
Structure of the Kidney. (a) Frontal section through the left kidney, showing major structures. (b) Shadow drawing to show the arrangement of the calyces and renal pelvis within the kidney. (c) Frontal section of kidney showing internal structures. (d) Urogram of left kidney showing calyces, renal pelvis, and ureter.

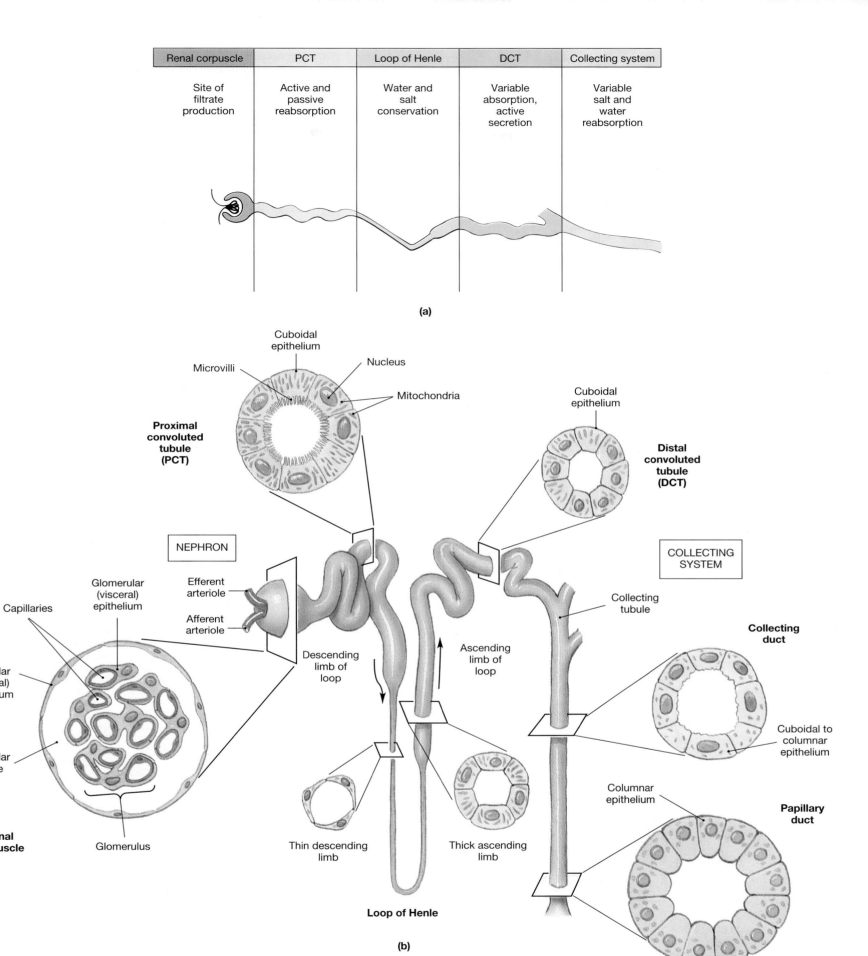

Renal corpuscle	PCT	Loop of Henle	DCT	Collecting system
Site of filtrate production	Active and passive reabsorption	Water and salt conservation	Variable absorption, active secretion	Variable salt and water reabsorption

(a)

Cuboidal epithelium

Microvilli

Nucleus

Mitochondria

Cuboidal epithelium

Proximal convoluted tubule (PCT)

Distal convoluted tubule (DCT)

NEPHRON

COLLECTING SYSTEM

Glomerular (visceral) epithelium

Efferent arteriole

Afferent arteriole

Capillaries

Collecting tubule

Collecting duct

Capsular (parietal) epithelium

Descending limb of loop

Ascending limb of loop

Capsular space

Cuboidal to columnar epithelium

Columnar epithelium

Renal corpuscle

Glomerulus

Thin descending limb

Thick ascending limb

Papillary duct

Loop of Henle

(b)

FIGURE 26-5
A Typical Nephron. (a) Diagrammatic view indicating major functions of each segment of the nephron and collecting system. (b) Histology of the nephron and collecting system.

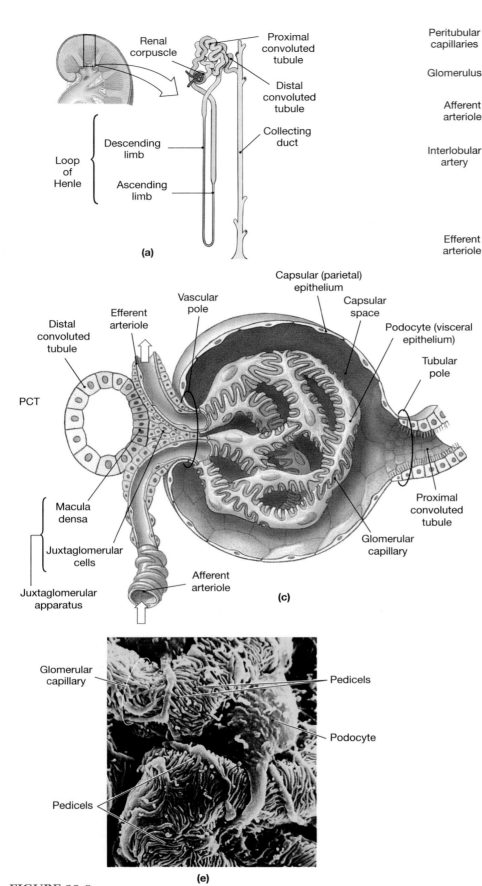

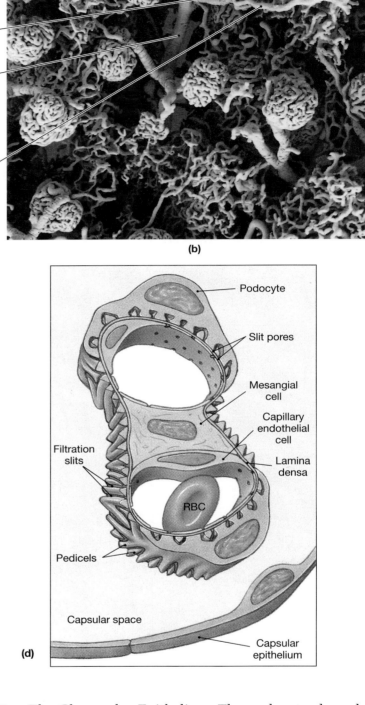

FIGURE 26-6
The Renal Corpuscle. (a) A juxtamedullary nephron. (b) SEM of several renal corpuscles, showing their three-dimensional structure. (SEM × 94) [Reproduced from R. G. Kessel and R. H. Kardon, *Tissues and Organs: A Text-Atlas of Scanning Electron Microscopy*, W. H. Freeman & Co., 1979]. (c) The renal corpuscle. (d) Diagrammatic view of the filtration apparatus. (e) Electron micrograph of the glomerular surface, showing individual podocytes and their processes. (SEM × 20,800)

670

3. *The Glomerular Epithelium*: The podocytes have long cellular processes that wrap around the outer surfaces of the basement membrane. These delicate "feet," or **pedicels** (PED-i-celz) (Figure 26-6d,e●) are separated by narrow spaces called **slit pores**. Because the slit pores are very small, the filtrate entering the capsular space consists of water with dissolved ions, small organic molecules, and few if any plasma proteins.

In addition to metabolic wastes, the filtrate contains other organic compounds such as glucose, free fatty acids, amino acids, and vitamins. These potentially useful materials are reabsorbed by the proximal convoluted tubule.

The Proximal Convoluted Tubule
(Figures 26-5b/26-6c/26-7a)

The entrance to the proximal convoluted tubule, or simply PCT, lies almost directly opposite the vascular pole of the glomerulus, at the **tubular pole** of the renal corpuscle (Figure 26-6c●). The lining of the PCT consists of a simple cuboidal epithelium whose exposed surfaces are blanketed with microvilli (Figures 26-5b and 26-7a●). These cells actively absorb organic nutrients, ions, and plasma proteins (if any) from the filtrate as it flows along the PCT. As these solutes are absorbed, osmotic forces pull water across the wall of the PCT and into the surrounding inter-stitial spaces. Absorption represents the primary function of the PCT; as tubular fluid passes along the PCT, the epithelial cells reabsorb virtually all of the organic nutrients and plasma proteins and 60% of the sodium ions, chloride ions, and water. The PCT also actively reabsorbs potassium, calcium, magnesium, bicarbonate, phosphate, and sulfate ions.

The Loop of Henle
(Figure 26-7d)

The proximal convoluted tubule ends at an acute bend that turns the renal tubule toward the medulla. This bend marks the

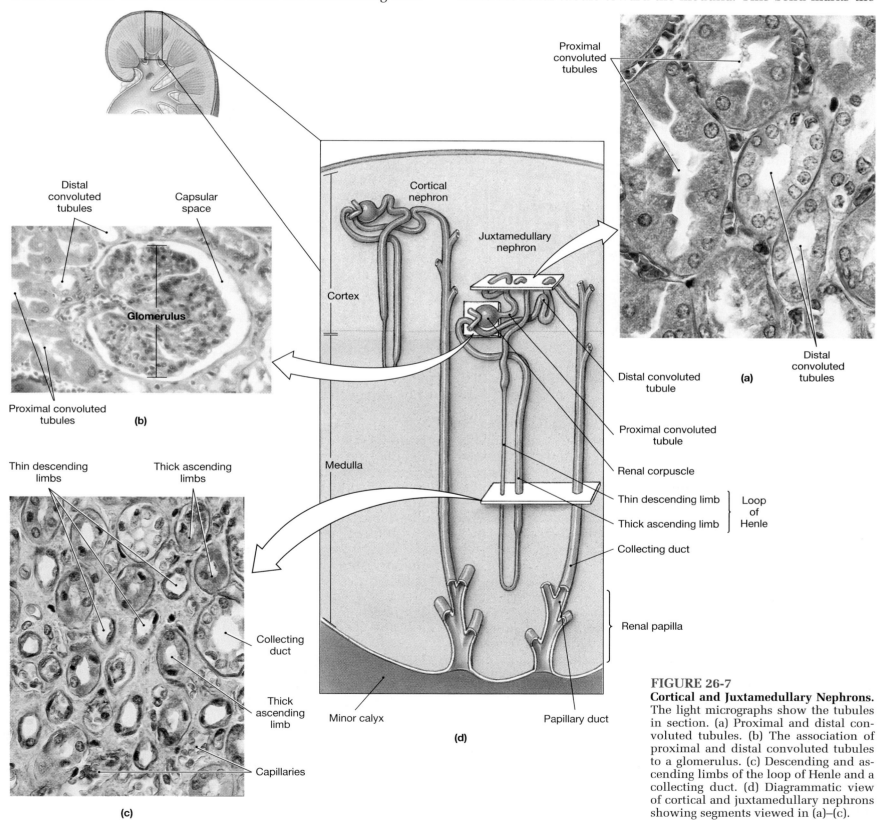

FIGURE 26-7
Cortical and Juxtamedullary Nephrons. The light micrographs show the tubules in section. (a) Proximal and distal convoluted tubules. (b) The association of proximal and distal convoluted tubules to a glomerulus. (c) Descending and ascending limbs of the loop of Henle and a collecting duct. (d) Diagrammatic view of cortical and juxtamedullary nephrons showing segments viewed in (a)–(c).

start of the loop of Henle (Figure 26-7d●). The loop of Henle can be divided into a **descending limb** and an **ascending limb**. The descending limb travels in the medulla toward the renal pelvis, and the ascending limb returns toward the cortex. Each limb contains a **thick segment** and a **thin segment**. (The terms *thick* and *thin* refer to the thickness of the surrounding epithelium, not to the diameter of the lumen.)

Thick segments are found closest to the cortex, whereas a thin squamous epithelium lines the deeper medullary portions. The thick ascending limb, which begins deep in the medulla, contains active transport mechanisms that pump sodium and chloride ions out of the tubular fluid. As a result of these transport activities, the medullary interstitial fluid contains an unusually high concentration of solutes. Near the base of the loop, in the deepest part of the medulla, the osmolarity of the interstitial fluid is roughly four times that of plasma (1200 mOsml vs. 300 mOsml). The thin descending and ascending limbs are freely permeable to water, but relatively impermeable to ions and other solutes. The high osmotic concentration of their surroundings results in an osmotic flow of water out of the nephron.

The net effect is that the loop of Henle reabsorbs an additional 25% of the water from the tubular fluid and an even higher percentage of the sodium and chloride ions. Reabsorption in the PCT and loop of Henle normally reclaims all of the organic nutrients, 85% of the water, and over 90% of the sodium and chloride ions. The remaining water, ions, and all of the organic wastes filtered at the glomerulus enter the distal convoluted tubule.

The Distal Convoluted Tubule
(Figures 26-6c/26-7)

The ascending limb of the loop of Henle ends where it forms a sharp angle that places the tubular wall in close contact with the glomerulus and its accompanying vessels. The distal convoluted tubule (DCT) begins at that bend. The initial portion of the DCT crosses the vascular pole of the renal corpuscle, passing between the afferent and efferent arterioles (Figure 26-6c●).

In sectional view (Figure 26-7●), the DCT differs from the PCT in that (1) the DCT has a smaller diameter, (2) the epithelial cells of the DCT lack microvilli, and (3) the boundaries between the epithelial cells in the DCT are distinct.

The distal convoluted tubule is an important site for (1) the active secretion of ions, acids, and other materials, and (2) the selective reabsorption of sodium ions from the tubular fluid. The sodium transport activities of the DCT are controlled by circulating levels of *aldosterone* secreted by the adrenal cortex. ∞ [p. 494]

THE JUXTAGLOMERULAR APPARATUS *(Figure 26-6c).* The epithelial cells adjacent to the vascular pole are taller than those seen elsewhere along the DCT, and their nuclei are clustered together. This region, detailed in Figure 26-6c●, is called the **macula densa** (MAK-ū-la DEN-sa). The cells of the macula densa are closely associated with unusual smooth muscle fibers in the wall of the afferent arteriole. These muscle fibers are known as **juxtaglomerular cells** (*juxta*, near). Together the macula densa and juxtaglomerular cells form the **juxtaglomerular apparatus**. The juxtaglomerular apparatus is an endocrine structure that secretes two hormones, *renin* and *erythropoietin*, that were described in Chapter 19. ∞ [p. 495] These hormones, released when renal blood pressure or blood flow decreases, elevate blood volume and blood pressure and restore normal rates of filtrate production.

The Collecting System
(Figures 26-5/26-7d)

The distal convoluted tubule, the last segment of the nephron, opens into the collecting system. The collecting system consists of *collecting tubules*, *collecting ducts*, and *papillary ducts* (Figure 26-5●, p. 669). Individual collecting tubules connect each nephron to a nearby collecting duct (Figure 26-7d●). Each collecting duct receives fluid from many collecting tubules, draining both cortical and juxtamedullary nephrons. Several collecting ducts converge to empty into the larger papillary duct that empties into a minor calyx. The epithelium lining the collecting system begins as simple cuboidal cells in the collecting tubules and converts to a columnar epithelium in the collecting and papillary ducts.

In addition to transporting fluid from the nephron to the renal pelvis, the collecting system makes final adjustments to its osmotic concentration and volume. The regulatory mechanism involves changing the permeability of the collecting ducts to water. This change is significant because the collecting ducts pass through the medulla, where the loop of Henle has established very high solute concentrations in the interstitial fluids. If collecting duct permeability is low, most of the tubular fluid reaching the collecting duct will enter the renal pelvis. If collecting duct permeability is high, there will be an osmotic flow of water into the medulla. A small amount of highly concentrated urine will then reach the renal pelvis. ADH (antidiuretic hormone) is the hormone responsible for controlling the permeability of the collecting system. ∞ [p. 488] The higher the levels of circulating ADH, the greater the amount of water reabsorbed, and the more concentrated the urine. ⸸ *Urine Composition* and *Hemodialysis [p. 774]*

The Blood Supply to the Kidneys
(Figures 22-17/26-7d/26-8/26-9)

The kidneys receive 20–25% of the total cardiac output. In normal individuals, about 1200 mℓ of blood flows through the kidneys each minute. Each kidney receives blood from a **renal artery** that originates along the lateral surface of the abdominal aorta near the level of the superior mesenteric artery (Figure 22-17●). ∞ [p. 561] As the renal artery enters the renal sinus, it provides blood to the **segmental arteries** (Figure 26-8●). Segmental arteries further divide into a series of **interlobar arteries** that radiate outward, penetrating the renal capsule and extending through the renal columns between the renal pyramids. The interlobar arteries supply blood to the **arcuate** (AR-kū-āt) **arteries** that arch along the boundary between the cortex and medulla of the kidney. Each arcuate artery gives rise to a number of **interlobular arteries** supplying portions of the adjacent renal lobe. Branching from each interlobular artery are numerous afferent arterioles.

Blood reaches the vascular pole of each glomerulus through an afferent arteriole and leaves in an efferent arteriole. Blood travels from the efferent arteriole to form a capillary plexus, known as the **peritubular capillaries,** which supply the proximal and distal convoluted tubules. Each efferent arteriole supplies blood to the peritubular capillaries that form a network around the PCT and DCT (Figure 26-8c,d●). The peritubular capillaries provide a route for the pickup or delivery of substances that are reabsorbed or secreted by these portions of the nephron. In juxtamedullary nephrons, the efferent arterioles

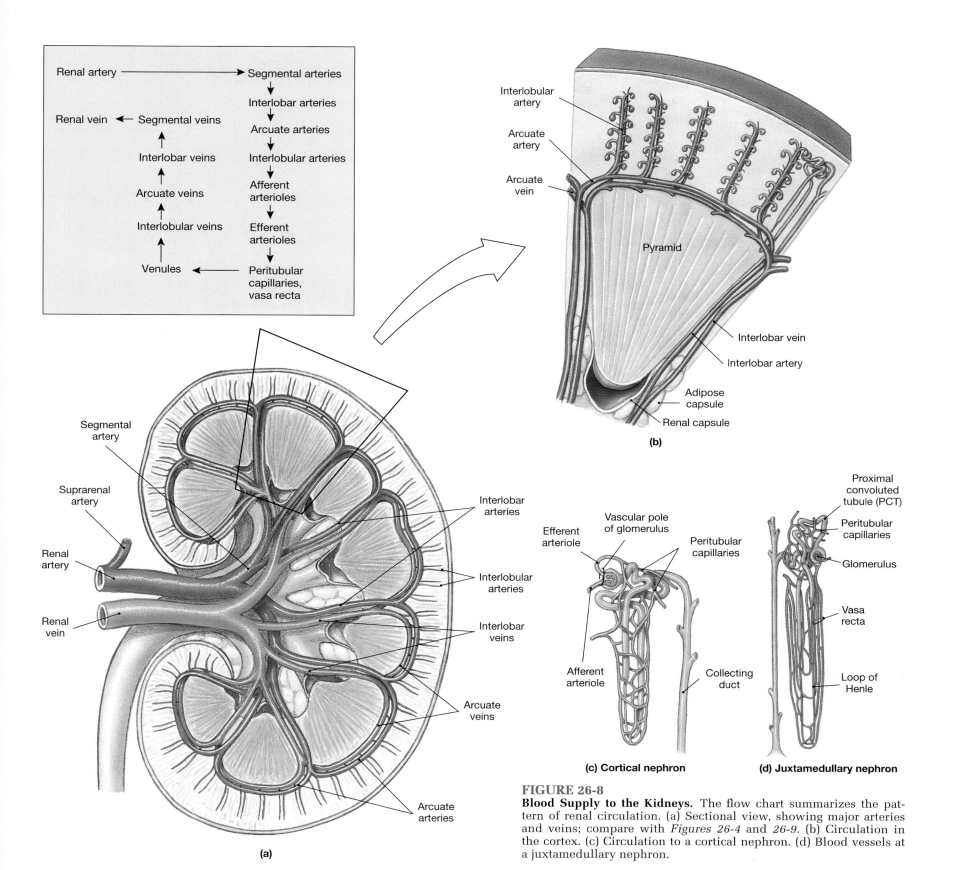

FIGURE 26-8
Blood Supply to the Kidneys. The flow chart summarizes the pattern of renal circulation. (a) Sectional view, showing major arteries and veins; compare with *Figures 26-4* and *26-9*. (b) Circulation in the cortex. (c) Circulation to a cortical nephron. (d) Blood vessels at a juxtamedullary nephron.

and peritubular capillaries are connected to a series of long, slender capillaries that accompany the loops of Henle into the medulla (Figure 26-7d●). These capillaries, known as the **vasa recta** (*rectus*, straight), absorb and transport solutes and water reabsorbed into the medulla from filtrate in the loops of Henle and collecting ducts. Under normal conditions the removal of solutes and water by the vasa recta precisely balances the rate of solute and water reabsorption in the medulla.

From the peritubular capillaries and vasa recta, blood enters a network of venules and small veins that converge on the **interlobular veins**. In a mirror image of the arterial distribution, the interlobular veins deliver blood to **arcuate veins** that empty into **interlobar veins**. The interlobar veins drain into the **segmental veins** that merge to form the **renal vein**. Many of the blood vessels just described are visible in the corrosion cast of the human kidneys and in the renal arteriogram shown in Figure 26-9●.

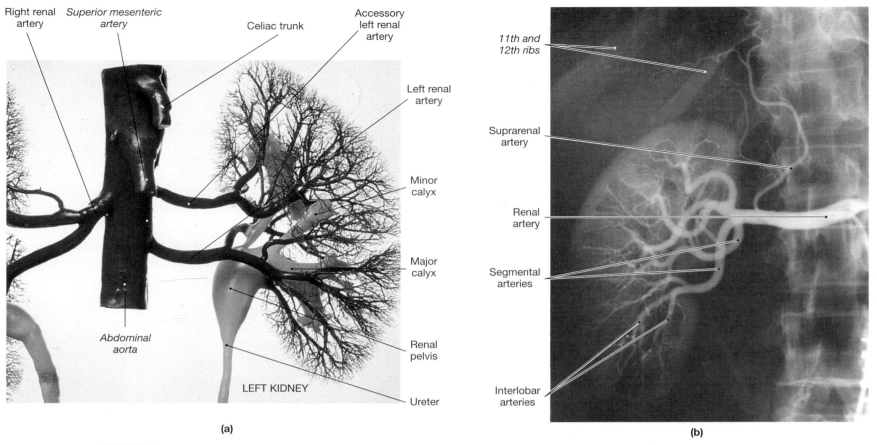

Right renal artery | *Superior mesenteric artery* | Celiac trunk | Accessory left renal artery

Left renal artery

Minor calyx

Major calyx

Abdominal aorta

Renal pelvis

LEFT KIDNEY

Ureter

(a)

11th and 12th ribs

Suprarenal artery

Renal artery

Segmental arteries

Interlobar arteries

(b)

FIGURE 26-9
Renal Vessels and Blood Flow. (a) Corrosion cast of the circulation and conducting passageways of the kidneys. This individual had an accessory renal artery delivering blood to the superior portion of the left kidney. Accessory renal arteries are relatively common; they may originate at the aorta or from the common or external iliac arteries. (b) A renal angiogram.

▌CLINICAL BRIEF
Examination of the Kidneys and Urinary System

The anatomy of the urinary system can be examined using a variety of sophisticated procedures. **Computerized tomography** (CT) scans can provide useful information concerning localized abnormalities (Figure 26-10a●).

The blood flow through the kidney can be checked by *angiography* (Figure 26-10b●), a procedure described in Chapter 21. ∞ [p. 533] Administering a radiopaque compound that will enter the urine permits the creation of a **pyelogram** (PĪ-e-lō-gram; *pyelos*, pelvis), or **IVP**, by taking an X-ray of the kidneys (Figure 26-9b●) or of the entire urinary system (Figure 26-11●). Pyelography permits detection of anatomically unusual kidney, ureter, or bladder structures and masses.

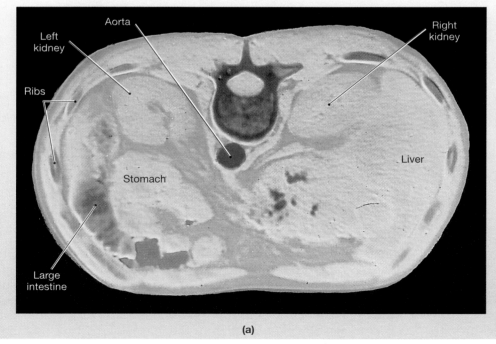

Left kidney | Aorta | Right kidney

Ribs

Liver

Stomach

Large intestine

(a)

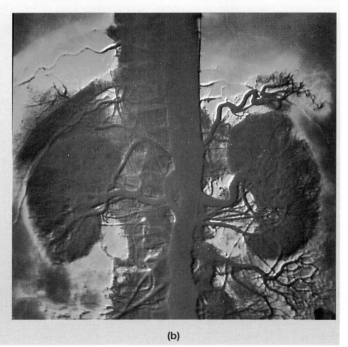

(b)

FIGURE 26-10
Images of the Urinary System. (a) A CT scan showing the position of the kidneys in a transverse section through the trunk. (b) An angiogram of the kidneys, showing the extensive blood supply.

Innervation of the Kidneys

Urine production in the kidneys is regulated in part through *autoregulation,* which involves reflexive changes in blood flow and filtration rates. Both hormonal and neural mechanisms can supplement or adjust the local responses. The kidneys and ureters are innervated by **renal nerves**. Most of the nerve fibers involved are sympathetic postganglionic fibers from the superior mesenteric ganglion. A renal nerve enters each kidney at the hilus and follows the tributaries of the renal arteries to reach individual nephrons. Known functions of sympathetic innervation include: (1) regulation of glomerular blood flow and pressure through innervation of afferent and efferent arterioles; (2) stimulation of renin release; and (3) direct stimulation of water and sodium ion reabsorption.

√ **Trace the path of a drop of blood from the renal artery to a glomerulus.**

√ **Now trace the route back to the renal vein.**

√ **Trace the path taken by filtrate in traveling from a glomerulus to a minor calyx.**

STRUCTURES FOR URINE TRANSPORT, STORAGE, AND ELIMINATION
(Figure 26-11)

Filtrate modification and urine production end when the fluid enters the minor calyx. The remaining parts of the urinary system (the *ureters*, *urinary bladder*, and *urethra*) are responsible for the transport, storage, and elimination of the urine. Figure 26-11●, a pyelogram of the urinary system, provides an orientation to the relative sizes and positions of these organs.

The minor and major calyces, the renal pelvis, the ureters, the urinary bladder, and the proximal portion of the urethra are lined by a *transitional epithelium* that can tolerate cycles of distension and contraction without damage. ∞ [p. 59]

The Ureters
(Figures 26-3/26-4/26-11/26-12a-c)

The ureters are a pair of muscular tubes that extend caudally from the kidneys for about 30 cm (12 in.) before reaching the urinary bladder. Each ureter begins at the funnel-shaped renal pelvis (Figures 26-4, p. 668, and 26-11●). As the ureters descend from the kidneys, they pass inferiorly and medially over the psoas muscles. The ureters are retroperitoneal, and they are firmly attached to the posterior abdominal wall (Figure 26-3●, pp. 666, 667). The paths taken by the ureters in men and women are somewhat different because of variations in the nature, size, and position of the reproductive organs (Figure 26-12a,b●). In the male, the base of the urinary bladder lies between the rectum and the symphysis pubis; in the female, the base of the urinary bladder sits inferior to the uterus and anterior to the vagina.

The ureters penetrate the posterior wall of the urinary bladder without entering the peritoneal cavity. They pass through the bladder wall at an oblique angle, and the **ureteral opening** is slitlike, rather than rounded (Figure 26-12c●). This shape helps to prevent backflow of urine toward the ureter and kidneys when the urinary bladder contracts.

Histology of the Ureters
(Figure 26-13a)

The wall of each ureter consists of three layers: (1) an inner mucosa covered by a transitional epithelium; (2) a middle muscular layer made up of longitudinal (inner) and circular (outer) bands of smooth muscle; and (3) an outer connective tissue layer that is continuous with the fibrous renal capsule and peritoneum (Figure 26-13a●). Starting at the kidney, about every half-minute, peristaltic contractions of the muscular wall serve to "milk" urine out of the renal pelvis and along the ureter to the bladder.

The Urinary Bladder
(Figure 26-12c,d)

The urinary bladder is a hollow muscular organ that functions as a temporary "storage reservoir" for urine. The dimensions of the urinary bladder vary, depending on the state of distension, but the full urinary bladder can contain about a liter of urine.

The superior surfaces of the urinary bladder are covered by a layer of peritoneum, and several peritoneal folds assist in stabilizing its position. The **middle** (or *median*) **umbilical ligament**, or **urachus** (Ū-ra-kus), extends from the anterior and superior border toward the umbilicus (Figure 26-12c●). The **lateral** (or *medial*) **umbilical ligaments** pass along the sides of the bladder and also reach the umbilicus. These fibrous cords contain the vestiges of the two *umbilical arteries* that supplied blood to the placenta dur-

FIGURE 26-11
A Pyelogram of the Urinary System. The X-ray was taken after the introduction of a radiopaque dye that was filtered into the urine.

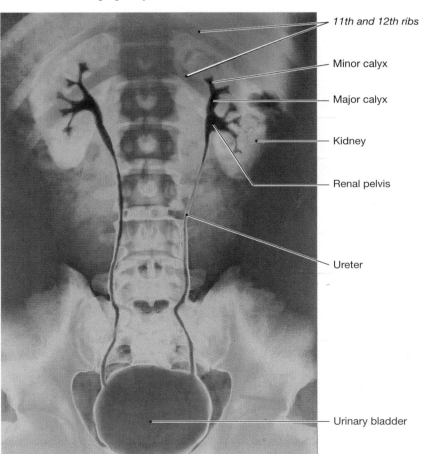

11th and 12th ribs

Minor calyx

Major calyx

Kidney

Renal pelvis

Ureter

Urinary bladder

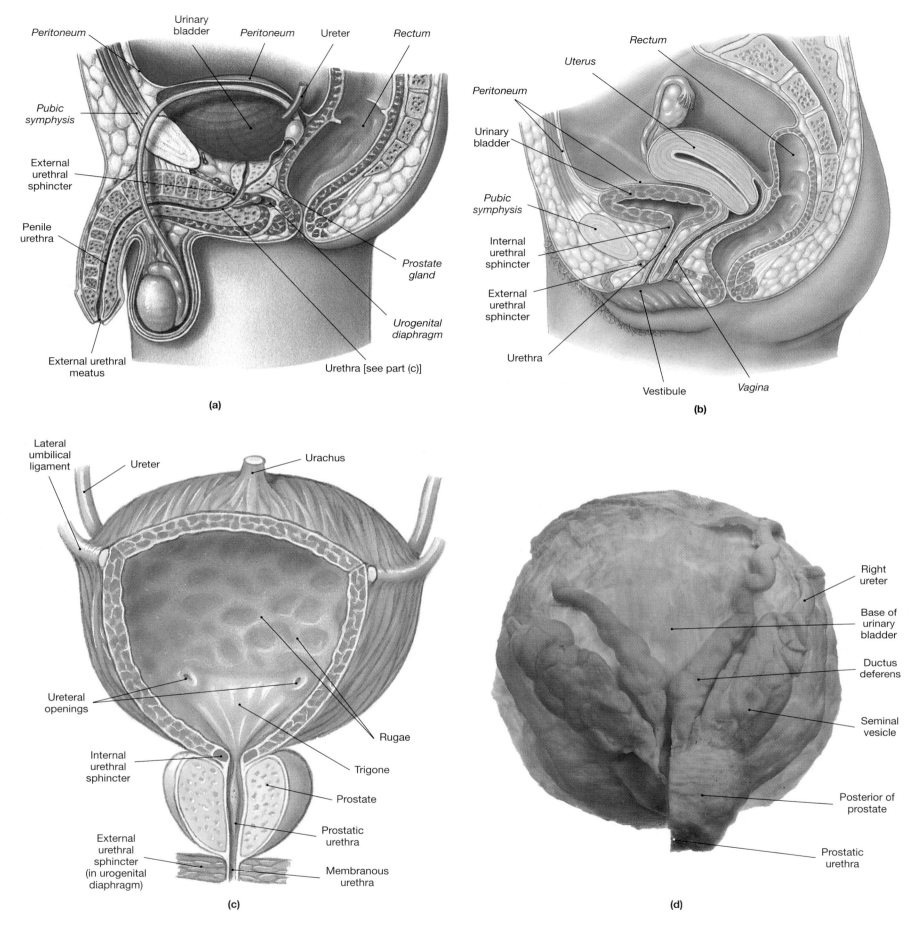

FIGURE 26-12
Organs Responsible for the Conduction and Storage of Urine. (a) Position of the ureter, urinary bladder, and urethra in the male. See also *Figure 26-3a*. (b) Position of the same organs in the female. (c) Anatomy of the urinary bladder. (d) The male urinary bladder and accessory reproductive structures as seen in posterior view.

ing embryonic and fetal development. ∞ [p. 655] The posterior, inferior, and anterior surfaces lie outside the peritoneal cavity, and in these areas tough ligamentous bands anchor the bladder to the pelvic and pubic bones.

In sectional view (Figure 26-12c,d●), the mucosa lining the urinary bladder is usually thrown into folds, or **rugae**, that disappear as the bladder fills. The triangular area bounded by the ureteral openings and the entrance to the urethra constitutes the **trigone** (TRĪ-gōn) of the urinary bladder. The mucosa here is smooth and very thick, and the trigone acts as a funnel that channels urine into the urethra when the urinary bladder contracts.

The urethral entrance lies at the apex of the trigone, at the most inferior point in the bladder. The region surrounding the urethral opening, known as the **neck** of the urinary bladder, contains a muscular **internal urethral sphincter**, or *sphincter vesicae*. The smooth muscle of the internal urethral sphincter provides involuntary control over the discharge of urine from the bladder.

Histology of the Urinary Bladder
(Figure 26-13b)

The wall of the bladder contains mucosa, submucosa, and muscularis layers (Figure 26-13b●). The muscularis layer con-

sists of inner and outer longitudinal smooth muscle layers, with a circular layer sandwiched between. Collectively, these layers form the powerful **detrusor** (dē-TROO-sor) muscle of the urinary bladder. Contraction of this muscle compresses the urinary bladder and expels its contents into the urethra. A layer of serosa covers the superior aspect of the urinary bladder.

■ CLINICAL BRIEF
Problems with the Conducting System

Local blockages of the collecting tubules, collecting ducts, or ureters may result from the formation of **casts**, small blood clots, epithelial cells, lipids, or other materials. Casts are often excreted in the urine and visible in microscopic analysis of urine samples. **Calculi** (KAL-kū-lī), or "kidney stones," form from calcium deposits, magnesium salts, or crystals of uric acid. This condition is called **nephrolithiasis** (nef-rō-li-THĪ-a-sis). The blockage of the urinary passage by a stone or other factors, such as external compression, results in **urinary obstruction**. Urinary obstruction is a painful and serious problem because it will reduce or eliminate filtration in the affected kidney.

Kidney stones are usually visible on an X-ray, and if peristalsis and fluid pressures are insufficient to dislodge them, they must be surgically removed or destroyed. One interesting nonsurgical procedure involves breaking kidney stones apart with a *lithotripter*, the same apparatus used to destroy gallstones.

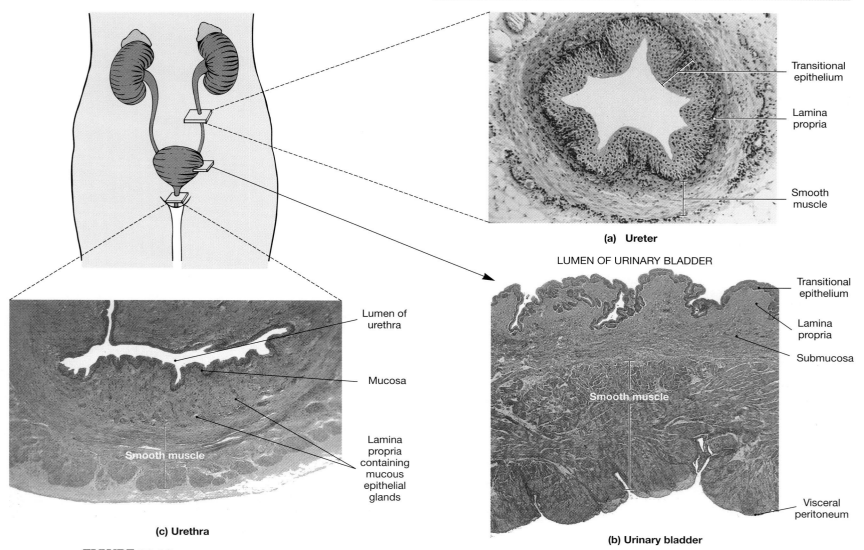

FIGURE 26-13
Histology of the Collecting and Transport Organs. (a) A ureter seen in transverse section. Note the thick layer of smooth muscle surrounding the lumen. (For a close-up of transitional epithelium, review *Figure 3-5c*.) (LM × 65) (b) The wall of the urinary bladder. (LM × 36) (c) A transverse section through the urethra. (LM × 61)

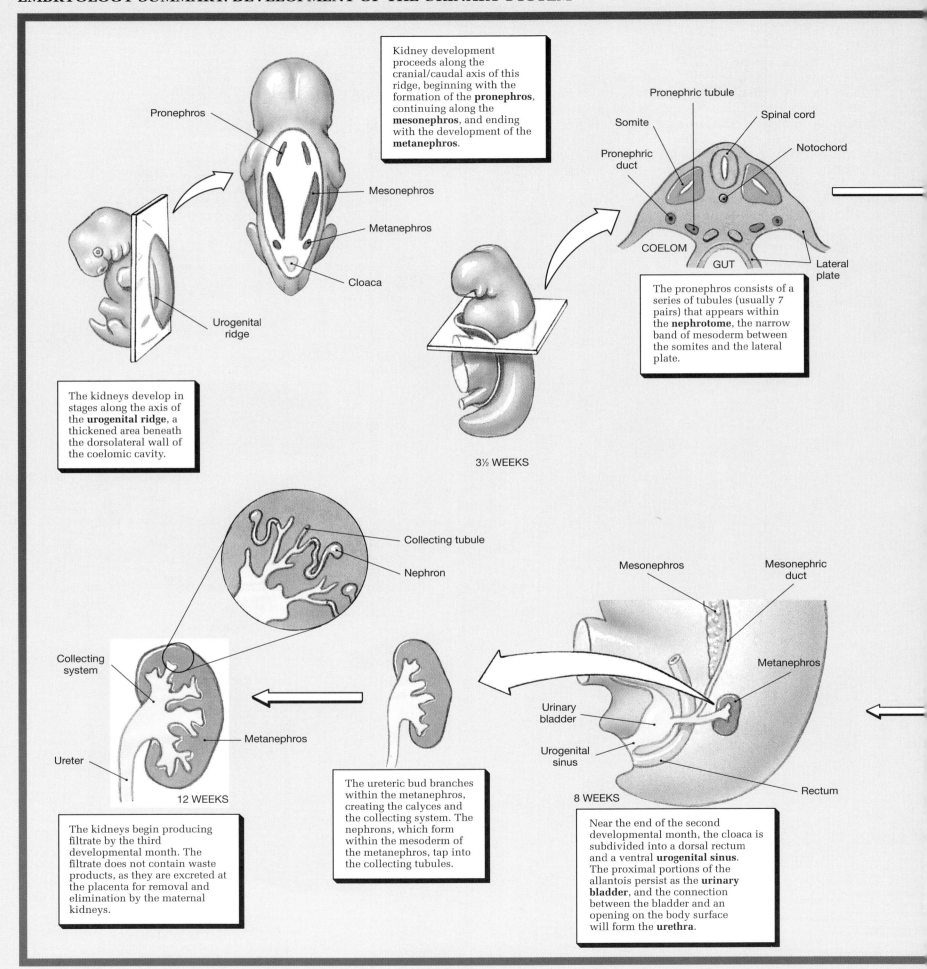

Kidney development proceeds along the cranial/caudal axis of this ridge, beginning with the formation of the **pronephros**, continuing along the **mesonephros**, and ending with the development of the **metanephros**.

Pronephros

Mesonephros

Metanephros

Cloaca

Urogenital ridge

The kidneys develop in stages along the axis of the **urogenital ridge**, a thickened area beneath the dorsolateral wall of the coelomic cavity.

3½ WEEKS

Pronephric tubule

Somite

Pronephric duct

Spinal cord

Notochord

COELOM

GUT

Lateral plate

The pronephros consists of a series of tubules (usually 7 pairs) that appears within the **nephrotome**, the narrow band of mesoderm between the somites and the lateral plate.

Collecting tubule

Nephron

Collecting system

Metanephros

Ureter

12 WEEKS

The kidneys begin producing filtrate by the third developmental month. The filtrate does not contain waste products, as they are excreted at the placenta for removal and elimination by the maternal kidneys.

The ureteric bud branches within the metanephros, creating the calyces and the collecting system. The nephrons, which form within the mesoderm of the metanephros, tap into the collecting tubules.

Mesonephros

Mesonephric duct

Metanephros

Urinary bladder

Urogenital sinus

Rectum

8 WEEKS

Near the end of the second developmental month, the cloaca is subdivided into a dorsal rectum and a ventral **urogenital sinus**. The proximal portions of the allantois persist as the **urinary bladder**, and the connection between the bladder and an opening on the body surface will form the **urethra**.

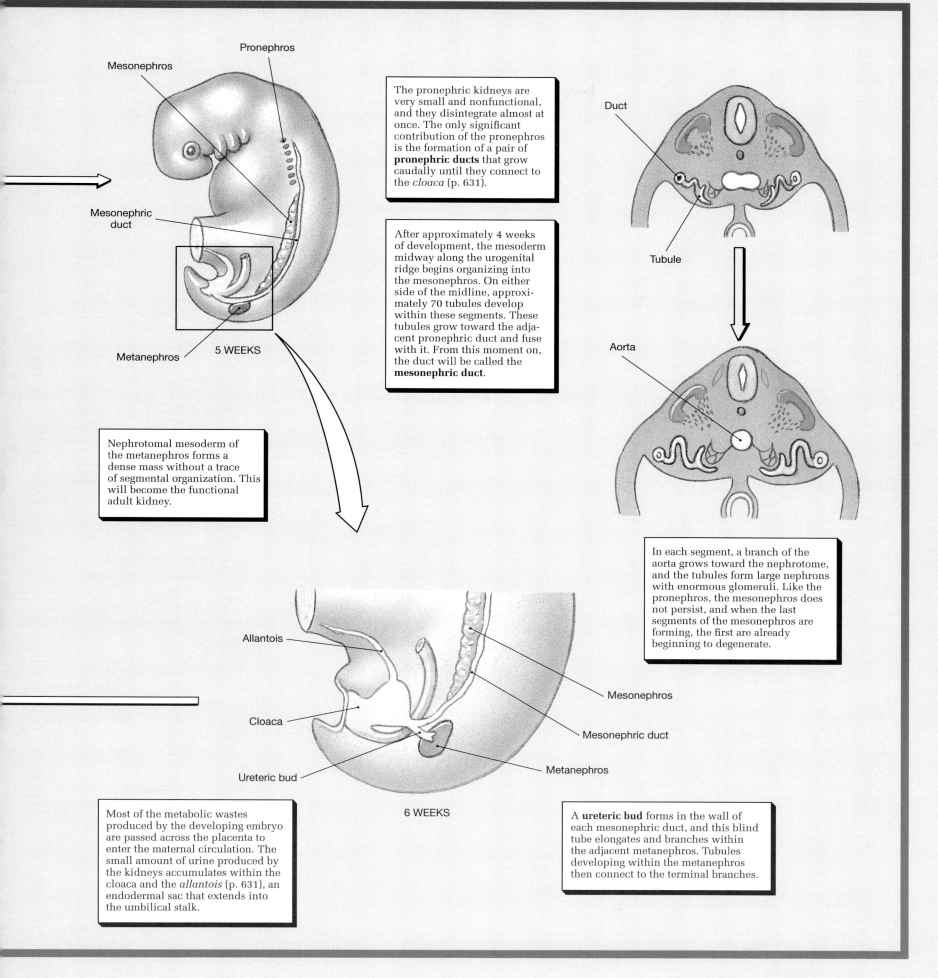

Mesonephros

Pronephros

Mesonephric duct

Metanephros

5 WEEKS

The pronephric kidneys are very small and nonfunctional, and they disintegrate almost at once. The only significant contribution of the pronephros is the formation of a pair of **pronephric ducts** that grow caudally until they connect to the *cloaca* (p. 631).

After approximately 4 weeks of development, the mesoderm midway along the urogenital ridge begins organizing into the mesonephros. On either side of the midline, approximately 70 tubules develop within these segments. These tubules grow toward the adjacent pronephric duct and fuse with it. From this moment on, the duct will be called the **mesonephric duct**.

Nephrotomal mesoderm of the metanephros forms a dense mass without a trace of segmental organization. This will become the functional adult kidney.

Duct

Tubule

Aorta

In each segment, a branch of the aorta grows toward the nephrotome, and the tubules form large nephrons with enormous glomeruli. Like the pronephros, the mesonephros does not persist, and when the last segments of the mesonephros are forming, the first are already beginning to degenerate.

Allantois

Cloaca

Ureteric bud

6 WEEKS

Mesonephros

Mesonephric duct

Metanephros

Most of the metabolic wastes produced by the developing embryo are passed across the placenta to enter the maternal circulation. The small amount of urine produced by the kidneys accumulates within the cloaca and the *allantois* (p. 631), an endodermal sac that extends into the umbilical stalk.

A **ureteric bud** forms in the wall of each mesonephric duct, and this blind tube elongates and branches within the adjacent metanephros. Tubules developing within the metanephros then connect to the terminal branches.

The Urethra
(Figure 26-12)

The urethra extends from the neck of the urinary bladder (Figure 26-12c●, p. 676) to the exterior. The female and male urethra differ in length and in function. In the female, the urethra is very short, extending 3–5 cm (1–1.5 in.) from the bladder to the vestibule (Figure 26-12b●). The external urethral opening, or **external urethral meatus**, is situated near the anterior wall of the vagina.

In the male, the urethra extends from the neck of the urinary bladder to the tip of the penis, a distance that may be 18–20 cm (7–8 in.). The male urethra can be subdivided into three portions (Figure 26-12a,c,d●, p. 676): (1) the *prostatic urethra*, (2) the *membranous urethra*, and (3) the *penile urethra*.

The **prostatic urethra** passes through the center of the prostate gland (Figure 26-12c●, p. 676). The **membranous urethra** includes the short segment that penetrates the urogenital diaphragm, the muscular floor of the pelvic cavity. The **penile** (PĒ-nīl) **urethra** extends from the distal border of the urogenital diaphragm to the external urethral meatus at the tip of the penis (Figure 26-12a●). The functional differences between these regions will be considered in Chapter 27.

In both sexes, as the urethra passes through the urogenital diaphragm, a circular band of skeletal muscle forms the **external urethral sphincter,** or *sphincter urethra*. The contractions of both the external and internal urethral sphincters are controlled by branches of the *hypogastric plexus*. ∞ [p. 439] Only the external urethral sphincter is under voluntary control, via the perineal branch of the pudendal nerve. ∞ [p. 353] The autonomic innervation of the external sphincter becomes important only if voluntary control is lacking, as in infants or adults after spinal cord injuries. (See the section on the micturition reflex.)

Histology of the Urethra
(Figure 26-13c)

The urethral lining consists of a stratified epithelium (Figure 26-13c●, p. 677) that varies from transitional at the neck of the urinary bladder, through stratified columnar at the midpoint, to stratified squamous near the external urethral meatus. The lamina propria is thick and elastic; the mucous membrane is thrown into longitudinal folds. Mucus-secreting cells are found in the epithelial pockets, and in the male the epithelial mucous glands may form tubules that extend into the lamina propria. Connective tissues of the lamina propria anchor the urethra to surrounding structures. In the female, the lamina propria contains an extensive network of

■CLINICAL BRIEF

Urinary Tract Infections

Urinary tract infections, or **UTIs**, result from the colonization of the urinary tract by bacterial or fungal invaders. The intestinal bacterium *Escherichia coli* is most often involved, and women are particularly susceptible to urinary tract infections because of the close proximity of the external urethral orifice to the anus. Sexual intercourse may also push bacteria into the urethra and, since the female urethra is relatively short, toward the urinary bladder.

The condition may be asymptomatic (without symptoms), but it can be detected by the presence of bacteria and blood cells in the urine. If inflammation of the urethral wall occurs, the condition may be termed **urethritis**, while inflammation of the lining of the bladder represents **cystitis**. Many infections affect both regions to some degree. Urination becomes painful, a symptom known as **dysuria** (dis-Ū-rē-a), and the bladder becomes tender and sensitive to pressure. Despite the discomfort produced, the individual feels the urge to urinate frequently. Urinary tract infections usually respond to antibiotic therapies, although subsequent reinfections may occur.

veins, and the entire complex is surrounded by concentric layers of smooth muscle.

The Micturition Reflex and Urination

Urine reaches the urinary bladder by the peristaltic contractions of the ureters. The process of urination is coordinated by the **micturition reflex**. Stretch receptors in the wall of the urinary bladder are stimulated as it fills with urine. Afferent fibers in the pelvic nerves carry the impulses generated to the sacral spinal cord. Their increased level of activity (1) facilitates parasympathetic motor neurons in the sacral spinal cord, (2) stimulates contraction of the urinary bladder, and (3) stimulates interneurons that relay sensations to the cerebral cortex. As a result, we become consciously aware of the fluid pressure in the urinary bladder. Voluntary urination involves relaxing the external sphincter and descending facilitation of the micturition reflex. When the external urethral sphincter relaxes, feedback through the autonomic nervous system relaxes the internal urethral sphincter. Tensing of the abdominal and expiratory muscles increases abdominal pressures and assists in compressing the urinary bladder. In the absence of voluntary relaxation of the external urethral sphincter, reflexive relaxation of both sphincters will eventually occur as the urinary bladder nears capacity. ✝ *Problems with the Micturition Reflex [p. 775]*

√ **An obstruction of the ureters by a kidney stone would interfere with the flow of urine between what two points?**

√ **Describe how urine in the urinary bladder is prevented from returning back to the kidney.**

AGING AND THE URINARY SYSTEM

In general, aging is associated with an increased incidence of kidney problems. Age-related changes in the urinary system include:

1. *A decline in the number of functional nephrons*: The total number of kidney nephrons drops by 30–40% between ages 25 and 85.

2. *A reduction in glomerular filtration*: This reduction results from decreased numbers of glomeruli, cumulative damage to the filtration apparatus in the remaining glomeruli, and reductions in renal blood flow.

3. *Reduced sensitivity to ADH*: With age the distal portions of the nephron and collecting system become less responsive to ADH. Less reabsorption of water and sodium ions occurs as a result; urination becomes more frequent, and daily fluid requirements increase.

4. *Problems with the micturition reflex*: Several factors are involved in such problems:

a. The sphincter muscles lose muscle tone and become less effective at voluntarily retaining urine. This loss of tone leads to problems with incontinence, often involving a slow leakage of urine.

b. The ability to control micturition is often lost after a stroke, Alzheimer's disease, or other CNS problems affecting the cerebral cortex or hypothalamus.

c. In males, **urinary retention** may develop secondary to chronic inflammation of the prostate gland. In this condition swelling and distortion of prostatic tissues compress the prostatic urethra, restricting or preventing the flow of urine.

Related Clinical Terms

hemodialysis (hē-mō-dī-AL-i-sis): A technique in which an artificial membrane is used to regulate the composition of the blood. ✝ *Hemodialysis [p. 774]*

pyelogram (PĪ-e-lō-gram): An image obtained by taking an X-ray of the kidneys after a radiopaque compound has been administered. *[p. 674]*

calculi (KAL-kū-lī): "Kidney stones" formed from calcium deposits, magnesium salts, or crystals of uric acid. *[p. 677]*

nephrolithiasis (nef-rō-li-THĪ-a-sis): The presence of kidney stones. *[p. 677]*

urinary obstruction: Blockage of the conducting system by a calculus or other factors. *[p. 677]*

urinary tract infection (UTI): Urinary tract infection caused by bacteria or fungi. *[p. 680]*

urethritis: Inflammation of the urethral wall. *[p. 680]*

cystitis: Inflammation of the urinary bladder lining. *[p. 680]*

dysuria (dis-Ū-rē-a): Painful urination. *[p. 680]*

incontinence (in-KON-ti-nens): An inability to voluntarily control urination. ✝ *Problems with the Micturition Reflex [p. 775]*

automatic bladder: A condition in which the micturition reflex remains intact, but the individual cannot prevent the reflexive emptying of the urinary bladder. ✝ *Problems with the Micturition Reflex [p. 775]*

CHAPTER SUMMARY AND REVIEW

STUDY OUTLINE

Related Key Terms

INTRODUCTION *[p. 663]*

1. The functions of the urinary system include: (1) eliminating organic waste products; (2) regulating plasma concentrations of ions; (3) regulating blood volume and pressure by adjusting the volume of water lost and releasing erythropoietin and renin; (4) helping to stabilize blood pH; and (5) conserving nutrients.

ORGANIZATION OF THE URINARY SYSTEM *[p. 663]*

1. The urinary system includes the **kidneys**, the **ureters**, the **urinary bladder**, and the **urethra**. The kidneys produce **urine** (a fluid containing water, ions, and soluble compounds); during **urination** urine is forced out of the body *(see Figure 26-1).*

micturition

THE KIDNEYS *[p. 663]*

Superficial Anatomy of the Kidney *[p. 665]*
1. The kidneys are located on either side of the vertebral column between the last thoracic and third lumbar vertebrae *(see Figures 26-1 and 26-2a).*
2. The position of the kidneys in the abdominal cavity is maintained by: (1) the overlying peritoneum, (2) contact with adjacent visceral organs, and (3) supporting connective tissues *(see Figures 26-2 and 26-3).*
3. The ureter and renal blood vessels are attached to the **hilus** of the kidney *(see Figure 26-3).*

renal capsule • **adipose capsule** • **renal fascia** • **floating kidney**

Sectional Anatomy of the Kidney *[p. 665]*
4. The kidney can be divided into an outer **cortex** and inner **medulla**. The medulla contains 6–18 **renal pyramids**, whose tips, or **papillae** project into the **renal sinus**. **Renal columns** separate adjacent pyramids. A **renal lobe** contains a renal pyramid, the overlying area of renal cortex, and adjacent tissues of the renal columns *(see Figure 26-4a,c).*
5. The ureter is continuous with the **renal pelvis**, which branches to form **major** and **minor calyces** *(see Figure 26-4b–d).*

Histology of the Kidney *[p. 665]*
6. The **nephron** (the basic functional unit in the kidney) consists of a **renal tubule** that empties into the **collecting system**. From the **renal corpuscle**, the **tubular fluid** travels through the **proximal convoluted tubule (PCT)**, the **loop of Henle**, and the **distal convoluted tubule (DCT)**. It then flows through the **collecting tubule**, **collecting duct**, and **papillary duct** to reach the minor calyx *(see Figure 26-5).*
7. Nephrons are responsible for: (1) production of filtrate, (2) reabsorption of organic nutrients, and (3) reabsorption of water and ions. The **capsular epithelium** lines the outer wall of the renal corpuscle. Blood arrives via the **afferent arteriole** and departs in the **efferent arteriole** *(see Figure 26-6b,c).*
8. At the **glomerulus**, **podocytes** cover the **lamina densa** of the capillaries that project into the **capsular space**. The **pedicels** of the podocytes are separated by narrow **slit pores** *(see Figure 26-6c–e).*

Bowman's capsule • **glomerular epithelium** • **vascular pole**

9. The proximal convoluted tubule (PCT) actively reabsorbs nutrients, plasma proteins, and electrolytes from the tubular fluid. The loop of Henle includes a **descending limb** and an **ascending limb**; each limb contains a **thick segment** and a **thin segment**. The ascending limb delivers fluid to the distal convoluted tubule (DCT), which actively secretes ions and reabsorbs sodium ions from the urine *(see Figures 26-5 to 26-7)*.

10. Roughly 85% of the nephrons are **cortical nephrons** found within the cortex; the **juxtamedullary nephrons** are closer to the medulla with their loops of Henle extending deep into the renal pyramids *(see Figures 26-7d/26-8c,d)*.

The Blood Supply to the Kidneys *[p. 672]*
11. The vasculature of the kidneys includes the **renal**, **segmental**, **interlobar**, **arcuate**, and **interlobular arteries**, and the **interlobar**, **arcuate**, **interlobular**, **segmental**, and **renal veins**. Blood travels from the efferent arteriole to the **peritubular capillaries** and the **vasa recta**. Diffusion occurs between the capillaries of the vasa recta and the tubular cells *(see Figures 26-8 and 26-9)*.

Innervation of the Kidneys *[p. 675]*
12. The kidneys and ureters are innervated by **renal nerves**. Sympathetic activation regulates glomerular blood flow and pressure, stimulates renin release, and accelerates sodium ion and water reabsorption.

STRUCTURES FOR URINE TRANSPORT, STORAGE, AND ELIMINATION *[p. 675]*

1. Tubular fluid modification and urine production end when the fluid enters the renal pelvis. The rest of the urinary system is responsible for transporting, storing, and eliminating the urine *(see Figure 26-11)*.

The Ureters *[p. 675]*
2. The ureters extend from the renal pelvis to the urinary bladder and are responsible for transporting urine to the bladder *(see Figures 26-3/26-4/26-11 to 26-13)*.

The Urinary Bladder *[p. 675]*
3. The bladder is stabilized by the **urachus or middle** (*median*) **umbilical ligament** and the **lateral** (*medial*) **umbilical ligaments**. Internal features include the **trigone**, the **neck**, and the **internal urethral sphincter**. The mucosal lining contains prominent **rugae** (folds). Contraction of the **detrusor** muscle compresses the bladder and expels the urine into the urethra *(see Figures 26-12/26-13)*.

The Urethra *[p. 680]*
4. In both sexes, as the urethra passes through the urogenital diaphragm, a circular band of skeletal muscles forms the **external urethral sphincter**, which is under voluntary control *(see Figures 26-12/26-13)*.

The Micturition Reflex and Urination *[p. 680]*
5. The process of urination is coordinated by the **micturition reflex**, which is initiated by stretch receptors in the bladder wall. Voluntary urination involves coupling this reflex with the voluntary relaxation of the external urethral sphincter.

AGING AND THE URINARY SYSTEM *[p. 680]*

1. Aging is usually associated with increased kidney problems. Age-related changes in the urinary system include: (1) a declining number of functional nephrons; (2) reduced glomerular filtration; (3) reduced sensitivity to ADH; (4) problems with the micturition reflex (**urinary retention** may develop in men whose prostate glands are inflamed).

Related Key Terms
tubular pole • macula densa • juxtaglomerular cells • juxtaglomerular apparatus

ureteral opening

external urethral meatus • prostatic urethra • membranous urethra • penile urethra

---

1 REVIEW OF CHAPTER OBJECTIVES

1. Describe the functions of the urinary system and its relation to other excretory organs.
2. Identify the components of the urinary system and their functions.
3. Describe the external surface features of the kidneys and their relationships to adjacent tissues and organs.
4. Identify the structure and function of features within the kidneys.
5. Describe the histological organization and structures of the nephron.
6. Identify the blood vessels that supply blood to the nephrons.
7. Identify the innervation of the kidneys and the effects of innervation on renal function.
8. Describe the location, gross anatomy, and microstructure of the ureters, urinary bladder, and urethra.
9. Identify the functions of the ureters, urinary bladder, and urethra.
10. Discuss the micturition reflex and its control.
11. Describe the effects of aging on the urinary system.

2 REVIEW OF CONCEPTS

1. Other than the elimination of organic waste from the body, what role does the urinary tract play in fluid homeostasis?
2. How does the kidney act to conserve water in the urine to prevent dehydration of the body?
3. How does the role of the perirenal fat pad differ from that of adipose tissues in other parts of the body?
4. What is the role of the renal cortex in filtration?
5. What is the physiological function of the nephrons?
6. What is the role of Bowman's capsule in the renal corpuscle?
7. What unique role does the structure of the podocytes play in filtration?
8. How does the filtrate change as it passes along the kidney tubules?
9. What is the role of the proximal convoluted tubule (PCT) in urine formation?
10. What effects do active transport mechanisms in the thick segment of the ascending limb of the loop of Henle have on the filtrate?

11. How does the juxtaglomerular apparatus regulate blood volume and pressure?
12. What is the role of the collecting system in concentrating the glomerular filtrate?
13. What is the function of the vasa recta?
14. How do the renal nerves affect the function of the kidney?
15. What functions of the urinary system are unique to the bladder?
16. Why do problems that cause retention of urine in the bladder have the potential to cause kidney damage?
17. How does the urethra in the male differ from that in the female?
18. What is the role of the micturition reflex in controlling bladder and urethral activity?
19. What physiological changes accompany the training that an infant undergoes when learning to control the timing of urine release?

3 CRITICAL THINKING AND CLINICAL APPLICATION QUESTIONS

1. Why can a blow to the back near the waist cause serious internal organ injury?
2. A person suffering from renal failure has been put on a diet restricted in water and salt, with a low protein content. What effects should this diet have on kidney function? If this dietary treatment is unsuccessful, what treatments can improve this person's health?
3. Among the most common medical assays is the analysis of urine composition. What kinds of information can be obtained from this test?
4. A young woman complains of pain and tenderness in the lower pelvic area, an urge to urinate frequently, and pain upon urination. Urinalysis shows she has blood and bacteria in her urine. What relatively common condition could she have, and why might she be particularly susceptible to this type of a problem?

27

The Reproductive System

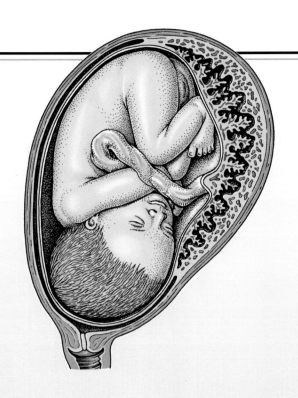

CHAPTER OUTLINE AND OBJECTIVES

An individual's life span can be measured in decades, but the human species has survived for hundreds of thousands of years through the activities of the reproductive system. The human reproductive system produces, stores, nourishes, and transports functional male and female reproductive cells, or **gametes** (GAM-ēts). The combination of the genetic material provided by a **sperm** from the father and an **ovum** (Ō-vum) from the mother occurs shortly after **fertilization**. Fertilization creates a **zygote** (ZĪ-gōt), a single cell whose divisions will, in approximately 9 months, produce an infant who will mature as part of the next generation. The reproductive system also produces sex hormones that affect the structure and function of all other systems.

This chapter will describe the structures and mechanisms involved with the production and maintenance of gametes and, in the female, the development and support of the developing embryo and fetus. Chapter 28, the last chapter in the text, will begin at fertilization and consider the process of development.

ORGANIZATION OF THE REPRODUCTIVE SYSTEM

The reproductive system includes:

- Reproductive organs, or **gonads** (GŌ-nadz), that produce gametes and hormones,
- Ducts that receive and transport the gametes,
- Accessory glands and organs that secrete fluids into these or other excretory ducts, and
- Perineal structures associated with the reproductive system. These perineal structures are collectively known as the **external genitalia** (jen-i-TĀ-lē-a).

The male and female reproductive systems are functionally quite different. In the adult male, the gonads, or **testes** (TES-tēz; singular *testis*), secrete *androgens*, principally testosterone, and produce one-half billion sperm per day. After storage, mature sperm travel along a lengthy duct system where they are mixed with the secretions of accessory glands, creating the mixture known as **semen** (SĒ-men). During **ejaculation** (e-jak-ū-LĀ-shun), the semen is expelled from the body.

The female gonads, or **ovaries**, typically produce only one mature ovum per month. This gamete travels along short **uterine tubes** (*oviducts*) that end in the muscular **uterus** (Ū-ter-us). A short passageway, the **vagina** (va-JĪ-na), connects the uterus with the exterior. During intercourse, male ejaculation introduces semen into the vagina, and the sperm cells ascend the female reproductive tract, where they may encounter one or more female gametes.

ANATOMY OF THE MALE REPRODUCTIVE SYSTEM
(Figure 27-1)

The principal structures of the male reproductive system are shown in Figure 27-1●. Proceeding from the testes, the sperm cells, or **spermatozoa** (sper-ma-tō-ZŌ-a), travel along the **epi-**didymis (ep-i-DID-i-mus), the **ductus deferens** (DUK-tus DEF-e-renz), or *vas deferens*, the **ejaculatory** (ē-JAK-ū-la-tō-rē) **duct**, and the **urethra** before leaving the body. Accessory organs, notably the **seminal** (SEM-i-nal) **vesicles**, the **prostate** (PROS-tāt) **gland**, and the **bulbourethral** (bul-bō-ū-RĒ-thral) **glands** secrete into the ejaculatory ducts and urethra. The external genitalia include the **scrotum** (SKRŌ-tum), which encloses the testes, and the **penis** (PĒ-nis), an erectile organ that surrounds the distal portions of the urethra.

The Testes
(Figures 27-1/27-2)

Each testis has the shape of a flattened oval roughly 5 cm (2 in.) long and 2.5 cm (1 in.) wide, with a weight of 10–15 g (0.35–0.53 oz). The testes hang within the scrotum, a fleshy pouch suspended inferior to the perineum and anterior to the anus. Note the location and relation of the testes in sagittal section (Figure 27-1●) and in frontal section (Figure 27-2●).

Descent of the Testes
(Figure 27-3)

During development, the testes form inside the body cavity adjacent to the kidneys. The relative positions of these organs change as the fetus enlarges, and the testes gradually move inferiorly and anteriorly toward the anterior abdominal wall. The **gubernaculum testis** is a ribbon of connective tissue fibers that extends from each testis to the posterior wall of a small, inferior pocket of the peritoneum. As growth proceeds, the gubernacula do not elongate, and the testes are held in position (Figure 27-3a●, p. 688). During the seventh developmental month (1) growth continues at a rapid pace, and (2) circulating hormones stimulate contraction of the gubernaculum testis. Over this period the relative position of the testes changes further, and they move through the abdominal musculature accompanied by small pockets of the peritoneal cavity. This process is known as the **descent of the testes** (Figure 27-3b●).

As it moves through the body wall, each testis is accompanied by the ductus deferens and the testicular blood vessels, nerves, and lymphatics. Together these structures form the body of the *spermatic cord*.

The Spermatic Cord
(Figure 27-2)

The **spermatic cords** consist of layers of fascia, tough connective tissue, and muscle enclosing the blood vessels, nerves, and lymphatics supplying the testes. Each spermatic cord begins at the deep inguinal ring, extends along the inguinal canal, exits at the superficial inguinal ring, and descends into the scrotum (Figure 27-2●). The spermatic cords form during the descent of the testes. ☨*Testicular Torsion [p. 775]*

Each spermatic cord contains the ductus deferens, the testicular artery, the **pampiniform plexus** (pam-PIN-i-form; *pampinus*, tendril + *forma*, form) of the testicular vein, and the **ilioinguinal** and **genitofemoral nerves** from the lumbar plexus.

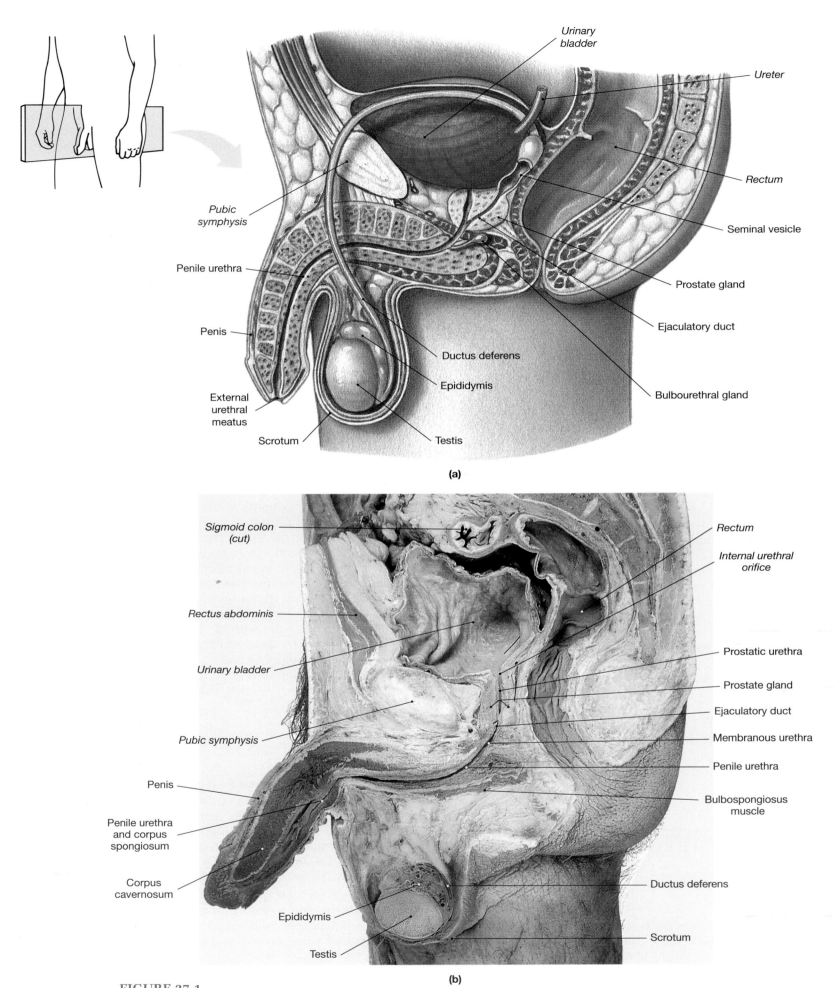

Urinary
bladder

Ureter

Pubic
symphysis

Rectum

Seminal vesicle

Penile urethra

Prostate gland

Penis

Ejaculatory duct

Ductus deferens

External
urethral
meatus

Epididymis

Bulbourethral gland

Scrotum

Testis

(a)

Sigmoid colon
(cut)

Rectum

Internal urethral
orifice

Rectus abdominis

Prostatic urethra

Urinary bladder

Prostate gland

Ejaculatory duct

Pubic symphysis

Membranous urethra

Penis

Penile urethra

Penile urethra
and corpus
spongiosum

Bulbospongiosus
muscle

Corpus
cavernosum

Ductus deferens

Epididymis

Scrotum

Testis

(b)

FIGURE 27-1
The Male Reproductive System, Sagittal View. (a) Diagrammatic view of the male reproductive system in sagittal section.
(b) Male reproductive system and relation to abdominal organs as seen in sagittal section.

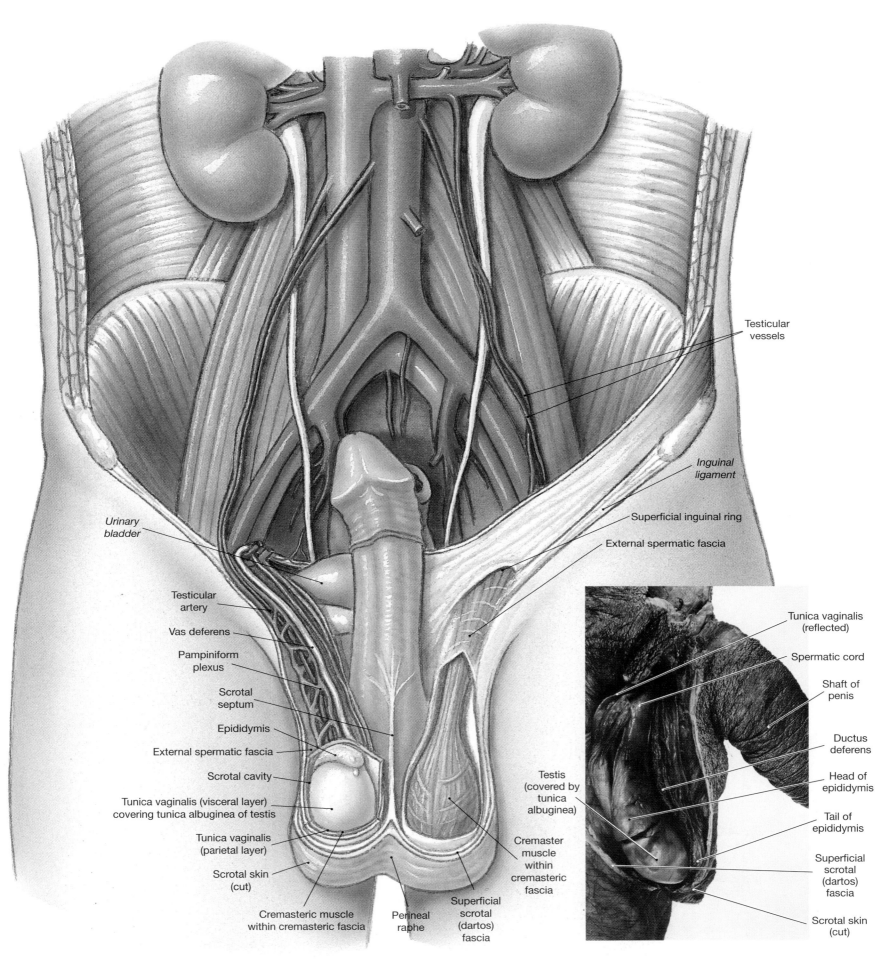

Testicular
vessels

Inguinal
ligament

Superficial inguinal ring

External spermatic fascia

Urinary
bladder

Testicular
artery

Vas deferens

Pampiniform
plexus

Scrotal
septum

Epididymis

External spermatic fascia

Scrotal cavity

Tunica vaginalis (visceral layer)
covering tunica albuginea of testis

Tunica vaginalis
(parietal layer)

Scrotal skin
(cut)

Cremasteric muscle
within cremasteric fascia

Perineal
raphe

Superficial
scrotal
(dartos)
fascia

Testis
(covered by
tunica
albuginea)

Cremaster
muscle
within
cremasteric
fascia

Tunica vaginalis
(reflected)

Spermatic cord

Shaft of
penis

Ductus
deferens

Head of
epididymis

Tail of
epididymis

Superficial
scrotal
(dartos)
fascia

Scrotal skin
(cut)

FIGURE 27-2
The Male Reproductive System, Frontal and Lateral View

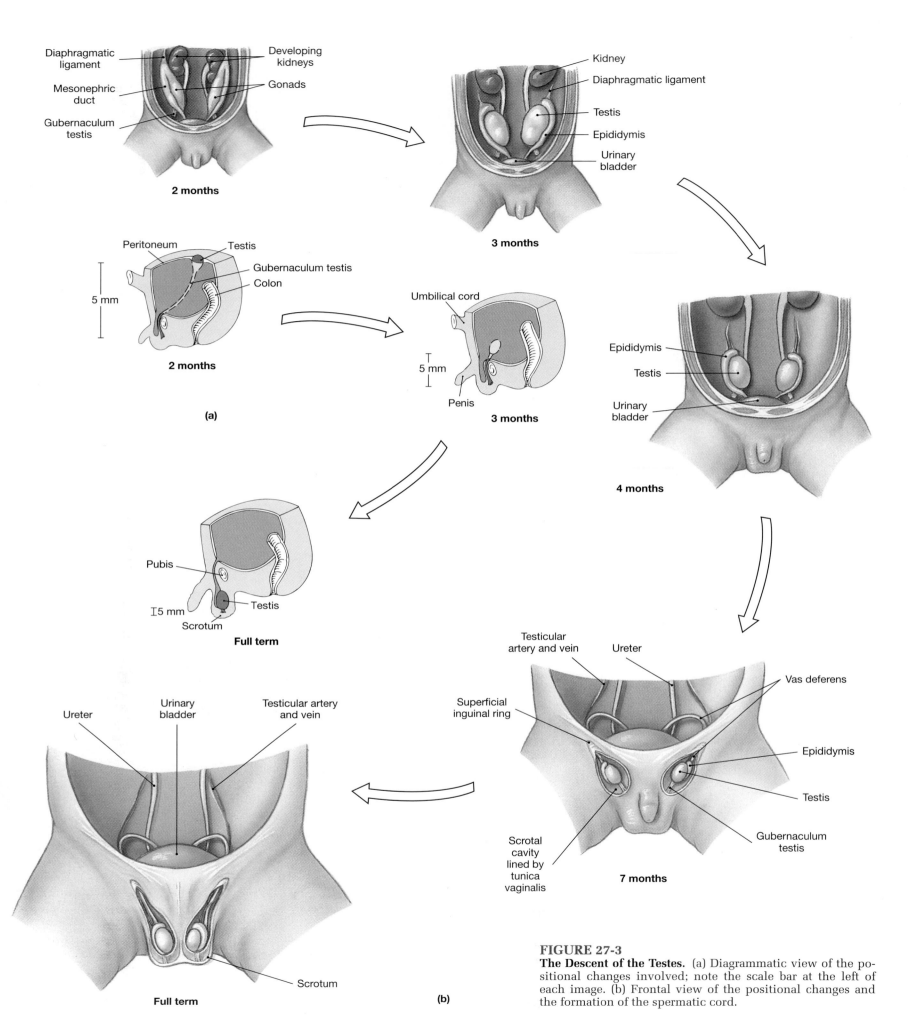

FIGURE 27-3

The Descent of the Testes. (a) Diagrammatic view of the positional changes involved; note the scale bar at the left of each image. (b) Frontal view of the positional changes and the formation of the spermatic cord.

Diaphragmatic ligament
Developing kidneys
Mesonephric duct
Gonads
Gubernaculum testis

2 months

Kidney
Diaphragmatic ligament
Testis
Epididymis
Urinary bladder

3 months

Peritoneum
Testis
Gubernaculum testis
Colon
5 mm

2 months

(a)

Umbilical cord
5 mm
Penis

3 months

Epididymis
Testis
Urinary bladder

4 months

Pubis
5 mm
Testis
Scrotum

Full term

Testicular artery and vein
Ureter
Superficial inguinal ring
Vas deferens
Epididymis
Testis
Scrotal cavity lined by tunica vaginalis
Gubernaculum testis

7 months

Ureter
Urinary bladder
Testicular artery and vein
Scrotum

Full term

(b)

The narrow canals linking the scrotal chambers with the peritoneal cavity are called the *inguinal canals.* These passageways usually close, but the persistence of the spermatic cords creates weak points in the abdominal wall that remain throughout life. As a result, *inguinal hernias,* discussed in Chapter 10, are relatively common in males. ∞ [p. 274] The inguinal canals in females are very small, containing only the ilioinguinal nerves and the round ligaments of the uterus. The abdominal wall is nearly intact, and inguinal hernias in women are very rare.

The Scrotum and the Position of the Testes
(Figures 27-2/27-4a)

The scrotum is divided internally into two separate chambers, and the partition between the two is marked by a raised thickening in the scrotal surface known as the **perineal raphe** (RĀ-fē) (Figures 27-2 and 27-4a●). Each testis lies in a separate compartment, or **scrotal cavity**, with a narrow space separating the inner surface of the scrotum from the outer surface of the testis. A serous membrane, the **tunica vaginalis**, lines the scrotal cavity and reduces friction between the opposing surfaces.

The scrotum consists of a thin layer of skin and the underlying superficial fascia. The dermis contains a layer of smooth muscle, the **dartos** (DAR-tōs), and tonic contraction of the dartos causes the characteristic wrinkling of the scrotal surface. A layer of skeletal muscle, the **cremaster** (kre-MAS-ter) **muscle**, lies beneath the dermis. Contraction of the cremaster tenses the scrotum and pulls the testes closer to the body. These contractions are controlled by the *cremasteric reflex* (Table 17-2, p. 441). Contraction occurs during sexual arousal and in response to changes in temperature. Normal sperm development in the testes requires temperatures around 1.1°C (2°F) lower than those found elsewhere in the body. The cremaster moves the testes away from or toward the body as needed to maintain acceptable testicular temperatures. When air or body temperatures rise, the cremaster relaxes and the testes then move away from the body. Cooling the scrotum, as when entering a cold swimming pool, results in cremasteric contractions that pull the testes closer to the body and keep testicular temperatures from falling.

The scrotum is richly supplied with sensory and motor nerves from the *hypogastric plexus* and branches of the *ilioinguinal nerves,* the *genitofemoral nerves,* and the *pudendal nerves.* ∞ [pp. 353, 354] The vascular supply to the scrotum includes the **internal pudendal arteries** (from the internal iliac arteries), the **external pudendal arteries** (from the femoral arteries), and the *cremasteric branch* of the **inferior epigastric arteries** (from the external iliac arteries). ∞ [pp. 560, 562] The names and distributions of the veins follow those of the arteries.

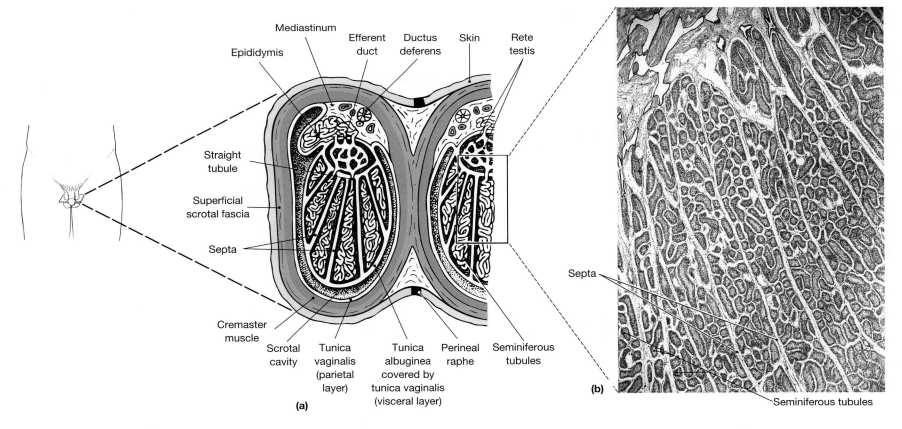

FIGURE 27-4
Structure of the Testes. (a) Diagrammatic sketch and anatomical relationships of the testes. (b) Photomicrograph showing general view of septa separating seminiferous tubules. (LM × 25)

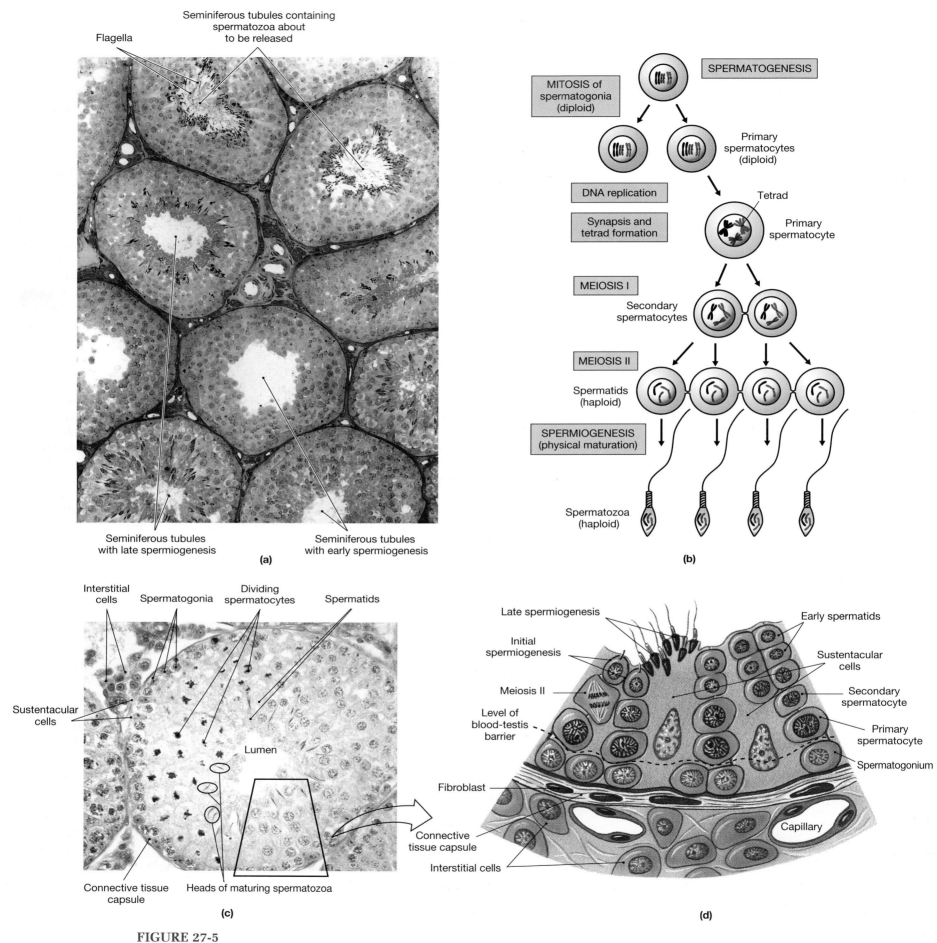

FIGURE 27-5

The Seminiferous Tubules. (a) Seminiferous tubules in sectional view. (b) Meiosis in the testes. (c) Spermatogenesis within one segment of a seminiferous tubule. (LM × 983) (d) Diagrammatic view of the cells in the wall of a seminiferous tubule.

Structure of the Testes
(Figure 27-4)

Beneath the serous membrane covering the testis lies the **tunica albuginea** (TŪ-ni-ka al-bū-JIN-ē-a), a dense layer of connective tissue rich in collagen fibers. The fibers of this network are continuous with those surrounding the adjacent epididymis. The collagen fibers of the tunica albuginea also extend into the substance of the testis, forming fibrous partitions, or *septa* (Figure 27-4●). These septa converge toward the **mediastinum** of the testis. The mediastinum supports the blood vessels and lymphatics supplying the testis and the ducts transporting sperm into the epididymis.

Histology of the Testes
(Figures 27-4/27-5/27-7a)

The septa subdivide the testis into a series of **lobules**. Roughly 800 slender, tightly coiled **seminiferous** (sem-in-IF-er-us) **tubules** are distributed among the lobules (Figure 27-4●). Each tubule averages around 80 cm (31 in.) in length, and a typical testis contains nearly one-half mile of seminiferous tubules. Sperm production occurs within these tubules.

Each seminiferous tubule has the form of a U connected to a single **straight tubule** that enters the mediastinum of the testis (Figure 27-7a●). Within the mediastinum, these tubules are extensively interconnected, forming a maze of passageways known as the **rete testis** (RĒ-tē; *rete*, net). Fifteen to twenty large **efferent ducts** connect the rete testis to the epididymis.

Because the seminiferous tubules are tightly coiled, histological preparations most often show them in transverse section. Each tubule is surrounded by a delicate capsule, and loose connective tissue fills the spaces between the tubules. Within those spaces are found numerous blood vessels and large **interstitial cells** (*cells of Leydig*) (Figure 27-5a,c,d●). Interstitial cells are responsible for the production of male sex hormones, called *androgens*. *Testosterone* is the most important androgen. (Testosterone and other sex hormones were introduced in Chapter 19.) ∞ [p. 499] Testosterone (1) stimulates spermatogenesis; (2) promotes the physical and functional maturation of spermatozoa; (3) maintains the accessory organs of the male reproductive tract; (4) determines secondary sexual characteristics such as the distribution of facial hair and adipose tissue, muscle mass, and total body size; (5) stimulates growth and metabolism throughout the body; and (6) influences brain development by stimulating sexual behaviors and sexual drive.

Testosterone production accelerates markedly at puberty, initiating sexual maturation and the appearance of secondary sexual characteristics.

Spermatogenesis and Meiosis
(Figure 27-5)

Sperm cells, or spermatozoa, are produced through the process of **spermatogenesis** (sper-ma-tō-JEN-e-sis). Spermatogenesis begins at the outermost layer of cells in the seminiferous tubules. Stem cells called **spermatogonia** (sper-mat-ō-GŌ-nē-a) divide mitotically throughout adult life. One of the daughter cells remains as an undifferentiated stem cell, while the other is pushed toward the lumen of the tubule. The latter cell differentiates into a **primary spermatocyte** (sper-MAT-ō-sīt) that prepares to begin *meiosis*, a form of cell division that produces gametes containing one-half the normal number of chromosomes.

Because they contain one member of each pair of chromosomes, gametes are called **haploid** (HAP-loid; *haplo*, single).

Meiosis (mī-Ō-sis) is a special form of cell division leading to the production of sperm or ova. Mitosis and meiosis differ significantly in terms of the events that take place in the nucleus. Mitosis involves a single division that produces two daughter cells, each containing 23 pairs of chromosomes. Meiosis involves a pair of divisions, and produces four gametes that each contain 23 individual chromosomes.

In the testes, the first step in meiosis is the division of a primary spermatocyte to produce a pair of **secondary spermatocytes**. Each secondary spermatocyte then divides, producing a pair of **spermatids** (SPER-ma-tidz). As a result, four spermatids are produced for every primary spermatocyte that enters meiosis (Figure 27-5b●). Spermatogonia, spermatocytes undergoing meiosis, and spermatids are depicted in Figure 27-5c,d●.

Spermatogenesis is stimulated by circulating FSH (follicle-stimulating hormone) and testosterone; testosterone is produced by interstitial cells of the testis in response to LH (luteinizing hormone). ∞ [p. 488]

Spermiogenesis
(Figure 27-5b,d)

Each spermatid matures into a single **spermatozoon** (sper-ma-tō-ZŌ-on), or sperm cell. This maturation process is called **spermiogenesis** (Figure 27-5b,d●). During spermiogenesis, spermatids are embedded within the cytoplasm of large **sustentacular** (sus-ten-TAK-ū-lar) **cells** (*Sertoli cells*). Sustentacular cells are attached to the tubular capsule and extend toward the lumen between the spermatocytes undergoing meiosis. As spermiogenesis proceeds, the spermatids gradually develop the appearance of mature spermatozoa. At *spermiation*, a spermatozoon loses its attachment to the sustentacular cell and enters the lumen of the seminiferous tubule. The entire process, from spermatogonial division to spermiation, takes approximately 9 weeks.

Sustentacular Cells
(Figure 27-5b,d)

Sustentacular cells have five important functions:

1. *Maintenance of the blood-testis barrier*: The seminiferous tubules are isolated from the general circulation by a **blood-testis barrier** (Figure 27-5d●) comparable to the blood-brain barrier. ∞ [p. 371] Tight junctions between extensions of sustentacular cells isolate the inner portions of the seminiferous tubule from the surrounding interstitial fluid. Transport across the sustentacular cells is tightly regulated so that conditions inside the tubule remain very stable. The lumen of a seminiferous tubule contains a fluid very different from interstitial fluid; for example, tubular fluid is high in androgens, estrogens, potassium, and amino acids. The blood-testis barrier is essential to preserving these differences. In addition, developing spermatozoa contain sperm-specific antigens in their cell membranes. These antigens, not found in somatic cell membranes, would be attacked by the immune system if the blood-testis barrier did not prevent their detection.

2. *Support of spermatogenesis*: Spermatogenesis depends on the stimulation of sustentacular cells by circulating FSH and testosterone. Stimulated sustentacular cells then in

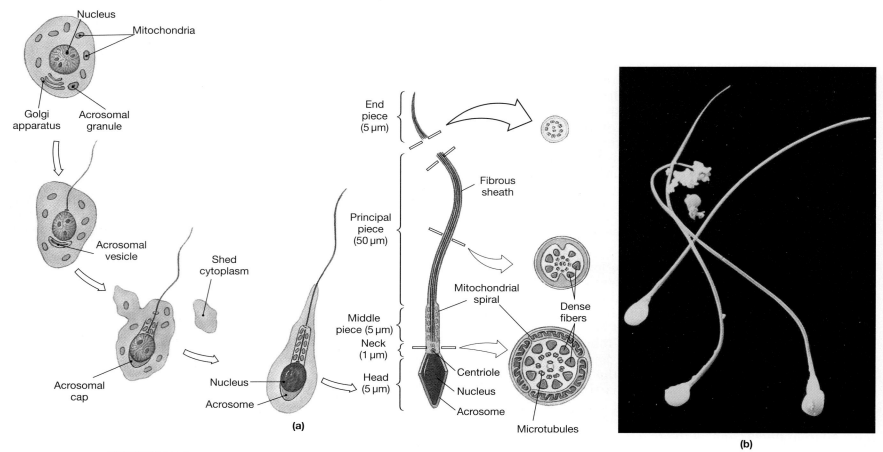

FIGURE 27-6
Spermiogenesis and Spermatozoon Structure. (a) Differentiation of a spermatid into a spermatozoon. (b) Human spermatozoa. (SEM × 1688)

some way promote the division of spermatogonia and the meiotic divisions of spermatocytes.

3. *Support of spermiogenesis*: Spermiogenesis requires the presence of sustentacular cells. These cells surround and enfold the spermatids, providing nutrients and chemical stimuli that promote their development.

4. *Secretion of inhibin*: Sustentacular cells secrete a hormone, **inhibin** (in-HIB-in). Inhibin, introduced in Chapter 19, depresses the pituitary production of follicle-stimulating hormone (FSH) and gonadotropin-releasing hormone (GnRH). ∞ [p. 499] The faster the rate of sperm production, the greater the amount of inhibin secreted.

5. *Secretion of androgen-binding protein*: **Androgen-binding protein (ABP)** binds androgens (primarily testosterone) in the fluid contents of the seminiferous tubules. This protein is thought to be important in elevating the concentration of androgens within the tubules and stimulating spermiogenesis.

Anatomy of a Spermatozoon
(Figure 27-6)

There are three distinct regions to each spermatozoon: the *head*, the *middle piece*, and the *tail* (Figure 27-6●).

- The **head** is a flattened oval containing densely packed chromosomes. The tip contains the **acrosomal** (ak-rō-

SŌ-mal) **cap**, a membranous compartment containing enzymes involved in the preliminary steps of fertilization.

- A short **neck** attaches the head to the **middle piece**. The neck contains both centrioles of the original spermatid. The microtubules of the distal centriole are continuous with those of the middle piece and tail. Mitochondria arranged in a spiral around the microtubules provide the energy needed to move the tail.

- The **tail** is the only example of a flagellum in the human body. ∞ [p. 36] A **flagellum** moves a cell from one place to another. Although cilia beat in a predictable, waving fashion, the flagellum of a spermatozoon has a complex, corkscrew motion. The microtubules of the flagellum are surrounded by a dense, fibrous sheath.

Unlike other, less specialized cells, a mature spermatozoon lacks an endoplasmic reticulum, Golgi apparatus, lysosomes, peroxisomes, inclusions, and many other intracellular structures. Because the cell does not contain glycogen or other energy reserves, it must absorb nutrients (primarily fructose) from the surrounding fluid.

The Male Reproductive Tract

The testes produce physically mature spermatozoa that are incapable of fertilizing an ovum. The other portions of the male reproductive system are concerned with the functional maturation, nourishment, storage, and transport of spermatozoa.

692

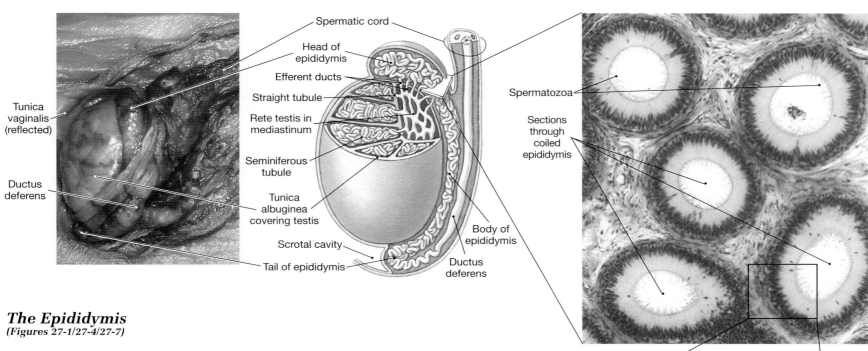

(a)

(b)

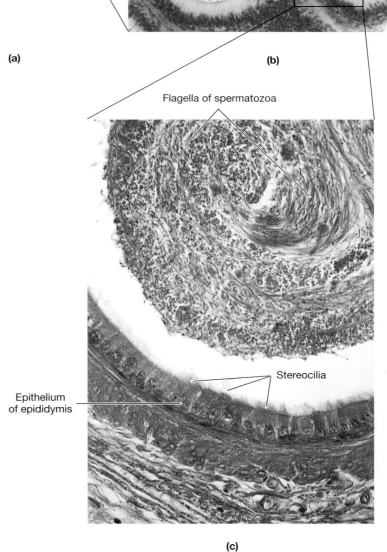

Flagella of spermatozoa

Stereocilia

Epithelium of epididymis

(c)

The Epididymis
(Figures 27-1/27-4/27-7)

Late in their development, the spermatozoa become detached from the sustentacular cells and lie within the lumen of the seminiferous tubule. Although they have most of the physical characteristics of mature sperm cells, they are still functionally immature and incapable of coordinated locomotion or fertilization. Fluid currents then transport the cell along the straight tubule, through the rete testis (Figure 27-7a●), and into the epididymis. The epididymis is lined by a distinctive simple columnar epithelium with long stereocilia (Figure 27-7b,c●).

The epididymis lies along the posterior border of the testis (Figures 27-1, p. 686, 27-4, p. 689, and 27-7a●). It has a firm texture and can be felt through the skin of the scrotum. The epididymis consists of a tubule almost 7 m (23 ft) long, coiled and twisted so as to take up very little space. The epididymis has a *head*, a *body*, and a *tail*.

- The superior **head** receives spermatozoa via the efferent ducts of the mediastinum of the testis.
- The **body** begins distal to the last efferent duct and extends inferiorly along the posterior margin of the testis.
- Near the inferior border of the testis, the number of convolutions decreases, marking the start of the **tail**. The tail recurves, and as it ascends, its histological organization changes until the stereocilia disappear and it becomes indistinguishable from that of the attached ductus deferens. The tail is the principal region involved with sperm storage.

The epididymis has the following functions:

1. It *monitors and adjusts the composition of the tubular fluid*. The columnar epithelial lining of the epididymis bears distinctive *stereocilia* (Figure 27-7b,c●) that increase the surface area available for absorption and secretion into the tubular fluid. ∞ [p. 58]

2. It *acts as a recycling center for damaged spermatozoa*. Cellular debris and damaged spermatozoa are absorbed, and the products of enzymatic breakdown are released into the surrounding interstitial fluids for pickup by the epididymal circulation.

3. It *stores spermatozoa and facilitates their functional maturation*. It takes about 2 weeks for a spermatozoon to pass through the epididymis, and during this period it

FIGURE 27-7
The Epididymis. (a) (Left) Appearance of the testis and epididymis on gross dissection. (Right) Diagrammatic view of the testis and epididymis, showing the sectional plane of (b). (b) A light micrograph showing the organization of tubules and the surrounding connective tissues. (LM × 49) (c) A micrograph showing epithelial characteristics, especially the elongate stereocilia characteristic of the epididymis. (LM × 1304)

completes its functional maturation. Although spermatozoa leaving the epididymis are mature, they remain immobile. To become active, motile, and fully functional, the spermatozoa must undergo **capacitation**, and the epididymis secretes a substance that prevents premature capacitation. Capacitation normally occurs in two steps: Spermatozoa become motile when mixed with secretions of the seminal vesicles, and they become capable of successful fertilization when exposed to conditions inside the female reproductive tract.

Transport along the epididymis involves some combination of fluid movement and peristaltic contractions of smooth muscle. After passing along the tail of the epididymis, the spermatozoa enter the ductus deferens.

The Ductus Deferens
(Figures 27-1/27-2/27-7a/27-8a,b)

The ductus deferens, or *vas deferens*, is 40–45 cm (16–18 in.) long. It begins at the tail of the epididymis (Figure 27-7a●) and ascends through the inguinal canal in the spermatic cord (Figure 27-2●, p. 687). Inside the abdominal cavity, the ductus deferens passes posteriorly, curving inferiorly along the lateral surface of the urinary bladder toward the superior and posterior margin of the prostate gland (Figure 27-1●, p. 686). Just before it reaches the prostate and seminal vesicles, the ductus deferens becomes enlarged, and the expanded portion is known as the **ampulla** (am-PYŪL-la) (Figure 27-8a●).

The wall of the ductus deferens contains a thick layer of smooth muscle (Figure 27-8b●). Peristaltic contractions in this layer propel spermatozoa and fluid along the duct, which is lined by a pseudostratified ciliated columnar epithelium. In addition to transporting sperm, the ductus deferens can store spermatozoa for several months. During this time the spermatozoa remain in a state of suspended animation, with low metabolic rates.

The junction of the ampulla with the base of the seminal vesicle marks the start of the **ejaculatory duct.** This relatively short passageway (2 cm, or less than 1 in.) penetrates the muscular wall of the prostate gland and empties into the urethra (Figure 27-1●, p. 686) near the ejaculatory duct from the other side.

The Urethra
(Figures 27-1/27-9)

The urethra of the male extends from the urinary bladder to the tip of the penis, a distance of 15–20 cm (6–8 in.). It is divided into *prostatic*, *membranous*, and *penile regions* (Figures 27-1, p. 686, and 27-9●, p. 697). These subdivisions were considered in Chapter 26. ∞ [p. 680] The urethra in the male is a passageway used by both the urinary and reproductive systems.

The Accessory Glands
(Figures 27-1/27-8a)

The fluids contributed by the seminiferous tubules and the epididymis account for only about 5% of the final volume of semen. The seminal fluid is a mixture of the secretions of many different glands, each with distinctive biochemical characteristics. Important glands include the *seminal vesicles*, the *prostate gland*, and the *bulbourethral glands* (Figures 27-1, p. 686, and 27-8a●). Major functions of these glands include: (1) activating the spermatozoa; (2) providing the nutrients spermatozoa need for motility; (3) propelling spermatozoa and fluids along the reproductive tract, primarily through peristaltic contractions; and (4) producing buffers that counteract the acidity of the urethral and vaginal contents.

The Seminal Vesicles
(Figures 27-1a/27-8a,c/27-9a)

The ductus deferens on each side ends at the junction between the ampulla and the duct draining the seminal vesicle. The seminal vesicles are embedded in connective tissue on either side of the midline, sandwiched between the posterior wall of the urinary bladder and the rectum. Each seminal vesicle is a tubular gland, with a total length of around 15 cm (6 in.). The body of the gland has many short side branches, and the entire assemblage is coiled and folded into a compact, tapered mass roughly 5 cm by 2.5 cm (2 in. by 1 in.). The location of the seminal vesicles can be seen in Figures 27-1a, p. 686 (lateral view), 27-8a (posterior view), and 27-9a (anterior view)●, p. 697.

The seminal vesicles are extremely active secretory glands, with an epithelial lining that contains extensive folds (Figure 27-8c●). The seminal vesicles contribute about 60% of the volume of semen, and although the vesicular fluid usually has the same osmotic concentration as blood plasma, the composition is quite different. In particular, the secretion of the seminal vesicles contains relatively high concentrations of fructose, which is easily metabolized by spermatozoa. Seminal fluid is discharged into the ductus deferens at *emission*, when peristaltic contractions are under way in the ductus deferens, seminal vesicles, and prostate gland. These contractions are under the control of the sympathetic nervous system. When mixed with the secretions of the seminal vesicles, previously inactive but functional spermatozoa begin beating their flagella, becoming highly mobile.

The Prostate Gland
(Figures 27-1/27-8a,e/27-9a)

The prostate gland is a small, muscular, rounded organ with a diameter of about 4 cm (1.6 in.). The prostate gland encircles the proximal portion of the urethra as it leaves the urinary bladder, as shown in Figures 27-1, p. 686 (sagittal view), 27-8a (posterior view), and 27-9a (anterior view)●, p. 697. The glandular tissue of the prostate consists of a cluster of 30–50 *compound tubuloalveolar glands* (Figure 27-8e●). ∞ [p. 62] These glands are surrounded and wrapped in a thick blanket of smooth muscle fibers.

The prostatic glands produce weakly acidic secretion, **prostatic fluid**, that contributes 20–30% of the volume of semen. In addition to several other compounds of uncertain significance, prostatic secretions contain **seminalplasmin** (sem-i-nal-PLAZ-min), an antibiotic that may help prevent urinary tract infections in males. These secretions are ejected into the prostatic urethra by peristaltic contractions of the muscular wall. ⊤*Prostate Cancer [p. 775]*

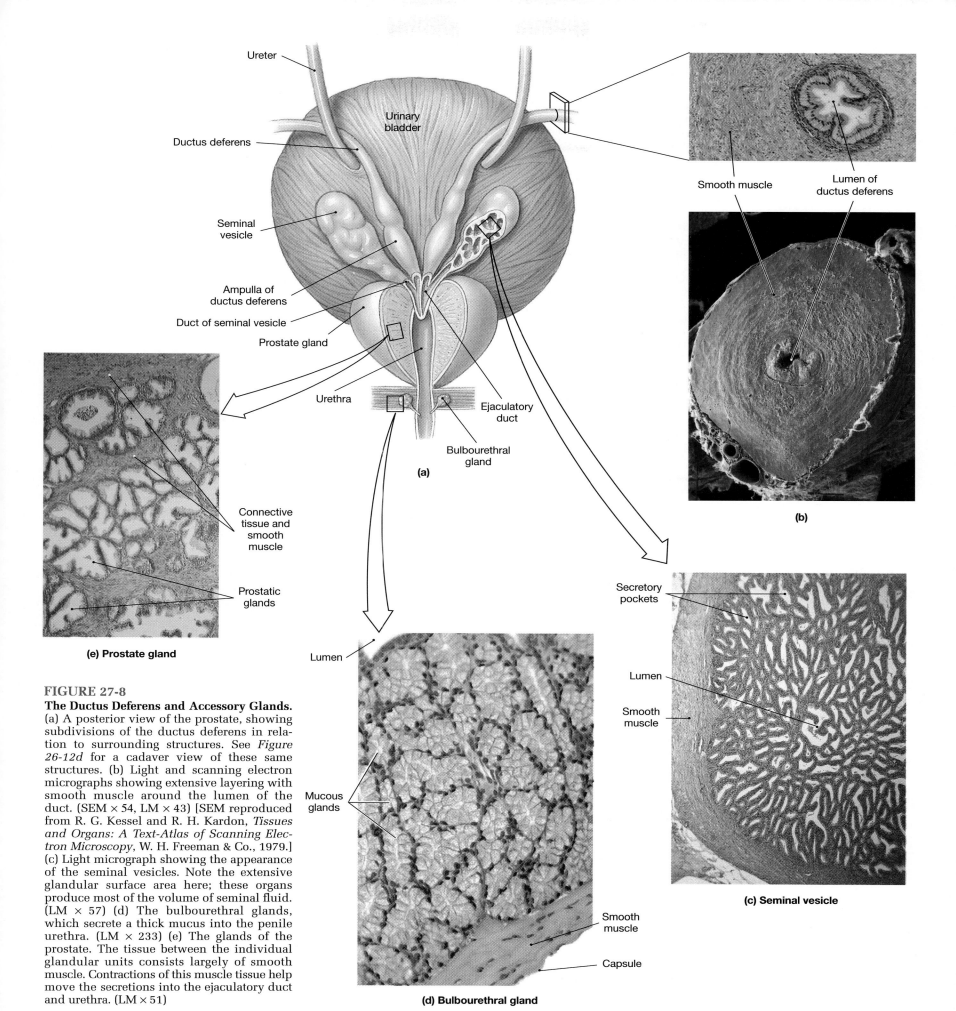

Ureter

Urinary bladder

Ductus deferens

Seminal vesicle

Smooth muscle

Lumen of ductus deferens

Ampulla of ductus deferens

Duct of seminal vesicle

Prostate gland

Urethra

Ejaculatory duct

Bulbourethral gland

(a)

(b)

Connective tissue and smooth muscle

Prostatic glands

(e) Prostate gland

Secretory pockets

Lumen

Smooth muscle

(c) Seminal vesicle

Lumen

Mucous glands

Smooth muscle

Capsule

(d) Bulbourethral gland

FIGURE 27-8

The Ductus Deferens and Accessory Glands.
(a) A posterior view of the prostate, showing subdivisions of the ductus deferens in relation to surrounding structures. See *Figure 26-12d* for a cadaver view of these same structures. (b) Light and scanning electron micrographs showing extensive layering with smooth muscle around the lumen of the duct. (SEM × 54, LM × 43) [SEM reproduced from R. G. Kessel and R. H. Kardon, *Tissues and Organs: A Text-Atlas of Scanning Electron Microscopy*, W. H. Freeman & Co., 1979.] (c) Light micrograph showing the appearance of the seminal vesicles. Note the extensive glandular surface area here; these organs produce most of the volume of seminal fluid. (LM × 57) (d) The bulbourethral glands, which secrete a thick mucus into the penile urethra. (LM × 233) (e) The glands of the prostate. The tissue between the individual glandular units consists largely of smooth muscle. Contractions of this muscle tissue help move the secretions into the ejaculatory duct and urethra. (LM × 51)

The Bulbourethral Glands
(Figures 27-1/27-8d/27-9a)

The paired bulbourethral glands, or *Cowper's glands*, are situated at the base of the penis, covered by the fascia of the urogenital diaphragm as seen in Figures 27-1, p. 686 (sagittal view), 27-8a (posterior view), and 27-9a (anterior view)●. The bulbourethral glands are round, with diameters approaching 10 mm (less than 0.5 in.). The duct of each gland travels alongside of the penile urethra for 3–4 cm (1.2–1.6 in.) before emptying into the urethral lumen. These are compound, tubuloalveolar mucous glands (Figure 27-8d●) that secrete a thick, sticky, alkaline mucus. This secretion helps to neutralize any urinary acids that may remain in the urethra and provides lubrication for the tip of the penis.

Semen

A typical ejaculation releases 2–5 mℓ of semen. This volume of fluid, called an **ejaculate**, contains:

1. *Spermatozoa*: A normal **sperm count** ranges from 20 to 100 million spermatozoa per milliliter.
2. *Seminal fluid*: **Seminal fluid**, the fluid component of semen, is a mixture of glandular secretions with a distinctive ionic and nutrient composition. In terms of total volume, seminal fluid contains the combined secretions of the seminal vesicles (60%), the prostate (30%), the sustentacular cells and epididymis (5%), and the bulbourethral glands (<5%).
3. *Enzymes*: Several important enzymes are present in the seminal fluid, including: (1) a protease that may help to dissolve mucous secretions in the vagina and (2) seminalplasmin, an antibiotic enzyme that kills a variety of bacteria including *Escherichia coli*.

The Penis
(Figures 27-1/27-2/27-9)

The penis is a tubular organ that contains the distal portion of the urethra (Figures 27-1, p. 686, and 27-2●, p. 687). It conducts urine to the exterior and introduces semen into the female vagina during sexual intercourse. The penis (Figure 27-9●) is divided into three regions:

- The **root** of the penis is the fixed portion that attaches the penis to the rami of the ischium. This connection occurs within the urogenital triangle immediately inferior to the pubic symphysis.
- The **body (shaft)** of the penis is the tubular, movable portion. Masses of *erectile tissue* are found within the body.
- The **glans** of the penis is the expanded distal end that surrounds the external urethral meatus.

The skin overlying the penis resembles that of the scrotum. The dermis contains a layer of smooth muscle, and the underlying loose connective tissue allows the thin skin to move without distorting underlying structures. The subcutaneous layer also contains superficial arteries, veins, and lymphatics.

A fold of skin, the **prepuce** (PRĒ-pūs), or *foreskin*, surrounds the tip of the penis. The prepuce attaches to the relatively narrow **neck** of the penis and continues over the glans that surrounds the **external urethral meatus**. There are no hair follicles on the opposing surfaces, but **preputial** (prē-PŪ-shal) **glands** in the skin of the neck and the inner surface of the prepuce secrete a waxy material known as **smegma** (SMEG-ma). Unfortunately, smegma can be an excellent nutrient source for bacteria. Mild inflammation and infections in this region are common, especially if the area is not washed thoroughly and frequently. One way of avoiding trouble is to perform a **circumcision** (ser-kum-SIZH-un) and surgically remove the prepuce. In Western societies (especially in the United States), this procedure is usually performed shortly after birth. Although controversial, continuation of this practice is supported by strong cultural biases and epidemiological evidence. Uncircumsised males have a higher incidence of urinary tract infections and are at greater risk for penile cancer than circumsised males.

Beneath the loose connective tissue, a dense network of elastic fibers encircles the internal structures of the penis. Most of the body of the penis consists of three cylindrical columns of **erectile tissue** (Figure 27-9b,c●). Erectile tissue consists of a three-dimensional maze of vascular channels incompletely separated by partitions of elastic connective tissue and smooth muscle fibers. In the resting state, the arterial branches are constricted and the muscular partitions are tense. This combination reduces blood flow into the erectile tissue. The smooth muscles in the arterial walls relax under parasympathetic stimulation. When the muscles relax: (1) the vessels dilate, (2) blood flow increases, (3) the vascular channels become engorged with blood, and (4) **erection** of the penis occurs. The flaccid penis hangs beneath the pelvic symphysis anterior to the scrotum, but during erection the penis stiffens and assumes a more upright position.

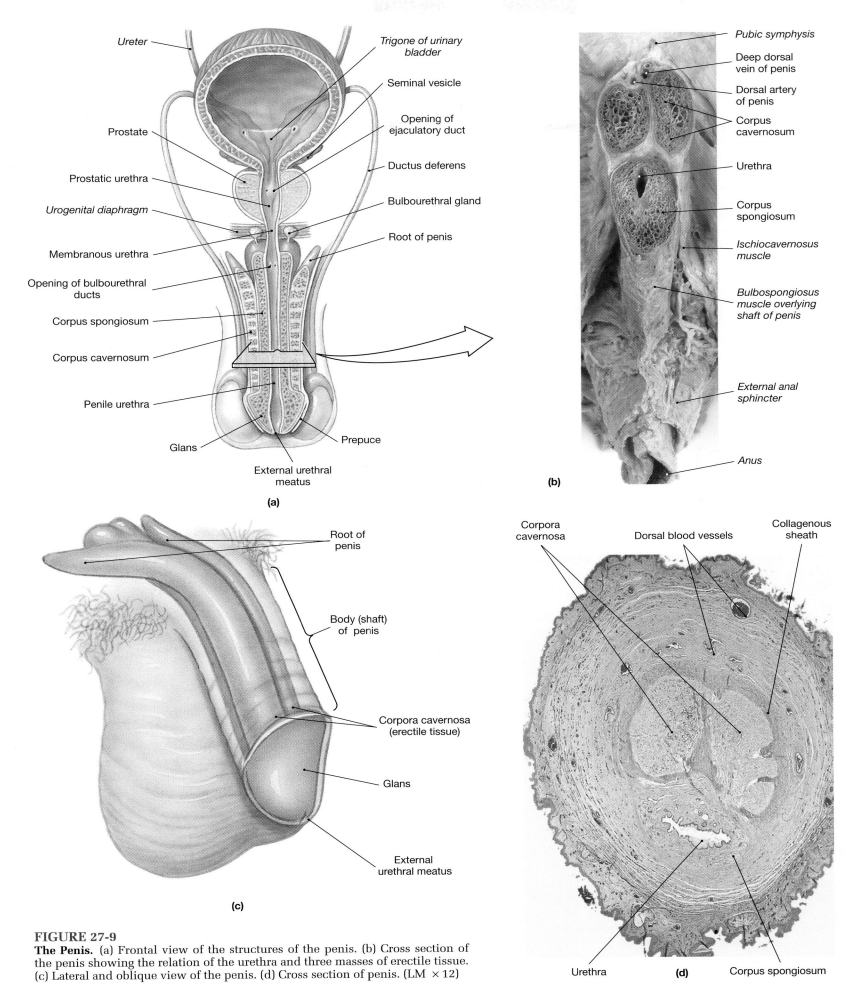

(a)

Ureter

Trigone of urinary bladder

Prostate

Seminal vesicle

Opening of ejaculatory duct

Prostatic urethra

Ductus deferens

Urogenital diaphragm

Bulbourethral gland

Membranous urethra

Root of penis

Opening of bulbourethral ducts

Corpus spongiosum

Corpus cavernosum

Penile urethra

Glans

Prepuce

External urethral meatus

(b)

Pubic symphysis

Deep dorsal vein of penis

Dorsal artery of penis

Corpus cavernosum

Urethra

Corpus spongiosum

Ischiocavernosus muscle

Bulbospongiosus muscle overlying shaft of penis

External anal sphincter

Anus

(c)

Root of penis

Body (shaft) of penis

Corpora cavernosa (erectile tissue)

Glans

External urethral meatus

(d)

Corpora cavernosa

Dorsal blood vessels

Collagenous sheath

Urethra

Corpus spongiosum

FIGURE 27-9
The Penis. (a) Frontal view of the structures of the penis. (b) Cross section of the penis showing the relation of the urethra and three masses of erectile tissue. (c) Lateral and oblique view of the penis. (d) Cross section of penis. (LM × 12)

On the anterior surface of the flaccid penis, the two cylindrical **corpora cavernosa** (KOR-pō-ra ka-ver-NŌ-sa) are separated by a thin septum and encircled by a dense collagenous sheath. At their bases the corpora cavernosa are bound to the ramus of the ischium by tough connective tissue ligaments, the **crura** ("legs"; singular *crus*) of the penis. The corpora cavernosa extend along the length of the penis as far as the neck, and the erectile tissue within each corpus cavernosum surrounds a central artery (Figure 27-9b,c●).

The **corpus spongiosum** (spon-jē-Ō-sum) surrounds the urethra. This erectile body extends from the superficial fascia of the urogenital diaphragm to the tip of the penis, where it expands to form the glans. The sheath surrounding the corpus spongiosum contains more elastic fibers than does that of the corpora cavernosa, and the erectile tissue contains a pair of arteries.

After erection has occurred, semen release involves a two-step process. During **emission** the sympathetic nervous system coordinates peristaltic contractions that in sequence sweep along the ductus deferens, the seminal vesicles, the prostate gland, and the bulbourethral glands. These contractions mix the fluid components of the semen within the male reproductive tract. Ejaculation then occurs, as powerful, rhythmic contractions begin in the *ischiocavernosus* and *bulbospongiosus* muscles of the pelvic floor. ∞ [pp. 270, 274] The ischiocavernosus muscles insert along the sides of the penis, and their contractions primarily serve to stiffen the organ. The bulbospongiosus wraps around the base of the penis, and its contraction pushes semen toward the external urethral orifice. These contractions are controlled by reflexes involving the lower lumbar and upper sacral segments of the spinal cord.

√ **On a warm day, would the cremaster muscle be contracted or relaxed? Why?**

√ **What effect would you predict a vasectomy to have on sperm production?**

ANATOMY OF THE FEMALE REPRODUCTIVE SYSTEM
(Figures 27-10/27-11/27-15)

A woman's reproductive system must not only produce functional gametes, but it must protect and support a developing embryo and nourish the newborn infant. The principal organs of the female reproductive system (Figure 27-10●) include the *ovaries*, the *uterine tubes* (*Fallopian tubes* or *oviducts*), the *uterus*, the *vagina*, and the components of the external genitalia. As in the male, a variety of accessory glands secrete into the reproductive tract.

The ovaries, uterine tubes, and uterus are enclosed within an extensive mesentery known as the **broad ligament** (Figures 27-11 and 27-15●). The uterine tubes run along the superior border of the broad ligament and open into the pelvic cavity lateral to the ovaries. The free edge of the broad ligament that attaches to each uterine tube is known as the **mesosalpinx** (mez-ō-SAL-pinks). A thickened fold of the broad ligament, the **mesovarium** (mez-ō-VAR-ē-um), supports and stabilizes the position of each ovary.

The broad ligament attaches to the sides and floor of the pelvic cavity, where it becomes continuous with the parietal peritoneum. The broad ligament thus subdivides the pelvic cavity. The pocket formed between the posterior wall of the uterus and the anterior surface of the colon is the **rectouterine** (rek-tō-Ū-te-rin) **pouch**, while that between the anterior wall of the uterus and the posterior wall of the urinary bladder is the **vesicouterine** (ves-i-kō-Ū-ter-in) **pouch**. These subdivisions are most apparent in sagittal section (Figure 27-10●).

Several other ligaments assist the broad ligament in supporting and stabilizing the position of the uterus and associated reproductive organs. These ligaments travel within the mesentery sheet of the broad ligament on the way to the ovaries or uterus. The broad ligament limits side-to-side movement and rotation, and the other ligaments (described with the ovaries) prevent superior-inferior movement.

The Ovaries
(Figures 27-10/27-11/27-15)

The **ovaries** are small organs located near the lateral walls of the pelvic cavity (Figures 27-10, 27-11, and 27-15●). These organs are responsible for the production of ova and the secretion of hormones. The position of each ovary is stabilized by the mesovarium and by a pair of supporting ligaments: the *ovarian ligament* and the *suspensory ligament*. The **ovarian ligament** extends from the uterus, near the attachment of the uterine tube, to the medial surface of the ovary. The **suspensory ligament** extends from the lateral surface of the ovary past the open end of the uterine tube to the pelvic wall. The major blood vessels, the **ovarian artery** and **ovarian vein**, travel to and from the ovary within the suspensory ligament. They extend through the mesovarium and are connected to the ovary at the **ovarian hilum** (Figure 27-11●).

A typical ovary measures approximately 5 cm by 2.5 cm (2 in. by 1 in.) and weighs 6–8 g (roughly 0.25 oz). It has a pink or yellowish coloration and a nodular consistency that resembles cottage cheese or lumpy oatmeal. The visceral peritoneum covering the surface of each ovary overlies a layer of dense connective tissue called the **tunica albuginea**. The tissues of the ovary can be divided into a superficial *cortex* and a deep *medulla*. The production of gametes occurs in the cortex.

The Ovarian Cycle and Oogenesis
(Figures 27-12/27-13)

Ovum production, or **oogenesis** (ō-ō-JEN-e-sis), occurs on a monthly basis, as part of the **ovarian cycle**. Ovum development occurs in specialized structures called **ovarian follicles** (ō-VAR-ē-an FOL-i-klz). Unlike the situation in the male gonads, the stem cells, or **oogonia** (ō-ō-GŌ-nē-a), of the female complete their mitotic divisions before birth. There are roughly 2 million **primary oocytes** (Ō-ō-sīts) at birth; by the time of puberty, that number has declined to around 400,000. The rest degenerate, a process called **atresia** (a-TRĒ-zē-a). The remaining primary oocytes are located in the outer portion of the ovarian cortex near the tunica albuginea, in clusters called **egg nests**. Each primary oocyte within an egg nest is surrounded by a simple squamous layer of follicular cells, and the combination is known as a **primordial** (prī-MOR-dē-al) **follicle**.

At puberty, rising levels of FSH trigger the start of the ovarian cycle, and each month thereafter some of the primordial

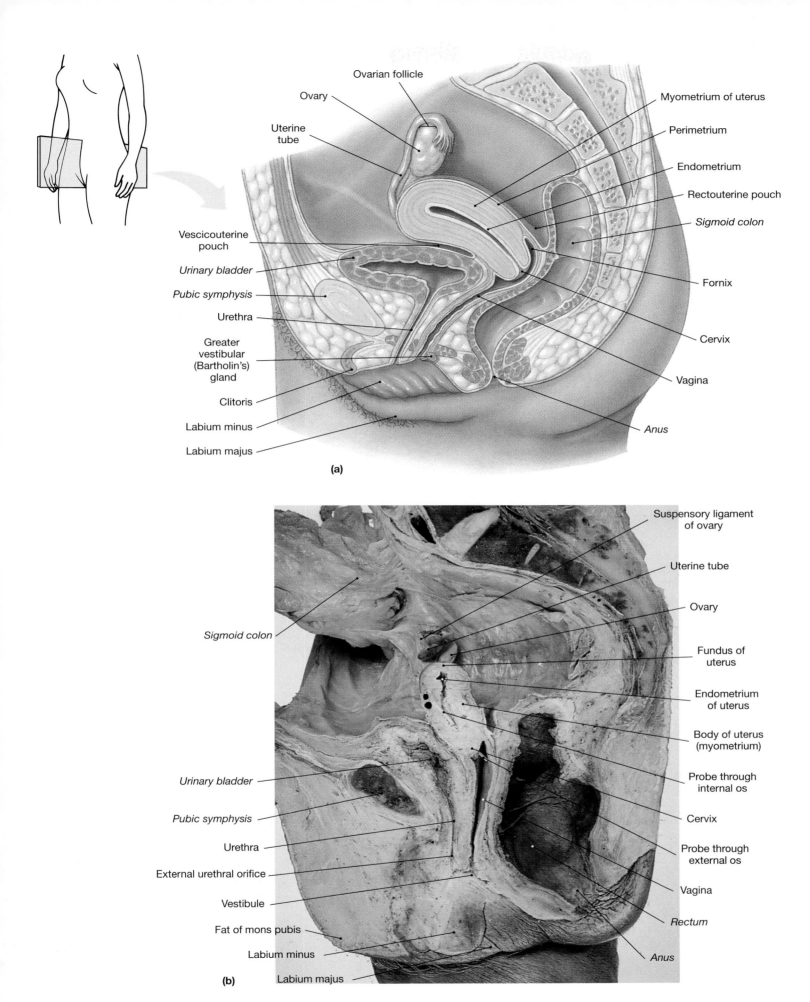

Ovarian follicle

Ovary

Uterine tube

Myometrium of uterus

Perimetrium

Endometrium

Rectouterine pouch

Sigmoid colon

Vescicouterine pouch

Urinary bladder

Pubic symphysis

Urethra

Greater vestibular (Bartholin's) gland

Clitoris

Labium minus

Labium majus

Fornix

Cervix

Vagina

Anus

(a)

Suspensory ligament of ovary

Uterine tube

Ovary

Fundus of uterus

Endometrium of uterus

Body of uterus (myometrium)

Probe through internal os

Cervix

Probe through external os

Vagina

Rectum

Anus

Sigmoid colon

Urinary bladder

Pubic symphysis

Urethra

External urethral orifice

Vestibule

Fat of mons pubis

Labium minus

(b)

Labium majus

FIGURE 27-10

The Female Reproductive System. (a) Diagrammatic view of the female reproductive system as viewed in sagittal section. (b) Female pelvis and perineum in sagittal section.

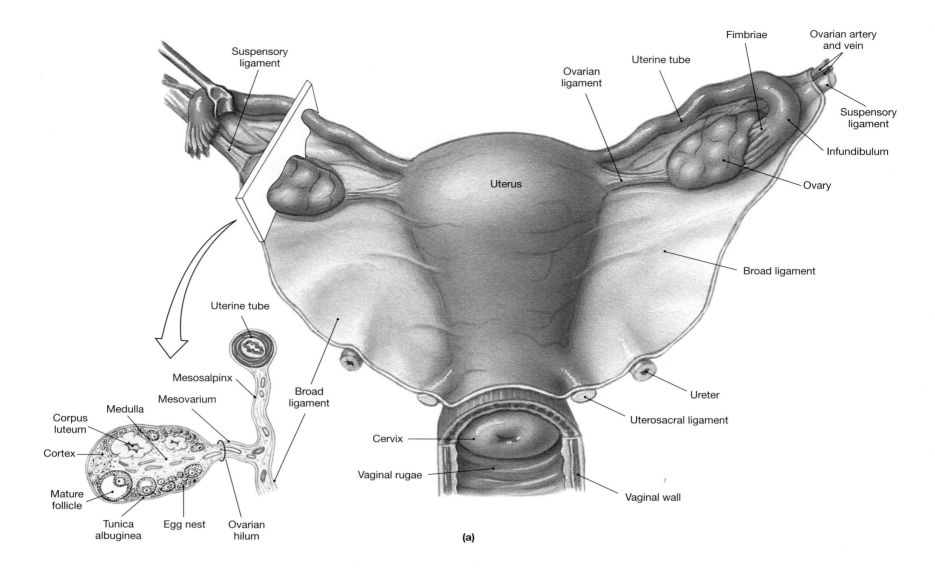

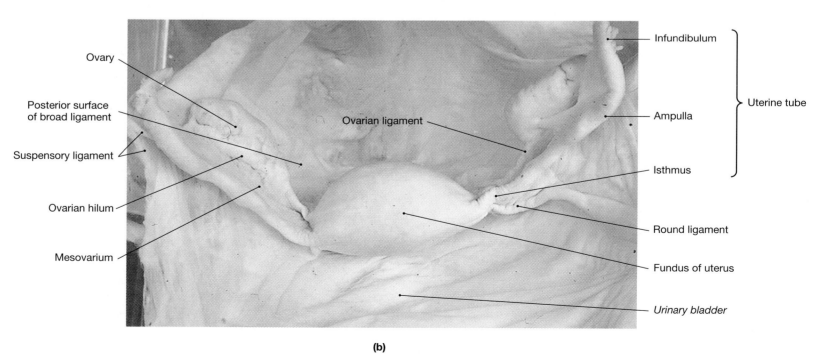

(b)

FIGURE 27-11

The Ovaries, Uterine Tubes, and Uterus. (a) Frontal view of the ovaries, uterine tubes, and uterus along with their supporting ligaments, isolated from the abdominopelvic cavity. (b) Superior and frontal view of isolated ovaries, uterine tubes, and uterus with supporting ligaments.

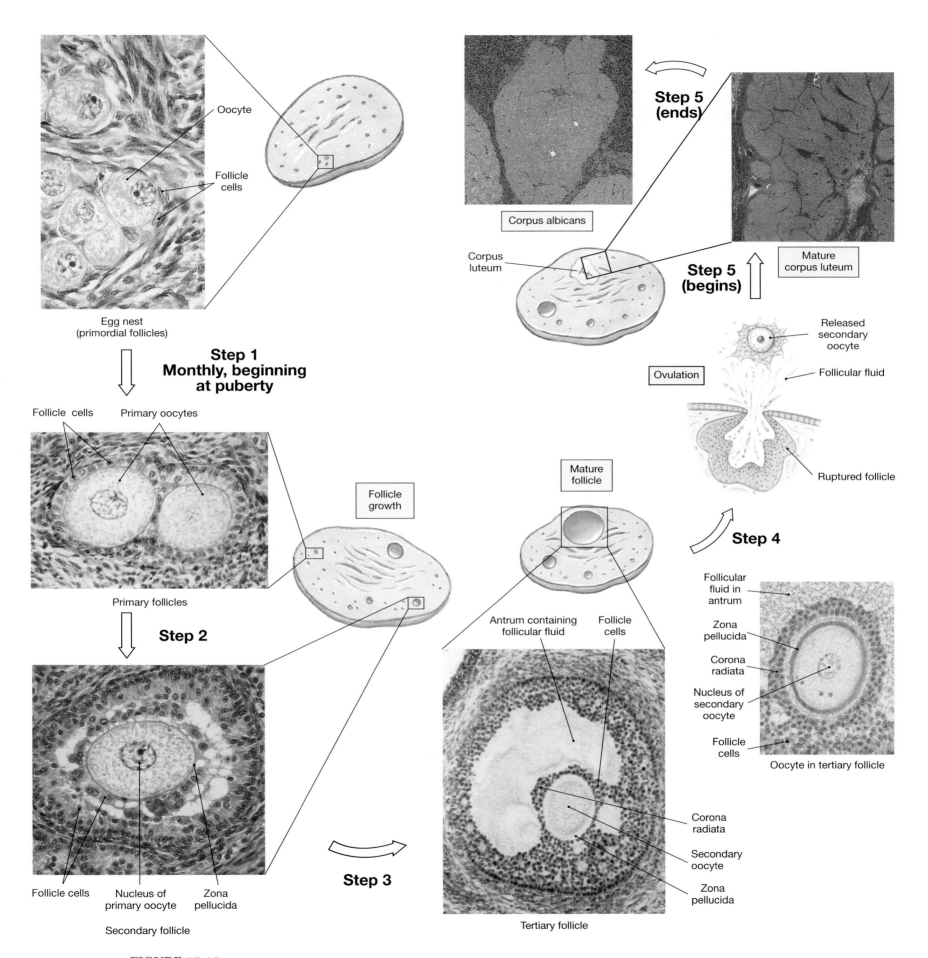

Oocyte

Follicle cells

Egg nest (primordial follicles)

Step 1
Monthly, beginning at puberty

Follicle cells   Primary oocytes

Primary follicles

Step 2

Follicle growth

Follicle cells   Nucleus of primary oocyte   Zona pellucida

Secondary follicle

Step 3

Mature follicle

Antrum containing follicular fluid   Follicle cells

Corona radiata

Secondary oocyte

Zona pellucida

Tertiary follicle

Step 5 (ends)

Corpus albicans

Corpus luteum

Mature corpus luteum

Step 5 (begins)

Released secondary oocyte

Follicular fluid

Ovulation

Ruptured follicle

Step 4

Follicular fluid in antrum

Zona pellucida

Corona radiata

Nucleus of secondary oocyte

Follicle cells

Oocyte in tertiary follicle

FIGURE 27-12
The Ovarian Cycle. Follicular development during the ovarian cycle. (Egg nest, LM × 1440; primary follicles, LM × 1092; secondary follicle, LM × 1052; tertiary follicle, LM × 136; corpus luteum, LM × 208)

follicles will be stimulated to undergo further development. Important steps of this cycle are shown in Figure 27-12● and summarized below.

Step 1: Formation of Primary Follicles.

The ovarian cycle begins as activated primordial follicles develop into **primary follicles**. In a primary follicle, the follicular cells enlarge and undergo repeated cell divisions. This division creates several layers of follicular cells around the oocyte. As the wall of the follicle thickens further, a space opens up between the developing oocyte and the innermost follicular cells. This region, which contains interdigitating microvilli from the follicle cells and the oocyte, is called the **zona pellucida** (ZŌ-na pel-LŪ-si-da). Follicular cells continually provide the developing oocyte with nutrients.

The conversion from primordial to primary follicles and subsequent follicular development occurs under FSH stimulation. As the follicular cells enlarge and multiply, they release steroid hormones, the *estrogens*. Small quantities of estrogens are also contributed by interstitial cells scattered within the ovarian cortex. The hormone **estradiol** (es-tra-DĪ-ol) is the most important estrogen, and it is the dominant hormone prior to ovulation. Estrogens (1) stimulate bone and muscle growth, (2) maintain female secondary sex characteristics, (3) affect CNS activity, including sex-related behaviors and drives, (4) maintain the function of the reproductive glands and organs, and (5) initiate repair and growth of the uterine lining.

Step 2: Formation of Secondary Follicles.

Although many primordial follicles develop into primary follicles, only a few will proceed to the next step. The transformation begins as the wall of the follicle thickens and the deeper follicular cells begin secreting small amounts of fluid. This **follicular fluid**, or *liquor folliculi*, accumulates in small pockets that gradually expand and separate the inner and outer layers of the follicle. At this stage the complex is known as a **secondary follicle**. Although the oocyte continues to grow at a slow pace, the follicle as a whole now enlarges rapidly because of this accumulation of fluid.

Step 3: Formation of a Tertiary Follicle.

Eight to ten days after the start of the ovarian cycle, the ovaries usually contain only a single secondary follicle destined for further development. By the tenth to the fourteenth day of the cycle, it has formed a **tertiary follicle**, or **mature Graafian** (GRAF-ē-an) **follicle**, roughly 15 mm in diameter. This complex spans the entire width of the cortex and stretches the ovarian capsule, creating a prominent bulge in the surface of the ovary. The oocyte projects into the expanded central chamber, or **antrum** (AN-trum), surrounded by a mass of follicular cells.

Until this time, the primary oocyte has been suspended in prophase of the first meiotic division. That division is now completed. Although the nuclear events during oogenesis are identical to those in spermatogenesis, the cytoplasm of the primary oocyte is not evenly distributed (Figure 27-13●). Instead of producing two secondary oocytes, the first meiotic division yields a **secondary oocyte** and a small, nonfunctional **polar body**. The secondary oocyte then proceeds to the metaphase stage of the second meiotic division; that division will not be completed unless fertilization occurs. Like the first division, the second meiotic division will produce an ovum and a nonfunctional polar body. Thus instead of producing four gametes, each of approximately equal size, oogenesis produces a single ovum containing most of the cytoplasm of the primary oocyte and polar bodies that are simply containers for the "extra" chromosomes.

Step 4: Ovulation.

As the time of egg release, or **ovulation** (ov-ū-LĀ-shun), approaches, the secondary oocyte and the surrounding follicular cells lose their connections with the follicular wall and drift free within the antrum. This event usually occurs on day 14 of a 28-day cycle. The follicular cells surrounding the oocyte are now known as the **corona radiata** (ko-RŌ-na rā-dē-A-ta). The distended follicular wall then ruptures, releasing the follicular contents, including the oocyte, into the pelvic cavity. The sticky follicular fluid usually keeps the corona radiata attached to the surface of the ovary, where direct contact with the *fimbriae* or fluid currents established by their ciliated epithelium can transfer the oocyte to the uterine tube.

The stimulus for ovulation is a sudden rise in LH levels that weakens the follicular wall. The surge in LH release coincides with, and is triggered by, a peak in estrogen levels as the tertiary follicle matures.

Step 5: Formation and Degeneration of the Corpus Luteum.

The empty follicle initially collapses, and ruptured vessels bleed into the lumen. The remaining follicular cells then invade the area, proliferating to create an endocrine structure known as

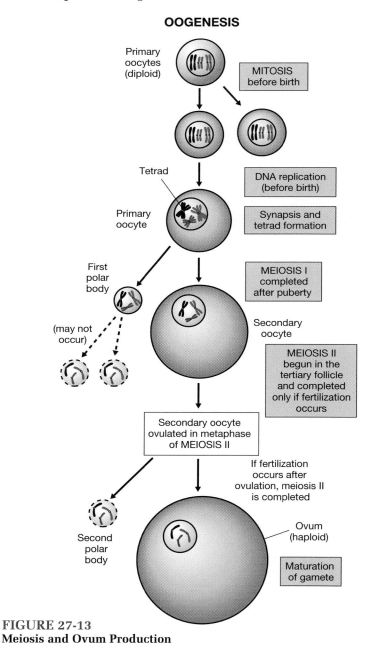

OOGENESIS

Primary oocytes (diploid)

MITOSIS before birth

Tetrad

DNA replication (before birth)

Primary oocyte

Synapsis and tetrad formation

First polar body

MEIOSIS I completed after puberty

(may not occur)

Secondary oocyte

MEIOSIS II begun in the tertiary follicle and completed only if fertilization occurs

Secondary oocyte ovulated in metaphase of MEIOSIS II

If fertilization occurs after ovulation, meiosis II is completed

Second polar body

Ovum (haploid)

Maturation of gamete

FIGURE 27-13
Meiosis and Ovum Production

the **corpus luteum** (LŪ-tē-um), named for its yellow color (*lutea*, yellow). This process occurs under LH stimulation.

The lipids contained in the corpus luteum are used to manufacture steroid hormones known as **progestins** (prō-JES-tinz), principally the steroid **progesterone** (prō-JES-ter-ōn). Although moderate amounts of estrogens are also secreted by the corpus luteum, progesterone is the principal hormone of the postovulatory period. Its primary function is to prepare the uterus for pregnancy.

Unless pregnancy occurs, the corpus luteum begins to degenerate roughly 12 days after ovulation. Progesterone and estrogen levels then fall markedly. Fibroblasts invade the nonfunctional corpus luteum, producing a knot of pale scar tissue called a **corpus albicans** (AL-bi-kanz). The disintegration, or *involution*, of the corpus luteum marks the end of the ovarian cycle. However, the decline in progesterone and estrogen levels stimulates **gonadotropin-releasing hormone (GnRH)** production at the hypothalamus. This hormone triggers a rise in FSH and LH production by the anterior pituitary gland, and the entire cycle begins again.

The hormonal changes involved with the ovarian cycle in turn affect the activities of other reproductive tissues and organs. At the uterus, the hormonal changes are responsible for the maintenance of the *uterine cycle*, discussed in a later section.

Age and Oogenesis

Although many primordial follicles may have developed into primary follicles, and several primary follicles converted to secondary follicles, usually only a single secondary oocyte will be released into the pelvic cavity at ovulation. The rest undergo atresia. At puberty there are approximately 200,000 primordial follicles in each ovary. Forty years later, few if any follicles remain, although only around 500 will have been ovulated over the interim.

The Uterine Tubes
(Figures 27-10/27-11/27-14/27-15c)

Each **uterine tube** is a hollow, muscular tube measuring roughly 13 cm (5 in.) in length. View Figures 27-10, p. 699 (sagittal view), 27-11, p. 700, 27-14 (anterior view), and 27-15c (interior view)● for location and structure. Each uterine tube is divided into three regions:

1. *The Infundibulum:* The end closest to the ovary forms an expanded funnel, or **infundibulum** (in-fun-DIB-ū-lum), with numerous fingerlike projections that extend into the pelvic cavity. The projections are called **fimbriae** (FIM-brē-ē). The inner surfaces of the infundibulum are lined with cilia that beat toward the entrance to the expanded initial segment of the uterine tube, the *ampulla*.
2. *The Ampulla:* The enlarged or expanded portion of the uterine tube that is connected to the infundibulum is known as the **ampulla**.
3. *The Isthmus:* The enlarged ampulla leads to the **isthmus** (IS-mus), a short segment adjacent to the uterine wall. The **intramural** (in-tra-MŪ-ral) **portion** of the uterine tube passes through the wall of the uterus and opens into the uterine cavity.

Histology of the Uterine Tube
(Figure 27-14)

The epithelium lining the uterine tube comprises ciliated and nonciliated columnar epithelial cells (Figure 27-14c●). The mucosa is surrounded by concentric layers of smooth muscle (Figure 27-14b●). Transport of the materials along the uterine tube involves a combination of ciliary movement and peristaltic contractions in the walls of the uterine tube. A few hours before ovulation, sympathetic and parasympathetic nerves from the hypogastric plexus "turn on" this beating pattern. The uterine tubes transport a secondary oocyte for final maturation and fertilization. It normally takes 3–4 days for an oocyte to travel from the infundibulum to the uterine chamber. *If fertilization is to occur, the secondary oocyte must encounter spermatozoa during the first 12–24 hours of its passage.* Fertilization typically occurs in the ampulla of the uterine tube.

Along with its transport function, the uterine tube also provides a rich, nutritive environment containing lipids and glycogen. This mixture provides nutrients to both spermatozoa and a developing pre-embryo. Unfertilized oocytes degenerate in the terminal portions of the uterine tubes or within the uterus.

■ CLINICAL BRIEF
Pelvic Inflammatory Disease (PID)

Pelvic inflammatory disease (PID) is a major cause of sterility in women. This condition, an infection of the uterine tubes, affects an estimated 850,000 women each year in the United States alone. Sexually transmitted pathogens are often involved, and as many as 50–80% of all first cases may be due to infection by *Neisseria gonorrhoeae*, the organism responsible for symptoms of **gonorrhea** (gon-ō-RĒ-a), a sexually transmitted disease discussed in the Clinical Issues appendix. PID may also result from invasion of the region by bacteria normally found within the vagina. Symptoms of pelvic inflammatory disease include fever, lower abdominal pain, and elevated white blood cell counts. In severe cases the infection may spread to other visceral organs or produce a generalized peritonitis.

Sexually active women in the 15–24 age group have the highest incidence of PID. Although oral contraceptive use decreases the risk of infection, the presence of an intrauterine device (IUD) may increase the risk by 1.4–7.3 times. Treatment with antibiotics may control the condition, but chronic abdominal pain may persist. In addition, damage and scarring of the uterine tubes may cause infertility by preventing the passage of a zygote to the uterus. Recently, another sexually transmitted bacterium, *Chlamydia*, has been identified as the probable cause of up to 50% of all cases of PID. Despite the fact that women with this infection may develop few if any symptoms, scarring of the uterine tubes may still produce infertility.

The Uterus
(Figures 27-10/27-11/27-15c)

The **uterus** (YŪ-ter-us) provides mechanical protection, nutritional support, and waste removal for the developing embryo (weeks 1–8) and fetus (from week 9 to delivery). In addition, contractions in the muscular wall of the uterus are important in ejecting the fetus at the time of birth. The position of the uterus within the pelvic cavity and its relation to other pelvic organs can be seen in different views in Figures 27-10, (sagittal), p. 699, 27-11, (anterior) p. 700, and 27-15c (interior)●.

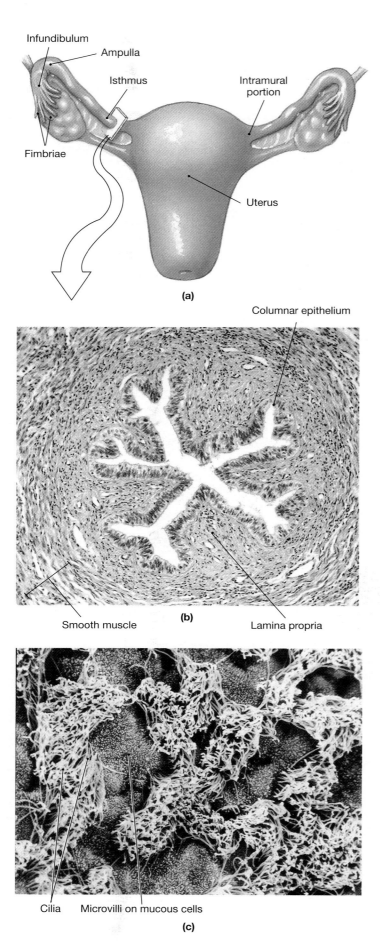

FIGURE 27-14
The Uterine Tubes. (a) Regions of the uterine tubes. (b) A sectional view of the isthmus. (LM × 122) (c) The ciliated lining of the uterine tube.

Infundibulum
Ampulla
Isthmus
Intramural portion
Fimbriae
Uterus

(a)

Columnar epithelium
Smooth muscle
Lamina propria

(b)

Cilia
Microvilli on mucous cells

(c)

The uterus is a small, pear-shaped organ about 7.5 cm (3 in.) long with a maximum diameter of 5 cm (2 in.). It weighs 30–40 g (1–1.4 oz). In its normal position, the uterus bends anteriorly near its base, a condition known as **anteflexion** (an-tē-FLEK-shun). In this position the body of the uterus lies across the superior and posterior surfaces of the urinary bladder (Figure 27-10●, p. 699). If instead, the uterus bends backward toward the sacrum, the condition is termed **retroflexion** (re-trō-FLEK-shun). Retroflexion, which occurs in about 20% of adult women, has no clinical significance.

Suspensory Ligaments of the Uterus
(Figure 27-15a)

In addition to the mesenteric sheet of the broad ligament, three pairs of suspensory ligaments stabilize the position of the uterus and limit its range of movement (Figure 27-15a●). The **uterosacral** (ū-te-rō-SĀ-kral) **ligaments** extend from the lateral surfaces of the uterus to the anterior face of the sacrum, keeping the body of the uterus from moving inferiorly and anteriorly. The **round ligaments** arise on the lateral margins of the uterus just inferior to the uterine tubes. They extend anteriorly, passing through the inguinal canal before ending in the connective tissues of the external genitalia. These ligaments primarily restrict posterior movement of the uterus. The **lateral** (*cardinal*) **ligaments** extend from the base of the uterus and vagina to the lateral walls of the pelvis. These ligaments also tend to prevent the inferior movement of the uterus. Additional mechanical support is provided by the skeletal muscles and fascia of the pelvic floor. ∞ [p. 270]

Internal Anatomy of the Uterus
(Figure 27-15b,c)

The uterus is divided into two anatomical regions (Figure 27-15c●): the *body* and the *cervix*. The uterine **body**, or *corpus*, is the largest division of the uterus. The **fundus** is the rounded portion of the body superior to the attachment of the uterine tubes. The body ends at a constriction known as the uterine **isthmus**. The **cervix** (SER-viks) is the inferior portion of the uterus that extends from the isthmus to the vagina.

The tubular cervix projects about 1.25 cm (0.5 in.) into the vagina. Within the vagina, the distal end of the cervix forms a curving surface surrounding the **external orifice** of the uterus, also known as the *cervical os* (*os*, opening or mouth). The cervical os leads into the **cervical canal**, a constricted passageway that opens into the **uterine cavity** of the body at the **internal orifice**, or *internal os* (Figure 27-15b,c●).

Each region of the uterus is supplied with branches of the uterine and ovarian arteries and veins. Numerous lymphatic vessels also supply each portion of the uterus. The uterus is innervated by autonomic fibers from the hypogastric plexus and sacral segments S_3 and S_4. Sensory afferents from the uterus enter the spinal cord via the eleventh and twelfth thoracic nerves.

The Uterine Wall
(Figure 27-15c)

The dimensions of the uterus are highly variable. In adult women of reproductive age who have not borne children, the uterine wall is about 1.5 cm (0.5 in.) thick. The uterine wall has an outer, muscular **myometrium** (mī-ō-MĒ-trē-um; *myo*-, muscle + *metra*, uterus) and an inner glandular **endometrium** (en-dō-

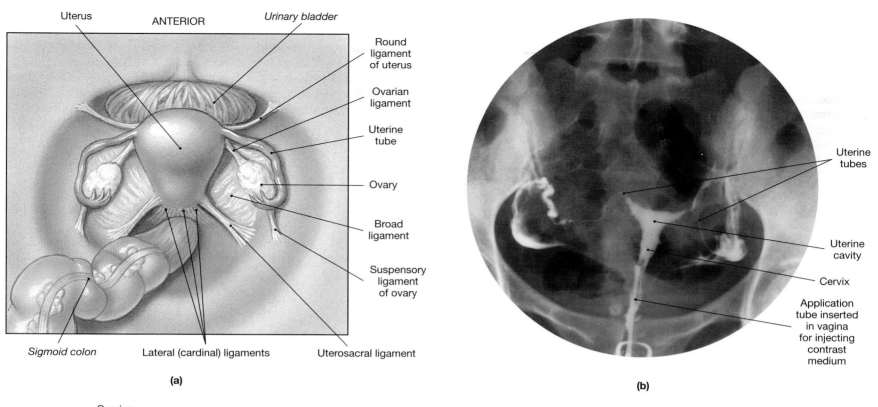

(a)

(b)

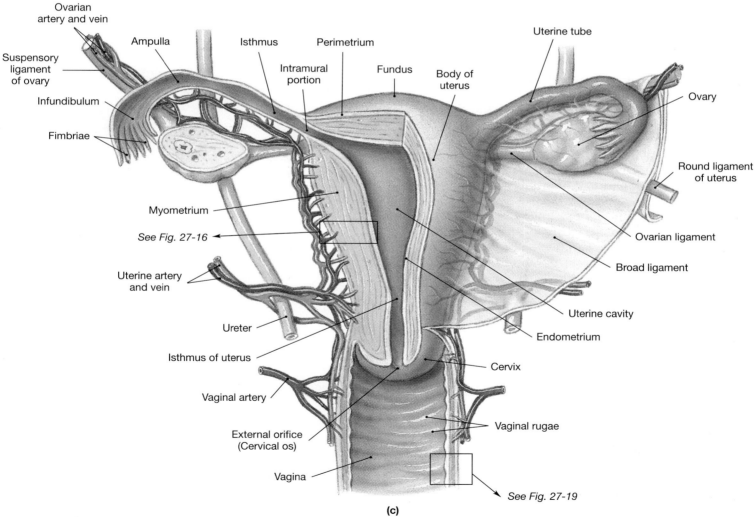

(c)

FIGURE 27-15

The Uterus. (a) As seen from above, the uterus and ligaments stabilizing its position in the pelvic cavity. (b) The uterine cavity and uterine lumen as seen in a hysterosalpingogram. (c) Details of uterus, uterine tube, and ovarian structure; for histological details, see *Figure 27-16*.

Cervical Cancer

Cervical cancer is the most common reproductive system cancer in women age 15–34. Roughly 13,000 new cases of invasive cervical cancer are diagnosed each year in the United States, and approximately 40% of them will eventually die of this condition. Another 50,000 patients are diagnosed with a less aggressive form of cervical cancer. (Other uterine tumors and cancers are discussed in the Clinical Issues appendix.) ⊤*Uterine Tumors and Cancers [p. 775]*

Most women with cervical cancer fail to develop symptoms until late in the disease. At that stage, vaginal bleeding, especially after intercourse, pelvic pain, and vaginal discharge may appear. Early detection is the key to reducing the mortality rate for cervical cancer. The standard screening test is the Pap smear, named for Dr. George Papanicolaou, an anatomist and cytologist. The cervical epithelium normally sheds its superficial cells, and a sample of cells scraped or brushed from the epithelial surface can be examined for abnormal or cancerous cells. The American Cancer Society recommends yearly Pap tests at ages 20 and 21, followed by smears at 1- to 3-year intervals until age 65.

The primary risk factor for this cancer is a history of multiple sexual partners. It appears likely that these cancers develop after viral infection by one of several different human papillomaviruses (HPV) that can be transmitted through sexual contact. ⊤*Sexually Transmitted Diseases [p. 776]*

Treatment of localized, noninvasive cervical cancer involves the removal of the affected portion of the uterus. Treatment of more advanced cancers typically involves a combination of radiation therapy, **hysterectomy** (complete or partial removal of the uterus), lymph node removal, and chemotherapy.

MĒ-trē-um), or *mucosa*. The fundus anterior and posterior surfaces of the uterine body are covered by a serous membrane continuous with the peritoneal lining. This incomplete serosal layer is called the **perimetrium** (Figure 27-15c●).

The endometrium contributes about 10% to the mass of the uterus. The glandular and vascular tissues of the endometrium support the physiological demands of the growing fetus. Vast numbers of uterine glands open onto the endometrial surface. These glands extend deep into the lamina propria almost all the way to the myometrium. Under the influence of estrogen, the uterine glands, blood vessels, and endothelium change with the various phases of the monthly *uterine cycle*.

The myometrium is the thickest portion of the uterine wall, and it forms almost 90% of the mass of the uterus. Smooth muscle in the myometrium is arranged into longitudinal, circular, and oblique layers. The smooth muscle tissue of the myometrium provides much of the force needed to move a large fetus out of the uterus and into the vagina.

Blood Supply to the Uterus

The uterus receives blood from branches of the **uterine arteries,** which arise from branches of the *internal iliac arteries,* and the *ovarian arteries,* which arise from the abdominal aorta inferior to the renal arteries. There are extensive interconnections among the arteries to the uterus. This arrangement helps ensure a reliable flow of blood to the organ, despite changes in position and the changes in uterine shape that accompany pregnancy.

Histology of the Uterus
(Figure 27-16)

The endometrium can be divided into a **functional zone,** the layer closest to the uterine cavity, and an outer **basilar zone** adjacent to

the myometrium. The functional zone contains most of the uterine glands and contributes most of the endometrial thickness. The basilar zone attaches the endometrium to the myometrium and contains the terminal branches of the tubular glands (Figure 27-16b●).

Within the myometrium, branches of the uterine arteries form **arcuate arteries** that encircle the endometrium. **Radial arteries** supply **straight arteries** that deliver blood to the basilar zone of the endometrium, and **spiral arteries** that supply the functional zone (Figure 27-16a●).

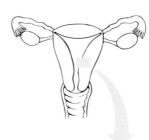

FIGURE 27-16
The Uterine Wall. (a) A sectional view of the uterine wall, showing the endometrial regions and the circulatory supply to the endometrium. (b) Basic histology of the uterine wall. (LM × 32)

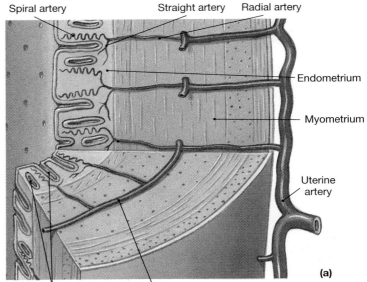

(a)

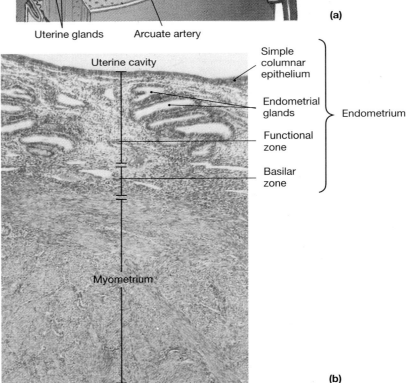

(b)

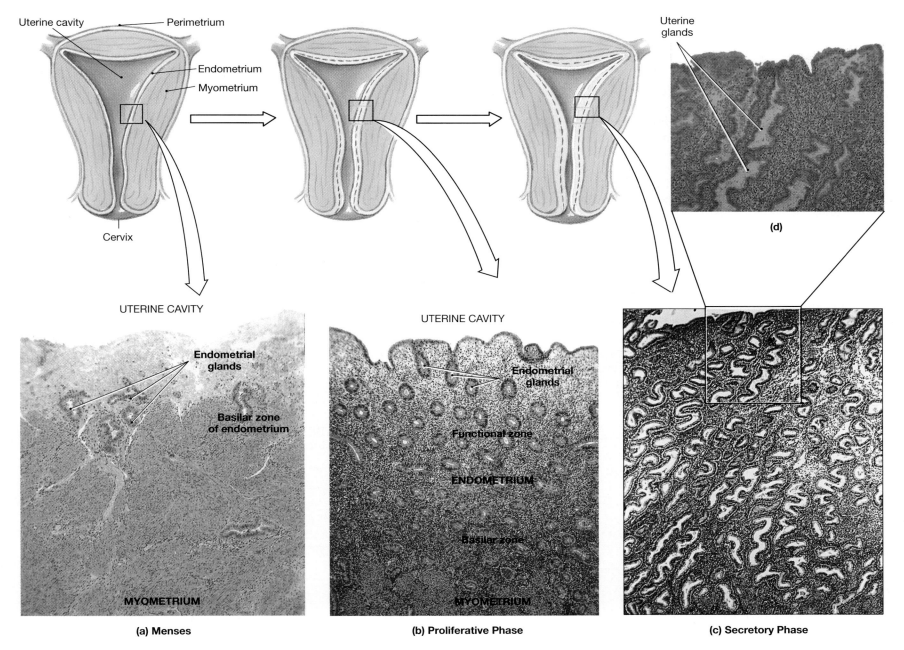

FIGURE 27-17
Histological Changes in the Uterine Cycle. These micrographs show the appearance of the endometrium at menstruation and during the proliferative phase and the secretory phase of the cycle. (a) Menses. (LM × 83) (b) Proliferative phase. (LM × 87) (c) Secretory phase. (LM × 66) (d) Detail of uterine glands. (LM × 64)

The structure of the basilar zone remains relatively constant over time, but that of the functional zone undergoes cyclical changes in response to sexual hormone levels. These alterations produce the characteristic histological features of the *uterine cycle.*
✝*Endometriosis [p. 776]*

The Uterine Cycle
(Figures 27-17/27-18)

The **uterine cycle,** or **menstrual** (MEN-stroo-al) **cycle,** averages 28 days in length, but it can range from 21 to 35 days in normal individuals. The cycle can be divided into three phases: (1) *menses,* (2) the *proliferative phase,* and (3) the *secretory phase.* The histological appearance of the endometrium during each phase is shown in Figure 27-17●. The phases occur in response to hormones associated with the regulation of the ovarian cycle (Figure 27-18●).

MENSES *(Figures 27-17a/27-18).*   The uterine cycle begins with the onset of **menses** (MEN-sēz), a period marked by the wholesale destruction of the functional zone of the endometrium. The arteries begin constricting, reducing blood flow to this region, and the secretory glands and tissues of the functional zone begin to die. Eventually the weakened arterial walls rupture, and blood pours into the connective tissues of the functional zone. Blood cells and degenerating tissues break away and enter the uterine lumen, to be lost by passage through the external orifice and into the vagina. This sloughing of tissue, which continues until the entire functional zone has been lost (Figure 27-17a●) is called **menstruation** (men-stroo-Ā-shun). Menstruation usually lasts from 1 to 7 days, and over this period roughly 35–50 mℓ of blood are lost. Painful menstruation, or **dysmenorrhea,** may result from uterine inflammation and contraction or from conditions involving adjacent pelvic structures.

Menses occurs when progestin and estrogen concentrations fall at the end of the ovarian cycle. It continues until the next group of ovarian follicles are developing and estrogen concentrations rise once again (Figure 27-18●).

THE PROLIFERATIVE PHASE *(Figures 27-17b/27-18).* The basilar zone, including the basal portions of the uterine glands, survives menses because its circulatory supply remains constant. In the days following the completion of menses, and under the influence of circulating estrogens, the epithelial cells of the glands multiply and spread across the endometrial surface, restoring the integrity of the uterine epithelium (Figure 27-17b●). Further growth and vascularization will result in the complete restoration of the functional zone. As this reorganization proceeds, the endometrium is said to be in the **proliferative phase**. Because this restoration occurs at the same time as the enlargement of primary and secondary follicles in the ovary, and under the influence of follicular hormones, it is also known as the *preovulatory phase* or the *follicular phase* of the uterine cycle (Figure 27-18●).

By the time ovulation occurs, the functional zone is several millimeters thick, and prominent mucous glands extend to the border with the basilar zone. At this time the endometrial glands are manufacturing a mucus rich in glycogen. The entire functional zone is highly vascularized, with small arteries spiraling toward the inner surface from larger trunks in the myometrium.

THE SECRETORY PHASE *(Figures 27-17c/27-18).* During the secretory phase of the uterine cycle, the endometrial glands enlarge, accelerating their rates of secretion, and the arteries elongate and spiral through the tissues of the functional zone as seen in Figure 27-17c●. This activity occurs under the combined stimulatory effects of progestins and estrogens from the corpus luteum. Because this phase begins at the time of ovulation, and it persists as long as the corpus luteum remains intact, it can also be called the *postovulatory phase*, or *luteal phase*, of the menstrual cycle (Figure 27-18●).

Secretory activities peak about 12 days after ovulation. Over the next day or two, the glandular activity declines, and the uterine cycle comes to a close as the corpus luteum stops producing stimulatory hormones. A new cycle then begins with the onset of menses and the disintegration of the functional zone. The secretory phase usually lasts 14 days. As a result, the date of ovulation can be determined after the fact by counting backward 14 days from the first day of menses.

MENARCHE AND MENOPAUSE. The uterine cycle of events begins with the **menarche** (me-NAR-kē), or first uterine cycle at puberty, typically at age 11–12. The cycles continue until age 45–50, when **menopause** (MEN-ō-paws), the last uterine cycle, occurs. Over the intervening 35 to 40 years, the regular appearance of uterine cycles will be interrupted only by unusual circumstances such as illness, stress, starvation, or pregnancy.

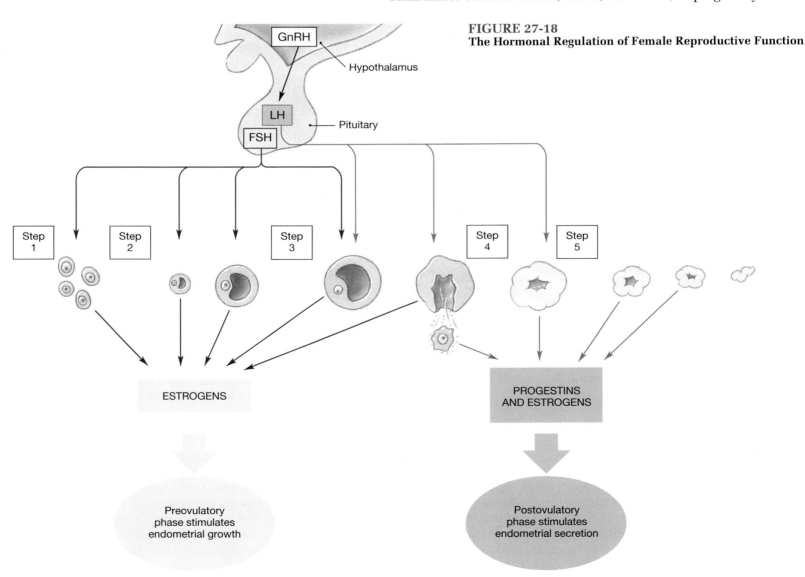

FIGURE 27-18
The Hormonal Regulation of Female Reproductive Function

The Vagina
(Figures 27-10/27-11a)

The **vagina** is an elastic, muscular tube extending from the cervix of the uterus to its external opening to outside the body, the *vestibule* (Figures 27-10, p. 699, and 27-11a●, p. 700). The vagina has an average length of 7.5–9 cm (3–3.5 in.), but because the vagina is highly distensible, its length and width are quite variable.

At the proximal end of the vagina, the cervix projects into the **vaginal canal**. The shallow recess surrounding the cervical protrusion is known as the **fornix** (FOR-niks). The vagina lies parallel to the rectum, and the two are in close contact posteriorly. Anteriorly, the urethra travels along the superior wall of the vagina as it travels from the urinary bladder to its opening into the vestibule. The primary blood supply of the vagina is via the **vaginal branches** of the internal iliac (or uterine) arteries and veins. Innervation is from the hypogastric plexus, sacral nerves S_2–S_4, and branches of the pudendal nerve.

The vagina has three major functions:

1. It serves as a passageway for the elimination of menstrual fluids;

2. It receives the penis during sexual intercourse and holds spermatozoa before they pass into the uterus;

3. In childbirth, it forms the lower portion of the birth canal through which the fetus passes during delivery.

■ CLINICAL BRIEF
Vaginitis

An infection of the vaginal canal, known as **vaginitis** (va-jin-Ī-tis), may be caused by fungal, bacterial, or parasitic organisms. In addition to any discomfort that may result, the condition may affect the survival of sperm and thereby reduce fertility. There are several different forms of vaginitis, and minor cases are relatively common. **Candidiasis** (kan-di-DĪ-a-sis) results from a fungal (yeast) infection. The organism responsible appears to be a normal component of the vaginal microbial population in 30–80% of normal women. Symptoms include itching and burning sensations, and a lumpy white discharge may also be produced. Topical antifungal medications are used to treat this condition.

Bacterial (nonspecific) vaginitis results from the combined action of several bacteria. The bacteria involved are normally present in about 30% of adult women. In this form of vaginitis, the vaginal discharge contains epithelial cells and large numbers of bacteria. Antibiotics are often effective in controlling this condition.

Trichomoniasis (trik-ō-mō-NĪ-a-sis) involves infection by a parasite, *Trichomonas vaginalis*, introduced by sexual contact with a carrier. Because it is a sexually transmitted disease, both partners must be treated to prevent reinfection.

A more serious vaginal infection by *Staphylococcus* bacteria is responsible for symptoms of **toxic shock syndrome (TSS)**. Symptoms include high fever, sore throat, vomiting and diarrhea, and a generalized rash. As the condition progresses, shock, respiratory distress, and kidney or liver failure may develop, and 10–15% of all cases prove fatal. Women have developed this condition while using tampons or vaginal sponges, and people of either sex may develop TSS after abrasion or burn injuries that promote bacterial infection.

Histology of the Vagina
(Figures 27-10b/27-15c/27-19)

In sectional view, the lumen of the vagina appears constricted, forming a rough H. The vaginal walls contain a network of blood vessels and layers of smooth muscle, and the lining is moistened by the secretions of the cervical glands and by the movement of water across the permeable epithelium. The vagina and vestibule are separated by an elastic epithelial fold, the **hymen** (HĪ-men), which may partially or completely block the entrance to the vagina. The two bulbocavernosus muscles pass on either side of the vaginal orifice, and their contractions constrict the entrance. These muscles cover the *vestibular bulbs*, masses of erectile tissue that pass on either side of the vaginal entrance. These are the equivalents of the corpora cavernosa of the male.

The vaginal lumen is lined by a stratified squamous epithelium (Figure 27-19●) that in the relaxed state is thrown into folds, called *rugae* (Figure 27-15c●, p. 705). The underlying lamina propria is thick and elastic, and it contains small blood vessels, nerves, and lymph nodes. The vaginal mucosa is surrounded by an elastic **muscularis layer**, with layers of smooth muscle fibers arranged in circular and longitudinal bundles continuous with the uterine myometrium. The portion of the vagina adjacent to the uterus has a serosal covering continuous with the pelvic peritoneum; over the rest of the vagina the muscularis layer is surrounded by an *adventitia* of fibrous connective tissue.

The vagina contains a normal population of resident bacteria, supported by the nutrients found in the cervical mucus. The metabolic activity of these bacteria creates an acid environment, which restricts the growth of many pathogenic organisms. An acid environment also inhibits sperm motility, and for this reason the buffers contained in seminal fluid are important to successful fertilization.

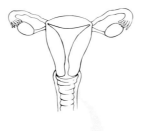

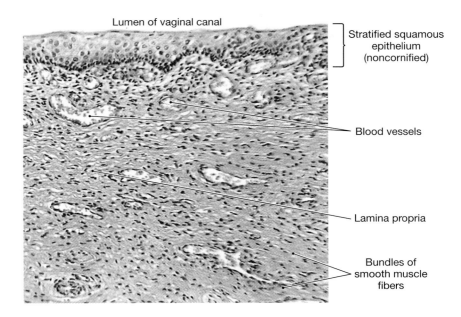

Lumen of vaginal canal

Stratified squamous epithelium (noncornified)

Blood vessels

Lamina propria

Bundles of smooth muscle fibers

FIGURE 27-19
Histology of the Vaginal Wall. (LM × 36)

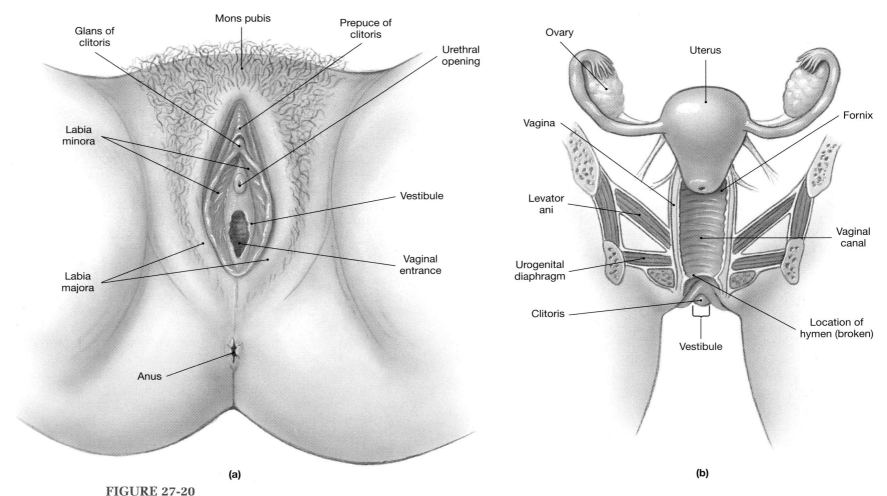

FIGURE 27-20
The Female External Genitalia. (a) An external view of the female perineum. (b) A sectional view showing the relative positions of the internal and external reproductive structures.

The External Genitalia
(Figures 27-10a/27-20)

The region enclosing the female external genitalia is usually called the **vulva** (VUL-va), or **pudendum** (pyū-DEN-dum) (Figure 27-20a●). The vagina opens into the **vestibule**, a central space bounded by the **labia minora** (LĀ-bē-a mi-NOR-a; singular *labium minus*). The labia minora are covered with a smooth, hairless skin. The urethra opens into the vestibule just anterior to the vaginal entrance. The **paraurethral glands,** or *Skene's glands,* discharge into the urethra near the external urethral orifice. Anterior to the urethral opening, the **clitoris** (KLI-tō-ris) projects into the vestibule. The clitoris is the female equivalent of the penis, derived from the same embryonic structures (see the Embryological Summary, pp. 711–713). Internally the clitoris contains erectile tissue comparable to the corpora cavernosa. The clitoris becomes engorged with blood during sexual arousal. A small erectile *glans* sits atop the organ, and extensions of the labia minora encircle the body of the clitoris, forming the **prepuce,** or *hood,* of the clitoris.

A variable number of small **lesser vestibular glands** discharge their secretions onto the exposed surface of the vestibule, keeping it moistened. During arousal, a pair of ducts discharges the secretions of the **greater vestibular glands** (*Bartholin's glands*) into the vestibule near the posterolateral margins of the vaginal entrance (Figure 27-10a●, p. 699). These mucous glands resemble the bulbourethral glands of the male.

The outer limits of the vulva are established by the *mons pubis* and the *labia majora*. The prominent bulge of the **mons pubis** is created by adipose tissue beneath the skin anterior to the pubic symphysis. Adipose tissue also accumulates within the fleshy **labia majora** (singular, *labium majus*), which encircle and partially conceal the labia minora and vestibular structures. The outer margins of the labia majora are covered with the same coarse hair that covers the mons pubis, but the inner surfaces are relatively hairless. Sebaceous glands and scattered apocrine sweat glands secrete onto the inner surface of the labia majora, moistening them and providing lubrication.

The Mammary Glands
(Figure 27-21)

At birth, the newborn infant cannot fend for itself, and several key systems have yet to complete their development. Over the initial period of adjustment to an independent existence, the infant gains nourishment from the milk secreted by the maternal **mammary glands**. Milk production, or **lactation** (lak-TĀ-shun), occurs in the mammary glands of the breasts, specialized accessory organs of the female reproductive system.

The mammary glands lie in the subcutaneous tissue of the **pectoral fat pad** beneath the skin of the chest (Figure 27-21a●, p. 714). Each breast bears a small conical projection, the **nipple**, where the ducts of underlying mammary glands open onto the body surface. The region of reddish-brown skin surrounding each nipple is known as the **areola** (a-RĒ-ō-la). Large sebaceous glands beneath the areolar surface give it a granular texture.

710

EMBRYOLOGY SUMMARY: DEVELOPMENT OF THE REPRODUCTIVE SYSTEM

EMBRYOLOGY SUMMARY: THE REPRODUCTIVE SYSTEM—INDIFFERENT STAGES

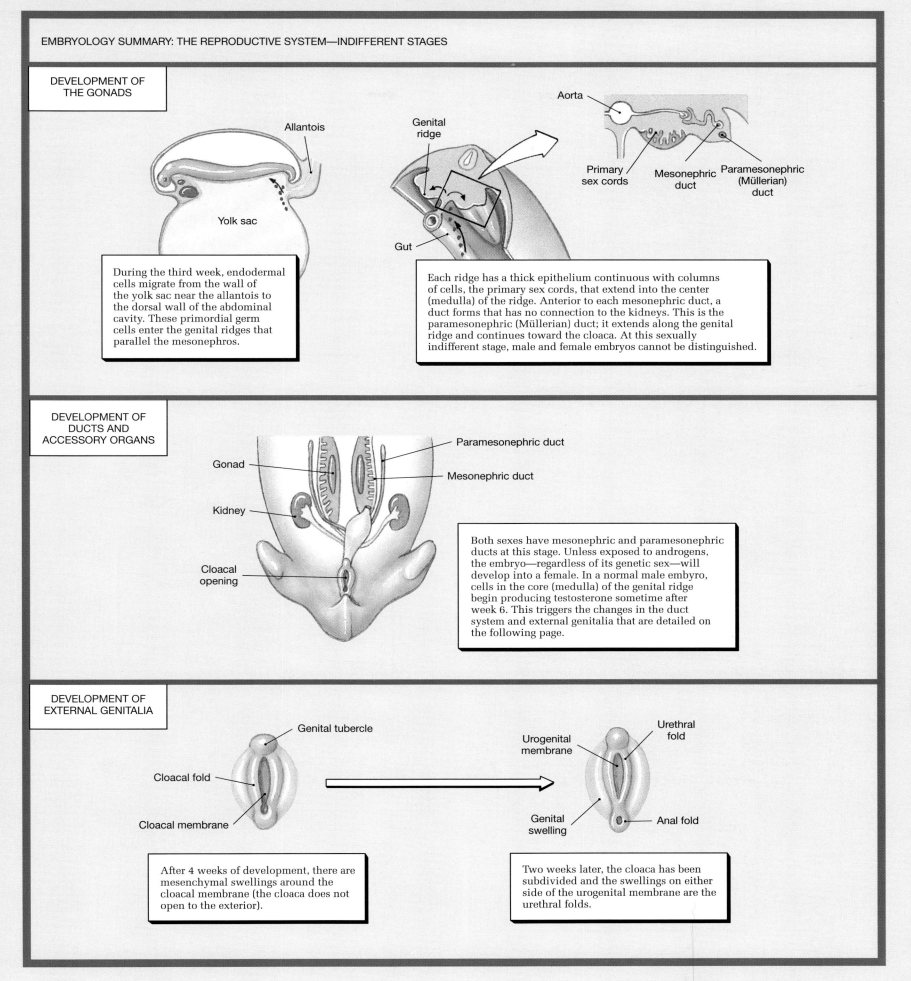

DEVELOPMENT OF THE GONADS

Allantois

Yolk sac

Genital ridge

Gut

Aorta

Primary sex cords

Mesonephric duct

Paramesonephric (Müllerian) duct

During the third week, endodermal cells migrate from the wall of the yolk sac near the allantois to the dorsal wall of the abdominal cavity. These primordial germ cells enter the genital ridges that parallel the mesonephros.

Each ridge has a thick epithelium continuous with columns of cells, the primary sex cords, that extend into the center (medulla) of the ridge. Anterior to each mesonephric duct, a duct forms that has no connection to the kidneys. This is the paramesonephric (Müllerian) duct; it extends along the genital ridge and continues toward the cloaca. At this sexually indifferent stage, male and female embryos cannot be distinguished.

DEVELOPMENT OF DUCTS AND ACCESSORY ORGANS

Gonad

Kidney

Cloacal opening

Paramesonephric duct

Mesonephric duct

Both sexes have mesonephric and paramesonephric ducts at this stage. Unless exposed to androgens, the embryo—regardless of its genetic sex—will develop into a female. In a normal male embyro, cells in the core (medulla) of the genital ridge begin producing testosterone sometime after week 6. This triggers the changes in the duct system and external genitalia that are detailed on the following page.

DEVELOPMENT OF EXTERNAL GENITALIA

Genital tubercle

Cloacal fold

Cloacal membrane

Urogenital membrane

Urethral fold

Genital swelling

Anal fold

After 4 weeks of development, there are mesenchymal swellings around the cloacal membrane (the cloaca does not open to the exterior).

Two weeks later, the cloaca has been subdivided and the swellings on either side of the urogenital membrane are the urethral folds.

EMBRYOLOGY SUMMARY: DEVELOPMENT OF THE REPRODUCTIVE SYSTEM

EMBRYOLOGY SUMMARY: MALE DEVELOPMENT

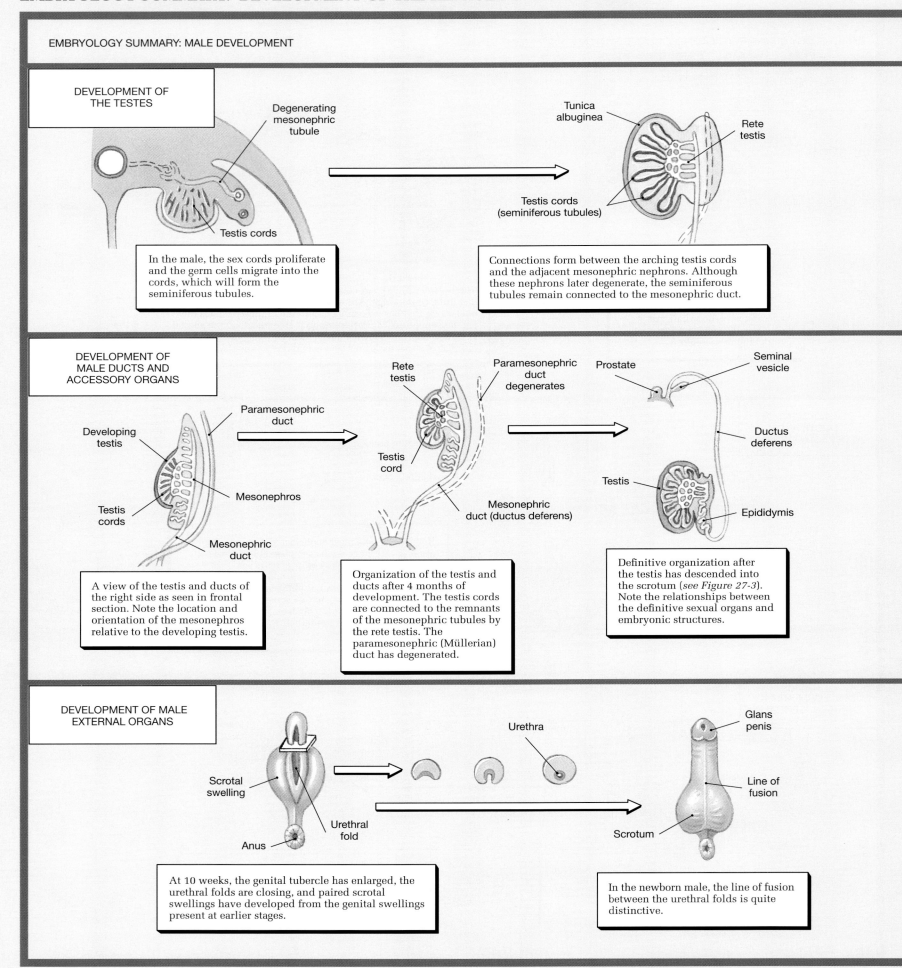

DEVELOPMENT OF THE TESTES

Degenerating mesonephric tubule

Testis cords

Tunica albuginea

Rete testis

Testis cords (seminiferous tubules)

In the male, the sex cords proliferate and the germ cells migrate into the cords, which will form the seminiferous tubules.

Connections form between the arching testis cords and the adjacent mesonephric nephrons. Although these nephrons later degenerate, the seminiferous tubules remain connected to the mesonephric duct.

DEVELOPMENT OF MALE DUCTS AND ACCESSORY ORGANS

Developing testis

Paramesonephric duct

Testis cords

Mesonephros

Mesonephric duct

Rete testis

Testis cord

Paramesonephric duct degenerates

Mesonephric duct (ductus deferens)

Prostate

Seminal vesicle

Ductus deferens

Testis

Epididymis

A view of the testis and ducts of the right side as seen in frontal section. Note the location and orientation of the mesonephros relative to the developing testis.

Organization of the testis and ducts after 4 months of development. The testis cords are connected to the remnants of the mesonephric tubules by the rete testis. The paramesonephric (Müllerian) duct has degenerated.

Definitive organization after the testis has descended into the scrotum (see Figure 27-3). Note the relationships between the definitive sexual organs and embryonic structures.

DEVELOPMENT OF MALE EXTERNAL ORGANS

Scrotal swelling

Urethral fold

Anus

Urethra

Glans penis

Line of fusion

Scrotum

At 10 weeks, the genital tubercle has enlarged, the urethral folds are closing, and paired scrotal swellings have developed from the genital swellings present at earlier stages.

In the newborn male, the line of fusion between the urethral folds is quite distinctive.

DEVELOPMENT OF THE OVARIES

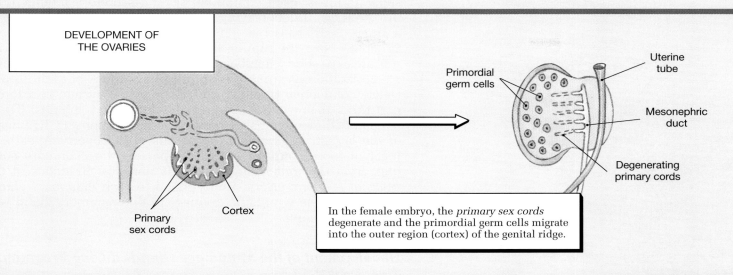

In the female embryo, the *primary sex cords* degenerate and the primordial germ cells migrate into the outer region (cortex) of the genital ridge.

DEVELOPMENT OF FEMALE DUCTS AND ACCESSORY ORGANS

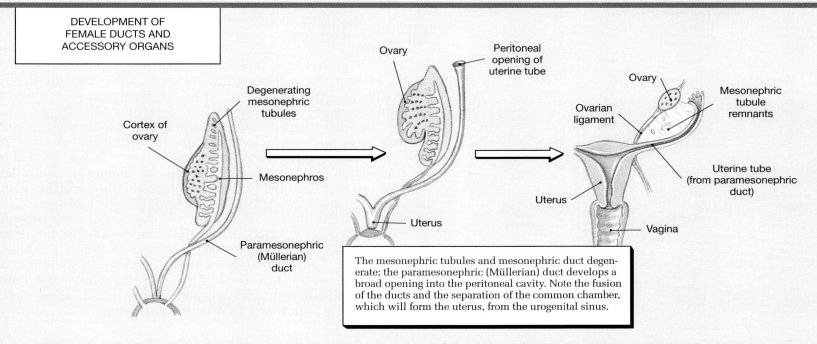

The mesonephric tubules and mesonephric duct degenerate; the paramesonephric (Müllerian) duct develops a broad opening into the peritoneal cavity. Note the fusion of the ducts and the separation of the common chamber, which will form the uterus, from the urogenital sinus.

DEVELOPMENT OF FEMALE EXTERNAL GENITALIA

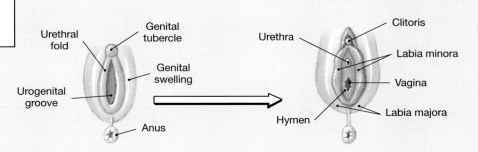

In the female, the urethral folds do not fuse; they develop into the labia minora. The genital swellings will form the labia majora. The genital tubercle, which in the male forms the glans of the penis, develops into the clitoris. The urethra opens to the exterior immediately posterior to the clitoris. The hymen remains as an elaboration of the urogenital membrane.

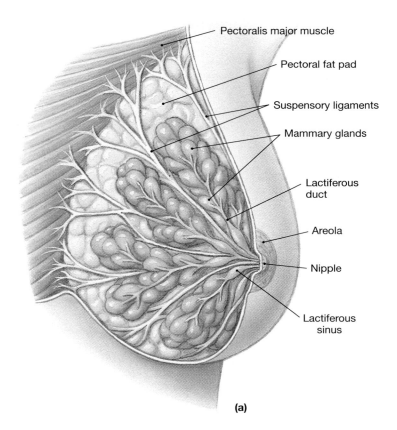

Pectoralis major muscle

Pectoral fat pad

Suspensory ligaments

Mammary glands

Lactiferous
duct

Areola

Nipple

Lactiferous
sinus

(a)

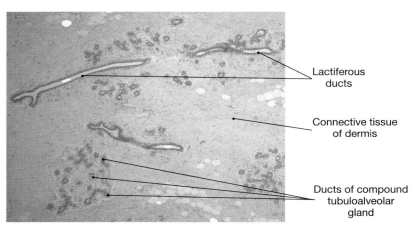

Lactiferous
ducts

Connective tissue
of dermis

Ducts of compound
tubuloalveolar
gland

(b) Resting mammary gland

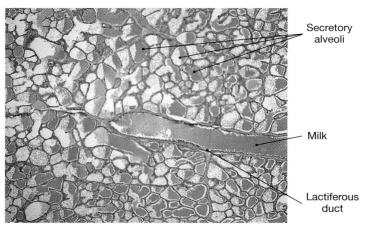

Secretory
alveoli

Milk

Lactiferous
duct

(c) Active mammary gland

FIGURE 27-21
The Mammary Glands of the Female Breast. (a) Gross anatomy of the breast. (b,c) Micrographs comparing the histological organization of (b) the resting (LM × 60) and (c) the active (LM × 131) mammary glands.

The glandular tissue of the breast consists of a number of separate lobes, each containing several secretory lobules (Figure 27-21b●). Ducts leaving the lobules converge, giving rise to a single **lactiferous** (lak-TIF-e-rus) **duct** in each lobule (Figure 27-21c●). Near the nipple, that lactiferous duct expands, forming an expanded chamber called a **lactiferous sinus**. Usually 15–20 lactiferous sinuses open onto the surface of each nipple. Dense connective tissue surrounds the duct system and forms partitions that extend between the lobes and lobules. These bands of connective tissue, known as the **suspensory ligaments of the breast**, originate in the dermis of the overlying skin. A layer of loose connective tissue separates the mammary complex from the underlying pectoralis muscles. Branches of the *internal thoracic artery* supply blood to each mammary gland. ∞ [p. 557] The lymphatic drainage of the mammary gland was detailed in Chapter 23. ∞ [p. 589]

Development of the Mammary Glands during Pregnancy (Figure 27-21b,c)

Figure 27-21b,c● compares the histological organization of the inactive and active mammary glands. The resting mammary gland is dominated by a duct system, rather than by active glandular cells. The size of the breast in a nonpregnant woman reflects primarily the amount of adipose tissue in the breast, rather than the amount of glandular tissue. The secretory apparatus does not develop until pregnancy occurs.

Further mammary gland development requires a combination of hormones, including *prolactin (PRL)* and *growth hormone (GH)* from the anterior pituitary gland. ∞ [p. 488] Under stimulation by these hormones, aided by **human placental lactogen** (LAK-tō-jen) **(HPL)** from the placenta, the mammary gland ducts become mitotically active, and gland cells begin to appear. By the end of the sixth month of pregnancy, the mammary glands are fully developed, and the gland begins producing secretions that are stored in the duct system. Milk is released when the infant begins to suck on the nipple. This stimulation causes the release of oxytocin at the posterior pituitary, and oxytocin triggers the contraction of smooth muscles in the walls of the lactiferous ducts and sinuses, ejecting milk.

√ **As the result of infections such as gonorrhea, scar tissue can block the lumen of each uterine tube. How would this blockage affect a woman's ability to conceive?**

√ **Would blockage of a single lactiferous sinus interfere with delivery of milk to the nipple? Explain.**

Pregnancy and the Female Reproductive System

If fertilization occurs, the zygote (fertilized egg) undergoes a series of cell divisions, forming a hollow ball of cells known as a **blastocyst** (BLAS-tō-sist). Upon arrival in the uterine cavity, the blastocyst initially obtains nutrients by absorbing the secretions of the uterine glands. Within a few days, it contacts the endometrial wall, erodes the epithelium, and buries itself in the endometrium. This process, known as **implantation** (im-plan-TĀ-shun), initiates the chain of events leading to the formation of a special organ, the **placenta** (pla-SEN-ta), that will support embryonic and fetal development over the next 9 months.

The placenta provides a medium for the transfer of dissolved gases, nutrients, and waste products between the fetal and maternal bloodstreams. It also acts as an endocrine organ, producing hormones. The hormone **human chorionic** (kō-rē-ON-ik) **gonadotropin (HCG)** appears in the maternal bloodstream soon after implantation has occurred. The presence of HCG in blood or urine samples provides a reliable indication of pregnancy. In function, HCG resembles LH, for in the presence of HCG, the corpus luteum does not degenerate. If it did, the pregnancy would end, for the functional zone of the endometrium would disintegrate.

In the presence of HCG, the corpus luteum persists for about 3 months before degenerating. Its departure does not trigger the return of menstrual periods, because by then the placenta is actively secreting both estrogen and progesterone. Over the following months, the corpus luteum and the placenta also synthesize two additional hormones: **relaxin**, which increases the flexibility of the pelvis during delivery and causes dilation of the cervix during birth, and *human placental lactogen*, which helps prepare the mammary glands for milk production. *†Conception Control [p. 776]*

AGING AND THE REPRODUCTIVE SYSTEM

The aging process affects the reproductive systems of men and women. The most striking age-related changes in the female reproductive system occur at menopause, while changes in the male reproductive system occur more gradually and over a longer period of time.

Menopause

Menopause is usually defined as the time that ovulation and menstruation cease. Menopause typically occurs at age 45–55, but in the years preceding it, the regularity of the ovarian and menstrual cycles gradually fades. **Premature menopause** occurs before age 40. A shortage of primordial follicles is the underlying cause of these developments. Menopause is accompanied by a sharp and sustained rise in the production of GnRH, FSH, and LH, while circulating concentrations of estrogen and progesterone decline. The decline in estrogen levels leads to reductions in the size of the uterus and breasts, accompanied by a thinning of the urethral and vaginal walls. The reduced estrogen concentrations have also been linked to the development of osteoporosis and a variety of cardiovascular and neural effects, including "hot flashes," anxiety, and depression.

The Male Climacteric

Changes in the male reproductive system occur more gradually, over a period known as the **male climacteric**. Circulating testosterone levels begin to decline between ages 50 and 60, coupled with increases in circulating levels of FSH and LH. Although sperm production continues (men can father children well into their eighties), there is a gradual reduction in sexual activity in older men, which may be linked to declining testosterone levels. Some clinicians are now tentatively suggesting the use of testosterone replacement therapy to enhance libido (sexual drive) in elderly men.

Related Clinical Terms

testicular torsion: Twisting of the spermatic cord resulting from rotation of the testis inside the scrotal cavity. *†Testicular Torsion [p. 775]*

orchiectomy (or-kē-EK-tō-mē)**:** Surgical removal of a testis. *†Testicular Torsion [p. 775]*

cryptorchidism (kript-OR-ki-dizm)**:** Failure of one or both testes to descend into the scrotum by the time of birth. *[p. 689]*

prostate cancer: A malignant, metastasizing cancer that is the second most common cause of cancer deaths in males. *†Prostate Cancer [p. 775]*

prostatectomy (pros-ta-TEK-tō-mē)**:** Surgical removal of the prostate gland. *†Prostate Cancer [p. 775]*

prostate-specific antigen (PSA): An antigen whose concentration in the blood increases in prostate cancer patients. *†Prostate Cancer [p. 775]*

pelvic inflammatory disease (PID): An infection of the uterine tubes. *[p. 703]*

gonorrhea (gon-ō-RĒ-a)**:** A sexually transmitted disease. *[p. 703]*

leiomyomas (lī-ō-mī-Ō-maz)**,** or **fibroids:** Benign myometrial tumors that are the most common tumors in women. *†Uterine Tumors and Cancers [p. 775]*

endometrial polyps: Benign epithelial tumors of the uterine lining. *†Uterine Tumors and Cancers [p. 775]*

sexually transmitted diseases (STDs): Diseases transferred from one individual to another primarily or exclusively through sexual contact. Examples include gonorrhea, syphilis, herpes, and AIDS. *†Sexually Transmitted Diseases [p. 776]*

endometriosis (en-dō-mē-trē-Ō-sis)**:** Growth of endometrial tissue outside of the uterus. *†Endometriosis [p. 776]*

vaginitis (va-jin-Ī-tis)**:** Infection of the vaginal canal by fungal or bacterial pathogens. *[p. 709]*

vasectomy (vaz-EK-to-mē)**:** Surgical removal of a segment of the ductus deferens, making it impossible for spermatozoa to reach the distal portions of the reproductive tract. *†Conception Control [p. 776]*

STUDY OUTLINE

INTRODUCTION [p. 685]

1. The human reproductive system produces, stores, nourishes, and transports functional **gametes** (reproductive cells). **Fertilization** is the fusion of a **sperm** from the father and an **ovum** from the mother to create a **zygote** (fertilized egg).

ORGANIZATION OF THE REPRODUCTIVE SYSTEM [p. 685]

1. The reproductive system includes **gonads**, ducts, accessory glands and organs, and the **external genitalia**.

2. In the male, the **testes** produce sperm, which are expelled from the body in **semen** during **ejaculation**. The **ovaries** (gonads) of a sexually mature female produce an egg that travels along **uterine tubes** to reach the **uterus**. The **vagina** connects the uterus with the exterior.

ANATOMY OF THE MALE REPRODUCTIVE SYSTEM [p. 685]

1. The **spermatozoa** travel along the **epididymis**, the **ductus deferens**, the **ejaculatory duct**, and the **urethra** before leaving the body. Accessory organs (notably the **seminal vesicles**, **prostate gland**, and **bulbourethral glands**) secrete into the ejaculatory ducts and urethra. The **scrotum** encloses the testes, and the **penis** is an erectile organ **(see Figures 27-1 to 27-9)**.

The Testes [p. 685]

2. The testes remain connected to internal structures via the **spermatic cords**. The **perineal raphe** marks the boundary between the two chambers in the scrotum **(see Figures 27-1/27-2)**.

3. The **dartos** muscle gives the scrotum a wrinkled appearance; the **cremaster muscle** pulls the testes closer to the body. The **tunica albuginea** surrounds each testis. Septa extend from the tunica albuginea to the **mediastinum**, creating a series of **lobules**. **Seminiferous tubules** within each lobule are the sites of sperm production. From there, sperm pass through a **straight tubule** to the **rete testis**. **Efferent ducts** connect the rete testis to the epididymis. Between the seminiferous tubules, there are **interstitial cells** that secrete sex hormones. Seminiferous tubules contain **spermatogonia**, stem cells involved in **spermatogenesis (see Figures 27-1 to 27-5)**.

gubernaculum testis • **scrotal cavity** • **tunica vaginalis** • **pudendal arteries** • **inferior epigastric arteries** • **primary spermatocyte** • **haploid** • **meiosis** • **secondary spermatocytes** • **spermatids** • **spermatozoon** • **spermiogenesis** • **sustentacular cells** • **blood-testis barrier** • **inhibin** • **androgen-binding protein (ABP)**

4. The **descent of the testes** through the inguinal canals occurs during development **(see Figure 27-3)**.

5. Layers of fascia, connective tissue, and muscle collectively form a sheath, the spermatic cord, which encloses the ductus deferens, testicular artery and vein, **pampiniform plexus**, and the **ilioinguinal** and **genitofemoral nerves (see Figure 27-2)**.

Anatomy of a Spermatozoon [p. 692]

6. Each spermatozoon has a **head**, **middle piece**, and **tail (see Figure 27-6)**.

acrosomal cap • **neck** • **flagellum**

The Male Reproductive Tract [p. 692]

7. From the testis, the spermatozoa enter the epididymis, an elongate tubule with **head**, **body**, and **tail** regions. The epididymis monitors and adjusts the composition of the tubular fluid and serves as a recycling center for damaged spermatozoa **(see Figures 27-1/27-4/27-7)**.

capacitation

8. The **ductus deferens** begins at the epididymis and passes through the inguinal canal as one component of the spermatic cord. Near the prostate, it enlarges to form the **ampulla**. The junction of the base of the seminal vesicle and the ampulla creates the **ejaculatory duct**, which empties into the urethra **(see Figures 27-1/27-7/27-8)**.

9. The urethra extends from the urinary bladder to the tip of the penis. It can be divided into three regions: the *prostatic urethra, membranous urethra,* and *penile urethra* **(see Figures 27-1/27-9)**.

The Accessory Glands [p. 694]

10. Each **seminal vesicle** is an active secretory gland that contributes about 60% of the volume of semen; its secretions contain fructose that is easily metabolized by spermatozoa **(see Figures 27-1/27-8/27-9)**.

11. The prostate gland secretes fluids that account for 20–30% of the volume of semen **(see Figures 27-1/27-8/27-9)**.

prostatic fluid • **seminalplasmin**

12. Alkaline mucus secreted by the bulbourethral glands has lubricating properties **(see Figures 27-1/27-8/27-9)**.

Semen [p. 696]

13. A typical ejaculation releases 2–5 mℓ of semen (an **ejaculate**), which contains a **sperm count** of 20 to 100 million sperm per milliliter.

seminal fluid

The Penis [p. 696]

14. The skin overlying the penis resembles that of the scrotum. Most of the body of the penis consists of three masses of **erectile tissue**. Beneath the superficial fascia, there are two **corpora cavernosa** and a single **corpus spongiosum** that surrounds the urethra. Dilation of the erectile tissue with blood produces an **erection (see Figures 27-1/27-2/27-9)**.

root • **body (shaft)** • **glans** • **prepuce** • **neck** • **external urethral meatus** • **preputial glands** • **smegma** • **circumcision** • **crura** • **emission**

ANATOMY OF THE FEMALE REPRODUCTIVE SYSTEM *[p. 698]*

Related Key Terms

1. Principal organs of the female reproductive system include the **ovaries**, **uterine tubes**, **uterus**, **vagina**, and external genitalia *(see Figures 27-10 to 27-21).*

2. The ovaries, uterine tubes, and uterus are enclosed within the **broad ligament** (an extensive mesentery). The **mesovarium** supports and stabilizes each ovary *(see Figures 27-10/27-11/27-15).*

mesosalpinx • rectouterine pouch • vesicouterine pouch

The Ovaries *[p. 698]*

3. The ovaries are held in position by the **ovarian ligament** and the **suspensory ligament**. The **ovarian artery** and **vein** enter the ovary at the **ovarian hilum**. Each ovary is covered by a **tunica albuginea** *(see Figures 27-10/27-11/27-15).*

4. Oogenesis (ovum production) occurs monthly in **ovarian follicles** as part of the **ovarian cycle**. Development proceeds through **primordial**, **primary**, **secondary**, and **tertiary (mature Graafian) follicles**. At **ovulation**, a **secondary oocyte** and the surrounding follicular walls of the **corona radiata** are released through the ruptured ovarian wall *(see Figure 27-12).*

oogonia • primary oocytes • atresia • egg nests • zona pellucida • estradiol • follicular fluid • antrum • polar body

5. The follicular cells remaining within the ovary form the **corpus luteum** that later degenerates into a **corpus albicans** of scar tissue *(see Figures 27-12/27-13).*

6. The hypothalamic secretion of **GnRH** triggers the pituitary secretion of FSH and the synthesis of LH. FSH initiates follicular development, and activated follicles and ovarian interstitial cells produce estrogens. **Progesterone**, one of the steroid hormones called **progestins**, is the principal hormone of the postovulatory period. Hormonal changes are responsible for the maintenance of the uterine cycle *(see Figure 27-18).*

The Uterine Tubes *[p. 703]*

7. Each uterine tube has an **infundibulum** with **fimbriae** (projections), an **ampulla**, an **isthmus**, and an **intramural portion** that opens into the uterine cavity. For fertilization to occur, the ovum must encounter spermatozoa during the first 12–24 hours of its passage from the infundibulum to the uterus *(see Figures 27-10/27-11/27-14/27-15).*

The Uterus *[p. 703]*

8. The uterus provides mechanical protection and nutritional support to the developing embryo. Normally the uterus bends anteriorly near its base (**anteflexion**). It is stabilized by the broad ligament, **uterosacral ligaments**, **round ligaments**, and the **lateral ligaments** *(see Figures 27-10/27-11/27-15).*

retroflexion

9. The gross divisions of the uterus include the **body**, **fundus**, **isthmus**, **cervix**, **external orifice**, **uterine cavity**, **cervical canal**, and **internal orifice**. The uterine wall can be histologically divided into an inner **endometrium**, a muscular **myometrium**, and a superficial **perimetrium** *(see Figures 27-15/27-16).*

basilar zone

10. A typical 28-day **uterine cycle** (**menstrual cycle**) begins with the onset of **menses** and the destruction of the **functional zone** of the endometrium. This process of **menstruation** continues from 1 to 7 days *(see Figures 27-17/27-18).*

dysmenorrhea

11. After menses, the **proliferative phase** begins and the functional zone undergoes repair and thickens. Menstrual activity begins at **menarche** and continues until **menopause** *(see Figures 27-17/27-18).*

The Vagina *[p. 709]*

12. The vagina is a muscular tube extending between the uterus and external genitalia. A thin epithelial fold, the **hymen**, partially blocks the entrance to the vagina *(see Figures 27-10/27-11/27-19).*

vaginal canal • fornix • vaginal branches • muscularis layer

The External Genitalia *[p. 710]*

13. The structures of the **vulva** (**pudendum**) include the **vestibule**, **labia minora**, **clitoris**, **prepuce** (hood), **labia majora**, and the **lesser** and **greater vestibular glands** *(see Figures 27-10/27-20).*

paraurethral glands • mons pubis

The Mammary Glands *[p. 710]*

14. The **mammary glands** lie in the subcutaneous layer beneath the skin of the chest. The glandular tissue of the breast consists of secretory lobules. Ducts leaving the lobules converge into a single **lactiferous duct** and expand near the nipple, forming a **lactiferous sinus**. The ducts of underlying mammary glands open onto the body surface at the **nipple** *(see Figure 27-21).*

pectoral fat pad • areola • suspensory ligaments of the breast

15. Branches of the *internal thoracic artery* supply blood to each breast. **Lactation** occurs in the mammary glands of the breasts for nourishing the newborn.

Pregnancy and the Female Reproductive System *[p. 714]*

16. If fertilization occurs, **implantation** of the **blastocyst** occurs in the endometrial wall. The **placenta** that develops produces several hormones that modify the ovarian and uterine cycles. **Human chorionic gonadotropin (HCG)** maintains the corpus luteum for several months. The placenta also produces **relaxin** and **human placental lactogen (HPL)**. HPL, GH, and PRL (prolactin) are primarily responsible for preparing the mammary glands for milk production.

AGING AND THE REPRODUCTIVE SYSTEM *[p. 715]*

Menopause *[p. 715]*

1. Menopause (the time that ovulation and menstruation cease) typically occurs around age 50. Production of GnRH, FSH, and LH rises, while circulating concentrations of estrogen and progestins decline.

premature menopause

The Male Climacteric *[p. 715]*

2. During the **male climacteric**, between ages 50 and 60, circulating testosterone levels decline, while levels of FSH and LH rise.

13. How are ovarian follicular types classified?

1 REVIEW OF CHAPTER OBJECTIVES

1. Describe the purpose of the reproductive system.
2. Identify the structures of the male and female reproductive systems, and summarize the functions of each structure.
3. Describe the gross anatomy and functions of the principal structures of the male reproductive system.
4. Describe the histological structure of the organs and accessory glands of the male reproductive system.
5. Discuss spermatogenesis and the storage and transport of sperm cells.
6. Describe the gross anatomy and functions of the principal organs of the female reproductive system.
7. Describe the histological structure of the organs and components of the female reproductive system.
8. Discuss oogenesis and the transport of ova.
9. Describe the ovarian and uterine cycles and the hormones that regulate and coordinate these cycles.
10. Describe the anatomical and hormonal changes that accompany pregnancy.
11. Discuss the changes in the reproductive system that occur at menopause and the male climacteric.

2 REVIEW OF CONCEPTS

1. What are the functions of the spermatic cord?
2. Why are the testes located in the scrotum rather than being intra-abdominal as are the ovaries?
3. What role do the sex hormones have in effecting the physical changes boys undergo at puberty?
4. How does the process of meiosis differ from mitosis?
5. What is the role of the sustentacular cells in sperm maturation?
6. What is the role of the epididymis in the process of sperm maturation and growth?
7. What is the role of the ductus deferens in sperm transport?
8. What is the role of the accessory reproductive glands in promoting normal sperm function?
9. Describe the structure of erectile tissue and the erectile mechanism of the penis.
10. What factors stabilize the positions of the organs of the female reproductive tract?
11. How does the production of ova during the life of a woman differ from the production of sperm during the life of a man?
12. What anatomical problem might result from a failure of the inguinal canal to close in a male?

13. How are ovarian follicular types classified?
14. What is the function of the corpus luteum?
15. Why do women undergo menopause typically in their early to mid-fifties?
16. How does the endometrium of the uterus change during the typical 28-day female reproductive cycle?
17. What determines the size and shape of the breasts?

3 CRITICAL THINKING AND CLINICAL APPLICATION QUESTIONS

1. A 61-year-old man complains of a burning sensation upon urination and an increased frequency of the urge to urinate. What might be the cause of the problem? If, instead of the symptoms described above, the man complained simply of having difficulty producing a flow of urine, what alternative problem might he be experiencing?
2. A newborn premature male baby has a scrotal sac of normal size and shape, but it is flaccid on the left. What might be the reason for this, and what can be done to correct it? If this condition remains uncorrected, what problem could it cause?
3. A young woman experiences severe abdominal cramping, nausea, and sweating, accompanied by irregular vaginal bleeding that has occurred intermittently for several days. After seeing a gynecologist and having a blood test and an ultrasound examination, the woman is admitted to the hospital, with a diagnosis of ectopic pregnancy. What might the blood test and ultrasound have shown to support this diagnosis?
4. A 24-year-old woman develops a fever, has lower abdominal pain, and a blood test reveals that she has an elevated white blood cell count. If she has PID, what accounts for the symptoms, and what could be the ultimate effect of this problem?
5. At an annual gynecological examination, a woman is given a Pap smear. What disease does this test detect, and how does it work?
6. Among the most common reproductive system problems women face are symptoms of vaginal itching and burning, often accompanied by a lumpy white discharge. What is this condition, what is the cause, and how is it treated?
7. A woman in her late thirties discovers a lump in her right breast upon self-examination. Although she knows that many breast lumps are benign, she immediately arranges to see a gynecologist. What kind of test will she be likely to have, and what treatment options exist?

28

Human Development

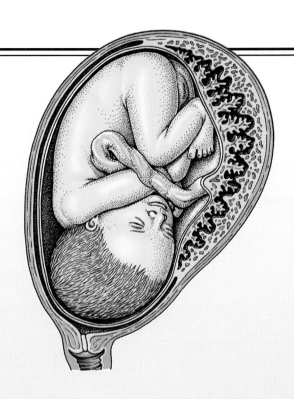

CHAPTER OUTLINE AND OBJECTIVES

The process of **development** is the gradual modification of anatomical structures during the period from conception to maturity. The changes are truly remarkable—what begins as a single cell slightly larger than the period at the end of this sentence becomes an individual whose body contains trillions of cells organized into tissues, organs, and organ systems. The creation of specialized cell types during development is called **differentiation**. Differentiation occurs through selective changes in genetic activity. A basic understanding of human development provides a framework for enhancing the understanding of anatomical structures. This discussion will focus on highlights of the developmental process rather than describing the events in great detail. (The development of specific systems has already been described in the Embryology Summaries in earlier chapters.)

AN OVERVIEW OF DEVELOPMENT

Development involves (1) the division and differentiation of cells and (2) changes that produce and modify anatomical structures. Development begins at fertilization, or **conception**, and can be divided into periods characterized by specific anatomical changes. **Embryology** (em-brē-OL-ō-jē) considers the developmental events that occur in the first 2 months after fertilization. **Fetal development** begins at the start of the ninth week and continues up to the time of birth. Embryological and fetal development are sometimes referred to collectively as **prenatal development**, which will be the primary focus of this chapter. **Postnatal development** commences at birth and continues to maturity. We will briefly consider the *neonatal period* that immediately follows delivery, but other aspects of childhood and adolescent development were considered in earlier chapters dealing with specific systems.

FERTILIZATION

Fertilization involves the fusion of two haploid gametes, producing a zygote containing the normal somatic number of chromosomes (46). ∞ [p. 691] The functional roles and contributions of the spermatozoon and the ovum are very different. The spermatozoon simply delivers the paternal chromosomes to the site of fertilization, but the ovum must provide all of the nourishment and genetic programming to support the embryonic development for nearly a week after conception. The volume of the ovum is therefore much greater than that of the spermatozoon.

Normal fertilization occurs in the ampulla of the uterine tube, usually within a day of ovulation. Over this period of time, the oocyte has traveled a few centimeters, but the spermatozoa must cover the distance between the vagina and the isthmus. The sperm arriving in the vagina are already motile, but they cannot fertilize an egg until they have undergone *capacitation* within the female reproductive tract. ∞ [p. 694]

Contractions of the uterine musculature and ciliary currents in the uterine tubes have been suggested as likely mechanisms for accelerating the movement of spermatozoa from the vagina to the fertilization site. The passage time may be from 30 minutes to 2 hours.

Even with transport assistance and available nutrients, this is not an easy passage. Of the 200 million spermatozoa introduced into the vagina from a typical ejaculate, only around 10,000 enter the uterine tube, and fewer than 100 actually reach the ampulla. A male with a sperm count below 20 million per milliliter is functionally **sterile** because too few spermatozoa will survive to reach the oocyte. One or two spermatozoa cannot accomplish fertilization, because of the condition of the oocyte at ovulation.

The Oocyte at Ovulation
(Figure 28-1)

Ovulation occurs before the completion of oocyte maturation, and the secondary oocyte leaving the follicle is in metaphase of the second meiotic division. Metabolic operations have also been discontinued, and the oocyte drifts in a sort of suspended animation, awaiting the stimulus for further development. If fertilization does not occur, the oocyte disintegrates without completing meiosis.

Fertilization is complicated by the fact that when the oocyte leaves the ovary, it is surrounded by a layer of follicle cells, the *corona radiata*. ∞ [p. 702] The events that follow are diagrammed in Figure 28-1●. These cells protect the gamete as it passes through the ruptured follicular wall and into the infundibulum of the uterine tube. Although the physical process of fertilization requires only a single sperm in contact with the oocyte membrane, that spermatozoon must first penetrate the corona radiata. The acrosomal cap contains **hyaluronidase** (hī-a-lūr-ON-a-dāz), an enzyme that breaks down the intercellular cement between adjacent follicle cells. At least a hundred spermatozoa must release hyaluronidase before the connections between the follicular cells break down enough to permit fertilization. No matter how many spermatozoa slip through the gap, only a single spermatozoon will accomplish fertilization and activate the oocyte. When that spermatozoon passes through the zona pellucida and contacts the secondary oocyte, their plasma membranes fuse, and the sperm enters the **ooplasm**, or cytoplasm of the oocyte. The process of membrane fusion is the trigger for **oocyte activation**. Oocyte activation involves a series of changes in the metabolic activity of the secondary oocyte. The metabolic rate of the oocyte rises suddenly, and changes in the cell membrane immediately prevent fertilization by additional sperm. One of the most dramatic changes is the completion of meiosis.

Pronucleus Formation and Amphimixis
(Figure 28-1)

After oocyte activation and the completion of meiosis, the nuclear material remaining within the ovum reorganizes as the **female pronucleus** (Figure 28-1●). While these changes are under way, the nucleus of the spermatozoon swells, becoming the **male pronucleus**. The male pronucleus then migrates toward the center of the cell, and the two pronuclei fuse in a process called **amphimixis** (am-fi-MIK-sis). Fertilization is now complete, with the formation of a **zygote** containing the normal complement of 46 chromosomes. The zygote now prepares to begin dividing; these mitotic divisions will ultimately produce billions of specialized cells. ⸸*Chromosomal Analysis [p. 777]*

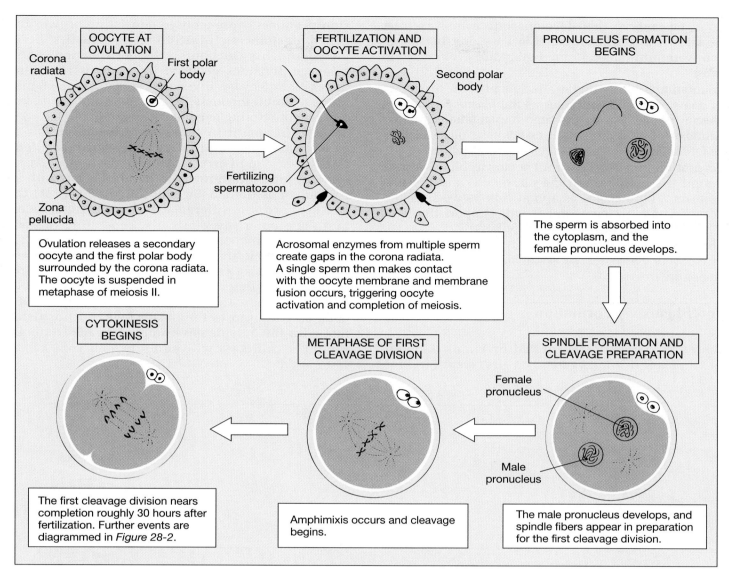

FIGURE 28-1
Fertilization and Preparation for Cleavage. During human fertilization the entire sperm enters the cytoplasm. In many other mammals and invertebrates, only the head of the sperm enters the ooplasm.

A Preview of Prenatal Development

The time spent in prenatal development is known as the period of **gestation** (jes-TĀ-shun). For convenience, the gestation period is usually considered as three integrated **trimesters**, each 3 months in duration:

- The **first trimester** is the period of embryonic and early fetal development. During this period the rudiments of all the major organ systems appear.
- In the **second trimester** the organs and organ systems complete most of their development. The body proportions change, and by the end of the second trimester the fetus looks distinctively human.
- The **third trimester** is characterized by rapid fetal growth. Early in the third trimester most of the major organ systems become fully functional, and an infant born 1 month or even 2 months prematurely has a reasonable chance of survival.

THE FIRST TRIMESTER

By the end of the first trimester (twelfth developmental week), the fetus is almost 75 mm (3 in.) long and weighs perhaps 14 g (0.5 oz). The events that occur in the first trimester are complex, and the first trimester is the most dangerous period in prenatal life. Only about 40% of conceptions produce embryos that survive the first trimester. For this reason pregnant women are usually warned to take great care to avoid drugs or other disruptive stresses during the first trimester, in the hopes of preventing an error in the delicate processes under way.

Many important and complex developmental events occur during the first trimester. We will focus attention on four general processes: *cleavage, implantation, placentation,* and *embryogenesis.*

    1.   Cleavage (KLĒ-vej) is a sequence of cell divisions that begins immediately after fertilization and ends at the first contact with the uterine wall. Over this period the zygote becomes a **pre-embryo** that develops into a multicellular complex known as a **blastocyst**. (Cleavage and blastocyst formation were introduced in the Embryology Summary in Chapter 3.) ∞ [p. 53]

2. Implantation begins with the attachment of the blastocyst to the endometrium and continues as the blastocyst invades the maternal tissues. During the time implantation is under way, a number of other important events take place that set the stage for the formation of vital embryonic structures.

3. Placentation (pla-sen-TĀ-shun) occurs as blood vessels form around the periphery of the blastocyst, and the **placenta** appears. The placenta is a vital link between maternal and embryonic systems, and it will support the fetus during the second and third trimesters.

4. Embryogenesis (em-brē-ō-JEN-e-sis) is the formation of a viable embryo. This process establishes the foundations for all major organ systems.

Cleavage and Blastocyst Formation
(Figure 28-2)

Cleavage (Figure 28-2●) is a series of cell divisions that subdivides the cytoplasm of the zygote. The first cleavage divi-

sion produces a pre-embryo consisting of two identical cells, called **blastomeres** (BLAS-tō-mērz). The first division is completed roughly 30 hours after fertilization, and subsequent cleavage divisions occur at intervals of 10–12 hours. During the initial cleavage divisions, all of the blastomeres undergo mitosis simultaneously, but as the number of blastomeres increases, the timing becomes less predictable.

At this point in time, the pre-embryo is a solid ball of cells, resembling a mulberry. This stage is called the **morula** (MOR-ū-la; "mulberry"). After 5 days of cleavage, the blastomeres form a hollow ball, the blastocyst, with an inner cavity known as the **blastocoele** (BLAS-tō-sēl). At this stage you can begin to see differences between the cells of the blastocyst. The outer layer of cells, separating the outside world from the blastocoele, is called the **trophoblast** (TRŌ-fō-blast). The function is implied by the name: *trophos*, food + *blast*, precursor. These cells will be responsible for providing food to the developing embryo. A second group of cells, the **inner cell mass**, lies clustered at one end of the blastocyst. These cells are exposed to the blastocoele, but insulated from contact with the outside environment by the trophoblast. In time the inner cell mass will form the embryo. ⚕ *Technology and the Treatment of Infertility [p. 777]*

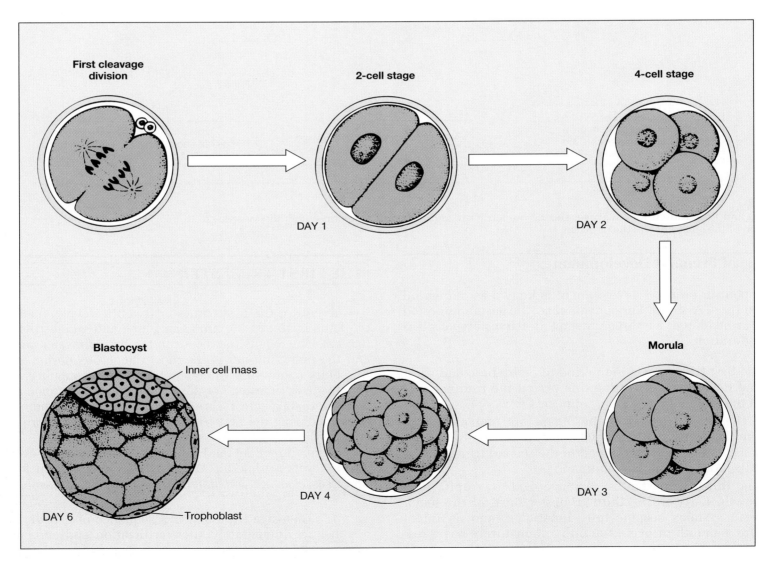

FIGURE 28-2
Cleavage and Blastocyst Formation

Implantation
(Figure 28-3)

At fertilization the zygote is still 4 days away from the uterus. It arrives in the uterine cavity as a morula, and over the next 2–3 days, blastocyst formation occurs. Over this period the active cells are gaining nutrients from the fluid within the uterine cavity. This fluid, rich in glycogen, is secreted by the endometrial glands. When fully formed, the blastocyst contacts the endometrium, usually in the fundus or body of the uterus, and implantation occurs. Stages in the implantation process are illustrated in Figure 28-3●.

Implantation begins as the surface of the blastocyst closest to the inner cell mass touches and adheres to the uterine lining

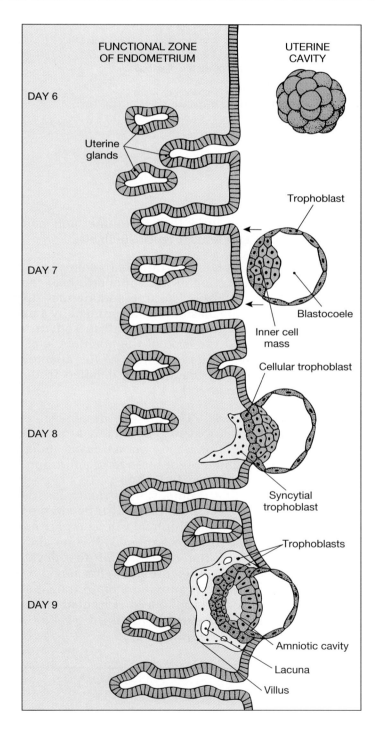

FIGURE 28-3
Stages in the Implantation Process

(see *Day 7*, Figure 28-3●). At the point of contact, the trophoblast cells divide rapidly, making the trophoblast several layers thick. Near the endometrial wall, the cell membranes separating the trophoblast cells disappear, creating a layer of cytoplasm containing multiple nuclei (*Day 8*). This outer layer, called the **syncytial** (sin-SISH-al) **trophoblast**, then erodes a path through the uterine epithelium, by secreting the enzyme *hyaluronidase*. This enzyme dissolves the intercellular cement between adjacent epithelial cells, just as the hyaluronidase released by spermatozoa dissolves the connections between cells of the corona radiata. At first this erosion creates a gap in the uterine lining, but the migration and divisions of epithelial cells soon repair the surface. When the repairs are completed, the blastocyst loses contact with the uterine cavity, and development occurs entirely within the *functional zone* of the endometrium. ∞ [p. 706]

As implantation proceeds, the syncytial trophoblast continues to enlarge and spread into the surrounding endometrium (*Day 9*). The digestion of uterine glands releases nutrients that are absorbed by the syncytial trophoblast and distributed by diffusion across the underlying **cellular trophoblast** to the inner cell mass. These nutrients provide the energy needed to support the early stages of embryo formation. Trophoblastic extensions grow around endometrial capillaries, and as the capillary walls are destroyed, maternal blood begins to percolate through trophoblastic channels known as **lacunae**. Fingerlike **villi** extend away from the trophoblast into the surrounding endometrium, and these extensions gradually increase in size and complexity as development proceeds. Over the next few days, the trophoblast begins breaking down larger endometrial veins and arteries, and blood flow through the lacunae accelerates.
☩ *Problems with the Implantation Process [p. 777]*

Formation of the Blastodisc
(Figures 28-3/28-4)

In the early blastocyst stage, the inner cell mass has little visible organization. But, by the time of implantation, the inner cell mass is separating from the trophoblast. The separation gradually increases, creating a fluid-filled chamber called the **amniotic** (am-nē-OT-ik) **cavity**. The amniotic cavity can be seen in *Day 9* of Figure 28-3●, and additional details from *Days 10–12* are shown in Figure 28-4●. At this stage the cells of the inner cell mass are organized into an oval sheet that is two cell layers thick. This oval, called a **blastodisc** (BLAS-tō-disk), initially consists of an epithelial layer, or **epiblast** (EP-i-blast), facing the amniotic cavity and an underlying **hypoblast** (HĪ-pō-blast) exposed to the fluid contents of the blastocoele.

Gastrulation and Germ Layer Formation
(Figure 28-4)

A few days later, a third layer begins forming through the process of **gastrulation** (gas-troo-LĀ-shun) (*Day 12*, Figure 28-4●). During gastrulation, cells in specific areas of the epiblast move toward the center of the blastodisc, toward a line known as the **primitive streak**. At the primitive streak, these migrating cells leave the surface and move between the epiblast and hypoblast. This movement creates three distinct embryonic layers with markedly different fates. Once gastrulation begins, the layer remaining in contact with the amniotic cavity is called the **ectoderm**, the hypoblast is known as the **endoderm**, and the intervening, poorly organized layer is the **mesoderm**. The formation of mesoderm and the devel-

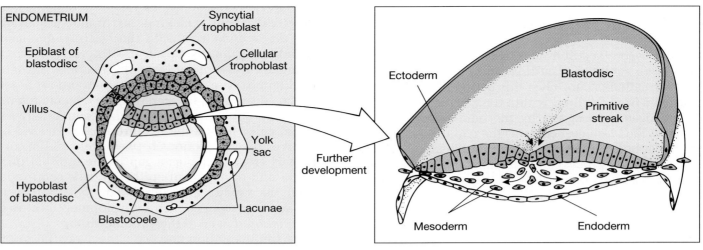

ENDOMETRIUM
Syncytial trophoblast
Epiblast of blastodisc
Cellular trophoblast
Villus
Yolk sac
Hypoblast of blastodisc
Blastocoele
Lacunae

Further development

Ectoderm
Blastodisc
Primitive streak
Mesoderm
Endoderm

Day 10: The blastodisc begins as two layers, an epiblast facing the amniotic cavity and the hypoblast exposed to the blastocoele. Migration of epiblast cells around the amniotic cavity is the first step in the formation of the amnion. Migration of hypoblast cells creates a sac that hangs below the blastodisc. This is the first step in yolk sac formation.

Day 12: Migration of epiblast cells into the region between epiblast and hypoblast gives the blastodisc a third layer. From the time this process, called gastrulation, begins, the epiblast is called *ectoderm,* the hypoblast *endoderm,* and the migrating cells *mesoderm.*

FIGURE 28-4
Blastodisc Organization and Gastrulation

opmental fates of these **germ layers** were introduced in an Embryology Summary in Chapter 3. ∞ [p. 83] Table 28-1 contains a more comprehensive listing of the contributions each germ layer makes to the body systems described in earlier chapters.

The Formation of Extraembryonic Membranes
(Figure 28-5)

Germ layers also participate in the formation of four **extraembryonic membranes**: the *yolk sac* (endoderm and mesoderm), the *amnion* (ectoderm and mesoderm), the *allantois* (endoderm and mesoderm), and the *chorion* (mesoderm and trophoblast). Although these membranes support embryonic and fetal development, they leave few traces of their existence in adult systems. Figure 28-5● shows representative stages in the development of the extraembryonic membranes.

THE YOLK SAC *(Figures 28-4/28-5a-c).* The first of the extraembryonic membranes to appear is the **yolk sac**. The yolk sac begins as migrating hypoblast cells spread out around the outer edges of the blastocoele to form a complete pouch suspended below the blastodisc. This pouch is already visible 10 days after fertilization (Figure 28-4●). As gastrulation proceeds, mesodermal cells migrate around this pouch and complete the formation of the yolk sac (Figure 28-5a-c●). Blood vessels soon appear within the mesoderm, and the yolk sac becomes an important site of blood cell formation.

THE AMNION *(Figure 28-5a,b,e).* The ectodermal layer also undergoes an expansion, and ectodermal cells spread over the inner surface of the amniotic cavity. Mesodermal cells soon follow, creating a second, outer layer (Figure 28-5a,b●). This combination of mesoderm and ectoderm is the **amnion** (AM-nē-on). As the embryo and later the fetus enlarges, this membrane continues to expand, increasing the size of the amniotic cavity. The

amnion encloses **amniotic fluid** that surrounds and cushions the developing embryo and fetus (Figure 28-5b,e●).

THE ALLANTOIS *(Figure 28-5b).* The third extraembryonic membrane begins as an outpocketing of the endoderm near the base of the yolk sac (Figure 28-5b●). The free endodermal tip then grows toward the wall of the blastocyst, surrounded by a mass of mesodermal cells. This sac of endoderm and mesoderm is the **allantois** (a-LAN-tō-is), and the base of the allantois later gives rise to the urinary bladder. (The formation of the allantois and its relationship to the urinary bladder was detailed in the Embryology Summary in Chapter 26.) ∞ [p. 679]

THE CHORION *(Figure 28-5a,b).* The mesoderm associated with the allantois spreads until it extends completely around the inside of the trophoblast, forming a mesodermal layer underneath the trophoblast. This combination of mesoderm and trophoblast is the **chorion** (KOR-ē-on) (Figure 28-5a,b●).

When implantation first occurs, the nutrients absorbed by the trophoblast can easily reach the blastodisc by simple diffusion. But as the embryo and the trophoblastic complex enlarge, the distance between the two increases and diffusion alone can no longer keep pace with the demands of the developing embryo. The chorion solves this problem, for blood vessels developing within the mesoderm provide a rapid transit system linking the embryo with the trophoblast. Circulation through those chorionic vessels begins early in the third week of development, when the heart starts beating.

Placentation
(Figure 28-5)

The appearance of blood vessels in the chorion is the first step in the creation of a functional placenta. By the third week of development (Figure 28-5b●), mesoderm extends along the core of

TABLE 28-1   The Fates of the Primary Germ Layers

Ectodermal Contributions

Integumentary system: epidermis, hair follicles and hairs, nails, and glands communicating with the skin (apocrine and merocrine sweat glands, mammary glands, and sebaceous glands)

Skeletal system: pharyngeal cartilages and their derivatives in the adult (portion of sphenoid bone, the auditory ossicles, the styloid processes of the temporal bones, the cornu and superior rim of the hyoid bone)*

Nervous system: all neural tissue, including brain and spinal cord

Endocrine system: pituitary gland and adrenal medullae

Respiratory system: mucous epithelium of nasal passageways

Digestive system: mucous epithelium of mouth and anus, salivary glands

Mesodermal Contributions

Skeletal system: all components except some pharyngeal derivatives

Muscular system: all components

Endocrine system: adrenal cortex, endocrine tissues of heart, kidneys, and gonads

Cardiovascular system: all components, including bone marrow

Lymphatic system: all components

Urinary system: the kidneys, including the nephrons and the initial portions of the collecting system

Reproductive system: the gonads and the adjacent portions of the duct systems

Miscellaneous: the lining of the body cavities (thoracic, pericardial, peritoneal) and the connective tissues supporting all organ systems

Endodermal Contributions

Endocrine system: thymus, thyroid, and pancreas

Respiratory system: respiratory epithelium (except nasal passageways) and associated mucous glands

Digestive system: mucous epithelium (except mouth and anus), exocrine glands (except salivary glands), liver, and pancreas

Urinary system: urinary bladder and distal portions of the duct system

Reproductive system: distal portions of the duct system, stem cells that produce gametes

*The neural crest is derived from ectoderm and contributes to the formation of the skull and the skeletal derivatives of the embryonic pharyngeal arches.

---

each of the trophoblastic villi, forming **chorionic villi** in contact with maternal tissues. These villi continue to enlarge and branch, forming an intricate network within the endometrium. Blood vessels continue to be eroded, and maternal blood slowly percolates through lacunae lined by syncytial trophoblast. Chorionic blood vessels pass close by, and exchange between the embryonic and maternal circulations occurs by diffusion across the syncytial and cellular trophoblast layers.

At first the entire blastocyst is surrounded by chorionic villi. The chorion continues to enlarge, expanding like a balloon within the endometrium, and by the fourth week, the embryo, amnion, and yolk sac are suspended within an expansive, fluid-filled chamber (Figure 28-5c●). The connection between the embryo and chorion, known as the **body stalk**, contains the distal portions of the allantois and blood vessels carrying blood to and from the placenta. The narrow connection between the endoderm of the embryo and the yolk sac is called the **yolk stalk**. (The formation of the yolk stalk and body stalk was detailed in the Embryology Summary in Chapter 25.) ∞ [p. 630]

The placenta does not continue to enlarge indefinitely. Regional differences in placental organization begin to develop as placental expansion creates a prominent bulge in the endometrial surface. This relatively thin portion of the endometrium, the **decidua capsularis** (dē-SID-ū-a kap-sū-LA-ris), no longer participates in nutrient exchange, and the chorionic villi disappear (Figure 28-5d●). Placental functions are now concentrated in a disc-shaped area situated in the deepest portion of the endometrium, a region called the **decidua basalis** (ba-SA-lis). The rest of the endometrium, which has no contact with the chorion, is called the **decidua parietalis** (pa-rī-e-TAL-is). As the end of the first trimester approaches, the fetus moves farther away from the placenta (Figure 28-5d,e●). It remains connected by the **umbilical cord**, or **umbilical stalk**, which contains the allantois, placental blood vessels, and the yolk stalk.

The developing fetus is totally dependent on maternal organ systems for nourishment, respiration, and waste removal. These functions must be performed by maternal systems in addition to their normal operations. For example, the mother must absorb enough oxygen, nutrients, and vitamins for herself and her fetus, and she must eliminate all of the generated wastes. Although this is not a burden over the initial weeks of gestation, the demands placed upon the mother become significant in subsequent trimesters, as the fetus grows larger. In practical terms the mother must breathe, eat, and excrete for two.

Placental Circulation
(Figure 28-6)

Figure 28-6a● diagrams circulation at the placenta near the end of the first trimester. Blood flows to the placenta through the paired **umbilical arteries** and returns in a single **umbilical vein**. The chorionic villi (Figure 28-6b●) provide the surface area for active and passive exchange between the fetal and maternal bloodstreams. As noted in Chapter 27, the placenta also synthesizes important hormones that affect maternal as well as embryonic tissues. ∞ [p. 715] Human chorionic gonadotropin production begins within a few days of implantation, and during the second and third trimesters the placenta also releases progesterone, estrogens, human placental lactogen, and relaxin. These hormones are synthesized and released into the maternal circulation by the trophoblast. ☩ *Problems with Placentation [p. 778]*

EMBRYOGENESIS
(Figures 28-5b,c/28-7)

Shortly after gastrulation begins, folding and differential growth of the embryonic disc produce a bulge that projects into the amniotic cavity (Figure 28-5b●). This projection is known as the **head fold**, and similar movements lead to the formation of a **tail fold** (Figure 28-5c●). The **embryo** is now physically as well as developmentally separated from the rest of the blastodisc and the extraembryonic membranes. The definitive orientation of the embryo can now be seen, complete with dorsal and ventral surfaces and left and right sides.

These changes in proportions and appearance that occur between the fourth developmental week and the end of the first trimester are visually summarized in Figure 28-7●, p. 730. The developmental history of structures labeled in Figure 28-7a,b● can be followed in more detail by referring to the Embryology Summaries in earlier chapters as noted in Table 28-2.

The first trimester is a critical period for development because events in the first 12 weeks establish the basis for organ formation, a process called **organogenesis**. Embryology Summaries in earlier chapters described major features of organogenesis in each organ system. Important developmental milestones are indicated in Table 28-2; those interested in additional details should refer to the Embryology Summaries cross-referenced in the table. ☩ *Teratogens and Abnormal Development [p. 778]*

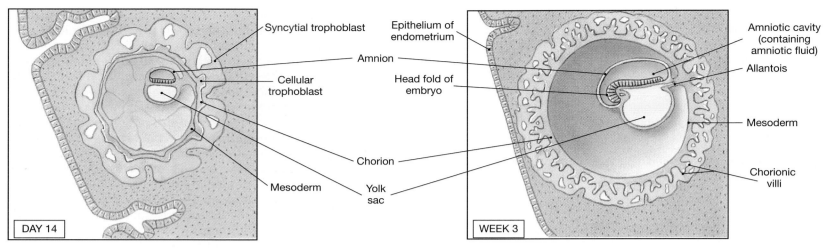

Syncytial trophoblast

Amnion

Cellular trophoblast

Mesoderm

Chorion

Yolk sac

DAY 14

Epithelium of endometrium

Head fold of embryo

Amniotic cavity (containing amniotic fluid)

Allantois

Mesoderm

Chorionic villi

WEEK 3

(a) Migration of mesoderm around the inner surface of the trophoblast creates the chorion. Mesodermal migration around the outside of the amniotic cavity, between the ectodermal cells and the trophoblast, creates the amnion. Mesodermal migration around the endodermal pouch below the blastodisc creates the definitive yolk sac.

(b) The embryonic disc bulges into the amniotic cavity at the head fold. The allantois, an endodermal extension surrounded by mesoderm, extends toward the trophoblast.

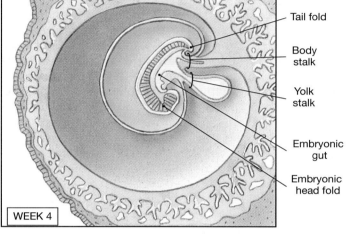

Tail fold

Body stalk

Yolk stalk

Embryonic gut

Embryonic head fold

WEEK 4

(c) The embryo now has a head fold and a tail fold. Constriction of the connection between the embryo and the surrounding trophoblast constricts the yolk stalk and body stalk.

Uterus

Myometrium

Decidua parietalis

Decidua capsularis

Decidua basalis

Umbilical stalk

Placenta

Chorionic villi

Yolk sac

Uterine lumen

WEEK 5

(d) The developing embryo and extraembryonic membranes bulge into the uterine cavity. The trophoblast pushing out into the uterine lumen remains covered by endometrium, but no longer participates in nutrient absorption and embryo support. The embryo moves away from the placenta, and the body stalk and yolk stalk fuse to form an umbilical stalk.

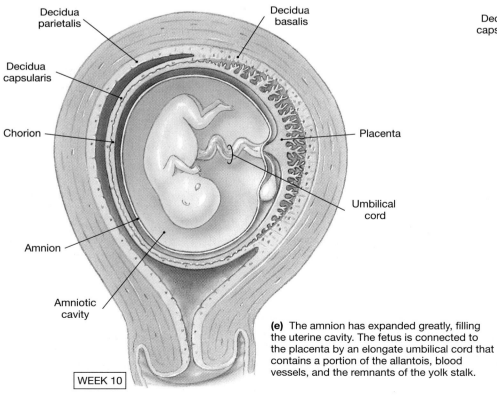

Decidua parietalis

Decidua capsularis

Chorion

Amnion

Amniotic cavity

Decidua basalis

Placenta

Umbilical cord

WEEK 10

(e) The amnion has expanded greatly, filling the uterine cavity. The fetus is connected to the placenta by an elongate umbilical cord that contains a portion of the allantois, blood vessels, and the remnants of the yolk stalk.

FIGURE 28-5
Embryonic Membranes and Placenta Formation

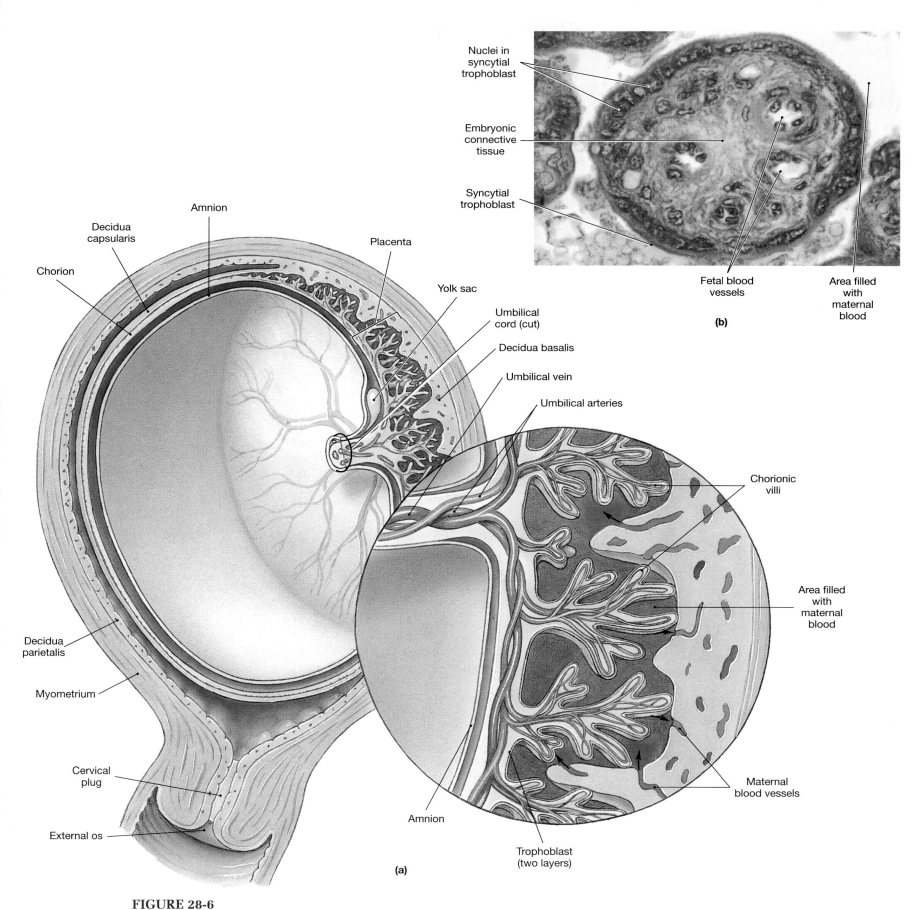

FIGURE 28-6
A Three-Dimensional View of Placental Structure. (a) For clarity the uterus is shown after the embryo has been removed and the umbilical cord cut. Blood flows into the placenta through ruptured maternal blood arteries. It then flows around chorionic villi that contain fetal blood vessels. Fetal blood arrives over paired umbilical arteries and leaves over a single umbilical vein. Maternal blood reenters the venous system of the mother through the broken walls of small uterine veins. Note that no actual mixing of maternal and fetal blood occurs. (b) A cross section through a chorionic villus, showing the syncytial trophoblast exposed to the maternal blood space. (LM × 2880)

TABLE 28-2 An Overview of Prenatal Development

Background Material
Ch. 3: Formation of Tissues (p. 53)
Development of Epithelia (p. 63)
Origins of Connective Tissues (p. 74)
Development of Organ Systems
(pp. 82–83)

Gestational Age (Months)	Size and Weight	Integumentary System	Skeletal System	Muscular System	Nervous System	Special Sense Organs
1	5 mm 0.02 g		(b) Somite formation	(b) Somite formation	(b) Neural tube formation	(b) Eye and ear formation
2	28 mm 2.7 g	(b) Nail beds, hair follicles, sweat glands	(b) Axial and appendicular cartilage formation	(c) Rudiments of axial musculature	(b) CNS, PNS organization, growth of cerebrum	(b) Taste buds, olfactory epithelium
3	78 mm 26 g	(b) Epidermal layers appear	(b) Ossification centers spreading	(c) Rudiments of appendicular musculature	(c) Basic spinal cord and brain structure	
4	133 mm 150 g	(b) Hair, sebaceous glands (c) Sweat glands	(b) Articulations (c) Facial and palatal organization	Fetus starts moving	(b) Rapid expansion of cerebrum	(c) Basic eye and ear structure (b) Peripheral receptor formation
5	185 mm 460 g	(b) Keratin production, nail production			(b) Myelination of spinal cord	
6	230 mm 823 g			(c) Perineal muscles	(b) CNS tract formation (c) Layering of cortex	
7	270 mm 1492 g	(b) Keratinization, nail formation, hair formation				(c) Eyelids open, retina sensitive to light Taste receptors functional
8	310 mm 2274 g		(b) Epiphyseal plate formation			
9	346 mm 2912 g					
Postnatal development		Hair changes in consistency and distribution	Formation and growth of epiphyseal plates continue	Muscle mass and control increase	Myelination, layering, CNS tract formation continue	
Chapter containing relevant Embryology Summary		4: Development of the Integumentary System (pp. 104–105)	6: Development of the Skull (pp. 156–157), Development of the Spinal Column (pp. 168–169) 7: Development of the Appendicular Skeleton (pp. 200–201)	10: Development of the Muscular System (pp. 272–273)	13: Introduction to the Development of the Nervous System (p. 325) 14: Development of the Spinal Cord and Spinal Nerves (pp. 356–357) 15: Development of the Brain and Cranial Nerves (pp. 402–403)	18: Development of Special Sense Organs (pp. 476–477)

Note: (b) = begin formation; (c) = complete formation.

Endocrine System	Cardiovascular and Lymphatic Systems	Respiratory System	Digestive System	Urinary System	Reproductive System
	(b) Heartbeat	(b) Trachea and lung formation	(b) Intestinal tract, liver, pancreas (c) Yolk sac	(c) Allantois	
(b) Thymus, thyroid, pituitary, adrenal glands	(c) Basic heart structure, major blood vessels, lymph nodes and ducts (b) Blood formation in liver	(b) Extensive bronchial branching into mediastinum (c) Diaphragm	(b) Intestinal subdivisions, villi, salivary glands	(b) Kidney formation (adult form)	(b) Mammary glands
(c) Thymus, thyroid gland	(b) Tonsils, blood formation in bone marrow		(c) Gallbladder, pancreas		(b) Definitive gonads, ducts, genitalia
	(b) Migration of lymphocytes to lymphatic organs, blood formation in spleen			(b) Degeneration of embryonic kidneys	
	(c) Tonsils	(c) Nostrils open	(c) Intestinal subdivisions		
(c) Adrenal glands	(c) Spleen, liver, bone marrow	(b) Alveolar formation	(c) Epithelial organization, glands		
(c) Pituitary gland			(c) Intestinal plicae		(b) Testes descend
		Complete pulmonary branching and alveolar formation		Complete nephron formation at birth	Descent complete at or near time of delivery
	Cardiovascular changes at birth; immune system becomes operative thereafter				
19: Development of the Endocrine System (pp. 496–497)	21: Development of the Heart (p. 538) 22: Development of the Circulatory System (pp. 574–575)	24: Development of the Respiratory System (pp. 618–619)	25: Development of the Digestive System (pp. 630–631)	26: Development of the Urinary System (pp. 678–679)	27: Development of the Reproductive System (pp. 711–713)

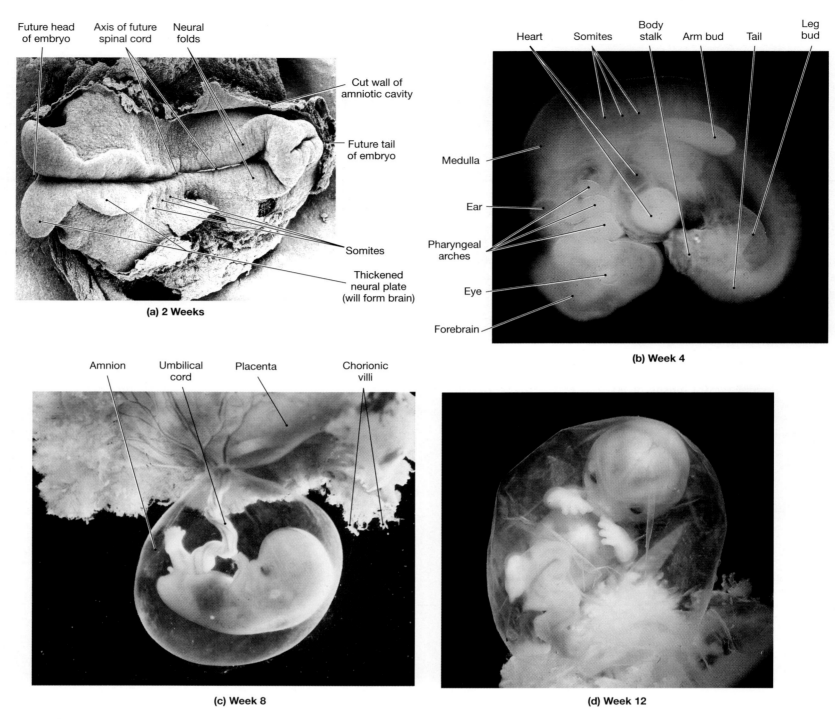

Future head of embryo   Axis of future spinal cord   Neural folds

Cut wall of amniotic cavity

Future tail of embryo

Somites

Thickened neural plate (will form brain)

(a) 2 Weeks

Heart   Somites   Body stalk   Arm bud   Tail   Leg bud

Medulla

Ear

Pharyngeal arches

Eye

Forebrain

(b) Week 4

Amnion   Umbilical cord   Placenta   Chorionic villi

(c) Week 8

(d) Week 12

FIGURE 28-7
The First Trimester. (a) An SEM of the superior surface of a 2-week embryo; neurulation (neural tube formation) is under way.
(b–d) Fiberoptic views of human embryos.

√ **What is the fate of the inner cell mass of the blastocyst?**

√ **Which extraembryonic membranes, if improperly developed, would affect the circulatory system?**

THE SECOND AND THIRD TRIMESTERS
(Figures 28-7 to 28-9)

By the start of the second trimester (Figure 28-7d●), the rudiments of all of the major organ systems have formed. Over the next 3 months, these systems will complete their functional development, and the fetus at that time will weigh around 0.64 kg (1.4 lb).

Over this period the fetus, encircled by the amnion, grows faster than the surrounding placenta. Soon the mesodermal outer covering of the amnion fuses with the inner lining of the chorion. Figure 28-8● shows a 4-month fetus as viewed with a fiberoptic endoscope and a 6-month fetus as seen in ultrasound.

During the third trimester, all of the organ systems become functional. The rate of growth begins to decrease, but in absolute terms this trimester sees the largest weight gain. In 3 months the fetus puts on around 2.6 kg (5.7 lb), reaching a full-term weight of somewhere near 3.2 kg (7 lb). Important events in organ system development in the second and third trimesters are detailed in the Embryology Summaries in earlier chapters, and highlights are noted in Table 28-2.

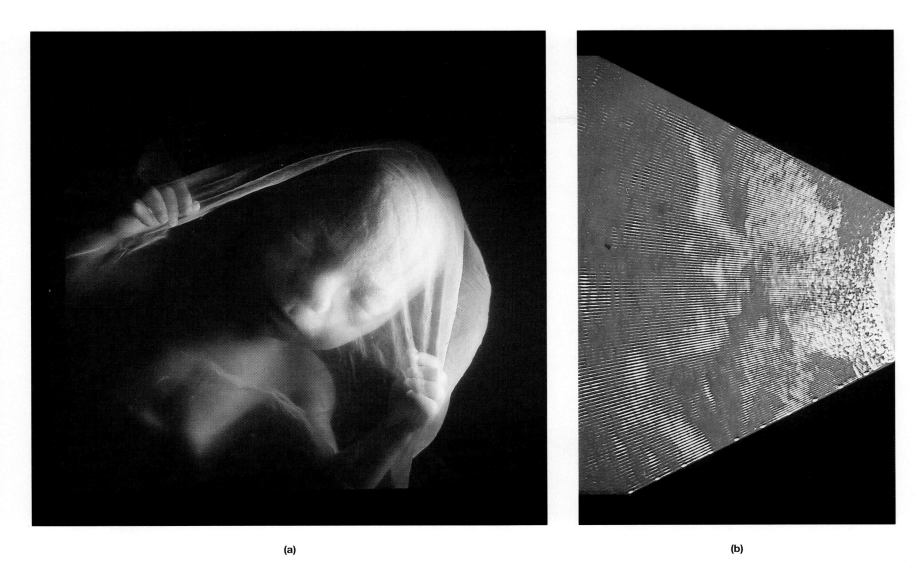

(a)   **(b)**

FIGURE 28-8
The Second and Third Trimesters. (a) A 4-month fetus seen through a fiberoptic endoscope. (b) Head of a 6-month fetus as seen using ultrasound equipment.

At the end of gestation, a typical uterus will undergo a tremendous increase in size. It will grow from 7.5 cm (3 in.) to 30 cm (12 in.) long and will contain almost 5 ℓ of fluid. The uterus and its contents weigh roughly 10 kg (22 lb). This remarkable expansion occurs through the enlargement and elongation of existing smooth muscle fibers. Figure 28-9● shows the position of the uterus, fetus, and placenta from 16 weeks to full term. When the pregnancy is at term, the uterus and fetus push many of the abdominal organs out of their normal positions (Figure 28-9c●).

LABOR AND DELIVERY

The goal of labor is the forcible expulsion of the fetus, a process known as **parturition** (par-tū-RISH-un), or birth. During true labor, each labor contraction begins near the top of the uterus and sweeps in a wave toward the cervix. These contractions are strong and occur at regular intervals. As parturition approaches, the contractions increase in force and frequency, changing the position of the fetus and moving it toward the cervical canal.

Stages of Labor
(Figure 28-10)

Labor is usually divided into three stages (Figure 28-10●): the *dilation stage*, the *expulsion stage*, and the *placental stage*.

The Dilation Stage
(Figure 28-10a)

The **dilation stage** (Figure 28-10a●) begins with the onset of true labor, as the cervix dilates and the fetus begins to slide down the cervical canal. This stage typically lasts 8 or more hours, but during this period the labor contractions occur at intervals of once every 10–30 minutes. Late in the process, the amnion usually ruptures, an event sometimes referred to as "having the water break."

The Expulsion Stage
(Figure 28-10b)

The **expulsion stage** (Figure 28-10b●) begins as the cervix dilates completely, pushed open by the approaching fetus. Expulsion continues until the fetus has completed its emergence from the vagina,

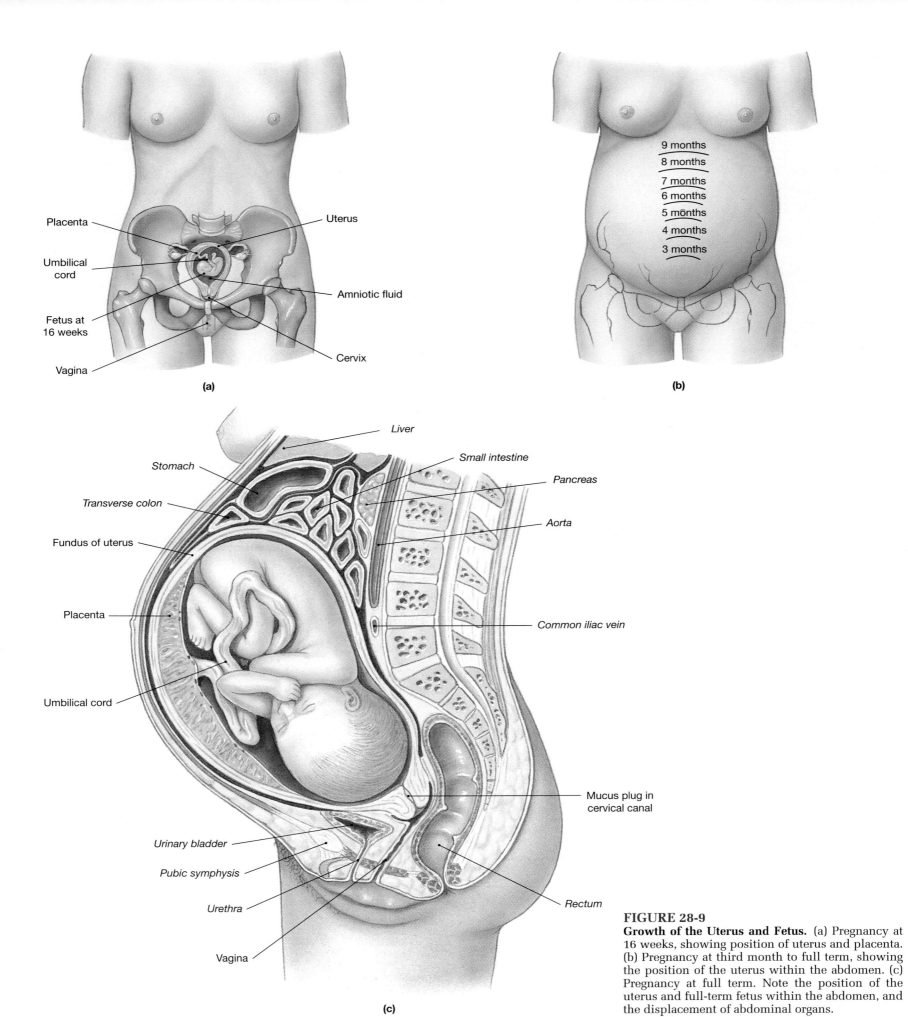

Placenta

Umbilical cord

Fetus at 16 weeks

Vagina

Uterus

Amniotic fluid

Cervix

(a)

9 months
8 months
7 months
6 months
5 months
4 months
3 months

(b)

Liver

Stomach

Transverse colon

Fundus of uterus

Placenta

Umbilical cord

Small intestine

Pancreas

Aorta

Common iliac vein

Mucus plug in cervical canal

Urinary bladder

Pubic symphysis

Urethra

Vagina

Rectum

(c)

FIGURE 28-9
Growth of the Uterus and Fetus. (a) Pregnancy at 16 weeks, showing position of uterus and placenta. (b) Pregnancy at third month to full term, showing the position of the uterus within the abdomen. (c) Pregnancy at full term. Note the position of the uterus and full-term fetus within the abdomen, and the displacement of abdominal organs.

732

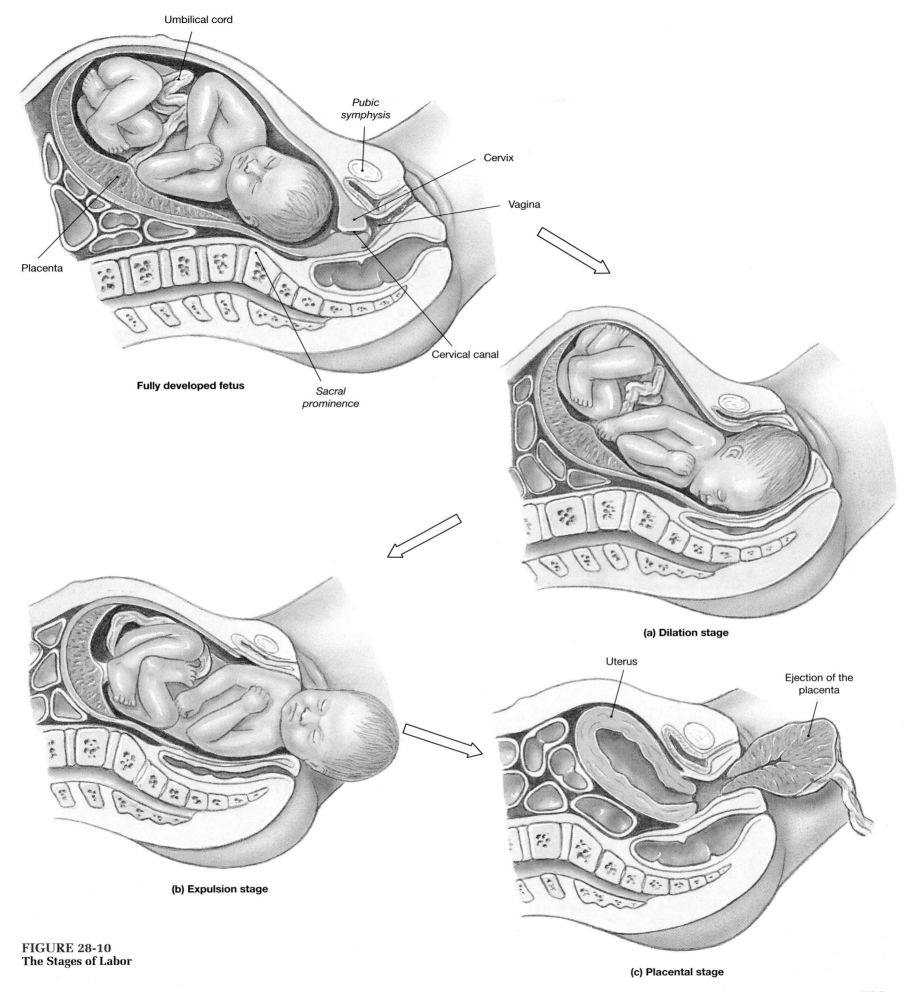

Umbilical cord

Pubic
symphysis

Cervix

Vagina

Placenta

Cervical canal

Fully developed fetus

Sacral
prominence

(a) Dilation stage

Uterus

Ejection of the
placenta

(b) Expulsion stage

(c) Placental stage

FIGURE 28-10
The Stages of Labor

a period usually lasting less than 2 hours. The arrival of the newborn infant into the outside world represents birth, or **delivery**.

If the vaginal canal is too small to permit the passage of the fetus, and there is acute danger of perineal tearing, the passageway may be temporarily enlarged by making an incision through the perineal musculature. After delivery this **episiotomy** (e-pē-zē-OT-ō-mē) can be repaired with sutures, a much simpler procedure than dealing with the bleeding and tissue damage associated with a potentially extensive perineal tear. If unexpected complications arise during the dilation or expulsion stages, the infant may be removed by **cesarean section**. In such cases an incision is made through the abdominal wall, and the uterus is opened just enough to allow passage of the infant's head. This procedure is performed during 15–25% of the deliveries in the United States. Efforts are now being made to reduce the frequency of both episiotomies and cesarean sections.

The Placental Stage
(Figure 28-10c)

During the **placental stage** of labor (Figure 28-10c●), the muscle tension builds in the walls of the partially empty uterus, and the organ gradually decreases in size. This uterine contraction tears the connections between the endometrium and the placenta. Usually within an hour after delivery, the placental stage ends with the ejection of the placenta, or "afterbirth." The disruption of the placenta is accompanied by a loss of blood (as much as 500–600 mℓ), but because the maternal blood volume has increased during pregnancy, the loss can be tolerated without difficulty.

■ **CLINICAL BRIEF**
Breech Births

In 3–4% of deliveries, the legs or buttocks of the fetus enter the vaginal canal first. Such deliveries are known as **breech births**. Risks to the infant are relatively higher in breech births because the umbilical cord may become constricted, and placental circulation cut off. Because the head is normally the widest part of the fetus, the cervix may dilate enough to pass the legs and body but not the head. Entrapment of the fetal head compresses the umbilical cord, prolongs delivery, and subjects the fetus to severe distress and potential damage. If the fetus cannot be repositioned manually, a cesarean section is usually performed.

Premature Labor

Premature labor occurs when true labor begins before the fetus has completed normal development. The chances of newborn survival are directly related to body weight at delivery. Even with massive supportive efforts, infants born weighing less than 400 g (14 oz) will not survive, primarily because the respiratory, cardiovascular, and urinary systems are unable to support life without the aid of maternal systems. As a result, the dividing line between *spontaneous abortion* and **immature delivery** is usually set at 500 g (17.6 oz), the normal weight near the end of the second trimester.

Infants delivered before completing 7 months of gestation (weight under 1 kg) have less than a 50:50 chance of survival, and most survivors suffer from severe developmental abnormalities. A **premature delivery** produces a newborn weighing over 1 kg (35.2 oz), and its chances of survival range from fair to excellent depending on the individual circumstances. ✝ *Complexity and Perfection: An Improbable Dream [p. 778]*

THE NEONATAL PERIOD

Developmental processes do not cease at delivery, for the newborn infant has few of the anatomical, functional, or physiological characteristics of the mature adult. The **neonatal period** extends from the moment of birth to 1 month thereafter. A variety of physiological and anatomical alterations occurs as the fetus completes the transition to the status of a newborn infant, or **neonate**. Before delivery, transfer of dissolved gases, nutrients, waste products, hormones, and immunoglobulins occurred across the placental interface. At birth, the newborn infant must become relatively self-sufficient, with the processes of respiration, digestion, and excretion performed by its own specialized organs and organ systems. The transition from fetus to neonate may be summarized as:

1. The lungs at birth are collapsed and filled with fluid, and filling them with air involves a massive and powerful inspiratory movement.

2. When the lungs expand, the pattern of cardiovascular circulation changes because of alterations in blood pressure and flow rates. The ductus arteriosus closes, isolating the pulmonary and systemic trunks, and the closure of the foramen ovale separates the atria of the heart, completing separation of the pulmonary and systemic circuits. These circulatory changes were discussed in Chapter 21. ∞ [p. 528]

3. Typical heart rates of 120–140 beats per minute and respiratory rates of 30 breaths per minute in neonates are normal, and considerably higher than those of adults.

4. Prior to birth, the digestive system remains relatively inactive, although it does accumulate a mixture of bile secretions, mucus, and epithelial cells. This collection of debris is excreted in the first few days of life. Over that period the newborn infant begins to nurse.

5. As waste products build up in the arterial blood, they are filtered into the urine at the kidneys. Glomerular filtration is normal, but the urine cannot be concentrated to any significant degree. As a result, urinary water losses are high, and neonatal fluid requirements are much greater than those of adults.

6. The neonate has little ability to control body temperature, particularly in the first few days after delivery. As the infant grows larger and increases the thickness of its insulating subcutaneous adipose "blanket," its metabolic rate also rises. Daily and even hourly alterations in body temperature continue throughout childhood.

■ **CLINICAL BRIEF**
Evaluating the Newborn Infant

Each newborn infant gets closely scrutinized after delivery. This procedure checks for the presence of anatomical and physiological abnormalities. It also provides baseline information useful in assessing postnatal development. In addition to general appearance, the pulse, respiratory rate, weight, length, and other physical dimensions are noted. Newborn infants are also screened for genetic and metabolic disorders, such as *phenylketonuria* (*PKU*) or congenital *hypothyroidism*.

The **Apgar rating** considers heart rate, respiratory rate, muscle tone, response to stimulation, and color at 1 and 5 minutes after birth. In each category the infant receives a score ranging from 0 (poor) to 2 (excellent), and the scores are then totaled. An infant's Apgar rating (0–10) has been shown to be an accurate predictor of newborn survival and the presence of neurological damage. For example, newborn infants with *cerebral palsy* usually have a low Apgar rating.

Related Clinical Terms

amniocentesis: An analysis of fetal cells taken from a sample of amniotic fluid. ⊤ *Chromosomal Analysis [p. 777]*

chorionic villi sampling: An analysis of cells collected from the chorionic villi during the first trimester. ⊤ *Chromosomal Analysis [p. 777]*

infertility: The inability to have children. ⊤ *Technology and the Treatment of Infertility [p. 777]*

in vitro fertilization: Fertilization outside the body, usually in a test tube or petri dish. ⊤ *Technology and the Treatment of Infertility [p. 777]*

gestational neoplasm: A tumor formed by undifferentiated, rapid growth of the syncytial trophoblast; if untreated, the neoplasm may become malignant. ⊤ *Problems with the Implantation Process [p. 777]*

ectopic pregnancy: A pregnancy in which implantation occurs somewhere other than the uterus. ⊤ *Problems with the Implantation Process [p. 777]*

placenta previa: A condition resulting from implantation in or near the cervix. ⊤ *Problems with Placentation [p. 778]*

abruptio placentae (a-BRUP-shē-ō pla-SEN-tē): Tearing of the placenta sometime after the fifth gestational month. ⊤ *Problems with Placentation [p. 778]*

teratogens (ter-AT-ō-jens): Stimuli that disrupt normal development by damaging cells, altering chromosome structure, or altering the chemical environment of the embryo. ⊤ *Teratogens and Abnormal Development [p. 778]*

fetal alcohol syndrome (FAS): A neonatal condition resulting from maternal alcohol consumption; characterized by developmental defects often involving the skeletal, nervous, and/or cardiovascular systems. ⊤ *Teratogens and Abnormal Development [p. 778]*

breech birth: A delivery wherein the legs or buttocks of the fetus enter the vaginal canal first. *[p. 734]*

congenital malformation: A severe structural abnormality, present at birth, that affects major systems. ⊤ *Complexity and Perfection: An Improbable Dream [p. 778]*

Apgar rating: A method of evaluating newborn infants; a test for developmental problems and neurological damage. *[p. 734]*

CHAPTER SUMMARY AND REVIEW

STUDY OUTLINE

Related Key Terms

INTRODUCTION *[p. 720]*

1. **Development** is the gradual modification of physical and physiological characteristics from conception to maturity. The creation of different cell types is **differentiation**.

AN OVERVIEW OF DEVELOPMENT *[p. 720]*

1. **Prenatal development** occurs before birth; **postnatal development** begins at birth and continues to maturity.

conception • embryology • fetal development

FERTILIZATION *[p. 720]*

1. Fertilization normally occurs in the uterine tube within a day after ovulation. Sperm cannot fertilize an egg until they have undergone *capacitation*.

sterile

The Oocyte at Ovulation *[p. 720]*
2. The acrosomal caps of the spermatozoa release **hyaluronidase**, an enzyme that separates cells of the corona radiata and exposes the oocyte membrane. When a single spermatozoon contacts that membrane, fertilization occurs and **oocyte activation** follows *(see Figure 28-1).*

ooplasm

Pronucleus Formation and Amphimixis *[p. 720]*
3. During activation the secondary oocyte completes meiosis, and **amphimixis** follows *(see Figure 28-1).*

female pronucleus • male pronucleus • zygote

A Preview of Prenatal Development *[p. 721]*
4. The 9-month **gestation** period can be divided into three **trimesters**.

THE FIRST TRIMESTER *[p. 721]*

1. **Cleavage** subdivides the cytoplasm of the zygote in a series of mitotic divisions; the zygote becomes a **blastocyst**. During **implantation** the blastocyst burrows into the uterine endometrium. **Placentation** occurs as blood vessels form around the blastocyst and the **placenta** appears.

pre-embryo • embryogenesis

Cleavage and Blastocyst Formation *[p. 722]*

2. The blastocyst consists of an outer **trophoblast** and an **inner cell mass** *(see Figure 28-2 and Table 28-1)*.

Implantation *[p. 723]*

3. As the trophoblast enlarges and spreads, maternal blood flows through open **lacunae**. After **gastrulation**, the **blastodisc** contains an embryo composed of **endoderm**, **ectoderm**, and an intervening **mesoderm**. These **germ layers** help form four **extraembryonic membranes**: the yolk sac, amnion, allantois, and chorion *(see Figures 28-3 to 28-5 and Table 28-1)*.

4. The **yolk sac** is an important site of blood cell formation. The **amnion** encloses fluid that surrounds and cushions the developing embryo. The base of the **allantois** later gives rise to the urinary bladder. Circulation within the vessels of the **chorion** provides a rapid transit system linking the embryo with the trophoblast *(see Figures 28-4/28-5)*.

Placentation *[p. 724]*

5. **Chorionic villi** extend outward into the maternal tissues, forming an intricate, branching network through which maternal blood flows. As development proceeds, the **umbilical cord** connects the fetus to the placenta. The trophoblast synthesizes HCG, estrogens, progestins, HPL, and relaxin *(see Figures 28-5/28-6)*.

EMBRYOGENESIS *[p. 725]*

1. The first trimester is critical for the **embryo**, because events in the first 12 weeks establish the basis for **organogenesis** (organ formation) *(see Figures 28-5b,c/28-7 and Table 28-2)*.

THE SECOND AND THIRD TRIMESTERS *[p. 730]*

1. In the second trimester, the organ systems near functional completion. During the third trimester, the organ systems become functional *(see Figures 28-7 to 28-9 and Table 28-2)*.

LABOR AND DELIVERY *[p. 731]*

1. The goal of labor is **parturition**, the forcible expulsion of the fetus.

Stages of Labor *[p. 731]*

2. Labor can be divided into three stages: **dilation stage**, **expulsion stage**, and **placental stage** *(see Figure 28-10)*.

Premature Labor *[p. 734]*

3. Complications of labor and delivery include breech birth and **premature delivery**.

THE NEONATAL PERIOD *[p. 734]*

1. The **neonatal period** extends from birth to 1 month of age.

2. In the transition from fetus to **neonate**, the respiratory, circulatory, digestive, and urinary systems begin functioning independently. The newborn must also begin thermoregulating.

Related Key Terms

blastomeres • morula • blastocoele

syncytial trophoblast • cellular trophoblast • villi • amniotic cavity • epiblast • hypoblast • primitive streak

amniotic fluid

body stalk • yolk stalk • decidua capsularis • decidua basalis • decidua parietalis • umbilical stalk • umbilical arteries • umbilical vein

head fold • tail fold

delivery • episiotomy • cesarean section

immature delivery

---

1 | REVIEW OF CHAPTER OBJECTIVES

1. Describe the process of fertilization.
2. Discuss the stages of embryonic and fetal development.
3. Describe the events that take place in the first trimester, and explain why they are vital to the survival of the embryo.
4. Describe the events that take place in embryogenesis.
5. Describe the events that take place in the second and third trimesters.
6. Explain the stages of labor and the events that occur immediately before and after delivery.
7. Identify the anatomical changes that occur as the fetus completes the transition to the newborn infant.

2 | REVIEW OF CONCEPTS

1. How does the process of differentiation differ from that of growth and development?
2. How does the role of the sperm differ from that of the ovum in the process of fertilization and embryogenesis?
3. Why must many spermatozoa arrive at the oocyte for fertilization to occur even though only one can fertilize the ovum?
4. Why is the embryo particularly susceptible to damage from alcohol and other drugs during the first trimester of pregnancy?
5. Of what significance to the process of development are the cells that form the trophoblast?
6. What is the role of the primitive streak in the process of gastrulation?
7. What different functions are served by the three layers of the embryo?
8. What is the importance of the yolk sac in embryonic development?
9. What is the function of the amnion to the developing embryo?
10. What is the function of the allantois?
11. What is the function of the mesoderm in the chorion?
12. What is the function of the chorionic villi in providing maternal nutrients to the fetus?
13. What is the function of the placenta in maintaining the fetus?
14. Why do women have difficulty breathing and have digestive and urinary discomfort in the last trimester of pregnancy?
15. Why is the head-first presentation of a baby during the birth process the easiest for a normal delivery?

CRITICAL THINKING AND CLINICAL APPLICATION QUESTIONS

1. As the age of prospective parents increases, particularly over the age of 40, it is more important for them to have prenatal testing to detect genetic abnormalities in the fetus, many of which increase in frequency as women grow older. What methods of determining the genetic composition of the fetus are available?

2. If the fetus tested by one of the sampling methods discussed above is found to have one or more genetic abnormalities, what problems could arise?

3. A couple has been attempting to have a child for more than a year, but conception has not occurred; therefore, the couple has sought medical assistance. In this case the problem has been determined to be with the husband. What kinds of fertility problems occur in males, and how may they be treated?

4. If the couple above discovered that the woman had the problem preventing conception, what might be the cause and how could it be treated?

5. A woman partway through the third trimester of an otherwise normal pregnancy begins to suffer intermittent bouts of bleeding, some of them severe. Placenta previa is suspected. What is this condition, and what treatment is provided?

APPENDIXES

SCANNING ATLAS

Page	Scan	Description
p. 741	1a,b,c	MRI scans of the brain, horizontal sections, in superior to inferior sequence. *See also Figures 6-4a (p. 136), and 6-11a (p. 145).*
	1d,e	MRI scans of the brain, parasagittal and sagittal sections. *See also Figures 6-4b (p. 136), 6-11a (p. 145), 15-2 (p. 367), 15-6 (p. 372), 15-12 (p. 380), and 15-15 (p. 385).*
p. 742	2a-d	MRI scans of the brain, coronal (frontal) sections, in anterior to posterior sequence. *See also Figures 6-13 (p. 148), 15-2 (p. 367), and 24-3 (p. 602).*
p. 743	3a	MRI scan of the cervical spine, sagittal section. *See also Figure 6-20 (p. 161).*
	3b	MRI scan of the spinal column, sagittal section. *See also Figure 6-18 (p. 158), 6-22 (p. 163), and 6-23 (p. 164).*
p. 743	4	MRI scan of the pelvis and hip joint, horizontal sections, in superior to inferior sequence. *See also Figures 8-12 (p. 222), 8-13 (p. 223), 8-14 (p. 295), 11-17 (p. 300).*
p. 744	5a,b	MRI scans of the knee joint, horizontal sections, in superior to inferior sequence. *See also Figures 8-14 (p. 223), 11-14 (p. 295), 11-15 (p. 298), 11-17 (p. 300), 11-18 (p. 300), 11-19 (p. 303), 22-18 (p. 562), 22-19 (p. 563), and 22-25 (p. 570).*
p. 744	6a,b	MRI scans of the knee joint, parasagittal sections, in lateral to medial sequence. *See also Figures 8-14 (p. 225), 11-14 (p. 295), 11-15 (p. 298), 11-17 (p. 300), 11-18 (p. 300), 11-19 (p. 303), 22-19 (p. 563), and 22-25 (p. 570).*
p. 745	7a,b	MRI scans of the knee joint, frontal sections, in posterior to anterior sequence. *See also Figures 8-14 (p. 225), 11-14 (p. 295), 11-15 (p. 298), 11-17 (p. 300), 11-18 (p. 300), 11-19 (p. 303), 22-18a (p. 562), 22-19 (p 563), and 22-25 (p. 570).*
p. 745	8a	MRI scan of the ankle joint, parasagittal section. *See also Figures 7-17d (p. 198), and 8-17 (p. 227).*
	8b	MRI scan of the ankle joint, frontal section. *See also Figures 8-16 (p. 226) and 11-22 (p. 227).*
pp. 746-747	9a-i	MRI scans of the trunk, horizontal (transverse) sections, in superior to inferior sequence.

Note to the Reader: Try to use the sequential images to enhance your understanding of these three-dimensional structures. To test yourself, try to label images 9c, 9e, 9g, and 9h, using your mental image and the labeled scans in that series as references.

Credits: MRI and CT scans courtesy of Dr. Eugene C. Wasson III, and staff of Maui Radiology Consultants, Maui Memorial Hospital.

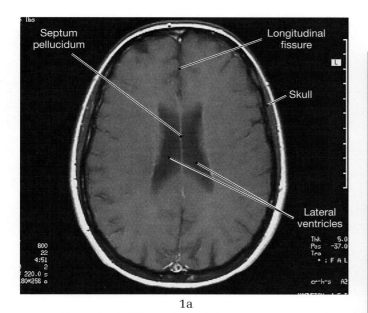

Septum pellucidum

Longitudinal fissure

Skull

Lateral ventricles

1a

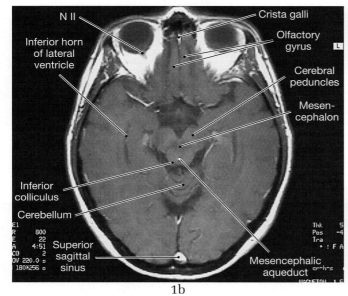

N II

Crista galli

Olfactory gyrus

Inferior horn of lateral ventricle

Cerebral peduncles

Mesencephalon

Inferior colliculus

Cerebellum

Superior sagittal sinus

Mesencephalic aqueduct

1b

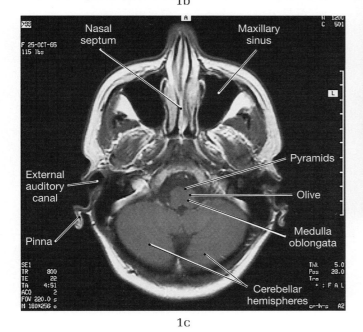

Nasal septum

Maxillary sinus

External auditory canal

Pyramids

Olive

Pinna

Medulla oblongata

Cerebellar hemispheres

1c

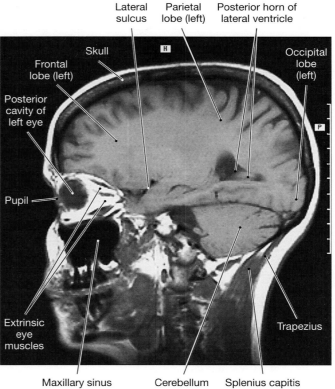

Lateral sulcus

Parietal lobe (left)

Posterior horn of lateral ventricle

Skull

Frontal lobe (left)

Occipital lobe (left)

Posterior cavity of left eye

Pupil

Extrinsic eye muscles

Maxillary sinus

Cerebellum

Splenius capitis

Trapezius

1d

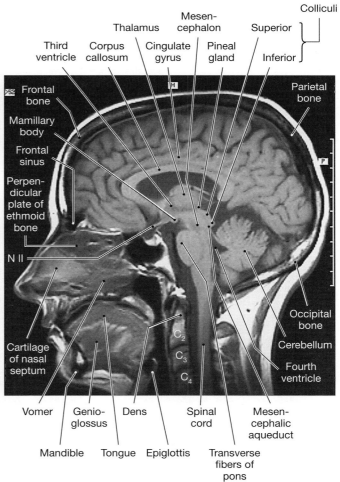

Thalamus

Mesencephalon

Colliculi

Third ventricle

Corpus callosum

Cingulate gyrus

Pineal gland

Superior

Inferior

Frontal bone

Parietal bone

Mamillary body

Frontal sinus

Perpendicular plate of ethmoid bone

N II

Occipital bone

Cerebellum

Cartilage of nasal septum

Fourth ventricle

Vomer

Genioglossus

Dens

Spinal cord

Mesencephalic aqueduct

Mandible

Tongue

Epiglottis

Transverse fibers of pons

C₂
C₃
C₄

1e

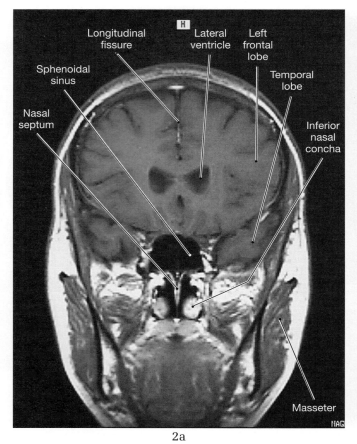

Sphenoidal sinus — Nasal septum — Longitudinal fissure — Lateral ventricle — Left frontal lobe — Temporal lobe — Inferior nasal concha — Masseter

2a

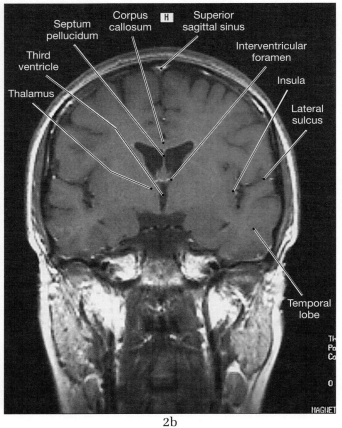

Septum pellucidum — Third ventricle — Thalamus — Corpus callosum — Superior sagittal sinus — Interventricular foramen — Insula — Lateral sulcus — Temporal lobe

2b

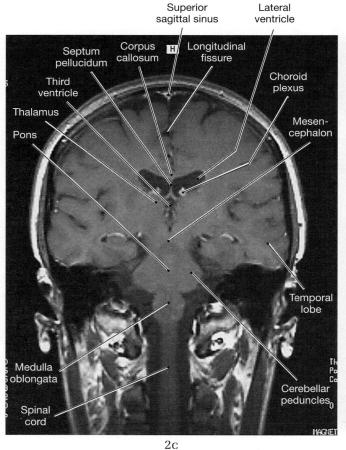

Septum pellucidum — Third ventricle — Thalamus — Pons — Corpus callosum — Superior sagittal sinus — Longitudinal fissure — Lateral ventricle — Choroid plexus — Mesencephalon — Temporal lobe — Medulla oblongata — Spinal cord — Cerebellar peduncles

2c

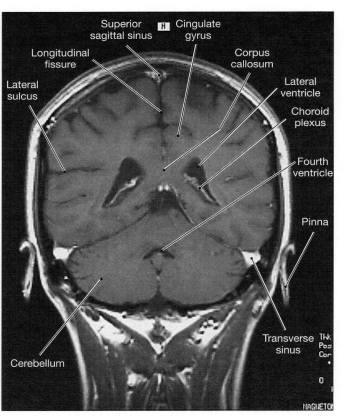

Longitudinal fissure — Lateral sulcus — Superior sagittal sinus — Cingulate gyrus — Corpus callosum — Lateral ventricle — Choroid plexus — Fourth ventricle — Pinna — Cerebellum — Transverse sinus

2d

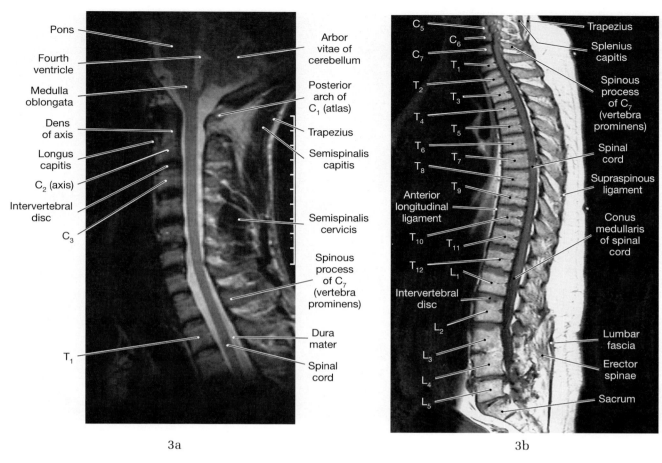

Pons

Fourth ventricle

Medulla oblongata

Dens of axis

Longus capitis

C_2 (axis)

Intervertebral disc

C_3

T_1

Arbor vitae of cerebellum

Posterior arch of C_1 (atlas)

Trapezius

Semispinalis capitis

Semispinalis cervicis

Spinous process of C_7 (vertebra prominens)

Dura mater

Spinal cord

3a

C_5

C_6

C_7

T_1

T_2

T_3

T_4

T_5

T_6

T_7

T_8

T_9

Anterior longitudinal ligament

T_{10}

T_{11}

T_{12}

L_1

Intervertebral disc

L_2

L_3

L_4

L_5

Trapezius

Splenius capitis

Spinous process of C_7 (vertebra prominens)

Spinal cord

Supraspinous ligament

Conus medullaris of spinal cord

Lumbar fascia

Erector spinae

Sacrum

3b

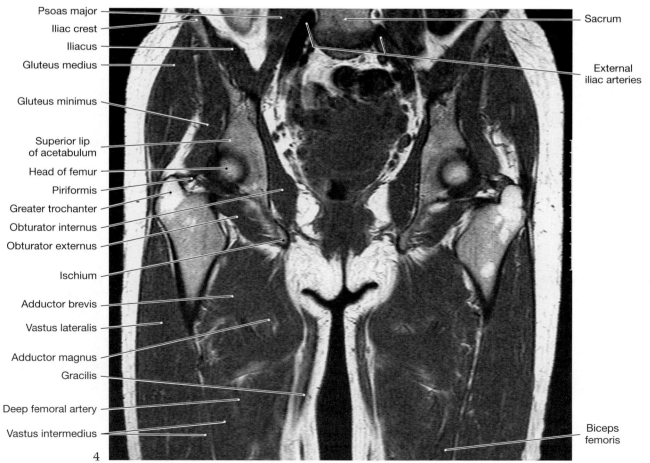

Psoas major

Iliac crest

Iliacus

Gluteus medius

Gluteus minimus

Superior lip of acetabulum

Head of femur

Piriformis

Greater trochanter

Obturator internus

Obturator externus

Ischium

Adductor brevis

Vastus lateralis

Adductor magnus

Gracilis

Deep femoral artery

Vastus intermedius

Sacrum

External iliac arteries

Biceps femoris

4

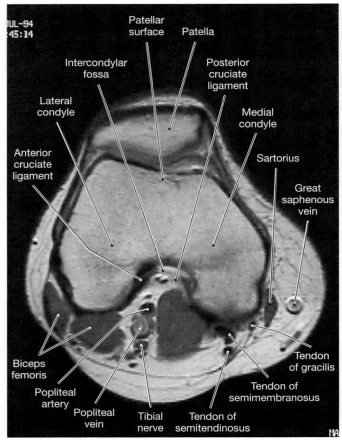

5a

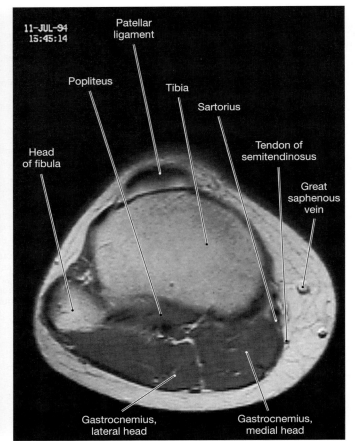

5b

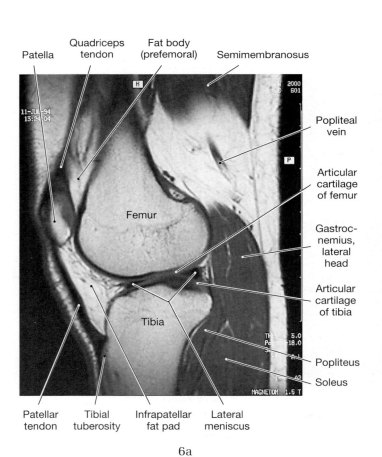

6a

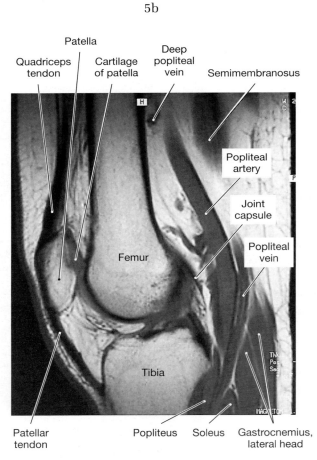

6b

744

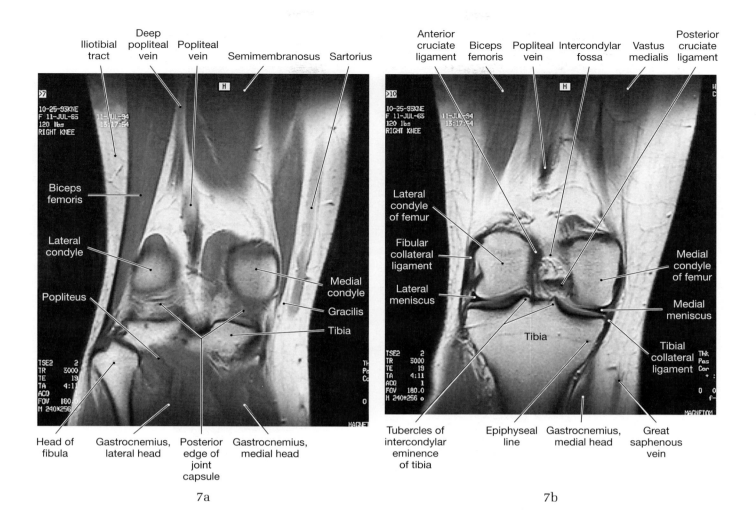

Iliotibial tract · Deep popliteal vein · Popliteal vein · Semimembranosus · Sartorius

Biceps femoris

Lateral condyle

Popliteus

Head of fibula · Gastrocnemius, lateral head · Posterior edge of joint capsule · Gastrocnemius, medial head

Medial condyle · Gracilis · Tibia

7a

Anterior cruciate ligament · Biceps femoris · Popliteal vein · Intercondylar fossa · Vastus medialis · Posterior cruciate ligament

Lateral condyle of femur

Fibular collateral ligament

Lateral meniscus

Tibia

Medial condyle of femur

Medial meniscus

Tibial collateral ligament

Tubercles of intercondylar eminence of tibia · Epiphyseal line · Gastrocnemius, medial head · Great saphenous vein

7b

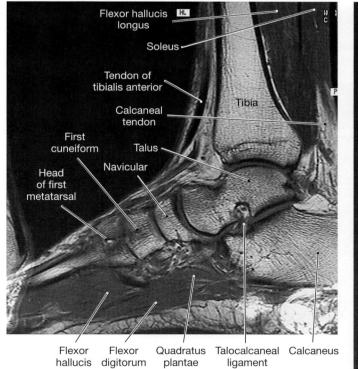

Flexor hallucis longus

Soleus

Tendon of tibialis anterior

Calcaneal tendon

First cuneiform

Head of first metatarsal

Talus

Navicular

Tibia

Flexor hallucis brevis · Flexor digitorum brevis · Quadratus plantae · Talocalcaneal ligament · Calcaneus

8a

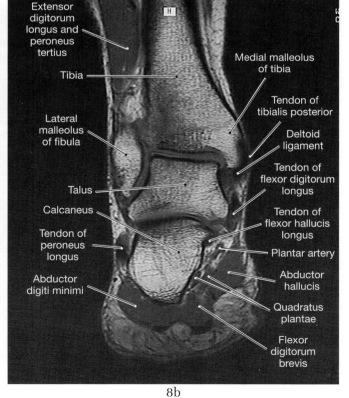

Extensor digitorum longus and peroneus tertius

Tibia

Lateral malleolus of fibula

Talus

Calcaneus

Tendon of peroneus longus

Abductor digiti minimi

Medial malleolus of tibia

Tendon of tibialis posterior

Deltoid ligament

Tendon of flexor digitorum longus

Tendon of flexor hallucis longus

Plantar artery

Abductor hallucis

Quadratus plantae

Flexor digitorum brevis

8b

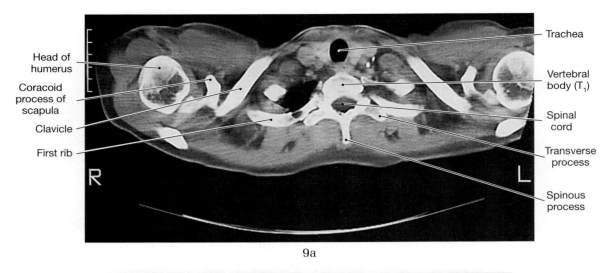

Head of
humerus

Coracoid
process of
scapula

Clavicle

First rib

Trachea

Vertebral
body (T$_1$)

Spinal
cord

Transverse
process

Spinous
process

R   L

9a

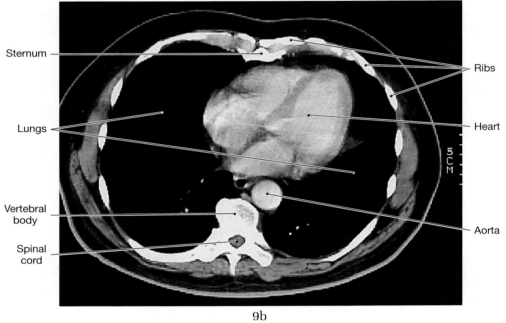

Sternum

Lungs

Vertebral
body

Spinal
cord

Ribs

Heart

Aorta

5 CM

9b

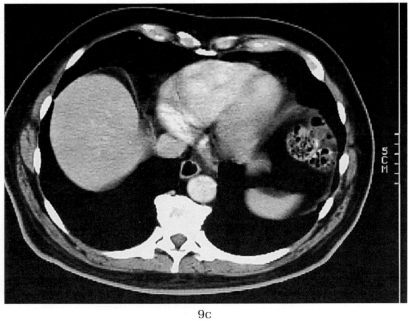

5 CM

9c

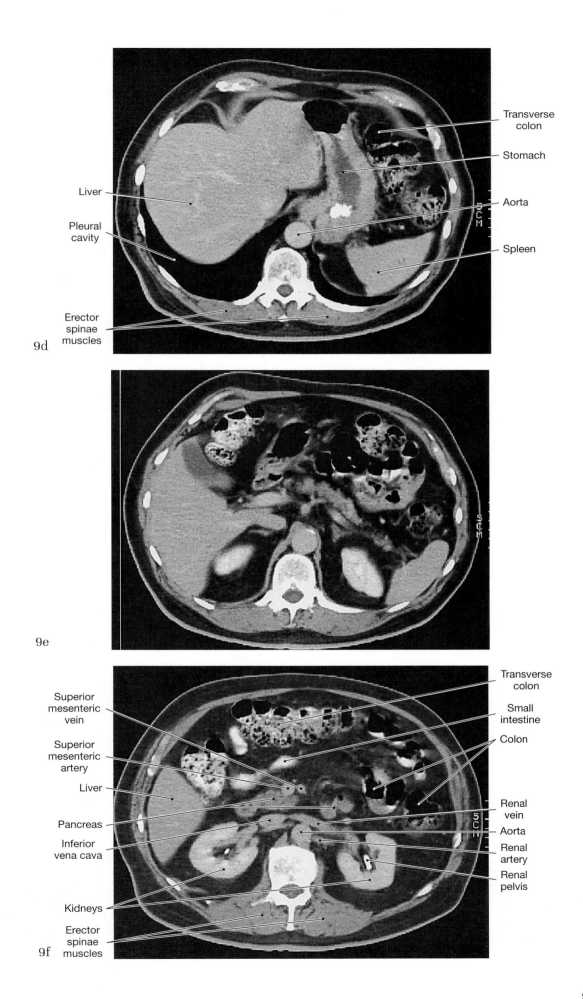

9d

Transverse colon

Stomach

Aorta

Liver

Pleural cavity

Spleen

Erector spinae muscles

9e

9f

Superior mesenteric vein

Superior mesenteric artery

Liver

Pancreas

Inferior vena cava

Kidneys

Erector spinae muscles

Transverse colon

Small intestine

Colon

Renal vein

Aorta

Renal artery

Renal pelvis

CLINICAL ISSUES

chapter 1
An Introduction to Anatomy

ANATOMY AND OBSERVATION *Page 15*

Technical innovation has affected modern medicine in many ways, but some of the most important diagnostic procedures have not changed significantly in a thousand years. The next time you visit a physician, pay attention to the way information is obtained. First, your chart is reviewed, giving the doctor information about past or preexisting problems. Then you are asked to describe the reasons for your visit.

As you talk, the doctor not only listens carefully, but also watches closely. This is really the first stage in physical assessment, a time for general questions such as, "Is the patient moving, speaking, and thinking normally?" The answers will later be integrated with the results of more precise observations.

A physical examination typically follows. There are four basic components to a physical examination:

1. **Inspection**: Inspection is careful observation. A general inspection involves examining body proportions, posture, and patterns of movements. Local inspection is the examination of sites or regions of suspected injury. Of the four components of the physical exam, inspection is often the most important because it provides the largest amount of useful information. Many diagnostic conclusions can be made on the basis of inspection alone; most skin conditions, for example, are identified in this way. A number of endocrine problems and inherited metabolic disorders can produce subtle changes in body proportions that are easily overlooked by the untrained eye.

2. **Palpation**: In palpation the physician uses hands and fingers to feel the body. This procedure provides information on skin texture and temperature, the presence of abnormal masses, the pattern of the pulse, and the location of tender spots. Once again, the procedure relies on an understanding of normal anatomy. A small, soft, lumpy mass in one spot is a salivary gland; in another location it could be a tumor. The importance of a tender spot can be recognized in diagnosis only if the observer knows what organs lie beneath it.

3. **Percussion**: Percussion is tapping with the fingers or hand to obtain information about the densities of underlying tissues. For example, the chest normally produces a hollow sound, because the lungs are filled with air. That sound changes in pneumonia, when the lungs contain large amounts of fluid. Of course, to get the clearest chest percussions, the fingers must be placed on the right spots.

4. **Auscultation**: Auscultation (aws-kul-TĀ-shun; *auscultare*, to listen) is listening to body sounds, usually using a stethoscope. This technique is particularly useful for checking the condition of the lungs during breathing. The wheezing sound heard in asthma is caused by constriction of the airways, and pneumonia produces a gurgling sound, indicating that fluid has accumulated in the lungs. Auscultation is also important in diagnosing heart conditions. Many cardiac problems affect the sound of the heartbeat or produce abnormal swirling sounds during blood flow.

The entire process of physical examination relies on one fact: The doctor already knows the superficial and deep anatomy of the human body. To detect the abnormal, you must first understand the normal.

chapter 2
The Cell

CAUSES OF CANCER *Page 48*

A relatively small number of cancers are actually inherited; 18 different types have been identified to date, including two forms of leukemia. Most cancers develop through the interaction of genetic and environmental factors. They will be discussed individually, although in practice it is difficult to separate the two completely.

Genetic Factors Two related genetic factors are involved in the development of cancer: *hereditary predisposition* and *oncogene activation*. An individual born with genes that increase the likelihood of cancer has a **hereditary predisposition** for the disease. Under these conditions a cancer is not guaranteed, but it is highly likely. The inherited genes usually affect tissue abilities to metabolize toxins, control mitosis and growth, perform repairs after injury, or identify and destroy abnormal tissue cells. As a result, body cells become more sensitive to local or environmental factors that would have little effect on normal tissues.

Cancers may also result from ideopathic somatic mutations that modify genes involved with cell growth, differentiation, or mitosis. As a result, an ordinary cell converts into a cancer cell. The modified genes are called **oncogenes** (ONG-kō-jēnz); the normal genes are called **proto-oncogenes**. Oncogene activation occurs through the alteration of normal somatic genes. Because these mutations do not affect reproductive cells, the cancers caused by active oncogenes are not inherited.

A proto-oncogene, like other genes, has a regulatory component (gene ON/OFF) and a structural component that contains the triplets that determine protein structure. Mutations in either portion of the gene may convert it to an active oncogene. A small mutation can accomplish this conversion; changing one nucleotide out of a chain of 5000 can convert a normal proto-oncogene to an active oncogene. In some cases, a viral infection can trigger activation of an oncogene. For example, one of the papilloma viruses appears to be responsible for many cancers of the cervix.

More than 50 oncogenes have now been identified. In addition, a group of anticancer genes has now been identified. These genes, called **tumor-suppressing genes (TSG)**, or **anti-oncogenes**, suppress mitosis and growth in normal cells. Mutations that alter these genes make oncogene activation more likely. TSG mutation has been suggested as important in promoting several cancers, including several blood cell cancers, breast cancer, and ovarian cancer.

Environmental Factors Many cancers can be directly or indirectly attributed to environmental factors called **carcinogens** (kar-SIN-ō-jenz). Carcinogens are ions or compounds that stimulate the conversion of a normal cell to a cancer cell. Some carcinogens are **mutagens**, compounds that damage DNA strands and sometimes cause chromosomal breakage. Radiation is an example of a mutagen that has carcinogenic effects.

There are many different carcinogens in the environment. Plants manufacture poisons that protect them from insects and other predators, and although their carcinogenic activities are relatively weak, many common spices, vegetables, and beverages contain compounds that can be carcinogenic if consumed in large quantities. Animal tissues may also store or concentrate toxins, and hazardous compounds of many kinds can be swallowed in contaminated meals. Cosmic radiation, X-rays, and other radiation sources can also cause cancer.

Specific carcinogens will affect only those cells capable of responding to that particular physical or chemical stimulus. The responses vary because differentiation produces cell types with specific sensitivities. For example, benzene can produce a cancer of the blood, cigarette smoke a lung cancer, and vinyl chloride a liver cancer. Very few stimuli can produce cancers throughout the body; radiation exposure is a notable exception. In general, cells undergoing mitosis are more likely to respond to chemical or radiational carcinogens than are cells in interphase. As a result, cancer rates are highest in epithelial tissues, where stem cell mitoses are occurring at a rapid rate, and relatively low in neural and muscle tissues, where mitoses do not normally occur.

chapter 3
The Tissue Level of Organization

INFANTS AND BROWN FAT *Page 68*

Infants' body temperatures are more unstable than those of adults. They cannot shiver, and they lose heat more rapidly because of their relatively large surface-to-volume ratios. The adipose tissue between the shoulder blades, around the neck, and possibly elsewhere in the upper body is different from most of the adipose tissue in the adult. It is highly vascularized, and the individual adipocytes contain numerous mitochondria. Together these characteristics give the tissue a deep, rich color responsible for the name **brown fat**. When stimulated by the nervous system, lipid breakdown accelerates; the cells do not capture the energy released, and it radiates into the surrounding tissues as heat. The circulating blood is warmed by this heat and also distributes it throughout the body. In this way an infant can accelerate metabolic heat generation by 100% very quickly. There is little if any brown fat in the adult; with larger body size, skeletal muscle mass, and insulation, shivering is significantly more effective in elevating body temperature.

LIPOSUCTION *Page 68*

Liposuction is a surgical procedure for the removal of unwanted adipose tissue. In liposuction, a small incision is made through the skin, and a tube is inserted into the underlying adipose tissue. Suction is then applied, and chunks of tissue containing adipocytes, other cells, fibers, and ground substance are removed.

This practice has received a lot of attention in the news, and many advertisements praise the technique as easy, safe, and effective. In fact, it is (1) not always easy, (2) sometimes dangerous, and (3) of limited effectiveness. The density of adipose tissue varies according to body location and from individual to individual, and adipose tissue is not always easy to suck through a tube. An anesthetic must be used to control pain, and this poses risks. Blood vessels are stretched and torn, and extensive bleeding can occur. Probably the most serious complication is the potential for infection after the liposuction treatment.

Finally, adipose tissue can repair itself, and adipocyte populations recover over time. The only way to ensure that fat lost through liposuction will not return is to adopt a lifestyle that includes a proper diet and adequate exercise. In fact, such a lifestyle could produce the same weight loss *without* liposuction, eliminating the surgical expense and risk.

CARTILAGES AND KNEE INJURIES *Page 72*

The knee is an extremely complex joint that contains both hyaline cartilage and fibrocartilage. The hyaline cartilage caps bony surfaces, while pads of fibrocartilage within the joint prevent contact when movements are under way. Many sports injuries involve tearing of the cartilage pads. This loss of cushioning places more strain on the cartilages within joints and leads to further joint damage. The articular cartilages are not only avascular, but they also lack a perichondrium. As a result, they heal even more slowly than other cartilages. Surgery usually produces only a temporary or incomplete repair. For this reason, most competitive sports have rules designed to reduce the number of knee injuries.

For example, in football "clipping" is outlawed because it produces stresses that can tear the knee cartilages.

Recent advances in tissue culture have enabled researchers to grow fibrocartilage in the laboratory. Chondrocytes removed from the knees of injured dogs are cultured in an artificial framework of collagen fibers. They eventually produce masses of fibrocartilage that can be inserted into the damaged joints. Over time the pads change shape and grow, restoring normal joint function. In the future this technique may be used to treat severe knee injuries in humans.

PROBLEMS WITH SEROUS MEMBRANES *Page 77*

If serous membranes are damaged, the production of transudate fluid may cease, and the parietal and visceral mesothelia will rub against one another, producing an abrasion. This mesothelial damage attracts fibroblasts, which migrate into the damaged area and bind the opposing membranes together with a network of collagen fibers. Although this binding reduces friction by limiting movement, the collagen fibers may compress blood vessels, nerves, or other vital structures in the region. These restrictive fibrous connections are called **adhesions**. Adhesions may occur as a result of surgery, infection, or other injuries that damage serous membranes.

Several clinical conditions, including infection and chronic irritation, can cause the abnormal buildup of fluid within one of the ventral body cavities. **Pleurisy**, or **pleuritis**, is an inflammation of the pleural cavities. At first the membranes become rather dry, and the opposing membranes may scratch against one another. This scratching produces a characteristic sound known as a **pleural rub**. Adhesions seldom form between the serous membranes of the pleural cavities. More frequently, continued inflammation and rubbing lead to a gradual increase in the production of fluid to levels well above normal. Fluid then accumulates in the pleural cavities, producing a condition known as **pleural effusion**. Pleural effusion can also be caused by heart conditions that elevate the pressure in blood vessels of the lungs. As fluids build up in the pleural cavities, the lungs are compressed, making it difficult to breathe. The combination of severe pleural effusion and heart disease can be lethal.

Pericarditis is an inflammation of the pericardium. This condition often leads to a **pericardial effusion**: an abnormal accumulation of fluid in the pericardial cavity. When sudden or severe, the fluid buildup can seriously reduce the efficiency of the heart and restrict blood flow through major vessels.

Peritonitis, an inflammation of the peritoneum, can occur after an infection of or injury to the peritoneal lining. Peritonitis is a potential complication of any surgical procedure that requires opening the peritoneal cavity. Liver disease, kidney disease, or heart failure can cause an increase in the rate of fluid movement through the peritoneal lining. The accumulation of fluid, called **ascites** (a-SĪ-tēz), creates a characteristic abdominal swelling. Distortion of internal organs by the contained fluid can result in a variety of symptoms; heartburn, indigestion, and low back pain are common complaints. The condition of heartburn will be discussed in greater detail in the digestive system.

TISSUE STRUCTURE AND DISEASE *Page 81*

Pathologists (pa-THOL-ō-jists) are physicians who specialize in the anatomical diagnosis of disease processes. In their analyses they synthesize anatomical and histological observations to determine the nature and severity of the disease.

Disease processes affect the histological organization of tissues and organs. As an example, consider the histological changes induced in the respiratory epithelium by one relatively common irritating stimulus, cigarette smoke. The first abnormality to be observed is **dysplasia** (dis-PLĀ-zē-a), a change in the normal shape, size, and organization of tissue cells. It is usually a response to chronic irritation or inflammation, and the changes are reversible.

Epithelia and connective tissues may undergo more radical changes in structure, caused by the division and differentiation of stem cells. **Metaplasia** (me-ta-PLĀ-zē-a) is a structural change that dramatically alters the character of the tissue. Metaplasia is also a reversible condition. During **anaplasia** (a-na-PLĀ-zē-a) tissue cells change size and shape, often becoming unusually large or abnormally small. In anaplasia tissue organization breaks down, and a tumor forms. Unlike dysplasia and metaplasia, anaplasia is irreversible.

There is no single, universally effective cure or preventive measure for cancer; there are too many separate causes, possible mechanisms, and individual differences. The goal of cancer treatment is to achieve **remission**. A tumor in remission either ceases to grow or decreases in size. Basically, the treatment of malignant tumors must accomplish one of the following to produce remission.

1. **Surgical removal or destruction of individual tumors**: Tumors that contain malignant cells can be surgically removed or destroyed by radiation, heat, or freezing. These techniques are very effective if the treatment is undertaken before metastasis has occurred. For this reason early detection is important in improving survival rates for all forms of cancer.

2. **Killing metastatic cells throughout the body**: This treatment is much more difficult and potentially dangerous, because healthy tissues are likely to be damaged at the same time. At present the most widely approved treatments are chemotherapy and radiation.

Chemotherapy involves the administration of drugs that will either kill the cancerous tissues or prevent mitotic divisions. These drugs often affect stem cells in normal tissues, and the side effects are often quite unpleasant. For example, because chemotherapy slows the regeneration and maintenance of epithelia of the skin and digestive tract, patients lose their hair and experience nausea and vomiting. Several drugs are often administered simultaneously or in sequence, because over time cancer cells can develop a resistance to a single drug. This strategy may also reduce the severity of any side effects. Chemotherapy is often used in the treatment of many kinds of metastatic cancer.

Hormone manipulation may offer a less toxic form of chemotherapy. The compounds used may suppress tumor growth or hormones that stimulate tumor growth. Examples include *Tamoxifen*, used to treat breast cancer, and *antigonadotropins* used to treat prostate cancer.

Localized radiation is often used to destroy cancers confined to a small region of the body. Massive doses of radiation are necessary if the cancer has metastasized widely. For example, in advanced cases of lymphoma, a cancer of the immune system, enough radiation is administered to kill all of the blood-forming cells in the body. After treatment, new blood cells must be provided by a *bone marrow transplant*.

An understanding of molecular mechanisms and cell biology is leading to new approaches that may revolutionize the treatment of cancer. One approach focuses on the fact that cancer cells are usually ignored by the immune system. In **immunotherapy**, substances are administered that help the immune system recognize and attack the cancer cells. More elaborate experimental procedures involve the creation of customized antibodies using gene-splicing techniques. The resulting antibodies are specifically designed to attack the tumor cells in one particular patient. Although this technique shows promise, it remains difficult, costly, and very labor-intensive.

Another new procedure builds upon immunotherapy. Standard chemotherapy drugs can be attached to antibodies that will bind to tumor cells and ignore normal tissues. In **boron neutron capture therapy (BNCT)** antibodies made to attack cancer cells are labeled with an isotope of boron (B). After these antibodies have been administered and have bound to cancer cells, the patient is irradiated with neutrons. These neutrons do not damage normal tissues. However, the boron atoms absorb neutrons and then release alpha particles (2 neutrons + 2 protons), radiation that kills the cancer cells quite effectively. Because alpha particles do not penetrate far through body tissues, they affect only the cancer cells that have been "tagged" with boron atoms. Other, normal tissues remain unaffected.

Incidence and Survival Rates A statistical profile of cancer incidence and survival rates in the United States is presented in Table A-1. Interestingly, the profile would be different for other countries, reflecting different environmental conditions, hereditary factors, exposure to infectious diseases, and cultural dietary preferences. Bladder cancer is common in Egypt, stomach cancer in Japan, and liver cancer in Africa.

Advances in chemotherapy, radiation procedures, and molecular biology have produced significant improvements in the survival rates for several types of cancer. The improved survival rates indicated in Table A-1, however, reflect advances not only in therapy, but also in early detection. Much of the credit goes to increased public awareness and concern about cancer. In general, the odds of survival increase markedly if the cancer is detected early, especially before it undergoes metastasis. Despite the variety of possible cancers, the American Cancer Society has identified seven "warning signs" that mean it's time to consult a physician. These warning signs are presented in Table A-2.

TABLE A-2   Seven Warning Signs of Cancer

Change in bowel or bladder habits
A sore that does not heal
Unusual bleeding or discharge
Thickening or lump in breast or elsewhere
Indigestion or difficulty in swallowing
Obvious change in wart or mole
Nagging cough or hoarseness

TABLE A-1   Cancer Incidence and Survival Rates in the United States

Site	Estimated New Cases (1996)	Estimated Deaths (1996)	Five-Year Survival Rates Diagnosis Date 1960–63	Five-Year Survival Rates Diagnosis Date 1983–89
DIGESTIVE TRACT				
Esophagus	12,300	11,200	2.5%	10%
Stomach	22,800	14,000	9.5%	17%
Colon and rectum	133,500	54,900	36.0%	59%
RESPIRATORY TRACT				
Lung and bronchi	177,000	158,700	6.5%	13%
URINARY TRACT				
Kidney and other urinary structures	30,600	12,000	37.5%	56%
Urinary bladder	52,900	11,700	38.5%	80%
REPRODUCTIVE SYSTEM				
Breast	185,700	44,560	54.0%	81%
Ovary	26,700	14,800	32.0%	40%
Testis	7,400	370	63.0%	93%
Prostate gland	317,100	41,400	42.5%	79%
NERVOUS SYSTEM	17,900	13,300	18.5%	26%
CIRCULATORY SYSTEM	52,200	56,200	14.0%[a]	39%[a]
SKIN (MELANOMA ONLY)	38,300	7,300	60.0%	84%

Data courtesy of the American Cancer Society.
[a]Lymphocytic leukemia only.

chapter 4

The Integumentary System

PSORIASIS AND XEROSIS *Page 91*

Excessive production of keratin is called **hyperkeratosis** (hī-per-ker-a-TŌ-sis). The effects are easily observed as calluses and corns. Calluses appear on thick-skinned areas exposed to mechanical stress, such as the palms of the hands or the heels. Corns are more localized and form on or between the toes.

In **psoriasis** (sō-RĪ-a-sis) the stratum germinativum becomes unusually active in specific areas, including the scalp, elbows, palms, soles, groin, and nails. Normally an individual stem cell divides once every 20 days, but in psoriasis it may divide every day-and-a-half. Keratinization is abnormal and often incomplete by the time the outer layers are shed. The affected areas appear to be covered with small scales that continually flake away. Most cases are painless and treatable.

Psoriasis appears in 20–30% of the individuals with an inherited tendency for the condition. Roughly 5% of the general U.S. population has psoriasis to some degree, often triggered by stress and anxiety.

Xerosis (zē-RŌ-sis), or "dry skin," is a common complaint of older persons and almost anyone living in an arid climate. Under these conditions the cell membranes deteriorate, and the stratum corneum becomes more a collection of scales than a unified covering. This condition can increase the rate of water loss by insensible perspiration by 75 times.

SKIN CANCERS *Page 93*

Almost everyone has several benign tumors of the skin; freckles and moles are examples. Skin cancers are the most common form of cancer, and the most common skin cancers are caused by prolonged exposure to sunlight.

A **basal cell carcinoma** is a malignant cancer that originates in the germinativum (basal) layer. This is the most common skin cancer, and roughly two-thirds of these cancers appear in areas subjected to chronic UV exposure. These carcinomas have recently been linked to an inherited gene.

Squamous cell carcinomas are less common, but almost totally restricted to areas of sun-exposed skin. Metastasis seldom occurs in squamous cell carcinomas and almost never in basal cell carcinomas, and most people survive these cancers. The usual treatment involves surgical removal of the tumor, and at least 95% of patients survive 5 years or more after treatment. (This statistic, the 5-year survival rate, is a common method of reporting long-term prognoses.)

Compared with these common and seldom life-threatening cancers, **malignant melanomas** (mel-a-NŌ-maz) are extremely dangerous. In this condition, cancerous melanocytes grow rapidly and metastasize throughout the lymphatic system. The outlook for long-term survival is dramatically different, depending on when the condition is diagnosed. If the condition is localized, the 5-year survival rate is 90%; if widespread, the survival rate drops to 14%.

Fair-skinned individuals who live in the tropics are most susceptible to all forms of skin cancer, because their melanocytes are unable to shield them from the ultraviolet radiation. Sun damage can be prevented by avoiding exposure to the sun during the middle hours of the day and by using a sunblock (not a tanning oil or sunscreen)—a practice that also delays the cosmetic problems of sagging and wrinkling. Everyone who expects to be out in the sun for any length of time should choose a broad-spectrum sunblock with a sun protection factor (SPF) of at least 15; blonds, redheads, and people with very fair skin are better off with a sun protection factor of 20 to 30. (One should also remember the risks before spending time in a tanning salon or tanning bed.)

The use of sunscreens has now become even more important as the ozone gas in the upper atmosphere is destroyed by our industrial emissions. Ozone absorbs UV before it reaches the earth's surface, and in doing so, it assists the melanocytes in preventing skin cancer. Australia, which is most affected by the depletion of the ozone layer near the South Pole (the "ozone hole"), is already reporting an increased incidence of skin cancers.

DERMATITIS *Page 94*

Dermatitis is an inflammation of the skin that involves primarily the papillary region of the dermis. The inflammation usually begins in a portion of the skin exposed to infection or irritated by chemicals, radiation, or mechanical stimuli. Dermatitis may cause no physical discomfort, or it may produce an annoying itch, as in poison ivy. Other forms of this condition can be quite painful, and the inflammation may spread rapidly across the entire integument. Some common forms of dermatitis are:

1. **Contact dermatitis** usually occurs in response to strong chemical irritants. It produces an itchy rash that may spread to other areas as scratching distributes the chemical agent; poison ivy is an example.
2. **Eczema** (EK-se-ma) is a dermatitis that can be triggered by temperature changes, fungus, chemical irritants, greases, detergents, or stress. Hereditary or environmental factors or both can encourage the development of eczema.
3. **Diaper rash** is a localized dermatitis caused by a combination of moisture, irritating chemicals from fecal or urinary wastes, and flourishing microorganisms.
4. **Urticaria** (ur-ti-KAR-ē-a), also known as **hives**, is an extensive allergic response to food, drugs, an insect bite, infection, stress, or other stimulus.

TUMORS IN THE DERMIS *Page 95*

Tumors seldom develop in the dermis, and those that do appear are usually benign. Two forms of benign tumors called **hemangiomas** may appear among dermal blood vessels during embryonic development. Viewed from the surface, these form prominent *birthmarks*. A **capillary hemangioma** involves capillaries of the papillary layer. It usually enlarges after birth, but subsequently fades and disappears. **Cavernous hemangiomas**, or "port-wine stains," affect larger vessels in the dermis. Such birthmarks usually last a lifetime.

BALDNESS AND HIRSUTISM *Page 99*

Two factors interact to determine baldness. A bald individual has a genetic susceptibility, but this tendency must be triggered by large quantities of male sex hormones. Many women carry the genetic background for baldness, but unless major hormonal abnormalities develop, as in certain endocrine tumors, nothing happens to their hair.

Male pattern baldness affects the top of the head first, only later reducing the hair density along the sides. Thus hair follicles can be removed from the sides and implanted on the top or front of the head, temporarily delaying a receding hairline. This procedure is rather expensive (thousands of dollars), and not every transplant is successful.

Alopecia areata (al-ō-PĒ-shē-ah ar-ē-A-ta) is a localized hair loss that can affect either sex. The cause is not known, and the severity of hair loss varies from case to case. This condition is associated with several disorders of the immune system; it has also been suggested that periods of stress may promote alopecia areata in individuals already genetically prone to baldness.

Skin conditions that affect follicles can contribute to hair loss. Baldness can also result from exposure to radiation or to many of the toxic (poisonous) drugs used in cancer therapy. Both radiation and anticancer drugs have their greatest effects on rapidly dividing cells, and thus tend to damage matrix cells.

Hairs are dead, keratinized structures, and no amount of oiling, shampooing, or dousing with kelp extracts, vitamins, or nutrients will influence either the exposed hair or the follicle buried in the dermis. Untested treatments for baldness were banned by the Food and Drug Administration in 1984. Minoxidil, a drug originally marketed for the control of blood pressure, appears to stimulate hair follicles when rubbed onto the scalp. It is now available on a nonprescription basis. Treatment involves applying a 2% solution to the scalp twice daily; after 4 months, over one-third of patients reported satisfactory results.

Hirsutism (HER-sūt-izm; *hirsutus*, bristly) refers to the growth of hair on women in patterns usually characteristic of men. Because considerable overlap exists between the two sexes in terms of normal hair distribution, and there are significant racial and genetic differences, the precise definition is more often a matter of personal taste than objective analysis. Age and sex hormones may play a role, for hairiness increases late in pregnancy, and menopause produces a change in body hair patterns. Severe hirsutism is often associated with abnormal androgen (male sex hormone) production, either in the ovaries or adrenal glands.

ACNE *Page 100*

Individuals with a genetic tendency toward **acne** have larger than average sebaceous glands, and when the ducts become blocked the secretions accumulate. Inflammation develops, and bacterial infection may occur. The condition usually surfaces at puberty, as production of sex hormones accelerates. The secretory output of the sebaceous glands may be further encouraged by anxiety, stress, physical exertion, certain foods, and drugs.

The visible signs of acne are called **comedos** (ko-ME-dōz). "Whiteheads" contain accumulated, stagnant secretions. "Blackheads" contain more solid material that has been invaded by bacteria. Although neither condition indicates the presence of dirt in the pores, washing may help to reduce superficial oiliness.

Acne usually fades after sex hormone concentrations stabilize. Peeling agents may help reduce inflammation and minimize scarring. In cases of severe acne, the most effective treatment usually involves the discouragement of bacteria by the administration of topical or systemic antibiotic drugs or topical or systemic vitamin A derivatives, such as tretinoin (Retin-A).

INFLAMMATION OF THE SKIN *Page 103*

The epidermis provides significant protection from mechanical and chemical hazards because of its thick cell layers and keratin covering. Although the surface of the skin normally contains a variety of microorganisms, most of them are harmless as long as they remain outside the stratum corneum. Penetration of the superficial layers may produce familiar conditions such as "athlete's foot" (a fungal infection), and warts and cold sores (both viral infections). To reach the underlying connective tissues, a bacterium must survive the bacteriocidal ingredients of sebum, avoid being flushed from the surface by the sweat gland secretions, penetrate the stratum corneum, squeeze between the junctional complexes of deeper layers, escape the Langerhans cells, and cross the basement membrane. Unfortunately, once there, the papillary layer of the dermis provides all the elements for the growth of microorganisms: warmth, darkness, moisture, and nutrients.

If the protective barriers are crossed, or if an injury breaks through the epidermis, mast cells within the dermis respond by triggering a powerful inflammatory response. Inflammation begins immediately after an injury and produces swelling, redness, heat, and pain. Inflammation in the skin is important in defending the body against serious injury and disease. Inflammation and regeneration are controlled at the tissue level. The two phases overlap; isolation establishes a framework that guides the cells responsible for reconstruction, and repairs are under way well before cleanup operations have ended.

Inflammation Many stimuli can produce inflammation, including impact, abrasion, distortion, chemical irritation, infection by pathogenic organisms (such as bacteria or viruses), or extreme temperatures (hot or cold). The common factor is that each of these stimuli either kills cells, damages fibers, or injures the tissue in some other way. These changes alter the chemical composition of the interstitial fluid: Damaged cells release prostaglandins, proteins, and potassium ions, and the injury itself may have introduced foreign proteins or pathogens.

Immediately after the injury, tissue conditions become even more abnormal. **Necrosis** (ne-KRŌ-sis) is the tissue degeneration that occurs after cells have been injured or destroyed. This process begins several hours after the initial event, and the damage is caused by lysosomal enzymes. Lysosomes break down through autolysis, releasing digestive enzymes that first destroy the injured cells and then attack surrounding tissues. The accumulation of debris, fluid, dead and dying cells, and other necrotic tissue components that may result is known as **pus**. Pus often forms at an infection site in the dermis. An accumulation of pus in an enclosed tissue space is called an **abscess**.

These tissue changes trigger the inflammatory response by releasing chemicals that stimulate mast cells. When an injury occurs that damages fibers and cells, the stimulated mast cells release chemicals (**histamine** and **heparin**). These chemicals stimulate local macrophages, affecting smooth muscle tissue and endothelial cells in the region. The combination of abnormal tissue conditions and chemicals released by mast cells stimulate almost at once local neurons. The pain sensations generated trigger immediate action (such as pulling the affected part of the body away from the source of the injury) that reduces the chances for further damage. Activation of mast cells results in three major changes:

1. Histamine relaxes the smooth muscle tissue in the vessel walls, and the vessels enlarge, or **dilate**. This dilation increases the blood flow through the tissue, giving the region a reddish color and making it warm to the touch. The increased blood flow accelerates the delivery of nutrients and oxygen and the removal of dissolved waste products and toxic chemicals. It also brings white blood cells to the region. These cells migrate into the injury site and assist in the defense and cleanup operations.

2. Histamine makes the endothelial cells of local capillaries more permeable. Plasma, including blood proteins, diffuses into the injured tissue, and the area becomes swollen. Once in the tissue, some of the blood proteins combine to form large, insoluble fibers, known as **fibrin**. The heparin released by mast cells prevents the formation of fibrin at the injury site, but a dense fibrous meshwork, called a **clot**, surrounds the damaged area. The clot walls off the inflamed region, slowing the spread of cellular debris or bacteria into adjacent tissues.

3. Stimulated macrophages and newly arrived white blood cells defend the tissue and perform cleanup operations. Fixed macrophages, free macrophages, monocytes, and microcytes phagocytize debris and bacteria. Lymphocytes convert to plasma cells, producing antibodies that attack pathogens and foreign proteins. Usually the combination of physical attack, through phagocytosis, and chemical attack, through antibodies, succeeds in cleaning up the region and eliminating the inflammatory stimulus.

Regeneration When damage extends through the epidermis and into the dermis, bleeding usually occurs. The fibrin clot, or **scab**, that forms at the surface temporarily restores the integrity of the epidermis and restricts the entry of additional microorganisms. Cells of the stratum germinativum undergo rapid divisions and begin to migrate along the sides of the wound in an attempt to replace the missing epidermal cells.

If the wound covers an extensive area or involves a region covered by thin skin, dermal repairs must be under way before epithelial cells can cover the surface. Fibroblast and mesenchymal cell divisions produce mobile cells that invade the deeper area of injury, migrating along fibrin strands. Endothelial cells of damaged blood vessels also begin to divide, and capillaries follow the fibroblasts, eventually uniting and providing a circulatory supply. The combination of fibrin clot, fibroblasts, and an extensive capillary network is called **granulation tissue**. Over time, the fibrin dissolves, and the number of capillaries declines. Fibroblast activity leads to the appearance of collagen fibers and typical ground substance.

While dermal repairs are in progress, **contraction** pulls the edges of the wound closer together. The mechanism of contraction is uncertain, but it is an essential part of the healing process when damage has been extensive. For example, after an amputation, over 90% of the exposed surface is covered by contraction of the wound edges, rather than by the divisions and migration of epithelial cells. Contraction distorts the adjacent surface, as if the skin were being stretched to cover the injury site. If contraction and epithelial cell migration cannot cover the wound, **skin grafts** may be required.

COMPLICATIONS OF INFLAMMATION *Page 103*

When bacteria invade the dermis, the production of bacterial toxins and the debris from dead cells powerfully stimulate the inflammation process. Pus is an accumulation of debris, fluid, dead and dying cells, and necrotic tissue components. Pus typically forms at the site of an infection in the dermis. When pus accumulates in an enclosed tissue space, the result is an abscess. In the skin, an abscess can form as pus builds up inside the fibrin clot that surrounds the injury site. If the cellular defenses succeed in destroying the invaders, the pus will be either absorbed or surrounded by a fibrous capsule, creating a cyst.

Erysipelas (er-i-SIP-e-las; *erythros*, red + *pella*, skin) is a widespread inflammation of the dermis caused by bacterial infection. If the inflammation spreads into the subcutaneous layer and deeper tissues, the condition is called **cellulitis** (sel-ū-LĪ-tis). Erysipelas and cellulitis develop when bacterial invaders break through the fibrin wall. The bacteria involved produce large quantities of hyaluronidase, an enzyme that liquefies the ground substance, and fibrinolysin, an enzyme that breaks down fibrin and prevents clot and abscess formation. These are serious conditions that require prompt antibiotic therapy.

An **ulcer** is a localized shedding of an epithelium. **Decubitus ulcers**, also known as "bedsores," may afflict bedridden or mobile patients with circulatory restrictions, especially when splints, casts, or bedding continually press against superficial blood vessels. Such sores most often affect the skin near joints or projecting bones, where the vessels are pressed against hard underlying structures. The chronic lack of circulation kills epidermal cells, removing a barrier to bacterial infection, and eventually the dermal tissues deteriorate as well. (A comparable necrosis will occur in any tissues deprived of adequate circulation.) Bedsores can be prevented or treated by frequent changes in body position that vary the pressures applied to specific blood vessels.

BURNS AND GRAFTS *Page 103*

Burns result from exposure of the skin to heat, radiation, electrical shock, or strong chemical agents. The severity of the burn reflects the depth of penetration and the total area affected. Table A-3 summarizes a classification of burns based on the depth of involvement.

TABLE A-3 A Classification of Burns		
Classification	*Damage Report*	*Appearance and Sensation*
First-degree burn	*Killed*: superficial cells of epidermis. *Injured*: deeper layers of epidermis, papillary dermis.	Inflamed, tender.
Second-degree burn	*Killed*: superficial and deeper cells of epidermis; dermis may be affected. *Injured*: damage may extend into reticular layer of the dermis, but many accessory structures unaffected.	Blisters, pain.
Third-degree burn	*Killed*: all epidermal and dermal cells. *Injured*: hypodermal and deeper tissues and organs.	Charred, no sensation at all.

First- and second-degree burns are also called **partial-thickness** burns because damage is restricted to the superficial layers of the skin. Accessory structures such as hair follicles and glands are usually unaffected. **Full-thickness burns**, or **third-degree burns**, destroy the epidermis and dermis, extending into subcutaneous tissues. These burns are actually less painful than second-degree burns, because sensory nerves are destroyed along with accessory structures, blood vessels, and other dermal components. Extensive third-degree burns cannot repair themselves, and the site remains exposed to potential infection.

Roughly 10,000 people die from burns each year in the United States. The larger the area burned, the more significant the effects on integumentary function. Burns that cover more than 20% of the skin surface represent serious threats to life because they affect the following functions:

Fluid and Electrolyte Balance Even areas with partial-thickness burns lose their effectiveness as barriers to fluid and electrolyte losses. In full-thickness burns, the rate of fluid loss through the skin may reach five times the normal level.

Thermoregulation Increased fluid loss means increased evaporative cooling. More energy must be expended to keep body temperature within acceptable limits.

Protection from Attack The epidermal surface, damp from uncontrolled fluid losses, encourages bacterial growth. If the skin is broken at a blister or the site of a third-degree burn, infection is likely. Widespread bacterial infection, or **sepsis** (*septikos*, rotting), is the leading cause of death in burn victims.

Because full-thickness burns cannot heal without aid, surgical procedures are necessary to encourage healing. In a **skin graft**, areas of intact skin are transplanted to cover the burn site. A **split-thickness graft** takes a shaving of the epidermis and superficial portions of the dermis. A **full-thickness graft** involves the epidermis and both layers of the dermis.

With the development of fluid replacement therapies, infection control methods, and grafting techniques, the recovery rate for severe burns has improved dramatically. At present, young patients with burns over 80% of the body have an approximately 50% chance of recovery.

Recent advances in cell culture techniques may improve survival rates further. It is now possible to remove a small section of skin and grow it under controlled laboratory conditions. Over time, the germinative cell divisions produce large sheets of epidermal cells that can then be used to cover the burn area. From initial samples the size of postage stamps, square yards of epidermis have been grown and transplanted onto body surfaces. Although questions remain concerning the strength and flexibility of the repairs, skin cultivation represents a substantial advance in the treatment of serious burns.

SCAR TISSUE FORMATION *Page 103*

What regulates the extent of scar tissue formation is not known. In some individuals, most often dark-skinned people, scar tissue formation may continue beyond the requirements of tissue repair. The result is a flattened mass of scar tissue that begins at the injury site and grows into the surrounding dermis. This thickened area of scar tissue, called a **keloid** (KĒ-loyd), is covered by a shiny, smooth epidermal surface. Keloids most often develop on the upper back, shoulders, anterior chest, and earlobes.

SYNTHETIC SKIN *Page 103*

Epidermal culturing can produce a new epithelial layer to cover the burn site. A second new procedure provides a model for dermal repairs that takes the place of normal granulation tissue. A special synthetic skin is used. The imitation has a plastic (silastic) epidermis and a dermis composed of collagen fibers and ground cartilage. The collagen fibers are taken from cow skin, and the cartilage from sharks. Over time, fibroblasts migrate among the collagen fibers and gradually replace the model framework with their own. The silastic epidermis is intended only as a temporary cover that will be replaced by either skin grafts or a cultured epidermal layer.

chapter 5

The Skeletal System: Osseous Tissue and Skeletal Structure

INHERITED ABNORMALITIES IN SKELETAL DEVELOPMENT *Page 118*

There are several inherited conditions that result in abnormal bone formation. Three examples are *osteogenesis imperfecta, Marfan's syndrome*, and *achondroplasia*.

Osteogenesis imperfecta (im-per-FEK-ta) is an inherited condition appearing in 1 individual in about 20,000, that affects the organization of collagen fibers. Osteoblast function is impaired, growth is abnormal, and the bones are very fragile, leading to progressive skeletal deformation and repeated fractures. The ligaments and tendons are very "loose," thereby permitting excessive movement at the joints.

Marfan's syndrome is also linked to defective collagen fiber production. This condition, which affects approximately 1 individual in 10,000, is due to a genetic defect that causes the production of an abnormal form of fibrillin, a protein important to normal connective tissue strength and elasticity. Because connective tissues are found in most organs, the effects of this defect are widespread. The most visible sign of Marfan's syndrome involves the skeleton; individuals with Marfan's syndrome are usually tall, with abnormally long limbs and fingers. But the most serious consequences involve the cardiovascular system.

Roughly 90% of the people with Marfan's syndrome have abnormal cardiovascular systems. Inside the heart, connective tissue supports the valves that ensure a one-way flow of blood. In Marfan's syndrome, these valves are often defective, and the heart works at reduced efficiency. Outside the heart, connective tissue that reinforces the walls of the aorta, a large blood vessel leaving the heart, may be too weak to resist the blood pressure. As a result, a bubble (aneurysm) forms in the vessel wall. If the bubble breaks, there is a sudden, fatal loss of blood.

Achondroplasia (a-kon-drō-PLĀ-zhē-a) is a condition resulting from abnormal epiphyseal activity. In this case the epiphyseal plates grow unusually slowly, and the individual develops short, stocky limbs. Although there are other skeletal abnormalities, the trunk is normal in size, and sexual and mental development remain unaffected. The adult will be an *achondroplastic dwarf*.

HYPEROSTOSIS *Page 119*

The excessive formation of bone is termed **hyperostosis** (hī-per-os-TŌ-sis). In **osteopetrosis** (os-tē-ō-pe-TRŌ-sis; *petros*, stone) the total mass of the skeleton gradually increases because of a decrease in osteoclast activity. Remodeling stops, and the shapes of the bones gradually change. Osteopetrosis in children produces a variety of skeletal deformities. The primary cause for this relatively rare condition is unknown.

In **acromegaly** (*akron*, extremity + *megale*, great) an excessive amount of growth hormone is released after puberty, when most of the epiphyseal plates have already closed. Cartilages and small bones respond to the hormone, however, resulting in abnormal growth at the hands, feet, lower jaw, skull, and clavicle.

OSTEOMALACIA *Page 120*

In the condition of **osteomalacia** (os-tē-ō-ma-LĀ-shē-ah; *malakia*, softness) bone size remains the same but the mineral content decreases, thereby softening the bones. In this condition the osteoblasts are working hard, but the matrix is not accumulating enough calcium salts. This condition can occur in adults or children whose diet contains inadequate levels of calcium or vitamin D.

OSTEOPOROSIS AND OTHER SKELETAL ABNORMALITIES ASSOCIATED WITH AGING *Page 121*

Osteoporosis (os-tē-ō-pōr-Ō-sis) exists when a reduction in bone mass has proceeded to the point that normal function is compromised. In typical osteoporosis, trabecular bone and the endosteal surfaces are eroded, so bone sizes and shapes remain normal but bone mass and strength decrease. Current statistics suggest that 29% of women between the ages of 45 and 79 can be considered osteoporotic. The increase in incidence after menopause has been linked to decreases in the production of the female sex hormones (estrogens). The incidence of osteoporosis in men of the same age is estimated at 18%.

In this condition the bones become very fragile. Fragility of the bones frequently leads to breakage, and subsequent healing occurs only slowly. Vertebrae may collapse, distorting the normal alignment of vertebrae and putting pressure on spinal nerves. The hunched-over posture that develops in older women with severe osteoporosis has been called a "widow's hump." Therapies that boost estrogen levels, dietary changes to elevate calcium levels in the blood, and exercise that stresses bones and stimulates osteoblast activity appear to slow but not completely prevent the development of osteoporosis. Treatment with calcitonin (by injection or nasal spray) has been also shown to slow the development of osteoporosis.

Osteoporosis can also develop as a secondary effect of many cancers. Cancers of the bone marrow, breast, or other tissues release a chemical known as **osteoclast-activating factor**. This compound increases both the number and activity of osteoclasts and produces a severe osteoporosis.

Infectious diseases that affect the skeletal system also become more common in older individuals.

- **Osteomyelitis** (os-tē-ō-mī-e-LĪ-tis; *myelos*, marrow) is a painful infection of a bone most often caused by bacteria. This condition, most common in people over 50 years of age, can lead to dangerous systemic infections.

- **Paget's disease**, also known as **osteitis deformans** (os-tē-Ī-tis de-FOR-manz), appears to be the result of a viral infection. This condition affects roughly 10% of the population over 70. Osteoclast activity accelerates, producing areas of acute osteoporosis, and osteoblasts produce abnormal matrix proteins. The result is a gradual deformation of the entire skeleton that is particularly noticeable in the skull.

chapter 6
The Skeletal System: Axial Division

THE THORACIC CAGE AND SURGICAL PROCEDURES *Page 166*

Surgery on the heart, lungs, or other organs in the thorax often involves entering the thoracic cavity. The mobility of the ribs and the cartilaginous connections with the sternum allow the ribs to be temporarily moved out of the way. Special rib-spreaders are used, which push them apart in much the same way that a jack lifts a car off the ground for a tire change. If more extensive access is required, the sternal cartilages can be cut and the entire sternum can be folded out of the way. Once replaced, the cartilages are reunited by scar tissue, and the ribs heal fairly rapidly.

After thoracic surgery, chest tubes may penetrate the thoracic wall to permit drainage of fluids. To install a **chest tube** or obtain a sample of pleural fluid, the wall of the thorax must be penetrated. This process, called **thoracentesis** (thō-ra-sen-TĒ-sis) or **thoracocentesis** (thō-ra-kō-sen-TĒ-sis; *kentesis*, perforating), involves the penetration of the thoracic wall along the superior border of one of the ribs. Penetration at this location avoids damaging vessels and nerves within the costal groove.

chapter 8
The Skeletal System: Articulations

RHEUMATISM, ARTHRITIS, AND SYNOVIAL FUNCTION *Page 207*

Proper synovial function depends on healthy articular cartilages. When an articular cartilage has been damaged, the matrix begins to break down, and the exposed cartilage changes from a slick, smooth gliding surface to a rough feltwork of bristly collagen fibers. This feltwork drastically increases friction, damaging the cartilage further. Eventually the central area of the articular cartilage may completely disappear, exposing the underlying bone. Fibroblasts are attracted to areas of friction, and they begin tying the opposing bones together with a network of collagen fibers. This network may later be converted to bone, locking the articulating elements into position. Such a bony fusion, called **ankylosis** (an-ke-LŌ-sis), eliminates the friction by making movement impossible. Degenerative changes may be caused by simply immobilizing a joint. When motion ceases, so does circulation of synovial fluid, and the cartilages begin to suffer. **Continuous passive motion (CPM)** of any injured joint appears to encourage the repair process by improving the circulation of synovial fluid. The movement is often performed by a physical therapist or machine during the recovery process.

Rheumatism (ROO-ma-tizm) is a general term that indicates pain and stiffness affecting the skeletal system, the muscular system, or both. There are several major forms of rheumatism. **Arthritis** (ar-THRĪ-tis) includes all of the rheumatic diseases that affect synovial joints. Arthritis always involves damage to the articular cartilages, but the specific cause may vary. For example, arthritis can result from bacterial or viral infection, injury to the joint, metabolic problems, or severe physical stresses.

The diseases of arthritis are usually considered as either **degenerative** or **inflammatory** in nature. Degenerative diseases begin at the articular cartilages, and modification of the underlying bone and inflammation of the joint occur secondarily. Inflammatory diseases start with the inflammation of synovial tissues, and damage later spreads to the articular surfaces. We will consider a single example of each type.

Osteoarthritis (os-tē-ō-ar-THRĪ-tis), also known as **degenerative arthritis**, or **degenerative joint disease (DJD)**, usually affects older individuals. In the U.S. population 25% of women and 15% of men over 60 years of age show signs of this disease. The condition seems to result from cumulative wear and tear on the joint surfaces. Some individuals, however, may have a genetic predisposition to develop osteoarthritis, for researchers have recently isolated a gene linked to the disease. This gene codes for an abnormal form of collagen that differs from the normal protein in only 1 of its 1000 amino acids.

Rheumatoid arthritis is an inflammatory condition that affects roughly 2.5% of the adult population. The cause is uncertain, although allergies, bacteria, viruses, and genetic factors have all been proposed. The synovial membrane becomes swollen and inflamed, a condition known as **synovitis** (sīn-ō-VĪ-tis). The cartilaginous matrix begins to break down, and the process accelerates as dying cartilage cells release lysosomal enzymes.

In their later stages, inflammatory and degenerative forms of arthritis produce an inflammation that spreads into the surrounding area. Ankylosis, common in the past when complete rest was routinely prescribed for arthritis patients, is rarely seen today. Regular exercise, physical therapy, and drugs that reduce inflammation, such as aspirin, or even low doses of the immunosuppressant *methotrexate*, can slow the progress of the disease. Surgical procedures can realign or redesign the affected joint, and in extreme cases involving the hip, knee, elbow, or shoulder, the defective joint can be replaced by an artificial one. Joint replacement has the advantage of eliminating the pain and restoring full range of motion. Prosthetic (artificial) joints are weaker than natural ones, but elderly people seldom stress them to their limits.

BURSITIS *Page 216*

When bursae become inflamed, causing pain in the affected area whenever the tendon or ligament moves, such a condition is termed **bursitis**. Inflammation can result from the friction associated with repetitive motion, pressure over the joint, irritation by chemical stimuli, infection, or trauma. Bursitis associated with repetitive motion often occurs at the shoulder; for example, golfers, pitchers, ceiling painters, and tennis players may develop bursitis, usually at the subscapular bursa. The most common pressure-related bursitis is a **bunion**. Bunions form over the base of the big toe as a result of the friction and distortion of the joint caused by tight shoes, especially those with pointed toes. There is chronic inflammation of the region, and as the wall of the bursa thickens, fluid builds up in the surrounding tissues. The result is a firm, tender nodule. There are special names for bursitis at other locations; the names of these conditions indicate the occupations most often associated with them. In "housemaid's knee," which accompanies prolonged kneeling, the affected bursa lies between the patella (*kneecap*) and the skin. The condition of "student's elbow" is a form of bursitis that can result from propping your head above a desk while reading your anatomy textbook.

Most of the symptoms of bursitis subside if the stimulus is removed, and a variety of anti-inflammatory drugs may help. In extreme cases the affected bursae can be surgically removed in a **bursectomy**.

HIP FRACTURES AND THE AGING PROCESS
Page 221

Hip fractures are most often suffered by individuals over 60 years of age, when osteoporosis has weakened the thigh bones. These injuries may be accompanied by dislocation of the hip or pelvic fractures. For individuals with osteoporosis, healing proceeds very slowly. In addition, the powerful muscles that surround the joint can easily prevent proper alignment of the bone fragments. Trochanteric fractures usually heal well, if the joint can be stabilized; steel frames, pins, screws, or some combination of those devices may be needed to preserve alignment and permit healing to proceed normally.

Severe fractures of the femoral neck have the highest complication rate of any fracture because the blood supply to the region is relatively delicate. The surgical procedures used to treat trochanteric fractures are often unsuccessful in stabilizing femoral neck fractures. If more complex pinning operations fail, the entire joint can be replaced. In this "total hip" procedure, the damaged portion of the femur is removed. An artificial portion of the femur, including the head and neck, is attached by a spike that extends into the marrow cavity of the shaft. Special cement is used to anchor it in place and to attach a new articular surface to the acetabulum.

A CASE STUDY: AVASCULAR NECROSIS
Page 221

Everyone agreed that Bo Jackson was an athletic phenomenon. At age 28 he was playing professional football with the L.A. Raiders and professional baseball for the Kansas City Royals—and starring in both sports. But on January 13, 1991, things changed dramatically when Bo was tackled near the sidelines in an NFL playoff game. The combination of pressure and twisting applied by the tackler and the tremendous power of Bo's thigh muscles produced an unusual injury, a fracture-dislocation of the hip. Although he experienced severe pain, the immediate damage to the femur and hip was limited; however, the complications of the injury were more damaging than the injury itself. The dislocation tore blood vessels in the capsule and along the femoral neck, where the capsular fibers attach. Two problems gradually developed as a result:

1. The mineral deposits in the bone of the pelvis and femur are turned over very rapidly, and osteocytes have high energy demands. A reduction in blood flow first injures and then kills them. When bone maintenance stops in the affected region, the matrix begins to break down. This process, called *avascular necrosis*, weakened and eroded Bo's femoral head.

2. The initial impact and damage to the articular cartilages was followed by joint immobility, and the articular chondrocytes became nutrient-starved. The combination resulted in the gradual loss of the articular cartilages of the femur and acetabulum.

After more than a year of rehabilitation, Bo Jackson had surgery to replace the damaged joint with an artificial hip. Joint replacement eliminates the pain and restores full range of motion. However, prosthetic (artificial) joints are weaker than natural ones. His long-term future in professional sports may be as a pinch hitter, a relatively low-stress position.

KNEE INJURIES *Page 225*

Athletes place tremendous stresses on their knees. Ordinarily, the medial and lateral menisci move as the femoral position changes. Placing a lot of weight on the knee while it is partially flexed can trap a meniscus between the tibia and femur, resulting in a break or tear in the cartilage. In the most common injury, the lateral surface of the lower leg is driven medially, tearing the medial meniscus. In addition to being quite painful, the torn cartilage may restrict movement at the joint. It can also lead to chronic problems and the development of a "trick knee," a knee that feels rather unstable. Sometimes the meniscus can be heard popping in and out of position when the knee is extended.

An **arthroscope** uses fiber optics to permit exploration of a joint without major surgery. Optical fibers are thin threads of glass or plastic that conduct light. The fibers can be bent around corners, so they can be introduced into a knee or other joint and moved around, enabling the physician to see what is going on inside the joint. If necessary, the apparatus can be modified to perform surgical modification of the joint at the same time. This procedure, called **arthroscopic surgery**, has greatly simplified the treatment of knee and other joint injuries. Ideally, small pieces of cartilage can be removed and the meniscus surgically trimmed. Large tears may require **meniscectomy**, the removal of the affected cartilage. New tissue-culturing techniques may someday permit the replacement of the meniscus.

An arthroscope cannot show the physician soft tissue details outside the joint cavity, and repeated arthroscopy eventually leads to the formation of scar tissue and other joint problems. Magnetic resonance imaging is a cost-effective and noninvasive method of viewing and examining, without injury, soft tissues around the joint.

Other knee injuries involve tearing one or more stabilizing ligaments or damaging the patella. *Torn ligaments* are often difficult to correct surgically, and healing is slow. The patella can be injured in a number of ways. If the lower leg is immobilized (as it might be in a football pileup, for example) while you try to straighten your leg, the muscles that extend the knee are powerful enough to pull the patella apart. Impacts to the anterior surface of the knee may also shatter the kneecap. Treatment involves surgical removal of fragments and repair of the tendons and ligaments, followed by immobilization of the joint. Total knee replacements are rarely performed on young people, but they are becoming increasingly common among elderly patients with severe arthritis.

PROBLEMS WITH THE ANKLE AND FOOT *Page 225*

The ankle and foot are subjected to a variety of stresses during normal daily activities. In a **sprain**, a ligament is stretched to the point where some of the collagen fibers are torn. The ligament remains functional, and the structure and stability of the joint is not affected. The most common cause of a sprained ankle is a forceful inversion of the foot that stretches the lateral ligament. An ice pack is usually required to reduce swelling, and with rest and support the ankle should heal in about 3 weeks.

In more serious incidents, the entire ligament may be torn apart, simply termed a *torn ligament*, or the connection between the ligament and the malleolus may be so strong that the bone breaks before the ligament. In general, a broken bone heals more quickly and effectively than does a torn ligament. A dislocation often accompanies such injuries.

In a **dancer's fracture** the proximal portion of the fifth metatarsal is broken. This fracture usually occurs while the body weight is being supported by the longitudinal arch. A sudden shift in weight from the medial portion of the arch to the lateral, less elastic border breaks the fifth metatarsal close to its articulation.

Individuals with abnormal arch develoment are more likely to suffer metatarsal injuries than individuals with normal arch development. In the condition of **flat feet**, one loses or never develops the longitudinal arch. *Fallen arches* may develop as tendons and ligaments stretch and become less elastic. Obese individuals or those who must constantly stand or walk on the job are likely candidates. Children have very mobile articulations and elastic ligaments, and they often have *flexible flat feet*. Their feet look flat only while they are standing, and the arch reappears when they stand on their toes or sit down. This condition usually disappears as growth continues.

Congenital talipes equinovarus (clubfoot), results from an inherited developmental abnormality that affects 2 in 1000 births. Boys are affected roughly twice as often as girls. One or both feet may be involved, and the condition may be mild, moderate, or severe. The underlying problem is abnormal muscle development (sometimes related to abnormal innervation) that distorts growing bones and joints. Usually the tibia, ankle, and foot are affected, and the feet are turned medially and inverted. The longitudinal arch is exaggerated, and if both feet are involved, the soles face one another. Prompt treatment with casts or other supports in infancy helps to alleviate the problem, and fewer than half of the cases require surgery. Kristi Yamaguchi, Olympic Gold Medalist in figure skating, was born with this condition.

chapter 9
The Muscular System: Skeletal Muscle Tissue

DUCHENNE'S MUSCULAR DYSTROPHY *Page 243*

The **muscular dystrophies** (DIS-trō-fēz) are congenital diseases that produce progressive muscle weakness and deterioration. One of the most common and best understood conditions is **Duchenne's muscular dystrophy (DMD)**.

This form of muscular dystrophy appears in childhood, often between the ages of 3 and 7. The condition generally affects only males. A progressive muscular weakness develops, and the individual usually dies before age 20 because of respiratory paralysis. Skeletal muscles are primarily affected, although for some reason the facial muscles continue to function normally. In later stages of the disease the facial muscles and cardiac muscle tissue may also become involved.

The skeletal muscle fibers in a DMD patient are structurally different from those of normal individuals. Abnormal membrane permeability, cholesterol content, rates of protein synthesis, and enzyme composition have been reported. DMD sufferers also lack a protein, called *dystrophin*, found in normal muscle fibers. Although the functions of this protein remain uncertain, researchers have recently identified and cloned the gene for dystrophin. Rats with DMD have been cured by insertion of this gene into their muscle fibers—a technique that may eventually be used to treat human patients.

The inheritance of DMD is sex-linked: Women carrying the defective genes are unaffected, but each of their male children will have a 50% chance of developing DMD. Now that the specific location of the gene has been identified, it is possible to determine whether a woman is carrying the defective gene or not.

MYASTHENIA GRAVIS *Page 243*

Myasthenia gravis (mī-as-THĒ-nē-a GRA-vis) is characterized by a general muscular weakness that is often most pronounced in the muscles of the arms, head, and chest. The first symptom is usually a weakness of the eye muscles and drooping eyelids. Facial muscles are often weak as well, and the individual develops a peculiar smile known as the "myasthenic snarl." As the disease progresses, pharyngeal weakness leads to problems with chewing and swallowing, and it becomes difficult to hold the head upright.

The muscles of the upper chest and arms are next to be affected. All of the voluntary muscles of the body may ultimately be involved. Severe myasthenia gravis produces respiratory paralysis, with a mortality rate of 5–10%. However, the disease does not always progress to such a life-threatening stage. For example, roughly 20% of patients experience eye problems but no other symptoms.

The condition results from a decrease in the number of ACh receptors on the motor end plate. Before the remaining receptors can be stimulated enough to trigger a strong contraction, the ACh molecules are destroyed by cholinesterase. As a result, muscular weakness develops.

The primary cause of myasthenia gravis appears to be a malfunction in the immune system. The body attacks the ACh receptors of the motor end plate as if they were foreign proteins. For unknown reasons, women are affected twice as often as men. Estimates of the incidence of this disease in the United States range from 2 to 10 cases per 100,000 population.

One approach to therapy involves the administration of drugs, such as *neostigmine*, that are termed *cholinesterase inhibitors*. These compounds are enzyme inhibitors that prevent the enzymatic breakdown of ACh. With cholinesterase activity reduced, the concentration of ACh at the synapse can rise enough to stimulate the surviving receptors and produce muscle contraction.

POLIO *Page 244*

Because skeletal muscles depend on their motor neurons for stimulation, disorders that affect the nervous system can have an indirect effect on the muscular system. The poliovirus attacks and kills motor neurons in the spinal cord and brain. As the neurons die, the dependent motor units first become paralyzed and then undergo atrophy. The resulting condition is called **polio**. In severe cases, paralysis of the respiratory muscles makes breathing impossible without the assistance of an "iron lung" or some comparable device that provides mechanical ventilation.

Polio has been almost completely eliminated from the U.S. population through a successful immunization program. In 1954 there were 18,000 new cases; there were 8 in 1976. Unfortunately, many parents refuse to immunize their children against the poliovirus on the assumption that the disease has been "conquered." This assumption is a mistake because (1) there is no cure for polio, (2) the virus remains in the environment, and (3) approximately 38% of children ages 1–4 have not been immunized. Although recent efforts by the World Health Organization have eliminated polio from the Western Hemisphere, it remains a threat in the rest of the world. With the rate of international travel increasing, a major epidemic could still occur.

chapter 10
The Muscular System: Muscle Organization and the Axial Musculature

HERNIAS *Page 270*

When the abdominal muscles contract forcefully, pressure in the abdominopelvic cavity can rise dramatically, and those pressures are applied to internal organs. If the individual exhales at the same time, the pressure is relieved, because the diaphragm can move upward as the lungs collapse. But during vigorous periods of isometric exercises or when lifting a weight while holding one's breath, pressure in the abdominopelvic cavity can rise to 1500 pounds per square inch, roughly 100 times normal pressures. Pressures this high can cause a variety of problems, among them the development of a *hernia*. A **hernia** develops when an organ protrudes through an abnormal opening. There are many types of hernias; at this time we will consider only *inguinal* (groin) *hernias* and *diaphragmatic hernias*.

During the sixth month of development, the male testes descend into the scrotum by passing through the abdominal wall at the **inguinal** (ING-gwuh-null) **canals**. Abdominal pressure actually assists in the descent process. In the adult male, the spermatic ducts and associated blood vessels penetrate the abdominal musculature at the inguinal canals on their way to the abdominal reproductive organs. In an **inguinal hernia**, the inguinal canal enlarges and abdominal contents such as a portion of the intestine (or more rarely the urinary bladder) are forced into the inguinal canal. If the herniated structures become trapped or twisted within the inguinal sac, surgery may be required to prevent serious complications. Inguinal hernias are not always caused by unusually high abdominal pressures. Injuries to the abdomen or inherited weakness or distensibility of the canal may have the same effect. Although a weak spot does exist at the site of the inguinal canal, females are much less likely to develop inguinal hernias.

The esophagus and major blood vessels pass through an opening in the muscular diaphragm. In a **diaphragmatic hernia** or *hiatal hernia* (hī-Ā-tal; *hiatus*, a gap or opening), abdominal organs slide into the thoracic cavity, most often through the **esophageal hiatus**, the opening used by the esophagus. The severity of the condition will depend on the location and size of the herniated organ(s). Hiatal hernias are actually very common, and most go unnoticed. Radiologists see them in about 30% of patients examined with barium contrast techniques. When clinical complications develop, they usually occur because abdominal organs that have pushed into the thoracic cavity are exerting pressure on structures or organs there. As is the case with inguinal hernias, a diaphragmatic hernia may result from congenital factors or from an injury that weakens or tears the diaphragmatic muscle.

chapter 11
The Muscular System: The Appendicular Musculature

SPORTS INJURIES *Page 283*

Many Americans participate in exercise programs and sports on a regular basis; more than 30 million Americans go jogging, and millions more participate in various amateur and professional sports. As a result, sports injuries are very common, and sports medicine has become an active area of professional and academic research interest.

Sports injuries affect amateurs and professionals alike. For example, in 1986 the approximately 1500 players in the National Football League reported 1500 injuries. At the amateur level, a 5-year study of college football players indicated that 73.5% experienced mild injuries, 21.5% moderate injuries, and 11.6% severe injuries in the course of their playing careers. Contact sports are not the only activities that show a significant injury rate; a study of 1650 joggers running at least 27 miles per week reported 1819 injuries in a single year.

Muscles and bones respond to increased use by enlarging and strengthening. Poorly conditioned individuals are therefore more likely to subject their bones and muscles to intolerable stresses. Training is also important in minimizing the use of antagonistic muscle groups and keeping joint movements within the intended ranges of motion. Planned warm-up exercises before athletic events stimulate circulation, improve muscular performance and control, and help prevent injuries to muscles, joints, and ligaments. Stretching exercises stimulate muscle circulation and help keep ligaments and joint capsules supple. Such conditioning extends the range of motion and prevents sprains and strains when sudden loads are applied.

Dietary planning can also be important in preventing injuries to muscles during endurance events, such as marathon running. Emphasis has often been placed on the importance of carbohydrates, leading to the practice of "carbohydrate loading" before a marathon. But muscles also utilize amino acids extensively while operating within aerobic limits, and an adequate diet must include both carbohydrates and proteins.

Improved playing conditions, equipment, and regulations also play a role in reducing the incidence of sports injuries. Jogging shoes, ankle or knee braces, helmets, and body padding are examples of equipment that can be effective. The substantial penalties now earned for "personal fouls" in contact sports have reduced the numbers of neck and knee injuries.

Several injuries common to those engaged in active sports may also affect nonathletes, although the primary causes may differ considerably. A partial listing of activity-related conditions would include the following:

Bone bruise: bleeding within the periosteum of a bone

Bursitis: inflammation of the bursae around one or more joints

Stress fractures: cracks or breaks in bones subjected to repeated stresses or trauma

Muscle cramps: prolonged, involuntary, and painful muscular contractions

Sprains: tears or breaks in ligaments or tendons

Strains: tears or breaks in muscles

Tendinitis: inflammation of the connective tissue surrounding a tendon

chapter 13
The Nervous System: Neural Tissue

AXOPLASMIC TRANSPORT AND DISEASE *Page 326*

Each synaptic knob contains rough endoplasmic reticulum, ribosomes, and mitochondria needed to synthesize the neurotransmitter released at the synapse. Much synthesis activity also occurs at the cell body: Additional neurotransmitter, enzymes, mitochondria, and lysosomes all form within the perikaryon. Many of these products are transported to the synaptic knobs along the length of the axon. This **axoplasmic transport** occurs along neurotubules, and the process requires adenosine triphosphate (ATP).

While some materials are traveling from the cell body to the periphery of the neuron, other substances are being transported to the cell body. This reverse transport process is called *retrograde flow*. Retrograde flow may deliver chemicals or debris that indicate damage to the axon or synaptic knobs. This flow will trigger an immediate change in the cell's functions.

Rabies is an acute disease of the central nervous system (CNS). The rabies virus can infect any mammal, wild or domestic, and with few exceptions the result is death within 3 weeks. Rabies is usually transmitted to people through the bite of a rabid animal. There are an estimated 15,000 cases of rabies each year worldwide, the majority of them the result of dog bites. Only about five of those cases, however, are diagnosed in the United States, and since dogs and cats in the United States are vaccinated against rabies, cases here are most commonly caused by the bites of raccoons, foxes, skunks, and bats.

Although these bites usually involve peripheral sites, such as the hand or foot, the symptoms are caused by CNS damage. The virus present at the injury site is absorbed by synaptic knobs in the region. It then gets a free ride to the CNS, courtesy of retrograde flow. Over the first few days following exposure, the individual experiences headache, fever, muscle pain, nausea, and vomiting. The victim then enters a phase marked by extreme excitability, with hallucinations, muscle spasms, and disorientation. There is difficulty in swallowing, and the accumulation of saliva makes the individual appear to be foaming at the mouth. Coma and death soon follow.

Preventive treatment, which must begin almost immediately after exposure, consists of injections containing antibodies against the rabies virus. This postexposure treatment may not be sufficient following massive infection, which can lead to death in as little as 4 days. Individuals such as veterinarians or field biologists who are at high risk for exposure often take a preexposure series of injections. These injections bolster the immune defenses and improve the effectiveness of the postexposure treatment. Without treatment, rabies infection is always fatal.

Rabies is perhaps the most dramatic example of a clinical condition directly related to axoplasmic flow. Many toxins, however, including heavy metals, some pathogenic bacteria, and other viruses, use this mechanism to enter the CNS.

DEMYELINATION DISORDERS *Page 331*

Demyelination is the progressive destruction of myelin sheaths in the CNS and peripheral nervous system (PNS). The result is a gradual loss of sensation and motor control that leaves affected regions numb and paralyzed. Many unrelated conditions can cause demyelination; four important examples of demyelinating disorders are heavy metal poisoning, diphtheria, multiple sclerosis, and Guillain-Barré syndrome. Multiple sclerosis is discussed in the appendix for Chapter 14; other demyelination disorders are considered here.

Heavy Metal Poisoning Chronic exposure to heavy metal ions, such as arsenic, lead, or mercury, can lead to glial cell damage and demyelination. As demyelination occurs, the affected axons deteriorate, and the condition becomes irreversible. Historians note several interesting examples of heavy metal poisoning with widespread impact. For example, lead contamination of drinking water has been cited as one factor in the decline of the Roman Empire. In the seventeenth century, the great physicist Sir Isaac Newton is thought to have suffered several episodes of physical illness and mental instability brought on by his use of mercury in chemical experiments. Well into the nineteenth century, mercury used in the preparation of felt presented a serious occupational hazard for those employed in the manufacture of stylish hats. Over time, mercury absorbed through the skin and across the lungs accumulated in the CNS, producing neurological damage that affected both physical and mental function. (This effect is the source of the expression "mad as a hatter.") More recently, Japanese fishermen working in Minamata Bay, Japan, collected and consumed seafood contaminated with mercury discharged from a nearby chemical plant. Levels of mercury in their systems gradually rose to the point that clinical symptoms appeared in hundreds of people. Making matters worse, mercury contamination of developing embryos caused severe, crippling birth defects.

Diphtheria Diphtheria (dif-THĒ-rē-a; *diphthera, membrane* + *-ia*, disease) is a disease that results from a bacterial infection of the respiratory tract. In addition to restricting airflow and sometimes damaging the respiratory surfaces, the bacteria produce a powerful toxin that injures the kidneys and adrenal glands, among other tissues. In the nervous system, diphtheria toxin damages Schwann cells and destroys myelin sheaths in the PNS. This demyelination leads to sensory and motor problems that may ultimately produce a fatal paralysis. The toxin also affects cardiac muscle cells, and heart enlargement and failure may occur. The fatality rate for untreated cases ranges from 35 to 90%, depending on the site of infection and the subspecies of bacterium. Because an effective vaccine exists, cases are relatively rare in countries with adequate health care. With political change and poverty causing a drop in immunization rates, Russia is experiencing a diphtheria epidemic that began early in 1993.

Guillain-Barré Syndrome Guillain-Barré syndrome is characterized by a progressive demyelination. Symptoms initially involve weakness of the legs, which spreads rapidly to muscles of the trunk and arms. These symptoms usually increase in intensity for 1–2 weeks before subsiding. The mortality rate is low (under 5%), but there may be some permanent loss of motor function. The cause is unknown, but because roughly two-thirds of Guillain-Barré patients develop symptoms within 2 months after a viral infection, it is suspected that the condition may result from a malfunction of the immune system. In severe cases, *plasmapheresis* has proven helpful. In this procedure the patient's blood is filtered, and the plasma and antibodies are discarded and replaced by donor plasma. High intravenous doses of gamma globulin has also helped.

GROWTH AND MYELINATION OF THE NERVOUS SYSTEM *Page 335*

The development of the nervous system continues for many years after birth. Nerve cells increase in number for the first year after delivery, and most of the important interconnections between neurons occur after birth, rather than before. Growth of the brain is completed by age 4, but the neurons have yet to be interconnected extensively. The myelination of axons may not be completed until early adolescence. The level of nervous system development limits mental and physical performance. For example, the degree of interconnection between neurons in the CNS affects intellectual abilities, and the myelination of axons improves coordination and control by decreasing the time between reception of a sensation and completion of a response.

chapter 14

The Nervous System: The Spinal Cord and Spinal Nerves

SPINAL MENINGITIS *Page 342*

Meningitis is the inflammation of the meningeal membranes. Meningitis often disrupts the normal circulatory and cerebrospinal fluid supplies, damaging or killing neurons and glial cells in the affected areas. Although the initial diagnosis may specify the meninges of the spinal cord (spinal meningitis) or brain (cerebral meningitis), in later stages the entire meningeal system is usually affected.

The warm, dark, nutrient-rich environment of the meninges provides ideal conditions for a variety of bacteria and viruses. Microorganisms that cause meningitis include those associated with middle ear infections, pneumonia, streptococcal ("strep"), staphylococcal ("staph"), or meningococcal infections, and tuberculosis. The malaria parasite can also cause a severe form of meningitis. These pathogens may gain access to the meninges by traveling within blood vessels or by entering at sites of vertebral or cranial injury. Headache, chills, high fever, disorientation, and rapid heart and respiratory rates appear as higher centers are affected. Without treatment, delirium, coma, convulsions, and death may follow within hours.

The mortality rate for viral meningitis ranges from 1 to 50% or higher, depending on the type of virus, the age and health of the patient, and other factors. There is no effective treatment for viral meningitis. Bacterial meningitis can be combated with antibiotics and the maintenance of proper fluid and electrolyte balance, given early diagnosis. The incidence of one form of meningitis caused by the bacteria *Hemophilus influenza* has been reduced by the vaccination of infants.

The most common clinical assessment involves checking for a "stiff neck" by asking the patient to touch chin to chest. Meningitis affecting the cervical portion of the spinal cord results in an increase in the muscle tone of the extensor muscles of the neck. So many motor units become activated that voluntary or involuntary flexion of the neck becomes painfully difficult if not impossible.

SPINAL ANESTHESIA *Page 342*

Local anesthetics introduced into the subarachnoid space of the spinal cord produce a temporary sensory and motor paralysis. The effects spread as the anesthetic diffuses along the cord, and precise control of the regional effects can be difficult to achieve. Problems with overdosing are seldom serious, because the diaphragmatic breathing muscles are controlled by upper cervical spinal nerves and most spinal anesthesia involves the thoracic and lumbar regions. Thus respiration continues even when the thoracic and abdominal segments have been paralyzed. Sometimes more precise control can be obtained by injecting the anesthetic into the epidural space, producing an **epidural block**. This technique has the advantage of affecting only the spinal nerves in the immediate area of the injection. Epidural anesthesia may be difficult to achieve, however, in the upper cervical, midthoracic, and lumbar regions, where the epidural space is extremely narrow. **Caudal anesthesia** involves the introduction of anesthetics into the epidural space of the sacrum. Injection at this site paralyzes lower abdominal and perineal structures. Epidural blocks in the lower lumbar or sacral regions or caudal anesthesia are typically used to control pain during childbirth.

TECHNOLOGY AND MOTOR PARALYSIS *Page 345*

If a peripheral axon is damaged but not displaced, normal function may eventually return as the cut stump grows across the injury site away from the soma and along its former path. The mechanics of this process were detailed at the close of Chapter 13. ⟨⟩ [p. 332]

In order for normal function to be restored, several things must happen. The severed ends must be relatively close together (1–2 mm), they must remain in proper alignment, and there must be no physical obstacles between them, such as the dense fibers of scar tissue. These conditions can be created in the laboratory, using experimental animals and individual axons or small axonal bundles. But in accidental injuries to peripheral nerves the edges are likely to be jagged, intervening segments may be lost entirely, and elastic contraction in the surrounding connective tissues may pull the cut ends apart and misalign them.

Until recently the surgical response would involve trimming the injured nerve ends, neatly sewing them together, and hoping for the best. This procedure was often unsuccessful, in part because scalpels do not produce a smoothly cut surface and the thousands of broken axons would never be perfectly aligned. Moreover, nerve axons are not highly elastic, so if a large segment of the nerve was removed, crushed, or otherwise destroyed there would be no way to bring the intact ends close enough to permit any regeneration. In such instances a **nerve graft** could be inserted, using a section from some other peripheral nerve. The functional results were even less likely to be wholly satisfactory, for the growing axonal tips had to find their way across not one but two gaps, and the chances for successful alignment were proportionately smaller. Nevertheless, any return of function was certainly better than none at all!

Two very different research strategies are now being pursued. One focuses on the physical and biochemical control of nerve regeneration, while the other looks for a more "high-tech" solution. An example of a biological approach that focuses on physical factors is the use of a synthetic sleeve to guide nerve growth. The sleeve is a tube with an outer layer of silicone around an inner layer of cowhide collagen bound to the proteoglycans from shark cartilage. Using this sleeve as a guide, axons can grow across gaps as large as 20 mm (0.75 in.). The procedure has yet to be tried on humans, and functional restoration is incomplete because proper alignment does not always occur. Experiments on other mammals have successfully reunited the cut sections of large peripheral nerves.

A second biological line of investigation involves the biochemical control of nerve growth and regeneration. Neurons are influenced by a combination of growth promoters and growth inhibitors. Damaged myelin sheaths apparently release an inhibitory factor that slows the repair process. Researchers have made a monoclonal antibody, *IN-1*, that will inactivate the inhibitory factor released in the damaged spinal cords of rats. They are also experimenting with the application of *nerve growth factor*, a protein in the CNS that is required for normal growth and development of many neural components. These treatments stimulate repairs, even in severed spinal cords. Prior to this research, regeneration of nerves in the CNS was thought to be impossible or highly unlikely.

In the meantime, other research teams are experimenting with the use of computers to stimulate specific muscles and muscle groups electrically. The technique is called **functional electrical stimulation**, or **FES**. This approach often involves implanting a network of wires beneath the skin with their tips in skeletal muscle tissue. The wires are connected to a computer that may be small enough to be worn at the waist. The wires deliver minute electrical stimuli to the muscles, depolarizing their membranes and causing contractions. With this equipment and lightweight braces, quadriplegics have walked several hundred yards and paraplegics several thousand.

Even more impressive results have been obtained using a network of wires woven into the fabric of close-fitting garments. This network provides the necessary stimulation without the complications and maintenance problems that accompany implanted wires. A paraplegic woman in a set of electronic "hot pants" completed several miles of the 1985 Honolulu Marathon, and more recently a paraplegic woman walked down the aisle at her wedding. A commercial version of the system may be on the market in the near future.

Such technological solutions can provide only a degree of motor control without accompanying sensation. Everyone would prefer a biological procedure that would restore the functional integrity of the nervous system. For now, however, computer-assisted programs such as FES can improve the quality of life for thousands of individuals. The 1995 spinal cord injury of actor Christopher Reeves has brought publicity and additional funds to this area of research.

MULTIPLE SCLEROSIS *Page 345*

Multiple sclerosis (skler-Ō-sis; *sklerosis*, hardness), or **MS**, is a disease characterized by recurrent incidents of demyelination affecting axons in the optic nerve, brain, and/or spinal cord. Common symptoms include partial loss of vision and problems with speech, balance, and general motor coordination. The initial symptoms appear as the result of myelin degeneration within the white matter of the lateral and poste-

rior columns of the spinal cord or along tracts within the brain. For example, spinal cord involvement may produce weakness, tingling sensations, and a loss of "position sense" for the arms or legs. During subsequent attacks the effects become more widespread, and the cumulative sensory and motor losses may eventually lead to a generalized muscular paralysis.

The time between incidents and the degree of recovery vary from case to case. In about one-third of all cases the disorder is progressive, and each incident leaves a greater degree of functional impairment. The average age at the first attack is 30–40; the incidence in women is 1.5 times that among men.

Recent evidence suggests that this condition may be linked to a defect in the immune system that causes it to attack myelin sheaths. MS patients have lymphocytes that do not respond normally to foreign proteins, and because several viral proteins have amino acid sequences similar to those of normal myelin, it has been proposed that MS results from a case of mistaken identity. For unknown reasons MS appears to be associated with cold and temperate climates. It has been suggested that individuals developing MS may have an inherited susceptibility to the virus that is exaggerated by environmental conditions. The yearly incidence in the United States averages around 50 cases for every 100,000 in the population. There is as yet no effective treatment, but in several cases treatment with interferon has slowed the progress of the disease.

REFLEXES AND DIAGNOSTIC TESTING *Page 360*

Many reflexes can be assessed through careful observation and the use of simple tools. The procedures are easy to perform, and the results can provide valuable information about damage to the spinal cord or spinal nerves. By testing a series of spinal and cranial reflexes, a physician can assess the function of sensory pathways and motor centers throughout the spinal cord and brain.

Neurologists test many different reflexes; only a few are so generally useful that physicians make them part of a standard physical examination. The *knee jerk* (patellar), *ankle jerk*, and *biceps* and *triceps reflexes* are stretch reflexes controlled by specific segments of the spinal cord. Testing these reflexes provides information about the corresponding spinal segments.

Stroking an infant's foot on the side of the sole produces a fanning of the toes known as the **Babinski sign**, or *positive Babinski reflex*. This response disappears as descending inhibitory synapses develop. In the adult, the same stimulus produces a curling of the toes, called a **plantar reflex** or *negative Babinski reflex*, after about a 1-second delay. If either the higher centers or the descending tracts are damaged, the Babinski sign will reappear.

The **abdominal reflex**, present in the normal adult, results from descending spinal facilitation. In this reflex, a light stroking of the skin produces a reflexive twitch in the abdominal muscles that moves the navel toward the stimulus. This reflex disappears following damage to descending tracts.

In **hyporeflexia** normal reflexes are present but weak. In **areflexia** (ā-re-FLEK-sē-a; *a*-, without) normal reflexes cannot be detected. Hyporeflexia or areflexia may indicate temporary or permanent damage to skeletal muscles, dorsal or ventral nerve roots, spinal nerves, the spinal cord, or the brain.

Hyperreflexia occurs when higher centers maintain a high degree of stimulation along the spinal cord. Under these conditions reflexes are easily triggered, and the responses may be grossly exaggerated. This effect can also result from spinal cord compression or diseases that target higher centers or descending tracts. One potential result of hyperreflexia is the appearance of alternating contractions in opposing muscles. When one muscle contracts, it stimulates the stretch receptors in the other. The stretch reflex then triggers a contraction in that muscle, and this contraction stretches receptors in the original muscle. This self-perpetuating sequence can be repeated indefinitely; it is called **clonus** (KLŌ-nus).

A more extreme hyperreflexia develops if the motor neurons of the spinal cord lose contact with higher centers. Often, following a severe spinal injury, the individual first experiences a temporary period of areflexia known as *spinal shock*. ∞ [p. 345] When the reflexes return, they respond in an exaggerated fashion, even to mild stimuli. The reflex contractions may occur in a series of intense muscle spasms potentially strong enough to break bones. In the **mass reflex** the entire spinal cord becomes hyperactive for several minutes, issuing exaggerated skeletal muscle and visceral motor commands.

chapter 15

The Nervous System: The Brain and Cranial Nerves

CRANIAL TRAUMA *Page 368*

Cranial trauma is a head injury resulting from violent contact with another object. Head injuries account for over half of the deaths attributed to trauma each year. There are roughly 8 million cases of cranial trauma annually, and over a million of these are major incidents involving intracranial hemorrhaging, **concussion**, contusion, or laceration of the brain. The characteristics of spinal concussion, contusion, and laceration are introduced in the Clinical Brief on *Epidural and Subdural Hemorrhages* on p. 368 of the text, and the same descriptions can be applied to injuries of the brain.

Concussions usually accompany even minor head injuries. A concussion involves a temporary loss of consciousness and some degree of amnesia. Physicians examine any concussed individual quite closely and may X-ray the skull to check for skull fractures or cranial bleeding. Mild concussions produce a brief interruption of consciousness and little memory loss. Severe concussions produce extended periods of unconsciousness and abnormal neurological functions. Severe concussions are often associated with contusions (bruises) or lacerations (tears); the possibilities for recovery vary depending on the areas affected. Extensive damage to the reticular formation may produce a permanent state of unconsciousness, while damage to the lower brain stem will usually prove fatal.

CEREBELLAR DYSFUNCTION *Page 387*

Cerebellar function may be permanently altered by trauma or a stroke, or temporarily by drugs such as alcohol. Such alterations can produce severe disturbances in motor control. **Ataxia** (*ataxia*, a lack of order) refers to the disturbance of balance that in severe cases leaves the individual unable to stand without assistance. Less severe conditions cause an obvious unsteadiness and irregular patterns of movement. The individual often watches his or her feet to see where they are going and controls ongoing movements by intense concentration and voluntary effort. Reaching for something becomes a major exertion, for the only information available must be gathered by sight or touch while the movement is taking place. Without the cerebellar ability to adjust movements while they are occurring, the individual becomes unable to anticipate the time course of a movement. Most often, a reaching movement ends with the hand overshooting the target. This inability to anticipate and stop a movement precisely is called **dysmetria** (dis-MET-rē-a; *dys-*, bad + *metron*, measure). When the person attempts to correct the situation, the hand usually overshoots again in the opposite direction, and then again. The hand continues to oscillate back and forth until either the object can be grasped or the attempt is abandoned. This oscillatory movement is known as an **intention tremor**.

Clinicians check for ataxia by watching an individual walk in a straight line; the usual test for dysmetria involves touching the tip of the index finger to the tip of the nose. Because many drugs impair cerebellar performance, the same tests are used by police officers to check drivers suspected of alcohol or other drug abuse.

chapter 16

The Nervous System: Pathways and Higher-Order Functions

TAY-SACHS DISEASE *Page 415*

Tay-Sachs disease is a genetic abnormality involving the metabolism of *gangliosides*, important components of nerve cell membranes. Victims lack an enzyme needed to break down one particular ganglioside, which accumulates within the lysosomes of CNS neurons and causes them to deteriorate. Affected infants seem normal at birth, but within 6 months neurological problems begin to appear. The progress of symptoms typically includes muscular weakness, blindness, seizures, and death, usually before age 4. No effective treatment exists, but prospective parents can be tested to determine whether they are carrying the gene responsible for this condition. The disorder is most prevalent in one ethnic group, the Ashkenazi Jews of Eastern Europe.

HUNTINGTON'S DISEASE *Page 419*

Huntington's disease (*Huntington's chorea*) is an inherited disease marked by a progressive deterioration of mental abilities and pronounced movement disorders. There are racial differences in the incidence of this condition; Caucasians have by far the highest incidence, 3–4 cases per million population.

In Huntington's disease the cerebral nuclei show degenerative changes, as do the frontal lobes of the cerebral cortex. The basic problem is the destruction of neurons secreting ACh and GABA (an inhibitory neurotransmitter) in the cerebral nuclei. The cause of this deterioration is not known. The first signs of the disease usually appear in early adulthood. As you would expect in view of the areas affected, the symptoms involve difficulties in performing voluntary and involuntary patterns of movement and a gradual decline in intellectual abilities leading eventually to dementia. Screening tests can now detect the presence of the gene for Huntington's disease, which is located on chromosome 4. However, no effective treatment is available.

chapter 18

The Nervous System: General and Special Senses

THE CONTROL OF PAIN *Page 448*

Pain management poses a number of problems for clinicians. Painful sensations can result from tissue damage or sensory nerve irritation; it may originate where it is perceived, be referred from another location, or represent a "false" signal generated along the sensory pathway. The treatment differs in each case, and an accurate diagnosis is an essential first step.

When pain results from tissue damage, the most effective solution is to stop the damage and end the stimulation. This solution is not always possible. Alternatively, the painful sensations can be suppressed at the injury site. Topical or locally injected anesthetics inactivate nociceptors in the immediate area. Aspirin and related analgesics reduce inflammation and suppress the release of irritating chemicals, such as enzymes or prostaglandins, in damaged tissues.

Pain can also be suppressed by inhibition of the pain pathway. Analgesics related to morphine reduce pain by mimicking the action of *endorphins*, pain-relieving chemicals released inside the CNS. Surgical steps can be taken to control severe pain, including: (1) the sensory innervation of an area can be destroyed by an electric current; (2) the dorsal roots carrying the painful sensations can be cut (a **rhizotomy**); (3) the ascending tracts can be severed (a **tractotomy**); or (4) thalamic or limbic centers can be stimulated or destroyed. These options, listed in order of increasing degree of effect, surgical complexity, and associated risk, are used only when other methods of pain control have failed to provide relief.

Many aspects of pain generation and control remain a mystery. Some patients experience a significant reduction in pain after receiving a nonfunctional medication. It has been suggested that this "placebo effect" results from endorphin release triggered by the expectation of pain relief. The Chinese technique of acupuncture to control pain has recently received considerable attention. Fine needles are inserted at specific locations and are either heated or twirled by the therapist. Several theories have been proposed to account for the positive effects, but none is widely accepted. The acupuncture points do not correspond to the distribution of any of the major peripheral nerves, and pain relief may come from the central release of endorphins. Developing a second source of pain or sensation may reduce the perception of pre-existing pain elsewhere in the body. Perhaps acupuncture and old-fashioned "mustard plastards" work in this way.

ASSESSMENT OF TACTILE SENSITIVITIES *Page 450*

Because tactile sensitivities can be affected by various pathological conditions or damage to neural pathways, mapping tactile responses can sometimes aid clinical assessment. Sensory losses with clear regional boundaries indicate trauma to spinal nerves. For example, sensory loss along a dermatomal boundary (see Figure 14–8●, p. 347) can permit a reasonably precise determination of the affected spinal nerve(s). More widespread sensory loss may indicate damage to ascending tracts in the spinal cord. **Anesthesia** implies a total loss of sensation; **hypesthesia**, a reduction in sensitivity; and **paresthesia**, abnormal sensations, such as the pins-and-needles sensation when an arm or leg "falls asleep" because of pressure on a peripheral nerve.

Regional sensitivity to light touch can be checked by gentle contact with a fingertip or a slender wisp of cotton. The **two-point discrimination test** provides a more detailed sensory map for tactile receptors. Two fine points of a drawing compass, bent paper clip, or other object are applied to the skin surface simultaneously. The subject then describes the contact. When the points fall within a single receptive field, the individual will report only one point of contact. A normal individual loses two-point discrimination at 1 mm (0.04 in.) on the surface of the tongue, at 2–3 mm (0.08–0.12 in.) on the lips, at 3–5 mm (0.12–0.20 in.) on the backs of the hands and feet, and at 4–7 cm (1.6–2.75 in.) over the general body surface.

Vibration receptors are tested by applying the base of a tuning fork to the skin. Damage to an individual spinal nerve produces insensitivity to vibration along the paths of the related sensory nerves. If the sensory loss results from spinal cord damage, the injury site can often be located by walking the tuning fork down the spinal column, and resting its base on the vertebral spines.

VERTIGO, DIZZINESS, AND MOTION SICKNESS *Page 461*

The term **vertigo** describes an inappropriate sense of motion. This description distinguishes it from "**dizziness**," a sensation of light-headedness and disorientation that often precedes a fainting spell. Vertigo can result from disturbances in central processing or abnormal conditions at the peripheral receptors.

Any event that sets endolymph into motion can stimulate the equilibrium receptors. Placing an ice pack in contact with the temporal bone or flushing the external auditory canal with cold water may chill the endolymph in the outermost portions of the labyrinth and establish a temperature-related circulation of fluid. A mild and temporary vertigo is the result. Consumption of excessive quantities of alcohol and exposure to certain drugs can also produce vertigo by changing the composition of the endolymph or disturbing the hair cells.

In **Ménière's disease**, distortion of the membranous labyrinth by high fluid pressures may rupture the membranous wall and mix endolymph and perilymph together. The receptors in the vestibule and semicircular canals then become highly stimulated, and the individual may be unable to start a voluntary movement because of the intense spinning or rolling sensations experienced. In addition to the vertigo, the victim may "hear" unusual sounds as the cochlear receptors are activated. Other causes of vertigo include viral infection of the vestibular nerve and damage to the vestibular nucleus or its tracts.

The exceedingly unpleasant symptoms of **motion sickness** include headache, sweating, flushing of the face, nausea, vomiting, and various changes in mental perspective. (Sufferers may go from a state of giddy excitement to almost suicidal despair in a matter of moments.) It has been suggested that the condition results when central processing stations, such as the mesencephalic tectum, receive conflicting sensory information. Sitting belowdecks on a moving boat or reading a magazine in a car or airplane often provides the necessary conditions. Your eyes report that your position in space is not changing, but your labyrinthine receptors detect every bump and roll. As a result, "seasick" sailors watch the horizon rather than their immediate surroundings, so that the eyes will provide visual confirmation of the movements detected by the inner ear. It is not known why some individuals are almost immune to motion sickness, whereas others find travel by boat or plane almost impossible. Even space shuttle astronauts are not immune. Roughly half of them develop acute space-sickness that can limit their activities for 1–2 days. NASA research continues and may soon help sailors as well as astronauts.

TESTING AND TREATING HEARING DEFICITS *Page 464*

In the most common hearing test, a subject listens to sounds of varying frequency and intensity generated at irregular intervals. A record is kept of the responses, and the graphed record, or **audiogram**, is compared with that of an individual with "normal" hearing.

Bone conduction tests are used to discriminate between conductive and nerve deafness. If you put your fingers in your ears and talk quietly, you can still hear yourself because the bones of the skull conduct the sound waves to the cochlea, bypassing the middle ear. In a bone conduction test the physician places a vibrating tuning fork against the skull. If the subject hears the sound of the tuning fork in contact with the skull, but not when held next to the auditory meatus, the problem must lie within the external or middle ear. If the subject remains unresponsive to either stimulus, the problem must be at the receptors or along the auditory pathway.

Several effective treatments exist for conductive deafness. A hearing aid overcomes the loss in sensitivity by simply increasing the intensity of stimulation. Surgery may repair the tympanic membrane or free damaged or immobilized ossicles. Artificial ossicles may also be implanted if the originals are damaged beyond repair.

There are few possible treatments for nerve deafness. Mild conditions may be overcome by the use of a hearing aid if some functional hair cells remain. In a **cochlear implant** a small battery-powered device is inserted beneath the skin behind the mastoid process. Small wires run through the round window to reach the cochlear nerve, and when the implant "hears" a sound it stimulates the nerve directly. Increasing the number of wires and varying their implantation sites make it possible to create a number of different frequency sensations. Those sensations do not approximate normal hearing because there is as yet no way to target the specific afferent fibers responsible for the perception of a particular sound. Instead, a random assortment of afferent fibers are stimulated, and the individual must learn to recognize the meaning and probable origin of the perceived sound. Recently, scientists have been able to stimulate regrowth of damaged hair cells in chickens and hamsters. If this procedure can be extended to humans, a major form of nerve deafness may be treated.

SCOTOMAS AND FLOATERS *Page 472*

Abnormal blind spots, or **scotomas**, that are fixed in position may result from compression of the optic nerve, damage to retinal photoreceptors, or central damage along the visual pathway. Scotomas are abnormal, permanent features of the visual field. Most readers will probably be more familiar with floaters, small spots that drift across the field of vision. Floaters are common, temporary phenomena that result from blood cells or cellular debris within the vitreous body. They can be detected by staring at a blank wall or a white sheet of paper.

chapter 19

The Endocrine System

DIABETES INSIPIDUS *Page 488*

There are several different forms of **diabetes**, all characterized by excessive urine production (**polyuria**). Although diabetes can be caused by physical damage to the kidneys, most forms are the result of endocrine abnormalities. The two most important forms are *diabetes insipidus*, considered here, and *diabetes mellitus*, considered later.

Diabetes insipidus develops when the posterior pituitary no longer releases adequate amounts of antidiuretic hormone (ADH). Water conservation at the kidneys is impaired, and excessive amounts of water are lost in the urine. As a result, an individual with diabetes insipidus is constantly thirsty, but the fluids consumed are not retained by the body. Mild cases may not require treatment, as long as fluid and electrolyte intake keep pace with urinary losses. In severe diabetes insipidus the fluid losses can reach 10 ℓ per day, and a fatal dehydration will occur unless treatment is provided. One innovative treatment method involves administering a synthetic form of ADH, *desmopressin acetate* (DDAVP) in a nasal spray. The drug enters the bloodstream after diffusing through the nasal epithelium.

THYROID GLAND DISORDERS *Page 491*

Normal production of thyroid hormones establishes the background rates of cellular metabolism. These hormones exert their primary effects on large, metabolically active tissues and organs, including skeletal muscles, the liver, and the kidneys. Inadequate production of thyroid hormones, or **hypothyroidism**, in an infant is marked by inadequate skeletal and nervous development and a metabolic rate as much as 40% below normal levels. This condition affects approximately 1 birth out of every 5000. Hypothyroidism beginning later in childhood will retard growth and delay puberty. Adults with this condition are lethargic and unable to adjust to cold temperatures. The symptoms, collectively known as **myxedema** (miks-e-DĒ-ma), include subcutaneous swelling, dry skin, hair loss, low body temperature, muscular weakness, and slowed reflexes.

Hypothyroidism usually results from some problem involving the thyroid gland rather than with pituitary production of thyroid-stimulating hormone (TSH). One useful testing procedure involves monitoring the rate of iodine uptake by the thyroid gland, which provides an indication of its functional capabilities. Thyroid hormone and TSH levels may also be analyzed. Treatment involves the administration of synthetic thyroid hormones to maintain normal blood concentrations.

A **goiter** is an enlargement of the thyroid gland. That enlargement may not necessarily indicate increased production of thyroid hormones, but merely an increase in follicular size. One form of goiter occurs if the thyroid fails to obtain enough iodine to meet its synthetic requirements. Under TSH stimulation, follicle cells produce large quantities of thyroglobulin, but they are unable to provide the iodine needed to create functional thyroid hormones. As a result, TSH levels rise further, and the thyroid gland begins to enlarge. Administering iodine may not solve the problem, for the sudden availability of iodine may produce symptoms of *hyperthyroidism* (see below) as the stored thyroglobulin becomes activated. The usual therapy involves the ingestion of thyroxine, which feeds back on the hypothalamus and pituitary to inhibit the production of TSH. Over time the resting thyroid may return to its normal size and functional capabilities.

Thyrotoxicosis, or **hyperthyroidism**, occurs when thyroid hormones are produced in excessive quantities. The metabolic rate climbs, and the skin becomes flushed and moist with perspiration. Blood pressure and heart rate increase, and the heartbeat may become irregular as circulatory demands escalate. The effects on the CNS make the individual restless, excitable, and subject to shifts in mood and emotional states. Despite the drive for increased activity, the subject has limited energy reserves and fatigues easily.

In **Graves' disease** excessive thyroid activity leads to goiter and the symptoms of hyperthyroidism. Protrusion of the eyes, or **exophthalmos** (eks-ahf-THAL-mos) may also appear, for unknown reasons. Graves' disease has a genetic basis and may be an autoimmune disorder. It affects women much more often than men. Treatment may involve the use of antithyroid drugs, surgical removal of portions of the glandular mass, or destruction of part of the gland by exposure to radioactive iodine. President and Mrs. Bush both had Graves' disease while in the White House. Mr. Bush developed heart arrhythmias, and Mrs. Bush suffered exophthalmos. No environmental cause was found despite an intensive investigation.

Hyperthyroidism may also result from thyroid tumors, inflammation, or immune system disorders. In extreme cases the individual's metabolic processes accelerate out of control. During a **thyrotoxic crisis** the subject experiences an extremely high fever, rapid heart rate, and the malfunctioning of a variety of physiological systems.

DISORDERS OF PARATHYROID FUNCTION *Page 491*

When the parathyroid gland secretes inadequate or excessive amounts of parathyroid thormone (PTH), calcium concentrations move outside of normal homeostatic limits. Inadequate PTH production, a condition called **hypoparathyroidism**, leads to low calcium concentrations in body fluids. The most obvious symptoms involve neural and muscle tissues, where calcium ions have important functions. The nervous system becomes more excitable, and the affected individual may experience hypocalcemic tetany, characterized by spasms in the muscles of the arms, hands, and face. Hypoparathyroidism may develop after neck surgery if the blood supply to these glands is restricted. In many other cases the primary cause of the condition is uncertain. Treatment is difficult because at present PTH can be obtained only by extraction from the blood of normal individuals. Thus PTH is extremely costly, and because supplies are very limited, PTH administration is not used to treat this condition, despite its probable effectiveness. As an alternative, a dietary combination of vitamin D and calcium can be used to elevate body fluid calcium concentrations. (Vitamin D stimulates the absorption of calcium ions across the lining of the digestive tract.)

DISORDERS OF THE ADRENAL CORTEX *Page 494*

Clinical problems related to the adrenal gland vary depending on which of the adrenal zones becomes involved. The conditions may result from changes in the functional capabilities of the adrenal cells (primary conditions) or disorders affecting the regulatory mechanisms (secondary conditions). In **hypoaldosteronism** the zona glomerulosa fails to produce enough aldosterone, usually because the kidneys are not releasing adequate amounts of renin. Affected individuals lose excessive amounts of water and sodium ions at the kidneys, and the water loss leads to low blood volume and pressure. Changes in electrolyte concentrations affect transmembrane potentials, eventually causing dysfunctions in neural and muscular tissues.

Hypersecretion of aldosterone results in the condition of **aldosteronism**. Under continued aldosterone stimulation, the kidneys retain sodium ions very effectively, but potassium ions are lost in large quantities. In response, potassium ions move out of the cells and into the interstitial fluids, only to be lost in turn. A crisis eventually develops when low extracellular potassium concentrations disrupt normal cardiac, neural, and kidney cell functions.

Addison's disease results from the inability of cells in the zona fasciculata to respond to ACTH. Affected individuals become weak and lose weight because of a combination of appetite loss and digestive tract malfunctions. They cannot adequately mobilize energy reserves, and their blood glucose concentrations fall sharply within hours after a meal. Stresses cannot be tolerated, and a minor infection or injury may lead to a sharp and fatal decline in blood pressure. A particularly interesting symptom is the increased melanin pigmentation in the skin. The ACTH molecule and the melanocyte-stimulating hormone (MSH) molecule are similar in structure, and at high concentrations ACTH stimulates the MSH receptors on melanocytes.

Cushing's disease results from overproduction of glucocorticoids. The symptoms resemble those of a protracted and exaggerated response to stress. Glucose metabolism is suppressed in most tissues, and although blood glucose levels climb, lipid reserves are mobilized, and peripheral proteins are broken down. Lipids and amino acids are mobilized in excess of the existing demand. The energy reserves are shuffled around, and the distribution of body fat changes. Adipose tissues in the cheeks and around the base of the neck become enlarged at the expense of other areas, producing a "moon-faced" appearance. The demand for amino acids falls most heavily on the skeletal muscles, which respond by breaking down their contractile proteins. This response reduces muscular strength and endurance.

The chronic administration of large doses of steroids can produce symptoms similar to those of Cushing's disease, but such treatment is usually avoided. Roughly 75% of cases result from an overproduction of ACTH, and afflicted individuals may also show changes in skin pigmentation. If the problem stems from a tumor within the adrenal cortex, the zona reticularis may also be affected. The overproduction of androgens then produces symptoms of the **androgenital syndrome**. In women this condition leads to the gradual development of male secondary sexual characteristics, including body and facial hair patterns, adipose tissue distribution, and muscular development. Tumors affecting the zona reticularis of males may sometimes result in the production of large quantities of estrogens. In this condition, **gynecomastia** (*gynaikos*, woman + *mastos*, breast), the individual develops female secondary sexual characteristics.

DISORDERS OF THE ADRENAL MEDULLA *Page 494*

A **pheochromocytoma** (fē-ō-krō-mō-sī-TŌ-mah) is a tumor that produces catecholamines in massive quantities. The tumor usually develops within the adrenal medulla, but it may also involve other sympathetic ganglia. The most dangerous symptoms are rapid and irregular heartbeat and high blood pressure; other symptoms include uneasiness, sweating, blurred vision, and headaches. This condition is rare, and surgical removal of the tumor is the most effective treatment.

DIABETES MELLITUS *Page 495*

Diabetes mellitus (mel-LĪ-tus; *mellitum*, honey) is characterized by glucose concentrations in the blood that are high enough to overwhelm the reabsorption capabilities of the kidneys. Glucose appears in the urine (**glycosuria**), and urine production becomes excessive (polyuria). Other metabolic products, such as fatty acids and other lipids, are also present in abnormal concentrations.

Diabetes mellitus may be caused by genetic abnormalities, pathological conditions, injuries, immune disorders, or hormonal imbalances. There are two major types of diabetes mellitus: *insulin-dependent* and *non-insulin-dependent*.

In **insulin-dependent diabetes mellitus (IDDM)**, or **Type I diabetes**, the primary cause is inadequate insulin production by the beta cells of the pancreatic islets. Type I diabetes most often appears in individuals under 40 years of age. Because it frequently appears in childhood, it has been called **juvenile-onset diabetes**.

When insulin concentrations decline, cells can no longer absorb glucose from their surroundings, and they remain glucose-starved despite the presence of adequate or even excessive amounts of glucose in the circulation. After a meal rich in glucose, blood concentrations may become so elevated that the kidney cells cannot reclaim all of the glucose molecules entering the urine. The high urinary concentration of glucose limits the ability of the kidney to conserve water, so the individual urinates frequently and may become dehydrated. The chronic dehydration leads to disturbances of neural function (blurred vision, tingling sensations, disorientation, fatigue) and muscle weakness.

Despite high blood concentrations, glucose cannot enter endocrine tissues, and the endocrine system responds as if glucose were in short supply. Alpha cells release glucagon, and glucocorticoid production accelerates. Peripheral tissues then break down lipids and proteins to obtain the energy needed to continue functioning. The breakdown of fatty acids leads to the generation of molecules called **ketone bodies**. These are metabolic acids whose accumulation can cause a dangerous reduction in blood pH. This condition, called **ketoacidosis**, often triggers vomiting, and in severe cases it can precede a fatal coma.

If the individual survives, an impossibility without insulin therapy, long-term treatment involves a combination of dietary control and administration of insulin, either by injection or by infusion using an **insulin pump**. The treatment is complicated by the fact that tissue glucose demands cycle up and down, depending on meals, physical activity, emotional state, stress, and other factors that are hard to assess or predict. It is therefore difficult to maintain stable and normal blood glucose levels over long periods of time. The best results to date have been achieved by monitoring blood glucose concentrations closely and administering insulin injections up to 5–7 times daily.

Non-insulin-dependent diabetes mellitus (NIDDM), or **Type II diabetes**, typically affects obese individuals over 40 years of age. Because of the age factor this condition is also called **maturity-onset diabetes**. Insulin levels are normal or elevated, but peripheral tissues no longer respond normally. Treatment consists of weight loss and dietary restrictions that may elevate insulin production and tissue response. Oral medications can help when weight loss and dietary restrictions fail.

These forms of diabetes mellitus may affect 1% of the U.S. population. Maturity-onset diabetes is roughly three times as common as insulin-dependent diabetes. Even with treatment, patients with diabetes mellitus often develop chronic medical problems. In general, these problems are related to circulatory system abnormalities. The most common examples include the following:

1. Vascular changes at the retina, including proliferation of capillaries and hemorrhaging, often cause disturbances of vision. This condition is called **diabetic retinopathy**.

2. Changes occur in the clarity of the lens, producing cataracts.

3. Small hemorrhages and inflammation at the kidneys cause degenerative changes that can lead to kidney failure. This condition is called **diabetic nephropathy**.

4. A variety of neural problems appear, including nerve palsies, paresthesias, and autonomic dysfunction. These disorders, collectively termed **diabetic neuropathy**, are probably related to disturbances in the blood supply to neural tissues.

5. Degenerative changes in cardiac circulation can lead to early heart attacks. For a given age group, heart attacks are 3–5 times more likely in diabetic individuals.

6. Other peripheral changes in the vascular system can disrupt normal circulation to the extremities. For example, a reduction in blood flow to the feet can lead to tissue death, ulceration, infection, and loss of toes or a major portion of one or both feet.

ENDOCRINE DISORDERS *Page 500*

Endocrine disorders fall into two basic categories: they reflect either abnormal hormone production (too much or too little) or abnormal cellular sensitivity (extremely sensitive or insensitive). The symptoms are interesting because they highlight the significance of normally "silent" hormonal contributions. An abbreviated summary is presented in Table A-4 on the next page.

chapter 20

The Cardiovascular System: Blood

BLOOD TESTS AND RBCs *Page 508*

Blood tests provide information about the general health of an individual, usually with a minimum of trouble and expense. Several common blood tests focus attention on red blood cells (RBCs), the most common and abundant cellular elements. This section describes several common examples.

Reticulocyte Count *Reticulocytes* are immature red blood cells that are still synthesizing hemoglobin. Most reticulocytes remain in the bone marrow until they complete their maturation, but some enter the circulation. Reticulocytes normally account for around 0.8% of the circulating erythrocyte population. Values above 1.5% or below 0.5% indicate that something is wrong with the rates of RBC survival or maturation.

Hematocrit (Hct) The hematocrit value is the percentage of whole blood occupied by cells. Normal adult hematocrits average 46 for men and 42 for women, with ranges of 40–54 for men and 37–47 for women. Because the numbers of RBCs far exceed those of white blood cells (WBCs), the hematocrit is normally used as a monitor for circulating erythrocytes. Although an abnormal hematocrit indicates that a problem exists, additional tests are needed to make a more definitive diagnosis. These procedures examine the size, age, abundance, and hemoglobin content of the erythrocytes.

Hemoglobin (Hb) Concentration The hemoglobin concentration test determines the amount of hemoglobin in the blood, expressed in grams per 100 mℓ (g/dℓ). Normal values range from 14 to 18 g/dℓ in males and 12 to 16 g/dℓ in females. The differences in hemoglobin concentration reflect the differences in hematocrit. For both sexes, a single RBC contains 27–33 picograms (pg) of hemoglobin. RBCs containing normal amounts of hemoglobin are termed **normochromic**, while **hyperchromic** and **hypochromic** indicate higher and lower than normal hemoglobin content, respectively.

RBC Count The normal number of RBCs per microliter of blood is approximately 5.2 million. Calculations based on the hematocrit, hemoglobin content, and RBC count can be used to develop a better picture of the condition of the RBCs. Values often reported in blood screens include mean corpuscular volume and hemoglobin concentration.

Mean Corpuscular Volume (MCV) The mean volume of an individual red blood cell, in cubic micrometers is calculated by dividing the volume of red cells per microliter by the RBC count, using the formula:

$$\text{MCV} = \frac{\text{Hct}}{\text{RBC count (in millions)}} \times 10$$

Normal values range from 82.2 to 100.6 μm^3. Using the values given above, the mean corpuscular volume would be:

$$\text{MCV} = \frac{46}{5.2} \times 10 = 88.5 \ \mu\text{m}^3$$

Cells of normal size are **normocytic**, while larger or smaller than normal RBCs are called **macrocytic** or **microcytic**, respectively.

TABLE A-4 Clinical Implications of Endocrine Malfunctions

Hormone	Underproduction Syndrome	Principal Symptoms	Overproduction Syndrome	Principal Symptoms
Growth hormone (GH)	Pituitary growth failure (p. 120)	Retarded growth, abnormal fat distribution, low blood glucose hours after a meal	Gigantism, acromegaly (p. 120)	Excessive growth
Antidiuretic hormone (ADH)	Diabetes insipidus ♆(p. 752)	Polyuria	SIADH (syndrome of inappropriate ADH secretion	Increased body water content and hyponatremia
Thyroxine (TX, T$_4$)	Myxedema, cretinism ♆(p. 753)	Low metabolic rate, body temperature; impaired physical and mental development	Graves' disease ♆(p. 753)	High metabolic rate, body temperature
Parathyroid hormone (PTH)	Hypoparathyroidism ♆(p. 753)	Muscular weakness, neurological problems, tetany due to low blood calcium concentrations	Hyperparathyroidism	Neurological, mental, muscular problems due to high blood calcium concentrations; weak and brittle bones
Insulin	Diabetes mellitus (Type I) ♆(p. 754)	High blood glucose, impaired glucose utilization, dependence on lipids for energy, glycosuria	Excess insulin production or administration	Low blood glucose levels, possibly causing coma
Mineralocorticoids (MC)	Hypoaldosteronism ♆(p. 753)	Polyuria, low blood volume, high blood potassium concentrations	Aldosteronism ♆(p. 753)	Increased body weight due to water retention, low blood potassium concentration
Glucocorticoids (GC)	Addison's disease ♆(p. 753)	Inability to tolerate stress, mobilize energy reserves, maintain normal blood glucose concentrations	Cushing's disease ♆(p. 753)	Excessive breakdown of tissue proteins and lipid reserves, impaired glucose metabolism
Epinephrine (E), norepinephrine (NE)	None identified		Pheochromocytoma ♆(p. 753)	High metabolic rate, body temperature, and heart rate; elevated blood glucose levels; other symptoms comparable to those of excessive autonomic stimulation
Estrogens (female)	Hypogonadism	Sterility, lack of secondary sexual characteristics	Androgenital syndrome ♆(p. 753)	Overproduction of androgens by zona reticularis of adrenal leads to masculinization
	Menopause	Cessation of ovulation	Precocious puberty	Early production of developing follicles and estrogen secretion
Androgens (male)	Hypogonadism, eunuchoidism	Sterility, lack of secondary sexual characteristics	Gynecomastia ♆(p. 753)	Abnormal production of estrogens, sometimes due to adrenal or interstitial cell tumors, leads to breast enlargement
			Precocious puberty (p. 494)	Early production of androgens, leading to premature physical development and behavioral changes

Mean Corpuscular Hemoglobin Concentration (MCHC) The amount of hemoglobin within a single RBC is expressed in picograms. Normal values range from 27 to 34 pg, calculated as:

$$\text{MCHC} = \frac{\text{Hb}}{\text{RBC count (in millions)}} \times 10$$

As an example, Table A-5 shows how this information can be used to distinguish among four major types of anemia.

TABLE A-5 RBC Tests and Anemias

Anemia type	Hct	Hb	Reticulocyte count	MCV	MCHC
Hemorrhagic	low	low	normal	normal	normal
Aplastic	low	low	very low	normal	normal
Iron deficiency	low	low	normal or low	low	low
Pernicious	low	low	very low	high	high

1. **Hemorrhagic anemia** results from severe bleeding. Erythrocytes are of normal size, each contains a normal amount of hemoglobin, and reticulocytes are present in normal concentrations, at least initially. Blood tests would therefore show a low hematocrit and low hemoglobin, but the MCV, MCHC, and reticulocyte counts would be normal.

2. In **aplastic** (ā-PLAS-tik) **anemia** the bone marrow fails to produce new red blood cells. The 1986 nuclear accident in Chernobyl (in the former U.S.S.R.) caused a number of cases of aplastic anemia. The condition is usually fatal unless surviving stem cells repopulate the marrow or a *bone marrow transplant* is performed. In aplastic anemia the circulating red blood cells are normal in all respects, but because new RBCs are not being produced the reticulocyte count is extremely low.

3. In **iron deficiency anemia**, normal hemoglobin synthesis cannot occur because iron reserves are inadequate. Developing red blood cells cannot synthesize functional hemoglobin, and as a result they are unusually small. A blood test therefore shows a low hematocrit, low hemoglobin content, low MCV, and low MCHC, but often a normal reticulocyte count.

4. Finally, in **pernicious** (per-NISH-us) **anemia** normal red blood cell maturation ceases because of an inadequate supply of vitamin B_{12}. Erythrocyte production declines, and the red blood cells are abnormally large and may develop a variety of bizarre shapes. Blood tests from a person with pernicious anemia indicate a low hematocrit with a very high MCV and a low reticulocyte count.

SICKLE CELL ANEMIA *Page 510*

Sickle cell anemia affects roughly 0.14% of the African-American population in the United States. This condition results from a change in the amino acid sequence in one of the globin chains of the hemoglobin molecule. When the blood contains an abundance of oxygen, the hemoglobin molecules and the erythrocytes that carry them appear normal. But when the defective hemoglobin gives up enough of its stored oxygen, the complex of hemoglobin molecules within the red blood cells changes shape, and the cells become stiff and markedly curved.

This "sickling" does not affect the oxygen-carrying capabilities of the erythrocytes, but it causes them to become more fragile and easily damaged. Moreover, an RBC that has folded to squeeze into a narrow capillary may lose its oxygen, change shape, and become stuck. A circulatory blockage results, and the nearby tissues become oxygen-starved. Symptoms of this condition include pain and damage to a variety of organs and systems, depending on the location of the obstructions. In addition, the trapped red blood cells eventually die and break down, producing a characteristic anemia. Transfusions of normal blood can temporarily prevent additional complications, and there are experimental drugs that can control or reduce sickling.

This disease is genetically associated with resistance to another disease. To develop sickle cell anemia an individual must have two copies of the sickle cell gene, one from each parent. If only one sickling gene is present, the individual is said to have the **sickling trait**. In such cases most of the hemoglobin is of the normal form, and the erythrocytes function normally. But the presence of the abnormal hemoglobin somehow gives the individual the ability to resist the parasitic infections that produce the symptoms of **malaria**. The malaria parasites enter the bloodstream when an individual is bitten by an infected mosquito. In a normal person the microorganisms then invade and reproduce within the erythrocytes. But when they enter an erythrocyte from a person with the sickling trait, the cell responds by sickling. Either the sickling itself kills the parasite, or the sickling attracts the attention of a phagocyte. In either event the individual remains unaffected by the disease, while normal individuals sicken and often die. Roughly 8% of the African-American population carries a single gene for sickle cell anemia; in parts of Africa where malaria is common the percentage is 3–4 times higher.

HEMOLYTIC DISEASE OF THE NEWBORN *Page 511*

Genes controlling the presence or absence of any agglutinogen in the erythrocyte membrane are provided by both parents, so a child may have a blood type different from that of either parent. During pregnancy, when fetal and maternal circulatory systems are closely intertwined, the mother's agglutinins may cross the placental barrier, attacking and destroying fetal red blood cells. The resulting condition is called **hemolytic disease of the newborn**, or **HDN**.

There are many forms of HDN, some so mild as to remain undetected, but those involving the Rh agglutinogen are potentially quite dangerous. Because HDN results from the passage of *maternal* agglutinins across the placental barrier, an Rh-positive mother (who lacks anti-Rh agglutinins) can carry an Rh-negative fetus without difficulty. Potential problems appear when an Rh-negative woman carries an Rh-positive fetus. Sensitization usually occurs at delivery, when bleeding occurs at the placenta and uterus and fetal blood enters the maternal circulation. This event can trigger the maternal production of anti-Rh agglutinins. Within 6 months of delivery, roughly 20% of Rh-negative mothers who carried Rh-positive children have become sensitized.

Because the anti-Rh agglutinins are not produced in significant amounts until after delivery, this first infant will not be affected. But if a second pregnancy occurs involving an Rh-positive fetus, the mother will respond by producing massive amounts of anti-Rh agglutinins. These agglutinins then enter the fetal circulation, destroying fetal red blood cells and producing a dangerous fetal anemia. The fetal demand for blood cells increases, and they begin leaving the bone marrow and entering the circulation before completing their development. Because these immature RBCs are called **erythroblasts**, the condition is known as **erythroblastosis fetalis** (e-rith-rō-blas-TŌ-sis fē-TAL-is). Without treatment, the fetus will probably die before delivery or shortly thereafter.

The problem can be avoided completely by preventing the maternal production of agglutinins by the administration of anti-Rh agglutinins (available under the name *RhoGam*) during and after delivery. These "foreign" agglutinins destroy any fetal red blood cells that cross the placental barrier. Thus there are no exposed agglutinogens to stimulate the maternal immune system, sensitization does not occur, and anti-Rh agglutinins are not produced. This relatively simple procedure has almost entirely prevented HDN mortality caused by Rh incompatibilities.

THE LEUKEMIAS *Page 513*

Leukocytosis with white blood cell counts of 100,000 $\mu\ell$ or more usually indicates the presence of some form of **leukemia** (loo-KĒ-mē-ah). Leukemias characterized by the presence of abnormal granulocytes or other cells of the bone marrow are called **myeloid**, and those involving lymphocytes are termed **lymphoid**. The first symptoms appear as immature and abnormal white blood cells appear in the circulation. As their numbers increase they travel via the circulation, invading tissues and organs throughout the body.

These cells are extremely active, and they require abnormally large amounts of energy. As in other cancers, invading leukemic cells gradually replace the normal cells, especially in the bone marrow. Red blood cell and platelet formation decline, with resulting anemia and impaired blood clotting, and untreated leukemias are invariably fatal.

Leukemias may be classified as *acute* (short and severe) or *chronic* (prolonged). Acute leukemias may be linked to radiation exposure, hereditary susceptibility, viral infections, or unknown causes. Chronic leukemias may be related to chromosomal abnormalities or immune system malfunctions. Survival in untreated acute leukemia averages about 3 months; individuals with chronic leukemia may survive for years.

Effective treatments exist for some forms of leukemia and not others. For example, when acute lymphoid leukemia is detected early, 85–90% of patients can be held in remission for 5 years or longer, but only 10–15% of patients with acute myeloid leukemia survive 5 years or more. The yearly mortality rate for leukemia (all types) in the United States has not declined appreciably in the past 30 years, remaining at around 6.8 per 100,000 population.

One option for treating acute leukemias is to perform a **bone marrow transplant**. In this procedure massive chemotherapy or radiation treatment is given, enough to kill all of the cancerous cells. Unfortunately, the patient's blood cells and stem cells in the bone marrow and other blood-forming tissues are also destroyed. The individual then receives an infusion of healthy bone marrow cells that repopulate the blood and marrow tissues.

If the bone marrow is extracted from another person (a **heterologous marrow transplant**), care must be taken to ensure that the blood types and tissue types are compatible. If they are not, the new lymphocytes may attack the patient's tissues, with potentially fatal results. Best results are obtained when the donor is a close relative. In an *autologous marrow* *transplant* bone marrow is removed from the patient, cleansed of cancer cells, and reintroduced after radiation or chemotherapy treatment. Although there are fewer complications, the preparation and cleansing of the marrow are technically difficult and time-consuming.

Bone marrow transplants are also performed to treat patients whose bone marrow has been destroyed by toxic chemicals or radiation. For example, heterologous transplants were used successfully in the former U.S.S.R. to treat survivors of the Chernobyl nuclear reactor accident in 1986.

PROBLEMS WITH THE CLOTTING PROCESS
Page 514

An overactive clotting system can cause problems as serious as those caused by an underactive one. The endothelial lining of blood vessels is very smooth and slick, but there are occasional bends and ripples where vessels branch or where adjacent epithelial cells interlock. If the clotting system becomes oversensitive, any of these irregular surfaces may break platelets apart or cause platelet aggregation. Each of these possibilities can have unpleasant consequences. When platelets break apart, clot formation begins in the circulation. These blood clots do not stick to the wall of the vessel but continue to drift around until plasmin digests them or they become stuck in a small blood vessel. A drifting blood clot is an example of an **embolus** (EM-bō-lus; *embolos*, plug). When an embolus becomes stuck in a blood vessel, it blocks circulation to the area downstream, killing the affected tissues. This condition is called an **embolism**. Embolus formation has been a recurring problem for individuals receiving artificial heart valves because clots form on the artificial surfaces. When these clots break free, they form emboli that may get stuck in capillaries in the brain, causing strokes. If an embolus forms in the venous system it will probably become lodged in one of the capillaries of the lungs, causing a condition known as a *pulmonary embolism*.

A **thrombus** (*thrombos*, clot) begins to form when platelets stick to the wall of an intact blood vessel. Often the platelets are attracted to areas called **plaques**, where endothelial and smooth muscle cells are storing large quantities of lipids. The blood clot gradually enlarges, projecting into the lumen of the vessel and reducing its diameter. Eventually the vessel may be completely blocked, or a large chunk of the clot may break off, creating an equally dangerous embolus.

Clinicians may attempt to prevent unwanted clotting by administering drugs that depress the clotting response or dissolve clots already present. Important anticoagulant drugs include **heparin** (HEP-a-rin), which inactivates thrombin, and **coumadin** (COO-ma-din), which depresses the synthesis of several clotting factors. Specific conditions, including pulmonary or venous embolisms and acute blockage of the coronary arteries can be treated by dissolving clots with the enzymes **streptokinase** (strep-tō-KĪ-nās), **urokinase** (ū-rō-KĪ-nās), or **tissue plasminogen activator** (t-PA). Aspirin, which stabilizes platelets, helps prevent thrombus formation and can prevent strokes and second heart attacks.

chapter 21

The Cardiovascular System: The Heart

THE CARDIOMYOPATHIES *Page 528*

The **cardiomyopathies** (kar-dē-ō-mī-OP-a-thēz) include an assortment of diseases with a common symptom: the progressive, irreversible degeneration of the myocardium. Cardiac muscle fibers are damaged and replaced by fibrous tissue, and the muscular walls of the heart become thin and weak. As muscle tone declines, the ventricular chambers become greatly enlarged. When the remaining fibers cannot develop enough force to maintain cardiac output, symptoms of heart failure develop.

Chronic alcoholism and coronary artery disease are probably the most common causes of cardiomyopathy in the United States. Infectious agents, including viruses, bacteria, fungi, and protozoans, can also produce cardiomyopathies. Diseases affecting neuromuscular performance, such as muscular dystrophy, can also damage cardiac muscle fibers, as can starva-

tion or chronic variations in the extracellular concentrations of calcium or potassium ions. Finally there are several inherited forms of cardiomyopathy, as well as a significant number of cases in which the primary cause cannot be determined.

Individuals suffering from severe cardiomyopathies may be considered as candidates for **heart transplants**. This surgery involves the complete removal of the weakened heart and its replacement by a heart taken from a suitable donor. To survive the surgery, the recipient must be in otherwise satisfactory health. Because the number of suitable donors is limited, the available hearts are usually assigned to individuals younger than age 50. Out of the 8000–10,000 U.S. patients each year suffering from potentially fatal cardiomyopathies, only around 1000 receive heart transplants. There is an 80–85% 1-year survival rate, and a 50–70% 5-year survival rate after successful transplantation. This rate is quite good, considering that these patients would have died if the transplant had not been performed.

Unfortunately, many individuals with cardiomyopathy who are initially selected for surgery succumb to the disease before a suitable donor becomes available. For this reason there continues to be considerable interest in the development of an artificial heart. One model, the Jarvik-7, had limited clinical use in the 1980s. Attempts to implant it on a permanent basis were unsuccessful, primarily because of formation of blood clots on the mechanical valves. When these clots broke free, they formed drifting emboli that plugged peripheral vessels, producing strokes, kidney failure, and other complications. In 1989 the federal government prohibited further experimental use of the Jarvik-7 as a permanent heart implant. Modified versions of this unit and others now under development may still be used to maintain transplant candidates while awaiting the arrival of a donor organ.

An interesting experimental procedure involves using skeletal muscle tissue to apply permanent patches to injured hearts or to build small accessory pumps. For example, one procedure involves freeing a portion of the latissimus dorsi muscle from the side. This flap, with its circulation intact, is moved into the thoracic cavity and folded to form a sling around the heart. A pacemaker is then used to stimulate the flap's contraction, and when it contracts it squeezes the heart and helps push blood into the major arteries. Such permanent patches are less stressful than heart transplants because (1) they leave the damaged heart in place, and (2) the transplanted tissue is taken from the same individual, so it will not be attacked by the immune system. Although preliminary results have been encouraging, few procedures have been performed, and the use of skeletal muscle patches remains an experimental concept rather than a recognized treatment for cardiomyopathy. A new treatment proposed instead of transplantation involves removing part of the dilated left ventricle. This apparently improves pumping efficiency significantly.

VALVULAR HEART DISEASE *Page 531*

Abnormalities in atrioventricular or semilunar valve function reduce the efficiency of the heart. Minor valve problems frequently go unnoticed, and some degree of regurgitation often occurs in otherwise normal individuals. When valve function deteriorates to the point that the heart cannot maintain adequate circulatory flow, symptoms of valvular heart disease appear. Congenital malformations may be responsible, but often the condition develops after *carditis*, an inflammation of the heart.

One common cause of carditis is *rheumatic* (roo-MA-tik) *fever*, an inflammatory condition that may develop after infection by streptococcal bacteria. Rheumatic fever most often affects children of ages 5–15; symptoms include high fever, joint pain and stiffness, and a distinctive full-body rash. Obvious symptoms usually persist for less than 6 weeks, although severe cases may linger for 6 months or more. The carditis that develops in 50–60% of patients often escapes detection, and scar tissue forms gradually in the myocardium and the heart valves. Valve condition deteriorates over time, and valve problems serious enough to affect cardiac function may not appear until 10–20 years after the initial infection.

Over the interim the affected valves become thickened and often calcified to some degree. This thickening narrows the opening guarded by the valves, producing a condition called **valvular stenosis** (ste-NŌ-sis; *stenos*, narrow). The resulting clinical disorder is known as *rheumatic heart disease* (*RHD*). The thickened cusps stiffen in a partially closed position, but the valves do not completely block the circulation, because the edges of the cusps are rough and irregular. Regurgitation may occur, and much of the blood pumped out of the heart may flow right back in. The ventricular musculature must work harder to maintain adequate circulation, and in severe cases it is not up to the task. The heart weakens, and peripheral tissues begin to suffer from oxygen and nutrient deprivation. This cycle ultimately leads to heart failure.

INFECTION AND INFLAMMATION OF THE HEART
Page 531

Many different microorganisms may infect heart tissue, leading to serious cardiac abnormalities. **Carditis** (kar-DĪ-tis) is a general term indicating inflammation of the heart. Clinical conditions resulting from cardiac infection are usually identified by the primary site of infection. For example, those affecting the endocardium produce symptoms of *endocarditis*. **Endocarditis** affects primarily the chordae tendineae and valves, and the mortality rate may reach 21–35%. The most severe complications result from the formation of blood clots on the damaged surfaces. These clots subsequently break free, entering the circulation as drifting emboli that may cause strokes, heart attacks, or kidney failure.

Bacteria, viruses, protozoans, and fungal pathogens that attack the myocardium produce **myocarditis**. The microorganisms implicated include those responsible for many of the conditions discussed elsewhere, including diphtheria, syphilis, and polio. The infected heart muscle becomes extremely sensitive, and the heart rate rises dramatically. Over time, abnormal contractions may appear, and these may prove fatal.

An equally great variety of pathogens can infect the pericardium, producing a condition called **pericarditis**. The inflamed pericardial layers rub against one another as the heart beats, producing a characteristic scratching sound. In addition, pericardial irritation and inflammation often results in an increased production of pericardial fluid. Fluid then collects in the pericardial sac, potentially restricting the movement of the heart. This condition is called **cardiac tamponade** (tam-po-NĀD). Treatment involves draining excess pericardial fluid and cutting a window in the restricting pericardium.

RHEUMATIC HEART DISEASE AND VALVULAR STENOSIS *Page 531*

Mitral stenosis and *aortic stenosis* are the most common forms of valvular heart disease. About 40% of patients with **rheumatic heart disease (RHD)** develop mitral stenosis, and two-thirds of them are women. The reason for the correlation between female gender and mitral stenosis is unknown. In **mitral stenosis** blood enters the left ventricle at a slower than normal rate, and when the ventricle contracts, blood flows back into the left atrium as well as into the aortic trunk. As a result, the left ventricle has to work much harder to maintain adequate systemic circulation. In severe cases of mitral stenosis the ventricular musculature is not up to the task. The heart weakens, and peripheral tissues begin to suffer from oxygen and nutrient deprivation. This weakened condition of the heart leads to **heart failure**.

Symptoms of **aortic stenosis** develop in roughly 25% of patients with RHD; 80% of these individuals are males. Symptoms of aortic stenosis are initially less severe than those of mitral stenosis. In this condition the left ventricle enlarges because of the narrow valvular opening and works harder just to maintain normal circulatory function. Clinical problems develop when the opening narrows to the point that adequate blood flow cannot occur. Symptoms then resemble those of mitral stenosis.

PROBLEMS WITH PACEMAKER FUNCTION *Page 533*

Symptoms of severe bradycardia (below 50 beats per minute) include weakness, fatigue, confusion, and loss of consciousness. Drug therapies are seldom helpful, but artificial pacemakers can be used with considerable success. Wires run to the atria, the ventricles, or both, depending on the nature of the problem, and the unit delivers small electrical pulses to stimulate the myocardium. Internal pacemakers are surgically implanted, batteries and all. These units last 7–8 years or more before another operation is required to change the battery. External pacemakers are used for temporary emergencies, such as immediately after cardiac surgery. Only the wires are implanted, and an external control box is worn on the belt.

There are over 50,000 artificial pacemakers in use at present. The simplest provide constant stimulation to the ventricles at rates of 70–80 per minute. More sophisticated pacemakers vary their rates to adjust to changing circulatory demands, as during exercise. Others are able to monitor cardiac activity and respond whenever the heart begins to function abnormally.

Tachycardia, usually defined as a heart rate of over 100 beats per minute, increases the workload on the heart. At very high heart rates cardiac performance suffers because the ventricles do not have enough time to refill with blood before the next contraction occurs. Chronic or acute incidents of tachycardia may be controlled by drugs that affect the permeability of pacemaker membranes or block the effects of sympathetic stimulation.

MONITORING THE LIVING HEART *Page 537*

Many techniques can be used to examine the structure and performance of the living heart. No single diagnostic procedure can provide the complete picture, so the tests used will vary depending on the suspected nature of the problem. A standard chest X-ray will show the basic size, shape, and orientation of the heart. Additional details require more specialized procedures to enhance the clarity of the images.

Because the heart is constantly moving, ordinary computerized tomography (CT) and magnetic resonance imaging (MRI) scans create blurred images. Special instruments and computers that generate images at high speed can be used to develop three-dimensional still or moving pictures of the heart as it beats. These procedures produce dramatic images, but the cost and complexity of the equipment have so far limited their use to major research institutions. Although PET scans can be used to diagnose disorders of coronary circulation (see Figure 21–9a,b●, p. 534), the cost limits the clinical use of this technology too.

Ultrasound analysis, called **echocardiography** (ek-ō-kar-dē-OG-ra-fē), provides images that lack the clarity of CT or MRI scans, but the equipment is relatively inexpensive and portable. Recent advances in data processing have made the images suitable for following details of cardiac contractions, and echocardiography is now an important diagnostic tool. Another relatively widespread approach involves preparation of a **coronary angiogram** (see Figure 21–8c●, p. 532). In this procedure, radiopaque dyes are injected into the coronary circulation, permitting X-ray analysis of coronary blood flow.

chapter 22

The Cardiovascular System: Vessels and Circulation

ANEURYSMS *Page 544*

An **aneurysm** (AN-yū-rizm) is a bulge in the weakened wall of a blood vessel, usually an artery. This bulge resembles a bubble in the wall of a tire, and like a bad tire, the affected artery may suffer a catastrophic blowout. The most dangerous aneurysms are those involving arteries of the brain, where they cause strokes, and of the aorta, where a blowout will cause fatal bleeding in a matter of minutes.

Aneurysms are most often caused by chronic high blood pressure. With age, the vessel walls become less elastic, and when a weak point develops the high arterial pressures distort the wall, creating an aneurysm. Unfortunately, because they are often painless, they are likely to go undetected. When aneurysms are found by ultrasound or other scanning procedures, the risk of rupture can sometimes be estimated on the basis of their size. For example, an aortic aneurysm larger than 6 cm has a 50:50 chance of rupturing in the next 10 years. Treatment often begins with the reduction of blood pressure by means of vasodilators or beta-blockers. An aneurysm in an accessible area, such as the abdomen, may be surgically removed and the vessel repaired.

Although high blood pressure is most often responsible for aneurysm formation, any trauma or infection that weakens vessel walls can lead to an aneurysm. In addition, at least some aortic aneurysms have been linked to inherited disorders, such as Marfan's syndrome (p. 754), that weaken connective tissues. Affected individuals have weak arterial walls, making them more likely to develop aortic aneurysms. It is not known whether genetic factors are involved in the development of other aneurysms.

PROBLEMS WITH VENOUS VALVE FUNCTION
Page 549

One of the consequences of aging is a loss of elasticity and resilience in connective tissues throughout the body. Blood vessels are no exception, and with age the walls of veins begin to sag. This change usually affects the superficial veins of the arms and legs first, because at these locations gravity opposes blood flow. The situation is aggravated by a lack of exercise or an occupation requiring long hours standing or sitting. Because there is no muscular activity to help keep the blood moving, venous blood pools on the proximal (heart) side of each valve. As the venous walls are distorted, the valves become less effective, and gravity can then pull blood back toward the capillaries. Normal blood flow is thereby further impeded, and the veins become grossly distended. These sagging, swollen vessels are called **varicose** (VAR-i-kōs) **veins**. Varicose veins of the leg are relatively harmless but unsightly; surgical procedures are sometimes used to remove or constrict the offending vessels.

Varicose veins are not limited to the extremities, and another common site involves a network of veins in the walls of the anus. Pressures within the abdominopelvic cavity rise dramatically when the abdominal muscles are tensed. Straining to force defecation can force blood into these veins, and repeated incidents leave them permanently distended. These distended veins, known as **hemorrhoids** (HEM-ō-roidz), can be uncomfortable and in severe cases extremely painful.

Hemorrhoids are often associated with pregnancy, because of changes in circulation and abdominal pressures. Minor cases can be treated by the topical application of drugs that promote contraction of smooth muscles within the venous walls. More severe cases may require the surgical removal or destruction of the distended veins.

chapter 23
The Lymphatic System

LYMPHEDEMA *Page 585*

Blockage of the lymphatic drainage from a limb produces **lymphedema** (lim-fe-DĒ-ma). In this condition, interstitial fluids accumulate, and the limb gradually becomes swollen and grossly distended. If the condition persists the connective tissues lose their elasticity, and the swelling becomes permanent. Lymphedema is painless, and the condition by itself does not pose a major threat to life. The danger comes from the continual risk of an uncontrolled infection developing in the affected area. Because the interstitial fluids are essentially stagnant, toxins and pathogens can accumulate and overwhelm the local defenses without fully activating the immune system in the rest of the body.

Temporary lymphedema of the feet and ankles may be caused by tight clothing constricting the lymphatic circulation or by prolonged standing or sitting. Elevating the feet and loosening clothing may eliminate the problem. Chronic lymphedema usually results as scar tissue forms in an area of injury. Trauma, infections, and surgical procedures are often implicated. In **filariasis** (fil-a-RĪ-a-sis), a parasitic nematode (roundworm) carried by mosquitoes forms massive colonies within lymphatic channels. Repeated scarring of the passageways eventually blocks lymphatic drainage and produces the extreme lymphedema with permanent distension of tissues known as **elephantiasis** (el-e-fan-TĪ-a-sis).

Therapy for chronic lymphedema consists of treating infections by the administration of antibiotics, and (when possible) reducing the swelling. One possible treatment involves the application of elastic wrappings that squeeze the tissue. This external compression elevates the hydrostatic pressure of the interstitial fluids and opposes the entry of additional fluid from the capillaries.

SCID *Page 587*

In **severe combined immunodeficiency disease (SCID)**, the individual fails to develop either cellular or humoral immunity. Lymphocyte populations are reduced, and normal B and T cells are not present. Patients with SCID are unable to provide an immune defense, and even a mild infection can prove fatal.

Total isolation offers protection at great cost and with severe restrictions on the individual's lifestyle. Bone marrow transplants from compatible donors, usually a close relative, have been used to colonize lymphatic tissues with functional lymphocytes.

The most famous SCID patient was the "bubble boy," David. David was kept in physical isolation until age 12, when he received a bone marrow transplant. Before the donor marrow cells established a functional immune system, David died of cancer. Although technology had protected him from external pathogens, without immune surveillance he had no defense against threats from within.

In 1990, genetic engineering techniques were used to insert normal genes into the lymphocytes of two children suffering from one form of SCID. The experiment was successful, and the children have regained and (thus far) retained apparently normal immune function.

AIDS *Page 587*

Acquired immune deficiency syndrome, or **AIDS**, is caused by a virus known as **human immunodeficiency virus (HIV)**, which selectively infects helper T cells. Infection of T cells by itself impairs the immune response, because these cells play a central role in coordinating cellular and humoral responses to antigens. To make matters worse, suppressor T cells are relatively unaffected by the virus, and over time the excess of suppressing factors "turns off" the normal immune response. Circulating antibody levels decline, cellular immunity is reduced, and the body is left without defenses against a wide variety of microbial invaders.

Because the immune function is so reduced, ordinarily harmless pathogens can initiate lethal infections, known as *opportunistic infections*. AIDS patients are especially prone to lung infections and pneumonia, often caused by *Pneumocystis carinii* or other fungi, and to a wide variety of bacterial, viral, and protozoan diseases. Because immune surveillance is also depressed, the risk of cancer increases. One of the most common cancers seen in AIDS patients, though very rare in normal individuals, is **Kaposi's sarcoma**, characterized by rapid cell division in endothelial cells of cutaneous blood vessels.

Infection with HIV occurs through intimate contact with the body fluids of infected individuals. Although all body fluids carry the virus, the major routes of transmission involve contact with blood, semen, or vaginal secretions. Most AIDS patients become infected through sexual contact with an HIV-infected person (who may *not necessarily* be suffering from the clinical symptoms of AIDS). The next largest group of patients consists of intravenous drug users who share contaminated needles. A relatively small number of individuals have become infected with the virus after receiving a transfusion of contaminated blood or blood products. Finally, an increasing number of infants are born with the disease, having acquired it from infected mothers prior to or at delivery.

The best defense against AIDS consists of avoiding sexual contact with infected individuals. All forms of sexual intercourse carry the potential risk of viral transmission. The use of synthetic condoms greatly reduces the chance of infection (although it does not completely eliminate it). Condoms that are not made of synthetic materials are effective in preventing pregnancy but do not block the passage of viruses.

Clinical symptoms of AIDS may not appear for 5–10 years or more after infection, and when they do appear they are often mild, consisting of *lymphadenopathy* and chronic but nonfatal infections. After a variable period of time, full-blown AIDS develops. AIDS is almost invariably fatal, and in the absence of a major clinical advance most of those who carry the virus will eventually die of the disease. Deaths in the United States have already climbed above 350,000, and estimates of the number of infected individuals range from 1 to 2 million. The numbers worldwide are even more frightening: the World Health Organization (1995) estimates that 20 million people may be infected.

Despite intensive efforts, a vaccine has yet to be developed that will provide immunity from HIV infection. However, the survival rate for AIDS patients has been steadily increasing because new drugs are available that slow the progress of the disease, and improved antibiotic therapies help combat secondary infections. This combination is extending the life span of patients while the search for more-effective treatment continues.

INFECTED LYMPHOID NODULES *Page 588*

The lymphocytes in a lymphoid nodule are not always able to destroy bacterial or viral invaders that have crossed the adjacent epithelium. When such pathogens become established in a lymphoid nodule, an infection develops. **Tonsillitis** is an infection of one of the tonsils, most often the pharyngeal tonsil. An individual with bacterial tonsillitis develops a high fever and leukocytosis (an abnormally high white blood cell count). The affected tonsil becomes swollen and inflamed, sometimes enlarging enough to partially block the entrance to the trachea. Breathing then becomes difficult, and in severe cases impossible. As the infection proceeds, abscesses develop within the tonsillar tissues, and the bacteria may enter the bloodstream by passing through the lymphatic capillaries and vessels to the venous system.

In the early stages, antibiotics may control the infection, but once abscesses have formed, the best treatment involves surgical drainage of the abscesses. **Tonsillectomy**, the removal of the tonsil, was once highly recommended and frequently performed to prevent recurring tonsillar infections. The procedure does reduce the incidence and severity of subsequent infections, but questions have been raised concerning the overall cost to the individual. The tonsils represent a first line of defense against bacterial invasion of the pharyngeal walls. If they are removed, bacteria may not be detected until a truly severe infection is well under way.

Appendicitis usually follows an erosion of the epithelial lining of the appendix. Several factors may be responsible for the initial ulceration, notably bacterial or viral pathogens. Bacteria that normally inhabit the lumen of the large intestine then cross the epithelium and enter the underlying tissues. Inflammation occurs, and the opening between the appendix and the rest of the intestinal tract may become constricted. Mucus secretion accelerates, and the organ becomes increasingly distended. Eventually the swollen and inflamed appendix may rupture or perforate. If it does, the bacteria will be released into the warm, dark, moist confines of the abdominopelvic cavity, where they can cause a life-threatening infection (peritonitis). The most effective treatment for appendicitis is the surgical removal of the organ, a procedure known as an **appendectomy**.

LYMPHOMAS *Page 589*

Lymphomas are malignant cancers consisting of abnormal lymphocytes or lymphocytic stem cells. Over 30,000 cases of lymphoma are diagnosed in the United States each year, and that number has been steadily increasing. There are many different types of lymphoma. One form, called **Hodgkin's disease (HD)**, accounts for roughly 40% of all lymphoma cases. Hodgkin's disease most often strikes individuals at ages 15–35 or those over age 50. The reason for this pattern of incidence is unknown; although the cause of the disease is uncertain, an infectious agent (probably a virus) is suspected. Other types are usually lumped together under the heading of **non-Hodgkin's lymphoma (NHL)**. They are extremely diverse, and in most cases the primary cause remains a mystery. At least some forms reflect a combination of inherited and environmental factors. For example, one form, called *Burkitt's lymphoma*, most often affects male children in Africa and New Guinea. The affected children have been infected with the *Epstein-Barr virus* (EBV).[1] The EBV infects B cells, but under normal circumstances the infected cells are destroyed by the immune system. EBV is widespread in the environment, and childhood exposure usually produces lasting immunity. Children developing Burkitt's lymphoma may have a genetic susceptibility to EBV infection; in addition, presence of another illness, such as malaria, may weaken their immune systems to the point that a lymphoma can develop. As improved antibiotic treatment prolongs AIDS patients' lives, more of them are developing NHL.

The first symptom associated with any lymphoma is usually a painless enlargement of lymph nodes. The involved nodes have a firm, rubbery texture. Because the nodes are painless, the condition is often overlooked until it has progressed to the point that secondary symptoms appear. For example, patients seeking help for recurrent fever, night sweats, gastrointestinal or respiratory problems, or weight loss may be unaware of any underlying lymph node changes. In the late stages of the disease, symptoms can include liver or spleen enlargement, CNS dysfunction, pneumonia, a variety of skin conditions, and anemia.

In planning treatment, clinicians consider the histological structure of the nodes and the stage of the disease. In examination of a biopsy, the structure of the node is described as nodular or diffuse. A nodular node retains a semblance of normal structure, with follicles and germinal cen-

---

[1]This highly variable virus is also responsible for *infectious mononucleosis* (discussed in *Disorders of the Spleen, p. 771*).

ters. In a diffuse node the interior of the node has changed, and follicular structure has broken down. In general, the nodular lymphomas progress more slowly than the diffuse forms, which tend to be more aggressive. On the other hand, the nodular lymphomas are more difficult to treat and are more likely to recur even after remission has been achieved.

The most important factor influencing treatment selection is the stage of the disease. Table A-6 includes a simplified staging classification for lymphomas. When a lymphoma is diagnosed early (stage I or II), localized therapies may be effective. For example, the cancerous node(s) may be surgically removed and the region(s) irradiated to kill residual cancer cells. Success rates are very high when a lymphoma is detected in these early stages. For Hodgkin's disease, localized radiation can produce remission lasting 10 years or more in over 90% of patients. Treatment of localized NHL is somewhat less effective. The 5-year remission rates average 60–80% for all types; success rates are higher in nodular forms than for diffuse forms.

Although these are encouraging results, it should be noted that few lymphoma patients are diagnosed while in the early stages of the disease. For example, only 10–15% of NHL patients are diagnosed at stages I or II. For lymphomas at stages III and IV, treatment most often involves chemotherapy. Combination chemotherapy, in which two or more drugs are administered simultaneously, is the most effective treatment. For Hodgkin's disease, a four-drug combination with the acronym MOPP (nitrogen mustard, Oncovin [vincristine], prednisone, and procarbazine) produces lasting remission in 80% of patients.

Bone marrow transplantation is a treatment option for acute, late-stage lymphoma. When suitable donor marrow is available, the patient receives whole-body irradiation, chemotherapy, or some combination of the two sufficient to kill tumor cells throughout the body. This treatment also destroys normal bone marrow cells. Donor bone marrow is then infused, and over the next 2 weeks the donor cells colonize the bone marrow and begin producing red blood cells, granulocytes, monocytes, and lymphocytes.

Potential complications of this treatment include the risk of infection and bleeding while the donor marrow is becoming established. The immune cells of the donor marrow may also attack the tissues of the recipient, a response called **graft versus host, or GVH, disease**. For a patient with stage I or II lymphomas, without bone marrow involvement, bone marrow can be removed and stored (frozen) for over 10 years. If other treatment options fail, or the patient comes out of remission at a later date, an **autologous marrow transplant** can be performed. This option eliminates the need for donor typing and the risk of GVH disease.

TABLE A-6   Cancer Staging in Lymphoma

Stage I: Involvement of a single node or region (or a single extranodal site)

Typical treatment: surgical removal and/or localized irradiation; in slowly progressing forms of NHL, treatment may be postponed indefinitely.

Stage II: Involvement of nodes in two or more regions (or an extranodal site and nodes in one or more regions) on the same side of the diaphragm

Typical treatment: surgical removal and localized irradiation that includes an extended area around the cancer site (the extended field).

Stage III: Involvement of lymph node regions on both sides of the diaphragm. This is a large category that is subdivided on the basis of the organs or regions involved. For example, in stage III the spleen contains cancer cells.

Typical treatment: combination chemotherapy, with or without radiation; radiation treatment may involve irradiating all of the thoracic and abdominal nodes plus the spleen (*total axial nodal irradiation*, or *TANI*).

Stage IV: Widespread involvement of extranodal tissues above and below the diaphragm

Treatment is highly variable, depending on the circumstances. Combination chemotherapy is always used; it may be combined with whole-body irradiation. The "last resort" treatment involves massive chemotherapy followed by a bone marrow transplant.

BREAST CANCER *Page 589*

Breast cancer is the primary cause of death for women between the ages of 35 and 45, but it actually becomes even more common after age 50. There were approximately 46,000 deaths in the United States from breast cancer in 1993, and nearly 180,000 new cases reported. An estimated 7% of women in the United States will develop breast cancer at some point in their lifetimes. The incidence is highest among Caucasians, somewhat lower in African Americans, and the lowest in Asians and Native Americans. Notable risk factors include (1) a family history of breast cancer, (2) a pregnancy after age 30, and (3) early menarche (first menstrual period) or late menopause (last menstrual period). Despite repeated studies (and rumors), there are no proven links between oral contraceptive use, estrogen therapy, breast feeding, fat consumption, or alcohol use and breast cancer. It appears likely that multiple factors are involved; most women never develop breast cancer, even women in families with a history of this disease.

Early detection of breast cancer is the key to reducing mortalities. *Most breast cancers are found through self-examination*, but the use of clinical screening techniques has increased in recent years. **Mammography** involves the use of X-rays to examine breast tissues; the radiation dosage can be restricted because only soft tissues must be penetrated. This procedure gives the clearest picture of conditions within the breast tissues. Ultrasound can provide some information, but the images lack the detail of standard mammograms. **Thermography** maps the surface temperatures on the skin of the breasts. Because cancer cells have abnormally high metabolic rates and increased vascularization, tumors are significantly warmer than the surrounding tissues. The heat can be detected with this technique, but unfortunately the results are subject to considerable variation.

For treatment to be successful the cancer must be identified while it is still relatively small and localized. Once it has grown larger than 2 cm (0.78 in.) the chances for long-term survival worsen. A poor prognosis also follows if the cancer cells have spread through the lymphatic system to the axillary lymph nodes. If the nodes are not yet involved, the chances of 5-year survival are about 82%, but if four or more nodes are involved the survival rate drops to 21%.

Treatment of breast cancer begins with the removal of the tumors and sampling of the axillary lymph nodes. Because the cancer cells usually begin spreading before the condition is diagnosed, surgical treatment involves the removal of part or all of the affected breast.

- In a **segmental mastectomy**, or "lumpectomy," only a portion of the breast is removed. This, combined with irradiation, is the preferred treatment for small, localized tumors.
- In a **total mastectomy**, the entire breast is removed, but other tissues are left intact.
- In **radical mastectomy**, an operation rarely performed in the 1990s, the pectoralis muscles, the breast, and the axillary lymph nodes are removed. In a *modified radical mastectomy*, the most common operation, the breast and nodes are removed but the muscular tissue remains intact.

A combination of chemotherapy, radiation treatments, and hormone treatments may be used to supplement the surgical procedures.

DISORDERS OF THE SPLEEN *Page 595*

An impact to the left side of the abdomen can distort or damage the spleen. Such injuries are known risks of contact sports, such as football or hockey, and more solitary athletic activities, such as skiing or sledding. The spleen tears so easily, however, that a seemingly minor blow to the side may rupture the capsule. The result is serious internal bleeding and eventual circulatory shock.

Because the spleen is relatively fragile, it is very difficult to repair surgically. (Sutures usually tear out before they have been tensed enough to stop the bleeding.) Treatment for a severely ruptured spleen involves its complete removal, a process called a **splenectomy** (splē-NEK-tō-mē).

The spleen responds like a lymph node to infection, inflammation, or invasion by cancer cells. The enlargement that follows is called **splenomegaly** (splē-nō-MEG-a-lē; *megas*, large), and splenic rupture may also occur under these conditions. One relatively common condition causing splenomegaly is **mononucleosis**. This condition, also known as the "kissing disease," results from infection by the Epstein-Barr virus (EBV). In addition to splenic enlargement, symptoms of acute mononucleosis include fever, sore throat, widespread swelling of lymph nodes, increased numbers of lymphocytes in the blood, and the presence of circulating antibodies to the virus. The condition most often affects young adults (age 15–25) in the spring or fall. Only the symptoms are treated, because no drugs are effective against this virus. The most dangerous aspect of the disease is the risk of rupturing the enlarged spleen, which becomes fragile. Patients are therefore cautioned against heavy exercise or other activities that increase abdominal pressures. If the spleen does rupture, severe hemorrhaging may occur; death will follow unless transfusion and an immediate splenectomy can be performed.

An individual whose spleen is missing (following splenectomy) or nonfunctional has **hyposplenism** (hī-pō-SPLĒN-ism). Hyposplenism usually does not pose a serious problem, but such individuals are prone to certain bacterial infections, and special immunizations may be recommended. In **hypersplenism** the spleen becomes overactive, and the increased phagocytic activities lead to anemia (low numbers of RBCs), leukopenia (low numbers of WBCs), and thrombocytopenia (low numbers of platelets). Splenectomy is the only known cure for hypersplenism.

IMMUNE DISORDERS *Page 595*

Because the immune response is so complex, there are many opportunities for things to go wrong. A great variety of clinical conditions may result from disorders of the immune functions associated with the lymphatic system. In an **immunodeficiency disease** either the immune system fails to develop normally or the immune response is blocked in some way. Examples of immunodeficiency diseases include **severe combined immunodeficiency disease (SCID)** and *acquired immune deficiency syndrome* (AIDS). SCID and AIDS are discussed in this appendix on page 769.

Autoimmune disorders develop when the lymphatic system mistakenly targets normal body cells and tissues. Lymphocytes usually recognize and ignore the antigens normally found in the body. The recognition system can malfunction, however, and when it does the activated B cells begin to manufacture antibodies against other cells and tissues. The trigger may be a reduction in suppressor T cell activity, excessive stimulation of helper T cells, tissue damage that releases large quantities of antigens, viral or bacterial toxins, or some combination of factors.

The symptoms produced depend on the identity of the antigen attacked by these misguided antibodies, called **autoantibodies**. Several conditions described in earlier chapters are autoimmune disorders. For example, *rheumatoid arthritis* occurs when autoantibodies form immune complexes within connective tissues, especially around the joints. [p. 207] Many other autoimmune disorders appear to be cases of mistaken identity. For example, proteins associated with the measles, Epstein-Barr, influenza, and other viruses contain amino acid sequences that are similar to those of myelin proteins. As a result, antibodies that target these viruses may also attack myelin sheaths. This mechanism accounts for the neurologic complications that sometimes follow a vaccination or viral infection.

Allergies are inappropriate or excessive immune responses to antigens. The sudden increase in cellular activity or antibody titers can have a number of unpleasant side effects. For example, neutrophils or cytotoxic T cells may destroy normal cells while attacking the antigen, or the antibody-antigen complex may trigger a massive inflammatory response. Antigens that trigger allergic reactions are often called **allergens**.

SYSTEMIC LUPUS ERYTHEMATOSUS *Page 595*

Systemic lupus erythematosus (LOO-pus e-rith-ē-ma-TŌ-sis), or **SLE**, appears to result from a generalized breakdown in the antigen recognition mechanism. An individual with SLE manufactures autoantibodies against nucleic acids, ribosomes, clotting factors, blood cells, platelets, and lymphocytes. The immune complexes form deposits in tissues, producing anemia, kidney damage, arthritis, and vascular inflammation. CNS function deteriorates with CNS vasculitis.

The most obvious sign of this condition is the presence of a butterfly-shaped rash of the face, centered over the bridge of the nose. SLE affects women nine times as often as men, and the U.S. incidence averages 2–3 cases per 100,000 population. There is no known cure, but almost 80% of SLE patients survive 5 years or more after diagnosis. Treatment consists of controlling the symptoms and depressing the immune response through administration of specialized drugs or corticosteroids.

chapter 24

The Respiratory System

OVERLOADING THE RESPIRATORY DEFENSES
Page 600

Large quantities of airborne particles may overload the respiratory defenses and produce a variety of different illnesses. The presence of irritants in the lining of the conducting passageways may provoke the formation of abscesses, and damage to the epithelium in the affected areas may allow the irritants to enter the surrounding tissues of the lung. The scar tissue that forms reduces the elasticity of the lung and may restrict airflow along the passageways. Irritants or foreign particles may also enter the lymphatics of the lung, producing inflammation of the regional lymph nodes. Chronic irritation and stimulation of the epithelium and its defenses cause changes in the epithelium that increase the likelihood of lung cancer. Severe symptoms of such disorders develop slowly and may take 20 years or more to appear. **Silicosis** (sil-i-KŌ-sis), produced by the inhalation of silica dust, **asbestosis** (as-bes-TŌ-sis), from the inhalation of asbestos fibers, and **anthracosis** (an-thra-KŌ-sis), the "black lung disease" of coal miners, are examples of conditions caused by overloading the respiratory defenses.

NOSEBLEEDS *Page 601*

The extensive vascularization of the nasal cavity and the relatively vulnerable position of the nose make a nosebleed, or **epistaxis** (ep-i-STAK-sis), relatively common. Bleeding usually involves vessels of the mucosa covering the anterior cartilaginous portion of the septum. Packing the vestibule with gauze or pinching the external nares together to squeeze the vessels against the septum will often control the bleeding until clotting occurs. More severe bleeding originating elsewhere in the nasal cavity may require packing the posterior portion of the cavity via the internal nares.

Epistaxis can result from any factor affecting the integrity of the epithelium or the underlying vessels. Examples would include trauma, such as a punch in the nose, drying, infections, allergies, or clotting disorders. Hypertension may also provoke a nosebleed by rupturing small vessels of the lamina propria.

DISORDERS OF THE LARYNX *Page 605*

Infection or inflammation of the larynx is known as **laryngitis** (lar-in-JĪ-tis). This condition often affects the vibrational qualities of the vocal cords; hoarseness is the most familiar symptom. Mild cases are temporary and seldom serious, but bacterial or viral infection of the epiglottis or upper trachea in children can be very dangerous because the swelling may close the glottis or trachea and cause suffocation. **Acute epiglottitis** (ep-i-glot-TĪ-tis) can develop relatively rapidly following a bacterial infection of the throat. Although most common in children, it does occur in adults, especially those with *Hodgkin's disease* or *leukemia*, two cancers described earlier. Serious inflammation and edema of the trachea can also occur in small children following infection with one of the parainfluenza viruses. The condition, **laryngotracheobronchitis**, is more commonly called **croup** (kroop).

BRONCHITIS *Page 609*

Bronchitis (brong-KĪ-tis) is an inflammation of the bronchial lining. The most characteristic symptom is the overproduction of mucus, which leads to frequent coughing. An estimated 20% of adult males have chronic bronchitis. This condition is most often related to cigarette smoking, but it can also result from other environmental irritants, such as chemical vapors. Over time the increased mucus production can block smaller airways and reduce respiratory efficiency. This condition is called **chronic airways obstruction** or *chronic obstructive pulmonary disease (COPD)*.

EXAMINING THE LIVING LUNG *Page 613*

A chest X-ray remains the standard diagnostic screening test for chest conditions. This procedure can detect abnormalities in lung structure including scar tissue formation, fluid accumulation, or distortion of the conducting passageways. CT scans show much greater definition of internal structures and can clarify the nature of abnormalities initially spotted on a chest X-ray. CT scans are particularly helpful in diagnosing and tracking the progression of lung cancers. Lung scans are made when radioactive tracers are injected or radioactive gases are inhaled. These procedures can detect abnormalities in airflow or pulmonary blood circulation.

Bronchoscopy (brong-KOS-kō-pē) involves the insertion of a fiberoptic bundle, or *bronchoscope*, a few millimeters in diameter into the trachea. The bundle, once inserted, can be steered along the conducting passageways to the level of the smaller bronchi. In addition to permitting direct visualization of bronchial structures, the bronchoscope can collect tissue or mucus samples from the respiratory tract. In **bronchography** (brong-KOG-ra-fē) a bronchoscope or catheter introduces a radiopaque material into the bronchi. This technique can permit detailed analysis of bronchial masses, such as tumors, or other obstructions along the bronchial tree (see Figure 24-10b●, p. 612).

RESPIRATORY DISTRESS SYNDROME (RDS)
Page 613

Surfactant cells begin producing surfactants at the end of the sixth fetal month. By the eighth month surfactant production has risen to the level required for normal respiratory function. **Neonatal respiratory distress syndrome (NRDS),** also known as *hyaline membrane disease (HMD)*, develops when surfactant amounts are inadequate. Although there are inherited forms of NRDS, the condition most often accompanies premature delivery.

In the absence of surfactants, the alveoli tend to collapse during exhalation, and although the conducting passageways remain open, the newborn infant must then inhale with extra force to reopen the alveoli on the next breath. In effect, every breath must approach the power of the first, and the infant rapidly becomes exhausted. Respiratory movements become progressively weaker, eventually the alveoli fail to expand, and gas exchange ceases.

One method of treatment involves assisting the infant by administering air under pressure, so that the alveoli are held open. This procedure, known as **positive end-expiratory pressure (PEEP)**, can keep the newborn alive until surfactant production increases to normal levels. Surfactant from other sources can also be provided; suitable surfactants can be extracted from cow lungs (*survanta*), obtained from the fluids that surround full-term infants, or synthesized using gene-splicing techniques (*exosurf*). These preparations are usually administered in the form of a fine mist of surfactant droplets. This method of delivery is called **nebulization**.

Surfactant abnormalities may also develop in adults, as the result of severe respiratory infections or other sources of pulmonary injury. Alveolar collapse follows, producing a condition known as **adult respiratory distress syndrome (ARDS)**. PEEP is often used in an attempt to maintain life until the underlying problem can be corrected, but at least 50–60% of ARDS cases result in fatalities.

TUBERCULOSIS *Page 613*

Tuberculosis (tū-ber-kū-LŌ-sis), or **TB**, results from a bacterial infection of the lungs, although other organs may be invaded as well. The bacteria, *Mycobacterium tuberculosis*, may colonize the respiratory passageways, the interstitial spaces, and/or the alveoli. Symptoms are variable, but usually include coughing and chest pain, with fever, night sweats, fatigue, and weight loss.

At the site of infection, macrophages and fibroblasts proceed to wall off the area, forming an abscess. If the scar tissue barricade fails, the bacteria move into the surrounding tissues and the process repeats itself. The resulting masses of fibrous tissue distort the conducting passageways, increasing resistance and decreasing airflow. In the alveoli, the attacked surfaces are destroyed. The combination reduces the vital capacity and the area available for gas exchange.

Treatment for TB is complex, because (1) the bacteria can spread to many different tissues, (2) they can develop a resistance to standard antibiotics relatively quickly, and (3) to be effective, treatment is prolonged. As a result, several drugs are used in combination over a period of 6–9 months. The most effective drugs now available include *isoniazid*, which interferes with bacterial replication, and *rifampin*, which blocks bacterial protein synthesis.

Tuberculosis is a major health problem throughout the world, but especially in underdeveloped countries. An estimated 2 billion people are infected at this time. There are 7–9 million cases diagnosed each year and 3 million deaths annually due to tuberculosis infection. The problem is much less severe in developed nations, such as the United States. However, tuberculosis was extremely common in the United States earlier in this century. An estimated 80% of the people born around the turn of the century became infected with tuberculosis during their lives. Although many were able to meet the bacterial challenge, it was still the number one cause of death in 1906. These statistics have been drastically altered with the arrival of antibiotics and techniques for early detection of infection. Between 1906 and 1984 the death rate fell from 2 deaths per 1000 population to 1.5 deaths per 100,000 population.

Tuberculosis today is unevenly distributed through the U.S. population, and is resurging as a public health problem. Several groups are at relatively high risk for infection. For example, Hispanics, African Americans, prison inmates, new immigrants, hospital employees, and individuals with immune disorders, such as AIDS patients, are more likely to be infected than other members of the population. Although at present only 2–5% of young American adults have been infected, the incidence and death rate have increased each year since 1984. Recently, strains resistant to many antibiotics have appeared, further complicating efforts to control the disease. An infectious disease that is airborne (spread through the air), potentially fatal, and threatens to become resistant to all current treatments has led to calls for supervised treatment of infected persons up to and including institutionalization, quarantine, and other forms of confinement until the patient is noninfectious. Interest in developing a more effective vaccine than the BCG vaccine currently used in Europe is increasing.

PNEUMONIA *Page 613*

Pneumonia (nū-MŌ-nē-a) is an infection of the lobules of the lung. As inflammation occurs within the lobules, respiratory function deteriorates as a result of fluid leakage into the alveoli and/or swelling and constriction of the respiratory bronchioles. When bacteria are involved, they are frequently species normally found in the mouth and pharynx that have somehow managed to evade the respiratory defenses. As a result, pneumonia becomes more likely when the respiratory defenses have been compromised by other factors, such as epithelial damage from smoking or the breakdown of the immune system in AIDS. The most common pneumonia that develops in AIDS patients results from infection by the fungus *Pneumocystis carinii*. These organisms are normally found in the alveoli, but in healthy individuals the respiratory defenses are able to prevent infection and tissue damage.

PULMONARY EMBOLISM *Page 614*

The lungs are the only organs that receive the entire cardiac output. Blood pressure in the pulmonary circuit is usually relatively low, and pulmonary vessels can easily become blocked by blood clots, fat masses, or air bubbles in the pulmonary arteries. Blockage of a vessel stops blood flow to all of the alveoli serviced by the obstructed vessel. This condition is called a **pulmonary embolism**. A condition described earlier, *venous thrombosis*, can promote development of a pulmonary embolism. If the blockage remains in place for several hours, the alveoli will permanently collapse. If the blockage occurs in a major pulmonary vessel, rather than a minor tributary, pulmonary resistance increases. This resistance places extra strain on the right ventricle, which may be unable to maintain cardiac output. Congestive heart failure may be the result.

PNEUMOTHORAX *Page 616*

Any injury to the thorax that penetrates the parietal pleura or damages the alveoli and the visceral pleura can allow air into the pleural cavity. This condition, called **pneumothorax** (nū-mō-THŌR-aks), breaks the fluid bond between the pleurae and allows the elastic fibers to contract. The result is called a collapsed lung, or **atelectasis** (at-e-LEK-ta-sis; *ateles*, imperfect + *ektasis*, expansion).

A pneumothorax may develop because of imperfections and air leakage in the walls of superficial alveoli. These problems often have a genetic basis. A sudden lung collapse, called a **spontaneous pneumothorax**, may be triggered by heavy exercise, coughing, or other activities that increase the pressures inside the alveoli of the lungs.

THORACENTESIS *Page 616*

The delicate pleural membranes can become inflamed because of chronic irritation or infection. Inflammation produces symptoms of pleuritis, or pleurisy. Chronic inflammation often causes a change in the permeability of the pleura, leading to a **pleural effusion** (an abnormal accumulation of fluid within the pleural cavities). When a pleural effusion is detected on an X-ray, samples of pleural fluid are obtained, using a long needle inserted between the ribs. This sampling procedure is called **thoracentesis** (or *thoracocentesis*). The fluid collected is usually checked for the presence of blood, white blood cells, and bacteria.

LUNG CANCER, SMOKING, AND DIET *Page 621*

Lung cancer, or *pleuropulmonary neoplasm*, is an aggressive class of malignancies originating in the bronchial passageways or alveoli. These cancers affect the epithelial cells lining conducting passageways, mucous glands, or alveoli. Symptoms usually do not appear until the condition has progressed to the point that the tumor masses are restricting airflow or compressing adjacent mediastinal structures. Chest pain, shortness of breath, a cough or wheeze, and weight loss are common symptoms. Treatment programs vary depending on the cellular organization of the tumor and whether metastasis (cancer cell migration) has occurred, but surgery, radiation, or chemotherapy may be involved.

Deaths from lung cancer were rare at the turn of the century, but there were 29,000 in 1956, 105,000 in 1978, and 153,000 in 1994. These figures continue to rise, with the number of diagnosed cases doubling every 15 years. Each year 22% of new cancers detected are lung cancers, and in 1994, 100,000 men and 72,000 women were diagnosed with this condition. Lung cancers now account for 35% of all cancer deaths, making this condition the primary cause of cancer death in the U.S. population. Despite advances in the treatment of other forms of cancer, the survival statistics for lung cancer have not changed significantly. Even with early detection the 5-year survival rates are only 30% (men) to 50% (women), and most lung cancer patients die within a year of diagnosis.

Detailed statistical and experimental evidence has shown that *85–90% of all lung cancers are the direct result of cigarette smoking*. Claims to the contrary are simply unjustified and insupportable. The data are far too extensive to detail here, but the incidence of lung cancer for nonsmokers is 3.4 per 100,000 population, while the incidence for smokers ranges from 59.3 per 100,000 for those burning between a half-pack and a pack per day, to 217.3 per 100,000 for those smoking one to two packs per day. Before 1970, this disease affected primarily middle-aged men, but as the number of women smokers has increased so has the number of women dying from lung cancer.

Smoking changes the quality of the inspired air, making it drier and contaminated with several carcinogenic compounds and particulate matter. The combination overloads the respiratory defenses and damages the epithelial cells throughout the respiratory system. The histological changes that follow were described in Chapter 3. ⬯ (p. 81) Whether lung cancer develops appears to be related to the total cumulative exposure to the carcinogenic stimuli. The more cigarettes smoked, the greater the risk, whether those cigarettes are smoked over a period of weeks or years. The histological changes induced by smoking are reversible, and a normal epithelium will return if the stimulus is removed. At the same time the statistical risks decline to significantly lower levels. Ten years after quitting, a former smoker stands only a 10% greater chance of developing lung cancer than a nonsmoker.

The fact that cigarette smoking often causes cancer is not surprising in view of the toxic chemicals contained in the smoke. What is surprising is that more smokers do not develop lung cancer. There is evidence that some smokers have a genetic predisposition for developing one form of lung cancer. Dietary factors may also play a role in preventing lung cancer, although the details are controversial. In terms of their influence on the risk of lung cancer, there is general agreement that (1) vitamins C and A have no effect, (2) beta carotene, an orange pigment, and other vegetable components reduce the risk, and (3) a high-cholesterol, high-fat diet increases the risk.

chapter 25

The Digestive System

ACHALASIA AND ESOPHAGITIS *Page 638*

In the condition known as **achalasia** (ak-a-LĀ-zē-a), a bolus of food from the mouth descends the esophagus relatively slowly and its arrival does not cause the opening of the lower esophageal sphincter. Materials then accumulate at the base of the esophagus like cars at a stop light. Secondary peristaltic waves may occur repeatedly, adding to the individual's discomfort. The most successful treatment involves cutting the circular muscle layer at the base of the esophagus or expanding a balloon in the lower esophagus until the muscle layer tears.

A weakened or permanently relaxed sphincter can cause inflammation of the esophagus, or **esophagitis** (ē-sof-a-JĪ-tis), as powerful gastric acids enter the lower esophagus. The esophageal epithelium has few defenses from acid and enzyme attack, and inflammation, epithelial erosion, and intense discomfort are the result. Occasional incidents of reflux, or backflow, from the stomach are responsible for the symptoms of "heartburn." This relatively common problem supports a multimillion dollar industry devoted to producing and promoting antacids.

STOMACH CANCER *Page 644*

Stomach cancer is one of the most common lethal cancers, responsible for roughly 15,000 deaths in the United States each year. Because the symptoms may resemble those of gastric ulcers, the condition may not be discovered in its early stages. Diagnosis usually involves X-rays of the stomach at various degrees of distension. The mucosa can also be visually inspected using a flexible instrument called a **gastroscope**. Attachments permit the collection of tissue samples for histological analysis.

Treatment of gastric cancer involves the surgical removal of part or all of the stomach. Even a **total gastrectomy** (gas-TREK-tō-mē) can be tolerated, because the stomach provides no essential digestive services other than the secretion of intrinsic factor for vitamin B$_{12}$ absorption.

GASTROENTERITIS *Page 648*

An irritation of the small intestine may lead to a series of powerful peristaltic contractions that eject the contents of the small intestine into the large intestine. An extremely powerful irritating stimulus will produce a "clean sweep" of the absorptive areas of the digestive tract. Vomiting clears the stomach, duodenum, and proximal jejunum, and peristaltic contractions evacuate the distal jejunum and ileum. Bacterial toxins, viral infections, and various poisons will sometimes produce these extensive gastrointestinal responses. Conditions affecting primarily the small intestine are usually referred to as **enteritis** (en-ter-Ī-tis) of one kind or another. If both vomiting and diarrhea are present, the term **gastroenteritis** (gas-trō-en-ter-Ī-tis) may be used instead.

DIVERTICULITIS AND COLITIS *Page 651*

In **diverticulosis** (dī-ver-tik-ū-LŌ-sis) pockets (*diverticula*) form in the mucosa, usually in the sigmoid colon. These get forced outward, probably by the pressures generated during defecation. If the pockets push through weak points in the muscularis externa, they form semi-isolated chambers that are subject to recurrent infection and inflammation. The infections cause pain and the occasional bleeding characteristic of **diverticulitis** (dī-ver-tik-ū-LĪ-tis). In severe cases the diverticula may perforate, setting the bacteria loose in the peritoneal cavity.

The general term **colitis** (kō-LĪ-tis) may be used to indicate a condition characterized by inflammation of the colon. The *irritable bowel syndrome* is characterized by diarrhea, constipation, or an alternation between the two. When constipation is the primary problem, this condition may be called a **spastic colon**, or *spastic colitis*. The irritable bowel syndrome may have a psychological component. Inflammatory bowel disease, or ulcerative colitis, involves chronic inflammation of the digestive tract, most often affecting the colon. The mucosa becomes inflamed and ulcerated, extensive areas of scar tissue develop, and colonic function deteriorates. Acute diarrhea, cramps, and bleeding are common symptoms. Treatment of severe inflammatory bowel disease may involve a **colectomy** (kō-LEK-tō-mē), the removal of all or a portion of the colon. If a large part or even all of the colon must be removed, normal connection with the anus cannot be maintained. Instead, the end of the intact digestive tube is sutured to the abdominal wall, and wastes then accumulate in a plastic pouch or sac attached to the opening. If the attachment involves the colon, the procedure is a **colostomy** (kō-LOS-tō-mē); if the ileum is involved it is an **ileostomy** (il-ē-OS-tō-mē).

CIRRHOSIS *Page 654*

The underlying problem in **cirrhosis** (sir-Ō-sis) appears to be the widespread destruction of hepatocytes by exposure to drugs (especially alcohol), viral infection, ischemia, or blockage of the hepatic ducts. Two processes are involved in producing the symptoms. Initially the damage to hepatocytes leads to the formation of extensive areas of scar tissue that branch throughout the liver. The surviving hepatocytes then undergo repeated cell divisions, but the fibrous tissue prevents the new hepatocytes from achieving a normal lobular arrangement. So the liver gradually converts from an organized assemblage of lobules to a fibrous aggregation of poorly functioning cell clusters.

chapter 26

The Urinary System

URINE COMPOSITION *Page 672*

Table A-7 presents typical values for the most important components of normal urine.

TABLE A-7   General Characteristics of Normal Urine
pH: 6.0 (range: 4.5–8)
Specific gravity: 1.003–1.030
Osmolarity: 855–1335 mOsm
Water content: 93–97%
Volume: 500–1500 mℓ/day
Color: clear yellow
Odor: varies depending on composition
Bacterial content: sterile

HEMODIALYSIS *Page 672*

If drugs, infusions, and dietary controls cannot stabilize the composition of the blood, more drastic measures are taken. In one technique, called **hemodialysis** (hē-mō-dī-AL-i-sis), an artificial membrane is used to regulate the composition of the blood. The basic principle involved in this process, called **dialysis**, involves passive diffusion across a semipermeable membrane. The patient's blood flows across an artificial *dialysis membrane* that contains pores large enough to permit the diffusion of small ions, but small enough to prevent the loss of plasma proteins. On the other side of the membrane flows a special **dialysis fluid**.

In effect, diffusion across the dialysis membrane takes the place of normal glomerular filtration, and the characteristics of the dialysis fluid ensure that important metabolites remain in the circulation, rather than diffusing across the membrane.

In practice, silastic tubes, called *shunts*, are inserted into a medium-sized artery and vein. (The usual location is in the forearm, although the lower leg is sometimes used.) When connected to the dialysis machine, the individual sits quietly while blood circulates from the arterial shunt, through the machine, and back via the venous shunt. Inside the machine, the blood flows across a dialyzing membrane, where diffusion occurs. For chronic dialysis, an artery is surgically connected to a vein in the forearm creating an arterio-venous anastomosis that serves as a biological shunt. During dialysis, two needles are inserted into the shunt and connected to the machine.

PROBLEMS WITH THE MICTURITION REFLEX
Page 680

Incontinence (in-KON-ti-nens) refers to an inability to voluntarily control urination. This condition may reflect damage to the CNS, the spinal cord, or the nerve supply to the bladder or external sphincter. Incontinence often accompanies the dementia of Alzheimer's disease, and it may also result from a stroke or spinal cord injury. In most cases the individual develops an **automatic bladder**. The micturition reflex remains intact, but voluntary control of the external sphincter is lost and the individual cannot prevent the reflexive emptying of the bladder.

Infants lack voluntary control over urination because the necessary corticospinal connections have yet to be established. Toilet training before age 2 usually involves training the parent to anticipate the timing of the reflex, rather than training the child to exert conscious control.

Childbirth can stretch and damage the sphincter muscles, and some women then develop **stress incontinence**. In this condition elevated intraabdominal pressures, such as during a cough or sneeze, can overwhelm the sphincter muscles, causing urine to leak out.

chapter 27

The Reproductive System

TESTICULAR TORSION *Page 685*

Because the testes are relatively loosely attached to the scrotal walls, they may become twisted within the scrotal cavity. This condition, called **testicular torsion**, usually occurs in children or adolescents. Symptoms include pain in the groin and inguinal region, local inflammation, and swelling of the scrotum on the affected side. Treatment involves prompt external or surgical manipulation of the testis to relieve the twisting. Because any kinks in the spermatic cord severely restrict the arterial supply to the testis, corrective measures must be taken within 4–6 hours, before the testicular tissues become permanently damaged. If the tissues are deprived of circulation for longer periods, the damage will be irreversible, and the affected testis may have to be removed. This surgical procedure is called an **orchidectomy** (ōr-ki-DEK-tō-mē); *orchis*, testis); the common term *castration* indicates that both testes have been removed.

PROSTATE CANCER *Page 694*

Prostate cancer is the second most common cancer in men, and it is the second most common cause of cancer deaths in males. Each year 130,000 cases are diagnosed in the United States, and there are approximately 34,000 deaths. Most patients are over age 65. There are racial differences in susceptibility that are poorly understood; the incidence is relatively high among African Americans and low among Asians.

Prostate cancer usually originates in one of the secretory glands, and as it progresses it produces a nodular lump or swelling on the prostatic surface. Palpation of the prostate gland through the rectal wall is the easiest diagnostic screening procedure, but transrectal prostatic ultrasound provides more detailed information.

If the condition is detected before the cancer cells have spread to other organs, the usual treatment is either localized radiation or the surgical removal of the prostate gland. This operation, called a **prostatectomy** (pros-ta-TEK-tō-mē), is often effective in controlling the condition, but undesirable side effects may include a loss of sexual function and urinary incontinence. Modified surgical procedures can reduce these risks and maintain normal sexual function in almost three out of four patients.

One recent screening method involves a blood test for **prostate-specific antigen (PSA)**. Elevated levels of this antigen, normally present in low concentrations, may indicate the presence of prostate cancer. This test is more sensitive than the serum enzyme assay previously used for screening purposes. The enzyme test for prostatic acid phosphatase detects prostate cancer in comparatively late stages of development.

Once metastasis is under way, and the lymphatic system, lungs, bone marrow, liver, or adrenal glands are involved, the survival rates are significantly lower. Potential treatments for metastatic prostate cancer include more intensive radiation dosage, hormonal manipulation, lymph node removal, and aggressive chemotherapy. Because the cancer cells are stimulated by testosterone, treatment may involve castration or hormones that depress gonadotropin-releasing hormone (GnRH) or luteinizing hormone (LH) production. Until recently the usual hormone selected was diethylstilbestrol (DES), an estrogen. There are two other drug options: (1) Drugs that mimic GnRH are given in high doses, producing a surge in LH production followed by a sharp decline to very low levels, presumably as the endocrine cells adapt to the excessive stimulation. (2) Drugs that block the action of androgens can be given. Several new drugs, including Flutamide, block the cytoplasmic receptors for testosterone and prevent stimulation of the cancer cells. Despite these interesting advances in treatment, however, the average survival time for patients diagnosed with advanced prostatic cancer is only 2.5 years.

UTERINE TUMORS AND CANCERS *Page 706*

Uterine tumors are the most common tumors in women. It has been estimated that 40% of women over age 50 have benign uterine tumors involving smooth muscle and connective tissue cells. These **leiomyomas** (lē-ō-mī-Ō-maz), or *fibroids*, are stimulated by estrogens, and they can grow quite large, reaching weights as great as 13.6 kg (30 lb) if left untreated. Occlusion of the uterine tubes, distortion of adjacent organs, and compression of blood vessels may lead to a variety of complications. In young women, observation or conservative treatment with drugs or restricted surgery may be utilized. In older women, a decision may be made to remove the entire uterus, a procedure termed a *hysterectomy*.

Benign epithelial tumors in the uterus are called **endometrial polyps**. Roughly 10% of women probably have polyps, but because of their small size and lack of symptoms the condition passes unnoticed. If bleeding occurs or if the polyps become excessively enlarged, they can be surgically removed.

Uterine cancers are less common, affecting approximately 11.9 per 100,000 women. In 1992 there were roughly 46,000 new cases reported in the United States and roughly 10,000 deaths. There are two types of uterine cancers, *endometrial* and *cervical*.

Endometrial cancer is an invasive cancer of the endometrial lining. Roughly 32,000 cases are reported each year in the United States, with approximately 5500 deaths. The condition most often affects women age 50–70. Estrogen therapy, used to treat osteoporosis in postmenopausal women (see Osteoporosis and Other Skeletal Abnormalities Associated with Aging on page 755) increases the risk of endometrial cancer by 2–10 times. Adding cyclic progesterone therapy to the estrogen therapy seems to reduce or eliminate this risk.

There is no satisfactory screening test for endometrial cancer. The most common symptom is irregular bleeding, and diagnosis typically involves examination of a biopsy of the endometrial lining by suction or scraping. The prognosis varies with the degree of metastasis. Treatment of early-stage endometrial cancer involves a hysterectomy followed by localized radiation therapy. In advanced stages, more aggressive radiation treatment is recommended. Chemotherapy has not proven to be very successful in treating endometrial cancers, and only 30–40% of patients with advanced endometrial cancer benefit from this approach.

Cervical cancer can be detected early by the Pap smear, and depending on how advanced it is, can be treated by local surgery, hysterectomy, and/or irradiation. In 1996 there were estimates of 15,700 new cases and 4900 deaths in the United States.

SEXUALLY TRANSMITTED DISEASES *Page 706*

Sexually transmitted diseases, or **STDs**, are transferred from individual to individual, usually or exclusively by sexual intercourse. A variety of bacterial, viral, and fungal infections are included in this category. At least two dozen different STDs are currently recognized. All are unpleasant. *Chlamydia* can cause pelvic inflammatory disease (PID) and infertility. ∞ (p. 703) Other types of STDs are quite dangerous, and a few, including AIDS, are deadly. Here we will discuss three of the most common sexually transmitted diseases: *gonorrhea, syphilis,* and *herpes.* AIDS was discussed earlier in Chapter 23 on page 769.

Gonorrhea The bacterium *Neisseria gonorrhoeae* is responsible for gonorrhea, one of the most common sexually transmitted diseases in the United States. Nearly 2 million cases are reported each year. These bacteria usually invade epithelial cells lining the male or female reproductive tract. In relatively rare cases they will also colonize the pharyngeal or rectal epithelium.

The symptoms of genital infection vary, depending on the sex of the individual concerned. It has been estimated that up to 80% of women infected with gonorrhea experience no symptoms, or symptoms so minor that medical treatment is thought to be unnecessary. As a result these individuals act as carriers, spreading the infection through their sexual contacts. An estimated 10–15% of women infected with gonorrhea experience more acute symptoms, because the bacteria invade the epithelia of the uterine tubes. This type of infection probably accounts for many of the cases of PID in the U.S. population; as many as 80,000 women may become infertile each year as the result of scar tissue formation along the uterine tubes after gonorrheal infections.

Diagnosis in males seldom poses as great a problem, for all but 20–30% of infected males develop symptoms recognized as requiring immediate medical attention. The urethral invasion is accompanied by pain on urination (*dysuria*) and often a viscous urethral discharge. A sample of the discharge can be cultured to permit positive identification of the organism involved. Treatment of gonorrhea involves the administration of antibiotics.

Syphilis Syphilis (SIF-i-lis) results from infection by the bacterium *Treponema pallidum.* Untreated syphilis can cause serious cardiovascular and neurological illness years after infection, or it can be spread to the fetus during pregnancy to produce congenital malformations. The annual reported incidence of this disease has now risen to roughly 15 cases per 100,000 population, the highest rate in 40 years. An equivalent or greater number probably went unrecognized or unreported.

Primary syphilis begins as the bacteria cross the mucous epithelium and enter the lymphatics and bloodstream. At the invasion site the bacteria multiply, and after an incubation period ranging from 1.5 to 6 weeks their activities produce a raised lesion, or **chancre** (SHANG-ker). This lesion remains for several weeks before fading away, even without treatment. In heterosexual men the chancre usually appears on the penis; in women it may develop on the labia, vagina, or cervix. Lymph nodes in the region usually enlarge and remain swollen even after the chancre has disappeared.

Symptoms of **secondary syphilis** appear roughly 6 weeks later. Secondary syphilis usually involves a diffuse, reddish skin rash. Like the chancre, the rash fades over a period of 2–6 weeks. These symptoms may be accompanied by fever, headaches, and uneasiness. The combination is so vague that the disease may easily be overlooked or diagnosed as something else entirely. In a few instances more serious complications such as meningitis, hepatitis, or arthritis may develop.

The individual then enters the **latent phase**. The duration of the latent phase varies widely. Fifty to seventy percent of untreated individuals with latent syphilis fail to develop the symptoms of **tertiary syphilis**, or late syphilis, although the bacterial pathogens remain within their tissues. Those destined to develop tertiary syphilis may do so 10 or more years after infection.

The most severe symptoms of tertiary syphilis involve the CNS and the cardiovascular system. Neurosyphilis may result from bacterial infection of the meninges or the tissues of the brain or spinal cord or both. **Tabes dorsalis** (TĀ-bēz dor-SAL-is) results from the invasion and demyelination of the posterior columns of the spinal cord and the sensory ganglia and nerves. In the cardiovascular system the disease affects the major vessels, leading to aortic stenosis, aneurysms, or calcification.

Equally disturbing are the effects of transmission from mother to fetus across the placenta. These cases of **congenital syphilis** are marked by infections of the developing bones and cartilages of the skeleton and progressive damage to the spleen, liver, bone marrow, and kidneys. The risk of transmission may be as high as 80–95%, so maternal blood testing is recommended early in pregnancy.

Treatment of syphilis involves the administration of penicillin or other antibiotics.

Herpes Virus **Genital herpes** results from infection by herpesviruses. Two different viruses are involved. Eighty to ninety percent of genital herpes cases are caused by a specific virus known as *HV-2* (herpes simplex virus Type 2), a virus usually associated with the genitalia. The remaining cases are caused by *HV-1,* the same virus responsible for cold sores on the mouth. Typically within a week of the initial infection the individual develops a number of painful, ulcerated lesions on the external genitalia. In women ulcers may also appear on the cervix. These gradually heal over the next 2–3 weeks. Recurring infections are common, although subsequent incidents are less severe.

Infection of the newborn infant during delivery with herpesviruses present in the vagina can lead to serious illness because the infant has few immunological defenses. Recent development of the antiviral agent *acyclovir* has helped treatment of initial infections.

ENDOMETRIOSIS *Page 707*

In **endometriosis** (en-dō-mē-trē-Ō-sis) an area of sloughed endometrial tissue reattaches elsewhere in the body and begins to grow outside the uterus. The severity of the condition depends on the size of the abnormal mass and its location. Abdominal pain, bleeding, pressure on adjacent structures, and infertility are common symptoms. As the island of endometrial tissue enlarges, the symptoms become more severe.

Diagnosis can usually be made by using a laparoscope inserted through a small opening in the abdominal wall. Using this device a physician can see the outer surfaces of the uterus and uterine tubes, the ovaries, and the lining of the pelvic cavity. Treatment may involve surgical removal of the endometrial masses or hormonal therapies that suppress menstruation.

CONCEPTION CONTROL *Page 715*

For physiological, logistical, financial, or emotional reasons most adults practice some form of conception control during their reproductive years. When the simplest and most obvious method, sexual abstinence, is unsatisfactory for some reason, another method of contraception must be used to avoid unwanted pregnancies.

Well over 50% of U.S. women age 15–44 are practicing some method of contraception; there are many different methods of contraception; only a few will be considered here.

Sterilization makes one unable to provide functional gametes for fertilization. Either sexual partner may be sterilized with the same net result. In a **vasectomy** (vaz-EK-tō-mē) a segment of the ductus deferens is removed, making it impossible for spermatozoa to pass from the epididymis to the distal portions of the reproductive tract. The surgery can be performed in a physician's office in a matter of minutes. After a 1-cm section is removed, the cut ends are usually tied shut, and scar tissue forms a permanent seal. Alternatively, the cut ends of the ductus deferens can be blocked with silicone plugs that can later be removed to restore fertility. After a vasectomy the man experiences normal sexual function, for the epididymal and testicular secretions normally account for only around 5% of the volume of the semen. Spermatozoa continue to develop, but they remain within the epididymis until they degenerate.

In the female the uterine tubes can be blocked through a surgical procedure known as a **tubal ligation**. Because the surgery involves entering the abdominopelvic cavity, complications are more likely than with the vasectomy. As in a vasectomy, attempts may be made to restore fertility after a tubal ligation.

Oral contraceptives manipulate the female hormonal cycle so that ovulation does not occur. Contraceptive pills use a combination of progestins and estrogens that prevents production of FSH and the resulting stimulation of ovarian follicles. There are now at least 20 brands of combination oral contraceptives available, and over 200 million women are using them worldwide. In the United States, 25% of women under age 45 use the combination pill to prevent conception.

The **condom**, also called a *prophylactic*, or "rubber," covers the body of the penis during intercourse and keeps spermatozoa from reaching the female reproductive tract. **Vaginal barriers** such as the *diaphragm, cervical cap*, or *vaginal sponge* rely on similar principles. A diaphragm, the most popular form of vaginal barrier in use at the moment, consists of a dome of latex rubber with a small metal hoop supporting the rim. Because vaginas vary in size, women choosing this method must be individually fitted. Before intercourse the diaphragm is inserted so that it covers the cervical os, and it is usually coated with a small amount of spermicidal jelly or cream, adding to the effectiveness of the barrier. The cervical cap is smaller and lacks the metal rim. It, too, must be fitted carefully, but unlike the diaphragm it may be left in place for several days. The vaginal sponge consists of a small synthetic sponge saturated with a *spermicide*, a sperm-killing foam or jelly.

An **intrauterine device (IUD)** consists of a small plastic loop or a T that can be inserted into the uterine chamber. The mechanism of action remains uncertain, but it is known that IUDs stimulate prostaglandin production in the uterus. The net result is an alteration in the chemical composition of uterine secretions, and the changes in the intrauterine environment lower the chances for fertilization and subsequent implantation. IUDs are in limited use today in the United States, but they remain popular in many other countries.

Vaginal rings and the **Norplant system** involve silicone rubber structures impregnated with progesterone hormones that diffuse at a constant rate. The vaginal rings are placed in the vagina, while the Norplant system involves the insertion of small rods beneath the skin. Both methods have undergone clinical trials, and the Norplant system has shown long-term effectiveness comparable to that of combined oral contraceptives, at lower progesterone levels. Although the Norplant system is now available by prescription, a relatively high cost has to date limited the use of this contraceptive method. An injectable progesterone that provides contraception for three months, and a progesterone-only pill that is taken daily are also available.

Male contraceptives are also under development. *Gossypol*, a yellow pigment extracted from cottonseed oil, produces a dramatic decline in sperm count and sperm motility after 2 months. It can be administered topically, as it is readily absorbed through the skin. Fertility usually returns within a year after treatment is discontinued.

chapter 28

Human Development

CHROMOSOMAL ANALYSIS *Page 720*

In **amniocentesis** a sample of amniotic fluid is removed and the fetal cells that it contains are analyzed. This procedure permits the identification of at least 20 different congenital conditions. The needle inserted to obtain a fluid sample is guided into position using ultrasound. Unfortunately, amniocentesis has two major drawbacks:

1. Because the sampling procedure represents a potential threat to the health of the fetus and mother, amniocentesis is performed only when there are known risk factors present. Examples of risk factors would include a family history for specific conditions, or in the case of Down syndrome, a maternal age over 35.

2. Sampling cannot safely be performed until the volume of amniotic fluid is large enough that the fetus will not be injured during the sampling process. The usual time for amniocentesis is at a gestational age of 14–15 weeks. It may take several weeks to obtain results once samples have been collected, and by the time the results are received, the option of therapeutic abortion may no longer be available.

A procedure called *chorionic villi sampling* analyzes cells collected from the villi during the first trimester. This procedure has a higher risk of miscarriage and misdiagnosis.

TECHNOLOGY AND THE TREATMENT OF INFERTILITY *Page 722*

Infertility, or the inability to have children, has been the focus of media attention for the past 10 years. An estimated 10–15% of U.S. marriages are infertile, and another 10% are unable to have as many children as desired. An infertile, or sterile, woman is unable to produce functional eggs and/or support a developing embryo. An infertile man is incapable of providing a sufficient number of motile spermatozoa for successful fertilization. Because sterility of either sexual partner will have the same result, diagnosis and treatment of infertility must involve evaluation of both sexual partners. Approximately 60% of infertility cases can be attributed to problems with the female reproductive system.

In cases of male infertility due to low sperm counts, semen from several ejaculates can be pooled, concentrated, and introduced into the female reproductive tract. This technique, known as *artificial insemination*, may lead to normal fertilization and pregnancy. If the husband cannot produce functional spermatozoa, viable spermatozoa can be obtained from a "sperm bank" that stores donor sperm cells.

When there are problems with the transport of the egg from the ovary to the uterine tube, due to scarring of the fimbriae or other problems, a procedure called **GIFT** (*gamete intrafallopian tube transfer*) can be used. In this process, mature oocytes are removed from the ovaries, placed in the uterine tubes, and exposed to high concentrations of spermatozoa from the husband or donor.

In the GIFT procedure, fertilization occurs in its normal location, within the uterine tube. This site is not essential, and fertilization can also take place in a test tube or Petri dish. This process is called **in vitro fertilization** (*vitro*, glass), or **IVF**. If a carefully controlled fluid environment is provided, early development will proceed normally. One variation on the GIFT procedure, called **ZIFT** (*zygote intrafallopian tube transfer*), exposes selected oocytes to spermatozoa outside the body and inserts zygotes or early pre-embryos, rather than mature oocytes, into the uterine tubes. Alternatively, the zygote can be maintained in an artificial environment through the first 2–3 days of development. This procedure is often selected if the uterine tubes are damaged or blocked. The pre-embryo is then placed directly into the uterus, rather than into one of the uterine tubes.

PROBLEMS WITH THE IMPLANTATION PROCESS *Page 723*

The trophoblast undergoes repeated nuclear divisions, shows extensive and rapid growth, has a very high demand for energy, invades and spreads through adjacent tissues, and fails to activate the maternal immune system. In short, the trophoblast has many of the characteristics of cancer cells. In an estimated 0.1% of pregnancies, a definitive placenta does not develop, and instead the syncytial trophoblast runs wild, forming a **gestational neoplasm** or **hydatidiform** (hī-da-TID-i-fōrm) **mole**. Prompt surgical removal of the mass is essential, sometimes followed by chemotherapy, for about 20% of hydatidiform moles will metastasize, invading other tissues with potentially fatal results.

Implantation usually occurs at the endometrial surface lining the uterine cavity. The precise location within the uterus varies, although most often implantation occurs in the body of the uterus. This is not an ironclad rule, and in an **ectopic pregnancy** implantation occurs somewhere other than within the uterus.

The incidence of ectopic pregnancies is approximately 0.6%. Women who douche regularly have a 4.4 times higher risk of experiencing an ectopic pregnancy than those who don't, presumably because the flushing action pushes the zygote away from the uterus. If the uterine tube has been scarred by a previous episode of pelvic inflammatory disease, there is also an increased risk of an ectopic pregnancy. Although implantation may occur within the peritoneal cavity, in the ovarian wall, or in the cervix, 95% of ectopic pregnancies involve implantation within a uterine tube. The tube cannot expand enough to accommodate the developing embryo, and it usually ruptures during the first trimester. At this time the hemorrhaging that occurs in the peritoneal cavity may be severe enough to pose a threat to the woman's life.

In a few instances the ruptured uterine tube releases the embryo with an intact umbilical cord, and further development can occur. About 5% of these **abdominal pregnancies** actually complete full-term development; normal birth cannot occur, but the infant can be surgically removed from the abdominopelvic cavity. Because abdominal pregnancies are possible, it has been suggested that men as well as women could act as surrogate mothers if a zygote were surgically implanted in the peritoneal wall. It is not clear how the endocrine, cardiovascular, nervous, and other systems of a man would respond to the stresses of pregnancy. The procedure has been tried successfully in mice, however, and experiments continue.

PROBLEMS WITH PLACENTATION *Page 725*

In a **placenta previa** (PRĒ-vē-a; "in the way") implantation occurs in or near the cervix. This condition causes problems as the growing placenta approaches the internal cervical orifice. In a total placenta previa the placenta actually extends across the internal orifice, while a partial placenta previa only partially blocks the os. The placenta in this condition is characterized by a rich fetal blood supply with erosion of maternal blood vessels within the endometrium. Where the placenta passes across the internal orifice the delicate complex hangs like an unsupported water balloon. As the pregnancy advances, even minor mechanical stresses can be enough to tear the placental tissues, leading to massive fetal and maternal hemorrhaging.

Without ultrasound scanning, most cases are not diagnosed until the seventh month of pregnancy, as the placenta reaches its full size. At this time the dilation of the cervical canal and the weight of the uterine contents are pushing against the placenta where it bridges the internal orifice. Minor, painless hemorrhaging usually appears as the first sign of the condition. The diagnosis can usually be confirmed by ultrasound scanning. Treatment in cases of total placenta previa usually involves bed rest for the mother until the fetus reaches a size at which cesarean delivery can be performed with a reasonable chance of neonatal (newborn) survival.

In an **abruptio placentae** (ab-RUP-shē-ō pla-SEN-tē) part or all of the placenta tears away from the uterine wall sometime after the fifth month of gestation. The bleeding into the uterine cavity and the pain that follows usually will be noted and reported, although in some cases the shifting placenta may block the passage of blood through the cervical canal. In severe cases the hemorrhaging leads to maternal anemia, shock, and kidney failure. Although maternal mortality is low, the fetal mortality rate from this condition ranges from 30 to 100%, depending on the severity of the hemorrhaging.

TERATOGENS AND ABNORMAL DEVELOPMENT *Page 725*

Teratogens (ter-AT-ō-jens) are stimuli that disrupt normal development by damaging cells, altering chromosome structure, or acting as abnormal inducers. **Teratology** (ter-a-TOL-ō-jē) is literally the "study of monsters," and it considers extensive departures from the pathways of normal development. Teratogens that affect the embryo in the first trimester will potentially disrupt cleavage, gastrulation, or neurulation. The embryonic survival rate will be low, and the survivors will usually have severe anatomical and physiological defects affecting all of the major organ systems. Errors introduced into the developmental process during the second and third trimesters will be more likely to affect specific organs or organ systems, for the major organizational patterns are already established. Nevertheless, the alterations reduce the chances for long-term survival.

Many powerful teratogens are encountered in everyday life. The location and severity of the resulting defects vary depending on the nature of the stimulus and the time of exposure. Radiation is a powerful teratogen that can affect all living cells. Even the X-rays used in diagnostic procedures can break chromosomes and produce developmental errors; thus nonionizing procedures such as ultrasound are used to track embryonic and fetal development. Fetal exposure to the microorganisms responsible for syphilis or rubella ("German measles") can also produce serious developmental abnormalities, including congenital heart defects, mental retardation, and deafness.

Even practices usually considered socially acceptable for adults may not be acceptable to the fetus. Two examples are consumption of alcohol and smoking. **Fetal alcohol syndrome (FAS)** occurs when maternal alcohol consumption produces developmental defects such as skeletal deformation, cardiovascular defects, and neurological disorders. Mortality rates can be as high as 17%, and the survivors are plagued by problems in later development. The most severe cases involve mothers who consume the alcohol content of at least 7 ounces of hard liquor, 10 beers, or several bottles of wine each day. But because the effects produced are directly related to the degree of exposure, there is probably no level of alcohol consumption that can be considered completely safe. Fetal alcohol syndrome is the number one cause of mental retardation in the United States today, affecting roughly 7500 infants each year.

Smoking presents another major risk to the developing fetus. In addition to introducing potentially harmful chemicals, such as nicotine, smoking lowers the oxygen content of maternal blood and reduces the amount of oxygen arriving at the placenta. The fetus carried by a smoking mother will not grow as rapidly as one carried by a nonsmoker, and smoking increases the risks of spontaneous abortion, prematurity, and fetal death. There is also a higher rate of infant mortality after delivery, and postnatal development can be adversely affected.

COMPLEXITY AND PERFECTION: AN IMPROBABLE DREAM *Page 734*

The expectation of prospective parents that every pregnancy will be idyllic and every baby a perfect specimen reflects deep-seated misconceptions about the nature of the developmental process. These misconceptions lead many to believe that when serious developmental errors occur someone or something is at fault, and blame might be assigned to maternal habits, such as smoking, alcohol consumption, improper diet, maternal exposure to toxins or prescription drugs, or the presence of other disruptive stimuli in the environment. The prosecution of women giving birth to severely impaired infants for "fetal abuse" (exposing a fetus to known or suspected risk factors) represents an extreme example of this philosophy.

Although environmental stimuli may indeed lead to developmental problems, such factors are only one component of a complex system normally subject to considerable variation. Even if every pregnant woman were packed in cotton fluff and locked in her room from conception to delivery, developmental accidents and errors would continue to appear with regularity.

Spontaneous mutations are the result of random errors in the replication process, and such incidents are relatively common. At least 10% of fertilizations produce zygotes with abnormal chromosomes, and because spontaneous mutations usually fail to produce visible defects, the actual number of mutations must be far larger. Most of the affected zygotes die before completing development, and only about 0.5% of newborn infants show chromosomal abnormalities resulting from spontaneous mutations.

Because of the nature of the regulatory mechanisms, prenatal development does not follow precise, predetermined pathways. For example, considerable variation exists in the pathways of blood vessels and nerves, because it doesn't matter how the blood or neural impulses get to their destination, as long as they do get there. If the variations fall outside acceptable limits, however, the embryo or fetus fails to complete development. Very small changes in heart structure may result in the death of a fetus, while large variations in venous distribution are extremely common and relatively harmless. Virtually everyone can be considered abnormal to some degree, because no one has characteristics that are statistically "average" in every respect. An estimated 20% of your genes differ from those found in the majority of the population, and minor defects such as extra nipples or birthmarks are quite common.

Current evidence suggests that as many as half of all conceptions produce zygotes that do not survive the cleavage stage. These disintegrate within the uterine tubes or uterine cavity, and because implantation never occurs there are no obvious signs of pregnancy. These instances of preimplantation mortality are often associated with chromosomal abnormalities. Of those embryos that implant, roughly 20% fail to complete 5 months of development, with an average survival time of 8 weeks. Severe problems affecting early embryogenesis or placenta formation are usually responsible.

Prenatal mortality tends to eliminate the most severely affected fetuses. Those with less extensive defects may survive, completing full-term gestation or arriving via premature delivery. **Congenital malformations** are structural abnormalities, present at birth, that affect major systems. Spina bifida, hydrocephaly, and Down syndrome are among the most common congenital malformations; these conditions were described in earlier chapters of the text and this appendix. The incidence of congenital malformations at birth averages around 6%, but only one-third of them are categorized as severe. Only 10% of these congenital problems can be attributed to environmental factors in the absence of chromosomal abnormalities or genetic factors, including a family history of similar or related defects.

Medical technology continues to improve our abilities to understand and manipulate physiological processes. Genetic analysis of potential parents may now provide estimates concerning the likelihood of specific problems, although the problems themselves remain outside our control. But even with a better understanding of the genetic mechanisms involved, it will probably never be possible to control every aspect of development and thereby prevent spontaneous abortions and congenital malformations. There are simply too many complex, interdependent steps involved in prenatal development, and malfunctions of some kind are statistically inevitable.

WEIGHTS AND MEASURES

Accurate descriptions of physical objects would be impossible without a precise method of reporting the pertinent data. Dimensions such as length and width are reported in standardized units of measurement, such as inches or centimeters. These values can be used to calculate the volume of an object, a measurement of the amount of space it fills. **Mass** is another important physical property. The mass of an object is determined by the amount of matter it contains; on earth the mass of an object determines its weight.

Most U.S. readers describe length and width in terms of inches, feet, or yards; volumes in pints, quarts, or gallons; and weights in ounces, pounds, or tons. These are units of the **U.S. system** of measurement. Table A-1 summarizes the familiar and unfamiliar terms used in the U.S. system. For reference purposes, this table also includes a definition of the "household units," popular in recipes and cookbooks. The U.S. system can be very difficult to work with, because there is no logical relationship between the various units. For example, there are 12 inches in a foot, 3 feet in a yard, and 1760 yards in a mile. Without a clear pattern of organization, converting feet to inches or miles to feet can be confusing and time-consuming. The relationships between ounces, pints, quarts and gallons, or ounces, pounds, and tons are no more logical.

In contrast, the **metric system** has a logical organization based on powers of 10, as indicated in Table A-2. For example, a **meter (m)** represents the basic unit for the measurement of size. When measuring larger objects, data can be reported in terms of **dekameters** (*deka,* ten), **hectometers** (*hekaton,* hundred), or **kilometers** (**km**; *chilioi,* thousand); for smaller objects, data can be reported in **decimeters** (**dm**; 0.1 m; *decem,* ten), **centimeters** (**cm** = 0.01 m; *centum,* hundred), **millimeters** (**mm** = 0.001 m; *mille,* thousand), and so forth. Notice that the same prefixes are used to report weights, based on the **gram (g)**, and volumes, based on the **liter (ℓ)**. This text reports data in metric units, usually with U.S. equivalents. You

---

TABLE A-1   The U.S. System of Measurement

Physical Property	Unit	Relationship to Other U.S. Units	Relationship to Household Units
Length	inch (in.)	1 in. = 0.083 ft	
	foot (ft)	1 ft = 12 in.	
		= 0.33 yd	
	yard (yd)	1 yd = 36 in.	
		= 3 ft	
	mile (mi)	1 mi = 5,280 ft	
		= 1,760 yd	
Volume	fluidram (fl dr)	1 fl dr = 0.125 fl oz	
	fluid ounce (fl oz)	1 fl oz = 8 fl dr	= 6 teaspoons (tsp)
		= 0.0625 pt	= 2 tablespoons (tbsp)
	pint (pt)	1 pt = 128 fl dr	= 32 tbsp
		= 16 fl oz	= 2 cups (c)
		= 0.5 qt	
	quart (qt)	1 qt = 256 fl dr	= 4 c
		= 32 fl oz	
		= 2 pt	
		= 0.25 gal	
	gallon (gal)	1 gal = 128 fl oz	
		= 8 pt	
		= 4 qt	
Mass			
	grain (gr)	1 gr = 0.002 oz	
	dram (dr)	1 dr = 27.3 gr	
		= 0.063 oz	
	ounce (oz)	1 oz = 437.5 gr	
		= 16 dr	
	pound (lb)	1 lb = 7000 gr	
		= 256 dr	
		= 16 oz	
	ton (t)	1 t = 2000 lb	

TABLE A-2 The Metric System of Measurement

Physical Property	Unit	Relationship to Standard Metric Units	Conversion to U.S. Units	
Length	nanometer (nm)	1 nm = 0.000000001 m (10^{-9})	= 4×10^{-8} in.	25,000,000 nm = 1 in.
	micrometer (µm)	1 µm = 0.000001 m (10^{-6})	= 4×10^{-5} in.	25,000 µm = 1 in.
	millimeter (mm)	1 mm = 0.001 m (10^{-3})	= 0.0394 in.	25.4 mm = 1 in.
	centimeter (cm)	1 cm = 0.01 m (10^{-2})	= 0.394 in.	2.54 cm = 1 in.
	decimeter (dm)	1 dm = 0.1 m (10^{-1})	= 3.94 in.	0.25 dm = 1 in.
	meter (m)	standard unit of length	= 39.4 in.	0.0254 m = 1 in.
			= 3.28 ft	0.3048 m = 1 ft
			= 1.09 yd	0.914 m = 1 yd
	dekameter (dam)	1 dam = 10 m		
	hectometer (hm)	1 hm = 100 m		
	kilometer (km)	1 km = 1000 m	= 3280 ft	
			= 1093 yd	
			= 0.62 mi	1.609 km = 1 mi
Volume	microliter (µℓ)	1 µℓ = 0.000001 ℓ (10^{-6}) = 1 cubic millimeter (mm^3)		
	milliliter (mℓ)	1 mℓ = 0.001 ℓ (10^{-3}) = 1 cubic centimeter $(cm^3$ or cc)	= 0.03 fl oz	5 mℓ = 1 tsp
				15 mℓ = 1 tbsp
				30 mℓ = 1 fl oz
	centiliter (cℓ)	1 cℓ = 0.01 ℓ (10^{-2})	= 0.34 fl oz	3 cℓ = 1 fl oz
	deciliter (dℓ)	1 dℓ = 0.1 ℓ (10^{-1})	= 3.38 fl oz	0.29 dℓ = 1 fl oz
	liter (ℓ)	standard unit of volume	= 33.8 fl oz	0.0295 ℓ = 1 fl oz
			= 2.11 pt	0.473 ℓ = 1 pt
			= 1.06 qt	0.946 ℓ = 1 qt
Mass	picogram (pg)	1 pg = 0.000000000001 g (10^{-12})		
	nanogram (ng)	1 ng = 0.000000001 g (10^{-9})		
	microgram (µg)	1 µg = 0.000001 g (10^{-6})	= 0.000015 gr	66,666 µg = 1 gr
	milligram (mg)	1 mg = 0.001 g (10^{-3})	= 0.015 gr	66.7 mg = 1 gr
	centigram (cg)	1 cg = 0.01 g (10^{-2})	= 0.15 gr	6.7 cg = 1 gr
	decigram (dg)	1 dg = 0.1 g (10^{-1})	= 1.5 gr	0.67 dg = 1 gr
	gram (g)	standard unit of mass	= 0.035 oz	28.35 g = 1 oz
			= 0.0022 lb	453.6 g = 1 lb
	dekagram (dag)	1 dag = 10 g		
	hectogram (hg)	1 hg = 100 g		
	kilogram (kg)	1 kg = 1000 g	= 2.2 lb	0.453 kg = 1 lb
	metric ton (kt)	1 kt = 1000 kg	= 1.1 t	
			= 2205 lb	0.907 kt = 1 t

Temperature	Centigrade		Fahrenheit	
Freezing point of pure water	0°		32°	
Normal body temperature	36.8°		98.6°	
Boiling point of pure water	100°		212°	
Conversion °C → °F:	°F = (1.8 × °C) + 32	°F → °C:	°C = (°F − 32) × 0.56	

should use this opportunity to become familiar with the metric system, because most technical sources report data only in metric units, and most of the rest of the world uses the metric system exclusively. Conversion factors are included in Table A-2.

Pharmacies at one time used the **apothecary system**, a relatively specialized system of measurement borrowed from England when America was still a colony. This system has been largely replaced by the metric system, and we will ignore apothecary units in this text. The apothecary system deals only with volumes and weights, and the volumetric units are comparable to those of the U.S. system. The two systems differ, however, in terms of the definitions of mass units.

The U.S. and metric systems also differ in their methods of reporting temperatures; in the United States, temperatures are usually reported in degrees Fahrenheit (°F), whereas scientific literature and individuals in most other countries report temperatures in degrees Centigrade or Celsius (°C). The relationship between temperatures in degrees Fahrenheit and those in degrees Centigrade has been indicated at the bottom of Table A-2.

ANSWERS TO CONCEPT CHECK QUESTIONS

CHAPTER 1

Page 6

1. A histologist investigates the structure and properties of tissues.
2. A gross anatomist investigates organ systems and their relationships to the body as a whole.

Page 19

1. The two eyes would be separated by a midsagittal section.
2. The body cavity inferior to the diaphragm is the abdominopelvic (or peritoneal) cavity.
3. The fall would affect your forearm.

CHAPTER 2

Page 37

1. The fingerlike projections on the surface of the intestinal cells are microvilli. They serve to increase the cells' surface area so that these cells can absorb nutrients more efficiently.
2. Since the flagellum is an organelle of locomotion, sperm cells lacking a flagellum would be unable to move. (As a result, they would probably be unable to reach and fertilize an egg.)

Page 43

1. The function of mitochondria is to produce energy for the cell in the form of ATP molecules. A large number of mitochondria in a cell would indicate a high demand for energy.
2. SER functions in the synthesis of lipids such as steroids. Ovaries and testes would be expected to have a great deal of SER because they produce large amounts of steroid hormones.

CHAPTER 3

Page 62

1. No. A simple squamous epithelium does not provide enough protection against infection, abrasion, and dehydration and is not found in the skin surface.
2. The process described is holocrine secretion.

Page 78

1. The tissue is probably deep fascia, a type of dense regular connective tissue.
2. Collagen fibers add strength to connective tissue. We would therefore expect vitamin C deficiency to result in the production of connective tissue that is weaker and more prone to damage.

Page 80

1. This is an example of a mucous membrane.
2. Since cardiac and skeletal muscle are both striated (banded), this must be smooth muscle tissue.
3. All of these regions are subject to mechanical trauma and abrasion—by food (pharynx and esophagus), feces (anus), and intercourse or childbirth (vagina).

CHAPTER 4

Page 91

1. Cells are constantly shed from the outer layers of the stratum corneum.
2. The splinter is lodged in the stratum granulosum.

Page 92

1. Keratin fibers develop within cells of the stratum granulosum. As keratin fibers are produced, these cells become thinner and flatter, and their cell membranes become thicker and less permeable. As these cells die, they form the densely packed layers of the stratum lucidum and stratum corneum.
2. Sanding the tips of one's fingers will not permanently remove fingerprints. Since the ridges of the fingerprints are formed in layers of the skin that are constantly regenerated, they will eventually reappear. The actual pattern of the ridges is determined by arrangement of tissue in the dermis, which is not affected by sanding.

Page 99

1. When the dermis is stretched excessively, the elastic fibers are overstretched and are not able to recoil. The skin then forms folds or wrinkles, called stretch marks, in the affected areas.
2. Contraction of the arrector pili muscles pulls the hair follicles erect, depressing the area at the base of the hair and making the surrounding skin appear higher. The result is known as "goose bumps" or "goose pimples."

3. Hair is a derivative of the epidermis, and if the epidermis were destroyed by the injury there would be no hair follicles to produce new hair.

Page 102

1. Sebaceous glands secrete an oily lipid (sebum) that coats hair shafts and epidermis, provides lubrication to the epidermis, and inhibits the growth of bacteria.
2. Apocrine sweat glands produce a secretion containing several kinds of organic compounds. Some of these have an odor, and others produce an odor when metabolized by skin bacteria. Deodorants are used to mask the odor of these secretions.

CHAPTER 5

Page 114

1. If the ratio of collagen to hydroxyapatite in a bone increased, the bone would be more flexible and less strong.
2. Concentric layers of bone around a central canal are indicative of an osteon or Haversian systems. Osteons make up compact bone. Since the ends (epiphyses) of long bones are primarily cancellous (spongy) bone, this sample most likely came from the shaft (diaphysis) of a long bone.
3. Since osteoclasts function in breaking down or demineralizing bone, the bone would have less mineral content and as a result it would be weaker.

Page 119

1. Long bones of the body, like the femur, have a plate of cartilage called the epiphyseal plate, that separates the epiphysis from the diaphysis as long as the bone is still growing lengthwise. An X-ray would indicate whether or not the epiphyseal plate was still present. If it was, then growth was still occurring, and if not the bone had reached its adult length.

Page 121

1. Bones increase in thickness in response to physical stress. One common type of stress that is applied to a bone is that produced by muscles. We would expect the bones of an athlete to be thicker after the addition of the extra muscle mass because of the greater stress that the muscle would apply to the bone.

CHAPTER 6

Page 144

1. Each internal jugular vein passes through an opening between the temporal bone and occipital bone.
2. The sella turcica contains the pituitary gland and is located in the sphenoid bone.
3. Nerve fibers to the olfactory bulb, which deals with the sense of smell, pass through the cribriform plate from the nasal cavity. If the cribriform plate failed to form, these sensory nerves could not reach the olfactory bulbs and the sense of smell (olfaction) would be lost.

Page 154

1. The paranasal sinuses function to make some of the skull bones lighter and to produce mucus.
2. The most powerful muscles that are involved in closing the mouth attach to the mandible at the coronoid process. A fracture of the coronoid process would make it difficult for these muscles to function properly and close the mouth.
3. Since many muscles that move the tongue and the larynx are attached to the hyoid bone, you would expect a person with a fractured hyoid bone to have difficulty moving their tongue and swallowing.

Page 167

1. The odontoid process is found on the second cervical vertebra, or axis, which is located in the neck.
2. Improper compression of the chest during CPR could and frequently does result in a fracture of the sternum or ribs.

CHAPTER 7

Page 181

1. The clavicle attaches the scapula to the sternum and thus restricts the scapula's range of movement. If the clavicle is broken, the scapula will have a greater range of movement and will be less stable.
2. The two rounded prominences on either side of the elbow are parts of the humerus (the lateral and medial epicondyles).
3. The radius is in a lateral position when the forearm is in the anatomical position.

Page 199

1. The three bones that make up the coxa are the ilium, ischium, and pubic bones.

2. Although the fibula is not part of the knee joint nor does it bear weight, it is an important point of attachment for many leg muscles. When the fibula is fractured, these muscles cannot function properly to move the leg and walking is difficult and painful. The fibula also helps stabilize the ankle joint.
3. Mark has most likely fractured the calcaneus (heel bone).

CHAPTER 8

Page 212
1. Originally, the joint is a type of syndesmosis. When the bones fuse, the bones along the suture represent a synostosis.
2. a. abduction
   b. supination
   c. flexion

Page 219
1. Since the subscapular bursa is located in the shoulder joint, an inflammation of this structure (bursitis) would be found in the tennis player. The condition is associated with repetitive motion that occurs at the shoulder, such as swinging a tennis racket. The jogger would be more at risk for injuries to the knee joint.
2. Mary has most likely fractured her ulna.

Page 225
1. The iliofemoral, pubofemoral, and ischiofemoral ligaments would all be found in the hip joint.
2. Damage to the menisci in the knee joint would result in a decrease in the joint's stability. The individual would have a harder time locking the knee in place while standing and would have to use muscle contractions to stabilize the joint. When standing for long periods, the muscles would fatigue and the knee would "give out." We would also expect the individual to experience pain.

CHAPTER 9

Page 239
1. Since tendons attach muscles to bones, severing the tendon would disconnect the muscle from the bone so that when the muscle contracted nothing would happen.
2. Skeletal muscle appears striated when viewed under the microscope because it is composed of the myofilaments actin and myosin, which are arranged in such a way as to produce a banded appearance in the muscle.

Page 243
1. During contraction the I band gets smaller, but the width of the A band remains constant.
2. Stimulation of a motor neuron triggers the release of chemicals at the neuromuscular junction, which alter the transmembrane potential of the sarcolemma. This change sweeps across the surface of the sarcolemma and into the T-tubules. The change in the transmembrane potential of the T-tubules triggers the release of calcium ions by the sarcoplasmic reticulum. This release initiates the contraction.

Page 244
1. A motor unit with 1500 fibers is most likely from a large muscle involved in powerful, gross body movement. Muscles that control fine and/or precise movements such as movement of the eye or the fingers have only a few fibers per motor unit, whereas muscles of the legs, for instance, that are involved in powerful contractions have hundreds of fibers per motor unit.

Page 245
1. The sprinter requires large amounts of energy for a relatively short burst of activity. To supply this demand for energy, the muscles switch to anaerobic metabolism. Anaerobic metabolism is not as efficient in producing energy as aerobic metabolism, and the process also produces acidic waste products. The combination of less energy and the waste products contributes to fatigue. Marathon runners, on the other hand, derive most of their energy from aerobic metabolism, which is more efficient and does not produce the level of waste products that anaerobic metabolism does.
2. Individuals who are naturally better at endurance types of activities such as cycling or marathon running have a higher percentage of slow-twitch muscle fibers, which are physiologically better adapted to this type of activity than the fast-twitch fibers, which are less vascular and fatigue faster.

CHAPTER 10

Page 253
1. A circular muscle, or sphincter.
2. The main antagonist of the biceps brachii muscle is the triceps brachii muscle.
3. This is a long muscle which flexes a digit.

Page 270
1. Biting and chewing.
2. Damage to the external intercostal muscles would interfere with the process of breathing.

3. A blow to the rectus abdominis would cause the muscle to contract forcefully, resulting in flexion of the torso. In other words, you would "double up."

CHAPTER 11

Page 291
1. When you shrug your shoulders you are contracting your levator scapulae muscles.
2. The rotator cuff muscles include the supraspinatus, infraspinatus, subscapularis, and teres minor. The tendons of these muscles help to enclose and stabilize the shoulder joint.
3. Injury to the flexor carpi ulnaris would impair the ability to flex and adduct the hand.

Page 301
1. Injury to the obturator muscle would interfere with your ability to rotate your thigh laterally.
2. The hamstring refers to a group of five muscles that collectively function in flexing the leg. These muscles are the biceps femoris, semimembranous, semitendinosus, gracilis, and sartorius.
3. The calcaneal tendon attaches the soleus and gastrocnemius muscles to the calcaneus (heel bone). When these muscles contract, they cause extension of the foot. A torn calcaneal tendon would make plantar flexion of the foot difficult, and the opposite action, dorsiflexion, would be more pronounced as a result of less antagonism from the soleus and gastrocnemius.

CHAPTER 13

Page 330
1. Sensory neurons of the peripheral nervous system are usually unipolar; thus this tissue is most likely associated with a sensory organ.
2. Microglial cells are small phagocytic cells that are found in increased number in damaged and diseased areas of the CNS.

Page 331
1. Cutting the axon of a neuron prevents the transmission of the nerve impulse along the length of the axon.
2. Because of saltatory conduction, myelinated fibers conduct action potentials much faster than nonmyelinated fibers, so the axon conducting at 50 m/s is myelinated.

CHAPTER 14

Page 343
1. The ventral root of spinal nerves is composed of visceral and somatic motor fibers. Damage to this root would interfere with motor function.
2. The cerebrospinal fluid that surrounds the spinal cord is found in the subarachnoid space, which lies beneath the arachnoid membrane and the pia mater.

Page 353
1. Since the poliovirus would be located in the somatomotor neurons, we would find it in the anterior gray horns of the spinal cord where the cell bodies of these neurons are located.
2. The dorsal rami of spinal nerves innervates the skin and muscles of the back. In this case we would expect the skin and muscles of the back of the neck and shoulders to be affected.
3. The phrenic nerves that innervate the diaphragm originate in the brachial plexus. Damage to this plexus or more specifically to the phrenic nerves would greatly interfere with the ability to breathe and possibly result in death.

CHAPTER 15

Page 381
1. If one of the interventricular foramina became blocked, cerebrospinal fluid would not be able to flow from the first or second ventricle to the third. Since cerebrospinal fluid would continue to be formed, the blocked ventricle would swell with fluid, a condition known as hydrocephalus.
2. The primary motor cortex is located in the precentral gyrus of the frontal lobe of the cerebrum.

Page 389
1. Changes in body temperature would stimulate the preoptic area of the hypothalamus, a division of the diencephalon.
2. The vermis and arbor vitae are structures associated with the cerebellum.
3. Even though the medulla oblongata is small, it contains many vital reflex centers, including those that control breathing and regulate the heart and blood pressure. Damage to the medulla oblongata can result in a cessation of breathing, or changes in heart rate and blood pressure that are incompatible with life.

Page 401
1. Damage to the vagus nerve (cranial nerve X) can result in death, since it has motor fibers to regulate breathing, heart rate, and blood pressure.
2. The cranial nerve responsible for tongue movements is the hypoglossal nerve.

8. Since the abducens nerve (cranial nerve VI) controls lateral movements of the eyes, we would expect an individual with damage to this nerve to be unable to move their eyes laterally.

CHAPTER 16

Page 417

1. The fasciculus gracilis in the posterior column of the spinal cord is responsible for carrying information about touch and pressure from the lower part of the body to the brain.
2. The anatomical basis for opposite side motor control is that crossing over (decussation) occurs, and the pyramidal motor fibers innervate lower motor neurons on the opposite side of the body.
3. The superior portion of the motor cortex exercises control over the hand, arm, and upper portion of the leg. An injury to this area would affect the ability to control the muscles in those regions of the body.

Page 422

1. An inability to comprehend the written or spoken word indicates a problem with the general interpretive area of the brain, which in most individuals is located in the left temporal lobe of the cerebrum.

Page 423

1. The reticular activating system (RAS) is responsible for rousing the cerebrum to a state of consciousness. If a sleeping individual's RAS were stimulated, she would certainly wake up.
2. The brain is affected by age. The common age-related anatomical changes in the CNS are: (1) the loss of cortical neurons; (2) decrease in blood flow to the brain due to the accumulation of fatty deposits; (3) decrease in the number and organization of dendritic branching and synaptic connections; and (4) abnormal accumulation of intercellular deposits, including lipofuscin, plaques, and neurofibrillary tangles.

CHAPTER 17

Page 436

1. The neurons that synapse in the collateral ganglia originate in the lower thoracic and upper lumbar portion of the spinal cord and pass through the chain ganglia to the collateral ganglia.
2. Blocking the beta receptors on cells would decrease or prevent sympathetic stimulation of those tissues. This would result in decreased heart rate and force of contraction, and relaxation of the smooth muscle in the walls of blood vessels. These changes would contribute to lowering a person's blood pressure.

CHAPTER 18

Page 455

1. Since nociceptors are pain receptors, if they were stimulated, you would perceive a painful sensation in your affected hand.
2. Proprioceptors relay information about limb position and movement to the central nervous system, especially the cerebellum. Lack of this information would result in uncoordinated movements and the individual would probably be unable to walk.

Page 465

1. Without the movement of the round window, the perilymph would be moved by the vibration of the stapes at the oval window, and there would be little or no perception of sound.
2. Loss of stereocilia (as a result of constant exposure to loud noises for instance) would reduce hearing sensitivity and could eventually result in deafness.

Page 473

1. Since the cornea is avascular, the cells of the cornea must obtain oxygen and nutrients from the tear fluid that passes over its surface.
2. The two structures most affected by an abnormally high intraocular pressure are: (1) the canal of Schlemm (the aqueous humor no longer has free access to this structure); and (2) the optic nerve (the nerve fibers of this structure are distorted, which affects visual perception).
3. An individual born without cones would be able to see only in black and white (monochromatic) and would have very poor visual acuity.

CHAPTER 19

Page 494

1. Most of the thyroid hormone in the blood is bound to proteins called thyroid-binding globulins. This represents a large reservoir of thyroxine that guards against rapid fluctuations in the level of this important hormone. Because there is such a large amount stored in this way, it takes several days to deplete the supply of hormone, even after the thyroid gland has been removed.

2. Removal of the parathyroid glands would result in a decrease in the blood levels of calcium ions. This could be counteracted by increasing the amount of vitamin D and calcium in the diet.

CHAPTER 20

Page 514

1. If a person with Type A blood received a transfusion of Type B blood, the red cells would clump or agglutinate, potentially blocking blood flow to various organs and tissues. When blood is not adequately supplied to tissues, these tissues may be damaged or destroyed.
2. In an infected cut we would expect to find a large number of neutrophils. Neutrophils are phagocytic white cells that are usually the first to arrive at the site of an injury and that specialize in dealing with infectious bacteria.

CHAPTER 21

Page 525

1. The most obvious characteristic that differentiates cardiac muscle tissue from skeletal muscle tissue is that cardiac muscle cells are small with a centrally placed nucleus, and they are connected to neighboring cells at intercalated disks.

Page 533

1. The pulmonary trunk.
2. When the ventricles begin to contract, they force the AV valves to close, which in turn pulls on the chordae tendineae, which then pull on the papillary muscles. The papillary muscles respond by contracting, counteracting the force that is pushing the valves upward.
3. If these cells were not functioning, the heart would still continue to beat but at a slower rate.
4. Damage to the left bundle branch would affect systolic and diastolic phases, so left ventricular contraction would not occur at an appropriate time. The ability to eject blood at a high pressure from the left ventricle is impaired.

Page 537

1. The cardiac centers of the medulla receive continuous information about CO_2, O_2, pH, and pressure via the sensory fibers of the vagus nerve. It is on the basis of this information that the cardiac centers adjust the cardiac output. In addition, the parasympathetic control of heart rate and force contraction would be lost.

CHAPTER 22

Page 550

1. The blood vessels are veins. Arteries and arterioles have a relatively large amount of smooth muscle tissue in a thick, well-developed tunica media.
2. Blood pressure in the arterial system pushes blood into the capillaries. Blood pressure on the venous side is very low, and other forces help keep the blood moving. Valves in the walls of venules and medium-sized veins permit blood flow in only one direction, toward the heart, preventing the backflow of blood toward the capillaries.

Page 571

1. The left subclavian artery is the branch of the aorta that sends blood to the left shoulder and upper limb.
2. Organs served by the celiac artery include the stomach, spleen, liver, and pancreas.

CHAPTER 23

Page 595

1. The thoracic duct drains lymph from the area beneath the diaphragm and the left side of the head and thorax. Most of the lymph enters the venous blood by way of this duct. A blockage of this duct would not only impair circulation of lymph through most of the body; it would also promote accumulation of fluid in the extremities (lymphedema).
2. During an infection, the lymphocytes and phagocytes in the lymph nodes in the affected region undergo cell division to better deal with the infectious agent. This increase in the number of cells in the node causes the node to become enlarged or swollen.

CHAPTER 24

Page 616

1. The tracheal cartilages are C-shaped to allow room for esophageal expansion when large portions of food or liquid are swallowed.
2. Chronic smoking damages the lining of the air passageways. Cilia are seared off the surface of the cells by the heat, and the large number of particles that escape filtering are trapped in the excess mucus that is secreted to protect the irritated lining. This combination of circumstances creates a situation in which there is a large amount of thick mucus that is difficult to clear from the passages. The cough reflex is an attempt to remove this material from the airways.

1. Since the rib penetrates the chest wall, the thoracic cavity will be damaged, as well as the inner pleura. Atmospheric air will then enter the pleural cavity. This space is normally at a lower pressure than the outside air, so when the air enters, the natural elasticity of the lungs will not be compensated and the lung will collapse. This condition is called a pneumothorax.
2. As a result of emphysema the larger air spaces and lack of elasticity will increase compliance.
3. Since the fluid produced in pneumonia takes up space that would normally be occupied by air, the vital capacity will be decreased.

CHAPTER 25

Page 633

1. The components of the mucosa are: (1) mucosal epithelium, depending upon location, may be simple or stratified; (2) lamina propria, loose connective tissue underlying the epithelium; and (3) muscularis mucosae, bands of smooth muscle fibers arranged in concentric layers. The mucosa of the digestive tract is an example of a mucous membrane, serving both absorptive and secretory functions.
2. Mesenteries provide an access route for the passage of blood vessels, nerves, and lymphatics to and from the digestive tract.
3. Peristalsis is waves of muscular contractions that move substances the length of the digestive tube. Segmentation activities churn and mix the contents of the small and large intestines but do not produce net movement in a particular direction.

Page 637

1. The oral cavity is lined by the oral mucosa, which is composed of nonkeratinized stratified squamous epithelium.
2. At mealtime, saliva lubricates the mouth and dissolves chemicals that stimulate the taste buds. Saliva contains the digestive enzyme salivary amylase, which begins the chemical breakdown of carbohydrates.

Page 639

1. Parasympathetic stimulation increases muscle tone and motility in the digestive tract. A drug that blocks this activity would decrease the rate of peristalsis.
2. The fauces is the opening between the oral cavity and the pharynx.
3. The process that is being described is deglutition, or swallowing.

Page 644

1. The vagus nerve contains parasympathetic motor fibers that can stimulate gastric secretions. This can occur even if food is not present in the stomach (cephalic phase of gastric digestion). Cutting the branches of the vagus that supply the stomach would prevent this type of secretion from occurring and decrease the chance of ulcer formation.

Page 648

1. The characteristic lining of the small intestine contains plicae circulares, which support intestinal villi. The villi are covered by a simple columnar epithelium whose plasma membrane free surface portions are specialized microvilli. This arrangement increases the total area for digestion and absorption to over 200 m².

2. A narrowing of the ileocecal valve would interfere with the flow of materials from the small intestine to the large intestine.

Page 657

1. Removal of the gallbladder due to gallstones or inflammation requires the surgical removal of fundus, body, and neck portions of the gallbladder and the adjacent cystic duct.
2. The condition of cystic fibrosis interferes with the digestion of sugars, starches, lipids, nucleic acids, and proteins.

CHAPTER 26

Page 675

1. The route blood must take from the renal artery to the glomerulus is: Renal artery → Segmental arteries → Interlobar arteries → Arcuate arteries → Interlobular arteries → Afferent arterioles → Glomerulus.
2. The route blood must take from the glomerulus to the renal vein is: Glomerulus → Efferent arterioles → Peritubular capillaries, vasa recta → Venules → Interlobular veins → Arcuate veins → Interlobar veins → Segmental veins → Renal vein.
3. Filtrate flows from glomerulus to a minor calyx via the following route: glomerulus → Descending limb of loop → Loop of Henle → Ascending limb of loop → Collecting tubule → Collecting duct → Papillary duct → Minor calyx.

Page 680

1. An obstruction of the ureters would interfere with the passage of urine from the renal pelvis to the urinary bladder.
2. The ureters penetrate the posterior wall of the urinary bladder at an oblique angle, terminating in a ureteral opening within the bladder. The ureteral opening appears as a slitlike opening, which prevents the backflow of urine toward the ureter and kidney.

CHAPTER 27

Page 698

1. The cremaster muscle as well as the dartos muscle would be relaxed on a warm day, so that the scrotal sac could descend away from the warmth of the body and cool the testes.
2. A vasectomy has no effect on spermatogenesis (sperm production).

Page 714

1. A blockage of the uterine tube would cause sterility.
2. Blockage of a single lactiferous sinus would not interfere with milk moving to the nipple because each breast usually has between 15 and 20 lactiferous sinuses.

CHAPTER 28

Page 730

1. The inner cell mass of the blastocyst eventually develops into the embryo.
2. Improper development of the yolk sac and allantois would affect the development and function of the circulatory system.

FOREIGN WORD ROOTS, PREFIXES, SUFFIXES, AND COMBINING FORMS

Many of the words we use in everyday English have their roots in other languages, particularly Greek and Latin. This is especially true for anatomical terms, many of which were introduced into the anatomical literature by Greek and Roman anatomists. This list includes some of the foreign word roots, prefixes, suffixes, and combining forms that are part of many of the biological and anatomical terms you will see in this text.

Each entry starts with the commonly encountered form or forms of the prefix, suffix, or combining form followed by the word root (shown in italics) with its English translation. One example is given to illustrate the use of the prefix, suffix, or combining form, but there are many others, and you will see them as you progress through the text.

a-, *a-,* without: avascular
ab-, *ab,* from: abduct
-ac, *-akos,* pertaining to: cardiac
ad-, *ad,* to, toward: adduct
aden-, adeno-, *adenos,* gland: adenoid
af-, *ad,* toward: afferent
-al, -alis, pertaining to: brachial
-algia, *algos,* pain: neuralgia
ana-, *ana,* up, back: anaphase
andro-, *andros,* male: androgen
angio-, *angeion,* vessel: angiogram
anti-, ant-, *anti,* against: antibiotic
apo-, *apo,* from: apocrine
arachn-, *arachne,* spider: arachnoid
arthro-, *arthros,* joint: arthroscopy
-asis, -asia, state, condition: homeostasis
astro-, *aster,* star: astrocyte
atel-, *ateles,* imperfect: atelectasis
baro-, *baros,* pressure: baroreceptor
bi-, *bi-,* two: bifurcate
blast-, -blast, *blastos,* precursor: blastocyst
brachi-, *brachium,* arm: brachiocephalic
brady-, *bradys,* slow: bradycardia
bronch-, *bronchus,* windpipe, airway: bronchial
cardi-, cardio-, -cardia, *kardia,* heart: cardiac
-centesis, *kentesis,* puncture: thoracocentesis
cerebro-, *cerebrum,* brain: cerebrospinal
chole-, *chole,* bile: cholecystitis
chondro-, *chondros,* cartilage: chondrocyte
chrom-, chromo-, *chroma,* color: chromatin
circum-, *circum,* around: circumduction
-clast, *klastos,* broken: osteoclast
coel-, -coel, *koila,* cavity: coelom
contra-, *contra,* against: contralateral
cranio-, *cranium,* skull: craniosacral
cribr-, *cribrum,* sieve: cribriform
-crine, *krinein,* to separate: endocrine
cyst-, -cyst, *kystis,* sac: blastocyst
desmo-, *desmos,* band: desmosome
di-, *dis,* twice: disaccharide
dia-, *dia,* through: diameter
diure-, *diourein,* to urinate: diuresis
dys-, *dys-,* painful: dysmenorrhea
-ectasis, *ektasis,* expansion: atelectasis
ecto-, *ektos,* outside: ectoderm
ef-, *ex,* away from: efferent
emmetro-, *emmetros,* in proper measure: emmetropia
encephalo-, *enkephalos,* brain: encephalitis
end-, endo-, *endos,* inside: endometrium
entero-, *enteron,* intestine: enteric
epi-, *epi,* on: epimysium
erythema-, *erythema,* flushed (skin): erythematosis
erythro-, *erythros,* red: erythrocyte
ex-, *ex,* out, away from: exocytosis
ferr-, *ferrum,* iron: transferrin
-gen, -genic, *gennan,* to produce: mutagen

genicula-, *geniculum,* kneelike structure: geniculate
genio-, *geneion,* chin: geniohyoid
glosso-, -glossus, *glossus,* tongue: hypoglossal
glyco-, *glykys,* sugar: glycogen
-gram, *gramma,* record: myogram
-graph, -graphia, *graphein,* to write, record: electroencephalograph
gyne-, gyno-, *gynaikos,* woman: gynecologist
hem-, hemato-, *haima,* blood: hemopoiesis
hemi-, *hemi-,* half: hemisphere
hepato-, *hepaticus,* liver: hepatocyte
hetero-, *heteros,* other: heterosexual
histo-, *histos,* tissue: histology
holo-, *holos,* entire: holocrine
homeo-, homo-, *homos,* same: homeostasis
hyal-, hyalo-, *hyalos,* glass: hyaline
hydro-, *hydros,* water: hydrolysis
hyo-, *hyoeides,* U-shaped: hyoid
hyper-, *hyper,* above: hyperpolarization
ili-, ilio-, *ilium:* iliac
infra-, *infra,* beneath: infraorbital
inter-, *inter,* between: interventricular
intra-, *intra,* within: intracapsular
ipsi-, *ipse,* itself: ipsilateral
iso-, *isos,* equal: isotonic
-itis, *-itis,* inflammation: dermatitis
karyo-, *karyon,* body: megakaryocyte
kerato-, *keros,* horn: keratin
kino-, -kinin, *kinein,* to move: bradykinin
lact-, lacto-, -lactin, *lac,* milk: prolactin
-lemma, *lemma,* husk: plasmalemma
leuko-, *leukos,* white: leukocyte
liga-, *ligare,* to bind together: ligase
lip-, lipo-, *lipos,* fat: lipoid
lyso-, -lysis, -lyze, *lysis,* dissolution: hydrolysis
mal-, *mal,* abnormal: malabsorption
mamilla-, *mamilla,* little breast: mamillary
mast-, masto-, *mastos,* breast: mastoid
mega-, *megas,* big: megakaryocyte
mero-, *meros,* part: merocrine
meso-, *mesos,* middle: mesoderm
meta-, *meta,* after, beyond: metaphase
mono-, *monos,* single: monocyte
morpho-, *morphe,* form: morphology
-mural, *murus,* wall: intramural
myelo-, *myelos,* marrow: myeloblast
myo-, *mys,* muscle: myofilament
natri-, *natrium,* sodium: natriuretic
neur-, neuro-, *neuron,* nerve: neuromuscular
oculo-, *oculus,* eye: oculomotor
oligo-, *oligos,* little, few: oligopeptide
-ology, *logos,* the study of: physiology
-oma, -oma, swelling: carcinoma
onco-, *onkos,* mass, tumor: oncology
-opia, *ops,* eye: optic
-osis, -osis, state, condition: neurosis
osteon, osteo-, *os,* bone: osteocyte
oto-, *otikos,* ear: otoconia

para-, *para,* beyond: paraplegia
patho-, -path, -pathy, *pathos,* disease: pathology
pedia-, *paidos,* child: pediatrician
peri-, *peri,* around: perineurium
-phasia, *phasis,* speech: aphasia
-phil, -philia, *philus,* love: hydrophilic
-phobe, -phobia, *phobos,* fear: hydrophobic
-phylaxis, *phylax,* a guard: prophylaxis
physio-, *physis,* nature: physiology
-plasia, *plasis,* formation: dysplasia
platy-, *platys,* flat: platysma
-plegia, *plege,* a blow, paralysis: paraplegia
-plexy, *plessein,* to strike: apoplexy
podo, *podon,* foot: podocyte
-poiesis, *poiesis,* making: hemopoiesis
poly-, *polys,* many: polysaccharide
presby-, *presbys,* old: presbyopia
pro-, *pro,* before: prophase
pterygo-, *pteryx,* wing: pterygoid
pulp-, *pulpa,* flesh: pulpitis
retro-, *retro,* backward: retroperitoneal
-rrhea, *rhein,* flow, discharge: amenorrhea
sarco-, *sarkos,* flesh: sarcomere
scler-, sclero-, *skleros,* hard: sclera
semi-, *semis,* half: semitendinosus
-septic, *septikos,* putrid: antiseptic
-sis, state or condition: metastasis
som-, -some, *soma,* body: somatic
spino-, *spina,* spine, vertebral column: spinodeltoid
-stomy, *stoma,* mouth, opening: colostomy
stylo-, *stylus,* stake, pole: styloid
sub-, *sub,* below; subcutaneous
syn-, *syn,* together: synthesis
tachy-, *tachys,* swift: tachycardia
telo-, *telos,* end: telophase
therm-, thermo-, *therme,* heat: thermoregulation
-tomy, *temnein,* to cut: appendectomy
trans-, *trans,* through: transudate
-trophic, -trophin, -trophy, *trophikos,* nourishing: adrenocorticotropic
tropho-, *trophe,* nutrition: trophoblast
tropo-, *tropikos,* turning: troponin
uro-, -uria, *ouron,* urine: glycosuria

EPONYMS IN COMMON USE

Eponym	Equivalent Terms	Individual Referenced
The cellular level of organization (Chapter 2)		
Golgi apparatus		Camillo Golgi (1844–1926), Italian histologist; shared Nobel Prize in 1906
Krebs cycle	Tricarboxylic or citric acid cycle	Hans Adolph Krebs (1900–1981), British biochemist; shared Nobel Prize in 1953
The skeletal system (Chapters 5–8)		
Colles' fracture		Abraham Colles (1773–1843), Irish surgeon
Haversian canals	Central canals	Clopton Havers (1650–1702), English anatomist and microscopist
Haversian systems	Osteons	
Pott's fracture		Percivall Pott (1714–1788), English surgeon
Volkmann's canals	Perforating canals	Alfred Wilhelm Volkmann (1800–1877), German surgeon
Wormian bones	Sutural bones	Olas Worm (1588–1654), Danish anatomist
The muscular system (Chapters 9–11)		
Achilles tendon	Calcaneal tendon	Achilles, hero of Greek mythology
Cori cycle		Carl Ferdinand Cori (1896–) and Gerty Theresa Cori (1896–1957), American biochemists; shared Nobel Prize in 1947
The nervous system (Chapters 13–17)		
Broca's center	Speech center	Pierre Paul Broca (1824–1880), French surgeon
Foramina of Luschka	Lateral foramina	Hubert von Luschka (1820–1875), German anatomist
Foramen of Magendie	Median foramen	François Magendie (1783–1855), French physiologist
Foramen of Munro	Interventricular foramen	John Cummings Munro (1858–1910), American surgeon
Nissl bodies		Franz Nissl (1860–1919), German neurologist
Purkinje cells		Johannes E. Purkinje (1787–1869), Czechoslovakian physiologist
Nodes of Ranvier		Louis Antoine Ranvier (1835–1922), French physiologist
Island of Reil	Insula	Johann Christian Reil (1759–1813), German anatomist
Fissure of Rolando	Central sulcus	Luigi Rolando (1773–1831), Italian anatomist
Schwann cells		Theodor Schwann (1810–1882), German anatomist
Aqueduct of Sylvius	Mesencephalic aqueduct	Jacobus Sylvius (Jacques Dubois, 1478–1555), French anatomist
Sylvian fissure	Lateral sulcus	Franciscus Sylvius (Franz de le Boë, 1614–1672), Dutch anatomist
Pons varolii	Pons	Costanzo Varolio (1543–1575), Italian anatomist
Sensory function (Chapter 18)		
Organ of Corti		Alfonso Corti (1822–1888), Italian anatomist
Eustachian tube	Auditory (pharyngotympanic) tube	Bartolomeo Eustachio (1520–1574), Italian anatomist
Golgi tendon organs	Tendon organs	*See* Golgi apparatus *under* The Cellular Level (Chapter 2)
Hertz (Hz)		Heinrich Hertz (1857–1894), German physicist
Meibomian glands		Heinrich Meibom (1638–1700), German anatomist
Corpuscles of Meissner		Georg Meissner (1829–1905), German physiologist
Merkel's discs		Friedrich Siegismund Merkel (1845–1919), German anatomist
Pacinian corpuscles		Filippo Pacini (1812–1883), Italian anatomist
Ruffini's corpuscles		Angelo Ruffini (1864–1929), Italian anatomist
Canal of Schlemm		Friedrich S. Schlemm (1795–1858), German anatomist
Glands of Zeis		Edward Zeis (1807–1868), German ophthalmologist
The endocrine system (Chapter 19)		
Islets of Langerhans	Pancreatic islets	Paul Langerhans (1847–1888), German pathologist

Eponym	Equivalent Terms	Individual Referenced
Interstitial cells of Leydig	Interstitial cells	Franz von Leydig (1821–1908), German anatomist
The cardiovascular system (Chapters 20–22)		
Bundle of His		Wilhelm His (1863–1934), German physician
Purkinje cells		See under The Nervous System (Chapters 13–17)
Starling's law		Ernest Henry Starling (1866–1927), English physiologist
Circle of Willis	Cerebral arterial circle	Thomas Willis (1621–1675), English physician
The lymphatic system (Chapter 23)		
Hassall's corpuscles		Arthur Hill Hassall (1817–1894), English physician
Kupffer cells		Karl Wilhelm Kupffer (1829–1902), German anatomist
Langerhans cells		See Islets of Langerhans under The Endocrine System (Chapter 19)
Peyer's patches	Aggregate lymphoid nodules	Johann Conrad Peyer (1653–1712), Swiss anatomist
The respiratory system (Chapter 24)		
Adam's apple	Laryngeal prominence of thyroid cartilage	Biblical reference
Bohr effect		Niels Bohr (1885–1962), Danish physicist; won 1922 Nobel Prize
Boyle's law		Robert Boyle (1621–1691), English physicist
Charles' law		Jacques Alexandre César Charles (1746–1823), French physicist
Dalton's law		John Dalton (1766–1844), English physicist
Henry's law		William Henry (1775–1837), English chemist
The digestive system (Chapter 25)		
Plexus of Auerbach	Myenteric plexus	Leopold Auerbach (1827–1897), German anatomist
Brunner's glands	Duodenal glands	Johann Conrad Brunner (1653–1727), Swiss anatomist
Kupffer cells	Stellate cells	See under The Lymphatic System (Chapter 23)
Crypts of Lieberkuhn	Intestinal crypts	Johann Nathaniel Lieberkuhn (1711–1756), German anatomist
Plexus of Meissner	Submucosal plexus	See Corpuscles of Meissner under Sensory Function (Chapter 18)
Sphincter of Oddi	Hepatopancreatic sphincter	Ruggero Oddi (1864–1913), Italian physician
Peyer's patches		See under The Lymphatic System (Chapter 23)
Duct of Santorini	Accessory pancreatic duct	Giovanni Domenico Santorini (1681–1737), Italian anatomist
Stensen's duct	Parotid duct	Niels Stensen (1638–1686), Danish physician/priest
Ampulla of Vater	Duodenal ampulla	Abraham Vater (1684–1751), German anatomist
Wharton's duct	Submandibular duct	Thomas Wharton (1614–1673), English physician
Foramen of Winslow	Epiploic foramen	Jacob Benignus Winslow (1669–1760), French anatomist
Duct of Wirsung	Pancreatic duct	Johann Georg Wirsung (1600–1643), German physician
The urinary system (Chapter 26)		
Bowman's capsule	Glomerular capsule	Sir William Bowman (1816–1892), English physician
Loop of Henle		Friedrich Gustav Jakob Henle (1809–1885), German histologist
The reproductive system (Chapters 27–28)		
Bartholin's glands	Greater vestibular glands	Casper Bartholin, Jr. (1655–1738), Danish anatomist
Cowper's glands	Bulbourethral glands	William Cowper (1666–1709), English surgeon
Fallopian tube	Uterine tube/oviduct	Gabriele Falloppio (1523–1562), Italian anatomist
Graafian follicle	Tertiary follicle	Reijnier de Graaf (1641–1673), Dutch physician
Interstitial cells of Leydig	Interstitial cells	See under The Endocrine System (Chapter 19)
Glands of Littre	Lesser vestibular glands	Alexis Littre (1658–1726), French surgeon
Sertoli cells	Sustentacular cells	Enrico Sertoli (1842–1910), Italian histologist

ABBREVIATIONS USED IN THIS TEXT

ACh	acetylcholine
AChE	acetylcholinesterase
ACTH	adrenocorticotropic hormone
ADH	antidiuretic hormone
ADP	adenosine diphosphate
AIDS	acquired immune deficiency syndrome
AMP	adenosine monophosphate
ANP	atrial natriuretic peptide
ANS	autonomic nervous system
ARC	AIDS-related complex
ARDS	adult respiratory distress syndrome
ATP	adenosine triphosphate
ATPase	adenosine triphosphatase
AV	atrioventricular
bpm	beats per minute
CAD	coronary artery disease
cAMP	cyclic AMP
CAPD	continuous ambulatory peritoneal dialysis
CCK	cholecystokinin
CHF	congestive heart failure
CNS	central nervous system
COPD	chronic obstructive pulmonary disease
CP	creatine phosphate
CPK or CK	creatine phosphokinase
CPM	continual passive motion
CPR	cardiopulmonary resuscitation
CRF	chronic renal failure
CRH	corticotropin-releasing hormone
CSF	cerebrospinal fluid; colony-stimulating factors
CT	computed tomography; calcitonin
CVA	cerebrovascular accident
CVS	cardiovascular system
DCT	distal convoluted tubule
DDST	Denver Developmental Screening Test
DJD	degenerative joint disease
DMD	Duchenne's muscular dystrophy
DNA	deoxyribonucleic acid
DOPA	dopamine
DPG	diphosphoglycerate
ECF	extracellular fluid
ECG	electrocardiogram
EDV	end-diastolic volume
EEG	electroencephalogram
EKG	electrocardiogram

ELISA	enzyme-linked immunosorbent assay
ESV	end-systolic volume
FES	functional electrical stimulation
FSH	follicle-stimulating hormone
GABA	gamma aminobutyric acid
GAS	general adaptation syndrome
GC	glucocorticoids
GH	growth hormone
GH-IH	growth hormone-inhibiting hormone
GH-RH	growth hormone-releasing hormone
GIP	glucose-dependent insulinotropic hormone
GnRH	gonadotropin-releasing hormone
GTP	guanosine triphosphate
Hb	hemoglobin
HCG	human chorionic gonadotropin
HDL	high-density lipoprotein
HDN	hemolytic disease of the newborn
HIV	human immunodeficiency virus
HLA	human leukocyte antigen
HMD	hyaline membrane disease
HPL	human placental lactogen
HR	heart rate
HTLV-III	human T cell lymphotropic virus, type III (= HIV)
Hz	Hertz
ICF	intracellular fluid
ICSH	interstitial cell-stimulating hormone, also called LH
IM	intramuscular
ISF	interstitial fluid
LDL	low-density lipoprotein
LH	luteinizing hormone
LM	light micrograph
MC	mineralocorticoid
MI	myocardial infarction
mm Hg	millimeters of mercury
mOsm	milliosmoles
MRI	magnetic resonance imaging
mRNA	messenger RNA
MS	multiple sclerosis
MSH	melanocyte-stimulating hormone
MSH-IH	melanocyte-stimulating hormone–inhibiting hormone
NE	norepinephrine
NRDS	neonatal respiratory distress syndrome

PAC	premature atrial contractions
PAT	paroxysmal atrial tachycardia
PCT	proximal convoluted tubule
PCV	packed cell volume
PEEP	positive end-expiratory pressure
PET	positron-emission tomography
PFC	perfluorochemical emulsions
PIH	prolactin-inhibiting hormone
PMN	polymorphonuclear leukocyte
PNS	peripheral nervous system
PRL	prolactin
PTH	parathyroid hormone
PVC	premature ventricular contraction
RAS	reticular activating system
RBC	red blood cell
REM	rapid eye movement
RER	rough endoplasmic reticulum
RHD	rheumatic heart disease
RNA	ribonucleic acid
rRNA	ribosomal RNA
SA	sinoatrial
SCID	severe combined immunodeficiency
SEM	scanning electron micrograph
SER	smooth endoplasmic reticulum
SIADH	syndrome of inappropriate ADH secretion
SIDS	sudden infant death syndrome
SLE	systemic lupus erythematosus
SV	stroke volume
SVC	superior vena cava
T_3	triiodothyronine
T_4	tetraiodothyronine, also called thyroxine
TB	tuberculosis
TBG	thyroid-binding globulin
TEM	transmission electron micrograph
TIA	transient ischemic attack
TRH	thyrotropin-releasing hormone
tRNA	transfer RNA
TSH	thyroid-stimulating hormone
TX	thyroxine
UTI	urinary tract infection
UV	ultraviolet
VF	ventricular fibrillation
VPRC	volume of packed red cells
VT	ventricular tachycardia
WBC	white blood cell

GLOSSARY OF KEY TERMS

abdomen: Region of trunk bounded by the diaphragm and pelvis.

abdominopelvic cavity: Portion of the ventral body cavity that contains abdominal and pelvic subdivisions.

abducens (ab-DŪ-senz): Cranial nerve VI; innervates the lateral rectus muscle of the eye.

abduction: Movement away from the midline.

abortion: Premature loss or expulsion of an embryo or fetus.

abscess: A localized collection of pus within a damaged tissue.

absorption: The active or passive uptake of gases, fluids, or solutes.

accommodation: Alteration in the curvature of the lens to focus an image on the retina; decrease in receptor sensitivity or perception following chronic stimulation.

acetabulum (a-se-TAB-ū-lum): Fossa on lateral aspect of pelvis that accommodates the head of the femur.

acetylcholine (ACh) (as-e-til-KŌ-lēn): Chemical neurotransmitter in the brain and PNS; dominant neurotransmitter in the PNS, released at neuromuscular junctions and synapses of the parasympathetic division.

acetylcholinesterase (AChE): Enzyme found in the synaptic cleft, bound to the postsynaptic membrane, and in tissue fluids; breaks down and inactivates ACh molecules.

achalasia (ak-a-LĀ-zē-a): Condition that develops when the lower esophageal sphincter fails to dilate, and ingested materials cannot enter the stomach.

Achilles tendon: Calcaneal tendon.

acid: A compound whose dissociation in solution releases a hydrogen ion and an anion; an acid solution has a pH below 7.0 and contains an excess of hydrogen ions.

acinus/acini (AS-i-nī): Histological term referring to a blind pocket, pouch, or sac.

acne: Condition characterized by inflammation of sebaceous glands and follicles; commonly affects adolescents and most often involves the face.

acoustic: Pertaining to sound or the sense of hearing.

acquired immune deficiency syndrome (AIDS): A disease caused by the **human immunodeficiency virus (HIV),** characterized by destruction of helper T cells and a resulting severe impairment of the immune response.

acromegaly: Condition caused by overproduction of growth hormone in the adult, characterized by thickening of bones and enlargement of cartilages and other soft tissues.

acromion (a-KRŌ-mē-on): Continuation of the scapular spine that projects above the capsule of the scapulohumeral joint.

acrosomal cap (ak-rō-SŌ-mal): Membranous sac at the tip of a sperm cell that contains hyaluronic acid.

actin: Protein component of microfilaments; forms thin filaments in skeletal muscles and produces contractions of all muscles through interaction with thick (myosin) filaments; *see* **sliding filament theory.**

action potential: A conducted change in the transmembrane potential of excitable cells, initiated by a change in the membrane permeability to sodium ions: *see also* **nerve impulse.**

active transport: The ATP-dependent absorption or excretion of solutes across a cell membrane.

acute: Sudden in onset, severe in intensity, and brief in duration.

adaptation: Alteration of pupillary size in response to changes in light intensity; in CNS often used as a synonym for accommodation: physiological responses that produce acclimatization.

Addison's disease: Condition resulting from hyposecretion of glucocorticoids, characterized by lethargy, weakness, hypotension, and increased skin pigmentation.

adduction: Movement toward the axis or midline of the body as viewed in the anatomical position.

adenine: A purine, one of the nitrogen bases in the nucleic acids RNA and DNA.

adenohypophysis (ad-e-nō-hī-POF-i-sis): The anterior portion of the pituitary gland, also called the **anterior pituitary** or the **pars distalis.**

adenoid: The pharyngeal tonsil.

adenosine triphosphate (ATP): A high-energy compound consisting of adenosine with three phosphate groups attached; the third is attached by a high-energy bond.

adhesion: Fusion of two mesenteric layers following damage or irritation of their opposing surfaces.

adipocyte (AD-i-pō-sīt): A fat cell.

adipose tissue: Loose connective tissue dominated by adipocytes.

adrenal cortex: Superficial portion of adrenal gland that produces steroid hormones.

adrenal gland: Small endocrine gland secreting steroids and catecholamines, located superior to each kidney.

adrenal medulla: Core of the adrenal gland; a modified sympathetic ganglion that secretes catecholamines into the blood following sympathetic activation.

adrenocortical hormone: Any of the steroids produced by the adrenal cortex.

adrenocorticotropic hormone (ACTH): Hormone that stimulates the production and secretion of glucocorticoids by the zona fasciculata of the adrenal cortex; released by the anterior pituitary in response to CRF.

adventitia (ad-ven-TISH-a): Superficial layer of connective tissue surrounding an internal organ; fibers are continuous with those of surrounding tissues, providing support and stabilization.

afferent: Toward.

afferent arteriole: An arteriole bringing blood to the glomerulus of the kidney.

afferent fiber: Axons carrying sensory information to the CNS.

agglutination (a-gloo-ti-NĀ-shun): Aggregation of red blood cells due to interactions between surface agglutinogens and plasma agglutinins.

agglutinins (a-GLOO-ti-ninz): Immunoglobulins in plasma that react with antigens on the surfaces of foreign red blood cells when donor and recipient differ in blood type.

agglutinogens (a-GLOO-tin-ō-jenz): Antigens on the surfaces of red blood cells whose presence and structure are genetically determined.

agonist: A muscle responsible for a specific movement.

agranular: Without granules; **agranular leukocytes** are monocytes and lymphocytes; the **agranular reticulum** is an intracellular organelle that synthesizes and stores carbohydrates and lipids.

AIDS: *see* **acquired immune deficiency syndrome.**

AIDS-related complex (ARC): Early symptoms of HIV infection, consisting chiefly of lymphadenopathy, fevers, and chronic nonfatal infections.

alba, albicans, albuginea (AL-bi-kanz) (al-bū-JIN-ē-a): White.

albinism: Absence of pigment in hair and skin caused by inability of body to produce melanin.

albumins (al-BŪ-mins): The smallest of the plasma proteins; function as transport proteins and important in contributing to plasma oncotic pressure.

aldosterone: A mineralocorticoid (steroid) produced by the zona glomerulosa of the adrenal cortex that stimulates sodium and water conservation at the kidneys; secreted in response to the presence of angiotensin II.

aldosteronism: Condition caused by the oversecretion of aldosterone, characterized by fluid retention, edema, and hypertension.

allantois (a-LAN-tō-is): One of the extraembryonic membranes; it provides vascularity to the chorion and is therefore essential to placenta formation; the proximal portion becomes the urinary bladder.

alpha cells: Cells in the pancreatic islets that secrete glucagon.

alpha receptors: Membrane receptors sensitive to norepinephrine or epinephrine; stimulation usually results in excitation of the target cell.

alveolar sac: An air-filled chamber that supplies air to several alveoli.

alveolus/alveoli (al-VĒ-ō-lī): Blind pockets at the end of the respiratory tree, lined by a simple squamous epithelium and surrounded by a capillary network; gas exchange with the blood occurs here.

Alzheimer's disease: Disorder resulting from degenerative changes in populations of neurons in the cerebrum, causing dementia characterized by problems with attention, short-term memory, and emotions.

amacrine cells (AM-a-krīn): Modified neurons in the retina that facilitate or inhibit communication between bipolar and ganglion cells.

amino acids: Organic compounds whose chemical structure can be summarized as $R-CHNH_2COOH$.

amnesia: Temporary or permanent memory loss.

amniocentesis: Sampling of amniotic fluid for analytical purposes; used to detect certain forms of genetic abnormalities.

amnion (AM-nē-on): One of the extraembryonic membranes; surrounds the developing embryo/fetus.

amniotic fluid (am-nē-OT-ik): Fluid that fills the amniotic cavity; provides cushioning and support for the embryo/fetus.

amphiarthrosis (am-fē-ar-THRŌ-sis): An articulation that permits a small degree of independent movement.

amphicytes (AM-fi-sīts): Supporting cells that surround neurons in the PNS; also called satellite cells.

amphimixis (am-fi-MIK-sis): The fusion of male and female pronuclei following fertilization.

ampulla/ampullae (am-PYŪL-la): A localized dilation in the lumen of a canal or passageway.

amygdala/amygdaloid nucleus (ah-MIG-da-loid): A cerebral nucleus that is a component of the limbic system and acts as an interface between that system, the cerebrum, and sensory systems.

amylase: An enzyme that breaks down polysaccharides, produced by the salivary glands and pancreas.

anabolism (a-NAB-ō-lizm): The synthesis of complex organic compounds from simpler precursors.

analgesia: Relief from pain.

anal triangle: The posterior subdivision of the perineum.

anaphase (AN-a-fāz): Mitotic stage in which the paired chromatids separate and move toward opposite ends of the spindle apparatus.

anastomosis (a-nas-tō-MŌ-sis): The joining of two tubes, usually referring to a connection between two peripheral vessels without an intervening capillary bed.

anatomical position: An anatomical reference position, the body viewed from the anterior surface with the palms facing forward; supine.

anatomy (a-NAT-ō-mē): The study of the structure of the body.

anaxonic neuron (an-ak-SON-ik): A CNS neuron that has many processes but no apparent axon.

androgen (AN-drō-jen): A steroid sex hormone produced primarily by the interstitial cells of the testis, and manufactured in small quantities by the adrenal cortex in either sex.

anemia (a-NĒ-mē-ah): Condition marked by a reduction in the hematocrit and/or hemoglobin content of the blood.

anencephaly (an-en-SEF-a-lē): Development defect characterized by incomplete development of cerebral hemispheres and cranium.

anesthesia: Total or partial loss of sensation from a region of the body.

aneurysm (AN-yū-rizm): A weakening and localized dilation in the wall of a blood vessel.

angiogram (AN-jē-ō-gram): An X-ray image of circulatory pathways.

angiography: X-ray examination of vessel distribution following the introduction of radiopaque substances into the bloodstream.

angiotensin I, II: Angiotensin II is a hormone that causes an elevation in systemic blood pressure, stimulates secretion of aldosterone, promotes thirst, and causes the release of ADH; a converting enzyme in the pulmonary capillaries, converts angiotensin I to angiotensin II.

angiotensinogen: Blood protein produced by the liver that is converted to angiotensin I by the enzyme renin.

ankyloglossia (ang-ki-lō-GLOS-ē-a): Condition characterized by an overly robust and restrictive lingual frenulum.

annulus (AN-ū-lus): A cartilage or bone shaped like a ring.

anorectal canal (ā-nō-REC-tal): The distal portion of the rectum that contains the rectal columns and ends at the anus.

anorexia nervosa: An eating disorder marked by a loss of appetite and pronounced weight loss.

anoxia (an-OK-sē-a): Tissue oxygen deprivation.

antagonist: A muscle that opposes the movement of an agonist.

antebrachium: The forearm.

anteflexion (an-te-FLEK-shun): Normal position of the uterus, with the superior surface bent forward.

anterior: On or near the front or ventral surface of the body.

anterior pituitary: *See* **pituitary gland.**

antibiotic: Chemical agent that selectively kills pathogenic microorganisms.

antibody (AN-ti-bod-ē): A globular protein produced by plasma cells that will bind to specific antigens and promote their destruction or removal from the body.

anticoagulant: Compound that slows or prevents clot formation by interfering with the clotting system.

antidiuretic hormone (ADH) (an-tī-dī-ū-RET-ik): Hormone synthesized in the hypothalamus and secreted at the posterior pituitary; causes water retention at the kidneys, and an elevation of blood pressure.

antigen: A substance capable of inducing the production of antibodies.

antrum (AN-trum): A chamber or pocket.

anuria (a-NŪ-rē-a): Cessation of urine production.

anus: External opening of the anorectal canal.

aorta: Large, elastic artery that carries blood away from the left ventricle, and into the systemic circuit.

aortic reflex: Baroreceptor reflex triggered by increased aortic pressures; leads to a reduction in cardiac output and a fall in systemic pressure.

apex (Ā-peks): A pointed tip, usually referring to a triangular object and positioned opposite a broad base.

Apgar: A test used to assess the neurological status of a newborn infant.

aphasia: Inability to speak.

apnea (AP-nē-a): Cessation of breathing.

apneustic center (ap-NŪ-stik): Respiratory center whose chronic activation would lead to apnea at full inspiration.

apocrine secretion: Mode of secretion where the glandular cell sheds portions of its cytoplasm.

aponeurosis/aponeuroses (ap-ō-nū-RŌ-sēz): A broad tendinous sheet that may serve as the origin or insertion of a skeletal muscle.

appendicitis: Inflammation of the appendix.

appendicular: Pertaining to the upper or lower limbs.

appendix: A blind tube connected to the cecum of the large intestine.

appositional growth: Enlargement by the addition of cartilage or bony matrix to the outer surface.

aqueous humor: Fluid similar to perilymph or CSF that fills the anterior chamber of the eye.

arachnoid (a-RAK-noid): The middle meninges that encloses CSF and protects the central nervous system.

arachnoid villi: Processes of the arachnoid that project into the superior sagittal sinus; sites where CSF enters the venous circulation.

arbor vitae: Central, branching mass of white matter inside the cerebellum.

arcuate (AR-kū-āt): Curving.

areflexia (ā-re-FLEK-sē-a): Absence of normal reflex responses to stimulation.

areola (a-RĒ-ō-la): Pigmented area that surrounds the nipple of a breast.

areolar: Containing minute spaces, as in areolar connective tissue.

arrector pili (ar-REK-tor PĪ-li): Smooth muscles whose contractions cause piloerection.

arrhythmias (a-RITH-mē-az): Abnormal patterns of cardiac contractions.

arteriole (ar-TĒ-rē-ōl): A small arterial branch that delivers blood to a capillary network.

artery: A blood vessel that carries blood away from the heart and toward a peripheral capillary.

arthritis (ar-THRĪ-tis): Inflammation of a joint.

arthroscope: Fiberoptic device intended for visualizing the interior of joints; may also be used for certain forms of joint surgery.

articular: Pertaining to a joint.

articular capsule: Dense collagen fiber sleeve that surrounds a joint and provides protection and stabilization.

articular cartilage: Cartilage pad that covers the surface of a bone inside a joint cavity.

articulation (ar-tik-ū-LĀ-shun): A joint; formation of words.

arytenoid cartilages (ar-i-TĒ-noid): A pair of small cartilages in the larynx.

ascending tract: A tract carrying information from the spinal cord to the brain.

ascites (a-SĪ-tēz): Overproduction and accumulation of peritoneal fluid.

aseptic: Free from pathogenic contamination.

asphyxia: Unconsciousness due to oxygen deprivation at the CNS.

aspirate: To remove or obtain by suction; to inhale.

association areas: Cortical areas of the cerebrum responsible for integration of sensory inputs and/or motor commands.

association neuron: *See* **interneuron.**

asthma (AZ-ma): Reversible constriction of smooth muscles around respiratory passageways frequently caused by an allergic response.

astigmatism: Visual disturbance due to an irregularity in the shape of the cornea.

astrocyte (AS-trō-sīt): One of the glial cells in the CNS.

atelectasis (at-e-LEK-ta-sis): Collapse of a lung or a portion of a lung.

atherosclerosis (ath-er-ō-skle-RŌ-sis): Formation of fatty plaques in the walls of arteries, leading to circulatory impairment.

atresia (a-TRĒ-zē-a): Closing of a cavity, or its incomplete development; used in the reproductive system to refer to the degeneration of developing ovarian follicles.

atria: Thin-walled chambers of the heart that receive venous blood from the pulmonary or systemic circuits.

atrial natriuretic peptide (nā-tre-ū-RET-ik): Hormone released by specialized atrial cardiocytes when they are stretched by an abnormally large venous return; promotes fluid loss and reductions in blood pressure and venous return.

atrioventricular (AV) node (ā-trē-ō-ven-TRIK-ū-lar): Specialized cardiocytes that relay the contractile stimulus to the bundle of His, the bundle branches, the Purkinje fibers, and the ventricular myocardium; located at the boundary between the atria and ventricles.

atrioventricular valve: One of the valves that prevent backflow into the atria during ventricular systole.

atrophy (AT-rō-fē): Wasting away of tissues from lack of use or nutritional abnormalities.

auditory: Pertaining to the sense of hearing.

auditory ossicles: The bones of the middle ear: malleus, incus, and stapes.

autoimmunity: Immune system sensitivity to normal cells and tissues, resulting in the production of autoantibodies.

autolysis: Destruction of a cell due to the rupture of lysosomal membranes in its cytoplasm.

automatic bladder: Reflex micturition following stimulation of stretch receptors in the bladder wall; seen in patients who have lost motor control of the lower body.

automaticity: Spontaneous depolarization to threshold, a characteristic of cardiac pacemaker cells.

autonomic ganglion: A collection of visceral motor neurons outside the CNS.

autonomic nerve: A peripheral nerve consisting of preganglionic or postganglionic autonomic fibers.

autonomic nervous system (ANS): Centers, nuclei, tracts, ganglia, and nerves involved in the unconscious regulation of visceral functions; includes components of the CNS and PNS.

autopsy: Detailed examination of a body after death, usually performed by a pathologist.

autoregulation: Alterations in activity that maintain homeostasis in direct response to changes in the local environment; does not require neural or endocrine control.

autosomal (aw-tō-SŌ-mal): Chromosomes other than the X or Y chromosomes that determine the genetic sex of an individual.

avascular (ā-VAS-kū-lar): Without blood vessels.

avulsion: An injury involving the violent tearing away of body tissues.

axilla: The armpit.

axolemma: The cell membrane of an axon, continuous with the cell membrane of the soma and dendrites and distinct from any glial cell coverings.

axon: Elongate extension of a neuron that conducts an action potential away from the soma and toward the synaptic terminals.

axon hillock: Portion of the neural soma adjacent to the initial segment.

axoplasm (AK-sō-plazm): Cytoplasm within an axon.

Babinski sign: Reflexive dorsiflexion of the toes following stroking of the plantar surface of the foot; positive reflex (Babinski sign) is normal up to age 1.5 years; thereafter a positive reflex indicates damage to descending tracts.

bacteria: Single-celled microorganisms, some pathogenic, that are common in the environment.

baroreception: Ability to detect changes in pressure.

baroreceptor reflex: A reflexive change in cardiac activity in response to changes in blood pressure.

baroreceptors (bar-ō-rē-SEP-torz): Receptors responsible for baroreception.

base: A compound whose dissociation releases a hydroxyl ion (OH⁻) or removes a hydrogen ion from the solution.

basement membrane: A layer of filaments and fibers that attach an epithelium to the underlying connective tissue.

basilar membrane: Membrane that supports the organ of Corti and separates the cochlear duct from the scala tympani in the inner ear.

basophils (BĀ-sō-filz): Circulating granulocytes (WBCs) similar in size and function to tissue mast cells.

B cells: Lymphocytes responsible for the production of antibodies, following their conversion to plasma cells.

benign: Not malignant.

beta cells: Cells of the pancreatic islets that secrete insulin in response to elevated blood sugar concentrations.

beta receptors: Membrane receptors sensitive to epinephrine; stimulation may result in excitation or inhibition of the target cell.

bicarbonate ions: HCO_3^-; anion components of the bicarbonate buffer system.

bicuspid (bī-KUS-pid): A sharp, conical tooth, also called a canine tooth.

bicuspid valve: The left AV valve, also known as the **mitral valve.**

bifurcate: To branch into two parts.

bile: Exocrine secretion of the liver that is stored in the gallbladder and ejected into the duodenum.

bile salts: Steroid derivatives in the bile, responsible for the emulsification of ingested lipids.

bilirubin (bil-ē-ROO-bin): A reddish pigment, a product of hemoglobin catabolism.

biopsy: The removal of a small sample of tissue for pathological analysis.

bipennate muscle: A muscle whose fibers are arranged on either side of a common tendon.

bladder: A muscular sac that distends as fluid is stored and whose contraction ejects the fluid at an appropriate time; used alone, the term usually refers to the urinary bladder.

blastocoele (BLAS-tō-sēl): Fluid-filled cavity within a blastocyst.

blastocyst (BLAS-tō-sist): Early stage in the developing embryo, consisting of an outer trophoblast and an inner cell mass.

blastodisc (BLAS-tō-disk): Later stage in the development of the inner cell mass; it includes the cells that will form the embryo.

blastomere (BLAS-tō-mēr): One of the cells in the morula, a collection of cells produced by the division of the zygote.

blood-brain barrier: Isolation of the CNS from the general circulation; primarily the result of astrocyte regulation of capillary permeabilities.

blood clot: A network of fibrin fibers and trapped blood cells.

blood pressure: A force exerted against the vascular walls by the blood, as the result of the push exerted by cardiac contraction and the elasticity of the vessel walls. It is usually measured along one of the muscular arteries, with systolic pressure measured during ventricular systole, and diastolic pressure during ventricular diastole.

blood-testis barrier: Isolation of the seminiferous tubules from the general circulation, due to the activities of the sustentacular (Sertoli) cells.

boil: An abscess of the skin, usually involving a sebaceous gland.

bolus: A compact mass; usually refers to compacted ingested material on its way to the stomach.

bone: *See* **osseous tissue.**

bowel: The intestinal tract.

brachial: Pertaining to the arm.

brachial plexus: Network formed by branches of spinal nerves C_5–T_1 en route to innervate the arm.

brachium: The arm.

bradycardia (brā-dē-KAR-dē-a): An abnormally slow heart rate.

brain stem: The brain minus the cerebrum and cerebellum.

brevis: Short.

Broca's center: The speech center of the brain, usually found on the neural cortex of the left cerebral hemisphere.

bronchial tree: The trachea, bronchi, and bronchioles.

bronchitis (brong-KĪ-tis): Inflammation of the bronchial passageways.

bronchoscope: A fiberoptic instrument used to examine the bronchial passageways.

bronchus/bronchi: One of the branches of the bronchial tree between the trachea and bronchioles.

buccal (BUK-al): Pertaining to the cheeks.

buffer: A compound that stabilizes the pH of a solution by removing or releasing hydrogen ions.

bulbar: Pertaining to the brain stem.

bulbourethral glands (bul-bō-ū-RĒ-thral): Mucous glands at the base of the penis that secrete into the penile urethra; also called Cowper's glands.

bundle branches: Specialized conducting cells in the ventricles that carry the contractile stimulus from the bundle of His to the Purkinje fibers.

bundle of His (hiss): Specialized conducting cells in the interventricular septum that carry the contracting stimulus from the AV node to the bundle branches and thence to the Purkinje fibers.

bursa: A small sac filled with synovial fluid that cushions adjacent structures and reduces friction.

bursectomy: The surgical removal of an inflamed bursa.

bursitis: Painful inflammation of one or more bursae.

cesarean section: Surgical delivery of an infant via an incision through the lower abdominal wall and uterus.

calcaneal tendon: Large tendon that inserts on the calcaneus; tension on this tendon produces plantar flexion of the foot; also called the **Achilles tendon.**

calcaneus (kal-KĀ-nē-us): The heelbone, the largest of the tarsal bones.

calcification: The deposition of calcium salts within a tissue.

calcitonin (kal-si-TŌ-nin): Hormone secreted by C cells of the thyroid when calcium ion concentrations are abnormally high; restores homeostasis by increasing the rate of bone deposition and the renal rate of calcium loss, and inhibiting calcium uptake at the digestive tract.

calculus/calculi (KAL-kū-lī): Concretions of insoluble materials that form within body fluids, especially the gallbladder, kidneys, or urinary bladder.

callus: A localized thickening of the epidermis due to chronic mechanical stresses; a thickened area that forms at the site of a bone break as part of the repair process.

calvaria (kal-VAR-ē-a): The skullcap, formed of the frontal, parietal, and occipital bones.

calyx/calyces (KĀL-i-sēz): A cup-shaped division of the renal pelvis.

canaliculi (kan-a-LIK-ū-lī): Microscopic passageways between cells; bile canaliculi carry bile to bile ducts in the liver; in bone, canaliculi permit the diffusion of nutrients and wastes to and from osteocytes.

cancellous bone (KAN-sel-us): Spongy bone, composed of a network of bony struts.

cancer: A malignant tumor that tends to undergo metastasis.

cannula: A tube that can be inserted into the body; often placed in blood vessels prior to transfusion or dialysis.

canthus, medial and lateral (KAN-thus): The angles formed at either corner of the eye between the upper and lower eyelids.

capacitation (ka-pas-i-TĀ-shun): Activation process that must occur before a spermatozoon can successfully fertilize an egg; occurs in the vagina following ejaculation.

capillaries: Small blood vessels, interposed between arterioles and venules, whose thin walls permit the diffusion of gases, nutrients, and wastes between the plasma and interstitial fluids.

capitulum (ka-PIT-ū-lum): General term for a small, elevated articular process; used to refer to the rounded distal surface of the humerus that articulates with the radial head.

caput: The head.

carbohydrase (kar-bō-HĪ-drāz): An enzyme that breaks down carbohydrate molecules.

carbohydrate (kar-bō-HĪ-drāt): Organic compound containing carbon, hydrogen, and oxygen in a ratio that approximates 1:2:1.

carbon dioxide: CO_2, a compound produced by the decarboxylation reactions of aerobic glycolysis.

carbonic anhydrase: An enzyme that catalyzes the reaction $H_2O + CO_2 \rightarrow H_2CO_3$; important in carbon dioxide transport, gastric acid secretion, and renal pH regulation.

carboxypeptidase (kar-bok-sē-PEP-ti-dāz): A protease that breaks down proteins and releases amino acids.

carcinogenic (kar-sin-ō-JEN-ik): Stimulating cancer formation in affected tissues.

cardia (KAR-dē-a): The area of the stomach surrounding its connection with the esophagus.

cardiac: Pertaining to the heart.

cardiac cycle: One complete heartbeat, including atrial and ventricular systole and diastole.

cardiac glands: Mucous glands characteristic of the cardia of the stomach.

cardiac output: The amount of blood ejected by the left ventricle each minute; normally about 5 liters.

cardiac tamponade: Compression of the heart due to fluid accumulation in the pericardial cavity.

cardiocyte (KAR-dē-ō-sīt): A cardiac muscle cell.

cardiomyopathy (kar-dē-ō-mī-OP-a-thē): A progressive disease characterized by damage to the cardiac muscle tissue.

cardiopulmonary resuscitation: Method of artificially maintaining respiratory and circulatory function.

cardiovascular: Pertaining to the heart, blood, and blood vessels.

cardiovascular centers: Poorly localized centers in the reticular formation of the medulla of the brain; includes cardioacceleratory, cardioinhibitory, and vasomotor centers.

cardium: The heart.

carina (ka-RĪ-na): A ridge on the inner surface of the base of the trachea that runs anteroposteriorly, between two primary bronchi.

carotene (KAR-ō-tēn): A yellow-orange pigment found in carrots and in green and orange leafy vegetables; a compound that the body can convert to vitamin A.

carotid artery: The principal artery of the neck, servicing cervical and cranial structures; one branch, the internal carotid, represents a major blood supply for the brain.

carotid body: A group of receptors adjacent to the carotid sinus that are sensitive to changes in the carbon dioxide levels, pH, and oxygen concentrations of the arterial blood.

carotid sinus: A dilated segment of the internal carotid artery whose walls contain baroreceptors sensitive to changes in blood pressure.

carotid sinus reflex: Reflexive changes in blood pressure that maintain homeostatic pressures at the carotid sinus, stabilizing blood flow to the brain.

carpus/carpal: The wrist.

cartilage: A connective tissue with a gelatinous matrix and an abundance of fibers.

castration: Removal of the testes, also called bilateral orchiectomy.

catabolism (ka-TAB-ō-lizm): The breakdown of complex organic molecules into simpler components, accompanied by the release of energy.

cataract: A reduction in lens transparency that causes visual impairment.

catecholamines (kat-e-KŌL-am-inz): Epinephrine, norepinephrine, and related compounds.

catheter (KATH-e-ter): Surgical instrument, a tube inserted into a body cavity or along a blood vessel or excretory passageway for the collection of body fluids, blood pressure monitoring, or the introduction of medications or radiographic dyes.

cauda equina (KAW-da ek-WĪ-na): Spinal nerve roots distal to the tip of the adult spinal cord; they extend caudally inside the vertebral canal en route to lumbar and sacral segments.

caudal/caudally: Closest to or toward the tail (coccyx).

caudate nucleus (KAW-dāt): One of the cerebral nuclei of the extrapyramidal system, involved with the unconscious control of muscular activity.

cavernous tissue: Erectile tissue that can be engorged with blood; found in the penis and clitoris.

cecum (SĒ-kum): An expanded pouch at the start of the large intestine.

cell: The smallest living unit in the human body.

cell-mediated immunity: Resistance to disease through the activities of sensitized T cells that destroy antigen-bearing cells by direct contact or through the release of lymphotoxins; also called cellular immunity.

cellulitis (sel-ū-LĪ-tis): Diffuse inflammation, usually involving areas of loose connective tissue, such as the subcutaneous layer.

cementum (se-MEN-tum): Bony material covering the root of a tooth, not shielded by a layer of enamel.

center of ossification: Site in a connective tissue where bone formation begins.

central canal: Longitudinal canal in the center of an osteon that contains blood vessels and nerves, also called the Haversian canal; a passageway along the longitudinal axis of the spinal cord that contains cerebrospinal fluid.

central nervous system (CNS): The brain and spinal cord.

central sulcus: Groove in the surface of a cerebral hemisphere, between the primary sensory and primary motor areas of the cortex.

centriole: A cylindrical intracellular organelle composed of nine groups of microtubules, three in each group; functions in mitosis or meiosis by forming the basis of the spindle apparatus.

centromere (SEN-trō-mēr): Localized region where two chromatids remain connected following chromosome replication; site of spindle fiber attachment.

centrosome: Region of cytoplasm containing a pair of centrioles oriented at right angles to one another.

centrum: The vertebral body.

cephalic: Pertaining to the head.

cerebellum (ser-e-BEL-um): Posterior portion of the metencephalon, containing the cerebellar hemispheres; includes the arbor vitae, cerebellar nuclei, and cerebellar cortex.

cerebral cortex: An extensive area of neural cortex covering the surfaces of the cerebral hemispheres.

cerebral hemispheres: Expanded portions of the cerebrum covered in neural cortex.

cerebral nuclei: Nuclei of the cerebrum that are important components of the extrapyramidal system.

cerebral palsy: Chronic condition resulting from damage to motor areas of the brain during development or at delivery.

cerebral peduncle: Mass of nerve fibers on the ventrolateral surface of the mesencephalon; contains ascending tracts that terminate in the thalamus and descending tracts that originate in the cerebral hemispheres.

cerebrospinal fluid: Fluid bathing the internal and external surfaces of the CNS; secreted by the choroid plexus.

cerebrovascular accident (CVA): A stroke; occlusion of a blood vessel supplying a portion of the brain, resulting in damage to the dependent neurons.

cerebrum (SER-e-brum): The largest portion of the brain, composed of the cerebral hemispheres; includes the cerebral cortex, the cerebral nuclei, and the internal capsule.

cerumen: Waxy secretion of integumentary glands along the external auditory canal.

ceruminous glands (se-RŪ-mi-nus): Integumentary glands that secrete cerumen.

cervical enlargement: Relative enlargement of the cervical portion of the spinal cord due to the abundance of CNS neurons involved with motor control of the arms.

cervix: The lower part of the uterus.

chalazion (kah-LĀ-zē-on): An inflammation and distension of a Meibomian gland on the eyelid; also called a sty.

chancre (SHANG-ker): A skin lesion that develops at the primary site of a syphilis infection.

charley horse: Soreness and stiffness in a strained muscle, usually involving the quadriceps group.

chemoreception: Detection of alterations in the concentrations of dissolved compounds or gases.

chemotaxis (kē-mō-TAK-sis): The attraction of phagocytic cells to the source of abnormal chemicals in tissue fluids.

chloride shift: Movement of plasma chloride ions into RBCs in exchange for bicarbonate ions generated by the intracellular dissociation of carbonic acid.

cholecystitis (kō-lē-sis-TĪ-tis): Inflammation of the gallbladder.

cholecystokinin (CCK) (kō-lē-sis-tō-KĪ-nin): Duodenal hormone that stimulates the contraction of the gallbladder and the secretion of enzymes by the exocrine pancreas; also called pancreozymin.

cholelithiasis (kō-lē-li-THĪ-a-sis): The formation or presence of gallstones.

cholesterol: A steroid component of cell membranes and a substrate for the synthesis of steroid hormones and bile salts.

choline: Chemical compound, a breakdown product or precursor of acetylcholine.

cholinergic synapse (kō-lin-ER-jik): Synapse where the presynaptic membrane releases ACh on stimulation.

cholinesterase (kō-li-NES-te-rās): Enzyme that breaks down and inactivates ACh.

chondrocyte (KON-drō-sīt): Cartilage cell.

chondroitin sulfate (kon-DROI-tin): The predominant proteoglycan in cartilage, responsible for the gelatinous consistency of the matrix.

chordae tendineae (KOR-dē TEN-di-nē-ē): Fibrous cords that brace the AV valves in the heart, stabilizing their position and preventing backflow during ventricular systole.

chorion/chorionic (KOR-ē-on) (kō-rē-ON-ik): An extraembryonic membrane, consisting of the trophoblast and underlying mesoderm, that forms the placenta.

choroid: Middle, vascular layer in the wall of the eye.

choroid plexus: The vascular complex in the roof of the third and fourth ventricles of the brain, responsible for CSF production.

chromatid (KRŌ-ma-tid): One complete copy of a DNA strand.

chromatin (KRŌ-ma-tin): Histological term referring to the grainy material visible in cell nuclei during interphase; the appearance of the DNA content of the nucleus when the chromosomes are uncoiled.

chromosomes: Dense structures, composed of tightly coiled DNA strands and associated histones, that become visible in the nucleus when a cell prepares to undergo mitosis or meiosis; normal human somatic cells contain 46 chromosomes apiece.

chronic: Habitual or long-term.

chylomicrons (kī-lō-MĪ-kronz): Relatively large droplets that may contain triglycerides, phospholipids, and cholesterol in association with proteins; synthesized and released by intestinal cells and transported to the venous blood via the lymphatic system.

chyme (kīm): A semifluid mixture of ingested food and digestive secretions that is found in the stomach as digestion proceeds.

chymotrypsin (kī-mō-TRIP-sin): A protease found in the small intestine.

chymotrypsinogen: Inactive proenzyme secreted by the pancreas that is subsequently converted to chymotrypsin.

ciliary body: A thickened region of the choroid that encircles the lens of the eye; it includes the ciliary muscle and the ciliary processes that support the suspensory ligaments of the lens.

cilium/cilia: A slender organelle that extends above the free surface of an epithelial cell, and usually undergoes cycles of movement; composed of a basal body and microtubules in a 9 × 2 array.

circulatory system: The network of blood vessels that are components of the cardiovascular system.

circumduction (sir-kum-DUK-shun): A movement at a synovial joint where the distal end of the bone describes a circle, but the shaft does not rotate.

circumvallate papilla (sir-kum-VAL-āt pa-PIL-la): One of the large, dome-shaped papillae on the dorsum of the tongue that form the V that separates the body of the tongue from the root.

cirrhosis (sir-RŌ-sis): A liver disorder characterized by the degeneration of hepatocytes and their replacement by connective tissue.

cisterna (sis-TUR-na): An expanded chamber.

clavicle (KLAV-i-kul): The collarbone.

cleavage (KLĒV-ij): Mitotic divisions that follow fertilization of the ovum and lead to the formation of a blastocyst.

cleavage lines: Stress lines in the skin that follow the orientation of major bundles of collagen fibers in the dermis.

climacteric: Age-related cessation of gametogenesis in the male or female due to reduced sex hormone production.

clitoris (KLI-to-ris): A small erectile organ of the female that is the developmental equivalent of the male penis.

clone: The production of genetically identical cells.

clonus (KLŌ-nus): Rapid cycles of muscular contraction and relaxation.

clot: A network of fibrin fibers and trapped blood cells; also called a **thrombus.**

clotting factors: Plasma proteins synthesized by the liver that are essential to the clotting response.

clotting response: Series of events that result in the formation of a clot.

coccygeal ligament: Fibrous extension of the dura mater and filum terminale; provides longitudinal stabilization to the spinal cord.

coccyx (KOK-siks): Terminal portion of the spinal column, consisting of relatively tiny, fused vertebrae.

cochlea (KOK-lē-a): Spiral portion of the bony labyrinth of the inner ear that surrounds the organ of hearing.

cochlear duct (KOK-lē-ar): Membranous tube within the cochlea that is filled with endolymph and contains the organ of Corti; also called the **scala media.**

codon (KŌ-don): A sequence of three nitrogen bases along an mRNA strand that will specify the location of a single amino acid in a peptide chain.

coelom (SĒ-lom): The ventral body cavity, lined by a serous membrane and subdivided during development into the pleural, pericardial, and abdominopelvic (peritoneal) cavities.

coenzymes (kō-EN-zīmz): Complex organic cofactors, usually structurally related to vitamins.

cofactors: Ions or molecules that must be attached to the active site before an enzyme can function; examples include mineral ions and several vitamins.

colectomy (kō-LEK-tō-mē): Surgical removal of part or all of the colon.

colitis: Inflammation of the colon.

collagen: Strong, insoluble protein fiber common in connective tissues.

collagenous tissues: (ko-LA-jin-us): Dense connective tissues in which collagen fibers are the dominant fiber type; includes dense regular and dense irregular connective tissues.

collateral ganglion (kō-LAT-er-al): A sympathetic ganglion situated in front of the spinal column and separate from the sympathetic chain.

Colles' fracture (KOL-lēz): Fracture of the distal end of the radius and possibly the ulna, with posterior and dorsal displacement of the distal bone fragments.

colliculus/colliculi (ko-LIK-ū-lus): A little mound; in the brain, used to refer to one of the cortical thickenings in the roof of the mesencephalon; the superior colliculus is associated with the visual system, and the inferior colliculi with the auditory system.

colon: The large intestine.

colonoscope (kō-LON-ō-skōp): A fiberoptic device for examining the interior of the colon.

colostomy (kō-LOS-tō-mē): The surgical connection of a portion of the colon to the body wall, sometimes performed after a colectomy to permit the discharge of fecal materials.

colostrum (kō-LOS-trum): Secretion of the mammary glands at the time of childbirth and for a few days thereafter; contains more protein and less fat than the milk secreted later.

coma (kō-ma): An unconscious state from which the individual cannot be aroused, even by strong stimuli.

comedo (kō-MĒ-dō): An inflamed sebaceous gland.

comminuted: Broken or crushed into small pieces.

commissure: A crossing over from one side to another.

common bile duct: Duct formed by the union of the cystic duct from the gallbladder and the bile ducts from the liver; terminates at the duodenal ampulla, where it meets the pancreatic duct.

common pathway: In the clotting response, the events that begin with the appearance of thromboplastin and end with the formation of a clot.

compact bone: Dense bone containing parallel osteons.

compensation curves: The cervical and lumbar curves that develop to center the body weight over the legs.

complement: 11 plasma proteins that interact in a chain reaction following exposure to activated antibodies on the surfaces of certain pathogens, and which promote cell lysis, phagocytosis, and other defense mechanisms.

compliance: The ability of certain organs to tolerate changes in volume; a property that reflects the presence of elastic fibers and smooth muscles.

compound: A molecule containing two or more elements in combination.

concentration: Amount (in grams) or number of atoms, ions, or molecules (in moles) per unit volume.

conception: Fertilization.

concha/conchae (KONG-kē): Three pairs of thin, scroll-like bones that project into the nasal cavities; the superior and medial conchae are part of the ethmoid, and the inferior are separate bones.

concussion: A violent blow or shock; loss of consciousness due to a violent blow to the head.

condyle: A rounded articular projection on the surface of a bone.

cone: Retinal photoreceptor responsible for color vision.

congenital (kon-JEN-i-tal): Already present at the birth of an individual.

congestive heart failure (CHF): Failure to maintain adequate cardiac output due to circulatory problems or myocardial damage.

conjunctiva (kon-junk-TĪ-va): A layer of stratified squamous epithelium that covers the inner surfaces of the lids and the anterior surface of the eye to the edges of the cornea.

conjunctivitis: Inflammation of the conjunctiva.

connective tissue: One of the four primary tissue types; provides a structural framework for the body that stabilizes the relative positions of the other tissue types; includes connective tissue proper, cartilage, bone, and blood; always has cell products, cells, and ground substance.

contractility: The ability to contract, possessed by skeletal, smooth, and cardiac muscle cells.

contracture: A permanent contraction of an entire muscle following the atrophy of individual muscle cells.

contralateral reflex: A reflex that affects the side of the body opposite to the stimulus.

conus medullaris: Conical tip of the spinal cord that gives rise to the filum terminale.

convergence: In the nervous system, the innervation of a single neuron by the axons from several neurons; this is most common along motor pathways.

coracoid process (KOR-a-koid): A hook-shaped process of the scapula that projects above the anterior surface of the capsule of the shoulder joint.

Cori cycle: Metabolic exchange of lactic acid from skeletal muscle for glucose from the liver; performed during the recovery period following muscular exertion.

cornea (KOR-nē-a): Transparent portion of the fibrous tunic of the anterior surface of the eye.

corniculate cartilages (kor-NIK-ū-lāt): A pair of small laryngeal cartilages.

cornification: The production of keratin by a stratified squamous epithelium; also called **keratinization.**

cornu: A horn.

corona radiata (ko-RŌ-na rā-dē-A-ta): A layer of follicle cells surrounding an oocyte at ovulation.

coronoid (KOR-ō-noid): Hooked or curved.

corpora quadrigemina (KOR-pō-ra quad-ri-JEM-i-na): The superior and inferior colliculi of the mesencephalic tectum (roof) in the brain.

corpus/corpora: Body.

corpus albicans: The scar tissue that remains after degeneration of the corpus luteum at the end of a uterine cycle.

corpus callosum: Bundle of axons linking centers in the left and right cerebral hemispheres.

corpus cavernosum (KOR-pus ka-ver-NŌ-sum): Masses of erectile tissue within the body of the penis (male) or clitoris (female).

corpus luteum (LOO-tē-um): Progestin-secreting mass of follicle cells that develops in the ovary after ovulation.

corpus spongiosum (spon-jē-Ō-sum): Mass of erectile tissue that surrounds the urethra in the penis, and expands distally to form the glans.

cortex: Outer layer or portion of an organ.

Corti, organ of: Receptor complex in the scala media of the cochlea that includes the inner and outer hair cells, supporting cells and structures, and the tectorial membrane; provides the sensation of hearing.

corticobulbar tracts (kor-ti-kō-BUL-bar): Descending tracts that carry information/commands from the cerebral cortex to nuclei and centers in the brain stem.

corticospinal tracts: Descending tracts that carry motor commands from the cerebral cortex to the anterior gray horns of the spinal cord.

corticosteroid: A steroid hormone produced by the adrenal cortex.

corticosterone (kor-ti-KOS-te-rōn): One of the corticosteroids secreted by the zona fasciculata of the adrenal cortex; a glucocorticoid.

corticotropin: *See* **adrenocorticotropic hormone (ACTH).**

corticotropin-releasing hormone (CRH): Releasing hormone secreted by the hypothalamus that stimulates secretion of ACTH by the anterior pituitary.

cortisol (KOR-ti-sol): One of the corticosteroids secreted by the zona fasciculata of the adrenal cortex; a glucocorticoid.

costa/costae: A rib.

cotransport: Membrane transport of a nutrient, such as glucose, in company with the movement of an ion, usually sodium; transport requires a carrier protein but does not involve direct ATP expenditure and can occur regardless of the concentration gradient for the nutrient.

countercurrent multiplication: Active transport between two limbs of a loop that contains a fluid moving in one direction; responsible for the concentration of the urine in the kidney tubules.

coxa/coxae: The bones of the hip.

cranial: Pertaining to the head.

cranial nerves: Peripheral nerves originating at the brain.

craniosacral division (krā-nē-ō-SAK-ral): *See* **parasympathetic division.**

craniostenosis (krā-nē-ō-sten-Ō-sis): Skull deformity caused by premature closure of the cranial sutures.

cranium: The braincase; the skull bones that surround the brain.

creatine: A nitrogenous compound synthesized in the body that can bind a high-energy phosphate and serve as an energy reserve.

creatine phosphate: A high-energy compound present in muscle cells; during muscular activity the phosphate group is donated to ADP, regenerating ATP.

creatinine: A breakdown product of creatine metabolism.

crenation: Cellular shrinkage due to an osmotic movement of water out of the cytoplasm.

cribriform plate: Portion of the ethmoid bone of the skull that contains the foramina used by the axons of olfactory receptors en route to the olfactory bulbs of the cerebrum.

cricoid cartilage (KRĪ-koid): Ring-shaped cartilage forming the inferior margin of the larynx.

crista/cristae: A ridge-shaped collection of hair cells in the ampulla of a semicircular canal; the crista and cupula form a receptor complex sensitive to movement along the plane of the canal.

cross-bridge: Myosin head that projects from the surface of a thick filament, and that can bind to an active site of a thin filament in the presence of calcium ions.

cruciate ligaments: A pair of intracapsular ligaments (anterior and posterior) in the knee.

cryptorchidism (kript-OR-ki-dizm): The failure of the testes to descend into the inguinal canal during late fetal development.

cryptorchid testis: An undescended testis that is in the abdominopelvic cavity rather than the scrotum.

cuneiform cartilages (kū-NĒ-i-form): A pair of small cartilages in the larynx.

cupula (KŪ-pū-la): A gelatinous mass that sits in the ampulla of a semicircular canal in the inner ear, and whose movement stimulates the hair cells of the crista.

Cushing's disease: Condition caused by oversecretion of adrenal steroids.

cuspids (KUS-pids): Conical upper teeth with sharp ridges, located in the upper jaw on either side posterior to the second incisor.

cutaneous membrane: The epidermis and papillary layer of the dermis.

cuticle: Layer of dead, cornified cells surrounding the shaft of a hair; for nails, *see* **eponychium.**

cutis: The skin.

cyanosis: Bluish coloration of the skin due to the presence of deoxygenated blood in vessels near the body surface.

cyst: A fibrous capsule containing fluid or other material.

796

cystic duct: A duct that carries bile between the gallbladder and the common bile duct.

cystitis: Inflammation of the urinary bladder.

cytokinesis (sī-tō-ki-NĒ-sis): The cytoplasmic movement that separates two daughter cells at the completion of mitosis.

cytology (sī-TOL-ō-jē): The study of cells.

cytoplasm: The material between the cell membrane and the nuclear membrane.

cytoskeleton: A network of microtubules and microfilaments in the cytoplasm.

cytosol: The fluid portion of the cytoplasm.

cytotoxic: Poisonous to living cells.

cytotoxic T cells: Lymphocytes of the cellular immune response that kill target cells by direct contact or through the secretion of lymphotoxins; also called killer T cells.

daughter cells: Genetically identical cells produced by mitosis.

decerebrate: Lacking a cerebrum.

decomposition reaction: A chemical reaction that breaks a molecule into smaller fragments.

decubitis ulcers: Ulcers that form where chronic pressure interrupts circulation to a portion of the skin.

decussate: To cross over to the opposite side, usually referring to the crossover of the pyramidal tracts on the ventral surface of the medulla oblongata.

defecation (def-e-KĀ-shun): The elimination of fecal wastes.

deglutition (deg-loo-TISH-un): Swallowing.

delta cell: A pancreatic islet cell that secretes somatostatin.

dementia: Loss of mental abilities.

demyelination: The loss of the myelin sheath of an axon, usually due to chemical or physical damage to Schwann cells or oligodendrocytes.

dendrite (DEN-drīt): A sensory process of a neuron.

denticulate ligaments: Supporting fibers that extend laterally from the surface of the spinal cord, tying the pia mater to the dura mater and providing lateral support for the spinal cord.

dentin (DEN-tin): Bonelike material that forms the body of a tooth; it differs from bone in lacking osteocytes and osteons.

deoxyribonucleic acid (DNA) (dē-ok-sē-rī-bō-nū-KLĀ-ik): DNA strand: a nucleic acid consisting of a chain of nucleotides containing the sugar deoxyribose and the nitrogen bases adenine, guanine, cytosine, and thymine. DNA molecule: two DNA strands wound in a double helix and held together by weak bonds between complementary nitrogen base pairs.

depolarization: A change in the transmembrane potential that moves it from a negative value toward 0 mV.

depression: Inferior (downward) movement of a body part.

dermatitis: Inflammation of the skin.

dermatome: A sensory region monitored by the dorsal rami of a single spinal segment.

dermis: The connective tissue layer beneath the epidermis of the skin.

desmosomes (DEZ-mō-sōmz): A cell junction consisting of a thin proteoglycan layer reinforced by a network of intermediate filaments that lock the two cells together.

detrusor muscle (dē-TROO-sor): Smooth muscle in the wall of the urinary bladder.

detumescence (dē-tū-MES-ens): Loss of a penile erection in the male.

development: Growth and the acquisition of increasing structural and functional complexity; includes the period from conception to maturity.

diabetes insipidus: Polyuria due to inadequate production of ADH.

diabetes mellitus (mel-LĪ-tus): Polyuria and glycosuria, most often due to inadequate production of insulin with resulting elevation of blood glucose levels.

dialysis: Diffusion between two solutions of differing solute concentrations across a semipermeable membrane containing pores that permit the passage of some solutes and not others.

diapedesis (dī-a-pe-DĒ-sis): Movement of white blood cells through the walls of blood vessels by migration between adjacent endothelial cells.

diaphragm (DĪ-a-fram): Any muscular partition; often used to refer to the respiratory muscle that separates the thoracic from the abdominopelvic cavities.

diaphysis (dī-AF-i-sis): The shaft of a long bone.

diarrhea (dī-a-RĒ-a): Abnormally frequent defecation, associated with the production of unusually fluid feces.

diarthrosis (dī-ar-THRŌ-sis): A synovial joint.

diastole (dī-AS-tō-lē): A period of relaxation; may refer to either the atria or the ventricles.

diastolic pressure: Pressure measured in the walls of a muscular artery when the left ventricle is in diastole.

diencephalon (dī-en-SEF-a-lon): A division of the brain that includes the epithalamus, thalamus, and hypothalamus.

differential count: The determination of the relative abundance of each type of white blood cell, based on a random sampling of 100 WBCs.

differentiation: The gradual appearance of characteristic cellular specializations during development, as the result of gene activation or repression.

diffusion: Passive molecular movement from an area of relatively high concentration to an area of relatively low concentration.

digestion: The chemical breakdown of ingested materials into simple molecules that can be absorbed by the cells of the digestive tract.

digestive system: The digestive tract and associated glands.

digestive tract: An internal passageway that begins at the mouth and ends at the anus.

dilate: To increase in diameter; to enlarge or expand.

diploe (DIP-lō-ē): A layer of spongy bone between the internal and external tables of a flat bone.

dislocation: Forceful displacement of an articulating bone to an abnormal position, usually accompanied by damage to tendons, ligaments, the articular capsule, or other structures.

distal: Movement away from the point of attachment or origin; for a limb, away from its attachment to the trunk.

distal convoluted tubule: Portion of the nephron closest to the collecting tubule and duct; an important site of active secretion.

diuresis: Fluid loss at the kidneys; the production of urine.

divergence: In neural tissue, the spread of excitation from one neuron to many neurons; an organizational pattern common along sensory pathways of the CNS.

diverticulitis (dī-ver-tik-ū-LĪ-tis): Inflammation of a diverticulum.

diverticulosis (dī-ver-tik-ū-LŌ-sis): The formation of diverticula.

diverticulum: A sac or pouch in the wall of the colon or other organ.

dizygotic twins (dī-zī-GOT-ik): Twins that result from the fertilization of two different ova.

dopamine (DŌ-pah-mēn): An important neurotransmitter in the CNS.

dorsal: Toward the back, posterior.

dorsal root ganglion: PNS ganglion containing the cell bodies of sensory neurons.

dorsiflexion: Elevation of the superior surface of the foot.

Down syndrome: A genetic abnormality resulting from the presence of three copies of chromosome 21; individuals with this condition have characteristic physical and intellectual deficits.

duct: A passageway that delivers exocrine secretions to an epithelial surface.

ductus arteriosus (ar-tē-rē-Ō-sus): Vascular connection between the pulmonary trunk and the aorta that functions throughout fetal life; normally closes at birth or shortly thereafter, and persists as the ligamentum arteriosum.

ductus deferens (DUK-tus DEF-e-renz): A passageway that carries sperm from the epididymis to the ejaculatory duct.

duodenal ampulla: Chamber that receives bile from the common bile duct and pancreatic secretions from the pancreatic duct.

duodenal glands: *See* submucosal glands.

duodenal papilla: Conical projection from the inner surface of the duodenum that contains the opening of the duodenal ampulla.

duodenum (dū-OD-e-num): The proximal 1 ft of the small intestine that contains short villi and submucosal glands.

dura mater (DŪ-ra MĀ-ter): Outermost component of the meninges that surround the brain and spinal cord.

dynamic equilibrium: Maintenance of normal body orientation as sudden changes in position (rotation, acceleration, etc.) occur.

dyslexia: Impaired ability to comprehend written words.

dysmenorrhea: Painful menstruation.

dysuria (dis-Ū-rē-a): Painful urination.

eccrine glands (EK-rin): Sweat glands of the skin that produce a watery secretion.

echocardiography (ek-ō-kar-dē-OG-ra-fē): Examination of the heart using modified ultrasound techniques.

ectoderm: One of the three primary germ layers; covers the surface of the embryo and gives rise to the nervous system, the epidermis and associated glands, and a variety of other structures.

ectopic (ek-TOP-ik): Outside of its normal location.

effector: A peripheral gland or muscle cell innervated by a motor neuron.

efferent: Away from.

efferent arteriole: An arteriole carrying blood away from the glomerulus of the kidney.

efferent fiber: An axon that carries impulses away from the CNS.

ejaculation (ē-jak-ū-LĀ-shun): The ejection of semen from the penis as the result of muscular contractions of the bulbocavernosus and ischiocavernosus muscles.

ejaculatory duct (ē-JAK-ū-la-tō-rē): Short ducts that pass within the walls of the prostate and connect the ductus deferens with the prostatic urethra.

elastase (ē-LAS-tāz): A pancreatic enzyme that breaks down elastin fibers.

elastin: Connective tissue fibers that stretch and rebound, providing elasticity to connective tissues.

electrocardiogram (ECG, EKG) (ē-lek-trō-KAR-dē-ō-gram): Graphic record of the electrical activities of the heart, as monitored at specific locations on the body surface.

electroencephalogram (EEG): Graphic record of the electrical activities of the brain.

electrolytes (ē-LEK-trō-līts): Soluble inorganic compounds whose ions will conduct an electric current in solution.

electron: One of the three fundamental particles; a subatomic particle that bears a negative charge and normally orbits around the protons of the nucleus.

eleidin (e-LĒ-i-din): A protein that forms as a precursor of keratin.

elephantiasis (el-e-fan-TĪ-a-sis): A lymphedema caused by infection and blockage of lymphatics by mosquito-borne parasites.

elevation: Movement in a superior, or upward, direction.

embolism (EM-bō-lizm): Obstruction or closure of a vessel by an embolus.

embolus (EM-bō-lus): An air bubble, fat globule, or blood clot drifting in the circulation.

embryo (EM-brē-ō): Developmental stage beginning at fertilization and ending at the start of the third developmental month.

embryogenesis (em-brē-ō-JEN-e-sis): The process of embryo formation.

embryology (em-brē-OL-ō-jē): The study of embryonic development, focusing on the first 2 months after fertilization.

emesis (EM-e-sis): Vomiting.

emmetropia: Normal vision.

emulsification (ē-mul-si-fi-KĀ-shun): The physical breakup of fats in the digestive tract, forming smaller droplets accessible to digestive enzymes; normally the result of mixing with bile salts.

enamel: Crystalline material similar in mineral composition to bone, but harder and without osteocytes, that covers the exposed surfaces of the teeth.

encephalitis: Inflammation of the brain.

endocarditis: Inflammation of the endocardium of the heart.

endocardium (en-dō-KAR-dē-um): The simple squamous epithelium that lines the heart and is continuous with the endothelium of the great vessels.

endochondral ossification (en-dō-KON-dral): The conversion of a cartilaginous model to bone, the characteristic mode of formation for skeletal elements other than the bones of the cranium, the clavicles, and sesamoid bones.

endocrine gland: A gland that secretes hormones into the blood.

endocrine system: The endocrine glands of the body.

endocytosis (en-dō-sī-TŌ-sis): The movement of relatively large volumes of extracellular material into the cytoplasm via the formation of a membranous vesicle at the cell surface; includes pinocytosis and phagocytosis.

endoderm: One of the three primary germ layers; the layer on the undersurface of the embryonic disc, that gives rise to the epithelia and glands of the digestive system, the respiratory system, and portions of the urinary system.

endogenous: Produced within the body.

endolymph (EN-dō-limf): Fluid contents of the membranous labyrinth (the saccule, utricle, semicircular canals, and cochlear duct) of the inner ear.

endometrial glands: Secretory glands of the endometrium.

endometrium (en-dō-MĒ-trē-um): The mucous membrane lining the uterus.

endomysium (en-dō-MĪS-ē-um): A delicate network of connective tissue fibers that surrounds individual muscle cells.

endoneurium: A delicate network of connective tissue fibers that surrounds individual nerve fibers.

endoplasmic reticulum (en-dō-PLAZ-mik re-TIK-ū-lum): A network of membranous channels in the cytoplasm of a cell that function in intracellular transport, synthesis, storage, packaging, and secretion.

endorphins (en-DOR-finz): Neuromodulators produced in the CNS that inhibit activity along pain pathways.

endosteum: An incomplete cellular lining found on the inner (medullary) surfaces of bones.

endothelium (en-dō-THĒ-lē-um): The simple squamous epithelium that lines blood and lymphatic vessels.

enkephalins (en-KEF-a-linz): Neuromodulators produced in the CNS that inhibit activity along pain pathways.

enteritis (en-ter-Ī-tis): Inflammation of the intestinal tract.

enterocrinin: A hormone secreted by the duodenal lining when exposed to acid chyme; stimulates the secretion of the duodenal glands.

enteroendocrine cells (en-ter-ō-EN-dō-krin): Endocrine cells scattered among the epithelial cells lining the digestive tract.

enterogastric reflex: Reflexive inhibition of gastric secretion initiated by the arrival of acid chyme in the small intestine.

enterohepatic circulation: Excretion of bile salts by the liver, followed by absorption of bile salts by intestinal cells for return to the liver via the hepatic portal vein.

enterokinase: An enzyme in the lumen of the small intestine that activates the proenzymes secreted by the pancreas.

enzyme: A protein that catalyzes a specific biochemical reaction.

eosinophils (ē-ō-SIN-ō-filz): A granulocyte (WBC) with a lobed nucleus and red-staining granules; participates in the immune response, and is especially important during allergic reactions.

ependyma (ep-EN-di-mah): Layer of cells lining the ventricles and central canal of the CNS.

epiblast (EP-i-blast): The layer of the inner cell mass facing the amniotic cavity prior to gastrulation.

epicardium: Serous membrane covering the outer surface of the heart; also called the visceral pericardium.

epidermis: The epithelium covering the surface of the skin.

epididymis (ep-i-DID-i-mus): Coiled duct that connects the rete testis to the ductus deferens; site of functional maturation of spermatozoa.

epidural block: Anesthesia caused by the elimination of sensory inputs from dorsal nerve roots following the introduction of drugs into appropriate regions of the epidural space.

epidural space: Space between the spinal dura mater and the walls of the vertebral foramen; contains blood vessels and adipose tissue; a frequent site of injection for regional anesthesia.

epiglottis (ep-i-GLOT-is): Blade-shaped flap of tissue, reinforced by cartilage, that is attached to the dorsal and superior surface of the thyroid cartilage; it folds over the entrance to the larynx during swallowing.

epimysium (ep-i-MĪS-ē-um): A dense investment of collagen fibers that surrounds a skeletal muscle, and is continuous with the tendons/aponeuroses of the muscle, and with the perimysium.

epineurium: A dense investment of collagen fibers that surrounds a peripheral nerve.

epiphyseal plate (e-pi-FI-sē-al): Cartilaginous region between the epiphysis and diaphysis of a growing bone.

epiphysis (e-PIF-i-sis): The end of a long bone.

epistaxis (ep-i-STAK-sis): Nosebleed.

epithelium (e-pi-THĒ-lē-um): One of the four primary tissue types; a layer of cells that forms a superficial covering or an internal lining of a body cavity or vessel.

eponychium (ep-ō-NIK-ē-um): A narrow zone of stratum corneum that extends across the surface of a nail at its exposed base; also called the **cuticle.**

equational division: The second meiotic division.

equilibrium (ē-kwi-LIB-rē-um): A dynamic state, where two opposing forces or processes are in balance.

erection: Stiffening of the penis prior to copulation due to the engorgement of the erectile tissues of the corpora cavernosa and the corpus spongiosum.

erythema (er-i-THĒ-ma): Redness and inflammation at the surface of the skin.

erythrocyte (e-RITH-rō-sīt): A red blood cell; an anucleate blood cell containing large quantities of hemoglobin.

erythrocytosis (e-rith-rō-sī-TŌ-sis): An abnormally large number of erythrocytes in the circulating blood.

erythropoiesis (e-rith-rō-poy-Ē-sis): Red blood cell formation.

erythropoietin (e-rith-rō-POI-e-tin): Hormone released by tissues, especially the kidneys, exposed to low oxygen concentrations; stimulates hematopoiesis in bone marrow.

Escherichia coli: Normal bacterial resident of the large intestine.

esophagus: A muscular tube that connects the pharynx to the stomach.

estradiol (es-tra-DĪ-ol): The primary estrogen secreted by ovarian follicles.

estrogens (ES-trō-jenz): The dominant sex hormones in females; notably estradiol.

eupnea (ŪP-nē-a): Normal quiet breathing.

eversion (ē-VER-shun): A turning outward.

excitable membranes: Membranes that conduct action potentials, a characteristic of muscle and nerve cells.

excretion: Elimination from the body.

exocrine gland: A gland that secretes onto the body surface or into a passageway connected to the exterior.

exocytosis (ek-sō-sī-TŌ-sis): The ejection of cytoplasmic materials by fusion of a membranous vesicle with the cell membrane.

expiration: Exhalation; breathing out.

expiratory reserve: The amount of additional air that can be voluntarily moved out of the respiratory tract after a normal tidal expiration.

extension: An increase in the angle between two articulating bones; the opposite of flexion.

extensor retinaculum (ret-i-NAK-ū-lum): A thickening of the fascia of the forearm at the wrist or the leg at the ankle, forming a band of dense connective tissue that holds extensor muscle tendons in place.

external auditory canal: Passageway in the temporal bone that leads from the external auditory meatus to the tympanic membrane of the inner ear.

external ear: The pinna, external auditory canal, and tympanic membrane.

external nares: The nostrils; the external openings into the nasal cavity.

exteroceptors: Sensory receptors in the skin, mucous membranes, and special sense organs that provide information about the external environment and our position within it.

extracellular fluid: All body fluid other than that contained within cells; includes plasma and interstitial fluid.

extraembryonic membranes: The yolk sac, amnion, chorion, and allantois.

extrafusal fibers: Contractile muscle fibers, as opposed to the sensory intrafusal fibers (muscle spindles).

extrapyramidal system: Nuclei and tracts associated with the involuntary control of muscular activity.

extremities: The limbs.

extrinsic pathway: Clotting pathway that begins with damage to blood vessels or surrounding tissues and ends with the formation of tissue thromboplastin.

fabella: A sesamoid bone often found in the gastrocnemius muscle just behind the knee.

facilitated diffusion: Passive movement of a substance across a cell membrane via a protein carrier.

facilitation: Depolarization of a neuron cell membrane toward threshold, or making the cell more sensitive to depolarizing stimuli.

falciform ligament (FAL-si-form): A sheet of mesentery that contains the ligamentum teres, the fibrous remains of the umbilical vein of the fetus.

falx (FALKS): Sickle-shaped.

falx cerebri (FALKS ser-Ē-brē): Curving sheet of dura mater that extends between the two cerebral hemispheres; encloses the superior sagittal sinus.

fascia (FASH-a): Connective tissue fibers, primarily collagenous, that form sheets or bands beneath the skin to attach, stabilize, enclose, and separate muscles and other internal organs.

fasciculus (fa-SIK-ū-lus): A small bundle, usually referring to a collection of nerve axons or muscle fibers.

fatty acids: Hydrocarbon chains ending in a carboxyl group.

fauces (FAW-sēz): The passage from the mouth to the pharynx, bounded by the palatal arches, the soft palate, and the uvula.

feces: Waste products eliminated by the digestive tract at the anus; contain indigestible residue, bacteria, mucus, and epithelial cells.

fenestra: An opening.

fenestrated (FEN-es-trāt-ed): Having multiple openings; used when referring to very permeable capillaries whose endothelial cells are penetrated by pores of varying sizes.

fertilization: Fusion of egg and sperm to form a zygote.

fetus: Developmental stage lasting from the start of the third developmental month to delivery.

fibrillation (fi-bril-Ā-shun): Uncoordinated contractions of individual muscle cells that impair or prevent normal function.

fibrin (FĪ-brin): Insoluble protein fibers that form the basic framework of a blood clot.

fibrinogen (fī-BRIN-ō-jen): Plasma proteins that can be converted, by the action of enzymes, into insoluble strands of fibrin that form the basis for a blood clot.

fibrinolysis (fī-brin-OL-i-sis): The breakdown of the fibrin strands of a blood clot by a proteolytic enzyme.

fibroblasts (FĪ-brō-blasts): Cells of connective tissue proper that are responsible for the production of extracellular fibers and the secretion of the organic compounds of the extracellular matrix.

fibrocartilage: Cartilage containing an abundance of collagen fibers; found around the edges of joints, in the intervertebral discs, the menisci of the knee, etc.

fibrous tunic: The outermost layer of the eye, composed of the sclera and cornea.

fibula (FIB-ū-la): The lateral, relatively small bone of the lower leg.

filariasis (fil-a-RĪ-a-sis): Condition resulting from infection by mosquito-borne parasites; may cause elephantiasis.

filiform papillae: Slender conical projections from the dorsal surface of the anterior two-thirds of the tongue.

filtrate: Fluid produced by filtration at a glomerulus in the kidney.

filtration: Movement of a fluid across a membrane whose pores restrict the passage of solutes on the basis of size.

filum terminale: A fibrous extension of the spinal cord that extends from the conus medullaris to the coccygeal ligament.

fimbriae (FIM-brē-ē): A fringe; used to describe the fingerlike processes that surround the entrance to the uterine tube.

fissure: An elongated groove or opening.

fistula: An abnormal passageway between two organs or from an internal organ or space to the body surface.

flaccid: Limp, soft, flabby; a muscle without muscle tone.

flagellum/flagella (fla-JEL-ah): An organelle structurally similar to a cilium, but used to propel a cell through a fluid.

flatus: Intestinal gas.

flexion (FLEK-shun): A movement that reduces the angle between two articulating bones; the opposite of extension.

flexor: A muscle that produces flexion.

flexor reflex: A reflex contraction of the flexor muscles of a limb in response to an unpleasant stimulus.

flexure: A bending.

fluoroscope: An instrument that permits the examination of the body with X-rays in real time, rather than via fixed images on photographic plates.

folia (FŌ-lē-ah): Leaflike folds; used in reference to the slender folds in the surface of the cerebellar cortex.

follicle (FOL-i-kl): A small secretory sac or gland.

follicle-stimulating hormone (FSH): A hormone secreted by the anterior pituitary; stimulates oogenesis (female) and spermatogenesis (male).

folliculitis (fō-lik-ū-LĪ-tis): Inflammation of a follicle, such as a hair follicle of the skin.

fontanel (fon-tah-NEL): A relatively soft, flexible, fibrous region between two flat bones in the developing skull.

foramen: An opening or passage through a bone.

forearm: Distal portion of the arm between the elbow and wrist.

forebrain: The cerebrum.

fornix (FOR-niks): An arch, or the space bounded by an arch; in the brain, an arching tract that connects the hippocampus with the mamillary bodies; in the eye, a slender pocket found where the epithelium of the ocular conjunctiva folds back upon itself as the palpebral conjunctiva.

fossa: A shallow depression or furrow in the surface of a bone.

fourth ventricle: An elongate ventricle of the metencephalon (pons and cerebellum) and the myelencephalon (medulla) of the brain; the roof contains a region of choroid plexus.

fovea (FŌ-vē-a): Portion of the retina providing the sharpest vision, with the highest concentration of cones; also called the **macula lutea.**

fracture: A break or crack in a bone.

frenulum (FREN-ū-lum): A bridle; *see* **lingual frenulum.**

frontal plane: A sectional plane that divides the body into anterior and posterior portions.

fructose: A hexose (simple sugar containing six carbons) found in foods and in semen.

fundus (FUN-dus): The base of an organ.

fungiform papillae: Mushroom-shaped papillae on the dorsal and dorsolateral surfaces of the tongue.

furuncle (FŪ-rung-kl): A boil, resulting from the invasion and inflammation of a hair follicle or sebaceous gland.

gallbladder: Pear-shaped reservoir for the bile secreted by the liver.

gametes (GAM-ēts): Reproductive cells (sperm or eggs) that contain one-half of the normal chromosome complement.

gametogenesis (ga-mē-tō-JEN-e-sis): The formation of gametes.

gamma aminobutyric acid (GABA) (GAM-ma a-MĒ-nō-bū-TIR-ik): A neurotransmitter of the CNS whose effects are usually inhibitory.

gamma motor neurons: Motor neurons that adjust the sensitivities of muscle spindles (intrafusal fibers).

ganglion/ganglia: A collection of nerve cell bodies outside of the CNS.

gap junctions: Connections between cells that permit electrical coupling.

gaster (GAS-ter): The stomach; the body or belly of a skeletal muscle.

gastrectomy (gas-TREK-tō-mē): Partial or total surgical removal of the stomach.

gastric: Pertaining to the stomach.

gastric glands: Tubular glands of the stomach whose cells produce acid, enzymes, intrinsic factor, and hormones.

gastrin (GAS-trin): Hormone produced by enteroendocrine cells of the stomach, when exposed to mechanical stimuli or vagal stimulation, and the duodenum, when exposed to chyme containing undigested proteins.

gastritis (gas-TRĪ-tis): Inflammation of the stomach.

gastroenteric reflex (gas-trō-en-TER-ik): An increase in peristalsis along the small intestine triggered by the arrival of food in the stomach.

gastroileal reflex (gas-trō-IL-ē-al): Peristaltic movements that shift materials from the ileum to the colon, triggered by the arrival of food in the stomach.

gastrointestinal (GI) tract: An internal passageway that begins at the mouth, ends at the anus, and is lined by a mucous membrane; also known as the **digestive tract.**

gastroscope: A fiberoptic instrument that permits visual inspection of the stomach lining.

gastrulation (gas-troo-LĀ-shun): The movement of cells of the inner cell mass that creates the three primary germ layers of the embryo.

gene: A portion of a DNA strand that functions as a hereditary unit and is found at a particular locus on a specific chromosome.

genetic engineering: Research and experiments involving the manipulation of the genetic makeup of an organism.

genetics: The study of mechanisms of heredity.

geniculate (je-NIK-ū-lāt): Like a little knee; the medial geniculates and the lateral geniculates are thalamic nuclei in the walls of the thalamus of the brain.

genitalia (jen-i-TĀ-lē-a): Reproductive organs.

genotype (JĒN-ō-tīp): The genetic complement of a particular individual.

germinal centers: Pale regions in the interior of lymphatic tissues or nodules, where mitoses are under way.

gestation (jes-TĀ-shun): The period of intrauterine development.

gingivae (JIN-ji-vē): The gums.

gingivitis: Inflammation of the gums.

gland: Cells that produce exocrine or endocrine secretions, derived from epithelia.

glans penis: Expanded tip of the penis that surrounds the urethra meatus; continuous with the corpus spongiosum.

glaucoma: Eye disorder characterized by rising intraocular pressures due to inadequate drainage of aqueous humor at the canal of Schlemm.

glenoid fossa: A rounded depression that forms the articular surface of the scapula at the shoulder joint.

glial cells (GLĒ-al): Supporting cells in the neural tissue of the CNS and PNS.

globular proteins: Proteins whose tertiary structure makes them rounded and compact.

globulins (GLOB-ū-lins): Globular plasma proteins with a variety of important functions.

glomerular capsule: Expanded initial portion of the nephron that surrounds the glomerulus.

glomerulonephritis (glō-mer-ū-lō-nef-RĪ-tis): Inflammation of the glomeruli of the kidneys.

glomerulus (glō-MER-ū-lus): A ball or knot; in the kidneys, a knot of capillaries that projects into the enlarged, proximal end of a nephron; the site where filtration occurs, the first step in the production of urine.

glossopharyngeal nerve (glos-ō-fa-RIN-jē-al): Cranial nerve IX.

glottis (GLOT-is): The passage from the pharynx to the larynx.

glucagon (GLOO-ka-gon): Hormone secreted by the alpha cells of the pancreatic islets; elevates blood glucose concentrations.

glucocorticoids (gloo-kō-KOR-ti-koyds): Hormones secreted by the zona fasciculata of the adrenal cortex to modify glucose metabolism; cortisol, cortisone, and corticosterone are important examples.

glucose (GLOO-kōs): A 6-carbon sugar, $C_6H_{12}O_6$, the preferred energy source for most cells and the only energy source for neurons under normal conditions.

glucose-dependent insulinotropic hormone (GIP): A duodenal hormone released when the arriving chyme contains large quantities of carbohydrates; triggers the secretion of insulin and a slowdown in gastric activity.

glycerides: Lipids composed of glycerol bound to 1–3 fatty acids.

glycogen (GLĪ-kō-jen): A polysaccharide that represents an important energy reserve; a polymer consisting of a long chain of glucose molecules.

glycolipids (glī-cō-LIP-idz): Compounds created by the combination of carbohydrate and lipid components.

glycoprotein (glī-kō-PRŌ-tēn): A compound containing a relatively small carbohydrate group attached to a large protein.

glycosuria (glī-cō-SŪ-rē-a): The presence of glucose in the urine.

goblet cell: A goblet-shaped, mucus-producing, unicellular gland found in certain epithelia of the digestive and respiratory tracts.

goiter: Enlargement of the thyroid gland.

Golgi apparatus (GOL-jē): Cellular organelle consisting of a series of membranous plates that give rise to lysosomes and secretory vesicles.

Golgi tendon organ (GOL-jē): *See* **tendon organ.**

gomphosis (gom-FŌ-sis): A fibrous synarthrosis that binds a tooth to the bone of the jaw; *see* **periodontal ligament.**

gonadotropic hormones: FSH and LH, hormones that stimulate gamete development and sex hormone secretion.

gonadotropin-releasing hormone (GnRH) (gō-nad-ō-TRŌ-pin): Hypothalamic releasing hormone that causes the secretion of FSH and LH by the anterior pituitary gland.

gonadotropins (gō-nad-ō-TRŌ-pins): Hormones that stimulate the gonads (testes or ovaries).

gonads (GŌ-nadz): Organs that produce gametes and hormones.

gout: Clinical condition resulting from elevated uric acid concentrations in the blood and peripheral tissues.

granulocytes (GRAN-ū-lō-sīts): White blood cells containing granules visible with the light microscope; includes eosinophils, basophils, and neutrophils; also called granular leukocytes.

gray matter: Areas in the central nervous system dominated by nerve cell bodies, glial cells, and unmyelinated axons.

gray ramus: A bundle of postganglionic sympathetic nerve fibers that go to a spinal nerve for distribution to effectors in the body wall, skin, and extremities.

greater omentum: A large fold of the dorsal mesentery of the stomach that hangs in front of the intestines.

greater vestibular glands: Mucous glands in the vaginal walls that secrete into the vestibule; the equivalent of the bulbourethral glands of the male.

greenstick fracture: A fracture most often affecting the long bones of young children.

groin: The inguinal region.

gross anatomy: The study of the structural features of the human body without the aid of a microscope.

growth hormone (GH): Anterior pituitary hormone that stimulates tissue growth and anabolism when nutrients are abundant, and restricts tissue glucose dependence when nutrients are in short supply.

gustation (gus-TĀ-shun): Taste.

gynecologists (gī-ne-KOL-ō-jists): Physicians specializing in the pathology of the female reproductive system.

gyrus (JĪ-rus): A prominent fold or ridge of neural cortex on the surfaces of the cerebral hemispheres.

hair: A keratinous strand produced by epithelial cells of the hair follicle.

hair cells: Sensory cells of the inner ear.

hair follicle: An accessory structure of the integument; a tube lined by a stratified squamous epithelium that begins at the surface of the skin and ends at the hair papilla.

hair root: A thickened, conical structure consisting of a connective tissue papilla and the overlying matrix, a layer of epithelial cells that produces the hair shaft.

hallux: The great toe.

haploid (HAP-loid): Possessing one-half of the normal number of chromosomes; a characteristic of gametes.

hard palate: The bony roof of the oral cavity, formed by the maxillary and palatine bones.

Hassall's corpuscles: Aggregations of epithelial cells in the thymus whose functions are unknown.

haustra (HAWS-tra): Saclike pouches along the length of the large intestine that result from tension in the taenia coli.

Haversian system: *See* **osteon.**

heart block: A cardiac arrhythmia due to conduction delays that affect communication between the atria and ventricles.

Heimlich maneuver (HĪM-lik): A technique for removing an airway blockage by external compression of the abdomen and forceful elevation of the diaphragm.

helper T cells: Lymphocytes (T cells) whose secretions and other activities coordinate the cellular and humoral immune responses.

hematocrit (hē-MAT-ō-krit): Percentage of the volume of whole blood contributed by cells; also called the packed cell volume (PCV) or the volume of packed red cells (VPRC).

hematologists (hē-ma-TOL-ō-jists): Specialists in disorders of the blood and blood-forming tissues.

hematoma: A tumor or swelling filled with blood.

hematuria (hē-ma-TŪ-rē-a): The presence of abnormal numbers of red blood cells in the urine.

heme (HĒM): A porphyrin ring containing a central iron atom that can reversibly bind oxygen molecules; a component of the hemoglobin molecule.

hemiplegia: Paralysis affecting one side of the body (arm, trunk, and leg).

hemocytoblasts: Stem cells whose divisions produce all of the various populations of blood cells.

hemodialysis (hē-mō-dī-AL-i-sis): Dialysis of the blood.

hemoglobin (hē-mō-GLŌ-bin): Protein composed of four globular subunits, each bound to a single molecule of heme; the protein found in red blood cells that gives them the ability to transport oxygen in the blood.

hemolysis: Breakdown (lysis) of red blood cells.

hemophilia (hē-mō-FIL-ē-a): A congenital condition resulting from the inadequate synthesis of one of the clotting factors.

hemopoiesis (hē-mō-poi-Ē-sis): Blood cell formation and differentiation.

hemorrhage: Blood loss.

hemorrhoids (HEM-ō-roidz): Swollen, varicose veins that protrude from the walls of the rectum and/or anorectal canal.

hemostasis: The cessation of bleeding.

hemothorax: The entry of blood into one of the pleural cavities.

heparin (HEP-a-rin): An anticoagulant released by activated basophils and mast cells.

hepatic duct: Duct carrying bile away from the liver lobes and toward the union with the cystic duct.

hepatic portal vein: Vessel that carries blood between the intestinal capillaries and the sinusoids of the liver.

hepatitis (hep-a-TĪ-tis): Inflammation of the liver, resulting from exposure to toxic chemicals, drugs, or viruses.

hepatocyte (he-PAT-ō-sīt): A liver cell.

hernia: The protrusion of a loop or portion of a visceral organ through the abdominopelvic wall or into the thoracic cavity.

herniated disc: Rupture of the connective tissue sheath of the nucleus pulposus of an intervertebral disc.

heterotopic: Ectopic; outside of its normal location.

heterozygous (het-er-ō-ZĪ-gus): Possessing two different alleles at corresponding loci on a chromosome pair; the individual's phenotype may be determined by one or both of the alleles.

hexose: A 6-carbon simple sugar.

hiatus (hī-Ā-tus): A gap, cleft, or opening.

hilus (HĪ-lus): A localized region where blood vessels, lymphatics, nerves, and/or other anatomical structures are attached to an organ.

hippocampus: A portion of the limbic system that is concerned with the organization and storage of memories.

hirsutism (HER-sut-izm): Excessive hair growth in women that follows the distribution pattern typical of adult males; sometimes caused by the overproduction of androgens.

histamine (HIS-ta-mēn): Chemical released by stimulated mast cells or basophils to initiate or enhance an inflammatory response.

histology (his-TOL-ō-jē): The study of tissues.

histones: Proteins associated with the DNA of the nucleus, and around which the DNA strands are wound.

holocrine secretion (HŌL-ō-krin): Form of exocrine secretion where the secretory cell becomes swollen with vesicles and then ruptures.

homeostasis (hō-mē-ō-STĀ-sis): The maintenance of a relatively constant internal environment.

homologous chromosomes (hō-MOL-ō-gus): The members of a chromosome pair, each containing the same gene loci.

homozygous (hō-mō-ZĪ-gus): Having the same gene for a particular character on two homologous chromosomes.

hormone: A compound secreted by one cell that travels through the circulatory system to affect the activities of cells in another portion of the body.

human chorionic gonadotropin (hCG): Placental hormone that maintains the corpus luteum for the first 3 months of pregnancy.

human immunodeficiency virus (HIV): The infectious agent that causes **acquired immune deficiency syndrome (AIDS).**

human leukocyte antigen (HLA): Antigens on cell surfaces important to foreign antigen recognition and that play a role in the coordination and activation of the immune response.

human placental lactogen (hPL): Placental hormone that stimulates the functional development of the mammary glands.

humoral immunity: Immunity resulting from the presence of circulating antibodies produced by plasma cells.

hyaline cartilage (HĪ-a-lin): The most common type of cartilage; the matrix contains collagen fibers. Examples include the connections between the ribs and sternum, the tracheal and bronchial cartilages, and synovial cartilages.

hyaluronic acid: A proteoglycan in the matrix of many connective tissues that gives the matrix a viscous consistency; also functions as intercellular cement.

hyaluronidase (hī-a-lūr-ON-a-dāz): An enzyme that breaks down hyaluronic acid; produced by some bacteria and found in the acrosomal cap of a sperm cell.

hydrocephalus: Condition resulting from excessive production or inadequate drainage of cerebrospinal fluid.

hydrostatic pressure: Fluid pressure.

hymen: A membrane that forms during development, covering the entrance to the vagina.

hypercapnia (hī-per-KAP-nē-a): High plasma carbon dioxide concentrations, often the result of hypoventilation or inadequate tissue perfusion.

hyperglycemia: Elevated plasma glucose concentrations.

hyperopia: The farsighted condition, characterized by an inability to focus on objects close by.

hyperplasia: Abnormal enlargement of an organ due to an increase in the number of cells.

hyperpnea (hī-perp-NĒ-a): Abnormal increases in the rate and depth of respiration.

hyperreflexia: Abnormally exaggerated reflex responses to stimulation.

hypersecretion: Overactivity of glands that produce exocrine or endocrine secretions.

hypertension: Abnormally high blood pressure.

hyperthermia: Excessively high body temperature.

hyperthyroidism: Excessive production of thyroid hormones.

hypertonic: When comparing two solutions, used to refer to the solution with the higher osmolarity.

hypertrophy (HĪ-per-trō-fē): Increase in the size of tissue without cell division.

hyperventilation (hī-per-ven-ti-LĀ-shun): A rate of respiration sufficient to reduce the plasma pCO_2 to levels below normal.

hypoblast (HĪ-pō-blast): The undersurface of the inner cell mass that faces the blastocoele of the early embryo.

hypocapnia: Abnormally low plasma pCO_2, usually the result of hyperventilation.

hypodermic needle: A needle inserted through the skin to introduce drugs into the subcutaneous layer.

hypodermis: The subcutaneous layer, a region of loose connective tissue also called the **superficial fascia.**

hypoesthesia: Abnormally decreased sensitivity to stimuli.

hypoglossal nerve: N XII, the cranial nerve responsible for the control of the muscles that move the tongue.

hyponychium (hī-pō-NIK-ē-um): A thickening in the epidermis beneath the free edge of a nail.

hypophyseal portal system (hī-pō-FIS-ē-al): Network of vessels that carry blood from capillaries in the hypothalamus to capillaries in the anterior pituitary gland (hypophysis).

hypophysis (hī-POF-i-sis): The anterior pituitary gland, which can be further subdivided into the pars distalis and the pars intermedia.

hyporeflexia: Abnormally depressed reflex responses to stimuli.

hyposecretion: Abnormally low rates of exocrine or endocrine secretion.

hypothalamus: The floor of the diencephalon; region of the brain containing centers involved with the unconscious regulation of visceral functions, emotions, drives, and the coordination of neural and endocrine functions.

hypotonic: When comparing two solutions, used to refer to the one with the lower osmolarity.

hypoventilation: A respiratory rate insufficient to keep plasma pCO_2 within normal levels.

hypovolemic (hī-pō-vō-LĒ-mik): An abnormally low blood volume.

hypoxia (hī-POKS-ē-a): Low tissue oxygen concentrations.

ileocecal valve (il-ē-ō-SĒ-kal): A fold of mucous membrane that guards the connection between the ileum and the cecum.

ileostomy (il-ē-OS-tō-mē): Surgical creation of an opening into the ileum; the opening created when the ileum is surgically attached to the abdominal wall.

ileum (IL-ē-um): The last 8 ft of the small intestine.

ilium (IL-ē-um): The largest of the three bones whose fusion creates a coxa.

immunity: Resistance to injuries and diseases caused by foreign compounds, toxins, and pathogens.

immunization: Developing immunity by the deliberate exposure to antigens under conditions that prevent the development of illness but stimulate the production of memory B cells.

immunoglobulin (i-mū-nō-GLOB-ū-lin): A circulating antibody.

implantation (im-plan-TĀ-shun): The erosion of a blastocyst into the uterine wall.

impotence: Inability to obtain or maintain an erection in the male.

incisors (in-SĪ-zerz): Two pairs of flattened, bladelike teeth located at the front of the dental arches in both the upper and lower jaws.

inclusions: Aggregations of insoluble pigments, nutrients, or other materials in the cytoplasm.

incontinence (in-KON-ti-nens): Inability to voluntarily control micturition (or defecation).

incus (IN-kus): The central auditory ossicle, situated between the malleus and the stapes in the middle ear cavity.

infarct: An area of dead cells resulting from an interruption of circulation.

infection: Invasion and colonization of body tissues by pathogenic organisms.

inferior: A directional reference meaning below.

inferior vena cava: The vein that carries blood from the parts of the body below the heart to the right auricle.

infertility: Inability to conceive.

inflammation: A nonspecific defense mechanism that operates at the tissue level, characterized by swelling, redness, warmth, pain, and some loss of function.

inflation reflex: A reflex mediated by the vagus nerve that prevents overexpansion of the lungs.

infundibulum (in-fun-DIB-ū-lum): A tapering, funnel-shaped structure; in the nervous system, refers to the connection between the pituitary gland and the hypothalamus; the infundibulum of the uterine tube is the entrance bounded by fimbriae that receives the ova at ovulation.

ingestion: The introduction of materials into the digestive tract via the mouth.

inguinal canal: A passage through the abdominal wall that marks the path of testicular descent, and that contains the testicular arteries, veins, and ductus deferens.

inguinal region: The area near the junction of the trunk and the thighs that contains the external genitalia.

inhibin (in-HIB-in): A hormone produced by the sustentacular cells that inhibits the pituitary secretion of FSH.

initial segment: The proximal portion of the axon, adjacent to the axon hillock, where an action potential first appears.

injection: Forcing of fluid into a body part or organ.

inner cell mass: Cells of the blastocyst that will form the body of the embryo.

inner ear: *See* **internal ear.**

innervation: The distribution of sensory and motor nerves to a specific region or organ.

insertion: Point of attachment of a muscle that is more movable.

inspiration: Inhalation; the movement of air into the respiratory system.

inspiratory reserve: The maximum amount of air that can be drawn into the lungs over and above the normal tidal volume.

insoluble: Incapable of dissolving in solution.

insomnia: Sleep disorder characterized by an inability to fall asleep.

insula (IN-sū-la): A region of the temporal lobe that is visible only after opening the lateral sulcus.

insulin (IN-sū-lin): Hormone secreted by the beta cells of the pancreatic islets; causes a reduction in plasma glucose concentrations.

integument (in-TEG-ū-ment): The skin and associated organs (hair, glands, receptors, etc.).

intercalated discs (in-TER-ka-lā-ted): Regions where adjacent cardiocytes interlock and where gap junctions permit electrical coupling between the cells.

intercellular cement: Proteoglycans, containing the polysaccharide hyaluronic acid, found between adjacent epithelial cells.

intercellular fluid: *See* **interstitial fluid.**

interdigitate: To interlock.

interferons (in-ter-FĒR-ons): Peptides released by virally infected cells, especially lymphocytes, that make other cells more resistant to viral infection and slow viral replication.

interleukins (in-ter-LOO-kins): Peptides released by activated monocytes and lymphocytes that assist in the coordination of the cellular and humoral immune responses.

internal capsule: Term given to the appearance of the white matter of the cerebral hemispheres on gross dissection of the brain.

internal ear: The membranous labyrinth that contains the organs of hearing and equilibrium.

internal nares: The entrance to the nasopharynx from the nasal cavity.

internal respiration: Diffusion of gases between the blood and interstitial fluid.

interneuron: An association neuron; neurons inside the CNS that are interposed between sensory and motor neurons.

interoceptors: Sensory receptors monitoring the functions and status of internal organs and systems.

interosseous membrane: Fibrous connective tissue membrane between the shafts of the tibia and fibula or the radius and ulna; an example of a fibrous amphiarthrosis.

interphase: Stage in the life of a cell during which the chromosomes are uncoiled and all normal cellular functions except mitosis are under way.

intersegmental reflex: A reflex that involves several segments of the spinal cord.

interstitial cell-stimulating hormone: An alternative name for LH in the male; stimulates androgen production by the interstitial cells of the testes.

interstitial fluid (in-ter-STISH-al): Fluid in the tissues that fills the spaces between cells.

interstitial growth: Form of cartilage growth through the growth, mitosis, and secretion of chondrocytes inside the matrix.

interventricular foramen: The opening that permits fluid movement between the lateral and third ventricles.

intervertebral disc: Fibrocartilage pad between the centra of successive vertebrae that acts as a shock absorber.

intestinal crypt: A tubular epithelial pocket lined by secretory cells and opening into the lumen of the digestive tract; also called an intestinal gland.

intestine: Tubular organ of the digestive tract.

intracellular fluid: The cytosol.

intrafusal fibers: Muscle spindle fibers.

intramembranous ossification (in-tra-MEM-bra-nus): The formation of bone within a connective tissue without the prior development of a cartilaginous model.

intramuscular injection: Injection of medication into the bulk of a skeletal muscle.

intraocular pressure: The hydrostatic pressure exerted by the aqueous humor of the eye.

intrapleural pressure: The pressure measured in a pleural cavity; also called the intrathoracic pressure.

intrapulmonary pressure (in-tra-PUL-mō-ner-ē): The pressure measured in an alveolus; also called the intra-alveolar pressure.

intrauterine: Within the uterus; used to refer to the period of prenatal development.

intrinsic factor: Glycoprotein secreted by the parietal cells of the stomach that facilitates the intestinal absorption of vitamin B_{12}.

inversion: A turning inward.

in vitro: Outside of the body, in an artificial environment.

in vivo: In the living body.

involuntary: Not under conscious control.

ion: An atom or molecule bearing a positive or negative charge due to the acceptance or donation of an electron.

ipsilateral: A reflex response affecting the same side as the stimulus.

iris: A contractile structure made up of smooth muscle that forms the colored portion of the eye.

ischemia (is-KĒ-mē-a): Inadequate blood supply to a region of the body.

ischium (IS-kē-um): One of the three bones whose fusion creates the coxa.

islets of Langerhans: *See* **pancreatic islets.**

isometric contraction: A muscular contraction characterized by rising tension production but no change in length.

isotonic: A solution having an osmolarity that does not result in water movement across cell membranes; of the same contractive strength.

isotonic contraction: A muscular contraction during which tension climbs and then remains stable as the muscle shortens.

isthmus (IS-mus): A narrow band of tissue connecting two larger masses.

jaundice (JAWN-dis): Condition characterized by yellowing of connective tissues due to elevated tissue bilirubin levels; usually associated with damage to the liver or biliary system.

jejunum (je-JŪ-num): The middle portion of the small intestine.

joint: An area where adjacent bones interact; an articulation.

juxtaglomerular apparatus: The macula densa and the juxtaglomerular cells; a complex responsible for the release of renin and erythropoietin.

juxtaglomerular cells: Modified smooth muscle cells in the walls of the afferent and efferent arterioles adjacent to the glomerulus and the macula densa.

juxtamedullary nephrons: The 15% of nephrons whose loops of Henle extend into the medulla; these nephrons are responsible for creating the osmotic gradient within the medulla.

karyotyping (KAR-ē-ō-tī-ping): The determination of the chromosomal characteristics of an individual or cell.

keratin (KER-a-tin): Tough, fibrous protein component of nails, hair, calluses, and the general integumentary surface.

keratinization (KER-a-tin-ī-zā-shun): The production of keratin by epithelial cells.

keratinized (KER-a-tin-īzd): Containing large quantities of keratin.

keratohyalin (ker-a-tō-HĪ-a-lin): A protein precursor of keratin.

ketone bodies: Keto acids produced during the catabolism of lipids and ketogenic amino acids; specifically acetone, acetoacetate, and beta-hydroxybutyrate.

kidney: A component of the urinary system; an organ functioning in the regulation of plasma composition, including the excretion of wastes and the maintenance of normal fluid and electrolyte balance.

killer T cells: *See* **cytotoxic T cells.**

Kupffer cells (KOOP-fer): Stellate cells of the liver; phagocytic cells of the liver sinusoids.

kyphosis (kī-FŌ-sis): Exaggerated thoracic curvature.

labia (LĀ-bē-a): Lips; labia majora and minora are components of the female external genitalia.

labrum: A lip or rim.

labyrinth: A maze of passageways; usually refers to the structures of the inner ear.

lacrimal gland (LAK-ri-mal): Tear gland on the dorsolateral surface of the eye.

lactase: An enzyme that breaks down a disaccharide (lactose) in milk.

lactation (lak-TĀ-shun): The production of milk by the mammary glands.

lacteal (LAK-tē-al): A terminal lymphatic within an intestinal villus.

lactic acid: Compound produced from pyruvic acid during glycolysis.

lactiferous duct (lak-TIF-e-rus): Duct draining one lobe of the mammary gland.

lactiferous sinus: An expanded portion of a lactiferous duct adjacent to the nipple of a breast.

lacuna (la-KŪ-nē): A small pit or cavity.

lambdoidal suture (lam-DOID-al): Synarthrotic articulation between the parietal and occipital bones of the cranium.

lamellae (la-MEL-lē): Concentric layers of bone within an osteon.

lamina (LA-min-a): A thin sheet or layer.

lamina propria (PRŌ-prē-a): A layer of loose connective tissue situated immediately beneath the epithelium of a mucous membrane.

laminectomy: Removal of the spinous processes of a vertebra to gain access and treat a herniated disc.

Langerhans cells (LAN-ger-hanz): Cells in the epithelium of the skin and digestive tract that participate in the immune response by presenting antigens to T cells.

laparoscope (LAP-a-rō-skōp): Fiberoptic instrument used to visualize the contents of the abdominopelvic cavity.

large intestine: The terminal portions of the intestinal tract, consisting of the colon, the rectum, and the anorectal canal.

laryngopharynx (la-ring-gō-FAR-inks): Division of the pharynx inferior to the epiglottis and superior to the esophagus.

larynx (LAR-inks): A complex cartilaginous structure that surrounds and protects the glottis and vocal cords; the superior margin is bound to the hyoid bone and the inferior margin is bound to the trachea.

lateral: Pertaining to the side.

lateral apertures: Openings in the roof of the fourth ventricle that permit the circulation of CSF into the subarachnoid space.

lateral ventricle: Fluid-filled chamber within one of the cerebral hemispheres.

laxatives: Compounds that promote defecation via increased peristalsis or an increase in the water content and volume of the feces.

lens: The transparent body lying behind the iris and pupil and in front of the vitreous humor.

lesion: A localized abnormality in tissue organization.

lesser omentum: A small pocket in the mesentery that connects the lesser curvature of the stomach to the liver.

leukemia (loo-KĒ-mē-ah): A malignant disease of the blood-forming tissues.

leukocyte (LOO-kō-sīt): A white blood cell.

leukocytosis (loo-kō-sī-TŌ-sis): Abnormally high numbers of circulating white blood cells.

leukopenia (loo-kō-PĒ-nē-ah): Abnormally low numbers of circulating white blood cells.

leukopoiesis (loo-kō-poy-Ē-sis): White blood cell formation.

ligament (LIG-a-ment): Dense band of connective tissue fibers that attach one bone to another.

ligamentum arteriosum: The fibrous strand found in the adult that represents the remains of the ductus arteriosus of the fetus.

ligamentum nuchae (NOO-kē): An elastic ligament that extends between the vertebra prominens and the external occipital crest.

ligamentum teres: The fibrous strand in the falciform ligament that represents the remains of the umbilical vein of the fetus.

ligate: To tie off.

limbic system (LIM-bik): Group of nuclei and centers in the cerebrum and diencephalon that are involved with emotional states, memories, and behavioral drives.

limbus (LIM-bus): The edge of the cornea, marked by the transition from the corneal epithelium to the ocular conjunctiva.

liminal stimulus: A stimulus sufficient to depolarize the transmembrane potential of an excitable membrane to threshold, and produce an action potential.

linea alba: Tendinous band that runs along the midline of the rectus abdominis.

lingual: Pertaining to the tongue.

lingual frenulum: An epithelial fold that attaches the inferior surface of the tongue to the floor of the mouth.

lipase (LI-pāz): A pancreatic enzyme that breaks down triglycerides.

lipemia (lip-Ē-mē-a): Elevated concentration of lipids in the circulation.

lipid: An organic compound containing carbons, hydrogens, and oxygens in a ratio that does not approximate 1:2:1; includes fats, oils, and waxes.

lipogenesis (lī-pō-JEN-e-sis): Synthesis of lipids from nonlipid precursors.

lipolysis: The catabolism of lipids as a source of energy.

lipoprotein (lī-pō-PRŌ-tēn): A compound containing a relatively small lipid bound to a protein.

liver: An organ of the digestive system with varied and vital functions that include the production of plasma proteins, the excretion of bile, the storage of energy reserves, the detoxification of poisons, and the interconversion of nutrients.

lobule (LOB-ūl): The basic organizational unit of the liver at the histological level.

loose connective tissue: A loosely organized, easily distorted connective tissue containing several different fiber types, a varied population of cells, and a viscous ground substance.

lordosis (lor-DŌ-sis): An exaggeration of the lumbar curvature.

lumbar: Pertaining to the lower back.

lumen: The central space within a duct or other internal passageway.

lungs: Paired organs of respiration, situated in the left and right pleural cavities.

luteinizing hormone (LH) (LOO-tē-in-ī-zing): Anterior pituitary hormone that in the female assists FSH in follicle stimulation, triggers ovulation, and promotes the maintenance and secretion of the endometrial glands; in the male, stimulates spermatogenesis; also known as **interstitial cell-stimulating hormone.**

luxation (luks-Ā-shun): Dislocation of a joint.

lymph: Fluid contents of lymphatic vessels, similar in composition to interstitial fluid.

lymphadenopathy (lim-fad-e-NOP-a-thē): Pathological enlargement of the lymph nodes.

lymphatics: Vessels of the lymphatic system.

lymphedema (lim-fe-DĒ-ma): Swelling of peripheral tissues due to excessive lymph production or inadequate drainage.

lymph nodes: Lymphoid organs that monitor the composition of lymph.

lymphocyte (LIM-fō-sīt): A cell of the lymphatic system that participates in the immune response.

lymphokines: Chemicals secreted by activated lymphocytes.

lymphopoiesis: The production of lymphocytes.

lymphotoxin (lim-fō-TOK-sin): A secretion of lymphocytes that kills the target cells.

lysis (LĪ-sis): The destruction of a cell through the rupture of its cell membrane.

lysosome (LĪ-sō-sōm): Intracellular vesicle containing digestive enzymes.

lysozyme: An enzyme present in some exocrine secretions that has antibiotic properties.

macrophage: A phagocytic cell of the monocyte-macrophage system.

macula (MAK-ū-la): A receptor complex in the saccule or utricle that responds to linear acceleration or gravity.

macula densa (MAK-ū-la DEN-sa): A group of specialized secretory cells in a portion of the distal convoluted tubule adjacent to the glomerulus and the juxtaglomerular cells; a component of the juxtaglomerular apparatus.

macula lutea (LOO-tē-a): The fovea.

malignant cancer: A form of cancer characterized by rapid cellular growth and the spread of cancer cells throughout the body.

malleus (MAL-ē-us): The first auditory ossicle, bound to the tympanic membrane and the incus.

mamillary bodies (MAM-i-lar-ē): Nuclei in the hypothalamus concerned with feeding reflexes and behaviors; a component of the limbic system.

mammary glands: Milk-producing glands of the female breast.

manubrium: The broad, roughly triangular, superior element of the sternum.

manus: The hand.

marrow: A tissue that fills the internal cavities in a bone; may be dominated by hemopoietic cells (red marrow) or adipose tissue (yellow marrow).

mass peristalsis: Powerful peristaltic contraction that moves fecal materials along the colon and into the rectum.

mass reflex: Hyperreflexia in an area innervated by spinal cord segments distal to an area of injury.

mast cell: A connective tissue cell that when stimulated releases histamine, serotonin, and heparin, initiating the inflammatory response.

mastectomy: Surgical removal of part or all of a mammary gland.

mastication (mas-ti-KĀ-shun): Chewing.

mastoid sinus: Air-filled spaces in the mastoid process of the temporal bone.

matrix: The ground substance of a connective tissue.

maxillary sinus (MAK-si-ler-ē): One of the paranasal sinuses; an air-filled chamber lined by a respiratory epithelium that is located in a maxillary bone and opens into the nasal cavity.

meatus (mē-Ā-tus): An opening or entrance into a passageway.

mechanoreception: Detection of mechanical stimuli, such as touch, pressure, or vibration.

medial: Toward the midline of the body.

mediastinum (mē-dē-as-TĪ-num): Central tissue mass that divides the thoracic cavity into two pleural cavities; includes the aorta and other great vessels, the esophagus, trachea, thymus, the pericardial cavity and heart, and a host of nerves, small vessels, and lymphatics.

medulla: Inner layer or core of an organ.

medulla oblongata: The most caudal of the five brain regions, also known as the **myelencephalon.**

medullary cavity: The space within a bone that contains the marrow.

medullary rhythmicity center: Center in the medulla oblongata that sets the background pace of respiration; includes inspiratory and expiratory centers.

megakaryocytes (meg-a-KAR-ē-ō-sīts): Bone marrow cells responsible for the formation of platelets.

meiosis (mī-Ō-sis): Cell division that produces gametes with half of the normal somatic chromosome complement.

Meissner's corpuscles: Touch receptors located within dermal papillae adjacent to the basement membrane of the epidermis.

melanin (MEL-a-nin): Yellow-brown pigment produced by the melanocytes of the skin.

melanocyte (me-LAN-ō-sīt): Specialized cell found in the deeper layers of the stratified squamous epithelium of the skin, responsible for the production of melanin.

melanocyte-stimulating hormone (MSH): Hormone of the pars intermedia of the anterior pituitary that stimulates melanin production.

melanomas (mel-a-NŌ-maz): Dangerous malignant skin cancers that involve melanocytes.

melatonin (mel-a-TŌ-nin): Hormone secreted by the pineal gland; inhibits secretion of MSH and gonadotropins.

membrane: Any sheet or partition; a layer consisting of an epithelium and the underlying connective tissue.

membrane flow: The movement of sections of membrane surface to and from the cell surface and components of the endoplasmic reticulum, the Golgi apparatus, and vesicles.

membranous labyrinth: Endolymph-filled tubes of the inner ear that enclose the receptors of the inner ear.

menarche (me-NAR-kē): The beginning of menstrual function.

meninges (men-IN-jēz): Three membranes that surround the surfaces of the CNS; the dura mater, the pia mater, and the arachnoid.

meningitis: Inflammation of the spinal or cranial meninges.

meniscectomy: Removal of a meniscus.

meniscus (men-IS-kus): A fibrocartilage pad between opposing surfaces in a joint.

menopause (MEN-ō-paws): The cessation of uterine cycles as a consequence of the aging process and exhaustion of viable follicles.

menses (MEN-sēz): The first menstrual period that normally occurs at puberty.

menstrual (MEN-stroo-al) **cycle:** See **uterine cycle.**

menstruation (men-stroo-Ā-shun): The sloughing of blood and endometrial tissue at menses.

Merkel's discs: Sensory nerve endings that contact special receptors called Merkel cells, located within the deeper layers of the epidermis.

merocrine (MER-ō-krin): A method of secretion where the cell ejects materials through exocytosis.

mesencephalic aqueduct: Passageway that connects the third ventricle (diencephalon) with the fourth ventricle (metencephalon).

mesencephalon (mez-en-SEF-a-lon): The midbrain.

mesenchyme (MEZ-en-kīm): Embryonic/fetal connective tissue.

mesentery (MEZ-en-ter-ē): A double layer of serous membrane that supports and stabilizes the position of an organ in the abdominopelvic cavity and provides a route for the associated blood vessels, nerves, and lymphatics.

mesoderm: The middle germ layer that lies between the ectoderm and endoderm of the embryo.

mesothelium (mez-ō-THĒ-lē-um): A simple squamous epithelium that lines one of the divisions of the ventral body cavity.

messenger RNA (mRNA): RNA formed at transcription to direct protein synthesis in the cytoplasm.

metabolism (me-TAB-ō-lizm): The sum of all of the biochemical processes under way within the human body at a given moment; includes anabolism and catabolism.

metabolites (me-TAB-ō-līts): Compounds produced in the body as the result of metabolic reactions.

metacarpals (met-a-KAR-pals): The five bones of the palm of the hand.

metalloproteins (me-tal-ō-PRŌ-tēnz): Plasma proteins that transport metal ions.

metaphase (MET-a-fāz): A stage of mitosis wherein the chromosomes line up along the equatorial plane of the cell.

metaphysis (me-TAF-i-sis): The region of a long bone between the epiphysis and diaphysis, corresponding to the location of the epiphyseal plate of the developing bone.

metarteriole (met-ar-TĒ-rē-ōl): A vessel that connects an arteriole to a venule and that provides blood to a capillary plexus.

metastasis (me-TAS-ta-sis): The spread of a disease from one organ to another.

metatarsal: One of the five bones of the foot that articulate with the tarsals (proximally) and the phalanges (distally).

metencephalon (met-en-SEF-a-lon): The pons and cerebellum of the brain.

micelle (mī-SEL): A spherical aggregation of bile salts, monoglycerides, and fatty acids in the lumen of the intestinal tract.

microcephaly (mī-krō-SEF-a-lē): An abnormally small cranium, due to premature closure of one or more fontanels.

microfilaments: Fine protein filaments visible with the electron microscope; components of the cytoskeleton.

microglia (mī-KROG-lē-a): Phagocytic glial cells in the CNS, derived from the monocytes of the blood.

microphages: Neutrophils and eosinophils.

microtubules: Microscopic tubules that are part of the cytoskeleton, and are found in cilia, flagella, the centrioles, and spindle fibers.

microvilli: Small, fingerlike extensions of the exposed cell membrane of an epithelial cell.

micturition (mik-tū-RI-shun): Urination.

midbrain: The mesencephalon.

middle ear: Space between the external and internal ear that contains auditory ossicles.

midsagittal plane: A plane passing through the midline of the body that divides it into left and right halves.

mineralocorticoids: Corticosteroids produced by the zona glomerulosa of the adrenal cortex; steroids such as aldosterone, that affect mineral metabolism.

miscarriage: Spontaneous abortion.

mitochondrion (mī-tō-KON-drē-on): An intracellular organelle responsible for generating most of the ATP required for cellular operations.

mitosis (mī-TŌ-sis): The division of a single cell that produces two identical daughter cells; the primary mechanism of tissue growth.

mitral valve (MĪ-tral): The left AV, or bicuspid, valve of the heart.

mixed gland: A gland that contains exocrine and endocrine cells, or an exocrine gland that produces serous and mucous secretions.

mixed nerve: A peripheral nerve that contains sensory and motor fibers.

modiolus (mō-DĪ-ō-lus): The bony central hub of the cochlea.

mole: A quantity of an element or compound having a mass in grams equal to its atomic or molecular weight.

molecular weight: The sum of the atomic weights of the atoms in a molecule.

molecule: A compound containing two or more atoms that are held together by chemical bonds.

monocytes (MON-ō-sīts): Phagocytic agranulocytes (white blood cells) in the circulating blood.

monoglyceride (mon-ō-GLI-se-rīd): A lipid consisting of a single fatty acid bound to a molecule of glycerol.

monokines: Secretions released by activated cells of the monocyte-macrophage system to coordinate various aspects of the immune response.

monosaccharide (mon-ō-SAK-ah-rīd): A simple sugar, such as glucose or ribose.

monosynaptic reflex: A reflex where the sensory afferent synapses directly on the motor efferent.

monozygotic twins: Twins produced through the splitting of a single fertilized egg (zygote).

morula (MOR-ū-la): A mulberry-shaped collection of cells produced through the mitotic divisions of a zygote.

motor unit: All of the muscle cells controlled by a single motor neuron.

mucins (MŪ-sins): Proteoglycans responsible for the lubricating properties of mucus.

mucosa (mū-KŌ-sa): A mucous membrane; the epithelium plus the lamina propria.

mucous: An adjective referring to the presence or production of mucus.

mucous membrane: *See* **mucosa.**

mucus: Lubricating secretion produced by unicellular and multicellular glands along the digestive, respiratory, urinary, and reproductive tracts.

multipennate muscle: A muscle whose internal fibers are organized around several different tendons.

multipolar neuron: A neuron with many dendrites and a single axon, the typical form of a motor neuron.

multiunit smooth muscle: Smooth muscle tissue whose muscle cells are innervated in motor units.

muriatic acid: Hydrochloric acid (HCl).

muscarinic receptors (mus-kar-IN-ik): Membrane receptors sensitive to acetylcholine (ACh) and to muscarine, a toxin produced by certain mushrooms; found at all parasympathetic neuroeffector junctions and at a few sympathetic neuroeffector junctions.

muscle: A contractile organ composed of muscle tissue, blood vessels, nerves, connective tissues, and lymphatics.

muscle tissue: A tissue characterized by the presence of cells capable of contraction; includes skeletal, cardiac, and smooth muscle tissue.

muscularis externa (mus-kū-LAR-is): Concentric layers of smooth muscle responsible for peristalsis.

muscularis mucosae: Layer of smooth muscle beneath the lamina propria responsible for moving the mucosal surface.

mutagens (MŪ-ta-jenz): Chemical agents that induce mutations and may be carcinogenic.

myalgia (mī-AL-jē-a): Muscle pain.

myasthenia gravis (mī-as-THĒ-nē-a GRA-vis): Muscular weakness due to a reduction in the number of ACh receptor sites on the sarcolemmal surface; suspected to be an autoimmune disorder.

myelencephalon (mī-el-en-SEF-a-lon): The medulla oblongata.

myelin (MĪ-e-lin): Insulating sheath around an axon consisting of multiple layers of glial cell membrane; significantly increases conduction rate along the axon.

myelination: The formation of myelin.

myeloid tissue: Tissue responsible for the production of red blood cells, granulocytes, monocytes, and platelets.

myenteric plexus (mī-en-TER-ik): Parasympathetic motor neurons and sympathetic postganglionic fibers located between the circular and longitudinal layers of the muscularis externa.

myocardial infarction (mī-ō-KAR-dē-al): Heart attack; damage to the heart muscle due to an interruption of regional coronary circulation.

myocarditis: Inflammation of the myocardium.

myocardium: The cardiac muscle tissue of the heart.

myofibrils: Organized collections of myofilaments in skeletal and cardiac muscle cells.

myofilaments: Fine protein filaments, composed of the proteins actin (thin filaments) and myosin (thick filaments).

myoglobin (MĪ-ō-GLŌ-bin): An oxygen-binding pigment especially common in slow skeletal and cardiac muscle fibers.

myogram: A recording of the tension produced by muscle fibers on stimulation.

myometrium (mī-ō-MĒ-trē-um): The thick layer of smooth muscle in the wall of the uterus.

myopia: Nearsightedness, an inability to accommodate for distant vision.

myosepta: Connective tissue partitions that separate adjacent skeletal muscles.

myosin: Protein component of the thick myofilaments.

myositis (mī-ō-SĪ-tis): Inflammation of muscle tissue.

nail: Keratinous structure produced by epithelial cells of the nail root.

narcolepsy: A sleep disorder characterized by falling asleep at inappropriate moments.

nares, external (NA-rēz): The entrance from the exterior to the nasal cavity.

nares, internal: The entrance from the nasal cavity to the nasopharynx.

nasal cavity: A chamber in the skull bounded by the internal and external nares.

nasolacrimal duct: Passageway that transports tears from the nasolacrimal sac to the nasal cavity.

nasolacrimal sac: Chamber that receives tears from the nasolacrimal ducts.

nasopharynx (nā-zō-FAR-inks): Region posterior to the internal nares, superior to the soft palate, and ending at the oropharynx.

necrosis (NEK-rō-sis): Death of cells or tissues from disease or injury.

negative feedback: Corrective mechanism that opposes or negates a variation from normal limits.

neonate: A newborn infant, or baby.

neoplasm: A tumor, or mass of abnormal tissue.

nephritis (nef-RĪ-tis): Inflammation of the kidney.

nephrolithiasis (nef-rō-li-THĪ-a-sis): Condition resulting from the formation of kidney stones.

nephron (NEF-ron): Basic functional unit of the kidney.

nerve impulse: An action potential in a nerve cell membrane.

neural cortex: An area where gray matter is found at the surface of the CNS.

neurilemma (nū-ri-LEM-ma): The outer surface of a glial cell that encircles an axon.

neuroeffector junction: A synapse between a motor neuron and a peripheral effector, such as a muscle cell or gland cell.

neurofibrils: Microfibrils in the cytoplasm of a neuron.

neurofilaments: Microfilaments in the cytoplasm of a neuron.

neuroglandular junction: A specific type of neuroeffector junction.

neuroglia (nū-ROG-lē-a): Nonneural cells of the CNS and PNS that support and protect the neurons.

neurohypophysis (nū-rō-hī-POF-i-sis): The posterior pituitary, or pars nervosa.

neuromuscular junction: A specific type of neuroeffector junction.

neuron (NŪ-ron): A nerve cell.

neurotransmitter: Chemical compound released by one neuron to affect the transmembrane potential of another.

neurotubules: Microtubules in the cytoplasm of a nerve cell.

neurulation: The embryological process responsible for the formation of the CNS.

neutropenia: An abnormally low number of neutrophils in the circulating blood.

neutrophil (NŪ-trō-fil): A phagocytic microphage (granulocyte, WBC) that is very numerous and usually the first of the mobile phagocytic cells to arrive at an area of injury or infection.

nicotinic receptors (nik-ō-TIN-ik): ACh receptors found on the surfaces of sympathetic and parasympathetic ganglion cells, that will also respond to the compound nicotine.

nipple: An elevated epithelial projection on the surface of the breast, containing the openings of the lactiferous sinuses.

Nissl bodies: The ribosomes, Golgi, RER, and mitochondria of the perikaryon of a typical nerve cell.

nitrogenous wastes: Organic waste products of metabolism that contain nitrogen, such as urea, uric acid, and creatinine.

NK cells (**n**atural **k**iller cells): Lymphocytes responsible for immune surveillance, the detection and destruction of cancer cells.

nociception (nō-sē-SEP-shun): Pain perception.

node of Ranvier: Area between adjacent glial cells where the myelin covering of an axon is incomplete.

nodose ganglion (NŌ-dōs): A sensory ganglion of cranial nerve X.

noradrenaline: Catecholamine secreted by the adrenal medulla, released at most sympathetic neuroeffector junctions, and at certain synapses inside the CNS; also called **norepinephrine.**

norepinephrine (nor-ep-i-NEF-rin): A catecholamine neurotransmitter in the PNS and CNS, and a hormone secreted by the adrenal medulla; also called **noradrenaline.**

normovolemic (nor-mō-vō-LĒ-mik): Having a normal blood volume.

nucleic acid (nū-KLĒ-ik): A polymer of nucleotides containing a pentose sugar, a phosphate group, and one of four nitrogenous bases that regulate the synthesis of proteins and make up the genetic material in cells.

nucleolus (nū-KLĒ-ō-lus): Dense region in the nucleus that represents the site of RNA synthesis.

nucleoplasm: Fluid content of the nucleus.

nucleoproteins: Proteins of the nucleus that are generally associated with the DNA.

nucleus: Cellular organelle that contains DNA, RNA, and proteins; a mass of gray matter in the CNS.

nucleus pulposus (pul-PŌ-sus): The gelatinous core of an intervertebral disc.

nutrient: An organic compound that can be broken down in the body to produce energy.

nystagmus: Involuntary, continual movement of the eyes as if to adjust to constant motion.

obesity: Body weight 10–20 percent above standard values as the result of body fat accumulation.

occlusal surface (o-KLŪ-sal): The opposing surfaces of the teeth that come into contact when processing food.

ocular: Pertaining to the eye.

oculomotor nerve (ok-ū-lō-MŌ-ter): Cranial nerve III, that controls the extrinsic oculomotor muscles other than the superior oblique and the lateral rectus.

olecranon: The proximal end of the ulna that forms the prominent point of the elbow.

olfaction: The sense of smell.

olfactory bulb (ol-FAK-tor-ē): Two olfactory nerves that lie beneath the frontal lobe of the cerebrum.

olfactory tract: Tract over which nerve impulses from the retina are transmitted between the optic chiasma and the thalamus.

oligodendrocytes (ol-i-gō-DEN-drō-sīts): CNS glial cells responsible for maintaining cellular organization in the gray matter and providing a myelin sheath in areas of white matter.

oncogene (ON-kō-jēn): A gene that can turn a normal cell into a cancer cell.

oncologists (on-KOL-ō-jists): Physicians specializing in the study and treatment of tumors.

oocyte (Ō-ō-sīt): A cell whose meiotic divisions will produce a single ovum and three polar bodies.

oogenesis (ō-ō-JEN-e-sis): Ovum production.

oogonia (ō-ō-GŌ-nē-a): Stem cells in the ovaries whose divisions give rise to oocytes.

oophorectomy (ō-of-ō-REK-tō-mē): Surgical removal of the ovaries.

oophoritis (ō-of-ō-RĪ-tis): Inflammation of the ovaries.

ooplasm: The cytoplasm of the ovum.

opsin: A protein, one structural component of the visual pigment rhodopsin.

opsonization: An effect of coating an object with antibodies; the attraction and enhancement of phagocytosis.

optic chiasma (OP-tik kī-AZ-ma): Crossing point of the optic nerves.

optic nerve: Nerve that carries signals from the eye to the optic chiasma.

ora serrata (Ō-ra ser-RA-ta): The anterior edge of the neural retina.

orbit: Bony cavity of the skull that contains the eyeball.

orchiectomy (or-kē-EK-tō-mē): Surgical removal of one or both testes.

orchitis: Inflammation of the testes.

organelle (or-gan-EL): An intracellular structure that performs a specific function or group of functions.

organic compound: A compound containing carbon, hydrogen, and usually oxygen.

organogenesis: The formation of organs during embryological and fetal development.

organs: Combinations of tissues that perform complex functions.

origin: Point of attachment of a muscle that is less movable.

oropharynx (or-ō-FAR-inks): The middle portion of the pharynx, bounded superiorly by the nasopharynx, anteriorly by the oral cavity, and inferiorly by the laryngopharynx.

osmolarity (oz-mō-LAR-i-tē): The total concentration of dissolved materials in a solution, regardless of their specific identities, expressed in terms of moles.

osmoreceptor: A receptor sensitive to changes in the osmolarity of the plasma.

osmosis (oz-MŌ-sis): The movement of water across a semipermeable membrane toward a solution containing a relatively high solute concentration.

osmotic pressure: The force of osmotic water movement; the pressure that must be applied to prevent osmotic movement across a membrane.

osseous tissue: A strong connective tissue containing specialized cells and a mineralized matrix of crystalline calcium phosphate and calcium carbonate.

ossicles: Small bones.

ossification: The formation of bone.

osteoblasts (OS-tē-ō-blasts): Cells that produce bone within connective tissue (intramembranous ossification) or cartilage (endochondral ossification); may differentiate into osteocytes.

osteoclast (OS-tē-ō-klast): A cell that dissolves the fibers and matrix of bone.

osteocyte (OS-tē-ō-sīt): A bone cell responsible for the maintenance and turnover of the mineral content of the surrounding bone.

osteogenesis (os-tē-ō-JEN-e-sis): Bone production.

osteogenic layer (os-tē-ō-JEN-ik): The inner, cellular layer of the periosteum that participates in bone growth and repair.

osteoid (OS-tē-oyd): The organic components of the bone matrix, produced by osteoblasts and osteocytes.

osteolysis (os-tē-OL-i-sis): The breakdown of the mineral matrix of bone.

osteon (OS-tē-on): The basic histological unit of compact bone, consisting of osteocytes organized around a central canal and separated by concentric lamellae.

osteopenia (os-tē-ō-PĒ-nē-a): The condition of inadequate bone production in the adult, leading to a loss in bone mass and strength.

osteoporosis (os-tē-ō-pōr-Ō-sis): A reduction in bone mass and strength sufficient to compromise normal bone function.

osteoprogenitor cells: Stem cells that give rise to osteoblasts.

otic: Pertaining to the ear.

otitis media: Inflammation of the middle ear cavity.

otoconia (otoliths) (ō-tō-KŌ-nē-a): Aggregations of calcium carbonate crystals in a gelatinous membrane that sits above one of the maculae of the vestibular apparatus.

oval window: Opening in the bony labyrinth where the stapes attaches to the membranous wall of the scala vestibuli.

ovarian cycle (ō-VAR-ē-an): Monthly cycle of gamete development in the ovaries, associated with cyclical changes in the production of sex hormones (estrogens and progestins).

ovary: Female reproductive gland.

ovulation (ōv-ū-LĀ-shun): The release of a secondary oocyte, surrounded by cells of the corona radiata, following the rupture of the wall of a tertiary follicle.

ovum/ova (Ō-vum): A gamete produced by the reproductive system of a female; an egg.

oxytocin (oks-i-TŌ-sin): Hormone produced by hypothalamic cells and secreted into capillaries at the posterior pituitary; stimulates smooth muscle contractions of the uterus or mammary glands in the female, but has no known function in males.

pacemaker cells: Cells of the SA node that set the pace of cardiac contraction.

Pacinian corpuscle (pa-SIN-ē-an): Receptor sensitive to vibration.

palate: Horizontal partition separating the oral cavity from the nasal cavity and nasopharynx; can be divided into an anterior bony (hard) palate and a posterior fleshy (soft) palate.

palatine: Pertaining to the palate.

palpate: To examine by touch.

palpebrae (pal-PĒ-brē): Eyelids.

pancreas: Digestive organ containing exocrine and endocrine tissues; exocrine portion secretes pancreatic juice; endocrine portion secretes hormones, including insulin and glucagon.

pancreatic duct: A tubular duct that carries pancreatic juice from the pancreas to the duodenum.

pancreatic islets: Aggregations of endocrine cells in the pancreas.

pancreatic juice: A mixture of buffers and digestive enzymes that is discharged into the duodenum under the stimulation of the enzymes secretin and cholecystokinin.

pancreatitis (pan-krē-a-TĪ-tis): Inflammation of the pancreas.

Papanicolaou (Pap) test: Test for the detection of malignancies of the female reproductive tract, especially the cervix and uterus.

papilla (pa-PIL-la): A small, conical projection.

paralysis: Loss of voluntary motor control over a portion of the body.

paranasal sinuses: Bony chambers lined by respiratory epithelium that open into the nasal cavity; includes the frontal, ethmoidal, sphenoid, and maxillary sinuses.

parasagittal: A section or plane that parallels the midsagittal plane but that does not pass along the midline.

parasympathetic division: One of the two divisions of the autonomic nervous system; also known as the craniosacral division; generally responsible for activities that conserve energy and lower the metabolic rate.

parathyroid glands: Four small glands embedded in the posterior surface of the thyroid; responsible for parathyroid hormone secretion.

parathyroid hormone: Hormone secreted by the parathyroid gland when plasma calcium levels fall below the normal range; causes increased osteoclast activity, increased intestinal calcium uptake, and decreased calcium ion loss at the kidneys.

parenchyma (par-ENG-ki-ma): The cells of a tissue or organ that are responsible for fulfilling its functional role.

paresthesia: Sensory abnormality that produces a tingling sensation.

parietal: Referring to the body wall or outer layer.

parietal cell: Cells of the gastric glands that secrete HCl and intrinsic factor.

Parkinson's disease: Progressive motor disorder due to degeneration of the cerebral nuclei.

parotid glands (pa-ROT-id): Large salivary glands that secrete a saliva containing high concentrations of salivary (alpha) amylase.

pars distalis (dis-TAL-is): The large, anterior portion of the anterior pituitary gland.

pars intermedia (in-ter-MĒ-dē-a): The portion of the anterior pituitary immediately adjacent to the posterior pituitary and the infundibulum.

pars nervosa: The posterior pituitary gland.

parturition (par-tū-RISH-un): Childbirth, delivery.

patella (pa-TEL-la): The sesamoid bone of the kneecap.

pathogenic: Disease-causing.

pathologist (pa-THOL-ō-jist): An M.D. specializing in the identification of diseases based on characteristic structural and functional changes in tissues and organs.

pedicel (PED-i-sel): A slender process of a podocyte that forms part of the filtration apparatus of the kidney glomerulus.

pedicles (PE-di-kls): Thick bony struts that connect the vertebral body with the articular and spinous processes.

pelvic cavity: Inferior subdivision of the abdominopelvic (peritoneal) cavity; encloses the urinary bladder, the sigmoid colon and rectum, and male or female reproductive organs.

pelvis: A bony complex created by the articulations between the coxae, the sacrum, and the coccyx.

penis (PĒ-nis): Component of the male external genitalia; a copulatory organ that surrounds the urethra, and that serves to introduce semen into the female vagina.

pepsin: Proteolytic enzyme secreted by the chief cells of the gastric glands in the stomach.

pepsinogen (pep-SIN-ō-jen): The inactive proenzyme that is secreted by chief cells of the gastric pits; after secretion it is converted to the proteolytic enzyme pepsin.

peptidases: Enzymes that split peptide bonds and release amino acids.

perforating canal: A passageway in compact bone that runs at right angles to the axes of the osteons, between the periosteum and endosteum.

perfusion: The blood flow through a tissue.

pericardial cavity (per-i-KAR-dē-al): The space between the parietal pericardium and the epicardium (visceral pericardium) that covers the outer surface of the heart.

pericarditis: Inflammation of the pericardium.

pericardium (per-i-KAR-dē-um): The fibrous sac that surrounds the heart, and whose inner, serous lining is continuous with the epicardium.

perichondrium (per-i-KON-drē-um): Layer that surrounds a cartilage, consisting of an outer fibrous and an inner cellular region.

perikaryon (per-i-KAR-ē-on): The cytoplasm that surrounds the nucleus in the soma of a nerve cell.

perilymph (PER-ē-limf): A fluid similar in composition to cerebrospinal fluid; found in the spaces between the bony labyrinth and the membranous labyrinth of the inner ear.

perimysium (per-i-MĪS-ē-um): Connective tissue partition that separates adjacent fasciculi in a skeletal muscle.

perineum (per-i-NĒ-um): The pelvic floor and associated structures.

perineurium: Connective tissue partition that separates adjacent bundles of nerve fibers in a peripheral nerve.

periodontal ligament (per-ē-ō-DON-tal): Collagen fibers that bind the cementum of a tooth to the periosteum of the surrounding alveolus.

periosteum (per-ē-OS-tē-um): Layer that surrounds a bone, consisting of an outer fibrous and inner cellular region.

peripheral nervous system (PNS): All neural tissue outside of the CNS.

peristalsis (per-i-STAL-sis): A wave of smooth muscle contractions that propels materials along the axis of a tube such as the digestive tract, the ureters, or the ductus deferens.

peritoneal cavity: *See* **abdominopelvic cavity.**

peritoneum (per-i-tō-NĒ-um): The serous membrane that lines the peritoneal (abdominopelvic) cavity.

peritonitis (per-i-tō-NĪ-tis): Inflammation of the peritoneum.

peritubular capillaries: A network of capillaries that surrounds the proximal and distal convoluted tubules of the kidneys.

permeability: Ease with which dissolved materials can cross a membrane; if freely permeable, any molecule can cross the membrane; if impermeable, nothing can cross; most biological membranes are selectively permeable.

peroxisome: A membranous vesicle containing enzymes that break down hydrogen peroxide (H_2O_2).

pes: The foot.

petrosal ganglion: Sensory ganglion of the glossopharyngeal nerve (N IX).

petrous: Stony, usually used to refer to the thickened portion of the temporal bone that encloses the inner ear.

Peyer's patches (PĪ-erz): Lymphatic nodules beneath the epithelium of the small intestine.

pH: The negative exponent of the hydrogen ion concentration, in moles per liter.

phagocyte: A cell that performs phagocytosis.

phagocytosis (fa-gō-sī-TŌ-sis): The engulfing of extracellular materials or pathogens; movement of extracellular materials into the cytoplasm by enclosure in a membranous vesicle.

phalanx/phalanges (fa-LAN-jēz): Digits; the bones of the fingers and toes.

pharmacology: The study of drugs, their physiological effects, and their clinical uses.

pharyngotympanic tube: A passageway that connects the nasopharynx with the middle ear cavity; also called the Eustachian or auditory tube.

pharynx (FAR-inks): The throat; a muscular passageway shared by the digestive and respiratory tracts.

phasic response: A pattern of response to stimulation by sensory neurons that are normally inactive; stimulation causes a burst of neural activity that ends when the stimulus either stops or stops changing in intensity.

phenotype (FĒN-ō-tīp): Physical characteristics that are genetically determined.

phonation (fō-NĀ-shun): Sound production at the larynx.

phosphate group: PO_4^{3-}.

phospholipid (fos-fō-LIP-id): An important membrane lipid whose structure includes hydrophilic and hydrophobic regions.

phosphorylation (fos-for-i-LĀ-shun): The addition of a phosphate group to a molecule.

photoreception: Sensitivity to light.

physiology (fiz-ē-OL-ō-jē): Literally the study of function; considers the ways living organisms perform vital activities.

pia mater: The tough, outer meningeal layer that surrounds the CNS.

pigment: A compound with a characteristic color.

piloerection: "Goosebumps" effect produced by the contraction of the arrector pili muscles of the skin.

pineal gland: Neural tissue in the posterior portion of the roof of the diencephalon, responsible for the secretion of melatonin.

pinealocytes (PIN-ē-a-lō-sīts): Secretory cells of the pineal gland.

pinna: The expanded, projecting portion of the external ear that surrounds the external auditory meatus.

pinocytosis (pin-ō-sī-TŌ-sis): The introduction of fluids into the cytoplasm by enclosing them in membranous vesicles at the cell surface.

pituitary gland: The "master gland," situated in the sella turcica of the sphenoid bone and connected to the hypothalamus by the infundibulum; includes the posterior pituitary (pars nervosa) and the anterior pituitary (pars intermedia and pars distalis).

placenta (pla-SENT-a): A complex structure in the uterine wall that permits diffusion between the fetal and maternal circulatory systems; also called the afterbirth.

placentation (pla-sen-TĀ-shun): Formation of a functional placenta following implantation of a blastocyst in the endometrium.

plantar: Referring to the sole of the foot.

plasma (PLAZ-mah): The fluid ground substance of whole blood; what remains after the cells have been removed from a sample of whole blood.

plasma cell: Activated B cells that secrete antibodies.

plasmalemma (plaz-ma-LEM-a): Cell membrane.

platelets (PLĀT-lets): Small packets of cytoplasm that contain enzymes important in the clotting response; manufactured in the bone marrow by cells called **megakaryocytes.**

pleura (PLOO-ra): The serous membrane lining the pleural cavities.

pleural cavities: Subdivisions of the thoracic cavity that contain the lungs.

pleuritis (ploor-Ī-tis): Inflammation of the pleura.

plexus (PLEK-sus): A complex interwoven network of peripheral nerves or blood vessels.

plica (PLĪ-ka): A permanent transverse fold in the wall of the small intestine.

pneumotaxic center (nū-mō-TAKS-ik): A center in the reticular formation of the pons that regulates the activities of the apneustic and respiratory rhythmicity centers to adjust the pace of respiration.

pneumothorax (nū-mō-THŌ-raks): The introduction of air into the pleural cavity.

podocyte (POD-ō-sīt): A cell whose processes surround the glomerular capillaries and assist in the filtration process.

polar body: A nonfunctional packet of cytoplasm containing chromosomes eliminated from an oocyte during meiosis.

pollex (POL-eks): The thumb.

polycythemia (po-lē-sī-THĒ-mē-a): An unusually high hematocrit due to the presence of excess numbers of formed elements, especially RBCs.

polymorph: Polymorphonuclear leukocyte; a neutrophil.

polypeptide: A chain of amino acids strung together by peptide bonds; those containing over 100 peptides are called proteins.

polysaccharide (pol-ē-SAK-ah-rīd): A complex sugar, such as glycogen or a starch.

polysynaptic reflex: A reflex with interneurons interposed between the sensory fiber and the motor neuron(s).

polyuria (pol-ē-Ū-rē-a): Excessive urine production.

pons: The portion of the metencephalon anterior to the cerebellum.

popliteal (pop-lit-Ē-al): Pertaining to the back of the knee.

porphyrins (POR-fi-rinz): Ring-shaped molecules that form the basis for important respiratory and metabolic pigments, including heme and the cytochromes.

porta hepatis: A region of mesentery between the duodenum and liver that contains the hepatic artery, the hepatic portal vein, and the common bile duct.

positive feedback: Mechanism that increases a deviation from normal limits following an initial stimulus.

postcentral gyrus: The primary sensory cortex, where touch, vibration, pain, temperature, and taste sensations arrive and are consciously perceived.

posterior: Toward the back; dorsal.

postganglionic neuron: An autonomic neuron in a peripheral ganglion, whose activities control peripheral effectors.

postovulatory phase: The secretory phase of the menstrual cycle.

precentral gyrus: The primary motor cortex on a cerebral hemisphere, located rostral to the central sulcus.

prefrontal cortex: Rostral portion of each cerebral hemisphere thought to be involved with higher intellectual functions, predictions, calculations, and so forth.

preganglionic neuron: Visceral motor neuron inside the CNS whose output controls one or more ganglionic motor neurons in the PNS.

premolars: Bicuspids; teeth with flattened occlusal surfaces located anterior to the molar teeth.

premotor cortex: Motor association area between the precentral gyrus and the prefrontal area.

preoptic nucleus: Hypothalamic nucleus that coordinates thermoregulatory activities.

preovulatory phase: A portion of the menstrual cycle; period of estrogen-induced repair of the functional zone of the endometrium through the growth and proliferation of epithelial cells in the glands not lost during menses.

prepuce (PRĒ-pūs): Loose fold of skin that surrounds the glans penis (males) or the clitoris (females).

preputial glands (prē-PŪ-shal): Glands on the inner surface of the prepuce that produce a viscous, odorous secretion, called **smegma.**

presbyopia: Farsightedness; an inability to accommodate for near vision.

prevertebral ganglion: *See* **collateral ganglion.**

prime mover: A muscle that performs a specific action.

proenzyme: An inactive enzyme secreted by an epithelial cell.

progesterone (prō-JES-ter-ōn): The most important progestin secreted by the corpus luteum following ovulation.

progestins (prō-JES-tinz): Steroid hormones structurally related to cholesterol.

prognosis: A prediction concerning the possibility or time course of recovery from a specific disease.

projection fibers: Axons carrying information from the thalamus to the cerebral cortex.

prolactin (prō-LAK-tin): Hormone that stimulates functional development of the mammary gland in females; a secretion of the anterior pituitary gland.

prolapse: The abnormal descent or protrusion of a portion of an organ, such as the vagina or anorectal canal.

proliferative phase: *See* **preovulatory phase.**

pronation (prō-NĀ-shun): Rotation of the forearm that makes the palm face posteriorly.

pronucleus: Enlarged egg or sperm nucleus that forms after fertilization but before amphimixis.

properdin: Complement factor that prolongs and enhances non-antibody-dependent complement binding to bacterial cell walls.

prophase (PRŌ-fāz): The initial phase of mitosis, characterized by the appearance of chromosomes, breakdown of the nuclear membrane, and formation of the spindle apparatus.

proprioception (prō-prē-ō-SEP-shun): Awareness of the positions of bones, joints, and muscles.

prostaglandin (pros-tah-GLAN-din): Lipoid secreted by one cell that alters the metabolic activities or sensitivities of adjacent cells; sometimes called "local hormones."

prostate gland (PROS-tāt): Accessory gland of the male reproductive tract, contributing roughly one-third of the volume of semen.

prostatectomy (pros-ta-TEK-tō-mē): Surgical removal of the prostate.

prostatitis (pros-ta-TĪ-tis): Inflammation of the prostate.

prosthesis: An artificial substitute for a body part.

protease: *See* **proteinase.**

protein: A large polypeptide with a complex structure.

proteinase: An enzyme that breaks down proteins into peptides and amino acids.

proteinuria (prō-tēn-ŪR-ē-a): Abnormal amounts of protein in the urine.

proteoglycan (prō-tē-ō-GLĪ-kan): Compound containing a large polysaccharide complex attached to a relatively small protein; examples include hyaluronic acid and chondroitin sulfate.

prothrombin: Circulating proenzyme of the common pathway of the clotting system; converted to thrombin by the enzyme thromboplastin.

proton: A fundamental particle bearing a positive charge.

protraction: Movement anteriorly in the horizontal plane.

proximal: Toward the attached base of an organ or structure.

proximal convoluted tubule: The portion of the nephron between Bowman's capsule and the loop of Henle; the major site of active reabsorption from the filtrate.

pruritis (prū-RĪ-tus): Itching.

pseudopodia (sū-dō-PŌ-dē-a): Temporary cytoplasmic extensions typical of mobile or phagocytic cells.

pseudostratified epithelium: An epithelium containing several layers of nuclei, but whose cells are all in contact with the underlying basement membrane.

psoriasis (sō-RĪ-a-sis): Skin condition characterized by excessive keratin production and the formation of dry, scaly patches on the body surface.

psychosomatic condition: An abnormal physiological state with a psychological origin.

puberty: Period of rapid growth, sexual maturation, and the appearance of secondary sexual characteristics; usually occurs at ages 10–15.

pubic symphysis: Fibrocartilaginous amphiarthrosis between the pubic bones of the coxae.

pubis (PŪ-bis): The anterior, inferior component of the coxa.

pudendum (pū-DEN-dum): The external genitalia.

pulmonary circuit: Blood vessels between the pulmonary semilunar valve of the right ventricle and the entrance to the left atrium; the blood circulation through the lungs.

pulmonary ventilation: Movement of air in and out of the lungs.

pulp cavity: Internal chamber in a tooth, containing blood vessels, lymphatics, nerves, and the cells that maintain the dentin.

pulpitis (pul-PĪ-tis): Inflammation of the tissues of the pulp cavity.

pupil: The opening in the center of the iris through which light enters the eye.

purine: An N compound with a ring-shaped structure; examples include adenine and guanine, two nitrogen bases common in nucleic acids.

Purkinje cell (pur-KIN-jē): Large, branching neuron of the cerebellar cortex.

Purkinje fibers: Specialized conducting cardiocytes in the ventricles.

pus: An accumulation of debris, fluid, dead and dying cells, and necrotic tissue.

putamen (pū-TĀ-men): Thalamic nucleus involved in the integration of sensory information prior to projection to the cerebral hemispheres.

P wave: Deflection of the ECG corresponding to atrial depolarization.

pyelogram (PĪ-el-ō-gram): A radiographic image of the kidneys and ureters.

pyelonephritis (pī-e-lō-nef-RĪ-tis): Inflammation of the kidneys.

pyloric sphincter (pī-LOR-ic): Sphincter of smooth muscle that regulates the passage of chyme from the stomach to the duodenum.

pylorus (pī-LOR-us): Gastric region between the body of the stomach and the duodenum; includes the pyloric sphincter.

pyrexia (pī-REK-sē-a): A fever.

pyrimidine: An N compound with a ring-shaped structure; examples include cytosine, thymine, and uracil, nitrogen bases common in nucleic acids.

pyruvic acid (pī-RŪ-vik): 3-carbon compound produced by glycolysis.

quadriplegia: Paralysis of the arms and legs.

radiodensity: Relative resistance to the passage of X-rays.

radiographic techniques: Methods of visualizing internal structures using various forms of radiational energy.

radiopaque: Having a relatively high radiodensity.

rami communicantes: Axon bundles that link the spinal nerves with the ganglia of the sympathetic chain.

ramus: A branch.

raphe (RĀ-fē): A seam.

receptor field: The area monitored by a single sensory receptor.

recessive gene: An allele that will affect the phenotype only when the individual is homozygous for that trait.

rectal columns: Longitudinal folds in the walls of the anorectal canal.

rectouterine pouch (rek-tō-Ū-te-rin): Peritoneal pocket between the anterior surface of the rectum and the posterior surface of the uterus.

rectum (REK-tum): The last 15 cm (6 in.) of the digestive tract.

rectus: Straight.

red blood cell: *See* **erythrocyte.**

reduction: The gain of hydrogen atoms or electrons, or the loss of an oxygen molecule.

reductional division: The first meiotic division, which reduces the chromosome number from 46 to 23.

reflex: A rapid, automatic response to a stimulus.

reflex arc: The receptor, sensory neuron, motor neuron, and effector involved in a particular reflex; interneurons may or may not be present, depending on the reflex considered.

refraction: The bending of light rays as they pass from one medium to another.

refractory period: Period between the initiation of an action potential and the restoration of the normal resting potential; over this period the membrane will not respond normally to stimulation.

relaxation phase: The period following a contraction when the tension in the muscle fiber returns to resting levels.

relaxin: Hormone that loosens the pubic symphysis; a hormone secreted by the placenta.

renal: Pertaining to the kidneys.

renal corpuscle: The initial portion of the nephron, consisting of an expanded chamber that encloses the glomerulus.

renin: Enzyme released by the juxtaglomerular cells when renal blood pressure or pO_2 declines; converts angiotensinogen to angiotensin I.

rennin: Gastric enzyme that breaks down milk proteins.

replication: Duplication.

repolarization: Movement of the transmembrane potential away from + mV values and toward the resting potential.

residual volume: Amount of air remaining in the lungs after maximum forced expiration.

respiration: Exchange of gases between living cells and the environment; includes pulmonary ventilation, external respiration, internal respiration, and cellular respiration.

respiratory minute volume: The amount of air moved in and out of the respiratory system each minute.

resting potential: The transmembrane potential of a normal cell under homeostatic conditions.

rete (RĒ-tē): An interwoven network of blood vessels or passageways.

reticular activating center: Mesencephalic portion of the reticular formation responsible for arousal and the maintenance of consciousness.

reticular formation: Diffuse network of gray matter that extends the entire length of the brain stem.

reticulocytes (re-TIK-ū-lō-sīts): The last stage in the maturation of red blood cells; normally the youngest red blood cells present in the blood.

reticulospinal tracts: Descending tracts that carry involuntary motor commands issued by neurons of the reticular formation.

retina: The innermost layer of the eye, lining the vitreous chamber; also known as the neural tunic.

retinene (RET-i-nēn): Visual pigment derived from vitamin A.

retraction: Movement posteriorly in the horizontal plane.

retroflexion (ret-rō-FLEK-shun): A posterior tilting of the uterus that has no clinical significance.

retrograde flow (RET-rō-grād): Transport of materials from the telodendria to the soma of a neuron.

retroperitoneal (re-trō-per-i-tō-NĒ-al): Situated behind or outside of the peritoneal cavity.

reverberation: Positive feedback along a chain of neurons, so that they remain active once stimulated.

rheumatism (ROO-ma-tizm): A condition characterized by pain in muscles, tendons, bones, or joints.

Rh factor: Agglutinogen that may be present (Rh-positive) or absent (Rh-negative) from the surfaces of red blood cells.

rhizotomy: Surgical transection of a dorsal root, usually performed to relieve pain.

rhodopsin (rō-DOP-sin): The visual pigment found in the membrane discs of the distal segments of rods.

rhythmicity center: Medullary center responsible for the basic pace of respiration; includes inspiratory and expiratory centers.

ribonucleic acid (RNA) (rī-bō-nū-KLĀ-ik): A nucleic acid consisting of a chain of nucleotides that contain the sugar ribose and the nitrogen bases adenine, guanine, cytosine, and uracil.

ribosome: An organelle containing rRNA and proteins, that is essential to mRNA translation and protein synthesis.

right lymphatic duct: Lymphatic vessel delivering lymph from the right side of the head, neck, and chest to the venous system via the right subclavian vein.

rigor mortis: Extended muscular contraction and rigidity that occurs after death, as the result of calcium ion release from the SR and the exhaustion of cytoplasmic ATP reserves.

rod: Photoreceptor responsible for vision under dimly lit conditions.

rostral: Toward the nose; used when referring to relative position inside the skull.

rough endoplasmic reticulum (RER): A membranous organelle that is a site of protein synthesis and storage.

rouleau/rouleaux (roo-LŌ): A stack of red blood cells.

round window: An opening in the bony labyrinth of the inner ear that exposes the membranous wall of the scala tympani to the air of the middle ear cavity.

rubrospinal tracts: Descending tracts that carry involuntary motor commands issued by the red nucleus of the mesencephalon.

Ruffini corpuscles (ru-FĒ-nē): Receptors sensitive to tension and stretch in the dermis of the skin.

rugae (ROO-gē): Mucosal folds in the lining of the empty stomach that disappear as gastric distension occurs.

saccule (SAK-ūl): A portion of the vestibular apparatus of the inner ear, responsible for static equilibrium.

sagittal plane: Sectional plane that divides the body into left and right portions.

salivatory nucleus (SAL-i-va-tōr-ē): Medullary nucleus that controls the secretory activities of the salivary glands.

saltatory conduction: Relatively rapid conduction of a nerve impulse between successive nodes of a myelinated axon.

sarcolemma: The cell membrane of a muscle cell.

sarcoma (sar-KŌ-ma): A tumor of connective tissues.

sarcomere: The smallest contractile unit of a striated muscle cell.

sarcoplasm: The cytoplasm of a muscle cell.

scala media: The central, endolymph-filled chamber of the inner ear; *see* **cochlear duct.**

scala tympani: The perilymph-filled chamber of the inner ear below the basilar membrane; pressure changes here distort the round window.

scala vestibuli: The perilymph-filled chamber of the inner ear above the vestibular membrane; pressure changes here result from distortions of the oval window.

scapula (SKAP-ū-la): The shoulder blade.

scar tissue: Thick, collagenous tissue that forms at an injury site.

Schlemm, canal of: Passageway that delivers aqueous humor from the anterior chamber of the eye to the venous circulation.

Schwann cells: Glial cells responsible for the neurilemma that surrounds axons in the PNS.

sciatica (sī-AT-i-ka): Pain resulting from compression of the roots of the sciatic nerve.

sciatic nerve (sī-AT-ik): Nerve innervating the posteromedial portions of the thigh and lower leg.

sclera (SKLER-a): The fibrous, outer layer of the eye forming the white area of the anterior surface; a portion of the fibrous tunic of the eye.

sclerosis: A hardening and thickening that often occurs secondary to tissue inflammation.

scoliosis (skō-lē-Ō-sis): An abnormal, exaggerated lateral curvature of the spine.

scrotum (SKRŌ-tum): Loose-fitting, fleshy pouch that encloses the testes of the male.

sebaceous glands (sē-BĀ-shus): Glands that secrete sebum, usually associated with hair follicles.

sebum (SĒ-bum): A waxy secretion that coats the surfaces of hairs.

secondary sex characteristics: Physical characteristics that appear at puberty in response to sex hormones, but that are not involved in the production of gametes.

secretin (sē-KRĒ-tin): Duodenal hormone that stimulates pancreatic buffer secretion and inhibits gastric activity.

semen (SĒ-men): Fluid ejaculate containing spermatozoa and the secretions of accessory glands of the male reproductive tract.

semicircular canals: Tubular components of the vestibular apparatus responsible for dynamic equilibrium.

semilunar valve: A three-cusped valve guarding the exit from one of the cardiac ventricles; includes the pulmonary and aortic valves.

seminal vesicles (SEM-i-nal): Glands of the male reproductive tract that produce roughly 60 percent of the volume of semen.

seminiferous tubules (se-mi-NIF-e-rus): Coiled tubules where sperm production occurs in the testis.

senescence: Aging.

septae (SEP-tē): Partitions that subdivide an organ.

serosa: *See* **serous membrane.**

serotonin (ser-ō-TŌ-nin): A neurotransmitter in the CNS; a compound that enhances inflammation, released by activated mast cells and basophils.

serous cell: A cell that produces a watery secretion containing high concentrations of enzymes.

serous membrane: A squamous epithelium and the underlying loose connective tissue; the lining of the pericardial, pleural, and peritoneal cavities.

serum: Blood plasma from which clotting agents have been removed.

sesamoid bone: A bone that forms in a tendon.

sigmoid colon (SIG-moid): The S-shaped 8-inch portion of the colon between the descending colon and the rectum.

sign: A clinical term for visible evidence of the presence of a disease.

simple epithelium: An epithelium containing a single layer of cells above the basement membrane.

sinus: A chamber or hollow in a tissue; a large, dilated vein.

sinusitis: Inflammation of a nasal sinus.

sinusoid (SĪ-nus-oid): An extensive network of vessels found in the liver, adrenal cortex, spleen, and pancreas; similar in histological structure to capillaries.

skeletal muscle: A contractile organ of the muscular system.

skeletal muscle tissue: Contractile tissue dominated by skeletal muscle fibers; characterized as striated, voluntary muscle.

sliding filament theory: The concept that a sarcomere shortens as the thick and thin filaments slide past one another.

small intestine: The duodenum, jejunum, and ileum; the digestive tract between the stomach and large intestine.

smegma (SMEG-ma): Secretion of the preputial glands of the penis or clitoris.

smooth endoplasmic reticulum: Membranous organelle where lipid and carbohydrate synthesis and storage occur.

smooth muscle tissue: Muscle tissue found in the walls of many visceral organs; characterized as nonstriated, involuntary muscle.

soft palate: Fleshy posterior extension of the hard palate, separating the nasopharynx from the oral cavity.

sole: The inferior surface of the foot.

solute: Material dissolved in a solution.

solution: A fluid containing dissolved materials.

solvent: The fluid component of a solution.

soma (SŌ-ma): Body.

somatic (sō-MAT-ik): Pertaining to the body.

somatic nervous system: System of nerve fibers that run from the central nervous system to the muscles of the skeleton.

somatomedins: Compounds stimulating tissue growth, released by the liver following GH secretion.

somatostatin: GH-IH, a hypothalamic regulatory hormone that inhibits GH secretion by the anterior pituitary.

somatotropin: Growth hormone, produced by the anterior pituitary in response to GH-RH.

sperm: *See* **spermatozoa.**

spermatic cord: Spermatic vessels, nerves, lymphatics, and the ductus deferens, extending between the testes and the proximal end of the inguinal canal.

spermatids (SPER-ma-tidz): The product of meiosis in the male, cells that differentiate into spermatozoa.

spermatocyte (sper-MAT-ō-sīt): Cells of the seminiferous tubules that are engaged in meiosis.

spermatogenesis (sper-ma-tō-JEN-e-sis): Sperm production.

spermatogonia (sper-ma-tō-GŌ-nē-a): Stem cells whose mitotic divisions give rise to other stem cells and spermatocytes.

spermatozoon/spermatozoa (sper-ma-tō-ZŌ-on)): A sperm cell, the male gamete.

spermicide: Compound toxic to sperm cells, sometimes used as a contraceptive method.

spermiogenesis: The process of spermatid differentiation that leads to the formation of physically mature spermatozoa.

sphincter (SFINK-ter): Muscular ring that contracts to close the entrance or exit of an internal passageway.

spina bifida (SPĪ-na BĪ-fi-da): A developmental abnormality in which the vertebral laminae fail to unite at the midline; the entire vertebral column and skull may be affected in severe cases.

spinal meninges (men-IN-jēz): Specialized membranes that line the vertebral canal and provide protection, stabilization, nutrition, and shock absorption to the spinal cord.

spinal nerve: One of 31 pairs of nerves that originate on the spinal cord from anterior and posterior roots.

spindle apparatus: A muscle spindle (intrafusal fiber) and its sensory and motor innervation.

spinocerebellar tracts: Ascending tracts carrying sensory information to the cerebellum.

spinothalamic tracts: Ascending tracts carrying poorly localized touch, pressure, pain, vibration, and temperature sensations to the thalamus.

spinous process: Prominent posterior projection of a vertebra, formed by the fusion of two laminae.

splanchnic nerves: Preganglionic (myelinated) sympathetic nerves that end in one of the collateral ganglia.

spleen: Lymphatic organ important for red blood cell phagocytosis, immune response, and lymphocyte production.

splenectomy (splē-NEK-tō-mē): Surgical removal of the spleen.

sprain: Forceful distortion of an articulation that produces damage to the capsule, ligaments, or tendons but not dislocation.

sputum (SPŪ-tum): Viscous mucus ejected from the mouth after transport to the pharynx by the mucus escalator of the respiratory tract.

squama: A broad, flat surface.

squamous (SKWĀ-mus): Flattened.

squamous epithelium: An epithelium whose superficial cells are flattened and platelike.

stapedius (stā-PĒ-dē-us): A muscle of the middle ear whose contraction tenses the auditory ossicles and reduces the forces transmitted to the oval window.

stapes (STĀ-pēz): The auditory ossicle attached to the tympanic membrane.

stenosis (ste-NŌ-sis): A constriction or narrowing of a passageway.

stereocilia: Elongate microvilli characteristic of the epithelium of the epididymis and portions of the ductus deferens.

steroid: A ring-shaped lipid structurally related to cholesterol.

stimulus: An environmental alteration that produces a change in cellular activities; often used to refer to events that alter the transmembrane potentials of excitable cells.

stratified: Containing several layers.

stratum (STRĀ-tum): Layer.

stratum corneum (KŌR-nē-um): Layers of flattened, dead, keratinized cells covering the epidermis of the skin.

stretch receptors: Sensory receptors that respond to stretching of the surrounding tissues.

stroma: The connective tissue framework of an organ, as distinguished from the functional cells (parenchyma) of that organ.

subarachnoid space: Meningeal space containing CSF; the area between the arachnoid membrane and the pia mater.

subclavian (sub-CLĀ-vē-an): Pertaining to the region under the clavicle.

subcutaneous layer: The layer of loose connective tissue below the dermis; also called the **hypodermis** or **superficial fascia.**

sublingual salivary glands (sub-LING-gwal): Mucus-secreting salivary glands situated under the tongue.

subluxation (sub-luks-Ā-shun): A partial dislocation of a joint.

submandibular salivary glands: Salivary glands nestled in depressions on the medial surfaces of the mandible; salivary glands that produce a mixture of mucins and enzymes (salivary amylase).

submucosa (sub-mū-KŌ-sa): Region between the muscularis mucosae and the muscularis externa.

submucosal glands: Mucous glands in the submucosa of the duodenum.

subserous fascia: Loose connective tissue layer beneath the serous membrane lining the ventral body cavity.

substantia nigra: A nucleus in the midbrain that is responsible for negative feedback control of the cerebral nuclei.

substrate: A participant (product or reactant) in an enzyme-catalyzed reaction.

sulcus (SUL-kus): A groove or furrow.

summation: Temporal or spatial addition of stimuli.

superficial fascia: *See* **subcutaneous layer.**

superior: Directional reference meaning above.

superior vena cava: The vein that carries blood from the parts of the body above the heart to the right atrium.

supination (su-pin-Ā-shun): Rotation of the forearm so that the palm faces anteriorly.

supine (sū-PĪN): Lying face up, with palms facing anteriorly.

suppressor T cells: Lymphocytes that inhibit B cell activation and plasma cell secretion of antibodies.

suprarenal gland (sū-pra-RĒ-nal): *See* **adrenal gland.**

surfactant (sur-FAK-tant): Lipid secretion that coats alveolar surfaces and prevents their collapse.

sustentacular cells (sus-ten-TAK-ū-lar): Supporting cells of the seminiferous tubules of the testis, responsible for the differentiation of spermatids, the maintenance of the blood-testis barrier, and the secretion of inhibin.

sutural bones: Irregular bones that form in fibrous tissue between the flat bones of the developing cranium; also called **Wormian bones.**

suture: Fibrous joint between flat bones of the skull.

sympathectomy (sim-path-EK-tō-mē): Transection of the sympathetic innervation to a region.

sympathetic division: Division of the autonomic nervous system responsible for "fight or flight" reactions; concerned primarily with the elevation of metabolic rate and increased alertness.

symphysis: A fibrous amphiarthrosis, such as those between adjacent vertebrae or between the pubic bones of the coxae.

symptom: Clinical term for an abnormality of function due to the presence of disease.

synapse (SIN-aps): Site of communication between a nerve cell and some other cell; if the other cell is not a neuron, the term neuroeffector junction is often used.

synarthrosis (sin-ar-THRŌ-sis): A joint that does not permit relative movement between the articulating elements.

synchondrosis (sin-kon-DRŌ-sis): A cartilaginous synarthrosis, such as the articulation between the epiphysis and diaphysis of a growing bone.

syncope (SIN-kō-pē): A sudden, transient loss of consciousness; a faint.

syncytial trophoblast (sin-SISH-al): Multinucleate cytoplasmic layer that covers the blastocyst; the layer responsible for uterine erosion and implantation.

syncytium: A multinucleate mass of cytoplasm, produced by the fusion of cells or repeated mitoses without cytokinesis.

syndesmosis (sin-dez-MŌ-sis): A fibrous amphiarthrosis.

syndrome: A discrete set of symptoms that occur together.

syneresis (si-NER-ē-sis): Clot retraction.

synergist (SIN-er-jist): A muscle that assists a prime mover in performing its primary action.

synostosis (sin-os-TŌ-sis): A synarthrosis formed through the fusion of the articulating elements.

synovial cavity (si-NŌ-vē-ul): Fluid-filled chamber in a diarthrodial joint.

synovial fluid: Substance secreted by synovial membranes that lubricates joints.

synovial membrane: An incomplete layer of fibroblasts confronting the synovial cavity, plus the underlying loose connective tissue.

synthesis (SIN-the-sis): Manufacture; anabolism.

system: An interacting group of organs that performs one or more specific functions.

systemic circuit: Vessels between the aortic semilunar valve and the entrance to the right atrium; the circulatory system other than vessels of the pulmonary circuit.

systole (SIS-tō-lē): The period of cardiac contraction.

systolic pressure: Peak arterial pressure measured during ventricular systole.

tachycardia (tak-ē-KAR-dē-a): An abnormally rapid heart rate.

tactile: Pertaining to the sense of touch.

taenia coli (TĒ-nē-a KŌ-lī): Three longitudinal bands of smooth muscle in the muscularis externa of the colon.

tarsus: The ankle.

T cells: Lymphocytes responsible for cellular immunity, and for the coordination and regulation of the immune response; includes regulatory T cells (helpers and suppressors) and cytotoxic (killer) T cells.

tears: Fluid secretions of the lacrimal glands that bathe the anterior surfaces of the eyes.

tectorial membrane (tek-TŌR-ē-al): Gelatinous membrane suspended over the hair cells of the organ of Corti.

tectospinal tracts: Descending extrapyramidal tracts carrying involuntary motor commands issued by the colliculi.

tectum: The roof of the mesencephalon of the brain.

telencephalon (tel-en-SEF-a-lon): The forebrain or cerebrum, including the cerebral hemispheres, the internal capsule, and the cerebral nuclei.

telodendria (te-lō-DEN-drē-a): Terminal axonal branches that end in synaptic knobs.

telophase (TEL-ō-fāz): The final stage of mitosis, characterized by the disappearance of the spindle apparatus, the reappearance of the nuclear membrane and the disappearance of the chromosomes, and the completion of cytokinesis.

temporal: Pertaining to time (temporal summation) or pertaining to the temples (temporal bone).

tendinitis: Painful inflammation of a tendon.

tendon: A collagenous band that connects a skeletal muscle to an element of the skeleton.

tendon organ: Receptor sensitive to tension in a tendon.

tentorium cerebelli (ten-TŌR-ē-um ser-e-BEL-ē): Dural partition that separates the cerebral hemispheres from the cerebellum.

teratogen (ter-AT-ō-jen): Stimulus that causes developmental defects.

teres: Long and round.

terminal: Toward the end.

tertiary follicle: A mature ovarian follicle, containing a large, fluid-filled chamber.

testes (TES-tēz): The male gonads, sites of gamete production and hormone secretion.

testosterone (tes-TOS-te-rōn): The principal androgen produced by the interstitial cells of the testes.

tetanic contraction: Sustained skeletal muscle contraction due to repeated stimulation at a frequency that prevents muscle relaxation.

tetanus: A tetanic contraction; also used to refer to a disease state resulting from the stimulation of muscle cells by bacterial toxins.

tetrad (TET-rad): Paired, duplicated chromosomes visible at the start of meiosis I.

tetraiodothyronine (tet-ra-ī-ō-dō-THĪ-rō-nēn): T_4, or thyroxine, a thyroid hormone.

thalamus: The walls of the diencephalon.

thalassemia (thal-ah-SĒ-mē-ah): A hereditary disorder affecting hemoglobin synthesis and producing anemia.

theory: A hypothesis that makes valid predictions, as demonstrated by evidence that is testable, unbiased, and repeatable.

therapy: Treatment of disease.

thermogenesis (ther-mō-JEN-e-sis): Heat production.

thermography: Diagnostic procedure involving the production of an infrared image.

thermoreception: Sensitivity to temperature changes.

thermoregulation: Homeostatic maintenance of body temperature.

thick filament: A myosin filament in a skeletal or cardiac muscle cell.

thin filament: An actin filament in a skeletal or cardiac muscle cell.

thoracoabdominal pump (thō-ra-kō-ab-DOM-i-nal): Changes in the intrapleural pressures during the respiratory cycle that assist the venous return to the heart.

thoracolumbar division (thō-ra-kō-LUM-bar): The sympathetic division of the ANS.

thorax: The chest.

threshold: The transmembrane potential at which an action potential begins.

thrombin (THROM-bin): Enzyme that converts fibrinogen to fibrin.

thrombocytes (THROM-bō-sīts): *See* **platelets.**

thrombocytopenia (throm-bō-sī-tō-PĒ-nē-ah): Abnormally low platelet count in the circulating blood.

thromboembolism (throm-bō-EM-bō-lizm): Occlusion of a blood vessel by a drifting blood clot.

thromboplastin: Enzyme that converts prothrombin to thrombin; enzyme formed by the intrinsic or extrinsic clotting pathways.

thrombus: A blood clot.

thymine: A pyrimidine found in DNA.

thymosin (THĪ-mō-sin): Thymic hormone essential to the development and differentiation of T cells.

thymus: Lymphatic organ, site of T cell formation.

thyroglobulin (thī-rō-GLOB-ū-lin): Circulating transport globulin that binds thyroid hormones.

thyroid gland: Endocrine gland whose lobes sit lateral to the thyroid cartilage of the larynx.

thyroid hormones: Thyroxine (T_4) and triiodothyronine (T_3), hormones of the thyroid gland; hormones that stimulate tissue metabolism, energy utilization, and growth.

thyroid-stimulating hormone (TSH): Anterior pituitary hormone that triggers the secretion of thyroid hormones by the thyroid gland.

thyroxine (TX) (thī-ROKS-in): A thyroid hormone (T_4).

tibia (TIB-ē-a): The large, medial bone of the leg.

tidal volume: The volume of air moved in and out of the lungs during a normal quiet respiratory cycle.

tissue: A collection of specialized cells and cell products that perform a specific function.

tonsil: A lymphatic nodule beneath the epithelium of the pharynx; includes the palatine, pharyngeal, and lingual tonsils.

topical: Applied to the body surface.

trabecula (tra-BEK-ū-la): A connective tissue partition that subdivides an organ.

trabeculae carneae (tra-BEK-ū-lē CAR-nē-ē): Muscular ridges projecting from the walls of the ventricles of the heart.

trachea (TRĀ-kē-a): The windpipe, an airway extending from the larynx to the primary bronchi.

tracheal ring: C-shaped supporting cartilage of the trachea.

tracheostomy (trā-kē-OS-tō-mē): Surgical opening of the anterior tracheal wall to permit airflow.

trachoma: An infectious disease of the conjunctiva and cornea.

tract: A bundle of axons inside the CNS.

tractotomy: The surgical transection of a tract, sometimes used to relieve pain.

transcription: The encoding of genetic instructions on a strand of mRNA.

transdermal medication: Administration of medication by absorption through the skin.

transection: To sever or cut in the transverse plane.

transfusion: Transfer of blood from a donor directly into the bloodstream of another person.

transient ischemic attack: A temporary loss of consciousness due to the occlusion of a small blood vessel in the brain.

translation: The process of peptide formation using the instructions carried by an mRNA strand.

transmembrane potential: The potential difference, in millivolts, measured across the cell membrane; a potential difference that results from the uneven distribution of positive and negative ions across a cell membrane.

transudate (TRANS-ū-dāt): Fluid that diffuses across a serous membrane and lubricates opposing surfaces.

treppe (TREP-ē): "Staircase" increase in tension production following repeated stimulation of a muscle, even though the muscle is allowed to complete each relaxation phase.

triad (liver): The combination of branches of the hepatic duct, the hepatic portal vein, and the hepatic artery, found at each corner of a liver lobule.

triad (muscle cell): The combination of a T tubule and two cisternae of the sarcoplasmic reticulum.

tricuspid valve (trī-KUS-pid): The right atrioventricular valve that prevents backflow of blood into the right atrium during ventricular systole.

trigeminal nerve (trī-JEM-i-nal): Cranial nerve V, responsible for providing sensory information from the lower portions of the face, including the upper and lower jaws, and delivering motor commands to the muscles of mastication.

triglyceride (trī-GLIS-e-rīd): A lipid composed of a molecule of glycerol attached to three fatty acids.

trigone (TRĪ-gōn): Triangular region of the bladder bounded by the exits of the ureters and the entrance to the urethra.

triiodothyronine: T_3, one of the thyroid hormones.

trisomy: The abnormal possession of three copies of a chromosome; trisomy 21 is responsible for the Down syndrome.

trochanters (trō-KAN-terz): Large processes near the head of the femur.

trochlea (TROK-lē-a): A pulley.

trochlear nerve (TROK-lē-ar): Cranial nerve IV, controlling the superior oblique muscle of the eye.

trophoblast (TRŌ-fō-blast): Superficial layer of the blastocyst that will be involved with implantation, hormone production, and placenta formation.

troponin/tropomyosin (trō-PŌ-nin) (trō-pō-MĪ-ō-sin): Proteins on the thin filaments that mask the active sites in the absence of free calcium ions.

trunk: The thoracic and abdominopelvic regions.

trypsin (TRIP-sin): One of the pancreatic proteases.

trypsinogen: The inactive proenzyme secreted by the pancreas and converted to trypsin in the duodenum.

T tubules: Transverse, tubular extensions of the sarcolemma that extend deep into the sarcoplasm to contact cisternae of the sarcoplasmic reticulum.

tuber cinereum (sin-Ē-re-um): Swelling in the floor of the hypothalamus at the attachment of the pituitary gland.

tuberculum (tū-BER-kū-lum): A small, localized elevation on a bony surface.

tuberosity: A large, roughened elevation on a bony surface.

tubulin: Protein subunit of microtubules.

tumor: A tissue mass formed by the abnormal growth and replication of cells.

tunica (TŪ-ni-ka): A layer or covering; in blood vessels: t. externa, the outermost layer of connective tissue fibers that stabilizes the position of the vessel; t. intima, the innermost layer, consisting of the endothelium plus an underlying elastic membrane; t. media, a middle layer containing collagen, elastin, and smooth muscle fibers in varying proportions.

turbinates: *See* **conchae.**

T wave: Deflection of the ECG corresponding to ventricular repolarization.

twitch: A single contraction/relaxation cycle in a skeletal muscle.

tympanic membrane (tim-PAN-ik): Membrane that separates the external auditory canal from the middle ear; membrane whose vibrations are transferred to the auditory ossicles and ultimately to the oval window; the "eardrum."

ulcer: An area of epithelial sloughing associated with damage to the underlying connective tissues and vasculature.

ultrasound: Diagnostic visualization procedure that uses high-frequency sound waves.

umbilical cord (um-BIL-i-kal): Connecting stalk between the fetus and the placenta; contains the allantois, the umbilical arteries, and the umbilical vein.

umbilicus: The navel.

unicellular gland: Goblet cell.

unipennate muscle: A muscle whose fibers are all arranged on one side of the tendon.

unipolar neuron: A sensory neuron whose soma lies in a dorsal root ganglion or a sensory ganglion of a cranial nerve.

unmyelinated axon: Axon whose neurilemma does not contain myelin, and where continuous conduction occurs.

urachus (Ū-ra-kus): The middle umbilical ligament.

uracil: One of the pyrimidines characteristic of RNA.

uremia (ū-RĒ-mē-a): Abnormal condition caused by impaired kidney function, characterized by the retention of wastes and the disruption of many other organ systems.

ureters (ū-RĒ-terz): Muscular tubes, lined by transitional epithelium, that carry urine from the renal pelvis to the urinary bladder.

urethra (ū-RĒ-thra): A muscular tube that carries urine from the urinary bladder to the exterior.

urethritis: Inflammation of the urethra.

urinalysis: Analysis of the physical and chemical characteristics of the urine.

urinary bladder: Muscular, distensible sac that stores urine prior to micturition.

urination: The voiding of urine; micturition.

uterine (menstrual) cycle: Cyclical changes in the uterine lining that occur in reproductive-age women. Each uterine cycle, which occurs in response to circulating hormones (*see* **ovarian cycle**), lasts 21–35 days.

uterus (Ū-ter-us): Muscular organ of the female reproductive tract where implantation, placenta formation, and fetal development occur.

utricle (Ū-tri-kl): The largest chamber of the vestibular apparatus; contains a macula important for static equilibrium.

uvea: The vascular tunic of the eye.

uvula (Ū-vū-la): A dangling, fleshy extension of the soft palate.

vagina (va-JĪ-na): A muscular tube extending between the uterus and the vestibule.

vagus nerve: N X, the cranial nerve responsible for most (75%) of the parasympathetic preganglionic output from the CNS.

varicose veins (VAR-i-kōs): Distended superficial veins.

vasa vasorum: Blood vessels that supply the walls of large arteries and veins.

vascular: Pertaining to blood vessels.

vascularity: The blood vessels in a tissue.

vascular spasm: Contraction of the wall of a blood vessel at an injury site, a process that may slow the rate of blood loss.

vasoconstriction: A reduction in the diameter of arterioles due to contraction of smooth muscles in the tunica media; an event that elevates peripheral resistance, and that may occur in response to local factors, through the action of hormones, or from stimulation of the vasomotor center.

vasodilation (vaz-ō-dī-LĀ-shun): An increase in the diameter of arterioles due to the relaxation of smooth muscles in the tunica media; an event that reduces peripheral resistance, and that may occur in response to local factors, through the action of hormones, or following decreased stimulation of the vasomotor center.

vasomotion: Alterations in the pattern of blood flow through a capillary bed in response to changes in the local environment.

vasomotor center: Medullary center whose stimulation produces vasoconstriction and an elevation in peripheral resistance.

vein: Blood vessel carrying blood from a capillary bed toward the heart.

venae cavae (VĒ-nē CĀ-vē): The major veins delivering systemic blood to the right atrium.

ventilation: Air movement in and out of the lungs.

ventilatory rate: The respiratory rate.

ventral: Pertaining to the anterior surface.

ventricle (VEN-tri-kl): One of the large, muscular pumping chambers of the heart that discharges blood into the pulmonary or systemic circuits.

ventricular folds: Mucosal folds in the laryngeal walls that do not play a role in sound production; the false vocal cords.

venules (VEN-ūlz): Thin-walled veins that receive blood from capillaries.

vermis (VER-mis): Midsagittal band of neural cortex on the surface of the cerebellum.

vertebral canal: Passageway that encloses the spinal cord, a tunnel bounded by the neural arches of adjacent vertebrae.

vertebral column: The cervical, thoracic, and lumbar vertebrae, the sacrum, and the coccyx.

vertebrochondral ribs: Ribs 8–10, false ribs connected to the sternum by shared cartilaginous bars.

vertebrosternal ribs: Ribs 1–7, true ribs connected to the sternum by individual cartilaginous bars.

vertigo: Dizziness.

vesicle: A membranous sac in the cytoplasm of a cell.

vestibular membrane: The membrane that separates the scala media from the scala vestibuli of the inner ear.

vestibular nucleus: Processing center for sensations arriving from the vestibular apparatus; located near the border between the pons and medulla.

vestibule (VES-ti-būl): A chamber; in the inner ear, the term refers to the utricle, saccule, and semicircular canals; also refers to (1) a region of the female external genitalia and (2) the space within the fleshy portion of the nose between the nostrils and the external nares.

vestibulospinal tracts: Descending tracts of the extrapyramidal system, carrying involuntary motor commands issued by the vestibular nucleus to stabilize the position of the head.

villus: A slender projection of the mucous membrane of the small intestine.

virus: A pathogenic microorganism.

viscera: Organs in the ventral body cavity.

visceral: Pertaining to viscera or their outer coverings.

visceral smooth muscle tissue: Smooth muscle tissue forming sheets or layers in the walls of visceral organs; the cells may not be innervated, and the layers often show automaticity (rhythmic contractions).

viscosity: The resistance to flow exhibited by a fluid, due to molecular interactions within the fluid.

viscous: Thick, syrupy.

vital capacity: The maximum amount of air that can be moved in or out of the respiratory system; the sum of the inspiratory reserve, the expiratory reserve, and the tidal volume.

vitamin: An essential organic nutrient that functions as a coenzyme in vital enzymatic reactions.

vitreous humor: Gelatinous mass in the vitreous chamber of the eye.

vocal folds: Folds in the laryngeal wall containing elastic ligaments whose tension can be voluntarily adjusted; the true vocal cords, responsible for phonation.

voluntary: Controlled by conscious thought processes.

vulva (VUL-va): The female pudendum (external genitalia).

Wallerian degeneration: Disintegration of an axon and its myelin sheath distal to an injury site.

white blood cells: Leukocytes; the granulocytes and agranulocytes of the blood.

white matter: Regions inside the CNS that are dominated by myelinated axons.

white ramus: A nerve bundle containing the myelinated preganglionic axons of sympathetic motor neurons en route to the sympathetic chain or a collateral ganglion.

Wormian bones: *See* **sutural bones.**

xiphoid process (ZĪ-foid): Slender, inferior extension of the sternum.

Y chromosome: The sex chromosome whose presence indicates that the individual is a genetic male.

yolk sac: One of the three extraembryonic membranes, composed of an inner layer of endoderm and an outer layer of mesoderm.

Zeis, glands of (ZĪS): Enlarged sebaceous glands on the free edges of the eyelids.

zona fasciculata (ZŌ-na fa-sik-ū-LA-ta): Region of the adrenal cortex responsible for glucocorticoid secretion.

zona glomerulosa (glō-mer-ū-LŌ-sa): Region of the adrenal cortex responsible for mineralocorticoid secretion.

zona pellucida (pel-LŪ-si-da): Region between a developing oocyte and the surrounding follicular cells of the ovary.

zona reticularis (re-tik-ū-LAR-is): Region of the adrenal cortex responsible for androgen secretion.

zygote (ZĪ-gōt): The fertilized ovum prior to the start of cleavage.

Index

Androgen-binding protein, 692
Androgens, 488, 691
    and adrenal gland, 494
Anemia, 508, 517
Anencephaly, 419, 424
Anesthesia, 478
Aneurysm, 573, 577
Angina pectoris, 534, 539
Angiography, 674
Angiotensin I, 495
Angiotensin II, 495
Angiotensinogen, 495
Angle
    mandible, 149
    ribs, 166
    scapula, 175
Angular motion, 211
Angular movements, 209, 210, 211
Ankle
    bones of, 198–99
    joint of, 225, 226
    ligaments of, 225
Ankyloglossia, 634
Ankylosis, 228
Annular ligaments, 219, 606
Anorectal canal, 650
Antagonist, muscles, 252
Antebrachial vein, 568
Anteflexion, 704
Anterior, 16
Anterior cardinal veins, 574
Anterior cerebral artery, 557
Anterior chamber of eye, 467
Anterior clinoid process, 143
Anterior commissure, 376
Anterior corticospinal tracts, 414
Anterior cranial fossa, 144
Anterior crest, 195
Anterior cruciate ligament, 225
Anterior gray horns, 344
Anterior inferior iliac spine, 186
Anterior interventricular branch, 531
Anterior interventricular sulcus, 525
Anterior lobes, 387
Anterior median fissure, 339
Anterior nucleus, 381, 384
Anterior pituitary, 488–89
Anterior section, 17
Anterior spinocerebellar tracts, 413
Anterior spinothalamic tracts, 412
Anterior surface of heart, 525
Anterior tibial artery, 562
Anterior tibial vein, 569
Anterior tubercles, 161
Anterior vertebral arch, 161
Anterior white columns, 345
Anterograde amnesia, 422
Anthracosis, 621
Antiangiogenesis factor, 75
Antibodies, 586
    functions of, 507
Antidiuretic hormone, 487
    function of, 487
Antigens
    in immune response, 586
    nature of, 585–86
Antrum, 702
Anulus fibrosus, 214
Anus, 650
Aorta
    abdominal aorta, 557, 560
    ascending, 528, 553
    descending, 528, 557, 560–64
    terminal segment of, 557
    thoracic aorta, 557

Aortic arch, 528, 553, 574
Aortic bodies, 450
Aortic hiatus, 585
Aortic semilunar valve, 528
Aortic sinus, 450, 528
Apex
    of heart, 525
    of lung, 607
    of nose, 601
    of sacrum, 165
Apgar rating, 734, 735
Aphasia, 420, 424
Apical foramen, 635
Apical ridge, 273
Aplastic anemia, 517
Apneustic center, 620
Apocrine secretion, 61
Apocrine sweat glands, 100
Aponeuroses, 70, 233, 235
Appendectomy, 595
Appendicitis, 595, 648
Appendicular muscular system, 254, 278
Appendicular skeleton, 130, 174–99
    development of, 200–201
    lower limb, 192–99
    pectoral girdle, 175–76
    pelvic girdle, 185–88
    upper limb, 180–84
Appetite, hypothalamic control, 384
Appositional growth, 72
Aqueduct of Sylvius, 368
Aqueous humor, 472
Arachnoid
    cranial meninges, 368
    spinal meninges, 342
Arachnoid trabeculae, 368
Arachnoid granulations, 368, 371
Arbor vitae, 387
Arches, of foot, 199
Arcuate arteries, 672
Arcuate fibers, 376
Arcuate line, 186
Arcuate veins, 673
Areflexia, 361
Areola, 710
Areolar tissue, 67–68
Arm. (See Upper limb)
Arrector pili muscles, 99
Arrhythmias, 537, 539
Arterial anastomosis, 547
Arteries, 544–45
    arterioles, 545
    axillary artery, 557
    basilar artery, 557
    brachial artery, 557
    bronchial arteries, 557
    carotid arteries, 557
    celiac artery, 557
    cerebral arteries, 557
    compared to veins, 544, 545, 564
    digital arteries, 557
    disorders of, 577
    dorsal arch, 564
    dorsalis pedis artery, 564
    elastic arteries, 544-45
    esophageal arteries, 557
    femoral artery, 562
    functions of, 521
    gastric artery, 560
    gonadal arteries, 560
    hepatic artery, 560
    iliac arteries, 560
    intercostal arteries, 557
    lumbar arteries, 560
    medium-sized arteries, 545

    mesenteric artery, 560
    ophthalmic artery, 557
    pericardial arteries, 557
    peroneal artery, 562
    phrenic arteries, 557
    plantar arteries, 564
    popliteal artery, 562
    posterior communicating arteries, 557
    pulmonary arteries, 553
    in pulmonary circulation, 553
    radial artery, 557
    renal arteries, 560
    splenic artery, 560
    subclavian arteries, 557
    superior phrenic arteries, 557
    suprarenal arteries, 560
    in systemic circulation, 553–64
    thoracic artery, 557
    tibial arteries, 562
    ulnar artery, 557
    umbilical arteries, 571
    vertebral artery, 557
    walls of, 544
Arterioles, 545
    pulmonary arterioles, 553
Arteriosclerosis, 548, 577
Arthritis, 228
Arthroscope, 228
Articular cartilage, 181, 207
Articular discs, 208
Articular facets, 160, 161, 166
Articular motion, 209
Articular processes, 159, 165
Articular surface, 186
Articular tubercle, 142
Articulations. (See Joints)
Artificial respiration, 620
Arytenoid cartilages, 603
Asbestosis, 621
Ascending aorta, 528, 553
Ascending colon, 633, 650
Ascending limb, 672
Ascending tracts, 345
Ascites, 84, 629
Association areas, 376
Association fibers, central white matter, 376
Association neurons, 330
Asthma, 613, 621
Astrocytes, 322–23, 332
Ataxia, 404
Atelectasis, 621
Atherosclerosis, 423, 548, 577
Athroscopic surgery, 228
Atlas, 161
ATP
    and mitochondria, 38
    in skeletal muscles, 235, 240
Atresia, 698
Atrial natriuretic peptide, functions of, 495
Atriovenous anastomoses, 547, 549
Atrioventricular bundle, 533
Atrioventricular node, 531, 533
Atrioventricular valve, 528
Atrium of heart, 521, 525, 528
    left, 528
    right, 525, 528
Atrophy of muscle, 224
Audiogram, 478
Audition. (See Hearing)
Auditory cortex, 376
Auditory ossicles, 142
Auditory reflex, 404
Auditory tube, 142
Auricle, 525
Auricular surface, 165

thymic hormones, 492, 587
thyroid hormones, 483
Horner's syndrome, 435, 442
Horns, of spinal cord, 344
Human body, levels of organization, 4–5
Human chorionic gonadotropin, 715, 725
Human development
　phases of, 720
　　pregnancy, 721–34
　　prenatal development, 720–31
Human placental lactogen, 714, 715, 725
Humeroulnar joint, 181
Humerus, 178–80, 180
Humoral immunity, 586
Hyaline cartilage, 72, 73
Hyaluronic acid, 44
Hyaluronidase, 720
Hydrocephalus, 373, 404
Hydrochloric acid, 642
Hydroxyapatite, 110
Hymen, 709
Hyoglossus, 262
Hyoid bone, 150
Hypaxial musculature, 273
Hyperextension, 211
Hyperostosis, 125
Hyperpnea, 616
Hyperreflexia, 361
Hypertension, 506
Hypertrophy, of muscle, 244
Hypervolemic blood volume, 506
Hypoblast, 723
Hypodermic needle, 96, 106
Hypodermis, 77, 96
Hypogastric plexus, 439, 680
Hypoglossal canals, 138
Hypoglossal nerve, 138, 262, 400
Hyponychium, 102
Hypophyseal artery, 487, 488
Hypophyseal fossa, 143
Hypophyseal portal system, 488
Hypophysis. (See Pituitary gland)
Hyporeflexia, 361
Hyposthesia, 478
Hypothalamus, 365, 371, 384
　and endocrine regulation, 484–85
　functions of, 365, 381, 384, 385
　and pituitary gland, 365
　structure of, 384
Hypothyroidism, 734
Hypovolemic blood volume, 506

I

I band, 239
Ileocecal valve, 648
Ileum, 645, 646
Iliac arteries, 560
Iliac crest, 186
Iliac fossa, 186
Iliac notch, 186
Iliac spine, 186
Iliac tuberosity, 186
Iliac vein, 569
Iliacus, 296
Iliocostalis group, spinal muscles,
　265, 266
Iliofemoral ligaments, 221
Ilioinguinal nerve, 685
Iliopectineal line, 186
Iliopsoas, 296
Iliotibial tract, 296
Ilium, 186
Immature delivery, 734

Immune system
　cellular immunity, 586
　disorders of, 595–96
　humoral immunity, 586
　and lymphocytes, 586
Immunodeficiency disease, 596
Immunoglobulins, 507, 586
Immunological surveillance, 586
Immunotherapy, 84
Impermeable membrane, 31
Implantation, of ovum, 714, 722, 723–24
Incisors, 636
Inclusions, 34
Incontinence, 681
Incus, 456, 457
Infarct, 536
Inferior, 16
Inferior angle, 175
Inferior articular facets, 161, 166
Inferior articular processes, 160
Inferior border of heart, 525
Inferior cerebellar peduncles, 387
Inferior colliculus, 384
Inferior demifacets, 164
Inferior extensor retinaculae, 297
Inferior ganglion, 398
Inferior hypophyseal artery, 487
Inferior iliac notch, 186
Inferior meatuses, 601
Inferior mesenteric artery, 560
Inferior mesenteric ganglion, 432
Inferior mesenteric plexus, 439
Inferior mesenteric vein, 571
Inferior nasal conchae, 601
Inferior nuchal lines, 138
Inferior oblique muscle, 260
Inferior orbital fissure, 146
Inferior phrenic artery, 492
Inferior ramus, 186
Inferior rectus muscle, 260
Inferior sagittal sinus, 368
Inferior section, 17
Inferior temporal lines, 138
Inferior thyroid artery, 489, 491
Inferior thyroid vein, 489
Inferior vena cava, 525, 569
Infertility, 735
Infraorbital foramen, 145
Infraspinous fossa, 176
Infundibulum, 384, 703
Ingestion, 626
Inguinal canals, 689
Inguinal hernia, 274, 689
Inguinal lymph nodes, 589
Inguinal region, lymphatic drainage of, 590
Inhibin, 692
　functions of, 499
Initial segment of axon, 326
Innate reflexes, 358
Inner cell mass, 53, 722
Inner ear, 142, 457, 459–63
Innervation, meaning of, 254
Innominate artery, 553
Insensible perspiration, 91
Insertion, of muscle, 252
Insula, 373
Insulin, functions of, 495
Insulin-dependent diabetes mellitus, 500
Integral proteins, 30
Integrative centers, cerebral hemispheres, 376
Integumentary system
　and aging, 95, 103
　blood supply to, 93, 95
　dermis, 94–95
　development of, 104–5

disorders of, 100, 106
epidermis, 90–93
functions of, 7
glands, 99–102
hair, 96–99
innervation of, 95
nails, 102
regulation of, 102–3
repair of, 103
subcutaneous layer, 77, 96
Intention tremor, 404
Interarticular crest, 166
Interatrial septum, 528, 538
Intercalated discs, 80
Intercarpal articulation, 219
Intercarpal joint, 219
Intercellular cement, 44
Intercondylar eminence, 195
Intercondylar fossa, 192
Intercostal arteries, 557
Intercostal muscles, 166, 268, 616
Intercostal veins, 569
Interlobar arteries, 672
Interlobular arteries, 672
Interlobular veins, 673
Intermediate fibers, 244–45
Intermediate filaments, 34
Intermediate hairs, 99
Intermediate junction, 44
Intermediate mass, 383
Internal acoustic canal, 142
Internal callus, 126
Internal capsule, 376, 383
Internal carotid artery, 557
Internal cerebral veins, 564
Internal elastic membrane, 544
Internal iliac artery, 560
Internal iliac veins, 569
Internal jugular vein, 568
Internal nares, 601
Internal oblique muscles, 268
Internal orifice, of uterus, 704
Internal root sheath, 96
Internal thoracic artery, 557
Internal thoracic vein, 569
Interneurons, 330, 332
Internodes, 324
Internus, 253
Interoceptors, 330, 447
Interosseous crest, 195
Interosseous membrane, 181, 195
Interosseus muscles, 297
Interphalangeal joints, 219, 225
Interphase, mitosis, 44
Interspinales, 265
Interstitial cells, 499, 691
Interstitial cell-stimulating hormone,
　function of, 488
Interstitial growth, 72
Interstitial lamellae, 113
Intertarsal joints, 225
Intertransversarii, 265
Intertrochanteric crest, 192
Intertrochanteric line, 192
Intertubercular groove, 180
Interventricular branch, 531
Interventricular foramen, 368
Interventricular septum, 528, 538
Interventricular sulcus, 525
Intervertebral discs, 214–15
　disorders of, 214, 215, 228
Intervertebral foramina, 160
Intestinal crypts, 646
Intestinal glands, 651
Intestinal juice, 648

Neuronal pools, 332, 334
    convergence, 332, 334
    divergence, 332, 334
    parallel processing, 334
    reverberation, 334
    serial processing, 332, 334
Neurons, 80, 326, 328–30
    and aging, 423
    anaxonic neurons, 329
    bipolar neurons, 329
    classification, 328–29
    interneurons, 330
    motor neurons, 330
    multipolar neurons, 329
    sensory neurons, 329–30, 410
    structure of, 322, 326, 328
    unipolar neurons, 329
Neurosecretions, 487
Neurotransmitters
    and adrenal medulla, 434
    and autonomic nervous system, 428, 430
    excitatory/inhibitory effects, 332
    and muscle contraction, 243
    and parasympathetic division, 437, 439
    release of, 326, 331–32
    and sympathetic nervous system, 435–36
Neurulation, 325
Neutrophils, 513
Newborn, neonatal period, 734
Nicotine, 439
Nicotinic receptors, 437
Nissl bodies, 326
Nitroglycerin, 534
NK cells, functions of, 586
Nociceptors, 447–48
Nodal cells, 533, 537
Nodes of Ranvier, 325
Nodose ganglia, 399
Non-insulin-dependent diabetes mellitus, 500
Nonstriated involuntary muscle, 80
Norepinephrine, 430, 434, 435, 533
    functions of, 495
Normocytic, 517
Normovolemic blood volume, 506, 517
Nose, 601
    muscles of, 257
    structure of, 601
Notch, 138
Notochord, 168
Nuchal lines, 138
Nuclear envelope, 38
Nuclear pores, 38
Nucleolus, 29
Nucleoplasm, 39
Nucleosome, 39
Nucleus, cell, 29, 38–40
Nucleus basalis, 422
Nucleus cuneatus, 389
Nucleus gracilis, 389
Nucleus pulposus, 214
Nucleus solitarius, 454
Nutrient artery, 119
Nutrient vein, 119
Nystagmus, 459

O

Oblique, 252
Oblique group, spinal muscles, 268–69
Oblique muscles, 260, 268
Obturator foramen, 186
Obturator groove, 186
Obturator muscles, 296
Occipital bone, 131, 138, 139

Occipital condyles, 138
Occipital crest, 138
Occipital lobe, 373
Occipitalis, 257
Occlusal surfaces, of teeth, 637
Ocular conjunctiva, 465
Oculomotor muscles, 260
Oculomotor nerve, 394
Odontoid process, 161
Olecranal joint, 181
Olecranon fossa, 180
Olecranon, 181
Olfaction (smell), 452–53
    olfactory discrimination, 453
    olfactory pathways, 452
    olfactory receptors, 452
Olfactory bulbs, 392
Olfactory cortex, 376
Olfactory epithelium, 452
Olfactory glands, 452
Olfactory nerve, 392
Olfactory organs, 450, 452
Olfactory receptors, 452
Olfactory region, 601
Olfactory tracts, 392
Oligodendrocytes, 324
Olivary nuclei, 389
Olives, 389
Omenta, 640, 642
Omentum, 633
Oncogene, 48
Oncologists, 84
Oocytes, 698, 702
    activation at ovulation, 720
Oogenesis, 698, 701–3
Oogonia, 698
Ooplasm, 720
Ophthalmic artery, 557
Ophthalmic branch, of trigeminal nerve, 395
Opposition, 212
Optic canal, 143
Optic chiasm, 384, 393
Optic cups, 476
Optic disc, 472
Optic groove, 143
Optic nerve, 393, 473
Optic stalks, 476
Optic tract, 384, 393
Optic vesicles, 476
Ora serrata, 470
Oral cavity, 633–37
    anatomy of, 633–34
    functions of, 633
    mastication, 637
    salivary glands, 634–35
    teeth, 635–37
    tongue, 634
Oral mucosa, 633–34
Orbicularis oris, 257
Orbital complex, 147, 149
Orbital fat, 467
Orbital fissure, 143, 146
Orbital rim, 145
Orbits, 147
Orchiectomy, 715
Organ of Corti, 461, 463
Organelles, 28, 34–43
    centrioles, 36
    cilia, 36–37
    endoplasmic reticulum, 40
    flagella, 37
    Golgi apparatus, 40–42
    intermediate filaments, 34
    lysosomes, 42
    membranous, 29, 38–43

microfilaments, 34
microtubules, 34, 36
microvilli, 36
mitochondria, 38
nonmembranous, 29, 34–38
nucleus, 38–40
peroxisomes, 42–43
ribosomes, 37–38
thick filaments, 34
Organisms, functions of, 5
Organ level, human body, 4
Organogenesis, 725
Organ systems
    cardiovascular system, 9
    development of, 82–83
    digestive system, 10
    endocrine system, 9
    integumentary system, 7
    lymphatic system, 10
    muscular system, 7
    nervous system, 8
    reproductive system, 12
    respiratory system, 10
    skeletal system, 7
    study of, 3
    urinary system, 10
Origin, of muscle, 252
Oropharynx, 601
Osmosis, 31
Osseous tissue. (See Bone)
Ossification, 115–16, 117
    endochondral, 116, 117
    intramembranous, 115–16
Ossification center, 115, 118
Osteoarthritis, 228
Osteoblasts, 74, 110
Osteoclast-activating factor, 126
Osteoclasts, 110
Osteocytes, 75, 110
Osteogenesis, 110
Osteogenesis imperfecta, 125
Osteoid, 110
Osteolysis, 110
Osteomalacia, 125
Osteomyelitis, 126
Osteon, 113
Osteopenia, 121, 126
Osteopetrosis, 125
Osteoporosis, 81, 121, 126
Osteoprogenitor cells, 112
Otic ganglia, 436
Otic placodes, 476
Otic vesicles, 476
Otitis media, 457
Otoconia, 460
Otoliths, 460
Oval window, 457, 459
Ovarian arteries, 560, 698
Ovarian cycle, 698
Ovarian follicles, 698
Ovarian hilum, 698
Ovarian ligament, 698
Ovarian vein, 698
Ovaries, 685, 698–703
    hormones of, 499
    ovarian cycle, 698, 701–3
    position of, 698
Ovulation, 702, 703, 720
Ovum
    and age, 703
    fertilization of, 703, 720–21
    production of, 698, 701–3
Oxygen, in blood, 506, 509, 510
Oxytocin, 487, 714
    functions of, 487

True pelvis, 186
True ribs, 166
True vocal cords, 605
Truncus arteriosus, 538
Tuber cinereum, 384
Tubercules, 161, 166, 180
Tuberculosis, 621
Tuberculum sellae, 143
Tubular glands, 62
Tubular pole, 671
Tubulin, 34
Tumors, 47, 81
    benign, 48
    growth of, 81
    malignant, 48
    primary and secondary, 81
Tumor-suppressing genes, 48
Tunica albuginea, 691, 698
Tunica externa, 544
Tunica interna, 544
Tunica media, 544
Tunica vaginalis, 689
Turbinate bones, 601
T wave, 536
Tympanic cavity, 142, 456
Tympanic duct, 461
Tympanic membrane, 142, 455
Tympanic reflex, 404
Type A blood, 510
Type AB blood, 510
Type B blood, 510
Type O blood, 510

U

Ulcers, 642
Ulna, 181, 182–83
Ulnar artery, 557
Ulnar collateral ligament, 219
Ulnar head, 181
Ulnar nerve, 350
Ulnar notch, 181
Ulnar vein, 568
Ultrasound, 21, 22, 23
Umbilical arteries, 571, 574–75, 675, 725
Umbilical cord, 725
Umbilical ligaments, 675
Umbilical stalk, 725
Umbilical vein, 571, 574–75, 725
Unicellular exocrine glands, 62
Unicellular glands, 62
Unipennate muscle, 250
Unipolar neurons, 329
Unmyelinated axons, 324
Upper limb, 180–84
    arteries of, 554–55
    carpals, 181
    hand, 181, 184
    humerus, 180
    lymphatic drainage of, 590
    muscles of, 280, 282–83, 284, 285, 286–90
    radius, 181, 182–83
    surface anatomy of, 315–16
    ulna, 181, 182–83
    veins of, 567, 568
Upper motor neuron, 413
Urachus, 675
Ureteral opening, 675
Ureteric bud, 679
Ureters, 675
    histology of, 675
    structure of, 675
Urethra, 680, 685
    histology of, 680

male, 694
    structure of, 680
Urethral fold, 711
Urethral meatus, 696
Urethral sphincter, 677, 680
Urethritis, 680, 681
Urinary bladder, 675, 677
    histology of, 677
    structure of, 675–77
Urinary obstruction, 677, 681
Urinary retention, 680
Urinary system
    and aging, 680
    development of, 678–79
    disorders of, 677, 680, 681
    functions of, 10
    kidneys, 663–75
    micturition reflex, 680
    ureters, 675
    urethra, 680
    urinary bladder, 675, 677
Urinary tract infections, 680, 681
Urination, process of, 680
Urine
    micturition reflex, 680
    production of, 663, 665, 667
    transport/storage of, 675–77, 680
Urogenital diaphragm, 270, 274
Urogenital groove, 714
Urogenital ridge, 678
Urogenital sinus, 678, 713
Urogenital triangle, 274
Urticaria, 106
Uterine cavity, 704
Uterine cycle, 703, 706, 707–8
    beginning and end of, 708
    menses, 707
    ovulation, 708
    proliferative phase, 708
    secretory phase, 708
Uterine tubes, 685, 703
    histology of, 703
    structure of, 703, 704
Uterosacral ligaments, 704
Uterus, 703–8
    histology of, 706–7
    internal anatomy of, 704
    position of, 703–4
    suspensory ligaments of, 703–6
    wall of, 704
Utricle, 459, 460
Uvea, 467
Uvula, 601, 634

V

Vagina, 685, 709–10
    disorders of, 709
    external genitalia, 710
    functions of, 709
    histology of, 709
Vaginal branches, 709
Vaginal canal, 709
Vaginitis, 709, 715
Vagus nerve, 399
Valves
    of heart, 529–30
    of lymphatic vessels, 583, 585
    of veins, 549
Valvular stenosis, 539
Varicose veins, 577
Vasa recta, 673
Vasa vasorum, 544
Vascular pole, 667

Vascular tunic, 467
Vas deferens, 694
Vasectomy, 715
Vasoconstriction, 544
Vasodilation, 544
Vasomotor center, 389
Vastus muscles, 297
Veins
    antebrachial vein, 568
    axillary vein, 568
    azygos vein, 569
    basilic vein, 568
    brachial vein, 568
    brachiocephalic vein, 569
    cephalic veins, 568
    cerebral veins, 564
    colic vein, 571
    compared to arteries, 544, 545, 564
    cubital vein, 568
    cystic vein, 571
    digital veins, 568
    disorders of, 577
    esophageal veins, 569
    facial veins, 568
    femoral vein, 569
    functions of, 521, 549
    gastric vein, 571
    gastroepiploic veins, 571
    gonadal veins, 569
    hemiazygos vein, 569
    hepatic veins, 569
    iliac vein, 569
    intercostal veins, 569
    jugular vein, 568
    large veins, 549
    lumbar veins, 569
    maxillary veins, 568
    medium-sized veins, 549
    mesenteric vein, 571
    peroneal vein, 569
    phrenic veins, 569
    plantar veins, 569
    popliteal vein, 569
    in pulmonary circulation, 553
    radial vein, 568
    rectal veins, 571
    renal veins, 569
    saphenous vein, 569
    splenic vein, 571
    subclavian vein, 568–69
    suprarenal veins, 569
    in systemic circulation, 564–71
    temporal veins, 568
    thoracic vein, 569
    tibial vein, 569
    ulnar vein, 568
    umbilical vein, 571
    venoconstriction, 550
    venous valves, 549
    venules, 549
    vertebral veins, 568, 569
    walls of, 544
Vellus hair, 99
Vena cava, 525
    inferior, 569
    superior, 525, 564, 568–69
    and systemic circulation, 564–69
Venoconstriction, 550
Venous reserve, 550
Venous valves, 549
Ventral, 16
Ventral body cavities, 18–19, 20
Ventral mesentery, 711
Ventral nuclei, 384
Ventral ramus, 345

Illustration Credits

PHOTOGRAPHS

To The Student

xxvi	Ward's Natural Science Establishment, Inc.
xxvii	Ralph Hutchings
xxxi (top)	Ralph Hutchings
xxxi (bottom)	Hinerfeld/Custom Medical Stock Photo

Chapter 1

1-14c	Ralph Hutchings
1-15a, b	Science Source/Photo Researchers, Inc.
1-16b	CNRI/Photo Researchers, Inc.
1-16c	Diagnostic Radiological Health Sciences Learning Lab
1-16d	Photo Researchers, Inc.

Chapter 2

2-1a	Frederic H. Martini
2-1b	Dennis Kunkel/CNRI/Phototake
2-8a	Fawcett/Hirokawa/Heuser/Science Source/Photo Researchers, Inc.
2-8b	M. Schliwa/Visuals Unlimited
2-11a	Dr. Don Fawcett/Photo Researchers, Inc.
2-12	CNRI/Science Source/Photo Researchers, Inc.
2-13a	Dr. Don W. Fawcett
2-13b	Biophoto Associates/Photo Researchers, Inc.
2-16a	Biophoto Associates/Photo Researchers, Inc.
2-17b	Dr. Birgit Satir, Yeshiva University
2-19e	Reproduced from M. G. Farquhar and J. E. Palade, "Junctional Complexes in Epithelia," *The Journal of Cell Biology,* 1963, vol. 17, p. 379.
2-22a,b,d–f	Ed Reschke/Peter Arnold, Inc.
2-22c	James Solliday/Biological Photo Service

Chapter 3

3-2b	Custom Medical Stock Photo
3-3c	Dr. C. P. LeBlond, McGill University
3-4a	Ward's Natural Science Establishment, Inc.
3-4b	Frederic H. Martini
3-5a	Michael J. Timmons
3-5b–d	Frederic H. Martini
3-6a–c	Frederic H. Martini
3-7a	Carolina Biological Supply/Phototake
3-8 (left)	Frederic H. Martini
3-8 (right)	CNRI/Science Photo Library/Photo Researchers, Inc.
3-9a	S. Elem/Visuals Unlimited
3-9b	Frederic H. Martini
3-11b	Ward's Natural Science Establishment, Inc.
3-12a,b	Lester V. Bergman & Associates, Inc.
3-13a	Science Source/Photo Researchers, Inc.
3-13b	Frederic H. Martini
3-13c	Ward's Natural Science Establishment, Inc.
3-14a	John D. Cunningham/Visuals Unlimited
3-14b	Bruce Iverson/Visuals Unlimited
3-14c	Frederic H. Martini
3-15a	Robert Brons/Biological Photo Service
3-15b	Science Source/Photo Researchers, Inc.
3-15c	Ed Reschke/Peter Arnold, Inc.
3-16	Frederic H. Martini
3-19a,b	G. W. Willis, MD/Biological Photo Service
3-19c	Frederic H. Martini
3-20	Frederic H. Martini

Chapter 4

4-3	John D. Cunningham/Visuals Unlimited
4-4b,c	Frederic H. Martini
4-7b	David Scharf Photography
4-9a	Elias-Paul's & Peters, Amenta, HISTOLOGY: AND HUMAN MICROANATOMY, 5/E, Puccin Publishers, 1987.
4-9b	Manfred Kage/Peter Arnold, Inc.
4-10b	John D. Cunningham/Visuals Unlimited
4-13	Frederic H. Martini
4-14a,b	Frederic H. Martini
4-16	C. Vergalee/Photo Researchers, Inc.

Chapter 5

5-2a	Ralph Hutchings
5-2b,c	Biophoto Associates/Photo Researchers, Inc.
5-3	Frederic H. Martini
5-4a,b	Frederic H. Martini
5-5a,b	Ralph Hutchings
5-8c	Frederic H. Martini
5-9a,b	Lester V. Bergman & Associates, Inc.
p. 122	Pott's fracture, comminuted fracture, and epiphyseal fracture: SIU/Visuals Unlimited
	Transverse fracture: Grace Moore/The Stock Shop, Inc./Medichrome
	Spiral fracture: SIU/Peter Arnold, Inc.
	Displaced fracture: Custom Medical Stock Photo
	Colles fracture: Scott Camazine/Photo Researchers, Inc.
	Greenstick fracture: Patricia Barber, RBP/Custom Medical Stock Photo
	Compression fracture: Lester V. Bergman & Associates, Inc.

Chapter 6

6-1	Ralph Hutchings
6-3a–e	Ralph Hutchings
6-4b	Ralph Hutchings
6-5b	Ralph Hutchings
6-6a–c	Ralph Hutchings
6-7a,b	Ralph Hutchings
6-8a	Michael J. Timmons
6-9c,d	Ralph Hutchings
6-11c	Ralph Hutchings
6-13d	Michael Siegfried, M.D., Lutheran General Hospital
6-14	Ralph Hutchings
6-15a	Ralph Hutchings
6-16b	Ralph Hutchings
6-17c	Michael J. Timmons
6-17d	Ralph Hutchings
6-18b	Ralph Hutchings
6-18c	Siemens Medical Systems, Inc.
6-20c	Ralph Hutchings
6-21a–d	Ralph Hutchings
6-22b,d	Ralph Hutchings
6-23b,d	Ralph Hutchings
6-23 (left)	Ralph Hutchings
6-24	Ralph Hutchings
6-25a,c	Ralph Hutchings

Chapter 7

7-1	Ralph Hutchings
7-2a	Ralph Hutchings
7-2b	Bates/Custom Medical Stock Photo
7-3a	Bates/Custom Medical Stock Photo
7-3b	Ralph Hutchings
7-5d–f	Ralph Hutchings
7-6b,d–f	Ralph Hutchings
7-7b,d	Ralph Hutchings
7-8d	Ralph Hutchings
7-9a	Ralph Hutchings
7-9b	Bates/Custom Medical Stock Photo
7-10a,b	Ralph Hutchings
7-11a,b	Ralph Hutchings
7-12d	University of Toronto
7-14a,b	Ralph Hutchings
7-15a,b	Ralph Hutchings
7-16a,b	Ralph Hutchings
7-17b	Ralph Hutchings
7-17d	Ralph Hutchings

Chapter 8

8-3	Ralph Hutchings
8-4	Ralph Hutchings
8-5	Ralph Hutchings
8-9d	Ralph Hutchings
8-10c	University of Toronto
8-10d	Patrick M. Timmons
8-11d	Patrick M. Timmons
8-12d	University of Toronto
8-13b	Ralph Hutchings
8-14b	Patrick M. Timmons
8-15c,d	University of Toronto
8-16c,e	Ralph Hutchings
8-17	Ralph Hutchings

Chapter 9

9-2a	Fred Hossler/Visuals Unlimited
9-2b	Dr. Don Fawcett
9-4b	Ward's Natural Science Establishment, Inc.
9-6a	Dr. Don Fawcett/Photo Researchers, Inc.
9-7a–c	J. J. Head/Carolina Biological Supply/Phototake
9-14a	Frederic H. Martini
9-14b	Cormack, D. (ed): HAM'S HISTOLOGY, 9th ed. Philadelphia: J. B. Lippincott, 1987. By permission.

Chapter 10

10-6	Mentor Networks Inc.
10-8	Ralph Hutchings
10-13b	Ralph Hutchings
10-15c	Ralph Hutchings
10-15d	Mentor Networks Inc.

Chapter 11

11-3	University of Toronto
11-6a,b	Mentor Networks Inc.
11-7a	Custom Medical Stock Photo
11-7c	Ralph Hutchings
11-8a	Mentor Networks Inc.
11-8c	Ralph Hutchings
11-10a	Michael J. Timmons and Patrick M. Timmons
11-10b	Ralph Hutchings
11-15c	Ralph Hutchings
11-16a	Ralph Hutchings
11-16b	Mentor Networks Inc.
11-17b	Ralph Hutchings
11-17c	Mentor Networks Inc.
11-18b	Ralph Hutchings
11-20c	Ralph Hutchings
11-21b	Ralph Hutchings
11-22c	Ralph Hutchings

Chapter 12

12-1a-c	Mentor Networks Inc.
12-2a,b	Mentor Networks Inc.
12-3a	Custom Medical Stock Photo
12-3b	Mentor Networks Inc.
12-4a,b	Mentor Networks Inc.
12-5a,b	Mentor Networks Inc.
12-6a	Custom Medical Stock Photo
12-6b,c	Mentor Networks Inc.
12-7a,b,d	Mentor Networks Inc.
12-7c	Custom Medical Stock Photo

Chapter 13

13-5a	Frederic H. Martini
13-6	Michael J. Timmons
13-7a	Biophoto Associates/Photo Researchers, Inc.
13-7b	Photo Researchers, Inc.
13-11c	David Scott/Phototake

Chapter 14

14-2a	Ralph Hutchings
14-2b	University of Toronto
14-3	Patrick M. Timmons
14-4a	Ralph Hutchings
14-4b	Hinerfeld/Custom Medical Stock Photo
14-5a	Michael J. Timmons
14-12	Ralph Hutchings
14-14a	Ralph Hutchings

Chapter 15

15-3c	Ralph Hutchings
15-7	Visuals Unlimited
15-8a	Ralph Hutchings
15-9a	Ralph Hutchings
15-11c	Michael J. Timmons
15-11d	Pat Lynch/Photo Researchers, Inc.
15-12a,b	Ralph Hutchings
15-13b,d	Ralph Hutchings
15-15b	Ralph Hutchings
15-16b	Ralph Hutchings
15-17a,b	Ralph Hutchings
15-17b (upper)	Ward's Natural Science Establishment, Inc.
15-20b	Ralph Hutchings

Chapter 17

17-6b	Ward's Natural Science Establishment, Inc.

Chapter 18

18-3c	Frederic H. Martini
18-5	Frederic H. Martini
18-7b	Frederic H. Martini
18-7c	G. W. Willis/Terraphotographics/Biological Photo Service
18-10b	Lennart Nilsson, BEHOLD MAN, Little, Brown, & Co., 1973.
18-14b	Lennart Nilsson, BEHOLD MAN, Little, Brown, & Co., 1973.

18-16c	Michael J. Timmons
18-16e	Ward's Natural Science Establishment, Inc.
18-16f	Prof. P. Motta, Dept. of Anatomy, University "La Sapienza," Rome/Science Photo Library/Photo Researchers, Inc.
18-18a	Ralph Hutchings
18-19d	Michael J. Timmons
18-19f	Ralph Hutchings
18-19g	University of Toronto
18-21a	Ed Reschke/Peter Arnold, Inc.
18-21c	Custom Medical Stock Photo
18-23	Ralph Hutchings

Chapter 19

19-4b	Manfred Kage/Peter Arnold, Inc.
19-7b,c	Frederic H. Martini
19-9b	Frederic H. Martini
19-9c	Frederic H. Martini
19-10c	Ward's Natural Science Establishment, Inc.
19-11a	Ward's Natural Science Establishment, Inc.
19-11b	Pfizer, Inc.
19-11c	Frederic H. Martini
19-12a,b	Lester V. Bergman & Associates, Inc.
19-12c	John Paul Kay/Peter Arnold, Inc.
19-12d	Custom Medical Stock Photo

Chapter 20

20-2a	David Scharf/Peter Arnold, Inc.
20-2b	Ed Reschke/Peter Arnold, Inc.
20-2c	Dennis Kunkel/CNRI/Phototake
20-5a–e	Ed Reschke/Peter Arnold, Inc.
20-6	Frederic H. Martini
20-7	Custom Medical Stock Photo

Chapter 21

21-2d	Ralph Hutchings
21-3c	Phototake
21-3d	Peter Arnold, Inc.
21-4b,c	Ralph Hutchings
21-5b	Lennart Nilsson, BEHOLD MAN, Little, Brown, & Co., 1973.
21-5c	Ralph Hutchings
21-6b	Ralph Hutchings
21-7c	Ralph Hutchings
21-8c	University of Toronto
21-8d	Ralph Hutchings
21-9a,b	DuPont Merck Pharmaceutical Company
21-9c	Peter Arnold, Inc.
21-11	Larry Mulvehill/Photo Researchers, Inc., Inc.

Chapter 22

22-1a,b	Michael J. Timmons
22-1c	Ward's Natural Science Establishment, Inc.
22-3a	Phototake
22-3b,c	BAILEY'S TEXTBOOK OF MICROSCOPIC ANATOMY by Kelly, Wood, & Enders. Copyright 1984 Williams & Wilkens.
22-4b	Biophoto Associates/Photo Researchers, Inc.
22-5 (top)	William Ober/Visuals Unlimited
22-5 (bottom)	B & B Photos/Custom Medical Stock Photo
22-9b	University of Toronto
22-11	Dr. E. L. Lansdown/University of Toronto
22-12b,c	Ralph Hutchings
22-13b	University of Toronto
22-14b	Ralph Hutchings
22-16	Ralph Hutchings
22-17b	University of Toronto
22-18b	Ralph Hutchings
22-19b	Ralph Hutchings

Chapter 23

23-3b	Frederic H. Martini
23-5	Ralph Hutchings
23-8a	Frederic H. Martini
23-8b	Biophoto Associates/Photo Researchers, Inc.
23-13	University of Toronto
23-14	University of Toronto
23-16c,d	Frederic H. Martini
23-17c	Frederic H. Martini

Chapter 24

24-2a	Frederic H. Martini
24-2b	Photo Researchers, Inc.
24-3b,c	Ralph Hutchings
24-5	Phototake
24-7c	John D. Cunningham/Visuals Unlimited
24-8a	Ralph Hutchings
24-10c	University of Toronto
24-10b,d	Ralph Hutchings
24-11b	Ward's Natural Science Establishment, Inc.
24-12c	TEXTBOOK OF HISTOLOGY, Bloom & Fawcett, 11/E. Copyright 1986, W.B. Saunders Publishers.

Chapter 25

25-2b	G. W. Willis, MD/Biological Photo Service
25-6b	Frederic H. Martini
25-9a	Alfred Pasieka/Peter Arnold, Inc.
25-9b	Astrid and Hanns-Frieder Michler/Science Photo Library/Photo Researchers, Inc.
25-10b,c	Dr. E. L. Lansdown/University of Toronto
25-11b	Ralph Hutchings
25-12d	Ward's Natural Science Establishment, Inc.
25-12e,f	Frederic H. Martini
25-15d	John D. Cunningham/Visuals Unlimited
2-15e (left)	Michael J. Timmons
25-15e (right)	G. W. Willis, MD/Biological Photo Service
25-16b	Ralph Hutchings
25-17b	Ralph Hutchings
25-18	Dr. E. L. Lansdown/University of Toronto
25-19b	Ward's Natural Science Establishment, Inc.
25-20b	Ralph Hutchings
25-21b	Ward's Natural Science Establishment, Inc.
25-21c	Michael J. Timmons
25-22c	University of Toronto
25-23c,d	Frederic H. Martini

Chapter 26

26-3b,c	Ralph Hutchings
26-4c	Ralph Hutchings
26-4d	Mentor Networks Inc.
26-6e	David M. Phillips/Visuals Unlimited
26-9a	Ralph Hutchings
26-9b	Dr. E. L. Lansdown/University of Toronto
26-10a	Photo Researchers, Inc.
26-10b	Science Photo Library/Photo Researchers, Inc.
26-11	Photo Researchers, Inc.
26-12d	Ralph Hutchings
26-13a	Ward's Natural Science Establishment, Inc.
26-13b	Frederic H. Martini
26-13c	Frederic H. Martini

Chapter 27

27-1b	Ralph Hutchings
27-2a	Ralph Hutchings

27-4b	Frederic H. Martini
27-5a	TEXTBOOK OF HISTOLOGY, Bloom & Fawcett, 11/E. Copyright 1986, W.B. Saunders Publishers.
27-5c	Ward's Natural Science Establishment, Inc.
27-6a	David M. Phillips/Visuals Unlimited
27-7a	Ralph Hutchings
27-7b	Frederic H. Martini
27-7c	Frederic H. Martini
27-8a	Ward's Natural Science Establishment, Inc.
27-8c–e	Frederic H. Martini
27-9b	Ralph Hutchings
27-9d	Ward's Natural Science Establishment, Inc.
27-10b	Ralph Hutchings
27-11b	Ralph Hutchings
27-12a–d	Frederic H. Martini
27-12f	G. W. Willis, MD/Biological Photo Service
27-12g	G. W. Willis, MD/Terraphotographics/Biological Photo Service
27-14b	Frederic H. Martini
27-14c	Custom Medical Stock Photo
27-15b	Dr. E. L. Lansdown/University of Toronto
27-16b	Ward's Natural Science Establishment, Inc.
27-17a,b,d	Frederic H. Martini
27-17c	Michael J. Timmons
27-19	Michael J. Timmons
27-21b	Fred E. Hossler/Visuals Unlimited
27-21c	Frederic H. Martini

Chapter 28

28-6b	Frederic H. Martini
28-7a	Dr. Arnold Tamarin
28-7b–d	A CHILD IS BORN, Dell Publishing Company. Copyright Boehringer Ingelheim International Gmbh, photo by Lennart Nilsson.
28-8a	A CHILD IS BORN, Dell Publishing Company. Copyright Boehringer Ingelheim International Gmbh, photo by Lennart Nilsson.
28-8b	Photo Researchers, Inc.

ART

All anatomical paintings and embryology summaries were rendered by William Ober and Claire Garrison, with the exception of the illustrations noted below.

Illustrations by Ron Ervin

2-16b, 5-12, 6-1a, 6-2, 6-3, 6-4a, 6-5a, 6-6c, 6-7, 6-8a and d, 6-12a–b, 6-13c, 6-15, 6-17a–b, 6-18a, 6-19, 6-22a, c, and e, 6-23a and c, 6-25b, 7-4, 7-6a and c, 7-7a and c, 7-11a, 7-13, 7-14, 7-16, 7-17, 8-8, 8-10, 8-12, 8-13, 8-14, 8-15, 10-7, 10-10, 10-11, 10-15, 10-17, 11-18, 11-19, 11-20, 11-21

Illustrations by Tina Sanders

6-13b, 14-6, 14-15, 15-5b, 15-16, 16-7, 17-1, 17-3, 17-4, 17-8, 17-11, 18-19a and c

Illustrations by MediVisuals

8-16

Illustrations by Craig Luce

24-3a and 27-21

All other line drawings rendered by Karen Noferi.